WITHDRAWN

ENCYCLOPEDIA
of HYDROLOGY
AND WATER
RESOURCES

Kluwer Academic Encyclopedia of Earth Sciences Series

ENCYCLOPEDIA OF HYDROLOGY AND WATER RESOURCES

Aim of the series

The Kluwer Academic *Encyclopedia of Earth Sciences Series* provides comprehensive and authoritative coverage of all the main areas in the Earth Sciences. Each volume comprises a focused and carefully chosen collection of contributions from leading names in the subject, with copious illustrations and detailed reference lists.

These books represent one of the world's leading reference resources for the Earth Sciences community. Previous volumes are being updated and new works published so that the volumes will continue to be essential reading for all practising and research geologists, teachers and students.

Series Editor

Rhodes W. Fairbridge has helped edit 16 encyclopedias in the Kluwer Academic *Encyclopedia of Earth Sciences Series* (previously Chapman & Hall), as well as authoring over 300 other scientific publications. During his career he has worked as a petroleum geologist in the Middle East, been a W.W.II intelligence officer in the SW Pacific and led expeditions to the Sahara, Arctic Canada, Arctic Scandinavia, Brazil and New Guinea. He is now Emeritus Professor of Geology at Columbia University.

Volume Editor

Reginald W. Herschy is the author of several other important hydrological works including *Streamflow Measurement*, Chapman & Hall (second edition, 1995). He is Chairman of the International Standards Organisation's Subcommittee on Hydrometry, and is an Honorary Research Fellow at the University of Bristol's Department of Geography.

Previous volumes in the series

Schwartz: *Encyclopedia of Beaches and Coastal Environments*, 1982
Finkl: *Encyclopedia of Applied Geology*, 1984
Oliver: *Encyclopedia of Climatology*, 1987
Finkl: *Encyclopedia of Field and General Geology*, 1988
Bowes: *Encyclopedia of Igneous and Metamorphic Petrology*, 1989
James: *Encyclopedia of Solid Earth Geophysics*, 1989

New and forthcoming volumes

Shirley: *Encyclopedia of Planetary Sciences*, 1997
Moores: *Encyclopedia of European and Asian Regional Geology*, 1997
Herschy: *Encyclopedia of Hydrology and Water Resources*, 1998
Alexander: *Encyclopedia of Environmental Science*
Marshal: *Encyclopedia of Geochemistry*
Gerrard: *Encyclopedia of Geomorphology*
Finkl: *Encyclopedia of Soil Science and Technology 2ed*

ENCYCLOPEDIA *of* HYDROLOGY AND WATER RESOURCES

edited by

REGINALD W. HERSCHY *and*
RHODES W. FAIRBRIDGE

KLUWER ACADEMIC PUBLISHERS

DORDRECHT / BOSTON / LONDON

Published by Kluwer Academic Publishers,
P.O. Box 17, 3300 AA Dordrecht, The Netherlands

Sold and distributed in North, Central and South America
by Kluwer Academic Publishhers,
101 Philip Drive, Norwell, MA 02061, U.S.A.

In all other countries, sold and distributed
by Kluwer Academic Publishers,
P.O. Box 322, 3300 AH Dordrecht, The Netherlands.

First edition 1998

© 1998 Kluwer Academic Publishers

Typeset in 8/8½pt Times by Photoprint, Torquay, Devon

Printed in Great Britain by The University Press, Cambridge

ISBN 0 412 74060 5

A legal record for this book is available from the British Library

Library of Congress Catalog Card Number: 98–72142

∞ Printed on permanent acid-free text paper, manufactured in
accordance with ANSI/NISO Z39.48–1992 and ANSI/NISO Z39.48–1984
(Permanence of Paper)

"The man of science is of no country; all mankind his countrymen"

James Smithson FRS
(Founder of the Smithsonian Institution, 1846)

Contents

Contributors

Deborah Anthony
College of Natural Resources
Department of Earth Resources
Colorado State University
Fort Collins, CO 80523, USA

V. Aravamuthan
Department of Civil Engineering
Louisiana State University
Baton Rouge, LA 70803–6405, USA

David Archer
2 Welburn Close
Ovingham, NE42 6BD, UK

A. John Arnfield
Department of Geography,
Ohio State University
103 Bricker Hall, 190 North Oval Mall
Columbus, OH 43210–1361, USA

Arthur J. Askew
World Meteorological Organization
CH 1211, Geneva 2, Switzerland

Nicholas G. Aumen
Okeechobee Systems Research Division
South Florida Water Management District
PO Box 24680
West Palm Beach, FL, 33416–4680, USA

William Back
United States Geological Survey
Reston, VA 22092, USA

Eric C. Barrett
Remote Sensing Centre, Department of Geography
University of Bristol
Bristol BS8 1SS, UK

Louis J. Battan (deceased)

Eric J. Best
Canberra College of Advanced Education,
PO Box 1, Belconnen
ACT 2616, Australia

Keith Beven
Institute of Environmental and Biological Sciences
Lancaster University
Lancaster LA1 4YQ, UK

Gary D. Bishop
US Environmental Protection Agency
200 SW 35th Street
Corvellis, OR, USA

Keith R. Boucher
Department of Geography
Loughborough University of Technology
Loughborough LE11 3TU, UK

F.C. Brassington
12 Culcheth Hall Drive, Culcheth
Warrington WA3 4PS, UK

Howard A. Bridgman
Department of Geography
University of Newcastle
Newcastle, NSW 2308, Australia

David J. Burdon
Food and Agricultural Organization of the United
  Nations
via DelleTerme di Caracella
Rome, Italy

M. Robbins Church
US Environmental Protection Agency
200 SW 35th Street
Corvellis, OR, USA

Jen-Hu Chang
Department of Geography
University of Hawaii, Porteus Hall
445, 2424 Maile Way
Honolulu, Hawaii 96822, USA

Malcolm K. Cleaveland
Department of Geography
University of Arkansas
Fayetteville, AR 72701, USA

Gerald A. Cole
Department of Zoology
Arizona State University
Tempe, AZ 85281, USA

J.A. Cole
3 Grangefield Way, Aldwick
Bognor Regis, PO21 4EG, UK

C.G. Collier
Department of Civil Engineering
Telford Research Institute
Salford University
Salford, M5 4WT, UK

Howard J. Critchfield
Office of the State Climatologist
Western Washington University
Bellingham, WA 98225, USA

Cui Dewei
Hydrological Service of Qinghai Province
82, Kunlun Road, Xining
People's Republic of China

William A. Dando
Department of Geography
University of North Dakota
Grand Forks, ND 58201, USA

John A. Day
609 North Cowls
McMinnville, OR 97128, USA

R.W. Durrenberger
6233 East Catalina Drive
Scottsdale, AZ 85251, USA

Philippe Dutartre
BRGM, BP 6009
45060 Orleans, Cedex 2, France

Val L. Eichenlaub
Department of Geography
Western Michigan University
Kalamazoo, MI 49008, USA

Clifford Embleton (deceased)

D.A. Ervine
Department of Civil Engineering
University of Glasgow, Oakfield Avenue
Glasgow, G12 8LT, UK

Rhodes W. Fairbridge
420 Riverside Drive, Apt 2-B
New York, NY 10025, USA

Thomas D. Fontaine
Department of Research
South Florida Water Management District
West Palm Beach, FL 33416–4680, USA

Peter G. Fookes
Winters Wood, 47 Crescent Road
Caterham, CR3 6LH, UK

Petre Gâştescu
Institute of Geography, Romanian Academy
Dimitrie Racovita, 12
Bucharest 70307, Romania

Lois D. George
Alabama Geological Survey, PO Drawer O
University, AL 35486, USA

David W. Gibbard
Kingsworthy, Winchester SO23 7QA, UK

Donald L. Gilman
W/NMC51, National Oceanic and Atmospheric
  Administration
National Weather Service
Washington, DC, USA

Brian Goodall
Department of Geography
University of Reading, Whiteknights
Reading, RG6 2AB, UK

Samuel N. Goward
Department of Geography, Social Sciences Building
University of Maryland
College Park, MD 20742, USA

Orman E. Granger
Department of Geography
University of California
Berkeley, CA 94720, USA

Colin Green
Flood Hazard Research Centre
Middlesex University, Queensway
Enfield, EN3 4SF, UK

John F. Griffiths
Texas A and M University College of Geosciences
College Station, TX 77843, USA

Pavel Ya. Groisman
Department of Geology and Geography
University of Massachusetts at Amherst
Amherst, MA 01003–0026, USA

John M. Grymes III
Southern Regional Climate Center
Department of Geography and Anthropology
Louisiana State University
Baton Rouge, LA 70803–4105, USA

James L. Guernsey
Department of Geography and Geology
Indiania State University
Terre Haute, IN 47809, USA

O.M. Hackett
US Geological Survey
Washington, DC 20242, USA

N.B. Harmancioglu
Department of Civil Engineering
Louisiana State University
Baton Rouge, LA 70803–6405, USA

Stuart A. Harris
Department of Geography
University of Calgary, Canada

Karl E. Havens
Okeechobee Systems Research Division
South Florida Water Management District
PO Box 24680
West Palm Beach, FL 33416–4680, USA

R.W. Herschy
2, Queensborough Drive
Reading, RG4 7JA, UK

P.G. Holland
21, Glencregagh Road
Belfast, BT8 4FZ, UK

Robert M. Hordon
Department of Geography
Rutgers University
New Brunswick, NJ 08903, USA

Alan Howard
Department of Geography
University of Reading, Whiteknights
Reading, RG6 2AB, UK

Denis A. Hughes
Institute for Water Research
Rhodes University
Grahamstown, PO Box 94, South Africa

Diane Ireland
14, Southlands, Kirkham
Preston PR4 2TR, UK

Jing Yueling
Hydrological Service of Qinghai Province
82 Kunlun Road, Xining
People's Republic of China

Charles R. Kolb (deceased)

Igho H. Kornblueh
Department of Physical Medicine
University of Pennsylvania
Philadelphia, PA 19146, USA

Zbigniew W. Kundzewicz
World Meteorological Organization
CH 1211, Geneva 2, Switzerland

Philip E. LaMoreaux
Philip E. LaMoreaux and Associates Inc.
PO Box 2310
Tuscaloosa, AL 35403, USA

Kamlesh P. Lulla
Department of Geography and Geology
Indiana State University
Terre Haute, IN 47809, USA

A.B. MacLeod
Department of Civil Engineering
University of Glasgow, Oakfield Avenue
Glasgow, G12 8LT, UK

T.J. Marsh
Institute of Hydrology
Wallingford, OX10 8BB, UK

Michael E. McCormick
Department of Ocean Engineering
US Naval Academy
Annapolis, MD 21402, USA

M.P. Mosley
Water and Soil Science Centre
Ministry of Works and Development
PO Box 1479
Christchurch, New Zealand

Benjamin Moulton
Department of Geography and Geology
Indiana State University
Terre Haute, IN 47809, USA

Robert A. Muller
Southern Region Climate Center
Department of Geography and Anthropology
Louisiana State University
Baton Rouge, LA 70803, USA

David H. Newsome
CNS Scientific & Engineering Services
5 Alderney Court, Montagu Street
Reading, RG1 4JN, UK

Sharon E. Nicholson
Department of Meteorology
Florida State University
404 Love Building 3034
Tallahasse, FL 32306, USA

John E. Oliver
Department of Geography, Geology and Anthropology
Indiana State University
Terre Haute, IN 47809, USA

Catherine Ottlé
CRPE, Centre Universitaire
10/12 Avenue de l'Europe
Velizy, 78140, France

G. Pender
Department of Civil Engineering
University of Glasgow, Oakfield Avenue
Glasgow, G12 8LT, UK

Troy L. Péwé
538 East Fairmont Drive
Tempe, AZ 85282, USA

J.R. Philip
Centre for Environmental Mechanics
CSIRO, GPO Box 821
Canberra, ACT 2601, Australia

Michael Price
PRIS, University of Reading
PO Box 227, Reading, RG6 2AB, UK

Peter Price
6 Aldford Close, Bromborough
Wirral, L63 OPT, UK

William H. Quinn
School of Oceanography
Oregon State University
Corvallis, OR 97331, USA

C.S. Ramage
Department of Meteorology
University of Hawaii
2525 Correa Road
Honolulu, HI 96822, USA

Günther Reichel
Gumpendorferstrasse 72/7
1060 Vienna, Austria

Gareth Roberts
Institute of Hydrology
Wallingford, OX10 8BB, UK

Mark Robinson
Institute of Hydrology
Wallingford, OX10 8BB, UK

G.G. Robson
British Standards Institution
389 Chiswick High Road
London, W4 4AL, UK

David T. Rudnick
Everglades Systems Research Division
South Florida Water Management District
PO Box 24680
West Palm Beach, FL 33416–4680, USA

George R. Rumney
3, Ridge Road
Groton Long Point, CT 06340, USA

Jean-Yves Scanvic
BRGM, PO Box 6009
Avenue de Concyr, Orleans
Cedex 2, 45060, France

Adrian E. Scheidegger
Institut für Geophysik
Technische Universität Wien
Gusshausstrasse 27–29
Vienna, A1040, Austria

Charlotte Schreiber
Department of Earth Science
Queens College
Flushing, NY 11367, USA

A.H. Schumann
Institute of Hydrology
Ruhr University Bochum
Universität Str. 150
44801 Bochum, Germany

Jonathan Shanklin
British Antarctic Survey
Cambridge, CB3 OET, UK

T.C. Sharma
School of Environmental Studies
PO Box 3900, Moi University
Eldoret, Kenya

Elizabeth M. Shaw
The Meadows, Castle Park, Hornby
Lancaster, LA2 8SB, UK

James R. Simpson
Westbank, New Ridley, Stocksfield
Northumberland, NE43 7RQ, UK

V.P. Singh
Department of Civil Engineering
Louisiana State University
Baton Rouge, LA 70803–6405, USA

J.H. Sircoulon
French National Committee (IAHS)
Orstom, 213 Rue Lafayette,
Paris, Cedex 75480, France

Alec J. Smith
Department of Geology
University College
London, UK

Keith Smith
School of Natural Sciences
University of Stirling
Stirling, FK9 4LA, UK

Stephen J. Stadler
Department of Geography
Oklahoma State University
Stillwater, OK 74078, USA

Ö. Starosolszky
VITUKI, Water Resources Research Centre PLC
1095 Budapest IX
Kvassay Jenö UT, Hungary

Donald Steila
Department of Geography and Earth Sciences
University of North Carolina
Unce Station
Charlotte, NC 28223, USA

R.E. Stevenson
Scripps Institute of Oceanography
La Jolla, CA 92037, USA

W.C. Swinbank (deceased)

Louis A. Toth
Department of Research
South Florida Water Management District
West Palm Beach, FL 33416–4680, USA

Philip Turton
16, Matlock Road
Reading, RG4 7BS, UK

United States Geological Survey
National Center, Mail Stop 415
Reston, VA 22092, USA

V.S. Vuglinsky
State Hydrological Institute
23 Second Line
19905 St Petersburg, Russia

Susan Walker
The Environment Agency
PO Box 12, Richard Fairclough House
Knutsford Road
Warrington, WA4 1HG, UK

Wayne M. Wendland
Climatology Section
Illinois State Water Survey
Champaign, IL, USA

Roger J.M. De Wiest
Department of Geological Engineering
Princeton University
Princeton, NJ 08540, USA

Ellen Wohl
College of Natural Resources
Department of Earth Resources
Colorado State University
Fort Collins, CO 80523, USA

World Meteorological Organization
CH 1211, Geneva 2, Switzerland

Robert L. Wright
Department of Geography
University of Sheffield
Sheffield, S10 2TN, UK

Thomas M. Yanosky
United States Geological Survey
National Center, Mailstop 415
Reston, VA 22092, USA

Masatoshi M. Yoshino
Institute of Geoscience
University of Tsukuba
Ibaraki, 305, Japan

Foreword

In my position over the last few years as Chief Hydrologist of the US Geological Survey, I have seen the increasing scarcity of the available water resources of the United States and those of the world. Users compete for a renewable resource that is becoming stressed in many locations. These stresses include demands for water for agricultural, industrial and domestic uses as well as demands for water in streams for purposes of preserving or enhancing habitat and water quality. Information is needed by those engineers and hydrologists and by informed lay persons in order better to utlilize and allocate water.

In the *Encyclopedia of Hydrology and Water Resources*, Dr Herschy has assembled the combined efforts of both practitioners and academics, who as world experts have provided some of the latest information in their fields. These diverse sources of knowledge have resulted in a reference work that should prove to be most useful as we move into the twenty-first century.

Of particular utility is the inclusion of water resources as well as the more traditional science of hydrology.

This encyclopedia covers many aspects of hydrology and water resources, including water pollution.

It is difficult for members of a busy general audience to quickly find up-to-date, summarized information on hydrology and water resources. Planners, policy makers, regulators and decision makers are the primary audience for this source of information.

In many parts of the world people face conflict and difficult choices as they consider how to manage their water resources. Many of these problem areas are given excellent coverage in the encyclopedia, which describes the situations in terms of water resources availability, water quality and management options.

With pleasure I recommend the *Encyclopedia of Hydrology and Water Resources* to those that need quick, authoritative information in these fields. This reference should prove particularly useful to the applied hydrologist and to those in academia.

Robert M. Hirsch
Chief Hydrologist, United States Geological Survey

Preface

The hydrological cycle is the source and substance of life on Earth. It took a thousand years to explain its concept and yet we still do not fully understand it. However we have come a long way from the concept of the early philosophers that the formation of rivers was due to sea water being forced somehow into huge caverns in the mountains.

In the world today, millions of people die each year because of poor water management and over 25 countries are still classified as water-scarce countries affecting human activity, health and growth. This encyclopedia does not pretend to embrace all aspects of the subject of hydrology and water resources; to do so would require several volumes. However, it contains an eclectic selection of numerous topics linked by the common thread of hydrology and water resources in the service of humanity. This volume is essentially a potpourri of miscellany that follow an applied theme where each topic may be considered an apophereta, something to be savored for the information it contains. Because it is not possible to consider every aspect of the subject in a reference work of this level and with a mandate so broad, the encyclopedia represents an attempt to provide an overview of diverse topics through a sample of a wide range of interrelated topics.

All entries are in alphabetical order with the length largely related to the importance of the topic. The reader may find that the same topic may be dealt with from various viewpoints in a number of different entries.

To find a particular topic, it is best to look first for that subject in its alphabetical context. Beyond that, the index and cross references will provide further guidance to the subject in question. The comprehensive index at the back of the volume will list, for a given term or name, every page in the book where that item appears. This is a fine way to find a subject presented in many different contexts. On the other hand, the cross references at the end of each entry act as a guide to other related entries. Wherever possible, extensive lists of references are appended to each entry. The opportunity was taken of examining previous encyclopedias in the series for suitable entries on hydrology and water resources and a selection of these was considered and updated where necessary. Due acknowledgement is therefore made to the authors and publishers for permission to publish these contributions in this volume. The encyclopedias utilized were:

Geomorphology, 1968. Reinhold Book Corporation, New York.

Geochemistry and Environmental Sciences, 1972. Dowden Hutchinson & Ross Inc, Stroudsburg, PA.

Applied Geology, 1984. Van Nostrand Reinhold Company, New York.

Climatology, 1987. Van Nostrand Reinhold Company, New York.

The creation of this encyclopedia, like others in the series, would not have been possible without the wholehearted cooperation of our contributors; there are over 114 of them in this volume and 17 different countries are represented.

Rhodes W. Fairbridge, Series Editor
Reginald W. Herschy, Volume Editor

Acknowledgements

The editors are deeply grateful to the following authors, publishers and organizations for granting permission to quote from books and papers published by them.

- Ashtec Europe Ltd, UK
- A.A. Balkema, Rotterdam
- C.V. Beadon, UK
- Asit K. Biswas, UK
- British Society of Dowsers, UK
- British Standards Institution, London
- Chapman & Hall, London
- Chartered Institution of Water and Environmental Management, London
- CRC Press Inc./Lewis Publishers, USA
- Department of the Environment, Water Directorate, UK
- J.C.I. Dooge, Ireland
- EARSeL, Paris
- Elsevier Science Ltd, Oxford, UK
- Environment Agency, UK
- The European Environment Agency; The Dobris Assessment, Copenhagen
- Sir Hugh Fish, UK
- Foundation for Water Research, UK
- Geological Society of America, Boulder, USA
- Institute of Hydrology/ODA, UK
- Institution of Civil Engineers, London
- International Association of Hydrological Sciences
- International Bottled Water Association, USA
- International Commission on Large Dams, Paris
- International Desalination Association, USA
- International Standards Organization, Geneva
- International Water Supply Association, London
- Smithsonian Institution, Washington, DC
- Sterling Publications, London
- UK Groundwater Forum, Wallingford, UK
- United Nations, New York
- United States Geological Survey, Reston, USA
- WaterAid, London
- John Wiley and Sons, New York
- World Bank, Washington, DC
- World Health Organization, Geneva
- World Meteorological Organization, Geneva

Selected international journals in hydrology and water resources

Advances in Water Resources Elsevier Science, Oxford, UK
Advances in Water Science PRC, Beijing
Agriculture and Forest Meteorology Elsevier Science, Amsterdam
Agricultural Water Management Elsevier Science, Amsterdam
Applied Hydrogeology (IAH) Verlag Heinz Hesse, Hanover, Germany
Aqua Blackwell, Oxford
Aqua Fennica Aqua Fennica, Finland
Aquatic Conservation John Wiley, Chichester, UK
Bulgarian Journal of Meteorology and Hydrology Bulgarian Academy of Science, Sofia, Bulgaria
Bulletin of The American Meteorological Society AMS Boston, USA
Bulletin de Liaison du Comité Interafricain d'Etudes Hydrauliques Orston, Paris
Ceres (FAO) FAO, Rome
Climate Monitor University of East Anglia, UK
Climate Change Kluwer, Dordrecht, Netherlands
Dams and Reservoirs British Dam Society of ICE, London
Danish Hydraulics DHI, Horsholm, Denmark
Deutsche Gewasser Kundliche Mitteilungen Schriftleitung, Koblenz
Earth Surface Processes and Landforms John Wiley, Chichester, UK
Ecological Monitoring Elsevier, Amsterdam
Environmental Science and Technology American Chemical Society, Ohio, USA
Environmetrics John Wiley, Chichester, UK
Geographical Journal Royal Geographical Society, London
Global Planetary Change Elsevier Science, Amsterdam
Groundwater Groundwater Pub. Co., Westerville, Ohio
Groundwater Monitoring & Remediation Groundwater Pub. Co., Westerville, Ohio
Houille Blanche Soc. Hydrotechnique de France, Paris
Hydrogeologie BRGM, Orleans, France
Hydrogeology Journal Verlag Heinz Heise, Hanover, Germany
Hydrological Processes John Wiley, Chichester, UK
Hydrological Sciences Journal IAHS, Wallingford, UK
Hydrology and Earth System Sciences EGS, Katlenburg-Lindau Germany
Hydroplus Levallois Perret, France
Hydrotitles Geosystems, Oxford, UK
ICID Journal ICID, New Delhi, India
International Journal of Climatology John Wiley, Chichester, UK
International Journal of Geographical Information Systems Taylor and Francis, London
International Journal on Hydropower and Dams Aqua-Media International, UK

International Journal of Remote Sensing Taylor and Francis, London
International Journal of Water Resources Development Carfax, Oxford, UK
International Water Power and Dam Construction: Wilmington Business Pub., UK
Irrigation Science Springer-Verlag, Heidelberg, Germany
Journal of American Water Resources Association AWRA, Hendon, VA, USA
Journal of Applied Ecology Blackwell Science, Oxford
Journal of Applied Hydrology (India) Association of Hydrology, Visakhapatnam, India
Journal of Applied Meteorology American Meteorological Society, Boston, MA, USA
Journal of Atmospheric Sciences American Meteorological Society, Boston, MA, USA
Journal of Climate American Meteorological Society, Boston, MA, USA
Journal of Contaminant Hydrology Elsevier Science, Amsterdam
Journal of Environmental Engineering (ASCE) American Society of Civil Engineers, New York
Journal of Hydraulic Research IAHR, Delft, Netherlands
Journal of Hydrology Elsevier Science, Amsterdam
Journal of Hydrology (New Zealand) New Zealand Hydrological Society, Wellington
Journal of Hydroscience and Hydraulic Engineering (Japan) Japanese Society of Civil Engineers, Tokyo
Journal of the Chartered Institution of Water and Environmental Management CIWEM, London
Journal of Irrigation and Drainage Engineering (ASCE) American Society of Civil Engineers, New York
Journal of Meteorology Artetech Pub. Co., UK
Journal of Water Resources Planning and Management (ASCE) Americas Society of Civil Engineers, New York
Land and Water International (Netherlands) NEDECO, The Hague
Monthly Weather Review American Meteorological Society, Boston, MA, USA
New Civil Engineer Thomas Telford (ICE), London
Nordic Hydrology Nordic Association for Hydrology, Lyngby
Proceedings of the Institution of Civil Engineers ICE, London
Rivers SEL and Associates, Colorado, USA
Russian Meteorology and Hydrology Allerton Press, New York
Soil Science Williams and Wilkins, Baltimore, USA
Soil Science Society of America Journal SSSA, Madison, WI, USA
Stochastic Hydrology and Hydraulics Springer International, Heidelberg, Germany
Transactions American Geophysical Union – EOS American Geophysical Union, Washington, DC

Transactions of the Institution of Engines of Australia, Civil Engineering Institution of Engineers of Australia, Barton

Water Air and Soil Pollution Kluwer, Netherlands

Water and Atmosphere NIWA, New Zealand

Water and Environment International Argus Business Media UK

Water International IAWQ, IL, USA

Water Quality International IAWQ, London

Water Research Pergamon Press

Water Resources Bulletin American Water Resources Association, VA, USA

Water Resources Journal (UNESCAP) United Nations, Bangkok

Water Resources Management Kluwer, Netherlands

Water Resources Research Amer. Geophysical Union, Washington, DC

Water SA South African Water Research Committee, Pretoria

Water Science and Technology (IAWQ) Elsevier Science, Oxford

Waterlines Intermediate Technology Pub, London

Waterway (UNESCO IHP) UNESCO, Paris

Weather Royal Meteorological, Society, UK

WMO Bulletin WMO, Geneva

World Climate News WMO, Geneva

World Water and Environmental Engineering Faversham House Group, UK

R.W. Herschy

Source

The Institute of Hydrology (UK) Library.

Units, symbols and conversion factors

Standard units are used for the measurement of length, area, volume, velocity, discharge, density, mass, pressure, gravitational acceleration and temperature. The Système International d'Unités (SI), as established by the International Organization for Standardization in 1960, is generally used throughout this volume, but the English system (customary units) is used where appropriate.

SI units

Historical note

The idea of decimal units of measurement was conceived by Simon Stévin (1548–1620), who also developed the even more important concept of decimal fractions. Decimal units were also considered in the early days of the French Académie des Sciences founded in 1666, but the adoption of decimal weights and measures was part of the general increase in administrative activity in Europe which followed the French Revolution. The statesman Talleyrand aimed at the establishment of a system of international decimal units of weights and measures 'à tous les temps, à tous les peuples'. On the advice of the scientists of his day, these were based on the meter as the unit of length and the gram as the unit of mass. The meter was intended to be one ten-millionth part of the distance from the North Pole to the equator at sea level, and passing through Paris; the gram was to be the mass of 1 cubic centimeter of water at its maximum density, at a temperature of approximately 4°C. A system of prefixes was developed to indicate powers of ten of the units, thus providing a flexible and convenient means of expression for a wide range of magnitudes and avoiding the need to use very large or very small numerical values. The prefixes also enabled the units of different sizes to be memorized with ease.

Although the decimal units were primarily devised as a benefit to industry and commerce, scientists soon realized their advantage and they were adopted in scientific and technical circles in continental Europe and in the UK. In 1832 Gauss arbitrarily chose length, mass and time as three 'fundamental' or base quantities and showed that magnetic flux could be calculated in terms of these; in 1851 Weber extended the theory to include all electrical quantities. During the period 1861–1867 the British Association Committee on Standards of Electrical Resistance adopted the electromagnetic units as outlined by Weber and later also adopted a system of 'practical' units (the ohm, ampere and joule), each of which was related by some power of ten to the corresponding unit of the fundamental system. In 1873 a further BA Committee recommended the general adoption by all scientists of the centimeter, the gram and the second as the units of measurement for the three 'fundamental' quantities, thus establishing the CGS system of units. In 1901 Professor Giorgi proposed that the base units be changed to the meter, the kilogram and the second (the MKS system) and to these he added an electrical base unit. The International Electrotechnical Commission adopted MKS in 1935 but the choice of the ampere as the electrical base unit was not made until 1950, giving the MKSA (or Giorgi) system covering both mechanics and electromagnetism.

Following the signing of the Metre Convention in Paris in 1875, the Conférence Générale des Poids et Mesures (CGPM) has been responsible for all international matters concerning the meter and the kilogram and with metrological studies in relation to these quantities; this responsibility was widened in 1921 to cover all units of measurement. The CGPM meets in Paris and under its authority to deal with scientific aspects of its work are the Comité International des Poids et Mesures (CIPM) assisted by various Consultative Committees, and the Bureau International des Poids et Mesures (BIPM). The laboratories of BIPM at Sèvres are the repository of the prototype kilogram and the former prototype meter. The kilogram is still defined in terms of the international prototype at Sèvres but the meter is now defined in terms of a specified number of wavelengths of a particular radiation of light.

At its tenth meeting, in 1954, the CGPM adopted a coherent system of units based on the four MKSA units together with the kelvin as the unit of temperature and the candela as the unit of luminous intensity. The eleventh CGPM in 1960 formally gave it the full title 'Système International d'Unités' for which the abbreviation is 'SI' in all languages. At the fourteenth CGPM, in 1971, the mole (symbol mol) was adopted as the unit for amount of substance and as the seventh base unit. Two additional special names (and symbols) were also adopted as follows: pascal (Pa) for newton per square meter and siemens (S) for reciprocal ohm. At the fifteenth CGPM, in 1975, two further special names (and symbols) were adopted: becquerel (Bq) for reciprocal second and gray (Gy) for joule per kilogram, as well as the additional prefixes peta (P) for 10^{15} and exa (E) for 10^{18}.

The International System of Units (SI)

This system includes three classes of units:

- base units;
- supplementary units;
- derived units.

Together these form the coherent system of SI units.

Base units

The SI is founded on the seven base units listed in Table 1.

Supplementary units

The CGPM has not yet classified certain units of the SI as either base units or derived units.

Table 1 Base units

Quantity	Name of base SI unit	Symbol
Length	meter	m
Mass	kilogram	kg
Time	second	s
Electric current	ampere	A
Thermodynamic temperature	kelvin	K
Amount of substance	mole	mol
Luminous intensity	candela	cd

Table 2 Supplementary units

Quantity	Name of supplementary SI unit	Symbol
Plane angle	radian	rad
Solid angle	steradian	sr

These units, listed in Table 2, are called 'supplementary units' and may be regarded either as base units or as derived units.

Derived units

Derived units are expressed in terms of base units and/or supplementary units by multiplication and division; for example, the SI unit for velocity is meter per second ($m\,s^{-1}$) and the SI unit for angular velocity is radian per second ($rad\,s^{-1}$).

For some of the derived SI units, special names and symbols exist; those approved by the CGPM are listed in Table 3.

It may sometimes be advantageous to express derived units in terms of other derived units having special names; for example, the SI unit for electric dipole moment is usually expressed as $C\,m$, but can be expressed as $A\,s\,m$.

Multiples of SI units

The prefixes given in Table 4 (SI prefixes) are used to form names and symbols of multiples (decimal multiples and submultiples) of the SI units.

The symbol of a prefix is considered to be combined with the unit symbol to which it is directly attached, forming with it a symbol for a new unit which can be provided with a positive or negative exponent and which can be combined with other unit symbols to form symbols for compound units. Note: because the name of the base unit kilogram already contains the SI prefix kilo, the names of decimal multiples and submultiples of the unit of mass are formed by adding appropriate prefixes to the word gram. Likewise, symbols for the

Table 4 Prefixes

Factor by which the unit is multiplied	Prefix Name	Prefix Symbol
10^{18}	exa	E
10^{15}	peta	P
10^{12}	tera	T
10^{9}	giga	G
10^{6}	mega	M
10^{3}	kilo	k
10^{2}	hecto	h
10^{1}	deca	da
10^{-1}	deci	d
10^{-2}	centi	c
10^{-3}	milli	m
10^{-6}	micro	μ
10^{-9}	nano	n
10^{-12}	pico	o
10^{-15}	femto	f
10^{-18}	atto	a

multiples and submultiples are formed by adding prefix symbols to the symbol g.
Examples of multiples of SI units include:

$$1\,cm^3 = (10^{-2}\,m)^3 = 10^{-6}\,m^3$$
$$1\,\mu s^{-1} = (10^{-6}\,s)^{-1} = 10^6\,s^{-1}$$
$$1\,mm^2\,s^{-1} = (10^{-3}\,m)^2\,s^{-1} = 10^{-6}\,m^2\,s^{-1}$$

Compound prefixes should not be used; for example, use nm (nanometer) and not 'mμm'.

Use of SI units and their multiples

The choice of the appropriate multiple (decimal multiple or submultiple) of an SI unit is governed by convenience, the multiple chosen for a particular application being the one which will lead to numerical values within a practical range.

The multiple can usually be chosen so that the numerical values will be between 0.1 and 1000, for example:

$1.2 \times 10^4\,N$	can be written as $12\,kN$
$0.003\,94\,m$	can be written as $3.94\,mm$
$1401\,Pa$	can be written as $1.401\,kPa$
$3.1 \times 10^{-8}\,s$	can be written as $31\,ns$

However, in a table of values for the same quantity or in a

Table 3 Derived units

Quantity	Name of derived SI unit	Symbol	Expressed in terms of base or supplementary SI units or in terms of other derived SI units
Frequency	hertz	Hz	$1\,Hz = 1\,s^{-1}$
Force	newton	N	$1\,N = 1\,kg\,m\,s^{-2}$
Pressure, stress	pascal	Pa	$1\,Pa = 1\,N\,m^{-2}$
Energy, work, quantity of heat	joule	J	$1\,J = 1\,N\,m$
Power	watt	W	$1\,W = 1\,J\,s^{-1}$
Electricity charge, quantity of electricity	coulomb	C	$1\,C = 1\,A\,s$
Electric potential, potential difference, electromotive force	volt	V	$1\,V = 1\,JC = 1\,WA^{-1}$
Electric capacitance	farad	F	$1\,F = 1\,CV^{-1}$
Electric resistance	ohm	Ω	$1\,\Omega = 1\,VA^{-1}$
Electric conductance	siemens	S	$1\,S = 1\,\Omega^{-1}$
Magnetic flux, flux of magnetic induction	weber	Wb	$1\,Wb = 1\,V\,S$
Magnetic flux density, magnetic induction	tesla	T	$1\,T = 1\,Wb\,m^{-2}$
Inductance	henry	H	$1\,H = 1\,Wb\,A^{-1}$
Luminous flux	lumen	lm	$1\,lm = 1\,cd\,sr$
Illuminance	lux	lx	$1\,lx = 1\,lm\,m^{-2}$
Activity (radioactive)	becquerel	Bq	$1\,Bq = 1\,s^{-1}$
Absorbed dose (of ionizing radiation)	gray	Gy	$1\,Gy = 1\,J\,kg^{-1}$

Table 5 Units of practical importance

Quantity	Name of unit	Unit symbol	Definition
Time	Minute	min	1 min = 60 s
	hour	h	1 h = 60 min
	day	d	1 d = 24 h
plane angle	degree	°	$1° = (\pi1/180)$ rad
	minute	'	$1' = (1/60)°$
	second	"	$1" = (1/60)'$
volume	liter	l	$1\,l = 1\,dm^3$
mass	tonne	t	$1\,t = 10^3$ kg

discussion of such values within a given context, it is generally better to use the same multiple for all items, even when some of the numerical values will be outside the range 0.1–1000. For certain quantities in particular applications, the same multiple is customarily used throughout; for example the millimeter is used for dimensions in most engineering drawings.

It is recommended that only one prefix should be used in forming a multiple of a compound SI unit.

Errors in calculations can be avoided more easily if all values are expressed in SI units, prefixes being replaced by powers of 10.

Rules for writing unit symbols
Unit symbols should be printed in roman (upright) type (irrespective of the type used in the rest of the text), should remain unaltered in the plural, should be written without a final full stop (period, point) and should be placed after the complete numerical value in the expression for a quantity, leaving a clear space between the numerical value and the unit symbol.

Unit symbols should be written in lowercase letters except that the first letter is written in upper case when the name of the unit is derived from a proper name. Example include:

m meter
s second
A ampere
Wb weber

A compound unit formed by multiplication of two or more units may be indicated in one of the following ways:

N·m, N.m, N m

Note that when using a unit symbol which coincides with the symbol for a prefix, special care should be taken to avoid confusion. The unit newton meter for torque should be written, for example, N m or m·N to avoid confusion with mN, the millinewton.

When a compound unit is formed by dividing one unit by another, this may be indicated in one of the following ways:

$\frac{m}{s}$, m/s or by writing the product of m and s^{-1}, for example as $m\,s^{-1}$.

In no case should more than one solidus (as in m/s) on the same line be included in such a combination unless parentheses are inserted to avoid all ambiguity. In complicated cases, negative powers or parentheses should be used.

Non-SI units which may be used together with the SI units and their multiples

There are certain units outside the SI which are nevertheless recognized by the CIPM as having to be retained either because of their practical importance (Table 5) or because of their use in specialized fields.

Prefixes given in Table 4 may be attached to many of the units given in Table 5; for example, milliter, ml; megaelectronvolt, MeV.

In a limited number of cases, compound units are formed with the units given in Table 5 together with SI units and their multiples; for example, kg/h; km/h.

Table 6 Basic units used in hydrology and water resources with conversion factors

Length	1 m	= 3.2808 ft
	1 ft	= 0.3048 m
	1 in	= 25.4 mm
	1 mile	= 1.6093 km
	1 km	= 0.6214 mile
	1 mm	= 0.0394 in
	1 cm	= 0.3937 in
Area	$1\,in^2$	$= 645.2\,mm^2$
	$1\,ft^2$	$= 0.0929\,m^2$
	$1\,mile^2$	$= 2.590\,km^2$
	$1\,mile^2$	= 259 ha
	1 acre	$= 4047\,m^2$
	1 acre	= 0.4047 ha
	1 ha	= 2.4710 acres
	$1\,km^2$	$= 0.3861\,mile^2$
	$1\,cm^2$	$= 0.1550\,in^2$
	$1\,m^2$	$= 10.763\,ft^2$
Volume	$1\,m^3$	= 1000 l
	$1\,ft^3$	= 28.32 l
	$1\,ft^3$	$= 0.02832\,m^3$
	$1\,ft^3$	= 6.23 gal
	$1\,ft^3$	= 7.4805 gal US
	1 gal	= 4.546 l
	1 gal	$= 0.00455\,m^3$
	1 gal	$= 0.1605\,ft^3$
	1 gal	= 1.201 gal US
	1 gal US	= 3.7851 l
	1 gal US	$= 0.003785\,m^3$
	$1\,cm^3$	$= 0.06102\,in^3$
	1 acre-ft	$= 1233\,m^3$
Velocity	$1\,m\,s^{-1}$	$= 3.2808\,ft\,s^{-1}$
	$1\,ft\,s^{-1}$	$= 0.3048\,m\,s^{-1}$
	$1\,mile\,h^{-1}$	$= 1.6093\,km\,h^{-1}$
	$1\,mile\,h^{-1}$	$= 0.447\,m\,s^{-1}$
Discharge (flow rate)	$1\,m^3\,s^{-1}$	= 86.400 Mld
	$1\,m^3\,s^{-1}$	$= 35.3147\,ft^3\,s^{-1}$
	$1\,m^3\,s^{-1}$	= 19.00526 Mgd
	1 Mld	$= 0.0115\,m^3\,s^{-1}$
	1 Mld	$= 11.574\,l\,s^{-1}$
	1 Mld	= 0.22 Mgd
	1 Mld	= 0.26 Mgd US
	$1\,ls^{-1}$	= 0.019 Mgd
	1 Mgd	= 4.546 Mld
	1 Mgd	$= 0.0526\,m^3\,s^{-1}$
	1 Mgd	$= 52.616\,l\,s^{-1}$
	1 Mgd	$= 4546\,m^3\,d^{-1}$
	$1\,ft^3\,s^{-1}$	$= 0.0283\,m^3\,s^{-1}$
	$1\,ft^3\,s^{-1}$	$= 28.32\,l\,s^{-1}$
	1 Mgd US	$= 3785\,m^3\,d^{-1}$
	1 Mgd US	$= 0.0438\,l\,m^3\,s^{-1}$
	$1\,ft^3\,s^{-1}\,mile^{-2}$	$= 0.01094\,m^3\,s^{-1}\,km^{-2}$
Density	Water	$= 1000\,kg\,m^{-3}$
Mass	1 lb	= 453.6 g
	1 lb	= 0.4536 kg
	1 ton	= 1016.05 kg
	1 ton	= 1.016 tonne
Pressure (head of water)	1 mm Hg	$= 1.333 \times 10^2$ Pa
Gravitational acceleration (g) 9.807 $m\,s^{-2}$		$= 32.175\,ft\,s^{-2}$

Temperature
(Fahrenheit temperature in °F) $= \frac{9}{5}$(Celsius temperature in °C) + 32

Temperature difference
(Temperature difference in °F) $= \frac{9}{5}$(Temperature difference in °C)

Table 7 Symbols, units and conversion factors (WMO)

I	II	III	IV	V	VI	VII
			Units		Conversion factor[a]	
Item	Element	Symbol	Recommended	Also in use		Remarks
1	Acceleration due to gravity	g	m s^{-2}	ft s^{-2}	0.305	ISO
2	Albedo	r	Expressed as a decimal			
3	Area (cross-sectional) drainage basin)	A	m^2	ft^2	0.0929	ISO
			km^2	acre	0.00405	ISO
				ha	0.01	
				mile^2	2.59	
4	Chemical quality		mg l^{-1}	ppm	~1	For dilute solutions
5	Chézy coefficient [$v(R_\mathrm{h}S)^{-1/2}$]	C	m$^{1/2}$ s^{-1}	ft$^{1/2}$ s^{-1}	0.552	ISO
6	Conveyance	K	m^3 s^{-1}	ft^3 s^{-1}	0.0283	ISO
7	Degree day	D	degree day	degree day	Conversion formula °C = 5/9 (°F − 32)	Col. IV is based on °C scale and Col. V on °F scale
8	Density	p	kg m^{-3}	lb ft^{-3}	16.0185	ISO
9	Depth, diameter, thickness	d	m	ft	0.305	ISO
			cm	in	2.54	
10	Discharge (river flow) (wells)	Q	m^3 s^{-1}	ft^3 s^{-1}	0.0283	ISO
		Q_we	l s^{-1}	gal (US) min^{-1}	0.063	
	(unit area QA^{-1}, or partial)	q	m^3 s^{-1} km^2	ft^3 s^{-1} mile^{-2}	0.0109	ISO
			l s^{-1} km^{-2}		10.9	
11	Drawdown	s	m	ft	0.305	
			cm		30.5	
12	Dynamic viscosity, (absolute)	η	N s m^{-2}			ISO Pa, s, kg m^{-1} s^{-1} also in use
13	Evaporation	E	mm	in	25.4	
14	Evapotranspiration	E_T	mm	in	25.4	
15	Froude number	Fr	Dimensionless number			ISO
16	Head, elevation	z	m	ft	0.305	ISO
17	Head, pressure	h_p	m	kg (force) cm^{-2}	10.00	
				lb (force) in^{-2}	0.705	
18	Head, static (water level)	h	cm	ft	30.05	ISO
			m		0.305	
	= $z + h_\mathrm{p}$	h				
19	Head, total	H	m	ft	0.305	ISO
	= $z + h_\mathrm{p} + h_\mathrm{v}$					
20	Head, velocity	h_v	cm	ft	30.5	
			m		0.305	
	= $v^2(2g)^{-1}$					
21	Hydraulic conductivity (permeability)	K	cm s^{-1}	m d^{-1}	0.00116	
				ft min^{-1}	0.508	
22	Hydraulic diffusity = TC_s^{-1}	D	cm^2 s^{-1}			
23	Hydraulic radius = AP_w^{-1}	R_h	m	ft	0.305	ISO
24	Ice thickness	d_g	cm	in	2.54	
25	Infiltration	f	mm	in	25.4	
26	Infiltration rate	I_f	mm h^{-1}	in h^{-1}	25.4	
27	Intrinsic permeability	k	10^{-8} cm^2	Darcy	0.987	
28	Kinematic viscosity	v	m^2 s^{-1}	ft^2 s^{-1}	0.0929	ISO
29	Length	l	cm	in	2.54	ISO
			m	ft	0.305	
			km	mile	1.609	
30	Manning's coefficient = $R_\mathrm{h}^{2/3} S^{1/2} v^{-1}$	n	s m$^{-1/3}$	s ft$^{-1/3}$	1.486	ISO $l/n = k$ roughness coefficient can also be used
31	Mass	m	kg	lb	0.454	ISO
			g	oz	28.35	
32	Porosity	n	%			α may also be used if needed
33	Precipitation	P	mm	in	25.4	
34	Precipitation intensity	Ip	mm h^{-1}	in h^{-1}	25.4	

Table 7 Continued

I	II	III	IV	V	VI	VII
			Units		Conversion factor[a]	
Item	Element	Symbol	Recommended	Also in use		Remarks
35	Pressure	p	Pa	hPa	100.0	See also Head, pressure
				mm Hg	133.3	
				in Hg	3386.0	
36	Radiation[b] (quantity of radiant energy per unit area)	R	J m^{-2}	ly	4.186×10^4	
37	Radiation intensity[b] (flux per unit area)	I_R	J m^{-2} s^{-1}	ly min^{-1}	697.6	
38	Radius of influence	r_2	m	ft	0.305	
39	Recession coefficient	C_r		Expressed as a decimal		
40	Relative humidity (moisture)	U	%			
41	Reynolds number	R_e		Dimensionless number		ISO
42	Runoff	R	mm	in	25.4	
43	Sediment concentration	c_s	kg m^{-3}	ppm	Depends on density	
44	Sediment discharge	Q_s	t d^{-1}	ton (US) d^{-1}	0.907	
45	Shear stress	τ	Pa			ISO
46	Slope (hydraulic, basin)	S		Dimensionless number		ISO
47	Snow cover	A_n	%			
48	Snow depth	d_n	cm	in	2.54	
49	Snow melt	M	mm	in	25.4	Normally expressed as daily
50	Soil moisture	U_s	% volume	% mass	Depends on density	
51	Soil moisture deficiency	U'_s	mm	in	25.4	
52	Specific capacity = $Q_{we}s^{-1}$	C_s	m^2 s^{-1}	ft^2 s^{-1}	0.0929	
53	Specific conductance	K	μS cm^{-1}			at $\theta = 25°C$
54	Specific yield	Y_s		Expressed as decimal		
55	Storage	S	m^3	ft^3	0.0283	
56	Storage coefficient (groundwater)	C_S		Expressed as a decimal		
57	Sunshine	n/N		Expressed as a decimal		Actual (n)/possible (N) hours
58	Surface tension	σ	N m^{-1}			ISO
59	Temperature	θ	°C	°F	Conversion formula °C $= \frac{5}{9}$ (°F − 32)	ISO t also in use
60	Total dissolved solids	m_d	mg l^{-1}	ppm	~1	For dilute solutions
61	Transmissivity	T	m^2 d^{-1}	ft^2 d^{-1}	0.0929	
62	Vapor pressure	e	Pa	hPa	100.0	
				mm Hg	133.3	
					3386.0	
63	Velocity (water)	v	m s^{-1}	ft s^{-1}	0.305	ISO
64	Volume	V	m^3	ft^3	0.0283	ISO
				acre ft	1230.0	
65	Water equivalent of snow	w_n	mm	in	25.4	
66	Weber number	W_e		Dimensionless number		
67	Wetted perimeter	P_w	m	ft	0.305	
68	Width (cross-section, basin)	b	m	ft	0.305	ISO
			km	mile	1.609	
69	Wind speed	u	m s^{-1}	km h^{-1}	0.278	
				mile h^{-1}	0.447	
				k_n (or kt)	0.514	
70	Activity (amount of radioactivity)	A	Bq (Becquerel)	Ci (Curie)	3.7×10^{10}	IAEA
71	Radiation fluence (or energy fluence)	F	J m^{-2}	erg cm^{-2}	10^3	IAEA
72	Radiation flux intensity (or energy flux intensity)	I	J m^{-2} s^{-1}	erg cm^{-2} s^{-1}	10^3	IAEA

NOTE: Where international symbols exist these have been used where appropriate and are indicated as ISO in the last column.

[a] Col. IV = Conversion factor (Col. VI) × Col. V.

[b] General terms. For detailed terminology and symbols see the WMO *Guide to Meterological Instruments and Methods of Observation* (WMO) No. 8).

Definitions of SI base units and supplementary units

Base units

Metre
The meter is the length equal to 1 650 763.73 wavelengths in vacuum of the radiation corresponding to the transition between the levels $2p_{10}$ and $5d_5$ of the ^{86}Kr atom.

Kilogram
The kilogram is the unit of mass; it is equal to the mass of the international prototype of the kilogram.

Second
The second is the duration of 9 192 631 770 periods of the radiation corresponding to the transition between the two hyperfine levels of the ground state of the ^{133}Cs atom.

Ampere
The ampere is that constant current which, if maintained in two straight parallel conductors of infinite length, of negligible circular cross-section, and placed 1 meter apart in vacuum, would produce between these conductors a force equal to 2×10^{-7} newton per meter of length.

Kelvin
The kelvin, unit of thermodynamic temperature, is the fraction 1/273.16 of the thermodynamic temperature of the triple point of water.

The Thirteenth CGPM (1967, Resolution 3) also decided that the kelvin and its symbol K should be used to express an interval or a difference of temperature. In addition to the thermodynamic temperature (symbol T) expressed in kelvins, use is also made of Celsius temperature (symbol t) defined by the equation $t = T - T_0$ where $T_0 = 273.15\,K$ by definition.

The Celsius temperature is in general expressed in degrees Celsius (symbol °C). The unit 'degree Celsius' is thus equal to the unit 'kelvin' and an interval or a difference of Celsius temperature may also be expressed in degrees Celsius.

Mole
The mole is the amount of substance of a system which contains as many elementary entities as there are atoms in 0.012 kilogram of ^{12}C. When the mole is used, the elementary entities must be specified and may be atoms, molecules, ions, electrons, other particles/or specified groups of such particles.

Candela
The candela is the luminous intensity, in the perpendicular direction, of a surface of 1/600 000 square meter of a black body at the temperature of freezing platinum under a pressure of 101 325 newtons per square meter.

Supplementary units

Radian
The radian is the angle between two radii of a circle which cut off on the circumference an arc equal in length to the radius.

Steradian
The steradian is the solid angle which, having its vertex in the center of a sphere, cuts off an area of the surface of the sphere equal to that of a square with sides of length equal to the radius of the sphere.

Units used in hydrology and water resources

Table 6 lists the basic units used in hydrology and water resources, together with conversion factors.

Table 8 Miscellaneous symbols

Item	Element	Symbol	Remarks
1	Concentration	c	ISO
2	Coefficient (in general)	C	ISO
3	Difference	Δ	ISO, values expressed in same units
4	Inflow	I	
5	Lag time	Δt	Various units
6	Load	L	
7	Number of (or rank)	m	ISO
8	Outflow	O	
9	Recharge	f	See Infiltration in Table 7
10	Total number	N	

Table 9 Recommended units appearing in Table 7

Item	Element	Symbol	Remarks
1	centimeter	cm	ISO
2	day	d	ISO
3	degree Celsius	°C	ISO
4	gram	g	ISO
5	hectare	ha	
6	hectopascal	hPa	ISO
7	hour	h	ISO
8	joule	J	ISO
9	kilogram	kg	ISO
10	kilometer	km	ISO
11	knot	kn, kt	
12	liter	l	ISO
13	meter	m	ISO
14	microsiemens	μS	
15	milligram	mg	ISO
16	millimeter	mm	ISO
17	minute	min	ISO
18	newton	N	ISO
19	parts per million	ppm	
20	pascal	Pa	ISO
21	percentage	%	
22	second	s	ISO
23	tonne (metric ton)	t	ISO
24	year	a	ISO
25	bequerel	Bq	IAEA

Recommended symbols, units and conversion factors

Table 7 lists symbols, units and conversion factors recommended by the WMO. Table 8, lists some miscellaneous symbols and Table 9 gives the recommended units appearing in Table 7.

Source

World Meteorological Organization, 1994. *Guide to Hydrological Practices*, 5th edn. WMO, Geneva.

Bibliography

BS 5555, 1993. *SI Units*. British Standards Institution, London.
Herschy, R.W., 1995. *Streamflow Measurement*, Chapman & Hall, London.
ISO 1000, 1992 SI units.

A

ACCESS TO AND ACCOUNTABILITY OF WATER RESOURCES

Few issues have a greater impact on our lives and on the life of the planet than the management of our most important natural resources: water. Today we have a new appreciation for the role of water in our lives, our economies and our ecosystem. Water is our lifeblood. Human beings, like other animals and plants, are made mostly of water. We need water to maintain basic health and sanitation. Some 8% of the world's freshwater supplies are used for this purpose. We need adequate supplies of water to feed ourselves. Agriculture accounts for some 63% of the world's use of freshwater (about 70% in the developing countries), and a third of the world's food crops are produced by irrigated agriculture. We also need water to develop and maintain vibrant economies. Industries use about a fifth of the world's freshwater supplies, often as a vital part of the production process. Factories use water for cooling, processing, and generating steam to run equipment and as a transporting agent. Finally, most animal and plant species depend on freshwater ecosystems, which are also important for maintaining regional weather patterns and even global climate.

In a world of some 6000 million people, more than one-third do not have safe drinking water and a quarter do not have sanitation (Figure A1). Some 50 000 deaths occur every day from waterborne diseases. As part of this water tragedy, 3 million children die every year under the age of five (or one child every 10 s). Only a fraction of the surface water of the Earth is of use to humanity, in fact about 0.3%. Such an important fraction, however, has never been precisely measured and no efficient management of water resources can be fully exploited until such an audit is seriously carried out. In most, if not all, developing countries where the demand for water is seriously overstretched, there are insufficient hydrometric stations to audit the situation.

The United Nations established the International Drinking Water Supply and Sanitation Decade (1981–1990) on the basis that 'All people have the right to have access to drinking water in quantity and of a quality equal to their basic needs'.

The aim, of course, was too ambitious and at the end of the decade, although some 700 million more people had been supplied with

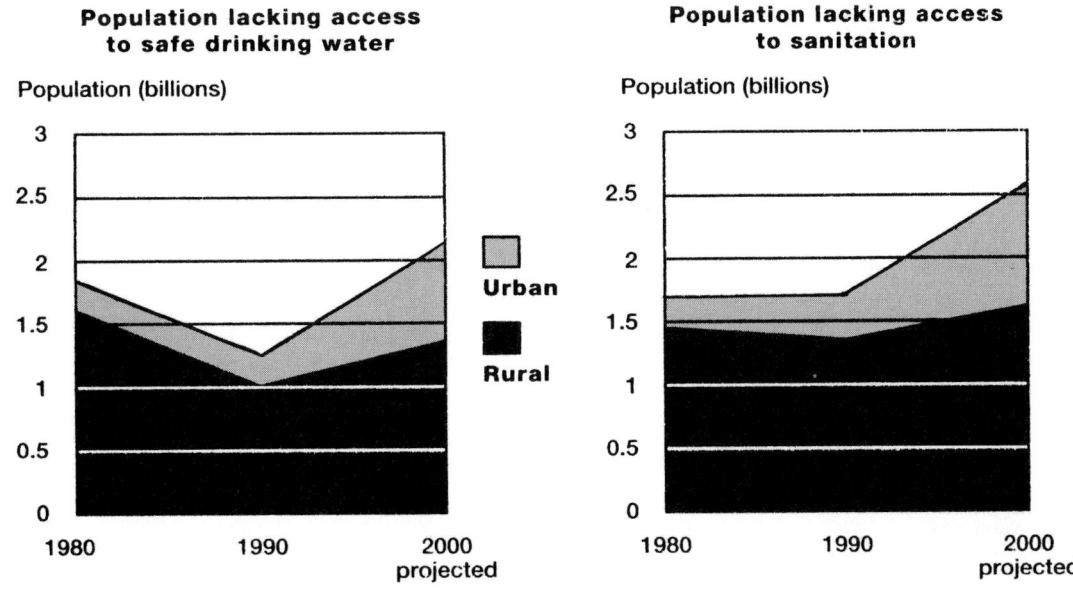

Figure A1 Population lacking access to (a) safe drinking water and (b) sanitation. (Source: *Sustaining Water: Population and the Futures of Renewable Water Supplies*, Population Action International, 1993.)

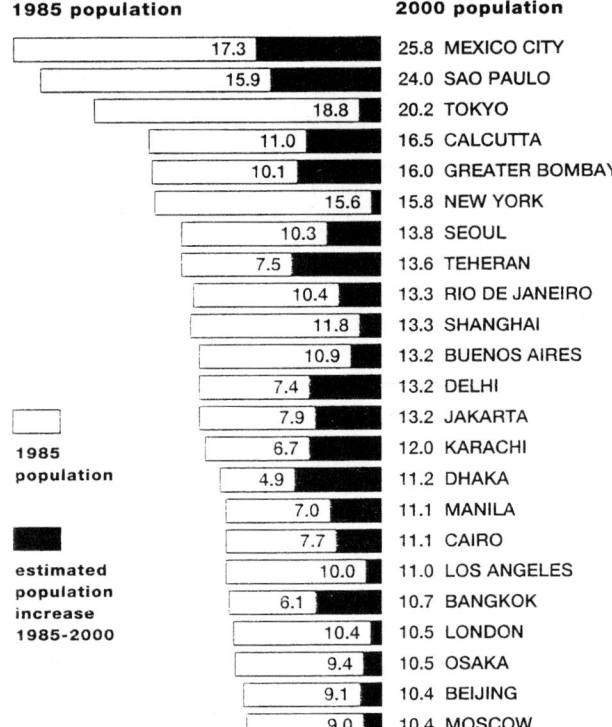

1985 population

2000 population

1985	2000	City
17.3	25.8	MEXICO CITY
15.9	24.0	SAO PAULO
18.8	20.2	TOKYO
11.0	16.5	CALCUTTA
10.1	16.0	GREATER BOMBAY
15.6	15.8	NEW YORK
10.3	13.8	SEOUL
7.5	13.6	TEHERAN
10.4	13.3	RIO DE JANEIRO
11.8	13.3	SHANGHAI
10.9	13.2	BUENOS AIRES
7.4	13.2	DELHI
7.9	13.2	JAKARTA
6.7	12.0	KARACHI
4.9	11.2	DHAKA
7.0	11.1	MANILA
7.7	11.1	CAIRO
10.0	11.0	LOS ANGELES
6.1	10.7	BANGKOK
10.4	10.5	LONDON
9.4	10.5	OSAKA
9.1	10.4	BEIJING
9.0	10.4	MOSCOW

1985 population

estimated population increase 1985–2000

Figure A2 The mega-cities. (Source: The World Environment 1972–1992, UNEP, UNFRA 1991.)

drinking water, the number without remained the same, due to the increase in population in the countries concerned.

The pace of urban expansion in the developing world is going to stretch planners to the limit, especially in the provision of water. In 1950 there were just two cities in the world with more than 8 million people: London and New York. By the year 2000 there will be 23 (Figure A2). The largest are already huge: Mexico City has now over 20 million people and faces the prospect of exhausting its water supply by the year 2000. Outright shortage of water is the first of many problems. History tells of ancient and not-so-ancient cities which drank up their surrounding water and perished – Babylon and Persepolis in the Middle East, and Fatehpur Sikri in India. In China at least 50 cities face acute shortages as the water table drops by 1–2 m a year. By the year 2000, Beijing will suffer a daily shortfall of some 0.5 million m³. Having overdrawn traditional surface and underground sources, cities such as Amman, Delhi, Santiago and Mexico City are pumping water from increasing distances and up increasing heights. In both Jakarta and Bangkok, excessive pumping of groundwater has led to intrusion of seawater into the aquifers causing land

subsidence. India, the world's second most populous country, will experience water shortage early in the twenty-first century.

Abundant as water may appear to be, we also have a new appreciation for how little freshwater there is on the Earth. Less than 3% of the world's water is freshwater, and most of this is in the ground, ice caps, and glaciers. Although enough precipitation falls each year on the land surface of the earth to cover the United States to a depth of 5 m or to fill all lakes, rivers and reservoirs 50 times over, about two-thirds of this evaporates back into the atmosphere, and more than half of what remains flows unused to the sea. Rainfall is also highly variable; the same area can experience droughts one year and floods the next. Withdrawals and the cost of recovery vary widely, as does the quality of water sources.

Indeed, water is critically scarce in many places. Generally, a country or region will experience periodic water stress when supplies fall below 1700 m³ per person per year. The global average annual supply of renewable freshwater is about 7400 m³ per person per year. However, 22 countries have renewable water resources of less than 1000 m³ per person, and 18 have more than 2000. By and large Latin America is best endowed, while the Middle East and North Africa is where water is most scarce (Table A1). By 2025 as many as 52 countries inhabited by some 3 billion people will be plagued by water stress or chronic water scarcity.

Issues of scarcity have put water at the top of the international political agenda. Agreement on access to water is an important part of the peace accords between Israel and its neighbors. A water treaty has also helped to maintain peace between India and Pakistan. However, water politics are not confined to historically conflicted or dry areas. Today, nearly 40% of the world's people live in more than 200 river basins that are shared by more than two countries. Even within countries, conflicts over water are often bitter. As populations and demand for limited supplies of water increase, interstate and international frictions over water can be expected to intensify.

R.W. Herschy

Bibliography

EARSeL, 1993. Report on the Hydrology and Water Resources Workshop, Dundee.
Herschy, R.W., 1995. *Introduction to EARSeL Hydrology Workshop*, Basel, A.A. Balkema, Rotterdam.
Serageldin, I., 1995. *Towards Sustainable management of Water Resources*, The World Bank.
WaterAid (ed.), 1994. *Mega-slums: The Coming Crisis*, WaterAid, London.

Cross references

Drinking water and sanitation
Environment priorities for development: water
Floods: world's maximum observed
Groundwater
Hydrological cycle
International rivers
Irrigated land area: world
Natural disasters

Table A1 Availability of water, by region (World Bank, 1992)

Region	Annual internal renewable water resources		Percentage of population living in countries with scarce annual per capita resources	
	Total (m³ × 1000)	Per capita (m³ × 1000)	Less than 1000 m³	1000–2000 m³
Sub-Saharan Africa	3.8	7.1	8.0	16.0
East Asia and the Pacific	9.3	5.3	<1.0	6.0
South Asia	4.9	4.2	0.0	0.0
Eastern Europe and former USSR	4.7	11.4	3.0	19.0
Other Europe	2.0	4.6	6.0	15.0
Middle East and North Africa	0.3	1.0	53.0	18.0
Latin America and the Caribbean	10.6	23.9	<1.0	4.0
Canada and the United States	5.4	19.4	0.0	0.0
World	40.9	7.7	4.0	8.0

Source: World Bank 1992

ACCURACY

Introduction

The accuracy or, more correctly, the error of a measurement of discharge may be defined as the difference between the measured flow and the true value. The true value of the flow is unknown and can only be ascertained (within close limits) by weighing or by volumetric measurement. An estimate of the true value has therefore to be made by calculating the uncertainty in the measurement, the uncertainty being defined as the range in which the true value is expected to lie expressed at the 95% confidence level (Figure A3). Although the error in a result is therefore, by definition, unknown, the uncertainty may be estimated if the distribution of the measured values about the true mean is known (Herschy, 1995).

The uncertainty in the measurement of an independent variable is normally estimated by taking N observations and calculating the standard deviation, where N is preferably at least 30. Using this procedure to calculate the uncertainty in a gauging for example, would require N consecutive measurements of discharge with different current meters at constant stage which is clearly impractical. An estimate of the true value has therefore to be made by examining all the various sources of error in the measurement.

In applying the theory of statistics to hydrological data it is assumed that the observations are independent random variables from a statistically uniform distribution. This ideal condition is seldom met in hydrometry or hydrology. For example, the measurement of velocity, by current meter, cannot be independent since the velocity itself is fluctuating with time due mainly to pulsations in flow. The measurement of stage at any particular moment in time is not strictly independent since the flow passing the reference gauge is continuous. Similarly other elements of the hydrological cycle (e.g. rainfall) are not independent. River flow is by nature, therefore, non-random, each hourly, daily, monthly and annual discharge being dependent on the previous hourly, daily, monthly and annual discharge. The cause of this is mainly the lag in the rainfall runoff relation, although in most catchments this may be an oversimplification. However, it is generally accepted by statisticians that the departure of hydrological data from the theoretical concept of errors is not serious provided care and attention are exercised in sampling techniques.

It should be stressed that the statistical analysis of hydrological data is only applicable if the field data have been obtained by acceptable

hydrometric principles and practices, i.e. to the relevant national or international standards. In this connection, a large responsibility is imposed on hydrometric observers who have to ensure that their equipment is all in good order, that the measurements are as precise as possible and that they are meticulously recorded. If the raw data are questionable for any reason it is the observer's duty to record this suspicion in the records. No amount of computer processing or statistical analysis will correct a wrong measurement. In addition, the observer should avoid getting into the position, for example, of producing a very precise measurement of the wrong water level, as may well happen under certain circumstances in the field. Statistical analysis is an aid to improving the presentation of the hydrometric data for the user's benefit, but the final quality of the data depends on the hydrologist.

Standard deviation

The most used statistical term in streamflow to estimate the uncertainty in a measurement is the standard deviation. Standard deviation is a measure of the dispersion or scatter of the observations about the arithmetic mean of the sample and is given by the equation

$$s_Y = \left[\frac{\sum\limits_{i=1}^{n} (Y_i - \bar{Y})^2}{N-1} \right]^{1/2} \tag{A1}$$

where s_Y is the standard deviation of the observations, $\bar{Y}$ is the arithmetic mean of the observations, Y_i, Y_i is the independent random observation of the variable Y and N is the number of observations.

In a linear regression, for example the stage–discharge curve, the term used is standard error of estimate which is numerically similar to standard deviation except that the linear regression replaces the arithmetic mean and $(N-1)$ is replaced by $(N-2)$. The standard error of estimate is given by the equation

$$S_e = \left(\frac{\sum d^2}{N-2} \right)^{1/2} \tag{A2}$$

where S_e is the standard error of estimate and d is the deviation of an observation from the computed value taken from the regression.

If a sample of streamflow data fits a bell-shaped symmetrical distribution, known as the normal distribution, then by statistical inference the dispersion of the observations about the mean is measured in standard deviations. Then, on average, 68% of the observations will lie within one standard deviation of the mean, 95% will lie within two standard deviations of the mean, and almost all (actually over 99%) will lie within three standard deviations of the mean (Figure A4). Generally the 95% level of confidence is used in hydrometry and provided the sample contains at least 30 observations the standard deviation is multiplied by 2. When the sample size is small, it is necessary to correct the statistical results that are based on a normal distribution by means of Student's t values. Student's t is a factor which compensates for the fact that the uncertainty in the standard deviation is large with small sample sizes. Student's t values at the 95% level are given in Table A2.

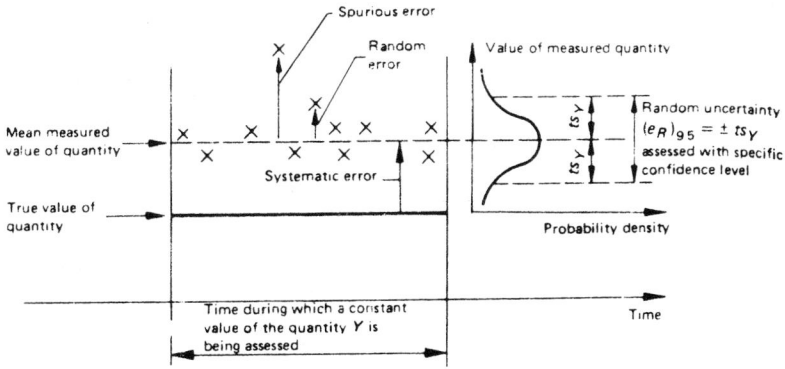

Figure A3 Basic statistical terms.

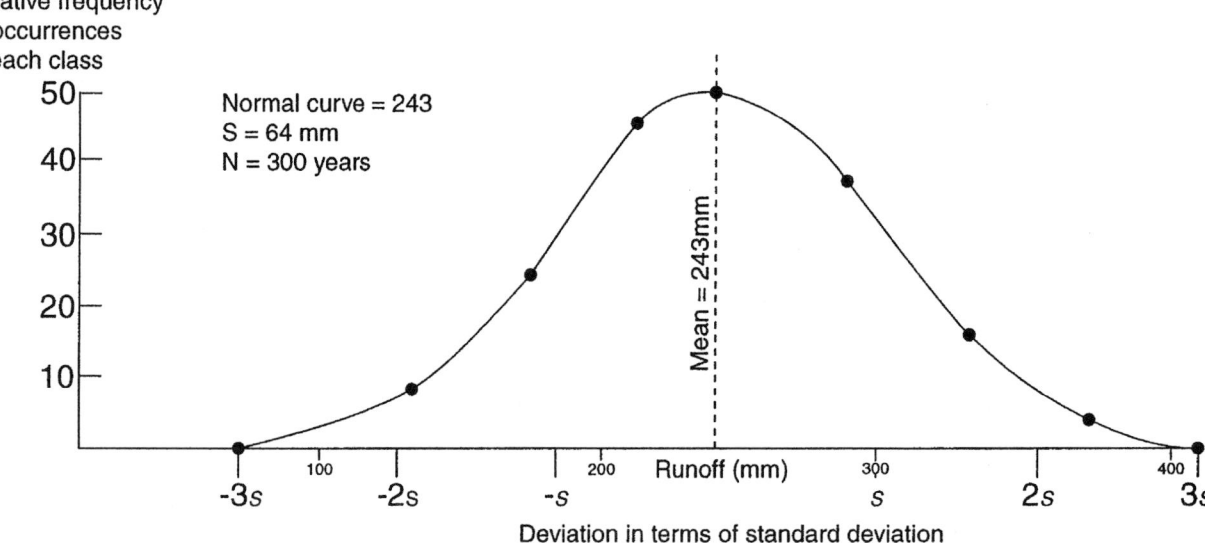

Figure A4 Normal distribution curve for annual runoff in River Thames at London for the period 1697–1996 (300 years). Values of runoff from 1697 to 1882 computed by mathematical model by P.C. Saxena, K.S. Rajagopalan and A. Choudhury.

Table A2 Values of Student's t at the 95% confidence level

Degrees of freedom	t
1	12.7
2	4.3
3	3.2
4	2.8
5	2.6
6	2.4
7	2.4
10	2.2
15	2.1
20	2.1
30	2.0
60	2.0
∞	1.96

Degrees of freedom for small samples

In a sample, the number of degrees of freedom is the sample size, N. When a statistic is calculated from the sample, the degrees of freedom associated with the statistic are reduced by one for every estimated parameter used in calculating the statistic. For example, from a sample of size N, $\bar{Y}$ is calculated and has N degrees of freedom, and the standard deviation is calculated from equation (A1) and has $(N-1)$ degrees of freedom because $\bar{Y}$ is used to calculate standard deviation. In calculating the standard error of estimate from a curve fit, the number of degrees of freedom which are lost is equal to the number of estimated coefficients for the curve; in the case of a stage–discharge relation C and n are the coefficients and 2 degrees of freedom are therefore lost and $(N-1)$ is replaced by $(N-2)$.

Percentage standard deviation

The standard deviation is normally expressed as a percentage in hydrometry by the following equation

$$s_Y = \frac{\left[\dfrac{\sum\limits_{i=1}^{n}(Y_i - \bar{Y})^2}{N-1}\right]^{1/2}}{\bar{Y}} \times 100 \qquad (A3)$$

Expressed in this form, the standard deviation is also known as the coefficient of variation (CV).

Standard deviation of the mean

The standard deviation of the mean, $s_{\bar{Y}}$, is an estimate of the uncertainty of the computed mean and expressed by the equation

$$s_{\bar{Y}} = \frac{s_Y}{\sqrt{N}} \qquad (A4)$$

Standard error of the mean relation

The standard error of the mean relation, S_{mr}, is similar to $s_{\bar{Y}}$ but refers to the uncertainty of the regression (for example, the stage–discharge curve). In this case, however, equation (A4) gives the value of S_{mr} at the centroid of the regression only and the value varies curvilinearly along the regression, being $s_Y/\sqrt{N}$ at the centre and a maximum at the extremes.

This is illustrated in the case of streamflow data in Figure A5 where the S_e limits are shown as parallel to the regression and the S_{mr} limits are shown as curves. Also shown in Figure A5 are the bell-shaped normal distributions for both S_e and S_{mr}. In theory the mean and standard deviation of the discharge observations are computed from samples of N discharges at M stages throughout the range (M and N both being at least 30) producing M normal distribution bell-shaped curves from MN discharge measurements. The standard deviation of the mean is computed from each of the M distributions and the points defining the mean, standard deviation and standard deviation of the mean for each of the M normal distributions joined to form the regression and the confidence limit curves of standard error of estimate and standard error of the mean. At the 95% level the standard error of estimate confidence limits will contain 95% of the discharges and the standard error of the mean confidence limits defines the uncertainty in the regression, the stage–discharge relation, that is the uncertainty of estimates of discharge from stage from the rating equation.

The application of standard deviation to streamflow data can be demonstrated by the following example. The annual runoff values in millimeters from the gauging station on the River Thames at London for the years 1697 to 1996, a total of 300 years (N) are shown in Figure A4 by means of a normal distribution.

Calculation gives the following values:

mean = 243 mm
standard deviation $S_Y = \pm 64$ mm
standard deviation of the mean

$$s_{\bar{Y}} = \frac{64}{\sqrt{(300)}} = \pm 4 \text{ mm}$$

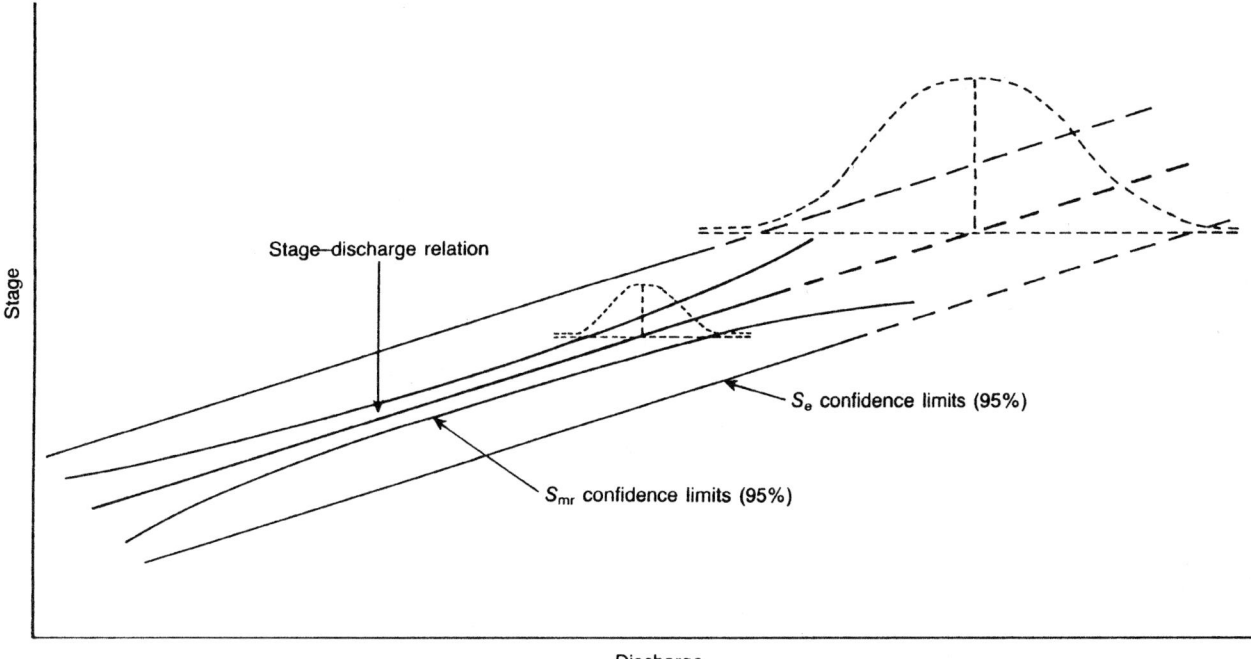

Figure A5 Illustration of confidence limits of a stage–discharge curve. The standard error of estimate confidence limits are straight and parallel on either side of regression and contain 95% of the current meter observations; the standard errors of the mean confidence limits are curved on either side of the regression, being narrowest at the center and maximum at the extremes; the stage–discharge relation will on average be within these limits for 95% of the time (19 years out of 20), assuming that there is no change in the hydraulic characteristics of the flow (e.g. change in control, change of datum).

At the 95% level

$$s_Y = \pm 128 \text{ mm}$$
$$s_{\bar{Y}} = \pm 8 \text{ mm}$$

From these results it can be stated that:

- The best estimate of the long-term annual runoff is the mean = 243 mm with an uncertainty of ±8 mm.
- The probability is that the mean in future will lie between 235 and 251 mm in 19 years out of 20.
- The uncertainty, or dispersion, of the annual runoff values about the mean is ±128 mm.

All we can conclude with certainty, however, is that the annual values, on average 19 years out of 20, will be within the confidence band 115–371 mm (243 − 128 mm and 243 + 128 mm), a 50% scatter on either side of the mean. We have more confidence in predicting the long-term annual mean, however, which on average, in 19 years out of 20, will be contained within the narrow confidence band of 235–251 mm.

Nature of errors

Errors of observation are usually grouped as random (or stochastic), systematic and spurious.

Random errors

Random errors are sometimes referred to as experimental errors and the observations deviate from the mean in accordance with the laws of chance such that the distribution usually approaches a normal distribution. They are the most important errors to be considered in streamflow (Figure A3).

Systematic errors

Systematic errors are those which cannot be reduced by increasing the number of observations if the instruments and equipment remain unchanged. In streamflow, systematic errors may be present in the water level recorder, in the reference gauge or datum, and in the current meter. These errors may be generally small but in some cases their effect may cause a systematic error in the stage–discharge

relation. It is also possible that the crest of a weir may be levelled-in incorrectly to the station datum, so producing a systematic error in head measurement which might have a serious effect on low values of discharge.

Spurious errors

These are human errors or instrument malfunction and cannot be statistically analysed. The observations must therefore be discarded (Figure A3).

It is sometimes difficult to distinguish between random and systematic errors in streamflow and some errors may be a combination of the two. For instance, where a group rating is used for current meters, each of the meters forming the group may have a plus or minus systematic error which is randomized to obtain the uncertainty in the group rating. A construction error in the length of a weir crest may be randomized by measuring the crest say N times after construction to obtain the mean and standard deviation, each measurement being independent of the others.

Theory of errors

If a quantity Q is a function of several measured quantities x, y, z, ..., the error in Q due to errors δx, δy, δz, ... in x, y, z, ..., respectively, is given by

$$\delta Q = \frac{\partial Q}{\partial x}\delta x + \frac{\partial Q}{\partial y}\delta y + \frac{\partial Q}{\partial z}\delta z + \cdots \quad (A5)$$

The first term in equation (A5), $(\partial Q/\partial x)\delta x$, is the error in Q due to an error δx in x only (i.e. corresponding to δy, δz, ..., all being zero). Similarly the second term $(\partial Q/\partial y)\delta y$ is the error in Q due to an error δy in y only. Squaring gives

$$\delta Q^2 = \left(\frac{\partial Q}{\partial x}\delta x\right)^2 + 2\frac{\partial Q}{\partial x}\frac{\partial Q}{\partial y}\delta x \delta y + \left(\frac{\partial Q}{\partial y}\delta y\right)^2 + \cdots \quad (A6)$$

Now the terms $(\partial Q/\partial x)(\partial Q/\partial y)\delta x \delta y$, etc, are covariance terms and, since they contain quantities which are as equally likely to be positive or negative, their algebraic sum may be conveniently taken as being either zero or else negligible as compared with the squared terms. Equation (A6) then becomes

$$\delta Q^2 = \left(\frac{\partial Q}{\partial x}\delta x\right)^2 + \left(\frac{\partial Q}{\partial y}\delta y\right)^2 + \left(\frac{\partial Q}{\partial z}\delta z\right)^2 + \cdots \tag{A7}$$

i.e. the error in Q, δQ, is the sum of the squares of the errors due to an error in each variable. Now

$$\frac{\partial Q}{\partial x} = yz, \quad \frac{\partial Q}{\partial y} = xz, \quad \frac{\partial Q}{\partial z} = xy, \tag{A8}$$

and

$$\delta Q = [(yz\delta z)^2 + (xz\delta y)^2 + (xy\delta z)^2 + \cdots]^{1/2} \tag{A9}$$

Dividing by $Q = xyz$

$$\frac{\delta Q}{Q} = \left[\left(\frac{yz}{xyz}\delta x\right)^2 + \left(\frac{xz}{xyz}\delta y\right)^2 + \left(\frac{xy}{xyz}\delta z\right)^2 + \cdots\right]^{1/2} \tag{A10}$$

and

$$\frac{\delta Q}{Q} = \left[\left(\frac{\delta x}{x}\right)^2 + \left(\frac{\delta y}{y}\right)^2 + \left(\frac{\delta z}{z}\right)^2 + \cdots\right]^{1/2} \tag{A11}$$

where $(\delta x)/x$, $(\delta y)/y$ and $(\delta z)/z$ are fractional values of the errors (standard deviations) in x, y and z, and if they are each multiplied by 100 they become percentage standard deviations. Let X_Q be the percentage standard deviation of Q and

X_x = percentage standard deviation of x
X_y = percentage standard deviation of y
X_z = percentage standard deviation of z

then

$$X_Q = \pm(X_x^2 + X_y^2 + X_z^2 + \cdots)^{1/2} \tag{A12}$$

which is generally referred to as the root-sum-square equation for the estimation of uncertainties. It is this equation that is employed to estimate the uncertainty in a current meter gauging.

Similarly, if

1. $$Q = \frac{x}{y}$$

then

$$X_Q = \pm(X_x^2 + X_y^2)^{1/2} \tag{A13}$$

2. $$Q = x \pm y$$

$$\delta Q^2 = \left(\frac{\partial Q}{\partial x}\delta x\right)^2 \pm \left(\frac{\partial Q}{\partial y}\delta y\right)^2$$

$$\delta Q = (\delta x^2 + \delta y^2)^{1/2}$$

$$X_Q = \frac{100\delta Q}{Q} = \frac{100}{Q}\left[\frac{x^2\delta x^2}{x^2} \pm \frac{y^2\delta y^2}{y^2}\right]^{1/2}$$

therefore

$$X_Q = \pm\frac{(x^2X_x^2 + y^2X_y^2)^{1/2}}{(x \pm y)} \tag{A14}$$

3. $$Q = Cbh^n \quad \text{(weir equation)}$$

then

$$X_Q = \pm(X_c^2 + X_b^2 + n^2X_h^2)^{1/2} \tag{A15}$$

which is the error equation for weirs, where C is the coefficient of discharge, b the length of crest, n the exponent of h (usually 3/2 for a weir and 5/2 for a V-notch) and h is the head over the weir.

Error equation

Velocity–area method

The general form of the working equation for computing discharge in the cross-section is

$$Q = \sum_{i=1}^{m} (b_i d_i \bar{v}_i) \tag{A16}$$

where Q is the total discharge in the cross-section, and b_i, d_i and $\bar{v}_i$ are the width, depth and mean velocity of the water in the ith of the m verticals or segments into which the cross-section is divided.

Equation (A16) assumes that a sufficient number of verticals have been taken in the cross-section but if this is not the case then equation (A16) should be multiplied by a factor F so that

$$Q = F\sum_{i=1}^{m} (b_i d_i \bar{v}_i) \tag{A17}$$

where $\bar{F}$ may be greater or less than unity. Equation (A17), therefore, requires to be optimized until sufficient verticals are employed so as to make F unity.

Let the following contributing random uncertainties, in making a single determination of discharge by current meter, be expressed as percentage standard deviations at the 95% confidence level:

X_b = uncertainty in width measurement;
X_d = uncertainty in depth measurement;
$X_{\bar{v}}$ = uncertainty in the mean velocity in the vertical;
X_Q = overall uncertainty in discharge;
X_m = uncertainty due to the limited number of verticals;

and let m be the number of verticals. Now

$$X_Q = \frac{\text{sum of the percentage errors in the segment discharges}}{\text{sum of the segment discharges}} \tag{A18}$$

Then

$$X_Q = \pm\left[\frac{\sum_{i=1}^{m} (b_i d_i \bar{v}_i)(Xb_i^2 + Xd_i^2 + X\bar{v}_i^2)^{1/2}}{\left(\sum_{i=1}^{m} b_i d_i \bar{v}_i\right)}\right] \tag{A19}$$

or

$$X_Q = \pm\left[\frac{\sum_{i=1}^{m} [(b_i d_i \bar{v}_i)^2(Xb_i^2 + Xd_i^2 + X\bar{v}_i^2)]}{\left(\sum_{i=1}^{m} b_i d_i \bar{v}_i\right)^2}\right]^{1/2} \tag{A20}$$

The random uncertainty, X_m, in using a limited number of verticals, has to be allowed for in equation (A20). In addition, the uncertainty in the mean velocity ($X_{\bar{v}}$) depends on the exposure time necessary to minimize the uncertainty due to pulsations in flow (X_e), the number of points taken in the vertical (X_p) and the uncertainty in the current meter rating (X_c). $X_{\bar{v}}^2$ should therefore be replaced by $(X_e^2 + X_p^2 + X_c^2)$ and the final equation now becomes

$$X_Q = \left[X_m^2 + \frac{\sum[(b_i d_i \bar{v}_i)^2(X_{b_i}^2 + X_{d_i}^2 + X_{e_i}^2 + X_{p_i}^2 + X_{c_i}^2)]}{\left(\sum_{i=1}^{m} b_i d_i \bar{v}_i\right)^2}\right]^{1/2} \tag{A21}$$

Now if the segment discharges ($b_i d_i \bar{v}_i$) are nearly equal and the random uncertainties X_{b_i} are nearly equal and of value X_b', and similarly for X_{d_i}, X_{e_i}, X_{p_i} and X_{c_i}, then

$$X_Q' = \pm\left[X_m^2 + \frac{1}{m}(X_b^2 + X_d^2 + X_e^2 + X_p^2 + X_c^2)\right]^{1/2} \tag{A22}$$

which is the simplified error equation. Note that the X' refers to a random uncertainty as distinct from a systematic uncertainty X'' – equation (A25).

Although the segment discharge or the contributing uncertainties in a gauging are seldom, if ever, equal, it has been found in practice that equation (A22) gives results which are not significantly different from those given by equation (A21). Where the verticals number 20 or more, the value of X_Q using either equation is generally less than 7%. The discrepancies therefore between the two equations become rather academic and it is clear that for routine gauging X_Q can be conveniently calculated from equation (A22). For special studies, however, which require more precise estimation of uncertainty X_Q, it is advisable to use equation (A21).

It will be evident that equations (A21) and (A22) can be employed to obtain a required value of X_Q by giving special consideration to the individual uncertainties in the equations. Individual uncertainties may usually be reduced by increasing the number of verticals, increasing the number of points in the verticals or, in the case of measuring structure, by reducing the uncertainty in the head measurement – equation (A15).

Systematic uncertainties may be combined by the root-sum-square method as in the case of random uncertainties unless the uncertainty is known or has been measured precisely, in which case it becomes a systematic error and should be added algebraically to the result. The uncertainty due to this source is then taken as zero. An example of this is a zero error E''_z in a gauge datum. If E''_z is not known, the uncertainties are taken as being random and the following equation is used:

$$X'_h = \pm\frac{100}{h}(E'^2_g + E'^2_z)^{1/2} \qquad (A23)$$

where X'_h is the percentage random uncertainty in head or stage measurement (95% level), E'_g is the random uncertainty in recorder reading (mm) and E'_z is the random uncertainty in gauge zero (mm).

If, however, E''_z is known then

$$X'_h = \frac{100}{h}(E'_g) \qquad (A24)$$

and the zero error E''_z is added algebraically to each reference gauge or recorder reading. If both the recorder and the gauge zero have known systematic errors then E''_g and E''_z should be added algebraically to each reference gauge reading.

In the velocity–area method the error equation for the systematic uncertainty X''_Q is

$$X''_Q = \pm(X''^2_b + X''^2_d + X''^2_c)^{1/2} \qquad (A25)$$

where X''_b is the percentage systematic uncertainty in the instrument measuring width, X''_d is the percentage systematic uncertainty in the instrument measuring depth and X''_c is the percentage systematic uncertainty in the current meter rating tank.

In weirs and flumes any systematic uncertainties are generally negligible as it is the practice to check carefully the dimensions after installation and amend any which do not conform to the design dimensions, thus randomizing any uncertainties in length of crest, width of throat, height of wing walls or divide walls, and gauge zero. Generally great care is exercised in levelling-in the crest level to the gauge zero. If the crest is level then these two values should be equal. Any departure, if it is measurable, is a systematic error; if it is not, the departure is estimated as a systematic uncertainty. If, in addition, the sign is unknown then the uncertainty has to be randomized [E'_z in equation (A23)]. This latter value is generally taken as ±3 mm. If a punched tape recorder or shaft encoder is used to record head or stage, E'_g is also taken as ±3 mm.

The overall random and systematic uncertainties in discharge are then combined by the root-sum-square method:

$$X_Q = \pm(X'^2_Q + X''^2_Q)^{1/2} \qquad (A26)$$

where X_Q is the overall uncertainty in a single determination of discharge.

Generally, systematic uncertainties are negligible compared with random uncertainties but it is important that they are investigated and if possible rectified.

The final presentation of the result of a single determination of discharge is then made by one of the following methods:

- discharge = $Q \pm X_Q$; random uncertainty = $\pm X'_Q$;
- discharge = Q; random uncertainty = $\pm X'_Q$; systematic uncertainty = $\pm X''_Q$.

Weirs and flumes

The error equation for the estimation of the uncertainty in a single determination of discharge for a weir, equation (A15), is:

if

$$Q = Cbh^n$$

then

$$X'_Q = \pm(X'^2_c + X'^2_b + n^2 X'^2_h)^{1/2} \qquad (A27)$$

where C is the coefficient of discharge; b is the length of crest; h is the gauged head; n is the exponent of h, usually $\frac{3}{2}$ for a weir and $\frac{5}{2}$ for a V-notch; X'_Q is the percentage random uncertainty in a single determination of discharge; X'_c is the percentage random uncertainty in the value of the coefficient of discharge; X'_b is the percentage random uncertainty in the measurement of the length of crest and X'_h is the percentage random uncertainty in the measurement of gauged head. All values of uncertainties are percentage standard deviations at the 95% level.

It is usually convenient to include any small systematic uncertainty components in the terms X'_c, X'_b and X'_h, in equation (A27). If the coefficient C, for example, has been established in the laboratory this is usually performed by a graphical relation of Q versus h. The uncertainty of the relation may therefore be taken as random, any systematic bias being due to systematic errors in the instrumentation. This should be insignificant in calibrations carried out in national laboratories (although it is possible for C to have a systematic error assigned to it when the calibration is applied in the field). It is also usual to investigate, on site, any suspicion of a systematic uncertainty in the gauge zero, or in the recorder, and correct these. If, however, an allowance is made in equation (A27) for zero error, E_z, this allowance should ensure that any systematic bias in the head measurement is included in X'_h.

An allowance of 0.1% in X'_b should more than compensate for any systematic error in the steel tape or other means for measuring the length of crest.

If, however, there is any suspicion of significant systematic errors in X_c, X_b or X_h the following equation is used to estimate the overall systematic uncertainty X''_Q

$$X''_Q = \pm(X''^2_c + X''^2_b + n^2 X''^2_h)^{1/2} \qquad (A28)$$

and the overall uncertainty calculated by the root-sum-square method as before [equation (A26)].

Other error propagation equations

In addition to the error propagation equations given in equations (A12)–(A15), the following equations are presented which have, or may have, application in estimating uncertainties in streamflow:

1.
$$v = \frac{1}{n}R^{2/3}S^{1/2} \qquad (A29)$$

$$X_v = \pm[X^2_n + \tfrac{4}{9}X^2_R + \tfrac{1}{4}X^2_s]^{1/2} \qquad (A30)$$

2.
$$v = kx^a y^b \qquad (A31)$$

$$X_v = \pm(a^2 X^2_x + b^2 X^2_y)^{1/2} \qquad (A32)$$

3.
$$v = \frac{1}{h} \qquad (A33)$$

$$X_v = \pm(X^2_h)^{1/2} \qquad (A34)$$

4.
$$W = Ax + By \qquad (A35)$$

$$X_w = \pm\frac{(A^2 x^2 X^2_x + B^2 y^2 X^2_y)^{1/2}}{(Ax + By)} \qquad (A36)$$

5.
$$W = Ax - By \qquad (A37)$$

$$X_w = \pm\frac{(A^2 x^2 X^2_x + B^2 y^2 X^2_y)^{1/2}}{(Ax - By)} \qquad (A38)$$

6.
$$Y = \frac{1}{t_1} - \frac{1}{t_2} \qquad (A39)$$

$$X_y = \pm\left[\left(\frac{t_1 t_2}{t_1 - t_2}\right)^2\left(\frac{X^2_{t_1}}{t^2_1} + \frac{X^2_{t_2}}{t^2_2}\right)\right]^{1/2} \qquad (A40)$$

Note: in the above equations X represents percentage standard deviation at the 95% level.

Uncertainty in the stage–discharge relation

The equation for the stage–discharge relation may be expressed in the general form

$$Q = C(h + a)^n \qquad (A41)$$

Table A3 Values required for calculation of S_e and S_{mr}

Observation no.	$h + a$	$\ln(h+a)(=x)$	$(x_i - \bar{x})^2$	Q_i	$\ln Q_i(=y_i)$	Q_c	$\ln Q_c(=y_c)$	$(y_i - y_c)^2$	$2S_{mr}$
1	0.157	−1.8515	1.8621	2.463	0.9014	2.323	0.8428	0.00342	1.97
2	0.158	−1.8452	1.8450	2.325	0.8437	2.345	0.8523	0.00007	1.96
3	0.188	−1.6713	1.4028	2.923	1.0726	3.060	1.1184	0.00209	1.80
4	0.192	−1.6502	1.3533	3.242	1.1726	3.160	1.506	0.00065	1.78
5	0.219	−1.5187	1.0646	3.481	1.3457	3.865	1.3520	0.00003	1.66
6	0.259	−1.3509	0.7465	4.995	1.6084	4.996	1.6086	0.00000	1.52
7	0.278	−1.2801	0.6292	5.410	1.6882	5.568	1.7170	0.00083	1.46
8	0.279	−1.2765	0.6235	5.422	1.6905	5.598	1.7224	0.00101	1.46
9	0.287	−1.2483	0.5797	5.883	1.7721	5.846	1.7658	0.00004	1.44
10	0.295	−1.2208	0.5386	6.154	1.8171	6.097	1.8078	0.00008	1.42
11	0.348	−1.0556	0.3234	7.376	1.9982	7.851	2.0606	0.00389	1.30
12	0.405	−0.9039	0.1739	9.832	2.2856	9.902	2.2927	0.00005	1.22
13	0.433	−0.8370	0.1226	11.321	2.4266	10.968	2.3950	0.00099	1.19
14	0.461	−0.7744	0.0826	12.372	2.5154	12.072	2.4909	0.00060	1.16
15	0.465	−0.7657	0.0777	11.825	2.4702	12.233	2.5041	0.00115	1.16
16	0.501	−0.6911	0.0417	13.826	2.6266	13.711	2.6182	0.00007	1.14
17	0.511	−0.6714	0.0340	14.102	2.6463	14.132	2.6484	0.00000	1.14
18	0.606	−0.5009	0.0002	19.020	2.9455	13.345	2.9094	0.00130	1.11
19	0.624	−0.4716	0.0002	19.970	2.9852	19.185	2.9541	0.00096	1.11
20	0.632	−0.4589	0.0008	20.280	3.0096	19.563	2.9736	0.00129	1.11
21	0.681	−0.3842	0.0105	21.204	3.0542	21.931	3.0879	0.00113	1.12
22	0.731	−0.3133	0.0301	23.996	3.1779	24.442	3.1963	0.00033	1.13
23	0.926	−0.0769	0.1681	36.242	3.5902	35.098	3.5581	0.00102	1.22
24	1.225	0.2029	0.4758	54.591	3.9999	53.855	3.9863	0.00018	1.38
25	1.411	0.3443	0.6909	67.327	4.2096	66.859	4.2026	0.00004	1.49
26	1.646	0.4983	0.9706	79.050	4.3701	84.631	4.4383	0.00465	1.62
27	1.895	0.6392	1.2681	110.783	4.7076	104.989	4.6538	0.00288	1.74
28	2.517	0.9231	1.9881	162.814	5.0926	162.095	5.0882	0.00001	2.02
29	3.150	1.1474	2.6709	227.600	5.4276	228.478	5.4314	0.00001	2.24
30	3.165	1.1522	2.6866	228.800	5.4328	230.145	5.4387	0.00003	2.25
31	3.191	1.1603	2.7133	228.500	5.4513	233.044	5.4512	0.00038	2.26
32	3.225	1.1709	2.7483	236.600	5.4664	236.854	5.4674	0.00000	2.27

$$\overline{\ln(h+a)} = -0.4869; \quad \Sigma(x_i - \bar{x})^2 = 27.9238; \quad \Sigma(y_i - y_c)^2 = 0.029\,18$$

$(y_i - y_c)^2 = (\ln Q_i - \ln Q_c)^2$
$(x_i - x^c)^2 = [\ln(h+a) - \overline{\ln(h+a)}]^2$
Discharge equation: $Q = 39.479(h - 0.115)^{1.53}$.

where C is a coefficient, h is the stage, a is a datum correction denoting the value of stage at zero flow in order to linearize the relation and n is an exponent usually in the range 1.5–2.5.

In order to estimate the uncertainty in the relation, equation (A41) is linearized by a logarithmic transformation of the form

$$\ln Q = \ln C + n \ln(h+a) \qquad (A42)$$

The procedure is then one of estimating S_e, the standard error of estimate and S_{mr}, the standard error of the mean (Figure A5).

S_e is calculated from

$$S_e = \left[\frac{\Sigma(\ln Q_i - \ln Q_c)^2}{N-2}\right]^{1/2} \qquad (A43)$$

where Q_i is the current meter observation and Q_c is the discharge taken from the rating curve corresponding to Q_i and $(h+a)$, where $Q_c = C(h+a)^n$.

Similarly, S_{mr} may be determined from

$$S_{mr} = \pm t S_e \left(\frac{1}{N} + \frac{[\ln(h+a) - \overline{\ln(h+a)}]^2}{\Sigma[\ln(h+a) - \overline{\ln(h+a)}]^2}\right)^{1/2} \times 100 \qquad (A44)$$

where t is Student's t correction for the sample size at the 95% confidence level for N gaugings and may be taken as 2 for 20 or more gaugings.

S_{mr} is calculated for each gauging on the stage–discharge curve at the relevant value of $(h+a)$. The limits will be curved, having a minimum value at the average value of $\ln(h+a)$. If the stage–discharge relation comprises one or more break points, S_e and S_{mr} are calculated for each segment and $(N-2)$ degrees of freedom are allowed for each segment. At least 20 current meter observations

should be available in each range before a statistically acceptable estimate can be made of S_e and S_{mr}. A typical stage–discharge relationship is tabulated in Table A3.

Substituting in equation (A43) for S_e from Table A3 gives:

$$S_e = \left(\frac{0.029}{30}\right)^{1/2} = 0.031$$

Therefore $t S_e = 2 \times 0.031 \times 100 = 6.2\%$ at the 95% level

This equation defines two parallel straight lines on either side of the stage–discharge curve and distant $2S_e$ (6.2%) from it, as shown in Figure A5. Therefore 95% of all the current meter observations, on average, will be contained within these limits. Similarly S_{mr} may be found from equation (A44) and substituting for S_{mr} from Table A3 gives

$$2S_{mr} = 6.2\left(0.031\,25 + \frac{[\ln(h - 0.115) + 0.4869]^2}{27.9238}\right)^{1/2}$$

The value of $2S_{mr}$ (95% level) for each current meter observation at stage $(h+a)$ may be so calculated and the results plotted on each side of the stage–discharge curve to give symmetrical limits with a minimum value as before at the average value of $\ln(h+a)$.

Therefore substituting for observation number 1 in Table A3 gives

$$2S_{mr} = 6.2\left(0.031\,25 + \frac{(-1.8515 + 0.4869)^2}{27.9238}\right)^{1/2} = 1.94\%$$

Similarly for observation number 18:

$$2S_{mr} = 6.2\left(0.031\,25 + \frac{(-0.5009 + 0.4869)^2)}{27.9238}\right)^{1/2} = 1.10\%$$

and for observation number 32:

$$2S_{mr} = 6.2\left(0.031\,25 + \frac{(1.1709 + 0.4869)^2)}{27.9238}\right)^{1/2} = 2.23$$

A summary for all the S_{mr} values so calculated is given in Table A3. The equation gives inner symmetrical curved limits on either side of the stage–discharge curve (as shown in Figure A5).

The above procedure should be satisfactory for most stage–discharge relations. If asymmetrical limits are preferred, the procedure is as follows:

$$S_{mr}\ \text{(upper 95\% confidence limit)} = 100(e^z - 1) \quad (A45)$$

$$S_{mr}\ \text{(lower 95\% confidence limit)} = 100(1 - e^{-z}) \quad (A46)$$

where z is the right-hand side of equation (A44) excluding the factor of 100.

Using the same example, therefore, for observation number 1 (where $z = 0.0194$), the upper confidence for S_{mr} becomes:

$$100(e^{0.0194} - 1) = 100(1.0196 - 1) = 1.96\%$$

and the lower confidence limit becomes

$$100\left(1 - \frac{1}{1.0196}\right) = 1.92\%.$$

As noted above, S_{mr} will have a minimum value when $\ln(h + a) = \overline{\ln(h + a)}$ and equation (14.46) reduces to

$$S_{mr} = \frac{tS_e}{\sqrt{N}} \quad (A47)$$

$$= \frac{6.2}{(32)^{1/2}} = \pm 1.1\%$$

which is the value given from equation (A44) for observation numbers 18–20 in Table A3 where $\ln(h + a)$ is approximately equal to $\overline{\ln(h + a)}$.

Uncertainty in the daily mean discharge

The value of discharge most commonly required for design and planning purposes is the daily mean discharge, which may be calculated by taking the average of the number of observations of discharge during the 24 h period.

The uncertainty in the daily mean discharge for a velocity–area station may be calculated from the following equation:

$$X_{dm} = \frac{\Sigma[(S_{mr}^2 + n^2X^2(h + a))^{1/2}Q_h]}{\Sigma Q_h} \quad (A48)$$

where X_{dm} is the uncertainty in the daily mean discharge (95% confidence level) and Q_h is the discharge corresponding to $(h + a)$.

The procedure is as follows:

1. Calculate $X(h + a)$ for each of the N values of discharge used to compute the daily mean, from equation (A23).
2. Calculate X_{dm} from equation (A48) using the appropriate value of S_{mr}.

A typical calculation for hourly values of discharge is given in Table A4.

The corresponding equation for a measuring structure is similar and may be expressed as follows:

$$X_{dm} = \frac{\Sigma[(X_c^2 + n^2X^2(h + a))^{1/2}Q_h]}{\Sigma Q_h} \quad (A49)$$

where X_c is the uncertainty in the coefficient of discharge. Note that X_b, the uncertainty in the length of crest (width of throat), has been neglected and note also that in this case the value of a is zero.

The uncertainty in the monthly mean and annual discharge may be estimated from the following equations:

Table A4 Typical computation for the uncertainty in the daily mean discharge using hourly values of discharge

Time	h (m)	$(h - 0.115)$ (m)	Q_h (m³ s⁻¹)	$X_{(h+a)}$ (%)	$2S_{mr}$ (%)	$[2S_{mr}^2 + n^2X^2(h + a)]^{1/2}Q_h$
0 900	1.225	1.110	46.314	0.4	1.3	64.84
1 000	1.565	1.450	69.707	0.3	1.5	115.53
1 100	1.971	1.856	101.699	0.2	1.8	183.06
1 200	2.293	2.178	129.906	0.2	1.9	246.82
1 300	2.520	2.405	151.186	0.2	2.0	302.37
1 400	2.670	2.565	165.850	0.2	2.0	331.70
1 500	2.789	2.674	177.814	0.2	2.0	355.63
1 600	2.872	2.767	186.328	0.2	2.1	391.29
1 700	2.929	2.814	192.255	0.2	2.1	403.74
1 800	2.981	2.876	197.717	0.1	2.1	415.21
1 900	3.034	2.929	203.339	0.1	2.2	447.34
2 000	3.067	2.952	206.867	0.1	2.2	455.11
2 100	3.082	2.967	208.478	0.1	2.2	458.65
2 200	3.065	2.950	206.653	0.1	2.2	454.64
2 300	3.026	2.911	202.487	0.1	2.2	445.47
2 400	2.975	2.860	197.084	0.1	2.1	413.88
0 100	2.915	2.800	190.793	0.2	2.1	400.66
0 200	2.845	2.730	183.543	0.2	2.1	385.44
0 300	2.747	2.632	173.558	0.2	2.0	347.11
0 400	2.628	2.513	161.697	0.2	2.0	323.39
0 500	2.495	2.380	147.788	0.2	2.0	297.57
0 600	2.365	2.250	136.543	0.2	1.9	259.41
0 700	2.257	2.142	126.635	0.2	1.9	240.61
0 800	2.164	2.049	118.320	0.2	1.8	212.98
		Σ	3883.56			
		Daily mean	161.815			Σ7952.45

$$X_{dm} = \frac{\Sigma[(2S_{mr}^2 + n^2X^2(h + a))^{1/2}Q_h]}{\Sigma Q_h}$$

$$= \frac{7952.45}{3883.56}$$

$$= \pm 2\%$$

Then daily mean discharge = 161.815 m³ s⁻¹ ± 2%.

$$X_{\text{mm}} = \frac{(X_{\text{dm}}Q_{\text{dm}})}{\Sigma Q_{\text{dm}}} \tag{A50}$$

and

$$X_{\text{aa}} = \frac{(X_{\text{mm}}Q_{\text{mm}})}{\Sigma Q_{\text{mm}}} \tag{A51}$$

where X_{mm} is the uncertainty in the monthly mean discharge (95% level), X_{aa} is the uncertainty in the annual discharge, Q_{dm} is the daily mean discharge, and Q_{mm} is the monthly mean discharge.

The percentage uncertainty in stage (or head) in the above equations may be found from equation (A23).

Systematic error in the stage–discharge curve

A systematic error in the stage–discharge curve has not been included in the above example because the curve is usually established using several different current meters, the inference being that any systematic error in the meters is randomized. However, the possible presence of a systematic error in the stage–discharge curve may require to be investigated. In effect the curve may shift along both the stage axis and the discharge axis, the resultant shift being maximum when stage and discharge shifts have opposite signs. The direction of shift, however, will be unknown and plus and minus values require to be assigned to both (say ±3 mm for stage and ±1% for discharge). The resultant of these two factors produces a bandwidth within which the curve will be expected to lie due to the systematic uncertainties alone. The analysis for S_{e} and S_{mr} is unaffected and is carried out independently. The result, however, should then be presented as

$$\text{discharge} = Q + S_{\text{mr}} + E_{\text{s}}$$

where E_{s} is the estimated systematic uncertainty.

An investigation of the reference gauge zero, however, would determine the sign and value of the stage shift, and if possible the error could be corrected on site or allowed for in plotting $(h + a)$ values. Any systematic error in the current meter rating tank, however, can only be found from a direct comparison of a selection of rating tanks by rating the same current meters in each.

Note that, by tradition, the dependable variable Q in the stage–discharge equation is plotted on the abscissa with stage on the ordinate. This procedure has no effect on the calculations or on the analysis of the uncertainties.

It should be noted that, in an ideal situation, the current meter observations would in fact fall on the stage–discharge curve, the stage–discharge relation being permanent, and S_{e} and S_{mr} both zero. The scatter about the curve experienced in practice, however, is due principally to the uncertainty in the current meter observations, the uncertainty in stage measurement, the instability of the station control, changing conditions in the channel due to scour or accretion and seasonal changes in the river regime. In view of these factors it is possible in some cases for S_{e} to be larger than the uncertainty in the current meter measurements $(X_{\bar{Q}})$. The current meter measurements may indeed be of high accuracy for the conditions prevailing at the time of measurement, but may be influenced by one or more of the above factors when plotted on the stage–discharge curve.

Estimation of length of record required at a streamflow station

The length of record required at a streamflow station to meet a specific uncertainty in the mean discharge may be calculated approximately from equation (A4):

$$s_{\bar{Y}} = \frac{ts_Y}{\sqrt{N}} \tag{A52}$$

where $s_{\bar{Y}}$ is taken as the specified uncertainty required in the mean monthly discharge or annual discharge as a percentage, s_Y is the standard deviation (95% level) of the monthly or annual discharge as a percentage [from equation (A3)], t is Student's t statistic and N is the number of years of record required to attain $s_{\bar{Y}}$.

If the number of years of record available is over 30, t is taken as having a value of 2; if the number of years of record available is less than 30 a Student's t multiplier is taken from Table A2.

Example

Table A5 shows 10 years of records of the monthly discharge at a gauging station together with the values of s_Y and $s_{\bar{Y}}$ calculated for each of the 12 months January–December. Values of s_Y have been multiplied by a Student's t factor of 2.2. From equation (A52):

$$N = \left(\frac{s_Y}{s_{\bar{Y}}}\right)^2 \tag{A53}$$

If a value of 10% is assigned to $s_{\bar{Y}}$ as being acceptable for the uncertainty in the mean monthly discharge, the number of years of record to attain this is approximately

$$N = \left(\frac{s_Y}{10}\right)^2 \tag{A54}$$

For January $s_Y = 100\%$, therefore

$$N = \left(\frac{100}{10}\right)^2 = 100 \text{ years}$$

For September $s_Y = 55\%$, therefore

$$N = \left(\frac{55}{10}\right)^2 = 30 \text{ years}$$

If the average value (71.75) is taken for s_Y for the period then

$$N = \left(\frac{71.75}{10}\right)^2 = 50 \text{ years}$$

It will be seen from equation (A54) that if a value of 5% is assigned to $s_{\bar{Y}}$, instead of 10%, then the above values of N are multiplied by 4. Whilst statistical methods of this nature applied to streamflow data need to be treated with caution because of the interdependence and skew nature of the data, the orders of magnitude of the above results are significant.

Table A5 Example of procedure for calculating the mean monthly discharge, the standard deviation (s_y) and the standard deviation of the mean ($s\bar{y}$) from the first 10 years of record at a gauging station

	Jan.	Feb.	Mar.	Apr.	May	June	July	Aug.	Sep.	Oct.	Nov.	Dec.
Gauged mean flow (m^3 s^{-1}):												
1963	1.175	1.200	1.975	2.614	2.492	2.012	1.658	1.453	1.206	1.089	1.414	1.558
1964	1.657	1.572	1.641	1.901	2.090	2.143	1.411	1.150	0.928	0.894	0.836	0.875
1965	0.922	0.893	0.893	0.845	0.783	0.743	0.686	0.624	0.681	0.684	0.756	1.009
1966	1.417	2.220	2.847	2.618	2.482	2.267	1.830	1.550	1.362	1.888	2.366	2.337
1967	2.540	2.842	3.575	3.337	2.615	2.016	1.577	1.341	1.135	1.225	1.665	1.891
1968	2.393	2.770	2.597	2.035	1.806	1.506	1.340	1.419	1.698	1.919	1.965	2.441
1969	3.409	3.616	2.285	2.972	2.520	2.138	1.895	1.559	1.274	1.555	0.894	0.980
1970	1.127	1.591	2.091	2.148	1.842	1.517	1.310	1.092	0.977	0.889	1.057	1.140
1971	1.403	2.371	2.522	2.461	2.238	2.433	2.358	2.047	1.486	1.426	1.265	1.218
1972	1.311	1.871	2.925	2.875	2.452	2.071	1.632	1.394	1.178	1.007	1.031	1.435
1963–1972 mean	1.735	2.095	2.435	2.381	2.132	1.885	1.570	1.363	1.203	1.218	1.325	1.488
$\pm s_y$, %	100	87	73	64	57	58	62	59	55	75	88	83
$\pm s_{\bar{y}}$, %	31	27	23	20	18	18	19	18	17	23	28	26

Table A6 Attainable uncertainties in a single measurement of discharge

Method	± Percentage uncertainty (95% level)
Current meter measurement	5
Floats	10–20
Slope – area	10–20
Fall – discharge	10–20
Dilution techniques	5
Thin-plate weir	2
Thin-plate V-notch	2
Triangular-profile (Crump) weir	5
Flat-V weir	5
Rectangular-profile weir	5
Round-nosed weir	5
Flumes	5
Moving boat	5
Ultrasonic	5
Electromagnetic	5–10

Uncertainties in individual methods

The uncertainties associated with each method of open channel flow depend on a number of factors, but the most important ones are

- hydraulic conditions of flow;
- measurement of head or stage;
- number of verticals taken in a current meter measurement;
- coefficient of discharge in the weirs and flumes methods;
- the stage – discharge relation in the velocity–area method;
- operation and maintenance of the station.

For good measurement practice carried out to ISO standards the attainable uncertainties in a single measurement of discharge may be taken as shown in Table A6.

The values of uncertainties in Table A6 may be considered attainable for average flow conditions but may require to be modified for conditions of extreme low flows or floods in accordance with error equations.

It has been shown that the uncertainty in the daily mean discharge, monthly mean discharge and annual discharge can be expected to be much better than the uncertainties shown in Table A6.

The examples given have been related specifically to streamflow measurement in order to demonstrate the application of statistics to hydrology. The principles and theory, however, apply to other variables in the hydrological cycle.

R.W. Herschy

Source

Herschy, R.W. (1995) *Streamflow Measurement*, 2nd edn, Chapman & Hall, London 524 pp.

Bibliography

Dymond, J.R. and Christian, R., 1982. Accuracy of discharge determined from a rating curve. *Hydrological Sciences Journal*, **27**(4), 493–504.

Herschy, R.W., 1978. Accuracy. In *Hydrometry: Principles and Practices* (ed. R.W. Herschy). John Wiley and Sons, Chichester, pp. 353–97.

Herschy, R.W., 1978. The accuracy of current meter measurements. *Proceedings of the Institution of Civil Engineers*, Part 2, 65 TN 187, pp. 431–7.

Herschy, R., 1993. The velocity area method. *Flow Measurement and Instrumentation*, **4**(1), 7–10.

Herschy, R., 1993. The stage – discharge relation. *Flow Measurement and Instrumentation*, **4**(1), 11–15.

Herschy, R.W., 1994. The analysis of uncertainties in the stage – discharge relation. *Flow Measurement and Instrumentation*, **5**(2), 188–190.

ISO 748, 1997. *Liquid Flow Measurement in Open Channels: Velocity – Area Methods*. ISO, Geneva, Switzerland.

ISO 1100/2, 1988. *Determination of the Stage – Discharge Relation*. ISO, Geneva, Switzerland.

ISO 5168, 1978. *Measurement of Fluid Flow: Estimation of Uncertainty of a Flow-rate Measurement*. ISO, Geneva, Switzerland.

ISO 7066/1, 1989. *Uncertainty in Linear Calibration Curves*. ISO, Geneva, Switzerland.

ISO 7066/2, 1988. *Uncertainty in Non-linear Calibration Curves*, ISO, Geneva, Switzerland.

ISO/TAG 4, 1995. *Guide on the Expression of Uncertainty in Measurement*. ISO, Geneva, Switzerland.

Lintrup, M., 1989. A new expression for the uncertainty of a current meter discharge measurement. *Nordic Hydrology*, **20**, 191–200.

Pelletier, P.M., 1987. Uncertainties in the determination of river discharge – a literature review. Eighth Canadian Hydrotechnical Conference, Montreal.

Cross reference

Water resources: dictionary of basic terms

ACCURACY OF HYDRODYNAMIC APPROXIMATIONS IN HYDROLOGY: NON-UNIFORM, STEADY FLOW

Errors of the kinematic wave and diffusion wave approximations are derived for non-uniform, time-independent cases of planar or channel flow under three types of boundary conditions: zero flow at the upstream end, and critical flow depth and zero depth gradient at the downstream end. The diffusion wave approximation is found to be accurate for $KF_0^2 \geq 5$, where K is the kinematic wave number and F_0 is the Froude number. However, in order for the kinematic wave approximation to be sufficiently accurate, KF_0^2 may have to be significantly greater than 5. The accuracy of the diffusion wave approximation is significantly influenced by the downstream boundary condition.

Introduction

In an overland flow a steady state is attained for constant rainfall of sufficiently long duration (greater than or equal to the time of concentration t_c), because the depth of flow at the outlet increases until equilibrium is reached. The same is true for flow in a channel subject to constant lateral inflow. For a channel receiving constant inflow at the upstream boundary, the flow at the downstream end would reach the equilibrium. Despite occurrences of steady state or time-independent flows, they have received much less attention in hydrology (Morris, 1978). The steady-state solution aids in understanding the nature of the water surface profile. It may help determine the condition for use of zero depth in place of zero influx at the upstream boundary. When rainfall duration is much greater than the time of equilibrium, steady-state water surface profiles are very useful.

A comprehensive discussion of steady-state flows using the diffusion wave (DW) approximation was undertaken by Govindaraju *et al.* (1988). They presented both numerical and analytical results for flux-type boundary conditions. The upstream boundary condition was one of zero inflow. Both the zero depth gradient and the critical depth downstream boundary conditions were investigated. For steep slopes, the upstream boundary condition of zero depth was found to be justifiable. It was shown that the critical depth condition at the downstream boundary was a stringent condition and might lead to problems for certain ranges of parameter values when seeking a numerical solution. Govindaraju *et al.* (1988) proposed an analytical approximation in the form of a cubic approximation of the DW model for the zero-depth gradient downstream boundary condition, which is accurate when $F_0^2 K$ is sufficiently small, where F_0 is the Froude number and K the kinematic wave number. Parlange *et al.* (1989) improved their approximation, using a Taylor series approximation, and showed that the improved approximation and the one by Govindaraju *et al.* (1988) would provide the upper and lower bounds of the exact solution. Most studies dealing with evaluating the adequacy of the kinematic wave (KW) or diffusion wave (DW) approximation have dealt with space–time dependent flows, and criteria have been derived as point values for given flow situations.

An explicit treatment of steady-state flows has not been included in these studies.

Pearson (1989) examined the criteria for using the KW approximation to the St Venant (SV) equations for shallow water flow. For steady-state one-dimensional flow over a plane he showed that for the condition of zero flow at the upstream boundary, the assumed steady-state upstream depth is generally non-zero, but is implicitly assumed to be zero in the kinematic wave approximation. By plotting contours of dimensionless steady-state upstream depth over a two-parameter plane, (F_0^2, KF_0^2), he obtained a new criterion for kinematic wave modeling as $K > 3 + 5/F_0^2$.

Parlange *et al.* (1990) were probably the first to undertake an investigation of errors in the KW and DW approximations by comparing their predictions with the exact numerical solution of the St Venant (SV) equations under steady-state conditions. It was shown that the two approximations could have significant errors even for critical flow and fairly large kinematic wave numbers. When the KW approximation was inaccurate, the improvement of the DW approximation was modest. Parlange *et al.* (1990) then proposed a more accurate approximation when the kinematic wave number was large. They then suggested splitting the solution of the SV equations in two regions, one near the downstream end of the plane and the other covering most of the plane. However, no explicit relations as a function of space between these criteria and errors resulting from the KW or DW approximation have yet been derived. As a result the actual amount of error of these approximations is usually not known. Furthermore, when carrying out hydrologic modeling it is not evident if the KW and DW approximations are valid for the entire length of the channel or a portion thereof. The objective of this article is to derive, under simplified conditions, errors for the KW and DW approximations as a function of space.

Dimensional formulation of shallow water wave (SWW) theory

The SWW theory can be described by some form of the SV equations. For flow over an infiltrating plane subject to uniform rainfall, these equations can be written in one-dimensional form on a unit width basis as:

Continuity equation:

$$\frac{\partial h}{\partial t} + \frac{\partial}{\partial x}(uh) = q - f \tag{A55}$$

Momentum equation:

$$\frac{\partial u}{\partial t} + \frac{\partial}{\partial x}\left(\frac{1}{2}u^2 + gh\right) = g(S_0 - S_f) - \frac{qu}{h} \tag{A56}$$

where h is the depth of flow (L), u is local mean velocity (L/T), q is uniform rainfall intensity (L/T), f is uniform infiltration rate (L/T), g is acceleration due to gravity, x is space coordinate in the direction of flow (L), t is time (T), S_0 is bed slope and S_f is frictional slope. Note $Q = uh$ is discharge [L^3/(TL)] per unit width. S_f can be approximated as

$$S_f = \beta \frac{u^2}{h} \tag{A57}$$

where β is some resistance parameter. If the Chézy relation is used then $\beta = g/C^2$, where C is Chézy's resistance parameter.

The dynamic wave (DYW) representation employs the full form of equations (A55) and (A56). The KW approximation is based on equation (A55) and equation (A56) with the left side omitted:

$$g(S_0 - S_f) - \frac{qu}{h} = 0 \tag{A58}$$

The DW approximation uses equation (A55) and equation (A56) with local and convective accelerations deleted:

$$g\frac{dh}{dx} = g(S_0 - S_f) - \frac{qu}{h} \tag{A59}$$

Analytical solutions of the SV equations or their variants in the KW and DW approximations are tractable only for simple cases. To that end, the time-independent flow is considered in this study. This case corresponds to the steady state condition.

Dimensionless formulation of shallow water wave theory

It is useful to express the SV equations in dimensionless form. To that end, the following normalizing quantiles are defined:

H_0 = normal depth of flow $Q_0 = q_{max}L$ at the end of the reach $x = L$
U_0 = normal velocity for $Q_0 = qL$
$Q_0 = U_0 H_0 = q_{max}L$
T = normalizing time = L/U_0 (A60)
q_{max} = maximum steady state q
F_0 = normalizing Froude number = $U_0/[(gH_0)^{0.5}]$

where L is the length of the channel. With the use of the above normalizing quantities, the following dimensionless quantities can be defined:

$$x_* = \frac{x}{L}, \quad t_* = \frac{t}{T} = \frac{tL}{U_0}, \quad h_* = \frac{h}{H_0}$$

$$U_* = \frac{u}{U_0}, \quad q_* = \frac{q}{q_{max}}, \quad Q_* = \frac{Q}{Q_0} = \frac{uh}{Q_0}$$

$$F_* = \frac{F}{F_0} = \frac{u_*}{(h_*)^{0.5}} \tag{A61}$$

Recall that the Froude number is

$$F = \frac{u}{(gh)^{0.5}} = \text{Froude number} \tag{A62}$$

and the kinematic wave number is

$$K = \frac{S_0 L}{F_0^2 H_0} \tag{A63}$$

With the introduction of the above normalizing quantities, equations (A55) and (A56) can be expressed in dimensionless form as

$$\frac{\partial h_*}{\partial t_*} + u_* \frac{\partial h_*}{\partial x_*} + h_* \frac{\partial u_*}{\partial x_*} = q_* - f_* \tag{A64}$$

$$\frac{\partial u_*}{\partial t_*} + u_* \frac{\partial u_*}{\partial x_*} + \frac{1}{F_0^2}\frac{\partial h_*}{\partial x_*} = K\left(1 - \frac{u_*^2}{h_*}\right) - \frac{u_* q_*}{h_*} \tag{A65}$$

The dimensionless form of equation (A58) is

$$K\left(1 - \frac{u_*^2}{h_*}\right) - \frac{u_* q_*}{h_*} = 0 \tag{A66}$$

and that of equation (A59) is

$$\frac{1}{F_0^2}\frac{\partial h_*}{\partial x_*} = K\left(1 - \frac{u_*^2}{h_*}\right) - \frac{u_* q_*}{h_*} \tag{A67}$$

Dimensionless time-independent flows

For time-independent flows, equation (A64) becomes

$$\frac{d}{dx_*}(u_* h_*) = q_* - f_* \tag{A68}$$

and equation (A68) becomes

$$u_* \frac{du_*}{dx_*} + \frac{1}{F_0^2}\frac{dh_*}{dx_*} = K\left(1 - \frac{u_*^2}{h_*}\right) - \frac{u_* q_*}{h_*} \tag{A69}$$

Equations (A68) and (A69) are the governing equations for the DYW representation for time-independent flows. The KW approximation is based on equation (A68) and equation (A69) with the left side deleted:

$$K\left(1 - \frac{u_*^2}{h_*}\right) - \frac{u_* q_*}{h_*} = 0 \tag{A70}$$

The DW approximation uses equation (A68) and equation (A69) with the convective acceleration term deleted:

$$\frac{1}{F_0^2}\frac{dh_*}{dx_*} = K\left(1 - \frac{u_*^2}{h_*}\right) - \frac{u_* q_*}{h_*} \tag{A71}$$

Equation (A70) can also be approximated by neglecting the momentum exchange term between lateral inflow and longitudinal channel flow as

$$u_*^2 = h_* \tag{A72}$$

Similarly, equation (A69) can be expressed as

$$u_* \frac{du_*}{dx_*} + \frac{1}{F_0^2} \frac{dh_*}{dx_*} = K\left(1 - \frac{u_*}{h_*}\right)^2 \tag{A73}$$

and equation (A71) as

$$\frac{1}{F_0^2} \frac{dh_*}{dx_*} = K\left(1 - \frac{u_*^2}{h_*}\right) \tag{A74}$$

Depending upon the presence of f_*, equation (A68) can be simplified. If $f_* = 0$, then

$$\frac{d}{dx_*}(u_* h_*) = q_* \tag{A75}$$

Dimensionless boundary conditions

Depending upon the type of flow, the boundary conditions have to be specified at the upstream boundary alone or at the upstream boundary as well as the downstream boundary. Two types of conditions at the upstream boundary are:

1. $u_*(0) = 0$, $h_*(0) = h_{*0}$ for a dry channel (A76)
2. $u_*(0) = u_*0$, $h_*(0) = h_{*0}$ for a wet channel (A77)

The first boundary condition, given by equation (A76), corresponds to zero flux at the boundary. It influences both the subcritical and supercritical flow outside zone A of the characteristic solution domain of the SV equations. The boundary condition given by equation (A77) corresponds to nonzero flux at the upstream boundary. The downstream boundary condition can be of two types:

1. The critical flow downstream boundary condition:

$$u_*(1) = \frac{h_*^{1/2}(1)}{F_0} \tag{A78}$$

This occurs when the channel ends at the steep bank of a river.
2. The zero-depth gradient downstream boundary condition:

$$\frac{dh_*(1)}{dx_*} = 0 \tag{A79}$$

For the supercritical flow:

$$u_*(1) > \frac{h_*^{1/2}(1)}{F_0} \tag{A80}$$

Both types of downstream boundary conditions can be employed in conjunction with appropriate upstream boundary condition.

Analysis of downstream boundary conditions for diffusion wave approximation when momentum exchange is zero

The boundary conditions are derived for two cases: (1) the momentum exchange term is neglected and (2) the momentum exchange term is retained. These will include the case when there is or there is no lateral inflow or lateral outflow. In each case, the upstream velocity is zero.

Equation (A68) gives

$$\frac{d}{dx_*}(u_* h_*) = q_{*1} \tag{A81}$$

and equation (A71) gives

$$\frac{dh_*}{dx_*} = KF_0^2\left(1 - \frac{u_*^2}{h_*}\right) \tag{A82}$$

or

$$u_* = h^{0.5}\left[1 - \frac{1}{KF_0^2}\frac{dh_*}{dx_*}\right]^{0.5} \tag{A83}$$

Substitution of equation (A83) in equation (A81) yields

$$\frac{d}{dx_*}\left\{h_*^{3/2}\left[1 - \frac{1}{KF_0^2}\frac{dh_*}{dx_*}\right]^{0.5}\right\} = q_{*1} \tag{A84}$$

Integration of equation (A84) from 0 to 1 yields

$$h_*^{3/2}(1)\left[1 - \frac{1}{KF_0^2}\frac{dh_*(1)}{dx_*}\right]^{0.5} - h_*^{3/2}(0) \times \left[1 - \frac{1}{KF_0^2}\frac{dh_*(0)}{dx_*}\right] = q_{*1} \tag{A85}$$

Equation (A85) is the general equation for boundary conditions. By substituting for $h_*(0)$ and $u_*(0)$ from equations (A76) and (A77), and using equations (A78), (A79) and (A80), different combinations of boundary conditions at the upstream and downstream ends can be obtained. The possible scenarios are: (1) upstream boundary condition of constant depth and zero discharge, (2) downstream boundary condition of critical flow depth and (3) zero depth gradient downstream boundary condition. The derivation of the above boundary conditions is illustrated in the ensuing discussion.

Critical flow depth downstream boundary condition
From equation (A67), $u_* q_*/h_* = 0$. If the upstream boundary condition is given by equation (A76), equation (A74) yields

$$\frac{dh_*(0)}{dx_*} = KF_0^2 \tag{A86}$$

The critical flow downstream boundary condition from equation (A74) with the substitution of equation (A78) is

$$\frac{dh_*(1)}{dx_*} = K(F_0^2 - 1) \tag{A87}$$

Substitution of equations (A86) and (A87) in equation (A85) yields

$$h_*^{3/2}(1)\left[1 - \frac{1}{KF_0^2}K(F_0^2 - 1)\right]^{0.5} = q_{*1} \tag{A88}$$

or

$$h_*(1) = F_0^{2/3} q_{*1}^{2/3} \tag{A89}$$

Zero depth gradient downstream boundary condition
Substitution of equation (A77) in equation (A82) yields

$$\frac{dh_*(0)}{dx_*} = KF_0^2 \tag{A90}$$

at the upstream end. Substitution of equations (A77) and (A79) in equation (A85) produces

$$h_*^{3/2}(1) = q_{*1} \tag{A91}$$

or

$$h_*(1) = q_{*1}^{2/3} \tag{A92}$$

General analysis for downstream boundary conditions
The value of h_* at $x_* = 1$ can be determined as follows. Equation (A82) can be written for u_* as

$$u_* = \left[h_*\left(1 - \frac{1}{KF_0^2}\frac{dh_*}{dx_*}\right)\right]^{0.5} \tag{A93}$$

Substitution of equation (A93) in equation (A68) with $\partial h_*/\partial t_* = 0$ yields

$$\frac{d}{dx_*}\left[h_*^{3/2}\left(1 - \varepsilon\frac{dh_*}{dx_*}\right)^{1/2}\right] = q_* - f_* \tag{A94}$$

where

$$\varepsilon = \frac{1}{KF_0^2}$$

Equation (A94) is integrated from $x_* = 0$ to $x_* = 1$ as

$$h_*^{3/2}(1)(1 - \varepsilon b)^{1/2} - h_*^{3/2}(0) \times (1 - \varepsilon a) = \int_0^1 [q_*(x_*) - f_*(x_*)]dx_* \tag{A95}$$

where

$$a = KF_0^2, \quad b = K(F_0^2 - 1) \qquad (A96)$$

Equation (A95) reduces to

$$h_*(1) = \left\{ (1 - \varepsilon b)^{-0.5} \int_0^1 \times [q_*(x_*) - f_*(x_*)] \mathrm{d}x_* \right\}^{2/3} \qquad (A97)$$

For the case $f_*(x_*) = 0$, $q_*(x_*) = q_* = 1$,

$$h_*(1) = (1 - \varepsilon b)^{-1/3} \qquad (A98)$$

The critical depth downstream boundary condition becomes

$$h_*(1) = F_0^{2/3} \qquad (A99)$$

and the zero depth gradient downstream boundary condition is

$$h_*(1) = 1 \qquad (A100)$$

Clearly, if $F_0 = 1$ then $b = 0$ for all K and the two downstream boundary conditions become identical. This implies that the steady-state depth profile is the same for either of the two downstream boundary conditions.

Analysis of downstream boundary conditions for diffusion wave approximation when momentum exchange is included

In this case the governing equations are equation (A81) and (A71). From equation (A71):

$$u_* = \frac{Kh_*}{q_*}\left(1 - \frac{u_*^2}{h_*}\right) - \frac{h_*}{F_0^2 q_*}\frac{\mathrm{d}h_*}{\mathrm{d}x_*} \qquad (A101)$$

Substitution of equation (A101) in equation (A81) leads to

$$\frac{\mathrm{d}}{\mathrm{d}x_*}\left[\frac{Kh_*^2}{q_*}\left(1 - \frac{u_*^2}{h_*}\right) - \frac{h_*^2}{F_0^2 q_*}\frac{\mathrm{d}h_*}{\mathrm{d}x_*}\right] = q_{*1} \qquad (A102)$$

Integration from $x_* = 0$ to 1:

$$\frac{Kh_*^2(1)}{q_*}\left[1 - \frac{u_*^2(1)}{h_*(1)}\right] - \frac{h_*^2(1)}{F_0^2 q_*}\frac{\mathrm{d}h_*(1)}{\mathrm{d}x_*}$$
$$- \frac{Kh_*^2(0)}{q}\left[1 - \frac{u_*^2(0)}{h_*(0)}\right] + \frac{h_*^2(0)}{F_0^2 q_*}\frac{\mathrm{d}h_*(0)}{\mathrm{d}x_*} = q_{*1} \qquad (A103)$$

Critical flow depth downstream boundary condition
At $x_* = 0$, $u_* = 0$, and equation (A86) applies. At $x_* = 1$, equation (A78) applies. With inclusion of these conditions, equation (A103) yields

$$\frac{\mathrm{d}h_*(1)}{\mathrm{d}x_*} = KF_0^2\left(1 - \frac{1}{F_0^2}\right) - \frac{F_0 q_*}{h_*^{0.5}(1)} \qquad (A104)$$

Substitution of equation (A104) in equation (A103) leads to

$$\frac{Kh_*^2(1)}{q_*}\left(1 - \frac{1}{F_0^2}\right) - \frac{h_*^2(1)}{F_0^2 q_*}\left[KF_0^2\left(1 - \frac{1}{F_0^2}\right) - \frac{F_0 q_*}{h_*^{0.5}(1)}\right]$$
$$- \frac{F_0 q_*}{h^{0.5}(1)} - \frac{K}{q_*}h_*^2(0) + \frac{h_*^2(0)}{F_0^2 q_*} \times KF_0^2 = q_{*1} \qquad (A105)$$

This simplifies to

$$h_*(1) = F_0^{2/3} q_{*1}^{2/3} \qquad (A106)$$

Zero depth gradient downstream boundary condition
With use of the upstream boundary condition at $x_* = 0$, $u_* = 0$, $\mathrm{d}h_*(0)/\mathrm{d}x_* = \mathrm{KF}_0^2$, and at the downstream boundary at $x_* = 1$, $\mathrm{d}h_*(1)/\mathrm{d}x_* = 0$, equation (A103) becomes

$$\frac{K}{q_*}h_*^2(1)\left[1 - \frac{u_*^2(1)}{h_*(1)}\right] - \frac{h_*^2(1)}{F_0^2 q_*}\frac{\mathrm{d}h_*(1)}{\mathrm{d}x_*} \times \frac{\mathrm{d}h_*(1)}{\mathrm{d}x_*} = q_{*1} \quad (A107)$$

or

$$\frac{K}{q_*}h_*^2(1)\left[1 - \frac{u_*^2(1)}{h_*(1)}\right] = q_{*1} \qquad (A108)$$

Substitution of equation (A76) in equation (A71) yields

$$\frac{\mathrm{d}h_*(1)}{\mathrm{d}x_*} = KF_0^2\left(1 - \frac{u_*^2(1)}{h_*(1)}\right) - \frac{F_0^2 q_* u_*(1)}{h_*(1)} \qquad (A109)$$

Therefore

$$KF_0^2\left(1 - \frac{u_*^2(1)}{h_*(1)}\right) - \frac{F_0^2 q_* u_*(1)}{h_*(1)} = 0 \qquad (A110)$$

This simplifies to a quadratic expression in u_* as

$$u_*^2(1) + \frac{q_*}{K}u_*(1) - h_* = 0 \qquad (A111)$$

Its solution is

$$u_*(1) = \frac{1}{2K}\left[- q_* + (q_*^2 + 4h_* K)^{0.5}\right] \qquad (A112)$$

Let $u_*(1) = C$. Then equation (108) becomes

$$\frac{K}{q_*}h_*^2(1)\left(1 - \frac{C_1^2}{h_*(1)}\right) = q_{*1} \qquad (A113)$$

or

$$h_*^2(1) - C_1^2 h_*(1) - \frac{q_* q_{*1}}{K} = 0 \qquad (A114)$$

Its solution is

$$h_*(1) = \frac{1}{2}\left[C_1^2 + \left(C_1^4 + \frac{4}{K}q_* q_{*1}\right)^{0.5}\right] \qquad (A115)$$

Analysis of downstream boundary conditions for dynamic wave representation with momentum exchange deleted

Recall the continuity equation:

$$\frac{\mathrm{d}}{\mathrm{d}x_*}(u_* h_*) = q_{*1} \qquad (A116)$$

and the momentum equation:

$$u_*\frac{\mathrm{d}u_*}{\mathrm{d}x_*} + \frac{1}{F_0^2}\frac{\mathrm{d}h_*}{\mathrm{d}x_*} = K\left(1 - \frac{u_*^2}{h_*}\right) \qquad (A117)$$

Equation (A117) can be written as

$$u_* = h_*^{0.5}\left[1 - \frac{u_*}{K}\frac{\mathrm{d}u_*}{\mathrm{d}x_*} - \frac{1}{KF_0^2}\frac{\mathrm{d}h_*}{\mathrm{d}x_*}\right]^{0.5} \qquad (A118)$$

Substitution of equation (A118) in equation (A116) yields

$$\frac{\mathrm{d}}{\mathrm{d}x_*}\left[h_*^{3/2}\left(1 - \frac{u_*}{K}\frac{\mathrm{d}u_*}{\mathrm{d}x_*} - \frac{1}{KF_0^2} \times \frac{\mathrm{d}h_*}{\mathrm{d}x_*}\right)^{0.5}\right] = q_{*1} \quad (A119)$$

Integrating equation (A119) from $x_* = 0$ to $x_* = 1$, we get

$$h_*^{3/2}(1)\left[1 - \frac{u_*(1)}{K}\frac{\mathrm{d}u_*(1)}{\mathrm{d}x_*} - \frac{1}{KF_0^2} \times \frac{\mathrm{d}h_*(1)}{\mathrm{d}x_*}\right]^{0.5} - h_*^{3/2}(0)$$
$$\times \left[1 - \frac{u_*(0)}{K} \times \frac{\mathrm{d}u_*(0)}{\mathrm{d}x_*} - \frac{1}{KF_0^2}\frac{\mathrm{d}h_*(0)}{\mathrm{d}x_*}\right] = q_{*1} \qquad (A120)$$

Recall that at $x_* = 0$, $u_* = 0$, and $\mathrm{d}h_*(0)/\mathrm{d}x_* = KF_0^2$. Therefore, equation (A120) becomes

$$h_*^{3/2}(1)\left[1 - \frac{u_*(1)}{K}\frac{\mathrm{d}u_*(1)}{\mathrm{d}x_*} - \frac{1}{KF_0^2} \times \frac{\mathrm{d}h_*(1)}{\mathrm{d}x_*}\right]^{0.5} = q_{*1} \quad (A121)$$

Critical flow depth downstream boundary condition
Recall that at $x_* = 1$, $u_*(1) = h_*^{0.5}(1)/F_0$. Therefore,

$$\frac{\mathrm{d}u_*(1)}{\mathrm{d}x_*} = \frac{1}{2F_0 h^{0.5}(1)}\frac{\mathrm{d}h_*(1)}{\mathrm{d}x_*} \qquad (A122)$$

Substitution of equation (A122) in equation (A121) yields

$$h_*^{3/2}(1)\left[1 - \frac{3}{2F_0^2 K}\frac{\mathrm{d}h_*(1)}{\mathrm{d}x_*}\right]^{0.5} = q_{*1} \qquad (A123)$$

From the momentum equation (A117), at $x_* = 1$,

$$\frac{dh_*(1)}{dx_*} = \frac{2}{3}K(F_0^2 - 1) \qquad (A124)$$

Inserting equation (A124) in equation (A123) we get

$$h_*(1) = F_*^{2/3} q_{*1}^{2/3} \qquad (A125)$$

Zero depth gradient downstream boundary condition
With use of the upstream conditions and $dh_*(1)/dx_* = 0$ at $x_* = 1$ in equation (A120), we get

$$h_*^3(1) - \frac{q_{*1}^2}{K}h_*(1) - q_{*1}^2 = 0 \qquad (A126)$$

An analytical solution of equation (A126) is wieldy. To that end, let

$$a_1 = -\frac{q_{*1}}{K}, \quad a_0 = -q_{*1}^2, \quad \text{and } a_2 = 0 \qquad (A127)$$

Therefore

$$p = \frac{1}{3}a_1 - \frac{1}{9}a_2^2 = -\frac{q_{*1}^2}{3K} \qquad (A128)$$

$$r = \frac{1}{6}(a_1 a_2 - 3a_0) - \frac{1}{27}a_2^3$$
$$= \frac{1}{6}(-3(-q_{*1}^2)) = \frac{1}{2}q_{*1}^2 \qquad (A129)$$

$$p_3 + r^2 = -\frac{q_{*1}^6}{27K^3} + \frac{q_{*1}^4}{4} \qquad (A130)$$

So $p^3 + r^2 > 0$, and we obtain one real root and a pair of complex conjugate roots. In equation (A130), $p^3 + r^2$ is always greater than zero for all q_{*1} values exceeding zero and K greater than $(4q_{*1}^2/27)^{1/3}$. The real root is given by

$$S_1 = [r + (p^3 + r^2)^{0.5}]^{1/3} \qquad (A131)$$
$$S_2 = [r - (p^3 + r^2)^{0.5}]^{1/3}$$

The root

$$h_*(1) = S_1 + S_2 \qquad (A132)$$

Analysis of downstream boundary conditions for dynamic wave representation with momentum exchange included

The momentum equation is

$$u_*\frac{du_*}{dx_*} + \frac{1}{F_0^2}\frac{dh_*}{dx_*} = K\left(1 - \frac{u_*^2}{h_*}\right) - \frac{u_* q_*}{h_*} \qquad (A133)$$

This gives

$$u_* = h_*^{0.5}\left[1 - \frac{u_*}{K}\frac{du_*}{dx_*} - \frac{1}{KF_0^2}\frac{dh_*}{dx_*} - \frac{u_* q_*}{Kh_*}\right]^{0.5} \qquad (A134)$$

Substitution of equation (A134) in equation (A116) yields

$$\frac{d}{dx_*}\left[h_*^{3/2}\left(1 - \frac{u_*}{K}\frac{du_*}{dx_*} - \frac{1}{KF_0^2}\frac{dh_*}{dx_*} - \frac{u_* q_*}{Kh_*}\right)^{0.5}\right] = q_{*1} \qquad (A135)$$

Integrating equation (A135) from $x_* = 0$ to $x_* = 1$;

$$h_*^{3/2}(1)\left[1 - \frac{u_*(1)}{K}\frac{du_*(1)}{dx_*} - \frac{1}{KF_0^2} \times \frac{dh_*(1)}{dx_*} - \frac{u_*(1)q_*}{Kh_*(1)}\right]^{0.5} - h_*^{3/2}(0)$$

$$\times \left[1 - \frac{u_*(0)}{K}\frac{du_*(0)}{dx_*} - \frac{1}{KF_0^2}\frac{dh_*(0)}{dx_*} - \frac{u_*(0)q_*}{Kh_*(0)}\right]^{0.5} = q_{*1} \qquad (A136)$$

Critical flow depth downstream boundary condition
Recall that at $x_* = 0$, $u_* = 0$, and at $x_* = 1$, $u_* = h_*^{0.5}/F_0$. From equation (A69) we obtain

$$\frac{dh_*(1)}{dx_*} = \frac{2}{3}K(F_0^2 - 1) - \frac{2}{3}\frac{F_0 q_*}{h_*^{0.5}} \qquad (A137)$$

Substitution of equation (A78) and use of the upstream boundary conditions in equation (A136) yield

$$h_*^{3/2}(1)\left[1 - \frac{3}{2KF_0^2}\frac{dh_*(1)}{dx_*} - \frac{q_*}{KF_0 h_*^{0.5}}\right]^{0.5} = q_{*1} \qquad (A138)$$

Insertion of equation (A137) in equation (A138) produces

$$h_*^{3/2}(1)\left[1 - \frac{3}{2KF_0^2}\left[\frac{2}{3}K(F_0^2-1) - \frac{2}{3}\frac{F_0 q_*}{h_*^{0.5}}\right] - \frac{q_*}{KF_0 h_*^{0.5}}\right]^{0.5} = q_{*1} \qquad (A139)$$

This simplifies to

$$h_*(1) = F_0^{2/3} q_{*1}^{2/3} \qquad (A140)$$

This gives the depth at the downstream.

Zero depth gradient downstream boundary condition
From the solution of the continuity equation,

$$u_*(1) = \frac{q_{*1}}{h_*(1)} \qquad (A141)$$

From equation (A69) with $dh_*(1)/dx_* = 0$ at $x_* = 1$,

$$\frac{du_*(1)}{dx_*} = \frac{q_{*1}}{h_*(1)} \qquad (A142)$$

Substitution of equations (A142) and (A141) and the upstream boundary conditions in equation (A136) gives

$$h_*^{3/2}(1)\left[\frac{Kh_*^2 - (q_{*1}^2 + q_{*1}q_*)}{Kh_*^2}\right]^{0.5} = q_{*1} \qquad (A143)$$

Squaring both sides and simplifying result in

$$h_*^3(1) - \frac{(q_{*1}^2 + q_{*1}q_*)}{K}h_*(1) - q_{*1}^2 = 0 \qquad (A144)$$

Types of scenarios

Depending upon the presence of lateral inflow and infiltration, four different scenarios can be considered:

1. $f = 0$, $q = q_0 = $ constant
   This includes the case $q = 0$.
2. $q = q_0 = $ constant
   $f = f_0 = $ constant
   This includes the case of $q = f = 0$.
3. $q - f = 0$, $q = q_0 = $ constant, $f = f_0 = $ constant
   This includes the case $q = f = 0$.
4. $q = 0$, $f = f_0 = $ constant
   This includes the case $f = 0$.

It may be noted that the scenario with $q = 0$ in equation (A75) or $q - f = 0$ in equation (A60) applies to the case of losing flow. The same would apply if $q - f < 0$.

Scenarios for determination of error

Error equations can be derived for the KW and DW approximations under the above-mentioned conditions for four different scenarios. To summarize, the following cases can be treated.

1. Scenario 1 ($q = q_0$, $f = 0$): equations (A75) and (A70) are the governing equations for the KW approximation, equations (A75) and (A71) for the DW approximation and equations (A75) and (A73) for the DYW representation, with the upstream boundary condition given by equation (A76).
2. Scenario 1 ($q = q_0$, $f = 0$): equations (A75) and (A70) are the governing equations for the KW approximation, equations (A75) and (A71) for the DW approximation and equations (A75) and (A69) for the DYW representation, with the upstream boundary condition given by equation (A69) and the downstream boundary condition given by equation (A79).
3. Scenario 1 ($q = q_0$, $f = 0$): equations (A75) and (A70) are the governing equations for the KW approximation, equations (A75) and (A71) for the DW approximation and equations (A75) and

(A69) for the DYW representation, with the upstream condition given by equation (A77) and the downstream boundary condition given by equation (A78).

4. Scenario 1 ($q = q_0$, $f = 0$): equations (A75) and (A72) are the governing equations for the KW approximation, equations (A75) and (A74) for the DW approximation and equations (A75) and (A73) for the DYW representation, with the upstream boundary condition given by equation (A77).

5. Scenario 2 ($q = q_0$, $f = 0$): equations (A75) and (A72) are the governing equations for the KW approximation, equations (A75) and (A73) for the DW approximation and equations (A75) and (A73) for the DYW representation, with the upstream boundary condition given by equation (A76) and the downstream boundary given by equation (A79).

6. Scenario 2 ($q = q_0$, $f = 0$): equations (A75) and (A72) are the governing equations for the KW approximation, equations (A75) and (A74) for the DW approximation and equations (A75) and (A73) for the DYW representation, with the downstream boundary given by equation (A78).

7. Scenario 2 ($q = q_0$, $f = f_0$): equations (A68) and (A72) are the governing equations for the KW approximation, equations (A68) and (A74) for the DW approximation and equations (A68) and (A73) for the DYW representation, with the upstream boundary condition given by equation (A77).

8. Scenario 2 ($q = q_0$, $f = f_0$): equations (A68) and (A72) are the governing equations for the KW approximation, equations (A68) and (A74) for the DW approximation and equations (A68) and (A73) for the DYW representation, with the upstream boundary condition given by equation (A77) and the downstream boundary condition given by equation (A79).

9. Scenario 2 ($q = q_0$, $f = f_0$): equations (A68) and (A72) are the governing equations for the KW approximation, equations (A68) and (A74) for the DW approximation and equations (A68) and (A73) for the DYW representation, with the downstream boundary condition given by equation (A78).

10. Scenario 2 ($q = q_0$, $f = f_0$): equations (A68) and (A70) are the governing equations for the KW approximation, equations (A68) and (A71) for the DW approximation and equations (A68) and (A69) for the DYW representation, with the upstream boundary condition given by equation (A77).

11. Scenario 2 ($q = q_0$, $f = f_0$): equations (A68) and (A70) are the governing equations for the KW approximation, equations (A68) and (A71) for the DW approximation and equations (A68) and (A69) for the DYW representation, with the upstream boundary condition given by equation (A76) and the downstream boundary condition given by equation (A79).

12. Scenario 2 ($q = q_0$, $f = f_0$): equations (A68) and (A70) are the governing equations for the KW approximation, equations (A68) and (A71) for the DW approximation and equations (A68) and (A69) for the DYW representation, with the upstream boundary condition given by equation (A76) and the downstream boundary condition given by equation (A78).

13. Scenario 3 ($q = q$, $f = f_0$): equations (A68) and (A72) are the governing equations for the KW approximation, equations (A68) and (A74) for the DW approximation, and equations (A68) and (A73) for the DYW representation, with the upstream boundary condition given by equation (A77).

14. Scenario 3, ($q = q_0$, $f = f_0$): equations (A68) and (A72) are the governing equations for the KW approximation, equations (A68) and (A74) for the DW approximation and equations (A68) and (A73) for the DYW representation, with the upstream boundary condition given by equation (A76) and the downstream boundary condition given by equation (A79).

15. Scenario 3 ($q = q_0$, $f = f_0$): equations (A68) and (A72) are the governing equations for the KW approximation, equations (A68) and (A74) for the DW approximation and equations (A68) and (A73) for the DYW representation, with the upstream boundary condition given by equation (A76) and the downstream boundary condition given by equation (A78).

16. Scenario 3 ($q = q_0$, $f = f_0$): equations (A68) and (A72) are the governing equations for the KW approximation, equations (A68) and (A74) for the DW approximation and equations (A68) and (A73) for the DYW representation, with the upstream boundary condition given by equation (A77).

17. Scenario 3 ($q = q_0$, f, $= f_0$): equations (A68) and (A72) are the governing equations for the KW approximation, equations (A68) and (A74) for the DW approximation and equations (A68) and

(A73) for the DYW representation, with the upstream boundary condition given by equation (A76) and the downstream boundary conditions given by equation (A79).

18. Scenario 3 ($q = q_0$, $f = f_0$): equations (A68) and (A72) are the governing equations for the KW approximation, equations (A68) and (A74) for the DW approximation and equations (A68) and (A73) for the DYW representation, with the upstream boundary condition given by equation (A76) and the downstream boundary conditions given by equation (A78).

19. Scenario 4 ($q = q_0$, $f = f_0$): equations (A68) and (A70) are the governing equations for the KW approximation, equations (A68) and (A71) for the DW approximation and equations (A68) and (A69) for the DYW representation, with the upstream boundary condition given by equation (A77).

20. Scenario 4 ($q = q_0$, $f = f_0$): equations (A68) and (A70) are the governing equations for the KW approximation, equations (A68) and (A71) for the DW approximation and equations (A68) and (A69) for the DYW representation, with the upstream boundary condition given by equation (A76); and the downstream boundary condition given by equation (A79).

21. Scenario 4 ($q = q_0$, $f = f_0$): equations (A68) and (A70) are the governing equations for the KW approximation, equations (A68) and (A71) for the DW approximation and equations (A68) and (A69) for the DYW representation, with the upstream boundary condition given by equation (A76) and the downstream boundary condition given by equation (A78).

These cases are summarized in Table A7 (Singh and Aravamuthan, 1992a–c).

Definition of error

The relative error E is defined (Singh *et al.*, 1993) as

$$E = \frac{S_K - S_D}{S_D} \tag{A145}$$

where S_K is the solution from the KW or DW approximation and S_D is the solution from the DYW representation. The solution can be either in terms of depth (h), velocity (u) or discharge (Q), i.e. $S\{u,h,Q\}$. Thus

$$E = \frac{u_K - u_D}{u_D}, \quad E = \frac{h_K - h_D}{h_D}, \quad E = \frac{Q_K - Q_D}{Q_D} \tag{A146}$$

where subscripts K and D correspond to the KW (or DW) and DYW solutions, respectively. The differential equation of E can be obtained by differentiating equation (A145) as

$$\frac{dE}{dx} = \frac{(E+1)}{S_K}\frac{dS_K}{dx} - \frac{(E+1)^2}{S_K}\frac{dS_D}{dx} \tag{A147}$$

$$= \frac{E+1}{S_K}\left[\frac{dS_K}{dx} - (E+1)\frac{dS_D}{dx}\right]$$

Note that

$$S_D = \frac{S_K}{E+1} \tag{A148}$$

Equation (A145) was used to define the error. The differential equation for error can, however, be defined without explicitly knowing S_D, so long as S_K is explicitly known.

Method of solution

The governing equations for DW and DYW representations with appropriate boundary conditions result in non-linear first-order ordinary differential equations. Analytical solutions of these equations do not seem tractable and hence the equations are integrated numerically using the fourth-order Runge–Kutta method. The DYW representation is assumed to be the correct solution and the spatial distributions of error in depths for the DW and KW approximations are calculated with respect to the DYW solution. The spatial variations of depth, velocity and error in depth are calculated for different values of lateral inflow, infiltration and upstream discharge, as well as for different boundary conditions. The equations are also solved with and without the momentum exchange term in the momentum equation.

Table A7 List of cases for derivation of error or the KW and DW approximations for steady, non-uniform flow

Case no.	Scenario no.	Governing equations			Boundary conditions			Lateral inflow q	Infiltration f
		KW approximation	DW approximation	DYW approximation	Upstream	Downstream (critical depth)	Downstream (zero-depth gradient)		
1	1	Equations A75 and A70	Equations A75 and A71	Equations A75 and A69	Equation A77	na	na	q_0	q
2	1	Equations A75 and A70	Equations A75 and A71	Equations A75 and A69	Equation A76	na	Equation A79	q_0	0
3	1	Equations A75 and A70	Equations A75 and A71	Equations A75 and A73	Equation A76	Equation A78	na	q_0	0
4	1	Equations A75 and A72	Equations A75 and A74	Equations A75 and A73	Equation A77	na	na	q_0	0
5	2	Equations A75 and A72	Equations A75 and A74	Equations A75 and A73	Equation A76	na	Equation A79	q_0	0
6	2	Equations A75 and A72	Equations A75 and A74	Equations A75 and A73	Equation A76	Equation A78	na	q_0	0
7	2	Equations A68 and A72	Equations A68 and A74	Equations A68 and A73	Equation A77	na	na	q_0	0
8	2	Equations A68 and A72	Equations A68 and A74	Equations A68 and A73	Equation A76	na	Equation A79	q_0	0
9	3	Equations A68 and A70	Equations A68 and A74	Equations A68 and A73	Equation A76	Equation A78	na	q_0	0
10	3	Equations A68 and A70	Equations A68 and A71	Equations A68 and A69	Equation A77	na	na	q_0	f_0
11	3	Equations A68 and A70	Equations A68 and A71	Equations A68 and A69	Equation A76	na	Equation A79	q_0	f_0
12	3	Equations A68 and A70	Equations A68 and A71	Equations A68 and A69	Equation A76	Equation A78	na	q_0	f_0
13	4	Equations A68 and A72	Equations A68 and A74	Equations A68 and A73	Equation A77	na	na	q_0	f_0
14	4	Equations A68 and A72	Equations A68 and A74	Equations A68 and A73	Equation A76	na	Equation A79	q_0	f_0
15	4	Equations A68 and A72	Equations A68 and A74	Equations A68 and A73	Equation A76	Equations A78	na	q_0	f_0
16	5	Equations A68 and A72	Equations A68 and A74	Equations A68 and A73	Equation A77	na	na	q_0	f_0
17	5	Equations A68 and A72	Equations A68 and A74	Equations A68 and A73	Equation A76	na	Equation A79	q_0	f_0
18	5	Equations A68 and A72	Equations A68 and A74	Equations A68 and A73	Equation A76	Equation A78	na	q_0	f_0
19	6	Equations A68 and A72	Equations A68 and A74	Equations A68 and A73	Equation A77	na	na	0	f_0
20	6	Equations A68 and A70	Equations A68 and A71	Equations A68 and A69	Equation A76	na	Equation A79	0	f_0
21	6	Equations A68 and A70	Equations A68 and A71	Equations A68 and A69	Equation A76	Equation A78	na	0	f_0

Error equations: non-zero depth at the upstream boundary

Kinematic wave solution

For the KW approximation with momentum exchange term retained, the governing equations are equations (A68) and (A72). Equation (A68) has the solution

$$u_* h_* = (q_* - f_*)x_* + a = q_{1*}x_* + a \quad (A149)$$

where a is the constant of integration and $q_{1*} = q_* - f_*$. Substituting equation (A149) in equation (A70) yields

$$Kh_*^3 - q_*(q_{1*}x_* + a)h_* - K(q_{1*}x_* + a)^2 = 0 \quad (A150)$$

If u_* is chosen as the dependent variable, we obtain

$$Ku_*^3 + u_*^2 q_* - K(q_{1*}x_* + a) = 0 \quad (A151)$$

When the momentum exchange term is neglected, equations (A150) and (A151) reduce to

$$h_* = (q_{1*}x_* + a)^{2/3} \quad (A152)$$

$$u_* = (q_{1*}x_* + a)^{1/3} \quad (A153)$$

The KW solution (dimensionless depth, velocity and discharge) was computed for different cases involving values of Froude number

$(F_0 = 0.1, 0.5$ and $1.0)$, kinematic wave number $(K = 3, 5, 10$ and $30)$, lateral inflow $q_* = 1.0$, zero discharge $a = 0$, at the upstream boundary and infiltration $f_* = 0.2$. The solution in terms of depth for a sample case $(F_0 = 0.5)$ is shown in Figure A6. In this case the solution was independent of K and F_0, and depended only on x_*. Although the depth and velocity varied non-linearly in space, the discharge varied linearly.

Diffusion wave solution

For the DW approximation with the momentum exchange retained, the governing equations are equations (A68) and (A71). Substituting equation (A149) into equation (A71) yields

$$\frac{dh_*}{dx_*} = KF_0^2\left(1 - \frac{(q_{1*}x_* + a)^2}{h_*^3}\right) - \frac{F_0^2(q_{1*}x_* + a)q_*}{h_*^2} \quad (A154)$$

If u_* is chosen as the dependent variable, we obtain

$$\frac{du_*}{dx_*} = [q_{1*}(q_{1*}x_* + a)\,u_* - KF_0^2(q_{1*}x_* + a)u_*^2 + KF_0^2 u_*^5 + q_* F_0^2 u_*^4]/[(q_{1*}x_* + a)^2] \quad (A155)$$

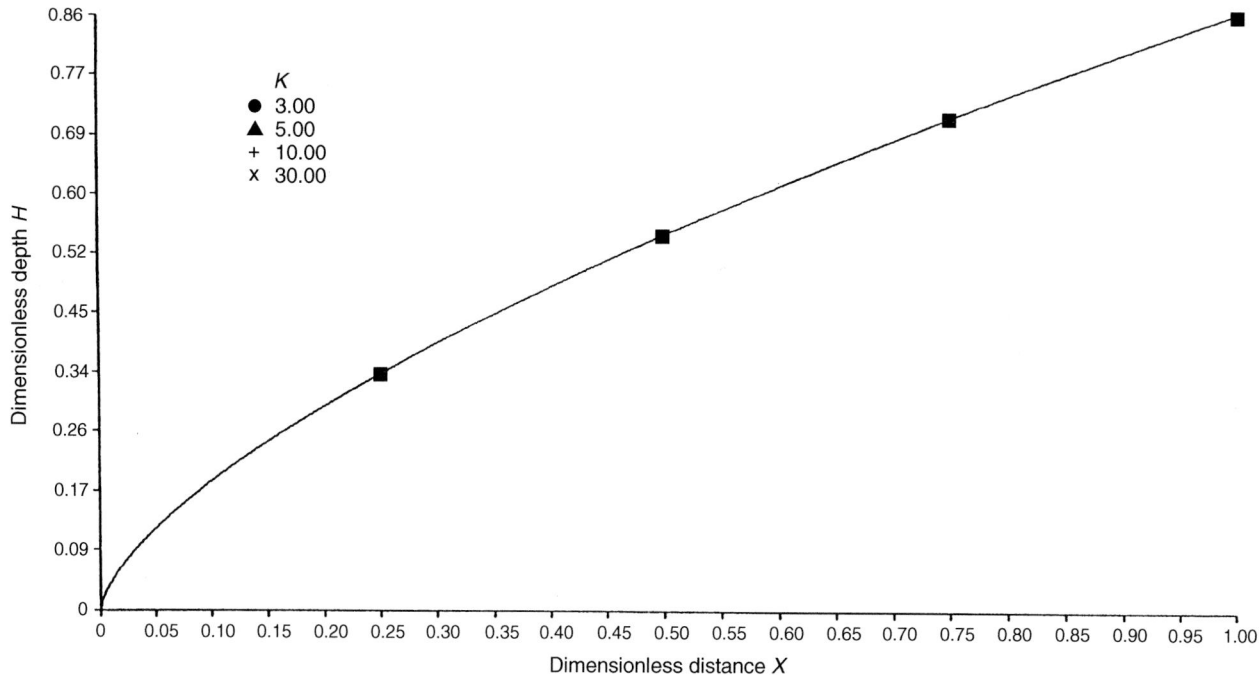

Figure A6 Dimensionless flow depth by the kinematic wave approximation as a function of dimensionless distance when the upstream velocity is zero, Froude number = 0.5, lateral inflow $q = 1.00$, infiltration $F = 0.20$ and K is variable.

When the momentum exchange term is neglected, equations (A154) and (A155) reduce to

$$\frac{dh_*}{dx_*} = KF_0^2 \left(1 - \frac{(q_{1*}x_* + a)^2}{h_*^3}\right) \tag{A156}$$

$$\frac{du_*}{dx_*} = [q_{1*}(q_{1*}x_* + a)u_* - KF_0^2(q_{1*}x_* + a)u_*^2 + KF_0^2u_*^5]/[(q_{1*}x_* + a)^2] \tag{A157}$$

An analytical solution of equation (A156) does not seem tractable; therefore, this equation was solved numerically using the fourth-order Runge–Kutta method. The DW solution (dimensionless depth, velocity and discharge) was computed for various values of K and F_0. For a sample case $F_0 = 0.5$, the solution in terms of dimensionless depth is shown in Figure A7. The solution is highly sensitive to K and F_0. For $F_0 = 0.5$, the flow depth at the upstream boundary significantly varied with K.

Dynamic wave solution

The governing equations for the DYW approximation with the momentum exchange term retained are given by equations (A68) and (A69).

Substituting equation (A149) into equation (A69) yields

$$\frac{dh_*}{dx_*} = \{F_0^2[Kh_*^3 - K(q_{1*}x_* + a)^2 - (q_{1*}x_* + a) \times q_*h_* - (q_{1*}x_* + a)q_{1*}h_*]\}/\{[h_*^3 - F_0^2(q_{1*}x_* + a)^2]\} \tag{A158}$$

If u_* is chosen as the dependent variable, we obtain

$$\frac{du_*}{dx_*} = [KF_0^2u_*^2(q_{1*}x_* + a) - KF_0^2u_*^5 - F_0^2u_*^4q_*]/[q_{1*}u_*(q_{1*}x_* + a) - (q_{1*}x_* + a)^2 + u_*^3F_0^2(q_{1*}x_* + a)] \tag{A159}$$

If the momentum exchange term is neglected, equations (A158) and (A159) reduce to

$$\frac{dh_*}{dx_*} = \{F_0^2[Kh_*^3 - K(q_{1*}x_* + a)^2 - (q_{1*}x_* + a)q_{1*}h_*]\}/[h_*^3 - F_0^2(q_{1*}x_* + a)^2] \tag{A160}$$

$$\frac{du_*}{dx_*} = [KF_0^2u_*^2(q_{1*}x_* + a) - KF_0^2u_*^5]/[q_{1*}u_*(q_{1*}x_* + a) - (q_{1*}x_* + a)^2 + u_*^3F_0^2(q_{1*}x_* + a)] \tag{A161}$$

An analytical solution to equation (A158) does not seem tractable; therefore, a numerical solution was obtained using the fourth-order Runge–Kutta method. The DYW solution (dimensionless flow depth, velocity and discharge) was computed for different cases. For a sample case ($F_0 = 0.5$) the solution in terms of dimensionless depth is shown in Figure A8. The solution was highly sensitive to K and F_0.

Error in KW approximation

For use of error equation (A147), h_K, dh_K/dx_*, h_D and dh_D/dx_* are needed. To that end,

$$h_K = (q_{1*}x_* + a)^{2/3}, \quad q_{1*} = q_* - f_* \tag{A162}$$

$$\frac{dh_K}{dh_*} = \frac{2q_{1*}}{3(q_{1*}x_* + a)^{1/3}} \tag{A163}$$

$$h_D = \frac{h_K}{(E + 1)} = \frac{(q_{1*}x_* + a)^{2/3}}{(E + 1)} \tag{A164}$$

$$\frac{dh_D}{dx_*} = \{F_0^2[K - K(E + 1)^3 - q_{1*}(q_{1*}x_* + a)^{-1/3} \times (E + 1)^2]\}/[1 - F_0^2(E + 1)^3] \tag{A165}$$

Substituting equations (A162)–(A165) into equation (A147) and a little algebraic manipulation yields

$$\frac{dE}{dx_*} = [2q_{1*}(E + 1)]/[3(q_{1*}x_* + a)] - \{(E + 1)^2F_0^2[K - K(E + 1)^3 - q_{1*}(q_{1*}x_* + a)^{-1/3}(E + 1)^2]\}/\{(q_{1*}x_* + a)^{2/3}[1 - F_0^2(E + 1)^3]\} \tag{A166}$$

Let $E_* = (E + 1)$ and $y = q_{1*}x_* + a$. Equation (A166) then becomes

$$\frac{dE_*}{dy} = (2E_*)/(3y) - \{E_*^2F_0^2[K - KE_*^3 - q_{1*}y^{-1/3}E_*^2]\}/\{q_{1*}y^{2/3}[1 - F_0^2E_*^3]\} \tag{A167}$$

This is the error differential equation for the KW approximation.

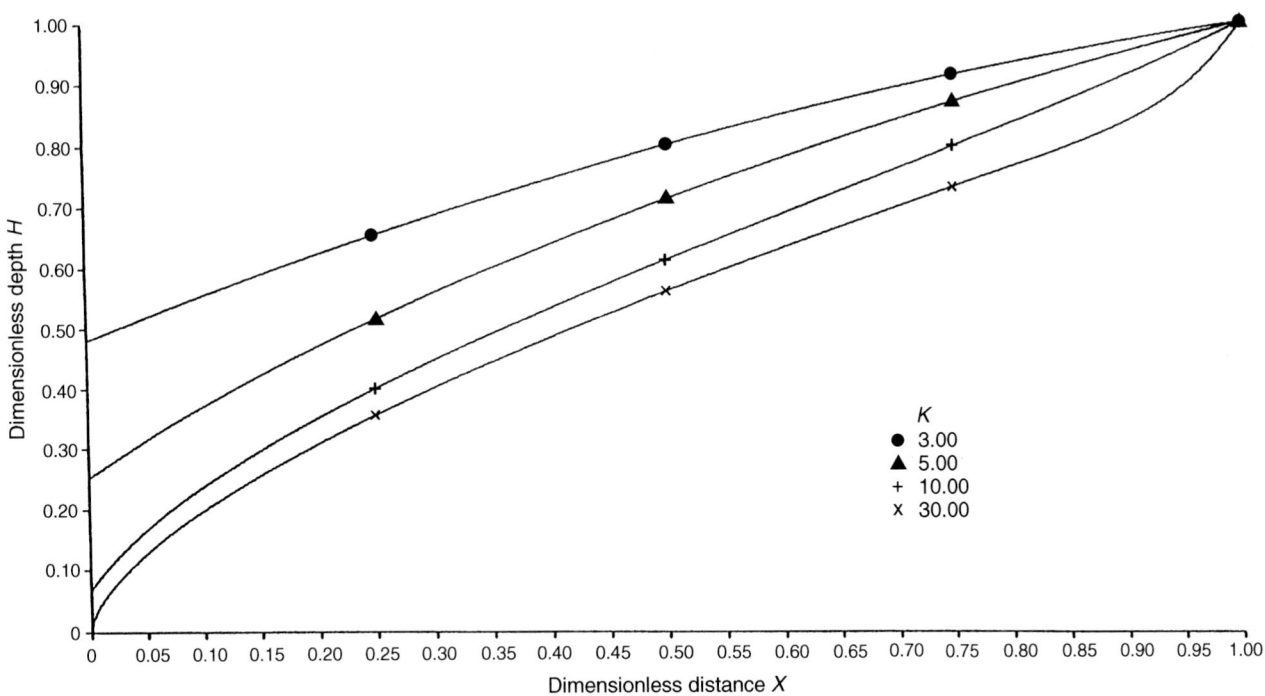

Figure A7 Dimensionless flow depth by the diffusion wave approximation as a function of dimensionless distance when the upstream velocity is zero, Froude number = 0.5, lateral inflow $q = 1.00$, infiltration $F = 0.20$ and K is variable. The upstream boundary condition is constant depth.

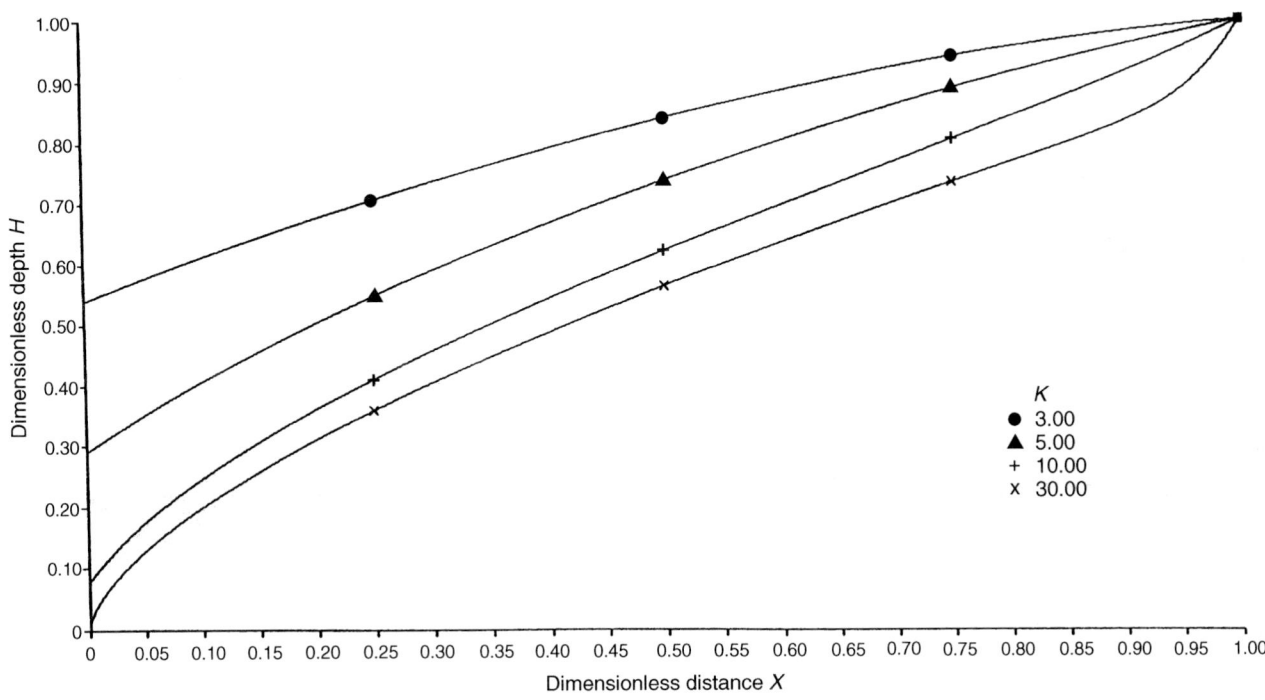

Figure A8 Dimensionless flow depth by the dynamic wave approximation as a function of dimensionless distance when the upstream velocity is zero, Froude number = 0.5, lateral inflow $q = 1.00$, infiltration $F = 0.20$ and K is variable. The upstream boundary condition is constant depth.

An explicit solution of equation (A167) is not possible. However, a numerical solution by the fourth-order Runge–Kutta method is relatively simple. The longitudinal variation of error in depth for a sample case $F_0 = 0.5$ is shown in Figure A9. The KW approximation

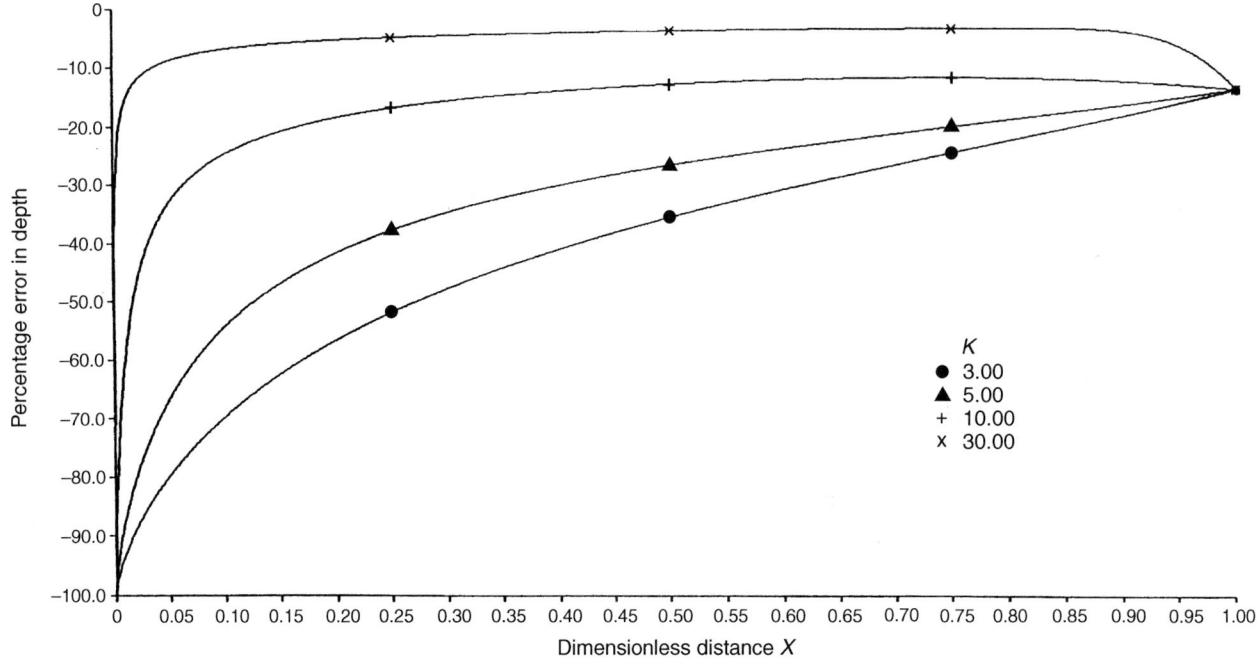

Figure A9 Error in the flow depth by the kinematic wave approximation as a function of dimensionless distance when the upstream velocity is zero, Froude number = 0.5, lateral inflow $q = 1.00$, infiltration $F = 0.20$ and K is variable. The upstream boundary condition is constant depth.

is a poor representation, especially for lower values of KF_0^2, even though for larger values of KF_0^2 the error declines significantly toward the downstream boundary.

Error in DW approximation

An explicit differential equation for error of the DW approximation does not seem tractable, but numerical computation of error is straightforward. The error variation for the flow depth in the longitudinal direction for a sample case $F_0 = 0.5$ is shown in Figure A10. As KF_0^2 increased, the error magnitude decreased. In general, the DW approximation was quite accurate, but its accuracy diminished for lower values of KF_0^2 toward the upstream boundary.

Error equations: zero depth gradient downstream boundary condition

Kinematic wave solution

The KW solution for h_* and u_* are obtained by substituting $a = 0$ (zero upstream discharge) in equations (A150)–(A153). The KW solution (dimensionless depth and velocity) for different values of K and $F_0 = 0.5$ is shown in Figure A6. The solution is independent of K and F_0, and depends only on x_*.

Diffusion wave solution

The DW solution for h_* and u_* are obtained by substituting $a = 0$ (zero upstream discharge) in equations (A154)–(A157). The DW solution (dimensionless depth and velocity) for different values of K and $F_0 = 0.5$ is shown in Figure A11. The solution is highly sensitive to K and F_0. For large values of KF_0^2 ($KF_0^2 = 7.5$), the upstream depth tends to zero.

Dynamic wave solution

The DYW solution for h_* and u_* is obtained by substituting $a = 0$ (zero upstream discharge) in equations (A158)–(A161). The DYW solution (dimensionless depth, velocity and discharge) for different values of K and $F_0 = 0.5$ is shown in Figure A12. The solution is highly sensitive to K and F_0. Whereas the depth and velocity vary

non-linearly with x_*, the discharge varies linearly. For $KF_0^2 \geq 5$, the upstream depth tends to vanish.

Error in KW approximation

The error differential equation is obtained by substituting $a = 0$ (zero upstream discharge) into equation (A166). The longitudinal variations of error in depth for a sample case $F_0 = 0.5$ is shown in Figure A13. The KW approximation gave error magnitudes of the order of 10% in the region $(0.05 < x_* < 1.0)$ for $KF_0^2 = 7.5$. The error magnitudes increase considerably for smaller values of KF_0^2 ($KF_0^2 = 0.75$) with error magnitudes greater than 50% for $x_* < 0.25$. This shows that the KW approximation is a poor representation for small values of KF_0^2

Error in DW approximation

An explicit expression for error in this case is not tractable, but its numerical computation is straightforward. The longitudinal variation of error in the flow depth is shown in Figure A14 for a sample case $F_0 = 0.5$. As KF_0^2 increases, the error magnitude decreases. In general the DW approximation was quite accurate, but its accuracy diminished for lower values of KF_0^2 and towards the upstream boundary.

Error equations: critical flow depth downstream boundary conditions

Kinematic wave solution

The KW solution for h_* and u_* are obtained by substituting $a = 0$ (zero upstream discharge) in equations (A150)–(A153). The KW solution (dimensionless depth, velocity and discharge) for different values of K and F_0^2 is shown in Figure A6. The discharge varies linearly, but the flow depth and velocity vary non-linearly. The solution is independent of K and F_0.

Diffusion wave solution

The DW solutions for h_* and u_* are obtained by substituting $a = 0$ (zero upstream discharge) in equations (A154)–(A157). The downstream boundary condition is given by equation (A78). The DW

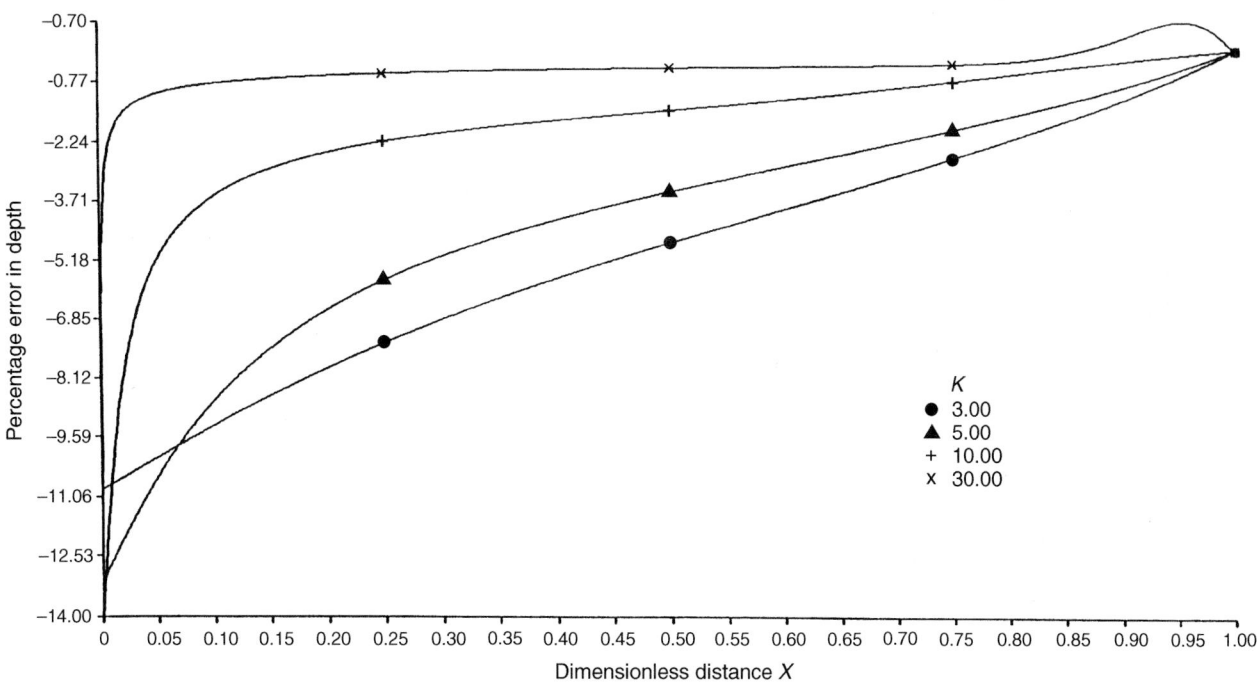

Figure A10 Error in the flow depth by the diffusion wave approximation as a function of dimensionless distance when the upstream velocity is zero, Froude number = 0.5, lateral inflow $q = 1.00$, infiltration $F = 0.20$ and K is variable. The upstream boundary condition is constant depth.

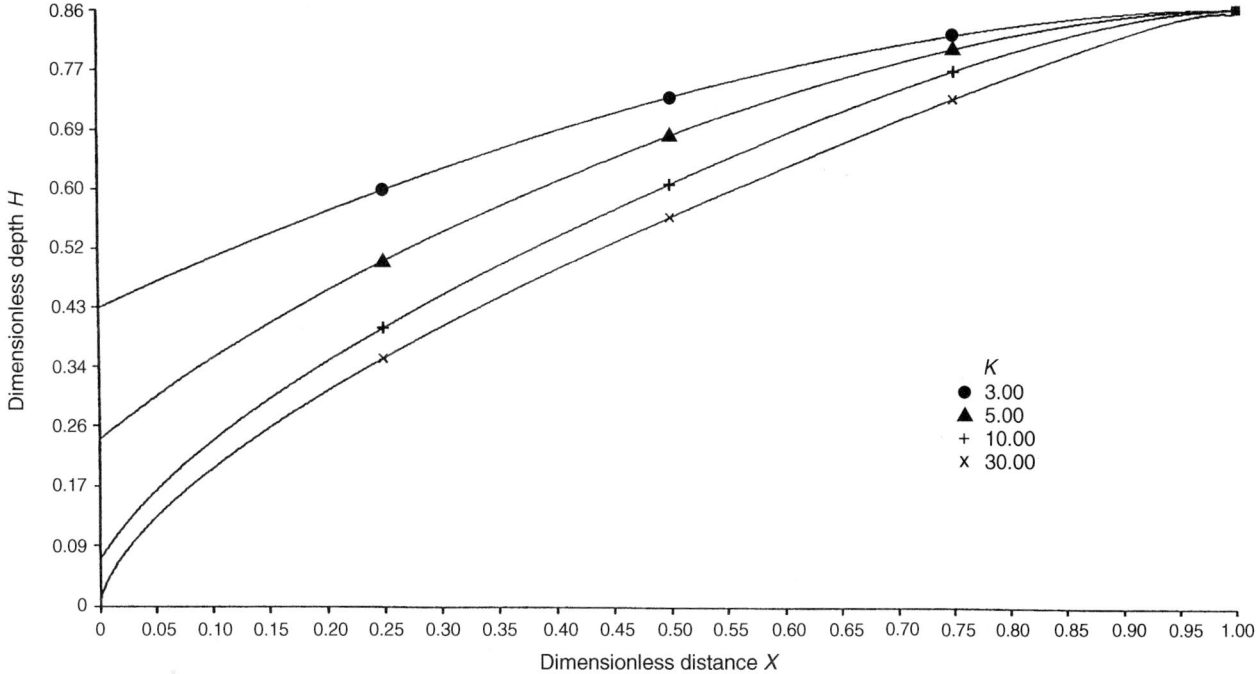

Figure A11 Dimensionless flow depth by the diffusion wave approximation as a function of dimensionless distance when the upstream velocity is zero, the downstream boundary condition is zero depth gradient, Froude number = 0.5, lateral inflow $q = 1.00$, infiltration $F = 0.20$ and K is variable.

solution (dimensionless depth and velocity) for different values of K and $F_0 = 0.5$ is shown in Figure A15. The solution is highly sensitive to K and F_0. The upstream depth tends to zero for $KF_0^2 \geq 5$. The solution changes significantly with F_0.

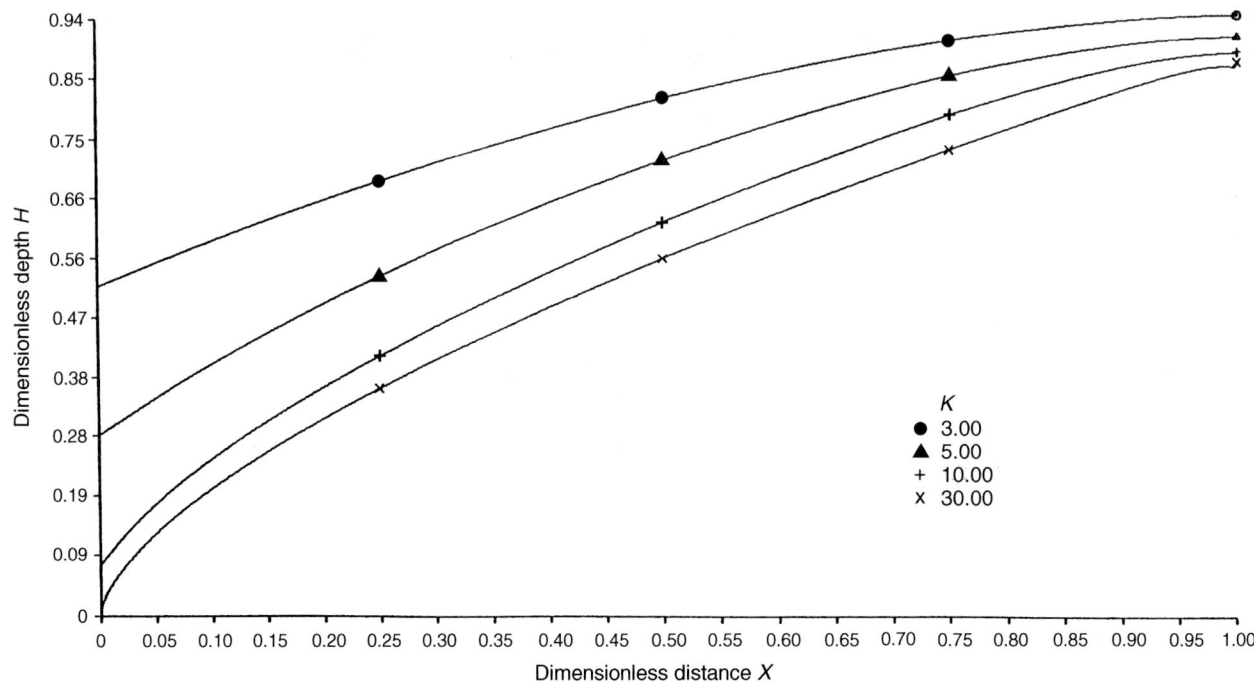

Figure A12 Dimensionless flow depth by the dynamic wave approximation as a function of dimensionless distance when the upstream velocity is zero, the downstream boundary condition is zero depth gradient, Froude number = 0.5, lateral inflow $q = 1.00$, infiltration $F = 0.20$ and K is variable.

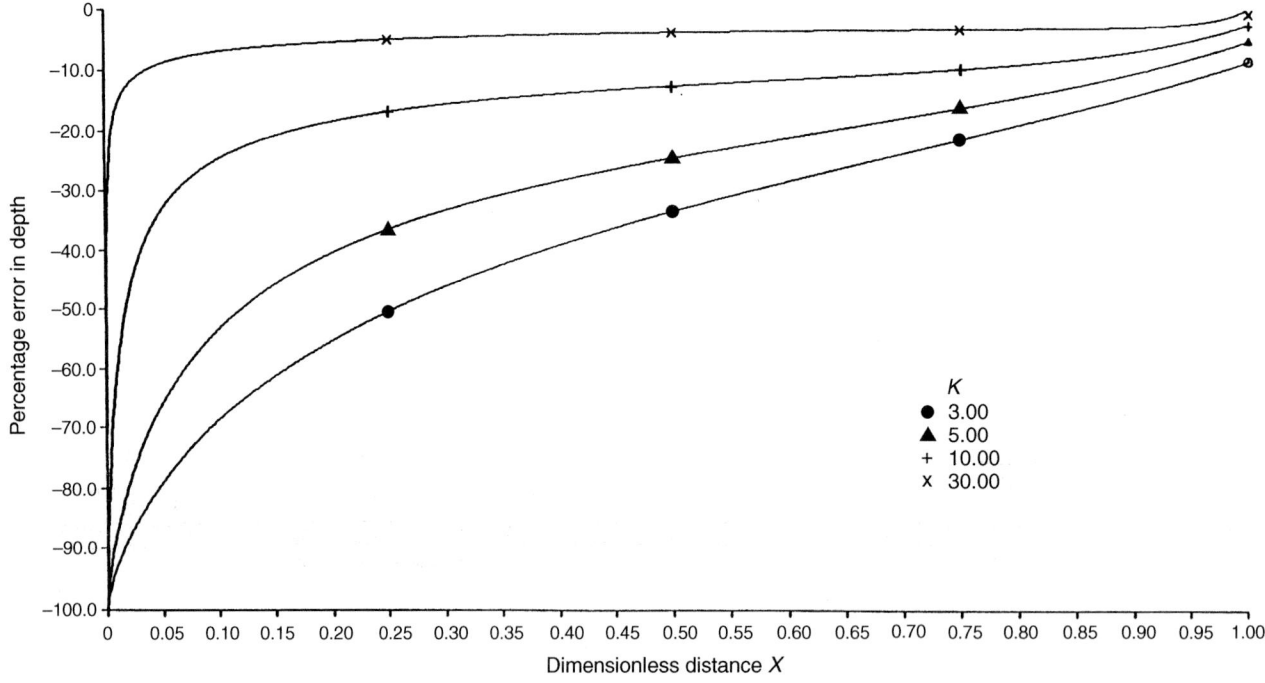

Figure A13 Error in the flow depth by the kinematic wave approximation as a function of dimensionless distance when the upstream velocity is zero, the downstream boundary is zero depth gradient, Froude number = 0.5, lateral inflow $q = 1.00$, infiltration $F = 0.20$ and K is variable.

Dynamic wave solution

The DYW solutions for h_* and u_* are obtained by substituting $a = 0$ (zero upstream discharge) in equations (A158)–(A161). The downstream boundary condition is given by equation (A78). The DYW

solution (dimensionless depth, velocity and discharge) for different values of K and $F_0 = 0.5$ is shown in Figure A16. The solution is highly sensitive to K and F_0. While the depth and velocity non-linearly vary with x_*, the discharge varies linearly. The upstream depth tends to zero for $KF_0^2 \geq 5$.

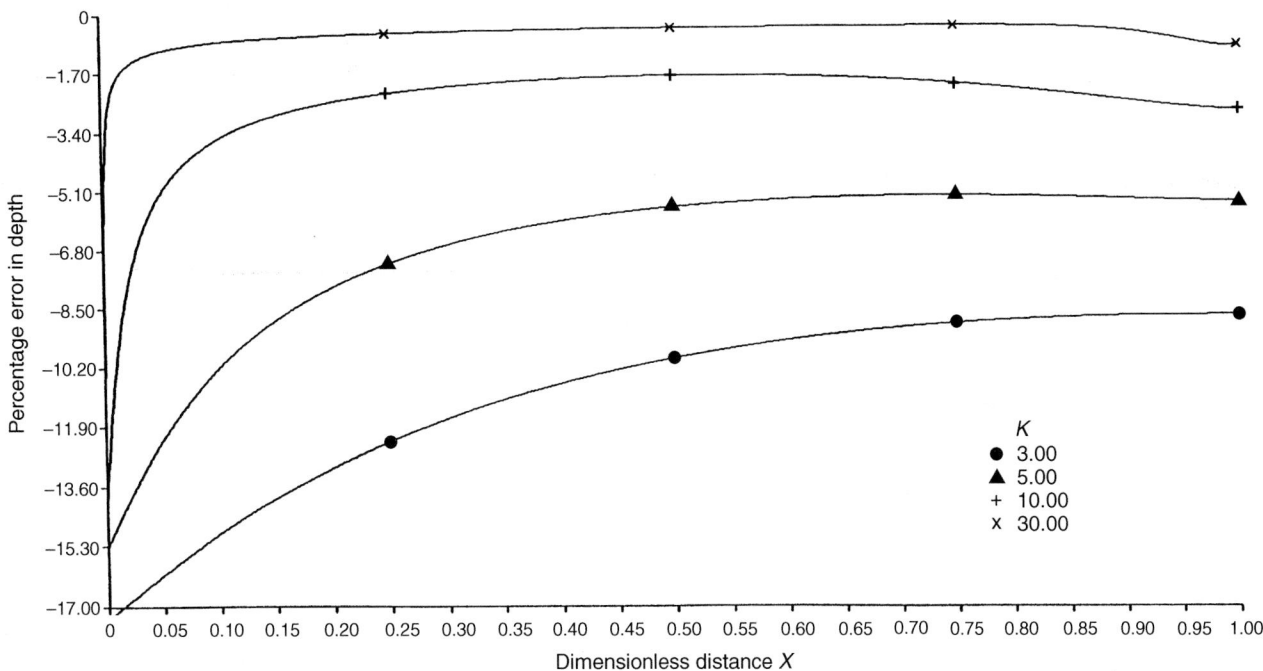

Figure A14 Error in the flow depth by the diffusion wave approximation as a function of dimensionless distance when the upstream velocity is zero, the downstream boundary is zero depth gradient, Froude number = 0.5, lateral inflow $q = 1.00$, infiltration $F = 0.20$ and K is variable.

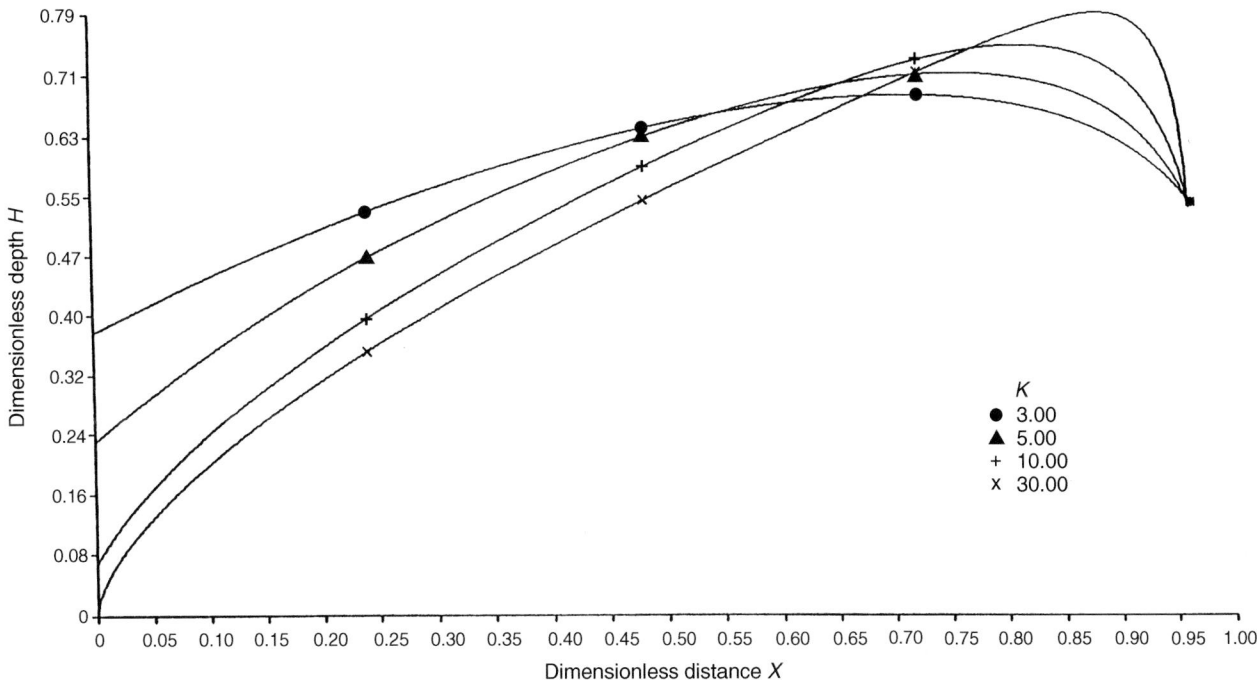

Figure A15 Dimensionless flow depth by the diffusion wave approximation as a function of dimensionless distance when the upstream velocity is zero, the downstream condition is critical flow, Froude number = 0.5, lateral inflow $q = 1.00$, infiltration $F = 0.20$ and K is variable.

Error in KW approximation

The error differential equation is obtained by substituting $a = 0$ (zero upstream discharge) into equation (A166). The longitudinal variation of error in depth for a sample case $F_0 = 0.5$ is shown in Figure A17. The KW approximation performed poorly at both the upstream and downstream ends of the channel with error magnitudes greater than

50% even for large values of KF_0^2. The KW approximation does not seem appropriate for this case and its usage is not recommended.

Error in DW approximation

The error of the DW approximation is computed numerically. The longitudinal variation of error in flow depth for a sample case $F_0 = 0.5$

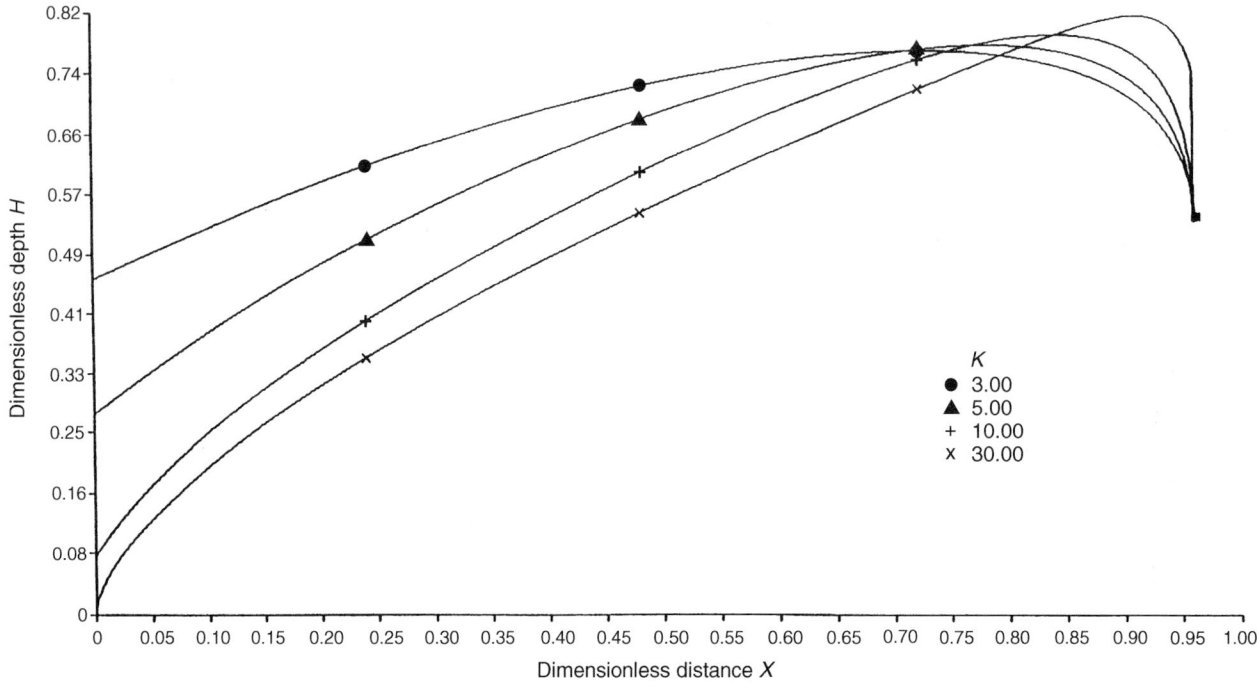

Figure A16 Dimensionless flow depth by the dynamic wave representation as a function of dimensionless distance when the upstream velocity is zero, the downstream condition is critical flow, Froude number = 0.5, and $K = 30.0$.

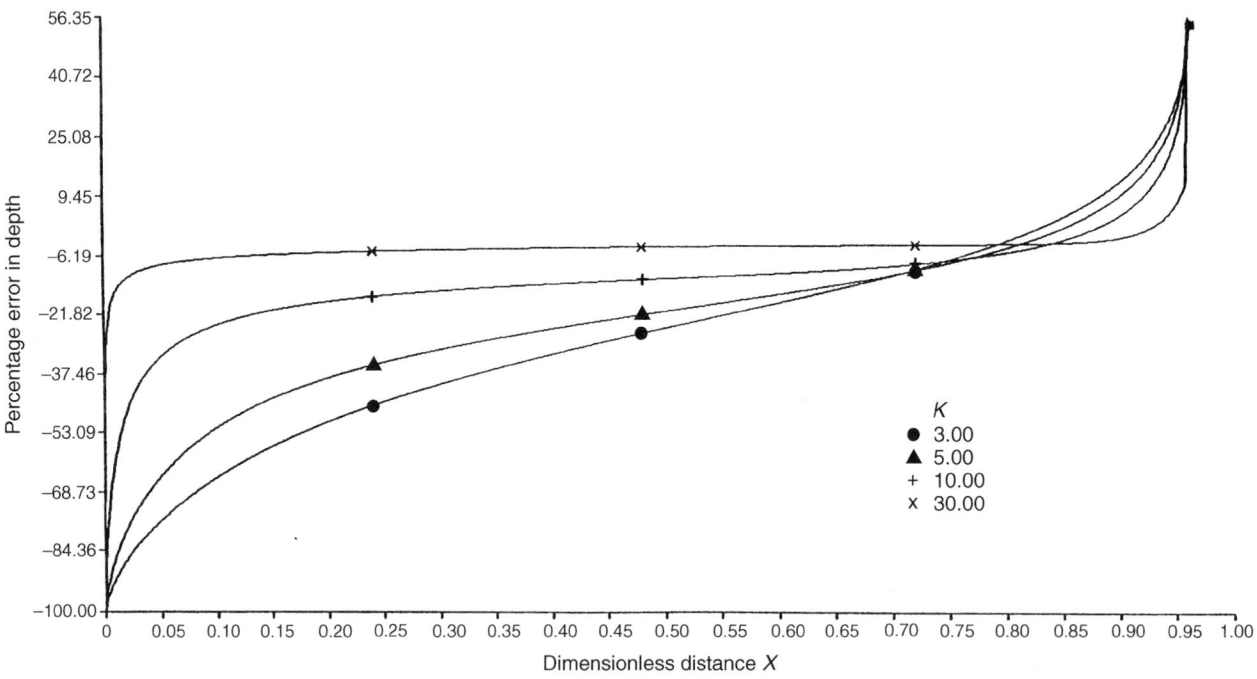

Figure A17 Error in the flow depth by the kinematic wave approximation as a function of dimensionless distance when the upstream velocity is zero, the downstream condition is critical flow, Froude number = 0.5, and $K = 30.0$.

is shown in Figure A18. As KF_0^2 increased, the error magnitude decreased. In general the DW approximation was quite accurate (error magnitude less than 2%) for $KF_0^2 = 7.5$ with an almost constant error profile. The error profile becomes steeper for small values of KF_0^2

with error magnitudes varying from 18% at the upstream boundary to about 14% at $x_* = 0.9$ for $KF_0^2 = 0.75$. If the above magnitudes of error are acceptable, then DW approximation is a reasonable representation.

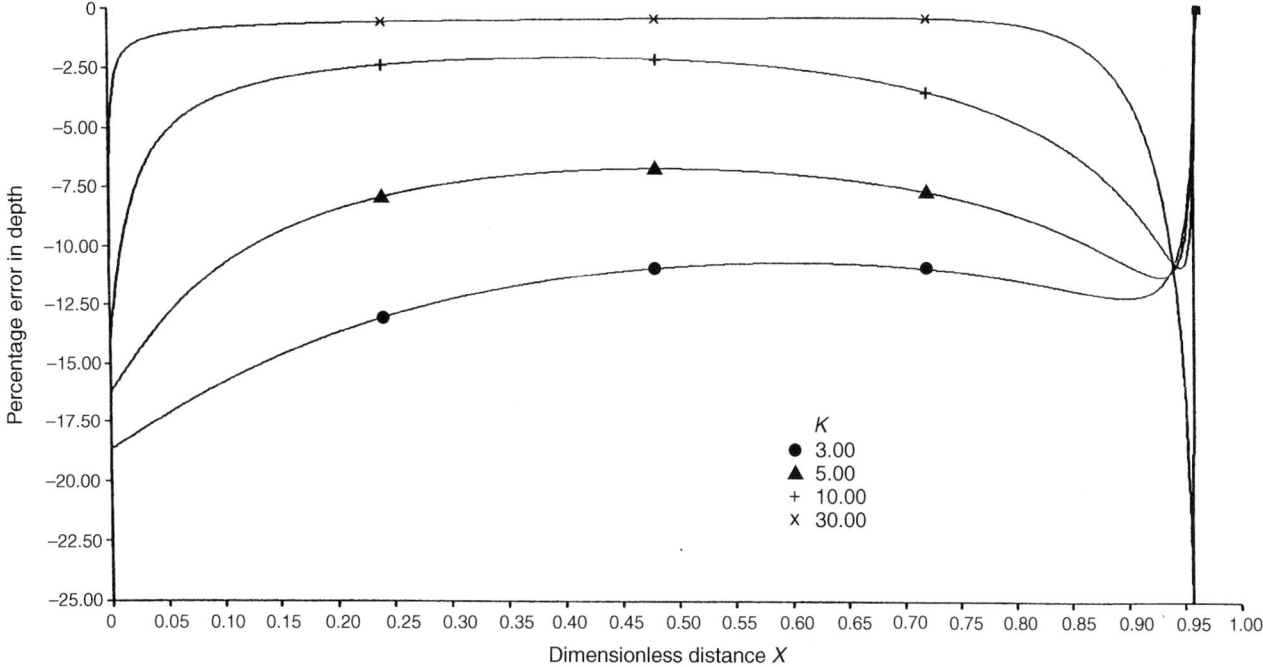

Figure A18 Error in the flow depth by the diffusion wave approximation as a function of dimensionless distance when the upstream velocity is zero, the downstream condition is critical flow, Froude number = 0.5, and $K = 30.0$.

Conclusions

The following conclusions can be drawn from this study.

1. When the upstream boundary conditions in terms of zero discharge and finite depth are known, the governing differential equation cannot be integrated directly from upstream to downstream. An iterative procedure must be followed by which the downstream depth, K and F_0 are varied until the upstream depth agrees with the given value. It follows that a solution may not be possible for some arbitrary upstream depth, K and F_0.

2. For upstream boundary condition of zero discharge and finite depth, the KW approximation was reasonable only for large values of KF_0^2 and was valid in the region $0.05 < x_* < 1.0$. The error profiles were flatter at almost a constant value of error ($\sim10.5\%$) for $KF_0^2 = 7.5$ and the profile became steeper as KF_0^2 decreased to 0.75. This is evident from Figure A9 where the error varies from almost 15% at the upstream boundary and greater than 50% at $x_* = 0.25$. This indicates that the KW approximation is a poor approximation for small values of KF_0^2.

   The DW approximation, on the other hand, gave good results with the magnitude of error less than 1% for $KF_0^2 = 7.5$ and varying from 1% at the upstream end to 11% at the downstream end for $KF_0^2 = 0.75$. For small values of KF_0^2, the DW approximation is better than the KW approximation. If the error magnitudes described above are acceptable, then the DW approximation is preferred to the KW approximation for small values of KF_0^2.

3. For the downstream boundary condition of zero depth gradient, the error profile for the KW approximation was flat for $KF_0^2 = 7.5$ with an error magnitude of the order of 15% in the region $0.05 < x_* < 1.0$. The error profile became steeper for small values of KF_0^2 ($KF_0^2 = 0.75$) with error magnitudes greater than 50% for $x_* < 0.25$. The DW approximation gave flatter error profiles when compared to the KW approximation. The errors were about 1% for $KF_0^2 = 7.5$ and varied from 10 to 17% for $KF_0^2 = 0.75$. This leads to the conclusion that the DW approximation is preferred to the KW approximation for small values of KF_0^2.

4. For the downstream boundary condition of critical depth, the KW approximation performed poorly at both the upstream and downstream ends ($x_* < 0.05$ and $x_* > 0.80$) for all values of KF_0^2. The error magnitude for $KF_0^2 = 7.5$ in the region ($0.05 < x_* < 0.80$) was about 6%. The error magnitudes for $KF_0^2 = 0.75$ varied from about 15% at $x_* = 0.80$ to about 65% at $x_* = 0.05$. The DW approximation was in good agreement with the DYW approximation with error less than 2% in the region ($0.05 < x_* < 0.80$). For small values of KF_0^2, the error magnitudes varied from about 18% at $x_* = 0.0$ to 15% at $x_* = 0.9$. The DW approximation is better than the KW approximation and should be preferred for smaller values of KF_0^2.

5. It was generally found that at the boundary farthest away from the boundary where the boundary condition was applied, the error always increased for both DW and KW approximations. This leads to the conclusion that if accurate solutions are needed in these regions and if error magnitudes indicated above are not acceptable, then the DYW representation should be used.

V.P. Singh and V. Aravamuthan

Bibliography

Govindaraju, R.S., Jones, S.E. and Kavvas, M.L., 1988. On the diffusion wave model for overland flow: 2. Steady state analysis. *Water Resources Research*, **24**(5), 745–754.

Morris, M., 1978. The effect of the small-slope approximation and lower boundary condition on the solutions of the Saint Venant equations. *Journal of Hydrology*, **40**, 31–47.

Parlange, J.Y., Hogarth, W., Sander, G.C. *et al.*, 1989. Comment on 'On the diffusion wave model for overland flow: 2. Steady state analysis,' by R.S. Govindaraju, S.E. Jones and M.L. Kavvas. *Water Resources Research*, **25**(8), 1923–1924.

Parlange, J.Y., Hogarth, W., Sander, G. *et al.*, 1990. Asymptotic expansion for steady-state overland flow. *Water Resources Research*, **26**(4), 579–583.

Pearson, C.P., 1989. One-dimensional flow over a plane: criteria for kinematic wave modeling. *Journal of Hydrology*, **111**, 39–48.

Singh, V.P. and Aravamuthan, V., 1992a. Errors in kinematic and diffusion wave approximations for time-independent flows: 1. Cases 1 to 6. Technical Report WRR20, Water Resources Program, Department of Civil Engineering, Louisiana State University, Baton Rouge, LA.

Singh, V.P. and Aravamuthan, V., 1992b. Errors in kinematic and diffusion wave approximations for time-independent flows: 2. Cases 7 to 13. Technical Report WRR21, Water Resources Program, Department of Civil Engineering, Louisiana State University, Baton Rouge, LA.

Singh, V.P. and Aravamuthan, V., 1992c. Errors in kinematic and diffusion wave approximations for time-independent flows: 3. Cases 14 to 19. Technical Report WRR22, Water Resources Program, Department of Civil Engineering, Louisiana State University, Baton Rouge, LA.

Singh, V.P., Aravamuthan, V. and Joseph, E.S., 1993. Accuracy of hydrodynamic models of flood discharge determinations. *Proceedings, International Conference on Environmental Management: Geo-Water and Engineering Aspects*, pp. 79–90, Wollongong, Australia.

Cross references

Accuracy
Accuracy of hydrodynamic approximations in hydrology: unsteady, uniform flow

ACCURACY OF HYDRODYNAMIC APPROXIMATIONS IN HYDROLOGY: UNSTEADY, UNIFORM FLOW

Unsteady cases of planar or channel flow are treated. The kinematic wave, diffusion wave and dynamic wave solutions are derived under different initial conditions and are parameterized through a dimensionless parameter γ. This parameter reflects the effect of initial depth of flow, channel bed slope, lateral inflow, infiltration and channel roughness. By comparing the kinematic wave and diffusion wave solutions with the dynamic wave solution, equations are derived in terms of γ for the error in the kinematic wave and diffusion wave approximations. These equations describe error as a function of time.

Introduction

Physically based models of overland flow, channel flow, surface irrigation and many other phenomena involving unsteady, free-surface open-channel flows are based on the shallow water wave (SWW) theory. These models are based either on the kinematic wave (KW) approximation (Lighthill and Whitham, 1955), diffusion wave approximation (DW), diffusion analogy (DA) or dynamic wave (DYW) representation. Lighthill and Whitham (1955) showed that in subcritical flow (at the Froude number – the ratio of inertial force to gravity force – less than one, appropriate to flood waves) the dynamic waves are rapidly attenuated and the kinematic waves become dominant. Using a dimensionless form of the St Venant (SV) equations, Woolhiser and Liggett (1967) obtained what is now referred to as the kinematic wave number, K, as a criterion for evaluating the adequacy of the KW approximation. The kinematic wave number reflects the effect of bed slope, channel length, normalizing depth and Froude number. For K greater than 20, the KW approximation was considered to be an accurate representation of the SV equations in modeling of overland flow. However, no relation between K and the error in the KW approximation was suggested. Morris and Woolhiser (1980) modified the above criterion with an explicit inclusion of Froude number, F_0, and showed, based on numerical experimentation, that $KF_0^2 \geq 5$ was a better indicator of the adequacy of the KW approximation. A relation between this criterion and the error resulting from the KW approximation was not derived, however.

Using a linear perturbation analysis, Ponce and Simons (1977) derived properties of the KW and DW approximations as well as DYW representations in modeling of open channel flows. They derived a spectrum showing the regions of the validity of the KW and DW approximations. Menendez and Norscini (1982) extended the work of Ponce and Simons by including the phase lag between the depth and velocity of flow. Their results were, however, similar to those of Ponce and Simons (1977). In another but similar study, Ponce *et al.* (1978), based on propagation characteristics of sinusoidal perturbation, derived criteria to evaluate the adequacy of the KW and DW approximations. Daluz Vieira (1983) compared solutions of the SV equations with those of the KW and DW approximations for a range of the values of F_0 and K, and defined the regions of validity of these approximations in the K–F_0 space.

Fread (1985) developed criteria for defining the range of application of the KW and DW approximations. These were based on an analysis of the magnitude of the normalized errors in the momentum equation due to the omission of certain terms. In a comprehensive study, Ferrick (1985) defined a group of dimensionless scaling parameters to establish the spectrum of river waves, with continuous transitions between wave types and subtypes. With the aid of these parameters he was able to discern when the KW and DW approximations would be valid.

In most of these studies, different types of criteria have clearly been established to evaluate the adequacy of the KW and/or DW approximations, but no explicit relations either in time or space between these criteria, and the errors resulting from these approximations have been derived yet. Furthermore, when conducting hydrologic modeling it is not evident if the KW and DW approximations are valid for the entire hydrograph or a portion thereof (Singh, 1992a,b,c, 1993a; Singh *et al.*, 1993). In other words, most of these criteria take on fixed point values for a given event. A historical perspective on these criteria was presented by Singh (1993, 1994a–c). Under simplified conditions, he (Singh, 1993, 1994a-c) derived error equations for the KW and DW approximations for unsteady, space-independent flow, which specify errors as a function of time. He considered nine different cases depending upon the inclusion or neglect of the lateral inflow or rainfall in the momentum equation. This article is based on his work and reports some additional results.

Shallow water wave (SWW) theory

The SWW theory can be described by some form of the SV equations. For flow over an infiltrating plane subject to uniform rainfall, these equations can be written in one-dimensional form on a unit width basis as:

Continuity equation:

$$\frac{\partial h}{\partial t} + \frac{\partial}{\partial x}(uh) = q - f \tag{A168}$$

Momentum equation:

$$\frac{\partial u}{\partial t} + \frac{\partial}{\partial x}\left(\frac{1}{2}u^2 + gh\right) = g(S_0 + S_f) - \frac{qu}{h} \tag{A169}$$

where h is the depth of flow (L), u is local mean velocity (L/T), q is uniform rainfall intensity (L/T), f is uniform infiltration rate (L/T), g is acceleration due to gravity, x is space coordinate in the direction of flow (L), t is time (T), S_0 is bed slope and S_f is frictional slope. Note $Q = uh$ is discharge (L^3/(TL)) per unit width. S_f can be approximated as

$$S_f = \beta \frac{u^2}{h} \tag{A170}$$

where β is some resistance parameter (T^2/L). If the Chézy relation is used for representing the friction then $\beta = g/C^2$, where C is Chézy's resistance parameter. Equation (A169) is based on accounting for inertial, pressure, gravity and frictional forces, as well as the momentum exchange between lateral inflow and overland (or channel) flow.

The DYW representation employs the full form of equations (A168) and (A169). The KW approximation is based on equation (A168) and equation (A169) with the left side omitted:

$$g(S_0 - S_f) - \frac{qu}{h} = 0 \tag{A171}$$

Equation (A171) neglects the effect of local and convective acceleration and the pressure gradient, but includes bed slope, frictional slope and the momentum exchange. The DW approximation uses equation (A168) and equation (A169) with local and convective acceleration deleted:

$$g\frac{\partial h}{\partial x} = g(S_0 - S_f) - \frac{qu}{h} \tag{A172}$$

Analytical solutions of the SV equations or their variants in the KW and DW approximations are tractable only for simple cases. To that end, the space-independent case is considered in this article. In this case the water surface is flat.

Space-independent flow

For space-independent (or uniform) flows, equation (A168) takes the form

$$\frac{dh}{dt} = q - f \tag{A173}$$

and equation (A169) becomes

$$\frac{dh}{dt} = g(S_0 - S_f) - \frac{qu}{h} \tag{A174}$$

The assumption of space independence implies omission of convective inertial and pressure forces in the momentum equation. These factors must be small by comparison with gravity, friction and local inertial forces to justify this assumption. Equations (A173) and (A174) are the governing equations for the DYW representation for spatially uniform flow. The KW approximation is based on equation (A173) and equation (A174) with the left side dropped:

$$g(S_0 - S_f) - \frac{qu}{h} = 0 \tag{A175}$$

The DW approximation uses equation (A173) as well as equation (A175). Therefore, for spatially uniform flow, the KW approximation is identical to the DW approximation. Equation (A175) can also be approximated by neglecting the momentum exchange between lateral inflow and longitudinal channel flow as

$$S_0 = S_f \tag{A176}$$

which can be expressed as equation (A170). Similarly, equation (A174) can be written as

$$\frac{dh}{dt} = g(S_0 - S_f) \tag{A177}$$

Depending upon the presence or absence of f, equation (A173) can also be simplified. If $f = 0$, then

$$\frac{dh}{dt} = q \tag{A178}$$

Although spatial independence seldom exists in surface flows, it does offer a useful first approximation and gives a lot of physical insight (Singh, 1992c; Singh *et al.*, 1993). Furthermore, when the SV equations are solved in the *x–t* plane, surface flow exhibits spatial independence during part of its rising limb or for part of the solution domain. On the other hand, the assumption of spatial independence can be likened to the concept underlying the systems approach so commonly employed in surface-water hydrology (Singh, 1988). In this approach the hydrological system is treated as lumped or as a black box.

Initial conditions

Two types of initial conditions can be assumed:

1. $h(0) = h_0, \; u(0) = u_0$ (A179a)
2. $u(0) = 0, \; h(0) = 0$ (A179b)

where h_0 is the initial depth of flow and u_0 is the initial flow velocity. Equation (A179a) implies that the channel is wet and has uniform flow at the beginning, whereas equation (A179b) considers the channel to be dry.

Types of scenarios

Depending upon the presence of lateral inflow and infiltration, four different scenarios can be considered:

1. $f = 0$, $q = q_0 = $ constant
   This includes the case q = 0
2. $q = q_0 = $ constant
   $f = f_0 = $ constant
   This includes the case $q = f = 0$.
3. $q - f = 0$, $q = q_0 = $ constant
   This includes the case $q = f = 0$.
4. $q = 0$, $f = f_0 = $ constant
   This includes the case f = 0.

It may be noted that the scenario with $q = 0$ in equation (A178) or $(q - f) = 0$ in equation (A173) applies to the recession hydrograph. The same applies if $(q - f) < 0$.

Definition of error

The relative error E can be defined as $E = (S_K - S_D)/S_D$, where S_K is the solution from the KW or DW approximation and S_D is the solution from the DYW representation. The subscripts K and D correspond to the KW (or DW) and DYW solutions, respectively. The solution can be either in terms of depth (h), velocity (u), or discharge (Q), i.e. $S = \{u, h, Q\}$. Thus the error differential equation is (Singh, 1993a):

$$E = \frac{u_K - u_D}{u_D}, \; E = \frac{h_K - h_D}{h_D} \text{ or } E = \frac{Q_K - Q_D}{Q_D} \tag{A180a}$$

$$\frac{dE}{dt} = \frac{(E + 1)}{S_K}\frac{dS_K}{dt} - \frac{(E + 1)^2}{S_K}\frac{dS_D}{dt} \tag{A180b}$$

Note that $S_D = S_K/(E + 1)$.

Scenarios for determination of error

Error equations can be derived for the KW and DW approximations under the above-mentioned initial conditions for four different scenarios. To summarize, the following cases can be treated:

1. Scenario 1: equations (A178) and (A175) are the governing equations for the KW approximation and equations (A178) and (A174) for the DYW representation, with the initial condition given by equation (A179a).
2. Scenario 1: equations (A178) and (A175) are the governing equations for the KW approximation and equations (A178) and (A174) for the DYW representation, with the initial condition given by equation (A179b).
3. Scenario 1: equations (A178) and (A176) are the governing equations for the KW approximation and equations (A178) and (A174) for the DYW representation, with the initial condition given by equation (A179a).
4. Scenario 1: equations (A178) and (A176) are the governing equations for the KW approximation and equations (A178) and (A174) for the DYW representation, with the initial condition given by equation (A179b).
5. Scenario 1: equations (A178) and (A176) are the governing equations for the KW approximation and equations (A178) and (A177) for the DYW representation, with the initial condition given by equation (A179a).
6. Scenario 1: equations (A178) and (A176) are the governing equations for the KW approximation and equations (A178) and (A177) for the DYW representation, with the initial condition given by equation (A179b).
7. Scenario 2: equations (A173) and (A175) are the governing equations for the KW approximation and equations (A173) and (A174) for the DYW representation, with the initial condition given by equation (A179a).
8. Scenario 2: equations (A173) and (A175) are the governing equations for the KW approximation and equations (A173) and (A174) for the DYW representation, with the initial condition given by equation (A179b).
9. Scenario 2: equations (A173) and (A176) are the governing equations for the KW approximation and equations (A173) and (A174) for the DYW representation, with the initial condition given by equation (A179a).
10. Scenario 2: equations (A173) and (A176) are the governing equations for the KW approximation and equations (A173) and (A174) for the DYW representation, with the initial condition given by equation (A179b).
11. Scenario 2: equations (A173) and (A176) are the governing equations for the KW approximation and equations (A173) and (A177) for the DYW representation, with the initial condition given by equation (A179a).
12. Scenario 2: equations (A173) and (A176) are the governing equations for the KW approximation and equations (A173) and (A177) for the DYW representation, with the initial condition given by equation (A179b).
13. Scenario 3: equation (A178) with $q = 0$ and equation (A175) are the governing equations for the KW approximation and equations

Table A8 List of cases for derivation of error for unsteady, uniform flow

| Case no. | Scenario no. | Governing equations | | Initial conditions | Lateral inflow q | Lateral outflow f |
		KW/DW approximation	DYW representation			
1	1	Equations A178 and A175	Equations A178 and A174	Equation 179a	q = constant	f = 0
2	1	Equations A178 and A175	Equations A178 and A174	Equation 179b	q = constant	f = 0
3	1	Equations A178 and A176	Equations A178 and A174	Equation 179a	q = constant	f = 0
4	1	Equations A178 and A176	Equations A178 and A174	Equation 179b	q = constant	f = 0
5	1	Equations A178 and A176	Equations A178 and A177	Equation 179a	q = constant	f = 0
6	1	Equations A178 and A176	Equations A178 and A177	Equation 179b	q = constant	f = 0
7	2	Equations A173 and A175	Equations A173 and A174	Equation 179a	q = constant	f = constant
8	2	Equations A173 and A175	Equations A173 and A174	Equation 179b	q = constant	f = constant
9	2	Equations A173 and A176	Equations A173 and A174	Equation 179a	q = constant	f = constant
10	2	Equations A173 and A176	Equations A173 and A174	Equation 179b	q = constant	f = constant
11	2	Equations A173 and A176	Equations A173 and A177	Equation 179a	q = constant	f = constant
12	2	Equations A173 and A176	Equations A173 and A177	Equation 179b	q = constant	f = constant
13	3	Equations A178 with $q=0$ and A175	Equations A178 with $q=0$ and A174	Equation 179a	q = 0	f = 0
14	3	Equations A178 with $q=0$ and A176	Equations A178 with $q=0$ and A174	Equation 179a	q = 0	f = 0
15	3	Equations A178 with $q=0$ and A176	Equations A178 with $q=0$ and A177	Equation 179a	q = 0	f = 0
16	4	Equations A173 with $q=0$ and A175	Equations A173 with $q=0$ and A174	Equation 179a	q = 0	f = constant
17	4	Equations A173 with $q=0$ and A176	Equations A173 with $q=0$ and A176	Equation 179a	q = 0	f = constant
18	4	Equations A173 with $q=0$ and A176	Equations A173 with $q=0$ and A177	Equation 179a	q = 0	f = constant

(A178) and (A174) for the DYW representation, with the initial condition given by equation (A179a).

14. Scenario 3: equation (A178) with $q = 0$ and equation (A176) are the governing equations for the KW approximation and equations (A178) and (A174) for the DYW representation, with the initial condition given by equation (A179a).

15. Scenario 3: equation (A178) with $q = 0$ and equation (A176) are the governing equations for the KW approximation and equations (A178) and (A177) for the DYW approximation, with the initial condition given by equation (A179a).

16. Scenario 4: equations (A173) with $q = 0$ and equation (A175) are the governing equations for the KW approximation and equation (A173) with $q = 0$ and equation (A174) for the DYW approximation with the initial condition given by equation (A179a).

17. Scenario 4: equation (A173) with $q = 0$ and equation (A176) are the governing equations for the KW approximation and equation (A173) with $q = 0$ and equation (A174) for the DYW approximation, with the initial condition given by equation (A179a).

18. Scenario 4: equation (A173) with $q = 0$ and equation (A176) are the governing equations for the KW approximation and equation (A173) with $q = 0$ and equation (A177) for the DYW approximation, with the initial condition given by equation (A179a).

The above cases are summarized in Table A8.

Error equations: use of non-zero initial conditions

Kinematic wave and diffusion wave solution

The governing equations are equations (A173) and (A176). Equation (A173) subject to equation (A179) has the solution:

$$h = h_0 + (q_0 - f_0)t \qquad (A181)$$

Equation (A181) states that the depth of flow increases linearly with time. In other words, the flow depth at any time is equal to the sum of the initial depth and the effective rainfall up to that time. For the kinematic wave (KW) approximation, equation (A176), in conjunction with equation (A170), yields

$$u = \left(\frac{S_0}{\beta}\right)^{0.5} h^{0.5} \qquad (A182)$$

It is convenient to define dimensionless time τ,

$$\tau = \frac{h}{h_0} = \frac{h_0 + (q_0 - f_0)t}{h_0}, \; \tau \geq 1 \qquad (A183)$$

where h_0 is the initial depth that acts as a normalizing depth. In terms of τ, equation (A181) becomes

$$h(\tau) = h_0\tau \qquad (A184)$$

and equation (A182) becomes

$$u = \left(\frac{S_0 h_0}{\beta}\right)^{0.5} \tau^{0.5} \qquad (A185)$$

The dimensionless flow depth $h_* = h/h_0$ and the dimensionless flow velocity $v = u/U$ become

$$h_* = \frac{h}{h_0} = \tau \qquad (A186)$$

$$v = \frac{u}{U} = \tau^{0.5} \qquad (A187)$$

where the normalizing velocity U is

$$U = \left(\frac{S_0 h_0}{\beta}\right)^{0.5} \qquad (A188)$$

The discharge Q can be expressed as

$$Q(t) = \left(\frac{S_0}{\beta}\right)^{0.5} [h_0 + q_* t]^{1.5} \qquad (A189)$$

Equation (A189) can be expressed in terms of τ as

$$Q(\tau) = \left(\frac{S_0}{\beta}\right)^{0.5} h_0^{1.5} \tau^{1.5} \qquad (A190)$$

The dimensionless discharge $Q_* = Q/Q_0$ can be expressed as

$$Q_*(\tau) = \tau^{1.5} \qquad (A191)$$

where the normalizing discharge Q_0 is

$$Q_0 = \left(\frac{S_0}{\beta}\right)^{0.5} h_0^{1.5} \qquad (A192)$$

The kinematic wave (KW) and diffusion wave (DW) solution is given by equations (A186) and (A187). The dimensionless velocity and dimensionless discharge are sketched in Figures A19 and A20, respectively. Clearly the velocity is independent of γ and increases parabolically with τ.

Dynamic wave solution

The governing equations are equations (A173) and (A177). Equation (A173) has the solution given by equation (A181). With the introduction of equation (A170), equation (A177) takes the form

$$\frac{du}{dt} = gS_0 - g\beta \frac{u^2}{h} \qquad (A193)$$

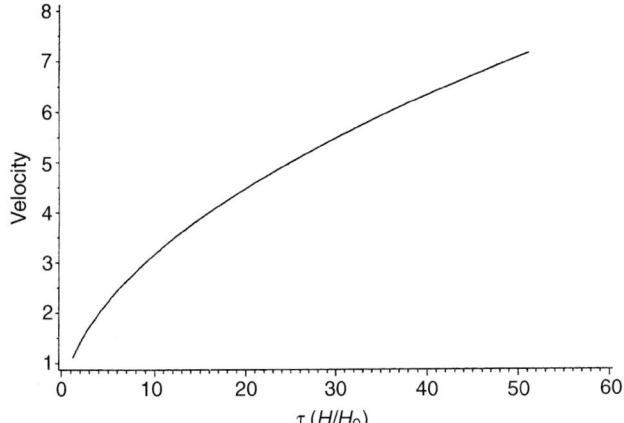

Figure A19 Dimensionless kinematic wave velocity as a function of dimensionless time for a space-independent case with lateral inflow neglected in the momentum equation, and $q = q_0$, $f = f_0$, $h(0) = h_0$ and $u(0) = u_0$.

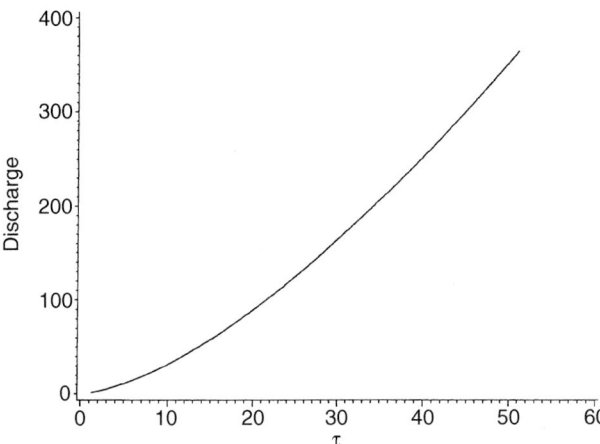

Figure A20 Dimensionless kinematic wave discharge as a function of dimensionless time for a space-independent case with lateral inflow neglected in the momentum equation, and $q = q_0$, $f = f_0$, $h(0) = h_0$ and $u(0) = u_0$.

In terms of τ, equation (A193) becomes

$$\frac{du}{d\tau} = \frac{gS_0h_0}{(q_0 - f_0)} - \frac{g\beta}{(q_0 - f_0)}\frac{u^2}{\tau} \tag{A194}$$

In dimensionless form, equation (A193) can be written as

$$\frac{dv}{d\tau} = \left(\frac{g}{q_0 - f_0}\right)(\beta h_0 S_0)^{0.5} - \left(\frac{g}{q_0 - f_0}\right)(\beta h_0 S_0)^{0.5}\frac{v^2}{\tau} \tag{A195}$$

If

$$\gamma = \frac{4g^2\beta S_0 h_0}{q_*^2}, \quad q_* = q_0 - f_0, \quad \Gamma = \gamma^{0.5} \tag{A196}$$

then

$$\frac{dv}{d\tau} = \frac{\gamma^{0.5}}{2} - \frac{\gamma^{0.5}}{2}\frac{v^2}{\tau} = \frac{\Gamma}{2} - \frac{\Gamma}{2}\frac{v^2}{\tau} \tag{A197}$$

The parameter γ reflects the effect of initial flow depth, bed slope, bed roughness, infiltration and lateral inflow. Equation (A197) has a form of the Riccati equation, and does not have an analytical solution.

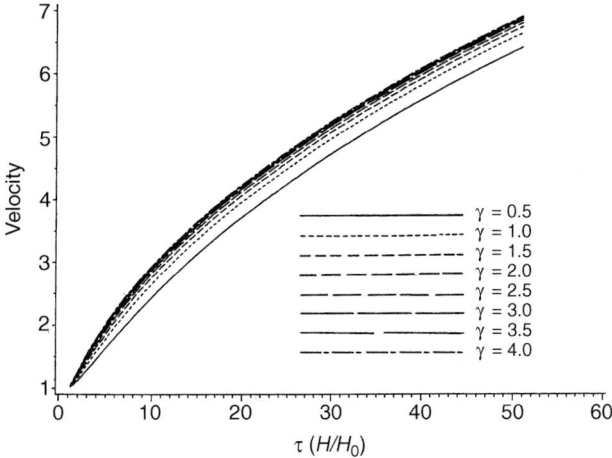

Figure A21 Dimensionless dynamic wave velocity as a function of dimensionless time for a space-independent case with lateral inflow neglected in the momentum equation, and $q = q_0$, $f = f_0$, $h(0) = h_0$ and $u(0) = u_0$.

Equation (A177) can be expressed in terms of Q as

$$\frac{dQ}{dt} = q_*\frac{Q}{(h_0 + q_*t)} + gS_0(h_0 + q_*t) - g\beta\frac{Q^2}{(h + q_*t)^2} \tag{A198}$$

In terms of τ, equation (A198) can be written as

$$\frac{dQ}{dt} = \frac{Q}{\tau} + \frac{h_0^2}{q_*}gS_0\tau - \frac{g\beta}{q_*h_0}\frac{Q^2}{\tau^2} \tag{A199}$$

In terms of dimensionless discharge $Q_* = Q/Q_0$, equation (A199) becomes

$$\frac{dQ_*}{d\tau} = \frac{Q_*}{\tau} + \frac{\gamma^{0.5}}{2}\tau - \frac{\gamma^{0.5}}{2}\frac{Q_*^2}{\tau^2} \tag{A200}$$

or

$$\frac{dQ_*}{d\tau} = \frac{Q_*}{\tau} + \frac{\Gamma}{2}\tau - \frac{\Gamma}{2}\frac{Q_*^2}{\tau^2} \tag{A201}$$

Equation (A197) is a special case of the Riccati equation. At $\tau = 1$, $dv/d\tau = 0$ and $v = 1$. Equation (A197) can be solved numerically by use of the fourth-order Runge–Kutta method. The dynamic wave solution is given by equation (A181) and the solution of equation (A197). The dimensionless velocity and dimensionless discharge are sketched as a function of τ for various values of γ and q_r, as shown in Figures A21 and A22. The velocity increases, for a fixed γ, with increasing τ. For a fixed τ, it also increases with increasing γ. For $\gamma \geq 1.5$ it is almost insensitive to γ. It is also independent of q_r.

Error in KW and DW approximations

With the substitution of equations (A187) and (A197) in equation (A180b), the error equation is found to be

$$\frac{dE}{d\tau} = C_0(\tau) + C_1(\gamma, \tau)E + C_2(\gamma, \tau)E^2, \quad E(1) = 0, \quad \tau \geq 1 \tag{A202}$$

where

$$C_0(\tau) = \frac{1}{2\tau} \tag{A203}$$

$$C_1(\gamma, \tau) = \frac{1}{2\tau}(1 - 2\gamma^{0.5}\tau^{0.5})$$

$$= \frac{1}{2\tau}(1 - 2\Gamma\tau^{0.5}) \tag{A204}$$

$$C_2(\gamma, \tau) = -\frac{\Gamma}{2\tau^{0.5}} \tag{A205}$$

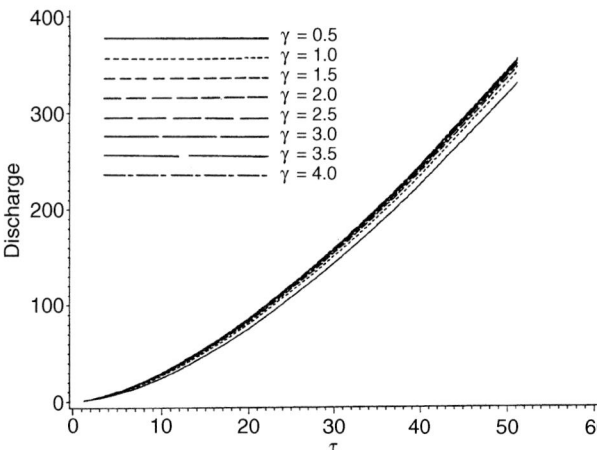

Figure A22 Dimensionless dynamic wave discharge as a function of dimensionless time for a space-independent case with lateral inflow neglected in the momentum equation, and $q = q_0$, $f = f_0$, $h(0) = h_0$ and $u(0) = u_0$.

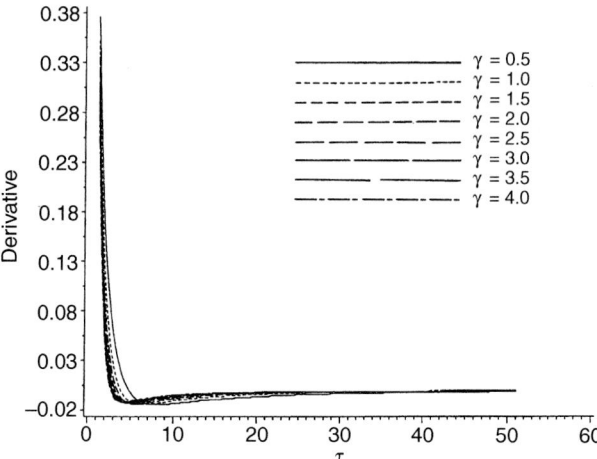

Figure A24 Error derivative as a function of dimensionless time for a space-independent case with lateral inflow neglected in the momentum equation, and $q = q_0$, $f = f_0$, $h(0) = h_0$ and $u(0) = u_0$.

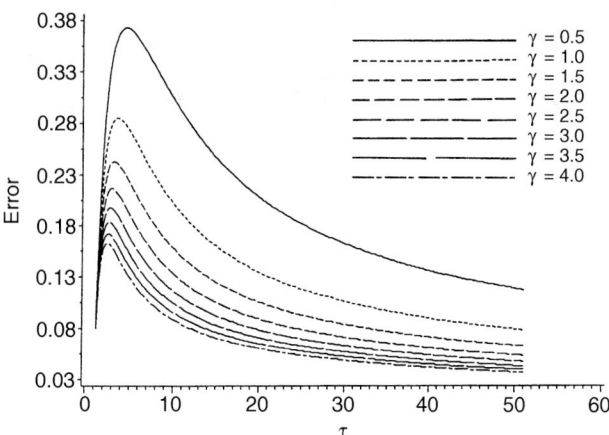

Figure A23 Error in the KW or DW approximation as a function of dimensionless time for a space-independent case with lateral inflow neglected in the momentum equation, and $q = q_0$, $f = f_0$, $h(0) = h_0$ and $u(0) = u_0$.

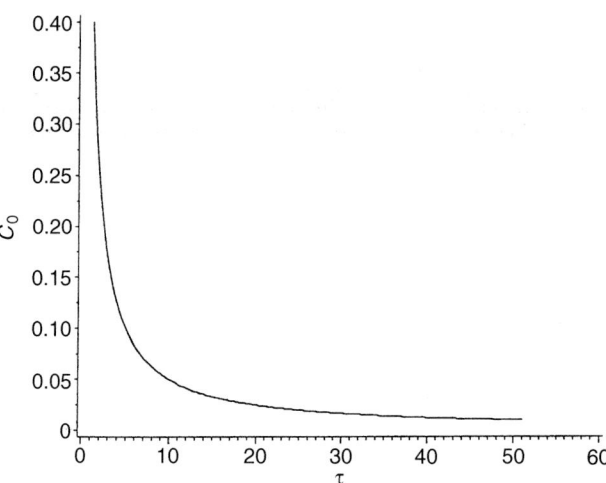

Figure A25 Coefficient C_0 as a function of dimensionless time for a space-independent case with lateral inflow neglected in the momentum equation, and $q = q_0$, $f = f_0$, $h(0) = h_0$ and $u(0) = u_0$.

Table A9 Maximum error and its time of occurrence

τ	Maximum error	γ
4.5	0.373	0.5
3.75	0.285	1.0
3.25	0.243	1.5
3.00	0.217	2.0
2.75	0.198	2.5
2.75	0.184	3.0
2.50	0.172	3.5
2.50	0.163	4.0

Equation (A202) is a Riccati equation and also holds for error in discharge. It can be solved by using the fourth-order Runge–Kutta method. At $\tau = 1$, $E(1) = 0$ and $dE(1)/d\tau = 0.5$. The distribution of error in the KW or DW solution is shown in Figure A23 for various values of γ. The distribution is highly skewed with a steep rise to a peak and an extended decline. Table A9 shows maximum error and its

time of occurrence. For a fixed value of τ, the error increases with decreasing γ. A graph of the error derivative is shown in Figure A24. For $\tau \geq 30$ it is almost independent of γ and τ. Even for $\tau \leq 30$ it is only mildly sensitive to γ. Coefficients C_0, C_1 and C_2 are plotted, as shown in Figures A25–A27. C_0 is positive, independent of γ and is inversely proportional to τ. C_1 is negative, and increases parabolically with the square root of τ. For a fixed value of τ, it increases with decreasing γ. C_2 is negative, inversely proportional to the square root of τ, and is indirectly proportional to γ. The variation of these coefficients can be further assessed by considering their derivatives.

Error equations: use of zero initial conditions

Kinematic wave and diffusion wave solution

Equation (A173), subject to equation (A179b), has the solution

$$h = (q_0 - f_0)t \qquad (A206)$$

From the KW approximation,

$$u = \left(\frac{S_0}{\beta}\right)^{0.5} h^{0.5} \qquad (A207)$$

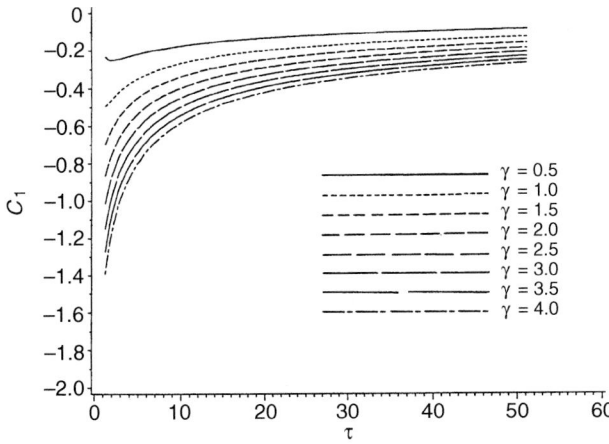

Figure A26 Coefficient C_1 as a function of dimensionless time for a space-independent case with lateral inflow neglected in the momentum equation, and $q = q_0$, $f = f_0$, $h(0) = h_0$ and $u(0) = u_0$.

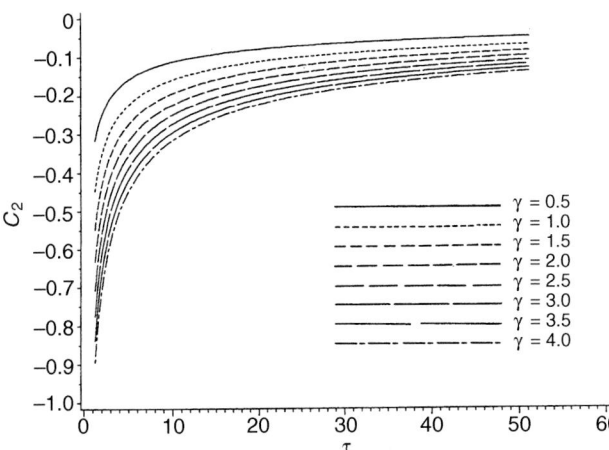

Figure A27 Coefficient C_2 as a function of dimensionless time for a space-independent case with lateral inflow neglected in the momentum equation, and $q = q_0$, $f = f_0$, $h(0) = h_0$ and $u(0) = u_0$.

In terms of a dimensionless parameter τ defined as

$$t = \frac{gt}{q_0 - f_0}, \quad \tau \geq 0 \tag{A208}$$

equation (A206) can be written as

$$h(\tau) = \frac{q_*^2}{g}\tau = h_0\tau, \quad q_* = q_0 - f_0, \quad h_0 = \frac{q_*^2}{g} \tag{A209}$$

and equation (A207) as

$$u = q_*\left(\frac{S_0}{g\beta}\right)^{0.5}\tau^{0.5} \tag{A210}$$

In dimensionless form, the flow depth $h_* = h/h_0$ is

$$h_* = \tau \tag{A211}$$

and flow velocity,

$$v = \frac{u}{U} = \tau^{0.5} \tag{A212}$$

where

$$U = q_*\left(\frac{S_0}{g\beta}\right)^{0.5} \tag{A213}$$

In terms of discharge Q,

$$Q(t) = \left(\frac{S_0}{\beta}\right)^{0.5}q_*^{1.5}\,t^{1.5} \tag{A214}$$

Equation (A214) can be expressed in terms of τ as

$$Q(\tau) = \left(\frac{S_0}{\beta}\right)^{0.5}\frac{q_*^3}{g^{1.5}}\tau^{1.5} \tag{A215}$$

The dimensionless discharge $Q_* = Q/Q_0$ can be expressed as

$$Q_*(\tau) = \tau^{1.5} \tag{A216}$$

where

$$Q_0 = \left(\frac{S_0}{\beta}\right)^{0.5}\frac{q_*^3}{g^{1.5}} \tag{A217}$$

The kinematic wave (KW) and diffusion wave (DW) solution is given by equations (A211) and (A212). The dimensionless velocity and dimensionless discharge have the appearance shown in Figures A19 and A20, respectively. The velocity is independent of γ and is directly proportional to the square root of τ.

Dynamic wave solution

Equation (A173) has the solution given by equation (A206). In terms of τ, equation (A193) becomes

$$\frac{du}{d\tau} = S_0 q_* - \beta\frac{g}{q_*}\frac{u^2}{\tau} \tag{A218}$$

In dimensionless form, equation (A218) becomes

$$\frac{dv}{d\tau} = (S_0 g\beta)^{0.5} - (S_0 g\beta)^{0.5}\frac{v^2}{\tau} \tag{A219}$$

If

$$\Gamma = 2(S_0 g\beta)^{0.5}, \quad \gamma = 4gS_0\beta, \quad \Gamma = \gamma^{0.5} \tag{A220}$$

then

$$\frac{dv}{d\tau} = \frac{\gamma}{2} - \frac{\gamma}{2}\frac{v^2}{\tau} \tag{A221}$$

The dimensionless parameter γ reflects the effect of bed slope and friction.

Equation (A177) can be expressed in terms of discharge Q as

$$\frac{dQ}{dt} = \frac{Q}{t} + gS_0 q_* t - \frac{g\beta}{q_*^2}\frac{Q^2}{t^2} \tag{A222}$$

In terms of τ, equation (A222) can be written as

$$\frac{dQ}{d\tau} = \frac{Q_*}{\tau} + S_0 h_0 q_*\tau - \frac{\beta q_*}{h_0^2}\frac{Q^2}{\tau^2} \tag{A223}$$

In terms of Q_*, equation (A222) can be expressed as

$$\frac{dQ_*}{d\tau} = \frac{Q_*}{\tau} + \frac{\gamma^{0.5}}{2}\tau - \frac{\gamma^{0.5}}{2}\frac{Q_*}{\tau^2} = \frac{Q_*}{\tau} + \Gamma\tau - \frac{\Gamma}{2}\frac{Q_*^2}{\tau^2} \tag{A224}$$

The dynamic wave (DYW) solution is given by equation (A211) and the solution of equation (A221). At $\tau = 0$, $v = 0$, and $dv/d\tau$ has a discontinuity which can be circumvented by use of forward differencing. Equation (A221) is solved by using the fourth-order Runge–Kutta method. The velocity is independent of q_t and increases with increasing τ. For a fixed value of τ it increases with γ, but becomes nearly independent of γ for $\gamma = 1.5$.

Errors in KW and DW approximations

Equations (212) and (221) are of the same form as equations (A187) and (A197), respectively. Therefore, the error equation for this case can be written as

$$\frac{dE}{d\tau} = C_0(\tau) + C_1(\gamma, \tau)E + C_2(\gamma, \tau)E^2, \quad E(0) = 0, \quad \tau \geq 0 \tag{A225}$$

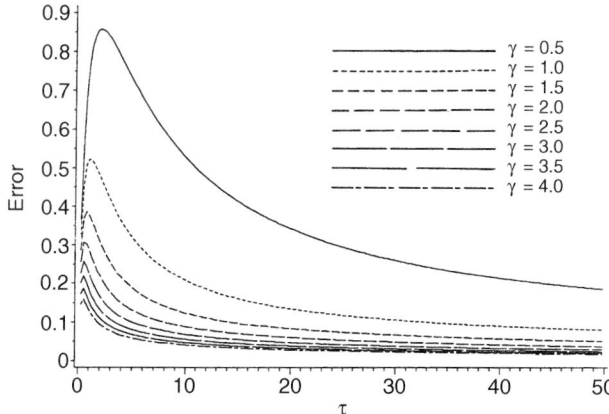

Figure A28 Error in the KW or DW approximation as a function of dimensionless time for a space-independent case with lateral inflow neglected in the momentum equation, and $q = q_0$, $f = f_0$, $h(0) = h_0$ and $u(0) = u_0$.

Table A10 Maximum error and its time of occurrence

τ	Maximum error	γ
2.0	0.857	0.5
1.00	0.522	1.0
0.75	0.386	1.5
0.50	0.307	2.0
0.50	0.258	2.5
0.50	0.219	3.0
0.50	0.187	3.5
0.50	0.160	4.0

where

$$C_0(\tau) = \frac{1}{2\tau} \tag{A226}$$

$$C_1(\gamma, \tau) = \frac{1}{2\tau}(1 - 2\gamma\tau^{0.5}) \tag{A227}$$

$$C_2(\gamma, \tau) = -\frac{\gamma}{2\tau^{0.5}} \tag{A228}$$

Equation (A225) is a Riccati equation, and also describes error in discharge. It has a discontinuity at $\tau = 0$. The discontinuity can be circumvented by use of forward differencing. Equation (A225) is solved by using the fourth-order Runge–Kutta equation.

The distribution of error in the KW or DW solution is sketched in Figure A28 for various values of γ. The maximum error and its time of occurrence are given in Table A10. The distribution is highly skewed with a steep rise to a peak and then an extended decline approaching a constant value for larger values of τ. The error increases with decreasing γ for a fixed value of τ. The derivative decreases steeply over a short range of τ, reaches a near-constant value for $\tau \geq 20$, and becomes independent of γ. C_0 is positive, independent of γ, and is inversely proportional to τ. C_1 may take on negative or positive values. For a fixed τ, C_1 increases with decreasing γ. C_2 is negative, inversely proportional to the square root of τ, and increases with decreasing γ.

Error equation for zero lateral inflow: use of non-zero initial conditions

In this case, $q = 0$ and $f = f_0$. Consequently, the only meaningful initial condition would be a finite depth, and a finite velocity, or a finite discharge. This corresponds to the case of a recession hydrograph.

Kinematic wave and diffusion wave solution

Equation (A173) takes the form

$$\frac{dh}{dt} = -f_0 \tag{A229}$$

which, subject to equation (A179a), has the solution:

$$h = h_0 - f_0 t \tag{A230}$$

It is convenient to define a dimensionless time parameter τ as

$$\tau = \frac{f_0 t}{h_0}, \; 0 \leq \tau \leq 1 \tag{A231}$$

Equation (A230) becomes

$$h = h_0(1 - \tau) = h_0\tau_*, \\ \tau_* = 1 - \tau, \; 0 \leq \tau_* \leq 1 \tag{A232a}$$

In dimensionless form the flow depth becomes

$$h_* = \frac{h}{h_0} = 1 - \tau \tag{A232b}$$

Equation (A176) takes the form

$$u = \left(\frac{S_0}{\beta}\right)^{0.5} h^{0.5} \tag{A233}$$

In terms of τ, equation (A233) becomes

$$u(\tau) = \left(\frac{S_0 h_0}{\beta}\right)^{0.5} \tau_*^{0.5}, \\ \tau_* = 1 - \tau, \; 0 \leq \tau \leq 1 \tag{A234}$$

In dimensionless terms, $v = u/U$,

$$v(\tau) = \tau_*^{0.5} = (1 - \tau)^{0.5} \tag{A235}$$

where

$$U = \left(\frac{S_0 h_0}{\beta}\right)^{0.5} \tag{A236}$$

The discharge can be expressed as

$$Q(\tau) = \left(\frac{S_0}{\beta}\right)^{0.5} h_0^{1.5} \tau_*^{1.5} \tag{A237}$$

The dimensionless discharge, $Q_* = Q/Q_0$ can be written as

$$Q_* = \tau_*^{1.5} \tag{A238}$$

where

$$Q_0 = \left(\frac{S_0}{\beta}\right)^{0.5} h_0^{1.5} \tag{A239}$$

The dimensionless velocity and discharge have the same appearance as shown in Figures A19 and A20.

Dynamic wave solution

The depth of flow is given by equation (A230) or (A232). Equation (A177), in terms of τ_*, can be written as

$$\frac{du}{d\tau_*} = -\frac{gS_0 h_0}{f_0} + \frac{g\beta}{f_0}\frac{u^2}{\tau_*} \tag{A240}$$

In dimensionless velocity, equation (A177) becomes

$$\frac{dv}{d\tau_*} = q_r\left[\frac{\gamma^{0.5}}{2} - \frac{\gamma^{0.5}}{2}\frac{v^2}{\tau_*}\right] \tag{A241}$$

where

$$\gamma = \frac{4g^2\beta S_0 h_0}{f_0^2} \text{ and } U = \left(\frac{S_0 h_0}{\beta}\right)^{0.5}, \\ v = \frac{u}{U}, \; q_r = \frac{q_0}{f_0} \tag{A242}$$

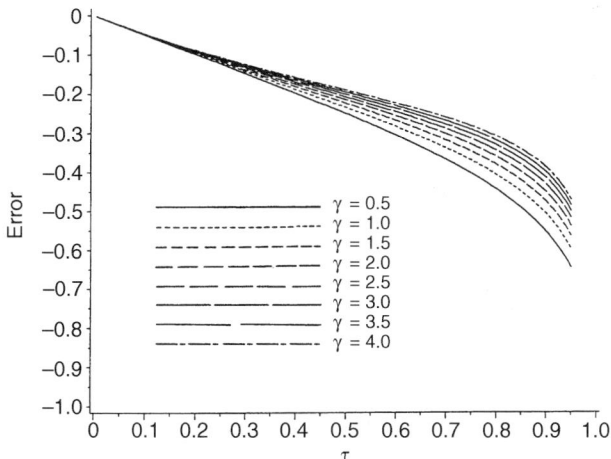

Figure A29 Error in the KW or DW approximation as a function of dimensionless time for a space-independent case with lateral inflow neglected in the momentum equation, and $q = 0$, $f = f_0$, $q_r = 1$, $h(0) = h_0$ and $f(0) = f_0$.

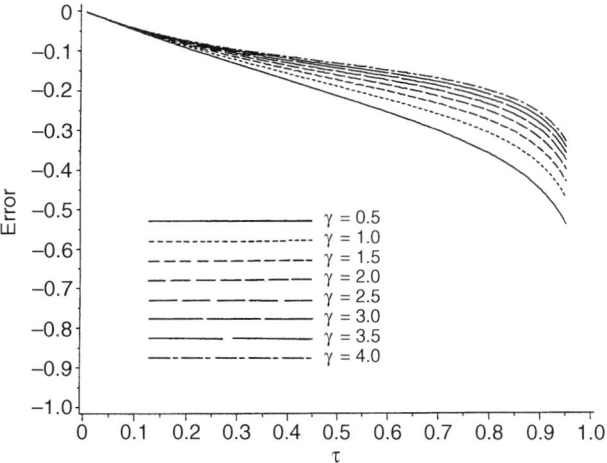

Figure A30 Error in the KW or DW approximation as a function of dimensionless time for a space-independent case with lateral inflow neglected in the momentum equation, and $q = 0$, $f = f_0$, $q_r = 2$, $h(0) = h_0$ and $f(0) = f_0$.

The dimensionless parameter γ combines the effects of friction, bed slope, initial flow depth and infiltration. Equation (A241) is a special form of the Riccati equation, which has to be solved numerically.

Equation (A177) can be expressed in terms of Q as

$$\frac{dQ}{d\tau_*} = \frac{Q}{\tau_*} - \frac{gS_0 h_0^2}{f_0}\tau_* - \frac{g\beta}{f_0 h_0^2}\frac{Q^2}{\tau_*^2} \qquad (A243)$$

In terms of dimensionless discharge, equation (A243) becomes

$$\frac{dQ_*}{d\tau_*} = \frac{Q_*}{\tau_*} + q_r\left[-\frac{\gamma^{0.5}}{2}\tau_* - \gamma^{0.5}\frac{Q_*^2}{\tau_*^2}\right] \qquad (A244)$$

The dimensionless dynamic wave velocity and discharge have the same appearance as in Figure A21.

Error in KW and DW approximations

The error differential equation is found to be similar to equation (A202):

$$\frac{dE}{d\tau_*} = C_0(\gamma, \tau_*) + C_1(\gamma, \tau_*)E + C_2(\gamma, \tau_*)E^2,$$

$$E(0) = 0, \; 0 \le \tau_* \le 1 \qquad (A245)$$

where

$$C_0(\gamma, \tau_*) = \frac{1}{2\tau_*} \qquad (A246)$$

$$C_1(\gamma, \tau_*) = \frac{1}{2\tau_*}\left(1 + 2q_r\gamma^{0.5}\tau_*^{0.5}\right) \qquad (A247)$$

$$C_2(\gamma, \tau_*) = +q_r\frac{\gamma^{0.5}}{\tau_*^{0.5}} \qquad (A248)$$

Equation (A245) is a Riccati equation and also holds for error in discharge. It can be solved numerically. Figures A29–A31 show error in the kinematic wave or diffusion wave approximation for various valus of γ.

Conclusions

The following conclusions can be drawn from this study. (1) The errors of the kinematic wave and diffusion-wave approximations follow a Riccati equation. (2) The error equations describe error as a function of time and are in terms of a dimensionless parameter γ. (3) The parameter γ combines the effects of the initial condition, lateral inflow, infiltration, friction and bed slope. (4) The errors significantly decline for $\gamma \ge 5$ and the kinematic wave and diffusion wave

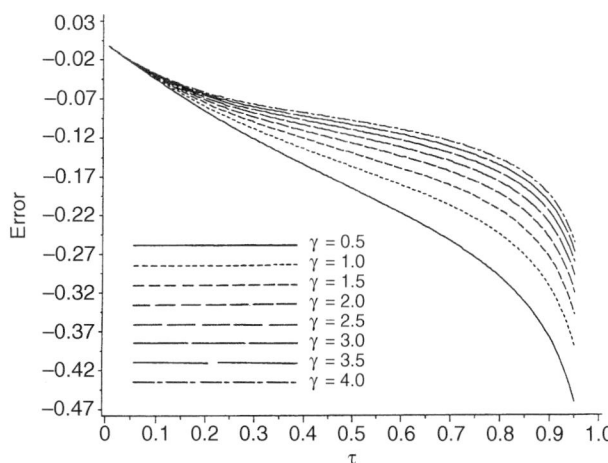

Figure A31 Error in the KW or DW approximation as a function of dimensionless time for a space-independent case with lateral inflow neglected in the momentum equation, and $q = 0$, $f = f_0$, $q_r = 3$, $h(0) = h_0$ and $f(0) = f_0$.

approximations become quite accurate. (5) For any value of γ, the errors also decline exponentially for dimensionless time greater than or equal to 10. (6) The definition of γ is influenced by the type of initial condition.

V.P. Singh

Bibliography

Daluz Vieira, J.H., 1983. Conditions governing the use of approximations for the Saint Venant equations for shallow water flow. *Journal of Hydrology*, **60**, 43–58.

Ferrick, M.G., 1985. Analysis of river wave types. *Water Resources Research*, **21**(2), 209–220.

Fread, D.L., 1985. Applicability criteria for kinematic and diffusion routing models. Laboratory of Hydrology, National Weather Service, NOAA, US Department of Commerce, Silver Spring, Maryland.

Lighthill, M.J. and Whitham, G.B., 1955. On kinematic waves: 1. flood movement in long rivers. *Proceedings of the Royal Society, London, Series A*, **229**, 281–316.

Menendez, A.N. and Norscini, R., 1982. Spectrum of shallow water waves: an analysis. *Journal of the Hydraulics Division, ASCE,* **108**(HY1), 75–93.

Morris, E.M. and Woolhiser, D.A., 1980. Unsteady, one dimensional flow over a plane: partial equilibrium and recession hydrographs. *Water Resources Research,* **16**(2), 355–360.

Ponce, V.M. and Simons, D.B., 1977. Shallow wave propagation in open channel flow. *Journal of the Hydraulics Division, ASCE,* **103**(HY12), 1461–1475.

Ponce, V.M., Li, R.M. and Simons, D.B., 1978. Applicability of kinematic and diffusion models. *Journal of the Hydraulics Division, ASCE,* **104**(HY3), 353–360.

Singh, V.P., 1988. *Hydrologic Systems: Vol. 1 – Rainfall–Runoff Modeling.* Prentice-Hall, Inc., Englewood Cliffs, NJ.

Singh, V.P., 1992a. Errors in kinematic wave and diffusion wave approximations for space-independent flows: 1. Cases 1 to 9. Technical Report WRR17, 155 pp., Water Resources Program, Department of Civil Engineering, Louisiana State University, Baton Rouge, LA.

Singh, V.P., 1992b. Errors in kinematic wave and diffusion wave approximations for space-independent flows: 2. Cases 10 to 19. Technical Report WRR18, pp. 156–329, Water Resources Program, Department of Civil Engineering, Louisiana State University, Baton Rouge, LA.

Singh, V.P., 1992c. Accuracy of hydrodynamic models of free-surface flows. *Proceedings, International Symposium on Hydrology of Mountainous Areas,* Shimla, India, pp. 331–352.

Singh, V.P., 1993. Quantifying the accuracy of hydrodynamic approximations for determination of flood discharges. *Journal of Indian Water Resources Society,* **13** (3–4), 172–185.

Singh, V.P., 1994a. Accuracy of kinematic-wave and diffusion-wave approximations for space-independent flows. *Hydrological Processes,* **8**(1), 45–62.

Singh, V.P., 1994b. Accuracy of kinematic-wave and diffusion-wave approximations for space-independent flows with lateral inflow neglected in the momentum equation. *Hydrological Processes,* **8**, 318–326.

Singh, V.P., 1994c. Derivation of errors of kinematic-wave and diffusion-wave approximations for space-independent flows. *Water Resources Management,* **8**, 57–62.

Singh, V.P. Aravamuthan, V. and Joseph, E.S., 1993. Accuracy of hydrodynamic models of flood-discharge determinations, in *Environmental Management: Geo-Water and Engineering Aspects* (eds R.N. Chowdhury and M. Sivakumar), Balkema, Rotterdam, pp. 79–89.

Woolhiser, D.A. and Liggett, J.A., 1967. Unsteady one-dimensional flow over a plane – the rising hydrograph. *Water Resources Research,* **3**, pp. 753–771.

Cross references

Accuracy
Accuracy of hydrodynamic approximations in hydrology: non-uniform, steady flow

ACIDIC DEPOSITION: ACIDIFICATION OF SURFACE WATERS

Acidic deposition has its origins primarily in emissions of oxides of sulfur (SO_2) and nitrogen (a mix of compounds usually denoted by NO_x) from the combustion of fossil fuels by electric utilities, factories and motor vehicles. A variety of chemicals (principally hydroxyl radicals, ozone or hydrogen peroxide) oxidize SO_2 and NO_x in the atmosphere to produce sulfuric and nitric acids. Although natural sources (e.g. volcanoes and forest fires) can also contribute to 'acid rain' artificial sources predominate in urban areas and can dominate the depositional chemistry of areas downwind.

Emitted SO_2 and NO_x can travel great distances before being deposited to the surface of the Earth. Deposition can occur in several ways: (1) wet deposition (i.e. precipitation), (2) dry deposition (e.g. as gases or particles) or (3) interception of fogs or cloud waters. Increasing forest canopy cover in natural watersheds effectively increases dry deposition and interception, with coniferous vegetation more effectively scavenging deposition than does deciduous vegetation. Increasing elevation increases both wet deposition (through

orographic effects) and dry deposition (through increased wind turbulence) (Turner *et al.*, 1990).

In some instances acidic deposition can cause *acidification* (loss of acid neutralizing capacity – ANC) of surface waters, sometimes to the point that lakes and streams become *acidic* (i.e., ANC $\leq \mu eq/l$) on either an episodic or chronic basis. Decreases in pH and increases in concentrations of monomeric aluminum compounds (leached from watershed soils) occur during severe surface-water acidification. Either of these chemical changes can prove toxic to aquatic organisms (Baker *et al.*, 1990). Although European scientists had known for centuries of the existence of acidic deposition, it was not until the 1960s that scientists in Scandinavia and the U.S. systematically began to note and call attention to the correlation between the acidity of rainfall and the apparent acidification of surface waters (see review by Bricker and Rice, 1993). For some time this association was primarily empirical, with the predominant mechanisms of cause and effect remaining a subject of much debate. In the early to mid-1980s a number of scientists published hypotheses of the mechanisms by which acidic deposition interacts in terrestrial systems leading to acidification of drainage waters (Seip, 1980; Galloway *et al.*, 1983; Krug and Frink, 1983; Reuss and Johnson, 1985, 1986). Despite the differences in emphasis favored by individual authors, the mechanisms proposed contained many elements either in common or directly complementary to one another. Today, the predominant controlling mechanisms seem reasonably well established and accepted (Reuss *et al.*, 1987; Turner *et al.*, 1990).

Mechanisms of action

When acidic deposition falls on watersheds containing highly weatherable minerals (e.g. carbonates) or well-buffered soils, the resulting effects on surface waters can be minimal or even undetectable. In watersheds with bedrock and minerals highly resistant to chemical weathering and with very poorly buffered soils (i.e. low cation exchange capacity and low base saturation), effects can be distinctly more pronounced.

The effects of acidic deposition on soil and surface water acidification can be divided usefully into two classes: intensity effects and capacity effects. Intensity effects refer to the immediate equilibrium reactions occurring between soils and soil solutions and in surface waters. Capacity effects refer to the longer-term changes in rates of flux of chemical constituents from soils to surface waters. The magnitude of intensity effects (e.g. soil solution pH) depend strongly on existing capacity conditions (e.g. the percentage of base saturation). This interdependence can be highly non-linear – a small change in capacity conditions can lead to large changes in intensity factors.

Intensity effects

The major anionic components of acidic deposition are sulfate and nitrate ions. In acidic deposition these anions are balanced primarily by ammonium and hydrogen ions. The ionic strength of such polluted rainfall can be considerably higher than that of rainfall in non-polluted areas. Such polluted deposition acts both to change the relative ionic composition of soil waters and to increase their ionic strength – with a variety of consequent interdependent effects.

First, the hydrogen ions from acidic deposition will increase the dissolution of primary or secondary minerals, causing increases in the concentrations of Al^{3+} in soil solution. Both the input H^+ and the newly dissolved Al^{3+} will engage in cation exchange reactions with the soil complex. So-called 'base' cations (i.e. Ca^{2+}, Mg^{2+}, Na^+, K^+) will be displaced from the exchange complex, as will Al^{3+}. Because of the nature of the exchange reactions, the exchange of Al^{3+} will be favored so that the soil solutions experience a shift towards the ion of higher valence. The Al^{3+} released in turn acts as a weak acid in solution leading to a lowering of soil solution pH. This combination of effects is known collectively as 'salt effect acidification' (Reuss and Johnson, 1985, 1986; Binkley *et al.*, 1988; Turner *et al.*, 1990). The more acidic the soil (i.e. the lower the base saturation) the greater will be this salt-effect acidification. Leaching of the resulting soil solutions to surface waters can lead to a progressive lowering (relative to pre-existing conditions – i.e. without acidic deposition) of surface water ANC and thus to surface water acidification.

Soil solutions unaffected by acidic deposition often can have low pH values as a result of a variety of naturally occurring processes. Usually, however, these solutions have at least slightly positive ANC. When such solutions leach to surface waters, CO_2 degasses from the solution (because it is supersaturated with respect to the atmosphere) and the pH rises. If ion loading from acidic deposition causes soil solution ANC to become negative then, when it leaches to surface waters, its pH will increase only minimally (if at all) (Reuss and Johnson, 1985, 1986). This can lead, in turn, to acidic surface waters – lakes and streams with ANC < 0.

Capacity effects

At the pH levels of soil solutions, the increasing hydrogen ions released due to shifts in exchange equilibria remain dissociated from the strong acid anions (i.e. NO_3^- and SO_4^{2-}) found in acidic deposition and thus base cations displaced from the soil exchange complex as the result of inputs of acidic deposition may leach from soils to surface waters. If renewal of available base cations (e.g. by weathering) cannot keep pace with this cation depletion, such leaching will lead to soil acidification – a progressive decrease in soil base saturation. As soil acidification progresses, the exchange complex and hence the resulting equilibrium soil solutions will become more and more dominated by H^+ and Al^{3+}. The ANC and pH of soil solutions and surface waters will continue to drop and the solution concentrations of toxic Al^{3+} will continue to rise. Because the increased leaching of basic and acidic cations is associated with increased leaching of sulfate and nitrate, some authors (e.g. Reuss et al., 1987) term the sulfate and nitrate anions 'carrier anions'. As the concentrations of sulfate and nitrate increase in leached soil solutions, so do the associated leaching of base cations and acids in soil solutions and the consequent acidification of soils and surface waters, respectively.

Both sulfur and nitrogen cycle in complex ways through terrestrial ecosystems. For sulfate the principal control on its mobility and soil solution concentrations appears to be inorganic adsorption on iron and aluminum hydrous oxides. Soils high in these hydrous oxides can adsorb significant amounts of sulfate relative to atmospheric loading from acidic deposition. This adsorption immobilizes (at least temporarily) the sulfate anion. It also releases hydroxide ions that help neutralize soil solution acidity. Soil sulfate adsorption is concentration dependent: as the amount of adsorbed sulfate increases, so does the equilibrium concentration of sulfate in soil solution. Thus, as loading of sulfate from acidic deposition continues, progressively less and less deposited sulfate is adsorbed and less and less soil acidity is neutralized. Concomitantly, soil solution sulfate concentrations increase as does the leaching of base cations and, increasingly, hydrogen and aluminum ions. The delays in the immediate soil and surface water acidification caused by this capacity effect vary according to a number of soil characteristics. For thin soils with low sulfate adsorption capacities the delays may be as short as years to a few decades (Reuss and Walthall, 1989; Church et al., 1992).

For nitrogen the principal mechanisms of immobilization in terrestrial systems appear to be biological (e.g. Aber et al., 1989). Compared to our knowledge of sulfur cycling in terrestrial systems, our understanding of integrated nitrogen cycling and especially rates of reaction is poor. During the 1980s, acidic deposition research and assessment activities focused on sulfur deposition and sulfate mobility (e.g. NAS, 1984; Church et al., 1992) rather than on considerations of nitrogen cycling and loss. This lack of concern with nitrogen deposition effects existed because it appeared that most forested ecosystems retained the vast majority of nitrogen deposited from the atmosphere – the principal retention mechanisms being a combination of vegetative uptake and soil microbial incorporation. More recently it has become clear that certain forests or forest ecosystems can approach or reach 'nitrogen saturation', which is defined as the availability of nitrogen in excess of biotic demand (Aber et al., 1989). The primary determinants of which forested systems approach or reach saturation appear to be forest age, site history (particularly timber removal or fire), atmospheric nitrogen deposition, and supply of other factors (e.g. moisture, phosphorus, carbon) that limit vegetative or microbial growth. As forest systems approach or reach nitrogen saturation, mineral nitrogen in the form of ammonium becomes more readily available in soils (both from mineralized organic matter and from ammonium inputs in acidic deposition) and soil nitrification increases dramatically (Aber et al., 1989). The results are (1) further soil acidification – nitrification of one mole of ammonium produces two moles of hydrogen ions (that will enter into the soil exchange reactions described earlier), and (2) increased soil solution nitrate concentrations (augmented by direct atmospheric deposition of nitrate). Consequently, increased leaching of nitrate

from these systems occurs which, as with sulfate, is associated with significant surface water acidification.

Hydrological flow paths

The pathways by which water flows through watersheds are important in determining the nature and extent of the chemical reactions that control the resulting solution composition. For example, soil solutions affected by acidic deposition which move only through shallow, poorly buffered soil horizons may result in drainage waters that are acidic. If the same solutions mix into deeper strata, however, they can be neutralized by larger volumes of more buffered groundwaters and can exit with positive ANC (Newton and April, 1982).

Episodic effects

Snowmelt events or heavy rainfalls can contribute acidic water directly to lakes and streams or can leach acidic water from saturated upper soil horizons leading to temporary surface-water acidification for periods ranging from hours to weeks. Such short term loss of ANC is termed 'episodic acidification'. Any temporary loss of surface water ANC whether caused by (1) mixing of more dilute meteoric waters with more concentrated surface waters, (2) flushing of naturally occurring organic acids from saturated soils or (3) flushing of acidic solutions from soils associated with mobile anions originating from atmospheric deposition qualifies as episodic acidification (Wigington et al., 1990, 1993). Thus many surface waters experience episodic acidification. In areas unaffected by acidic deposition, the first two scenarios dominate the observed episodes. Such episodes do not lead to conditions of aluminum toxicity, nor is it likely that the lowered pH values associated with flushing of organic acids significantly affect the native stream organisms that have evolved to live in these environments. In areas receiving acidic deposition, however, previously occurring natural episodes of lowered ANC can be intensified dramatically by acidification processes so that chemical changes (e.g. lowered pH, increased soluble aluminum species) can become more frequent and much more severe, leading to previously unknown toxic conditions for lake and stream organisms.

In-lake and in-stream neutralization reactions

A variety of processes can occur in lakes to affect ANC. Principal among these processes are sulfate reduction, nitrate assimilation, denitrification and cation exchange with sediments. Of these, microbial sulfate reduction occurring in lake sediments is usually the most important. In situations where watershed to lake ratios are small or where for other reasons (e.g. seepage lakes) the lake retention time is long (e.g. >2 years), in-lake ANC production can become important (or even dominant) relative to ANC contributions from the watershed (Cook and Schindler, 1983). For most drainage lakes of concern in the USA, however, watershed reactions and contributions dominate lake ANC values (Shaffer et al., 1988; Shaffer and Church, 1989). Because of relatively short residence times of drainage waters in individual stream segments, in-stream contributions to ANC are usually minimal (Turner et al., 1990).

M. Robbins Church

Bibliography

Aber, J.D., K.J. Nadelhoffer, P. Steudler and J.M. Melillo, 1989. Nitrogen saturation in northern forest ecosystems. BioScience, 39, 378–386.

Aber, J.D., J.M. Melillo, K.J. Nadelhoffer et al., 1991. Factors controlling nitrogen cycling and nitrogen saturation in northern temperate forest ecosystems. Ecol. Appl., 1, 305–315.

Baker, J.P., D.P. Bernard, M.J. Sale and S.W. Christensen, 1990. Biological Effects of Changes in Surface Water Acid-Base Chemistry, NAPAP Report 13. Acidic Deposition: State of Science and Technology, National Acid Precipitation Assessment Program, Washington, DC, 381 pp.

Binkley, D., C.T. Driscoll, H.L. Allen et al., 1988. Acid Deposition and Forest Soils. Springer-Verlag, New York, 149 pp.

Bricker, O.P., and K.C. Rice (1993) Acid rain, Ann. Rev. Earth Planet. Sci., 21, 151–174.

Church, M.R., P.W. Shaffer, K.W. Thornton et al., 1992. Direct/ Delayed Response Project: Future Effects of Long-Term Sulfur

Deposition on Stream Chemistry in the Mid-Appalachian Region of the Eastern United States. US Environmental Protection Agency, EPA/600/R-92/186, NTIS PB92–232370/AS, Washington, DC, 384 pp.

Cook, R.B. and D.W. Schindler, 1983. The biogeochemistry of sulfur in an experimentally acidified lake. *Ecol. Bull.* (Stockholm), **35**, 115–127.

Galloway, J.N., S.A. Norton and M.R. Church, 1983. Freshwater acidification from atmospheric deposition of sulfuric acid: A conceptual model. *Environ. Sci. Technol.*, **17**, 541A–545A.

Krug, E.C. and C.R. Frink, 1983. Acid rain on acid soil: A new perspective. *Science*, **221**, 520–525.

NAS (National Academy of Sciences), 1984. Acid deposition: processes of lake acidification. Summary of a discussion. National Research Council Commission on Physical Sciences, Mathematics, and Resources. Environmental Studies Board, Panel on Processes of Lake Acidification. National Academy Press, Washington, DC, 11 pp.

Newton, R.M. and R. April, 1982. Surficial geologic controls on the sensitivity of two Adirondack lakes to acidification. *N. E. Environ. Sci.*, **1**, 143–150.

Reuss, J.O., Cosby, B.J. and R.F. Wright, 1987. Chemical processes governing soil and water acidification. *Nature*, **329**, 27–32.

Reuss, J.O. and D.W. Johnson, 1985. Effect of soil processes on the acidification of water by acid deposition. *J. Environ. Qual.*, **14**, 26–31.

Reuss, J.O. and D.W. Johnson, 1986. *Acid Deposition and Acidification of Soils and Waters.* Ecol. Series, Vol. 59. Springer-Verlag, New York, 119 pp.

Reuss, J.O. and P.M. Walthall, 1989. Soil reaction and acidic deposition, in *Acidic Precipitation: Volume 4, Soils, Aquatic Processes, and Lake Acidification* (eds S.A. Norton, S.E. Lindberg, and A.L. Page), Springer-Verlag, New York, pp. 1–33.

Seip, H.M., 1980. Acidification of fresh waters: sources and mechanisms, in *Ecological impact of Acid Precipitation*, Proc. Int. Conf. SNSF Project: Acid Precipitation – Effects on Forest and Fish (eds D. Drablos and A. Tollan), The Norwegian Interdisciplinary Research Programme, Oslo, Norway, pp. 358–366.

Shaffer, P.W. and M.R. Church, 1989. Terrestrial and in-lake contributions to alkalinity budgets of drainage lakes: an assessment of regional differences. *Can. J. Fish. Aquat. Sci.*, **46**, 509–515.

Shaffer, P.W., R.P. Hooper, K.N. Eshleman and M.R. Church, 1988. Watershed vs. in-lake alkalinity generation: a comparison of rates using input-output studies, *Wat., Air, Soil Pollut.*, **39**, 263–273.

Stoddard, J.L., 1994. Long-term changes in watershed retention of nitrogen: Its causes and aquatic consequences, in *Environmental Chemistry of Lakes and Reservoirs*, Advances in Chemistry Series No. 237 (ed, L.A. Baker), American Chemical Society, Washington, DC, pp. 223–284.

Turner, R.S., R.B. Cook, H. Van Miegroet *et al.*, 1990. *Watershed and Lake Processes Affecting Surface Water Acid-Base Chemistry*, NAPAP Report 10. Acidic Deposition: State of Science and Technology. National Acid Precipitation Assessment Program, Washington, DC, 167 pp.

Wigington, P.J., Jr, T.D. Davies, M. Tranter and K.N. Eshleman, 1990. *Episodic Acidification of Surface Waters Due to Acidic Deposition*, NAPAP Report 12. Acidic Deposition: State of Science and Technology. National Acid Precipitation Assessment Program, Washington, DC, 200 pp.

Wigington, P.J., Jr, J.P. Baker, D.R. DeWalle *et al.*, 1993. *Episodic Acidification of Streams in the Northeastern United States: Chemical and Biological Results of the Episodic Response Project.* EPA/600/R-93/190, US Environmental Protection Agency, Washington, D.C., 337 pp.

Acknowledgement

The information in this document has been funded wholly (or in part) by the US Environmental Protection Agency. It has been subjected to the Agency's peer and administrative review, and it has been approved for publication as an EPA document. Mention of trade names or commercial products does not constitute endorsement or recommendation for use.

Cross reference

Acid rain

ACID RAIN

Acid rain has become one of the major environmental problems facing industrialized countries, Much has been written on the subject, often with political rhetoric and emotional speculation providing the theme. Despite this, evidence points to the existence of a problem that will be of increasing concern in the future.

Assessing the nature of acid rain and its potential impact requires background information concerning the concept of acidity and ways in which atmospheric processes influence the potential for the deposition of an acid by precipitation processes.

Acidity

The acidity of a substance is associated with the relative abundance of free hydrogen ions (H^+) when that substance is in a water solution. Acidity is measured on a logarithmic pH scale where a value of 7 indicates neutrality; decreasing values on the scale indicate increasing acidity and increasing values represent alkalinity. The pH scale with representative examples is shown in Table A11.

Absolutely pure water has a pH of 7 but if left standing in clean air its pH will decrease to near 5.6, a result of the absorption of atmospheric carbon dioxide to form weak carbonic acid. Most rainwater will have this pH value; acid rain is thus considered to exist when the pH is less than 5.6. For this increased acidity to occur, the atmosphere must contain chemicals to provide an acid source. The atmosphere also must have the ability to deposit the material at the surface. This deposition is completed through the process of atmospheric cleansing.

Atmospheric cleansing

The main process that results in acid deposition is the removal of chemicals through precipitation – the 'wet deposition' process. Atmospheric impurities are incorporated in the entire precipitation process, starting with cloud droplet formation, and are deposited as part of the eventual precipitation. Cleansing can also occur through 'dry deposition', which implies that substances in the atmosphere are deposited through gravitational settling. Large particles settle under their own weight; very small particles may be carried far from their source eventually to coagulate or polymerize until they are large enough to fall under their own weight. Such particles often are returned to earth through the wet deposition process.

The wet and dry depositional processes do not cover all aspects of deposition. Other forms include fog droplet interception (especially marked in high-altitude forests), dew, frost, and rime icing.

The distribution of wet-deposition acid in North America is shown in Fig. A32. The eastern part of the continent, centered on the northeastern USA and southeastern Canada, clearly has a higher acidity than the remainder of the area. In part, this distribution reflects the conditions in the drier west where contributions of alkaline substances to the atmosphere alter the chemistry of precipitation.

Acid rain sources

Acid rain primarily results from the release of sulfur oxides (SO) and nitrogen oxides (NO) into the air by industrial and transportation sources. These oxides are transformed into sulfuric acid and nitric acid through oxidation and hydrolysis. Table A12 provides a

Table A11 Sample pH values

Example	pH[a]
Neutral solution	7.0
Natural rainfall	5.6
Lethal to most fish	4.5
Average rainfall (NE USA)	4.4
Acidity of tomato juice	4.3
Rainfall at Toronto, Canada (February 1979)	3.5
Acidity of lemon juice	2.2
Rainfall at Mount Mitchell, NC (July 1986)	2.2
Battery acid	1.1

[a] For each unit decrease, acidity increases by a factor of 10.

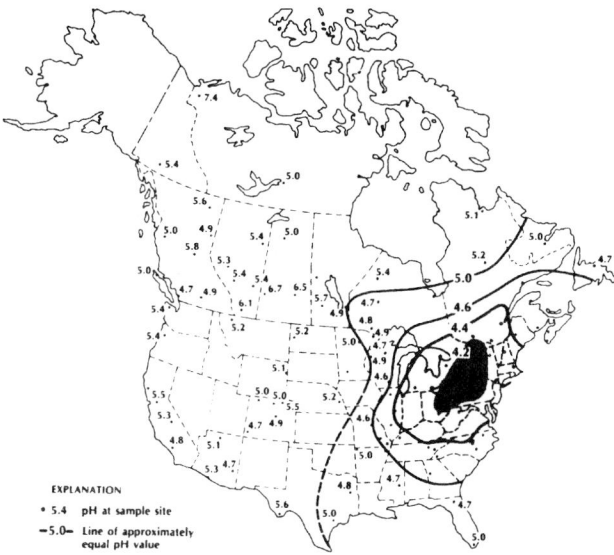

Figure A32 Distribution of acidity of rain in North America (wet deposition only) (after National Academy of Science, 1983).

Table A12 Chemicals that influence the acidity of rainfall

Ions in deposition	Source and relative significance	
	Natural	Human activities
Sulfate (SO)	Swamps	Power plants
	Volcanoes[c]	Industrial processes[a]
	Oceans	Smelters
Nitrate (NO)	Lightning	Industrial processes
	Soil biological activity[c]	Transportation[a]
Chloride (Cl)	Oceans[c]	Industrial processes[c]
	Land surface	Road salt
Ammonium (NH)	Biological processes	Industrial processes
	Animals[b]	Agriculture
Calcium (Ca)	Land surface[b]	Industrial processes[c]
		Agriculture
Sodium (Na)	Oceans	Road salt[c]
	Land surface[c]	
Magnesium (Mg)		Industrial processes
Potassium (K)	Land surface[c]	Agriculture

[a] Major significance.
[b] Moderate significance.
[c] Minor importance.

summary of all the major chemicals that influence the acidity of rain, with both natural and artificial sources identified. As noted above, significant in the listing is the relative importance of artificial sources of SO and NO. These are classed as of 'major significance' suggesting that acid rain is a direct result of human pollution of the atmosphere.

Integral to this conclusion is understanding the transport of air pollutants responsible for acid rain to locations currently experiencing increased acid deposition. That acidity is increasing is well illustrated in the time comparison shown in Figure A33.

Given that pollutants can return to earth almost immediately or remain aloft for a week or longer, then atmospheric circulation patterns and prevailing winds are significant in understanding the

distribution of acid rain. Preliminary analysis suggests that some 30% of the total sulfur compounds deposited over the eastern United States originate from site sources more than 500 km away from the region. Another one-third comes from sources between 200 and 500 km, with the remainder being derived from sources less than 200 km from the deposition site.

Location of the actual source of chemicals contributing to acid rain is a distinct scientific problem, which is compounded by the political and economic ramifications involved. For example, there has been much controversy over the relative role of Midwestern industries and power plants using high-sulfur coal, thereby promoting acid rain in the northeastern United States. Similarly, the role played by Scandinavian acid rain resulting from SO and NO emissions in the UK is of

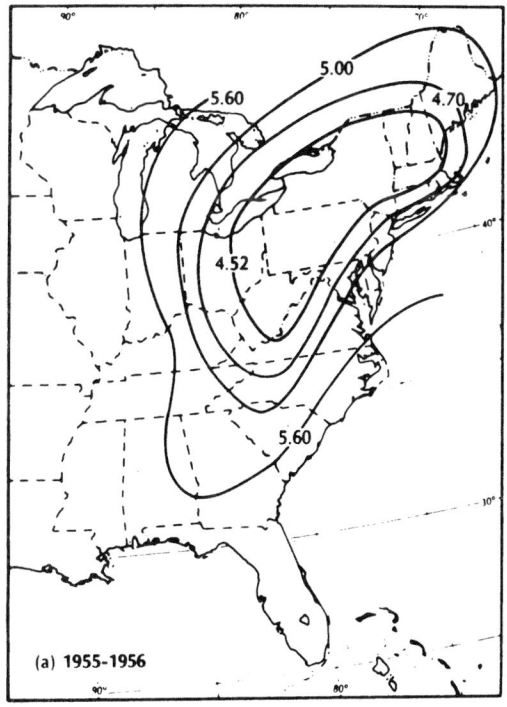

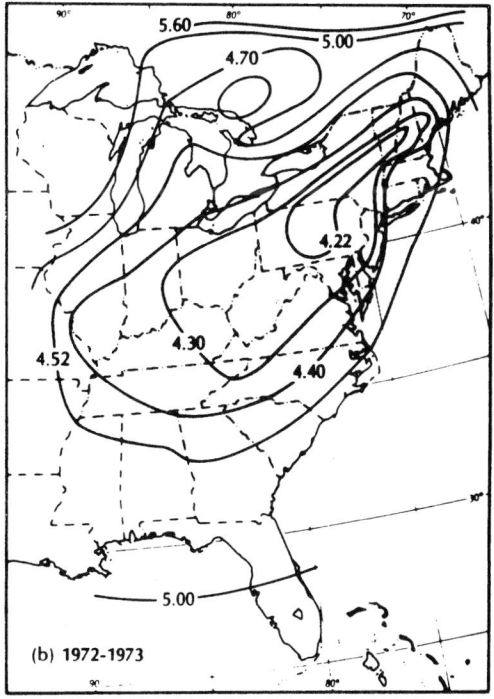

Figure A33 A comparison of rainfall acidity in the eastern United States for two time periods.

concern. Until definitive answers concerning the transport of pollutants are derived, it might prove difficult to get industry to invest large sums of money to rectify the situation, particularly when it could lead to loss of earnings for many people.

Effects of acid rain

In an account of the potential impacts of acid rain, Miller (1984) succinctly outlines the problem. He notes that some 30 years ago researchers in Scandinavia and the United States found that there was an increase in acidity and a decrease in fish population in certain lakes and streams. At first the correlation was attributed to acid rain; however, further investigation showed that the relationship was much more complex, since differences in surrounding soils and vegetation supply chemical ions that also modify acidity. Continued research shows that although this is the case, the addition of acid rain to sensitive lakes and streams does have an adverse effect.

Acid rain is also thought to have a detrimental effect on forests. Trees may be injured by the direct action of acid on leaves or by changes in soil chemistry caused by the acid solution. Such injuries may be responsible for the decline of several species of trees in Europe and North America over the past 30 or so years. Further research has also suggested that agricultural productivity may be adversely affected by acid rain.

Buildings and monuments also are affected by acid rain. Chemical reactions resulting from deposition of both wet and dry deposition have been identified as a cause of extensive damage in historical buildings in Europe. Unfortunately, many famous ancient structures are involved. In Greece, for example, statues carved in the fifth century BC were deteriorating so rapidly in recent times that they have been removed and replaced by fiberglass copies. In India, the famed Taj Mahal which is made of marble, is deteriorating through exposure to acid rain. In addition to damage to stone structures, acid also weakens other building materials, particularly iron, steel, zinc and paint, causing the lifetimes of structures to be shortened appreciably.

The future

It is clear that acid rain is an environmental hazard in many parts of the world. The phenomenon is basically associated with the industrialized regions with well-documented effects in Europe and North America. It can be anticipated, however, that the impacts of acid rain will be seen in the other high producers of SO_2, the former USSR and China.

Decreases in the occurrence of acid rain will occur when sulfur and nitrogen emissions into the atmosphere are reduced, a goal that can only be met through economic and political actions. The former requires appropriate alternatives in energy production and transportation, the latter requires national and international agreement. In 1979 an attempt to meet the problem of the transport of air pollutants across political boundaries was the Convention on Long-Range Transboundary Air Pollution, a weak but important accord between European and North American countries. Subsequently, actions such as the 1990 amendment of the 1970 Clean Air Act in the United States and the 1991 protocol of the UN Economic Commission for Europe call for reduced SO and NO emissions. It is clear, however, that much remains to be completed before the impacts of acid rain are negated.

John E. Oliver

Bibliography

Howells, G., 1990. *Acid Rain and Acid Water*. London: Ellis Horwood.
Miller, J.M., 1984. Acid rain, *Weatherwise*, **37**, 232–239.
National Academy of Science, 1983. *Acid Deposition: Atmospheric Processes in Eastern North America*. Washington, DC: National Academy Press.
Park, C., 1989. *Acid Rain: New Approach to an Old Problem*. New York: Routledge.
Peters, N.E., R.A. Schroeder and D.E. Troutman, 1982. *Temporal Trends in The Acidity of Precipitation and Surface Waters of New York*, US Geol. Survey Water Supply Paper 2188. Washington, DC: Government Printing Office.
Wisniewski, J. and J.D. Kinsman, 1982. An overview of acid rain monitoring activities in North America, *Am. Meteorol. Soc. Bull.* **63**, 598–618.

Cross references

Acidic deposition: acidification of surface water
Precipitation

ACTIVATED SLUDGE PROCESS: HISTORICAL

In 1914 the activated sludge process was first demonstrated by Edward Arden and William T. Lockett at the Davyhulme Sewage Works in Manchester, UK (see Activated sludge process; Figure A34). Like others they were trying to improve the performance of the newly invented biological filter on which an organic gel capable of purifying effluent could grow. Arden and Lockett showed that a floc built up from sewage solids served the same purpose as the filter film and could be used over and over again. Air bubbling through the floc ensured that it mixed with the sewage. They called the process the activated sludge for reference purposes failing a better term and it promised startling improvements in performance cutting treatment time from days to hours. Such was the superiority of the new process over existing ones that it spread rapidly around the world by the Activated Sludge Company which after several changes in name has finally emerged as Water Engineering, a subsidiary of North West Water – and Davyhulme is now North West Water's largest works.

R.W. Herschy

Bibliography

UK Water Services Association, 1994. A sewage celebration, *Water Bulletin*.

Cross reference

Activated sludge process

ACTIVATED SLUDGE PROCESS

The most efficient method for reducing the organic content of dilute organic wastes is by aerobic biological treatment in activated sludge systems, percolating filters, rotating biological contactors (RBCs) and natural and aerated oxidation ponds, all of which use the same basic biochemical processes to effect treatment. This entry describes the activated sludge process and reviews biological treatment with particular reference to the process.

In the basic process, wastewater is aerated in a tank with a seed of microorganisms (the activated sludge, often referred to as the mixed liquor suspended solids, MLSS), which oxidizes part of the organic matter to water and carbon dioxide, thus liberating energy. This energy is then used to synthesize more microorganisms. Hence the waste organic substances have a dual role: to serve as a source of carbon for growth and as a source of energy. The overall biochemical reactions are shown diagrammatically in Figures A35 and A36, where waste organic material is referred to as COHNS (carbon, oxygen, hydrogen, nitrogen, sulphur), derived from the symbols of the main chemical elements in organic matter. It is essential that the physical and chemical environment is satisfactory for the growth of microorganisms, e.g. suitable temperature and pH and the absence of toxic or inhibitory substances. In addition, the waste must contain all the nutrients and growth factors required by the microorganisms. After treatment the MLSS pass to a settlement tank where the active sludge settles and is then recycled (return sludge) to the inlet of aeration tank as the seed for the incoming untreated wastewater. From time to time active sludge is wasted from the process in order to control the concentration of MLSS in the aeration tank. The layout of the basic process, which typically produces a very high quality of final effluent, is shown in Figure A37.

Historical development of the process

In 1914 Arden and Lockett (1914) described fill-and-draw laboratory experiments on the aeration of sewage which led to the development

Figure A34 Plaque to commemorate the discovery of the activated sludge process in 1914.

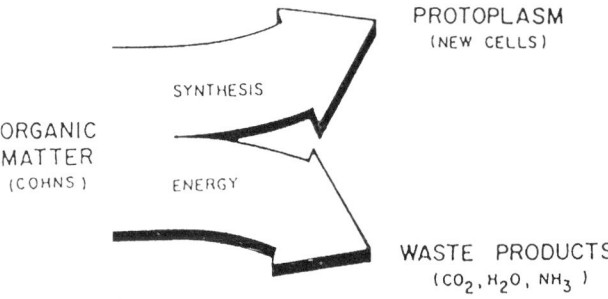

Figure A35 Aerobic metabolism of organic matter.

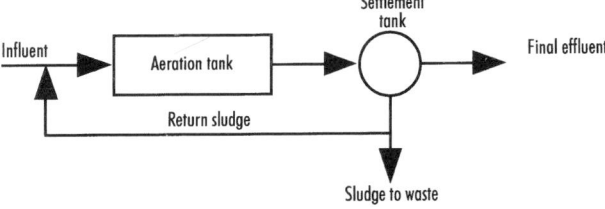

Figure A37 Layout of basic process.

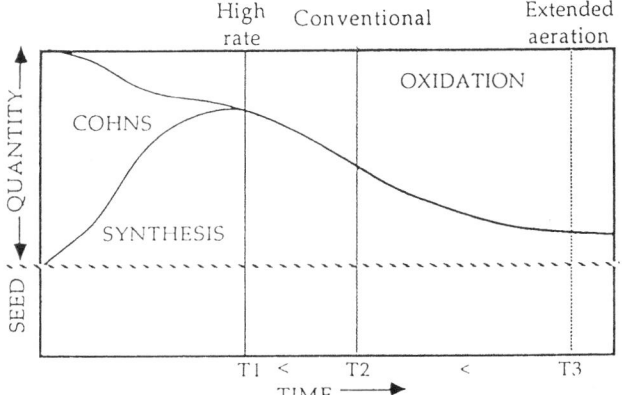

Figure A36 Effect of treatment time on oxidation and synthesis.

of the activated sludge process. During 1914 the work was transferred to a large-scale continuous-flow pilot plant in the Manchester (Davyhulme) sewage treatment works and the first full-scale plant was installed at Worcester in 1916. Since then it has been used extensively throughout the world.

Initially there was no clear understanding on how the process worked and the early literature contains many articles arguing the pros and cons of physical versus biological action. However, by 1930 the biological theory was well established due to the work of many research workers, e.g. Seizer (1928) in Germany, and Buswell and Long (1929) in the USA. The process has been the subject of countless studies in the field and in the laboratory to gain a better understanding of it, and to assess the value of modifications to the basic system.

Initially uniform aeration was applied throughout the whole aeration tank, but when it was appreciated that the basic or conventional process was characterized by rapidly diminishing oxygen requirements as treatment progresses, tapered aeration was proposed for continuous flow plants. Tapered aeration requires that the aeration facilities along the tank be sized to satisfy the anticipated oxygen demand with a consequent saving in power, i.e. the intensity of aeration is reduced along the length of the aeration tank (Figure A38).

Another modification of the basic process which overcomes any problem of high oxygen demand at the inlet to the aeration tank is the complete mix, activated sludge process. In this system the incoming

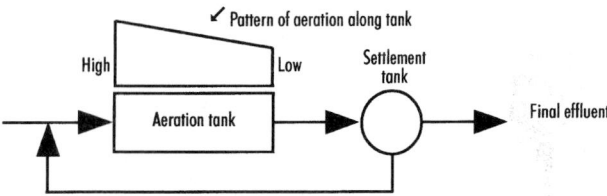

Figure A38 Tapered aeration.

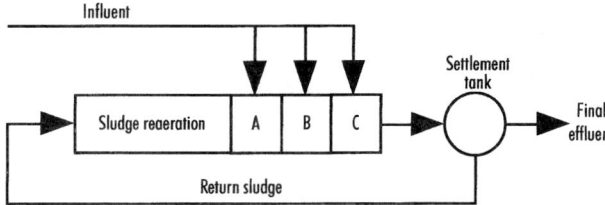

Figure A39 Step aeration.

wastewater is mixed immediately with the whole contents of the aeration tank, which ensures that high oxygen demands are distributed throughout the aeration tank. Similarly, potentially toxic and inhibitory concentrations of substances are diluted in the whole contents of the tank.

To overcome problems in the basic process, Gould (1942) proposed step aeration, which is shown diagrammatically in Figure A39. The first tank is reserved for reaerating the return sludge in order to regenerate its adsorptive properties; activated sludge has the ability to adsorb quickly relatively large amounts of colloidal and finely divided organic matter which, after a period of time, is metabolized by the microorganisms in the sludge. Wastewater is then added stepwise, e.g. one-third each at the beginning of tanks A, B and C in a three-tank system. Step aeration distributes the demand for oxygen during treatment and minimizes the possibility of oxygen deficiency at the inlet to the aeration tank. It also affords a greater degree of dilution of toxic and inhibitory substances in the incoming wastes. However, although high-quality final effluents can be achieved with respect to the removal of biological oxygen demand (BOD) and suspended solids (SS), it is not usually possible to obtain complete nitrification in step aeration systems unless long treatment times are used.

Until 1943 the conventional activated sludge process was used to produce final effluents of high quality with respect to BOD and SS. However, Setter and Edwards (1943, 1944) showed that, when appropriate, the process could be designed to produce reliable effluents of intermediate quality, i.e. by reducing the MLSS to about 500 mg l^{-1} and with retention times of about 2 h. This modification of the basic process is known as high-rate, or modified aeration, and typically removes 65–70% of the applied BOD and SS. It requires a minimum of retention time and aeration capacity, but nitrification is not possible with this modification of the activated sludge process.

Ulrich and Smith (1951, 1957) showed that the great adsorptive properties of a well-conditioned activated sludge could be used to advantage in another modification of the basic process called the biosorption or contact stabilization process (Figure A40). In this modification, reaerated activated sludge is brought into contact with unsettled wastewater and aerated for a period of about 30 min, during which time up to 90% of the BOD may be adsorbed and absorbed by

the active sludge. The MLSS are then settled and the sludge is aerated to regenerate its adsorptive properties before being mixed with more untreated wastewater.

More recent modifications of the activated sludge process have been developed to effect complete nitrification followed by denitrification, and for the biological removal of phosphate.

Oxygen is essential to all activated sludge systems and can be introduced into the aeration tank using (1) coarse- or fine-bubble, air diffusion systems; (2) surface, turbine-type aerators rotating about the vertical axis, and surface, brush-type aerators rotating about the horizontal axis, both of which produce turbulence and spray which allows the fine droplets to dissolve oxygen from the air; and (3) pure oxygen from cryogenic plants or from pressure-swing absorption (PSA) units. The use of pure oxygen can take place in covered aeration tanks, e.g. the UNOX system, or by side-stream oxygen injection in open tanks, e.g. the VITOX system. The use of pure oxygen has resulted in plants designed to operate with more than 20 000 mg l^{-1} MLSS, concentrations at least five times those typically used in air, activated sludge systems. Brush aerators can be used in conventional tanks, or in ditches and channels with a 'race track' layout, for example as used by Pasveer (1959) in The Netherlands.

Basic microbiology and design considerations of the process

For any particular waste the oxidation of a unit quantity of organic matter will liberate a specific quantity of energy which, for a given population of microorganisms, will be used to synthesize a specific mass of cellular material, i.e. linear relationships should exist between the oxidation of waste organic material (oxygen use) and the production of biological solids. When the waste matter has been metabolized and no further synthesis can occur, the microorganisms undergo endogenous respiration and oxidize their cellular material in order to obtain energy. Again, a linear relationship should exist between the oxidation of a unit quantity of cellular material (oxygen use) and the decrease in the mass of biological solids. This is indicated below using an empirical formula derived from chemical analyses to describe typical cellular material:

$$C_7H_{11}NO_3 + 7.5O_2 \rightarrow 7CO_2 + 4H_2O + NH_3$$
cellular material

Theoretical and actual curves describing these relationships are shown in Figs A41 and A42.

The species of microorganisms which predominate in a given environment will be influenced by the nature of the organic matter and nutrients present in the wastewater. Microbial species have diverse physiological properties and, because synthesized biological solids are wasted regularly from the treatment process, nutrients removed from the waste will typically not be recycled for reuse, so they must be present in the untreated waste. This is the fundamental reason for aerobic biological treatment: soluble and colloidal organic matter, or BOD, is converted to settleable biological solids which are recovered for further treatment and/or disposal. From the foregoing it is obvious that in aerobic treatment systems there are at least five important design considerations.

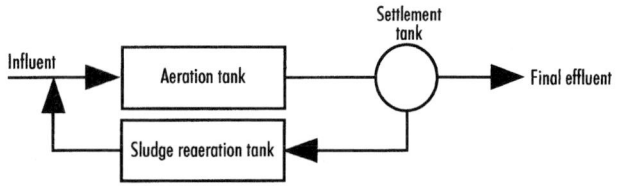

Figure A40 Layout of biosorption process.

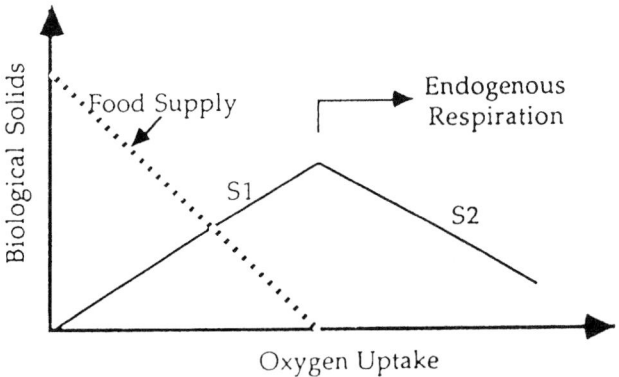

Figure A41 Theoretical relationship between synthesis and uptake.

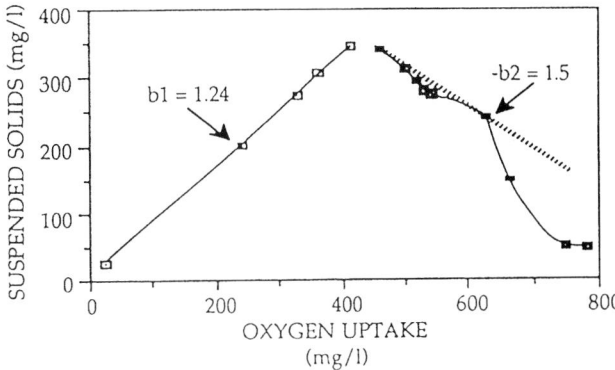

Figure A42 Suspended solids and oxygen uptake data for glucose with *Pseudomonas fluorescens* seed.

1. The quantity of organic matter to be removed. This requires reliable sampling and analyses of all wastewaters to be treated and the selection of an appropriate method of treatment which will produce the desired quality of final effluent.
2. The quantity of oxygen required to satisfy the metabolic requirements of the treatment microorganisms. This is a function of the quantity of pollution to be treated and the food to microorganism ratio (F/M), or the 'biochemical position,' of the treatment process.
3. The correct quantities and types of microorganisms to effect satisfactory treatment. This requirement is affected by the ability of the microbes to acclimatize to the organic matter in the waste, and by the nature of the treatment system which must ensure that sufficient numbers of microorganisms are present for an adequate time to metabolize the pollution. In this respect, highly variable wastewater characteristics can cause problems.
4. The quantity of secondary sludge for disposal. This quantity is dependent on the F/M of the treatment process and inert debris present in the raw wastewater.
5. The provision and maintenance of a satisfactory physical and chemical environment for the microorganisms. In this respect there must be strict control over the discharge of anything which may be harmful to the microbes, and the treatment plant may have to be protected from adverse weather conditions.

Sludge loading rate (F/M)

The major factor governing the performance of a treatment plant is the amount of organic matter a unit quantity of microorganisms will oxidize in unit time under a given set of conditions. The amount of organic matter to be treated in unit time is the design BOD load, expressed in kilograms applied per day. In the absence of a practicable method for measuring viable microbes in a system, the total weight of activated sludge in the aeration tank, i.e. the MLSS, is used on the assumption that there is a relationship between total weight of MLSS and the proportion which is viable microorganisms. The sludge loading rate, or F/M, is usually expressed as kilograms of applied BOD per day per kilogram of MLSS.

While it appears that an infinite number of sludge loading rates can be selected, for practical purposes designs usually fall into one of three categories: high-rate systems where the F/M is greater than about 0.35; conventional systems with F/M values in the range 0.20–0.25; and extended aeration systems where the F/M is less than 0.15. The F/M affects a number of factors in the treatment system as outlined below.

Sludge production and sludge age

The production of new sludge is mainly controlled by the F/M, and in a treatment system at equilibrium the daily production of sludge determines the sludge age. Sludge production can be expressed as a percentage increase in the total quantity of sludge in the treatment system, e.g. 10%, and the reciprocal of this is the sludge age, i.e. 10 days.

The actual daily quantity of sludge which must be wasted from a system is affected by several factors in addition to the F/M, for example biological solids lost in the final effluent, predators in the

Table A13 Relationship between growth, rate of production and sludge age for different values of F/M

F/M	Growth (a)	Rate of production (%)	Sludge age (days)
0.1	0.52	5.2	19.2
0.2	0.65	13.0	7.7
0.3	0.75	22.5	4.4
0.4	0.83	33.2	3.0

treatment system and non-degradable debris in the untreated waste. An empirical formula for calculating the weight of sludge produced per unit weight of applied BOD in UK temperatures is:

$$a = k + (F/M)^{0.5}$$

where a = weight of sludge produced per unit of applied BOD, and k = a constant to compensate for non-degradable debris (= 0.2 for typical settled sewage). This forumula has been used to calculate the relationships given in Table A13.

Removal of BOD

The sludge loading rate is one of the factors which affects the quality of the final effluent, but there is only an approximate correlation between F/M and BOD removed. When the F/M is below about 0.3, up to 95%, or more, of the BOD should be removed, but with higher sludge loadings the quality of the effluent will progressively deteriorate. This is accounted for by the fact that with high F/M values the sludge age is relatively short and the ecology of the sludge is immature, i.e. there are a limited number of microbial species present. In addition, with high sludge loading rates, diffuse microbial growths occur which do not readily flocculate and settle, i.e. these are 'high-energy' systems. Processes with F/M values lower than 0.3 have longer sludge ages, are more mature, have lower energy levels and typically flocculate readily. In addition, mature sludges usually maintain relatively large numbers of predator, or grazing, microorganisms which control the numbers of free swimming microbes in the system and, hence, the quality of the final effluent.

Oxygen requirements

The oxygen required during treatment is largely dependent on sludge age and nitrification, which is discussed later. The sludge loading rate affects the oxygen requirements of the treatment process in three ways: (1) as discussed previously, part of the applied BOD is oxidized to carbon dioxide and water; (2) a proportion of the biological solids synthesised from the applied BOD will be oxidized during endogenous respiration, how much being a function of the solids retention time in the system, or F/M; and (3) dissolved and colloidal organic matter adsorbed and absorbed by the activated sludge may be wasted from the system before it is metabolized. In this latter respect, the longer a sludge is retained in the process (low F/M values), the greater will be the amount of adsorbed and absorbed organic matter that is oxidized, which requires oxygen. Conversely, with high loading rates a large amount of surplus sludge is synthesized, which requires that a greater proportion of the synthesised biological solids must be wasted from the system and with it adsorbed and absorbed organic matter.

Vosloo (1973) compared and published data from several sources on the relationship between the sludge loading rate and the oxygen required for carbonaceous oxidation only, and in practice these have been used to good effect (Table A14).

The diurnal variations in applied BOD will result in varying oxygen demands, and if aeration is applied at an average rate over the day there would be periods when the dissolved oxygen (DO) level would fall to zero, with possibly adverse effects on process performance. At other times the demand for oxygen would be lower than average, thereby causing unnecessarily high DO levels which waste energy. The DO level should be 1.5–2 mg l^{-1}, but in practice upper and lower levels of 3 and 0.5 mg l^{-1} are commonly adopted for minimum and peak oxygen demand periods, respectively. Automatic control is effected using DO electrodes. When nitrification is required, additional oxygen must be supplied in the ratio of 4.5 kg O_2 per kg of nitrogen to be oxidized.

Microbial growth, treatment of specific organic compounds and nitrification

The species of microorganisms which predominate in a given treatment plant are influenced by the nature of the waste to be treated,

Table A14 Relationship between sludge loading rate and oxygen required for carbonaceous oxidation only

F/M kg kg^{-1}	Sludge age (days)	Oxygen required [kg (kg BOD)$^{-1}$	Ratio of maximum to mean rate of use	Maximum use [kg (kg BOD)$^{-1}$]
≤0.1	≥20	1.60	1.5	2.40
0.15	10.0	1.38	1.6	2.20
0.20	6.7	1.22	1.7	2.07
0.25	5.0	1.10	1.8	1.98
0.30	4.0	1.00	1.9	1.90
0.40	2.9	0.88	2.0	1.76
0.60	1.8	0.74	2.2	1.63

Table A15 Calculated critical sludge loading rates above which nitrification will not occur

Temperature (°C)	Growth rate constant (percentage increase per day)	Critical sludge age (days)	Critical F/M (kg kg^{-1})
5	5.42	18.5	0.10
10	9.87	10.1	0.16
15	18.00	5.6	0.25
20	32.80	3.0	0.40

the availability of nutrients and growth factors, physical and chemical environment, and sludge age. The significance of the latter is often not fully appreciated, yet it has a major influence on the treatment of specific organic compounds and nitrification; it is, perhaps, the most important factor to be taken into account when considering the treatability of a waste.

In a given environment different microorganisms grow at different rates, i.e. each microbe has a specific growth rate, or doubling time. In activated sludge systems the microorganisms increase in number by normal growth, but are reduced in number when excess sludge is wasted from the system. When a particular treatment plant is in equilibrium, the amount of synthesized sludge wasted from the system must be equal to the amount of new sludge which is produced and is, therefore, a function of F/M. This being so, it follows that if the rate of wasting sludge is greater than the specific growth rate of a microorganism, then the microbe cannot be present in any significant numbers in the activated sludge. This is of particular importance when the breakdown of specific organic compounds is required and where this can only be achieved by certain microorganisms with low rates of growth. In high F/M systems the rate of sludge production is high, i.e. the sludge age is low, and consequently only microorganisms with relatively short generation times will be present in significant numbers in the sludge. Downing and coworkers (1964a,b) showed the same was true for nitrification, while Downing and Knowles (1967) stated that in mixtures of activated sludge and domestic sewage, with pH in the range 7–8 and temperatures between 5 and 25°C, the growth rate of the nitrifying microorganism, *Nitrosomonas*, could be calculated using the following formula:

$$k = 0.18\ e^{0.12(t-15)}$$

i.e. the growth rate constant k increases by about 13% for each 1°C rise in temperature. Using this formula the calculated critical sludge loading rates above which nitrification will not occur have been calculated as shown in Table A15.

Nutrient requirements

Basically, aerobic biological systems are used to stabilize the carbonaceous content of wastewaters, which must be nutritionally balanced if there is to be unrestricted growth of the microorganisms. The microorganisms derive not only energy and food from the waste, but also the nutrients for growth and the constituent units of enzymes, coenzymes, enzyme activators and cell protoplasm without which the biochemical reactions would be impaired or inhibited.

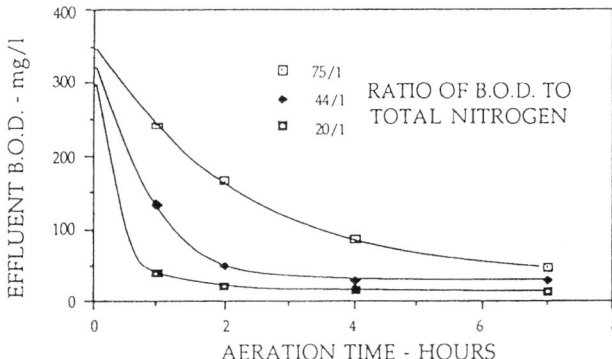

Figure A43 Effect of nitrogen deficiency on the treatment of trade was by the activated sludge process.

Nitrogen and phosphorus are required for the synthesis of proteins, energy-transfer enzymes, etc., and for conventional treatment systems must be present in the ratios:

BOD : available N : available P = 100 : 5 : 1.

If these nutrients are not present in sufficient quantities the rate of treatment will be reduced as shown in Fig A43.

The six elements of C, O, H, N, P (phosphorus) and S account for about 95% of the dry weight of microbial cells, but many other elements are known to be present in the remaining fraction and studies have shown that some of these must be present in macro or micro quantities for the satisfactory growth of microorganisms. It is known that potassium, magnesium, calcium, sodium, iron, manganese, cobalt, copper, nickel, zinc, molybdenum and vanadium are essential trace nutrients required by many bacteria. Beveridge and Doyle (1989) and Simpson *et al.* (1991) proposed that a lack of trace metals is a major cause of bulking activated sludge.

Settleability and return sludge

The settleability of activated sludge is usually reported in terms of the stirred sludge volume index (SSVI), which is the volume occupied by 1 g (dry weight) of mixed liquor suspended solids (MLSS) after settling for 30 min while being gently stirred. SSVIs are used in the design and operation of activated sludge systems: (1) to size the return activated sludge (RAS) system, (2) to control the level of MLSS in the aeration tank, and (3) as an indicator of the condition of the sludge with respect to settleability, i.e. it is widely accepted that sludges with SSVI values of 125 ml g^{-1} or less are satisfactory, whereas bulking sludges have SSVI values above 150 mg l^{-1}.

When a process design is being prepared the desired F/M is selected, e.g. 0.25 (kg BOD kg MLSS)$^{-1}$, together with concentration for the MLSS, e.g. 3 000 mg l^{-1}. Knowing the design quantity of BOD to be applied to the aeration tank, e.g. 3900 kg day^{-1}, the size of the tank can be calculated as follows:

$$M = \text{applied BOD/sludge loading rate}$$
$$= 3900/0.25$$
$$= 15\,600\ \text{kg MLSS}$$

Therefore

volume of aeration tank = kg MLSS × 1000/concentration of MLSS in mg l^{-1}
= 15 600 × 1000/3000
= 5200 m^3

In order to maintain the desired F/M in the aeration tank, assumptions have to be made regarding the concentration of suspended solids in the return activated sludge (RASSS). For steadystate conditions in the aeration tank, the mass of solids entering the final settlement tank must be equal to the mass of solids returned to the aeration tank and this is achieved using the following formula (see also Figure A44).

$$RY = (Q + R)X$$

from which

$$R = XQ/(Y - X)$$

where Q = flow to be treated (m^3 day^{-1}), X = concentration of MLSS

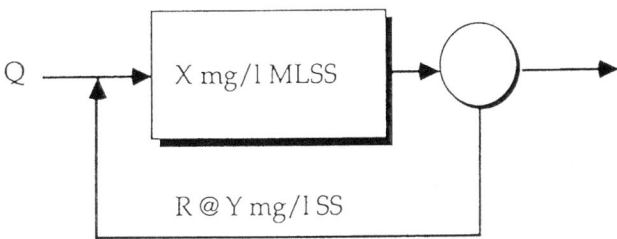

Figure A44 Sludge recirculation.

(mg l^{-1}), Y = concentration of RASSS (mg l^{-1}) and R = rate of sludge recirculation (m^3 day^{-1}).

Unless pre-design pilot plant studies have been carried out, Y will not be known, so the RAS system must be designed for a range of RASSS concentrations which may occur and the consequent different rates of sludge recirculation (R) which will be required. Even when Y is known, the RAS system should be designed to allow maximum flexibility in the operation of the treatment plant in order to accommodate changes in the SSVI, together with long-term variations in the applied BOD and/or flow to be treated. It is not uncommon to install RAS facilities which will permit variable sludge rates from 25 to 150% of the design inflow to the treatment plant.

When operating a treatment plant, SSVI measurements can be helpful in deciding R, the rate of sludge return. In the SSVI test the average concentration of solids in the settled sludge layer can be expressed as SSVI $\times$ 10^{-6} mg l^{-1}, which will be the RASSS concentration if the sludge settles in the final tanks in the same way as it does in the SSVI test. In practice this often occurs, but in a particular treatment plant it may not be the case because of the nature of the sludge, the geometry of the final tank and the sludge retention time in the final tank. Nevertheless, at such treatment plants it is often possible to develop a factor which corrects for local circumstances. The value of using SSV values to assist in calculating the sludge return rate cannot be under-emphasised.

The regular measurement of the SSVI is also very useful for monitoring the condition of a sludge. In this respect the trend of results is important, rather than any specific value, and SSVI values increasing above 150 mg l^{-1} should be regarded as the possible onset of sludge bulking problems.

In conventional treatment plants the design of final settlement tanks is usually based on the maximum rate of flow applied to the tank, ignoring recirculation. With the advent of pure, or enriched, oxygen activated sludge systems, which may be designed and operated with MLSS of 20 000 mg l^{-1} or more and SSVI values of 35 or less, the size of the final tanks should be determined by the maximum solids flux applied to the tanks. For a particular waste, pilot plant studies may be required to develop the relationships between the rates of settlement of different concentrations of MLSS.

The design and operation of aerobic biological treatment plants should take into account all of the variables which occur in the nature of wastewaters and ensure that nothing will adversely affect the proper growth of the treatment microorganisms. The activated sludge system is the most flexible method of treatment, taking into account the variations which occur in the nature of wastewaters, the biochemistry of wastewater treatment and the degree of treatment required.

James R. Simpson

Bibliography

Arden, E. and Lockett, W.T., 1914. Experiments on the oxidation of sewage without the aid of filters. *J. Soc. Chem. Ind.*, **33**, 523.
Beveridge, T.J. and Doyle, R.J., 1989. *Metal Ions and Bacteria*. John Wiley & Sons, New York.
Buswell, A.M. and Long, H.L., 1929. Microbiology and theory of activated sludge. *J. Am. Water Works Assn.*, **10**, 309.
Downing, A.L., Painter, H.A. and Knowles, G., 1964a. Nitrification in the activated sludge process. *J. Proc. Inst. Sew. Purif.*, (2), 130–158.
Downing, A.L. and Hopwood, A.P., 1964b. Some observations on the kinetics of nitrifying activated sludge plants. *Schweiz. Z. Hydrol.*, **26**, 271.
Downing, A.L. and Knowles, G., 1967. Population dynamics in biological treatment plants. *Proc. 3rd Int. Conf. Wat. Pollut. Res.*, (Munich 1966), **2**, 117.
Gould, R.H., 1942. Operating experiences in New York City. *Sew. Works J.*, **14**, 1, 70.
Pasveer, A., 1959. A contribution to the development in activated-sludge treatment. *J. Inst. Sewage Purif.*, (4), 436.
Seizer, A., 1928. Research on the mechanism of the activated-sludge process. *Gesundh.-Ing.* (Germany), **51**, 253, 278.
Setter, L.R. and Edwards, G.P., 1943, 1944. Modified sewage aeration. *Sew. Works*, **15**(4), 629 (1943), **16**(2), 278 (1944).
Simpson, J.R., List, E. and Dunbar, J.M., 1991. Bulking sludge: a theory and successful case histories. *J.I.W.E.M.*, **5**, 302–311.
Ulrich, A.H. and Smith, M.W., 1951, 1957. The biosorption process of sewage and waste treatment. *Sew. Ind. Wastes*, **23**(10), 1048 (1951), **29**(4), 400 (1957).
Vosloo, P.B.B., 1973. Oxygen requirements in the activated sludge process. *Wat. Pollut. Control*, **72**, 209–212.

Cross references

Activated sludge process: historical
Sewage treatment: general introduction
Sewage treatment processes

AFRICA: CLIMATE

Africa covers an area of more than 30 million km^2 and is second in size only to Asia. Of all continents, it is the most symmetrically located with regard to the equator, and this is reflected in its climatic zonation. The coastline is remarkably smooth and the continent has been called a giant plateau, since there is a relative absence of very pronounced topography, although some high mountains exist, especially in the East African region (Kilimanjaro, 5894 m; Mount Kenya, 5199 m; and the Ruwenzoris, 5120 m). Lake Victoria, astride the equator, covers an area of 70 000 km^2 and is exceeded in size only by Lake Superior among the world's fresh water lakes.

We will note below how certain evidence can be used to reconstruct the early climate of Africa, but climatic observations really only began with the European explorers of the late eighteenth and nineteenth centuries. Then, in the last few decades of the nineteenth century, meteorological services were formed that began systematic observations at a network of stations. In most cases the meteorological service followed the meteorological practices of the colonial or governing power and this characteristic has tended to persist, even after independence was obtained. A number of the countries have suffered from internal disturbance since independence, a fact often leading to a hiatus in the records. Nevertheless, the standard of observation generally has remained high at the first-order or synoptic stations, but care must be taken when using data from many of the cooperative or second-order stations.

Weather controls

As in other regions of the globe, the pattern of solar radiation is the major control of climate. However, the nature of the air–surface interface and topography also play important roles. Although ocean currents help to determine the climate of some narrow bands of land, it is by appreciating the nature of the air masses reaching a region that one can begin to understand the observed climatic pattern.

Most of the time the continent is affected by tropical air masses, often maritime (moist) in nature, but in certain areas and during certain months they can be of continental (dry) origin. At the extreme latitudinal boundaries of the continent, in the littoral region of the Mediterranean Sea, and in the area around and east of Cape Town, the effects of polar air masses cannot be ignored at the time of low sun. The terms high sun (summer) and low sun (winter) will be used here because in the tropics they are preferable to the thermally meaningless terms of summer and winter.

Temperature, at a particular station, is a rather conservative element with a relatively small annual range, and wind speeds are normally low compared to areas in higher latitudes. Precipitation, mostly rainfall, is the significant feature of the African climate. For rainfall to occur two criteria must be met – an adequate amount of water vapor within the atmosphere and the initiation of a cooling mechanism.

The cooling mechanism is usually obtained through the ascent of a large parcel of air. Such uplifting is generally due either to topography or to horizontal convergence of the air, which is simply the coming together of air parcels or masses. The ways in which such horizontal convergence can occur over Africa are detailed in Johnson and Morth (1960).

There are four important phenomena that determine rainfall amounts and patterns over the tropical continents: (1) the intertropical convergence zone (ITCZ), (2) the equatorial trough (ET), (3) easterly waves, and (4) tropical cyclones. The latter two play relatively minor roles over Africa.

The ITCZ, defined as a surface discontinuity separating the trade winds of the two hemispheres, can be identified readily on climatic charts, but it is not easily found on daily weather maps. The confluence of convergence of the usually relatively moist air masses leads to rainfall patterns that reflect the seasonal migration of the sun, with a time lag.

The equatorial trough (ET), the zonal pressure minimum, is detected up to an average height of 500 mbar (5500 m) with a mean position near the equator at that height. At lower levels there is evidence of a pronounced shift in location with season.

Weather situations

To set the stage for an appreciation of the various climatic patterns experienced, it is helpful to understand the weather situations dominant during certain months.

January

A broad low-pressure region is noted north of the equator with only light winds in evidence. In the upper air the divergent northeasterlies act to suppress rainfall. The surface position of the ET is north of the rain belt in Central Africa whereas, in the southern sector, upper level troughs cause heavy rains over the Angolan plateau. Frontal activity brings rain to the North African coast.

April

There has been a movement of the ET northward from the January situation. In West Africa the ET becomes identified more easily on the daily surface charts. Thompson (1965) considers the rainfall now to be the result of many complex interplays among synoptic processes and dismisses the concept of a continous zonal belt of rainfall moving northward.

July

The position of the surface ET is now at about its furthest north, near 20°N. There is a meridional (longitudinal) pressure gradient extending from the high-pressure belt of the southern hemisphere to the intense heat lows of Arabia and North Africa. In East Africa the topography leads to periods of convergence above the 700 mbar (3000 m) level, giving rise to the wettest month, while there is subsidence at 850 mbar (1500 m), where little rainfall is reported.

October

The rain belt is now moving southward, while a new trough begins to develop over Somalia and the Arabian Sea. Like April, this is a transitional month between the extremes of January and July.

Continental patterns of important climatic elements

The best method of identifying analogous climatic zones is to consider aspects of each of the important elements separately and then to combine them to obtain the overall picture. The elements selected here are temperature, precipitation, humidity and radiation.

Temperature

The mean annual temperature range (MATR), the difference between the mean temperatures of the hottest and coldest months, is of small magnitude over most of the continent, being less than 6°C over about half the continent. Its minimum value is 1.4°C at Barumbu in northernmost Zaire whereas the greatest is 23–24°C in parts of the Algerian Sahara. The dependence of the MATR on the continentality of the station, as well as its latitude, is shown in Table A16.

The mean annual diurnal temperature range is extremely dependent on continentality, as shown in Figure A45. Nearly all the coastal

Table A16 Mean annual temperature range (MATR)

	Latitude E	Longitude N	MATR (°C)
Port Harcourt	7°01'	4°46'	2.5
Lokoja	6°44'	7°48'	3.6
Kano	8°32'	12°32'	5.8
Zinder	9°00'	13°48'	9.4
Agadez	7°59'	16°59'	13.3
Tamanrasset	5°31'	22°42'	16.7
Ouargla	5°20'	31°54'	23.3
Biskra	5°44'	34°51'	22.7
Constantine	6°37'	36°22'	18.3
Philippeville	6°54'	36°52'	14.2

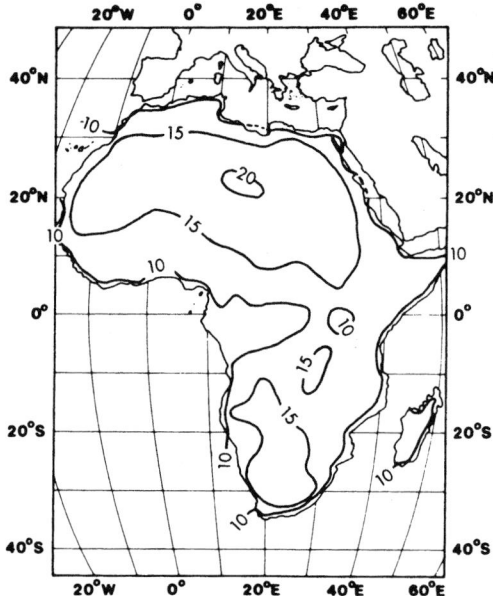

Figure A45 The mean diurnal temperature range (°C).

regions exhibit values of below 10°C, whereas in the central Sahara the range reaches 20°C, one of the highest values for any region of the world.

Actually, the best measure of the temperature variation is the highest mean monthly maximum temperature (H) and the lowest mean monthly minimum temperature (L). The patterns of these two variables are given in Figures A46 and A47. Figure A46 shows that it is only north of the equator where values exceeded 35°C and only in parts of the foggy coastal strip of southwestern Africa where values less than 20°C were reported. In Figure A47 the effects of elevation and latitude are more evident than those of continentality. Values below 5°C are unusual and only at high altitudes (over 1000–1500 m) in Algeria, Morocco and South Africa does L go below 0°C. The largest value is 26°C noted at Dallol, Ethiopia. The mean annual temperature variation (MATV), defined as H − L, which is depicted in Figure A48, is a combined measure of both annual and diurnal range and shows a relationship with both latitude and continentality. Again, the maximum values (above 30°C) occur in drier areas, the greatest being at Adrar in western Algeria, with 42°C. As would be expected, the equatorial littoral yields the lowest values, reaching only 8°C in Liberia and Sierra Leone.

Precipitation

Because both the ITCZ and the polar front lows exhibit large spatial movements, a basic seasonal pattern of precipitation can be identified on the continent. However, the complexities introduced by topography, upper-air conditions, ocean currents and inland lakes, among

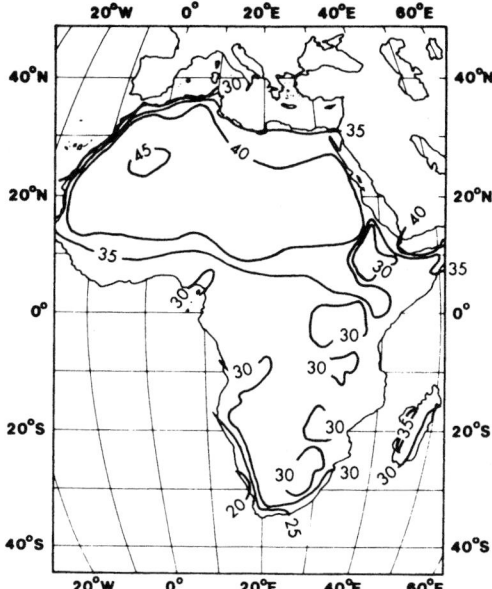

Figure A46 The highest mean monthly temperature (°C).

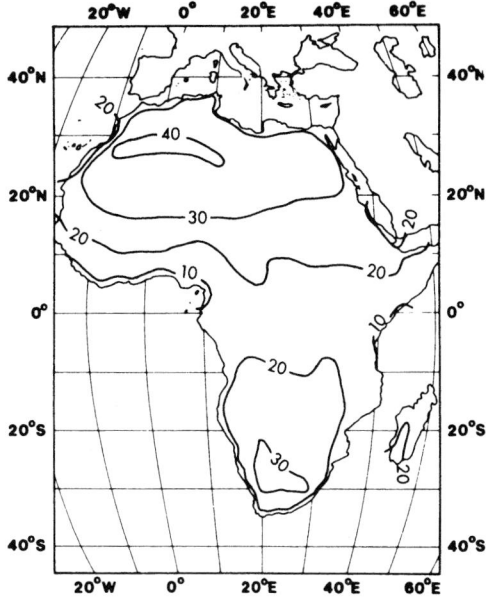

Figure A48 The difference between the highest mean monthly maximum temperature and the lowest mean monthly temperature (°C).

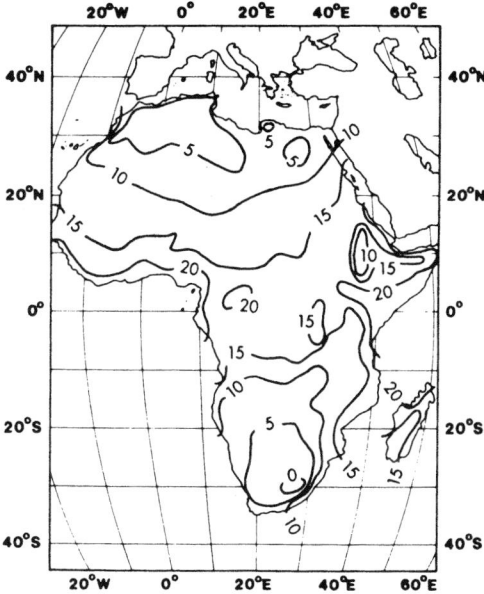

Figure A47 The lowest mean monthly temperature (°C).

others, make the detailed pattern extremely complicated. An example of this is given by Griffiths (1972) for East Africa, whereby over-laying the spatial patterns of mean monthly precipitation, 52 separate regions with 30 different rainfall seasons evolve.

Over most of the continent the seasonal distribution of precipitation exhibits the single significant maximum pattern, such as shown in Table A17. The season of maximum amount is generally around the time of high sun. In Figure A48 those regions in which three consecutive months receive at least 50% of the annual rainfall are shown. Only in the eastern sector of the Mediterranean coast is there a maximum at the time of low sun, but amounts involved are very small.

In the central belt of the continent most stations exhibit some degree of double maxima. However, the areas in which there is a really significant double swing during the year are quite small (Figure

A49). For this illustration, significance is defined as occurring when the difference between the secondary maximum and secondary minimum exceeds 5% of the annual mean (see example of Lagos in Table A17), and this criterion limits the regions to just two. The sector in the Horn of Africa is mostly semiarid, except around Nairobi, Kenya. For stations with annual mean rainfall of over 1000 mm, Kitui, Kenya, is unique; its mean monthly totals (mm) being 41, 24, 118, 244, 56, 5, 3, 5, 0, 82, 304 and 143, giving a 22% swing.

The average annual rainfall totals show a wide range (roughly 0–10 000 mm), exhibiting a decrease away from the equatorial regions to reach a minimum around 20–30° latitude, then showing a slight increase (Figure A50). Since snow and hail amounts are generally small, all precipitation can be considered as rain. Never-theless, snowfalls have been recorded in the Sahara, as far south as 15°N. Some falls have been quite heavy and reference to Dubief (1959, 1963) will give fuller details and some interesting photo-graphs.

A distinctive feature of tropical rainfall is its large variability, interpreted as the difference within monthly, seasonal and annual totals. The station of Makindu, Kenya, is outstanding in this respect. Although its mean annual value is 610 mm, it has recorded as low as 67 mm and as high as 1964 mm of precipitation. On the other hand, April, its wettest month (111 mm average), has had amounts ranging from 822 mm, which exceeds the annual mean, to 0 mm. In Figure A51 a measure of annual rainfall fluctuations is depicted. Use is made of the relative variability statistic, V_r, defined as mean deviation/mean:

$$V_r = \Sigma(X_i - \bar{X})/\Sigma X_i$$

where X_i is individual yearly amounts and $\bar{X}$ is the yearly mean. Values of V_r show dependence on the mean, $\bar{X}$, so data for 500 stations were used to compute the expected value of V_r as a function of $\bar{X}$, called $V_r(\bar{X})$, and comparing V_r for the station with its corresponding $V_r(\bar{X})$. Differences are given as a percentage of $V_r(\bar{X})$.

The great variability in certain areas of the continent can be illustrated further by two examples. Quseir, Egypt, has received 33 mm in a day – 11 times its average annual total; Lobito, Angola, had 536 mm of rain in one day – over 1.5 times its annual average fall of 330 mm.

Hail is not a common phenomenon on the continent, especially on the coast of the tropical regions. However, Maputo, Mozambique, experienced a very heavy fall in October 1977 that did considerable damage. Few places have more than five incidences annually, but a region around Kericho, Kenya, reports as many as 80 hailstorms per year (Frisby and Sansom, 1967).

Table A17 Mean monthly rainfall amounts (mm) for selected stations

	Jan.	Feb.	Mar.	Apr.	May	June	July	Aug.	Sept.	Oct.	Nov.	Dec.	Annual
						Single maximum							
Algiers, Algeria	116	76	57	65	36	14	2	4	27	84	93	117	641
Kano, Nigeria	0	1	2	8	71	119	209	311	137	14	1	0	873
Mbeya, Tanzania	199	165	161	116	17	1	1	1	3	15	52	152	883
Pretoria, South Africa	117	101	78	46	25	9	8	6	25	63	110	120	708
Wau, Sudan	0	4	20	69	132	170	199	234	179	130	8	0	1145
						Double maxima							
Lagos, Nigeria	40	57	100	115	215	336	150	59	214	222	77	41	1625

Note: double maxima significance $(222 - 59)/1625 = 10\%$. See text.

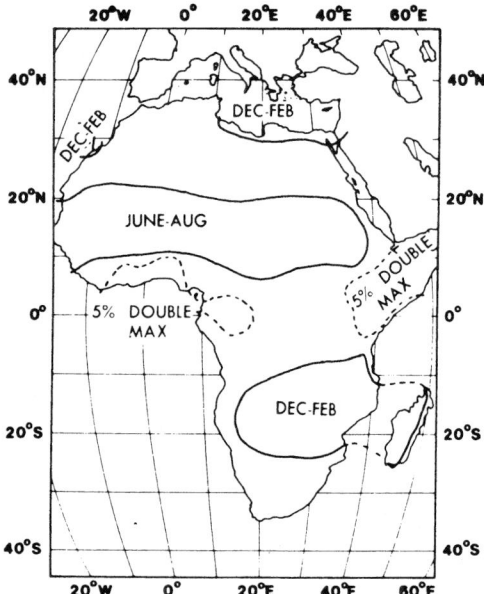

Figure A49 The 3-month period of maximum precipitation and those areas with a significant double maximum distribution.

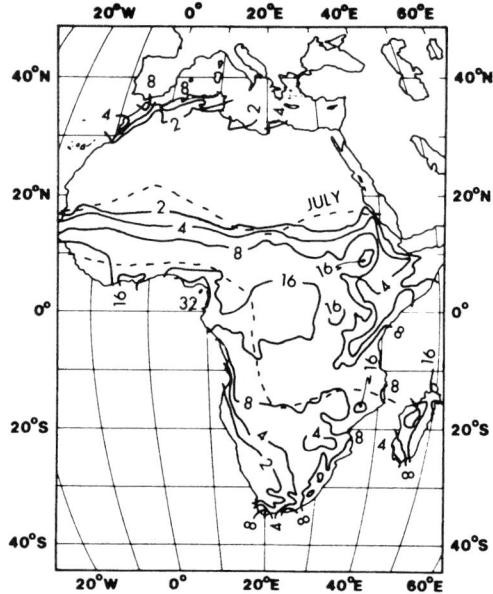

Figure A50 Mean annual precipitation (mm).

Thunderstorm days are frequent with over 20% of the continent reporting in excess of 100 annually. This band of 100 occurrences stretches from about Sierra Leone across the central area as far as Lakes Victoria and Malawi. There are a few locations where convective instability leads to annual values of more than 200, with Kampala, Uganda, 242; Bukavu, Zaire Republic, 221; and Calabar, Nigeria, 216 holding the top places.

Humidity

Relative humidity, as an expression of the atmospheric moisture condition, can be rather misleading because its impact on human comfort is dependent on the air temperature occurring at the same time. For this reason it is preferable to use the dew point temperature as an indicator since this shows little diurnal variation and can be related more readily to human comfort. Values in excess of around 21°C can be considered very sultry, and this isopleth is indicated by a thick line in Figures A52 and A53. Some scientists consider dew points above 18°C uncomfortable. With this threshold about 30% of the continent falls into this category in January and 25% in July.

Along the humid and hot coastal regions of Africa the trade winds and/or sea breezes provide reasonably comfortable conditions, contrary to most people's concepts of the humid tropics. When there is little wind, as is often the case inland, in cities or in wooded areas, the situation is quite enervating. For a good discussion of and information on human comfort conditions consult Terjung (1967).

Radiation

Africa extends from 38°N to 35°S, so that the annual fluctuation of solar radiation at the top of the atmosphere is small compared with that in higher latitudes. Mean annual global radiation (solar radiation measured on a horizontal plane at the surface) varies from nearly 600 ly day^{-1} in the Sahara–Nubia area to something less than 400 ly day^{-1} around Gabon, the Algerian coast and East London, South Africa.

Climatic zones

Using the findings of the earlier sections, it is possible to identify eight important climatic zones in Africa: (1) tropical wet; (2) tropical, short dry spell; (3) tropical, long dry spell; (4) tropical desert; (5) tropical highland; (6) subtropical desert; (7) subtropical, summer rain; and (8) subtropical, winter rain.

In addition, smaller zones of subtropical, uniform rain and subtropical highland can be found. The eight major zones are shown in Figure A54. For these purposes 'tropical' designates that the mean temperature of each month is 18°C or greater; 'desert' occurs when the mean annual rainfall (cm) is less than $16 + 0.9\,\bar{T}$, where $\bar{T}$ is the mean annual temperature (°C); and 'highland' is where the altitude causes the region to be classified in a different thermal zone from what it would be if at sea level.

The tropical wet climate (Kisangani, Table A18) exhibits some rain in all months. Temperatures are uniformly high all year round and the conditions are very enervating, although sea breezes and/or trade winds can reduce the stress along the coast. The tropical rain forest is found in abundance in this zone.

Surrounding the first zone is the tropical, short dry spell climate

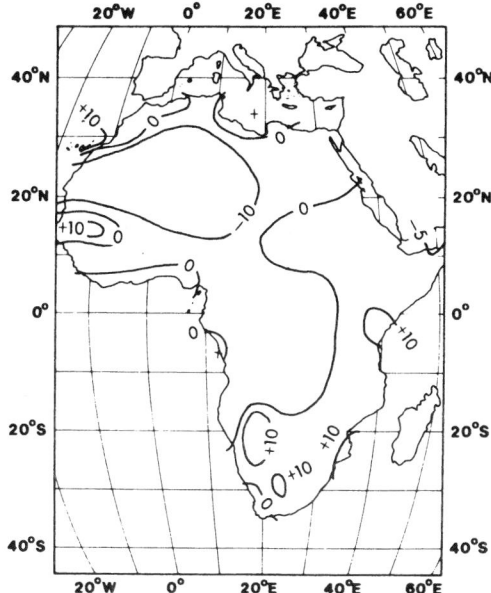

Figure A51 Variation of annual precipitation values in per cent (see text for details).

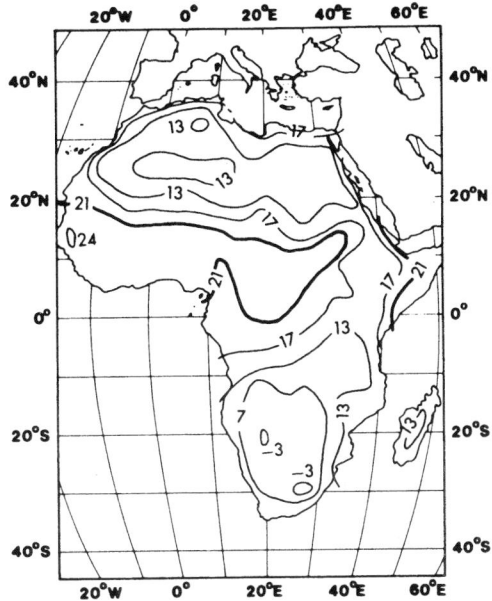

Figure A53 Mean July dewpoint temperature (°C).

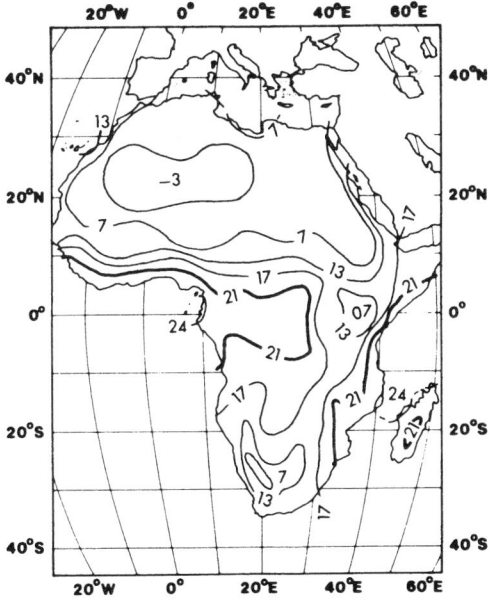

Figure A52 Mean January dewpoint temperature (°C).

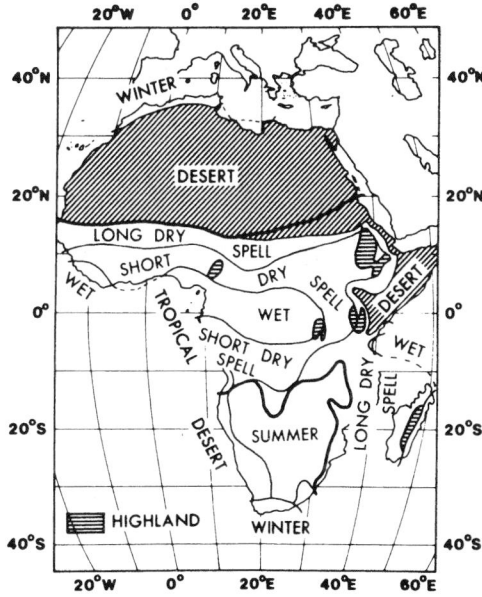

Figure A54 A simple climatic classification. Thick lines (N and S) indicate tropical boundaries (see text for details).

(Kinshasa, Table A18). Here a period of three to five dry months is experienced. Precipitation and temperature are still high, but the annual temperature range tends to be larger than in the wet climate. Vegetation changes from forest near the boundary with the previous zone to deciduous woodland on the drier side although, because this is an important climate for agriculture, much clearing has taken place.

The tropical, long dry spell climate (Niamey, Table A18) is on the equatorial side of the desert regions and has low rainfall for at least 6 months. Rainfall amounts are less than in the two zones discussed above and temperatures show a much larger seasonal swing. The area is susceptible to drought and at such times the often marginal agriculture suffers tremendously. Vegetation is normally savanna and scrubland. The northern belt is referred to as the Sahel, a region in

which famine has afflicted millions of inhabitants. The extreme variability of rainfall amount and frequency is a characteristic of the zone (Todorov, 1984).

Tropical desert climates (Obbia, Table A18) are not common, the biggest region being in the Horn of Africa where the prevailing winds, NE at low sun and SW at high sun, ensure that very few moist air masses reach the area.

The tropical highlands climate (Nairobi, Table A18) offers relief from the tropical heat, as well as a decrease in absolute humidity. Precipitation amounts can change quite rapidly in short distances as exposure to prevailing winds plays a dominant role. Generally, there is an increase in annual amount with height up to a belt of maximum rainfall, often around 2000 m or more, but changing according to the direction of slope. If the elevation is high enough (over about 5500 m)

Table A18 Monthly temperature and precipitation data for representative stations

	Jan.	Feb.	Mar.	Apr.	May	June	July	Aug.	Sept.	Oct.	Nov.	Dec.	Average temperature or total precipitation for year
Kisangani, Congo, D.R.: 0°26'N, 25°14'E, 410 m													
Mean maximum temperature (°C)	31.1	31.1	31.1	31.1	30.6	30.0	28.9	28.3	29.4	30.0	29.4	30.0	30.0
Mean minimum temperature (°C)	20.6	20.6	20.6	21.1	20.6	20.6	19.4	20.0	20.0	20.0	20.0	20.0	20.6
Precipitation (mm)	53	84	178	157	137	114	132	165	183	218	198	84	1703
Kinshasa, Congo, D.R.: 4°20'S, 15°18'E, 324 m													
Mean maximum temperature (°C)	30.6	31.1	31.7	31.7	31.1	28.9	27.2	28.9	30.6	31.1	30.6	30.0	30.0
Mean minimum temperature (°C)	21.1	21.7	21.7	21.7	21.7	19.4	17.8	18.4	20.0	21.1	21.7	21.1	20.6
Precipitation (mm)	135	145	196	196	157	8	3	3	30	119	221	142	1355
Niamey, Niger: 13°31'N, 2°06'E, 215 m													
Mean maximum temperature (°C)	33.9	36.7	40.6	42.2	41.1	38.3	34.4	31.7	33.9	38.3	38.3	34.4	36.7
Mean minimum temperature (°C)	14.4	17.2	21.7	25.0	26.7	25.0	23.3	22.8	22.8	23.3	18.3	15.0	21.1
Precipitation (mm)	0	2	5	8	33	81	132	183	91	13	1	0	549
Obbia, Somalia: 5°20'N, 48°31'E, 15 m													
Mean maximum temperature (°C)	29.4	30.6	32.2	33.9	31.7	29.4	28.3	28.9	39.4	30.0	31.7	30.6	30.6
Mean minimum temperature (°C)	22.2	23.3	24.4	25.5	25.0	23.9	22.2	22.2	22.8	23.3	23.3	22.8	23.3
Precipitation (mm)	12	0	8	21	33	0	1	1	2	38	25	25	166
Nairobi, Kenya: 1°16'S, 36°48'E, 1820 m													
Mean maximum temperature (°C)	25.0	26.1	25.0	23.9	22.2	21.1	20.6	21.1	23.9	24.4	23.3	23.3	23.3
Mean minimum temperature (°C)	12.2	12.8	13.9	14.4	13.3	11.7	10.6	11.1	11.1	12.8	13.3	12.8	12.8
Precipitation (mm)	38	64	124	410	157	46	15	23	30	53	109	86	1155
Wadi Halfa, Sudan: 21°55'N, 31°20'E, 125 m													
Mean maximum temperature (°C)	23.9	26.1	31.1	36.7	40.0	41.1	41.1	40.6	38.3	36.7	30.6	25.6	34.4
Mean minimum temperature (°C)	7.8	8.9	12.2	16.7	21.1	23.3	23.3	23.9	22.2	19.4	14.4	9.4	16.7
Precipitation (mm)	T[a]	T	T	T	T	0	T	T	T	T	T	0	1
Harare, Zimbabwe: 17°50'S, 31°08'E, 1403 m													
Mean maximum temperature (°C)	25.6	25.6	25.6	25.6	23.3	21.1	21.1	23.3	26.1	28.3	27.2	26.1	25.0
Mean minimum temperature (°C)	15.6	15.6	14.4	12.8	9.4	6.7	6.7	8.3	11.7	14.4	15.6	15.6	12.2
Precipitation (mm)	196	178	117	28	13	2	1	2	5	28	97	163	828
Cape Town, South Africa: 33°54'S, 18°32'E, 17 m													
Mean maximum temperature (°C)	25.6	26.1	25.0	22.2	19.4	18.3	17.2	17.8	18.3	21.1	22.8	24.4	21.7
Mean minimum temperature (°C)	15.6	15.6	14.4	11.7	9.4	7.8	7.2	7.8	9.4	11.1	12.8	14.4	11.7
Precipitation (mm)	15	8	18	48	79	84	89	66	43	30	18	10	508

[a] = trace

the region is permanently snowcapped. This transition from sea level to snowfield means that many vegetation belts are identifiable on the slopes.

The subtropical desert climate (Wadi Halfa, Table A18) is the most extensive of all zones on the continent. Summer temperatures in the Sahara are among the highest in the world, although they are not quite as great as in Namibia. Due to the low relative humidity and clear skies, diurnal temperature ranges can be extreme, with values in excess of 20°C often being reported. As may be expected, radiation and sunshine amounts are extremely large. Vegetation, while sparse, springs to life after any brief shower.

The subtropical, summer rain climate (Harare, Table A18) is found mainly in the southern plateau. Precipitation usually is so concentrated that about half the annual total falls in 3 months. Winters are generally very pleasant and comfortable.

The subtropical winter rain (or mediterranean) climate (Cape Town, Table A18) is found at the extremities of the continent. These areas can experience extremely hot and dusty winds in summer from their adjacent deserts, but from fall to spring conditions are ideal. Vegetation is xerophytic, able to withstand the long dry spell.

Table A19 lists some climatic extremes for the continent. It is interesting to note how many of these are also world record extremes. Even in the precipitation class only two stations, Waialeale, Hawaii, with 1455 mm and Cherrapunji, India, with 10 820 mm, exceed

Ureka's total. At another site near Cherrapunji a 5-year mean of 12 650 mm has been reported.

Past climates

There has been relatively little study of past climates in Africa compared with studies of Europe and North America. However, it is known that the continent has occupied very different latitudes from that in which it is presently situated due to tectonic plate movements. In Figure A55 the latitudinal changes in the positions of three points on the continent are shown. From this alone it can be appreciated that in the period 450–200 million years before present (Ma BP) the Cape Town site, occupying a position within the Antarctic Circle, must have had a very cold climate, whereas now it is in relatively the same latitude as it was 500 Ma BP. The Central Saharan site has shown an almost steady progression from 80°S and it is likely that around 300 Ma BP it was also semiarid to arid, but from 200 to 50 Ma BP it was quite wet and humid. The East African location has not shown such extreme latitudinal variation but, nevertheless, must have experienced midlatitude and subtropical climates before reaching its present equatorial situation.

Some remark must be made concerning climate changes in the Saharan area. There have been many papers on this subject and the consensus of opinion is that from 20 000 to 12 000 years ago there

Table A19 Some climatic extremes for Africa

	Temperature	
Absolute maximum	58°C[a]	Azizia, Libya (13 Sept. 1922)
Highest mean monthly maximum	47°C[a]	Bou-Bernous, Algeria (July)
Highest mean monthly	39°C[a]	Bou-Bernous, Algeria (July)
Highest mean annual	35°C[a]	Dallol, Ethiopia
Highest mean monthly minimum	32°C[a]	Dallol, Ethiopia
Highest mean of coldest month	31°C[a]	Dallol, Ethiopia
Highest absolute minimum	21°C[a]	Dallol, Ethiopia
Absolute minimum	−24°C (11°F)	Ifrane, Morocco (11 February 1935)
	Precipitation	
Highest mean annual	10 450 mm	Ureka, Equat, Guinea
	10 300 mm	Debundscha, Cameroons
Lowest mean annual	0.5 mm	Wadi Halfa, Sudan
	Miscellaneous	
Highest Average dew point	29°C	Assab, Ethiopia (June afternoons)
Highest mean annual sunshine	4300+ h	Wadi Halfa, Sudan
Highest hourly radiation	113 langleys[a]	Malange, Angola
	112 langleys[a]	Windhoek, Namibia

[a] World record

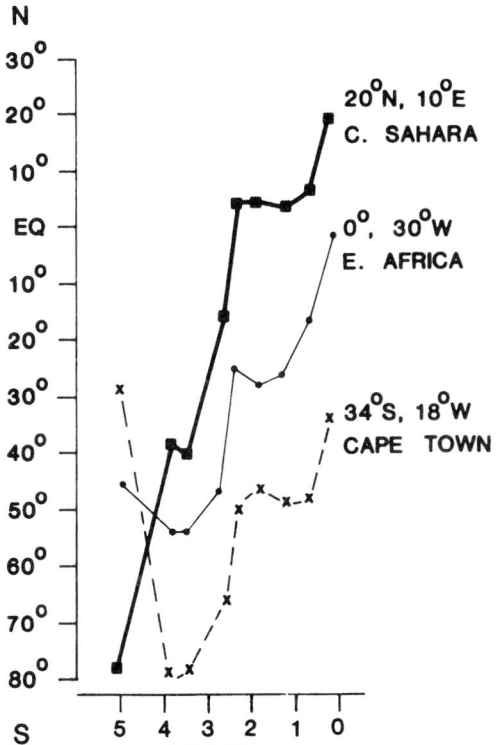

Figure A55 Latitudinal changes at three locations during the past 500 million years. (After Newell, 1974.)

was great aridity and the desert advanced southward. Following this period there were some very moist periods, while over the past 2 000 years the rainfall has declined sufficiently to make the agriculture practiced in the time of Roman occupation no longer feasible (Carpenter, 1969). Murphey (1951) claimed that the cause is basically artificial and, even today, the growth of the desert must be attributed in great measure to anthropogenic influences. It is interesting to note that there are reports of ice on the Nile in the ninth and eleventh centuries (Oliver and Fairbridge, 1987).

For eastern Africa a more recent study (Hastenrath, 1984) suggests that there was a distinct retreat or disappearance of glaciers around

11 000–15 000 years ago, the deglaciation beginning at lower altitudes (*ca.* 3000 m). The most detailed studies of the historical climatology of Africa have been published by Nicholson (1976, 1978) and Nicholson and Flohn (1980). Nicholson finds, in the times of anomalous climate and climatic discontinuities, reasonable correlation between the sub-Saharan area and that of southern Africa. She identifies anomalous weather patterns in the 1680s and 1830s and a major rainfall change around 1800. Apparently the nineteenth century had greater snowfall than the twentieth century.

In the period 1870–1895, both the Sahara and eastern Africa had above average rainfall, after which drier conditions set in and by the mid-1910s severe droughts were common in much of the tropics and subtropics. In the 1920s and 1930s there were indications of wetter conditions – Nile discharge up 35%, Lake Chad depth up 50%, and Sierra Leone reporting a third more rainfall than in the late nineteenth century.

Studies of African climate during the last 100 years or so are made problematic because of the vast areas for which the periods of record are very short. It is true that some temperature and precipitation measurements exist from the first half of the nineteenth century, such as in Tripoli and western Africa, but in general few reliable records exist before about 1890. Exceptions to this would include the island of Mauritius which has an almost unbroken record since around 1851.

A special project of the Global Climate Laboratory, part of the National Climatic Data Center, located in Asheville, North Carolina, is concerned with locating, extracting and digitizing data of monthly mean maximum and minimum temperatures and precipitation amount. This is now nearing completion and it is estimated that about 300 stations will yield sufficient data for the study of climatic fluctuations. In a few countries, including Egypt, Nigeria and South Africa, there are enough stations to allow a regional investigation.

John F. Griffiths

Bibliography

Carpenter, R., 1969. Climate and history, *Horizon*, **11**(2), 48.
Dubief, J., 1959. *Le Climat du Sahara*, Vol. 1. Algiers: Univeristy of Algeria.
Dubief, J., 1963. *Le Climat du Sahara*, Vol. 2. Algiers: University of Algeria.
Frisby, E.M. and H.W. Sansom, 1967. *Hail Incidence in the Tropics*, U.S. Army Electronics Command, Ecom. 02105-F.
Griffiths, J.F. (ed.), 1972. Climates of Africa, in *World Survey of Climatology*, Vol. 10. Amsterdam: Elsevier.
Hastenrath, S., 1984. *The Glaciers of Equatorial East Africa*. Dordrecht: Reidel.
Johnson, D.H. and H.T. Morth, 1960. Forecasting research in East Africa, in *Tropical Meteorology in Africa*, D.J. Bargman (ed.). Nairobi: Munitalp Foundation, pp. 56–137.

Murphey, R., 1951. The decline of North Africa since the Roman occupation, *Assoc. Am. Geographers Annals*, **41**(2), 116–132.

Newell, R.E., 1974. The earth's climatic history, *Technol. Rev.*, pp. 30–45.

Nicholson, S.E., 1976. A climatic chronology for Africa: synthesis of geological, historical, and meteorological information and data. Unpublished PhD dissertation. Madison: University of Wisconsin.

Nicholson, S.E., 1978. Climatic variations in the Sahel and other African regions during the past five centuries, *J. Arid Environments*, **1**, 3–24.

Nicholson, S.E. and H. Flohn, 1980. African environmental and climatic changes and the general atmospheric circulation in Late Pleistocene and Holocene, *Climate Change*, **2**, 313–348.

Oliver, J.E. and Fairbridge, R.W., 1987. *The Encyclopedia of Climatology*. New York: Van Nostrand Reinhold, 305–323.

Terjung, W.H., 1967. The geographical application of some physio-climatic indices to Africa, *Internat. J. Biometeorol.*, **11**(1), 5–19.

Thompson, B.W., 1965. *The Climate of Africa*. London: Oxford University Press.

Todorov, A.V., 1984. The changing rainfall regions and the 'normals' used for its assessment, *J. Climate and Appl. Meteorol.*, **24**, 97–107.

Cross references

Atmospheric processes associated with water in the atmosphere
Climate and climate change
Climate data: sources

AGROCLIMATOLOGY

Agroclimatology is concerned with the interaction between climatological and hydrological factors, on the one hand, and agriculture in the broad sense, including animal husbandry and forestry, on the other. Its aim is to apply climatological information for the purpose of improving farming practices and increasing agricultural productivity in quantity and in quality. Because climatic effects on animal adaptation and forestry are less complicated and less amenable to manipulation than those on crops, research in agroclimatology has been concentrated largely in the latter area.

Agroclimatology and agrometeorology share nearly the same aim – scope and methodology. However, in their application the latter tends to emphasize weather forecasting in dealing with daily problems, whereas the former is more concerned with the use of mean data as a guide to long-range planning.

Problems in agroclimatology may be grouped into four major categories:

- land use planning, i.e. selecting ecologically suitable and economically profitable crops and farming systems for a given region or selecting the best sites for a particular farming enterprise within a large area;
- agronomic practices, e.g. agricultural calendar (date of planting and harvesting), irrigation, fertilizer application, mulching, fallow, shading, mixed cropping and the use of machinery;
- crop-weather relationship, i.e. crop development from germination to vegetative growth, flowering and finally to yield formation in relation to crop growth as affected by photosynthesis, respiration, translocation, transpiration and other physiological processes;
- protective measures against frost, hail, food, drought, soil erosion, wind damage and weather-related problems of pests and diseases.

To investigate and solve the many and varied problems, agroclimatologists rely primarily on four types of research: (1) response of crop growth and development to one or more climatic factors, often carried out by experiments in a controlled environment by plant physiologists, (2) micrometeorological measurements of radiation profile, evapotranspiration and other parameters and their effects on various physical and/or physiological processes, mostly carried out in agricultural experimental stations, (3) regional survey of the various climatic factors, particularly those related to energy budget and water balance, and (4) synoptic and long-range weather forecasting.

Agroclimatological stations require more instrumentation and a denser network than those designed for weather forecasting. Except for a barometer, a visibility meter and upper air observation equipment, a complete agroclimatological station will usually have all the instruments for measuring air temperature, precipitation, wind and humidity. In addition, measurements of the various components of radiation, soil temperature, free water evaporation, evapotranspiration, soil moisture, and dew and fog drips are taken. Such stations are usually maintained by agricultural research institutions.

Climate as a determinant to crop distribution

All crops have their critical and optimum climatic requirements. In the 1940s, Nuttonson (1947) undertook a series of studies to define 'climatic analogues' for rice, wheat and other grain crops. He defined climatic analogs as 'areas sufficiently alike with respect to certain major weather characteristics that techniques and materials development for one area have applications and chance of success when transplanted to its climatic counterpart.' Nuttonson used temperature, dates of frosts and monthly rainfall as his basic climatic criteria.

A concept dating back over 200 years that is still widely used by Russian and Chinese scientists is the degree day, or the cumulative mean daily temperature above a certain minimum threshold. This concept holds that the completion of a crop life cycle requires a certain minimum amount of heat expressed as degree days. For instance, turnips will ripen when the cumulated degree days above 0°C have reached a value of 800–1000.

Nuttonson's climatic analogues and the degree day concept are crude indicators of the ecological limits imposed by extreme climatic conditions for crops whose distribution may range from the tropics to very high latitudes. They are of limited value in agricultural land-use planning as they fail to consider optimum climatic conditions for various crops.

Although pineapple will grow in areas where the mean temperature ranges from 20 to 27°C, the optimum temperature for the entire growing season is 23–24°C with a diurnal temperature range of about 10°C. This information has been used by Neild and Boshell (1976) as one of the criteria for selecting potential production areas.

For the various varieties of common beans, distribution is largely determined by temperature during the flowering period. Nearly all bean-growing regions of the world have a temperature between 17.5 and 25°C during the flowering period, with 55% of the areas falling within the narrow range of 20.0–22.5°C (Centro Internacional de Agricultura Tropical, 1978).

For some crops, night temperature may pose a more stringent restriction on distribution than mean daily temperature. For instance, the optimum night temperatures for flowering in macadamia nuts are 15–18°C. At night temperatures of 21°C or higher, such as in the humid tropics, racemes will not form to set fruits. Plants differ in their response to day length or photoperiod. Some plants are day length-neutral; however, others require a certain day length for flowering, tuberization and other developmental processes to occur.

The distribution of soybean cultivars is closely related to day length. Soybean flowers in the field only when the days are shortened to a critical value. Soybean will remain vegetative almost indefinitely if the days are long enough, or flower in less than a month if the days are short. Because of their response to photoperiod, most soybean varieties are adapted for full-season growth in a band usually no wider than 5 degrees of latitude. It is difficult to breed soybean cultivars that are insensitive to day length; the few genotypes that have been developed for the day length conditions in the tropics are of poor agronomic quality.

Flowering and tuberization in sweet potato are favored by natural day length of 11–13 h. During the summer at latitudes higher than 35°N, flowering is prohibited by long days. In long days of more than 18 h, branch length and leaf expansion are promoted at the expense of tuber yield (McDavid and Alamu, 1980). Yams also require a short day length for tuberization.

Most onions require long days to set bulbs. Even with the recently developed short-day varieties, bulbing starts very early and bulbs are small at day lengths of 12 h 30 min or shorter (Eavis and de Jeffers, 1970). This imposes a severe restriction on the expansion of onion cultivation in the tropics.

In the absence of irrigation, rainfall to some extent determines crop distribution as well as dates of planting and harvesting for some annuals. The minimum and optimum water needs of crops have been the subjects of many investigations. For instance, the International Rice Research Institute (1976) has considered a monthly rainfall of 100 mm to be the minimum for upland rice cultivation. This has been used as a criterion for evaluating the possibility of expanding upland rice cultivation in several Southeast Asian countries.

The length of the wet months is often known as the hydrological growing season in the tropics. Both cacao and oil palm require a hydrological growing season of 9 months in a year. In Africa, Lauer (1952) has shown that the boundary between the wet and dry savannas coincides with the isoline of a 5-month dry season.

Some crops cannot tolerate excessive rainfall that leads to water-logged conditions; others may be highly susceptible to wind damage. Both these factors are important considerations in the site selection for banana plantations.

Weather-related pests and diseases, which are particularly prevalent in tropical countries, also limit crop distribution. In some cases the incidence of pests and diseases may be affected by a change in humidity, wind movement or formation of dew, which are subject to microscale variations induced by topography, exposure, ground cover and the like. The demise of Fordlandia rubber plantation in Brazil in the 1940s and coffee plantations in Ceylon in 1879, and the shift of United Fruit Company's banana plantation in Costa Rica from the humid Caribbean lowlands to the subhumid Pacific coast in the 1930s were all caused by pest and disease problems.

Photosynthesis

The organic matter of higher plants originates from photosynthesis in which the chlorophyll utilizes solar radiation to produce carbo-hydrates from water and carbon dioxide. At the molecular level it takes eight photons of sunlight to liberate a carbon atom from a carbon dioxide molecule. However, this theoretical quantum efficiency of photosynthesis cannot be obtained in crop plants. Loomis and Williams (1963) have estimated that the maximum productivity of a crop surface receiving 500 langleys of solar radiation per day, which is the approximate summer value in the Corn Belt of the United States, is 71 (g carbohydrate) cm^{-2}. This calculation assumes complete light usage by the plant except for a portion of the radiation reflected by the surface, inactive absorption by the crop, and respiration loss. The maximum observed crop growth rate is about 80% of this theoretical maximum as light wastage in the field is inevitable.

A more realistic approach to simulate crop photosynthesis is to consider the response of leaves to radiation, the distribution of radiation in successive layers of the canopy, and the changing radiation intensity throughout the day. The studies by Monteith (1965) and de Wit (1965) are well-known examples of such an attempt.

For most plant leaves, photosynthesis is only slightly affected by temperature within the range 10–35°C, but is largely determined by the intensity of photosynthetically active radiation that coincides with the visible light of the solar spectrum. The photosynthetic rates usually increase with light intensity to a point, known as the saturation light intensity, and then level off. The saturation light intensities of leaves of major crops vary from about one-fifth to full sunlight. For a very few crops photosynthetic rates may decline at high light intensities. For example, Kumar and Tieszen (1980) have shown that as the leaf temperature of coffee reaches 25°C under high radiation intensity, mesophyll conductance decreases, which in turn reduces the intake of CO_2. Consequently, coffee grown under shade often possesses substantially higher photosynthetic rates.

Flowering plants may be grouped into C_3 plants whose first stable product of the photosynthetic reactions is the three-carbon compound, phosphoglyceric acid, and C_4 plants whose primary products are the four-carbon acids, malic and aspartic. Most agricultural crops belong to the former group; only sugarcane, maize and sorghum are C_4 plants, which also include many tropical Gramineae. In general, C_4 plants have a higher saturation light intensity and can utilize CO_2 at a lower atmospheric concentration level than C_3 plants (Monteith, 1978).

Although the top leaves of a plant may be light saturated during the greater part of the day, the lower leaves under shade may not receive sufficient light to maintain maximum photosynthetic rate. Light distribution within the canopy depends largely on the arrangement and orientation of leaves and to a lesser extent on their shape and structure. In general, light extinction from the top to the bottom leaves is very small in trees that usually spread out to a great height. In herbaceous crops the best arrangement is to have vertical top leaves changing gradually to a horizontal position at the bottom. A crop with such a leaf arrangement may be twice as productive as a similar crop that has horizontal top leaves.

Part of the photosynthetic product is lost in respiration that takes place in all organs and usually increases linearly with temperature, except at very high readings. The respiration rate of C_3 plants proceeds day and night. However, C_4 plants do not respire in the presence of light, which partly accounts for their high growth rate.

The ratio between the leaves and other non-photosynthesizing organs in a plant is an important determinant of the relative magnitude of photosynthesis and respiration. In mature trees the branches and trunks are so bulky that leaf assimilates are nearly equal to respiration loss. In most annual grain crops respiration accounts for about 25% of the gross photosynthetic rate in temperate climates and about 35% in the tropics. The proportion of non-photosynthesizing organs is least in root crops, which do not need long stems to support seeds as in grain crops (Coursey and Haynes, 1970). This is a major reason for the extremely high yield in some root crops such as cassava.

To maintain the maximum photosynthetic rate for the plant as a whole, the lowest leaves should have a photosynthetic rate equal to or slightly in excess of the respiration rate. This is known as the compensation point, which is determined by the radiation and temperature regime of a place as well as the saturation light intensity, the leaf area index (the leaf area subtended per unit area of land), and the light distribution within the canopy. When the lowest leaves are maintained at the compensation point, the plant may be said to have the optimum leaf area index. Other factors being equal, in a hot and cloudy climate the optimum leaf area index is lower than in a climate where radiation intensity is high but the temperature is not too excessive. This concept has been one of the bases for the breeding of the so-called 'miracle rice', which is shorter and less leafy than local varieties in the Philippines.

Assuming given values of light saturation intensity, light distribution within the canopy and temporal change of solar radiation, agronomists and agroclimatologists have presented programs for computing photosynthetic rates of different crops. The results indicate that for the same amount of solar radiation, the photosynthetic rate increases with day length. According to Monteith (1965), with a daily solar radiation of 400 langleys, the photosynthetic rate increases by some 15% from a day length of 12 h to 16 h. Thus, crop yields in high latitudes often exceed those in the tropics.

For crops whose economic yield consists of only a fraction of the total organic matter, factors other than photosynthesis may be important. The highest harvest index, which is the ratio between the economic yield and total organic matter, is often obtained at a leaf area index considerably below the optimum for maximum photosynthesis. For example, the harvest index of cassava decreases when the leaf index reaches a modest value of 3.5 (Centro International de Agricultura Tropical, 1979). Similarly, the productivity of fruit bunches in oil palm reaches a maximum at a leaf area index of 3 and declines sharply at higher values (Rees, 1963). For many crops it is important to maintain a proper balance between the source (photosynthates) and the sink (seeds, tubers or fruits) through pruning, proper spacing or plant breeding.

Yield component analysis

For cereal crops the relationship between weather and crop yields is far more complicated than the mechanistic calculation of photosynthesis would suggest. It is necessary to adopt a phenological approach to analyze the yield components. For instance, the yield of rice may be expressed by:

Amount of brown rice produced per unit area = total number of spikelets per unit area × ratio of ripening × average weight per grain

The total number of spikelets per unit area may be further divided into two components, i.e. number of ears per unit area times the number of spikelets per ear. The number of ears is determined during the early stage up to 10 days before maximum tillering. Cold damage during that stage may reduce ear number. The number of spikelets per ear is determined during the period from about 32 to 5 days before heading. The ideal climatic condition would be high radiation intensity and a mean daily temperature of about 22°C (Yoshida and Parao, 1976). As temperature increases above 22°C, the number of spikelets decreases. A similar adverse effect of high temperature has been observed in rye, wheat and barley.

The length of the ripening period is inversely related to the mean daily temperature. For a given locality the variation of ripening period from one year to another may be insignificant. However, the length of ripening period decreases from about 65 days in northern Japan to 30 days in the Philippines. The shortened ripening period of rice and other cereal crops is now generally recognized as a major reason for their low yields in the tropics.

Low night temperatures during the ripening period are beneficial in increasing grain weight with better quality. However, the weight of rice grain does not vary greatly because it is rigidly enclosed by the outer and inner glumes whose size has been determined during the period before anthesis. Among the cereals, wheat and barley have greater variations in grain size and weight.

It is clear from the analysis of yield components that different climatic factors exert varying effects during successive stages of crop development. Studies for all grain crops seem to indicate that the effect of solar radiation is greatest during and before the flowering period and that high temperature during the ripening period is unfavorable. The formula derived by the International Rice Research Institute for estimating rice yield potential in the tropics is illustrative (Yoshida, 1977):

$$Y = S(278 - 7.07t) \times 0.86 \times 18.1 \times 10^{-5}$$

where Y is yield in tons per hectare, S is solar radiation in $cal\,cm^{-2}\,day^{-1}$, t is temperature in $°C$ during 25 days before flowering, 0.86 is the average filled grain percentage, 18.1 is the average weight of 1000 grains weight and 10^{-5} is a correction factor.

Evapotranspiration

Evapotranspiration is the upward flux of water vapor from crop field to the atmosphere. When the canopy is fully developed and the supply of water adequate, evapotranspiration is at the maximum or potential level. For nearly all agricultural crops, stomata are fully open in the absence of water deficit. One notable exception is pineapple whose stomata are closed during the day and open at night (Ekern, 1965). The daytime closure of stomata impedes the flux of water vapor, reducing the potential evapotranspiration of pineapple to about one-third of ordinary crops.

The rate of potential evapotranspiration is primarily determined by atmospheric conditions. Evapotranspiration requires a supply of energy to convert water into the vapor form as well as a mechanism to remove that vapor from the interface to the free atmosphere. Since the net radiation energy and the turbulent mixing tend to increase with plant height, tall crops may have a potential evapotranspiration rate some 20–25% higher than short grass.

When the supply of water to the plant is less than adequate, the evapotranspiration rate falls short of the potential. The reduction varies greatly with plant species depending to a large extent on the response of stomata to soil moisture deficit.

Evapotranspiration can be accurately measured by a weighing lysimeter, which is rarely used except in expensive experimental work handled by well-trained personnel. Based on a comparison of lysimeter and climatological observations, a number of methods have been developed for the estimation of potential evapotranspiration. They include the use of evaporimeters such as evaporation pan and atmometers, the use of net radiation as an estimator, and the formulae proposed by Penman (1948), Thornthwaite (1948), Blaney and Criddle (1950) and others. Discussion of the advantages and disadvantages of the various methods can be found in Chang (1968) and Rosenberg (1974).

The concept of potential evapotranspiration serves as a useful reference point. The actual evapotranspiration rate of a crop during any stage of development may differ somewhat from this reference level because of incomplete ground cover or the senescence of the crop. The ratio between the actual water needs of a crop at any given stage and the theoretical potential evapotranspiration is known as the crop coefficient. Based on numerous experimental studies by various investigators, Doorenbos and Pruitt (1977) have summarized the average values of crop coefficient during successive stages of development for 36 crops.

Water balance

The water balance in the soil can be expressed by the following equation:

rainfall + irrigation water = changes in soil moisture + evapotranspiration + percolation + runoff

A bookkeeping procedure is often used by agroclimatologists to estimate several items in this equation. In addition to measurements of rainfall and irrigation water and an estimate of potential evapotranspiration by an appropriate method, the bookkeeping method requires a knowledge of the water-holding capacity of the soil to a depth that includes about 90% of the roots. As the soil moisture content gradually decreases, the actual evapotranspiration rate may proceed at the potential rate or may decrease progressively. Thus different soil moisture depletion curves may be adopted in the computation.

The climatological water balance computation not only gives an estimate of the moisture content in the soil, but also the amount of water deficit that is required to satisfy fully the water need of the plants as well as surplus water, which includes percolation and surface runoff. The water balance computation can be carried out at different time intervals and has numerous practical applications in both daily agronomic practices and in long-range water resources and land-use planning.

Water and yield relationship

Irrigation practices for maximum crop production require knowledge of the potential evapotranspiration and of the effects of water deficit on crop growth at different stages of development. These effects are complicated and vary greatly from one crop to another. However, some generalizations may be briefly stated according to four types of crops (Fischer and Hagan, 1965).

- Where yield is a major part of the vegetative growth, such as pasture, vegetable and stem fiber crops, maximum yield is usually obtained when water application is equal to potential evapotranspiration. The ratio between potential and actual evapotranspiration, known as the R index (Yao, 1969), is often linearly related to the ratio between maximum and actual yield during the period when the ground cover is nearly complete.
- Where yield is a chemical constituent occurring as a small fraction of growth such as quinine and rubber, slight water stress may be desirable. Loomis (1953) has pointed out that in some plants a reduction in growth may accelerate the differentiation process. Thus a water deficit that reduces the growth of the guayule plant before harvest by 20% may actually increase the rubber yield by as much as 45% (Wadleigh et al., 1946).
- Where yield is a carbohydrate-rich storage organ other than a fruit, such as potato, sugarcane and sugar beet, water stress during the harvest period has also been known to increase the sugar content.
- Where yield is a reproductive organ, either as flowers or fruits (including grain and other seed crops), the relationship between water and yield is most complex. Water deficits at different stages of crop development exert varying effects on the final yields. In general, water deficits during the flowering period are most critical.

Hiler and Clark (1971) have introduced the concept of crop susceptible factor as a numerical measure of the yield loss for a given stress day at any given growth stage. Subsequently, Hiler et al. (1974) have presented values of susceptible factors for various crops as a guide for irrigation practice.

A similar concept is the yield response factor proposed by Doorenbos and Kassam (1979). The yield response factor relates relative yield decrease to relative evapotranspiration deficit for different stages of crop development. Their tabulation of yield response factors for 23 crops from various sources provides a ready and invaluable reference.

Protective measures

Natural hazards such as frost, hail, flood, drought and strong wind may all create havoc in agriculture. Because such natural hazards are also of concern to other fields of human endeavor, most studies of extreme weather events appear in journals and monographs dealing with climatology in general rather than being restricted to the field of agroclimatology. Furthermore, the effects of adverse weather events on agricultural activities are difficult to generalize. Many important studies on this subject have been reviewed by Hurst and Rumney (1971).

Only studies of the shelterbelt and the relationship between climate and pests and diseases fall clearly within the field of agroclimatology. The shelterbelt, or windbreaks, affects temperature, precipitation, snow cover, evaporation, net radiation and wind in a complicated manner. The numerous studies, often with seemingly conflicting results, have been summarized in a World Meterorological Organization technical report by the Working Group of the Commission for Agricultural Meteorology (1964).

The relationship between climate and pests and diseases in agriculture has also been a major concern of the Commission for Agricultural Meteorology of the World Meteorological Organization. Several technical reports have been prepared dealing with the potato blight, Colorado potato beetle, desert locust, Japanese beetle and the like. The critical factor may be soil temperature, soil moisture, wind ventilation or the formation of dew. In some cases the incidence of pests and diseases vary greatly within a short distance because of micro- or topoclimatic variations.

Jen-Hu Chang

Bibliography

Blaney, H.F. and W.D. Criddle, 1950. *Determining Water Requirements in Irrigated Areas from Climatological Data*, Soil Conservation Service, Tech. Publ. 96. Washington, DC: US Department of Agriculture.

Centro Internacional de Agricultura Tropical, 1978. *Annual Report*. Cali, Colombia.

Centro Internacional de Agricultura Tropical, 1979. *Annual Report*, Cali, Colombia.

Chang, J.H., 1968. *Climate and Agriculture*. Chicago: Aldine.

Coursey, D.G. and P.H. Haynes, 1970. Root crops and their potential as food in the tropics, *World Crops*, **22**, 261–265.

de Wit, C.T., 1965. Photosynthesis of leaf canopies, *Agric. Res. Rep.*, **663**, 1–57.

Doorenbos, J. and A.H. Kassam, 1979. *Yield Response to Water*, Irrigation and Drainage Paper No. 33. Rome: FAO.

Doorenbos, J. and W. Pruitt, 1977. *Crop Water Requirements*, Irrigation and Drainage Paper No. 24. Rome: FAO.

Eavis, B. and W.C. de Jeffers, 1970. *Onions from a Possibility to a Reality in Barbados*, Bull. No. 2. Barbados: Ministry of Agriculture.

Ekern, P.C., Jr, 1965. Evapotranspiration of pineapple in Hawaii, *Plant Physiol.*, **40**, 736–739.

Fischer, R.A., and R.M. Hagan, 1965. Plant water relations, irrigation management and crop yield, *Experimental Agric.*, **1**, 161–177.

Hiler, E.A., and R.N. Clark, 1971. Stress day index to characterize effects of water stress on crop yield, *Am. Soc. Agric. Eng. Trans.*, **14**, 757–761.

Hiler, E.A., T.A. Howell, R.B. Lewis and R.P. Boos, 1974. Irrigation timing by the stress day index method, *Am. Soc. Agric. Eng. Trans.*, **17**, 393–398.

Hurst, G.W. and R.P. Rumney, 1971. *Protection of Plants against Adverse Weather*, Tech. Note No. 118. Geneva: World Meteorological Organization.

International Rice Research Institute, 1976. *Annual Report*. Los Banos, the Philippines.

Kumar, D. and L.L. Tieszen, 1980. Photosynthesis in *Coffea arabica*. II. Effects of water stress, *Experimental Agric.*, **16**, 21–27.

Lauer, W., 1952. Humide und aride Jahreszeihung qu den Vegetationsgurteln, *Bonner Geog. Abh.*, **9**, 15–98.

Loomis, R.S. and W.A. Williams, 1963. Maximum crop productivity: an estimate, *Crop Science*, **3**, 67–72.

Loomis, W.E., 1953. *Growth and Differentiation in Plants*. Ames: Iowa State University Press.

McDavid, C.R. and S. Alamu, 1980. Effect of daylength on the growth and development of whole plants and rooted leaves of sweet potato (*Ipomoea batatas*), *Trop. Agric.* (Trinidad), **57**, 113–119.

Monteith, J.L., 1965. Light distribution and photosynthesis in field crops, *Bot. Ann., New Series*, **9**, 17–37.

Monteith, J.L., 1978. Reassessment of a maximum growth rates for C_3 and C_4 crops, *Experimental Agric.*, **14**, 1–5.

Neild, R.E. and F. Boshell, 1976. An agroclimatic procedure and survey of the pineapple production potential of Columbia, *Agric. Meteorol.*, **17**, 18–92.

Nuttonson, M.Y., 1947. Ecological crop geography of the Ukraine and agroclimatic analogues in North America, *Inst. Crop Ecol.*, **1**, 1–24.

Penman, H.L., 1948. Natural evaporation from open water, bare soil, and grass, *Royal Soc. Proc.*, Ser. A, **193**, 120–145.

Rees, A.R., 1963. Relationship between crop growth rate and leaf area index in the oil palm, *Nature*, **197**, 63–64.

Rosenberg, N.J., 1974. *Microclimate: Biological Environment*. New York: Wiley.

Thornthwaite, C.W., 1948. An approach toward a rational classification of climate, *Geog. Rev.*, **38**, 55–94.

Wadleigh, C.H., H.G. Gauch and O.C. Magistad, 1946. *Growth and Rubber Accumulation in Guayale as Conditioned by Soil Salinity and Irrigation Regime*, Tech. Bull. 925. Washington, DC: US Department of Agriculture.

Working Group of the Commission for Agricultural Meteorology, 1964. *Windbreaks and Shelterbelts*, Tech. Note, No. 59. Geneva: World Meteorological Organization.

Yao, Y.M., 1969. The R index for plant water requirement, *Agric. Meteorol.*, **6**, 259–273.

Yoshida, S., 1977. Rice, in *Ecophysiology of Tropical Crops*, P. de T. Alvim, and T.T. Kozlowski (eds). New York: Academic Press, pp. 57–87.

Yoshida, S. and F.T. Parao, 1976. Climatic influence on yield and yield components of lowland rice in the tropics, in *Climate and Rice*. Los Banos, Philippines: International Rice Research Institute, pp. 471–494.

Cross references

Evaporation measurement
Evapotranspiration
Water budget analysis

ALBEDO AND REFLECTIVITY

Albedo is the percentage of solar radiation reflected by an object. The term is derived from the Latin *albus*, white. A pure white object would reflect all radiation that impinges on it and have an albedo of 100%. A pure black object would absorb all radiation and have an albedo of 0%. Bright Earth features such as clouds, fresh snow and ice have albedos that range from 50 to 95%. Forests, fresh asphalt and dark soils have albedos between 5 and 20%. Table A20 presents representative albedos for a variety of objects. Knowledge of albedo is important because absorbed solar radiation increases the amount of energy available to the Earth's surface and atmosphere, whereas reflected radiation returns to space (Houghton *et al.*, 1996).

Appreciation of the relation between albedo and climate extends historically to at least classical Greek times. The principles and theories by which albedo and reflectivity may be explained and measured were first formulated by P. Bouguer and J. Lambert in the eighteenth century, but accurate measurements did not begin until the early twentieth century (Fritz and Rigby, 1957). The work of early investigators, including A. Ångström, C. Dorno, N. Katlin, F. Götz, H. Kimball and others, rapidly developed an extensive body of knowledge concerning albedos that is still drawn on today (see annotated bibliography by Fritz and Rigby, 1957). One of the more interesting approaches to early observations of the Earth's planetary albedo employed measurements of earthshine and sunshine on the Moon (Danjon, 1936, cited in Fritz and Rigby, 1957). Recent use of aircraft and spacecraft as observing platforms have significantly expanded albedo studies (Barrett, 1992).

Reflectivity

Reflectivity is the capacity of an object to reflect solar radiation. It is described as a function of radiation wavelength and is determined by the physical composition of the object. The adjective 'spectral' is frequently used in conjunction with reflectivity to indicate that reflectivity varies as a function of wavelength in the radiation spectrum for most objects. Representative spectral reflectivity measurements for common Earth surface features are given in Figure A56. Soil reflectivity in general increases monotonically with increasing wavelength to about 1.3 μm and then decreases with sharp dips at 1.4 and 1.9 μm because of absorption by soil water. Living green vegetation reflectivity is low in the visible portion of the spectrum (0.4–0.7 μm) as a result of chlorophyll absorption, high in the near infrared (0.7–1.3 μm) because of internal cell structure, and decreases past 1.3 μm in a manner similar to soils due to absorption by water within leaves. Snow is highly reflective in the visible and decreases to low values in the infrared, again as a result of water absorption. Water reflects little radiation in any portion of the spectrum when solar elevation is high (Gurney *et al.*, 1993).

These spectral reflectivity curves vary significantly for each surface type as a function of physical condition and composition. Figure A57

Table A20 Albedos for selected objects (Sellers, 1965)

Water surfaces		
Winter:	0° latitude	6
	30° latitude	9
	60° latitude	21
Summer:	0° latitude	6
	30° latitude	6
	60° latitude	7
Bare areas and soils		
Snow, fresh-fallen		75–95
Snow, several days old		40–70
Ice, sea		30–40
Sand dune, dry		35–45
Sand dune, wet		20–30
Soil, dark		5–15
Soil, moist gray		10–20
Soil, dry clay or gray		20–35
Soil, dry light sand		25–45
Concrete, dry		17–27
Road, black top		5–10
Natural surfaces		
Desert		25–30
Savanna, dry season		25–30
Savanna, wet season		15–20
Chaparral		15–20
Meadows, green		10–20
Forest, deciduous		10–20
Forest, coniferous		5–15
Tundra		15–20
Crops		15–25
Cloud overcast		
Cumuliform		70–90
Stratus (150–300 m thick)		59–84
Altostratus		39–59
Cirrostratus		44–50
Planets		
Earth		34–42
Jupiter		73
Mars		16
Mercury		5.6
Moon		6.7
Neptune		84
Pluto		14
Saturn		76
Uranus		93
Venus		76
Human skin		
Blond		43–45
Brunette		35
Dark		16–22

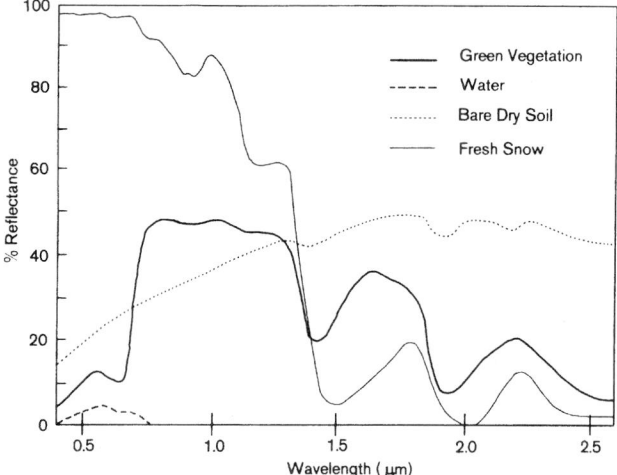

Figure A56 Spectral reflectance of selected Earth surface features.

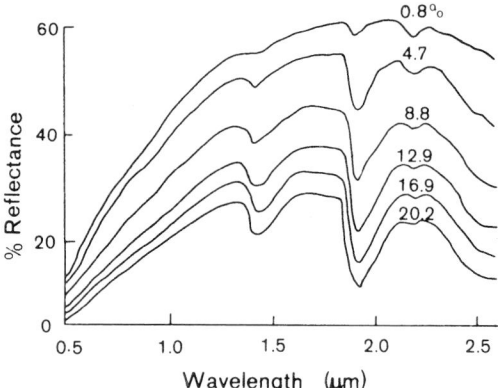

Figure A57 Variations of silty loam spectral reflectance as a function of water content (percentage water content shown by each plot). (After Kondrotyev, 1973.)

presents the variation of a silty loam soil reflectivity due to changes in moisture content. Soil reflectivity varies because of variations in moisture content, particle size, organic matter content, surface roughness and mineral composition. Vegetation reflectivity varies with percentage ground cover, canopy geometry, leaf size and area, and plant growth stage. Snow reflectivity varies with crystal size, compaction, age and liquid water content. Water reflectivity is affected by turbidity, depth and phytoplankton concentrations. Also, because water in its pure form is a dielectric, its albedo increases as the angle of incidence of radiation decreases. Water albedo is lowest when the sun is near zenith and increases to near 100% when the sun is near the horizon. (For further discussion, see Chapter 4, Kondrotyev, 1973; Chapter 8, Miller, 1981.) Other factors may affect the reflectivity of these surfaces and other materials such as rocks, and artificial materials (e.g. asphalt and concrete) also display unique spectral reflectivity patterns that are not fully known. Intensive research is underway to better understand the spectral reflectivity properties of the Earth as a result of the use of multispectral imaging systems for Earth observations (Barrett, 1992).

Relation of reflectivity to albedo

Albedo is the integrated product of incident solar radiation spectral composition and the spectral reflectivity of the object. Outside the atmosphere, solar radiation spectral composition is relatively constant, peaking at 0.5 µm, decreasing rapidly at shorter wavelengths to small amounts at 0.2 µm, and decreasing less rapidly at longer wavelengths to small amounts at about 4.0 µm. The atmosphere selectively absorbs and scatters solar radiation. As a result, at the Earth's surface, the spectral composition of solar radiation varies significantly as a function of atmospheric conditions (e.g. clouds, water vapor and dust) and solar elevation (Robinson, 1966). The majority of albedo measurements have been carried out under clear-sky, high sun elevation conditions (Table A20). Under cloudy conditions, radiation is predominately in the visible spectrum. This decreases the albedos of soils and vegetation but increases snow albedo (Miller, 1981). When atmospheric turbidity is high or the sun is low in the sky, the spectral distribution of solar radiation shifts to the red and infrared portion of the spectrum. Soil and vegetation albedos increase and snow albedo decreases (Kondrotyev, 1973). This variability points out the need to know both the spectral reflectivity of objects and the spectral composition of incident radiation in order to evaluate Earth albedos.

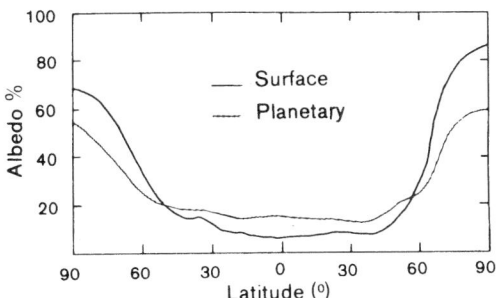

Figure A58 Latitudinal variations in surface and planetary albedo. (After Hummel and Reck, 1979.)

Surface and planetary albedos

Two global albedo measurements are of general interest to climatologists: surface and planetary albedos. Surface albedo is the ratio of incident to reflected radiation at the interface between the atmosphere and the Earth's land and water areas. Almost 75% of all solar energy absorbed by the Earth is absorbed at this interface; the remainder is absorbed in the atmosphere (Sellers, 1965). Any change in surface albedo will alter climate by significantly changing the amount of solar energy absorbed by the planet. Several studies have evaluated the Earth's surface albedo with resultant estimates ranging between 13 and 17% (Sellers, 1965; Kondrotyev, 1973; Hummel and Reck, 1979; Briegleb and Ramanathan, 1980). This range of results is suggestive of current limitations in knowledge of the distribution and reflectivity of Earth surface features.

Planetary albedo is the ratio between incident and reflected radiation at the top of the atmosphere. It includes the effects of reflection from the atmosphere, particularly clouds, and surface albedo. Only about 6% of the incident radiation is reflected by scattering in the atmosphere but, on average, 24% of incident radiation is reflected by clouds (Sellers, 1965). Changes in atmospheric turbidity or cloud cover can alter climate by changing the amount of solar radiation that reaches the Earth's surface. Studies of the Earth's planetary albedo have been carried out over the last 60 years (Barrett, 1992). Barrett notes that estimates have progressively decreased from 50% in early studies to current estimates between 30% and 35%, based on satellite estimates. He suggests that this trend is due to improved knowledge of global cloud cover. However, the possibility of interannual and longer-term variations in planetary albedo should not be overlooked.

Geographic patterns

Both surface and planetary albedos increase with distance from the equator (Figure A58). Surface albedo shows a slight minimum at the equator because of dense evergreen forest and a secondary minimum at 40°S latitude because of the large extent of open ocean at this latitude. Increases at 25–30° N and S result from the presence of subtropical deserts with relatively high albedos (>40%). At higher latitudes the seasonal or permanent occurrence of snow cover and sea ice raises average albedos, which are in excess of 60% in the polar regions. The latitudinal patterns of planetary albedo are less extreme than the surface trends. Two factors affect this difference. Tropical and midlatitude land areas that are vegetated are in latitudes of frequent cloud occurrence. Contrasts between vegetated and desert latitudes are thus less apparent in the planetary figures. In addition, interactions between surface albedo and atmospheric scattering and absorption tend to reduce albedo differences between the tropics and the poles. Where surface albedo is high, the refected radiation passes back and forth through the atmosphere many times, increasing absorption both at the surface and in the atmosphere. Over regions of low albedo, scattering in the atmosphere increases planetary reflectance when compared to surface albedos (Hummel and Reck, 1979).

Measurements

Traditionally, two pyranometers, one pointed toward the sky, the other toward the surface, have been used to measure albedos. Pyranometers are solar radiation measurement devices that respond thermally to record the amount of radiation incident on the device. They are designed to absorb all wavelengths of solar radiation equally (Sellers, 1965). In the late 1960s, photoelectric detectors, such as silicon cells, became widely used for albedo and spectral reflectance measurements (Dirnhirm, 1968). The advantage of photon detectors is that they respond quickly to changes in incident radiation and thus permit high-resolution measurements, particularly from aircraft and spacecraft (Barrett, 1992). One limitation of photon detectors is that they are sensitive to restricted spectral ranges. Silicon, for example, only senses wavelengths between 0.5 and 1.0 μm. Measurements must be either compensated for in those portions of the solar spectrum not observed or two or more different detectors must be used. However, ease of use in the field and in aircraft and spacecraft has significantly increased their use for albedo measurements (Gurney *et al.*, 1993; Shine *et al.*, 1996).

Human effects on Earth's albedo

Recently investigators have suggested that human modifications of the Earth's surface may be altering the planet's albedo. Otterman (1977) showed that overgrazing in desert regions can increase surface albedo by as much as 20%. Charney (1975) estimates that such changes may suppress rainfall, which would enhance the process of 'desertification' that is occurring in sub-Saharan Africa. More recently, Sagan *et al.* (1979) have proposed that extensive deforestation in tropical rainforests may significantly increase surface albedo and result in major climate changes. However, measurements of Brazilian tropical soils reflectivity do not support their suggestions showing overall albedos of 20% or less as a result of high iron oxide content (Stoner and Baumgardner, 1981). More complete knowledge of the diverse albedos of Earth materials is needed to assess accurately human effects on albedo. Continued expansion of human activities that alter the planet, including urbanization, agriculture and forestry, point to the need for better understanding of possible effects on albedo.

Samuel N. Goward

Bibliography

Barrett, E.C., 1992. *Introduction to Environmental Remote Sensing*. London: Chapman & Hall.
Briegleb, B. and V. Ramanathan, 1980. Spectral and diurnal variations in clear sky planetary albedo, *J. Appl. Meteorol.*, **21**, 1160–1171.
Charney, J., 1975. Dynamics of deserts and drought in the Sahel, *Q. J. Royal Meteorol. Soc.*, **101**, 193–202.
Dirnhirm, I., 1968. On the use of silicon cells in meteorological radiation studies. *J. Appl. Meteorol.*, **7**, 702–707.
Fritz, S. and M. Rigby, 1957. Selective annotated bibliography on albedo, *Meteorol. Abstr. Bibliogr.*, **8**, 952–998.
Gurney, R.J., J.L. Foster and C.L. Parkinson, 1993. Global vegetation mapping, in *The Atlas of Satellite Observations*, Cambridge: Cambridge University Press, 301–312.
Houghton, J.J., G.J. Jenkins, and J.J. Ephraums, 1996. Radiation forcing of climate, in *Climate Change, The IPCC Scientific Assessment*. Cambridge: Cambridge University Press, 47–68.
Hummel, J.R. and R.A. Reck, 1979. A global surface albedo model, *J. Appl. Meteorol.*, **18**, 239–253.
Kondrotyev, K. Ya. (ed.), 1973. *Radiation Characteristics of the Atmosphere and the Earth's Surface*, V. Pondit (trans.). New Delhi: Amerind.
Miller, D.H., 1981. *Energy at the Surface of the Earth*. International Geophysics Series, Vol. 27. New York: Academic Press.
Otterman, J., 1977. Anthropogenic impact on albedo of the Earth, *Climatic Change*, **1**, 137–155.
Robinson, N., 1966. *Solar Radiation*. New York: Elsevier.
Sagan, C., O.B. Toon and J.B. Pollock, 1979. Anthropogenic albedo changes and the Earth's climate, *Science*, **206**(4425), 1363–1368.
Sellers, W.D., 1965. *Physical Climatology*. Chicago: University of Chicago Press.
Shine, K.P., R.G. Derwent, D.J. Wuebbles and J.J. Morcrette, 1996. Albedo, in *The Encyclopedia of Climate and Weather*, S.H. Schneider (ed.), Oxford: Oxford University Press.
Stoner, E.R. and M.F. Baumgardner, 1981. Characteristic variations in reflectance of surface soils, *Soil Sci. Soc. Am. J.*, **45**, 1161–1165.

ALGAL GROWTH ON LAKES

Excessive blue-green algal growth on lakes is caused by progressive fertilization of the water and has become a common problem in many countries. The toxic genera are normally *Oscillatoria, Aphanezemnon, Microcystis* and *Anabaena*. These are single-celled algae which have the means to control their position in the water column by gas vacuoles, and they exist mainly in colonies, some of which resemble grass cuttings. Blue-green algae orientate by the sun. A scum is formed on the surface and as the algae eventually die, and their cell structure breaks down, internal chemicals are released, often of a toxic nature. Attempts have been made to control populations by chemical and physical means which have concentrated mainly on the decrease of phosphates and nitrates. Some blue-green algae, however, can convert atmospheric nitrogen into nitrates which can be used for organic synthesis, hence nitrate control has little or no effect (UK WSA, 1994). Phosphates can be controlled within water systems, but blue-green algae can store sufficient phosphates within their bodies to allow the population to increase significantly without external sources of phosphates being required. One popular tourist lake in Carinthia, Austria, was physically emptied to remove blue-green algae because chemicals had no effect.

It would seem from work done in the UK that temperature and sunshine need to be above average for scum formation to occur. Ideally the condition should be anticyclonic with light winds and low rainfall. Long periods of high temperatures, sunshine and a stable water column (where temperature decreases with depth) have the effect of increasing the algal population.

In healthy fresh or coastal water, algae share nutrients with rooted macrophytes. Algae have only a short life span yet represent the main food source for zooplankton in the aquatic food chain. Zooplankton plays a key role in this aquatic ecosystem as it responds effectively to variations in algal production through opportunist population dynamics.

The algal biomass density is dependent on algal production and algal removal through sedimentation and grazing by zooplankton or other secondary producers. In a healthy aquatic ecosystem there is a balance between production and consumption within each link in the food chain. Accumulation of nutrients inhibits phytoplankton dynamics while macrophytes also shelter zooplankton and young fish protecting them from predators. Zooplankton controls short-term variation in phytoplankton dynamics in fresh and coastal water (Collins, 1995).

Eutrophication problems are caused by a failure of the ecosystems to stabilize and control algal growth through competition with macrophytes and zooplankton grazing.

R.W. Herschy

Bibliography

Collins, M.D., 1995. Is weather the single most important factor controlling the development of blue-green algae scum? *Weather*, (Reading UK), **50**(6).
UK Water Services Association, 1994. Water Bulletin 632, 25 November.

ALLUVIAL VALLEY ENGINEERING

An alluvial valley is a gently sloping plain consisting of alluvium. It is normally delimited by uplands on either side that rise above the level of the valley to varying heights. More specifically, the alluvial valley of a given stream is that portion of its alluvial plain upstream from its deltaic plain. An alluvial valley is considered to be a more or less balanced system. Sediment supplied to it from the surrounding upland and its tributary alluvial valleys is gradually traded downstream until it is eventually deposited in the deltaic plain. Typically, there is neither net accumulation nor net removal of sediment from the alluvial valley. The volume of material deposited within the valley tends to about equal the amount of material removed from it and carried to the delta.

Alluvial valleys vary widely in width, and the thickness of alluvium within a given valley ranges from the depth of maximum scour during floods to many times the depth of scour. Regional subsidence, changes in base level caused by local obstructions, faulting and related factors significantly affect the thickness of alluvium. The most prominent base-level change affecting thickness of alluvium was the Late Wisconsinan drop in sea level. Nearly all the larger river valleys of the world were entrenched during the last drop in sea level. Where data are available, they almost always indicate that the thickness of the alluvium through which the larger rivers flow is considerably greater than the present depth of scour of the river.

Sediments within valleys almost always grade upward from coarse at the base to finer materials at the valley surface. Deviations from this generality can usually be attributed to remnants of older materials left within the valley from previous periods of alluviation, the entrance of tributaries into the valley, ponding within tributaries caused by rapid aggradation of the master stream, or tectonic influences such as faulting.

In temperate and humid areas of the world, alluvial valley fill consists exclusively of alluvium. In arid regions, however, significant thicknesses of eolian sands and silts are often intercalated with the valley fill (Glennie, 1970). Stream beds are often dry for long periods in such areas, and wind tends to rework previously deposited alluvium significantly. Because of this, landforms and depositional processes common to most river valleys vary to some extent from those in more humid areas. The same is true to a somewhat lesser extent in arctic regions, where permafrost and extended periods during which streams are frozen interject conditions that are geomorphically and depositionally unique to the arctic (see Permafrost).

The great majority of the world's river valleys are characterized by landforms produced by braided streams, meandering streams or a combination of the two. The alluvial valley of the lower Mississippi River belongs to the last category and is one of the world's more intensively studied river valleys. Most of the environments of deposition within the Mississippi Valley were formed by meandering streams; however, within the valley large areas of an older, slightly higher surface exist which reflect a time when the stream was braided. The classification of Mississippi Valley sediments used here is essentially that followed by Fisk (1944) and Kolb *et al.* (1968). For useful variations to this classification, particularly as such classification might apply to streams considerably smaller than the Mississippi, see Schumm (1977) and Happ (1971).

An understanding of the geological processes involved in the formation of the various environments of deposition within an alluvial valley, of the lithological and physical characteristics of these environments, and of their distribution in plan and profile are of importance in engineering, petroleum geology, environmental science, law and many other disciplines. The reconstruction of ancient petroliferous sedimentary environments based on a knowledge of the landforms and lithological characteristics found within present-day alluvial valleys has become of ever-increasing importance in oil exploration. The development of agriculture, housing and industry within valleys is often predicated on geological factors, as are environmental or flood control considerations, which just as often tend to discourage or modify such developments. Problems involving property rights and boundaries along rivers more and more frequently involve the expertise of the geologist and the fluvial morphologist.

Of particular importance is the application of geological parameters or techniques in the prediction of river behavior in the search for aggregate in areas where aggregate sources are sometimes scarce, and in resolving foundation problems. The value of geological studies cannot be overemphasized in locating and designing engineering structures within river valleys and in economizing on soil borings and laboratory testing programs. The major types of depositional environments within river valleys produce predictable sequences of soils, each having a fairly well-defined range of engineering properties.

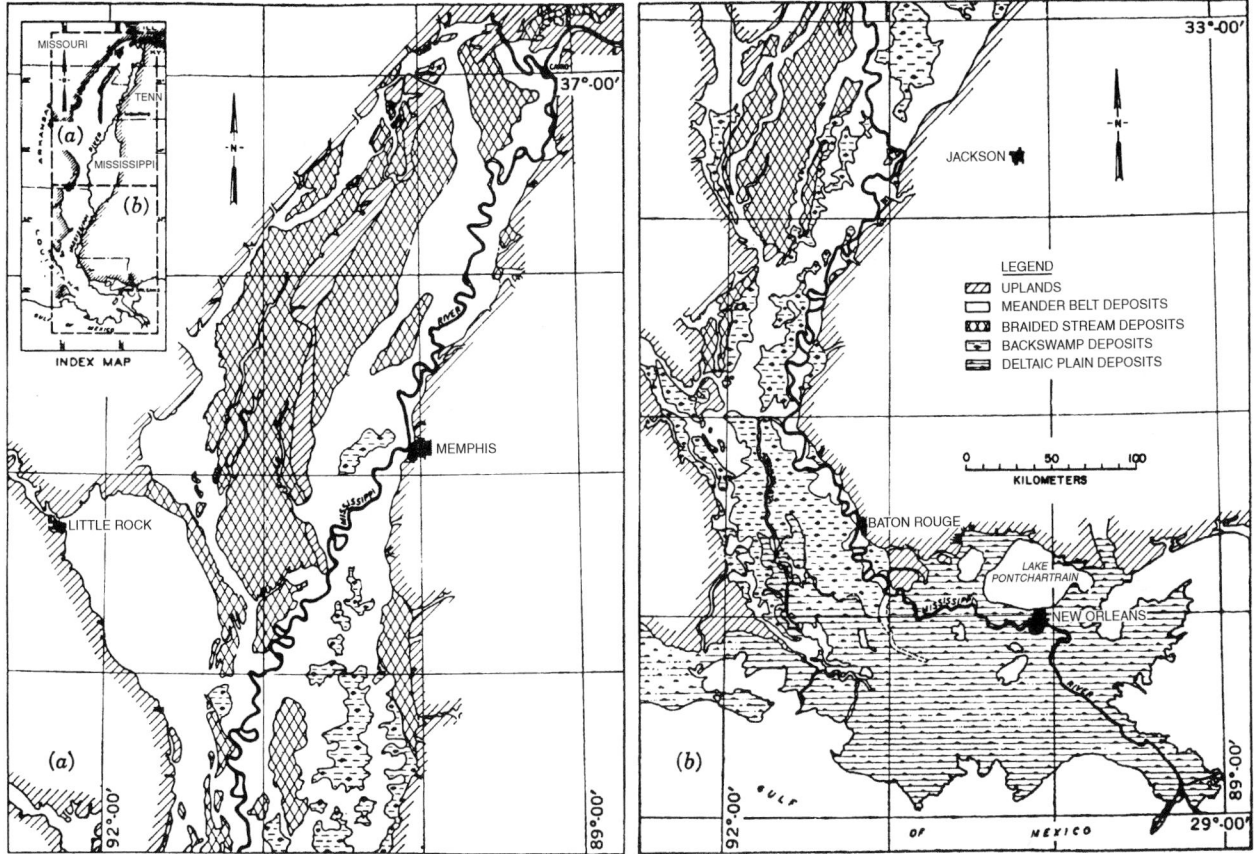

Figure A59 Distribution of depositional types in the lower Mississippi Valley. (From Kolb and Shockley, 1959.)

Alluvial valley of the lower Mississippi River

The alluvial valley of the Mississippi was profoundly affected by sea-level fluctuations during the Pleistocene. The latest drop in sea level is generally believed to have reached its maximum about 17 000 years ago. This drop caused the scouring of an entrenched valley beneath the present flood-plain surface. There is controversy about the effect of this relatively short-lived drop in sea level on the sediments forming the alluvial valley. Some contend that sediments of the previous interglacial deposits were removed entirely and that the valley was entrenched to new depths. Others believe that large remnants of the mid-Wisconsinan interglacial deposits remain at depths beneath more recent Holocene deposits. Still others hypothesize that entrenchment of the mid-Wisconsinan interglacial deposits extended no farther upstream than the latitude of Baton Rouge, Louisiana.

For our purposes it is sufficient to characterize the present alluvial valley as consisting of a substratum of sand and gravel – whatever its age – overlain by a topstratum of finer-grained materials. The substratum, the massive sand and gravel sequence that underlies the topstratum and overlies older Tertiary and Cretaceous strata, is treated as a single unit.

Figure A59 is a generalized map showing the distribution of the major depositional types that form the lower Mississippi alluvial plain. The alluvial valley, by definition, does not include the deltaic plain. Figure A60 shows the environments of deposition in plan and profile within a small portion of the valley in the vicinity of Greenville, Mississippi.

The substratum

The substratum consists of a wedge of coarse-grained material laid down during the earlier stages of the filling of the entrenched valley of the Mississippi River. Figure A60 illustrates the thickness and general nature of the substratum deposits at a centrally located site in the valley. The unit is composed predominantly of clean sand and gravel, with the material normally becoming coarser with depth. Cobbles up to 10 cm in diameter are sometimes encountered near the base of the unit. Occasional lenses of clay, sandy silt or silty sand are also found, but they are rare and discontinuous. Thicknesses of the substratum and the depth to the top of the substratum generally increase in a down-valley direction. The substratum is often encountered at depths as shallow as 3 m in the northern part of the valley and may average only 15 m in thickness. The depth to the substratum near Baton Rouge, Louisiana, on the other hand, is as much as 40 m and the thickness of the substratum may be more than 100 m.

Although a shallow depth to firm substratum sand may be desirable from a foundation standpoint, the relatively high permeability of the substratum may often result in an expensive dewatering operation if deep excavations are involved. Pressure-relief wells are often installed to minimize undesirable uplift pressures in cases in which the bottom of the excavation is in clays overlying substratum sands at shallow depths. Where the excavation bottom is in the substratum, the problem of keeping the excavation free of water may be sizeable. Permeabilities of the substratum generally range from 400 to 2000×10^{-4} cm s^{-1}, with the larger values associated with the deeper and coarser part of the deposit.

Water tables within the valley are near the surface; thus the highly permeable substratum provides a convenient and important aquifer for agricultural, industrial, municipal and individual use. The volume of the alluvial valley substratum is enormous. Kolb (1961) estimated that the amount of storage in this aquifer equals the total flow of the Mississippi for 2.5 years. Wells 70 m deep yield 15 000–18 000 l min^{-1}.

The substratum is also an important source of concrete aggregate and base-course fill. Dredging from gravel bars is common along the river, and numerous pits within the flood plain exploit substratum sands and gravels that occur at shallow depths. Even more important

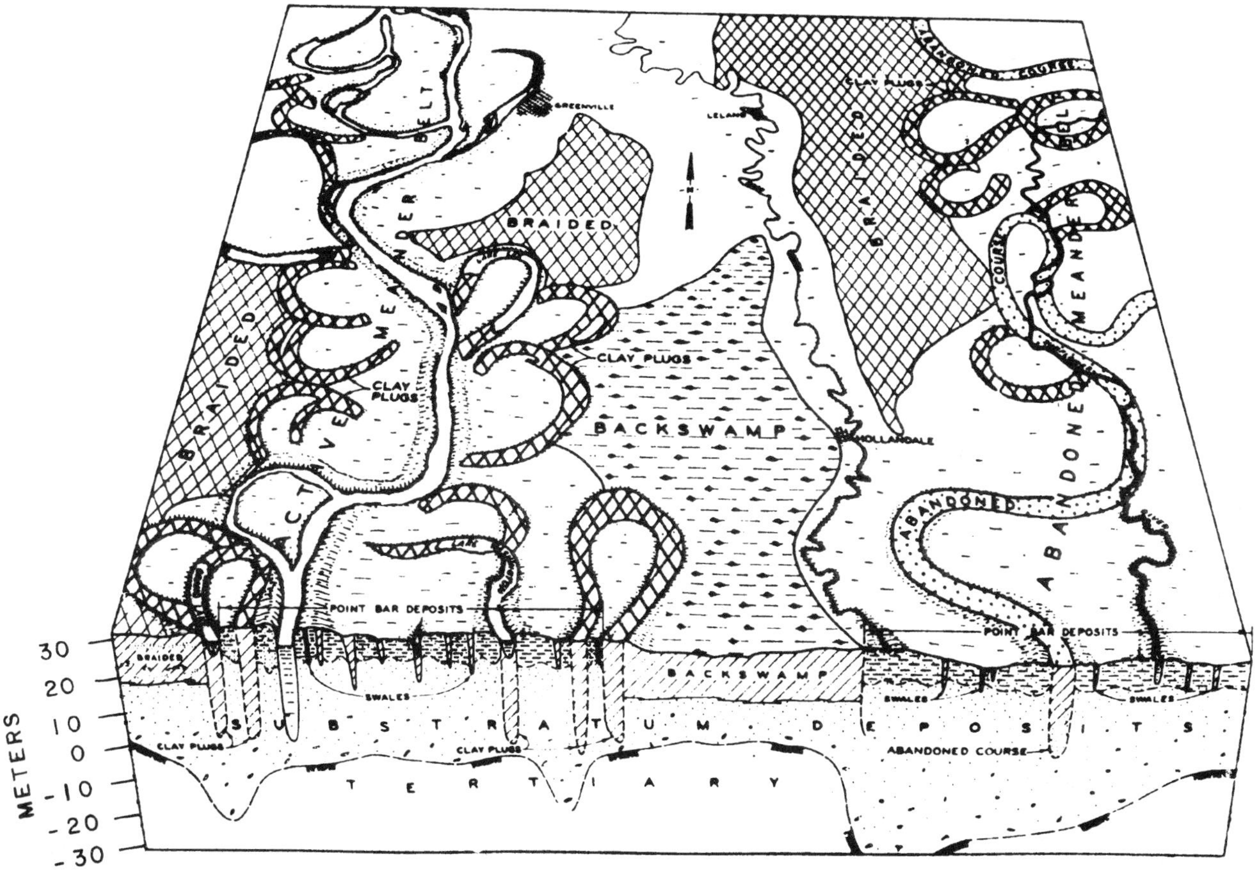

METERS

Figure A60 Major environments of deposition in the vicinity of Greenville, Mississippi. (From Kolb and Shockley, 1959.)

as an aggregate source are substratum sands and gravels in the alluvial terraces that border the valley.

Natural levees

Natural levees (Figure A61) are low ridges that flank both sides of streams which periodically overflow their banks. Since the coarsest and greatest quantities of sediment are deposited closest to the stream channels, the natural levees are highest and thickest in these areas and gradually thin away from the channels. In general, the greater the distance from the stream, the greater the percentage of the finer-grained sediments. Minute drainage channels trending at right angles to the parent stream (down the backslope of the levees) are rather common and tend to fill with fine sand. Major crevasses through the

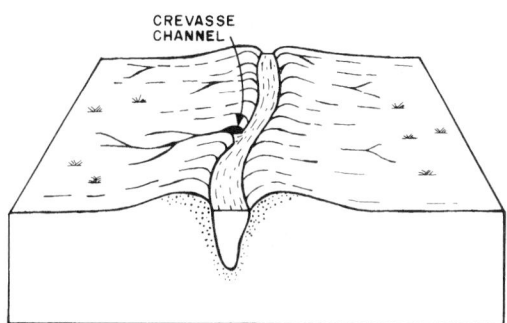

CREVASSE CHANNEL

Figure A61 Natural levee environment. Most of the natural levees along the Mississippi are now topped by artificial levees from 10 to 12 m high. Natural levees attain crest heights of 3–4 m above adjacent backswamp areas and may be 4 km in width.

levee sometimes form during floods and often carve surprisingly deep channels, which tend to fill with some of the coarsest material carried in suspension by the river.

Natural levees attain crest heights of 3–4 m above adjacent backswamp areas and may be 4 km wide. They typically consist of stiff to very stiff, brown to grayish brown silts, silty clays and clays that exhibit moderate to high degrees of oxidation. Natural water contents of the deposits are typically low, and organic matter is seldom present except in the form of roots. Early roads through the Mississippi Valley typically were sited on these higher, well-drained features. As highway technology advanced, however, roads were built as often as not without regard to these natural ridges. As a result, many of the older roads built on natural levees have performed well over the years, while more modern highways have posed serious problems, particularly where they cross low-strength, highly compressible clays characteristic of some of the other alluvial valley environments.

Alluvial aprons

Alluvial aprons (Figure A62) are combinations of alluvial and colluvial deposits that overlie flood-plain deposits along the valley walls and along the sides of upland remnants within the valley. Typically, symmetrical alluvial fans are present at the mouths of streams that drain the uplands. When these streams are closely spaced, the fans coalesce to form alluvial aprons. When the streams are more widely spaced, the fans are separated and the intervening portions of the aprons are composed mainly of sediments that have washed down from the uplands or that have been moved downslope by soil creep.

Alluvial aprons are common, particularly along the high eastern valley walls of the Mississippi alluvial valley. They occur but are less pronounced along the lower western valley walls. They are best developed near the mouths of the small streams that enter from the uplands and particularly where they overlie backswamp deposits and thus have not been affected by migration of the river. Alluvial apron

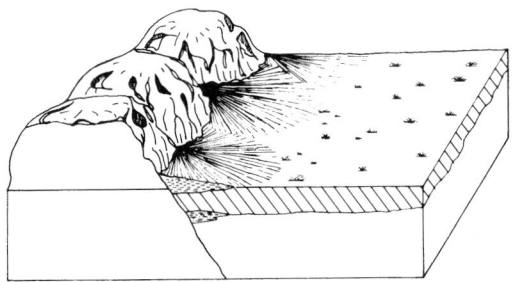

Figure A62 Alluvial apron environment. Aprons are particularly well developed along the loess-covered eastern valley wall of the lower Mississippi Valley.

widths of more than 3 km are common in the Yazoo Basin, for example, and elevations of 4–6 m above the flood-plain level occur near the valley edge. Borings made on alluvial aprons encounter soils that reflect the composition of the materials in the uplands. Along the eastern valley wall of the Mississippi Valley, the thick loess of the uplands is the predominant source material. Thus clayey silt, silt and fine sand are its typical constituents. Of interest to the engineering geologist and the environmental geologist is the accelerated rate at which streams in the loessial uplands are being entrenched. The result is an accelerated rate of growth of the alluvial aprons that these streams build as they drop their loads and build alluvial aprons at the valley borders. Exploitation of timber and landuse practices in the uplands are causing significant problems not only in the uplands but where alluvial aprons flank the valley borders.

Braided stream deposits

Braided stream deposits (Figure A63) consist of the sediments that were laid down by rapidly shifting, aggrading streams during the earlier stages of valley alluviation. The braided stream deposits were formed by shallow, anastomosing, ancestral streams of the Ohio, the Mississippi, the Arkansas and smaller streams emerging from the uplands adjacent to the entrenched valley. The great majority of the thick substratum deposits that underlie the valley and extend to depths of 100 m or more were laid down by braided streams. By the time alluviation within the valley has reached its present level, however, fines were being deposited and braided channels were gradually consolidating into single meandering streams, particularly in the southern portions of the valley. As meander belts became established and were extended up-valley, valley gradients were reduced and valley surfaces were built to levels slightly lower than the older braided stream surfaces. Consequently, these surfaces in the Mississippi Valley often stand as low terraces above the level of the meander belt portions of the valley, as shown on Figure A63.

Braided stream topstratum covers a fairly significant area within the lower Mississippi Valley (Figure A59). It consists of light gray to tan clays, silts and well-graded fine sands on the order of 3–10 m thick. Both water and organic contents of the sediments are low. The average grain size of the sediments increases toward the northern end of the basin.

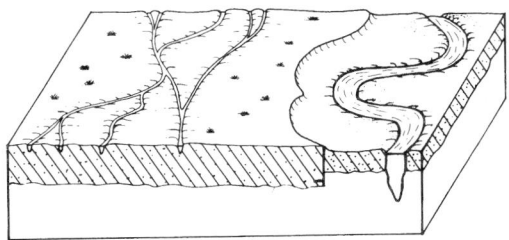

Figure A63 Braided stream environment. Sediments laid down by a shallow, aggrading network of streams during earlier phases of Mississippi Valley development. These now stand as low terraces within the valley above the level of those environments of deposition associated with meandering streams.

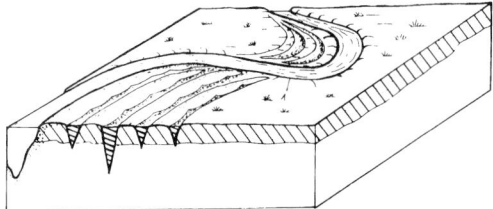

Figure A64 Point bar environment. Sediments laid down on the insides of the river bend as the stream meanders. They consist of an alternating series of swales (clay bodies) and ridges (silty and sandy silts) overlying substratum at shallow depths.

Point bar deposits

Point bar deposits (Figure A64) consist of sediment laid down on the insides of river bends as a result of the meandering of the stream. Although the deposits extend to a depth equal to the deepest portion, or thalweg, of the parent stream, only the uppermost, fine-grained portion is considered part of the topstratum. Within the point bar topstratum, there are two types of deposits: silty and sandy elongate bar deposits, or ridges, which are laid down during high stages of the stream, and silty and clayey deposits in arcuate depressions, or swales, which were laid down during falling river stages. Characteristically, the ridges and swales form an alternating series, the configuration of which conforms to the curvature of the migrating channel and indicates the direction and extent of meandering.

Point bar deposits are most widespread along the present course of the Mississippi River and along the abandoned courses of the river. Because of successive occupations of certain meander belts by streams of different sizes, complex patterns of ridge-and-swale topography are common. Point bar deposits consist of tan to gray clays, clayey silts, silts and fine sands in the ridges, and soft, gray clays and silty clays in the swales. Excluding the larger swales, which occasionally may be filled with clays over 20 m thick, the topstratum varies from 6 to 12 m thick. Both water and organic contents are high in the swale deposits, whereas they are both commonly low in the ridge deposits.

A common and potentially hazardous phenomenon associated with flooding of the Mississippi River is seepage beneath the levees and the formation of sand boils, particularly in the thin topstratum deposit characteristic of point bar ridges. Sand boils consist of sand carried by seepage forces to the surface on the landward side of levees. These features often form conical mounds with water – sometimes muddy water – issuing from the top of the mound. Although limited underseepage and through-seepage of the levees are generally acceptable, seepage beneath levees in the form of sand boils indicates active piping and poses a threat to the safety of the levee. Comprehensive studies of these features (Mansur et al., 1956; Kolb, 1976) have shown that the disposition of the various environments of deposition, the juxtaposition of pervious versus impervious floodplain deposits beneath the levee, and the angle at which such bodies are crossed by the overlying levees are controlling factors in localizing sand boils. Figure A65 shows the effect of elongate swales and channel fill deposits where these clayey deposits pass beneath a levee at an angle. Seepage is often heaviest and boil formation most marked within the acute angle. The clay body tends to concentrate seepage in the pervious ridge areas where the geometry of the levees vis-à-vis the trend of the swales resembles that shown in Figure A65. Note that boils also tend to form adjacent to the swale within the obtuse angle formed by the swale and the levee. Such seepage is generally less pronounced, however. Also of interest in this figure is the development of boils where a sand-filled crevasse channel lies beneath the levee; in this instance, a crevasse channel developed through natural levee deposits overlying backswamp clay.

Abandoned channels

Abandoned channels (Figure A66) – or clay plugs, as filled abandoned channels are commonly called – are segments of stream channels formed when the stream shortens its course. The abandoned segment may consist of an entire meander loop when the river cuts directly across the narrow neck between two converging arms of a loop (a neck cutoff), or it may be a portion of a loop formed when a

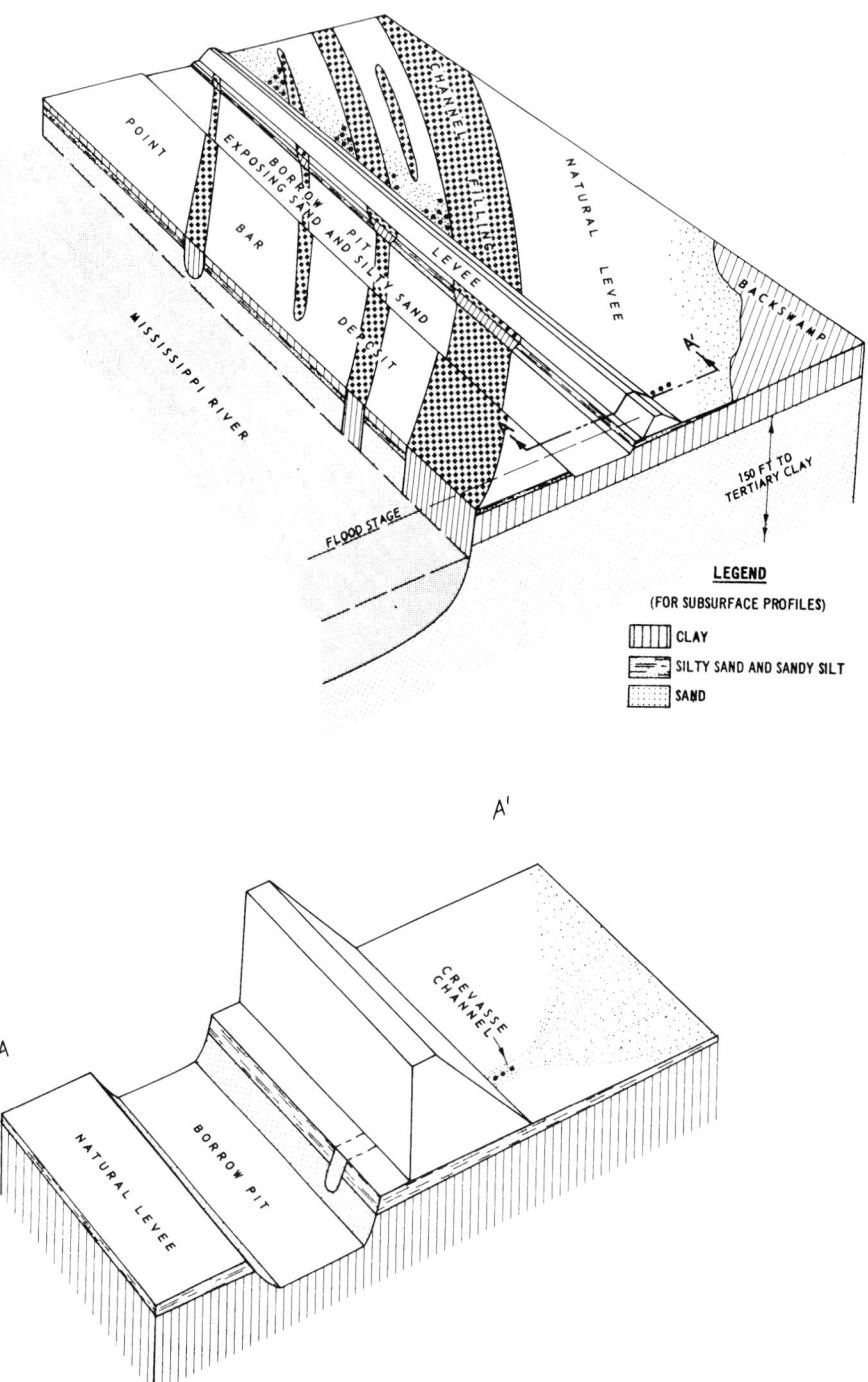

Figure A65 Sand boils (shown with asterisks) and seepage (shown with a dot pattern) that formed during the 1937 Mississippi flood. This is an example of only one of thousands of such phenomena that occurred on the landward side of the levees during the flood. A special case is illustrated in the expanded section shown along A–A'. Here a well-developed, semi-pervious natural levee deposit lies between backswamp clays and the artificial levee. In such instances seepage may occur in the extreme landward portions of the natural levee and boils may form in old natural levee crevasses backfilled with sand. (From Kolb, 1976.)

stream occupies a large point bar swale during a flood stage and abandons the outer portion of the loop (a chute cutoff).

Deposits filling abandoned channels are predominantly clay and silty clay. Shortly after cut-off, a wedge of sand fills both arms of the abandoned channel at the point of cut-off, and soon thereafter an oxbow lake is formed. The only material deposited within the oxbow lake is that from periodic overbank flow from the river. As the channel migrates away from the point of cut-off, only the finest materials find their way to the oxbow lake. Consequently, clays may fill these abandoned channels to the depth of the former active channel; clay deposits from 30 to 50 m thick are not uncommon. Abandoned channels of the Mississippi are usually 8–15 km long (following the loop) and 500–1500 m wide (channel width).

Clays settling within deep oxbow lakes are characteristically high

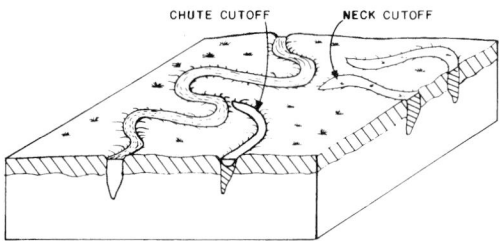

Figure A66 Abandoned channel environment. Abandoned segments of a loop formed when a stream shortens its course. These form lunate or horseshoe-shaped lakes that eventually fill with clays – clay plugs.

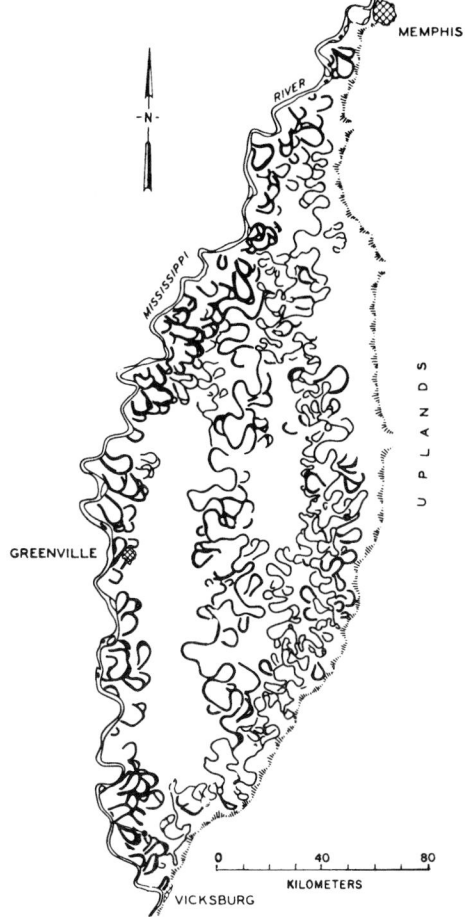

Figure A67 Abandoned channel and abandoned course deposits in the Yazoo Basin. Literally thousands of these massive clay bodies are found within the Mississippi alluvial valley. (From Kolb *et al.*, 1968.)

in water content. Because plants grow within the lake area only when it is filled or nearly filled with sediment, organic content is low except in the upper 3 m of the clay body. Clay plugs are among the deepest clay bodies found in the alluvial valley. They are relatively unconsolidated and tend to exhibit high compressibility and low strength. Special design and construction measures are usually necessary where levees, highways and other engineering structures cross abandoned channel deposits. Their distribution often vitally affects the location of drainage and navigation channels. In addition, these channel deposits have a significant effect on river migration. Wherever these deep clay bodies occur along the river, they act as 'hard' points that radically alter the normal rate and direction of stream migration and greatly influence revetment placement and levee location (Fisk, 1947). Valid predictions of the direction in which the river may migrate during the life of an installation is of prime importance to the engineering geologist entrusted with site location along the banks of the Mississippi. Bridges, pipelines, overbank flow structures, docks, levees and factories anxious to utilize cheap transportation facilities provided by the river are examples of the numerous installations that are affected by the stability of the river.

Of equal importance has been the application of fluvial morphology to the ubiquitous problem of boundary lines between states, countries and private properties that border on the Mississippi and its tributaries within the alluvial valley. Because the boundaries between states often follow the thalweg of the river, changes in the thalweg radically affect these boundaries. Simplistically speaking, rapid changes (avulsions) in the thalweg caused by a chute or a neck cutoff fix the boundary along the thalweg of the abandoned channel. Thus enclaves of Louisiana within Mississippi, for example, and of Mississippi within Arkansas are found along the Mississippi and its tributaries. Here, a portion of one state may be entirely surrounded by another. Conversely, state boundaries change as the thalweg shifts, if the shift is gradual. Huge portions of states adjoining the river are carved away each year as the river shifts and are added to the state across the river. Fluvial morphologists are thus often called on to testify in court cases involving riparian boundaries.

Abandoned channels formed by meandering streams of all sizes are numerous in the lower Mississippi Valley. Figure A67 shows the major clay plugs mapped in that portion of the alluvial valley known as the Yazoo Basin, a football-shaped flatland with Vicksburg, Mississippi, at its southern end and Memphis, Tennessee, at its northern end. The figure includes abandoned courses (discussed below), but most of the units delineated are abandoned channels or clay plugs.

Abandoned courses

Abandoned courses (Figure A68) are lengthy segments of river abandoned when the stream forms a new course across the flood plain. The abandoned course, varying from a few kilometers (but always more than one meander loop) up to hundreds of kilometers in length, gradually fills with sediment and is often occupied by a smaller or underfit stream. Studies of these features suggest that the old course fills with a wedge of sand, thickest where the new course diverges from the old, gradually thinning downstream. In many cases the smaller stream meanders within the confines of the larger meander belt and destroys surficial segments of the original abandoned course. In other cases the smaller stream delineates the extent of the abandoned course where there are no other indications of its presence.

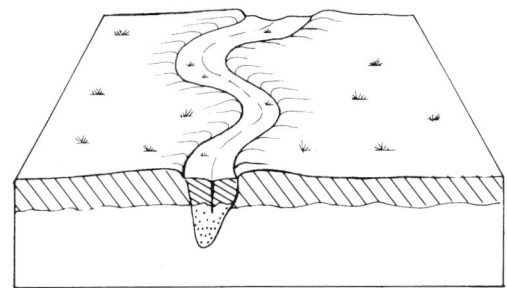

Figure A68 Abandoned course environment. Length segments of the river are abandoned when the stream chooses a new course across the floodplain. These features tend to fill with a wedge of sand which is thickest at the point of diversion and thins downstream.

Many of the surface expressions of Mississippi River abandoned courses have been destroyed by the meandering of smaller streams. The short, isolated segments that are recognized are similar in size and shape to abandoned channels. Abandoned courses of smaller streams are numerous throughout the valley. Unlike abandoned channels or clay plugs, topstratum deposits in the former are relatively thin, particularly in the immediate area of the point of diversion and from some distance downstream from this point (otherwise, the topstratum of the abandoned courses resembles those of the abandoned channels, i.e. highly compressible clays and silty clays).

Based on available data (Fisk *et al.*, 1952), it appears that the abandonment of one Mississippi River course and the development of another takes place over a fairly lengthy period (up to a century), and also that such a traumatic change in the river happens only once in 500 years or so. As such, it might seem that such changes are so rare as to be of little practical consequence during the life of a project. It becomes very significant, however, if this change is about to occur. This is true of the Mississippi. The Atchafalaya River (Figure A59) leaves the Mississippi north of Baton Rouge, Louisiana, and diverts water through a direct, markedly shorter route to the Gulf than the present channel, which flows past Baton Rouge and New Orleans. The Mississippi is on the verge of abandoning the latter course for the former and undoubtedly would have done so by now had not extensive and expensive control structures been built in the early 1950s to prevent it. The consequences of such a major diversion of the river are monumental. Some (Kolb, 1980) argue that the future of the southern part of the alluvial valley would be better served if the river were not shackled with artificial and costly engineering restraints and were permitted to choose the shorter path to the sea it would have chosen under natural conditions. Comprehensive studies might well conclude that a gradual, controlled diversion down the Atchafalaya, with necessary safeguards to ensure an adequate amount of fresh water down the present channel south of the point of diversion, might be a viable option.

Backswamp deposits

Backswamp deposits (Figure A69) consist of fine-grained sediments laid down in shallow ponded areas during periods of stream flooding. The coarser material from overbank flow is dropped near the stream to form natural levees. The finer material settles slowly in low-lying areas as the ponded water gradually drains off, seeps into the ground, or evaporates. Backswamp areas typically have very low relief, and a distinctive, complex drainage pattern develops in which the channels alternately serve as tributaries and distributaries at different times of the annual flood cycle.

Backswamp deposits are widespread within the alluvial valley and increase in areal extent in the wider portions of the valley, particularly in the southern portion. Deposits are continuous over areas as large as 500 km², and their thickness generally increases toward the south, where deposits as thick as 30 m are found. Varying thicknesses of natural levee or alluvial apron deposits often overlie backswamp deposits. Between 3 and 4 m of backswamp deposits, in turn, often overlie other valley deposits such as point bars and abandoned channels.

Soft to stiff, gray to dark gray-brown clays and silty clays are typical of backswamp deposits, and occasional thin layers of silt or sand may be found, but more than 80% of the deposit normally consists of clay. Organic matter in the form of disseminated particles, peat layers and large wood fragments are common. Because backswamp clays are exposed to desiccation and oxidation after each flood, they often develop strength properties considerably higher than similarly fine materials that settle out in an aqueous environment, such as the oxbow lakes that are the sites for abandoned channel

deposits. This preconsolidated strength is apparent not only in laboratory measurements of shear strengths but is also reflected in the lower water content of these sediments when compared with abandoned-channel and abandoned-course topstratum.

Clays are generally added to the backswamp in increments ranging from paper-thin to several centimeters thick. On drying between floods and after local precipitation, the clay shrinks and thousands of small cubical fragments often form, fragments that can be scooped up by the handful from the dried backswamp surface. The size and hardness of these dry clay pellets give rise to the term 'buckshot' clay. When wet, the extremely fine clay becomes maddeningly viscous and sticky and is referred to locally as 'gumbo'.

Because the backswamp deposits typically contain variable quantities of organic materials, they are sometimes used as a source for lightweight aggregate. On firing, organic particles in the clay form gases that expand and result in a light, porous, durable clinker that is much in demand in the southern part of the valley where aggregate is scarce and expensive and where lightweight aggregate is often used in high-rise buildings to impose as small a load as possible on soils that are characteristically low in bearing strength.

Summary

Alluvial valleys are gently sloping plains consisting of alluvium delimited on either side by uplands rising to varying heights. The principal depositional types are those associated with (1) braided streams and (2) meandering streams. The major alluvial valleys of the world fall in the latter category. The alluvial valley of the Mississippi River is characterized chiefly by depositional landforms associated with meandering channels; however, it also contains remnants of surfaces left by an older, braided Mississippi River.

Principal depositional types forming surficial deposits in the Mississippi Valley are natural levees, alluvial aprons, braided streams, point bars, abandoned channels, abandoned courses and backswamps. Each is characterized by a predictable sequence of soils, with predictable thicknessess and lateral limits, and having fairly well-defined ranges of engineering properties. A knowledge of the geological processes forming each depositional type and the distribution of these types has proved invaluable in locating and designing engineering structures in the valley and in economizing on soil borings and laboratory testing programs.

An understanding of the fluvial processes in alluvial valleys and the depositional types that result have numerous applications. Among them are (1) predictions of future meandering and migration of the river based on past migratory history; (2) age determinations for archeological sites on abandoned streams; (3) reconstruction of ancient petroliferous sedimentary environments based on present alluvial valley stratigraphy; (4) siting industrial, housing and agricultural developments having the least environmental impact; (5) resolving boundary disputes along rivers among states, counties and individuals; (6) locating sources and assessing resources for sand and gravel supplies; and (7) solving a multitude of problems associated with foundation stability beneath locks, dams, drainage facilities, and similar engineering structures associated with flood control and general valley development.

Charles R. Kolb

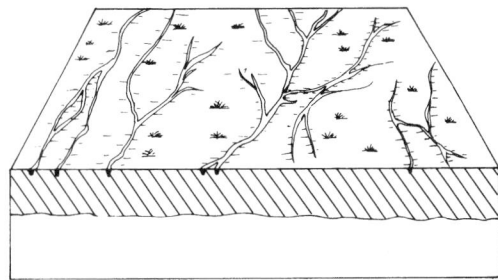

Figure A69 Backswamp environment. Fine-grained sediments are laid down in shallow ponded areas during floods. Because backswamp clays are exposed to desiccation and oxidation after each flood, they tend to develop a preconsolidation strength considerably higher than similar fine materials in the abandoned channel, abandoned course and swale environments.

Bibliography

Fisk, H.N., 1944. *Geological Investigation of the Alluvial Valley of the Lower Mississippi River.* Vicksburg, MI: Mississippi River Commission, 78p.

Fisk, H.N., 1947. *Fine-grained Alluvial Deposits and Their Effects on Mississippi River Activity,* 2 vols. Vicksburg, MI: US Army Corps of Engineers Waterways Experiment Station, 82p.

Fisk, H.N., Kolb, C.R. and Wilbert, L.J., 1952. *Geological Investigation of the Atchafalaya Basin and the Problem of Mississippi River Diversion,* 2 vols. Vicksburg, MI: US Army Corps of Engineers Waterways Experiment Station.

Glennie, K.W., 1970. *Desert Sedimentary Environments.* Developments in Sedimentology, **14**. Amsterdam: Elsevier, 222 pp.

Happ, S.C., 1971. Genetic classification of valley sediment deposits, *Am. Soc. Civ. Eng. Proc., J. Hydraulics Div.*, **97**, 43–53.

Kolb, C.R., 1961. Geology of the Lower Mississippi River Valley and its economic aspects, in James E. Noblin, Jr. (ed.), *Our Nuclear Future.* Jackson, MI: Mississippi Industrial and Technological Research Commission, 129–146.

Kolb, C.R., 1976. Geologic control of sand boils along Mississippi River levees, in D.R. Coates (ed.), *Geomorphology and Engineering*. Stroudsburg, PA: Dowden, Hutchinson and Ross, 99–113.

Kolb, C.R., 1980. Should we permit Mississippi–Atchafalaya diversion? *Gulf Coast Assoc. Geol. Soc. Trans.*, **30**, 145–150.

Kolb, C.R. and Shockley, X.X., 1959. Engineering geology of the Mississippi Valley, *Am. Soc. Civ. Eng. Trans.*, **124**, 633.

Kolb, C.R., Steinriede, W.B., Jr, Krinitzsky, E.L. *et al.*, 1968. Geological investigation of the Yazoo Basin, Lower Mississippi Valley, *US Army Corps Engineers, Waterways Expt. Sta. Tech. Rept. 3–480*, 160 pp.

Mansur, C.I., Kaufmann, R.I. and Schultz, J.R., 1956. Investigation of underseepage and its control: lower Miss. River Levees, *US Army Corps Engineers Waterways Expt. Sta. Tech. Memo 3–242*, 421 pp., appendixes and 241 plates.

Schumm, S.A., 1977. *The Fluvial System*. New York: Wiley, 338 pp.

Cross references

Channelization and bank stabilization
Dams
Deltaic plains
Floods
Reservoir capacity: water resources

AMAZON RIVER

The River Amazon is the world's largest, in terms of discharge, with a mean discharge of 220 000 m^3 s^{-1} or 6930 km^3 year^{-1}, which is ten times that of the Mississippi and five times that of the next largest, the Congo (Zaire). The drainage area of the Amazon basin is 6 915 000 km^2, nearly double that of the Congo (Zaire). It is a truly equatorial river, flowing generally from the northern and central Andes (Colombia, Ecuador, Peru and Bolivia) in an ENE direction to debouch in the Atlantic Ocean, exactly on the equator, while major tributaries join it from the left (coming from the Precambrian plateaus of the Guyanas and Venezuela) and from the right (coming from rather similar formations in central Brazil, the Mato Grosso and Caatinga; Table A21).

Occupying 38% of the total area of South America, the Amazon Basin receives *every day* 52% of the precipitation for the whole continent. This amounts to 2150 mm or 14 900 km^3 daily. Of this daily precipitation, 11 500 km^3 is carried in as clouds from the Atlantic Ocean and may be explained by the convergence of both the northerly and southerly trade wind systems in the intertropical convergence zone (ITCZ). Besides these rains, which have two peak seasons, 3400 km^3 (23%) are recycled by direct evaporation within the basin. The climate, in the Köppen system, ranges from Af to Aw, or tropical rainforest to tropical savanna with a daily convective overturn (Oliver and Fairbridge, 1987). Precipitation decreases markedly from west to east (where drought areas sometimes develop).

Table A21 The Amazon's tributaries (USSR Committee for IHD, 1978)

Name	Drainage area (km^2)	Discharge (m^3 s^{-1})	Discharge (km^3 year^{-1})
Ucayali	375 000	12 600	397
Maranon	350 000	15 600	491
Napo	106 000	6 500	205
Javari	91 000	4 260	134
Isa	123 600	7 160	225
Jerua	224 000	9 100	286
Japura	282 000	17 900	564
Purus	365 000	12 600	397
Rio Negro	691 000	29 300	923
Madeira	1 391 000	30 500	961
Tapajos	487 000	15 500	488
Xingu	513 000	16 000	504
Tocantins	770 000	16 800	529

The seasonal variance of the Amazon discharge is considerable and during the flood season the banks are inundated to create vast ephemeral lakes. Near Obidos, the 'dry' season flow (e.g. November) is about 72 000 m^3 s^{-1} and mean water depth is 41 m, compared with wet season flow (May–August) at up to 230 000 m^3 s^{-1} with water depth 7 m higher at 48 m. The interannual variance of the total discharge is also very large. Over a long time span the range is of the order of ±10%.

Geologically the Amazon Basin is constrained by a framework of three very distinctive rock types (Bigarella, in Fairbridge, 1975): (1) the Andes, consisting of Cenozoic volcanics and poorly consolidated sedimentary rocks, with core regions of acid igneous rocks (granitic family) and metamorphic types, of Mesozoic and Paleozoic age; (2) the plateaus of the Brazilian craton, which are dominated by nearly flat-lying Proterozoic quartzites and shales (dating mostly around 1–2 billion years); and (3) the Amazonian trough, which is marked by grabens and downwarped basins trending ENE–WSW, and dating back to the Paleozoic (up to 500 million years). There is a youthful infilling of over 4000 m in places of Tertiary deposits, largely sandstones and shales. Near Nova Olinda on the Purus River there is some petroleum accumulation.

Prior to the South Atlantic break-up of Gondwanaland which began here about 120 million years ago the ancestral Amazon trough partly lay in West Africa and was open to the west. Hand in hand with the break-up and subsequent sea-floor spreading in the South Atlantic, subduction began on the western margin of the South American plate and the Andes began to form. Initially their erosive products began to fill a foredeep trough along their eastern flank, now occupied by the upper Amazon and its tributaries. It appears that the mountains only became prominent in the late Tertiary and have been rising steadily (1–2 mm year^{-1}) during 2 million years or so.

The glacial oscillations (in 100 000 year cycles) of the Quaternary Period served to lower the sea level during each glacial maximum by about 135 m, so that transported sediment (partly abraded from the bed of the Amazon and its local tributaries) was projected directly to the edge of the Atlantic continental slope, and the rate of sedimentation offshore was up to five times higher than it is today. The mineralogy of the sediments (dominated by unweathered feldspars) shows that it was generated under hyper-arid climatic conditions (Damuth and Fairbridge, 1970). Wide areas of quartz sandplains in the tropical cratons are interpreted as podzolic residuals transported by glacial-age winds. The eolian accumulations are partly in now-vegetated dunes (Fairbridge and Finkl, 1984). Patches of tropical rainforest survived in so-called 'refugia' but large parts of the Brazilian cratonic region became savanna or subdesertic. The present humid conditions returned about 10 000 years ago, since when the glacial-regime thalweg has refilled with youthful sediments.

The above remarks are highly relevant with respect to the amount of water in storage in the saturated Quaternary sands and late Tertiary sandstones of the Amazon Trough. However, relatively little exploratory drilling has taken place, so that estimates of that paleohydrologic storage are not yet feasible. There may well be offshore freshwater seepage beyond the delta front. The flood period surface flow is so strong that in some seasons fresh water can be sampled at the sea surface up to 100 Km offshore.

Geochemically, the Amazon waters are rather unusual (Gibbs, 1967). The river carries about 232 × 10^{12} g year^{-1} in solution, and 499 × 10^{12} g year^{-1} in solids, to furnish a total of 116 × 10^6 g km^2 year^{-1}. It is calculated that 21.7% of the erosion is by solution. The dissolved load of the main flow is the highest in the world, but some of the tributaries, such as the Xingu, have among the lowest. The reason is the bedrock and topography of each catchment: the relatively erodible or soft materials and high relief of the Andes, in contrast to the flat quartzite plateaus that characterize the Xingu sources, some of which are so low in dissolved components that they are almost like distilled water and the pH may be as low as 4. Some 82% of the Amazon basin (mostly in Brazil) is of low relief and only 18% is mountainous.

Most of the low-relief part was until recent decades also rainforest which provided an almost complete protection against mechanical erosion. However, the late twentieth century uncontrolled population explosion in Brazil has led to land hunger and extensive deforestation. Unfortunately, not all of the newly cleared properties are properly cared for. The soil is largely lateritic, a kaolinitic clay often rich in colloidal silica and iron; under forest cover it is soft and malleable, but on exposure to the sun, the colloids dehydrate and the soil becomes a duricrust (ferricrete and silcrete) of extreme hardness. This process is irreversible, so that the rock-like crust becomes completely useless for agriculture, although the duricrust has a limited value for highway and foundation construction.

Rhodes W. Fairbridge

Bibliography

Damuth, A.E. and Fairbridge, R.W., 1970. Equatorial Atlantic deep-sea arkosic sands and ice-age aridity in tropical South America. *Bulletin of the Geological Society of America*, **81**, 189–206.

Fairbridge, R.W. (ed.), 1975. *The Encyclopedia of World Regional Geology, Part I: Western Hemisphere*. Stroudsburg: Dowden, Hutchinson & Ross, 704 pp.

Fairbridge, R.W. and Finkl, C.W., Jr, 1984. Tropical stone lines and podzolized sand plains as paleoclimatic indicators for weathered cratons. *Quarternary Science Reviews*, **3**, 41–71.

Gibbs, R.J., 1967. The geochemistry of the Amazon River System: Part I. The factors that control the salinity and the composition and concentration of the suspended solids, *Geol. Soc. Am. Bull.*, **78**, 1203–1232.

Oliver, J.E. and Fairbridge, R.W. (eds), 1987. *The Encyclopedia of Climatology*. New York: Van Nostrand Reinhold, 986 pp.

USSR Committee for I.H.D., 1978. *World Water Balance and Water Resources of the Earth*. Moscow and Paris: UNESCO, 663 pp.

Cross references

Hydrogeology
Rain
Rivers

ANTARCTIC OZONE HOLE

Introduction

The discovery of a spring-time depletion of the ozone layer above the Antarctic by British Antarctic Survey scientists in 1985 demonstrated that our understanding of the atmosphere is far from complete. This article explains how the ozone hole forms and answers some common questions.

Antarctica is a huge continent over 50 times the size of Great Britain. It rises from sea level at the coast to over 3000 m inland. Most of the research stations are at coastal sites for ease of access and transport of provisions by ship. Nations carrying out research on the ozone hole in the Antarctic include Argentina, Finland, France, Germany, Great Britain, Italy, Japan, New Zealand, Russia, the Ukraine and the USA (Figure A70). Ozone measurements in the Antarctic started in the mid-1950s and the longest continuous records are from Faraday (Vernadsky) and Halley.

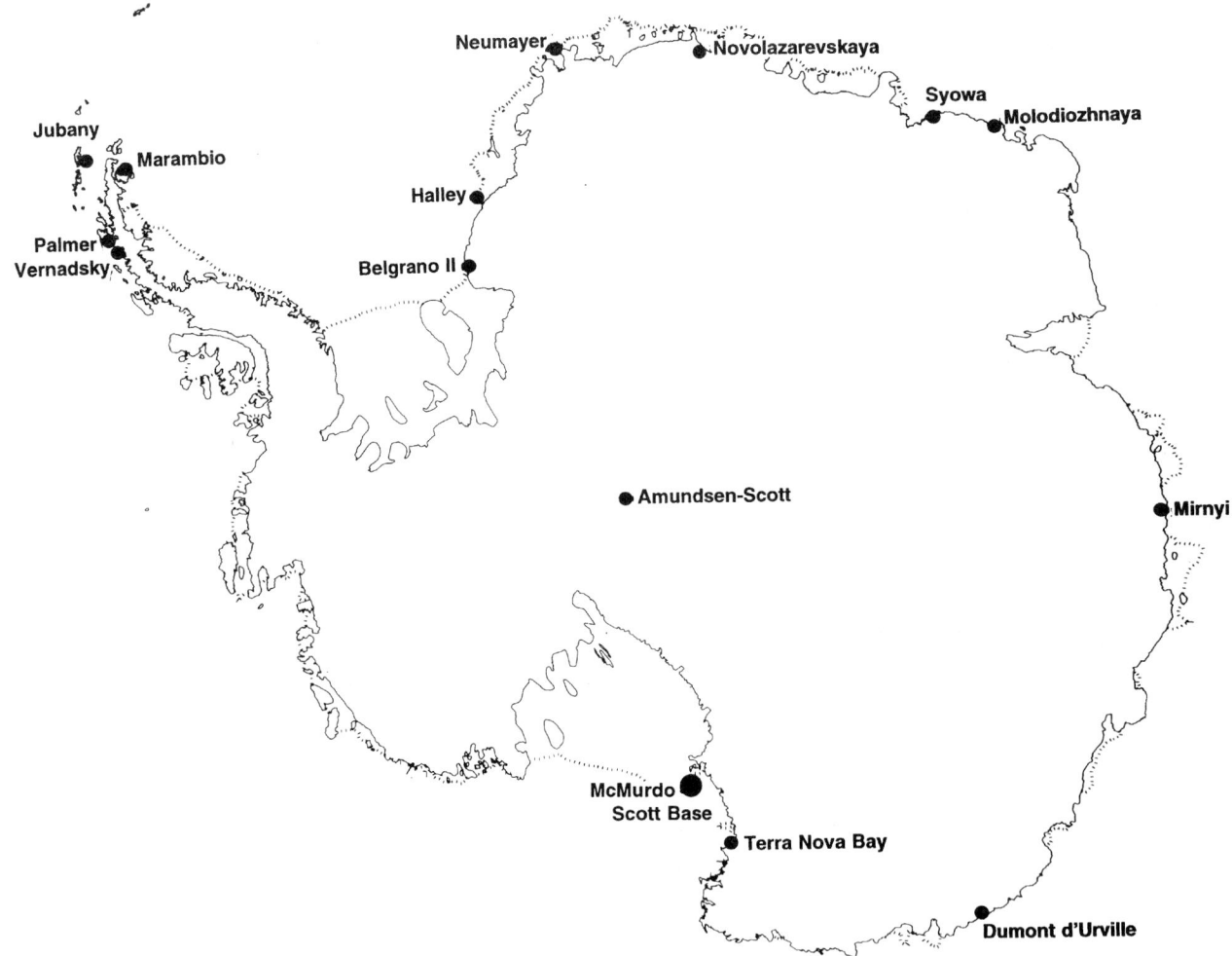

Figure A70 Map showing Antarctica and sites mentioned in the text.

Radiation from the Sun

The visible surface of the Sun is at a temperature of no more than 6000 K, but parts of its atmosphere have a temperature of over 1 000 000 K. It emits radiation in all regions of the electromagnetic spectrum but overall its spectrum resembles that of a black body at 5800 K. We can divide the spectrum into visible light with a wavelength between 400 and 750 nm (which we can see), infrared (which has a longer wavelength than visible light, for example the heat from an electric fire) and ultraviolet with a wavelength between 240 and 400 nm. Approximately 7% of the solar energy is emitted in the ultraviolet, 41% in the visible and 52% in the infrared. The ultraviolet part of the spectrum can be further divided into UV-A, UV-B and UV-C. UV-A lies between 400 and 320 nm and gives rise to a sun-tan and aging of the skin. UV-B lies between 320 and 280 nm and is the damaging part of the spectrum. UV-C lies between 280 and 240 nm and is totally absorbed in the atmosphere before it can reach the ground. The Dobson spectrophotometer (Figure A71) uses wavelengths between 305 and 340 nm to measure the amount of ozone in the atmosphere.

Ozone (O_3) in the stratosphere is created by the action of photons of ultraviolet light (hf), with wavelength shorter than 242 nm, on oxygen (O_2). A third body (M), which may be an oxygen or nitrogen molecule is needed to carry away the excess kinetic energy. Ozone is mainly created above regions where the sun is high in the sky.

$$O_2 + hf \rightarrow O + O$$
$$O_2 + O + M \rightarrow O_3 + M$$

Ozone can be destroyed by the action of light with a wavelength shorter than 1180 nm.

$$O_3 + hf \rightarrow O_2 + O$$
$$O + O_3 + M \rightarrow O_2 + M$$

These two reactions sequences, together with many other more complex ones, balance to give an average of around 300 DU of ozone.

Ozone in the atmosphere

Our atmosphere consists of nitrogen (78%) and oxygen (21%) and a small amount of other gases. Ozone (a molecule with three oxygen atoms) is quite a rare gas: even in the ozone layer there is less than one ozone molecule for every 100 000 molecules of air. Since ozone is a toxic, irritating gas, it is fortunate that its concentration at the Earth's surface is much lower than in the ozone layer. In the polluted air from large cities, ozone may be present in higher concentrations and this can cause severe health problems. In its proper place higher in the atmosphere, ozone provides a safety screen against harmful ultraviolet light from the Sun, which can cause sunburn, skin cancers and cataracts. The ozone layer also controls the temperature structure of the upper atmosphere because it can absorb radiant energy from the Sun.

Ozone has been measured from Halley and Vernadsky Research Stations (Vernadsky was the British station Faraday until 1996) for over 35 years using a Dobson spectrophotometer. This instrument compares the intensities of two wavelengths of ultraviolet light from the sun. One wavelength is strongly absorbed by ozone and the other is only weakly absorbed. Once the instrument has been calibrated, the ratio of the intensities tells us how much ozone there is in the atmosphere. Because the instrument normally uses the sun as a source of ultraviolet light it is not possible to make regular measurements of ozone during the dark Antarctic winter.

The amount of ozone in the atmosphere is measured in milli-atmocentimeters (Dobson Units, DU) and a typical measurement is about 300 DU. This means that if you took all the ozone in a vertical column above the instrument and brought it down to sea level it would form a layer just 3 mm thick. However, the ozone is not actually confined to a narrow layer in the atmosphere, but is spread throughout the atmosphere, mostly between 10 and 50 km above the ground. The graphs in Figure A72 show where it is to be found both before and after the formation of the ozone hole.

The data up to the mid-1970s show that ozone at Halley (Figure A73) was typically around 300 DU through the winter until late October (spring in the Antarctic). It then rose rapidly to about 400 DU by the beginning of December and slowly declined to reach 300 DU again by March. Since the 1980s, however, a seasonal decrease or 'hole' has appeared in the Antarctic ozone layer each spring, with values at Halley now dropping below 100 DU. Data from Faraday

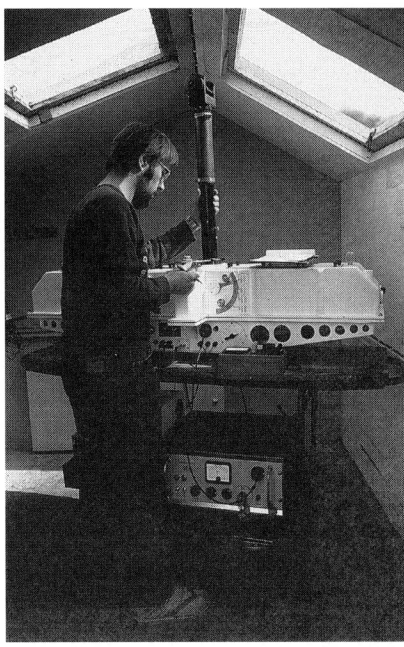

Figure A71 The Dobson spectrophotometer, which is used to measure ozone.

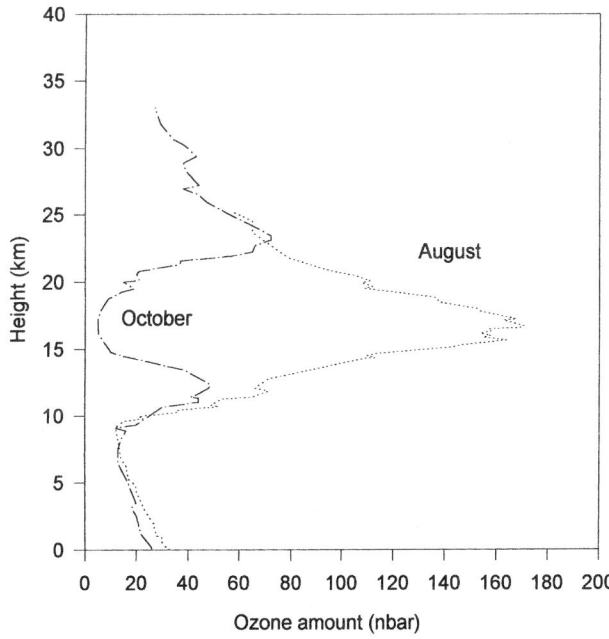

Figure A72 Vertical profiles of ozone at Halley, before (15 August) and during (13 October) the 1987 Antarctic ozone hole.

show a similar general picture (Figure A74), but there is a much greater day-to-day variation than at Halley, with a periodic fluctuation in the amount of ozone. The mean October ozone values at Halley show a rapid decline (Figure A75), which is just beginning to level out.

The ozone is being destroyed because of the release of chlorofluorocarbons (CFCs), mostly in the northern hemisphere. These spread throughout the world and diffuse into the stratosphere, where they are broken down by catalytic reactions to release chlorine. A catalytic reaction is one where a chemical is needed for the reaction to

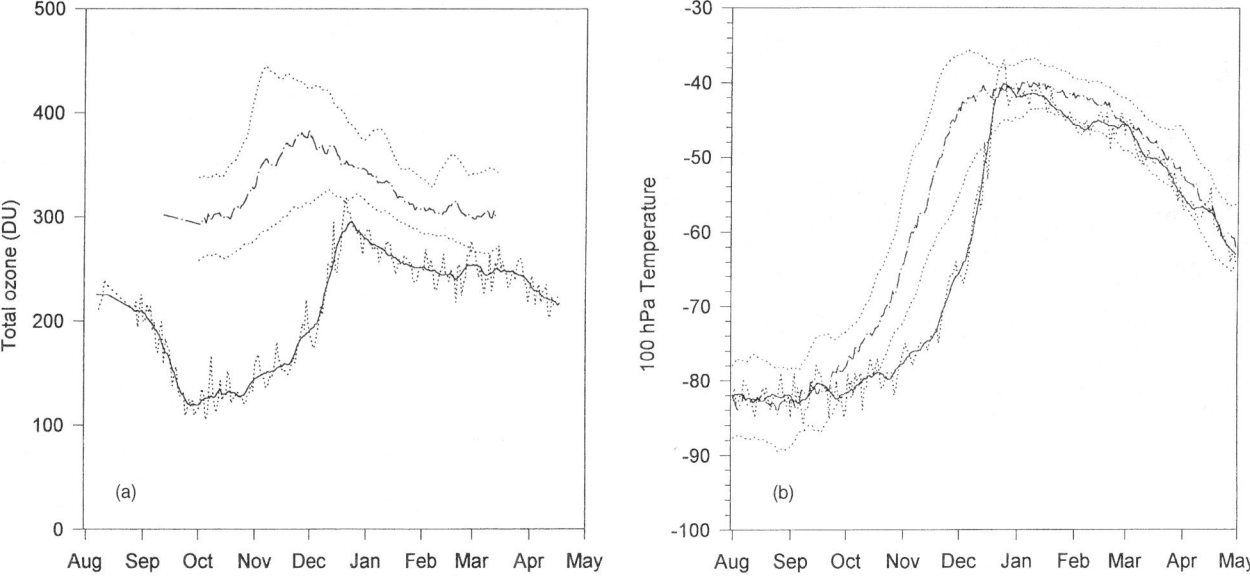

Figure A73 Halley ozone (a) and temperature (b) observations for 1995–1996. The upper lines show the historical mean and range of values before the ozone hole developed.

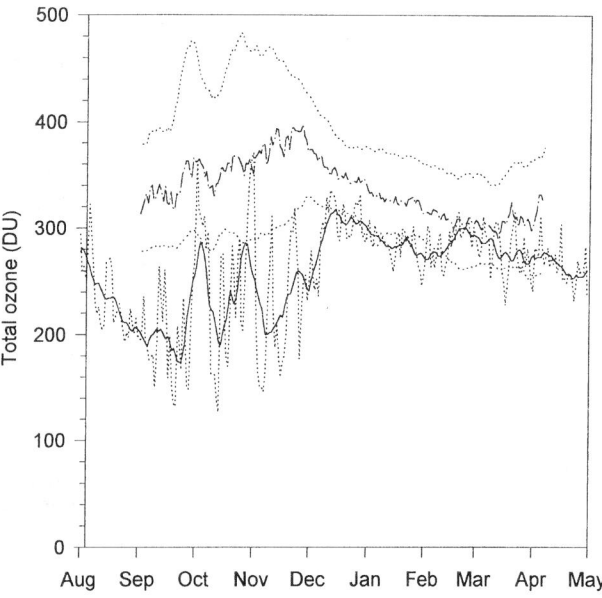

Figure A74 Faraday ozone observations for 1995–1996.

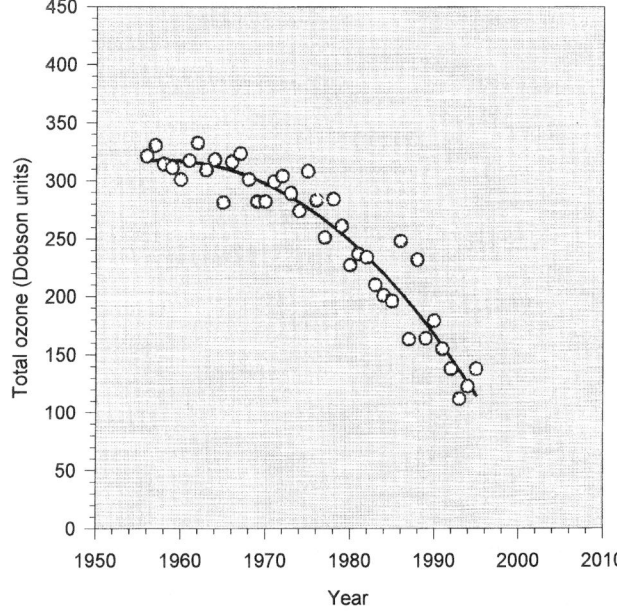

Figure A75 The decline in total ozone amount during October at Halley.

take place, but the chemical is not affected by the reaction. The simplest catalytic cycle which destroys ozone involves chlorine (Cl) and has chlorine monoxide (ClO) as an intermediary:

$$Cl + O_3 \rightarrow ClO + O_2$$
$$ClO + O \rightarrow Cl + O_2$$
$$\text{Net } \overline{Cl + O_3 + O \rightarrow Cl + 2O_2}$$

The same amount of chlorine is present before and after the reaction. The Antarctic ozone hole forms when more complex variations of this reaction take place during the early spring. Bromine (Br), a gas similar to chlorine, can also take part in the catalytic cycles.

During the Antarctic winter a strong westerly circulation around the continent, known as the circumpolar vortex, builds up in the stratosphere. This effectively cuts off the interior and allows it to cool, with temperatures falling below –80°C at 17 km (Figure A73). Thin clouds form (Figure A76), which enable reactions with gases which contain chlorine to take place. When the sun returns in the spring, the chlorine is able to take part in complex catalytic chemical reactions which destroy ozone and create the ozone hole. When the stratosphere warms up again during the late spring and summer, these reactions cease, the circumpolar vortex breaks down and the ozone hole disappears as fresh ozone is brought in.

Unlike Antarctica, which is a continent surrounded by oceans, the Arctic is an ocean surrounded by mountainous continents. This means that the stratospheric circulation is much more irregular, and the temperature does not fall as low as it does in the Antarctic. Stratospheric clouds are therefore less common, which prevents the formation of a deep ozone hole over the Arctic.

Every day at Halley a balloon is launched (Figure A77), carrying meteorological instruments. The instrument package signals back the temperature, humidity and pressure to an altitude of over 20 km. The

Figure A76 Mother-of-pearl clouds, a localized form of stratospheric cloud found near mountain ranges.

Figure A77 A balloon launch at Halley.

temperature near the height of the ozone maximum (about 17 km) is strongly linked with the total amount of ozone in the atmosphere, and this gives us additional information on the state of the ozone layer. The graph (Figure A73) shows that temperatures at 17 km during the 1995 Antarctic spring remained cold for much longer than normal. That year had the longest-lasting ozone hole yet observed.

The vertical distribution of ozone in the atmosphere is measured by launching a balloon which carries a chemical sensor, known as an ozone sonde. This uses the reaction between ozone and a solution of potassium iodide to generate an electric current. The strength of the current modulates a signal which is transmitted back to the ground. These measurements show that the normal maximum in the ozone layer at 17 km is replaced by a deep minimum during the spring ozone hole (Figure A72). Most of the ozone between 14 and 20 km altitude is destroyed, but when the circumpolar vortex breaks down the hole fills in.

From space a more complete picture can be seen. Several satellites carry sensors which make global measurements of atmospheric ozone and other trace gases. Maps made from these measurements clearly show the 'hole' as it forms in the spring and subsequently recovers in

the summer. They also show why the ozone amount at Faraday varies in a regular manner: as the air rotates around the pole, the station is sometimes inside the vortex in ozone-poor air and sometimes outside in ozone-rich air.

A major international study of the Antarctic ozone hole took place during the southern spring of 1987. Many flights were made in specially equipped high-altitude research aircraft from Punta Arenas in southern Chile, along the Antarctic Peninsula and towards the South Pole. Lidar (laser radar) and microwave experiments from McMurdo and Amundsen-Scott stations also probed the ozone layer. The measurements showed conclusively that chlorine plays a major role in forming the ozone hole. In the northern spring of 1989 similar studies took place over the Arctic. More recent studies from the Space Shuttle and satellites specially designed to study the atmosphere show that CFCs are the source of the chlorine and the clear link between it and ozone depletion.

The US has set up a UV Irradiance Monitoring Network at unpolluted sites throughout the world and data from the network clearly show that increases in UV are associated with decreases in ozone. Some recent studies have linked a decrease in phytoplankton production in the marginal ice zone of the Southern Ocean with the decrease in ozone associated with the Antarctic ozone hole. Other studies on the biological effects of ozone depletion are still in progress.

Some answers to questions

When was the ozone hole discovered?

The discovery of the annual depletion of ozone above the Antarctic was first announced in a paper by Joe Farman, Brian Gardiner and Jonathan Shanklin which appeared in *Nature* in May 1985. Later, NASA scientists reanalyzed their satellite data and found that the whole of the Antarctic was affected.

How long has the Antarctic ozone layer been studied?

Ozone was first measured from British Antarctic stations during the International Geophysical Year of 1957–1958. It was originally studied because of its influence on the temperature structure of the atmosphere, and also as a tracer for the circulation of stratospheric air. In the 1970s ozone became the focus of attention as a possible indicator of long-term changes in the atmosphere. Scientists realized that ozone might be affected by the increasing concentration in the atmosphere of anthropogenic gases such as nitric oxide and CFCs. In 1995 Paul Crutzen, Mario Molina and Sherwood Rowland received the Nobel prize for this pioneering work.

Will the ozone hole get bigger?

The depth of the Antarctic ozone hole has been steadily increasing, but it is unlikely to become much deeper than in 1993. During October of that year most of the ozone between 12 and 20 km altitude disappeared. The ozone destruction process has so far been confined to this height range, probably because the air above and below is too warm for stratospheric clouds to form. The area affected by ozone depletion has also increased since the beginning of the effect in the 1970s, but is limited by the strong circumpolar circulation and cannot normally extend beyond 55 S. The 1994 hole was the largest yet observed, covering over 23 million km². The shape of the hole is continually changing; sometimes it is circular and at other times very elongated. The duration of the hole has increased as well, because with less ozone present there is less warming of the stratosphere and the circumpolar circulation is more stable and lasts longer. Major volcanic eruptions, such as Mt Pinatubo in the Philippines which erupted in 1991, may put material which contributes to ozone depletion into the stratosphere.

Does the ozone hole affect the rest of the world?

At the moment, catastrophic ozone depletion is only seen in the Antarctic during the spring, but surrounding areas experience lowered ozone levels as the ozone hole decays at the end of the spring. As the ozone hole rotates, it may extend over populated areas for a short while when it is very elongated. For example it covered the tip of South America and the Falkland Islands for over a week in October 1994.

Figure A78 Spectacular Mother-of-pearl clouds over Cambridge, UK in February 1996.

Limited ozone depletion can occur above the Arctic, but at present it is confined to parts of the region and only lasts for a few days at a time. A continent-sized ozone hole might have appeared if CFC releases had continued at the high rates of the mid-1980s. In mid-February 1996 a spectacular display of stratospheric clouds was seen over Britain (Figure A78) and a fortnight later a small ozone hole passed over the country. Elsewhere in the northern hemisphere, stratospheric ozone amounts over temperate latitudes have fallen by 5–10% during the winter.

Is the ozone hole dangerous to scientists in Antarctica?

The intensity of ultraviolet light at Antarctic stations during the time of the ozone hole is not much more than on a tropical beach at midday. This is because in the tropics the light comes straight down through the atmosphere, whereas in the Antarctic the sun is much lower in the sky so that light takes a much longer path through the atmosphere. Nevertheless, it has always been very important to take precautions against sunburn in the Antarctic, because sunlight is also reflected from the snow surface. Recent studies suggest that some skin cancers are caused by a two-stage process: sunburn as a child causes a DNA mutation and further sunburn when older activates the cancer, which may appear many years after the original exposure.

How does ultraviolet damage living things?

All living cells, whether microbes, plants or animals, contain a complex molecule called DNA which carries the genetic code. This is the set of instructions which describes the structure and biochemistry of an organism. Unfortunately, DNA readily absorbs high-energy UV-B radiation and becomes damaged so that the instructions cannot be read properly. If the amount of UV-B entering the cell increases (as during the ozone hole), the risk of damage also increases and may result in malfunction or death of the organism. Some Antarctic organisms such as algae, lichens and mosses also contain a pigment called chlorophyll. This absorbs visible light as the energy source of photosynthesis for making organic compounds. Chlorophyll also absorbs UV-B light, so that the system becomes bleached and non-functional. Even enzymes and other proteins are damaged by this high-energy radiation. Living organisms therefore have to protect themselves from UV-B. Humans can cover their skin with artificial sunscreens, but natural protection systems have also evolved. Many microbes, plants and other animals synthesize protective pigments. Our skin cells synthesize brown melanin to protect against sunburn (which is caused by UV-B radiation), and so do Antarctic lichens on rocks near the edge of the polar ice-cap. A variety of sunscreen pigments are produced by Antarctic organisms on land, in fresh-water and in the sea. That is why exposed, snow-free rocks are often covered with bright orange and yellow lichens. Some lichens and microbes even live inside translucent rocks to shelter from high radiation levels and desiccating winds!

Does the greenhouse effect cause the ozone hole?

The greenhouse effect (producing global warming) and ozone depletion are two separate problems; however, there are links between them. Warming at the Earth's surface is caused by certain gases in the atmosphere which can trap energy from the Sun. An increase in the amount of these gases produces an increase in the surface temperature. The largest increase is in carbon dioxide from burning coal, oil, gas and forests, but other gases such as methane (from cattle and rice fields) play a part. A link with ozone depletion is that CFCs are gases which also contribute to the greenhouse effect.

A further link is that although the greenhouse effect warms the surface, it allows the higher atmosphere, where ozone is present, to cool. This means that more stratospheric clouds may form and so make the ozone hole worse.

Even if the problem of ozone depletion is solved, global warming will still remain. It will cause a rise in sea-level and change the regions where crops can be grown. The issue will be harder to tackle than ozone depletion, but is one which concerns everyone on our planet.

What is the Montreal Protocol?

The Montreal Protocol is an international agreement which was drawn up in September 1987. It originally aimed to halve the use of CFCs by 1999. However, reviews of the protocol held in 1990 in London and 1992 in Copenhagen imposed more stringent controls, so that all production of CFCs, CCl_4 and halons should cease by the year 2000. Many countries have even agreed to stop using CFCs before this deadline. Production of other ozone-depleting gases is to stop in the early years of the twenty-first century. Unfortunately the ozone hole will not immediately disappear because CFCs are such stable gases that they will remain in the atmosphere for decades after release.

Independent reviews by panels of scientists (e.g. the UK stratospheric Ozone Review Group reports) present conclusive evidence that CFCs are still increasing in the atmosphere and that chlorine from them is also increasing and is responsible for ozone depletion. Thanks to the provisions of the Montreal protocol and its subsequent amendments, the level of ozone depleting gases in the atmosphere will start dropping by the end of the 1990s.

Where were CFCs used?

CFCs were used in a wide variety of products. Due to public pressure the use of CFCs by the aerosol industry declined rapidly. The other major uses were 'foam blowing' for upholstery padding, freezer linings, fast-food cartons, cavity wall insulation, and as the fluid in refrigeration and air conditioning systems. CFCs were also used as solvents in industrial and electronic cleaning processes. Halons (bromofluorocarbons) were extensively used in fire extinguishing systems. CFCs were introduced because they were generally odorless, non-toxic, stable, non-flammable and compressible substances. It was their high stability which allowed them to get into the stratosphere where they were broken down to release active chlorine. The two simplest CFCs are CFC11 and 12, which have the chemical formulae $CFCl_3$ and CF_2Cl_2.

How can we mend the ozone hole?

The only way to mend the ozone hole is to stop releasing CFCs and other ozone depleting gases into the atmosphere. The restrictions of the Montreal Protocol and its extensions are helping to do this.

Many other suggestions have been made for mending the ozone hole, but these turn out to be difficult or dangerous to put into practice and do not solve the problem. Examples of suggestions include the use of aircraft or balloons. Supersonic transport planes could take ozone up to the Antarctic stratosphere. Unfortunately this would need many Concorde flights a day and the extra ozone would immediately start to be destroyed by the catalytic reactions. In addition each flight would consume many tonnes of fuel and the exhaust gases would further change the chemical make-up of the stratosphere and contribute to the greenhouse effect. Balloons carrying ozonizers could be used in place of planes, but it would need over 100 billion balloons, which would create an even bigger environmental problem!

Further reading

Journals which frequently have material on the ozone hole include:

- *Nature* (Original paper: Large losses of total ozone in Antarctica reveal seasonal ClO$_x$/NO$_x$ interaction, J.C. Farman, B.G. Gardiner and J.D. Shanklin, **315**, 207–210, 16 May 1985; Ten year update: Continued decline of total ozone over Halley, Antarctica, since 1985, A.E. Jones and J.D. Shanklin, **376**, 409–411, 3 August 1995).
- *New Scientist* (Reviews: pp. 50–54, 12 November 1987; Inside Science No. 9, pp. 1–4, 5 May 1988).
- *Science* (Reviews: **239**, 1489–1491, 1988; **241**, 785–786, 1988; **256**, pp. 342–349, 1992; **260**, 523–526, 1993).
- *Scientific American* (Articles: **258**(1), 20–26, January 1988; **264**, 40–47, June 1991)

The annual reports of the Stratospheric Ozone Review Group (HMSO) contain up-to-date reviews of the present understanding of ozone depletion.

There is much information about the ozone hole on the Internet. FAQs on the ozone hole are held in the newsgroups sci.environment, sci.answers and news.answers. The World Wide Web has several sites which carry images and news on the ozone hole and pointers to these are contained within the British Antarctic Survey web pages at http:\\www.bas.ac.uk.

Jonathan Shanklin

Acknowledgements

Thanks are due to Brian Gardiner, Anna Jones and David Wynn-Williams for their reviews and contributions to this article.

Cross references

Global warming
Greenhouse effect

ANTECEDENT PRECIPITATION

The amount of precipitation that has occurred prior to a single storm event can have a significant impact on the nature of the response of runoff to the rainfall during the event and partly accounts for the non-linearity in runoff response to similar storm rainfalls. This is particularly true in catchments where soil water or groundwater storage plays an important role in the generation of runoff and the principle underlies the 'variable source area' theory of runoff generation (Hewlett and Hibbert, 1967). In this context, antecedent precipitation becomes a surrogate for an estimation of the moisture status of the catchment prior to a storm event. Any approach that attempts such an estimation needs to account for the various processes that determine what proportion of the precipitation falling n days previously remains within the catchment as a contribution to current soil moisture. A relatively simple approach adopted by many authors is to use a decay or recession function of the form:

$$\text{API} = \sum_{i=1}^{N} [DR_i K^i]$$

where API is the antecedent precipitation index, DR_i is the daily rainfall i days before the storm, N is the total number of antecedent days over which to base the API, and K is the recession constant (< 1.0).

Alternative and more complex approaches use continuous soil moisture budgeting techniques where the processes of rainfall infiltration, subsurface drainage and evapotranspiration loss are estimated more directly. In the UK the Meteorological Office use a soil moisture budgeting technique to estimate regional values of soil moisture deficiencies continuously (Meteorological Office, 1981).

Antecedent precipitation indices have been found to be useful in various hydrological and water resource estimation approaches but have been used mostly in single event type flood simulation and estimation models and specifically in attempts to develop improved approaches to design flood estimation. Hughes (1984) made use of the above equation to estimate the moisture storage status at the beginning of a storm event within an isolated event simulation model.

Various total antecedent periods (N in the above equation) were tested but no clear indication of the optimum period to use emerged and the final choice of 20 days was largely arbitrary. Packman and Kidd (1980), amongst others, have questioned the validity of the assumption of probability equivalence between storm rainfall depth and resulting peak runoff which is inherent in the design storm concept. They have demonstrated that the probability distribution of peak flows is sensitive to the choice of design storm characteristics as well as antecedent precipitation characteristics. Hughes (1986) discusses an attempt to define the joint probability structure of storm rainfall depth and an antecedent precipitation index, while Schmidt and Schulze (1987) describe how the catchment moisture status has been incorporated into a South African design flood model based on the USDA SCS (United States Department of Agriculture's Soil Conservation Service) model.

Antecedent precipitation is also frequently used in simple models relating monthly or annual streamflow volumes to rainfall through some kind of regression equation having the form:

$$Q_t = a_1 P_t + a_2 P_{t-1} + a_1 P_{t-2} + a_3$$

where Q_t is runoff volume in the current month or year, $P_{t,t-1}$, etc. are precipitation in the current and previous months or years, and a_1–a_3 are parameters or coefficients of the equation.

Mostert *et al.* (1993) report on the importance of the impact of a high rainfall season on the vegetation cover of Namibian catchments and how this can cause a reduction in the amount of runoff in following years. This process was incorporated into a relatively simple monthly time series simulation model through the use of a component based on the previous 3 years seasonal rainfall totals.

Denis A. Hughes

Bibliography

Hewlett, J.D and A.R. Hibbert, 1967. Factors affecting the response of small watersheds to precipitation in humid areas. In: W.E. Sopper and H.W. Lull (editors), *Int. Symp. on Forest Hydrology*, Pergamon, Oxford, 275–290.

Hughes, D.A., 1984. An isolated event model based upon direct runoff calculations using an implicit source area concept. *Hydrological Sciences J.*, **29**(3), 311–325.

Hughes, D.A., 1986. Analysis of extreme rainfalls and antecedent catchment moisture using the bivariate Normal distribution, in *Multivariate Analysis of Hydrologic Processes*, Proc. 4th Int. Hydrology Symposium, July 1985, Fort Collins, Colorado, 519–530.

Meteorological Office, 1981. The Meteorological Office Rainfall and Evaporation Calculation System MORECS. Hydrological memorandum 45.

Packman, J.C. and C.H.R. Kidd, 1980. A logical approach to the design storm concept. *Water Resour. Res.*, **16**(6), 994–1000.

Schmidt, E.J. and R.E. Schulze, 1987. Flood volume and peak discharge from small catchments in Southern Africa, based on the SCS technique. Univ. of Natal, Pietermaritzburg, Dept. of Agric. Eng., ACRU Report No. 24, 163 pp.

Mostert, A.C., R.S. McKenzie and S.E. Crerar, 1993. A rainfall/runoff model for ephemeral rivers in an arid or semi-arid environment. *6th South African National Hydrological Symposium*, Pietermaritzburg, Sept. 1993, 219–224.

Cross references

Floods
Flood studies worldwide
Precipitation

AQUIFER

An aquifer is a lithological unit or combination of such units which has appreciably greater water transmissibility than adjacent units. It stores and transmits water, commonly recoverable in economically usable quantities. Impervious layers or beds of very low permeability which bound an aquifer are termed confining or aquicludes (e.g. the two shale horizons confining a porous sandstone). Any rock of low porosity and permeability is termed an aquitard. An aquitard is one so impervious that it neither stores nor transmits water (Shaw, 1994).

The yield or capacity of an aquifer is termed its storage coefficient, which is defined as the volume of water released or taken into storage per unit surface area, per unit change in the component of head normal to the surface of the aquifer (Chow, 1964). In the case of an unconfined aquifer, the storage coefficient is equal to the specific yield (Price, 1996).

Charlotte Schreiber

Bibliography

Chow, Ven Te, 1964. *Handbook of Applied Hydrology*, McGraw-Hill, New York, 1445pp.
Price, M., 1996. *Introducing Groundwater*, Second edition Chapman & Hall, London.
Shaw, E.M., 1994. *Hydrology in Practice*, Second edition, Chapman & Hall, London.

Cross references

Groundwater
Hydrogeology
Hydrology

ARAL SEA

During the past 30 years the Aral Sea has shrunk from some 66 900 km^2 in 1960 to 32 300 km^2 in area today (see under Lakes for location). During this time the depth has reduced by 16 m. By the year 2000 it is predicted that the area will shrink further to about 23 000 km^2 and the depth will drop by a further 3 m. The reason is an increasing demand for irrigation from the rivers Amu Dar'ya and Syr Dar'ya, both of which flow into the sea. The average annual inflow to the sea from a drainage area of 1.8×10^6 km^2 was estimated for 1981–1992 as 6.8 km^3 compared with 55 km^3 for the years 1911–1960, or 12%.

The difference between the total average annual flow of these two rivers, 103 and 112 km^3, and the 55 km^3 inflow is apparently due to substantial natural flow losses as the rivers cross the two Central Asian deserts, the Kyzylkum and Karakum, and pass through the deltas of the two rivers, as well as by consumptive use for irrigation. Irrigation increased between the years 1960 and 1980 from 5×10^6 to 6×10^6 h$_a$ but consumptive water use increased from 45 km^3 to 90 km^3.

In 1987 the Aral Sea divided into the large sea and the small sea and these have developed as separate water bodies with a human-made dike in 1992 separating the two seas. This prevented water loss from the small to the large sea. The average salinity has dropped from 10 g l^{-1} in 1960 to about 35 g l^{-1} for the large sea and 25 g l^{-1} for the small sea.

In 1960 the Aral Sea was the world's fourth largest lake and, in spite of its salinity, was populated by substantial numbers of freshwater fish sustaining a large fishing industry.

In 1991, in order to arrest and subsequently improve the situation, the State Commission of the Aral Sea (consisting of five republics of the Commonwealth of Independent States – Uzbekistan, Kazakhstan, Kyrgyzstan, Turkmenistan and Tadzhikistan) proposed a 20 year program to be implemented in three stages, 1991–1995, 1996–2000 and 2001–2010. The main objective of the program is that by 2010 the lake level would be raised by about 2 m to 41 m above sea level (in 1960 the lake level was 53 m); the area of the lake would be increased to 42 500 km^2; its volume increased to 430 km^3 (in 1960 the volume of the lake was 1100 km^3) and salinity would be reduced to 25 g l^{-1}, enough to support salt-tolerant fish. The crisis has stirred international attention and the tragedy was discussed at the 46th General Assembly of the United Nations in 1991. Subsequently, several UN organizations and the World Bank became involved in an Environment Assistance Plan to be implemented over a period of 10 years at an estimated cost of $50 million. However, hydrologists agree that the key issues of health, medical improvement (mainly from harmful soda dust), supply of clean drinking water (pesticides pollute the drinking water) and reconstruction of the cotton-supporting irrigation system to save water will take decades to implement fully and cost more than $1000 million.

R.W. Herschy

Bibliography

The World Bank, 1993. *The Aral Sea Crisis: proposed framework of activities.*
Micklin, P.P., 1994. The Aral Sea Problem, *Proc. Inst. Civ. Engns. Civ. Engng.*, **102**, Aug., 114–121, Paper 10154.
Robertson, J.O., 1995. Private communication with author.

Cross references

Irrigated land areas: world
Lakes
Limnology

ARID CLIMATES

More than one-third of the Earth's surface lacks sufficient moisture to support a continuous cover of vegetation, and in the dryer portions of the arid zone, vast areas are without vegetation. However, on the margins of the deserts, in the steppe and savannah regions of the world, grasslands support abundant herds of domestic and wild animals, and humans have introduced domesticated plants. It is along the desert margins that humans have tested nature by encroaching on areas best left in their natural state. One of the problems in assigning precise boundaries to the lands of the arid realm is that slight changes of climate in such fragile environments may result in widespread changes in nature and in human–environment relationships. Conversely, changes that humans make in the environment may result in changes in climate. Thus the boundaries of the arid realm cannot be delineated by narrow lines drawn on a map – they represent broad zones in which xeric vegetation types merge gradually with those of the more humid climatic types that surround them.

What is aridity?

Although many different environmental factors are involved in the explanation of why arid conditions exist, the essential feature of all arid lands is the small amount of precipitation that falls. However, just what quantity of precipitation is needed to make an area humid and what other environmental factors should be considered in defining arid conditions is not readily apparent. If precipitation evaporates rapidly after falling, a much greater amount of moisture is required for the growth of plants than if precipitation evaporates slowly. Soil conditions, temperature, wind and solar radiation are all factors that govern evaporation and growth rates and enter into the equations that govern the efficiency of rainfall.

Scholars have been seeking a definition of aridity on which they can agree for a long time. Because of the complexity of the problem, many argue for a simple solution. Some say that we should use an arbitrary value of rainfall such as 500 mm to mark the boundary between arid and humid zones. Others, accepting the idea of moisture deficiencies associated with varying environmental conditions but opting for a simple concept, say that aridity occurs when potential evaporation exceeds precipitation.

Perhaps the three best-known efforts at identifying the arid zones of the world were those of Köppen (1931), Thornthwaite (1931, 1948), and Meigs (1953).

Köppen

In his classification system, Köppen devised a method for the identification of aridity that took into account the season of occurrence of precipitation. Rainfall occurring during a hot summer is less effective than the same amount of precipitation falling in winter. To systematize this concept, Köppen devised formulae that can be used to determine if a place is arid, semiarid or humid depending on temperature, precipitation and season of occurrence of precipitation. The humid–arid boundary was defined as follows (P in millimeters and T in Celsius):

- winter concentration: $P = 2T$, 70% or more of precipitation falls during six winter months;
- summer concentration: $P = 2 (T\ 14)$, 70% or more of precipitation falls in six summer months;
- even distribution: $P = 2 (T\ 7)$, no season of maximum precipitation.

Thornthwaite

Thornthwaite's classification is based on precipitation effectiveness (P/E), temperature efficiency (T/E) and seasonal distribution of precipitation effectiveness. His system is more complex than Köppen's, and the element that delimits the arid zone – precipitation effectiveness – is based on the sum of 12 monthly values determined from a formula in which temperature is the important variable. Thornthwaite says that by comparing P/E with actual precipitation one can come up with a moisture index that may be used to delimit climatic zones. The arid zones have values below −20.

Meigs

Despite the limitation of Thornthwaite's system, his moisture index forms the basis for the classification system of the arid lands most widely in use today. Meigs, under the auspices of UNESCO's Advisory Committee on Arid Zone Research, devised a system that delimited the arid zone and identified climatic differences within the zone. He identified arid climates (A) as those in which there is not enough precipitation for crop production. The semiarid realm (S) has sufficient precipitation for certain types of crops, and grass is the most abundant vegetation type. In addition, Meigs identified an extremely arid zone (E) on the basis of a dry period of 12 months or more in which no precipitation fell. (Thornthwaite's moisture index values of −20, −40 and −57 can be used to make specific determinations of the boundaries of these zones.)

Meigs followed Köppen in recognizing the importance of the season of precipitation and Thornthwaite in stressing heat as a factor essential for the growth and ripening of plants. Where water is available for irrigation in arid lands, temperature becomes a more important factor than moisture in the climate of a place. Thus, Meigs's maps of arid climate types include a number of subdivisions where the first digit represents the coldest month and the second digit the warmest month; 03 indicates a winter month with temperatures below 0°C and 3 indicates a summer month with average temperatures of 20–30°C.

Characteristics of the climates of arid lands

Besides aridity, the arid lands of the world have other unique climatic characteristics that should be noted. Along the cool desert coastal zones, persistent fogs occur that provide moisture to the plants growing there. In other parts of the desert, dew forms and may be captured for human use. Abundant energy in the form of sunshine and wind are available. Climatic factors may be a boon or a bane, but they can never be ignored in the desert milieu.

Moisture

Life in the arid lands is attuned to arid conditions. However, droughts affect these regions with regularity. A failure of seasonal rains results in scarcities of feed for livestock and dependable sources of moisture for people. People living near irrigated oases face a diminished flow of water in the streams and springs used for irrigation. In nearby mountains less water is available to generate electricity for homes and factories. Drought is perhaps the greatest environmental hazard to life in the arid zone, leading to death, starvation and migration from the area.

Floods also lead to death and destruction of property in arid lands. In those regions where rain is an infrequent visitor, unusually heavy downpours disrupt the regular pattern of life. In primitive agricultural societies, floods destroy irrigation systems and an entire season's crops are lost. In urban environments, flooding is responsible for loss of life and property and the disruption of transportation systems. In those arid lands reached by tropical cyclones, the areas of heavy precipitation associated with these storms is extensive. Such storms are of particular significance in North America and Australia. In other arid lands, unusually heavy precipitation generally is associated with isolated thunderstorms that produce flash floods in desert washes.

Most of the precipitation associated with the midlatitude storms of winter is light to moderate in intensity and becomes a hazard only when it continues for some time or occurs in the form of snow. Snowfall is not an exceptional event in the subtropical desert regions, but it does cause difficulties when it occurs. Livestock are unable to get food and water and may succumb to the low temperatures accompanying the snow because they are unable to move about

freely. This limitation of freedom of movement also applies to people. Streets and highways become clogged because snow-clearing equipment is unavailable (Arnon, 1992).

In desert areas where virtually no rain occurs, dew and fog drip represent the only forms of atmospheric moisture available. In the coastal deserts of Peru and Chile, moisture obtained from the fogs that move inland across the coast ranges sustain a moderately dense shrub forest. Because of the great amount of radiational cooling that occurs in the deserts, the formation of dew is a frequent occurrence. In the days of Roman occupation of North Africa, large rock piles above a cistern were used to collect dew for domestic water supply. Dew is an important contributor to the water budgets of many desert regions.

Perhaps the greatest amount of research into moisture sources in the arid regions has been directed at problems relating to the liability, frequency, duration and intensity of precipitation. Generalizations about these matters are difficult because of the different kinds of atmospheric conditions that prevail in different parts of the world. For all of the arid lands it is quite safe to say that precipitation occurs infrequently and its dependability is low. In those areas where winter storms are dominant, precipitation amounts are light to moderate, and storms generally endure for half a day or less. In those areas subject to surges of tropical air where thunderstorms occur, intensities near the centers of the storms are high, but duration is short. When tropical cyclones invade the arid lands, intensities are great and the storms persist for several days.

Moisture and the lack of it is the climatic factor of most concern to those living in the arid realm, but temperature conditions are also important.

Sunshine and temperature

One of the greatest resources of the arid lands is sunshine. The areas on Earth that receive the greatest amounts of solar energy are located in the centers of the arid zones. Because of the drying and stabilizing effects of the descending air masses of the Hadley cells, insolation penetrates freely through the atmosphere to the Earth's surface. This energy has long been used to evaporate sea water to make salt, and in solar stills to produce fresh water. Today, the energy is being used to generate electricity, to pump ground water for irrigation and to heat water for domestic and commercial uses.

The effect of the clear skies on temperature is also readily apparent. Temperature ranges of 15–22°C are common and objects in the shade are appreciably cooler than those in the sun. Some of the highest temperatures ever recorded have been measured in the arid lands. These high temperatures have been the result of warm dry winds flowing down the slopes of mountains and being heated adiabatically to temperatures higher than those normally recorded. The highest temperature (58°C) was recorded at El Azizia about 40 km south of Tripoli (Kendrew, 1961). However, temperatures in the tropical deserts generally are lower than this but are usually above 40°C each day during the summer months. Night-time temperatures are normally 16–18°C lower. In winter, freezing temperatures at or near the surface are not unusual at night but, as the sun rises higher in the sky, temperatures rise to comfortable levels. Arid lands in higher latitudes experience temperature conditions comparable to those of adjacent humid environments. Advection of warm and cold air into these areas is more frequent and radiational cooling and heating of less significance.

Wind and dust

Wind and dust are inseparable in the arid lands. Air movements of any consequence pick up particles from the desert floor and transport them to other parts of the region. Small eddies of wind (dust devils) carry spirals of dust aloft across the landscape, appearing and disappearing at short intervals during the warmer parts of the day. Desert thunderstorms appear as dark menacing masses, laden with dust in the cold roll or air that precedes the main part of the storm. Frequently, dust and sand are the only elements that fall from such storms. Larger masses of air that accompany frontal passages pick up large quantities of dust and transport it great distances. One such wind, the sirocco, carries dust from the Sahara across the Mediterranean Sea at heights up to 3 km, occasionally falling as red

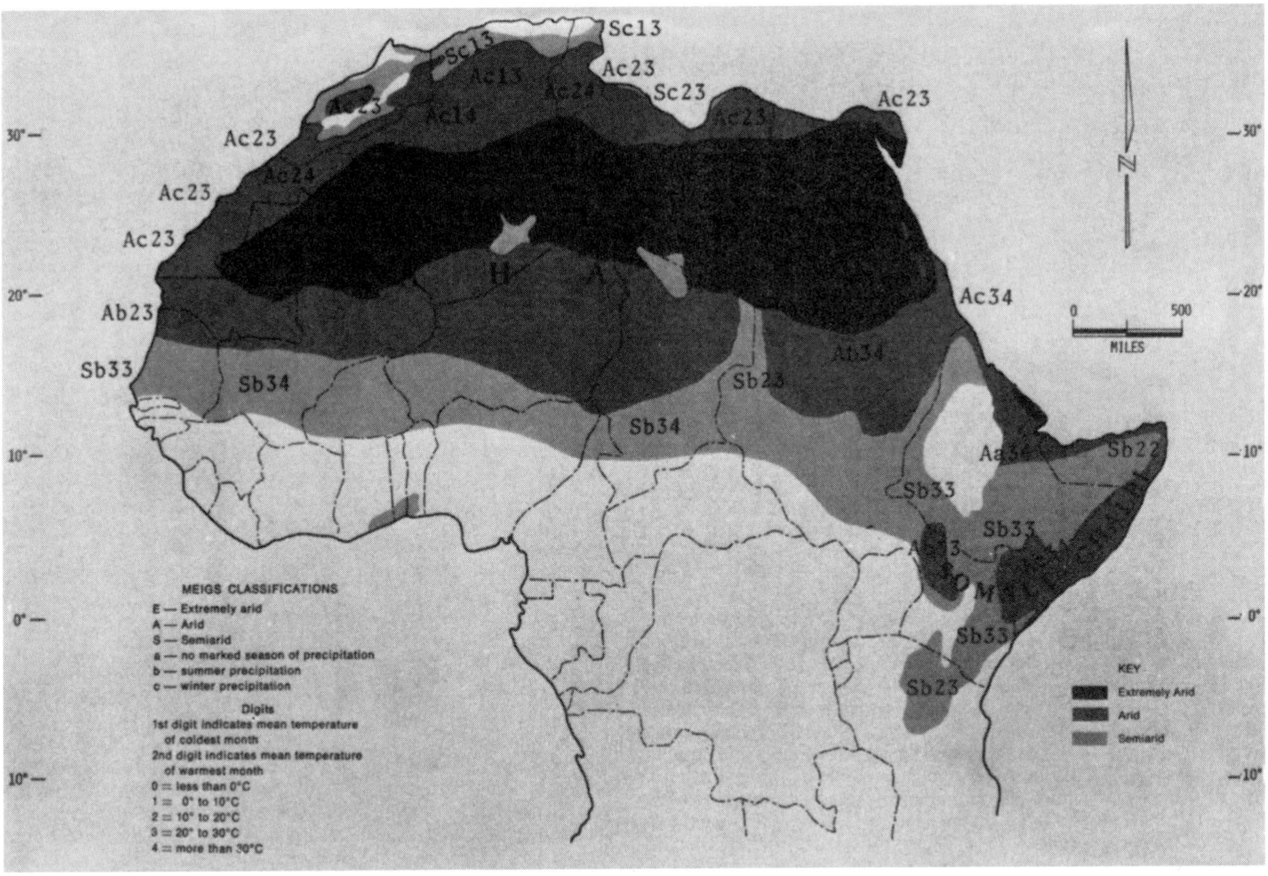

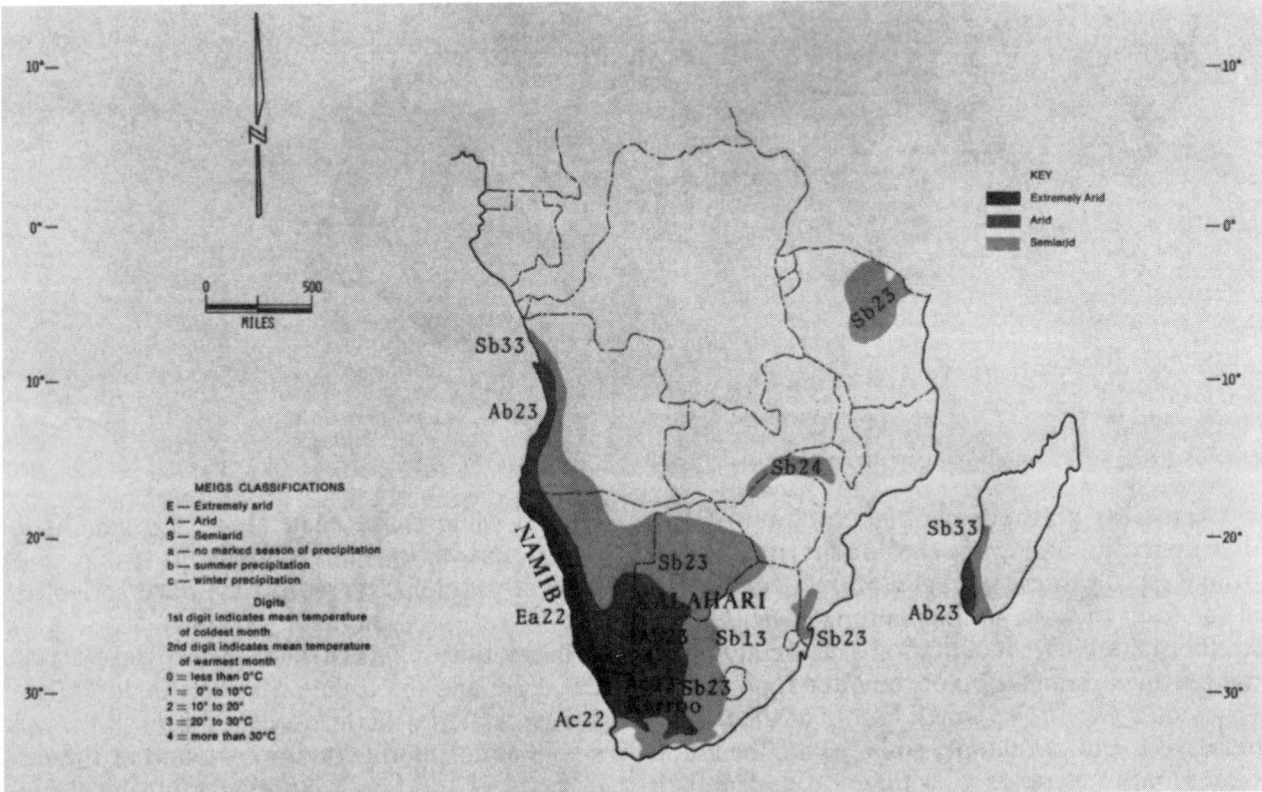

Figure A79 Arid lands of northern Africa. (From McGinnies *et al.*, 1968.)

Figure A80 Arid lands of southern Africa. (From McGinnies *et al.*, 1968.)

rain in southern Europe. The harmattan, another wind out of North Africa, occasionally carries yellowish dust from the Sahara across the Atlantic to the Caribbean (Kendrew, 1961). Dust is a nuisance and a hazard, entering the lungs to cause disease and affecting transport. On American highways severe accidents occur during dust storms; in primitive societies all travel ceases.

Arid zones of the world

Africa

The Sahara is a vast desert region that lies within the subtropical high-pressure belt (Figure A79). Across its northern margins, cyclonic

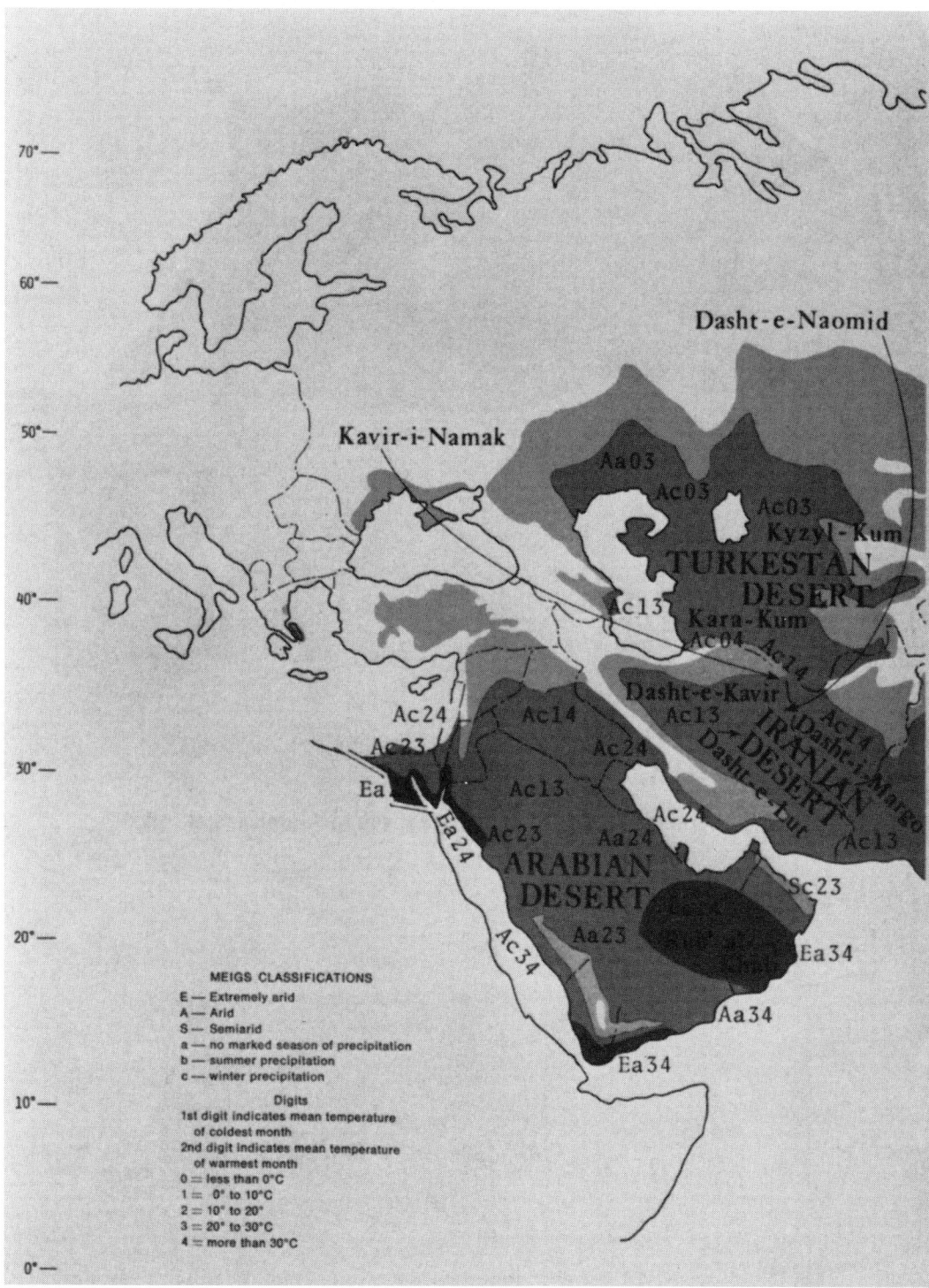

Figure A81 Arid lands of Asia: (a) western portion (From McGinnies *et al.*, 1968.)

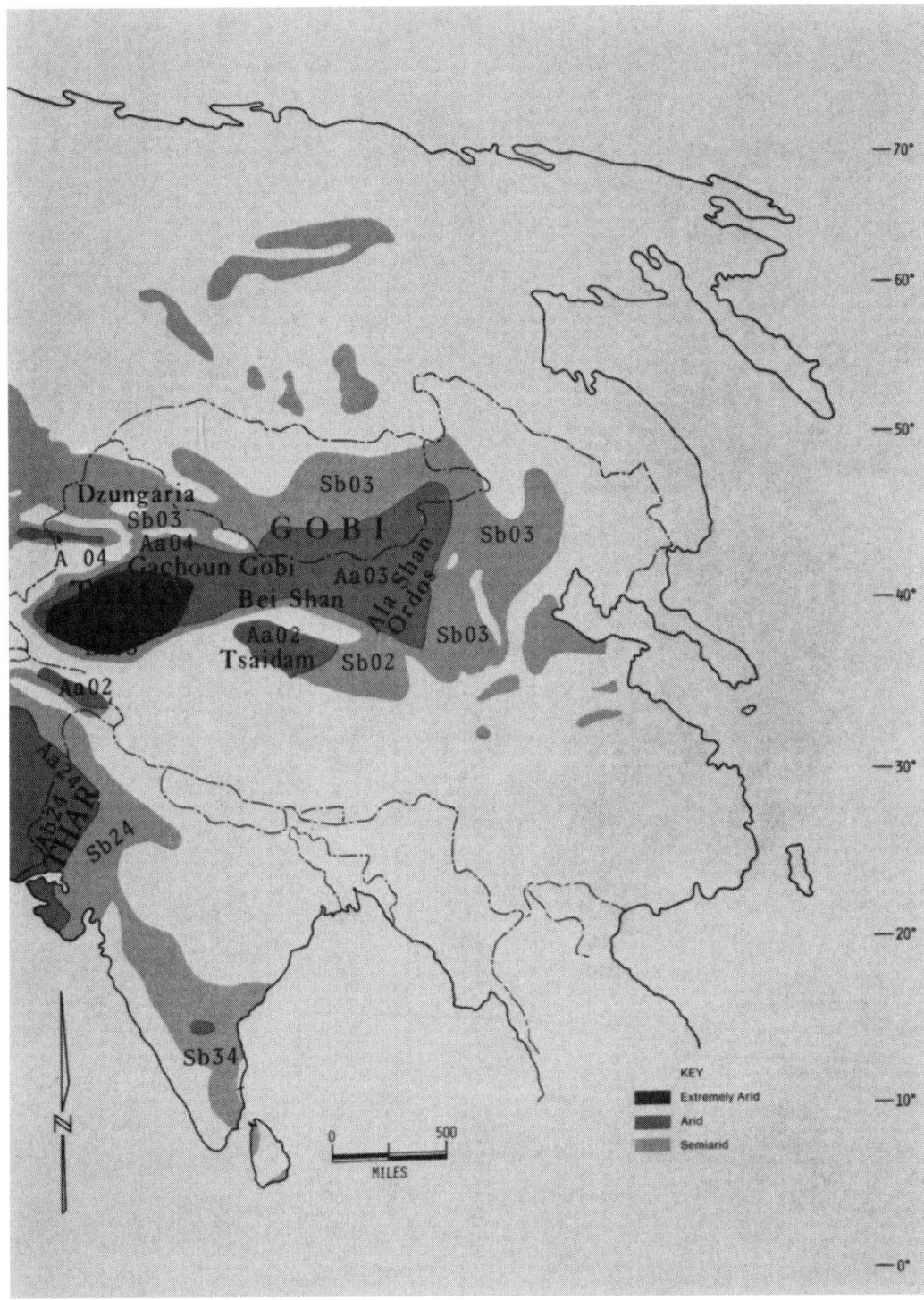

Figure A81 Arid lands of Asia: (b) eastern portion. (From McGinnies *et al.*, 1968.)

storms that penetrate and cross the Mediterranean Sea from west to east drop precipitation, occasionally in the form of snow in the winter season. On the southern margins, precipitation is associated with the northward migration of the intertropical convergence zone. In the Sahel, failure of these rains to arrive results in calamity and death to the unfortunate native peoples who have expanded their territories and their populations in the years when rainfall values have been above normal. In the desert, dust storms associated with strong winds

present problems for both sedentary agriculture and for migratory nomadic groups.

In southwest Africa the Namib Desert along the coast and the Kalahari Desert inland (Figure A80) are associated with the descending air masses and stable atmospheric conditions found at the eastern end of the South Atlantic subtropical high-pressure area. Portions of the Namib are extremely arid and devoid of vegetation. Stability of the air mass in the region is enhanced by cold upwelling water found

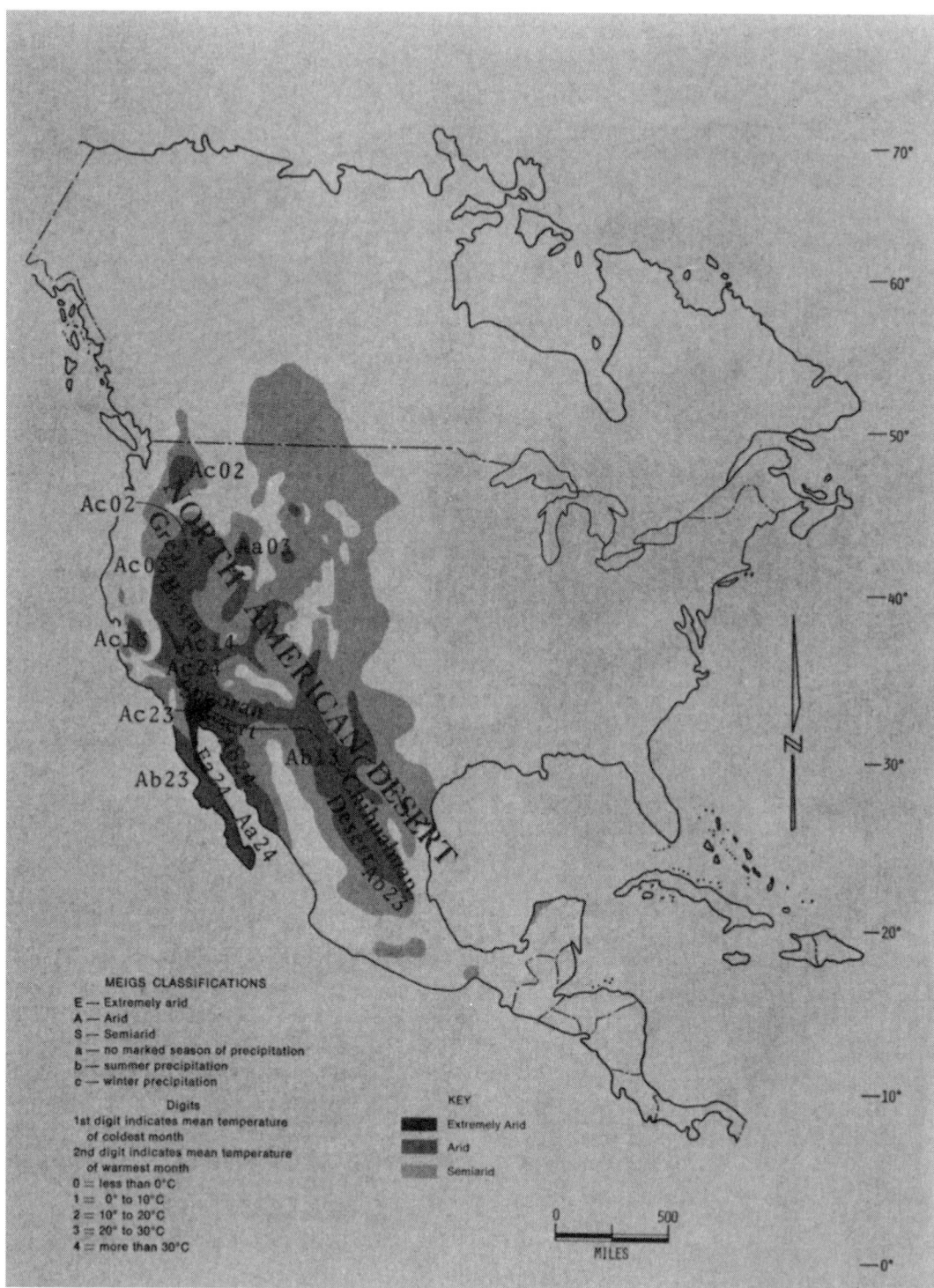

Figure A82 Arid lands of North America. (From McGinnies *et al.*, 1968.)

on the coastal side of the Benguela current. The interior desert, the Kalahari, receives more precipitation in the form of thunderstorms and is covered by a scrub forest.

Asia

The Arabian, Iranian and Thar deserts (Figure A81) fall under the influence of the subtropical high-pressure area of the northern hemisphere but are also located far from the principal source of moisture in the storms that cross the area. In the winter season, cyclonic storms that originate over the Atlantic Ocean or Mediterranean Sea pass through the area. However, only small amounts of moisture are left by the time the storms reach these interior locations. Additionally, mountains and high pressure tend to block their movements and divert the paths of the storms to the north.

The Turkestan, Takla-Makan and Gobi desert regions lie at the interior of the Eurasian continent, remote from sources of oceanic

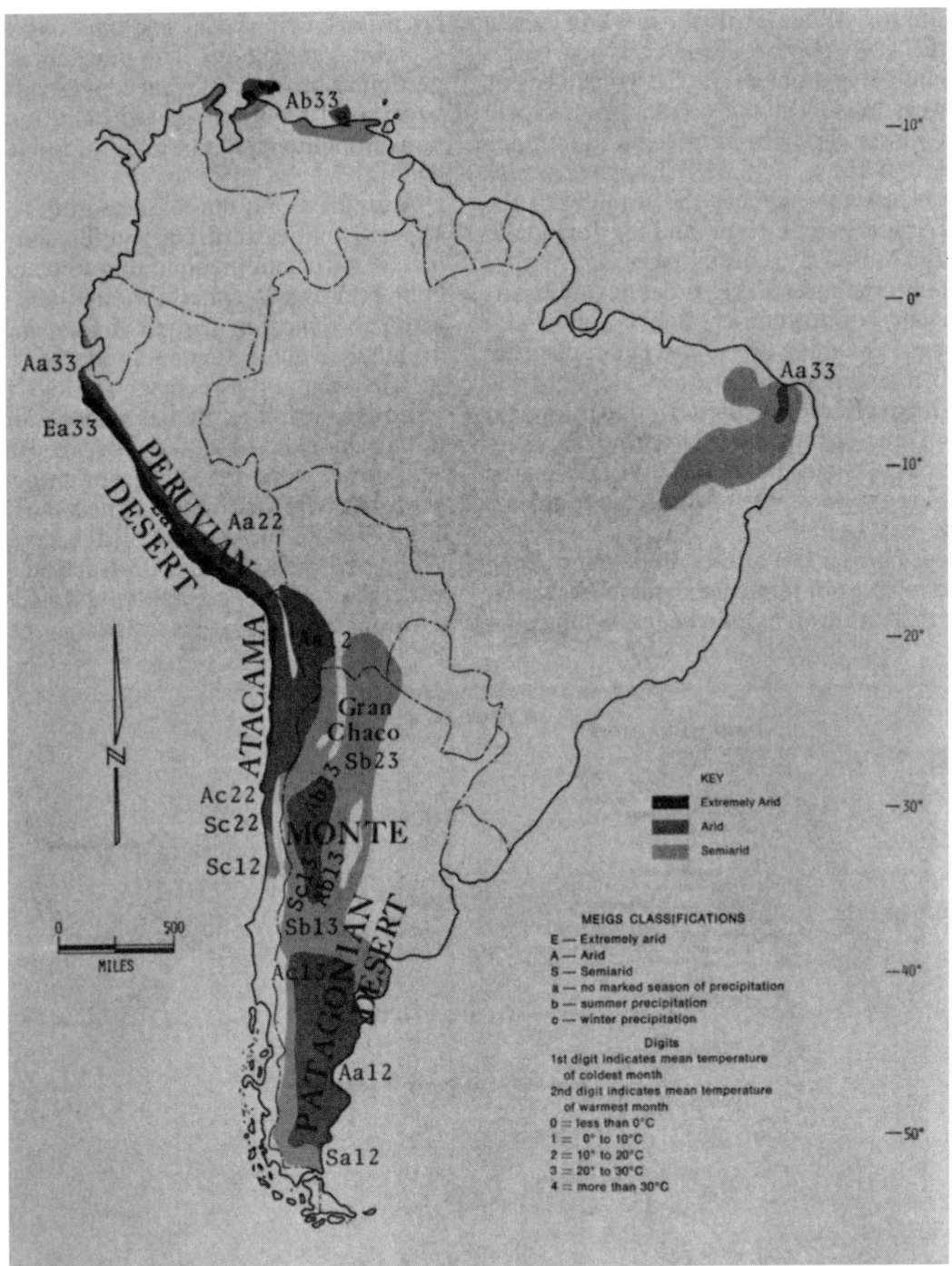

Figure A83 Arid lands of South America. (From McGinnies *et al.*, 1968.)

moisture and shielded from tropical air masses by gigantic mountain systems. In winter the Asiatic High blocks the movement of cyclonic storms across the region. In the summer the moist stream of air associated with the summer monsoon is diverted around the southeastern corner of Asia by the mountain masses of Pakistan, India, China and Malaysia.

North America

The arid zone of North America (Figure A82) comes under the influence of the belt of subtropical pressure, portions of it lie behind mountain barriers and much of it is remote from sources of moisture.

The driest portion, the Sonoran Desert, is dominated by the Pacific High for most of the year. In winter, occasional cyclonic storms penetrate the region, bringing small amounts of precipitation, occasionally in the form of snow. However, summer is the season of maximum precipitation. Afternoon thunderstorms associated with surges of tropical air may drop copious amounts of moisture on limited areas causing temporary flooding. By far the heaviest precipitation and the most serious flooding comes with tropical storms

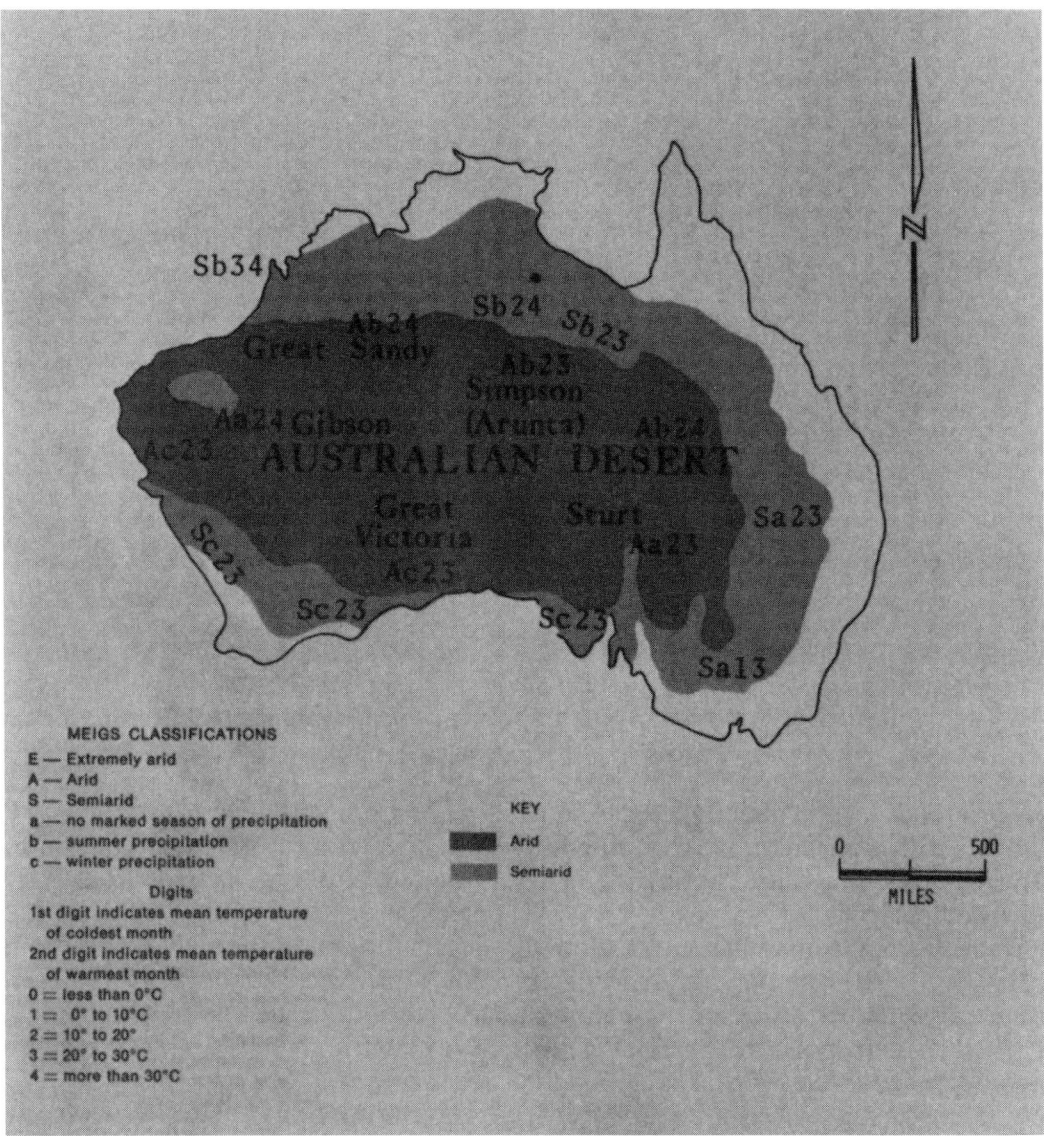

Figure A84 Arid lands of Australia. (From McGinnies *et al.*, 1968.)

that migrate into the area from the southeast Pacific Ocean in late summer and early fall.

In the Great Basin, maximum amounts of precipitation are associated with the cyclonic storms of winter which cross the area. However, these storms have lost most of their moisture in crossing the Sierra–Cascade barrier, and rainfall is generally light. Snow, falling in the surrounding mountains, represents a major source of water used in the region. In the summer season, thunderstorms forming in tongues of moist tropical air occasionally produce heavy rain and runoff.

The Chihuahua Desert of the Rio Grande Valley and north-central Mexico lies in an area that is protected from air masses from the Pacific and Atlantic oceans by mountain ranges. Aloft, stable descending air of the Hadley cell limits the formation of convective storms. Only in summer, when tropical hurricanes and easterly waves from the Caribbean and Gulf of Mexico penetrate the area, do heavy rains occur.

South America

The Peruvian and Atacama deserts (Figure A83) of the west coast of South America are among the driest areas on Earth. On the north, the intertropical convergence zone only occasionally penetrates more than several degrees south of the equator. When it does, a warm ocean current – El Niño – appears offshore and heavy rains cause

innumerable problems for the irrigated oases of northwest Peru. On the south, the cyclonic storms of winter rarely bring precipitation much farther north than 32°S. Offshore, the cold Humboldt current enhances the stability of the air in the South Pacific High, and fog frequently blankets the hills of the coastal zone.

On the eastern flanks of the Andes lie the deserts and grasslands of Argentina, Bolivia and Paraguay. Storms, sweeping out of the Pacific Ocean, drop their moisture on the western slopes of the Andes. The desiccated air flowing downslope on the eastern side of the Andes is warmed by compression and contributes to the aridity of the region by absorbing the available moisture as it crosses the plains.

Two unusual zones of aridity are located along the northern coast of Venezuela and in northeast Brazil. Both of these areas have been the subject of intensive investigations and are considered to be somewhat anomalous. The Brazilian arid zone, in particular, has been of great concern to that nation because of recurrent drought and forced migrations from the region caused by the lack of food and water (Hinman and Hinman, 1992).

Australia

Only the coastal portions of northern, eastern, southeastern and southwestern Australia receive sufficient precipitation to be classified as humid (Figure A84). Over half the continent is a desert, with subtropical and tropical grasslands occupying much of the remainder. In the southern hemisphere summer the intertropical front dips southward into the continent, occasionally penetrating far into the desert interior. Runoff from heavy rains flood desert washes and creates vast playa lakes. Tropical cyclones affect the Queensland coast, and in the northwest, the dreaded willy-willies cause extensive destruction to coastal settlements. Winter precipitation is associated with the troughs of low pressure that lie between the migratory anticyclones that encircle the globe north of Antarctica. Cold fronts force moist marine air to rise over the low hills and mountains of the coastal zones. The easterly flow of air on the back sides of the migratory anticyclones is forced to rise along the slopes of the Great Dividing Range to produce precipitation on the eastern coast of New South Wales and Queensland.

Within the Great Australian Desert, differences in the appearance of the landscape resulting from variation in environmental factors are reflected in local names for that portion of the desert. The Great Sand, Simpson, Gibson, Great Victoria and Sturt deserts reflect the variety of landscapes to be observed within the continent.

On the margins of the desert, livestock grazing predominates, but in the better soils and more humid locations the raising of wheat has been important. Where water is available, irrigation agriculture is practiced.

Change of climate

Questions about changes in climate have particular significance for those living in the arid lands of the world. Here, small changes in climate produce large swings in the capability of humans to survive. The question of the relative significance of nature and of humans in producing changes in the environment is nowhere more debated (see *Desertification*). Whether large settlements in the desert can long endure is an open question. The difficulties of survival in a fragile environment where water supply and water use are evenly balanced are well known to all. However, programs for coping with disasters in the arid realm are not well developed.

R.W. Durrenberger

Bibliography

Amiran, D. and A. Wilson (eds), 1973. *Coastal Deserts, Their Natural and Human Environments*. Tucson: University of Arizona Press.
Arnon, I., 1992. *Agriculture in Dry Lands: Principles and Practices*, Amsterdam: Elsevier, 979 pp.
Bailey, H.P., 1981. Climatic features of deserts, in *Water in Desert Ecosystems*, D.D. Evans and J.L. Thames, (eds). Stroudsburg, PA: Dowden, Hutchinson & Ross, pp. 13–41.
Borisov, A.A., 1965. *Climates of the USSR*. London: Oliver and Boyd.
Fogel, M.M., 1981. Precipitation in the desert, in *Water in Desert Ecosystems*, D.D. Evans and J.L. Thames, (eds). Stroudsburg, PA: Dowden, Hutchinson & Ross, pp. 219–234.
Hinman, C.W. and Hinman, J.W., 1992. *The Plight and Promise of Arid Land Agriculture*, Cambridge University Press.
Kendrew, W.G., 1961. *Climates of the Continents*. Oxford: Clarendon Press.
Köppen, W., 1931. *Grundriss der Klimakunde*, 2nd edn. Berlin: Walter de Gruyter.
McGinnies, W.G., B.J. Goldman and P. Paylore (eds), 1968. *Deserts of the World*. Tucson: University of Arizona Press.
Meigs, P., 1953. World distribution of arid and semi-arid homoclimates, in *Reviews of Research on Arid Zone Hydrology*. Paris: UNESCO, Arid Zone Research, pp. 203–210.
Thompson, R.D., 1975. *The Climatology of the Arid World*, Geog. Paper No. 35. Reading, England: University of Reading.
Thornthwaite, C.W., 1931. The climates of North America according to a new classification, *Geog. Rev.*, **21**, 633–655.
Thornthwaite, C.W., 1948. An approach toward a rational classification of climate, *Geog. Rev.*, **38**, 55–94.
Thornthwaite, C.W., and F.K. Hare, 1965. The loss of water to the air, *Meteorol. Monogr.*, **6**(28), 163–180.

Cross references

ARIDITY INDICES

Aridity indices are quantitative indicators of the degree of water deficiency present at a given location. A variety of aridity indices have been formulated, although the term aridity index specifically refers to the 1948 work of Thornthwaite. Aridity indexes have been applied at continental and subcontinental levels and are most commonly related to distributions of natural vegetation and crops. Critical values of the indices have been derived from observed vegetation boundaries. For instance, Köppen's 1918 classification defines the desert–steppe boundary as the 200 mm annual isohyet in regions where there is no seasonality of rainfall and the mean annual temperature is 5–10°C.

The formulation of aridity indices is not straightforward due to the nature of aridity. First, aridity is a function of the interplay between rainfall, temperature and evaporation. Use of mean annual rainfall as an index of aridity ignores the importance of temperature and evaporation. Aridity indexes that have gained widespread acceptance directly or indirectly take into account all three factors. Second, the arid regions generally have been recognized as having a paucity of climatological data. Given the temporal variability of precipitation inherent in arid regions, the lack of climatological data has been detrimental in attempts to quantitatively define the boundaries of aridity. Third, aridity indices must be considered from the standpoint of their eventual use. For example, the 1968 US Army World Desert Classification defines aridity with respect to military operations; application to world vegetation patterns would be inappropriate. A particular aridity index may serve several purposes, but no one index is appropriate for all uses. However, aridity indices are often mathematically related and to some extent have been used interchangeably on a global scale.

Identification of the arid zones of the Earth has roots that can be traced two millenia. Classical Greek thought identified the latitudinally controlled torrid, temperate and frigid zones of the world. Implicit in their thought was the concept that the torrid, low-latitude climates were arid. Not until long-term instrumental records and reliable world vegetation maps became available could true aridity indices be developed. Thus aridity indices are a product of the twentieth century. Table A22 outlines the major developments regarding aridity indices. For additional information, see Dzerdzeevskii (1958), Hare (1977) and International Crops Research Institute for the Semi-Arid Tropics (1980).

In 1900 Köppen originally qualitatively classified as arid those places that had desert vegetation. V.V. Dokutchaev in 1900 and A. Penck in 1910 qualitatively defined arid regions as places where annual evaporation exceeds precipitation. In 1905 both E.N. Transeau and G.N. Vyssotsky quantified this relationship. Yet this approach was not totally satisfactory because of the lack of reliable, worldwide evaporation measurements. Köppen's influential series of climatic classifications used mean annual temperature and precipitation combinations to define arid climates (1919, 1936). In a similar vein, W. Lang's 1920 Rain Factor Index was a ratio between mean annual precipitation and mean annual temperature. Lang's index, and a modified version devised by E. de Martonne in 1925, were widely used because their data requirements were minimal. However, their approach was limited in that the seasonality of temperature and precipitation were not addressed.

A. Meyer's 1926 Precipitation–Saturation Deficit Ratio was an attempt to obviate the need for dependable evaporation data. Meyer assumed the evaporation rate to be a function of the saturation deficit (saturation vapor pressure minus actual vapor pressure at a particular temperature). The Precipitation–Saturation Deficit Ratio was calculated from long-term temperature precipitation and relative humidity data and was found to be more reliable than temperature/precipitation-based indexes. Data availability limited the application of Meyer's ratio in that relative humidity data generally were not as available as were temperature and precipitation records.

Thornthwaite's work had an immense influence on the quantitative calculation of aridity. His Precipitation Effectiveness Index of 1931 is computed as ten times the sum of the monthly precipitation to evaporation ratio at a given location. Of practical importance was his

Table A22 Selected summary of aridity indexes

Year	Author	Remarks	Formula
1900	W. Köppen	*Xerophytic* (arid and semiarid) climates qualitatively defined through presence of vegetative types. No formula used	
1900	V.V. Dokutchaev	Defined aridity through comparison of annual precipitation with annual evaporation from a water surface. No formula used	
1905	E.N. Transeau	Used ratio of annual precipitation to evaporation to describe aridity. Along with Vyssotsky, the first quantitative aridity index	$\dfrac{P}{E}$
1905	G.N. Vyssotsky	Used ratio of annual precipitation to evaporation to describe aridity. Along with Transeau, the first quantitative aridity index	$\dfrac{P}{E}$
1910	A. Penck	Defined aridity through comparison of annual precipitation with annual evaporation from a water surface. An attempt to relate climate to landforms. No formula used	
1911	E.M. Oldekop	Precipitation compared with potential evaporation. E computed by multiplying the saturation deficit of the air by a coefficient of proportionality	$\dfrac{P}{E}$
1918	W. Köppen	Arbitrary climatic boundaries based on presumed vegetation boundaries. For example, desert and steppe were partitioned by 200 mm annual isohyet in areas where the mean annual temperature was 5–10°C; they were separated by the 320 mm isohyet where the mean annual temperature was 25°C. No formula used	
1920	W. Lang	Rain Factor. Mean annual precipitation (mm) and mean annual temperature (°C) compared	$\dfrac{P}{T}$
1922	W. Köppen	Precipitation compared formula at right. Several revisions of Köppen's scheme were formulated by the author himself and by others	$2(T + 7)$
1926	E. de Martonne	Index of Aridity. A modification of Lang's Rain Factor Index	$\dfrac{P}{T + 10}$
1926	A. Meyer	Absolute saturation deficit (mm of mercury) replaces evaporation	$\dfrac{P}{D}$
1928	E. Reichel	Inserts the number of days with precipitation (N) in the formula of deMartonne	$\dfrac{NP}{T + 10}$
1931	C.W. Thornthwaite	Precipitation Effectiveness. Monthly precipitation to evaporation ratios determined, summed, and multiplied by 10 to eliminate fraction (where n is an individual month, and T is the mean monthly temperature). For stations where evaporation data were not available, a formula using only precipitation and temperature data was provided. (Note: Formulae use English units)	$\left\{ \sum_{n=1}^{12} \dfrac{P_n}{E_n} \right\} 10$ $\left\{ \sum_{n=1}^{12} 115 \dfrac{P_n}{T_n} - 10 \right\}^{\frac{9}{10}}$
1932	V.B. Shostakovitch	t is the mean temperature during the growing period	$\dfrac{P}{t10}$
1933	L. Emberger	An attempt to incorporate the effect of the seasonality of temperature on aridity. M is the mean maximum temperature of the warmest month and m is the minimum temperature of the coldest month	$\dfrac{100P}{(M + m)(M - m)}$
1934	W. Gorozynski	Aridity Coefficient. C is the cosecant of latitude, T_r is the difference between the means of the hottest and coldest months, and P_r is the difference between the greatest and least annual precipitation totals over 50 years. The coefficient increases with increasing aridity with its maximum value near 100. (Note: formula uses English units)	$C\,T_r\,P_r$
1937	G.T. Selianinov	Effectiveness of precipitation in the growing season. Only mean month temperatures above 10°C are summed	$\dfrac{P10}{\sum_{n=1}^{12} T_n}$
1941	N.N. Ivanova	Calculation of a precipitation 'potential' evaporation ratio using the formula at right where t is the mean monthly temperature and a is the mean monthly relative humidity	$\dfrac{P}{E}$ $E = 0.0018(25 + t)^2$ $(100 - a)$

Table A22 Continued

Year	Author	Remarks	Formula
1942	E. de Martonne	Modification of earlier work incorporating a representation of the temperature (T_d) and precipitation of the driest month (P_d)	$\dfrac{\dfrac{P}{T+10}+\dfrac{12P}{T_d+10}}{2}$
1947	N.V. Bova	Inclusion of soil moisture conditions in a precipitation/temperature ratio. H is the initial moisture content of the soil	$\dfrac{H+P}{\sum\limits_{n=1}^{12} T_n}$
1948	C.W. Thornthwaite	Represents a water balance approach to aridity where I_h is the Humidity Index, s is the surplus moisture in the humid season, n is the water deficiency in the dry season, I_a is the Aridity Index, and I_m is the Moisture Index. Later modifications were made to this work	$I_h = 100s/n$ $I_a = 100d/n$ $I_m = I_h - 0.6\,I_a$ $I_m = \dfrac{100s - 60d}{n}$
1948	V.P. Popov	Index of Aridity. Σg is the annual effective precipitation, $t - t'$ is the mean annual wet-bulb depression and r is a factor based on day length	$\dfrac{\Sigma g}{2.4(t - t')r}$
1949	J.A. Prescott	Refinement of earlier formulae using precipitation and saturation deficit. (Note: this method uses English units)	$\dfrac{P}{0.7D}$
1950	A.A. Skvortsov	E_a, actual evaporation, compared to E_{st}, 'standard' evaporation measured from a water surface	$\dfrac{E_a}{E_{st}}$
1951	R. Capot-Rey	P and T refer to the mean annual precipitation and evaporation while p and t refer to the precipitation and evaporation of the wettest month	$\dfrac{100\dfrac{P}{e}+12\dfrac{P}{e}}{2}$
1951	M.I. Budyko	Radiational Index of Dryness. R is the mean annual net radiation and L is the latent heat of vaporization for water. This is the first index using a radiation balance approach	$\dfrac{R}{LP}$
1952	S.J. Kostin	Precipitation versus potential evapotranspiration for the same period	$\dfrac{P}{PE}$
1953	P. Meigs	Use of Thornthwaite's Moisture Index to classify and map the dry lands of the earth. No formula used	
1955	H. Gaussen	Classification based on the duration and severity of dry months. A dry month is defined by the conditions at right. Other factors considered by Gaussen's definition of aridity include number of rain days, humidity, mist and dew	$P \leq 2T$
1957	F.R. Bharucha and G.Y. Shanbhag	A reuse of the P/E index using the formula at right. E is the mean 24 h evaporation in inches, B is the mean wind velocity in miles per hour, h is the mean relative humidity in percent and e is the mean vapor pressure in inches of mercury	$E = (1.465 - 0.0186B)$ $(0.44 + 0.11BW)\dfrac{100}{h} - 1\ e$
1960	P. Meigs	Revision of 1953 maps. This work has become the most widely used identification of the world's arid regions	
1962	V.M. Meher-Homji	Index of Aridity–Humidity. S is the 'precipitation quantity factor' and X is the length of the day period	$S + X$
1965	C. Troll	Defined arid climates on the basis of number of months that the expression at right holds true	$P > PE$
1967	C.C. Wallen	Interannual variability (V_I), where n is a particular year in a series of N years. The second equation is an empirical one that describes the arid margin of dryland farming	$V_I = \dfrac{100\Sigma(P_n - 1 - P_n)}{P(N - 1)}$
1967	J. Cocheme and P. Franquin	Matches water availability to a crop's growth cycle through a comparison of the values of precipitation and evapotranspiration (ET). Can be used for any growing period	P vs. ET $V_I = 0.07\,\bar{P} + 22$
1968	US Army	World Desert Classification. Based on rainy days per month with a rainy day defined as any day with greater than 0.1 in of precipitation. Months are categorized in 4 categories by the categorization of the cumulative number of wet months in a year. Used with respect to personnel and military equipment	

Table A22 Continued

Year	Author	Remarks	Formula
1969	H. Lettau	Approaches aridity from the standpoint of surface energy and moisture fluxes. B is the Bowen Ratio (cf. Bowen Ratio) and C is the annual water surplus divided by the precipitation. The formula is equivalent to Budyko's Radiational Index of Dryness	$(1 + B)(1 - C)$
1970	W.K. Sly	The ratio of growing season precipitation to total amount of water required by the crop if lack of water is not to limit production. P is the growing season precipitation, SM is the start of the growing season and IR is the calculated irrigation requirement during the growing season	$\dfrac{P}{P + SM + IR}$
1971	G.H. Hargreaves	Moisture Available Index (MAI). For a specified period, the ratio of the monthly rainfall total expected with a 75% probability to the estimated potential evapotranspiration. Values of 1.00 to 0.00 were considered increasingly moisture deficient	$\dfrac{P_\mathrm{p}}{PE}$
1979	UNESCO	Map of the World Distribution of Arid Regions. Based on the ratio of precipitation to evapotranspiration with evapotranspiration being determined by Penman's method. This work was intended to replace Meig's 1960 work	$\dfrac{P}{ET}$
1980	R.P. Sarker and B.C. Biwas	Modification of MAI to consider weekly periods, various levels of rainfall total probabilities so that P_A is the assured rainfall of a period and PE is the potential evapotranspiration for the same period	$\dfrac{P_\mathrm{A}}{PE}$

Dates given are first appearance in the literature. All formulae are in metric units unless otherwise noted. Symbols have been modified from original sources for purposes of comparison.

accompanying empirical formula for deriving the Precipitation Effectiveness Index for stations recording only mean monthly temperature and precipitation. In 1948 and in subsequent revisions of his climatic classification, Thornthwaite employed the Aridity Index, which relates annual moisture deficit to annual potential evapotranspiration (see Water budget analysis). Weighted by 0.6 and subtracted from Thornthwaite's Humidity Index, the Aridity Index is a component of Thornthwaite's Index of Moisture. On the basis of the Index of Moisture, Thornthwaite categorized the world into nine moisture zones ranging from arid to perhumid. Evapotranspiration prominently figured into Thornthwaite's indices, yet it was measured at only a handful of sites worldwide. So, Thornthwaite devised a formula to estimate evapotranspiration through the use of a station's latitude and temperature. His concepts have gained wide use because of the simplicity of their data requirements and their general agreement with world vegetation patterns. However, some engineers and agriculturalists have criticized his methods as too general for use in specific applications. The formulae have been found to produce unreliable results in certain tropical locales.

Budyko (1951) offered a new approach by considering the heat and water balance equations of the Earth's surface. His Radiational Index of Dryness was the ratio of the mean annual net radiation (i.e. the radiation balance) to the product of the mean annual precipitation multiplied by the latent heat of vaporization for water. The warm dry conditions synonymous with arid regions are well characterized by Budyko's index. In practical terms, the Radiational Index of Dryness is the number of times the net radiative energy income at the surface can evaporate the mean annual precipitation. Although a number of writers have preferred Budyko's method of calculating aridity, a major limitation in application is the lack of long-term radiation records at many observation stations.

Other indices have tended to be refinements and hybridizations of the above notions. Of recent interest has been the use of aridity indices to define the agricultural boundary between arid and semiarid climates. UNESCO, FAO and WMO are in accord that the boundary should be drawn where lack of water makes dryland farming impossible. Thus aridity indices are gaining increased importance in the planning of water supplies for crops.

Meigs's maps (1953, 1960) have been the most widely cited classification of aridity. Meigs used Thornthwaite's Moisture Index to define aridity. The 1:2 500 000 *Map of the World Distribution of Arid Regions* (UNESCO, 1979) has been produced to refine Meigs's maps. Although the 1979 map continues use of the ratio of precipitation to evapotranspiration, evapotranspiration is calculated by the more-favored Penman method. By this definition approximately one-third of the world's continental surface can be classified as having some degree of aridity.

Drought indices

Percent of normal

The percent of normal is one of the simplest measurements of rainfall. It is calculated by dividing actual rainfall by long-term average rainfall, usually a 30-year mean, and multiplying by 100 to yield a percentage which can be calculated for a variety of time scales ranging from a month to a group of months or season or year. Analyses using the criterion of percent of normal are effective when used for a season, and the percent of normal precipitation for June–August 1993 highlighted the record flooding during that summer in the Midwest United States (National Drought Mitigation Center (NDMC), 1996).

Deciles

Another drought monitoring technique is the arrangement of monthly precipitation data into deciles, developed by Gibbs and Maher (1967). The technique divides the distribution of occurrences over a longterm record into sections for each 10% of the distribution, called **deciles**. The first decile is the rainfall amount not exceeded by the lowest 10% of the precipitation occurrences. The second decile is the precipitation amount not exceeded by the lowest 20% of occurrences. These deciles continue until the rainfall amount identified by the tenth decile is the largest precipitation amount within the longterm record. By definition, the fifth decile is the median, and it is the precipitation amount not exceeded by 50% of the occurrences over the period of record. The deciles are grouped into five classifications, as shown in Table A23.

Table A23 Decile classification for dry and wet periods

Decile	Percentage range	Category
1–2	Lowest 20%	Much below normal
3–4	Next lowest 20%	Below normal
5–6	Middle 20%	Near normal
7–8	Next highest 20%	Above normal
9–10	Highest 20%	Much above normal

Table A24 Classification for dry and wet periods (PDSI)

PDSI value	Drought category
4.00 or more	Extremely wet
3.00–3.99	Very wet
2.00–2.99	Moderately wet
1.00–1.99	Slightly wet
0.50–0.99	Incipient wet spell
0.49–−0.49	Near normal
−0.50–−0.99	Incipient dry spell
−1.00–−1.99	Mild drought
−2.00–−2.99	Moderate drought
−3.00–−3.99	Severe drought
−4.00 or less	Extreme drought

Table A25 SPI values

SPI value	Drought category	Time in category
0 to −0.99	Mild drought	24%
−1.00–−1.49	Moderate drought	9.2%
−1.50–−1.99	Severe drought	4.4%
−2.00 or less	Extreme drought	2.3%

Palmer Drought Severity Index (PDSI)

Palmer (1965) developed an index to measure the departure of the moisture supply in the water balance equation, taking into account more than just the precipitation deficit at specific locations. The object of the PDSI was to provide a measurement of moisture conditions that were standardized so that comparisons using the index could be made between locations and between months. The PDSI is a meteorological drought index and responds to weather conditions that have been abnormally dry or abnormally wet (NDMC, 1996). When the conditions change from dry to normal or wet, for example, the drought as measured by PDSI ends without taking into account streamflow, lake and reservoir levels and other long-term hydrological impacts (Karl and Knight, 1985). The calculation of PDSI is based on precipitation and temperature data, as well as the local available water content (AWC) of the soil. The inputs provide all the basic terms required to determine the water balance equation, including evapotranspiration, soil moisture loss from the surface layer, soil recharge and runoff.

The PDSI varies between about −6.0 and +6.0 and the index is calculated on a monthly basis in the United States. A long-term archive exists for every climate division with the National Climate Data Center since 1895. In addition, weekly Palmer Index values are calculated for the climate divisions during every growing season and published. Weekly Palmer Index maps are also available on the World Wide Web (WWW). Table A24 shows the PDSI classification for dry and wet periods.

Surface Water Supply Index (SWSI)

The SWSI was developed by Shafer and Dezman (1982) to complement the Palmer Index for moisture conditions across the state of Colorado and designed to be an indicator of surface water conditions including snow accumulation, which the Palmer Index did not. The objective of the SWSI was to incorporate both hydrological and climatological features into a single index resembling the Palmer Index for each major river basin in the state of Colorado (NDMC, 1996). Four inputs are required: snowpack, streamflow, precipitation and reservoir storage (winter), and during the summer, streamflow replaces snow in the equation. The procedure involves the collection of monthly data of precipitation stations, reservoirs and snowpack or streamflow over the catchment, each component summed and normalized using frequency analyses determined from a long-term data set. The probability of non-exceedance is determined for each component based on the frequency analyses (NDMC, 1996). This procedure permits comparisons of the probabilities to be made between the components, each component having a weight assigned to it depending on its typical contribution to the surface water within the catchment. These weighted components are added together to determine a SWSI value representing the whole catchment. Like the Palmer Index, the SWSI is centered around zero and has a range between −4.2 and +4.2.

Standard Precipitation Index (SPI)

McKee et al. (1993) developed the SPI to quantify the precipitation deficit for multiple time scales which reflect the impact of drought on the availability of the different water resources. Soil moisture conditions respond to precipitation anomalies on a relatively short scale, while groundwater, streamflow and reservoir storage reflect the longer-term precipitation anomalies and for these reasons the authors originally calculated the SPI for 3, 6, 12, 24 and 48 month time scales.

The SPI is calculated by taking the difference of the precipitation from the mean for a particular time scale, and dividing by the standard deviation, but because the precipitation is not normally distributed for time scales shorter than 12 months, an adjustment is made which allows the SPI to become normally distributed. Thus the mean SPI for a time scale and a location is zero and the standard deviation is one. This is an advantage because the SPI is normalized so that the wetter and drier climates can be represented in the same way (NDMC, 1996). Wet periods can also be monitored using the SPI.

McKee et al. (1993) used the classification system shown in Table A25 to define drought intensities resulting from the SPI. A drought event occurs any time the SPI is continually negative and reaches an intensity when the SPI is −1.0 or less. The event ends when the SPI becomes positive.

Crop Moisture Index (CMI)

The CMI uses a meteorological approach to monitor week to week crop conditions. Developed by Palmer (1968) from calculations of his PDSI and designed to evaluate short-term moisture conditions across major crop producing regions, it is based on the mean temperature and total precipitation for each week within a United States climate division. However, because it is designed to monitor short-term moisture conditions impacting a developing crop, the CMI is not a good long-term drought monitoring tool (NDMC, 1996).

National Rainfall Index (RI)

The RI was developed by Gommes and Petrassi (1994) to characterize recent precipitation patterns across Africa. It is calculated for each country by taking a national annual precipitation average weighted according to the long-term precipitation averages of all the individual stations, and the RI therefore permits comparisons to be made between years and between countries.

Dependable Rain (DR)

Le Houerou et al. (1993) applied yet another monitoring approach to the African continent concerned with the concept of dependable rains, which they define as the amount of rainfall that occurs in four of every five years (statistically not consecutively). In Africa the relation of the DR to the mean is complex and reflects the characteristics of annual precipitation across the continent. Near the Sahara the DR is about 40–50% of the annual mean, while in the 700–800 mm rainfall zone the DR is about 80% of the mean (NDMC, 1996).

Stephen J. Stadler

Bibliography

Budyko, M.I., 1951. O. Klimaticheskikh Factorakh Stoka (On climatic factors and runoff), *Problemyfiz. Geog.*, **16**, 41–48.

Dzerdeevskii, B.L., 1958. On some climatological problems and microclimatological studies of arid and semiarid regions in the U.S.S.R., in *Climatology and Microclimatology: Proceedings of the Canberra Symposium*. Paris: UNESCO, Arid Zone Research, pp. 315–323.

Gommes, R. and Petrassi, F., 1994. *Rainfall variability and drought in Sub-Saharan Africa since 1960*, Agrometeorology Series Working Paper No. 9. Rome: Food and Agriculture Organization.

Hare, F.K., 1977. Climate and desertification, in *Desertification: Its Causes and Consequences*, Secretariat of the United Nations Conference on Desertification (ed.). Oxford: Pergamon Press, pp. 63–168.

International Crops Research Institute for the SemiArid Tropics, 1980. *Climatic Classification: A Consultants Meeting*. Patancheru, India: IRCISAT,

Karl, T.R. and Knight, R.W., 1985. *Atlas of Monthly Palmer Hydrological Drought Indices (1931–1983) for the Contiguous United States*, Historical Climatology Series 3–7. Ashville, NC: National Climatic Data Center,

Köppen, W., 1918. Klassification der Klimate nach Tempertur, Niederschlag and Jahreslauf, *Petermanns Geog. Mitt.*, **64**, 193–203, 243–248.

Köppen, W., 1936. Das Geographische System der Klimate, in *Handbuch der Klimatologie*, vol. 1, pt. C, W. Köppen and R. Geiger (eds). Berlin: Gebrüder Borntröger.

Le Houerou, H.N., Popov G.F. and See, L., 1993. *Agro-bioclimatic Classification of Africa*, Agrometeorology Series Working Paper No. 6. Rome: Food and Agriculture Organization.

McKee, T.B., Doesken, N.J. and Kleist, J., 1995. The relationship of drought frequency and duration to time scales, *Preprints, 8th Conference on Applied Climatology*, 17–22 January, Anaheim, CA, pp. 179–184.

Meigs, P., 1953. *World Distribution of Arid and Semiarid Homoclimates*, Arid Zone Programme, vol. 1. Paris: UNESCO, pp. 203–210.

Meigs, P., 1960. *Distribution of Arid Homoclimates: Eastern Hemisphere: Western Hemisphere*. United Nations Maps No. 392 and No. 393, Revision 1. Paris: UNESCO.

National Drought Mitigation Center (NDMC), 1996. *Drought Indices*, NDMC, USA.

Palmer, W.C., 1965. *Meteorological Drought*, Research Paper No. 45. Washington, DC: US Department of Commerce Weather Bureau,

Palmer, W.C., 1968. Keeping track of crop moisture conditions, nationwide: the new Crop Moisture Index, Weatherwise, **21**, 156–161.

Shafer, B.A. and Dezman, A., 1982. Development of a Surface Water Supply Index (SWSI) to assess the severity of drought conditions in snowpack runoff areas, *Proc. of the Western Snow Conference*, pp. 164–175.

Thornthwaite, C.W., 1931. The climates of North America, *Geog. Rev.* **21**(3), 633–655.

Thornthwaite, C.W., 1948. An approach toward a rational classification of climate, *Geog. Rev.*, **38**(1), 55–94.

UNESCO, 1979. *Map of the World Distribution of Arid Regions*, MAB Tech. Note 7. Paris: UNESCO.

Cross references

Arid climates
Arid lands
Desertification
Drought
Water balance
Water budget analysis

ARID LANDS

Deserts are dry areas of sparse or nonexistent vegetation that comprise more than one-third of the Earth's land surface if semiarid regions are included. The general term **desert** usually refers to the hot, dry regions of the world. About 5% of the Earth's land surface can be classified as hot and extremely arid, and about 15% as hot and arid. These areas owe their existence largely to meteorological causes, being located along the Earth's two great subtropical belts of minimal rainfall or far away from centers of rainfall. Because the development and form of their ground conditions arise from past and present climates, a definition of a hot desert area from an engineering viewpoint is essentially related to climate (Somerville, 1996).

Attempts to delineate boundaries of arid zones have resulted in the development of a number of indices of aridity. For example, boundaries could be established in terms of mean annual temperature and mean annual precipitation. Indices of aridity, however, generally suffer from several disadvantages. In particular it is questionable whether mean annual values of precipitation indicate sufficient similarities between two regions to let them be placed in the same category. For example, if each of two regions of similar temperatures had 250 mm of rainfall annually but one experienced about 20 mm of

rain per month and the other only three large storms per year, their vegetation, drainage, soil and other characteristics would be significantly different. Even though aridity is a measure of dryness, the best index is probably one based on the water balance of the area, that is, the difference between the moisture received and the moisture lost. Moisture received is chiefly in the form of precipitation, whereas losses are from evaporation, runoff and seepage. There are several difficulties in any water balance evaluation, and this is particularly true of the vast arid areas with limited data and vegetation cover. Thornthwaite (1948) devised various indices that are closely related to water balance. His Aridity Index, for example, is related to the potential total evaporation from a continually damp area covered in vegetation, to the length of the day and to the temperature. Thornthwaite's index is easy to apply, with the use of nomograms: Meigs (1953) chose to use it in the preparation of his small-scale maps depicting the world's arid regions. Figure A85 shows Meigs's subdivision of the world's arid lands by climatic zones. Hot deserts, which occupy the great majority of the area, comprise arid and extremely arid areas.

Ground conditions of significance to engineering

For areas as large as those depicted by Figure A85, it is extremely difficult, if not actually misleading, to try to rationalize ground conditions for engineering purposes. Yet in some significant respects it is reasonable to do so because of a certain unity of conditions imposed by the overall climatic regime. The arid climate tends to produce particular forms of erosion with a dominance of mechanical weathering on high land which supplies coarse debris, which is subsequently transported by stream- or sheetflood to low land. There is also a transport of fine sediments by wind, in addition to a general upward leaching and surface precipitation of salts by evaporation. The ground is therefore commonly saline and covered with granular alluvial sediments, usually without humus.

The principal engineering problems associated with these desert conditions include (1) unstable terrain, such as wind-blown silt (loess) and sand (sheets, drifts and dunes), (2) aggressive salty ground, such as sabkhas, salinas, salt playas and some duricrusts, (3) unsuitable construction materials, such as some silts, sands, weak carbonate sediments and some duricrusts, (4) rapid erosion and deposition, by wind and floods – especially flash floods – and debris flows. Other problems occur but are generally similar to those found in temperate regions. For background discussions see Geological Society of London (1978) and Hopgood (1996).

As an aid to engineering feasibility evaluation of hot desert ground conditions, and Middle East conditions in particular, Fookes and Knill (1969) proposed a simple model based on 'mountain and piedmont plain' terrain and natural desert processes. Four zones were recognized, each with different desert surface characteristics and with different engineering behavior. An important part of the concept is that the width of the mountain (or hill) and the plain can vary from only a few hundred meters to tens or even hundreds of kilometers, the width of the individual zones stretching or shortening commensurately.

The overwhelming majority of engineering soils comprising the zones are granular (Foth, 1990). Their grading is related to the zone by reference to the distance from the mountains supplying the erosion debris: the farther from the mountains, the finer the deposits. Gravity and water transport the particles from the mountains into the plains where wind and water move the finer fractions around. In general, the grading of materials indicates their geotechnical engineering characteristics and their Casagrande group symbol classification (Foth, 1990). The base-leveled central plains are mainly composed of sands and silts. The general level of the water table is such that the capillary rise often reaches ground level; thus a high water table is the single most significant factor affecting ground engineering problems. Further discussion of the ground engineering conditions outlined here are also given in Fookes and Higginbottom (1980), Epps (1980) and Oweis and Bowman (1981).

Zone I: mountains

Even though mechanical weathering – which involves the splitting, exfoliation and crumbling of rocks – is dominant here, chemical weathering in the presence of moisture from dew and occasional rain or snow, although slow, also plays an important role. By decomposition and solution, rocks that would otherwise successfully resist the

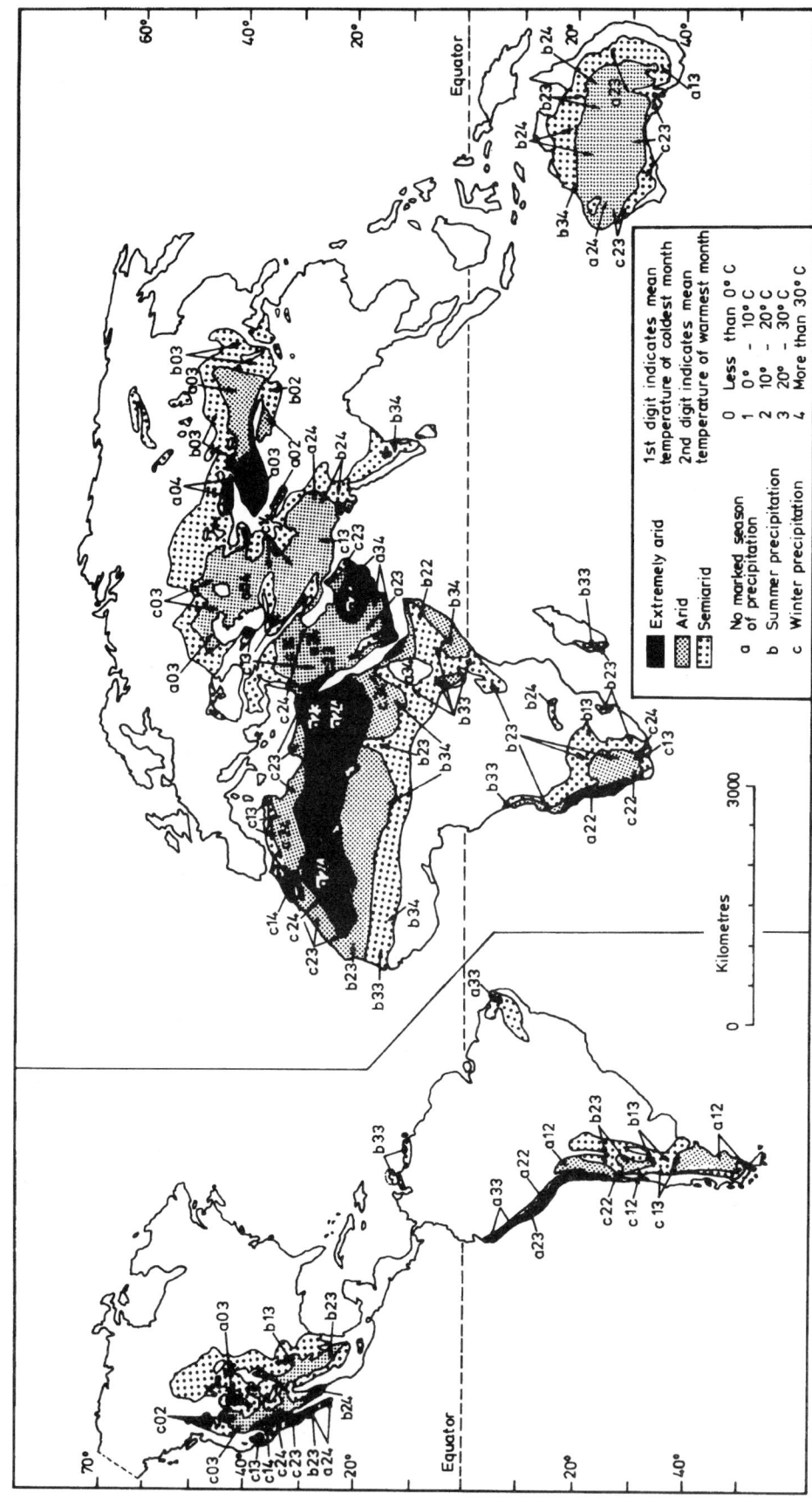

Figure A85 Extremely arid, arid and semiarid climatic zones of the world. (After Meigs, 1953.)

stresses set up by temperature changes are gradually weakened until they are shattered. Minute quantities of dissolved matter are brought to the surface by evaporation. Commonly the loose salts are blown away, but oxides of iron, accompanied by traces of manganese and other oxides, form a red, brown or black surface film that is known as desert varnish.

General problems associated with engineering works in zone I include major landslips (Wiseman et al., 1970) and karst topography (limestone surface solution features and subsurface cavities), while lesser slips and slope failures occur in any situation with unfavorable combinations of discontinuities, such as joints and faults (Fookes and Sweeney, 1976; Schuster and Krizek, 1978; Hoek and Bray, 1981). Within the mountains, the size of erosional debris ranges from poorly sorted medium angular gravel to very large boulders. Hazards occur from streamflooding and from talus slope movements, especially in semiarid areas (Figure A85). Duricrusts occur extensively in some areas of limestone terrain, such as the Near East, the Middle East and Australia (Figure A85). Pre-Tertiary hardrocks in semiarid mountain areas – for example, the Mediterranean and the Ethiopian highlands – may have residual soils developed on them. In areas marginal to wetter climates, such as the eastern Mediterranean, terra rossa soils may have developed on limestones (Chapman, 1971, 1974).

Zone II: aprons and sediments

When the apron consists of rock, it may be covered by a thin mantle of sand and gravel deposits, which sometimes hide an irregularly eroded, or even a terraced, rock surface. The engineering performance of the bedrock is directly related to its rock type, which may have a duricrusted surface with a leached or softer underlying zone.

Gravel fans are, however, generally the most extensive form of apron, especially in parts of the Middle East and southwestern United States. They are almost entirely composed of angular particles, boulders, and cobbles on the upper slopes grading to fine gravels downslope. The gravel fans occur in rough layers, which reflect deposition in times of flood and, being reasonably compact, they usually have good load-bearing characteristics, except where occasional silt or clay layers or debris flow materials occur. These materials have a fairly high permeability and sometimes form aquifers. They generally have a good borrow potential, but one should check their chemistry in addition to conducting the usual mechanical–physical properties tests when considering them for use in concrete. Such terrains are subject to flash floods, which is a particularly important consideration in road design.

Zone III: alluvial plains

Alluvial plains mainly consist of extensive splays of fine gravels and sands. Two common types of alluvial plain include the sandy–stony and silty–stony deserts, which respectively represent deposition from stream flow and overbank sheet flow. Such deposits generally provide good foundation conditions because the sands and fine gravels are waterlaid and reasonably dense. They also make good borrow for fill and aggregates, depending on their grading. Pavements composed of a single layer of single-sized stone may also occur extensively in this zone (and zones II and IV). These pavements can be scraped to supply aggregate, but their removal may expose underlying finer material that is susceptible to erosion by wind and water. Alluvial plains are subject to sheet- and streamflood hazards.

Zone IV: base-leveled plains

These surfaces are composed of wind-blown silts and sands, which are frequently modified by subsequent flooding or marine action. Perhaps the most extensive of all the zones, base-leveled plains are very common in arid areas (Figure A85), especially the Near and Middle East and parts of Australia. In areas where there is a high water table and where aggressive salty conditions occur, load bearing and other engineering performance is reduced, requiring dewatering in excavations and tanking of foundations. In areas of uniform or thin-bedded, uncemented sands, permeabilities are similar to those generally associated with grading, typically ranging from $k = 10^{-1}$ to 10^{-3} or 10^{-4} cm s^{-1}. Layered deposits are often cemented by carbonate and therefore their overall permeabilities tend toward 10^{-4} to 10^{-6} cm s^{-1}. Until more experience is gained in high water table locations, foundations for major works should be inspected by deep pits. Successful large dewatering operations can usually be conducted by well point systems. In layered deposits, difficulties may be expected with dredging or pile driving through thin, cemented layers.

Because many of the unconsolidated deposits have a large, finely granular component that is often not bound by clay or cement, erosion by wind or water is common and therefore protection may be required. Similarly, filter protection against the migration of fines may be necessary in underground and surface drainage systems and also in dewatering systems. Because metal filter screens readily corrode in certain ground waters, filter media constructed of synthetic fibers are widely used.

Most potential engineering problems are associated with this zone, namely clays, silts, windblown sands, salty soils and duricrusts.

Clays

These fine materials are uncommon, except near the coast and in areas of clay plains in Australia. The clays are usually calcareous and normally fall in the CI–CH range in the Casagrande classification. Some may have marked shrink–swell characteristics, which can cause problems for shallow foundations. Coastal clays are usually consolidated, but many have a desiccated crust; their strength ranges from soft to stiff. Some residual clays occur in the marginal highlands, especially in the semiarid areas, where chemical weathering probably occurred during pluvial periods or may even be occurring today. Special engineering problems are associated with estuarine clays, such as the plastic muds in the Nile delta.

Silts

Hot desert silts may have been blown in from higher latitudes or produced at least in part, within the desert system. Some silts may be loesses, which are potentially metastable, that is, they collapse on wetting when under load because the individual grains are not packed in a dense configuration. This condition also exists in some fine sands. When a higher load than any previous loading by nature is applied, the normal consolidation curve is followed. If the loess is wetted under its new load, however, a sudden collapse may occur. Loess areas are also susceptible to piping erosion and underground drainage channels (Fookes and Best, 1969). When loess is reworked and deposited from water, it will generally have the properties of conventional water-laid alluvial silt deposits.

Windblown sands

An eolian dune is a mound of windblown sand; the smallest may be only a meter or so high and may cover only 10 m^2, whereas the largest may be over 40 m high and may cover several square kilometers. Sand sheets in distinction, are large areas of gently undulating sandy surfaces with low relief. True dunes cover extensive areas mainly in the big sand seas, but they also occur in isolated areas on hard surfaces. Partly vegetated dunes occur in semiarid regions on the margins of deserts and in moist coastal areas (Pye and Tsoar, 1990).

Dune sands commonly have median grain diameters between 0.2 and 0.4 mm and range between extremes of 0.1 and 0.7 mm. It is of note that most dunes are well sorted and samples of sand from one dune usually have particles of similar size and rounded shape. Sand sheets, on the other hand, are poorly sorted and bimodal, comprising both coarse (0.6 mm) and fine (0.1 mm) grains. Dunes frequently have poor load-bearing capacities (loose to medium-dense) and can be difficult to compact as fill. Sheet sands have slightly better bearing capacity and exhibit compaction characteristics similar to those that have been reworked, transported and deposited by water. The latter are quite common around desert margins, in coastal areas, and in semi-humid climates.

Mobile dunes may blow over roads or buildings moving at a rate of several meters per month. Methods for controlling drifting sand include removal of dunes, realigning routes or structure, and stabilization by oiling, fencing, planting or paving.

Salty soils

Groundwaters in the Near and Middle East, parts of the southwestern United States and in Africa are frequently saline due to the presence of salts dissolved from the local bedrock or from seawater. Salts may occur at the soil surface in the form of salty crusts and elsewhere as windblown particles of sand or silt. Figure A86 shows common relationships between groundwater, capillary rise and the ground surface (Cooke, 1981). Where the capillary water does not reach the ground surface, its depth can be inferred from characteristic desiccation ground patterns (Neal, 1969). Salty capillary moisture reaching

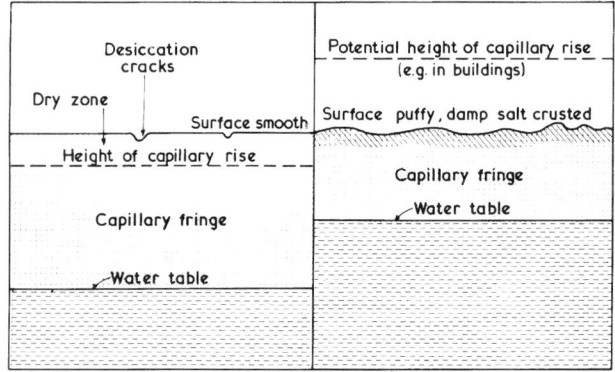

Figure A86 Schematic section showing low and high water tables, capillary rise and related surface featu.•s in zone IV sands. (After Brunsden et al., 1976.)

the ground surface usually produces dry, caked surfaces, whereas groundwater near the surface produces a damp, puffy surface. All these conditions depend on local circumstances, but in general an intuitive feel for salty ground can usually be obtained rather quickly.

Natural salt surfaces have a variety of names, depending on location and country (Neal, 1969). With much simplification these have been reduced for engineering purposes to sabkha (coastal salt marsh), playa (an ephemeral lake flat), salt playa (a playa with a salty surface due to evaporation of salty lake waters) and salina (local depression with high salt water table and attendant formation of salt crusts). The local salt regime and chemical conditions are frequently complex and often vary with the season so that each engineering site must be investigated separately.

Salts may actually help bind unsealed roads, but they can severely damage sealed roads. Building and engineering foundations may require tanking because of salt attack on masonry or concrete. Salts may also contaminate aggregate sources (Ellis and Russell, 1973; Netterberg et al., 1974; Fookes and Collis, 1976; Fookes and French, 1977). Reclaimed land adjacent to coasts, or in any high water table situation, may become aggressively saline where capillary moisture reaches the surface and evaporation takes place. Salt weathering is perhaps the most common form of weathering in salty lowland areas.

Duricrusts

Duricrusts occur in a variety of forms commonly with a hardened surface ranging from millimeters to tens and hundreds of centimeters thick, often with a leached, cavernous, porous or friable zone underneath. The term 'duricrust' was first applied in Australia to denote a surface or near-surface hardened accumulation of silica, alumina or iron oxides in varying proportions. Admixtures of other substances also occur, and the term is now applied by extension to encrusting layers of calcium or magnesium carbonate, gypsum and salt, the latter two being common products of desert climates. For engineering purposes, those duricrusts ending in '-crete', such as calcrete ($CaCO_3$) and silcrete (SiO_2), indicate hardened surfaces. Those ending in 'crust' – gypcrust ($CaSO_4 \cdot 2H_2O$) and salcrust (NaCl), for example – are softer accumulations usually occurring in areas of centrally draining deserts (salt playas or salinas) or coastal sabkhas. Sands may be locally cemented with carbonate to form cap rock or miliolite, especially near the coast. Mixtures of nodular calcrete, calcrete fragments, and drifted sand form the so-called desert fill. Massive development of calcrete on limestone rocks, common in the semiarid highland areas of the Middle East, probably represents a pedogenic relic.

Surface inspection of the different types of cretes can be quite misleading because the thickness of surface hardening will vary depending on the local circumstances, as will the leached or otherwise altered zone underneath. Each site should therefore be investigated by drilling if a knowledge of subsurface characteristics is required for engineering purposes. Indurated crusts make suitable aggregate borrow, but the chemical properties must also be evaluated. A depth profile of chemical and physical properties for a surface-altered

limestone bedrock may show, for example, an increase of gypsum and halite towards the surface, making it unacceptable for crushing as concrete aggregate; a decrease in porosity and an increase in unconfined compressive strength, however, make it otherwise attractive (Fookes and Higginbottom, 1980).

Site investigation

Where there is adequate exposure of the terrain, walkover surveys, inspection of air photos, engineering geology and geomorphological mapping are economical and successful techniques. In addition to basic mapping and air photo interpretation for feasibility and site investigation planning, maps can be prepared for specific requirements (*Quart. J. Eng. Geol.*, 1972), such as sand dune migration, potential salt hazard, potential flood hazard, borrow locations, and route and urban planning. Various forms of color air photography and satellite imagery are also valuable aids (Doornkamp, *et al.*, 1980).

Site investigation contracts for desert conditions should generally include pitting as a main tool of the investigation because many boring techniques tend to lose fines and break up large stones. There is also a need to develop improved techniques of boring, sampling and evaluating the predominantly granular desert soils (Fookes and Higginbottom, 1975). The durability of natural and artificial materials in salty or wet environments must be estimated if appropriate to the works. The nature of the ground chemistry and seasonal changes in the water table should be established, as they are important parameters in the planning stages of engineering works in arid environments.

Descriptions and classification systems (e.g. *Quart. J. Eng. Geol.*, 1972) generally do not adequately cover the range of calcareous soils and rocks found in many desert areas, especially the Near and Middle East. Accordingly, a pilot carbonate soil and rock classification system has been proposed by Fookes and Higginbottom (1975).

Peter G. Fookes

Bibliography

Brunsden, D., Doornkamp, J.C. and Jones, D.K.C. (eds), 1976. Geomorphology and superficial materials, *Bahrain Surface Materials Resources Survey*, 4, 1–124.

Chapman, R.W., 1971. Climatic changes and the evolution of landforms in the Eastern Province of Saudi Arabia, *Geol. Soc. Am. Bull.*, **82**, 2713–2728.

Chapman, R.W., 1974. Calcareous duricrust in Al-Hasa. Saudi Arabia, *Geol. Soc. Am. Bull.*, **85**, 119–130.

Cooke, R.U., 1981. Salt weathering in deserts, *Geol. Assoc. Proc.*, **92**, 1–16.

Doornkamp, J.C., Brunsden, D. and Jones, D.K.C. (eds), 1980. *Geology, Geomorphology and Pedology of Bahrain*. Norwich, UK: Geo Abstracts Ltd., 443 pp.

Ellis, C.I. and Russell, R.B.C., 1973. The use of salt-laden soils (sabkha) for low cost roads. *Department of Environment TRRL Paper PA 78/74.* 1–26.

Epps, R.J., 1980. Geotechnical practice and ground conditions in coastal regions of the United Arab Emirates, *Ground Eng.*, **13**, 12–25.

Fookes, P.G. and Best, R. 1969. Consolidation characteristics of some late Pleistocence, periglacial metastable soils in East Kent, *Quart. J. Eng. Geol.*, **2**, 103–128.

Fookes, P.G. and Collis, L., 1976. Cracking and the Middle East, *Concrete*, **10**, 14–19.

Fookes, P.G. and French, W.J., 1977. Soluble salt damage to surfaced roads in the Middle East. *Highway Engineer*, **24**, 10–20.

Fookes, P.G. and Higginbottom, I.E., 1975. The classification and description of near-shore carbonate sediments for engineering purposes, *Geotechnique*, **25**, 406–411.

Fookes, P.G. and Higginbottom, I.E., 1980. Some problems of construction aggregates in desert areas with particular reference to the Arabian Peninsula: 1) Occurrence and special characteristics. 2) Investigation, production and quality control, *Proc. Inst. Civil Eng.* **68**(1), 39–90.

Fookes, P.G. and Knill, J.L., 1969. The application of engineering geology to the regional development of Northern and Central Iran, *Eng. Geol. Internat. Jr.*, **3**, 81–120.

Fookes, P.G., and Sweeney, M., 1976. Stabilization and control of local rock falls and degrading rock slopes, *Quart. Jr. Eng. Geol.*, **9**, 37–55.

Foth, H.D., 1990.. *Fundamentals of Soil Science*, New York.

Geological Society of London, 1978. Proceedings of the conference on engineering problems associated with ground conditions in the Middle East, *Quart. Jr. Eng. Geol.*, **11**, 1–112.

Hoek, E. and Bray, J.W., 1981. *Rock Slope Engineering*, 3rd edn. London: Institution of Mining and Metallurgy, 360 pp.

Hopgood, J., 1996. Erosion and weathering, in *The Encyclopedia of Climate and Weather*. Cambridge: Cambridge University Press.

Meigs, P., 1953. World distribution of arid and semi-arid homo-climates, in *Reviews of Research of Arid Zone Hydrology*, Paris: UNESCO, 203–209.

Neal, J.T., 1969. Playa variation, in W.G. McGinnies and B.J. Goldman, eds, *Arid Lands in Perspective*. Tucson: University of Arizona Press, and Washington, DC: American Association for the Advancement of Science, pp. 13–44.

Netterberg, F., Blight, G.C., Theron, P.F. and Marais, G.P., 1974. Salt damage to roads with bases of crusher-run Witwatersrand quartzite, *Proc. 2nd Conf. Asphalt Pavements S. Africa*, Durban, Session 7, pp. 134–153.

Oweis, I. and Bowman, J., 1981. Geotechnical considerations for construction in Saudi Arabia, *Am. Soc. Civil Eng. Proc., J. Geotech. Eng. Div.*, **107**, 319–338.

Pye, K. and Tsoar, H., 1990. *Aeolian Sand and Sand Dunes*, London.

Quart. J. Eng. Geol., 1972. The preparation of maps and plans in terms of engineering geology, **5**, 295–382.

Schuster, R.L. and Krizek, R.J. (eds), 1978. *Landslides Analysis and Control*, Special Report 176. Washington, DC: National Academy of Sciences, 234 pp.

Somerville, R.C.J., 1996. Climate and weather, in *The Encyclopedia of Climate and Weather*. Cambridge: Cambridge University Press.

Thornthwaite, C.W., 1948. An approach toward a rational classification of climate, *Geog. Rev.*, **38**, 55–94.

Wiseman, G., Hayati, G., Frydman, S. *et al.*, 1970. A study of a landslide in Galilee, Israel, *Proc. 1st Int. Congr. Int. Assoc. Eng. Geology*, Paris, **1**, 50–61.

Cross references

Arid climates
Aridity indices
Arid zone hydrology
Deltaic plains
Desertification
Deserts
Evaporation: measurement

ARID ZONE HYDROLOGY

In a review paper on the hydrological characteristics of arid zones, McMahon (1979) defines arid areas as those with average annual precipitation less than 500 mm and average annual potential evapotranspiration greater than 800 mm. The main areas of the world that fall into this zone are large parts of Australia and southern Africa, the western seaboard of South America, the southwest and central west parts of North America and a broad band covering the whole of North Africa, the Middle East and central Asia. These areas cover more than a quarter of the Earth's land surface and include many developing countries.

Apart from generally low rainfall, high evaporative loss and consequently low runoff, arid areas are also characterized by high degrees of both spatial and temporal variability. The spatial variability is largely a consequence of the convective rainfall mechanisms that prevail in arid areas where storm cells develop and decay rapidly over relatively small areas. Table A26 illustrates the temporal variability of arid areas using values for the coefficients of variation (C_v = standard deviation/mean) and skewness (C_s, the shape of the probability distribution) of annual streamflow volumes taken from McMahon (1979) and Görgens and Hughes (1982). The table demonstrates that annual variability in arid zones can be as much as double that of the continental values, even where the arid zone variability is relatively low, as in the case of North America. The annual variability of arid zone rainfall is commonly much lower.

The high skewness values illustrate the importance of extremes in the time series of flow data derived from arid areas and the problems associated with using mean values as indicators of 'average flow conditions'. In fact the concept of 'average conditions' has very little meaning in arid areas. Monthly flow data for 17 years from the Mosetse River (1026 km^2) in Botswana reveal that two of the highest monthly flow volumes represent 281 and 159% of the mean annual runoff. Streamflow data from other Botswana catchments illustrate similar patterns with peak monthly values within 15 to 30 year records representing between 120 and 240% of mean annual values. Similar figures for moderate-sized catchments (200–1500 km^2) in central and western Zimbabwe (semiarid with 500 to 700 mm of rainfall) range from 67 to 134%, while values for Malawian catchments (>900 mm of rainfall) range from 27 to 58%.

The simplest measure of low-flow conditions in arid areas is the percentage of time that streamflow ceases. McMahon's (1979) paper suggests that the flow regimes of Australian, Eastern Mediterranean and Southern African arid rivers (zero annual flows occurring 3.8, 4.5 and 1.6% of the time, respectively) are much less reliable than their arid North American counterparts, where annual cease-to-flow conditions do not seem to occur. For the 30 streamflow data sets analyzed by Görgens and Hughes (1982) for arid South Africa, zero-flow months represented between about 13 and 86% of the record with a mean of some 37%. On a daily basis, the same South African catchments experience zero flows for greater than 90% of the time.

Short-term variability of hydrological processes in arid areas is also important and can be illustrated by the nature of the response of a 100 km^2 catchment in the Walnut Gulch basin area of Arizona, USA. During a single event in August 1971, flow progressed from zero at 1931 h to over 70 m^3 s^{-1} 4 minutes later, reaching a peak of 107 m^3 s^{-1} at 1947 h. Flows of over 50 m^3 m^3 s^{-1} were sustained for 1 h, while 2 h later the flow had reduced to less than 1 m^3 s^{-1}. Such events are typical of this and similar semiarid catchments.

Arid zone hydrological processes

The generally low vegetation cover and high rainfall intensities experienced in arid zones contribute to runoff processes that appear to be dominated by an excess of rainfall over infiltration rates (Hortonian runoff) or through saturation excess on areas where very thin soils exist. Runoff generation may be enhanced by sealing of the soil surface due to raindrop impact or salt encrustation reducing infiltration rates to very low values. However, while small scale hillslope processes tend to promote the generation of runoff, other factors operating at larger scales tend to limit the amount of streamflow that occurs at the outlets of moderate to large catchments. Where deeper soils and better vegetation cover exist in valley bottom situations, infiltration rates are generally higher and runoff generated as sheet or rill flow on slopes can partially or totally reinfiltrate before reaching the channel. The spatial variability in rainfall amounts and intensities

Table A26 Comparison of variability criteria based on annual streamflow values for continental or humid regions and arid regions of North America, Australia and South Africa

	Global total – arid zones	North America[a]		Australia[a]		South Africa[b]	
		Continent	Arid zone	Continent	Arid zone	Humid	Arid
No. of catchments	72	–	22	–	16	28	17
Average mean annual run off (mm)	30	–	30	–	21	348	32
Average annual C_v	0.99	0.34	0.65	0.67	1.27	0.57	1.14
Average annual C_s	1.80	–	1.60	1.10	2.20	–	2.00

Sources: [a] McMahon (1979), [b] Görgens and Hughes (1982)

also suggests that runoff is not generated throughout the catchment at the same time. Coupled with the losses that can occur within the channel itself, the implication is that rainfall amounts have to be quite high, or more generally widespread than usual, for runoff to survive and be experienced in higher-order channel systems. At least some of the confusion about the hydrological processes that prevail in arid zones must be attributed to the fact that they have been studied at different scales by different hydrologists. It should be recognized that observations of quite frequent occurrences of runoff generated on hillslopes from high-intensity rainfall are not incompatible with observations that the general occurrence of streamflow is rare.

The losses that occur within channels include seepage into underlying alluvium (Lane, 1983; Walters, 1990), recharge to groundwater through fractures in bedrock channel beds and evaporative losses from large rivers. Preliminary estimates (McKenzie *et al.*, 1993) of the evaporative losses from the 1377 km of channel of the Orange River, which passes through the semiarid Northern Cape Province of South Africa, are as high as $800 \times 10^6 \, \mathrm{m^3 \, year^{-1}}$. This represents a similar volume of water to that used for irrigation along the same reach of river.

High evapotranspiration rates and low antecedent soil water states suggest that in areas of moderate to deep soils direct recharge of groundwater through saturation of the soil profile is a rare event (Lloyd, 1986). Recharge is more likely to occur in areas of thin to no soil cover. Other indirect recharge mechanisms are likely to be dependent upon the concentration of surface water by runoff before significant amounts of recharge take place. This is particularly true for underlying material which does not include structural units of high conductivity (Peck, 1979). The conclusion is that recharge and channel transmission losses are strongly related, both where channels are underlain by permeable alluvial material or by fractured rock. Groundwater recharge events are therefore likely to occur almost as infrequently as streamflow events and, although the mechanisms might be reasonably well understood, quantitative estimates of the various elements are difficult to obtain. Lloyd (1986) suggests that measurement errors preclude obtaining reliable estimates of the flood volume differences between two gauging stations and that the infiltration is subject to high evaporative losses. The presence of fossil groundwater can confuse hydrogeological investigations in arid areas. The existence of well-established groundwater gradients would normally indicate some recharge, but storage depletion of ancient recharge mounds has also been cited as a possible cause (Lloyd, 1986).

One of the major problems with quantifying hydrological processes in arid areas is that the relative harshness of the environment, low population and consequent poor communication systems make it difficult to establish instrumentation networks. The high degree of variability also means that the instruments used have to be able to measure over wider ranges than in humid situations and that longer periods of time and denser networks are required before representative observations (in time and space) can be obtained. High stream velocities frequently occur with high sediment loads and large quantities of debris on flood wave fronts. All of these contribute to the need for specialized monitoring equipment which can be expensive and difficult to maintain. It is no accident that of the many experimental research catchments that do exist, or have existed in the past, very few are located within the arid zones of the world. Perhaps the best known, and those with the longest detailed records, are the catchments in the Walnut Gulch basin near Tombstone, Arizona. Similarly, the density of national hydrometric networks decreases dramatically in the arid zones. While the arid zone of Australia covers 75% of the continent, the density of streamflow measuring stations with over 15 years of data is one per $350\,000 \, \mathrm{km^2}$ compared with the equivalent figure of $3200 \, \mathrm{km^2}$ for the humid regions (McMahon, 1979).

The application of hydrological estimation methods to arid zones is therefore hampered by the lack of understanding of some of the processes involved and a lack of sufficient data to quantify the extreme spatial and temporal variability of inputs and responses. Many of the estimation techniques have also been developed for temperate regions and do not account for some of the processes or data characteristics of arid zones. For example, few generalized deterministic models are able to account satisfactorily for the processes of channel transmission loss and the high variability and frequency of zero flow conditions present problems for stochastic time series models. Even simple water balance methods have not proved to be reliable in arid areas, largely because the measurement errors associated with some components are an order of magnitude greater than other critical components.

Water resource development and exploitation

The high variability of arid zone streamflow places a serious constraint on the extent to which surfacewater resources can be exploited. Larger reservoir storage capacities are required for similar regulated yields than in humid areas. McMahon (1979) indicates that while the maximum potential streamflow regulation in humid areas may approach 90%, the equivalent figure for the more variable arid regimes is likely to be closer to 10%. Higher reservoir capacities, the likelihood of low storage levels prior to major events and relatively high sediment loads during individual runoff events all contribute to high sediment trap efficiencies and shorten the lifespan of artificial impoundments in arid areas or contribute to costs through the need to raise dam walls. Although there is no real evidence to suggest that arid lands have naturally higher sediment yields than humid areas, agricultural practices and poorly planned land uses have contributed to increased sediment loads in many arid zone rivers. Arid areas also take a long time to recover after disturbance and the annual sediment load is concentrated in fewer events than in humid areas, suggesting that sediment loads during individual events will be greater.

The lack of naturally occurring reliable sources of water and the perceived need to promote economic development has frequently led to the establishment of irrigation schemes adjacent to arid zone rivers which are supplied by water from upstream impoundments or imported through interbasin transfer schemes. The river channels are often used as natural conduits for water supply and also receive the return flow from drainage. This means that the flow regimes of such rivers are drastically altered from ephemeral and highly intermittent to permanently or seasonally flowing. Such changes can affect the riverine ecology and have been known to result in increases in the incidence of animal or human ailments through the creation of new habitats enjoyed by the organisms or insects that spread diseases.

High evapotranspiration rates and infrequent mobilization of subsurface stored water lead to the concentration of salts within soil water and groundwater, leading to poor water quality, which can deteriorate even further during long periods of storage in impoundments. Irrigation schemes on alluvial valley bottom areas can also contribute to a serious deterioration in downstream water quality as the return flow leaches out salts concentrated in the irrigated soils.

The paucity of available surfacewater supplies has prompted many countries to move towards exploiting groundwater for both domestic and agricultural water supply. In rural areas with dispersed populations, groundwater often represents the only economically feasible option for water supply. However, the relatively low rates and high variability of groundwater recharge can severely limit the potential of groundwater resource exploitation. Referring to African groundwater development, Foster (1984) identifies three principal problems:

- the difficulty of successfully siting boreholes or wells as a result of geohydrological inhomogeneity;
- the estimation of recharge and exploitable storage;
- the occurrence of groundwater with poor water quality characteristics.

Water quality problems associated with poor groundwater management practices abound in arid areas. The recirculation of initially good-quality groundwater for irrigation purposes can lead to serious salinization problems and suggests that shallow aquifers cannot be used for both abstractions and sinks. In Australia, deforestation of semiarid recharge areas at the end of the nineteenth century resulted in increased recharge and groundwater levels rising to the surface in the more arid areas down gradient. The high evaporation rates in these areas has led to the creation of surface salinization and the laying waste of millions of hectares of land (Lloyd, 1986).

Artificial groundwater recharge using excess surface water stored in recharge basins represents a method of improving the exploitation potential of water resources in arid areas. However, suitable aquifer situations where infiltration rates are high enough for recharge rates within the basin to exceed evaporation losses are not always present. Such approaches may be appropriate in areas with highly permeable aquifers where the topography permits the construction of substantial recharge reservoirs.

Denis A. Hughes

Bibliography

Foster, S.S.D., 1984. African groundwater development – the challenges for hydrogeological science. *Challenges in African Hydrology and Water Resources*, Proc. Harare Symp., July 1984, IAHS Publ. No. 144, pp. 3–12.

Görgens, A.H.M. and D.A. Hughes, 1982. Synthesis of streamflow information relating to the semi-arid Karoo Biome of South Africa. *South African J. of Science*, **78**(2), 58–68.

Lane, L.J., 1983. Transmission Losses, Washington; DC: US Dept. of Agriculture.

Lloyd, J.W., 1986. A review of aridity and groundwater. *Hydrol. Processes*, **1**, 63–78.

McKenzie, R.S., C. Roth and F. Stoffberg, 1993. Orange River Losses, in *Proc. of the 6th South African National Hydrological Symposium*, University of Natal, Pietermaritzburg, Sept. 1993, pp. 351–358.

McMahon, T.A., 1979. Hydrological characteristics of arid zones, in *Symp. on the Hydrology of Areas of low Precipitation*, IAHS Publ. No. 128, pp. 105–123, Proc. Canberra Symposium, Dec. 1979.

Peck, A.J., 1979. Groundwater recharge and loss, in *Symp. on the Hydrology of Areas of Low Precipitation*, IAHS Publ. No. 128, pp. 361–370, Proc. Canberra Symposium, Dec. 1979.

Walters, M.O., 1990. Transmission losses in arid region. *J. Hydrol. Eng.*, **116**(1), 129–138.

Cross references

Alluvial valley engineering
Arid climates
Aridity indices
Arid lands
Desertification
Deserts

ASIA: CLIMATE

Asia covers one-third of the Earth's surface and almost every known climate occurs on this continent. It is the largest and most climatically continental, has the highest average elevation above sea level, and is wettest, coldest and physically most diverse of all continents. Other than climate, few common denominators unite this vast land mass (Tsuchiya, 1964; Dando, 1983a,b; Oliver, 1984).

Climates in Asia are subject to land influences as opposed to maritime influences, and seasonal variations in temperature and moisture are extreme. Asia's climates differ from region to region because of variations in the amount, intensity and spatial distribution of solar energy, temperature, humidity and precipitation, atmospheric pressure and winds and storms, but there is unity in Asia's climatic diversity that is provided by the monsoon effect. Latitudinal differences in the amount of solar energy received provide the basic climatic control for Asia, extending from well above the Arctic Circle to near the equator. Differences in heating and cooling between high and low latitudes and between snow- and vegetation-covered surfaces produce regional atmospheric pressure contrasts that induce air movement (wind). Seasonal land and sea winds caused by atmospheric pressure reversals, from one season to another over land and water, produce the monsoon effect. Also, vast segments of Asia are thousands of kilometers from the ameliorating effect of warm ocean currents and from the moist rain-bearing air masses; other areas face the frozen Arctic or are located where mountain barriers inhibit advection of moisture. Using the Köppen climatic classification (Fig A87), Asia may be divided into three major climatic realms: Boreal Asia in the north (Dfb, Dfc, Dwa, Dwb, Dwd and ET), Desert Asia in the west and center (BS and BW) and Monsoon Asia in the south and east (Af, Aw, Cfa, Cfb, Cs and Cw). Boreal Asia is the largest and is the dominant cold season, wind-generating climatic realm. A complex set of interactive controls determines the climates of Asia. Concomitantly, many multifaceted physiogeographic features influence local climates and give distinctive regional climatic character to places. To understand Asia's climates, one must initially read simplified generalizations, then detailed professional publications written in a multitude of languages.

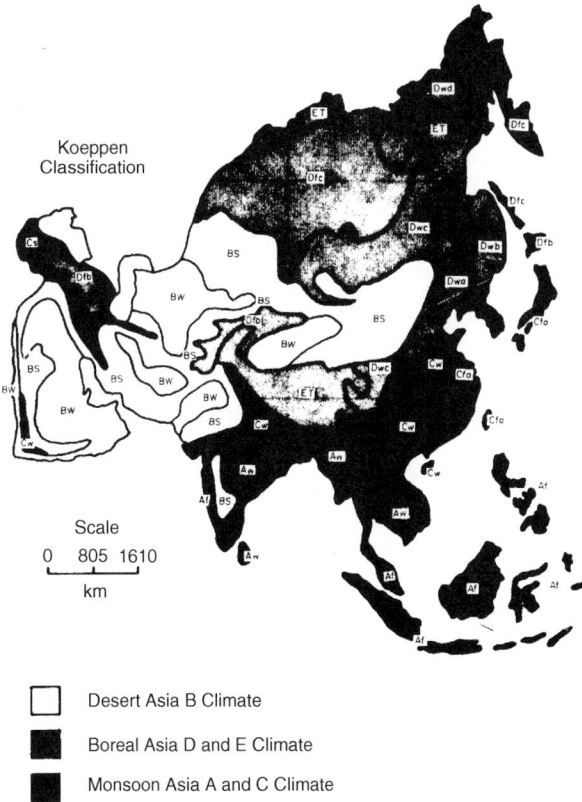

Koeppen
Classification

Scale

0   805 1610

km

☐ Desert Asia B Climate

■ Boreal Asia D and E Climate

■ Monsoon Asia A and C Climate

Figure A87 Asian climates.

Synoptic circulation

Synoptic climatology is a valuable non-quantitative approach to climatic description, emphasizing non-periodic weather episodes associated with transient weather disturbances and distinctive circulation patterns at various scales and duration. However, in a continent as large and as climatically complex as Asia, there is currently no comprehensive world climatological scheme based on weather regimes available that could be used to describe the dynamics of weather. As a consequence, statistical averages of climatic elements and synoptic circulation are employed in this study of Asia's climates as the method of climatic description.

Radiation component

The unequal distribution of solar radiation over Asia is the primary factor in its multifaceted climatic genesis. Asia's great latitudinal spread determines the intensity and duration of solar radiation. The annual march of solar radiation is determined by the angle at which the sun's rays strike the surface. In Asia's tropical belt, the sun remains high with little seasonal variation, accounting for continuous warm-to-hot year-round temperatures. In Asia's midlatitudes, solar radiation receipts exhibit a strong seasonal maximum and a strong seasonal minimum that are reflected in greater seasonal variations in temperature than in the tropical belt. In Asia's high latitudes there is a period of limited to no solar radiation received at the surface of the Earth, then a period of almost continuous solar radiation, resulting in a season with extremely low temperatures in the winter or low-sun period. Total solar radiation received at the Earth's land–sea surface during the entire year in kcal cm^{-2} year^{-1} ranges from 140–180 in Asia's tropical belt, through 160–120 in Asia's midlatitudes to 100–60 in Asia's high latitudes. Maximim solar radiation received at the Earth's surface occurs in the steppe and desert regions of southwest and west-central Asia. Transformation of available solar radiation is an essential ingredient in the process that produces Asia's climate, particularly temperature ranges (Borisov, 1965).

Moisture component

Asia's vast land mass, high average elevation and position within the westerly belt of planetary winds, combined with great variations in solar radiation, temperature and pressure cells produce a wide range in moisture variables such as specific humidity, relative humidity, absolute humidity, dew point temperature and precipitable water. Air masses from maritime tropical oceans and seas are the primary source of Asia's moisture. Asia is bordered on three sides by oceans: the Indian in the south, the Pacific in the east, and the Arctic in the north, along with the Red, Mediterranean, Black and Caspian seas on the west. Rate of evaporation from water and land surfaces depends on air temperature, wind and surface conditions, thus the potential for evaporation declines from tropical latitudes. Zonal average annual evaporation potential ranges from about 140 cm and 20°N to 20 cm at 60°N; specific humidity in grams per kilogram varies from 14 at 20°N to 5 at 60°N. Large amounts of evaporated water in the atmosphere and large amounts of latent and sensible heat along the east coast of Asia make this segment of the Earth's surface an important breeding ground for tropical disturbances and storms. Asia's moisture component is best understood as the circulation of water by a number of successive physical processes that take place over continental and ocean surfaces, i.e. evaporation, cloud formation, precipitation, moisture transfer and surface runoff. The moisture component is an important climate-forming factor associated with climatic elements such as precipitation, evaporation, cloudiness, fog, humidity and continentality (Meigs, 1953; Sulakvelidze, 1969).

General circulation features

The basic circulation features that give regional character to Asia's climates are the movement of air masses, transformation of air mass properties and interactions between air masses along fronts. Some typical characteristics of air masses and their source regions are as follows (Arakawa, 1969; Takakashi and Arakawa, 1981: Lydolph, 1977).

- Continental arctic (winter) air masses are stable, intensely cold (−55°C to −35°C), extremely dry (specific humidity is approximately 0.05–0.20 g kg⁻¹), with inversions; cA air masses enter Boreal Asia from the Arctic seas and at times penetrate to the Pacific Ocean; they are associated with clear skies, low temperatures and dense air.
- Continental polar (winter) air masses are very stable with strong temperature inversions, cold (−35 to −20°C) and dry (specific humidity is approximately 0.2–0.6 g kg⁻¹); they spread north and south of the 'great ridge' of high pressure in Siberia and Outer Mongolia at approximately 50°N and dominate the weather of the entire continent: generally, cP air masses produce clear, cold, and cloudless weather.
- Maritime tropical (winter) air masses are, for the most part, unable to penetrate eastern Asia in winter due to the intensity of the great Siberian high-pressure cell (Figure A88); mT air masses that influence the weather in southwestern Asia in winter circulate around the eastern edge of the Azores-Bermuda high-pressure cell where upper level subsistence is strong; mT air masses are cool, dry, and stable in winter.
- Maritime polar (winter) air masses have difficulty penetrating deep into the Asian continent in winter because the Siberian high pressure cell and strong outflowing winds in eastern Asia inhibit mP incursions; mP air masses are basically conditionally stable, cool (0–10°C) and moist (specific humidity is approximately 3–8 g kg⁻¹); they have significant influence on the weather of maritime eastern Siberia, Manchuria and Korea.
- Continental polar (summer) air masses are, in general, stable or conditionally stable, cool (5–15°C) and somewhat moist (specific humidity is approximately 4–9 g kg⁻¹); eastern and southern Asia are so dominated by mT air in summer that cP air contributes little to the weather of that region.
- Maritime polar (summer) air masses contribute to the summer weather of Manchuria, eastern Siberia and Japan and are in most cases stable to conditionally stable, mild (2–14°C), and humid (specific humidity is approximately 5–10 g kg⁻¹); conflicts between mP and mT air masses in summer lead to the formation of a semistationary front in northeastern Asia with associated overcast and drizzly weather.
- Maritime tropical (summer) air masses are initially conditionally stable and very moist (specific humidity is approximately 15–

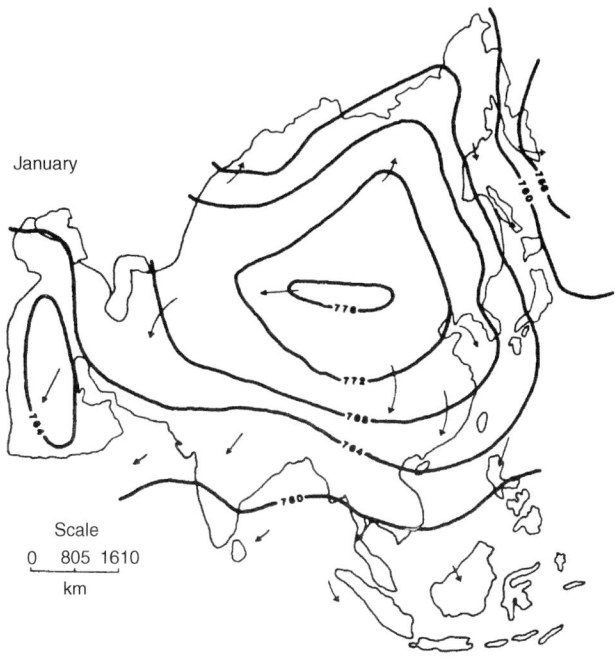

January

Scale

0   805 1610

km

—— ₇₉₀ —— Average atmospheric pressure in mm

Figure A88 Atmospheric pressure and winds in January.

20 g kg⁻¹); mT air masses provide moisture for the summer monsoon, and in spring invading mT air meets cP air in central and southern China producing very active cyclogenesis; mT air masses dominate the weather of eastern and southern Asia during the high-sun period.

- Continental tropical (summer) air masses develop over the Sahara and southwestern Asia in summer and are conditionally stable, dusty, very hot (30–40°C) and hold moisture (specific humidity is approximately 5–10 g kg⁻¹); cT air flows north and northwest in summer into western Asia, advecting heat and absorbing moisture en route, in addition to contributing a characteristic opalescent haze to the local climates.

Asian air masses show great contrasts in temperature, moisture and density; they are a manifestation of tremendous seasonal differences within source regions and are seasonally very pronounced. At any particular site, properties of air masses depend not only on the nature of the source region but also on the modification the air mass experiences en route from the source region. Air mass modifications en route are of great importance in determining the nature of weather associated with an air mass (Arakawa, 1937; Tu, 1939).

Siberian high-pressure cell

General atmospheric circulation over Asia is controlled by centers or cells of high or low pressure whose axes generally are east–west and whose pressure centers vary drastically from winter to summer. In winter an extensive, well-developed high-pressure cell, centered over Mongolia, dominates the weather over most of eastern and southern Asia (Figure A88). A thermally induced high, this pressure cell is very shallow and cannot be traced at the 850 mbar level. Triangular in shape, the apex of the pressure cell extends west to the Caspian Sea, and its base is anchored in the northeast near the Verkhoyansk Mountains of Siberia and in the southeast near the Chin Ling Mountains of eastern central China. This Siberian high effectively blocks penetration of moisture-bearing, moderating maritime air masses in winter. Acting as a wedge, forcing air masses to skirt northeasterly from the Black Sea across much of northern Siberia and to flow Southerly along the Kamchatka Peninsula towards northern Japan, this intense high generates continental, land-trajectory, dry and cool, low-level air masses that surge from the north and northeast to the south-west across all of southern and eastern Asia. This high,

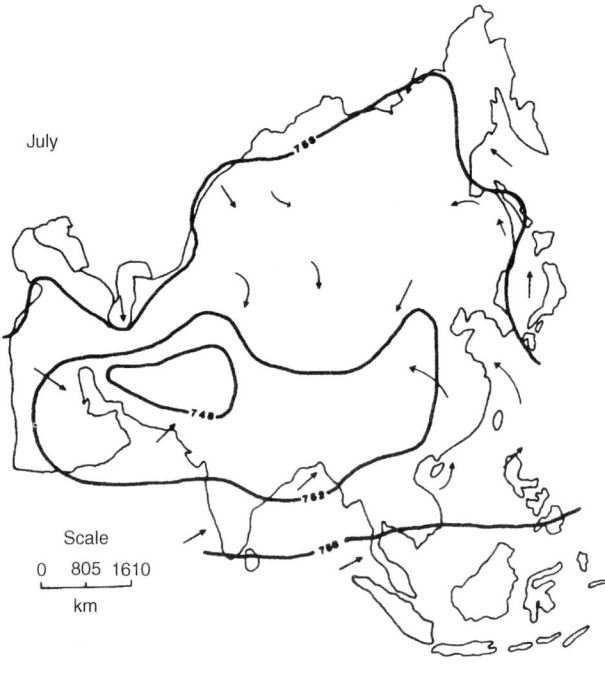

July

Scale

0   805   1610

km

——7oo—— Average atmospheric pressure in mm

Figure A89 Atmospheric pressure in July.

whose core and area of highest pressure is focused on Lake Baikal, pulsates in intensity and at times breaks into smaller high-pressure cells of less intensity. February marks the height of the Siberian high's dominance of the winter circulation over Asia, although a similar pattern, depicted on the January map, exists from November to March. In April the Siberian high weakens and shifts its center westward into a position over northeastern Russian Central Asia. It then dissipates in May, when the weather map of Asia begins to be dominated by an intensive, thermally induced low-pressure cell over southwestern Asia that is focused on the tip of the Arabian Peninsula, the Iranian Plateau and the Thar Desert of Pakistan (Figure A89) (Davydova *et al.*, 1966).

Southwest Asian low
In summer southern Asia's weather is dominated by a large, deep, thermally induced low-pressure cell that extends from the Arabian Peninsula to Central China and from central India to Russian Central Asia, centered approximately 30°N and 70°E. A complex cell, the Southwest Asian Low experiences east–west locational oscillations and occasional intense pressure deepening. Cyclonic circulation can be traced from the 700 mbar level and at the 500 mbar level to the tropopause; the circulation is dominated by anticyclonic circulation. Active large-scale subsidence associated with high-level anticyclonic activity produces intense middle tropospheric warming. In summer this low has the effect of interrupting the subtropical high-pressure system in the northern hemisphere by dividing the globe-girdling zonal band into two distinct large oceanic cells. One intense depression, the Southwest Asian low, induces a radical change in prevailing winds and storm tracts during the high-sun period. Air masses from the stable eastern end of the Azores–Bermuda high-pressure cell skirt this low from a northerly to northwesterly direction across the eastern rim of the Arabian Peninsula; less intense air masses from the northern quadrant of the Azores–Bermuda High sweep eastward across Turkey and into the northern extremities of Russian Central Asia. Air masses and storms spawned under the unstable western quadrant of the Hawaiian High sweep from the south and southeast in a northerly trajectory across Japan, extreme eastern China and the Russian Maritime Provinces. But the most constant and climatically significant air masses and storms are advected to this intense low-pressure cell because bent southwesterly trade winds spawned from a semi-permanent high-pressure cell over the Indian

Ocean in the southern hemisphere. In India and eastern Asia this modification of the general planetary wind system in summer constitutes the Asian monsoon (Chang, 1967).

Frontal dynamics and jet streams
The size and latitudinal position of the Asiatic land mass, the presence of numerous mountain ranges and highland areas in various directional alignments, complicated air mass characteristics and boundary surfaces, combined with the insular and peninsular characteristics of extreme southern and eastern Asia, give rise to seasonally complex frontal dynamics and jet-stream patterns. Lithospheric and hydrospheric energy-managing characteristics, along with seasonal variations and transformations of solar radiation, lead to the formation of weather-generating air masses along with modification of their properties as they are advected from source regions and interaction between the air masses along fronts. Frontal zones and wind systems conform to the location and circulation of air masses.

In winter four major zones of cyclonic activity are distinguished over Asia. One zone is located along the Asiatic arctic front, well above the Arctic Circle in northern Siberia, and extends along the shores of the continent. This front fluctuates greatly and, at times, arctic air masses penetrate to the Black Sea, southwestern Asia and eastern central Asia polar front, which develops in winter over the Mediterranean Sea and extends to the Caspian Sea. The third zone is along the East Asian polar front, which aligns itself in a northeasterly path from extreme southern Asia toward Japan. A fourth and final zone of cyclonic activity in winter is the east–west aligned South Asian intertropical front located near 10°S and traversing Java. Along these winter frontal zones, and moving in various directions and at varying speeds, depressions and anticyclones impart to the climates of Asia that special character by which one area differs from another.

During summer three major zones of pronounced cyclonic activity can be identified. The first is the southward-displaced Asiatic arctic front, which at times extends east–west across northwestern Siberia along the 70°N parallel. A second zone of pronounced cyclonic activity is the Asian polar front, which normally extends from the eastern tip of Lake Balkhash in Russian Central Asia, over Mongolia to the northernmost bend of the Amur River along the 50°N parallel. This oscillating, and at times southward-dipping, Asiatic polar front zone has been called the 'barometric backbone of Asia' in summer or the 'great ridge,' because a large number of anticyclones are observed each year between 50 and 55°N. The third major zone is along the South Asian intertropical front. Extending from the south-central portion of the Arabian Peninsula, across Pakistan and northern India, and over China at approximately 25°N, the South Asian intertropical front is well defined in some locations, but in others it is weak or absent. In this belt, convergence of surface winds results in large-scale lifting of warm, humid, relatively unstable air, producing numerous weak, rain-generating disturbances (Trewartha, 1981).

Pronounced horizontal variations in temperature, humidity and stability are found along major frontal zones and because frontal zones are areas of steep horizontal temperature gradients, there are usually strong winds aloft. Strong, narrow jet-stream currents, thousands of kilometers long, hundreds of kilometers wide and several kilometers deep, are concentrated along a nearly horizontal axis in the upper troposphere or stratosphere, producing strong vertical and lateral shearing action. Three distinct jet systems, the polar front jet, the subtropical jet and the tropical easterly jet, have a major impact on weather and climate in Asia. Latitudinal location of these jet systems, especially the polar jet, shifts considerably from day to day and from season to season, often following a meandering course. However, in general, the polar jet gives rise to storms and cyclones in the middle latitudes of Asia, the subtropical jet, noted for a predominant subsidence motion, gives rise to fair weather, and the tropical easterly jet is closely associated with the Indian monsoon. The polar front jet and the subtropical jet are most intense on the eastern margin of the continent and are most well developed during winter and early spring. Jet-stream wind speeds of up to 500 km hr^{-1} have been encountered over Japan. The subtropical jet, at times in winter, forms three planetary waves with ridges over eastern China and Japan. It frequently merges with the polar front jet, producing excessively strong jet-wind speeds. During summer the subtropical jet loses its intensity and is mapped only occasionally (Reiter, 1967).

Asian monsoon
The Asian monsoon is a gigantic multifactored, multifaceted weather system composed of diverse heat and moisture cycles that are

intimately related to local topography, modified air masses, quasi-stationary troughs and ridges, and jet streams. Weather over most of eastern and southern Asia during winter is dominated by the Siberian high-pressure cell and outflowing continental air masses (Figure A87). Low-level air flow is mainly from the north and is cold, dry and stable. Successive waves of cold northeasterlies commence in late September and early October, progressing farther and farther southward until they reach the southern China coast by late November or early December. In the higher midlatitudes during the low-sun period, a steep north–south pressure gradient persists. Southwestern Asia and the Indian subcontinent are protected from the cold Siberian air by blocking mountain ranges and highlands because cold fronts extend only 2000–3000 m or so above sea level. Cold fronts are frequently accompanied by gales that persist for days; humidity and temperatures are low. During March and April the Siberian high gradually weakens and incursions of moist maritime tropical air from the south and east replace the cold-to-cool northeasterlies, producing widespread stratus clouds, fog and drizzle that may persist for days (Das, 1968).

In summer a combination of the deep and elongated Southwest Asian low pressure cell, extending from the Arabian Peninsula to China (Figure A89), and the South Asian intertropical front, reaching its maximum poleward displacement, sets the stage for a marked seasonal reversal of air flow. Warm, moist and conditionally unstable southwesterly maritime air masses from relatively cooler oceanic source regions eventually overcome blocking atmospheric conditions, stream across India and flow northwestward over continental southern Asia, China and Japan. Considerable convective activity develops over land, and heavy showers and thunderstorms contribute a large portion of the summer rainfall maximum of the region. The characteristics, attributes, onset, and duration of the monsoon are site specific. In all cases, low-level wind patterns and resultant summer weather are complicated by topography. Distance from air mass source regions, moisture content of air masses, orographic barriers and atmospheric disturbances associated with cyclonic or inter-air mass convective activity determine distribution and quantity of precipitation. There is a pronounced difference between East Asian and South Asian monsoonal weather. The East Asian winter monsoon is much stronger than the South Asian winter monsoon, and the East Asian summer monsoon is much weaker than the South Asian summer monsoon (Kurashima, 1968; Trewartha and Horn, 1980).

Climatic elements

Climates vary from place to place in Asia because of differences in quantity, intensity and spatial dynamics of critical weather and climatic elements – specifically, temperature and precipitation along with atmospheric disturbances. Climatic elements are given regional distribution through the impact of climatic controls such as solar radiation and latitude, land mass and water body contrasts, high- or low-pressure cells, winds and air masses, storms, mountain barriers, altitudes above sea level and cold or warm ocean currents. Climatic controls working in various combinations, interactions, reactions and intensities induce modifications in hypothetical or short-term temperature and precipitation averages and produce those aspects of weather that give distinct character to climatic regions in addition to climatic change (Masatoshi and Urushibara, 1981).

Temperature

Great latitudinal differences ranging from south of the equator to above the Arctic Circle, a huge land mass, numerous mountain barriers to air mass flow, high elevation and distance from the ameliorating effects of warm ocean currents produce a wide range in mean annual energy budgets, average temperatures and temperature extremes (Figure A90). Northern Asia, specifically Siberia, is climatically isolated from moist tropical air masses and records the greatest mean annual temperature range on Earth. At Verkhoyansk, in the valley of the Yana River, January temperature averages −49°C, July averages +15°C, and the absolute temperature range is 103°C. Oymyakon, located in the same physiogeographic region, records an absolute temperature range of 104°C. Siberian winter low temperatures are proverbial and, for most of Siberia, the January mean annual temperature is less than −25°C. In contrast, much of southwestern Asia is subject to excessive heat (Figure A91). Summer temperatures in the interior lowlands of the Arabian Peninsula and Iran are so extreme that they rank among the hottest areas of the world. Mean daily temperatures in the Arabian Desert approach 36°C

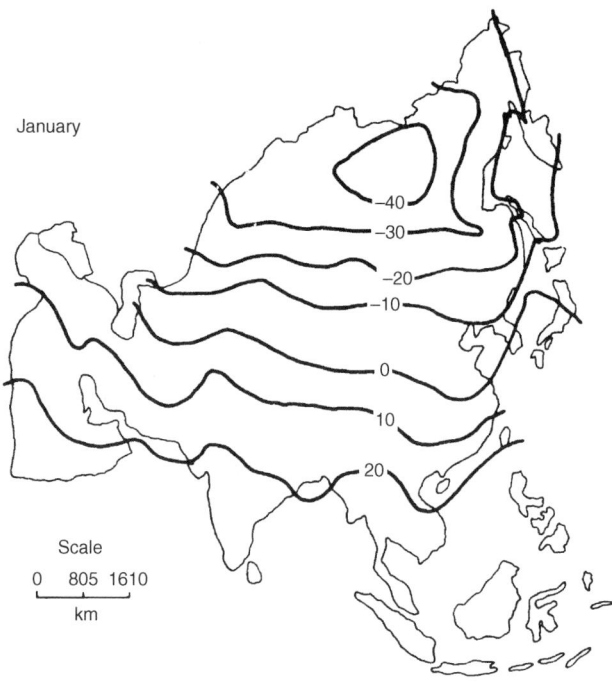

January

Scale
0   805 1610
         km

——10—— Temperature in Centigrade

Figure A90 Atmospheric temperature in January.

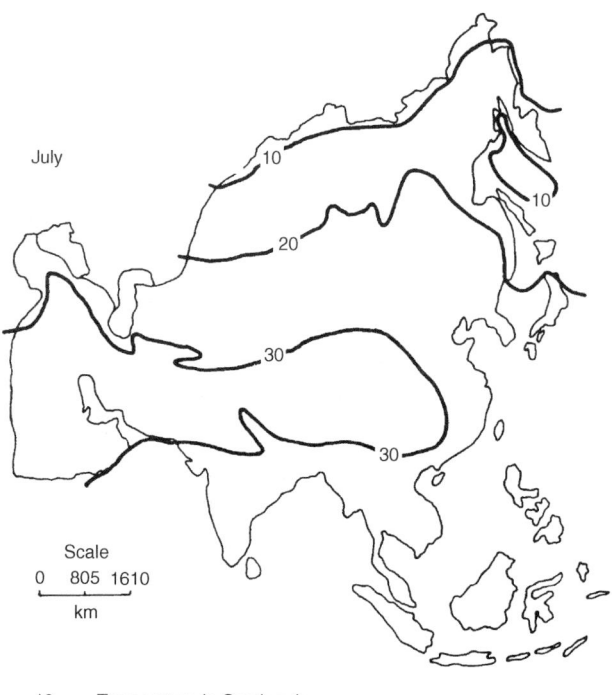

July

Scale
0   805 1610
         km

——10—— Temperature in Centigrade

Figure A91 Average temperature in July.

in July, and in the Iranian valleys of Khuzestan and Luristan daily maxima often exceed 44°C, with temperatures in excess of 50°C reported for Abadan. Yet Singapore, located about 150 km north of the equator, has an annual mean temperature of 27°C, with January averaging 26°C and July averaging 28°C; here there is very little temperature change from one month to another and the annual range is only 2°C. Asia's spatial temperature differences and ranges are greatest in winter, least in summer and more extreme in all seasons within the landlocked core rather than in the southern and eastern maritime periphery. Winter isotherms reflect the influence of ameliorating ocean currents, mountain barriers, altitude above sea level, high-pressure cells and solar radiation. Summer isotherms reflect solar radiation and latitude along with altitude (Miller *et al.*, 1983).

Precipitation

Mountain barriers, altitude, storms, and unusual winds combine to provide Asia with a most uneven distribution of moisture from the atmosphere. In general, precipitation receipts increase from north to south and from southwest to southeast (Figure A92). Throughout most of Siberia, annual average precipitation is scarcely 250 mm and in Russian Central Asia, the Tarim Basin, Gobi Desert, Plateau of Iran and the Arabian Peninsula annual average precipitation is scarcely 100 mm. Some of the driest areas are in the southwest with Eilat, Israel, averaging only 27 mm annual precipitation, Hail on the Arabian Peninsula only 17 mm annually and Tabriz, Iran, only 7 mm annually. On the other hand, the north eastern sector of the Indian subcontinent is considered by many scholars to be the wettest region on the surface of the Earth. Here the South Asian summer monsoon precipitates moisture in a veritable deluge. At Cherripunji the mean annual precipitation is 11 419 mm, with a 24 h maximum of 974 mm. Most of the precipitation is received from May to September, and the monthly average is 5 mm in December and 2 922 mm in June. The seasonal distribution of precipitation is of great importance because the summer heat and continental character of Asia affect precipitation effectiveness and thermal efficiency. In insular and island-studded southern, southeastern and eastern Asia, where there is limited or no frost, rainfall is relatively evenly distributed throughout the year (Af, Cfa and Dfb). Interior and rain-shadow locations in southern and eastern Asia experience a distinct dry season in the lowsun period or winter (Aw, Cw, Dwa and Dwb). Mediterranean Sea-influenced western Asia and the coastal regions of the Black and Caspian seas receive most of their precipitation in winter, at least three times as much precipitation in the wettest winter month as the driest month of summer (Cs). Precipitation in the dry realm of southwestern Asia, Russian Central Asia, the Tarim Basin and the Gobi Desert is minimal, erratic and uncertain as to time of fall, but most of the precipitation is received in summer from violent convective showers (BW) or a combination of violent convective showers and frontal activity (BS). In the subarctic and arctic areas of Siberia (Dwc, Dwd and ET), summer is the season of maximum temperatures, highest specific humidity, deepest penetration of maritime air masses under the influence of the summer monsoon, and is the period of maximum precipitation. Asia's central dry realm, aligned in a southwest–northeast orientation from the Red Sea coast of the Arabian Peninsula to the Plateau of Mongolia, separates Asia's southern and eastern warm-to-hot, wet belt from Asia's northern and northeastern cool-to-cold, limited precipitation belt (Griffiths and Driscoll, 1982).

Storms

Disturbances of the atmosphere marked by strong winds, dense clouds, heavy rain, sleet, hail or snow, or by combinations of two or more of these, are usually intense in Asia, for there exist at any one time great variations in local heat balances, water balances and air temperatures. Spatial and temporal differences in temperature, pressure, humidity and latitude give rise to an endless variety of types and patterns of local storms. Five categories of storms, induced by variations in air masses and atmospheric perturbations, give regional character to Asia's climates, i.e. wind storms, dust and snow storms, thunderstorms, tornadoes and typhoons.

Wind, dust and snow storms

Wind conveys heat, moisture and light-fugitive materials from one location to another and, in part, determines the motions of minor atmospheric perturbations and cyclonic storms. The magnitude of winds and their thermal–moisture attributes, rather than their direction, make them a significant climatic factor. Owing to its variety in relief and exposure, Asia is subject to numerous local winds and their associated weather. Ubiquitous are the warm, dry, gusty, downslope föhn-type winds generated, in most cases, by passing atmospheric disturbances in highland or mountainous areas. In Soviet Central Asia and Siberia, these föhn-like winds experienced along the Caspian Sea locally are called germich; in Uzbekistan, afghanetz; in Tadzhikistan, harmsil; in and near the great Fergana Valley, ursatevskiy and kastek; along the Russian-Chinese border at the Dzhungarian Gate, east of Lake Balkhash, evgey; in Iran, samoon; and on the islands of Indonesia, bohorok and kumbang. Also, for each season of the weather year, representative winds of local significance have been identified and named.

- Summer. A hot, strong, constant northerly summer wind, carrying considerable dust and obscuring the atmosphere, is called meltemi in Turkey; shamal, in Iraq; karaburan, in the Tarim Basin of Sinkiang and northwestern China; chang, in Turkmenia; and seistan or *winds of one hundred days*, in Iran. A hot, dry, southerly wind blowing off the sun-scorched deserts of southwestern Asia is known regionally as a khamsin and simoon. At times, on the southern edge of a modified arctic air mass that advects into southern Siberia and Russian Central Asia, hot, dry moisture-absorbing sukhovei winds eliminate local cloud cover, permit intense solar radiation to strike the Earth, and reduce relative humidity at the surface to a very low value both day and night. Also, a type of hotseason squall, usually accompanied by violent thunderstorms, heavy rain and hail, designated as a nor'wester, is experienced on the northern plains of the Indian subcontinent and Myanmar.
- Fall. Cold and often very dry northerly or northeasterly winds, preceded by a cold front, blow with great strength and violent gusts down from the mountains and high plateaus of northern Siberia and the Caucasus. Bora is the term used to identify this type of wind in the region between the Black and Caspian seas, and sarma and kharanka are bora-type winds in the Lake Baikal region of southern Siberia.
- Winter. Very cold northerly or northeasterly gale-force winds, often blowing at temperatures below −20°C and accompanied by falling or drifting snow, are generated from the back side of winter depressions in Siberia and Russian Central Asia. Sensation of cold

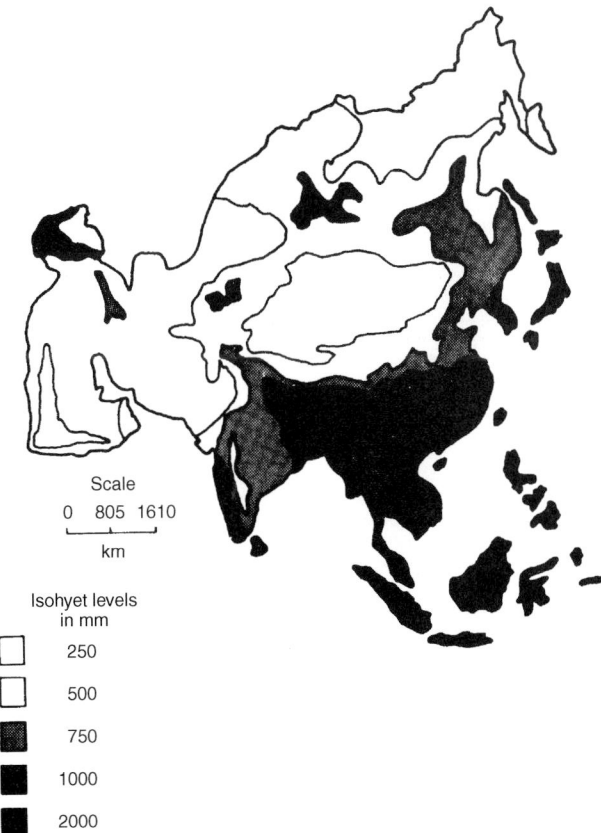

Scale
0   805   1610
——————
km

Isohyet levels
in mm

☐   250
☐   500
▨   750
■   1000
■   2000

Figure A92 Asian precipitation.

temperatures is increased by the low wind-chill factor associated with a buran or a purga, and the break in the comparative calm associated with the Siberian high-pressure cell.

- Spring. Continental depressions, passing eastward over central China and the Yellow Sea in spring and early summer, bring heavy overcasts, high humidity and rain to China and Japan. Bai-u, mai-yu, or plum rains, as they are called, are an extended period of unstable weather caused by stagnation of the polar front.

Winds that are repeated are given a medley of names by those who experience them. Such winds occur in a multiplicity of temporal and spatial scales, and the largest and most pronounced are a major contributing factor to the broad aspects of Asia's climate. Still, even the smallest wind, dust and snow storms are significant for they are responsible for day-to-day weather events (Noveck, 1959; Critchfield, 1983).

Thunderstorms and tornadoes

In Asia thunderstorms reach their maximum development over lowlands within the equatorial trough region. Each year areas of peninsular and insular southern Asia experience 180 or more days of such storms. These thunderstorms form in a conditionally unstable atmosphere when the measured lapse rate is greater than the saturated lapse rate and less than the dry adiabatic lapse rate. Most thunderstorms are ordinary convective cumulonimbus clouds within maritime tropical air masses that produce localized precipitation. Air mass thunderstorms, randomly scattered, are primarily initiated by daytime solar heating of land surfaces: frontal and orographic thunderstorms have distinct patterns and movements because they are triggered in a place or zone where unstable air is forced upward. Occasionally, thunderstorms modify their internal structures, grow larger and eventually become severe thunderstorms. Severe thunderstorms may produce hail, strong surface winds and tornadoes. At times they interact in a synergistic manner to produce intense and sometimes violent thundersqualls (Yoshimura, 1971; Magono, 1980).

The average number of days with thunderstorms and monthly distribution of thunderstorms is a complex matter. Malaya and Sumatra record a thunderstorm more than every other day – over 180 days with thunderstorms. Convective thunderstorms develop more frequently in these two places because of strong insolation and low wind velocities. Clouds of convective thunderstorms reach elevations of 12 000 m or more, and rainfall associated with these cells is short in duration, intense and localized. Few thunderstorms (less than five per year) are observed over the tundra regions of Siberia, the hill and desert regions of Jordan, the southeastern quadrant of the Arabian Peninsula, southern Iran, Afghanistan and Pakistan. Thunderstorm activity is a summer phenomenon in China, much of Russian Central Asia, Singkiang, Mongolia, the Maritime Provinces of the Russian Far East and Siberia. Annual thunderstorm frequency over Lebanon, Jordan, Israel and Kuwait is approximately 20 per year; they commence during the fall transitional period, are more prevalent in winter, and decline in number during the spring transitional period. Maximum thunderstorm activity over most of the Indian subcontinent and the Arabian Peninsula occurs in spring.

Remotely sensed data gathered from satellites and analyzed with the aid of computers have enabled research meteorologists and climatologists to identify, map and track a new type of thunderstorm, i.e. thunderstorm complexes. A typical Asian thunderstorm forms during the period of maximum daily surface heat and dissipates following precipitation at night when the energy from the sun is no longer reaching the Earth's surface. In contrast, thunderstorm complexes initially form during the heat of the day and grow to maximum size at night. By 0200 or 0300 h old cells and newly formed cells merge into nearly circular configurations hundreds of times larger than individual thunderstorms. Thunderstorm complexes or 'mega storms' may produce intense and widespread rainfall or hail, leading to flash flooding and crop destruction, violent wind and, at times, nocturnal tornadoes (Fujita, 1983).

The most destructive spin-off of a thunderstorm is the tornado, an extremely violent rotating column of air that descends from a thunderstorm's cloud base, causing destruction along a narrow track. An Asian tornado, usually a funnel-shaped cloud with upward-spiraling winds of terrific velocity travels in a straight track 150–500 m wide and several kilometers long at speeds ranging from 50 to 100 km h^{-1}. In general, three things are necessary to create a tornado: convergence, rotation and wind shear. Convergence occurs when air rises and two large air masses of different temperature and moisture

content collide. Rotation of the air is imparted by the Earth turning on its axis – the Coriolis effect. Wind shear, a strong shift in wind speed and direction, when combined with the thunderstorm's updraft, produces a whirling column of air. Jet streams are important because they supply an additional source of energy. Many tornado outbreaks and jet-stream migrations cause the areas of greatest tornado activity to shift as surface temperatures decrease during fall and winter months (Eagleman, 1985).

As the whirling mass of unstable air gains force, a rotating column of white condensation is formed at the base of the cloud. Dirt and debris sucked into this whirlwind darken the column, and as it reaches the ground it becomes a roaring swirl of deadly force. A notable feature of the climate of Asia is the relative infrequency of tornadoes, particularly in comparison with central and eastern North America at similar latitudes. Although records are inadequate, there are sufficient data to conclude that tornadoes occur on the order of once every 3–5 years in the northern Caspian Sea area and Russian Central Asia during May, June and July; once or twice each year in the northeastern part of the Indian subcontinent during March, April and May; two to five or more annually in China and Japan during August and September; and at least one every 4 years in the Philippines, Indonesia and Malaya. Tornadic storms occur in many other areas of Asia (many more than listed in the official data) and are given local names or designations.

Tropical cyclones

One of the most powerful and destructive storms is the tropical cyclone. Loosely referred to as any pressure depression (near 1000 mbar) originating above warm oceans in tropical regions, tropical cyclones are an important feature of the weather and climate of southern and eastern Asia, particularly from July to October. Different words are used worldwide to describe these tropical storms: typhoon in the western Pacific; baquios or baruio in the Philippines; tropical cyclone in the Indian Ocean; willy-willys in Australia; hurricane in the eastern Pacific and Atlantic; cordonazo in Mexico; and taino, in Haiti. The Royal Observatory of Hong Kong classifies tropical cyclones into four categories (Hsu 1982): (1) tropical depression, with maximum sustained winds of less than 18 m s^{-1}; (2) tropical storm, with maximum sustained winds between 18 and 25 m s^{-1}; (3) severe tropical storm, with maximum sustained winds between 25 and 33 m s^{-1}, and (4) typhoon, with maximum sustained winds of 33 m s^{-1} or more.

Although the organization and development of tropical cyclones are not fully understood and are under intensive study, their formation is associated with warm ocean surfaces of not less than 27°C located between 5 and 10°N, light-to-calm initial winds, and waves or troughs of low pressure deeply embedded in easterly wind streams, converging into an unstable atmosphere zone (the ITCZ). Large quantities of latent heat, released through condensation, are converged and transferred to higher levels, deepening the pressure center and intensifying the storm. An almost circular storm of extremely low pressure, into which winds spiral with great speed, is formed. Such storms never originate over land but do penetrate the margins of the Asian continent. Asian tropical cyclones travel slowly at speeds of 16–48 km h^{-1} and cut a destructive storm path 80–160 km wide; winds in the wall-cloud area achieve speeds in excess of 200 km h^{-1}. Strong winds and heavy rainfall are associated with the passage of a tropical cyclone over water and land. Storm tracks vary annually and no two recorded tracks have been exactly the same. Despite all irregularities, most tropical cyclones tend to move westward, then poleward, and finally turn eastward towards higher latitudes under the influence of both internal circulation and external steering currents, penetrating into the belt of westerly winds. This awesome tropical storm contributes between 25 and 50% of annual precipitation in many tropical weather stations. Flooding, destructive wind force and storm surge are responsible for much property damage and many human casualties (Kutzbach, 1979).

Tropical cyclone occurrence is restricted to specific seasons depending on the geographical location of the storm-affected region. In a 33-year period (1946–1978) the Royal Observatory of Hong Kong recorded 31 severe tropical storms per year, with 83% of the annual total recorded between June and November. Approximately 50% of these tropical storms attained typhoon intensity. Storm frequency in the Bay of Bengal varies widely throughout the year, but, in recent years, the majority of all cyclonic disturbances that developed into severe cyclonic storms did so in May, October or

November, with four to six tropical cyclones recorded each year. Frequency of depressions and storms in the Arabian Sea is much less than in the Bay of Bengal. Almost every summer some part of Asia is affected by severe tropical storms or tropical cyclones. From mid-November to April, very few tropical storms pass over the coasts of China and Korea, but in the warm July–October period, numerous tropical depressions, tropical storms, severe tropical storms and tropical cyclones (typhoons) are experienced. In the 1884–1955 period at least 438 tropical cyclones crossed various sections of the Chinese and Korean coasts. Japan's tropical cyclone season begins in June and ends in November, reaching its peak in September. In the 1918–1947 period 85 typhoons were reported in Japan. The tropical cyclone season is slightly longer in the Philippines, extending from June to December. Almost 90% of all tropical cyclones in the 1948–1962 period were noted in the summer and fall seasons. Tropical cyclones usually weaken over land, and few penetrate and persist more than 500 km inland. The rise in sea level, when combined with high tides, accounts for more damage and loss of life in Asia than violent wind (Riehl, 1979).

Conclusions

The climatic base of Asia is broadly similar in major areas of dense population. Heat and moisture variability or periodic inadequacy have been contributing factors to the marked concentration of settlements in river valleys or sites where irrigation water is available. A combination of physical circumstances and human ingenuity, in general, has produced a highly intensive Asian agricultural system. Agriculture, with some exceptions, is still largely subsistence rather than commercial. Nowhere in the world do so many people depend on subsistence agriculture, is the agricultural population so dense over such a large area or is food or famine so related to atmospheric conditions. Asia's food problem is one of the most critical contemporary issues, and optimization of agricultural output depends not only on reducing agricultural hazards such as storms, droughts, floods, hail, frost, insect pests and disease, but also on achieving the full potential of climatic energy and moisture available (Bryson and Murry, 1977).

Human-induced, local, microclimatic changes and long-term alteration of Asia's present macroclimatic regionalization could plunge this continent into absolute chaos. Crops currently are grown in areas with climates only marginally acceptable to plants. An annual change in temperature of only a few degrees Celsius would alter the climates of Asia sufficiently to render marginal agricultural regions unacceptable for food production and would wreak havoc on local food supplies. Moreover, a combination of human activities that modify climatic elements with natural causes of climatic change could lead to more frequent or more severe changes in Asia's three major climatic realms. For those struggling to secure a meager existence from a hectare or two of land in Asia, any climatic change would disrupt lifestyle; humans and human institutions are adjusted to precisely the weather that prevails (Budyko, 1977; Dando, 1980; Yoshino, 1984).

William A. Dando

Bibliography

Arakawa, H., 1937. The air masses of Japan, *Am. Meteorol. Soc. Bull.*, **18**, 407–410.
Arakawa, H. (ed.), 1969. *Climates of Northern and Eastern Asia*. World Survey of Climatology, Vol. 8. Amsterdam: Elsevier.
Borisov, A.A., 1965. *Climates of the USSR*, R.A. Ledward (trans.). Chicago: Aldine.
Bryson, R. and T. Murry, 1977. *Climates of Hunger*. Madison: University of Wisconsin Press.
Budyko, M.I., 1977. *Climatic Changes*, American Geophysical Union (trans.). Washington, DC: American Geophysical Union.
Chang, J., 1967. The Indian summer monsoon, *Geog. Rev.*, **57**, 372–96.
Critchfield, H., 1983. *General Climatology*. Englewood Cliffs, NJ: Prentice-Hall.
Dando, W., 1980. *The Geography of Famine*. London: Edward Arnold.
Dando, W., 1983a. *An Introduction to China's Diverse Physical Geographic Base*. Grand Forks, ND: US Dept. of Education, International Understanding Program.
Dando, W., 1983b. Famine in China, 1959–1961: some geographical insights, in *China in Readjustment*, C.K. Leung and S.S.K. Chin (eds). Hong Kong: University of Hong Kong Press.
Das, P., 1968. *The Monsoons*. New York: St Martin's Press.
Davydova, M., A. Kamenskii, N. Nekiukova and G. Tushinskii, (1966) *Fizicheskaia Geografiia SSSR*. Moscow: Prosveshchenie.
Eagleman, J.R., 1985. *Meteorology*. Belmont, California: Wadsworth.
Fujita, T., 1983. On the trail of twisters, *University of Chicago Mag.*, **75**, 6–13.
Griffiths, J. and D. Driscoll, 1982. *Survey of Climatology*. Columbus, Ohio: Charles E. Merrill.
Hsu, S.I., 1982. *Tropical Cyclone in the Western North Pacific*, Occasional Paper No. 24. Hong Kong: The Chinese University of Hong Kong, Department of Geography.
Kurashima, A., 1968. Studies on the winter and summer monsoon in East Asia based on dynamic concept. *Geophys. Mag.*, **34**(2), 145–235.
Kutzbach, G., 1979. *The Thermal Theory of Cyclones*. Boston: American Meteorological Society.
Lydolph, P.E., 1977. *Climates of the Soviet Union*. World Survey of Climatology, Vol. 7. Amsterdam: Elsevier.
Magono, C., 1980. *Thunderstorms*. Amsterdam: Elsevier.
Masatoshi, M. and R. Urushibara, 1981. Regionality of climatic change in East Asia, *GeoJournal*, **5**, 121–132.
Meigs, P., 1953. World distribution of arid and semiarid homo-climates, in *Reviews of Research on Arid Zone Hydrology*. Paris: UNESCO, pp. 203–209.
Miller, A., J. Thompson, R. Peterson and D. Haragan, 1983. *Elements of Meterology*, 4th edn. Columbus, Ohio: Charles E. Merrill.
Oliver, J., 1984. *Climatology*. Columbus, Ohio: Charles E. Merrill.
Noveck, S., 1959. *Russian–English Glossary of Physics of Fluids and Meterology*. New York: Interlanguage Dictionaries.
Reiter, E., (1967) *Jet Streams*. Garden City, NY: Anchor Books.
Riehl, H., 1979. *Climate and Weather in the Tropics*. London: Academic Press.
Sulakvelidze, G., 1969. *Rainstorms and Hail*, I. Schechtman (trans.). Jerusalem: Israel Program for Scientific Translations.
Takahashi, K. and H. Arakawa, 1981. *Climates of Southern and Western Asia*. World Survey of Climatology. Vol. 9. Amsterdam: Elsevier.
Trewartha, G.T., 1981. *The Earth's Problem Climates*, 2nd edn. Madison: The University of Wisconsin Press.
Trewartha, G. and L. Horn, 1980. *An Introduction to Climate*. New York: McGraw-Hill.
Tsuchiya, I., 1964. *The Climate of Asia*. World Climatology, Vol. 1. Tokyo: Kokon Shoin.
Tu, C., 1939. Chinese air mass properties, *Q. J. Royal Meteorol. Soc.*, **65**, 33–51.
Yoshimura, M., 1971. Regionality of secular variation in precipitation over monsoon Asia and its relation to general circulation, in *Water Balance of Monsoon Asia*, M. Yoshino (ed.). Tokyo: University of Tokyo Press, pp. 195–215.
Yoshino, M. (ed.), 1984. *Climate and Agricultural Land Use in Monsoon Asia*. Tokyo: University of Tokyo.

Cross references

Climate and climate change
Climate data: sources

ATMOSPHERE

The Earth's atmosphere comprises a mixture of gases; about 78% nitrogen, 21% oxygen with other gases, carbon dioxide, hydrogen, argon and other inert gases, making up the other 1%. The atmosphere also contains a variable quantity of water vapor, which is central in various weather phenomena, and dust, specks of carbon, spores, salt particles and other chemicals, particularly in the upper atmosphere.

Three-quarters of the atmosphere lies within 10 km of the Earth's surface, 10% within about 15 km and about 97% within 25 km. Most weather phenomena occur in the lowest 10 km, and more than half of the water vapor resides in the lowest 3 km.

On average, the temperature decreases with height in the lowest 10–16 km or so, in the **troposphere**. The rate of decrease, or lapse rate, depends upon whether the air is dry, in which case the temperature decreases at the rate of $10°C\,km^{-1}$, or saturated, in which case the temperature decreases at the rate of $5°C\,km^{-1}$. From the top of the troposphere, the **tropopause**, the temperature begins to increase slightly through the **stratosphere**, the top of which is around an altitude of 50 km. Above this level is the **mesosphere**, a region where the temperature decreases again with height. An altitude of 80 km marks the approximate beginning of the region known as the **thermosphere** in which the temperature increases with height to the edge of the atmosphere.

The movement of water from the oceans to the atmosphere and back to the oceans, sometimes via the land, is known as the **hydrological (or water) cycle**. Atmospheric processes involved in this cycle are the concern of hydrometeorology.

C.G. Collier

Cross references

Atmospheric processes
Atmospheric water vapour
Hydrological cycle

ATMOSPHERIC PROCESSES ASSOCIATED WITH WATER IN THE ATMOSPHERE

Equation of state for a perfect gas

The pressure (P), specific volume (V, volume per gram), and temperature (T) of a perfect gas are related by the equation of state:

$$PV = \frac{R^*}{m} T \qquad (A249)$$

where m is the molecular weight of the gas and R^* is known as the universal gas constant, 8.3144×10^7 erg g^{-1} K^{-1}. This equation reflects the observations of experiments with gases embodied in Boyles' Law i.e. $PV = $ constant at constant temperature, Charles' law, i.e. any gas at a given pressure expands uniformly with temperature, and Avagadro's hypothesis which states that at normal temperature and pressure (273.15 K and 1013 mbar) the molecular volume is the same for all gases, namely 22.415×10^3 cm^3. These three laws, as well as being true for a perfect gas, are nearly true for atmospheric gases.

Dalton's law states that in a mixture of gases, the total pressure is equal to the sum of the pressures which would be exerted by each gas if it filled the volume under consideration at the same temperature. Hence a mixture of gases, such as the atmosphere, behaves like a single gas provided

$$R = \frac{\Sigma m_i R_i}{\Sigma m_i} \qquad (A250)$$

where m_i is the molecular weight of the ith gas, and R_i is the corresponding individual gas constant. Equation (A250) may be evaluated for dry air giving $R_d = 2.8704 \times 10^6$ erg g^{-1} K^{-1}. For atmospheric air, which is usually a mixture of dry air and a variable proportion of water vapor, the gas constant will be a function of the water vapor content.

First law of thermodynamics

The first law of thermodynamics states that the energy of a system in a given state relative to a fixed normal state (usually at 0°C, 1 atmosphere pressure, and at rest in a given position) is equal to the algebraic sum of the mechanical equivalent of all the effects produced outside the system, when it passes in any way from the given state to the normal state. This energy is independent of the manner of the transformation.

The law is derived from two facts: (1) heat is a form of energy (Joule's law) and (2) energy is conserved. Conservation of energy (Q) may be expressed by the following.

$$dQ = dU + dW \qquad (A251)$$

where U is the internal energy of the gas and W is the work done by the gas in moving from one state to another. Generally, equation (A251) is expressed in terms of unit mass of gas.

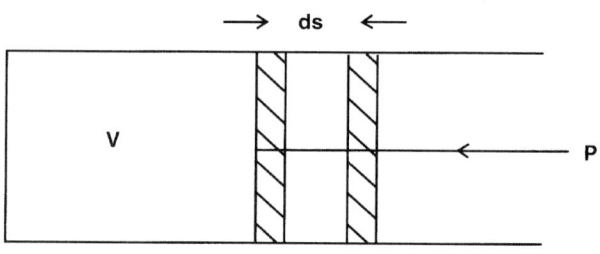

Frictionless moveable piston

Figure A93 Expanding gas pushing a piston.

$$dq = du + dw \qquad (A252)$$

Consider the work done by a unit mass of gas, dw, in expanding from V to $V + dV$ by pushing a frictionless, movable piston of area A (Figure A93). The work done against a pressure, p, is $ds\,pA = pdV$. Hence, equation (A252) may be rewritten as

$$dq = pdV + du \qquad (A253)$$

The internal energy is a function of V and T, that is,

$$du = \left(\frac{\partial U}{\partial T}\right)_V dT + \left(\frac{\partial U}{\partial V}\right)_T dV \qquad (A254)$$

However, for a perfect gas $(\partial U/\partial V)_T = 0$, and we may define the specific heat at constant volume (C_V) as $(\partial U/\partial T)_V$. Equation (A253) then becomes

$$dq = pdV + C_V dT \qquad (A255)$$

This is sometimes called the energy equation. Using equation (A249),

$$dq = (C_V + R)dT - Vdp \qquad (A256)$$

and considering a change at constant pressure, then $dp = 0$ and

$$dq = C_p dT = (C_V + R)dT \qquad (A257)$$

where C_p is the specific heat at constant pressure. Therefore

$$R = C_p - C_V \qquad (A258)$$

In the atmosphere we treat C_p and Cv as constant with values derived experimentally of 0.240 cal g^{-1} $°C^{-1}$ (10.04×10^6 erg g $°C^{-1}$) and 0.171 cal g^{-1} $°C^{-1}$ (7.17×10^6 erg g^{-1} $°C^{-1}$) respectively. The value of R is 0.0686 cal g^{-1} $°C^{-1}$ (2.8704×10^6 erg g^{-1} $°C^{-1}$), and we define $\gamma = C_p/C_V = 1.4$.

Dry adiabatic lapse rate

The atmosphere receives heat from the Sun or the surface of the Earth (radiation), and from processes involving friction, condensation, evaporation or turbulence. Hence, provided no condensation or evaporation occurs, atmospheric processes may be regarded as adiabatic, that is, involving no exchange of heat between that part of the atmosphere under consideration and its surroundings. Considering equations (A255) and (A256) for an adiabatic process, $dq = 0$ and therefore

$$pdV + C_V dT = 0, \; vdp - C_p dT = 0 \qquad (A259)$$

If the atmospheric process is reversible as well as adiabatic then, using equation (A249) and integrating,

$$TV^{R/C_V} = \text{constant} \qquad (A260)$$

$$Tp^{-K} = \text{constant} \qquad (A261)$$

$$pV^\gamma = \text{constant} \qquad (A262)$$

where $K = R/C_p$. Equation (A261) is known as Poisson's equation, and taking $p = 1000$ hectapascals (millibars) then we may define potential temperature, θ, as follows:

$$\theta = T \left(\frac{1000}{P}\right)^K \qquad (A263)$$

Therefore potential temperature is a constant for a dry adiabatic (no condensation or evaporation) process. From equation (A261):

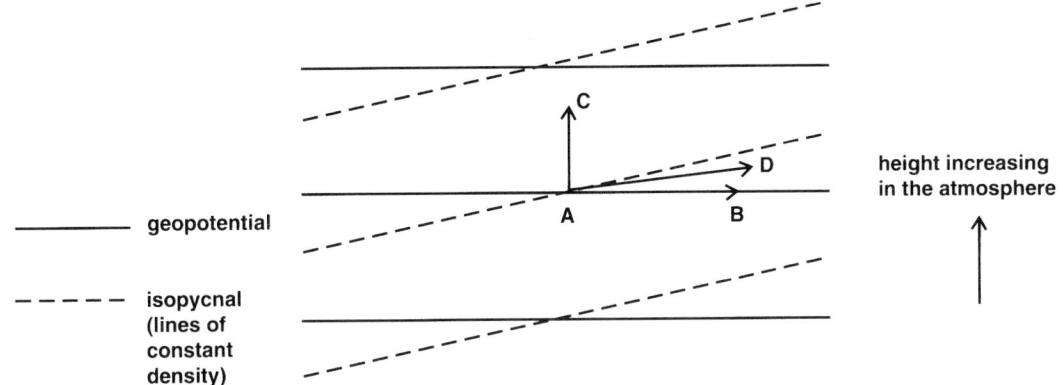

Figure A94 Sloping convection.

$$\log T - K \log p = \text{constant} \tag{A264}$$

Differentiating with respect to height Z, then

$$\frac{\partial T}{\partial Z} = K \frac{T}{p} \frac{\partial p}{\partial Z} \tag{A265}$$

It may be shown from the equations of motion for the atmosphere that, to a good approximation,

$$\frac{\partial p}{\partial Z} = -\rho g \tag{A266}$$

where ρ is the atmospheric density and g is the acceleration due to gravity. This is known as the hydrostatic equation. Substitution of equation (A266) into equation (A265) gives

$$\frac{\partial T}{\partial Z} = \frac{kT}{p}(-\rho g) \tag{A267}$$

and using equation (A249) then

$$\frac{\partial T}{\partial Z} = -\frac{g}{C_p} \gamma d = -9.76°\text{C km}^{-1} \tag{A268}$$

where d is the distance from the ground in kilometers. Therefore if a parcel of air rises dry adiabatically, then its temperature will fall at the rate of about $10°\text{C km}^{-1}$. This is the dry adiabatic lapse rate (DALR).

Saturated adiabatic lapse rate

For a sample of moist air in which no evaporation or condensation occurs, then equation (A268) may be used. However, when condensation occurs and the resulting water falls out of the sample, then the mass of the sample changes, and heat is lost with the fall out of the water. This is known as a pseudo-adiabatic process. If all the condensed water remains in the sample, then the process is of course reversible. In the atmosphere conditions are usually such that some, but not all, of the condensed water falls out of any sample of moist air.

Using equations (A249) and (A255), it may be shown that for saturated air which is lifted slightly,

$$\frac{\partial T}{\partial Z} = \gamma_s = \frac{\gamma_d\left(1 + \dfrac{Lx_s}{R_d T}\right)}{\left(1 + \dfrac{L^2 x_s}{R' C_p T^2}\right)} \tag{A269}$$

where γ_s is the saturated adiabatic lapse rate (SALR) and x_s is the saturated humidity mixing ratio, the mass of water vapor present in the moist air measured per gram of dry air when the moist air is saturated. Although γ_s varies with temperature and pressure, a typical value in the atmosphere is $-5.0°\text{C km}^{-1}$.

Stability and convection in the atmosphere

Moist air can become saturated, and hence produce precipitation, by movement upwards in the atmosphere. Consider a sample, referred to

as a parcel, of moist air for which the pressure in the parcel is the same as that of its environment. Assuming that the parcel can move vertically without disturbing the environment and does not mix with its environment, it can be shown from the equations of motion for the atmosphere that

$$\frac{dW}{dt} = -\frac{1}{p}\frac{\partial p}{\partial z} - g = \frac{(T - T')}{T'} g \tag{A270}$$

where W is the vertical velocity of the parcel, t is time, T is the temperature of the parcel and T' is the temperature of the environment. Making the further assumption that the movement is adiabatic, then

$$\frac{dW}{dt} = \frac{(T_0 - T'_0)}{T'_0} g + \frac{(\gamma - \gamma_a)}{T'_0} gz \tag{A271}$$

where T_0 and T'_0 are the initial temperature of the parcel and the environment, respectively, γ is the environmental lapse rate, Z is the vertical coordinate and γ_a is the appropriate adiabatic lapse rate, being γ_d if the parcel is unsaturated and γ_s if the parcel is saturated. If the temperature is constant in the horizontal then $T_0 = T'_0$ and

$$\frac{dW}{dt} = \frac{(\gamma - \gamma_a)}{T'_0} gz \tag{A272}$$

The atmosphere is regarded as stable, neutral or unstable if dW/dt is <0, 0 or >0. If $\gamma_s < \gamma < \gamma_d$ then the atmosphere is conditionally unstable, whereas if $\gamma < \gamma_s$ the atmosphere is absolutely stable, or if $\gamma > \gamma_d$ the atmosphere is absolutely unstable. Hence if moist air is lifted by some means it may become saturated and hence unstable, and may then continue to rise without any external force being applied.

This release of instability in the atmosphere, or convection, is of course much more complex than this, involving entrainment of environmental air into the rising air, and the effect of descending air on the rising air. Indeed, convection is not necessarily a vertical process. Consider the exchange of a parcel of air from A to B in Figure A94. The potential energy (energy arising from moving a parcel to a greater height against gravity) of the system does not change, and movement to C requires energy to raise the centre of gravity of the parcel upwards as in the movement of the parcel of air discussed so far.

However, movement to D could cause the centre of gravity of the parcel to be lowered, causing energy to be released. A full explanation of this type of convection known as 'sloping' or baroclinic convection is beyond the scope of this section, but it provides the basic mechanism for the development of large-scale atmospheric weather systems which produce most of the precipitation in midlatitudes.

C.G. Collier

Cross references

Atmosphere
Atmospheric water vapor

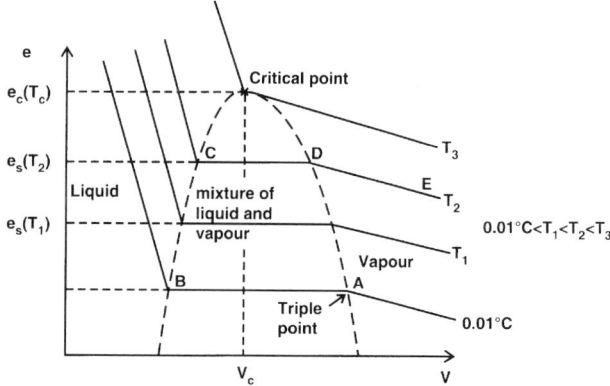

Figure A95 Pressure–volume graph for water.

ATMOSPHERIC WATER VAPOR

If we compress a sample of air containing water vapor at a temperature greater than 0°C, then the water vapour pressure (e) will increase until water begins to condense from the sample. From this point the water vapor pressure is constant, and is referred to as the saturated vapor pressure (e_s), which is a function of temperature. Eventually a very large increase of pressure is required to produce a small change in the volume of the liquid. This process is shown in Figure A95.

If the temperature was 0.01°C, then we find that the horizontal line AB in Figure A95 represents a mixture of water vapor, water and ice. In dealing with unsaturated water vapor (line DE in Figure A95), it is sufficiently accurate to treat water vapor as a perfect gas with a gas constant (R') equal to $R_d/0.622$.

When a change of phase occurs (vapor to liquid; liquid to solid; vapor to solid known as sublimation), heat is released or must be absorbed. The amount of heat required to transform 1 g of water to vapor at a constant temperature is defined as the latent heat of evaporation, and has a value of 597.3 cal g^{-1}. Similarly the latent heat of sublimation (ice to vapor, 677.0 cal g^{-1}), and the latent heat of fusion (ice to water, 79.7 cal g^{-1}) are defined. The relationship between the equilibrium pressure to the temperature during a phase change is given by the Clausius–Clapeyron equation:

$$\frac{de_s}{dT} = \frac{L_{12}}{T(V_2 - V_1)} \qquad (A273)$$

where L is the latent heat, and the subscripts refer to a phase change from a state defined by 1 to that defined by 2. Considering equation (A273) for the change of phase from ice to water, it may be shown that it takes about 135 atmospheres pressure to lower the melting point of ice to −1°C. Therefore in the atmosphere the melting point is strictly 0°C. However, water can remain as a liquid at temperatures well below 0°C. In these circumstances a sample of moist air can be supersaturated with respect to ice, but unsaturated with respect to supercooled water. If two surfaces, one of water and one of ice, come into contact with the moist air, then water will evaporate, but ice will grow by deposition of the water vapor. This occurs often in clouds at temperatures below 0°C, and is important in the formation of precipitation.

C.G. Collier

Cross references

Atmosphere
Atmospheric processes associated with water in the atmosphere

AUSTRALIA: CLIMATE AND WATER RESOURCES

Introduction

Australia's climate depends on four major factors. First, the size and shape of Australia determine how the general circulation affects the

continent, and what modifications the continent makes to the air passing over it. Australia has an area of 7 682 300 km², stretching from 10°41'S to 43°39'S and 113°09'E to 153°39'E at the extremes. Its greater longitudinal distance (4000 km) allows considerable airmass modification and a strong degree of continentality. Its limited latitudinal span allows a relatively moderate change in climate on a seasonal basis compared to larger continents such as North America.

Second, Australia's topography does little to obstruct air moving over the continent. The average altitude of the land mass is only 300 m, with 87% of the continent less than 500 m and 99.5% less than 1000 m. There are only three major topographic features: the eastern mountains and plateaus which provide the highest elevations on the continent; the interior lowlands; and the western plateau, which is generally between 300 and 600 m altitude. Only the eastern mountains have any major influence on synoptic flow.

Third, Australia is completely surrounded by water, which precludes any influences from unmodified polar air from the Antarctic. The major influence of the ocean is coastal, particularly affecting temperature, humidity and precipitation. However, neither the west nor the east coast has a significant upwelling of cold water similar to the west coasts of Africa or South America. Australia's climate can be generally described as moderately continental with some ocean influences.

Fourth, Australia's geographical location places it under the global subtropical high-pressure zone. Much of Australia is under the influence of dry subsiding air and can be considered arid or semiarid, although not as dry as the west coasts of the other southern hemisphere continents. Particularly in summer, Australia experiences high sun angles and intense solar radiation under mostly clear skies. With the exception of Tasmania and much of the east coast, most of the continent receives over 3000 h of sunshine per year, almost 70% of that possible (Bureau of Meteorology, 1988). Central and western Australia receive more than 3500 h or in January greater than 2700 J cm^{-2} d^{-1}. Solar radiation decreases zonally to 1890 J cm^{-2} d^{-1} in Darwin and meridionally to 2300 J cm^{-2} d^{-} on the east coast. In July the solar radiation pattern is much more zonal, decreasing from about 1900 J cm^{-2} d^{-1} on the north coast to 630 J cm^{-2} d^{-1} in Tasmania.

Synoptic circulation patterns

In general, Australia is affected by midlatitude westerlies on the southern fringe, tropical convergence on the northern fringe in summer, and stable subsiding air under the subtropical anticyclone belt in between. The chief determinant of the windfield (Figure A96) over Australia is the subtropical anticyclone belt. In winter prevailing wind directions over the northern half of the continent are southeasterly and are dry and cloud free. Afternoon northeasterly winds occur on the Queensland coast, and afternoon westerly winds on the north western Western Australian coast. The southern half of the continent is dominated by northwesterly winds but has a significant component of southwesterly and southerly winds associated with the passage of cold fronts and midlatitude depressions. Particularly in the morning, along the coast from Brisbane to Adelaide, shallow local drainage flows from the interior highlands occur which strongly affect the surface wind distribution. Western Australia poleward of 25°S has prevailing westerlies associated with synoptic anticyclones and frontal movements from the Indian Ocean.

In summer the pattern is more complex. In the north a dynamic low creates prevailing westerly winds which do not extend far into the continent. From the center to the east coast of Queensland southeasterly winds, on the equatorial side of high-pressure systems, continue to dominate. Northeasterly sea breezes along the east coast provide a significant proportion of the airflow. In most of South Australia, south to southeasterly winds prevail, except for coastal southwesterly winds north of Perth. New South Wales and Victoria have dominant winds from the south or southeast quarter.

The synoptic circulation patterns can best be described by comparing summer and winter seasons.

Winter (May–October) (Figure A97)

The subtropical anticyclone belt covers the northern two-thirds of the continent. Dry, stable southeasterly trades dominate Australia north of 20°S, with greater than 50% frequency. This air originates over the

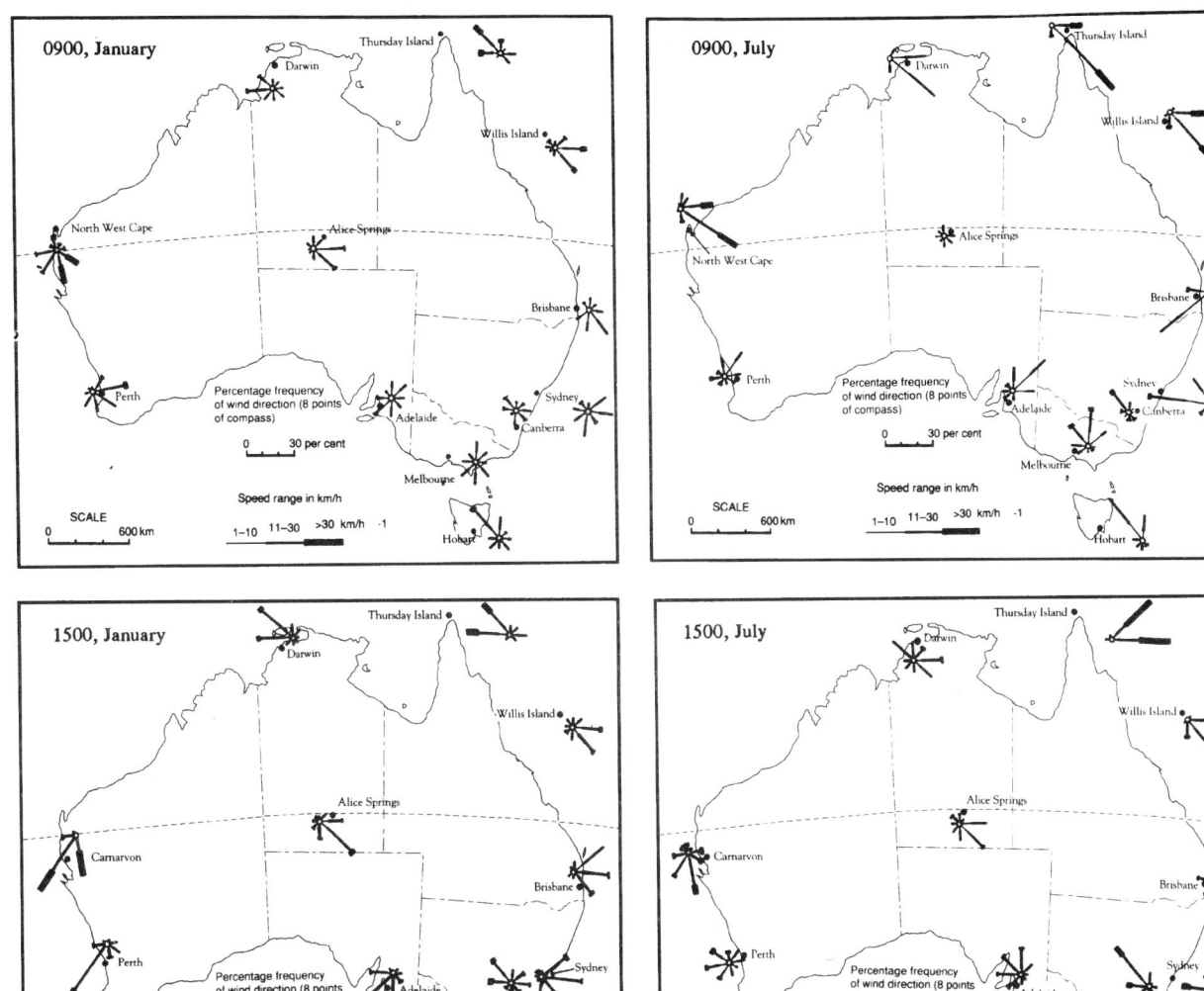

Figure A96 Representative monthly wind frequencies at 0900 and 1500 h in summer and winter for various stations.Prevailing winds are emphasised rather than wind frequency detail. (Source: Collis and Whitaker, 1996.)

Pacific, often loses some of its moisture due to orographic precipitation on the eastern highlands, and cools and stabilizes as it passes over the continental interior.

Frontal activity
South of 20°S the continent is under the influence of alternating anticyclones and cold fronts, associated with the northern edge of the midlatitude westerlies. Troughs and fronts from these systems extend into southern Australia (Figure A98), separating migratory anticyclones in 4–8 day intervals. These fronts move at an average speed of 35 km h^{-1} and can be divided into two types (Sturman and Tapper, 1996). Ana-fronts are more active and are associated with convection in the frontal zone. Kata-fronts are less active and are linked to subsiding air from anticyclones.

The impact of these fronts on the main continent is not usually strong and only in spring extends into the northern half of the country (Smith *et al.*, 1995). The major extratropical cyclonic belt is well south of the main continent, near 55°S, but a subsidiary belt between 30 and 40°S in winter is particularly active between 110–120° and 155–165°E (Oliver, 1979). Tasmania is most severely affected, being

closest in location to the midlatitude depressions. There is a striking absence of warm fronts because the frontal systems form well to the west of Australia and the warm front has swung south before the system reaches the continent.

Precipitation associated with winter frontal systems often occurs from the Northwest Australian cloud band (Sturman and Tapper 1996), which can stretch for distances of up to 5000 km between the Indian Ocean and south to southeastern Australia. A source of moisture and deep convection, the cloud band can bring heavy rainfall. Eighty per cent of these cloud bands occur between April and September, with an average of about two per month.

Occasionally, secondary depressions or cut-off lows moving towards low latitudes can initiate severe storm situations. Cold outbreaks from the Southern Ocean region clash with warmer, moist air from the Indian or Pacific Ocean. Development is better on the southeastern coast of Australia rather than on the southwestern coast. A slow-moving deep low-pressure system results, bringing strong winds and heavy precipitation for periods of 0.5–10 days (Tapper and Hurry, 1993). Such systems are more likely to develop in spring or fall (October–November or April–May) because the thermal gradient

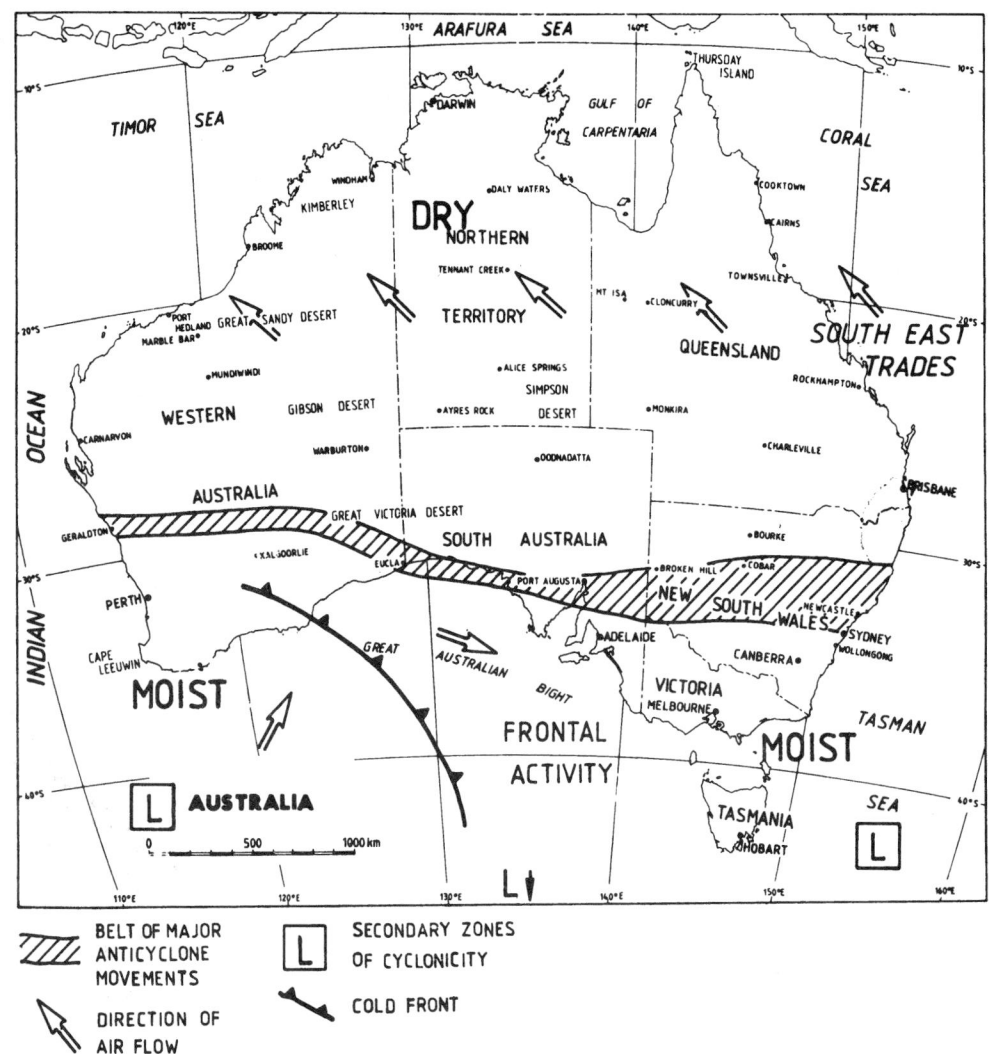

Figure A97 Simplified synoptic climatology of Australia, July surface circulation. (Source: Wendland and MacDonald, 1985; Bridgman, 1987.)

between air masses is strongest. The cause is either the remnants of a tropical storm, or intense cyclonic development in an easterly dip in the isobaric pattern. Important influences are the orientation of the eastern highlands, a strong sea-surface temperature gradient, and a upper level cold pool (Bridgman, 1986; Holland *et al.*, 1987). The most famous of these storms is the 'Sygna' storm at Newcastle which occurred on 26 May 1974, causing considerable local damage along the coast between Wollongong and Taree, but no loss of life.

The interaction of the Indian Ocean air and the cooler maritime air south of the continent brings winter rains to south and southwest Australia. In early winter these rains can be extensive and are related to midtropical convection and the strength of the westerlies. In late winter, more showery precipitation occurs associated with topographic and coastal convergence influences, and not with the general circulation (Allan and Haylock, 1993).

Winter anticyclones
Anticyclones dominate much of the continent in winter. On the average, 40 anticyclones move over Australia in a year, with slightly less in fall–winter than spring–summer. Anticyclones enter the continent at about 115°E and 27°S on the west coast, dip to 35°S in south Australia, and exit about 30°S off the east coast (Figure A97). The mean anticyclonic track is farthest north in June–July except on the east coast of the continent, where the lag in ocean heating defers the maximum northward extent to August–September (Gentilli, 1971; Tapper and Hurry, 1993).

Over the continent the strength and persistence of the anticyclones can be intensified by wintertime cooling of the land. Anticyclones may remain stationary in the interior for several days, particularly if the high develops a meridional (longitudinal) orientation. This creates blocking, and is much more common east of 150°E, in the Tasman Sea. Eastward movement of weather systems is retarded and the convective motion associated with depressions is countered (Baines, 1990; Sturman and Tapper 1996). Their duration is up to 12 days, bringing dry, fine weather.

The presence of anticyclones in the winter enhances the effects of local topography on the continent, encouraging meso-scale inversions and cold air drainage to occur. It is not unusual to see night-time floodplain temperatures 4–5°C cooler than those on the adjoining hilltop, caused by the downslope movement of cold air, and widespread frost and fog.

Subtropical jet stream
The trajectory and speed of movement of both anticyclones and fronts are strongly influenced by the subtropical jet stream, which is strongest over the southern part of the continent near 200–300 mbar (10 000 m altitude) in the winter (Bridgman, 1987; Sturman and Tapper, 1996). The mean latitude of the jet is 26–32°S, but varies considerably from season to season and year to year. In winter its mean core speed is 70 m s^{-1} and northerly meandering from about 25°S is most apparent. In summer the jet is much weaker (30 m s^{-1}), its mean axis position has moved to 31°S, and more extensive

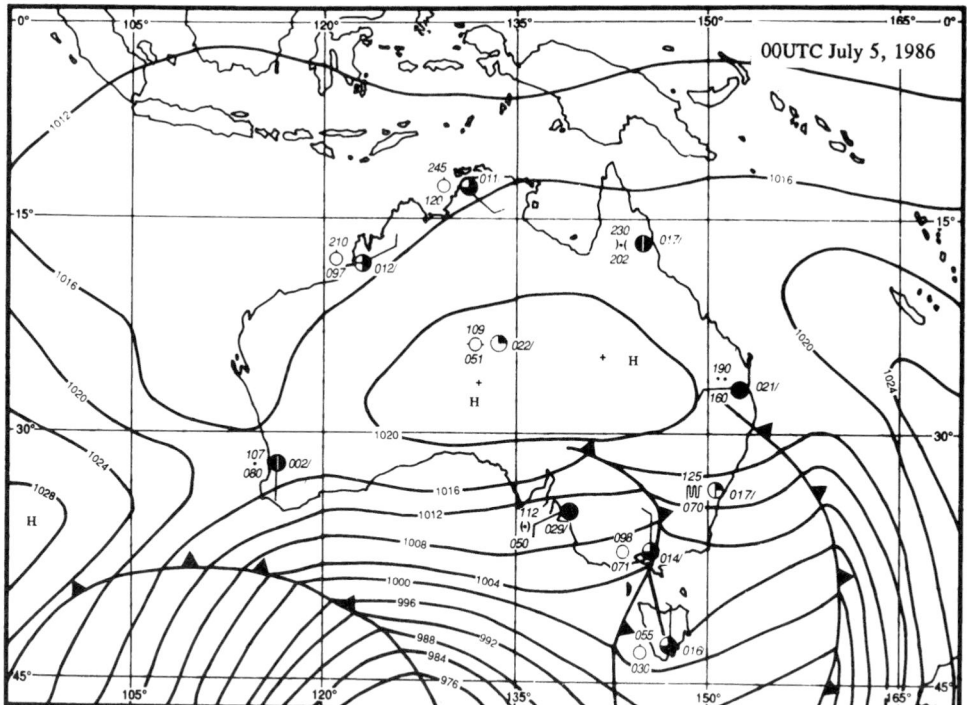

Figure A98 Representative synoptic meteorological situation in winter. Cold fronts parade across the southern third of the continent bringing cool cloudy weather and a light to moderate rainfall. Dry, cool conditions dominate the interior, with overnight frosts likely in many locations. (Source: Tapper and Hurry, 1993.)

southward departures occur. The jet is much less variable in direction and speed in winter.

The jet controls several significant features of Australia's synoptic pattern during the winter:

- It determines the strength of the zonal (west to east) flow depending on how much it meanders, controlling how fast synoptic systems move across the continent. Minor fluctuations in zonal flow influence the eastern Australia–Tasman Sea area more strongly than in Western Australia. Weakest jets occur in conjunction with blocking anticyclones. Split jet streams and jet fingers may also occur, mainly over eastern Australia.
- The jet is associated with the formation of upper air troughs and depressions which, when linked with a lower-level unstable airflow, can bring considerable rain to large areas of the southern half of the continent. Such troughs also provide the major medium for tropical moisture to reach the southern parts of the continent through the northwest cloud band. The strongest jets are also associated with strong cyclogenesis in the west Tasman Sea. It is not unusual in winter on the southeast coast for a shallow surface anticyclone with surface pressures near 1020 mbar to exist with steady rain falling through the system from a mid-tropospheric trough.
- The jet has a significant influence on rainfall systems occurring in late fall that are independent of those in the tropics and subtropics. When the jet is located between 20 and 25°S, these storms can bring strong rain and flooding to parts of the continental interior. They appear erratically in the general area from Geraldton to Port Hedland in Western Australia to Carpenteria to northeast New South Wales in eastern Australia, providing secondary and sometimes primary rainfall maxima for some individual stations (Bridgman, 1987).
- Anticyclonic movements, and the west to east or north to south elongation of anticyclones, relate strongly to the area of upper tropospheric convergence on the equatorial side of the jet. If the jet meanders considerably across the continent, the anticyclones will be elongated longitudinally and will be slow moving. If the jet has a strong zonal (east–west) component, the anticyclones will be zonally elongated and may move across the continent with considerable speed.

Summer (November–April) (Figure A99)

Summer anticyclones

In summer the circulation systems are 5–8° farther south, changing the climatic regimes affecting the northern and southern edge of the continent. Subsiding air from the Subtropical High-Pressure Belt, now located poleward of Australia between 35 and 45°S, covers most of the continent except for the northern tropical fringe and southern Tasmania. The descending air, originating at 10 000 m, warms adiabatically and subdivides into a series of travelling anticyclones (5–6 days periodicity) separated by troughs (Gentilli, 1971). The longitudinal extent of these anticyclones (Figure A100) is of the order of 2000–3000 km and the latitudinal extent is 1000–2000 km. The counter clockwise circulation around the high during its passage ensures dry air from the hot interior moves coastward, limiting rainfall periods to troughs between anticyclonic systems or to onshore easterlies along the east coast of the continent. The pressure in the center of the anticyclonic cells usually reaches 1030 mbar, and falls to about 1005 mbar in the trough between.

The frequency of anticyclone occurrence in the main belt (near 35°S and 125–135°E) is about 210 per year. In the Tasman Sea, at 165–170°E, another center of activity occurs, with a frequency of 194 per year. These anticyclones can be blocking systems.

The anticyclones channel heat into the Pilbara area, in Western Australia, creating a large semi-dynamic heat low (120 days per year). The hot, dry, gusty winds moving coastward from the interior have a local name, brickfielder, signifying hot, parched conditions.

A typical weekly synoptic pattern for the southern half of Australia might be cooler southerly winds in the east side of a high, changing to easterly, northeasterly and then northwesterly winds as the high passes. Along the coasts, temperatures might warm from 22–24°C in the cooler Pacific maritime air to 36–38°C or higher in the continental air from the interior. The situation remains extremely hot and dry until the trough, or cool change between anticyclones, brings relief.

Summer cool change

In Western Australia the cool change is weak, often dry, and is mainly a change in temperature and wind direction (Reeder and Smith, 1992). It forms in the Cape Leeuwin–Kalgoorlie area as an incursion of Indian Ocean air, and is mainly controlled by the warm to hot air

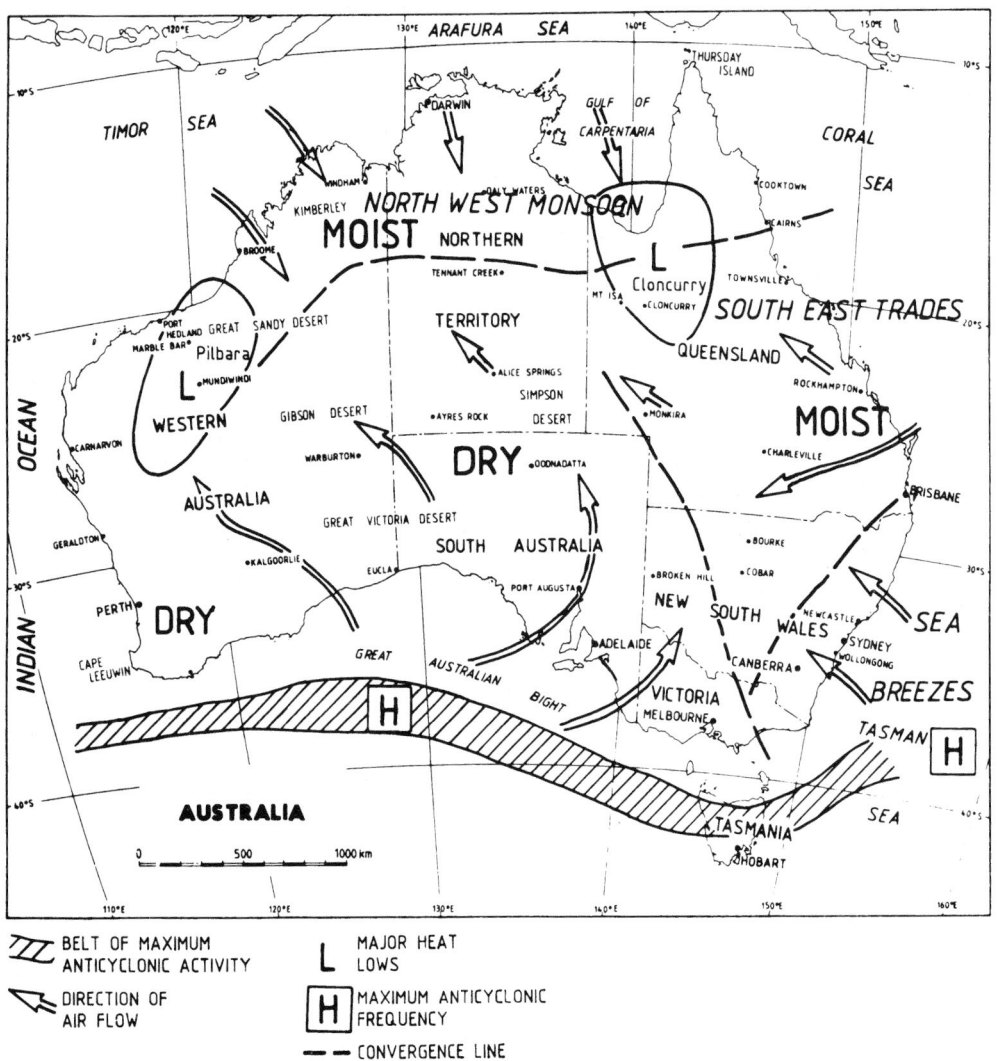

Figure A99 Simplified synoptic climatology of Australia, January surface circulation. (Source: Wendland and MacDonald, 1985; Bridgman, 1987.)

ahead, rather than the weak, cooler air behind. The change will often be sudden, bringing relief from exceedingly hot conditions, dropping temperatures by 10–15°C in the space of less than half an hour.

Cool changes reaching southeast Australia are often much stronger, with gusty winds and occasional rain. Between 1974 and 1983, 95 of these southerly busters were reported at Sydney Airport (Colquhoun *et al.*, 1985). They occur between November and March, and are channeled by the topography of the eastern highlands (Reeder and Smith, 1992; McInnis, 1993). As the buster moves across Victoria and southern New South Wales, the topography causes an S-shaped bend in the trough as friction from the mountains slows down the western side of the front. Since the depth of the change is only about 3–5 km, this change in orientation is not reflected at 500 mbar, but the buster may be associated with weak troughs at this level. About 62% of the busters occur between 1300 and 2100 h, since their speed of movement strength is enhanced by a sea breeze (Colquhoun *et al.*, 1985). Land breezes in the early morning retard buster movement. About half the cool changes are dry, but the change may be heralded by a roll cloud, gusty, squally winds (gusts up to 30 m s^{-1}) and thunderstorms. At passage, a temperature drop of 10°C in a few minutes and up to 22°C in 24 h may occur.

On occasion, cool changes are associated with prefrontal troughs, which can bring severe weather, heavy precipitation and flash floods. From 1957 to 1990 between November and March, 69 meso-scale severe storms occurred in the Sydney region, between November and

March (Speer and Geerts, 1994). The majority occurred between 1200 and 1800 h.

On a regional scale, other circulation patterns which are important are the west coast trough (Western Australia), the sea breeze, and various convergence lines, creating cloud patterns in standing waves, such as the 'morning glory'. The west coast trough develops between two strong highs centred south of the continent, as a dip in the isobaric pattern. The trough encourages hot, strong northeasterly winds from the interior to the coast, which continue until the trough moves inland and weakens (Tapper and Hurry, 1993).

The sea breeze brings cooling to the coast on summer days, as well as on occasional winter afternoons (Abbs and Physick, 1992). Established under clear skies and weak synoptic airflow, through land–water temperature contrasts of more than 7–8°C, the sea breeze penetrates 80–90 km inland on average, depending on topography. A well-developed sea breeze may blow for 2–7 h, depending on distance from the shore. Sea breezes are more strongly developed on the western and southwestern coasts than on the east coast. Bringing temperature relief of up to 15°C on a hot day, these breezes occur with daily frequencies of about 60% on the coast in summer. They often carry appropriate local names such as the Fremantle Doctor (Western Australia).

The morning glory is a spectacular wave structure that appears over the Gulf of Carpenteria just ahead of the wet season in Australia's north. It heralds strong wind squalls, long, narrow roll clouds and a pressure jump, but precipitation is rare (Collis and Whitaker, 1996).

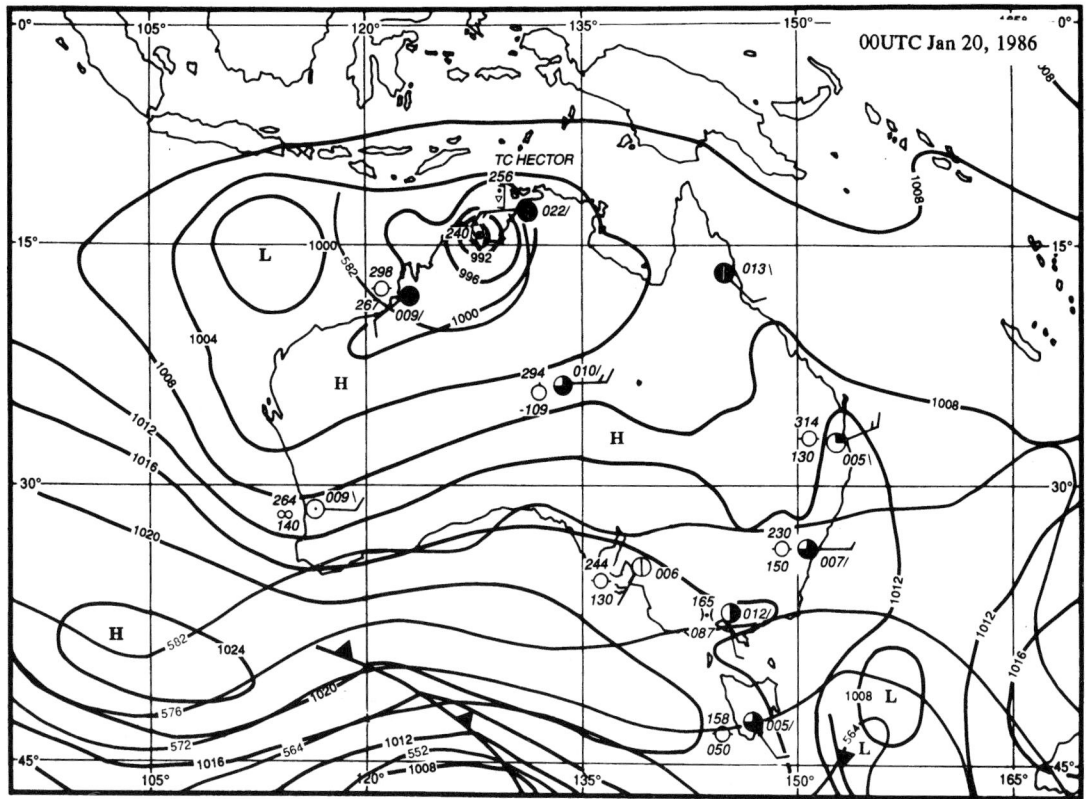

Figure A100 Representative synoptic meteorological situation in summer. A cool change separates an elongated high pressure ridge. Hot NW winds from the continental interior invade SW Australia. On-shore SE winds affect the E coast bringing cool, cloudy, and showery conditions. A tropical cycle (Hector) enters the continent near Darwin. (Source: Tapper and Hurry, 1993.)

Northern Australia rainy season, or monsoon

In northern Australia the shift of the subtropical high-pressure belts and subtropical jet allows a dynamically active low to form over the northern part of the continent (Figure A99) centred in the Cloncurry, Queensland area (maximum frequency 154 days per year). The result is a moist northwesterly air flow which meets the southeast trades along a line approximately from Port Hedland (Western Australia) to Cairns (Queensland), which is commonly described as a monsoon (Joseph *et al.*, 1991; Suppiah, 1992). Table A27 presents the range of dates of onset of the monsoon. The table illustrates the high variability in monsoon dynamics, dependent on the intertropical convergence zone (ITCZ) and its controlling features. There is a strong correlation with the previous season's monsoon rain in India, with below-normal rain creating a delay in monsoon onset in Australia. The monsoon season generally lasts from late December to mid-March.

The monsoonal flow generally extends to about 15°S, and is better developed east of 100°E (Sturman and Tapper, 1996). Its onset can be abrupt, producing violent thunderstorms and tropical depressions during active phases. During this time, convection reaches 400 hPa, close to the tropical tropopause. Rainfall is heaviest along the coast, decreasing quite rapidly inland. About 20% of the monsoon period consists of breaks, where moist humid air in the northwesterly airstream is not triggered into precipitation. The equatorial trough

becomes shallow, northwesterly flow weak, and mid-tropospheric easterly winds also weak (Sturman and Tapper 1996).

To the south and southeast of the monsoonal trough, southeasterly winds dominate the eastern third of the continent. Low-level moisture from the Pacific Ocean is, to a large part, expended on the eastern highlands, but often reaches the interior. Convective and orographic activity results in rain through instability and thunderstorms, occasionally enhanced by an upper level trough. The best rains occur when a mid-tropospheric trough extends southward from central Queensland, interacting with the unstable easterly flow (Tapper and Hurry, 1993).

Tropical cyclones

Australia is the only continent with the same incidence of *tropical cyclones* reaching either the west or east coast, due to the availability of warm seas to enhance formation and to the shape of the coastline. About 95% of tropical cyclones form in the latitudinal zone 9–19°S, in the shear line between the monsoon westerlies and the trade winds (Sturman and Tapper 1996). The tropical cyclones season occurs between late November and early May (Suppiah, 1992). On average, eight to ten cyclones form off the Australian coast every year, but there is considerable variation in frequency. On average about three reach northeast Queensland each year, two to three reach Western Australia, and one reaches the Northern Territory (Figure A101). Landfall most often occurs around Port Hedland on the west coast, near Townsville on the east coast, and Wyndham on the northeast coast of Western Australia.

The development of tropical cyclones is not the same on each coast (Oliver, 1979). Cyclones developing in the Timor or Arafura Sea off the north and northwest coast tend to travel west or WSW over ocean for long trajectories. They then recurve south or southeast to cross the northwest or west coast of the continent. In the first half of the cyclone season it is more probable that cyclones will cross the central part of the west coast, with higher probabilities shifting to the southern sector in the second half of the season. More cyclones cross

Table A27 Range of dates of onset and length of the North Australian summer monsoon (Suppiah, 1992)

	Minimum	Mean	Maximum	Standard deviation (days)
Onset	Nov. 23	Dec. 24	Jan. 27	15
Finish	Jan. 1	Mar. 7	Apr. 6	18
Length	11	74	119	25

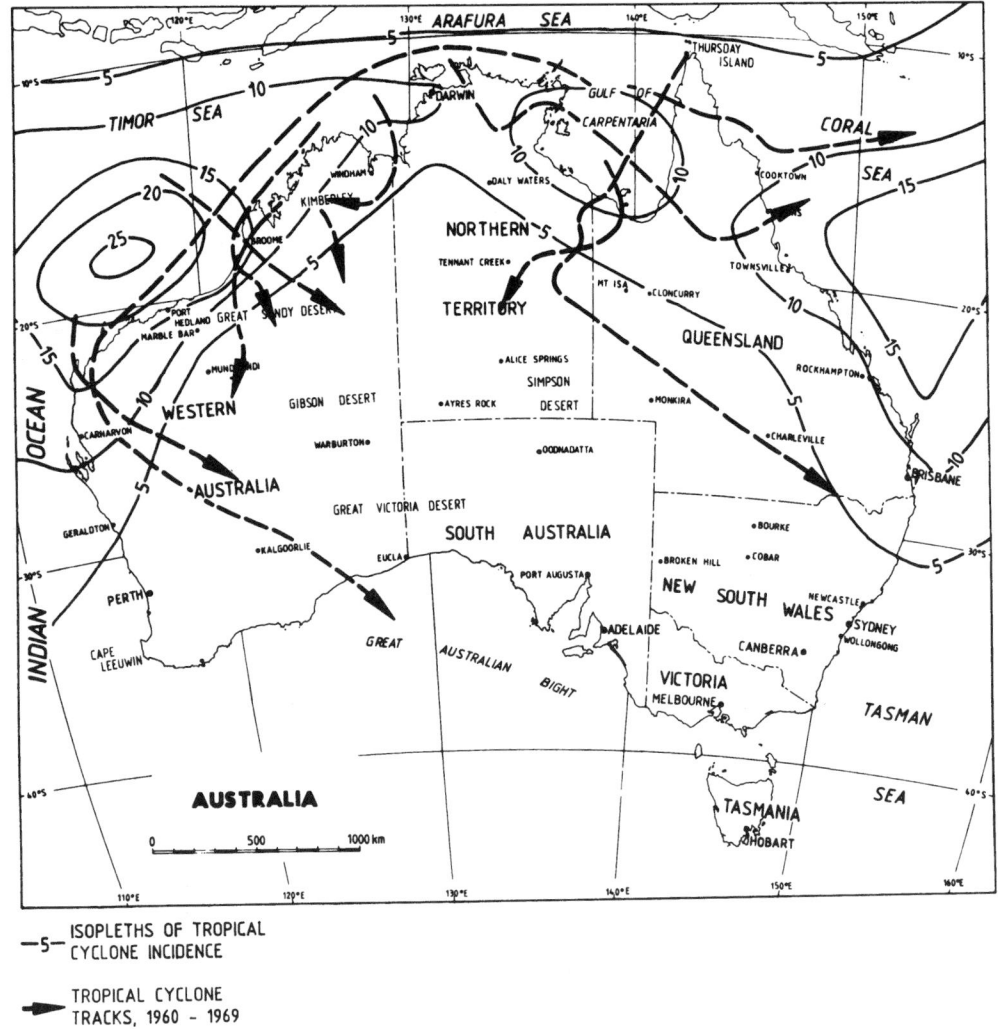

Figure A101 Average decadal incidence of tropical cyclone formation. Tropical cyclone tracks crossing the coast are shown to indicate representative storm paths. (After Bridgman, 1987; Collis and Whitaker, 1996.)

the north coast in April or May at the end of the season. Off Australia's northeast coast, tropical cyclones have little time to develop over the ocean before reaching landfall. February and March are the months of highest frequency, particularly between 15 and 20°S. The timing of the strongest tropical cyclones is linked to the active phases of the monsoon (Suppiah, 1992).

Tropical cyclones bring high winds and heavy rain to both coasts with considerable destruction potential. Flooding in much of central Queensland and in Western Australia is inevitable if the cyclone moves inland. Much of the area is sparsely inhabited, however, and rains from tropical cyclones are often the main source of moisture for crops and grassland. Only on rare occasions does a tropical cyclone strike a heavily inhabited area. The most famous Australian cyclone is Tracy which passed through Darwin on Christmas evening, 1974 (Collis and Whitaker, 1996), and destroyed more than half the city.

Links to El Niño/Southern Oscillation
Tropical cyclones, the monsoonal flow and precipitation in eastern Australia in general is very strongly influenced by the El Niño/Southern Oscillation (ENSO) circulation variations (Allen, 1988; Drosdowsky and Williams, 1991; Joseph *et al.*, 1991), and the associated changes in sea-surface temperature, especially in spring and early summer. There is a significant correlation between the Southern Oscillation index (SOI) and variations in precipitation, cloud cover and the diurnal range of temperature. Lower SOI brings less rain (and often drought), lower cloud amounts and greater diurnal

temperature ranges. A positive SOI brings higher than average rainfall, deeper incursions of the monsoon into central Australia and potential floods. Correlations (r) between precipitation and SOI reach 0.6 in western New South Wales, and are generally 0.4 over much of the eastern half of Australia (Sturman and Tapper, 1996). Correlations diminish greatly towards the southwest of the continent.

Climatic elements

Tables A28 and A29 present means of various climatic elements for January and July and some climatic extremes. Two elements are of major concern to Australia, temperature and moisture.

Temperature

The range of annual average air temperature is from 28°C on the Kimberly coast (northwest Western Australia) to about 4°C in the alpine mountain areas (>1500 m) in southeast Australia. In January (summer) the average daily minimum is virtually equivalent to the annual average air temperature. In July (winter) minima decrease from 21°C in the Darwin area to −20°C in the southeastern Australian mountains. Particularly in winter, the minima are tempered on the coast by the ocean. Local cold air drainage and topography eliminates any regularity of pattern in the eastern highlands.

Maximum temperatures are much more important. Australia is known for its heat extremes rather than its cold (Bureau of Meteorology 1988a,b, 1989). Between 1939 and 1968, heatwaves

Table A28 Mean climatic data for selected Australian capital cities, January and July (after Bridgman, 1987; Bureau of Meteorology, 1988a)

City	Latitude/ Longitude	MSL altitude (m)	Years of record	Pressure (mbar)	Wind speed (m s⁻¹)	Prevailing wind direction 9 a.m.	Prevailing wind direction 3 p.m.	Evaporation (mm)	Daily cloud (eighths)	Max. temperature (°C)	Min temperature (°C)	Mean temperature (°C)	Hours of sunshine	9 a.m. relative humidity	9 a.m. vapour pressure	Rainfall (mm) (days)	Fog days	Solar energy (MJ m⁻² d⁻¹)
Perth, WA	31°57′S 115°51′E	19.5	42															
January				1012.6	4.9	E	SSW	280	2.3	31.5	16.7	23.5	10.5	51	14.8	7(3)	0.2	27.3
July				1018.8	3.9	NNE	W	58	4.5	17.7	9.0	13.2	5.3	76	10.9	164(18)	1.6	9.4
Darwin, NT	12°25′S 130°52′E	31.0	45															
January				1006.2	2.6	W	NW	225	5.9	31.7	24.7	28.6	5.9	81	31.1	409(21)	0.0	18.4
July				1012.8	3.4	SE	E	229	1.3	30.3	19.2	25.1	9.8	62	17.6	1(1)	1.1	19.3
Sydney, NSW	33°56′S 151°10′E	6.0	47															
January				1012.7	3.4	NE	NE	217	4.7	25.7	18.5	22.0	7.2	68	18.8	102(12)	0.3	22.5
July				1018.5	3.2	W	WSW	95	3.5	16.0	7.9	11.8	6.2	74	9.6	100(10)	2.1	10.4
Canberra, ACT	35°19′S 149°12′E	571.0	47															
January				1012.0	1.8	NW	NW	251	4.1	27.7	12.9	20.3	8.9	60	13.1	60(8)	1.0	25.9
July				1020.2	1.2	NW	NW	54	4.4	11.0	−0.3	5.4	5.2	84	6.6	39(10)	7.9	9.5
Melbourne, Vic	37°41′S 144°51′E	132	16															
January				1012.8	3.6	S	S	228	4.1	26.0	13.4	19.9	8.1	68	13.1	42(8)	0.1	24.9
July				1202.2	3.6	N	N	47	5.2	12.8	4.9	9.5	3.7	81	8.9	33(14)	4.3	6.3
Hobart, Tas	42°55′S 147°20′E	4.0	28															
January				1010.6	3.5	NNW	SW	167	5.0	18.6	11.5	16.5	7.9	58	11.0	69(14)	0.3	23.2
July				1014.0	3.0	NNW	NNW	26	4.8	11.1	6.1	7.9	4.3	78	7.6	95(21)	1.4	5.5
Alice Springs, NT	23°49′S 133°53′E	545	38															
January				1007.0	NDª	ESE	SE	397	2.3	36.6	22.2	28.0	11.0	36	12.8	43(ND)	ND	26.7
July				1018.0	ND	ESE	SE	121	0.7	19.3	4.5	12.0	9.3	61	7.0	11(ND)	ND	16.0

ª ND = not determined.

Table A29 Extreme weather events in Australia

Weather event	Location	Period	Value
Rainfall			
Hourly total	Deeral Qld	13 Mar. 1936	330 mm
Daily total	Beerwah, Qld	3 Feb. 1893	907 mm
Monthly total	Bellanden Kerr, Qld	Jan. 1979	5387 mm
Annual total	Bellanden Kerr, Qld	1979	11251 mm
Highest annual mean	Babinda, Qld	32 years	4537 mm
Lowest annual mean	Troudaninna, SA	42 years	105 mm
Temperature			
Highest maximum	Cloncurry, Qld	16 Jan. 1889	53.1°C
Lowest minimum	Charlotte's Pass, NSW	29 Jun. 1994	−23°C
Longest heat wave[a]	Marble Bar, WA	30 Oct–7 April 1924	161 days
Maximum annual average	Wyndham, WA	–	35.5°C
Wind			
Maximum gust	Mardie, WA	19 Feb. 1975	259 km h^{-1}

[a] Defined as the number of days in a row the maximum temperature exceeded 37.8°C. Data courtesy of the National Climate Centre, Australian Bureau of Meteorology.

consisting of many hot, dry days killed 1326 persons. In January, 35°C is exceeded over most of the interior and 40°C regularly in northwestern Western Australia. In towns in this area, such as Marble Bar, maxima exceed 40°C for several weeks at a time. Maxima drop to less than 20°C in Tasmania and in the higher altitudes of the southeast Australian mountains. The coastal sea breezes create maximum temperature gradients in all directions from the interior, with the south and east coasts having maxima in the mid and upper 20s and the north and west coasts in the lower 30s. The maximum for single days under one synoptic system has reached 47.6°C in Adelaide, 46°C Melbourne and 45.3°C in Sydney (11–14 January 1939).

In July, except for the eastern highlands, maximum temperatures follow a latitudinal pattern, decreasing from 30°C near Darwin and on the Cape York Peninsula to 12°C on the south and Tasmanian coasts. July maxima in the mountains do not usually reach 5°C.

The sun and seasonal extent of cloud cover determines the month of highest temperature. Most of northern Australia, equatorward of 20°S, has highest maximum temperatures in November just before the onset of the monsoon, with some areas near Darwin and in the northwest having the highest maximum in October. Just south of this area, to a line from Geraldton (WA) to Tennant Creek (NT) to Cooktown (Qld), the highest maxima occur in December. Most of the interior and the northeast coast has highest maxima in January, and the west and east coasts in February, due to the lag in ocean temperatures.

The diurnal temperature range is least on the north coast (3–4°C in summer, up to 8°C in winter) and greatest in the interior, near Alice Springs (>8°C in summer, >12°C in winter).

A combination of hot, dry climatic conditions and strong winds from the centre to the coast can create serious bushfire problems, particularly in late spring and early summer. Bushfire behavior is controlled by fuel availability, topography, pressure tendency, wind direction and their interaction between individual fires. Cumulative antecedent rainfall, or the total rainfall over several previous months or seasons, correlates strongly with bushfire frequency (Love and Downey, 1986). The region of highest fire hazard is the coastal zone of eastern Victoria and southern New South Wales where, due to the regular availability of good fuel grown during wet periods, large bushfires during dry periods occur every third year. In an area roughly from southeast South Australia through Victoria and western New South Wales to the south Queensland coast, including southern Tasmania near Hobart, one big fire occurs approximately every 10 years. Recent severe fires include the Ash Wednesday fires in Victoria and South Australia in February 1983, where 75 people were killed, more than 2000 houses were destroyed and property damage was $430 million (Tapper and Hurry, 1993); and the east coast fires in the Sydney–Newcastle (NSW) area in January 1994, which caused minimal damage to occupied land but considerable destruction to bushland and the Royal National Park, south of Sydney.

Bushfires are enhanced when a blocking summer anticyclone occurs in the Tasman Sea and there is slow approach of a trough or depression from the southwest. The approach of the trough encourages strong, hot, dry, northwesterly winds out of the continental interior, creating havoc conditions for large bushfires. Control depends on the timing of the southerly change as the trough passes through and the ability of the bushfire brigades to back-burn appropriately for the wind shift. The occurrence of such situations is enhanced during low SOI periods.

Moisture (Figure A102)

Australia is a dry continent with certain localized exceptions. Vapor pressure and atmospheric water vapor do not change much diurnally (or in the interior seasonally) except along the coast or where there is a significant rainy season. Potential evaporation regularly overshadows precipitation. About 50% of the continent has a median rainfall less than 300 mm year^{-1}, and 80% less than 600 mm year^{-1}. Highest rainfall occurs on the northeast and north coasts, reaching 3800 mm year^{-1} in the eastern highlands, and lowest in the Simpson Desert (<150 mm year^{-1}) (Bureau of Meteorology, 1988). About 75% of the continent has an annual evaporation (class A pans) of greater than 2500 mm year^{-1}. In the central and northwest sections of the continent, potential evaporation reaches 4500 mm year^{-1}, more than 20 times the annual rainfall.

Rain days per year decrease from over 100 around Darwin and 120 or more on the northeast Queensland coast to less than 20 in the interior. In the area affected by the midlatitude westerlies, for example southwest Tasmania, more than 150 rain days per year can occur.

Precipitation can be strongly seasonal, depending on location. Five main precipitation regimes dominate the continent (Bureau of Meteorology, 1988a,b 1989) allowing an analysis by climatic zones (Figure A102):

- In Northern Australia, the monsoonal season (October to April) is markedly wet and the winter very dry. Most stations have more than 20 times the rainfall in summer compared to winter. The northeast Queensland Coast has the highest annual rainfall in the country.
- In southeast Queensland and NE New South Wales, summer is wet and winter is dry, but the difference between seasons is much less marked than under monsoon conditions. There is a tendency for a secondary peak in midwinter.
- In southeast Australia, including most of Victoria, eastern New South Wales and Tasmania, a relatively uniform precipitation regime exists. Victoria and Tasmania receive slightly more rain in winter than summer, and New South Wales the reverse, with the difference becoming greater closer to the equator.
- Southwestern Western Australia has a wet winter (May to October) and a dry summer, the Mediterranean climate caused by the seasonal shifting of the subtropical high-pressure Belt and the midlatitude westerlies.

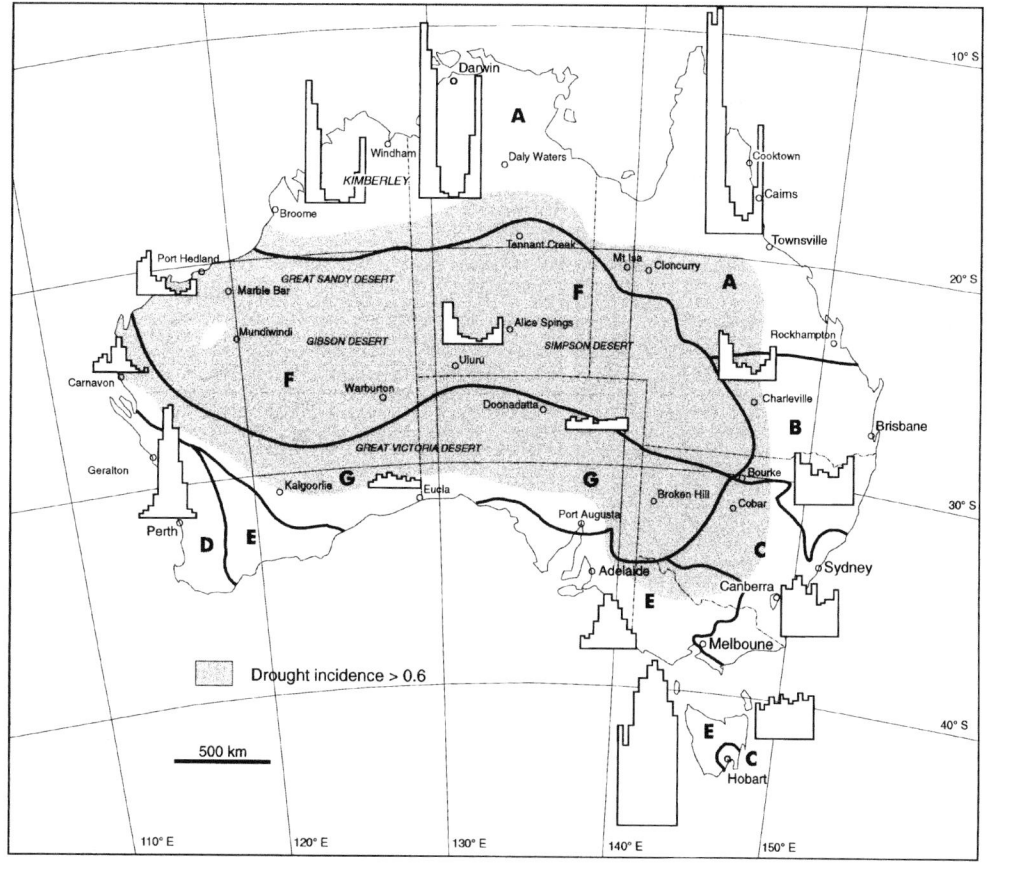
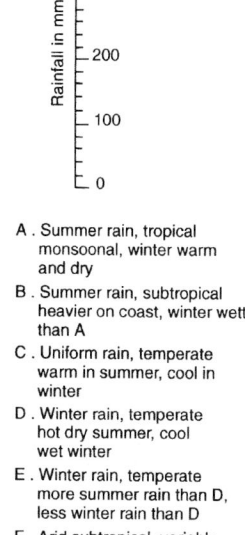

Figure A102 Representative monthly precipitation histograms (January begins at the left) for various parts of Australia. The data are averaged for various meteorological districts and not representative of single stations. Included are climatic districts based on temperature and precipitation. The region of high to excessive drought incidence is shaded. (Source: Bridgman, 1987.)

● More than half the continent, from northwestern Western Australia to the Great Australian Bight is semiarid to arid, with a weak summer seasonal rainfall distribution in the west and north. This is the area of greatest rainfall variability, with heaviest rains associated with occasional tropical cyclones and severe local thunderstorms.

On a more local scale, severe thunderstorms associated with fronts and other convective triggers create flooding and damage. Some of the strongest thunderstorms spawn tornadoes, particularly under extremely unstable conditions in the mid and upper troposphere. Tornadoes occur especially in the uplands of eastern Queensland and in the southwest of Western Australia (Sturman and Tapper, 1996). The frequency of tornadoes may be 100–200 per year (Oliver, 1979) over the continent, occurring mainly outside the dry interior. In the northern half of the country, tornadoes are relatively rare and only develop in summer, as a result of strong local convection related to the wet season or sometimes associated with tropical cyclones. In the southern half, tornadoes in summer occur under convective activity associated with weak lows or troughs. In winter a secondary maximum occurs associated with cold fronts emanating from deep depressions in the midlatitude westerlies.

Tornado formation is strongly influenced by upper atmospheric air flow, the subtropical jet, and upper air trough location. Most occur within 6° latitude of the jet stream or within 14° longitude of an upper air trough. Diurnally tornadoes occur most often between 1530 and 1830 h. Winds occasionally exceed 50 m s⁻¹.

Intensive surface heating under dry conditions can also create dust devils, known colloquially as willy-willys. Synoptic wind speeds must be less than 5 m s⁻¹, turbulence very strong and lapse rates much greater than the dry adiabatic rate. Maximum wind gusts can reach 19.5 m s⁻¹ (Hess and Spillane, 1990).

Snow, associated with winter cold fronts passing over elevated areas, is mainly confined to the mountains in southeast Australia, particularly just south of Canberra, and on Tasmania's Central Plateau and uplands. Frost depends as much on topography and continentality as latitude, but is most prevalent in Tasmania, where on the Central Plateau it can occur on 300 days per year. Throughout much of the moist continental interior, occasional frost days occur between April and October, particularly in areas prone to cold air drainage.

The lack of consistent rainfall over most of the continent creates one of Australia's most serious climatic problems, drought. Drought can be defined as a severe water shortage depending on the amount of water needed, or by the failure of rains at specific places. Where seasonal rainfall is critical, drought is defined as the failure of the wet season.

The Australian Bureau of Meteorology defines an index of drought to establish the concept of drought potential: 50th percentile of rainfall minus 10th percentile divided by the 30th percentile. High indexes (>0.6) mean a strong potential for drought. More than two-thirds of Australia has high to extreme drought indices (Figure A102). The only areas escaping are the north, south and east coast and Tasmania. Table A30 shows that drought, along with flood and tropical cyclones, create the highest cost from weather disasters in the country. The table emphasizes the importance of extreme weather to Australia.

Water resources

The dryness of the continent creates major limitations to Australia's water resources. Tables A31 and A32 present a summary of ground and surfacewater resources by state. The major source of groundwater is the Great Artesian Basin, which covers a significant proportion of inland Queensland and northern New South Wales. Surfacewater flow from permanent rivers and streams has been estimated at 397 × 10¹²

Table A30 Annual average cost of weather related disasters in Australia (Sturman and Tapper, 1996)

Type of disaster	Annual cost ($ million, 1989)	Percentage of cost
Drought	303	24
Bushfire	68	5
Storm	202	16
Flood	386	31
Tropical cyclone	258	21
Other	33	3
Total	1250	100

Table A31 Groundwater resources of Australia, by state (Castles, 1996)

State/territory	Groundwater resources (1×10^9)		
	Area of aquifer (km^2)	Fresh usable	Total usable
New South Wales	595 900	881	2 180
Victoria	103 700	469	862
Queensland	1 174 800	1760	2 840
South Australia	486 100	102	1 210
Western Australia	2 622 000	578	2 740
Tasmania	7 240	47	124
Northern Territory	236 700	994	4 420
Australia	5 226 400	4831	14 376

l year^{-1} (Castle, 1996), of which about 25% is considered useable for agriculture and other human activities. At 12%, Australia's runoff is far lower than that of the other continents, which range from 38 to 57%. Runoff in the north is confined mainly to the rainy season during summer. In the south there is a small maximum in winter.

Attempts to regulate or control surface flows have had only been partially successful. Problems of irregular rainfall, long drought periods, floods and very high evaporation interfere with the control process. In addition, overuse and overcontrol of surface water, particularly for agriculture, has created major salinity, sedimentation and nutrient pollution problems in many of the major river systems. This is particularly apparent in south-central New South Wales and northern Victoria, where the damage to the Murray–Darling river system is extreme.

Near-future climate change

The climate is varying continuously, and there are concerns about near-future changes associated with greenhouse warming in Australia. National attention in the form of a series of reports (*Climate Change*, 1994), focusing on the results of Bureau of Meteorology network measurements, and global and regional climate modeling suggest the following scenario is likely for Australia by the year 2030.

Higher temperatures in general across the continent are probable, with those areas north of 25°S warming by 0–1.5°; those areas south of 25°S by 0.5–2.0°C; and the regions more than 200 km inland from the coasts by 0.5–2.5°C. The extra heat capacity of the oceans will delay and minimize warming along the coast. Most of the change will most likely occur in the minimum temperature, decreasing the overall daily temperature range. Evidence that this has already been occurring comes from Queensland (Lough, 1995) and other parts of the country. Warmer minimum temperatures would decrease the number of days below 0°C (by up to six per year) and less snow cover would exist. However, an increase in the number of extremely hot days in summer is expected, particularly in the desert areas of Western Australia.

Rainfall zones are likely to shift southward, following a shift in global circulation patterns towards the poles. Summer rainfall is expected to be up to 20% greater over the continent. Winter rainfall may decrease by up to 10% in central Australia and increase by up to 10% in Tasmania. Changes in winter in southwest and southeast Australia are unclear, but there has been a downward trend in rainfall in the southwest since the mid-1960s (Allen and Haylock, 1993), which may continue. The climate record already shows an increase in Australian cloud cover during this century (Jones, 1991). There may be greater frequencies of high rainfall events, and decreased numbers of low rainfall events, as well as increased frequencies of drought, bushfires and fine weather (Sturman and Tapper, 1996).

Readers are referred to the following bibliography for further information on the climate and water resources of Australia. Earlier references are listed in Bridgman (1987).

Howard A. Bridgman

Bibliography

Abbs, D.J. and Physick, W.L., 1992. Sea-breeze observations and modelling, *Aust. Meteorol. Mag.*, **41**, 7–20.

Allen, R.J., 1988. El Nino Southern Oscillation influences in the Australasian region, *Prog. Phys. Geog.*, **12**, 313–48.

Allen, R.J. and Haylock, M.R., 1993. Circulation features associated with winter rainfall decrease in southwestern Australia, *J. Climate*, **6**, 1356–67.

Baines, P.G., 1990. What's interesting and different about Australian meteorology?, *Aust. Meteorol. Mag.*, **38**, 123–146.

Bridgman, H.A., 1986. The Sygna Storm at Newcastle – 12 years later, *Meteorol. Aust.*, **3**, 10–16.

Bridgman, H.A., 1987. Australia Climates of, in *Encyclopedia of Climatology*, ed. J. Oliver. New York: Elsevier.

Bureau of Meteorology, 1988a. *Climatic Atlas of Australia*. Canberra: Department of Administrative Services, 67 pp.

Bureau of Meteorology, 1988b. *Climatic Averages, Australia*. Brisbane: Watson Ferguson.

Bureau of Meteorology, 1989. *Climate of Australia*. Canberra: Australian Government Publishing Service.

Castles, I., 1996. *Year Book Australia 1995*. Canberra: Australian Bureau of Statistics.

Climate Change, 1994. National Report Under the Framework Convention of Climate Change. Canberra: Australian Government Publications Service.

Collis, K. and Whitaker, R., 1996. *The Australian Weather Book*. Sydney: National Book Distributors.

Colquhoun, J., Shepherd, D., Couleman, C. *et al.*, 1985. The southerly

Table A32 Surfacewater resources of Australia by state (Castles, 1996)

State	Area (km^2)	Surfacewater resources (1×10^9)			
		Mean annual runoff	Mean annual outflow	Usable resources	Developed
NSW	802 000	42 400	37 200	16 900	7 970
Vic	228 000	19 200	18 800	9 810	5 990
Qld	1 730 000	159 000	158 000	32 700	3 840
SA	984 00	2 120	1 250	384	124
WA	2 520 000	39 900	39 700	11 700	2 340
Tas	68 200	52 900	52 900	10 900	1 020
NT	1 350 000	81 200	79 200	17 700	59
Australia	7 680 000[a]	397 000	387 600	100 000	21 500

[a] Surfacewater totals rounded.

buster of southeastern Australia: an orographically forced cold front, *Mon. Wea. Rev.*, **113**, 2090–2107.

Drosdowsky, W. and Williams, M., 1991. The Southern Oscillation in the Australian Region. Part I: anomalies at the extremes of the oscillation, *J. Climate*, **4**, 619–638.

Gentilli, J., 1971. Climates of Australia, in *World Survey of Climatology*, Vol. 1 ed. H. Landsberg. Amsterdam: Elsevier, chapters 4–7, pp. 35–210.

Gentilli, J., 1972. *Australian Climatic Patterns*. Adelaide: Thomas Nelson, 285 pp.

Gentilli, J., 1988. Climatology, *Aust. Geog. Studies*, **26**, 21–44.

Hess, G.D. and Spillane, K.T., 1990. Characteristics of dust devils in Australia, *J. Appl. Meteorol.*, **29**, 498–507.

Holland, G.J., Lynch, A.H. and Leske, L.M., 1987. Australian east-coast cyclones. Part I synoptic overview and case study, *Mon. Wea. Rev.*, **115**, 3024–36.

Jones, P.A., 1991. Historical records of cloud cover and climate for Australia, *Aust. Meteorol. Mag.*, **39**, 181–190.

Joseph, P.V., Liebmann, B. and Hindon, H.H., 1991. Interannual variability of the Australian summer monsoon onset: possible influences of the Indian summer monsoon and El Nino, *J. Climate*, **4**, 529–538.

Lough, J.M., 1995. Temperature variations in a tropical–subtropical environment: Queensland, Australia 1910–1987, *Int. J. Climatol.*, **15**, 77–96.

Love, G. and Downey, A., 1986. A prediction of bushfires in central Australia, *Aust. Meteorol. Mag.*, **34**, 93–102.

McInnis, K.L., 1993. Australian southerly busters, part III: the physical mechanism and synoptic conditions contributing to development, *Mon. Wea. Rev.*, **121**, 3261–3281.

Oliver, J., 1979. Wind and storm hazards in Australia, in *Natural Hazards in Australia*, eds R. Heathcote and B. Thom. Canberra: Australian Academy of Sciences, pp. 119–142.

Reeder, M.J. and Smith, R.K., 1992. Australian spring and summer cold fronts, *Aust. Meteorol. Mag.*, **41**, 101–123.

Smith, R.K., Reeder, M.J., Tapper, N.J. and Christie, D.R., 1995. Central Australian cold fronts, *Mon. Wea. Rev.*, **123**, 16–28.

Speer, M. and Geerts, B., 1994. A synoptic-mesoalpha-scale climatology of flash-floods in the Sydney metropolitan area, *Aust. Meteor. Mag.*, **43**, 87–103.

Sturman, A.P. and Tapper, N.J., 1996. *The Weather and Climate of Australia and New Zealand*. Melbourne: Oxford University Press.

Suppiah, R., 1992. The Australian summer monsoon: a review, *Prog. Phys. Geog.*, **16**, 283–318.

Tapper, N. and Hurry, L., 1993. *Australia's Weather Patterns An Introductory Guide*. Mt Waverly (Victoria): Dellastra.

Wendland, W. and McDonald, N., 1985. Mean airstreams of Australia, *Aust. Geog. Stud.*, **23**, 28–37.

Cross references

Climate and climate change
Precipitation

AUSTRALIAN FLOOD STUDIES

The needs of the engineering profession for design flood estimation were first served in 1958 by *Australian Rainfall and Runoff* published by The Institution of Engineers Australia. The third edition of 1987 subtitled *A Guide to Flood Estimation* draws on the experience of the intervening years to reassess design philosophy, to give sound practical advice to non-specialists in hydrology and to incorporate new techniques founded on the extended records of rainfall and runoff.

The sources of streamflow and rainfall data are the Commonwealth Bureau of Meteorology, the Commonwealth Scientific and Industrial Research Organisation and the several pertinent authorities of each State and Territory. The areal extent of Australia covers many climatic regions with a wide range of weather types, so that different hydrological techniques of analysis are required to provide the optimum design flood values according to the various conditions and to the availability of data.

Rainfall

The basic data come from daily rain gauges and recording rain gauges or pluviometers with chart recorders. Statistical and meteorological analyses using algebraic and graphical procedures on the available records enabled regional differences to be incorporated into mapped durations and frequencies from which specific intensity–frequency–duration design curves can be evolved for a required location. Intensity–frequency–duration rainfall design curves extend from 5 min to 72 h with an average recurrence interval from 1 to 100 years.

For determining the temporal patterns of rainstorms used in the unit hydrograph method for determining flood hydrographs, Australia is divided into eight zones. For each of the eight zones, patterns have been derived for 20 durations ranging from 10 min to 72 h.

Design rainfall excess

The evaluation of the excess rainfall forming the flood hydrograph has to take into account the storm losses dependent on the initial wetness of the catchment and the storm rainfall pattern. From studies of infiltration and runoff processes, models of rainfall excess are explained and based on limited catchment records; models are recommended for the different states and territories.

Rural catchments

Of particular interest is the estimation of peak flows for small catchments (<24 km^2) and medium-sized catchments (250–1000 km^2), since nearly half of the national average expenditure is applied to schemes in these areas governed by estimated design floods. Recommended methods are given for locations where no data are available for the problem sites, as follows.

- The rational method with the difficult runoff coefficient and time of concentration being obtained from regionally derived values.
- Regional flood frequency methods obtained from flood frequency analyses of all catchment data in the region with sufficient records.
- Envelope curves of recorded floods from regional plots of maximum recorded floods at gauging stations versus catchment areas. This is a useful method of obtaining initial estimates and providing a check of other methods.
- Synthetic unit hydrographs and runoff routing where regional parameter values are available.
- Arbitrary methods, such as the use of empirical formulae based on overseas studies in different climatic regions and incorporating coefficients resulting from foreign data, are not to be recommended.

A comparable but fuller chapter in *Australian Rainfall and Runoff* deals with urban stormwater drainage. The wide range of methods used for determining design flows include the familiar hydrological methods but also include a number of hydraulic models to match urban conditions. An asset in urban areas is the more readily available records of rainfall and streamflow to establish reliable estimates of specific flood flows. Even so, design rules established in one urban area may not be applicable at the other side of the continent.

Flood estimation

Where there are good quality records of rainfall and streamflow, the classical methods of hydrological analysis can be used to derive requisite design floods.

- The unit hydrograph method. Following the introduction of the concept of the unit hydrograph by the American engineer Sherman in 1932, this method of hydrological analysis has been studied and expanded by many scholars and practitioners. The unit hydrograph is explained in detail in most hydrological textbooks. Its application in Australian conditions has been well documented, and where ungauged catchments are involved the modeling of synthetic unit hydrographs using catchment characteristics has been demonstrated.
- Runoff routing methods. Runoff routing involves the routing of storm rainfall excess through a model representing the catchment storage. The resultant direct runoff hydrograph must have added to it separately estimated baseflow to complete the total flood hydrograph. A variety of representations of catchment storage are

applied successfully in Australia where significant models have been developed. The nature of the storage–discharge relationship influences the different model types and can be taken as linear, $S = Kq$, or in a non-linear form, $S = kq^m$, where S is storage, q is discharge, k is a coefficient and m is an exponential coefficient. The temporary storm storage may be considered distributed along drainage channels and routed downstream using standard methods or concentrated in a single storage or a cascade of equal storages at the catchment outlet. Examples of such models are the synthetic unit hydrograph with concentrated storage linear at the outlet and the Nash model with a cascade of equal storages (not so satisfactory in Australia).

Alternatively, the storage may be represented by modeling the areal drainage network of the catchment. Network models include the Australian Rainfall Runoff Model (with a computer package RORB), in which the storages are represented between nodes on the rivers at the centroids of subcatchments, and the Watershed Bounded Network Model (WBNM), a more detailed development of RORB in which the storages model accounts for geomorphological effects within each subarea.

Several other rainfall routing models are presented and the methods of calibration of model parameters are described. The application of continuous simulation models for estimating design floods is also mentioned.

The statistical analyses of recorded flood flows are fully explained with worked examples of different techniques. The selection of design floods for particular engineering projects incorporates the consideration of capital costs. A careful appraisal is made of the choice of flood estimation method and the estimation of large floods, including the probable maximum flood accounts for the different storm types of the varying climatic regimes.

Australian Rainfall and Runoff combines the functions of a theoretical text and a practicing engineer's manual and is an invaluable hydrological contribution to Australian engineers.

Elizabeth M. Shaw

Bibliography

Pilgrim, D.H. (ed.), 1987. *Australian Rainfall and Runoff, A Guide to Flood Estimation*, The Institution of Engineers, Australia, 374 pp.

Cross references

Flood estimation: methods for developing countries
Floods
Flood studies worldwide
Floods, world's maximum observed

B

BERNOULLI ENERGY EQUATION

An equation ascribed to the Swiss mathematician Daniel Bernoulli (1700–1782), based on the principle of the conservation of energy (that is, 'energy cannot be destroyed') can be applied, *inter alia*, to parallel or gradually varied flow in an open channel with a small bed slope. It is probable that the equation was formulated first by Leonard Euler (1707–1783) and popularized later by Julius Weisbach in the middle of the nineteenth century. Considering flow through an open channel cross-section, the total energy upstream and downstream of that cross section is a constant value, although the component energy values may be different upstream from those downstream. The total energy components are 'position' energy (the elevation of the channel bed, above a horizontal datum), the pressure head and the velocity head, and the equation is:

$$z_1 + d_1 + \frac{\bar{v}_1^2}{2g} = z_2 + d_2 + \frac{\bar{v}_2^2}{2g} = \text{constant}$$

where z is the bed elevation above the horizontal datum in meters, d is the depth of water (m), $\bar{v}$ is the mean velocity of the water through a cross-section (m s^{-1}), g is the acceleration due to gravity (acceleration of free fall) at the specific geographical location (m s^{-2}), and subscripts 1 and 2 refer to the upstream and the downstream cross sections used for calculation purposes.

P.G. Holland

Bibliography

BS 3680, Part 1, 1991. *Glossary of Terms*, British Standards Institution, London.

Cross references

Energy head
Hydrologists (600 BC–AD 1900)
Water resources, dictionary of basic terms

BIOSPHERE

The three outermost spheres of planet Earth are the atmosphere, hydrosphere and lithosphere. Each is an essentially inorganic physical continuum, albeit diversified. The term biosphere is employed in two ways: (1) all the area that is occupied by or favorable for occupation by living organisms or (2) all the living organisms inhabiting the planet Earth. For the latter, a more definitive expression, the collective term *biota* is generally better, meaning a total population, usually coupled with an adjective such as 'planetary' to distinguish it from regional population units. Thus it is usual to employ 'biosphere' in its

dimensional sense, i.e. the physical space occupied by the Earth's biota. This space is a zone of variable dimensions superimposed on parts of the atmosphere, hydrosphere and lithosphere. [Care should be taken with the two accepted meanings of the term 'terrestrial', which in astronomy distinguishes anything relative to planet Earth, whereas in biology and geology in this connection it refers to land biotas or processes as distinct from aquatic or marine equivalents. Geologists often prefer the term 'continental' (whether belonging to either a continent or an island), as in a 'continental sediment' that is deposited in either a lake, river bed or dune (thus either aquatic or eolian), as distinguished from a 'marine sediment' that is deposited in the ocean.]

In general the biosphere is most restricted in the highest latitudes and displays its greatest width in the tropics and equatorial latitudes. Subject to organic adaptations, the entire hydrosphere is accessible for components of the Earth's biota. For the atmosphere, only the lower part is normally inhabitable, for reasons of both temperature and pressure. Quite exceptionally, migrating birds have been observed at an elevation of nearly 10 000 m, flying over Mt Everest. In contrast, only the thinnest, upper layer of the lithosphere is accessible to components of the biota, in the soil and pervious sediments.

Thus the biosphere can be physically defined as coinciding with the entire hydrosphere, together with the lowest 10 km of the atmosphere and the uppermost 1–10 m of the lithosphere. It is the expanded ecotope (habitat) of the entire planetary biota.

Rhodes W. Fairbridge

Bibliography

Henderson, L.J., 1913. *The Fitness of the Environment*. New York: Macmillan (paperback reprint: Beacon Press, Boston).
Odum, E.P., 1971. *Fundamentals of Ecology*, 3rd edn. Philadelphia: Saunders.
Odum, H.T., 1971. *Environment, Power, and Society*. New York: Wiley-Interscience.
Strahler, A.N. and Strahler, A.H., 1974. *Introduction to Environmental Science*. Santa Barbara, CA: Hamilton.

Cross reference

Atmosphere

BLACK SEA ENVIRONMENT

The Black Sea is fed by a basin of more than 2 million km^2, covering parts of 17 countries in Central and Eastern Europe, the former Soviet Union and Turkey. It receives the inflows of several major rivers, including the Danube, Don, Dnieper and Dniester. As an almost fully enclosed water body, the Black Sea is especially vulnerable to changes in the quantity and quality of inflows from these rivers. The

Don and Dnieper, in particular, have been highly developed for irrigation and other purposes through a chain of reservoirs.

Increasing pollutant loads from these rivers – especially the nutrients nitrogen and phosphorus – have led to algal blooms and the destruction of important nursery areas for fish. In addition, damming of the major rivers for navigation, flood control, water supply and, above all, for irrigation, has considerably altered the seasonal flow patterns of these rivers. The damming has also decreased the total inflow to the Black Sea, resulting in an increase in salinity in critical coastal and estuarine areas, especially in the Sea of Azov, which creates further problems for fish breeding. The overall result is a 90% decline in the once-productive Black Sea fishery over the last 30 years.

With assistance from the Global Environment Facility, the six Black Sea countries (Bulgaria, Georgia, Romania, Russia, Turkey and Ukraine) have begun a regional program to analyze the causes of observed environmental degradation and to propose solutions. Actions in the basin to regulate fertilizer use and to control point sources of pollution are expected to result in reductions of nutrient inflows. Pilot projects are proposed to restore fish production under the new salinity conditions. Under the World Bank-supported Environmental Management Project for Russia, a study of the Lower Don Basin will investigate ways to alter the operating rules for the major reservoirs to promote greater fish regeneration downstream.

Given the size of the problem and the importance of these reservoirs in the agricultural economies of Ukraine and Russia, it would be unrealistic to expect dramatic changes. Nevertheless, recognition of the problem and the development of mechanisms for regional cooperation now make progress much more likely.

R.W. Herschy

Source

The World Bank, 1994. *World Development Report 1994.*

BOTTLED WATER

The consumption and sale of bottled water is increasing annually and in the United States some 2500 million US gallons (2100 million UK gallons) were sold in 1994 at a total cost of about $3000 million. In the UK some 785 million litres (170 million UK gallons) were sold in 1995 at an estimated cost of $500 million to the consumer. The increase appears to be at the rate of more than 10% per annum. Bottled waters, whether they are 'still' or 'sparkling', have to conform to strict regulations which set the same quality standards as for tap water. No water should contain harmful bacteria and regulations in the UK give maximum bacteria levels for the first 12 h after bottling. After this, any increase should not be greater than what would normally be expected.

The International Bottled Water Association (IBWA) rules are more detailed in that they define explicitly the various terminology used by the manufacturers and state rules for practice and operational, source and product monitoring. They also give standards for labeling requirements and list some 200 contaminants for which maximum permitted levels in $mg\,l^{-1}$ are presented.

In a detailed consumer investigation of bottled water, *Which* (1995) examined over 50 bottled waters available in the UK. The average price per liter for still waters was found to be 34 pence (51 cents) and the average price per liter for sparkling water was 44 p (66 ¢) with a wide variation for both. *Which* compared these prices with a price of 0.07 p (0.16 ¢) for tap water.

Which concluded that all the samples tested had low mineral content and most contained high levels of bacteria, although the levels found were not harmful. According to *Which*, of the millions of tests carried out by UK water companies in 1993, over 98% met standards for bacteria, chemical composition, appearance and taste.

R.W. Herschy

Source

Zenith Research International, 1996. Private communication.

Bibliography

International Bottled Water Association, 1995. Model Bottled Water Regulation.
Which, 1995. Troubled Waters.

C

CASPIAN SEA

Details of the area and depth of the Caspian Sea – the world's largest lake – are given under Lakes. The sea, however, has been gradually declining in size as a result of the diversions from the rivers Volga and Ural, posing a serious threat to fisheries dominated by the sturgeon catch both in Russia and Iran. After years of receding levels, however, the sea level has been rising, resulting in demand from littoral dwellers for action to be taken with respect to their homes and work. The sea level is at −27 m and for many years the former USSR considered building a canal to join the Black Sea with the Caspian, over the mountains, some 700 km long. This remains only a proposal but hydrologists have argued that such a plan would upset the ecology of the Caspian sea, which at present has a salinity of about 11 g l^{-1} compared with 33 g l^{-1} for the ocean, the Black Sea being somewhat less.

Oil, fisheries and shipping are the three largest industries in the Caspian Sea and predominate all others.

R.W. Herschy

Source

Robertson, J.O., 1995. Personal communication.

Bibliography

Fairbridge, R.W., (ed.) 1968. *The Encyclopedia of Geomorphology.* Reinhold, 109–116.

Cross reference

Lakes

CENTRAL AMERICA AND WEST INDIES: CLIMATE

Located almost entirely within the tropics, Central America has great geographic diversity, a diversity that is reflected in local climatic conditions. Despite this diversity, the broad climatic patterns can be related to the basic planetary controls, namely the seasonally shifting subtropical highs of the Atlantic and Pacific Oceans and their associated winds.

The atmospheric circulation pattern is such that easterly winds prevail in all seasons. Exposure to these trade winds results in generally moist climates, although some water deficiencies occur in winter and spring. These dry seasons are due to the fact that in summer and fall the easterly trade winds are deep, whereas in winter and spring inversions that limit the precipitation process in the trade winds are lower and more frequent.

Given the low latitude location and the dominance of trade winds, climates of the area are all grouped as tropical, being either Af, Am or Aw in the Köppen classification. The major variation in this widespread pattern occurs in modification by altitude.

Mainland Central America

To Central American and Mexican inhabitants, there are three major climates in Central America and tropical Mexico. These three major climates are loosely identified by their altitude limits and relative temperatures as tierra caliente, tierra templada and tierra fria. No exact limits are understood by the application of these terms, and they are used with purely relative meaning that differs from place to place and, in many localities, from person to person. In Guatemala, McBryde (1942) among others, has attempted to establish meaningful temperature–elevation relationships among these terms, assigning very general limits of altitude as follows: tierra caliente, sea level to 1000 m; tierra templada, 1000–1900 m; and tierra fria, 1900–4200 m.

Northwestward from Panama, mean annual temperature ranges in tropical situations under 1524 m increase, although always within high limits, from 1°C in Panama to more than 11°C north of the Tropic of Cancer. The increasing range arises chiefly from lower temperatures in the cool season from December to April.

Seasons

In central America the climatic year consists of two more or less distinct seasons, a wet season principally from May to November and a dry season from December to April, although in many instances this dry season experiences considerable rain. March is usually the driest month. The rainy season is punctuated by two maxima: a first usually in May or June and a second usually in October, although delayed until November in Panama. These maxima are related to the altitude of the sun, lagging its zenithal position by about 1 month in the early part of the wet period and by nearly 2 months toward the end. Between these two peaks, rains subside in a sufficiently striking manner to have inspired in many localities the term canicula (dog days), intervals of diminished rain that occur for several weeks at any time from June through August.

Areas facing the Atlantic basin (the Gulf of Mexico and the Caribbean) are under the dominant control of the northeast trades throughout the year and are less sunny, cloudier, more humid, receive more precipitation and experience a shorter dry season than those facing the Pacific. The contrast shows up most plainly during the winter season of the northern hemisphere.

Precipitation

Precipitation is the primary key to climatic differentiation in tropical Central America and Mexico (Figure C1).

The easterly flow of the trades governs most of the weather, and the weather in general is distinguished by much sunshine, interspersed in the rainy season by showers and thundershowers. Cyclonic disturbances are relatively frequent along both coasts but seldom appear inland. They may be ill defined or may appear as well-developed circulations. They may move very slowly or may intensify into

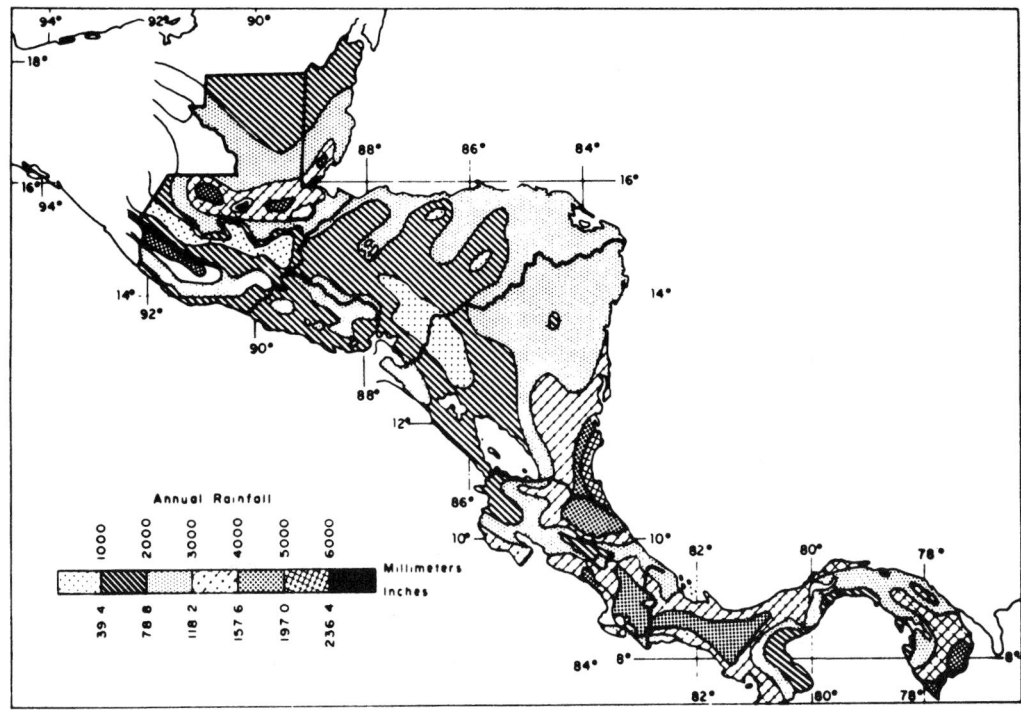

Figure C1 Mean annual rainfall (in inches) in Central America. (After Portig, 1965.)

Table C1 Selected climatic data of Cristobal (Colon), Panama (latitude 9°21′N; longitude 79°55′W; elevation 12 m)

Month	Mean sea-level pressure (mbar)	Temperature (°C)				Precipitation (mm)		
		Daily mean	Mean daily range	Extremes				
				Max.	Min	Mean	Mean	Max. in 24 h
Jan.	1011.5	26.8	6.4	31.7	17.8	70	85	77
Feb.	1010.8	26.8	6.1	32.8	17.8	38	40	113
Mar.	1010.8	27.0	6.1	33.9	17.8	40	42	106
Apr.	1010.5	27.2	6.1	33.9	18.9	94	105	157
May	1009.6	26.8	6.7	33.9	21.1	315	324	154
June	1010.1	26.4	7.2	35.0	21.1	314	307	137
July	1010.2	26.6	6.4	33.9	21.1	389	406	166
Aug.	1010.3	26.6	6.4	37.2	20.0	384	387	135
Sept.	1009.7	25.6	7.8	35.0	20.0	321	321	87
Oct.	1010.0	26.5	7.8	33.9	20.0	432	397	173
Nov.	1009.9	26.1	7.2	33.9	20.0	646	569	343
Dec.	1010.6	26.4	6.4	33.9	20.0	381	302	130
Annual	1010.3	26.6	6.7	34.1	18.1	3424	3285	343

tropical storms of hurricane strength. The slow-moving systems tend to remain in the Caribbean.

Rain appears to be derived from three main disturbances: steady rain from largely stratified clouds (temporals), thunderstorms and trade-wind showers without thunder. Hail is very rare and snow has been observed only as far south as the mountainous heights of southern Guatemala, where a number of elevations exceed 4000 m above sea level.

Hurricanes rarely affect the southern Central American countries but occasionally deliver destructive winds and flood-producing rains to eastern Nicaragua and northwestward along both coasts of Mexico, including the Yucatan Peninsula.

Normal yearly rains supporting the variety of tropical climates in Middle America vary widely, from the rainforest excesses of more than 5970 mm in southeastern Nicaragua to less than 635 mm toward

the poleward limits of scrub woodland near the Mexican coasts. Sample data are provided in Tables C1–C6.

Climate–vegetation relationships

The rainforest climate from Colombia to Nicaragua is represented by Cristobal at the Caribbean entrance to the Panama Canal. Cristobal (elevation 11 m) lies in a belt of tropical rainforest edging the Caribbean coast from Colombia to northern Nicaragua (Figure C2). Dominant winds throughout the year are northerly. Cristobal's normal temperatures range from 27°C in November to 28°C in April, and its mean annual rainfall is 3300 mm. Eight months (May–December) normally receive more than 254 mm of rain per month, and the first maximum is attained in July (400 mm), the second in November (570 mm). Rain falls in this period on from 22 to 26 days per month,

Table C2 Selected climatic data of San José, Costa Rica (latitude 9°56'N; longitude 84°08'W; elevation 1120 m)

Month	Mean sea-level pressure (mbar)	Temperature (°C) Daily mean	Extremes Max.	Extremes Min.	Precipitation (mm) Mean	Precipitation (mm) Max. in 24 h	Relative humidity (%)	Days with precipitation ≥0.1 mm
Jan.	1010.7	19.0	30.5	9.4	8	12	80	3
Feb.	1010.1	19.3	31.1	10.5	5	8	80	1
Mar.	1009.0	20.3	32.7	10.0	10	12	80	2
Apr.	1008.6	21.0	31.6	11.7	37	25	79	7
May	1008.1	21.4	31.0	12.2	244	83	84	19
June	1008.3	21.2	33.2	13.9	284	88	86	22
July	1008.9	20.6	28.9	12.2	230	123	86	23
Aug.	1008.8	20.8	29.4	13.3	233	60	85	24
Sept.	1008.4	20.9	30.0	13.3	342	60	86	24
Oct.	1008.7	20.6	29.4	12.7	333	75	88	25
Nov.	1009.5	19.9	28.7	11.1	172	75	84	14
Dec.	1010.6	19.3	30.5	9.4	46	22	82	6
Annual	1009.1	20.4	27.8	11.6	1944	123	83	170

Table C3 Selected climatic data of San Salvador (latitude 13°43'N; longitude 89°12'W; elevation 700 m)

Month	Mean stationary pressure (mbar)	Temperature (°C) Daily mean	Mean daily range	Extremes Max.	Extremes Min.	Precipitation (mm) Mean	Precipitation (mm) Max. in 24 h
Jan.	934.0	22.1	13.3	29.2	15.9	5	18
Feb.	934.0	22.4	15.0	30.8	15.8	3	24
Mar.	933.3	23.5	15.0	32.0	17.0	8	32
Apr.	933.1	24.2	13.6	32.1	18.5	60	88
May	932.8	23.7	12.0	30.7	18.7	190	120
June	933.1	23.1	10.6	29.3	18.7	322	205
July	933.7	22.9	11.5	29.6	18.1	304	95
Aug.	933.5	23.0	11.6	29.9	18.3	297	165
Sept.	932.8	22.5	10.7	29.1	18.4	325	175
Oct.	932.7	22.4	10.0	28.3	18.3	220	170
Nov.	933.3	22.0	11.2	28.3	17.1	35	54
Dec.	933.7	22.0	12.1	28.4	16.3	7	46
Annual	933.3	22.8	12.2	29.8	17.6	1775	205

Table C4 Selected climatic data of Fort-de-France, Martinique (latitude 14°35'N; longitude 61°12'W; elevation 144 m)

Month	Mean stationary pressure (mbar)	Temperature (°C) Daily mean	Mean daily range	Extremes Max.	Extremes Min.	Precipitation (mm) Mean	Precipitation (mm) Max. in 24 h
Jan.	1014.9	23.5	5.5	30.2	17.8	96	45
Feb.	1014.6	23.5	5.6	30.0	17.3	68	63
Mar.	1014.3	24.0	6.1	32.6	18.6	58	42
Apr.	1014.0	24.7	6.1	32.8	18.9	82	78
May	1013.7	25.4	5.8	32.5	19.9	126	123
June	1014.7	25.7	5.2	31.5	20.0	160	76
July	1014.4	25.6	5.0	31.3	19.5	214	71
Aug.	1014.1	26.0	5.5	33.0	20.3	227	103
Sept.	1013.0	25.9	5.7	32.9	20.6	232	238
Oct.	1012.7	25.6	5.6	32.0	20.2	221	152
Nov.	1012.7	25.2	5.4	31.8	19.8	230	133
Dec.	1013.2	24.2	5.1	31.0	17.4	126	86
Annual	1013.8	24.9	5.6	31.8	19.2	1840	238

Table C5 Selected climatic data of Managua, Nicaragua (latitude 12°08'N; longitude 86°11'W; elevation 56 m)

Month	Mean stationary pressure (mbar)	Temperature (°C)				Precipitation (mm)	
		Daily mean	Mean daily range	Extremes		Mean	Max. in 24 h
				Max.	Min.		
Jan.	1005.5	26.3	10.6	31.0	20.4	4	9
Feb.	1005.1	27.2	11.5	32.1	20.6	1	2
Mar.	1004.3	28.6	11.9	33.6	21.7	5	29
Apr.	1004.0	29.3	11.7	34.3	22.6	6	23
May	1003.9	29.4	10.6	34.0	23.4	76	92
June	1004.0	27.2	8.4	31.4	23.0	296	119
July	1004.8	26.9	8.3	30.9	22.6	134	89
Aug.	1004.1	27.2	9.0	31.4	22.4	130	60
Sept.	1003.5	26.9	9.1	31.3	22.2	182	119
Oct.	1003.3	26.5	8.7	30.8	22.1	243	108
Nov.	1004.1	26.3	9.7	30.6	20.9	59	45
Dec.	1005.1	26.1	10.8	30.8	20.0	6	9
Annual	1004.3	27.3	10.0	31.8	21.8	1142	119

Table C6 Selected climatic data of Curacao (latitude 12°12'N; longitude 68°58'W; elevation 8 m)

Month	Mean stationary pressure (mbar)	Temperature (°C)				Precipitation (mm)	
		Daily mean	Mean daily range	Extremes		Mean	Max. in 24 h
				Max.	Min.		
Jan.	1013.0	26.2	5	30.9	19.0	68	75
Feb.	1013.1	26.1	5	31.1	19.9	31	24
Mar.	1012.4	26.5	5	32.2	20.0	14	33
Apr.	1011.8	27.1	5	33.4	21.4	12	33
May	1011.6	27.7	5	35.6	22.2	18	43
June	1012.3	28.0	5	32.9	21.0	26	98
July	1012.7	28.0	6	33.8	22.6	34	42
Aug.	1011.7	28.4	6	34.9	20.4	48	64
Sept.	1010.8	28.8	6	35.8	20.4	31	91
Oct.	1010.3	28.3	6	35.2	20.0	67	119
Nov.	1010.8	27.8	5	33.5	20.0	98	157
Dec.	1011.9	26.9	5	32.7	20.1	85	104
Annual	1011.9	27.5	5	35.8	19.0	532	157

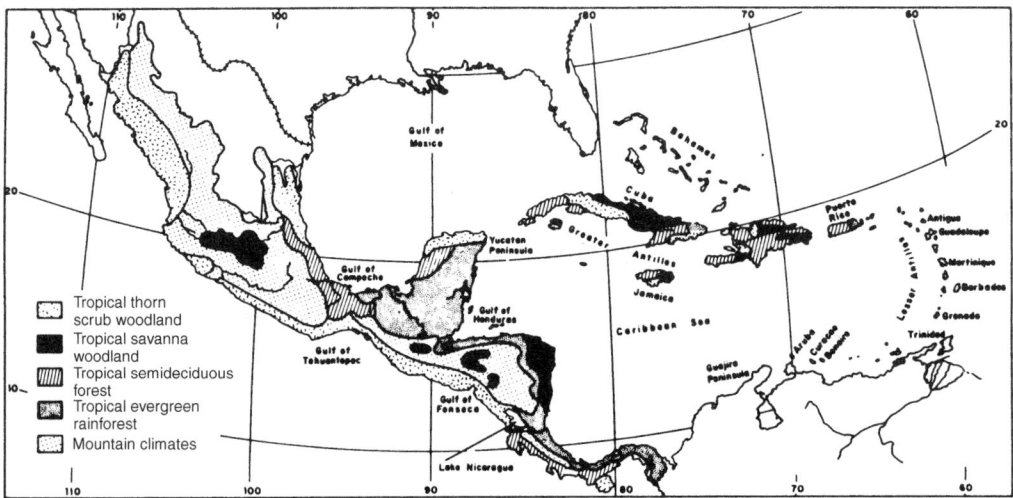

Figure C2 Vegetation regions of Middle America.

mean cloud cover is between 70 and 80% and relative humidity averages between 82 and 86%. An average of 87 thunderstorms occur from May through to October, out of a total of 99 for the year. The four dry months together, from January through April, average 260 mm, February and March each recording 38 mm, falling on 12–15 days per month. Cloud cover decreases but remains high at between 50 and 60%, and relative humidity averages 77%.

Tropical evergreen rainforest extends westward from Panama into western Costa Rica where the low mountains are heavily mantled with forest vegetation which thrives under normal rainfall exceeding 2540 mm that is adequately distributed throughout the year. Continuing northwestward into the Nicaraguan lowland, however, a rapid change occurs as the heavier forest cover gives way to more open savanna and thorn scrub woodland. From near Lake Nicaragua northwestward along the Pacific coast in an almost continuous distribution, the drier savanna and scrub woodland reaches beyond the Tropic of Cancer in Mexico. Reaching inland for a variable distance of between 16 and 32 km from the sea as far west as the Isthmus of Tehuantepec, the drier, more open woodlands spread farther inland beyond that point to 160 km or more from the sea along the Pacific sierras of Mexico. Diminishing rainfall amounts concentrated in an increasingly shorter season and a lengthening of the dry season are chiefly responsible for these changes.

From Lake Nicaragua northwestward across El Salvador to southeastern Guatemala, mean annual rainfall averages between 1650 and 1900 mm concentrated in a 6-month rainy season, interrupted by a dry season of 4–5 months when monthly amounts average less than 13 mm. From southeastern Guatemala to southern Chiapas in Mexico is a narrow belt of much heavier rainfall where annual amounts exceed 2540 mm and the dry period is only 3–4 months long. Here, semideciduous to selva vegetation is supported by the more abundant rain at elevations from 760 to 1350 m. From here to the Isthmus of Tehuantepec, annual amounts diminish again to between 1400 and 1650 mm. From the low saddle of Tehuantepec westward some 480 km to Acapulco, amounts lessen still further to between 1270 and 1520 mm, occurring in a 5-month wet season, and the dry period is 5–6 months long. Northwestward from Acapulco, annual totals average between 890 and 1020 mm, still falling in a 5-month wet season, and the dry period is also 5–6 months in length, each month averaging less than 13 mm of normal rain.

West Indies and Caribbean

Over the island region of the West Indies the overriding equability of the climate is chiefly modified by variations in elevation and esposure to the prevailing easterlies. This is particularly evident on the larger islands of Cuba, Jamaica, Hispaniola and Puerto Rico – the Greater Antilles.

For the West Indies there is no summer and winter but simply a change from wet season to a somewhat drier season. At stations almost anywhere on the smaller islands and along the coasts of the larger ones, temperatures vary within a mean daily range of 5.5–8.2°C, fluctuating between the low to upper 20s from December through April and between the mid 20s and lower 30s from May through November. In the interior of the larger islands, the mean daily range increases to 11°C or more from about 20°C to the low 30s or higher. Here the thermometer may exceed 35°C from time to time, and less frequently may reach over 38°C during the rainy season, but extremely high values are rare for most of the West Indies.

Frost is very unusual even at higher elevations, although cold waves from North America (the norther or el norte) may lower thermometer readings to below 10°C. The lowest value recorded in Puerto Rico has been 4.4°C at Aibonito in March. Extreme minima in most of the West Indies have seldom dropped below 12.7°C.

A general statement on the overall distribution of normal rainfall over the West Indies must be accepted with caution in view of the many contrasts among the islands in terms of size, shape, configuration, exposure to the trades, and geographical arrangement. Ignoring the influence of relief for the moment, the following pattern may be noted. In the low-lying Bahamas, mean annual rainfall ranges from more than 1500 mm among the northern group (Green Turtle Cay, 1633 mm annually in an 8-year period) to less than 650 mm among the southernmost group (Duncan Town, 610 mm annually over 11 years; Mathew Town, Great Inagua Island, 632 mm annually over 5 years). Nassau on New Providence Island in the northern group has a 59-year average of 1300 mm. Over the large islands of the Greater Antilles, values vary too widely according to the influence of major

relief features and other factors to warrant a generalization, but many stations report between 1100 mm and 1500 mm of mean yearly rain. Farther east the northern group of the Lesser Antilles have recorded normal amounts somewhat lower than these. This holds true as far as Antigua. However, beginning with Monserrat, normal yearly totals increase to between 1800 and 2300 mm. From here to Grenada, a span of some 650 km, a curving arc of small, mountainous volcanic islands lies athwart the main westward current of the steady trades. South of Grenada, mean annual rainfall again diminishes, and in Trinidad amounts are generally less than 1500 mm. The lower island of Barbados, about 160 km east of the archipelago, having elevations of less than 380 m, is an exception among the Lesser Antilles, with a yearly normal of 1250 mm. A dry belt in the southern Bahamas lies between the wetter northern group and the well-watered conditions prevailing over most of the West Indies. A still drier zone exists along the southern margins of the Caribbean in coastal Venezuela and Colombia and among the offshore islands.

Among the West Indies, elevation and exposure to or shelter from the dominant trades are the principal factors in normal rainfall distribution from place to place in each island. The orographic influence is a major cause of rainfall excess and deficiency, particularly during the height of the rainy season from May through November. It is in this period that waves in the easterlies occur most frequently. No general rule may be applied to all the West Indies as far as height–amount relationships are concerned. From the available records it appears likely that well-exposed heights in many instances receive over 5000 mm of rainfall annually, an amount far above the normal values at lower elevations.

Storms

Severe tropical cyclones and storms, originating as waves in the easterlies and occasionally gaining sufficient intensity to become hurricanes are a source of concern throughout the West Indies during the latter part of the rainy season. They ordinarily skirt the north coast of South America and the islands of Trinidad and Tobago, but elsewhere they pose a serious threat of property damage and loss of life. They also account for some of the excessive rains that have fallen in brief periods on the islands.

Cold-front rains are often excessive in the less rainy period of December–April and add a significant quotient to the normals for stations in the Greater Antilles, although northers as such seldom extend their influence effectively beyond the Virgin Islands and Antigua, east of Puerto Rico.

George R. Rumney and John E. Oliver

Bibliography
Lydolph, P.E., 1985. *The Climate of the Earth*. Totowa, NJ: Rowman and Allanheld.
McBryde, F.W., 1942. Studies in Guatemalan meteorology, *Am. Meteorol. Soc. Bull.*, **23**, 254–263.
Portig, W.H., 1965. Central American rainfall, *Geog. Rev.*, **55**, 68–90.
Rumney, G.R., 1968. *Climatology and the World's Climates*. New York: Macmillan.
Schwerdtfeger, W. (ed.), 1976. *Climates of Central and South America*, Vol. 12, *World Survey of Climatology*. New York: Elsevier.

Cross references

Climate and climate change
Climate data: sources

CHANNELIZATION AND BANK STABILIZATION

Channelization involves human modification and control of natural, existing waterways, usually to permit or promote economic development or to protect already established urban, agricultural and industrial developments. A specific channelization project may be undertaken for one or more of a number of reasons: (1) for flood control, (2) to drain wetlands, (3) to improve navigation and (4) to prevent bank erosion and channel migration, and thus to protect neighboring property. There has been much controversy about whether channelization, especially over the long term, is effective, and whether harmful effects may exceed the benefits.

Aims of channelization

Flood damage

Flood damage to structures, crops and so on amounts to about $1 billion annually in the United States alone. In addition, flooding causes widespread disruption of human activity, and may result in great loss of life. Floods can be controlled and abated by a number of means, including flood-control dams, levees and floodways, as well as channelization. Frequently, a comprehensive program will utilize all available measures to maximize protection. Channelization of a waterway aims to increase the hydraulic efficiently of the channel so that flood waters from upstream may pass as rapidly as possible through the reach. The boundary roughness of the channel is reduced by smoothing the channel perimeter and removing obstacles to flow such as trees, and the form roughness may also be reduced by realigning the channel to produce a straight or smoothly sinuous course (Acheson, 1968). By making, the channel more hydraulically efficient, the cross-sectional area of the water is reduced, and hence water depths and the chance of overbank flow are decreased.

Drainage

While channelization for flood control aims to deal with excess water from upstream of the affected reach, channelization for drainage is undertaken to remove excess water from the land in the immediate vicinity of the waterway. It is often extremely difficult, however, to distinguish between the flood control and land drainage effects of a particular project, and both benefits are frequently realized. Channelization for drainage has the same immediate aim as for flood control – to increase the hydraulic efficiency of a waterway, facilitate the evacuation of excess water and lower water levels. The water table in adjacent lands will thus be lowered, permitting agricultural development of or construction on former wetlands.

Navigation improvement

Channelization projects designed to improve navigation facilities aim to provide a navigable, shoal-free channel that will be easily negotiated by ships. The waterway must hence be trained to follow a desirable course, aligned to produce smoothly sinuous bends which to obviate the need for frequent dredging, should be self-maintaining. By constricting the channel width, flow velocities are increased and bed scour promoted. This helps maintain the desired channel depth throughout the channelized reach, although dredging may still be necessary.

Bank protection

A frequent aim of channel modification is to prevent bank caving and channel migration, which may destroy valuable farmland and threaten buildings, levees, and other structures. On a more limited scale, channelization of a stretch of river may be undertaken in conjunction with, for example, construction or relocation of a highway, in which the desired alignment encroaches on the channel. Channelization may also improve wildlife habitat, by preventing destructive bank erosion and channel migration (reviewed by Stevens *et al.* 1975b, pp. 558–561). For example, Milkovic and Petersen (1975) have described methods under test along the Sacramento River to improve riverine environments, by preventing erosion of the basins between channel and levee, and White (1975) described a project on a Wisconsin stream undertaken specifically to improve fish habitat.

A given channelization project may well provide benefits in all the preceding categories, and projects are increasingly undertaken for multipurpose water resource development. In turn, channelization may be only one aspect of river basin regulation, and may be associated with the construction of levees and flood control dams, land treatment for erosion control, and so on.

Channelization is usually undertaken on river flood plains, valley bottomlands and coastal plains, except where urban flood control or protection of structures against erosion is desired. The English fenlands or the Rhine and Scheldte deltas in the Low Countries – all highly productive and populous agricultural regions – are good examples of low-lying, gently sloping areas in which channelization

has provided substantial drainage, floor control, and navigation benefits. Channelization is largely restricted to the highly developed, economically advanced nations of the world, and especially to the North American and European continents.

Although many channelization projects have been on a large scale, data summarized by Little (1973), and Acheson (1968) suggest that projects in the United States and New Zealand, for example, tend to be quite small. Apart from small-scale channel modification carried out by individual landowners for drainage purposes, 65% of the projects completed by the US Army Corps of Engineers before 1972 were less than 8 km long, and 10% of the approved Soil Conservation Service projects averaged 6.6 km in modified channel length (Little, 1973). Nevertheless, between 1940 and 1970, the Corps of Engineers and the Soil Conservation Service modified over 55 000 km of river.

Channelization techniques

Channel modification may be accomplished in a number of ways, several of which may be used in a given project. Two basic references to channelization techniques, in New Zealand and the United States respectively, are Acheson (1968) and Winkley (1972).

Clearing and snagging

This procedure consists of removing large obstacles such as trees and rocks from the channel. Such obstacles retard flow by increasing turbulence and energy dissipation, and catch floating debris that exacerbates the problem. Removal of obstacles achieves a modest improvement in hydraulic efficiency and channel capacity, speeds evacuation of water and lowers water levels. Clearing and snagging is thus of particular importance in flood control or drainage projects, especially in small upstream channels. With bank grading it is also a prerequisite to bank stabilization and protection, discussed below.

Channel excavation

Greater improvement in the flow capacity of a waterway than that provided by clearing and snagging may be obtained by channel excavation. Two broad types of excavation may be identified: conventional and dredging.

Conventional
Equipment such as draglines, power shovels and bulldozers work from the bank of the waterway to widen and deepen the channel, and to form a more hydraulically efficient cross-section. To minimize costs and problems due to sedimentation downstream, this type of work is done as far as possible 'in the dry,' during periods of low flow. Conventional excavation is especially suitable for relatively small waterways, or for wide, braided rivers, in which shallow depths permit excavation of both bed and banks. It may also be used for smoothing and grading the upper banks of larger rivers, usually to prepare for bank stabilization with revetment.

The aim of conventional channel excavation is to increase flow capacity and to develop a more efficient channel shape. The trapezoid is the most common shape for unlined channels; although it is hydraulically efficient, it is an unstable shape for alluvial channels. Side slopes are dictated by the stability of the material through which the channel is cut; permissible side slopes for stability in various materials are listed in Table C7. The rectangle, with vertical side slopes, is a special case of the trapezoid, and is commonly used for channels built of stable materials, such as masonry.

Table C7 Suitable side slopes for channels built in various materials (from Chow, 1959)

Material	Side slope
Rock	Nearly vertical
Muck and peat soils	$\frac{1}{4}$:1
Stiff clay or earth with concrete lining	$\frac{1}{2}$:1 to 1:1
Earth with stone lining, or earth for large channels	1:1
Firm clay or earth for small ditches	$1\frac{1}{2}$:1
Loose, sandy earth	2:1
Sandy loam or porous clay	3:1

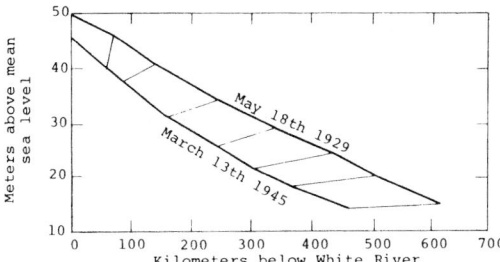

Figure C3 Effect of cut-offs on Lower Mississippi River surface profiles. Both curves are on rising stage at 43 000 m³ s⁻¹; upper is before and lower is after cut-offs. Lines between curves connect same locations on river before and after cut-offs. (After Carey, 1966.)

Dredging

There are three broad types of dredge – dipper, ladder and suction (Huston, 1967). A dipper dredge is merely a floating power shovel, and would be restricted to work in shallow water. Ladder dredges have an endless chain of buckets, which bring the bottom material to the surface and discharge it onto a conveyor. Suction dredges pick up the bottom material and water in suction pipes, and discharge the slurry via a spoil pipe supported by floats to the desired spoil area. Rather than being merely dumped, the spoil should be used to supplement other channelization operations, such as the closing of chutes at the heads of point bars or islands.

Channel cuts

A logical extension to enlarging existing channels is to excavate completely new sections of channel. Natural rivers have meandering courses, which have been regarded as undesirable for blood control and navigation purposes. Removing and bypassing bends by excavating channel cuts reduces resistance to flow due to form roughness and reduces the distance water must travel. For example, the cutoffs on the Mississippi River between Memphis and Baton Rouge described by Matthes (1948) shortened the river by 270 km from an original length of 1095 km. This shortening resulted in an appreciable lowering of flood stages on the river (Figure C3). Cutoffs also bring benefits to navigation and bank protection. Elimination of sharp meander bends simplifies negotiation of the waterway by ships, and reduces journey time. The caving banks in the cutoff bends are bypassed, and properly designed cutoffs may substantially reduce the lengths of eroding banks that must be maintained.

Realignment

An existing channel can be realigned by training the flow into the desired alignment with structures such as dikes and jetties. Whether for navigation or flood control purposes, the function of river-training structures is to persuade the flow into a smoothly sinuous alignment that is hydraulically more efficient than the natural channel. The channel is constructed so that widths are reduced and depths increased, thus promoting bed scour, which prevents deposition and makes the channel more or less self-maintaining. In other situations the training structures are intended to move the main flow away from the bank lines to prevent bank caving and channel migration.

A common type of realignment structure is the stone-fill dike. Its principal function is to direct the flow away from the bank; angling the dike upstream, downstream or normal to the bank accomplishes this purpose in any given location. Two configurations using stone dikes that deserve specific mention are the vane dike and the L-head dike system. Vane dikes are angled about 10° to the bank in a downstream direction, and generate less eddying and scouring than dikes constructed normal to the bank. The L-head dike includes a section of dike extending downstream from the main dike and parallel to the flow. When the L-heads close about half the gap between dikes in the dike field, they promote deposition between the dikes, decrease scour at the ends of the dikes and provide bank protection. Both dike configurations may be included in navigation channel or bank protection projects.

Other types of river-training structures are retards and jetty fields. Retards are permeable devices, such as timber piles, which are placed parallel to river banks to decrease flow velocities and prevent erosion.

Jetty fields are intended to train the main stream into a selected alignment, reduce flow velocities along the banks and eliminate erosion. They may be constructed of jacks, typified by those used in the middle Rio Grande to stabilize the braided channel and protect levees and adjacent areas. The unit consists of three 5 m long steel angles placed at right angles to each other, bolted together at the center and reinforced with wire. The Jacks are then connected with cables to form a jetty line. Two types of jetty lines are used in a jetty field: diversion lines are placed along the desired location of the channel, and retard lines are placed at an angle to the diversion lines and spaced between 40 and 80 m apart. Other types of structures are available, and many variations and combinations have been used.

Bank protection and stabilization

The river-training structures discussed in the preceding section provide one means of protecting banks against erosion, by moving high-velocity flows away from the bank line. They may be said to provide intermittent protection; continuous protection is provided by revetment, in which the entire bank is covered with some type of erosion-resistant material. High flow velocities are permitted against the bank line, but the revetment increases the resistance of the bank to errosion.

The simplest form of bank protection, at least of the upper banks, is *vegetation*. The foliage reduces flow velocities at the soil surface and the roots bind the bank material, thus reducing or eliminating erosion. Vegetation needs periodic maintenance to prevent deterioration of the channel; bank slopes must therefore be sufficiently gentle to permit the use of mowing machinery.

The standard practice in New Zealand has been to place willow poles along the bankline to be protected, wire them together and anchor them with cable, then bulldoze river gravel over the toe of the slope. The willows rapidly sprout and, after repeated layering (cutting of branches that then resprout), provide a dense line of trees along the bank. Because willows may spread so readily beyond the location where they are needed and because of difficulties with supply of appropriate poles, this type of work is being partly supplanted by the use of rock riprap (Acheson, 1968).

Riprap is perhaps the most common type of bank protection; it consists of a layer of rock fragments, preferably with a smooth size gradation and with an angular shape. The specifications for riprap on the upper banks of the Mississippi require a 25 cm (±5 cm) layer of rock with individual particles weighing between 6 and 25 lb, with an approximate gradation as follows (1 lb = 0.45 kg):

75–125 lb: 10% max.
25–74 lb: 40–60%
6–24 lb: 20–40%
<6 lb: 15% max.

With a good gradation of sizes, the interstices between the larger rocks are filled with the smaller sizes; interlocking is enhanced when the rock fragments are angular. Riprap may be used for the whole bank on smaller rivers, and for the upper banks only on rivers such as the Mississippi. When adequate riprap sizes are not available, rocks of cobble size may be placed in wire mesh mats or baskets and laid along the bank to provide protection.

Continuous bank protection that is fabricated in large sections and sunk against the underwater bank is termed mattress. The most effective yet designed is the articulated concrete mattress, formed of 20 concrete blocks spaced on a continuous wire mesh reinforcing. The sections are assembled on a barge, and lowered to the bottom as the barge moves out from the bank. Because of the scale of operation necessary, articulated concrete mattresses have been used solely on the lower Mississippi, on which over 1200 km have been laid. Other types of revetment in use are woven willow brush and woven lumber mattresses, which have been used on various rivers such as the Missouri, Arkansas and Red rivers. The lumber mattress consists of a mat of 10 × 2.5 cm boards woven together and sunk to the bottom with the aid of stone-filled cribs. Several other types of fabricated bank protection have been used; they have in common, apart from their ability to protect the bank from erosion, a degree of flexibility that permits them to conform to bank contours and to adjust to any undercutting and caving that may occur at the toe of the bank.

The ultimate in bank protection is complete lining of the channel with masonry or concrete, as has been done with the Los Angeles River, for example. Lining substantially reduces channel roughness, improves hydraulic efficiency and reduces flow stages, and completely eliminates bank erosion. It also permits vertical sidewalls, which substantially reduce the area covered by the channel, but which also represent a potential safety hazard. Only in urban areas where land values are high and structures are in close proximity to the channel can complete lining be economically justified.

Effects of channelization

Channelization has brought major benefits, both direct and indirect, to agriculture, transportation and other sectors of the economy. Little's report to the Council on Environmental Quality (1973) concluded that the direct benefits of the channelization projects that it studied were generally somewhat conservatively stated; the report has been severely criticized, however. Jahn and Trefethen (1972) considered that, for many projects, benefits are overstated and costs are understated. Since the late 1960s there has been recognition that there are significant environmental and other costs that must be considered in project evaluations.

In addition, it appears that the benefits of channelization may be only temporary, unless continued maintenance is undertaken. Thus, for example, bank vegetation must be cut back on channelized reaches to prevent deterioration of the channel and loss of capacity, while on the Mississippi River continual dredging is necessary to maintain the navigation channel. Moreover, there is uncertainty about the long-term effects of channelization. For example, there is some concern that the beneficial flood control effects of middle Mississippi River channelization are being lost, and that the situation may in fact be exacerbated in the future (e.g. Stevens et al., 1975a; Belt, 1975).

Several costs of channelization have been identified.

- Channelization causes the loss of large numbers of different, and increasingly rare, habitats for plant, animal and fish species. Drainage of wetlands and bottomlands represents an obvious loss of habitat, and an unavoidable loss given that drainage is a major justification for a large proportion of projects. In addition, excavated channels, with their wide, shallow flows, bare or eroding banks, and high sediment loads, are less ecologically productive than are natural rivers, while the excavation of cut-offs frequently leaves the bypassed bends stagnant and silted, and of limited value for fish and wildlife. Duvet et al. (1976) concluded that channelization had no long-term deleterious effects on forage fish species and benthic macroinvertebrates, but reduced trout populations by removing overhead cover (overhanging banks and vegetation) and deep pools. Examples of 'ecological disasters' in which the natural stream ecosystem has been almost completely disrupted – such as the attempt at channelization for flood mitigation in Crow Creek, Tennessee – are numerous, but measures are available to mitigate the environmental effects of channel modification. For example, Keller (1975) suggested that the bed of a channelized reach should be excavated to leave a winding low-water channel with the characteristic of a natural river. The Tennessee Valley Authority modified Bear Creek in such a way that the original channel was relatively untouched (Jahn and Trefethen, 1972). Meanders were cut off by a shallow grassed channel to take flood flows, but the natural channel was permitted to carry a normal flow at other times, so that fishery values were maintained. Mifkovic and Petersen (1975) listed a number of bank protection techniques, tested on the Sacramento River, which protect levees against bank erosion with minimum disruption of riparian habitat and aesthetic values. They concluded that bank protection can protect and preserve environmental values, given proper planning.
- Channelization frequently results in severe erosion and downstream sedimentation. For example, a section of the Blackwater River in Missouri was shortened in 1910 from 53.6 to 29 km, with an increase in gradient from 1.67 to 3.1 m km^{-1} (Emerson, 1971). The area of the original excavated cross-section was 38 m^2, but because of severe erosion induced by the increased slope and flow velocities, cross-sectional areas now range between 160 and 484 m^2. Channelization enabled the utilization of new floodplain land, but also caused erosional loss of farmland, and necessitated expensive bridge renewal and repair. Bird's (1979) case study of the Lang Lang River, Australia, described unintended impacts of channelization that are, if anything, even more disastrous. Koloseus (1972) suggested that these effects can be predicted, using analytical procedures developed by hydraulic engineers. To prevent such effects, measures such as installation of drop sills may be adopted, as is already done in many cases by the US Soil Conservation Service.

- There may also be effects on reaches and water bodies downstream of a channelized stretch of river. On the Blackwater River, for example, channelization was followed by sedimentation and increased flooding downstream. If flood waters or drainage water are rapidly passed through a hydraulically efficient, channelized section of waterway, flood peaks in unchannelized reaches downstream will be increased. Thus additional channelization may be required to deal with the negative effects of the original work. Channelization may also cause a deterioration in water quality, by shortening the time period in which the water can be purified. Channelization on the Kissimmee River, Florida, has led to severe deterioration of the quality of the waters of Lake Okeechobee, because flood waters flow straight to the lake, rather than spilling onto, and being filtered by, the floodplain.
- Many people regard channelized waterways as aesthetically objectionable, especially when maintained to limit vegetation regrowth.

Because of these and other environmentally undesirable effects, channelization is an extremely controversial subject in the USA, although it is less so in other countries. The main targets for criticism are projects designed for flood control and drainage; projects for erosion control and navigation do not seem to have so many unintended side effects. Many engineers now believe that channelization is in many cases the least desirable course of action to take to achieve a specific aim (Schoof, 1980). For flood control purposes, for example, it is pointed out that natural river flood plains provide ready-made reservoirs to reduce flood peaks, and that channelization to prevent inundation induces more problems downstream than it solves. Alternative measures such as floodplain zoning or floodway construction are being considered; the floodplain must obviously be maintained under a land-use system, such as pasture, which is not adversely affected by periodic inundation. There seems less chance of resolving to everyone's satisfaction the conflicts over channelization for drainage of wetlands. Both the US Soil Conservation Service and the US Army Corps of Engineers have severely cut back their rural channelization programs for flood mitigation and drainage, but in many situations channelization is justified. Given adequate planning, design, execution and maintenance, as required under Public Law 566, many effects of channelization, such as induced erosion and loss of visual amenity, may be mitigated or avoided. It is necessary for channelization to be viewed as just one means of achieving a desired end, and should be used when other methods are not appropriate.

M.P. Mosley

Bibliography

Acheson, A.R., 1968. *River Control and Drainage in New Zealand*. Wellington, NZ: New Zealand Ministry of Works, 296 pp.

Belt, C.B., Jr, 1975. The 1973 flood and man's constriction of the Mississippi River, *Science*, **189**, 681–684.

Bird, J.F., 1979. Geomorphological implications of flood control measures, Lang Lang River, Victoria, *Australian Geog. Studies*, **17**, 169–183.

Carey, W.C., 1966. Comprehensive river stabilization, *Am. Soc. Civil Eng. Proc., J. Waterways and Harbors Div.*, **92**(WW1), 87–108.

Chow, V.T., 1959. *Open Channel Hydraulics*. New York: McGraw-Hill, 680 p.

Duvet, W.A., Volkmar, R.D., Specht, W.L. and Johnson, F.W., 1976. Environmental impact of stream channelization, *Water Resources Bull.*, **12**, 799–812.

Emerson, J.W., 1971. Channelization: a case study, *Science*, **173**, 325–326.

Huston, J., 1967. Dredging fundamentals, *Am. Soc. Civil Eng. Proc., J. Waterways and Harbors Div.*, **93**(WW1), 45–69.

Jahn, L.R., and Trefethen, J.B., 1972. Placing channel modification in perspective in S.C. Csallany, T.G. McLaughlin, and W.D. Striffer (eds), *Watersheds in Transition, Symposium Proceedings*. Urbana, IL: American Water Resources Association, pp. 15–21.

Keller, E.A., 1975. Channelization: a search for a better way, *Geology*, **3**, 246–248.

Koloseus, H.J., 1972. Channel changes, *Civil Eng.*, **42**(2), 46.

Little, A.D., 1973. *Report on Channel Modifications: Report to the Council on Environmental Quality*, Vol. 1. Washington, DC: US Government Printing Office, 394 pp.

Matthes, G.H., 1948. Mississippi River cutoffs, *Am. Soc. Civil Eng. Trans.*, **113**, 1–15.

Mifkovic, C.S., and Petersen, M.S., 1975. Environmental aspects Sacramento – bank protection: *Am. Soc. Civil Eng. Proc., J. Hydraulics Div.*, **101**(HY5), 543–555.

Schoof, R., 1980. Environmental impact of channel modification, *Water Resources Bull.*, **16**, 697–701.

Stevens, M.A., Simons, D.B. and Schumm, S.A., 1975a. Man-induced changes of middle Mississippi River, *Am. Soc. Civil Eng. Proc., J. Waterways, Harbors and Coastal Engineering Div.*, **101**(WW2), 119–133.

Stevens, M.A., Simons, D.B. and Richardson, E.V., 1975b. Non-equilibrium river form, *Am. Soc. Civil Eng. Proc., J. Hydraulics Div.*, **101**(HY5), 557–566.

White, R.J., 1975. Trout population responses to streamflow fluctuations and habitat management in Big Roche-a-Cri Creek, Wisconsin, *Verh. Internat. Verein. Limnologie*, **19**, 2469–2477.

Winkley, B.R., 1972. Practical aspects of river regulation and control, in H.W. Shen (ed.), *River Mechanics*, Vol. 1. Fort Collins, CO: H.W. Shen, pp. 1–79.

Cross references

Alluvial valley engineering
Flow measurement: new technology
Hydromechanics
River engineering
Urban hydrology

CHÉZY FORMULA

This formula was developed by the French engineer, Antoine Chézy, during 1769 and verified by experiments on the Courpalet Canal and the River Seine in France. The basic formula is

$$v = c(r_h s)^{\frac{1}{2}}$$

where v is the water velocity (m s^{-1}), c is Chézy's C, a factor of flow resistance (non-dimensional), r_h is the hydraulic radius (m) and S is the slope of the hydraulic energy line (non-dimensional).

In this form it applies to uniform flow in an open channel. A number of hydraulicians have adapted Chézy's work for applications to open channel flow, notable amongst whom are H. Bazin of France who proposed the Bazin formula in 1897, E. Ganguillet and W.R. Kutter of Switzerland who proposed the G–K formula in 1869, Robert Manning of Ireland who presented the Manning formula at a meeting of the Institution of Civil Engineers of Ireland in 1889, N.N. Pavlovskii of the USSR who published the Pavlovskii formula in various forms (initially in 1925), Ralph W. Powell of the USA who proposed a Powell formula in a logarithmic form in 1950 and J.C. Stevens of the USA who used the Chézy formula as a basis of a number of prognoses concerning open channel flow in the 1930s and mid-1940s. Descriptions of these various applications would be lengthy and complicated, and the reader is referred to any of the many relevant standard textbooks for details. It is sufficient to emphasise the importance of understanding the concept of the original Chézy formula when using the various applications of it.

P.G. Holland

Bibliography

BS 3680. Part 1, 1991. *Glossary of Terms*. British Standards Institution, London.

Cross references

Hydrologists (600 BC–AD 1900)
Water resources: dictionary of basic terms

CHILGROVE HOUSE WELL, UK

The series of water level measurements for the Chilgrove House well, which penetrates the fissured Chalk and Upper Greensand aquifer of southern England, began in 1836 and is thought to be the longest continuous record of groundwater level variations in the world. It provides a unique historical perspective against which to assess contemporary changes in groundwater levels. Responsibility for monitoring the level at Chilgrove now rests with the Environment Agency and the full historical record is held on the National Groundwater Level Archive maintained by the British Geological Survey.

Chilgrove House is located on the Upper Chalk outcrops of the South Downs in a typically 'dry valley' in the headwaters of the River Lavant which drains through Chichester to the English Channel. The original well was a shaft about 1 m in diameter excavated to a depth of 41.15 m. During the notable drought of 1854–1855 the well was found to be silted and was cleaned out and deepened by 2.6 m. Further siltation was discovered during the 1933–1934 drought and, following a second cleaning out, a borehole (114 mm diameter) was sunk to extend the overall depth to 62.03 m.

The large annual range in groundwater levels – which can exceed 40 m – together with the very limited impact of groundwater abstraction on the natural variability of the water table make the Chilgrove well particularly suitable both for monitoring changes in groundwater resources and identifying short- and longer-term climatic influences on recharge rates.

Rainfall at the Chilgrove site is fairly evenly distributed through the year and the average annual total is around 800 mm of which about 60% is lost to evaporation. Evaporation losses are concentrated in the April–September period, imposing a marked seasonality on the rate of aquifer replenishment, 80% normally occurring over the 5 months beginning in November. Following dry, hot summers soil moisture deficits may persist well into the winter, greatly restricting the window of opportunity for aquifer recharge. As a consequence little or no recovery may be registered during notably dry winters. Such was the case in 1933–1934, 1975–1976 and 1991–92, and as the droughts continued the water table declined to approach its natural base level by the end of the ensuing summer.

Although the historical series is characterized by sustained periods with groundwater levels above, or below, the seasonal average, no overall trend can be identified. Steep falls and rapid recoveries (when saturation of the Chalk matrix may lag behind the levels in the well) are common. Following a notably rapid recovery in levels at the end of the 1976 drought, the water table followed the normal seasonal pattern, albeit generally above average, over the ensuing 12 years. Thereafter levels exhibited a volatility unmatched in the first 150 years of the series. Early in 1988, and again in 1990, levels climbed to their highest in 30 years but the period-of-record minima was 'equalled' late in 1989 and 1990, both exceptionally warm years. The water table then remained depressed until late in 1992 when a further notable recovery began (Figure C4). This culminated in a 38 m rise over the 3 months beginning in October 1993. Sustained rainfall on a saturated catchment eventually produced artesian conditions at Chilgrove – the well overflowed throughout an 18-day period. Such conditions have occurred rarely in the past but correlation with a nearby well (Compton, which also has a record exceeding 100 years) suggests that only in 1852 have the recent conditions been closely replicated. Excessive groundwater outflows were a major factor in causing the Lavant to exceed its previous highest flow by a wide

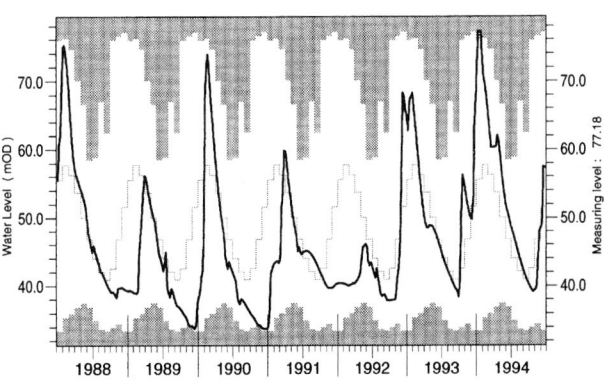

Figure C4 Water table levels in the Chilgrove House well, 1988–1994.

margin and to inflict over $7.5 million damage on Chichester in early 1994.

T.J. Marsh

Bibliography

Thomson, D.H., 1938. A 100 year's record of rainfall and water levels in the Chalk at Chilgrove, West Sussex. *J. Inst. Water. Eng.*, **10**(3), 193–201.
Monkhouse, R.A. *et al.*, 1990. Long-term hydrograph of groundwater levels in the Chilgrove House well in the Chalk of southern England. Wallchart series, British Geological Survey, NERC.
Marsh, T.J. *et al.*, 1994. *The 1988–92 Drought*. Hydrological data UK series, Institute of Hydrology, Wallingford, UK, 80 pp.

Cross references

Groundwater
Water table

CLIMATE AND CLIMATE CHANGE

It is widely recognized that water resources, together with food and energy, have the largest dependence on changes of climate. Part of the reason for this is that many water resources development projects are designed to function for around 100 years or even more in some cases. Therefore any changes in climate which may effect design parameters could endanger the long-term viability of the project, or even endanger lives through increased likelihood of sudden structural failures.

Definitions of climate

Climate may be defined as the sequence of weather which is experienced at a given locality. This sequence is usually taken to extend for between 50 and 100 years, the average life span of a human being. Hare (1985) considers a number of idealized time series of some variables such as air temperature or humidity, Figure C5. In this

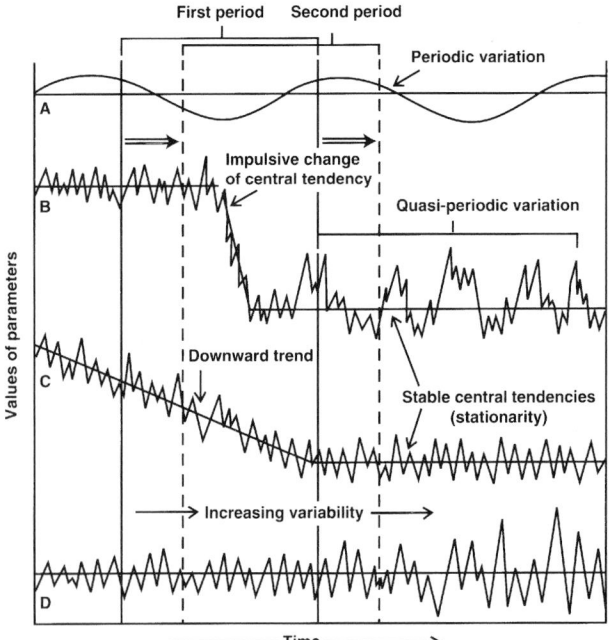

Figure C5 Idealized time series (curves A–D) of a representative parameter of a climatic element that is continuous in time (such as temperature or pressure). Vertical bars indicate arbitrary averaging or integrating periods (usually 30 years) which are recalculated each decade (see dashed bars). (After Hare, 1985.)

figure curves A to D illustrate modes of variation typical of these types of time series as follows:

- Curve A: a periodic variation about a mean value or central tendency. Such variations as these are rare in atmospheric time series, except for daily or annual solar parameters.
- Curve B: short-term variations typical of atmospheric time series. The series changes almost impulsively in a short period to a new regime within which quasi-periodic variations occur.
- Curve C: a series in which a downward trend occurs up to a particular point, thereafter the series is stationary.
- Curve D: a series with a constant central tendency, but the short-term variations increase in amplitude as time progresses.

In order to define a particular climate, a number of time series for several parameters are analyzed over a long period. In general, 30-year periods, updated every decade, are used, as shown in Figure C5 by the annotation first and second periods.

Climatic noise is that part of the variance of climate attributable to short-term weather changes, and climatic variability is the manner of variation of the climatic parameters within the typical averaging period (Hare, 1985). Hence the climate of an averaging period must be defined in terms of the central tendency and estimates of the variability. Climatic change occurs when the differences between successive averaging periods exceed the differences that can be accounted for by noise. Short-term changes lasting a few decades are referred to as climatic fluctuations.

Climate may be regarded as one element in a global system which includes other geophysical and biogeochemical cycles. Proposals for interdisciplinary studies of the total system have been made (Roederer, 1985), which aim to clarify various main linkages, classified (Hare, 1985) as:

- the natural energy system, the main element of which is the flux of solar energy, but which includes terrestrial and atmospheric radiation, heat fluxes in the soil and ocean, and energy transformations in the oceans and atmosphere.
- the hydrological cycle (q.v.).
- The carbon cycle, describing the exchange of carbon between atmosphere, oceans, biosphere and the solid earth (this cycle is of major importance to climate, particularly aspects of atmospheric carbon dioxide);
- other biogeochemical cycles, such as the exchange of nitrogen and chlorofluoromethanes, and of both synthetic and volcanic aerosols.

Perturbations of these linkages affect climate, but it is often difficult to separate out cause and effect. It is only with the help of numerical models that observed climatic change can be analyzed.

Evidence of climatic change

Instrumental records have existed in only a few places for as long as 200–300 years, although in China temperature records have been extended for several thousand years (Chu, 1973; Yeh and Fu, 1985). Hence our knowledge of past climate depends largely on analyses of phenomena such as tree rings, fossil and ice cores. Table C8 provides a summary of the data sources giving information on past climate. These data have enabled the climate of the last 1 million years, known as the Pleistocene Period, to be documented for the northern hemisphere, as shown in Figure C6.

Eight alterations between cold glacial epochs and relatively warm interglacial epochs occurring at intervals of about 100 000 years are evident. Large fluctuations in the northern hemisphere ice sheets and sea-level variations of around 100 m occurred with these transitions. About 18 000 years before the present, extensive areas of the northern hemisphere were covered with continental ice sheets, the sea level had dropped by about 120 m and the sea surface temperatures in the North Atlantic had dropped by up to 10°C.

Since the last interglacial maximum about 6000 years ago there has been a gradual cooling, as shown in Figure C7a, which ceased in the late eighteenth century, although from around 1940 cooling again occurred (Figure C7b,c). This central tendency was interrupted by the warm period of the Middle Ages from about AD 1150 to 1350, and the period from about AD 1500 to 1850 known as the 'Little Ice Age'.

One method of analyzing time series, known as power spectrum analysis, can provide evidence of periodicity, provided account is taken of any smoothing, filtering or removal of long-term trends. Figure C8 shows the power spectrum of the Manley record of

Table C8 Characteristics of paleoclimatic data sources (after Mason, 1976)

Proxy data source	Variable measured	Continuity of evidence	Potential geographical coverage	Period open to study (years BP)	Minimum sampling interval (years)	Usual dating accuracy (years)	Climatic inference
Layered ice cores	Oxygen isotope concentration, thickness (short cores)	Continuous	Antarctica, Greenland	10 000	1–10	±1–100	Temperature, accumulation
	Oxygen isotope concentration (long cores)	Continuous	Antarctica, Greenland	10 000+	Variable	Variable	Temperature
Tree rings	Ring-width anomaly, density, isotopic composition	Continuous	Midlatitude and high-latitude continents	1000 (common) 8000 (rare)	1	±1	Temperature, runoff, precipitation, soil moisture
Fossil pollen	Pollen-type concentration (varved core)	Continuous	Midlatitude continents	12 000	1–10	±10	Temperature, precipitation, soil moisture
	Pollen-type concentration (normal core)	Continuous	50°S to 70°N	12 000 (common) 200 000 (rare)	200	±5%	Temperature, precipitation, soil moisture
Mountain glaciers	Terminal positions	Episodic	45°S to 70°N	40 000	–	±5%	Extent of mountain glacier
Ice Sheets	Terminal positions	Episodic	Midlatitude to high latitude	25 000 (common) 1 000 000 (rare)	–	Variable	Area of ice sheet
Ancient soils	Soil type	Episodic	Lower and mid-latitude	1 000 000	200	±5%	Temperature, precipitation drainage
Closed-basin lakes	Lake level	Episodic	Midlatitudes	50 000	1–100 (variable)	±5% ±1	Evaporation, runoff, precipitation, temperature
Lake sediments	Varve thickness	Continuous	Midlatitudes	5000	1	±5%	Temperature, precipitation
Ocean sediments (common deep-sea cores, 2–5 cm (1000 years)$^{-1}$	Ash and sand accumulation rates	Continuous	Global ocean (outside red clay areas)	200 000	500+		Wind direction
	Fossil plankton composition	Continuous	Global ocean (outside red clay areas)	200 000	500+	±5%	Sea surface temperature, surface salinity, sea ice extent
	Isotopic composition of planktonic fossils, benthic fossils, mineralogical composition	Continuous	Global ocean (above CaCO$_3$ composition level)	200 000	500+	±5%	Surface temperature, global volume, bottom temperature and bottom water flux, bottom water chemistry
(Rare cores > 10 cm (1000 years)$^{-1}$	As above	Continuous	Along continental margins	10 000+	20	±5%	As above
(Cores < 2 cm (1000 years)$^{-1}$	As above	Continuous	Global ocean	1 000 000+	1000+	±5%	As above
Marine shorelines	Coastal features, reef growth	Episodic	Stable coasts, oceanic islands	400 000	–	±5%	Sea level, ice volume

temperatures for Central England with the long-term trend removed; the larger the area beneath the peaks, the more significant the peaks. Significant peaks are evident at 2.1, 5.2, 7.6, 14.5, 23 and 76 years. The first peak is associated with the quasi-biennial oscillation, the 3.1-year peak and perhaps the 5.2 and 7.6 year peaks could result from the Southern Oscillation (Philander, 1983), and the 23-year peak may be associated with the double sunspot cycle. However, the number of sunspots oscillates on roughly an 11-year cycle, but this periodicity is not particularly significant in Figure C8, although other analyses (e.g. King, 1973) have indicated the significance of this

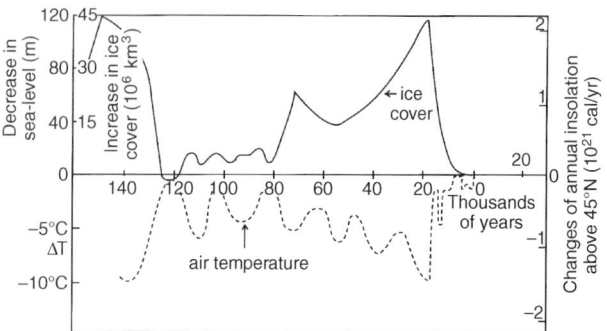

Figure C6 Changes in global ice cover and northern hemisphere air temperature north of 45°N over the last 150 000 years. (After Mason, 1976.)

frequency. This illustrates the dangers of drawing conclusions from limited or smoothed data sets, which may or may not reveal a range of periodicities.

The periodicities discussed so far relate to short-term climatic fluctuations. Further fluctuations may also be introduced by human activities, for example in increasing atmospheric carbon dioxide, mainly by burning fossil fuels (Figure C9), or releasing pollutants. These effects may manifest themselves as either atmospheric warming, as is thought to be the case for carbon dioxide (for review see Bach, 1976), or cooling as may be the case of aerosol input to the stratosphere produced by volcanic activity (Lamb, 1977; Kondratyev et al., 1985). However, the mechanisms are not yet fully understood and require detailed modeling before the total effects on climate can be defined under all circumstances. There is little evidence at present that the variability of weather is changing (Ratcliffe et al., 1978).

Local changes in the air–sea interaction produce further short-term climatic fluctuations. Global sea surface temperature (SST) anomalies, one example being the El Niño, a current of warm water that occurs off the coast of South America in certain years and may be associated with the Southern Oscillation, are thought to affect atmospheric systems (Namias and Cayan, 1981). Similar associations may explain rainfall deficiencies in the Sahel (sub-Saharan North Africa); (Folland et al., 1986). These types of relationships are sometimes referred to as teleconnections.

Longer-term climatic change is identified by power spectrum analysis of paleoclimatological records for the last million years (Figure C10). Significant peaks occur at periods of around 100 000, 40 000 and 20 000 years. Changes on these time scales are responsible for major changes of world climate.

Causes of climatic change

The causes of climatic change lie in the linkages listed above. However, the natural energy system is thought to control major changes of climate, with the other cycles producing fluctuations within the major changes, or perhaps accelerating or decelerating these changes. Each system is examined separately here although, as we shall see, their combined effects can only be assessed using a global numerical model.

Natural energy system

This system is concerned not only with solar energy, but also with the whole spectrum of terrestrial and atmospheric radiation, and energy transfers within both the oceans and the atmosphere. Climatic changes depend upon modification to the fluxes of energy and the amount of energy stored in the atmosphere, the earth and the oceans.

Changes in the oceans, manifest as sea surface temperature anomalies, generate climatic fluctuations, as we have noted in the previous section. However, whilst these influences are extremely important, particularly for long-range (periods of about a month ahead) forecasting (Gilchrist, 1986), it is less certain how much the oceans influence climatic change.

Changes in solar input and long-term changes in solar insolation seem to offer one explanation of some of the observed periodicities.

Wetherald and Manabe (1975) suggest that a 2% increase in the solar constant (total solar radiation) would produce a rise of 3°C in the mean global surface temperature. A decrease of 2%, however, would produce an average temperature drop of 4.3°C. It was also concluded that a 6% change (from −4% to +2%) in the solar constant would produce a 27% increase in the area-mean rates of precipitation. Changes such as these would not be uniform from equator to pole because of changes in snow cover and in albedo (reflection coefficient). Indeed, changes in the solar constant of this magnitude have not occurred in the last few hundred years, and although there is some evidence of atmospheric periodicities which coincide with the sunspot cycle which reflect in small changes of the solar constant, the relationship is still not uniformly accepted (for a review see Bonnet, 1985).

Changes of solar radiation are also brought about on very long time scales by changes in the orbit of the Earth about the Sun. The eccentricity of the orbit has a periodicity of 9600 years, changes in the obliquity of the axis of the Earth have a periodicity of 40 000 years, and changes due to the precession of longitude of the perihelion, equivalent to the precession of the solstices and equinoxes, have a periodicity of 21 000 years (Mason, 1976). All these periodicities appear in the power spectrum analysis of the oxygen isotope content of fossil plankton (Figure C10). The changes of incident solar radiation over the last million years, which are implied by this orbital geometry, were calculated as a function of season and latitude by Milankovitch (1930, 1938).

Mason (1976) compared the variations in annual insolation received northwards of 45°N due to changes in the orbit of the Earth, with changes of air temperature and ice cover over the last 150 000 years (Figure C11). There is a close correlation between the major advances and recessions of the ice and the 40 000 year cycle of variations in solar insolation; the maxima of the ice cover are nearly coincident with the minima in the radiation received at about 45°N. The oceans play a significant part in this process as they act as stores of heat which may be released as the sea ice melts, so aiding further melting. This type of feedback process is essential to explain the rather rapid transition from an ice age which cannot be explained entirely by the increased solar radiation. Changes in albedo caused by either increased or decreased snow cover may also generate complex feedback mechanisms.

Hydrological cycle

Although, as we shall discuss later, climatic change may have significant effects upon the hydrological cycle, the converse is also true. Changes in the vegetation, brought about by natural events, such as mountain building, earthquakes, volcanoes or the climate itself, modify evaporation and transpiration. However, the activities of humanity have had by far the largest impact upon natural vegetation. It has been suggested (Flohn, 1973) that over the last 8000 years about 11% of the land area has been converted to arable land, and 31% of the forest land is not as it originally was. The loss of forests is particularly important since they play a major part in the interception and re-evaporation of rain water (transpiration).

As well as modifying evaporation and transpiration, changes of land use result in changes of albedo which, particularly in areas around the edges of deserts, may cause instabilities in local climate. Charney (1975) suggested that a reduction of vegetation, with a consequent increase in albedo, in the Sahel region on the southern edge of the Sahara would lead to a perpetuation of arid conditions. It was shown, by numerical experiment, that increasing the albedo from 14 to 35% had the effect of decreasing the rainfall in the Sahel by about 40% during the rainy season.

Reductions of vegetation may result from human activities or reductions of soil moisture. Soil moisture content may also directly affect evaporation, and consequently the proportion of the net radiation available as latent heat. Walker and Rowntree (1977) demonstrated that deficiencies in soil moisture could, without necessarily significantly reducing the existing vegetation, cause deserts to persist. Anomalies of soil moisture may also propagate and affect wider areas (Rowntree and Bolton, 1983).

Changes in soil moisture may be brought about by irrigation, which results in increased evaporation. However, irrigation actually produces a rise of global temperature as the water causes a decrease in the reflection of incoming solar radiation. Budyko (1974) estimated that present-day irrigation generates an increase of the Earth's mean surface temperature of about 0.07°C.

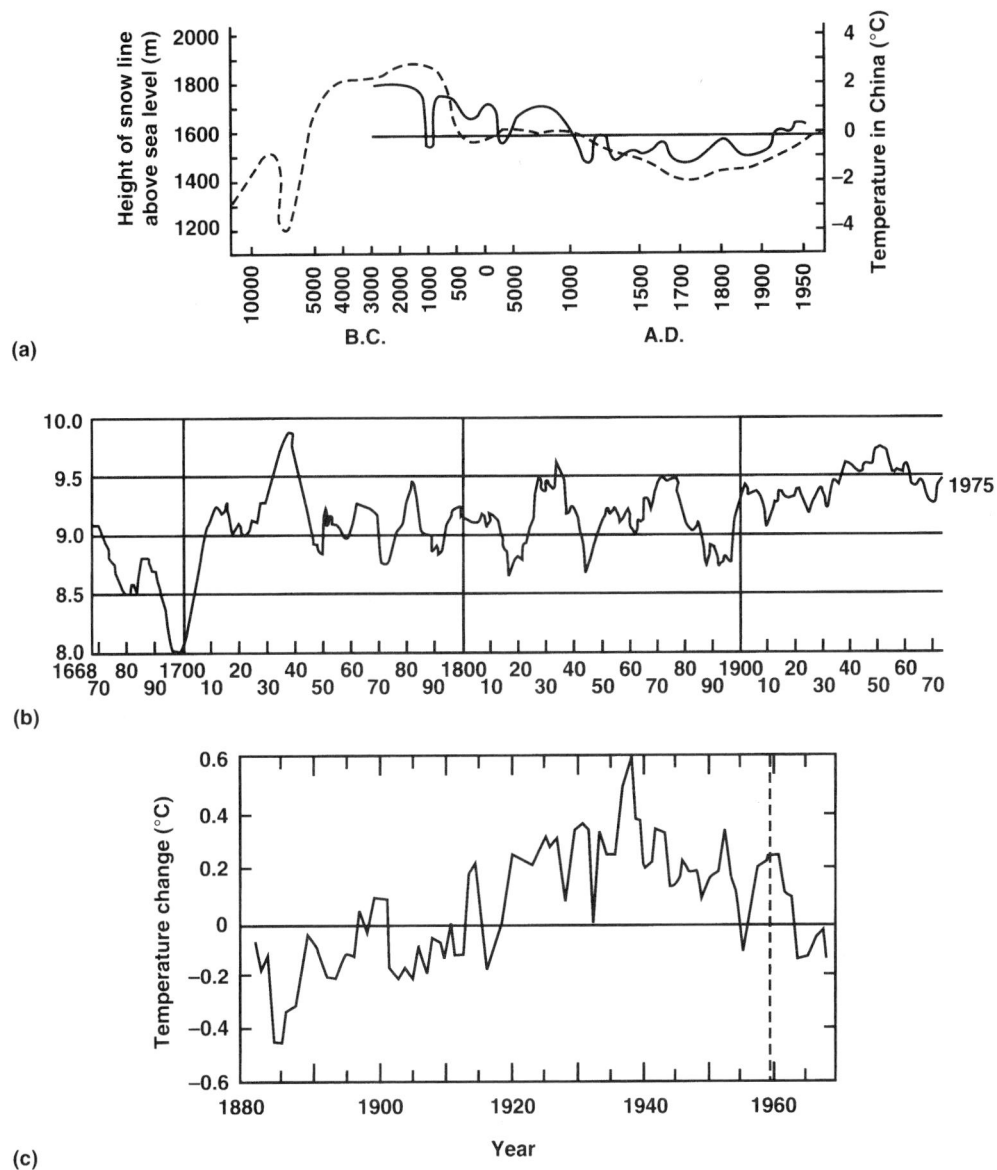

Figure C7 (a) Temperature curve in China during the past 5000 years (after Chu, 1973), (b) 10-year running means of Central England temperatures from 1650 to 1975 (after Manley, 1974) and (c) recorded changes in the annual mean temperature of the northern hemisphere since 1880 (after Budyko, 1969).

Increased use of water by humans for irrigation, domestic and industrial processes increases the rate at which water passes through the hydrological cycle over land. Whilst this may not in itself lead to climatic changes, it is likely to cause a redistribution of water resources, particularly if large-scale water transfers are involved (Shiklomanov, 1985), decreasing runoff, depleting groundwater resources and increasing the amounts of water in the oceans and the polar ice fields. It is these changes which could affect climate.

Carbon cycle

Carbon dioxide (CO_2) is a very small constituent of the atmosphere (0.03%), but it transmits short-wavelength radiation from the Sun, yet absorbs a portion of the longer wavelength radiation emitted by the Earth, and this property causes it to have a profound effect on climate. The transmission of short-wavelength radiation and the absorption of long-wavelength radiation is called the *greenhouse effect*.

The biosphere extracts carbon dioxide from the atmosphere in spring and summer, and returns carbon dioxide to the atmosphere during the rest of the year. Likewise, carbon dioxide is transferred between the atmosphere and the oceans. In addition to these natural circulations, humanity is injecting carbon dioxide into the atmosphere through the burning of fossil fuels, the cutting and burning of forest and the conversion of land for agricultural or urban development. A steady increase in atmospheric carbon dioxide has been measured, (Figure C9) although the size of the yearly increase is only about half of what would be expected if all the carbon dioxide produced by burning fossil fuel remained in the atmosphere. It is thought that most of the rest of the carbon dioxide, about 35% (Crane and Liss, 1985), is absorbed by the oceans, the seawater reacting with the carbon dioxide to form a solution of carbonic acid (H_2CO_3). Figure C12 shows schematically the fluxes of carbon in the carbon cycle. Since it is thought that the surface layer of the oceans can take up only about 10% of the carbon dioxide injected into the atmosphere each year, it would appear that water deeper in the oceans must be absorbing significant amounts. The mechanism by which this is accomplished is not yet fully understood.

The effects of the steady increase in atmospheric carbon dioxide have been studied using numerical models (e.g. Schneider, 1975;

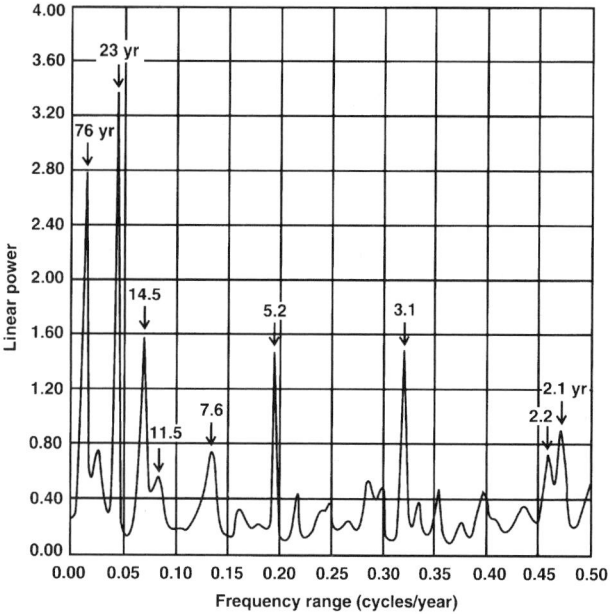

Figure C8 Power spectrum of the Manley record of Central England temperatures with the long-term trend removed. (After Mason, 1976.)

important changes in global climate. However, increased warming could produce increased cloudiness which would reduce solar radiation causing cooling, and so one effect could negate another. These complex interactions need to be studied in depth before the effects of increasing carbon dioxide can be understood properly.

Other biogeochemical cycles

A variety of particles of various types are found in the atmosphere. These particles are either anthropogenic or are produced by natural processes, such as from sea salt spray, windblown dust, forest fires, meteoric debris or volcanic emissions. Anthropogenic pollutants include chlorofluoromethanes (gases used in spray-cans) and other halocarbons, and sulfur compounds from the burning of fossil fuels (Bolin, 1979). Nuclear explosions, such as that of the Chernobyl accident in 1986 (ApSimon and Wilson 1986), may generate unusual pollutants.

The anthropogenic pollutants and oxides of nitrogen released by high-flying aircraft (and perhaps from fertilizers) may interfere with the natural processes occurring in the atmospheric ozone layer (e.g. Mitchell, 1983). Ozone is an unstable form of oxygen, and is constantly created and destroyed by incoming ultraviolet solar radiation (Crutzen and Andreae, 1985). These processes cause the atmosphere to be heated, and prevent much of the ultraviolet radiation from reaching the surface of the Earth. Changes in the ozone layer brought about by interactions with anthropogenic pollutants could modify the heating in the stratosphere, or cause a health hazard to humans by increasing the amount of ultraviolet radiation reaching the Earth's surface. Such effects are not yet monitored well enough for their impact on climate to be assessed.

Very small particles (aerosols) are also injected into the atmosphere either from industrial activities or from natural processes. The amount of aerosol affects visibility and turgidity (the ability to transmit solar radiation; e.g. Mass and Schneider, 1977; Brinkman and McGregor, 1983), and could have a long-term effect on global temperatures (e.g. Lamb, 1970; Schneider and Mass, 1975). Figure C13 shows the mean residence times of aerosols in different layers of the atmosphere.

Manabe and Wetherald, 1975, 1980; Mitchell, 1983). It has been found that a doubling of carbon dioxide would lead to an average global temperature increase of about 2°C in 70 years, although the temperature increase in the polar regions might be three to five times larger. Such changes, if confirmed, are significant, and could lead to

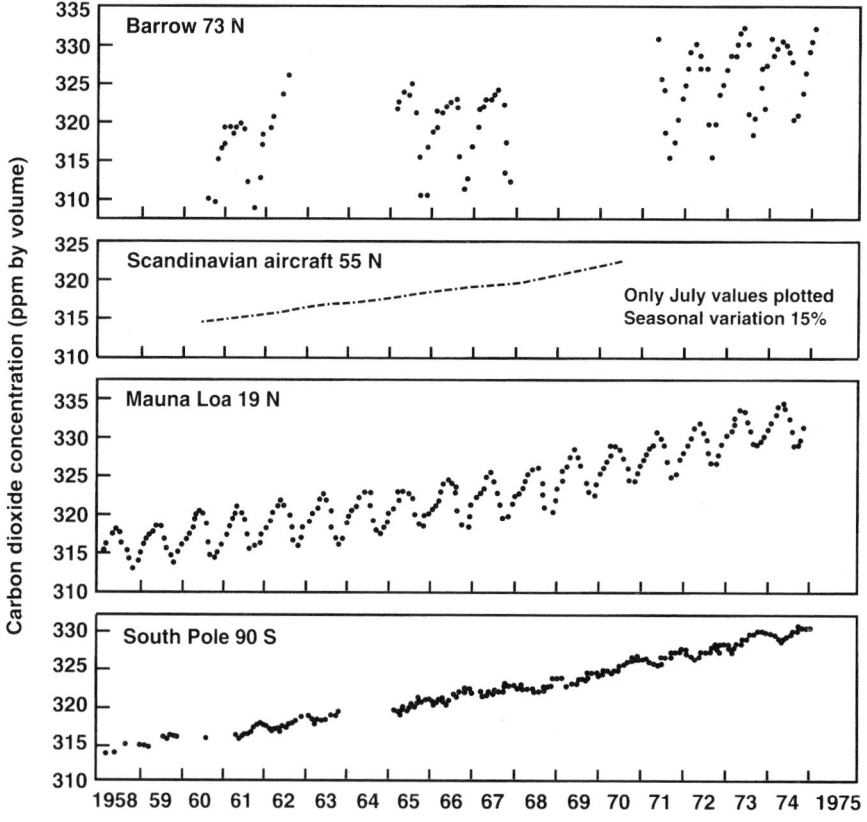

Figure C9 Changes in the atmospheric concentration of carbon dioxide. (After Rotty and Weinberg, 1977.)

Period in thousands of years

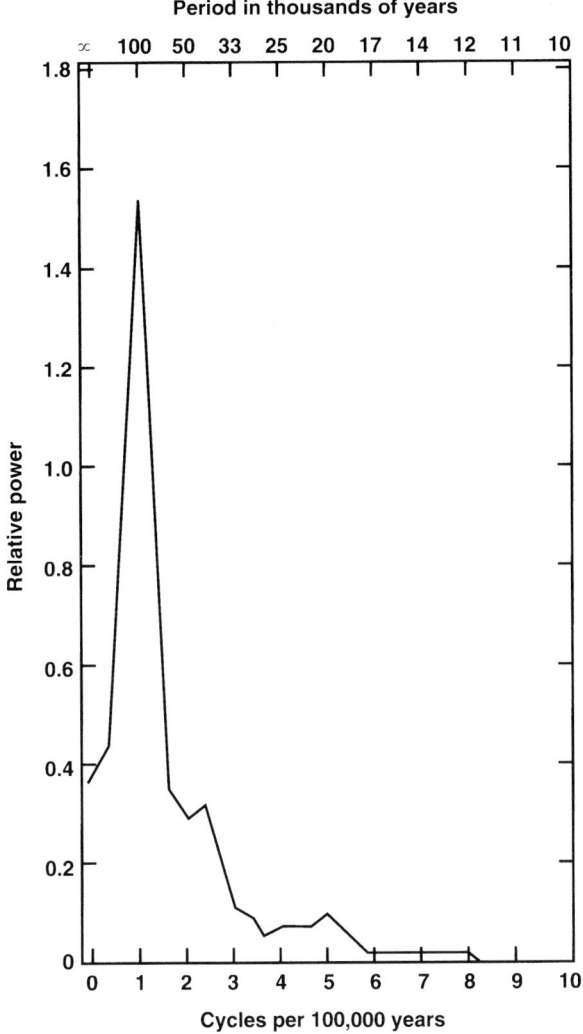

Figure C10 Power spectrum of a time series of observations of the oxygen isotope content in a deep-sea core from the equatorial Pacific which indicates fluctuations in global ice volume over the last 600 000 years. (Source: US National Academy of Sciences, 1975.)

Material close to the surface falls out in a matter of days, but material reaching the stratosphere may remain for several years. Volcanoes cause material to be placed directly in the stratosphere (Cronin, 1971), as do nuclear explosions either deliberate or accidentally. Such occurrences could have profound effects on the climate over many years, although it remains unclear the extent to which climatic effects, such as the so-called 'nuclear winter', would really occur (Golding *et al.*, 1986).

Modeling climatic change

The processes which may cause climatic change must be studied using numerical models, since the observations are often confusing, and it is often difficult to separate cause and effect. These models range from integral parameter, one-dimensional models to full three-dimensional representations of the global atmosphere. Much can be gained from simplistic models. For example the mean temperature of outgoing radiation from the Earth, T_r, may be estimated from,

$$\epsilon\sigma T_r^4 = q = \tfrac{1}{4}I_o(1 - A) \qquad (C1)$$

where I_o is the solar constant, A is the mean planetary albedo (0.28 derived from satellite data), ϵ is the emissivity and σ is the Stefan–Boltzmann constant ($\sim 5.67 \times 10^{-8}$ W m^{-2} K^{-4}. Equation (C1) gives

$T_r \sim 256$K, a mean temperature which exists in the atmosphere at a height of about 5 km.

In order to estimate the atmospheric circulation, at least the mean difference between the surface air temperature at the equator and the poles, ΔT, must be added to equation (1). Hence, the total (over the depth of the atmosphere) meridional heat flux produced by air movements is assumed to balance the heat losses due to radiation:

$$C_p M U \frac{\Delta T}{\Pi R/2} = \epsilon\sigma T_r^4 \qquad (C2)$$

where C_p is the specific heat at constant pressure, R is the radius of the Earth, M is the mass of a unit column of air and U is the typical atmospheric wind velocity, around 10 m s^{-1}. From equation (C2) ΔT can be evaluated. This approach to analyzing the circulation of a planetary atmosphere was developed by Golitsyn (1970, 1973). However, the real atmosphere of the Earth deviates considerably from the state of radiative equilibrium described by equation (C1). Nevertheless, it is possible to represent aspects of climate by these simple models and carry out limited experiments to investigate the effects of changing the values of particular parameters (Monin, 1986).

So far we have discussed integral parameter climate models, but more exact formulations of the vertical structure of the atmosphere also provide a useful framework for studies of climatic change. In a column of air a balance may be assumed between heat fluxes due to turbulent and radiative vertical heat exchange:

$$\frac{d}{dz}\lambda\frac{dT}{dz} + \alpha\xi_{ab}(F_d + F_u - 2f\sigma T^4 + \gamma S) = 0 \qquad (C3)$$

where z is the vertical coordinate upwards from the Earth's surface; λ is the vertical turbulent heat conductivity; α is the coefficient of absorption of longwave radiation; ξ_{ab} is the density of absorbing substances; F_d and F_u are the downward (atmosphere) and upward Earth's surface and atmosphere fluxes of long-wave radiation; T is blackbody temperature; f is a correction factor to allow for the fact that the substances are not perfect black bodies; S is the downward flux of short-wave solar radiation; and γ is the ratio of the absorption coefficient for short-wave and long-wave radiation. Equation (C3) may be solved with equations for the fluxes of radiative energy and appropriate boundary conditions. Whilst these one-dimensional models produce results which fit the thermal structure of the atmosphere quite well, the neglect of the divergence of the heat fluxes caused by largescale atmospheric motions restricts their applicability.

This problem can be partially overcome by constructing a model from the zonally averaged hydrostatic equations and the continuity equation as follows:

$$\frac{\partial\bar{p}}{\partial z} = -g\zeta \qquad (C4)$$

$$\frac{\partial\bar{V}_z}{\partial z} + \frac{1}{R\sin\theta}\frac{\partial\bar{V}_{\theta\sin\theta}}{\partial\theta} = 0$$

where $\bar{p}$ is the pressure, z is the vertical coordinate, ζ is the velocity, g is the acceleration due to gravity, R is the radius of the Earth, θ is the colatitude and $\bar{V}$ is the zonally meaned velocity component. These models can reproduce quite well the observed zonal temperature and velocity fields, together with the meridional circulation and the energy cycle (Stone, 1972; Kirichkov, 1978). Consequently they have been used to study aspects of climatic change (Stone, 1973).

In spite of the success of one-dimensional and zonal models in representing atmospheric processes, the complex interactions of these processes and their concomitant effect upon climate require the formulation of full three-dimensional atmospheric models. Such a model is constructed from the equations of motion, continuity and radiative transfer, and must include realistic representations of earth–atmosphere and ocean–atmosphere interactions. One example, which includes the hydrological cycle, has been described by Manabe *et al.* (1965; see also Corby *et al.*, 1972).

A rationale for the use of large-scale numerical models for the study of climate and climatic change has been given by Gilchrist (1978). Controlled experiments cannot be carried out on the atmosphere, and the interactions of systems on a global scale are so complex that making the assumptions of simpler models is not comprehensive enough to produce reliable results. The results of

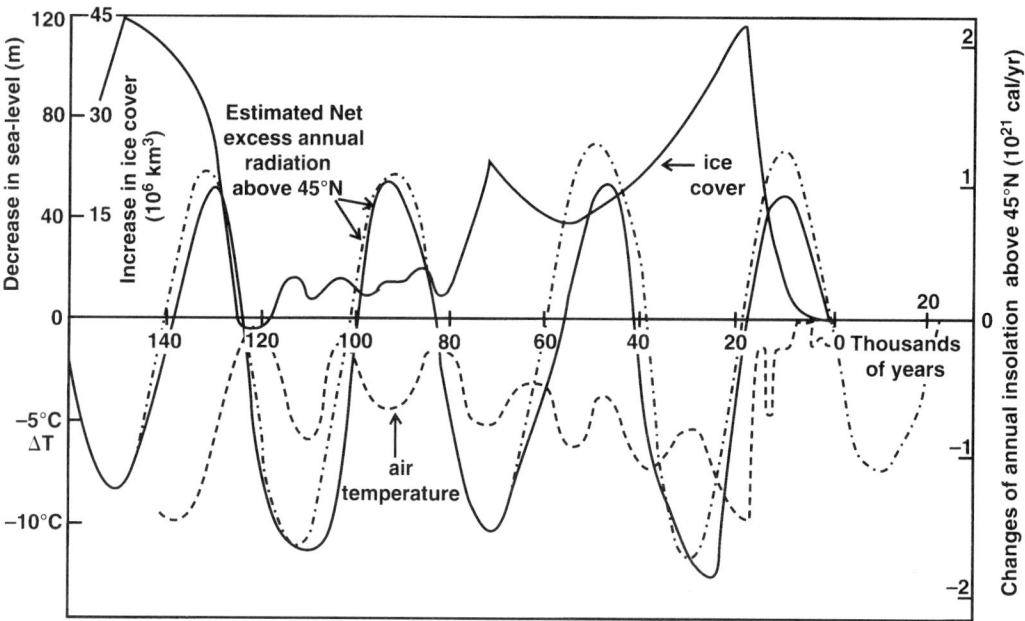

Figure C11 Change in global ice cover, northern hemisphere air temperatures (as in Figure C6) and total insolation north of 45°N over the last 150 000 years. (After Mason, 1976.)

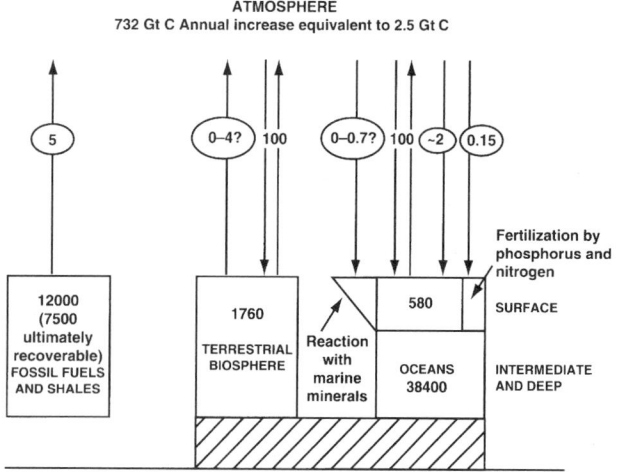

Figure C12 Fluxes of carbon between the atmosphere, the biosphere and the oceans. The sizes of the reservoirs are in gigatonnes (10^9 tonnes) of carbon, and the fluxes are in gigatonnes of carbon per year. Synthetic fluxes are circled. (After Crane and Liss, 1985.)

large-scale experiments with these models are impressive (e.g. Kutzbach and Street-Perrott, 1985; Rind, 1986; Chervin, 1986) and tend to support the theories of climatic change discussed above. The data produced by integrations of these models enable the changes (compared with present conditions) produced in mean global climate to be reproduced when conditions of the ice ages, (18 000 years before the present), Mesozoic Period (65 million years before the present) and doubling the level of carbon dioxide in the atmosphere are reproduced (e.g. Rind, 1986).

In recent years work has intensified, partly as a result of mounting concern about the impact that increases of greenhouse gases in the atmosphere might have. In 1990 the Intergovernmental Panel on Climate Change (IPCC) presented an assessment of the information relevant to climate change issues (IPCC, 1990). One of the main questions was 'How much do we expect the climate to change and how quickly?' In formulating an answer to this, some assumptions on the future rate of emission of the anthropogenic greenhouse gases

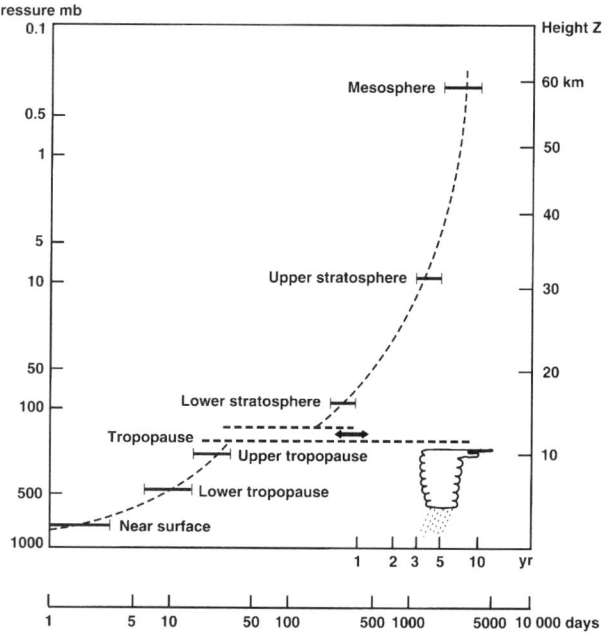

Figure C13 Mean residence times of aerosols in different layers. (After Flohn, 1973.)

needed to be made, and this led to the development of a number of emission 'scenarios'.

Such emission scenarios are used as input to general circulation models (GCMs). These models are the best tools we have with which to determine the likely climatic response to a given change in the atmospheric concentrations of greenhouse gases since they synthesize our knowledge of the physical and dynamical processes of the climate system and allow for many of the complex interactions between the various components. However, as pointed out in the (1990) IPCC report '. . . in their current state of development, the descriptions of many of the processes involved are comparatively crude'. It has therefore been a matter of high priority to improve the GCMs.

Recently the limitations of the work described in the IPCC Report have been addressed (Hadley Centre, 1992). For day-to-day weather forecasting the sea surface temperature distribution may be considered as fixed since it varies only slowly on that time scale. Such a simplification is not possible for climate prediction studies. On the annual and decadal time scales, changes in ocean temperature cannot be ignored, indeed they are central to the problem. Thus for climate prediction studies, changes in the ocean circulation, which govern the sea surface temperature distribution, must be modeled. This makes it necessary to conduct multi-decadal experiments with coupled atmosphere–ocean general circulation models (AOGCMs).

Unfortunately AOGCMs require very large supercomputing facilities. Until a few years ago these were not generally available. In consequence, most of the IPCC (1990) results were based on models which, in order to reduce the computational requirements, had only very simple ocean components. In most cases the ocean was represented by a shallow layer of water typically only a few tens of meters thick. It was sufficient to represent the seasonal interaction between ocean and atmosphere but not adequate to reproduce the thermal inertia and the internal circulations of the real oceans.

There are a number of detrimental consequences to this, in particular the model that (pseudo-)oceans respond rapidly to atmospheric changes, typically in less than a decade, whereas in the real world the response takes from decades to centuries. Thus these models cannot simulate the transient behavior of the real climate system as it adjusts to changes in the concentration of greenhouse gases. They can only simulate the eventual long-term 'equilibrium' state.

In the last few years computer technology has improved considerably. Sufficient computing power is now becoming available both to represent the full depth of the ocean within AOGCMs and to carry out experiments for periods of typically 100 years. GCMs can now simulate both the ocean dynamics and their thermal inertia, and hence the transient response of the climate system.

Current experimental evidence suggests that, with doubled atmospheric carbon dioxide, the following broadscale changes of climate are likely.

- A global mean surface warming of about 3–4°C, and more confidently between 1.5 and 4.5°C. Most of the larger warmings are obtained by the recent models, and are likely to have resulted from the more comprehensive treatments of clouds.
- A stratospheric cooling of between 3 and 5°C.
- An increase in warming with height in the tropical troposphere.
- Enhanced warming at the surface in high latitudes in winter. In the models the magnitude of this feature depends on the treatment of sea-ice and albedo, and the resulting feedbacks.
- Generally greater precipitation in equatorial regions and in middle and high latitudes, and a tendency to decreased precipitation in the tropics away from the equator.

Such changes as these will have an impact on water resources and flood management. For example, over a wide range of UK catchments a 10% increase in winter precipitation, a 2% increase in wet snow and a 10% increase in summer potential evaporation would lead to an increase in total annual runoff of around 5%. However in drier areas of lowland Britain, the seasonal pattern of change shows flows 8% higher during winter and 4% lower in summer. Much more work needs to be undertaken to assess the reliability of these assessments and the changes in climate being predicted.

C.G. Collier

Bibliography

ApSimon, H. and Wilson, J., 1986. Tracking the cloud from Chernobyl, *New Scientist*, 17 July, 42–45.

Bach, W., 1976. Global air pollution and climatic change, *Rev. Geophys. Space Phys.*, **14**, 429–474.

Bolin, B., 1979. Global ecology and man, in *Proc. World Climate Conf.*, WMO Publ. No 537, WMO, Geneva, pp. 88–111.

Bonnet, R.M., 1985. Solar terrestrial relations, in *Global Change*, Proc. Symp sponsored by Int. Council Sci. Unions (ICSU), 20th Gen. Ass., Ottawa, 25 September 1984, eds T.F. Malone and J.G. Redeemer, Cambridge Univ. Press, pp. 397–419.

Brinkman, A.W. and McGregor, J., 1983. Solar radiation in dense Saharan aerosol in Northern Nigeria, *Quart. J. R. Meteorol. Soc.*, **109**, 831–847.

Budyko, M.I., 1969. The effect of solar radiation variations on the climate of the earth, *Tellus*, **21**, 611–619.

Budyko, M.I., 1974. *Climate and Life*, Academic Press, New York.

Charney, J.G., 1975. Dynamics of deserts and drought in the Sahel, *Quart. J. R. Meteorol. Soc.*, **101**, 193–202.

Chervin, R.M., 1986. Interannual variability and seasonal climate predictability, *J. Atmos. Sci.*, **43**(3), 233–51.

Chu, K., 1973. A preliminary study of the climatic fluctuations during the last 5000 years in China, *Scientia Sinica*, **6**, 226–256.

Corby, G.A., Gilchrist, A. and Newson, R.L., 1972. A general circulation model of the atmosphere suitable for long period integrations, *Quart. J. R. Meteorol. Soc.*, **98**, 809–832.

Crane, A. and Liss, P., 1985. Carbon dioxide, climate and the sea, *New Scientist*, 21 November, 50–54.

Cronin, J.F., 1971. Recent volcanism and the stratosphere, *Science*, **172**, 847–849.

Crutzen, P.J. and Andreae, M.O., 1985. Atmospheric chemistry, in *Global Change*, Proc. Symp sponsored by Int. Council Sci. Unions (ICSU), 20th Gen. Ass., Ottawa, 25 September 1984, eds T.F. Malone and J.G. Roederer, Cambridge Univ. Press, pp. 75–113.

Flohn, H., 1973. Globale energiebilanz und klimaschwankungen, in *Bonner Meteorologische Abhandlungen, Westdeutscher Verlag*, pp. 75–117.

Folland, C.K., Palmer, T.N. and Parker, D.E., 1986. Sahel rainfall and worldwide sea temperatures, 1901–85, *Nature*, **320**, 17 April, 602–607.

Gilchrist, A., 1978. Numerical simulation of climate and climatic change, *Nature*, **276**(5686), 342–345.

Gilchrist, A., 1986. Long-range forecasting, *Quart. J. R. Meteorol. Soc.*, **112**, 567–592.

Golding, B.W., Goldsmith, P., Machin, N.A. and Slingo, A., 1986. Importance of local mesoscale factors in any assessment of nuclear winter, *Nature*, **319**(6051), 301–303.

Golitsyn, G.S., 1970. Similarity theory for large-scale motions of planetary atmospheres, *Dokl. Akad. Nauk SSSR*, **190**(2), 323–326.

Golitsyn, G.S., 1973. *An Introduction to the Dynamics of Planetary Atmospheres*. L. Gidrometeoizdat, 104 pp.

Hadley Centre, 1992. *The Hadley Centre Climate Change Experiment*, August, Hadley Centre for Climate Prediction & Research, Bracknell, UK, 20 pp.

Hare, F.K., 1985. Climatic variability and change, in *Climate Impact Assessment*, eds R.W. Kates, J.H. Ausubel and M. Berberian, John Wiley and Sons., New York 37–68.

IPCC, 1990. *Climate Change. The IPCC Scientific Assessment*, ed. J.T. Houghton, G.J. Jenkins and J.J. Ephraums, Cambridge Univ. Press, Cambridge, 365 pp.

King, J.W., 1973. Solar radiation changes and the weather, *Nature*, **245**, 443–446.

Kirichkov, S.E., 1978. Numerical experiments with a zonal atmospheric circulation model, *Izv Akad. Nauk. SSSR, Fiz. Atmos. Okeana*, **14**(7), 691–702.

Kondratyev, K.Ya., Moskalenko, N.I., Parzhin, S.N. and Skvortsova, S.Ya., 1985. Volcanic activity and climates of the Earth and Mars, *Geofisica Internacional*, **24**(2), 217–243.

Kutzbach, J.E. and Street-Perrott, F.A., 1985. Milankovitch forcing of fluctuations in the level of tropical lakes from 18 to 0kyr BP, *Nature*, **317**, 12 Sept., 130–134.

Lamb, H.H., 1970. Volcanic dust in the atmosphere; with a chronology and assessment of its meteorological significance, *Phil. Trans. R. Soc London.*, **A266**, 425–533.

Lamb, H.H., 1971. *Climate: Present, Past, and Future*, Vol. II, *Climatic History and Future*, Methuen, London; Barnes and Noble, New York, 835 pp.

Manabe, S. and Wetherald, R.T., 1975. The effect of doubling the CO_2 concentration on the climate of a general circulation model, *J. Atmos. Sci.*, **32**, 3–15.

Manabe, S. and Wetherald, R.T., 1980. On the distribution of climatic change resulting from an increase in CO_2 content of the atmosphere, *J. Atmos. Sci.*, **37**, 99–118.

Manabe, S., Smagorinsky, J. and Strickler R.F., 1965. Simulated climatology of a general circulation model with a hydrologic cycle, *Mon. Wea. Rev.*, **93**(12) 769–798.

Manley, G., 1974. Central England temperatures: monthly means 1659 to 1973, *Quart. J. R. Meteorol. Soc.*, **100**, 389–405.

Mason, B.J., 1976. Towards the understanding and prediction of climatic variations, *Quart. J. R. Meteorol. Soc.*, **102**, 473–498.

Mass, C. and Schneider, S.H., 1977. Statistical evidence on the influence of sunspots and volcanic dust on long term temperature records, *J. Atmos. Sci.*, **33**, 1995–2004.

Milankovitch, M., 1930. The astronomical theory of climate, in *Handbuch der Klimatologie I*, ed. Teil A. Koppen and Geiger, Berlin.

Milankovitch, M., 1938. Neue Ergebrusse der astronomiechen Theorie der Klimaschwankungen, *Bull. Ac. Sci. Math. Nat.*, No. 4, 41.

Mitchell, J.F.B., 1983. The seasonal response of a general circulation model to changes in CO_2 and sea temperatures, *Quart. J. R. Meteorol. Soc.*, **109**, 113–152.

Mitchell, J.F.B., 1984. The effect of pollutants on global climate, *Meteorol. Mag.*, **113**(1338), 1–16

Monin, A.S., 1986. *An Introduction to the Theory of Climate*, D. Reidel, Dordrecht, Atmospheric Sciences Library Series, 261 pp.

Namias, J. and Cayan, D.R., 1981. Large-scale air sea interactions and short-period climatic fluctuations, *Science*, **214**, 869–876.

Philander, S.G.H., 1983. El Nino southern oscillation phenomenon, *Nature*, **302**, 295–301.

Ratcliffe, R.A.S., Weller, J. and Collison, P., 1978. Variability in the frequency of unusual weather over approximately the last century, *Quart. J. R. Meteorol. Soc.*, **104**, 243–256.

Rind, D., 1986. The dynamics of warm and cold climates, *J. Atmos. Sci.*, **43**(1), 3–24.

Roederer, J.G., 1985. The proposed International Geosphere-Biosphere Program: some special requirements for disciplinary coverage and program design, in *Global Change*, Proc. Symp. sponsored by Int. Council Sci. Unions (ICSU), 20th Gen Ass., Ottawa, 25 Sept 1984, eds T.F. Malone and J.G. Roederer, Cambridge Univ. Press, pp. 1–19.

Rotty, R.M. and Weinberg, A.M., 1977. How long is coal's future?, *Climatic Change*, **1**, 45–57.

Rowntree, P.R. and Bolton, J.A., 1983. Simulation of the atmospheric response to soil moisture anomalies over Europe, *Quart. J. R. Meteorol. Soc.*, **109**, 501–526.

Schneider, S.H., 1975. On the carbon dioxide climate confusion, *J. Atmos. Sci.*, **62**, 2060–2066.

Schneider, S.H. and Mass, C., 1975. Volcanic dust, sunspots and temperature trends, *Science*, **190**, 741–746.

Schiklomanov, I.A., 1985. Large scale water transfers, in *Facets of Hydrology*, Vol. II, ed. J.C. Rodda, John Wiley & Sons, Chichester, 345–387.

Stone, P.H., 1972. A simplified radiative-dynamic model for the static stability of rotating atmospheres, *J. Atm. Sci.*, **29**(3), 405–418.

Stone, P.H., 1973. The effect of large-scale eddies on climatic change, *J. Atmos. Sci.*, **30**(4), 521–529.

US National Academy of Sciences, 1975. *Understanding Climatic Change*, NAS, Washington, DC, 239 pp.

Walker, J. and Rowntree, P.R., 1977. The effect of soil moisture on circulation and rainfall in a tropical model, *Quart. J. R. Meterol. Soc.*, **103**, 29–46.

Wetherald, R.T. and Manabe, S., 1975. The effects of changing the solar-constant on the climate of a general circulation model, *J. Atmos. Sci.*, **32**, 2044.

Wetherald, R.T. and Manabe, S., 1979. *Proceedings of the World Climate Conference*, World Meterological Organization Geneva, Publ. No. 537, 791 pp.

Yeh, T. and Fu, C., 1985. Climatic change – a global and multidisciplinary theme, in *Global Change*, Proc. Symp. sponsored by Int. Council Sci. Union (ICSU), 20th. Gen. Ass., Ottawa, 25 Sept. 1984, eds T.F. Malone and J.G. Roederer, Cambridge Univ. Press, pp. 127–145.

Cross references

Climate and climate change
Climatic data
Climate forecasting: monthly and seasonal

CLIMATE CHANGE AND ANCIENT CIVILIZATION

Climatic change during the postglacial period

The Quaternary Period is characterized as a glacial age. The coldest period of the last glaciation was about 18 000 BP, when the margins of ice sheets and glaciers began to retreat in northern Europe, North America and regions of the Eurasian continent. The end of the ice age brought great changes in the landscape, due not only to the retreat of ice sheets and glaciers but also to the rise in sea level caused by melted ice. The rise in sea level occurred not only in the postglacial period but also between 17 000 and 14 000 BP and between 10 000 and 7 000 BP. The latter phase was drastic in northern Europe. Worldwide rise in sea level from the last glaciation to the present is estimated to be 40–100 m in total (Maunder, 1992).

The Hypsithermal

The last glacial age ended about 10 000 BP, after which the climate became warmer and reached a peak period called the Climatic Optimum or the Hypsithermal that lasted from 6000 to 4500 BP. Lamb (1982a) reconstructed average latitudes of the lowest and highest air-pressure axes at sea level in the European sector of the northern hemisphere. In winter, during the Hypsithermal, the dry belt under the subtropical high pressure was located at 40–45°N (around 30°N today) and the wet belt under the subpolar low at around 57–58°N (about 50°N today), as shown in Figure C14. On the other hand, in summer the dry belt under the subtropical high was located at around 50°N during the Hypsithermal (around 33–35°N today); the wet belt under the subpolar low has not fluctuated greatly since 8000 BP. Circulation patterns over Africa of 8000 BP were reconstructed by Messerli (1980), as shown in Figure C15. The difference between Hypsithermal and present-day circulation patterns is strikingly obvious in this illustration.

In most parts of the world the climate between 7000 BP or earlier and 5000 BP was warmer by 1–3°C than it is today (Harding, 1982; Flohn and Fantechi, 1982). The middle latitudes enjoyed a warm climate and the intertropical convergence zone (ITCZ) shifted northward in the northern hemisphere (Suzuki, 1975). One result of this shift was that the arid regions from the Sahel to northwestern India, which today lie in the subtropical high-pressure zone (STHP) area, experienced a moister regime south of the STHP during the Hypsithermal. Europe experienced a warm, moist climate, and large beech and oak forests covered England and the Scandinavian Peninsula. Overall mean temperatures in Europe and North America seem to have been up to 2°C higher than at present (Lamb *et al.*, 1966, Lamb, 1977). In East Asia it was 5–8°C colder 10 000 BP but 2–3°C warmer 5000–6000 BP than today (Yoshino and Urushibara, 1978).

In the middle latitudes, glaciers on the mountains almost disappeared during the Hypsithermal. Ice in the Arctic Sea melted away and the ice sheets and glaciers in Greenland and Antarctica shrunk in this period. The minimum reconstructed 1,000–500 mbar layer thickness of 5150–5200 gpm (geopotential meters) was found over Arctic North America during the last ice age (22 000–19 000 BP), but the

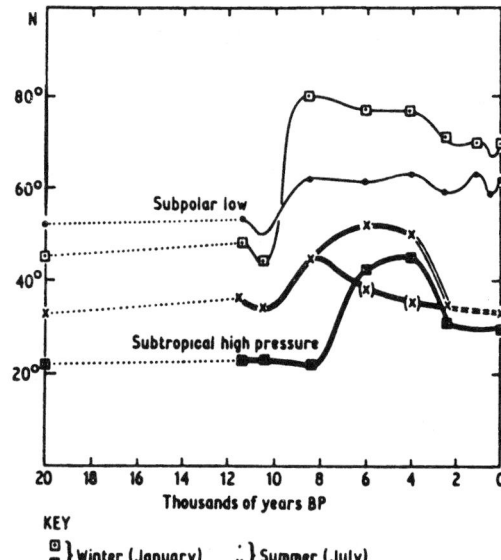

Figure C14 Estimated average latitudes of the lowest and highest barometric pressure at sea level in the European sector of the northern hemisphere since the latter part of the last ice age. (After Lamb, 1982a.)

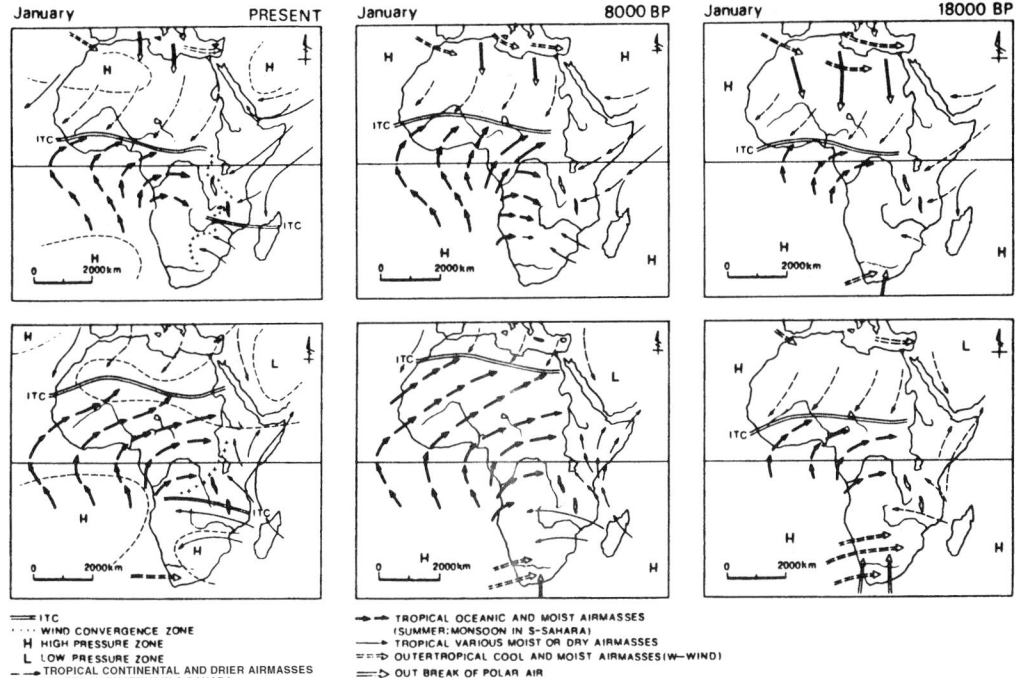

Figure C15 Circulation patterns in Africa at present, 8000 BP and 18 000 BP. (After Messerli, 1980.)

minimum thickness was 5400–5450 gpm over the Arctic Sea during the Hypsithermal (Lamb, 1974). By 6000 BP thermal gradients had weakened and the area of steep gradient shifted north of its current position. The great cold trough had broadened, weakened and shifted eastward, out over the Atlantic, in summer. It appears that the trough was located farther east than where it had been over North America in winter at 8500 BP (Lamb *et al.*, 1966). In short, the circulation became weaker on the whole, more zonally oriented, with its action centers located farther north. The North American cold trough was displaced eastward and the wavelength was increased, which resulted in the spread of westerlies at surface level.

Post-Hypsithermal

After the Hypsithermal, the climate deteriorated again. Large forests disappeared from England and Scandinavia during 3500–3000 BP. This tendency was observed not only in Europe but in all parts of the world. Suzuki (1978) thoroughly reviewed world climate tendencies dating back to 3500 BP and concluded that there is evidence of a sharp decrease in air temperature around this time in the Mediterranean region, East Asia, Australia, North America and South America. Of course, the decrease in temperature did not occur at the same time all over the world. It can be summarized that in 3500–3000 BP temperature dropped by 2–3°C from the maximum of the Hypsithermal period. In the Near East this period was characterized by a mild dryness and in northwestern India a sharp dryness. Such severe desertification at the zone around 25–35°N was caused by the northward shift of the Subtropical High-Pressure Zone in this period. The Polar Frontal Zone also shifted poleward and wet conditions prevailed along both zones. There are other indications that the climate became slightly wet in the tropical zone during this period, which may have been caused by the intensification of tropical westerlies. In short, the temperature decrease and the change in wet and dry conditions caused by the shifting of frontal zones and the Subtropical High-Pressure Zone were striking features in the period 3500–3000 BP.

By 2500 BP the renewed cooling trend of climates in the north had steepened the thermal gradient again, resulting in stronger circulation, longer wavelength and greater penetration of the westerlies over Europe (Lamb *et al.*, 1966). Around 2850 BP, glaciers in northern Europe and the Alps advanced suddenly and rivers were frozen in many parts of Europe due to the cold climate. East Asia was also cold during this period. This cold period, which continued until about AD

350, was generally wetter than at present. The water surface of inland lakes in Asia and North America reached its highest level. The climate then recovered and a warmer period occurred during the period AD 400–1200.

Ancient civilizations

After the climax stage of the last glaciation, the climate became warmer and reached its peak of the Hypsithermal as mentioned above. Until 12 000 BP, for instance, human inhabitants in the northern area in the Near East had lived in caves and hunted wild game in mountain areas. With the climatic change to warmer conditions, they came down to open living sites in the foothills where the ground was more favorable for cultivation (Wright, 1968). The northward shift in the vegetation zone was seen in Europe. In accordance with the invasion of woodland-type vegetation, insects, birds and fish extended their ranges.

During the warmest postglacial time, between 6500 and 5000 BP, the Sahara was moist. Animals and humans could roam about and cross what is now the world's greatest desert. The moist condition of the Sahara is supported by the study of the change in water levels of Lake Chad and other lakes (Messerli, 1980). Although the levels fluctuated severely in the beginning of the postglacial time, they reached their highest level about 8000 BP, with a second peak about 5000 BP, as shown in Figure C16.

Cooling first occurred in higher latitudes after about 5500 BP. Especially after 4800 BP the moist regime began to decline. The drier conditions confined human settlement and animals to oases and river valleys such as the Nile, Mesopotamia, Indus and Hwang-Ho (Yellow River). In this period, summer temperatures were 1–3°C higher than today's but winter temperatures were variable. The drying tendencies from about 5500 to 4800 BP onward seem to have been related to a climatic development of hemispheric, and probably global, extent (Lamb, 1982a).

Butzer (1966) also suggested that the Sahara recorded a moister subpluvial period between *ca.* 5500 and 2350 BC (with interruption by one or more dry spells) and a hyperarid oscillation between 2350 and 870 BC.

The dry and wet conditions differed from region to region during the Hypsithermal as shown in Figure C14. The regional change in cultural practices during the postglacial period is closely related to the climatic change, mainly due to the change of animals available. Irwin-William and Haynes (1970) reported a change in Paleo-Indian cultural

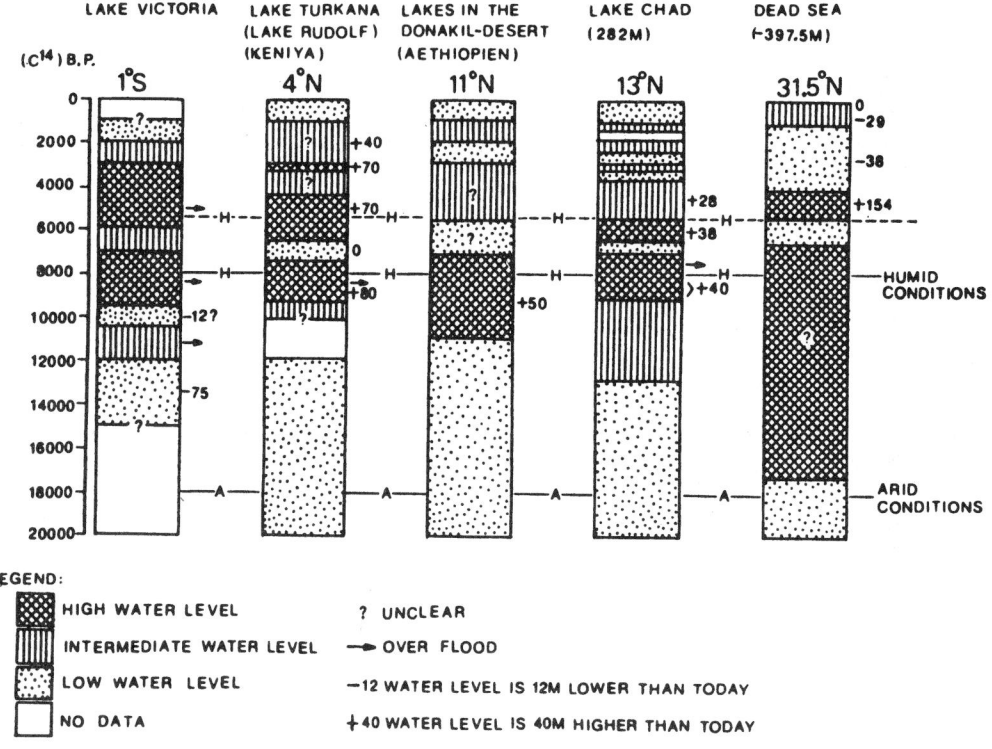

Figure C16 Water levels in lakes in Africa and the Dead Sea since 20 000 BP. (After Messerli, 1980.)

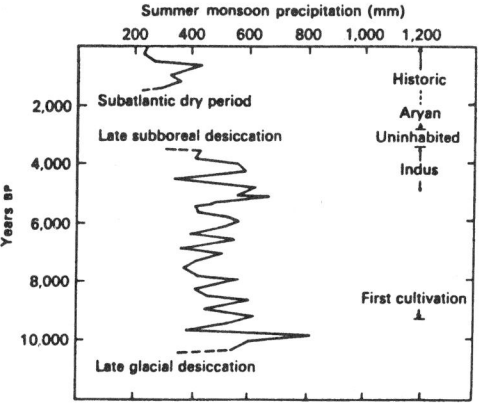

Figure C17 Reconstructed summer monsoon rainfall in Rajasthan, India, since 10 800 BP. (After Bryson, 1978.)

areas in the southwestern USA since 11 500 BP. After the last glacial period the dryness increased gradually and the Hypsithermal period was the driest in the Paleo-Indian region. Accordingly, their cultural area retreated owing to the decrease in the number of bison. Then, in the period between 5000 and 4500 BP, a wet condition pervaded this region and a dramatic increase in population took place.

Cities of Harappa and Mohenjodaro in the Indus valley in north-western India, which is now located in the Thar desert, flourished between about 4500 and 3700 BP. The decline of the Indus civilization can be attributed to increasing drought, as shown in Figure C17.

In China, during the Hypsithermal period, the overall average temperature may have been about 2°C warmer and the midwinter temperature about 5°C higher than today. This enabled the cultures of the Yanshao and Yin-Hsu periods to flourish, because there were rich subtropical fauna and flora concentrated in old cities such as Sian and Anyang (Chu, 1973). It is assumed that at the beginning of the Chou Dynasty (1066–256 BC), the climate was warm enough to grow bamboo extensively in the Hwang-Ho valley. The cultivation was soon destroyed, however; great droughts followed as well as freezing up of the Yangtze River region.

The first Chinese Neolithic agricultural civilization developed in the northern China plain. The sudden advance to a Bronze Age culture and the cultivation with irrigation of wheat and millet seem to suggest some contact with the European culture across central Asia, possibly during a moist period.

In Japan the Hypsithermal corresponded with the early and middle Johmon (Jomon) period. The Johmon culture seems to have been established by the people who came from the north during the cooler period prior to the Johmon period, and the people who came from the west during the warmer Johmon period. After about 4000 BP, the climate tended to deteriorate; it was warm but unstable. In the second half of the latter stage of the Johmon period, cooler climate covered northern Japan: the wetter region retreated to southwestern Japan due to the southward shift of the location of the Polar Frontal Zone during the warm season. There is evidence that the Johmon culture developed in the east first, then migrated to the southwest. This may be attributed to the shifting of wetter regions in Japan as mentioned above.

The Yayoi period in Japan began about 2500 BP, when the climate became cooler. This cooler period coincided with the above-mentioned cool period of the Chou Dynasty in China. There was a short warm stage between 80 and 400 BC. The air temperature in the Yayoi period seems to have been 1.0–1.5°C lower than today. In northern Japan it was cooler and wetter than today, with its peak in about 2500 BP (Yasuda, 1978). Paddy rice cultivation spread from southwestern to northeastern and northern Japan and established the basis of Japanese culture today, even though the climatic conditions were not friendly in this period. It is interesting to note that even among the different cultures in different periods there are similar phenomena, namely the coincidence in timing of the rise of a new culture with the decline of the prevailing favorable climate, as in the timing of the rise of the stone age culture in the Old World and the decline of the Hypsithermal.

The question now is, why were there no flourishing civilizations in the valleys of other great rivers such as the Ganges, the Yangtze and the Mississippi, as there were in the Nile, Hwang Ho, etc. At least one reason for this lack can be attributed to the climate. It seems that the drier conditions in the less inhabited valleys during the Hypsithermal did not experience the passage of ITCZ or polar frontal zones in this period.

Summarizing the relationship between the beginning of ancient civilizations and their climates, we may conclude that the ancient major civilizations began around 5000 BP – this was the time after the peak of the Hypsithermal. Strictly speaking, civilization began in the valleys along great rivers located in the marginal regions that started to experience cool and dry conditions. The rise of Egypt and the organized cultivation of the Nile valley by the use of early floods for irrigation may have been the necessary response to the increased food demand in a newly habitable region. In other words, civilization came about through the need to organize irrigation systems to produce food for the increased population, while the refugees presumably provided slave labor (Suzuki, 1979; Lamb, 1982a). The disruption of established ways, which the climatic events caused, provided the challenge and stimulus for undertaking deliberate cultivation and invention of new tools (Frisinger, 1977).

The decline of ancient civilization took place around 3500 BP, which was a turning point in postglacial climatic history. It is obvious that the ancient civilizations were not able to continue in the valleys along great rivers because increasingly dry climates caused limited crop production.

The agricultural areas had been expanded by 3500 BP and agricultural production was influenced greatly by the decreased rainfall, for example a decrease in rainfall of several hundred millimeters in the Indus valley. Outside the agricultural areas, people had to shift regions to live in accordance with the retreat and advance of grassland. This shifting was the cause of decline in the ancient civilizations and the migration of peoples.

The Mediterranean world in Roman times (Rome was founded in 753 BC) had a cooler climate with more winter rains (primarily from 600 to 200 BC). This period was one of great fertility for Greece, northern Africa and the Carthaginian and later Roman croplands. For a few hundred years there was a warming tendency and increasing dry weather until AD 400.

During the days of early Babylonia and Egypt there were two great inventions. One was the art of fashioning iron into tools with a cutting edge and the other was the building of seagoing ships. The first invention meant that humans could now live quite comfortably in colder or more humid climates than those of Babylonia or Egypt. The second, ship building, can be considered in relation to forest management. During the greatest days of Greece, the contrast between wet winters and dry summers was less marked than today. Such conditions were profitable for the growth of forests.

Viewpoints

There are several viewpoints on the relationship between climate and civilization (Pittock et al., 1978). In particular, the rise and fall of ancient civilizations seems to have been closely related to the change in climate. The ancient civilizations were built up on the basis of hunting and nomadism, which depended on the distribution of vegetation and animals. The flourishing of the ancient civilizations was supported by the agricultural production and a definite form of organized village life.

Huntington (1945) wrote on the role of climate in developing civilizations in the Babylonian and Mesopotamian regions. He pointed out rightly that 'the real problem [in developing civilizations] is to determine the exact nature of the [climatic] influence, its magnitude, and the extent to which its favorable and unfavorable aspects have counteracted one another.' Because of his general deterministic treatment of the relationship between climate and human activities, including civilization, his descriptions are not accepted by many people today. However, at least as far as his account of ancient civilization is concerned, his description is a sophisticated one. As pointed out by Spate (1952) and Oliver (1973), the historian Toynbee presented many similar ideas, although in a much more literary fashion.

It seems that the second half of the twentieth century is a time to discuss how climatic change may have influenced history (Manley, 1958; Carpenter, 1966; Lamb, 1968; Claiborn, 1970; Le Roy Ladurie, 1971; Singh, 1971; Chu, 1973; Dansgaard et al., 1975; Bryson, 1978; Suzuki, 1979; Wigley et al., 1981; Lamb, 1982a; Houghton et al., 1990). Ranging from determinism to probabilism, and possibilism to voluntarism, there can be various viewpoints on climate–civilization relationships (Oliver, 1973). However, at the present stage, it can be

concluded that even though the Holocene fluctuations are of relatively small magnitude, they have been sufficient to trigger cultural change in marginal situations. As Bryson (1978) wrote, cultural change includes changes in the economic base, such as agriculture, hunting-gathering and herding. How humanity responded to cultural change, i.e., the in situ modification of lifestyle, migration or literal disappearance of the people, must be determined locally (Jones, 1990).

Masatoshi M. Yoshino

Bibliography

Bryson, R.A., 1974. World Climate and World Food System III: The Lessons of Climatic History, Institute of Environmental Studies, Rep. 27. Madison: University of Wisconsin.

Bryson, R.A., 1978. Cultural economic and climatic records, in Climatic Change and Variability, A.B. Pittock et al. (eds). Cambridge: Cambridge University Press, pp. 316–327.

Butzer, K.W., 1966. Climate changes in the arid zones of Africa during early to mid-Holocene times, in World Climate from 8000 to 0 BC. London: Royal Meteorological Society, pp. 72–83.

Carpenter, R., 1966. Discontinuity in Greek Civilization. Cambridge: Cambridge University Press.

Chu, Ko-chen, 1973. A preliminary study of the climatic fluctuations during the last 5,000 years in China, Sci. Sinica, 16(2), 226–256.

Claiborn, R., 1970. Climate, Man, and History. New York: Norton.

Dansgaard, W., S.J. Johnson, N. Reeh et al., 1975. Climatic changes, Norsemen and modern man, Nature, 255, 24–28.

Flohn, H. and R. Fantechi, 1982. The Climate of Europe: Past, Present and Future. Dordrecht: Reidel.

Flohn, H. and S.E. Nicholson, 1979. Climatic fluctuations in the arid belt of the 'Old World' since the last glacial maximum: Possible causes and future implications, in Palaeoecology of Africa. Cape Town: Balkema.

Frisinger, H.H., 1977. The History of Meteorology to 1800, New York: Science History Publications.

Fukui, E., 1977. Climatic fluctuations, past and present, in Climate of Japan, E. Fukui (ed.). Amsterdam: Elsevier, pp. 261–305.

Harding, A., 1982. Climatic Change in Later Prehistory. Edinburgh: Edinburgh University Press.

Houghton, J.T., G.T. Jenkins and J.J. Ephraums, 1990. Climate Change, The Intergovermental Panel on Climate Change (IPCC). Cambridge: Cambridge University Press.

Huntington, E., 1945. Mainsprings of Civilization. New York: Wiley.

Irwin-William, C. and C.V. Haynes, 1970. Climatic change and early populations dynamics in the southwestern United States, Quaternary Research, 1, 59–71.

Jones, P.D., 1990. The climate of the last 1000 years, Endeavour, Elsevier.

Lamb, H.H., 1968. Climatic changes during the course of early Greek history, Antiquity, 42, 231–233.

Lamb, H.H., 1974. Climates and circulation regimes developed over the Northern Hemisphere during and since the last ice age, in Physical and Dynamic Climatology. Geneva: World Meteorological Organization, pp. 233–261.

Lamb, H.H., 1977. Climate: Present, Past and Future, Vol. 2. London: Methuen.

Lamb, H.H., 1982a. Climate, History and the Modern World. London: Methuen.

Lamb, H.H., 1982b. Reconstruction of the course of postglacial climate over the world, in Climatic Change in Later Prehistory. A. Harding (ed.). Edinburgh: Edinburgh University Press, pp. 11–32.

Lamb, H.H., R.P.W. Lewis and A. Woodroffe, 1966. Atmospheric circulation and the main climatic variables between 8000 and 0 BC: Meteorological evidence, in World Climate from 8000 to 0 BC. London: Royal Meteorological Society, pp. 174–217.

Le Roy Ladurie, E.L., 1971. History and climate, in Economy and Society in Early Modern Europe: Essays from Annales, P. Burke (ed.). London: Routledge Kegan Paul, pp. 134–169.

Manley, G., 1958. The revival of climatic determinism, Geog. Rev., 48, 98–105.

Manuel, D., 1996. Climate and weather: history and study, in Encyclopedia of Climate and Weather, S.H. Schneider (ed.). Oxford: Oxford University Press.

Messerli, B., 1980. Die afridanischen Hochgebirge und die Klimageschichte Afrikas in den letzten 20 000 Jahren, in Das Klima,

Analysen und Modelle, Geschichte und Zukunft, H. Oeschger *et al.* (eds.). Berlin: Springer-Verlag, pp. 64–90.

Nicholson, S.E. and H. Flohn, 1980. African environmental and climatic changes and the general atmospheric circulation in Late Pleistocene and Holocene, *Clim. Change* **2**, 313–348.

Oliver, J.E., 1973. *Climate and Man's Environment. An Introduction to Applied Climatology.* New York: Wiley.

Pittock, A.B., L.A. Frakes, D. Jenssen *et al.*, 1978. The effect of climatic change and variability on mankind, in *Climatic Change and Variability*, A.B. Pittock *et al.* (eds). Cambridge: Cambridge University Press, pp. 294–297.

Singh, G., 1971. The Indus Valley culture seen in the context of postglacial climatic and ecological studies in Northwest India, *Archaeol. and Phys. Anthropol. Oceana* **6**, 177–189.

Spate, O.H.K., 1952. Toynbee and Huntington: a study in determinism, *Geog. J.*, **118**, 406–428.

Suzuki, H., 1975. World precipitation, present and Hypsithermal, *University of Tokyo Dept. of Geog. Bull.*, **4** (in Japanese).

Suzuki, H., 1978. Kikô to bummei (Climate and civilization), *Kikô to Ningen Series, Asakura Shoten, Tokyo*, **4** (in Japanese).

Suzuki, H., 1979. 3,500 years ago. *University of Tokyo Dept. of Geog. Bull.*, **10**, 43–58.

Wigley, T.M.L., M.J. Ingram and C. Farmer, 1981. *Climate and History: Studies in Past Climates and Their Impact on Man.* Cambridge: Cambridge University Press.

Wright, H.E., 1968. Natural environment of early food production north of Mesopotamia, *Science*, **161**, 334–339.

Yasuda, Y., 1978. Prehistoric environment in Japan. *Tohoku University (Sendai), Sci. Rep., 7th Ser.*, **28**(2), 117–281.

Yoshino, M.M. and K. Urushibara, 1978. Climatic change and fluctuation in South and Southeast Asia, *Climatol. Notes*, **21**, 1–48.

Yoshino, M.M., 1979. Kôhyôki no kikô (Climate of postglacial period), in *Kikô hendô (Climatic change).* G. Yamamoto and Taiki Kankyo no Kagaku (eds). Tokyo: University of Tokyo Press, pp. 16–31 (in Japanese).

Cross references

Climate and climate change
Desertification
Tree rings in hydrological studies

CLIMATE CHANGE AND THE GREENHOUSE EFFECT

A greenhouse effect has always existed, keeping the Earth warmer than it would be without an atmosphere. What is popularly referred to today as the 'greenhouse effect' is really the anthropogenically enhanced greenhouse effect by which an extra warming of the surface and the lower atmosphere is produced, leading to disturbances in the geosphere/biosphere system and, notably, an increase in the mean global surface temperature and in the mean sea level.

The natural presence of 'radiatively active' gases in the atmosphere is essential for life: they trap heat in the lower atmosphere, thus creating – like a greenhouse – an environment which is far warmer (by about 33°C) inside than out. By increasing the concentration of greenhouse gases, however, additional infrared radiation, which otherwise would have been lost to space, is absorbed in the lower atmosphere and the Earth's radiation balance is upset. This energy is re-emitted in all directions, a large portion being sent back to the Earth's surface or elsewhere in the troposphere. This yields a radiative imbalance which can be restored only through a warming of the troposphere. On the other hand, the enhanced greenhouse effect will also cool the upper layers of the atmosphere (i.e. the stratosphere, and above 25–70 km).

Greenhouse gases in the atmosphere have increased since pre-industrial times by an amount that is radiatively equivalent to about a 50% increase in carbon dioxide (CO_2), although CO_2 itself has risen by about 25%; other gases have made up the rest. The global emissions of most greenhouse gases are expected to rise in the next decade: CO_2 at 0.5% year^{-1}, methane (CH_4) at 0.9% year^{-1}, and nitrous oxide (N_2O) at about 0.3% year^{-1}. In contrast, CFC emissions are expected to decrease to near zero around the year 2000 following internationally agreed phase-out measures. As a result of these trace gas trends, an effective doubling of greenhouse gas concentrations

(commonly measured as 'CO_2-equivalents') is expected around 2030.

There has been much scientific activity aimed at identifying clues of the enhanced greenhouse effect by searching for trends in climatic and hydrological records (mainly temperature, precipitation, glaciers, runoff, freezing/melting data and sea level). Conclusions have so far been mainly tentative because definite signals from an enhanced greenhouse effect are weak and/or smaller than natural fluctuations, most of the homogeneous and comparable records, if any, are too short, and regional differences have added to the confusion. Our partial understanding of multimedia mechanisms involved has also been a limitation to making a safe forecast of future trends, although this has fostered large international research efforts such as the World Climate Research Programme [WCRP, of the World Meteorological Organization (WMO), the International Council of Scientific Unions (ICSU) and the Intergovernmental Oceanographic Commission (IOC) of UNESCO] and the International Geosphere–Biosphere Programme (IGBP of ICSU), or international expert assessments such as under the Intergovernmental Panel on Climate Change (IPCC of WMO and UNEP).

At first glance, uncertainties about the distribution and timing of consequences of climate change may make some people feel that the cost of rapid implementation of actions intended to curb the anthropogenic emissions responsible might be prohibitive. However, regardless of what happens to climate, application of relevant measures would bring its own benefits, such as longer availability of fossil fuels and other natural resources, as well as more rational and efficient energy use. Additionally, consequences such as stronger and more frequent storms, higher seas and less water in rivers are so far-reaching that taking this 'precautionary' approach should be rewarding. This approach is also known as the 'no regrets' policy, because of the argument that preventing climate change would be far better than attempting to cure it, despite the uncertainties of how fast and how soon it might be happening.

Although Europe's land mass covers less than 10% of the world's total area and its population is some 15% of the world population, European countries contribute a significant amount to the global emissions of greenhouse gases and other substances which bring about large-scale changes in the atmosphere. Conversely, climate change may greatly affect Europe's 680 million people and its natural environment.

The causes

Human exploitation of the world's natural resources, primarily fossil fuels – coal, oil and natural gas – but also biomass resources through agricultural and forestry practices, result in the release each year of over 20 000 million tonnes of carbon dioxide (CO_2) into the atmosphere. In addition, other atmospheric gases, notably methane (CH_4), nitrous oxide (N_2O) and chlorofluorocarbons (CFCs), are being released as a result of human activities. All these gases together with ozone (O_3) and water vapour are referred to as 'greenhouse gases', that is, they are relatively transparent to incoming short-wave radiation from the Sun, but absorb long-wave infrared radiation emitted from the Earth (Figure C18), and consequently entrap heat in the atmosphere. Ozone in the lower atmosphere (troposphere) is not emitted directly but is formed from anthropogenic emissions of nitrogen oxides (NO_x), volatile organic compounds (VOCs), methane and carbon monoxide. Particles (e.g., sulfate) emitted into the atmosphere by anthropogenic and natural sources (mainly volcanoes) can also affect climate because they can reflect and absorb radiation.

Various physical and dynamic processes combine to redistribute heat horizontally and vertically in the atmosphere and the oceans. At the same time, they trigger various feedback effects that can either enhance or dampen (reduce) the initial trends. Most of these feedback processes are connected to the formation of clouds as well as to biospheric and oceanic responses.

Soils are an important reservoir for carbon; an estimate of the global soil carbon pool is 1500 gigatonnes, or twice the atmospheric pool, excluding carbon in organic soils. Globally, the amount of carbon in decaying plant litter and soil organic matter may exceed the amount of carbon in living biomass by a factor of two or three. Soils also contribute to the natural fluxes of greenhouse gases; the contribution of the whole terrestrial biota (soil, vegetation and fauna) is 30% for CO_2, 70% for CH_4 and 90% for N_2O.

Recent estimates suggest that 2 gigatonnes of carbon are absorbed

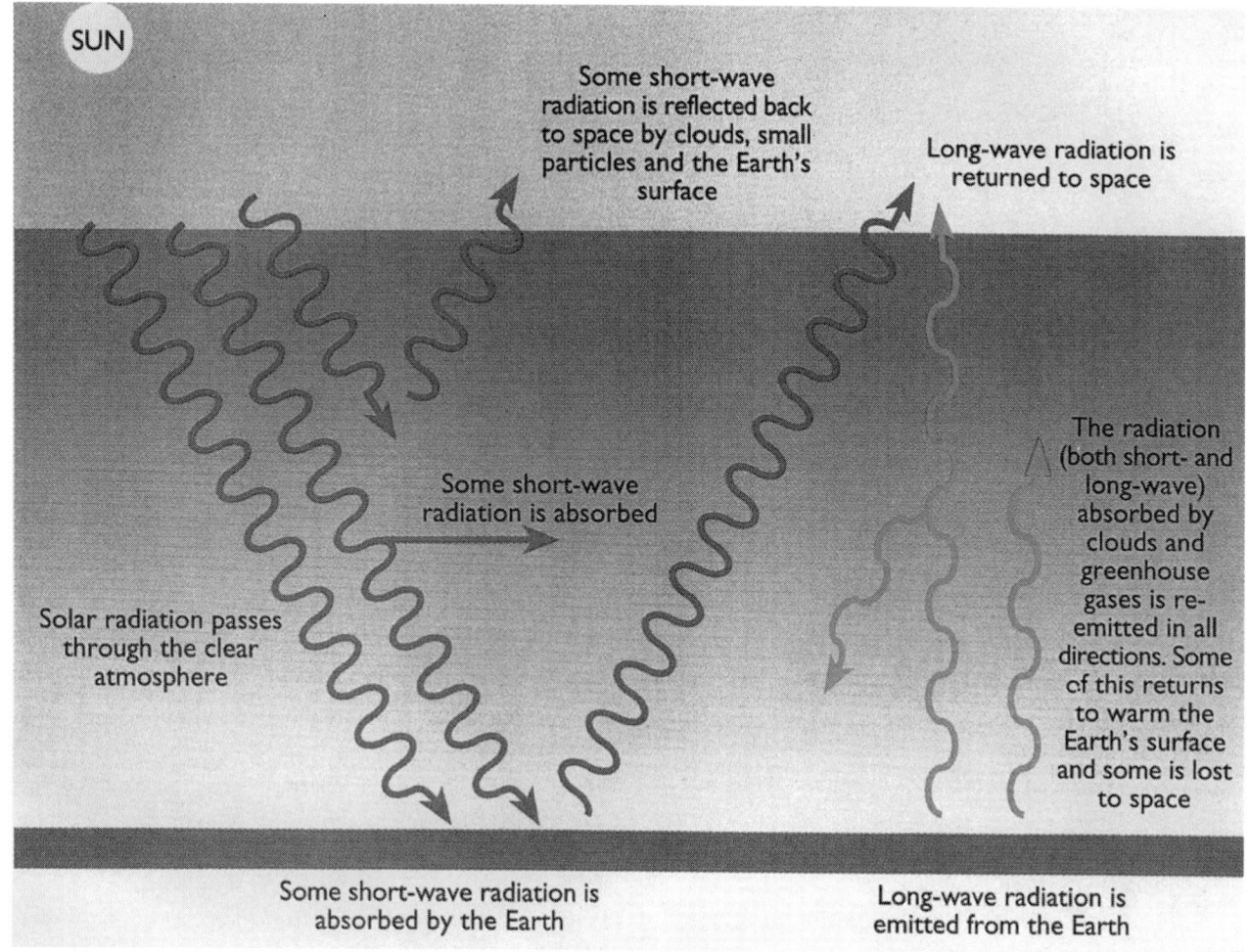

Figure C18 The greenhouse effect. (Source: IPCC, 1992.)

by the oceans each year to replace the carbon taken by phytoplankton from seawater and subsequently deposited on the sea bed when the phytoplankton die.

The lifetime of greenhouse gases in the atmosphere is determined by their sources and sinks in the oceans, atmosphere and biosphere. Carbon dioxide, CFCs and nitrous oxide are removed only slowly from the atmosphere and hence, following a change in emissions, their atmospheric concentrations take decades or centuries to adjust fully. Even if all anthropogenic emissions of CO_2 had been halted in the year 1993, about half of the increase in carbon dioxide concentration caused by human activities would still be observable in the year 2100.

The concept of relative global warming potentials (GWPs) has been developed as an index of the relative radiative effect (and hence potential climate effect) of equal emissions of each of the greenhouse gases, to take into account the differing time that they remain in the atmosphere and their different absorption properties. The GWP defines the time-integrated warming effect due to an instantaneous release of unit mass (1 kg) of a given gas in today's atmosphere, relative to that of carbon dioxide. The relative contributions will change over time, and a period of 20 years will indicate the response to emission change in the short term. Anthropogenic carbon dioxide emissions are currently responsible for more than half of the enhanced greenhouse effect (Figure C19). However, the GWP concept does not allow for the comparison of the relative effect of different greenhouse gases in terms of their contribution to climate change *per se*, such as may be measured in terms of the increase in the global mean surface temperature or sea-level rise. This comparison can only be made with the use of complete atmospheric models, and simple comparative indices are now being developed.

During the past century, the concentration of the most important greenhouse gases, and notably that of carbon dioxide, has been increasing steadily, showing that anthropogenic emissions have exceeded the natural capacity for their removal either through absorption at the Earth's surface or through chemical reactions in the atmosphere. Figure C20 illustrates the global CO_2, N_2O, CH_4 and CFC-11 concentrations since 1750.

The use of fossil fuels for energy production and transport is the most important source of global CO_2 emissions. Emissions due to coal are about twice those of natural gas for the same amount of energy (oil-related emissions are about 1.5 times as much). Present conversion of tropical forests to agriculture and pasture releases up to 15% of global emissions (approximately 7000 million tonnes of carbon in 1991).

Currently, European forests are a net sink for CO_2 as a result of afforestation and probably because of a CO_2 fertilization and nitrogen fertilization effect. In most cases the conversion of native forests to plantations will have resulted in a net loss of carbon to the atmosphere, including the carbon stored in timber products (Cannell *et al.*, 1992).

Ordinary agricultural practices also play a role in the problem of climate change. The yearly burning of biomass releases an amount of carbon, in the forms of carbon dioxide, carbon monoxide and methane, which is smaller than, but of the same order of magnitude as, the emissions from the burning of fossil fuels. The carbon released from biomass burning, however, like that from the decay of fallen leaves in temperate forests, does not correspond to a net release in the long term, since the same amount of carbon is removed from the atmosphere in the following growing season. Indeed, the oxidation of carbon at the higher temperature of burning tends to generate a higher

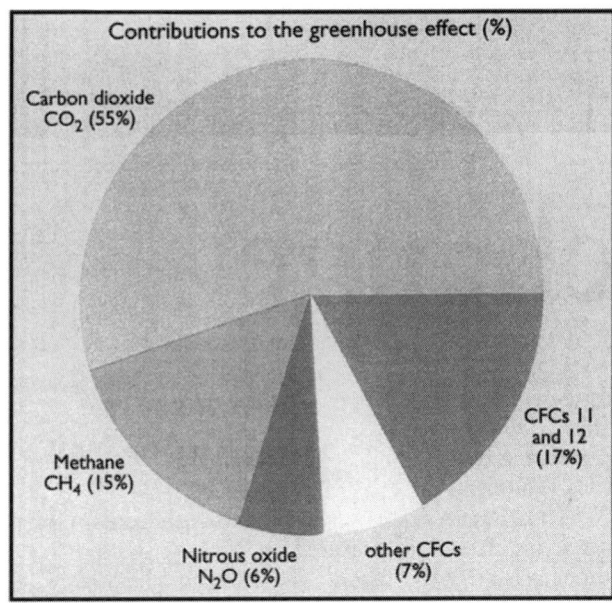

Figure C19 Current contributions of the different greenhouse gases to the enhanced greenhouse effect. (Source: IPCC, 1990.)

percentage of carbon dioxide, which produces a smaller enhancement of the greenhouse effect, molecule by molecule. The burning of biomass, on the other hand, does affect the chemistry of the troposphere, notably by increasing the ozone concentration, which may also have local pollution effects.

Scientific assessments of global climate change have generally considered only long-lived greenhouse gases merged into one equivalent surrogate (CO_2), which is assumed to double homogeneously by the middle of the twenty-first century. The IPCC reports (IPCC, 1990, 1992) mentioned the greenhouse property of ozone, but made no quantitative assessment of its warming potential. In effect, because ozone does not last long in the atmosphere, this assessment is more difficult since both temporal and spatial distribution should be taken into account. A recent estimate (Marenco *et al.*, 1994) computes a relative radiative forcing for ozone at least 1200 times higher than for CO_2, that is, much higher than the other greenhouse gases except CFCs. With its current concentrations, ozone would thus contribute 22% of the global warming in the northern hemisphere and 13% in the southern hemisphere.

The increase of methane concentration is correlated largely with increasing population, about 60% of global emissions being associated with human activities. Methane from fossil origins is released by the exploitation of coal, oil (if it is not used or flared) and natural gases, and from distribution systems. As a product of the anaerobic digestion of organic material, it is also released from landfills, wetlands, and rice paddies and by fermentation in the rumen of ruminants. Because of the potency of methane in the atmosphere and its relatively short lifetime, stabilizing CH_4 concentrations will have a substantial impact in reducing potential warming.

Nitrous oxide emissions are produced mostly by denitrification processes in oxygen-free environments with a high nitrate load such as soils and sediments in polluted waterbodies. N_2O is also released in limited quantities by the use of fossil fuels.

Carbon monoxide plays significant roles in controlling the chemistry of ozone production and hydroxyl radical (OH) destruction in the lower atmosphere. It directly affects the oxidizing capacity of the lower atmosphere and thus influences the concentrations of other important trace gases such as methane, methylchloroform and the HCFCs.

Global anthropogenic emissions of sulfur gases have increased by about a factor of three during the past century, leading to increased

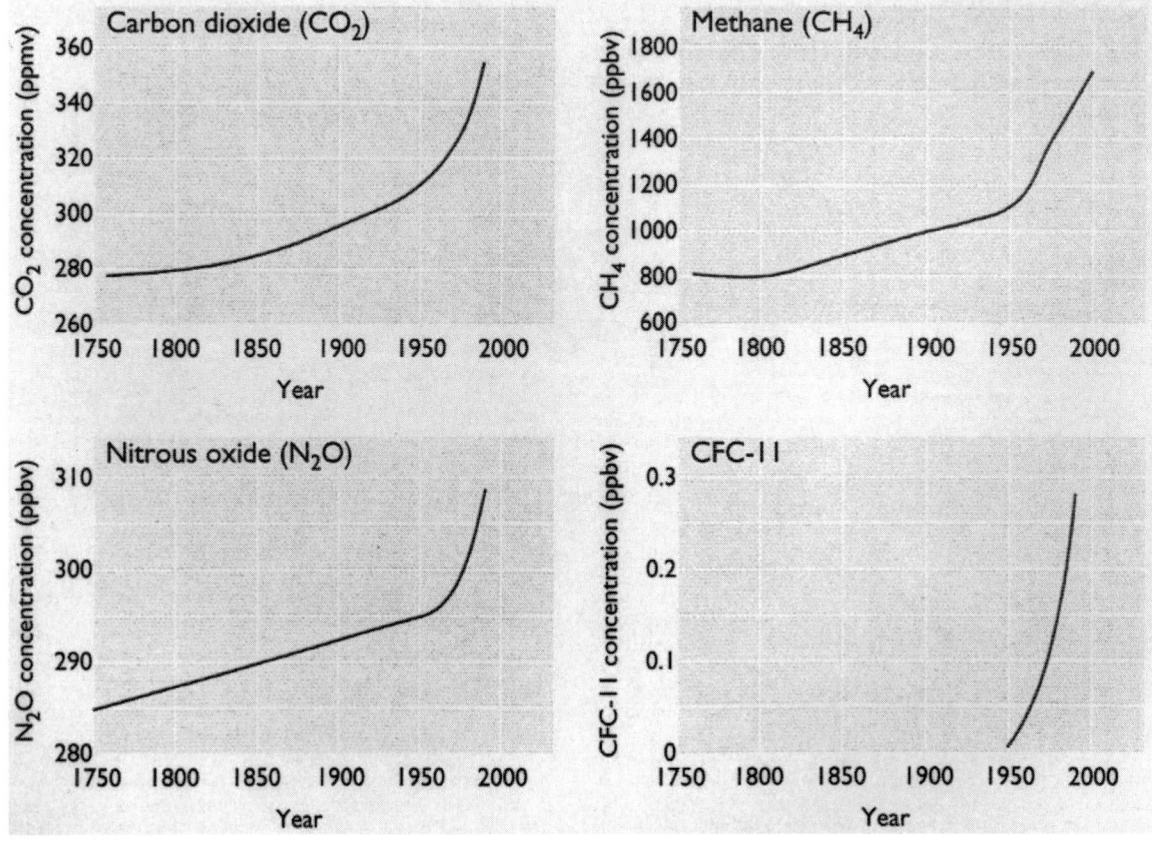

Figure C20 Concentrations of the greenhouse gases CO_2, N_2O, CH_4 and CFC-11 in the atmosphere since 1750. (Source: IPCC, 1990.)

Table C9 Best estimates of climate change projections over the next 50 years (Schneider, 1996)

Indicators	Annual average change	Distribution of changes					Confidence of projection	
		Regional average	Change in seasonality	Interannual variability	Significant transients		Global average	Regional average
Temperature	+2 to +5°C	−3 to +10°C	Yes	Down?	Yes		High	Medium
Sea level	+10 to +100 cm	−	No	?	Unlikely		High	Medium
Precipitation	+7 to 15%	−20 to +20%	Yes	Up	Yes		High	Low
Direct solar radiation	−10 to +10%	−30 to +30%	Yes	?	Possible		Low	Low
Evapotranspiration	+5 to +10%	−10 to +10%	Yes	?	Possible		High	Low
Soil moisture	?	−50 to +50%	Yes	?	Yes		?	Medium
Runoff	Increase	−50 to +50%	Yes	?	Yes		Medium	Low
Severe storms	?	?	?	?	Yes		?	?

sulfate aerosol concentrations, mainly in the northern hemisphere. Sulfate aerosols can affect the climate directly, by increasing the reflection of incoming solar radiation back to space in cloud-free air and, indirectly, by providing additional cloud condensation nuclei. Model calculations show that the increase in sulfate aerosol concentration reaches a factor of 100 over northern Europe in winter.

Consequences of climatic changes

The primary consequence of increased radiation absorption is climatic change or, in terms of the most common climate indicators, rapid changes of surface temperature – about 0.3°C per decade (IPCC, 1992) – and precipitation changes. The consequent expected sea-level rise (3–10 cm per decade) and changes in hydrological and vegetation patterns may have serious effects on society, leading to high risks and to substantial costs. Current models predict average global changes, but the impacts of climate change will be most keenly felt on a regional and local scale where uncertainties are larger (Table C9).

To evaluate the consequences of climate change, it is common to take as a reference case the doubling of CO_2 (or an equivalent amount of total greenhouse gases) in the atmosphere relative to its pre-industrial level (before the years 1750–1800). The IPCC business-as-usual scenario (which is considered a high emission scenario) suggests that this will occur by 2025. Under the IPCC low emission scenario, doubling will take place around 2060.

Changes in climate patterns

General circulation model estimates of the increase in the global annual average temperature at the surface of the Earth vary from 1.5 to 4.5°C, given an increase in greenhouse gases equivalent to a doubling of the pre-industrial CO_2 concentration. The current 'best' estimate is 2.5°C (IPCC, 1990). The average global increase in precipitation and evaporation is estimated at 3–15%. A recent reassessment of the IPCC 1990 climatic trends, which also takes into account the effects of CO_2 fertilization, feedback from stratospheric ozone depletion and the counteracting radiative effects of sulfate aerosols, yields new projections for radiative forcing of climate and for changes in global mean temperature and precipitation (Wigley and Raper, 1992). Changes in temperature and sea level are predicted to be less severe than those estimated previously, but are still far beyond (four to five times) the limits of natural variability.

The range of possible rise in temperature is determined largely by the uncertainty about the feedback processes. While the timing and extent of warming on a global basis is uncertain, still more indefinite is the change of climate at the regional level in Europe and elsewhere. The spatial resolution of models being still too coarse, regional estimates are less accurate than global ones. Models, however, consistently predict that temperature increases will be greater in the high latitudes than low latitudes. Hence the extent of temperature increases in northern Europe are expected to be larger on average than those in the Mediterranean regions. Using the new technique of nesting computer models, improved estimates of possible temperature changes in Europe are becoming available.

The IPCC collected best model guesses as to the change in climate for southern Europe from pre-industrial times to 2030, under the assumption that average global warming would be 1.8°C by that year. Temperature was estimated to be 2°C warmer in winter and 2–3°C warmer in summer, precipitation 0–10% greater in winter and 5–15% lower in summer, with a −5 to +5% change in soil moisture in winter

and −15 to −25% change in summer. The ranges in these values indicate the disagreement between the three different general circulation models used for these calculations. Although the models are relatively consistent, in some cases they do not agree on the direction of change.

Higher temperatures will yield more evaporation, resulting in more precipitation. This will intensify the hydrological cycles, with unknown consequences. Sudden regional or global climate changes following changes in the location and/or intensity of air and water flows cannot therefore be excluded.

Assessing climate impacts

Many studies on specific climate change impacts have not used a common framework, and some of their underlying assumptions may differ. As yet, there has been no comprehensive, systematic study of climate impacts in Europe. However, preliminary results from a still unpublished study of the European Commission (performed by the Climate Research Unit, East Anglia, UK; Environmental Resources Management Limited, London, UK; and the National Institute of Public Health and Environmental Protection (RIVM), Bilthoven, The Netherlands) indicate that the possible changes in the EU due to global warming would be increased agricultural yields and tourism, and reduced heating requirements. However, the study also points out that the monetary costs of protecting coastlines against sea-level rise are likely to far outweigh benefits.

Sea-level rise

A rise in the global mean sea level will be one important impact of global warming. It will result from

- the thermal expansion of sea water;
- melting of mountain glaciers;
- melting of inland ice caps, as in Greenland;
- changes in the Antarctic ice sheet.

According to IPCC best estimate scenarios, the global mean sea level would be about 22 cm higher than today by 2050 and around 50 cm higher by 2100 (the low and high scenario estimates give 15 and 90 cm respectively for 2100). A rise of 10–20 cm over the last 100 years has been observed, due mainly to oceanic thermal expansion and retreating mountain glaciers.

The consequent environmental effects are many and varied: permanent inundation of low-lying land; increased frequency of temporary flooding from high tides or storm surges; changes in rates of beach, dune or cliff erosion; salinization of estuaries, groundwater and surfacewater supplies, wetland ecosystems or agricultural soils; and effects on river hydrology, including inland flooding, through changes in river gradients.

Hydrological processes

Among the greatest potential impacts of climatic change will be the effects on the hydrological cycle. An increase in the incidence of extremes, such as floods and droughts, would cause increased frequency and severity of disasters, while changes in precipitation, evapotranspiration and soil moisture would strongly alter agricultural practice and water management systems, and lead to severe land degradation. Forecasts of future changes of water resources attributable to human-induced climate changes are still uncertain because of the unreliability of regional precipitation predictions. However, they do indicate that such changes would be fairly significant, even in the case of slight climate modification, and that not only river runoff but also water demand might be affected. For instance, a temperature rise of 1–2°C and a 10% drop in rainfall would reduce river runoff in arid regions by 4–70% (Shiklomanov, 1991).

Various case studies point out possible modifications in runoff and other processes that might occur if there were a doubling of CO_2. For example, a study of three catchments in Belgium (Bultot, 1988) indicated that winter floods may become more frequent in catchments that have low infiltration rates and unsteady base flows; however, in basins having high infiltration rates and more steady baseflow there might be a beneficial increase in base flow throughout the year. In a study of seven Norwegian stream basins, climate changes brought about by a doubling of CO_2 are estimated to increase runoff in mountainous catchments, while decreasing it in lowland basins because of greater evapotranspiration. The seasonal pattern of runoff will change markedly, particularly in catchments of intermediate elevation. Winter runoff will increase significantly, while summer runoff declines. Flooding will occur more frequently in the fall and winter, and the overall amount of flood damage is expected to increase. The net effect of changes in runoff patterns may also lead to a small increase in Norwegian hydropower production.

Preliminary model results have shown an increase in runoff (up to a doubling) of rivers discharging into the Arctic Ocean. Since freshwater runoff into the Arctic Ocean is a significant fraction of its total water mass, any substantial modification of the input of water as envisioned in model scenarios may change important thermal and salinity characteristics. This in turn could significantly modify currents, ice cover and atmospheric circulation. Changes in river flow also modify the delivery of dissolved constituents and particulates to the Arctic Ocean. Increases in nutrient inputs could increase net primary production and thereby sequester excess atmospheric CO_2.

Results of an investigation into the effects of global warming on river flows in the UK are given by Arnell and Reynard (1996). Twenty-one catchments were investigated under climate change scenarios based on the UK Climate Change Impacts Review Group (CCIRG, 1991). A large difference was found for various scenarios. The wettest scenario implied a general increase in runoff whilst the driest would result in a reduction of up to 30%, the largest percentage change being in the south. Under most scenarios the range of flows would be greater, with the higher flows in winter and lower flows in summer. However, there was considerable uncertainty in the magnitude of change in river flows due largely to differences between climate change scenarios.

Risks to ecosystems

Alteration of the hydrological cycle, with changes in runoff and moisture availability, will influence patterns of sedimentation and erosion and the recycling of organic matter and nutrients. These will in turn influence plant productivity, competition between species and biodiversity. The linkages are sometimes more subtle with, for example, changed temperature and precipitation conditions combining to favor the outbreak of plant diseases or insect plagues. These complex linkages between different factors, together with uncertainties in predicted water availability, make it difficult to predict changes in natural ecosystems towards new equilibria. Moreover, changes may take place so quickly that plants have no time to spread to new habitats by natural dispersion mechanisms and the actual vegetation response could therefore differ markedly from equilibrium predictions.

Climate changes will influence ecosystems and agriculture since temperature and evapotranspiration/precipitation determine the major biogeographical zones (such as northern boreal, temperate, Mediterranean). Large regional differences may arise due to small changes in one of the critical factors, such as temperature, precipitation, evaporation and soil structure, yielding changes in vegetation structure and the range of species of a given zone. The potential impacts of climate change on agriculture may have dramatic consequences since Europe, as a main exporter of agricultural products, contributes significantly to the nutritional balance of non-European countries. This possibility for a major disturbance in agricultural trades should be seen also in the context of a rapidly increasing world population. A study of the potential range of maize-growing areas in Europe found that a 1°C increase in mean annual temperature in Europe would translate into a northward shift of approximately 200–350 km in Western Europe, and 250–400 km in Eastern Europe. New areas of

potential maize production would be opened up in southern England, The Netherlands, Belgium, northern Germany and northern Poland. A mean annual increase of 4°C would move the boundary up into northern Russia and central Fennoscandia. However, this study did not take into account possible shifts in water availability, possible changes in the range of agricultural pests and diseases, or other factors that might limit agropotential. Indeed, the IPCC estimated under their business-as-usual scenarios that Mediterranean countries which already depend heavily on irrigation would have 15–25% less soil moisture in summer.

Although extensions of agricultural areas northward might prove to be beneficial, the same pressure on natural ecosystems to extend or shift their range may endanger these ecosystems. This is because of human or natural barriers that may prevent the migration of ecosystems as their climatic zones shift. The rate of change is also important because, in less than 100 years, Europe's bioclimatic zones may shift hundreds of kilometres in latitude (or hundreds of meters in latitude in high Alpine regions), and it is generally accepted that this will place considerable stress on terrestrial ecosystems. In some cases it can be expected that flora and fauna will be unable to migrate fast enough to survive.

Terrestrial ecosystems are thought to play an important role in determining regional and global climate; two examples of this are in Amazonia, where destruction of the tropical rainforest leads to warmer and drier conditions, and the Siberian forests, which represent 40 000 million tonnes of stored carbon (i.e., an amount equivalent to half of the forests of Amazonia). Boreal forest ecosystems may also affect climate: as temperatures rise, the amount of continental and oceanic snow and ice is reduced, so the land and ocean surfaces absorb greater amounts of solar radiation, reinforcing the warming in a 'snow–ice–albedo' feedback which results in large climate sensitivity to radiative forcings. This sensitivity is moderated, however, by the presence of trees in northern latitudes, which mask the high reflectance of snow, leading to warmer winter temperatures than if trees were not present. Results from the National Centre for Atmospheric Research (NCAR) global climate model show that the boreal forest warms both winter and summer air temperatures, relative to simulations in which the forest is replaced with bare ground or tundra vegetation (Bonan et al., 1992). This suggests that future redistributions of boreal forest and tundra vegetation (for example due to extensive logging in Russia/Siberia, or the influence of global warming) could initiate important climate feedbacks, which could also extend to lower latitudes. The position of the tree line distinguishing boreal forest from tundra vegetation has altered in response to past climate changes and is likely to change with the warmer climate caused by increased atmospheric CO_2 concentrations. Results from the NCAR coupled model raise the possibility of considerable climate changes caused merely by redistribution of boreal forest and tundra ecosystems. The decrease in snow-covered land surface albedo caused by the northward migration of boreal forest into tundra zones in response to climate warming may produce further warming.

Plants are the essential basis of all terrestrial life. Any significant variation in the productivity and composition of plant life would initiate a cascade of changes affecting animal life. At first glance, elevated CO_2 levels might seem an agricultural blessing by enhancing plant growth. This 'CO_2 fertilization effect', as it is called, is expected to be particularly pronounced if plants have plentiful supplies of nutrients, light and water. The CO_2 fertilization effect also promises to provide a buffer for concerns about global warming by drawing more CO_2 from the atmosphere. However, recent studies (Bazzaz and Fajer, 1992) suggest that such assumptions about the benefits of a world replete with CO_2 may be overstated. An isolated case of a plant's positive response to increased CO_2 levels does not necessarily translate into increased growth for entire plant communities. Farming will also be adversely affected by the differential response of plants. Important crops, such as maize and sugarcane, may experience yield reductions because of the increased growth of weeds.

Even the notion that plants will serve as sinks to absorb ever-mounting levels of CO_2 is questionable. Increased competition between plants will also diminish the CO_2 enhancement not only of natural ecosystems such as meadows and forests, but also of artificial ecosystems such as farms. Agricultural yields will improve in a CO_2-rich future only at the cost of large quantities of fertilizers, pesticides and water from irrigation.

Other organisms which depend on threatened plant species for food, shelter or mating sites may also become endangered. A strong reduction of species diversity would, in turn, undermine the integrity of natural ecosystems. Because individual plant and animal species supply a wealth of essential industrial, agricultural and medicinal products, the loss of diversity will have pervasive environmental and economic consequences. Changes in the nutritional quality of plant leaves could lower herbivore and predator populations within their habitats. If insect herbivores suffer population reductions in a world abundant with CO_2, many predators will have less prey. Some predatory insects, for example, feast on other insect pests that damage certain crops. Plant development and flowering times may be also altered unpredictably by elevated CO_2, thus disrupting pollination.

Shifts in climate and vegetation could also significantly affect future animal distribution, abundance and survival. Rapid rates of change, especially in combination with the existence of artificial urban and agricultural barriers, may affect the ability of many species to relocate to areas which are climatically and ecologically more favorable. Endangered or rare species would be particularly vulnerable to rapid change, especially if their distributions are spatially restricted and their niche width is narrow. Rapid climatic change therefore becomes a threat for current biodiversity.

Land degradation

The resulting change in atmospheric composition will affect soil biological processes and provoke, for instance, an increase in biomass and changes in the composition of vegetation and organic matter. These biological transformations will have an unforeseeable impact on soils. For instance, there is some concern about anticipated effects on soil carbon storage in tundra boreal peatland. On the other hand, human interference in the land cover brings about changes in the rainfall regime, in the hydrology of large river systems and in the Earth's albedo, all of which play a major role in the surface energy balance.

In the long term (over 50–100 years) a substantial increase in temperature and changes in rainfall patterns (perhaps slightly wetter winters) in Europe could modify land use. Whole ecosystems may shift northwards, but this consequence of climate change will undoubtedly be modified by land-use policy (Cannell et al. 1992). Land-use changes can have such important impacts that soil carbon decreases in the top 1 m by 40–60% when forest or grassland is converted to cropland, and by 25–35% when forest is converted to grassland. Overall, warmer temperatures are thought to result in more arid conditions. Increased aridity in Southern Europe could lead to a chain of events in the soil, beginning with the breakdown in supply of organic matter, continuing with salt accumulation and the formation of surface crusts, and ultimately ending in severe land degradation.

Loss of organic matter may result from removal of crops without the return of any material to compensate. It may also result from alteration of soil drainage, or tillage that accelerates oxidation of organic matter. Peaty soils tend to suffer oxidation of organic matter if drained, and can shrink alarmingly as a consequence. In drier climates, the loss of soil organic matter generally leads to a reduction in retention of soil moisture and a decline in vegetation cover, crops or natural plants, which in turn leads to increased erosion. A positive feedback can thus arise, as organic matter/moisture in soil falls, plant cover declines and thus the renewal of organic matter in soil is further reduced. But for the extrapolation of data and making predictions, the underlying processes, which are largely microbial, need to be understood.

R.W. Herschy

Source

Stanners, D. and Bordeau, P. (eds), 1993. *Europe's Environment; The Dobřiš Assessment*, European Environment Agency, Luxembourg.

Bibliography

Arnell, N.W. and Reynard, N.S., 1996. The effects of climate change due to global warming on river flows in Great Britain, *Journal of Hydrology*, **183**, 397.

Bazzaz, F.A. and Fajer, E.D., 1992. Plant life in a CO_2-rich world, *Scientific American*, 18–24.

Bonan, G.B., Pollard, D. and Thompson, S.L., 1992. Effects of boreal forest vegetation on global climate, *Nature*, **359**, 716–718.

Bultot, F., 1988. Land surface management and its impact on climate change and water balance, in Cottenie and Teller (eds) *SCOPE report on Belgian Research on Global Change*, IGBP.

Cannel, M.G.R., Dewar, R.C. and Thornley, J.H.M., 1992. Carbon flux and storage in European forest, in Teller, Mathy and Jeffers (eds), *Responses of Forest Ecosystems to Environmental Change*, Commission of the European Community, EUR 13902, Elsevier, London, pp. 256–271.

CCIRG (Climate Change Impacts Review Group), 1991. *The potential impacts of climate change in the United Kingdom*, HMSO, London.

Chandler, M.A., 1996. Global warming, in *The Encyclopedia of Climate and Weather* (ed. S.H. Schneider), Oxford University Press, Oxford.

IPCC, 1990. *Climate change: the IPCC scientific assessment.* Houghton, J.T., Jenkins, G. and Ephraum, J.J. (eds), Cambridge University Press, Cambridge.

IPCC, 1992. *The supplementary report to the IPCC scientific assessment of climate change 1992*, Houghton, J.T., Callander, B.A. and Varney, S.K. (eds), Cambridge University Press, Cambridge.

Marenco, A., Gouget, H., Nedelec, P. *et al.*, 1994. Evidence of long term increase in tropospheric ozone at mid-latitudes from Pic du Midi data series: positive radiative forcing, *Journal of Geophysical Research*, **99**(D8), 16617–16632.

Shiklomanov, I.A., 1991. The world's water resources, *IHD/IHP 25 years International Symposium*, 15–17 March 1990, UNESCO, Paris, pp. 93–126.

Schneider, S.H., 1990. Prudent planning for a warmer planet, *New Scientist*, 17 Nov, 48–51.

Wigley, T.M. and Raper, S.C.B., 1992. Implications for climate and sea level of revised IPCC emissions scenarios, *Nature*, **357**, 293–300.

Cross references

Climate and climate change
Greenhouse effect

CLIMATE DATA: SOURCES

Ideally, an understanding of the nature and intended use of climatic data precedes attempts to find a source. However, it is often necessary to strike a compromise between apparent needs and availability of data, although for many purposes useful data can be derived by statistical inference based on physical principles. Basic criteria for selection of an appropriate source are the specific variables or derivations required, geographical coverage, time scale and continuity, acceptable levels of accuracy or approximation, statistical complexity and format. The relative significance of these criteria varies among the potential uses in climatological research, technical applications or narrative descriptions. Once the objectives of a data search are established, it is usually possible to decide whether to consult an original source, such as an observational record, or to seek guidance from data services.

Agency sources of climatic data

Organizations that administer the collection and processing of climatic data offer primary orientation to archives, data banks or publications. They are also the best sources of current information on availability, alternative formats and costs. The international agency for coordination of worldwide climatic records is the World Meteorological Organization (WMO), which has its Secretariat at Geneva, Switzerland. The WMO publication *Meteorological Services of the World* is a directory of approximately 150 national members, including addresses of branches responsible for climatology. It also lists related academic and research institutions in several countries.

Both national and international scientific groups cooperate to maintain collections of data pertaining to climate. Principal centers are:

World Data Center A for Meteorology, National Climatic Data Center, Federal Building, Asheville, North Carolina 28801, USA. Telephone: 704–271–4994. Fax: 704–271–4246. E-mail: internet wdca@ncdc.noaa.gov

World Data Center B for Hydrometeorological Information, 6 Korolov St, Obninsk, Kaluga Region 249020, Russia.

World Data Center D for Meteorology, National Meterological Center, State Meteorological Administration, 46 Baishiqiao Rd, Western Suburb, Beijing 100081, China. Telephone 8312277 ext. 2615.

World Meteorological Organization, Case Postale 2300, CH-1211 Geneva 2, Switzerland.

World Data Center A: Glaciology, National Snow and Ice Data Center, University of Colorado, Boulder, Colorado 80309, USA.

World Data Center A: Oceanography, National Oceanic and Atmospheric Administration, Washington, DC 20235, USA.

World Data Center A: Solar–Terrestrial Physics National Geophysical Data Center, National Oceanic and Atmospheric Administration, 325 Broadway, Boulder, Colorado 80303, USA.

World Ozone Data Centre, Atmospheric Environment Service, 4905 Dufferin Street, Downsview, Ontario M3H 5T4, Canada.

Sunspot Index Data Center, 3 Avenue Circulaire, B1180 Brussels, Belgium.

The official repository for records generated by US government weather services is the National Climatic Data Center, Federal Building, Asheville, NC 28801, USA. The center publishes indexes as well as summaries of national and world climatic data and has facilities to provide electronic communication, computerized data, microfilm, or duplicate copies of original records. Certain early historical materials, mainly predating creation of the US Weather Bureau, are held by the National Archives and Records Service in Washington, DC.

Other data services sponsored by the US government are:

Data Services Division, Center for Environmental Assessment Services, 3300 Whitehaven Street, NW, Washington, DC 20235, USA.

Environmental Science Information Center, Environmental Satellite, Data and Information Services/National Oceanic and Atmospheric Administration, 6009 Executive Boulevard, Rockville, Maryland 20852, USA.

Satellite Data Services Division, National Oceanic and Atmospheric Administration, World Weather Building, Washington, DC 20233, USA.

The British Meteorological Office, London Road, Bracknell, Berkshire RG12 2SZ, UK, publishes compilations of climatic data for the United Kingdom and summaries for selected world stations.

In Canada and the United States regional data centers are maintained by the provinces and states in cooperation with federal agencies. Canadian regional climate offices operate in conjunction with the Canadian Climate Centre, 4905 Dufferin Street, Downsview, Ontario M3H 5T4. Addresses of state and regional climate centers in the United States may be obtained from the National Climatic Center.

A growing number of government regulatory agencies, academic institutions, research organizations and private companies collect local or regional data for a variety of purposes, mainly entailing technical applications. These potential sources usually can be identified by government offices and often have data that supplement official records. Formats vary from instrument charts or computer tapes to written records and published summaries.

Published guides to climatic data sources

Besides the obvious utility of general archive inventories or bibliographies, special climatic data catalogs and guides assist in the identification of sources. The WMO periodically revises its list of *Publications of the World Meteorological Organization*, which includes catalogs of meteorological data compilations.

For a detailed description of world data centers and their holdings see *Fourth Consolidated Guide to International Data Exchange Through the World Data Centres*, 1979, issued and periodically revised by the Secretariat of the International Council of Scientific Unions Panel on World Data Centers, National Academy of Sciences, 2101 Constitution Avenue, NW, Washington, DC 20418, USA.

Sources and characteristics of unpublished data collections held in Canada and the United States are catalogued in *The Interim Climate Data Inventory: A Quick Reference to Selected Climate Data*, 1980,

compiled by C.F. Ropelewski, M.C. Predoehl and M. Platto and published by the Center for Environmental Assessment Services. This inventory initiated a project for comprehensive indexing of specialized data sets for more than 100 atmospheric and surface parameters, including inferred (proxy) and land-use data. Its contents are also accessible through an interactive database system managed by the Center for Environmental Assessment Services.

Both published and unpublished data for the United States and selected world networks are described and illustrated with sample formats in *Selective Guide to Climatic Data Sources. Key to Meteorological Documentation No. 4.11*, 1969, Asheville, NC: National Weather Records Center, 90 pp. Revised editions prepared by K.D. Butson and W.L. Hatch under the same title were published by the National Climatic Data Center in 1979 and 1983.

Other useful guides to US data sources are:

NOAA Products and Services of the National Weather Service, National Environmental Satellite Service, Environmental Data Service and Environmental Research Laboratories, 1977, Rockville, Maryland: National Oceanic and Atmospheric Administration, 4 vols.

Ownbey, J.W. (ed.), 1980, *Guide to Standard Weather Summaries and Climatic Services*, Asheville, North Carolina: Naval Oceanography Command Detachment, 211 pp.

Products and Services Guide, 1996, National Climatic Data Center, Asheville, North Carolina: Climate Services Division, 52 pp.

Among indexes prepared and revised by the National Climatic Data Center to meet special needs are *Index of Original Surface Weather Records (Hourly, Synoptic and Autographic)* for stations in each state and *Index of Surface Marine Climatic Data Products. SOLMET Volume 1 – User's Manual, Hourly Solar Radiation-Surface Meteorological Observations*, 1979, indexes observed and derived solar radiation data for US stations. *STAR Tabulations Master List* identifies stations for which unpublished air pollution potential data have been prepared. *Index – Summarized Wind Data*, 1977, was compiled by M.J. Changery, W.T. Hodge, and J.V. Ramsdell and issued jointly by the National Climatic Data Center and Battelle Pacific Northwest Laboratories, Richland, Washington.

Published climatic data

Atlases and maps

Most world atlases and many national atlases contain maps that show the mean distribution of major climatic elements, usually including temperature, precipitation, barometric pressure and wind patterns. While affording regional generalizations, such maps often require interpolation of values for a particular place. The smaller the map scale, the greater is the probable error, especially for sites having locally anomalous environments. Specialized climatic atlases ordinarily incorporate more detail and may have supporting statistical tables. The following selection suggests their geographic and topical range.

Brooks, C.F., A.J. Connor et al., 1936. *Climatic Maps of North America*. Cambridge, MA: Harvard University Press, 26 maps.

Climatic Atlas of North and Central America, Vol. I, *Maps of Mean Temperature and Precipitation*, 1979. Geneva: World Meteorological Organization, 28 sheets.

Climatic Atlas of the United States, 1968. Washington, DC: Environmental Science Services Administration, Environmental Data Service, 80 pp.

Climatological Atlas of the British Isles, 1952. London: British Meteorological Office, 139 pp.

Diaz, H.F., 1980. *Atlas of Mean Winter Temperature Departures from the Long-Term Mean over the Contiguous United States 1895–1979*. Asheville, NC: National Climatic Center, 88 pp.

Fletcher, R.J. and G.S. Young, 1976. *Climate of Arctic Canada in Maps*. Edmonton: University of Alberta, 46 maps.

Hastenrath, S. and P.J. Lamb, 1977. *Climatic Atlas of the Tropical Atlantic and Eastern Pacific Oceans*. Madison: University of Wisconsin Press, 97 charts.

Hastenrath, S. and P.J. Lamb, 1979. *Climatic Atlas of the Indian Ocean*. Madison: University of Wisconsin Press, 2 vols.

Hoffmann, J.A.J., 1975. *Climatic Atlas of South America.* Geneva: World Meteorological Organization.
Landsberg, H.E., H. Lippmann, Kh. Paffen and C. Troll, 1969. *World Maps of Climatology,* 2nd edn. New York: Springer-Verlag, 5 maps.
Steinhauser, F. ed., 1970. *Climatic Atlas of Europe.* Geneva: UNESCO/World Meteorological Organization.
Taylor, J.A. and R.A. Yates, 1958. *British Weather in Maps.* London: Macmillan, 256 pp.
Thomas, M.K., 1953. *Climatological Atlas of Canada.* Ottawa: National Research Council, Canada, 256 pp.
US Navy Marine Climatic Atlas of the World, 1955–1959. Washington, DC: Chief of Naval Operations, 8 vols.
US Navy Marine Climatic Atlas of the World, revised 1974–1981. Asheville, NC: Naval Weather Service Detachment, 8 vols.
Visher, S.S., 1954. *Climatic Atlas of the United States.* Cambridge, MA: Harvard University Press, 403 pp.
Walter, H., E. Harnickell and D. Mueller-Dombois, 1975. *Climate-Diagram Maps of the Individual Continents and the Ecological Climatic Regions of the Earth.* Berlin: Springer-Verlag, 36 pp.

Monographs and books

An extensive treatment of world climates appears in *World Survey of Climatology,* H.E. Landsberg, Chief Editor, published as a series of 15 volumes beginning in 1969 by American Elsevier Publishing Company, Inc., New York. Volumes 1–3 are concerned with *General Climatology;* volume 4 considers *Climate of the Free Atmosphere;* and volumes 5–15 treat climates of major world divisions, including polar regions and the oceans.

Explanatory descriptions, tables and maps appear in a variety of additional climatology monographs and textbooks. Representative publications in this category are:

Bailey, H.P., 1966. *The Climate of Southern California.* Berkeley: University of California Press, 87 pp.
Baldwin, J.L., 1973. *Climates of the United States.* Washington, DC: National Oceanic and Atmospheric Administration, 113 pp.
Borisov, A.A., 1965. *Climates of the USSR.* Chicago: Aldine, 255 pp.
Boucher, K., 1975. *Global Climate.* New York: Halsted Press, 326 pp.
Chandler, T.J., 1965. *The Climate of London.* London: Hutchinson, 292 pp.
Chandler, T.J. and S. Gregory (eds.), 1976. *The Climate of the British Isles.* New York: Longman, 390 pp.
Climate of Andhra Pradesh, 1976. New Delhi: Government of India Meteorological Department, 143 pp.
Climate and Man, 1941 Yearbook of Agriculture, 1941. Washington, DC: US Department of Agriculture, 1248 pp.
Fukui, E. (ed.), 1977. *The Climate of Japan.* Tokyo: Kodansha, 317 pp.
Garnier, B.J., 1958. *The Climate of New Zealand.* London: Edward Arnold, 191 pp.
Green, C.R. and W.D. Sellers, 1964. *Arizona Climate.* Tucson: University of Arizona Press, 503 pp.
Hare, F.K. and M.K. Thomas, 1979. *Climate of Canada,* 2nd edn. New York: Wiley, 230 pp.
Haurwitz, B. and J.M. Austin, 1944. *Climatology.* New York: McGraw-Hill, 410 pp.
Hutchinson, P., 1974. *The Climate of Zambia.* Lusaka: Zambia Geographical Association.
Kendrew, W.G., 1961. *The Climates of the Continents,* 5th edn. Oxford: Clarendon Press, 608 pp.
Koeppe, C.E., 1931. *The Canadian Climate.* Bloomington, IN: McKnight and McKnight.
Köppen, W.P. and R. Geiger, 1930–1939. *Handbuch der Klimatologie.* Berlin: Gebrüder Borntraeger, 6 vols.
Lebedev, A.N. (ed.), 1970. *The Climate of Africa,* Vol. 1. Jerusalem: Israel Program for Scientific Translations, 482 pp.
Ludlum, D.M., 1982. *The American Weather Book.* Boston: Houghton Mifflin, 296 pp.
Lydolph, P.E., 1985. *The Climate of the Earth.* Totowa, Rowman and Alanheld, 386 pp.

Maxwell, J.B., 1980. *The Climate of the Canadian Arctic Islands and Adjacent Waters.* Hull, Quebec: Canadian Government Publishing Centre, 531 pp.
Ojo, O., 1977. *The Climates of West Africa.* London: Heinemann, 218 pp.
Ooi, J.-B. and C.L. Sien, 1974. *The Climate of West Malaysia and Singapore.* Singapore: Oxford University Press, 262 pp.
Papadakis, J., 1975. *Climates of the World and Their Potentialities.* Buenos Aires: published by author, 200 pp.
Phillips, D.W. and G.A. McKay (eds), 1981. *Canadian Climate in Review.* Ottawa: Canadian Government Publishing Centre, 104 pp.
Riehl, H., 1979. *Climate and Weather in the Tropics.* New York: Academic Press, 611 pp.
Rudloff, W., 1981. *World-Climates, with Tables of Climatic Data and Practical Suggestions.* Stuttgart: Wissenschaftliche Verlagsgesellschaft, 632 pp.
Rumney, G.R., 1968. *Climatology and the World's Climates.* New York: Macmillan, 658 pp.
Visher, S.S., 1944. *Climate of Indiana.* Bloomington, IN.: Indiana University, 511 pp.
Thompson, B.W., 1965. *The Climate of Africa.* London: Oxford University Press, 151 pp. and 132 maps.

Compilations and statistical summaries

The standard references for world climatic tables are in the series *World Weather Records.* Data from the earliest available dates through 1940 were published as *Miscellaneous Collections,* volume 79, 1927; volume 90, 1934; and volume 105, 1947, by the Smithsonian Institution, Washington, DC. The US Weather Bureau continued the series with a single volume for the decade 1941–1950 and six volumes for 1951–1960. The first of six additional volumes for 1961–1970 was published by the National Climatic Center in 1979. Under the sponsorship of WMO the National Climatic Center issues a continuing summary of surface and upper-air data for world stations entitled *Monthly Climatic Data for the World.*

Since first publication in 1958 the British Meteorological Office has periodically revised *Tables of Temperature, Relative Humidity and Precipitation for the World* in six parts, covering major divisions of the world.

Average Climatic Water Balance Data of the Continents, containing water budget tables for an extensive world network, was published in eight parts from 1962 to 1965 by C.W. Thornthwaite Associates, Laboratory of Climatology, Elmer, New Jersey. Water budget data also are found in A. Baumgartner and E. Reichel, *The World Water Balance, Mean Annual Global Continental and Maritime Precipitation, Evaporation and Runoff,* 1975, published by Elsevier Scientific Publishing Company, Amsterdam.

Member nations of the WMO prepare a wide variety of monthly, annual and occasional summaries for their respective areas. Current indexes and lists can be obtained from the appropriate governmental service agency. Major series issued by the National Climatic Data Center for US stations are *Climatological Data, National Summary,* monthly and annual; *Climatological Data* (by state), monthly and annual; *Local Climatological Data* (by station), monthy; *Hourly Precipitation Data* (by state), monthly; *Solar Radiation Data, Monthly Summary* (1977–1980); and *Storm Data,* monthly. Long-term data summaries appear in series such as *Climates of the States, Climatography of the United States,* and *Decennial Census of United States Climate.*

Unpublished records

Government meteorological centers and affiliated service agencies are the definitive sources of information on geographic coverage, time series, formats and availability of unpublished official climatic data. Local observers do not consistently maintain file copies of station records, which commonly are forwarded to central archives, especially if the station is part of an official network. In addition to paper or microfilm copies, formats of unpublished records may include instrument charts, telemetry printouts and various presentations of computerized data. Many holding centers can provide current indexes of unpublished data and advise users on available formats. Access to records held in the private sector ordinarily requires prior arrangement with the responsible agent.

Howard J. Critchfield

Cross references

Climate and climate change
Climate change and the greenhouse effect

CLIMATE FORECASTING: MONTHLY AND SEASONAL

Although theoretically guided projections of future states of the climate over a variety of time spans are now being tried, regularly issued forecasts by tested methods are confined to the immediately forthcoming month or season. They are often called long-range weather forecasts instead of climatic forecasts. The semantic ambiguity is a natural consequence of the fact that weather events shade into and overlap with climatic phenomenon on these time scales.

The predictibility of daily weather events fades to low levels by 5 or 6 days, but a 5-day average can be predicted with some – rather variable – skill over the range 6–10 days in advance. That product is called a medium-range forecast. Direct calculation with a dynamical model of the atmosphere underlies these short and medium-range forecasts. On the monthly or seasonal time scale, few predictions attempt to go beyond providing information about average temperature, total precipitation or other simple statistical characteristics of the weather. Systematic testing of the usefulness of the largest computational models of the atmosphere and oceans in the long-range forecast process has now become feasible, thanks to the growth of computer power, but the work is only beginning and its outcome remains uncertain.

Long-range forecasts may be aimed at specific locations (such as cities), geographical regions, nations or even a major part of the globe, depending on the user's requirements. A single area average is sometimes given for regions not exceeding a few thousand square kilometers, but a forecast map showing spatial variations is needed for larger areas.

A few of the forecasting methods now in use draw only on local measurements for their input data, but most employ data from a much larger area than the predictions will cover, for it is known that local weather can be affected within a few days by meteorological conditions thousands of kilometers away.

Current practice

According to a recent survey, long-range forecasts of some kind are being made by the national weather services of the United States, Canada, UK, France, Germany, The Netherlands, Portugal, Russia, East Germany, Poland, Czechoslovakia, Hungary, Bulgaria, Yugoslavia, India, Pakistan, China, Hong Kong, Burma, Indonesia, South Korea, Japan, Tunisia, Libya, Jordan, Madagascar, Jamaica, Guyana and Peru. In addition, private individuals and groups in some of these countries produce independent forecasts. The dissemination of forecasts varies greatly in mode and breadth from one nation to the next; some are published in full, others are restricted to a few official uses.

Methods

Because the long-range forecast problem has obdurately resisted all attempts at solution by quantitative physical theory, current methods of prediction – the modest survivors from a long history of generally fruitless research – are largely empirical in nature. The principal methods are as follows.

1. Statistical regression analysis and related techniques, based on time-lagged correlations between one or more predictor variables and the predictand, or on the autocorrelations of the predictand.
2. Contingency tables, in which the whole possible range of the predictand is divided into a few broad classes, the predictors also classified, and the frequency of occurrence found for each predictand class contingent on the prior occurrence of each possible combination of predictor classes.
3. Forward extrapolation of apparent periodicities or trends in the predictand record.
4. Kinematic extrapolation of recent rates of movement or of intensification of features on maps of the predictand.
5. Selection from the past of one or more *analogs* to the present situation, i.e. closely similar cases. The prediction may follow either the average or the most frequent subsequent development of the analog cases.
6. Forecaster's judgment, based on experience.
7. Forecaster's judgment, based on qualitative physical reasoning from current theoretical concepts.

Each of the above methods has often been abused, and each requires very cautious application to produce useful results. Only methods 1 and 2 are wholly objective, although 3, 4 and 5 can be made so. Most of the forecast services use a combination of several.

Accuracy

Predictions are said to have positive 'skill' if, on average, they show greater accuracy than would a series of control forecasts generated by random drawing from the climatological frequency distribution of the predictand. Forecasts generally are not considered useful unless they also maintain some margin of accuracy above another set of control forecasts generated by simple persistence (repetition) of the preceding value of the predictand.

Because of the great variety of ways in which long-range forecasts are expressed and of the uses to which they are put, it is difficult to rate their skill adequately or fairly with single numbers. If one must attempt it, nevertheless, the simplest common denominator is perhaps a score based on an even two-way choice, for instance above or below normal. Reduced to these terms, monthly and seasonal temperature forecasts in temperate latitudes would be right about 60% on average, while precipitation forecasts would rate less than 55%. All of these scores are higher than could be obtained by predicting values of the previous period to persist unchanged. A contributing reason for the lower scores for precipitation is the fact that rainfall occurs in more complex and broken spatial patterns than temperature, i.e. local precipitation totals over a period are sensitive to the paths and intensities of individual storms to a far greater degree than are average temperatures. It now appears that tropical rainfall may be significantly more predictable than extratropical rainfall, at least on the seasonal time scale. The strong interaction between ocean and atmosphere in the tropics produces this effect, but consequences of the interaction outside the tropics are less predictable (Rasmusson and Wallace, 1983; Shukla and Paolino, 1983; World Meteorological Organization, 1983).

Estimates of accuracy can be conveyed to forecast users most helpfully by expressing forecasts in terms of probabilities. This practice, long established in short-range weather forecasting, is just beginning in monthly and seasonal prediction. The US forecasts show isolines of probability on the maps, allowing forecasters to express variations in confidence according to place, season and occasion.

Physical theory

Some physical concepts, such as the length of stable waves in the westerly winds aloft that circle the globe in middle latitudes, the cooling effects of anomalous snow cover, and the contribution of unusually warm sea surface temperatures to increased cyclogenesis, are employed to a degree in long-range forecasting, qualitatively. However, an accepted quantitative physical theory of long-range forecasting cannot be said to exist.

There does exist a family of mathematical models of the global atmosphere, incorporating (with various simplifications and omissions) the differential equations for the conservation of momentum, energy and mass, plus the equation of state. Given initial conditions and boundary conditions, one can integrate a set of these equations numerically with respect to time, obtaining a step-by-step (or iterative) sequence of predictions. This kind of direct application of classical fluid dynamics and thermodynamics is being used successfully in short-range and medium-range forecasting. With the addition of radiative, latent heating and frictional processes, and after many time iterations, one obtains a dynamical prediction of certain aspects of climate, such as the normal wind and temperature fields, and the normal transport and transformations of momentum and energy.

The models of the general circulation, or some further development from them, would seem to provide natural generators of long-range forecasts. They are still incomplete in a physical sense, coarse in spatial resolution and subject to some cumulative computational errors, but these are deficiencies that are being overcome by experience and faster computers. More difficult problems arise with respect to the boundary and initial conditions. Some of the boundary conditions cannot be specified in advance, but may depend in part on the instantaneous or recent internal state of the atmosphere; good examples are the dependence of the radiation balance on cloudiness and on snow cover. Such variable feedback effects require very careful modeling if the computed system is to remain stable and produce realistic results. In addition, even the most sophisticated model must begin with initial conditions, which cannot be measured everywhere or with absolute accuracy. This inevitable incompleteness of knowledge of the state of the atmosphere at any one moment will cause a dynamical prediction beginning at that moment to lose all of its accuracy with respect to daily evolution after a week or two. The practical consequence of this result is the empirically long-recognized requirement that long-range forecasts deal only with gross statistics of the weather, such as average values and probability distributions.

The quest for physical insight into climatic phenomena of rather short time scales is now attracting many researchers (Nicholls, 1980; Madden, 1981; Namias and Cayan, 1981; Shukla, 1981; Simmons, 1982; Miyakoda *et al.*, 1983; Rasmusson and Wallace, 1983; Simmons *et al.*, 1983; World Meteorological Organization, 1983, 1984; Blackman *et al.*, 1984). Our best hopes for improving predictive models and predictions depend on the success of this quest.

Donald L. Gilman

Bibliography

Blackman, M.L., Y.-H. Lee, J.M. Wallace and H.-H. Hsu, 1984. Time variation of 500 mb height fluctuations with long, intermediate and short time scales as deduced from lag-correlation statistics, *J. Atmos. Sci.*, **41**, 981–991.
Gutzler, D.S. and J. Shukla, 1984. Analogs in the wintertime 500 mb height field, *J. Atmos. Sci.*, **41**, 177–189.
Madden, R.A., 1981. A quantitative approach to long-range prediction, *J. Geophys. Res.*, **86**, 9817–9825.
Miyakoda, K., T. Gordon, R. Caverly, *et al.*, 1983. Simulation of a blocking event in January 1977, *Monthly Weather Rev.*, **111**, 846–869.
Namias, J. and D.R. Cayan, 1981. Large-scale air–sea interactions and short-period climatic fluctuations, *Science*, **214**, 869–876.
Nicholls, N., 1980. Long-range weather forecasting: Value, status and prospects, *Rev. Geophysics Space Physics*, **18**, 771–778.
Rasmusson, E.M. and J.M. Wallace, 1983. Meteorological aspects of the El Nino/Southern Oscillation, *Science*, **222**, 1195–1202.
Shukla, J., 1981. Dynamical prediction of monthly means, *J. Atmos. Sci.*, **38**, 2547–2572.
Shukla, J. and D.A. Paolino, 1983. The Southern Oscillation and long range forecasting of the summer monsoon rainfall over India, *Monthly Weather Rev.*, **111**, 1830–1837.
Simmons, A.J., 1982. The forcing of stationary wave motion by tropical diabatic heating, *Q. J. Royal Meteorol. Soc.*, **108**, 503–534.
Simmons, A.J., J.M. Wallace and G.W. Branstator, 1983, Barotropic wave propagation and instability, and atmospheric teleconnection patterns, *J. Atmos. Sci.*, **40**, 1363–1392.
World Meteorological Organization, 1983. *Proceedings of the WMO/CAS Expert Study Meeting on Long-Range Forecasting*. WMO Long-range Forecasting Res. Pub. Ser. No. 1. Geneva: World Meteorological Organization.
World Meteorological Organization, 1984. *Long-range Weather Forecasting: Recent Research*, WMO Long-range Forecasting Res. Pub. Ser., No. 3. Geneva: World Meteorological Organization.

Cross references

Climate and climate change
Climate data: sources

CLOUDS (CLOUD SEEDING)

Humanity seeks to change the weather and climate to alleviate storm damage, but also to secure the food supply for an ever-growing world population (List, 1987). In practice this means modifying the tracks of hurricanes, preventing the occurrence of damaging hail or heavy rain, or increasing rainfall and snowfall in areas where this can lead to increased runoff for irrigation and water supply generally. All these

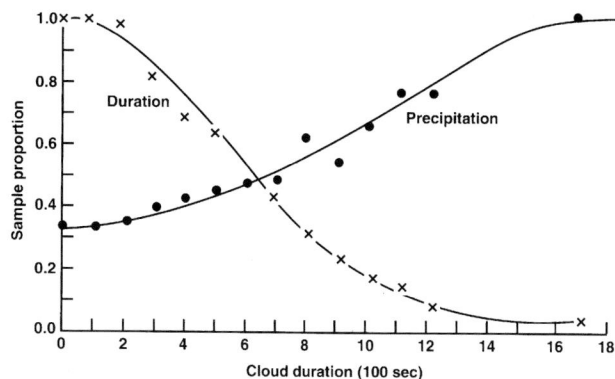

Figure C21 Probability that the supercooled portion of cumulus clouds will last for at least as long the period on the abscissa, and the probability that natural precipitation will develop in clouds of corresponding lifetime. (From Braham, 1986.)

measures are based upon the understanding of the process of cloud and precipitation formation.

Theoretical basis

It is impractical to modify weather by competing directly with the energies prevailing in the atmosphere, except locally (WMO, 1986). Best (1957) points out that to evaporate 1 km² of a cloud 600 m deep containing 0.1 g m^{-3} of liquid water would require 1.5×10^{18} erg. This calculation neglects the additional heat required to raise the temperature of the air to accommodate the extra water vapour thus produced.

Weather modification techniques attempt to use points of instability in weather systems, whereby a relatively small disturbance will have substantial impact on the natural evolution. The aim is to reduce evaporative loss from the natural system, thus providing more water for the production of precipitation. The production of latent heat will generate buoyancy and release potential instability, thereby causing convection. To produce these effects it is necessary either to provide artificial nuclei or to chill the air by means of dry ice (frozen carbon dioxide) pellets or other chilling substances. Usually silver iodide particles are used to provide suitable artificial nuclei. These particles become active as freezing nuclei at $-5°C$. Other materials used include lead iodide, organic substances such as metaldehyde and, for warm clouds, hydroscopic materials such as common salt. The number of effective nuclei varies markedly with temperature.

In general, it is believed that clouds are inefficient precipitation producers because of a lack of natural ice nuclei. If ice-nucleating material is introduced to convective clouds, then more of the cloud water will be converted to ice. Since ice crystals will grow more rapidly than water droplets, they eventually become of a sufficient size to fall out of the cloud to the ground as rain or snow depending upon the height of the melting level. Unfortunately, from the point of view of assessing the effectiveness of seeding, most cumulus clouds do not last long enough for precipitation embryos induced by seeding to attain sizes capable of reaching the ground. Those embryos which do reach the ground also have a high probability of forming natural precipitation anyway, as shown in Figure C21.

Assessment

Most methods of evaluating cloud-seeding experiments in current use are based upon a comparison of similar treated events, either historical or concurrent (Williams and Elliott, 1985). However, the high variability of precipitation makes the definition of incremental increases of the same magnitude as the variability quite difficult. In spite of this, some limited success is claimed, particularly in Israel, and work is continuing in several countries. In the USA experiments are being assessed by the National Oceanic and Atmospheric Administration (NOAA) in six states; Arizona, Utah, Nevada, North Dakota, Illinois and Texas (Reinking, 1992). So far the results are inconclusive. However, it is anticipated that the assimilation of high-resolution Doppler weather radar (WSR-88D) and wind profiler data into mesoscale cloud models will permit highly detailed real-time predictions of cloud and precipitation development with more accurate estimates of the transport and dispersion of seeding plumes.

C.G. Collier

Bibliography

Best, A.C., 1957. *Physics in Meteorology*, Pitman, London, 159 pp.
Braham, R.R., 1986. The cloud physics of weather modification, *WMO Bulletin*, **35**, 3, 215–222.
List, R., 1987. The need for weather and climate modification, *Proc. Int. Conf. on Agromet.*, Casena, 8–9 October, ed. F. Prodi, F. Rossi and G. Cristoferi, Fondazione Cesena Agricultura, pp. 61–86.
Reinking, R.F., 1992. The NOAA Federal/State cooperative program in atmospheric modification research: a new era in science responsive to regional and national water resource issues, *Preprints, Symp. on Planned and Inadvertent Weather Modification*, AMS, Boston, MA, 136–144.
Williams, M.C. and Elliott, R.D., 1985. Weather modification, in *Facets of Hydrology*, Vol II, ed. J.C. Rodda, John Wiley, Chichester, pp. 99–129.
WMO, 1986. Review of the present status of weather modification. Statement adopted by the Executive Council at its Thirty-Seventh Session, *WMO Bulletin*, **35**, 2, 140–144

Cross references

Climate and climate change
Climate forecasting: monthly and seasonal

COLEBROOK–WHITE EQUATION

The Colebrook–White equation is a complex equation applied to flow in pipes and used often in drainage design, especially in the design of stormwater sewers. The user is advised to use the *Tables for the Hydraulic Design of Pipes* published by the (UK) Hydraulics Research Station (now HRL plc) through Her Majesty's Stationery Office (publication author P. Ackers). The use of the tables includes a decision upon the type of pipe to be used, with a consequential assessment of the internal roughness coefficient for the pipe material.

The Colebrook–White equation for the mean velocity in pipes is:

$$\bar{v} = -\sqrt{32 g_n R_h S}\, \log_{10}\left(\frac{k}{14.8 R_h} + \frac{1.255\gamma}{R_h \sqrt{32 g_n R_h S}}\right)$$

where g_n is the acceleration due to gravity, k is the roughness height, R_h is the hydraulic radius, S is the energy gradient; $\bar{v}$ is the mean velocity and γ is the kinematic viscosity of the fluid.

P.G. Holland

Bibliography

BS 3680, Part 1, 1991. *Glossary of Terms*: British Standards Institution, London.

Cross reference

Water resources: dictionary of basic terms

COMPUTER MODELS

The advent of highspeed digital computers during the 1960s and 1970s contributed greatly to the development of computer-based models in hydrology and water resource analysis. Such models typically deal with large data sets consisting of time series of hydrological and water resource variables (rainfall, runoff, evaporation, water demand and use, etc.) which were almost impossible to analyze before the availability of digital computers. With the arrival of cheaper, more readily accessible, personal computer hardware and the inevitable increases in their power over the 15 years since the early 1980s there has been an associated increase in the number and range of computer models. Today, just about every aspect of hydrology and water resource planning and management has been reduced to a computer model by someone in some part of the world. The range of available computer models can be broadly classified into several groups according to their main areas of application:

- models for estimation of historical events and time series;
- models for prediction and forecasting future events;
- models for planning and design;
- models for operational control and management;
- models for teaching and academic research purposes.

Computer models have been developed to simulate atmospheric circulation systems, catchment hydrological processes, channel and estuarine hydraulic processes, groundwater movement, catchment sediment and water quality dynamics. They have also been developed to simulate the behavior of rivers and reservoirs under defined water abstraction patterns, to design and manage urban water reticulation, stormwater and effluent removal systems, for scheduling irrigation water applications and as a basis for a wide range of decision support systems in water resource management. Systems models of the hydrology and water resource exploitation of complete drainage basins are frequently used for integrated water resource planning and management. These 'models' often represent combinations of several smaller subcomponent models. Models for operational control are used for flood forecasting and water supply management and make use of 'real time' data inputs of the condition of the system being modeled so that short-term future predictions can be made as a basis for decision making. Teaching and research models are often developed in association with detailed experiments of hydrological processes in the field and can be the tools whereby advances are made in the development of more practically orientated models.

The emphasis today is often on the 'packaging' of the software and the ease with which users can apply the models to solve problems. Many of the hydrological and water resource analyses that form part of the models require large amounts of data to be processed and some software developers have provided interfaces with powerful database, GIS (geographical information systems) and DTM (digital terrain model) software. Such approaches encourage model users to take advantage of the best available information, which may be neglected if it is too difficult to incorporate into the model.

While there is little doubt that the wide range of models available has made an important contribution to the development of hydrology as a science and to the practical solution of water resource problems, certain reservations have been expressed by some hydrologists. Inevitably, any computer model is a simplification of reality and some of the reservations relate to making the models readily available to users who are less than familiar with the implications of such simplifications and the limitations of the model results. The implication is that the results from computer models should be carefully interpreted and not simply accepted as the 'truth' which requires no validation. The opinion has also been expressed that research into model development has proceeded at the expense of research into the understanding of how hydrological processes operate. Thus James (1991) refers to the science of hydrology as the 'supply side' and the application of hydrology as the 'demand side'. He argues that too much of the development of hydrological science has been driven by the demand for practical technologies rather than the search for scientific understanding. One implication is that many of the available computer models are poor reflections of reality, because they were developed in the absence of an adequate understanding of that reality. To a certain extent the increase in the power of the available computer hardware has tended to offset this, and it is now more practical to include detailed and distributed formulations of hydrological processes into computer models (Abbott *et al.*, 1986).

Denis A. Hughes

Bibliography

Abbott, M.B., J.C. Bathurst, J.A. Cunge *et al.*, 1986. An introduction to the European Hydrological System – Système Hydrologique Européen, 'SHE', i: History and philosophy of a physically-based, distributed modelling system. *J. Hydrol.*, **87**, 45–59.
James, L.D., 1991. Hydrology: infusing science into a demand-driven art, in D.S. Bowles and P.E. O'Connell (eds), *Recent Advances in the Modelling of Hydrologic Systems*, Dordrecht: Kluwer Academic, 31–43.

Cross references

Modelling of water resources systems
Model predictions: uncertainty
Models: parameter extermination

CONVEYANCE

The water-carrying capacity of an open channel expressed in terms of discharge and the bed slope:

$$K = \frac{Q}{\sqrt{S}}$$

where K is the conveyance (sometimes known as the conveyance factor), ($\text{m}^3\,\text{s}^{-1}$), Q is the discharge ($\text{m}^3\,\text{s}^{-1}$) and S is the bed slope.

P.G. Holland

Bibliography

BS 3680, Part 3A, 1980. *Velocity Area Methods*, British Standards Institution, London.

Cross reference

Water resources: dictionary of basic terms

CORRELATION COEFFICIENT

The correlation coefficient is a measure of the degree of linear association between two variables which have a bivariate normal distribution. When the correlation coefficient is positive one variable tends to increase as the other increases, when it is negative one variable tends to decrease as the other increases. The correlation coefficient will be computed as the normalized covariance of two variables x, y:

$$r = \frac{\text{cov}(x,\ y)}{s_x s_y} = \frac{\Sigma(x_i - \bar{x}) \cdot (y_i - \bar{y})}{\sqrt{[\Sigma(x_i - \bar{x})^2\ \Sigma(y_i - \bar{y})^2]}}$$

$$= \frac{n\,\Sigma x_i y_i - \Sigma x_i \Sigma y_i}{\sqrt{[n\Sigma x_i^2 - (\Sigma x_i)^2]\,[n\Sigma y_i^2 - (\Sigma y_i)^2]}}$$

Its value lies between $+1$ or -1. If the absolute value of r is 1 the linear relationship between both variables is perfect. If r is zero the variables are independent or uncorrelated. The correlation coefficient measures the degree of relationship in a statistical sense. It does not in any way guarantee the existence of a cause-and-effect relationship. The correlation coefficient is unaffected by linear transformations of one or both variables. The degree of a relationship between two variables expressed by the correlation coefficient can be tested for its statistical significance. The null hypothesis ($r = 0$) is that the variable y is independent of the variable x. This hypothesis will be rejected if the absolute value of the test statistic t is greater than the quantile of Student's t-distribution with $(n - 2)$ degrees of freedom and a probability of exceedence of $\alpha/2$ where n is the number data pairs used to estimate r and α is the level of significance. The test statistic t is computed with the relationship

$$t = \frac{r\sqrt{(n - 2)}}{\sqrt{(1 - r^2)}}$$

A.H. Schumann

Cross reference

Water resources: dictionary of basic terms

CURRENT METERING

Current metering is the process of using one or more current meters to assess the velocity component of the discharge determination for water in an open channel or in a pipe. The current meter can be used in a variety of ways; the more usual forms of use are described in the following paragraphs.

From a boat

This method is used where an open channel is too wide for a cableway installation or too deep for wading. The current meter can be suspended from a hand-held line but, generally, greater accuracy and instrument stability is achieved by the use of a wading rod (see By wading, below). There are safety considerations to be taken into account, especially at high velocities, and great care has to be taken to assess the exact location of the boat in relation to the configuration of the open channel.

From a bridge

Although a bridge can provide convenient, ready-made access to an open channel, the bridge abutments and especially any bridge piers can distort the velocity profile of a stream to the extent that the velocity measurements taken are not entirely representative of the general stream velocity. However, this form of use is advantageous

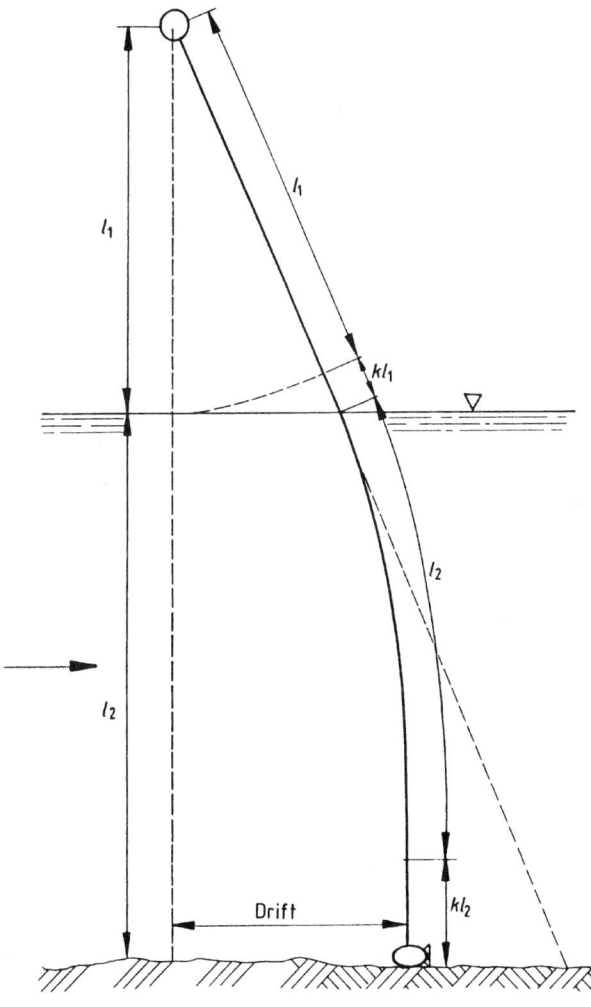

Key

kl_1 is the *air line correction*;

kl_2 is the *wet line correction*;

→ is the direction of the flow;

∇ is the level of the water.

Figure C22 Sounding line corrections.

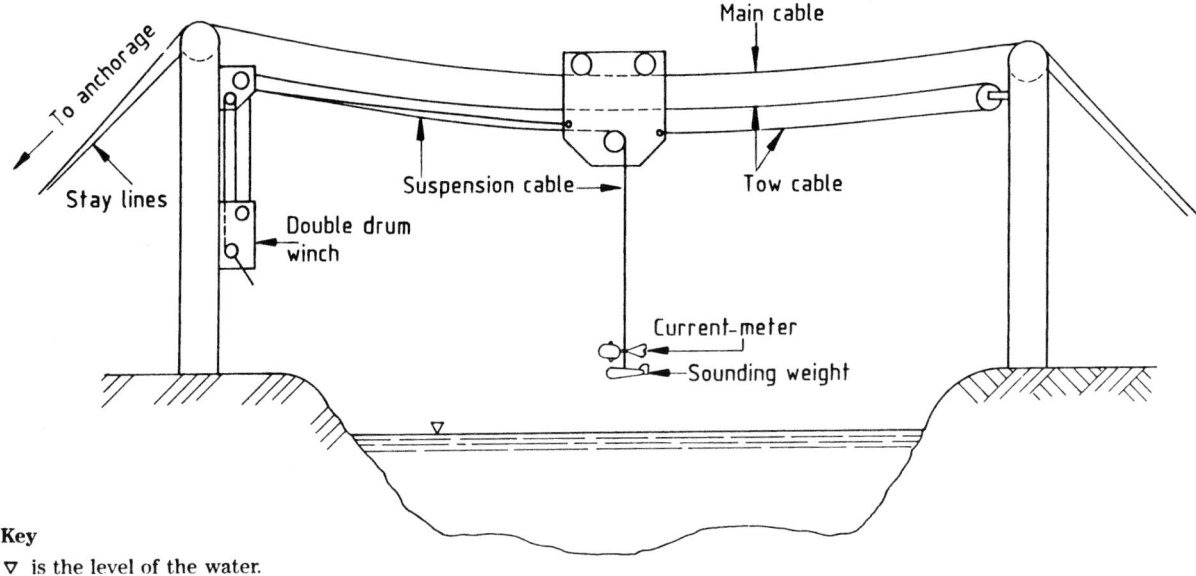

Key

▽ is the level of the water.

Figure C23 Cableway system: unmanned instrument carriage.

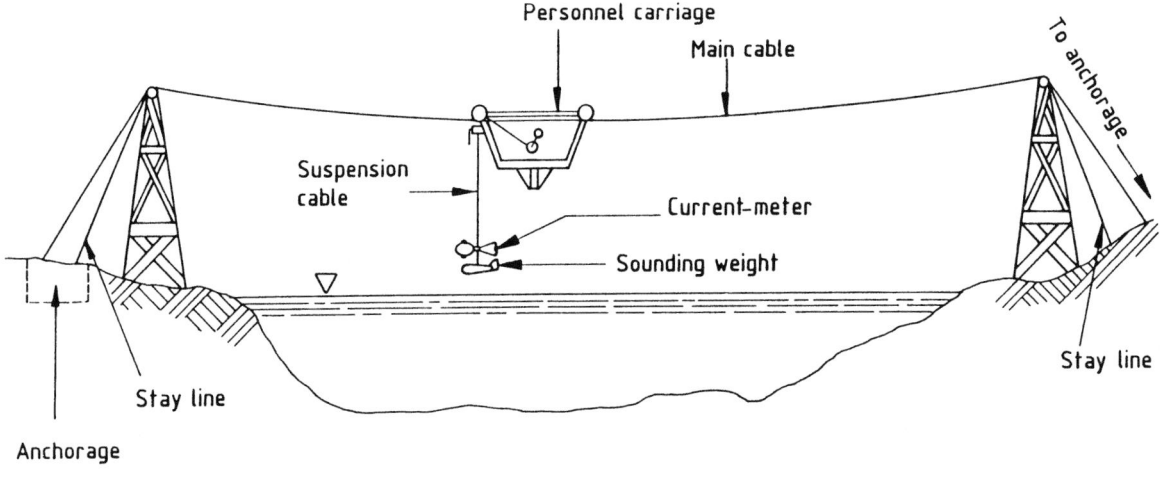

Key

▽ is the level of the water.

Figure C24 Cableway system: manned instrument carriage.

when wading is not practicable or when there is no good cableway site. It is particularly useful, for exploratory *ad hoc* assessments and for determining the general values of flood velocities. There are various types of small gantry-type cranes in use for suspending the current meter cable vertically by means of a sounding line at a short distance out from, and clear of, a bridge parapet, but a similar suspension can be achieved by hand-held means. The usual suspension mode is for measuring on the downstream side of the bridge, but at high stream velocities the current meter can drift downstream to the extent that the sounding line is no longer vertical but adopts a catenary profile. As the length of the sounding line is the effective means of determining the vertical position of the current meter, the real length of a catenary sounding line is longer than the true vertical depth of the current meter below the origin of the sounding line. Consequently, air line and wet line corrections have to be made, as shown in Figure C22.

From a cableway

This is a common means of suspending a current meter, generally at a permanent location, into an open channel. Also, there are forms of temporary, portable, demountable cableways in use for less permanent locations. A cableway is installed mainly where long-term routine measurements are needed at locations where the water in an open channel is deep or the stream velocity therein is too high for wading in safety. The essentials of the two types of cableway in use generally are shown idealistically in Figures C23 and C24, but actual designs vary according to manufacturer. Whatever type is used, the winch mechanisms must incorporate means of assessing the vertical depths at which the current meter is suspended and the horizontal distances of the current meter from some on-bank benchmark. For wide open channels it may be necessary to correct the distances measured to take into account the cable sag, caused by self-weight and the weight of the instrument carriage.

a.

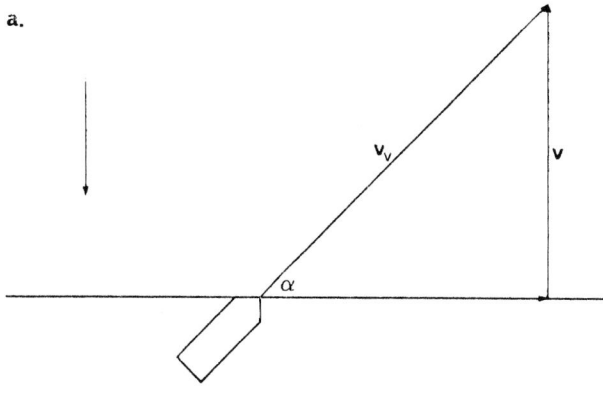

b.

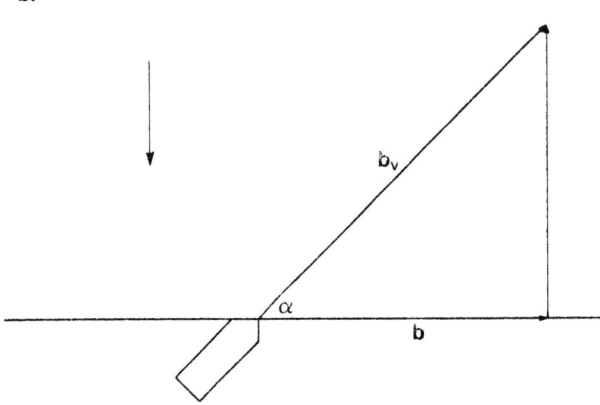

Figure C25 General diagrams of velocity vectors. (a) River velocity computation at sampling stations: α, angle of current meter relative to section line; v_v combined boat and river velocity; v, river velocity; $v = v_v \sin \alpha$. (b) Computation of segment width between sampling stations: α, as above; b_v relative distance of travel; b, actual distance of travel; $b = b_v \cos \alpha$. (Courtesy of Environment Canada.)

From a moving boat

This method is used for very large and wide open channels whereby the discharge is determined from a boat traversing the stream along the line of the measuring cross-section so that the stream velocity can be measured continuously by current meter, with simultaneous measurements of the depth of water and of the distance travelled by the boat. The gathered data is transferred to vector diagrams, as shown in Figure C25, for the computation of stream velocities as components of the discharge determination.

By wading

This can be the most accurate method of measurement of stream velocity in open channels, but its use, generally, for safety reasons, is restricted to water depths not exceeding upper thigh height, water velocities permitting the wader to stand still, and channel bed properties sufficiently firm to avoid the wader sinking unduly into the bed material. The current meter is fixed to a hand-held rod and there is an electrical circuit between the current meter and a pulse counter which records the number of rotations of the current meter rotor over a measured time span, from which the stream velocity can be deduced.

Where the water surface is covered by an ice sheet which cannot be broken easily, special precautions have to be taken and special instrument assemblies utilized to enable a current meter to be used in association with accurate means of assessing the depth of water below the ice.

P.G. Holland

Bibliography

BS 3680, Part 1, 1991. *Glossary of Terms*, British Standard Institution, London.

Cross reference

Water resources: dictionary of basic terms

CYANOBACTERIA (BLUE-GREEN ALGAE)

Cyanobacteria (blue-green algae) occur naturally in freshwater environments including lakes, reservoirs, ponds and slow-flowing rivers and estuaries. Under enhanced conditions the cyanobacteria may proliferate and become an unsightly and often poisonous bloom. According to Reynolds (1987) cyanobacterial blooms have done more to give eutrophication its bad name than any other consequence of lake enrichment. The most common bloom-forming genera of cyanobacteria include *Microcystis*, *Anabaena*, *Nodularia* and *Oscillatoria*.

Biological characteristics

Cyanobacteria are photosynthetic organisms and as such require light (400–700 nm wavelength) and a supply of inorganic carbon to grow. The photosynthetic rate is also affected by nutrient level (particularly phosphorus), water temperature and water chemistry. For *Microcystis*, the maximum photosynthetic rate of 28.9×10^{-6} (mol C) (mol cell C)$^{-1}$ s^{-1} at 20°C is achieved at a photon flux of 753×10^{-6} mol photon m^2 s^{-1} (Reynolds, 1990). Certain cyanobacteria such as *Oscillatoria agardhii* are able to increase their photosynthetic efficiency in persistent low light (Reynolds, 1990). The optimum water temperature range for cyanobacterial growth is 25–35°C (Reynolds and Walsby, 1975); blooms generally occur in hard water (Pearsall, 1932) within a pH range of 7.5–9.0 (Kratz and Myers, 1955; Brock, 1973).

The maximum growth rate of *Microcystis* at 20°C is 5.5×10^{-6} s^{-1} (Reynolds, 1990). If photosynthetic production is greater than this figure, then surplus carbohydrate is stored as ballast. An excess carbon production of 9×10^{-6} (mol C) (mol cell C)$^{-1}$ s^{-1} will increase cell density by 1.62×10^{-3} kg m^3 s^{-1}. In darkness, stored carbohydrate is used to meet respiratory demand [0.55×10^{-6} (mol C) (mol cell C)$^{-1}$ s^{-1}] resulting in density loss. A resultant characteristic of bloom-forming cyanobacteria is that their density varies diurnally above and below that of water (998.2 kg m^{-3} at 20°C) resulting in oscillatory vertical migration patterns. If sufficient density is lost then the cyanobacteria may migrate to the lake surface and form a concentrated bloom at or near the shoreline. This process is aided by the presence of rigid gas vesicles that give the cyanobacteria additional buoyancy. In some genera such as *Anabaena*, collapse (due to turgor pressure) and light-mediated synthesis of gas vesicles also contributes significantly to density change (Walsby, 1988).

Potential problems

A surface bloom of cyanobacteria may persist if there is a shortage of carbon; without carbon, photosynthesis cannot occur, meaning that 'ballast' accumulation is not possible and the bloom does not sink. Under these conditions, or if the bloom has been transported by wind currents to the shoreline, the bloom may become accessible to water body users. This is potentially dangerous because about 70% of blooms release toxins that can kill animals and cause illness in humans. The problems and occurrence of toxic cyanobacterial blooms were reviewed by, for example, NRA (1990), Lawton and Codd (1991) and Howard (1994). In 1990 and 1991, six dogs died at Loch Insh, Scotland, after ingesting toxin-contaminated water. Observations suggested that the dogs actually chose to drink from the bloom-affected water rather than from clean water located nearby – an apparent 'fatal attraction' to cyanobacteria (Codd *et al.*, 1992). Previously two soldiers had become seriously ill in 1989 after canoeing in Rudyard Lake, England, where a large *Microcystis* bloom was present. By 1989, 16 European countries had reported incidents involving toxic cyanobacteria (NRA, 1990). In Australia, Soong *et al.* (1992) described illness associated with exposure to toxic *Nodularia* blooms in Lake Alexandrina and Lake Albert in 1991. Such occurrences are not isolated geographically; it is a worldwide water quality problem that may increase as climatic change makes more water environments suitable for cyanobacterial growth.

The long-term amelioration of the problem will require integrated catchment management policies to reduce eutrophication. Artificial mixing devices installed in water bodies can reduce growth by altering the underwater light environment; in some cases however this may promote accelerated growth of diatoms and green algae. Short term solutions such as chemical algicide should be avoided as application may be detrimental to desirable flora and fauna and may cause a large pulse of growth-promoting nutrient and toxin to be released at once.

Alan Howard

Bibliography

Brock, T.D., 1973. Lower pH limit for the existence of blue-green algae: evolutionary and ecological implications, *Science*, **179**, 480–483.

Codd, G.A., Edwards, C., Beattie, K.A. *et al.*, 1992. Fatal attraction to cyanobacteria?, *Nature*, **359**, 110–111.

Howard, A., 1994. Problem cyanobacterial blooms: explanation and simulation modelling, *Transactions of the Institute of British Geographers*, **19**, 213–224.

Kratz, W.A. and J. Myers, 1955. Nutrition and growth of several blue-green algae, *American Journal of Botany*, **42**, 282–287.

Lawton, L.A. and G.A. Codd, 1991. Cyanobacterial (blue-green algal) toxins and their significance in UK and European waters, *Journal of the Institution of Water and Environmental Management*, **5**, 460–465.

NRA (National Rivers Authority), 1990. *Toxic Blue-green Algae*, Water Quality Series No. 2. London: National Rivers Authority.

Pearsall, W.H., 1932. Phytoplankton in the English Lakes II. The composition of the phytoplankton in relation to dissolved substances, *Journal of Ecology*, **20**, 241–262.

Reynolds, C.S., 1987. Cyanobacterial water blooms, in *Advances in Botanical Research*, Vol. 13 (ed. J.A. Callow). London: Academic Press, pp. 68–145.

Reynolds, C.S., 1990. Temporal scales of variability in pelagic environments and the response of phytoplankton, *Freshwater Biology*, **23**, 25–53.

Reynolds, C.S. and A.E. Walsby, 1975. Water blooms, *Biological Reviews*, **50**, 437–481.

Soong, F.S., Maynard, E., Kirke, K. and C. Luke, 1992. Illness associated with blue-green algae, *Medical Journal of Australia*, **156**, 67.

Walsby, A.E., 1988. Mechanisms of buoyancy regulation by planktonic cyanobacteria with gas vesicles, in *The Cyanobacteria* (eds P. Fay and C. Van Baalen). Amsterdam: Elsevier, pp. 377–392.

Cross references

Algal growth on lakes
Black Sea environment

D

DAMS

A dam is an engineering structure constructed across a valley or natural depression to create a water storage reservoir. Such reservoirs are required for three main purposes: (1) provision of a dependable water supply for domestic and/or irrigation use, (2) flood mitigation and (3) generation of electric power.

In providing a water supply, the reservoir storage is filled during the periods of above-average streamflow, thus ensuring a steady supply of water during periods of little or no streamflow. For flood mitigation, the storage reservoir is kept nearly empty during drought and periods of low rainfall, so that when the flood-generating rainstorms occur, the storage volume available in the reservoir provides a buffer against severe flooding in the river valley downstream of the dam. For power generation, the storage reservoir provides a head of water upstream of the dam, and the potential energy of this water is converted first to kinetic energy by passing the water through turbines, and then to electrical energy by generators.

A large dam has two essential requirements. First, it must be reasonably watertight. Therefore, either the dam is constructed of impermeable material (e.g. concrete), or it incorporates an impermeable membrane in its structure (e.g.) an earth core. Also, the dam foundations must be made watertight by grouting or other means if necessary. Second, the dam must be stable. Movement and deformation of the dam and its foundations cannot be eliminated, but they must be predicted and allowed for in the design.

Because of these requirements, the location and design of dams are invariably influenced to some extent by geological features. In many cases, geological factors such as foundation conditions and the proximity of construction materials are of overriding importance in determining the type of dam constructed at a given site. It therefore follows that a detailed knowledge of the geology of a prospective dam site and its environs is necessary before an informed decision can be made on the most suitable dam design and the estimated cost of construction.

The most important basic geological data required are the distribution and nature of the various rock types present in the area, the weathering profile and details of the structural geology. These data are obtained by a site investigation program that uses a wide range of data-gathering techniques, such as outcrop mapping, water pressure testing, geophysical surveys, joint surveys and laboratory testing of rock samples. The geological and physical data so obtained are then integrated to form a geomechanical model of the site, which provides the design engineer with a reasonably realistic and quantitative basis on which to design the dam and its associated structures. This process of collecting relevant geological data and presenting them in a form useful to the engineer is the main function of the engineering geologist (Attewell and Farmer, 1976; Bell 1980; Leggett, 1962; Paige, 1950).

Types of dams

In designing a dam for a particular site, the engineer has several basic types of dams to choose from (Thomas, 1976). Figure D1 summarizes the layout and significant characteristics of these basic dam types.

At some dam sites the most economical design has been a composite of two or more basic dam types. One particular type of composite concrete dam is the multiple arch design, which consists of several cylindrical arches supported by buttresses. This design is well suited to sites with geologically variable foundations – the buttresses are located on strong parts of the foundations, while the arches are located so that they bridge weak zones in the foundations.

Dam foundations

The foundations of a dam have to support the weight force of the dam, plus a significant component of the force that the reservoir of water exerts on the upstream side of the dam. For the dam types illustrated in Figure D1, there is a progressive decrease in the area of foundations, for a given dam height, from the earth dam (largest area) to the double-curvature arch dam (smallest area). This means that the bearing pressure (i.e. force per unit area) that must be supported by the foundations progressively increases from a minimum for an earth dam to a maximum for the arch dam; in other words the sequence of dam types from (1) to (6) in Figure D1 requires progressively stronger foundations. The overall strength of dam foundations is determined by the detailed geological characteristics of the area; it therefore follows that foundation geology at a proposed dam site is an important factor in deciding the most appropriate type of dam for the site (Walters, 1971; Wahlstrom, 1974).

Construction materials

Dams are constructed from large volumes of naturally occurring earth materials. For example, an earth-cored rockfill dam requires regolith with certain physical properties for the earth core, sand and gravel for filter zones and broken rock for rockfill; a concrete dam requires suitable broken rock or gravel for concrete aggregate, together with sand and cement. These construction materials must be obtained from as close to the dam site as possible if the dam is to be economically feasible. Therefore, the location and cost of extraction of construction materials constitute another important factor in determining the type of dam to be built. For instance, a site may have highly weathered bedrock that appears to be suitable only for the foundations of an earth dam. If there is little suitable earth material close to the site, however, it may well be cheaper to excavate the foundations to a depth suitable for a concrete dam than to transport earth material over a long distance to the site.

The location of adequate quantities of suitable construction materials is determined very much by the bedrock geology and the geological history of the area. It is therefore fairly obvious that a search for construction materials should be based on detailed geological mapping – both bedrock and surficial – of the area around

CROSS SECTION   PLAN   SALIENT FEATURES

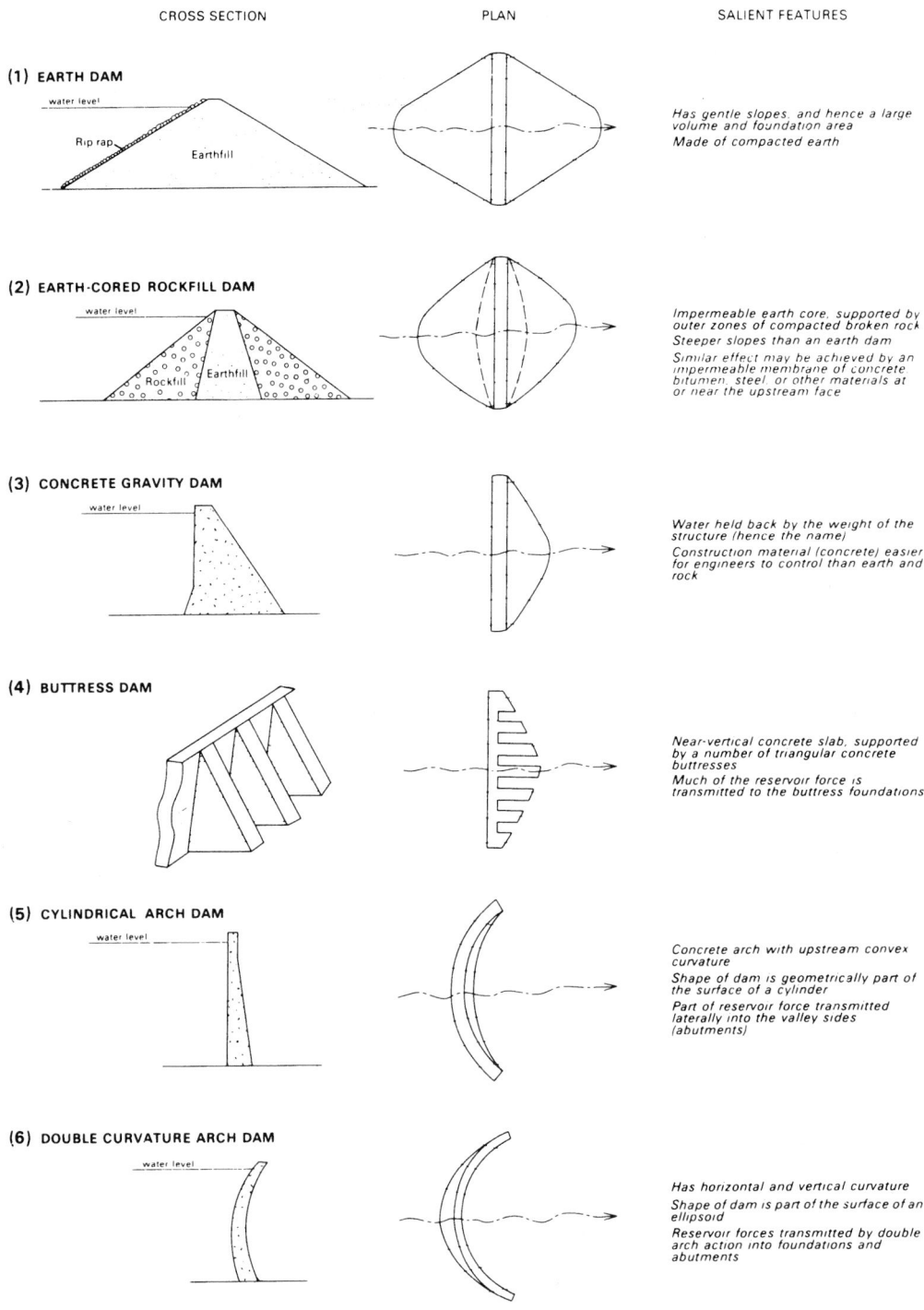

(1) EARTH DAM

Has gentle slopes, and hence a large
volume and foundation area
Made of compacted earth

(2) EARTH-CORED ROCKFILL DAM

Impermeable earth core, supported by
outer zones of compacted broken rock
Steeper slopes than an earth dam
Similar effect may be achieved by an
impermeable membrane of concrete
bitumen, steel, or other materials at
or near the upstream face

(3) CONCRETE GRAVITY DAM

Water held back by the weight of the
structure (hence the name)
Construction material (concrete) easier
for engineers to control than earth and
rock

(4) BUTTRESS DAM

Near-vertical concrete slab, supported
by a number of triangular concrete
buttresses
Much of the reservoir force is
transmitted to the buttress foundations

(5) CYLINDRICAL ARCH DAM

Concrete arch with upstream convex
curvature
Shape of dam is geometrically part of
the surface of a cylinder
Part of reservoir force transmitted
laterally into the valley sides
(abutments)

(6) DOUBLE CURVATURE ARCH DAM

Has horizontal and vertical curvature
Shape of dam is part of the surface of an
ellipsoid
Reservoir forces transmitted by double
arch action into foundations and
abutments

Figure D1 Basic types of dam design. (From Best, 1981.)

the dam site. However, the cost of extraction is based not only on the volume of the particular required material that is available, but also on the geometry of the deposit and its relationship to the ground surface. For instance, a quarry developed in a particular rock sequence to supply concrete aggregate will invariably involve the unavoidable extraction of a substantial volume of unsuitable material because of the geological variability (often caused by folding and faulting) of the quarry area. The expense of extracting and dumping such unusable material is unavoidable, and can be minimized only by a thorough

geological investigation of alternative sources of construction materials so that those finally selected are the most economical for dam construction.

Some idea of the volumes of construction materials required for large dams can be obtained from some dam statistics. The largest dam (by volume) yet constructed is Tarbela Dam in Pakistan, which is an earth dam containing $122 \times 10^6 \, \text{m}^3$ of materials. Several large dams have also been constructed in North America, such as those at Fort Peck, Montana ($96 \times 10^6 \, \text{m}^3$), Oahe, South Dakota ($70 \times 10^6 \, \text{m}^3$),

Gardiner, Saskatchewan, Canada (66×10^6 m³) and Oroville, California (60×10^6 m³). The largest concrete dam in the world is Sayany in the former USSR (9.1×10^6 m³), followed by Grande Coulee Dam in the United States (8.5×10^6 m³).

Choice of dam type

Once a site has been selected for a dam consideration has to be given to deciding which type of dam is most suited to the site. Some dam types may be impracticable because of the topography of the site; for instance, the valley may be too wide for the construction of an arch dam, or the valley sides may be too steep for an earth dam. However at any site several types of dam should be considered, and estimates of quantities of materials, cost of materials, amount of foundation excavation, amount and type of foundation treatment, cost of river diversion and spillway construction and so on, should be prepared for each type. The final choice should be the dam with the lowest estimated construction cost. In general, three factors control this final decision: (1) the topography of the dam site and reservoir area, (2) the strength and variability of the foundations and (3) the availability and suitability of construction materials. These factors are largely controlled by the geological structure and history of the site, and an informed decision requires a great deal of geological data analysis and interpretation, particularly for the second and third factors, presented in a manner that the engineer can use in design calculations.

The influence of topography and foundation geology on the selection of dam type is particularly evident where composite dams are constructed. For example, when studying the site for the Aviemore Dam in New Zealand, investigators discovered a fault beneath the river valley, with good dam foundations on one side of the fault and poor foundations on the other. Accordingly, it was decided to construct a concrete gravity dam, incorporating the power station, on that part of the site with good foundations, and an earth-cored rockfill dam on the poor foundations with the two dam types abutting against each other in the middle of the valley. Another example is the Rio Torto Dam in Italy, which has a concrete arch dam across the main river valley and a buttress dam and an earth dam section farther up the valley sides.

Site investigation for dams

The progress from an initial proposal to dam a river to the completion of construction of a dam is marked by a series of investigations of the site, using a wide range of techniques such as surveying and various engineering tests. However, the major component by far of a dam site investigation is the geological investigation of the site, the reservoir area and prospective sources of construction materials. The importance of geology in dam construction is evident in just about all technical reports prepared during the investigation of major dams; many case histories have been published in engineering and geological journals and conference proceedings.

The ultimate aim of a site investigation is the identification of a geomechanical model, with quantitative parameters, which represents the engineering properties of the site, and on which the design of the dam can be based. Intermediate stages in this process are the development of (1) a geological model, which describes the lithology, geological structure, weathering and geological history of the site; followed by (2) an engineering geological model, in which the physical properties of the geological features are described in some semi-quantitative manner (e.g. rock materials properties; description of joints; orientation, spacing and continuity analysis of joints). From the engineering geological model, quantification of physical properties using field and laboratory tests (often accompanied by some reasonable simplification of the geological model) produces the geomechanical model.

Phases of investigation

The geological investigation of a large dam is usually carried out in several phases associated with the sequence of engineering decisions to be made. The following is a brief description of the general sequence of dam site investigations.

Preliminary or reconnaissance investigation

At this stage, alternative locations are being considered for the dam, and the objective is to provide sufficient data to select the best (cheapest) site for dam construction.

Feasibility investigation

Once the preferred site has been selected, investigations are carried out to confirm the economic feasibility of the project at that site. Alternative types of structures are evaluated at this stage (e.g. concrete dam versus rockfill dam), and the investigation should provide sufficient data to allow comparative designs and cost estimates to be prepared.

Design investigation

Once the general type of engineering structure has been determined, still more information is required for the detailed design of the structure. For instance, it may have been decided that a rockfill dam should be constructed, but the spillway and diversion tunnel have yet to be specifically located. The data from this phase are used to complete the detailed design of the structure, to work out quantity estimates (such as volumes of rock excavation and construction materials) and to draw up the tender documents. This is a critical stage, as all geological conditions to be encountered during construction should be predicted and addressed in the design documents.

Construction investigation

In any major project, conditions actually encountered may necessitate changes in the proposed design – one hopes that these will be minor and result in little extra cost for the project. However, to ensure that variations from the anticipated geological conditions are detected as soon as possible, a systematic and progressive record of the geology exposed as construction progresses must be maintained. If conditions are very much as predicted, then little or no investigation will be required, but if a major unexpected geological problem arises, a proper investigation using a wide range of techniques may be necessary. The important feature of this stage is the maintenance of detailed, up-to-date records of the 'as-constructed' geology of the entire site.

Post-construction investigation

Once the structure or project becomes operational, regular inspections should be undertaken to monitor the behavior of the structure and to detect any unexpected changes that could affect the stability or efficient operation of the structure (e.g. blocking of groundwater drains, instability of rock or soil slopes on deterioration of exposed rock). If a problem does occur, a properly planned site investigation may be required to evaluate the problem and provide a solution. The plans of the 'as-constructed' site geology are invaluable in evaluating post-construction problems.

Site investigation techniques

During site investigations, information is required from the surface and near-surface bedrock, from the rock at depth, and on the nature and distribution of surficial materials. The following are brief details of the main techniques used to obtain this information. (For more detailed descriptions, see Attewell and Farmer, 1976; Bell, 1980; Best and Hill, 1967.)

Geological mapping

The main component of all site investigations is detailed large-scale mapping of natural and artificial exposures of bedrock (and of surficial materials where appropriate). The scale and detail of mapping depends to some extent on the phase of the investigation and the geological complexity of the site, but scales of between 1:100 and 1:500 are common. It is usually essential to use a detailed and accurate topographic map as a base for such mapping, to enable three-dimensional analysis of geological structures. One very important point about this mapping is that factual data should be kept separate from interpretive data. The geological map should show as much information as it is possible to record from the field observations, and should include as much quantitative (e.g. orientation) data as possible. Geological inferences and interpretation must inevitably be made from these basic data, but should be presented on separate, clearly labeled, interpretive plans and sections. Special engineering geological maps and sections are often prepared to present the geological information in the most appropriate manner (UNESCO, 1976).

Trenching and pitting

Dam sites rarely have anything like 100% exposure of bedrock; 10–25% is much more common. At such sites there will be very large gaps in the geological map where soil and scree deposits occur. Once the outcrop geology map has been prepared, critical areas of covered bedrock can be selected for exposure by bulldozing away the overburden. Apart from extending the knowledge of bedrock geology, bulldozed trenches may also be used to provide information on soil and weathering profiles, and to assess the excavation characteristics of selected areas.

Excavation of pits with a mechanical backhoe is done extensively in the mapping and evaluation of prospective sources of construction materials such as core material, sand and gravel. The information obtained by mapping the surficial deposits exposed in these pits enables a three-dimensional picture of their distribution to be drawn, which is essential for the economic evaluation of alternative sources of construction materials.

Sluicing

At geologically complex sites, narrow strips of exposed bedrock from trenching are sometimes not sufficient to provide the required information; larger areas of foundations need to be exposed. If the overburden thickness is not too great most of the overburden can be bulldozed away and the bedrock can then be cleaned with high-pressure water jets. While this technique may appear costly, the large amounts of geological information obtained and the increased reliability of the geological interpretation may well justify the expense. Another important advantage of this technique is that it makes available large areas of the bedrock for inspection by prospective tenders; consequently, the prices for foundation excavation and treatment will be more realistic than if the bedrock were visible only in drill cores and isolated pits and trenches.

Diamond core drilling

This is the most common way to obtain information and samples from depth at an engineering site. It is an expensive technique, so great care needs to be taken in planning drill holes to obtain the maximum amount of geological information with the minimum of drilling. A 50 mm diameter core is commonly used for site investigations; it should be obtained with a special core barrel incorporating a split inner tube capable of hydraulic ejection at the end of each drilling run. The drilling must be of high quality, and 100% core recovery should be obtained in all but very bad ground. Care should be taken to recover all seams and crushed zones of rock, as these are much more important than the sticks of core obtained from fresh rock. The driller's skill and motivation are the most important factors in diamond drilling for engineering site investigation (provided, of course, that the right equipment is used), as it is essential that the core recovered is as undisturbed as possible by the drilling process.

Water pressure testing (lugeon testing)

Quantitative data on the leakage properties of rock can be obtained by conducting water pressure tests in drill holes. Packers are used to isolate a section of the drill hole, water is pumped into the section under a known pressure, and the rate of water leakage into the rock is measured. Leakage rates are measured for a range of water pressures, and after appropriate corrections are made to the field data, the results give a good indication of the water tightness and degree of open jointing in the rock at depth.

Exploratory shafts or adits

Where an evaluation of the rock foundations at some distance below the ground surface is necessary for engineering design criteria (e.g. the abutments of an arch dam), direct access to these areas must be furnished by shafts or adits. Careful excavation techniques should be used in the areas of interest, as excessive blasting will give an incorrect picture of the rock properties. Detailed geological mapping of the walls, floor and roof of the excavations is carried out to provide a three-dimensional picture of the rock mass properties.

Geophysical surveys

Seismic and resistivity surveys are frequently carried out during dam site investigations, as they provide very useful data that complement the information from diamond drilling. They are also useful in assessing 'average' geological conditions, and can give information on rock mass properties that diamond drilling cannot provide. They cannot, however, be relied on to evaluate particular geological features (such as an individual clay seam or a narrow shear zone), and they are no replacement for a drill hole in identifying what the rock material is at a particular point some distance below ground surface.

The seismic refraction method is the most common geophysical technique used in engineering site investigations; special equipment has been specifically developed for such investigations. The information obtained enables useful extrapolation and interpolation of drillhole data, such as depth of weathering and tightness of jointing, and also provides values for certain physical properties of the rock mass (such as moduli of elasticity).

Laboratory testing

The quantification of rock material properties for the engineering geological and geomechanical models is provided by laboratory tests on drill-core specimens. Parameters commonly determined include density, porosity, compressive strength, shear strength, Young's modulus (both by direct tests and by geophysical tests) and Poisson's ratio. Some physical properties of defects in the rock – such as the friction characteristics of joints – can also be determined on core samples.

Field testing

Some important rock foundation properties such as shear strength, friction characteristics and modulus of elasticity can be determined by *in situ* field tests. They usually give much more reliable results than equivalent laboratory tests because the rock is tested in its natural environment. Also, the volume of the tested sample is much larger than laboratory test specimens, and so is much more likely to be representative of the rock mass. On the other hand, *in situ* field tests are very expensive, and even on major projects only a small number of tests can be afforded. It is therefore important that the test sites be selected with great care or unrepresentative data may be obtained.

Another category of field testing includes trial tests of engineering procedures that need to be evaluated before construction commences. For example, a trial blast may be carried out at a proposed quarry site to confirm that suitable rock for construction materials is obtainable. Other such field tests include trial grouting, trial ripping, trial loading, trial rockbolting and trial compaction tests.

Importance of geology in dam construction

When designing a large dam, the engineer has two prime goals: (1) the dam must be stable and successfully fulfill its function throughout its projected life, and (2) the dam and its associated structures must be constructed as economically as possible. Of course there is a dilemma here, because the two objectives act against each other; ensuring stability by overdesign increases the cost, while cost-cutting methods, if taken to extremes, will lead to an unsafe structure.

Stability of dams

On a worldwide scale, it is clear that the objective of constructing stable dams is not always achieved. During the 1900–1965 period, for example, about 1% of the 9000 large dams in service throughout the world have failed, and another 2% have suffered serious accidents; significantly, in more than half these incidents, the failure or damage could be related to geological causes (Stapledon, 1976). Brief details of some failures illustrate the magnitude of the problem.

- Malpasset Dam, France – 60 m high double-curvature arch dam that failed in 1959, causing 400 deaths. The cause of failure was uplift and sliding of a section of the rock foundations (Jaeger, 1963; Londe, 1967).
- Vajont Dam, Italy – a 265 m high double-curvature arch dam (at the time, the highest dam in the world). In 1963 some 260 million m^3 of rock slid into the reservoir just upstream of the dam, and over 2000 lives were lost in the flood caused by the displaced water (Muller, 1963). Because the dam itself did not give way, this is considered a reservoir failure rather than a dam failure.
- Baldwin Hills Storage Basin, USA – failed in 1963, after 12$\frac{1}{2}$ years of service, with the loss of five lives and \$15 million in damages. The cause of failure was subsidence and erosion due to movement along fault planes below the structure (James, 1968).

- Teton Dam, USA – a 92 m high earth dam that failed in 1976, with the loss of 11 lives and damage estimated at about $400 million. The investigating committee concluded that failure was probably caused by a combination of inadequate sealing of bedrock in a critical part of the dam foundations, and inadequate protection of the core material against internal erosion (US Department of the Interior, 1977).

It might be expected that progressive advances in dam design and construction techniques would result in a lower incidence of failures. This does not appear to be the case, however, for two main reasons.

First, with any technological advance there are always likely to be unforeseen factors that can produce unexpected problems. For example, when the Malpasset Dam was constructed, drainage of foundations to reduce hydrostatic pressures downstream of arch dams was not considered necessary (Malpasset was one of the first dams constructed in the double-curvature thin arch design). The subsequent failure was caused by hydrostatic uplift in the foundations.

Second, most of the obvious and easy dam sites around the world have now been utilized. This means that future dam construction will be necessary at progressively more difficult and geologically complex dam sites, which increases the probability of foundation problems. It is therefore clear that if dam failures and accidents are to be minimized in the future, the role of geology must be maintained or enhanced during the investigation, design and construction of dams.

Economic construction of dams

Hundreds of dams have cost much more than the original contract price; unexpected geological features, many of which could have been identified by a more thorough investigation, are often the cause of expensive problems during construction. Even today, too many dam sites are underinvestigated, and the resultant extra cost of construction is far greater than the additional expense that would have been necessary to carry out a thorough investigation. The problem is, of course, that the assessment of what constitutes a reasonable investigation has to be done well before the scheduled commencement of construction, when there is little information available. This is compounded by the fact that an investigation that is perfectly adequate for a simple site could well be quite inadequate for a geologically complex site. It is extremely difficult to plan site investigations that are adequate without being wasteful of resources, but it is clear that, on a worldwide scale, the pendulum needs to swing towards more thorough pre-construction investigations.

Eric Best

Bibliography

Attwell, P.B. and Farmer, I.W., 1976. *Principles of Engineering Geology*. London: Chapman & Hall, 1045 pp.
Bell, F.G., 1980. *Engineering Geology and Geotechnics*. London: Newnes-Butterworths, 497 pp.
Best, E.J., 1981. The influence of geology on the location, design and construction of water supply dams in the Canberra area, *Bur. Mineral Resources J. Australian Geology and Geophysics*, 6(2), 161–179.
Best, E.J. and Hill, J.K., 1967. Site investigation techniques used at Corin Dam site, Cotter River, ACT, *Proceedings of the 5th Australia–New Zealand Conference on Soil Mechanics and Foundation Engineering*, Australia: Institution of Engineers, 1–8.
Jaeger, C., 1963. The Malpasset report, *Water Power*, 15(2), 55–61.
James, L.B., 1968. Failure of Baldwin Hills Reservoir, Los Angeles, California, *Geol. Soc. Am. Eng. Geology Case Histories*, 6, 1–11.
Legget, R.F., 1962. *Geology and Engineering*. New York: McGraw-Hill, 884 pp.
Londe, P., 1967. Panel discussion, *Proceedings of the 1st Congress of the International Society of Rock Mechanics*, Vol. 3, Madrid: Editorial Blume, 449–453.
Muller, L., 1963. The rock slide in the Vajont valley, *Rock Mechanics Eng. Geology*, 2, 149–212.
Paige, S. (ed.), 1950. *Application of Geology to Engineering Practice*. Berkey Volume, New York: Geological Society of America, 327 pp.
Stapeldon, D.H., 1976. Geological hazards and water storage, *Bull. Internat. Assoc. Eng. Geol.* 14, 249–262.
Thomas, H.H., 1976. *The Engineering of Large Dams*. 2 vols. London: Wiley, 777 pp.
UNESCO, 1976. *Engineering Geological Maps – A Guide to Their Preparation*. Paris: Unesco Press, 79 pp.
US Department of Interior, 1977. *Failure of Teton Dam: A Report of Findings*. Washington, DC: US Government Printing Office, 107 pp. plus 638 pp. appendixes.
Wahlstrom, E.E., 1974. *Dams, Dam Foundations and Reservoir Sites*. Amsterdam: Elsevier, 278 pp.
Walters, R.C.S., 1971. *Dam Geology*. London: Butterworths, 470 pp.
World Register of Large Dams, 1988. The International Commission on Large Dams, Paris.
Wright, C.E., 1994. *UK Reservoir Failures and Safety Legislation, Dams and Reservoirs*, October. 4 pp.

Cross references

Dams: failure
Dams: world

DAMS: FAILURE

Consequences of dam failure

On 11 and 12 August 1979 the failure of the Macchu II Dam in the state of Gujarat, India produced a disastrous flood. Spillway design allowed for 5415 m^3 s^{-1} (presumably designed to withstand the probable maximum flood or PMF) with a 2.73 m freeboard. Flooding resulted from 520 mm of rainfall in 21 h. The dam overtopped and the wing embankments were destroyed almost immediately. The dam break caused an 8–10 m high flood wave down the valley killing 2000 people in the town of Morvi alone 9 km downstream (Anon, 1979). Following this disaster, the PMF was re-evaluated at 20 925 cumecs and a decision made to rebuild the dam. When the new design study was nearing completion, an even heavier cyclone crossed the catchment producing 700 mm of rainfall in a single day. As a result, the PMF was again re-evaluated, this time to 26 420 cumecs, almost five times the initial estimate (Pessoa, 1989).

The Dale Dyke Dam was built in 1858 to supply water to Sheffield. It was a (then) impressive 28.9 m high and 381 m long embankment dam. In 1864 the outlet valves were closed to raise the water level in the Bradfield reservoir. In the evening of the 11 March, a crack appeared in the air face which gradually widened and later the dam gave way. Nearly one-quarter of the embankment collapsed and an estimated 91 million liters of water flooded towards Sheffield. Some parts of the city flooded to a depth of 2.8 m; approximately 250 people were killed and approximately 800 houses destroyed.

At least 60 years later in Wales, in 1925, the failure of Eigiau Dam and subsequent failure of the Coedty Dam just downstream as a result caused the village of Dolgarrog to be swamped and 16 people were killed. The Eigiau Dam failure was due to foundation seepage leading to piping blowout. This was a bad year for British dams with three failures and one near failure. At least 60 catastrophic dam failures (Wright, 1994) have occurred in the UK (Binnie and Partners, 1986) of which four are due to overtopping. Despite recent advances in the field of dam construction and design, no structure can be made completely failsafe.

Gruner (1963) states that approximately one-third of all dam failures worldwide are due to inadequate spillway design. Thomas (1979) describes the findings from different countries since 1933 and gives statistics on the causes of failure of different types of dams around the world. The general consensus is that between one-quarter and one-third of all dam failures are due to overtopping. In 1973 the International Commission on Large Dams (ICOLD) produced a report on the failure and accidents to large dams (ICOLD, 1973).

Legislation in the UK

Recent dam failures throughout the world have focused attention on dam design guidelines. Great Britain was probably the first country to pass legislation which states that dams should be subject to periodic scrutiny (Smith, 1972). In 1930 the Reservoirs (Safety Provision) Act was passed. Essentially the Act states that all reservoirs exceeding 5 million gallons above adjacent natural ground level in capacity must be inspected at least once every 10 years and that all dams should be designed by qualified engineers. However, catastrophic (breach with an escape of water) failures continued to occur with over ten recorded between 1960 and 1971 with at least a further 12 overtopping events during the same period (Binnie and Partners, 1988). The legislation was strenthened by the Reservoirs Act 1975 (which required continual supervision between inspections).

World legislation

Legislation and safety criteria for dams vary quite significantly throughout the world. Many countries regulate the abstraction rates from the river or reservoir with little or no reference to the safety of the dam. This is particularly true for many arid or semiarid regions. In countries such as Canada, dam safety and river pollution are controlled by the provinces with no Federal control (Thomas, 1979). Some countries do not have any regulations on dam construction and safety, preferring instead to adopt the guidelines of other countries, without enforcing them. This is true in Denmark, for example (Pessoa, 1989).

Dams and reservoirs in the UK

There are some 2600 dams in Great Britain whose reservoirs are covered by the Reservoirs Act 1975 (BRE, 1994). Almost 80% of the dams are embankment type (BRE, 1994), but probably more interesting is that 70% of the dams covered by the Act were constructed before 1931 (Binnie and Partners, 1988), many of which are located in upland areas above densely populated areas.

British reservoirs are grouped into four hazard categories (ICE, 1989):

- Category A: reservoirs where a breach will endanger lives in a community.
- Category B: reservoirs where a breach
  - may endanger lives not in a community;
  - will result in extensive damage.
- Category C: reservoirs where a breach will pose negligible risk to life and cause limited damage.
- Category D: Special cases where no loss of life can be foreseen as a result of a breach and very limited additional flood damage will be caused.

PMF versus frequency analysis

Before detailed design and construction can begin on a reservoir and dam, consideration must be given to the meteorology and hydrology of the proposed area in which the construction is to take place. Once the site has been selected and provisionally approved, there ensues a more detailed study of the catchment.

Spillway design requires knowledge of the expected flood peak for the catchment. There are basically two approaches to this problem. Wallis and Wood (1985) for example, argue that extreme flood estimation should be carried out solely by some form of frequency analysis. The benefits of frequency analysis for the economics of structures is well known and its use and approach has improved considerably over the years, yet it is still basically a method of curve fitting and extrapolation whose errors will be amplified by sampling errors and choice of distribution. The type of distribution chosen should be considered very carefully.

The concept of PMF has been around for at least 80 years (Riedel, 1977). In the US the PMF was known as the Maximum Probable Flood until the mid-1950s. The US Corps of Engineers define the PMF as 'the flood that may be expected from the most severe combination of critical meteorologic and hydrologic conditions that are reasonably possible in a region' (Corps of Engineers, 1975). In dam design this is the inflow hydrograph.

Politically as well as psychologically, it is beneficial to avoid using the word 'risk' which is inherent in the probabilistic approach to dam and reservoir design. The statistical approach does not mean that the degree of risk is automatically increased, but implies that the flood of magnitude q has a probability of occurrence of p so that the $p - dp$ probability flood is of larger magnitude. It is UK practice to postulate an upper limit which can reasonably be expected to occur. The concept of PMF is used in British dam and reservoir safety for structures where failure could result in extensive damage and loss of life.

Existing dams and the need for continuing research

As mentioned above, 70% of dams covered by the Reservoirs Act were constructed before 1931. As improved methods for calculation of the design inflow hydrograph are developed, the inspecting engineer has to verify that existing structures still pass the safety criteria.

For example the Stocks Dam, situated on the River Hodder, a tributary of the River Ribble in West Yorkshire, UK, was completed in 1932. The 33 m high dam comprises an embankment with a puddle-clay core supported by shoulders of boulder clay and protected by stone pitching on the upstream face.

Following the Dunsop storm of 8 August 1967, where a peak flood greatly exceeding the then current flood standard occurred, the Fylde Water Board commissioned an investigation into the adequacy of the spillway at the Stocks Dam. The study was subsequently extended to cover the heightening of the spillway and dam to provide additional storage in the reservoir. In the investigation, the Dunsop storm was transposed over the Stocks catchment. The study concluded that the dam should be heightened and the spillway redesigned (Seddon, 1971). This was subsequently carried out, and a new spillway provided.

Continuing research into extreme flood and rainfall estimates is essential and should be considered and compared to previous estimates used for designing and modifying existing dams and reservoirs as well as in the design and construction of new structures.

R.W. Herschy

Source

B.N. Austin, I.D. Cluckie, C.G. Collier and P.J. Hardaker, 1995. *Radar-based estimation of probable maximum precipitation and flood.* Report prepared for the Water Directorate, Department of Environment, United Kingdom, by METSTAR Consultants, Meteorological Office and the Telford Institute, Department of Civil Engineering, University of Salford.

Bibliography

Anon., 1979. International news note, *Water Power and Dam Construction*, October, 3.

Binnie and Partners, 1986. Modes of dam failure and flooding and flood damage following dam failure. Contract report to the Department of the Environment, (Contract No. PECD 7/7/184).

Binnie and Partners, 1988. Estimation of flood damage following potential dam failure: Guidelines. Contract report to the Department of the Environment (Contract No PECD 7/7/259).

Bobee, B., G. Cavadias, F. Ashkar *et al.*, 1993. Towards a systematic approach to comparing distributions used in flood frequency analysis, *J. Hydrology*, **142**, 121–136.

BRE, 1994. *Register of British Dams*, Building Research Establishment.

Corps of Engineers, 1975. Recommended Guidelines for Safety Inspection of Dams, in *National Program of Inspection of Dams*, Appendix D, Washington, DC.

Gruner, E., 1963. *Dam Disasters*. Proceedings of the ICE, Paper No. 6648.

ICOLD, 1973. *Lessons from Dam Incidents* (in English and French), Paris, International Commission on Large Dams, pp. 201.

ICE, 1989. *Floods and Reservoir Safety: An Engineering Guide*, Institution of Civil Engineers, London pp. 56.

Pessoa, M.L., 1989. The Hydrological Utilisation of Radar Derived Rainfall Data in Modelling the Behavior of Severe Storms and Extreme Floods, PhD Thesis, Dept. of Civil Eng., University of Birmingham.

Riedl, J.T., 1977. Assessing the probable maximum flood, *Water Power and Dam Construction*, December, 29–34.

Smith, N., 1972. *A History of Dams*. Citadel Press, Inc., New York.

Seddon, B.T., 1971. *Spillway investigations for Stocks Dam*. Proceedings of the ICE, Paper No. 7374.

Thomas, H.H., 1979. *The Engineering of Large Dams*, Parts I and II. John Wiley & Sons, Chichester.

Wallis, J.R. and E.F. Wood, 1985. Relative accuracy of Log Pearson III Procedures, *J. Hydraulic Eng.*, **111**(7), 1043–1056.

Wright, C.E., 1994. UK reservoir failures and safety legislation, *Dams and Reservoirs*, October, pp. 4.

Cross references

Dams
Lakes

DAMS: WORLD

Records of the world's dams are kept by the International Commission on Large Dams (ICOLD) in Paris. In 1950 there were 5196 dams in service in the world; in 1986 (the latest figures available) there were 36 235, and Table D1 gives the number of dams by continents. It shows that of this total, China had 18 595 dams in 1986 compared to only eight in 1950 (only dams over 15 m in height are recorded). About 80% of the China dams are less than 30 m in height and are practically all embankment dams.

Table D1 Number of dams by continent

Continents	1950		1982		1986	
Europe	1323	25.11%	3961	11.26%	4215	11.63%
Asia	1562	29.65%	22 789	64.80%	23 286	64.26%
America	2099	39.84%	7303	20.77%	7479	20.64%
Africa	133	2.52%	665	1.89%	763	2.11%
Australasia	151	2.87%	448	1.27%	492	1.36%
Total	5268	100.00%	35 166	100.00%	36 235	100.00%
China	8	0.15%	18 595	52.88%	18 820	51.94%

Table D4 World's largest reservoirs in terms of capacity

Capacity (10^6 m^3)	Name	Country	Year
204 800	Owen Falls[a]	Uganda	1954
169 000	Bratsk	Former USSR	1964
162 000	High Aswan	Egypt	1970
160 368	Kariba	Zimbabwe/Zambia	1959
147 960	Akosombo	Ghana	1965
141 851	Daniel Johnson	Canada	1968
135 000	Guri	Venezuela	1986
73 300	Krasnovarsk	Former USSR	1967
70 309	W A C Bennett	Canada	1967
68 400	Zeya	Former USSR	1978
63 000	Cahora Bassa	Mozambique	1974
61 715	La Grande 2 Barrage	Canada	1978
60 020	La Grande 3 Barrage	Canada	1981
59 300	Ust-Ilim	Former USSR	1977
58 200	Boguchany	Former USSR	C[b]
58 000	Kuibyshev	Former USSR	1955
54 400	Serra da Mesa	Brazil	C
53 790	Caniapiscau Barrage KA 3	Canada	1980
49 800	Bukhtarma	Former USSR	1960
48 700	Atatürk	Turkey	C
46 000	Irkutsk	Former USSR	1956

Table D4 Continued

Capacity (10^6 m^3)	Name	Country	Year
45 500	Tucurui	Brazil	1984
35 900	Vilyui	Former USSR	1967
35 400	Sanmenxia	China	1960
34 852	Hoover	USA	1936
34 100	Sobradinho	Brazil	1979
33 304	Glen Canyon	USA	1966
32 203	Skins Lane No. 1	Canada	1953
31 790	Jenpeg	Canada	1975
31 500	Volgograd	Former USSR	1958
31 300	Sayano-Shushensk	Former USSR	C
30 600	Keban	Turkey	1974
29 959	Iroquois	Canada	1958
29 000	Itaipu	Brazil	1982
29 000	Loma de la Lata	Argentina	1977
28 973	Churchill Falls (GR-1)	Canada	1971
28 370	Missi Falls Control	Canada	1976
28 100	Kapchagay	Former USSR	1970
27 920	Garrison	USA	1953
27 675	Kossou	Ivory Coast	1972
27 433	Oahe	USA	1958
26 000	Razza Dyke	Iraq	1970
25 400	Rybinsk	Former USSR	1941
24 700	Longyangxia	China	C
24 700	Mica	Canada	1972
24 000	Tsimlyansk	Former USSR	1952
23 700	Kenney	Canada	1952
23 500	Ust-Khantaika	Former USSR	1970
23 400	Shuikou	China	C
22 950	Furnas	Brazil	1963
22 119	Fort Peck	USA	1937
21 626	Xinanjiang	China	1960
21 166	Ilha Solteira	Brazil	1973
21 000	Yacyreta	Argentina	C

[a] This capacity is not fully obtained by a dam; the major part of it is the natural capacity of a lake; Owen Falls is not the greatest artificial lake.
[b] Under construction in 1986.

Table D5 Distribution of dams by types

TE/ER	PG	VA	CB	MV	Total
28 845	3 953	1 527	337	136	34 798
82.9%	11.3%	4.4%	1%	0.4%	100%

Key: TE, earth dam; ER, rockfill dam; PG, gravity dam; CB, buttress dam; VA, arch dam; MV, multiarch dam.

Table D2 shows the number of dams in each country and Table D3 gives a list of the world's highest dams. Table D4 shows a list of the world's largest reservoirs in terms of capacity and Table D5 shows the distribution of dams by types (1982).

R.W. Herschy

Source

International Commission on Large Dams (ICOLD) *World Register of Large Dams*, 1984, updated 1988, Paris.

Table D2 Number of dams by country

Dams ≥ 15 m		Dams ≥ 30 m		Dams ≥ 60 m		Dams ≥ 100 m		Dams ≥ 150 m		Dams ≥ 200 m	
1. China	18 820	1. China	2287	1. USA	319	1. USA	65	1. USA	18	1. USA	4
2. USA	5459	2. USA	1185	2. Japan	233	2. Japan	47	2. Switzerland	9	2. Switzerland	4
3. Japan	2228	3. Japan	741	3. Spain	132	3. Spain	32	3. Japan	6	3. Former USSR	4
4. India	1137	4. Spain	381	4. China	117	4. Switzerland	25	4. Canada	6	4. Canada	2
5. Spain	737	5. Italy	274	5. Italy	88	5. Italy	20	5. Former USSR	5	5. Iran (Islamic Rep. of)	2
6. Korea	690	6. India	220	6. France	57	6. France	15	6. Italy	5	6. Austria	1
7. Canada	608	7. Australia	179	7. India	48	7. Canada	13	7. Austria	4	7. Colombia	1
8. Great Britain	535	8. Mexico	168	8. Switzerland	47	8. Former USSR	13	8. France	4	8. Spain	1
9. Brazil	516	9. France	167	9. Canada	45	9. China	13	9. Turkey	4	9. India	1
10. Mexico	503	10. Canada	162	10. Australia	45	10. India	12	10. Brazil	3	10. Italy	1
11. France	468	11. Brazil	148	11. Mexico	42	11. Mexico	11	11. Iran (Islamic Rep. of)	3	11. Mexico	1
12. South Africa	452	12. Great Britain	130	12. Former USSR	38	12. Austria	10	12. Australia	3	12. Turkey	1
13. Italy	440	13. South Africa	127	13. Brazil	34	13. Australia	10	13. Spain	2	13. Yugoslavia	1
14. Australia	409	14. Former USSR	101	14. Turkey	32	14. Argentina	9	14. Colombia	2	14. Philippines	1
15. Norway	245	15. Norway	89	15. Yugoslavia	23	15. Brazil	8	15. Venezuela	2	15. Honduras	1
16. Germany (FR)	191	16. Turkey	88	16. Portugal	20	16. Yugoslavia	8	16. India	2		
17. Czechoslovakia	146	17. Switzerland	84	17. Venezuela	20	17. Venezuela	8	17. Albania	1		
18. Switzerland	144	18. Yugoslavia	77	18. Romania	20	18. Turkey	7	18. China	1		
19. Sweden	141	19. Austria	68	19. Argentina	17	19. Iran (Islamic Rep. of)	7	19. Ecuador	1		
20. Romania	133	20. Argentina	63	20. Germany (FR)	16	20. Romania	6	20. Greece	1		
21. Former USSR	132	21. Germany (FR)	60	21. Austria	15	21. Colombia	5	21. Honduras	1		
22. Austria	123	22. Romania	53	22. Norway	14	22. Philippines	5	22. Malaysia	1		
23. Yugoslavia	123	23. Korea (Rep. of)	52	23. Algeria	13	23. Greece	4	23. Mexico	1		
24. Bulgaria	108	24. Portugal	52	24. Morocco	13	24. Portugal	4	24. Paraguay	1		
25. Turkey	103	25. Czechoslovakia	52	25. South Africa	12	25. Thailand	4	25. Philippines	1		
26. Albania	98	26. Albania	48	26. Colombia	12	26. Korea (Rep. of)	4	26. Romania	1		
Total	34689		7054		1472		365		88		26

Table D3 World's highest dams

Height above lowest foundation (m)	Type[a]	Name	Country	Year
335	TE/ER	Rogun	Former USSR	C[b]
300	TE	Nurek	Former USSR	1980
285	PG	Grande Dixence	Switzerland	1961
272	VA	Inguri	Former USSR	1980
262	VA	Vajont	Italy	1961
261	ER	Manuel Moreno Torres (Chicoasén)	Mexico	1980
261	ER/TE	Tehri	India	C
253	ER/TE	Kishau	India	C
245	VA	Ertan	China	C
243	ER	Guavio	Colombia	C
242	VA/PG[c]	Sayano-Shushensk	Former USSR	C
242	TE	Mica	Canada	1972
237	ER	Chivor	Colombia	1975
237	VA	Mauvoisin	Switzerland	1957
234	VA	El Cajon	Honduras	1984
233	VA	Chirkey	Former USSR	1978
230	TE	Oroville	USA	1968
226	PG	Bhakra	India	1963
221	VA/PG[c]	Hoover	USA	1936
220	VA	Contra	Switzerland	1965
220	VA	Mratinje	Yugoslavia	1976
219	PG	Dworshak	USA	1973
216	VA	Glen Canyon	USA	1966
215	PG	Toktogul	Former USSR	1978
214	MV	Daniel Johnson	Canada	1968
210	ER	San Roque	Philippines	1985
208	VA	Luzzone	Switzerland	1963
207	ER/PG	Keban	Turkey	1974
203	VA	Dez	Iran (Islamic Rep. of)	1962
202	VA	Almendra	Spain	1970
201	VA	Khudoni	Former USSR	C
200	VA	Karoun	Iran (Islamic Rep. of)	1975
200	VA	Kölnbrein	Austria	1977
196	PG/ER/TE	Itaipu	Brazil/Paraguay	1982
195	ER	Altinkaya	Turkey	C
194	VA	New Bullard's Bar	USA	1970
192	PG	Lakhwar	India	C
191	ER	New Melones	USA	1979
186	VA	Kurobe	Japan	1964
186	TE	Swift	USA	1958
186	VA	Zillergründl	Austria	1986
185	VA	Mossyrock	USA	1968
185	VA	Oymapinar	Turkey	1984
184	ER	Atatürk	Turkey	C
183	PG	Shasta	USA	1945
183	TE	WAC Bennett	Canada	1967
180	VA	Amir Kabir	Iran (Islamic Rep. of)	1964
180	ER	Dartmouth	Australia	1979
180	VA	Emosson	Switzerland	1974
180	VA	Tehchi	Taiwan	1974
180	VA	Tignes	France	1952
176	ER	Takase	Japan	1978
175	ER	Ayvacik	Turkey	1981
175	PG	Lijiaxia	China	C
175	PG/ER	Revelstoke	Canada	1983
174	PG	Alpe-Gera	Italy	1964
173	TE	Don Pedro	USA	1971
173	VA	Karakaya	Turkey	C
172	VA	Hungry Horse	USA	1953
172	PG	Longyangxia	China	C
172	TE	Thissavros	Greece	C
171	VA	Cahora Bassa	Mozambique	1974
170	VA	Daniel Palacios	Ecuador	1982

Table D3 Continued

Height above lowest foundation (m)	Type[a]	Name	Country	Year
169	VA	Idukki	India	1974
168	ER	Charvak	Former USSR	1977
168	ER	Gura Apelor	Romania	C
168	ER	La Grande 2 Barrage	Canada	1978
168	PG	Grand Coulee	USA	1942
168	TE	Miller Ash Pond	USA	1980
167	ER	Fierze	Albania	1978
166	VA	Dongfeng	China	C
166	ER	Thomson	Australia	1985
166	VA	Vidraru	Romania	1965
165	TE	Kremasta	Greece	1965
165	VA	Ross	USA	1949
165	PG	Wujiangdu	China	1982
165	ER	Maroun	Iran (Islamic Rep. of)	C
164	TE	Trinity	USA	1962
163	PG	Piedra del Aguila	Argentina	C
163	PG	Sardar Sarovar	India	C
162	ER/PG	Guri	Venezuela	1986
162	ER	Talbingo	Australia	1971
160	ER	Foz do Areia	Brazil	1980
160	TE/ER	Grand-Maison	France	1984
160	ER	Salvajina	Colombia	1985
160	ER/TE	Theim Dam Ranjit	India	C
160	VA	Yellowtail	USA	1966
160	PG	Sengwon	D.P.R. Korea	C
158	ER	Canales	Spain	C
158	TE	Yacambu	Venezuela	1986
158	ER	Cougar	USA	1964
158	ER	Emboracacção	Brazil	1982
158	VA	Gökcekaya	Turkey	1972
158	ER	Naramata	Japan	C
157	VA	Dongjiang	China	C
157	PG	Okutadami	Japan	1961
157	VA	Speccheri	Italy	1957
156	PG	Sakuma	Japan	1956
156	VA/TE	Zeuzier	Switzerland	1957
155	ER	Goescheneralp	Switzerland	1960
155	VA	Monteynard	France	1962
155	VA	Nagawado	Japan	1969
155	VA	Place Moulin	Italy	1965
155	ER	Kenyir	Malaysia	1985
154	VA/PG	Bhumibol	Thailand	1964
154	ER	Tedorigawa	Japan	1979
153	VA	Curnera	Switzerland	1967
153	VA	Flaming Gorge	USA	1964
153	ER	Gepatsch	Austria	1965
153	VA	Santa Giustina	Italy	1950
151	PG	Dorna	Spain	C
151	ER	Menzelet	Turkey	C
151	VA	Zervreila	Switzerland	1957
151	VA	Geheyin	China	C
150	VA	Canelles	Spain	1960
150	ER	Finstertal	Austria	1980
150	VA/CB	Roselend	France	1961
150	TE	Big Horn	Canada	1972

		In operation	Under construction
	Member countries	89[d]	27
Dams ≥ 150 m	Non-members	2[e]	
	Total	91	

[a] TE, earth dam; ER, rockfill dam; PG, gravity dam; VA, arch dam; MV, multiarch dam.
[b] Under construction in 1986.
[c] Arch-gravity dam.
[d] 90 with Itaipu for Brazil and Paraguay.
[e] Taiwan, Mozambique.

DANUBE RIVER: HYDROLOGY AND GEOGRAPHY

The Danube has been given a number of different names: Danubius in Latin, Istros in Greek (the lower), Dunăre in Romanian, Donau in German, Dunaj in Slovakian, Duna in Hungarian, Dunav in Serbian and Bulgarian and Duna in Russian.

The Danube is the second largest (805 300 km^2) and longest (2860 km) river in Europe after the Volga. Its source area lies in the central-western part of Europe, in the Schwartzwald, where two of its tributaries spring from – the Breg and the Brigach – bringing their waters together at Donaueschingen. Thence the Danube crosses central Europe to Budapest, then the Pannonian Depression down to the junction with the Drava (Drau), finally to pierce the Carpathian range at the Iron Gate and form a defile. It represents the southern border between the Romanian Plain and the Prebalkan Tableland. From Călăraşi (Romania) and Silistra (Bulgaria) to the Black Sea, the Danube skirts round the Dobrogea Plateau and its mountains, forming a delta.

Drainage basin

The Danube basin covers 8% of the continent of Europe, its waters flowing through the territory of 13 states (Germany, Switzerland, Austria, Czech Republic, Slovakia, Hungary, Slovenia, Croatia, the Federal Republic of Yugoslavia, Romania, Bulgaria, the Republic of Moldova and the Ukraine).

The process of formation of the river bed and valley of the Danube ended in the Late Pliocene and the Early Quaternry as the result of the successive drainage, with time, of several huge lakes from the Vienna, Pannonian and Pontic basins, themselves separated from the vast Sarmatian Sea by the uplifting of the Alps, the Carpathians and the Stara Planina Mountains.

The points of epigenetic or catchment penetration show up today in the well-known gates or defiles (Devin, Iron Gate), dividing the river course into three distinct sections: the upper or Alpine, the middle or Pannonian and the lower or Pontic (Walachian or Romanian).

Upper course

The upper course, 1060 km in length, extends from the source to the Devin Gate. After Brig, it joins the Brigach at Donaueschingen. In the Jurassic limestones of the Swabo-Franconian Jura Mountains the Danube loses some 5 m^3 s^{-1} of its flow which escapes through the underground karst to Tuttligen, in the Neckar drainage basin, a tributary of the Rhine.

The Danube flows between the Hercynian Swabo-Franconian Jura range, on the left, and the Prealpine heights, on the right, into the Münich Basin at Ulm, from where its channel becomes navigable. The major right bank tributaries (the Riss, Iller, Günz, Mindel, Lech, Isar, Traun and Enns) issue from the northern slopes of the Alps, accounting for the alpine discharge of this sector. Most of these tributaries drain many lakes situated in glacial, moraine-barred valleys (the Forggen, Ammer, Starnberger, Walchen, Tegern, Chiem, Atter, Mond, Hallstätter and Traun). Some of the important left bank tributaries are the Altmühl, the Naab and the Regen, the last one joining at Regensburg. The Ludwigs Canal, built on the Altmühl and the Main (a tributary of the Rhine), connects the Danube to the Rhine and the Black Sea with the North Sea. The tributary with the highest discharge rate in this sector is the Inn (810 m^3 s^{-1}), which at this point exceeds that of the Danube (660 m^3 s^{-1}). As the riverbed slope varies from 0.6 to 0.9%, streamflow speeds increase to 1–3.5 m s^{-1}. A landscape of gorges or defiles occurs wherever the Danube crosses mountain summits or runs through hard rock, e.g. at Neuburg, Kalheim, Wachau, Bisamberg (Kahlenberg, the cataracts of Ardagger), Persenbeug, Strudel and Wirbel. The Grein–Ybbs defile downstream of the junction with the Enns, near the town of Linz, is particularly picturesque. Defiles and cracks in slopes increase the hydropower potential (hydropower stations at Lochenstein (140 MW), Aschach (258 MW), Ottensheim, Linz, etc.).

In Vienna the Danube divides into three streams: Donau Kanal, Alte Donau, which is abandoned but still enjoyed by holidaymakers, and the Danube proper, used for navigation.

Before passing through the Devin Gate, the Danube receives a left bank tributary, the Czech Morova (average discharge 62 m^3 s^{-1}), forming a 100 km long border between Austria and Slovakia.

Middle course

The middle course from the Devin Gate to Baziaş (Romania) (725 km long) and the portion which narrows at Devin represents the passage of the river from the Vienna Basin to the Pannonian Basin, through the smaller Carpathian range (Male Karpati). A few kilometers farther downstream of the Devin Gate, and the point where the Danube enters Slovakian territory, lies Bratislava, Slovakia's capital. Between Bratislava and Komarno (Komaron on the Hungarian side), the river forms two arms – the Danube itself on the right, forming the border between Slovakia and Hungary, and the smaller Danube (Mali Dunaj) on the left, forming a 90 km long island called Inland Delta by the Slovaks. It is in this section that the two countries built a hydropower station at Gabcikova, subsequently contested by Hungary when the Slovaks commissioned part of it.

From the Smaller Carpathians (Devin Gate) and the Mid-Hungarian Mountains (Visegrad Defile), the Danube passes through the Kiss Alföld (Smaller Plain) in the south, and the Slovak Plain in the north. Some of the main left bank tributaries, originating from the Tatra Mountains, are the Vah and the Nitra (discharge rate 139 m^3 s^{-1}), the Hron (82 m^3 s^{-1}) and the Ipoly, which forms a 150 km long border between Hungary and Slovakia. An important right bank tributary is the Raaba (96 m^3 s^{-1}). After leaving the Visegrad Defile, the Danube takes a N–S orientation, flowing along 275 km through the middle of Hungary. Budapest, its capital, lies on either side of the Danube (Buda, on the right, named after hills of the same name, and Pesta on the left, in a somewhat higher floodplain). The two sides of the city are connected by numerous bridges both old and new. The place where Budapestans go for recreation and leisure is the Margareta Island. On the southern periphery of Budapest the Danube divides into two arms: the river proper on the right, and Soroksari-Duna on the left with the Csepel Island lying in between. Further on, the riverbed slope gets smaller (0.05%) and the Danube meanders through the Pannonian Plain, leaving behind many abandoned arms, which are called 'morotva' (oxbow), on both sides. From the southern section of the plain, the Danube receives its largest tributaries, substantially increasing its flow. First is the Drava (670 m^3 s^{-1}), then the Tisza (814 m^3 s^{-1}) and the Sava (1460 m^3 s^{-1}), the last one running into the Danube at Belgrade, the capital of the Federal Republic of Yugoslavia.

From its junction with the Sava down to Baziaş, where the lower course begins, the Danube receives three major left bank tributaries, all issuing from Romanian territory: the Tamis (Timiş), close to Belgrade, the Karas (Caraş) and the Nera (on the frontier between Romania and the Federal Republic of Yugoslavia). There is one right bank tributary, the Serbian Morava, with a high discharge value of 210 m^3 s^{-1}.

Lower course

The lower course, 1075 km long, forms Romania's natural border with the Federal Republic of Yugoslavia, Bulgaria, the Republic of Moldova and the Ukraine. This reach boasts the longest and most beautiful defile, namely the Iron Gate (144 km) and the most striking geographic river braiding (Drobeta–Turnu Severin to Călăraşi, 566 km). It is a large floodplain and forming two islands (Ialomiţa and Brăila, 195 km). This is the reach of maritime navigation and of the Danube Delta (Brăila–Sulina, 170 km).

The Iron Gate Defile, between Baziaş and Gura Văii, narrows down in some sections and becomes wider in others, where small basins are formed. The first narrowed section is at Coronini, where large floods (e.g. in 1897) formed a temporary upstream lake, suggestively called Mare Album, extending beyond Belgrade. Similar conditions led to the formation of the Moldova Veche islet. The tectonic limestone zone at Coronini, known as Babacalin Rock, constitutes an obstacle for navigation, likewise Cozia, Doica, Islazi, Tahtalia, Vrani, Vlas, Iuţi and Pregrada. A first Iron Gate dam was built at Pregrada and Iuţi Rocks. The Iron Gate hydropower station, shared jointly by Romania and the Federal Republic of Yugoslavia, was commissioned in 1971, and has operated at full capacity (2100 MW) ever since. The construction of the dam at Gura Văii and the formation of a storage lake flooded the Adakeleh Island and some Roman remains. A second hydropower station (Iron Gate II), downstream of the Iron Gate, was also built jointly with the Federal Republic of Yugoslavia.

In this section, the largest Danube tributaries lie on the left bank in Romanian territory (the Jiu, the Olt and the Argeş); although right bank tributaries from Yugoslavia and Bulgaria are more numerous, they are much smaller (the Timoc, Ogosta, Iskar, Vit, Osam, Iantra and the Lom). In 1954 a road and rail bridge was built between Giurgiu (Romania) and Ruse (Bulgaria). Between Călăraşi and Brăila, the Danube and its many arms encompass the floodplain itself, rich in lakes and backwaters but frequently flooded, and suggestively named Balta Ialomiţei (Borcei) and Balta Brăilei ('balta' means floodplain). Both are diked and the land is used for farming.

An impressive bridge for rail traffic, the longest in Europe, was built between 1890 and 1895 by the engineer Anghel Saligny. It was a remarkable feat for that time, spanning the river from Feteşti to Cernavodă. A second bridge for road and rail traffic was commissioned in 1987. Not far downstream, where the river forms a single channel, the road bridge, built in 1970, is the longest of the Danube bridges (length 1450 m, of which 750 m are suspended over the Danube). The last reach of the river is from the city-port of Brăila to the Black Sea. It is known as the maritime sector, because high-tonnage sea vessels enter the Sulina Arm (where engineering works have been completed) and have navigated up to Brăila since early in the twentieth century, a distance of 170 km.

In this reach the Danube receives two of its largest left bank tributaries, the Siret and the Pruth. Engineering works have facilitated the navigation of vessels as far as the delta (Ceatal Chilia), with the consequent flourishing of the city-ports of Brăila, Galaţi (Romania) and Reni (Ukraine). The river also forms the border between Romania, the Republic of Moldova and the Ukraine. This is the section of the Danube Delta which extends between the Chilia arm in the north (117 km), Tulcea arm (19 km) and the Sfântul Gheorghe arm in the south (109 km), an area of 4152 km^2. Eighty-two per cent of this territory (3446 km^2) lies in Romania. The perpetual territorial evolution of this geographical area gives the Danube a discharge of 6473 m^3 s^{-1} and 58 million tonnes per year of sediment. The delta was and still is one of Europe's unique faunal and floral sites. Despite massive human pressure on this environment for agricultural, forestry and fish-farming purposes (until 1989), the delta still preserves many areas in a natural condition. In 1990 the Danube delta, together with the Razim–Sinoie Lake complex in the south and the marine waters down to the 20 m isobath were declared a biosphere reserve, benefitting from legislation, management and administration.

Discharge regime

In the upper course, the Danube regime is determined by its alpine tributaries increasing the flow in June. The middle and the lower courses of the river are under the influence of the Drava and the Sava, which add to the flow in spring (April–May), depleting the volume in the fall (September–October). In winter and summer, discharge is moderate.

Average discharge

There is a progressive downstream increase in the average annual discharge – 1470 m^3 s^{-1} at Passau, after the junction with the Inn, 1920 m^3 s^{-1} in Vienna, 2350 m^3 s^{-1} in Budapest and 5300 m^3 s^{-1} after the junction with the Drava, the Tisza and the Sava. On reaching the delta, flows reach 6473 m^3 s^{-1}, through the contribution of its lower-section tributaries. Within the delta itself, the discharge values are unevenly distributed between the three arms of the Danube: Chilia (58%), Sfântul Gheorghe (23.2%) and Sulina (18.8%). The same proportion holds for the discharge of 58 million tonnes of sediment per year, which causes the secondary delta of the Chilia arm (in Ukraine) to advance into the sea at an annual rate of about 40–80 m.

Maximum discharge

Maximum discharge is recorded in spring and occasionally in summer, when flows are high: 15 100 m^3 s^{-1} at Orşova in April 1940, 15 900 m^3 s^{-1} at Olteniţa in May 1942 and 15 500 m^3 s^{-1} at Ceatal Chilia in July 1970.

Minimum discharge

Minimum discharge occurs in the fall and sometimes in winter: 1250 m^3 s^{-1} at Orşova in January 1954, 1450 m^3 s^{-1} at Olteniţa in January 1964 and 1350 m^3 s^{-1} at Ceatal Chilia in October 1921. As the climate is temperate continental, blocks of ice float down the river in the lower reaches from December until the beginning of March and, in particularly heavy winters, the ice bridge may last for 45–50

days, with traffic obstructions at Zimnicea, Călăraşi, Topalu and Cotu Pisicii, with sections closed to navigation during certain intervals.

Water pollution

Water pollution is moderate, despite the quantities of waste emptied into the river by the large cites on its banks (Vienna, Bratislava, Budapest and Belgrade), which has increased the quantity of water pollutants. However, the river's impressive self-purification ability enables it to recover in the lower reaches, despite total dissolved solids (TDS) values of up to 350–400 mg l^{-1}.

History of the River Danube

The Danube is a navigable waterway of significant importance and since ancient times it has helped form links between the populations inhabiting its banks. Traces of settlements date back thousands of years as people were attracted by fertile floodplains and terraces, the wildlife in willow forests and the wealth of fish populating the river itself and its many great lakes. Unfortunately, this was not always the case and there were times, spanning some 500 years, when the lower reaches divided the peoples. The Romans turned the river into a political frontier, easily defended because of the fierce resistance of the population, the Dacians, in particular. Therefore, the Empire set up a fleet on the Danube, built strategic roads and bridges – one at Drobeta-Turnu, Severin Trajan's Bridge, constructed by Appolodorus from Damascus, and another at Celei, near Corabia. The walls at Cazane still preserve some stone inscriptions known by the name of Tabula Trajana and Tabula Domitiana, marking the construction of the Roman road on the right bank of the river under Emperor Trajan (AD 98–117). Traces of Greek and Roman strongholds erected on this bank, many in Dobrogea, confirm the intense activity experienced in the lower reaches during ancient times. Herodotus of Halicarnas (484–425 BC), the author of a nine-volume work, entitled *Histories*, tells about the inroads made by Darius I, the emperor of the Persians, presumably chasing the Scitians as far as Isaccea on the Danube (514 BC), which is an indication of the river having also been navigable in those days. In the period of migrations, the Danube became a gateway to the Balkans. The fourteenth century ruler of Walachia, Mircea the Old, strengthened his cities on the Danube: Drîstor (Silistra), Giurgiu and Turnu Măgurele.

Management of navigation

The expansion of the Ottoman Empire and the establishment of some Turkish dominions on the river (Turnu, Giurgiu and Brăila) put limitations to navigation on the Danube, the situation being solved only in 1829 by the Treaty of Adrianople. The European Danube Commission (EDC), set up in 1856, was assigned the task of management of navigation on the river and the undertaking of corrective works to this end. In addition to the riparian states, Great Britain and France also enjoyed membership. The Commission discharged its duties until 1948, when the Belgrade Convention legislated the rights of riparian states. Under the terms of that Convention, the Romanian sector of the Danube fell under the control of the Fluviatile Administration of the Lower Danube (FALD), with central offices in Galaţi, a body belonging to the Danube Commission located in Budapest. This form of organization stimulated the development of several city-ports, which besides trading and transport functions started building and repairing river barges. Plans for connecting the Danube to the Rhine and the North Sea, respectively, go back to the reign of Charlemagne (AD 793). They were aimed at extending the transport by canal between the Altmühl (near Treuchtlingen) and the Main, and some traces of those works are still visible at Karlsgraben. Eventually, the two rivers were connected when the Ludwigs–Donau–Main Canal was built (1836–1845). It stretches 177 km from Kahlheim on the Altmühl to Banberg on the Main. The Canal was operated until 1945 but a new connection was made in 1959. The Main–Danube Canal, commissioned in 1992, is 55 m wide at water level, 31 m at the bottom, and 4–4.5 m deep, with 12 m wide × 190 m long spillways. It allows the passage of 90 m long vessels (1500 dwt) and of tugs (up to 3000 dwt).

Another goal was to shorten the distance to the Black Sea by having a canal built on the lower reaches of the Danube, between Cernavodă and Constanţa. Its main axis (Cernavoda–Basarabi–Agigea) was built between 1976 and 1984; the Basarabi (Poarta Albă)–Năvodari section was finished in 1988. In this way, the distance to the sea is about 400 km less. The Danube-Black Sea Canal is 64.2 km long (the Poarta Albă–Năvodari branch is 30 km long), 70–80 m wide at water level and 7–7.5 m at the bottom, allowing the passage of vessels of 5000 dwt, and 6 m draught at a speed of 8–9 km h^{-1}, and of double-barge convoys of 3000 dwt, each up to 296 m long and 23 m wide.

Peter Gâştescu

Bibliography

Gâştescu, P., 1990. *Fluviile Terrei*, Ed. Sport-turism, Bucharest.
Gâştescu, P., 1983. *Geografia României*, Vol. I, *Geografia fizică*, Ed. Academiei Române, Bucharest.

Cross references

Danube River: development
Lakes
Razim–Sinoie lake complex, Romania
Rivers

DANUBE RIVER: DEVELOPMENT

The river basin of the Danube includes glacier-covered mountains, mountain chains of intermediate height covered by forest, highlands and uplands, table lands, plateaus with deeply carved river valleys, and wide plains and depressions. Due to the elongated shape of the river basin in the west–east direction and diverse relief features, the climatic conditions are variable. With respect to climate, the catchment extends from the western regions of the upper Danube, with a strong Atlantic influence, to the eastern territories affected by continental climate. In the Drava (Drau) and Sava basins the climate is influenced by the Mediterranean Sea. Average annual air temperature ranges from −6.2 to +12°C.

The hydrological regime is substantially influenced by precipitation. Average annual precipitation fluctuates within the range from 3000 mm in the high mountains to 400 mm in the delta region. The estimated mean annual evaporation is between 100 and 700 mm.

The hydrographical stream system is characterized by about 120 major tributaries and several thousand lakes including Attersee, Neusiedler See (Fertő), Balaton and Wörther See.

Extensive hydrological investigations of the water regime started in the second half of the eighteenth century and the early nineteenth century. National hydrological services were established before the end of the nineteenth century.

Mean annual specific runoff varies from 19.4 to 4.3 l s^{-1} km^{-2} at the main gauging stations on the main river and the tributaries and a graph showing the minimum, average and maximum discharges along the river from source to mouth is shown in Figure D2.

The development of river structures and engineering works on the Danube is connected with the population expansion, economic development and culture in this area. The first training works were constructed during the period of the Roman Empire.

International cooperation of Danube countries concerning navigation is based on agreements since 1856. The recent 'Danube Commission' was founded in 1948 in Belgrade, and its headquarter is in Budapest. All common problems concerning navigation, water management and water works are dealt with, for example recommendations for the waterway (depth, width, curvature, slope, sizes of shiplocks, discharge capacity) from Regensburg to the Black Sea, including some tributaries (the Drava, Tisza, Sava and Prut).

The proposal to construct a navigable waterway connecting the rivers Rhine, Main and the Danube dates back to ancient times (AD 793). The new Main–Danube Canal was constructed between 1959 and 1991. The 204 km long sector between Regensburg and Bamberg is divided into 18 reservoirs over a 103 km length. In 1984 a 64 km Danube–Black Sea canal was completed in Romania, by shortening the waterway by 370 km. Several river barrages have been constructed in Germany and Austria, and at the Iron Gate by Romania and Yugoslavia.

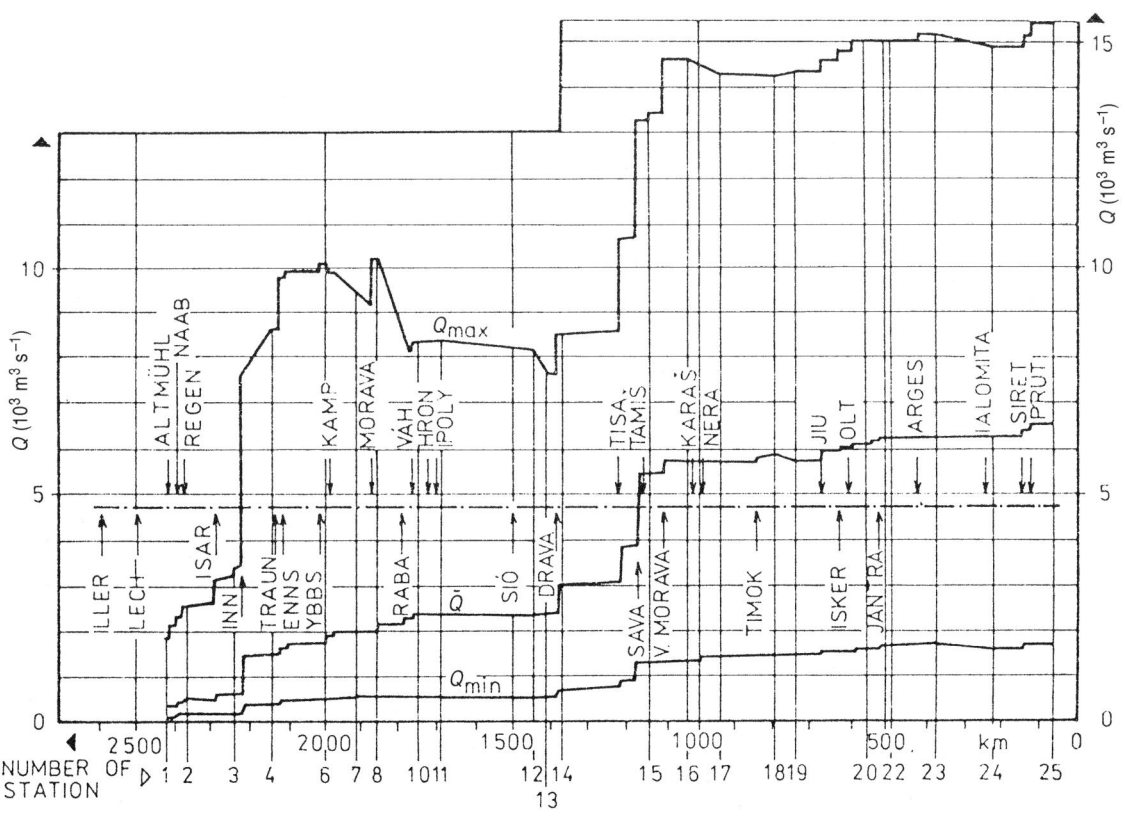

Figure D2 Minimum, average and maximum discharges along the River Danube from source to mouth.

River regulation involves artificial control of natural flow to reduce flood peaks. Control measures include building embankments along the river to confine flood waters, dredging to deepen the channel, alleviating scouring, increasing the gradient of the river by cutting across river bends, shortening the course, and removal of shoals. The construction of hydraulic works has still not been completed.

Although the water quality of the Danube is still acceptable, local pollution can be observed at several places and incidental water pollution, particularly due to navigation, has been frequently observed. In order to protect Danube water quantity and quality, a convention was signed in June 1994 in Sofia and the Danube Environmental Programme was launched as the joint efforts of riparian countries, the World Bank, Global Environmental Facility, UNDP and the European Commission.

Ö. Starosolsky

Bibliography

Horvárth, S., 1979. *Ice Conditions on the Danube* (in Hungarian). VITUKI, Budapest,

Kresser, W., 1973. *Die Donau und ihre Hydrologie Wasser und Energiewirtschaft*, No. 3/4, Vienna.

Pardé, M. *et al.*, 1965. Le Danube. *La Houille Blanche*, Grenoble.

Regionale Zusammenarbeit der Donauländer, 1993. *Schwebstoff- und Geschieberegime der Donau.* UNESCO, IHP.

Stancik, A., Jovanovic, S. *et al.*, 1988. *Danube. Hydrology of the River.* Publishing House Priroda, Bratislava.

Starosolszky, Ö., 1992. Flood control in the Danube countries – a case study. *Proceedings of the NATO ASI on Coping with Floods.* Applied Sciences, Vol. 257, Erice.

Tőry, K., 1972. *The Danube and its regulation* (in Hungarian). Akadémiai Kiadó, Budapest.

Cross references

Danube River: hydrology and geography
Lakes
Razir–Sinoie lake complex, Romania
Rivers

DATA PROCESSING IN HYDROLOGY

Data processing in hydrology refers to the several stages of manipulation that measurements of hydrological processes, taken in the field, are subjected to before they can be efficiently stored and/or made available for analysis. It therefore represents the essential interface between data collection and data use. For the purposes of this definition, data processing does not include all the common forms of analysis that hydrological data are subjected to in order to derive estimates of means, probabilities, extremes, etc.

The basis of a great deal of research and practice in hydrology lies in the data that are collected from field instrumentation in order to quantify temporal and spatial variations in processes. Data collection networks vary from national, operated by state hydrometric agencies for general water resource availability determination, through to local, operated largely by research organizations for very specific purposes. The former frequently involve a great number of replications of gauges measuring basic hydrological variables (rainfall, evaporation, streamflow, groundwater levels, etc.) and operate over long periods of time so that historical trends can be recorded. The latter usually involve fewer measurement points, but with a higher spatial density and tend to have shorter life spans that last for specific experimental periods. Because of the costs and effort involved in collecting these data and because of their high volume, it is important that adequate data processing approaches are used to ensure that their full value can be realized at a later stage when the data are analyzed. All too often, the time and effort spent collecting data in the field is wasted through the lack of an efficient and robust data processing system.

Some of the important issues related to data processing are

data encoding and digitization;
error checking, quality control and data integrity;
calibration accuracy and stability;
data collection versus data storage resolution;
data storage and retrieval system.

Data encoding and digitization

There are many different ways in which process measurements can be recorded in the field, and the first stage of most data processing systems is to encode these measurements into digital form. The simplest form of field data collection is to use hand-written field sheets, and this is still the standard approach for many national hydrometric agencies, particularly with respect to rainfall, evaporation and other meteorological variables. These data are then usually manually typed into a database. Older types of automated recording devices use paper strip charts and pen traces. These data can be encoded manually or by digitizers connected to computers, both of which represent quite lengthy procedures if the network consists of a large number of gauges. Punched paper tape recording systems have been used for many years, but have largely been superseded by magnetic tape or solid-state recording media. All of these have the advantage of being able to transfer the data to a computer database rapidly and efficiently and facilitate the collection of data at finer time resolutions than is possible with manual systems. They do, however, suffer from the problem that no hardcopy version of the field data is available and if the data are lost or damaged once transferred, they cannot be recovered.

Some specific applications in hydrology and water resource management require field data to be collected and processed in 'real time'. Modern telemetry systems using instruments connected to telephone lines, radio and satellite links allow data recorded at field stations to be interrogated or downloaded to computers at any time.

Error checking, quality control and data integrity

A browse through many countries' national archives of hydrological data will reveal the importance of these items. Typical problems include missing data, accumulated data, data recorded on the wrong day or at the wrong time and obvious instrument failure. In many cases the databases do not include flags to indicate the likely accuracy or quality of the data and it is almost impossible to express any degree of confidence in the data set as a whole. Similarly, with long records of historical data it becomes an impossible task to examine the raw data and carry out error checking procedures at a time when the data are being analyzed. It is therefore essential that error checking and quality control are carried out at the time that the data are collected or initially processed for long-term storage.

The type of error checking procedures that can be carried out include cross-correlation of data from nearby gauges to identify anomalies and periodic checking of the instruments against standards. It is also important to check for errors introduced during data processing (see Calibration accuracy below).

Calibration accuracy and stability

Many measurements obtained from field instruments require processing through calibration tables or equations to derive the required process values. Perhaps the most commonly known example is a stage–discharge relationship which converts measurements of water level behind a weir or flume, or within a stable channel section, to discharges or flow rates. If water levels are observed within natural channel sections, it is often necessary to check the calibration after large flood events when bed scour, deposition or bank erosion may have occurred. In other cases, the instruments themselves require some form of calibration to ensure that the measurements are accurate. One example is tipping bucket raingauges, which may occasionally require adjustment to ensure that each tip of the bucket represents a defined or known depth of rainfall.

Data collection versus data storage resolution

The time resolution used to collect and store hydrological data is of great importance given the variable rates of change of hydrological processes in different climates. To analyze rainfall intensities in temperate climates, a resolution of about 1 h may be adequate for most purposes, but in semiarid and arid climates, short-term intensity variations can be much greater and recording intervals of minutes may be necessary. Similar comments can be made with respect to streamflow, where variations of greater than $10 \text{ m}^3 \text{ s}^{-1}$ can occur in minutes in small- to medium-sized semiarid catchments. Observations are frequently made at intervals that are shorter than required (hourly, for example) for the immediate purpose of the gauge network. This is particularly true when sophisticated automatic recording devices (electronic data loggers, for example) are used. However, it is also frequent practice to condense the detailed observations into longer interval (daily, for example) summaries for long-term database storage and discard the original data after error checking. Given the relative cost of collecting the data compared with that of data storage, this approach is something of a false economy and raw data should always be stored when possible. Discarding the original data precludes further error checking at a later stage, as well as the possibility that a future user may require the higher-resolution data. The same applies to the storage of the original, pre-calibration raw data.

Data storage and retrieval system

Where a large number of measurement stations form part of the same database, as in the case of a national archive, the methods of data storage and data retrieval assume great importance if the value of the data is to be realized. Fortunately, modern database storage, enquiry and retrieval software is readily available and there is very little excuse for not implementing efficient systems. The main components of an adequate system should allow a user to be able to identify easily the type of measurement, the instrument location details, the length of record, the extent of missing data and quality of the data. It should also allow the data to be extracted for later use in various formats. One of the most important aspects of any database system is the need to make frequent backups in case of failure of the primary storage system.

In conclusion, it should be recognized that the pressures on available water resources are increasing throughout the world and that the need for information is increasing. There have been many recent developments in the techniques that are available to analyze hydrological data and information. However, many of these rely upon accurate data inputs and have overtaken the supply of available historical data. There is very little that can be done about collecting data from past events, but it is important that this lesson is learned and that all data collected in the future are securely stored using systems that permit rapid and efficient retrieval.

Denis A. Hughes

Cross references

Accuracy
Rain
Streamflow measurement

DELTAIC PLAINS

Definitions and types

A deltaic plain consists of active or abandoned deltas, which are either overlapping or contiguous to one another. A delta is a relatively flat area at the mouth of a river or a river system in which sediment load is deposited and distributed. That portion of a drainage basin within which the sediment load is traded or carried in transit is referred to as an alluvial valley (see Alluvial valley engineering). The alluvial valley merges downstream with the deltaic plain, often where the main stream channel branches, or has branched in the past, into multiple distributaries. The deltaic plain does not necessarily begin downstream from its most upstream distributary or its most upstream abandoned distributary (Carter and Woodroffe, 1994). In the case of the Mississippi River, the most upstream distributary at present is the Atchafalaya River (Figure D3). However, much of the area through which the Atchafalaya now flows consists of fluviatile backswamp deposits. Moreover, individual abandoned Mississippi River deltas have been mapped in some detail, and only the stippled area shown on Figure D3 contains such deltaic masses. Thus landforms and environments of deposition characteristics of deltas identify deltaic plain fluviatile landforms and environments of deposition are characteristic of and identify an alluvial valley.

This concept is useful in describing those characteristics of the deltaic plain of interest to disciplines as disparate as fluvial morphology, environmental geology, hydrology and river engineering. It is of particular importance to the engineering geologist and the soil mechanics engineer concerned with the distribution and strength parameters of the notoriously weak soils that support the levees, docks, industrial structures, urban developments, harbors and similar installations that are built to develop and make deltaic plains habitable. The problems associated with such development and the cost involved in resolving them are often the primary reasons why many of the world's larger deltaic plains are undeveloped or under-developed (Selley, 1988).

The deltaic plains of the world are as varied as the river systems that give them birth. Table D6 lists 46 of the world's rivers in order of their average discharge. Note that although the Amazon ranks first in average discharge, it ranks third in yearly suspended load, and that the Hwang-Ho (Yellow River), which ranks first in yearly suspended load, ranks twenty-ninth in average discharge. Obviously, the discharge of a given river and the amount of sediment it carries to the sea are important parameters in the distribution of environments of deposition and their associated sediment types within the growing deltaic plain. Other parameters of equal or greater importance include the depth of water into which the delta is being built (expressed in Table D6 in terms of the slope of the offshore areas), the magnitude of the tides, the climate and the offshore wave action and currents. These factors, together with the tectonics and geometry of the receiving basin and interrelated parameters, are used by Coleman (1981) and Coleman and Wright (1975) to describe and classify world deltas. A somewhat similar group of factors is proposed by Morgan (1970), who divides the four factors controlling and influencing delta formation into (1) river regime, mainly particle size and quantity of material transported by a river to its delta and variations in these properties during seasonal fluctuations in flow; (2) coastal processes, essentially the influence of waves, tides and currents on the seaward margins of the delta; (3) structural behavior and the relation of sea level to the depositional site; and (4) climatic factors, particularly those that affect vegetation within the delta (Pye, 1994).

Galloway (1975) proposes the more simplified threefold classification shown in Figure D4, based on whether a delta is fluvially dominated, tidally dominated or principally affected by wave energy. The modern Mississippi is fluvially dominated with characteristically high discharge and sediment load. Elongate land areas tend to form along a limited number of distributaries. Lobate deltas like the Ebro, the Danube and abandoned deltas of the Mississippi are considered typical of deltas influenced chiefly by fluvial and wave processes. In tidally dominated rivers such as the Ganges–Brahmaputra and the Colorado, estuarine deltas are typical and distributaries with flaring mouths and many mid-channel bars are common. Where wind and wave processes are influenced by more moderate tidal action, barrier beaches and offshore sandy tidal flats are typical, such as in the Copper and the Niger rivers. Cuspate deltas are thought to be characteristic of rivers, such as the Sao Francisco, where wave forces are predominant.

Despite the number and variety of factors that affect delta formation, some depositional processes and environments are common to all deltas. One of the more intensively studied of the world's deltaic plains is that of the Mississippi. As indicated earlier, the present Mississippi delta is a good example of a fluvially dominated system, a system affected, moreover, by a negligible tidal range and by only moderate wave effects. It is commonly described as a birdfoot delta for it has a few well-developed distributaries separated by wide, shallow bays. The distributaries extend seaward in elongate projections like a bird's foot. Most of the abandoned deltas of the Mississippi, on the other hand, formed lobate deltas, often described as horsetail deltas, where numerous distributaries were occupied and abandoned as the deltas built seaward as huge fans. No bays of consequence occurred between distributaries. This is attributed to the fact that the ancient Mississippi deltas were built rapidly across a shallow continental shelf with water depth ranging from a few meters to generally less than 30 m. The modern Mississippi, in contrast, is dumping its load at the very edge of the continental shelf into water that was 150–200 m deep in the recent past. Figure D5 shows the major Mississippi River deltas that were built and abandoned in the recent geological past. Figure D6B illustrates the types of sediments associated with the present delta and contrasts those in Figure D6A with the sedimentary units associated with past Mississippi deltas.

The soils that form the deltaic plain of the Mississippi have been extensively studied by geologists and engineers, particularly during the past half-century. Tens of thousands of shallow borings were made and comprehensive reports and papers published. (Some of the more notable of these are listed as references at the end of this entry.) Although differences exist between environments and associated

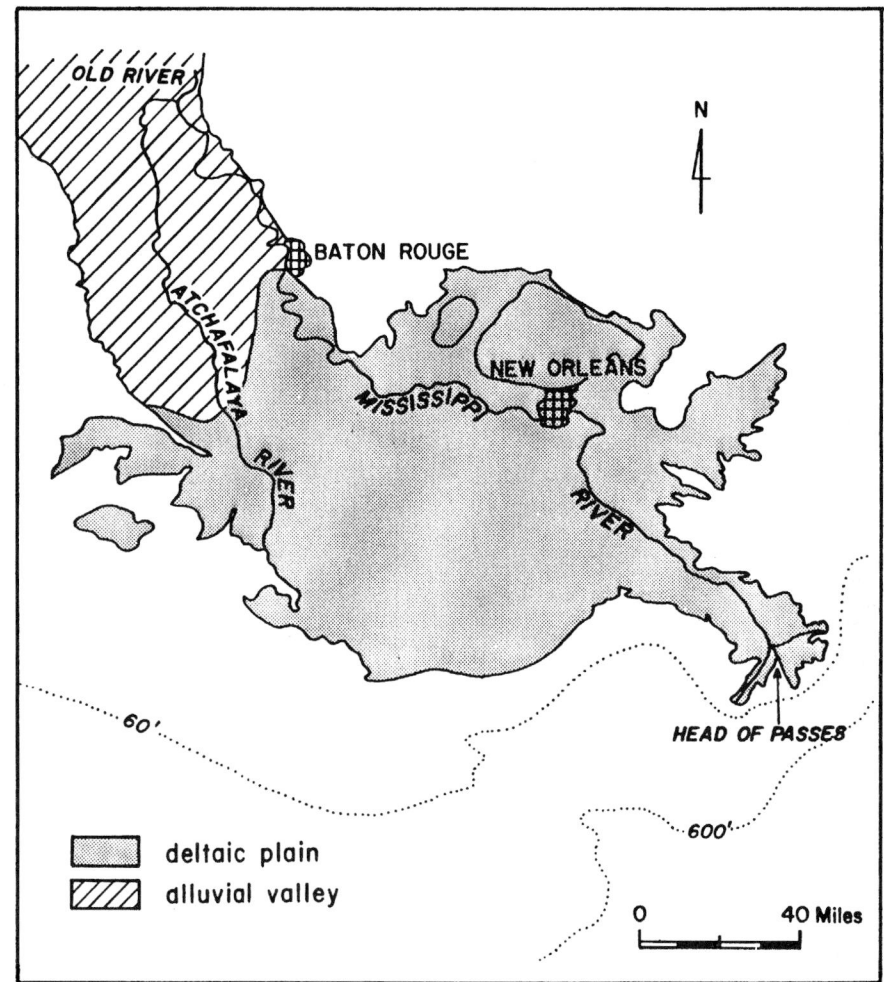

Figure D3 Mississippi River deltaic plain. The boundary between the deltaic plain and upstream fluviatile deposits of the alluvial valley is approximate.

soil units characteristic of the Mississippi deltaic plain and other deltas, the similarities between other deltas and the Mississippi are remarkable.

The following discussion emphasizes the situation found in the Mississippi deltaic plain. Occasional comments are included where landforms or their constituent soils are known to differ significantly from those of the Mississippi. Sources for data on the Mississippi are principally Kolb and Van Lopik (1958), Fisk *et al.* (1954) and Bernard and LeBlanc (1965). Data contrasting the Mississippi with other world deltas were based on many references, chief among them Allen (1965) for the Niger; Andel (1967) for the Orinoco; Arnborg, Walker and Peippo (1962) for the Colville; Coleman (1981) for the Klang, the Ord, the Burdekin, the Sao Francisco and the Senegal; Kolb and Dornbusch (1975) for the Mekong; Naidu and Mowatt (1975) for the Colville; Nelson (1970) for the Po; and Rodolfo (1975) for the Irrawaddy. The references cited at the end of this entry contain additional sources. For more extensive bibliographies on deltas, refer to Coleman (1981).

Pre-deltaic plain deposits

Deltaic plain deposits of world rivers are underlain by a variety of soil and rock types, the depth to and the nature of which are often extremely important parameters in areas where good foundations for heavy structures are at a premium. Deltaic plain soils of the Mekong, the Colorado and the Burdekin, for example, overlie and are flanked in places by ancient rock strata. The deltaic plain of the Ebro is flanked and underlain by Mesozoic and Pliocene sedimentary units in

some areas. Deltaic plain deposits of the Mississippi, probably in common with most of the world's deltaic plains, overlie Pleistocene deposits. These deposits normally consist of soils that are non-indurated but that have been subjected to tens of thousands of years of consolidation, desiccation, oxidation and erosion during the last Wisconsinan drop in sea level. Multistoried buildings, major highways, harbors and so on are usually founded on pilings driven to this horizon. As shown by the slanted lines in Figure D5, the Pleistocene flanks the deltaic plain on the west and north. The irregularly eroded surface slopes gradually seaward beneath Lake Pontchartrain, and beneath New Orleans it averages about 20 m below the surface. It is readily recognized in borings. The colors of the overlying Holocene deltaic soils are generally gray to blue-gray. The upper portion of the Pleistocene, in contrast, is characteristically oxidized to a yellow or orange color. The Pleistocene is also marked by a distinctive stiffening in soil consistency and soil strength, a decrease in water content and the occurrence of calcareous concretions. Strengths commonly reach 144 kPa.

Deltaic plain deposits

Classification

The Mississippi deltaic plain consists of the present delta and at least six additional deltas, the oldest of which was abandoned sometime after sea level reached its present stand about 5,000 years ago (Figure D5). Environments of deposition are divided into (1) fluvial sediments deposited along the major active or abandoned streams that traverse

Table D6 Data on selected world deltas (after Inman and Nordstrom, 1971; Coleman, 1981)

River (location)	Average discharge ($m^3 s^{-1}$)	Rank yearly suspended load	Climate	Offshore slope (%)	Tidal range (m)	Average wave power ($\times 10^7$ ergs^{-1})
1. Amazon (Brazil)	149 736	3	Tropical rainforest	0.5	4.9	0.193
2. Congo (Zaire)	40 441	21	Subtropical steppe		1.70	
3. Ganges–Brahmaputra (Bangladesh)	34 500	2	Tropical rainforest	1.5	3.6	0.586
4. Orinoco (Venezuela)	25 200	19	Tropical rainforest	2.8	1.8	
5. Yangtze-Kiang (China)	22 231	8	Humid subtropical	0.013	3.7	0.127
6. Yenisey (Former USSR)	17 190	34	Tundra		0.40	
7. Lena (Former USSR)	15 661		Tundra	0.8	0.21	
8. Mississippi (USA)	15 631	6	Humid subtropical	7.0	0.43	0.034
9. Mekong (Vietnam)	14 168	4	Tropical savanna	4.3	2.6	
10. St. Lawrence (Canada)	14 160	35	Humid continental			
11. Parana (Argentina)	12 658	17	Tropical rainforest	0.3	0.64	
12. Ob (Former USSR)	12 631	31	Subarctic		0.70	
13. Irrawaddy (Myanmar)	12 558	10	Tropical rainforest	1.4	2.7	0.193
14. Tocantins (Brazil)	10 110		Tropical rainforest		4.30	
15. Amur (Former USSR)	9 714	23	Subarctic		2.30	
16. Niger (Nigeria)	8 769	22	Tropical rainforest	6.2	1.4	2.007
17. Mackenzie (Canada)	8 532	20	Tundra	1.9	0.34	
18. Volga (Former USSR)	7 736	27	Steppe		0	
19. Magdalena (Colombia)	7 500	11	Tropical savanna	36.0	1.1	206.25
20. Colombia (USA)	7 278	25	Marine west coast			
21. Zambezi (Mozambique)	7 164	16	Tropical savanna		4.00	
22. Danube (Romania)	6 250		Humid continental	14.1	0	0.034
23. Yukon (USA)	5 154	18	Subarctic		1.20	
24. Indus (Pakistan)	4 274	9	Tropical desert	9.6	2.6	14.15
25. Red (Vietnam)	3 913	14	Humid subtropical	3.7	1.9	
26. Sao Francisco (Brazil)	3 420		Tropical savanna	11.2	1.86	30.415
27. Pechora (Former USSR)	3 362		Tundra	6.2	0.73	
28. Godavari (India)	3 180		Tropical savanna	12.8	1.2	
29. Hwang Ho (China)	2 571	1	Humid continental	1.5	1.13	0.218
30. Rhine (Netherlands)	2 223	36	Marine west coast		1.7	
31. Rhone (France)	1 528	24	Mediterranean		0.20	
32. Po (Italy)	1 484		Humid subtropical	5.6	0.73	
33. Nile (Egypt)	1 480	15	Desert	7.3	0.43	10.25
34. Dneiper (Former USSR)	1 370		Steppe	4.4	0	
35. Shatt al Arab (Iran and Iraq)	1 300	5	Desert	0.470	2.5	
36. Klang (Malaysia)	1 100		Tropical rainforest	4.1	4.2	0.218
37. Senegal (Senegal)	867		Steppe	16.0	1.22	112.42
38. Chao Phraya (Thailand)	831	33	Tropical savanna	0.6	2.4	0.736
39. Colorado (USA)	595	12	Desert		8.2	
40. Ebro (Spain)	552		Mediterranean	36.0	0	0.155
41. Colville (Alaska)	491		Tundra	2.1	0.21	0.001
42. Burdekin (Australia)	475		Humid subtropical	9.2	2.2	6.414
43. Murray (Australia)	400	25	Mediterranean		2.80	
44. Tana (Kenya)	172		Tropical savanna	0.032	2.9	
45. Ord (Australia)	166		Tropical savanna	3.9	5.8	1.062
46. Rio Grande (USA)	142	28	Steppe			

Note: values of the average discharge may, in some cases, differ from those given in Table D6, depending on the location of the gauging station and the period over which the average discharge is calculated.

the deltaic plain; (2) fluvial-marine deposits laid down off the mouths of the deltas as they advance; (3) organic and minor inorganic deposits formed *in situ* or carried short distances and redeposited in a paludal environment; and (4) marine deposits resulting from reworking of the deltaic deposits by tides, wind and waves.

The prevalence and the types of landforms within each of these groups vary from delta to delta. *Fluvial* environments and fluvial-marine environments predominate in deltas such as the Mississippi with high sediment input, high discharge, minor reworking of sediments by waves, and low tidal ranges. Paludal environments become increasingly important in deltaic plains in humid tropical areas such as the Mekong, the Chao Phraya, and the Orinoco. They form far smaller portions of the deltaic mass in the drier climates. Marine environments are more important volumetrically and more varied where wave and tidal energy is great and fluvial dominance is less pronounced. Evaporites form major constituents of such deltaic plains in the drier climates. Reworking of marine

environments by the wind is also an important factor in the drier climates.

Fluvial deposits

Fluvial environments in the Mississippi deltaic plain are conveniently devided into (1) natural levees, (2) point and lateral bars, and (3) abandoned courses and distributaries. Natural levees form slightly elevated areas flanking streams within the deltaic plain. They are the most conspicuous highs on otherwise strikingly level plains. Nearly all major inhabited and cultivated areas in the Mississippi deltaic plain are located on them. The levees are formed by the deposition of the coarsest sediments carried in suspension by flood waters that top the river banks. Figure D7 is a schematic representation of typical crest elevations, widths, thicknesses and soil types forming the natural levees from Donaldsonville, Lousiana, to Head of Passes (Figure D5). Soil strengths of natural levee deposits are typically high, with cohesive strengths ranging between 183 and 575 kPa. Desiccation and

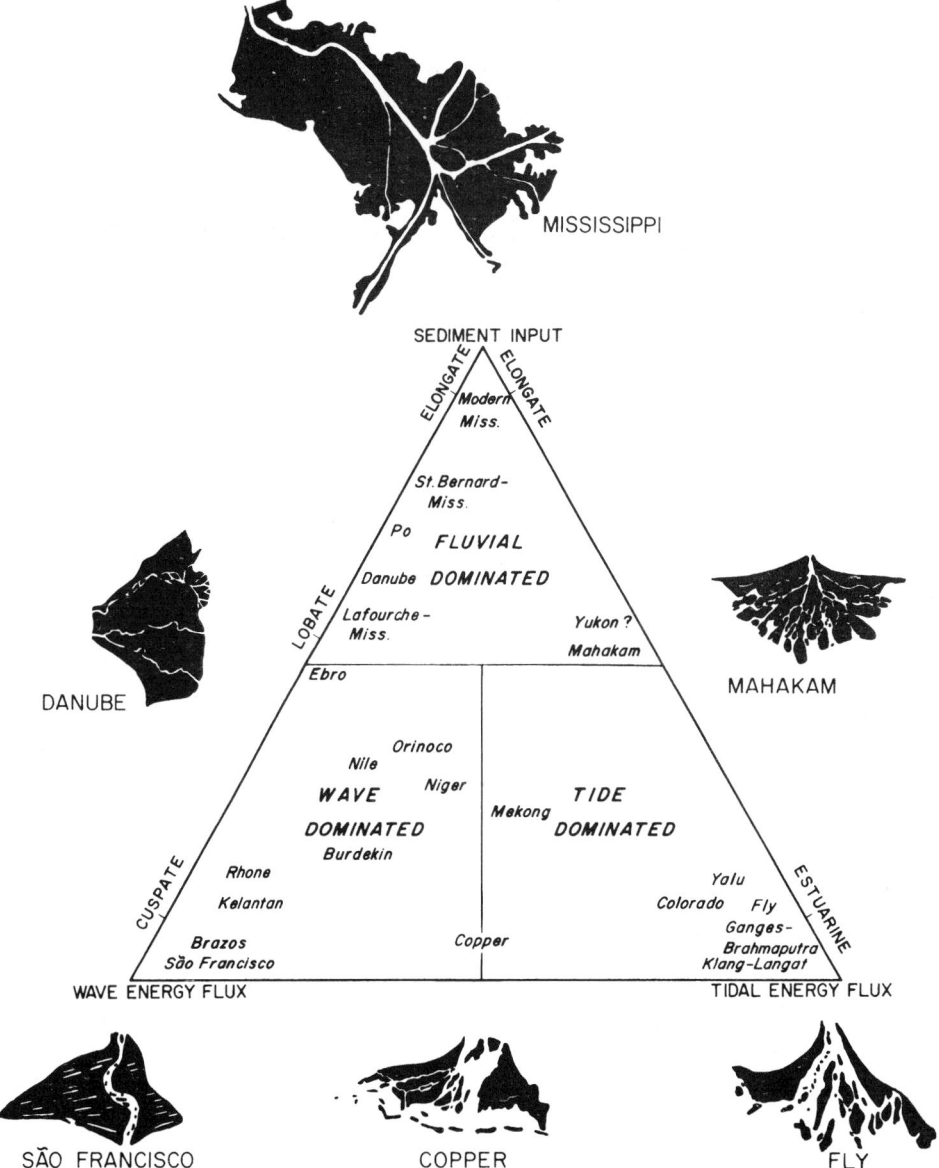

Figure D4 Schematic diagram illustrating a threefold division of deltas into fluvial-dominated, wave-dominated and tide-dominated types. (Based on Galloway, 1975.)

oxidation undoubtedly account for their high strengths. Natural levees are common to most streams transiting deltaic plains. Their crest heights and widths are generally proportional to the size – that is, the discharge and suspended load – of the streams they flank. In streams with high tidal ranges, such as the Ord, the Colorado, the Ganges–Brahmaputra and the Klang, they are only poorly developed. Arctic rivers such as the Colville and the Ob are essentially devoid of landforms recognizable as natural levees in their deltaic plains.

Point bars or lateral bars are formed by meandering and lateral migration of the river or its distributaries. These environments normally contain the coarsest material carried by the river or moved along the bottom as bed load. Whereas point bar deposits make up more than 75% of the deposits forming the Mississippi alluvial valley upstream from the deltaic plain, they become progressively less prevalent along the river downstream from Baton Rouge; downstream from New Orleans, almost all accretions along the sides of the channel consist of occasional lateral bars. The term point bar is normally applied to accretions within a bend; lateral bar is used for the more or less straight accretionary segments that form on either side of the master stream or its larger distributaries. Samples from hundreds of borings in point bar deposits in the portion of the Mississippi deltaic plain between Donaldsonville and the Gulf consist principally of poorly graded fine sand. The upper portions of the deposits vary from clay to silty sand with a characteristic increase in grain size with depth. Depths to sand are highly variable, and soil types change rapidly, both horizontally and vertically. Only the basal one-third to one-half of the point bar deposits can be expected to be fairly clean, fine-grained sand. Point and lateral bars are common along present and abandoned channels of many of the world's deltaic plains. Many rivers, however, such as the Mekong, Sao Francisco, Klang, Burdekin and Ganges–Brahmaputra, typically develop middle bars and islands rather than the point or lateral bars common along the Mississippi.

Abandoned courses and distributaries, although mapped in some detail in the Mississippi deltaic plain, are only poorly known and delineated in most of the world's deltaic plains. Exceptions include such well-documented river deltas as the Po, the Rhone, the Niger and the Colorado. In the Mississippi deltaic plain, sediments filling

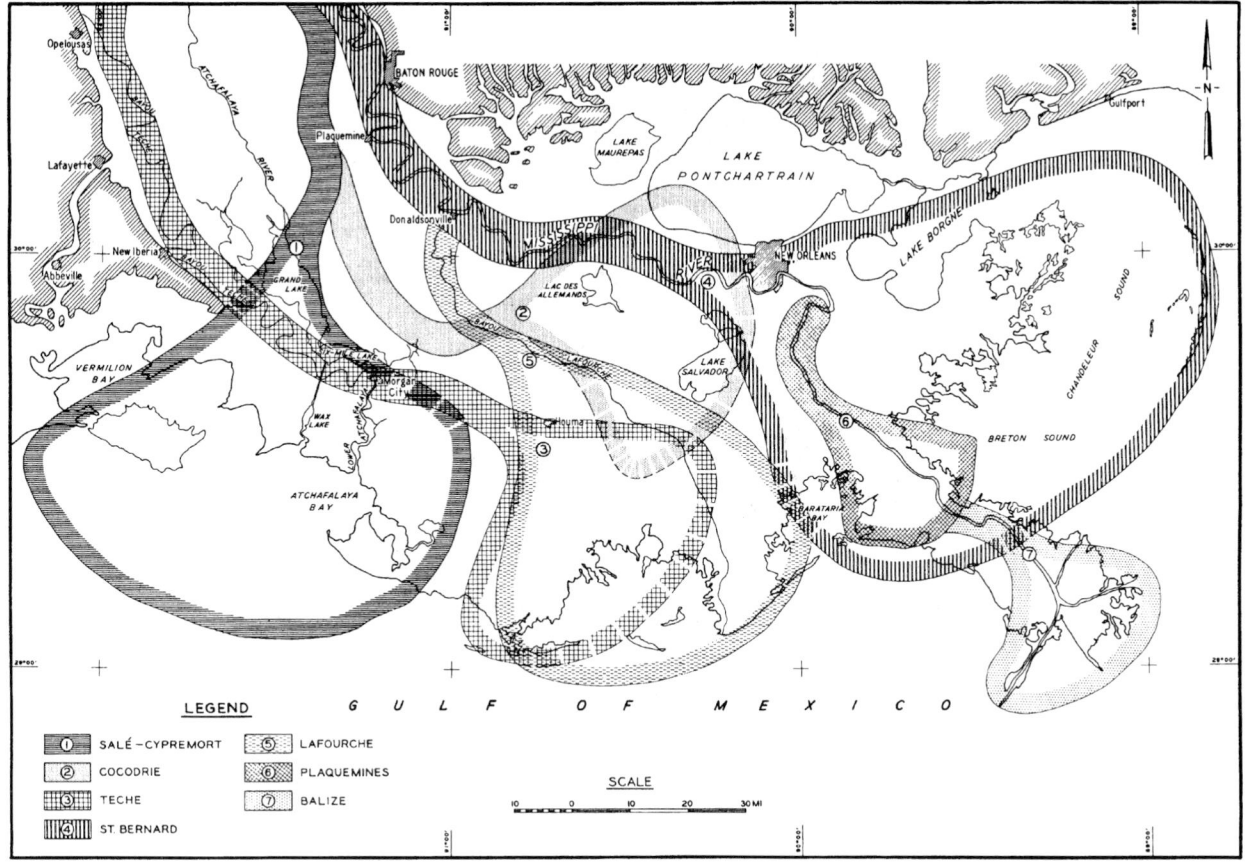

Figure D5 Mississippi River deltas. (From Kolb and Van Lopik, 1958.)

abandoned courses and distributaries form distinctive ribbons of relatively coarse sediments within the deltaic mass, sandy bodies that are comparatively small volumetrically but are of considerable significance to the engineering geologist. Abandoned course deposits fill the main channel left by the river when it is diverted at some point upstream into a new, shorter course to the sea (Figure D8). As shown in Figure D5, six such diversions of the Mississippi occurred in the recent geological past. Each course is now filled with distinctive sediments to depths and widths as great as the former channel −45–50 m deep, for example, and from 0.18 to 1.2 km wide. Fisk *et al.* (1952) showed that abandonment of a course is a slow process, but that a critical phase of the diversion process occurs when about 50% of the river's flow is diverted through the new channel. The former course is then plugged with sand just downstream from the point of diversion, and the new channel rapidly enlarges to take the entire flow. After abandonment, only high water is capable of breaching the sand wedge at the head of the abandoned course and sandy materials are distributed for some distance downstream from the point of diversion. The resulting body of sediment typically consists of a wedge of sand, gradually thinning downstream, overlain by a complementary clay wedge thinning upstream.

Abandoned distributaries are integral parts of deltaic advance, and hundreds, perhaps thousands, of narrow bands of distinctive sediments filling abandoned distributaries attest to the importance of this environment of deposition in the Mississippi deltaic plain. They range from a few meters of organic fill material – more logically considered part of the ubiquitous marsh deposits that flank or overlie them – to wedges of sandy organic soils 20 m or more thick. In deltas characterized by high tides and high sand loads, such as the Colorado, the Ord and the Burdekin, abandoned distributaries are typically filled with coarse clastics rather than with organic soils. Both ebb and flow tides sweep sand into the abandoned courses and distributaries, and thick, coarse-grained fill is not uncommon.

Fluvial-marine deposits

Three environments of deposition, each forming distinct lithological entities, characterize the advance of a delta: the prodelta, the intradelta and the interdistributary. The prodelta consists of the first terrigenous sediments introduced into the depositional area. The basal portions of the prodelta sediments are deposited in an essentially marine environment, but as the subaerial delta advances, more and more fluvial deposits are introduced and very fine clays with many marine shells give way to leaner clays and silts, and shells progressively begin to make up less of the unit (Perillo, 1995). Mississippi prodelta deposits are about 80% clay. Kolb and Kaufman (1967) have mapped thick wedges of prodelta clay that underlie the deltaic plain of the Mississippi and the immediate offshore areas. Strengths vary between 96 and 335 kPa. Thicknesses of the deposits range from a few meters to as much as 120 m. Oomkens (1970) records a maximum thickness of 50 m of prodelta clays in the Rhone delta. Naidu and Mowatt (1975) describe a prodelta sequence off the Colville in Alaska that differs basically both in process and in lithology from the prodelta of temperate and tropical areas. The Colville and other Arctic rivers are icebound for most of the year. At the spring break-up, water begins to issue from 'strudels' (cracks and holes in the ice), carrying with it silts, sands and sometimes gravel from the river bottom and spreading them out onto the more-or-less intact ice shelf that covers the delta. Essentially two prodelta facies result from rafting and melting of the ice. Inside the 20 m isobath, the deposits consist of a heterogeneous mixture ranging in grain size from clay to gravel; outside the isobath, the prodelta consists of very poorly sorted gravelly mud with molluscan and echinoderm remains.

Associated with and essentially a part of the prodelta clay environment are mud flats and mud lumps. Mud flats are principally prodelta clays which, instead of being deposited in a broad fan seaward of the delta, are swept by littoral and wind currents onto sheltered areas along the coast (Hardisty, 1990). They sometimes become established as huge masses of organic clays overgrown with

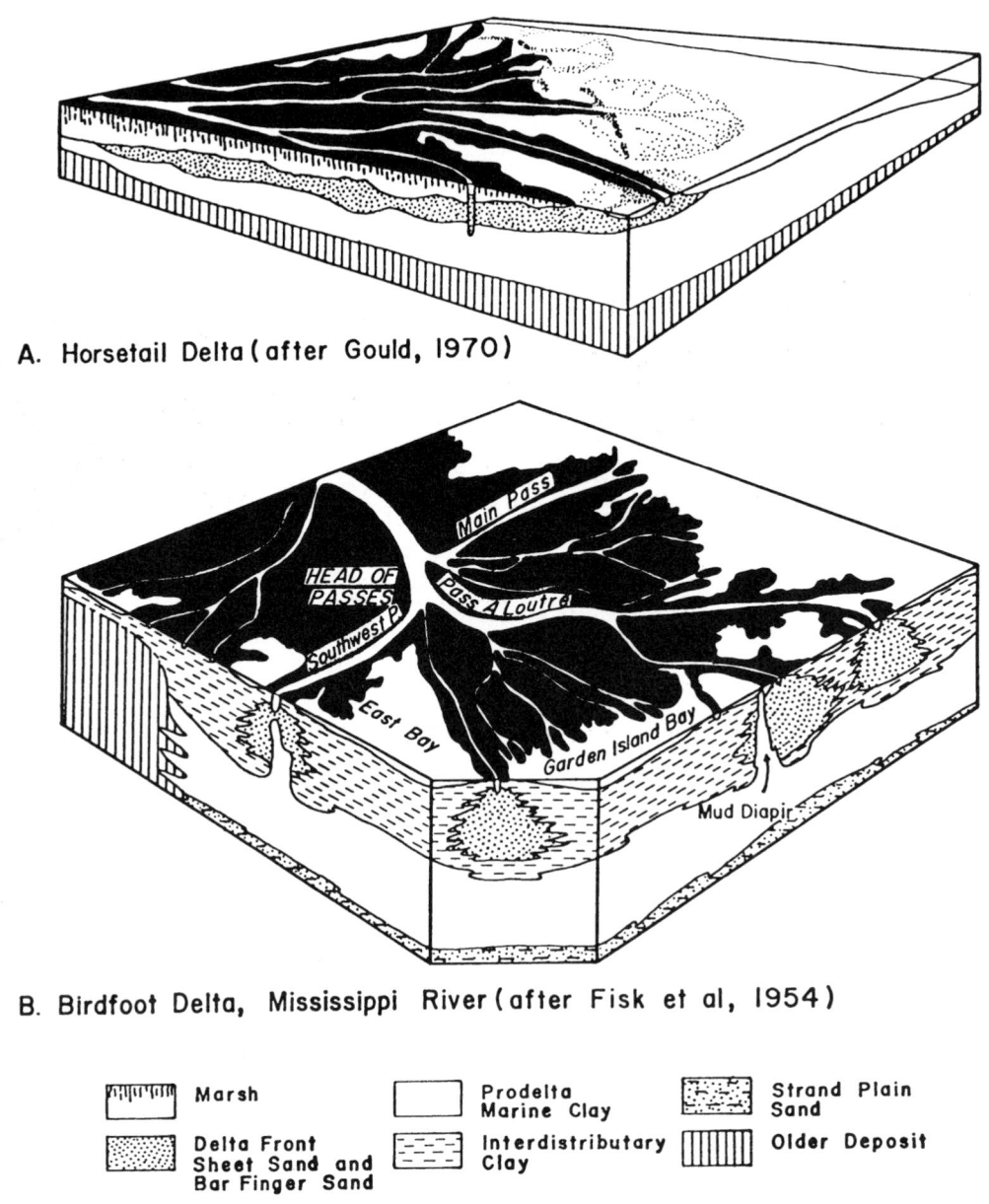

A. Horsetail Delta (after Gould, 1970)

B. Birdfoot Delta, Mississippi River (after Fisk et al, 1954)

Marsh		Prodelta Marine Clay		Strand Plain Sand	
Delta Front Sheet Sand and Bar Finger Sand		Interdistributary Clay		Older Deposit	

Figure D6 Ancient and modern Mississippi River delta types. (After Gould, 1970.)

marsh vegetation. Mud flats are common in many world deltas, for example, the Colorado, the Irrawaddy and the Klang. Mud lumps are prodeltaic clays that have worked their way upward through overlying coarser deltaic deposits – sometimes more than 100 m – as fingers of clay that reach the surface as mud islands within or immediately seaward of the Mississippi subaerial delta. Extensive studies, such as those by Morgan *et al.* (1963), detail these curious features that characterize the advance of the modern Mississippi delta. They probably occur in other world deltas (perhaps the Orinoco), but they have not been well documented.

Intradelta deposits are the coarser sediments associated with deltaic advance. As stream velocity decreases at the mouth of a distributary, the greater part of its load is deposited in distributary mouth bars. Sediments accumulate on the bar crest or on the seaward side of the bar. They also top underwater and subaerial natural levees on either side of a distributary as the delta builds seaward. Thus an irregular wedge of clastic sediments builds on either side and below the distributary as it advances (Colella and Prior, 1990). The mass sinks into the underlying prodeltaic clay and, in the Mississippi, distinctive

sandy units, sometimes 50 m or more thick, are formed beneath and parallel to the major distributaries. Often called bar fingers (Fisk *et al.*, 1954), they are evident in Figure D6B. Where many distributaries form channels in close proximity to one another, the coarse deposits tend to coalesce to form a single unit called a delta front sand sheet, also shown in Figure D6B. Delta front sand sheets have been reported from many world deltas. The Mekong, Colorado, Klang, Niger, Nile and Rhone are but a few of the deltas where delta front sand sheets have formed. The sand units that form as the deltas advance in some of these deltas (the Klang, the Colorado and the Ord, for example) are reworked by strong tidal currents into elongate bars that extend seaward, principally underwater, in the offshore areas for several kilometers. These bars shift with the fluvial and tidal cycles and can range up to tens of meters in height. Coleman (1981) cites a height of 10–22 m for such tidal bars on the Ord. Lengths vary from 1 to 15 km. These bars result from convergence of flood- and ebb-dominated bedload transport. On being buried during deltaic advance, these features form significant sand bodies in the deltaic mass. Variations of this process occur in wave-dominated deltas where the

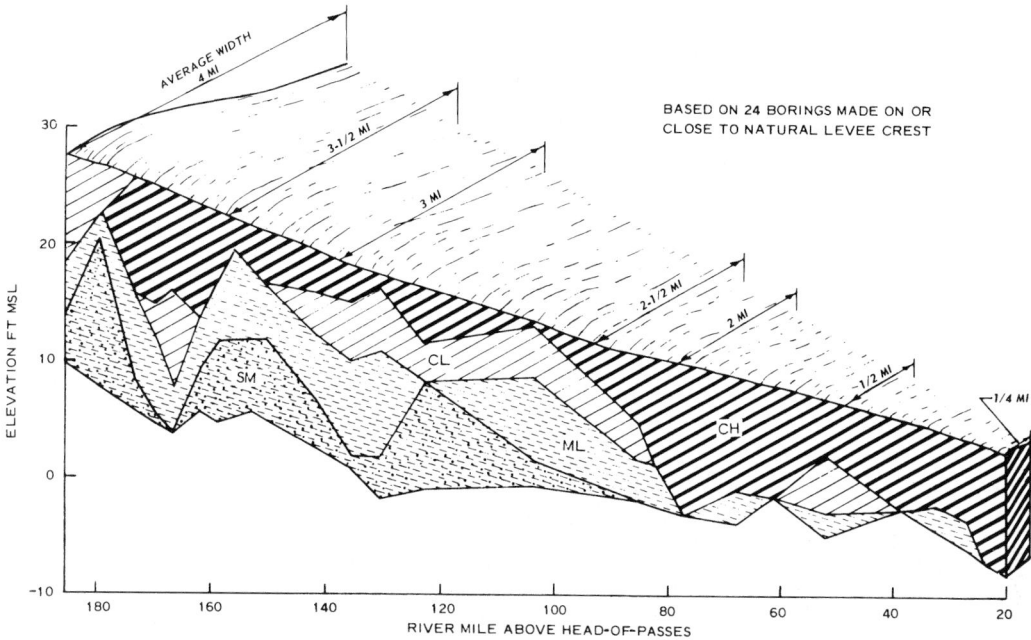

Figure D7 Longitudinal slope, width, thickness and general composition (soil type) of natural levee deposits from Donaldsville, Louisiana, to Head of Passes. (From Kolb, 1962.)

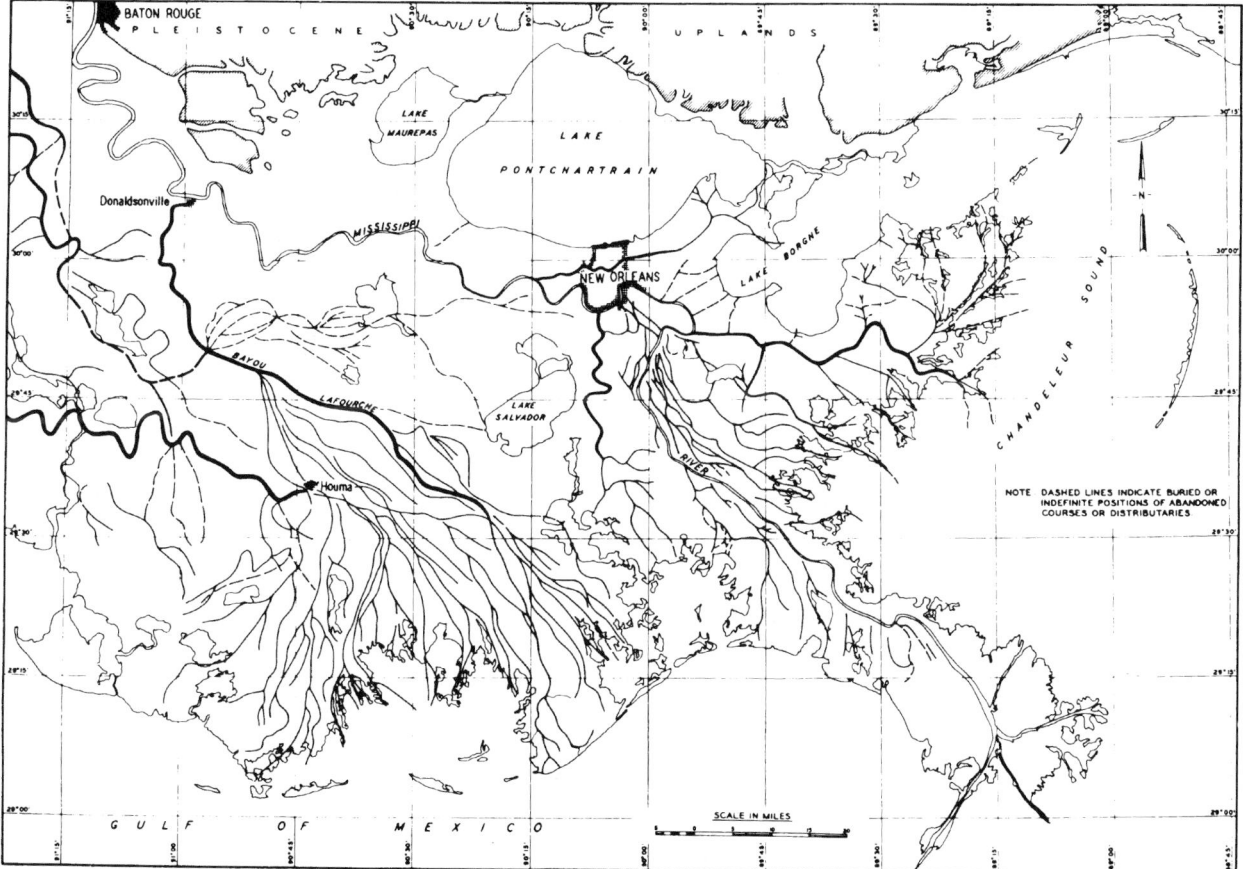

Figure D8 Distribution of abandoned courses and distributaries in the Mississippi deltaic plain. (From Kolb and Van Lopik, 1958)

sandy bars arrange themselves at some angle to the direction of delta advance in the offshore area; some eventually parallel the shoreline. Where they reach the surface, the bars from alongshore beaches or barrier beaches (Davis, 1993).

Studies indicate that nearly 85% of the materials within the Mississippi bar fingers and the delta front sand sheets (that is, within the intradelta environment) consists of silt or fine sand. Clay and organic clay occur in moderate amounts. From a geotechnical standpoint, these units tend to flank the paths of abandoned courses and distributaries irregularly and to grade laterally sometimes into poorly consolidated clays which form between the more widely spaced distributaries, the interdistributary environment.

Considerable thicknesses of interdistributary clays are deposited between the present Mississippi River distributaries. They form discrete bodies that grade downward into prodeltaic clays and upward into the richly organic clays of swamp or marsh deposits. The line of demarcation between the interdistributary and overlying marsh is indistinct. A true marsh or swamp forms when the watery area between distributaries or flanking the main channel has shallowed sufficiently to support vegetative growth. Interdistributary deposits are characteristically low-strength, high-water-content organic clays, some of the softest and least consolidated of the materials forming the deltaic mass. Interdistributary deposits in the more arid deltaic plains such as the Ord, the Shatt al Arab, the Senegal and the Indus, are commonly interfingered with evaporites. Tidal flats larger than 100 km^2 in area with gypsiferous and saline sediments 7 m thick are reported from the Ord.

Paludal deposits

Paludal deposits are largely *in situ* organic sediments. The predominant paludal environments are swamps and marshes, areas half land and half water which seldom rise more than 0.5 m above mean gulf level in the Mississippi delta. Complexly intermixed with these environments are clayey lacustrine and tidal channel deposits. The formation of lakes and tidal channels is an important part of marsh development in the Mississippi deltaic plain. The majority of these features, however, are shallow, insignificant water bodies in their original state that, when abandoned, leave behind sediments that for all practical geotechnical purposes can be classified as marsh. Nevertheless, a few are of such size and depth that they form deep channels or lakes during their active stages and, on being abandoned, fill with fairly massive bodies of organic clay.

Deposits filling abandoned tidal channels or left by migrating tidal channels become volumetrically important in deltas characterized by moderate to high tides and significant amounts of fluvially introduced sediments. Coleman (1981) describes tidal channels in the Klang completely filled with sand. Allen (1965, 1970) describes these features in the Niger:

> There is a major reticulate pattern of interconnected meandering tidal creeks which surround flats whose platform-like upper surfaces lie between tide lines. Cutoff meander loops are not uncommon and many of the creeks give evidence of capture by others; evidently the creeks are unstable, shifting their position in a manner similiar to river channels. . . . Each meander loop encloses a tidal point-bar. The bar deposits become finer grained upwards but at most levels muddy as well as sandy deposits can be found. The creeks reach a maximum depth of about 20 m. . . . The lower deposits consist of thick layers of fine- to coarse-grained sand in alternation with gray and black silty clays and clayey silts rich in finely divided organic matter and drifted debris.

The marsh forms flat, pervasive expanses of grasses and sedges which, together with the swamp environment, covers about 90% of the Mississipi deltaic plain. The vegetation of the marsh grows in close formation, providing a relatively firm surface underfoot, or the grasses may grow in tufts with mud and water between them. In some areas flotant, a mat of floating marsh, overlies black organic muck that has the consistency of thick gravy. Foundation problems in such areas are manifold. Interstate highways crossing them are often built on concrete pilings and have among the highest costs per kilometer for highways anywhere in the United States. Subsidence, the distance from the zone of active wave attack, and complex biological factors affect the distribution of land and water. Peats, organic oozes and humus are formed as the marsh plants die and are covered by water. Normal subaerial oxidation processes are limited, decay is largely due to anaerobic bacteria, and in stagnant water thick deposits consisting almost entirely of organic materials 20 m or more deep are formed.

Variable amounts of inorganic materials are found in the marsh deposits, the greatest amounts in what is termed fresh water marsh in areas subject to repeated inundation by flood flows or other freshwater sources. In the New Orleans area, swamp and marsh deposits underlie large portions of the expanding city, and subsidence of the marsh due to groundwater withdrawal, compaction and other reasons causes a host of foundation problems. Areas once above sea level now lie at elevations some 8 m below sea level and 15 m below the top of massive artificial levees that protect the city from recurrent floods on the Mississippi. Marsh deposits are common to most world deltas but are fairly unimportant in the drier climatic zones. In the Arctic they take on characteristics and landforms unique to the Arctic environment. Arnborg *et al.* (1962) describe patterned ground in the Colville delta, wind-oriented lakes, and ice-wedge polygons that often contain clastic subsurface materials intercalated with the polygon ice. Pingos, thick masses of perennial ice, rise above the surface of the deltaic plain at irregular intervals in mounds a few hundred meters in diameter at their bases and many meters high.

Swamps are distinguished from marshes principally because of the dense growth of trees on the former and their absence on the latter (Bird, 1993). This is reflected in swamp deposits by the typical occurrence of partially decayed stumps and trunks of trees. The organic content of swamp deposits is high but is generally lower than that of marsh deposits. Mangrove swamps are essentially saltwater swamps. Only found in a few isolated instances in the Mississippi deltaic plain and other temperate-zone deltas, mangrove swamps are of significant areal and geotechnical importance in the humid tropical deltas of the world. Mangrove swamps (e.g. of the Irrawaddy, Mekong, Orinoco, Godavari and Niger) are described as impenetrable jungles periodically inundated by the tides. The closely packed profuse vegetation, with its exposed intricate root systems, tends to trap clastic sediments brought in by flooding from fluvial sources as well as marine tidal sources. These sediments, mingled with finely divided organic matter and layers of drifted debris, form black organic muds and peats. These, in turn, are intercalated with the organic refuse left by decay of the mangrove jungle so that the mangrove swamps are sometimes built on organic platforms to elevations more than 1 m higher than the banks of the tidal and distributary channels that traverse them. Information about the thickness and variability of the strata typical of mangrove swamps is meager; it is known, however, that the deltaic plain of the Klang contains 18 m of peat.

Marine deposits

The three principal environments of marine deposition in the Mississippi deltaic plain are (1) bay-sound and the closely related nearshore gulf, (2) reef and (3) beach. Sedimentation in these environments takes place exclusively under marine conditions; as such, it might be argued that they are not truly deltaic plain deposits. However, large portions of deltaic plains consist of marine deposit that formed beneath, at the margins, or in the paths of deltaic advance or retreat (Whateley and Pickering, 1989). In the case of the Mississippi, significant volumes of marine sedimentation are intercalated with the deltaic plain. Of the Holocene deposits in Louisiana, only the Chenier Plain of western Louisiana is considered entirely marine and is not considered part of the deltaic plain.

Nearshore gulf and bay-sound deposits are relatively coarse, shelly sediments that are difficult to distinguish lithologically when encountered in borings. Environmentally they are distinctive in that nearshore gulf deposits are laid down at the borders of the open ocean, while bay-sound deposits are laid down in relatively quiet waters protected by barrier beaches and carpet the floors of bays and sounds. The nearshore gulf deposits are generally sandier than their bay-sound counterparts. They form a continuous blanket seaward of the bays and sounds except where interrupted by active deposition at the mouth of the Mississippi. Nearshore gulf deposits in the Mississippi deltaic plain are associated with a transgressive sea. Thus each abandoned delta is characterized by a variable thickness of such deposits as these deltas subsided and succumbed to wave attack. One of the complicating factors in delineating these buried basal sandy deposits in borings is their frequent merging in the Mississippi deltaic plain with 'strand plain' sands and shells at the base of the deltaic deposits immediately overlying the Pleistocene. Here a fairly persistent shell-and-sand horizon formed as the sea level rose in late Pleistocene and early Holocene times. Where identifiable, this horizon is more aptly referred to as the strand plain, not an integral part of the deltaic plain deposits. Not much is mentioned in the literature concerning near-shore or bay-sound deposits in other deltas. That comparable deposits exist is certain, however, particularly where deltaic plains are subsiding and undergoing attrition through wave attack.

Reefs are shell units that are undoubtedly associated with many deltaic plains, particularly those in tropical and subtropical areas; they have not, however, been widely reported in the literature (Pye and Lancaster, 1993). The only reef-forming mollusc of importance in the Mississippi deltaic plain is the oyster, *Crassostrea virginica*. Oysters typically build reefs in nearshore areas where regular influxes of fresh water bring about a mean salinity range of 10–30%, and where the mean temperature ranges between 10 and 25°C. Reefs of varying areas occur fairly extensively throughout the immediate offshore areas of southern Louisiana, particularly in the more sheltered portions of the bay-sound environment. The thickness of the reef deposits averages between 1 and 5 m; some are many kilometers in length. They are widely used for building roads and as aggregate in an area where coarse material suitable for such purposes is at a premium. Locating and exploiting shell reefs buried within the Mississippi deltaic plain has become big business in southern Louisiana, where natural material coarser than fine sand is virtually nonexistent.

From the standpoint of their engineering significance, beaches associated with the deltaic plain of southeastern Louisiana fall into two classes: sand beaches and shell beaches. Sand beaches are by far the more common in fluvially and sediment-dominated deltas such as the Mississippi. Shell beaches in southern Louisiana form along the inner margins of bays and sounds and on islands within these water bodies. Waves typically fracture the shells, and such deposits can be readily distinguished in soil borings from the more intact remains typical of reefs. Sand beaches in southern Louisiana consist almost entirely of fine sand and range in height from 1 to 5 m above sea level. They compress the softer deltaic materials at their bases, and the sand bodies sometimes reach thicknesses of 13 m or more. As the various deltas wax and wane, new sets of beaches are formed and subsequently become buried within the deltaic mass. Although not nearly so valuable as the coarser shell deposits of the reef and shell beach environments, these sand units are often sought and exploited as permeable fill and for surcharging and compacting areas underlain by marsh peats and highly organic clays.

Compared with the majority of deltaic plains, sand beaches are only poorly developed within and flanking the Mississippi deltaic plain. Stranded beach ridges reach far inland within the Mekong delta, for example, and form conspicuous elongate highs. Low sand beaches with extensive spits and eolian dunes typify the Ebro, and the Niger is characterized by nearly continuous sand beaches along the shoreline of its deltaic plain (Pye and Lancaster, 1993). The Nile and the Senegal have conspicuous broad, high sand beaches and barrier beaches with well-developed windblown dunes. The Sao Francisco is characterized by extremely large eolian dunes attaining elevations in excess of 22 m. Beneath the dunes and inland on the delta plain, broad sandy beach ridges, plastered one against another, form the major landforms within the entire deltaic plain (Coleman, 1981).

Summary

There are marked differences between geotechnically significant deltaic environments in the Mississippi and other world deltas; however, these differences are not so great that the framework developed for the Mississippi cannot be applied to other deltas. The grouping of deltaic environments into fluvial, fluvial-marine, paludal and marine types can be applied to any deltaic plain (Carter and Woodroffe, 1994). The principal differences in environmental types of geotechnical significance between the Mississippi and other world deltas are due chiefly to (1) discharge of the stream(s) forming the delta, (2) the amount and nature of sediment load, (3) the depth of water into which the deltaic material is being deposited, (4) the magnitude of the tides and offshore wave action and currents and (5) the climate.

Charles R. Kolb

Bibliography

Allen, J.R.L., 1965. Late Quaternary Niger Delta and adjacent areas, *Am. Assoc. Petroleum Geol. Bull.*, **49**, 547–600.
Allen, J.R.L., 1970. Sediments of the modern Niger Delta, in J.P. Morgan (ed.), *Deltaic Sedimentation Modern and Ancient, Soc. Econ. Paleontologists and Mineralogists Spec. Pub.* **15**, 138–151.

Andel, T.H., 1967. The Orinoco Delta, *J. Sed. Petrology*, **37**(2), 297–310.
Arnborg, L.E., Walke, H.J. and Peippo, J., 1962. *Suspended Load in the Colville River, Alaska*. Baton Rouge: Coastal Studies Institute, Louisiana State University, Rept. 54, 131–144.
Bernard, H.A., and LeBlanc, R.J., 1965. Resume of the Quaternary geology of the northwestern Gulf of Mexico province, in W.E. Wright and D.G. Frey (eds), *The Quaternary of the United States: Review Volume for the VII Congress of the International Association for Quaternary Research*, Princeton, NJ: Princeton University Press, 137–186.
Bird, E.C., 1993. *Submerging Coasts*, New York: Wiley, 184 pp.
Carter, R.W.G., and Woodroffe, C.D. (eds), 1994. *Coastal Evolution*. Cambridge: Cambridge University Press, 517 pp.
Colella, A. and Prior, D.B. (eds), 1990. *Coarse Grained Deltas*. Oxford: Blackwell Scientific Publications, 357 pp.
Coleman, J.M., 1981. *Deltas – Processes of Deposition and Models for Exploration*, 2nd edn. Minneapolis: Burgess, 124 pp.
Coleman, J.M. and Wright, L.D., 1975. Modern river deltas: variability of processes and sand bodies, in M.L. Broussard (ed.), *Deltas, Models for Exploration*, 2nd edn. Houston, TX: Houston Geological Society, 99–150.
Davis, R.A., 1993. *The Evolving Coast*. New York: Scientific American Library, 231 pp.
Fisk, H.N., Kolb, C.R. and Wilbert, L.J., 1952. *Geological Investigation of the Atchafalaya Basin and Problems of Mississippi River Diversion*. Vicksburg, MS: US Army Corps of Engineers, Mississippi River Commission, 145 pp.
Fisk, H.N., McFarlan, E., Jr, Kolb, C.R. and Wilbert, L.J., Jr, 1954. Sedimentary framework of the modern Mississippi Delta, *J. Sed. Petrology*, **24**, 76–99.
Galloway, W.E., 1975. Process framework for describing the morphological and stratigraphic evolution of deltaic depositional systems, in M.L. Broussard (ed.), *Deltas, Models for Exploration*, 2nd edn. Houston, TX: Houston Geological Society, 87–98.
Gould, H.R., 1970. The Mississippi Delta complex, in J.P. Morgan (ed.), *Deltaic Sedimentation: Modern and Ancient, Soc. Econ. Paleontologists and Mineralogists Spec. Pub.* **15**, 3–30.
Hardistry, 1990. *Beaches, Form and Process*. London: Unwin Hyman, 324 pp.
Inman, D.L. and Nordstrom, C.E., 1971. On the tectonic and morphologic classification of coasts, *J. Geology*, **79**, 1–21.
Kolb, C.R., 1962. Engineering soils bordering the Mississippi from Donaldsonville to the Gulf, *US Army Corps Engineers, Waterways Expt. Sta. Misc. Paper 3–481*.
Kolb, C.R. and Dornbusch, W.K., 1975. The Mississippi and Mekong deltas – a comparison, in M.L. Broussard (ed.), *Deltas, Models for Exploration*, 2nd edn. Houston, TX: Houston Geological Society, 193–207.
Kolb, C.R., and Van Lopik, J.R., 1958. Geology of Mississippi River deltaic plain, southeastern Louisiana, *US Army Corps Engineers, Waterways Expt. Sta. Tech. Rept. 3–483 and 3–484*, 2 vols.
Morgan, J.P., 1970. Depositional processes and products in the deltaic environment, in J.P. Morgan (ed.), *Deltaic Sedimentation: Modern and Ancient, Soc. Econ. Paleontologists and Mineralogists, Spec. Pub. 15*.
Morgan, J.P., Coleman, J.M. and Gagliano, S.M., 1963. *Mudlumps at the Mouth of South Pass, Mississippi River: Sedimentology, Paleontology, Structure, Origin, and Related Deltaic Processes*. Baton Rouge: Coastal Studies Institute, Louisiana State University, Coastal Studies Series No. 10, 190 pp.
Naidu, A.S. and Mowatt, T.C., 1975. Depositional environments and sediment characteristics of the Colville and adjacent deltas, northern arctic Alaska, in M.L. Broussard (ed.), *Deltas, Models for Exploration*, 2nd edn. Houston, TX: Houston Geological Society, 268–283.
Nelson, B.W., 1970. Hydrology, sediments dispersal, and recent historical development of the Po River Delta, Italy, in J.P. Morgan (ed.), *Deltaic Sedimentation: Modern and Ancient, Soc. Econ. Paleontologists and Mineralogists, Spec. Pub. 15*, 152–184.
Oomkens, E., 1970. Depositional sequences and sand distribution in the post-glacial Rhone Delta complex, in J.P. Morgan (ed.), *Deltaic Sedimentation: Modern and Ancient, Soc. Econ. Paleontologists and Mineralogists Spec. Pub.*, 15, 198–212.
Perillo, G.M.E. (ed.), 1995. *Geomorphology of Estuaries*. Amsterdam: Elsevier, 471 pp.

Pye, K. (ed.), (1994). *Weathering and Deltas: sediment Transport and Depositional Processes.* Oxford: Blackwell, 397 pp.

Pye, K. and Lancaster, N. (eds), 1993. *Aeolian Sediments, Ancient and Modern.* Oxford: Blackwell, 167 pp.

Rodolfo, K.S., 1975. The Irrawaddy Delta: tertiary setting and modern offshore sedimentation, in M.L. Broussard (ed.), *Deltas, Models for Exploration*, 2nd edn. Houston, TX: Houston Geological Society, 339–356.

Selley, R.C., 1988. *Applied Sedimentology.* London: Academic Press, 446 pp.

Whateley, M.K.G. and Pickering, K.T. (eds), 1989. *Deltas, Sites and Traps for Fossil fuels.* Oxford: Blackwell Scientific Publications, 360 pp.

Cross references

Alluvial valley engineering
Maritime zones
River engineering
Rivers

DENSITY FOR A MINIMUM NETWORK OF HYDROLOGICAL STATIONS

The minimum network is one that will avoid serious deficiencies in developing and managing water resources on a scale commensurate with the overall level of economic development and environmental needs of the country. It should be developed as rapidly as possible, incorporating existing stations, as appropriate. In other words, such a network will provide the framework for expansion to meet the information needs of specific water uses.

The concept of network density is intended to serve as a general guideline if specific guidance is lacking. As such, the design densities must be adjusted to reflect actual socio-economic and physio-climatic conditions. Computer-based mathematical analysis techniques should also be applied, where data are available, to optimize the network density required to satisfy specific needs. For example, the network analysis using generalized least squares (NAUGLS) developed by the US Geological Survey (Moss and Tasker, 1991) offers a promising approach for optimizing the stream gauges in a basic network for regional information.

In the following sections, minimum densities of various types of hydrological stations are recommended for different climatic and geographic zones (WMO, 1992). It is impossible to define a sufficient number of zones to represent the complete variety of hydrological conditions. A limited number of larger zones have been defined in a somewhat arbitrary manner.

The simplest and most precise criterion for the classification of zones would be on the basis of the areal and seasonal variation of rainfall. Each country could present a good map of annual precipitation and a minimum network would be developed from this. However, this would not help the various countries that need a network most as they have very few prior records, and the establishment of a good precipitation map is impossible. One must consider, as a special category, the countries with very irregular rainfall distribution. It is not advisable to base the classification on this one characteristic.

Population density also affects network design. It is almost impossible to install and operate, in a satisfactory way, a number of stations where population is sparse. For example, to set up more than two gauges on a catchment of 1000 km², when the population of the area is only 100 people, is almost impossible, especially if this population is not permanent. Besides, it is difficult even to find observers in thinly populated areas where access is poor. Sparsely settled zones, in general, coincide with various climatic extremes, such as arid regions, polar regions or tropical forests. The use of totalizers (storage gauges) is recommended in such cases because they need little maintenance and infrequent visits.

On the other extreme, densely populated urban areas need a very dense raingauge network for both temporal and spatial resolution of storms and for design, management, and real-time control of the storm–drainage systems and for other engineering applications.

From these considerations, some general rules have been adopted for the definition of density norms. Six types of physiographic regions have been defined for minimum networks:

- Coastal;
- Mountainous;
- Interior plains;
- Hilly/undulating;
- Small islands (surface areas less than 500 km²);
- Polar/arid.

For the last type of region, it is necessary to group together the areas in which it does not seem currently possible to achieve completely acceptable densities because of sparse population, poor development of communications facilities, or for other economic reasons.

Minimum densities for climatological stations

The following kinds of data are collected at a climatological station in the basic network: precipitation, snow survey and evaporation. It is understood here that evaporation- or snow-measuring stations, particularly the former, will generally measure temperature, humidity and wind because these meteorological elements affect evaporation and melting.

Precipitation stations

The minimum densities for precipitation stations are provided in Table D7. These densities are not applicable to the great deserts (Sahara, Gobi, Arabian, etc.) and great ice fields (Antarctic, Greenland and the Arctic islands) that have no organized hydrographic networks. In these regions, precipitation is not studied by raingauge networks of standard type, but by special stations and methods of observation.

If one follows certain principles of installation and use, the small number of stations in the minimum network can furnish the most immediate needs. In general, precipitation gauges should be as uniformly distributed as is consistent with practical needs for data and the location of volunteer observers. In mountainous regions, attention must be given to vertical zonality by using storage gauges to measure precipitation at high altitudes. Snow surveys may be used to supplement the network, but they should not be counted as part of the network.

The minimum network should consist of three kinds of gauges:

- Standard gauges. These gauges are read daily for quantity. Besides daily depth of precipitation, observations of snowfall, the depth of snow on the ground and the state of the weather are to be made at each standard precipitation station;
- Recorders. In developing networks, it is advisable to aim to have at least 10% of such stations in cold climates. The greatest density of recording stations should be achieved in those areas subject to intense, short-duration rainfalls. Such stations will provide valuable information on the intensity, distribution and duration of precipitation.

For urban areas where the time resolution needed for rainfall measurements is of the order of 1–2 min, special attention should be paid to the time synchronization of the raingauges. For reliable measurements, tipping bucket raingauges with an electronic memory (or another computer readable medium) are recommended.

In assigning priorities to locations for recording raingauge installations, the following types of areas should be given priority: urban areas (population in excess of 10 000) where extensive

Table D7 Recommended minimum densities of precipitation stations

Physiographic unit	Minimum densities per station (area in km² per station)	
	Non-recording	Recording
Coastal	900	9 000
Mountainous	250	2 500
Interior plains	575	5 750
Hilly/undulating	575	5 750
Small islands	25	250
Urban areas		10–20
Polar/arid	10 000	100 000

drainage systems are likely to be constructed river basins in which major river control systems are anticipated or in operation, large areas inadequately covered by the existing network, and special research projects.

- Storage gauges (totalizers). In sparsely settled or remote regions, such as in desert or mountainous terrain, storage gauges may be used. These gauges are read monthly, seasonally or whenever it is possible to inspect the stations.

Location of precipitation gauges relative to stream-gauging stations

To ensure that precipitation data are available for extending stream-flow records, for flood forecasting purposes, or for hydrological analysis, coordination of the locations of the precipitation gauges with respect to those of the stream gauges is of great importance. Precipitation gauges should be located so that basin precipitation can be estimated for each stream-gauging station. These will usually be located at or near the stream gauge and in the upper part of the gauged drainage basin. A precipitation gauge should be located at the site of the stream gauge only if the observations will be representative of the general area. There can be cases in which it is desirable to locate the precipitation gauge some distance away from the stream gauge, as for instance when the stream gauge is in a narrow, deep valley.

Snow surveys

Where applicable, observations of snowfall, water equivalent of snow and depth of snow on the ground should be made at all precipitation stations in the minimum network.

The water equivalent of snow at the time of maximum accumulation is an indication of total seasonal precipitation in regions where winter thaws and winter snow melt are insignificant. In such regions, surveys of the snow cover on selected courses may be useful in estimating seasonal precipitation at points where the normal observations are unavailable. Such snow-cover surveys will also provide useful information for river forecasting and flood studies.

Snow cover surveys are conducted by special teams equipped with simple instruments for sampling the accumulated snow and for determining its depth and water equivalent. The number of the snow courses, their locations and lengths will depend upon the topography of the catchments and the purposes for which the data are being collected. The full range of elevation and the types of exposure and vegetation cover in the area of interest should be considered in selecting representative courses. It has been suggested that one course for 2000–3000 km^2 is a reasonably good density for less homogeneous regions, and one course for 5000 km^2 in homogeneous and plain areas. However, each case must be considered on its own merits, and these generalities must not be applied indiscriminately.

In the early stages of network development, snow-cover surveys will usually be made only once a year, near the expected time of maximum accumulation. It will be desirable, later on, to extend the operation to include surveys at regular intervals throughout the snowfall season. As soon as it becomes feasible, the snow cover observations should be augmented by observations of related meteorological factors, such as radiation, soil temperature and wind velocity.

Evaporation stations

Evaporation can be estimated indirectly in the water budget, energy budget and aerodynamic approaches, and directly by extrapolation from pan measurements. An evaporation station consists of a pan of standard national designs where daily observations of evaporation are made, together with daily observations of precipitation, maximum and minimum water and air temperatures, wind movement, and relative humidity or dewpoint temperature. The norms recommended for a minimum network of evaporation stations within areas of uniform physiography are given in Table D8.

Evaporation plays an important role for long-term studies of the water regime of lakes and reservoirs and for water management. In such cases, the number and distribution of evaporation stations are determined according to the area and configuration of the lakes and the climatic region or regions involved.

Table D8 Minimum density of evaporation stations

Physiographic unit	Minimum density per station (area in km^2 per station)
Coastal	50 000
Mountainous	50 000
Interior plains	50 000
Hilly/undulating	50 000
Small islands	50 000
Polar/arid	100 000

Minimum densities for hydrometric stations

Streamflow stations

The main objective of the stream-gauging network is to obtain information on the availability of surfacewater resources, their geographical distribution and their variability in time. Magnitude and frequency of floods and droughts are of particular importance in this regard.

The minimum densities for streamflow stations are given in Table D9. These norms are not applicable to the great deserts with no defined stream networks (such as the Sahara, Gobi, Arabian and Korakorum deserts) and great ice fields (Antarctic, Greenland and Arctic islands).

In general, a sufficient number of streamflow stations should be located along the main stems of large streams to permit interpolation of discharge between the stations. The specific location of these stations should be governed by topographic and climatic considerations. If the difference in flow between two points on the same river is not greater than the limit of error of measurement at the station, then an additional station is unjustified. In this context it must also be stressed that the discharge of a small tributary cannot be determined accurately by substracting the flows at two main-stream gauging stations which bracket the mouth of the tributary. Where the tributary flow is of special interest in such a case, a station on the tributary will be required. It will usually take its place as a secondary station in the minimum network. The streamflow stations may be interspersed with stage stations.

Wherever possible, the base stations should be located on streams with natural regimes. Where this is impractical, it may be necessary to establish additional stations on canals or reservoirs to obtain the necessary data to reconstruct the natural flows at the base stations. Computed flows past hydroelectric plants or control dams may be useful for this purpose, but provisions will have to be made for calibration of the control structures and turbines and for the periodic checking of such calibrations during the life of the plants.

Stations should be located on the lower reaches of the major rivers of the country, immediately above the river mouths (usually above tidal influence), or where the rivers cross borders. Stations should also be located where rivers issue from mountains and above the points of withdrawal for irrigation water. Other hydrometric stations are situated at points, such as where the discharge varies to a considerable extent, below the points of entry of the major tributaries, at the outlets from lakes, and at those locations where large structures are likely to be built.

To ensure adequate sampling, there should be at least as many gauging stations on small streams as on the main streams. However, for small streams, a sampling procedure becomes necessary as it is impracticable to establish gauging stations on all of them. The

Table D9 Recommended minimum densities of streamflow stations

Physiographic unit	Minimum density per station (area in km^2 per station)
Coastal	2 750
Mountainous	1 000
Interior plains	1 875
Hilly/undulating	1 875
Small islands	300
Polar/arid	20 000

discharge of small rivers is strongly influenced by local factors. In highly developed regions, where even the smallest watercourses are economically important, network deficiencies are keenly felt even on streams draining areas as small as 10 km^2.

Stations should be installed to gauge the runoff in different geologic and topographic environments. Because runoff varies greatly with elevation in mountains, the basic network stations must be located in such a way that they can, more or less evenly, serve all parts of a mountainous area, from the foothills to the higher regions. Account should be taken of the varying exposure of slopes, which is of great significance in rough terrain. Similarly, consideration should be given to stations in districts containing numerous lakes, whose influence can be determined only through the installation of additional stations.

River stages

Stage (height of water surface) is observed at all stream-gauging stations to determine discharge. There are places where additional observations of water level only are needed as part of a minimum network:

- at all major cities along rivers, river stages are used for flood forecasting, water supply, and transportation purposes;
- on major rivers, at points between stream-gauging stations, records of river stage may be used for flood routing and forecasting purposes.

Lake and reservoir stages

Stage, temperature, surge, salinity, ice formation, etc., should be observed at lake and reservoir stations. Stations should be established on lakes and reservoirs with surface areas greater than 100 km^2. As in the case of rivers, the network should sample some smaller lakes and reservoirs as well.

Sediment discharge and sedimentation

Sediment stations may be designed either to measure total sediment discharge to the ocean or to measure the erosion, transport and deposition of sediment within a country, basin, etc. In designing a minimum network, emphasis should be placed on erosion, transport and deposition of sediment within a country. An optimum network would contain a sediment station at the mouth of each important river discharging into the sea.

Sediment transport by rivers is a major problem in arid regions, particularly in those regions underlain by friable soils and in mountainous regions where, for engineering applications, the amount of sediment loads should be known.

Although the densities given in Table D10 serve as guides in considering a basic network, the designer must be forewarned that sediment transport data are much more expensive to collect than other hydrological records. Consequently, great care must be exercised in selecting the number and location of sediment transport stations. Emphasis should be placed on those areas where erosion is known to be severe. After a few years of experience, it may be desirable to discontinue sediment measurements at those stations where sediment transport no longer appears to be of importance.

Sediment transport data may be supplemented by surveys of sediment trapped in lakes or reservoirs. Echo sounding devices are useful for this purpose. However, information obtained in this way is not considered a substitute for sediment transport measurements at river stations.

Table D10 Recommended minimum densities for sediment stations

Physiographic unit	Minimum density per station (area in km^2 per station)
Coastal	18 300
Mountainous	6 700
Interior plains	12 500
Hilly/undulating	12 500
Small islands	2 000
Polar/arid	200 000

Table D11 Recommended minimum densities for water quality stations

Physiographic unit	Minimum density per station (area in km^2 per station)
Coastal	55 000
Mountainous	20 00
Interior plains	37 500
Hilly/undulating	47 500
Small islands	6 000
Polar/arid	200 000

Water quality stations

The usefulness of a water supply depends, to a large degree, on its chemical quality. Observations of chemical quality, for the purposes of this contribution, consist of periodic sampling of water at stream-gauging stations and analyses of the common chemical constituents.

The number of sampling points in a river depends on the hydrology and the water uses. The greater the water quality fluctuation, the greater the frequency of measurement required. In humid regions, where concentrations of dissolved matter are low, fewer observations are needed than in dry climates, where concentrations, particularly of critical ions such as sodium, may be high.

As a minimum network, records of water quality should be obtained at the densities shown in Table D11.

Water temperature

The temperature of water should be measured and recorded each time a hydrometric station is visited to measure discharge or to obtain a sample of the water. The time of day of the measurement should also be recorded. At stations where daily stage observations are made, temperature observations should also be made daily. These observations, whose cost is negligible, may provide data which are useful in studies of aquatic life, pollution, ice formation, sources of cooling water for industry, temperature effects on sediment transport, or solubility of mineral constituents.

Source

World Meteorological Organization, 1994. Guide to Hydrological Practices, 5th edn, WMO Geneva.

Bibliography

Moss, M.E. and Tasker, G.D., 1991. An intercomparison of hydrological network design technologies. *Hydrological Science Journal*, **36**, 209–221.
World Meteorological Organization, 1992. *Proceedings of the International Workshop on Network Design Practices*. 11–15 November 1991, Koblenz, Germany.

Cross references

Evaporation: measurement
Evapotranspiration
Hydrological yearbooks
Rain
Streamflow measurement

DESALINATION

The sun and wind turn very salty seawater into very pure water vapor, all the sodium, calcium and magnesium compounds being left behind. Similarly, in cold regions, when salt water solidifies relatively pure ice is formed. The technique of turning seawater into fresh water is called desalination, perhaps an illogical term since the process is one of removing the water and leaving the salt behind.

Desalination, using simple evaporation and condensation of heated seawater, may have been used by the Greeks and Romans two thousand years ago to provide emergency water supplies. Desalination has been used in ships since steam propulsion became common, but

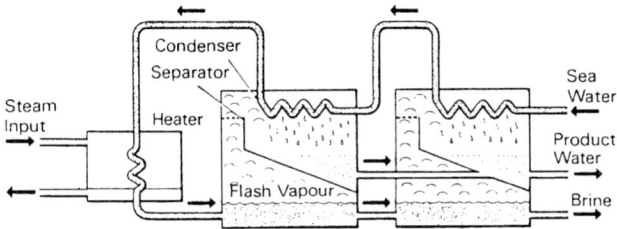

Figure D9 Multi-stage flash distillation (UKAEA, 1970).

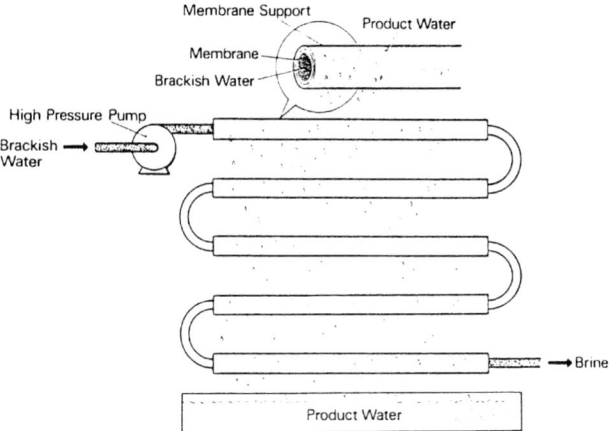

Figure D10 Reverse osmosis (UKAEA, 1970).

experiments were undertaken in the British Navy on board sailing ships at the time of Samuel Pepys (1633–1703).

Desalination has already made tremendous contributions to the development of countries in arid zones of the world and is now playing a part in the water supply of non-arid countries.

'Desalting' has come to mean all treatment of water in the context of desalination, that is, fresh water, river water, brackish water, seawater and wastewater, by means of thermal (evaporation) or membrane (reverse osmosis, nanofiltration and electrodialysis).

The method of desalination is to some extent dependent on the degree of salinity in the water and there should be a source of energy. Seawater contains about 35 000 ppm salts (3.5%) and brackish water about 0.5% salts.

In general, distillation and evaporation are more suitable for use with high-salinity water while methods of desalination using a membrane such as reverse osmosis or electrodialysis are more suited to lower-salinity water because of the cost of operation. The choice of desalination, therefore, in any given situation depends largely on the economics of the various processes including the chemical types of salts present in the water, e.g. large amounts of magnesium and calcium compounds – chlorides, sulfates, carbonates and hydrogencarbonates.

Methods

The two most common methods are as follows.

Multi-stage flash distillation (MSF)

The overall principle of MSF is simply that of evaporating saline water and obtaining pure water by condensation (Figure D9). Heated brine is introduced in open channel flow into a chamber under reduced pressure. Some of the water evaporates immediately (flashes) and is condensed on tubes cooled by the feed water flowing towards the steam-heated heat input section. A series of such chambers at progressively reduced pressure forms the plant. The feed seawater increases in temperature as it passes through the condenser tubes – in the opposite direction to the flashing flow – towards the heat input section. The condensed steam is collected on trays below the condensers and pumped out of the plant as water.

Reverse osmosis

The method depends on the properties of a semi-permeable membrane which, when used to separate water from a salt solution, allows fresh water to pass into the brine compartment under the influence of osmotic pressure (Figure D10). If a pressure in excess of this value is applied to the salt solution, fresh water will pass from the brine into the water compartments.

Thermally, the only energy required is that to pump the feed water to the osmotic pressure. In practice, higher pressures must be used in order to have a useful volume of water pass through a unit area of membrane per day.

Distribution of plants

The total capacity of plants worldwide either in service or under construction, is nearly 20 million m³ day⁻¹ of which about 65% is seawater desalination. About 70% of all desalination units are located in Middle East countries (approx 8.5 million m³ day⁻¹). In the USA, brackish water desalination is increasingly being used in Florida and southern states with a capacity of some 2.7 million m³ day⁻¹. Other areas include islands in the Mediterranean, especially Malta, the Balearics and the Canary Islands, and in southern Spain. African

Table D12 Total number of desalination plants according to capacity

Capacity (10^6 m³ day⁻¹)	Number of units
0–0.5	3 250
0.5–1.0	2 000
1.0–1.3	2 000
1.3–1.8	500
1.8–2.2	500
2.2–2.7	500
2.7–3.1	500
3.1–3.6	250
3.6–4.0	250
4.0–4.5	250
Total	10 000

Table D13 Distribution of types of plant

Type	%
Multistage flash (MSF)	60
Reverse osmosis (RO)	25
Multiple effect distillers (MED)	5
Electrodialysis (ED)	5
Vapour compression (VC)	5

countries have less than 5% of the total. Table D12 gives the number of plants, or units, in million m³/d and Table D13 gives the approximate distribution of different types of plant.

Research

Research on the desalination of seawater, funded by the European Union, is being continued notably by a team from the Mediterranean countries of Cyprus, Greece, Italy, Jordan and Portugal, under the leadership of the Higher Technical Institute in Nicosia. Two pilot solar- and wind-powered desalination plants are planned over the next 3 years (1997–2000), one in Greece and one in Jordan. It is envisaged that the two plants will each desalt about 3 m³ day⁻¹ at the end of the study. A hybrid solar- and wind-powered desalination plant has also been erected in Tenerife near El Medano, supported by the European Union, with an average freshwater production of 630 l h⁻¹ (World Water, 1997).

Costs

The cost of desalinated water is higher than that of existing supplies. Present-day 'natural' supplies cost on average less than $1 per m³

(UK) whereas the cost of desalinated water can vary between \$1 and \$2 per m³. The lower cost of 'natural' water is due mainly to the capital costs of reservoir construction, pumping stations and aqueducts which have been amortized and do not appear in the price structure. For a realistic comparison, the actual cost of development of new water supplies should be made including the cost of reservoir construction, aquifers, dams, treatment works, pipelines and land. If calculated on this basis, it is estimated that such new supplies would cost between \$1.5 and \$2.25 per m³ (UK).

R.W. Herschy

Bibliography

IDA News, May, 1996. The latest inventory report, IDA, MA., USA.

Temperley, T.G., 1997. Desal solution leaves shortage gloom at sea, *Water Services*, **101**(1214), 8–9.

United Kingdom Atomic Energy Authority and Central Office of Information, 1970. *Desalination and its role in water supply*, HMSO, London

Wade, N. and Callister, K., 1997. Desalination; the state of the art, CIWEM, **11**(2), London, 87–97.

Wangnick, K., 1987. *The 1986 world market of desalting plants*, The International Desalting Association (IDA), MA., USA.

World Water, 1997. **20**(5), Faversham House Group, London.

Cross references

Water resources
World water balance

DESERT HYDROLOGY

Water in the desert

The terms water and desert seem somewhat incongruous, but water is a very active force in creating and sculpting the desert environment. Its patterns of availability in time and space control the biota of deserts. Thus the surface hydrology of deserts is a significant aspect of the environment.

The water supply in desert regions can be characterized as endogenous, originating within the desert, or exogenous, originating externally in more humid environments. The endogenous sources are stream flow and surface or subsurface reservoirs. These reservoirs include water which has accumulated in sands (which have high infiltration, but dry out only at the surface), aquifers in alluvial fans or occasionally in limestone rocks (karst features), playas and freshwater accumulations along some coasts. A much greater volume of water is exogenous, originating from groundwater and rivers.

The water availability in a given dryland region is therefore a function not only of the local climate, but also of various geographical factors. These include soil type, slope and topography, the nature of the drainage system, proximity to groundwater (depth of the water table) and proximity to seasonal or exotic streams. The driest areas tend to be the playas and rocky slopes. Favorable environments include highlands, where rainfall is higher than on the plains; depressions and the bases of slopes or inselbergs, where runoff collects; and oases, which may be spring-fed from groundwater. The water table tends to be elevated under sand dunes, allowing dunes to support plant growth. The dry beds of exogenous streams are also moist environments where lush vegetation often thrives. Desert aquifers provide another source of water.

Overview of surface hydrology

The hydrological regime in arid regions differs fundamentally from that in wetter regions. The differences result from numerous factors, but three are basic: the relative sparseness of the vegetation cover, the small amount of rainfall and its episodic and localized nature (see later). Vegetation utilizes water and facilitates infiltration; it also returns moisture to the atmosphere by evapotranspiration. As a result, a vegetation cover reduces immediate runoff but promotes storage and delayed runoff. It therefore moderates the influence of extreme precipitation events on streamflow. The low rainfall reduces the generation of runoff, and hence stream discharge. Its intermittency in time and space are strongly reflected in streamflow, which is also episodic or strongly seasonal and usually spatially intermittent and variable.

A summary of the hydrology of deserts is difficult, because the desert environment is so diverse. An overview is facilitated through the use of a classic physiographic framework distinguishing shield-and-platform deserts and mountain-and-basin deserts on the basis of topography and structure (Cooke and Warren, 1975; Mabbutt, 1979). The latter are areas of high relief superimposed on low plains. These deserts are often in cold-winter, midlatitude locations such as the cordillera of the Americas and eastern Asia. The shield-and-platform deserts are broad plains of low relief, covered with stone surfaces, sand seas or finer materials. They tend to be areas of warm continental climates, like the Sahara or Australian deserts.

Most desert environments represent a point on a transitional continuum with these two types at the extreme ends (Figure D11). The continuum begins with the desert uplands, areas of high relief and exposed bedrock where erosion is dominant and the desert watershed originates. The remaining settings, in a downslope sequence, are the piedmont, the stony deserts, the lake basin and the sand deserts. The piedmont is a more gentle slope, marked by an abrupt change of gradient from the uplands. It is an area of both erosion and deposition. The flatter, lowland environments are largely depositional. This sequence is evident in all deserts, but is geographically compressed in the mountain-and-basin deserts. In contrast, the piedmont and upland settings are markedly diminished in the shield-and-platform deserts.

Each environment is functionally related to those which are upslope and downslope, with desert rivers and floodplains linking the uplands with the piedmont and lowlands. The uplands control the runoff and the supply of detritus which reach the piedmont and the plains below.

Along this continuum of physiographic settings are associated sequences of aridity, drainage patterns, energy supply and relative dominance of the forces of wind and water. There is a general increase of aridity downslope, largely because of the orographic enhancement of rainfall in the uplands. Correspondingly, there is change in the nature of the drainage from connected, branching systems with distinct channels in the uplands to disorganized and disintegrated networks with sparse, thin channels of flow in the lower regions.

The importance of fluvial activity wanes downslope, as eolian activity increases; the desert lakes often mark the termination of desert drainage and the transition from the fluvial to the eolian regime. In regimes of erosion and weathering in the uplands, the energy for geomorphological processes derives from insolation and the potential energy of water under the influence of gravity (Cooke and Warren, 1975). On the hillslopes the potential energy of water is of primary importance, with a secondary role played by the potential energy of loose soil (e.g. sand blast and dust abrasion). In the drainage channels, energy is almost exclusively supplied by water. In the playas the energy which sculpts the surface is supplied by the sun and wind. Thus the various settings of the desert environment have a close relationship to atmospheric processes.

Desert uplands and slope

The upland landscape of the desert consists of steep and rugged exposed rocks and debris-covered slopes (Figure D12). This segment of the landscape, where erosion is the dominant process, produces most of the runoff. At the top is a relatively flat exposed surface (waxing slope), abutting a steep cliff or free-face; here movement is gravity controlled. Below is the debris-mantled slope. In steeper portions debris is coarse and moves downslope by mass movement (debris-controlled slopes), but on gentler slopes the debris is finer and moved by slope wash (wash-controlled or rain-washed slopes). The debris usually forms only a thin layer; these are slopes of transportation where there is an equilibrium between a supply of material from the cliff face above and removal downslope by gravity and water. In the uplands flow is unconfined or overland; water moves through shallow and impermanent rills rather than permanent channels. On the rugged slopes below, drainage becomes organized and water flows through deep channels.

The piedmont is a gentler slope, marked by an abrupt change of gradient from the debris-mantled slopes above it. The greater the aridity, the more pronounced is the junction of the piedmont and debris slope, and the lower the gradient of the piedmont. The

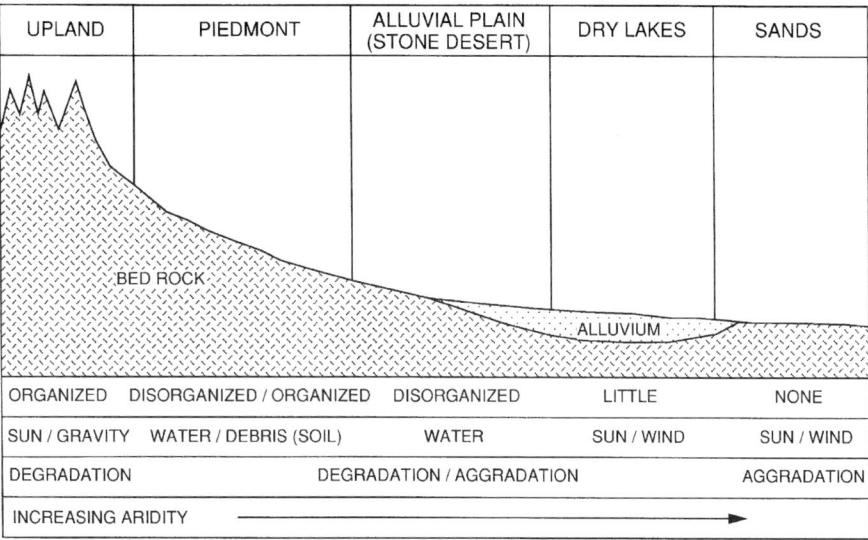

Figure D11 Physiographic framework of deserts: generalized schematic of the relations between desert uplands and lowlands and surface hydrology: drainage patterns, geomorphic energy sources, geomorphic processes, moisture regime.

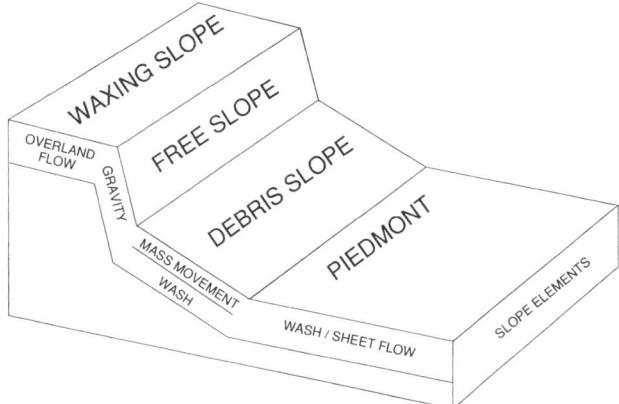

Figure D12 Schematic illustration of the relationships between desert slope elements and mass movement.

piedmont intervenes between the connected upland drainage patterns and the disintegrated drainage of the plains below. Few permanent drainage channels continue from the hills through the piedmont. At the transition between degradation on the slopes and aggradation on the plains, the landforms of the piedmont may be either erosional or depositional. The former are termed pediments, plains cut in bedrock where drainage takes the form of networks of tiny distributary rills, representing diminishing channels of runoff. In areas of stone pavements, they are obstructed or their development hindered and drainage is through shallow and sandy channels instead. The depositional forms are alluvial fans and bajadas. In general, pediments and fans are rarely found together; pediments are common in areas of lower relief, fans in areas of higher relief.

Alluvial fans are cone-shaped bodies of alluvial detritus formed at the outlet of a mountain valley or canyon. These are formed when a wadi leaves the confines of a mountain channel and emerges on the plains below, sometimes the consequence of a short-lived flood. The energy of the stream is dissipated by surface friction, but the unconfined water also spreads or diverges; both factors lead to deceleration of the flow. The low velocity cannot support the suspended load and material is deposited on the surface. The fan is characterized by a radial distribution of drainage, with feeder channels and gullies. The fans may be a few hundred meters to tens of kilometres in extent. Where the fans are compound or coalesce to form unstructured alluvial plains, the term bajada is applied.

Desert lowlands

Below the piedmont are the lowlands or plains; these consist of stony deserts, river beds and floodplains, sparsely vegetated flatlands, desert lake basins and ultimately sand deserts. The stone deserts have quite varied surfaces. They may include flat plains of closely packed stones or solid rock surfaces. The term hamada is used for boulder-rich terrain and tablelands; the term reg (serir in the Sahara) is applied to pavements of finer materials. The stony deserts also include regions of hard and inpenetrable duricrusts, formed as a result of restricted leaching in deserts. Some, like laterites (iron-rich crusts) and silcretes (crusts of siliceous materials like sands and quartz grains) tend to be relics of past climates.

The desert lowlands are traditionally considered to be largely depositional environments. Research in arid and semiarid regions has altered this idea, replacing it with the 'erosion cell' model. The landscape is a complex and dynamic mosaic with simultaneous processes of erosion and deposition governed by the spatial variations in hydraulic and aerodynamic resistances to water and wind erosion. These variations are a consequence of vegetation, soil characteristics and local variations in slope. Pickup and Chewings (1986) describe this as a set of overlaid and interspersed 'erosion cells', each consisting of a source zone, a transfer zone and a sink. Erosion, in the source zone leads to deposition in the sink, via the transfer zone. The dispersion and deposition of the soil and nutrients alter the patterns of vegetation growth. The restructuring of the surface by these processes modifies the surface resistances, thus altering the loci of the deposition and erosion. Thus the environment changes over relatively short periods of time. In some cases the processes of removal were assumed to be desertification, rather than stages in this dynamic equilibrium.

Scattered over the desert lowlands are lake basins occupying the lowest areas of the desert drainage network or lying beyond its termination. The most common term for these basins is playa but they are also called sebkhas or chotts in North Africa, dry lakes in North America and pans in Southern Africa. The nature of these basins is quite diverse even within an individual desert. The playa sediments are generally fine grained and saline, a characteristic which demonstrates the role of evaporation in playa formation. These vary in size and number, ranging from a few square meters to thousands of square kilometers. The largest is Lake Eyre in Australia, occupying an area of 9300 km^2. Their occurrence is quite frequent; for example, over 1000 playas have been recorded in North Africa. However, because they are generally quite small, playas represent only about 1% of desert surfaces. Many playas, especially those of the Sahara and western United States, are relics of past climates, such as the pluvial periods of the Pleistocene.

These lake basins differ widely with respect to the amount of water they contain and length of time they contain water. Most of these lake basins remain dry except for brief periods of seasonal or ephemeral flooding. Water supply results from surface runoff, direct precipitation or groundwater discharge. The basin contains water whenever the precipitation plus surfacewater supply exceeds evaporation. The drying cycle is generally dominant, with surface runoff and precipitation being removed by infiltration and evaporation. Groundwater is discharged by evaporation and by transpiration from plants, such as the phreatophytes that often thrive on the playas. During the dry period erosion is intense.

The hydrologic regime of the playas is largely dependent on the relative importance of surface water inflow (precipitation and runoff) and groundwater discharge. Moist playas are largely dependent on groundwater, dry playas on surface sources. The greater the importance of groundwater, the more reliable the moisture supply and the longer the moist period in the basin. Those entirely dependent on groundwater are often perennial. The variability of water on the playa surface is lower when groundwater supplements the surface inflow. The dry playas, which are primarily dependent on episodic rainfall, are often closed basins (those without an outlet). These strongly reflect local or regional rainfall fluctuations. The response time to rainfall events is dependent on the ratio of the area of the drainage basin to that of the playa surface. In many cases the response to rainfall is almost immediate because the surface is inpenetrable hardpan or the surface clays may expand and seal when water is absorbed. In such cases a brief shower can bring a rapid and spectacular inundation of the surface. For larger playas, however, the response time may be 2–4 years.

Precipitation regime

Precipitation in deserts tends to be episodic and localized, with aridity erratically interrupted by torrential rains. It is highly variable in both time and space. Annual totals are determined by a small number of rain events, during which rainfall is usually of high intensity and short duration. A typical rain might last for a few minutes or hours. These characteristics are valid for most deserts, but particularly for those in subtropical latitudes, where the precipitation regime is convective.

In the higher latitudes, especially those with a winter rainfall regime, precipitation may be in the form of persistent drizzle. In Mauritania, for example, where such a situation is termed 'heug' weather, the drizzle may persist for several days. In the coastal deserts of South America, this drizzle is termed 'garua'. Such events are generally associated with frontal systems.

Rainfall is the dominant form of precipitation in most deserts. The exception is some coastal deserts, where fog-water is a more significant source of moisture. In the Namib, for example, water condensed from the frequent fogs is several times greater than rainfall at many locations. Hail and thunderstorms are uncommon in deserts. Snow occurs occasionally in some deserts, particularly those in higher-latitudes, but even the northern Sahara receives snow every few years.

The world's deserts are quite diverse with respect to the amount and seasonality of precipitation. Areas of the Sahara, Namib and Peru–Atacama deserts are so dry than many years pass without a drop of rain. Rainless stretches of 10–14 years have been recorded at locations such as Swakopmund, Namibia, and Lima, Peru, in the coastal deserts of Southern Africa and South America. Often, when rain does fall, the mean annual rainfall will fall within hours or days. At Lima, where the mean annual is 46 mm, 1524 mm fell during one storm in 1925. In the Sahara at Nouakchott, Mauritania, over twice the annual mean, or 249 mm, fell on one day; at Tamanrasset, Algeria, the mean annual rainfall of 48 mm once fell in 1 h.

Few deserts are as dry as those. In the Sonoran, Mojave, Iranian and Arabian deserts and in the Takla-Makan, mean annual rainfall is of the order of tens of millimeters in their driest cores. It exceeds 80–100 mm everywhere in the Thar, Gobi, Patagonian and Australian deserts and in the Kalahari. In contrast, it scarcely falls below 200 mm in the Chihuahuan Desert and Great Basin of North America.

The seasonality of rainfall can also be quite diverse. By some definitions, a true desert is a location with no preferred or reliable season of rainfall. This is, however, too simple. Subtropical deserts are generally the transition between the summer rainfall regime of the tropics and the winter rainfall regime of the Mediterranean-type climates. Hence, the slight rain that falls tends to be in summer on the equatorward side, in winter on the poleward side. Generally, however, the rare rains can occur in nearly any month. The patterns are much more complex in the midlatitude deserts, especially the rain-shadow deserts.

Perhaps the best generalization that can be made is that winter rainfall tends to be more effective than summer rainfall. For this reason, the mean annual rainfall associated with the tropical margin of a desert such as the Sahara may be significantly higher than that marking the poleward border.

Rainfall in deserts is highly variable in time (Figure D13). The coefficient of variation of annual rainfall is generally over 50% in deserts, compared to 5–10% in humid regions. At Nouadibou, Mauritania, with mean annual rainfall of 32 mm, annual totals have ranged from 0–301 mm; on one day in 1909 140 mm fell. At Biskra, Algeria (mean annual rainfall of 140 mm), annual totals range from 32 to 638 mm.

Desert rainfall is also localized in space because most occurs during a small number of storms which are limited in spatial scale. Studies by Sharon (1981) in the Namib Desert have shown, for example, that the typical convection cell is 10–50 km and that cells are typically spaced at preferred distances of 30–40 or 80–100 km. This means that for individual storms, rainfall at stations just 2–3 km apart can differ by factors of 10–20. The monthly rainfall at two locations a few kilometers apart might be quite different. For these reasons, desert rain events were thought to be completely random and localized occurrences. Satellites have shown, however, that although this is true

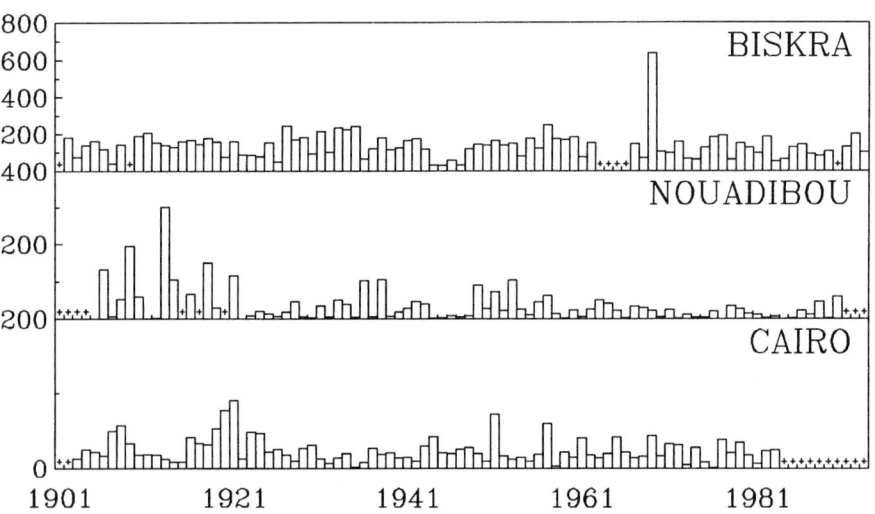

Figure D13 Annual rainfall (mm) for three desert towns in Africa.

of individual rain cells, these cells tend to occur in organized, large-scale meteorological disturbances.

An interesting example of a desert storm occurred in September 1969 in Algeria and Tunisia. At Biskra, Algeria, where the mean annual rainfall is 140 mm, this storm brought 299 mm in September 1969 (210 mm of it in 2 days). The same system brought nearly 800 mm to Sidi bou Zid, Tunisia, during September and October, months in which the mean rainfall is on the order of 10–20 mm.

Runoff and streamflow

The primary factors which influence the amount and nature of runoff are the proportions of vegetation cover and bare ground and the intensity, duration, amount and seasonal distribution of rainfall. Other factors include the slope of the terrain, soil type and condition, and antecedent soil moisture. In dryland regions the most important factor is generally rainfall intensity, with more intense events producing more runoff.

Runoff generation is relatively low in desert lowlands, where the sands and stone pavements promote infiltration, clay pans inhibit it and drainage is disorganized. In desert uplands, where much rock is exposed and rainfall is often more abundant and more intense than on the plains, runoff generation can be quite high. Because the bare rock on desert hillslopes limits absorption, rainwash can occur at a relatively low threshold, rainfall totals as low as 5 mm and intensities as low as 0.5 mm min^{-1} can produce rainwash.

The nature of runoff in arid regions contrasts with that in humid regions. Runoff takes place at the surface (overland flow), in the subsurface (throughflow) and from groundwater (base flow). The dominant form of runoff in humid regions is throughflow, i.e. lateral flow of water which has infiltrated the upper soil layers. In dryland regions, overland flow (lateral surface flow of water directly intercepted from rainfall) is dominant because soils and mantles are too thin to maintain throughflow and the water table is generally too deep to supply streams. The overland flow is primarily Hortonian, resulting from rainfall intensity exceeding infiltration capacity, but saturation overland flow (resulting from saturated soils) can occur locally. Evidence of the predominance of Hortonian flow in dryland regions is the greater dependence of runoff on rainfall intensity than on rainfall duration.

The response of streams to runoff depends not only on the total amount of runoff generated, but also on the path the runoff follows from the watershed to the stream and on the characteristics of the catchment. Groundwater is the slowest and least direct, but most reliable, source of streamflow. The greater the proportion of rainfall which infiltrates the soil, the lower the peak discharge and the more evenly distributed in time the streamflow. Overland flow is generally

the fastest and most direct, and this produces the quickest response in stream discharge.

The low infiltration and the high proportion of overland flow in arid regions promote high and immediate runoff generation, and a quick and peaked response in streamflow (Figure D14). These are typical characteristics of stream flow in dryland regions. Where rainfall is episodic, streamflow is also typically brief and infrequent. Floods at Nahal Yael, in the Gulf of Aquaba (Schick and Sharon, 1974) peak within 5–20 min and usually last just 1–4 h (the longest persisted for 24 h).

The response of streams in the desert uplands is particularly rapid and intense; runoff and even floods can result from very little rainfall. Thresholds for initiating streamflow were found to be 7.5 mm of rainfall with a mean intensity of 0.5 mm min^{-1} in the Sinai and 5 mm with an intensity of 0.5 mm min^{-1} in the Hoggar of the Sahara (Mabbutt, 1979). Catastrophic floods occur in the White Mountains of California and Nevada with as little as 100 mm of rainfall. In general, a much higher proportion of rainfall in the uplands ultimately becomes runoff than in the lowlands (Mabbutt, 1979). An example from Nahal Yael gives 35–60% in the head of the catchment, but as little as 5–10% for the watershed as a whole.

In the lowlands the streamflow is moderated by the basin it must traverse; in large basins there can be a considerably lagged response. The nature of surface materials and drainage in the lowlands,

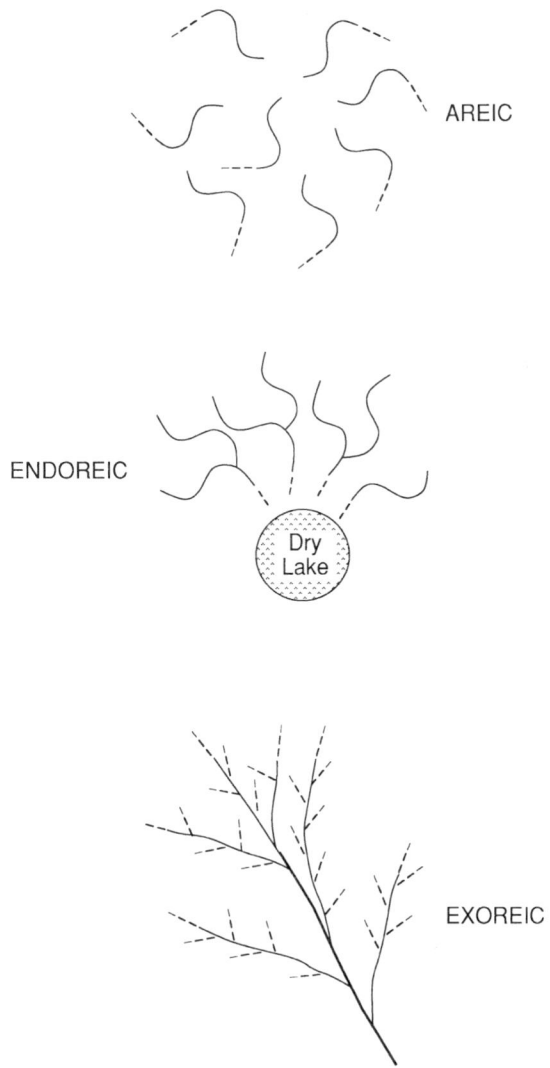

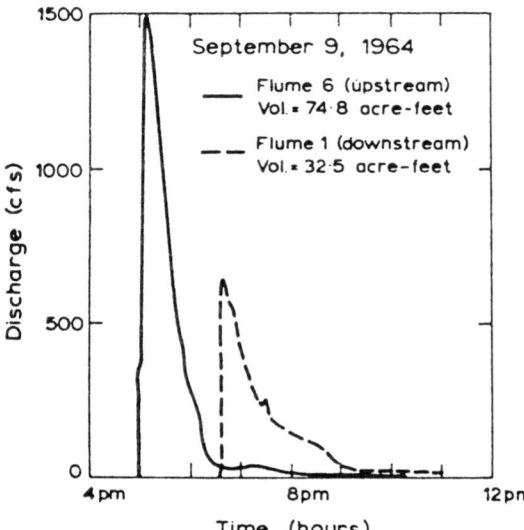

Figure D14 Typical stream hydrograph for streams in arid regions. Transmission losses for a flood in Walnut Gulch, Arizona, represented by the hydrographs for two flumes 10.9 km apart in the ephemeral stream channel.

Figure D15 Schematic illustration of three types of drainage organization. Areic (also arheic) is drainage devoid of systematic organization; endoreic refers to inland or internal drainage; exoreic refers to seawards-directed drainage.

however, is not conducive to high runoff generation; often localized runoff will occur, but will be lost to infiltration or evaporation before reaching stream channels. Thus another characteristic of dryland streamflow is that the total discharge and peak flow progressively diminish downstream. The frequency of flooding likewise diminishes downstream.

A number of studies have examined various hydrological characteristics of streamflow in arid and semiarid regions, their interrelationships and how they compare with flows in humid regions. Studies in the Walnut Gulch catchment of the southwestern United States showed that total runoff correlated best with total precipitation, that peak runoff was best correlated with rainfall intensity, flow duration with watershed length, and that lag time was best correlated with watershed area (Schreibner and Kincaid, 1967; Osborn and Lane, 1969). These same relationships have been established in other dryland regions.

Fluvial and Eolian processes

Water is an active agent in sculpting the desert; in desert regions it is particularly effective because flow intensity is high and unbridled by deep channels and it spreads across land which is generally devoid of vegetation. Eolian or wind forces also play a role, but the energy is proportional to velocity and mass, hence flowing water is more effective than wind and carries coarser, more abrasive materials.

The size of the particles which can be mobilized and held in suspension in water depends on the velocity of flow. Particles will be mobilized when the drag force of the flowing water exceeds the forces bonding the particle to the surface. One bonding agent is soil moisture; erodibility is inversely proportional to soil moisture. Another is sorting of grains. The presence of non-erodible roughness elements (such as rocks and pebbles) also reduces erodibility by absorbing some of the force of the wind. Vegetation cover also plays a role, since it both bonds the soil surface and decreases the velocity of water flowing over the surface.

Many of the characteristics of the dryland environment are conducive to the mobilization and transport of surface particles. The sparse vegetation and the bare, often sandy soil increase the erodibility of the surface. The low soil moisture content also makes the surface prone to erosion and the low surface roughness of bare ground increases surface drag.

The transport of material by water is analogous to that by air, with the same forces involved. It is much more effective, however, because water is denser than air and the pressures involved are therefore greater. The three primary mechanisms are rainsplash, overland flow (sheet or slope wash) and gullying (creation of channels when erodibility or erosion potential change). The latter two are lateral movements and hence related to velocity of flow. The effectiveness of rainsplash is a function of the kinetic energy of raindrops. This in turn is related to their terminal velocities (and hence size) and their number (a function of rainfall intensity). Low intensity rains have small drop size, hence low kinetic energy and low erosion potential. Generally, a threshold intensity of rainfall must be exceeded before it is erosive (Jackson, 1989).

Wind and water act collectively to erode the landscape. The rate of erosion, especially that by wind, is difficult to assess. However, the factors which promote erosion are well understood, and their relationship to regional climate has been established in a number of studies. Studies of sediment yield suggest that erosion rates vary widely in the drylands, from nil to over 300 m^3 km^{-2} year^{-1}. These figures neglect wind erosion and dissolved sediment load, which can both be considerable in arid regions (Cooke and Warren, 1975).

Erosion is traditionally estimated from the universal soil loss equation, which relates total erosion by rainfall to six variables:

$$E_r = f(R, K, L, S, C, P)$$

namely rainfall erosivity (R), soil erodibility (K), length of slope (L), steepness of slope (S), and C and P are factors involving land management and erosion control. Numerous forms of this equation exist. Unfortunately the functional relationships of the variables are not well established, particulary for arid climates.

An analogous equation for wind erosion (Woodruff and Siddoway, 1965) takes the form

$$E_w = f(I, K, C, L, V)$$

where the relevant variables are soil erodibility by wind (I), surface roughness (K), local climate (C), a length (L) related to field length downwind, and vegetative cover (V). The climate term includes winds, precipitation and evaporation (and hence, implicitly, soil moisture).

Despite the strong erosion potential in deserts, total erosion (wind plus water) appears to be relatively small. Erosion from arid lands

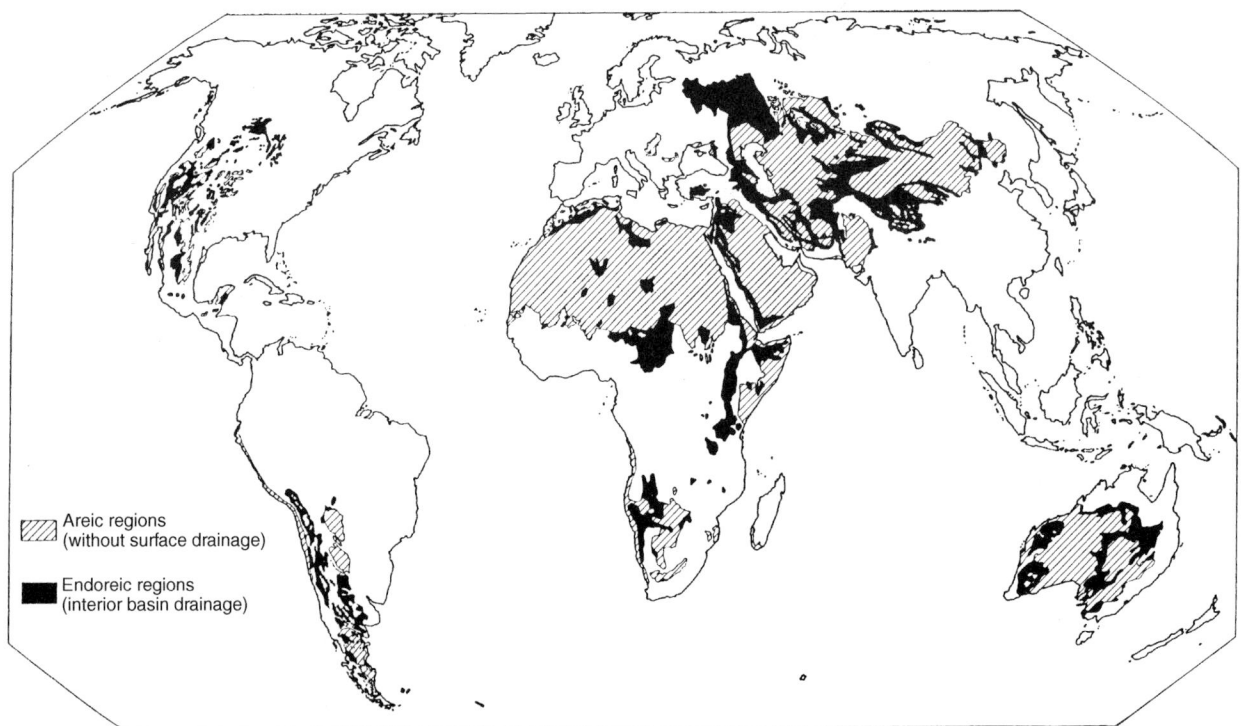

Figure D16 Location of areas of areic and endoreic drainage.

Figure D17 The 'erosion cell' as visualized in the Australian landscape. Deposition forms the cluster of vegetation in the upper right, which enhances the forces of erosion in the surrounding lower land. As the cluster grows, it becomes prone to wind erosion, creating a dynamic mosaic in the environment.

accounted for only about 4% of total world erosion. Such figures are meaningless, however, without reference to the watershed in deserts (Mabbutt, 1979), since runoff varies markedly between the upland head of the catchment and the desert plains. In general the rate of erosion on the uplands is about two to five times greater than on the plains.

The relatively small amount of erosion in deserts is clearly demonstrated by the classic work of Langbein and Schumm (1958), illustrating the relationship between mean annual sediment yield and mean annual 'effective rainfall' (the amount of rainfall required to produce a given amount of runoff under specified temperature conditions). Sediment yield is extremely low in arid regions; it is greatest in semiarid regions where annual rainfall is between about 250 and 600 mm, ranges which are generally associated with semiarid climates. In these regions, rainfall is more frequent than in deserts and the surface often contains more erodible material.

Desert streams

Streams fall into three classes: perennial, ephemeral and intermittent. Perennial rivers flow year-round; ephemeral rivers flow for only short periods, seasonally or in response to episodic rainfall. Intermittent rivers are discontinuous in space; often much of the river will dry up, leaving only ponds in the depressions in the channel. Endogenous streams in arid regions are predominantly ephemeral, but intermittent streams become more common with increasing rainfall. The water level in desert streams is highly localized in time and spatially, reflecting the characteristics of the dryland rainfall regime. Most of the perennial rivers in desert regions are exogenous (also called exotic or allogenic). Since they tend to originate in subhumid regions, these rivers, such as the Nile or Niger, tend to have a strongly seasonal regime. Water is well below the level of the stream bank much of the year and is accessible only with pumps.

In general, exogenous rivers draining into dryland regions become smaller downstream, because they are constantly losing water through evaporation and infiltration, and their response to episodic floods becomes increasingly moderated. This stands in sharp contrast to rivers of humid regions. Sediment load, on the other hand, increases downstream as the stream erodes the surface.

The ephemeral streams, like the exogenous perennial rivers, have permanent channels with steep sides; these are generally termed wadis or arroyos, but have numerous local names. Their discharge initially increases downstream, as the water collects in the channel, but at some distance downstream discharge diminishes and, in most cases, ends long before the flow can reach the sea.

The drainage networks in arid and semiarid regions have distinctive characteristics which distinguish them from drainage in humid regions (Figure D15). The main difference is that they are generally endoreic,

Figure D18 The runoff from intense desert storms often forms pools on the surface, which eventually evaporate but provide temporary water supply.

Figure D19 Vegetation grows in the desert 'vleis' of southern Africa, small depressions or flat pans in which runoff accumulates.

draining into internal basins rather than reaching the ocean, or areic, with no integrated network of surface flow. Streams in humid regions have exoreic drainage, eventually reaching the sea. Endoreic drainage is advantageous in that it concentrates runoff in small areas, and thereby reduces water loss through evaporation. The degree of endoreism is clearly a function of rainfall; the global distribution of areas of endoreic and areic drainage corresponds closely to the distribution of deserts (Figure D16). Another characteristic of dryland

Figure D20 Fog rills in stones in the Namib Desert. Fog water condenses on the stones and drips downward, dissolving minerals in the stones.

Figure D23 Life thrives in desert oases, such as Palm Springs, California.

Figure D21 Desert streams, like this one in the Kalahari, are intermittent but powerful, capable of carrying large stones and boulders.

Figure D24 This dry wadi in Australia is lined by large trees utilizing water remaining deep in the stream bed.

Figure D22 Desert pans are subject to rapid flooding. The Bonneville Salt Flats of Utah become a vast lake after intense rains.

Figure D25 Dense, closed-canopy forest thriving in the bed of the Kuiseb River, Namibia, which floods about once in 10 years.

Figure D26 Dunes retain a water reservoir which promotes rapid vegetation growth in wet years.

Figure D27 The ancient city of Shifta, settled around the first century BC in Israel, supported a dense population via sophisticated techniques of water harvesting.

drainage is that the actual area of the drainage basin is often considerably larger than its effective area (that where runoff originates) because the basins were formed during pluvial periods of the past.

The streams of dryland regions also have a different relationship to the water table than those of humid regions. They are less dependent on groundwater, being sustained primarily by overland flow. Consequently, streams in dry climates lose water through seepage and recharge the groundwater below. These are termed influent streams, in contrast to the effluent streams of moist climates which receive flow from the groundwater. Thus, near the exogenous streams of arid regions, the water table is locally determined by the river, and vegetation growth will be extremely lush along the stream banks or even in its dry bed. In contrast, the water table determines the streamflow in humid regions.

In general, the streams in dryland regions commence in the uplands of the basin as dispersed, overland flow, but the flow becomes organized on the hillslopes, where it moves through deep channels. These are often referred to as wadis or arroyos, especially if the sides of the channel are particularly steep. The channels then tend to disintegrate and the network becomes dispersed and disorganized in the lower elevations and on sand and clay plains. Drainage density is thus high on the hillslopes, where it is proportional to the degree of aridity. It is also inversely related to the percentage of bare ground, since vegetation absorbs the flow and reduces runoff. There are few drainage channels on the plains because the ground soaks up the little rain that falls there and the flow from the uplands, which is too meager to reach the sea, is lost through infiltration. This is especially true for sand, which has a high infiltration capacity, but clays will swell and absorb water, likewise leaving little available for runoff and streamflow.

Figures D17–D27 show certain features of desert hydrology. These include the 'erosion cell' in Australian landscape; runoff from intense desert storms; vegetation growth in the desert 'vleis' of South Africa; fog rills in stones in the Namib desert; desert streams; desert pans; desert oases; Australian dry Wadi; dense forest; dunes; and a view of the ancient city of Shifta.

Sharon E. Nicholson

Bibliography

Chorley, R.J., 1969. *Introduction to Geographical Hydrology*, Methuen., London, 206 pp.

Cooke, R.U. and A. Warren, 1975. *Geomorphology in Deserts*, Batsford, London, 394 pp.

Dunne, T. and L.B. Leopold, 1978. *Water in Environmental Planning*, W.H. Freeman and Co., New York, 818 pp.

Jackson, I.J., 1989. *Climate, Water and Agriculture in the Tropics*, Longman Scientific and Technical, London, 377 pp.

Langbein, W.B. and S.A. Schumm, 1958: Yield of sediment in relation to mean annual precipitation. *Trans. Am. Geophys. Union*, **39**, 1076–1084.

Mabbutt, J.A., 1979. *Desert Landforms*, MIT Press, Cambridge, MA, 340 pp.

Osborn, H.B. and L. Lane, 1969. Precipitation–runoff relations for very small semiarid rangeland watersheds. *Wat. Resour. Res.*, **5**, 419–425.

Pickup, G. and V.H. Chewings, 1986. Mapping and forecasting soil erosion patterns from Landsat on a microcomputer-based image processing facility. *Aust. Rangel. J.*, **8**, 57–62.

Shick, A.P. and D. Sharon, 1974. *Geomorphology and Climatology of Arid Watersheds*. US Army Eur. Res. Office Tech. Paper DA JA-72-C-3874. Dept. of Geography, University of Jerusalem.

Schreibner, H.A. and D.R. Kincaid, 1967. Regression models for predicting on-site runoff from short-duration convective storms. *Wat. Resour. Res.*, **3**, 389–395.

Sharon, D., 1981. The distribution of rainfall in space and time in the Namib desert. *J. Climatol.*, **1**, 69–75.

Woodruff, N.P. and F.H. Siddoway, 1965. A wind erosion equation. *Proc. Soil Sci. Soc. Am.*, **29**, 602–608.

Cross references

Aridity indices
Arid lands
Arid zone hydrology
Desertification

DESERTIFICATION

Introduction and overview

A common definition of desertification is 'the expansion of desert-like conditions and landscapes to areas where they should not occur climatically' (Graetz, 1991). The term evokes an image of the 'advancing desert', a living environment becoming sterile and barren, but this is not an accurate picture. Desertification is a process of land degradation and its impact may be confined to relatively small scales. Desertification has been most severe in drylands, arid and semiarid regions, and most of these lands are at risk, if not properly managed. However, the extent of affected lands has been greatly exaggerated and the issue of desertification has become a controversial one, suffering from a dearth of data and rigorous scientific study (Mainguet, 1991; Graetz, 1991; Hellden, 1991).

The concept of good lands becoming deserts has resurged from time to time since the beginning of the twentieth century, but the most intense interest in the topic came on the heels of the 1970s drought in Sahelian West Africa. Some scientists, such as the late meteorologist Jule Charney (1975), claimed that the drought had been a result of desertification in the region. In 1977 the United Nations announced that globally some 35 million km^2 of land had been 'desertified' and that perhaps 35% of the Earth's land surface was at risk of undergoing similar changes. With the sounding of that alarm, desertification became the focus of much scientific, political and institutional attention. Nevertheless, we did not achieve any understanding of the process nor any accurate assessment of the extent and degree of desertification globally.

As these gaps in our knowledge became recognized, the topic of desertification became quite controversial. Land degradation has clearly affected many regions, including the African Sahel, but it is also apparent that the problem had been exaggerated by early studies. Nonetheless, studies of desertification have established many important ideas, particularly concerning the arid and semiarid environments. One is that people, climate and the environment are intricately linked in a relatively delicate balance; the feedbacks between these components are required to maintain a stable system. The increased human pressure on the land in recent decades may threaten the balance by making the impact of climatic fluctuations more severe. The Earth's dryland environments, which support some 20% of its population, are at risk of severe degradation. Prudent strategies of land management are needed to maintain these lands.

Definition of desertification

Part of the reason for our lack of understanding of desertification is the lack of any agreed upon definition of the term. Some 100 vastly different definitions exist. The emphasis ranges from human impact on the land to economic impact, landforms and vegetation and even phenomena on geological time scales.

Most definitions emphasize the concept of diminished land productivity and they restrict it to cases where the cause is human and not climatic. The best known definition is that of the United Nations (1980): 'the diminution or destruction of the biological potential of land [which] can lead to desert-like conditions'. Unfortunately, most of the evidence of desertification derives from locations and time periods which were simultaneously affected by either drought or long-term declines in rainfall. Consequently, there is disagreement concerning the causes and processes of degradation, the extent to which the changes are natural or anthropogenic, the amount of land affected or at risk, and the reversibility of desertification.

Mainguet (1991) proposes a more encompassing definition: 'Desertification, revealed by drought, is caused by human activities in which the carrying capacity of land is exceeded; it proceeds by exacerbated natural or man-induced mechanisms, and is made manifest by intricate steps of vegetation and soil deterioration which result, in human terms, in an irreversible decrease or destruction of the biological potential of the land and its ability to support population.'

Desiccation and desertification: examples from the recent past

Throughout the twentieth century there have been numerous cases of dryland regions becoming increasingly desiccated. As with recent claims of desertification, these incidents were surrounded by controversy concerning human versus climatic causes. A notable example is the belief, just after the turn of the twentieth century, that the Kalahari had been progressively drying up, a claim which led to the creation of a drought investigation committee to evaluate the situation. A proposed solution was a grandiose flooding scheme, the creation of Lake Kalahari, to bring back the waning rains (Schwarz, 1920).

The concept of the encroaching Sahara goes well back in time, to papers noting the desiccation of the Senegal River and nearby wells (Bovill, 1921) and the degradation of the forests in northern Nigeria, eastern Mali and southern Niger (Stebbing, 1935). These events, at the time, were attributed to human activities.

Landscape degradation has also affected huge regions of Australia. Severe degradation occurred in Australia at the end of the nineteenth century. The number of sheep in New South Wales declined from 13 million in 1890 to 4–5 million in 1900 (Graetz, 1991).

These three situations in Australia and West and Southern Africa have two commonalities; they were all roughly contemporaneous and the observed trends paralleled a climatic desiccation which affected nearly all the global tropics but was particularly apparent in Australia and Africa. The decline of the waters and forests in the Sahel, as well as the 'drying up' of the Kalahari, followed a two-decade decline in rainfall. Thus, in these examples, landscape degradation is accompanied by climatic perturbations. Likewise, farmland was ruined and soil was eroded during the Dust Bowl days of the 1930s in the Great Plains of the USA during the worst drought conditions on record in the region. These examples illustrate the inseparability of climate and desertification.

This tandem of events occurred again with the drought which ravaged Sahelian West Africa in the late 1960s and early 1970s. The drought brought a recent surge of interest in desertification. During the most severe conditions, ca. 1969–1975 and ca. 1982 to the present, reportedly a million people starved, 40–50% of the population of domestic stock perished and millions of people took refuge in camps and urban areas and became dependent on external food aid (Graetz, 1991). However, it is virtually impossible to separate the impact of drought from that of desertification. The case for desertification was enhanced by aerial and satellite photos showing evidence of large-scale human impact on the land. International borders separating grazed and ungrazed land were vividly seen from space: Afghanistan–Soviet Union, Namibia–Botswana, Sinai–Negev. More heavily grazed areas appeared as brighter spots. Fenced 'exclosures', i.e. huge fenced ranches where the extent of grazing was rigidly controlled, were likewise visible as dark patches in aerial photos.

Causes of desertification

Many general causes have been identified, with the roots of desertification lying in societal changes such as increasing population, sedentarization of the indigenous nomadic peoples, the breakdown of traditional market and livelihood systems, innovation of new and inappropriate technology in the affected regions, and in general bad strategies of land management. Associated with these changes are growing livestock numbers, cultivation in marginal areas, intensive irrigation and deforestation.

In West Africa, where desertification is considered to be intense, much of the problem has been related to government policies and its effects on the pastoral societies of the Sahel. Incentives were given for growing cash crops, at the expense of subsistence agriculture, and cultivation expanded into the southern Sahelian fringes traditionally left for dry season pastures. The result was overuse or 'over-cultivation' of the land.

Also, the sedentarization of nomads was encouraged for a variety of reasons, such as centralized government control of education and health, and their traditional migration routes became blocked by national borders. Their impact became more localized, instead of spread out over the vast region as they migrated. Their animals clustered around the few available wells, overgrazing the surrounding land; this had dramatic effects in the face of the drought which prevailed in the 1960s and 1970s. These policies were exasercbated by an increase in population of both pastoralists and cattle by about 3% per year, a result of better access to medical and veterinary services (Lamprey, 1975).

These changes alter the land's topography, vegetation and soils. Topsoil is eroded away and sometimes vast amounts are washed away to produce huge gullies. The soil's texture, organic matter and nutrient contents are changed in ways which reduce its fertility. Poor irrigation and management practices lead to salinization and waterlogging of the soil. The land cover becomes more barren or nutrient-rich and diverse species are replaced by vegetation of poorer quality. The carrying capacity of the land is dramatically reduced.

Figure D28 Biomass burning in the savanna of Botswana. Slash and burn agriculture is implicated as a cause of desertification.

Figure D30 Many savanna peoples are turning from traditional building materials to wood construction, leading to deforestation in many areas.

Figure D29 Firewood is commonly used for cooking in the developing world. Gathering wood for fuel can reduce the productivity of the land by reducing the vegetation cover and making the soil more erosion-prone.

Figure D31 Grazing can degrade the environment because it is often concentrated around the few wells, which provide water supply.

Processes and physical manifestations of desertification

The processes of desertification include removal of vegetation, erosion and compaction of soils, excessive water consumption, irrigation, inappropriate use of agricultural technology and other examples of poor land management practices. Natural vegetation is cleared for agriculture; savanna lands, fields and pastures are burned at the end of the dry season (Figure D28); wood is gathered for firewood (Figure D29), charcoal and building supplies (Figure D30); and animals overgraze grasslands (Figure D31). This generally leaves the surface more barren, and more susceptible to wind and water erosion. The erosion removes the organic topsoil, fine materials and sometimes the whole soil surface. In many arid regions the use of tractors and other technologies further enhance the erodibility. Animals trample the soil surface, compacting and aggregating materials in ways which hinder drainage and infiltration. When irrigated land is improperly drained, salt and other chemical residues are left behind.

Desertification is manifested as changes in the vegetation cover, soil and surface topography. Soil structure is altered, materials are leached, nutrients and organics are lost. Impervious horizons, such as lateritic crusts, may form. Vast amounts of soil are removed by water and wind; gullies and badlands form. In many areas the fertility of soils is reduced by salinization or alkalinization. Water supplies also become enriched in salt. Bare ground is exposed. Sand dunes are built or mobilized, often encroaching upon vegetation. The amount of vegetation or primary productivity may be reduced or vegetation composition may change, with rich grasses, forbs and herbs replaced

by less palatable shrubs. The spatial distribution of vegetation and water supply is unfavorably altered, the landscape becoming more heterogeneous and resources contracted (Schlesinger *et al.*, 1990).

Overall, desertification represents a severe disturbance in an ecosystem that has otherwise been stabilized over time. Many scientists are concerned that the process can destabilize the environment in such a way that the process dramatically accelerates. In other words, desertification might be self-accelerating (Graetz, 1991; Schlesinger *et al.*, 1990). If this is the case, the process might be irreversible.

Desertification and climate

In 1975 a well-known meteorologist at MIT, Jule Charney, wrote a classic but highly controversial paper suggesting that such desertification at the hands of humans had caused the drought which was occurring in the Sahel. His mechanism was based on the bright appearance of overgrazed land on satellite and aerial photos: the degradation had exposed highly reflective soil which increased the albedo of the land, thereby returning more of the solar radiation back to space. His model showed that this would, overall, reduce the net radiation over the Sahara and adjacent Sahel; to compensate for this loss of energy and associated cooling, the subsidence of air (which produces warming) would be enhanced over the region. The subsidence would further suppress rainfall, providing a positive feedback mechanism by which droughts could be self-accelerating, or which could even produce drought.

Charney's theory that desertification caused the Sahel drought of the 1970s has never been widely accepted. However, the underlying concept – that changes of the land surface can significantly influence weather and climate – has received increasing attention from the scientific community. Subsequently a number of field studies were conducted to assess the impact of desertification and other land surface changes on the atmosphere.

The first studies emphasized albedo changes. Field measurements showed that the soil in the overgrazed Sinai reflects 46% of the solar radiation, compared to 25% in the protected area of the Negev (Otterman, 1981). Correspondingly, ground temperatures in the Sinai were shown to be up to 5°C cooler than on the darker, Negev side (Otterman and Tucker, 1985). Similar differences in albedo are evident between the American Sonoran Desert and the more heavily grazed Mexican side of the border (Bryant et al., 1990). However, in contrast to the situation in the Sinai/Negev region, temperatures are generally 2–4°C lower on the ungrazed, darker side. These temperature differences appear to have an impact on soil moisture and cloudiness, although no changes in precipitation have been demonstrated.

Despite these observations, the albedo theory is losing support. A number of studies from field observations, aerial or satelllite photos and vegetation information have shown that the albedo changes which occur are much smaller than Charney's theory requires and are inconsistent with the changes of rainfall and vegetation which occur (e.g. Courel et al., 1984; Gornitz, 1985). The effects of changes of evapotranspiration and soil moisture appear to be more important, as well as the increased dust generated by desertification or drought (Nicholson, 1988; Lare and Nicholson, 1993).

Overall, observational evidence of the climatic impact of desertification is more difficult to obtain. However, numerical models of the effect of vegetation patterns on the atmosphere support the idea that desertification can potentially influence weather and climate (Avissar and Pielke, 1989; Xue and Shukla, 1993).

Prevention and remedies

Mainguet (1991) sketches a number of ways that desertification can be prevented or reversed. Many of these have been long-established methods of soil, vegetation and water conservation. Some are indigenous strategies going back at least hundreds of years in arid and semiarid regions.

Vegetation can recover naturally or artifically. The unknown is the length of time required. Natural recovery means exclusion of humans and animals from a region, allowing a field to lie fallow. The degree and rapidity of recovery depends on the degree of aridity, soil condition and degree and extent of degradation. In the Sahel, recovery in productivity has been shown to take place within 1–2 years.

Artificial recovery requires active management techniques. Examples are terracing, plowing, improved water distribution techniques (e.g. drip irrigation), water harvesting and fertilization. Controlled grazing helps to implement these. The introduction of plant species which increase forage quality, replace species lost to desertification or provide resistance to drought is another general strategy. Vegetation is also introduced to create wind breaks or to prevent soil loss.

Many of these methods have been used in arid lands to increase productivity. When carefully managed they can allow the desert to bloom or the human-induced desert to recover its full potential. Most must be created and adapted to specific regions; when they are not, they may instead enhance the land degradation.

Sharon E. Nicholson

Bibliography

Avissar, R. and R.A. Pielke, 1989. A parameterization of heterogeneous land surface for atmospheric numerical models and its impact on regional meteorology, *Monthly Weather Review*, **117**, 2113–2136.

Bovill, E.W., 1921. The encroachment of the Sahara on the Sudan, *Journal of the African Society*, London.

Bryant, N.A., Johnson, L.F., Brazel, A.J. *et al.*, 1990. Measuring the effect of overgrazing in the Sonoran Desert, *Climatic Change*, **17**, 243–264.

Charney, J.G., 1975. Dynamics of deserts and drought in Sahel, *Quarterly Journal of the Royal Meteorological Society*, **101**, 193–202.

Courel, M., Kandel, R., and Rasool, S., 1984. Surface albedo and the Sahel drought, *Nature*, **307**, 528–538.

Gornitz, V., 1985. A survey of anthropogenic vegetation changes in West Africa during the last century – climatic implications, *Climatic Change*, **7**, 285–325.

Graetz, R.D., 1991. *Desertification: A Tale of Two Feedbacks, Ecosystem Experiments* (H.A. Mooney *et al.*, eds), New York: Wiley, pp. 59–87.

Hellden, U., 1991. Desertification – time for an assessment?, *Ambio*, **20**, 372–383.

Lamprey, H., 1975. Report on the Desert Encroachment Reconnaissance in Northern Sudan, Khartoum: National Council for Research/Ministry of Agriculture, Food and Natural Resources.

Lare, A.R. and Nicholson, S.E., 1994. Contrasting conditions of surface water balance in wet years and dry years as a possible land surface–atmosphere feedback mechanism in the West African Sahel, *Journal of Climate*, **7**, 653–668.

Mainguet, M., 1991. *Desertification, Natural Background and Human Mismanagement*, Berlin: Springer-Verlag.

Nicholson, S.E., 1988. Land surface–atmosphere interaction: physical processes and surface changes and their impact, *Progress in Physical Geography*, **12**, 36–65.

Otterman, J., 1981. Satellite and field studies of man's impact on the surface in arid regions, *Tellus*, **33**, 68–77.

Ottterman, J. and Tucker, C.J., 1985. Satellite measurements of surface albedo and temperatures in semi-desert, *Journal of Climate and Applied Meteorology*, **24**, 228–234.

Schlesinger, W.H., Reynolds, J.F., Cunningham, G.L. *et al.*, 1990. Biological feedbacks in global desertification, *Science*, **247**, 1043–1048.

Schwarz, E.H.L., 1920. *The Kalahari or Thirstland Redemption*, Capetown: T. Maskew Miller.

Stebbing, E.P., 1935. The encroaching Sahara, *Geographical Journal*, **86**, 509–510.

Tucker, C.J., Dregne, H.E. and Newcomb, W.W., 1991. Expansion and contraction of the Sahara desert from 1980 to 1990, *Science*, **253**, 299–301.

United Nations, 1980. *Desertification* (M.K. Biswas and A.K. Biswas, eds), Oxford: Pergamon Press.

Xue, Y. and Shukla, J., 1993. The influence of land surface properties on Sahel climate: Part I. Desertification, *Journal of Climate*, **6**, 2232–2245.

Cross references

Aridity indices
Arid lands
Arid zone hydrology
Deserts

DESERTS

Introduction

The common perception of a desert is dry region with little or no vegetation cover and inhospitable conditions for life. While some aspects of this image apply to many deserts, these environments are characterized by a remarkable diversity in vegetation, surface materials and climate and most teem with life well adapted to the environmental conditions. Also, modern technology has made it possible for deserts to sustain large populations and even crops.

The definition of a desert depends on the aspect of environment of principal concern, such as vegetation, terrain, culture, climate or resources. Hence there is no simple way to define or delineate desert environments. Even climatic boundaries of deserts, and therefore geographic boundaries, are hard to establish because there is a gradual and continual transition between arid and semiarid environments. Peveril Meigs, whose classification of arid lands is that most widely used today, states that many environmental traits distinguish deserts, but the essential one is scarcity of precipitation.

In basic climatic terms, a desert can be defined as an area which receives little or no rainfall and experiences no season of the year in which rain regularly occurs. However, rainfall alone insufficiently defines climatic boundaries of deserts because it varies dramatically from year to year and because aridity is not determined solely by rainfall. What distinguishes deserts from non-deserts is the amount of moisture available to the ecosystem, or the difference between the rainfall a region receives and that lost through evaporation. According to estimates based on this concept, true deserts make up roughly 20% of the Earth's land surface and another 15% is semiarid.

Deserts are concentrated in Asia, Africa and Australia. The largest expanse of arid land lies in an almost continuous area stretching nearly half a hemisphere across northern Africa and Asia, an inhospitable barrier separating most of Europe from southeastern Asia. Africa and Asia respectively contain 37% and 34% of the world's arid lands. The principal deserts of the world include (Figure D32) the Kalahari–Namib, Somali–Chalbi and Sahara on the African continent; the Arabian, Iranian and Turkestan deserts of the Middle East; the Thar, Takla-Makan and Gobi deserts of Asia; the Monte–Patagonian and Atacama–Peruvian Deserts of South America; the Australian Desert; and the North American Desert, including the Great Basin and the Sonoran, Mojave and Chihuahuan deserts. Vast tracts of semiarid land, such as the American Great Plains and the African Sahel, border each of these.

The deserts shown in Figure D32 tend to be situated in subtropical latitudes or in the lee of major mountain ranges, as the factors which promote aridity are common to these locations (see later). Deserts are also found at high altitudes and high latitudes, but these differ in two basic respects: the low rainfall is primarily a result of cold conditions and evapotranspiration is inherently low and of minor importance in determining water availability.

Overall the world's deserts are extremely diverse. This diversity results from the complexity of climatic conditions which produce them and the varying surface and terrain features, and the vast variety of vegetation thriving in these regions.

Classification of deserts

Although sparse or negligible rainfall is a common characteristic of all deserts, aridity is better defined by moisture availability. This is a residual in the balance between water availability through precipitation and water loss via evapotranspiration. The latter quantity represents the maximum amount of water which could be evaporated by solar energy or transpired by plants under conditions of constant moisture supply. Potential evapotranspiration can be as high as 3000 mm in some regions. In deserts, however, water is limited and hence this maximum is never achieved. Nevertheless, this parameter can be considered as 'water demand' and climatic characteristics such as temperature or incoming radiation are used to approximate this demand. Rainfall, in turn, represents water availability.

Water demand, whether assessed by net radiation, temperature or potential evapotranspiration, varies greatly from one region to another. It tends to decrease with increasing latitude and it is higher in summer than winter. Therefore the amount of rainfall at the desert margin tends to decrease with increasing latitude, and it is higher for deserts with summer rainfall. Within a geographical region, water demand is more constant and desert boundaries are often delineated regionally on the basis of mean annual rainfall. In Australia, for example, the generally accepted limit for the desert is 400–500 mm of rainfall in the north and 250 mm in the more temperate climate of the southern border.

A comparison between supply and demand forms the basis of most definitions of deserts. The best-known classification scheme for arid lands, that of Peveril Meigs, assesses 'demand' by way of potential evapotranspiration; others use temperature. The classification scheme of the Russian climatologist Budyko represents it by the net radiation received at some location, i.e., the difference between the net heat gain from the sun and the net heat loss by radiation from the ground. All of these assessments are in reasonable agreement on global space scales, but the standard is generally Meigs's system, which provides the basis for the maps in Figure D32.

The Budyko classification has some advantages, in that in allows for a simple quantitative comparison of the degree of aridity of various deserts and its aridity index is readily interpreted physically. The index is a parameter termed the 'dryness ratio'. To produce it, net radiation is converted to energy units and the index is defined as the ratio of mean annual rainfall to the amount of rainfall which this energy suffices to evaporate. In Budyko's classification, the boundary between deserts and semiarid regions is where the ratio is 3 (Figure D33). This ratio is as high as 200 in the central Sahara, but is generally below 10 in the North American deserts.

Causes of aridity

The formation of rainfall requires moisture supply, unstable atmospheric conditions and ascending air. Ascent can be produced by extreme heating of the surface, by a convergent pattern of air flow, or orographically forced by mountain barriers. Aridity is promoted by the opposite conditions: a lack of atmospheric moisture, stable air, subsidence (i.e. descending air motion) and a divergent pattern of air flow (i.e. air streams spreading apart from each other). Another cause is distance from the main tracks of major weather systems.

Except in polar and high-altitude deserts, a lack of sufficient moisture is generally not the overriding cause of the arid climate. The case of the Sahara Desert clearly demonstrates this. In summer the atmosphere above the Sahara is enriched with about as much moisture as that over the wetter regions of the southeastern United States. Likewise, the Namib Desert of southwestern Africa is a very humid environment, although it is one of the driest (i.e. most rainless) deserts on Earth.

The other factors, stable and subsiding air and divergent flow, are common to two meteorological situations linked to the occurrence of dry climates: the semi-permanent high-pressure systems which prevail in the subtropical latitudes and the rainshadows on the leeward sides of mountain barriers. These are the most common locations of deserts.

Descending currents are particularly effective in promoting aridity. They originate at high altitudes, where the air is dry, and as they sink they undergo intense compressional heating. Consequently, the air is very hot and dry when it reaches the surface. The warming also produces a temperature inversion (temperature increasing with elevation), which stabilizes the lower atmosphere and suppresses the uplift required to produce clouds and precipitation. Air descends in the core of high-pressure cells because of the pattern of surface air flow; it descends also in the lee of mountains.

Near the high-pressure cells, aridity is further enhanced by a temperature inversion produced by the decending air and by cold water on the eastern flanks of the highs. The occurrence of this unusually cold water is linked to the surface wind flow around the high, which produces the upwelling of cold water and advects cold polar water into relatively low latitudes. Along the coast of South America, the upwelled water is as cold as 12°C, compared to over 20°C in the mid-ocean.

Both the subtropical highs and mountain barriers further promote aridity by blocking the passage of major weather systems. Mountains also act as barriers to the inland penetration of moist, maritime air masses. The transformation of air masses crossing the mountains produces a 'rainshadow'. Air approaching a mountain range is forced to rise, promoting cloud formation and rainfall on the windward side and the peaks. The air which reaches the leeward side has dropped its moisture at higher elevations and is relatively dry. As it sinks in the lee, compressional heating accentuates its dryness.

Most of the world's largest deserts are under the influence of the subtropical high: the Australian Desert, the Peruvian–Atacama Desert of South America, the southwestern United States, the Namib Desert and Kalahari of Southern Africa, and the Sahara. Aridity is strongest on the western sides of continents in these latitudes, where the influence of the highs is greatest and where the coastal deserts lie.

The deserts of the midlatitudes are generally located in the lee of major mountain ranges: the Patagonian Desert of South America, many of the deserts of the Middle East and Asia, and the intermontane region of the western United States. Topography also plays a role in the Somali–Chalbi Desert of Ethiopia and the Horn of Africa and, to some extent, the Thar Desert of India.

In most cases, however, complementary factors also play a role in producing aridity. For example, the influence of the subtropical highs cannot suffice to explain the full extent of the Sahara and its continuation into Arabia and the Middle East. There, neither the cold-season temperate latitude weather systems to the north nor the warm-season tropical disturbances on its equatorward side penetrate to its interior. The deserts of Asia and the Middle East are far from prevailing storm tracks. The latter two regions are also quite distant from any moisture sources. In the Somali–Cholbi Desert, one factor is local patterns of wind flow, jet streams lying near the surface. A coastal, low-level jet stream enhances the aridity of the Peruvian Desert, the most latitudinally extensive desert in the world.

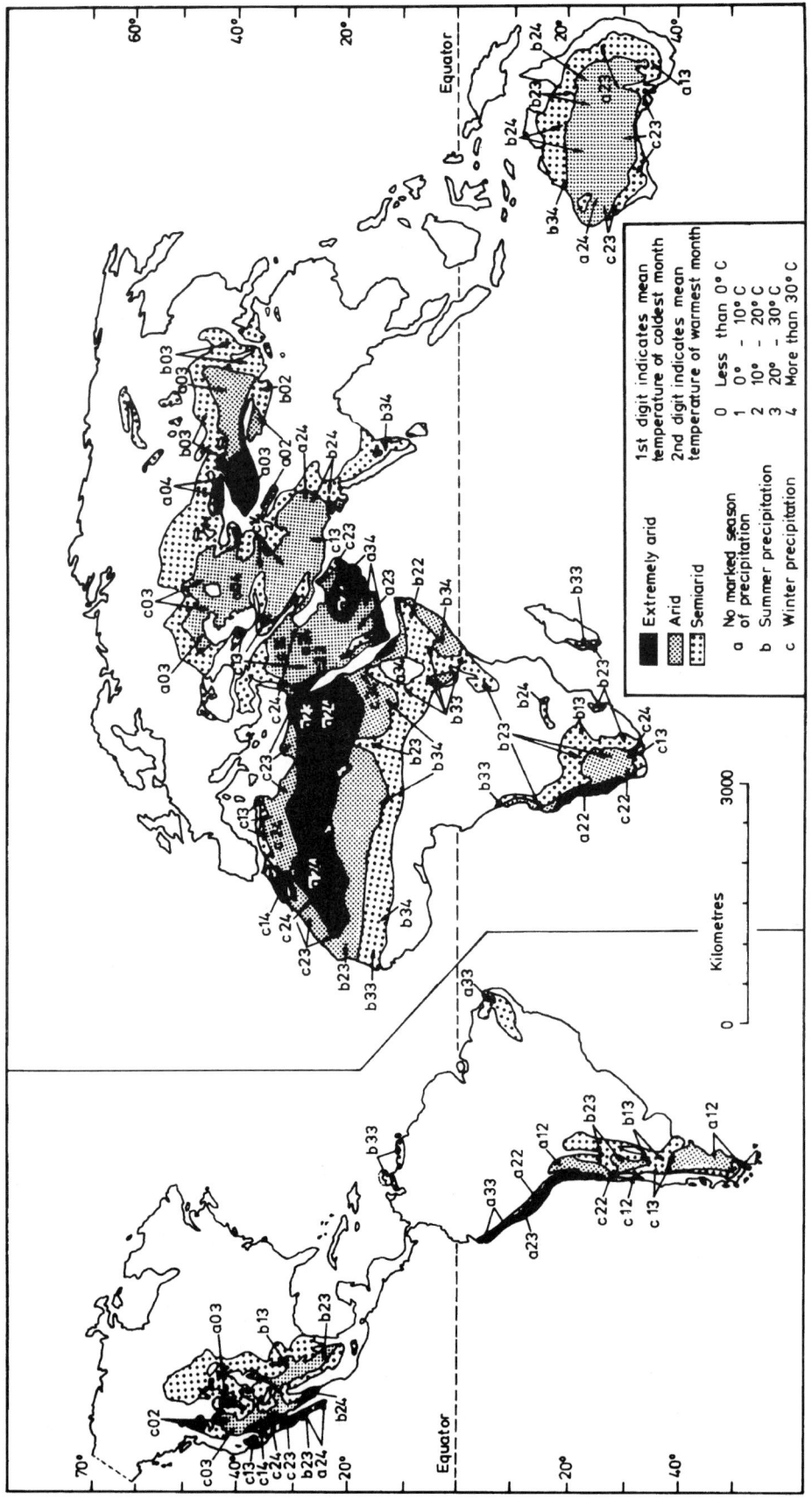

Figure D32 Location of the world's major deserts, according to the Meig's classification. (After Meigs, 1953.)

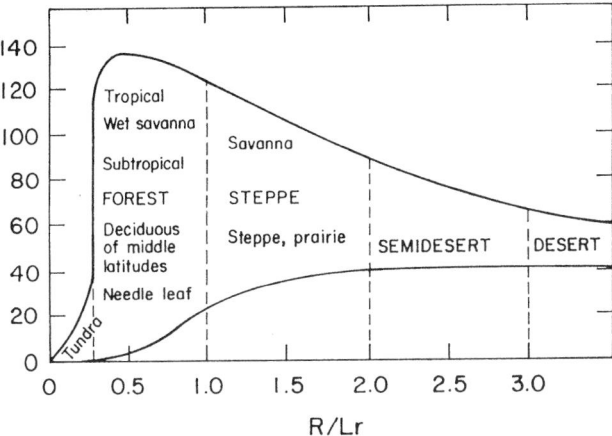

Figure D33 Budyko's concept of geobotanical zonality: the principal ecosystems as a function of net radiation and the dryness ratio (ratio between annual net radiation and annual average precipitation multiplied by the latent heat of condensation).

Figure D35 Exposed barren rock is typical of many barren landscapes. In the photograph, the rock desert begins to the north of the Kuiseb River, the bed of which is lined by dense stands of trees in the foreground. To the south of the river lie dune fields.

Thus the causes of desert climates are quite complex and often regional influences must be taken into account. The complexity and the regionalization of these factors gives a nearly unique climatic identity to each of the Earth's major dryland regions, despite their many common climatic characteristics.

The desert surface (Figures D34–D38)

The vision of the sandy desert is applicable to only a portion of the world's arid lands. Other surface types include bedrock (often termed hammada), stone pavements (termed reg and serir), depositional flats and desert crusts. Sand fields and dunes are common in the Sahara and in parts of the Australian Desert, but the Syrian and Gobi deserts are largely hammada and reg types. As the desert gives way to semiarid landscape, soils become more common and widespread.

The desert surface also contains hydrological networks carrying the flow of ephemeral stream channels and exotic streams and other landscape features. The latter commonly include alluvial fans deposited by streams, pediments (eroded bedrock platforms at the foot of hillslopes), and inselbergs (isolated, steep-sided hills void of vegetation), mesas and badlands. These landforms and surface types represent the combined forces of weathering, which breaks down materials, wind and water erosion, which transports and removes finer materials, and depositional processes (see Desert hydrology).

Stone pavements are fields of coarse material and rock fragments resting among or on finer materials such as sand, silt and clay. These are commonly produced by the process of deflation of surface

Figure D36 The rocky desert north of the Kuiseb is home to many ostriches. These animals regulate their temperature through posture, using raised feathers as a heat shield, raised wings to expose the body to cool breezes and conserve water by exhaling unsaturated air.

Figure D34 Shield and platform desert in the Namib. The uplands in the background provide runoff for the flat floodplain below.

Figure D37 Badlands are erosional features of rock deserts, as in Bryce Canyon, Utah.

Figure D38 Flat pans, or dry lake beds, dot the desert floor. This one is western Australia is filled with a bright-colored evaporite deposit, strongly contrasting with the sparse ground cover.

Figure D39 Dwarf shrubs of the Mojave desert of California.

sediments: finer materials are progressively transported by wind and occasional water currents, leaving behind the coarser residue. In deserts, of course, the sediment field must be a product of ancient streamflow and probably a more humid climate.

Sand surfaces are depositional: these particles are bounced by the wind and pushed along, though never really suspended, and they are left behind when the force of the wind decreases. Deposition by water produces other surface features, such as alluvial fans and desert flats (also termed sébkhas, pans and playas). The desert flats contain crystalline materials, such as chlorides, carbonates and sulfates (e.g. rock salt, gypsum, lime and sodium salts), which are left behind during repeated cycles of inflow and evaporation.

Smaller-scale cycles of moistening and evaporation also produce the desert crusts. Vertical water movement, which is preferentially upward in dry regions, leaves crusts of hard materials such as gypsum, lime, laterites (iron-rich material) and silicates. Although these can accumulate in lower soil horizons, they are commonly at or near the surface.

Sand dunes are an interesting feature of the desert which reflects the patterns of prevailing winds. These occupy less than 20% of the surface area of the world's deserts. They cover about 28% of the Sahara but only 1% of the desert surface in North America. A hierarchy of dune types and variety of dune forms exists, based on size, dune morphology, and relationship to the sand-transporting winds (see Sand dunes).

Classically, two types of dunes are distinguished: longitudinal (or linear) dunes, which are rougly parallel to the prevailing winds, and transverse dunes, which are aligned roughly normal to the wind. In reality a great variety and complexity of dune forms exist. The simpler dune forms result from simple wind regimes, i.e. one or two prevailing directions. As the complexity of the prevailing winds increases, so does the complexity of dunes. The availability of surface material can also influence the form of dunes.

Desert vegetation (Figures D39–D44)

Many deserts sustain a rich vegetation cover, which can include grasses, trees and shrubs (although fewer in number than in the wetter environments). In the true deserts arboreal species (trees and tall shrubs) are rare, as are perennial grasses. The dominant vegetation in most deserts includes dwarf and low shrubs (such as *Artemisia* or sagebrush) and succulents (cacti and euphorbia). Annual grasses may be present in the years with abnormally high rainfall.

The world's deserts differ greatly in the amount of biomass, productivity, characteristic species, variety and richness of these species, the percentage surface cover, and size and variety of life forms. This partly reflects the diversity of desert climates, but vegetation is even more diverse because it depends also on physical and evolutionary factors, such as topography, surface materials, local hydrology, climatic change or continental drift. Distinctive characteristics of desert vegetation do emerge, however, because plants must

Figure D40 Joshua tree of the Mojave of southern California.

Figure D41 Hummock grasses of the Australian desert.

Figure D42 Tussock grasses of the Namib desert.

Figure D43 Succulents are a major landscape element in the Sonoran desert of Arizona.

adapt to conditions of low and variable moisture, extreme temperature and high salinity common to most deserts.

Moisture availability is the principal determinant of the type and amount of vegetation because water is central to the growth process of plants. This factor represents the interplay of rainfall, evaporation and the underlying ground surface. Thus the environmental characteristics to which the flora adapt include the amount and rhythm of rainfall, atmospheric moisture, solar radiation and surface temperature, ground chemistry and soil moisture conditions. The last factor is strongly dependent on soil type, slope and topography, the types and characteristics of surface materials, the nature of the drainage system, and the proximity to groundwater, the water table, seasonal streams and exotic streams.

The most general characteristics of desert vegetation are relative scarcity and its variability in time and space. Deserts contain more bare ground than other environments, this characteristic serving to reduce the competition for soil moisture. Ground cover generally varies from about 0 to 50%. Many desert species have extensive root systems to utilize water well below the surface and a low ratio of above- to below-ground biomass, a characteristic that reduces water loss via transpiration. Plant cover is generally concentrated in wetter niches, such as stream beds. Most plants are annuals or ephemerals, so that growth is seasonal in nature. Both coverage and biomass fluctuate in response to varying moisture during the year or over longer periods.

Desert vegetation consists of two general classes, based on the primary method of resisting moisture stress: perennials which avoid drought, and annuals or ephemerals, which evade it. The latter grow profusely during periods of favorable precipitation and produce seeds which lie dormant during drier episodes, until wetter conditions return. The perennials include phreatophytes, which develop long tap roots penetrating downward to the water table, and xerophytes, which adapt to low water supply and high salinity.

Phreatophytes often seek wetter habitats, near stream channels and wadis, springs or lakeshores. Examples are the date palm, tamarisk and mesquite. Roots often extend 10–15 m or more below ground, but for the mesquite 50 m or more is not unusual.

The xerophytes include many dwarfed, woody species, with reduced water demand, and succulents which resist drought by storing water in their leaves, roots and stems. The succulents include the cacti of the American deserts and the Old World euphorbia. Other xerophytes tolerate drought, surviving long periods of dehydration in a vegetative stage, which gives them a brown and dead appearance.

An additional category of dryland vegetation is the halophytes or salt-tolerant species, such as saltbushes (*Atriplex* sp.). Some species can tolerate high salt concentrations in cell sap; others exclude salt from moisture intake or excrete it when high concentrations are reached.

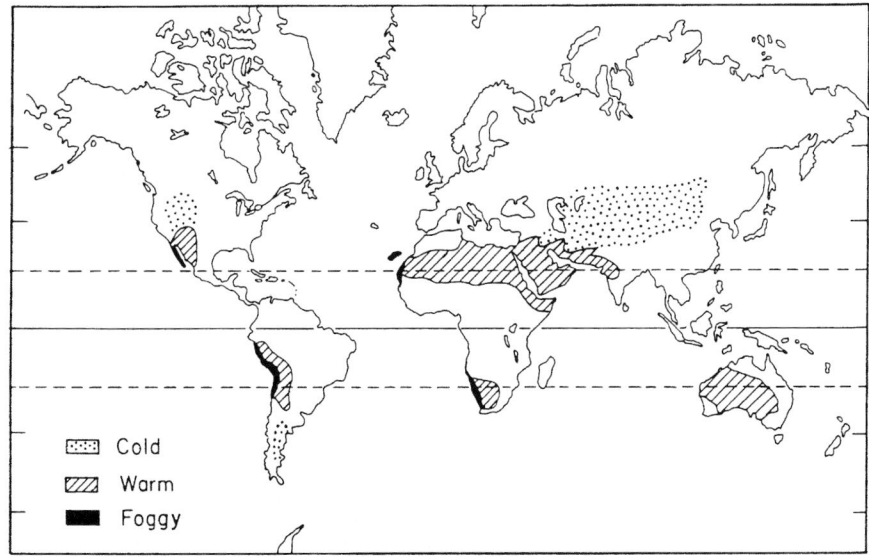

Figure D44 Classification of desert regions into cold, hot and foggy deserts.

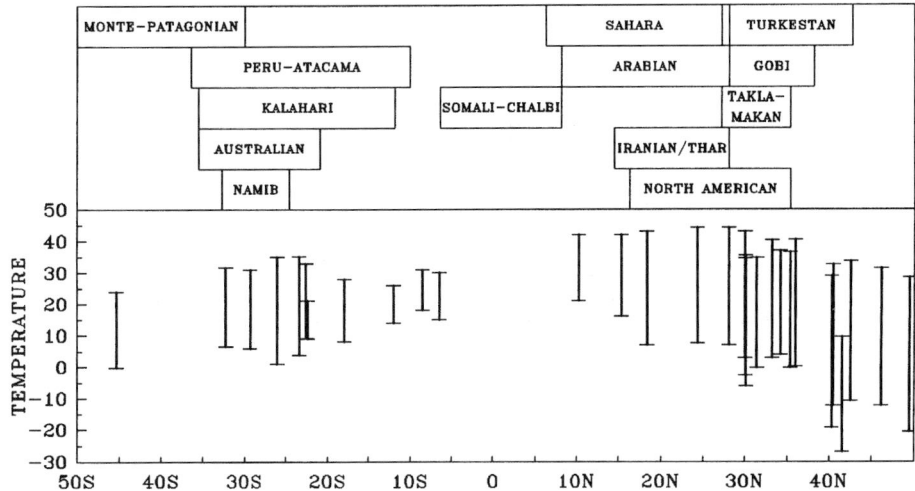

Figure D45 Annual temperature range versus latitude for arid and semi-arid regions. Bars extend from mean temperature of the warmest month to mean temperature of the coldest month. Note the smaller temperature range in the southern hemisphere, as consequence of the large ratio of water to land.

Diversity and extremes of desert climates

Most desert environments share several common climatic characteristics. These include meager rainfall which is highly variable in time and space, and often concentrated in very small, localized areas; low atmospheric relative humidity; an extreme thermal environment with large fluctuations of temperature between day and night and during the year; and generally low cloudiness and high insolation. Individual deserts differ markedly with respect to the amount of rainfall and the degree of aridity. The thermal characteristics, especially mean annual temperature and its seasonal variations, are determined to a large extent by latitude.

Consequently, several types of deserts are distinguished on the basis of climate: warm, cold-winter and foggy or coastal deserts (Figure D44). In the warm deserts of the lower latitudes, temperatures are high all year round and freezing is rare or absent. These may occasionally receive incursions of frigid air from higher latitudes, producing frost or even snow. The midlatitude deserts generally experience cold winters, often with seasonally occurring freezing temperatures. Both are characterized by low relative humidity throughout the year.

In contrast, coastal deserts are relatively humid and the temperatures relatively mild in both winter and summer, a consequence of the influence of the coastal waters. Compared to inland deserts, both the diurnal and annual temperature ranges are reduced. Also, they tend to be frequently cloud covered and fog is a frequent occurrence.

The rainfall conditions in deserts are quite diverse. Some, such as the Sahara, have virtually rainless sectors with less then 10 mm of rainfall annually. Rainless stretches of 10–14 years have been recorded in parts of the Sahara and at locations such as Swakopmund, Namibia, and Lima, Peru, in the coastal deserts of Southern Africa and South America. Few deserts are this dry. In the Sonoran, Mojave, Iran and Arabian deserts and in the Takla-Makan, mean annual rainfall is on the order of tens of millimeters in their driest cores. It exceeds 80–100 mm everywhere in the Thar, Gobi, Patagonian and Australian deserts and in the Kalahari. In contrast, it scarcely falls below 200 mm in the Chihuahuan Desert and Great Basin of North America.

The temperature conditions are also diverse because they depend on several factors, including latitude, elevation, distance from the coast and the degree of aridity (Figure D45). Except in coastal deserts, mean daily maximum temperatures during the summer months are generally of the order of 30–45°C. The mean daily winter minima are usually above −10°C, except in the interior of Asia, where they may be as low as −20 or −30°C. There is a general tendency for annual temperature range to increase with latitude, varying from about 15°C in low-latitude deserts to 20–35°C in the cold-winter deserts. The mean diurnal range is likewise diverse, varying from about 4°C in coastal deserts to 22°C in high latitudes.

These mean conditions of rainfall and temperature do not fully illustrate the extreme range of thermal conditions experienced in deserts. In Australia and Asia the absolute maximum air temperatures are of the order of 48–50°C, but temperatures of 57 and 58°C have been recorded in the Sahara, and 57°C at Death Valley, California. Surface temperature commonly fluctuates by over 30°C between day and night in the Kara Kum Desert of Asia and by over 40°C in parts of the North American deserts. Some deserts of the former Soviet Union experience absolute temperature ranges in excess of over 100°C, with air temperatures as low as −58°C having been recorded.

Ground temperatures are even more extreme, with temperatures in excess of 70°C having been recorded at several locations. At Port Sudan on the Red Sea, a sand temperature of 83.5°C was recorded. The deserts also experience extreme cold: at Repetek, in the Kara Kum Desert of Asia, where a sand temperature of 79.4°C was once recorded, the ground temperature can drop to about −40°C.

The precipitation regime is similarly extreme. In some areas, many years pass without a drop of rain, but when rain does fall, it is often torrential. Often the mean annual rainfall can fall within hours or days. At Lima, where the mean annual is 46 mm, 1524 mm fell during one storm in 1925. In the Sahara at Nouakchott, Mauritania, over twice the annual mean, 249 mm, fell on one day; at Tamanrasset, Algeria, the mean annual rainfall of 48 mm once fell in 1 h.

Thermal regime

The subtropical location of many deserts and the scant cloud cover of most produce a regime of strong solar insolation and high surface temperatures. Temperature is enhanced by the dryness of the soil and the lack of a dense vegetation cover to absorb and redistribute the solar radiation near the ground. The heat is concentrated at the surface because the dry ground transports little heat to deeper layers of soils; temperature decreases rapidly with depth into the ground and with height above the surface. In the Kara Kum Desert of Asia the daytime temperature typically drops over 15°C within the first few centimeters and 20–30°C within the first 10 cm below the surface. The air temperature can drop by more than 10°C within the first 10 cm above the surface.

The concentration of heat at the surface means that there is no subsurface thermal reservoir. As a consequence of this and the sparse vegetation cover, the surface cools extremely rapidly and efficiently at night. The generally clear skies and the dryness of the air near the ground allow most of the heat accumulated by day to escape to the upper atmosphere by night (90% compared to 50% in humid regions). This accentuates the night-time cooling.

The result is a large range of both ground and air temperature. At Khartoum, Sudan, at 15°N latitude in the eastern Sahara, the daily air temperature range is of the order of 18°C in the dry season and 11°C in the wet season. At Alice Springs, at 23°S in central Australia,

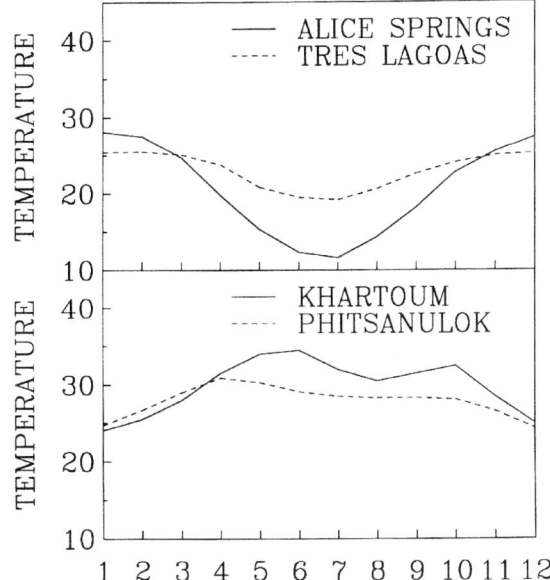

Figure D46 Mean monthly temperature at pairs of stations at comparable latitudes but differing precipitation regimes. Alice Springs and Khartoum are desert locations; Tres Lagoas, Brazil, and Phitsanulok, Thailand, are regions of tropical forests.

the diurnal range is about 15°C. The daily fluctuations at humid stations of comparable latitude would be about half as great during the wet season. At a station in the Sahara, during the course of a day the air temperature fell from a daytime maximum exceeding 37°C to −1°C at night. At Death Valley, California, a daily range of 41°C was observed in August, 1891; at Tucson, Arizona, the record is 56°C.

The lack of a surface heat reservoir in deserts tends to lead also to a high annual range of temperature. To a large extent, the annual range is dependent on latitude, increasing with latitude as seasonal contrasts become important. Nevertheless, the annual temperature range at a desert location will tend to be greater than that at a humid location at a comparable latitude. The annual ranges for Alice Springs (24°S) and Khartoum (16°N) are 16 and 10°C, respectively, compared with approximately 6°C at Tres Lagoas, Brazil, and Phitsanulok, Thailand, at similar latitudes but with humid climates (Figure D46). In deserts near the equator, the annual range is often much less, so that the daily temperature range is several times larger than the annual range.

Hydrological regime

Much of the hydrological character of a desert is dependent on its latitudinal location. Tropical rainfall prevails at low latitudes, such as in the southern Sahara or the northern sector of the Australian Desert. A midlatitude rainfall regime prevails at these latitudes, such as the poleward sectors of these deserts or in the deserts of North America and much of Asia. The tropical and midlatitude regimes differ with respect to both the nature and seasonality of the precipitation and, consequently, with respect to the loss of moisture through runoff and evapotranspiration.

Tropical rainfall is produced by convective processes (i.e. cloud formation related to localized surface heating and dynamic processes). In the midlatitudes, most rainfall (especially that which occurs in winter) is linked to large-scale warm or cold fronts. Convective rainfall tends to be much more localized and intense than frontal rainfall. As a result, in areas where convective rainfall prevails, whether deserts or humid regions, rainfall is highly variable in space, confined to raincells of the order of 10–50 km, or less. The monthly rainfall at two locations a few kilometers apart might be quite different. For this reason, rainfall in deserts was often thought to be a completely random and localized occurrence. Satellites have shown, however, that although this is true of individual raincells, these cells tend to occur in organized, large-scale meteorological disturbances.

Convective rainfall tends to be of both short duration and high intensity. A typical rain might last for a few minutes or hours, compared to days for frontal rainfall. The high intensity of the convective rainfall also affects its effectiveness, since the ground quickly becomes saturated and much of the rainfall is lost to runoff. Because there are few drainage channels in deserts, the runoff may form a thin sheet of water, transforming the barren ground to an enormous lake within minutes. This runoff produces flash flooding in many deserts, especially when rain falls over barren rock at higher elevations. As little as a few millimeters of rain can produce flow in dry desert wadis; the flow may emerge suddenly, peak rapidly, but last only a few hours. Less than 100 mm can produce catastrophic floods. In the Tadmait Plateau of the Sahara, a rain of approximately 16 mm generated a flood with peak discharge of about 1600 m³ s⁻¹.

Not only the rainfall in individual storms, but also the mean distribution in a desert, can be a mosaic; this is particularly true in the rainshadow deserts. In the state of Washington, in the western United States, mean annual rainfall varies from 3000 mm on the peaks of the Cascades to less than 200 mm in the leeside valleys only 40 km away. Isolated mountains, such as Tibesti or the Hoggar in the Sahara, dramatically enhance rainfall. Peaks may routinely receive 100 mm or more per year, compared to a few millimeters in the surrounding regions.

Although by some definitions a true desert is a location without a regular rainy season, in most desert regions there is a preference for concentration in either the warm or cool season (Figure D47). The subtropical deserts represent a transition from tropical, summer rainfall, which prevails along their equatorward margins, to mid-latitude, winter rainfall on their poleward borders. Both the seasonality and amount of rainfall decrease towards their center.

In the cold-winter deserts of the mid-latitudes, fewer generalizations can be made about the seasonal occurrence of precipitation. For example, in the western Great Plains of the United States, precipitation is concentrated in summer, but the desert Southwest tends to receive both summer and winter rainfall, with greater aridity in the transition seasons. In some deserts of the Middle East, the maximum is in spring. In most midlatitude deserts, summer rainfall has the characteristics of tropical rain, falling in intense but brief bursts which promote high runoff. Winter rainfall is usually likely to be due to large frontal systems, of low to moderate intensity but persisting for long periods. For these reasons, and because potential evapotranspiration is higher in summer than winter, summer precipitation is less effective for vegetation growth.

Snow occasionally falls in the poleward margins of the subtropical deserts. At oases such as Ouarghla and Ghardaia in the northern Sahara, snow falls as often as once in ten years. At Laghouat snow fell nearly every year towards the end of the nineteenth century and now occurs every 2–5 years. The traditional housing is not meant to withstand such occurrences, which may lead to the collapse of roofs. In the higher latitudes, snow is much more common, especially in deserts where precipitation is concentrated in the winter months. These include the Takla-Makan and Iranian deserts.

The distribution of rainfall in a desert is erratic in both space and time, especially in regions of tropical rainfall (Figure D48). The amount of rainfall varies tremendously from year to year; many years may pass without a drop. Usually rain occurs only a few days within the year and most of the rain which falls occurs within short periods, sometimes as briefly as a few hours. Several times the mean annual rainfall may occur within one day. In Helwan, Egypt, where the mean annual rainfall is about 20 mm, seven storms produced a quarter of the rain which fell during an entire 20-year period. At Nouadibou, Mauritania, with mean annual rainfall of 32 mm, annual totals have ranged from 0–301 mm; on one day in 1909 140 mm fell. At Biskra, (Algeria) (mean annual rainfall of 140 mm), annual totals range from 32 to 638 mm, but 299 mm fell in September of 1969 (210 mm of it in 2 days). The same storm system brought nearly 800 mm to Sidi bou Zid, Tunisia during September and October, months in which the mean rainfall is of the order of 10–20 mm.

Coastal deserts

Deserts are common along the western coasts of continents in the subtropical latitudes, where high pressure prevails. These deserts have certain climatic characteristics, such as a high frequency of fog, which distinguish them from other desert regions. Their origin lies in three main factors: the aridifying influence of the subtropical high-pressure cells over the oceans and adjacent coasts; the cold water which exists

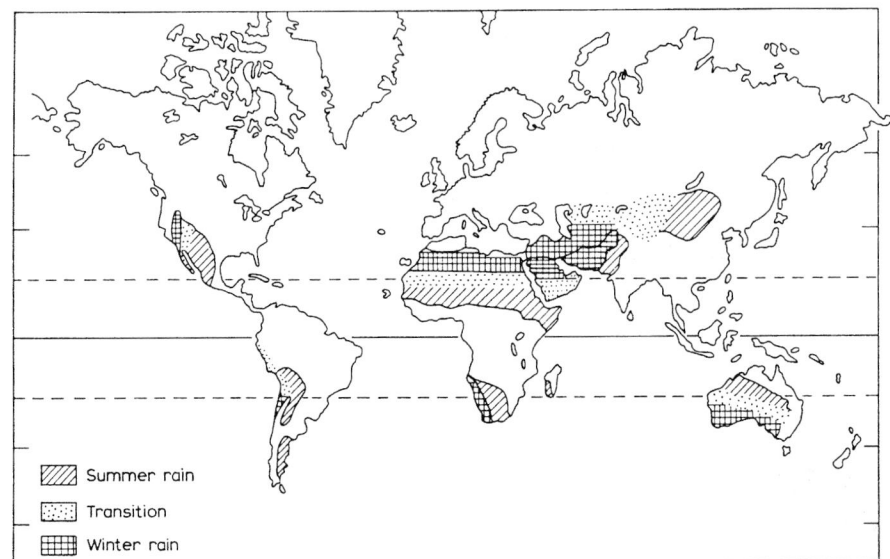

Figure D47 Areas of hot deserts with summer rain, winter rain and rainfall during the transition seasons.

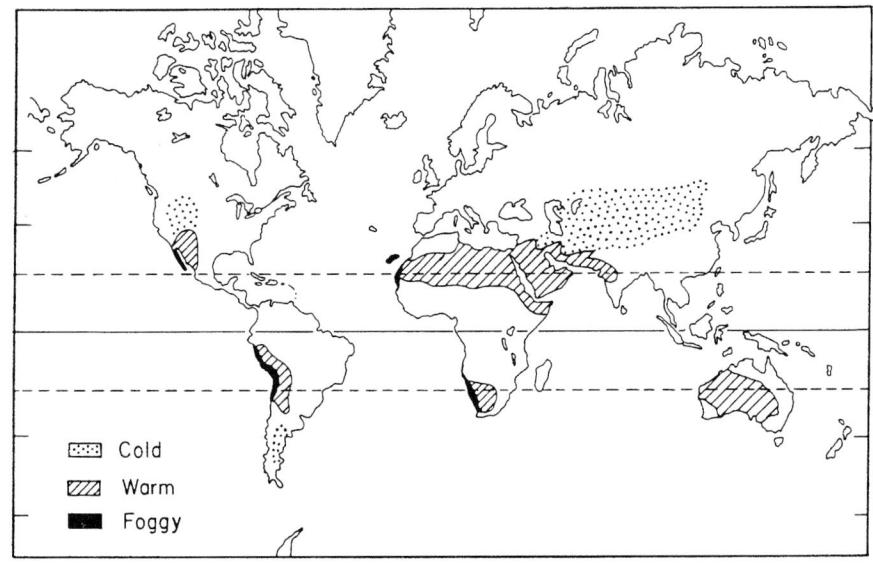

Figure D48 Mean annual rainfall (mm) at Biskra, Algeria, 1901–1990. Plus signs indicate no data.

along these coasts; and a frictional effect of the shoreline on the coastal winds. Aridity is greatest at the coast, with rainfall increasing inland. Many local factors serve to accentuate the coastal aridity, such as near-surface jet streams constrained by coastal mountain chains.

The extent of the coastal deserts ranges from 25° of latitude for the Peruvian–Atacama Desert of western South America to less than 5° of latitude for western Australia. The latitude of their equatorward margin, with summer rainfall, is highly variable, but the poleward border with winter rainfall is in all cases at about 25° N or S of the equator. In those of Africa and South America, rainfall approaches zero in the arid core.

The climatic conditions of coastal deserts are less extreme than those of inland deserts. Many of the coastal deserts are relatively moist environments, with a high frequency of fog and relatively high atmospheric humidity. In some, more precipitation is received as fog-water than as rainfall. The temperature regime is generally quite moderate, since the water, with its high thermal capacity, dampens both diurnal and annual fluctuations. In the Namib, for example, the

daily and annual range are about 6°C at the coast, compared to 15–20°C just 100 km inland.

These more moderate temperatures, together with the moisture provided by the fogs, create a more favorable environment for plants and animals than interior deserts. In the Peru–Atacama and Namib deserts some plant and animal species have special adaptations which allow them to utilize fog-water. Certain plants can absorb water directly through the leaf surface. A species of beetle in the Namib builds trenches on the dunes to trap water as it condenses on the sand. Another 'basks' on the dune with its head pointed downward and body to the wind, causing fog droplets to condense on its back and roll downward into its mouth.

Many coastal deserts are affected by a phenomenon called El Niño (q.v.) a major change of temperature and wind patterns in the Pacific that has global climatic consequences. During El Niño years the cold water along the South America desert coast disappears, establishing conditions which promote intense rainfall. In March and April of 1965, an El Niño brought 600 mm of rainfall to areas of coastal Peru which receive on average about 80 mm during those months. The El Niño often renders similar changes along the coasts of southern California, southwestern Africa and in other coastal deserts.

Winds, sand and dust

The desert surface conditions, barren ground and extreme heat interact with the prevailing regional scale winds to produce a local wind regime which is strong and turbulent. The sparse vegetation cover means little surface friction to dissipate wind near the ground, so mean wind speeds are quite high in deserts. In the afternoon steady gale-force winds may prevail for hours, but at night, when the surface cools and the air becomes stable, winds are often calm. The desert winds are hot and dry and often blow from the desert to surrounding regions of more humid climate. Some, like the harmattan of the Sahara, may provide welcome relief from stifling humidity.

The hot desert ground and rapid temperature drop above the ground produce unstable conditions which create gusty, turbulent winds. These are very effective in lifting particles from the surface. These winds are reflected in the patterns of sand dunes and surface erosion. They also produce sand and dust storms and smaller and shorter-lived dust devils. The heavy sand does not stay aloft very long, but the smaller dust particles do. The dust layer over the Sahara, for example, extends to over 5 km in altitude, producing vivid red colors in the clouds at this height.

When a dust storm occurs in a desert, its effects can be instant and dramatic. Within minutes bright sunshine changes to an eerie, dusk-like ambience of red-brown haze and the temperature can drop more than 15°C. One particular type of dust storm, called a haboob in North Africa and the southwestern United States, originates as a strong, turbulent downdraft in a thunderstorm. The dust is kicked up by what is called a density current, cold air which sinks to the ground from high altitude. As it hits the surface, it spreads laterally, churning up dust in violent gusts that may exceed 100 km h^{-1}. The visibility can rapidly approach zero, producing near darkness in mid-afternoon.

Microclimates in the desert (Figures D49–D51)

There are a number of microenvironments, or habitats, in deserts which offer shelter from the harsh conditions. It is here where life thrives. Some, such as parts of sand dunes and desert plants, modulate the thermal extremes. Others, such as oases and riverine environments, offer more favorable moisture conditions. Even dry riverbeds can support a virtual forest. Favorable habitats can also be dictated by topography or soils; the arid surface conditions are accentuated by stone pavements but moderated in lowlands near high relief. On the other hand, desert depressions, such as the Qattara depressions of

Figure D50 Many desert plants have long, sharp thorns to ward off grazing animals.

Figure D51 The *Welwitschia*, an unusual plant of the Namib desert, absorbs the fog water through its leaves. The long, flat, low surfaces of its leaves promote condensation. The relatively moderate microclimate of the plant provides a refuge for insects and small animals.

Egypt or the Chott el Jerid of Tunisia, are the hottest places on Earth. At Tozeur, in the latter, the mean daily maximum temperature exceeds 42°C in June and July.

Even an environment as small as a dune contains a number of microhabitats, as each part of the dune is affected differently by sun and wind. The side facing the morning sun will heat up first and by 0800 h may be 10°C warmer than elsewhere. In the afternoon the windy dune crest may be the coolest location by more than 10°C. Characteristic plant and animal communities reside at different parts of the dunes and in the interdune areas.

Water accumulates deep inside sand dunes, because the surface dries out first. This deep water reservoir helps to support plant life. Moisture also varies among the dune habitats. In coastal deserts, for example, where sea breezes bring fogs inland, moisture is most efficiently captured where the wind is strongest.

Temperature and surface moisture are also moderated by plant cover. The temperature in litter underneath and within the leaves of a *Welwitschia* plant may be as much as 20–30°C cooler than on the surrounding, exposed ground surface. The soil moisture content can be twice as high as on the exposed ground. Insects and small animals take refuge within the plant cover.

Climatic change (Figures D52-D54)

The Earth's deserts have undergone numerous climatic fluctuations on time scales of tens to tens of thousands of years. The fluctuations in

Figure D49 White, loose clothing is traditional attire of many of the desert peoples; it reflects the sun's rays and catches the desert breezes. The camel deals with the low moisture supply by retaining water in its humps.

Figure D52 Inselbergs, such as Ayers Rock, Australia, are common features of many deserts.

Figure D54 The coastal deserts are very different environments, with high relative humidity and moderate temperatures. Most lie along cold-water coasts, where the nutrients in upwelled water attract fish, upon which flocks of shore birds feed.

Figure D53 In some deserts the day–night temperature range is so extreme that the expansion and contraction of rocks forms huge cracks, as rock layers peel away from the cooler interior.

these regions are roughly synchronous with major changes over the Earth as a whole. Those of higher latitudes have experienced fluctuations of both temperature and rainfall, but the major changes in the low-latitude deserts involve mainly rainfall. In recent times, generally only the semiarid margins have been affected, but over the last 30 000 years the global extent of deserts has changed markedly.

During the last glacial maximum, about 18 000 BP, the low-latitude deserts generally expanded, with the Sahara advancing nearly 10° towards the equator and its dunes overriding the lakes and rivers south of its previous border. The expansion of the arid zone into the tropics was so complete that the rainforests nearly vanished, their species taking refuge in a few highland habitats. At the same time 'pluvial', or humid, conditions prevailed in many of the midlatitude deserts. Great lakes covered much of the current states of Utah, Nevada and California. Many dried up at the end of the ice age, about 10 000 years ago, and they left behind huge beds of fine sediments, such as the Bonneville Salt Flats of Utah, where the ultra-smooth surface has permitted a vehicle to set a land-speed record of 763 mph.

The conditions which commenced about 10 000 BP are in stark contrast, with a low-latitude 'pluvial' period reaching a maximum some 5 000 BP. Then, the low-latitude deserts were generally reduced to their hyperarid cores. There is little evidence anywhere of active sand dunes at that time. In parts of the Sahara, Neolithic man herded cattle and animals grazed on savanna vegetation; fish hooks un-

covered in archaeological sites attest to the presence of lakes and human occupation in now-hyperarid regions.

A more recent period in which aridity has waxed and waned was around the Middle Ages, some 600–1200 BP. Considered to be a global warm epoch, the core of this period was one of wetter conditions along the margins of many deserts. This was probably the case in the Peruvian Desert, the southwestern United States, Australia, the Mediterranean, the Middle East and the southern Sahara. Major civilizations, such as the Mali Empire, thrived in presently semiarid regions of Africa. Caravan routes traversed now-waterless plains of the Sahara and several towns flourished along these routes. On the other hand, drier conditions prevailed in some deserts and semi-desert regions of higher latitudes, such as the Great Plains of the central United States.

In recent centuries, similar fluctuations of climate have affected desert regions, although less severely. These are an inherent characteristic of the desert environment, and one to which life in the deserts has adapted.

Sharon E. Nicholson

Bibliography

Cooke, R.U. and A. Warren, 1975. *Geomorphology in Deserts*. B.T. Batsford Ltd., London.
Evenari, M., I. Noy-Meir and D.W. Goodall (eds), 1985. *Hot Deserts and Arid Shrublands. Ecosystems of the World*, **12**, Elsevier, Amsterdam.
Goudie, A. and J. Wilkinson, 1977. *The Warm Desert Environment*. Cambridge University Press, Cambridge.
Meigs, P., 1953. World distribution of arid and semi-arid homo-climates, in *Reviews of Research on Arid Zone Hydrology*. Arid Zone Research, UNESCO, Paris, pp. 203–210.
Nicholson, S.E., 1995. *Dryland Climatology*. Oxford University Press, Oxford.
Seely, M.K. (ed.), 1994. *Deserts. The Illustrated Library of the Earth*. Weldon-Owen, Sydney.
West, N.E. (ed.), 1985. *Temperate Deserts and Semi-Deserts. Ecosystems of the World*, **5**, Elsevier, Amsterdam.

Cross references

Arid climates
Aridity indices
Arid lands
Arid zone hydrology
Desert hydrology
Desertification

DESIGN FLOOD ASSESSMENT

A major objective in water management is to see that excess water from extreme flood events is controlled so as to minimize distress and hardship to the population and damage to the environment. Both the river engineer and the water resources engineer need the skills of the hydrologist to evaluate flood flows.

The frequency of occurrence of floods of different magnitude can be estimated by a variety of methods depending on the availability of hydrometric data. A logical procedure for obtaining flood flows for selected return periods is given by the UK *Flood Studies Report* (NERC, 1975). However, before the publication of this report, experienced design engineers adopted a variety of other techniques, some hydraulic and some hydrological in derivation. Several of the major flood alleviation schemes in the UK were designed from analyses of insufficient basic hydrological measurements and decisions on the frequency (return period) of a particular flood magnitude have often been made from very scanty information.

The actual choice of a design flood magnitude with its assessed return period depends both on the expected life of an engineering scheme and on the degree of protection required. In the limit, flood defences should never fail, but the cost of providing complete protection could be prohibitive. The consideration of the chance of failure is fundamental in the design of reservoir spillways, since impounding dams should never be overtopped or the structure allowed to fail. Any future urban effects on a design flood should also be considered in preparing a scheme for flood protection. The final choice of a scheme based on flood magnitude and return period and the associated costs and benefits, is made by the client, the authority commissioning the scheme. However, the engineer must assess the benefits of the protection and weigh these against the estimated costs for each of a range of schemes.

Protection of valuable inner city properties and dense housing areas near a river may merit a 1 in 200 year scheme with the authorities being obliged to meet the cost. For land drainage works incorporating flood banks, sluices and weirs, and estimated life of 75 years is considered reasonable, whereas channel regrading should last 50 years (Nixon, 1966). In making the decisions, local authorities are much influenced by the severity of any recent major flood that has just occurred. Thus the floods of 1947 due to snow melt in eastern England, estimated to be a 75-year event, governed the design of such major undertakings as the Great Ouse Flood Protection Scheme; in southwest England, the floods of 1960 were the yardstick against which many schemes were measured, to be replaced more recently by 1979 inundations. In some locations, a 30-year flood might be recommended for the design, but less vulnerable sites may only be worth a one in 5 year protection. At this lower end of the design flood range, the degree of protection for settlements and developed land joins the recommended standards for agricultural land. In the UK, these range from very high potential horticultural land with no more than 1 in 100 year frequency to very low potential grazing land of a floodplain which could be inundated every year.

Elizabeth M. Shaw

Bibliography

Natural Environment Research Council (NERC), 1975. *Flood Studies Report*, 5 vols.
Nixon, M., 1966. Economic evaluation of land drainage works. Chapter 4 in *River Engineering and Water Conservation Works*, eds Thorn, R.B., Butterworths, 520 pp.
Shaw, E.M., 1994. *Hydrology in Practice*, Chapman & Hall, London.

Cross references

Floods
Floods: largest in the USA, China and the world
Flood studies worldwide
Floods, world, maximum observed

DEW

Dew is any water that is condensed onto the ground, rocks, grass and so on. It generally forms at night due to radiational cooling (most effective on clear, still, cool nights), which causes the temperature of the air to fall below its dewpoint. If the temperature is below freezing, hoar frost will form, but if the temperature drops below freezing after the dew has formed the result is white dew. Dew is especially effective when the ground layer of air has a high relative humidity, as along river valleys, near swamps, etc. Dew is often the only form of moisture available to plants and animals in extreme deserts.

An optical effect of dew is known as heiligenschein or Cellini's halo, brought about by an early morning sun producing a shadow over the observer's head. Reflection from the dew-covered surface produces a 'saintly' halo, said to have been described first by Cellini, who found it a divine sign.

Rhodes W. Fairbridge and John E. Oliver

Bibliography

Gendezelman, S.D., 1980. *The Science and Wonders of the Atmosphere*. New York: Wiley.
Huschke, R.E., 1959. *Glossary of Meteorology*. Boston: American Meteorological Society.

Cross references

Dewpoint
Dew ponds

DEWPOINT

Dewpoint is the temperature at which saturation of the air occurs, i.e. the temperature at which the observed vapor pressure is equal to the saturation vapor pressure. If a parcel of air is hypothetically held at constant pressure and vapor content, the temperature at which it must be cooled to reach saturation is called its dewpoint or, if below 0°C, its frost point. Provided that any change in temperature of a parcel of air occurs without change in pressure or vapor content, the dewpoint will always be constant, i.e. conservative. On the other hand, if the parcel rises adiabatically, the dewpoint is quasi-conservative, because the dewpoint of moist air drops only at about one-fifth the rate of the dry adiabatic lapse rate.

A dewpoint hygrometer may be used to determine surface dewpoint. A refrigerated metal surface is brought in contact with air, causing condensation when it is slightly below the temperature of the thermodynamic dewpoint. More usually dewpoint is simply determined with a psychrometer (wet and dry bulb thermometers), used with appropriate tables (Table D14).

Rhodes W. Fairbridge and John E. Oliver

Bibliography

Wallace, J.M. and P.V. Hobbs, 1977. *Atmospheric Science: An Introductory Survey*. New York: Academic Press.

Cross references

Dew
Dew ponds

DEW PONDS

Dew ponds are artificial, saucer-shaped depressions commonly found on upland surfaces in northwest Europe, most notably in southern England. They are used to provide convenient small-scale watering sources for sheep and cattle in otherwise dry areas. Large-scale domestication of these animals began in Africa during the late Paleolithic, and following the end of the last ice age animal husbandry accompanied human immigration into northwest Europe during the Mesolithic (around 8000 BC) and especially in the Neolithic (5000–2000 BC). During the nineteenth century, however, the widespread extension of piped water supplies caused dew ponds to fall into disuse, although in the late twentieth century a sociological trend towards a simplification of life systems has led to the revival of the old facilities (see below).

Table D14 Dewpoint temperature (1000 mbar)

Columns 1–22 give the Wet-bulb depression ($T_d - T_w$).

Dry-bulb temperature (°C)	Saturation vapor pressure (mbar)	1	2	3	4	5	6	7	8	9	10	11	12	13	14	15	16	17	18	19	20	21	22
−20	1.2540	−33																					
−18	1.4877	−28																					
−16	1.7597	−24																					
−14	2.0755	−21	−36																				
−12	2.4409	−18	−28																				
−10	2.8627	−14	−22																				
−8	3.3484	−12	−18	−29																			
−6	3.9061	−10	−14	−22																			
−4	4.5451	−7	−11	−17	−29																		
−2	5.2753	−5	−8	−13	−20																		
0	6.1078	−3	−6	−9	−15	−24																	
2	7.0547	−1	−3	−6	−11	−17																	
4	8.1294	1	−1	−4	−7	−11	−19																
6	9.3465	4	1	−1	−4	−7	−13	−21															
8	10.722	6	3	1	−2	−5	−9	−14															
10	12.272	8	6	4	1	−2	−5	−9	−14	−28													
12	14.017	10	8	6	4	1	−2	−5	−9	−16													
14	15.977	12	11	9	6	4	1	−2	−5	−10	−17												
16	18.173	14	13	11	9	7	4	1	−1	−6	−10	−17											
18	20.630	16	15	13	11	9	7	4	2	−2	−5	−10	−19										
20	23.373	19	17	15	14	12	10	7	4	2	−2	−5	−10	−19									
22	26.430	21	19	17	16	14	12	10	8	5	3	−1	−5	−10	−19								
24	29.831	23	21	20	18	16	14	12	10	8	6	2	−1	−5	−10	−18							
26	33.608	25	23	22	20	18	17	15	13	11	9	6	3	0	−4	−9	−18						
28	37.796	27	25	24	22	21	19	17	16	14	11	9	7	4	1	−3	−9	−16					
30	42.430	29	27	26	24	23	21	19	18	16	14	12	10	8	5	1	−2	−8	−15				
32	47.551	31	29	28	27	25	24	22	21	19	17	15	13	11	8	5	2	−2	−7	−14			
34	53.200	33	31	30	29	27	26	24	23	21	20	18	16	14	12	9	6	3	−1	−5	−12	−29	
36	59.422	35	33	32	31	29	28	27	25	24	22	20	19	17	15	13	10	7	4	0	−4	−10	
38	66.264	37	35	34	33	32	30	29	28	26	25	23	21	19	17	15	13	11	8	5	1	−3	−9
40	73.777	39	37	36	35	34	32	31	30	28	27	25	24	22	20	18	16	14	12	9	6	2	−2

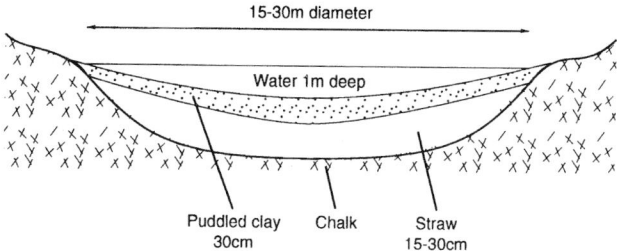

15-30m diameter

Water 1m deep

Puddled clay   Chalk   Straw
30cm   15-30cm

Figure D55 Cross-section of a typical dew pond.

Construction

The ponds are normally saucer shaped, about 16 m in diameter, and 1 m deep at the center usually with a substantial catchment of some 30 m diameter. To be classed as a dew pond, except for its own natural catchment, there should be no ingress of water from springs, channels or any form of drainage into the pond. The essential ingredient of the dew pond is a completely waterproof layer of puddled clay, usually of about 30 cm thick, over a 30 cm layer of straw (Figure D55). The clay is not allowed to dry out and crack and hence break the seal. At the turn of the twentieth century the puddling of the clay was often done by oxen but today puddling is performed by caterpillar tractor. Otherwise, construction methods have changed little over the centuries except that, when a pond is restored, a butyl rubber or polythene liner is sometimes used between the straw and the puddled clay layer (Beldon, 1997; Larkin, 1997).

Restoration of ancient dew ponds in southern England is in progress today, with the support of various environmental groups, notably the Sussex Downs Conservation Board and the Brighton Environmental Services. Using old maps, these organizations inspected over 230 possible sites where dew ponds were known to have existed at one time. From the survey, only 68 sites proved to be genuine dew ponds, of which 40 still held water but only 14 were being used for stock watering (Carpenter, 1995) Figure D56 shows two of a selection of the 30 dew ponds restored to date, and Figure D57 shows diagrammatically the method of restoration generally in

use today. This figure, in the form of a notice, and giving some notes on dew ponds, is displayed for the public's information throughout the S Downs at selected dew ponds.

Although Figure D55 shows one method of construction of a dew pond, and although it is basically the method adopted in the restoration of ponds today, it has not been the only one in the past, and over the last two centuries there have been several variations, depending mainly on the part of the country in which the dew ponds were constructed. Martin (1915) shows that straw and puddled clay were used in one position or another and often in a series of layers of (from the bottom) straw, clay, straw, clay, straw, clay (giving six layers); clay, straw, chalk rubble; quicklime, clay, quicklime, straw, broken chalk; chalk puddle only; straw, clay, stones; lime tempered clay, straw, loose rubble or broken chalk.

Historical

It is believed that dew ponds date from prehistoric times, possibly from the Neolithic or even Palaeolithic periods. In both of these periods, the water table on the chalk of the Sussex Downs was probably much the same as it is today. The ponds were to be found high up in the downs where stock required to be watered rather than exposing them to a long trek down to the valleys below.

Gilbert White, the naturalist, mentions the ponds in his writings in 1776 when he noted that

the little ponds on the chalk hills were not known to fail even when ones in the vales were dried up. White wrote 'Now we have many such little round ponds in the district [Selborne]; and one in particular on our sheep-down . . . [contains] perhaps not more than two or three hundred hogsheads of water [300 hogsheads = 73 000 l] yet never is known to fail, though it affords drink for 300 or 400 sheep and for at least twenty head of large cattle besides; and in spite of evaporation from sun and wind, and perpetual consumption by cattle, yet [these ponds] constantly maintain a moderate share of water, without overflowing in the wettest seasons, as they would do if supplied by springs. By my journal of May 1775, it appears that the small and even considerable ponds in the valleys are now dried up, while the small ponds in the very tops of hills are but little affected. Can the difference be accounted for from evaporation alone, which is certainly more prevalent in

Figure D56 Dew pond near Brighton on the Sussex Downs, UK.

Figure D56 (continued) Dew pond near Brighton on the Sussex Downs, UK.

bottoms? or, rather, have not those elevated ponds some unnoticed recruits, which in the night-time counterbalance the waste of the day, without which the cattle alone must soon exhaust them.

Clearly White was intrigued by the ponds and did he too think of dew supplementing them by night, but was too professional to say so?

Theory

Dew consists of drops of water, the condensation of atmospheric water vapor on any surface where the temperature has fallen below the dew point (q.v.). The latter is defined as the temperature at which the saturation pressure equals the vapor pressure. If the ambient temperature is below 0°C, it becomes the frost point. Dew is often the only source of water available for plants and animals in desert regions. In northwest Europe, dry areas exist, especially on the Upper Cretaceous chalk formation (widespread in southern England, northern France and Germany) which is homogeneously porous and consequently well drained, even though there is abundant rainfall (750–1000 mm). Thus there is a genuine need for dew ponds. However, the term is based on a misleading deduction, dating from antiquity. Walford (1924) reported it as 'the great dewpond myth'. Alas, the maximum experimental dew accumulation is only 50 mm year^{-1}. Earlier experiments noted by Pugsley (1939) were carried out by Slade (1877), Hubbard and Hubbard (1904) and by Martin (1915), but all led to the same conclusion. Nevertheless, the term enjoys the priority of antiquity and is therefore permanently established.

The explanation is that in a region of annual precipitation up to 1000 mm, rather evenly distributed throughout the year as a rule, and where the annual evaporation is only 450 mm, there should be an abundant water supply except in major drought cycles (as in 1994–1997, the driest in centuries). There seems scope, however, for further research. Slade (1877) notes that 'shepherds reported that some dewy nights had added an inch to the water measurement of the afternoon before', which may well be a little propaganda encouraged by the pond makers. Nevertheless, the meteorological setting is significant. In a region of prevailing westerlies, the often nearly saturated humid air mass is forced to rise 200–300 m as it encounters the Downs,

where the air temperature may fall below the dewpoint and a 'scotch mist' may result.

R.W. Herschy

Bibliography

Beckett, A., 1923. *The Spirit of the Downs*, Chapter. XX, Methuen, London.

Belden, P., 1997. Private communication and discussion (Sussex Downs Conservation Board).

Bilham, E.G., 1938. *The Climate of the British Isles*, Macmillan, London.

Burton, L.C., 1956. The dew ponds of Sussex, *The Angler*, **8**.

Carpenter, G., 1995. Oases of the Downs, *The Countryman* **100**(6), London.

Hubbard, A.J. and Hubbard, G., 1904. *Neolithic Dewponds and Cattle Ways.*

Larkin, D., 1997. Private communication and discussion (Brighton and Hove Environmental Services).

Martin, E.A., 1915. *Dewponds: History, Observation and Experiment*, Werner Laurie, London.

Miall, L.C., 1900. The British Association Report, Bradford.

Phillip, A., 1935. Dewponds of Sussex, *Country Life*, London.

Pugsley, A.J., 1939. Dewponds in fable and fact, *Country Life*, London.

Slade, H.P., 1877. *A Short Practical Treatise on Dewponds*, Spon, London.

Toms, H.S., 1926. *Rock Pond*, Standean, The Downland Post.

Walford, E., 1924. The great dewpond myth, in *Discovery*, Vol. 5, 245 pp.

Wenlock, C., 1939. Dewponds of the Sussex Downs, *The Sussex County Magazine*.

White, Gilbert, 1775. *Natural History of Selborne*.

Cross references

Evaporation: measurement
Evapotranspiration
Groundwater
Precipitation

DEW PONDS

Dew Ponds are traditional hill ponds used for watering sheep before there was piped water and troughs.

Dew Ponds are downland oases providing an important habitat for wildlife as well as being significant landscape features.

The South Downs are made of porous chalk so, in order to hold water, a layer of non porous / impervious clay is 'puddled' into a chalk depression.

A modern technique for restoring Dew Ponds is to put a liner underneath the clay. The liner holds the water should the clay crack during hot and dry weather.

Folklore has provided the name Dew Pond (and in some areas - Mist Pond) for these hill ponds. Although dew and mist do contribute a small amount of water, the vast majority comes from rainfall. They retain water in all but the driest summers because water evaporation is much less than annual rainfall.

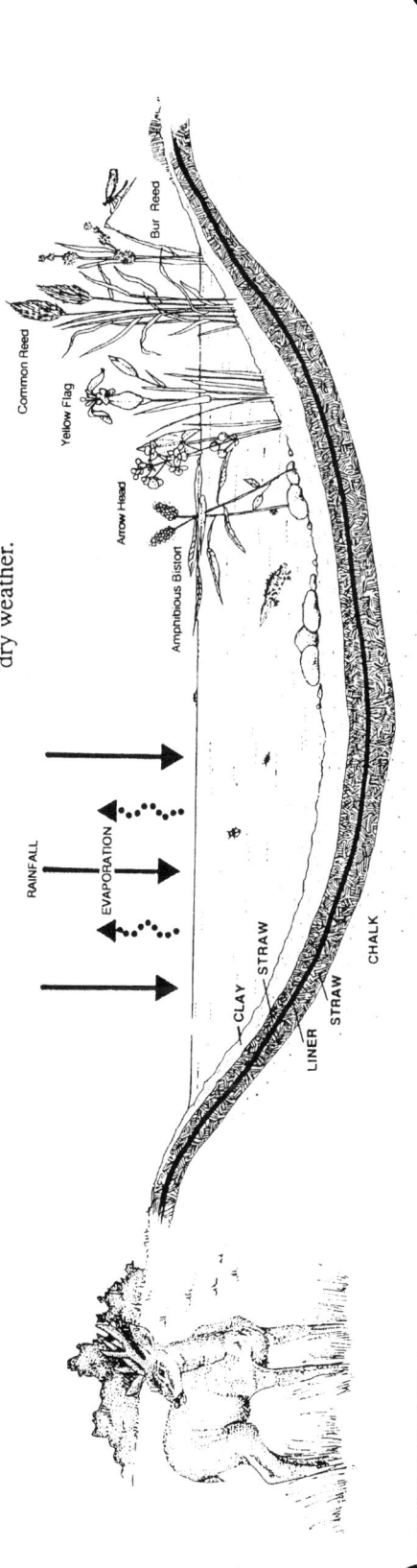

RAINFALL

EVAPORATION

CLAY
STRAW
LINER
STRAW
CHALK

Common Reed

Bur Reed

Yellow Flag

Arrow Head

Amphibious Bistort

Figure D57 Copy of notice for public information placed at dew ponds showing generally the method of restoration. (Courtesy of Sussex Downs Conservation Board.)

DRAINAGE

Artificial drainage is an integral part of agricultural intensification in many parts of the world. The primary concern of agricultural drainage engineers is the control of water table conditions within an area of crop production. The installation of artificial drains to remove 'excess' water from the soil confers a number of benefits to farmers on wet land. It promotes the breakdown of organic matter in the soil and the uptake of nutrients, plant growth is more uniform over a field, animals and crops are less susceptible to disease and pest attack, less mechanical power is needed for tillage operations, and the greater bearing strength of the soil increases the time during the year when heavy machinery can be used on the land. Worldwide, about 12% of the cultivated area has artificial drainage which is between one-third and a half of the land that would benefit from drainage (Smedema and Ocks, 1995).

The different types of drainage systems may be broadly divided into surface drainage systems, subsurface or groundwater drainage systems and the main drainage channel systems that convey the water away.

Surface drainage

Surface drainage systems comprise open ditches and land grading, and are intended to gather excess water from the surface of the land and conduct it to a collector drain. This is the oldest and simplest type of drainage. The excess may result from the intensity of the rain exceeding the infiltration capacity of the soil, or may be due to the soil being fully wetted up, leaving no further scope for infiltration. At the field scale, the collector drains are essentially carriers of excess water, often being widely spaced and their position being determined by topography; they are not intended to have a significant drawdown effect on the water table within the fields. An example of this type of drainage is the field boundary ditch that often surrounded small fields in the nineteenth century. The grading of land may be intended to remove surface depressional hollows and to provide a uniform slope, and includes the plowing of land into raised bedding systems (sometimes known as ridge-and-furrow) to provide elevated, drier cropping sites.

Subsurface drainage

Subsurface drainage involves the installation of underground pipes and hence the removal of excess water by subsurface and groundwater means.

Shallow drainage is normally confined to poorly permeable soils in which shallow flow can occur through the more permeable topsoil. In the tropics, shallow drainage systems are based upon raised beds and shallow ditches, whilst in Europe buried pipes are often installed in heavy clay soils in conjunction with artificial soil loosening (mole drains or subsoiling); in such circumstances, a combination of shallow flow through the loosened soil and groundwater flow through the undisturbed subsoil occurs. A gravel backfill over the pipes assists downward flow through the soil profile to the pipes.

Groundwater drainage is generally confined to relatively permeable soils and is usually based upon systems of pipes or ditches, with modern practice increasingly making use of perforated plastic pipes in place of the clay tile drains which were formerly used. True groundwater flow occurs when the water table rises above the level of the drain, thus creating the elevation head needed to generate flow towards it.

There is also interaction between surface and subsurface drainage (the two may be used in conjunction). Surface and subsurface flows have been modeled, for example, using the soil water management model DRAINMOD (Skaggs, 1980).

The distribution of subsurface drained land is shown for Europe in Figure D58. The greatest intensity of drainage in Europe is concentrated in the north around the Baltic and the North Sea. This is largely due to a combination of climatic conditions (rainfall and temperature) and the presence of glacially derived clay soils, which can result in prolonged waterlogging (Green, 1980). In the Mediterranean countries waterlogging is rarely a problem and there is much less need for drainage. Such national drainage statistics are not only subject to differences in methods of collection and accuracy; they also hide regional variations within individual countries. Thus drainage in the Scandinavian countries is largely confined to their southern margins,

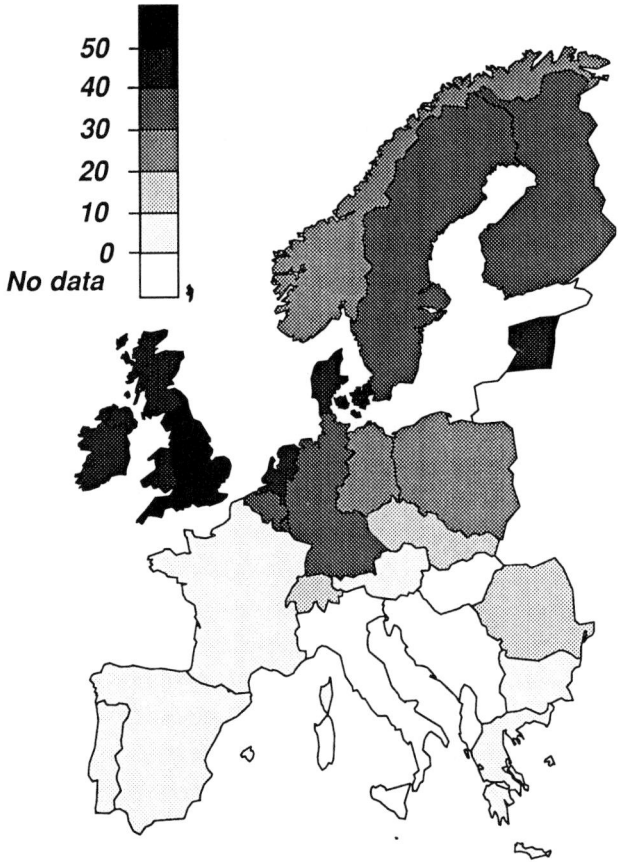

Figure D58 Percentage of agricultural land in Europe with subsurface drainage.

and in the UK to the most productive agricultural land in the east of the country. The detailed distribution of drainage depends on a combination of physical factors such as rainfall, evaporation, soil type and topography, as well as on social factors including land tenure, the size of holdings and government policy.

The UK has a long history of involvement with drainage and was in fact the originator of modern field drainage (Van der Beken, 1987). It is the most extensively drained country in Europe, possibly in the world, and for many years this was actively encouraged by the government for strategic reasons (the country is a large net importer of food) through systems of grants to farmers. It is also probably unique in the detail of information about field drainage, due to the records kept by a centralized system of government support to farmers for making improvements to their land. For 40 years after World War II statistics were recorded for every drainage scheme carried out (Figure D59).

Improvements to main channels

Main drainage describes the system of collectors comprising either closed pipes, or more commonly open channels, which are used to convey the water away from an area. The channels may form part of a natural river system or may be artificially constructed, in which case they would normally have a prismatic type of cross-section.

Environmental impacts

As concern about environmental impacts continues to grow, it is increasingly important to be aware that agricultural effects may extend beyond the boundary of a field or a farm. Thus drainage may cause a drying of surrounding land and will inevitably affect the pattern of water flows from the land and into the receiving watercourses. The downstream impacts of farm drainage have been

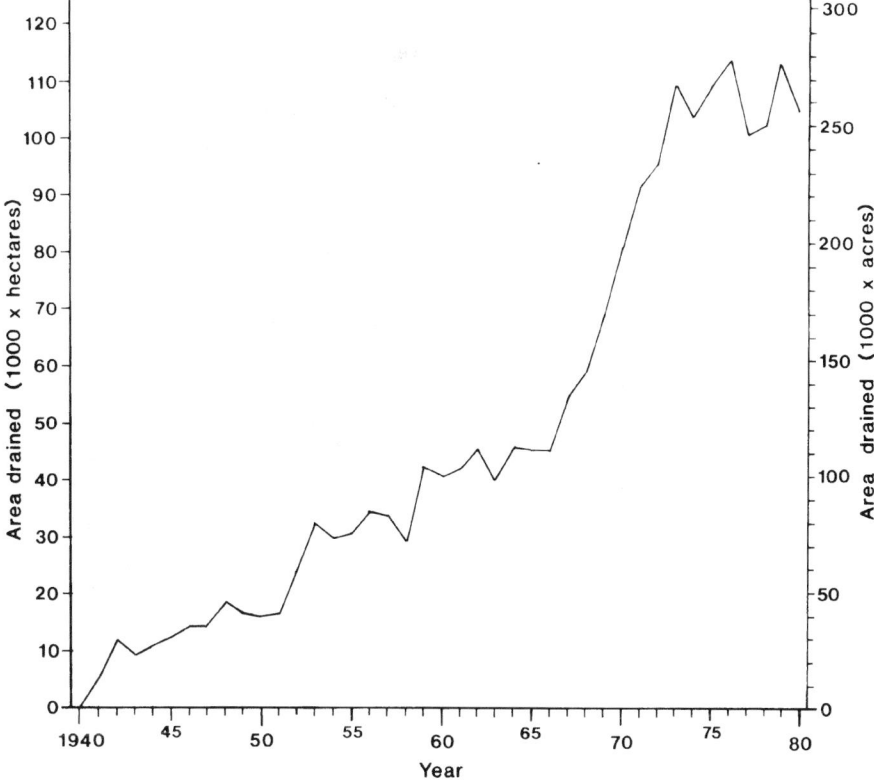

Figure D59 Rate of field drainage in England 1939–1980. (From Robinson and Armstrong, 1988.)

the source of much uninformed speculation, but recent research has reached the following broad conclusions.

Surface drainage increases peak flows by reducing surface ponding. The effect is similar in manner to that of main channel drainage where peak flows farther downstream are increased because of the reduction of overbank storage on the floodplain.

The impact of subsurface drainage depends upon soil wetness. At naturally wet sites, drainage increases the available soil water storage capacity, thereby reducing peak storm flows, whereas at drier locations artificial drainage increases peak flows as a result of the shorter flow paths and steeper hydraulic gradients.

The effect of drainage on low flows varies with the type of drainage. In the case of surface drainage the effect would be to reduce low flows by the removal of excess water which would otherwise have to drain away slowly through the soil, supporting baseflow. The result of subsurface drainage is to increase low flows, provided that the artificial drainage system removes water from a greater thickness of soil profile than the natural drainage system. Otherwise, if the drains are no deeper than the original channels, the reverse may occur and the flow in streams will cease sooner in dry weather.

Concerns about the damage to stream ecology resulting from the dredging and straightening of main channels have resulted in a move to more environmentally friendly design with less emphasis on linear channels and less destruction of river bed and bankside habitats.

Mark Robinson

Bibliography

Armstrong, A.C., 1978. *A digest of drainage statistics.* FDEU Technical Report 78/7, MAFF, London, UK.

Green, F.H.W., 1980. Current field drainage in northern and western Europe. *J. Environ. Management,* **10**, 149–153.

Robinson, M. and Armstrong, A.C., 1988. The extent of agricultural field drainage in England and Wales, 1971–80. *Trans. Inst. British Geographers,* **13**, 19–28.

Robinson, M. and Rycroft, D.W., 1996. The impact of drainage on streamflow, in van Schilfgaarde J. and Skaggs, R.W. (eds), *Agricultural Drainage.* American Society of Agronomy, Madison, Chapter 23.

Skaggs, R.W., 1980. DRAINMOD Reference Report: *Methods for design and evaluation of drainage water management systems for soils with high water tables.* Soil Conservation Service, US Department of Agriculture, South Technical Center, Fort Worth, Texas.

Smedema, L.K. and Ocks, W.J., 1995. The state of land drainage in the world. *National Seminar on Subsurface Drainage,* 24–26 May 1995. Jaipur, India, Vol II, pp. 1–15.

Smedema, L.K. and Rycroft, D.W., 1983. *Land Drainage.* Batsford, London.

Van Der Becken, A., 1987. The development of the theory and practice of land drainage in the nineteenth century, in Wunderlich, W.O. and Prins, J.E. (eds), *Water for the Future.* A.A. Balkema, Rotterdam, pp. 91–99.

Cross references

Groundwater
Irrigation and drainage

DRAWDOWN, CONE OF DEPRESSION

Drawdown is a term applied to the maximum lowering of the groundwater table caused by pumping or artesian flow (Figure D60). It is measured as the difference between the initial level of water in a well before pumping, and the static, or stabilized, level of water after a long period of pumping. The static level is achieved when the flow into the well from the aquifer is equal to the rate of withdrawal.

The drawdown may be determined, for an unconfined aquifer, from the following equation (after Todd, 1980):

$$Q = \frac{\pi K(D^2 - d^2)}{\ln(R/r)}$$

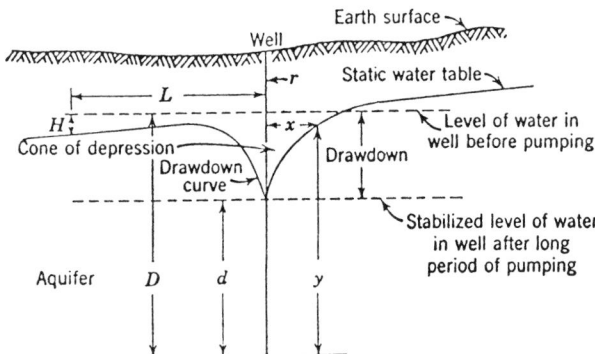

Figure D60 Diagrammatic representation of a well penetrating an unconfined aquifer. Pumping has lowered the water table locally to develop a cone of depression around the well Key: r = radius of well in metres or feet; d = thickness of water in well (in metres or feet measured from the base of the aquifer being pumped; x and y = coordinates of any point on the drawdown curve; D = thickness of aquifer below static water level (in metres or feet); H = difference in head between two manometers in path of fluid movement; L = distance along the path of movement between positions of manometers.

where Q is the flow in (m³ s⁻¹); K is the hydraulic conductivity in (m³ s⁻¹); d is the thickness of water in the well when pumping is in progress, measured from the base of aquifer being pumped (m); D is the thickness of the aquifer below the static water level measured to the base of the well when entirely in the aquifer, or to the base of the aquifer if the well is deeper (m); R is the radius of base of cone of depression (m), and r is the radius of the well (m). Note: the equation is applicable in either metric or fps units

Within the aquifer (q.v.), immediately around a well, the upper boundary of the water is lowered as water is pumped out (Figure D60). The water surface boundary, described by an inverted cone, is known as the cone of depression, which theoretically extends outward from the well to the limits of the water-bearing bed. The radial distance out from the well over which this water boundary is actually lowered, below that of the static water table, is termed the radius of influence, and is approximately 400 m. This must be taken into account for the proper spacing of wells (Boulton, 1954).

Charlotte Schreiber

Bibliography

Boulton, N.S., 1954. The drawdown of the water table under non steady conditions near a pumped well in an unconfined formation, *Inst. Civ. Eng. Proc.*, London, **3**(3), 564–579.
Todd, D.K., 1980. *Groundwater Hydrology*, John Wiley & Sons, New York.

Cross references

Aquifer
Chilgrove House well
Groundwater
Water table

DRINKING WATER AND SANITATION

The access to drinking water and sanitation in 148 countries world-wide is shown in Table D15.

R.W. Herschy

Source

World Development Report, 1994. The World Bank.

Cross references

Access to and accountability of water resources
Environment priorities for development: water

Groundwater
Sanitation and clean water
Water availability and river water quality
Water balance
Water resources: Europe
World water balance

DROUGHT

Drought has been defined in many ways, including (1) a period of rainfall deficiency, (2) a relative state of forest flammability, (3) occurring when a specific agricultural crop or pasture yields less than expected amounts, (4) denoting a critical level of soil moisture or groundwater depletion, and (5) poetically as a 'valley of rain deficiency in the broad sweep of time and weather.' Current theory identifies the phenomenon as a temporary negative deviation in environmental moisture status (EMS).

Drought identification: early developments

During the first decade of the twentieth century, the US Weather Bureau identified drought as occurring during any period of 21 or more days with rainfall 30% or more below normal. Tannehill (1947) cites the test results of this criteria as it was applied to the District of Columbia, wherein 62 cases of drought were recorded within a 33-year period. Detailed examination of these drought events revealed that most were not perceived by individuals affected by them as significant dry periods. It was also learned that ample or heavy precipitation frequently preceded the drought episode, that soil moisture reserves during the events were often at high levels – sufficient to support vegetation's water demands – and that drought identification measures based solely on precipitation were inadequate. Numerous researchers subsequently have shown that an area's moisture status is constituted by more elements than precipitation receipts alone.

Thornthwaite (1948), Penman (1948, 1956), and Holdridge (1962, 1967) stressed the importance of considering moisture demands that are associated with a region's energy regime, i.e. potential evapo-transpiration, when evaluating its water balance. Chang (1968) and Keetch and Byram (1968) have shown soil moisture status also to be critical in the determination of EMS, that the level at which soil moisture is present antecedent to a precipitation deficiency can serve as a limiting factor on drought intensity, and that soil moisture reserves at a point in time serve as a cumulative index of water transfers between the Earth and its atmosphere. Early attempts at incorporating these parameters in drought evaluation have been described by Ventskevich, (1961) who detailed use of the Selyaninov hydrothermal coefficient (GTK) and Kulik soil water approach.

The GTK appraised drought by uniting, with a single formula, both precipitation and temperature parameters, the latter being used as a convenient means of approximating evapotranspiration:

$$GTK = \frac{(\Sigma \text{ precipitation} \times 10)}{(\Sigma \text{ temperature} > 10)}$$

where precipitation is expressed in millimeters and temperature in degrees Celsius. Drought identification is theoretically derived from the ratio when the coefficient falls below 1.0. Selyaninov's methodology was criticized because of three primary defects: (1) soil moisture reserves were ignored, (2) total precipitation, including that which comprised runoff, was included in the assessment, and (3) the estimate of evaporability was unsubstantiated.

Kulik, on the other hand, defined drought based on only soil moisture reserves, considering it to be a period of time during which plant-available water in the upper 20 cm of the solum is less than 20 mm for at least 10 days. A major drawback to Kulik's approach is the soil moisture requirement of 20 mm or more in the upper 20 cm for drought not to exist. Under this assumption, locales with porous soils would always experience drought as would arid climatic zones. Even though porous soils may lack plant-available water, the climatic milieu need not be experiencing dry conditions. Soils nearby may actually be above field capacity due to recent precipitation. The methodology appears better suited to identifying soil and/or climatic aridity than it does to detecting significant deviations in soil moisture status.

Table D15 Access to drinking water and sanitation (percentage of population)

Country	Access to safe drinking water							Access to sanitation						
	Total			Urban		Rural		Total			Urban		Rural	
	1970	1980	1990	1980	1990	1980	1990	1970	1980	1990	1980	1990	1980	1990
Low income economies														
1 Mozambique[a]	–	–	22	–	44	–	17	–	–	21	–	61	–	11
2 Ethiopia[a]	6	–	18	–	70	–	11	12	–	17	–	97	–	7
3 Tanzania[a]	13	–	52	–	75	–	46	–	–	77	–	76	–	77
4 Sierra Leone	12	14	39	50	80	2	20	–	12	39	31	55	6	31
5 Nepal	2	11	37	83	66	7	34	1	2	6	16	34	1	3
6 Uganda	22	11	33	45	60	8	30	78	13	60	40[b]	32	10	60
7 Bhutan	–	7	34	50	60	5	30	–	–	43	–	70	–	37
8 Burundi	–	23	46	90	92	20	43	–	35	19	40	64	35	16
9 Malawi[a]	–	41	51	77	66	37	49	–	83	–	100	–	81	–
10 Bangladesh	45	39	78	26	39	40	89	6	3	12	21	40	1	4
11 Chad	27	–	57	–	–	–	–1	–	–	–	–	–	–	–
12 Guinea-Bissau[a]	–	10	25	18	18	8	27	–	15	21	21	30	13	18
13 Madagascar[a]	11	21	21	80	62	7	10	–	2	–	9	–	–	–
14 Laos	48	21	28	21	47	12	25	–	4	11	11	30	3	8
15 Rwanda	67	55	69	48	84	55	67	–	51	23	60	88	50	17
16 Niger	20	33	53	41	98	32	45	–	7	14	36	71	3	4
17 Burkina Faso[a]	12	31	70	27	44	31	70	–	7	7	38[b]	35	5	5
18 India	17	42	73	77	86	31	69	18	7	14	27	44	1	3
19 Kenya	15	26	49	85	–	15	–	49	30	–	89	–	19	–
20 Mali	–	6	11	37	41	0	4	–	14	24	79	81	0	10
21 Nigeria	–	36	42	60	100	30[b]	22	–	–	28	–	80	–	11
22 Nicaragua	35	39	55	91	76	10	21	18	18	35	–	–	–	–
23 Togo[a]	17	38	70	70	100	31	61	–	13	22	24	42	10	16
24 Benin	29	18	55	26	73	15	43	14	16	45	48	60	4	35
25 Central African Republic	–	–	24	–	19	–	26	–	–	46	–	45	–	46
26 Pakistan	21	35	55	72	82	20	42	–	13	25	42	53	2	12
27 Ghana	35	45	–	72	63	33	–	55	26	61	47	63	17	60
28 China[a]	–	–	72	–	87	–	68	–	–	85	–	100	–	81
29 Tajikistan	–	–	–[c]	–	–	–	–	–	–	–	–	–	–	–
30 Guinea	–	15	52	69	100	2	37	11	11	–	54	–	1	0
31 Mauritania	17	–	66	80	–	85	–	–	1	–	5	–	–	–
32 Sri Lanka	21	28	60	65	80	18	55	65	67	50	80	68	63	45
33 Zimbabwe	–	–	84	–	95	–	80	–	–	40	–	95	–	22
34 Honduras	34	59	64	50	85	40	48	24	31	62	40	89	26	42
35 Lesotho[a]	3	15	47	37	59	11	45	11	14	21	13	14	14	23
36 Egypt, Arab Republic	93	84	90	88	95	64	86	–	26	50	45	80	10	26
37 Indonesia	3	23	34	35	35	19	33	13	23	45	29	79	21	30
38 Myanmar	18	21	74	38	79	15	72	36	20	22	38	50	15	13
39 Somalia[a]	15	32	36	60[b]	50	20	29	–	17	17	45[b]	41	5	5
40 Sudan[a]	19	51	34	–	90	31[b]	20[b]	–	12	12	63[b]	40	0	5
41 Yemen, Republic	14	24	–	100	–	18	–	–	–	–	–	–	–	–
42 Zambia[a]	37	46	59	65	76	32	43	17	70	55	100[b]	77	48[b]	34
Middle-income economies														
Lower–middle income														
43 Ivory Coast	44	–	69	–	57	–	80	–	–	91	–	81	–	100
44 Bolivia	33	36	53	69	76	10	30	13	19	26	37	38	4	14
45 Azerbaijan	–	–	–[c]	–	–	–	–	–	–	–	–	–	–	–
46 Philippines	36	45	81	65	93	43	72	58	72	70	81	79	67	63
47 Armenia	–	–	–[c]	–	–	–	–	–	–	–	–	–	–	–
48 Senegal	–	43	44	33	65	25	26	–	3	47	5	57	2	38
49 Cameroon	32	–	44	–	42	–	45	–	–	–	–	–	–	–
50 Kyrgyzstan Republic	–	–	–[c]	–	–	–	–	–	–	–	–	–	–	–
51 Georgia	–	–	–[c]	–	–	–	–	–	–	–	–	–	–	–
52 Uzbekistan	–	–	–[c]	–	–	–	–	–	–	–	–	–	–	–
53 Papua New Guinea	70	16	33	55	94	10	20	14	15	56	96	57	3	56
54 Peru	35	50	53	68	68	21	24	36	37	58	57	76	0	20
55 Guatemala	38	46	62	89	92	18	43	22	30	60	45	72	20	52
56 Congo Republic[a]	27	20	38	36	92	3[b]	2	–	6	–	17	–	0	2
57 Morocco	51	–	56	100	100	–	18	30	–	–	–	100	–	–
58 Dominican Republic	37	60	68	85	82	33	45	57	15	87	25	95	4	75
59 Ecuador	34	50	54	82	63	16	44	–	26	48	39	56	14	38
60 Jordan	77	86	99	100	100	65	97	–	70	100	94	100	34	100

Table D15 Continued

Country	Access to safe drinking water							Access to sanitation						
	Total			Urban		Rural		Total			Urban		Rural	
	1970	1980	1990	1980	1990	1980	1990	1970	1980	1990	1980	1990	1980	1990
61 Romania[a]	–	–	95	–	100	-	90	–	–	97	–	100	–	95
62 El Salvador	40	50	47	67	87	40	15	37	47	59	80	85	26	38
63 Turkmenistan	–	–	–[c]	–	–	–	–	–	–	–	–	–	–	–
64 Moldova	–	–	–[c]	–	–	–	–	–	–	–	–	–	–	–
65 Lithuania	–	–	–[c]	–	–	–	–	–	–	–	–	–	–	–
66 Bulgaria[a]	–	–	90	–	100	–	96	–	–	100	–	100	–	100
67 Colombia	63	86	86	–	87	79	82	50	66	64	100	84	4	18
68 Jamaica[a]	62	51	72	–	95	–	46	94	–	–	–	14	–	–
69 Paraguay	11	21	–	39	61	10	–	–	92	47	95	31	89	60
70 Namibia	–	–	47	–	90	–	37	–	–	13	–	24	–	11
71 Kazakhstan	–	–	–[c]	–	–	–	–	–	–	–	–	–	–	–
72 Tunisia[a]	49	60	70	100	100	17	31	63	55	47	100	71	–	15
73 Ukraine	–	–	–[c]	–	–	–	–	–	–	–	–	–	–	–
74 Algeria	–	–	–	–	–	–	–	10	–	–	–	–	–	–
75 Thailand	17	63	77	65	–	63	85	–	45	–	64	–	41	86
76 Poland[a]	–	–	89	–	94	–	82	–	–	100	–	100	–	100
77 Latvia	–	–	–[c]	–	–	–	–	–	–	–	–	–	–	–
78 Slovak Republic	–	–	–	–	–	–	–	–	–	–	–	–	–	–
79 Costa Rica[a]	74	90	92	100	100	68	84	53	87	96	93	100	82	93
80 Turkey[a]	53	76	84	95	100	62	70	–	–	92	56	95	–	90
81 Iran, Islamic Republic	35	66	89	82	100	50	75	74	–	71	–	100	–	35
82 Panama[a]	69	81	84	100	100	65	66	73	45	85	62	100	28	68
83 Czech Republic	–	–	–	–	–	–	–	–	–	–	–	–	–	–
84 Russian Federation	–	–	–[c]	–	–	–	–	–	–	–	–	–	–	–
85 Chile[a]	56	84	87	100	100	17	21	29	–	85	99	100	–	6
86 Albania	–	–	97	–	100	–	95	–	–	100	–	100	–	100
87 Mongolia	–	–	80	–	100	–	58	–	–	75	–	100	–	47
88 Syrian Arab Republic[a]	71	74	79	98	91	54	68	–	50	63	74	72[b]	28	55
Upper–middle income														
89 South Africa	–	–	–[c]	–	–	–	–	–	–	–	–	–	–	–
90 Mauritius	61	99	95	100	100	98	92	78	94	94	100	92	90	96
91 Estonia	–	–	–[c]	–	–	–	–	–	–	–	–	–	–	–
92 Brazil	55	72	87	80	95	51	61	55	21	72	32	84	–	32
93 Botswana	29	–	90	–	100	–	88	–	–	88	–	100	–	85
94 Malaysia	29	63	78	90	96	49	66	57	70	94	100	94	55	94
95 Venezuela	75	86	92	91	–	50	36	45	87	–	90	–	70	72
96 Belarus	–	–	c	–	–	–	–	–	–	–	–	–	–	–
97 Hungary[a]	–	–	98	–	100	–	95	–	–	100	–	100	–	100
98 Uruguay	92	81	95	96	100	2	–	78	59	–	59	–	60	–
99 Mexico	54	73	89	64	94	43	–	23	38	–	51	85	12	–
100 Trinidad and Tobago	96	97	96	100	100	93	88	–	92	98	95	100	88	92
101 Gabon[a]	–	–	66	–	90	–	50	–	–	–	–	–	–	–
102 Argentina[a]	56	54	64	65	73	17	17	85	79	89	89	100	32	29
103 Oman[a]	–	–	46	–	87	–	42	–	–	40	–	100	–	34
104 Slovenia	–	–	–	–	–	–	–	–	–	–	–	–	–	–
105 Puerto Rico	–	–	–	–	–	–	–	–	–	–	–	–	–	–
106 Korea, Republic	58	75	93	86	100	61	76	–	–	90	–	–	–	–
107 Greece[a]	–	–	98	–	100	–	95	–	–	98	–	100	–	95
108 Portugal[a]	–	–	92	–	97	–	90	–	–	97	–	100	–	95
109 Saudi Arabia[a]	49	90	93	92	100	87	95	29	70	81	81	100	50	30
High income economies														
110 Ireland	–	96	100	–	100	–	100	–	94	100	–	100	–	100
111 New Zealand	–	–	97	–	100	–	82	–	–	–	–	–	–	88
112 Israel[c]	–	96	100	–	100	–	97	–	–	99	–	99	–	95
113 Spain	82	90	100	–	100	–	100	–	90	100	–	100	–	100
114 Hong Kong[c]	–	100	100	100	100	95	96	–	94	88	100	90	–	50
115 Singapore[c]	–	100	100	100	100	–	–	–	80	–	80	97	–	–
116 Australia	99	–	–	–	100	–	100	–	–	–	–	100	–	100
117 United Kingdom	99	99	100	–	100	–	100	–	85	100	–	100	–	100
118 Italy	85	90	100	–	100	–	100	–	99	100	–	100	–	100
119 Netherlands	99	100	100	–	100	–	100	–	100	100	–	100	–	100
120 Canada	96	98	100	–	100	–	100	–	–	–	–	–	–	–

Table D15 Continued

| | Access to safe drinking water | | | | | | | Access to sanitation | | | | | | |
| | Total | | | Urban | | Rural | | Total | | | Urban | | Rural | |
Country	1970	1980	1990	1980	1990	1980	1990	1970	1980	1990	1980	1990	1980	1990
121 Belgium	95	98	100	–	100	–	100	–	99	100	–	100	–	100
122 Finland	53	70	96	–	99	–	90	–	72	100	–	100	–	100
123 United Arab Emirates[c]	–	92	100	95	100	81	100	–	80	95	93	100	22	77
124 France	92	98	100	–	100	–	100	–	85	100	–	100	–	100
125 Austria	–	80	100	–	100	–	100	–	85	100	–	100	–	100
126 Germany	–	100	100	–	100	–	100	–	–	100	–	100	–	100
127 United States	–	100	–	–	–	–	–	–	98	–	–	–	–	–
128 Norway	98	–	100	–	100	–	100	–	85	100	–	100	–	100
129 Denmark	90	100	100	–	100	–	100	–	100	100	–	100	–	100
130 Sweden	78	86	100	–	100	–	100	–	85	100	–	100	–	100
131 Japan	–	–	96	–	100	–	85	–	–	–	–	–	–	–
132 Switzerland	97	98	100	–	100	–	100	–	85	100	–	100	–	100
Selected economies not included in main tables														
Angola	–	26	40	85	73	10	20	–	20	22	40	25	15	20
Barbados	98	99	100	100	100	28	100	–	–	100	–	100	–	100
Cyprus	100	100	100	100	100	100	100	–	100	97	100[b]	96	100	100
Fiji	37	77	80	94	96	66	69	–	70	75	85	91	60	65
Gambia, The	12	–	77	85	100	–	48	–	–	67	–	100	–	27
Guyana	75	72	79	–	100	60	71	100	86	85	100	97	80	81
Haiti	–	19	41	48	56	8	35	–	–	25	–	44	–	17
Iceland	–	–	100	–	100	–	100	–	–	100	–	100	–	100
Iraq	51	–	77	–	93	–	41	48	–	–	–	96	–	–
Kuwait	51	87	–	–	100	–	–	–	–	–	–	100	–	–
Liberia[a]	15	–	50	–	93	16	22	16	–	6	–	4	–	8
Luxembourg	–	–	100	–	100	–	100	–	–	100	–	100	–	100
Malta	–	100	100	100	100	100	100	–	97	100	100	100	84	100
Suriname[a]	–	88	68	–	82	79[b]	56	–	–	49	–	64	–	36
Swaziland[a]	–	–	31	–	100	–	7	–	–	45	–	100	–	25
Congo, D.R.	11	–	39	–	68	–	24	6	–	23	–	46	–	11

[a] 1990 data refer to 1988; World Resources Institute 1992.
[b] World Resources Institute 1992.
[c] Economies classified by the United Nations or otherwise regarded by their authorities as developing.

Kulik's definition raises a serious question regarding the misuse of drought terminology. Can soil moisture deficiency or climatic aridity be equated with drought? Thornthwaite (1963) referred to arid climatic regimes as regions of perennial drought. Seasonal drought has been used to describe the cyclically low rainfall periods of the tropical savanna, Mediterranean and subtropical monsoon climates. Further, McBryde (1982) qualitatively defined the term normal seasonal drought as the normality of periodic low rainfall regimes and the term abnormal drought as dry periods within them, the latter having damaging environmental consequences. Steila (1983) points out that all the above terms are subjective, in opposition to drought theory, and that their use should be discouraged. They imply regularity and permanence to a phenomenon commonly recognized (and professionally identified) as temporary in duration and probabilistically uncommon. Furthermore, none of the foregoing can account for actual drought occurrence without redefinition of the descriptive term.

Drought identification: recent

Van Bavel and Carriker, in addressing the likelihood of drought return, devised a methodology for determining drought probability. Their concern focuses on agricultural drought, which they define as 'a condition in which sufficient soil moisture is not available in the root zone for plant growth and development' (Van Bavel and Carriker, 1957, p. 5). Four major environmental parameters are evaluated: rainfall, evapotranspiration, available soil moisture, and the moisture requirements of plants.

The system consists of combining daily rainfall data with estimates of daily water losses through evapotranspiration in order to determine the number of times soil moisture storage is reduced to zero in relation to its maximum capacity. Measurement begins on March 1st and runs to October 31st, assuming the soil has maximum retention at the beginning of the period. Evapotranspiration is computed with the Penman (1956) formula. When the base amount of moisture in the soil is exhausted, successive days without replenishment are recorded as drought days. Using the foregoing procedure, frequency tabulations are determined and probabilities derived for droughts of varying lengths.

Knowledge of the probable occurrence of moisture deficient periods during the growing season enhances the farm manager's opportunity to make sound economic decisions, e.g. whether or not to invest in irrigation equipment. There are limitations, however, associated with the use of the Van Bavel–Carriker approach: (1) drought is equated with moisture deficiency; (2) maximum soil moisture availability is assumed to exist on March 1st (if a dry winter precedes the growing season, a drought event would be approached much more rapidly than if that period were wet); (3) potential for winter drought and drawdown of regional water supplies is ignored: and (4) an amount of precipitation as low as 0.25 mm would break a drought period, an unrealistic assumption.

In an alternative approach by Keetch and Byram (1968) the US Department of Agriculture Forest Service developed an index of drought for use by fire control managers. Based on an 8 in (200 mm) soil moisture storage capacity, the Drought Index (DI) is expressed in hundredths of an inch of soil moisture depletion, ranging from 0 to 800. Zero represents no moisture deficiency, whereas 800 indicates absolute drought. Stages of drought intensity are represented as follows: incipient stage (no drought): 0–99; stage 1: 100–199; stage 2: 200–299; stage 3: 300–399; stage 4: 400–499; stage 5: 500–599; stage 6: 600–699; stage 7: 700–800.

The theory and framework of the DI are based on the following assumptions.

● The rate of moisture loss in a forested area will depend on the density of the vegetation cover in the area. In turn, the density of the vegetation cover and, consequently, its transpiring capacity are a function of the mean annual rainfall. Furthermore, the vegetation will use most of the available moisture.

- The vegetation–rainfall relation is approximated by an exponential curve in which the rate of moisture removal is a function of the mean annual rainfall. Therefore, the rate decreases with decreasing density of vegetation, hence with decreasing mean annual rainfall.
- The rate of moisture loss from soil is determined by evapotranspiration relations.
- The depletion of soil moisture with time is approximated by an exponential curve form in which wilting point moisture is used as the lowest moisture level. Thus, the expected rate of drop in soil moisture to the wilting point, under similar conditions, is directly proportional to the amount of available water in the soil layer at a given time.
- The depth of the soil layer wherein the drought events occur is such that the soil has 8 in (200 mm) of available plant water.

Computation of the DI is based on a daily water-budgeting procedure whereby the drought factor (which indicates degree of drying and relates to the flammability of organic fuels) is balanced with precipitation and soil moisture. The basic structure of the DI technique is rooted in traditional evapotranspiration calculations and is theoretically sound. However, it is limited by the range of its soil moisture holding capacity. While 8 in depletion can never be reached due to the exponential rate of moisture removal, a drought at the beginning of stage 7 would, although severe be much less intense than one nearing the exhaustion limit. If the soils are depleted of moisture, or nearly so, drought may have reached its maximum limit in terms of crops and native vegetation, but a further moisture deficit most certainly would be reflected in the continued drop of stream levels and groundwater supplies. Thus the index portrays what it is intended to: a measure of flammability that could create forest fires. It is not useful for portraying drought as a measure of total environmental stress and therefore has limited application.

Drought identification: current

Current drought theory stems largely from Palmer (1965), who established criteria for EMS parameters on the basis of indices referred to as climatically appropriate fore existing conditions (CAFEC) and defined drought as:

An interval of time, generally of the order of months or years in duration, during which the actual moisture supply at a given place rather consistently falls short of the climatically expected or climatically appropriate moisture supply.

To compute Palmer's Drought Index (DI) requires the determination of CAFEC parameters for evapotranspiration (ET), soil moisture loss ($\hat{L}$), soil moisture recharge ($\hat{R}$), runoff ($\hat{RO}$), and precipitation ($\hat{P}$). CAFEC symbols are denoted with a circumflex ($\hat{N}$). All of the CAFEC values, except for precipitation, are calculated by taking the ratio of long-term mean measures to long-term potential measures and multiplying this value by the parameter value occurring during a specific month. For example, in order to compute CAFEC evapotranspiration (ET), the following procedure is followed:

1. The coefficient of evapotranspiration is computed by the equation $k_e = \overline{ET}/\overline{PE}$, where k_e = coefficient of evapotranspiration, $\overline{ET}$ = mean evapotranspiration, and $\overline{PE}$ = mean potential evapotranspiration.
2. The computed ET for a given month is multiplied by k_e yielding ET.

Upon determination of each of the CAFEC parameters, excluding precipitation, it is possible to compute CAFEC precipitation from the following equation, which is based on the concept of supply and demand:

$$\hat{P} = \hat{ET} + \hat{R} + \hat{RO} - L$$

A departure of actual monthly precipitation from CAFEC – derived precipitation is obtained by subtracting the latter from the former: $d = P - \hat{P}$ where d is the monthly departure, P = monthly precipitation, and $\hat{P}$ = CAFEC precipitation for that month. The d value is used in conjunction with a climatic characteristic, or weighting factor, to then determine a monthly anomaly index: $z = d \times K$, where z = monthly anomaly index, d = monthly departure and K = weighting factor. The final quantitative drought measure is established by

Table D16 Palmer drought indices

Moisture status class	Drought index
Extremely wet	≥ 4.00
Very wet	3.00 to 3.99
Moderately wet	2.00 to 2.99
Slightly wet	1.00 to 1.99
Incipient wet spell	0.50 to 0.99
Near normal	0.49 to -0.49
Incipient dry spell	-0.50 to -0.99
Mild drought	-1.00 to -1.99
Moderate drought	-2.00 to -2.99
Severe drought	-3.00 to -2.99
Extreme drought	≤ -4.00

evaluating the degree of change in a monthly z value relative to those derived for the preceding and following month, and it also determines the moisture status class for a given month according to the criteria in Table D16.

Only the barest essentials of the Palmer drought identification technique are revealed in the foregoing summary. Serious students of drought are encouraged to study his monograph, *Meteorological Drought*. The approach integrates all recognized biospheric EMS variables except those that are atmospherically diffused (e.g. water vapor) and provides the first acceptable procedure for evaluating the role of potential evapotranspiration and soil moisture in either intensifying or ameliorating drought status. The system has been widely applied in measuring EMS changes within the United States. Even so, research has shown its moisture anomaly indices to be insufficiently sensitive to identify many drought periods that are both of short duration and capable of reducing agricultural productivity. Havens (1969, p. 2) states that Palmer's index 'appeared to be a very good indicator of drought but was slow to react to changes in the drought situation. It normally took about three or more months of abnormally low streamflow before the PDI [Palmer Drought Index] reacted significantly.' Steila (1971) and Sanders (1972) independently noted that Palmer's index has a persistent lag characteristic that at times presents an indication of environmental moisture status not actually occurring. They also point out that the analytic procedure is too complicated for the layman and produces dimensionless parameters of drought status that cannot be equated with other known environmental moisture variables.

Shear and Steila (1974) proposed an alternative means of using water budget analysis to identify EMS anomalies. The procedure, like Palmer's, accounts for precipitation, potential evapotranspiration and soil moisture, but yields moisture status departure values that are expressed in the same units as precipitation, i.e. they are areally applicable water-depth measures having equivalent meteorological significance in diverse climatic realms, a feature valuable for comparative analysis of regional EMS departures from normality. It has been established that this technique is sensitive to short-term changes in moisture status, as is evident in both crop and hydrological response, and is simple to apply when compared with the Palmer system (Steila, 1971, 1972; Sanders, 1972; Shear and Steila, 1974).

Drought indices are derived using the Shear–Steila definition by computing departure of actual from normal *EMS*. The procedure involves:

1. Calculation of the normal water balance with the accounting system described by Thornthwaite and Mather (1957). Table D17 is an example of the accounting methodology applied to the West Tennessee climatic division. Data input include only mean monthly temperature and precipitation values and assume a regional average soil moisture storage component of 6 in (150 mm).
2. Mean monthly EMS ($\overline{MS}$) is derived by the equation:

$$MS = \overline{AW} - \overline{PE}, \tag{D1}$$

where AW = mean monthly available water and PE = mean monthly potential evapotranspiration, and wherein

$$\overline{AW} = \overline{ST_0} + \overline{PPT} \tag{D2}$$

where ST_0 = mean monthly soil moisture storage at the beginning of a month, and $\overline{PPT}$ = monthly precipitation. Substituting, it is possible to derive

$$\overline{MS} = (\overline{ST_0} + \overline{PPT}) - \overline{PE} \tag{D3}$$

Table D17 Thornthwaite water balance – West Tennessee climatic division

	Jan.	Feb.	Mar.	Apr.	May	Jun.	Jul.	Aug.	Sep.	Oct.	Nov.	Dec.
Temperature (°F)	40.30	42.90	50.10	60.50	69.00	77.40	80.20	79.50	72.70	62.00	49.40	41.80
Heat index	0.88	1.34	2.88	5.73	8.50	11.59	12.69	12.41	9.82	6.19	2.71	1.14
Duration of sunlight	0.87	0.85	1.03	1.10	1.21	1.22	1.24	1.16	1.03	0.97	0.86	0.84
Unadjusted potential evapotranspiration	0.27	0.43	1.02	2.20	3.44	4.87	5.40	5.26	4.04	2.41	0.95	0.36
Adjusted potential evapotranspiration	0.23	0.36	1.05	2.43	4.16	5.94	6.69	6.10	4.17	2.33	0.82	0.30
Precipitation	5.96	4.65	5.28	4.42	4.16	3.96	4.04	3.05	3.30	2.85	4.27	4.47
Precipitation potential evapotranspiration	5.73	4.29	4.23	1.99	0.00	−1.99	−2.65	−3.05	−0.87	0.52	3.45	4.17
Accumulated water loss	0.00	0.00	0.00	0.00	0.00	−1.99	−4.64	−7.69	−8.56	−6.74	−0.68	0.00
Storage	6.00	6.00	6.00	6.00	6.00	4.29	2.74	1.62	1.39	1.91	5.36	6.00
Storage change	0.00	0.00	0.00	0.00	0.00	−1.71	−1.55	−1.12	−0.23	0.52	3.45	0.64
Actual evapotranspiration	0.23	0.36	1.05	2.43	4.16	5.67	5.59	4.17	3.53	2.33	0.82	0.30
Surplus	5.73	4.29	4.23	1.99	0.00	0.00	0.00	0.00	0.00	0.00	0.00	3.53
Deficiency	0.00	0.00	0.00	0.00	0.00	−0.27	−1.10	−1.93	−0.64	0.00	0.00	0.00

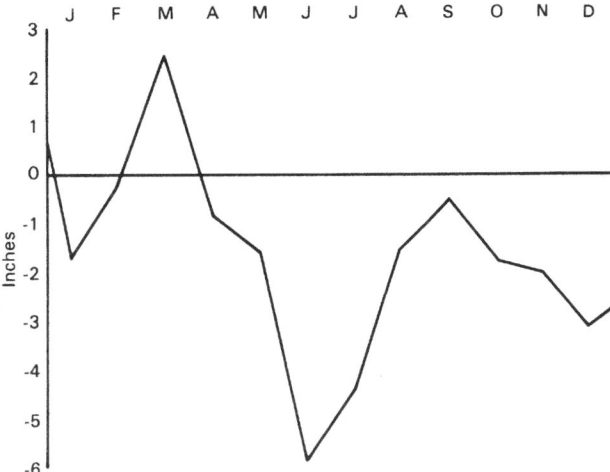

Figure D61 Monthly drought indices for the Western Texas climatic division.

Table D18 Drought classification

Environmental moisture status class	Drought index (in)
Above normal	≥1.00
Near normal	0.99 to −0.99
Mild drought	−1.00 to −1.99
Moderate drought	−2.00 to −2.99
Severe drought	−3.00 to −3.99
Extreme drought	≤−4.00

or its equivalent

$$\overline{MS} = \overline{ST_0} + (\overline{PPT} - \overline{PE}) \qquad (D4)$$

The same monthly value may also be obtained by

$$\overline{MS} = \overline{ST} + \bar{s}\bar{d} \qquad (D5)$$

where $\overline{ST}$ = mean soil storage at the termination of the month, $\bar{s}$ = mean monthly surplus and $\bar{d}$ = mean monthly deficit. (Note: surplus and deficiency cannot occur simultaneously; hence, either the former is added or the latter subtracted from the soil storage factor.) Equations (D1), (D3), (D4), and (D5) yield the same mathematical results.

3. Actual monthly EMS *(MS)* is determined by the same procedure outlined in 2 above, except that observed temperature and precipitation measures are incorporated into the equation:

$$MS = ST + sd$$

4. Departure of *MS* from $\overline{MS}$ provides a measure of EMS deviation from normality that is expressed as an index of drought *(DI)*:

$$DI = MS - \overline{MS}$$

Figure D61 is a historical trace of EMS departures. Intensity of drought, similar to the Palmer approach, may be identified by degree of departure indexes from normality. Table D18 lists drought classes.

Drought prediction

Drought identification should be evaluated relative to climatic fluctuations established for millenia. Available meteorologic observations do not afford climate analysts such luxury, however, as record continuity seldom exists prior to 100 or 150 years before the present. Consequently, long-range EMS changes are indirectly derived (National Science Board, 1972) from interpretation of oceanic and lake sediments, tree ring and pollen analysis, and historical documents. Based on 350 years of dendrochronological data, Mitchell *et al.* (1978) suggest that a 22-year drought rhythm associated with the Hale solar cycle can be identified in the western United States. In particular, the recorded large-scale moisture deficit events of the 1890s, 1910s and 1950s, and the spatially limited dryness of the 1970s within the Great Plains (Warrick, 1980) lend support to the concept of cyclically recurring moisture deficiencies within that specific geographic region. Similar rhythmic patterns defining water stress periods for other areas are not as identifiable as is the western US cycle. This suggests that controlling mechanisms in the Earth–atmosphere system are determining a regional climatic moist–dry cycle, wherein the dry intervals are expected periods of climatic aridity and possibly do not constitute drought. If the latter is indeed occurring, it requires re-evaluation based on negative EMS departures derived from EMS associated with multiple dry-cycle phases.

Eddy and Cooter (1978), through frequency simulation, propose that the 1930s western Kansas moisture deficiency has a return interval of 360 years. Considered relative to the 22-year rhythm, the likelihood of return of a dry phase of the cycle being as severe would be 0.163, well within recognized moisture status departures that identify drought status (Steila, 1983). It is improbable that all other dry-cycle phases on the Great Plains would likewise be classified as drought events when appropriately evaluated.

Cyclic moisture deficiency is temporally predictable, whereas drought is randomly distributed in both time and space. As might be expected, when all EMS departures are analyzed (both positive and negative), they approximate a normal distribution, with the majority of cases clustering about the mean. Drought, in general, is recognized as being pronounced when negative departures statistically approximate or are less than −σ (standard deviation; Steila, 1983). Utilizing

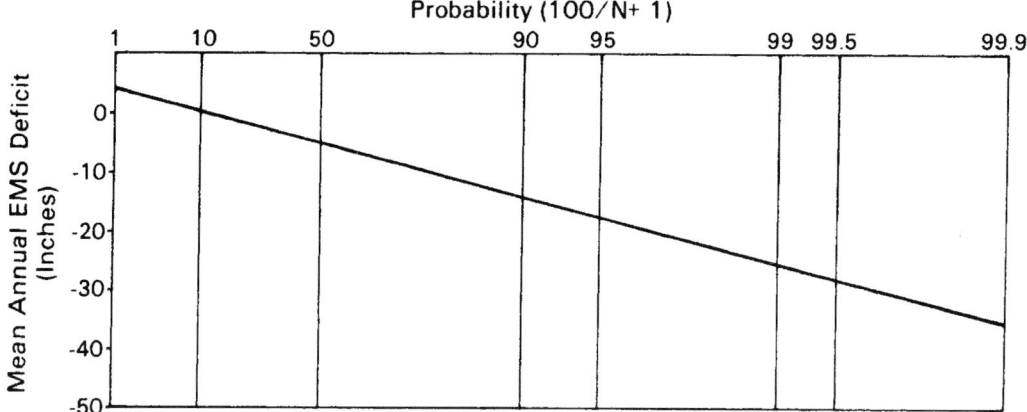

Figure D62 Annual negative environmental moisture state totals for the Northwestern Arizona climatic division.

extreme-event probability techniques, it is possible to determine meaningful recurrence intervals for accumulated annual moisture departures of specified intensity. To illustrate, Figure D62 is a plot of total annual negative EMS for northwestern Arizona. The curve demonstrates data normality for negative years and provides probability estimates for recurrence periods. This approach offers promise for portraying characteristics of annual moisture deficits but does not easily lend itself to comparative analysis of drought events that are not confined to calendar years.

Causes of drought

Virtually all droughts occur when an area is under the control of slow and prevailing subsiding motions of air masses. Frequently, the source of the air is from continental interiors where available moisture for diffusion into the atmosphere is limited. Under these conditions rainfall potential is low for three primary reasons: (1) the air's humidity status is initially low; (2) descending air experiences adiabatic heating, thus its relative humidity declines and moisture-holding capability increases; and (3) air motion is generally unfavorable for condensation processes. At the same time, reduced cloudiness provides for greater sunshine receipts, therefore increasing potential evapotranspiration (PE) demands and enhancing soil moisture loss.

Persistence of areas of subsidence is linked to planetary waves in the upper-level westerlies, wherein long-lived anomalies are superimposed on the general circulation pattern. The anomalies are quasi-stationary or change in position very slowly (National Science Board, 1972), and they determine whether air subsidence or lifting will dominate surface weather. When the duration and intensity of these anomalies result in regional subsidence over an extended time period, drought is likely to prevail.

Future advances in determining the likelihood of drought occurence appear to lie in identifying processes that generate large-scale anomalies in the high-altitude circulation patterns. Midlatitude upper-atmospheric flow patterns have been shown by Bjerknes (1969) to respond readily to oceanic–atmospheric interactions in equatorial latitudes. Specifically significant to distorting flow patterns and creating regional anomalies in the midlatitudes are variations in the transfer of heat and moisture from the ocean and the liberation of heat via condensation in the air of the tropics. More research in this area is needed.

Once drought has been established within a region, it appears to persist and expand into adjacent areas (National Science Board, 1972). This appears to be related to positive feedback mechanisms. For example, desiccation of soil seems to influence subsequent air circulation and available moisture for downwind precipitation. Simultaneously, large-scale interactions that create abnormal wind systems can produce variations in surface temperature, which in itself promotes further development of the abnormal circulation pattern. Much has been learned about drought-producing factors, yet too little

is known at the present time to identify comfortably the phenomenon's recurrence.

Other considerations

Climatologists generally accept the notion that drought is not solely related to precipitation receipts, but is rather a function of moisture supply, available water in storage and evapotranspiration demand. Climatologic literature contains abundant references to drought events, their relative intensity and environmental consequences. Minimal research attention, however has been directed towards identifying the role of individual drought components in contributing to the intensification or amelioration of the phenomenon.

Steila (1971, 1983) evaluated the monthly and annual contribution of potential evapotranspiration (PE) as a percentage of total drought status. Findings for the 1936 drought in the southeastern United States indicate that above-normal PE accounts for as little as 1% to as much as 75% of drought status during given months, the largest percentages normally being attributed to months of minor negative EMS departures. The annual drought contribution of PE averaged 14–15% of total yearly drought status. Although these findings suggest that cumulative, negative precipitation and soil moisture storage departures from normal may dominate the drought status of humid subtropical climates, the results are too tenuous to apply to other climatic regimes. The separate drought contributions of precipitation and soil moisture storage have not yet been evaluated in the professional literature. This remains an area in need of considerable research attention.

Can drought occur during periods of surplus EMS? Many qualitative definitions of drought refer to the phenomenon as a 'state of dryness' (e.g. McBryde, 1982). Quantitative research, however, clearly identifies periods wherein an environment has surplus moisture availability and yet, at the same time, total water supplies are appreciably below those normally associated with the same region. Under such conditions the area is not dry, but it is unquestionably below its expected norm with respect to water supply (Palmer, 1965; Steila, 1971; Sanders, 1972). In short, there is drought in regard to expected EMS. Petterson (1980) substantiated the consequences of these statements in her analysis of the 1977 drought in the southeastern United States. In several parts of the study region, water budget analysis and hydrological data indicated surplus water to be available during the winter of 1977. Precipitation receipts, however, were well below normal. Although these environments were not dry, they received too little rainfall to adequately recharge groundwater and soil reserves to normal levels. The latter factor proved detrimental to crops planted later that spring. Indeed, it can be seen that drought can occur during periods of surplus EMS.

Summary

Drought identification and evaluation procedures have slowly evolved from simplistic approaches that considered the phenomenon to be a rainfall deficiency, to problem-specific models of limited applicability, and finally to sophisticated techniques that quantitatively appraise total environmental moisture status and its deviation from normality. As such, they have provided a means by which the relative severity of a given drought occurrence can be assessed.

The regional causes for drought development are well known. What are unknown are the global dynamics that generate regional circulation patterns capable of initiating drought conditions, thus placing great limitations on prediction.

Drought has affected all of humankind. Some feel its impacts directly, for example, the farmers whose livelihoods are endangered as their crops wither and die in the field. Others experience indirect effects, such as increased cost of agricultural products. The large scale of most drought events and their relation to global circulation patterns will pose a limit to attempts at controlling their occurrence. Nonetheless, continued research into specific causes and into prediction of future drought events will aid in development of strategies to cope with the phenomenon and to utilize environmental resources better.

Donald Steila

Bibliography

Bjerknes, J., 1969. Atmospheric telecommunications from the equatorial Pacific, *Monthly Weather Rev.*, **97**, 163–172.

Chang, J.-H., 1968. *Climate and Agriculture: An Ecologic Survey.* Chicago: Aldine.

Eddy, A., and E. Cooter, 1978. *A Drought Probability Model for the USA Northern Plains.* Norman: Department of Meteorology, University of Oklahoma.

Havens, A.V., 1969. *Economic Impact of Drought on Water Systems in Passaic River Basin, N.J.* New Brunswick, NJ: New Jersey Agricultural Experiment Station.

Holdridge, L.R., 1962. The determination of atmospheric water movements, *Ecology*, **43**, 1–9.

Holdridge, L.R., 1967. *Life Zone Ecology.* San Jose, Costa Rica: Tropical Science Center.

Keech, J.J. and G.M. Byram, 1968. *A Drought Index for Forest Fire Control.* Asheville, NC: Southeastern Forest Experiment Station.

McBryde, W.F., 1982. Drought as a seasonal phenomenon, *Prof. Geographer*, **34**, 347.

Mitchell, J.M., C.W. Stockton and D.M. Meko, 1978. Evidence of a 22-year rhythm of drought in the western United States related to the Hale Solar Cycle since the 17th century, *Proceedings, Symposium/Workshop on Solar – Terrestrial Influences on Weather and Climate,* July 24–28. Columbus: Ohio State University.

National Science Board, 1972. Drought, the causes and nature of drought and its prediction, in *Patterns and Perspectives in Environmental Science.* Washington, DC: National Science Foundation, pp. 165–68.

Palmer, W.C., 1965. *Meteorological Drought.* Washington, DC: US Government Printing Office.

Penman, H.C., 1948. Natural evaporation from open water, bare soil, and grass, *Royal Soc. (London) Proc.*, **A193**, 120–145.

Penman, H.C., 1956. Estimating evaporation, *Am. Geophys. Union Trans.*, **37**(1), 43–46.

Petterson, L.M., 1980. An analysis of the drought of 1977 in the southeastern United States. M.A. Thesis. Greenville, NC: East Carolina University.

Sanders, C.G., 1972. A comparison of the Palmer method and the Shear method of drought determination in North Carolina and South Carolina. M.A. Thesis. Athens, GA: The University of Georgia.

Shear, J.A. and D. Steila, 1974. The assessment of drought intensity by a new index, *Southeastern Geographer*, **13**(1), 195–201.

Steila, D., 1971. Drought analysis in four southern states by a new index. Ph.D. Diss. Athens, GA: The University of Georgia.

Steila, D., 1972. *Drought in Arizona.* Tucson: College of Business and Public Administration, The University of Arizona.

Steila, D., 1983. Quantitative versus qualitative drought identification, *Prof. Geographer*, **35**, 192–194.

Tannehill, I.R., 1947. *Drought: Its Causes and Effects.* Princeton: Princeton University Press.

Thornthwaite, C.W., 1948. An approach toward a rational classification of climate, *Geog. Rev.*, **38**, 55–94.

Thornthwaite, C.W., 1963. Drought, *Encyclopedia Britannica*, Vol. 7, pp. 699–701.

Thornthwaite, C.W. and J.R. Mather, 1957. *Instructions and Tables for Computing Potential Evapotranspiration and the Water Balance.* Centerton: Laboratory of Climatology, Drexel Institute of Technology.

Van Bavel, C.H.M. and J.R. Carriker, 1957. *Agricultural Drought in Georgia.* Athens: Georgia Agricultural Experiment Station.

Ventskevich, G.Z., 1961. *Agrometeorology.* Jerusalem: Israel Program for Scientific Translations.

Warrick, R.A., 1980. Drought in the Great Plains: A case study of research on climate and society in the USA, in *Climatic Constraints and Human Activities*, J. Ausubel and A.K. Biswas (eds). Oxford: Pergamon Press, pp. 93–123.

Cross references

DROUGHT IN WESTERN EUROPE, 1988–1992

Background

Western Europe has a largely temperate climate grading towards more continental conditions to the east and a more Mediterranean climate to the south. In the latter areas the summers are relatively dry but generally rainfall is, on average, well distributed throughout the year. As a consequence of the regular passage of Atlantic frontal systems – augmented by convectional storms during the summer half-year – sustained rainfall deficiencies are relatively rare in Europe. Long-term rainfall records for the UK show a number of interesting perturbations but no overall trend. Other Western European rainfall records display similar characteristics (Thomsen, 1993), but some indicate a slight trend towards increased precipitation over the last 40 years, mostly attributable to wetter winter seasons.

Total water abstractions in Europe presently amount to around 15% of renewable resources but rise to 70% in Belgium. Reservoir and river flow offtakes are the major sources in much of the UK but, for Europe as a whole, groundwater is the principal supply source, accounting for about 75% of all abstractions. Water resource problems are most manifest in regions of low rainfall and high population density. Some, like the Netherlands and the Ruhr benefit – via the Rhine – from Alpine runoff; others have to rely principally on indigenous resources. The English lowlands, with a population exceeding 20 million and average annual rainfall of less than 700 mm, present a combination of water resources and demand that is inherently vulnerable to year-on-year variations in runoff and recharge.

Following a decade when, despite rising water demand, water resources throughout Western Europe had not been seriously stressed, the 1988–1992 period was characterized by protracted drought conditions which stimulated reviews of water management policies in a number of countries. These reviews were given greater impetus by the search for practical and scientifically based sustainable development options.

Spatial and temporal variations in drought intensity were large (Table D19) but the impact on agriculture was significant and widespread; hydrological stress was also evident in environmental and water resources terms. Low river flows and depleted reservoir stocks restricted irrigation over large parts of Europe, and in Spain, where in the south, some hydropower stations were forced to shut down in 1992, the unequal distribution of water triggered protracted inter-regional political conflict. From the fall of 1992 the focus of the drought shifted towards Mediterranean Europe as sustained heavy rainfall gradually transformed the hydrological picture in much of Western Europe.

Fortunately, the drought provided many useful insights into both the scale and scope of the water resources and environmental problems caused by long-term rainfall deficiencies and the strategies needed to combat them (Marsh *et al.*, 1994). The development of innovative, and more environmentally sympathetic, management strategies will be of particular significance if the extraordinary weather patterns recently experienced become a more familiar feature of our climate in the future.

Table D19 1988–1992 rainfalls for selected European sites

	Rainfall	1988	1989	1990	1991	1992
Bergen	mm	2325	2844	2980	2399	2935
	% LTA[a]	117	144	151	121	148
Copenhagen	mm	669	532	612	594	576
	%LTA	105	83	96	93	90
De Bilt	mm	887	661	716	649	918
	%LTA	110	82	88	80	113
Madrid	mm	418	561	328	289	376
	%LTA	92	124	72	64	83
Nuremburg	mm	808	519	581	518	541
	%LTA	126	81	91	81	85
Paris	mm	757	601	493	512	503
	%LTA	120	96	78	81	80

[a] %LTA = percentage of 1941–1970 average.

Synoptic context

Over the 4-year period beginning in the spring of 1988 most rain-bearing frontal systems followed a relatively northerly track remote from the English lowlands and the North European Plain. As a result the normal west-to-east rainfall gradient was greatly accentuated. This was well demonstrated by the abundant precipitation in Norway coupled with large deficiencies in, for example, central Germany and in Great Britain, where nationwide rainfall totals were close to normal but northwest Scotland was very wet whilst the lowlands of England were exceptionally dry. The drought was punctuated by a number of wet interludes, but the hydrological impact of this unusual and very persistent disturbance to the normal rainfall distribution was reinforced by the abnormally high temperatures which characterized much of the period (and encouraged high rates of evaporative loss).

A homogeneous temperature record exists for central England from 1659 (Manley, 1974). The 1988–1992 period is the warmest 5-year sequence in the series by a considerable margin. Consequently, evaporation rates remained well above average for lengthy periods, especially in 1989 and 1990 when potential evaporation losses in the English lowlands were more typical of parts of southern Europe. Throughout much of southern and eastern Britain, and parts of Western Europe, the record evaporation demands produced persistently dry lowland soils which, by robbing the rainfall of much of its effectiveness, served to exacerbate drought conditions significantly.

Whilst in the more mountainous maritime regions fringing the Atlantic, evaporation accounts for less than a quarter of the annual rainfall, this proportion rises to over 70% in much of the European lowlands. An important consequence of high evaporation rates and very dry soils during 1988–92 was that a 15–20% rainfall deficiency reduced aquifer recharge and runoff rates by 50–80%. Commonly, this major depletion in resource occurred in regions where concentrations of population, commercial activity and intensive agriculture generate the greatest demand for water.

Rainfall

Spatial and temporal variations in rainfall deficiency were large over the full compass of the drought, but in England the drought could be traced back to March 1988 and a general intensification throughout much of Western Europe could be recognized following the winter of 1989–1990. For England and Wales as a whole, the drought achieved its greatest severity over the period beginning in March 1990. Notwithstanding a relatively wet spring in 1992, the 28-month rainfall total up to and including June 1992 is eclipsed only by the minima established during the prolonged droughts of the mid-1850s and late 1780s (Marsh et al., 1994). Over the longest time spans the drought was markedly more severe in the eastern lowlands of England.

As important as the deficiencies themselves was the fact that 1988–1989 was commonly the first of a cluster of dry winters. For Copenhagen, the November 1988 to February 1989 rainfall total was the second lowest in 40 years and the three subsequent November–February periods also produced below average rainfall, although in some of these years, winter rainfall deficiencies were partially counterbalanced by a wet, early spring. Similar rainfall patterns occurred in De Bilt, Netherlands, and Paris, where none of the individual November-February periods were exceptionally dry but

1988–1989, 1990–1991 and 1991–1992 each rank amongst the driest ten such sequences since 1950.

From Denmark to Spain, the drought intensified again at the beginning of 1992 but generally declined in severity thereafter, although not as briskly as in the UK. The August–November period in 1992 was wet in some regions and produced substantial reductions in rainfall deficiencies. In others, the Paris Basin for example, the drought abated at a slower pace but, by the spring of 1993, overall deficiencies had moderated by comparison with those obtaining 12 months previously (Merillon and Scherer, 1993). In southern Spain, however, water resource problems continued well into 1993 (and thereafter) with demand restrictions operating in, for example, Seville and Cadiz. Southern Europe excepted, the winter of 1993–1994 was wet heralding a period of abundant rainfall which produced several severe flood episodes.

River flow

Most of Western Europe experienced runoff deficits between 1988 and 1992. The timing and intensity of the maximum deficiencies varied considerably across Europe. In some basins the deficits were worse during 1991; the annual flow for the Elbe, Germany, for example, was only 51% of the long-term average. In others the drought was most extreme during 1990. The rapid change from low to high and back to low flow conditions observed in large parts of Britain during 1989–1990 also characterised many continental catchments but the greater geographical spread of the major river systems provided more opportunity for compensating runoff (e.g. from the Alps) to support low flows elsewhere.

The severity of the drought is best indexed by runoff accumulations in the 2–4 year time frames. Average flows over the 48-month period to May 1992 were the lowest on record on the River Ebro, Spain, and ranks third most severe for the Rhine (in a 60-year series). Runoff for the Weser, Germany, was similarly depressed but more extreme conditions were experienced farther to the west. The low flow statistics for many rivers in eastern and southern England were largely redefined over the 1989–1992 period. In part this reflects the limited length of most gauging station records – relatively few exceed 40 years. A lengthy historical perspective is provided by the flow record for the River Thames. This suggests that only during the 1901–1903 and 1933–1935 droughts have lower 24-month flows occurred during the twentieth century and the significance of these historical minima is certainly exaggerated by the tendency of low flows to be underestimated prior to the major refurbishment of the Teddington Weir gauging station in 1951.

Depressed runoff rates over an extended period were associated with a shrinkage in the stream network that is without modern parallel in large parts of Western Europe. The corresponding loss of amenity and aquatic habitat was considerable. For example, by late 1990 over 3000 km of rivers had dried up in southern France and, as the groundwater contribution to summer and fall flows continued to diminish in the English lowlands, headwaters remained dry for several years. Over much of Western Europe the environmental impact was exacerbated in those catchments where groundwater pumping, often over many years, had steadily reduced river flows and caused the headwater sources to migrate downstream. In lowland England this provided impetus to a major lowflow alleviation program introduced by the National Rivers Authority, aimed at restoring a healthy aquatic environment in some 40 rivers (Figure D63).

Groundwater

The limited rainfall – in the winter particularly – and exceptionally warm conditions caused the normal seasonal variation in groundwater levels to be barely identifiable over much of Western Europe during the 1988–1992 period. In some aquifers, weak or nonexistent winter recoveries allowed erratic water table recessions to continue over 3–5 years such that by the middle of 1992 water tables had often fallen below period-of-record minima (Figure D64).

In the UK the regions where the long-term drought achieved its greatest severity coincide broadly with those areas where groundwater is the major source of water supply. In much of the eastern lowlands of England a cluster of three or four winters with modest aquifer replenishment separated by extended groundwater level recessions provide the background to the very depressed water tables in the summer of 1992. On the basis of a sparse monitoring network (for the pre-1950 period), it appears that overall groundwater resources for

Figure D63 The River Ver in 1991 (dry), an example of the combined impact of drought and long-term groundwater abstraction.

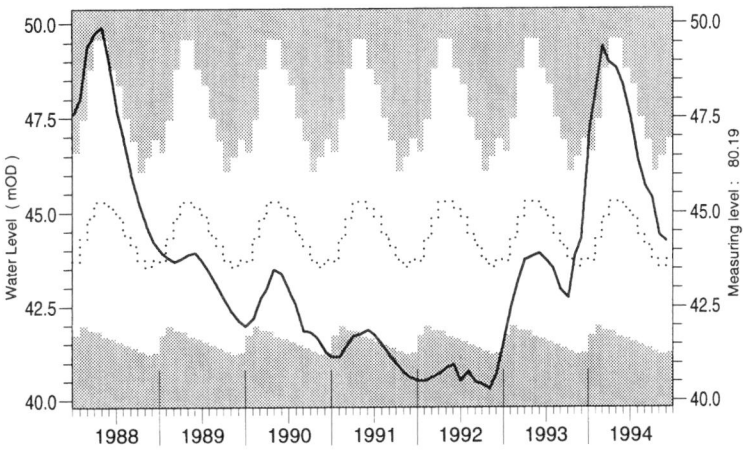

Figure D64 Groundwater levels at the Washpit Farm borehole, 1988–1992.

England and Wales had declined to their lowest since at least the turn of the twentieth century. Notable declines in water tables were also reported through the major aquifers of northern France and the North European Plain. Over the 2 years from the late summer of 1990 the failure of shallow wells was commonplace and reduced yields from public supply boreholes were widely reported.

By the spring of 1992, the water supply outlook was very fragile in much of Western Europe; groundwater droughts tend to be protracted events with substantial replenishment required to return depressed water tables to even average levels. However, the return of a predominantly westerly airflow in the autumn of 1992 – carrying a series of active frontal systems – produced abundant aquifer recharge, in the west especially, and very brisk recoveries in groundwater resources over the ensuing two winters (Institute of Hydrology, 1994).

Drought impact

The impact of a drought depends on the interaction of hydrological conditions, water resource availability and the pressure of local demand. The biggest impacts of the 1988–1992 drought in Europe have been in areas with the greatest pressures on resources, and especially in those areas with high irrigation demands; these are not necessarily the same as the areas which registered the most severe hydrological drought. In the English lowlands, demand management measures – mostly hosepipe bans – were on occasions (such as the spring of 1990) introduced when overall resources remained healthy but peak demands overstretched the supply systems. Such circumstances created a number of public relations difficulties for the water industry and emphasized the need for wider education regarding drought and its effects. The drought provided a timely reminder of the vulnerability of Western Europe to long-term rainfall deficiencies and the need for innovative, and more environmentally sensitive, water management procedures to combat their impact in the future.

T.J. Marsh

Bibliography

Cannell, M.G.R. and Pitcairn, S.E.R. (eds), 1993. *Impacts of the Mild Winters and Hot Summers 1988–90*. Dept. of the Environment, HMSO.

Institute of Hydrology, 1993. *1992 Yearbook*, Hydrological data UK series, Institute of Hydrology, Wallingford, UK, 175 pp.

Manley, G., 1974. Central England temperatures: monthly means 1659 to 1973. *Quart. J. Roy. Meteorol Soc.*, **100**, 389–405.

Marsh, T.J., Monkhouse, R.A., Arnell, N.W. *et al.*, 1994. *The Drought*, Hydrological data UK series. Institute of Hydrology, Wallingford, UK, 80 pp.

Merillon, Y and Scherer, J.C., 1993. La secheresse de 1992. *La Houille Blanche*, **8**, 559–50.

Thomsen, R., 1993. Future droughts, water shortages in parts of western Europe. *Trans. AGU*, **74**, 164–165.

Cross references

Drought
Drought management
Droughts

DROUGHT MANAGEMENT

Introduction

Droughts in the developed world are often not significant in global terms, nor do they pose a threat to life as they do in many parts of the world. However, in those countries where adequate quantities of water are taken for granted, even relatively minor events can be taken very seriously by both the public and the media. This is particularly the case where restrictions in water use affect social and economic activities. With increasing interest worldwide in ideas of sustainability the impact of droughts on ecology and amenity is also being increasingly appreciated.

Mawdsley *et al.* (1994) have shown that droughts can be classified either as an environmental drought or a water supply drought. An environmental drought is caused by a shortage of rain often compounded by high evaporation. Its severity can be assessed by hydrometeorological indicators which measure direct impact on the hydrological cycle, such as rainfall, temperature and evapotranspiration. An environmental drought is significant primarily for those uses directly affected by a lack of water such as ecology, fisheries, agriculture and horticulture and abstractions which take water directly for small surface supplies which lack storage.

On the other hand, a water supply drought may reflect as much a lack of resilience or mismanagement of a water resources system as a lack of rainfall. For those affected by a water supply drought, typically domestic and industrial consumers, they are concerned about reduction in reliability of supplies, whether this is caused by a 'natural' event or by the management regime of the water resources network. However, the options to manage a drought clearly depend on whether the drought is in an environmental or water supply category. Many droughts will have both environmental and water supply impacts.

In the UK the National Rivers Authority, a non-departmental government body, is an environmental regulator charged with the regulation, management and protection of water resources. As such it is required to manage both environmental and water supply droughts and in its regulatory role it is required to ensure that in the absence of an environmental drought, a water supply drought has not been caused by either a lack or mismanagement of water resources infrastructure. On the other hand, water service companies which provide water and sewerage services are likely only to be interested in managing water supply droughts.

Options for managing droughts

There are two main categories of drought management measures – those which increase supply availability and those which reduce demand. Supply enhancement measures typically involve the temporary relaxation of statutory conditions on abstraction on a site-specific basis, for example, allowing abstraction from unlicenced sources or allowing greater volumes of abstraction than that allowed under abstraction licences. Relaxation of river flow or groundwater levels constraints designed to protect the environment from the adverse impact of an abstraction may be temporarily allowed. Compensation water releases from a reservoir, which normally protect the rights of downstream abstractors or meet 'in river' needs, may be reduced.

Such relaxation of statutory conditions, intended to protect the environment and other abstractions, would generally be seen as a last resort by an environmental regulator. However, it is generally recognized that in a severe drought some supply enhancement could be necessary. The skill is in determining where such actions should be targeted in order to minimize the detrimental impact on the environment.

In contrast to the site-specific nature of supply enhancement measures, demand management measures can often be applied across a wider geographical area or even a region. Within the UK, demand typically increases in the early stages of a drought as consumers use more water, for example to water gardens.

Opportunities exist to reduce demand on a water resources system, either by reducing the use of water by consumers or by various water supply network management measures. Reduction in consumer consumption can be encouraged by publicity and appeals for domestic and industrial consumers to save water. Under UK legislation, consumers can be restricted in using water for garden watering or car washing by the imposition of a ban on the use of hosepipes. Legal powers are available to ban the use of water for non-essential uses, for example in commercial car washes. The National Rivers Authority is able to restrict abstraction for spray irrigation purposes. Such powers, because of their impact on livelihood, are only introduced when a drought becomes severe. Ultimately, rota cuts in supply or standpipes could be introduced if the droughts are sufficiently serious.

It is often possible to reduce the demand on a water resource system using network management measures. These include the rescheduling of non-essential maintenance and the reduction of in-pipe pressure to reduce both customer use and leakage. Opportunities often exist to re-zone supplies in order to equalize the risks to supply by transferring water from a water resources system with a low risk of failure to a water supply zone served by a water resource system experiencing a higher level of risk. Most major water resource systems are now sufficiently interconnected to allow this sort of flexibility on a regional scale.

Mawdsley *et al.* (1994) showed that during the UK drought of 1989–1992 there were 37 orders to ban the non-essential use of water in different areas of England and Wales. This compares with 157 temporary relaxations of statutory abstraction conditions.

Operating policies and contingency plans

Whilst droughts by their nature are rare events, the need to take precautionary action occurs much more frequently on occasions when an incipient drought does not fully develop. Experience has shown (Howard *et al.*, 1994) that by including drought containment measures in routine operating policies the implications of potential increase in water resource system operating costs and detrimental impacts on the environment and risks to supplies can be minimized. If such measures are not included in routine operating policies, the opportunity to take early precautionary action is lost, leading to higher risks to supplies and environmental damage later in the year with consequent increase in costs. Typically, operating policies are designed to protect against a repetition of drought of specified reliability. This may be expressed in probabilistic terms, e.g. a 1 in 50 year or 1 in 100 year event, or may relate to protection against the worst event in the historic runoff record.

Nonetheless, it is essential to have contingency plans in place should unforeseen circumstances arise or should the developing drought turn into an event more severe than that against which the operating policy is designed to protect.

Contingency plans should normally be based on experiences gained during previous droughts. They may take the form of a detailed decision diagram which indicates what management action should be undertaken at different stages in a drought. Alternatively, the contingency plan may be a simple list of supply enhancement or demand management measures and under what circumstances these actions should be taken. By developing these management action plans outside the pressure of managing a drought event, ad hoc and ill thought through responses to a drought should be avoided.

Monitoring

In the build-up to an incipient drought and during the management of the drought event proper, it is essential that extensive monitoring of the water resources system takes place. Typically, such monitoring begins with an assessment of the severity of the potential drought which is developing by calculating the likelihood of occurrence of the rainfall or runoff event which has occurred until now. This often takes the form of a probabilistic analysis of cumulative rainfall or runoff records or comparison of the current event with known critical droughts in the historic records. Reservoir storages or groundwater levels are sometimes used in such assessments, but are often difficult to use directly because of the impact of abstraction practices on the records. In carrying out assessment of drought severity, difficulties arise in determining both the starting date of the event and the duration which should be considered. Ideally this should normally relate to the time of year when storage in the water resource system would typically begin to drawdown and to the critical period of the resource.

Further monitoring and analysis can be helpful in the actual management of the drought in order to assess the scale and nature of management action that would be appropriate. Such assessment techniques have been extensively described by Walker *et al.* (1993) in the context of water supply systems and by Walker and Smithers (1995) who describe the value of such techniques in the management of water resource systems both to protect the environment and to maintain water supplies. Risk assessment can be carried out for reservoir-based resource systems. These take into account the anticipated inflow under different runoff conditions and the current volume of water in storage. Assessments can then be made of the pattern of storage and hence the risks to supply at different rates of abstraction, and the abstraction and compensation which the water resource system can support under different runoff conditions. Similar analysis can also be used to establish the chance of refill, and when return to normal operation can be anticipated.

These types of analysis are not suitable for either short critical-period sources or for river abstractions. For these situations the pattern of recession of either river flows or lake or reservoir levels can be compared with those of previous critical droughts. These can be modified to take into account different abstraction rates.

These analyses can be used to establish the demand management or supply enhancement measures which are most appropriate for the circumstances in order to manage the drought so as best to protect the interests of both the environment and water supply.

Susan Walker

Bibliography

Howard, K.W., Smithers, H., Walker, S. and Wyatt, T., 1994. Evaluating the impact of drought measures within complex water resources systems, in *Advances in Water Resources Technology and Management*, eds Tsakiris, G. and Santos, M.A. Balkema, Rotterdam.
Mawdsley, J.A., Petts, G. and Walker, S., 1994. *Assessment of Drought Severity*. Occasional Paper No. 3, British Hydrological Society, Wallingford.
Walker, S., Jowitt, P.W. and Bunch, A.H., 1993. Development of a decision support system for drought management within North West Water, *J Instn Water and Environ. Management*, **7**(3), 295–303.
Walker, S. and Smithers, H.A., 1995. Recent advances in drought management with particular reference to NW England. IAHS Publ. No. 231, pp. 107–116.

Cross references

Drought
Drought in Western Europe, 1988–1992
Droughts

DROUGHTS

Introduction

W.C. Palmer (1965), a noted authority in the study of droughts, has said 'drought means various things to various people depending on their specific interest. To the farmers drought means a shortage of moisture in the root-zone of his crops. To the hydrologist it suggests below average water levels in streams, lakes, reservoirs, and the like. To the economist it means a shortage which affects the established economy'. In scientific terms, a drought can be defined as a shortage in the availablity of natural waters with respect to normal for a place and time. The natural waters can be in the form of precipitation, streamflow, groundwater level, lake level or soil moisture and are termed drought variables.

The behavior of the drought variable can be studied on an annual, monthly or daily scale, i.e. over the time period that the aforesaid variables are averaged or totaled. The most commonly used time unit in drought analysis is the year followed by the month (Bonacci, 1993). Although the use of year as a time unit is too large, it can be used successfully to abstract the information on the regional behavior of the drought. The monthly time unit seems to be more appropriate for monitoring drought effects in agriculture, water supply and groundwater levels. For studying the behavior of short-term droughts within the time scale of a year, the time unit of a day has also been used (Gupta and Duckstein, 1975; Smart, 1983; Zelenhasic and Salvai, 1987; Sharma, 1995).

Based on the variable selected, droughts can be classified as meteorological, hydrological, agricultural or socio-economic (Wilhite and Glantz, 1985). Meteorological drought refers to lack of precipitation as to cause the extended dry period. Hydrological droughts are defined in terms of below-normal streamflow, depleted surface or groundwater storage. Agricultural drought is associated with soil water which is insufficient to support crop growth. Socio-economic drought relates to periods of low water supply (from natural and artificial sources) which affects society's productive or consumptive activities in economic terms. The most studied droughts as of today are meteorological and hydrological. Relatively less published information exists on agricultural and socio-economic types of drought. The causes of a drought are poorly understood and some studies tend to link them to El Niño/Southern Oscillation (ENSO) events and sea surface temperatures (Li and Makarau, 1994).

Drought parameters

The important parameters quantifying a drought are (1) duration (2) severity (3) location in absolute time, i.e. its initiation and termination, and (4) areal coverage. At times, a parameter, namely magnitude, is also used, which is merely the ratio of the severity to duration (Dracup *et al.*, 1980a). The concept of the above parameters can be best illustrated using Figure D65. The most basic element for deriving the above parameters is the truncation or threshold level. The truncation level specifies the demand level of the water in terms of some statistic of the drought variable and serves to divide the time series of the drought variable into 'deficit' and 'surplus' sections. In practice the selection of the truncation level is not arbitrary but is a function of the nature of the drought being studied. Dracup *et al.* (1980a) defined the truncation level by the following expression:

$$x_0 = x_m - e\sigma = x_m(1 - e \cdot C_v) \tag{D6}$$

where x_0 is the truncation level, x_m is the mean of series, σ is the standard deviation, e is an elective scaling factor and C_v is the coefficient of variation. If e is zero, then $x_0 = x_m$, i.e. the reference (truncation) level for drought identification is the mean of the time series of a drought variable such as annual rainfall or runoff. If one recognizes the 90% of the mean annual rainfall or runoff having the coefficient of variation of 0.40 to be the truncation level, then $e = (1 - 0.90)/0.40 = 0.25$. The other corollary of equation (D6) is that when the time series of the drought variable is standardized (mean = 0 and standard deviation = 1) then the truncation level (designated by Z_0) is zero for the mean annual level and -0.25 for 90% of the mean annual level ($C_v = 0.40$).

Diverse definitions of drought severity

Although the terms duration and areal coverage have been defined in nearly consistent manner in the literature, considerable disagreement

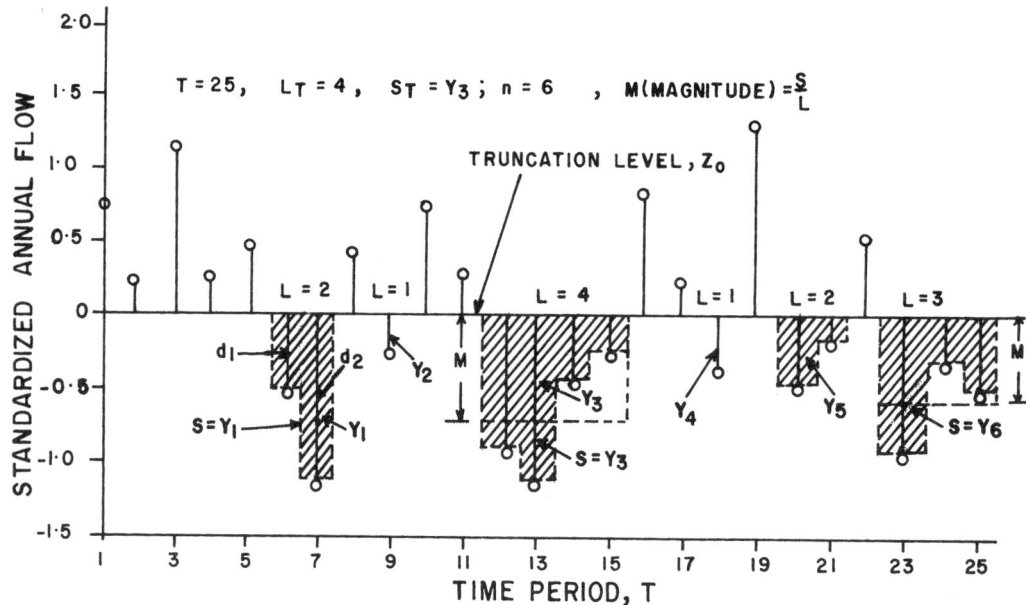

Figure D65 Definition sketch of drought parameters.

exists in defining the term severity. Yevjevich (1967) and Dracup *et al.* (1980a) define the severity as the cumulative shortage or deficit sum with reference to the defined truncation level, and therefore severity has the unit of millimeters or cubic meters. This connotation of the severity has been followed in the context of hydrological droughts. On the other hand, in the literature of a meteorological drought, the severity has not been defined in the aforesaid units of the depth of water shortage but through some form of index. Some initial index of the drought severity originates from Australia through the analyses of dry spells using monthly rainfall records, such as Foley's (1957) drought severity index based on residual mass curve technique. Other more objective indices such as decile range (Gibbs and Maher, 1967) and the standardized indices (Gibbs, 1975) were introduced in subsequent years.

In the technique of residual mass curve analysis, departures are taken from the monthly average, reduced to the proportions of the average annual rainfall to overcome seasonal variability and these proportions are summed. The steepness of the fall of the residual mass curve provides an index of the severity. In using decile ranges the monthly, seasonal or annual rainfall records are arranged in ascending order and then divided into ten equal parts or deciles. The decile range in each class is then found. The next step is to scan through the unranked data and classify the rainfall for each year into the decile range in which it falls using the decile limit. The decile range in which a particular year occurs gives some indication of its departure from average. Decile range 1 suggests abnormally dry conditions whereas decile range 10 indicates abnormally wet conditions. This procedure has formed the basis of the Australian Drought Watch System.

In the standardized index method (Gibbs, 1975), the monthly, seasonal or annual rainfall sequences are standardized using the mean (truncation level) and standard deviation of the individual time unit. The resulting standardized (z_i) sequence fluctuates around zero with the standard deviation of unity. Droughts have been categorized as mild if $-0.5 < z_i \leq -0.2$, moderate if $-0.8 < z_i \leq -0.5$ and severe if $z_i \leq -0.8$. No drought conditions are regarded to exist for $z_i > -0.2$, at least for agricultural purposes (Downing *et al.*, 1984).

The concept of cumulative departures or residuals from monthly averages was extended by Herbst *et al.* (1966) in South Africa. They realized that the main problem in analyzing droughts is in separating their occurrence in the record, i.e. defining their occurrence in the record, even if the long-term monthly means are regarded as truncation levels. The problem of identification of the points of onset and termination of drought on the time scale is compounded by the carryover effect in the monthly rainfall sequences; that is, the benefit to vegetation of above-average rainfall persists for some time after the

rain actually fell due to the storage of moisture in the soil, and conversely, the recovery of vegetation after a drought is not immediate, the deleterious effect persisting for some time after adequate rain has fallen. The analysis was preceded by first computing the effective rainfall through the use of a weighting factor W_t as follows:

$$E_t = R_t - [R_{t-1} - TR_{t-1}]W_t \qquad (D7)$$

and

$$W_t = 0.1 \left[1 + TR_t \sum_t^{12} TR_t/12 \right] \qquad (D8)$$

in which E_t = effective rainfall, R_t = actual measured rainfall, TR_t = truncation level, i.e. mean monthly rainfall (designated by m_t) and W_t = weighting factor, and all quantities refer to a particular month t ($t = 1$–12).

The important parameter for locating the onset and termination of a drought and thus defining the duration (j, months) is the mean monthly deficit (MMD). The parameter was calculated for each of the 12 months from the differences of 12 months of actual rainfall and mean monthly rains (truncation levels) for the entire period. Serious students of drought are encouraged to study the papers by Herbst *et al.* (1966) and Mohan and Rangacharya (1991) to understand fully the procedure to determine the duration of a drought spell (j). The drought severity index was computed by multiplying the drought intensity index Y by the duration of drought spell j. The drought intensity index DI is defined as

$$DI = \left[\sum_{t=1}^{j} [(E_t - TR_t) - (MMD)_t]/\sum_{t=1}^{j} (MMD)_t \right] \qquad (D9)$$

The drought severity index is equivalent to $DI.j$ and has the dimension of time as DI is a dimensionless quantity. In general the high intensity reflects the low percentage of the average rainfall over the drought spell. The Herbst method was applied to identify droughts in the UK and results were found to be encouraging (Shaw, 1984). Shaw further notes that the technique for defining droughts from rainfall records is objective and useful as the rainfall data are widely available. Furthermore, the nucleus of the drought lies in the rainfall, as it is the primary input to any source of water. The drought severity indices can further be extended for mapping the regional droughts.

Although the Herbst method has performed satisfactorily in South Africa and the UK, some difficulties were experienced in India (Mohan and Rangacharya, 1991), where the monthly variability of rainfall and streamflow sequences is quite high (coefficients of variation 0.37–2.60) due to the highly seasonal pattern of rainfall.

Table D20 Drought classification by Palmer's Drought Severity Index

Palmer's Drought Severity Index	Degree of drought
PDSI < −4.0	Extremely dry
−4.0 < PDSI ≤ −3.0	Severely dry
−3.0 < PDSI ≤ −2.0	Moderately dry
−2.0 < PDSI ≤ −1.0	Mildly dry
−1.0 ≤ PDSI < +1.0	Near normal
+1.0 ≤ PDSI < +2.0	Mildly wet
+2.0 ≤ PDSI < +3.0	Moderately wet
+3.0 ≤ PDSI < +4.0	Severely wet
+4.0 ≤ PDSI < +4.0	Extremely wet

Table D21 Drought classification by the method proposed by Shear and Steila (Steila, 1987)

Environmental moisture status class	Drought index (in)
Above normal	≥1.00
Near normal	0.99 to −0.99
Mild drought	−1.00 to −1.99
Moderate drought	−2.00 to −2.99
Severe drought	−3.00 to −3.99
Extreme drought	≤ −4.00

They suggested, therefore, that for seasonal rainfall and streamflow conditions the truncation levels should be modified as

$$TR_t = m_t - \sigma_t^2/m_t \qquad (D10)$$

where m_t = mean monthly rainfall or runoff for the month, t, and σ_t = standard deviation for the same month. It should be noted that when the monthly values show small deviations, the truncation levels tend to approach the mean monthly values as proposed by Herbst *et al.* (1966).

One of the best-known indices of the drought severity is that proposed by Palmer (1965) from the US Weather Bureau. The index is being extensively and routinely used in the USA and many other countries of the world for agricultural planning, forecasting crop yields and other related purposes. The index is based on a definition of drought as a prolonged period of abnormal moisture deficiency, and the drought severity is a function of moisture demand and supply. Drought is identified as the phenomenon with temporary negative deviation in environmental moisture status (EMS). The scale of severity of drought designated as Palmer's Drought Severity Index (PDSI) is given in Table D20. The PDSI is computed by using observed precipitation and temperature data and a moisture accounting procedure (Palmer, 1965). The PDSI values of various areas in the USA are routinely published in the weekly weather and crop bulletins.

Although Palmer's method of drought identification has been widely acclaimed, it has attracted some criticism. Steila (1987) reports that Palmer's index has a persistent lag characteristic that at times presents an indication of environmental moisture status not actually occurring. It is also pointed out that the analytical procedure is too complicated for the layman and produces the parameters of the environmental moisture status in terms of non-dimensional quantities. Generally, the hydrological balance equation expresses the environmental moisture status in the dimensions of depth (mm). Steila presents an alternative means of using water budget analysis to identify the anomalies in the environmental moisture status. The procedure, similar to Palmer's, accounts for precipitation, potential evapotranspiration and soil moisture but yields moisture status departure in units of precipitation, a feature highly desirable for expressing the drought severity index in depth units. It is claimed that the proposed technique is sensitive to short-term changes in moisture status and simple to apply when compared to Palmer's system. Drought indexes are derived by computing a departure of actual environmental moisture status from the normal counterparts and are therefore in depth units. The drought classification by this method is shown in Table D21.

Analysis of extreme drought duration and severity

One of the applications of drought analysis using streamflow records (hydrological drought) is to plan water storage for drought periods. Such planning for the design of water storage systems requires knowledge of the longest drought and the largest severity for the desired return period. The droughts lasting for the longest duration and experiencing the largest severity have been designated as critical droughts in the hydrological literature (Askew *et al.*, 1971; Sen, 1980). The analysis of critical droughts has been carried out using the techniques of stochastic simulation, i.e. chiefly by Monte Carlo methods of data generation (Askew *et al.*, 1971; Frick *et al.*, 1990) and/or analytical models of stochastic theory. One theory which has almost invariably played the central role in analytical formulations is the theory of runs, i.e. run length representing the drought duration

and the run sum the drought severity (Millan and Yevjevich, 1971; Guerraro and Yevjevich, 1975; Sen, 1980, 1989; Chander *et al.*, 1981; Zelenhasic and Salvai, 1987; Sharma 1994, 1995). A significant deviation from the theory of runs has been employed by Lee *et al.* (1986) where droughts have been studied using the notions of reliability theory, renewal processes and hazard functions. Due to the prominence of the run theory in the analysis of hydrological droughts, a brief description of the theory is presented below.

If a standardized time series z_i; is truncated at a level z_0, then the events of surpluses ($z_i > z_0$) and deficits ($z_i \leq z_0$) would emerge along the time axis (Figure D65). The truncation level can be assigned its probability level as $q = P(z_i \leq z_0)$ where q is the probability of drought corresponding to the truncation level z_0 and $P(\ldots)$ represents the notation of probability. Any uninterrupted sequence of deficits can be regarded as a drought length (duration) equal to the number of deficits in the sequence, designated by L ($L = 1, 2, 3, \ldots, j$). Each drought duration is associated with a deficit sum, i.e. the sum of the individual deficits in the successive epochs of the spell (Figure D65) and is designated as S. This deficit sum S is treated as drought severity. If the data sequence follows a normal probability distribution (or can be normalized) then, in view of the central limit theorem, the deficit sum or the drought severity can be regarded as following truncated normal distribution (Sen, 1980). Because of the property of normality, the drought probability quantile q and truncation level z_0 are related and, for a given z_0, q can be obtained from the standard normal curve or table. Values of q are 0.5, 0.42, 0.34, 0.27, 0.21 and 0.16 for z_0 equal to 0, −0.2, −0.4, −0.6, −0.8 and −1.0.

One can expect n ($n = 0, 1, 2, 3, \ldots, i$) drought spells (runs) over a period of T years and correspondingly there will be n values of severities. Each drought spell has a length $L = 1, 2, 3, 4, \ldots, j$ years and severity $S = Y_1, Y_2, Y_3, \ldots, Y$. A designer is interested in the longest value of L designated as L_t and the largest value of S designated as S_T. The period of T means a sample size of T ($T = 10, 20, \ldots, 100$) years and can be regarded as being equivalent to a return period of T years (Horn, 1989) for the aforesaid largest values.

The probabilistic relationships for L_T and S_T can be obtained by applying the theorem of extremes of a random number of random variables, as advocated by Todorovic and Woolhiser (1975). Pursuing the work of Sen (1980), the following expressions can be written for L_T and S_T.

$$P(L_T \leq j) = \exp[-Tq(1-r)(1-P(L \leq j))] \qquad (D11)$$

$$P(S_T \leq Y) = \exp[-Tq(1-r)(1-P(S \leq Y))] \qquad (D12)$$

where q is the probability quantile defined earlier, and r is the conditional probability of any year being a drought year given that the past year is also a drought year. One major constituent of equations (D11) and (D12) is that the number of drought occurrences (n) during the period of T years follows the Poisson probability law. A simplification of equation (D11) can yield $E(L_T)$:

$$E(L_T) = \sum_{j=1}^{T} j \cdot P(L_T = j) \qquad (D13)$$

Drought lengths (L) have been found to follow the geometric probability distribution with good agreement (Sen, 1980; Sharma, 1994), hence $P(L_T = j)$ can be evaluated from the following expression:

$$P(L_T = j) = \exp[-Tq(1-r)r^{j-1}][\exp(Tq(1-r)^2 r^{j-1}) - 1] \qquad (D14)$$

The conditional probability r is related to the first-order serial correlation coefficient from the following equation (Sen, 1977):

$$r = q + \frac{1}{2\pi q} \int_0^\rho [\exp(-0.5z_0^2(1 + \tau))] \cdot [(1 - \tau^2)^{-\frac{1}{2}}] d\tau \quad (D15)$$

where ρ is the first-order serial correlation coefficient in the sequences of the drought variable and τ is the dummy variable of integration. Values of r for various values of ρ and q can be evaluated through a numerical integration procedure. Therefore evaluation of the length of the longest drought over a period of T years is conveniently found using equations (D13)–(D15).

In order to evaluate $E(S_T)$, equation (D12) can be simplified to

$$E(S_T) = \int_0^\infty (S_T) \cdot f(S_T) dS_T \text{ i.e. mean} = \int_0^\infty x f(x) dx \quad (D16)$$

where $f(S_T)$ is the probability density function (pdf) of S_T [and $f(x)$ is the pdf of some variable x], which is not fixed because the pdf of S in equation (D12) is not known. Equation (D16) can therefore be solved numerically by first evaluating $P(S \le Y)$ in equation (D12). If S is assumed to be normally distributed then

$$P(S \le Y) = \frac{1}{\sqrt{(2\pi)}\sigma_s} \int_0^Y \exp\left[-\frac{1}{2}\left(\frac{S - \mu_s}{\sigma_s}\right)^2\right] dS \quad (D17)$$

where μ_s and σ_s are the mean and standard deviation of the deficit sums. The severity S is made of several deficits in succession. If the deficit (d) is regarded to have a truncated normal distribution with mean μ_d and standard deviation σ_d, then μ_s and σ_s can be estimated approximately from (Kotz and Neumann, 1963)

$$\mu_s = k\mu_d \quad (D18)$$

$$\sigma_d^2 = k\sigma_d^2\left[\frac{1 + \rho}{1 - \rho} - \frac{2\rho(1 - \rho^k)}{k(1 - \rho)^2}\right] \quad (D19)$$

The expression for μ_d and σ_d can be calculated from (Sharma, 1994)

$$\mu_d = -\frac{\exp(-0.5z_0^2)}{q\sqrt{(2\pi)}} - z_0 \text{ (absolute)} \quad (D20)$$

$$\sigma_d^2 = 1 - \frac{z_0 \cdot \exp(-0.5z_0^2)}{q\sqrt{(2\pi)}} - \frac{\exp(-z_0^2)}{q^2 \cdot 2\pi} \quad (D21)$$

Sharma further arrived at the optimal value of k after extensive simulation experiments:

$$k = 0.5E(L_T) + 0.5\left(\frac{1}{1 - r}\right). \quad (D22)$$

Note that the evaluation of $E(S_T)$ is involved as it requires the numerical integration of several functions. Furthermore equation (D22) implicitly confirms that $E(S_T)$ is strongly correlated with $E(L_T)$, as cited in the hydrological literature (Sen, 1977; Dracup et al., 1980b; Lee et al., 1986). It should be borne in mind that for the random occurrence of drought spells, $r = q$ in the above equations. The values of $E(L_T)$ and $E(S_T)$ for various return periods, truncation levels and autocorrelations can be computed and are shown in Figure D66. Two points become evident from these plots: (1) the drought duration and severity increases as the dependence level in the drought variable increases and (2) the ratio of $E(S_T)$ to $E(L_T)$ is about 1 for a truncation level z_0 between -0.2 and 0 for return periods from 1 to 500 years and ρ from 0 to 0.7. For simplicity in the design equations, one can assume $E(S_T) \sim E(L_T)$ since the estimation of $E(L_T)$ is far simpler than that of $E(S_T)$.

It was also found that the relationship $E(L_T) \sim E(S_T)$ is valid for two parameter lognormal and gamma pdf's of the drought variable. However, in the computation of $E(L_T)$ the values of q and r at the desired truncation level will differ from those applicable to the normal pdf. For example, at the mean level of truncation, z_0 for lognormal and gamma variates ($C_v = 0.4$ and $\rho = 0$) will take on the values equal to 0.19 and 0.13 (as against $z_0 = 0$ for the normal variate). The corresponding values of q (= r values) from the normal probability tables are 0.58 and 0.55 (as against 0.5 for the normal pdf). These values of z_0 and q should be used for estimation of $E(L_T)$ in the relationships indicated above for lognormal and gamma pdf's. When the drought sequence is autocorrelated, say $\rho = 0.5$, $q = 0.58$, $r =$

0.73 for a lognormal pdf and 0.55 and 0.71 for a gamma pdf, respectively, as against 0.5 and 0.67 for a normal pdf.

The analysis of S_T in equations (D12)–(D22) is in the standardized form. The actual largest drought severity (expected value), designated as D_T, can be written as (Horn, 1989).

$$D_T = E(S_T)\sigma_x = E(S_T)C_v x_m \quad (D23)$$

where x_m is the mean drought severity index. Equation (D23) can be manipulated as

$$D_T = x_m - x_m + E(S_T)\sigma_x = x_m + F_T\sigma_x \quad (D24)$$

where F_T can be termed as the drought frequency factor and can be written in the form

$$F_T = [E(S_T) - C_v^{-1}] \quad (D25)$$

since for the design truncation levels ($z_0 = 0$ to -0.2). $E(S_T) \sim E(L_T)$, therefore equation (D25) can be written as

$$F_T = [E(L_T) - C_v^{-1}] \quad (D26)$$

It can be seen that equation (D24) is analogous to the flood frequency formula

$$Q_T = Q_m + K_T\sigma_Q$$

commonly cited in hydrological texts, where Q_T is the flood magnitude, say in $m^3 s^{-1}$, for the return period of T years, Q_m and σ_Q are the mean and standard deviation of a flood sequence Q, such as an annual maximum flood series, and K_T is a frequency factor.

In an earlier development, Millan and Yevjevich (1971) discovered that the distribution of L_T and S_T can be represented by the lognormal probability functions. With the lognormal approximation, the mean and median of the logarithms of data are assumed to coincide and the antilogarithms of the means of these distributions provide reasonable estimates of median values of L_T and S_T. The expressions for the natural logs of the median values of L_T and S_T designated as L_{Tm} and S_{Tm} were obtained as follows:

$$\ln(L_{Tm}) = 1.275 + 0.9024 \ln(q) + 0.2703 \ln(T) + 0.00156 \ln(\gamma + 0.0001) + 0.0237 \ln(\rho + 0.0001) \quad (D27)$$

$$\ln(S_{Tm}) = 1.1336 + 1.1876 \ln(q) + 0.3046 \ln(T) - 0.0162 \ln(\gamma + 0.0001) + 0.0273 \ln(\rho + 0.0001) \quad (D28)$$

in which γ is the coefficient of skewness and other terms are defined earlier. The median value of the actual longest severity designated by D_{Tm} can be written as

$$D_{Tm} = \sigma_x S_{Tm} \quad (D29)$$

The most significant independent variables in equation (D27) and (D28) are q (a measure of truncation level) and return period (T), followed by lag 1 serial correlation coefficient. This is due to the fact that a large serial correlation implies 'persistence' in drought events, with an increased likelihood of multiple year droughts. Skewness has the smallest effect, at least on the length of the drought spell, a fact which has been corroborated by Sen (1980) and Chander et al. (1981). The value of L_{Tm} and S_{Tm} can also be calculated using the theoretical equations (D11)–(D12), which will essentially converge to $E(L_T)$ and $E(S_T)$ in view of the strong assumption of normality implicit in them.

Identification of persistence in drought occurrences

One important element in the functions for estimation of the largest drought severity and the longest drought length is the identification of the persistence and the parameterization of ρ and r. The persistence can be identified by the parametric approach in which the coefficient of lag 1 serial correlation can be estimated from the series of drought variables and checked for its significance (Chander et al., 1981). For a standardized z_i sequence of drought variables, Sen (1978) has proposed a method based on run analysis in which an estimate of lag 1 serial correlation coefficient ρ can be obtained from the following relationship:

$$\rho = \sum_{i=2}^n z_i z_{i-1} \sum_{i=1}^n z_i^2 \quad (D30)$$

in which case the z_i or (original x_i) series is truncated at the median level and the conditional probability of any year being a wet year given the previous year is also wet (designated by pp) is computed by a numerical counting procedure. An estimate of the lag 1 serial correlation coefficient is obtained from the relationship

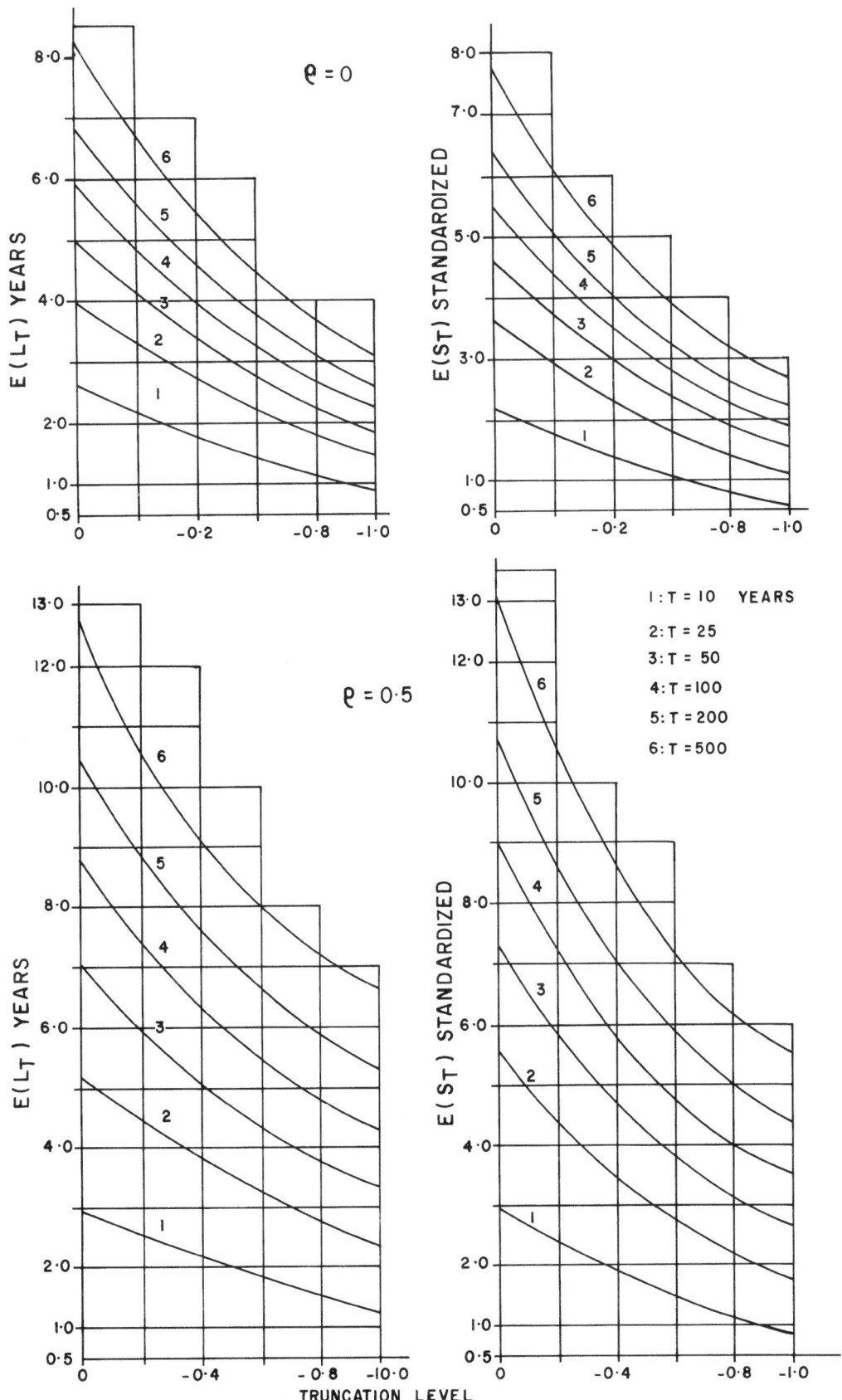

Figure D66 Prediction of the largest values of drought parameters.

$$\rho = \sin \pi \ (pp - 0.5) \qquad (D31)$$

For a random series of drought variables, ρ can be regarded as being approximately normally distributed with a mean of zero and standard error of $1/\sqrt{n}$ (n being the sample size).

The other method for identifying the persistence is based on evaluation of the simple and transitional probability matrices for a lag 1 Markov process. The transitional probability matrix for a lag 1 Markov process can be written as

$$\begin{bmatrix} pp & 1 - pp \\ 1 - r & r \end{bmatrix} \qquad (D32)$$

Under the assumption of the random occurrences of drought spells, the above matrix takes the form

$$\begin{bmatrix} 1 - q & q \\ 1 - q & q \end{bmatrix} \qquad (D33)$$

The above matrices equations (D32) and (D33) can be tested for equivalence using a chi-square test criterion (Medhi, 1982). The critical value of the chi-square statistic is 3.84 at a 5% level of significance and one degree of freedom for inferring the randomness against the Markovian structure of persistence in the drought spells. The estimates of q, pp and r are obtained by a simple counting procedure from the historic data (Sharma, 1994).

At times, a simple non-parametric test known as run test (Miller and Freund, 1985) can also be used to test the randomness of drought spells. The test however does not quantify the value of r, should the drought spells behave in a Markovian fashion. Beran and Rodier (1985) have noted that the persistence phenomenon becomes more apparent when regional data rather than the individual station data are analyzed.

Regional analysis and estimation of droughts

In general in the past, the regional behavior of droughts has been studied first by analyzing the point behavior (i.e. analyses of point rainfall or streamflow data) and then mapping the relevant parameters over a region or country. For instance, isoline maps of 1 in 10, 1 in 50, 1 in 100 or 1 in 500 year frequency droughts at the defined truncation levels or the drought severity indices as those of Palmer or Herbst et al. can be drawn. Shaw (1984) presents a map of the Herbst drought severity index for England and Wales in the UK. This type of presentation is quite useful for understanding by the general public and by politicians. Some attempts at the regional analysis of droughts through stochastic approaches are underway and an excellent review has been made by Rossi et al. (1992). One important tool which is indispensible in the regional analysis is the multiple regression algorithm (Paulson et al., 1985; Mimikou et al., 1993) which involves drought parameters, geomorphic parameters and climatic parameters for the development of regressional equations.

Drought prediction and forecasting

The behavior of droughts in the frequency domain is well studied as has been described in the foregoing sections, and a satisfactory level of methodology is available for planning and designing the water resources systems for meeting the shortages during the drought periods. However, the forecasting aspect which is more important, from the point of view of drought preparedness and early warning, is still fraught with great difficulty. On a short time scale such as months or seasons, it has at times been possible to give some indications of the inception of drought and also of its termination (Beran and Rodier, 1985) but on the long-term basis such possibilities are remote. Short-term forecasts essentially employ:

- linear regression models using weather variables such as air pressure, air and sea temperatures, wind variables, even sunspot number or some other cyclic variable;
- teleconnections, i.e. links between sea surface temperature and inland weather, wind in East Africa and the monsoon in India, location of the ITCZ (intertropical convergence zone) and jet stream, El Niño, etc.;
- time series forecasting methods employing the notions of Box and Jenkins ARIMA (autoregressive integrated moving average) models;
- recession rates of streamflow hydrographs and other water bodies;

- regression models involving river flows, soil moisture, rainfall, air temperature, groundwater levels or some snow-related parameters.

In general, short-term forecasts are expressed probabilistically, i.e. with confidence limits attached to them.

Conclusions

Droughts are natural disasters which occur frequently, affect large areas, cause great loss to agricultural production and reduction in water resources, and inflict adverse environmental impacts such as desertification. The indirect effects of drought often extend beyond the reduction of agricultural output, at times leading to famine, the spread of refugees, social instability and even death. The most important elements of the drought from the point of view of water resource planning and design are the longest duration and the largest severity for a desired return period with reference to some pre-determined demand level designated as the truncation level. The rainfall or runoff sequences constitute the variables for the analysis of the duration and severity. The truncation level is taken to be between 0 and -0.2 in terms of the standardized value. Severity in the context of hydrologic drought is defined as the cumulative deficit (mm or m^3), unlike meteorological droughts in which it may be expressed through indexes. The analysis of drought duration and severity has been performed successfully using the theory of runs. Drought spells (deficit runs) can be represented by the Poisson law, the drought duration (run length) by the geometric law and the severity (run sum) by the truncated normal law of probability theory. Droughts may evolve randomly or follow a Markovian persistence in response with the persistence in the sequences of drought variable being represented by a lag 1 serial correlation coefficient. The aforesaid laws of stochastic theory when applied to runs can be used to estimate the mean or median values of the longest duration and largest severity for the desired return period. The design value of the drought severity can be expressed through a formula similar to the flood frequency formula

$$Q_T = Q_m + K_T \sigma_Q$$

commonly cited in hydrologic texts. The regional behavior of droughts can be assesed by mapping suitable parameters through point analyses. The forecasting techniques for droughts on the long-term basis are almost nonexistent. However on a short-term basis some probabilistic forecasts can be made.

T.C. Sharma

Bibliography

Askew, A.J., W.G. Yeh and W.A. Hall, 1971. A comparative study of critical drought simulation, Wat. Resour. Res., 7(1), 52–62.
Beran, M.A. and J.A. Rodier, 1985. Hydrological Aspects of Drought, UNESCO-WMO, Studies and Reports in Hydrology, 39, 149 pp.
Bonacci, O., 1993. Hydrological identification of drought, Hydrological Processes, 7, 249–262.
Chander, S., N.S. Kambo, S.K. Spolia and A. Kumar, 1981. Analysis of surplus and deficit using runs, J. Hydrol. 49, 193–208.
Downing, T.E., K.W. Gitu, C.M. Kamau and J. Barton, 1984. Drought in Kenya, in T.E. Downing, K.W. Gitu and C.M. Kamau (eds), Coping with Drought in Kenya: National and Local Strategies, Lynne Reinner Publishers, London.
Dracup, J.A., K.S. Lee and E.G. Paulson Jr, 1980a. On definition of droughts, Wat. Resour. Res., 16(2), 289–296.
Dracup, J.A., K.S. Lee and E.G. Paulson Jr, 1980b. On the statistical characteristics of drought events, Wat. Resour. Res., 16(2), 297–302.
Foley, J.C., 1957. Droughts in Australia, Bureau of Meteorology Bulletin No.43, Melbourne.
Frick, D.M., D.Bode and J.D. Salas, 1990. Effect of drought on urban water supplies, I: drought analysis, J. Hyd. Eng., ASCE, 116(6), 733–753.
Gibbs, W.J. and J.V. Maher, 1967. Rainfall Deciles as Drought Indicators, Bureau of Meteorology Bulletin No. 45, Melbourne.
Gibbs, W.J., 1975. Drought, its definition, delineation and effects, Drought Special Environmental Report No.5, WMO, Geneva, 40 pp.
Guerraro-Salazar, P. and V. Yevjevich, 1975. Analysis of Drought Characteristics by the Theory of Runs, Hydrol. Pap. 80, Colorado State University, Fort Collins, USA.

Gupta, V.K., and L. Duckstein, 1975. A stochastic analysis of extreme droughts, *Wat. Resour. Res.*, **11**(2), 221–228.

Herbst, P.H., P.B. Bredenkamp and H.M.G. Barker, 1966. A technique for the evaluation of drought from rainfall data, *J. Hydrol.*, **4**, 264–272.

Horn, D.H., 1989. Characteristics and spatial variability of droughts in Idaho, *J. Irrig. Drain. Eng., ASCE*, **115**(1), 111–123.

Kotz, S. and Neumann, J., 1963. On the distribution of precipitation amounts for periods of increasing length, *J. Geophys. Res.*, **68**(12), 3635–3640.

Lee, K.S., J. Sadeghipour and J.A. Dracup, 1986. An approach for frequency analysis of multiyear drought durations, *Wat. Resour. Res.*, **22**(5), 655–662.

Li, K. and A. Makarau, 1994. *Drought and Desertification*, Report to World Climate Application Programme 28, WMO/TD No. 605, 68 pp.

Medhi, J., 1982. *Stochastic Processes*, Wiley Eastern Limited, New Delhi, India.

Millan, J. and V. Yevjevich, 1971. *Probabilities of Observed Droughts*, Hydrol. Pap. 50, Colorado State University, Fort Collins, USA.

Miller, I. and J.E. Freund, 1985. *Probability and Statistics for Engineers*, Prentice-Hall International, Englewood Cliffs, New Jersey.

Mimikou, M.A., Y.S. Kouvopoulos and P.S. Hadjissavva, 1993. Analysis of multi-year droughts in Greece, *Wat. Resour. Dev.*, **9**(3), 281–291.

Mohan, S. and N.C.V. Rangacharya, 1991. A modified method of drought identification, *Hydrol. Sci. J.*, **36**(1–2), 11–21.

Palmer, W.C., 1965. *Meteorological Drought*, Research Paper No. 45, US Weather Bureau, Washington, DC.

Paulson, E.G., J. Sadeghipour and J.A. Dracup, 1985. Regional frequency analysis of multiyear droughts using watershed and climatic information, *J. Hydrol.*, **77**, 57–76.

Rossi, G., M. Benedini, G. Tsakiris and S. Giakoumakis, 1992. On regional drought estimation and analysis, *Wat. Resour. Management*, **6**, 249–277.

Sen, Z., 1977. Run-sums of annual flow series, *J. Hydrol.*, **35**, 311–324.

Sen, Z., 1978. Autorun analysis of hydrologic time series, *J. Hydrol.*, **36**, 75–88.

Sen, Z. 1980. Statistical analysis of hydrologic critical droughts, *J. Hydraul. Div., ASCE*, **106**(HY1), 99–115.

Sen, Z. 1989. The theory of runs with application to drought prediction – comment, *J. Hydrol.*, **110**, 383–391.

Sharma, T.C., 1990. Stochastic features of drought in Kenya, East Africa, in K.W. Hipel (ed.), *Stochastic and Statistical Methods in Hydrology and Environmental Engineering*, Kluwer Academic, Vol. 1, 125–137.

Sharma, T.C. 1995. A Markov–Weibull model of the Kenyan longest dry spells and the largest rain sums, *J. Hydrol.*

Sharma, T.C. 1996. Estimation of drought severity for independent and dependent hydrological series. *Water Resources Management.*

Sharma, T.C. 1996. Prediction of multiyear extremal droughts. Conference Proceedings Int. Conf. in Water Resources and Environment Research, Kyoto University, Japan. Oct. 28–31, 1996.

Shaw, E.M., 1984. *Hydrology in Practice.*, Van Nostrand Reinhold, UK.

Smart, G.M., 1983. Drought analysis and soil moisture prediction, *J. Irrig. Drain. Eng., ASCE*, **109**(2), 251–261.

Steila, D. 1987. Drought, in J.E. Oliver and R.W. Fairbridge eds, *The Encyclopedia of Climatology*, Van Nostrand Reinhold, New York, pp. 388–395.

Todorovic, P. and D. A. Woolhiser, 1975. A stochastic model of n day precipitation, *J. Appl. Meteorol*, **14**, 17–24.

Wilhite, D.A. and M.H. Glantz, 1985. Understanding the drought phenomenon: the role of definitions. *Water International*, **10**, 111–120.

Yevjevich, V. 1967. *An Objective Approach to Definition and Investigations of Continental Hydrological Droughts*. Hydrology Paper 23, Colorado State University, Fort Collins, Colorado.

Zelenhasic, E. and A. Salvai, 1987. A method of streamflow drought analysis, *Water Resour. Res.*, **23**(1), 156–168.

Cross references

Drought
Drought management

E

ECHO SOUNDING

This is the process of operating an instrument called an echo sounder which uses the reflection of an acoustic signal from an object to determine the distance of the object to the instrument. In open channels this is achieved vertically, almost invariably by placing the echo sounder in a boat, in order to determine the water depths as the echo sounder traverses the water surface of the open channel. The echo sounder can be installed in a purpose-built location in the boat, or attached to the underside of the boat's hull, or suspended over the side of the boat by means of a small gantry. The depth readout can be digital, analog or display visual. The accuracy of the reflective signal varies with the composition of the channel bed and poor accuracy pertains when the bed material is fine silt or when there is significant weed growth. The inaccuracies are less significant, in percentage terms, in deep water than when the water is shallow. The effective use of the depth data obtained is heavily dependent upon the accuracy to which the boat's positioning is recorded, relative to on-bank datum points.

P.G. Holland

Bibliography

BS 3680, Part 1, 1991. *Glossary of Terms*, British Standards Institution, London.

Cross reference

Water resources: dictionary of basic terms

EL NIÑO

Originally, El Niño referred to the warm current that sets southward each year along the coasts of southern Ecuador and northern Peru during the southern hemisphere summer when the southeast trade winds are weakest. It was named El Niño (the child) by devout inhabitants of this region in reference to the 'Christ Child,' since it ordinarily sets in shortly after Christmas. The movement southward (to 4–6° south of the equator) of the zone of discontinuity between this thin, warm, low-salinity equatorial water layer and the cool Peru current water is a regular southern hemisphere summer phenomenon (Figure E1). However, in exceptional years, when the seasonal decrease in strength of the southeast trades is abnormally large, the warm current penetrates much farther south than usual along the Peruvian coast, occasionally past Callao at 12°S. Through common usage over recent years, the term El Niño is now reserved for the exceptional year phenomenon by most publishers and environmental scientists.

After learning more about El Niño and the important Peruvian anchoveta fishery it so seriously affected, the definition has been

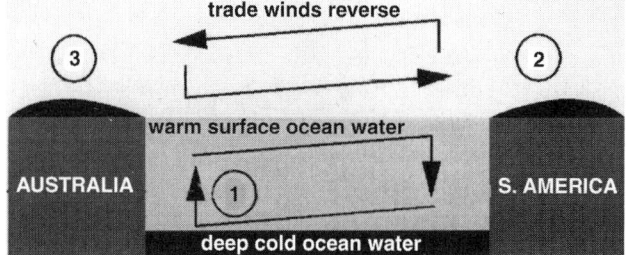

Figure E1 El Niño: changing seawater temperature. (Source: Roth, 1997.)

altered considerably. It not only involves the thin southward flowing equatorial surface water layer, but also an influx of waters from the west and northwest beneath this surface layer. The invading thin surface layer, which has a significantly lower salinity than the subtropical surface water farther to the west of the Peruvian coast, is nutrient depleted unlike the cool, highly productive Peruvian current and its coastal upwelled waters which usually prevail along the Peruvian coast. These infrequent invasions ordinarily set in during the southern hemisphere summer when sea temperatures are at a seasonal high, but they may at times set in well into the fall and the effects may persist for a year or more. Additional symptoms of the stronger El Niño, some or all of which may be noted, are torrential downpours, flood and erosion in the normally arid coastal lowlands of northern Peru, red tide, invasion by tropical nekton, and mass mortality of various marine organisms, including guano birds, sometimes with subsequent decomposition, and release of hydrogen sulfide (Wooster, 1960). El Niño occurs at irregular intervals and may appear two years in succession and then not reappear for another 3–12 years [refers to the moderate and strong El Niño categories of Quinn *et al.* (1978) which seriously affect the fishery]. Particularly strong El Niños occurred in 1864, 1877–1878, 1884, 1891, 1899–1900, 1925–1926, 1941, 1957–1958, 1972–1973 and 1982–1983. The fisheries along the northern Chilean coast, like those off Peru, are sensitive to environmental disturbances (Caviedes, 1981), and the Southern Oscillation-related climatic changes that spawn El Niño seriously affect them (Quinn and Neal, 1983).

El Niño is essentially a regional symptom of the large-scale ocean–atmosphere fluctuation, known as the Southern Oscillation (Berlage, 1957). Southern Oscillation indices (differences in sea level atmospheric pressure between sites located in the South Pacific subtropical high-pressure region and the Indonesian equatorial low-pressure region) are often used to represent the Southern Oscillation (Quinn, 1974; Quinn *et al.*, 1978). They can be used to monitor and also to obtain an outlook on the Southern Oscillation-related short-term climatic changes that occur over the equatorial Pacific, the coastal and offshore regions of western South America, and the Indonesian–Australian regions. Although many different indices have been used for these purposes, the Easter–Darwin, Totegegie–Darwin, Rapa–Darwin and Tahiti–Darwin indices are particularly effective for following and assessing these climatic developments. In order to emphasize the interannual changes, the regular annual cycle can be largely eliminated by subtracting the long-term average or normal monthly pressure values from the individual monthly values. Plots of the filtered data (smoothed by using a 3 month running mean or triple 6 month running mean filter) are then used to detect, identify and evaluate the changes that occur (Quinn *et al.*, 1978). It appears that most of the large short-term climatic changes and their characteristic current and weather patterns over the lower latitudes of the Pacific Ocean are associated with either the El Niño (low index) or anti-El Niño (high index) phases of the Southern Oscillation. At times, a broader connotation, El Niño type, is used when describing the Southern Oscillation-related events; in this way, one can account for events that evolve in a similar manner (associated with falling and low Southern Oscillation indices) but vary in timing, intensity and extent (Quinn *et al.*, 1978).

In recent years the definition of El Niño has been modified by this realization that it is a regional manifestation of the large-scale ocean–atmosphere circulation fluctuation and that it is brought about by relaxation from a prolonged period of strong southeast trades and equatorial easterlies (represented by rising and high Southern Oscillation indices; Quinn, 1974; Wyrtki, 1975). The magnitude of the southeast trade relaxation (as indicated by falling and low indices) and its timing in relation to the regular regional seasonal relaxation tends to determine the strength of the resulting El Niño. During the period of prolonged strong southeast trades and equatorial easterlies, the South Equatorial Current is intensified, coinciding with an east to west build-up in sea level and an accumulation of warm water in the western tropical Pacific and, as soon as the wind stress relaxes, the accumulated water flows eastward, probably in the form of an internal equatorial Kelvin wave (Wyrtki, 1975). This wave leads to the accumulation of warm equatorial undercurrent water off Ecuador and Peru and to a depression of the usually shallow thermocline there. In addition to the generation of internal Kelvin waves and Rossby waves, as discussed by Hurlburt *et al.* (1976) and McCreary (1976), it is assumed that the eastward flowing currents (i.e. the North Equatorial Countercurrent, South Equatorial Countercurrent and Equatorial Undercurrent) are intensified and the westward-flowing South Equatorial Current is weakened when the relaxation occurs (Wyrtki *et al.*, 1976). Hydrographic data off the coasts of Ecuador and Peru confirm the thermal structure depression and poleward spreading during El Niño (Enfield, 1981) and, although coastal upwelling may

continue, it is from this accumulated warm water above the base of the thermocline. This too causes coastal surface waters to be much warmer and less productive than water from the usual Peruvian current source.

The foregoing hypothesis concerning the evolution of an El Niño is based in general on the developments that have taken place since 1950, and in particular on the more recent El Niños (e.g. 1957–1958, 1965–1966, 1972–1973 and 1976–1977) when more data were available. Also, in agreement with the age-old definitions, it was believed that the stronger developments initially set in during the southern hemisphere summer or early fall, and that their initial onset generally precedes the occurrence of many of the other Southern Oscillation-related climatic changes with which El Niño is associated (e.g. the occurrence of anomalously warm water in the central equatorial Pacific, anomalously heavy rainfall in the central and western equatorial Pacific, Indonesian–Australian droughts and abnormally heavy rainfall in subtropical Chile). It was also generally believed that the onset of any large-scale interannual stimulus in southern hemisphere winter would lead to a weak or negligible event as far as the coastal regions of southern Ecuador and Peru were concerned. However, circumstances surrounding the 1982–1983 El Niño have by now cast considerable doubt on several assumptions and beliefs included in the foregoing hypothesis with regard to the course of changes leading up to and following the El Niño phenomenon. Perhaps in the past we have placed too much emphasis on the more specific developmental aspects of El Niño and not enough emphasis on the larger-scale climatic changes with which it is associated.

The 1982–1983 El Niño was an exceptionally strong one, yet the time that it set in, the sequence of other associated changes over the lower latitudes of the Indo-Pacific area and the characteristic trends in the more directly related atmospheric and oceanic variables differ greatly from both the older traditional viewpoints and the more modern hypothesis derived from the detailed studies of prior El Niño events, for example the following.

● This El Niño was not preceded by an extended period of high Southern Oscillation indexes. In fact, due to a long-term climatic shift (Quinn and Zopf, 1984), index anomalies remained relatively low for about 8.5 years following April 1976 (see the Easter–Darwin plot in Figure E2). Also, there was no prolonged period of strong southeast trades and equatorial Pacific easterlies preceding this development, and likewise, no prior anomalously large rises in sea level in the western equatorial Pacific tide gauge records.

● Although it was a strong El Niño, it did not set in near Christmas nor during the southern hemisphere summer or early fall seasons.

Figure E2 Three-month running mean plots of anomalies of the difference in sea level atmospheric pressure (mbar) between Easter Island (27°10'S, 109°26'W) and Darwin, Australia (12°26'S, 130°52'E), and 3-month running mean plot of sea surface temperature anomalies (°C) for Chimbote, Peru (9°10'S, 78°31'W).

The low pre-event index anomaly peak occurred during January 1982 (in the southern hemisphere summer), and there was a large and rapid fall in the index and index anomalies (Figure E2) during the following months. Positive sea surface temperature anomalies started showing up along the coast of Peru during the June–October period (Figure E2), becoming high at most Peruvian coastal stations from Callao northward by October 1982. Therefore, it set in during the southern hemisphere winter–early spring, totally out of phase with the traditional onset time.

- The onset of El Niño along the coast of Peru did not precede the occurrence of anomalously warm waters in the central equatorial Pacific, the onset of heavy rainfall in the central and western equatorial Pacific, the occurrence of drought conditions over Indonesia and Australia, or the abnormally heavy Chilean subtropical rainfall, as it usually does. In this case the El Niño and these other associated large-scale changes set in at about the same time; in fact, most of the other changes were noted to set in earlier than El Niño.

One of the consistent features of the 1982–1983 El Niño development was the magnitude of the Southern Oscillation index fall from the small peak to the extremely deep trough. For the 3-month running mean of the Easter-Darwin index, this fall was 11.1 mbar to a low point in August 1982: the index anomalies reached a minimum in early 1983 after a fall of 8.8 mbar (Figure E2). These figures are representative of a substantial El Niño development. Also, the equatorial winds, instead of relaxing from strong easterlies to weak easterlies or near-calm conditions, changed from mostly weak easterlies to mostly westerlies over the western and central equatorial Pacific during the latter half of 1982. Although the prolonged anomalously large build-ups in sea level in the western equatorial Pacific prior to this event did not occur, there were abnormally large falls from June 1982 on (Halpern, 1983).

Canby (1984) provides a comprehensive report on the global ramifications of the 1982–1983 El Niño. Evidence indicates that it was probably the strongest El Niño on record. Quinn and Zopf (1984) theorize that the unusual strength and widespread nature of this event were the result of a strong interannual fluctuation (a manifestation of the Southern Oscillation) taking place at almost the same time that the cumulative effects of a large-scale, long-term climatic change (Quinn and Neal, 1983) were coming to a head.

After the 1982–1983 El Niño conditions ran their course and subsided and available developmental findings were gathered, it became apparent that existing concepts and hypotheses concerning the sequence of changes leading up to and following El Niño must be modified somewhat. Although the time of onset for the 1982–1983 El Niño differed greatly from the norm, it resembled some unusual cases of the past. For example, the strong 1941 El Niño set in during late 1940 although its activity peaked in early 1941: however, the difference between the 1940–1941 and the 1982–1983 cases is that there was a large pre-event peak in the Southern Oscillation indices in 1938 prior to the 1939–1941 activity (Quinn and Zopf, 1984). It now appears that El Niño can set in at almost any time of the year, and developmental characteristics and spatial distribution of related activity can vary considerably from event to event.

We must also realize that originally the El Niño occurrences and their intensities were to some extent determined by their effects on the Peruvian anchoveta fishery, and it was this anchoveta fishery that made Peru the leading fish-producing nation from the late 1960s until 1971 (Idyll, 1973). The record catches of 1970–1971 followed by the strong 1972–1973 El Niño led to a precipitous drop in Peruvian fishmeal production through 1973 (Figure E3), from which there has never been a full recovery. After the 1976–1977 El Niño the anchoveta became a much smaller portion of the total fish catch in both the Peruvian and northern Chilean fisheries, where they depended more on the contributions of sardines, mackerel and other fish for their fishmeal sources (Quinn and Neal, 1983). Peru fell from its position of leading fishmeal producer in the Fishmeal Exporters Organization to a number two position behind Chile by 1980 (Figure E3).

One of the main problems in understanding El Niño has been the lack of data. However, following the 1983 disaster, the World Meteorological Organization conducted a research program and, as a result, the US National Oceanic and Atmospheric Administration (NOAA) installed a series of buoys referred to as the TOGA TAO Array across the tropical Pacific to monitor winds and ocean temperature to a depth of 500 m (Roth, 1997). Japan plans to extend the TOGA TAO Array into the Western Pacific and Indian Ocean and

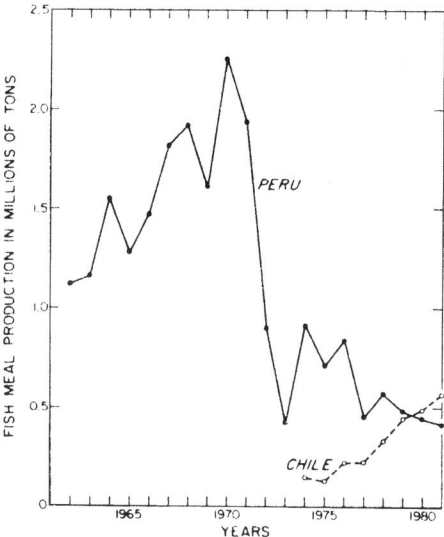

Figure E3 The Peruvian fishmeal production for 1962–1981 and Chilean fishmeal production for 1974–1981 in millions of tonnes as obtained from the National Marine Fisheries Service. (From Quinn and Neal, 1983.)

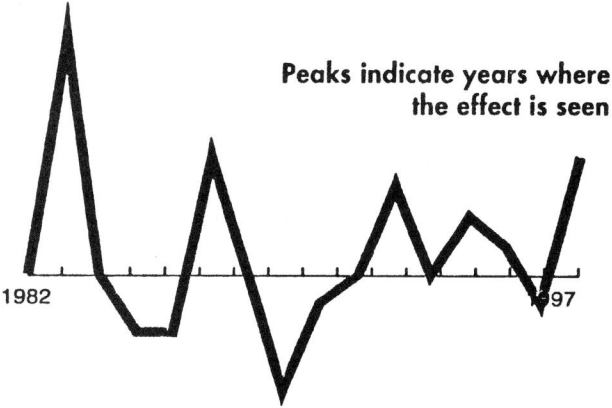

Figure E4 The El Niño effect off western South America. (Source: Roth, 1997.)

further research is being carried out under the WMO's World Climate Research Programme. Although this daily data is a significant step forward, the array cannot measure all the variables to be keyed into a complex model and resort is made to producing forecasts based on probability to create a model of the climate system that evolves. As a result of the research, the El Niño of 1997–1998 will be the most closely observed in history and, as well as the stationary buoys, satellites and ships will also be used in data gathering (Nash, 1997). Until recently, El Niño came more or less periodically every 2–7 years (Figure E4) but in the early 1990s several El Niños appeared in a row, one after the other. However, in 1997, after dying down in 1995 and 1996, it arrived earlier than usual, in April, instead of its normal onset in December, and temperatures rose with surprising speed. The extent of its swath over the equatorial ocean is some 10 000 km.

William H. Quinn

Bibliography

Berlage, H.P. ,1957. Fluctuations of the general atmospheric circulation of more than one year, their nature and prognostic value, *K. Ned. Meteorol. Inst. Meded. Verh.*, **69**.

Canby, T.Y. (ed.), 1984. El Niño's ill wind. *Natl Geographic*, **165**(2), 144–183.

Caviedes, C.N., 1981. The impact of El Niño on the development of the Chilean fisheries, in *Resource Management and Environmental Uncertainty: Lessons from Coastal Upwelling Fisheries*. M.H. Glantz and J.D. Thompson (eds.). New York: Wiley-Interscience, pp. 354–368.

Enfield, D.B., 1981. El Niño: Pacific eastern boundary response to interannual forcing, in *Resource Management and Environmental Uncertainty: Lessons from Coastal Upwelling Fisheries*. M.H. Glantz and J.D. Thompson (eds.). New York: Wiley-Interscience, pp. 213–254.

Halpern, D. (ed.), 1983. *1982 Equatorial Pacific Warm Event. Tropical Ocean-Atmosphere Newsletter*, no. 16. Seattle, WA: University of Washington.

Hurlbert, H.E., J.C. Kindel and J.J. O'Brien, 1976. A numerical simulation of the onset of El Niño, *J. Phys. Oceanog.*, **6**, 621–631.

Idyll, C.P., 1973. The anchovy crisis, *Sci. American* **228**(6), 22–29.

McCreary, J.P., 1976. Eastern tropical ocean responses to changing wind systems: With application to El Niño, *J. Phys. Oceanog.* **6**, 632–645.

Nash, J.M., 1997. Is it El Niño of the century?, *Time*, **150**(7), New York.

Quinn, W.H., 1974. Monitoring and predicting El Niño invasions. *J. Appl. Meteorol.*, **13**, 825–830.

Quinn, W.H. and V.T. Neal, 1983. Long-term variations in the Southern Oscillation, El Niño, and Chilean subtropical rainfall, *Fishery Bull.*, **81**(2), 363–374.

Quinn, W.H. and D. Zopf, 1984. The unusual intensity of the 1982–1983 ENSO event, in *Tropical Ocean Atmosphere Newsletter*, no. 26, D. Halpern (ed.), Seattle, WA: University of Washington, pp. 17–20.

Quinn, W.H., D.O. Zopf, K.S. Short and R.T.W. Kuo Yang, 1978. Historical trends and statistics of the Southern Oscillation, El Niño, and Indonesian droughts, *Fishery Bull.*, **76**(3), 663–678.

Roth, S., 1997. Winds of Frightening Change, *Water Bulletin*, No. 762, Water Services Association, London.

Wooster, W.S., 1960. El Niño, *California Cooperative Oceanic Fisheries Investigations*. Rep. no. 7, Marine Research Committee. Department of Fish and Game, State of California, pp. 43–45.

Wyrtki, K., 1975. El Niño – The dynamic response of the equatorial Pacific to atmospheric forcing, *J. Phys. Oceanog.*, **5**, 572–584.

Wyrtki, K., E. Stroup, W. Patzert, *et al.*, 1976. Predicting and observing El Niño, *Science*, **191**(4225), 343–346.

Cross references

Australia: climate and water resources
Climate and climate change
Droughts

ENERGY HEAD

The sum of the elevation of a water surface in an open channel above the channel bed (or above the crest of a structure placed in the channel) – effectively the water depth – and the velocity head, based on the mean water velocity in the channel is:

$$H = h + \left(\frac{\alpha \bar{v}^2}{2g} \right)$$

where H is the energy head, h is the head of water level, $\bar{v}$ is the mean water velocity, α is the energy correction factor and g is the acceleration due to gravity.

The energy head is known also as the total head and the energy correction factor is known also as the Coriolis energy coefficient. For most practical applications in open channels, the Coriolis energy coefficient can be assumed to be unity, although its real value can vary between 1.03 and 1.36 (see also Bernoulli energy equation).

P.G. Holland

Bibliography

BS 3680, Part 1, 1991. *Glossary of Terms*, British Standards Institution, London.

Cross reference

Water resources: dictionary of basic terms

ENTROPY IN ENVIRONMENTAL AND WATER RESOURCES

Introduction

The term entropy as a scientific concept was initially used in thermodynamics as early as the 1850s by Clasius. Later in 1877, Boltzmann provided a probabilistic interpretation of the concept within the context of statistical mechanics. The explicit relationship between entropy and probability was developed in the early 1900s by Planck. Finally, Shannon (1948a,b) used the concept to present an economical description of the properties of long sequences of symbols, and applied the results to a number of basic problems in coding theory and data transmission. With his remarkable contributions, Shannon developed the basis of modern information theory. Later, Jaynes (1957a,b) re-evaluated the method of maximum entropy and applied it to a variety of problems involving the determination of unknown parameters from incomplete data (Papoulis, 1991).

Since the pioneering work of Shannon (1948a,b), much attention has been focused on the use of entropy and energy dissipation rate relationships in environmental and water resources engineering. The energy dissipation relationships are derived from well-established thermodynamic principles of entropy creation. The entropy concept, based in both thermodynamics and statistical mechanics, has been shown to be capable of playing a significant role in a wide spectrum of scientific areas.

Entropy is a measure of the degree of uncertainty or disorder associated with a system. Indirectly, it also reflects the information content of space–time measurements. These features of entropy have been mathematically formulated in the theory of entropy by Shannon (1948a,b) and the principle of maximum entropy by Jaynes (1957a,b). The works of Shannon and Jaynes form the basis for a wide range of applications of entropy in hydrology and water resources, covering frequency analyses, parameter estimation, catchment modeling, flow forecasting, spectral analysis, assessment of model performance, design of hydrometric networks, data acquisition systems, modeling of sediment yield and pollutant loading, reliability of water distribution systems, and so on (Singh and Fiorentino, 1992).

Engineering decisions are frequently made with less than perfect information. Such decisions may often be based on experience, professional judgement, rules of thumb, crude analyses, safety factors or probabilistic methods. Usually, decision making under uncertainty tends to be relatively conservative. Although probabilistic methods allow for a more explicit and quantitative accounting of uncertainty, their major difficulty stems from availability of limited data. The entropy concept enables determination of the least biased probability distributions with limited knowledge and data.

On the other hand, the energy dissipation rate relationships have been utilized to rationalize the evolution of fluvial geomorphological systems in general and stream channels in particular (Yang, 1971a). Fundamental to these relationships has been the use of analogies between thermodynamic concepts of heat evolved and system temperature and the postulated geomorphic equivalents of potential energy change and elevation. Although there are reservations about some of the mathematical and empirical development applying thermodynamic principles in water resources (Kennedy *et al.*, 1964; Nordin, 1977; Davy and Davies, 1979), the validity of the analogies is, in general, accepted (Scheidegger, 1964).

Concepts and definitions

The concept of entropy

The scientific literature provides a multitude of definitions of entropy. Different definitions may be appropriate for different uses. Of particular importance, however, is the distinction between entropy as an objective measure of some property of a system, and entropy as a subjective concept for use as a model-building tool to maximize the use of available information. When entropy is used as a subjective concept, it is associated not with the system itself but with the information about the system that is known. Wilson (1970) has presented four different ways to view the concept of entropy:

- entropy as a measure of system property (such as order and disorder, reversibility and irreversibility, complexity and simplicity, etc.);
- entropy as probability for measurement of information, uncertainty or probability;
- entropy as a statistic of a probability distribution for measurement of information or uncertainty;
- entropy as the negative of a Bayesian log-likelihood function for measure of information.

The first is an objective view of entropy and the last three are subjective views (Singh and Fiorentino, 1992).

Entropy as a measure of system property
Boltzmann considered that entropy is the measure of disorder in a system and further that the degree of disorder is essentially the amount of uncertainty associated with particular microscopic states. Boltzmann's definition described entropy in probabilistic terms and constituted the basis for statistical thermodynamics (Harmancioglu *et al.*, 1992a).

Boltzmann (1872) used entropy as a measure of the degree of ignorance as to the true state of a thermodynamic system:

$$H = k \log p \qquad (E1)$$

where H is the entropy, p is the probability of system state, and k is the Boltzmann constant.

According to the second law of thermodynamics, the entropy of a closed system always either remains constant or increases. Mathematically,

$$dH \geq 0 \qquad (E2)$$

where dH is the change in entropy. This law asserts that the system cannot receive more in energy than the amount of external work done. Conversely, the system cannot supply more energy to its environment in the form of work than it has energy available. A system can be construed as any volume of space selected for scientific analysis. A system can be small or quite big. An example of a small system is the laboratory watershed or the experimental flume, and an example of a large system is the entire Mississippi River basin. A closed system is a very special kind of system which is thermally insulated from the rest of the environment. It is hard to think of a closed hydrologic system in the thermodynamic sense. Nevertheless, a confined aquifer over a small time scale would be a good example. A natural watershed over a short time period would be another acceptable example. In general, the entropy of a closed system is higher than when it is made 'open'. In other words, the disorder of a closed system is greater than when it is open, and the entropy increases as the disorder of the system increases. The hydrological cycle is a closed system at the global scale in a hydrological sense, but it is an open system at the watershed scale. By analogy with a thermodynamic system, it would be reasonable to say that the entropy of the hydrological cycle at the global scale is greater than the entropy of the hydrological cycle at the watershed scale (Singh and Fiorentino, 1992).

The relation of entropy to disorder is hydrologically quite meaningful. We consider an example of a water–vapor system. Water vapors in the atmosphere are quite disordered. When the vapors change to liquid form (rain) or solid state (snow and hail), the changed state has a much higher degree of order or a much lower degree of disorder. Ice crystals or snowflakes have a beautiful order. The entropy of a water–vapor system is much higher when it is vapor than when it is water or snow. As another example, we consider a river basin consisting of a number of sub-basins. Let us consider the uppermost first-order watershed that has a certain amount of disorder. As we move down from this watershed to a second-order watershed that encompasses the first-order watershed, the disorder increases. This means that the entropy increases. One may, somewhat presumptuously, postulate that on average the entropy increases with the watershed order in a river network.

The concept of entropy or disorder can also be related to the concept of space–time scales of hydrological processes. As an example, we will consider a rainfall event whose continuous time observations are available. If rainfall intensity is plotted against time as a histo-

gram, the shape of the histogram depends on the time scale chosen for the plot, for the averaging of the rainfall intensity entirely rests on the time scale. If the time scale is, say, 1 min, the histogram may have several peaks and may exhibit great disorder. If the time scale is increased to 1 h, the resulting histogram may have only one peak and will certainly display a higher degree of order and lower entropy. If the time scale is increased to the duration of the rainfall event, the resulting histogram will simply plot as a rectangular response. Clearly then, as the time scale increases, the observed rainfall data exhibit greater order and lower entropy. In general, the entropy of a hydrological system decreases with increasing time scale or decreases with decreasing time variability. This means that entropy can be a measure of the temporal variability of a hydrological system (Singh and Fiorentino, 1992).

As regards the spatial scale, the entropy of a hydrological system can be thought to decrease with increasing homogeneity of the system. In general, as the spatial scale increases, homogeneity decreases. Thus, one can postulate that the entropy of a hydrological system decreases with decreasing spatial scale and can, in turn, be considered as a measure of the heterogeneity of the system. In general, spatial variability increases with size or area of a watershed. Then, larger watersheds can be said to possess higher entropy than smaller watersheds.

Entropy, uncertainty, information and probability
Shannon (1948a,b) adopted Boltzmann's definition with the idea that the entropy concept could be used to measure other types of disorder besides those of thermodynamic microscopic states. Through his significant contributions to communications theory (later known as information theory), Shannon showed that entropy describes the amount of uncertainty in any probability distribution. Thus the entropy concept can be used as a measure of uncertainty and indirectly as a measure of information in probabilistic terms. Basically what Shannon defined is 'informational entropy' (Harmancioglu *et al.*, 1992a).

Similar to the role of thermodynamic entropy in physics, informational entropy has found a wide area of application in various different fields, including water resources engineering. In these applications, uncertainties associated with (or information conveyed by) systems of concern have been measured by the probabilistic definition of entropy.

To describe the entropy concept in a probabilistic sense, let us consider a random variable whose behavior is described by a probability distribution. There is some uncertainty associated with this distribution and, for that matter, with any distribution used to describe the random variable. The concept of entropy provides a quantitative measure of this uncertainty.

Every finite scheme describes a state on uncertainty. We may perform an experiment to investigate the nature of the random variable. Through experimental observations, we have the actual outcome, and we know only the probabilities of possible outcomes. The information thus gained can be regarded as equal to the amount of uncertainty removed from realization of the random variable that existed before the experiment. In this sense, entropy is a measure of information provided by realization of the probability distribution. Naturally, the amount of uncertainty is different in different schemes.

As an example, we may consider a system with two possible outcomes: rain (A_1) with probability p_1, and no rain (A_2) with probability p_2 on any day. Based on any experimental analysis or any other method, the following is obtained:

$$\begin{array}{cc} \text{Scheme 1} & \text{Scheme 2} \\ \begin{pmatrix} A_1 & A_2 \\ 0.5 & 0.5 \end{pmatrix} & \begin{pmatrix} A_1 & A_2 \\ 0.9 & 0.1 \end{pmatrix} \end{array}$$

The first scheme obviously represents much more uncertainty than the second. In the second scheme, the result of the experiment is almost certainly A_1. In the first case it is prudent to refrain from making any predictions. Entropy is a measure of the uncertainty associated with each scheme; it is higher for the first scheme than for the second scheme (Singh and Fiorentino, 1992).

To express the uncertainty mathematically, one may follow Shannon's (1948a,b) definition of entropy as a measure of uncertainty. If S is a system of events, $E_1, E_2, \ldots, E_n$ and $P(E_k) = p_k$ is the probability of the kth event recurring, then the entropy of the system is

$$H(S) = \sum_{k=1}^{n} p_k \ln p_k \qquad (E3)$$

with

$$\sum_{k=1}^{n} p_k = 1$$

In the above, $H(S)$ is the entropy of the probability distribution $P = (p_1, p_2, \ldots, p_n)$. By making use of equation (E3), the uncertainty of the two schemes that was just alluded to can now be quantified. Accordingly, for the first scheme,

$$H(S) = -0.5 \ln 0.5 - 0.5 \ln 0.5 = 0.693$$

and for the second scheme,

$$H(S) = -0.9 \ln 0.9 - 0.1 \ln 0.1 = 0.135$$

Clearly, the first scheme is about fourfold as uncertain as scheme 2 (Singh and Fiorentino, 1992).

Entropy and probability distribution
Jaynes (1957a,b) states:

> Just as in applied statistics the crux of a problem is often the devising of some method of sampling that avoids bias, our problem is that of finding a probability assignment which avoids bias while agreeing with whatever information is given. The great advance provided by information theory lies in the discovery that here is a unique, unambiguous criterion for the 'amount of uncertainty' represented by a discrete probability distribution, which agrees with our intuitive notions that a broad distribution represents more uncertainty than does a sharply peaked one, and satisfies all other conditions which make it reasonable.

This measure of uncertainty was discovered by Shannon (1948a,b) and is the entropy of the probability distribution (Shannon and Weaver, 1949). Khinchin (1957) showed that this is a unique and unambiguous measure of uncertainty. Jaynes (1957a,b) further states:

> It is now evident how to solve our problem; in making inferences on the basis of partial information we must use that probability distribution which has a maximum entropy subject to whatever is known. This is the only unbiased assumption we can make; to use any other would amount to arbitrary assumption of information which by hypothesis we do not have.

Thus entropy not only provides a measure of the uncertainty of a probability distribution, but also allows, through maximization, derivation of the least-biased probability distribution subject to the information about the random variable (Singh and Fiorentino, 1992).

Entropy and maximum likelihood function
In Bayesian statistics, we deal with degrees of belief and compute probability distributions which are compatible with the available information. Intuitively, it would then seem that the Bayesian methods of statistical inference and entropy are related to each other. For a random variable X, which can take values $x_1, x_2, \ldots, x_n$, with probability distribution $P(X, \theta)$ where θ is the parameter, the likelihood function $L(x, \theta)$ is:

$$L(x, \theta) = \prod_{i=1}^{n} p(x, \theta)$$

It is often convenient to work with the log-likelihood function

$$\ln L(x, \theta) = \sum_{i=1}^{n} \ln p(x_i, \theta) \qquad (E5)$$

When the sample size n increases, then the value of θ, obtained by maximizing $\ln L(x, \theta)$ will, under appropriate conditions, tend to the true value. By invoking the law of large numbers,

$$\lim_{n \to \infty} [1/n \ln L(x, \theta)] = E[\ln p(x_i, \theta)] \qquad (E6)$$

where $E[.]$ is the expectation of $[.]$ and

$$E[\ln p(x_i, \theta)] = \sum_{k=1}^{n} p(x_i, \theta) \ln p(x_i, \theta)$$

$$= \sum_{k=1}^{n} p_i \ln p_i \qquad (E7)$$

Equations (E3) and (E7) show the relation between the Shannon entropy and the maximum likelihood. This comparison implies that $p(x, \theta)$, derived by maximization of entropy, is the one that minimizes the likelihood function. Stated another way, the probability distribution that maximizes entropy makes the weakest assumption which is consistent with the available information. Thus the Bayesian maximum likelihood procedure and entropy-maximizing procedure can be complementary to each other (Singh and Fiorentino, 1992).

Entropy as a measure of information

Entropy as defined in information theory
In information theory the concept of and the measure for 'information content' has been derived from statistical and probabilistic principles. The theory analyzes the statistical structure of a series of numbers, signs or symbols that make up a communication signal, without considering at all their kind, meaning, value or any other subjective characteristics. The term 'information content' here refers to the capability of signals to create communication, and the basic problem is the generation of correct communication by sending a sufficient amount of signals, leading neither to any loss nor to repetition of information (Shannon and Weaver, 1949; Khinchin, 1957).

The basic principles of information theory were developed by Shannon, who considered the transmission of signals through a communication channel to be a stochastic process. Shannon expressed the concept of information as 'entropy' since his mathematical formula for the concept is similar to the entropy function defined in statistical mechanics (Shannon and Weaver, 1949; Harmancioglu, 1980, 1981).

According to Shannon, information is attained only when there is uncertainty about an event. This uncertainty points out the presence of alternative results the event may assume and the action of making selections among them. Alternatives with a high probability of occurrence convey little information and vice versa. Here the probability of occurrence of a certain alternative is the measure of uncertainty or the degree of expectedness of a sign, symbol or number. It is this uncertainty that Shannon refers to as 'entropy'. When such uncertainty is removed, the result is information. Therefore, the information gained is indirectly measured as the amount of uncertainty or of entropy.

Shannon defines the concept of 'entropy' objectively in mathematical terms as the expected value of the probabilities of alternative values an event may assume. Once the statistical structure of a process is known, its entropy can be computed and expressed in specific units (as bits per symbol as Shannon defined). According to Shannon's formulation, the entropy function always assumes positive values which, in addition to other characteristics mentioned, makes it readily acceptable as an objective criterion for measuring the information content of any statistical process. Thus the concept has found a wide area of application in various fields of science (Harmancioglu et al., 1992a).

A short historical perspective
In a series of contributions, Shannon (1948a,b) developed the mathematical theory of entropy. Later, Jaynes (1957a,b) formulated the principle of maximum entropy (POME) and applied it in statistical physics (Jaynes, 1961, 1982). The works of Shannon and Jaynes opened up this new area of research and provided the major impetus for applications of entropy in various areas of science and technology (Singh and Rajagopal, 1987). These areas cover communications, economics, thermodynamics, psychology, hydrology, statistical mechanics, reservoir engineering, turbulence, structural reliability, optimization and decision theory, and landscape evolution, apart from linguistics and social sciences (Harmancioglu, 1980, 1981).

Leopold and Langbein (1962) were perhaps the first to have applied the concept of entropy in geomorphology and landscape evolution. Equations for longitudinal profiles for the most probable state of the river were derived by defining the energy distribution in the river. Scheidegger (1967) used entropy to develop a thermodynamic analogy for river meandering, Yang (1971a,b) applied it to study stream morphology. The laws of average stream fall and the minimum rate of energy expenditure governing the formation of natural streams were derived. Davy and Davies (1979) examined the thermodynamic basis of this concept as applied to fluvial morphology, and concluded that the use of entropy in analysis of stream behavior and sediment transport was of dubious validity. Combining with Horton's and Scheidegger's stream ordering systems, Sharp (1970) used entropy to determine optimum sampling methods for pollution in a river. Paulson and Garrison (1973) applied it to measure the areal concentration of water oriented industry in the Tennessee Valley region.

The first examples for the application of the entropy concept in hydrology date back to the early 1970s. Sonuga (1972, 1976) used the principle of maximum entropy in flood frequency analysis and rainfall–runoff relationships. Jowitt (1979) discussed the properties and problems associated with this concept in parameter estimation for the extreme type I distribution. The principle of maximum entropy refers to the maximization of Shannon's entropy function to estimate the unknown probability distribution of a certain hydrological process in the case of inadequate data. Panu and Unny (1979) used entropy in feature extraction and developed the methods of pattern recognition in hydrology. Amorocho and Espildora (1973) applied entropy to derive an objective criterion based on marginal entropy, conditional entropy and transinformation to assess the uncertainty of the Stanford watershed model in simulating streamflows from a basin in California for which historical records were available. Their results showed the value and limitations of this concept in assessing model performance.

As may be observed from the above, the concept of entropy has a very short background in hydrology and water resources, and hydrological studies based on entropy have been relatively few. However, their results were promising and therefore called for future research. Recently, the concept has found versatile uses in water resources research.

Definitions

Shannon (1948a) and later Jaynes (1957a) defined entropy as expectation of information or, conversely, as a measure of uncertainty. As described above, if S is a system of discrete events E_1, E_2, ..., E_n, and $p(E_k) = p_k$ is the probability of kth event recurring, then the entropy of the system is

$$H(S) = -\sum_{k=1}^{n} p_k \ln p_k \qquad (E8)$$

with

$$\sum_{k=1}^{n} p_k = 1$$

This discrete representation is extended to the continuous case. If $x \in (-\infty, \infty)$ and $y \in (-\infty, \infty)$ are two random variables, then the marginal entropy $H(x)$ and joint entropy $H(x, y)$ can be defined as

$$H(x) = \int_{-\infty}^{+\infty} f(x) \ln f(x) \, dx \qquad (E9)$$

with

$$\int_{-\infty}^{+\infty} f(x) \, dx = 1$$

$$H(x, y) = \int_{-\infty}^{+\infty} \int_{-\infty}^{+\infty} f(x, y) \ln f(x, y) \, dx \, dy \qquad (E10)$$

where $f(x)$ is the probability density function (pdf) of random variable x, and $f(x, y)$ is the joint pdf of x and y. Here f of x is not the same as f of y. If y is conditioned on x then the conditional entropy is defined as

$$H(y/x) = -\int_{-\infty}^{+\infty} \int_{-\infty}^{+\infty} f(x, y) \ln f(x/y) \, dx \, dy \qquad (E11)$$

Equation (E11) can be related to the joint entropy in equation (E10) and the marginal entropy in equation (E9) as

$$H(y|x) = H(x, y) - H(x) \qquad (E12)$$

and

$$H(y|x) \leq H(y) \qquad (E13)$$

If x and y are independent, then

$$H(x, y) = H(x) + H(y) \qquad (E14)$$

Therefore, equation (E10) will be bounded by equation (E14):

$$H(x, y) < H(x) + H(y) \qquad (E15)$$

In other words, the joint entropy of dependent x and y will be less than the joint entropy of independent x and y. The difference between these entropies defines transinformation $T(x, y)$ as

$$T(x, y) = H(x) + H(y) - H(x, y) \qquad (E16)$$

or

$$T(x, y) = \int_{-\infty}^{+\infty} \int_{-\infty}^{+\infty} f(x, y) \ln \{f(x, y)/[f(x)f(y)]\} dx \, dy \qquad (E17)$$

Transinformation represents the amount of information common to both stochastically dependent x and y. In other words, stochastic dependence reduces the uncertainty by this amount. For the independent x and y, $T(x, y) = 0$. Equation (E16) can also be expressed, using equation (E12), as

$$T(x, y) = H(x) - H(y|x) \qquad (E18)$$

Cast differently,

$$H(y) = H(y|x) + T(x, y) \qquad (E19)$$

Equation (E19) expresses the marginal entropy as the sum of uncertainty in y reduced by knowledge of x and the uncertainty still remaining (Harmancioglu, 1980, 1981; Singh and Rajagopal, 1987).

Principle of maximum entropy

Another important characteristic of entropy is that it allows assignment of probabilities (derivation of pdf) to a given system of events based on prior knowledge. Jaynes (1957b) formalized this characteristic into what is termed the principle of maximum entropy (POME). According to POME, the derived pdf is minimally prejudiced and makes maximum use of the given information (Singh and Rajagopal, 1987).

In search of an appropriate probability distribution for a given random variable, one has to maximize entropy. In practice, however, it is common that some information is available on the random variable. The selected probability distribution must be consistent with the given information. From all such distributions, one should choose the distribution that has the highest entropy. To that end, Jaynes (1957a,b) developed what is called the principle of maximum entropy (POME). According to POME, the minimally prejudiced assignment of probabilities is that which maximizes entropy subject to the given entropy, i.e., POME takes into account all of the given information and at the same time avoids consideration of any information that is not given.

If there is no information available on the random variable, then all the outcomes are equally likely, that is, $p_i = 1/n$, $i = 1, 2, ..., n$. Using Jensen's inequality or the method of Lagrange multipliers, it can be shown that Shannon's measure of entropy is maximum in this case and may indeed serve as an upper bound of entropy for all cases involving some information. This is consistent with Laplace's principle of insufficient reason, which states that all outcomes should be considered equally likely, unless there is information to the contrary. In the event that some information is given, the outcomes will not be equally likely. In that case POME accomplishes as uniform or as broad a distribution as possible subject to the given information. The maximum entropy in the presence of some information will be less than the maximum entropy in the absence of that information. The difference between these two maximum entropies may be regarded as a measure of the bias due to the given information. Maximizing entropy means minimizing this bias. That is why POME is said to yield the minimally biased or prejudiced assignment of probabilities. POME may also be called the principle of minimum bias or minimum prejudice. The probability distribution, obtained by application of POME, is variously called the maximum entropy probability distribution (MEPD), the most uncertain probability distribution (MUPD), the most uniform probability disribution (MUPD), the most likely probability disribution (MLPD), the least biased probability distribution (LBPD) or the most random probability distribution (MRPD; Singh, 1992).

Information on a random variable may be available in many ways such as moments (expectations), bounds, point values, incomplete means and probabilities. In entropy literature this information is expressed as constraints. The basis of the information may be intuition, experience, experimental observations, etc. When the information is available in the form of a prior probability distribution A expressed as

$$A = (a_1, a_2, a_3, \ldots, a_n), \sum_{i=1}^{n} a_i = 1 \qquad (E20)$$

we define another measure of entropy, called Bayesian entropy, as

$$B(P, A) = B(p_1, p_2, p_3, \ldots, p_n) = -\sum_{i=1}^{n} p_i \ln \left[\frac{p_i}{a_i(a_i)_{min}} \right] \quad (E21)$$

$$a_i = 0, \; i = 1, 2, \ldots, n$$

where

$$(a_i)_{min} = \min (a_1, a_2, a_3, \ldots, a_n) \qquad (E22)$$

Equation (E21) can be written as

$$B(P, A) = -\sum_{i=1}^{n} p_i \ln \left[\frac{p_i}{a_i} \right] - \ln(a_i)_{min} \qquad (E23)$$

The summation term in equation (E23) vanishes if $p_i = a_i$ for all i. Thus the Bayesian entropy is maximum when $p_i = a_i$ for all i and is minimum when the outcome with minimum a priori probability is certain to occur, i.e. $(a_i)_{min} = 1$. The Bayesian entropy specializes into the Shannon entropy if the prior distribution is uniform (Singh, 1992).

Concentration theorem

The concentration theorem was formulated by Jaynel (1979). It shows the spread of lower entropies around the maximum entropy. For the marginal entropy $H(x)$ of a random variable x with a probability density function $f(x)$, the entropy is in the range given by some class $C: H_{max}, - \Delta H \leq H(X) \leq H_{max}$, where H_{max} is given by POME. For m constraints, n prohibited for N observed different realizations, the concentration of these probabilities near the upper bound is given by the concentration theorem: Asymptotically, $2NAH$ is distributed over the class C as K^2 (chi-square) with $n - m - 1$ degrees of freedom (DOF), independently of the nature of the constraints. Denoting the critical K^2 for k DOF at 100% significance level as $K_c^2(P)$, ΔH is given in terms of the upper tail areas $1 - F$ as $K_c^2(1 - F) = 2N\Delta H$, where F is the cummulative distribution function.

Types of entropy

The preceding discussion shows that entropy indeed measures uncertainty of a probability distribution. Shannon's entropy given in equation (E3) is one measure. Other measures of uncertainty have also been proposed, including Renyi's entropy (Renyi, 1961), generalized Renyi's entropy (Aczel and Daroczy, 1975; Kapur, 1968), generalized gamma-entropy (Behara and Chawla, 1974), and others due to Havrada and Chavrat (1967), and Kapur (1967, 1986). Khinchin (1957) showed that among all the measures of entropy, the Shannon entropy possesses the most desirable attributes and is the most useful and most natural. It is, therefore, used most commonly in a wide range of areas. Other entropic measures are useful but to a limited extent (Singh and Fiorentino, 1992).

Entropy of more than two variables

In the multivariate case, the total entropy of M stochastically independent variables X_m ($m = 1, \ldots, M$) is (Harmancioglu, 1981; Harmancioglu and Alpaslan, 1992)

$$H(X_1, X_2, X_3, \ldots, X_M) = \sum_{m=1}^{M} H(X_m) \qquad (E24)$$

where $H(X_m)$ represents the marginal entropy of each variable X_m in the form of

$$H(X_m) = K \sum_{n=1}^{N} p(x_n) \log [1/p(x_n)] \qquad (E25)$$

with $K = 1$ if $H(X_m)$ is expressed in napiers for logarithms to the base e. Equation (E25) defines the entropy of a discrete random variable X_m

with N elementary events of probability $P_n = p(x_n)$ ($n = 1, \ldots, N$) (Shannon and Weaver, 1949). For continuous density functions, $p(x_n)$ is approximated as $[f(x_n).\Delta x]$ for small Δx, where $f(x_n)$ is the relative class frequency and Δx is the length of class intervals (Amorocho and Espildora, 1973). Then the marginal entropy for an assumed density function $f(x_n)$ is

$$H(X_m; \Delta x) = \int_{-\infty}^{+\infty} f(x) \log [1/f(x)]dx + \log[1/\Delta x] \qquad (E26)$$

In the above, the selection of Δx becomes a crucial decision as it affects the values of entropy (Harmancioglu and Alpaslan, 1992; Harmancioglu et al., 1986).

If significant stochastic dependence occurs between M variables, the total entropy has to be expressed in terms of conditional entropies $H(X_m/X_1, \ldots, X_m)$ added to the marginal entropy of one of the variables (Harmancioglu, 1981; Topsoe, 1974):

$$H(X_1, X_2, \ldots, X_M) = H(X_1) + \sum_{m=2}^{M} H(X_m|X_1, \ldots, X_{m-1}) \quad (E27)$$

Since entropy is a function of the probability distribution of a process, the multivariate joint and conditional probability distribution functions of M variables need to be determined to compute the above entropies (Harmancioglu, 1981):

$$H(X_1, X_2, \ldots, X_m) =$$
$$\int_{-\infty}^{+\infty} \int_{-\infty}^{+\infty} f(x_1, \ldots, x_m) \cdot \log f(x_1, \ldots, x_m) \cdot dx_1 \; dx_2 \ldots dx_m \quad (E28)$$

$$H(X_m|X_1, \ldots, X_{m-1}) =$$
$$\int_{-\infty}^{+\infty} \int_{-\infty}^{+\infty} f(x_1, \ldots, x_m) \cdot \log f(x_m|x_1, \ldots, x_{m-1}) \cdot dx_1 \; dx_2 \ldots dx_m \quad (E29)$$

The common information between M variables, or the so-called transinformation, $T(X_1, \ldots, X_m)$, can be computed as the difference between the total entropy of equation (E24) and the joint entropy of equation (E28). It may also be expressed as the difference between the marginal entropy $H(X_m)$ and the conditional entropy of equation (E29). It follows from the above that the stochastic dependence between multi-variables causes their marginal entropies and the joint entropy to be decreased. This feature of the entropy concept has been used in the spatial design of hydrometric data collection networks to select appropriate numbers and locations in order to avoid redundant information (Harmancioglu and Alpaslan, 1992).

An important step in the computation of any kind of entropy is to determine the type of probability distribution function which best fits the analyzed processes. If a multivariate normal distribution is assumed, the joint entropy of X (the vector of M variables) is obtained as (Harmancioglu, 1981; Harmancioglu and Alpaslan, 1992)

$$H(X) = (M/2) \ln 2 + \frac{1}{2} \ln|C| + M/2 - M \ln \Delta x \qquad (E30)$$

where M is the number of variables and $|C|$ is the determinant of the covariance matrix C. Equation (E30) gives a single value for the entropy of M variables. If logarithms of observed values are evaluated by the above formula, the same equation can be used for lognormally distributed variables. The calculation of conditional entropies in the multivariate case can also be realized by equation (E30) as the difference between two joint entropies. For example, the conditional entropy of variable X with respect to two other variables Y and Z can be determined as:

$$H(X|Y, Z) = H(X, Y, Z) - H(Y, Z) \qquad (E31)$$

Uses of entropy

Several applications of entropy have been reported in hydrological, environmental and water resources literature. The entropy concept can be used in a quite general way for the following purposes: (1) derivation of hypotheses, (2) interpretation of theories, (3) system dynamics and (4) quantification of information or loss of information (Singh and Fiorentino, 1992).

Derivation of hypotheses
The POME-based method gives rise to the best estimate of a probability distribution taking account of all the available information. If the available information is changed then the estimate will also

be changed. The methodology is very general and applicable to analysis of different systems, and the range of possible application deriving hypotheses (or estimating the form of system models) is very wide indeed. For a wide class of problems marked by their complexity, the methodology provides a set of rules for model construction compatible with the given information, and an internal consistency not easily achievable otherwise.

Interpretation of theories
There exists a multitude of models, hypotheses or theories which are constructed without use of the entropy concept. If these models can also be derived using POME, then the model parameters can be expressed in terms of the constraints or the Lagrange multipliers. In this manner the model parameters can be physically interpreted, and so also the theory underlying the model. Consider, for example, the case of the gamma distribution, which is frequently used in hydrology. An application of POME to this distribution shows that the expected value of the random variable and the expected value of the log-transformed random variable constitute sufficient statistics for derivation of the gamma distribution. Thus the two parameters of the gamma distribution can be interpreted in terms of these physically meaningful statistics.

System dynamics
Entropy arises from the second law of thermodynamics. In most studies in physics and other fields, a system is assumed to move from one equilibrium position to another, and only the initial and final states are considered. In many physical systems, the relaxation time – the time taken by the system to return to a new equilibrium position after the disturbance – is quite short, and a comparative static model may not be unduly unreasonable. For other systems, however, the relaxation time is long and transient effects are important. The entropy concept can be applied to incorporate system dynamics in model building (Tomlin, 1969, 1970).

Information or loss of information
Entropy is the expected information, and this concept is applied to design data acquisition and gathering systems. Theil (1967) has used the entropy concept to investigate the loss of information when certain kinds of category aggregation arise. In hydrology it can be used to compute the loss of information due to grouping of supposedly similar watersheds.

Application in water resources

General

It appears that the use of the entropy concept in water resources has been along two lines. The first area of application comprises basically river morphology and river hydraulics, whereas the second one covers diverse problems in statistical hydrology. One may consider specific issues such as sediment transport or pollution transport under a separate group. However, the way entropy is defined in these problems eventually foresees their inclusion in either the first or second area of applications (Harmancioglu et al., 1992a).

The investigations in the first group rely on the thermodynamic basis of entropy and are basically non-probabilistic in nature. The pioneering works of Leopold and Langbein (1962) and Yang (1971a,b), followed by Scheidegger (1967), may be considered within this group. Their studies in fluvial morphology foresaw the analogies between (1) heat energy in a thermodynamic system and potential energy in a stream system and (2) temperature in a thermodynamic system and elevation in a stream system. Later, however, Davy and Davies (1979) analyzed the validity of these analogies and concluded that the 'application to stream systems of certain laws governing the behavior of entropy in a thermodynamic system has been shown to be strictly illegitimate, since the character of the stream system and the processes occurring within it lie outside the range of application of the principles governing entropy behavior. Similarly, any analogous principles can find no support from their thermodynamic counterparts in the stream system' (Davy and Davies, 1979). Obviously, the application of the entropy concept in fluvial geomorphology has become controversial and requires further research, considering the results obtained so far.

The second area of entropy application in water resources is purely probabilistic in nature and comprises the analysis of a wide range of problems in statistical hydrology. A common feature of all these

applications is that the entropy concept is used as a measure of uncertainty associated with random processes, probability distributions, models and model parameters. Entropy, as a quantitative measure of uncertainty, does prove to have its merits and usefulness in investigations on uncertainty problems, since 'uncertainty' is one of the oldest and the most intriguing problems in hydrology and water resources engineering. This is basically the reason why informational entropy is more widely used in water resources than thermodynamic entropy.

The studies along this second line are based on Shannon's definition of informational entropy. Jaynes also used entropy in a probabilistic context, but rather in a reversed form of Shannon's entropy. As discussed above, he basically considered the problem of inferring a probability distribution function which would have maximum entropy and developed the principle of maximum entropy (POME). A large number of applications of the entropy concept in water resources rely on POME, especially when partial or incomplete information exists about a problem. By means of POME, one doesn't select a probability distribution by merely considering the partial information at hand. Rather, the probability distribution which maximizes Shannon's entropy is chosen as it is subject to partial information. This means that the 'probability distribution which results from this constrained maximization process will then be one which introduces minimum bias into the probability estimation process' (Templeman, 1989). Thus Jaynes's POME has provided a useful means of avoiding bias in problem solving with incomplete information. This significant feature of the principle was readily accepted in water resources engineering to solve not only problems of distribution fitting, but also other more practical cases where the basic obstacle is a lack of sufficient information.

In the following sections, uses of both Shannon's and Jaynes's entropy principles in water resources are summarized with respect to different fields of application.

Assessment of uncertainty

Within a general context, the entropy concept is used in water resources to assess uncertainties associated with

- hydrological variables;
- hydrological systems and their models;
- parameters of probability distribution functions.

In this case, either Shannon's or Jaynes's entropy principle, whichever is applicable, is used to define in quantitative units the uncertainty (or indirectly the information) contained in the above three cases. As such, entropy constitutes a unique means of expressing uncertainty in objective terms.

The basic advantage of the entropy concept here is that it measures information in specific units, i.e. decibels, bits and napiers, so that one can assess information objectively. This probably is the most significant feature of entropy since one of the major problems in hydrology and water resources has been the lack of a precise and quantitative definition of information (Harmancioglu et al., 1992a).

Flood frequency analysis: parameter estimation and derivation of frequency distributions

POME has also been employed for estimating parameters of frequency distributions used for analysis of floods. Essentially, POME enables derivation of a probability distribution subject to a given set of constraints. The only parameters that the distribution has are those that are expressed in terms of constraints. In a way, the POME-based distributions are parameter free. Conversely, POME can be applied to estimate the parameters of a given probability distribution. A number of studies have been carried out on parameter estimation by the POME-based method and its comparison with other parameter estimation methods (Singh and Fiorentino, 1992). Jowitt (1979) discussed the properties and problems associated with application of POME to estimation of parameters of the extreme value type I distribution. Singh and Jain (1985), Jain and Singh (1986), and Arora and Singh (1987a,b, 1989) compared this method of parameter estimation with the methods of moments (MOM), probability-weighted moments (PWM), mixed moments (MOMM), maximum likelihood estimate (MLE), incomplete means (IM) and least squares (MOLS), and found it comparable to the method of maximum likelihood estimate and better than five other methods. Singh and Singh (1985a) applied the POME-based method to estimate the parameters of the gamma distribution, and found it comparable to the methods of moments, cummulants, maximum likelihood estimate and least squares. Singh and Singh (1985b, 1985c, 1988), and Arora and Singh (1989) compared the POME-based method with the method of moments [direct (MOMD), indirect (MOMI) and mixed (MOMM)] and the maximum likelihood estimate for the Pearson type III and log-Pearson type III distributions. The POME-based method was comparable to some methods or better than others. Singh (1987) estimated, using POME, the parameters of the extreme value type III distribution and found the POME-based method better than the methods of moments and maximum likelihood estimate. In other comparative studies, the POME-based method was found to be superior to or as good as the best of the other methods. Specifically, this was true in the case of two-component extreme value distribution (Fiorentino et al., 1987a,b), three-parameter lognormal distribution (Singh et al. 1989, 1990a; Singh and Singh, 1987) and the Weibull distribution (Singh et al., 1990b).

Singh and Rajagopal (1987) used POME to develop a new parameter estimation method, called the parameter-space expansion method, which is applicable to any probability distribution with any number of parameters that is expressible in direct form. They applied this method to estimate parameters of the Weibull and extreme value type III distributions. This method was either as good as or superior to the methods of moments and maximum likelihood estimate.

Derivation of functional relationships

The principle of maximum entropy can be used to develop functional relationships between two or more variables, such as rainfall and runoff, sediment yield and runoff, or water quality and runoff. In such cases, entropy provides a means of deriving stochastic relations, taking into account both stochastic and deterministic characteristics (Singh and Rajagopal, 1987). By maximizing the conditional entropy subject to certain constraints, the probability distribution of a dependent variable (like sediment yield) can be obtained by conditioning it on the probability distribution of the independent variable (like runoff). Such a distribution results in a minimally prejudiced assignment of probabilities on the basis of given information. Singh and Krstanovic (1987) used this procedure to model sediment yield from upland watersheds and Sonuga (1976) to model the rainfall–runoff relationship. The results of these studies showed that such an approach can be used in practice very easily, especially in cases where the data available are minimum. However, more research is called for to delineate its advantages over other methods and models (Singh and Krstanovic, 1987).

Measuring the information content of random processes

Water resources engineers and scientists in general have always searched for objective criteria to define how much information their observations or experiments reveal about the events they analyze in the decision making process. Since the validity of decisions depends on how sufficiently the observations reflect the real event that takes place in nature, the basis for water resources planning studies lies in the data that are observed and the amount of information they convey. Therefore, an objective criterion is required to measure their information content. Among its few alternatives, the entropy concept seems to be one of the most promising criteria so far developed, as revealed through its characteristics clearly defined in the principles of the information theory. Uslu and Tanriover (1979) have analyzed the information content and further the transfer of information between the observations of two streamgauging stations, both by the entropy concept and by Fisher's (Fisher, 1966) reciprocal of the variance criterion, as measures of information content. Their conclusion was that the latter criterion is appropriate to measure a priori information and that entropy actually measures a posteriori information. In their study, Uslu and Tanriover (1979) calculated marginal and conditional entropies in the case of continuous variables for which p_k of equation (E8) is approximated by $f(x_k)\Delta x$, $f(x_k)$ being the relative class frequency and Δx the length of class intervals. Here, the choice of Δx is significant since marginal and conditional entropies attain values relative to the level of uncertainty defined by Δx. Amorocho and Espildora (1973) earlier discussed the same problem associated with entropy calculations, where uncertainties measured by $H(x)$ and $H(y/x)$ decrease as Δx increases. Harmancioglu (1980, 1981) extended the use of the entropy concept to the multivariate case and developed mathematical definitions for the entropy of multi-variables, both dependent and independent. Particularly for a single random variable, she analyzed the effect of

serial dependence upon the information content of an observed variable and discussed that such an approach could be used to determine optimum sampling intervals with respect to time. In the multivariate case, the effects of both serial and cross-correlations were considered. The mathematical expressions thereby developed were proposed by Harmancioglu (1980) as convenient methods of determining optimum sampling intervals with respect to both time and space.

There are also studies which use the entropy concept to assess the stochastic structure of hydrological data. For example, Harmancioglu (1980, 1981) showed that the serial dependence structure of a hydrological series can be evaluated by entropy to determine the required order of dependence models. Padmanabhan and Rao (1988) used the maximum entropy spectral analysis (MESA) of hydrological data for determination of periodicities. They claim that MESA is an alternative method which does not show the disadvantages of conventional methods.

Evaluation of information transfer between hydrological processes

Harmancioglu and Yevejevich (1985, 1987) defined the concepts of transferred and transferable information, the former being measured by the classical correlation coefficient r and the latter by entropy measures. The entropy measure used in this context is transinformation T between two or more variables, computed by equation (E16). Here, the marginal entropies $H(x)$ and $H(y)$ and the joint entropy $H(x, y)$ are estimated from samples by replacing the probabilities in equations (E9) and (E10) by the corresponding frequencies (Harmancioglu, 1981). The transinformation T_o computed as such represents the upper limit of transferable information between the processes. The measure for this limit is also expressed by the informational correlation coefficient R_o which is basically a function of transinformation:

$$R_o = [1 - \exp(-2T_o)]^{0.5} \qquad (E32)$$

R_o is a dimensionless measure of stochastic dependence with $0 < R_o < 1.0$ (Linfoot, 1957; Harmancioglu et al., 1986).

The computation of the transferable information is free of any assumptions regarding the probability distributions of variables and the form of transfer function (basically of a regression-type function). On the other hand, the classical correlation coefficient r measures the amount of information transferred between variables under specified assumptions (like linearity and normality). Thus, when it is compared to the upper limit of transferable information R_o, one can make inferences as to the goodness of information transfer by regression and that of the underlying assumptions. A similar inference can be made by defining the amount (in per cent) of transferred information by the ratio T/T_o, where T is computed for a particular type of distribution and transfer function. In this case, T can be expressed as a function of the correlation coefficient r for an assumed type of regression equation (Harmancioglu et al., 1986). Harmancioglu and Yevjevich (1985, 1987) have applied these concepts to analyze the transfer of information among river points.

Transinformation T_o, as the upper limit of common information between two processes, also represents the level of dependence (or association) between their variables. This feature of transinformation can be used effectively to select among water quality variables to be observed (Harmancioglu and Yevjevich, 1986; Harmancioglu et al., 1986). Similar investigations were performed on scarce data from a polluted basin in Turkey (Harmancioglu et al., 1987). Harmancioglu and Baran (1989) used the same approach to evaluate recharge systems of river basins with different flow regimes.

The results of all these aforementioned studies show that the use of entropy coefficients as the measure of transferable information and correlation coefficients as the measure of transferred information definitely deserves further analysis and development (Yevjevich, 1987). The basic difficulty in the computation of T_o and R_o values has been the selection of the number of class intervals (NCI) in frequency analyses as both entropy measures increase with an increase of NCI. However, since the variation of entropy values within a range of selected NCI is much smaller than the variation of r values corresponding to the transferred information, the problem of entropy information coefficients of transferable information depending on the size of class intervals does not preclude its use even in the case of continuous random variables (Harmancioglu and Yevjevich, 1987).

Assessment of recharge systems for a river basin

Concepts of transferable and transferred information can be used to evaluate different inputs to a basin which produce the output runoff. Harmancioglu and Baran (1989) used such an approach to attain the maximum information about runoff at a point along a river. Their study involved the analysis of information transfer between runoff–runoff and precipitation–runoff processes in basins with different recharge systems.

These investigations were demonstrated on four case studies in Turkey, where each basin is dominated by different inputs (rainfall, snowfall, karstic spring contributions or mixed) to produce runoff. The purpose there was to delineate the typical features of each recharge system that need to be considered in the process of information transfer. Such an approach is also claimed to serve in recognizing the basic properties of the recharge systems and in evaluating input-output relations. The results of this study showed within-the-year variation of transferred and transferable information to reflect how each basin responds to seasonally varying inputs. Particularly in karstic basins, neither runoff–runoff nor precipitation–runoff regressions were found to be sufficient in transferring information especially during the non-rainy seasons, such as sub-basins with different degrees of karstification respond diversely to input precipitation. The problem here is a multivariate one where both cross and serial correlations are taken into account.

Cetiner (1988) performed a similar study for karstic basins to investigate their specific features in the process of information transfer.

Evaluation of data acquisition systems

Hydrological data are collected in both time and space. Optimum sampling intervals are required to evaluate the efficiency of the data acquisition systems. Harmancioglu (1980, 1981) emphasized the potential characteristic of the entropy concept in delineation of optimum sampling intervals in data collection systems, both with respect to time and space. Using streamflow data, she developed the mathematical formulation of the entropy concept when significant serial correlations occur for a single hydrological variable and thereby analyzed the statistical structure of the process itself by using the entropy concept. She employed this approach in the determination of optimum sampling frequencies with respect to time. Next, she developed the mathematical formulations for the entropy concept in the case of multi-variables, considering the complex situations of uncertainty that exist when both the serial and the cross-correlations between variables are taken into account. She used these considerations to delineate optimum sampling intervals with respect to space. Harmancioglu (1980, 1981) applied the above procedure to daily streamflow observations of flow at the Orenkoy and Yapilar gauging stations in the Esencay river basin in Turkey and examined the effect of serial dependence and cross-correlations upon marginal and conditional entropies of streamflows at the two stations. She found that for winter months, the amount of remaining uncertainty was high. For such months, the amounts of repeated information, or transinformations, between successive observations were very low. Hence it was necessary to observe the runoff on a daily basis, and observations made at intervals greater than a day will cause loss of information. The amount of repeated information between the two stations increases in the summer months to values as high as 38%, which means that there still remains a significant percentage (62%) of the uncertainty of the Yapilar flows although the flows at Orenkoy are known. The transinformation reduces to the order of 5% in the winter months, so that 95% of the monthly marginal entropy still remains at Yapilar. Thus observations at both stations are required.

In another study, Harmancioglu (1984) applied the entropy concept to determine the optimum sampling intervals in time for NH_4^+ concentrations (in ppm) in water for data observed at 40 min intervals at Choisy-le-Roi, Paris. Later, Harmancioglu and Yevjevich extended the use of the method for purposes of spatial design in case of streamflow gauging stations by defining transferred and transferable amounts of information (Harmancioglu and Yevjevich, 1985, 1987; Yevjevich, 1987; Harmancioglu and Baran, 1989). Along similar lines, they used the same concept for water quality data, basically for purposes of selecting variables to be sampled (Harmancioglu et al., 1986; Harmancioglu and Yevjevich, 1986).

Caselton and Husain (1980) and Husain (1989) used the entropy concept to estimate regional hydrological uncertainty and information at both gauged and ungauged grids in a basin, using rainfall data.

Their results show that the entropy method presents a convenient means of evaluating an optimum spatial design with respect to both the numbers and the locations of gauging stations.

Harmancioglu et al. (1992b) and Harmancioglu and Alpaslan (1992) have used the entropy concept for the design of water quality monitoring networks. They have realized temporal, spatial and combined temporal–spatial design criteria based on entropy. The results are highly promising as the benefits of a monitoring network are defined quantitatively in terms of information gain measured by entropy.

Krstanovic and Singh (1988a,b) applied the entropy approach to space and time evaluation of rainfall networks in Louisiana. Space and time dependencies amongst raingauges were examined by autocovariance and cross-covariance matrices. Using POME, multivariate normal distributions were derived. The joint and conditional entropies and transinformation were computed. The reduction or gain of information at a particular gauge was used as a parameter to decide whether to retain or eliminate that raingauge. Finally, the lines of equal information (or isoinformation contours) were constructed.

The entropy measures of information were further applied by Krstanovic and Singh (1993a,b) to rainfall network design and by Goulter and Kusmulyono (1993) to prediction of water quality at discontinued water quality monitoring stations in Australia. Similar considerations were used for design of data collection systems (Singh and Krstanovic, 1986; Harmancioglu, 1984). In these studies the entropy concept has been shown to hold significant potential as an objective criterion which can be used in both spatial and temporal design of networks. Although the method has not been yet utilized for actual design in practice, the results of current research indicate that it is a promising technique and that it may be developed into a powerful tool for design purposes.

Assessment of model performance

Another potential characteristic of the entropy concept with respect to hydrology is that it represents an objective criterion fairly well and has proved to give successful results in the comparison between various mathematical models developed for a certain hydrological process, in the selection of the most appropriate model, and in the evaluation of the degree of completeness and efficiency of a certain model to represent a natural phenomenon, which actually are some of the major problems of synthetic hydrology.

The specific features of the transinformation concept have led to its use in the evaluation of model performance and selection of the best model to represent a hydrological process or system. Accordingly, the model that produces the highest transinformation between observed and simulated data is considered to produce the best fit. Amorocho and Espildora (1973) initiated such an approach and showed the limitations and merits of using the entropy criterion in model evaluation. They have also discussed the selection of class interval size Δx with respect to the accuracy of entropy calculations. Further, they pointed out that 'the concepts involved in the measure of uncertainty require that the probability frequency distributions of the outcomes of a process be bounded . . . In practical applications the unbounded frequency distributions should be truncated, thus they will define a finite region of uncertainty' (Amorocho and Espildora, 1973).

Later, Chapman (1986) extended the original use of the method by Amorocho and Espildora to evaluate the reduction of uncertainty in hydrological data due to application of a model. He proposed a complementary approach to overcome the limitations of the technique. Particularly for the selection of Δx, he claimed that one should use proportional rather than fixed class intervals. He gave general equations for the proportional class interval and solved them for assumed lognormal and gamma distributions and extended to data series with zero values. He proposed a more general criterion of model performance to be the ratio of the transinformation to the marginal entropy of observed data.

Along similar lines, Baran and Harmancioglu (1990) compared the goodness of three models in representing the recession period flows at a streamgauging station where a significant portion of runoff is made up of karstic spring effluents. The results of this study have led to further investigations on the problem of class interval size, which are currently continued.

Optimization and decision support systems

Recently, Templeman (1989) considered the extension of POME to problems of optimization and decision support systems where the major problem is incomplete information. He proposed that a water supply network can be analyzed by the entropy concept to calculate 'the least biased assignment of pipe flow rates' for a looped network where information required to determine head losses is not available.

Templeman (1989) further suggested that POME can be satisfactorily used in optimization problems. In this case, 'maximizing the entropy of the Pareto multipliers in a multi-criteria optimization problem generates the solution of the minimax optimization problem'. Maximum entropy therefore forms the link between multi-criteria and minimax optimization. Within this respect, Templeman calls for further research as the concept of entropy is highly promising for the solution of optimization problems in civil engineering.

Assessment of regional information on floods

Although the entropy principle has been used in the form of POME to infer distributions of flood events, it may also be employed as Shannon's informational entropy to assess the regional information about floods within a basin. To this end, the total uncertainty about flood events observed by a number of streamgauging stations in a basin can be described as the joint entropy of multi-variables, with each variable representing observations at particular space points. Next, the contribution of each station to the reduction of this total regional uncertainty may be assessed by considering the joint entropy of different combinations of stations. Then the combination which produces maximum reduction in the total uncertainty of the basin about flood events can be accepted as the required system of stations to characterize the floods at a regional scale. If certain stations do not contribute significantly to uncertainty reduction, their observations may be considered redundant with respect to basin-wide information gathering. On the contrary, the existing stations may prove to be insufficient in producing the required information about flood events. Then, new space points may be considered as potential sites of observation to reduce the total regional uncertainty. Such an approach is similar to that used in the design of hydrometric data networks. However, specific features of floods, e.g. the dependence or independence structure of variables, distribution functions, etc. need to be considered. The applicability of the above approach is demonstrated by Harmancioglu (1994a) in a Turkish river basin where coping with floods is a significant problem.

Streamflow forecasting

Currently, three types of entropy forecast models are available. The first type of model is the Bayesian entropy model (BEM) combining POME and Bayesian processors of forecast (BPF). This model is general, and is not specifically developed for hydrological forecasting. The second type of entropy model is based on the entropy minimax approach and has been applied to long-term annual forecasting of droughts using seven rainfall stations in northern California (Christensen, 1981; Eilbert and Christensen, 1983). The entropy minimax approach partitions the test area into disjointed subareas or 'patterns' to optimize their information content. The precipitation index, computed as weighted yearly average precipitation of all available stations in a given (128 year) period, was used to define their information content. This approach gave promising results and exhibited an ability to predict long-term precipitation within three probability limits (78%, 61%, 41%). Predictions were made for 49 years of data, and were validated at 95% confidence level.

Spectral analysis is used to identify significant periodicities in time series. MESA, maximum entropy spectral analysis, introduced by Burg (1975), is a powerful method, with three degree resolutions. It uses the autoregressive process with maximum order (all-pole process) for extrapolation of the autocorrelation function (acf) using POME. This aspect of the method can be used advantageously to develop time series forecasting models. The maximum entropy criterion, used in MESA, imposes the fewest constraints on the unknown time series by maximizing its randomness, thus producing a minimally biased solution. Time series analysis, spectral analysis and entropy are interrelated. The entropy concept is closely related to some time series models such as the AR models (Van den Boss, 1971) and spectral analysis in MESA (Burg, 1975). The benefits of MESA

in hydrology have been discussed by Krstanovic and Singh (1987), and Padmanabhan and Rao (1988).

The third type of entropy model is based on spectral analysis. Krstanovic and Singh (1987) employed MESA for Spring Creek in Louisiana. They computed maximum entropy spectra for various flood characteristics, and compared them with the observed spectra. In a similar study, Padmanabhan and Rao (1988) applied MESA to some rainfall and streamflow time series data from India.

In a series of two papers, Krstanovic and Singh (1991a,b) developed a univariate streamflow forecasting model, using POME via MESA. The model was developed from MESA by employing the extended autocorrelation function (acf) subject to certain constraints. The model parameters are Lagrange multipliers, which were expressed in normalized form convenient to describe the model behavior. The model uses acf to perform forecasting; the stronger acf weights the greater portion of known streamflow, and the weaker acf weights only the most recent values. The forecasting equations in the model were developed for three types of streamflow forecasting: forward, backward and intermittent forecasting. The forecasting equations were compared with respect to forecast accuracy. The second paper in the series verifies the entropy-based univariate model for long-term forecasting on five rivers from different regions of the world. The results of the model are compared with the corresponding results of ARIMA and state–space model. The Lagrange multipliers of the univariate model are found similar to autocorrelation coefficients of the ARIMA model. Forecasts by ARIMA and univariate models were comparable for periodic streamflow, but for forecasting of highly variable streamflows the univariate model was found to be superior (Krstanovic and Singh, 1991a,b).

Using Burg's maximum entropy spectral analysis (MESA), Krstanovic and Singh (1993a,b) developed a real-time flood forecasting model. Fundamental to MESA is the extension of autocovariance and cross-covariance matrices describing the correlations within and between rainfall and runoff series. These matrices are used to derive the model forecasting equations with and without feedback. The MESA-based model was found to be similar to a bivariate autoregressive model and under certain conditions the two models became equivalent. The model was verified on five watersheds from different regions of the world, wherein the sampling time and the forecast lead time varied from several minutes to a day. The model was superior to a state–space model for all events where it was difficult to obtain prior information about model parameters.

Application in environmental resources

Design of water quality monitoring networks

Application of the entropy concept to network design
The design of water quality monitoring networks is still a controversial issue for a number of reasons. First, there are difficulties in the selection of temporal and spatial sampling frequencies, the variables to be monitored and the sampling duration. Second, benefits of monitoring cannot be defined in quantitative terms for reliable benefit–cost analyses. In each case, there are no definite criteria established yet to solve these two problems. The entropy principle can be effectively used to develop such criteria on the basis of quantitatively expressed information expectations and information availability. This approach is justified in the sense that a monitoring network is basically an information system. In fact, investigations on application of the entropy principle in network design for water quality monitoring have revealed promising results, particularly in the selection of technical design features such as monitoring sites, time frequencies, variables to be sampled and sampling duration (Harmancioglu and Alpaslan, 1992; Alpaslan et al., 1992; Harmancioglu et al., 1992b,c, 1994).

Entropy is a measure of the degree of uncertainty of random hydrological processes. Since the reduction of uncertainty by means of making observations is equal to the amount of information gained, the entropy criterion indirectly measures the information content of a given series of data (Harmancioglu, 1981). According to the entropy concept as defined in communication (or information) theory, the term 'information content' refers to the capability of signals to create communication. The basic problem is the generation of correct communication by sending a sufficient number of signals, leading neither to loss nor to repetition of information (Shannon and Weaver, 1949).

Each sample collected actually represents a signal from the natural system which has to be deciphered so that the uncertainty about the real system is reduced. Application of engineering principles to this problem calls for a minimum number of signals to be received to obtain the maximum amount of information. Redundant information does not help reduce the uncertainty further, it only increases the costs of obtaining the data. These considerations represent the essence of the field of communications and hold equally true for hydrological data sampling, which is essentially communicating with the natural system. On the basis of this analogy, a methodology based on the entropy concept of information theory has been proposed for the design of hydrological data networks. The basic characteristic of entropy as used in this context is that it is able to represent quantitative measures of 'information'. As a data collection network is basically an information system, this characteristic is the essential feature required in a monitoring network (Alpaslan *et al.*, 1992; Harmancioglu *et al.*, 1992b).

The definitions of entropy given in information theory (Shannon and Weaver, 1949) to describe the uncertainty of a single variable can be extended to the case of multiple variables as described above. In this case the stochastic dependence between two processes causes their total entropy and the marginal entropy of each process to be decreased. The same is true for dependent multi-variables (Harmancioglu, 1981). This feature of the entropy concept can be used in the spatial design of monitoring stations to select appropriate numbers and locations so as to avoid redundant information. The investigation of spatial frequencies requires the assessment of reduction in the joint entropy of two or more variables [computed by equation (E28)] due to the presence of stochastic dependence between them. In this case, the reduction is equivalent to the redundant information [transinformation to be computed as the difference between equations (E28) and (E29)] in the series of the same water quality variable observed at different sites. The optimum combinations of required sampling sites are selected so as to minimize the total transinformation of M stations. On the other hand, the marginal entropy of a single process that is serially correlated is less than the uncertainty it would contain if it were independent. In this case, serial dependence acts to reduce marginal entropy and causes a gain in information (Harmancioglu, 1981). This feature of the entropy concept is suitable for use in the temporal design of sampling stations. The analysis of temporal frequencies by the entropy method is based on the assessment of reduction in the marginal entropy of a process due to the presence of serial dependence. Such a reduction, if any, is equivalent to the redundant information of successive measurements. To this end, one may consider a water quality variable X being measured at an interval Δt yielding x_i, $i = 0, 1, 2, \ldots, N$, where N is the sample size. The marginal entropy of the sample is computed with an assumed or known distribution of X. Also computed is the conditional entropy. The intervals between measurements can be analyzed to determine whether there is any repetition of information between successive values. The sample is then divided into subsamples x_{i-k}, $k = 0, 1, 2, \ldots, m$, where $k =$ lag, $m << N$, in order to get a better idea of serial dependence. The problem of determining the optimum sampling interval reduces to determining the uncertainty that remains in x_i when the values x_{i-k}, $k = 0, 1, 2, \ldots, m$ are known. This calls for computing the uncertainty $H(x_i|x_{i-k})$. The analysis can be applied to increasing time intervals (Harmancioglu and Alpaslan, 1992).

With respect to water quality in particular, the entropy principle can be used to evaluate five basic features of a monitoring network: temporal frequency, spatial orientation, combined temporal/spatial frequencies, variables sampled and sampling duration. The third feature represents an optimum solution with respect to both the time and space dimensions, considering that an increase in efforts in one dimension may lead to a decrease in those in the other dimension (Harmancioglu and Alpaslan, 1992). To determine the variables to be sampled, the method can be employed, not to select from a large list of variables but to reduce their number by investigating information transfer between the variables (Harmancioglu *et al.*, 1986, 1992a,b; Harmancioglu and Yevjevich, 1986). Assessment of sampling duration may be approached in a number of ways. If station discontinuance is the matter of concern, decisions may be made in an approach similar to that applied in spatial orientation. The problem is much simpler when a sampling site is evaluated for the redundancy of information it produces in the time domain. If no new information is obtained by continuous measurements, sampling may be stopped permanently or temporarily (Harmancioglu, 1994b).

Advantages of the entropy method in design of monitoring networks

Basic role of the entropy method

The primary role of the entropy concept in the design of water quality monitoring networks must be disclosed before discussing its advantages. The studies carried out so far show that the method works quite well for the assessment of an existing network. It appears as a potential technique when applied to cases where a decision must be made to remove existing observation sites, and/or reduce the frequency of observations, and/or terminate the sampling program. The method may also be used to select the numbers and locations of new sampling stations as well as to reduce the number of variables to be sampled (Harmancioglu and Alpaslan, 1992).

On the other hand, the entropy method cannot be employed to initiate a network; that is, it cannot be used for design purposes unless a priori collected data are available. This is true for any other statistical technique that is used to design and evaluate a monitoring network. In fact, the design process is an iterative procedure initiated by the selection of preliminary sampling sites and frequencies. This selection has to be made essentially by non-statistical approaches. After a certain amount of data is collected, initial decisions are evaluated and revised by statistical methods. It is through this iterative process of modifying decisions that the entropy principle works well. Its major advantage is that such iterations are realized by quantifying the network efficiency and cost-effectiveness parameters for each decision made.

Measure of information and usefulness of data

One of the valuable aspects of the entropy concept as used in network design is its ability to provide a precise definition of 'information' in tangible terms. This definition expresses information in specific units (i.e. napiers, decibels or bits) so that it constitutes a completely quantitative measure. At this point it is important to note the distinction between the two terms 'data' and 'information'. The term 'data' represents a series of numerical figures which constitute a means of communication with nature; what these data actually communicate to us is 'information'. This distinction means that availability of data is not a sufficient condition for availability of information unless those data have utility, and the term 'information' describes this utility or usefulness of data. Among the various definitions of information proposed to date, the entropy measure appears to be the only one that gives credence to the relevance or utility of data.

The value of data can also be expressed in quantitative terms since it is measured by the amount of information the data convey. This observation implies that monitoring benefits may eventually be assessed on the basis of quantitative measures rather than indirect descriptions of information. In comparison with the current methods, the entropy method develops a clearer and more meaningful picture of data utility versus cost (or information versus cost) trade-offs. This advantage occurs because both the information and the costs can be measured in terms of quantitative units. For example, if cost considerations require data to be collected less frequently, the entropy measure describes quantitatively how much information would be risked by increasing the sampling intervals (Harmancioglu, 1984; Harmancioglu and Alpaslan, 1992). By such an approach, it is possible to express how many bits of information would be lost against a certain decrease in costs (or in monetary measures). Similarly, it is possible to define unit costs of monitoring in terms of the number of dollars per bit of information.

Network efficiency and flexibility

Efficiency is related to objectives of monitoring in that the latter delineates 'information expected' from monitoring and the former describes 'information produced' by a network. The 'information produced' is a function of the technical features of a network related to the variables sampled, spatial and temporal sampling frequencies and the duration of sampling. It is plausible then to define efficiency as the 'informativeness' of a network. If the design of the network is such that this information is maximized, then the requirement of efficiency is satisfied. The entropy theory can be used to test whether the supplied information is optimal or not, thereby ensuring system efficiency (Harmancioglu and Alpaslan, 1992). A network, once designed and in operation, has to be evaluated for efficiency, particularly if the monitoring objectives have been changed or revised. The entropy method may again be used to assess the data collected to determine how much information is conveyed by the network under present conditions. If revisions and modifications are made, their contribution to an increase in information can be measured by the same method. Within this respect, the entropy theory also serves to maintain flexibility in the network since each decision regarding the technical features can be assessed on objective grounds.

Information-based design strategy

The entropy theory may be used to set up an information-based design strategy. As noted earlier, the approach for developing the design strategy for efficient and cost-effective networks encompasses two steps: (1) delineation of design considerations and (2) technical design of the network. The first step is to define the objectives of monitoring and information needs associated with each objective. The entropy method can be employed basically for the second step to be combined with cost considerations to realize both informativeness and cost effectiveness (Harmancioglu *et al.*, 1992c). This two-stage process can permit the design procedures to be developed to match the information expected from monitoring. Such an approach covers both the 'demand' (objectives of monitoring) and the 'reaction' (monitoring practices) parts of the problem in an integrated fashion. Both parts can then be defined in terms of 'information' as 'information needed' and 'information supplied'. The efficiency and effectiveness of the network can be realized by matching these two aspects. The 'demand' part of the design problem can be addressed by specifying the information expected of each objective of monitoring. The 'reaction' portion of the problem covers the more specific questions of any design procedure, such as the selection of variables to be sampled and the selection of temporal and spatial frequencies. Solution of the problems associated with this 'reaction' step requires (1) an extraction information from available data and (2) a transfer of information among water quality variables with respect to time and space. These two steps are shown to be effectively accomplished by entropy-based measures (Harmancioglu *et al.*, 1986; Harmancioglu and Singh, 1991).

Cost effectiveness

A major difficulty underlying both the design and evaluation of monitoring systems is the lack of an objective criterion to assess the cost effectiveness of the network. In this assessment, costs are relatively easy to estimate, but benefits are often described indirectly in terms of other parameters, using optimization techniques, Bayesian decision theory or regression methods (Schilperoort *et al.*, 1982; Tirsch and Male, 1984). Thus a realistic evaluation of benefit/cost considerations cannot be achieved since benefits are not directly quantified. Actually, benefits of monitoring can only be measured by means of the information conveyed by collected data; that is, they are a function of the value or worth of data. The concept of entropy can also be used to quantify the benefits of monitoring since it describes the utility of data. Here, benefits of monitoring are expressed as the information supplied which is quantified in tangible units by entropy measures. Cost effectiveness can be evaluated by comparing costs of monitoring versus information gained via monitoring. The issue is then an optimization problem to maximize the amount of information (benefits of monitoring) while minimizing the accruing costs. The technical features of design can then be evaluated with respect to cost effectiveness (Harmancioglu and Alpaslan, 1992).

Space–time sampling frequencies and selection of variables

The entropy method measures the information content of available data (extraction of information) and assesses the goodness of information transfer between temporal or spatial data points (transfer of information). These two functions constitute the basis of the solution to the design problems of what, where, when and how long to observe. Such a solution is based on the maximization of information transfer between variables, space points and time points, respectively. The amount of information transfer used in such an analysis can be measured by entropy in specific units. The selection of each technical design factor can be evaluated again by means of entropy to define the amount of information conveyed by the data collected by each of the selected monitoring procedures. These evaluations may eventually provide the ability to make quantitatively based rational decisions on how long a gauge should be operated (Alpaslan *et al.*, 1992).

Harmancioglu and Alpaslan (1992) demonstrated the applicability of the entropy method in assessing the efficiency and the benefits of

an existing water quality monitoring network with respect to temporal, spatial and combined temporal–spatial design features. They described the effect of each feature upon network efficiency and cost effectiveness by entropy-based measures. For example, the effect of extending the sampling interval from monthly to bimonthly measurements for three variables investigated was found to lead to significant losses of information for dissolved oxygen, chloride and electrical conductivity. Here, the selection of an appropriate sampling interval is made by assessing how much information the decision maker would risk versus given costs of monitoring. A similar evaluation can be made with respect to the number and location of required sampling sites, where changes in rates of information gain are investigated with respect to the number of stations in the network. Harmancioglu and Alpaslan (1992) further combined both the spatial and temporal frequencies to assess the variation of information with respect to both space and time dimensions. The results of these analyses have shown the applicability of the entropy concept in network assessment.

Limitations of the entropy method in network design

The above-mentioned advantages of the entropy principle indicate that it is a promising method in water quality monitoring network design problems because it permits quantitative assessment of efficiency and benefit/cost parameters. However, some limitations of the method must also be noted for further investigations on entropy theory.

As the situation holds true for the majority of statistical techniques, a sound evaluation of network features by the entropy method requires the availability of sufficient and reliable data. Applications with inadequate data often cause numerical difficulties and hence unreliable results. For example, when assessing spatial and temporal frequencies in the multivariate case, the major numerical difficulty is related to the properties of the covariance matrix of equation (E30) (Harmancioglu and Alpaslan, 1992). When the determinant of the matrix is too small, entropy measures cannot be determined reliably since the matrix becomes ill conditioned. This often occurs when the available sample sizes are very small.

On the other hand, the question with respect to data availability is 'how many data points would be considered sufficient'. For example, Goulter and Kusmulyono (1993) claim that the entropy principle can be used to make 'sensible inferences about water quality conditions' but that sufficient data are not available for a reliable assessment. The major difficulty here arises from the nature of water quality data, which are often sporadically observed for short periods of time. With such 'messy' data, application of the entropy method poses problems both in numerical computations and in evaluation of the results. Particularly, it is difficult to determine when a data record can be considered sufficient.

With respect to the temporal design problem, all evaluations are based on the temporal frequencies of available data so that, again, the method inevitably appears to be data dependent. At present, it appears to be difficult to assess smaller time intervals than what is available. However, the problem of decreasing the sampling intervals may also be investigated by the entropy concept provided that the available monthly data are reliably disaggregated into short interval series. This aspect of entropy applications has to be investigated in future research.

Another important point in entropy applications is that the method requires the assumption of a valid distribution type. The major difficulty occurs here when different values of the entropy function are obtained for different probability distribution functions assumed for the same variable. On the other hand, the entropy method works quite well with multivariate normal and lognormal distributions. The mathematical definition of entropy is easily developed for other skewed distributions in bivariate cases. However, the computational procedure becomes much more difficult when their multivariate distributions are considered. When such distributions are transformed to normal, then uncertainties in parameters need to be assessed.

Another problem that has to be considered in future research is the mathematical definition of entropy concepts for continuous variables. Shannon's basic definition of entropy is developed for a discrete random variable, and the extension of this definition to the continuous case entails the problem of selecting the discretizing class intervals x to approximate probabilities with class frequencies. Different measures of entropy vary with x such that each selected value of x constitutes a different base level or scale for measuring uncertainty. Consequently, the same variable investigated assumes different values of entropy for each selected value of x. It may even take on negative values which contradict the positivity property of the entropy function in theory. A possible procedure to overcome the above deficiency may be to define confidence levels for entropy measures, particularly for transinformation. This is still an issue that needs to be investigated. Some researchers propose the use of a function $m(x)$ such that the marginal entropy of a continuous variable is expressed as

$$H(x) = - \int_{-\infty}^{+\infty} f(x) \ln [f(x)/m(x)] \, dx \qquad (E33)$$

'where $m(x)$ is an "invariant measure" function, proportional to the limiting density of discrete points' (Jaynes, 1983). The approach seems to be statistically justified; however, it is still uncertain what the $m(x)$ function may represent in reality. Jaynes (1983) also discusses that it can be an *a priori* probability distribution function, but there are then controversies over the choice of an a priori distribution so that the problem still remains unsolved.

One last problem that needs to be investigated in future research is the development of a quantifiable relationship between monitoring objectives and technical design features in terms of the entropy function. As stated earlier, an information-based design strategy requires the delineation of data needs or information expectations. To ensure network efficiency, 'information supplied' and 'information expected' must be expressed in quantifiable terms by the entropy concept. At the current level of research, if one considers that the most significant objective of monitoring is the determination of changes in water quality, then the entropy principle does show such changes with respect to time and space. However, future research has to focus on the quantification of information needs for specific objectives (e.g. trend detection or compliance) by means of entropy measures (Harmancioglu *et al.*, 1994).

Water quality modeling

Singh and Krstanovic (1987) derived a stochastic model for sediment yield by using POME. The model described the probability distribution of sediment yield conditioned on the probability distribution of runoff volume. The model distribution parameters were determined from prior information about the runoff volume and sediment yield such as their means and covariance. In another study, Singh and Krstanovic (1988) used POME to develop a stochastic–deterministic model for modeling water quality constituents (phosphorus). The stochastic component was based on POME, and the deterministic component on a power function. The stochastic component was fitted to the residuals obtained by subtracting the observed values from the values predicted by the deterministic component. The results of both studies indicated that POME was a viable tool for stochastically modeling these hydrologic processes (Singh and Fiorentino, 1992).

Assessment of treatment plant efficiencies

Tai and Goda (1980, 1985) define 'thermodynamic efficiency' on the basis of thermodynamic entropy and show that the efficiency of a water treatment system can be described by the rate of decrease in the entropy of polluted water. They also relate the reduction in thermodynamic entropy to the entropy of discrete information conveyed by the output process of a treatment plant (TP).

The informational entropy concept may also be effectively used in environmental engineering system design and operation, where the processes encountered are random by nature. A particular area of application within this respect covers water and wastewater treatment plants. The input and output processes of a TP fluctuate significantly with time. This variability is often insufficiently recognized in both the design and operation of treatment systems. Alpaslan (1994) proposed the use of entropy measures of information in the assessment of input/output uncertainty of TP. Such measures help to identify how variable or how steady the inputs or the outputs are so that both the design parameters and the operational efficiency of a TP can be evaluated. According to the approach applied, the operational efficiency of a TP is assessed by means of the 'dynamic efficiency index' (DEI) which represents the rate of reduction in the uncertainty of inputs to produce minimum amount of entropy or uncertainty in the outputs. In essence, two requirements are foreseen for an effective and reliable treatment system: (1) the highest reduction in input uncertainty must be obtained, and (2) the outputs must be insensitive to the inputs; that is, the condition of independence must be satisfied. The entropy measures, as proposed by Alpaslan (1994), can be

effectively used to assess whether these two requirements are met by the operation of the TP.

Alpaslan (1994) defined the 'dynamic efficiency index, DEI' as the efficiency level that produces maximum reduction in the uncertainty $H(X_i)$ of the inputs to arrive at a minimum amount of uncertainty $H(X_e)$ in the outputs. In the ideal case, the TP is expected to produce outputs that comply with an effluent standard, the value of which is generally constant. In entropy terms, this means that the requirement is $H(X_e) = 0$. In practice, though, the outputs will fluctuate around the standard, and the TP is expected to reduce the variability of the output (effluent) so that such fluctuations are kept below the standard value. In entropy terms again, this indicates that the uncertainty of $H(X_e)$ should be minimum or that it should approach zero. Then the performance of the TP can be evaluated by means of the 'dynamic efficiency index, DEI' which measures the maximum reduction in $H(X_i)$ so that it approaches zero or in practical terms minimum $H(X_e)$ value. Such a measure can be expressed in per cent as

$$\text{DEI} = [H(X_i) - H(X_e)]/H(X_i \times 100\% \qquad (E34)$$

The term 'dynamic' is used here to indicate that efficiency is expressed on the basis of variability of input/output processes by using entropy measures of uncertainty. In essence, the above definition involves the first requirement for efficient treatment, namely that the difference between input/output variability must be maximized. This difference is actually an indicator of treatment capacity such that if it reflects low efficiency, then the process or design parameters of the TP may have to be changed to increase the DEI. Furthermore, calculation of the DEI for different values of different process parameters can help to identify those parameters which significantly affect the treatment system. The TP may be considered sensitive to those values of particular parameters which lead to maximum reductions in $H(X_i)$. Then, such parameters will need to be more strictly observed throughout monitoring procedures for both the design and the operation of the TP.

The above approach to assessment of efficiency appears to comply with the thermodynamic efficiency definition given by Tai and Goda (1980, 1985). As mentioned earlier, their description of efficiency refers to the decrease in the thermodynamic entropy of polluted water, where the treated media move from a state of disorder to order. The terms 'order' and 'disorder' are analogous in both the thermodynamic and the informational system considered. In the former, 'disorder' refers to thermodynamic disorder or pollution, which can be measured by the thermodynamic entropy of the system. In the latter, 'disorder' indicates high variability in the system, again quantified by entropy measures, albeit in an informational context. Accordingly, the two efficiency definitions, one given by Tai and Goda (1980, 1985) and the other by Alpaslan (1994), are similar in concept; the major difference between them is that the former is given in a thermodynamic framework, whereas the latter is presented on a probabilistic basis.

The second requirement for effective treatment is recognized as the insensitivity of effluents with respect to the inputs; that is, the correlation between the inputs and the outputs is required to be a minimum for a reliable treatment system. Such a requirement may also be considered as an indicator of TP efficiency. Entropy measures can again be employed to investigate the relationship between input/output processes. In this case, conditional entropies in the form of $H(X_e/X_i)$ have to be computed. The condition $H(X_e|X_i) = H(X_e)$ indicates that the outputs are independent of the inputs and, consequently, that the TP is effective in processing the inputs. Otherwise, if inputs and outputs are found to be correlated with $H(X_e|X_i) = H(X_e)$, this implies that the effluents are sensitive to the inputs and that the treatment system fails to transform the inputs effectively.

Another entropy measure of correlation between the input and the output processes is transinformation $T(X_i, X_e)$. If transinformation between the two processes is zero, this indicates that they are independent of each other. In the case of complete dependence, $T(X_i, X_e)$ will be as high as the marginal entropy of one of the processes.

The methodology described above is applied by Alpaslan (1994) to the case of Seka Dalaman Paper Factory treatment plant in Turkey, for which input and output data on BOD (biochemical oxygen demand), COD (chemical oxygen demand) and TSS (total suspended solids) concentrations are available. These results show that the treatment processes applied result in TSS having the highest reduction in input uncertainty and COD having the lowest reduction. This feature is also reflected by the dynamic efficiency index which reaches the highest value for TSS and the lowest for COD. It is interesting to note that when the classical efficiency measure in the form of

$$E = (S_i - S_e)/S_i \qquad (E35)$$

(with E being efficiency, and S_i and S_e, the random inputs and outputs, respectively, of the TP) gives values of the order of 96, 91 and 75%, respectively, for TSS, BOD and COD, the efficiencies defined by entropy measures on a probabilistic basis result in the respective values of 50, 36 and 25%. Although the two types of efficiencies described do not achieve similar values, their relative values for each variable were found to be of the same order; that is, both types of efficiencies are the highest for TSS and the lowest for COD with BOD in between. Accordingly, Alpaslan (1994) concluded that the output TSS of the treatment process is insensitive to the inputs and that the operation of the TP can be considered to be reliable in the case of TSS. The level of dependence increases slightly for BOD and COD, indicating some correlation between the inputs and the outputs. For these two variables, the conditional entropies are close to but not equal to the marginal entropies of the output processes. Thus Alpaslan inferred that the reliability of the TP decreases for BOD and COD.

In his study, Alpaslan (1994) also claimed that the definition of TP efficiency on the basis of entropy measures has further advantages, the most significant one being in sensitivity analyses of process parameters. Such parameters in biological treatment, for instance, may be maximum specific growth rate, decay rate and yield coefficient, the values of which must be selected for the design of TP. The TP system is either sensitive or insensitive to these design parameters so that its efficiency is eventually affected by them. The effects of these parameters are already recognized; however, the degree of uncertainty they convey to the system are not well quantified. Values of these parameters for design purposes are either taken from literature or determined by laboratory analyses. Outputs from a simulation model of a TP may be observed with respect to different parameter values and in calculating the DEI for the system for each case. The TP system may be considered sensitive to those values of parameters which lead to maximum reductions in $H(X_i)$. Then those parameters will need to be more strictly observed throughout the data collection procedures for the design as well as for the operation of TP. Alpaslan (1994) claimed here that further investigations are needed on the subject, particularly in disclosing the relationship between the informational entropy measures and the dynamic (physical, thermodynamic, etc.) processes of treatment.

Physical basis of entropy

Relation between thermodynamic, Shannon and Boltzmann entropies

The concept of entropy has its origins in classical thermodynamics and is commonly known as 'thermodynamic entropy' in relation with the second law of thermodynamics. As described in the previous sections, such a non-probabilistic definition of entropy has been used widely in physical sciences, including hydrology and water resources.

According to the second law of thermodynamics, or the so-called entropy law, the amount of disorder (entropy) in any closed conservative thermodynamic system tends to a maximum. As such, 'classical thermodynamics ... is concerned only with the macroscopic states of matter, i.e., with experimentally observable properties such as temperature, pressure, volume, etc. Clasius defined entropy in this non-probabilistic context as a function of these macroscopic quantities' (Templeman, 1989). This is what we know as 'thermodynamic entropy'.

Later, Boltzmann gave a new definition for entropy by analyzing microscopic states of a thermodynamic system. His purpose was to infer eventually on the nature of a macroscopic state which was made up of possible combinations of various microscopic states. According to Boltzmann, 'the macroscopic maximum entropy state corresponded to a thermodynamic configuration which could be realized by the maximum number of different micro-states' (Templeman, 1989). Boltzmann considered that entropy is the measure of disorder in a system and further that the degree of disorder is essentially the amount of uncertainty associated with particular microscopic states. Boltzmann's definition described entropy in probabilistic terms and constituted the basis for statistical thermodynamics (Harmancioglu *et al.*, 1992a).

Boltzmann's definition of entropy as a measure of disorder in a system was given in prohabilistic terms and constituted the basis for statistical thermodynamics. Later, Shannon followed up on Boltzmann's definition, claiming that the entropy concept could be used to measure disorder in systems other than thermodynamic ones. Shannon's entropy is what we know as 'informational entropy', which measures uncertainty (or indirectly information) about random processes.

Entropy as a phenomenological descriptor

Derivation of velocity distributions

Hydraulic applications of entropy and POME have encompassed derivations of distributions of velocity, shear stress and suspended sediment concentration, and estimation of diffusion coefficient for pollutant transport in open channel flows. Fundamental to these applications has been the derivation of velocity distributions.

There are presently two basic methods and a recently proposed method to obtain a time-averaged horizontal velocity profile: the logarithmic distribution law, the power law and the two constraint entropy methods by Chiu (1988). The entropy method uses POME to maximize the information content of the data. The entropy method produces four integral equations in four unknowns. These equations are derived from the physical contraints on the system (i.e. probability, continuity, momentum and energy) and are not solvable by exact analytical means except for the two constraint cases (Chiu, 1988). The constraints Chiu used were the probability and continuity constraints. He employed this method (Chiu, 1988, 1989) to derive a velocity distribution or the horizontal velocity in a wide open channel with uniform flow.

Following Chiu (1988, 1989), Barbe *et al.* (1991) derived an approximate velocity distribution by employing POME subject to three constraints. The results of application to field data showed that this approximation was reasonably accurate. The approximation enabled an analytical expression for the velocity distribution. By using these POME-based velocity distributions, it is now possible to relate the coefficients of empirical velocity distributions to the constraints based on the laws of conservation of mass, momentum and energy.

Derivation of geomorphological laws

Since the classical work of Leopold and Langbein (1962), much attention has been focused on the use of thermodynamic basis of entropy concepts in analyzing the behavior of streams. Two thermodynamic principles have been applied to streams. The first principle is that the most probable state of a system is such that its entropy is a maximum. The second is the principle of minimum entropy production rate. The justification for the use of these principles is based on the analogies between (1) heat energy in a thermodynamic system and potential energy in a stream system, and (2) temperature in a thermodynamic system and elevation in a stream system (Scheidegger, 1964). Davy and Davies (1979), however, questioned these analogies.

Leopold and Langbein (1962) derived equations for the longitudinal profiles of rivers that were mathematically comparable to those observed in the field. Yang (1971a) derived two laws which govern the formation of all stream systems. The first is the law of average stream fall, where it is stated that under the dynamic equilibrium condition the ratio of average fall between any two different-order streams in the same river basin is unity. This law provides a measure of the maturity of the stream system and indicates whether the stream should aggrade or degrade in the future (Singh and Fiorentino, 1992). The second law is the law of least rate of energy expenditure, which is the basis of further work on fluvial morphology and sediment transport (Yang, 1971b, 1972). According to this law, during its evolution towards its equilibrium condition a natural stream chooses its course of flow in such a manner that the rate of potential energy expenditure per unit mass of water along this course is a minimum. This minimum value depends on the external constraints imposed on the stream. Using these laws, Yang computed streambed profiles which were in agreement with the observed data.

Kapoor (1990) investigated into spatial uniformity of power and the altitudinal geometry of river networks. The power or potential energy imparted to the water per unit time between any two points was considered as proportional to the product of discharge and the elevation difference between those two points. This product of flow and the elevation drop of a river reach is approximately equal to the energy that the stream dissipates in doing work to transport water and sediment, and as heat to overcome friction during its flow. Kapoor (1990) hypothesized that river networks adjust their geometries in the altitudinal space to achieve a state of maximum spatial uniformity of power. He mathematically showed that the concept of uniformity was linked to minimum variance and maximum information–theoretical entropy. This is in accord with Yang's (1971b) law of average stream fall. The maximum spatial uniformity of power was found to be the link between river network geometry and flows (Singh and Fiorentino, 1992).

Fiorentino *et al.* (1993) explored the connection between entropy and potential energy to analyze the morphological characteristics of a drainage basin under the assumption that the only information available on a drainage basin is its mean elevation. The mean elevation is found to be linearly related to the entropy of the drainage basin. This relation leads to a linear relation between the mean elevation of a subnetwork and the logarithm of its topological diameter. Furthermore, the relation between the fall in elevation from the source to the outlet of the main channel and the entropy of its drainage basin is found to be linear, and so also is the case between the elevation of a node and the logarithm of its distance from the sounce. When the drainage basin is ordered in accordance with the Horton–Strahler ordering scheme, a linear relation is obtained between the drainage basin entropy and the basin order. This relation can be characterized as a measure of the basin network complexity. The basin entropy is found to be linearly related to the logarithm of the magnitude of the basin network. Thus, this relation leads to a non-linear relation between the network diameters and magnitude, where the exponent is found to be related to the fractal dimension of the drainage network. Also, the exponent of the power law relating the channel slope to the network magnitude is related to the fractal dimension of the network. Fiorentiono *et al.* verified these relationships on three drainage basins in Southern Italy.

Distribution of groundwater potential

Barbe *et al.* (1994) employed POME to derive a probability distribution for the piezometric head *h* in groundwater flow. Two cases of one-dimensional steady flow for confined and unconfined aquifers were considered. For the case of confined flow, no constraint was used, whereas for the case of unconfined flow the conservation of mass was the constraint. For homogeneous, isotropic and confined flow, the probability distribution of the piezometric head is a uniform distribution. The piezometric head is found to be a linear function of the distance. If the cross-sectional area and the hydraulic conductivity of the confined aquifer are not constant, a constraint based on the conservation of mass will have to be defined and that would lead to a different probability distribution.

For homogeneous, isotropic and unconfined flow, the probability distribution of the piezometric head is a non-linear function. The solution for *h* is found to be the same as given by the Dupuit–Forchheimer (DF) approximation. The advantage of using POME is that it permits a probabilistic interpretation of aquifer behavior for a given set of aquifer properties.

Evaluation of the applicability of entropy in water resources

It follows from the previous sections that the informational entropy principle can be effectively used to solve versatile problems in water resources engineering. 'Uncertainty' and 'information' are the most significant, yet the least clarified, concepts in water resources. The decision making process is highly dependent on the evaluation of available information about processes investigated. However, there has been no quantitative measure developed yet to assess uncertainty or information in objective terms. Thus the two concepts are often described subjectively or indirectly in terms of other variables or parameters. Under these conditions, entropy appears to be a highly attractive concept since it directly defines uncertainty and information. Furthermore, it may be described as the only technique that measures these two factors in quantitative units. Expressing uncertainty or information in specific units is basically the most significant contribution of entropy to investigations in water resources engineering. This feature leads to more reliable and objective evaluations within the decision-making process, especially when inferences are to be based on incomplete or inadequate information.

On the other hand, there are still some difficulties at the mathematical level which need to be solved before entropy can be widely accepted as a principal technique in water resources engineering. The major problem is the controversy associated with the mathematical definition of entropy for continuous distribution functions. Shannon's entropy as given in equation (E8) is originally formulated for discrete variables. Shannon extended this expression to the continuous case by simply replacing the summations with integral operations as in equation (E9), which is not mathematically justified. Thus, the expressions (E9)–(E11) are not valid under the assumptions initially made in defining entropy for the discrete case. What researchers have proposed for solving this problem is to approximate the discrete probabilities p_k by $f(x)\Delta x$, where $f(x)$ is the relative class frequency and Δx the size of class intervals. Under these conditions the selection of Δx becomes a crucial problem, as stated in earlier sections. The significance of Δx is that each specified class interval size gives a different reference level of zero uncertainty with respect to which the calculated marginal or conditional entropies are measured. In this case, these entropies become relative to the horizontal axis and change in value as Δx changes. The unfavorable result here is that the marginal uncertainty of a random process may assume different values depending on the selection of Δx, so that one cannot evaluate the degree of uncertainty within the process. In certain cases the marginal entropy of a variable even becomes negative (Harmancioglu, 1980, 1981), a situation which contradicts Shannon's definition of entropy. The only parameter that seems to be independent of Δx is transinformation T; however, it was shown by Harmancioglu *et al.* (1986) that it is also affected by the selection of Δx intervals.

Apart from the major problem discussed in the preceding paragraphs, there are further mathematical difficulties associated with the definition of entropy in case of (1) probability distribution functions other than the normal and lognormal pdf's; (2) extension of the definition for multivariate normal pdf to other distribution functions in the multivariate case. The problem of attaining a different value of entropy for the same variable under assumptions of different pdf's also remains unresolved.

Another significant step that has to be accomplished for the development of the entropy concept in water resources is to verify its advantages over other classical techniques. We need to answer questions such as whether, in comparison with the maximum likelihood estimation procedure, one would prefer to use POME in estimating parameters of a distribution, or why one would preferably use R_0 or T_0 to assess information transfer between random variables when the classical correlation coefficient r can do the same thing. These questions need to be answered so that the entropy method can be readily accepted as a valid and reliable technique in hydrology and water resources engineering.

N.B. Harmancioglu and V.P. Singh

Bibliography

Aczel, J. and Daroczy, Z., 1975. *On Measures of Information and Their Characterization.* Academic Press, New York.

Alpaslan, N., 1994. Assessment of treatment plant efficiencies by the entropy principle. In *Time Series Analysis in Hydrology and Environmental Engineering, Proceedings of the International Conference on Stochastic and Statistical Methods in Hydrology and Environmental Engineering* (eds K.W. Hipel, A.I. McLeod, U.S. Panu and V.P. Singh), Kluwer Academic Publishers, Dordrecht Water Science and Technology Library, Vol. 10/3, pp. 177–190.

Alpaslan, N., Harmancioglu, N.B. and V.P. Singh, 1992. The role of the entropy concept in design and evaluation of water quality monitoring networks, *Entropy and Energy Dissipation in Water Resources* in (eds V.P. Singh and M. Fiorentino), Dordecht, Kluwer Academic Publishers, Water Science and Technology Library, pp. 261–282.

Amorocho, J. and Espildora, B., 1973. Entropy in the assessment of uncertainty of hydrologic systems and models. *Water Resources Research,* **9**(6), 1551–1522.

Arora, K. and Singh, V.P., 1987a. On statistical intercomparison of EV1 estimators by Monte Carlo simulation. *Advances in Water Resources,* **10**(2), 87–107.

Arora, K. and Singh, V.P., 1987b. An Evaluation of Seven Methods for Estimating Parameters of the EVI Distribution, in *Hydrologic*

Frequency Modeling (ed. V.P. Singh), D. Reidel, Boston, pp. 383–394.

Arora, K. and Singh, V.P., 1989. A comparative evaluation of the estimates of log-Pearson type (LP) 3 distribution. *Journal of Hydrology*, **105**, 19–37.

Baran, T. and Harmancioglu, N.B., 1990. Assessment of mathematical models with exponential functions describing karstic spring discharges. *UKAM, IAHS & IAH, International Symposium and Field Seminar on Hydrogeologic Processes in Karst Terrains*, Session X on Modeling, Antalya, Turkey, October 1990, 15pp. + figures.

Barbe, D.E., Cruise, J.F. and Singh, V.P., 1991. Solution of the three-constraint entropy-based velocity distribution. *Journal of Hydraulic Engineering, ASCE*, **117**(10), 1389–1396.

Barbe, D.E., Cruise, J.F. and Singh, V.P., 1994. *Derivation of a Distribution for the Piezometric Head in Groundwater Flow Using Entropy in Stochastic and Statistical Methods in Hydrology*, Vol. 2, (ed. K.W. Hipel), Kluwer Academic Publishers, Dordrecht, pp. 151–161.

Behara, M. and Chawla, J.S., 1974. Generalized gamma-entropy. *Selecta, Statistica Canadiana*, **2**, 15–38.

Boltzmann, L., 1872. Weitere Studien uber das Warmegleichgewich unter Gasmolekulen. *K. Acad. Wiss. (Wein) Sitzb., II Abt.*, **66**, 275.

Burg, J.P., 1975. Maximum Spectral Analysis. PhD Dissertation, Stanford University, Palo Alto, California.

Caselton, W.F. and Husain, T., 1980. Hydrologic networks: information transmission, *Journal of Water Resources Planning and Management Division, ASCE*, **106**(WR2), 503–529.

Cetiner, A., 1988. *Hydrologic Information Transfer in River Basins Fed by Karstic Spring Effluents* (in Turkish). Izmir, Dokuz Eylul University, Institute of Technological Sciences, Civil Engineering Department, M.Sc. Thesis in Hydrology and Hydraulic Structures, No. 21.

Chapman, T.G., 1986. Entropy as a measure of hydrologic data uncertainty. *Journal of Hydrology*, **85**, 111–126.

Chiu, C.L., 1988. Entropy and 2D-velocity distribution in open channels. *Journal of Hydraulic Engineering, ASCE*, **114**(7), 738–756.

Chiu, C.L., 1989. Velocity distribution in open channel flow. *Journal of Hydraulic Engineering, ASCE*, **115**(5), 576–594.

Christensen, R.A., 1981. An exploratory application of entropy minimax to weather prediction: estimating the likelihood of multi-year droughts in California, in *Entropy Minimax Sourcebook*, Vol. IV: *Applications* (ed. R.A. Christensen), Entropy Limited, Lincoln, Massachusetts, pp. 495–544.

Davy, B.W. and Davies, T.R.H., 1979. Entropy concepts in fluvial geomorphology: a reevaluation. *Water Resources Research*, **15**(1), 103–106.

Eilbert, R.F. and Christensen, R.A., 1983. Performance of the entropy hydrological forecasts for California water years 1948–1977. *Journal of Climate and Applied Meteorology*, **22**, 1654–1657.

Fisher, R.A., 1966. *Design of Experiments*, 8th edn, Oliver and Boyd, Edinburgh, 248 pp.

Fiorentino, M., Claps, P. and Singh, V.P., 1993. An entropy-based morphological analysis of river basin networks. *Water Resources Research*, **29**(4), 1215–1224.

Fiorentino, M., Singh, V.P. and Arora, K., 1987a. On the two-component extreme-value distribution and its point and regional estimators, in *Regional Flood Frequency Analysis* (ed. V.P. Singh), D. Reidel Publishing Co., Boston, pp. 252–272.

Fiorentino, M., Arora, K. and Singh, V.P., 1987b. The two-component extreme-value distribution for food frequency analysis: another look and derivation of a new estimation method. *Stochastic Hydrology and Hydraulics*, **1**, 199–208.

Goulter, I. and Kusmulyono, A., 1993. Entropy theory to identify water quality violators in environmental management, in *Geo-Water and Engineering Aspects* (eds R. Chowdhury and M. Sivakumar), A.A. Balkema, Rotterdam, pp. 149–154.

Harmancioglu, N., 1980. *Measuring the Information Content of Hydrological Processes by the Entropy Concept* (in Turkish). PhD Thesis in Hydrology and Hydraulic Structures, No. 4, Ege University, Faculty of Engineering, 164 pp.

Harmancioglu, N., 1981. Measuring the information content of hydrological processes by the entropy concept. Centennial of Ataturk's Birth, *Journal of the Civil Engineering Faculty of Ege University*, 13–38.

Harmancioglu, N., 1984. Entropy concept as used in determination of optimum sampling intervals, in *Proceedings of Hydrosoft '84, International Conference on Hydraulic Engineering Software*, Portoroz, Yugoslavia, pp. 96–102.

Harmancioglu, N., 1994a. Assessment of information and uncertainty related to floods, in *Coping with Floods* (eds G. Rossi, N.B. Harmancioglu and V. Yevjevich), Kluwer Academic Publishers, NATO-ASI Series, Series E, Vol. 257, pp. 171–184.

Harmancioglu, N., 1994b. An entropy-based approach to station discontinuance, in *Time Series Analysis in Hydrology and Environmental Engineering, Proceedings of the International Conference on Stochastic and Statistical Methods in Hydrology and Environmental Engineering* (eds K.W. Hipel, A.I. McLeod, U.S. Panu and V.P. Singh), Kluwer Academic Publishers, Water Science and Technology Library, Vol. 10/3, pp. 163–176.

Harmancioglu, N.B. and Alpaslan, N., 1992. Water quality monitoring network design: a problem of multi-objective decision making, *AWRA, Water Resources Bulletin*, Special Issue on Multiple-Objective Decision Making in Water Resources, **28**, pp. 1–14.

Harmancioglu, N. and Baran, T., 1989. Effects of recharge systems on hydrologic information transfer along rivers. *IAHS, Proceedings of the Third Scientific Assembly – New Directions for Surface Water Modeling*, IAHS Publ. 181, pp. 223–233.

Harmancioglu, N.B. and Singh, V.P., 1991. An information-based approach to monitoring and evaluation of water quality data, in *Advances in Water Resources Technology* (ed. G. Tsakiris). Proc. of the European Conference ECOWARM, A.A. Balkema, Rotterdam pp. 377–386.

Harmancioglu, N.B. and Yevjevich, V., 1985. Transfer of hydrologic information along rivers partially fed by karstified limestones, in *Proceedings of International Symposium on Karst Water Resources*, Ankara, IAHS Publ. 161, pp. 161–171.

Harmancioglu, N.B. and Yevjevich, V., 1986. *Transfer of Information Among Water Quality Variables of the Potomac River, Phase III: Transferable and Transferred Information*. Report to DC Water Resources Research Center of the University of the District of Columbia, Washington, DC, June 1986, 81 p.

Harmancioglu, N.B. and Yevjevich, V., 1987. Transfer of hydrologic information among river points. *Journal of Hydrology*, **91**, 103–118.

Harmancioglu, N.B., Yevjevich, V. and Obeysekera, J.T.B., 1986. Measures of information transfer between variables. *Proceedings of Fourth International Hydrological Symposium on Multivariate Analysis of Hydrologic Processes*, (eds H.W. Shen *et al.*), pp. 481–499.

Harmancioglu, N., Ozer, A. and Alpaslan, N., 1987. Evaluation of Water Quality data (in Turkish), in *Ankara, Chamber of Civil Engineers of Turkey, IX. Technical Congress Proceedings*, November 16–20, 1987, Vol. II, pp. 113–129.

Harmancioglu, N.B., Singh, V.P. and Alpaslan, N., 1992a. Versatile uses of the entropy concept in water resources, in *Entropy and Energy Dissipation in Water Resources* (eds V.P. Singh and M. Fiorentino), Kluwer Academic Publishers, Dordrecht, Water Science and Technology Library, pp. 91–117.

Harmancioglu, N.B., Alpaslan, N. and Singh, V.P., 1992b. Application of the entropy concept in design of water quality monitoring networks, in *Entropy and Energy Dissipation in Water Resources*, (eds V.P. Singh and M. Fiorentino), Kluwer Academic Publishers, Dordrecht, Water Science and Technology Library, pp. 261–283.

Harmancioglu, N.B., Singh, V.P. and Alpaslan, N., 1992c. Design of water quality monitoring networks, in *Geomechanics and Water Engineering in Environmental Management*, (ed. R.N. Chowdhury), A.A. Balkema, Rotterdam, Ch. 8, pp. 267–296.

Harmancioglu, N.B., Alpaslan, N. and Singh, V.P., 1994. Assessment of the entropy principle as applied to water quality monitoring network design, in *Time Series Analysis in Hydrology and Environmental Engineering, Proceedings of the International Conference on Stochastic and Statistical Methods in Hydrology and Environmental Engineering* (eds K.W. Hipel, A.I. McLeod, U.S. Panu and V.P. Singh), Kluwer Academic Publishers, Water Science and Technology Library, Vol. 10/3, pp. 135–148.

Havrada, J.H. and Charvat, F., 1967. Quantification methods of classificatory processes: concept of structural entropy. *Kybernatica*, **3**, 30–35.

Husain, T., 1989. Hydrologic uncertainty measure and network design. *Water Resources Bulletin*, **25**(3), 527–534.

Jain, D. and Singh, V.P., 1986. Estimating parameters of EVI distribution of flood frequency Analysis. *Water Resources Bulletin,* **23**(1), 59–71.

Jaynes, E.T., 1957a. Information theory and statistical mechanics I. *Phys. Rev.,* **106**, 620–630.

Jaynes, E.T., 1957b. Information theory and statistical mechanics II. *Phys. Rev.,* **108**, 171–190.

Jaynes, E.T., 1961. *Probability Theory in Science and Engineering,* McGraw-Hill, New York.

Jaynes, E.T., 1979. Concentration of distributions at entropy maxima. in *E.T. Jaynes: Papers on Probability, Statistics and Statistical Physics* (ed. R.D. Rosenkratz), D. Reidel, Boston, pp. 315–335.

Jaynes, E.T., 1982. On the rationale of entropy methods. *Proceedings of IEEE,* **70**(19), 939–959.

Jaynes, E.T., 1983. *Papers on Probability, Statistics and Statistical Physics* (ed. R.D. Rosenkrantz), D. Reidel, Dordrecht, *Studies in Epistemology, Logic, Methodology and Philosophy of Science,* Vol. 158.

Jowitt, P.W., 1979. The extreme value type-1 distribution and the principle of maximum entropy. *Journal of Hydrology,* **42**, 23–38.

Kapoor, V., 1990. Spatial uniformity of power and the altitudinal geometry of river networks. *Water Resources Research,* **26**(10), 2303–2310.

Kapur, J.N., 1967. Generalised entropies of order a and type b. *Mathematics Seminar,* **4**, 79–94.

Kapur, J.N., 1968. On information of order a and b. *Proceedings of the Indian Academy of Science,* **48A**, 65–76.

Kapur, J.N., 1986. Four families of measures of entropy. *Indian Journal of Pure and Applied Mathematics,* **17**(4), 429–449.

Kennedy, J.F., Richardson, P.D. and Sutera, S.P., 1964. Discussion of 'Geometry of river channels' by W.R. Langbein. *Journal of the Hydraulics Division, Proceedings of ASCE,* **90**(HY6), 332–347.

Khinchin, A.I., 1957. *Mathematical Foundations of Information Theory,* Dover Publ., New York, 120 pp.

Krstanovic, P.F. and Singh, V.P., 1987. A multivariate stochastic flood analysis using entropy, in *Hydrologic Frequency Modeling* (ed. V.P. Singh), D. Reidel, Dordrecht, pp. 515–540.

Krstanovic, P.F. and Singh, V.P., 1988a. *Application of Entropy Theory to Multivariate Hydrologic Analysis,* Vol. 2. Technical Report WRR9, Water Resources Program, Dept. of Civil Engineering, Louisiana State University, Baton Rouge, LA, pp. 271–557.

Krstanovic, P.F. and Singh, V.P., 1988b. *Application of Entropy Theory to Multivariate Hydrologic Analysis,* Vol. 1, Technical Report WRR9, Water Resources Program, Dept. of Civil Engineering, Louisiana State University, Baton Rouge, LA, pp. 1–271.

Krstanovic, P.F. and Sing, V.P., 1991a. A univariate model for long-term streamflow forecasting: 1. Development. *Stochastic Hydrology and Hydraulics,* **5**, 173–188.

Krstanovic, P.F. and Singh, V.P., 1991b. A univariate model for long-term streamflow forecasting: 1. Application. *Stochastic Hydrology and Hydraulics,* **5**, 189–205.

Krstanovic, P.F. and Singh, V.P., 1993a. Evaluation of rainfall networks using entropy: I. Theoretical development. *Water Resources Management,* **6**, 279–293.

Krstanovic, P.F. and Singh, V.P., 1993b. Evaluation of rainfall networks using entropy: II. Application. *Water Resources Management,* **6**, 295–314.

Krstanovic, P.F. and Singh, V.P., 1993c. A real-time flood forecasting model based on maximum entropy spectral analysis: 1. Development. *Water Resources Management,* **7**(2), 109–130.

Krstanovic, P.F. and Singh, V.P., 1993d. A real-time flood forecasting model based on maximum entropy spectral analysis: II. Application. *Water Resources Management,* **7**(2), 131–152.

Leopold, L.B. and Langbein, W.B., 1962. *The Concept of Entropy in Landscape Evaluation.* USGS Prof. Paper 500-A, pp. A1–A20.

Linfoot, E.H., 1957. An information measure of correlation. *Information and Control,* **1**, 85–89.

Nordin, C.F., 1977. Discussion of 'Applicability of Unit Stream Power Equation' by C.T. Yang and J.B. Stall. *Journal of the Hydraulics Division, Proceedings of ASCE,* **103**(HY2), 209–211.

Padmanabhan, G. and Rao, A.R., 1988. Maximum entropy spectral analysis of hydrologic data. *Water Resources Research,* **24**(9), 1519–1534.

Panu, U.S. and Unny, T.E., 1979. Entropy concept in feature extraction and hydrologic time series analysis, in *Proceedings Third International Hydrology Symposium,* Colorado State University, Fort Collins, Colorado, pp. 100–115.

Papoulis, A., 1991. *Probability, Random Variables and Stochastic Processes.* McGraw-Hill, New York. Ch. 15.

Paulson, A.S. and Garrison, C.B., 1973. Entropy as a measure of the areal concentration of water oriented industry. *Water Resources Research,* **9**(2), 263–269.

Renyi, A., 1961. On measures of entropy and information in *Proceedings of the 4th Berkeley Symposium on Mathematical Statistics and Probability,* Vol. I, pp. 547–561.

Scheidegger, A.E., 1964. Some implications of statistical mechanics in geomorphology. *IAHS bulletin,* **9**(1), pp. 12–16.

Scheidegger, A.E., 1967. A thermodynamic analogy for meander systems. *Water Resources Research,* **3**(4), 1041–1046.

Schilperoot, T., Groot, S., Wetering, B.G.M. and Dijkman, F., 1982. *Optimization of the Sampling Frequency of Water Quality Monitoring Networks,* Waterloopkundig Laboratorium Delft, Hydraulics Lab, Delft, The Netherlands.

Shannon, C.E., 1948a. A mathematical theory of communications, I and II. *Bell System Tech. Journal,* **27**, 379–423.

Shannon, C.E., 1948b. A mathematical theory of communication, III and IV. *Bell System Tech. Journal,* **27**, 623–656.

Shannon, C.E. and Weaver, W., 1949. *The Mathematical Theory of Communication,* The University of Illinois Press, Urbana, Illinois.

Sharp, W.E., 1970. Stream orders as a measure of sample source uncertainty. *Water Resources Research,* **6**(3), 919–926.

Singh, V.P., 1987. On application of the Weibull distribution in hydrology. *Water Resources Management,* **1**, 33–43.

Singh, V.P., 1992. Entropy-based probability distributions for modeling of environmental and biological systems, in *Structuring Biological Systems* (ed. S.S. Iyengar), *CRC Press,* Ch. 6., pp. 167–208.

Singh, V.P. and Fiorentino, M., 1992. A historical perspective of entropy applications in water resources, in *Entropy and Energy Dissipation in Water Resources,* (eds V.P. Singh and M. Fiorentino), Kluwer Academic Publishers, Dordrecht, Water Science and Technology Library, pp. 21–61.

Singh, V.P. and Jain, D., 1985. Comparing methods of parameter estimation for EVI distribution for flood frequency analysis, in *Proceedings of the Vth World Congress on Water Resources,* Brussels, Belgium, pp. 1119–1132.

Singh, V.P. and Krstanovic, P.F., 1986. Space design of rainfall networks using entropy, in *Proceedings of the International Conference on Water Resources Needs and Planning in Drought Prone Areas,* Khartoum, Sudan, pp. 173–188.

Singh, V.P. and Krstanovic, P.F., 1987. A stochastic model for sediment yield using the principle of maximum entropy. *Water Resources Research,* **23**(5), 781–793.

Singh, V.P. and Krstanovic, P.F., 1988. A stochastic model for water quality constituents, in *Proceedings of the 6th APD-IAHR Congress,* Kyoto, Japan.

Singh, V.P. and Rajagopal, A.K., 1987. Some recent advances in the application of the principle of maximum entropy (POME) in hydrology, in *Water for the Future* (eds J.C. Rodda and N.C. Matalas), Proceedings of the Rome Symposium, April 1987, IAHS Publication No. 164, pp. 353–364.

Singh, V.P. and Singh, K., 1985a. Derivation of the gamma distribution by using the principle of maximum entropy. *Water Resources Bulletin,* **21**(6), 941–962.

Singh, V.P. and Singh, K., 1985b. Derivation of the Pearson type (PT) III distribution by using the principle of maximum entropy (POME). *Journal of Hydrology,* **80**, 197–214.

Singh, V.P. and Singh, K., 1985c. Pearson type III distribution and the principle of maximum entropy, in *Proceedings of the Vth World Congress on Water Resources,* Brussels, Vol. 3, pp. 1133–1146.

Singh, V.P. and Singh, K., 1987. Parameter estimation for TPLN distribution for flood frequency analysis. *Water Resources Bulletin,* **23**(6), 1185–1191.

Singh, V.P. and Singh, K., 1988. Parameter estimation for log-Pearson type III distribution by POME. *Journal of Hydraulic Engineering,* **114**(3), 112–122.

Singh, V.P., Cruise, J.F. and Ma, M., 1989. *A Comparative Evaluation of the Estimators of the Two Distributions by Monte Carlo Simulation Method.* Technical Report WRR13, Water Resources Program, Dept. of Civil Eng., Louisiana State University, Baton Rouge, LA, 126 pp.

Singh, V.P., Cruise, J.F. and Ma, M., 1990a. A comparative evaluation of the estimators of the three parameter lognormal distribution by Monte Carlo simulation. *Computational Statistics and Data Analysis*, **10**, 71–85.

Singh, V.P., Cruise, J.F. and Ma, M., 1990b. A comparative evaluation of the estimators of the Weibull distribution by Monte Carlo simulation. *Journal of Statistical Computation and Simulation*, **36**, 229–241.

Sonuga, J.O., 1972. Principle of maximum entropy in hydrological frequency analysis. *Journal of Hydrology*, **17**, 177–191.

Sonuga, J.O., 1976. Entropy principle applied to rainfall–runoff process. *Journal of Hydrology*, **30**, 81–94.

Tai, S. and Goda, T., 1980. Water quality assessment using the theory of entropy, in (ed. M.J. Stiff) *River Pollution Control*, USA, Ellis Horwood, Ch. 21, pp. 319–330.

Tai, S. and Goda, T., 1985. Entropy analysis of water and wastewater treatment processes, *International Journal of Environmental Studies*, **25**, 13–21.

Templeman, A.B., 1989. *Entropy and Civil Engineering Optimization*. NATO/ASI on Optimization and Decision Support Systems in Civil Engineering, Edinburgh, June 1989, 17 p.

Theil, H., 1970. *Economics and Information Theory*. North Holland, Amsterdam.

Tirsch, F.S. and Male, J.W., 1984. River basin water quality monitoring network design, in *Options for Reaching Water Quality Goals*, Proceedings of 20th Annual Conference of AWRA (ed. J.M. Schad), AWRA Publications, pp. 149–156.

Tomlin, S.G., 1969. A kinetic theory of distribution and similar problems. Environment and Planning, **1**, 221.

Tomlin, S.G., 1970. Time-dependent traffic distribution. *Transportation Research*, **4**(1).

Topsoe, F., 1974. *Informationstheorie*. B.G. Teubner, Stuttgart, 88 pp.

Uslu, O. and A. Tanriover, 1979. Measuring the information content of hydrological processes (in Turkish), in *Proceedings of the First National Congress on Hydrology*, Istanbul, Nov. 1979, pp. 437–443.

Van den Boss, A., 1971. Alternative interpretation of maximum entropy spectral analysis. *IEEE Transactions on Information Theory*, **IT-17**, 493–494.

Wilson, A.G., 1970. The use of the concept of entropy in system modelling. *Operational Research Quarterly*, **21**(2), 247–265.

Yang, G.T., 1971a. Potential energy and stream morphology. *Water Resources Research*, **7**(2), 311–322.

Yang, G.T., 1971b. On river meanders. *Journal of Hydrology*, **13**, 231–253.

Yang, G.T., 1972. Unit stream power and sediment transport. *Journal of Hydraulics Division, Proceedings of ASCE*, **98**(HY10), 1805–1826.

Yevjevich, V., 1987. Stochastic models in hydrology. *Stochastic Hydrology and Hydraulics*, **1**, 17–36.

Cross reference

Accuracy

ENVIRONMENTAL PRIORITIES FOR DEVELOPMENT: WATER

The health of hundreds of millions of people is threatened by contaminated drinking water, particulates in city air and smoky indoor air caused by the use of cooking fuels such as dung and wood. Productivity of natural resources is being lost in many parts of the world because of the overuse and pollution of renewable resources – soils, water, forests and the like. Amenities provided by the natural world, such as the enjoyment of an unpolluted vista or the satisfaction that a species is being protected from extinction, are being lost as habitats are degraded or converted to other uses. Because the interaction of various pollutants with other human and natural factors may be hard to predict, some environmental problems may entail losses in all three areas: health, productivity and amenity.

The priorities that developing countries set for their own environments will not necessarily be those that people in richer countries might want them to adopt. Thus, although some cultures in poor countries may value their natural heritage strongly, most developing country governments are likely to give lower priority to amenity damage as long as basic human needs remain unmet.

National priorities vary. In Sub-Saharan Africa, for example, contaminated drinking water and poor sanitation contribute to infectious and parasitic diseases that account for more than 62% of all deaths – twice the level found in Latin America and 12 times the level in industrial countries. Higher-income countries have virtually eliminated these waterborne health risks.

Water

Access to safe water remains an urgent human need in many countries. Part of the problem is contamination; tremendous human suffering is caused by diseases that are largely conquered when adequate water supply and sewerage systems are installed. The problem is compounded in some places by growing water scarcity, which makes it difficult to meet increasing demand except at escalating cost.

The most widespread contamination of water is from disease-bearing human wastes, usually detected by measuring fecal coliform levels. Human wastes pose great health risks for the many people who are compelled to drink and wash in untreated water from rivers and ponds. Data from UNEP's Global Environment Monitoring System (GEMS) demonstrate the enormous problem of such contamination, with poor and deteriorating surfacewater quality in many countries. Water pollution from human wastes matters less in countries that can afford to treat all water supplies, and it can in principle be reversed with adequate investment in treatment systems. But water quality has continued to deteriorate even in some high-income countries.

The capacity of rivers to support aquatic life is decreased when the decomposition of pollutants lowers the amount of oxygen dissolved in the water. Unlike fecal contamination, oxygen loss does not threaten health directly, but its effects on fisheries may be economically important. Human sewage and agro-industrial effluent are the main causes of this problem; nutrient runoff in agricultural areas with intensive fertilizer use is another contributor. Although inadequate levels of dissolved oxygen tend to affect shorter lengths of rivers than does fecal contamination, a sample of GEMS monitoring sites in the mid-1980s found that 12% had dissolved oxygen levels low enough to endanger fish populations. The problem was worst where rivers passed through larger cities or industrial centers. In China, only five of 15 river stretches sampled near large cities were capable of supporting fish. High-income countries have seen some improvement over the past decade. Middle-income countries have, on average, shown no change, and low-income ones show continued deterioration (Figure E5).

Where industry, mining and the use of agricultural chemicals are expanding, rivers become contaminated with toxic chemicals and with heavy metals such as lead and mercury. These pollutants are hard to remove from drinking water with standard purification facilities. They may accumulate in shellfish and fish, which may be eaten by people who do not realize that the food is contaminated. In a sample of fish and shellfish caught in Jakarta Bay, Indonesia, 44% exceeded WHO guidelines for lead, 38% those for mercury and 76% those for cadmium. In Malaysia, after it was found that lead levels in 12 rivers frequently exceeded the national standard for safe drinking water, the country began monitoring rivers for heavy metals. During the 1980s lead also worsened or became a problem for the first time in some rivers in Brazil (Paraíba and Guandu), Korea (Han) and Turkey (Sakarya).

As surface water near towns and cities becomes increasingly polluted and costly to purify, public water utilities and other urban water users have turned to groundwater as a potential source of a cheaper and safer supply. Monitoring of groundwater for contamination has lagged behind monitoring of surface water, but that is beginning to change as in many places groundwater, too, is becoming polluted. It is often more important to prevent contamination of groundwater than of surface water. Aquifers do not have the self-cleansing capacity of rivers and, once polluted, are difficult and costly to clean.

One of the principal origins of groundwater pollution is seepage from the improper use and disposal of heavy metals, synthetic chemicals and other hazardous wastes. In Latin America, for instance, the quantity of such compounds reaching groundwater from waste dumps appears to be doubling every 15 years. Sometimes industrial effluents are discharged directly into groundwater. In coastal areas, overpumping causes salt water to infiltrate freshwater aquifers. In some towns contamination occurs because of lack of sewerage systems or poor maintenance of septic tanks. Where intensive

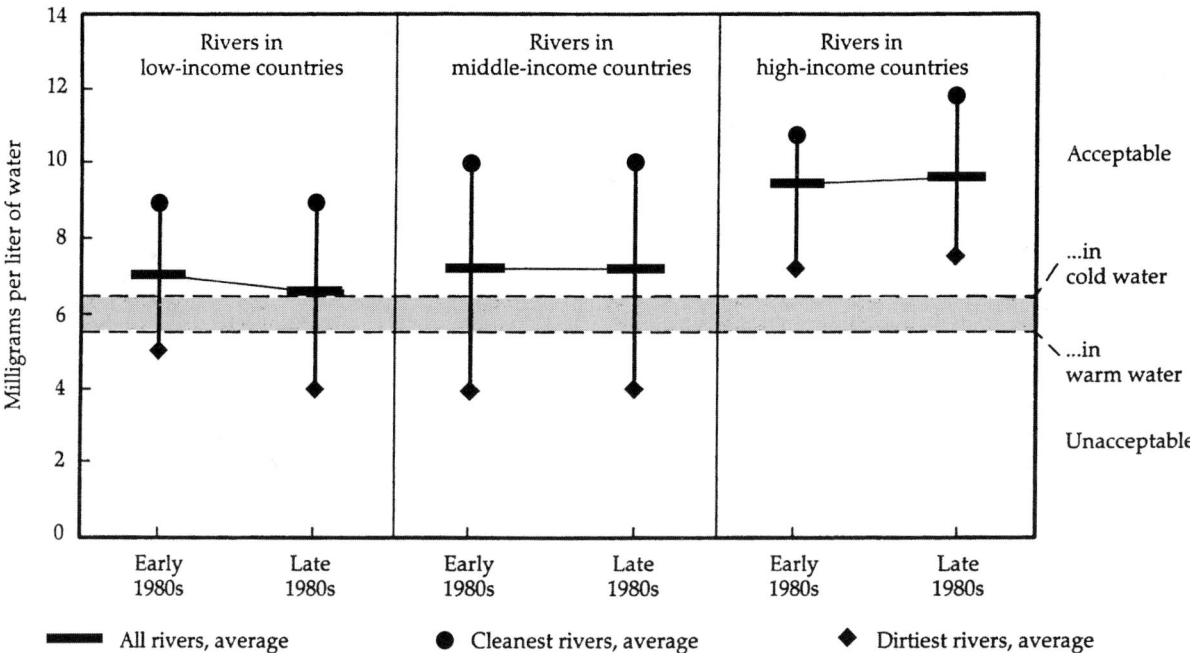

Figure E5 Dissolved oxygen in rivers: levels and trends across country income groups. Data are for 20 sites in low income countries, 31 sites in middle-income countries and 17 sites in high-income countries. 'Cleanest rivers' and 'dirtiest rivers' are the first and last quartiles of sites when ranked by water quality. Periods of time series differ somewhat by site. US Environmental Protection Agency water standards for supporting aquatic life are used as the criteria for acceptability.

agriculture relies on chemical inputs combined with irrigation, the chemicals often leach into groundwater.

Water quality has continued to deteriorate despite substantial progress in bringing sanitation services to the world's population. Little has been done to extend the treatment of human sewage. The replacement of septic tank systems with piped sewerage systems greatly reduces the risks of groundwater pollution but leads to increased pollution of surface water unless the sewage is treated. Yet in Latin America as little as 2% of sewage receives any treatment. Moreover, despite the expansion of sanitation services, the absolute number of people in urban areas without access to these services is thought to have grown by more than 70 million in the 1980s, and more than 1.7 billion people worldwide are without access (Figure E6).

Access to uncontaminated water has barely kept pace with population growth. Official WHO figures suggest that between 1980 and 1990 more than 1.6 billion additional people were provided with access to water of reasonable quality. In fact, however, many of those who officially have access still drink polluted water. At least 170 million people in urban areas still lack a source of potable water near their homes, and in rural areas, although access has increased rapidly in the past decade, more than 855 million are still without safe water (Figure E6).

It is the poor – the woman in Niamey drawing water from an open sewage channel or the Bangladeshi child washing household utensils in a pool also used as a latrine – who bear the brunt of risks from contaminated water. The differences in access to safe water by income exist both within and across countries. The gap in access between lower- and higher-income countries has narrowed only slightly, and within countries inequities continue to be striking. For example, a family in the top fifth income group in Peru, the Dominican Republic or Ghana is, respectively, three, six and 12 times more likely to have a house connection than a family in the bottom fifth income group in those countries. The rural poor are more likely to rely directly on rivers, lakes and unprotected shallow wells for their water needs and are least able to bear the cost of simple preventive measures such as boiling water to make it safe for drinking. In many cities in developing countries, poor households in neighborhoods unserved by the municipal water system buy water from private vendors, typically at prices several times greater than the charges for households with municipal hook-ups.

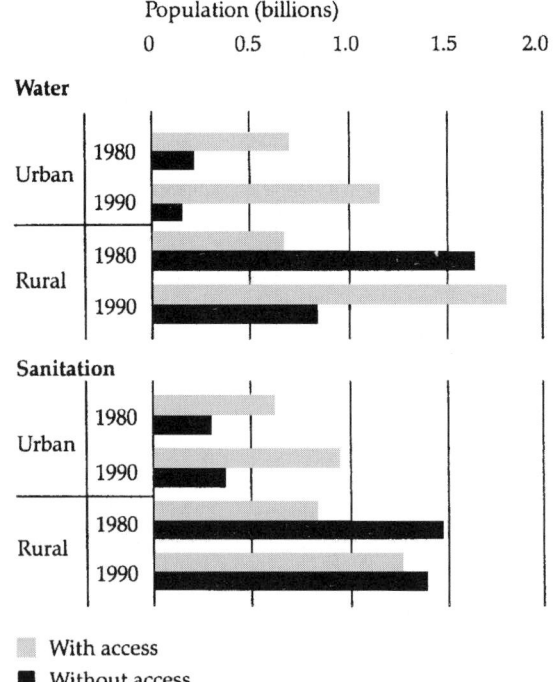

Figure E6 Access to safe water and adequate sanitation in developing countries, 1980 and 1990. (Source: World Health Organization data.)

Water scarcity

Globally, fresh water is abundant. Each year an average of more than 7000 m^3 per capita enters rivers and aquifers. It does not always arrive where and when it is needed. Twenty-two countries already have

Table E1 Availability of water, by region

Region[a]	Annual internal renewable water resources		Percentage of population living in countries with scarce annual per capita resources	
	Total (10^3 km^3)	Per capita (10^3 m^3)	1000 m^3	1000–2000 m^3
Sub-Saharan Africa	3.8	7.1	8	16
East Asia and the Pacific	9.3	5.3	<1	6
South Asia	4.9	4.2	0	0
Eastern Europe and former USSR	4.7	11.4	3	19
Other Europe	2.0	4.6	6	15
Middle East and North Africa	0.3	1.0	53	18
Latin America and the Caribbean	10.6	23.9	<1	4
Canada and United States	5.4	19.4	0	0
World	40.9	7.7	4	8

[a] Regional groups include high-income economies. Sub-Saharan Africa includes South Africa.
Source: World Resources Institute data; World Bank data.

renewable water resources of less than 1000 m^3 per capita – a level commonly taken to indicate that water scarcity is a severe constraint. An additional 18 countries have less than 2000 m^3 per capita on average, dangerously little in years of short rainfall. Most of the countries with limited renewable water resources are in the Middle East, North Africa and Sub-Saharan Africa, the regions where populations are growing fastest (Table E1). Elsewhere, water scarcity is less of a problem at the national level, but it is nevertheless severe in certain watersheds of northern China, west and south India and Mexico.

Water scarcity is often a regional problem. More than 200 river systems, draining over half of the planet's land area, are shared by two or more countries. Overpumping of groundwater aquifers that stretch under political borders also injects international politics into the management of water scarcity.

When water is scarce, countries may sometimes have to make awkward choices between quantity and quality. As river flows decline, effluents are less diluted. In countries with inadequate effluent treatment, water quality can often be improved only if supplies from dams are used to maintain flows for dilution rather than for other economic uses. Often, the disparate agencies involved in water management cannot agree on trade-offs between quantity and quality.

In many countries water scarcity is becoming an increasing constraint, not just on household provision but on economic activity in general. Downstream cities can become so short of water as it is drawn off upstream that their industries are seasonally forced to curtail operations. That, indeed, has become routine during dry months in the Indonesian regional capital of Surabaya. As industry, irrigation and population expand, so do the economic and environmental costs of investing in additional water supply. There is growing awareness of the need to integrate the management of water demand from the different sectors of the economy.

Health effects

The use of polluted waters for drinking and bathing is one of the principal pathways for infection by diseases that kill millions and sicken more than a billion people each year. Diseases such as typhoid and cholera are carried in infected drinking water; others are spread when people wash themselves in contaminated water. Because of their effect on human welfare and economic growth, deficient water supplies and sanitation pose the most serious environmental problems that face developing countries today. Consider first the consequences for health.

The direct impact of waterborne diseases is huge, especially for children and the poor (who are most at risk). Unsafe water is implicated in many cases of diarrheal diseases which, as a group, kill more than 3 million people, mostly children, and cause about 900 million episodes of illness each year. At any one time more than 900 million people are afflicted with roundworm infection and 200 million with schistosomiasis. Many of these conditions have large indirect health effects – frequent diarrhea, for instance, can leave a child vulnerable to illness and death from other causes.

A key question is what the reduction in this burden of disease and death would be if water and sanitation were improved. This is not a

Table E2 Effects of improved water and sanitation on sickness

Disease	Millions of people affected by illness	Median reduction attributable to improvement (%)
Diarrhea	900[a]	22
Roundworm	900	28
Guinea worm	4	76
Schistosomiasis	200	73

[a] Refers to number of cases per year.

Table E3 Effects of water supply and sanitation improvements on morbidity from diarrhea

Type of improvement	Median reduction in morbidity (%)
Quality of water	16
Availability of water	25
Quality and availability of water	37
Disposal of excreta	22

simple question to answer, or one on which all epidemiologists agree. Too little is known about how risks and diseases are distributed and interact with each other, and uncertainty remains over the extent to which modest changes in infrastructure account for longterm health improvements. But some impression can be gained from a recent comprehensive review by the US Agency for International Development (USAID), which summarized the findings from about 100 studies of the health impact of improvements in water supplies and sanitation (Table E2). Most of the interventions studied were improvements in the quality or availability of water or in the disposal of excreta. The review showed that the effects of these improvements are large, with median reductions ranging from 22% for diarrhea to 76% for guinea worm. It also showed that environmental improvements have a greater impact on mortality than on illness, with median reductions of 60% in deaths from diarrheal diseases. A companion WHO analysis of the largest group of health impact studies – those on the effect of water and sanitation on diarrheal diseases – suggests that the effects of making several kinds of improvements at the same time (say, in the quality and availability of water) are roughly additive (Table E3). Project experience shows that the gains are reinforced by educating mothers and improving hygiene.

Taking these studies as a guideline, it is possible to make a rough estimate of the effects of providing access to safe water and adequate sanitation to all who currently lack it. If the health risks of these people were reduced by the levels shown in Table E2, then there would be

- 2 million fewer deaths from diarrhea each year among children under 5 years of age (as an indication of magnitudes, about 10 million infants die each year in developing countries from all causes);

- 200 million fewer episodes of diarrheal illness annually;
- 300 million fewer people with roundworm infection;
- 150 million fewer people with schistosomiasis;
- 2 million fewer people infected with guinea worm.

Other effects

The costs of water pollution include the damage it does to fisheries, which provide the main source of protein in many countries, and to the livelihoods of many rural people. For instance, pollution of coastal waters in northern China is implicated, along with overfishing, in a sharp drop in prawn and shellfish harvests. Heavy silt loads aggravated by land development and logging are reducing coastal coral and the fish populations that feed and breed in it, as in Bacuit Bay in Palawan, the Philippines. Fish are often contaminated by sewage and toxic substances that make them unfit for human consumption. Sewage contamination of seafood is thought responsible for a serious outbreak of hepatitis A in Shanghai and for the recent spread of cholera in Peru.

Excessive water withdrawal contributes to other environmental problems. In addition to displacing people and flooding farmland, damming rivers for reservoirs alters the mix of fresh and salt water in estuaries, influences coastal stability by affecting sedimentation, and transforms fisheries by changing spawning grounds and river hydrology. When groundwater is drawn off at a rate faster than the rate of natural recharge, the water table falls. In China's northern provinces, where ten large cities rely on groundwater for their basic water supply, water tables have been dropping – by as much as 1 m per year in wells serving Beijing, Xian and Tianjin. In the southern Indian state of Tamil Nadu a decade of heavy pumping has brought about a drop of more than 25 m in the water table. The costs are often substantial and go beyond the additional costs of pumping from greater depths and replacing shallow wells with deep tubewells. Coastal aquifers can become saline, and land subsidence can compact underground aquifers and permanently reduce their capacity to recharge themselves. Sewers and roads may also be harmed, as has happened in Mexico City and Bangkok.

R.W. Herschy

Source

The World Bank, 1992. *World Development Report*.

Cross references

Water availability and river water quality
Water resources: Europe
Water resources: quality assessment
Water quality for drinking: WHO guidelines

ESTUARINE HYDROLOGY

An estuary is the wide mouth of a river, or arm of the sea, where the tide meets the river current, or flows and ebbs. It may also be defined as 'a body of water in which the river water mixes with and measurably dilutes sea water' (Ketchum, 1951). These definitions do not overlap completely, because a lagoon connected with the sea may also be affected by the tide.

Some scientists prefer to describe the environment in terms of the salinity of the water (saline, brackish or fresh), but saline water is not restricted to marginal marine areas. Such a description does not consider, therefore, the most characteristic aspect of the estuarine environment – that it is a region of steep and variable gradients in the environmental conditions (Figure E7).

If the physiography of estuaries is a sole consideration, they can be defined as 'bodies of water bordered by and partly cut off from the ocean by land masses that were originally shaped by nonmarine agencies. They are usually perpendicular to the coastline and most of them occupy the drowned mouths of stream valleys and are, therefore,

usually considered as evidence of submergence' (Emery and Stevenson, 1957; see also Fairbridge, 1980).

Classification

There is no system which is universally used to classify estuaries. A broad classification separates normal (or positive) estuaries, in which freshwater inflow exceeds evaporation, from inverse (hypersaline) estuaries in which evaporation exceeds freshwater inflow. Neutral estuaries are those in which neither evaporation nor river discharge dominates.

Estuaries along most coastlines have been formed partly by the subsidence of the land mass and partly by the rise in sea level. These embayments are usually elongate indentures of the coastline with rivers flowing in from the landward ends. Deep estuaries are known as rias. In eastern North America most estuaries are shallow with irregular, or dendritic, shorelines and are normal estuaries.

Along the Gulf Coast of the United States, marine processes have built a series of barrier islands parallel to the coastline. Most of the islands extend across the mouths of estuaries, forming a lagoon and decreasing the width of the estuarine entrance to the open sea.

The exchange of water, in such cases, between the estuary and the open sea is modified by the intervening lagoon in which evaporation may exceed freshwater inflow. The waters in the estuary, then, have salinities higher than normal as a result of the exchange with the lagoonal water.

Water characteristics and circulation

The important feature in an estuary is the intermixing of seawater with the fresh water from land drainage. This interaction usually produces a variation, both horizontal and vertical, in the salinity of estuarine waters. In normal estuaries, salinities range from nearly zero at the river's mouth, to approximately 30‰ at the seaward extremity. In addition, there is generally an increase in salinity with depth (Wiley, 1976).

An inverse estuary also has greater salinities at depth, but the highest salinities are at the head of the embayment rather than at the mouth. There may be a difference of several parts per thousand between the salinity at the head and that of normal seawater.

Temperature

The water in estuaries, and especially that overlying the tidal flats, is relatively thin, so it follows the variations in temperature of the atmosphere more closely than does the water of the open sea. The water is much colder in winter and warmer in summer than is the sea. The diurnal variation is also greater than in the sea.

There are pronounced variations in water temperature with depth. During the winter, the water is cold and nearly isothermal at all depths. In some instances, in response to cold weather, the surface water may become a degree or so colder than the deeper water. During the summer, solar radiation and minor wind mixing produce a high temperature at the surface with less change at depth. The difference between surface and bottom temperatures may also be influenced by warm or cold river water flowing over the dense seawater.

Where evaporation exceeds river inflow, the summer surface water may become so saline as to sink to the floor of the estuary and flow out of the entrance beneath the incoming seawater. The temperature of the deep water may then be higher than that at the surface.

Circulation

Water movements in estuaries result mainly from the interaction of tides, river flow and wind. The tides and river flow are usually the dominant factors. In Gulf Coast estuaries, where the tide range is small and river discharge is at times negligible, wind-induced currents are most important.

Stable estuaries

In normal estuaries the distribution of temperature and salinity, and the circulation pattern, are controlled almost exclusively by the tide range and the river inflow. Tidal currents tend to produce turbulent mixing of the river water and the seawater. However, the low-density fresh water above the seawater results in a stable vertical stratification

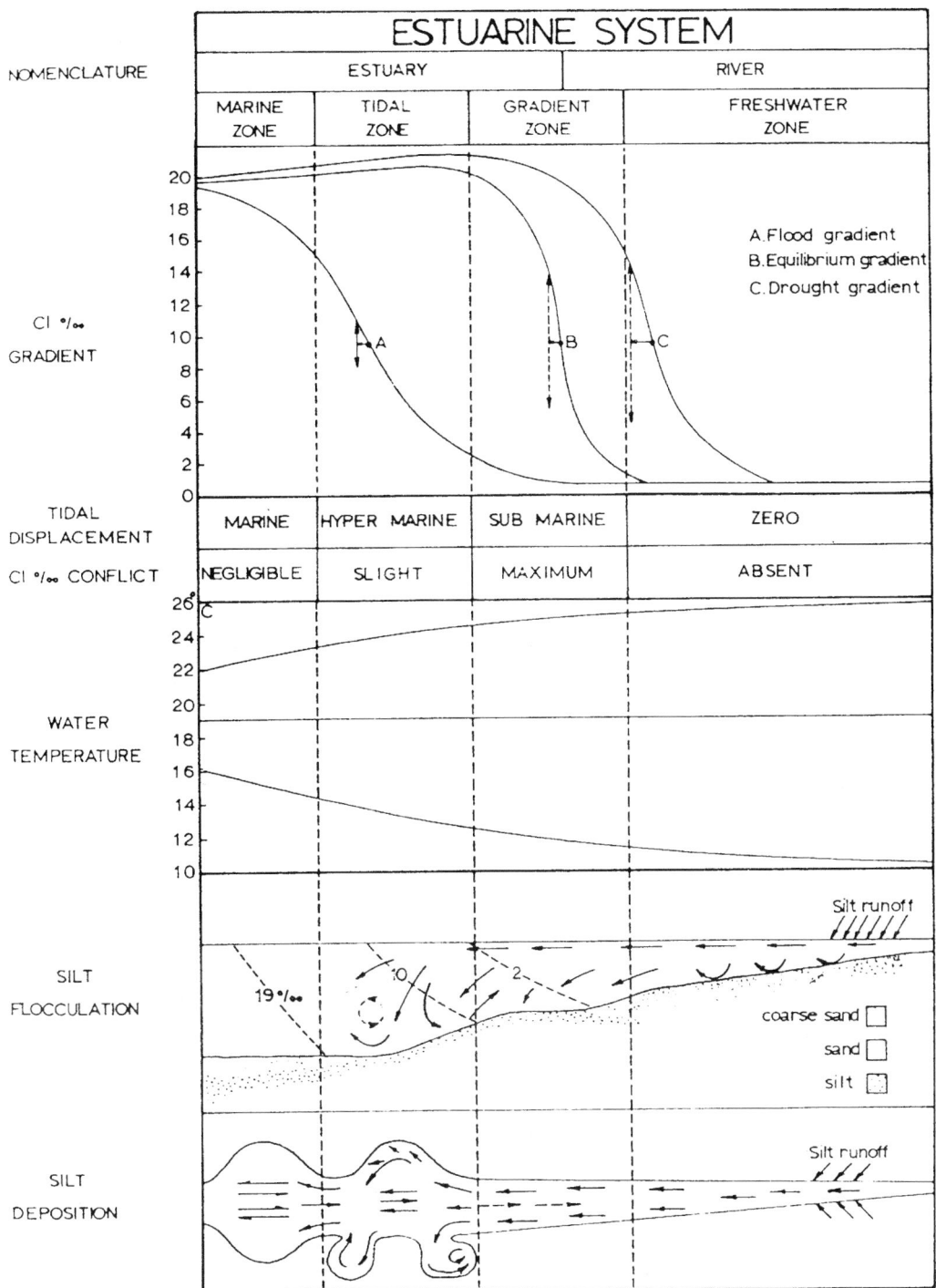

Figure E7 Zonal features of a composite Australian system. (Modified from Rochford, 1951.)

which resists mixing. As a consequence, the relative magnitudes of the river discharge and the tidal flow are significant in controlling the physical structure of the water in the estuary.

Where the river flow is large in relation to the tidal exchange, the seawater enters the estuary as a saltwater 'wedge' along the bottom. However, there is frictional drag between the overlying fresh water, the saltwater wedge and the bottom. The relative velocities of the seaward-flowing fresh water and the intruding saltwater wedge control the magnitude of the friction factor. Thus the actual position of the wedge is closely dependent on the volume of the river flow. When the volume of river discharge is great, the wedge extends only a short distance into the estuary and, of course, vice versa (Pritchard, 1952).

The salinity of the saltwater wedge remains similar to that of the open sea because there is only minor mixing with the seaward-flowing fresh water. However, at the interface between the two types of water, waves form and sometimes intrude into the surface water. Thus, the salt content in the upper layers increases slightly as the water moves seaward. Even so, throughout the estuary, a sharp salinity gradient exists between the two water layers.

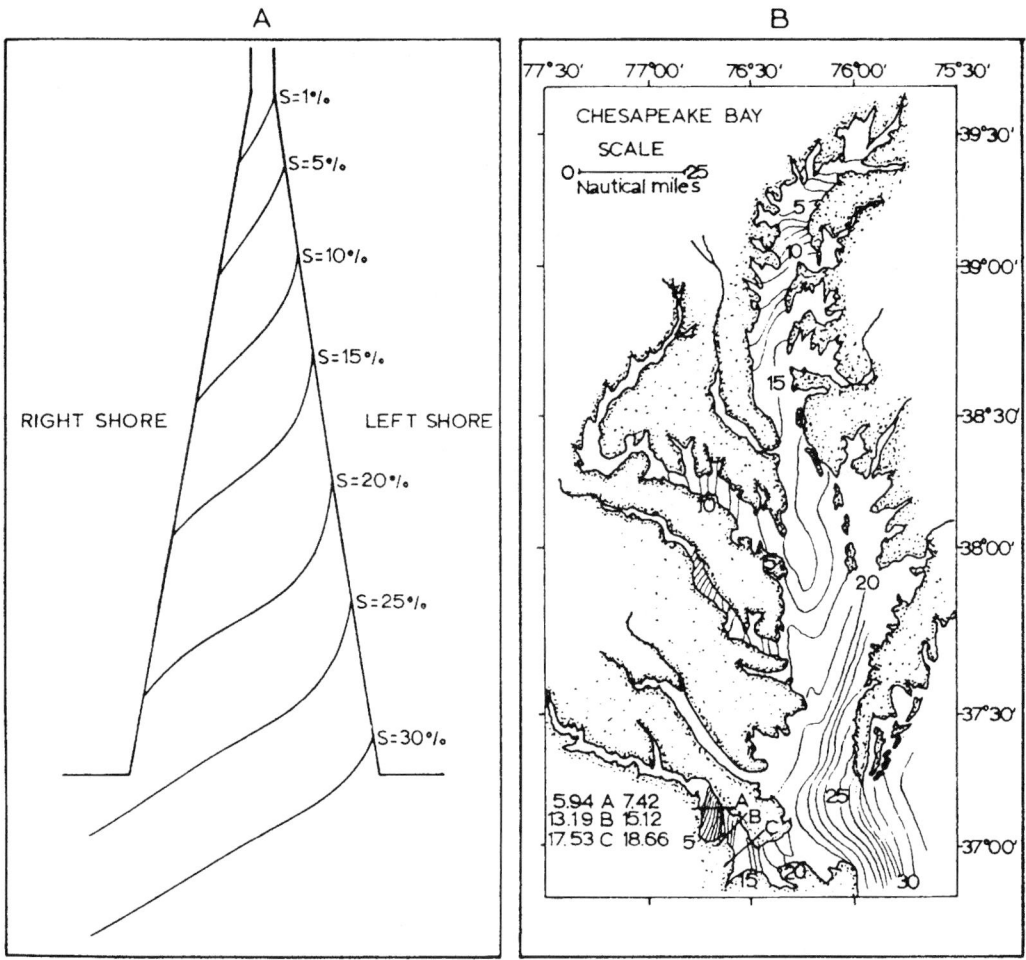

Figure E8 Hypothetical and actual isohalines in a northern hemisphere estuary. (Modified from Pritchard, 1952.)

The loss of salt water from the wedge to the upper layer is compensated by a flow of water from the sea. The exchange from below takes place all along the upper interface of the wedge. As a result there is a flow directed upstream at all positions within the wedge. The landward-moving water in the salt wedge is minor, however, and of the two, the seaward flow of surface water is far greater.

Partly mixed estuaries
Where tidal movements are great as compared to the volume of river discharge, mixing between the seawater and fresh water is sufficient to destroy sharp interfaces. The salt wedge, in such cases, does not exist as an identifiable feature, but a transition layer of definitely increasing salinity does occur. In such an estuary, however, the salinity in both the upper and lower layers decreases toward the head of the estuary.

The chief cause of currents in estuaries in which the waters are partly mixed is the tide. As in the stratified estuary, there is a net water movement superimposed on the tidal currents – a net seaward flow at the surface and a net flow toward the head in the deeper layers. These water motions are not as well defined as in a stratified estuary, and there is no sharp current interface. The flow from the deeper layers toward the surface decreases towards the head of the embayment. The volume rate of seaward flow increases, therefore, towards the mouth (Pritchard, 1952).

Mixed estuaries
In wholly mixed estuaries, the movements induced by the tide are far greater than those produced by the river inflow. The waters are

completely mixed and are isohaline from the surface to the bottom. At all depths, the salinity decreases from the mouth to the head.

In such estuaries, the outward flowing water is deflected to the right, in the northern hemisphere, because of Earth rotation (Figure E8). Thus, in wide estuaries, the salinity is less on the right side (looking towards the estuarine mouth) than on the left. A net seaward flow exists along the right side and a net landward flow on the left. Water also moves laterally across the estuary from the left to the right side, resulting in horizontal mixing.

In narrow, well-mixed estuaries, mixing induced by tidal action may be great enough to eliminate any lateral salinity gradient. There is a net seaward flow in all waters and the only difference in salinity is the normal decrease towards the head.

Estuaries bordered by lagoons
Along coastal regions where barrier islands extend across the mouths of estuaries, the water bodies are usually so shallow that mixing by winds is sufficient to produce homogeneous water. Tidal currents are only significant through the inlets between the barrier islands. The total volume of water which flows in and out is relatively small. As a consequence, the rise and fall of the tide and tidal currents are minor within the estuary, and the most significant currents are from wind action.

There is, necessarily, a net flow of water out of these shallow estuaries sufficient to remove the water added by freshwater discharge. The large cross-sectional area of the estuary and the dampening effect of the coastal lagoon reduce the flow so that it is normally not directly measurable. The net motion may be completely modified by wind action to the extent that high water and constant, net inflow may occur during times when prevailing winds blow up the estuary. Strong winds blowing from the land reverse this effect and result in extremely low water levels in the estuary and the extrusion of estuarine waters many kilometers to sea.

Carbon dioxide and oxygen

The oxygen content of estuarine waters reaches a minimum in the pre-dawn hours and a maximum in the later daylight hours in response to the respiration of plants and animals and photosynthesis by the plants. This general trend is complicated by the tidal regime and the river discharge, but annual means indicate such maxima and minima exist (Stevenson and Emery, 1958).

The range of oxygen concentrations is greatest in shallow water where the volume of organisms is greatest. Ranges of from nearly zero at sunrise to 260% of saturation have been measured in areas where marsh grasses are concentrated. In the main body of estuarine water, oxygen values usually lie between 80 and 120% of saturation.

A close approach to the variation of carbon dioxide is that of the pH because, basically, the higher the content of carbon dioxide the lower is the pH. Thus the photosynthetic process of plants should tend to make the pH low at night and high during the day. The range is usually not great with the maximum measured at slightly more than one pH unit (7.5–8.6; Stevenson and Emery, 1958).

There is also a seasonal cycle of carbon dioxide and oxygen. Greater ranges occur in the summer in response to increased plant growth. Increased river flow dilutes the water and reduces the carbon dioxide.

The oxygen concentration usually decreases with depth. Where fresh water flows over sea-water, there is only a slow replacement of bottom water from the open sea which may already be low in oxygen. Accordingly, the bottom water may become increasingly more stagnant and less hospitable to oxygen-consuming life. Hypersaline estuaries may have highly oxygenated waters throughout their depth because the bottom water has only recently sunk from the surface.

Seiches in estuaries

In some of the larger estuaries, the periodic flooding by the tides is supplemented by seiches, long stationary waves. The simplest seiche is one whose node is at the mouth of the estuary and the antinode near the head. The period of the seiche is controlled by the length and depth of the body of water, and where its natural period nearly coincides with that of the tide, as at the Bay of Fundy, a great fluctuation in sea level occurs (about 15 m). In most estuaries the seiche is only a few centimeters and is obscured by the much greater tidal amplitude.

Waves

Wind waves are small in estuaries because of the short fetch and the shallow water. They usually cause little erosion although, when the tide is high, waves may stir up the muddy sediment on tidal flats. Waves may transport some sand and, because the largest waves come across the widest part of the estuary, they form sand spits pointing upstream in tidal channels.

A tidal bore (a wave of translation) is common in narrow estuaries and tidal channels. As a result of the shape of the entrance and bottom friction, the flooding tide is held back for a time until the water finally rushes up the channel as a steep wall of water. Bores may be from a few centimeters to several meters in height and move at velocities as great as 18 km h^{-1}. The character of a tidal bore is determined, in part, by the river discharge which must be sufficient to hold back the tide for a period of time.

Estuarine sediments

The inorganic sediments of estuaries are derived from inflowing rivers bordering sea cliffs, the sea floor outside the estuary and the reworked deposits of tidal flats and marshes along the shores. Regardless of the source, much reworking of sediment occurs within estuaries. Erosion, too, is evident from the migration of tidal channels and the muddy color of the water when no river inflow is taking place. Some estuaries have entrances narrow enough so that tidal currents scour the bottom locally, leaving rocky or gravelly bottoms. The prevailing condition, however, must be one of deposition, and the average rate of deposition is greater than that of the open sea.

Distribution of grain sizes.

The coarsest sediment in most estuaries is on the barrier or bay-mouth bar, and consists of sand and cobbles. Generally, this material is too coarse to have been transported across the tidal flats, but is derived from erosion of a sea cliff, then transported and deposited by longshore currents and waves. The excellent sorting and absence of much silt and clay may result from the turbulence of the waves.

The flat portions of the floors of estuaries that are deeper than about 6 m are usually covered by sediment which becomes progressively finer with depth of water. A smooth concentric pattern of sediments may occur, ranging from sand along the shore to fine mud at depth. Such a distribution occurs only where the bottom is relatively flat and current conditions are mild. In estuaries where the deeper areas are extremely irregular, mud occurs only in depression and coarse sediments characterize the shallower bottoms.

The sediment distribution in shallow areas, mostly the tidal flats, is more complex but usually follows a systematic pattern. Most of the flow of water is confined to well-defined channels which slowly migrate over the tidal flat (as shown by remapping at intervals of several years). The velocity of the water is such that the finer grains are swept out, leaving the coarse sediment in the channel. The areas between the channels consist of poorly sorted mud which becomes finer with distance from the tidal channels. Probably most, but not all, of the reworking of sediment in estuaries takes place on the shallow flats where the ebbing and flooding currents erode and redeposit the sediment.

Organic constituents

The sediments contain the remains of all phyla of animals and much plant debris. Even though the remains become scattered by scavenging, decomposition and diagenesis, the organisms still have enriched the sediments in organic matter, calcium carbonate, silica, nutrients and other constituents.

Sediments in estuaries located in areas where precipitation exceeds evaporation have organic nitrogen contents from 0.2 to 0.6%, and sediments in hypersaline areas below 0.2%. The percentage used to differentiate between these areas, or 0.2% organic nitrogen, corresponds to about 1.7% organic carbon, or 2.9% total organic matter. Phosphorus is also abundant in sediments of normal environments, ranging from 0.1 to 0.4% (Bayley, 1995).

Calcium carbonate is variable because of the presence or absence of shells and because of solution induced by acidic conditions. In coastal bays in temperate and arctic regions, calcium carbonate ranges between 0 and 6%, whereas in bays of tropical regions it is 10–47%. Seasonally blocked estuaries in subtropical regions are marked by stagnation, the bacterial liberation of H_2S and iron sulfide minerals.

Manner of deposition

In the tidal section of a freshwater river, a transition takes place and the distribution of sediments may be quite variable and confused. When the estuary proper is reached, there is some admixture of sea salts, and where the net upstream flow in lower layers occurs, there is a distinct change in sediment distribution. Finer sediments tend to be deposited in the channel (the reverse of conditions commonly found in river channels). In most streams the bulk of suspended material probably is silt which is deposited directly out of suspension. Clay sizes, however, may be deposited through flocculation. The clays then fall to the deeper floors of the estuaries. Sediment may also travel down rivers at or near the surface in large floating floccules containing organic debris. When these settle to the bottom or are stranded by lowering water level, they are held by capillary action. Near the mouth of the estuary, coarser sediment are again found in the channel as a result of wave action and because much of the silt load has already been deposited in the channel farther upstream.

In ecology, the organic habitats and ecosystems are known as **lentic**, i.e. characterized by moving waters (Fairbridge and Marsh, 1998).

R.E. Stevenson

Bibliography

Arons, A.B. and Stommel, H., 1961. A mixing-length theory of tidal flushing, *Trans. Am. Geophys. Union*, **32**(3), 419–421.

Bayley, P.B., 1995. Understanding large river floodplain ecosystems, *Bioscience*, **45**(3), 153–158.

Fairbridge, R.W., 1980. The estuary: its definition and geodynamic cycle, in Olausson, E. and Cato, I. (eds), *Chemistry and Biochemistry of Estuaries*, Wiley, New York.

Fairbridge, R.W. and Marsh, 1998. Lentic and lotic ecosystems, in Alexander, D.E. (ed.), *The Encyclopedia of Environmental Science*, Chapman and Hall, London.

Ketchum, B.H., 1951. The flushing of tidal estuaries, *Sewage Ind. Wastes*, **23**(2), 198–209.

Lauff, G.H. (ed.), (1967) *Estuaries*, Washington, D.C., AAAS., Publ. 83, 757 pp.

Pritchard, D., 1952. Salinity distribution and circulation in the Chesapeake Bay estuarine system, *J. Marine Res.*, **11**(2), 106–123.

Pritchard, D.W., 1955. Estuarine circulation patterns, *Proc. Am. Soc. Civil Eng.*, **81**(717).

Rochford, D.J., 1951. Studies in Australian estuarine hydrography. I. Introduction and comparative features. *Australian J. Marine Freshwater Res.*, **2**(1).

Stevenson, R.E. and Emery, K.O. (1958) Marshlands at Newport Bay, California, *Allan Hancock Foundation Publ. Occas. Paper*, **20**.

Wiley, M. (ed.), 1976. *Estuarine Processes*, Academic Press, New York, 2 vols, 588 pp, 444 pp.

Cross references

Evaporation: measurement
Evapotranspiration
Saltwater wedge
Seiche
Suspended sediment monitoring: use of Doppler current profiles
Water resources: dictionary of basic terms

EUROPE: CLIMATE, A HYDROLOGICAL PERSPECTIVE

Factors affecting climate

The climatic domain of the continent of Europe extends from the mid-Atlantic islands of the Azores (29°W) across the open North European Plain to the Ural Mountains of the Russian Federation (60°E) and from the islands of the Mediterranean Sea in the south (35°N) to the islands of the Barents Sea which lie well within the Arctic Circle (80°N). The total area is about 15 million km² of which about two-thirds is land. Europe, along with its islands and peninsulas, is the most maritime of the continents. Large areas to the west of the 10°E longitude (Hamburg–Milan) experience a temperate oceanic climate where extremes of temperature and rainfall are infrequently experienced. The chief factors promoting this comparatively unique assemblage of climates are

- the predominant west-to-east movement of weather systems;
- the extensive area of abnormally warm surface waters of the North Atlantic;
- the virtual absence of north–south aligned mountain ranges between latitudes 45 and 60°N that might otherwise alter the nature of the westerly flow;
- the presence of large inland seas such as the warm Mediterranean, the cooler Baltic and the smaller, less warm, Caspian and Black seas.

Elements of the climate of Europe

Energy balance

In common with other midlatitude regions, Europe experiences a net deficit in radiation except at its southern margins. Values derived from satellite information for the Earth–atmosphere system indicate that annual deficits increase from -20 W m^{-2} over central Italy to around -80 W m^{-2} over southern Scandinavia and northern Norway (Raschke *et al.*, 1973). The top of atmosphere satellite-derived values for January for both cloudy and clear skies are lower than this and are about -100 W m^{-2} for southern Italy, falling to -190 W m^{-2} over

Scandinavia. The deficit cannot be attributed principally to the excessive loss of heat from the land surface, though this loss may be significant in cold snowy winters over central and eastern Europe. It is the radiative cooling at mid-tropospheric levels which is also significant. This is at a maximum around latitude 60°N, and it is in this region that maximum cooling takes place in winter (Figure E9). The effects of this cooling on Eurpean climates are profound. In the long term, such cooling produces latitudinal zones of temperature gradient in the atmosphere. These baroclinic zones are associated with high winds and disturbed weather conditions that so often characterize the climate north of the latitude of the Alps. Compensation for the deficit in radiant energy is effected by the transfer of sensible and latent heat from lower latitudes associated with cyclonic disturbances (midlatitude, frontal depressions).

The incoming solar radiation provides a major source of energy which helps to drive the hydrological cycle. The radiant energy is used to heat the surface of the land and adjacent oceans resulting through small-scale turbulence, in the heating (H) of the lowest layer of the atmosphere. The energy may also be used in evaporation (LE). The Bowen ratio β (H/LE) expresses the pathway of energy transfer and is of importance in climate and hydrological studies. Typical graphs are presented for grass near Munich (Figure E10). They show β values of 1.0 in winter and 0.3 in summer (Kessler, 1985). In general, temperate forests and grassland may be expected to have β values of between 0.4 and 0.8 (Oke, 1987) but this value increases significantly under summer drought conditions when radiant energy is transformed primarily into sensible heat (H). This will enhance the drought effect and diminish the role of the water vapor component in the hydrological cycle. In winter, frozen ground and snow cover also significantly affect the value of the Bowen ratio.

The hydrological cycle is driven by the energy input, especially net radiation. Values of surface net radiation are not easily measured from satellites. They are mainly dependent on the latitude and cloudiness of locations. Table E4 provides some general values for two locations in Europe. Interannual variations of about 3 MJ m^{-2} in these values have been measured.

Temperature

Average and accumulated temperatures
The influence of the warm waters of the Atlantic Ocean are most clearly seen in the winter pattern of surface temperatures. Average January values of 7°C in lowland locations of western Ireland decline steadily eastward to around -10°C in the vicinity of Moscow, leading to a meridional pattern between latitudes 45 and 60°N. In summer, the heating of the continental area produces a more zonal thermal pattern with isotherms tilted WSW–ENE (Figure E11). Thus the 17°C isotherm for July run from along the north coast of France across to the southern Baltic Sea towards Leningrad. In northern European Russia, summers, though warm, are short and are constantly threatened by outbreaks of cold air of Arctic origin. A measure of the increasing warmth from north to south through European Russia is provided by the extent to which summer daily air temperatures exceed 10°C. Budyko (1974) found that accumulated daily temperature values of 750 (temperature >10°C) occur at latitude 70°N, increase to 2000 at 55°N and exceed 3000 around the Black Sea at 45°N. On a somewhat different basis, Smith (1976) produced degree-day values over 10°C for England for the six summer months using empirical formulae. In the cloudier climate of the British Isles, values might be expected to be lower. They range from below 500 (temperature >10°C) in northern England (55°N) and approach 1000 in the Channel Islands (49°N). Using this method the equivalent value for Moscow would be 722. Analysis of temperature data by Gerasimov (1964) shows that average diurnal temperatures rise above 10°C by 1 July along the coastlands of the Barents Sea and by 21 April in the Crimea. Similar dates in autumn are 11 August and the 15 October, respectively, when temperatures fall below 10°C. These dates also give an indication of the duration of the growing season. Alternatively, the frost-free period may be used.

Frost-free period
The average length of the frost-free period on the open plains of European Russia increased from 75 days in the north to around 200 along the northern shore of the Black Sea. Further west, across the European Plain, the length of the frost-free period shows much greater variation (Lednicky, 1985). Over lowland France, for example, values

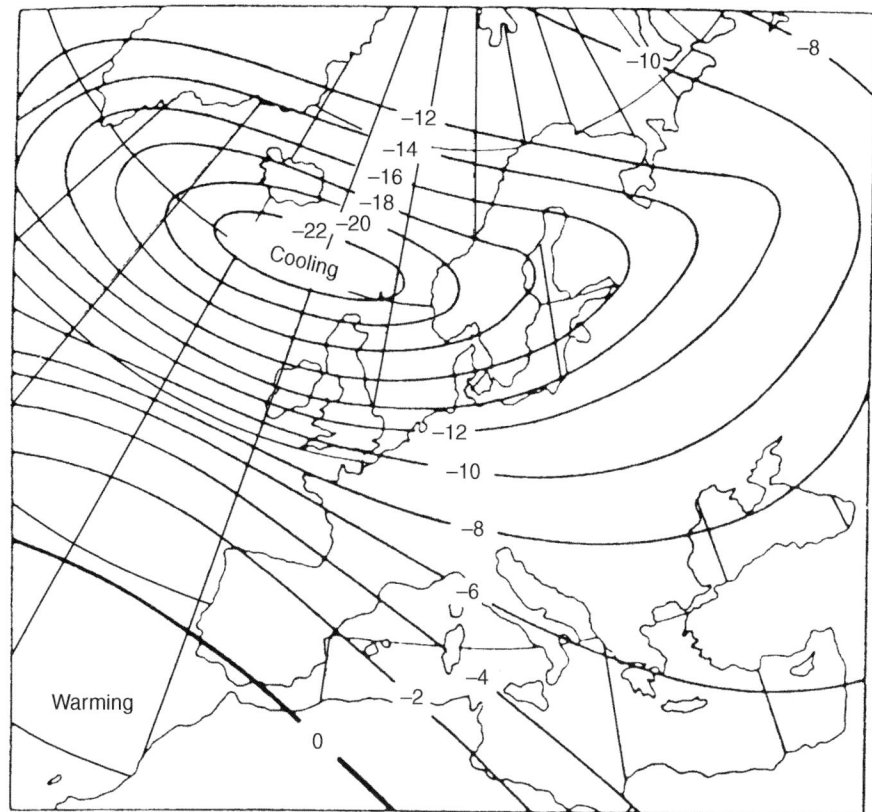

Figure E9 Change in the heat content of the troposphere from mid-November through to mid-December in W m^{-2}. Average values are for 1955–1959. Values for maximum cooling represent 7.1% of the incoming radiation at solar noon for 60°N at end of November at the top of the atmosphere.

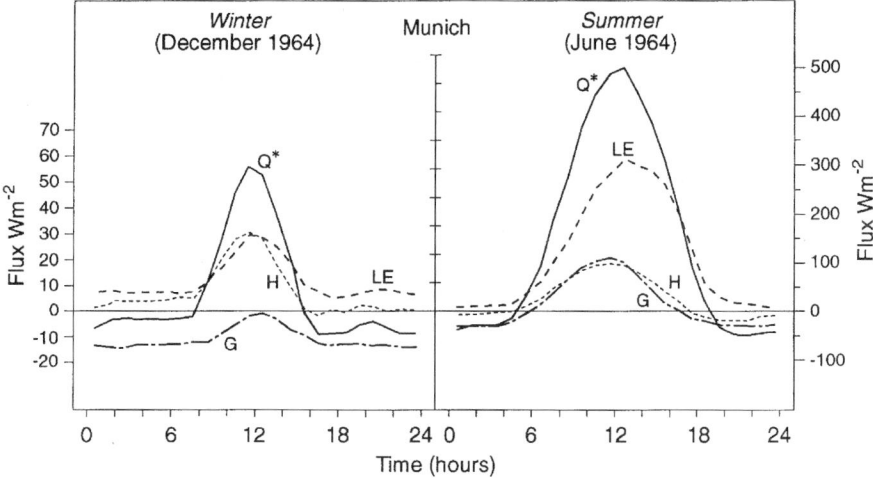

Figure E10 Mean diurnal variation of components of the heat balance. Q^* is net radiation, H and LE are fluxes of sensible and latent heat and G is the soil heat flux for a grass surface near Munich, 48°16'N 11°41'E for December and June 1964. (After Kessler, 1985.)

range from 160 days in Alsace in eastern France to 240 days along the coastlands of the Bay of Biscay (Garnier, 1954). In contrast, the determination of the frost-free period in mountainous countries such as Austria is largely a function of elevation. In the Austrian Alps the frost-free season at 800 m above mean sea level lasts for about 155 days per year. At a height of 1600 m this is reduced to 104 days, whereas above 2500 m spells of below freezing temperatures may be

expected even at the height of summer. In the Iberian Peninsula the frost-free period is governed both by altitude and proximity to the sea. Much of Spain experiences less than 250 days frost-free. Only the extreme southeast and southwest of the peninsula are entirely free of frost during most years. An empirical relationship has been established between air and ground frosts by Hulme *et al.* (1995) and a complete data set constructed for Europe.

Table E4 Mean daily net radiation values for selected stations in Europe (MJ m^{-2}) (after Johannessen, 1970)

Location	Lat. (°N)	Mar.	June	Sept.	Dec.
Trondheim	63°25'	−2.4	9.4	1.9	−4.4
Copenhagen	55°41'	2.3	11.5	4.1	−1.2

Extreme values

Extreme values of temperature tend to reflect the local controls of topography as well as distance from the North Atlantic Ocean and adjacent seas. Low temperatures show much greater variation in time and space than high temperatures. Very dry, cold air, often of Arctic origin, with little wind is required for temperatures to fall below −20°C. Such an occurrence is far more common in winter over countries such as Finland, where temperatures below −30°C are recorded nearly every winter. The lowest temperature indicated in the standard climatic tables is −51°C at Syktyvkar (61°40'N, 50°51'E) in northeast European Russia. Moscow, along with many stations in northern Russia and Finland, has experienced temperatures of −40°C at measuring screen height. The penetration of cold air into parts of eastern Europe also means that very low temperatures have occurred at relatively low latitudes, as may be seen from the following examples: Iasi, Rumania, at 47°N, −30°C; Sofia, Bulgaria, 42°30'N, −27.5°C. There were also reports of −26°C from several locations in Thrace, Greece, in December 1972. Only the southernmost part of Spain and a part of the Mediterranean island of Crete have absolute minima at or above 0°C. Comparatively maritime countries such as Spain experience frost in severe winters. Madrid recorded −10.1°C sometime during the period 1901–1930 and Rome, Italy has reported −7.4°C. Over the British Isles, minima of −10°C have been recorded along the western seaboard. Inland values of −25°C are very rare, but occurred in the Dee Valley of the Grampians in Scotland and in the West Midlands of England during the winters of 1995–1996 and 1981–1982, respectively.

In contrast, there is little variation in absolute maxima. Over much of northern Europe temperatures occasionally climb into the low 30s (Jenkinson, 1985), while over the remainder of Europe temperatures between 35 and 39°C have occurred. Apart from isolated events, temperatures above 40°C have only been recorded south of latitude 42°N. Absolute maximums of 45°C (Palermo, Sicily) and 47°C (Malaga, Spain) testify to occasional advection northwards across the Mediterranean of very warm desert air from the Sahara. The searing heat accompanying the scirocco and Levante winds is well known.

Continentality

This is an indicator of the diurnal and seasonal range in temperature. Various indices have been formulated to express the degree of continentality over Europe. In a general sense, the indices make use of the range of the mean temperature between the coldest and warmest months of the year and is known as the annual range of mean temperature. The range increases from the Atlantic coastlands eastward across Europe into Russia (Figure E12). A relatively simple index has been proposed by Tsenker (in Borisov, 1965) who proposed that $K = A/\phi$ where K is the index, A is the annual range and ϕ the latitude. Similar values are obtained from Conrad's index where $K = [1.7A/\sin(\phi + 10)] − 14$ (Conrad, 1946). More complex formulae appear in Russian literature, an example being that of Ivanov, who included a term for diurnal range of temperature and an additional term for relative humidity deficit. Using Conrad's equation, a value of 4 is obtained for western Ireland, 12.5 for London, 24 for Berlin and 40 for Moscow. Indices of over 30 are also found in parts of the Mediterranean countries, especially where inversions of temperature occur in winter, such as Milan (32). The index may be calculated for individual years and this may indicate the degree of moderating

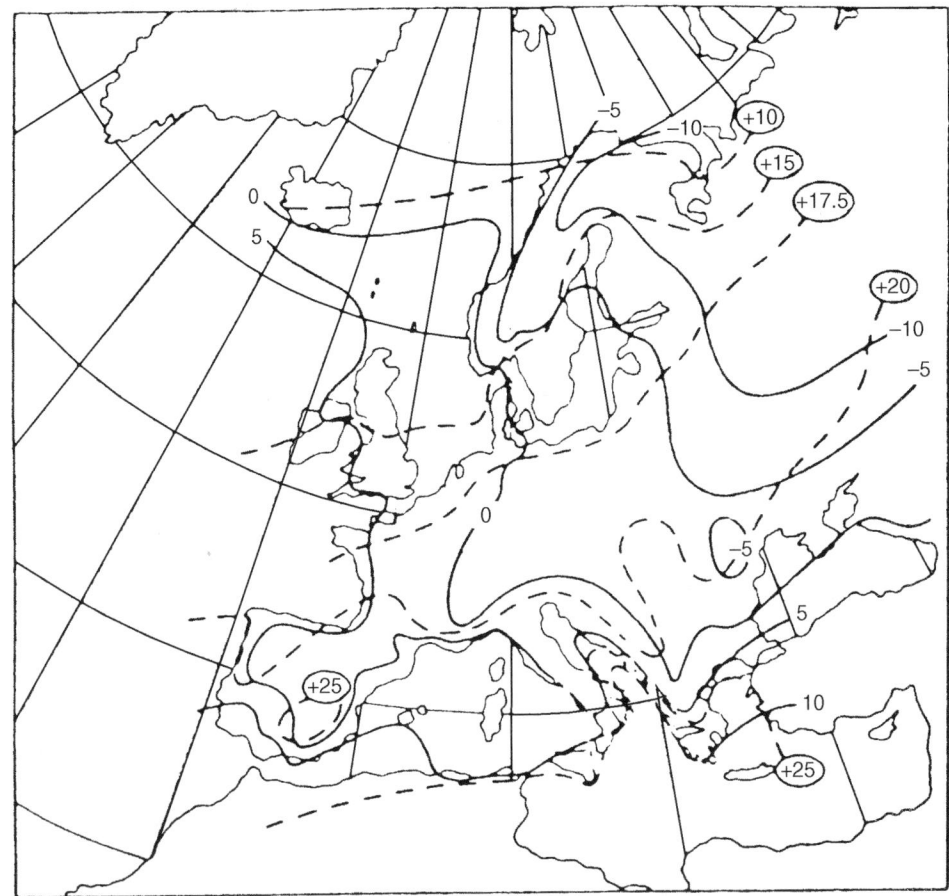

Figure E11 Mean temperatures for January (solid lines) and July (dashed lines) in °C for the WMO data period 1951–1980.

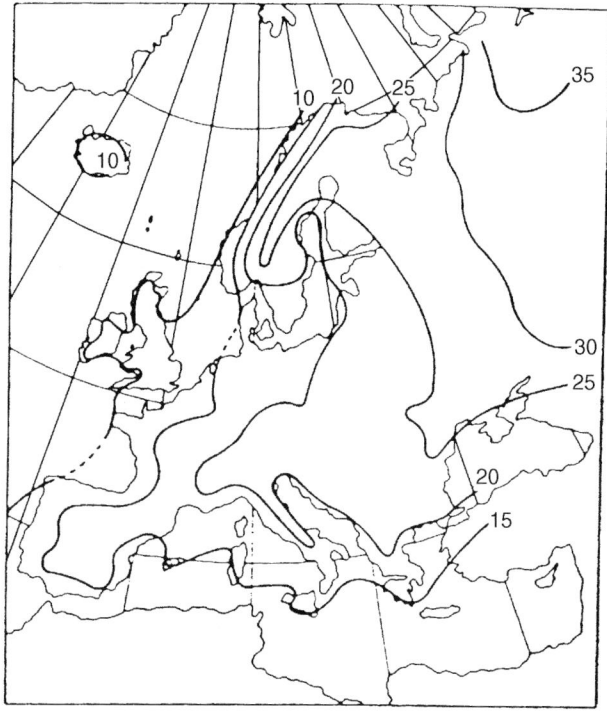

Figure E12 The annual range of mean monthly air temperatures in °C for the WMO data period 1951–1980.

Atlantic oceanic influence in any one year. In the context of the northern hemisphere atmospheric circulation, a place may experience an increase in continentality if anticyclones are more dominant than expected. Thus Conrad's index for central England rose to 15 in 1995 when high pressure dominated the summer.

Trends in temperature
Trends in land surface temperatures have been published by Houghton *et al.* (1990) as derived from P.D. Jones' data set. Figure E13 shows temperature trends using a data set covering most of Europe. Despite marked oscillations in the temperature series since 1920, a general upward trend is discernible. This may be compared with the area farther north where a marked cooling took place in the first half of the twentieth century. Data for individual countries do not necessarily follow this dominant trend (Türkes, 1995). Recent seasonal variations within Europe are shown in Figure E14 where the

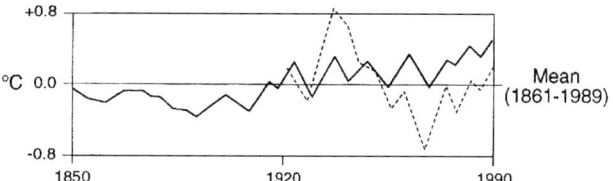

Figure E13 Annual surface land temperature anomaly variations from the long-term average (1861–1989) for midlatitude Europe shown in full line (40–60°N: 0–60°E) and for high-latitude Europe (60–80°N; 0–60°E) in dashed line. (Extracted from Houghton *et al.*, 1990.)

winter decade 1981–1990 was warmer over most of Europe, especially east of 25°E, when compared with the standard 30-year period 1951–1980. Summers in the 1980s were warmer across western Europe but close to average elsewhere (Houghton *et al.*, 1992). Trends in temperature between the surface and 9 km over Europe (tropospheric temperatures at 300–1000 hPa) are more difficult to identify and depend upon the frequency of warm and cold pools of air stagnating over Europe and the position of the main upper westerly flow. If, as seems likely, European temperatures reflect the general trend over all continents in the midlatitude belt of 30–60°N, a significant rise of 0.9 K occurred after 1984 with the largest warming taking place in winter where temperatures rose by 1 K between the late 1970s and the late 1980s, ending the decade 0.6 K higher than during the reference period of the 1950s (Weber, 1995). The effect on the regional hydrological cycle is not so clearly seen (Loaiciga *et al.*, 1996).

Moisture fluxes

Increasing interest is being shown in the role of moisture advection in the regional hydrological cycle. There are five main routes open to low-level entry of moisture into central Europe. These are in order of importance:

1. southwesterly entrance from the Bay of Biscay across France, southern England and Belgium;
2. northwesterly entrance across the North Sea and Denmark;
3. southerly, from the western Mediterranean and around the western side of the Alps;
4. southeasterly, northwards across the Aegean into the Ukraine and sometimes northwest into the Hungarian Basin;
5. northern entrance from the Arctic across Finland.

Moisture flux into Europe is very variable in time and space and is mostly associated with the passage of warm sectors of depressions which pass northeast near to or across the British Isles. The conveyor belt of such low-pressure systems is chiefly responsible for increases in the moisture content, indicated by low-level mixing ratios of

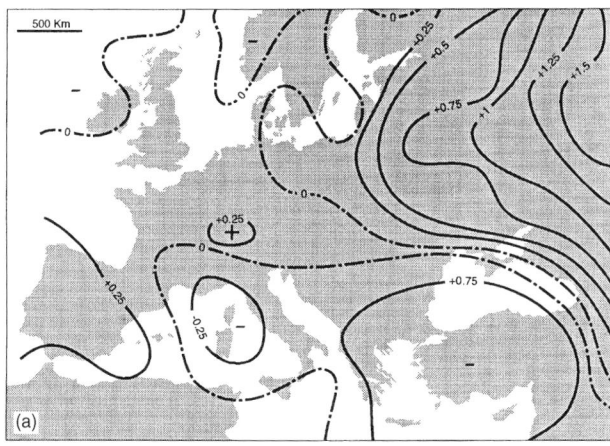

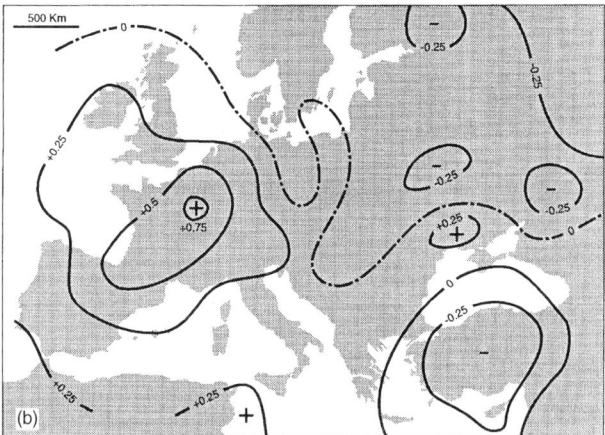

Figure E14 European surface (land and sea) temperature anomalies for the decade 1981–1990 compared to the WMO standard period 1951–80. (a) Winter months: December–February, (b) summer months: June–August. (Extracted from Houghton *et al.*, 1992.)

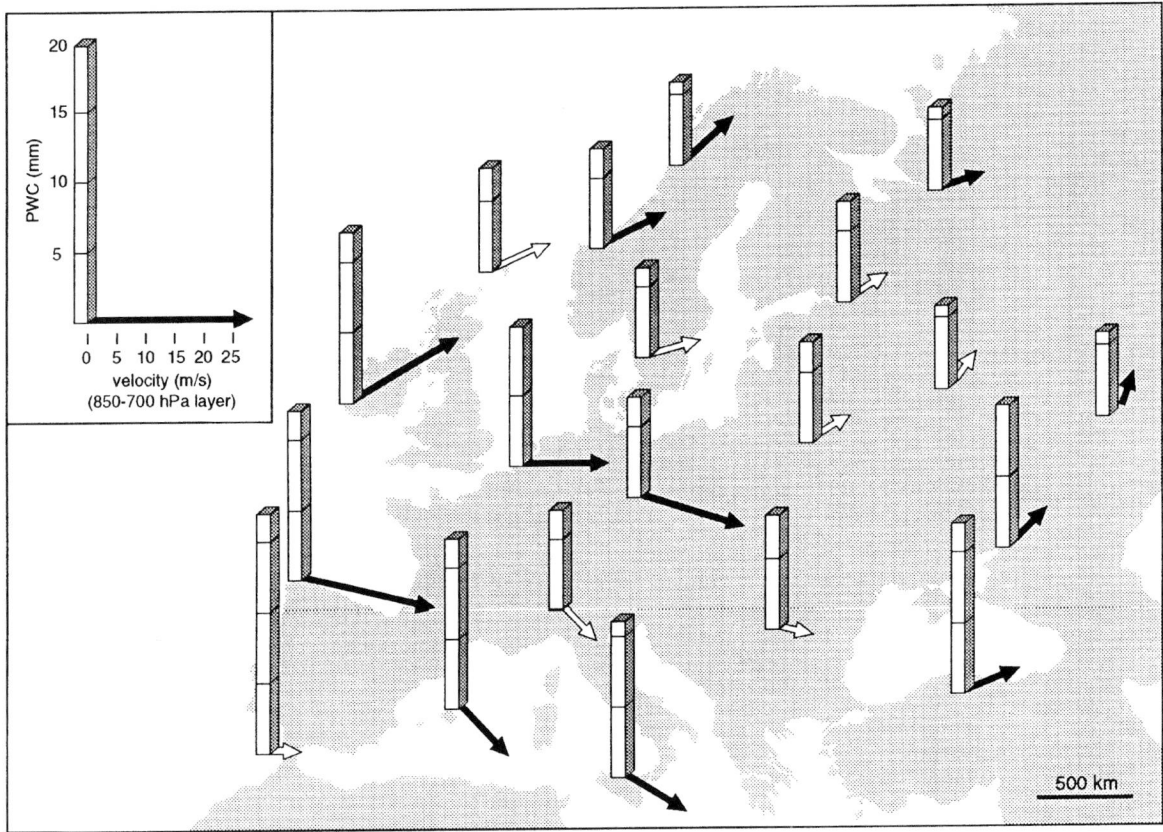

Figure E15 Sample mean winter precipitable water content (PWC) for selected stations across Europe (January 1995). Mean vector wind directions and speeds are also indicated for the 850–700 hPa layer (1.5–3 km) where most moisture advection takes place. Open arrows indicate a steadiness factor of less than 50% (data from MCDW: NOAA). The steadiness factor is defined as the ratio of mean vector wind speed to mean scalar wind speed.

between 7 and 10 g kg^{-1}, sometimes associated with daily precipitation rates of about 25 mm. Figure E15 shows sample water vapor content expressed as precipitable water content (PWC) for a typical winter month. The mean vector wind direction and strength between about 1.5 and 3 km are also shown. As expected, most of the moisture content is being advected eastward across the western seaboard of Europe. Over central Europe there is net divergence leading to a fairly uniform PWC of around 6 mm – half of that over the western seaboard open to Atlantic influences. The flow also becomes far more variable in direction as it encounters the cold anticyclone of continental Asia. Precipitation mechanisms also deplete the PWC over central Europe.

Precipitation

This vital component of the hydrological cycle is highly variable in time and space. The uplift required to produce precipitation is likely to occur either due to low-level convergence, especially along frontal zones, or to instability in the lower part of the atmosphere. Frontal ascent of 5–10 cm s^{-1} tends to produce widespread, moderate rain, whilst convection with core ascent values of 30 m s^{-1} within a storm cell is likely to lead to locally heavy precipitation. The storage of water in upland reservoirs also relies on precipitation inception or enhancement of these two mechanisms due to uplift over hills and mountains. Certain types of synoptic weather situation produce the bulk of such precipitation, notably the uplift of moist air within the warm sectors of depressions (cyclones). The synoptic origins and intensities of precipitation over the English Midlands have been studied by Matthews (1972). Sometimes generally wet years over lowland areas fail to provide sufficient upland precipitation to replenish reservoirs if warm sectors are either too stable (upper level convergence) or possess inadequate precipitable water content.

Annual precipitation

Values of annual lowland rainfall show little spatial variation in the zone stretching from the English Midlands through to the Urals of Russia (Figure E16). The passage of fronts ensures moderate precipitation across the European Plain, excessive falls being rare and usually associated with slow-moving disturbances. A decline from 640 to 500 mm can be observed eastward across southern England (Atkinson and Smithson, 1976) but values rise again across the Low Countries – evidence of the ease with which moist Atlantic air penetrates across Europe and the diminishing effect of the ridge from the Azores anticyclone. A few areas such as Bydgoszcz (northwest of Warsaw) and coastal regions of Lithuania record annual values less than 500 mm.

As a general rule, annual rainfall over the European lowlands declines both northward toward the Arctic and southward towards the Mediterranean where Almeria in southeast Spain receives only 200 mm in an average year – the driest place in Europe (Escardo, 1970). Many other regions of Spain also experienced inadequate rainfall during the decade 1986–1995. Everywhere the existence of hills, mountains, enclosed basins and peninsulas complicates this simple pattern. Precipitation over mountain regions is enhanced by uplift of airmasses on windward slopes producing isohyet gradients. Using over 6500 rainfall stations over Britain, Bleasdale and Chan (1972) produced the regression relationship between annual mean precipitation and altitude of $R = 714 + 2.42\,H$ where R is the annual mean precipitation and H is the height in meters. In fact, every region and every year possess unique constants a and b in a regression equation which makes hydrological forecasting difficult.

Highest annual precipitation values are found in mountainous areas where valleys are open to the southwest or westerly airflow. A few localities have recorded in excess of 5000 mm mean annual precipitation for the period 1931–1960, notably south of Nordfjord, Norway, and Sprinkling Tarn, English Lake District, and possibly inland from

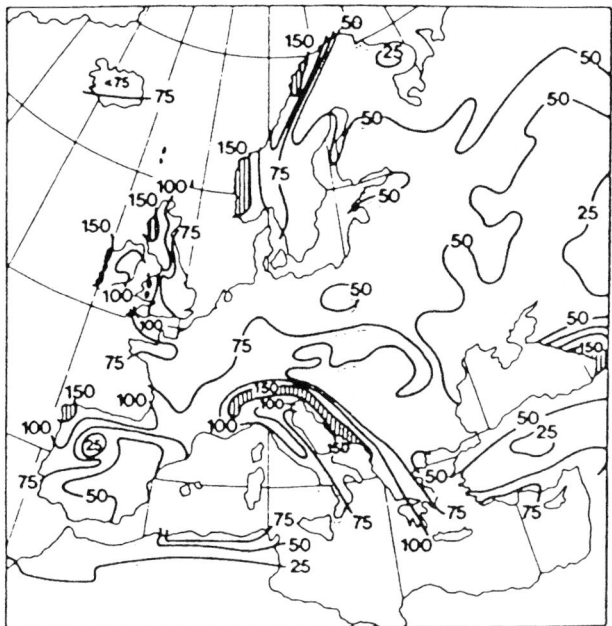

Figure E16 Mean annual precipitation in centimetres.

Boka Kotorska near the Montenegro–Albanian border. A value of 4000 mm may possibly be accepted at Monchsgrat in the Bernese Alps, although heavy winter snow makes measurement difficult. It is probable that one or two localities in the Bavarian Alps, Germany, and in the Sierra de Gredos, west of Madrid, Spain, receive 3000 mm, whereas 2000 mm would seem feasible for the highest parts of the Central Massif and Jura mountains in France. Longterm variations in annual precipitation have been analyzed for central Europe by Brazdil *et al.* (1985). They discovered a rising trend (1880–1980) for parts of Germany and a falling trend across Hungary, with other areas showing complex patterns sometimes, as in Moravia, with inter-decadal swings of 50 mm.

Summer half-year pattern
The percentage of rain falling in the summer half of the year (April–September) during the period 1861–1970 has been analyzed by Tabony (1981) who found that most of Europe north of 55°N and east of 5°E received on average more than 50% with the Hungarian Plain reaching 65%. This may chiefly be accounted for by the relatively high summer temperatures over much of central and eastern Europe which lead to enhanced convection. This shower activity may be fairly random if the pressure gradients are weak, or it may be organized in association with cold fronts as incursions of cool air of maritime polar origin sweep southeast into central Europe. Violent

thunderstorms accompanied by squally winds are characteristic of the mountain regions such as the Alps. The hydrological usefulness of heavier summer rainfall is offset by higher evaporation.

Seasonal pattern
Invasions of moist air across Europe are most frequent when the circulation is vigorous and depressions are moving quickly east across Scotland and Scandinavia into the Baltic region. Such a situation is most likely to occur in early winter and is least likely to occur in spring when anticyclones usually are present over some part of Europe. In fact, over much of Europe the percentage of annual precipitation falling in spring is remarkably constant, lying between 16 and 21% (Table E5). A winter maximum of rainfall might well be expected to occur fairly widely. That this is not so indicates that many other factors influence the seasonal distribution. As may be seen from five of the stations in Table E5, a summer (June–August) maximum occurs over eastern England and central, eastern and northern Europe, except the coastlands open to Atlantic influence. A zone of fall maximum occurs from Portugal through western France, central Britain, the coastlands of the Low Countries and Norway. This maximum is associated with relatively high sea surface temperatures leading to upward transfer of water vapor into the boundary layer. The moisture may then be ingested into the circulation of midlatitude depressions (cyclones) that become more vigorous and active from September to November.

The Mediterranean region with its mountains, peninsulas, islands and inland sea presents a more complex pattern of seasonal precipitation (Figure E17). From a hydrological standpoint, it is more convenient to state the driest period of the year, which is summer, when extensions of the subtropical high pressure lie over the region. During the remainder of the year, although frontal depressions can sometimes be identified, many disturbances are ill defined and rather transitory. Moreover, their activity in terms of producing rain varies on both a seasonal and annual basis. It is generally true that a fall maximum is more likely in the western and northern Mediterranean, probably associated with the transitory 'western mid-level trough' (Jacobeit, 1987) and that a winter maximum is more evident in the southern and eastern Mediterranean, as at Athens, Greece (Table E5). A bimodal maximum (spring, fall) is a feature of much of Spain, at least when the tropospheric westerly flow is relatively strong.

Comprehensive analysis of seasonal rainfall frequency in the Mediterranean was carried out by Reichel and Huttary (in Trewartha, 1981). More recent investigations have employed indices and statistical analyses in attempts to describe present rainfall patterns and to search for regular periodicities (Fukui, 1966; Tabony, 1981). Attempts have also been made to use principal components analysis to identify rainfall patterns across the Mediterranean (Goossens, 1985). It should also be noted that there is a tendency for the general characterisiics of a regime to change over periods of 30 years (Tabony, 1981). Thus the fall may be considered 'wet' in one decade, spring in the next.

Duration of heavy falls
As noted above, convective storms contribute significant amounts to summer rainfall over much of Europe. Extreme maximum precipitation over periods of less than 24 h is probably highest for lowland

Table E5 Seasonal precipitation as a percentage of the annual precipitation (1931–1960)

	Latitude		Winter (DJF) (%)	Spring (MAM) (%)	Summer (JJA) (%)	Fall (SON) (%)	Annual Total (mm)
Northern Europe							
Bergen (west coast)	60	23	27	17	22	34	1958
Stockholm (central)	59	17	22	16	33	29	555
Arkhangelsk (northeast)	64	40	18	18	33	31	539
Central Europe							
Valentia (west coast)	51	56	31	19	20	30	1398
Brussels (western)	51	51	26	17	29	27	817
Warsaw (central)	52	13	19	19	39	22	471
Moscow (eastern)	55	45	17	21	37	25	575
Southern Europe							
Madrid (Iberia)	40	25	28	30	12	30	436
Rome (central)	41	54	35	20	6	39	749
Athens (eastern)	37	58	42	21	7	30	402

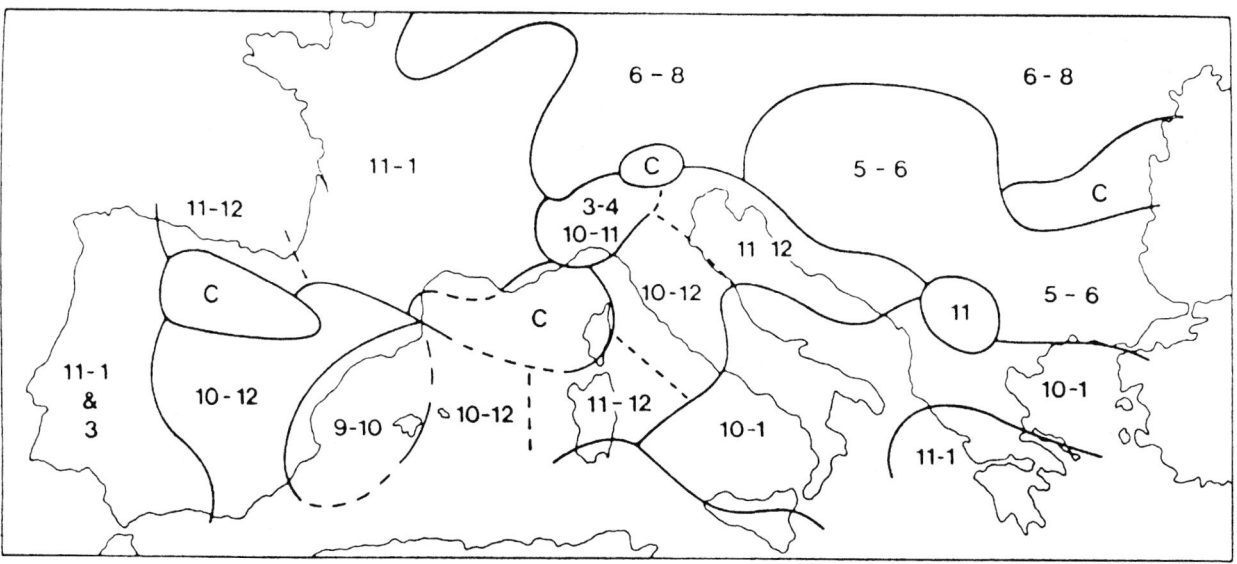

Figure E17 Mid-twentieth century seasonal distribution of rainfall. The figure shows months of maximum rainfall over southern Europe based on World Meteorological Organization Tables 1931–1960. Months are numbered from January (1) through to December (12). Complex regimes with no distinct seasonal maximum are indicated by 'C'.

Table E6 Data relating to two notable rainfall events with return periods in excess of 100 years (after Doneaud, 1971; Bleasdale, 1975)

Isohyet (mm)	Area enclosed (km^2)	Rainfall (10^9 m^3)
Central and northern Romania, 12–14 May 1970, duration 48 h		
25	50 000	2.5
100	6 000	0.66
Scottish Highlands, 26–27 March 1968, duration 36 h		
100	12 500	1.96
200	2 700	0.64

stations where thunderstorms dominate. Continuous heavy rain for periods in excess of 24 h depends on steady inflow and uplift of moisture over a region. An inspection of available data shows that rare events producing 250 mm seem a possibility over much of coastal western and southern Europe. Inland, intense storms yielding about 200 mm have also been recorded. Available moisture for intense thunderstorms diminishes eastward and northward. Records show the occurrence over European Russia of maximum 24 h falls of around 100 mm. Less intense but continuous rainfall may extend over much larger areas, producing devastating floods usually contributed to in winter and spring by snowmelt (Boucher, 1972, Fink *et al.*, 1996). Such widespread rainfall is usually a combination of three components: (1) uplift of moist air over mountains, (2) vigorous wave activity along a quasi-stationary frontal zone and (3) convective instability. In such instances over 600 million m^3 of rain may fall over a limited area (Table E6). The estimation of frequency and duration of rainfall events may be expressed in terms of return periods based on statistical analysis. This method of presenting rainfall data is widely used in hydrological applications. An example for a maritime location near Exeter in southwest England and open to Atlantic influences is shown in Figure E18. The straight lines approximately describe about 99% of rainfall events. Extreme events with return periods in excess of 20 years seem to possess their own unique distribution (Clark, 1991). From a hydrological standpoint such extreme events cannot easily be estimated since they invariably depend upon a unique combination of meteorological elements. For much of lowland Europe away from maritime influences, the slope of the upper lines is about half that shown in Figure E18 with 24 h maximum values being around 50–60 mm. However, intense short-duration storms will raise the position of the lower lines.

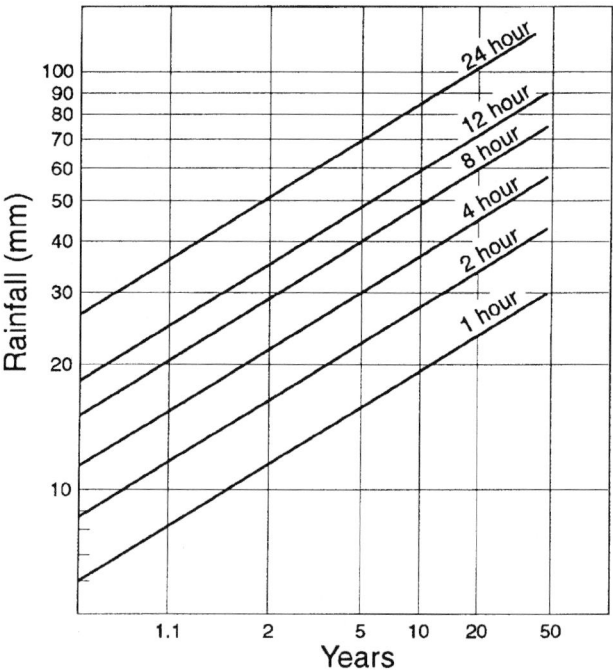

Figure E18 Rainfall frequency-duration lines for six selected hour periods for Hemyock (30 km NE from Exeter, SW England) in a region subject to heavy rainfall. (Adapted from Clark, 1991.)

Thunderstorms and related phenomena

Over much of continental Europe, thunderstorms are typically a summer phenomenon though super-cell storms are very rare. Within European Russia, about 50% of the thunderstorms are associated with cold fronts, 22% with warm fronts and 28% with convergence within airmasses (Borisov, 1965). Thunderstorm frequency increases with the rise in surface temperatures in early summer and reaches a maximum in June in parts of southern European Russia, in July over much of central and northern Europe, including northern Italy, in August in parts of Sweden, in September and October in the western Mediterranean and in winter over southern Italy as well as parts of northwestern Britain.

The occurrence of hail is most closely associated with 'airmass' thunderstorms rather than frontal storms, especially where these develop in hilly or mountainous regions such as the Alps. Extensive damage may occur. Small-scale tornadoes occasionally accompany storms (Meaden, 1985), but nowhere have they been reported to attain the magnitude and frequency of those in the United States. Over western parts of Europe they may be associated with cold fronts at any time of the year. Their tracks rarely exceed 1 km. There have been fairly frequent sightings of water spouts over the western and central Mediterranean, mostly during thundery weather.

Snowfall
The relatively high sea surface temperatures in winter mean that snowfall in winter is variable over low ground in western Europe. The snow cover persists for more than 7 days only in the coldest winters. As a general rule, monthly average temperatures must lie below 1°C for snow cover to persist. The number of individual days with complete snow cover increases northeast across the European continental plain from 5 days in the Midlands of England to around 40 days over central Poland, 135 in the Moscow region and up to 200 along the Arctic fringe. The mean maximum depth of snow cover increases from 20 cm in southwest Russia to about 75 cm on the lowlands to the west of the Urals. Most mountain areas are liable to receive heavy snowfall, although valley aspect and the character of individual winters play an important part in determining accumulated values. Prolonged lowland snowfall exceeding 25 cm over western Europe is only likely on the cold side of an active frontal trough. Such a situation leads to very cold continental air being drawn west over northern Germany towards central Britain, while humid air to the south is forced to ascend over the cold easterlies. Snow enhancement occurs if small disturbances run east, typically along the English Channel into central Germany. Polar lows occasionally bring blizzards to Scotland as in the winter of 1995–1996, but their development depends on unstable cyclonic disturbances developing in cold northerly airflows crossing relatively warm ocean waters.

Drought

Lack of precipitation is a normal part of the seasonal climatic cycle within the Mediterranean in summer. An appropriate index is used to describe the rainless period. Aridity indices, such as those of De Martonne, have traditionally been employed by Italian and French climatologists to describe summer drought (Pinna, 1957). The existence of a rainless period is closely associated with mid-tropospheric subsidence. Therefore the existence and maintenance of summer drought depend on the strength and position of the Azores anticyclone and its eastward extension. When this anticyclone is well developed, drought is complete and may last for up to 4 months in southern parts of Spain, Italy and Greece. Occasionally during summers, such as those of 1959, 1976 and 1995, the belt of high pressure may be

located farther north over northern France and the North Sea, bringing drought to areas usually well supplied with precipitation and leading to summer rainfall over the Mediterranean (Perry, 1976). Annual variations in precipitation are of great concern, not only to countries such as Spain and Portugal, but also to the grain-growing areas of Poland and Russia where the failure of adequate snowmelt and spring rain may seriously reduce yields.

Severe droughts are associated with rainfall deficiencies of 50% over periods of 6 months or more. The impact of such droughts is greatest if they occur during the soil moisture recharge period (Hounam et al., 1975; Marsh and Turton, 1996). As elsewhere, the intensity of the drought is increased by high winds, low humidity and high temperatures. Areas likely to suffer such conditions lie along the southern border of Europe and across the steppes of the Ukraine and Kazakhstan. In the latter area the steppe climate is characterized by an extreme annual range of temperature approaching 80°C, frequent droughts and desiccating, dust-laden winds known as sukhovei. During the period 1880–1950, serious drought occurred every 3.5 years on average (Poliarus, in Borisov, 1965). The effect of one year's drought is magnified by the tendency of one drought year to follow another. Such was the case over much of European Russia from 1889 to 1892, and in southern Spain from 1917 to 1919.

Within the midlatitude belt of Europe, the occurrence of drought is closely associated with the position, height and persistence of ridges in the mid-level airflow of the atmosphere. There is some evidence of persistence in airflow patterns over a number of years. For this reason, periods of a decade may show general deficits in precipitation. The droughts of the early 1970s in European Russia prompted a number of studies (Buchinskiy, 1975; Chistyakova, 1975). In an earlier investigation, Davitaya (1958) found that droughts over the former southern USSR reflected shifts in the mid-tropospheric ridge–trough pattern. A thorough investigation by Rauner (1980) into the simultaneous recurrence of drought in different regions of Europe revealed that drought affected the whole of the grain zone of the former USSR on at least five separate occasions during the period 1891–1975. On a regional basis, droughts were found to have affected the Ukraine in 43 out of 106 years (1870–1976) and the Volga region, 1000 km farther east, in 41 out of 96 years (1880–1976) of which 19 years were common to both regions. There have been 20 years when drought was recorded simultaneously in European Russia and in western Europe during the period 1700–1976, three of which occurred consecutively from 1747–1749. There is also some evidence that periods of drought in Europe became more frequent during the latter part of the twentieth century (Marsh and Turton, 1996), though whether this was part of a more permanent change in the climatic pattern has been difficult to establish.

Water balance
Figure E19 depicts the annual atmospheric component of the hydrological cycle as assessed per 5° latitude band for the 'peripheral' and

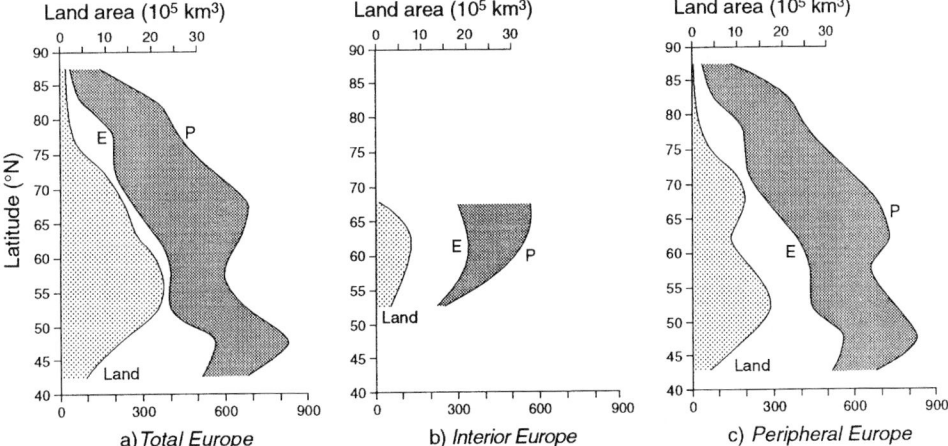

a) *Total Europe*   b) *Interior Europe*   c) *Peripheral Europe*

Figure E19 Atmospheric water balance for Europe; precipitation (P) and evaporation (E) shown in solid lines; land area dashed line. (a) Total area of Europe, (b) interior Europe and (c) peripheral Europe. Geographical area covered: East Border: Jugarski Peninsula, Ural Mountains, Ural/Embra Divide, north edge of Caspian Sea, Kuma, Manych Separation, Sea of Azov, Kerch Strait. Included: Azores, Madeira Island, Faroes, Jan Mayen, Bear Island, Spitzbergen, Franz Joseph Land, Novaya Zemlya, Vaygach Island.

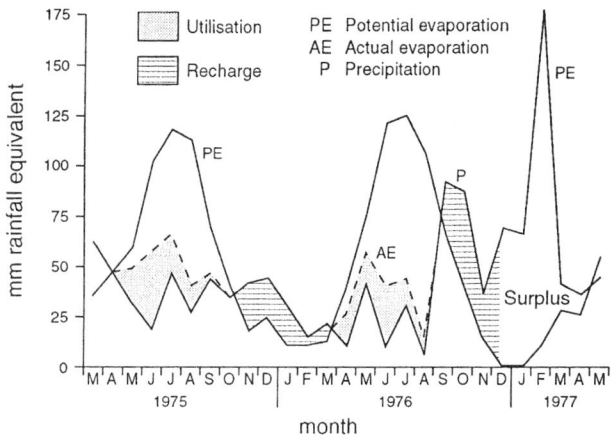

Figure E20 Soil moisture deficits during the prolonged drought period 1975–1976 at Sutton Bonington in the English Midlands using the Thornthwaite method.

'core' areas of Europe. The shaded area between 'P' and 'E' represents the atmospheric moisture deficit which, if multiplied by the land area (dashed line) provides an indication of the amount of net moisture content that must be imported into Europe. At ground level and on a seasonal basis, the hydrological significance of recharge of soil moisture and aquifers during winter is of great importance. Deficits of 11–13 cm may accumulate during the summer, even in the more humid western parts of Europe, as in 1976 and 1995. The least satisfactory situation is where winter recharge of moisture is incomplete, placing considerable stress on water supplies in the following year. This is shown in Figure E20 for a location in central England (Boucher, 1996) where the different characteristics of the climatic water budget may be seen using the Thornthwaite method.

Synoptic climatology

It is sometimes desirable to investigate aspects of regional climate and hydrological impact from a generic standpoint. The study of the frequency of weather types and pressure patterns, known as synoptic climatology, provides such an insight.

Weather types

A detailed survey of this approach is to be found in Barry and Perry's text *Synoptic Climatology* (1973). The authors indicate that European climate may be analyzed using either a static or a dynamic (kinematic) framework. In essence, this means that weather elements are synthesized into 'types' that correspond to recognized features in the circulation such as troughs and ridges. Alternatively, the basic flow structure and stability of the air may be described. British research (Lamb, 1972) has favored a somewhat less rigorous scheme than a number of German and Russian climatologists (Hess and Brezowsky, 1969; Chubukov, 1977). Undoubtedly the work by Baur (1931, 1944, 1947, 1963), in Germany, greatly influenced classification of weather systems. He examined large-scale circulation patterns over Europe and the nearby eastern North Atlantic, and searched for patterns that persisted for several days due to quasi-stationary waves in the mid-tropospheric flow. Baur called these periods of similar flow 'Grosswetter' as against the weather characteristics of individual days (see references in Barry and Perry, 1973). Flow patterns (Grosswetterlagen) were identified on the basis of the strength or weakness of the zonal flow, along with indicators specifying the nature of isobaric curvature (closed, open, cyclonic, anticyclonic, etc.). Such schemes are best applied to the zone bounded by latitudes 45–60°N. On the northern and southern margins of Europe, weakness in the mid-level flow at 500 hPa makes analysis more problematic when using this method. More recently, an attempt has been made by Fraedrich (1990) to correlate Grosswetter with climatic variations in other parts of the world and, in particular, the ENSO (El Niño/Southern Oscillation) events in the Pacific Ocean.

The subjectivity involved in determining the category of a flow pattern has led to analysis using various indices. The most widely used is a form of zonal index employing values of pressure difference

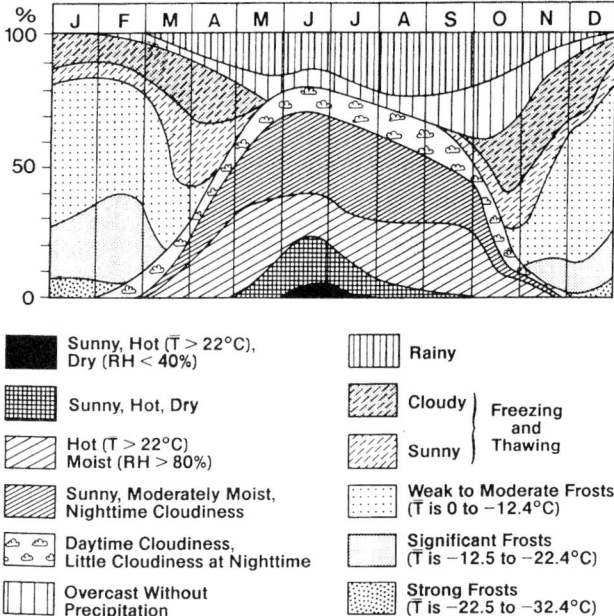

Figure E21 Frequency of weather types for Moscow. T̄ is mean daily temperature and RH is mean relative humidity. (After Gerasimov, 1964.)

across the westerly flow. Some attempt has been made, however, to link 'types of weather' with numerical indices. Murray and Lewis (1966) proposed a PSCM index represented by four components of flow: progressive, southerly, cyclonic and meridional. This index can be employed on a daily basis, although its correlation with individual climatic elements such as maximum temperature cannot be expected to be statistically significant.

Despite the limitations already mentioned, three examples of the annual regime of weather types are given. The first is for Moscow and portrays a frequency analysis by Gerasimov (1964) which employs temperature, cloud cover and relative humidity as criteria (Figure E21). The second example is for Kew to the west of London, England and shows two interpretations, that of Belasco (1952) in Figure E22a employing an airmass classification, based on airflow types similar to those identified by the Bergen School of meteorologists. The third which is of greater interest to hydrologists, is that of Lamb (1972) employing a weather-type analysis of precipitation for Southampton in Figure E22b, showing that cyclonic situations contributed about 30% throughout the year whilst the 'westerly' category dominated the winter months. Using Lamb's data, Wilby (1995) has indicated that there has been a long-term annual decline in westerlies over England since 1900 (Figure E23). On the other hand, long-term seasonal changes in weather types over continental Europe have been analyzed by Bardossy and Caspary (1990), who have shown that 'zonal' westerly airflow types have recently become more prominent in the winter halfyear, except for November (Figure E24). The 1981–1992 record for regions of the British Isles shows declining cyclonicity and increasing westerly types in the majority of months when compared with the 1951–1990 period (Mayes, 1994).

The term singularities has been applied by Lamb (1964) to periods of distinctive weather lasting between 7 and 14 days that have a tendency to recur in a number of years at about the same time of year. Trewartha (1981) has associated this calendar of weather episodes with the Grosswetterlagen of German climatologists. For a certain region of Europe a calendar sequence of probable types of weather may be assembled. Singularities probably represent the climatic impact of adjustments to the general circulation over the Atlantic and European sectors. These in turn are a response to radiative forcing factors at the surface – particularly eastern North Atlantic sea surface temperatures – and in the troposphere above. A singularity, such as an anticyclonic spell, should be regarded as a statement of statistical probability of a particular circulation pattern occurring during a given period of the climatic record. An example of such a calendar is presented in Table E7, which refers chiefly to the British Isles. Most

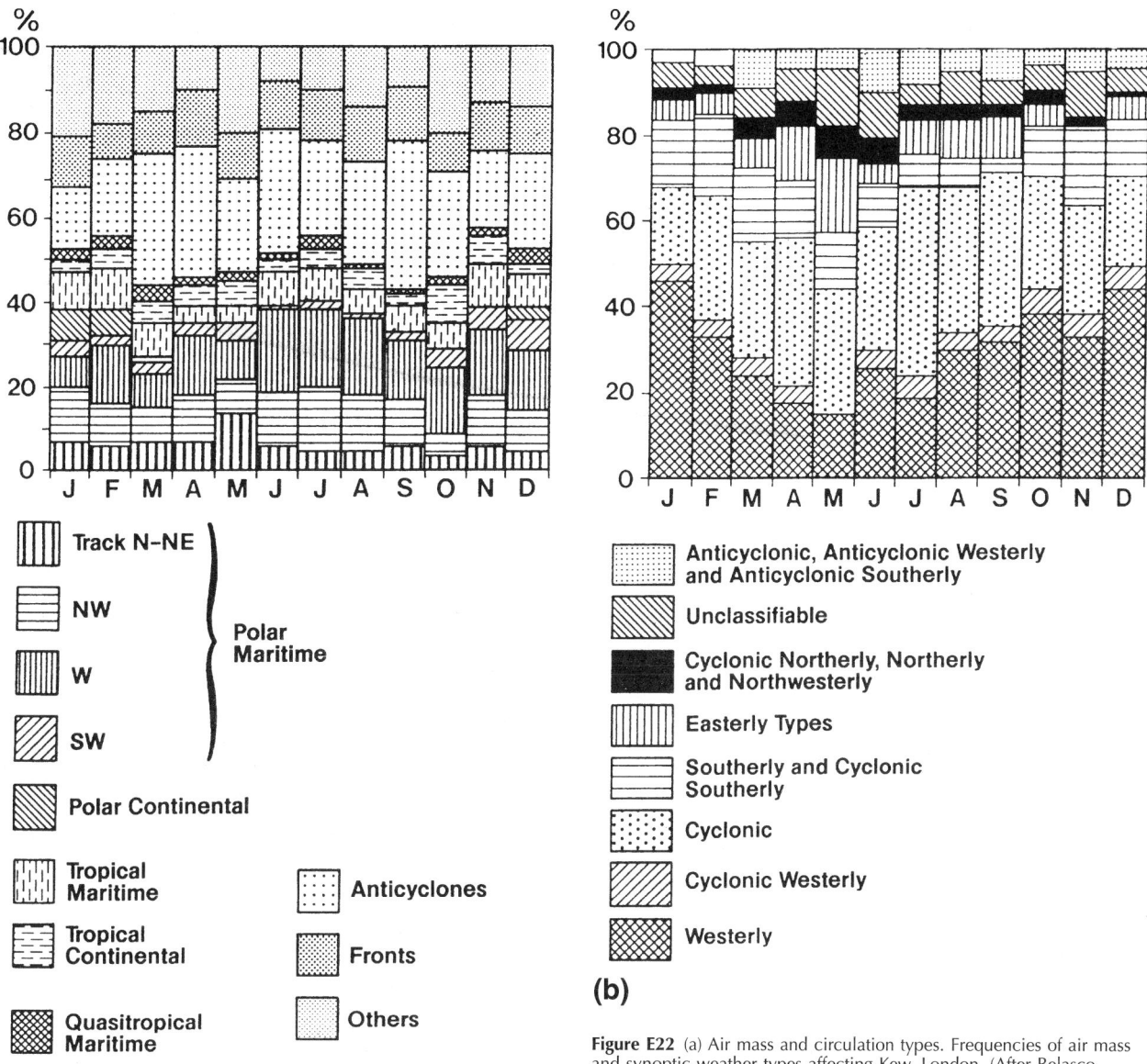

Figure E22 (a) Air mass and circulation types. Frequencies of air mass and synoptic weather types affecting Kew, London. (After Belasco, 1952.) (b) Air mass and circulation types. Contribution of H.H. Lamb's weather types to monthly mean precipitations at Southampton, central-southern England, 1921–1950. (After Barry and Perry, 1973.)

singularities shown in the table also occur in the neighboring continent but statistical significance diminishes eastwards. Periods of mean anticyclonic activity over western Europe are likely to be associated with cyclonic activity over eastern Europe under 'ridge-and-trough' patterns (Figure E26a) as happened in the summer of 1995. It can also be seen from Table E7 that there is a tendency for periods of cyclonic activity to alternate with anticyclonic spells. The singularity table is not a climatic forecasting tool. For reasons not well understood, certain singularities become statistically significant for a number of years in the record before declining, even disappearing, only to re-emerge in a later decade (Murray, 1993). It is interesting to postulate that such changes form part of a climatic signal in the atmospheric system, indicating fundamental adjustments within the hemispheric circulation (Lamb, 1982).

Pressure patterns at the surface

The mean centers of both anticyclonic and cyclonic activity lie outside of Europe over the Azores Islands, west of Portugal, and southwest of Iceland, respectively. When the westerly flow is well established, ridges and troughs move eastward away from these

centres across central Europe. These may be associated with closed isobaric highs and lows at the surface. On the average monthly mean pressure charts for 1951–1966 (Meteorological Office, 1975), the shape of the North Atlantic low-pressure center changes, being at its most intense in December and January at 997 hPa (Figure E25a); thereafter, it declines until July when there is no easily identifiable center of low pressure close to Europe. The Azores anticyclone appears weakest in March (1020 hPa) and lies equatorward of 30°N. It reasserts its dominance and is at its most extensive in July (1025 hPa) with ridges extending east and northeast (Figure E25b). It has been noted that July is also the wettest month over much of Europe, indicating that breakdowns in the high pressure are frequent in some summers or that troughs dominate downstream.

Pressure changes at 700 hPa (1000 m), mid-twentieth century

Surface pressure features no longer appeared at this level on the maps of 5-day mean heights for 1951–1965 (Wahl and Lahey, 1969). Instead, the circumpolar vortex was much more evident, the winter circulation being dominated by a ridge extending north with its axis to

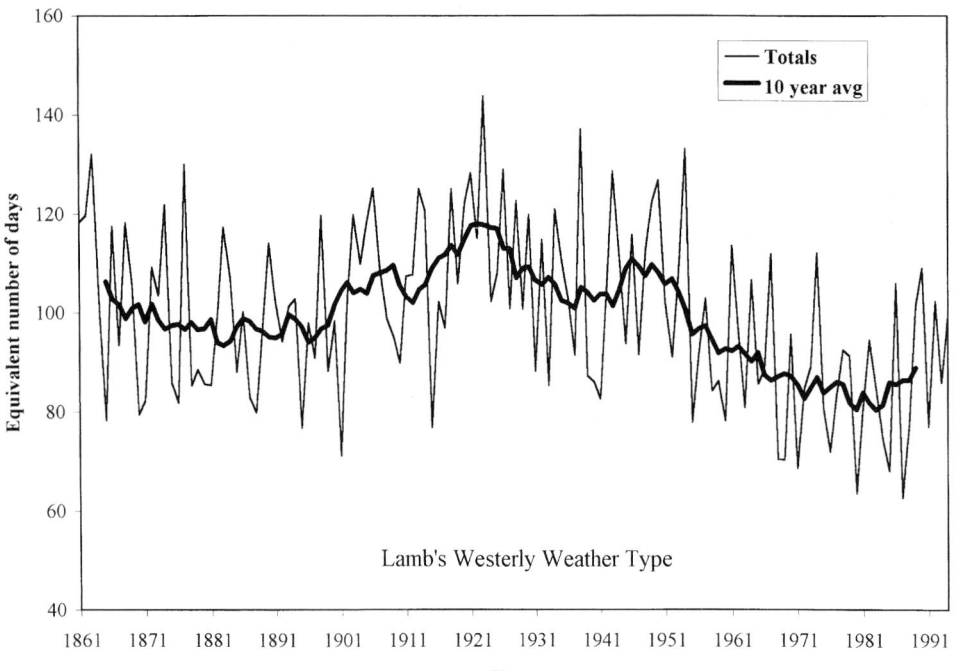

Figure E23 Annual values and 10-year moving average (RM) of the frequency of Hubert H. Lamb's westerly type flow for the period 1861–1994 as defined by the Climatic Research Unit (CRU), employing weighted day equivalent values. The 10-year RM is plotted mid-point. (Data from CRU/UEA, Norwich, UK.)

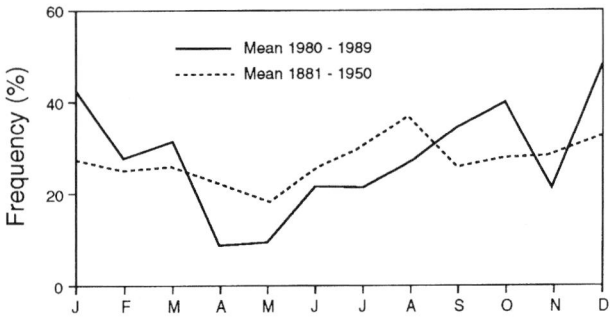

Figure E24 Monthly frequency of the 'zonal' westerly types in the European airflow classification, 1881–1950 (solid line) and 1980–1989 (broken line). (After Bardossy and Caspary, 1990.)

the west of Spain and the British Isles, while a trough lay over eastern Europe (Figure E26a). The trough persisted through to mid-May before disappearing and being replaced at higher latitudes by a weaker trough further upstream at about 10°W (Figure E26b). This, in turn, disappeared and by the end of August the whole of central and northern Europe was dominated by a westsouthwest air flow at this level. Troughs on these 5-day average maps did not reappear again until the beginning of November. Even some of these features were difficult to locate at higher levels in the troposphere. An analysis of the average height of the 500 hPa surface at about 5.2 km for January revealed a weak ridge somewhat east of the British Isles with an attendant broad trough feature over European Russia (Figure E27a). By July the mean pattern around 5 km in the mid-trosphere was approximately zonal with the suggestion of a trough over eastern Europe (25°E) as indicated in (Figure E27b). Tabular presentation of this approach is provided in Tables E8 and E9, which show the meridional mean pressure gradient at the surface and at 500 hPa. The hemispheric flow is known to be much weaker in July than in January

Table E7 Abbreviated calendar of singularities (after Lamb, 1964)

Dates	Singularity	Probability of occurrence over British Isles	Comments for Germany
11/22–12/10	Early winter storms	>50% (1)	Zonal westerlies
12/17–12/24	Continental anticyclones	25%	>60%
12/26–1/12	Storms of mid-winter	Westerly 60%	Westerlies 72%
1/19–1/25	Continental anticyclones	25% (5)	>75%
1/27–2/4	Renewed winter storms	Variable (1)	Precipitation peak 70%
2/7–2/22	Anticyclones	25%	>60%
2/26–3/9	Cold spell	50%	No regular features
3/12–3/22	Anticyclones	35% (1)	>60%
3/30–4/15	Atlantic depressions	Westerly 40% (5)	Rainfall peak
4/16–5/20	Spring northerlies	25%	3 days 75% (1)
6/10–6/30	Increase in westerlies	Westerly 52% by 20th	Westerlies 80%
7/15–8/15	Unstable cyclonic	30–35%	Westerlies >80%
8/17–9/2	End of summer	Westerly 50%	Rainfall peak
9/6–9/19	Anticyclones	30–40%	>70%
9/21–9/30	Atlantic storms	25%	Anticyclones >70%
9/30–10/15	Anticyclones	Peak 40% (5)	Part >70%
10/24–11/13	Late fall rains	Westerly 35%	Anticyclone >60%
11/15–11/24	Anticyclones	30% (1)	Anticyclones >70%

Notes: statistically tested figures exceeding the level of 5% probability of occurrence are indicated by (5) and 1% probability by (1) usually for part of the period indicated in column 1. Length of the record in total was from 1873 to 1961. Some figures refer to parts of this period.

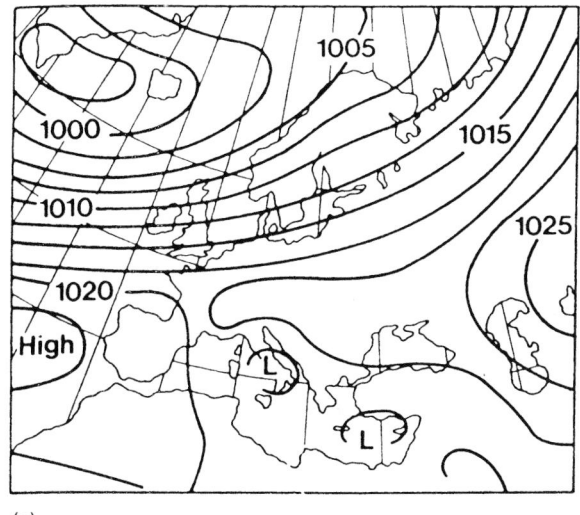

(a)

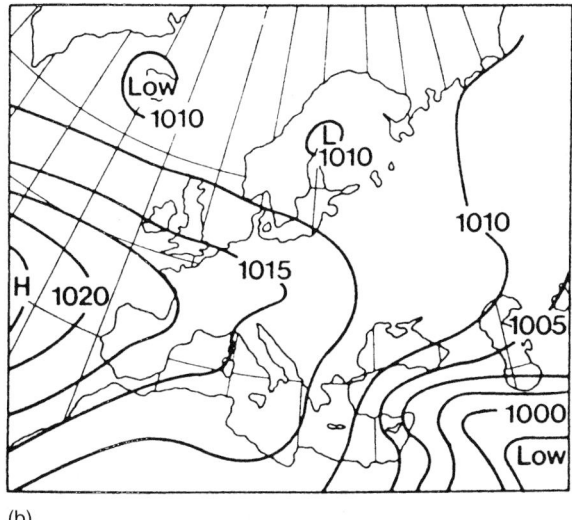

(b)

Figure E25 Mid-twentieth century average sea level pressure in hPa for (A) January and (B) July (after Meteorological Office, 1975).

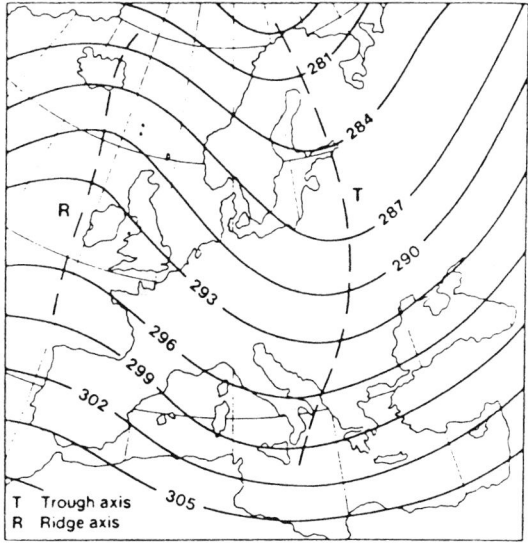

(a)

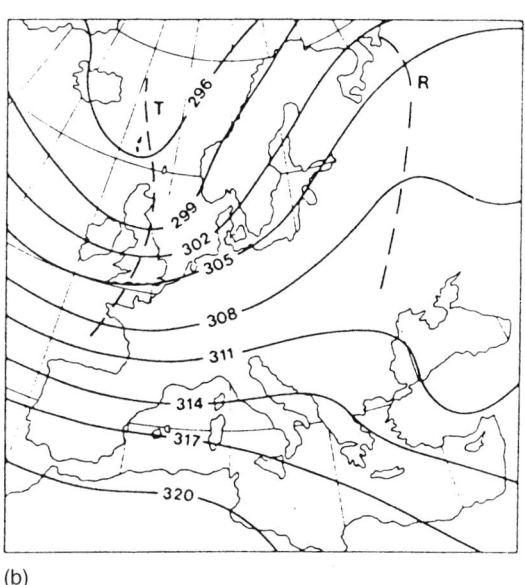

(b)

Figure E26 Mid-twentieth century average height of the 700 hPa surface in geopotential decameters for (A) mid-February and (B) mid-July. Main troughs (T) and ridges (R) are indicated.

and this is evident at 20°W, but at the prime meridian (°0) there is very little difference in this gradient between summer and winter at 500 Pa. This again emphasizes the degree of penetration of the mid-level westerlies across Europe during most summers.

Jet flow across Europe

Mean airflow over Europe in the troposphere is predominantly from the west between high pressure to the south and low pressure to the north. Imposed on this pattern are waves of large amplitude having a wavelength of several thousand kilometers. Changes in amplitude and movement downstream produce an ever-varying pattern of airflow. Embedded in this flow lie narrow zones of fast-moving air known as jet streams that extend downstream for several thousand kilometers. They may be about 400 km wide, and are associated with air mass boundaries in the vicinity of the polar front. They appear and disappear with the changes in the circulation pattern. Within the core of the Polar front jet, winds may exceed values of 60 m s^{-1} over Europe. The jet flows are also important from a hydrological standpoint since large-scale moisture advection and ascent of air in the vicinity of the polar front often gives rise to moderate and potentially widespread rainfall events.

Blocking patterns

One of the most frequent interruptions to the westerly flow over Europe results from blocking patterns developing either upstream in the mid-North Atlantic or over Europe. The presence of a large anticyclone, persisting on average for 16 days, causes disruption and bifurcation of the mid-level flow. Blocking may be dominant in some years, such as 1995, or only weakly developed in others. The effect of the blocking on the distribution of precipitation is most evident with large-scale anomalies arising. Abnormally low temperatures may also result if blocking occurs in winter, as in February 1986. Much attention has been given to these features in the literature. Rex (1950) has provided a comprehensive review of the climatic data relating to blocking over Europe. He indicated that the frequency of blocking action by quasi-stationary anticyclones in the eastern North Atlantic sector rises from below 20% of all days in September to over 40% in April, thereafter declining sharply. A low zonal index accompanied by meridional flow is therefore a characteristic of spring

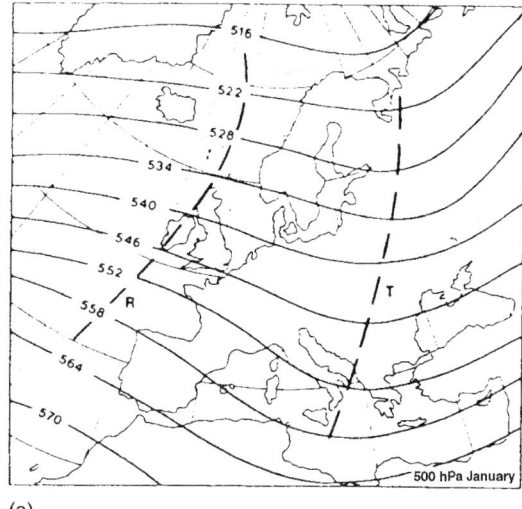

(a)

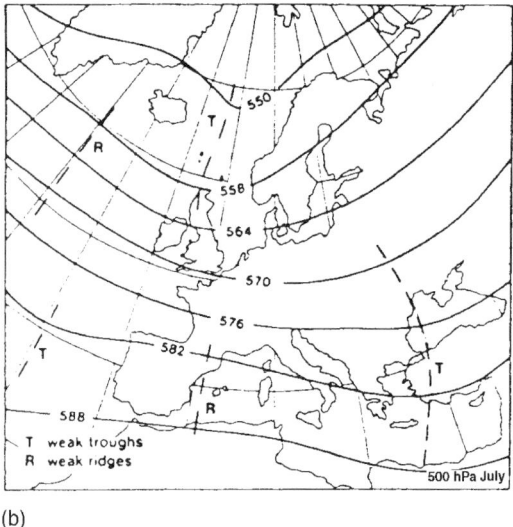

(b)

Figure E27 Mid-twentieth century average height of the 500 hPa surface in geopotential decameters for (A) January and (B) July (1951–1966).

Table E8 Meridional mean pressure difference at 40°N to 60°N in hPa at the surface

	20°W	0	20°E	40°E
January	15.2	9.0	3.7	3.0
April	11.2	2.5	-1.0	-4.0
July	11.5	6.2	2.1	-0.3
October	14.5	7.0	3.7	3.0

Note: positive values represent high pressure at 40°N and low pressure at 60°N.

Table E9 Meridional mean difference in the height of the 500 hPa level in geopotential meters between 40°N and 60°N, representing an index of intensity of westerly flow; High values represent a high zonal index

	20°W	0	20°E	40°E
January	140	82	68	76
July	92	86	66	60

Note: positive values indicate westerly flow.

Obstruction to westerly flow in the lower part of the troposphere (below 3 km) in winter may also be the result of the development and persistence of a cold anticyclone over snow-covered Siberia. The westward extension of this 'high' into European Russia causes continental air to flow around the southern margin of the anticyclone in winter (see Lamb's winter severity index, 1969). On average, airflow associated with the Siberian anticyclone is found on 24 days in January over the lower Volga Plain, decreasing to 12 days in the vicinity of Leningrad (Lydolph, 1977). Occasionally, this bitterly cold and initially dry air crosses the Low Countries and the North Sea into England. The hydrological impact is likely to be greatest at the end of such a spell, particularly if rain falls on snowfields and frozen ground. Extensive flooding occurs, as in England in 1947. Individual cases of blocking have been studied by Namias (1964) and others, whereas more recent attempts at numerical stimulation are reported by Davies (1978) and Tibaldi and Ji (1983).

Surface anticyclones and depressions

Anticyclones

Throughout the year 'cells' break away from the semi-permanent Azores high and travel slowly northeast, often close to the English Channel and either across the European Plain or central Europe into southern European Russia. Two to four new anticyclones a year tend to form along this same axis. Another favorite center for anticyclogenesis lies over Scandinavia, whilst high values over the Mediterranean represent the relocation and development of the Azores high. Annual mean speeds of eastward (progressive) moving highs are of the order of 6° of longitude per day at 55°N. Somewhat surprisingly, anticyclones moving westward (retrogressing) away from Europe travelled at 7° per day on average (Meteorological Office, 1975) and accounted for 40% of mobile anticyclones. There was little significant seasonal pattern evident. Figure E28 shows the total number of days in which the center of a high was located within a 5° × 5° grid. Counts were made daily.

over western Europe. This has been confirmed in a seasonal study undertaken by Knox and Hay (1985) for the period 1946–1978. The results for the eastern North Atlantic and Europe are shown in Table E10. According to Bulinskaya (in Lydolph, 1977), anticylonic activity is most frequent both in summer and winter over European Russia between latitudes 45 and 50°N which Borisov calls 'the great ridge', although Russian climatologists do not use the term blocking to describe these anticyclones. Lamb's analysis (1969) of European blocking covered a period stretching back to AD 1100 in which he identified a quasi-50-year alternating pattern of westerlies and blocking anticyclones.

Table E10 Frequency of occurrence of blocking per 10° of longitude from the southernmost latitude to 75°N for the period 1946–1978. Blocking was defined in terms of geopotential decameters (1 gpdm approx = 10 m) anomalies persisting for more than 6 days (Knox and Hay, 1985)

	20°W	10°W	0°	10°E	20°E	30°E	40°E	50°E	60°E
Winter	26	35	28	33	28	21	30	29	38
Spring	41	45	30	41	40	30	39	39	38
Summer	33	38	40	37	26	28	26	32	37
Fall	42	40	40	32	27	27	37	37	38

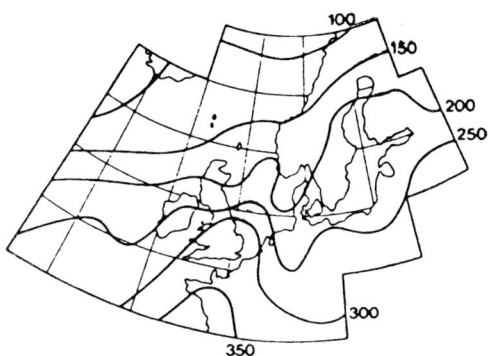

Figure E28 Total number of days with anticylonic centers at 1230 UT 1899–1938 (after Meteorological Office, 1975.)

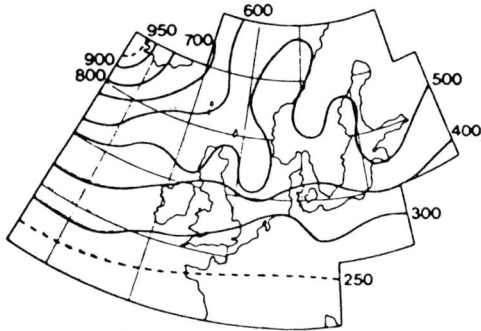

Figure E29 Total number of days with low-pressure centers at 1230 UT, 1899–1938, adjusted to unit area size (after Meteorological Office, 1975.)

Midlatitude depressions

[It should be noted that, in European literature, the term cyclone is usually reserved for tropical storms. Midlatitude cyclonic disturbances are referred to as depressions or 'lows'.] Frontal depressions affect much of Europe north of 45°N throughout the year and parts of the Mediterranean during the winter half. A wide variety of non-frontal depressions also occur and include heat lows (thunder lows), polar lows (Businger, 1985), complex lows and lee depressions (Lanzinger *et al.*, 1990). Most of the latter are associated with infrequent synoptic weather patterns. Major depressions that develop central pressure below 980 hPa originate over the western North Atlantic and attain maximum intensity before reaching land. This is evident from Figure E29 which shows the diminishing influence of low pressure southeastward into central Europe. A feature of great importance over the sea areas bordering northwestern Europe is the ability of some lows to deepen suddenly, creating high winds and storm surges – the so-called 'bombs' (Sanders and Gyakum, 1980). Very occasionally, such deepening leads to hurricane-force winds sweeping inland across parts of Western Europe (Burt and Mansfield, 1988).

Figure E30 shows the areas most likely to experience cyclogenesis. It should be noted that the map does not convey any realistic idea of the development, movement and intensity of the lows. For example, the center over northern Italy reflects the formation of many shallow but active lee depressions (Zenone and Lecce in Wallén, 1977). Other 'centers' are less geographically fixed and do not appear on all map interpretations of the data (Borisov, 1970; Trewartha, 1981).

The movement of depressions over the northern hemisphere has been investigated by Klein (1957) and others and is shown in Figure E31. The continuous lines denote main tracks and the dashed lines indicate less frequent routes. The main Icelandic low is renewed by depressions moving into the area from the mid-North Atlantic Ocean.

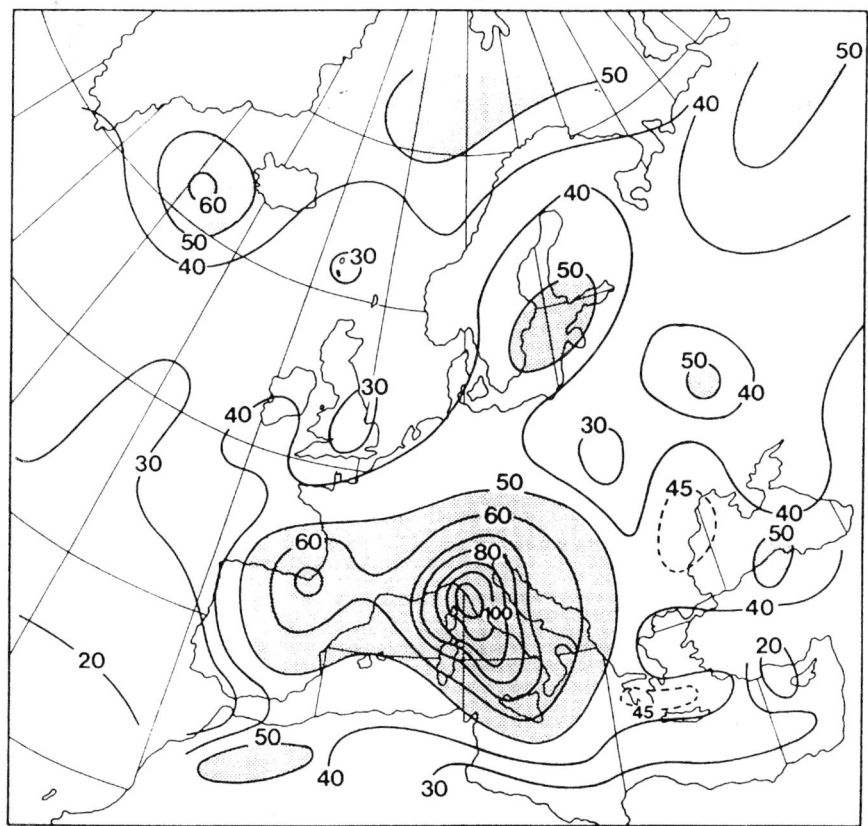

Figure E30 Twenty-year frequency of cyclogenesis, 1909–1914 and 1924–1937 (Sources: various.)

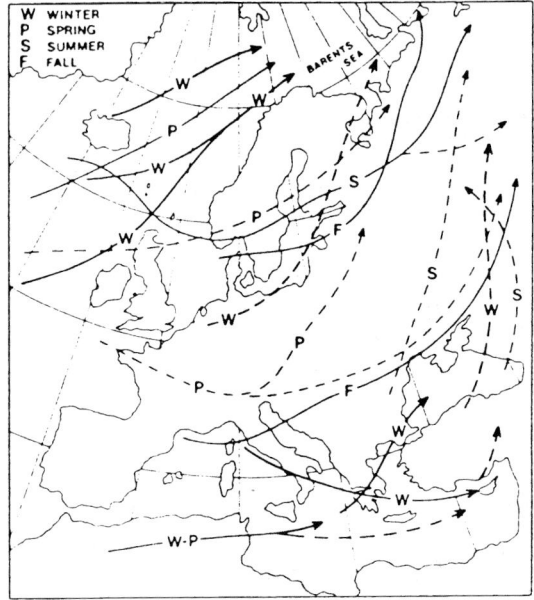

Figure E31 Main tracks of travelling depression (after Klein, 1957).

Tracks across northwestern Europe are usually followed by subsidiary lows circulating around the southern side of the parent depression, although some do break away beneath the southwesterly jet and travel across the Barents Sea. Most European lows either follow the northern or southern boundary of the continent in winter, whereas in summer a central route across southern Scandinavia and the Baltic Sea has been established. Over European Russia, tracks are difficult to identify in winter when high pressure predominates, but low pressure is more in evidence in summer. This is particular so in central European Russia – a region that stretches from the western borders with Poland eastward towards the southern Urals. In this region there are twice as many cyclonic centers in summer as in winter. Some of these lows will be associated with fronts and outbreaks of cool polar air. Others are more easily identified with showery, thundery outbreaks of rain. A few originate from the Black Sea area.

Depressions in the Mediterranean

It is unusual for the whole of the Mediterranean, which extends 3700 km eastward from Gibraltar, to be free of low pressure areas at any one time, except in summer. Depressions forming within the Mediterranean usually do so in response to quasi-stationary anti-cyclones farther north over central or western Europe. Other factors contributing to cyclogenesis in the Mediterranean are the presence of a baroclinic or frontal zone, lee effects in strong airflows and instability in airmasses (Meteorological Office, 1975). Analysis of upper-air charts shows that depressions travel in the direction of the flow at 500 hPa, whereas thunderstorms are more likely to follow the flow pattern at 700 hPa. Perry (1981) has also drawn attention to the role of sea surface temperature anomalies in promoting cyclonic activity within the Mediterranean. In addition, despite the inhospitable environment of the North African desert, a number of disturbances originate from this area, subsequently moving northeast across Italy or into the eastern Mediterranean (Boucher, 1975; Prezerakos, 1985).

Cold pools

During periods of low zonal index and well-developed meridional flow – especially when blocking occurs in the eastern North Atlantic section – masses of cold air may be transported southward over Europe and may become entirely surrounded within the troposphere by relatively warm air. The severing of the tips of these tongues of cold air produces tropospheric cold pools. When such an event occurs over the western Mediterranean, it becomes the center for instability, thunderstorms and abnormal precipitation (Boucher, 1982). On some occasions, central pressures may fall as low as 980 hPa, as in late January 1986. These events may be accompanied by cyclogenesis in

the Genoan area instigated by air flowing around the southern French Alps. This aspect was thoroughly investigated under the WMO experimental program ALPEX-SOP in 1982 (Lanzinger *et al.*, 1990).

Conclusion

The ever-changing pattern of the climate of western Europe provided the stimulus for the development of the science of meteorology in the early part of the twentieth century – notably the establishment of the Bergen School founded in 1918 by the Norwegian physicist Vilhelm Bjerknes (Jewell, 1981). This day-to-day variability has posed problems for climatologists who seek to establish order out of seeming chaos (Volkert, 1985; Lamb, 1985). The emphasis on circulation, weather types and Grosswetterlagen has provided a valuable tool for hydrologists investigating variations in precipitation yields (Sweeney and O'Hare, 1992) and coupled synoptic–hydrological modeling (Wilby *et al.*, 1994). Regional impacts resulting from global warming trends indicate the direction of future studies (Wilby, 1995).

Keith R. Boucher

Bibliography

Atkinson, B and Smithson, P., 1976. Precipitation, in Chandler, T.J. and Gregory, S. (eds) *The Climate of the British Isles*, 129–182.

Bardossy, A. and Caspary, H.J., 1990. Detection of climate change in Europe by analysing European atmospheric circulation patterns from 1881 to 1989. *Theor. Appl. Climatol.*, **42**, 155–167.

Barry, R.G. and Perry, A.H., 1973. *Synoptic Climatology*. London: Methuen.

Baur, F., 1931. Die Formen der atmosphärischen Zirkulation in der gemässigten Zone. *Gerlands Beitr. Geophys.*, **34**, 264–309.

Baur, F., 1944. Über die grundsätzliche Möglichkeit langfristiger Witterungsvorhersagen. *Ann. Hydrogr. Mait. Met. (Hamburg)*, **72**, 15–25.

Baur, F., 1947. *Musterbeispiele europäischer Grosswerterlagen*. Wiesbaden: Deterich.

Baur, F., 1963. *Grosswetterkunde und langfristige Witterungsvorhersage*. Frankfurt-am-Main: Akad. Verlagsgesellschaft.

Belasco, J.E., 1952. Characteristics of air masses over the British Isles. *Geophys. Mem.*, **11**(87), 1–34.

Bleasdale, A., 1975, 1968. An outstanding year for multiple events with exceptionally heavy and widespread rainfall, in *British Rainfall 1968*. London: Her Majesty's Stationary Office, pp. 223–231.

Borisov, A.A., 1965. *Climates of the USSR*. London: Oliver and Boyd.

Borisov, A.A., 1970. *Klimatografiya Sovetskogo Soyruza*. Leningrad: Leningrad University.

Boucher, K., 1972. The Rumanian flood disaster of May 1970. *Weather*, **27**(2), 55–62.

Boucher, K., 1975. *Global Climate*. London: English Universities Press, 326 pp.

Boucher, K., 1982. Tropospheric cold pool development over the Western Mediterranean, in *Proceedings of the First Hellenic–British Climatological Congress*, Athens, 1980, pp. 139–154.

Boucher, K., 1996. Measurement and instrumentation, in Wilby, R. (ed.) *Contemporary Hydrology*. Chichester; UK: Wiley, Ch. 4.

Brázdil, Šamaj, F., Valovič, Š., 1985. Variations of spatialannual precipitation sums in Central Europe in the period 1881–1980. *J. Climatol.*, **5**, 617–631.

Buchinsky, I. Ye., 1975. Droughts in the Ukraine (in Russian). *Moscow Vses Geog Obsch. Izv.*, **107**, 207–213.

Budyko, M.I., 1974. *Climate and Life*. New York: Academic Press.

Burt, S.D. and Mansfield, D.A., 1988. The Great Storm. *Weather*, **43**(3), 90–110.

Businger, S., 1985. The synoptic climatology of polar low outbreaks. *Tellus*, **37A**, 419–432.

Chistyakova, Ye. A., 1975. April precipitation deficits and surpluses in the southern half of Soviet Europe (in Russian). *Leningrad Gidromet. Nauc Issled. Tsent SSSR T. Vyp*, **166**, 17–29.

Chubukov, L.A., 1977. Graphical methods of representing climatic weather types. *Moscow: Akad. Nauk Mezhd. Geog. Kom.*, **2**, 81–88.

Clark, C., 1991. A four-parameter model for the estimation of rainfall frequency in south-west England. *Meteorological Magazine*, **120**(1423), 21–28.

Conrad, V., 1946. Usual formulas of continentality and their limits of validity *Am Geophys. Union Trans*, **27**, 663–664.

Davies, D.R., 1978. Blocking anticyclones. *Weather*, **33**(1), 30–32.

Davitaya, F.F., 1958. Studies of drought and sukhovei (in Russian). *Izv. Akad. Nauk. SSSR Ser. Geog.*, **5**, 131–136.

Doneaud, A., 1971. Meteorological factors causing Romanian floods in 1970. *World Meteorol. Organ. Bull.*, **20**(1), 28–31.

Escardo, A.L., 1970. The climate of the Iberian peninsula, in Wallén, C.C. (ed), *Climate of Northern and Western Europe*, Vol. 5, World Survey of Climatology. Amsterdam: Elsevier, Ch. 5.

Fink, A. *et al.*, 1996. Aspects of the January 1995 flood in Germany. Weather, **51**(2), 34–39.

Fraedrich, K., 1990. European Grosswetter during the warm and cold extremes of the El Niño/Southern Oscillation. *Int. J. Climatol.*, **10**, 21–31.

Fukui, E., 1966. Numerical expression for the development of the Mediterranean climate with its regional varieties. *Tokyo Geog. Papers*, **10**, 149–173.

Garnier, M., 1954. Contribution à l'étude des gelées en France. *Le Météorologie*, **35**, 369–378.

Gerasimov, I.P. (ed.), 1964. *Fiziko-Geograficheskiy Atlas Mira*. Moscow: Academic Naukita.

Goossens, C., 1985. Principal component analysis of Mediterranean rainfall. *J. Climatol.*, **5**, 379–388.

Hess, P. and Brezowsky, H., 1969. Katalog der Grosswetterlagen Europas. *Ber. Dtsch. (Offenbach)*, **15**(113), 1–56.

Houghton J.T. *et al.* (eds), 1990. *Climate Change*: The IPCC Scientific Assessment. Cambridge.

Houghton J.T. *et al.* (eds), 1992. *Climate Change 1992*. The Supplementary Report to the IPCC Scientific Assessment, Cambridge.

Hounam, C.E. *et al.*, 1975. *Drought and Agriculture*, Tech. Note No. 138. Geneva: World Meteorological Organization.

Hulme, M. *et al.*, 1995. Construction of a 1961–1990 European climatology for climate change modelling and impact applications. *Int. J. Climatol.*, **15**, 1333–1363.

Jacobeit, J., 1987. Variations of trough positions and precipitation patterns in the Mediterranean area. *J Climatol.*, **7**, 453–476.

Jenkinson, A.F., 1985. Hot spells in central England. *Weather*, **40**, 127–128.

Jewell, R., 1981. *Tor Bergeron's First Year in the Bergen School*. Basel: Birkhäuser Verlag.

Johannessen, T.W., 1970, The Climate of Scandinavia, in Wàllen, C.C. (ed.) *Climates of Northern and Western Europe*, Vol. 5, *World Survey of Climatology*. Amsterdam: Elsevier, Ch. 2.

Kessler, A., 1985. Heat balance climatology, in Essenwanger, O.M. (ed.) *General Climatology, Vol. 1A, World Survey of Climatology*. Amsterdam: Elsevier.

Klein, W.H., 1957. *Principal Tracks and Mean Frequencies of Cyclones and Anticyclones in the Northern Hemisphere*, Res. Paper No. 4. Washington DC.: US Weather Bureau.

Knox, J.L. and Hay, J.E., 1985. Blocking signatures in the Northern Hemisphere: frequency distribution and interpretation. *J. Climatol.*, **5**, 1–16.

Lamb, H.H., 1964. *The English Climate*, London: English Universities Press.

Lamb, H.H., 1969. Climatic fluctuations, 5 in Flohn, H. (ed.) *General Climatology* 2, Vol 2, *World Survey of Climatology*. Amsterdam: Elsevier, Ch. 5.

Lamb, H.H., 1972. British Isles weather types and a register of the daily sequence of circulation patterns 1861–1971. *Geophys. Mem.*, **16**(116), 1–85.

Lamb, H.H., 1982. *Climate, History and the Modern World*. London: Methuen.

Lamb, H.H., 1985. *Climate and its variability in the North Sea – Northeast Atlantic region*. Stavanger: North Sea.

Lanzinger, A. *et al.* (eds), 1990. *Alpex Atlas: case studies of ALPEX-SOP cyclones in the western Mediterranean*. Inst. Met. and Geophysics, Univ. of Innsbruck, Austria, 334 pp.

Lednicky, V., 1985. Climatological characteristics of a summer period index. *Hydromet. Ustav: Met. Zpravy*, **38**, 94–95.

Loaiciga, H.A. *et al.*, 1996. Global warming and the hydrological cycle. *J. of Hydrology*, **174**, 83–127.

Lydolph, P.E., 1977. *Climates of the Soviet Union*. London: Oliver & Boyd.

Marsh, T.J. and Turton, P.S., 1996. The 1995 drought – a water resources perspective. *Weather*, **51**(2), 46–53.

Matthews, R.P., 1972. Variation of precipitation intensity with synoptic type over the Midlands. *Weather*, **27**(2), 63–72.

Mayes, J., 1994. Recent changes in the monthly distribution of regional weather types in the British Isles. *Weather*, **49**(5), 156–162.

Meaden, G.T. *et al.* (eds), 1985. Proceedings of the First Conference on Tornadoes, Waterspouts, Wind-devils and Severe Storm Phenomena. *J. Meteorol.*, **10**(100), 178–248.

Meteorological Office, Great Britain, 1975. *Weather in Home Waters*, Vol II, Part I. London: Her Majesty's Stationery Office.

Murray, B. and Lewis, R.P.W., 1966. Some aspects of the synoptic climatology of the British Isles as measured by simple indices *Meteorol. Mag.*, **95**, 193–203.

Murray, R., 1993. Bias in southerly synoptic types in decade 1981–90 over the British Isles. *Weather*, **48**, 152–54.

Namias, J., 1964. Seasonal persistence and recurrence of European blocking during 1958–1960. *Tellus*, **16**, 394–407.

Perry, A.H., 1976. Mediterranean downpours in 1976, *J. Meteorol.*, **2**(13), 10–11.

Perry, A.H., 1981. Mediterranean climate – a synoptic reappraisal. *Prog. Phys. Geog.*, **5**(1), 107–113.

Pinna, M., 1957. La carta dell'indice di aridità par l'Italia, *Attidel XVII Congresso Geografico Italiano*.

Prezerakos, N.G., 1985. The northwest African depressions affecting the south Balkans, *J. Climatol.*, **5**, 643–654.

Raschke, E., Vonderhaar, T.H., Bandeen, W.R. and Pasternak, M., 1973. The radiation balance of the earth–atmosphere system during 1969–70 from Nimbus 3 measurements. *J. Atmos. Sci.*, **30**, 341–364.

Rauner, Yu. L., 1980. The synchronous recurrence of droughts in the grain growing regions of the Northern Hemisphere, *Soviet Geog.*, **21**(3), 159–179.

Rex, D.F., 1950. Blocking action in the middle troposphere and its effects upon regional climate. *Tellus*, **2**, 196–211; 275–301.

Sanders, F. and Gyakum, J.R., 1980. Synoptic–dynamic climatology of the 'bomb'. *Mon. Weather Rev.*, **114**, 1589–1606.

Smith, L.P., 1976. *The Agricultural Climate of England and Wales*, MAFF Tech. Bull. Series. London: Her Majesty's Stationery Office.

Sweeney, J.C. and O'Hare, G.P., 1992. Geographical variations in precipitation yields and circulation types in Britain and Ireland. *Trans. Inst. British Geog.*, **17**, 448–63.

Tabony, R.C., 1981. A principal component and spectral analysis of European rainfall, *J. Climatol.*, **1**, 283–294.

Tibaldi, S. and Ji, L.R., 1983. On the effect of model resolution on numerical simulation of blocking. *Tellus*, **35A**(1), 28–38.

Trewartha, G.T., 1981. *The Earth's Problem Climates*, 2nd edn. Madison: University of Wisconsin Press.

Türkes, M., 1995. Variations and trends in annual mean air temperatures in Turkey with respect to climatic variability. *Int. J. Climatol.*, **15**, 557–569.

Volkert, H., 1985. On the mesoscale variability of meteorological fields – the example of Southern Bavaria. *Contri. Atmos. Phys.*, **58**, 498–516.

Wahl, E.W. and Lahey, J.F., 1969. *A 700 mb Atlas for the Northern Hemisphere*. Madison: University of Wisconsin Press.

Wallén, C.C., 1977. *Climates of Central and Southern Europe*, Vol. 6, *World Survey of Climatology*. Amsterdam: Elsevier.

Weber, G.R., 1995. Seasonal and regional variations of tropospheric temperatures in the Northern Hemisphere 1976–1990, *Int. J. Climatol.*, **15**, 259–274.

Wilby, R.L., Greenfield, B. and Glenny, C., 1994. A coupled synoptic–hydrological model for climate change impact assessment. *J. Hydrology*, **153**, 265–290.

Wilby, R.L., 1995. Greenhouse hydrology, *Prog. Phys. Geog.*, **19**(3), 351–369.

Cross references

EVAPORATION: MEASUREMENT

Natural evaporation is the process whereby water at the Earth's surface, either in liquid or solid form, is converted to vapor and transferred into the atmosphere. It is therefore the reverse component to precipitation in the global water cycle and, in total over the whole surface of the Earth, the two must balance on the average. Furthermore, since the storage capacity of the atmosphere for water in all phases is equivalent to only a few centimeters depth of liquid water, whereas the average evaporation for the whole Earth approaches 100 cm year^{-1}, this balance must hold over comparatively short intervals of time.

It should, however, be stressed that the balance is achieved only on a global reckoning, the world patterns of evaporation and precipitation showing gross differences. The consequent transfer of water vapor from the source of evaporation to the area of precipitation, with the associated transfer of latent heat, provides the most important single factor in the overall redistribution of heat over the Earth's surface. It is for this reason that knowledge of the distribution of evaporation over the whole Earth is basic to an understanding of the general circulation of the atmosphere, while the economic importance of evaporation over the continents is being increasingly recognized as water usage grows rapidly. Over the oceans, the latter aspect is, of course, of no consequence, but the combined study of evaporation and precipitation in these regions will lead to increased knowledge of drift currents in and between the oceans, which are already known to be considerable.

Measurement of evaporation

The measurement of evaporation presents difficult problems, the difficulty varying with the nature and extent of the evaporating surface. Here we shall discuss briefly the principal methods that have been developed to meet different requirements, indicating the applicability and limitations of each.

Hydrological balance

When all other factors contributing to the hydrological or water balance for a given area are known, evaporation may be estimated as a residual. The hydrological balance may be expressed as

$$P = R + S + E$$

P being precipitation, R the surface runoff of water, S the seepage of water into the ground, which may increase the local storage or percolate to deeper levels and be removed as underground currents, and E the evaporation. Adequate measurement of P and R presents problems in sampling over the area in question. S is obviously a difficult term to assess, although in certain circumstances (when it is known from the local geological structure that underground currents out of the area do not exist) it can, over a sufficient time interval, be assumed to be zero. This method of estimating evaporation must evidently be very approximate in general, although in certain favorable circumstances, for example catchment areas of easy topographical form, results of acceptable accuracy may be obtained.

In certain studies, e.g. in agricultural research, the gravimetric method of determining evaporation is of prime importance. This technique, which is a specific refinement of the hydrological balance method, consists of isolating in an impermeable container *in situ* a soil block of appropriate size, whose water balance is known in terms of water applied and water lost by successive weighings of the block. This method is capable of high accuracy and, depending on the precision of the weighing equipment, can be used for the measurement of evaporation over periods less than 1 h when the evaporation rate is high. There are, however, obvious limitations to the size of such installations and the method is unsuitable for the study of evaporation when the terrain carries irregular cover.

Soil moisture profile

Other methods of measuring the water loss from soil by evaporation may be mentioned briefly. They are all based on the determination of the vertical distribution of moisture in the soil, the 'soil moisture profile,' and its change with time, and include measurement of the moisture content by sampling and weighing; measurement of electrical conductivity of materials placed in the soil, whose conductivity is moisture dependent (gypsum, fiberglass, nylon, etc.); the measurement of the thermal conductivity of soil, which is also moisture

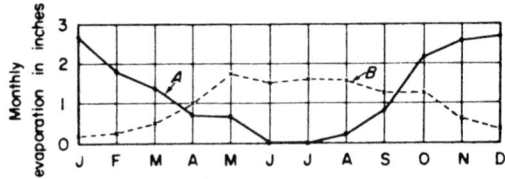

Figure E32 Estimated monthly evaporation rates for (A) Lake Superior and (B) nearby shallow lake. (From Wisler and Braker, 1959.)

dependent; and the determination of the scattering of neutrons by the hydrogen atoms in water. Each is subject to limitations of one sort or another (for details and reference to original work, see review by Deacon *et al.*, 1958).

Pan evaporation

The measurement of water loss from pans and tanks has occupied a central position in experimental studies of evaporation. The results from such observations are usually presented by means of a formula due originally to Dalton (Shaw, 1993):

$$E = C(e_w - e_a)$$

where E is the rate of evaporation, e_w is the saturation water vapor pressure at the surface temperature of the water, e_a is the vapor pressure in the air, and C is a factor that incorporates the effects of wind speed, barometric pressure and other variables such as exposure. Much of this work was originally carried out in the United States, in particular by Rohwer (1931) at Fort Collins, Colorado, Rohwer summarized the results of his studies, made from a tank 3 feet (0.9 m) square, in the formula

$$E = 0.771(1.465 - 0.0186B)$$
$$(0.44 + 0.118w)(e_w - e_a)$$

where E is measured in inches per day; B (the barometric pressure), e_w and e_a, in inches of mercury; and w (the wind speed near the ground), in miles per hour.

Such measurements, if they can be related to water losses from larger water bodies such as dams and reservoirs and from freely transpiring vegetation, are of obvious economic importance. What is wanted, ideally, is a single conversion factor (pan coefficient) applicable at all times under any conditions. Consideration of the basic dependence of pan evaporation on the size, design, and exposure of the pan and its comparatively rapid response to changing atmospheric conditions (whereas the larger body of water will be more or less thermally inert depending on its size) emphasizes the difficulty of achieving this factor. This is clearly brought out in the results of evaporation studies over Lake Superior. Figure E32 shows the monthly evaporation rates from Lake Superior and a nearby small shallow lake, both calculated from Rohwer's formula. The physical reasons for the curves being 6 months out of phase are obviously connected with the relative thermal inertia of the two lakes. This case is admittedly extreme, but it does emphasize the fact that each water body will have its own pan coefficient, and that this is likely to vary with pan or tank type and season. Nevertheless, such instruments, particularly if they are standardized, have a useful role to play in assessing the comparative evaporative needs of different regions, even though their quantitative indications must be interpreted with caution.

Other methods

A direct method of measuring evaporation has been developed in recent years, which stems from the recognition of evaporation as a process of turbulent diffusion. It can be shown that the product, averaged over any time interval, of the instantaneous values of vertical air velocity and specific humidity near the evaporating surface represents the average evaporation for the period. Instrumentation has been designed to measure the fine structure of the air in respect of these properties (Deacon *et al.*, 1958) and the technique has been verified in the field. This method has the advantage that it can be used for the measurement of evaporation from any type of uniform surface and is subject to none of the restrictions that limit many other techniques. Though up to the present time its use has been confined to research studies of evaporation, it is capable of development for routine use (Shaw, 1993).

Evaporation in the climatic context

Each of the above methods finds its particular use in specific studies of evaporation over relatively small areas, the interest being chiefly in the economy of water storages and in the water needs for agriculture. None is suitable in the larger context of the delineation of evaporation on the climatic scale, although the hydrological balance method in favorable circumstances provides an exception to both of these statements.

For such surveys, where it is sought to estimate evaporation in terms of climatological variables, there are two principal approaches. In the first, which is usually known as the aerodynamic method, the rate of evaporation E is related to the difference in water vapor content of the air at two levels above the evaporating surface and some function $f(u)$ of the wind speed in the diffusion formula

$$E = f(u)(q_1 - q_2)$$

where q_1 and q_2 represent the specific humidity of the air at the levels z_1 and z_2.

The quantity $f(u)$ is determined empirically, or may be derived from the theory of atmospheric turbulence as

$$f(u) = \frac{\rho k^2 u_2 (1 - u_1/u_2)}{(\ln z_2/z_1)}$$

where u_1 and u_2 are the wind speeds at the levels z_1 and z_2, ρ is the air density and k is von Karman's constant (~ 0.4). It must be stressed that this formulation for $f(u)$ is valid only when the air is in or near neutral equilibrium. However, over much of the Earth's oceanic surface, the assumption of neutral stability is justified, and then u_1/u_2 has a constant value so $f(u)$ reduces to a constant factor times the wind speed, and

$$E = A(q_1 - q_2)u$$

For the empirical determination of $f(u)$, it is necessary for the evaporation to be known independently, either from consideration of the water balance or from energy conservation. In the latter method, which can be used in its own right for the determination of evaporation and is the second principal approach referred to above, the latent heat expended on evaporation can be estimated as a residual when all other components in the energy balance at the evaporating surface are known. This balance is represented as

$$R_N = H + LE + S$$

where R_N is the net radiant energy reaching the surface, which can be estimated from empirical formulas in terms of climatological variables or can be measured; H is the sensible heat conduction from the surface to the air; L is the latent heat of vaporization of water; and S is the rate of heat storage below the evaporating surface. Over land, S is a relatively small component and, over long periods, may be neglected. However, over the oceans this is not so, and for this reason among others, the aerodynamic method has been preferred over these regions as we shall see below.

The mechanism of transfer of sensible heat into the air is similar to that of water vapor, and so the problem of determining evaporation from the energy balance resolves into that of correctly apportioning the available energy $R_N - S$ (or, nearly enough, R_N) between sensible and latent heat transfer. For this purpose the Bowen ratio technique is available. In this method, identity of the transfer mechanisms of heat and water vapor is assumed, and the ratio of the transfer of sensible heat H to latent heat LE (the Bowen ratio) may be written

$$\frac{H}{LE} = \frac{c_p(T_1 - T_2)}{L(q_1 - q_2)} = \beta$$

where T_1 and T_2 are the temperatures of the air at the levels z_1 and z_2, and c_p is the specific heat of air at constant pressure.

Hence, from the equation representing energy balance, E may be expressed as

$$E = \frac{R_N}{L(1 + \beta)}$$

Although each method may, in principle, be used for the measurement of evaporation, each requires great accuracy in the determination of the small differences in temperature and specific humidity that usually occur. The instrumentation for achieving this on a routine network basis is not yet available. Over the oceans, however, the difficulty is relieved by the fact that differences in temperature and vapor pressure between the surface and the air are relatively large, and a knowledge of the surface temperature automatically provides the value of the vapor pressure there. Over the land, no such simplification is generally available. Furthermore, departures from neutrality that call into question the assumptions of equality of transfer coefficients, referred to above, are much more pronounced over the land than over sea. For reasons such as these, evaporation surveys over land areas have, up to the present, had to rely largely on estimates made from the hydrological balance and, being subject to the inherent errors of this method, can at best be only very approximately correct.

One of the most extensive surveys of evaporation yet published is that due to Budyko (1956). Evaporation is estimated by the aerodynamic method, the appropriate transfer coefficient being determined from a variety of evidence including energy balance, studies of lake evaporation, etc. The formula used by Budyko is

$$E = 2.4 \times 10^{-6} u(q_s - q_a)$$

where E is in g cm^{-2} s^{-1}, u is the wind speed in cm s^{-1} at the usual height for shipboard observation, and q_s and q_a are the specific humidity of the air at the sea surface and at shipboard level,

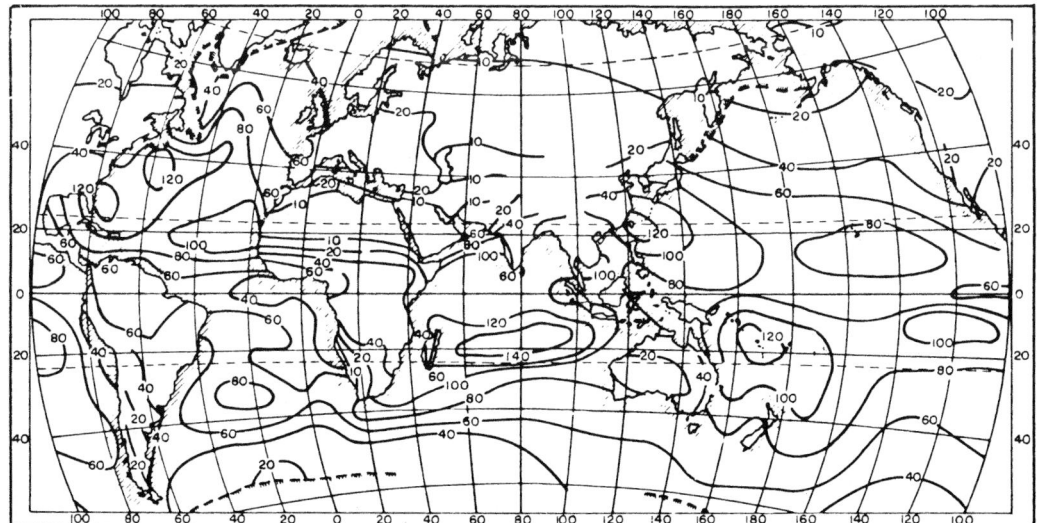

Figure E33 Mean annual evaporation rates for the Earth (kcal cm^{-2}). (From Budyko, 1956.)

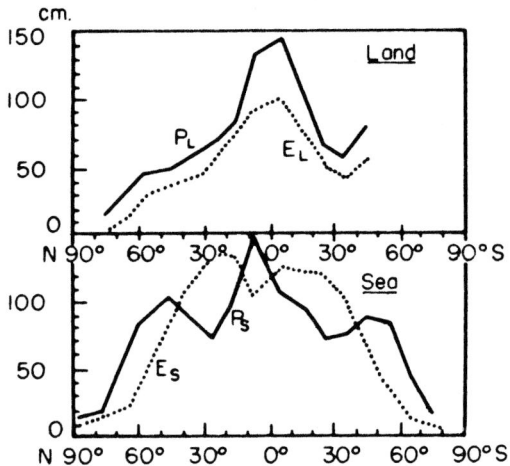

Figure E34 Mean annual precipitation (P) and evaporation (E) for land and ocean surfaces (cm).

Table E11 Average annual rates of evaporation for oceans and continents (cm year⁻¹) (after Budyko, 1956)

Oceans		Continents	
Atlantic	104	Europe	36
Indian	138	Asia	39
Pacific	114	North America	40
Arctic	12	South America	86
		Africa	51
		Australia	41

respectively: q_s is assumed to correspond to saturation at the sea surface temperature.

Since the world totals of evaporation and precipitation must balance, and since the estimates of either must be made on grossly inadequate observations, there is considerable latitude for interpreting and reconciling the observations of precipitation from coastal and island stations with estimates of evaporation from formulae whose coefficients are only approximately known and that necessarily use observational material whose quality may often be questionable. Observations of wind speed, temperature, and humidity can only properly be made on shipboard if special precautions are taken, and the measurement of sea surface temperature is another source of error.

Although estimates of evaporation show a variation in excess of 20%, and thus emphasize the approximate knowledge of the level of oceanic evaporation, there is nevertheless reasonably good agreement in the evaporation pattern on the broad scale. The distribution shown in Figure E33 is taken from Budyko's work. A notable feature is the distribution of evaporation rates in the subtropical high atmospheric pressure belts, with high rates over oceanic areas reaching a maximum in the central Indian Ocean interrupted by low values over land areas. The equatorial minimum reflects the comparatively high cloudiness of that region. The marked contrast in the precipitation–evaporation relationship between land and ocean surfaces is brought out in Figure E34. Over the land areas, where evaporation relies largely on local precipitation, the patterns of the two quantities are in good agreement, in strong distinction to the precipitation minima and evaporation maxima over the oceans in the trade wind belts.

Table E11 presents final estimates of average annual rates of evaporation from the oceans and the continents, based on the work of Budyko, 1956. While the figures shown in the table are perhaps the best available at the present time, their necessarily approximate nature must be borne in mind.

W.C. Swinbank

Bibliography

Budyko, M.I., 1956. *The Heat Balance of the Earth's Surface*. N.A. Stepanova (trans.). Washington, DC: Office of Technical Services. US Department of Commerce.

Deacon, E.L., C.H.B. Priestley and W.C. Swinbank, 1958. Evaporation and the water balance, in *Climatology, Reviews of Research*. Arid Zones Res. No. 10. Paris: UNESCO.

Rohwer, C., 1931. *Evaporation from Free Water Surfaces*, US Dept. of Agriculture Tech. Bull. 271. Washington, DC, 96 pp.

Shaw, E., 1993. *Hydrology in Practice*. London: Chapman & Hall.

Wisler, C.O. and E.F. Braker, 1959. *The Heat Balance of the Earth's Surface*. N.A. Stepanova (trans.). Washington, DC: Office of Technical Services, US Department of Commerce.

Cross references

Evapotranspiration
Hydroclimatology
Water budget analysis

EVAPOTRANSPIRATION

Moisture is returned directly to the atmosphere through a number of processes. The change in state from solid or liquid form to gaseous water vapor comprises the process of evaporation and sublimation. Evaporation occurs when input of energy onto an evaporating surface causes water molecules to pass from that surface to the atmosphere; this will occur when the vapor pressure of the air is below its saturation value. The rate of evaporation is governed by the state of a number of variables including water vapor, temperature and air motion, and several formulae are available to determine the rate at which it occurs. Perhaps the oldest of these is the Dalton equation given as

$$E_o = (e_s - e)f(u)$$

where e_s is the vapor pressure of the evaporating surface, e is the vapor pressure at some height above that surface, and $f(u)$ is a function of the horizontal wind speed. Since these parameters vary widely over the Earth's surface, the rates of evaporation vary enormously.

Transpiration

Just as moisture is returned to the atmosphere through evaporation, it is also returned by the process of transpiration. Transpiration is the term applied to the loss of moisture by plants, such losses occurring through stomata on the exposed leaf surface, with the rate of loss of water being controlled by guard cells on the leaf. During the day these are usually open to expose the moist leaf interior, with resulting high transpiration; at night they are generally closed. Some question exists concerning the role of transpiration in plant growth and development, but it is generally agreed that transpiration rates control the movement of moisture through plants, and this is obviously related to the transport of materials through the plant. The relative amounts of moisture lost through evaporation and transpiration obviously vary appreciably, depending on the nature of the ground surface. But as shown in Table E12, transpiration rates are often significantly higher than evaporation rates over densely vegetated areas.

While the study of both evaporation and transpiration is significant in itself, it is convenient to treat them as a single process in applied climatic studies. The loss of water through the combined process is termed evapotranspiration.

Evapotranspiration measurement

The rate at which evapotranspiration occurs depends on both meteorological and botanical characteristics, but it can be assumed that under a given set of conditions, an upper limit will exist. Clearly, to satisfy this maximum rate there must be sufficient moisture available. If moisture is in limited supply, the loss of water will be lower than the maximum rate. It is therefore necessary to recognize two evapotranspiration rates. Potential evapotranspiration is the maximum amount of water lost, assuming all moisture requirements can be met; actual evapotranspiration is the observed amount that, if moisture is limited, will be lower than the potential rate.

Problems exist in determining both the evaporation rate and the transpiration rate. It follows then that the estimation of evapo-

Table E12 Evaporation and transpiration losses for selected forest types (data from Ward, 1967)

Forest	Forest age (Years)	Evaporation (e) (mm/growing season)	Transpiration (T) + Interception (I) (mm/growing season)	Ratio $\dfrac{E}{(T+I)}$
Aspen	20	78	314	1:4.0
Aspen	60	84	282	1:3.4
Scots pine	20	48	363	1:7.6
Scots pine	60	87	340	1:3.9

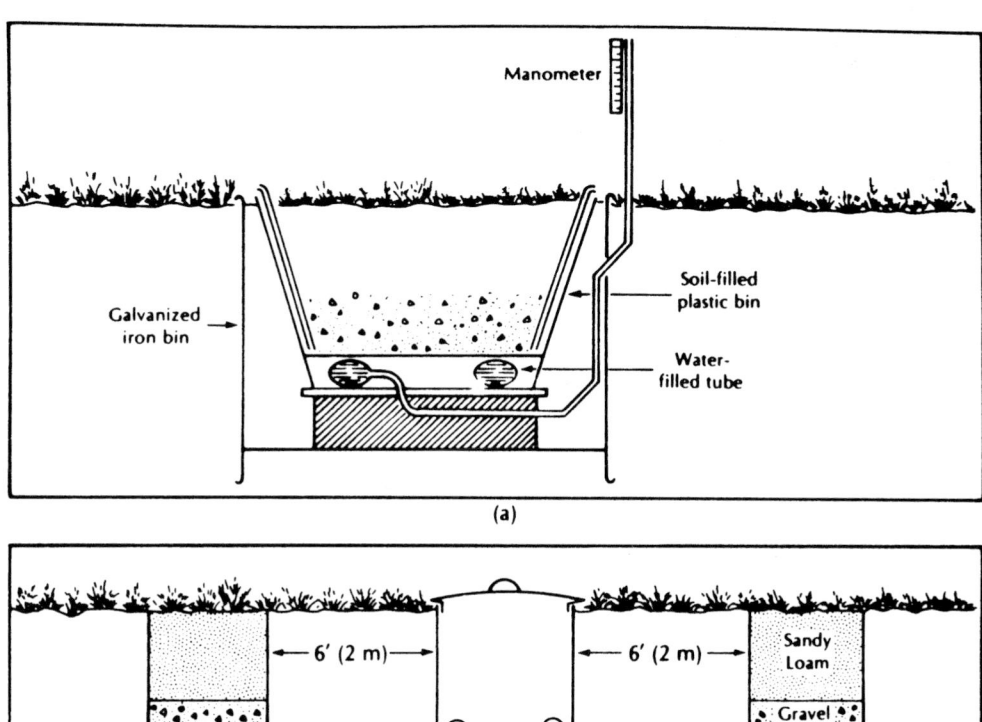

(a)

(b)

Figure E35 (a) Section through a simple weighing lysimeter. (b) Section through an evapotranspirometer. (From Oliver, 1973.)

transpiration rates is equally problematic. In measuring the potential rate, water can be added at known quantities and its disposal can be determined more readily. A number of instruments have been designed to measure both rates (see examples in Figure E35). The evapotranspirometer, since it is measuring potential evapotranspiration, gives fairly reliable results. The example shown consists of a number of tanks that are filled with a soil of similar constituency to that of the surrounding area and covered by a continuous vegetation cover. Since all of the moisture entering the system is accounted for by evapotranspiration or by percolation into the collecting jars, then the moisture consumed by evapotranspiration can be calculated. The measurement of the actual rate is open to larger experimental error. An example of the type of instrument used, in this case the floating lysimeter, is shown in Figure E35.

The emplacement, location and operation of lysimeters require great care and expenditure. Clearly, they must be large enough to provide a representative sample of a given area and deep enough so that they do not significantly alter the natural profile of soil moisture. Such instruments are rarely found outside of experimental stations

and thus provide standards and supplements to evapotranspiration rates derived using other methods.

With the lack of observed data, investigators have turned to other methods of estimating evapotranspiration losses. Essentially, these concern analysis of the variables that influence moisture losses; derived formulae include such factors as solar radiation, temperature, plant cover, and wind speed. Three approaches are available: the empiric in which derived equations are based on observation, the aerodynamic where the physics of the atmospheric processes responsible for evapotranspiration are evaluated, and the energy budget approach, which estimates the amount of energy available to cause moisture transfer back to the air. Numerous formulae have been derived using each approach. Empirical methods are represented by the work of Thornthwaite (1948), Blaney and Criddle (1950) and Makkink (1957). The aerodynamic approach is illustrated by the method of Thornthwaite and Holzman (1942), whereas the energy budget approach has been investigated by Bowen (1926), Budyko (1956) and Penman (1963). To demonstrate the methodology used, two methods are discussed below.

Estimating methods

Penman's method uses the combined influences of turbulent transfer and the energy budget. His derived equation for the determination of evapotranspiration uses vapor pressure, net radiation and the drying power of air at a given temperature. To derive this formula, Penman had first to obtain an expression that allowed determination of both net radiation and the drying power of air. Such a formula is necessarily quite complex, as is the expression for the evaluation of the drying power of air. Even so, and as is shown in Table E13,

---

Table E13 Estimation of evapotranspiration using Penman's method (Oliver, 1973, p. 54)

Formula:

$$E = \frac{(\Delta H + \gamma Ea)}{(\Delta + \gamma)} \tag{1}$$

where E = evaporation, mm per day;

δ = the slope of saturation vapor pressure versus temperature curve at air temperature T, in millibars per degree C;

H = the heat budget term (see below);

γ = the psychrometric constant;

Ea = the drying power of the air (see below).

Step 1   Calculation of H

H = (incoming shortwave radiation) (reflection coefficient) − (net outward long-wave radiation)

$$H = Ra(I - r) (0.18 + 0.55[n/N]) - \sigma Ta^4(0.56 - 0.092 \sqrt{ed}) (0.10 + 0.90[n/N]) \tag{2}$$

where Ra = calculated maximum solar radiation reaching earth in absence of atmosphere. Expressed in evaporation units and available from prepared tables (e.g. Brunt);

r = reflection coefficient of surface;

n/N = ratio of actual to possible hours of bright sunshine;

σTa^4 = blackbody function, Ta in degrees Kelvin. Available from prepared tables or $\sigma = 2.01 \times 10^{-9}$ mm day^{-1}

ed = mean vapor pressure.

Step 2   Calculation of Ea

$$Ea = 0.35(I + u/100)(ea - ed) \tag{3}$$

where u = wind speed at height of 2 m in miles day^{-1}

ea = saturation vapor pressure at temperature T

Step 3   Calculation of E

Substitute values for H, Ea, Δ, and γ into Equation 1.

Note that this provides evaporation from a free water surface. For evaporation from a vegetated area an empirical constant (ratio of E to ET for that area) is introduced.

$$ET = fE$$

where f = empirical coefficient, which varies over space and time.

Example

Data for North Carolina, month of June:

Mean monthly temperature	= 75.9°F (24.4°C)
Mean monthly relative humidity	= 72%
Mean monthly sunshine	= 65%
Mean monthly wind speed at 2 m	= 67.7 miles day^{-1}
Mean monthly extraterrestrial radiation (in evaporation units)	= 16.93 mm day^{-1}
Reflection coefficient	= 0.05
f value for area	= 0.7

Step 1

$H = Ra(I - r) (0.18 + 0.55[n/N]) - \sigma Ta^4(0.56 - 0.092 \sqrt{ed}) (0.10 + 0.90[n/N])$

$Ra = 16.93$; $r = 0.05$; $n/N = 0.65$; $ed = 16.6$ mm; that is, at T of 75.9°F ea is equal to 23.0 mm, so at RH 0.65%, $ed = 16.6$.

$H = 16.93(I - 0.05) (0.18 + 0.55.0.65) - 15.64(0.56 - 0.092.4.07) (0.10 + 0.90.0.65)$

$H = 6.64$ mm day^{-1}

Step 2

$$Ea = 0.35(1 + u/100) (eq - ed)$$
$$u = 67.7, ea = 23.0, ed = 16.6$$
$$Ea = 0.35(1 + 67.7/100) (23.0 - 16.6)$$
$$Ea = 3.71 \text{ mm day}^{-1}$$

Step 3

$$E = \frac{(\Delta H + 0.27\, Ea)}{(\Delta + 0.27)}$$

$\Delta = 0.77$ mm; $H = 6.64$; $Ea = 3.71$

$$E = \frac{(0.77 + 0.27.3.71)}{(0.77 + 0.27)} = 5.88 \text{ mm day}^{-1}$$

$$ET = fE$$

where $f = 0.7$; $E = 5.88$

$ET = 0.7 \times 5.88$

$ET = 4.12$ mm day^{-1}

Table E14 Methods for estimating potential evapotranspiration from short grass at Aspendale. Australia (after Sellers, 1965)

Month	T (°C)	$ET\, g^{a}$ (mm day^{-1})	Estimated potential evapotranspiration (mm/day^{-1})				
			Budyko and Penman	McIlroy	Thornthwaite	Blaney and Criddle	$0.2T^{b}$
January	23.3	7.76	6.86	7.57	4.47	4.91	4.66
February	21.1	5.62	5.44	6.12	3.51	4.34	4.22
March	19.6	4.21	3.76	4.22	2.80	3.82	3.92
April	17.2	2.94	2.76	3.16	2.04	3.24	3.44
May	12.6	1.31	1.48	1.67	1.09	2.52	2.52
June	10.9	0.99	1.01	1.19	0.80	2.26	2.18
July	10.0	0.93	1.17	1.34	0.75	2.25	2.00
August	11.0	1.37	1.60	1.76	0.89	2.55	2.20
September	13.0	2.30	2.64	2.88	1.50	3.02	2.60
October	15.8	4.07	4.00	4.28	2.09	3.64	3.16
November	17.8	5.26	4.76	5.27	2.73	4.17	3.56
December	10.1	5.99	5.99	6.56	3.47	4.62	4.02
Annual total		1296	1260	1398	793	1257	1170

[a] ET_g based upon observed values of seven lysimeters.
[b] A simple empiric statement, where T is mean monthly temperature, which appears to provide a reasonably good estimate.

Table E15 Estimation of evapotranspiration using Thornthwaite's method (Oliver, 1973, p. 58)

Formula:

$$E = 1.6(10T/l)^{a}$$

where E = monthly potential evapotranspiration in cm;
T = mean monthly temperature (°C);
l = heat index; the sum of 12 monthly i values. Constant for a given location;
a = constant, a function of I.

Step 1 Calculation of I
Obtained by summing prepared tables of monthly i values. The extract below provides an example.

Monthly i values—monthly mean temperature

T (°C)	0.0	0.1	0.2	0.3	0.4	0.5	0.6	0.7	0.8	0.9
22	9.42	9.49	9.55	9.62	9.68	9.75	9.82	9.88	9.95	10.01
23	10.08	10.15	10.21	10.28	10.35	10.41	10.48	10.55	10.62	10.68
24	10.75	10.82	10.89	10.95	11.02	11.09	11.16	11.23	11.30	11.37
25	11.44	11.50	11.57	11.64	11.71	11.78	11.85	11.92	11.99	12.06

Step 2 Calculation of unadjusted PE

Derived from a prepared nomogram. On the I scale plot the I value derived in step 1. Connect this with a straight line to the point of convergence. Use the constructed line to read off unadjusted potential evapotranspiration.

Step 3 Calculation of adjusted PE

Values derived from the nomogram are adjusted for day and month length. Extract of the table used for this is given below. it shows the mean possible duration of sunlight for given latitudes expressed in units of 30 days of 12 h each. Multiply by correction factor given in the table.

Lat. (°N)	J	F	M	A	M	J	J	A	S	O	N	D
39	0.85	0.84	1.03	1.11	1.23	1.24	1.26	1.18	1.04	0.96	0.84	0.82
40	0.84	0.83	1.03	1.11	1.24	1.25	1.27	1.18	1.04	0.96	0.83	0.81
41	0.83	0.83	1.03	1.11	1.25	1.26	1.27	1.19	1.04	0.96	0.82	0.80

Example

Monthly Temperatures (°C) for Seabrook, New Jersey

J	F	M	A	M	J	J	A	S	O	N	D
0.9	1.2	5.9	11.3	17.5	22.3	24.7	23.7	20.2	14.0	7.6	2.3

Table E15 Continued

Step 1

For each month refer to table for *i* value. For example, June 22.3 = 9.62; July 24.7 = 11.23

J	F	M	A	M	J	J	A	S	O	N	D
0.7	1.2	1.29	3.44	6.66	9.62	11.23	10.55	8.28	4.75	1.89	0.31

$$I = \Sigma i = 58.21$$

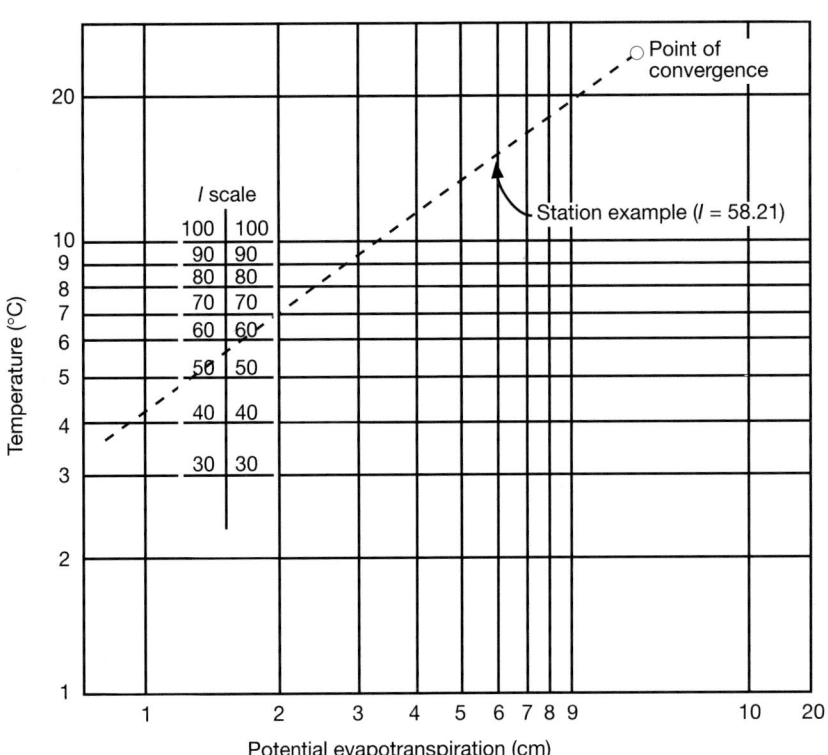

Step 2   Locate *i* = 58.21 on nomogram. Join to point of convergence. Read off unadjusted *ET* values.

J	F	M	A	M	J	J	A	S	O	N	D
0.1	0.2	1.6	4.1	7.5	10.6	12.2	11.5	9.3	5.5	2.3	0.4

Step 3   Calculate monthly (adjusted) potential evapotranspiration. Using latitude of station, multiply each month by correction factor.

J	F	M	A	M	J	J	A	S	O	N	D
(0.1)	(0.2)	1.6	4.6	9.2	13.1	15.4	13.6	9.7	5.3	1.9	0.3

Each gives the potential transpiration for each month. The sum = 75.0, gives the annual potential evapotranspiration in centimeters.

working with the system literally involves simple arithmetic and observance of set techniques.

The formula was derived through research in the UK, but results obtained from its use appear to hold true for many other parts of the world. As shown in Table E14, it appears to provide fairly reliable results for the Australian region in which it was tested. Investigators looking into the agricultural potential of most regions would do well to consider the Penman approach in estimating moisture requirements.

Thornthwaite's method is probably the best known and most widely used for estimating potential evapotranspiration in the United States. Working in the eastern part of the country, Thornthwaite devised a formula that essentially is based on the availability of temperature data. His method (outlined by example in Table E15) uses mean monthly temperature and an empiric heat index, which is itself an exponential function of temperature, as inputs. The derived unadjusted potential evapotranspiration is corrected by using actual daylight hours and number of days in the month in question.

While the method is widely used, it has been criticized. Perhaps the fault most often cited is the fact that temperature, which is the major variable used in the system, is not the best indicator of evapotranspiration rates; radiation values probably provide a more precise guide. Chang (1959) gives a number of examples where, because of the time lag between incoming radiation and temperature maxima, the Thornthwaite method gives imprecise results. Furthermore, since the formula was based on lysimeter data observed in watersheds in the eastern United States, the method does not always give good results elsewhere in the world. Criticism is also made of the fact that Thornthwaite assumes that evapotranspiration ceases at temperatures below 0°C.

Despite such criticism, there is little doubt that the Thornthwaite method is a useful and valuable approach, particularly when monthly data are used (shorter-term results are more questionable). Its value is enhanced since the only data required to estimate evapotranspiration are temperatures, and these variables are readily available for many stations throughout the world. The numerous publications by the Thornthwaite Associates (in 1950 and following years) are also an asset to its application.

Other estimates, in which actual plant types are considered, are also available. The method proposed by Blaney and Criddle (1950) uses the equation

$$U = KF = kf$$

where U = the consumptive use (evapotranspiration) in inches, F = the sum of the monthly consumptive use factors (the sum of f, the product of mean monthly temperature and monthly percentage of daytime hours) and K = the empiric coefficient for the plant in question (k being the monthly coefficient). Estimates of the plant coefficient (K) are derived from observed data for crops in arid and semiarid regions. Typical values are alfalfa 0.85, corn 0.80, citrus trees 0.06 and rice 1.20. When used in humid regions the coefficient values need to be decreased by 10%.

Papadakis (1966) has used evapotranspiration under different climatic regimes to formulate a climate classification based on agricultural potential. Although the classification he proposes is somewhat complex, it contains much that is interesting regarding climate and water needs. For example, in evaluating evapotranspiration he uses saturation vapor pressure corresponding to monthly temperatures. This is given by

$$E = 0.5625 \, (e_{\text{ma}} - e_{\text{d}})$$

where E is the monthly potential evapotranspiration in centimeters, e_{ma} is the saturation vapor pressure corresponding to the average daily maximum temperature, and e_{d} is vapor pressure corresponding to the mean monthly temperature.

To facilitate use of the formula a prepared table is used, from which values are derived from a slightly modified formula. In this table, the expresion $e_{\text{ml}} - 2$ is substituted for e_{d}. The new expression corresponds to the vapor pressure equivalent to the average daily minimum temperature minus 2; Papadakis bases this on the observation that the 'normal' difference between the average daily minimum temperature and the dewpoint is 2°C. As an example of the use of this method, Papadakis gives the case of Salvador, Brazil. In July the average maximum temperature is 29.9°C, the average minimum 23.2°C with the corresponding vapor pressures (multiplied by 0.5625) 238 and 142 mm, respectively. This gives a monthly potential evapotranspiration of 96 mm.

In tests (as indicated in Table E15) of various estimates of potential evapotranspiration, the Papadakis formula appears to hold up well.

John E. Oliver

Bibliography

Blaney, H.F. and W.D. Criddle, 1950. *Determining Water Requirements in Irrigated Areas from Climatological Data*. Soil Conservation Service Tech. Pub. No. 96. Washington, DC: US Department of Agriculture.
Bowen, I.S., 1926. The ratio of heat losses by conduction and by evaporation from any water surface, *Phys. Rev.*, **27**, 779–787.
Budyko, M.I., 1956. *The Heat Balance of the Earth's Surface*, N.A. Stepanova (trans.). Washington, DC: Office of Technical Services, US Department of Commerce.
Chang, J.-Hu., 1959. An evaluation of the 1948 Thornthwaite classification. *Assoc. Am. Geog. Annals*, **49**, 24–30.
Makkink, G.F., 1957. Testing the Penman formula by means of lysimeters, *J; Inst. Water Eng.*, **11**, 277–288.
Oliver, J.E., 1973. *Climate and Man's Environment*. New York: Wiley.
Papadakis, J., 1966. Climates of the world and their agricultural potentialities. Published by author, Buenos Aires.
Penman, H.L., 1963. *Vegetation and Hydrology*. Commonwealth Bureau of Soils Tech. Comm. No. 53. Farnham Royal, UK: Commonwealth Agricultural Bureaux.
Sellers, W.D., 1965. *Physical Climatology*. Chicago: University of Chicago Press.
Thornthwaite, C.W., 1948. An approach toward a rational classification of climate, *Geog. Rev.*, **38**, 55–94.
Thornthwaite, C.W. and B. Holzman, 1942. *Measurements of Evaporation from Land and Water Surfaces*, Tech. Bull. No. 817. Washington, DC: US Dept. of Agriculture.
Thornthwaite Associates, 1963–1964. *Average Climatic Water Balance Data of the Continents*, Vols 16 and 17, *Publications in Climatology*. Centerton, NJ.
Ward, R.C., 1967. *Principles of Hydrology*. New York: McGraw-Hill.

Cross reference

Evaporation: measurement

EVERGLADES, FLORIDA, USA

Introduction

The 'river of grass' label often attributed to the Everglades (Douglas, 1945) still holds true in many parts of this impressive subtropical wetland system, and is clearly visible to those who have flown over south Florida. From the ground, the river of grass is also apparent but the patchiness of the landscape is less easily detected. In reality, the Everglades landscape contains a multitude of wetland types including sawgrass marshes, sloughs, marl- and peat-based wet prairies, tree islands, pinelands and, at its southernmost extreme, mangroves and the Florida Bay estuary. Before major human influence, the Everglades landscape was even more diverse and included custard apple swamps, short-hydroperiod wet prairies and cypress strands (Davis et al., 1994).

Access to this wetland system is difficult. Airboat travel is the norm, and aside from the limited geographical area covered by tourist concession airboats, few venture deep into the interior wetlands. Once there, however, the visitor is welcomed by a plethora of intriguing biological utterances, magnificent but greatly diminished wading bird populations (Ogden, 1994), occasional bellowing alligators, sometimes searing heat and, depending on the area and season, biting and sucking insects of impressive numbers and noise.

At one time, the Everglades covered about 1.2 million ha, extending south from Lake Okeechobee to the peninsular tip of Florida, east to the coastal ridge (with occasional connections to the sea through areas known as the transverse glades), and west to the Immokalee Ridge (roughly the border of the Big Cypress National Preserve). An area of 283 290 ha immediately south of Lake Okeechobee that was once sawgrass marsh is now, however, dominated by agricultural uses (primarily sugarcane). Additional areas along the eastern border of the Everglades that were once cypress stands are now urban areas. This loss of about half of the original Everglades wetland area has resulted in a loss of habitat diversity, including that required for wading bird foraging, a reduction of total areal productivity and a loss of three major vegetation communities.

Formation of the Everglades occurred largely in recent geological time, about 5000 years ago. Water historically flowed slowly in a southerly direction from Lake Okeechobee to Florida Bay, creating conditions favorable for development of an underlying peat deposit that is one of the worlds largest. In the late nineteenth century the construction of canals was initiated to drain the Everglades to make it suitable for agriculture. About a century later, thousands of kilometers of canals and levees had been built, providing not only land for agriculture, but also accomplishing flood control and water supply objectives. What resulted from these canals and the large pumps that move water through them, collectively know as the Central and Southern Florida (C&SF) project, was lowered water tables, altered hydropatterns and transport of vast quantities of freshwater to the ocean that would have naturally flowed down the Shark River and Taylor Sloughs into Florida Bay (Light and Dineen, 1994). A system once characterized by hydrological sheetflow had become a highly compartmentalized and regulated system.

Excess nutrient runoff (eutrophication) from the Everglades Agricultural Area (EAA) has contributed to shifts of vegetation communities in the Water Conservation Areas (WCAs) from sawgrass to cattail. Nutrient runoff from the EAA has resulted from fertilizer application and through oxidation of the peat soils due to agricultural water management practices. Oxidation and other factors have led to an average annual soil subsidence rate of approximately 2.5 cm. In some locations, this translates to soil losses of more than 1.5 cm in thickness. Eutrophication and changed hydropatterns have altered the extent and coverage of naturally occurring fires, and have provided areas suitable for successful invasion of exotic species such as *Melaleuca* and *Casaurina*. *Melaleuca* presents a significant problem because of its potential to spread, its ability to outcompete native vegetation and its high evapotranspiration potential. The latter could have a significant impact on the water budget of South Florida and the ability to satisfy water demands by urban, agricultural and natural systems.

Although there is little disagreement that the Everglades have been regrettably altered during the last century, there is considerable discussion about what a restored system should look like. What vegetation patterns existed before human influence? Were they constant or constantly changing? What hydropattern and nutrient levels would be conducive to restore the landscape? Because half of the Everglades area has been lost to agricultural and urban uses, should the restoration goal be to reproduce the original Everglades in half the space? Alternatively, should restoration efforts return the Everglades that remain to what that particular area was historically, without attempting to restore short hydroperiod wetlands that existed on the edges of the original Everglades? In addition, given that Florida Bay historically received water from an area twice the size it is now, can sufficient water be sent to the Bay through half of the original Everglades without adversely affecting upstream hydro-pattern restoration efforts?

Determining how to accomplish an Everglades restoration, while sustaining agricultural and urban systems of South Florida will be the subject of intense debate and the focus of major scientific and engineering programs. A first step already underway requires implementation of best management practices (BMP) in the EAA and construction of about 16 000 ha of ecologically engineered wetlands called Stormwater Treatment Areas (STAs) for reduction of EAA phosphorus runoff. The combination of BMPs and STAs, as well as future nutrient reduction technologies, are intended to achieve reductions in phosphorus loads and concentrations that will protect the Everglades from imbalances of flora and fauna. Additional restoration efforts will focus on identifying historical hydropattern needs of the Everglades, and developing engineering solutions to achieve the needed water distribution, depth and timing of delivery.

Thomas D. Fontaine

Bibliography

Davis, S.M., L.H. Gunderson, W.A. Park *et al.*, 1994. Landscape dimension, composition, and function in a changing Everglades ecosystem, in S.M. Davis and J.C. Ogden, (eds), *Everglades: The Ecosystem and its Restoration*, St Lucie Press, 826 pp.
Douglas, M.S., 1947. *The Everglades: River of Grass* (revised edn., 1988), Pineapple Press, Sarasota, FL. 448 pp.
Light, S. and W. Dineen, 1994. Water control in the Everglades: A historical perspective, in S.M. Davis and J.C. Ogden, (eds), *Everglades: The Ecosystem and its Restoration*, St Lucie Press, pp. 47–84.
Ogden, J.C., 1994. A comparison of wading bird nesting colony dynamics (1931–1946 and 1974–1989) as an indication of ecosystem conditions in the southern Everglades, pp. 533–570, in S.M. Davis and J.C. Ogden, (eds), *Everglades: The Ecosystem and its Restoration*, St Lucie Press, pp. 533–570.

Cross reference

Okeechobee Lake, Florida, USA: human impacts, research and lake restoration

EXPERIMENTAL BASIN

Catchments or drainage basins are also known as watersheds (especially in the USA). Basin studies are sometimes divided into representative and experimental catchments, as originally conceived in the International Hydrological Decade (IHD). Representative basins are chosen to be 'typical' of a particular combination of factors

such as land use, geology, topography etc, but have been subject to criticism that many of them are only representative of themselves. In contrast, experimental catchments are distinguished by being subject to some deliberate modification (e.g. urbanization or cutting down an existing forest), so that the impact of the change on basin dynamics (usually streamflow) may be studied. However, the distinction is often blurred. Thus, a so-called representative basin which may be completely forested is not unchanging through time. Even without deliberate modification (e.g. clear felling), the trees will grow over time and may alter the catchment behavior, in addition to which there may be a sudden devastating (natural) deforestation due to insect attack or wildfire. Similarly, two representative basins under different land use may be used to make a comparison of flows, and so constitute a paired catchment experiment, subject to the assumption that other characteristics influencing streamflow – such as soils and topography – are similar.

In practice, the term experimental basin is now commonly applied to any instrumented basin where scientific research into hydrological phenomena is being undertaken.

Basin studies may be broadly grouped into three main types.

• Correlation studies in which the streamflow of different basins is compared. The simplest example is of two basins, selected to be as similar as possible in all respects except vegetation, so that all differences in their streamflow are attributed to land use. This approach has the great advantage of providing results immediately, but suffers from the fundamental problem that no two catchments are ever completely identical (soil, topography, geology, shape, etc). There has been increasing recognition of the limitations of such interbasin comparisons when inevitably other factors that may be of hydrological importance differ too. Thus some of the observed differences – but how much is not known – will be due to factors other than vegetation cover. A more complex approach, which makes some allowance for the effects of other variables is a multiple basin study statistically relating the values of a flow parameter (such as mean annual flow) to these parameters (land-use cover, soils, climate, etc). This provides a crude empirical prediction of flow response, but in practice provides little understanding of the processes involved, and so potentially can actually lead to misunderstanding.

• Single-basin studies offer a more direct approach. Before undergoing a land cover change the streamflow behavior of a basin is related to meteorological variables (for example, in a rainfall runoff model). This model is then used to predict the pattern of flows that would have subsequently occurred in the absence of the land cover change; these are compared with the flows actually observed. Since the same catchment is studied under different vegetation covers, this approach avoids the problem of differences in other physical factors encountered with the statistical approach. There is, however, the problem of climatic variability; it may be that the weather in the calibration period before the vegetation change was very different to that in the period afterwards.

• Paired-basin experimental studies are often preferred in practice, and combine the correlation and single-basin experimental approaches. Two similar catchments are studied for a calibration period to allow for any systematic differences in flow (due, for example, to differences in geology or topography) to be quantified. Then one catchment (the experimental basin) is subject to a change, and the other (the control catchment) remains unchanged. Comparison of the measured flows from the two basins before and after the vegetation change allows the effects of climatic variability to be identified.

Worldwide there have been many hundreds if not thousands of experimental basin studies, of which the following are some of the best known.

In 1674 Pierre Perrault (q.v.) published the results of probably the world's first basin study in which he compared annual rainfall and flow measurements in the catchment of the Seine upstream of Aignay-le-Duc (approximately 93 km^2 in area Figures E36–E38). This demonstrated, for the first time, that rainfall was sufficient to account for streamflow and helped overturn earlier notions that the flow of rivers was generated by condensation of dew within underground caverns.

In 1900, following a series of disastrous floods in the Alps two small basins (about 0.6 km^2 each) were established in the Bernese Emmental region of Switzerland in what was probably the first land-use study. One was mainly forest whilst the other was largely pasture

Figure E36 Source of the River Seine.

Figure E37 Aisey sur Seine.

Figure E38 River Seine at Moulin de la Forge (near Orret).

and used for cattle. Direct comparisons were made between the flows from the basins to answer questions whether forestry helped reduce flooding (and if tree felling could increase it). The forested basin had lower peak flows and annual streamflow; it also supported the highest dry weather baseflows. Subsequent work has suggested that differences between the basins' outflows might reflect soil differences as well as vegetational ones.

The Wagon Wheel Gap experiment in southern Colorado, USA, was the first true experimental basin study (in the sense that it involved deliberate catchment modification). Two small forested catchments (< 1 km^2) were established in a steep, mountainous area. After an 8-year calibration period the forest was removed on one catchment and measurements were continued for a further 7 years, over which it was noted that the cutting increased annual flow (relative to the unchanged 'control' catchment).

At Coweeta in the eastern USA, approximately 60 years of data have been collected into the relationship between land treatment and water resources, making it the oldest continuously operating catchment study in the world. In rugged landscape with hardwood forest in the southern Appalachian Mountains, some 25 small catchments have been studied including afforestation and deforestation. Initially a number of the subcatchments were subject to various treatments of forest cover conversions. The study has subsequently broadened to include more environmental studies, and to cover water quality including acid rain.

The Coweeta forested watershed inspired the ecosystem research at many other sites including Hubbard Brook in the northern Appalachian Highlands of New Hampshire, USA, set up in 1955. The central aim of that study was to determine land management methods for controlling streamflow extremes. Since the 1960s it has also become the site of some of the most intensive water quality studies.

The Plynlimon catchment study in the uplands of mid-Wales compares the water balance of two adjacent catchments, one under grassland and the other mainly under forest. In addition, there have been numerous field-based process studies to provide the understanding and quantification necessary to model the different processes under the two land uses and very importantly to be able to extrapolate the results to other areas.

The Krofdorf experiment in Germany is a good example of a sophisticated multiple paired catchment approach. Four small forested catchments were instrumented and after a 10-year calibration period the trees on two basins were cut. Subsequent changes in runoff and water quality from these basins were assessed by comparison of the observed flows with those predicted using statistical relationships derived with the unaltered basins during the calibration period. This study was of particular interest since two control basins were used and they yielded somewhat different predictions.

Mark Robinson

Bibliography

Bates, C.G. and Henry, A.J., 1928. Forest and streamflow experiment at Wagon Wheel Gap, Colorado. *Monthly Weather Review*, Supplement 30, 1–79, US Dept. Agriculture, Weather Bureau.

Bormann, F.H. and Likens, G.F., 1979. *Pattern and process in a forested ecosystem.* Springer-Verlag, New York, 253 pp.

Brechtel, H.M. and Fuehrer, H.-W., 1991. *Water yield control in beech forest – a paired watershed study in the Krofdorf forest research area.* Int. Assoc. Hydrol. Sci., Publ. 204, 477–484.

Kirby, C., Newson, M.D. and Gilman, K., 1991. *Plynlimon research: the first two decades.* Report No 109, Institute of Hydrology, Wallingford, UK, 188 pp.

Rodda, J.C., 1976. Basin studies, In Rodda, J.C. (ed.) *Facets in Hydrology.* John Wiley and Sons, Chichester, 257–297.

Swank, W.T. and Crossley, D.A., 1988. *Forest ecology and hydrology at Coweeta.* Ecological Studies No 66. Springer-Verlag, New York, 469 pp.

Whitehead, P.G. and Robinson, M., 1993. Experimental basin studies – an international and historical perspective of forest impacts. *J. Hydrology*, **145**, 217–230.

Cross references

Water movement in unsaturated soils
Water resources

F

FLOAT

A float is a natural or artificial body that is supported in the water by buoyance forces and used in open channels to determine water velocities. A surface float determines surface velocities, a subsurface float determines velocities below the water surface, a double float is used to determine velocities at a predetermined depth but which has an attached surface float to indicate the position in the water of the sub-surface component, and a rod float or a velocity rod determines velocities at a pre-determined water depth (Figure F1). Float operations can be affected by wind blowing across the water surface. The surface velocity is seldom an indicator of the mean velocity in an open channel: a rough interpretation is that the mean velocity is 0.7 times the surface velocity. Surface floats need not be sophisticated: oranges or table tennis balls can be just as good as purpose-made floats.

P.G. Holland

Bibliography

BS 3680, Part 1, 1991. *Glossary of Terms*, British Standards Institution, London.

Cross reference

Water resources: dictionary of basic terms

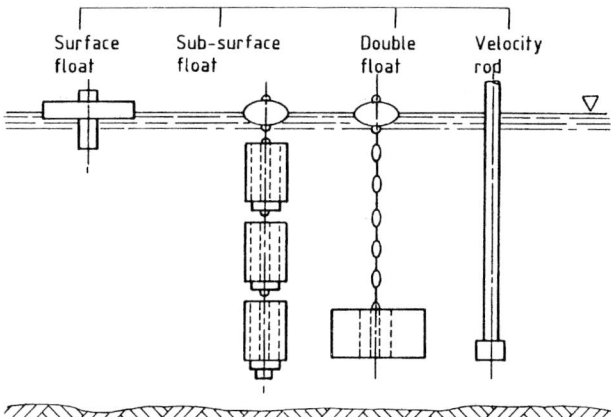

Key

▽ is the level of the water.

Figure F1 Types of float.

FLOOD ESTIMATION: METHODS FOR DEVELOPING COUNTRIES

An investigation into flood estimation in developing countries by the UK Institute of Hydrology for the UK Overseas Development Administration developed procedures suitable for use in a wide variety of smaller projects and for preliminary analysis of major works. The procedure can be used at sites which have no observed streamflow data.

When a long streamflow record is available, estimates of floods of a specified return period can be conveniently performed by normal flood frequency methods by plotting flood magnitude against return period in years. However, it is usually the case, unfortunately, that no data or only a limited period of data is available.

The index used in the study is the mean annual flood (MAF), which at a streamflow station is the mean of the series of the maximum flood peaks, occurring each year of the record. At ungauged sites, or those with short records, the MAF was estimated from catchment area and in some cases by a number of physical or climatic characteristics of the catchment which allowed the MAF to be estimated. Such equations for MAF were developed from which the instantaneous flood peak flow (Q) could be estimated for return periods up to 500 or 1000 years. Figure F2 shows the countries considered in the investigation and Table F1 gives the MAF equations developed for each region. Table F2 gives the predicted values of Q/MAF for return periods of 20, 100 and 500 years.

Figure F3a,b gives the resulting flood frequency curves with Q/MAF plotted against return periods. For any value of MAF, therefore, Q can be estimated.

List of symbols and abbreviations used in the Figures and Tables

AAR	catchment average rainfall (mm)
APBAR	mean annual maximum catchment 1-day rainfall (mm)
AREA	catchment area (km^2)
f.s.e.e	factorial standard error of estimate (from a regression equation)
GEV	general extreme-value
LAKE	the proportion of the catchment area which is upstream of lakes or reservoirs (if the total surface area of lakes is less than 1% of the catchment controlled, LAKE = 0)
MAF	mean annual flood, the mean of a series of annual maximum flood peaks (m^3 s^{-1}).
PADDY	proportion of catchment under paddy rice (value 0–1)
q	Q/MAF
Q	instantaneous flood peak (m^3 s^{-1})
r^2	coefficient of determination of a regression equation
SLOPE	the slope from the catchment outlet to the highest point above the end of the longest stream in the catchment (m km^{-1})

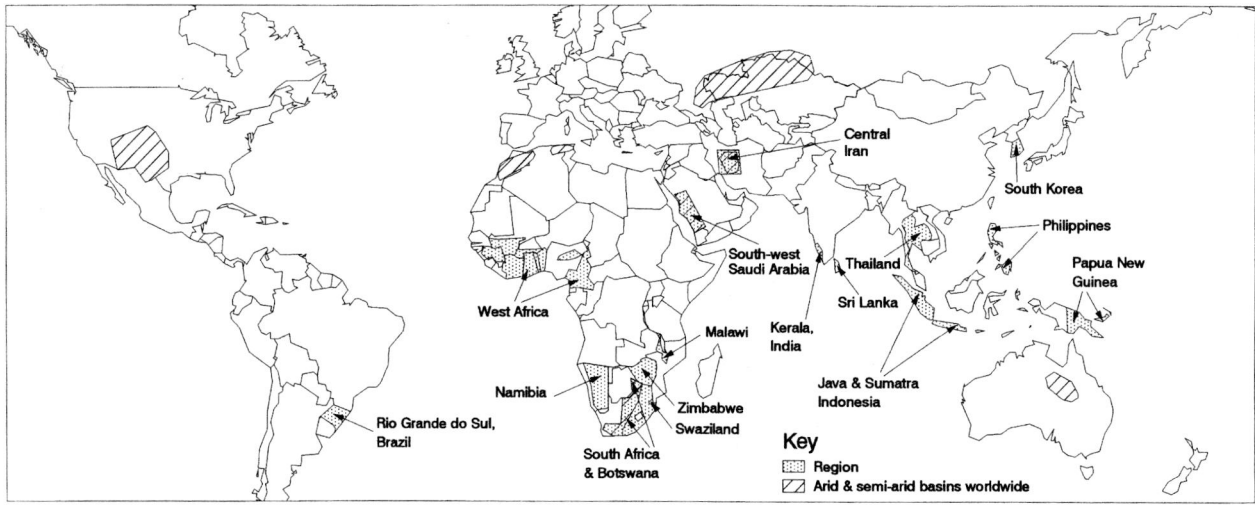

Figure F2 Countries and regions studied in flood estimation study.

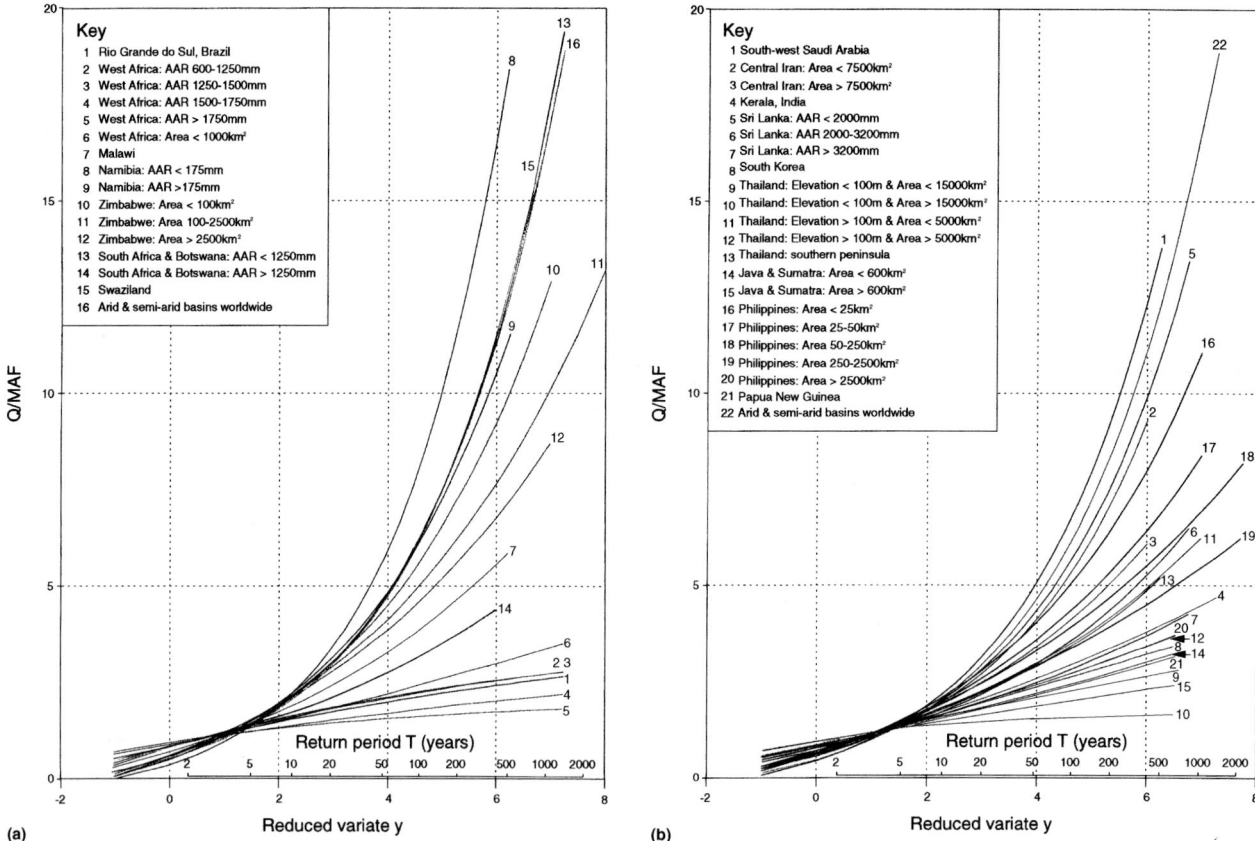

Figure F3 Comparison of regional flood frequency curves.

Table F1 Summary of recommended MAF prediction equations

Country or grouping	Prediction equation	r^2	f.s.e.e.
Rio Grande do Sul, Brazil (59)	$MAF = 8.75 \times 10^{-5}\ AREA^{0.987}\ S1085^{0.419}\ AAR^{1.017}$	0.913	1.49
West Africa			
>8°W (35)	$MAF = 7.86 \times 10^{-9}\ AREA^{0.933}\ AAR^{2.260}$	0.910	1.38
8°W to 2°W (86)	$MAF = 4.22 \times 10^{12}\ AREA^{0.807}\ AAR^{3.378}$	0.905	1.60
2°W to 4°E (41)	$MAF = 7.34 \times 10^{-7}\ AREA^{0.747}\ AAR^{1.887}$	0.856	1.58
9°E to 16°10'E and >8°N (16)	$MAF = 3.87 \times 10^{-6}\ AREA^{0.335}\ AAR^{2.308}$	0.819	1.54
9°E to 16°10'E and <8°N (46)	$MAF = 2.80 \times 10^{-10}\ AREA^{0.929}\ AAR^{2.652}$	0.943	1.44
Malawi (28)	$MAF = 2.89\ AREA^{0.553}\ STMFRQ^{0.360}$	0.381	2.39
Namibia (40)	$MAF = 2.63\ AREA^{0.460}$	0.651	1.92
Zimbabwe (234)	$MAF = 1.46\ AREA^{0.665}$	0.836	1.87
South Africa and Botswana (109)	$MAF = 6.97\ AREA^{0.450}$	0.542	2.19
	$MAF = 0.0964\ AREA^{0.515}\ AAR^{0.587}$	0.593	2.10
Swaziland (38)	$MAF = 2.93\ AREA^{0.570}$	0.657	1.76
Southwest Saudi Arabia (28)	$MAF = 0.0625\ AREA^{0.578}\ AAR^{0.727}$	0.452	2.41
Central Iran (24)	$MAF = 4.09 \times 10^{-4}\ AREA^{0.618}\ AAR^{1.362}$	0.694	2.21
Kerala, India (75)	$MAF = 5.14\ AREA^{0.722}$	0.613	2.04
Sri Lanka (69)	$MAF = 0.0285\ AREA^{0.670}\ AAR^{0.688}$	0.790	1.49
South Korea			
Area < 1000 (9)	$MAF = 1.71 \times 10^{-4}\ AREA^{0.680}\ AAR^{1.545}$	0.767	1.59
Area > 1000 (24)	$MAF = 2.50 \times 10^{-3}\ AREA^{0.646}\ AAR^{1.288}\ (1 + PADDY)^{-0.186}$	0.830	1.36
Thailand			
Main part (106)	$MAF = 2.56\ AREA^{0.625}$	0.729	1.91
S Peninsula (16)	$MAF = 1.23\ AREA^{0.841}$	0.818	2.05
Java and Sumatra, Indonesia (110)	$MAF = 8.20 \times 10^{-6}\ AREA^{0.852}\ APBAR^{2.640}$	0.881	1.61
	$MAF = 8.00 \times 10^{-6}\ AREA^{v}\ APBAR^{2.445}\ SLOPE^{0.117}\ (1 + LAKE)^{-0.85}$ where $v = 1.02 - 0.0275\ \log(AREA)$	0.889	1.59
Philippines			
Regions 1–2 (49)	$MAF = 15.3\ AREA^{0.623}$	0.675	1.92
Regions 3–8 (222)	$MAF = 11.7\ AREA^{0.616}$	0.638	2.10
Regions 9–12 (62)	$MAF = 11.5\ AREA^{0.502}$	0.459	2.61
Papua New Guinea (29)	$MAF = 6.08\ AREA^{0.676}$	0.918	1.58
Arid and semiarid basins worldwide (162)	$MAF = 1.87\ AREA^{0.578}$	0.55	2.88
	$MAF = 0.172\ AREA^{0.573}\ AAR^{0.416}$	0.57	2.85

Table F2 Summary of recommended regional flood frequency curves

Country or grouping	No. stations	No. years	GEV parameters			Predicted floods		
			u	α	k	q_{20}	q_{100}	q_{500}
Rio Grande do Sul, Brazil	57	1209	0.830	0.348	0.0959	1.73	2.12	2.46
West Africa								
AAR 600–1250	53	1034	0.806	0.424	0.1360	1.84	2.26	2.59
AAR 1250–1500	51	795	0.813	0.390	0.1095	1.80	2.22	2.57
AAR 1500–1750	70	1286	0.881	0.234	0.0756	1.50	1.79	2.04
AAR > 1750	27	487	0.908	0.219	0.1826	1.41	1.59	1.72
Area < 1000	26	304	0.804	0.314	−0.0437	1.80	2.41	3.05
Malawi	28	509	0.655	0.422	−0.1968	2.36	3.81	5.80
Namibia								
AAR < 175	9	100	0.336	0.448	−0.4834	3.30	7.97	18.09
AAR > 175	37	510	0.448	0.513	−0.3391	3.08	6.14	11.39
Zimbabwe								
Area < 100	53	954	0.486	0.516	−0.3018	2.97	5.63	9.93
Area 100–2500	139	2575	0.527	0.541	−0.2332	2.85	4.99	8.09
Area > 2500	42	737	0.562	0.534	−0.1996	2.73	4.59	7.13
South Africa and Botswana								
AAR < 1250	101	3808	0.470	0.430	−0.4039	2.94	6.23	12.50
AAR < 1250	8	233	0.733	0.343	−0.1710	2.06	3.13	4.53
Swaziland	38	756	0.485	0.410	−0.4128	2.87	6.12	12.39
Southwest Saudi Arabia	30	378	0.427	0.459	−0.4094	3.09	6.67	13.57
Central Iran								
Area < 7500	16	198	0.559	0.376	−0.3806	2.63	5.27	10.10
Area > 7500	9	145	0.636	0.419	−0.2307	2.42	4.07	6.43
Kerala, India	76	1171	0.747	0.370	−0.0991	2.02	2.90	3.92
Sri Lanka								
AAR < 2000	17	360	0.525	0.404	−0.3818	2.76	5.59	10.81
AAR 2000–3200	29	699	0.703	0.330	−0.2486	2.15	3.54	5.59
AAR > 3200	23	595	0.773	0.311	−0.1358	1.91	2.76	3.81
South Korea	24	542	0.775	0.373	−0.0256	1.93	2.60	3.29
Thailand[b]								
Group 1	18	290	0.828	0.310	0.0233	1.72	2.18	2.62
Group 2	13	284	0.919	0.243	0.3128	1.39	1.51	1.58
Group 3	54	942	0.691	0.388	−0.1835	2.22	3.50	5.19
Group 4	24	496	0.780	0.330	−0.0829	1.89	2.63	3.47
S Peninsula	16	284	0.708	0.352	−0.2050	2.15	3.40	5.13
Java and Sumatra, Indonesia								
Area < 600	47	541	0.812	0.290	−0.0671	1.77	2.37	3.05
Area > 600	48	468	0.866	0.239	0.0175	1.56	1.92	2.27
Philippines								
Area < 25	47	887	0.558	0.450	−0.2941	2.69	4.95	8.54
Area 25–50	37	646	0.603	0.466	−0.2206	2.56	4.32	6.81
Area 50–250	127	2208	0.641	0.457	−0.1752	2.42	3.88	5.79
Area 250–2500	104	1762	0.696	0.422	−0.1276	2.22	3.34	4.70
Area > 2500	18	243	0.768	0.356	−0.0715	1.94	2.70	3.55
Papua New Guinea	50	450	0.818	0.280	−0.0682	1.74	2.33	2.98
Arid and semi-arid basins worldwide								
All stations	162	3637	0.476	0.428	−0.4003	2.92	6.15	12.28

[a] Values in brackets = number of catchments studied.

[b] In Thailand, groups 1–4 are in the main part of the country, while the southern peninsula is treated separately. The groups are defined as follows: Group 1: elevation < 100, area < 15 000; group 2: elevation < 100, area > 15 000; group 3: elevation > 100, area < 5000; group 4: elevation > 100, area > 5000.

STMFRQ the number of stream junctions (as shown on a 1:50 000 scale map) divided by the catchment area (junctions km^{-2})

S1085 the slope of the longest stream in the catchment between 10 and 85% point (m km^{-1})

R.W. Herschy

Source

Meigh, J. 1995. *Regional Flood Estimation Methods for Developing Countries:* Report to the Overseas Development Administration. The UK Institute of Hydrology, ODA Report 95/1.

Cross references

Flood frequency analysis
Flood hazard management
Floods
Floods, river and multistage channels
Floods studies worldwide
Floods: worlds' maximum observed

FLOOD FREQUENCY ANALYSIS

Introduction

Flood frequency analysis is the means by which flood discharge magnitude (Q) is related to the probability of its being equaled or exceeded in any year or to its frequency of recurrence or return period (T). The return period and recurrence interval (terms which are used interchangeably) are used to indicate the long-term average interval between floods of a given magnitude. The return period and exceedence probability are reciprocals.

Frequency analysis is most commonly applied to peak instantaneous discharges, but may also be applied to daily mean flow or to a volume over a specified duration.

Analysis is carried out on an observed historic record of river flow with the aim of assessing future probabilities of exceedence. The flood record is regarded as if it were a random sample from a homogeneous population of floods with the assumption that such a record provides a reasonable approximation of the 'true' probability distribution which generated the historic record. It is also usually assumed that there will be no temporal change in the underlying statistics due to climatic change or to land-use changes although, on occasions, records may be adjusted to a common set of basin land-use conditions.

Frequency analysis of a gauged record is carried out by estimating the parameters of a selected probability distribution from recorded data. A large number of probability distributions and methods of application have been used and recommended for interpolation or extrapolation, but there is no convincing theoretical basis for specifying a single distribution either for universal or for a particular application.

Annual maximum and peaks over a threshold (POT) analysis

There are two methods of extracting and arranging the flood data from the continuous flow hydrograph. The annual maximum series draws a single maximum value from each calendar year or water year of record so that the number of data points equals the number of years of record, eg. $Q1$, $Q2$, and $Q3$ in Figure F4. Such a data set excludes some peaks which, although the second highest in a given year, are much larger than the largest in other years. The partial duration series in contrast, includes all independent peaks over a specified threshold discharge and is thus sometimes referred to as the **peaks** over a threshold (POT) series. On Figure F4 it includes q1 to q6 in addition to Q1 to Q3. The threshold is usually selected to include up to an average of four events per year. The annual exceedence series is a special POT series where the threshold is chosen such that there are N exceedences in N years of record.

A return period using the annual maximum series is the mean time interval between years containing a flood of given magnitude or greater, and is thus limited to assessment of return periods of more than 1 year. The POT series can be used to assess return periods of more frequently occurring events as, for example, may be required for in-river works, and is considered preferable for floods up to a 5-year return period. At higher return periods the difference between the two series becomes small. Langbein (1949) gives a relationship between the two series as follows:

$$T_e = \frac{1}{\ln T_m - \ln (T_m - 1)} \qquad (F1)$$

where T_m and T_e are the return periods of the annual maximum series and the POT series, respectively.

The relationship assumes that the intervals between events are randomly distributed and follow a Poisson process, and that events are independent. In practice, although the assumption of independence is usually (though not invariably) acceptable for annual maximum exceedences, it is perceived as a problem for the POT series where successive floods above a threshold may be affected by the same weather system or the second flood may be conditioned by catchment wetness and channel storage produced by the first. Various empirical procedures have been adopted to exclude such events by specifying a time separation between peaks dependent on basin area and/or by an intervening decline in discharges (e.g. Cunnane, 1979; Water Resources Council, 1982). In spite of these restrictions, POT event independence cannot be assured and equation (F1) thus overestimates T_e (Archer, 1981a; Page and McElroy, 1981).

Requirements for flood frequency analysis

Before launching into a description of statistical analysis it is appropriate to consider the use to which such analysis is put and the hydrological and engineering context in which it is applied.

Flood frequency analysis was developed in the first instance in response to the need for information for the safe and economical design of engineering structures either for the conveyance of flood

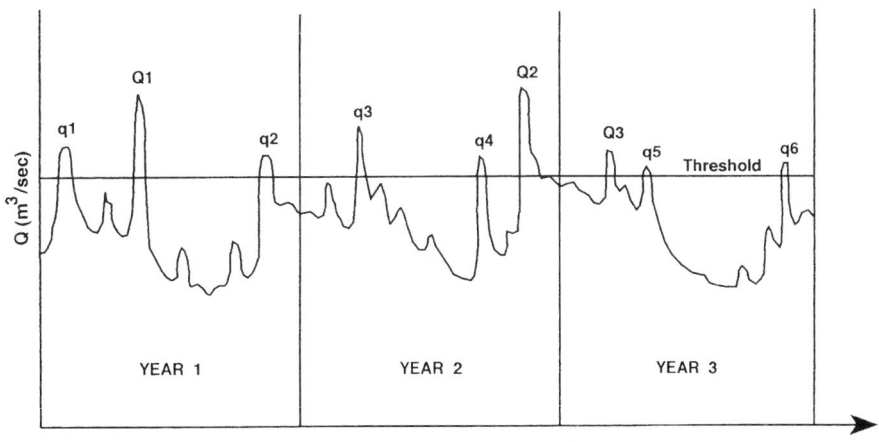

Figure F4 Extraction of flood peaks for annual maximum and peaks over a threshold (POT) series analysis.

Table F3 Levels of service based on flood return periods as specified by Northumbrian Water Authority, 1979

	Return period (yrs)
Non-agricultural land (residential)	40
Non-agricultural land (commercial/industrial)	25
Agricultural land	
Grade 1 (area > 10 acres: 4 ha)	15
Grade 2 (area > 100 acres: 40 ha)	10
Grade 3 (area > 100 acres: 40 ha)	10
Frequency of flooding shall not be greater than indicated return periods	

flows (bridges, culverts, diversion channels, reservoir spillways) or for the protection or mitigation from flooding of land and property (embankments and walls). Earliest application of techniques was in the United States (e.g. Fuller, 1914; Foster, 1924; Hazen, 1930). Ultimately the engineer needed to set a suitable level for the top of embankments or a discharge to pass through a waterway.

The decision on the acceptable risk associated with such discharges and levels is largely independent of flood frequency analysis and is based on a balance between the cost of protection and the consequences of the design level or discharge being exceeded over the expected life of the scheme. In some instances such risk is expressed directly in terms of the return period. An example is shown in Table F3 for standards applied by Northumbrian Water Authority (1979) in northeast England.

The decision on protection levels for major schemes is now more frequently based on a full cost–benefit analysis covering a range of schemes and design levels. Benefits of protection are assessed using relationships developed between river stage and the cost of damage. Detailed guidance on such assessment is provided, for example, by Penning–Rowsell and Chatterton (1977) and MAFF (1993).

Apart from the design of structures, flood frequency analysis is used for planning purposes and the designation of land-use categories based on flood zoning with respect to vulnerability. It is also used in the appraisal of the rarity of an observed event.

Most requirements are for probability over the full year, but for the risk to agricultural land or for the scheduling of river maintenance works, assessment of seasonal risks over a specified short period may be needed.

Hydrological and engineering context

The application of statistical flood frequency analysis to a flood record forms only one component of an engineering design calculation, each stage of which may involve some degree of extrapolation and hence approximation. Typically four stages are involved.

Flood flow measurement and extrapolation

Whilst standards can be set for accuracy of flow measurement in 'normal' flows, few stations are rated accurately up to the highest observed level. Gauging structures become non-modular and are bypassed, blocked or damaged. Current meter gauging becomes hazardous due to high velocities and debris and sediment entrainment. Overbank flows are difficult to measure, while channel cross-section may change due to scour. Level measurement instruments may cease to function owing to sediment blockage, physical range limitation or submergence. It may be impractical to reach the station in flood conditions.

It is not uncommon to find an extrapolated rating to be in error by more than 25% following application of improved techniques or successful gauging at a higher stage. Potter and Walker (1981) demonstrate how discontinuous measurement error can distort the apparent form of the flood frequency curve.

Extrapolation of flood frequency curve

Data length is often far short of the required design return period. Standard errors of estimation for the extrapolated flood frequency can be calculated for the selected distribution and fitting method, but errors can also arise from an inappropriate choice of distribution or

non-stationariness of the sample or population. The inclusion of historical flood flows can usefully extend the record but discharges are subject to greater measurement errors.

Spatial extrapolation

Data are rarely available exactly at the site of interest and spatial extrapolation is required from one or more gauging stations. For a station on the same river reach the effect of ungauged inflow between station and design site must be assessed. For ungauged tributaries and rivers, alternative methods of regionalization may be adopted. In addition to statistical estimation errors, errors may also arise from inhomogeneity of flood frequency characteristics within regional data sets or from inadequate representation of the design site.

Hydraulic extrapolation

To calculate a design flood level at the site of interest, a discharge–stage relationship must be estimated. The design site, unlike the gauging station, does not have the benefit of a structure or current meter observed rating to extrapolate to the design discharge, and estimation is primarily achieved on the basis of the hydraulic characteristics of the reach, with uncertainty arising from such parameter estimation.

Much theoretical research has been directed to the analysis of a data set for a single site for flood frequency extrapolation, implying the accuracy of the data and the adequacy of the analysis for design. Hydrological practitioners are obliged to take a broader and more balanced view.

Independence assumption

The ultimate customer for flood frequency analysis is the householder whose property is at risk above a given threshold. The fact that successive events are hydrologically or statistically dependent is irrelevant to the householder, if they are so separated as to duplicate the loss. It is independence of effect which is paramount. If the assumptions of the frequency model are violated by the data, it is the model rather than the data which requires amendment.

Flood frequency analysis of a gauged record

Whilst observing the cautions noted above, flood frequency analysis for a single site or as part of a process of regionalization is still necessary. The standard procedure is to assume the frequency distribution that flood events are likely to follow and to evaluate the parameters of the selected distribution using a selected probability plotting position and fitting method. Thus for a single site there are three choices to be made between

- annual maximum or peaks over a threshold analysis;
- frequency distributions;
- fitting methods.

Once these choices have been made the application of the procedure is objective.

Annual maximum analysis

Annual maximum analysis is generally preferred to peaks over a threshold analysis if the required return period is greater than 5 years and especially if more than 10 years of record are available (Sutcliffe, 1978). The discussion below concentrates on the annual maximum series. Recent statistical developments in the use of the peaks over a threshold series are outlined in Naden (1992).

Frequency distributions

The frequency of floods from a sample of annual maxima can be represented by a relative frequency diagram where the number of annual maximum floods in each discharge class can be plotted against the discharge (Figure F5a). In the limit, where the population is large and discharge class interval is small, the distribution tends towards a smooth curve and the frequency can be interpreted as a probability (Figure F5b). The relationship between probability and discharge is then described by a probability distribution and the total area under the distribution is equal to 1.0. The probability distribution may be converted to a cumulative exceedence probability distribution by summing incremental areas of $Q = X_i$, proceeding from the upper tail (Figure F5c).

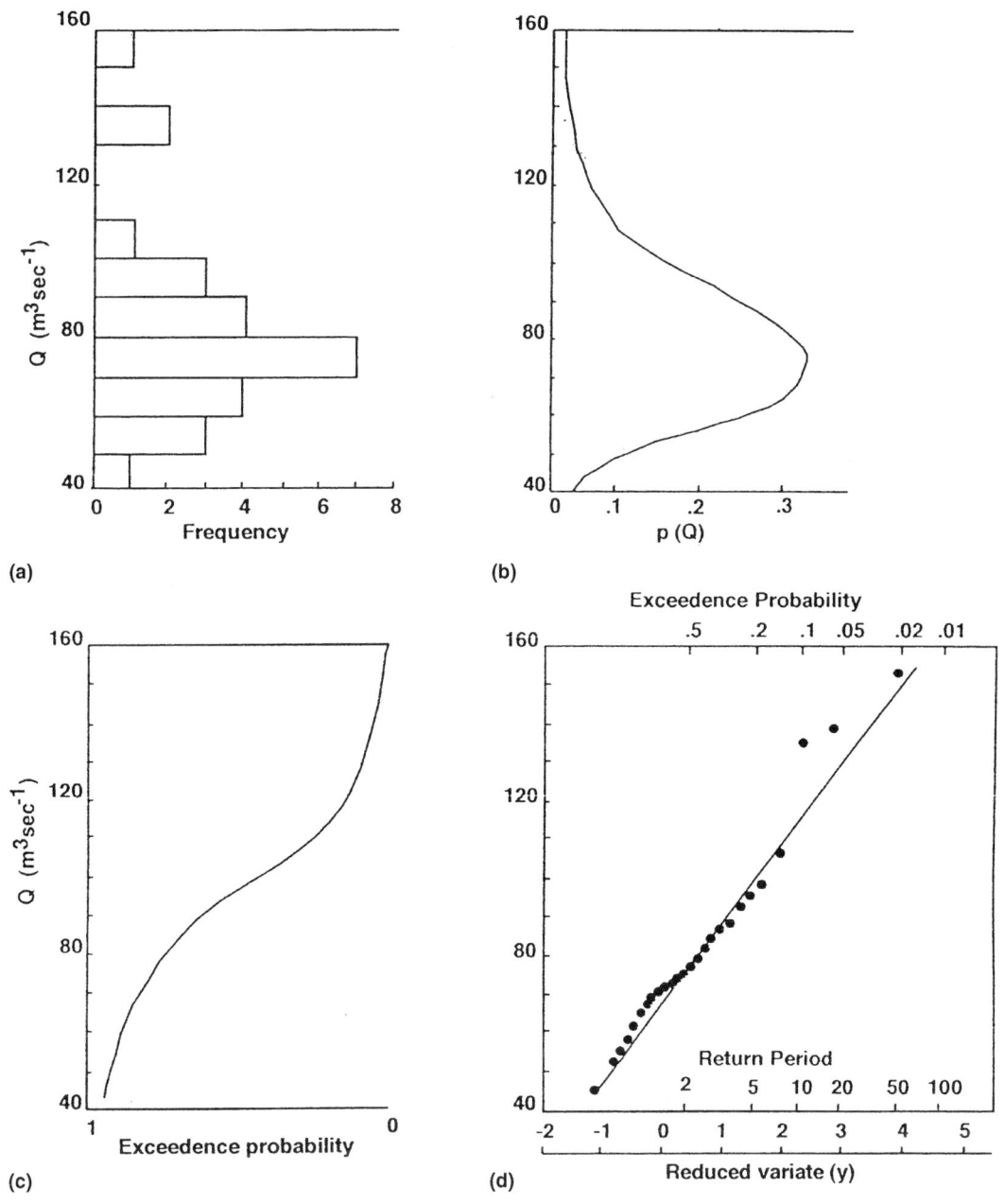

Figure F5 Annual maximum peak flow estimation. (a) Frequency diagram, (b) probability distribution, (c) cumulative probability with arithmetic probability scale and (d) flood frequency using EV1 (Gumbel) plot.

The cumulative frequency can then be linearized for specific distributions by expressing the probability as a reduced variate (y). The probability (or return period) scale is then marked alongside the y scale, giving the more familiar form of the flood frequency plot (Figure F5d). Plotting in this way permits ready graphical inspection of the success in fitting a particular distribution to the data and is widely (though unwisely) used as a basis for extrapolating well beyond the range of the data.

Each distribution type is characterized by its own formula but the precise form for a particular population is defined by distribution constants or parameters of the distribution. Thus for the well-known normal distribution defined by:

$$f(x) = \frac{1}{\sigma\sqrt{(2\pi)}} e^{(-x-\mu)^2/2\sigma^2} \qquad (F2)$$

the parameter μ is the mean value of the variate and σ is the standard deviation. More generally the parameter μ locates the position of the population variate on the x axis and is thus referred to as a location parameter, σ controls the spread of the variate on either side of $x = \mu$ and is thus referred to as a scale parameter. Different distributions have their own location and scale parameters which are not necessarily μ and σ. Two-parameter distributions commonly used for flood frequency include the lognormal and the Gumbel, also referred to as extreme value type 1 (EV1).

Greater flexibility in the fitting of distributions can be achieved by the use of a third parameter, most often a shape parameter which controls the skewness or asymmetry of the distribution. The general extreme value (GEV) (Jenkinson 1955), Pearson type III and log-Pearson type III are three-parameter distributions which are often applied to flood series.

The numerical fitting of a theoretical frequency distribution to a flood data set consists of the determination of the distribution parameters which then define the fitted line on the flood frequency plot. The flood discharge for any specified return period can then be obtained by graphical inspection or by calculation from the distribution parameters. Thus for the Gumbel EV1 distribution whose distribution function has the form

$$F(x) = \exp\left[-\exp\left(-\frac{x-u}{\alpha}\right)\right] \quad (F3)$$

where u is a location parameter, α is a scale parameter and skew is fixed at 1.14, and the Gumbel reduced variate (y_t) is calculated from

$$y_t = -\ln[-\ln(1-1/T)] \quad (F4)$$

then the magnitude of an event with return period T can be estimated from

$$X_T = u + \alpha y_T \quad (F5)$$

Distribution functions and the characteristics of commonly used distributions are described in detail in the *Flood Studies Report* (NERC, 1975) and Cunnane (1989).

Method of fitting

Fitting may be done either graphically or numerically. Numerical methods are generally preferred because of their objectivity although different fitting procedures lead to different curves and hence to different estimates of the T-year flood. There are several methods in common use including the method of moments, probability weighted moments, and maximum likelihood.

By the method of moments the moments of the distribution such as mean, variance and skewness are computed from the data, and in turn used to estimate distribution parameters. Thus for the EV1 distribution where the sample mean $\hat{\mu}_Q$ and variance $\hat{\sigma}^2_Q$ are estimated from the data, the corresponding estimates of the u and α in equation (F3) are

$$u = \hat{\mu}_Q - 0.5772\alpha$$
$$\alpha = \frac{\sqrt{6}\hat{\sigma}_Q}{\pi} \quad (F6)$$

The fitted line is then defined by equation (F5) in which α and u are substituted.

Derived moments are sensitive to small changes in data values especially those which are unusually high (outliers) in the upper tail of the distribution. Probability weighted moments, introduced to hydrology by Greenwood *et al.* (1979) limit the effect of outliers by giving them a lower probability weighting.

According to the method of maximum likelihood, a probability density function, expressed as $f(x, \theta)$ where θ represents parameters to be estimated, the parameters which give the largest value of the likelihood function:

$$L = f(x_1/\theta) f(x_2/\theta) \ldots f(x_n/\theta) \quad (F7)$$

are assumed to be the most likely to have generated the observed sample. They are thus the maximum likelihood estimates of the distribution parameters. Hand calculation of the maximum likelihood methods is impractical as they require iteration. However computer programs for flood frequency analysis using a range of fitting methods and distributions are now widely available and hand calculation is rarely necessary.

Data preparation and probability plotting positions

Preparation of data for analysis is illustrated in Table F4 for the gauged record at Rutherford Bridge on the River Greta, an 86.1 km^2 basin in the Pennine hills in northeast England. The record period is 1960 to 1985 ($N = 26$). Annual maximum discharges are extracted, arranged in descending order of magnitude and allocated a rank.

To display the data graphically, the next stage is to chose a plotting position on the probability axis for an event of given rank i in a sample of N years. Historically the Weibull (1939) formula $p(x) = i/(N+1)$ gained acceptance. However, Cunnane (1978) showed that the Weibull formula gives a biased estimate of the magnitude with exceedence probability p. Although unbiased plotting position formulae can be found for a particular distribution, the Gringorten (1963) formula:

Table F4 Preparation of annual maximum flood data for frequency analysis: River Greata at Rutherford Bridge, A, excluding and B, including 1986 flood

Q (m^3 s^{-1})	Rank, i A	Rank, i B	Gringorten return period plotting position A	Gringorten return period plotting position B	Reduced variate, y_i A	Reduced variate, y_i B
210.4		1		48.43		3.87
118.0	1	2	46.64	17.38	3.83	2.82
113.2	2	3	16.74	10.59	2.79	2.31
109.2	3	4	10.20	7.62	2.27	1.96
95.4	4	5	7.34	5.95	1.92	1.69
93.8	5	6	5.73	4.88	1.65	1.47
92.1	6	7	4.70	4.13	1.43	1.28
90.5	7	8	3.98	3.59	1.24	1.12
86.6	8	9	3.46	3.17	1.07	0.97
83.7	9	10	3.05	2.84	0.92	0.83
81.0	10	11	2.73	2.57	0.79	0.71
77.4	11	12	2.47	2.35	0.66	0.59
73.6	12	13	2.26	2.16	0.54	0.48
71.6	13	14	2.08	2.00	0.42	0.37
70.7	14	15	1.98	1.86	0.31	0.26
68.4	15	16	1.79	1.74	0.20	0.16
67.3	16	17	1.68	1.64	0.10	0.06
65.9	17	18	1.58	1.54	−0.01	−0.04
65.1	18	19	1.49	1.46	−1.11	−0.14
60.5	19	20	1.41	1.39	−0.22	−0.24
58.4	20	21	1.34	1.32	−0.32	−0.35
57.8	21	22	1.27	1.26	−0.44	−0.46
52.4	22	23	1.21	1.20	−0.56	−0.58
51.7	23	24	1.16	1.15	−0.69	−0.71
49.5	24	25	1.11	1.10	−0.84	−0.86
49.5	25	26	1.06	1.06	−1.04	−1.05
49.2	26	27	1.02	1.02	−1.35	−1.36

$$p(X_i) = \frac{i - 0.44}{N + 0.12} \quad (F8)$$

which is the appropriate form for the Gumbel (EV1) distribution, can also be used with little loss of precision for other distributions, and it is now widely used. Plotting position formulae differ most for the highest ranked value.

Mathematical fitting by the methods outlined above does not depend on the choice of plotting position formula but plotting permits graphical inspection to be made of the match of fitted distribution and observed data. On the other hand, graphical fit by eye or other methods is strongly affected by the choice of plotting position. Even using unbiased plotting positions, graphical fit by eye has received disapproval because of its obvious subjectivity. However, as Cunnane (1975) points out, 'the fact that some distribution has got to be assumed, is often overlooked, and the choice between distributions must always have an element of subjectivity in it, notwithstanding the use of goodness of fit tests. Graphical methods should not be lightly dismissed; a degree of difficulty tends to generate veneration for a method while simplicity tends to breed the familiar contempt'. More recently, Klemes (1986) has strongly opposed the sophistication of numerical fitting methods and Bardsley (1994) has defended the use of graphical fitting.

Comparison of estimates: Rutherford Bridge 1960–1985

Conclusions might be drawn from a comparison of the effect of choice of distribution and fitting method on the T-year estimate for the River Greta at Rutherford Bridge (Table F5).

- Using the data set for 1960–1985, distributions give quite consistent estimates up to a record length N and, even at $2N$ and $4N$, the difference between estimates is satisfactorily small. In this instance the best fit three-parameter distribution differs little from the two-parameter best fit (shape parameter near its two-parameter constant value).
- Estimates by different fitting methods are sensibly similar for the

Table F5 Comparison of *T*-year flood estimates by a range of distributions and fitting methods: River Greta at Rutherford Bridge, 1960–1985

Distribution	Fitting method	Return period (years)						
		2	5	10	25	50	100	200
EV1	MOM	72	90	101	116	127	138	149
EV1	PWM	72	91	103	119	131	142	154
EV1	MLE	72	90	102	117	128	139	150
GEV	PWM	72	91	103	118	129	139	149
GEV	MLE	72	90	102	117	128	139	150
LN2	MOM	73	91	102	115	125	134	143
LN2	MLE	73	91	102	115	125	134	143
LN3	MOM	73	91	102	114	122	130	137
LN3	PWM	72	91	103	118	128	139	149
LN3	MLE	71	90	105	123	138	154	170
P3	MOM	73	91	102	114	122	130	137
LP3	MOM	72	90	102	117	128	138	149

EV1 = extreme value type 1 (Gumbel); GEV = general extreme value;
LN2 = two Parameter lognormal; LN3 = three Parameter lognormal;
P3 = Pearson type 3; LP3 = Log-Pearson type 3.
MOM = methods of moments; PWM = probability weighted moments;
MLE = maximum likelihood estimate.

Table F6 Comparison of *T*-year flood estimates by a range of distributions and fitting methods: River Greta at Rutherford Bridge, 1960–1986

Distribution	Fitting method	Return period (years)						
		2	5	10	25	50	100	200
EV1	MOM	75	104	123	147	165	183	200
EV1	PWM	75	101	118	140	156	172	187
EV1	MLE	75	97	111	129	143	156	170
GEV	PWM	71	96	116	148	176	209	248
GEV	MLE	71	95	117	152	185	226	276
LN2	MOM	74	103	123	147	166	185	204
LN2	MLE	76	100	115	135	149	163	177
LN3	MOM	72	99	120	149	173	198	226
LN3	MLE	71	97	118	149	175	204	235
P3	MOM	69	97	120	152	178	204	231
LP3	MOM	72	97	117	148	175	206	242

EV1 = extreme value type 1 (Gumbel); GEV = general extreme value;
LN2 = two Parameter lognormal; LN3 = three Parameter lognormal;
P3 = Pearson type 3; LP3 = Log-Pearson type 3.
MOM = methods of moments; PWM = probability weighted moments;
MLE = maximum likelihood estimate.

same distribution except for the three-parameter lognormal where there are 13% and 18% differences between moments and maximum likelihood at 50- and 100-year return periods.

- Nevertheless, beyond 2*N* (50 years) the difference between methods can make a difference of a factor of 2 in the estimated return period. Thus a discharge of 130 m³ s⁻¹ could represent a 100-year flood by LN3-MOM or P3-MOM and less than a 50-year event by EV1-PWM or LN3-MLE.

The data set for Rutherford Bridge 1960–1985 is particularly well behaved. However, in spite of apparent consistency, the dangers of design based on a single station record, even if 25 years in length, can be readily illustrated by the addition of a single year of record in which an exceptional flood occurred.

Problem of outliers – Rutherford Bridge 1960–1986

In August 1986 a flood event, known as Hurricane Charlie, gave a peak discharge 1.8 times the previous highest in the record. Revised plotting positions and reduced variates for the extended record are shown in Table F4 and revised *T*-year estimates are shown in Table F6.

- The estimated return period of the observed 1986 flood, based on distributions fitted to the 1960–1985 record, ranged from 1000 years to 0.6 million years with a median of 16 500 years.
- New *T*-year estimates of discharge are much higher over most of the return period range. Whilst the 100-year flood estimate increases by an average of over 30%, the 5-year event also increases by 9%. A discharge of 130 m³ s⁻¹ becomes less than a 20-year return period by most estimates.
- Distributions and the fitting methods respond quite differently to the additional flood, with three-parameter distributions tending to bend more to accomodating the new point and thus giving much higher extrapolated flows than two-parameter distributions. For two-parameter distributions *T*-year estimates are higher by MOM than by MLE. However, estimates by GEV–MLE gave the highest of all extrapolations. No stable solution was available for LN3 by MLE.

In the light of such variations and in the absence of knowledge of the true population (or if it exists), a choice of distribution with lowest standard errors has little commonsense support. The use of sample-based standard errors and confidence limits is of limited value.

The flood of 1986 arose from a tropical hurricane, transformed as it crossed the Atlantic Ocean. In this example the storm generated a flood which plots as an outlier on the frequency graph, but many such events are indistinguishable in magnitude from the general population. Labeling as Hurricane Charlie highlights the questionable assumption of population homogeneity. There is nothing inherent in the data or in the statistical procedures use to judge the frequency of

such events, but the allocation of a return period of more than 1000 years based on the 25-year record seems not only doubtful but dangerously presumptive for design.

How then can one avoid the dangers of lurking outliers and improve upon single-station estimates, and gain confidence in those estimates? Three basic types of procedure are available:

- the incorporation of historical flood information;
- regionalization of flood statistics;
- generation of flood frequency from rainfall.

Incorporation of historical flood information

Gauged records rarely exceed 50 years but historical information on flooding may extend back over many centuries. Chinese historical flood records (Chen *et al.*, 1975) are perhaps the most comprehensive, with flood descriptions and water level markers up to 2000 years old. Glos and Krause (1967) have described historical data for major rivers in Central and Eastern Europe and Thompson *et al.* (1964) in New England. Archer (1992) has compiled a detailed historical flood record for the rivers of northeast England. Historical data suffer from two notable disadvantages – they may be of low accuracy and they may be reported inconsistently.

Retrospective assessment of discharges based on flood markers may be straightforward if the river channel and the stage–discharge control have remained stable. Where there have been changes, as is often the case, unusual combinations of historical and hydraulic analysis may be needed (e.g. Archer, 1993). Hosking and Wallis (1986) show by computer simulations that even where a single historical flood magnitude is subject to an error of 40%, it can still improve the accuracy of a short gauged record.

Historical flood information is often based on flood marks on walls and bridges or on the occurrence of flooding of a building (especially public buildings such as churches and schools where written records are maintained). Such data indicate the level of floods which have risen above a threshold during some historic period and are known as censored samples. For the analysis of censored samples it is assumed that all floods above the threshold have been marked and that floods in intervening years, for which no marks exist, have failed to reach the fixed point. Where it is not possible to guarantee this condition, it may be preferable to raise the threshold and to eliminate some lower hisorical floods. Muliiple thresholds covering different segments of the historical record are possible (Hirsch and Stedinger, 1987).

The method of maximum likelihood is now favored for the estimation of the distribution parameters from a censored sample (Condie and Lee, 1982; Stedinger and Cohn, 1986). Leese (1973), who first applied censored sample theory to flood frequency analysis, gives an example of the River Avon at Bath in southwest England for a gauged record from 1940 to 1968 combined with ten historical floods dating back to 1865.

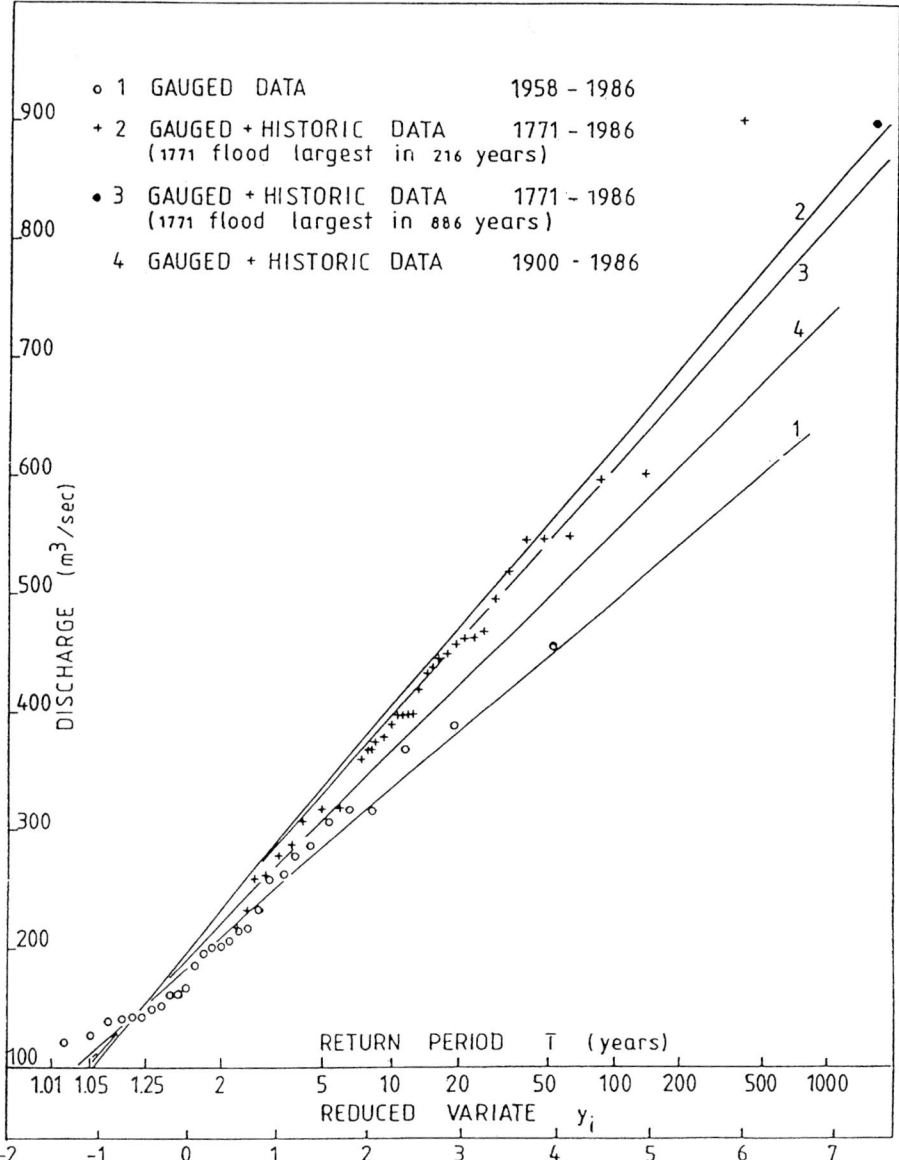

Figure F6 Comparison of annual maximum flood frequency plots (EV1-MOM) for the River Wear at Durham, northeast England, including and excluding historic data. (After Archer, 1987.)

The use of a longer record typically available from historical information (200–500 years) highlights the problem according to Klemeš (1993) that, the longer a hydrological series, the lower the credibility of the assumption of stationariness. For basins little affected by land-use change, the difference in quantile estimates with historical data included or excluded may indeed be illustrations of the impact of climate variability. However, from a practical point of view, the record is also likely to be more representative of the climate variability of a future period of similar length than the short gauged record. Figure F6 from Archer (1987) shows a striking difference between flood frequency plots for gauged and historic periods on the River Wear at Durham, and also demonstrates the effects of alternative historic periods relevant to the largest known flood which occurred in 1771.

Regional flood analysis

Regionalization of flood statistics may be employed not only to limit the influence of sampling variability at a single site with a short record, but also as a means of estimation for ungauged sites. Regional analysis is most commonly based on an index flood, the mean annual flood ($\bar{Q}$), and the use of a dimensionless flood frequency curve or growth curve relating flood magnitude of given return period Q_T to $\bar{Q}$. Dimensionless flood series are pooled and a regional average curve is fitted to the combined data.

Cunnane (1988) summarizes the variety of ways in which this plotting is carried out. Simple methods include the averaging of $Q/\bar{Q}$ from different sites, within grouped ranges of the reduced variate y, for instance -2.0 to -1.5, -1.5 to -1.0, etc., and plotting against the midpoint of the range (NERC, 1975). More sophisticated procedures include the fitting of a five-parameter Wakeby distribution by probability weighted moments (Houghton, 1978). Historical flood data may be incorporated in the analysis.

For reapplication to a particular site, $\bar{Q}$ may be determined from the site record if available, or from basin characteristics, and used to rescale the $Q_T/\bar{Q}$ relationship. Such methods were first used by Dalrymple (1960) and have since been widely used for national flood estimation schemes (NERC, 1975; McKerchar and Pearson 1989).

The principal difficulty with regional flood frequency analysis is in the identification of a homogeneous region (Wiltshire, 1986). All sites

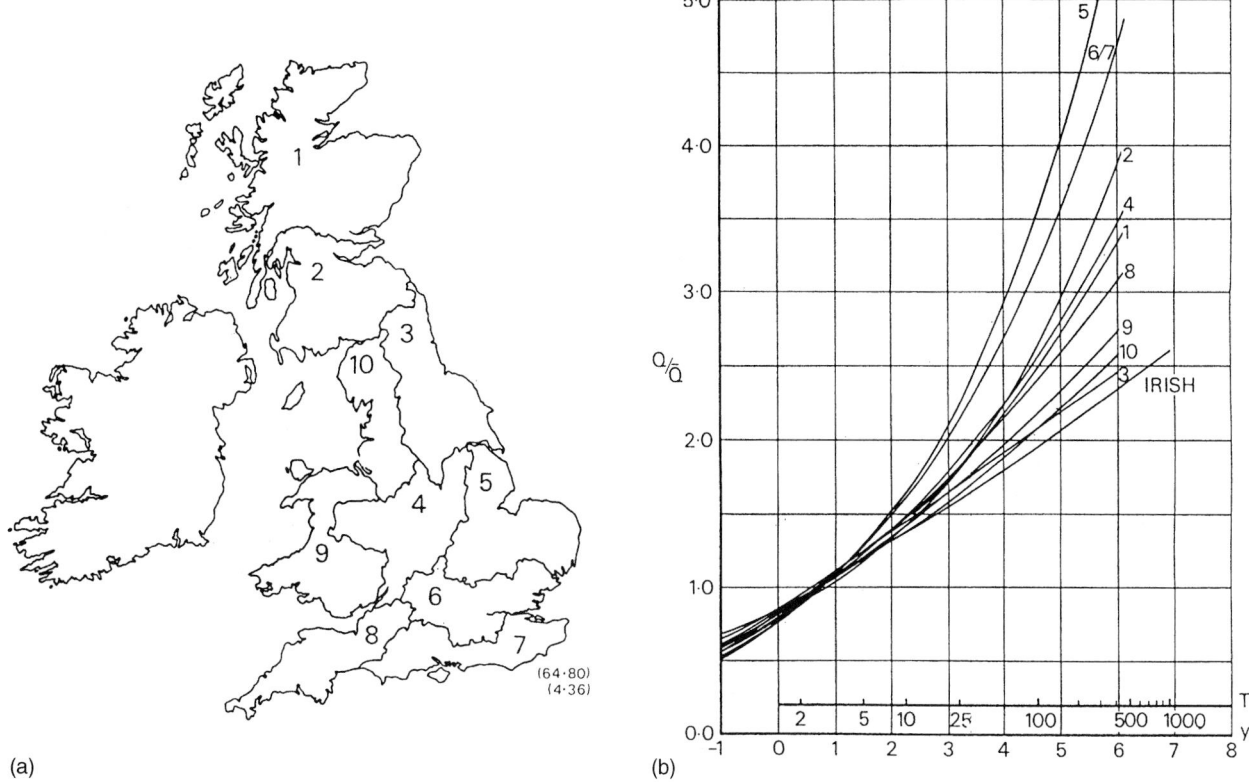

Figure F7 (a) British flood regions and, (b) standardized regional flood growth curves. (After NERC, 1975.)

within such regions are assumed to have a common form of distribution with identical parameter values, and that any observed departures from this form are the result of sampling variation. The greater the homogeneity of a region, the greater the gain in using regional instead of single-site estimation.

Such regions may be geographically defined in advance by physiographic or administrative boundaries (Figure F7a; NERC, 1975), by classification of statistics of basin flood frequency distribution such as coefficients of variation (C_V) and skewness (C_S) (Moseley, 1981) or by classification of drainage basin characteristics, for example by cluster analysis (Acreman and Sinclair, 1986).

Tests may then be applied to determine whether the records in a group differ from one another by amounts that could not reasonably be expected by chance, including graphical methods (Dalrymple, 1960) and likelihood ratio tests (Acreman and Sinclair, 1986). However, Cunnane (1988) notes that the power of such tests with currently available amounts of hydrological data is low. 'It is not possible to say with absolute certainty whether the flood frequency behaviour of any one of M sites in a region is inconsistent with that of the other $(M-1)$ sites, nor is it possible to divide, with great assurance, a large group of catchments into homogeneous subgroups.' Departures from homogeneity are not thought to invalidate the benefits of using regional estimates over single-site estimates.

Generation of flood frequency from rainfall statistics

Rainfall records are of greater length and have wider coverage than peak flow statistics on a worldwide basis. Since precipitation (including snow) generates floods, it seems logical to derive flood frequency statistics from extreme rainfall statistics, with the potential for assessing not only the peak flow but also the shape of the flood hydrograph. Whilst such methods have been studied for some decades (Paulhus and Miller, 1957), they have received less attention than the direct analysis of flood flow.

The 'gradex' method proposed by Guillot and Duband (1967) is one such procedure and has been widely used in Europe for the estimation of extreme floods. The method assumes that, above a reference return period, the frequency distribution of extreme rainfall

and flood volume run parallel. The success of the method is inherently difficult to verify, but Reed (1994) has drawn attention to underlying weaknesses in the absence of regionalization of rainfall statistics and the lack of flexibility in the chosen EV1 distribution to reflect the skewness exhibited by rainfall extremes. In addition, the method takes no account of the effect of non-linear catchment response on flood frequency in the transformation from rainfall to runoff. Particularly important is the role of floodplain storage (Archer and Kelway, 1987) as illustrated below.

A more complex approach was attempted by NERC (1975) in the synthesis of a design flood hydrograph from rainfall. Using 1500 events from 140 basins they optimized the combination of rainfall return period, duration and profile, in conjunction with a unit hydrograph/losses model, to synthesize a flood peak and hydrograph of given return period. Rainfall statistics were regionalized.

The method achieves about the same level of success in matching the synthesized flood frequency curve to those from gauged records as the regionalized statistical method (Archer, 1987). However, it is interesting that the generated flood frequency curve, like the gradex method, runs virtually parallel with rainfall frequency. Neither catchment wetness nor channel non-linearities have any significant influence on simulated flood growth (C_V or C_S). The method is insufficiently flexible to match the range of C_S values typical of flood frequency curves, which is wider than for rainfall frequency curves.

Flood frequency and physical hydrology

Klemes (1986, 1993) has severely criticized the 'mathematistry' (sic) of statistical flood estimation and has advocated an approach which is more dependent on the physical hydrology of the basin. A series of annual maximum floods is clearly not a random sample from a homogeneous population. Events arise from rainfall and snowmelt events of different origins, magnitudes and rates of occurrence. Rainfalls are transformed by catchment conditions, especially of soil moisture status, and further transformed by channel and flood plain storage operating differently over different ranges of discharge. The analysis of flood peaks as if they were drawn from a homogeneous population, even where historical data and regionalization have been

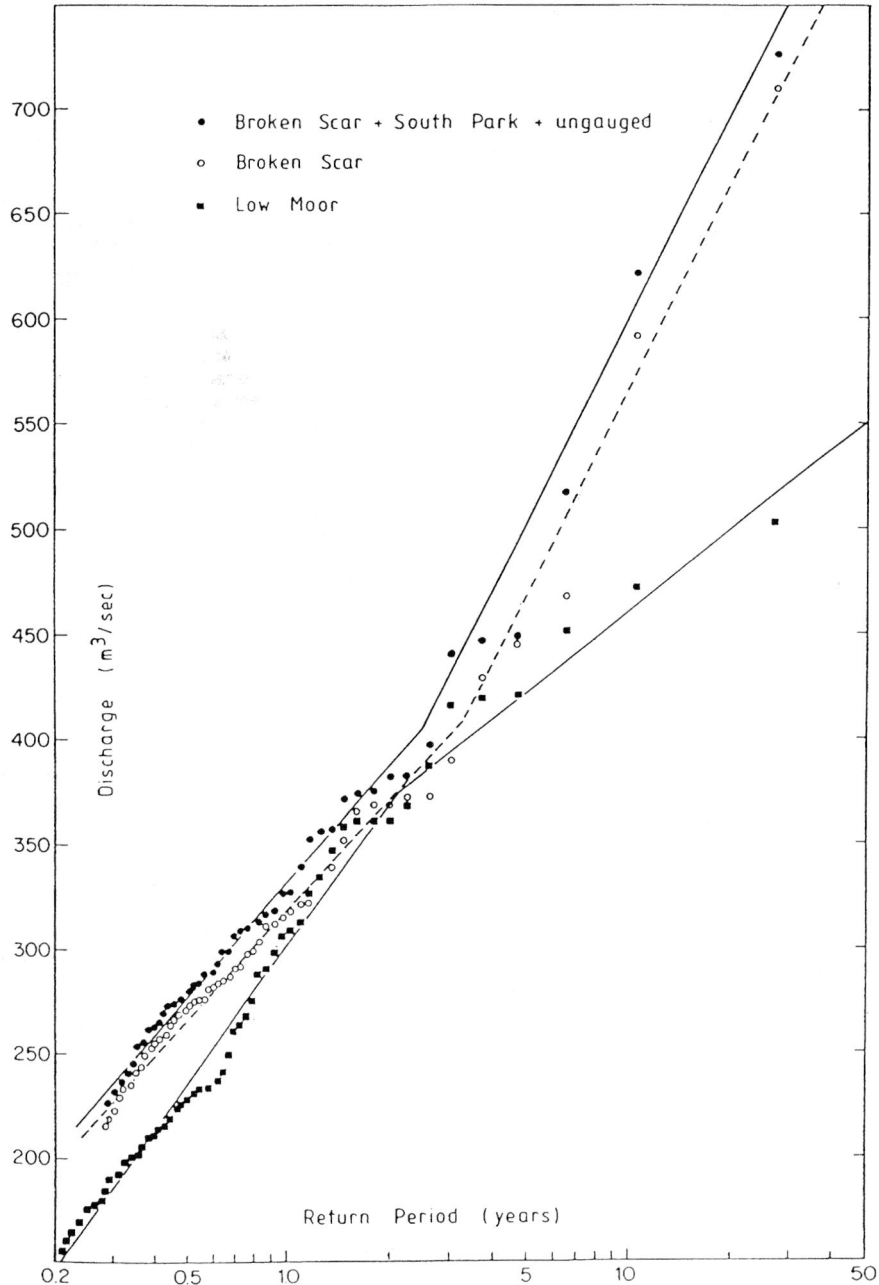

Figure F8 The impact of channel and flood plain storage on the transformation of an upstream (Broken Scar) to a downstream (Low Moor) flood frequency curve using POT series analysis on the River Tees in northeast England. (After Archer and Kelway, 1987).

incorporated, can only lead so far in our understanding of flood event generation. There is a need to unpick the component influences, attempt to establish their frequencies and dependences and then to restitch physically feasible combinations to estimate the flood frequency distribution.

One physical influence which is well understood in flood routing but widely ignored in flood frequency analysis is the attenuation of peak flow resulting from channel, floodplain and lake storage. Archer (1981a) described floodplain effects on flood frequency in northeast England and in 1988 illustrated them for a lowland reach of the River Tees with significant overbank storage and little lateral inflow, using the same record period for both stations (Figure F8). The downstream flood growth curve has much smaller C_V than the upstream one and a break-of-slope where the overbank storage commences.

The finding in the United States that the coefficient of variation (C_V) varied inversely with basin area (Benson, 1962; Riggs, 1973) which led to the temporary abandonment of index flood methods, may be largely due to increasing downstream channel and floodplain attenuation.

The influence of lake attenuation is well illustrated by Acreman and Sinclair (1986), whose analysis of Scottish basins isolated a group of stations with significant lake (loch) storage, which were represented by a relatively flat EV3 distribution. Similarly Sutcliffe (1981) found that the flattest regional curve in an international set was from the lake district of southeast Finland.

An opposing influence of reservoirs may apply where a large reservoir is drawn down for much of the year. Most annual maxima are represented by releases and trivial overflows, but a few events are generated by major storms coinciding with a full reservoir. The resulting flood frequency curve is steep and highly skewed.

This is an artificial case but it represents a more general condition on dry catchments where soil storages must generally be replenished before overflow occurs, but the largest events occur when storm rainfall coincides with soils at or above field capacity. The resulting flood frequency distribution, which combines the frequencies of storm rainfall and percentage runoff, has a much higher C_V than on wet upland basins where storm rainfall is the dominant control and the range of percentage runoff is much lower. Archer (1981a) illustrates the systematic increase in C_V from wetter to drier basins in northeast England. A measure of flood growth $Q_{30}/\bar{Q}$ was found to be inversely correlated with basin mean annual rainfall, with regression significant at 0.1% level. On a national scale the slopes of ten regional flood frequency curves in the British Isles increase from the wetter north and west to the drier southeast (Figure F7b; NERC, 1975).

It has been argued that such regional variations in flood frequency characteristics could also arise from sampling variability from a single homogeneous population and this might indeed be possible. However, it seems illogical to reject the evidence for physical influences on flood frequency on the basis of low-powered statistical tests.

Seasonality: the key to understanding

Perhaps the most convincing demonstration of the controlling influence of catchment wetness on flood frequency is in the seasonal distribution of flood occurrence and its relationship to seasonality in catchment wetness and storm rainfall. Archer (1981b) shows that on drier basins in northeast England, monthly storm rainfall exceedence probabilities are highest in summer from July to September, coinciding with the season of lowest flood risk. Dry basin conditions normally override the effects of greater rainfall intensity. There is a shift towards increasing summer flood risk at higher return periods and on wetter catchments, but only on the wettest upland catchment does summer flood risk exceed winter risk. Archer (1981b) also illustrates a practical method for the assessment of seasonal flood risk based on regionalized monthly probabilities of occurrence.

Flood frequency analysis: the way ahead?

We have come far enough downstream on the floodwave of statistical hydrology. It is time to row ashore, perhaps against the current and concentrate effort on the physical basis of flood generation and effect on flood frequency through the study of basin and channel controls. In the meantime, existing methods of regionalization and historical extension are to be preferred to the sole use of a gauged record, even if of moderate length.

David Archer

Bibliography

Acreman, M.C. and Sinclair, C.D., 1986. Classification of drainage basins according to their physical characteristics; an application to flood frequency in Scotland. *J. Hydrol.*, **84**, 365–380.

Archer, D.R., 1981a. A catchment approach to flood estimation. *J. Instn Wat. Eng. Sci.*, **35**(3), 275–289, and communication, **35**(6) 528–535.

Archer, D.R., 1981b. Seasonality of flooding and the assessment of flood risk. *Proc. Instn. Civ. Eng., Pt 2*, **70**, 1023–1035.

Archer, D.R., 1987. Improvement in flood estimates using historical flood information on the River Wear at Durham. *British Hydrol. Soc., First National Symp.*, Hull, 5.1–5.9.

Archer, D.R., 1989. Flood wave attenuation due to channel and floodplain storage and effects on flood frequency, in *Floods – Hydrological, Sedimentological and Geomorphological Implications* (eds Bevan, K. and Carling, P), John Wiley, Chichester, pp. 37–46.

Archer, D.R., 1992. *Land of Singing Waters; The Rivers and Great Floods of Northumbria*. Spredden Press, Stocksfield.

Archer, D.R. and Kelway, P.S., 1987. A computer system for flood estimation and its use in evaluating the Flood Studies rainfall runoff method. *Proc. Instn Civ. Eng., Pt.* 2, **83**, 601–612.

Archer, N., 1993. Discharge estimate for Britain's largest flood: River Tyne, 17th November, 1771. *British Hydrol Soc., Fourth National Symp.*, Cardiff, 4.1–4.6.

Bardsley, W.E., 1994. Against objective statistical analysis of hydrological extremes. *J. Hydrol.*, **162**, 429–431.

Benson, M.A., 1962. *Evolution of methods for evaluating the occurrence of floods*. US Geol. Surv. Water Supply Pap. 1580A.

Chen, J.Q., Ye, Y-Y. and Tan, W.Y., 1975. The important role of historical flood data in the estimation of spillway design floods. *Scientia Sinica*, **28**(5), 669–680.

Condie, R. and Lee, K., 1982. Flood frequency analysis with historic information. *J. Hydrol.*, **58**, 47–61.

Cunnane, C., 1975. Flood estimation by statistical methods, in *Flood Studies Conference*, Inst. of Civil Engineers, London, pp. 43–46.

Cunnane, C., 1978. Unbiased plotting positions – a review. *J. Hydrol.*, **37**, 205–222.

Cunnane, C., 1979. A note on the Poisson assumption in partial duration series models. *Water Resources Res.*, **15**(2), 489–494.

Cunnane, C., 1988. Methods and merits of regional flood frequency analysis. *J. Hydrol.*, **100**, 269–290.

Cunnane, C., 1989. *Statistical distributions for flood frequency analysis*. World Met. Org., Operational Hydrol. Rep. 33.

Dalrymple, T., 1960. *Flood frequency methods*. US Geol. Surv. Water Supply Pap. 1543A, 11–51.

Foster, H.A., 1924. Theoretical frequency curves and their application to engineering. *Trans. Am. Soc. Civ. Eng.*, **87**, 142–173.

Fuller, W.E., 1914. Flood flows. *Trans. Am. Soc. Civ. Eng.*, **77**, 564–617.

Glos, E. and Krause, R., 1967. Estimating the accuracy of statistical flood values by means of long-term discharge records and historical data. *Proc. Symp. on Floods and their Computation*, Leningrad, 144–151.

Greenwood, J.A., Landwehr, J.M., Matalas, N.C. and Wallis, J.R., 1979. Probability weighted moments: definition and relation to parameters of distributions expressible in inverse form. *Water Resources Res.*, **15**(5), 1049–1054.

Gringorten, I.I., 1963. A probability rule for extreme probability paper. *J. Geophys. Res.*, **68**(3), 813–814.

Guillot, P. and Duband, D., 1967. La méthode du gradex pour le calcul de la probabilité des crues à partir des pluies. AISH. Publ. 84, 560–569.

Hazen, A., 1930. *Flood Flows*. John Wiley, New York.

Hirsch, R.M. and Stedinger, J.R., 1987. Plotting positions for historical floods and their precision. *Water Resources Res.*, **23**(4), 715–720.

Hosking, J.R.M. and Wallis, J.R., 1986. The value of historical data in flood frequency analysis. *Water Resources Res.*, **22**(11), 1606–1612.

Houghton, J.C., 1978. Birth of a parent: the Wakeby distribution for modelling flood flows. *Water Resources Res.*, **14**(6), 1105–1109.

Klemes, V., 1986. Dilettantism in hydrology: transition or destiny. *Water Resources Res.*, **22**(9), 1775–1885.

Klemes, V., 1993. Probability of extreme hydrometeorological events, a different approach. In *Proc. of Symp. on Extreme Hydrological Events: Precipitation, Floods and Droughts*, Yokohama, IAHS Publ. 213, 167–176.

Langbein, W.G., 1949. Annual floods and the partial duration series. *Trans. Am. Geophys. Un.*, **30**, 879–881.

Leese, M.N., 1973. The use of censored data in estimating *T*-year floods. *Proc. UNESCO/WMO/IASH Symp. on Design of Water Resources Projects with Inadequate Data*, Madrid, 236–247.

McKerchar, A.I. and Pearson, C.P., 1989. *Flood Frequency in New Zealand*. Hydrol. Centre Publ. 20, Christchurch.

Ministry of Agriculture, Fisheries and Food (MAFF), 1993. *Flood and Coastal Defence*. Project Appraisal Guidance Note, MAFF, London.

Moseley, M.P., 1981. Delimitation of New Zealand hydrological regions. *J. Hydrol.*, **49**, 173–192.

Naden, P.S., 1992. Analysis and use of peaks-over-threshold data in flood estimation, in *Floods and Flood Management* (ed. A.J. Saul), Kluwer Academic Publ., Dordrecht, 131–143.

Natural Environment Research Council (NERC), 1975. *Flood Studies Report*, 5 vols., NERC, London.

Northumbrian Water Authority, 1979. *Target Levels of Service for Land Drainage*, Internal Report.

Page, K.J. and McElroy, L., 1981. Comparison of annual and partial duration series floods on the Murrumbidgee River. *Water Resources Bull.*, **17**(12), 286–289.

Paulhus, J.C.H. and Miller, J.F., 1957. Flood frequencies derived from rainfall data. *J. Hyd. Div. ASCE.*, 83.

Penning-Rowsell, E.C. and Chatterton, J.B., 1977. *The benefits of flood alleviation: A manual of assessment techniques*. Saxon House.

Potter, K.W. and Walker, J.F., 1981. A model of discontinuous measurement error and its effect on the probability distribution of flood discharge measurements. *Water Resources Res.*, **17**(5), 1505–1509.

Reed, D.W., 1994. On the gradex method of estimating extreme floods. *Dams and Reservoirs*, 17–19.

Riggs, H.C., 1973. *Regional analysis of streamflow characteristics*. US Geol. Surv. Water Resources Invest. Tech., Book 4.

Stedinger, J.R. and Cohn, T.A., 1986. Flood frequency analysis with historical and palaeoflood information. *Water Resources Res.*, **22**(5), 785–793.

Sutcliffe, J.V., 1978. *Methods of Flood estimation, a guide to the Flood Studies Report*. Report No. 49, Institute of Hydrology, Wallingford, UK.

Sutcliffe, J.V., 1981. Use of flood studies report overseas, in *ICE Flood Studies Report – Five Years On*. Thomas Telford, London, 7–10.

Thomson, M.T., Gannon, W.B., Thomas, M.P. and Hayes, G.S., 1964. *Historical floods in New England*. US Geol. Surv. Water Supply Pap., 1779-M.

Water Resources Council, 1982. *Guidelines for determining flood flow frequency*. Bull. 178, Hydrol. Comm., Washington, DC.

Weibull, W., 1939. *A Statistical Theory of Strength of Materials*. Ing. Vet. Ak. Handl., Stockholm, 151.

Wiltshire, S.E., 1986. Identification of homogeneous regions for flood frequency analysis. *J. Hydrol.*, **84**, 287–302.

Cross references

Flood hazard management
Floods
Flood studies worldwide
Floods, river and multi-stage channels
Floods: world's maximum observed

FLOOD HAZARD MANAGEMENT

What is the problem?

The objective in flood management is to maximize the efficiency of the use of flood-prone land and to maximize the economic productivity of this land, after the costs of flood damages and flood alleviation are taken into account. There are thus two interrelated issues: should development intensity on the flood-prone land be increased; and is it desirable to intervene to reduce flood losses? Counterintuitively, rising flood losses can consequently be a sign of increased economic efficiency (Green *et al.*, 1993), as well as simply the consequence of economic growth.

The desirability of intensifying development on flood-prone land depends upon the costs, financial and environmental, of development elsewhere. Where these are high, then it can be efficient to encourage development on flood-prone land. In many countries, development can take place nowhere except on flood-prone land. In delta areas, such as the Netherlands and Bangladesh, there is essentially no other land. In mountainous countries, such as Japan and Italy, there is little land to develop other than valley floors and coastal fringes. In neither case is it sensible to call for development always to be located away from the floodplains; instead, it is a question of what development should be encouraged on flood-prone land and what development should be directed elsewhere.

Population pressures and urbanization are forcing the development of marginal land, such as the char lands in Bangladesh and steep slopes around many cities. These more marginal lands are typically settled by the poorest groups in society who are thus exposed to the greatest risks (Hewitt, 1983). At the same time, floodplains are environmentally rich areas (Marchand and Toornstraf, 1986) and in traditional agricultural communities, the agricultural pattern will have adapted to the flood regime and may be dependent upon it (Marchand, 1987). Thus, in many parts of the world, some floods provide benefits to agricultural communities, it being the extreme events which cause problems.

Those living on floodplains, particularly the inhabitants of the floodplains of India, Bangladesh and China, have adapted over the years to the flood risk both in terms of agricultural practices and in minimizing the potential losses. Thus the problem in flood hazard management is to enable the public to be able to manage the hazard more effectively than they could if they had to rely solely upon their own resources.

Flood problems in agricultural areas are also usually associated with either irrigation problems in a dry season or a drainage problem (Hunting Technical Services, 1991). Intervention in such circumstances can do more harm than good if it worsens the drainage or irrigation problem, disrupts the pattern of land usage of which the floodplain forms part or, for example, damages fisheries (Drijver and Marchand, 1985). When the total productivity of the floodplain is considered, ill-judged interventions may reduce flood losses only at the cost of reducing overall productivity and reducing economic efficiency.

Conversely, there are no countervailing benefits from floods in urban areas, but the rivers themselves are valuable as recreational resources. In the nineteenth and early twentieth century, rivers in cities were treated at best as convenient transport routes. In this period, rivers were routinely canalized as part of flood defence works, usually destroying their ecological and recreational value. However, in recent years the recreational and amenity value of rivers has been increasingly recognized (Bureau of Water Resources, 1985) and river corridors are being restored to a more natural state in order to increase their environmental and amenity value (Gardiner, 1991).

Consequently, flood hazard management must be situated in the broader context of integrated catchment planning rather than treated in isolation, and floods must be regarded as only one of the many issues involved in the appropriate management of a catchment (Newson, 1992).

Making things worse

The risk of flooding is rarely stable: climatic change is likely to result in flood problems in some areas becoming worse either because of changes in precipitation or of runoff (Riebsame, 1989). Similarly, development itself increases the proportion of land surface which is impermeable, thus increasing both the rate and quantity of runoff (Bureau of Water Resources, 1985). Again, groundwater abstraction in some areas, such as Bangkok and Shanghai, has resulted in subsidence, thus exacerbating an existing flooding problem.

Vulnerability

Vulnerability is generally defined as the interaction between the hazard and resources available to those at risk to adapt to or cope with that hazard (Blaikie *et al.*, 1994). Thus it has two connected elements: the challenge posed by the hazard, and the capacity to respond to that challenge by those at risk and by society as a whole. By intervention it is sought either to reduce the challenge or to strengthen the capacity to cope, thus increasing the public's ability to successfully manage the flood risk. It is also necessary to distinguish between the vulnerability of those directly at risk and the vulnerability of the society of which those people are part. The support offered to flood victims from those outside of the flood area is an important contribution to the victims' coping resources; equally, the effects of a flood can extend beyond the boundaries of the flooded area.

The challenge posed by a flood depends upon its characteristics. If the flood rises rapidly, then there is little time for those at risk to take any action to mitigate the effects. Rapid rates of rise are also associated with high flow velocities, which pose a significant risk to life. The threat is mainly as a result of the structural failure of buildings in which people have taken refuge, this failure occurring in consequence of the pressure of water (Sangrey, 1975), battering damage from waterborne debris or the undermining of the buildings' foundations.

Whilst rapid rates of rise are associated with small, steep catchments, the greatest potential problems are usually associated with the failure of artificial structures including dams, levees and structures such as bridges which act as dams or throttles until failure. The greatest loss of life is usually found in those situations and on delta areas exposed to cyclones, hurricanes or other tropical storms (DeKay and McClelland, 1993).

The time of the year at which the flood occurs, in areas with pronounced seasons, affects the extent of the challenge. Ice floods cause particular problems (Inland Waters Directorate, 1985) and floods in cold seasons multiply the problems of recovery when people struggle to repair and replace damaged properties whilst living in cold, damp buildings.

Coping capacity is enhanced by accurate expectations about the nature of flooding and knowledge of what to do in a flood. Experience can provide such knowledge, so that floods are seen as a routine rather than a threatening event, but people can be trapped by their experience and come to expect that all future floods will be like those that they have experienced in the past. If an extreme event develops, then experience may lead people to prepare for the wrong event, to seek to protect their property when they should evacuate. The intervention challenge is thus to enable people to learn about events which they have not experienced and to provide them with information as to what to do when such an event is threatened.

Experience of flooding generally results in local residents making adaptations prior to a flood, for example house raising or flood proofing (Paul, 1984). The adaptation of buildings is a particularly important adaptation because they can act to protect both the occupants and their possessions. Other forms of adaptive behaviour often include informal flood forecasting strategies (Emergency Management Australia, 1995).

Coping capacity depends significantly upon the wealth of those affected, those with high incomes, savings and insurance being better able to prepare for a flood and to recover after a flood (Green and Penning-Rowsell, 1986). Other household characteristics, such as the age profile of the household, and the kinship and friendship linkages are important in recovery; disruption of kinship links both decreases the household's ability to recover and increases the stress of the flood itself (Drabek, 1986). It is generally believed that social support helps recovery, but there is no substantive evidence that this is the case although flood victims value such help (Ketteridge and Green, 1994).

Buildings mediate between the flood and the contents and occupants of the building. In addition, damage to the buildings also delays the resumption of normal usage of the building. The extent of the damage caused to buildings and their contents largely depends upon the depth of flooding (Penning-Rowsell and Fordham, 1994). The materials carried by the flood may cause direct damage, debris causing battering damage, and salt damaging many materials. The sediment deposited by retreating floodwaters (River Bureau, undated) may substantially increase the difficulties of cleaning up after the flood. The duration of the flood does not generally increase significantly the physical damages caused by a flood, but the losses of production obviously continue throughout the period of flooding as well as for some time afterwards (Parker et al., 1987).

Advanced economies are typically more susceptible to flood than are simpler economies. Advanced technologies suffer more damage and generally require replacement rather than repair; for example, electronic equipment suffers more damage from flooding than electro-mechanical plant (Parker et al., 1987). In addition, because of economies of scale, advanced economies are characterized by the concentration of capacity into a few large plants; disruption of such a plant can substantially reduce the nation's production of that product. Similarly, the shift to 'just-in-time' manufacturing, the minimization of stocks of raw materials and parts, makes such plants extremely susceptible to disruption of transport.

Floods can cause a significant health hazard. In some countries, flooded areas can form breeding areas for insects and other disease vectors (Beinin, 1985; Parker and Thompson, 1991). Disruption of water supplies can reduce accessibility to potable water for drinking and hygiene, and the contamination of flood waters by fecal material spreads disease vectors (Wellings, 1983). Although in developed countries alternative methods of supplying potable water are usually successful in substantially reducing the risk of waterborne diseases, the proportion of the population who report some health effects is usually in the range of 40–70% of households (Parker et al., 1987). Many of these health effects are believed to be a consequence of the stress resulting both from the flood itself and from the process of recovery from the flood. This stress itself is generally considered by flood victims to be one of the worst consequences of flooding. Having to leave home, particularly if the family has to split up, is also disruptive and households may require months if not years to recover from the effects of a flood (Drabek, 1986). In general, these non-monetary impacts of flooding are regarded by flood victims as worse than the financial losses they experienced (Green and Penning-Rowsell, 1986); although financial losses are not involved, the losses are nevertheless economic losses, but ones which are more difficult to evaluate (Penning-Rowsell et al., 1992).

In many areas where freshwater flooding is prevalent, agricultural systems are typically highly adapted to and dependent upon the annual flood both to bring water for irrigation and sometimes nutrients (Brammer, 1990), although floods can also leave deposits of sand or other material which reduce soil fertility. In such societies it is the extreme floods, rather than the frequent floods, which overwhelm the coping capacity of the local communities. The relationship between the flood season and growing seasons here is critical; this general relationship is much less true of temperate climates than of tropical climates. In agricultural areas it is thus livestock rather than crops which experience the greatest losses in extreme floods.

Table F7 is a checklist to screen areas for vulnerability to flooding.

Why intervene?

Government intervention is, therefore, appropriate under a number of circumstances. First, if the floodplain occupants have inadequate information about the nature and extent of the risk of flooding that they face and the most effective adaptations to adopt, then governments may intervene to given them such information. Second, the challenge may be beyond that in which the individuals can mobilize resources with which to cope. Third, the individual adaptations may interact in such a way as make the flood risk greater elsewhere; floodproofed properties reduce the damage to those properties but may block flood flows and reduce floodplain storage capacity, so possibly making flooding worse elsewhere.

It may also be more efficient to direct or attract development to flood-prone areas than to allow development to take place in those areas in which risk-averse developers would otherwise locate. The environmental damage caused by development elsewhere, or the costs of providing infrastructural support, may be such that the most efficient location for development is the floodplain.

Lastly, there may be economies of scale from those forms of intervention which rely upon government action compared to those interventions which individuals or groups of individuals can undertake on their own. For example, if controlling runoff would be more efficient than floodproofing, then only government is likely to be able to ensure that source control is undertaken. Similarly, a structural flood alleviation scheme will be more efficient than floodproofing individual properties once development intensity exceeds a certain level.

The overall objective is not to reduce floods, since these are a natural part of the hydrological cycle, but to have manageable flood events.

What are the options?

A rather artificial distinction has come to be made between 'structural' solutions and 'non-structural' solutions (Table F8), and to regard them as alternatives rather than as complements. The enthusiasm for non-structural solutions arose from their neglect in the first half of the century in favor of sole reliance upon structural solutions. A more balanced view is now possible.

The different options either reduce the challenge faced or aid the coping process, and not all of the options are equally applicable in any given situation. As Table F8 indicates, the 'structural' versus 'non-structural' distinction is almost identical to a distinction between strategies which reduce the challenge and those which enhance coping capability.

The use of a portfolio of strategies is also preferable because of the risk of failure of any single strategy. This risk of failure arises for two reasons. Firstly, there is always the possibility of a more extreme flood event occurring than the maximum with which a particular strategy can cope. Second, the reliability of the different strategies is often questionable; flood warning systems are often very unreliable (Smith, 1986) and embankments fail (Adnan, 1991). Part of the process of evaluating a possible intervention strategy should be an assessment of what will happen when it fails.

It is therefore necessary to consider the effect of the intervention strategy across the entire spectrum of flood events in that area. In highly adapted agricultural areas it is the extreme events which cause the problem, but the positive and negative effects of the intervention on frequent events must also be considered. Conversely, whilst interventions in urban areas are usually targeted at reducing the impact of frequent floods, the effects of those interventions on more extreme events must also be considered. The failure of a levee can

Table F7 Checklist for screening vulnerability to flooding

- Will a flood last for an extended period (more than a day)?
- Will the depth of flooding exceed waist deep?
- Will the flood deposit sediment and pollutants (including salt)?
- Will the expected velocity of flow be high enough to be impassible on foot?
- Are there any existing flood protection structures which may fail? What is the probability of failure?
- Are there any structures which may hold back the flood but then fail (e.g. bridges)? What is the probability of failure? Are there any other throttles?
- Is the time to peak less than 12 h?
- Have there been any developments in the catchment which will increase runoff (e.g. urban intensification, deforestation)?
- Is there a history of flooding? When was the last significant flood? Are local residents and administrations aware of the risk of flooding?
- Has there been any new development on the floodplain since the last major flood?
- Does flolooding occur in winter or the growing season?
- Is any part of the floodplain a natural depression which will act as a reservoir once flooded?
- Does any part of the floodplain lie behind an embankment, intended for flood protection or other purposes (e.g. a railway embankment)? How extreme must a flood be before this fails or is overtopped?

Are there any of the following on the flood plain:
- Major chemical hazards (water treatment works included)?
- Toxic waste disposal sites?
- Any areas of single storey housing? Any areas of mobile homes, caravans, tents or chalets?
- Any concentrations of elderly people?
- Any occupied basements?
- Have any of the properties been floodproofed or raised above some flood levels?
- Are there any natural refuge areas?
- Are there many possible evacuation routes which the residents can take from the flood area?
- Are buildings likely to collapse in a flood?
- Any main railway, metro/underground stations or signal boxes?
- Underground car parks?
- Any power stations or high-voltage electricity switching stations?
- Any main telephone exchanges?
- Any water treatment works?
- Any sewage treatment works?
- Any pumping stations (gas, water, sewerage, oil, etc.)?
- Will more than 10% of the capacity of the local road network be cut?
- Will more than 25% of the road links across the river be cut?
- Is there any significant water transport at present?
- Will any bridges act as dams or weirs and is there a risk of collapse?
- Will any main railway lines be affected?
- Will any road underpasses be affected?
- Are there any large factories?
- Any warehousing or open air storage areas including vehicle parks?
- Any accident and emergency hospitals?
- Any headquarters for police or other emergency services?
- Are there many livestock?
- Are local residents relatively wealthy, covered by flood insurance or a government loss compensation scheme?

Table F8 Intervention options

	Reduce challange	Enhance coping response
Structural		
	Runoff control	
	Artificial washlands	
	Flood storage reservoirs	
	Channel improvements	
	Levees/embankments	
Non-structural		
		Flood proofing
		Flood insurance
		Loss compensation
		Information
		Flood warnings
		Refuge areas and evacuation
	Floodplain zoning	Counselling and emergency help
	House raising	Property acquisition

result in worse flooding than if there had been no levee constructed (Penning-Rowsell *et al.*, 1992).

Since the flood which poses the least challenge is generally one which is slow rising, shallow, slow in velocity, deposits the least sediment and is short in duration, the ideal intervention shifts the existing floods to follow this pattern. However, these characteristics are inherently mutually contradictory and trade-offs have to be made.

Equally, considered across the entire spectrum of possible flood events, those options which attenuate the effects of floods across the entire spectrum of events are preferable to those which are effectively designed to fail at a given return period. Thus improvements to storage capacity or channel capacity will usually be, in principle, preferable to embankments or levees since, if the channel capacity is increased, for example, then there will be less water out of bank for all possible flood events. Conversely, if an embankment fails or is overtopped, the resultant flooding can be more severe than if there was no embankment (Penning-Rowsell *et al.*, 1992).

Decreasing the quantity and more particularly the rate of runoff, runoff control, in order to extend the time to peak gives more time for coping actions, including emergency works, to be made effective. It also tends to reduce the velocities of flows. Since urbanization typically increases the rate and quantity of runoff by increasing the proportion of impermeable area, source control is particularly appropriate within urban areas. Strategies adopted in Sydney, Bordeaux (Madiec, 1992) and elsewhere to reduce the rate and volume of runoff include infiltration areas, swales and detention ponds.

In non-urban areas, high rates of runoff are often accompanied by soil loss (Blaikie, 1985) with the resultant loss of soil fertility, a loss which may be more serious in economic terms than any change in flood damage. Deforestation is also cited as a cause of increased flood risk.

River improvements over the last 100 years have often resulted in a loss of natural storage; recently, there has been a new emphasis on the recovery of such flood storage by the construction of engineered wetlands, the use of 'winter' and 'summer' embankments, and washlands (Nature Conservancy Council, 1983). In part, this is a result of the new recognition of the important environmental and amenity values of such areas (Gardiner, 1991) as well as their advantages in flood alleviation.

Dams range in scale from large-scale structures to local structures, sometimes involving inflatable materials, or wire mesh or brushwood frameworks. Large-scale dams, because of their cost, often need to be managed for multiple purposes including hydroelectric generation and storage for irrigation. Such multipurpose management involves compromises since the flow storage requirements for the different purposes rarely coincide. Dam failure often results in catastrophic loss of life and significant economic losses (DeKay and McClelland, 1993).

Channel improvements, at their worst, involve straight trapezoid concrete channels which minimize the resistance to flow. Multi-stage channels, which maintain a sweetening flow during low flow periods, are now commonly employed.

Levees, embankments and walls are solutions which are often necessary when space precludes any other solution. In many urban areas, development has taken place up to the banks of the rivers so that channel improvements, for example, would require expensive land acquisition.

Development in a floodplain is sometimes termed floodplain encroachment; this label is dangerous since it presupposes that development should not intrude upon a floodplain. Instead, a decision has to be taken as to what development is appropriate for the floodplain, and what flood alleviation strategies to adopt for that development. Floodplain zoning, the restriction of some types of development to lower risk areas of the floodplain, and requirements that buildings are floodproofed, is an approach which is increasingly used (Ericksen, 1986). It may be coupled to the provision of subsidized flood insurance, or limitations on compensation for flood losses to areas complying with the zoning restrictions.

Floodproofing has been widely adopted by people occupying floodplains in the past. Buildings can either be raised on mounds or stilts, or the external walls (and floors) can be made to act as a waterproof membrane. The principle can also be extended to groups of buildings; the buildings or the boundary walls of those properties on the periphery of a development may be floodproofed, along with the entrances of the streets, so as to protect all of the buildings in the development. Because openings have to be closed in the event of a flood, boundary floodproofing depends upon the operation of a reliable system of flood warnings.

Flood insurance is intended to fulfill two purposes; first, the usual purpose of insurance, that of loss sharing both over time and between people so that the infrequent but catastrophic loss is converted to a small but constant payment. The more frequent the event which is to be insured against, the nearer the loss and the actuarially fair insurance premium will be in value, but insurance premiums will always exceed the actuarially fair sum because of the need for the insurer to recover administration costs and make a profit. One consequence is therefore that if a structural scheme can be justified on economic grounds, then this will always be more efficient than an insurance strategy (Green *et al.*, 1993).

At the same time, it is often also intended to increase economic efficiency by bringing home to occupiers of floodplains the risks which they are running by charging insurance premiums which reflect the flood risk. Consequently, it is necessarily dependent upon a system of floodplain mapping and zoning. Floodplain zoning requires floodplain mapping and thus the definition of the boundaries of the area at risk. Since floods can vary in magnitude up to the probable maximum flood, these boundaries are always to some extent arbitrary, it merely being the case that those outside of the boundaries are at a lower risk of flooding. Governments often pay compensation for some forms of flood losses and increasingly what are labeled as flood insurance schemes are compensation schemes where the expertise of commercial companies is used to assess flood losses and deliver compensation.

The public may have inadequate information as to the best ways of responding both before and during a flood and also after a flood. Governments may usefully determine what information the public considers that is needed and make that information readily available to the public (Emergency Management Australia, 1995).

Flood warnings are then a specific form of information. It is necessary to draw a distinction between a flood forecast and a flood warning. A forecast is a technical prediction of some flood event at some place at some time in the future. That forecast is only turned into a warning if a message reaches those who will have to cope with that flood in time for them to respond effectively to that challenge. Whilst attention has focused upon improvements to the forecasting component, for example, by use of weather radar, it is the conversion of forecasts into warnings which is typically the weakest link in the chain (Parker and Neal, 1990).

The reliability of a flood warning system can be expressed by the following formula (CNS Scientific and Engineering Services, 1991):

$$P_f \cdot P_d \cdot P_i \cdot P_a \cdot P_c$$

where P_f is the probability that an accurate forecast is made, P_d is the probability that the forecast is disseminated, P_i is the probability that a member of the individual household will be available to be warned, P_a is the probability that the individual household is physically able to respond to the warning, and P_c is the probability that the individual knows how to respond effectively.

P_d can be further broken down into the probabilities associated with each of the chain of actions necessary for a forecast to be turned into a warning. In each case, the probabilities are time specific; that is, they vary over time. However, the probabilities vary in different ways with time; the longer the lead time, the less reliable the forecast but the greater is the likelihood that the forecast will be disseminated and that there will be someone available to be warned, and that they can respond effectively. The increase in the probability that those at risk will be warned in time must be traded off against the reduction in the accuracy of the warning that also results from an increase in lead time. Typically, the reliability of a warning system is rather low and is highest when there is recent experience of flooding.

Two of the concerns often expressed about warnings are that those at risk may 'panic' and of the 'cry wolf syndrome'; that if warnings are given which are not then followed by a flood, the credibility of future warnings will be diminished. For these reasons, the process of converting a forecast into a warning is often delayed until such time that there is a high probability of a flood occurring. In consequence, the conversion process is often unsuccessful, the message failing to reach those who need it (Parker and Neal, 1990). To be effective, flood warnings must also have a number of characteristics (Emergency Management Australia, 1995); in particular, they must contain advice as what to do.

Given sufficient warning, emergency protection works and floodproofing of individual properties can be successful. Alternatively, it may be necessary to help those on the floodplain to evacuate. In either case, the likelihood of success is increased by the existence of an existing and tested system of multi-hazard emergency planning (Perry, 1985).

Successful evacuation depends upon the residents having somewhere to evacuate to. Therefore, a traditional intervention strategy which is being reinvented is the construction of refuge areas. Raised mounds were a traditional feature of Zeeland in the Netherlands within the coastal dike rings; house raising is commonplace in many parts of the world. Raised cyclone shelters have been constructed in Bangladesh, and road and railway embankments, as well as flood embankments, are now intended to act as refuge areas, having been traditionally informally used for such purposes. Such refuge areas should be multipurpose, being used for other purposes, such as schools or hospitals, in normal times.

Psychological counseling of disaster victims has begun to be introduced in some countries (National Institute for Mental Health, 1979) in an attempt to help victims recover from the trauma of the disaster. Adopting this response depends upon the availability of a sufficient number of trained counselors from outside of the affected area. Therefore, the option of developing community self-help through existing social organisations in the areas likely to be affected may be more practical. Friends, relatives and NGOs are the primary support to flood victims, supplemented by state emergency and other services. Whilst important in demonstrating social concern to the victims, such help can be poorly targeted in terms of the sort of help which flood victims believe would be most useful to them (Ketteridge and Green, 1994).

Where a very few properties are at a high risk, it is sometimes appropriate to purchase those properties, by property acquisition, and demolish the property prior to allowing only new development which is not very susceptible to flooding on that land. It is often very expensive, for example, to reduce the risk of flooding of property to sewage overflows in storms where this overflow is due to intensified development upstream. In this case it is often cheaper to buy and demolish the affected property than to protect it against flooding.

Cost recovery

Structural flood alleviation schemes have conventionally been undertaken by one of two means: either by associations of landowners, as a form of common property resource management, the Wattenschapen of the Netherlands being the most famous example (Wagret, 1967), or through the state. In the former case, the common property resource associations have devised various taxation strategies to raise the capital and maintenance costs of structures.

Where the state has intervened, the costs have instead been borne by general taxation rather than by those who directly benefit by the scheme; thus, those at risk are effectively subsidized by the general population unless the costs are recharged to the people who are protected by the scheme. It may also lead to the development of floodplain land in the expectation that the state will then construct a flood alleviation scheme, thus encouraging the development of floodplain land. It is necessary to plan the development of floodplains rather than to respond to development pressures.

In other instances, where flood problems are created or worsened by development which changes runoff, the economically efficient solution would be to charge according to the change in runoff. Few such schemes have been introduced and where they have the rationale behind their use has been an equity argument of recovering the costs of surfacewater drainage according to the load generated from a piece of land.

Problems in intervention

In the 1960s the term 'river improvement' was often used to describe interventions which in practice turned rivers into canals devoid of environmental and amenity value. The concern now is to undertake interventions which maintain or add environmental and amenity; for example, by the use of 'two-stage channels' (Nature Conservancy Council 1983) or the use of 'winter' and 'summer' embankments.

The problems of geomorphological instability following interventions are now recognized (Brookes 1988), along with the problems and costs of channel maintenance.

How to decide

Benefit/cost analysis is extensively used in deciding whether to provide flood alleviation and what level of protection to provide (Penning-Rowsell and Chatterton, 1977). The capitalized costs of constructing and maintaining the proposed scheme are compared to the expected value of the resulting reduction in flood losses, where the expected value of the resulting reduction is calculated by plotting the loss from a flood event against the exceedance probability for that event and calculating the area under the resulting loss probability curve.

An intervention strategy is likely to have many other impacts as well as those upon the efficiency of use of the floodplain. These impacts may include effects upon income distribution, the role of women and the environment. A full assessment of the impacts should be conducted. Therefore, multi-criteria analysis (Nijkamp et al., 1990) is a second technique which has been adopted in some countries, the different options being compared against a selected range of objectives which may include environmental impacts as well as economic efficiency. Environmental assessment is routinely undertaken of individual schemes. With increasing emphasis on catchment management (Newson 1992), and flood alleviation schemes which are made up of many small, localised improvements, strategic environmental assessments (Therivel et al. 1992) are beginning to be used to select between alternative programmes of improvements.

Along with catchment management, there is increasing emphasis on public consultation at a minimum and, increasingly, public participation (McPherson and McGarry, 1987). The emphasis is consequently shifting from what to do to how to decide what to do. At one extreme, the successful implementation of projects is more likely if the public are involved; embankments in Bangladesh have been breached by those who believe either that they are causing waterlogging or shifting the floodwaters onto their land (Ahmad, 1989). At the other extreme, there is a long tradition of community self-help in flood protection which is now being revived (Maskrey, 1989). Moreover, those affected by flooding are often those most able to identify the reasons why flooding is a problem and to be aware of the trade-offs necessarily involved in any intervention strategy.

Methods of enabling public participation are embryonic, the difficulties being both to ensure that the participation process is not captured by special interest groups and that the public have sufficient information so as to realistically be able to participate. Experiments have been undertaken on using geographic information systems and multimedia computer systems to enable the people to select and structure the information which they require (Penning-Rowsell and Fordham, 1994). Some forms of multi-criteria analysis are designed to enable multi-party decision making (von Winterfeldt and Edwards, 1986), since the choice between the options will often involve conflicts of interest. Environmental mediation procedures (Brown et al., 1995) are consequently being developed to enable such conflicts to be resolved.

Conclusions

The objective in flood hazard management is to maximize the efficiency of use of flood-prone land. To do this, it is necessary to integrate flood hazard management into the broader context of catchment management both in terms of land management and water management. Successful hazard management increases the public's ability to manage the flood hazard, either by reducing the challenge that floods present or by strengthening the public's capacity to respond effectively to those challenges.

It is likely that a mixture of strategies will prove to be the most effective way of achieving these ends, those of manageable floods. In choosing between the strategies it is necessary to compare the effectiveness of the options against all the possible flood events rather than only across the most frequent events. Benefit/cost analysis is a useful way of analyzing and comparing the different impacts of management options, not only in terms of reducing flood losses and increasing productivity but also in their effects on the environment. Public participation, rather than consultation, perhaps through a catchment management committee, is increasingly part of the process of not only selecting the management strategies to adopt but also the long-term process of catchment management.

Colin Green

Bibliography

Adnan, S., 1991. Floods, people and the environment: a critical review of flood protection measures in Bangladesh, *Grass Roots*, **1**(1), 227–48.

Ahmad, M. (ed.), 1989. *Flood in Bangladesh*, Dhaka: Community Development Library.

Beinin, L., 1985. *Medical Consequences of Natural Disasters*, New York: Springer-Verlag.

Blaikie, P., 1985. *The Political Economy of Soil Erosion in Developing Countries*, London: Longmans.

Blaikie, P., Cannon, T., Davis, J. and Wisner, B., 1994. *At Risk: Natural Hazards: People's Vulnerability and Disasters*, London: Routledge.

Brammer, H., 1990. Floods in Bangladesh 2, flood mitigation and environmental aspects, *Geographical Journal*, **156**, 158–165.

Brookes, A., 1988. *Perspectives for Environmental Improvement*, Chichester: John Wiley.

Brown, V., Smith, D.I., Wiseman, R. and Handmer, J., 1995. *Risks and Opportunities: Managing Environmental Conflict and Change*, London: Earthscan.

Bureau of Water Resources, 1985. *Stream Corridor Management: A Basic Reference Manual*, Albany, NY: New York State Department of Environmental Conservation.

CNS Scientific and Engineering Services, 1991. *The Benefit–Cost of Hydrometric Data – River Flow Gauging*, Marlow: Foundation for Water Research.

DeKay, M.L. and McClelland, J.H., 1993. Predicting loss of life in cases of down failure and flash flood, *Risk Analysis*, (2), 193–204.

Drabek, T.E., 1986. *An Inventory of Sociological Findings – Human System Responses to Disaster*, Berlin: Springer-Verlag.

Drijver, A. and Marchand, M., 1985. *Taming the Floods: environmental aspects of floodplain development in Africa*, Leiden: Centre for Environmental Studies, University of Leiden.

Emergency Management Australia, 1995. *Flooding Warning: An Australian Guide*, Mount Macedon: Emergency Management Australia.

Ericksen, N.J., 1986. *Creating Flood Disasters?*, Water and Soil Miscellaneous Publication No. 77, Wellington: National Water and Soil Conservation Authority.

Gardiner, J.L. (ed.), 1991. *River Projects and Conservation: A Manual for Holistic Appraisal*, Chichester: John Wiley.

Green, C.H. and Penning-Rowsell, E.C., 1986. Evaluating the intangible benefits and costs of a flood alleviation proposal, *Journal of the Institution of Water Engineers and Scientists*, **40**(3), 229–248.

Green, C.H., Parker, D.J. and Penning-Rowsell, E.C., 1993. Designing for failure, in Merriman, P.A. and Browitt, C.W.A. (eds), *Natural Disasters: Protecting Vulnerable Communities*, London: Thomas Telford.

Hewitt, K. (ed.), 1983. *Interpretation of Calamity from the Viewpoint of Human Ecology*, London: Allen and Unwin.

Hunting Technical Services/Sanyu Consultants, 1991. *Bangladesh Flood Action Plan: FAP12 FCD/1 Agricultural Study*, Report to Ministry of Irrigation, Water Development and Flood Control, Dhaka: Hunting Technical Services.

Inland Waters Directorate, 1985. *Federal Guidelines for the National Flood Damage Reduction Program*, Environment Canada.

Ketteridge, A.M. and Green, C.H., 1994. *EUROFLOOD: Full Flood Impacts*, Technical Annex, Enfield: Flood Hazard Research Centre.

Madiec, H., 1992. A means of fighting pollution in urban storm water overflow: compensating techniques, the experience in Bordeaux (France), paper given at CONFLO 92, Oxford.

Marchand, M., 1987. The productivity of African floodplains, *International Journal of Environmental Studies*, **29**, 201–211.

Marchand, M. and Toornstraf, H., 1986. *Ecological Guidelines for River Basin Development*, cml Report No. 28, Leiden: Centrum Voor Milieukunde, Rijksuniversiteit Leiden.

Maskrey, A., 1989. *Disaster Mitigation: A Community-based Approach*, Oxford: Oxfam.

McPherson, H.J. and McGarry, M.G., 1987. User participation and implementation strategies in water and sanitation projects, *Water Resources Development*, **3**(1), 23–30.

National Institute for Mental Health, 1979. *Crisis Intervention Programs for Disaster Victims in Smaller Communities*, Washington DC: US Department of Health and Human Services.

Nature Conservancy Council, 1983. *Nature Conservation and River Engineering*. Shrewsbury: Nature Conservancy Council.

Newson, M., 1992. *Land Water and Development: River Basin Systems and their Management*, London: Routledge.

Nijkamp P., Rietveld, P. and Voogd, H., 1990. *Multicriteria Evaluation in Physical Planning*, Amsterdam: North Holland.

Parker, D.J. and Neal, J., 1990. Evaluating the performance of flood warning systems, in Handmer, J. and Penning-Rowsell, E.C. (eds) *Hazards and the Communication of Risk*, Aldershot: Gower.

Parker, D.J. and Thompson, P.M., 1991. Floods and tropical storms, in World Health Organisation, *The Challenge of African Disasters*, Addis Ababa: Panafrican Centre for Emergency Preparedness and Response.

Parker, D.J., Green, C.H. and Thompson, P.M., 1987. *Urban Flood Protection Benefits: A Project Appraisal Guide*, Aldershot: Gower.

Paul, B.K., 1984. Perception of and adjustment to floods in Jamuna floodplain, Bangladesh, *Human Ecology*, **12**, 3–19.

Penning-Rowsell, E.C. and Chatterton, J.B., 1977. *The Benefits of Flood Alleviation: A Manual of Assessment Techniques*, Aldershot: Gower.

Penning-Rowsell, E.C. and Fordham, M. (eds), 1994. *Floods Across Europe*, London: Middlesex University Press.

Penning-Rowsell, E.C., Green, C.H., Thompson, P.M. *et al.*, 1992. *The Economics of Coastal Management: A Manual of Assessment Techniques*, London: Belhaven.

Perry, R.W., 1985. *Comprehensive Emergency Management: Evacuating Threatened Populations*, London: JAI.

Riebsame, W.E., 1989. *Assessing the Social Implications of Climate Fluctuations: A Guide to Climate Impact Studies*, Nairobi: UNEP.

River Bureau, undated. *Rivers in Japan and other countries*, Tokyo: Ministry of Construction.

Sangrey, D.A., 1975. *Evaluating the impact of structurally interrupted flood plain flows*, Technical Report No. 98, Cornell University.

Smith, D.I., 1986. Cost-effectiveness of flood warnings, in Smith, D.I. and Handmer, J.W. (eds), *Flood Warning in Australia*, Canberra: Centre for Resource and Environmental Studies, Australian National University.

Therivel, R. Wilson, E. Thompson, S. *et al.*, 1992. *Strategic Environmental Assessment*, London: Earthscan.

von Winterfeldt, D. and Edwards, W., 1986. *Decision Analysis and Behavioral Research*, Cambridge: Cambridge University Press.

Wagret, P., 1967. *Polderlands*, London: Methuen.

Wellings, F.M., 1983. Public health research needs, in Schicht, A. and Semonin, R.G. (eds), *A Plan for Research on Floods and their Solution in the United States*, Illinois: Illinois State Water Survey.

Cross references

Flood frequency analysis
Floods
Floods, river and multi-stage channels
Flood studies worldwide
Floods; world's maximum observed

FLOODS

Floods are among the most damaging of natural hazards, and indeed Newson (1975) cites work that lists them first in a world table of natural disasters during the 20 years after 1947. They are most common on river floodplains and over alluvial fans. Thus the frequency of overbank flow in floodplains is often in the range of 6 months to 2 years, with 1.5 years a common value (Cooke and Doornkamp, 1974). Likewise, flooding is a natural feature on alluvial fans that occur chiefly on the flanks of uplands in semiarid areas, forming where the ephemeral flows of mountain streams debouch onto the adjoining footslopes and plains. Although flooding is a serious hazard in humid regions, it can be devastating also in semiarid regions, where high rates of runoff following storms over sparsely vegetated slopes produce widespread, costly flood damage down-valley (Bigger, 1974; Rantz, 1975). Recurring floods are also typical in coastal and estuarine zones (Ward, 1978).

Damage from flooding is increasing in many areas despite expensive efforts to reduce it (Cooke and Doornkamp, 1974; US Water Resources Council, 1974). This seems related to the increasing use of drainage basins and floodplains, commonly made possible by partial protection works, which created major changes in the hydrological system. Urbanization leads to especially rapid change, replacing permeable by impermeable surfaces and a natural system of channels by storm sewers and other drains. The effects on flood hydrograph parameters are well documented, producing increases in peak discharge values and decreases in time between peaks, accompanied by rapid changes in channel geometry and sediment yield, for example (Leopold, 1974; Newson, 1975; Fox, 1976; Lazaro, 1979).

The nature and occurrence of floods are governed by diverse factors, including rainfall characteristics, intrinsic properties of the drainage catchment, and land use in the catchment. Because of the complex interactions involved, much work has been directed toward predicting floods by identifying the most important variables that are relatively simply derived. The term is used here to cover both flood prediction (for engineering design purposes) and flood forecasting (for warning purposes), although there are major distinctions between them (Ward, 1978). Geomorphological variables figure prominently in this work. Hence, using the drainage basin as the basic study unit, numerous investigations have related flood characteristics to such variables as catchment area, drainage density and main channel slope. Analysis of the characteristics of floods themselves is important in planning preventive measures. Thus the statistical probability of flood events of a given magnitude can be quantified through magnitude and/or frequency analyses, which provide a yardstick by which the hazard and the cost of preventing it can be assessed. Similarly, flood hydrographs showing peak flow, total runoff, rate of change of discharge and flood-to-peak interval, for example, have direct application in planning flood prevention. Rainfall analyses are also basic to flood analyses, because heavy rainfall is the major cause of flooding, and raingauging commonly has both a longer duration and a denser network than river gauging.

In many drainage basins, however, both raingauge networks and rivergauging are inadequate, so planning for flood prediction and management must be based on the study of catchment characteristics. Common approaches to estimating river flows in this case include precise surveys of channel cross-sections, of flood water and surface slope as evidenced by debris lines, and of hydraulic roughness. Instantaneous peak discharge has been assessed by the 'slope–area' method, and simple 'flood formulae' have been developed to relate the maximum peak flow at a site to the catchment area, with a coefficient and exponents being fixed for the area in question (Newson, 1975; Ward, 1978). This has led to the development of many versions of the 'rational formula,' which uses a figure for rainfall intensity as well as catchment area together with a dimensionless constant, the 'runoff coefficient.' Equations used for calculating time of concentration have been based on catchment area, average slope, and straight-line distance from outflow to remotest divide.

The application of the rational formula depends on the selection of an appropriate value for the runoff coefficient. Theoretically this coefficient incorporates all factors controlling the proportion of rain that becomes runoff, including surface configuration vegetation, infiltration rate, soil storage capacity and drainage patterns. But all these data are rarely available, so many coefficient values place emphasis on the two most important and most readily identified features, landforms and land use. Slope gradient is a major consideration, usually in association with the nature of the surface, surface storage and degree of saturation (Gray, 1970; Hudson, 1971).

As Hudson explains, the advantage of the rational formula is that it can always be used for estimating maximum runoff rates no matter how little recorded information is available. Hence it is of value in areas lacking rainfall or hydrologic records. However, several methods have been evolved for estimating rates of runoff where more detailed information is available. An approach used by the US Soil Conservation Service considers the effect of four principal factors in the catchment, two of which are geomorphological – 'relief' and surface storage – and the others soil infiltration and vegetal cover. Each factor has four categories ranging from 'low' to 'extreme' as runoff-producing characteristics, and with corresponding numerical 'weightings' of 10 to 40 for relief and 5 to 20 for the other variables. The sum of the four weightings is the 'watershed characteristic,'

which, by reference to prepared tables or charts, is used to predict peak flow values.

Hudson describes a similar method developed for use in some African conditions and further refinements where the effect of management practices is added. The US Geological Survey and others have taken this catchment analysis further to obtain multiple regression expressions for mean annual flood and a variety of catchment characteristics. Values obtained for specific flood recurrence intervals can then be used in engineering design. Orsborn (1976) discusses such regression models and relatively sophisticated 'hydrogeological' methods whereby drainage basin properties, especially geomorphological characteristics, are analyzed to identify parameters having the best correlations with gauged streamflows so that flows for ungauged streams can be predicted using their basin parameters and the gauged correlations (Wandle 1977; Petton, 1988).

Geomorphological considerations are also important in more elaborate approaches as applied, for example, by the US Department of Agriculture's Hydrograph Laboratory (Holtan and Lopez, 1971). Here a model has been evolved to provide a mathematical continuum from drainage divide to catchment outlet. The framework for the model consists of dividing a catchment into 'hydrological response zones' corresponding to the three major geomorphological divisions: uplands (i.e. interfluve crests), hillsides and bottomlands. Soils within each zone are grouped according to land capability classes and infiltration, with evaporation and overland flow computed for each zone using published data and conventional formulae.

Input to the model consists of a continuous record of rainfall weighted to represent the catchment. Vegetation, land use and soil properties are considered in computing evapotranspiration and infiltration. Detailed geomorphological characteristics also receive emphasis. For instance, the model incorporates 'depression storage' of the surface water that is held until dissipated by infiltration, which is extremely important where techniques such as terracing and contour furrowing are practiced. Rainfall in excess of infiltration is routed in the model across each hydrologic response zone and cascaded, subject to further infiltration, to the channel across subsequent zones at lower elevations. The equation for computing overland flow includes a coefficient dependent on roughness, length and degree of slope; channel flows and subsequent return flows are routed by simultaneous solutions of this equation and a storage function. In this way, flow from different parts of the catchment may be routed separately through watershed storage and then summed to predict watershed outflow. Ward (1978) outlines a number of other computer simulation models incorporating such drainage basin, or hydrogeological, characteristics and meteorological inputs.

A central problem in predicting storm runoff is to identify the transfer function that converts a given hyetograph of effective water input on a catchment into the response hydrograph. This function can be developed through mathematical representation of the instantaneous unit hydrograph (Dingman, 1994), and the widely used geomorphological instantaneous unit hydrograph which combines unit hydrograph theory with geomorphic parameters, including drainage network characteristics (Rodriguez-Iturbe and Valdes, 1979; Patton, 1988; Gupta and Mesa, 1988).

Modeling of catchment water flow and runoff prediction is now focused upon physically based 'distributed' models which simulate the spatial distribution of catchment properties, unlike earlier 'lumped' models whereby catchments are characterized by single parameters. In distributed models a catchment is treated as a dynamic system, firstly by replicating catchment processes in terms of the physical mechanisms which produce 'event responses' in water transfers by, for example, overland flow, channel flow, unsaturated subsurface flow and saturated flow (Dingman, 1994; Fawcett et al., 1995). Accurate representation of catchment behavior requires these mechanisms to be viewed in relation to the environmental conditions under which they occur and which determine the distribution of flow pathways. Such conditions are commonly linked to topography (Beven and Kirkby, 1979; Bernier, 1985; Troendle, 1985; Moore and Foster, 1990; O'Loughlin, 1990; Wood et al., 1990). Second, process dynamics are portrayed in terms of temporal variability – for instance, in the extent of runoff-contributing source areas (Pearce et al., 1986; Beven, 1987), and in the spatial structure of the drainage network (Wharton, 1994).

Of fundamental concern in all such representations of process systems are the problems of calibration and validation of models (Fawcett et al., 1995).

Four main approaches have been developed to help control

flooding: (1) watershed management, (2) protective structures, (3) land-use control in flood plains and (4) calculated risk and absorption of loss (Cooke and Doornkamp, 1974; US Water Resources Council, 1974; US Army Corps of Engineers, 1975; Noble, 1976; Ward, 1978; Dunne 1988).

Human activities in river catchments, including vegetation clearance and improper agricultural methods, have greatly increased the dangers of flooding in urban or other areas down-valley. Consequently, land-use management practices, termed watershed management, are necessary to reduce and delay runoff to rivers in these catchments. They include the construction of terraces, contour furrowing and contour strip cropping, crop rotation and reforestation. Whereas these measures reduce peak flows from many smaller storms, they may have little effect on large floods. Consequently, protective structures have been designed with the object of reducing overflow in the floodplain and ensuring the relatively harmless passage of water through the threatened areas. They mostly involve the construction of storm channels and reservoirs to divert or retard excessive runoff, levees or floodwalls to confine flood flow, various bank protection structures, and channel modifications to increase the river's water transmitting capability. These latter include straightening, steepening, widening, or deepening channels, locally lining them with concrete, and using dikes to direct flow into desired alignments

The third approach mentioned, land-use control in flood plains, often referred to as floodplain regulation, does not attempt to reduce or eliminate flooding but rather to minimize its damaging effects by controlling the use made of flood-prone areas, especially the type and location of buildings. It includes designation of floodways, zoning prohibitions, subdivision regulations, building codes and other ordinances.

The final approach is that of calculated risk and absorption of loss. Here property owners anticipate that they will have to bear little loss, or temporarily evacuate to prevent greater loss, or rely on some form of public relief subsidy or insurance to meet their losses. The extent of such losses is strongly influenced by the availability of flood forecasting and warning services, and by floodproofing, or adjusting buildings and their contents to withstand flooding.

Applied geomorphology figures prominently in these control measures. Thus the two most common approaches, watershed management and the building of protective structures along the river, are largely applied geomorphology – albeit mostly practiced by engineers and agriculturalists – in that they involve building new landforms, ranging from levees and diversion channels to dams and terraces, or modifying existing ones by deepening and straightening channels, contour furrowing, and so forth. Likewise, floodplain regulation is a geomorphological exercise in terms of defining floodways, identifying the extent and likely frequency of previous floods, areas of potential inundation, and tracts of varying degree of risk that are thus suited for different subdivision and building purposes. Furthermore, flood prediction and associated flood frequency reports and inundation maps have major geomorphological components, as noted earlier.

The factors influencing the selection of remedies to deal with flooding vary markedly from place to place. As Newson (1975) explains by reference to UK examples, there may be little choice as to the methods employed because local circumstances of geomorphology and urban conditions may preclude the application of many measures. Alternatively, costly construction can be avoided where the risks are less great. In this latter case, a system of public adjustment to flooding could be based on flood danger warnings, which requires some idea of the typical flood hydrograph for the particular catchment, to time the flood's arrival downstream and its duration at sites of risk. Flooding problems are extremely complex in many areas, however, and it is difficult to find permanent solutions, as in southern California (Rantz, 1975). Because the population in this region is growing rapidly, there is great pressure on the land for building and other development purposes. But floods here are of tremendous violence, and so their control necessitates a variety of interrelated remedies, including construction of flood-water reservoirs and debris basins; diversion of flood waters onto areas where sediment can be deposited and excess water can percolate underground; realigning, enlarging and paving permanent channels to accommodate excess runoff; and retarding erosion and surface runoff in catchments by means of various slope and channel modifications.

Despite the complexities of resolving problems with flood control, certain general principles are evident. In the first place, satisfactory results can be achieved only if the geomorphological process systems within the flood-prone basin are properly understood and allowed for in the design of control measures. Second, watershed management and protective structures in the danger areas down-valley should be regarded as complementary rather than alternative approaches. Third, management of floodplain use must be a continuing priority together with the devising of systems to regulate streamflows for all beneficial purposes, not only flood control but also water use and conservation generally. The complementary nature of the main flood control methods is again apparent here: reservoirs, which are among the most expensive measures, are advantageous because flood prevention becomes part of general water resource planning by checking excess runoff at or near its source in the uplands. Thus 'regulating reservoirs' and 'trans-basin aqueducts' have been chosen as a key element of water resource development in England and Wales (Newson, 1975). These general principles, and especially the need to understand the dynamics of geomorphological process systems, are well illustrated in examples given by Rantz (1975), Noble (1976), Kolb (1976), Keller (1976) Palmer (1976) and Dunne (1988).

Landslides

The term landslide is used here to refer to all types of rapid mass movements. Landslides, like flooding, regularly cause major disasters with much loss of life, and are frequently recurring hazards that are expensive in terms of the physical damage they produce. They may be the actual agent of destruction, as when masses of debris sweep over settlements, or the initial cause, as in Norwegian fjords when the resultant waves destroy shoreline villages, and as in the worst dam disaster in history, which killed 2600 people at the Vaiont Dam in Italy in 1963 (Cooke and Doornkamp, 1974; Morton and Streitz, 1975; Kiersch, 1975). In many areas landslides occur regularly as a continuing sequence of natural events. Elsewhere slopes may be comparatively stable until landslides are triggered by disturbance factors that are either relatively infrequent, such as earthquakes, or new developments, such as landuse changes. In particular, landslides are an increasing hazard in areas of urban development, chiefly because slopes that appear naturally stable may become unstable if moisture conditions, loadings or gradients are changed through such developments.

Landslides are related to gravity-produced shearing stresses, which increase with slope gradient, and slope height and unit weight of the underlying materials. Surficial processes of differential volume change – such as freezing and thawing, and shrinking and swelling – produce further shearing stresses. Landslides occur, therefore, when the shear stress forces creating downslope movement exceed the shear strength of the materials. Hence for each kind of material, varying in terms of bulk density, angle of internal friction, and cohesion, there is a limiting slope gradient above which rapid mass movement will occur from time to time, and below which the slope is relatively stable. The steepness of a slope influences not only the susceptibility to sliding but also the volume of the slide. In engineering work it has also been found that the time lag between slope excavation and subsequent failure in clays, for instance, is related to slope steepness. 'External' geomorphological relationships are also most important. Thus shear stress may be increased through the removal, by natural erosion or artificial excavations, of 'supporting' material along the 'toe' of a slope, or by the accumulation of talus in the higher parts. Similarly, shear strength may be decreased by increased pore-water pressure due to human-induced changes in surficial water systems through reservoir construction, producing higher groundwater tables, or landuse practices resulting in increased run-on. Therefore the location of landsliding is strongly influenced not only by the form and composition of slopes, but also by their geomorphological setting.

Geomorphology has valuable applications in both predicting areas of potential landsliding and controlling the causes of landslides. Thus Cooke and Doornkamp (1974) compiled a checklist whereby, with reference to readily observable features of slope form and composition, any slope unit in an area of potential landsliding could be given a stability rating on a scale, ranging from stability through increasing degrees of potential instability to failure. Inherited features can also be important because potentially unstable relict landforms, which developed under conditions no longer operative, are quite common and can be reactivated during excavations for road work, for instance. However, they can often be identified by geomorphological ground survey and air photo interpretation because even these ancient landslides commonly have distinctive slope forms within and around them. Leggett's (1973) case studies illustrate such diverse geomorphological considerations in areas prone to landsliding. They demonstrate that the toe of a landslide is a critical location, because it requires little interference with this distal margin to reduce the length of the arc on which are acting two major forces: the weight of the mass of material that is tending to move downslope, and the shear strength that provides the resistance to movement. Predictably, therefore, many valley-side slopes above eroding streams or road cuts are especially prone to landsliding.

Bailey (1973) illustrated the importance of environmental interrelationships in landsliding and the value of applied geomorphology in elucidating them. He identified the connections between fluvial processes and slope stability in a study of landslide hazards in Wyoming, and found that the soil and rock mantles of slopes were in precarious balance with the physical environment, and road construction and logging activities disturbed that balance. He identified the location of unstable areas, the associated environmental factors and processes, and the limitations to management practices. Active landslides, inactive landslides and active taluses were recorded on a landslide hazard map, and guidelines for logging and road construction were described in each case. Oversteepened slopes characterized by active or inactive debris avalanches and mudflows, and potential landslide areas where no apparent movement had yet taken place, were also differentiated and similar guidelines laid down. Slope gradient was recognized as a characterizing feature for mass movements of all types, and was positively correlated with road-related slides.

The extent and importance of applied geomorphology in the engineer's dealing with landslide problems are seen in the manual produced by the US Highway Research Board's Committee on Landslide Investigations (Eckel, 1958). Procedures used in the identification and interpretation of both actual and potential slides are largely based on geomorphological considerations. Thus a checklist for use by engineers to recognize different kinds of active or recently active landslides depends on detailed examination of the form, microrelief and surface composition of predefined sets of geomorphological units for both 'stable parts surrounding the slide' and 'parts that have moved' (Ritchie, 1958, pp. 56–57). In identifying areas of potential slides, 'special attention should be given to the slopes, changes in slope, and their relationship to the different materials involved' (Ritchie, 1958, p. 51). Geomorphological interpretation of airphotos plays a major role. This is because landform differences can be easily identified on aerial photographs and because of their close bedrock relationships, as well as slope–process associations, which are important in delimiting areas of actual and potential landslides. Each type of landform, distinguished by a specific kind of geological material, overburden and topographic expression 'poses relatively distinct problems for the engineer, particularly from the standpoint of landslide susceptibility' (Ta Liang and Belcher, 1958, p. 71). In more detailed airphoto interpretation, individual landslides are indicated by 'the sharp line of break at the scarp; the hummocky topography of the sliding mass below it; the elongated, undrained depressions in the mass; and the abrupt differences in vegetative and tonal characteristics between the landslide and the adjoining stable slopes' (Ta Liang and Belcher, 1958, p. 72). Air photo clues to sites of potential landsliding include banks undercut by streams, steep slopes, drainage lines on higher ground contributing seepage waters downslope, and seepage depressions.

For such reasons, detailed geomorphological mapping provides an ideal framework for planning geotechnical or soil engineering studies prior to highway design, for example. This approach was applied by Brunsden et al. (1975) (see also Cooke and Doornkamp, 1974), who demonstrated that mapping of slopes that are potentially unstable, or have already failed, is useful in the planning stages of engineering projects. Routes can be aligned to avoid these slopes, or appropriate remedial measures can be designed and costed to enable them to be crossed. Such mapping also facilitates planning the deployment of costly research effort by identifying potentially unstable slopes in which field sampling and laboratory analysis should be concentrated, with less intensive coverage of relatively stable surfaces. The geomorphological basis of terrain evaluation for landslide investigations is demonstrated also by Schuster and Krizek (1978).

Most methods for controlling landslides have been devised by engineers in constructing roads, buildings and other structures. They have been developed because of the widespread occurrence of slopes that become unstable once there is human disruption of their natural state. Moreover, the cuttings and embankments of many 'artificial'

landforms are inherently unstable without such control methods. Four main groups of methods are employed, other than avoiding the danger area: excavation, drainage, restraining structures and miscellaneous methods (Eckel, 1958). To be effective, they depend on the application of geomorphological knowledge in identifying the critical causative factors of landsliding in a particular locality, and how best to modify these to induce relative stability. They involve the design and construction of new stable slopes (examined in detail by Schuster and Krizek, 1978) or the modification of the original ones and their surficial water systems.

Preventive measures in excavation consist of designing slopes, varying according to the material, whose gradients and drainage will minimize sliding. Such slopes may be produced by removing part of the original head or upper part of the slide, by grading the slope to a gentler inclination, or by cutting benches. Particular attention has to be given to surficial water conditions because increases in pore-water pressure are a common trigger mechanism in landsliding. Appropriate measures include draining the slide and creating diversion channels upslope. This may involve construction of paved ditches, installation of flumes or conduits, and paving or bituminous treatment of slopes, in order to minimize the possibility of water from run-on or runoff percolating the unstable area. Major landform modifications may be necessary, such as regrading overloaded slopes at the head of a slide, building structures where support is needed, or diverting channels undercutting the toe of the slope. Many restraining structures are concentrated at the critical toe of a slide, including buttresses to provide added weight and thus to increase the resistance to movement. Among miscellaneous methods, artificial cementation has been used to increase the shear strength of the soil, and partial removal of the toe of a landslide can be a temporary expedient to protect a structure until more permanent safeguards can be provided.

These methods for controlling slope instability are among the more widely used of traditional engineering applications of geomorphology because, in Leighton's words 'Slope, including its underlying material, is the most important geologic and engineering element in hillside (urban) development' (Leighton, 1974, p. 207). He illustrates this by citing examples both of predicting the occurrence of landslides and of treating areas of actual or potential movement – unstable slopes being made stable by resculpturing and 'grading'. Other case studies described by Leighton (1976) underline the crucial role of geomorphology in the identification, interpretation and engineering control of landslide problems.

Robert L. Wright

Bibliography

Askochensky, A.N., 1973. Basic trends and methods of water control in the arid zones of the Soviet Union, in D.R. Coates (ed.), *Environmental Geomorphology and Landscape Conservation, III: Non-Urban Regions*. Stroudsburg, PA: Dowden, Hutchinson & Ross, pp. 207–216.
Bailey, R.G., 1973. Forest land use implications, in D.R. Coates (ed.), *Environmental Geomorphology and Landscape Conservation, III: Non-Urban Regions*. Stroudsburg, PA: Dowden, Hutchinson & Ross, pp. 388–413.
Bennett, H.H. and Chapline, W.R., 1973. Soil erosion – a national-menace. Part I: Some aspects of the wastage caused by soil erosion, in D.R. Coates (ed.), *Environmental Geomorphology and Landscape Conservation, III: Non-Urban Regions*. Stroudsburg, PA: Dowden, Hutchinson & Ross, pp. 57–83.
Bernier, P.Y., 1985. Variable source areas and storm-flow generation: an update of the concept and simulation effort, *J. Hydrol.*, **79**, 195–213.
Beven, K.J., 1987. Towards the use of catchment geomorphology in flood frequency predictions, *Earth Surf. Process. Landf.*, **12**, 69–82.
Beven, K.J. and Kirkby, M.J., 1979. A physically-based variable contributing area model of basin hydrology, *Hydrol. Sci. Bull.*, **24**, 43–69.
Bigger, R., 1974. The flood problem, in D.R. Coates (ed.), *Environmental Geomorphology and Landscape Conservation, II: Urban Areas*. Stroudsburg, PA: Dowden, Hutchinson & Ross, pp. 187–196.
Brunsden, D., Doornkamp, J.C., Fookes, P.G. *et al.*, 1975. Large scale geomorphological mapping and highway engineering design, *Quart. J. Eng. Geology*, **8**, 227–253.

Carson, M.A. and Kirkby, M.J., 1972. *Hillslope Form and Process.* Cambridge, England: Cambridge University Press, 475 pp.

Cooke, R.U. and Doornkamp, J.C., 1974. *Geomorphology in Environmental Management.* Oxford: Clarendon Press, 413 pp.

Currey, D.T., 1977. The role of applied geomorphology in irrigation and groundwater studies, in J.R. Hails (ed.), *Applied Geomorphology.* Amsterdam: Elsevier, 51–83.

Dingman, S.L., 1994. *Physical Hydrology.* Englewood Cliffs, NJ: Prentice Hall, 575 pp.

Dunne, T., 1988. Geomorphologic contributions to flood-control planning, in V.R. Baker, R.C. Kochel and P.C. Patton (eds), *Flood Geomorphology,* New York: Wiley, pp. 421–438.

Eckel, E.B. (ed.), 1958. Landslides and engineering practice, *Highway Research Board Spec. Rep. 29,* Washington D.C., NAS-NRC Publication 544, 232 pp.

Fawcett, K.R., Anderson, M.G., Bates, P.D. *et al.,* 1995. The importance of internal validation in the assessment of physically based distributed models, *Trans. Inst. Br. Georg.,* **20,** 248–265.

Food and Agriculture Organization, 1965. Soil erosion by water – some measures for its control on cultivated lands, *Agricultural Development Paper 87,* 284 pp.

Foose, R.M. and Hess, P.W., 1976. Scientific and engineering parameters in planning and development of a landfill site in Pennsylvania, in D.R. Coates (ed.), *Geomorphology and Engineering,* Stroudsburg, PA: Dowden, Hutchinson & Ross, pp. 289–312.

Fox, H.L., 1976. The urbanizing river: a case study in the Maryland piedmont, in D.R. Coates (ed.), *Geomorphology and Engineering.* Stroudsburg, PA: Dowden, Hutchinson & Ross, pp. 245–271.

Freeman, V.M., 1974. Water spreading as practiced by the Santa Clara water-conservation district, Ventura County, California, in D.R. Coates (ed.), *Environmental Geomorphology and Landscape Conservation, II: Urban Areas.* Stroudsburg, PA: Dowden, Hutchinson & Ross, pp. 111–117.

Glenn, L.C., 1973. Denudation and erosion in the southern Appalachian region and the Monongahela basin, in D.R. Coates (ed.), *Environmental Geomorphology and Landscape Conservation, III: Non-Urban Regions.* Stroudsburg, PA: Dowden, Hutchinson & Ross, pp. 36–56.

Gray, D.M., 1970. *Handbook on the Principles of Hydrology.* New York: Water Information Center, 591 p.

Gupta, V.K. and Mesa, O.J., 1988. Runoff generation and hydrologic response via channel network geomorphology – recent progress and open problems, *J. Hydrol.,* **102,** 3–28.

Holtan, H.N. and Lopez, N.C., 1971. USDAHL-70 Model of Watershed Hydrology, *US. Dept. Agriculture Research Service Tech. Bull. 1435,* 84p.

Hudson, N., 1971. *Soil Conservation,* London: Batsford, 320 pp.

Jepson, H.G., 1973. Prevention and control of gullies, in D.R. Coates (ed.), *Environmental Geomorphology and Landscape Conservation, III: Non-Urban Regions.* Stroudsburg, PA: Dowden, Hutchinson & Ross, pp. 283–298.

Keller, E.A., 1976. Channelization: environmental, geomorphic, and engineering aspects, in D.R. Coates (ed.), *Geomorphology and Engineering.* Stroudsburg, PA: Dowden, Hutchinson & Ross, pp. 115–140.

Kiersch, G.A., 1975. The Vaiont reservoir disaster, G.D. McKenzie and R.O. Utgard (eds), *Man and His Physical Environment: Readings in Environmental Geology,* 2nd edn. Minneapolis: Burgess, pp. 71–75.

Kirkby, M.J. and Morgan, R.P.C. (eds), 1980. *Soil Erosion.* Chichester, England: Wiley, 312 pp.

Kolb, C.R., 1976. Geologic control of sand boils along Mississippi river levees, in D.R. Coates (ed.), *Geomorphology and Engineering.* Stroudsburg, PA: Dowden, Hutchinson & Ross, pp. 99–113.

Lazaro, T.R., 1979. *Urban Hydrology.* Ann Arbor, MI: Ann Arbor Science, 249 pp.

Leggett, R.F., 1973. *Cities and Geology.* New York: McGraw-Hill, 624 pp.

Leighton, F.B., 1974. Landslides and hillside development, in D.R. Coates (ed.), *Environmental Geomorphology and Landscape Conservation, II: Urban Areas.* Stroudsburg, PA: Dowden, Hutchinson & Ross, pp. 206–223.

Leighton, F.B., 1976. Geomorphology and engineering control of landslides, in D.R. Coates (ed.), *Geomorphology and Engineering,* Stroudsburg, PA: Dowden, Hutchinson & Ross, pp. 273–287.

Leopold, L.B., 1974. Hydrology for urban land planning, in D.R. Coates (ed.), *Environmental Geomorphology and Landscape Conservation, II: Urban Areas.* Stroudsburg, PA: Dowden, Hutchinson & Ross, pp. 69–86.

Moore, I.D. and Foster, G.R., 1990. Hydraulics and overland flow, in M.G. Anderson and T.P. Burt (eds), *Process Studies in Hillslope Hydrology,* Chichester, UK: Wiley, pp. 215–254.

Morton, D.M. and Streitz, R., 1975. Landslides, in G.D. McKenzie and R.O. Utgard (eds), *Man and His Physical Environment: Readings in Environmental Geology,* 2nd edn. Minneapolis: Burgess, pp. 58–60.

Newson, M.D., 1975. *Flooding and Flood Hazard in the United Kingdom.* Oxford: Oxford University Press, 60 pp.

Noble, C.C., 1976. The Mississippi River flood of 1973, in D.R. Coates (ed.), *Geomorphology and Engineering.* Stroudsburg, PA: Dowden, Hutchinson & Ross, pp. 79–98.

O'Loughlin, E.M., 1990. Perspectives on hillslope research, in M.G. Anderson and T.P. Burt (eds), *Process Studies in Hillslope Hydrology,* Chichester, UK: Wiley, pp. 501–516.

Orsborn, J.F., 1976. Drainage basin characteristics applied to hydraulic design and water resources management, in D.R. Coates (ed.), *Geomorphology and Engineering.* Stroudsburg, PA: Dowden, Hutchinson & Ross, pp. 141–171.

Palmer, L., 1976. River management criteria for Oregon and Washington, in D.R. Coates (ed.), *Geomorphology and Engineering.* Stroudsburg, PA: Dowden, Hutchinson & Ross, pp. 329–346.

Patton, P.C., 1988. Drainage basin morphometry and floods, in V.R. Baker, R.C. Kochel and P.C. Patton (eds), *Flood Geomorphology,* New York: Wiley, pp. 51–64.

Pearce, A.J., Stewart, M.K. and Sklash, M.G., 1986. Storm runoff generation in humid headwater catchments. 1, Where does the water come from?, *Water Resour. Res.,* **22,** 1263–1272.

Rantz, S.E., 1975. Urban sprawl and flooding in southern California, in G.D. McKenzie and R.O. Utgard (eds), *Man and His Physical Environment: Readings in Environmental Geology.* 2nd edn. Minneapolis: Burgess, pp. 45–52.

Ritchie, A.M., 1958. Recognition and identification of landslides, in E.B. Eckel (ed.), *Landslides and Engineering Practice.* Highway Research Board Spec. Rept. 29. Washington DC: NAS-NRC Publication 544, pp. 48–68.

Rodriguez-Iturbe, I. and Valdes, J.B., 1979. The geomorphic structure of hydrologic response. *Water Resour. Res.,* **15,** 1409–1420.

Schneider, W.J., 1975. Hydrologic implications of solidwaste disposal, in G.D. McKenzie and R.O. Utgard (eds.), *Man and His Physical Environment: Readings in Environmental Geology.* 2nd edn. Minneapolis: Burgess, pp. 125–134.

Schuster, R.L. and Krizek, R.J. (eds), 1978. *Landslides Analysis and Control.* Washington, DC: National Academy of Sciences, 234 pp.

Schwab, G.O., Frevert, R.K., Edminster, T.W. and Barnes, K.K., 1966. *Soil and Water Conservation Engineering,* 2nd edn. New York: Wiley, 683 pp.

Smith, D.D. and W.H., Wischmeier, 1962. Rainfall erosion, *Advances in Agronomy,* **14,** 109–148.

Stall, J.B., 1973. Man's role in affecting the sedimentation of streams and reservoirs, in D.R. Coates (ed.), *Environmental Geomorphology and Landscape Conservation, III: Non-Urban Regions.* Stroudsburg, PA: Dowden, Hutchinson & Ross, 103–119.

Stallings, J.H., 1957. *Soil Conservation.* Englewood Cliffs, NJ: Prentice-Hall, 575 pp.

Strahler, A.N., 1973. The nature of induced erosion and aggradation, in D.R. Coates (ed.), *Environmental Geomorphology and Landscape Conservation, III: Non-Urban Regions.* Stroudsburg, PA: Dowden Hutchinson & Ross, pp. 18–35.

Ta Liang and Belcher, D.J., 1958. Airphoto interpretation, in E.B. Eckel (ed.), *Landslides and Engineering Practice.* Highway Research Board Spec. Rept. 29. Washington DC: NAS-NRC Publication 544, pp. 69–92.

Troendle, C.A., 1985. Variable source area models, in M.G. Anderson and T.P. Burt (eds), *Hydrological Forecasting,* Chichester, UK: Wiley, pp. 347–403.

U.S. Army Corps of Engineers, 1975. Guidelines for reducing flood damages, in G.D. McKenzie and R.O. Utgard (eds), *Man and His Physical Environment: Readings in Environmental Geology,* 2nd edn. Minneapolis: Burgess, pp. 53–57.

US Water Resources Council, 1974. Floods and flood damages, in D.R. Coates (ed.), *Environmental Geomorphology and Landscape*

Conservation, II: Urban Areas. Stroudsburg, PA: Dowden, Hutchinson & Ross, pp. 158–167.

Wandle, S.W., 1977. Estimating the magnitude and frequency of floods on natural-flow streams in Massachusetts. Boston, MA: US Geological Survey, *Water Resources Investigations*, 77–139.

Ward, R., 1978. *Floods: A Geographical Perspective.* London: Macmillan, 244 pp.

Wharton, G., 1994. Progress in the use of drainage network indices for rainfall–runoff modelling and runoff prediction, *Prog. Phys. Geog.*, **18**, 539–557.

Withers, B. and Vipond, S., 1974. *Irrigation Design and Practice.* London: Batsford, 306 pp.

Wood, E.F. Sivaplan, M. and Beven, K., 1990. Similarity and scale in catchment storm response. *Rev. Geophys.*, **28**, 1–18.

Cross references

Alluvial valley engineering
Flood frequency analysis
Floods
Flood studies worldwide
Floods, river and multi-stage channels
Floods: world's maximum observed

FLOODS: LARGEST IN THE USA, CHINA AND THE WORLD

Tables F9–F11 list the largest rainfall–runoff floods for a given drainage area in the USA, China and the world derived from the latest publications available (Costa, 1987; Rodier and Roche, 1984). The USA list is confirmed, as far as is possible, by personal communication with Costa (1997) as floods not having been exceeded to date. The floods from all three tables are plotted in Figure F9 showing an envelope curve of the USA data (Costa, 1987) and in Figure F10 by recession lines through all the data. It is assumed that the data in Tables F9–F11 are derived from rainfall–runoff floods and not from the effect of landslides, dam failures and ice jams, etc. It can be seen that there is similarity between the three sets of data, possibly suggesting that the data may have come from the same population. It is also believed that many, if not most of the floods have either been

Table F10 The largest rainfall–runoff floods in the People's Republic of China (Costa, 1987)

Flood no.	Province, county	Date	Drainage area (km^2)	Discharge (m^3 s^{-1})
1	Neimonggol, Baotou	1900	39.4	1 640
2	Guangdong, Baisha	1894	75.3	3 420
3	Lianing, Yixian	1930	97.2	4 000
4	Lianing, Yixian	1930	154	5 320
5	Lianing, Suizhong	1894	171	7 000
6	Taiwan, Nantou	1979	259	7 780
7	Guangdong, Yaxian	1946	644	10 700
8	Henan, Fangcheng	1896	746	11 300
9	Henan, Biyang	1975	760	13 000
10	Anhui, Shucheng	1853	1 110	12 100
11	Taiwan, Taizhong	1959	1 980	18 300
12	Shandong, Yishui	1730	2 278	17 500
13	Zhejiang, Qingtian	1912	3 255	19 200
14	Guangdong, Changjing	1887	4 634	28 300
15	Shandong, Linyi	1730	10 315	30 000
16	Zhejiang, Qingtian	1912	13 500	30 400
17	Shaanxi, Ankang	1583	38 700	36 000
18	Lianing, Kuandian	1888	55 420	44 600
19	Hubei, Guanghua	1935	95 220	57 200
20	Hubei, Zhongxiang	1935	140 340	57 900
21	Guangxi, Wuzhou	1915	329 710	54 500
22	Sichuan, Nanxi	1520	639 230	65 600
23	Sichuan, Chongqing	1870	866 560	100 000
24	Sichuan, Wanxian	1870	974 880	114 800
25	Hubei, Yichang	1870	1 005 500	110 000

indirectly measured from the slope–area method or by extrapolation of the stage–discharge curve. The flood, (No. 9), in Table F11 was caused by a typhoon and was measured by the Manning equation using a cross-sectional area of 1510 m^3, a hydraulic radius of 11.53 m, a water surface slope of 0.008077, a resistance coefficient (*n*) of 0.067 and a mean velocity of 6.9 m s^{-1} (Costa, 1987), the discharge

Table F9 The largest rainfall-runoff floods in the United States (Costa, 1987)

Flood no.	Stream	Date	Drainage area (km^2)	Discharge (m^3 s^{-1})	Maximum rain intensity
1	Praire Creek trib. near Tyler, TX[a]	8 May 1936	0.0065	0.31	?
2	Unnamed trib. in Brushy and Sandy Ck. basins near Risel, TX – SW18 (SCS expt. watershed)[a]	31 October 1940	0.0124	0.65	?
3	Brawley Wash trib. near Tucson, AZ[a]	26 September 1962	0.021	1.95	178 mm (48 h)$^{-1}$
4	Castle Creek Trib. no 2 near Rockford, SD	28 July 1955	0.05	2.8	127 mm (2 h)$^{-1}$
5	Cimarron Ck. trib. near Cimarron, NM	5 June 1958	0.13	9.54	115 mm (24 h)$^{-1}$
6	Wenatchee River trib. near Monitor, WA	25 August 1956	0.39	25.6	?
7	Lahontan Reservoir trib. no. 3 near Silver Springs, NV	20 July 1971	0.57	47.6	?
8	Little Pinto Ck. trib. near Newcastle, UT	11 August 1964	0.78	74.5	?
9	Humboldt River trib. near Rye Patch, NV	31 May 1973	2.20	251	127 mm (1 h)$^{-1}$
10	Lane Canyon near Nolin, OR	26 July 1965	13.1	807	?
11	Meyers Canyon near Mitchell, OR	13 July 1956	32.9	1 540	102 mm (2 h)$^{-1}$
12	Bronco Ck. near Wikieup, AZ	18 August 1971	49.2	2 080	89 mm (0.75 h)$^{-1}$
13	South Fork Wailua River near Lihue, Kauai	5 April 1963	58.0	2 472	?
14	Eldorado Canyon at Nelson Landing, NV	14 September 1974	59.3	2 152	152 mm (1 h)$^{-1}$
15	North Fork Hubbard Ck. near Albany, TX	4 August 1978	102	2 920	738 mm (24 h)$^{-1}$
16	Jimmy Camp Ck. near Fountain, CO	17 June 1965	141	3 510	203 mm (6 h)$^{-1}$
17	Mailtrail Ck. near Loma Alta, TX	24 June 1948	195	4 810	?
18	Seco Ck. 18 km above D'Hanis, TX	31 May 1935	368	6 510	610 mm (3.5)h^{-1}
19	West Nueces River near Kickapoo Spgs., TX	14 June 1935	1 041	16 425	305 mm (24 h)$^{-1}$
20	Eel River at Scotia, CA	23 December 1964	8 063	21 300	315 mm (24 h)$^{-1}$
21	Susquehanna River at Conowingo, MD	24 June 1972	70 189	32 000	376 mm (24 h)$^{-1}$
22	Ohio River at Metropolis, IL	1 February 1937	525 770	52 392	>559 mm (31 d)$^{-1}$
23	Mississippi River near Arkansas City, AR	20 April 1927	2 928 513	70 007	?

[a] Data lost or measurement not seen.

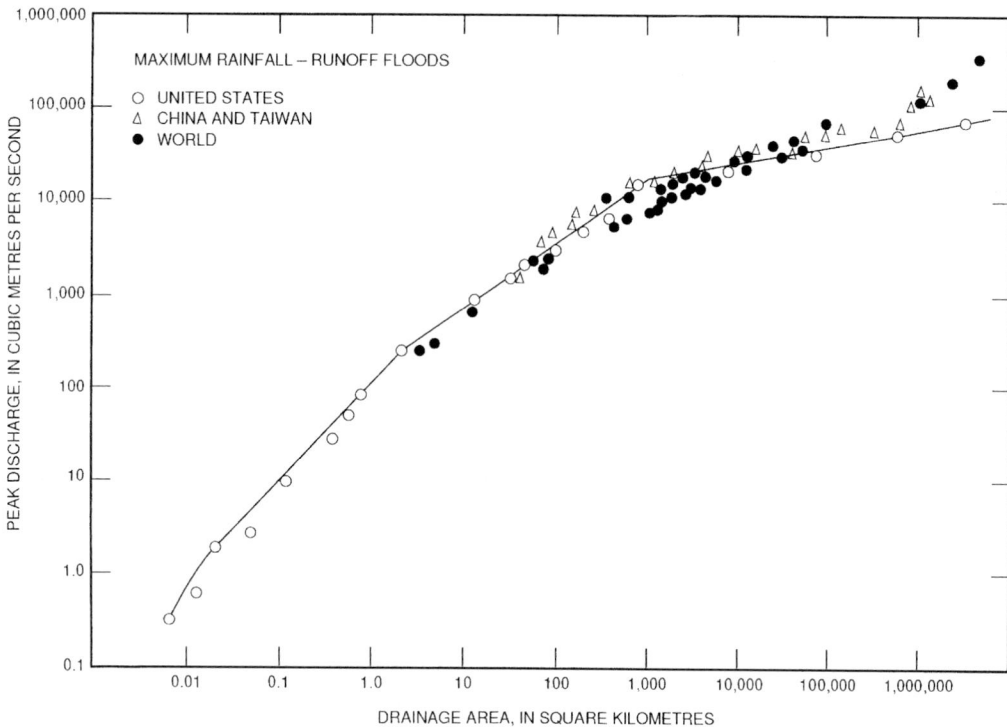

Figure F9 Envelope curve of the maximum rainfall-runoff floods measured in the United States, compared with the maximum rainfall-runoff floods from China and from the world. Data from Tables F9–F11. (After Costa, 1987.)

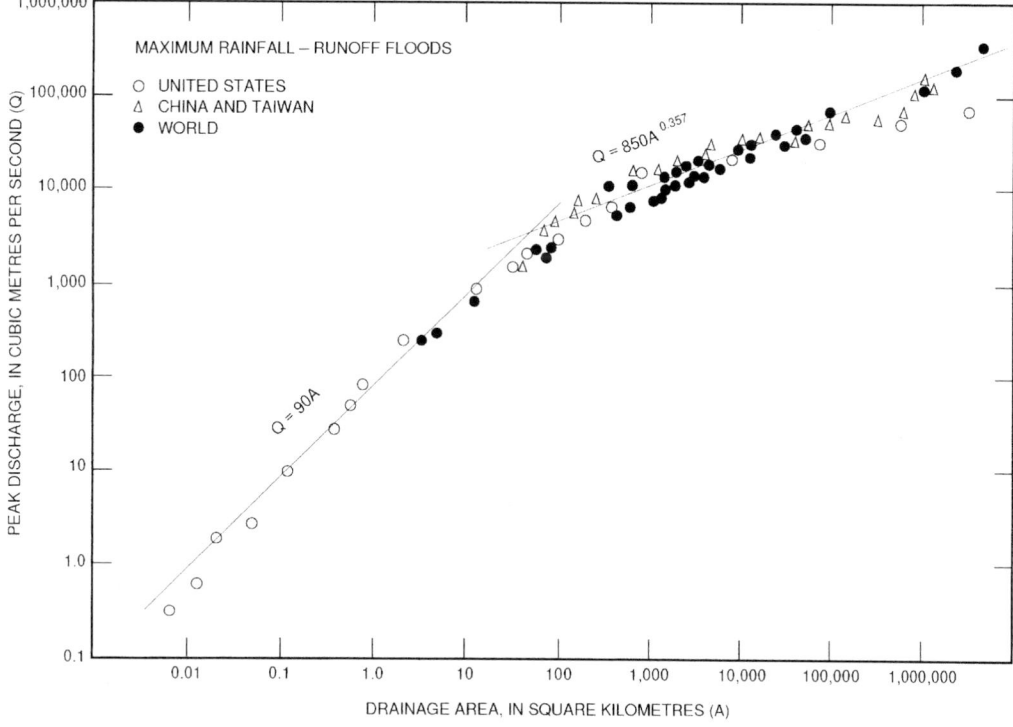

Figure F10 Recession curves drawn through observations.

Table F11 Maximum rainfall–runoff floods in the world (Rodier and Roche, 1984)

Flood no.	Stream	Date	Drainage area (km²)	Discharge (m³ s⁻¹)
1	San Rafael, California, USA	16 January 1973	3.2	250
2	Little San Gorgonio, California, USA	25 February 1969	4.5	311
3	Halawa, Hawaii, USA	4 February 1965	12	762
4	S.F. Wailua, Hawaii, USA	15 April 1963	58	2 470
5	Buey, Cuba	7 October 1963	73	2 060
6	Papenoo, Tahiti (France)	12 April 1983	78	2 200
7	San Bartolo, Mexico	30 September 1976	81	3 000
8	Quinnè, New Caledonia (France)	8 March 1975	143	4 000
9	Quaiéme, New Caledonia (France)	24 December 1981	330	10 400
10	Yaté, New Caledonia (France)	25 December 1981	435	5 700
11	Little Nemaha, Nebraska, USA	9 May 1950	549	6 370
12	Haast, New Zealand	2 December 1979	1 020	7 690
13	Mid. Fork American, California, USA	23 December 1964	1 360	8 780
14	Cithuatlan, Mexico	27 October 1959	1 370	13 500
15	Pioneer, Australia	23 January 1918	1 490	9 840
16	Hualien, Taiwan (China)	1973	1 500	11 900
17	Niyodo, Japan	9 August 1963	1 560	13 510
18	Kiso, Japan	June 1961	1 680	11 150
19	West Nueces, Texas USA	14 June 1935	1 800	15 600
20	Machhu, India	11 August 1979	1 900	15 600
21	Tamshui, Taiwan (China)	11 September 1963	2 110	16 700
22	Shingu, Japan	26 September 1959	2 350	19 025
23	Pedernales, Texas, USA	11 September 1952	2 450	12 500
24	Daeryorggang, North Korea	12 August 1975	3 020	13 500
25	Yoshino, Japan	9 September 1974	3 750	14 470
26	Cagayan, Philippines	1959	4 245	17 550
27	Tone, Japan	15 September 1947	5 110	16 900
28	Nueces, Texas, USA	14 June 1935	5 504	17 400
29	Eel, California, USA	23 December 1964	8 060	21 300
30	Pecos, Texas, USA	28 June 1954	9 100	26 800
31	Betsiboka, Madagascar	4 March 1927	11 800	22 000
32	Toedonggang, North Korea	29 August 1967	12 175	29 000
33	Han, South Korea	18 July 1925	23,880	37 000
34	Jhelum, Pakistan	1929	29 000	31 100
35	Hanjiang, China	1583	41 400	40 000
36	Mangoky, Madagascar	5 February 1933	50 000	38 000
37	Narmada, India	6 September 1970	88 000	69 400
38	Changjiang, China	20 July 1870	1 010 000	110 000
39	Lena, Russia	8 June 1967	2 430 000	189 000
40	Amazonas, Brazil	June 1953	4 640 000	370 000

measured being 10 400 m⁻³ s⁻¹. The rainfall recorded was 1692 mm in 24 h

R.W. Herschy

Bibliography

Costa, J.E., 1987. A comparison of the largest rainfall–runoff floods in the United States with those of the People's Republic of China and the world. *Journal of Hydrology*, **96**, 101–115.
Rodier, J.A. and Roche, M., 1984. World catalogue of maximum observed floods, IAHS, Publ. 143, 354 pp.

Cross references

Floods estimation for developing countries
Flood frequency analysis
Floods
Flood studies worldwide
Floods, river and multi-stage channels
Floods: the world's maximum observed

FLOOD STUDIES FOR THE BRITISH ISLES

At the instigation of the Institution of Civil Engineers and with the financial support of the Natural Environment Research Council (NERC) of the government Department of Education and Science, a special team of hydrologists was set up at the Institute of Hydrology. Working in collaboration with colleagues in the Meteorological Office, the Hydraulics Research Station and the Irish Department of Public Works, they assembled all available precipitation and river flow data in the British Isles and abstracted the relevant information on maximum rainfalls and river peak discharges. From statistical analyses of the data and using well proven hydrological techniques and their developments, they produced a methodology for estimating maximum possible floods and of the flood discharges for selected return periods for all sizes of rivers in the British Isles.

The report of this thorough and rewarding study, *The Flood Studies Report* (FSR), was published in five volumes (NERC, 1975)

Vol. I   *Hydrological Studies* (570 pp.)
Vol. II   *Meteorological Studies* (91 pp.)
Vol. III *Flood Routing Studies* (85 pp.)
Vol. IV   *Hydrological Data* (549 pp.)
Vol. V   *Maps* (24)

The first three volumes are scholarly texts that have been acclaimed internationally and they provide a valued schematic example to hydrologists in the developing countries. Experience in the application of the methods of the FSR by hydrologists and engineers in the UK has led to further studies and some improvements in detail. The five volumes form a basic manual for estimating a design flood and they warrant a place in all design offices concerned with schemes in water engineering. From the experience gained in using the FSR techniques and applying increased data sets, a series of Flood Studies Supplementary Reports (FSSR) have been published by the Institute of Hydrology. Modifications and updating of some of the equations have been made from the FSSRs.

The recommended procedure for evaluating the frequency of a given flood magnitude for a particular river depends on the availability of data and the amount of discharge detail required. A good long record of over 25 years of annual maximum flows would provide, by extreme-value statistical analysis, satisfactory estimates of floods of return periods up to say 500 years. A requirement for the detailed shape of a flood hydrograph from only 2 or 3 years' data, or even none at all, would need estimates to be found by a combination of statistical and deterministic approaches. The main deterministic technique used in the FSR is the unit hydrograph method of relating rainfall to streamflow.

For advising on a design flood flow for a scheme, it is necessary to make estimates of flood magnitudes for a selection of return periods by more than one method and then to compare results. Especially when there are only limited data, gross margins of error can be avoided by comparing design floods evaluated by different techniques. An outline of the various procedures that may be followed according to data availability is described in general terms.

Statistical methods

(a) River flow records over 25 years

The peak flows are abstracted for each year of record, and this annual maximum series is fitted by the extreme-value EVI (Gumbel) distribution. Estimated values of flood discharges, Q_T, for any return period T, can then be obtained. In practice, for river flood protection works, return periods of over 500 years are rarely required.

(b) Records over 25 years and 10–25 years

The mean annual flood, $\bar{Q}$, is estimated from the annual maximum series. Then the required $\bar{Q}_T$ is taken from a $Q/\bar{Q}$ plot against return period for an appropriate region of the UK. Eleven curves were derived from all the assembled records, the regions being defined by comparable records which have been combined together to give common $Q/\bar{Q}$ relationships. For the long (>25 years) period records this estimate of $\bar{Q}_T$ can be compared to the direct estimate from the fitted EVI distribution in (a).

(c) Records of 3–10 years

Estimates of $\bar{Q}$ should be obtained by three methods:

1. Use of the peak-over-threshold (POT) series. This method now needs a more detailed specification of the selected threshold and of independence of peaks. The number of exceedances per year is assumed to follow a Poisson distribution whose parameter λ is estimated by:

$$\hat{\lambda} = M/N$$

where M is the number of exceedances in N years of record.

The peak magnitudes q_i are treated as an exponential distribution to give an estimate of the parameter β by:

$$\hat{\beta} = \bar{q} - \bar{q}_0 = \sum_{I=1}^{M} (q_i - q_0)/M$$

where $\bar{q}$ is the average peak and q_0 is the threshold. Then $\bar{Q}$ may be estimated from:

$$\bar{Q} = q_0 + \hat{\beta}\ln\hat{\lambda} + 0.5772\,\hat{\beta}$$

2. The short period record is extended by taking the monthly maximum flow peaks, deriving regression equations with comparable data from neighboring long-term stations, then using the regression relationships to estimate more peaks at the short-period station from the long-term records. An estimated value of $\bar{Q}$ is then calculated. This method needs care and competence in statistical methods.

3. The value of the mean annual flood, $\bar{Q}$, can also be calculated from a multiplicative equation relating $\bar{Q}$ to several catchment characteristics. The parameters of the equation were obtained by multiple regression between the assembled peak flows and the appropriate characteristics of each catchment with satisfactory records. Thus

for the short-period station, the catchment area particulars are obtained from relevant maps, and the mean annual flood is calculated from an equation of the form.

$$\bar{Q} = 0.0201\ \text{AREA}^{0.94}\ \text{STMFRQ}^{0.27}\ \text{S1085}^{0.16}\ \text{SOIL}^{1.23}$$
$$\text{RSMD}^{1.03}\ (1 + \text{LAKE})^{-0.85}$$

where AREA is in km^2, STMFRQ is stream frequency in junctions per km^2, S1085 is the stream slope between 10 and 85% of length in m km^{-1}, SOIL is an index determined from five soil types, RSMD is the net 1 day 5-year rainfall in mm and LAKE is an index of lake area as proportion of total area.

For detailed explanations and determining the mapped variables, the reader is referred to Vol. I of the *Flood Studies Report* (NERC, 1975).

When a value of $\bar{Q}$ has been determined by averaging the estimates by the three methods, then required values of Q_T are obtained from the appropriate regional curve of $Q/\bar{Q}$.

No records, ungauged catchments

For UK catchments the above regression equation method (c, 3) can be used to derive an estimate of $\bar{Q}$ using the appropriate regional equation for $\bar{Q}$ and the catchment characteristics measured on topographical maps and read from the specially compiled maps of soil and rainfall indices in the FSR. Then the Q_T values are taken from the corresponding regional curve of $Q/\bar{Q}$. However, before deciding on a design storm for a flood protection scheme, a river gauging station should be established at a suitable site so that some records may be obtained both to add confidence to this crude $\bar{Q}$ estimate and to enable alternative Q_T values to be determined by the unit hydrograph method.

Elizabeth M. Shaw

Bibliography

Natural Environment Research Council (NERC), 1975. *Flood Studies Report*. In five volumes.

Shaw, E.M., 1994. *Hydrology in Practice*, Chapman & Hall, London.

Cross references

Flood estimation: methods for developing countries
Flood frequency analysis
Flood hazard management
Floods
Floods: largest in USA, China and the world

FLOOD STUDIES WORLDWIDE

Since the publication of the *Flood Studies Report* (FSR), the hydrologists of the UK Institute of Hydrology have applied the FSR techniques to engineering problems in many countries. In evaluating the flood discharges for required return periods, the computational method used is usually determined by the availability of data. Where hydrometric measurements are limited, estimates are obtained by more than one method and engineering judgement is required in deciding on design values to recommend to clients.

Currently, research is being undertaken on the statistical analysis of long series of annual maximum floods from all over the world and, wherever there are sufficient records, regional growth curves are being developed as defined in Flood studies for the British Isles. At present records from 70 countries have been analyzed and it is hoped to expand the study when further data series have been assembled.

The series of annual maximum flood peaks for each station were converted to a non-dimensional form (q) by dividing each peak (Q) by the mean annual flood ($\bar{Q}$) so that comparisons between stations within a region could be made. The analyses of the q values in a region were combined, and regional flood frequency curves obtained by fitting the data to the general extreme-value (GEV) distribution chosen for its flexibility in accommodating the conditions of different climates and terrain.

Table F12 Selected results from European flood studies

	GEV parameters			$q(T)$ values		
	u	δ	k	50	100	50
Czechoslovakia	0.717	0.356	−0.1821	2.74	3.28	4.82
Denmark	0.814	0.302	−0.0368	2.08	2.33	2.92
Hungary and adjacent parts of Yugoslavia	0.793	0.281	−0.1389	2.25	2.61	3.57

The probability of an annual maximum $q_i \leqslant q$ is given by

$$F(q) = \exp\left[-\left(1 - \frac{k(q-u)}{\alpha}\right)^{1/k}\right]$$

with three parameters u, α and k.

The predicted values of q are provided for return periods (T) of 50 100 and 500 years from which Q_T can be determined:

$$Q_T = \bar{Q} \times q_T$$

given that a mean annual flood value is available. Table F12 gives results from three European Countries (see also Flood estimation: methods for developing countries).

The shapes of the growth curves are influenced by climate and by catchment area and within each region local differences would be expected to produce divergent results. The present results provide only a guideline for planning and pre-feasibility studies at ungauged sites or in areas deficient in good-quality data. More confident results are expected from further studies.

Elizabeth M. Shaw

Source

Shaw, E.M., 1994. *Hydrology in Practice*, Chapman & Hall, London.

Cross references

Flood estimation: methods for developing countries
Flood frequency analysis
Flood hazard management
Floods
Floods: largest in USA, China and the world
Flood studies for the British Isles
Floods, rivers and multi-stage channels
Floods: the world's maximum observed

FLOODS, RIVERS AND MULTI-STAGE CHANNELS

Flooding: an overview

Society's perception is that the number of river flooding events appears to have increased in recent years. This perception is ill founded and is probably due to the more extended coverage given to flooding by the media nowadays. In fact, it is impossible to prove conclusively whether the amount of rainfall has actually increased over the years. Some regions report greater rainfall than in previous years and others less. What is clear, however, is that urbanization itself has exacerbated river flooding problems. Humans exhibit a clear tendency to reside on floodplains, adjacent to rivers. The problem with this custom is that the floodplain is an integral part of the river, which typically experiences inundation once or twice per year. Undaunted by this fact, in many instances, we have actually built our cities and towns in the middle of these rivers. Such a strategy assumes that we are able to protect ourselves from these floods when they occur and a myriad of solutions have been proposed, mainly by engineers, with mixed success. The floods of the Mississippi (1993) and in Bangladesh (1989) are a testament to the fact that complete protection from river floods has not yet been achieved, despite expenditure in terms of billions of dollars.

A very contentious proposal at this stage of our history is that river engineers, when using traditional methods to address problems of river flooding, have on balance made matters worse! Not only can

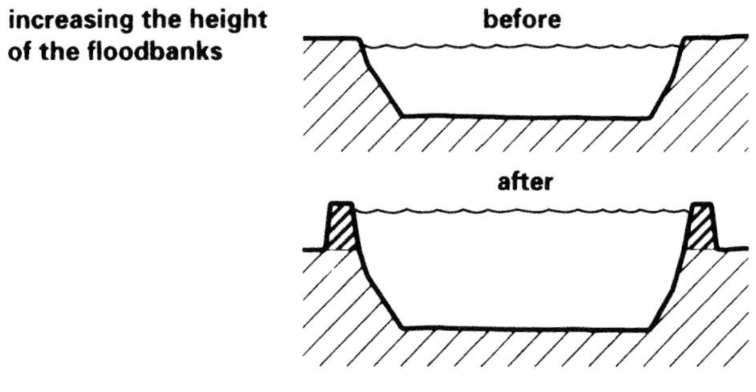

Figure F11 Flood banks at the edge of a river (From RSPB, 1994.)

certain flood defence measures offer little or no extra flood defence, but they can even contribute to destroying river habitats and the river environment. Often, according to Purseglove (1989), natural rivers have been engineered into 'more efficient drains', with very little consideration given to the resulting environmental impact.

Engineers who in the past were preoccupied with simply 'draining away' flood waters as quickly as possible from a particular site often just transferred even greater flood discharges downstream. These downstream areas, unaltered by any engineering works but receiving increased volumes of flood water, tended to suffer much worse flood damage than before. These simple draining techniques include resectioning, altering the cross-sectional shape of an existing river channel; realignment, changing the course of a river and bank erection; and building flood banks adjacent to the river's main channel to contain any rise in water levels. Some of the disadvantages associated with these traditional methods of river management are listed below.

- Resectioning. In upland rivers, resectioning can cause instability as the river attempts to re-establish its equilibrium in terms of width, depth and slope. Sometimes, in lowland rivers, increased amounts of sediment will result which can impair capacity without dredging. Instream and bankside species of flora and fauna may also be destroyed.
- Realignment. In upland and lowland rivers, realignment can cause severe instability, resulting in bed scour and deposition. If the river is straightened, then areas of slow-flowing water (often located at the knuckles of the bends in a meandering section) will be lost, destroying the habitat of various species of flora and fauna.
- Bank erection. Figure F11 illustrates these banks (levees) built at the edge of a river to contain any rise in water levels. A flood would formerly have transferred part of its energy and sediment to adjacent floodplains. The floodplains would have acted as temporary storage areas for the passing flood thus reducing the peak water levels in the river. Following the construction of flood banks, the connection between the river and floodplain is broken, wetlands are slowly destroyed, and the new 'improved' river is left with greater erosive energy. A clear case of humanity working against nature!

A further assessment of these traditional methods may be found in Hey (1994), in which he categorizes the methods of river management and elaborates upon their detrimental effects.

Environmentally sensitive river management

For many years, the precepts underlying the engineering world's river management strategy have been predominantly concerned with controlling and utilizing the water available from any particular river system. In recent years, these narrowly focused objectives have been measurably tempered by society's increased awareness and resulting

section x - y

before

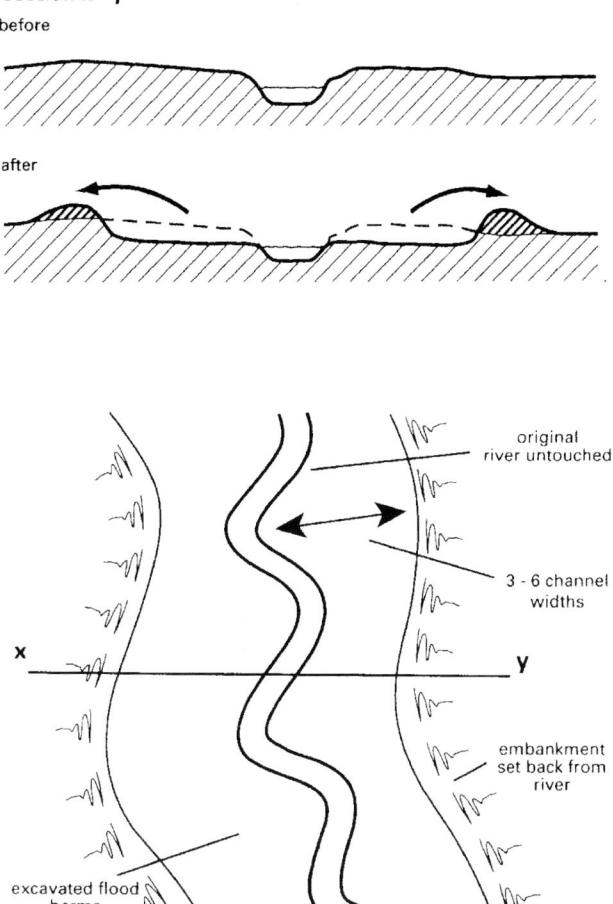

Figure F12 A typical two-stage channel (From RSPB, 1994.)

the natural geomorphology and floral and faunal diversity of a river system.

Currently, one of the better options available to limit the environmental impact of flood alleviation methods, whilst achieving maximized hydraulic performance is the multistage channel. Further corroboration of this opinion can be found in Hey *et al.* (1994) and Reeve and Bettess (1990).

These multistage channels are created when an artificial limit is imposed on a floodplain by the building of new flood banks which are often set about three to six channel widths from the river, as shown in Figure F12. There will, of course, still be situations in which rectilinear channel design will be the most appropriate solution, mainly when there is a limitation of space. However, even if there is space limitation, partial use of a multistage channel may still prove to be the optimal solution.

Sellin and Giles (1988) describe a successful project, in which the detailed stages from the preliminary model studies to the final full-scale implementation of a multistage channel are chronicled. Over a period of 6 years, Sellin's team from Bristol University studied the River Roding, Essex. The final alignment is shown in Figure F13. They used scale models to establish the optimal design for a multi-stage channel proposed to alleviate flooding in this river. They compared the hydraulic performance of their scale models and the full-scale construction and even gave details about the habitat and ecology of the final development, including a discussion about how the management of the floodplain (for example, cutting the grass) affected the performance of the channel. Their reporting of the many stages of the study forms an excellent guide for any similar projects in the future. However, the need to build physical models to enable the design of multistage channels will hopefully become less important with the advent of experimentally and theoretically derived calculation procedures which are intended to be sufficient to enable the design of these multistage channels. The viability of the calculation procedures currently available is assessed later in this article. It is important to note that the final design for the River Roding utilized a two-stage channel (a specific type of multistage channel). The distinction between two-stage and multistage channels is explained in the next section.

Multistage and two-stage channels

Sellin and Giles (1988) divide multistage channels into two distinct types.

- Retained floodplain channels are those in which the natural floodplain is retained as the upper channel with the flood banks constructed somewhere on the floodplain (Figure F14).
- Bermed floodplain channels are ones which are constructed by excavation of a wholly artificial berm to form the upper channel (this is usually below the natural floodplain level). The flood banks are placed at the extremities of the berms (Figure F14).

demand for the implementation of environmentally sensitive solutions to flood alleviation problems. A tangible dividend has become available, encouraging engineers to consider conservation and enhancement of the environment as one of their important design criteria. The consequence of such a switch in emphasis is that, nowadays, most of the favored solutions to flood alleviation problems tend to maintain

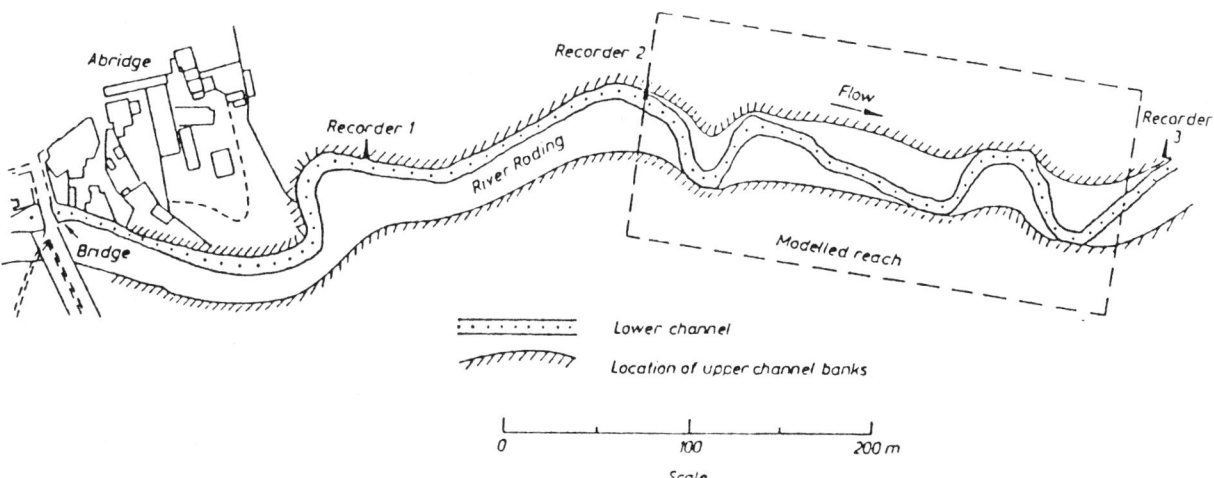

Figure F13 Two-stage channel for the River Loding, Essex. (From Sellin and Giles, 1988.)

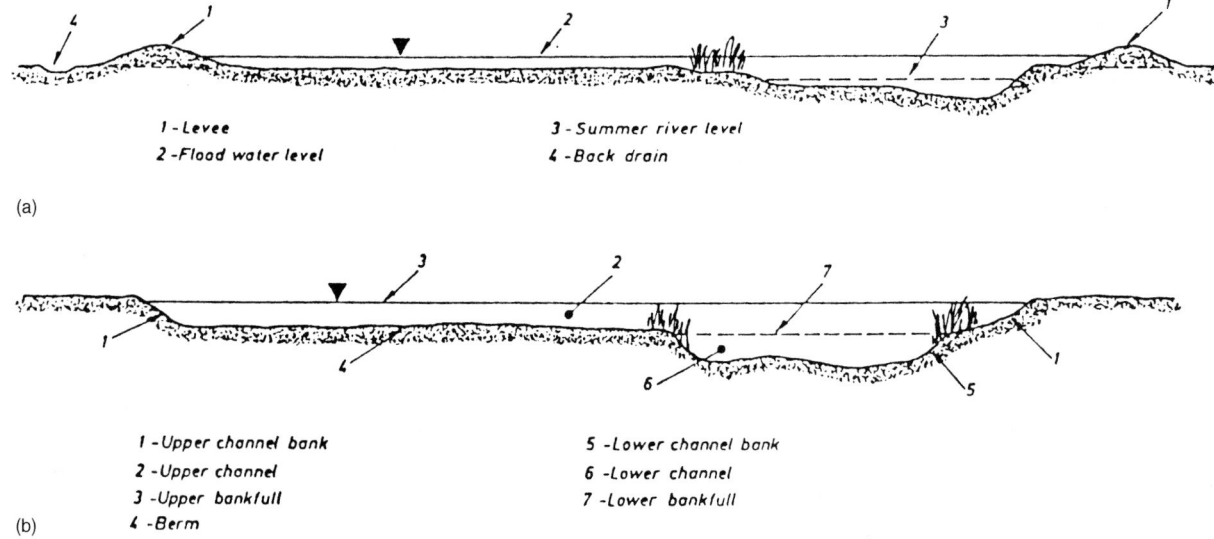

(a)

(b)

Figure F14 Types of multi-stage channel. (From Sellin and Giles, 1988.)

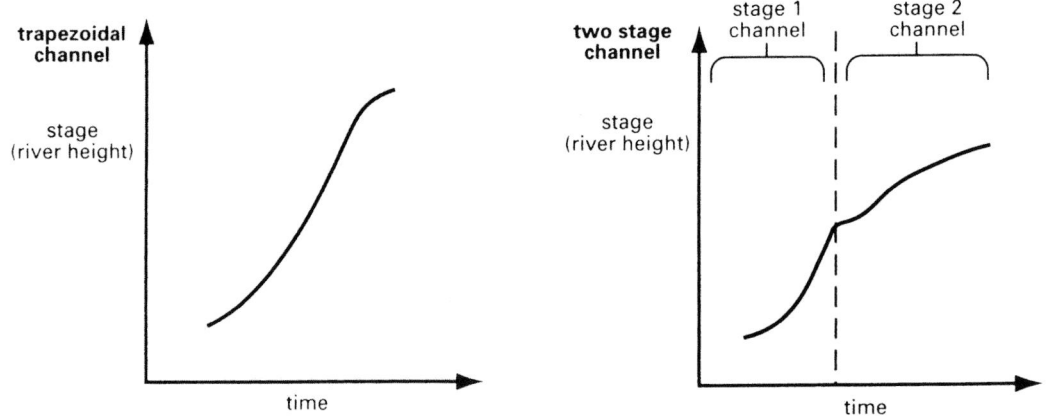

Figure F15 Hydrographs: trapezoid versus two-stage channels. (From RSPB, 1994.)

Retained floodplain channels are undeniably the preferred choice based on environmental considerations. Their natural stability and the maintenance of the endemic instream and bankside habitats is an automatic by-product of the original channel being retained. If the space limitations coupled with the possible difficulties of managing the floodplains still allow the required hydraulic performance to be met, then this type of channel would certainly be the first-choice solution to any flood alleviation problem.

However, with the everyday reality of financial, spatial and demographic restrictions that will be encountered in most design processes, the compromise solution of a multistage bermed channel will perhaps yield more desirable results. It will be easier to predict hydraulic performance for multistage channels because the final channel cross-section will be configured exactly to the designers requirements. Second, it will still be possible to maintain a varied floodplain habitat. The bermed channel solutions have the potential to be more flexible and therefore will be implemented more often than the retained channel options.

These multistage channels, with their excavated berms, may be made up of a number of floodplain levels or tiers. However, for ease of construction, a specific example of a multistage channel is often chosen as the ideal solution to a river management problem; a two-stage channel. This type of channel has floodplain berms, which may be sloping, but have only one tier.

Section x–y in Figure F12 shows the transformation of an original river system into a two-stage channel which will carry lower flows in its original, untouched main channel and with higher flows inundating the berms at more regular intervals. River flooding is now confined to the compound two-stage channel, which is wide enough to cause attenuation of the flood peak (water levels and discharge rates) as shown schematically in Figure F15.

The new river corridor will of course occupy more land than a conventional engineered channel, and therefore will be more problematic to construct in urban areas. The lack of space may prevent a two-stage channel from being chosen, although a berm cut on one side of the river only is an improvement on radical surgery to the original river. An example of this is given in Figure F16 for the case of the River Ray in Oxfordshire, UK.

Conveyance in two-stage channels

The key question needing answering prior to specifying the configuration of a two-stage or multistage channel is: what is the predicted total discharge which can be carried by the compound channel, or alternatively, for a given flood discharge rate, what size and dimensions should a compound channel be?

This is an intriguing question, the answer of which is very complicated to derive. The original river can be defined in terms of cross-sectional shape, bed slope, bed roughness characteristics, bed

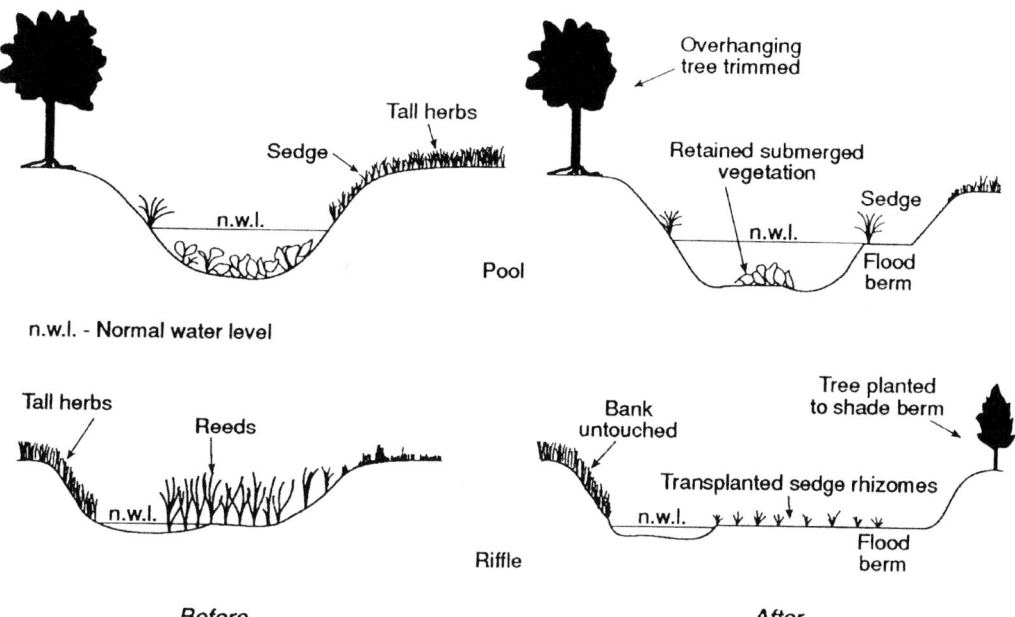

n.w.l. - Normal water level

Before   **After**

Figure F16 River Ray in Oxfordshire. (From Brookes, 1988.)

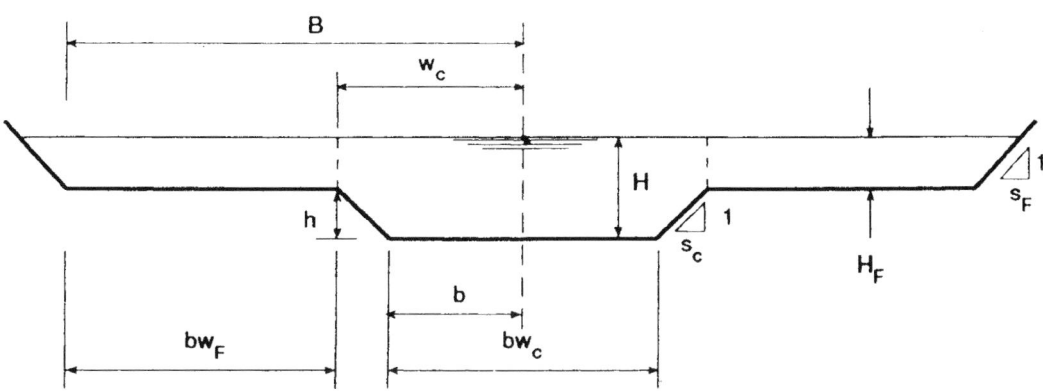

Figure F17 Idealized cross-section of straight two-stage channel. (From Ackers, 1991.)

forms, and bend/meander configuration (plan-form geometry). Meanwhile, the upper channel is likely to be much straighter, to have a steeper bed slope, a wider aspect ratio (width/depth) and a very different bed roughness in the form of grass, trees and hedgerows. Hence we have two separate channels mixing as one. The complexity of the three-dimensional mixing is almost intractable. The typical cross-sectional shape of a two-stage channel is illustrated in Figure F17.

Some of the popular methods used presently to predict the conveyance capacity of meandering channels flowing overbank are presented in the rest of this article. It is interesting to note that currently more research has been completed at Glasgow, Aberdeen and Bristol Universities as part of the British government's EPSRC Series B extension program to investigate further overbank flooding in meandering channels. The final aim of this project was to develop an improved and fully comprehensive design method to predict the conveyance capacity of any meandering channel flowing overbank. The resulting design manual was published in 1997.

Presently, the conventional approach to determine the conveyance capacity of a two-stage channel is to perform separate calculations for the main channel and floodplain flows. This is facilitated through the expedient use of imaginary vertical subdivisions at the junction as shown in Figure F17.

When the bed friction coefficient for the main channel alone and the floodplain alone are determined, then a theoretical discharge can be computed combining both subsections and based on bed friction losses only:

$$Q_{\text{friction}} = A_c\left(\frac{8gR_cS_c}{f_c}\right)^{1/2} + A_f\left(\frac{8gR_fS_f}{f_F}\right)^{1/2} \qquad (F1)$$

The definition of each symbol is given in the notation section.

It is now known that the estimate of discharge from equation (F1) is a gross overestimate of how much a compound cross-section can convey. This is because the intense interaction, mixing and energy losses other than bed friction have been ignored. The interaction between the main channel and floodplain flows is very important and in certain circumstances can reduce the combined compound flow rate to under 50% of that estimated from equation (F1).

Research program into two-stage channels

Hydraulic research into compound channels was born 40–50 years ago. Early investigators discovered the intense interaction between river channel and floodplain flow.

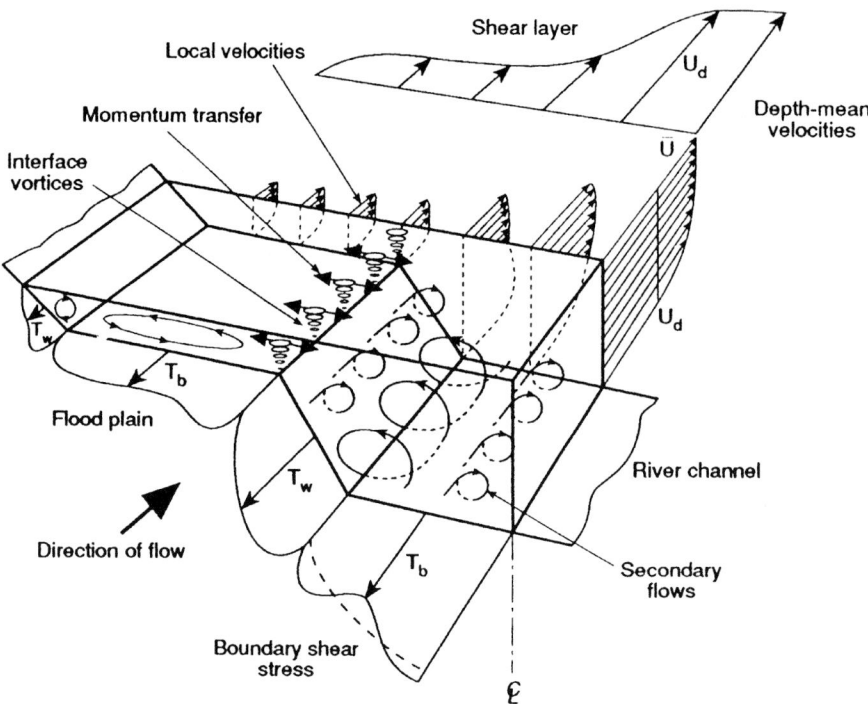

Figure F18 Hydraulic aspects of overbank flow. (From Knight, 1989.)

Large vortices in plan exist at the channel–floodplain junction, these dissipate turbulent energy and transfer momentum from the main channel to the floodplain.

A schematic diagram by Knight (1989) of the same physical processes is given in Figure F18 showing interface vortices, secondary current development in the deeper river section, the distribution of boundary shear stress and depth-averaged velocity distribution across the river channel/floodplain width. A classic shear layer develops.

The physical process is more complex for the meandering compound case, as sketched by Ervine et al. (1993; Figure F19). Floodplain flow approaches the main channel flow in the crossover region between bends. An intense mixing occurs, with part of the floodplain flow entrained down into the main channel, and the remainder crosses over to the opposite floodplain. The shearing process drives a large-scale secondary cell which grows in strength and size approaching the next bend. When this cell arrives at the next bend, its rotational sense is in the opposite direction to bend secondary cells for inbank flows. This cell cannot be sustained. It decays rapidly just beyond the bend and is accompanied by a vigorous expulsion of flow from the main channel out onto the floodplain. For smooth floodplains, the flow expulsion has elements akin to side-weir flow, with raised water levels in the main channel before overspilling, and lower water levels on the floodplain just downstream of the overspilling. The area experiencing the maximum levels of bed shear stress is indicated in Figure F20.

Further investigation of the overall bed shear stress distribution and associated flow mechanisms will enable engineers to develop an understanding of the scouring and deposition patterns in meandering multistage channels.

A major research program into compound flows and two-stage channels has been continuing in the UK since 1985. The Engineering and Physical Sciences Research Council (EPSRC) provided funding for a Flood Channel Facility (FCF) at Hydraulics Research, Wallingford.

The Facility is 56 m long and 10 m wide, carrying discharges in excess of 1 m³ s⁻¹. In the period 1986–1989, research centered on straight and slightly skewed compound flows. A skewed flow is shown in Figure F21. From 1989 to 1992, work moved to meandering compound flows (Figure F22) with two major sinuosities tested, namely 1.374 and 2.04. The final test series (1994–1998) is involved with sediment transport and the stability of two-stage channels.

The third phase is intended to look at the processes of rivers that provide floodplains with sediment, and the geomorphological effects of flows in excess of bank full level.

Discharge estimates for straight compound channels

Several methods have been proposed for calculating stage and discharge in straight compound channels. The first is a simple numerical model developed by Wark et al. (1991), the lateral distribution method, and the second, a hand-calculation methodology developed by Ackers (1991), is simply known as the Ackers method.

Lateral distribution method

The authors of this method start with the Reynolds equations for three-dimensional flow. These can be reduced for the case of straight, uniform, steady flow in depth-averaged form to be

$$gDS_{xf} + \frac{B_s f/q/q}{8D^2} - \frac{d}{dy}\left(\frac{dq}{dy}\right) = 0 \qquad \text{(F2)}$$

Each parameter is defined in the notation section.

The third term in Equation (F2) represents the lateral shear stress in the compound channel. This term can be evaluated using the eddy viscosity concept, determined from a knowledge of lateral eddy viscosity, ν_t. ν_t can be derived by chosing appropriate values for the non-dimensional eddy viscosity (NEV, λ) and by utilizing the following definition:

$$\nu_t = \lambda U_* D \qquad \text{(F3)}$$

Wark et al. (1990) found λ to have mean values between 0.2 to 0.3 for model studies in the Flood Channel Facility, Wallingford, the lower value representing rough floodplains and the higher value smooth floodplains. They recommended, however, a standard value of 0.16 for compound flows based on a fuller analysis of model studies and real river data.

The most significant aspect of this work concerns the fact that when bed friction is known for the deeper river channel and the floodplains/berms, then a lateral distribution of discharge can be simply computed.

A typical result for the River Severn, UK is shown in Figure F23 comparing measured values with those computed from the LDM method. Wark et al. (1991) demonstrated that the LDM method

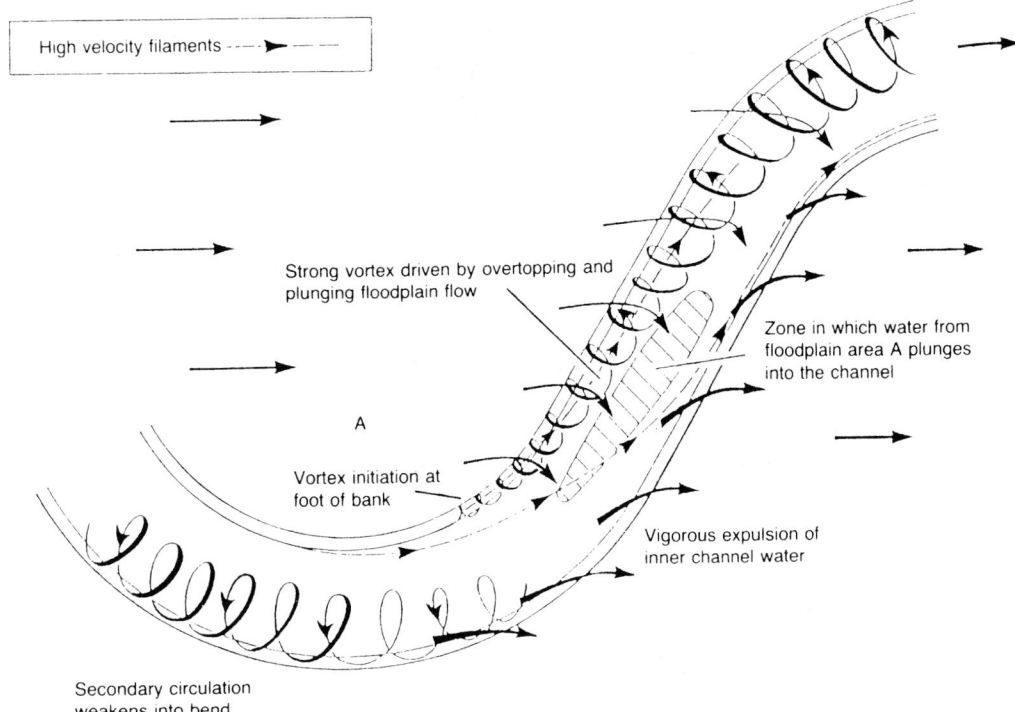

Figure F19 Boundary shear stress in straight two-stage channels. (From Ervine, 1993.)

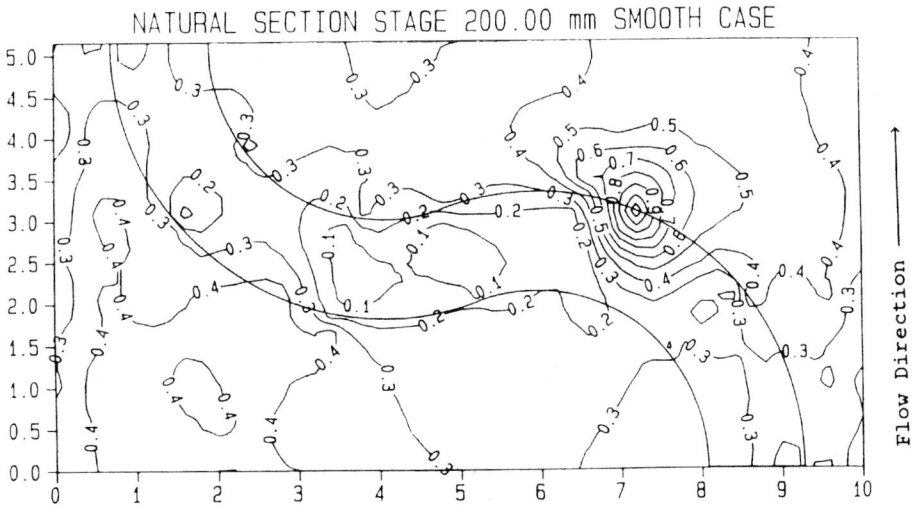

Figure F20 Boundary shear stress in meandering two-stage channels. (From Lorena, 1992.)

performs better than any of the traditional 'sum of segments/zones methods'. However, they also admit that, in common with all the other procedures that are being used currently, accuracy is most dependent upon the choice of values for bed roughness being accurate.

Ackers method (1991)

Ackers developed a method of computing discharge in a compound channel, when the channel is straight, or skewed at angles up to 10% to the floodplain direction.

The method can be carried out as a series of hand calculations and has been calibrated from data taken from the Flood Channel Facility, Wallingford, UK. The method has also been applied to real river channels with some success.

An idealized compound cross-section is shown in Figure F17 and is subdivided into three zones by vertical subdivisions at the junctions of the main channel and floodplain on either side. The basic discharges in each zone are computed from a knowledge of bed friction only and then corrected for compound flow using the Ackers (1991) equations.

Figure F21 Layout of skew channel flume. (From Sellin, 1990.)

Figure F22 Layout of meandering channel flume. (From Lorena, 1990.)

Ackers identified four main regions of flow behavior depending on the relative compound flow depth. Figure F24 shows a plot of the variation of DISADF (discharge adjustment factor) for a straight compound flow with smooth floodplains. DISADF is essentially the ratio of the actual compound discharge to the discharge estimated from bed friction alone.

The correction factor DISADF varies significantly in regions 1 to 3, until at large relative flow depths DISADF values are equal to the

Stage = 6.087 m AOD

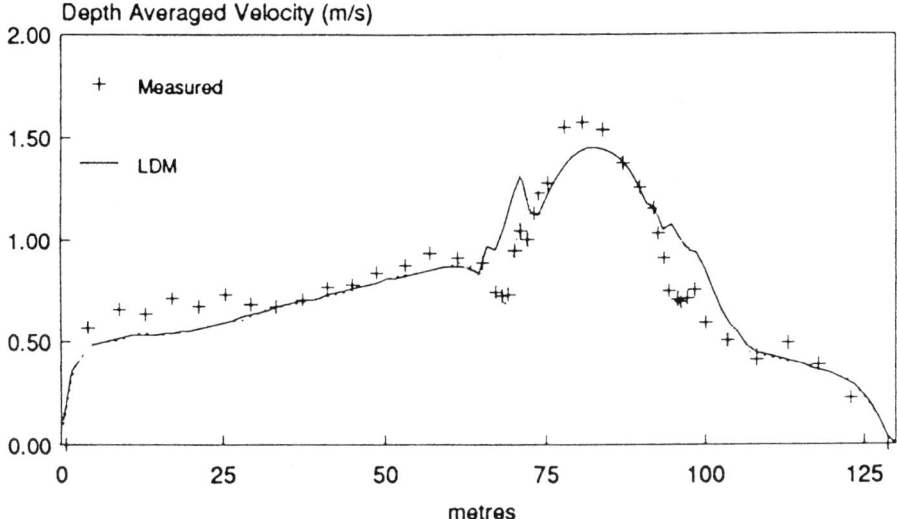

Figure F23 Typical velocity distribution in River Severn. (From Wark, 1991.)

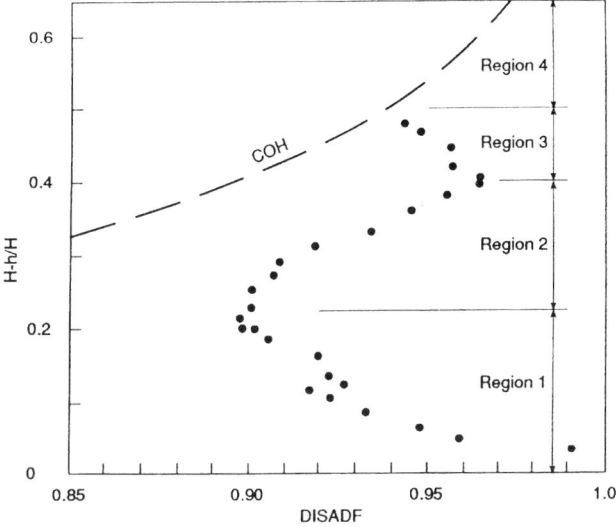

Figure F24 Relative depth versus DISADF for smooth floodplains. (From Ackers, 1991.)

compound channel coherence COH, which is defined by Ackers (1991) and is a measure of the departure of compound flow behaviour from that of a single channel.

A step-by-step design guide to the Ackers method is given in James *et al.* (1994).

Discharge estimates for meandering compound channels

It is clear that most lowland rivers prone to flooding are predominantly sinuous or meandering. A measure of this is known as sinuosity, and is usually defined as the ratio of the total curved length along a river channel divided by the straight direct route along the valley. Sinuosities of less than 1.2 are considered low, whereas those greater than 1.7 or 1.8 are considered very large.

When a meandering or sinuous river overflows its banks, or a two-stage channel is created with a meandering river, the resultant compound flow is complex. Ervine *et al.* (1993) have investigated at least seven parameters which influence the interaction between the flood-plain and main channel flows, and hence the conveyance of the compound cross-section. These parameters include

- the sinuosity of the main river channel;
- the relative flow depth between the floodplains and main channel;
- the relative bed roughness between the floodplains and main channel;
- the width of the meander belt width relative to the total flooded width;
- the lateral slope of the floodplain/berms towards the main channel;
- the aspect ratio, cross-sectional shape and side slopes of the river channel;
- the Reynolds Number, *Re*, or scale of model tests and field tests.

Such complexity creates problems in estimating the actual discharge in a compound cross-section, compared with that calculated accounting for bed friction losses only. These two values can be different by a factor of two, underlining the importance of research into meandering compound channels.

Figure F25 shows a plot of the discharge correction factor F_* needed to compute the actual compound flow rate, compared with that calculated using bed friction only. The correction factor F_* is plotted with the sinuosity of the main river channel and for a range of relative flow depths.

Figure F25 demonstrates clearly that, for high sinuosity compound flows, if the only energy loss mechanism assumed to be acting is that of bed friction, then the actual discharge may be as low as 50% of that predicted using simple 'sum of zones' theory. One alternative method for predicting discharge in meandering compound channels was outlined by James and Wark (1992).

Their method subdivides the compound cross-section into four zones as shown in Fig. F26. Zone 1 is in the main river channel below bankfull level, zone 2 is above bankfull level, extending laterally to the edge of the meander belt width and zone 3 is above bankfull level on either side of the main channel.

The James and Wark (1992) method uses correction factors for each zone and computes the total compound discharge by summing over the four zones in Fig. F26. A step-by-step guide is given in James *et al.* (1994) claiming an accuracy level for this method to within 5%.

Conclusion

There is no doubt that river engineers have made mistakes with river training and flood alleviation. The sensitive environmental option has often been placed near the bottom of a list of options. This is now changing rapidly in the wake of global environmental concern and the gradual destruction of our natural waterways.

One environmentally sensitive method of dealing with river floods is the creation of a multistage channel, or two-stage channel in its simplest form. This encourages compound flow, allows the river channel to meander in its original natural mode, allows some

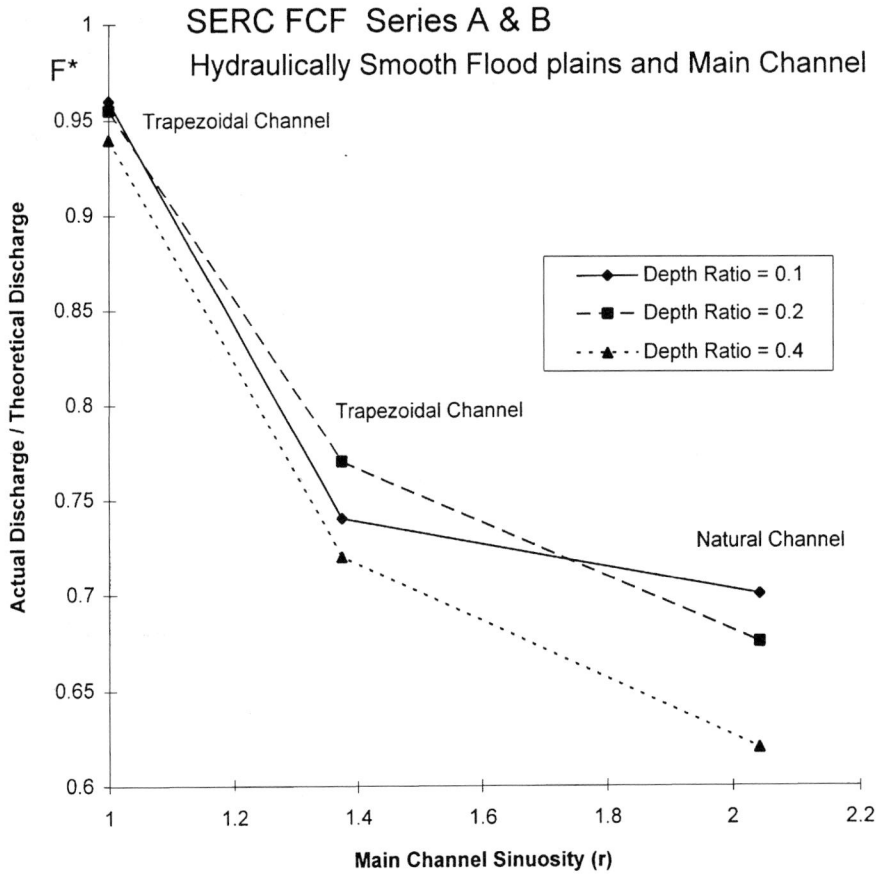

Figure F25 F^* versus sinuosity for multiple depth ratios.

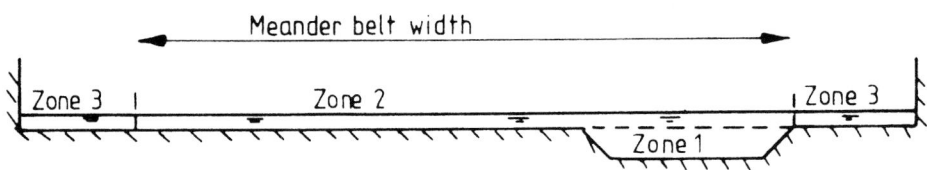

Figure F26 Cross-section subdivision into three zones. (From James and Wark, 1992.)

overbank flow onto side berms and still provides attenuation to flood peaks.

Following a research program in the UK, engineers can now design two-stage channels from the viewpoint of estimating stage and discharge. This can be done with some accuracy for straight, skewed and meandering compound flows, using design methods by Ackers (1991) and James and Wark (1992).

One area of uncertainty over two-stage channels concerns long-term stability and sediment transport matters. A specifically designed two-stage channel may be straight or meandered. If it is straight, and berms on either side have been lowered to form the upper stage, then it is likely that the side berms will be subject to deposition of sediment, especially in areas of lower velocity. It is not known where sediments will be deposited, whether the berms will simply silt up and return to their original level, or what distribution of sediment sizes will return to the floodplains/berms.

For the case of meandering compound flows the outlook is even more uncertain. Figure F20 by Lorena, shows the distribution of boundary shear stress in a meandering two-stage channel. There exists a 'hot-spot' of shear stress activity, producing shear values ten times those in the crossover region of the main channel. Such a distribution of shear is likely to cut out a type of flood relief channel taking almost

a straight line between alternate bends. There is likely to be increased deposition in the center of the floodplain and at the crossover region of the main channel. The two-stage channel may change its configuration, and it may be unstable!

Notation

$Q_{friction}$	Discharge rate (bed friction losses only)
q	Discharge rate per unit width
A_c, A_f	Cross-sectional area of main channel or floodplain
R_c, R_f	Hydraulic radius of main channel or floodplain
g	Gravitational constant
S_c, S_f	Longitudinal bed slope of main channel or floodplain
S_{xf}	Longitudinal friction slope of main channel or floodplain
f_c, f_f	Darcy–Weisbach friction factor of main channel or floodplain
D	Local depth of flow
y	Coordinate direction across flow direction
B_s	Ratio of arbitary sloping area to its horizontal projection
B	Top width of main channel
λ	Non-dimensional eddy viscosity (NEV)

U_* Local shear velocity
v_t Lateral eddy viscosity

D.A. Ervine and A.B. MacLeod

Bibliography

Ackers, P., 1991. *The hydraulic design of straight compound channels.* Report SR 281, Hydraulics Research, Wallingford, UK.

Brookes, A., 1988. *Channelized Rivers: Perspectives for environmental management.* John Wiley and Sons, Chichester.

Elliot, S.C. and Sellin, R.H.J., 1990. SERC flood channel facility skewed flow experiments. *J. Hyd. Res. IAHR,* **28**(2), 127–214.

Ervine, D.A., Willetts, B.B., Sellin, R.H.J., Lorena, 1993. Factors affecting conveyance in meandering compound flows. *Proc. ASCE, J. Hyd. Eng.,* **119**(12).

Ervine, D.A., Sellin, R.H.J. and Willetts, B.B., 1994. Large flow structures in meandering compound channels, in *Proc. 2nd Int. Conf. on River Flood Hydraulics* (eds W.R. White and J. Watts). John Wiley & Sons, Chichester, pp. 459–470.

Hey, R.D., 1994. *Environmentally sensitive River Engineering.* The River Handbook II, Blackwell, Oxford.

James, C.S. and Wark, J.B., 1992. *Hydraulics manual for meandering compound channels.* Final Report EX2006, Hydraulics Research, Wallingford, UK.

James, C.S., Wark, J.B. and Ackers, P., 1994. *Design of straight and meandering compound channels,* NRA R&D Report 13, RR, Wallingford, UK.

Knight, D.W., 1989. Hydraulics of flood channels, in *Floods; Hydrological, Sedimentological and Geomorphological Implications* (eds K Beven and P Carling). John Wiley and Sons, Chichester.

Purseglove, J., 1989. *Taming the Flood. A history and natural history of rivers and wetlands.* Oxford University Press (with Channel 4 Television Co.), Oxford, UK.

Reeve, C.E. and Bettess, R., 1990. Hydraulic performance of environmentally acceptable channels, in *Proc. Int. Conf. on River Flood Hydraulics,* Wallingford, UK, John Wiley and Sons, Chichester, pp. 279–287.

RSPB, NRA and RSNC, 1994. *The New Rivers and Wildlife handbook.* Royal Society for the Protection of Birds.

Sellin, R.H.J. and Giles, A., 1988. *Two stage channel flow.* University of Bristol, Department of Civil Engineering.

Wark, J.B., Samuels, P.G. and Ervine, D.A., 1990. A practical method of estimating velocity and discharge in compound channels in *Proc. Int. Conf. on River Flood Hydraulics,* Wallingford, UK. John Wiley and Sons, Chichester, pp. 163–172.

Wark, J.B., Slade, J.E. and Ramsbottom, D.M., 1991. *Flood discharge assessment by the lateral distribution method,* Report SR277, Hydraulics Research, Wallingford, UK.

Cross references

Flood estimation: methods for developing countries
Flood frequency analysis
Flood hazard management
Floods
Floods: largest in USA, China and the world
Flood studies worldwide
Floods, world's maximum observed

FLOODS: WORLD'S MAXIMUM OBSERVED

The knowledge of exceptionally large floods is essential to the hydrological and civil engineering solution to many problems in water resources management. In 1984 Rodier and Roche published their *World Catalogue of Maximum Observed Floods* (IAHS, 1984). The work was a contribution to the International Hydrological Programme of UNESCO and involved the collection of flood information from 95 countries. The catalogue contains 1400 large floods of which 38 of the world's maximum floods, in relation to catchment, are plotted in Figure F27. The straight-line relation gives an equation:

$$Q = 500A^{0.43}$$

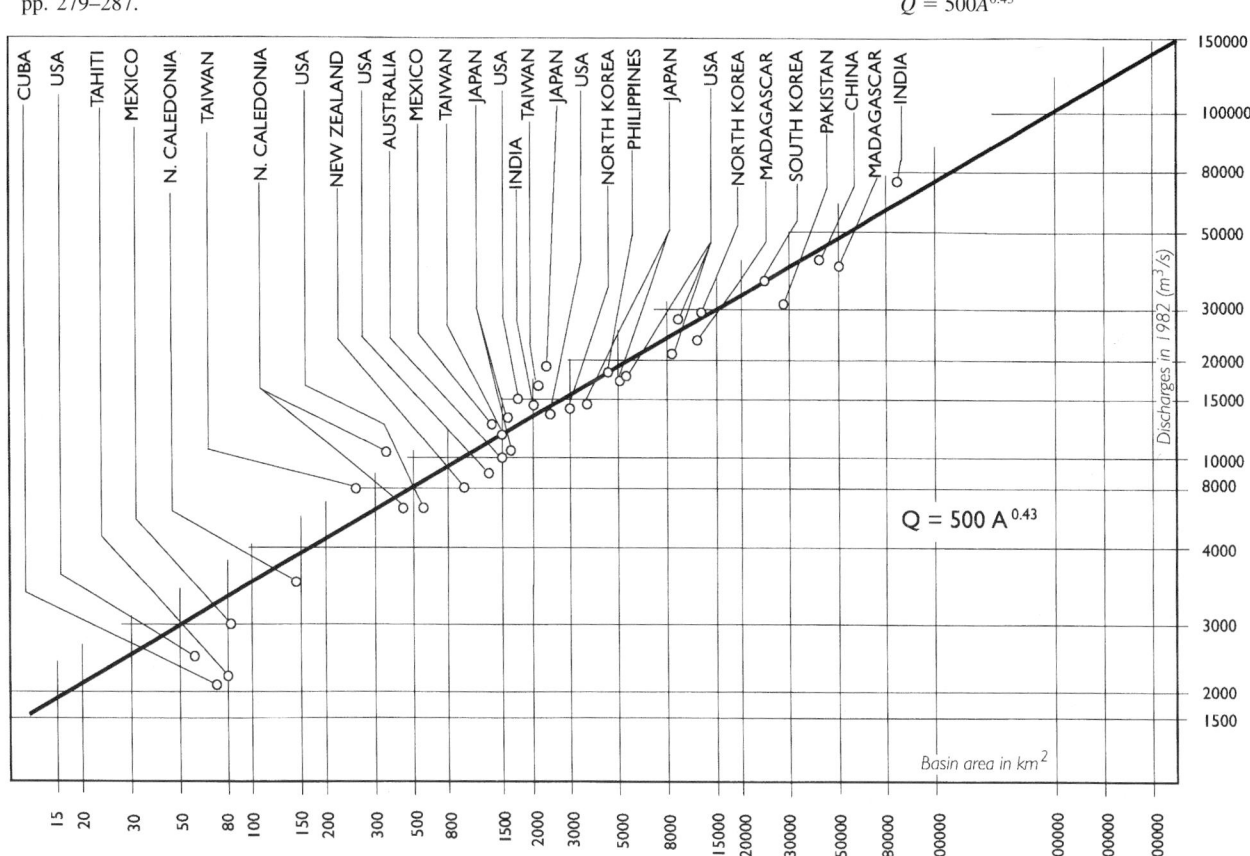

Figure F27 Maximum flood discharges in the world plotted against catchment area on log paper. (After Rodier and Roche, 1984.)

where Q is the discharge in $m^3 s^{-1}$ and A is the catchment area in km^2.

The highest discharges ever measured, with catchments, are:

- The Amazon at Obidos: 370 000 $m^3 s^{-1}$ in 1953 (4 640 000 km^2);
  250 000 $m^3 s^{-1}$ in 1963;
  250 000 $m^3 s^{-1}$ in 1972 (moving boat measurement, not included in catalog);
  239 600 $m^3 s^{-1}$ in 1976;
- The Lena at Kusur: 189 000 $m^3 s^{-1}$ in 1967 (2 430 000 km^2);
- The Yangtze at Yitchang: 110 000 $m^3 s^{-1}$ in 1870 (1 010 000 km^2).

The oldest maximum flood records are:

1342	The Main at Frankfurt Osthaven	(4 000 $m^3 s^{-1}$)	(24 765 km^2)
1501	The Danube at Vienna	(14 000 $m^3 s^{-1}$)	(101 700 km^2)
1583	The Hanjiang at Ankang	(40 000 $m^3 s^{-1}$)	(41 400 km^2)
1651	The Isère at Grenoble	(2500 $m^3 s^{-1}$)	(5720 km^2)
1658	The Seine at Paris	(2500 $m^3 s^{-1}$)	(44 300 km^2)

and the longest continuous record of discharge is that of the Seine at Paris, starting in 1732.

R.W. Herschy

Bibliography

Rodier, J.A. and Roche, M., 1984. *World Catalogue of Maximum Observed Floods*, IAHS, Wallingford UK.

Cross references

Flood estimation: methods for developing countries
Flood frequency analysis
Flood hazard measurement
Floods: largest in USA, China and the world
Flood studies worldwide
Floods, river and multi-stage channels

FLORIDA BAY: STATUS AND RESTORATION

Florida Bay characteristics

Florida Bay is an estuary located between the southern tip of the Florida Peninsula and the Florida Keys. This estuary is large and shallow, with an area of approximately 2200 km^2 and an average depth of about 1 m. About 80% of Florida Bay's area is within the boundary of Everglades National Park. Fresh water flows into the estuary along its 60 km long northern boundary with the Everglades wetland. Taylor Slough and the C-111 canal, in the eastern Everglades, are the primary sources of this freshwater flow. These waters enter the northeastern Bay through about 20 creeks (McIvor et al., 1994) and also diffusely enter the bay as Everglades sheet flow and perhaps as ground water seepage. Additionally, fresh water may enter the Bay indirectly from the Shark River Slough, which flows through the Everglades into the Gulf of Mexico. Some of this fresh water enters the Bay along its western boundary with the Gulf. Currently, quantitive estimates of fresh water flow entering these major sources are available. During the 1980s, the volume of water flowing into C-111, Taylor Slough and Shark River Sough was about $2 \times 10^8 m^3 year^{-1}$, $1 \times 10^8 m^3 year^{-1}$ and $7 \times 10^8 m^3 year^{-1}$, respectively (Light and Dineen 1994). However, no quantitative estimates of fresh water flowing from these sources into the bay are available.

The circulation patterns of water within the bay and exchange of these waters with the Gulf of Mexico and the Atlantic Ocean are also unknown. Preliminary evidence indicates that there is a net easterly flow of water into the bay from the Gulf and a net southerly flow of water from the bay through passes in the Florida Keys (Smith, 1994; Wang et al., 1994). Water circulation within the bay is highly restricted by a complex maze of mud banks, which are composed of biogenic carbonate. These banks partition the bay into numerous basins and dampen the relatively low tidal energy of the Gulf of Mexico. In the northeastern bay, tidal ranges are only about 1 cm (Wang et al., 1994). Thus circulation appears to be more influenced by wind than tides.

A consequence of the poor flushing of Florida Bay by marine waters is that a wide range of salinity occurs. With the high solar energy of southern Florida, evaporation rates generally exceed rainfall rates (Schomer and Drew, 1982). Thus, during periods of drought, or with insufficient inflow of water from the Everglades, Florida Bay becomes hypersaline. Salinity as high as 70‰ has been measured over the last 30 years (McIvor et al., 1994). In contrast, during periods of peak flow from the Everglades, such as following Tropical Storm Gordon in the fall of 1994, salinities can drop drastically. Between August and November 1994, the salinity dropped from hypersaline levels (35 to 50‰) to a range from 25‰ in the central bay to near zero along the north coast of the bay (unpublished data).

The shallow depths and the carbonate muds of Florida Bay have many ecological consequences. Because of light penetrating through the shallow water column to the bottom of the bay, primary production is dominated by seagrass beds that cover much of the bay (Zieman et al., 1989). These beds not only provide habitat and a food base for fauna, but also stabilize the bay's muddy sediments and remove water column nutrients, minimizing sediment resuspension and phytoplankton growth and thus maintaining water clarity. The chemical characteristics of the bay's carbonate muds also influence the bay's ecological structure and function. Because phosphorus availability is thought to limit pelagic and benthic primary production in Florida Bay (Fourqurean et al., 1992, 1993), and carbonate sediments can bind inorganic phosphorus (DeKanel and Morse, 1978), productivity may be minimized and water clarity maximized by carbonate–phosphorus interactions.

Ecological change and water resources

The history of Florida Bay has been marked by drastic and rapid change. The bay is geologically young; as recently as 5000 years ago, the current area of the bay was above sea level and part of the Everglades (Davies and Cohen, 1989, Wanless et al., 1994). This wetland was flooded by the sea during a period of rapid sea-level rise between 5000 and 3000 years ago (Wanless et al., 1994). Little is known about the ecological history of the bay, but based on remnants of seagrasses within the sediments (Wanless and Tagett, 1989), it is clear that seagrasses have been an important part of the bay throughout its history.

In the twentieth century, the pace of ecological change has accelerated because of human perturbation. One of the most important anthropogenic perturbations has been the construction of canals that have drained much of the original Everglades and diverted large quantities of fresh water away from Florida Bay and towards the Atlantic and Gulf coasts (see Light and Dineen, 1994, for details). By 1931, five major canals diverted water to the sea and the Tamiami Trail, which crossed the Everglades and disrupted water flow, was completed. During the 1950s and 1960s, canals were constructed between the eastern Tamiami Trail and northeastern Florida Bay. These canals provide flood protection and water for the Miami and Homestead areas, but drain water from the eastern Everglades toward Biscayne Bay.

Other major anthropogenic perturbations of the Florida Bay ecosystem that occurred in this century include construction of the Flagler railroad through the Florida Keys, locally increased nutrient input from the development of the Keys, regionally increased input of nutrients from the west Florida coast to the Gulf of Mexico and eventually to Florida Bay (Lapointe et al., 1994), and increased fishing pressure. Based on changes in the stable isotope record in Florida Bay corals near the Keys, Swart et al. (1994) have hypothesized that the Flagler railway construction may have been particularly important. During construction, passes between the islands were filled and water exchange between the bay and the Florida Straits decreased. With decreased flushing, organic matter and nutrients have accumulated, resulting in the increased eutrophication of the bay.

Coincident with these human activities, the ecological structure of Florida Bay appears to be changing. The most dramatic change has been the mass mortality of seagrass in the bay (Robblee et al., 1991) that has occurred since 1987. The area affected by this die-off was initially reported to be 40 km^2, but more recent observations indicate that as much as 400 km^2, or about 25% of the Bay's seagrass coverage in the early 1980s (Zieman et al., 1989), has been affected (McIvor et al., 1994). After seagrasses started dying, Florida Bay waters became more turbid. This turbidity has been due not only to increased sediment resuspension because of the absence of seagrass root binding, but also to phytoplankton blooms that have been sustained since 1992 (unpublished data). Other ecological changes include declines in the catch of pink shrimp, lobsters and finfish, as well as the mass mortality of sponges (McIvor et al., 1994).

The cause of these changes is uncertain. Several hypotheses have been presented regarding the cause of the seagrass die-off. These include the physiological stress associated with periods of elevated salinity and high-temperature, sulfide toxicity, a pathogenic infection and anthropogenic eutrophication leading to decreased light availability for seagrasses (Robblee et al., 1991; Durako et al., 1993; Carlson et al., 1994; Durako and Kuss, 1994, Tomasko and Lapointe, 1994; Zieman et al., 1994). One possible scenario, presented by Zieman et al. (1994), is that the diversion of fresh water from the Bay stabilized the salinity regime of the Bay at more marine to hypersaline levels. These conditions favored the growth of turtle grass (Thalassia testudinum), which grew to an unsustainable biomass prior to the die-off. Several factors then probably triggered the crash of this species in the bay (Durako et al. 1993).

The cause of phytoplankton blooms in Florida Bay likewise is uncertain. While nutrients from decomposing seagrass certainly have helped to sustain these blooms, the 4-year delay between the maximum seagrass mortality and bloom initiation indicates that more complex ecosystem processes may be involved. Blooms potentially may be stimulated by increased export of nutrients from the Everglades and mangrove zone adjacent to the bay during recent years, which had high rainfall. Blooms also may be stimulated by sedimentary nutrients if the phosphorus binding capacity of the carbonate mud has decreased. Mechanisms that could affect phosphorus–carbonate interactions include the coating of carbonate particles by dissolved organic matter (McGlathery et al. 1992), perhaps from decaying seagrass, and changes in ionic strength coincident with changing freshwater inflow.

In order successfully to restore the Florida Bay ecosystem, there must be a more complete understanding of the historical characteristics and variability of the bay, the extent to which current characteristics differ from historical characteristics, and the mechanisms that caused these changes. Current Florida Bay restoration plans, as mandated recently by 1994 Florida State law, entail modifying the quantity, distribution, timing and quality of water deliveries to Florida Bay. Nevertheless, the link between past water management practices in South Florida and ecological change in Florida Bay is unclear. Environmental research must elucidate the nature of this link and the nature of other interacting causes of change, in order effectively and efficiently to guide environmental management.

David T. Rudnick

Bibliography

Carlson, P.R., L.A. Yarbro and T.R. Barber, 1994. Relationship of sediment sulfide to mortality of Thalassia testudinum in Florida Bay. Bull. Mar. Sci., 54, 733–74.
Davies, T.D. and A.D. Cohen, 1989. Composition and significance of the peat deposits of Florida Bay. Bull. Mar. Sci., 44, 387–398.
DeKanel, J. and J.W. Morse, 1978. The chemistry of orthophosphate uptake from seawater on to calcite and aragonite. Geochim. Cosmochim. Acta, 42, 1335–1340.
Durako, M.J. and K.M. Kuss, 1994. Effects of Labyrinthula infection on the photosynthetic capacity of Thalassia testudinum. Bull. Mar. Sci., 54, 727–732.
Durako, M.J., T.R. Barber, J.B.C. Bugden et al., 1993. Seagrass die-off in Florida Bay, in Proceedings of the 1992 Gulf of Mexico Symposium, F.J. Webb (ed.), US EPA, Tarpon Springs, FL, pp. 14–15.
Fourqurean, J.W., J.C. Zieman and G.V.N. Powell, 1992. Phosphorus limitation of primary production in Florida Bay: evidence from the C:N:P ratios of the dominant seagrass Thalassia testudinum. Limnol. Oceanogr., 37, 162–171.
Fourqurean, J.W., R.D. Jones and J.C. Zieman, 1993. Processes influencing water column nutrient characteristics and phosphorus limitation of phytoplankton biomass in Florida Bay, FL, USA: inferences from spatial distributions. Est. Coast. Shelf Sci., 36, 295–314.
Lapointe, B.E., D.A. Tomasko and W.R. Matzie, 1994. Eutrophication and trophic state classification of seagrass communities in the Florida Keys. Bull. Mar. Sci., 54, 696–717.
Light, S.S. and J.W. Dineen, 1994. Water control in the Everglades: a historical perspective, in Everglades: The Ecosystem and Its Restoration, S.M. Davis and J.C. Ogden (eds), St Lucie Press, Delray Beach, FL, pp. 47–84.
McGlathery, K.J., R.W. Howarth and R. Marino, 1992. Nutrient limitation of the macroalga. Penicillus capitatus, associated with subtropical seagrass meadows in Bermuda. Estuaries 15, 18–25.
McIvor, C.C., J.A. Ley and R.D. Bjork, 1994. Changes in freshwater inflow from the Everglades to Florida Bay including effects on biota and biotic processes: a review. in Everglades: The Ecosystem and Its Restoration, S.M. Davies and J.C. Ogden (eds), St Lucie Press, Delray Beach, FL, pp. 117–146.
Robblee, M.B., T.R. Barber, P.R. Carlson, Jr et al., 1991. Mass mortality of the tropical seagrass Thalassia testudinum in Florida Bay (USA). Mar. Ecol. Prog. Ser., 71, 297–299.
Schomer, N.S. and R.D. Drew, 1982. An ecological characterization of the lower Everglades, Florida Bay, and the Florida Keys. US Fish and Wildlife Service, Office of Biological Services, Washington, DC, WS/OBS-82/58.1, 246 pp.
Smith, N.P., 1994. Long-term Gulf-to-Atlantic transport through tidal channels in the Florida Keys. Bull. Mar. Sci., 54, 602–609.
Swart, P.K., P. Kramer, J.J. Leder et al., 1994. A 120 year record of natural and anthropogenic variations in Florida Bay based on oxygen and carbon isotopic variations in a coral Solenastrea bornoni. Bull. Mar. Sci., 54, 1085.
Tomasko, D.A. and B.E. Lapointe, 1994. An alternative hypothesis for Florida Bay seagrass die-off. Bull. Mar. Sci., 54, 1086.
Wang, J.D., J. van de Kreeke, N. Krishnan and D. Smith, 1994. Wind and tide response in Florida Bay. Bull. Mar. Sci., 54, 579–601.
Wanless, H.R. and M.G. Tagett, 1989. Origin, growth, and evolution of carbonate mudbanks in Florida Bay. Bull. Mar. Sci. 44, 454–489.
Wanless, H.R., R.W. Parkinson and L.P. Tedesco, 1994. Sea level control on stability of Everglades wetlands, in Everglades: The Ecosystem and Its Restoration, S.M. Davis and J.C. Ogden (eds), St Lucie Press, Delray Beach, FL, pp. 199–223.
Zieman, J.C., J.W. Fourqurean and R.L. Iverson, 1989. Distribution, abundance and productivity of seagrasses and macroalgae in Florida Bay. Bull. Mar. Sci., 44, 292–311.
Zieman, J.C., R. Davis, J.W. Fourqurean and M.B. Robblee, 1994. The role of climate in Florida Bay seagrass dieoff. Bull. Mar. Sci., 54, 1088.

Cross reference

Okeechobee Lake, Florida, USA: human impacts, research and lake restoration

FLOW MEASUREMENT: NEW TECHNOLOGY

Introduction

Volumetric flow measurement is an important process factor in the water industries. Data is required by public and private bodies for various reasons such as resource planning, abstraction/discharge control, storm flow control and control of flow to treatment works.

Different methods are used throughout the water and waste water industries to measure flow. Three methods are discussed utilizing non-intrusive velocity–area (V–A) techniques:

doppler ultrasonics;
transit time ultrasonics;
electromagnetic technology.

The applicability of these techniques to different flow measurement problems is illustrated from the author's experience of using particular instruments of each type in a variety of applications in the water and waste industries.

V–A methods may be defined as the simultaneous measurement of velocity and the cross-sectional area of the flowing water to derive flow by multiplying the two. In an open channel or a part-filled pipe, measurement of cross-sectional area usually reduces to measuring depth.

Ultrasonic doppler technology

Principle of operation

The method is based on the principle named after Christian Doppler who discovered the phenomenon in 1843. Sound reflected from a

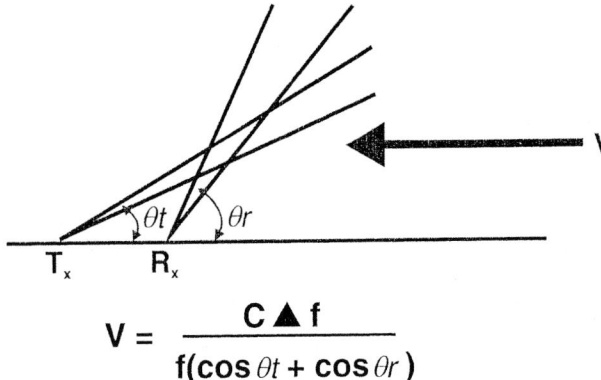

$$V = \frac{C \blacktriangle f}{f(\cos \theta t + \cos \theta r)}$$

Figure F28 The Doppler principle.

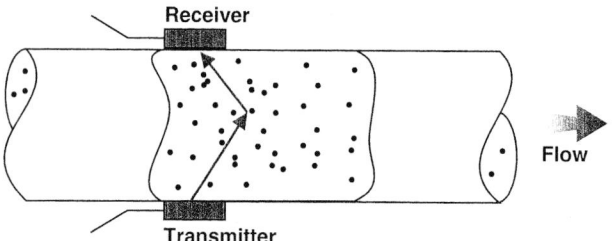

Figure F29 The Doppler technique in a closed conduit.

moving object is perceived as being of a different frequency by a stationary observer. If the reflector is moving towards the observer the frequency becomes higher and vice versa. The principle is used for the measurement of fluid flow where the source of sound and the receiver (the observer) are ultrasonic transducers and the reflectors are particles being carried in the moving fluid.

The transmitter and receiver have conical fields and particle velocity data is obtained from the volume of overlap.

The frequency shift

$$df = v(\cos \theta_t + \cos \theta_r)f/c$$

where θ_t is the angle of the transmission cone to the flow direction, θ_r that of the receiver cone and f and c are the frequency and velocity of sound, respectively (Figure F28).

Full closed pipes
In full closed pipes the usual practice is to locate the transducers on either side of the pipe as in Figure F29. In some instruments they are actually clamped on the outside of the pipe, the sound being transmitted through the walls. The velocity profile is fairly predictable provided certain installation constraints are observed and the velocity measured in the overlapping volume bears a known relationship to the mean velocity. The cross-sectional area is fixed since the pipe is full, so the velocity measurement is proportional to flow.

Open channels
In open channels the limited sensing volume produces errors in relating the measured velocity to the mean because of the variations in velocity profile. If the transmitter and receiver are brought together, the sensing volume extends throughout the liquid to the limit of penetration of the beams (Figure F30).

A major limitation of first-generation open channel Doppler flowmeters is that signals from particles close to the probe are of greater amplitude and dominate the velocity spectrum, causing a bias which is dependent on particle density.

More modern designs overcome this defect to a large extent by two methods. One of these removes the amplitude component of the signals, e.g. by spectral analysis. Provided the returned signal is detectable there is no bias towards stronger signals from local reflectors. The Peek Mainstream is an example of an instrument of this type. The other method looks at signals received at certain times

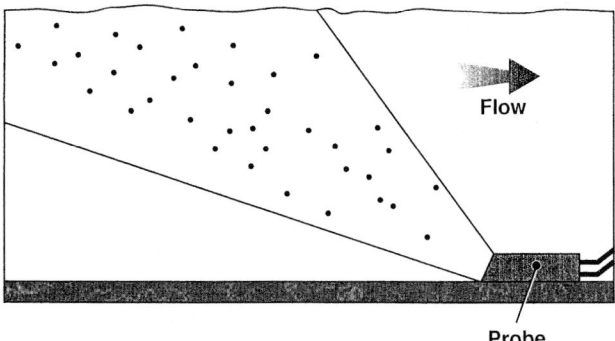

Figure F30 The Doppler technique in an open channel.

after transmission to define how far away the reflectors are which are responsible for them. This method is called range-gated or time-gated Doppler.

Areas of application

The Doppler method is most suitable for channels or pipes up to about 2 m and relies on the presence of reflecting particles. In fact a very low density of particles is sufficient, but the method will not work in perfectly clean water.

Doppler is best suited to liquids which are dirty or are aerated. Some instruments may also be used on sludges with up to about 10% solids. The relatively low cost, low power and ease of installation make this type of equipment particularly well suited to survey work and temporary installations.

Technology limitations

Attention should be paid to the shape of the transmission field relative to the channel dimensions and depth range for open channels. Failure to do so will lead to errors in flow calculation resulting from measurement of velocity which relates poorly to the mean channel velocity. In practice, at the present state of technology this limits the maximum size of channel or pipe to 2–3 m, the minimum length to 5–10 diameters and minimum depth to 50 mm or 10% of the width.

The sensor may be located at the bottom or side of an open channel, but in the case of complex cross-sectional shapes it might be difficult to avoid areas where the beam will not reach.

Doppler equipment is not suitable for very clean liquids. Clamp-on technology may give poor results on certain types of pipe (e.g. copper or copper–nickel alloy) or where the inside becomes coated.

Ultrasonic transit time technology

Principle of operation

Ultrasonic transit time flow meters use one or more pairs of transducers mounted on opposite sides of the pipe or channel. A pulse of ultrasonic sound is transmitted from each of a pair of transducers and received by the opposite transducer along a path at an angle (θ) to the flow direction (Figure F31).

The apparent velocity of sound upstream is less than that downstream by an amount equal to the average water velocity along the path resolved through the angle θ between the path and the flow direction. By measuring the times of flight, the water velocities are calculated by the gauge along the following lines:

$$t_1 = \frac{L}{(c + V_p)} \quad t_2 = \frac{L}{(c - V_p)}$$

where L is the length of ultrasonic path, c is the velocity of sound in water, V_p is the mean water velocity resolved in path direction and, t_1 and t_2 are flight path times.

These equations combine to give

$$V_p = \frac{L}{2}\left(\frac{1}{t_1} - \frac{1}{t_2}\right)$$

which is independent of c. Also

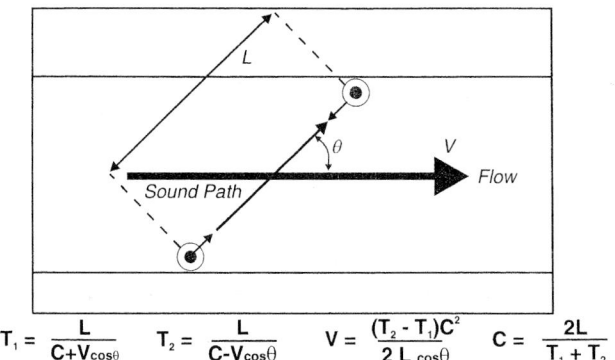

$$T_1 = \frac{L}{C+V\cos\theta} \qquad T_2 = \frac{L}{C-V\cos\theta} \qquad V = \frac{(T_2 - T_1)C^2}{2L\cos\theta} \qquad C = \frac{2L}{T_1 + T_2}$$

Figure F31 Ultrasonic transit time: cross-section.

$$c = \frac{L}{2}\left(\frac{1}{t_1} + \frac{1}{t_2}\right)$$

which is used for depth measurement. Hence the average velocity of the water V_1 is

$$V_1 = \frac{V_p}{\cos\theta}$$

Full closed pipes

In full closed pipes the velocity profile is usually fairly predictable, provided certain installation constraints are observed and the velocity is measured by a single ultrasonic path where velocity is proportional to the mean velocity. The cross-sectional area is fixed since the pipe is full, so the velocity measurement is proportional to flow. The transducers may be mounted internally, or in some instruments clamped on the outside of the pipe, the sound being transmitted through the walls.

In cases where the velocity profile is expected to vary, more than one path is used to give a better estimate of mean velocity. In practice this is not possible for small pipes.

Open channels

For open channels (or when there is a free surface in a part-filled pipe) and the water level varies, the velocity is usually measured in a similar manner at several water levels between the surface and the bottom and the flow computed in these 'slices' of the river. Each 'slice' is of a known thickness provided that a vertical path or ancillary depth gauge is used. The total discharge or volume flow is found by summing these separate measurements. Where the velocity profile is predictable, it is still possible to obtain an accurate flow figure with only one or two paths.

Sometimes, particularly in rivers, the measurement of θ is difficult. Skew flow may be present at different depths and this effect may not be constant. By employing paths in a crossed configuration, errors as a result of skew flow are virtually canceled out by the corresponding errors from the path in the opposite direction.

Areas of application

This technology is suitable for a large range of sizes of channels or pipes by correct selection of transducers (e.g. ISO 6416). It has been used on pipes of 10 or 20 mm diameter up to rivers of hundreds of meters wide. It is best suited to fairly clean water since the ultrasonic paths are easily obstructed (see below).

Technology limitations

Anything which might obstruct, attenuate, scatter or refract the ultrasonic beams will affect the efficiency or actually stop a flowmeter which uses this technology, including

- suspended solids;
- entrained bubbles of air or other gases;
- physical obstructions;
- weed growth in a river;
- inhomogeneity which could cause refraction; in practice gradients of temperature or salinity are particularly troublesome.

The method is not suitable for open channels which are shallow compared with the width because a minimum depth is required for a path to operate (ISO 6416). Clamp-on instruments are not suitable for some pipe materials (see above).

Electromagnetic gauges

Principle of operation

When a conductor moves in a magnetic field a voltage is generated in the conductor. In 1832 Faraday realized that this effect could be used to measure liquid flow, but it was over 100 years before technology allowed a practical demonstration.

This principle is commonly encountered in liquid flow measurement for relatively small pipes or point current metering, but has been developed further to measure flows in rivers, channels and large sewer pipes. This method is described in ISO 9213 but is briefly outlined below.

Faraday's law of electromagnetic induction relates the induced e.m.f. to the length of a conductor moving in a magnetic field by the equation:

$$E = Tvb$$

where E is the e.m.f. generated normal to the direction of movement (V), T is the magnetic field intensity (T), v is the velocity of conductor (m s^{-1}) and b is the length of conductor (m). When applied to a flow meter the conductor is the water or fluid flowing in a channel or pipe of width b.

In practice, most channels and river beds will have some significant electrical conductivity which can affect the signal predicted from the equation above. To reduce this attenuation and to reduce electrical interference, an insulated membrane or lining is installed.

In practice the magnetic field will not be uniform and this, coupled with the above effect means the flow formula is complex and has a general form for a given channel

$$Q = f(E, d)$$

where Q is flow and d is depth.

A low-frequency vertical magnetic field is created by an electromagnetic coil, which is typically 200 turns and carrying an alternating current of between 3 and 5 A. The coil is buried below the river bed or bridged above the channel or pipe to generate a magnetic field across the full channel width (Figures F32 and F33).

For river and open channel applications, strip electrodes at the sides of the channel pick up the induced signal, which is then amplified and extracted from background noise by a detector operating synchronously with the alternating field. This is the flow signal relating to the liquid velocity. The gauge processor is programmed with the geometric parameters of the channel and computes the liquid flow by

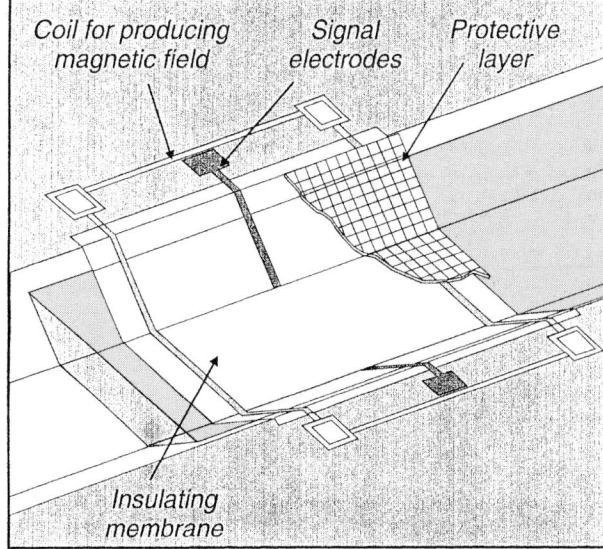

Figure F32 Electromagnetic technique using a buried coil.

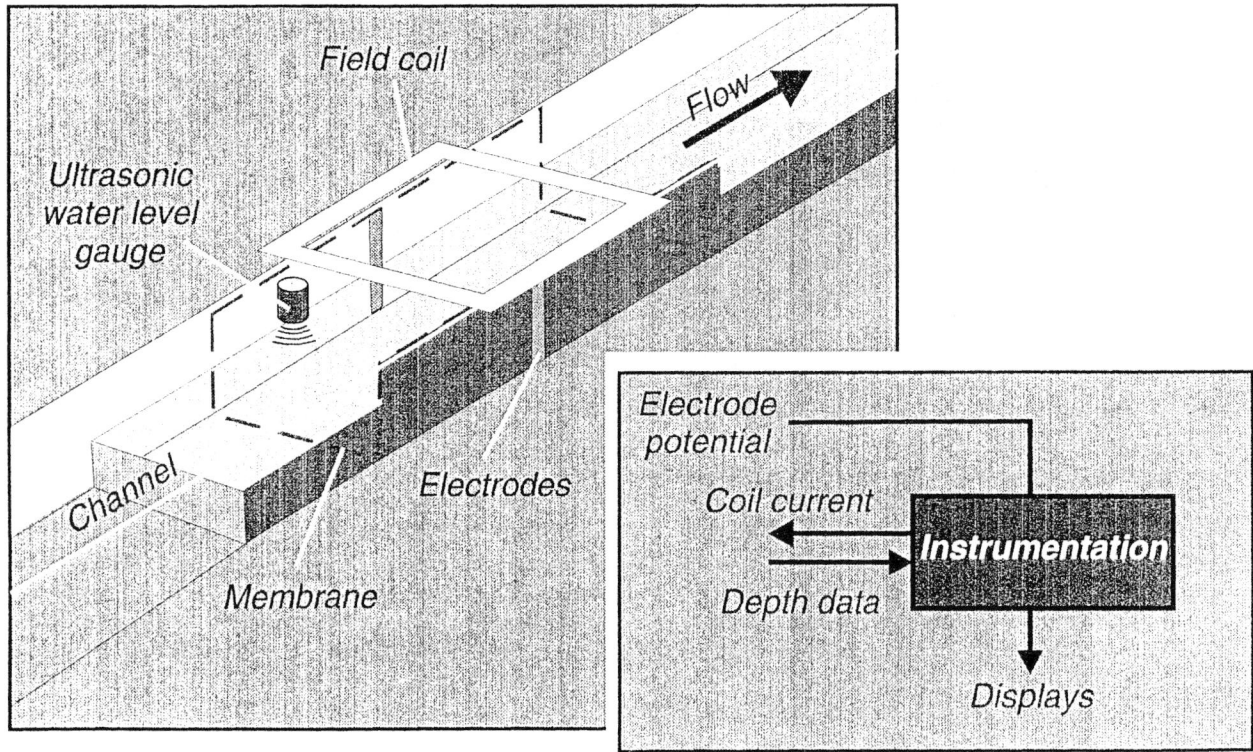

Figure F33 Electromagnetic technique using a coil above the channel.

taking the output signal from the electrodes and a depth measurement from an auxiliary depth gauge.

For part-filled, fully enclosed pipes, a glass-reinforced plastic (GRP) or coated concrete pipe section is installed to form the gauged section of the pipeline. Strip electrodes are mounted on the inside of the pipe and the coil positioned above, or preferably a pair of coils positioned above and below.

Areas of application

The method is very robust and tolerant of bubbles, suspended solids and siltation, and thus is most suitable for part-filled pipes or open channels with dirty water or sewage and small rivers with silt or weed problems.

The effective integration over the whole channel means that a shorter straight length is usable than most methods. Installations with straight approach lengths only two or three times the width have been successful.

Technology limitations

- it is difficult to produce an accurate flow formula without calibration;
- the cost of installation increases rapidly as the size of channel increases, making it prohibitively expensive for larger rivers (the largest known is 30 m wide).
- the civil cost of retro-fitting, e.g. to existing sewers, is very high.

Examples of applications

Deephams WwTW, UK (Figure F34)

User
Thames Water Utilities.

Application
Continuous flow measurement of discharge from works into River Lee. The water level in the channel is affected by water backing up when the River Lee water level is high. This does not affect the operation of the gauge which can measure reverse flows. No special

Figure F34 Continuous measurement of discharge of treated wastewater from sewage works using an ultrasonic gauge.

civil work is required for the installation, which was carried out without interrupting the flow.

Process data

Fluid:	Water (treated effluent from sewage works)
Flow rate:	$1–10 \, \text{m}^3 \, \text{s}^{-1}$
Depth:	0.5–2 m
Channel:	Concrete rectangular channel, 3 m wide

Configuration
- Two in-line ultrasonic paths measuring water velocity at heights of 0.245 and 0.645 m. Length of ultrasonic paths is 5.44 m at an angle of 33° to the flow direction.
- One upward-looking depth transducer.
- Transducer frequency 1.0 MHz.

Figure F35 Measurement of flow to works by an electromagnetic gauge.

Outputs
- Local display of flow, depth and velocity of water from each path.
- Isolated analog (4–20 mA) outputs of flow and depth.
- Pulsed output of flow.
- Totalizer.

Mansfield Sewage Treatment Works (Figure F35)

The gauge is installed on the input to the works and is used to control penstocks which limit the flow to treatment. Excess flow is sent to storm tanks and pumped back when the input drops.

Site
Flow to works, Mansfield WRW

Customer
Severn Trent Water, East Division

Process data
Liquid: Sewage and stormwater
Flow rate: 0.1–1.7 m^3 s^{-1}
Depth: 0.1–1.2 m

Gauge Type
- Electromagnetic open-channel flow gauge.
- Coils located inside 100 mm GRP trunking in the form of a 2000 × 2000 mm square.
- Number of turns = 300.
- Coil positioned above channel.
- Depth control by Hycontrol (with additional 4–20 mA output).
- Performance typically ±5%.

Outputs
- Variable pulse to data logger for flow.
- Analog (4–20 mA) of flow for telemetry and chart recorder.
- Analog (4–20 mA) of flow for Penstock control.
- Pulse for totalizer for 'total flow' reading
- Settable flow figure alarm.

Sewer surveys (Figure F36)

This type of instrument uses the Doppler principle and is widely used for survey work in part-filled sewer pipes. The tolerance of the dirty conditions and the simple single combined sensor make it particularly suitable for this type of work. The low power consumption allows operation on internal batteries and the large data storage capacity means that prolonged recording periods can be used.

Muscat Pumping Station (Figure F37)

This is another example of a temporary monitoring application, this time of pumped groundwater abstraction in Oman. Due to the low rainfall, it is important that a fine balance is struck between the natural

Figure F36 Flowmeter giving velocity and depth used for sewer survey work.

replenishment of the aquifers and the rate of abstraction. Over-pumping can lead to saline contamination. The flowmeter is used periodically to monitor these rates.

Site
One of the many farms being monitored on the coast near Muscat, Oman.

Customer
Ministry of Water Resources, Oman.

Process data
Groundwater flow rates 10–100 l s^{-1} in pipes of 50–150 mm diameter of carbon steel.

Chickenhall Sewage Treatment Works (Figure F38)

In this case, clamp-on technology was used because of the lower cost of installation. The requirement here is to measure storm flow when the pipe becomes full. Due to the hydraulic condition, the pipe can only discharge when it is full. A level detector is used to detect this condition and switch on the flowmeter.

Site
Chickenhall Sewage Treatment Works.

Customer
McDowells Consulting Engineers.

Process data
Raw sewage in cast iron–bitumen pipe.

Conclusions

Traditional methods of open-channel flow measurement rely on level measurement in conjunction with rated sections or weir or flume structures. Often these methods are not suitable or will not operate over the full range.

There are several techniques available for flow measurement using velocity × area methods which overcome these limitations. Each technique has its strengths and limitations and it is important to select the most appropriate one for each application.

David W. Gibbard

Figure F37 Portable clamp-on ultrasonic transit time flowmeter, Muscat, Oman.

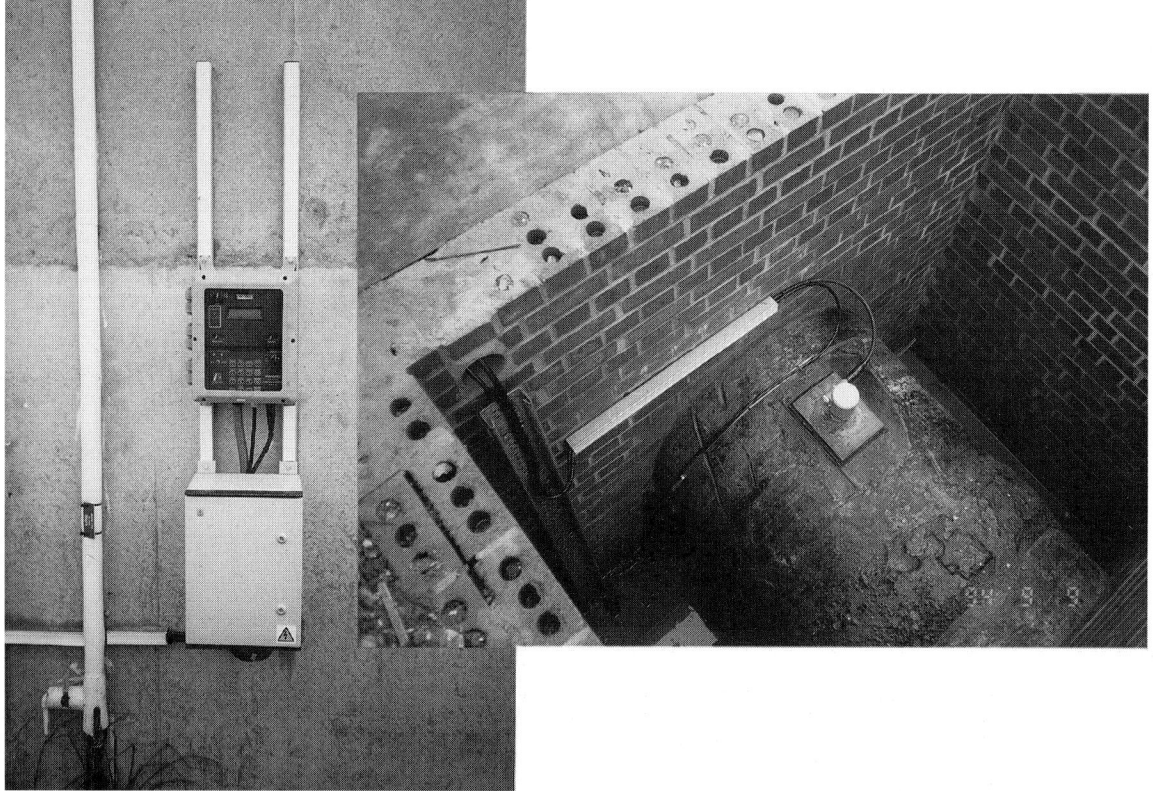

Figure F38 Dedicated polysonics Doppler flowmeter at Chickenhall Sewage Treatment Works.

Bibliography

ISO 6416:1992. *Measurement of liquid flow in open channels. Measurement of discharge by the ultrasonic (acoustic) method.*
ISO 9213:1992. *Measurement of total discharge in open channels. Electromagnetic method using a full-channel-width coil.*

Cross references

Gauge
Gauging station
Streamflow measurement
Water resources: dictionary of basic terms

FLOW THROUGH WEIRS, FLUMES, ORIFICES, SLUICES AND PIPES

The discharge equation for measuring weirs and flumes is:

$$Q = \left(\frac{2}{3}\right)^{3/2} C_d b \sqrt{(g)} H^{3/2} \qquad \text{(F4)}$$

where Q is the discharge (m^3 s^{-1}), C_d is the discharge coefficient, b is the width, length of crest (m), g is the acceleration due to gravity (9.81 m s^{-1}) and H is the total head (m).

Since the total head H cannot be measured in practice, an iterative procedure is necessary to compute the discharge from equation (F4). To avoid this, the discharge equation can, however, be presented as

$$Q = \left(\frac{2}{3}\right)^{3/2} C_d C_v b \sqrt{(g)} h^{3/2} \qquad \text{(F5)}$$

which is the basic equation for measuring structures where C_v is the dimensionless coefficient of velocity allowing for the velocity of approach, and C_d is the coefficient of discharge. Values of C_v for measuring structures may be obtained from Table F13 where C_v has been plotted against $(C_d bh)/A$, where A is the cross-sectional area of flow at the head measurement section.

Thin-plate weirs

Rectangular thin-plate weir

A diagrammatic illustration of the basic weir form is shown in Figure F39. This particular form of the thin-plate rectangular weir is often referred to as a rectangular-notch weir or 'contracted' weir, so called because the nappe is contracted. When $b/B = 1$, or when the width of the notch is equal to the width of the channel, the weir is termed a full-width rectangular thin-plate weir or sometimes a 'suppressed' weir because the nappe does not have side contractions.

Rectangular-notch weir

The equation for discharge for a rectangular-notch (contracted) weir avoiding the calculation of coefficients, is (in metric units)

$$Q = 0.554\left(1 - 0.0035\frac{h}{p}\right)(b + 0.0025)\sqrt{(g)}(h + 0.001)^{3/2} \quad \text{(F6)}$$

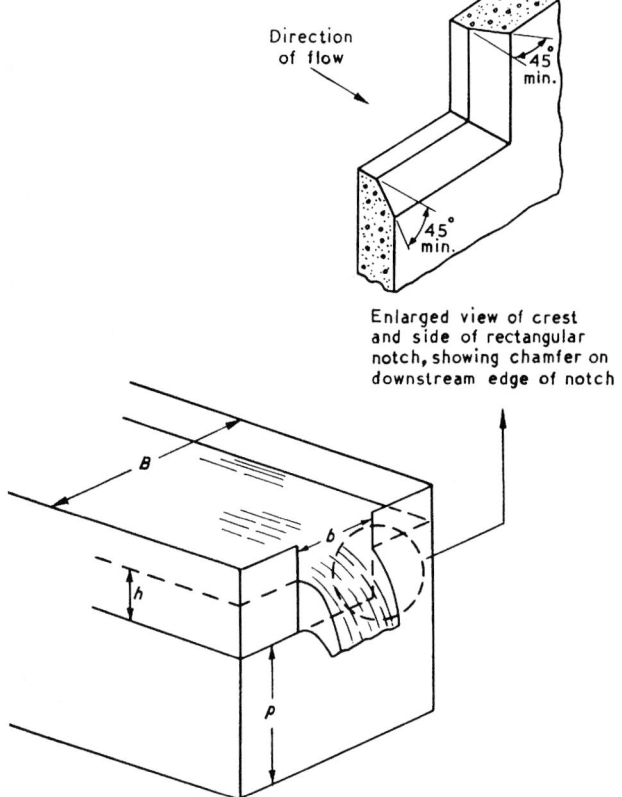

Enlarged view of crest and side of rectangular notch, showing chamfer on downstream edge of notch

Figure F39 Diagrammatic illustration of thin-plate weir.

Provided the respective limitations are observed, this equation should give flows within about 2%.

The limitations on the use of equation (F4) are:

- h should not be less than 0.03 m;
- b should not be less than 0.15 m;
- P should not be less than 0.10 m;
- h/P should not exceed 2.0;
- $(B - b)/2$ should not be less than 0.10 m;
- the head measurement section should be located at a distance of 4–5 times the maximum head upstream from the weir.

Full-width rectangular thin-plate weir

The Hydraulics Research (UK) equation for a full width rectangular thin plate weir is (in metric units)

$$Q = 0.564\left(1 + \frac{0.150h}{P}\right)b\sqrt{(g)}(h + 0.001)^{3/2} \qquad \text{(F7)}$$

Practical limitations on the use of the equation are:

Table F13 Coefficient of approach velocity C_v for values of $C_d bh/A$, where A is the area of cross-sectional flow at the head measuring section, h is the gauged head and P is the height of the weir (or flume)

$C_d bh$	0.00	0.01	0.02	0.03	0.04	0.05	0.06	0.07	0.08	0.09
0.1	1.003	1.004	1.004	1.005	1.006	1.006	1.007	1.008	1.008	1.009
0.2	1.010	1.011	1.012	1.013	1.014	1.015	1.016	1.018	1.019	1.020
0.3	1.021	1.023	1.024	1.026	1.028	1.030	1.032	1.034	1.036	1.038
0.4	1.040	1.042	1.044	1.046	1.049	1.051	1.054	1.056	1.059	1.061
0.5	1.064	1.067	1.070	1.073	1.076	1.080	1.082	1.086	1.090	1.093
0.6	1.097	1.101	1.105	1.110	1.115	1.120	1.125	1.130	1.135	1.140
0.7	1.144	1.150	1.156	1.163	1.170	1.177	1.184	1.192	1.200	1.208
0.8	1.218	1.226	1.236	1.246	1.225					

- *h/P* should not exceed 2.5;
- *h* should not be less than 0.02 m;
- *b* should not be less than 0.20 m;
- *P* should be not less than 0.15 m;
- the head measurement section should be located at a distance upstream from the weir of 2.5*P* to 3*P*.

Triangular (V-notch) thin-plate weir

A diagrammatic illustration of the basic weir form is shown in Figure F40. The three sizes of V-notches commonly used are (Figure F41):

1. A 90° notch in which the dimension across the top is twice the vertical depth (tan θ/2 = 1). This is the most common type of V-notch.
2. A half 90° notch (θ = 53°8') in which the dimension across the top is equal to the vertical depth (tan θ/2 = 0.5).
3. A quarter 90° notch (θ = 28°4') in which the dimension across the top is half the vertical depth (tan θ/2 = 0.25).

Notches (2) and (3) above nominally deliver a half and a quarter of the discharge, respectively, of the 90° notch.

The BSI equation of discharge is

$$Q = \frac{8}{15}\sqrt{(2g)}C_\mathrm{d}\tan\frac{\theta}{2}h^{5/2} \qquad (F8)$$

and the experimentally determined values of C_d, may be found from Figure F42.

Practical limitations applicable to the use of equation (F8) are:

- *h/P* should not exceed 0.4;
- *h/B* should not exceed 0.2;

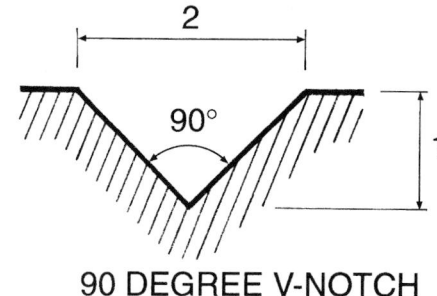

90 DEGREE V-NOTCH

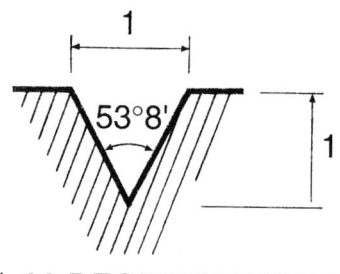

¹/₂ 90 DEGREE V-NOTCH

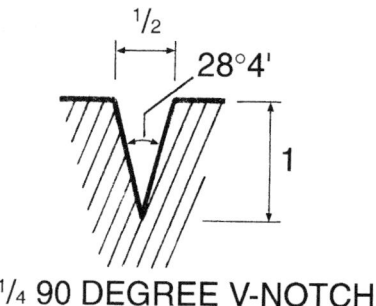

¹/₄ 90 DEGREE V-NOTCH

Figure F41 Commonly used V-notches.

- *h* should be between 0.05 and 0.38 m;
- *P* should be not less than 0.45 m;
- *B* should be not less than 1 m;
- the head measurement section should be located at a distance upstream of the notch of 4–5 times the maximum head.

It should be noted that the maximum head to be used with equation (F8) is 0.38 m. If larger heads are to be measured, a loss of accuracy in the discharge measurement has to be accepted. There is, however, insufficient experimental data available to give guidance on this aspect. However, if a 90° notch is used, and this is the most common form, the maximum head to be gauged may be increased provided the *h/P* ratio is within the range 0.2 to 2.0, *P* is not less than 0.09 m and *P/B* is between 0.10 and 1.0. When *B* is large compared with *P* (*P/B* = 0.1), the coefficient C_d is substantially constant and equal to 0.578.

For the above conditions for the 90° notch, *h* in equation (F8) is replaced by (*h* + 0.001) m. This adjustment compensates for the combined effects of viscosity and surface tension and may be an advisable adjustment at very low heads but becomes unnecessary at large heads.

Simplified discharge equations for design or spot measurements

For design purposes or for spot measurements, the following discharge equations may be used.

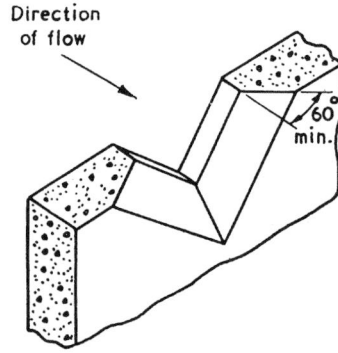

Enlarged view of V notch, showing chamfer on down-stream edge of notch

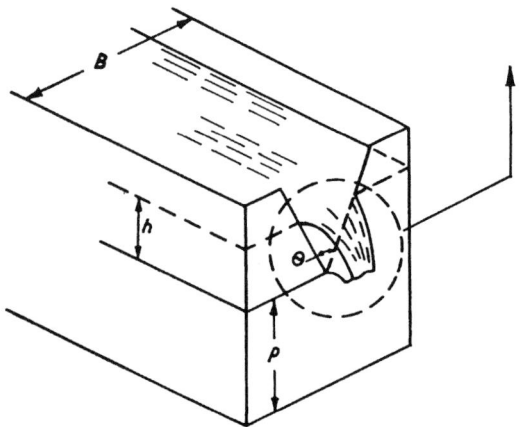

Figure F40 Diagrammatic illustration of V-notch thin-plate weir.

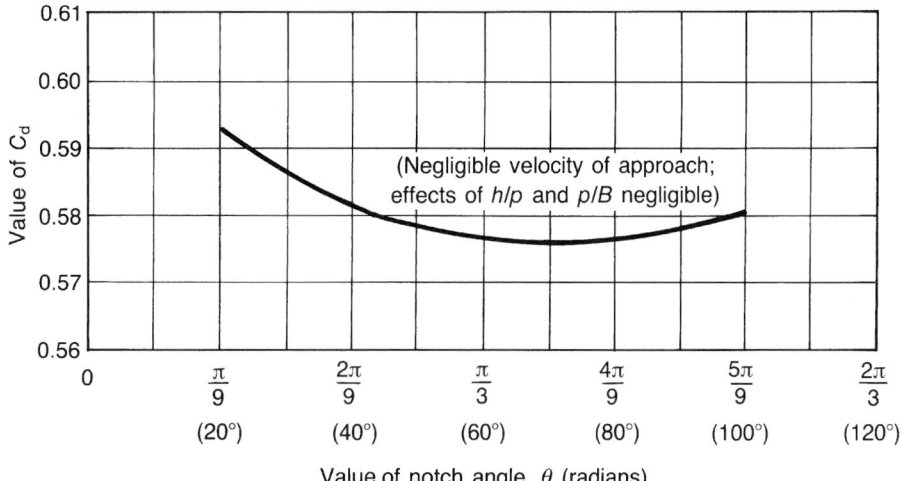

Figure F42 V-notch thin-plate weir coefficient of discharge C_d related to notch angle θ.

Rectangular notch (contracted thin-plate weir):

$$Q = 1.73bh^{3/2} \text{ m}^3 \text{ s}^{-1} \tag{F9}$$

Full-width rectangular thin-plate weir:

$$Q = 1.766(1 + 0.150h/p)bh^{3/2} \tag{F10}$$

V-notch thin-plate weir (90°):

$$Q = 1.365h^{5/2} \text{ m}^3 \text{ s}^{-1} \tag{F11}$$

V-notch thin-plate weir ($\frac{1}{2}$ 90°):

$$Q = 0.682h^{5/2} \text{ m}^3 \text{ s}^{-1} \tag{F12}$$

V-notch thin-plate weir ($\frac{1}{4}$ 90°):

$$Q = 0.347h^{5/2} \text{ m}^3 \text{ s}^{-1} \tag{F13}$$

Broad-crested weirs

Triangular profile (Crump) weir

This form of weir is the one most used in UK rivers over the past 40 years and has given good operational service providing reliable records. A diagrammatic illustration of the basic weir form is shown in Figure F43. The weir has a slope of 1:2 (one vertically to two

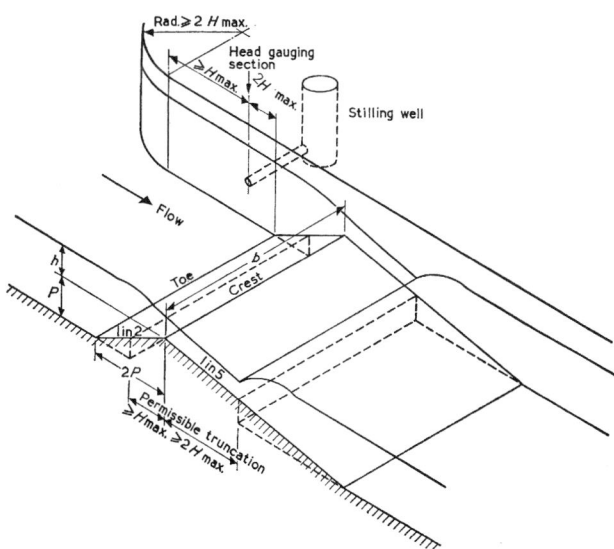

Figure F43 Diagrammatic illustration of triangular-profile (Crump) weir.

horizontally) on the upstream face and 1:5 on the downstream face, a geometry that tests have shown to give an essentially constant coefficient of discharge and a high modular limit (75%).

The equation of discharge is

$$Q = C_v C_d \sqrt{(g)} bh^{3/2} \tag{F14}$$

or

$$Q = 0.633 C_v \sqrt{(g)} bh^{3/2} \tag{F15}$$

In terms of total head, the discharge equation becomes

$$Q = 0.633 \sqrt{(g)} bH^{3/2} \tag{F16}$$

Note:

1. In the form of equation (F5), equation (F15) would become

$$Q = \left(\frac{2}{3}\right)^{3/2} \times 1.16 C_v \sqrt{(g)} bh^{3/2} \tag{F17}$$

since

$$\left(\frac{2}{3}\right)^{3/2} \times 1.16 = 0.633$$

2. When computing C_v from Table F13 the coefficient 1.16 should be used to calculate $(C_d bh)/A$.

3. Since $A = b(h + P)$ at the head measuring section

$$\frac{C_d bh}{A} \text{ becomes } C_d \frac{h}{h + P} = 1.16 \frac{h}{h + P}$$

This assumes that the width of river at the head measuring section is the same as the width of the weir (Figure F43).

Practical limitations applicable to the use of equations (F15) and (F16) are:

- h should not be less than 0.03 m (for a crest of smooth metal or equivalent);
- h should not be less than 0.06 m (for a crest of fine concrete or equivalent);
- P should not be less than 0.06 m;
- b should not be less than 0.3 m;
- h/P should not exceed 3.0;
- b/h should not be less than 2.0;
- the Froude number should not be greater than 0.5;
- the head measuring section should be located at a distance of twice the maximum head ($2H_{max}$) from the crest-line of the weir;
- h_2/h should not be greater than 0.75, where h_2 is the downstream gauged head (modular, or free flow, limit).

Rectangular-profile weir

Of all the precalibrated broad-crested weirs in operational use today there has probably been more research carried out on the rectangular

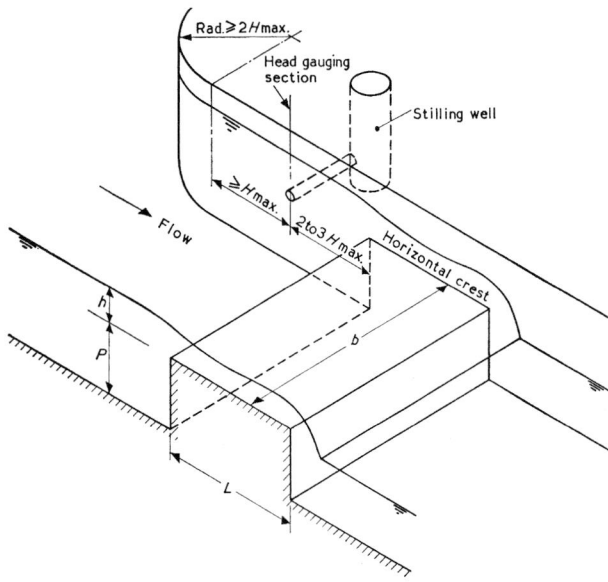

Figure F44 Diagrammatic illustration of rectangular broad-crested weir.

profile weir than on any other. This has not been because of the merit of the weir as a gauging structure but rather due to the hydraulic considerations in the variable coefficient range. Indeed, reported installations of the weir have been few compared to other types of weir. However, there are many existing rectangular profile weirs which are used for operational purposes such as irrigation devices or for compensation water measurement from reservoirs which were built before the International Standard on the weir was published. Some of these weirs might conform, even approximately, to the limitations given later although a degradation in the uncertainty in discharge has to be accepted. The weir is one of the easiest to construct in the field, the main requisite being that it has to have sharp right-angle corners. The main disadvantages of the weir are that silt and debris collect behind the structure and it has a low modular limit. A diagrammatic illustration of the weir form is shown in Figure F44.

The equation of discharge is

$$Q = \left(\frac{2}{3}\right)^{3/2} C_d C_v \sqrt{(g)} bh^{3/2} \tag{F18}$$

The coefficient of discharge C_d has a constant value of 0.86 in the range

$$\frac{h}{P} \leqslant 0.5$$

and

$$\frac{h}{L} \leqslant 0.3$$

where L (m) is the length of the weir in the direction of flow. The coefficient of velocity C_v is obtained from Table F13.

Other practical limitations for the use of equation (F18) are:

- h should not be less than 0.06 m;
- b should not be less than 0.3 m;
- P should not be less than 0.15 m;
- L/P should not be less than 0.15 nor greater than 7.0.

Flumes

General

A flume is a flow measurement device which is formed by a constriction in the channel. The constriction can be a narrowing in the channel or a hump, or both. A flume with a hump in the invert has a

discharge equation identical to that of the broad-crested weir. An advantage of the flume over a weir is its capacity to transport sediment.

Rectangular-throated flume

The rectangular flume consists of a constriction of rectangular cross-section symmetrically disposed with respect to the approach channel. There are three types: (1) with side contractions only, (2) with a bottom contraction (or hump) only and (3) with both side and bottom contractions. A diagrammatic illustration of the rectangular flume is shown in Figure F45.

The discharge equation is

$$Q = \left(\frac{2}{3}\right)^{3/2} C_v C_d b \sqrt{(g)} h^{3/2} \tag{F19}$$

where

$$C_d = \left(1 - \frac{0.006L}{b}\right)\left(1 - \frac{0.003L}{h}\right)^{3/2}$$

L is the length of flume throat and b is the width of throat.

C_v is found from Table F13 taking the cross-sectional area of flow, A, as

$$A = B(h + P) \tag{F20}$$

where B is the width of the approach channel and P is the height of the hump (with no hump, $P = 0$). Practical limitations on the use of the above discharge equation (F19) are:

- b should not be less than 0.10 m;
- h/b should not be more than 3;
- h/L should not be more than 0.50 but may be permitted to increase to 0.70 with an additional uncertainty in the coefficient of discharge of 2%;
- h should not be more than 2 m;
- h should not be less than 0.05 m;
- $(bh)/[B(h + P)]$ should not be greater than 0.7;
- the head measurement section is located at a distance of 3–4 times h_{max} upstream of the leading edge of the entrance transition.

Simplified discharge equation for design or spot measurements
For design purposes or for spot measurements, the following discharge equation may be used:

$$Q = 1.8bh^{3/2} \text{ m}^3 \text{ s}^{-1} \tag{F21}$$

Parshall flume

The Parshall flume has a rectangular cross-section and comprises three main parts: a converging inlet section with a level floor, a throat section with a downward sloping floor and a diverging outlet section with an upward sloping floor.

The control section of the flume is not located in the throat as in the previous flumes described, but near the end of the level floor, or crest, in the converging section. Because of this, Parshall flumes are considered as short-throated flumes. Laboratory calibration has been carried out in the modular, free-flow range but because of their low modular (submergence) ratio the Parshall flume is sometimes used with a tapping for measuring the downstream water level. A diagrammatic illustration of the Parshall flume is shown in Figure F46.

The discharge equation is

$$Q = Kh^u \tag{F22}$$

where K is a dimensional factor which is a function of the throat width b. The power u varies between 1.522 and 1.600.

Table F14 summarizes the dimensions for the range of Parshall flumes normally used and Table F15 gives details of the equations of discharge, discharge range for flumes and the head range.

Practical limitations for the application of equation (F22) are:

- h should not be less than 0.015 m;
- h should not be more than 1.83 m;
- the head measurement section (c) for free flow is located at $(b/3) + 0.813$ m (or $\frac{2}{3}A$) upstream from the downstream end of the horizontal crest. Note that this distance is measured along the wall and not axially (Figure F46).

Dimensions to be strictly followed are given in Figure F46 and Table F14.

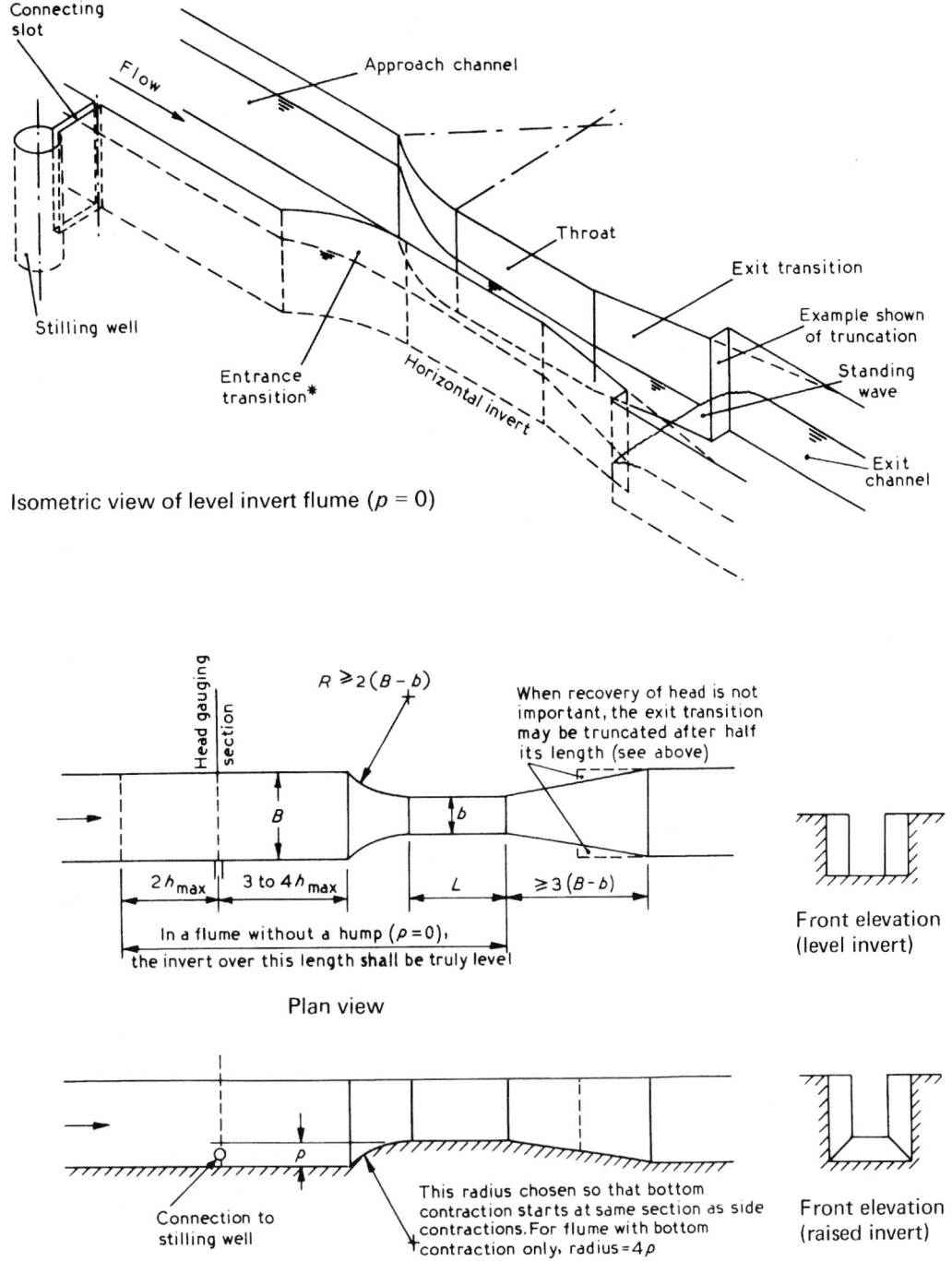

Isometric view of level invert flume ($p = 0$)

Plan view

Longitudinal section of flume with raised invert (hump)

Figure F45 Diagrammatic illustration of rectangular-throated flume.

Round-crested weirs

The overflow spillways of dams and barrages may have crest sections of circular or parabolic form and can often be adapted for flow measurement (Figure F47). The advantage of the parabolic crest is that, for a given discharge, the minimum possible head is required over a given length of crest.

The general equation of discharge for these weirs is

$$Q = Cbh^{3/2} \text{ m}^3 \text{ s}^{-1} \qquad (F23)$$

where

$$C \text{ for circular weirs} = 2.03(h/R)^{0.07} \qquad (F24)$$

and

$$C \text{ for parabolic wiers} = 1.86h^{0.1} \qquad (F25)$$

where R is the radius of the circular crest. The equations of discharge are then

circular weirs: $Q = 2.03(h/R)^{0.07}bh^{3/2} \text{ m}^3 \text{ s}^{-1}$ (F26)

parabolic weirs: $Q = 1.86bh^{1.6} \text{ m}^3 \text{ s}^{-1}$ (F27)

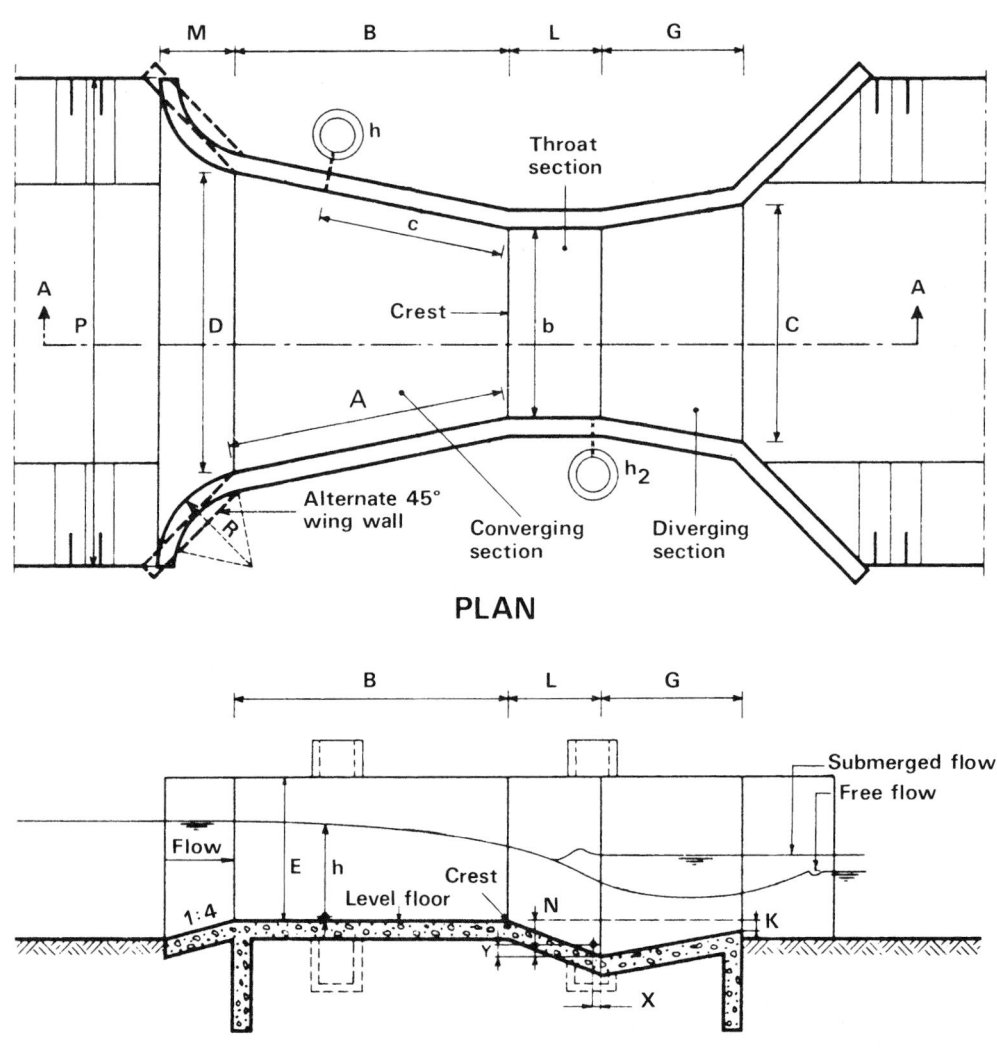

Figure F46 Diagrammatic illustration of Parshall flume.

The limits of application for the above equations are (approximately):

- h should not be less than 0.05 m;
- h/p should not be less than 3;
- h/P_2 should be less than 1.5 (where P_2 is the height of the weir above the downstream channel);
- $\dfrac{b}{h}$ should not be less than 2;
- the modular limit $h/_2h_1 = 0.3$ (where h_2 is the downstream head above crest level);
- the upstream head, h, should be measured at a distance 2–3 times h_{max} upstream from the weir face.

It should be noted that the discharge coefficient C in equations (F23), (F24) and (F25) has dimensions of the square root of gravity acceleration; that is, it includes g. Many discharge equations for non-standard weirs have in the past been presented in this form and in imperial units. To convert the equation

$$Q = Cbh^{3/2}$$

where b and h are in feet, Q is in ft^3 s^{-1} and C is in dimensions of the square root of gravity acceleration, to the metric equation

$$Q = Cbh^{3/2}$$

where b and h are in meters, Q is in m^3 s^{-1} and C is in dimensions of the square root of gravity acceleration, the imperial equation is multiplied by 0.552.

Note: the above equations of discharge are valid for weirs with vertical upstream faces.

Orifices and sluices

Orifices (Figure F48) are normally used in waterworks or in irrigation works to measure small discharges, but can also be designed to provide constant discharge under constant head. They are generally circular, rectangular or U-shaped. The general equation for flow through orifices is

$$Q = CAv \; \text{m}^3 \, \text{s}^{-1} \tag{F28}$$

where C is the coefficient of discharge, A is the cross-sectional area of the orifice (m^2) and v is the average velocity through the orifice (m s^{-1}). Then

$$Q = CA\sqrt{(2gh)} \; \text{m}^3 \, \text{s}^{-1} \tag{F29}$$

where h is the head above the center of the orifice (m). For large rectangular orifices under low heads

$$Q = \int dq = Cb\sqrt{(2g)}\int_{h_2}^{h_1} h^{1/2} dh \; \text{m}^3 \, \text{s}^{-1} \tag{F30}$$

Table F14 Dimensions of standard Parshall flumes. Note flume sizes 0.076–2.44 m have approach aprons rising at 1:4 slope and the following entrance roundings: 0.076–0.228 m, radius 0.4 m; 0.30–0.90 m, radius 0.51 m; 1.2–2.4 m, radius 0.60 m

Widths			Axial lengths			Wall depth	Vertical distance below crest		Converging		Gauge points	
												h2
Size throat width b (m)	Upstream end D (m)	Downstream end C (m)	Converging section B (m)	Throat section L (m)	Diverging section G (m)	Converging section E (m)	Dip at throat N (m)	Lower end of flume K (m)	Wall length[a] A (m)	h distance upstream of crest[b] c (m)	X (m)	Y (m)
0.025	0.167	0.093	0.357	0.076	0.204	0.153–0.229	0.029	0.019	0.363	0.241	0.008	0.013
0.051	0.213	0.135	0.405	0.114	0.253	0.153–0.253	0.043	0.022	0.415	0.277	0.016	0.025
0.076	0.259	0.178	0.457	0.152	0.30	0.305–0.610	0.057	0.025	0.466	0.311	0.025	0.038
0.152	0.396	0.393	0.610	0.30	0.46	0.61	0.114	0.076	0.719	0.415	0.051	0.076
0.229	0.573	0.381	0.862	0.30	0.46	0.76	0.114	0.076	0.878	0.588	0.051	0.076
0.305	0.844	0.610	1.34	0.61	0.91	0.91	0.228	0.076	1.37	0.914	0.051	0.076
0.457	1.02	0.762	1.42	0.61	0.91	0.91	0.228	0.076	1.45	0.966	0.051	0.076
0.610	1.21	0.914	1.50	0.61	0.91	0.91	0.228	0.076	1.52	1.01	0.051	0.076
0.914	1.57	1.22	1.64	0.61	0.91	0.91	0.228	0.076	1.68	1.12	0.051	0.076
1.22	1.93	1.52	1.79	0.61	0.91	0.91	0.228	0.076	1.83	1.22	0.051	0.076
1.52	2.30	1.83	1.94	0.61	0.91	0.91	0.228	0.076	1.98	1.32	0.051	0.076
1.83	2.67	2.13	2.09	0.61	0.91	0.91	0.228	0.076	2.13	1.42	0.051	0.076
2.13	3.03	2.44	2.24	0.61	0.91	0.91	0.228	0.076	2.29	1.52	0.051	0.076
2.44	3.40	2.74	2.39	0.61	0.91	0.91	0.228	0.076	2.44	1.62	0.051	0.076
3.05	4.75	3.66	4.27	0.91	1.83	1.22	0.34	0.152	2.74	1.83		
3.66	5.61	4.47	4.88	0.91	2.44	1.52	0.34	0.152	3.05	2.03		
4.57	7.62	5.59	7.62	1.22	3.05	1.83	0.46	0.229	3.50	2.34		
6.10	9.14	7.31	7.62	1.83	3.66	2.13	0.68	0.31	4.27	2.84		
7.62	10.67	8.94	7.62	1.83	3.96	2.13	0.68	0.31	5.03	3.35		
9.14	12.31	10.57	7.92	1.83	4.27	2.13	0.68	0.31	5.79	3.86		
12.19	15.48	13.82	8.23	1.83	4.88	2.13	0.68	0.31	7.31	4.88		
15.24	18.53	17.27	8.23	1.83	6.10	2.13	0.68	0.31	8.84	5.89		

[a] For sizes 0.3–2.4 m, $A = b/2 + 1.2$ m.
[b] h is located 2/3 A distance from crest for all sizes, distance is wall length, not axial. For symbols see Figure F46.

Table F15 Discharge characteristics of Parshall flumes

Throat width b (m)	Discharge range (m³ s⁻¹ × 10⁻³)		Equation $Q = Kh^u$ (m³ s⁻¹)	Head range (m)		Modular limit h_2/h
	Minimum	Maximum		Minimum	Maximum	
0.025	0.09	5.4	$0.0604h^{1.55}$	0.015	0.21	0.50
0.051	0.18	13.2	$0.1207h^{1.55}$	0.015	0.24	0.50
0.076	0.77	32.1	$0.1771h^{1.55}$	0.03	0.33	0.50
0.152	1.50	111	$0.3812h^{1.58}$	0.03	0.45	0.60
0.229	2.50	251	$0.5354h^{1.53}$	0.03	0.61	0.60
0.305	3.32	457	$0.6909h^{1.522}$	0.03	0.76	0.70
0.457	4.80	695	$1.056h^{1.538}$	0.03	0.76	0.70
0.610	12.1	937	$1.428h^{1.550}$	0.046	0.76	0.70
0.914	17.6	1427	$2.184h^{1.566}$	0.046	0.76	0.70
1.219	35.8	1923	$2.953h^{1.578}$	0.06	0.76	0.70
1.524	44.1	2424	$3.732h^{1.587}$	0.06	0.76	0.70
1.829	74.1	2929	$4.519h^{1.595}$	0.076	0.76	0.70
2.134	85.8	3438	$5.312h^{1.601}$	0.076	0.76	0.70
2.438	97.2	3949	$6.112h^{1.607}$	0.076	0.76	0.70
		in m³ s⁻¹				
3.048	0.16	8.28	$7.463h^{1.60}$	0.09	1.07	0.80
3.658	0.19	14.68	$8.859h^{1.60}$	0.09	1.37	0.80
4.572	0.23	25.04	$10.96h^{1.60}$	0.09	1.67	0.80
6.096	0.31	37.97	$14.45h^{1.60}$	0.09	1.83	0.80
7.620	0.38	47.14	$17.94h^{1.60}$	0.09	1.83	0.80
9.144	0.46	56.33	$21.44h^{1.60}$	0.09	1.83	0.80
12.192	0.60	74.70	$28.43h^{1.60}$	0.09	1.83	0.80
15.240	0.75	93.04	$35.41h^{1.69}$	0.09	1.83	0.80

Note: each of the discharges in the upper half of the table requires to be divided by 10^3.

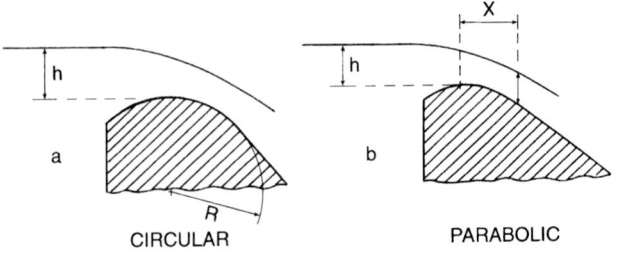

Figure F47 Round-crested weirs, (a) circular and (b) parabolic, being sections of spillways of dams and often adapted for flow measurement.

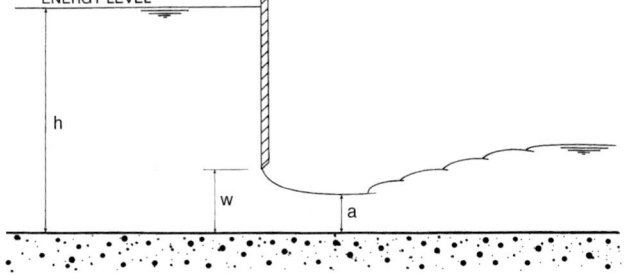

Figure F49 Detail of an undershot (sluice gate).

$$= \frac{2}{3}Cb\sqrt{(2g)}(h_1^{3/2} - h_2^{3/2}) \qquad \text{(F31)}$$

where h_1 is the distance from the water surface to the lower edge of the orifice and h_2 is the distance from the water surface to the top edge of the orifice.

In practice, however, equation (F29) is generally used for all orifices under free-flow conditions.

Sluice gates

The free discharge below a sluice gate (Figure F49) is given by the following equation:

$$Q = CA\sqrt{[2g(h - a)]} \text{ m}^3 \text{ s}^{-1} \qquad \text{(F32)}$$

where a is the water level immediately below the sluice gate or taken as equal to nw where h is the upstream water level, w is the gate opening and n is the coefficient of contraction, which may be taken as 0.61.

Submerged orifices

The basic discharge equation for submerged orifices and sluice gates (Figure F50) is

$$Q = CA\sqrt{[2g(h_1 - h_2)]} \text{ m}^3 \text{ s}^{-1} \qquad \text{(F33)}$$

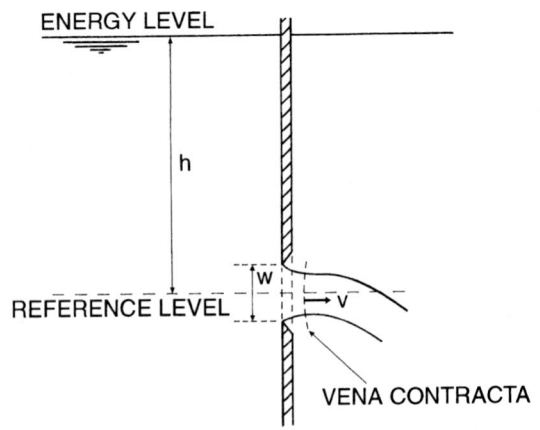

Figure F48 Detail of a freely discharging orifice.

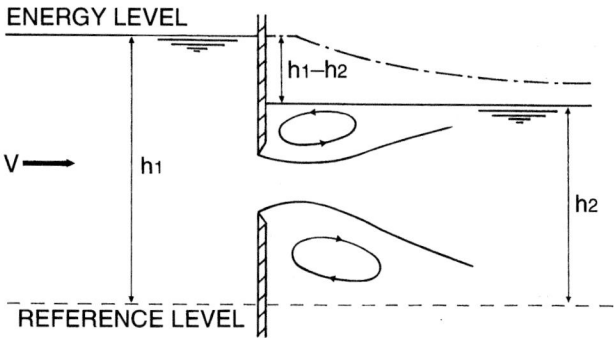

Figure F50 Detail of a submerged orifice.

where $(h_1 - h_2)$ is the differential head across the orifice or sluice gate, or the difference between the upstream and downstream water levels.

Discharge coefficients

Little recent research has been done on discharge coefficients of orifices and sluice gates but past work indicates the following values with uncertainties of about 5%:

● orifices and sluice gates under free-flow conditions 0.62;
● orifices and sluice gates under submerged flow conditions 0.60.

It is advisable, however, that until further research is carried out, these values should be checked by current meter in individual cases wherever possible.

Limits of application for orifices

● The upstream edge of the orifice should be sharp and smooth in accordance with the profile shown in Figure F50.
● The upstream face of the orifice should be vertical.
● For rectangular orifices the top and bottom edges should be horizontal and the sides should be vertical.
● The distance from the edge of the orifice to the bed and sides of the approach and tailwater channels should be greater than twice the smallest dimension of the orifice or in the case of circular orifices not less than the radius of the orifice.
● In order to neglect velocity of approach, the wetted cross-sectional area of the approach channel where head is measured should be not less than 10 times the area of the orifice.
● In the case of submerged orifices the differential head across the orifice should be not less than 0.03 m.

Flow in pipes

Closed conduit flowmeters

Hydrologists are often required to audit closed-conduit flowrate in which venturi meters or orifice plate meters, etc., are employed to measure flow in pipes running full. These meters are based on the principle that when water passes through a contraction in a pipe, it accelerates. The resulting increase in kinetic energy is balanced by a decrease in the static pressure at that point in the pipe and the pressure drop caused by the contraction is proportional to the square of the flow rate for a given flowmeter.

The general relation between flow rate and pressure drop for a differential pressure meter is:

$$Q = \frac{Ca\sqrt{(2gh)}}{(1 - a^2/A^2)^{1/2}} \text{ m}^3 \text{ s}^{-1}$$ (F34)

or

$$Q = CE\pi d^2/4 \sqrt{(2gh)} \text{ m}^3 \text{ s}^{-1}$$ (F35)

where C is the coefficient of discharge and

$$E = \frac{D^2}{\sqrt{(D^4 - d^4)}}$$ (F36)

where a is the cross-sectional area of the contraction (m²), A is the cross-sectional area of the pipe (m²), d is the diameter of the contraction (m), D is the diameter of the pipe (m) and h is the head

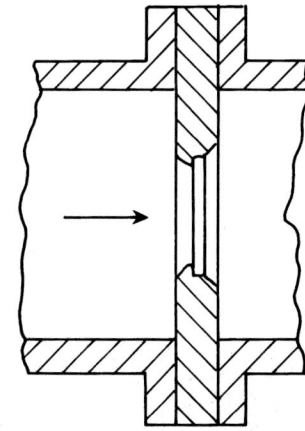

Figure F51 Orifice plate meter.

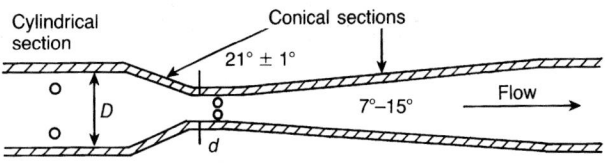

Figure F52 Venturi tube meter.

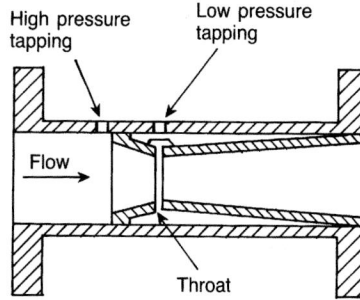

Figure F53 Dall tube meter.

difference between contraction and adjacent pipe (differential pressure in meters of water).

It should be noted that the value of C is a function of Reynolds number [equation (F38)] but normally is constant above a Reynolds number of about 3×10^5 based on the diameter of the contraction.

The most common types of flowmeters for water measurement are the square-edged orifice plate and the venturi tube (Figures F51 and F52). Typical values of the coefficient of discharge for these are 0.6 and 0.98, respectively.

The orifice plate is simply a plate with a hole in it (usually concentric) and installed transversely in a pipe with differential pressure tappings before and after the plate. There should be at least four tappings in each plane of pressure measurement, distributed evenly around the pipe. This applies to all types of differential pressure meters so that the pressure distribution is constant across each of the two planes of pressure measurement.

Venturi meters consist of a cylindrical 'throat' section preceded by a short contraction and followed by a longer expansion to allow pressure recovery. They are more expensive to make and install than an orifice plate but pressure recovery ensures a much lower head loss than with an orifice plate.

The Dall tube meter (Figure F53) was invented to ensure a low head loss. It consists of two cones each with a substantial included angle between which is a circumferential slot. The abrupt change of boundary contour results in a flow curvature which increases the

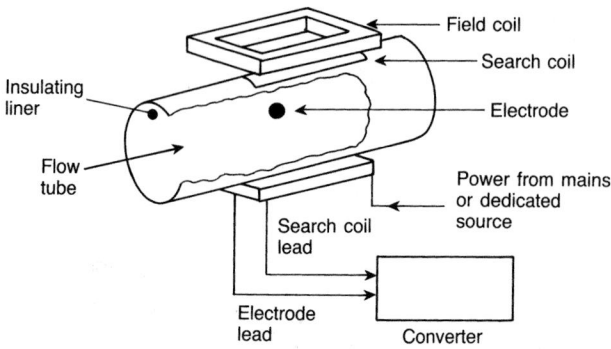

Figure F54 Electromagnetic flowmeter ('Magflow' meter).

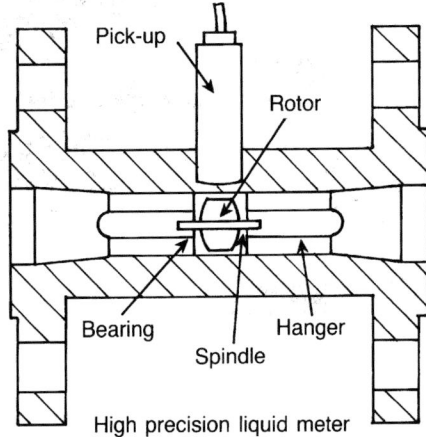

Figure F55 Turbine meter.

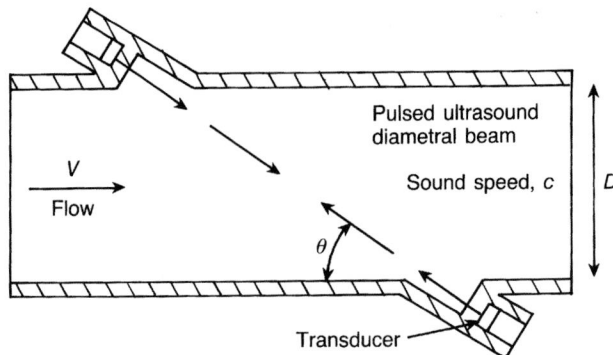

Figure F56 Ultrasonic pipe flowmeter.

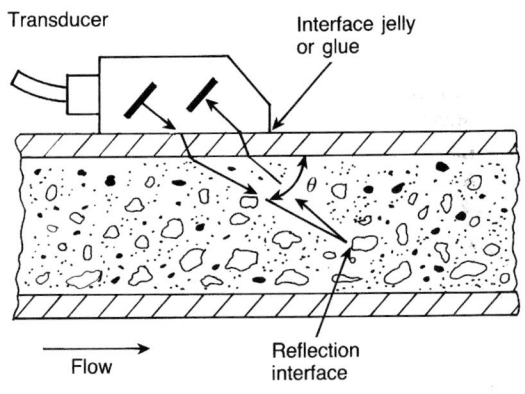

Figure F57 Ultrasonic Doppler flowmeter.

differential head produced, and the sudden reduction in cross-sectional area at the upstream pressure tapping gives a local pressure increase which also augments the pressure differences.

Manometers are most commonly used to measure differential pressure. Care should be exercised to ensure the leads to the manometer are free from air bubbles, and frequent 'bleeding' of the leads to eliminate bubbles is essential. If a U-tube manometer is used to determine the differential pressure, the effective density of the manometer liquid is mercury density minus water density.

Differential meters have a small range (turndown ratio) and because of the square-law relation between flow rate and differential pressure, a 3:1 flow range is about the most which can be measured without changing the manometers or pressure transducers.

Other types of flowmeter include the electromagnetic flowmeter (Figure F54), commonly referred to as the 'Magflow' meter, which is becoming more popular because of its minimal head loss and large range of flow (typically 30:1) and is less sensitive to pipework installation effects. These meters become increasingly more expensive than differential pressure meters as the pipe diameter increases. Frequent checking of the meter calibration is advisable if the highest accuracy is to be maintained. The principle of the electromagnetic meter is the same as that of the open channel gauge.

In the turbine flowmeter (Figure F55), a free-spinning rotor is mounted axially in the pipe and a magnetic pick-up is employed to measure the speed of rotation which is proportional to flow rate except at very low flows. Turbine meters have a range of about 10:1 and have a much lower head loss than differential pressure meters but higher than electromagnetic meters.

Ultrasonic pipe flowmeters (Figure F56) operate on the same principle as the open channel gauge and in recent years have become attractive as an option. They have no resistance to flow, are virtually independent of viscosity and the cost for large pipes of 0.5 m and above is little different from that for small pipes; also the accuracy improves as the pipe size increases. The configuration of the meter in its simplest form

has one pair of transducers across a diameter but more sophisticated meters have four pairs of transducers located across four chords which do not pass through the centre of the pipe centre line. The transmissions in this case are averaged or weighted to give the mean velocity. More recent versions employ a clamp-on meter which has the advantage that flowrate can be measured in existing pipes without drilling or cutting but only one path having one pair of transducers is used, the path passing through the center of the pipe.

The best ultrasonic meters use a 'time-of-flight' principle but some meters may employ a Doppler principle which relies on a single beam being reflected from particles moving with the flow (Figure F57). The frequency with which these reflected signals are received is then a function of the frequency of the transmitted beam, the velocity of the water and the speed of the ultrasound. This system is less accurate than the 'time-of-flight' method which is mostly used in commercial meters.

The flowrate accuracy attainable with closed-conduit meters is very much higher than that achieved in open-channel gauges; meters manufactured and installed in accordance with national or international standards have an uncertainty of about 1.5% but a meter having an individual calibration by an absolute method can achieve an uncertainty of as low as 0.5%. It is essential, however, that the conditions under which the meter is used are identical to those under which it was calibrated.

Discharge through unmetered pipes

Flowing full

The discharge of unmetered pipes flowing full is based on the head loss along the pipeline. The head loss is referred to as the 'head loss due to friction', h_f, and the energy gradient is given as $S = h_f/L$ where L is the length of pipeline. The flow through full pipes is a problem often encountered in hydrometry and several formulae are available for the estimation of discharge. The most common discharge equation is the Darcy–Weisbach equation for turbulent flow:

$$h_f = \frac{4fLv^2}{2gD} \text{ (m)} \tag{F37}$$

from which v can be determined. In equation (F37) f is a non-dimensional coefficient dependent on the relative roughness and Reynolds number Re where

$$Re = \frac{vD}{v} \tag{F38}$$

In equations (F37) and (F38) D is the pipe diameter (m) and in equation (F38) v is the kinematic viscosity ($m^2\,s^{-1}$). It should be noted that equation (F37) is now usually replaced by

$$h_f = \frac{\lambda Lv^2}{2gD} \text{ (m)} \tag{F39}$$

where $\lambda = 4f$. For laminar flow ($Re \leqslant 2000$)

$$\lambda = \frac{64}{Re} \tag{F40}$$

Values of λ for steady uniform turbulent flow can be found from charts such as the *Moody diagram* or the *Charts for the hydraulic design of channels and pipes* by the UK Hydraulics Research Ltd. (respectively). Values of λ may also be determined from the Colebrook–White equation for full pipes:

$$\frac{1}{\lambda} = -2 \log\left(\frac{k}{3.7D} + \frac{2.51}{Re\sqrt{\lambda}}\right) \tag{F41}$$

where k is the effective roughness size of the pipe wall, and by combining equations (F39) and (F40) the following equation for v is obtained:

$$v = -2\sqrt{(2gDS)} \log\left[\frac{k}{3.7D} + \frac{2.51v}{D\sqrt{(2gDS)}}\right]. \tag{F42}$$

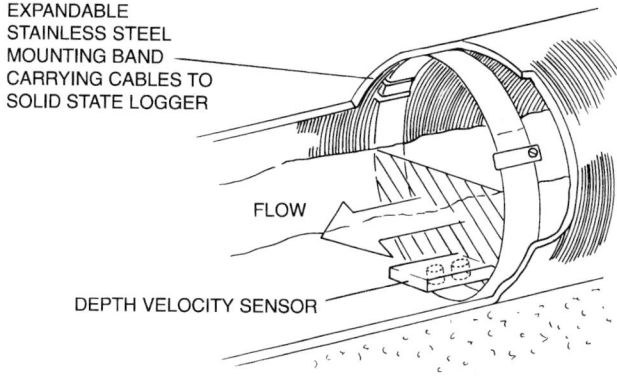

EXPANDABLE
STAINLESS STEEL
MOUNTING BAND
CARRYING CABLES TO
SOLID STATE LOGGER

FLOW

DEPTH VELOCITY SENSOR

Figure F58 Diagrammatic sketch of a depth–velocity module to measure flow in partially filled pipes.

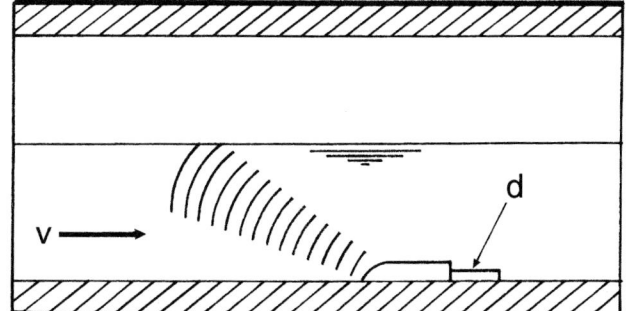

v

d

Figure F59 Diagrammatic sketch of a Doppler–velocity module and pressure transducer depth module to measure flow in partially filled pipes.

Partially full pipes

The Colebrook–White equation may be used to estimate flow in partially filled pipes. By replacing D in equation (F42) by $4R$ the equation becomes:

$$v = -\sqrt{(32gRS)} \log\left[\frac{k}{14.8R} + \frac{1.255v}{R\sqrt{(32gRS)}}\right] \tag{F43}$$

where R is the hydraulic radius $= A/P$, where A is the cross-sectional area of flow (m^2) and P is the wetted perimeter (m). It is evident that in sewers the value of k will have higher values than for clean pipes. Recommended values can be found in *Charts for the hydraulic design of channels and pipes* by UK Hydraulics Research Ltd.

New methods of measuring flow in sewers, channels and culverts

A number of instrument manufacturers have developed new systems for measuring flow in sewers, channels and culverts by means of a sensor installed in the invert of the pipe or culvert (or on the bed of a channel). This sensor is designed to measure both depth of flow and velocity and to combine these two components by the velocity–area method to provide an estimate of flowrate (Figures F58 and F59). The depth component of the sensor is normally a pressure transducer and the velocity element may operate on the electromagnetic, Doppler (Figure F55), or acoustic principle. For systems which measure velocity only to compute flow, area is keyed into the software from a laptop PC.

R.W. Herschy

Source

Herschy, R.W. (1995). *Streamflow Measurement*, Chapman & Hall, London, 524 pp.

Bibliography

Ackers, P., White, W.R., Perkins, J.A. and Harrison, A.J.M., 1978. *Weirs and Flumes for Flow Measurement.* John Wiley and Sons, Chichester.

Addison, H., 1940. *Hydraulic Measurements.* Chapman & Hall, London.

Baker, R.C., 1989. *Flow Measurement – An Introductory Guide.* MEP, London.

Bos, M.G., 1976. *Discharge Measurement Structures.* Publication No. 161, Delft Hydraulics Laboratory, Delft.

BS 5792, 1980. *Specification for Electromagnetic Flow Meters* (no ISO equivalent). HMSO, London.

BS 7405, 1991. *Guide to the Selection and Application of Flowmeters* (no ISO equivalent). HMSO, London.

BS 3680, Part 4D, 1981. *Methods of Liquid Flow in Open Channels – Compound Gauging Structures.* HMSO, London.

Clemmens, A.J., Bos, M.J. and Replogle, J.A., 1993. *Flume-design and Calibration of Long Throated Measuring Flumes.* ILRI Pub. 54, Wageningen, The Netherlands.

Douglas, J.F., Gasiorek, J.M. and Swaffield, J.A., 1979. *Fluid Mechanics.* Pitman, London.

Featherstone, R.E. and Nalluri, C., 1988. *Civil Engineering Hydraulics.* BSP Professional Books, Oxford.

Herschy, R.W., White, W.R. and Whitehead, E., 1977. *The Design of Crump Weirs.* Technical Note No. 8, Department of the Environment (Water Data Unit), London.

Horton, R.E., 1907. *Weir Experiments, Coefficients and Formulas.* US Geological Survey Water Supply and Irrigation Paper 200.

Hydraulics Research Ltd, 1993. *Charts for the Hydraulic Design of Channels and Pipes*, 5th edn. Wallingford, Oxfordshire, UK.

ISO 5167/1 (BS 1042/1.1), 1992. *Square Edged Orifice Plates, Nozzles, and Venturi Tubes.* ISO, Geneva, Switzerland.

ISO 9104 (BS 7526), 1991. *Evaluating the Performance of Electromagnetic Flow Meters.* ISO, Geneva, Switzerland.

ISO 1438/1, 1980. *Liquid Flow Measurement in Open Channels: Thin Plate Weirs.* ISO, Geneva, Switzerland.

ISO 3846, 1989. *Liquid Flow Measurement in Open Channels: Free Overfall Weirs of Finite Crest Width (Rectangular Broad Crested Weirs).* ISO, Geneva, Switzerland.

ISO 4359, 1983. *Liquid Flow Measurement in Open Channels: Flumes.* ISO, Geneva, Switzerland.

ISO 4360, 1984. *Liquid Flow Measurement in Open Channels: Triangular Profile Weirs (Crump).* ISO, Geneva, Switzerland.

ISO 4374, 1989. *Liquid Flow Measurement in Open Channels: Round Nose Horizontal Weirs.* ISO, Geneva, Switzerland.

ISO 4377, 1989. *Liquid Flow Measurement in Open Channels: Flat V Weirs.* ISO, Geneva, Switzerland.

ISO 8368, 1985. *Liquid Flow Measurement in Open Channels: Guidelines for the Selection of Flow Gauging Structures.* ISO, Geneva, Switzerland.

Lewit, E.H., 1947. *Hydraulics and the Mechanics of Fluids.* Pitman, London.

Linford, A., 1949. *Flow Measurement and Meters.* E & FN Spon, London.

Parker, P.A.M., 1949. *The Control of Water.* Routledge and Kegan Paul, London.

Cross references

Accuracy
Bernoulli's energy equation
Chézy formula
Colebrook–White equation
Flow measurement: new technology
Current metering

Energy head
Float
Flow measurement: new technology
Flume
Froude number
Gauge
Gauging station
Manning formula
Reynolds number
Stage–discharge relation
Stream flow measurement
Water resources: dictionary of basic terms
Weir: flow measurement

FLUME

A flume is an artificial open channel with clearly specified shape and dimensions that may be used for the measurement of flow. It is necessary for the shape and the dimensions to remain constant so, almost invariably, flumes are constructed in concrete, usually to a standardized design according to type. There are proprietry designs of flumes which are factory preformed out of rigid materials, such as

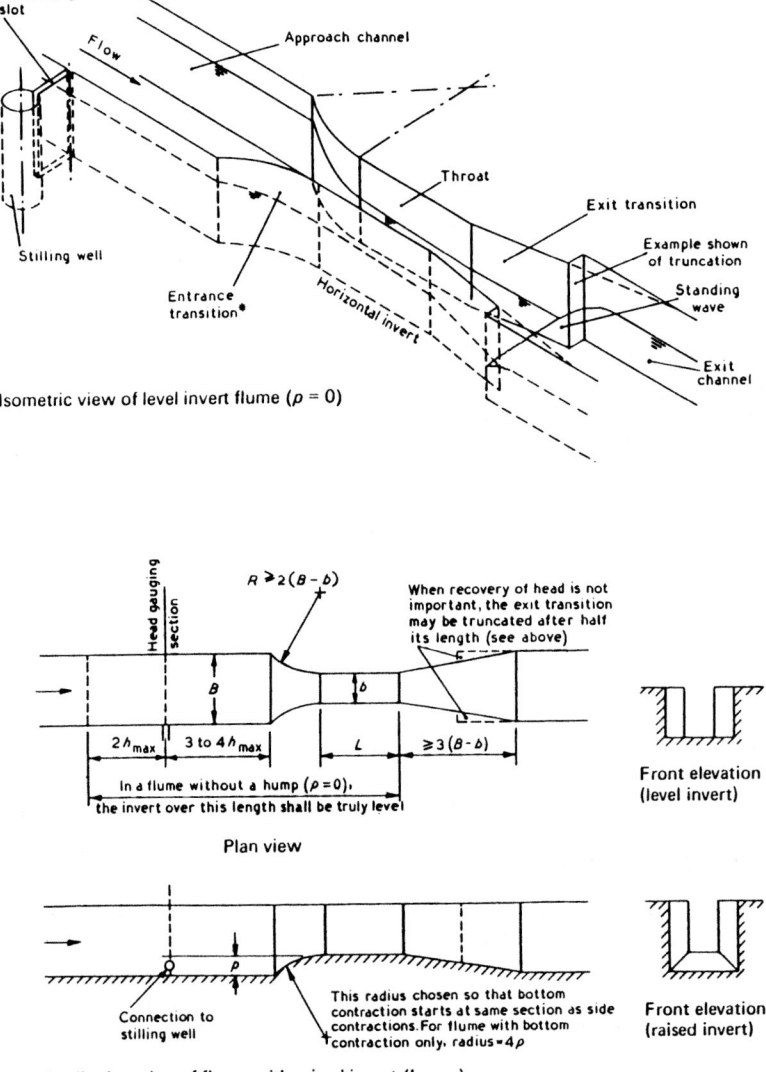

Figure F60 Diagrammatic illustration of rectangular-throated flume.

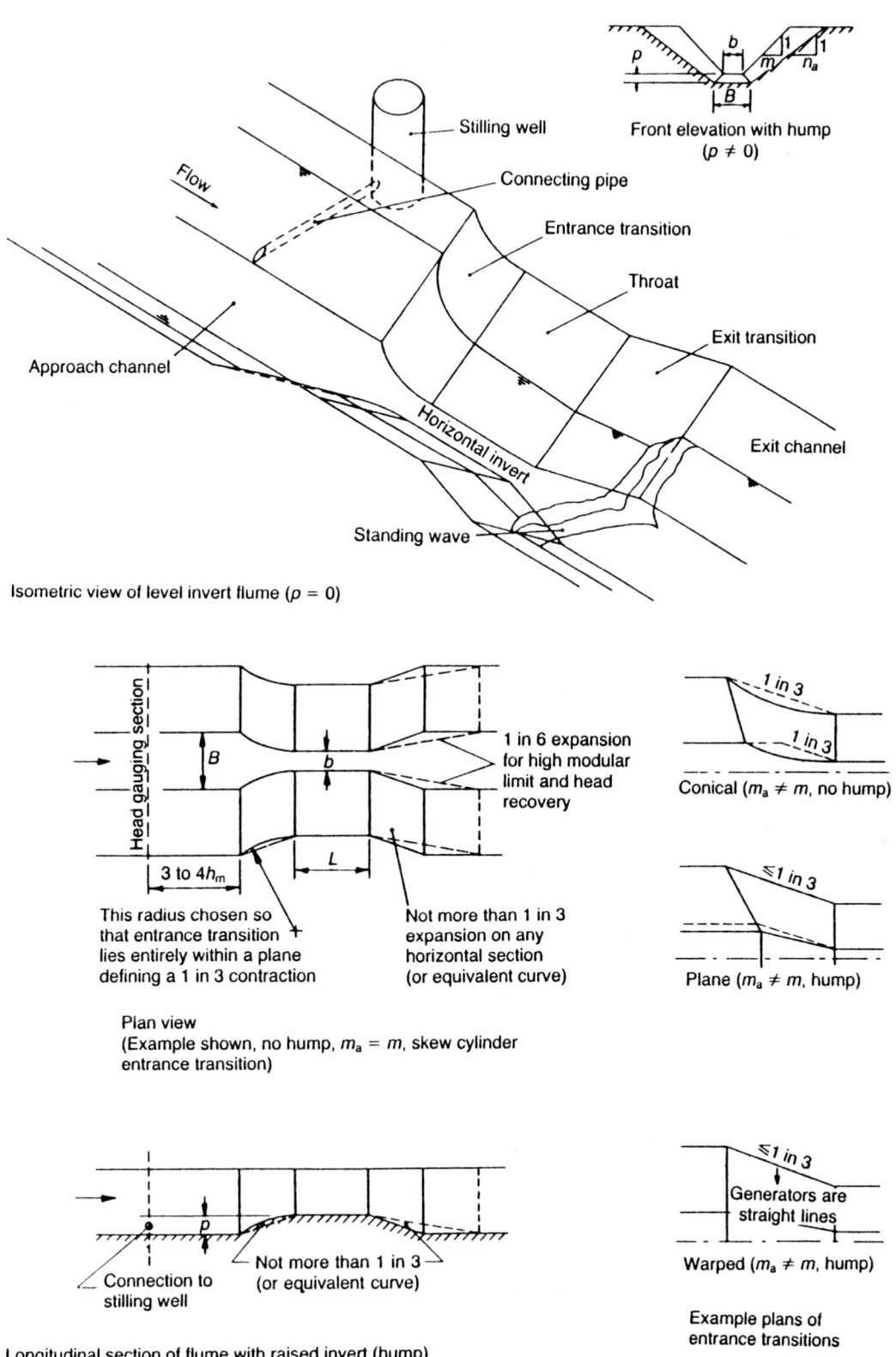

Figure F61 Diagrammatic illustration of trapezoid-throated flume.

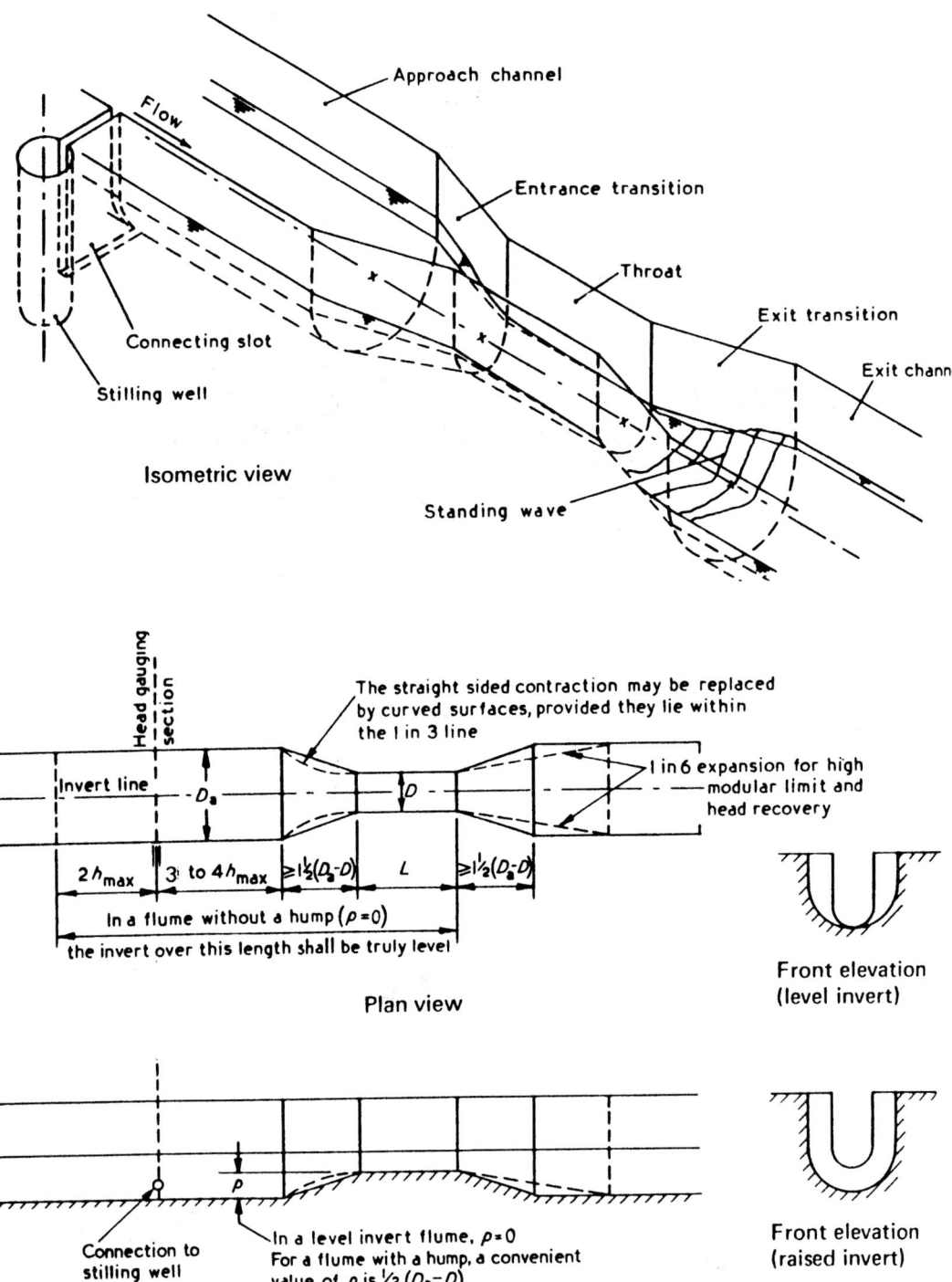

Isometric view

The straight sided contraction may be replaced
by curved surfaces, provided they lie within
the 1 in 3 line

Head gauging section

Invert line — D_a —

D

1 in 6 expansion for high
modular limit and
head recovery

$2h_{max}$ 3 to $4h_{max}$ $\geq 1\frac{1}{2}(D_a-D)$ L $\geq 1\frac{1}{2}(D_a-D)$

In a flume without a hump ($\rho=0$)
the invert over this length shall be truly level

Plan view

Front elevation
(level invert)

Connection to
stilling well

ρ

In a level invert flume, $\rho=0$
For a flume with a hump, a convenient
value of ρ is $\frac{1}{2}(D_a-D)$

Front elevation
(raised invert)

Longitudinal section of flume with raised invert (hump)

Figure F62 Diagrammatic illustration of U-throated flume.

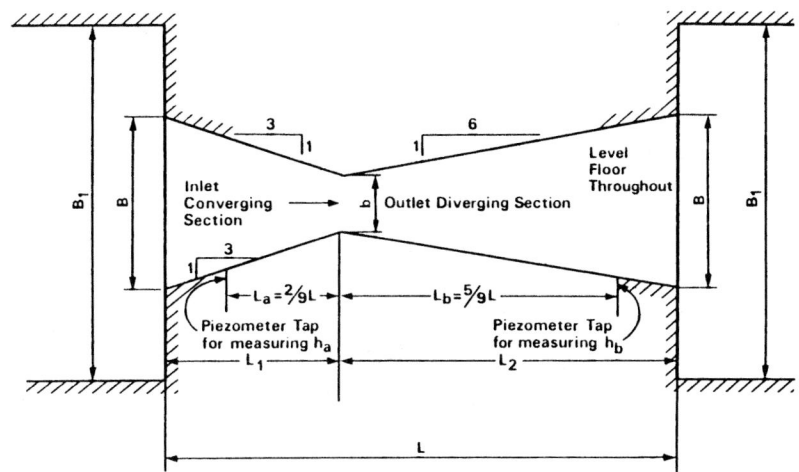

Figure F63 Diagrammatic illustration of cut-throated flume.

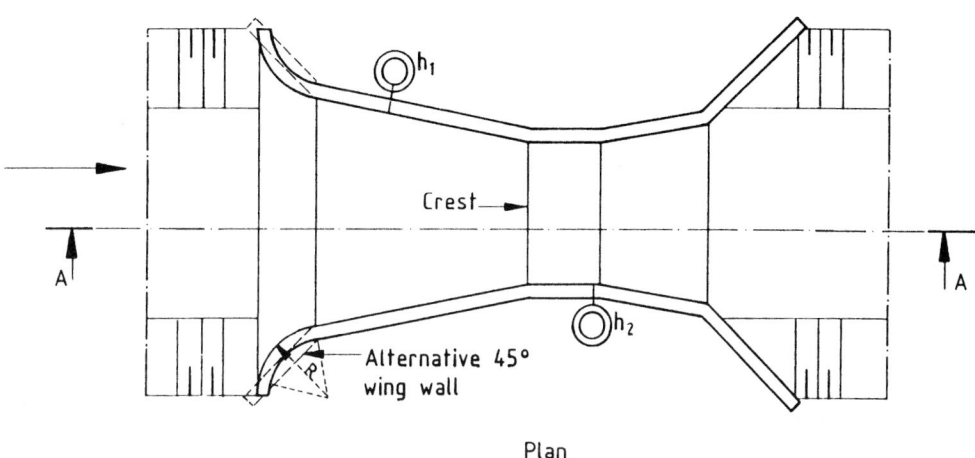

Plan

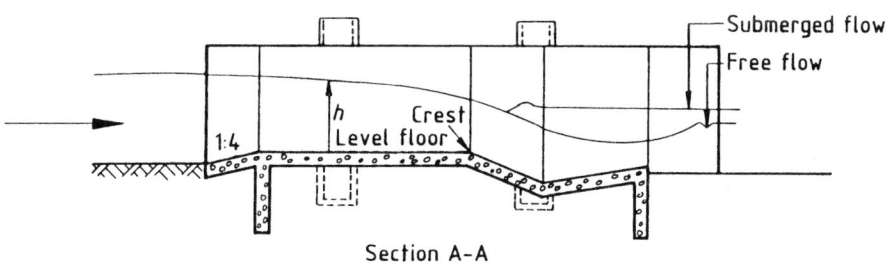

Section A–A

Key

h is the head of *liquid level*;

h_1 and h_2 are locations for measurements of head of *liquid level*;

→ is the direction of the flow.

Figure F64 Diagrammatic illustration of Parshall flume.

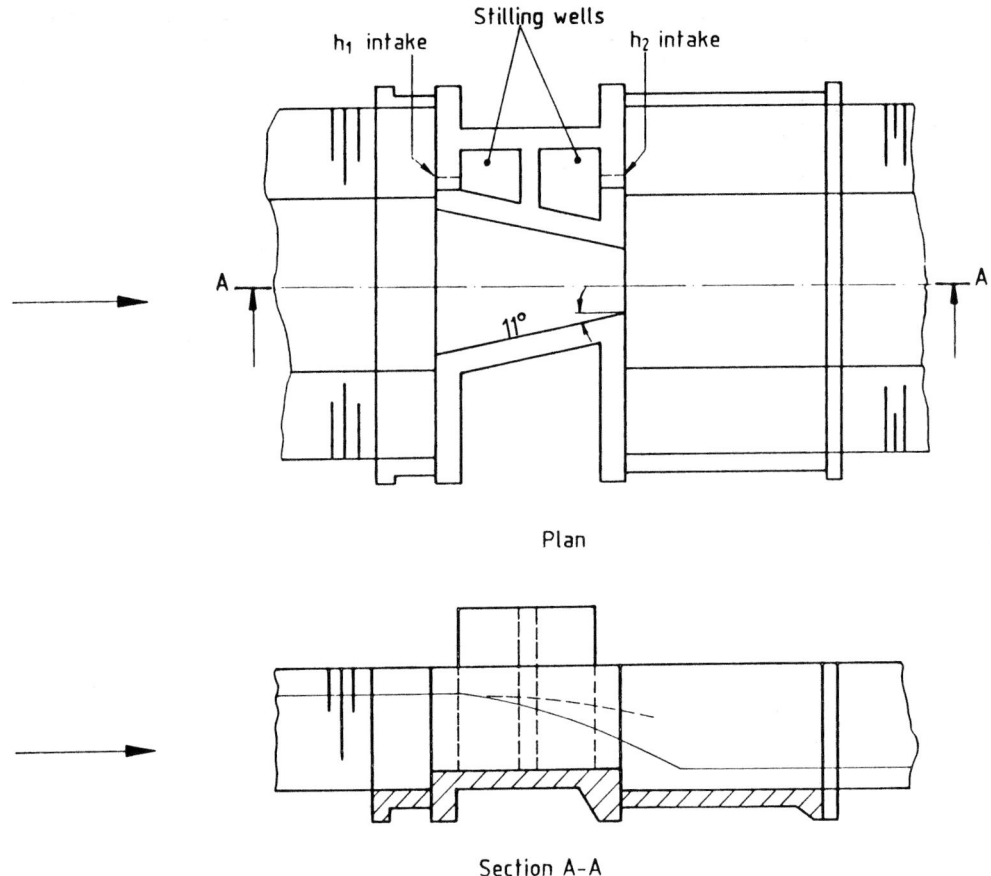

Plan

Section A-A

Key

h_1 and h_2 are locations for measurements of head of *liquid level*.

Figure F65 Diagrammatic illustration of Saniiri flume.

fiberglass, and inserted into a previously rigidly lined open channel. The basic equation for calculating open channel discharges through flumes is built up from critical depth theory, augmented by experimental data, whereby the design of the flume creates critical flow through it, giving rise to a minimum value of the specific energy for each given discharge. A general flume equation takes the form:

$$Q = \left(\frac{2}{3}\right)^{3/2} Cb\sqrt{g}H^{3/2} \text{ m}^3 \text{ s}^{-1}$$

where Q is the discharge (m³ s⁻¹), b is the effective width of the flume throat (m), g is the acceleration due to gravity (acceleration of free fall) at the flume location (m s⁻²), C is the discharge coefficient and H is the effective total head (m).

The general flume equation is adapted, as necessary, for the various types of flume. The main types of long-throated flumes in use are:

- rectangular-throated (Figure F60)
- trapezoidal-throated (Figure F61)
- u-throated (round-bottomed) (Figure F62)

The main types of short-throated flumes are:

- cut-throat (Figure F63)
- Parshall (Figure F64)
- Saniiri (Figure F65)

P.G. Holland

Source

BS 3680, Part 1, 1991. *Glossary of terms*, British Standards Institution, London.

ISO 748, 1996. *Velocity Area Methods*, International Standards Organization, Geneva.

Bibliography

BS 3860, 1981. Part 4C. *Flumes*, British Standards Institution, London.
ISO 4359, 1983. *Flumes*. International Standards Organization, Geneva.
Herschy, R.W., 1995. *Streamflow Measurement*, 2nd Edn. Chapman & Hall, London.

Cross reference

Water resources: dictionary of basic terms: flow through weirs, flumes, orifices, sluices and pipes

FOG AND MIST

Fog

Fog is a stratus cloud that lies on, or very close to, the surface of the Earth. The horizontal visibility in fog is reduced to less than 1 km, according to international definition.

Fog is an aggregate of very small water droplets in a size range of 10–50 μm, typically in concentrations of 10–100 cm⁻³. The air in fog usually feels wet because the humidity is very high, often but not necessarily above 95%, and the observer is experiencing contact with many small droplets.

The atmosphere always contains an adequate number of condensation nuclei, though in certain circumstances there will be differing mixes of types of nuclei, e.g. along an ocean shoreline there will be a greater than normal number of salt nuclei, which are highly water-loving, or hygroscopic. The haziness of beaches with breaking surf is accounted for by the fact that salt particles start to take on water at relative humidities as low as 60–70%.

A critical relative humidity beyond which condensation will be initiated, and fog forms, can be achieved in four ways: (1) addition of water vapour to the volume of space in question; (2) cooling of the volume by contact with a colder surface; (3) cooling by infrared radiation from the volume itself; (4) expansional cooling due to ascent of the airmass. Of these four ways, the first two are of major importance in the formation of fog.

Fogs may be broadly classified as those that form within airmasses and those that form at the boundaries between different airmasses, i.e. in conjunction with fronts. In outline form:

I. Airmass fogs
   A. Advection types
      1. Transport of warm air over a cold surface
         (a) Land and sea breeze fog
         (b) Sea fog
      2. Transport of cold air over a warmer, wet surface
         (a) Steam fog
         (b) Arctic 'sea smoke'
   B. Radiation types
      1. Ground fog (ice fog if particles are ice crystals)
      2. High inversion fog
   C. Expansional cooling fog (upslope fog)
   D. Combinations of A, B and C
II. Frontal fogs
   A. Prefrontal warm front fog
   B. Frontal passage fog

Advection implies primarily horizontal transport, though vertical transport may be significantly present in certain circumstances. When relatively warm air is carried over a cooler surface, its temperature is lowered by contact cooling (conduction) and the relative humidity rises. After the critical value is exceeded, condensation on the cloud condensation nuclei (CCN) begins and fog forms. The cooling of the surface layers is carried to higher layers by turbulent mixing, which results from wind drag over the surface. If the air becomes slightly unstable because of this mixing the fog may have a low ceiling, as is common with the California coastal stratus.

When cold, dry air moves over a warmer, wet surface, rapid evaporation takes place and saturates the colder air. The resultant condensation is the steam fog (steam smoke) seen in the Arctic Ocean areas.

Radiation fog results from a different set of conditions. If skies are clear and the air is relatively still, nocturnal cooling of the ground by escaping terrestrial radiation chills the surface layers of air. Fog forms at ground level when the air temperature is lowered to the dewpoint. Slight turbulent mixing increases the depth of cooling and the thickness of the fog. Ground fog tends to form in the late night or early morning hours; it then 'burns off' in the later morning hours when solar shortwave radiation penetrates the shallow cloud, warms the ground, lowers the relative humidity and causes the water droplets to evaporate.

Upslope fog forms when a stable airmass moves slowly up over higher terrain, cools by expansion, and finds its temperature lowered to the dewpoint. This fog is characteristic of the Western Plains states in the United States.

Frontal fogs form primarily when additional water vapour from evaporating precipitation elevates the dewpoint temperature, and evaporative cooling lowers the temperature of the air. Such fogs are most characteristic of warm fronts.

Mist

According to international definition, mist consists of an aggregate of microscopic sized droplets ($\sim$10 μm) producing a thin, grayish veil over the landscape, reducing visibility to a lesser extent than fog. Mist is intermediate in all respects between damp haze and fog.

However, in the United States the term mist has come to have a popular usage of a hydrometeor that is intermediate between fog and drizzle. Oregon mist or Scotch mist are terms used to describe the occurrences of very light, 'misty' precipitation. Mist particles range in size from 50 to 500 μm, the latter large enough to fall from the cloud.

Trees and other objects such as grasses collect moisture from drifting heavy fog or mist, as sometimes occurs in the Redwood Forests in northern California. Fog-drip can collect as much as 0.8 mm of water in a single night – the equivalent of a light shower. This phenomenon prevents excessive aridity in the coastal forests during the rainless California summers.

John A. Day

Bibliography

Day, J.A. and G. Sternes, 1970. *Climate and Weather*. Reading, MA: Addison-Wesley.
Huschke, R.E. (ed.), 1959. *Glossary of Meteorology*. Boston: American Meteorological Society.
Schaefer, V.J. and J.A. Day, 1981. *A Field Guide to the Atmosphere*. Boston: Houghton Mifflin.

Cross references

Climate data: sources
Clouds (cloud seeding)
Dew ponds
Precipitation

FREQUENCY ANALYSIS

Use of frequency analysis in hydrology

The occurrence of many extreme events in hydrology cannot be forecasted on the basis of deterministic information with the sufficient skill and lead time as those decisions which are sensitive to their occurrence. In such cases, a probabilistical approach is required in order to incorporate the effects of such phenomena into decisions. If the occurrences can be assumed to be independent in time, i.e., the timing and magnitude of an event bears no relation to preceding events, then frequency analysis can be used to describe the likelihood of any one or a combination of events over the time horizon of a decision. Hydrological phenomena that are commonly described by frequency analysis are storm precipitation and annual flood maxima.

Frequency analysis can be conducted either graphically or mathematically. In the graphical approach, the historical observations of the variable of interest are ordered in increasing or decreasing magnitude, and a graph of the magnitudes of the events versus an estimate of their frequency of exceedance, or recurrence interval, is plotted. A smooth curve is then fitted through the plotted points to describe the probability of any particular event's future occurrence. Special graph paper is available that can be used to attempt to depict the smooth curve as a straight line.

The mathematical approach to frequency analysis relies on the assumption of a specific mathematical description, known as a probability distribution, to define the equivalent of the smooth curve of the graphical approach. The parameters of the probability distribution are defined as functions of the statistics of the hydrological observations.

Statistical series and return periods

In probabilistical analysis, a series is a convenient sequence of data, such as hourly, daily, seasonal, or annual observations of a hydrological variable. If the record of these observations contains all the events that occurred within a given period, the series is called a complete duration series (Shaw, 1964). For convenience, the record often contains only events of magnitude above a preselected base. Such a series is called a partial duration series. A series that contains only the event with the largest magnitude that occurred in each year is called an annual maximum series.

The use of the annual maximum series is very common in probabilistical analysis for two reasons. The first is for convenience, as most data are processed in such a way that the annual series is readily available. The second is that there is a theoretical basis for extrapolating annual series data beyond the range of observation, but with partial series data, such theory is lacking. A reason for the absence of statistical theory for the partial duration series is the lack of independence of events that might follow one another in close sequence.

Table F16 Corresponding return periods for annual and partial series

Partial series	Annual series
0.50	1.16
1.00	1.58
1.45	2.00
2.00	2.54
5.00	5.52
10.00	10.50

A limitation of annual series data is that each year is represented by only one event. The second highest event in a particular year may be higher than the highest in some other years, yet it would not be contained in the series. Accordingly, an event of a given magnitude would have a different frequency of occurrence for each of the two series.

The complete duration series may be required for the stochastic approach in which independence is not required. It may also serve for a probabilistic analysis of data from arid regions where the events are rare and almost independent.

The return period, T_r of a given event is the average number of years within which the event is expected to be equalled or exceeded only once. The event that, in expectation, will be equalled or exceeded every N years is the N-year event, X_{Tr}. Both terms refer to the expected average frequency of occurrence of an event over a long period of years. The return period is equal to the reciprocal of the probability of exceedance in a single year.

For return periods exceeding 10 years, the differences in return periods between the annual and partial series is inconsequential. Table F16 presents factors for conversions between the two series.

Mathematical approach to frequency analysis

Probability distributions used in hydrology

Probability distributions are used in a wide variety of hydrological studies, e.g., water resources studies, studies of extreme high and low flows, droughts, reservoir volumes, rainfall quantities, and in time-series models. The principal distributions used in hydrology are listed in Table F17. Their mathematical definitions are given in the *WMO Statistical Distributions for Flood Frequency Analysis* (WMO, 1989).

Annual totals, such as flow volumes or rainfall depths, tend to be distributed normally or almost so because of the central limit theorem of statistics. Monthly and weekly totals are less symmetric, display a definite skewness (mostly positive) and cannot usually be modelled by the normal distribution.

Table F17 Probability distributions used in hydrology

Name	Acronym
Normal	(N)
Lognormal	(LN)
Pearson type 3	(P3)
Extreme value type 1	(EV1)
Extreme value type 2	(EV2)
Extreme value type 3	(EV3)
Three-parameter gamma distribution	
Gamma	(G)
Log-Pearson type 3	(LP3)
General extreme value	(GEV)
Weibull	
Wakeby	(WAK)
Boughton	
Two-component EV	(TCEV)
Log-logistic	(LLG)
Generalized logistic	(GLC)

Annual extremes (high or low) and peaks over a threshold have positively skewed distributions. The part of a sample that lies near the mean of the distribution can often be described well by a variety of distributions. However, the individual distributions can differ significantly and very noticeably from one another in the values estimated for large return periods. Since hydraulic design is often based on estimates of large recurrence-interval events, it is important to be able to determine them as accurately as possible. Hence, the choice of distribution is very important for such cases. The choice of distributions is discussed in the WMO *Statistical Distributions for Flood Frequency Analysis* (WMO 1989), which includes a discussion on the methods available for choosing between distributions and how these choices are dependent on a number of technical issues, such as the character of hydrological data and the method of parameter estimation.

Parameter estimation

In addition to the consideration of the choice of distribution, the method of parameter estimation used with it may have an effect on the outcome. Traditionally, the method of ordinary moments (MOM) has been popular in hydrology even though it has been recognized as statistically inefficient in comparison to the method of maximum likelihood (ML). The method of probability-weighted moments (PWMs), introduced by Greenwood *et al.* (1979) is, in many cases, convenient to apply, and it has been found by Hosking *et al.* (1985) to be comparable with ML in its statistical properties for sample sizes which are normally encountered in hydrology.

A more recent methodology employing L-moment statistics (Hosking, 1990) shows considerable improvement over the more conventional maximum likelihood or method at moments techniques. Applications of this regionalized technique are beginning to be reported in the analysis of extreme-value data.

Homogeneity of data

The homogeneity of hydrological data is the requirement for a valid statistical application. There are many reasons why data series may not be homogeneous, for example:

- A time series of maximum discharges may contain both snow melt and rainfall discharges;
- A time series may contain discharges formed both before construction of a hydraulic structure in undisturbed conditions and after construction when the runoff regime is controlled; or
- A time series may contain discharges that include mixtures of systematic and random errors.

Data homogeneity may also be disturbed by the anthropogenic changes of climate as well.

A detailed analysis of the data is the most effective method of evaluating data homogeneity. The methods of analysis are usually based on plotting different types of runoff dependencies upon runoff-producing factors (physical and mathematical) to discover causes of a disturbance of homogeneity. The following types of time series reconstructions are possible when non-homogeneity is established and when its causes are discovered:

- Non-homogeneous data are corrected to homogeneous conditions (recovery of natural runoff, computation of empirical frequencies, etc.);
- The record is subdivided into a number of homogeneous samples (water discharges produced by mud flows, maximum rainfall discharges, runoff availability and absence, etc.); and
- Known systematic errors are corrected and spurious data are deleted from the record.

Source

World Meteorological Organization, 1994. *Guide to Hydrological Practices*, 5th edn, WMO, Geneva.

Bibliography

Greenwood, J.A., Landwehr, J.M., Matalas, N.C. and Wallis, J.R., 1979. Probability weighted moments: definition and relation to parameters of several distributions expressible in inverse form. *Water Resources Research*, **15**(5), 1049–1054.
Hosking, J.R.M., Wallis, J.R. and Wood, E.F., 1985. Estimation of

the generalized extreme-value distribution by the method of probability-weighted moments. *Technometrics*, **27**(3), 257–261.

Hosking, J.R.M., 1990. L-Moments: analysis and estimation of distributions using linear combinations of order statistics. *Journal of the Royal Statistical Society B*, **51**(3).

Shaw, T.T., 1964. Frequency analysis. *Handbook of Applied Hydrology* (V.T. Chow, ed.), Section 8-I, McGraw-Hill, New York.

World Meteorological Organization (WMO), 1989. *Statistical Distributions for Flood Frequency Analysis* (C. Cunnane). Operational Hydrology Report No. 33, WMO-No. 718, Geneva.

Cross reference

Flood frequency analysis

FROST

Frost is a solid phase of water. As a mineral it crystallizes according to the hexagonal system featuring six-sided plates, needles, clusters and columns. The crystals are subject to innumerable variations of twinning and dendritic or arborescent growth. None of the variations are unusual in the mineral world, but the melting point of water limits observations, whereas similar crystallization can be found in commonly seen metallic and non-metallic minerals. The complex crystals are the object of a variety of platitudinal visual imitations in the forms of snow crystals drawn by artists and in reality as seen by frost on the windows.

From the climatologists' point of view, the frost of specific interest is hoar frost or white frost that accumulates on surfaces in places with appropriate temperatures. The white color is a product of small air bubbles in the ice, cutting down on the transparency of the ice crystal, the reaction of poor sky light and the lack of sunshine. At the critical temperature at which frost forms, sunshine would prove disastrous in a very short time. The 'hoar' implies a gray or grayish tone often seen on objects covered with frost. Hoar frost or the veneer of ice crystals is indicative of three conditions:

- surfaces on which the frost forms must be 0°C or below;
- the surrounding air is saturated at 0°C or slightly below;
- nuclei are present so that the process of sublimation can take place.

Each of these three conditions is part of the natural environment from time to time.

Surfaces suitable for the accumulation of frost are cooled by outward radiation and advection, so that the ambient temperature of 0°C or less is reached. The temperature of the atmosphere has already been reduced to the near-freezing condition by normal processes. The loss of heat to or below the freezing level by the atmosphere and surface objects in dry air and without reaching saturation creates a frost known as black frost. Vegetation, when exposed to freezing conditions, upon thawing will turn dark or black. The water that is part of the cell structure of the plant solidifies, cell walls burst and the plant materials deteriorate.

Saturated air at the appropriate temperature begins to give up its moisture. If this is in the range above 0°C, the product is dew and, should the temperature drop, the dew particles solidify to coat the object with a veneer of ice, which is amorphous. Thus the saturation-level temperature is most critical in the formation of frost.

Nuclei are as essential for crystal formation as they are for dew, raindrop or cloud formation. The surfaces on which frost accumulates depend on the presence of these nuclei. Nuclei can be dust particles, irregularities in the configuration of leaves, plant hairs or a host of other features. It is difficult to conceive of a situation where an adequate number of nuclei would not be present.

Although the literature is sprinkled with terms related to frost such as frost-wedging, ice-wedging, frost-buckling and others implying solid ice, under scrutiny many of these terms are not frost at all but gelification of water as a liquid. Innumerable citations imply that ice formed by sublimation in pore space and microcellular space in rocks, soils and other material should be classified as frost. Again it is probably interstitial and interfacial water that has crystallized or solidified. There is no recognition in the literature that water vapor is found free in cellular or pore space of such consequence that sublimation could occur to form frost to the extent of creating expansion of material and the destruction of, for example, plant cells.

Protoplasm in plants functions at varying temperatures for each plant within restricted ranges. Frost, as defined here, implies near 0°C temperatures, but some plants cease protoplasm functioning at temperatures higher or lower than 0°C. Thus a plant may suffer 'chilling' injury but not as a result of frost or freezing.

It is conceivable in certain types of cavities and caverns that frost accumulates over a period of time and a repetition of the conditions may permit consolidation of this frost into ice masses. The condition is simulated in large cold-storage or deep-freeze rooms. Thus, large and small cavities opened to the atmosphere and near the surface experience frost activities. The results are such phenomena as ice-filled sink holes or ice caves.

Whatever the process of formation or product, frost is a symptom of a climatic situation in which temperatures have been reduced through radiation or advection to the freezing stage. Thus frost, particularly hoar frost, is a symptom of or forerunner to plant or crop loss.

Because frost can freeze plant tissue, which marks the end of growth for the plant, it is significant that this can occur at either end of the growing season. Since destruction is involved, it is evident that frost is a natural hazard along with other climatic hazards such as excessive precipitation, tornadoes and hail. Most citations to frost overlook this hazard factor, perhaps because it is not as spectacular or forceful as high winds, high water, large hailstones or heavy snowfalls. The agriculturist could well proclaim the facts by playing the game and trying to produce a crop that fits in between the last killing frost of the growing season in spring and the first killing frost at the end of the growing season in fall. In doing so, the farmer recognizes frost as a hazard.

The accumulated experiences of frost as a hazard leads to a considerable discussion in many climatology texts of techniques for frost or freeze prevention. These are generally beyond what the botanist and geneticist have already done to produce plants that mature in shorter periods of time, or by the climatologist who has defined the perimeters of frost incidence for any particular area. Indeed the greatest incidence of activity in frost prevention is in the marginal or fringe areas where, for reasons of possible economic gain, the farmer or orchardist is willing to take a chance on planting in the hope of being able to produce a crop.

Frost damage prevention techniques can be classified into several groups by procedures and practices.

- Identification of the critical and perhaps typical physical site of low or high frost incidence in a general region. These can be defined as wind shelters, topographic favorable regions, or shore position, etc.
- Development of physical equipment such as fans, heaters, brushes, sprinklers and plant shields to modify temperatures or reduce radiation.

Whatever method or combination of methods is chosen to prevent destruction by frost, the choice is usually to modify temperatures a few degrees usually not more than 4 or 5°C. The techniques involve reducing radiation, improving wind circulation, discouraging sublimation or creating a fog or smoke cover. The physical modification of the environment in the immediate situation of a frost or freeze hazard does not represent all that can be done.

Long before agriculture is considered for a particular area or crop, a climatic record or history would reveal the probabilities of frost or freeze incidence and the chance of success if the correct plans and methods are used. Additional terrain analysis helps to identify local microclimatic conditions in basins, alluvial fans, slopes, plains or other topographic conditions. Peri-shore or valley sites could and should be evaluated.

In the competition for the early harvest–high price crop, farmers for many years have tried to protect young plants with shields of plastic or paper. Commercial outlets have provided the devices in large numbers. As is true of other procedures, the shield protects within limited temperature ranges. The shield in addition may speed growth and thus 'harden' the plant against chill.

Another shield device is the drowning or flooding of crops, especially cranberries, to prevent low temperatures from damaging tender berries near the end of the growing season.

Thus frost is (1) a form of mineral, (2) a climatic condition, and (3) an economic situation.

In the literature, in addition to terms already mentioned, are such terms as frost-free period (growing season), frost density, killing frost, frost warning, frost circles, frost zones, mush frost, late frost and early frost. Each of these terms have limited interpretations, the use of which will reflect the science with which it is associated.

Benjamin Moulton

Bibliography

Critchfield, H.J., 1974. *General Climatology*, 3rd edn. Englewood Cliffs, NJ: Prentice-Hall.
Geiger, R., 1966. *The Climate Near the Ground*, rev. ed. Cambridge, MA: Harvard University Press.
Schaefer, V.J., 1964. Preparation of permanent replicas of snow, ice, and frost, *Weatherwise*, **17**, 278–287.

Cross references

Agroclimatology
Climate data: sources
Dewpoint
Dew pond

FROUDE NUMBER (*Fr*)

This is a dimensionless parameter devised by W. Froude (1810–1879) to represent the ratio between the inertia and gravity forces in a liquid. In open channels the number is given by the expression:

$$F_r = \frac{\bar{V}}{\sqrt{Dg}}$$

where *Fr* is the Froude number, $\bar{V}$ is the mean velocity of the water, g is the acceleration due to gravity (acceleration of free fall) at the geographical location in the channel and D is the mean depth of water at the same location.

A number of other scientists have proposed similar dimensionless ratios, including Th. Rehbock, Boris A. Bakhmeteff, J. Boussinesq, F.V.A.E. Engel, J.C. Stevens and C.J. Posey, but the Froude number has enjoyed general acceptance because of the simplicity of measuring the number's components in open-channel hydrometry. A Froude number of unity represents **critical flow**, with less than unity and more than unity representing **subcritical flow** and **supercritical flow**, respectively.

P.G. Holland

Bibliography

BS 3680, Part 1, 1991. *Glossary of Terms*. British Standards Institution, London.

Cross reference

Water resources: dictionary of basic terms

G

GAUGE (GAGE)

A gauge is the device installed at a gauging station in an open channel for measuring the water surface relative to a datum. Gauges can take several forms but the following are the commonest in use.

- Vertical gauge. A graduated vertical scale, fixed to a staff or a structure, against which may be read the water level. Also known as a staff gauge.
- Inclined gauge. A gauge on a slope, generally graduated directly to indicate vertical heights. Also known as a ramp gauge. Inclined gauges are used when the banks of an open channel have a significant shallow slope. Their use in these circumstances reduces the incidence of debris accumulating against the gauge (so causing waves and observational inaccuracies) and also permits a greater accuracy of reading.
- Float gauge. A gauge consisting essentially of a float that rides on the water surface, usually within a stilling well, and rises or falls with the water surface. Usually the rise and fall is transmitted to a recording or an indicating device. A float gauge is not to be confused with a float used for determining water velocities.
- Point gauge. A water level measurement gauge, the essential element of which is a pointed rod that is lowered until it touches the water surface. The instant when the part touches the water surface often is indicated by an electrical device displaying a light signal or a sonic noise. A point gauge can take the form of a fixture to a vertical lined bank (Figure G1) but with the pointer touching the surface, not protruding into the water, or it can be attached to a hand-held or instrument-held steel measuring tape for reading the vertical distance of the water surface below a known datum.

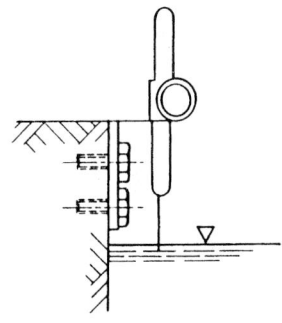

Key

▽ is the level of the water.

Figure G1 Point gauge.

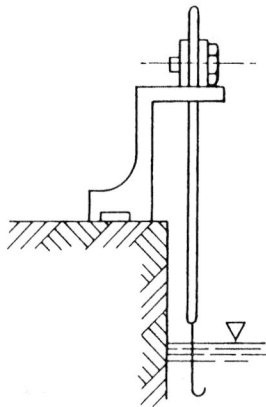

Key

▽ is the level of the water.

Figure G2 Hook gauge.

- Hook gauge. A water level measurement gauge, the essential element of which is an inverted pointed hook that, after immersion, is raised until it touches the water surface. Its mode of use is similar to that of the point gauge (Figure G2) but the alternative of using it with a measuring tape is not commonly employed.
- Crest stage gauge. A gauge, usually vertical, used to indicate the flood peak stage in an open channel.
- Wire weight gauge. A graduated gauge connected to a weighted wire or chain, which is lowered to make contact with the surface of the liquid. (Note: the contact with the liquid is determined visually.)

P.G. Holland

Bibliography

BS 3680, Part 1, 1991. *Glossary of Terms*. British Standards Institution, London.

Cross reference

Water resources: dictionary of basic terms

GAUGING STATION

In hydrometry, this is the most important location in an open channel and it is the site at which systematic measurements of water level or discharge or both are made. Gauging stations are at their most accurate where the profile of the measuring section, or of the measuring reach, is stable. There may be some form of control of the water level, either by means of the natural channel bed profile, such as a rock outcrop across the width of the channel, or by structural means in the form of a weir or a flume. The profile of the channel, and the form of control, decrees the actual composition of the gauging station but there is always a means of measuring the water level, either continuously or by manual observation, and if there is no structure, or a weir or a flume which cannot accommodate the full range of the discharges, there has to be some means of measuring water velocities.

The water level measurement is by means of a gauge and it is usual for a recording gauge to be installed within a recorder house built over a stilling well. This well comprises an annular ring, of 1 m diameter or more, set vertically into the channel bank with one or two horizontal open-ended small-bore pipes taken from it into the open channel at a level at or near to the bed level of the channel. The recorder is erected over the top of the well with the result that the water level in the well, which is the same as the mean water level in the channel, is devoid of any wind-induced or other surface turbulence and an accurate determination of the water level is obtained. There are circumstances when a stilling well is not needed and an example of such a gauging station is shown (Figure G3), located at the Todd River, Alice Springs, Northern Territory, Australia. This river dries up for many months of the year but, when the rains do come, there are enormous floods for which it is necessary to record the flood peaks for property insurance and general record purposes.

Figure G3 A gauging station at the Todd River bed, Alice Springs, Australia.

For certain types of gauging station there may be a need for two water level measurement locations, either to determine the slope of the water surface through a reach or to measure the water level upstream of and downstream of a structure. Such a gauging station is known as a twin-gauge station. Where there is only one water level measuring location, the gauging station is known as a single-gauge station.

The measurement of complementary velocities at a gauging station, in order to determine discharges is achieved by current metering.

P.G. Holland

Bibliography

BS 3680, Part 1, 1991. *Glossary of Terms*, British Standards Institution, London

Cross references

Gauge
Streamflow measurement
Water resources: dictionary of basic terms

GHYBEN–HERZBERG THEORY

Saltwater encroachment or intrusion is the shoreward movement of water from a sea or ocean into confined or unconfined coastal aquifers and the subsequent displacement of fresh water from these aquifers.

The hydrostatic equilibrium between immiscible freshwater and saltwater bodies in contact with each other along a certain interface was studied first by Ghyben and Herzberg (De Wiest, 1965). The equation for the depth of the interface is (Figure G4)

$$z_s = \frac{\rho_f}{\rho_s - \rho_f} z_w \tag{G1}$$

where ρ_f is the density of fresh water, ρ_s is the density of salt water, z_w is the height of the freshwater table above mean sea level, and z_s is the depth of the interface below sea level. This equation is in good agreement with measurements made in the field indicating that for every meter of fresh water above mean sea level, the thickness of the freshwater lens resting on the salt water was about 40 m. The limitations of the hydrostatic theory are obvious: if both fluids were truly in static condition, the water table would have zero slope and the interface would become horizontal, with fresh water overlying salt water by mere density difference. Furthermore, fresh water is in a continuous state of motion due to changes in the water table, for example because of replenishment, evaporation and discharge. It has been recognized for a long time that fresh water seeps into the ocean above sea level. Such water was tapped in earlier times as potable water for use on sea-going vessels. The escape of fresh water below sea level was not considered either in the Ghyben–Herzberg theory.

Hubbert, pointing at the dynamic rather than the hydrostatic equilibrium of the freshwater/saltwater interface, showed the discrepancy between the actual depth to salt water and the depth as calculated by the Ghyben–Herzberg formula for flow conditions near the shore line (Figure G5).

Hubbert's concepts were confirmed and extended by Luscynski (De

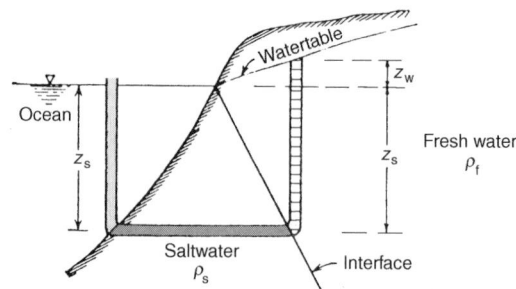

Figure G4 Saltwater intrusion according to the Ghyben–Herzberg theory. (Courtesy R.J.M. de Wiest.)

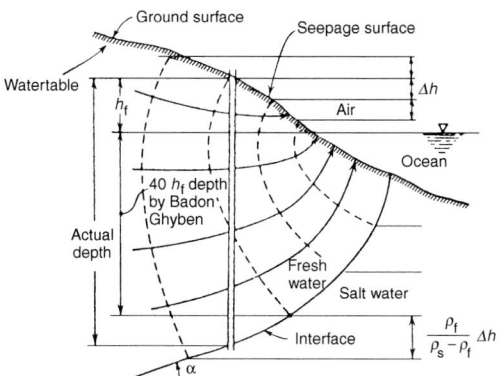

Figure G5 Discrepancy between actual depth to salt water and depth calculated by Ghyben–Herzberg theory. (Courtesy M.K. Hubbert.)

Wiest, 1965) who took into account the existence of a zone of dispersion between fresh water and salt water. Contingent upon the reading of water levels in some observation wells, Luscyznski's work allows for the computation of a three-dimensional velocity picture in a medium where the salt content of the water gradually varies. This picture may be constructed regardless of the often complicated boundary conditions created by the geological nature of the aquifers which preclude a complete analytical solution of the problem. The value of Lusczynski's work resides in its practical application.

Roger J.M. De Wiest

Bibliography

Davis, S.N. and De Wiest, R.J.M., 1966. *Hydrogeology*, New York, John Wiley & Sons, 463 pp.

De Wiest, R.J.M., 1965. *Geohydrology*, New York, John Wiley & Sons, 366 pp.

De Wiest, R.J.M., 1969. *Flow through Porous Media*, New York, Academic Press, 530 pp.

Hubbert, M.K., 1940. The theory of groundwater motion, *J. Geol.*, **48**(8), 785–944.

Meinzer, O., 1942. *Hydrology*, New York, Dover, 712 pp.

Muskat, M., 1937. *The Flow of Homogeneous Fluids Through Porous Media*, New York, McGraw-Hill Book Co., 763 pp.

Todd, D.K., 1959. *Ground Water Hydrology*, New York, John Wiley & Sons, 336 pp.

Tolman, C.F., 1937. *Ground Water*, New York, McGraw-Hill Book Co.

Cross references

Groundwater
Hydrology
Hydrology: coastal terrain
Water table

GLOBAL POSITIONING SYSTEM (GPS): OVERVIEW

The Navstar Global Positioning System (GPS), is a US Department of Defense space-based radio navigation system that provides continuous, all-weather, global navigation capability, and is available to all worldwide users free of charge.

GPS is the most ambitious global navigation system ever attempted. Extremely accurate three-dimensional position and velocity accuracy is achieved through high-technology user equipment which tracks the satellites. For real-time navigation, the user equipment tracks four satellites to solve for four unknowns, longitude, latitude, altitude and time. If only three satellites are visible, altitude must be entered into the user equipment to get the three unknowns of longitude, latitude and time (Ashtech Europe, 1996).

Background

The use of heavenly bodies (i.e. the Sun, the Moon, stars and planets) for purposes of navigation was started centuries ago and was used extensively by the early Portugese navigators to explore this planet. These early explorers made 'position fixes' by combining the known positions of the heavenly bodies with on-board position measurements made with an instrument called an 'astrolabe'. The results of this simple technique were sufficiently accurate to allow navigators to find their approximate positions even when far from land. In the eighteenth century, the sextant, compass and star and Sun tables were integrated with the clock to improve navigation performance significantly. Later, a technological breakthrough in the use of radio signal direction-finding for navigation produced significant advances in the accuracy of position fixing.

The use of artificial Earth satellites for purposes of navigation originated with Sputnik I in October 1957. Satellite navigation combined the methods of celestial navigation, as used by the early explorers, with those of radio navigation to achieve systems having revolutionary improvements in accuracy and performance. The fundamental difference between artificial satellite navigation using radio signals and other radio navigation methods is simply the geometry. Space offers the opportunity for line-of-sight signal propagation over vast areas of the world, so the usual trade-off of less accuracy for greater range is not involved. Also, since satellite signals penetrate the ionosphere rather than being reflected by it, difficulties encountered with 'sky waves' are eliminated. Artificial Earth satellites are obviously desirable platforms from which to provide navigational services, but these advantages have been gained at the price of increased sophistication. In the 40 years since Sputnik 1, space technology has generated a positioning system which provides users with position accuracies of a few meters or so, velocity measurements to within a 0.1 m s^{-1} and time readings within a few billionths of a second.

The development of Transit I, the first navigation satellite system, was triggered by observations made on signals from the first Sputnik. Officially begun in December 1958, Transit I resulted in a worldwide navigation system which has been in continuous operation since January 1964. Transit basically allowed a calculation perpendicular to the orbit plane. Fixes would require measurements from two satellites separated in time. Transit was scheduled to phase out in 1995.

It became obvious that a global navigation satellite system had much to offer in terms of accurate all-weather, continuous, worldwide navigation capability. Consequently, the Defense Department established requirements for a DOD tri-service worldwide navigation satellite system. Between 1967–1969 preliminary concept formulations and system design studies were conducted by the US Air Force for such a system, which was designated System 621B. As a result of these efforts, combined with mission analyses and parametric studies, a space-based navigation system was developed that called for 20 satellites, deployed in synchronous orbits, whose ground tracks formed four 'eggbeater-shaped clusters' extending to 60° N and S latitudes. Satellite tracking and control was to be maintained from ground stations in the continental United States through inter-satellite links. This inter-satellite tracking approach minimized the vulnerability of the system to physical attack on ground stations. System 621B was designed to make direct, simultaneous range measurements from at least three satellites and instantly compute a position fix at the intersection of three spheres with centers at the satellites. Simultaneous range measurements from a fourth satellite eliminated the need for synchronization of satellite user clocks, since the time bias could be calculated in the navigation solution, which consisted of using four measurements to solve for the three unknown positions and one time.

Concurrently with the USAF space-based navigation studies, the Naval Research Laboratory (NRL) conceived the idea of a timing/navigation satellite system (TIMATION). Development of the TIMATION system was to consist of two phases. Phase I (TIMATION I, II and III) was initiated primarily as a technology effort to investigate the behavior of high-stability crystal oscillators in low-altitude orbits and to verify the TIMATION technique. Phase II involved the development and deployment of the operational system. The TIMATION concept involved making direct range measurements from the satellite to the user, with time-delay readings being taken each minute during a satellite pass. The direct range measurements were made by making phase measurements on several side tones modulated on a carrier signal. TIMATION I and II, which were launched in 1967 and 1969, performed precise time transfer, navigation and geodesy experiments and transmitted both side-tone-ranging (STR) and pseudo-random noise (PRN) signals. TIMATION satellites were used to conduct navigation and time transfer experiments. TIMATION III, subsequently identified as the NTS-1 (GPS Navigation Technology Satellite 1), was launched in mid-1974 with the first spaceborne atomic clock. As a result of these studies, a TIMATION global navigation system was proposed by the Navy, using 21–27 satellites in medium (8 h) orbits with both STR and PRN signals.

The USAF system 621B and the Navy TIMATION system were both candidates for the DOD Navigation Satellite System. Budgetary constraints would not permit the deployment of two independent systems. The compromise configuration consisted essentially of orbits proposed by the Navy to permit evolutionary deployment and the signal structure and frequencies proposed by the USAF for maximum user performance and a new satellite control approach for minimum cost. The program resulting from this composite effort is the Navstar Global Positioning System.

The GPS program was approved in 1973 for a demonstration phase at a minimum cost. Launch vehicles came from inactivated Atlas missiles which had been in the wheat fields in Kansas. Six satellites were approved and funded to provide the demonstration. Satellite control was through a dedicated control station. Military development of the first user equipment proceeded. The accuracy test results met or exceeded all expectations. As a result, production of 28 satellites was authorized and the contract awarded in 1982 to build the satellites and launch them on the shuttle. The Shuttle accident occurred just prior to the launching of the first production satellite. As a consequence, about $1 billion was appropriated to procure the Delta IIs to launch the satellites and a 3-year delay in the deployment of the full constellation resulted.

The GPS system

The current baseline satellite orbital configuration consists of 21 primary satellites in 55° inclined (relative to the equator) circular 12 h orbits to transmit navigational signals. The six planes of the operational constellation are 60° apart in longitude. The continuous four-satellite global coverage is provided by placing four unequally spaced satellites in each of six orbit planes. The satellites are positioned so that any three satellite outages caused by failure or an unhealthy satellite will still give the best system results. Thus the baseline 21 primary satellite constellation has three active spare satellites in orbit for the military Air Force Space Command commander to keep 21 satellites operational at all times.

The satellites radiate two spread-spectrum pseudorandom noise (PRN) radio signals. The navigation message, which consists of onboard clock and satellite ephemeris information, is modulated onto the PRN sequence. The navigation signals are transmitted on two frequencies L1 (1575.42 MHz) and L2 (1227.6 MHz). Both are coherently derived from a highly stable onboard atomic clock.

The control segment consists of a master control station located at Falcon AFB near Colorado Springs, CO, USA, five monitor stations and three ground antennas. The radiometric data from the satellites are tracked by the monitor stations. Accurate ephemeris and clock parameters are estimated by extensive data processing at the master control station. Predicted ephemeris and clock information in the form of navigation messages is periodically transmitted (uplinked) to the satellites from the ground antennas for later transmission to users. The control segment is also responsible for maintaining the health of the satellites.

The user equipment has an antenna, receiver, signal processing and data processing capabilities. The satellite-transmitted radio signal is received by the equipment knowing the signal PRN code, obtaining pseudo-range data and demodulating the navigation message. Data from four satellites allow the user state vector (consisting of position, velocity and time) to be computed.

R.W. Herschy

Source and bibliography

Ashtech Europe Ltd, 1996. *The Global Positioning System (GPS) Overview*, Ashtech Europe Ltd, UK.

Cross reference

Remote scanning

GLOBAL WARMING

The world's climate has never remained stationary. However, increased emissions of greenhouse gases are believed to cause the trapping of the Sun's heat and a small increase in world temperature.

It is today generally agreed that some 6 billion tonnes of carbon in the form of carbon dioxide are emitted into the atmosphere annually as a result of the burning of fossil fuels. Carbon dioxide concentrations are estimated to account for 57% of the greenhouse effect, while 14% is due to methane and 5% due to nitrous oxide as a result of the use of nitrogen fertilizers. It is forecast that this may double in the next 50 years, which could lead to global warming of 1–5°C over the next century, having catastrophic effects on the Earth's ecosystems. Already there are regions which are experiencing the consequences of global warming where, for example, glaciers are shrinking.

Models show that by the year 2045 average temperatures for the south of England will increase by 2°C. Changes in air temperature will also change river water temperature estimated in the UK by 1–2°C and would have considerable significance for the ecology of streams. The increased temperature and changed climate circulation will result in a change in precipitation. Winter precipitation in southern England would increase by 10–15% and summer precipitation would increase in northern England and Scotland, but with a decrease in southern England of about 10%.

The increased temperature would result in an increase in evapotranspiration and hence less moisture in the soil. Winter stream flows would increase and summer flows reduce by as much as 20%. Single-season reservoirs where there is appreciable summer inflow could have less yield, and those reservoirs whose yield is based on small inflows during drought conditions would be drawn down more frequently. In agriculture there could be severe reductions in food production.

R.W. Herschy

Bibliography

Binnie, C.J.A. Presidential address, 1995. The Chartered Institution of Water and Environmental Management. *Newsweek*, Nov. 6, 1995.

Cross references

Antarctic ozone hole
Climate and climate change
Climate change and ancient civilization
Climate data: sources
Greenhouse effect

GREENHOUSE EFFECT: GENERAL

The term greenhouse effect has frequently been attributed in the media and in the scientific literature as the principal mechanism of global climate change. Its name stems from the phenomenon of local warm-up observed in greenhouses with glass roofs and walls used for horticulture. Incident solar radiation penetrates through the transparent glass of a greenhouse which is partly opaque to the longer-wave energy reradiated from the Earth's surface. A part of this infrared radiation is trapped in the glass of the greenhouse and warms up the closed volume. A similar effect is produced by polyatomic molecules in the atmosphere, which absorb part of infrared radiation emitted by the Earth's surface which otherwise would have escaped to the outer space. This mechanism warms up the Earth's surface and the lower atmosphere while cooling the upper atmosphere.

The principal greenhouse gases are water vapor, carbon dioxide, methane, nitrous oxide and freons. Although the volumes of the last three gases in the atmosphere are far lower than those of the carbon dioxide, their collective contribution to the greenhouse effect is comparable to that of carbon dioxide. Methane as a greenhouse gas is some 25 times stronger than carbon dioxide and freons are as much as 10 000 times stronger.

The greenhouse effect is the principal mechanism maintaining the Earth's climate within the range suitable for sustaining life. Without it the global mean annual temperature would be 33°C lower. However, the present concern is that the atmospheric concentration of greenhouse gases will grow further causing stronger warming which could damage the adaptation of nations to definite and stationary climatic conditions.

The increase of anthropogenic emissions of carbon dioxide (increased combustion of fossil fuels and deforestation) and some other greenhouse gases (such as methane and freons) and of their concentration in the atmosphere have been observed. For example, the atmospheric concentration of carbon dioxide has grown by over a quarter in the last two centuries. Over the last decades the atmospheric concentration of carbon dioxide has grown at the mean rate of 1.4 ppm per year, which corresponds to the addition of 3.3 gigatonnes of carbon to the total mass of carbon dioxide already present in the atmosphere. In the case of CFCs, i.e. the family of chemicals produced synthetically by humans, the natural background concentration was equal to zero.

The issue of climate change has raised much recent interest and dispute. Although a global temperature rise of ~0.6°C has been observed in the last 100 years, this rise is not uniform. There have been episodes of stronger growth of Earth surface temperature observed in 1890s and 1920–1930s, while in 1940s a cooler period started, which turned into a warming tendency only after mid-1970s. Yet translating these tendencies into impacts on water resources is at present subject to high uncertainty. No general and unambigous evidence of change has been deciphered in hydrological data.

Zbigniew W. Kundzewicz

Cross references

Antarctic ozone hole
Climate and climate change
Greenhouse effect

GREENHOUSE EFFECT

The so-called greenhouse effect is the mechanism whereby the equilibrium magnitude of the Earth's surface is maintained at a higher temperature than that needed to balance the solar radiation absorbed by the Earth–atmosphere system. On average, the atmosphere is more transparent to solar radiation than to terrestrial radiation. Only a small proportion of the terrestrial radiation emitted by the Earth's surface passes directly to space; most is absorbed by the atmosphere. In accordance with Kirchoff's law, the atmosphere emits terrestrial radiation but does so at a lower temperature than the surface. To achieve an energy balance, the whole system (surface and atmosphere) must maintain a higher temperature than would have been the case without terrestrial radiation absorption in the atmosphere.

The term 'greenhouse effect' derives from the mistaken view that higher temperatures within greenhouses are maintained by an analogous process resulting from the glass cover's high transparency to solar radiation and high absorptivity of terrestrial radiation. In fact, this effect is but a small factor in producing the elevated greenhouse temperature, which arises primarily from restriction of vertical diffusion of heated air within the structure. The alternative expression 'atmospheric effect' was proposed (Fleagle and Businger, 1963) but has failed to gain currency.

The concept of planetary equilibrium temperature is an important one in understanding the greenhouse phenomenon. The Earth must maintain a mean temperature at which its terrestrial radiation emission per unit area equals the solar radiation absorbed per unit area. If these two radiant flux densities are not equal, the Earth–atmosphere system will warm (if solar input exceeds terrestrial output) or cool (if output exceeds input), until a new temperature is attained at which radiation balance is achieved.

The solar constant for the Earth is 1353 W m^{-2}. This is intercepted by the cross-sectional area of the planet, assumed to be spherical; hence the flux intercepted is $1353 \times \pi R^2$ watts, where R is the radius of the Earth. The average flux density for the total surface area of the planet is obtained by dividing the above flux by the area of the planetary surface, $4\pi R^2$. Hence the mean solar irradiance of the Earth is $1353/4 \approx 338$ W m^{-2}. Of this amount, about 28% is backscattered or reflected to space by the gases of the Earth's atmosphere, clouds, aerosols and the surface. Thus, the *absorbed* solar radiation, per unit

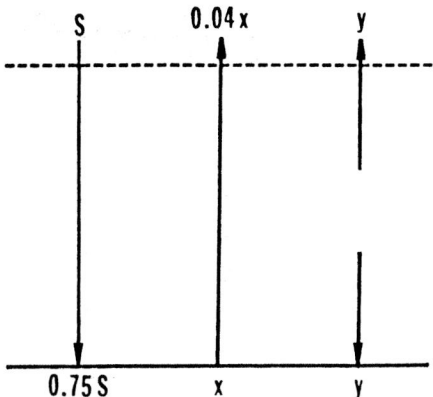

Figure G6 Simplified energy budget of the earth–atmosphere system.

area of surface, is $(1 - 0.28) \times 338 \sim 243$ W m^{-2}. If the Earth had no atmosphere (or had an atmosphere that was transparent to solar and terrestrial radiation), each square meter of the surface would, on average, emit 243 W m^{-2} to maintain an equilibrium with the solar input and hence a stable climate. The Stefan–Boltzmann law shows that the surface, treated as a black body, must exhibit a mean temperature of 255.9 K (-17.3°C) to achieve this balance. This is about 32°C cooler than the observed mean temperature of the Earth's surface, which is near 15°C, and represents a considerably less hospitable global climate than actually exists.

The surface emits terrestrial radiation in response to solar heating. Only a very small proportion of this radiation escapes directly to space through the 'atmospheric window.' The remainder is absorbed by water vapor, carbon dioxide, ozone, clouds, aerosols and some trace gases. By virtue of Kirchoff's law, these substances in the atmosphere also emit terrestrial radiation, both upwards and downwards. This radiation is, in turn, absorbed by the radiatively active constituents of the atmosphere and by the surface, with only a small proportion of the upward emission reaching space directly, primarily in the peripheral regions of the gaseous absorption bands and in the 'atmospheric window.' This process involves a complex series of radiative exchanges among different levels in the atmosphere. In the present context, these exchanges produce further components of terrestrial radiation loss to space emanating from various elevations in the atmosphere that augment the loss from the surface. Since most of this radiation is emitted within the troposphere, in which temperature decreases with height, these additional components of terrestrial radiation loss originate from bodies cooler than the surface. Hence the irradiance emerging from the top of the atmosphere is smaller than that leaving the surface. Radiative balance can be achieved only if the loss from the total Earth–atmosphere system equals 243 W m^{-2}, a requirement that necessitates higher temperatures at the surface (and throughout the troposphere) than would be the case with no atmosphere.

This may be illustrated quantitatively using the following highly simplified example (based on Wallace and Hobbs, 1977), in which the atmosphere is represented as a single layer possessing a uniform temperature (Figure G6). S is the mean available solar radiation flux density (243 W m^{-2}), x is the terrestrial radiation emission from the Earth's surface and y is that from the atmosphere. If the absorptivities of the atmospheric layer for solar and terrestrial radiation are 25 and 96%, respectively, the balance for the Earth's surface may be written

$$0.75S + y = x$$

and that for the whole system is

$$S = 0.04x + y$$

Solving simultaneously,

$$x = 1.68S \text{ and } y = 0.93S$$

For the Earth's surface, using the Stefan–Boltzmann law,

$$\sigma T^4 = x = 1.68S = 1.68 \times 243 = 408 \text{ W m}^{-2}$$

giving $T = 291.3$ K $= 18.1°C$. For the atmosphere, using also Kirchoff's Law,

$$0.96\sigma T^4 = y = 0.93S = 0.93 \times 243 = 226 \text{ W m}^{-2}$$

giving $T = 253.8$ $K = -19.4°C$.

Clearly, these results are only approximate and the estimate of the Earth's surface temperature is 3°C too high. This error results from use of an oversimplified model of the energy budget, particularly the neglect of convective losses of sensible and latent heat from the surface and the use of a single-layer atmosphere emitting equal quantities of radiation in both directions. Nevertheless, this calculation serves to illustrate the significance of the greenhouse effect that has produced, in this case, an increase of 35°C in the equilibrium surface temperature of the Earth, while maintaining energy balance at the upper boundary of the atmosphere.

Since the classical work of Chamberlin and Arrhenius at the beginning of the twentieth century (Schneider, 1975), climatologists and others have speculated on the significance of changing atmospheric composition on the efficiency of the greenhouse effect and, hence, on global climate. Attention has been focused primarily on changes in atmospheric carbon dioxide concentration brought about by the combustion of fossil fuels and the clearing of natural vegetation, although models of the radiative effects of other trace gases have also revealed possible effects (Wang et al., 1976). Estimates of the expected increase in atmospheric carbon dioxide concentration vary, depending on the assumptions made regarding alternate reservoirs for the additional carbon, but increases of the order of 20–25% during the second half of the twentieth century are generally accepted (Schneider, 1975). Some workers have suggested that by the third or fourth decade of the twenty-first century human activity will have doubled the pre-industrial revolution concentration (Schneider, 1975).

Numerous attempts have been made to evaluate the impact of such a change on the greenhouse effect and to determine the resulting temperature changes within the Earth-atmosphere system. Schneider (1975) summarized much of the early work (primarily one- and two-dimensional radiation budget and radiative–convective models) and concluded that the best estimate of the effect of doubling carbon dioxide concentration on surface temperature was an increase of between 1.5°C and 3.0°C. More recent investigations, using three-dimensional general circulation models (Manabe and Stouffer, 1980; Manabe and Wetherald, 1980; Mitchell, 1983) have tended to substantiate expected temperature increases in the vicinity of 3°C, with the likelihood of greater warming at high latitudes, over the continents, and in the winter half of the year. Such numerical experiments also have shown cooling in the stratosphere and an increase in the vigor of the hydrological cycle.

Important work in recent years has led to a better understanding of the greenhouse effect and to climate change. This work has been performed by the Intergovernment Panel on Climate Change (IPCC) (Houghton et al., 1990, 1996; Tegart et al. 1991; Watson et al., 1996). As a result of this detailed work which comprised a series of working groups on most, if not all, aspects of the subject including scientific assesssment, impacts, adaptions and mitigations, estimates were presented of key greenhouse gases affected by human activity (Table G1) and estimates of climate change for five regions of the world by

Table G1 A summary of key greenhouse gases affected by human activities (Watson et al., 1996)

	CO_2	CH_4	N_2O	CFC-12	HCFC-22 (a CFC substitute)	CF_4 (a perfluorocarbon)
Pre-industrial concentration	280 ppmv	700 ppbv	275 ppbv	Zero	Zero	Zero
Concentration in 1992	355 ppmv	1714 ppbv	311 ppbv	503 pptv[a]	105 pptv	70 pptv
Recent rate of concentration change per year (over 1980s)	1.5 ppmv/year^{-1} 0.4% year^{-1}	13 ppbv/year 0.8% year^{-1}	0.75 ppbv/year 0.25% year^{-1-1}	18–20 pptv/year 4% year^{-1-1}	7–8 pptv/year 7% year^{-1-1}	1.1–1.3 pptv/year^{-1} 2% year^{-1-1}
Atmospheric lifetime (years)	(50–200)[b]	(12–17)[c]	120	102	13.3	50 000

[a] 1 pptv = 1 part per trillion (million million) by volume.
[b] No single lifetime for CO_2 can be defined because of the different rates of uptake by different sink processes.
[c] This has been defined as an adjustment time which takes into account the indirect effect of methane on its own lifetime.

Table G2 Estimates for changes by 2030

Central North America (35–50°N, 85–105°W)	The warming varies from 2 to 4°C in winter and 2 to 3°C in summer. Precipitation increases range from 0 to 15% in winter whereas there are decreases of 5–10% in summer. Soil moisture decreases in summer by 15–20%
Southern Asia (5–30°N, 70–105°E)	The warming varies from 1 to 2°C throughout the year. Precipitation changes little in winter and generally increases throughout the region by 5–15% in summer. Summer soil moisture increases by 5–10%
Sahel (10°–20°N, 20°W–40°E)	The warming ranges from 1 to 3°C. Area mean precipitation increases and area mean soil moisture decreases marginally in summer. However, throughout the region, there are areas of both increase and decrease in both parameters throughout the region
Southern Europe (35–50°N, 10°W–45°E)	The warming is about 2°C in winter and varies from 2 to 3°C in summer. There is some indication of increased precipitation in winter, but summer precipitation decreases by t–15%, and summer soil moisture by 15–25%
Australia (12–45°S 110–115°E)	The warming ranges from 1 to 2°C in summer and is about 2°C in winter. Summer precipitation increases by around 10%, but the models do not produce consistent estimates of the changes in soil moisture. The area averages hide large variations at the subcontinental level.

[a] IPCC business-as-usual scenario; changes from pre-industrial. The numbers given are based on high-resolution models, scaled to be consistent with our best estimate of global mean warming of 1.8°C by 2030. For values consistent with other estimates of global temperature rise, the numbers should be reduced by 30% for the low estimate or increased by 50% for the high estimate. Precipitation estimates are also scaled in a similar way. Confidence in these regional estimates is low.

Table G3 Estimated percentage changes in mean annual runoff at three sites in the UK in response to changes in temperature and precipitation[a]

Temp. change (°C)	Change in annual precipitation					
	−10%	0%	10%	20%	30%	40%

(a) Initial mean annual precipitation, 1000 mm; initial mean annual temperature, 8°C; initial estimated mean annual runoff, 530 mm						
1	−21.3	−4.8	12.3	29.7	47.4	65.3
2	−25.8	−9.6	7.1	24.2	41.7	59.4
3	−30.4	−14.7	1.7	18.5	35.8	53.3
4	−35.1	−19.8	−3.8	12.7	29.6	46.8
5	−39.8	−25.0	−9.5	6.6	23.2	40.1

(b) Initial mean annual precipitation, 1000 mm; initial mean annual temperature, 10°C; initial estimated mean annual runoff, 479 mm						
1	−23.0	−5.6	12.6	31.2	50.2	69.6
2	−28.1	−11.2	6.4	24.7	43.4	62.5
3	−33.3	−17.0	0.2	18.0	36.3	55.1
4	−38.5	−22.8	−6.2	11.1	29.0	47.4
5	−43.7	−28.7	−12.7	4.1	21.5	39.5

(c) Initial mean annual precipitation, 800 mm; initial mean annual temperature, 10°C; initial estimated mean annual runoff, 311 mm						
1	−26.1	−7.0	13.3	34.4	56.3	78.8
2	−32.2	−14.0	5.5	26.0	47.3	69.3
3	−38.3	−20.9	−2.3	17.4	38.1	59.5
4	−44.2	−27.8	−10.1	8.9	28.8	49.5
5	−49.9	−34.6	−17.8	0.3	19.4	39.4

[a] Based on the Turc formula (Beran and Arnell, 1989).

the year 2030 (Table G2). An example for the UK is given in Table G3.

A. John Arnfield

Bibliography

Beran, M. and Arnell, N., 1989. *Effect of climate change on quantitative aspects of United Kingdom water resources*, Institute of Hydrology, Wallingford, 93 pp.

Fleagle, R.G. and J.A. Businger, 1963. *An Introduction to Atmospheric Physics*. New York: Academic Press.

Houghton, J.T., Jenkins, G.J. and Ephraums, J.J., 1990. *Climate Change, Scientific Assessment*, IPCC. Cambridge: Cambridge University Press.

Houghton, J.T., Meira Filho, L.G., Callander, B.A. *et al.*, 1996. *Climate Change, 1995, The science of climate change*, IPCC. Cambridge: Cambridge University Press.

Manabe, S. and R.J. Stouffer, 1980. Sensivity of a global climate model to an increase of CO_2 concentration in the atmosphere. *J. Geophys. Research*, **85**, 5529–5554.

Manabe, S. and R.T. Wetherald, 1980. On the distribution of climate change resulting from a change in the CO_2 content of the atmosphere, *J. Atmos. Sci.* **37**(1), 99–118.

Matthews, W.H., W.W. Kellogg and G.D. Robinson, 1971. *Man's Impact on the Climate*. Cambridge, MA: MIT Press.

Mitchell, J.F.B., 1983. The seasonal response of a general circulation model to changes in CO_2 and sea temperatures, *Royal Meteorol. Soc. Quart. J.*, **109**(459), 113–152.

Schneider, S.H., 1975. On the carbon dioxide-climate confusion, *J. Atmos. Sci.*, **32**(11); 2060–2066.

Tegart, W.J.McG., Sheldon, G.W., Griffiths, D.C., 1990. *Climate Change, Impacts Assessment*, IPCC. Canberra: Government of Australia.

Wallace, J.M. and P.V. Hobbs, 1977. *Atmospheric Science: An Introductory Survey*. New York: Academic Press.

Wang, W.C., Y.L. Yung, A.A. Lacis *et al.*, 1976. Greenhouse effects due to man-made perturbations of trace gases, *Science*, **194**, 685–690.

Watson, R.T., Zinyowera, M.C., Moss, R.H., Dokken, D.J., 1996. *Climate Change, 1995, Impacts, Adaptions and Mitigations*. Cambridge: Cambridge University Press.

Cross references

Antarctic ozone hole
Atmosphere
Climate and climate change
Climate change and the greenhouse effect
Greenhouse effect: general

GROUNDWATER

Groundwater occurs in the natural openings in the soils and rocks of the Earth's crust. These spaces range from large caverns, which may have dimensions of the order of tens or hundreds of meters, to pore spaces which may be less than 1 μ m in size between the grains or crystals of rocks. In the soil and near-surface rocks, it is usual for smaller pores to be filled or partly filled with water which is at a pressure less than that of the atmosphere, and for larger pores to be mainly filled with air; this region is called the unsaturated zone. At greater depths all the pores will be completely filled with water which is at a pressure greater than atmospheric pressure; this region is called the saturated zone, and the surface separating the two zones is the water table (Figure G7) Some definitions of groundwater include all the water in pore spaces in rocks, but most restrict the term to water in the saturated zone, i.e. below the water table.

Some rocks and soils are more porous than others, and so will be better stores of water. More importantly, the pores in some rocks are larger or better connected, so that water can move easily through these materials, which are said to be permeable. Rocks that are porous and permeable enough to store and transmit water in useful quantities are called aquifers.

After the oceans, groundwater probably represents the largest store of water on Earth. The distribution of groundwater within the Earth's crust is controlled by the distribution of porosity, whereas the circulation of water is controlled by the distribution of permeability. Both of these properties tend to decline with depth below the Earth's surface, the pores being closed by the pressure exerted by the overlying rocks, so that at depths of a few kilometers the rocks usually contain little free water, and this water is scarcely moving. At greater depths, the rocks begin to flow and all pore spaces are in effect closed. The 'Superdeep' well drilled in the Kola Peninsula in Russia encountered water moving in fracture systems to a depth of 9000 m; below this, water was still present to 12 000 m, but the fractures containing it appeared to be isolated and discontinuous.

Thus groundwater extends over such an interval of the Earth's crust that in the upper part of the saturated zone it may be intimately related to soil water, whereas in the lower part of the saturated zone it may be involved in tectonic processes.

It is generally believed that groundwater can originate in three ways. The most important of these from the human standpoint is meteoric water, which is derived from precipitation and infiltration within the hydrological cycle. Meteoric water accounts for most of the groundwater that is abstracted for human use or which discharges naturally from aquifers to provide river baseflow. This water enters the soil, and moves downwards through the unsaturated zone of the underlying aquifer until it reaches the water table. The water table rises in response to infiltration until at some locations it reaches the ground surface. This will usually happen at the lowest points on the ground surface, normally in river valleys. At these points, groundwater flows from the aquifer as springs or seepages (Figure G8).

Groundwater can also originate as water trapped in sedimentary rocks at the time of their deposition. Some of this entrapped water will be expelled from the pore spaces in the sediments as they are buried and compacted under the weight of the sediments deposited on top of them, but some may remain for millions of years. This entrapped water is termed connate water, and it is usually regarded as water that has been removed from the hydrological cycle for a long (in geological terms) time. Because many sedimentary rocks are of marine origin, much connate water is saline, but not all saline water is connate; meteoric water may circulate to great depths, dissolving minerals in the rock formations through which it moves, and itself becoming highly mineralized.

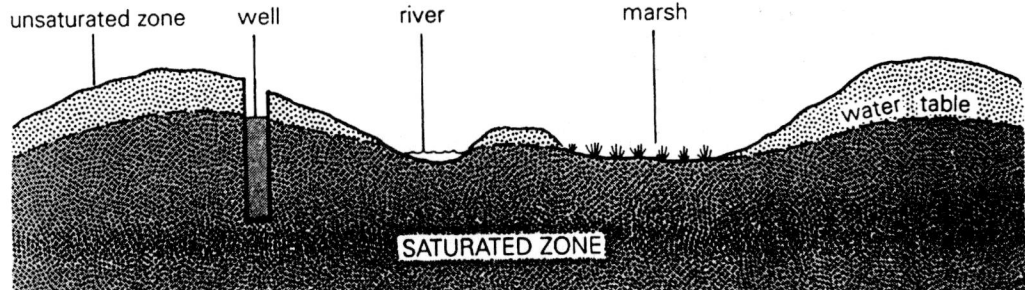

Figure G7 The water table. In general the water table is not flat; it rises and falls with the ground surface but in a subdued way, so that it is deeper beneath hills and shallower beneath valleys. It may even coincide with the ground surface. If it does, we can easily tell because the ground will be wet and marshy.

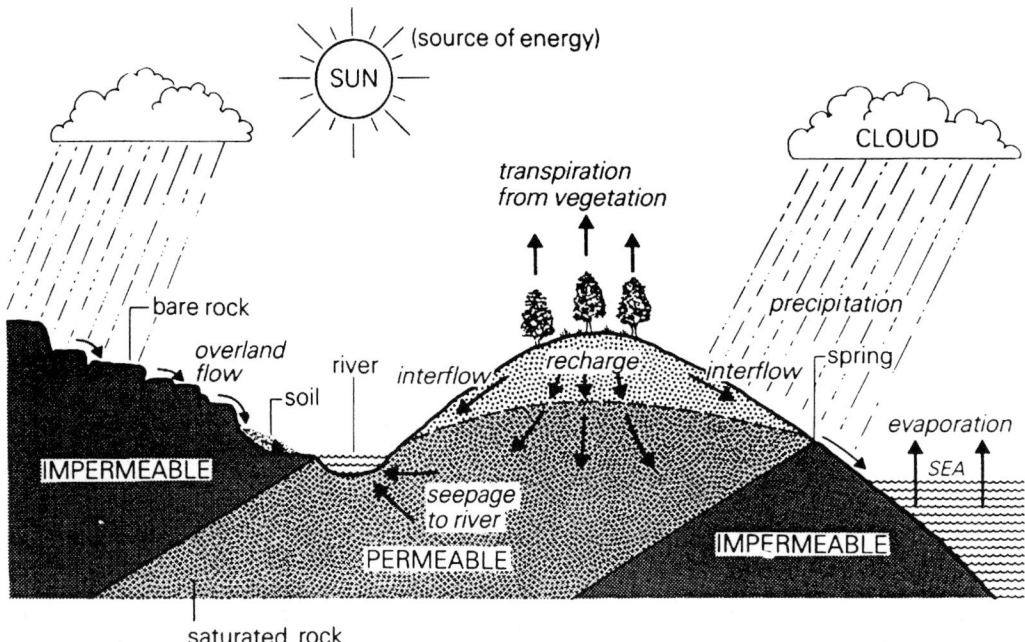

Figure G8 Diagrammatic representation of the water cycle.

There is a third possible origin for groundwater. Juvenile water is the name given to the water believed to be derived from igneous processes within the Earth, and which can contribute unusual constituents to the meteoric groundwater which it joins. According to strict definition juvenile water has never previously taken part in the hydrological cycle; one theory is that all the Earth's water originated at one time as juvenile water. Others point out that juvenile water is indistinguishable from meteoric water that has penetrated to great depths and become intimately associated with igneous processes; it is therefore possible that much or all of this supposed 'juvenile' water is really meteoric in origin. It is also possible that some of the water released during igneous events originates as oceanic or connate water trapped within rocks carried beneath the Earth's crust, as crustal plates collide and sink beneath each other at subduction zones.

Virtually all rocks have some permeability, but the permeability of some rocks is so low that they can be regarded as effectively impermeable. If a layer of impermeable rock occurs beneath an aquifer, it will restrict the movement of groundwater downwards from the aquifer. If part of the aquifer is also overlain by one of these layers, movement of water from the aquifer becomes so restricted that the groundwater in the aquifer becomes confined under pressure. A well drilled into the confined part of the aquifer will encounter groundwater under pressure, so that the water will rise up inside the well and may overflow at the surface.

The movement of groundwater is described by Darcy's law, which relates specific discharge (volume rate of flow through a unit cross-sectional area perpendicular to the flow direction) to the potential gradient or hydraulic gradient and to the hydraulic conductivity of the porous medium (the rock or soil). The hydraulic conductivity depends both on the permeability of the porous medium through which flow is occurring and on the kinematic viscosity of the fluid. In a medium which is isotropic, i.e. has the same permeability in every direction at any point, the direction of flow will be parallel to the hydraulic gradient; if the soil or rock is anisotropic, the direction of flow will not in general be parallel to the hydraulic gradient, but will be between the hydraulic gradient and the direction in the formation in which the permeability is greatest.

The speed of flow is determined by the specific discharge and the porosity through which flow takes place (the kinematic porosity). Other things being equal, the lower the kinematic porosity, the faster will be the speed of flow.

The quality of groundwater depends on the quality of the input water, including the amount of material dissolved in rainfall, and the processes that take place to alter that initial quality. Groundwater generally moves very slowly through rocks, and is in intimate contact with their constituents for long periods of time, giving ample opportunity for changes to take place. The processes include

● dissolution of minerals, including the products of weathering;

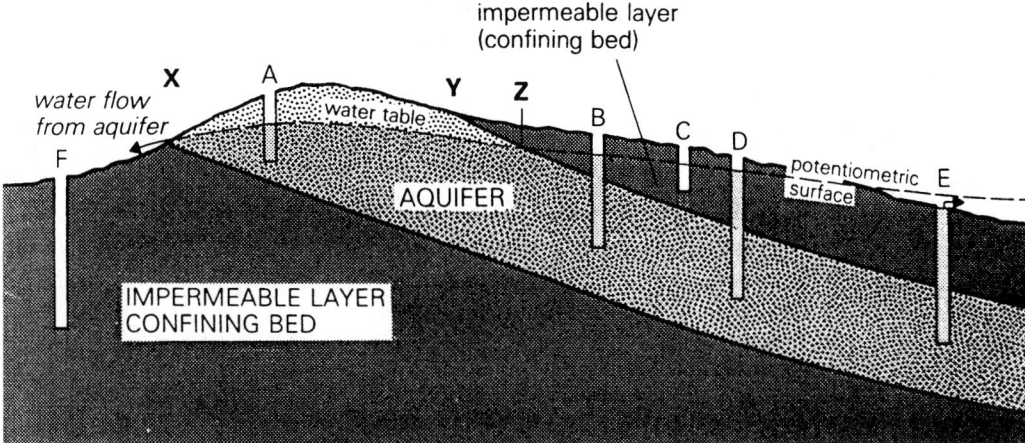

Figure G9 A confined aquifer and its potentiometric surface. Between X and Z the aquifer is unconfined and has a water table; to the right of Z the aquifer is confined and has a potentiometric surface. Wells A, B, D and E enter permeable material and strike water. Wells C and F are in impenetrable materials which will water only very slowly.

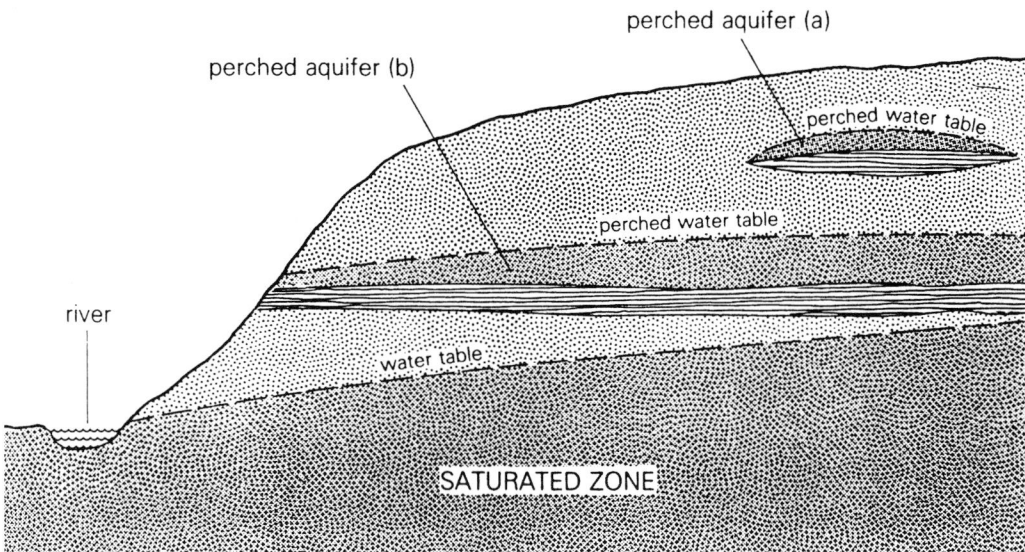

Figure G10 Common occurrences of perched aquifers. Perched aquifers: (a) caused by impermeable material of limited extent; (b) occurring where an impermeable bed intersects a valley side some way above the river level. Situation (b) is a more common cause of perched aquifers of moderate extent than situation (a) – in the latter the perched aquifer will probably exist only after a period of infiltration.

- concentration of dissolved constituents by evaporation and other processes, including ultrafiltration as water is forced through poorly permeable layers;
- microbial processes.

In addition, groundwater can be contaminated by human activity.

Aquifer

The term aquifer is used to describe rocks that are porous enough to store significant quantities of groundwater and sufficiently permeable to allow that water to flow at economically useful rates to wells or springs (Meinzer, 1923). A more recent definition (Lohman *et al.*, 1972) states that an 'aquifer is a formation, group of formations, or part of a formation that contains sufficient saturated permeable material to yield significant quantities of water to wells or springs'.

The term 'formation' means a distinct lithological unit that is capable of being traced for a significant distance. In this context it does not imply any specific mode of origin. Sedimentary, igneous and metamorphic rocks can all form productive and extensive aquifers (e.g. the volcanic rocks of the Columbia Plateau region of the

northwestern United States). Although Meinzer did not explicitly say so, it is clear from a later paper (Meinzer, 1945) that he intended the term aquifer to include the unsaturated part of the permeable unit as well as the saturated material below the water table, and this convention has generally been followed.

Aquifers are frequently divided into two main classes: those with an unsaturated zone and those in which all the pore spaces of the permeable material are filled with water throughout the full thickness of the aquifer. The first group are usually referred to as unconfined or water-table aquifers, although they are sometimes called phreatic aquifers (from the Greek *phrear* for well). The second group are referred to as confined aquifers, because all the groundwater within them is confined, at a pressure greater than that of the atmosphere, between less permeable layers (Figure G9). Confined aquifers are also called artesian aquifers, after the Latin name *Artesium* for the province of Artois in northern France where the condition was formerly noteworthy. In reality, the distinction between confined and unconfined aquifers is rather artificial, since many aquifers have an unconfined portion at their outcrop and a confined portion where they dip below less permeable strata (Figure G9).

A well sunk into an unconfined aquifer will encounter inflowing water when it reaches the water table, and this will be the approximate level at which water will stand in the well. A well sunk into a confined aquifer will encounter little or no water until it has penetrated the confining layer above the aquifer; when it reaches the top of the aquifer, water will rise up in the well to the level of an imaginary surface called the potentiometric surface. If this surface is above ground level, water will flow naturally from the well.

In an unconfined aquifer water is released from storage by drainage of the void space, the water that drains out being replaced by air. In a confined aquifer water is released from storage by elastic processes; as the head in the aquifer is reduced, the water expands and the framework of the aquifer contracts, the relative importance of the two processes depending on the compressibility of the framework of the aquifer. The reduction in volume of the aquifer, particularly if it contains or is overlain by compressible material such as clays or silts, can lead to lowering of the ground surface. Such subsidence following groundwater withdrawal has caused major problems in places including the Central Valley of California, Venice, Mexico City, Bangkok, Shanghai and parts of Taiwan.

Sometimes a layer of material of low permeability occurs above the water table in the unconfined region of an aquifer, and is in turn overlain by permeable material. Infiltrating water is held up by this layer to form a saturated lens, usually of limited extent, above the saturated zone of the aquifer proper (Figure G10). Such an occurrence is called a perched aquifer, and the upper limit of the saturated material is called a perched water table. Perched aquifers do not usually make large or reliable sources of water supply, and it sometimes happens that the act of drilling or deepening a well in a perched aquifer penetrates the impermeable layer and allows perched water to drain through to the aquifer beneath.

Michael Price

Bibliography

Meinzer, O.E., 1923. *Outline of ground-water hydrology, with definitions.* US Geological Survey Water-Supply Paper 494.
Meinzer, O.E., 1945. Problems of the perennial yield of artesian aquifers. *Economic Geology,* **40**(3), 159–163.
Lohman, S.W. *et al.*, 1972, *Definitions of selected ground-water terms – revisions and conceptual refinements.* US Geological Survey Water-Supply Paper 1988.

Cross references

Aquifer
Groundwater: UK
Hydrogeology
Hydrology: subsurface waters
Water resources
Water table
World water balance

GROUNDWATER: UK

Groundwater use and importance

Groundwater provides about one-third of the water abstracted for public supplies in England and Wales 11% in Northern Ireland and 3% in Scotland. In terms of overall water abstracted for all purposes, which is considerably greater than that abstracted for public supply and includes non-consumptive uses such as cooling, groundwater accounts for less than 15% of the total. This difference in use reflects the better quality of groundwater compared with surface water and thus its importance as a public supply source. The relative amounts of total groundwater and surface water abstracted are shown in Figure G11. In contrast, Figure G12 shows the distribution of groundwater and surfacewater abstractions for public water supply by region. This figure illustrates the wide regional variation in the use of groundwater for public supply, with relatively minor use in some regions (for example Welsh, Northumbria, South West), while in others groundwater is a very important source of public supply (Thames, Severn–Trent, Anglian, Wessex and Southern).

In most regions the greatest proportion is used for public water supply, and this is particularly true for central and southeast England.

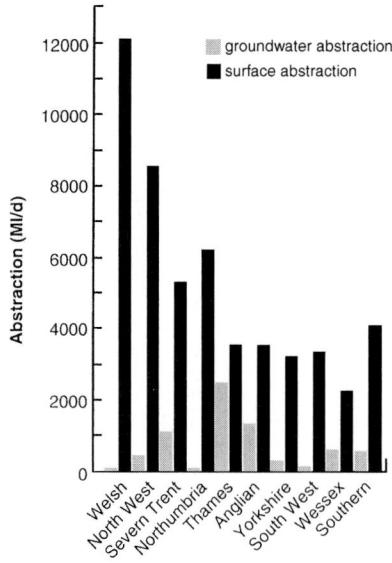

Figure G11 Total groundwater and surface water abstraction in England and Wales for the year 1989–1990.

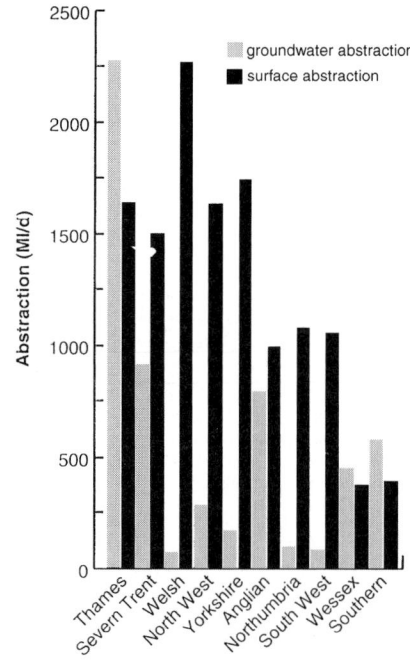

Figure G12 Groundwater and surface water abstraction for public water supply in England and Wales for the year 1989–1990.

Industrial and agricultural purposes tend to dominate non-public supply use. It is generally recognized that local groundwater development is a least-cost option for meeting increasing water demand, at typically £0.1–0.5 million ($0.15–0.75 million) per Ml day^{-1}, a figure in line with demand management gains and direct river abstraction. However, in some cases the environmental costs may significantly alter the economics of groundwater development, if the impacts on baseflows and wetlands are considered.

Table G4 Aquifers in the UK

Aquifer	Lithology	Flow type[a]	Importance[b]
Superficial deposits	Thin and local alluvial deposits; thicker and more extensive glacial gravels	I	●●
Crag	Sands (Quaternary) and shelly sands (Pilocene)	I	●●
Eocene sands	Sands, clays and pebble beds	I	●
Chalk	Thick, soft, white microporous clastic limestone, fractured in upper part	F	●●●●[c]
Lower Greensand	Variable sands and sandstones (commonly glauconitic) with clays and sandy limestones	I	●●●[c]
Hastings Beds	Alternating sand–clay sequence; sands, silts and clays	M	●●
Spilsby Sandstone	Medium-grained sands	I	●●
Portland and Purbeck Beds	Limestones, shales, sandstones and evaporites	F	●
Corallian	Massive limestones and marls, limestones and calcareous sandstones in N England; fractured	F	●●●
Middle Jurassic limestones (Great and Inferior Oolites)	S England: massive fractured oolitic limestones with clays and marls. Lincolnshire Inferior Oolite: oolitic shelly limestones with calcareous sandstones and cementstones. Well fractured	F	●●●
Lower Jurassic sands (Bridport and Yeovil Sands)	Sands with variable clay	I	●●
Marlstone Rock	Thin limestones and ironstones	F	●
Permo-Triassic sandstones	Thick layered heterogeneous sandstones and conglomerates, variably fractured	M	●●●●[c]
Magnesian Limestone	Massive dolomitic and reef limestones with marls, sandstones and breccias; well fractured	F	●●●[c]
Coal Measures	Thick fractured sandstones, alternating with mudstones and coals	M	●●
Millstone Grit	Thick sandstones, grit, mudstones and shales, variably fractured	M	●●
Carboniferous Limestone	Massive fractured karstic limestone	F	●●●[c]
Devonian, Sandstone (Old Red Sandstone)	Sandstones, marls and (particularly in Scotland) conglomerates	M/F	●●

[a] I, intergranular flow; F, fracture flow; M, mixed flow.
[b] ● Aquifer of relatively little importance; ●● aquifer providing useful local supplies; ●●● aquifer important locally; ●●●● aquifer of great national importance.
[c] Important aquifers.

Groundwater resources

The principal control on the occurrence of groundwater in the UK is the distribution of the relatively impermeable ancient (Precambrian and Paleozoic) rocks to the north and the west, and the general restriction of the more permeable, younger (Mesozoic and Paleogene) strata to the south and east of England. Thus, in broad terms, upland Britain or the regions to the north and west such as Northern Ireland, Scotland, Wales and Cornwall do not possess major groundwater resources of regional scale, although available groundwater resources are locally very important and they are often extensively used for private domestic water supplies.

Table G4 lists the aquifers occurring in the UK, with a rating of their importance. The principal aquifers of the UK lie in post-Carboniferous rocks and comprise the Chalk, the Middle Jurassic limestones, the Lower Cretaceous sandstones and the Permo-Triassic sandstones. Of these, the Chalk and Permo-Triassic sandstone aquifers are the most extensive, the Chalk underlying much of southeast England, and the Permo-Triassic aquifer outcropping extensively in the Midlands and northwest England. Figure G13 gives the outcrop areas of the aquifers in England and Wales.

The relative replenishment and abstraction of groundwater by aquifer in England and Wales is shown in Figure G14 (1977 data). The figure clearly shows the dominance of the Chalk and Permo-Triassic sandstones as groundwater sources. It also suggests that replenishment substantially exceeds use.

However, the figures conceal the fact that not all replenishment is available for abstraction, since groundwater contributions to rivers must be maintained, and in some coastal regions seaward flow of groundwater must be permitted to prevent saline intrusion. In England and Wales, the Environment Agency has drawn up a priority list of 40 locations where unacceptably low river flows are considered to be caused by excessive authorized abstractions, rather than drought. In Scotland and Northern Ireland, groundwater abstraction licences are not currently required. Competition for resources is occurring in some aquifer units, notably the Sherwood Sandstone Group in the Belfast area and the Dumfries Permian basin. Only one case of derogation has yet been considered and that was resolved by recourse to the basic planning law.

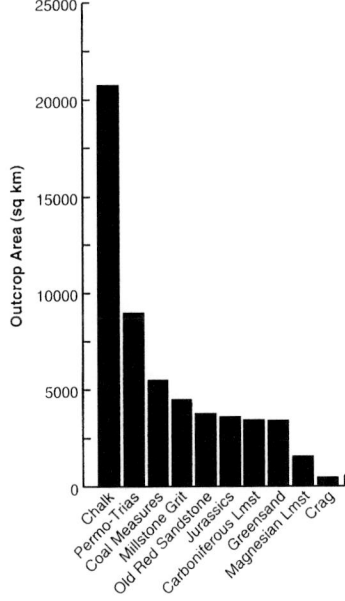

Figure G13 Outcrop areas of the English and Welsh aquifers.

Groundwater quality

Quality is a more complex variable in groundwaters than in surface waters. The relatively slow movement of water through the ground means that residence times are long, giving ample scope for interaction between the water and the material of the containing aquifer. As well as being more complex, quality is more difficult to assess in groundwater because of its relative inaccessibility.

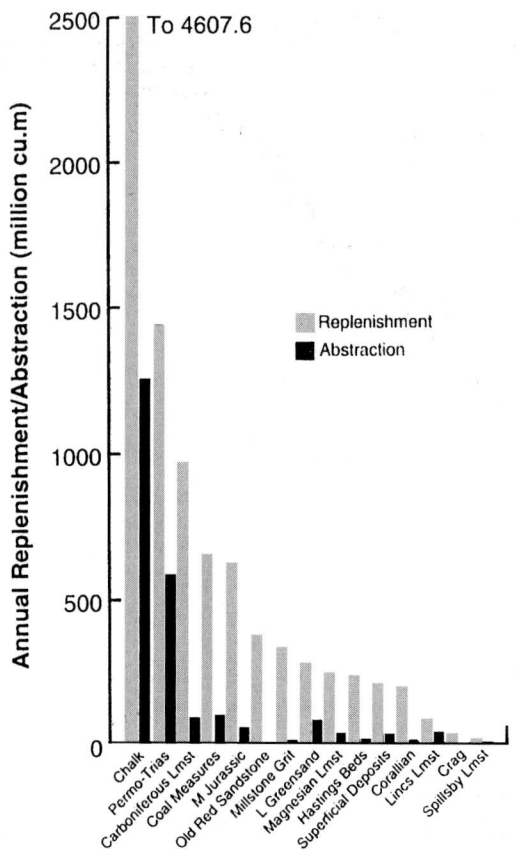

Figure G14 Replenishment and abstraction figures of groundwater by aquifer in England and Wales.

The natural quality of the groundwater which is typically exploited from shallow aquifers is almost invariably good, with low concentrations of major ions throughout most of the aquifers of the country. There is typically an evolutionary process of cation exchange and oxygen removal down gradient (down-dip), as groundwater moves away from the recharge area. This eventually leads to reducing conditions deep in the confined zone of an aquifer, which can produce concentrations of metals which are problematic for groundwater supplies. Much of the groundwater put into supply in southern or eastern England is hard water from carbonate aquifers. The carbonate nature of these aquifers provides the buffering capacity which moderates the influence of acidic deposition, at least in the areas where groundwater is most heavily utilized. Acidic deposition is, therefore, only an important groundwater quality concern for some of the smaller, shallower aquifers of parts of upland Britain.

Most groundwater quality problems result from anthropogenic activity. The most important problems are a result of contamination by nitrate and pesticides, largely from agricultural activity, and trace organic compounds and other pollutants arising from urban and industrial development. Other groundwater quality issues, such as acid mine drainage and saline intrusion of coastal aquifers, may have significant impacts but these are typically of more restricted extent, affecting relatively small volumes of groundwater.

The most widespread, and probably the most well-documented, groundwater quality issue is that of nitrate. Since the mid-1970s, attention has focused on steadily rising groundwater nitrate concentrations in the areas of eastern and central England which receive the lowest rainfall and support the most intensive cereal cultivation. Although the rate of increase may have declined as a result of the agricultural measures introduced to lessen the risk of nitrate leaching, nevertheless significant areas of the Chalk, Lincolnshire Limestone and Sherwood Sandstone aquifers now contain groundwater with relatively high nitrate concentrations.

The Environment Agency has assembled a large volume of historical nitrate data and, as a result, 72 such Nitrate Vulnerable Zones (NVZs) covering a total area of 650 000 hectares were proposed in 1994 for England and Wales. Nitrate is a less problematic quality consideration in Scotland and Northern Ireland because a combination of less intensive farming and significantly higher infiltration produces generally lower nitrate concentrations in groundwater. Nevertheless, there are two NVZs in Scotland and one anticipated in Northern Ireland.

Other groundwater quality issues have become apparent only more recently, as a combination of stricter legislation and improved analytical techniques allow trace organic compounds to be determined routinely. Recent research employing surveys of groundwater quality beneath several of our major cities has indicated the extent of contamination by trace organic compounds arising from current and, especially, past industrial activities. A wide range of compounds has been observed, and it can be anticipated that the legacy of past industrial development will be increasingly seen in groundwater, as it is in contaminated land. Today's contaminated land is tomorrow's groundwater pollution problem.

The same considerations of tightening legislation and improved analytical techniques apply to pesticides. Hardly any knowledge of pesticide occurrence in groundwater existed before the mid-1980s, and some of the earliest results must be treated with caution because confirmatory determinations were not performed. Routine analysis of pesticides in groundwater at detection limits appropriate to assessing the situation with respect to a maximum admissable concentration (as specified in legislation) of 0.1 $\mu g\, l^{-1}$ dates back only 4–5 years. Thus the current situation is one of 'the more one looks, the more one finds'. While the current data do not provide an overall national picture, it is clear that low concentrations of pesticide residues, in the range 0.05–0.15 $\mu g\, l^{-1}$, do occur regularly in many groundwater sources. Concentrations of over 1 $\mu g\, l^{-1}$, are, however, extremely rare.

It is also apparent from the existing data that the most commonly encountered compounds are the triazine herbicides, which have been widely used in non-agricultural situations as general weedkillers. Much less information exists for Scotland and Northern Ireland, although it can be anticipated that pesticides are likely to be less of a quality issue because of the dilution effect caused by relatively high effective rainfall. To date, hardly any information exists about the presence of pesticide metabolites, many of which may be as toxic and persistent as the parent compound.

Groundwater regulation

Although groundwater has been exploited in the UK country for millennia, it has been managed for less than 50 years. A major new regulatory dimension was added through membership of the European Union. Directives adopted by the European Union normally impose a duty on the Member States to show compliance within a given time limit. However, a European Union Directive is not effective within a Member State until it is implemented by the Government of that State. This is achieved using the national systems of law and administration, resulting in new or amended Acts or Regulations.

Table G5 shows the European Council (EC) Directives which affect groundwater management. Generally, European Water Directives fall into two categories. The first defines acceptable water quality for particular purposes (e.g. Directive 80/778/EEC). The second category relates to the quality of potentially polluting discharges, and includes Directives 76/464/EEC and 80/68/EEC. More recently the European Commission's Fourth Action Programme on the Environment introduced a new source-directed approach to pollution problems, and this has led to Directives 91/271/EEC and 91/676/EEC. Other Directives may relate to groundwater, but only indirectly; for example Directive 86/278/EEC (Use of Sewage Sludge in Agriculture) is aimed at protecting the environment from contamination resulting from uncontrolled applications of sewage sludge to agricultural land.

Figure G15 shows diagrammatically the different perspective of the regulators and the suppliers.

Aquifer properties and their consequences

The relative importance of the issues affecting groundwater will vary greatly with hydrogeological environment and aquifer type. The behavior of groundwater is to a large extent governed by the

Table G5 EC Directives relevant to groundwater

Directive	Main purpose
Directive 76/464/EEC Dangerous Substances to Water	To protect the aquatic environment from discharge of dangerous substances. Covered inland surface waters, estuaries, coastal waters and groundwater. Provided framework for control of 'Black' List I and 'Gray' List II substances
Directive 80/68/EEC Groundwater	To protect groundwater against pollution caused by specified dangerous substances. Defined List I (most dangerous) and List II (other dangerous) substances. EC member states obliged to prevent List I and limit amounts of List II substances reaching groundwater
Directive 80/778/EEC Quality of Water for Human Consumption	Defines quality standards (for some 60 substances) and monitoring requirements. Annex I sets out maximum permissible concentration and guide levels for 62 parameters and minimum required concentration levels for four parameters
Directive 9/271/EEC Urban Waste Water Treatment	To encourage the reuse of water and the development of improved waste management. Lays down minimum requirements for treatment of urban waste water and disposal of sludge
Directive 91/676/EEC Protection of Waters Against Pollution Caused by Nitrate from Agricultural Sources	To encourage agricultural practices which are environmentally beneficial, and in particular to reduce at source contamination by nitrate from agricultural practices. Restrictions placed on fertilizer use, maximum permissible nitrate level in drinking water set; zones vulnerable to nitrate pollution to be identified

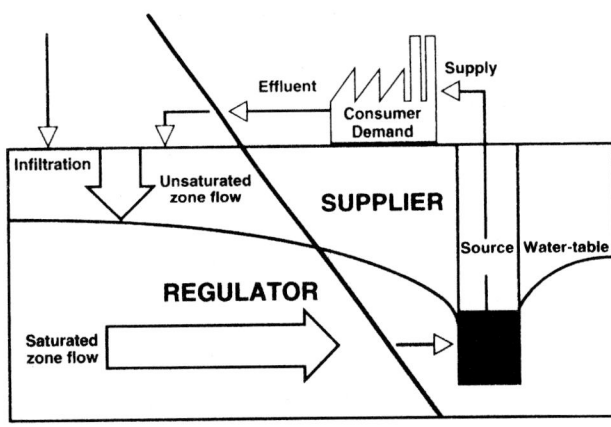

Figure G15 Different perspectives of the regulators and suppliers.

properties of the host aquifer. Table G6 shows the relative aquifer properties of the main aquifers in the UK. The permeability is a measure of the ability to transmit water (flow through unit area in unit time); porosity is the proportion of void space in the rock; and drainable porosity is that part of the porosity that will drain (a function of pore connectivity and size).

The Chalk is the principal aquifer of the UK, with boreholes commonly yielding several million liters per day, sometimes exceeding 10 Ml day^{-1} from large-diameter wells, some with adit systems. The matrix of the Chalk is too fine-grained to possess a significant permeability; it is only an aquifer as a result of open water-bearing fractures. These fractures, and therefore the zones of high permeability, are not uniformly distributed throughout the Chalk, either with depth or geographically. Open fractures tend to be restricted to the upper few tens of meters, and therefore, although the Chalk may be several hundred meters thick in places, the productive thickness of the aquifer may be as little as 50 m. Also, there is considerable areal variability in transmissivity (permeability × aquifer thickness) with values varying over a few kilometers by an order of magnitude or more. These variations are thought to be associated with topography, with valleys often having higher values than interfluves. The porosity of the Chalk is generally very high, but the pore spaces are mostly too small to be involved with flow, and the drainable porosity of the aquifer is low. Atypically, the Chalk at outcrops in Northern Ireland is karstified with several large sinks and risings in the coastal area of County Antrim.

The aquifer properties of the Permo-Triassic sandstones are controlled by a combination of factors, including the lithology of the material (for example, grain size and type), the degree of cementation, and the extent of fracturing. These factors, and therefore the

properties of the aquifer, are often both complex and varied, both laterally and with depth. In broad terms, however, the aquifer often has high permeability and high porosity, much of which is drainable. Well yields from the Permo-Triassic sandstones are variable, ranging up to 10 Ml day^{-1} in the Midlands, around 1 Ml day^{-1} in the north of England, Scotland and Northern Ireland, and less in the Cheshire Plain.

In the Middle Jurassic limestones, groundwater tends to move in discrete fractures and permeability is only meaningful for sufficiently large blocks of rock. This difficulty increases in the Permian Magnesian Limestone and in the Carboniferous Limestone, where most of the flow is in large discrete conduits. The porosities of these aquifers are generally small, and this can substantially affect borehole yields. In particular the Jurassic limestones drain rapidly and high yields are only found where the limestones are confined, where very high yields may be obtained. Yields from the Magnesian Limestone are dependent upon wells intersecting productive fractures, although several megaliters per day are possible, and in the Carboniferous Limestone drilling is very speculative, with yields depending entirely on the size of fracture encountered.

The flow characteristics of the major aquifers are shown schematically in Figure G16. Flow through fractures is a very important (and relatively little understood) characteristic of most of the aquifers in the UK and has profound consequences for the prediction of their behavior, particularly with regard to groundwater transport processes. Figure G16 illustrates the point that while fracture flow is unimportant in minor unconsolidated aquifers, it is important to varying degrees in nearly all the major aquifers. Although fractures are generally of limited significance in the Lower Greensand, they can be very significant in the Permo-Triassic sandstones, and are fundamental to flow in the Chalk, Jurassic limestones and the Carboniferous Limestone. Figure G16 also illustrates in a general way fracture size for various aquifers. Thus a trend in the limestone aquifers is shown, ranging from the Chalk, where fractures are generally narrow (although there are many exceptions) through the Jurassic and Magnesian Limestones to the largely karstic Carboniferous Limestone, in which flow is predominantly in large conduits. By plotting the flow characteristics of the aquifers in this way a trend of increasing flow unpredictability can be shown, from matrix-dominated, relatively predictable aquifers such as unconsolidated sands, to the Carboniferous Limestone, where both paths of groundwater flow and borehole success are very uncertain.

If the properties of an ideal aquifer are well known, it is possible by using simple models to predict groundwater flows in response to pumping. Thus, for example, protection zones, drawn as lines of constant travel time may be drawn around a pumping well. Figure G17 shows such generalized lines of constant travel time for a well pumping in different aquifers using simplified assumptions concerning the properties of the aquifers. Thus after a certain period of pumping, the model predicts that a particle will have reached the well from a much greater distance in the Chalk than in the Permo-Triassic sandstones. Other models based on generalized aquifer properties can

Table G6 General aquifer properties of major UK aquifers

Aquifer	Average permeability	Porosity	Drainable porosity	Notes
Chalk	●●●●	●●●●	●	Aquifer good in valleys, poor on hills
Lower Greensand	●●●	●●●●	●●●	
Jurassic limestones	●●●●	●●●	●●	Permeability unpredictable
Permo-Triassic sandstones	●●●	●●●	●●●	
Magnesian Limestone	●●●	●●	●●	Permeability very unpredictable
Carboniferous Limestone	●●●	●	●	Permeability extremely unpredictable

●●●● very high   ●●● high   ●● medium   ● low

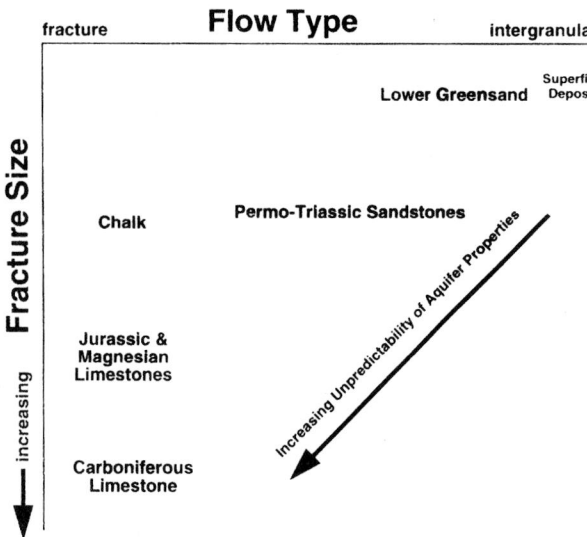

Figure G16 Flow characteristics of major UK aquifers.

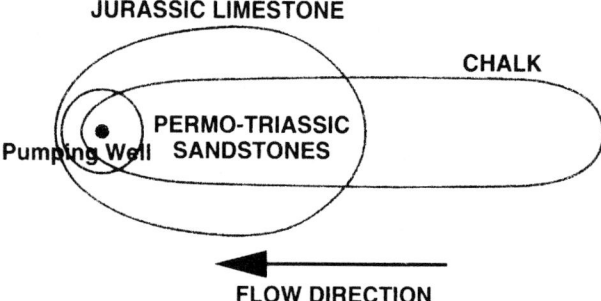

Figure G17 Generalized shape of groundwater protection zones (of constant travel time) for various aquifers.

be used to show that in an aquifer with high transmissivity and low storage such as the Chalk, pumping influences will spread much more rapidly than in an aquifer with high storage, such as an unfractured porous sandstone.

It must be recognized that reality is much more complex than can be portrayed by simple models. For example the shapes of constant travel time shown in Figure G17 rely on the aquifers being homogeneous, and this is not the case. In particular, the effect of fracturing in UK aquifers renders models subject to varying degrees of uncertainty.

Hydrogeological and basic processes

Understanding many of the groundwater issues involves a detailed knowledge of groundwater flow and the types of chemical reactions that occur in aquifers. This understanding of scientific principles and processes is particularly necessary for groundwater as its remoteness and the cost of access (e.g. for measuring or sampling) means that deductive and predictive analysis (i.e. inference) is much more important than for surface water, where extensive measurement can be readily made.

Fundamental advances in basic science such as in statistics, numerical analysis, geology, analytical chemistry and instrumentation, geophysics and microbiology can all have important repercussions in groundwater science. These advances may, for example, be in terms of new theoretical approaches or of advances in monitoring. However, knowledge of the underlying processes alone is not sufficient to provide practical solutions to problems. These processes must be combined with some kind of model which links the processes together in a meaningful way and with data and databases which define certain key properties or parameters of the particular issue being considered. Finally, there must be an overall approach or methodology (e.g. risk analysis) which defines the way in which these components are brought together to solve a particular problem.

The most complex of the elements are the hydrogeological processes, which include a broad range of applied scientific topics. The key hydrogeological processes are described in Table G7. It is important to emphasize that the processes all interrelate, and to some extent overlap; however, each is sufficiently important in its own right to deserve separate attention.

The various hydrogeological processes, which are identified above, can themselves be subdivided further into a number of different basic scientific processes. These basic processes along with their associated data requirements, are listed in Table G8. For example, understanding saturated flow needs data on the spatial distribution of permeability and porosity. Reactive transport requires knowledge of the major geochemical processes such as acid–base reactions, redox reactions and sorption as well as of a great deal of fundamental chemical thermodynamic data.

This results in a hierarchical approach to the science underpinning the resolution of groundwater issues, leading from the fundamentals of physics, chemistry and biology through to the hydrogeological processes. There is a two-way flow of information between these areas of science: the basic sciences feed the applied sciences with new ideas while the applied sciences such as hydrogeology stimulate new solutions to practical problems from the basic sciences.

Models

Modeling is not an end in itself, it is part of a wider process of decision making. There are a variety of types of model designed for different end purposes. These range from management models, which aim to provide guidance to management about the consequences of various course of action, to research models, in which the main purpose is to aid our conceptualization and understanding of the processes taking place so that extrapolation can be made to new and unknown situations. An increasingly important use of models is in risk assessment which attempts to use scientific understanding of an issue and the consequences of various courses of action in terms of the risks posed. Models are not a substitute for data collection or human decision-making. Models can range from simple operational 'rules of thumb' to sophisticated research models aimed at increasing the understanding of the underlying processes.

Table G7 Hydrogeological processes

Soil processes	Groundwater recharge enters the unsaturated zone throught the soil. The soil and vegetation are important in determining the amount and timing of groundwater recharge and its quality. The soil forms an important barrier against groundwater pollution but contaminated soils can also be a potential long-term source of polluted groundwater. The soil contains a large number of microorganisms which carry out important transformations, e.g. organic nitrogen to nitrate, degradation of toxic organic substances, consumption of oxygen and production of carbon dioxide
Unsaturated zone flow and transport	The unsaturated zone of UK aquifers is often 10–50 m deep. Since the rate of flow of water through the unsaturated zone of many UK aquifers is in the order of 0.5–1.5 m year^{-1}, the unsaturated zone provides an important delay between input of a contaminant at the soil surface and when it finally reaches the water table. Although the unsaturated zone contains less organic matter and fewer microorganisms than the soil, its adsorbent properties and capacity for degradation may provide an additional barrier against many contaminants. Aquitards can significantly affect recharge and contaminant migration to the water table
Saturated zone flow and transport	Understanding the rate and direction of groundwater flow in an aquifer is a key part of most hydrogeological investigations. Frequently this information is derived from a computer model. Such models require a detailed understanding of the physical properties of the aquifer, preferably in 3–D. Although the saturated zone of aquifers can be more than 100 m deep, the hydraulically active part is often concentrated in the top 50 m. Modelling flow within an aquifer often has to be based on sparse field observations and inadequate characterization of the aquifer heterogeneity. Flow predictions based on such models therefore have a considerable degree of uncertainty. Predicting the movement of chemicals in aquifers is even more demanding. Aquitards have a strong influence on flow directions and contaminant pathways
Reactive transport	Many, but not all, chemicals are sorbed on the aquifer matrix or undergo chemical reaction during transport, so delaying their arrival at a groundwater pumping station. In the case of some organics, such as pesticides, this delay may be such that all of the chemical has had time to degrade by the time it reaches the pumping station. Quantifying the extent and rate of such reactions is critical in understanding the long-term fate of many groundwater pollutants. It is one of the most active areas of current groundwater research
Microbiology	Subsurface microbiology is a relatively new science but there is now ample evidence to show that a wide range of microorganisms are present and probably active in UK aquifers. However, it is difficult to quantify exactly how active they presently are. Microorganisms are responsible for a number of important transformations, including most redox reactions, such as organic carbon oxidation, denitrification and methane production. The *in situ* microorganisms can sometimes be stimulated to increase the rate of breakdown of a biodegradable groundwater contaminant. Such bioremediation can be quite efficient in shallow aquifers
Fracture flow	The major UK aquifers are all fractured to some extent, with fracture flow dominant in some. Such fractures can lead to preferential flow of water and dissolved chemicals, and can lead to fast pollutant transport. The yield of pumping boreholes is often mostly derived from the rapid flow of water through fractures. The importance and extent of fracture flow is often difficult to model as it is very sensitive to the assumed fracture sizes, spacing and distribution
Multiphase flow	Petrol and many organic solvents do not mix with water and so flow through aquifers as a separate phase. Since petrol is lighter than water, it will tend to accumulate close to the water table. Many solvents are denser than water and so will tend to sink to the bottom of an aquifer, often very rapidly. Most of these organic substances are slightly soluble in water and so also get transported with the groundwater flow. They are also volatile and will partition into the gas phase. Knowledge of the flow of such chemicals is important for understanding their fate in aquifers and in designing clean-up programs. Modelling such flow is difficult
Gas exchange	Below the soil there is relatively little exchange of gases with the atmosphere. Some major gases, such as oxygen and carbon dioxide, are important in determining the geochemical environment of an aquifer. Some trace gases, such as radon and methane, can be potentially dangerous. Others such as the various nitrogen gases can provide clues to the biological transformations of nitrogen species taking place. At contaminated sites, the transport of organic vapors may be important in pollutant migration and remediation. Gases move by diffusion and convection both in the gas phase and in the dissolved phase

Sustainability

The concept of sustainability is often employed with regard to groundwater management and development and in relation to the needs of society and the environment. This concept is open to wide interpretation, however; it is not clearly defined and can be used to justify a diverse range of actions (Table G9).

In general, the concept of sustainable water resources management incorporates ideas of costs and benefits to present and future generations. If activities which influence the groundwater system, such as pumping and pollution, can be managed in such a way that the overall welfare of society is increased without sacrificing the interests of future generations, then that method of management can be considered sustainable. Welfare might be derived from a broad array of water uses, ranging from domestic, agricultural and industrial to the maintenance of ecosystems. However, sustainable management does not mean that everything must remain the same, with the existing quality and quantity of water resources remaining intact. Rather, it is

the value to society of different water uses that are to be maintained or increased. These values might include economic productivity, human health and environmental quality.

In addition, there is a scale question: welfare benefits to a small group could bring welfare benefits or disbenefits to society as a whole (for example, as could be argued one way or another in the cases of maintaining agricultural communities or maintaining headwater streamflows for recreational fishing); similarly preventing environmental benefits or disbenefits at a larger scale (for example, as could be argued either way for interbasin transfers). Clarity over the meaning of sustainability in groundwater management is obviously important and should be sought; there are trade-offs to be considered and economic questions to be asked.

Economic questions

In addition to understanding the science of the processes underpinning the groundwater issues, it is clear that a significant part of the solution

Table G8 Characterization of hydrogeological processes

Hydrogeological process	Basic processes, parameters, models and associated data requirements
Soil processes	Evaporation and recharge, nutrient uptake, mineralization of organic matter, humics, ion exchange and sorption, heterogeneity, acid–base equilibria, CO_2 production, mineral weathering, formation of secondary minerals, contaminated soil characterization, bioavailability and toxicity, clean-up, hyperaccumulator plants
Unsaturated flow	Preferential flow, non-Darcian flow, transfer functions, stochastic hydrology, geostatistics and transport, groundwater–surfacewater interactions, unsaturated hydraulic conductivity, diffusivity, hysteresis, aquitard behavior
Saturated flow and transport	Permeability, porosity, dynamic porosity, dynamic storage, spatial variability, geological control, flowpaths and mixing, density-driven mixing, geochemistry, connate waters, groundwater 'age', isotopes, diffusion and dispersion, equivalent porous medium flow models, double porosity and double permeability models, formation factors, aquitard behavior
Reactive transport	Acid–base and redox reactions, mineral solubility and precipitation, solid solutions, surface chemistry, facilitated transport, rate-limited sorption, rate-limited dissolution, thermodynamic databases, partition coefficients, surface binding constants and site densities, structure–activity relationships, heterogeneity and non-linear sorption, humics, isotopes, diffusion coefficients and tortuosity, dispersion coefficients, hydrolysis
Microbiology	Biodegradation and transformation, redox reactions, diffusion, reaction pathways, microbial biomass, microbial adsorption, preferential flow, rate constants, temperature dependence, microcosm studies, remediation
Fracture flow	Fracture size, spacing, aperture and orientation distributions, fracture surface characteristics, flow pathways, tracer tests, breakthrough curves, fracture chemistry, fracture models, network models, channel models
Multiphase flow	Wettability, dissolution, volatilization, residual saturation, solubility, densities, geophysics, relative permeability
Gas exchange	Gas and vapor diffusion, convection (gas venting), Henry's law partition coefficients, gas density gradients, non-ideal transport and rate-limited sorption

Table G9 Possible elements of sustainable groundwater development

Possible management goal	To ensure that the benefits to different users provided by a groundwater/aquifer system will meet present objectives of society without compromising the ability of the system to meet future objectives	
Possible objectives of society	Economic productivity and efficiency	('wealth creation')
	Equity	
	Human health	('quality of life')
	Environmental protection	
Possible conflicts and trade-offs	Interests of present versus future generations	
	Groundwater development for one use vs another	
	Abstraction for supply vs environmental needs	
	Groundwater protection vs unrestricted economic activity	
	Lowest price water to consumers vs. environmental protection and full cost pricing	

to some of the groundwater issues will lie in answering questions of economics and the political economy. In the case of groundwater, little good conceptual work has been done, either in the UK or elsewhere, which applies new concepts of environmental and institutional economics to the following.

- Groundwater resources degradation (both in terms of quantity and quality) caused by anthropogenic activity. Degradation does not refer to changes in quantity or quality caused by anthropogenic activity, since these changes may be beneficial (lowered water levels enhancing recharge or reducing flood risk; scavenger pumping improving quality), but could be defined as a reduction in the value (in constant terms) of stock as a consequence of changes in quantity or quality.
- The consequent impacts of degraded or exploited but undegraded groundwater resources on society and the ecosystem. Impacts could be physical (e.g. subsidence, or reduced surface flows), financial/economic (e.g. increased costs of supply), or environmental (e.g. reduced biodiversity). Impacts may also be intergenerational (e.g. the loss of the physical aquifer fabric as a natural resource asset, or the loss of an ecological resource).

The economics of groundwater development are thus intricately linked with the viability of groundwater management strategies and the enforceability of pollution protection and abstraction of policies,

and trade-offs are apparent. For example, instruments of an abstraction policy (e.g. regulatory, such as licencing, or economic, such as pricing) will generally increase the costs faced by users, while bringing environmental benefits. What are these costs and benefits and what is the consumer's 'willingness to pay'? Protection policies have significant impacts on agricultural practices and industrial processes, possibly increasing investment and operating costs and thus reducing profitability or increasing consumer prices. An important analytical question, needing inputs from environmental economic analysis, which would pose significant theoretical problems, is whether to protect upstream (i.e. in the catchment) or to treat downstream (i.e. at the borehole).

A further important area for economic analysis is that of the costs of groundwater clean-up and the value of benefits, both to public supplies and to the environment.

One overall emerging theme is the essentially ethical conflict over the 'condition' at which our environment should be preserved. Our environment has been 'engineered' for centuries, with (for example) meadows, woodlands, hedgerows and a 'built' environment that bear little resemblance to a 'pristine' landscape. Groundwater flowing into surface water plays its part in sustaining the environment that we know. Maintaining that status quo – or restoring it to what we perceive to be an earlier status quo – is clearly one, not unreasonable, goal. But we still choose to pick those historical conditions that we would prefer to avoid: high water levels below London's central business district; the return to waterlogged land in coalfield areas which have been arable farmland for over a century due to dewatering by mine pumping; flood risks in formerly flashy rivers. The questions of engineering perennial rather than ephemeral flows, or sustaining vulnerable wetlands, are fraught with the dilemmas of sustainable development, discussed above. There is, of course, the option, increasingly adopted, of artificially sustaining important environmental assets, through compensation pumping, thus engineering both substantial groundwater resource development and environmental conservation. This, however, also raises ethical questions and potential conflicts, not least in public perceptions.

Public perceptions appear to be, to some extent, bipolar, with concern for rising costs of water on the one hand and increasing concern at damage to the water environment on the other. Perceptions of risk are possibly characterized, in part at least, by the growing use of bottled water for drinking. There appears to be generally very little public understanding of groundwater flow and quality, and this is often compounded by poor representation in the media – often inferring 'underground rivers'. Enhanced public awareness of groundwater and its role in our daily lives and the landscape, and public participation in decision making on some of the more intractable ethical questions, could make a significant contribution to sustainable groundwater management.

Quantity: the storage, conjunctive use and compensation questions

The UK has relatively abundant water resources but they are seasonally distributed and rapidly lost to the sea. Thus the problem is largely one of meeting peak demands during times of low rainfall and low river flows, through providing storage close to centers of demand. The groundwater quantity questions revolve around the effective management of groundwater storage to optimize water resources development while minimizing environmental effects, particularly on sensitive river flows and wetlands. Some perceive groundwater pumping to be the cause of many problems and that greater investments in surface reservoirs to capture abundant river flows would be a solution, despite the considerable potential storage of UK aquifers. This general view of groundwater is relatively widespread. In part this represents a significant shift over the last three decades, as in the 1960s and 1970s use of groundwater storage was actively promoted, due to widespread opposition to new reservoirs.

Pioneering work in the 1960s and 1970s led by the Water Resources Board (WRB), demonstrated the benefits that could be derived from conjunctive use of surface water and groundwater, which exploits groundwater storage for river regulation to meet peak demand downstream and also for providing compensation flows for environmental purposes. The WRB work also revealed many problems, particularly in the Chalk. The Chalk typically has a low storage coefficient and a high transmissivity, causing pumping effects to be propagated rapidly, which may then affect stream flows. Conjunctive use locations in the Chalk therefore need to be selected where storage coefficients are unusually high or where boreholes can be distant from the river, although this increases conveyance costs and reduces benefit. Triassic sandstones have a lower transmissivity and higher storage coefficient than the Chalk. This has enabled the successful Shropshire Groundwater Scheme. Development of conjunctive use schemes are now a demonstrated and accepted water resources management strategy. Nevertheless, existing schemes are little used, partly because the new institutional framework for water means that the water industry is primarily interested in river regulation for ensuring downstream supplies (e.g. for pumped storage reservoirs) and the Environment Agency's interest is primarily in river regulation to satisfy environmental objectives. Conjunctive use means taking both into account, and institutional and financial mechanisms need to be established on a case by case basis, as they have been in the Shropshire Groundwater Scheme, where the Environment Agency is the 'operator' for both environmental purposes and for the water companies. In addition, due to relatively high operating costs, schemes are regarded primarily as insurance policies (with low capital cost), only to be used when needed.

A further strategy which has been investigated in the UK for several decades is that of augmentation of groundwater storage through artificial recharge. The most significant scheme in operation is the Lee Valley/Enfield–Haringey Artificial Recharge Scheme in North London, where surplus treated mains water is injected into the dewatered Chalk aquifer and the Lower London Tertiary sands, which are overlain by the London Clay. The scheme is designed to provide strategic drought storage and demonstrates that imaginative management of groundwater, including artificially replenishing the reservoir in periods when supply exceeds demand, can provide a sustainable resource with minimum environmental impacts. However, in recent strategic planning, artificial recharge of groundwater has been excluded as an option for development, largely on the grounds of a perceived lack of suitable geological structures. Nevertheless, it is recognized that further studies are needed to identify additional favorable sites.

In the 1960s the Water Resources Board initiated an extensive series of field investigations of the feasibility of artificially recharging the principal British aquifers by means of basins and wells, and no practical problems were encountered. In addition, some artificial recharge in the UK has focused on replenishing freshwater aquifers with tertiary treated water or the disposal of wastewater through seepage lagoons, channels and drains. Recharge in the latter case is incidental, as the key objective is wastewater disposal.

In the light of the recent experience gained in the United States with aquifer storage recovery (ASR), studies may be expanded to include non-potable aquifers as potential groundwater reservoirs (see below). Attention may be focused on areas with the greatest potential, not only from the hydrogeological point of view but also where an ASR scheme could meet a genuine demand. Areas likely to benefit most from this type of scheme, assuming appropriate geological structures exist, would include coastal holiday areas which have difficulty in meeting peak summer demands.

Aquifer storage recovery in the USA

In the United States the concept of artificial recharge of groundwater has recently been developed to include the use of non-potable or saline aquifers for storage of water (Pyne, 1995). The injected water (generally treated to potable standards to avoid clogging problems) forms a lens of good-quality water within the saline water body for later recovery. This development greatly increases the potential for using artificial recharge, or aquifer storage recovery (ASR). The main objective of using ASR is to manage water resources by use of aquifer storage as a strategic reservoir to meet variations in demand, almost irrespective of native groundwater quality, using almost no land area and having minimal environmental impact.

Operation of a typical ASR installation is designed to smooth out annual variability by recharging aquifers during periods of low demand and recovering the water during periods of high demand. When a non-potable or saline aquifer is used as the ASR reservoir, the saline water has to be flushed from the fractures and pore spaces, a process that requires several cycles for the scheme to become fully operational. With each cycle, the percentage of water of acceptable quality recovered progressively approaches the quality of the recharge water.

The number of operational ASR schemes installed in the United States has increased dramatically since 1991 (from 10 to 20 with 40 more in development) as the technology and methodologies have become established and accepted. Schemes have been installed in a wide variety of hydrogeological environments and have recovery capacities varying from 2 to 385 Ml day^{-1}. The capital costs of several schemes that have been implemented in the United States range from \$53 000 to \$4.5.5 000 per Ml day^{-1}. These costs are not directly comparable with British costs but provide a useful indication of scale. Similarly, a comparison of costs for expansion of five schemes in the United States show ASR schemes to be between 5 and 43% of the cost of a scheme not using ASR.

Groundwater research

- Although actual figures are not available, the total spent on research that specifically targets potable groundwater as its primary objective in the UK by all parties could be less than £1.5 million (\$2.25 million) per year and is unlikely to exceed £3 million (\$4.5 million); public sector spend is less than £1 million (\$1.5 million), for a resource that
  - provides supplies to about 20 million people in the UK;
  - provides all bottled water in the UK;
  - is a crucial strategic store of water;
  - maintains river flows and wetlands.

- Groundwater sales in England and Wales for public water supply (after treatment and reticulation) are about 2000 million $m^3\,year^{-1}$, with an annual sales value of, say, £1.2 billion ($1.8 billion); groundwater abstraction for industry, agriculture and fish farming yields a further 500 million $m^3\,year^{-1}$.
- Private, unregulated supplies (almost all groundwater) are estimated to withdraw an estimated 50 million $m^3\,year^{-1}$ to supply about 5000 000 people, at a notional value of £30 million ($45 million).
- Bottled water sales in the UK in 1995 (all groundwater) were about 785 million $l\,year^{-1}$, with a sales value of about £360 million ($500 million).
- Industry expenditure on treatment of contaminated groundwater is very high, and during the period 1990–1991 to 1993–1994 was around £2.5 billion ($3.75 billion) at 1993–1994 prices; and on new treatment to remove pesticides from water put into supply was £122 million ($183 million) in 1992–1993, and in 1993–1994 and 1994–1995 companies were planning to spend £170 million ($255 million) and £207 million ($310 million), respectively. A high proportion of expenditures on pesticide and nitrate treatment was on groundwaters. These costs do not include recurrent spends.

R.W. Herschy

Source

Grey, D.R.C., Kinniburgh, D.G., Barker, J.A. and Bloomfield, J.P., 1995. *Groundwater in the UK. A Strategic Study. Issues and Research Needs*. UK Groundwater Forum Report FR/GF1. Foundation for Water Research, Marlow, UK.

Bibliography

Pyne, R.D.G., 1995. *Groundwater Recharge and Wells: A Guide to Aquifer Storage Recovery*. Lewis Publishers, USA.

Cross references

Bottled water
Chilgrove House well, UK
Groundwater
Hydrodynamics: porous media
Hydrogeology
Hydrological cycle
Hydrology: subsurface waters
Trafalgar Square borehole, London, UK
Water resources
World water balance

H

HYDROCLIMATOLOGY

Hydroclimatology was defined by Langbein (1967) as the 'study of the influence of climate upon the waters of the land.' It includes hydrometeorology as well as the surface and near-surface water processes of evaporation, runoff, groundwater recharge and interception. The total hydrological cycle, then, is the basis for a discussion of hydroclimatology.

The water budget equation based on conservation of mass is given as:

$$P = R + ET + U + S_s + S_g + I, \qquad (H1)$$

where P is precipitation, R is streamflow, ET is evapotranspiration, U is the subsurface underflow, S_s is the change in soil moisture, S_g is the change in groundwater storage and I is interception. When the equation is evaluated using annual mean data, the change in groundwater storage, soil moisture and subsurface underflow are assumed to be zero; and if the interception is sufficiently small,

$$P = R + ET \qquad (H2)$$

Although the components of the total budget are easy to identify, quantification of each of the components over various time scales is always difficult, and sometimes impossible to measure or to estimate correctly.

Measurement of water budget components

Precipitation

Measurement of precipitation is typically accomplished by catching the rain or snow in a flat-bottomed, vertically sided canister, and simply measuring the depth, either after each precipitation event or continuously by an automatic sensor and recorder. Catchment of solid precipitation (particularly under windy conditions) often underestimates, in that snow is easily blown over the gauge opening rather than being captured by it. Liquid precipitation and the water equivalent of freezing precipitation, too, are subject to depletion by wind, though not to the extent of snowfall.

A different, but equally important problem, relates to the measurement of the spatial distribution of rainfall. Mid- and high-latitude precipitation in summer, and at low latitudes throughout the year, are convective in nature, i.e. precipitation emanating from cumulus clouds. Individual clouds typically exhibit a horizontal extent of only a few kilometers, therefore the spatial distribution of precipitation, even over flat terrain or the ocean, can be quite variable, since the mean separation of gauges is of the order of 50 km (in the USA). In the extreme case, rainfall at the rate of a few tens of millimeters per hour can occur in one location with no precipitation occurring only a few kilometers distant. Rain gauge densities sufficient to measure small scale anomalies of, say, 1 km scale have only been maintained for research, rather than for operational networks. Changnon (1979) describes 14 meso networks of rain gauges in Illinois operated by the Illinois State Water Survey during the last four decades. The networks ranged from 0.03 km^2 (18 gauges) to 3800 km^2 (250 gauges). The former was operated during 20 warm seasons, the latter for four. Some of the precipitation anomaly patterns are discussed by Changnon and Vogel (1981). Spatial correlation of rainfall is directly related to latitude, and greater during the cold season. For example, summer rainfall for sites with 5 km separation typically correlate about +0.2 to +0.3 in Florida, and about +0.4 to +0.6 in Illinois (Jones and Wendland, 1984). Winter precipitation in mid- and high latitudes is much more spatially continuous (of the order of hundreds of kilometers), being a product of large-scale cyclones with attendant stratus clouds.

The spatial distribution of precipitation can be determined from existing raingauge networks, with their limitations, or can be inferred from radar imagery. Radar echo intensity be calibrated with appropriately placed raingauges within the area of the radar scan. Currently, the decrease in precipitation data quality from gauges to radars to satellites is substantial; however, the Doppler radars currently being installed at a few hundred sites in the USA exhibit substantially improved resolution over former radars, so that improved precipitation distribution and quantity estimates from radar should become available within the foreseeable future.

Surface runoff

Runoff is composed of several potential components: direct runoff from rainfall, and indirect, lagged runoff from snow- and icemelt, and runoff contributed for interflow if re-entering a stream. Each of the above can lag the precipitation event by substantially different intervals, depending on the rate of the liquid precipitation event, the slope and aspect of the surface, soil moisture content and the presence of ground frost. Indirect runoff depends on the depth of snowpack, direct solar intensity and the magnitude and duration of temperatures greater than 0°C. Interestingly, infiltration of liquid precipitation or snow melt into essentially frozen soil under a snow cover can represent in excess of 95% of the precipitation (Price et al., 1978). Runoff from surface ice and snow is substantially reduced if ablation occurs via sublimation during clear sunny days with temperatures less than 0°C.

The rate of runoff largely determines the degree of erosion from the underlying surface, although erosion also depends on surface cover, slope, aspect, etc. Sediment yield in a stream (including eroded material) has been found to be highly related to the coefficient of variation (standard deviation/mean) of precipitation (Harlin, 1978). Spring runoff from snowpack often contributes to riverine flooding, an especially serious problem when deep snowpack melts rapidly due to very warm, prolonged temperatures over still-frozen ground.

Evapotranspiration

The rate of evaporation from a surface into the air is a function of the strength of an energy source, the vapor pressure gradient between the soil or vegetation surfaces and that of the ambient air, and wind speed.

Evaporation is the total moisture loss through the soil–air interface, and that which transpires through the vegetation.

In the United States, evaporation is measured by the water loss from an evaporation pan (25 cm deep, 1.2 m diameter) initially filled with water to 5 cm from the rim. Because of turbulence to air flow caused by the structure itself, and heat transfer through the sides of the container, pan evaporation tends to be greater than that over an open water surface (Budyko, 1974). This difference from actual evaporation can be reduced by sinking the pan so that the evaporating surface is near the level of the undisturbed ground surface (World Meteorological Organization, 1971). Reasonable estimates of evaporation from a cropped surface can be obtained from a lysimeter (a container several meters in diameter and 1–2 m deep), filled with soil and growing vegetation. Water loss is determined by noting weight loss of the lysimeter. That water loss can vary greatly is seen from the observation that evapotranspiration over a pine forest was found to exceed that from a free-water surface by about 10% over six warm seasons (Holmes and Wronski, 1981).

Estimates of evaporation and evapotranspiration can be calculated. For example, Thornthwaite (1948) presented an equation to calculate potential evapotranspiration (evapotranspiration from an unlimited supply of water) as a function of daily mean air temperature and day length. Penman (1963) related evapotranspiration to vapor pressure, wind speed, surface temperature and the short-wave radiation available at the surface. Estimates using Penman's method in Illinois are typically greater than those calculated by the Thornthwaite equation (Jones, 1966). Monteith (1965) suggests an improved evapotranspiration calculation, whereby evapotranspiration is determined as a function of energy (usually radiation) temperature, humidity, wind speed, leaf resistance and the number and distribution of stomata. Bavel (1966) derived a relationship between potential evapotranspiration and net radiation, ambient air properties and surface roughness. Priestley and Taylor (1972) demonstrated that the latent and sensible heat fluxes could be closely estimated using parameterizations of energy over land and of bulk aerodynamic type over the sea. The latter three relationships, based on a greater number of physically related independent variables, yield good estimates of evapotranspiration under varying conditions. However, the greater number of independent variables limit their use to sites where sufficient observations are available, i.e. primarily micrometeorological research sites.

On a larger scale, Rasmusson (1966) presented the spatial distribution of mean monthly water vapor transport over North America, determined from 2 years of twice-daily aerological data. These analyses exhibit a reasonable pattern of flux divergence and inferred evapotranspiration.

Groundwater

Groundwater includes two components: subsurface underflow and groundwater storage. These two components are difficult to measure separately and are therefore often treated as one quantity and evaluated as a residual in the water budget equation, i.e. groundwater recharge, that quantity representing the change to subsurface underflow and groundwater storage. Rehm et al. (1982) review methods to estimate groundwater recharge: calculations based on hydrograph records, Darcy flux calculations (where vertical flow is a function of permeability, hydraulic gradient and porosity), and flow partitioning from an analysis of groundwater flow. Two examples of the first method are found in Schicht and Walton (1961) and O'Hearn and Gibb (1980).

Soil moisture

Soil moisture is moisture in the unsaturated zone, originating from surface water that enters the soil through percolation. Soil moisture can be measured by physically removing a soil column, weighing it, heating to expel water and weighing again. The weight difference is attributed to water from the soil mass. Obviously, disturbing the soil precludes further measurement at the same site. In situ measurements of soil moisture can repeatedly be made with a calibrated neutron probe, which is lowered in a pipe into the soil, obtaining measurements at desired intervals. Belding et al. (1995) have suggested using a time-domain reflectometer device to measure soil moisture. A helical coil is buried in place, and through appropriate circuitry, soil moisture measurements at discrete intervals can be obtained repeatedly.

Models have been developed to calculate the change in soil moisture from some earlier time, that time often being saturated conditions in early spring. Soil moisture estimates are updated as a function of temperature and precipitation. Dale et al. (1982) describe a budget technique that estimates soil moisture under different field and drainage conditions.

Interception

Interception is that quantity of precipitation that impinges on vegetation and other surfaces above the soil, some of which can be either assimilated into the interception medium or evaporated prior to reaching the surface. In long-term determinations, interception is usually assumed to be minimal. However, interception can be a significant moisture source to vegetation. For example, in coastal Peru where precipitation is minimal, vegetation obtains moisture by interception from surface-based clouds that drift through the vegetation. Cloud droplets settle on the leaves, some of which drop to the soil surface and are available to the tree roots. Interception in an Australian plantation forest with closed canopy was found to be 1.8 mm by Holmes and Wronski (1981).

Worldwide distribution of evaporation and runoff ratios

Two parameters which are helpful in delineating hydroclimatological regions are evaporation ratio and runoff ratio. Evaporation ratio is the energy consumed by evaporation (E) during a given time interval expressed as a percent of the total net radiation (R_n) received during the same interval. Runoff ratio is the total runoff (RO) during a given time interval expressed as a percent of total precipitation (P) accumulated during the same interval. The raw data required for these ratios (i.e. heat expended for mean annual evaporation, net radiation, and precipitation) were obtained from Barry (1969), Budyko (1956), and Geiger (undated), respectively. The mean annual distribution of the runoff is given in Sellers (1967) and Budyko (1974). Additional heat budget detail from tropical oceans was determined from data presented by Hastenrath and Lamb (1978, 1979). Runoff was calculated from estimates of precipitation minus evaporation. Resolution of detail is limited to distances of about 1000 km.

Evaporation ratio

Budyko (1956) and Sellers (1967) have discussed the concept of evaporation ratio. The ratio, i.e. the energy required to evaporate the total amount of water as a percentage of the available net radiation during a given time, differentiates between those areas where evaporation can be (but need not be) totally supported by net radiation (indicated by evaporation ratios of less than 100%) and those areas where the energy required for evaporation comes, in part, from sensible energy sources, such as atmospheric advection and the Earth's surface, indicated by values greater than 100%.

Ratios greater than 100% dominate the midlatitude oceans of both hemispheres. There are relatively large ratios along the western oceanic margins, where the sea surface is relatively warm, and diminished ratios over eastern oceans. Maximum values in excess of 250% are found over the Gulf Stream. The tropical oceans and eastern margins of the Atlantic and Pacific oceans typically exhibit evaporation ratios of 50–60%.

Evaporation ratios of less than 50% dominate the continents, particularly the southwestern United States, southern South America, Africa, southwestern and central Asia, and Australia. Sizeable areas with evaporation ratios less than 10% are found over northern Africa and interior Australia, where minimal precipitation limits evaporation. Maximum ratios are found over northern hemisphere continents poleward of about 60°N, exhibiting values in excess of 150%. Midlatitude continental ratios are between 50 and 100%. For the continents as a whole, net radiation is sufficient to support estimated evaporation.

Evaporation and net radiation estimates are less accurate over oceans and high-latitude continents, as are therefore the evaporation ratios. However, it is unlikely that the estimates are sufficiently in error to change significantly the patterns of the evaporation ratio.

Runoff ratio

Budyko (1956) and Sellers (1967) also have discussed this ratio, i.e. the runoff expressed as a percentage of mean annual precipitation. Substantial areas of the continents exhibit ratios near zero, including northern and western North America, east-central South America, Australia, North Africa and northern Eurasia. Runoff is exceedingly meager in these areas. The remainder of the continents typically exhibit ratios between 30 and 50% except southern South America and western India (>75%), and southeastern Africa (>100%). Only in these latter two areas does runoff exceed precipitation, at least at this scale of analysis.

Subtropical oceans typically exhibit effective precipitation ratios less than zero, due to the considerable evaporation under subtropical anticyclones. The ratios tend to be slightly positive near the equator and over high-latitude oceans, where cloud cover tends to dominate due to the intertropical convergence and the subpolar cyclones, respectively. Lowest effective precipitation ratios are found over the eastern margins of the southern hemisphere oceans. Again, because evaporation and precipitation estimates are less likely to be accurate over oceans and high latitudes, runoff and effective precipitation ratios are more suspect in those areas.

Human influences on hydroclimate

Purposeful and inadvertent impacts have been exerted on the hydroclimate in limited areas and times by human activity. The effect of large metropolitan areas on the urban and downstream precipitation distribution has been demonstrated by data from, and downstream of, St Louis Missouri, by Changnon et al. (1981) and other large cities.

Reducing water loss by evaporation from lakes and evapotranspiration from vegetation by the addition of monomolecular films to the surfaces has been shown to be feasible (Roberts, 1961).

Atmospheric carbon dioxide concentrations have been increasing since observations began about 100 years ago, and are expected to continue into the future, due to burning of fossil fuels and extensive forest clearance. Carbon dioxide plays a prominent role in the greenhouse effect, i.e. warming the lower atmosphere by reducing terrestrial long-wave radiation losses to space. It is anticipated that the present atmospheric carbon dioxide concentration will double by the mid-twenty-first century, resulting in mean annual warming 1.5–3.0°C (Schneider, 1975). This warming may decrease atmospheric stability, affecting precipitation, and will further modify the hydrologic budget since relative humidities are expected to remain constant, with the concomitant increase in absolute humidity (Watts, 1980).

The damming and channeling of the Mississippi River to improve navigation and flood control has changed the river's flow and sediment carrying capacity. The change imposed on discharge and residence time of water on its journey to the Gulf of Mexico has had an impact on the local hydrological budget. Reversing the direction of flow of the Snowy River in southeastern Australia modified local water budgets by draining regions that formerly received runoff and delivering water to others that formerly exported runoff.

On a larger and speculative scale, Borisov (1973) discusses changes that would result from diverting warm Pacific water to the Arctic Ocean, thereby freeing Arctic Ocean coastal areas from ice during much of the year. In addition to improving shipping opportunities for currently ice-locked ports during much of the year, the open ocean may lead to increased precipitation over nearby continental regions. This would increase runoff of fresh water to the Arctic Ocean, modifying the distribution of density and thus impacting ocean circulation.

The hydrological budget may also be modified by purposeful cloud seeding, in which a nucleating agent is injected into a cloud. These agents promote ice crystal formation, either by cooling cloud temperatures or by providing a crystalline structure conducive to ice crystal formation. Positive results in some (but not all) controlled seeding experiments has been statistically demonstrated (briefly reviewed by Kerr, 1982).

Conclusions

The components of the water budget exhibit spatial and temporal patterns over the surface of the Earth in response to various climate forcing functions and surficial constraints, for example, components of atmospheric general circulation, atmospheric stability and moisture, radiation, topography, surface albedo and surface water. As any one component changes, so do one or more components of the hydrological budget.

The occurrence and frequency of unusual hydrological events prompts questions of climate change and what the future may hold. Recent examples of such events include the intensity and frequency of droughts in northern Africa. Lamb (1985) has shown that a Sub-Saharan drought persisted from 1968 through to 1985, a relatively long string of moisture-deficient years relative to the ca. 45-year record. Areas of the upper Midwest have experienced several substantial variations in precipitation recently, notably the drought of the 1988 growing season and the floods of the summer of 1993. The drought exhibited a recurrence interval of some 40 years, whereas the 1993 flood exceeded the 100-year magnitude. The occurrence of extreme magnitude events, or several such events in a relatively short time may reflect natural variance, or may signal changes from one stable climate regime to another.

That climate and hydrological responses change with time is clearly shown from the paleoclimatic record. A particularly interesting example is an inferred reduced hydrological cycle during the last glacial maximum suggested by increased atmospheric dust present at that time, as found in Antarctic glacial cores (Yung et al., 1996).

Wayne M. Wendland

Bibliography

Barry, R.G., 1969. The world hydrologic cycle, in Water, Earth and Man, R.J. Chorley (ed.). London: Methuen, pp. 11–29.

Bavel, C.H.M. van, 1966. Potential evaporation: The combination concept and its experimental verification, Water Resources Research, 2, 455–467.

Belding, M.J., Hollinger, S.E. and Peppler, R.A., 1995. Development of new soil temperature and soil moisture profilers for the Illinois Climate Network. Preprint, 11th International Conf. On Interactive Information and Processing Systems for Meteorology, Oceanography and Hydrology. 75th American Meteorological Society Annual Meeting. Boston, MA, pp. 363–364.

Borisov, P., 1973. Can Man Change the Climate? Moscow: Progress Publishers.

Budyko, M.I., 1956. The Heat Balance of the Earth's Surface, N.A. Stepanova (trans.). Washington, DC: Office of Technical Services, US Dept. of Commerce.

Budyko, M.I., 1974. Climate and Life, D.H. Miller (trans.). New York: Academic Press.

Changnon, S.A., 1979. The Illinois climate center, Am. Meteorol. Soc. Bull., 60, 1157–1164.

Changnon, S.A. and Vogel, J.L., 1981. Hydroclimatological characteristics of isolated severe rainstorms, Water Resources Research, 17, 1694–1700.

Changnon, S.A., Semonin, R.G., Auer, A.H. et al., 1981. METRO-MEX: A review and summary, Meteorol. Mons., 18, 1–181.

Dale, R.F., Nelson, W.L., Scheeringa, K.L. et al., 1982. Generalization and testing of a soil moisture budget for different drainage conditions, J. Appl. Meteorol., 21, 1417–1426.

Geiger, R., undated. Mean annual precipitation (chart). Darmstadt, Germany: Justus Perthes Publications.

Harlin, J.M., 1978. Reservoir sedimentation as a function of precipitation variability, Water Resources Bull., 14, 1457–1465.

Hastenrath, S. and Lamb, P.J., 1978. Heat Budget Atlas of the Tropical Atlantic and Eastern Pacific Oceans. Madison: University of Wisconsin Press.

Hastenrath, S. and Lamb, P.J., 1979. Climatic Atlas of the Indian Ocean, pt II. Madison: University of Wisconsin Press.

Holmes, J.W. and Wronski, E.B., 1981. The influence of plant communities upon the hydrology of catchments, Agricult. Water Manage., 4, 19–34.

Jones, D.M.A., 1966. Variability of Evapotranspiration in Illinois. Circular 89. Champaign, IL: Illinois State Water Survey.

Jones, D.M.A. and Wendland, W.M., 1984. Some statistics of instantaneous precipitation, J. Climatol. and Appl. Climatol., 23, 1273–1285.

Kerr, R.A., 1982. Test fails to confirm cloud seeding effect, Science, 217, 234–236.

Lamb, P.J., 1985. Rainfall in Subsaharan West Africa during 1941–1983, Zeit. für Gletscherkunde und Glazialgeologie, 21, 131–139.

Langbein, W.G., 1967. Hydroclimate in The Encyclopedia of Atmospheric Sciences and Astrogeology, R.W. Fairbridge (ed.). New York: Reinhold, pp. 447–451.

Monteith, J.L., 1965. Evaporation and environment, Symposium of the Society for Experimental Biology, 19, 205–234.

O'Hearn, M. and Gibb, J.P., 1980. *Groundwater Discharge to Illinois Streams*, Contract Rep. 246. Champaign, IL: Illinois State Water Survey.

Penman, H.L., 1963. *Vegetation and Hydrology*, Tech. Comm. No. 53. Farnham, England: Commonwealth Bureau of Soils, Commonwealth Agricultural Bureau.

Price, A.G., Hendrie, L.K. and Dunne, T., 1978. Controls on the production of snowmelt runoff, in *Modeling of Snow Cover Runoff*, S.C. Colbeck and M. Ray (eds). Hanover, NH: US Army Cold Regions Research and Engineering Laboratory, pp. 257–268.

Priestley, C.H.B. and Taylor, R.J., 1972. On the assessment of surface heat flux and evaporation using large-scale parameters, *Monthly Weather Rev.*, **100**, 81–92.

Rasmusson, E.M., 1966. *Atmospheric Water Vapor Transport and the Hydrology of North America*, Report A. Cambridge, MA: Massachusetts Institute of Technology, Department of Meteorology.

Rehm, B.W., Morgan, S.R. and Groenwold, G.H., 1982. National groundwater recharge in an upland area of central North Dakota, U.S.A., *J. Hydrology*, **59**, 293–314.

Roberts, W.J., 1961. Reduction of transpiration, *J. Geophys. Research*, **66**, 3309–3312.

Schicht, R.J. and Walton, W.C., 1961. *Hydrologic Budgets for Three Small Watersheds in Illinois*, Rep. of Investigation 40. Champaign, IL: Illinois State Water Survey.

Schneider, S., 1975. On the carbon dioxide–climate confusion, *J. Atmos. Sci.*, **32**, 2060–2066.

Sellers, W.D., 1967. *Physical Climatology*. Chicago: University of Chicago Press.

Thornthwaite, C.W., 1948. Approach toward a rational classification of climate, *Geog. Rev.*, **38**, 55–94.

Watts, R.G., 1980. Climate models and CO_2-induced climatic changes, *Clim. Change*, **2**, 387–408.

World Meteorological Organization, 1971. *Guide to Meteorological Instruments and Observing Practices*, 4th edn. Geneva: World Meteorological Organization.

Yung, Y.L., Lee, T., Wang, C.-H. and Shieh, Y.-T., 1996. Dust: A diagnostic of the hydrologic cycle during the last glacial maximum, *Science*, **271**, 962–963.

Cross references

Acid rain
Evaporation: measurement
Evapotranspiration
Hydrology
Precipitation
Semiarid regions
Water resources

HYDRODYNAMICS: POROUS MEDIA

The study of flow through porous media is of importance in connection with many geological applications. Such diversified fields as soil mechanics, ground water hydrology and petroleum engineering rely heavily on it as basic to their individual problems.

Porous media

Porous media are solid bodies that contain 'pores,' small void spaces, which are distributed more or less frequently throughout the material. The problem of complete geometric characterization of a porous medium has not yet been solved. One is able only to define some geometric parameters of a porous medium that are based on averages. The first of these is the porosity, P, which is equal to the average ratio of the void volume to the bulk volume of the porous medium. The second is the specific surface area, S, which is the average ratio of internal surface to the bulk volume of the porous medium.

Darcy's law

More than a century ago, Henri Darcy (1856) made some experiments to investigate the flow of water through the sand filters of the water purification plant at Dijon, France. The following law can be deduced from Darcy's experiments:

$$q = - (k/\mu) \, (\mathrm{grad} \; p - \rho g)$$

where q is the filtration velocity vector (the quotient of the total sum of flow vectors per unit of time through an infinitesimal cross-section of porous medium divided by that cross-section; its magnitude is q), μ is the viscosity of the fluid, ρ is the density of the fluid, and g is the gravity vector (of magnitude g and direction downward). Finally, k has been termed the 'permeability' of the porous medium.

To apply the basic equations of flow through porous media to practical cases, one has to add a continuity condition

$$-P(\partial \rho / \partial t) = \mathrm{div} \; (\rho q)$$

and an equation of state for the fluid

$$\rho = \rho(p)$$

where, in addition to the symbols already defined, t is time. As is evident, this system of equations is non-linear, which makes it difficult to obtain solutions of it. Fortunately, however, it is often possible to linearize the system of basic equations to a heat conductivity equation for which many solutions are known (Carslaw and Jaeger, 1959).

A peculiar problem occurs if one studies gravity flow with a free surface in a porous medium. The free surface is an equipressure surface: any streamline having one point in common with it must lie entirely within it. If the free surface intersects an 'open' surface of the porous medium, then a surface of seepage will be formed below the line of intersection. The problem of finding the free surface, even under steady-state conditions, is thus a problem with a floating boundary condition. Various analytical methods have been adapted to this end; the most comprehensive survey is probably that in a monograph by Polubarinova-Kochina (1962).

Limitations of Darcy's law

Darcy's law is valid only in a certain seepage velocity domain, outside which more general flow equations must be used to describe the flow correctly. To characterize this seepage velocity domain, it is customary to introduce a Reynolds number (Re), as follows:

$$Re = q\rho\delta/\mu$$

where, in addition to the symbols defined earlier, the quantity δ is a microscopic diameter associated with the porous medium (pore diameter). The universal critical Reynolds number, beyond which Darcy's law is no longer valid, ranges between 0.1 and 75. Forchheimer suggested in 1901 (Scheidegger, 1974) that Darcy's law should be modified for high-flow velocities by including a second-order term in the velocity, which leads to (in a linear system):

$$\partial p/\partial x = aq + bq^2$$

Solutions of the preceding high-velocity flow equations for particular cases are difficult to obtain. Notwithstanding the difficulties, Engelund discussed some possibilities in 1953 (Scheidegger, 1974).

A breakdown of Darcy's law also occurs in gases at low pressures, due to the various molecular effects that may come into prominence in rarefied gas dynamics. Thus, Knudsen (slip) flow or even molecular flow may occur.

Theoretical models of porous media

The flow laws just discussed are essentially empirical laws, deduced from a series of experiments. However, since the fluids flowing through a porous medium can in most instances be regarded as ordinary Newtonian, that is, viscous fluids, it should be expected that their motion can be described by obtaining a suitable solution of the Navier–Stokes equations. This could be done if it were possible to formulate the boundary conditions of the problem correctly, stating that the fluid must stick to the walls of the pores. Unfortunately, porous media are of an extremely complex nature; thus it is entirely impossible to treat the flow of a fluid confined within the pore space in any manner that could claim to be microscopically exact. This seems to preclude forever the understanding of such things as Darcy's law from a microscopic standpoint.

A way out of this difficulty is to represent an actual porous medium by something that the human brain can comprehend. One therefore makes gross simplifications in a porous medium to be able to formulate the boundary conditions and to integrate the basic Navier–Stokes equations. Such simplified versions of porous media are called models.

The simplest model consists of a bundle of parallel capillaries of circular cross-section, all with the same diameter. A serious drawback of such a model is that all the pores are supposed to go from one face of a porous medium right through to the other. This is evidently a picture far removed from actuality.

The opposite extreme picture would be obtained by assuming that the pore space is lined up serially, so that each particle of fluid would have to enter at one pinhole at one side of a porous medium and travel through very tortuous channels through all the pores, and then emerge at only one pinhole at the other face of the medium. Obviously, this picture is just as unreal as a parallel type model; a realistic model lies somewhere between the two extremes.

A different type of model has been suggested by Emersleben in 1925 (Scheidegger, 1974). It may be noted that, in the models just discussed, the porous medium has been visualized essentially as a piece of solid material with holes in it. Emersleben took the opposite view, in that he visualized the porous medium as a fluid-filled space with a few obstacles in it. The drag exercised by all the obstacles on the fluid, then, represents the resistance of the medium to flow. It turns out that this type of model satisfactorily describes the flow through highly porous media, such as an agglomeration of fibers.

All these models are unsatisfactory on general grounds, because they attempt to describe disordered porous media by well-ordered models. Disorder can be taken into account by introducing the concepts of fractal geometry. Thus it has been suggested to consider a porous medium as a Sierpinski carpet of a certain fractal dimension (e.g. Turcotte, 1992, p. 169 ff.). In this fashion, an explanation of the sudden onset of permeability with increasing porosity in certain classes of porous media (e.g. fractured limestones and regular sandstones) is obtained. However, for the description of the flow dynamics in a porous medium, recourse has to be taken to statistical mechanics (Scheidegger, 1954, 1974). This leads to a statistical treatment of the hydrodynamics in porous media, in which either the flow paths of the individual fluid particles or the flow channels in a porous medium are assumed to be randomly distributed. The main prediction is that of occurrence of mechanical dispersion in porous media by which individual fluid particles become intermixed, not by the molecular diffusive motion but by the splitting up and rejoining of flow channels.

Displacement processes

Of particular interest in connection with geological applications is the theory of displacement processes in porous media, which are important in the production of oil from underground strata, in the exploitation of groundwater resources and in the assessment of the spread of pollutants in the latter.

Displacement occurs in a different fashion, depending on whether the two fluids involved are miscible with each other or not. Turning first to immiscible displacement, we note that a theory can be obtained by writing Darcy's law for each phase, introducing a 'relative permeability' k (as a fraction of total permeability K) to the phase (i) that is assumed to be a function of saturation only. This, in conjunction with the usual continuity equations and equations of state for both flowing phases, leads to a complete system of differential equations that describes the displacement processes. Unfortunately, it is non-linear, and solutions have been obtained only for the one-dimensional case.

Turning to miscible displacement, we note that there are no fronts; the predominant effect is a progressive blurring of the interface between the two fluids, which has been called dispersion. It is described by a diffusivity equation referring to a moving coordinate system connected with the overall motion of the mixture of the two fluids. Solutions have been obtained for the linear case. The mechanical dispersion of the two fluids is caused by the inter-connections of the flow channels. This shows that the statistical models discussed earlier have a direct application to miscible displacement theory (Scheidegger, 1974).

The above treatments of the displacement processes assume that the latter are stable. However, already a very elementary theory shows that this may not be the case. In such an elementary theory, no mixing between the two flowing phases is assumed to take place at all. The two phases are confined to domains whose extent changes with time. Each fluid moves within its own domain according to Darcy's law. The solutions of the corresponding differential equations show that instabilities may result if a fluid is displaced from a porous medium by a less viscous one. In this case, fingering is liable to occur, which means that the displacement front becomes unstable: fingers form that

shoot at relatively great speed through the porous medium. On a more sophisticated basis, the process is amenable to treatment by system theory: taking the mean position of the displacement front as a reference, any 'regular' displacement is a quasi-stationary state of the system at the 'edge of chaos'; if it becomes unstable, the deviations from it are fractal. Thus the range goes from a perfectly stable displacement process [viscosity ratio (of displacing fluid/displaced fluid) > 1] through a quasi-stable to an unstable fingering state, until complete chaos (represented by dispersion, or 'diffusion') is reached. The general sequence of states referred to above has been observed in nature and has been modeled on computers many times (e.g. Kauffman, 1993); it seems, however, that it is a universal law of nature (Scheidegger, 1995, 1996).

Adrian E. Scheidegger

Bibliography

Carslaw, H.S. and Jaeger, J.C., 1959. *Conduction of Heat in Solids.* Oxford: Clarendon Press, 510 pp.

Darcy, H., 1856. *Les Fontaines Publiques de la Ville de Dijon.* Paris: V. Dalmont, 647 p.

Kauffman, S.A., 1993. *The Origins of Order.* Oxford: Oxford University Press, 709 pp.

Polubarinova-Kochina, P.Ya., 1962. *Teoriya dvizheniya grunto-vykh vod.* Moscow: Gosudarstv. Izdatelstvo Tekhnicheskoi-Teoreticheskoi Literatury, 676 p (English translation by R. De Wiest, Princeton, University Press).

Scheidegger, A.E., 1954. Statistical hydrodynamics in porous media, *J. Appl. Physics,* **25**, 994–1001.

Scheidegger, A.E., 1974. *The Physics of Flow Through Porous Media,* 3rd edn. Toronto: University of Toronto Press, 373 pp.

Scheidegger, A.E., 1995. Order at the edge of chaos in geophysics. *Abstracts IUGG XXI Assembly,* **1**, A–11, UA51A–10.

Scheidegger, A.E., 1996. Ordnung am Rande des Chaos: Ein neues Naturgesetz. *Oesterr. Z. Vermessung & Geoinf.,* **84**, No. 1.

Turcotte, D.L., 1992. *Fractals and Chaos in Geology and Geophysics.* Cambridge: University Press, 221 pp.

Cross references

Drainage
Hydrology
Hydromechanics
Soil and water management
Water: categories
Water movement in unsaturated soils

HYDRO-ECOLOGY: PHABSIM

PHABSIM (physical habitat simulation) is a hydro-ecological model, developed in the United States, which enables the assessment of impacts caused by changing flow regimes, or channel geometry, on the available habitat for selected species.

PHABSIM has been used widely in the United States with applications in over 38 states. In recent years its application has been extended with studies taking place in at least 13 other countries, including eight in Europe. Developed by the US Fish and Wildlife Service, PHABSIM allows the simulation of the relation between streamflow and available physical habitat as defined by water depth and velocity, substrate and available cover. It contains a number of hydraulic models which predict values of depth and velocity at selected simulation discharges. These models require calibration using flow data collected in the field at one or more calibration discharges. Observations of substrate and cover are also recorded, using an appropriate coding system, and are assumed to be independent of discharge. Once calibrated to a study site, the model can simulate values of microhabitat variables over the full range of discharges within the river reach in question.

The simulated values of the microhabitat variables modeled within PHABSIM are combined with data which describe the relative suitability of those variables for the selected target species life stages as shown in Figure H1. These data are termed 'habitat suitability

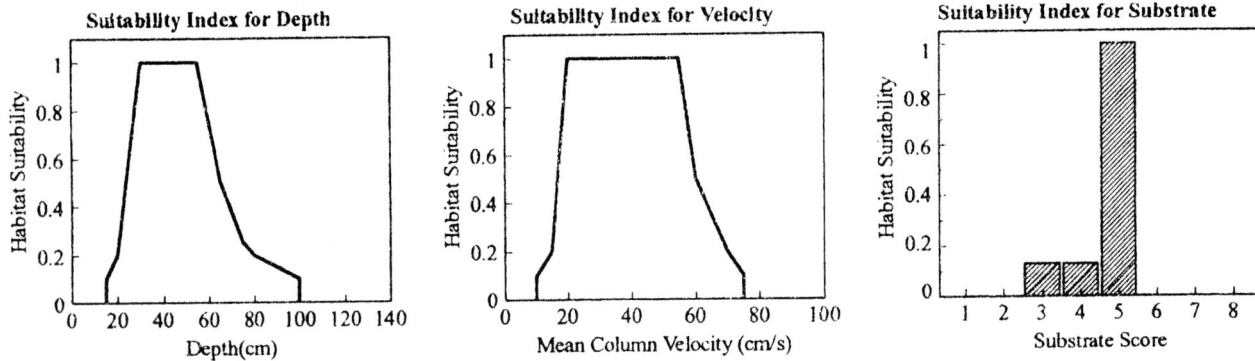

Figure H1 Examples of habitat suitability indices for fry/juvenile brown trout, based on observations by staff of NRA South Western Region, UK.

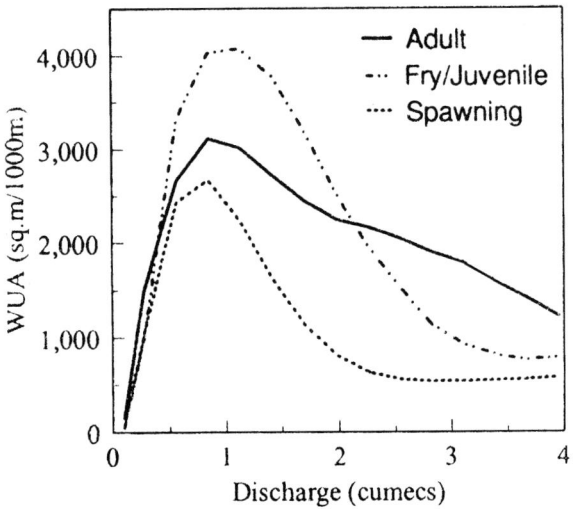

Figure H2 Weighted usable area versus discharge for brown trout in the River Allen, UK.

indices' and may be derived from expert opinion, existing literature or by field sampling techniques such as electro-fishing or snorkelling. The PHABSIM habitat models which combine these two sets of information produce results showing the available habitat within the study reach, expressed as weighted usable area (WUA) versus discharge. An example set of results is shown in Figure H2.

The WUA–discharge relationship can then be used to derive habitat time series from observed and simulated flow series. This enables the analysis of the frequency of the available habitat under existing and predicted flow regimes and can be carried out on a seasonal basis as required. The example habitat duration curves (Figure H3), for the River Allen, show values for both the 'historical' (including the effects of the historical groundwater abstraction regime) and 'naturalized' (with the effects of the abstraction removed) flow regimes.

R.W. Herschy

Bibliography

Johnson, I.N. and Law, F.M., 1995. Computer models for quantifying the hydro-ecology of British rivers. Institution of Water and Environmental Management, **9**(3).

Hydro-Ecology, 1995. *Phabsim Newsletter*, 4 pp. The Institute of Hydrology, UK.

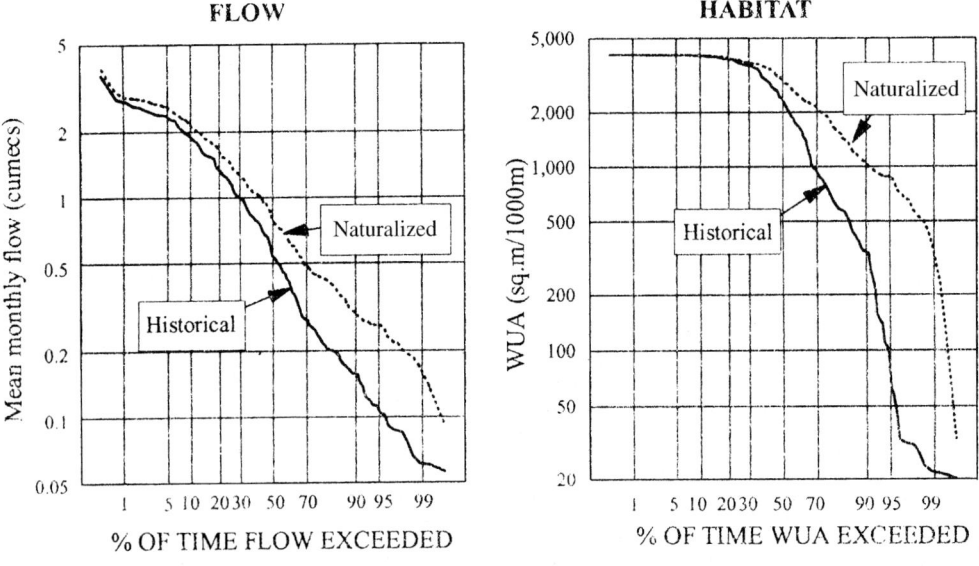

Figure H3 Example of flow and habitat duration curves (River Allen, UK, downstream study site).

HYDROGEOLOGY

Hydrogeology is defined as the science which studies the occurrence, distribution, movement and chemical composition of water beneath the surface of the Earth based on a knowledge of geology (Maxey, 1964). This recognition of the importance of geological factors, and the inclusion of the rocks within the scope of investigations, contrasts hydrogeology with other forms of groundwater studies (Davis and DeWeist, 1966: Todd, 1980). The closely related discipline of groundwater hydrology or geohydrology embraces the study of groundwater but without the same emphasis on the geological aspects. These latter terms tend to be used to cover specialist aspects of groundwater studies such as those undertaken by hydrochemists, hydraulic engineers and applied mathematicians (Downing and Jones, 1985). Resulting from the complexity of the subject, hydrogeological projects usually require the integration of a wide range of specialist skills, including those of geology, hydraulics, chemistry, geophysics, civil engineering and mathematics (Downing and Wilkinson, 1991; Headworth and Skinner, 1986). This requirement for a diversity of expertise is illustrated by the broad scope of standard hydrogeological text books, for example Brandon (1986), Davis and DeWeist (1966), Driscoll (1986), Erdélyi and Gálfi (1988) and Todd (1980). In practice, therefore, the fine distinctions between the different groundwater sciences are often blurred.

Groundwater forms one part of the hydrological cycle, from which it follows that hydrogeology may also be regarded as a branch of the hydrological sciences. The close relationship between the various components of the hydrological cycle means that in order to study a groundwater system it is necessary to consider the rest of the cycle including rainfall, evaporation and surfacewater flows. The other important features which require field assessment or measurement are the rock types and their hydraulic properties. It is particularly important to know the permeability (or more specifically the hydraulic conductivity) of all the rocks in the study area.

In both sedimentary rocks such as sandstone, and unconsolidated sediments such as sand and gravel, groundwater is contained in and moves through the pore spaces between individual grains. Fissure systems in solid rocks can significantly increase the hydraulic conductivity of the rock mass and are often important in determining the successful yield of a well. In crystalline aquifers such as some limestone and igneous rocks, groundwater flow takes place through fissures and very little moves through the body of the rock itself. Often the significant zones of groundwater flow are in the near-surface weathered materials or in broken ground associated with faulting. The Basement Complex which occurs over much of the African continent is a good example of the importance of weathering and fissure systems to groundwater resources development. Some geological materials do not transmit groundwater at significant rates, while others only permit small quantities to flow through them. Although these poorly permeable materials do not transmit much water, they play a major role in controlling the movement of water through aquifers.

The hydrological cycle is described in detail elsewhere. In essence, however, a groundwater system consists of rainfall recharge percolating into the ground, reaching the water table and then flowing through rocks of varying hydraulic conductivity towards natural discharge points. The rate at which water flows through the system depends upon the rainfall, evaporation rates, the geological conditions and many other factors. A hydrogeological investigation of necessity must address each part of the total system.

Historical development

The development of an understanding of the occurrence and movement of groundwater has its roots in antiquity with ancient peoples attempting to explain spring flows, through to more modern times with the scientific developments starting during the eighteenth and nineteenth centuries and continuing to the present day. There are a number of accounts of these early developments in standard textbooks (Davis and DeWeist, 1966; Todd, 1980) based on historical reviews. In early times it was generally believed that the water discharging from springs could not originate from rainfall as the rainfall was held to be inadequate and the ground too hard to permit water flow. Ancient Greek philosophers thought that springs were formed by seawater flowing through special channels which extended beneath mountain ranges with the flowing water becoming purified before

emerging at springs. These ideas were still held until at least into the seventeenth century. The French physicist Edmé Mariotte (*ca.* 1620–1684) carried out a number of hydrological experiments and was the first person to propose that rainwater infiltrating into the ground forms the source of spring water (Todd, 1980).

Hydrogeology as a science developed from the application of geological principles in the search for optimum locations for the construction of new wells or other abstraction systems. The foundations of geology as a science were laid down largely during the eighteenth century and their practical application in groundwater studies quickly followed. For example, the English geologist William Smith (1769–1839) acted as a consultant during the early part of the nineteenth century, advising on a variety of applied geological problems including the siting of wells at Scarborough, UK, and elsewhere (Sheppard, 1917).

It was recognized at an early stage that it was important to be able to quantify the flow of water through geological materials. During the middle of the eighteenth century Henri Darcy (1803–1858), a French water engineer carried out experiments at Dijon to define the flow of water through sand filters. The resultant publication (Darcy, 1856) defined for the first time the relationship between several measurable variables and the permeability of the sand. This mathematical relationship is known as Darcy's law and forms the basis of all studies of flow through porous media. Dupuit (1863) was the first to develop a formula to describe the flow of water into a well based on Darcy's law. Many other workers have since added to the methods of analyzing the flow of groundwater to a well and through geological materials in general. Initially the formulae were restricted to the steady-state situation where it is assumed that water is continuously pumped from a well at constant rate. A great advance was made by Theis (1935) who introduced an equation for the non-steady flow towards a well. Because the volume of groundwater storage is depleted as water levels decline during the pumping period, this method allowed the storage coefficient to be calculated from field data for the first time. Other significant contributions to the development of analytical methods for non-steady-state pumping conditions were made by Boulton (1951, 1963) and Jacob (1940, 1963). Since that time a large number of analytical methods have been developed to enable pumping test data to be analyzed in most geological and hydrogeological circumstances, and in any configuration of well construction (Kruseman and DeRidder, 1991).

The ultimate practical application of mathematics and computer methods in hydrogeology is the development of models which simulate the groundwater flow system (Rushton and Redshaw, 1979; Bear and Verruijt, 1987). The established general equations of groundwater flow are used to simulate the relationship of the hydrogeological variables at a large number of locations across an aquifer. The massive computational power of modern computers is used to perform these calculations rapidly, thereby making such techniques a practical tool for hydrogeological assessment. Although the purpose of developing a groundwater model is often to simulate various conditions of flow or pumping to make predictions for planning purposes, groundwater models have a more fundamental application in data validation. Models are capable of handling extensive data sets and allow the relationships between each variable to be tested. This enables the quality of the data on each variable to be assessed, in addition to testing elements of the conceptual understanding which has been developed for that particular aquifer system. This latter application of groundwater models has proved to be an invaluable tool in many hydrogeological studies (Rushton, 1984).

During the early development of groundwater modeling techniques, computer code was specially written for each application. A modern trend, however, is for standard computer codes to be applied and adapted. This has the advantage that it is easier for simulations to be carried out to the same standards and that by having many thousands of users the mistakes in each program have been discovered. In addition, as groundwater models are being used by environmental regulators it is important that each modeled situation is treated in the same way and hence standard model code has been selected to provide this consistency. One of the most commonly used groundwater model computer codes is MODFLOW (McDonald and Harbaugh, 1988), which was originally developed by the US Geological Survey.

Hydrological investigations

Groundwater studies vary widely in their scope depending on the technical objectives of the project and the hydrogeological setting. At one end of the scale they may comprise no more than a desk study drawing on the available information for an area, and at the other they require extensive field work and observations extending over several years (Headworth and Skinner, 1986).

A distinguishing feature of hydrogeology is its heavy dependence on field observation and measurement. In any study the first step is to develop a conceptual model (or general understanding) of the groundwater flow system and the interrelationships with the geology. Field data are then collected which enable this conceptual model to be refined systematically and quantified as required (Brassington, 1988; Downing and Jones, 1985). There are several phases to such studies, although both the amount of effort put into each one, and the other in which they are carried out, are often varied to meet the needs of a particular set of circumstances.

Over the past two decades there has been a significant development in the availability of sophisticated field equipment for measuring groundwater levels, taking samples and measuring chemical parameters. The tendency has been for an increased use of electronic and computer-based systems in taking these measurements, collecting the raw field data and in data storage and processing.

A desk study is an essential prerequisite to any hydrogeological field investigation. It comprises the assembly of all the available relevant information, thereby defining the additional data to be collected in later phases of the work. The desk study includes an examination of published geological information to identify potential aquifers from their lithologies; topographic maps and aerial photographs are used to identify spring lines and groundwater discharge points; groundwater level records define the groundwater flow directions; and abstraction records define the extent of the current level of resource development and help the understanding of the changes in the flow system which may have occurred. Information is also needed on the rainfall and evaporation losses to define the extent of annual recharge.

Once a desk study has been completed, an initial reconnaissance is usually carried out. It may be necessary to carry out additional geological mapping and the information gained from the geological and topographical maps must be verified. The reconnaissance also forms the start of the survey to locate the abstraction wells and identify where groundwater level measurements can be taken. An essential part of this reconnaissance is to identify the type of information needed to complete an understanding of the groundwater system, thereby planning the exploration phase of the investigation.

At the start of the investigation a database of the available data is developed which includes groundwater level records, abstraction rates, spring flows and water quality data. New data are added as they are obtained and compared with the conceptual model developed at the desk study stage. An increasingly more detailed understanding of the groundwater system is built up in this way, and is used to plan the remaining stages of the investigation. The continual re-evaluation of data is an important feature of groundwater investigations because the data available are usually sparse.

Once the initial reconnaissance of an area has been completed, a monitoring program is established to record groundwater levels, the flow of springs and streams and abstraction rates. Additionally, rainfall and evaporation measurements are usually required together with water chemistry information. The more detailed information collected during this phase is used in conjunction with the background data collected by monitoring organizations. Computer-based databases are usually employed to manage the data sets assembled during hydrogeological studies.

Further information is frequently required on the aquifer geology and hydraulic properties to enable a comprehensive understanding of the local hydrogeology to be achieved. Such data can only be obtained from exploratory drilling, geophysics and conducting pumping tests on suitable existing or newly drilled boreholes, to assess the aquifer's hydraulic characteristics.

Once the extent of an aquifer has been established and its boundaries identified, it should be possible to quantify the volumes of water which are passing through the groundwater system. The amount of recharge can be assessed using information about rainfall and evaporation. Discharges from the aquifer can be estimated from spring flow measurements, stream gauges and the amount of water pumped from local wells. The water inputs and outflows from the aquifer are compared in a water balance exercise. This provides confidence in the overall assessment of recharge and to the conceptual understanding of the groundwater flow system. An apparent imbalance is generally indicative of the need for further field data or a refinement in the conceptual model. This stage of the investigation may also incorporate the use of computer models both to validate the available data and predict changes in the system.

Current and future issues

Although hydrogeological theory largely developed in the pursuit of water resources to be developed, it has been extended into a number of related fields, particularly over the last quarter of the twentieth century (Downing and Wilkinson, 1991). As the fields in which hydrogeology has been applied have grown in significance since the 1960s, the involvement of other disciplines in groundwater studies has also increased. Hydrogeological programs have become more complicated and require the overall coordination of many strands by the hydrogeologist, who is now often cast in a management role.

The development of groundwater resources remains important in many parts of the world where the rapidly increasing population maintains a growing demand for water supplies. Where possible, groundwater tends to be developed in preference to surface sources because wells can often be located in convenient positions in relation to villages, and groundwater tends to have a better and more reliable quality. Hydrogeologists from the developed countries have assisted in groundwater development in the developing world for several decades. There is now a general change in emphasis in the way these projects are carried out with an increasing involvement from the local nationals who are being trained in hydrogeological skills and techniques.

During the past decade the environmental impact of groundwater abstraction on the other components of the water environment have been recognized by environmental regulators in a number of countries (Brassington, 1992). Hydrogeologists have been employed in assessing the extent to which gradually increasing groundwater development has reduced river flows and in designing schemes to rectify the environmental damage caused by techniques such as augmenting river flows, replacing some sources and river-bed lining. Reductions in groundwater abstraction, generally in older urban centers, has caused a rise in the water table, or groundwater rebound. An international review of examples of rising groundwater levels shows that significant reductions in industrial abstraction have occurred in many countries including the USA, the UK, Germany, Japan and Denmark (Wilkinson and Brassington, 1991). In these situations the long-term local control on the water table provided by abstraction has been removed, allowing the natural groundwater levels to be restored and increasing the resources available for new development. In some instances, problems are being caused by a threat to the stability of building foundations and other engineering structures or in the mobilization of pollutants as the rising water levels inundate contaminated sites and landfills. Other causes of rising water levels were identified, such as leaking water-supply pipes and sewers, and over-irrigation.

During the 1970s it was realized by governments in a number of countries that various industrial activities appeared to be causing a long-term contamination of groundwater resources. The slow rate of groundwater flow compared to surface water means that once pollutants have reached a groundwater body it is likely to remain contaminated for a very long period before the contaminants are flushed out. The disposal of industrial and domestic wastes and the use of chemicals in agriculture were the two areas first studied. The recognition that landfills are potential point sources of pollution led to a series of government-funded research projects and a steady improvement in landfill practice in many countries. The main areas of future interest in waste disposal are centered on waste minimization and recycling. Concern is also raised regarding the potential impact on groundwater from older landfills where the standards applied at the time of their operation are now regarded as inadequate. There is also interest in the recovery of energy from the generation of methane during the biodegradation of the waste materials.

Studies of agrochemicals first concentrated on nitrogenous fertilizers and later on the use of chemical pesticides to kill unwanted insects and plants. (Parker *et al.*, 1991). The research into the processes which are responsible for the widespread increases in concentrations of nitrate in groundwater has lead to control measures being put in place in Member States of the European Union and elsewhere. The studies of pesticides in groundwater, however, is more technically difficult because of the wide range of chemicals in common use, the low level of concentrations permitted for these chemicals in drinking water and the difficulties in taking samples without cross-contamination (Lawrence and Foster, 1987). As a result, the research programs into the distribution of these materials are likely to be an important part of hydrological work over the next decade or so.

Land contaminated by industrial process was identified as a major source of environmental pollution during the 1980s. The highest profile for this work was in the USA where the Federal Government faced the problem of water pollution emanating from former industrial activities on sites which are now under a different ownership. These concerns led in part to the formation of the US Environmental Protection Agency and federal funding for research into clean-up technologies. In many instances the contaminants are leached from the near-surface environment to reach the underlying water table. Where soluble, the contaminants dissolve in the groundwater or otherwise may float on top and are then dispersed from the site via the groundwater flow system. Initially it was thought that these contaminated sites should be completely cleaned up. However, the experience gained in attempting to restore sites has shown that in most cases, although large reductions in contaminant concentration can be achieved, it is very expensive and technically extremely difficult to remove it all.

A number of formerly common industrial practices have been identified as causes of groundwater contamination (Lloyd *et al.*, 1991). These include fallout from chimneys, casual surface disposal of wastes, disposal of solvents and sludges into pits, the use of old quarries as factory waste disposal sites and the use of abandoned wells and mine shafts for waste disposal. Groundwater contamination in urban areas also results from small-scale leakage from sewers giving rise to high nitrate levels and accidental spillage.

Groundwater contamination from industrial and agricultural sources has resulted in governments establishing a legislative framework for environmental regulators, such as the US Environmental Protection Agency and the National Rivers Authority in England and Wales, forming groundwater protection policies. Regulatory bodies also monitor activities to establish whether or not a deterioration in groundwater quality is taking place. It has become necessary for baseline surveys to be completed from which to monitor changes and to set out scientifically and statistically robust complex methodologies for reviewing the monitoring data (Gibbons, 1994).

The development of the nuclear power industry which took place from the early 1950s produced a significant quantity of radioactive waste. These materials range from highly radioactive materials such as spent fuel to slightly contaminated protective clothing used by operatives in nuclear installations. An important feature of these waste materials is that they will remain radioactive for a significantly long period. Many countries have considered methods for the disposal of radioactive waste by deep burial. As the main potential agent for the migration of the radionuclides from the waste is via groundwater flow systems, hydrogeologists are likely to continue to play a central role in this activity. Such studies have added to the scientific literature, particularly on the hydraulic conductivity of poorly permeable materials and deep-seated groundwater flow mechanisms.

The development of geothermal energy resources began in the early part of the twentieth century, although most took place in the latter half (Tomasson and Smárason, 1985). Geothermal energy is generally extracted by pumping superheated groundwater from rocks in geothermal zones or by injecting water into hot dry rocks via specially constructed boreholes. The essential involvement of water in the energy extraction process means that hydrogeologists have played a central role in geothermal energy research programs (Downing *et al.*, 1991).

Groundwater has long been recognized as a significant agent in geotechnical engineering considerations. For example, slope stability is often controlled by groundwater flows which destabilize the materials and lubricate the slip face. Hydrogeologists are increasingly working with engineering geologists.

Over the past decade a development in the planning control activities of local and central government bodies is the requirement for a comprehensive environmental impact assessment for all major new schemes. Where the proposed new development has a groundwater aspect, hydrogeologists are having an increasing role in the preparation of these impact assessment statements. New water resources development is an obvious example of where hydrogeological

assessments are needed, but the requirement is equally strong in other activities such as mining and quarrying, and the construction of new pipelines and roads (Brassington, 1991).

The future role of hydrogeology is likely to be a continuation of the development and management of finite groundwater resources, particularly in the developing world where there is still a basic requirement for a clean reliable water supply. The second main thrust is also likely to be a continuation of an existing involvement in the investigation and management of contaminated sites. Hydrogeologists will also continue to play a central role in aspects of environmental impact assessment, industrial and radioactive waste disposal and the exploitation of geothermal energy.

F.C. Brassington

Bibliography

Bear, J. and A. Verruijt, 1987. *Modeling Groundwater Flow and Pollution*. D. Reidel Publishing Company, Dordrecht, 414 pp.

Boulton, N.S., 1951. The flow pattern near a gravity well in a uniform water baring medium. *Journal Inst. Civil Engineers, London*, **36**, 534–550.

Boulton, N.S., 1963. Analysis of data from non-equilibrium pumping tests allowing for delayed yield from storage. *Proc. Inst. Civil Engineers, London*, **26**, 469–482.

Brandon, T.W., 1986. *Groundwater: Occurrence, Development and Protection*. Institution of Water Engineers and Scientists, London, 615 pp.

Brassington, F.C., 1988. *Field Hydrogeology*. John Wiley & Sons, Chichester, 175 pp.

Brassington, F.C., 1991. Construction causes hidden chaos. *Geoscientist*, **14**, 8–11.

Brassington, F.C., 1992. Present and future use of groundwater resources. *Modern Geology*, **16**, 363–374.

Darcy, H., 1856. Les fontaines publiques de la ville de Dijon. Victor Dalmont, Paris, 647 pp.

Davis, S.N. and R.J.M. DeWeist, 1966. *Hydrogeology*. John Wiley and Sons, New York, 463 pp.

Downing, R.A., 1993. Groundwater resources, their development and management in the UK: an historical perspective. *Quarterly Journal of Engineering Geology*, **26**, 335–358.

Downing, R.A. and G.P. Jones, 1985. Hydrogeology – some essential facets, in *Hydrogeology in the Service of Man*. Memoirs of the 18th Congress of the International Association of Hydrogeologists, Cambridge, pp. 1–16.

Downing, R.A., R.H. Parker and D.A. Gray, 1991. Geothermal energy in the United Kingdom, in *Applied Groundwater Hydrogeology*, R.A. Downing and W.B. Wilkinson (eds). Clarendon Press, Oxford, pp. 283–301.

Downing, R.A. and W.B. Wilkinson (eds), 1991. *Applied Groundwater Hydrogeology*. Clarendon Press, Oxford, 340 pp.

Driscoll, F.G., 1986. *Groundwater and Wells*, 2nd edn. Johnson Division, St Paul, 1089 pp.

Dupuit, J., 1863. *Etudes Théoriques et Pratiques sur le Mouvement des Eaux dans Découverts et à Travers les Terrains Perméables*, 2nd edn. Dunod, Paris, 304 pp.

Erdélyi, M. and J. Gálfi, 1988. *Surface and Subsurface Mapping in Hydrogeology*. John Wiley and Sons, Chichester (Akadémiai Kiadó, Budapest), 384 pp.

Gibbons, R.D., 1994. *Statistical Methods for Groundwater Monitoring*. John Wiley and Sons, New York. 286 pp.

Headworth, H.G. and A.C. Skinner, 1986. Hydrogeological Investigations, in *Groundwater: Occurrence, Development and Protection*, T.W. Brandon (ed.). Institution of Water Engineers and Scientists, London, pp. 229–269.

Jacob, C.E., 1940. On the flow of water in an elastic artesian aquifer. *Trans. Am. Geophysical Union*, **72**, 574–586.

Jacob, C.E., 1963. Correction of drawdowns caused by a pumped well tapping less than the full thickness of an aquifer, in *Methods of determining permeability, transmissibility and drawdown*, R. Bentall (ed.), US Geological Survey Water Supply Paper 1536–1, pp. 272–282

Kruseman, G.P. and N.A. De Ridder, 1991. *Analysis and Evaluation of Pumping Test Data*. International Institute for Land Reclamation and Improvement, Wageninen. 377 pp.

Lawrence, A.R. and S.S.D. Foster, 1987. *The pollution threat from agricultural pesticides and industrial solvents: a comparative review in relation to British aquifers*. Hydrogeology Report No. 897/2, British Geological Survey, Keyworth.

Lloyd, J.W., G.M. Williams, S.S.D. Foster *et al.*, 1991. Urban and industrial groundwater pollution, in *Applied Groundwater Hydrogeology*, R.A. Downing and W.B. Wilkinson (eds). Clarendon Press, Oxford, pp. 134–148.

Maxey, G.B., 1964. Hydrogeology, in *Handbook of Applied Hydrology*, Ven Te Chow (ed.). McGraw-Hill, New York, pp. 4.1–4.36.

McDonald, M.G. and A.W. Harbaugh, 1988. A modular three-dimensional finite-difference ground-water flow model in *Techniques of Water Resources Investigations of the United States Geological Survey*, Book 6, *Modeling Techniques*. US Geological Survey, Washington, DC, Chapter A1.

Parker, J.M., C.P. Young and P.J. Chilton, 1991. Rural and agricultural pollution of groundwater, in *Applied Groundwater Hydrogeology*, R.A. Downing and W.B. Wilkinson (eds). Clarendon Press, Oxford. pp. 149–163.

Rushton, K.R., 1984. *Groundwater: a hidden resource*. Inaugural lecture, University of Birmingham, 17 pp.

Rushton, K.R. and S.C. Redshaw, 1979. *Seepage and Groundwater Flow*. John Wiley & Sons, Chichester, 332 pp.

Sheppard, T., 1917. William Smith, his maps and memoirs. *Proc. Yorkshire Geological Society*, **19**, 75–253.

Theis, C.V., 1935. The relation between the lowering of the piezometric surface and the rate and duration of discharge of a well using groundwater storage. *Trans. Am. Geophysical Union*, **16**, 519–524.

Todd, D.K., 1980. *Groundwater Hydrology*, 2nd edn. John Wiley & Sons, New York, 535 pp.

Tómasson, J. and Ó.B. Smárason, 1985. Developments in geothermal energy, in *Hydrogeology in the Service of Man*. Memoirs of the 18th Congress of the International Association of Hydrogeologists, Cambridge, pp. 189–211.

Wilkinson W.B. and F.C. Brassington, 1991. Rising groundwater levels – an international problem, in *Applied Groundwater Hydrogeology*, R.A. Downing and W.B. Wilkinson (eds). Clarendon Press, Oxford, pp. 35–53.

Cross references

Groundwater
Groundwater: UK
Hydrogeology: history in USA
Hydrology
Hydrology: subsurface waters

HYDROGEOLOGY: HISTORY IN USA

People have observed the occurrence and movement of water in streams, springs and lakes for thousands of years. Engineering structures ranged from those as simple as an irrigation ditch to elaborate stone aqueducts to transfer water from areas of supply of areas of demand. Even so, the study of hydrology is a young science with most of its development occurring in the twentieth century. It is generally accepted that the scientific study of groundwater began in 1856 with a publication by Henri Darcy outlining the general equation of flow known throughout the world as Darcy's law. This is the basic concept on which all principles of the physics of groundwater, that is physical hydrogeology, is based. The principles of chemical hydrogeology are based on concepts of law of mass action, chemical thermodynamics, kinetics and organic chemistry. Along with Darcy's work the other two revolutionary developments in the science of hydrogeology were (1) the transient well hydraulics analysis of C.V. Theis in 1935 and (2) the introduction of large digital computers in the early 1960s. Darcy provided the empirical law and Theis provided the methodology for measurement of *in situ* hydrological properties of geologic formations and prediction of the response of groundwater systems to pumping. Digital computers provide the means for assessment for groundwater resources on a regional scale within the context of the full hydrological cycle.

Hydrogeology, as its name implies, is historically rooted in both geology and hydrology. Books explaining the historical development of these sciences include Adams (1954), who traced the historical development of geological sciences, Biswas (1970), who did the same for the hydrological sciences, Rouse and Ince (1957), who discussed much of hydrological interest in their book on the history of hydraulics, and Freeze and Back (1983), who selected benchmark papers and explained their relevance in a historical context. The first three provide a wealth of detail from the contributions of the early Egyptians, and the Greek and Roman philosophers up through the emergence of the Earth sciences in Europe during the eighteenth and nineteenth centuries. Several shorter articles on the history of hydrogeology are by Baker and Horton (1936) discussing the ideas of the early philosophers, Ferris and Sayre (1955), tracing the development of quantitative groundwater hydrology from Darcy's time through to 1950, and Bredehoeft *et al.* (1982), outlining the growth of ideas with respect to regional circulation of groundwater. Chow (1964) provides a concise discussion of the history of hydrology. He divided the time since 1800 into four periods. The period of modernization (1800–1900) is described as the grand era of experimental hydrology; the period of empiricism (1900–1930) is noted as a time when modern quantitative hydrology was still immature. The period of rationalization (1930–1950) saw the emergence of many great hydrologists who used rational analysis rather than empiricism to solve hydrological problems, and the period of theorization (1950–present) has seen the innovative application of mathematical analysis to hydrology. Using this classification, Darcy's experiments fall into the period of modernization. Theis' contributions belong to the period of rationalization, and the growth of computer simulation technology is a feature of the period of theorization.

The chemical aspects of hydrogeology developed quite differently from the physical aspects. Although well construction techniques were practiced by many early civilizations, great demands on groundwater resources and the need for improved engineering techniques and well drilling equipment did not develop until after Darcy's experiment in 1856. Therefore, by the time extensive groundwater exploration was necessary, Darcy's law provided the integrating concept and provided the rationale for exploration, well spacing and pumping regimes.

Demands were significantly different for the chemical aspects in that people needed to know the chemical character of water – springs, lakes and rivers – for diverse reasons. As each situation developed it was approached empirically with little theory available for extrapolation from previous experience. Some of the early chemical analyses (late eighteenth century) were of water from mineral spas used as health resorts. Eventually, it became imperative that engineers knew the chemical character of water (middle and late nineteenth century) available for both stationary and locomotive steam engines. The difficult and complex problems of boiler-scale formation and removal could not be economically resolved by engineering techniques of that time. It was therefore decidedly advantageous to avoid the use of water that would cause scale and corrosion problems in boilers. Later, during the early twentieth century, knowledge of chemistry of water was required to select appropriate sources for municipal, agricultural and industrial uses. This diversity of applications resulted in the multifaceted study of chemical hydrogeology. The early work of chemists was to provide descriptions and chemical characterizations of surface water and groundwater, justified economically by determining the fitness of the water for any intended use. The fundamental scientific need was to develop new analytical techniques and revise standard techniques giving the necessary precision for the desired descriptions.

These early studies of chemistry of water became part of hydrogeology in the middle of the twentieth century when chemical data were incorporated into a hydrological or geological analysis of drainage basins. The coupling of chemical descriptions of water with local and regional flow systems within the geologic framework of aquifers demonstrated the existence of relationships that required explanations, if the spatial distribution of chemical constituents within groundwater regimes were to be understood. Investigation of the problem of saltwater encroachment into coastal aquifers stimulated development of both physical and chemical aspects of the science of hydrogeology and afforded early training for many future leading scientists.

The evolutionary phase of chemical hydrogeology ended in the 1950s. Many of the important chemical reactions had been identified and shown to exert major controls on the chemical character of groundwater; such reactions as ion exchange, mineral solution and sulfate reduction were well understood. Sanitary engineers understood the concept of mineral equilibria in the sense that some waters are corrosive and capable of dissolving minerals and plumbing systems, whereas other waters could precipitate minerals within the pipes.

Also, biologists and limnologists had made great progress in understanding oxidation–reduction reactions as an environmental control.

A great deal of understanding was gained in the late 1950s and during the 1960s. Two major texts were Garrels (1960) and Hem (1959); although Hem's was not marketed as a textbook, it is used extensively as one throughout the scientific world. Other publications included application of redox potential to natural environments; the use of ^{14}C for groundwater studies, the isotopic variation of meteoric water, the concept of hydrochemical facies; the first application of chemical thermodynamics to groundwater, the cyclic flow of salt water in coastal aquifers; ion-exchange characteristics of minerals, and the role of geological membranes in the development of anomalous head values and the origin of brines. Chemical thermodynamics was first applied to low-temperature aqueous systems by Garrels and colleagues in working on the exploration and genesis of uranium deposits in the Colorado Plateau. These fundamental approaches led to the concept of mass balance and the use of equilibrium models and mass transfer concepts. These concepts are the basis for all geochemical studies of groundwater currently being undertaken. For example, mineral equilibrium studies have now been combined with isotopic analyses within the geological and hydrological framework to determine sources of water; the origin, fate and behavior of chemical constituents; and controlling reactions and groundwater flow rates.

All these ideas and applications are contributing to our predictive capabilities by integration with groundwater flow models through the solute transport equation. The solute transport equation includes two mass transport components – advection and hydrodynamic dispersion – and one mass transfer component, or reactive term, which includes the effects of all chemical reactions. Solute concentration is a function of mineral solubility, flowpath and residence time. Essentially all investigations of groundwater chemistry are based to some extent on the transport equation, either in a formal mathematical manner or by intuitive consideration of the equation parameters. The transport equation indicates that movement of solutes in subsurface systems is controlled by three parameters: groundwater velocity, hydrodynamic dispersion and the reaction term.

William Back

Bibliography

Adams, F.D., 1954. *The Birth and Development of the Geological Sciences*, Dover, New York, 506 pp.

Back, W. and Freeze, R.A. (eds), 1983. *Chemical Hydrogeology, Benchmark Papers in Geology*, Vol. 73, Hutchinson Ross, Stroudsburg, PA, 432 pp.

Baker, M.N. and Horton, R.E., 1936. Historical development of ideas regarding the origin of springs and groundwater, *Am. Geophys. Union Trans.*, **17**, pp. 395–400.

Biswas, A., 1970. *History of Hydrology*, North-Holland, Amsterdam, 336 pp.

Bredehoeft, J.D., Back, W. and Hanshaw, B.B., 1982. Regional ground water flow concepts in the United States: historical perspective, in *Proceedings of Symposium on Recent Trends in Hydrogeology*, Geological Society of America Special Paper 189, pp. 297–316.

Chow, V.T., 1964. Hydrology and its development, in *Handbook of Applied Hydrology*, V.T. Chow, ed., McGraw-Hill, New York, pp. 1–21.

Darcy, H., 1856. *Les fontaines publiques de la ville de Dijon*, Victor Dalmont, Paris.

Ferris, J.G. and Sayre, A.N., 1955. The quantitative approach to ground-water investigations, *Econ. Geology*, 50th anniv. issue, 714–747.

Freeze, R.A. and Cherry, J.A., 1972. *Groundwater*, Prentice-Hall, Englewood Cliffs, NJ, 604 pp.

Freeze, R.A. and Back, W., 1983. *Physical Hydrogeology, Benchmark Papers in Geology*, Vol. 72, Hutchinson Ross, Stroudsburg, PA, 448 p.

Garrels, R.M., 1960. *Mineral Equilibria at Low Temperature and Pressure*, Harper & Bros., New York, 254 pp.

Hem, J.D., 1959. *Study and Interpretation of the Chemical Characteristics of Natural Waters, US Geological Survey Water-Supply Paper 1473*, 363 pp.

Pourbaix, M.J.N., 1949. *Thermodynamics of Dilute Aqueous Solutions*, Edward Arnold, London, 136 pp.

Rouse, H. and Ince, S., 1957. *History of Hydraulics*, Iowa State University, Institute of Hydraulic Research, Ames, IA, 269 pp.

Schoeller, H., 1955. Geochimie des eaux souterraines, *Inst. Francais Petrole Rev.*, **10**(3), 181–213; **10**(4), 219–246.

Cross references

Groundwater
Groundwater: UK
Hydrogeology
Hydrology

HYDROLOGICAL CYCLE

General concepts

The hydrological cycle (or water cycle) is the never-ending circulation of water and water vapor over the entire Earth (Chow, 1964; Voskrensky, in UNESCO, 1978). This circulation penetrates the three parts of the total Earth system (Emiliani, 1995): the atmosphere (the gaseous envelope above the hydrosphere), the hydrosphere (the water covering the surface of the Earth), and the lithosphere (the solid rock beneath the hydrosphere). Solar energy and gravity provide the energy for the circulation.

The idea of a hydrological cycle was expressed at least two centuries BC in the Bible (Ecclesiastes), but it took another 1000 years before the concept was accepted and expressed by several natural philosophers in ancient Greece and in Western Europe during the Middle Ages. Nevertheless, it is still surprising that it was not until the eighteenth century that several geologists in France and Britain made the discovery that rivers were actually responsible for cutting their own valleys.

The hydrological cycle has no beginning or end. Water is evaporated from the oceans and land, with the former providing the largest amounts. The evaporated water is carried into the atmosphere, usually drifting tens to hundreds of kilometers before being returned to the earth as rain, snow, hail or sleet (McDonald, 1962). This precipitated water may be intercepted and transpired by plants, may run over the ground surface and into streams, or may infiltrate into the ground. A considerable part of the intercepted and transpired water and the surface runoff returns to the air by evaporation. The infiltrated water may seep down to deeper zones of the Earth, forming groundwater storage which may later flow out to streams as runoff and finally evaporate into the atmosphere to complete the hydrological cycle.

Thus the hydrological cycle is generally described in terms of precipitation (P), infiltration (I), evaporation (E), transpiration (T), surface runoff (R) and groundwater flow (G). On planet Earth it behaves as a closed system, but rates are regionally, seasonally and interannually variable, as are storage times. The latter range from the scale of days, to a few million years (stored as glacial ice), up to hundreds of millions of years (buried in sediments).

The quantity of water going through the hydrological cycle during a given period for an area can be evaluated by the hydrological equation (or continuity equation):

$$I - O = \Delta S$$

where I is the total inflow of surface runoff, groundwater and total precipitation; O is the total outflow, which includes evapotranspiration, and subsurface and surface runoff from the area; and ΔS is the change in storage in the various forms of retention and interception.

A quantitative approach to the hydrological cycle may be found in Peixoto and Oort (1992), which also contains a useful summary of the variability of climate. That variability has been particularly felt during the twentieth century in North Africa, notably the Sahel (see discussion in Leroux, 1996; Tardy, 1986).

The hydrological cycle may be illustrated qualitatively (Figure H4), descriptively (Figure H5 and H6) and quantitatively (Figure H7).

Magnitude of the hydrological cycle

Table H1 shows that each year approximately 500 000 km³ of water is evaporated from the Earth's surface. Of this amount, the oceans account for 428 000 km³ (85%), and inland water bodies and wet soils provide the remaining 72 000 km³ (15%) (see World water balance).

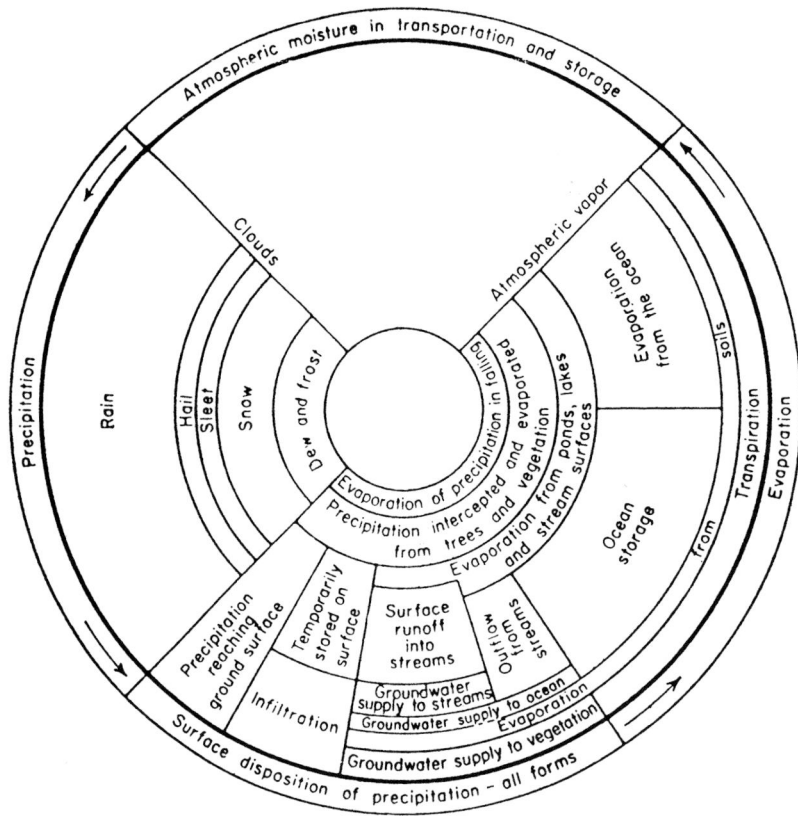

Figure H4 The hydrological cycle: a qualitative representation. (Source: Horton, 1931.)

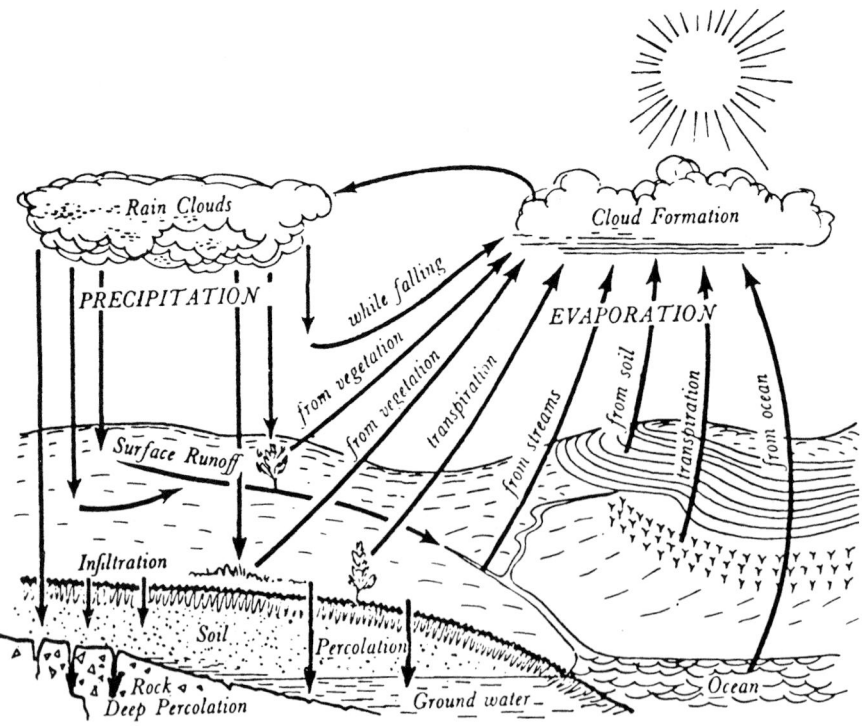

Figure H5 The hydrological cycle: a descriptive representation. (Source: Ackermann *et al.*, 1955.)

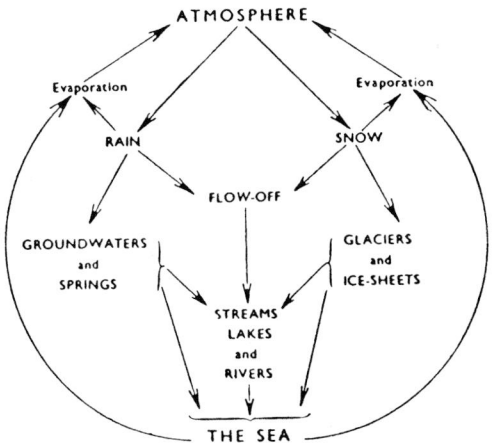

Figure H6 The circulation of meteoric water. In addition to the main sources of evaporation indicated here, it should be noted that evaporation takes place from all exposed surfaces of water and ice (e.g. lakes, rivers, glaciers and ice sheets) and also from the soil and from plants and animals. Part of the water that ascends from the depths by way of volcanoes reaches the surface for the first time; such water is called juvenile water to distinguish it from the meteoric water already present in the hydrosphere and other outer zones of the Earth. (Note: the word runoff is generally used rather than flow-off. Source: Holmes, 1965.)

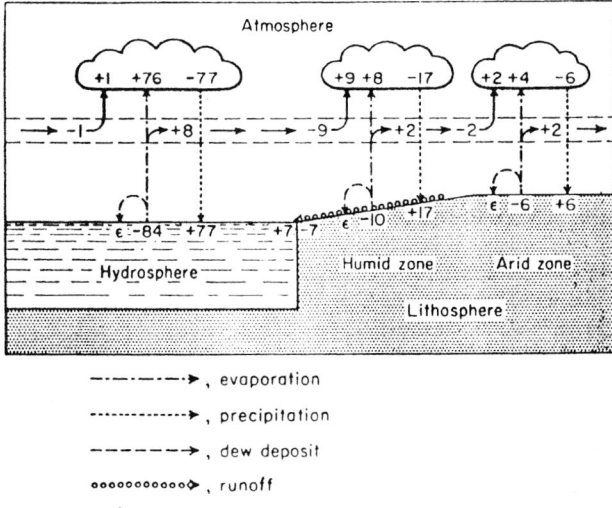

—·—·—·—▶, evaporation

·············▶, precipitation

— — — — — ▶, dew deposit

ooooooooooo◇, runoff

removal from and addition to
horizontal advection of water vapor

ε, values less than 0.5 relative unit.

Figure H7 The hydrological cycle: a quantitative representation. 100 relative units = 85.7 g cm^{-2} year^{-1} or 85.7 cm mean annual precipitation for the world. (Source: Lettau, 1954.)

Most of the inland evaporation occurs into relatively dry air masses. Much of the water evaporated from the oceans is transported by maritime air masses (which can hold considerably more water vapor than continental air masses) to the continents, where total precipitation amounts to 108 000 km^3 year^{-1}. This is enough water to cover the entire state of Texas (692 408 km^2) to a depth of 150 m. Of the 108 000 km^3 of water precipitated, 40 000 km^3 (37%) returns to the sea as runoff to balance the excess of precipitation over inland evaporation (UNESCO, 1978; Leroux, 1996; Table H1).

Reichel (1952) calculates that the mean annual precipitation for the world is 864 mm, which is balanced by a comparable amount of

Table H1 Water cycle, quantities of precipitation, evaporation and runoff (source: US Geological Survey)

	Volume (km^3)	Percentage of total water
Annual evaporation		
From oceans	430 000	0.025
From land	70 000	0.005
Total	500 000	0.031
Annual precipitation		
On oceans	390 000	0.024
On land	110 000	0.007
Total	500 000	0.031
Annual runoff to oceans		
From rivers and ice caps	38 000	0.003
Groundwater outflow to oceans	2 000	0.0001
Total	40 000	0.0031

Notes:
1. Evaporation is a measure of total water participating annually in the water cycle.
2. The groundwater outflow to oceans is set arbitrarily to about 5% of surface runoff.

evaporation. It is estimated that 97% of all the water in the world, or over 1350 × 10^6 km^3 is contained within the oceans. If the Earth were a uniform sphere, this volume of water would cover the Earth to a depth of 244 m (Wolman, 1962).

The total volume of fresh water on the Earth is estimated at 35 × 10^6 km^3, approximately distributed as follows (%) (UNESCO, 1978; see also Table H2):

Polar ice and glaciers	75
Groundwater between 762 and 3810 m	14
Groundwater less than 762 m	11
Lakes	0.3
Soil moisture	0.06
Atmosphere	0.035
Streams	0.03

Note that these figures are stationary estimates of distribution. Although huge amounts of water pass through the atmosphere, the water content is relatively small at any given moment.

The average annual precipitation over the continental United States would amount to 762 mm if it were spread evenly. However, topographic configurations and patterns of atmospheric circulation result in uneven distribution of precipitation, ranging from a few tens of millimeters in the arid southwest to over 250 cm in parts of the Pacific Northwest. The 17 western states receive only 25% of the total precipitation but contain 60% of the land area.

The 76 cm of water for the United States represents 56 000 km^3 year^{-1} or 1.63 × 10^{13} l day^{-1}. Of this amount, 54.6 cm or 71.7% is returned to the atmosphere by the processes of evapotranspiration. The remaining 21.6 cm or 28.3%, becomes surface and groundwater runoff into the oceans (Robinove, 1963).

Role of the hydrologic cycle

If the atmosphere and the earth are considered as separate entities, radiation and conduction fail to provide balanced heat budgets, as the Earth's surface has a net gain and the free atmosphere a net loss. The link between the gain and loss is the hydrological cycle.

Some of the heat absorbed by the Earth's surface is expended in evaporation, and therefore transferred to latent heat (the quantity of heat absorbed or emitted without change in temperature during a change of state of a unit mass of a material, 'hidden heat'), which is later realized as sensible heat (the heat added to a body when its temperature is changed) and released to the atmosphere when the vapor condenses to clouds. Evaporation is high where relatively cool air sweeps over warmer oceans. The highest evaporation values found in the northern hemisphere occur in the Atlantic and Pacific trade wind belts south of 30°N. High evaporation values also occur over the northwestern Pacific and North Atlantic oceans during the northern

Table H2 Distribution of the world's estimated water supply (data from UNESCO, 1978; and other sources)

Location	Surface area (km²)	Water volume (km³)	Percentage of total water
Surface water:			
Freshwater lakes	1 236 000	91 000	0.009
Saline lakes and inland seas	822 000	85 000	0.008
Average (instantaneous) in stream channels		1 250	0.0001
Subsurface water: soil moisture and vadose water	130 000 000	67 000	0.005
Groundwater 1 km deep	130 000 000	4 200 000	0.3
Groundwater deep lying	130 000 000	4 200 000	0.3
Total water on land areas	130 000 000	8 600 000	0.6
Icecaps and glaciers	16 000 000	24 000 000	2.1
Atmosphere (at sea level)	510 000 000	13 000	0.001
World oceans	361 000 000	1 340 000 000	97
Total (rounded)		1 400 000 000	100

It should be noted that the above figures are estimates with an unknown uncertainty, which may be as much as 20%. Tables of estimates of the world's water supply may vary acccording to how the estimates were arrived at and in this Encyclopedia, although several tables and figures are presented by different authors which are in fairly close agreement with each other, accurate figures will only be available when sufficient measurements are made.

Table H3 Oceanic and land-drainage areas (UNESCO, 1978)

	Regional Area (10⁶ km²)	Water Area (10⁶ km²)	Runoff Area (10⁶ km²)	River Discharge (10³ km³)	Precipitation (10³ km³)	Evaporation (10³ km³)
Pacific	182.6	178.7	24.8	14.1	26.2	11.6
Atlantic	92.7	91.7	50.7	19.8	57.3	34.1
Indian	77.0	76.2	20.9	5.6	15.8	9.1
Arctic	18.5	14.7	22.4	5.1	11.2	6.0
Total runoff, etc.			119.0	44.7	110.0	60.9

Notes:
1. The world ocean is here deemed to consist of only four units. Marginal seas (e.g. Mediterranean) are attached to nearest ocean. The 'Southern Ocean' and 'Antarctic Ocean', recognized in ecological studies, are identifiable as 'sectors' of the major oceans.
2. The boundaries of oceans, determined by the International Hydrographic Bureau, are summarized in Fairbridge, R.W. (ed.), 1996: *The Encyclopedia of Oceanography* (New York).
3. Endoreic (internal drainage) areas of the continents amount to 30.2×10^6 km², receiving 8.7×10^3 precipitation, and providing 8.1×10^3 km³ evaporation.
4. The world ocean covers 361.3×10^6 km², with a volume of 1338.5×10^6 km³.
5. The total land area (including islands) is 149×10^6 km², 20% of which is marked by endoreic drainage. The Atlantic Ocean receives 34% of the runoff, the Pacific 17%, the Arctic 15% and the Indian Ocean 14%.
6. Evaporation annually from the total surface of the globe is 577 000 km³, of which 505 000 km³ is from the ocean and 72 000 km³ is from the land. Precipitation returns 458 000 km³ to the ocean and 47 000 km³ is transferred to the land in atmospheric circulation.

hemisphere winter when cold, dry continental air masses (cP and cA) move over warmer waters (Petterssen, 1964).

The average life of water vapor molecules in air varies from an hour to several days. Latent heat is usually liberated far from the regions where evaporation occurred. This is particularly true of evaporation in the trade wind belts, which supply much of the vapor that eventually precipitates in middle and high latitudes. Thus the circulation of water is a key part of heat transfer from low to high latitudes and from oceans to continents (Petterssen, 1964).

Return of water to the oceans

In spite of the relatively uniform pattern of evaporation in the various latitudinal belts of the ocean, there is a marked regional imbalance in the return flow of water to the oceans. The explanation lies in the concentration of major rivers (Amazon, Mississippi, Congo, Niger, St Lawrence, Danube, Po, Nile and Rhine) which drain into the Atlantic Ocean and its marginal seas (Gulf of Mexico, Black Sea). In contrast, the Pacific has only a limited number of major discharge outlets (Yangtze, Hwang-Ho, Yukon, Columbia, and Colorado; see Rivers). Table H3 provides further evidence that the Atlantic not only drains the largest portion of the Earth's land surface, but has the highest proportion of land area draining into the ocean area.

As water is evaporated from the ocean, it becomes desalinated, but storm-generated wave bubbles carry salt into clouds, thereby contaminating the rainfall ('cyclic salts'). Both anthropogenic and natural contaminants modify the quality of the water at every stage of the cycle (Camp, 1963). Anthropogenic carbon dioxide has certainly been

rising dramatically in the twentieth century, and model projections indicate that this should force a small increase in the intensity of the hydrological cycle (National Research Council, 1983).

Robert M. Hordon

Bibliography

Ackermann, W.C., Colman, E.A. and Ogrosky, H.O., 1955. 'From ocean to sky to land to ocean,' in *Water*, Yearbook Agr., (US Dept. Agriculture), pp. 41–51.
Allaby, M., 1992. *Water: Its Global Nature*. New York: Facts on File, 208 pp.
Barry, R.G. and Chorley, R.J., 1987. *Atmosphere, Weather, and Climate*. London: Methuen, 460 pp.
Berner, E.K. and R.A. Berner, 1996. *Global Environment: Water, Air, and Geochemical Cycles*. Upper Saddle River, New Jersey: Prentice Hall, 376 pp.
Camp, T.R., 1963. *Water and Its Impurities*. New York: Reinhold Publ., 355 pp.
Chow, V.T., 1964. Hydrology and its Development, in *Handbook of Applied Hydrology*. New York: McGraw-Hill, pp. 1–1 to 1–22.
Emiliani, C., 1995. *Planet Earth*. Cambridge: University Press, 2nd edn., 718 pp.
Falkenmark, M., 1989. *Hydrological Phenomena in Geosphere-Biosphere Interactions: Outlooks to Past, Present and Future*, Wallingford, Oxfordshire, UK: International Association of Hydrological Sciences (IASH) in cooperation with the International

Institute for Hydraulic and Environmental Engineering, 81 pp., IASH Monographs and Reports No. 1.

Holmes, A., 1965. *Principles of Physical Geology*, New York: Ronald Press, rev. edn., 1288 pp.

Horton, R.E., 1931. The field, scope and status of the science of hydrology, *Trans. Am. Geophys. Union*, **12**, 189–202.

Leroux, M., 1996. *La Dynamique du Temps et du Climat*. Paris: Masson, 310 pp.

Lettau, H., 1954. A study of the mass, momentum and energy budget of the atmosphere, *Arch. Meteorol. Geophys. Bioklimatol., Ser. A*, **7**, 131–153.

Livingston, D.A., 1963. *Chemical composition of rivers and lakes*, US Geol. Surv. Professional Paper 440-G.

McDonald, J.E., 1962. The evaporation–precipitation fallacy, *Weather*, London, **17**(5), 168–170, 172–177.

Moore, J.W., 1989. *Balancing the Needs of Water Use*. New York: Springer-Verlag, 267 pp.

National Research Council, 1983. *Changing Climates: Report of a Carbon Dioxide Assessment Committee*. Washington, DC: National Academy Press.

Peixoto, J.P and Oort, A.H., 1992. *Physics of Climate*. New York: American Institute of Physics, 520 pp.

Petterssen, S., 1964. Meteorology, in *Handbook of Applied Hydrology*. New York: McGraw-Hill, pp. 3–1 to 3–39.

Reichel, E., 1952. Der Stand des Verdunstungs problems (The status of the evaporation problem), *Ber. Deut. Wetterdienst*, Bad Kissingen, **35**, 155.

Robinove, C.J., 1963. What's Happening to Water, *Smithsonian Inst. Ann. Rept.*, 1962, 375–389.

Tardy, Y., 1986. *Le Cycle de l'Eau: Climats, Paleoclimats et Geochimie Globale*. Paris: Masson, 338 pp.

UNESCO, 1978. *World Water Balance and Water Resources of the Earth*. Paris: UNESCO, (Prepared by the USSR Committee for the International Hydrological Decade), 663 pp.

Viessman, W. Jr, G.L. Lewis and J.W. Knapp, 1989. *Introduction to Hydrology*, 3rd edn. New York: Harper & Row, 780 pp.

Wetzel, R.G., 1983. *Limnology*, 2nd edn. Philadephia: Saunders, 767 pp.

Wolman, A., 1962. *Water Resources, A Report to the Committee on Natural Resources of the National Academy of Sciences – National Research Council*, Natl Acad. Sci. Natl Res. Council, Publ. 1000-B.

Cross references

Evaporation: measurement
Groundwater
Hydrology
Lakes
Precipitation
Rain
Rivers
Water on Earth and other planets
Water allocation and use
Water balance
Water budget analysis
Water resources
Water use
World water balance

HYDROLOGICAL MAPPING

Introduction

Mapping is a concise way to display and summarize the spatial distribution of hydrological variables, such as the elements of the hydrological balance (i.e. precipitation, evapotranspiration and runoff). Hydrological maps can be an effective communication tool for scientific, educational and socio-political purposes and thus can play an important role in many of the environmental and social problems (e.g. acid precipitation and climate change) that our world faces today (McKay, 1976).

Hydrological maps are a relatively recent development (probably <150 years) and can take many forms (McKay and Thomas, 1971; Robinson, 1971). For gauged, measured or estimated data having a continuous scale of measurement the two most common types are (1) a point map, or (2) an isoline map when data are interpolated/ extrapolated over the Earth's surface. This paper deals mainly with the second type of map, isoline maps. For a more comprehensive examination of the myriad types of hydrological maps, see *Hydrological Maps: a contribution to the international hydrologic decade* (UNESCO/WMO, 1977).

Precipitation

Precipitation is the amount of water that falls or precipitates on the Earth's surface in a liquid or solid form. Typically, precipitation is measured by catching rain or snow in a cylinder and either (1) determining the depth periodically or after each event, or (2) measuring the rainfall on a continuous basis via weighing or a tipping bucket-type device. Measurement of precipitation is problematic in that large errors can occur due to operator error, wind and turbulence affecting the catch, snow blowing in or out of the orifice, and the lack or undercatch of occult precipitation (e.g. fog drip or rime, Dingman *et al.*, 1988). These errors are usually ignored in most precipitation mapping (McKay and Thomas, 1971).

In 1839 the Danish cartographer Olsen produced the earliest known map with precipitation as its theme. In 1841 Berghaus produced, for his *Physikalischer Atlas*, a map showing actual precipitation values and using isoline symbology. After the 1840s precipitation maps became fairly common (Robinson, 1982). By at least the early twentieth century many investigators recognized the strong relationship between elevation and precipitation and they began to create precipitation isoline maps using graphical studies of the relationship of precipitation to elevation, or distance from a coast or mountain crest (e.g. Lee, 1911). With the advent of digital cartography and more sophisticated interpolation techniques, a plethora of interpolation options has become available. Examples of these methods include simple trend surface analysis (e.g. Hughes, 1982), inverse distance weighting, and kriging (e.g. Dingman *et al.*, 1988; Phillips *et al.*, 1992). Tabios and Salas (1985) have compared many of these and other techniques).

Although many precipitation maps are still being made by hand, researchers are relying more and more on automated interpolation procedures. Recent developments in automated precipitation mapping include the use of thin-plate spline methods (Hutchinson and Bischof, 1983; Hutchinson, 1993) and the creation of sophisticated statistical–topographic models such as the precipitation–elevation regressions on independent slopes model (PRISM) (Daly *et al.*, 1994; see Figure H8 for a PRISM-based map). This model shows great promise for increasing the accuracy of precipitation maps. With the growing use of computers and easy access to a wide variety of interpolation algorithms, the number of automated maps produced is likely to increase. One concern regarding such mapping is that not all available automated contouring methods take into account the complex interaction of scale, elevation, slope and aspect that affects precipitation; thus maps with gross errors can sometimes result.

Evapotranspiration

Evapotranspiration is the amount of water that is evaporated from the Earth's surface and transpired by plants. Evapotranspiration can be expressed in two forms, actual and potential. Actual evapotranspiration is, as the name implies, the amount of water actually evaporated and transpired. Potential evapotranspiration is the amount that would be evaporated and transpired from a vegetated surface unlimited by the supply of water. Potential evapotranspiration is usually calculated by formulae (e.g. the temperature-based method of Thornthwaite, 1948). Actual evapotranspiration is measured by lysimeter, or for longer time periods (annual or longer) it can be determined by the water balance, i.e. evapotranspiration equals precipitation minus runoff, where these values are known (e.g. Kittredge, 1938). When the water balance approach is used, evapotranspiration is often referred to as 'water loss'. Most maps of evapotranspiration use isoline symbolism.

An early, and possibly the first, example of the mapping of evapotranspiration (water loss) in the United States was the work of Kittredge (1938). He mapped water loss for the USA, based on previous maps of runoff and precipitation. Williams *et al.* (1940) calculated the water balance for about 200 water basins in the eastern USA and produced a map of evapotranspiration from the results. Using Thornthwaite's temperature-based formula for estimating potential evapotranspiration, Thornthwaite (1948) and Thornthwaite

LONG–TERM PRECIPITATION
PRISM METHOD

Precipitation (Inches)

- □ < 20
- ▨ 20 – 40
- ▨ 40 – 60
- ▨ 60 – 80
- ■ > 80

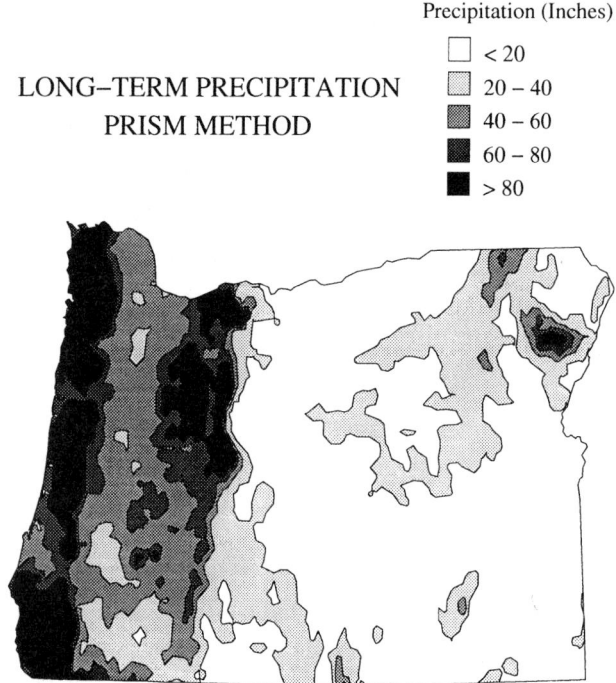

Figure H8 PRISM-based long-term (1961–1990) precipitation map for the state of Oregon, USA. Contour interval: 20 in (For further information on PRISM contact George Taylor, Oregon Climate Service, (503) 737–2 5707, oregon@ats.orst.edu) (Source: C. Daly, personal communication, 1994.)

LONG–TERM EVAPOTRANSPIRATION
MANUAL METHOD

Evapotranspiration (cm)

- □ no data
- ▨ < 40
- ▨ 40 – 50
- ▨ 50 – 60
- ■ > 60

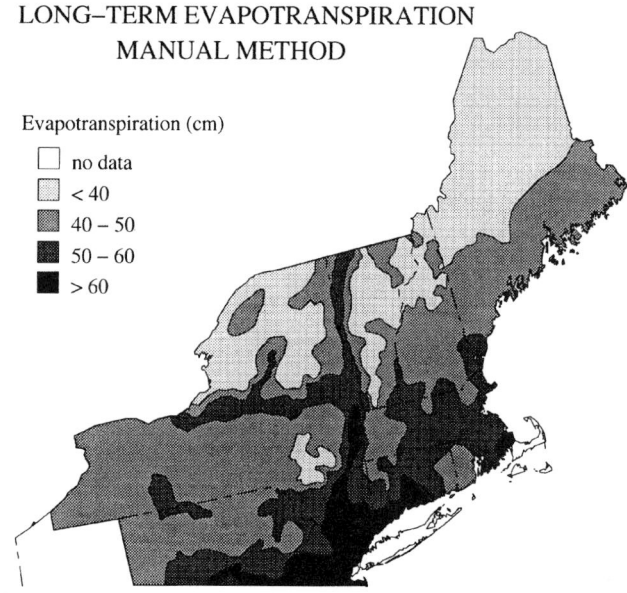

Figure H9 Manual method long-term (1951–1980) evapotranspiration for the northeastern United States. Contour interval: 10 cm. (Source: Church *et al.*, 1995.)

LONG–TERM EVAPOTRANSPIRATION
AUTOMATED METHOD

Evapotranspiration (cm)

- □ no data
- ▨ < 40
- ▨ 40 – 50
- ▨ 50 – 60
- ■ > 60

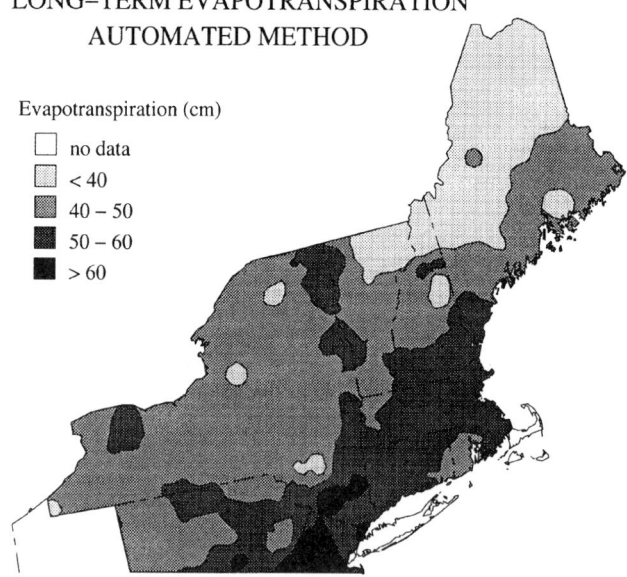

Figure H10 Automated method long-term (1951–1980) evapotranspiration for the northeastern United States. Contour interval: 10 cm. (Source: Church *et al.*, 1995.)

et al. (1958) produced maps for the USA and eastern North America respectively. Gurnell (1981) provides a summary of some other evapotranspiration maps made between 1950 and 1980.

Recent maps of evapotranspiration include the work of Stancik and Jovanovic (1988) (mentioned by Domokos and Sass, 1990), who based their map of evapotranspiration of the Danube Basin on empirical formulae. Milly (1994) produced maps of potential and actual evapotranspiration for the eastern USA based on a modified version of the Thornthwaite (1948) formula and the water balance, respectively. Church *et al.* (1995) used the water balance to produce maps of actual evapotranspiration, by both manual and automated methods, for the northeastern USA (Figures H9 and H10). Mapping based on remotely sensed data is also under development (e.g. Running *et al.*, 1989). Such mapping shows promise for increasing the accuracy and speed with which evapotranspiration maps are produced.

Runoff

Runoff depth (often referred to as just runoff) is computed by dividing the volume of water discharged by a stream or river over a selected period (e.g. 1 year) by the area of the contributing watershed. This runoff depth can be visualized as being the residual of precipitation after the demands of evapotranspiration have been met. Maps of runoff (i.e. runoff depth) use isolines and are usually for annual or longer time periods. Runoff maps can be useful in the evaluation of water resources, such as the estimation of flow from ungauged streams, and for other scientific and informational purposes.

Runoff maps have been produced in the United States almost since gauging began (around 1890). Langbein *et al.* (1949) give a brief history of early American runoff mapping.

Most recent runoff maps can be placed into two categories with regards to their production: manual and automated. Manual maps are of two types, those relying on gauge data and subjective consideration of local precipitation patterns and other geographic considerations (e.g. Krug *et al.*, 1990; Domokos and Sass, 1990), and those that employ empirical estimates as well as actual runoff values (e.g. Thornthwaite *et al.*, 1958; Liebscher, 1972). Figure H11 shows the manually produced map of Krug *et al.* (1990).

Automated methods to create runoff maps employ computer algorithms to interpolate estimated runoff (or estimated and gauged runoff together) to create the runoff surface. Solomon *et al.* (1968) developed a raster-based automated system in which they used various physical and climatic factors to create a regression formula. They then applied this formula across a grid surface to produce a runoff map. Foyster (1975) employed a similar methodology but based her method on a water balance formula developed by Penman (1950). More recently, Bishop and Church (1992, 1995) have examined methods that employ both gauged runoff and estimated runoff at precipitation stations. Through various methods (regression formulae, estimates of evapotranspiration, or runoff/precipitation

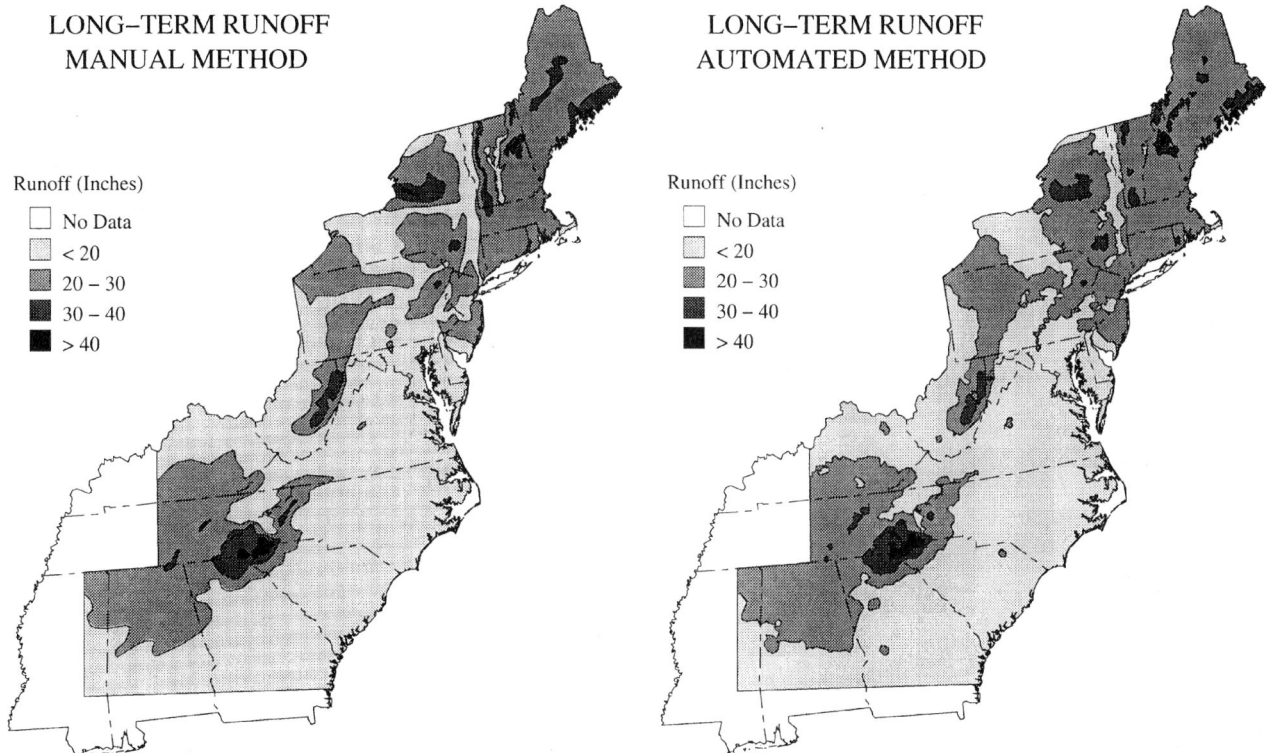

LONG–TERM RUNOFF MANUAL METHOD

Runoff (Inches)
- ☐ No Data
- < 20
- 20 – 30
- 30 – 40
- > 40

LONG–TERM RUNOFF AUTOMATED METHOD

Runoff (Inches)
- ☐ No Data
- < 20
- 20 – 30
- 30 – 40
- > 40

Figure H11 Manual method long-term (1951–1980) runoff map for the eastern United States. Contour interval: 10 in. (Source: Krug *et al.*, 1990.)

Figure H12 Automated method long-term (1951–1980) runoff map for the eastern United States. Contour interval: 10 in. (Source: Bishop And Church, 1995.)

ratios) they used estimates of runoff, along with actual gauged data, in a simple interpolation routine to produce maps of runoff. Figure H12 shows an example of one of these maps. Bishop and Church (1992, 1995) found that maps made with some of their methods provide estimates of runoff as accurate (or more so) than estimates derived from manually produced maps, e.g. that of Krug *et al.* (1990).

Ratios

Ratios of elements of the water balance include Runoff/Precipitation (*R/P*) and Evapotranspiration/Precipitation (*ET/P*). These ratios express the proportion of precipitation, for a given time period, that results in runoff or is consumed in evapotranspiration; for example, an *R/P* value of zero represents an area where evapotranspiration consumes all precipitation.

The mapping of ratios of elements of the water balance are relatively rare in the literature. Examples include the mapping of consumptive use as a percentage of precipitation (*ET/P* × 100) for the continental USA by Hidore (1966), a map of R/P for Great Britain by Ward (1967), the mapping of quickflow (from hydrograph separation) to precipitation (a variant of *R/P*) for the eastern United States by Woodruff and Hewlett (1970), and a map of *R/P* for Poland by Dynowska (1971) (presented in UNESCO/WMO, 1977). More recent examples include a global map of *R/P* × 100 produced by Wendland (1987), a map of *R/P* for the Danube Basin by Domokos and Sass (1990), maps of *R/P* for the eastern USA by Milly (1994), and maps of R/P for the northeast United States produced by Church *et al.* (1995).

Conclusions

Hydrological mapping is currently in a state of transition/evolution from time-consuming manual-based methods to more automated procedures. As long as developers and users are aware of the limits of spatial interpolation/extrapolation of hydrological variables, the increasing use of automation should lead to the production of more accurate and easily reproducible maps.

Gary D. Bishop and M. Robbins Church

Bibliography

Bishop, G.D. and Church, M.R., 1992. Automated approaches for regional runoff mapping in the northeastern United States, *J. Hydrol.*, **138**, 361–383.

Bishop, G.D. and Church, M.R., 1995. Mapping long-term regional runoff in the eastern United States using automated approaches, *J. Hydrol.*, **169**, 189–207.

Church, M.R., Bishop, G.D. and Cassell, D.L., 1995. Maps of regional evapotranspiration and runoff/precipitation ratios in the northeast United States, *J. Hydrol.*, **168**, 283–298.

Daly, C., Neilson, R.P. and Phillips, D.L., 1994. A statistical–topographic model for mapping climatological precipitation over mountainous terrain, *J. Appl. Meteorol.*, **33**, 140–158.

Dingman, S.L., Seely-Reynolds, D.M. and Reynolds III, R.C., 1988. Application of kriging to estimating mean annual precipitation in a region of orographic influence. *Water Resour. Bull.*, **24**, 329–339.

Domokos, M. and Sass, J., 1990. Long-term water balances for subcatchments and partial national areas in the Danube basin, *J. Hydrol.*, **112**, 267–292.

Dynowska, I., 1971. Typy rezimow rzecznych w Polsce [Types of river regimes in Poland]. *Zesz. Nauk UJ, 268, Prace Geograf. z.28,* Prace Inst. Geograf., z.50.

Foyster, A.M., 1975. Mapping runoff by the grid square technique, *Nord. Hydrol.*, **6**, 207–221.

Gurnell, A.M., 1981. Mapping potential evapotranspiration: the smooth interpolation of isolines with a low density station network, *Appl. Geog.*, **1**(3), 167–183.

Hidore, J., 1966. Regional variations in natural water consumption in the conterminous United States, *J. Hydrol.*, **4**, 79–90.

Hughes, D.A., 1982. The relationship between mean annual rainfall and physiographic variables applied to a coastal region of southern Africa, *South African Geog. J.*, **64**(1), 41–50.

Hutchinson, M.F. and Bischof, R.J., 1983. A new method for estimating the spatial distribution of mean seasonal and annual rainfall applied to the Hunter Valley, New South Wales, *Australian Meteorol. Mag.*, **31**(3), 179–184.

Hutchinson, M.F., 1993. Interpolating rainfall means – getting the temporal statistics correct, in *Proceedings of the 2nd International Conference/Workshop on Integrating Geographic Information Systems and Environmental Modeling*, Sept. 26–30, 1993 Breckenridge, CO, National Center for Geographic Information and Analysis, Santa Barbara, CA.

Kittredge, J. Jr, 1938. The magnitude and regional distribution of water losses influenced by vegetation, *J. For.*, **36**, 775–778.

Krug, W.R., Gebert, W.A., Graczyk, D.J. *et al.*, 1990. *Map of Mean Annual Runoff for the Northeastern, Southeastern and Mid-Atlantic United States, Water Years 1951–1980*, US Geological Survey Water Resources Investigation Rep. 88–4904, US Geological Survey, Madison, WI, 11 pp.

Langbein, W.B. *et al.*, 1949. *Annual Runoff in the United States*, US Geological Survey Circular 52, Dept. of the Interior, Washington, DC, 14 pp.

Lee, C.H., 1911. Precipitation and Altitude in the Sierra, *Monthly Weather Review*, **39**, 1092–1099.

Liebscher, H., 1972. A method for runoff-mapping from precipitation and air temperature data, in *World water balance, Proc. of the Reading Symp., Reading, 15–23 July 1970*, Vol. 1. UNESCO/IASH/WMO, Gentbrugge, Belgium, pp. 115–121.

McKay, G.A. and Thomas, M.K., 1971. Mapping of climatological elements, *Canadian Cartographer*, **8**(1), 27–40.

McKay, G.A., 1976. Hydrological Mapping, in *Facets of Hydrology*, J.C. Rodda (ed.). London: John Wiley and Sons, pp 1–36.

Milly, P.C.D., 1994. Climate, soil water storage, and the average annual water balance, *Water Resour. Res.*, **30**(7), 2143–2156.

Penman, H.L., 1950. The water balance of the Stour catchment area, *J. Inst. Water Eng.*, **4**, 457–469.

Phillips, D.L., Dolph, J. and Marks, D., 1992. A comparison of geostatistical procedures for spatial analysis of precipitation in mountainous terrain, *Agricultural and Forest Meteorol.*, **58**, 119–141.

Robinson, A.H., 1971. The genealogy of the isopleth, *Cartographic J.*, **8**, 49–53.

Robinson, A.H., 1982. *Early Thematic Mapping in the History of Cartography*, Univ. of Chicago Press, Chicago, IL, 266 pp.

Running, S.W., Nemani, R.R., Peterson, D.L. *et al.*, 1989. Mapping regional forest evapotranspiration and photosynthesis by coupling satellite data with ecosystem simulation, *Ecology*, **70**(4), 1090–1101.

Solomon, S.I., Denouvilliez, J.P., Chant, E.J. *et al.*, 1968. The use of a square grid system for computer estimation of precipitation, temperature, and runoff, *Water Resour. Res.*, **4**, 919–929.

Stancik, A. and Jovanovic, S. (eds), 1988. *Hydrology of the River Danube*, Publishing House Priroda, Bratislava.

Tabios, G.Q. III and Salas, J.D., 1985. A comparative analysis of techniques for spatial interpolation of precipitation, *Water Resour. Res.*, **21**(3), 365–380.

Thornthwaite, C.W., 1948. An approach towards a rational classification of climate, *Geographical Review*, **38**(1), 55–94.

Thornthwaite, C.W., Mathers, J.R. and Carter, D.B., 1958. *3 Water Balance Maps of Eastern North America* (with a forward by E.A. Ackerman). Resources for the Future, Washington, DC, 47 pp.

UNESCO/WMO, 1977. *Hydrological Maps: a contribution to the international hydrological decade*. Studies and Reports in Hydrology No. 20. Geneva: UNESCO/WMO, 204 pp.

Ward, R.C., 1967. *Principles of Hydrology*, McGraw-Hill, New York, NY.

Williams, G.R. *et al.*, 1940. *Natural Water Loss in Selected Drainage Basins*, US Geological Survey Water Supply Paper 846, Washington, DC: US Government Printing Office.

Wendland, W.M., 1987. Hydroclimatology, in *Encyclopedia of Climatology*, Encyclopedia of Earth Sciences, Vol. XI. J.E. Oliver and R.W. Fairbridge (eds). New York, NY: Van Nostrand Reinhold, pp. 497–502.

Woodruff, J.F. and Hewlett, J.D., 1970. Predicting and mapping the average hydrologic response for the eastern United States, *Water Resour. Res.*, **6**, 1312–1326.

Acknowledgements

The information in this document has been funded wholly (or in part) by the US Environmental Protection Agency. It has been subjected to the Agency's peer and administrative review, and it has been approved for publication as an EPA document. Mention of trade names or commercial products does not constitute endorsement or recommendation for use.

Cross references

Evapotranspiration
Precipitation
Water balance
Water budget analysis

HYDROLOGICAL SERVICES

Functions of hydrological services

Accurate information on the condition and trend of a country's water resources – surface and subsurface, quantity and quality – is required for economic and social development and for the maintenance of environmental quality. Uses of water resources information are many and varied. Almost every sector of a nation's economy uses water information for planning, development or operational purposes. Water is of inestimable value to all countries, and as competition for water increases, water information grows in value. Because the cost of government programs must be properly justified, it is becoming very important to demonstrate the benefits of hydrological information (WMO, 1990). Ratios of benefit to cost of up to 40:1 have been cited (that is, the value of the information is 40 times its cost of collection). Benefit–cost ratios in the range 5 to 10 seem to be generally plausible, with values of 9.3 and 6.4 being found in studies in Canada and Australia (Acres Consulting Services, 1977; Australian Water Resources Council, 1988). The assessment of the UK streamgauging network found a benefit/cost ratio of between 1.2 (lowest) and 7 (highest) depending on how the benefits were quantified (CNS Scientific and Engineering Services, 1989). Regardless of the actual numerical values, water managers in all countries subscribe to the view that water information is a cost-effective programme and is a prerequisite for wise water management.

Uses of hydrological information

The primary role of a hydrological service, or equivalent agency, is to provide information to decision makers on the status and trends of a country's water resources. Such information may be required for:

● assessing a country's water resources (quantity, quality, distribution in time and space), the potential for water-related development, and the ability to supply actual or foreseeable demands;
● planning, designing, and operating water projects;
● assessing the environmental, economic and social impacts of water resource management practices, existing and proposed, and adopting sound policies and strategies;
● assessing the impacts on water resources of other non-water sector activities, such as urbanization or forest harvesting;
● providing security for people and property against water-related hazards, particularly floods and droughts.

In general, a hydrological service provides the necessary information for water resources assessment, which is defined (UNESCO, WMO, 1988; WMO/UNESCO, 1991) as 'the determination of the sources, extent, dependability, and quality of water resources, upon which is based an evaluation of the possibilities for their utilization and control.' With the growing recognition of such issues as global climatic change and the environmental impacts of urbanization, there is an increasing emphasis upon reliable water information as the foundation for sustainable development and management of water resources. This implies that future generations, as well as the present one, will continue to enjoy adequate and available water supplies to meet their social, environmental and economic needs. A hydrometric program designed purely for currently defined needs may be inadequate in the long-term.

Functions and responsibilities of a hydrological service

Water resources information may be required at a single, specific place, such as at a proposed dam site, or across an entire region, for example along a proposed highway route that crosses numerous watercourses. In the first case, information may be economically collected at the single site or in the catchment area upstream. Such information might be called use-specific. In the second case, it would be impracticable to collect information at every river crossing. General purpose data, representative of the whole region, must then be collected at a few locations, and a means must be provided of transferring the information to other sites for which no data are available. To achieve this capability, a basic network of observation stations will be required. The principal characteristic of the data obtained from such a network is that they may be used for a variety of unforeseen purposes. They are representative of the hydrology of the area, and they must be collected to standards that are able to meet the reasonable requirements of any likely user.

To meet these requirements, a hydrological service must

- establish the requirements of existing or possible future users of water resources information;
- define the standards (accuracy, precision, timeliness, accessibility, etc.) of the data that are implied by those requirements;
- design and establish hydrometric networks to measure the various types of data required. Both use-specific and basic networks, that may be complementary or even overlapping, may be needed;
- develop methods for transferring information from measurement sites to other locations in the region for which it is representative;
- collect data and maintain quality control of the data collection process by inspection of field installations and field practice;
- process and archive data and maintain control of the quality and security of the archived data;
- make the data accessible to users, when, where and in the form they require, including
  - dissemination of hydrological forecasts and warnings;
  - publication of yearbooks of basic data in paper, microfiche or computer compatible (CD-ROM, floppy disk, etc.) form;
  - preparation of reports on water resources in which data are comprehensively analyzed. This may include media such as hydrological atlases or databases in geographical information systems; informative or educational material for use by the general public, the news media and schools;
  - information for project design including the frequencies of streamflow extremes;
- inform potential users of the information that is available and assist them to make the best use of it;
- develop new technology and carry out research into hydrological and related processes to assist the user in interpreting and understanding the data;
- develop staff (training) and other functions related to quality assurance, such as the preparation of instruction manuals and assessment or acceptance testing of new instrumentation;
- ensure coordination with other agencies that acquire water-related or other relevant information, such as hydrogeological, water use, topographic, land use or climatic information.

A simplified schematic flow chart is given in Figure H13.

The hydrological service may carry out these functions as a service to a particular client, such as a power company, perhaps on contract. On the other hand, it might provide an entirely public service, using funds provided from general taxation, because its products are of value to the public at large. Whichever is the case, considerable emphasis must be placed upon communicating with the users, both to determine their requirements and to ensure that the hydrological service's products are readily accessible and used to the greatest extent possible. Increasingly, natural resources are managed in an integrated fashion, which requires that a variety of different types of data be available – hydrological, geological, topographic, land use, socio-economic (e.g. water use), and so on. Rapidly evolving computer technology facilitates this, but often outstrips the ability of organizations to collaborate and exchange information.

Types of data required

Many classifications have been proposed for the uses of hydrological information (Rodda and Flanders, 1985). The areas of application identified by hydrological services in just three countries, Canada,

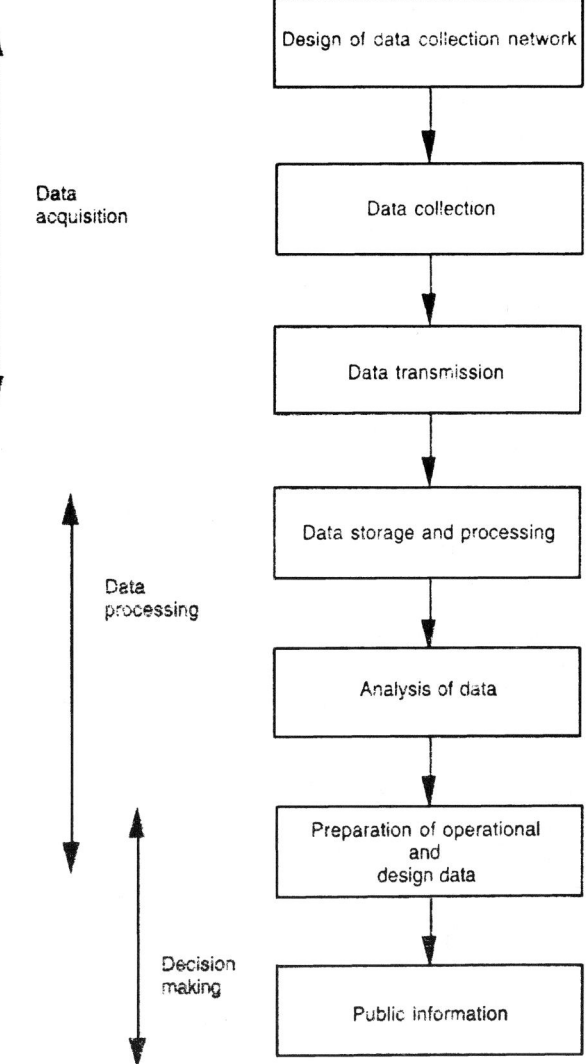

Figure H13 Activities of a hydrological service.

Australia and the United States (Acres Consulting Services, 1977; Australian Water Resources Council, 1988; Fontaine *et al.*, 1984), indicate the diversity of uses for streamflow data alone, and of course other types of hydrological data have additional applications. In a rather novel approach to classification, the UNESCO/WMO *Water Resource Assessment Activities – Handbook for National Evaluation* (1988) recognized a number of types of water resource projects that required hydrological information. A similar tabulation with a more conventional definition of water information sectors is provided by the Australian Water Resources Council (Australian Water Resources Council, 1988).

For basic water resources assessment purposes, the major elements of the hydrological system that must be considered can be classified as inflows, storages, and outflows (Figure H14). In many cases, other types of data would be required, including data on groundwater levels and water quality, water use [consumption, irrigation return flows, non-consumptive uses, such as the biological oxygen demand (BOD) of wastes disposed to a watercourse, etc.], and non-hydrological data, such as the intensity of use for recreation or bathing, the quantity of fish harvested from a watercourse, and so on.

Together these imply a vast range of water-related data and information that the hydrological service and other related agencies may need to supply. Different levels of economic and social development, the sensitivity of the natural environment to disturbance by human activity, and the nature of the physical environment itself (climate, topography, the abundance or otherwise of water, and so

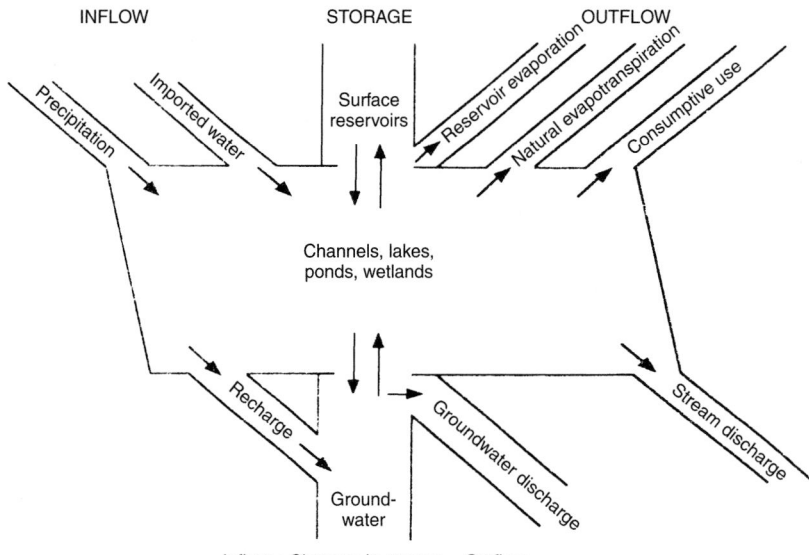

Figure H14 Diagram showing major elements of a hydrological system need for a water budget of a typical river basin in a subhumid region.

on), all determine the level of information required. One framework that has been proposed (UNESCO, 1988) is the transition from ecological orientation to construction orientation and finally to resource management orientation. In each phase, different types of information are required, depending on the number and types of decision that are to be made. In the first phase, society essentially adapts to the environment, including the natural hydrological regime. In the second, the water resource is increasingly exploited, but is still abundant relative to demand. Decision making tends to focus on means of exploiting the resource by construction of dams, irrigation schemes, and so on. The main requirement for information relates to the statistics of spatial and temporal variability of the water resource. In the third phase the resource is no longer relatively abundant. Human activity is itself markedly influencing (usually in a negative way) the size and quality of the resource. Decision making is increasingly focused on regulating demand and supply to allocate efficiently a valuable resource amongst competing users. Hence information is required, not just on the status of the water resource, but also on usage and the impact of that usage.

The range of possible alternative decisions increases through the three phases. Hence the need for more and different types of information increases. This implies a progressive evolution in the role of a hydrological service of a given country and also that hydrological services in different countries may have very different requirements placed upon them. Nevertheless, providing information on water quantity – total volumes, year to year variability, extreme-values – is perhaps the basic activity for most hydrological services. Water quality is of rapidly growing importance in many countries because of its significance for consumptive uses (domestic water supply, industrial and agricultural purposes), in-stream uses (fish resources, aquaculture, recreational and bathing use) and because of environmental concerns (eutrophication of lakes, damage to natural freshwater and estuarine ecosystems).

Real-time forecasting of extremes

The preceding sections have emphasized the water resources assessment role of hydrological services, which requires long, continuous records and an eye on future needs, as well as present-day management. However, a major requirement in many countries is the provision of forecasts and warnings of extreme hydrological events, principally floods, droughts, storm surges and avalanche hazards. Many of these events are both weather and water related, so that forecasts are commonly provided in cooperation with the national meteorological service.

Although the information needed to provide forecasts and warnings may ostensibly be similar to that required for water resources assessment, (i.e. precipitation intensities, water levels, and so forth),

in fact the specific requirements are very different. In forecasting, the greatest need is for information to be timely, easily understood and reliable, so that rapid decisions can be made with confidence. High precision, continuous data collection or conformity to scientifically based sampling designs may be less important. The disparity between data requirements for resource assessment and forecasting/warning may present significant practical difficulties to a hydrological service charged with both responsibilities and may require virtually separate instrumentation, data transmission systems and data dissemination procedures.

Organization of hydrological and meteorological services

The manner in which hydrological services are organized varies widely from country to country as a function of such factors as the governmental and political system, the size of the country, its stage of economic development, the physical environment and the particular information needs of the country. Information needs evolve with time, which implies that the most appropriate form of organization for providing information may also change with time.

There are four main patterns of organization (Rodda and Flanders, 1985):

● a combined hydrological and meteorological service that is a distinct arm of a central government;
● a separate hydrological service, most likely as part of a central government department that has the primary responsibility for water;
● no single hydrological service, and responsibility for acquiring water resources information and for other aspects of operational hydrology is shared by several central government departments;
● various aspects of operational hydrology are the responsibility of several or many specialized agencies that operate at the site, river basin, provincial or regional government level, probably with the coordination by a central government body.

Combinations of these patterns may be found in some countries. Recently, there has been a trend towards commercialization of hydrological services and institutes, as governments require them to become more market driven. In the extreme case, this could result in establishing a hydrological service or institute as a private sector company from which the government and other clients purchase the service that they require on contract.

It is a common practice for the national meteorological service to provide weather forecasts and a wide range of water-related meteorological and climatological data for operational hydrology. Often, however, data collection networks operated by meteorological services focus on airports and urban areas because of the strong historical relationship between meteorology and aviation and the location of observers. Hence, hydrological services, whether at national, state or local level, are likely to supplement data obtained from meteorological services with data that they themselves collect. Additional telemetered raingauges or river-monitoring stations are sited in unpopulated headwater areas to provide early flood warning or supplementary data collection networks required for detailed water resources assessment on a full catchment basis.

The history of water resources development in many countries, in which the initial emphasis was on exploitation for hydroelectricity or irrigation or on flood control, has led to a situation where hydrology and water resources assessment is the responsibility of government departments, such as the Ministry of Energy or the Ministry of Agriculture and Fisheries. Indeed, frequently, several such departments have developed their own programmes of water resources assessment and management. Many countries, as a consequence, have several data collection networks and several hydrological archives. There are, in such circumstances, dangers of duplication of effort, inconsistent standards of data collection, incompatible data processing and archiving, and a failure to make full use of all available information or outright competition for resources.

Almost all countries recognize the need for the coordination of agencies with water-related responsibilities, and many have established coordinating bodies at the central government level. There are excellent examples where such coordination has been successful, but coordination and collaboration require a considerable amount of communication and hard work to be wholly effective. So there are also many countries where arrangements have not been effective. Some of the most successful examples of coordination may be found in international river basins, where all countries have a common interest in standardizing their data acquisition techniques, facilitating communication, and so on.

In principle, the most effective arrangement would be for all water-related activities to be located in a single agency. However, in practice, countries with fragmented hydrological services may assess and manage their water resources as effectively as others with a centralized service. While there seems to have been a recent trend for increased coordination or centralization of water-related functions, some countries have been moving in the opposite direction and relegating responsibilities as much as possible to the local level. The principal need is for an unhindered flow of information to be achieved from the data provider to the data user. Arrangements to achieve this – a ministry of water resources, an inter-agency coordinating committee or a water resources council with responsibility for national oversight or simple day-to-day contact – can be varied to suit the particular circumstances. Several publications (Rodda and Flanders, 1985; Godwin et al., 1990; WMO, 1977) provide examples and guidance on arrangements for hydrological services.

Source

World Meteorological Organization, 1994. *Guide to Hydrological Practices*, 5th edn, WMO, Geneva.

Bibliography

Acres Consulting Services, 1977. Economic evaluation of hydrometric data. *Report to the Department of Fisheries and Environment*, Ottawa.
Australian Water Resources Council, 1988. *The Importance of Surface Water Resources Data to Australia*. Water Management Series 16, Australian Government Publishing Service, Canberra.
CNS Scientific and Engineering Services, 1989. *The benefit–cost of hydrometric data – River flow gauging*, Dept. of the Environment, pub. by FWR Marlow, UK.
Fontaine, R.A., Moss, M.E., Smith, J.A. and Thomas, W.O., 1984. *Cost effectiveness of the stream-gauging program in Maine: a prototype for nationwide implementation. US Geological Survey Water-Supply Paper 2244*, Reston, Virginia.
Godwin, R.B., Foxworthy, B.L. and Vladimirov, V.A., 1990. *Guidelines for water resource assessments of river basins. Technical Documents in Hydrology*, IHP-III Project 9.2, UNESCO, Paris.
Rodda, J.C. and Flanders, A.F., 1985. *The Organization of Hydrological Services: Facets of Hydrology*. Volume 2, Chapter 14, Wiley, New York.
United Nations Educational, Scientific and Cultural Organization/ World Meteorological Organization, 1988. *Water Resource Assessment Activities – Handbook for National Evaluation*.
World Meteorological Organization, 1977. *Casebook of Examples of Organization and Operation of Hydrological Services*. Operational Hydrology Report No. 9, WMO No. 461, Geneva.
World Meteorological Organization, 1990. *Economic and Social Benefits of Meteorological and Hydrological Services*. Proceedings of the Technical Conference, Geneva, 26–30 March 1990, WMO No. 733, Geneva.
World Meteorological Organization/United Nations Educational, Scientific and Cultural Organization, 1991. *Progress in the implementation of the Mar del Plata Action Plan and a strategy for the 1990s. Report on Water Resources Assessment*.

Cross references

Hydrological yearbooks
Water agencies: national and international

HYDROLOGICAL YEARBOOKS

The collation, normally on a national basis, of hydrometeorological data to provide a continuing series of snapshots of hydrological conditions in readily accessible publications has a long history. Prior to the 1970s the production of hydrological yearbooks was taken as one tangible sign that hydrometric data acquisition and dissemination had come of age. Then, as one country after another developed computer-based river flow and groundwater level archives supported by relatively sophisticated data retrieval services, the need for yearbooks was increasingly questioned. The versatility of the new archiving systems seemed to provide a more efficient and effective means of ensuring that the resources devoted to flow measurement and hydrometric data processing were fully capitalized on. Commonly, the projected benefits associated with the introduction of computer technology resulted in an increasing number of countries abandoning yearbook publication.

A trend away from yearbooks is entirely understandable; many national archiving sytems are impressively user friendly, have a proven pedigree and clearly stimulate a broader usage of hydrological data. Often, however, the focus of attention when reviewing more traditional approaches to data handling has been on what is technically feasible rather than what is likely to be deliverable. Given the often challenging field conditions, the inadequacies of the gauging station network and the limited funds and expertise available to most responsible national agencies, it is essential to determine what is realistically sustainable.

In this context the assumed diminishing benefits, both direct and indirect, of yearbook production deserve re-examination. Yearbooks, together with the associated activities that ensure the timely placing of validated data into the public domain, can perform a number of functions as yet not fully addressed by systems conceived as rendering them redundant.

The principal aims of yearbook publications vary from country to country, but commonly the objectives will include some, or all, of the following:

* to catalog available water resources and flood information;
* to provide a representative snapshot of water resources and establish a series of benchmarks at a time when documentation of the impact of climatic variability is vital;
* to serve as an archive in itself and provide a safeguard against the loss or destruction of original data;
* to serve as an introduction to the national hydrological information service and a gateway to complementary data sets;
* to help motivate field and office staff;
* to provide tangible evidence of a return on substantial investment in hydrometric networks;
* to promote discipline and concern for accuracy within the national data acquisition system;
* to provide a clear incentive to keep archives up to date and a focus for a regular hydrometric audit linked to quality assurance targets;
* to increase awareness of hydrological and water resources issues and encourage the wider use of hydrological data;
* to promote hydrological research through the ready availability of hydrometric and other relevant data.

Although many individual countries acknowledge yearbooks as both a source and a backup to computer systems, the most important of the above objectives is, arguably, the discipline imposed by the need to place good-quality data in the hands of a user community in a timely manner (Marsh, 1995). The exposure of contemporary data to early user scrutiny brings benefits across the full range of hydrometric activities. It serves to motivate data suppliers, helps ensure that field personnel do not remain remote from the end product of their endeavors and, crucially, focuses attention on the need to maximize the utility of archived data. The absence of such a discipline can, and does, lead to data being left unprocessed until a compelling need for them is identified – only for validation to be all but impossible at a remove of many years.

Over the last decade, concern has grown that hydrological data collection (and analysis) is not keeping pace with present water development and management needs, not to mention the expected new demands associated with the design and implementation of practical sustainable development options (avoidance of conflict here). In such circumstances, yearbooks can provide the means both of harnessing better the hydrological monitoring that is undertaken and to help determine network development and data processing priorities. By making the yearbook a central component in regular reviews of network effectiveness and data utility, the publication becomes a driving force behind the development of improved water resources assessments and, ultimately, water management procedures. As importantly, gaps in records and sequences of anomalous data – which can undermine the value of even the most sophisticated computer-based system – can be identified and rectified; from the user's viewpoint this is an especially valuable adjunct to the routine publication of data.

Yearbooks also increasingly serve as shop windows for a range of related data sets and services thereby increasing accessibility to much-needed relevant hydrological spatial information. Unlike the meteorological service in many countries, hydrological services often involve several national or regional agencies, and a well-designed yearbook series helps improve collaboration between agencies and increase user awareness of data availability – the latter can still be surprisingly limited, even in developed countries, especially where stewardship of the national hydrometric archives has passed successively to a number of different agencies.

Apart from the central need to feature basic data, there is little point in being prescriptive with regard to the style and contents of yearbooks. The preferred format will vary greatly according to circumstances. Some smaller countries may wish to continue, or initiate, the publication of yearbooks embracing comprehensive sets of daily mean flows. By contrast, larger countries with access to sophisticated computing facilities may prefer to publish subsets of representative data together with annual catalogs of what is available in supporting archives. A comprehensive approach is of particular importance in those parts of the world where data are being lost faster than they are being effectively archived. This will help to redress a situation where in some regions, parts of Africa for example, more data are available outside the continent than in. Additionally, the use of data acquisition and handling procedures appropriate for yearbook production will themselves help countries to exploit more fully the powerful analytical packages now becoming generally available.

Yearbooks cannot compete with the data accessibility offered by modern computer networks or the data delivery capabilities of, for example, optical disks or televisual links. Nonetheless, the continuing value of yearbooks – perhaps in a different guise to those familiar a decade ago – emerges more clearly when the entire data acquisition/dissemination enterprise is reviewed and attention is directed to the overall information output of individual hydrometric networks. It may well be that the long-term demise of printed yearbooks is inevitable, but to condemn them to a premature death is to misunderstand their importance at a critical time.

T.J. Marsh

Bibliography

Marsh, T.J., 1995. Yearbooks – dinosaurs or dynamos. World Meteorological Organization Bulletin, **44**(1), 60–63.

Cross references

Hydrological services
Water agencies: national and international

HYDROLOGISTS (600 BC–AD 1900)

The following provides short biographical notes of philosophers, scientists and engineers involved in the development of hydrology.

Leone Battista Alberti (1404–1472)

The origin of rivers and springs had still not been solved by the dawn of the fifteenth century and consequently the concept of a correct hydrological cycle was a major ambition of all learned people in solving the motion of water. The Italian Alberti was such a person and his views, although corresponding with the thoughts of the period, made substantial progress in his contribution to solving the hydrological problem by his treatise *Ten Books on Architecture*, a major portion of which was devoted to water. In it he covered all aspects of the subject known at the time but accepted that he had no new explanation to present except that he agreed with the general belief that rivers originated from rain. His ideas about groundwater did not contribute to learning although the subject was covered in detail in his treatise. He concluded that if a well was dug, water would be found only if the well was sunk to the level of the nearest river.

Georgius Agricola (Georg Bauer 1494–1555)

Agricola was quite explicit about the origin of groundwater as he viewed it, but he believed, as did Aristotle, that the creation of water within the Earth was from the condensation of vapor. The cause of vaporization was primarily from subterranean heat from burning bitumen. Agricola maintained that groundwater came from infiltration and condensation of subterranean steam generated by heating deeply percolated surface water. Therefore both Agricola and Aristotle believed in the subterranean generation of water, and the percolation of seawater landward was justified by the fact that wells dug near the shore yielded saline water. Agricola believed in the use of divining rods for locating water (Figure H15), but then so did Robert Boyle (1626–1691), one of the founders of the Royal Society of London.

Anaxagoras of Clazomenae (500–428 BC)

An Ionian philospher, his explanation for the regular rise of the Nile turned out to be almost the correct one. The sea, he believed, was produced by the waters of the Earth and from the rivers which flow into it; rivers depend on the rain and on the waters within the Earth which was hollow and had water in its cavities. The Nile rises in summer owing to the water that comes down from snows in Ethiopia was his philosophy.

Figure H15 Agricola's concept of the use of divining rods. (From Agricola, *De re Metallica*.)

Figure H16 Aristotle.

Anaximander of Miletos (610–545 BC)

A hydrological philospher, he was a contemporary of Thales and influenced by him. He believed that life originated in water. With continuous evaporation he believed that land emerged where once there was an all-engulfing sea and that eventually the sea would dry up. He believed that precipitation was due to moisture being drawn up from the Earth by the Sun.

Aristotle (385–322 BC) (Figure H16)

The Aristotelian concept of the universe was similar to that of Plato and Pythagoras. He believed in five elements and each element had its appropriate place in the universe. He was the first to write a book on meteorology entitled *Meteorologica*. His concept of the hydrological cycle was that rainfall resulted from a great amount of vapor and that heavy showers occurred as a cloud of water vapor descended into warm air. He did not agree with Plato that a subterranean reservoir existed from which all rivers originated. He contended that the origin of rivers came from rainfall and percolation, subterranean condensation of air into water and condensation of vapors rising from some source (not stated).

Like Hippocrates he believed that only 'light' water was evaporated by the Sun and argued that the sea must persist for ever because as the Sun approaches water, it draws it up, and as the Sun recedes, the water comes down as rain, and as long as this arrangement continued the sea would never dry up.

Thomas Barker (1716–1800)?

Barker was the brother-in-law of Gilbert White and is credited for carrying out the longest record of rainfall measurement in the eighteenth century using the same gauge (1736–1796). White himself was an observer for 8 years from 1779 to 1786.

John Frederic La Trobe Bateman (1810–1889)

Bateman was apprenticed as a pupil engineer at the early age of 15, but became one of the most eminent water engineers of the nineteenth century. His works both in the UK and abroad were numerous, most involving the design and construction of water supply schemes, the best known being Glasgow, Manchester and Buenos Aires. He was invited by many countries either in an advisory capacity or in the design and preparation of water supply schemes. He had also a great interest in rainfall measurement which he used to good effect in reservoir design, there being very few, if any, river flow gauges at that time. Other works he carried out involved flood protection and river

improvement. He also designed harbor and dock works in the UK and Ireland. In 1869 he was engaged in the design of a tunnel or tube crossing of the English Channel, but that had to wait for another 100 years!

Henri Emile Bazin (1829–1917)

Bazin joined Darcy in Dijon in 1854 and both carried out major laboratory hydraulic research experiments. Investigation of velocity distribution in open channels was performed by a Pilot tube made by Darcy and their experiments showed that the maximum velocity V_{max} in natural rivers and open channels occurred at the surface, and found that the velocity distribution at the center of a wide open channel where side effects were negligible, velocity v at depth d was

$$\frac{V_{max} - v}{H} = 20 \left(\frac{d}{H}\right)^2$$

where d and H (total depth) are in meters.

In narrow open channels having a width less than 5 times the depth they found the maximum velocity to occur 'somewhat' below the surface. Darcy and Bazin used a channel nearly 600 m long which had rectangular, trapezoidal, triangular and semicircular sections with various types of lining. They also carried out experiments on flow through orifices and conduits and produced equations for velocity in the form

$$RS = \left(a + \frac{b}{R}\right) V^2$$

where R is the hydraulic radius and S is the surface slope.

Bazin's equation of 1897 for C in the Chézy equation expressed in fps units was

$$C = \frac{157.6}{1 + \dfrac{Y}{\sqrt{R}}}$$

where Y is the rugosity factor.

Bazin was probably best known in hydrometry for his discharge equation for the flow over thin-plate weirs:

$$Q = \left(3.25 + \frac{0.08}{D}\right)\left[1 + 0.55\left(\frac{D}{D+p}\right)^2\right]LD^{1.5}$$

where D is the head, L is the length of the weir and p is the height of the sill of the weir (dimensions in feet).

The shortened Bazin equation is

$$Q = \left[3.41 + 1.69\left(\frac{D}{D+P}\right)\right]LD^{1.5} \text{ (in fps units)}$$

Nathaniel Beardmore (1816–1872)

The first work on hydrology in the English language was Beardmore's *Manual of Hydrology* published in 1862. Previously he had published in 1850 a book on *Hydraulic Tables*. Beardmore was responsible for the planning and design of railways, harbors, bridges and drainage and waterworks, and advised on the water supply of cities such as Edinburgh, Glasgow, Moscow and Odessa. Although he did not put forward new hydrological principles, nevertheless he was very much involved in the practice of hydrology and water resources.

Venerable Bede (AD 674–735)

The Venerable Bede was an English theologian and historian who compiled a summary of the causes of the inundation of the Nile. He was the first Englishman to write about weather and arguably was the founder of English meteorology. He contributed to concepts of the hydrological cycle but only conceded that seawater was salty and could not be raised by the suns rays, but rain, rivers and lakes were not salty. He also proposed a theory about the origin of the Nile.

Jacques Besson (1500–1580)?

A professor at Orleans, Besson's small 85-page book on water gave a clear and correct explanation of the hydrological cycle. He believed that water was evaporated by the sun and came down as rainfall which was sufficient to sustain river flow and springs. He maintained that in this context evaporation equalled rainfall, and considered that the salinity of the sea required no explanation since it had been salty since the beginning of time.

It should be pointed out that Besson was a contemporary of Palissy (see below) and each may have been aware of the other's work. The clear understanding of the hydrological cycle for the first time may therefore be ascribed to both.

Daniel Bernoulli (1700–1782)

The Bernoulli family were prolific mathematicians, but so far as hydrology, or hydrometry, is concerned, it is Daniel who was most interested. Daniel Bernoulli taught mathematics at St Petersburg from 1725 to 1732 but returned to his native Basle to teach anatomy, physics and botany. Bernoulli, in his analysis of the pressure–velocity relation, stated the principle that the sum of potential and kinetic energies of a freely falling body is constant and his main objective was to establish a relation between pressure and velocity. It is not known who first solved the relationship but certainly Bernoulli was one of the first. The 'Bernoulli equation', as it is known today, was not in fact his work; it was concerned with potential and kinetic energy and did not include the effect of pressure. Nevertheless, Bernoulli was one of the pioneers and directed much needed attention to the pressure–velocity relation.

Al Biruni (973–1048)

The origin and reason for the artesian wells of Modena had been the subject of conjecture and discussion for many years prior to AD 1000. Biruni, however, presented for the first time the correct explanation of the nature of these wells.

Benedetto Castelli (1577–1644) (Figure H17)

Castelli was a student of Galileo and taught mathematics at the universities of Rome and Pisa, and was ordered by Pope Urban VIII to apply his thoughts to the flow in rivers. Galileo had admitted to his student that he, Galileo, knew more about the planets than he did on river flow. Probably his major contribution to hydrometry was that he contended that previous work had ignored the velocity and, for the first time, corrected the erroneous discharge equation to $Q = AV$ where Q is discharge, A is cross-sectional area and V is velocity.

Figure H17 Benedetto Castelli (1577–1644).

Figure H18 Antoine Chézy (1718–1798).

Castelli used floats to measure the velocity, the practice which was followed by practically all engineers of the time (floats are still included in International Standards on flow measurement).

Geronimo Cardano (1501–1576)

Cardano was a prolific writer. It is argued that he plagiarized freely from Leonardo's manuscripts. He reduced the traditional four elements to three, omitting five, and believed that the Earth was like a sponge and full of subterranean water and that the proportion of land exceeded that of water. Water, he maintained, remained on the Earth's surface because there was not enough room for it inside. During this period the terms air and vapor were synonymous and Cardano was confident that the main source of water supply for rivers came from the conversion of vapor into water.

Antoine Chézy (1718–1798) (Figure H18)

Chézy's interest in hydrometry started when he became involved in the water supply of Paris when Perronet was director of the Ecole des Ponts et Chaussées. An additional supply was required for Paris and Chézy and Perronet considered that the River Yvette would be a suitable source. Chézy designed the channel required to carry the water to Paris, taking into consideration length, slope and cross-section. It was probably this work which led to Chézy's experiments with discharge in channels and eventually the Chézy discharge equation in 1775. The equation for velocity apparently lay in Chézy's reports undiscovered until Clemens Herschel found it among the files of the Ponts et Chaussées and later had it published. To prove his equation, Chézy carried out field work on the Courpalet canal and the Seine (see also Darcy below).

John Dalton (1766–1844)

Dalton was engaged in rainfall and evaporation measurements and determined the evaporation at a site near Manchester from 1795. A cylindrical vessel of tinned iron 10 in (25 cm) in diameter and 3 ft (0.9 m) deep was used. Two pipes were connected to the vessel, one at the bottom, the other 1 in (2.5 cm) from the top. The vessel was filled for a few centimeters with gravel and sand and the remainder with soil and put into a hole in the ground, the surrounding space being filled with earth except on one side, for the convenience of putting bottles to the two pipes. Water was added to saturate the earth. A regular record was kept of the quantities of rain which ran off from the surface of the earth through the upper pipe and the quantity that percolated through the same to the bottom pipe. Rainfall was measured by a vessel having the same dimensions as the one used for evaporation measurements.

Dalton assumed that evaporation − rainfall = quantity of water in the two bottles and concluded that the annual evaporation was 25 in (60 cm), the evaporation increases with rainfall but not proportionally and there was no difference between evaporation from bare earth and grass.

In 1802 Dalton presented a theory of vapor pressure which provided a basis for determining the evaporation from water surfaces in the form:

$$E = C(e_w - e_a)$$

where E is the rate of evaporation in inches per day, C is a coefficient, e_w is the maximum vapor pressure (in mercury) and e_a is the actual vapor pressure (in mercury).

The method is still used today with slight modifications for wind and/or temperature. John Dalton and Edmund Halley are arguably recognized as the founders of British hydrology.

Henri Philibert Gaspard Darcy (1803–1858)

Darcy's design and construction of the water supply to Dijon not only included hydrometric concepts, but his report on the project published in 1856 contained the historical background and legal aspects of water law. As a result of his work in Dijon he was retained by the City of Brussels on a similar contract. For his contribution to hydrometry he was ably assisted by Bazin, who joined Darcy's staff in Dijon in 1854, and between them they proceeded to carry out major laboratory research sponsored by the French government and the book *Recherches Hydraulique* was published in 1865 under the joint authorship of Darcy and Bazin. Darcy's empirical formula for the purification of water by filtration through sand:

$$Q = \frac{KA(H + L)}{L}$$

where L and A are the length and cross-section area, K is a constant and H is the head of water above the sample, still bears his name (see also Bazin, above).

Charles Augustin de Coulomb (1736–1806)

The basic theory of all nineteenth century discharge equations was expressed in a paper in 1800 by Coulomb. From his investigations he concluded that the resistance could be represented by a function containing two terms, one of which varied with the first power of the velocity and the other with its second power. Coulomb's law of resistance to flow in open channels was tested later by Girard in 1803, De Prony in 1804 and Eyetelwein in 1850 and accordingly modified by the latter to $V = 50.9\sqrt{(RS)}$ and by Lahmeyer in 1845 to $V = 49.87\ V^{\frac{1}{4}}\sqrt{(RS)}$ where R is the hydraulic radius and S is the surface slope. In 1846 De Saint-Venant suggested the following equation with tables for quick computation $V = 60\sqrt{(RS)}$. All of these equations have been reduced to the Chézy-type equation (see also Chézy, above).

Democritus of Abdera (460–357 BC)?

Democritus claimed that, as a physical philosopher, he wandered over a larger part of the world than any other person of the time and observed climates, land and water. He claimed that the snow melts and flows away in the northern parts during the summer solstice, forming clouds from the vapours. The etesian wind drove the clouds to the south and Egypt, giving rise to storms which fill the lakes and the Nile (hinting that the Nile had its source in rain-fed lakes of central Africa).

René Descartes (1596–1650)

Descartes considered that, like others, seawater diffuses through a series of subterranean channels in various directions until it reaches large caverns at the base of the mountains and there the water evaporates due to the heat of the Earth's interior and the salt is left behind. The water is formed from the condensed vapor by the low temperature at the top of the vaults and emerges as streamflow. He did not consider, however, what happened to the enormous deposits of salt left behind. Nevertheless Descartes' theory was dominant for nearly two centuries.

Henry De Saumarez (1715)?

The current meter, as we know it today, was developed from ships' logs for locating the longitude of ships at sea. De Saumarez from the Island of Guernsey entered a British Government competition in 1714 which rewarded the most acceptable method for discovering the longitude at sea. De Saumarez did not win the competition but with his 1715 model, which he tested in 1720 on the Thames estuary, he produced probably the first published velocity data. He used a vertical-axis meter, or ships' log, in his first design but also a horizontal-axis meter for use in weedy rivers.

Mathew D. Dobson (?)

Dobson was actually a medical practitioner but maintained probably the first record of rainfall, evaporation and temperature for four consecutive years (1772–1775) in Liverpool in two tin vessels, one for rainfall the other for evaporation. J.C. Rodda has since estimated the evaporation of the same period as Dobson's from the equation $E = 0.17T − 7.18$ where E is the monthly pan evaporation in inches and T is the monthly mean maximum temperature in °F.

Dobson's method of rainfall measurement accorded with today's standards of measurement; most raingauges of the time, and indeed later, were placed on the roofs of buildings.

Pierre Louis Georges du Buat (1738–1809)

Du Buat was a contemporary of Chézy and developed a discharge equation for open channel velocity similar to that of Chézy. The impact of the equation was enormous, although it was more cumbersome but based on the same Chézy principles – V proportional to RS. Although he did not take into account the surface tension, he recognized the boundary layer influence on flow. This was probably the beginning of the present-day boundary layer theory.

Ecclesiastes (250 BC)

'All the rivers run into the sea; yet the sea is not full; unto the place from whence the rivers came, thither they return again' (*Ecclesiastes*, Chapter 1, verse 7).

This verse is perhaps the first recognition of the hydrological cycle. The missing link, of course, between the sea and the returning of the rivers to the sea (loss plus evaporation) was not understood. It would not be for nearly another 1500 years. However, the practice of hydrology and water resources was recorded as far back as 3200 BC in the reign of King Scorpion and in 3000 BC when King Menes is reported to have dammed the Nile and diverted its course. In 2850 BC the Sadd el Kafara dam failed, and the origin of the Indus Valley water supply and drainage system dates from 2750 BC. Chinese water resources dates back to 2200 BC and the Nile and Red Sea were joined by a navigational canal in 1950 BC.

Eratosthenes (276–194/192 BC)

Eratosthenes was the chief librarian at Alexandria and drew a reasonably accurate map of the Nile as far as Khartoum, and hinted that the equatorial lakes were the source of the river. The inundation, or flooding, of the river, he suggested, was simply because of the rains. He hinted that Aristotle had previously suggested the same theory.

Michael Faraday (1791–1867)

Faraday had little schooling but was fascinated by anything to do with science. He attended some lectures given by Humphry Davy at the Royal Institution in London and persuaded Davy to take him on as a laboratory assistant. Such was his dedication that he proceeded to be the first to demonstrate the principle of electromagnetic induction and showed how this principle could be used, first to generate electricity and then to convert it from energy into motion. From these discoveries came the generator and the electric motor. His activities in flow measurement came in 1832 when he noticed that when the motion of water flowing in a river cuts the vertical component of the Earth's magnetic field, an electromotive force (e.m.f.) is induced in the water which can be sensed by electrodes. He found this emf was directly proportional to the average velocity in the river. Faraday, to prove his theory, carried out experiments on the River Thames in

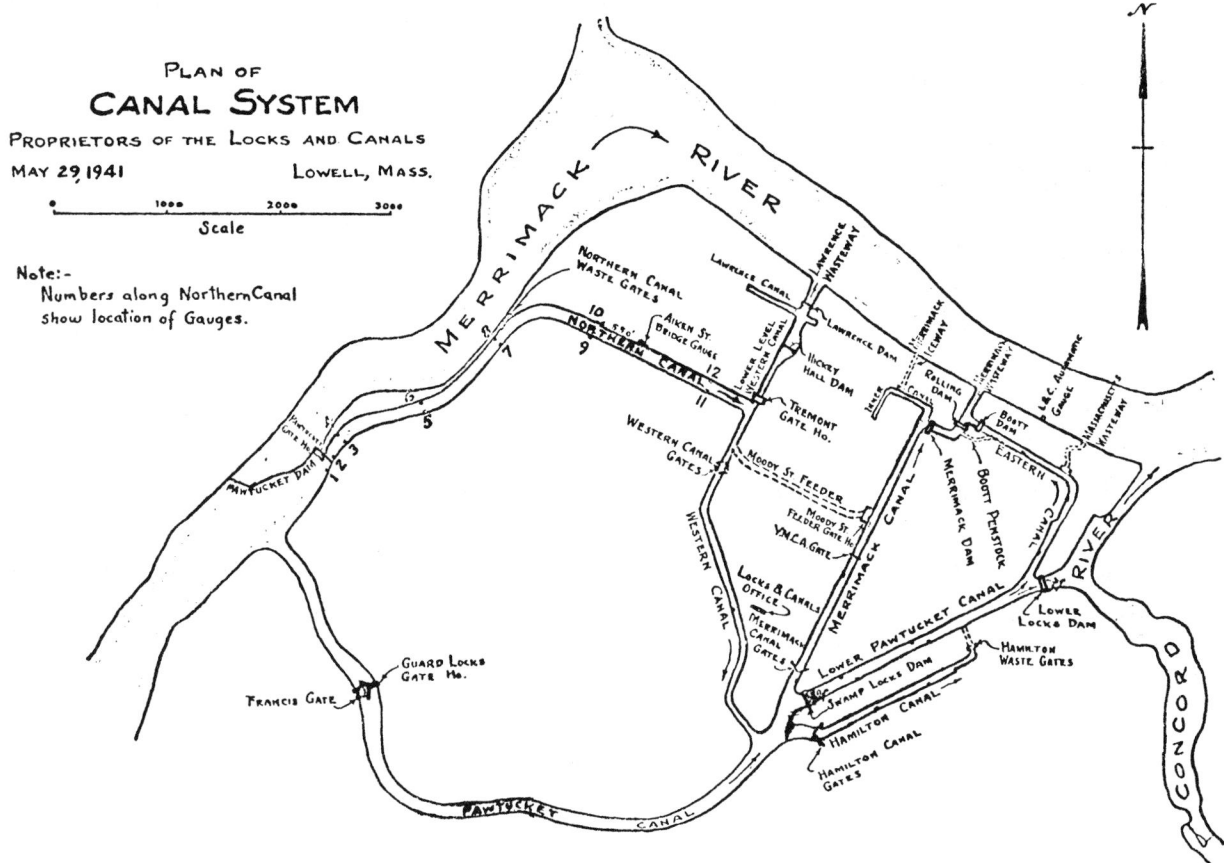

Figure H19 The canals of Lowell, Massachusetts.

London when he suspended a cable across Westminster Bridge and at each end suspended an electrode in the river flow. In the center of the cable he fixed a galvanometer and related the galvanometer deflection to the flow. It was not until 1953, however, that the method was again taken up when it was used to measure the flow through the Straits of Dover using one of the existing telephone cables to sense the emf.

In 1968 the UK Water Resources Board investigated the principle using a coil installed in the bed of the river to produce a vertical magnetic field. The method is now commonly used for the flow in rivers as well as for pipes, and electromagnetic current meters have also been successfully developed.

Giovanni Fontana da Meli (1540–1614)

Fontana's contribution to hydrometry concerned his investigations into flood estimation of the Tiber at Rome and the flood of December 1598 in particular. His treatise, dedicated to Pope Clement VIII, was first published in 1599 and reprinted again in 1640 at the recommendation of Benedetto Castelli. The method used was the $Q = A$ concept, so after 1500 years no improvement had taken place in flow measurement. Nevertheless, the conclusion of Fontana to relieve the flooding of Rome was to design channel improvement methods.

Fontana also designed fountains and obelisks and the extension of the Vatican. He was eventually deprived of his office by Clement VIII and moved to Naples where he designed the royal palace (1592).

James B. Francis (1815–1892)

The experiments of J.B. Francis on thin-plate weirs were made between 1848 and 1852, mostly at the lower locks of Pawtucket canal leading from Concord River past Lowell Dam to slack water of the Merrimack River, Massachusetts (see Figure H19). The experiments were made to determine the exponent n in the weir equation

$$Q = CLH^n$$

where C is the discharge coefficient, L is the length of the weir crest and H is the head over the weir. Francis found n to be equal to 1.47 but adopted the value 1.5 or 3/2. From numerous tests at Lowell, Francis determined the value of C to be 3.33 and his discharge equation became, allowing for the velocity of approach:

$$Q = 3.33L[(D + h)^{3/2} - h^{3/2}] \text{ (in fps units)}$$

where $h = v^2/2g$, and v was found by successive approximations and D was the head over the weir.

Also involved at Lowell in flow measurement were J.F. Baldwin (1782–1862), George Washington Whistler (1800–1849), and C.S. Storrow (1809–1904).

Paolo Frisi (1727–1784)

Frisi was Professor of Mathematics at the University of Milan and a member of the major scientific societies. His patrons included Maria Theresa, Catherina II and Joseph II. In the field of hydrometry his influence was considerable and plans of all major water works were submitted to him for comment. His book on the regulation of river flow and torrents had a major influence on the subject of hydrometry, not only at home but also abroad, and the British Government had it translated so that it could be made available to British engineers engaged in irrigation and river regulation in India. In his book he discussed the work on open channel flow by Castelli, Viviani, Zendrini, Manfredi, Poleni, Grandi and Guglielmini.

Sextus Julius Frontinus (AD 35?–104)

At one time Governor of Britain, Frontinus received his appointment as Commissioner of water works of Rome from Emperor Nerva in about AD 97. His book *Deaquis urbis Romae, libre II* contained a wealth of information on water supply systems used by the Romans during the first century AD. He had no clear understanding that it was necessary for velocity to be measured in order to calculate discharge

discharge appeared to persist throughout the Roman period with the exception of Hero who suggested the equation of discharge as $Q = AV$. However Frontinus did consider the discharge with a given head as $Q = A\sqrt{(2gh)}$.

Frontinus's concepts of flow probably had some influence on the design of the Claudia and Anio Norus aqueducts of Rome. Table H4 gives details of the Roman aqueducts as they existed while under the jurisdiction of Frontinus.

Alfonse Fteley (1837–1903) and Frederic Pike Stearns (1851–1919) (Figures H21 and H22)

Both Fteley and Stearns were directly involved in the development of hydrology in the nineteenth century, but their most important contribution to hydrometry was their involvement in weir flow measurement. Indeed, one of the earliest references to the rectangular-profile broad-crested weir is that of Fteley and Stearns in their historic paper of 1883. They attempted to devise a method of correcting the head over a rectangular profile weir to give equivalent head over a thin-plate weir. Thus when flashboards or stop-planks were installed, the observed water levels could be interpreted in terms of the reasonably well-established data on thin-plate weirs. They varied the length of the crest in the direction of flow from 50 to 250 mm in their laboratory tests and the results showed variations with crest length. However, the equation which they developed was too complex for everyday use and fell into obscurity. Fteley and Stearns proposed the following discharge equation for thin-plate weirs:

$$Q = 3.31LH^{1.5} + 0.007L \text{ (in fps units)}$$

In working on the Boston water supply in 1873, Fteley decided he needed a current meter and developed one with the aid of Stearns. The Fteley and Stearns current meter remained in use for over 25 years. A model is now in the Smithsonian (Figure H23).

Galileo Galilei (1564–1642)

Galileo did not contribute directly to hydrology, although he carried out experiments on Aristotles' doctrines and proved that some of them were incorrect. His greatest contribution was possibly his teaching and both Benedetto Castelli and Evangelista Torricelli were his students.

Emile Oscar Ganguillet (1818–1894) and Wilhelm Rudolph Kutter (1818–1888) (Figure H24)

The Swiss involvement in open channel flow was probably initiated by Ganguillet and Kutter who carried out a series of hydraulic

Figure H20 (a) The aqueducts of Rome, restored. The Anio Novus and the Claudia in the left foreground, combined in a single structure; the Marcia, Tepula and Julia also borne by one single arcade, on the right. (Source: painting by Zeno Dima, courtesy of Deutches Museum Munich.) (b) The Roman aqueduct at Segovia, Spain, 1996.

in open channels, as had been presented by his teacher Hero of Alexandria. In designing the Roman aqueducts (Figure H20), it was the practice to ensure an adequate slope and it is unlikely that the designers related slope to cross-sectional area. It is presumed that they adopted a process of trial and error. Frontinus, however, seemed to have a concept of the relation of head to velocity. Herschel, in researching Frontinus's work, noted that the unit of measurement used was the *quinaria*, which was the area of a pipe of $1\frac{1}{4}$ *digits* in diameter (one *quinaria* was equivalent to 19 000 or 22 500 l per 24 h, ±7500–11 000 l according to site conditions and aqueduct design). It is believed that the practice of using area only in the calculation of

Table H4 Aqueducts under Frontinus' charge

Name	Builder	Date built	Source	Length (miles)	Size of aqueduct (ft)[a]	Quality of water	Elevation of delivery above Tiber wharves (ft)	Number of delivery tanks	Amount in quinariae		
									Available water	Used in city	Used outside the city
Appia	Claudius	312 BC	Spring	10.29	2.0 × 6.0	Excellent	28	20	704	699	5
Anio Vetus	Dentator	272–269 BC	River	39.55	2.5 × 7.0	Turbid	84	35	1 610	1 102	508
Marcia	Marcius	144–140 BC	Spring	56.73	4.6 × 9.0	Excellent	125	51	1 935	1 098	837
Tepula	Caepio and Longinus	125 BC	Spring	11.00	2.0 × 3.5	Warmest	128	14	445	331	114
Julia	Agrippa	33 BC	Spring	14.19	1.5 × 5.0	Excellent	133	17	803	597	206
Virgo	Agrippa	19 BC	Spring	12.97	2.2 × 5.0	Excellent	35	18	2 504	2 304	200
Alsietina	Augustus	AD 10	Lake	20.39	5.8 × 8.7	Not palatable	–	–	392	–	392
Claudia[b]	Caligula and Claudius	AD 38–52	Spring	43.34	3.0 × 6.2	Excellent	158	92	5 625	3 824	1 801
Anio Novus[b]	Caligula and Claudius	AD 38–52	River	53.98	4.3 × 9.0	Turbid	158				
Total				262.44				247	14 018	9 955	4 063

[a] The size of channels vary from place to place and hence dimensions are only approximate.
[b] The Anio Novus (upper) and the Claudia (lower) form a double-decked aqueduct.

Figure H21 Alphonse Fteley (1837–1903).

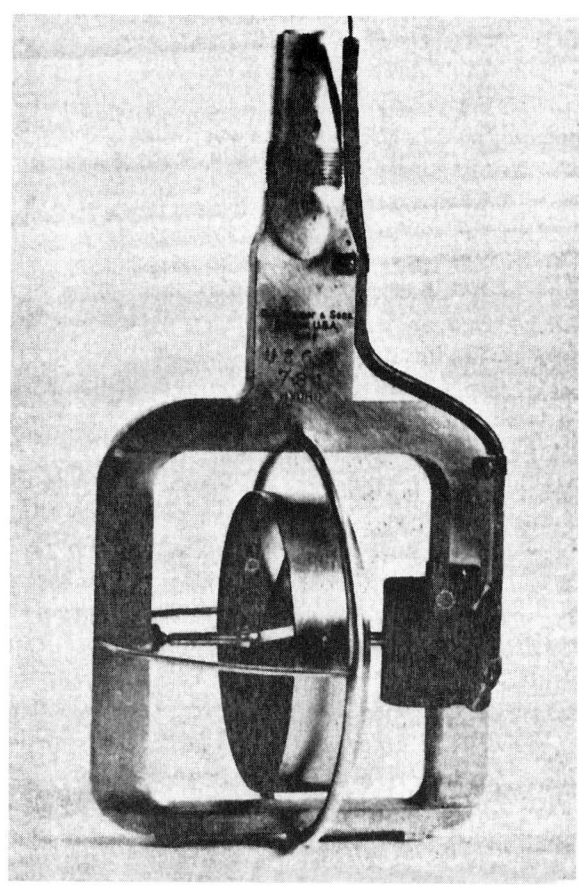

Figure H23 A Fteley–Stearns current meter manufactured by C.L. Berger & Sons, Boston. (NMHT 330410, Smithsonian photo 75409.)

Figure H22 Frederic Pike Stearns (1851–1919).

experiments in the Swiss Alps. The former was the chief engineer of the Department of Public Works at Bern and Kutter was a member of his staff. The Swiss hydrometric service is still located in Bern, one of the foremost in the world. Ganguillet and Kutter were invited by Humphreys and Abbot to check the validity of their formula based on C in the Chézy equation. They concluded that it was only valid for rivers with gentle slopes and that the Bazin equation was as inapplicable to the Mississippi as that of Humphreys and Abbot was

Figure H24 Emile Oscar Ganguillet (1818–1894).

to channels with steep slopes. It should be noted that Thrupp in Vol. 171 of the *Proceedings of the Institution of Civil Engineers* stated that for slopes of between 1/10 000 and 1/100 000, there was a transition for the value of velocity in the Chézy equation.

Ganguillet and Kutter's formula for the value of *C* in the Chézy equation found wide acceptance worldwide. In metric units it is:

$$C = \frac{23 + \dfrac{0.00155}{S} + \dfrac{1}{n}}{1 + \dfrac{n}{\sqrt{R}}\left(23 + \dfrac{0.00155}{S}\right)}$$

n is a coefficient of rugosity depending on the nature of the bed and sides of the river or channel and became known as 'Kutter's *n*' (also used in Manning's equation).

The Chézy equation for velocity was used extensively for channel design, and Ganguillet and Kutter's formula for *C* was an immediate success. It should be noted however that later not all water engineers or hydrologists viewed the equation with having the necessary accuracy in rivers. Indeed, P.A. Morley Parker in his classic book on the *Control of Water* (1949) considered that the Chézy equation was suitable only for artificial channels of regular section and when applied to rivers it was merely an interpolated formula of very limited range.

Nathan C. Grover (1868–1956)

Grover was Chief Hydraulic Engineer of the United States Geological Survey from 1903 to 1939. The Geological Survey was created in 1879 and under Grover and Powell's direction (J.W. Powell was the Director of the Survey) the survey made a major contribution to world hydrology which continues today. As a result of the Embudo Camp, New Mexico, on the Rio Grande in 1888 there began systematic work in collecting records of streamflow data for studying the water resources of the United States. Classic papers of the period of Grover's term of office were E.C. Murphy's Water Supply Paper No. 64 in 1902, on *Accuracy of streamflow measurement*, his *Hydrographic manual* with J.C. Hoyt and G.B. Hollister in 1904 (Water Supply Paper No. 94) and R.E. Horton's Water Supply Paper (revised) of 1907 on *Weir experiments, coefficients and formulas* (Water Supply Paper No. 200).

Giovanni Domenico Guglielmini (1656–1710) (Figure H25)

Guglielmini made a considerable contribution to international hydrometry and became the holder of the first Chair in Hydrometry in the University of Bologna in 1692 as well as City and District Water Engineer. Probably the first book on hydrometry was that published

Figure H25 Domenico Guglielmini (1655–1710).

Figure H26 Edmund Halley (1656–1742).

by him in 1697. His experiments were more precise than either Torricelle's or Maggiotte's and he was convinced that flow through an orifice, sluice or weir was proportional to *h*.

Guglielmini also carried out velocity distribution experiments in open channels and concluded that the parabolic distribution of velocities with zero at the surface would be valid only in the case of a perfect fluid.

Herodotus (484–425 BC)

Hydrological phenomena were included in the many aspects Herodotus was interested in. He explains the theories as to why the Nile's inundation began at the beginning of the summer solstice. He was also fascinated by the flow of the Danube (Ister River), and noted the contrast with the flow of the Nile which overflowed its banks with regularity, while the Danube maintained the same stage during the summer as in winter, and he endeavoured to explain the reason. Herodotus found that the Upper Nile actually flowed in the same direction as the Danube – west to east – and this idea was partially correct.

Edmund Halley (1656–1742) (Figure H26)

The major British contributions to hydrology during the seventeenth and early eighteenth centuries were the evaporation experiments of Edmund Halley and the raingauges of Sir Christopher Wren, Robert Hook and John Ray. Halley also conducted the first British expedition to study Antarctic icebergs and penguins. He was appointed Astronomer Royal in 1720, Secretary of the Royal Society in 1713 and a member of the French Academy of Sciences in 1729.

His results on evaporation experiments were reported in four papers to the *Philosophical Transactions of the Royal Society* between 1687 and 1715. He is reputed to have been the first to explain the concept of evaporation from the oceans and concluded that enough evaporation took place from the oceans to more than replenish all rivers and springs by rainfall. He demonstrated, therefore, that the landward phase of the hydrological cycle is essentially atmospheric. Halley and John Dalton are arguably recognized as the founders of British hydrology.

Johann Georg Gustav Hellmann (1854–1939)

Hellmann made considerable efforts to systematize precipitation measurements, as well as being one of the leading meteorologists of his time. The Hellmann gauge, named after him, is today one of the world's most used raingauges amounting to 20% of the 150 000 raingauges in use worldwide.

Figure H27 The world's first valid streamflow measurement as described by Hero of Alexandria, first century AD. (Drawn by A.H. Frazier.)

Hero of Alexandria (1st century AD)

The calculation of discharge from $Q = AV$ was first enunciated by Hero of Alexandria. His two most important works are *Pneumatica* and *Dioptra* and in the former he describes the application of the syphon. The latter book is on land surveying.

In hydrometry his most important contribution is that of calculating the discharge of a spring, explaining for the first time that discharge depended on both velocity and cross-sectional area (Figure H27). However, his concept was not accepted by either Vitruvius or Frontinus, or others.

It was not until Castelli, some 1500 years later, that it was finally recognized that discharge was equal to area times velocity.

Clemens Herschel (1842–1930) (Figure H28)

Through his association with James Francis at Lowell Massachusetts he developed a keen interest in hydraulics and hydrometry, and although he could not claim to have actually invented the principle of the Venturi flume or the Venturi meter, their use in flow measurement is credited, with good reason, to Herschel. Indeed it could be argued that he was one of the most influential hydrometric engineers of all time. The Venturi meter for closed conduit flow was entirely developed by Herschel and it is still one of the most common devices in use today, having an uncertainty of 1% or less (Figure H29). Herschel took photographs of the Frontinus Manuscript at the

Figure H28 Clemens Herschel (1842–1930).

Figure H30 Ralph Leroy Parshall (1881–1959).

Montecassino monastery and located the last Chézy work on flow measurement and translated both into English.

The Parshall flume (see Flume), developed in the United States in 1920, by Leroy Parshall (1881–1959) (Figure H30), was based on Herschel's Venturi flume. It is still used extensively, particularly in the United States. Herschel attended many professional meetings on hydrometry and his views were sought after by engineers of the time. One such meeting to which both he and Robert E. Horton contributed was held at the Institution of Civil Engineers in 1913 when E. Sandeman introduced his paper on the flow of the River Derwent in England using the Venturi-type flume (*Min. Proc. Inst. Civ. Eng.*, Vol 194).

Hippocrates (460–400? BC)

Hippocrates was recognized as the father of medicine, but he had also some definite ideas regarding the constitution of water and thought it comprised of two parts, one being thin, light and clear, the other thick, turbid and dark colored.

He also performed experiments to show that the thinner part of the water could be eliminated by evaporation.

Robert Hooke (1635–1703)

Hooke was the author of Hooke's law of elasticity and was the first Curator of experiments at the Royal Society of London and an inventive genius. Several writers have ranked him second only to Leonardo da Vinci. His current meter was developed in 1663 for use both as a ships' log and a river current meter.

Andrew Atkinson Humphreys (1810–1883) and Henry Larcom Abbot (1831–1927) (Figures H31, H32)

Humphreys and Abbot were both West Point graduates and were responsible for the Mississippi delta survey of 1851 to 1860. The importance of this survey was not that it was the first river survey ever carried out but that it was by far the most detailed water resources survey up to that time. The report covered 610 pages and, published in 1861, was translated into many languages. It contained considerable aspects of hydrometry and hydraulics which were partly original and partly compiled. They used double floats for the measurement of velocity and noted that for the same stage the discharge could change by as much as 20% and selected mean values for the establishment of the stage–discharge relation. Presumably this may have been the first

(a)

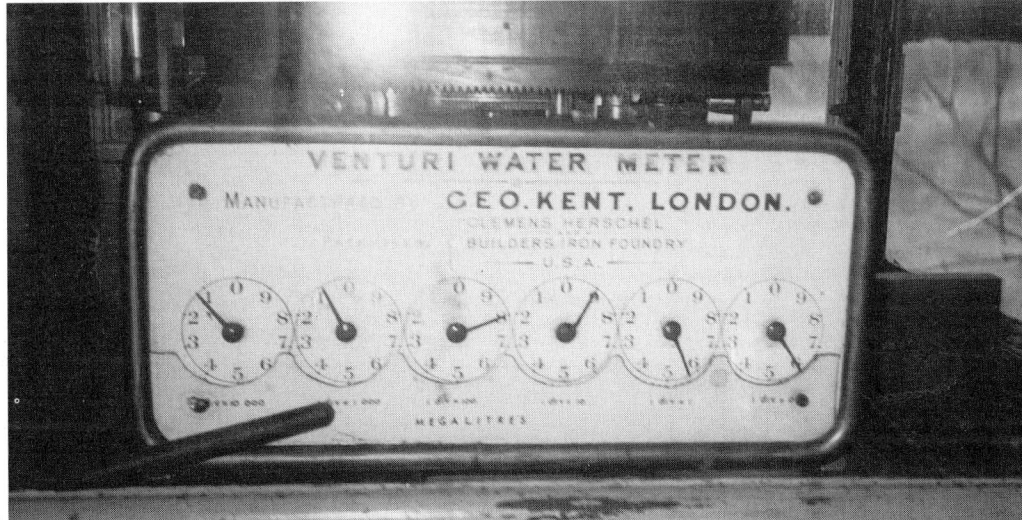

(b)

(c)

Figure H29 Herschel Venturi meter: (a) 9 in Venturi tube,
(b) recorder. (c) Clemens Herschel's first commercial Venturi tube.

Figure H31 Andrew Atkinson Humphreys (1810–1883).

Figure H33 Athanasius Kircher (1602–1680).

Figure H32 Henry Larcom Abbot (1831–1927).

Figure H34 Kircher's view of a cavern being supplied with seawater.

indication of loop ratings in the stage–discharge curve. Humphreys and Abbot developed their own discharge equation, presumably based on Chézy and Du Buat, but it did not include a roughness term and the equation failed to gain acceptance. It was studied by Ganguillet and Kutter at Humphreys and Abbot's request. Although Humphreys and Abbot preferred double floats for measuring velocity, they did employ one current meter – the Saxton meter – in their study of the Mississippi River, but only to a limited extent. They maintained for many years that the use of double floats was by far the best method of velocity measurement and that current meters were simply 'pretty toys having contributed to retard progress in river hydraulics'.

Athanasius Kircher (1602–1680) (Figure H33)

Kircher wrote several treatises while professor at Würzburg, one of which included the subterranean world including a section on the origin of rivers and springs. He was unable, however, to free himself from the concept that rivers received their supply of water from the sea and that great caverns existed below mountains. Rivers flowed out of these caverns so that the water could be used for supply and irrigation as well as for navigation.

The problem he faced, however, as others did before him, was the connection from the sea to the caverns and the raising of the water to a higher level. The part he got right was that the rivers ultimately return to the sea. Elaborate diagrams were produced by Kircher to show how the seawater was returned to the mountain caverns (Figure H34).

Steponas Kolupaila (1892–1964)

During the course of its highly successful operations, the Ott firm (Figures H35–H37) collaborated with many distinguished European hydraulic engineers and often incorporated their suggestions in their

Figure H35 Albert Ott (1847–1895).

Figure H36 Herman Ott.

Figure H37 Ludwig Ott.

each of the major instruments on which he had contributed improvements. The meter case, containing a current meter with his component propellor, carries a brass plate engraving and is now in the Smithsonian Institution.

Leonardo da Vinci (1452–1519) (Figure H38)

Leonardo was a genius. He was a sculptor, painter and architect. He gave hydrology and hydraulics a tremendous boost and was a prolific writer, and by far the bulk of his notes were devoted to hydrology and hydraulics. It is said that he had a correct understanding of the hydrological cycle, and had a clear concept of the origin of rivers and

instruments: Steponas Kolupaila was one such hydraulic engineer who made a major contribution to both current meter design and open channel flow physics. The most important aspect of his 4 years with Ott in designing horizontal-axis propellor-type meters was the development of the component propellor meter which measured the axial component of the oblique velocity.

When Kolupaila left Ott to become Professor of Civil Engineering at the University of Notre Dame, the Ott firm presented him with one

Figure H38 Leonardo da Vinci (1452–1519).

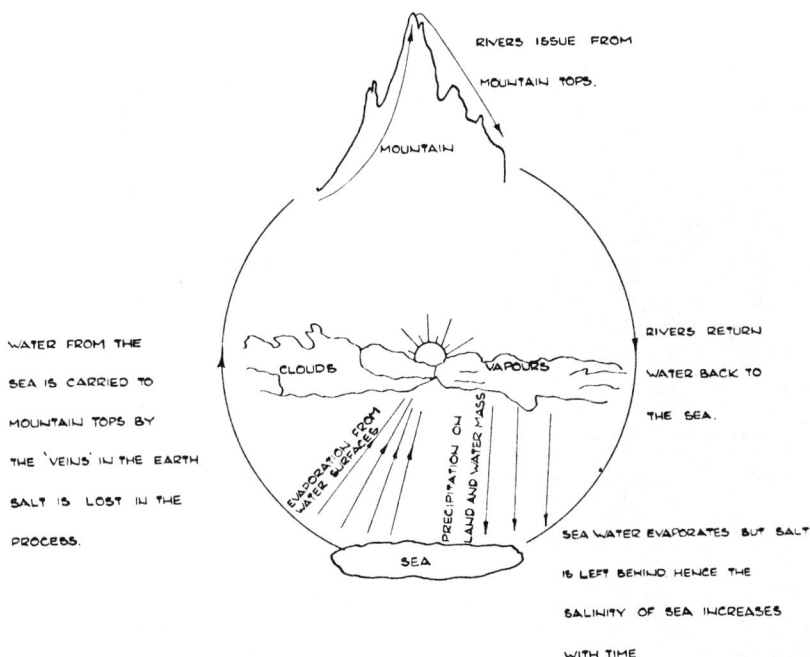

Figure H39 Leonardo's concept of the hydrological cycle.

Figure H40 Leonardo conducting experiments on the velocity distribution in streams. (Drawn by A.H. Frazier.)

springs (see Figure H39), although his theories were not entirely accepted with those currently held at the time.

Leonardo had a better understanding of open channel flow than any of his predecessors or contemporaries and his theory was based on experiments in the field. Had he introduced a time element instead of distance in his theory of motion he would have postulated one of the fundamental laws of motion, namely velocity = acceleration × time, nearly 100 years before Galileo. Figure H40 shows an artist's impression of how he conducted his observations of flow measurement in the field. He studied river bed configuration and carried out observations on velocity distribution in open channels both by the use of models and field work. He also noted that the salinity of the sea was the result of rivers discharging their salt content into the oceans and that salinity gradually increased through evaporation. He also put forward theories of irrigation, drainage and flood control.

William Newsham McClean (1874–1968)

McClean became a pupil of Sir John Wolfe Barry and Brunel. For 2 years he was on St John's staff on the construction of Barry Docks,

then until 1900 he was on the Middlesbrough Dock extension. The Boer War interrupted his civil engineering work and he eventually retired from the army with the rank of Captain. For the rest of his career he was referred to as Captain McClean. He set up in Westminster as a consulting engineer and from 1919 onwards he formed River Flow Records. It was as an authority on the measurement of water resources that McClean was to make his chief contribution to engineering and science.

Indeed, as time went on he gave himself increasingly to the service of pure and applied science, due perhaps to the influence of his father, Frank McClean FRS, and his grandfather, J.R. McClean MP, FRS, President of the Institution of Civil Engineers from 1864 to 1865. As early as 1912 McClean had begun to demonstrate river flow measurement by means of the current meter, a technique almost unknown in Britain at the time. He began by measuring the flows of the rivers of the Great Glen of Scotland and almost entirely at his own expense he continued this work only with the interruption of World War II until he retired in 1950.

McClean continued the generous patronage of his family to both science and art and funded engineers and others to help him gauge, record and publish Scottish river flows for many years without any return, reward or recognition save the admiration and affection of his fellow engineers. There were also other pioneers of flow measurement in Britain which included such names as Allard, Clark, Crump, Inglis, Jameson, Kershaw, Sanderman and Wright, to name just a few, and who were McClean's contemporaries. Perhaps, however, McClean's most important contribution to hydrology and water resources in Britain came when he became Secretary of the British Association for the Advancement of Science's committee which prepared an appeal to the Prime Minister, Ramsay Macdonald in 1934 to set up a survey of the water resources of Great Britain. Although the memorandum of appeal was signed by the Presidents of the Institution of Civil Engineers and British Association, it is believed that preliminary informal discussions were held between the Prime Minister and McClean, who lived near each other in the north of Scotland. As a result of the appeal, the Inland Water Survey was formed in 1935, and its work is continued today by the Environment Agency in England and Wales and by the Scottish Environment and Protection Agency in Scotland.

Robert Manning (1816–1897) (Figure H41)

Robert Manning joined the Irish Office of Public Works of Ireland in 1846 as a clerk but eventually worked himself up to become Chief

Figure H41 Robert Manning (1816–1897).

Engineer responsible for planning, design and construction of harbors, navigation and drainage. He is renowned, however, for the establishment of the Manning equation for the calculation of average velocity in open channels:

$$V = \frac{1.49}{n} R^{2/3} S^{1/2} \text{ (in fps units)} \tag{H1}$$

or

$$V = \frac{1}{n} R^{2/3} S^{1/2} \text{ (in metric units)} \tag{H2}$$

In reconstructing the way in which Manning arrived at the monomial formula which carries his name, it is necessary to look behind the order of presentation in Manning's paper of 1889 and retrace the steps through a careful analysis of the text. Luckily the methodical nature of his presentation allows this to be done.

In accordance with his view of the usefulness of previous formulae as a basis of a preliminary analysis, Manning first made a comparison of a selection of formulae and commented that 'this work has occupied the author's scant leisure for more than four years'. He goes on to say in his paper of 1889:

To investigate separately each of the numerous formulae which have been published during the last century would be a tedious task, for on examination many, if not most of them, will be found to differ only in form. Seven of the best known of them were therefore selected, and for easy comparison were reduced to the same units – viz. the metre for length and the second for time. The safe assumptions were made that each author of a formula had framed it so as to reproduce as nearly as he could the experimental velocities on which it was founded, and that, therefore, by taking the mean results of all of them an approximation to the truth might be arrived at.

The seven formulae chosen by Manning were those of Du Buat (1786), Eytelwein (1814), Weisbach (1845), St Venant (1851), Neville (1860), Darcy and Bazin (1865) and Ganguillet and Kutter (1869). The exact form of the equation used in each case was given in an appendix to the paper (Manning, 1889 p. 192).

The comparison of the formulae was made by tabulating for a given slope the calculated velocity for each formula for values of R between 0.25 and 30 m. The mean of the seven calculated velocities was taken as an approximation to the truth and was tabulated and plotted. A

Table H5 Comparison of Manning's formula with those of others

Authority	Number of experiments	Number differing from formula [equation (H4)] more than 7%
Bazin	104	12
Kutter	40	5
Revy	1	0
Ftely and Stearns	15	0
Humphrey and Abbott	10	8
Total	170	25

formula was then sought 'that would closely represent the mean curve formed by the co-ordinates of R and V'. The tabulation for $S = 0.0001$ of both the calculated velocities and the ratio of the calculated velocities to the mean for the seven formulae are reproduced in his paper. The compilation of these tables, without the aid of a calculator or even a slide rule, must have been tedious and time consuming.

The first formula found by Manning to fit the mean V–R relationship was

$$V = 32 \sqrt{[RS(1 + R^{1/3})]} \tag{H3}$$

which Manning described as 'entirely empirical'. He goes on: 'The second was found on the assumption that the exponent of S was constant and equal to the square root of that function.' If then it was possible to represent the velocity by a monomial equation such as Chezy's, it should take the form

$$V = C \cdot S^{1/2} R^X \tag{H4}$$

This was found to be the case, and for the mean value of the velocities the equation was found approximately to be

$$V = 46 \, S^{1/2} R^{4/7} \tag{H5}$$

Manning comments that 'numerous empirical equations closely approximating to that curve may be found'. He was evidently satisfied that the form of equation was sufficiently in line with hydraulic principles, simple enough to calibrate and use, and sufficiently close to the mean result of experimenters to warrant a closer analysis.

After this preliminary analysis, Manning accepted the form of equation (H4) as promising and turned to an analysis of those experimental results in which he had the greatest confidence. In his own words: 'In order to obtain the value of the exponent x from a more extended range of observations, the author availed himself of Bazin's experiments on artificial channels, specially constructed for experimental purposes, with the following results.'

Mean value of $x = 0.6623$
Mean value of $C = 103$

Further comparison was made with 170 experiments, summarized in Table H5. It is interesting to report that the Manning papers are now in the keeping of Professor J.C.I. Dooge in Dublin presented to him in 1956 by Manning's great-granddaughter Mrs Pamela Smyth (née Manning).

Edmé Mariotte (1620–1684)

Mariotte did not belong to the school of water scientists which had developed in Italy towards the latter half of the seventeenth century. Nevertheless he was one of the few outstanding men of the time and arguably the most eminent hydrologist of the late seventeenth century. His work on hydraulics and hydrostatics presented to the Academy of Sciences in 1669 had a major impact on hydraulics and hydrology. He demonstrated clearly by experimental investigation that rainfall was the source of river flow and springs and also presented the first acceptable measurements of velocity in open channels. His treatise on the flow of water and other fluids was published after his death in 1686 and translated into English in 1718. It was a classic of its time.

Mariotte used interconnected floats to show that the velocity at the bottom of the river was less than that at the surface and used a mean velocity of two-thirds of the surface velocity for his calculations of discharge. He measured the discharge of the Seine at Pont Royal and found it to be 3333 ft³ s⁻¹ (94.4 m³ s⁻¹) which he said was about one-sixth of the total precipitation. Mariotte also measured the rainfall at

Dijon for 3 years to show that rainfall was adequate to supply the flow in rivers. It was this rainfall information he used in the Seine work.

For his investigations into river flow and precipitation, it could be argued that Mariotte was the founder of hydrometry.

James Mulvaney (1822–1892)

James Mulvaney was the younger brother of the Commissioner of Drainage William T. Mulvaney and has been credited rightly with the measurement of flood discharges in Ireland using self-registering rainfall and flood gauges and was probably the first hydrologist to understand the concept of time of concentration. Irish engineers were foremost in flood flow estimation during the period 1842 to 1847 and based their procedure on anticipated maximum rainfall and runoff factors. By 1847 a national flood equation as follows could be used:

$$Q = 2.52 \text{ CIA}$$

where Q is the design discharge in $ft^3 \, min^{-1}$, C is the runoff factor, I is the maximum daily rainfall (1.5–2 in) and A is the catchment area in acres.

Malvaney was the originator of the method and pointed out that to determine more accurate values of the above equation, long-period data of both rainfall and streamflow would be necessary. It may well have been the case that the dedication to flood flow estimation by Irish engineers encouraged J.E. Nash to develop the unit hydrograph in the early 1950s.

Bernard Palissy (1499–1590)? (Figure H42)

Palissy was a land surveyor and potter and perfected a technique for making enameled pottery. He was also interested in geology, botany, agriculture, chemistry, zoology, minerology and hydrology. Up to his time very few scientists or philosophers had a correct understanding of the hydrological cycle. Although it was claimed that he was a plagiarist and indeed was familiar with the work of Vitrunius and Roman predecessors, he stated categorically that rivers and springs could not have any other source than from rainfall. Being accused of contradicting the philosophers past and present did not worry him as he had confidence in his theory. He refuted the age-old belief that rivers originated directly from seawater or from air that had been converted into water. Palissy also enunciated the theory and practice of artesian wells, recharge of wells, lag and change of stage in rivers. He also considered the principle of forestation for the prevention of soil erosion and the design of 'fountains' for domestic water supply. Leonardo and Palissy were the outstanding figures of their times in the field of hydrology and water resources.

Figure H42 Bernard Palissy (1499–1590).

Figure H43 Henry de Pitot (1695–1771).

Henry de Pitot (1695–1771) (Figure H43)

Pitot became superintendent of the Canal du Midi in Languedoc in 1740 and was concerned with the design and construction of flood control works, bridges, aqueducts and land drainage. His major claim to fame, however, was in his invention of the Pitot tube. He reviewed the velocity distribution in open channel flow in a paper of 1732 and put forward the concept of frictional resistance of the bottom velocities. He deplored the use of floats in the measurements of velocities and gave his reasons which are still appropriate today. He said that all these problems could be overcome by his new instrument which had the advantage that the use of the instrument was as simple as plunging a stick into the water. The Pitot tube is still used today in both open channel and closed conduit flow measurement and an international (ISO) standard covers its use.

Plato (428–348 BC) (Figure H44)

Plato founded the Academy in Athens in 387 BC, which was a milestone in the history of science and philosophy. He accepted the four basic elements of matter, fire, air, water and earth as first postulated by the Greek philosopher Empedocles of Agrigentum, and added a fifth which was taken to be 'heaven'. Plato, in his explanation on the origin of rivers and springs, maintained that under the Tartarus concept there were numerous interconnected perforations and passages in the Earth's interior. He imagined a huge subterranean reservoir which he called Tartarus. Socrates credited Plato with describing the first pluvial concept of the hydrological cycle. Plato in his concept of the flow of water explained that from the Tartarus the water flowed by a circuitous route forming rivers, lakes, seas and springs and back to Tartarus.

Plato was also interested in water law. Anyone was permitted to draw water from a common stream on his land as long as it did not cut off the flow of a private stream. If individuals intentionally polluted or wasted water of a stream or reservoir, they would be required to pay damages equal to the value of the loss.

Pierre Perrault (1608–1680)

Along with Mariotte and Halley, Perrault (q.v.) was arguably recognized as one of the founders of the science of hydrology as we know it today. His book on the origin of springs was a classic of the time and in it he went into great detail to prove the concept of Aristotle (and others) of the hydrological cycle to be incorrect. Perrault showed by his study of the Seine that there was sufficient

Figure H44 Plato.

Figure H45 William Gunn Price (1853–1928).

annual rain to cause the river to flow for 1 year and 'serve losses such as feeding of trees, plants, grasses and evaporation'. He concluded that the concept applied to all rivers of the world and that more hydrometric measurements were needed to prove his theory, but according to his measurements (in units of pouces, muids and toises) only about one-sixth of the rain that fell found its way into the rivers.

Marquis Giovanni Poleni (1683–1761)

Poleni became professor of astronomy at Padua at the age of 26 and later professor of physics and mathematics. He served in the field of flood control and water engineering. In his treatise of 1717 he analyzed the flow of water through a rectangular opening which extended to a free surface (thinplate weir), and found the discharge equation:

$$Q = \frac{2}{3}Cb \sqrt{(2g)}h^{3/2}$$

where C is the discharge coefficient, b is the width, g is the acceleration due to gravity and h is the total head.

Robert Plot (1640–1696)

Plot was a Fellow of the Royal Society of London and contributed investigations into the origin of springs and rivers and was perhaps the first to prepare a classification of rivers. Nevertheless he was a member of the school who believed that salt water ascended to the mountain tops through capillary action, ridding it of salt in the process.

William Gunn Price (1853–1928) (Figure H45)

Between 1879 and 1896 Price was an assistant engineer with the Mississippi River Commission engaged in measuring flows of the Mississippi, Ohio and Missouri Rivers. While working in these rivers he conceived the design of his first current meter. He had, in the past, used the 'Herschel' meter, having a horizontal shaft in these rivers, but in January 1882 wrote to Nathan C. Grover then Chief Hydraulic Engineer to the USGS (see Grover above) stating that he was dissatisfied with the meter since there was no means for excluding water from its bearings. Also the Ellis meter, apparently the only other meter available, had the same problem with the vertical shaft bearings. He thereupon decided to develop his own current meter by using inverted cup bearings. By employing four mechanics, 'the best mechanics in town', the meter was 'completed next day' and used to measure the great Mississippi flood of that year (1882). The original

meter of 1882 is now in the Smithsonian Institution. Price was probably the foremost US authority on current meters. The Geological Survey has probably owned more current meters than any other organization in the world and over 95% of them have been of the Price type.

John Ray (1627–1705)

Ray was probably the first to move from the concept of the subterranean abyss of water storage to an almost correct theory of the hydrological cycle. He believed that the Sun attracted 'vapors' from the earth and sea and that the wind was responsible for driving the vapors from the sea toward the land where the vapor fell as rain. The rivers derived their supply of water from rain and a vast quantity of flow was carried down into the sea, and the circulation of water was thus completed.

Osborne Reynolds (1842–1912)

Reynolds was a physicist and chemist and gave his name to the Reynolds number (q.v.) proposed by him in 1883 for characterizing the type of flow in a pipe or conduit flowing full where the resistance to motion depends on the viscosity of the liquid and the influence of inertia. It is the product of the mean velocity and the diameter or depth of flow divided by the kinematic viscosity of the liquid.

Santorio Santorio (1561–1630)

During the eighteenth century, velocity measurement devices have been credited to Henry Pitot and Reinhard Woltman, but the earliest current meter was actually developed by the Italian physician Santorio. He designed it, however, for medical reasons in order to lull his patients to sleep by the gentle soft noise of quietly falling water. He very soon realized, however, that it had other uses, such as for the measurement of the velocity of water, but he made no effort to convert the data to velocities, the water impacting on a flat plate, no doubt making the conversion difficult.

Joseph Saxton (1799–1873)

Saxton was the first and outstanding American to develop a current meter. He designed it in London while working in the Adelaide Gallery (in the Strand) for Joseph Cubitt, son of the prominent civil engineer William Cubitt (1785–1861). In 1836 the gallery published details of the meter (Figure H46).

Saxton made field measurements with the meter in 1834 in the River Colne at Denham Point in accordance with a plan to improve

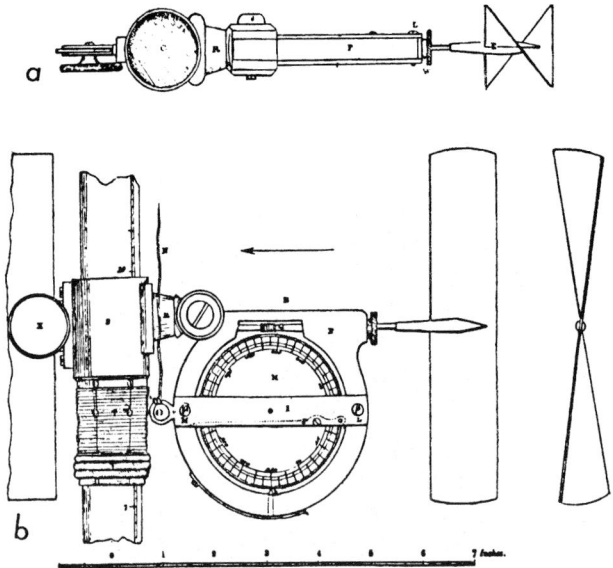

Figure H46 Joseph Saxton's water current meter: (a) top view and (b) side and end views. (From Adelaide Gallery, *Description of the Current Meter.*)

London's water supply. The meter was illustrated in Elliot's catalogue in the 1850s together with the following description: 'Elliot's (i.e. Saxton's) current meter is a most useful instrument for ascertaining velocities either a few inches below the surface, at the bottom, or any depth between'. The Proprietors of Locks and Canals of Lowell, Massachusetts, were among those who acquired one of Elliot's models of Saxton's meters and in 1956 they contributed it to the Smithsonian Institution.

Saxton's current meter was the only current meter used in Humphreys and Abbot's report on the Mississippi River. Over 99% of the other observations were made by double floats. In actual fact, however, current meters were not accepted without controversy even by hydraulic engineers, and Humphreys and Abbot stated that double floats were superior to any device for measuring velocities.

Gaius Plinius Secundus (AD 27–79)

Pliny drew his views on meteorology (and hydrology) from the writings of Lucretius and Seneca and was confident that the freezing of rain was the cause of both hail and snow. His thoughts on the origin of springs and rivers were also similar to those of Lucretius, Seneca and contemporaries.

Lucius Annaeous Seneca (4 BC–AD 65) (Figure H47)

Seneca, although Spanish, carried out most of his work in Rome and is best known for his treatise *Quaestiones naturales* which was of major interest to hydrology. In it he discusses hail and snow. Seneca also considered the inundations of the Nile and presented some thoughts on the origin of rivers and springs and underground water.

John Smeaton (1724–1792)

Smeaton, the builder of the famous Eddystone lighthouse, was one of the major contributors to water science in England. He was responsible for the design of various harbors and drainage works and carried out extensive tests on scale models particularly on water wheels and windmills. His current meter consisted of a rotor designed by himself and which is now in the Smithsonian Institution.

William Smith (1769–1839)

Smith is often credited with being the father of English geology, his contribution to hydrology being in groundwater, although his interest also lay in civil engineering and water resources in particular. He was directly involved in the water supply to Scarborough which involved both groundwater abstraction and open channel flow and in the design of canals and locks.

Figure H47 Lucius Annaeus Seneca (4 BC–AD 65).

George James Symons (1838–1900)

Symons spent some 40 years of his life coordinating the rainfall observations throughout Britain and it was because of his efforts that the first volume of *English Rainfall* was published in 1860–1861. It has been published regularly ever since as *British Rainfall*. Hydrologists owe their gratitude to Symons for his rationalization of rainfall records. Later Glasspoole (1897–1981) continued the work started by Symons.

Thomas Telford (1757–1834)

Telford, the father of civil engineering in Britain, was born near Langholm in Scotland. His father was a shepherd who died a few months after Telford was born. Telford left school at the age of 14, where he could only have learned the three Rs, to become a herdsman in the valley of the River Esk. However, his intelligence and love of reading attracted notice. He became an apprentice stone mason in Langholm and his 'mark' can still be seen under the bridge over the Esk in that border town. He moved to Edinburgh, where he worked on the building of the new town on the north side of Princes Street, and made a study of architecture before moving to London. There he was employed as a hewer in the building of Somerset House, later moving to Portsmouth where, as foreman, he supervised the building of the substantial Commissioner's house in Portsmouth Dockyard. It was there that he first became interested in docks and harbors. It was not until a few years later, however, that he became interested in water engineering when he was made Surveyor of Public Works for Shropshire and then engineer and architect for the Ellesmere Canal to link the rivers Mersey, Dee and Severn. His achievements in building the canal were unparalleled in engineering and marked by great originality. The two aqueducts at Chirk and Pont-y-Cysylte were, and still are, works of art, the latter having 19 arches over a length of some 300 m with the aqueduct in a cast iron trough 3.5 m wide at a height over the River Dee of 36 m (Figure H48). The construction started in 1795 and the canal was open to traffic in 1805. Later canals included the Göta canal linking the Baltic with the North Sea, a canal distance of 88 km but a waterway length of 193 km. For his work on the Göta Canal, Telford received a knighthood from the King of Sweden. He also designed and built the Caledonian Canal in Scotland, the canal having a length of 96 km with 28 locks. Apart from his network of canals, still in use today mainly for pleasure craft and water conveyance, during his lifetime Telford constructed 1200 bridges, many hundreds of kilometers of roads, and many docks and harbors.

However, as great a contribution as any that Telford made to engineering was his fostering of the Institution of Civil Engineers and agreeing to become the first President in 1820. It was through his influence that a Royal Charter was granted by George IV in 1828. Telford died in 1834 at 24 Abingdon Street, Westminster, and had the unique distinction of being buried in Westminster Abbey. A few years ago the new town of Telford was established in Shropshire, the county where he had his first commission as a civil engineer.

Figure H48 Pont-y-Cysylte aqueduct over the Dee at Llangollen, UK.

Thales of Miletos (624–584 BC)

Thales was a hydrological philosopher fascinated by the regular inundation of the River Nile. Measurements of Nile floods, however, were made at least 1000 years earlier, because agricultural taxes were based on them. Some of the high water marks were cut into the bedrock at the first and second cataracts. Regular annual flood figures have been kept since the beginning of the Muslim era in 612 AD. Thales believed in the common belief that Egypt had been created by the Nile. He was the first Greek astronomer and mathematician and acknowledged as one of the Seven Wise Men of the ancient times. Thales believed as others did that the Earth floated on water like a flat disc and that water was the primary matter.

Theophrastus (370–287 BC)?

The original work of Theophrastus on meteorology is lost, although an abstract exists in Arabic which seems to indicate that his concepts of the hydrological cycle followed somewhat the convention of the time. It can be argued, however, that he later recognized that the concept did not answer all the questions of the cycle.

It is perhaps interesting to note that around this time the first measurements of rainfall were being made in India by Kautilya, who was Chancellor of Exchequer and who based land tax on the amount of precipitation the land received.

Adolph Thiem (1836–1908)

Probably the most notable contribution to groundwater in Germany was made by Adolph Thiem. Thiem was City Engineer of Dresden and made major theoretical analysis of problems causing the flow towards gravity wells, artesian wells and filter galleries. By making the necessary assumptions, he derived the same expressions as Dupuit had derived for gravity and artesian wells. In his paper published in 1870, a classic in the field of groundwater, he also considered partially penetrating gravity wells and non-steady seepage. He also attempted to measure the velocity of groundwater flow.

Lucretius Carus Titus (96–55 BC)?

Lucretius believed that moisture rose from everything, especially from the sea, and when the winds blew they drove the clouds across the sea. The clouds picked up the moisture and lost their water content in the form of rain. However, he did not agree that rivers originated from rain but from the filtration of seawater. He also had views on the origin and rise of the Nile.

Evangelista Torricelli (1608–1647)

It was Torricelli who first stated that velocity was related to head and presented for the first time the equation $V = \surd(2gh)$. However, he was not able to evaluate g and it was left to Christian Huygens (1629–1695) to determine g in 1673. The equation was not used, however, until the two Bernoullis (father and son) did so in 1748. Others, however, worked on the velocity equation including Maggiotti, MacLaurin, Poleni, Newton, Guglielmini, Grandi, Mariotte and Michelotti.

Richard Townley (1629–1707)

Townley is credited with making the first continuous measurements of rainfall in Britain, starting in 1677. The gauge was fixed to the roof of his house, Townley Hall in Lancashire.

Antonio Valisnieri (1661–1730)

Valisnieri was president of the University of Padua and published a treatise in 1715 on the origin of rivers based on his personal observations in the Alps. He was one of the outstanding figures of hydrology of his time and found no indication of seawater being forced out of the mountain tops, a common belief of the time. He concluded, as a result of his observations, that melting snow and rainfall provided the necessary supply of water to all springs, lakes and rivers. He also discovered, with the aid of local shepherds, that hidden subterranean channels took water down to Modena, which became the source of the famous Modena artesian wells which had received considerable speculation in the past.

Bernhardus Varenius (1622–1650)

Varenius was probably the most famous geographer of his time and influenced geography for more than a century. His concept of the origin of rivers and streams was a compromise of those of predecessors and contemporaries. He assumed that the seawater hypothesis was correct – rivers discharge into the sea, and since the sea does not increase in volume, the seawater has to be carried to the source of the rivers. He maintained that salt springs were supplied by conduits from the sea.

Giovanni Battista Venturi (1746–1822) (Figure H49)

Venturi was professor of natural philosophy at the University of Modena and later at Pavia and was a noted civil engineer. His hydraulic experiments on the motion of fluids was carried out in the physical laboratory of the University of Modena. The results were

Figure H51 Vitruvius' method for locating water.

Figure H49 Giovanni Battista Venturi (1764–1822).

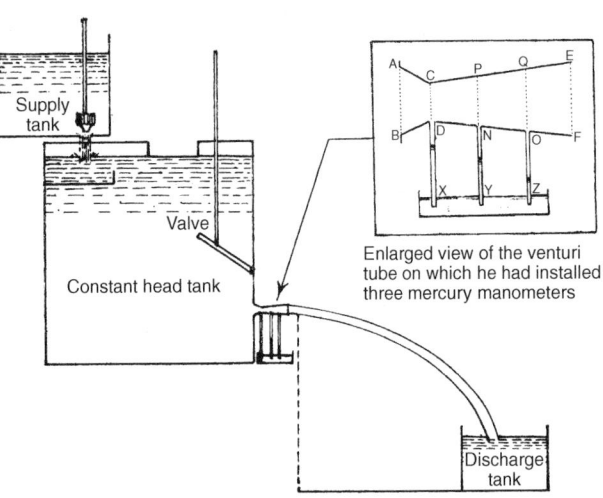

Figure H50 Schematic representation of equipment used by Giovanni Battista Venturi to test how rapidly the various types of outlets would permit filling of the lowermost (discharge) tank. The 'winner' (insert) was ultimately known as the Venturi tube. (Adapted from Venturi, *Experimental Inquiries*.)

published in Paris in 1797. His investigations included the use of various types of measuring structure in open channels to determine the discharge, one of which was a flume which later became known, through Clemens Herschel, as the Venturi flume and Venturi meter (Figure H50).

Vitruvius (first century BC)

Born in Italy, his treatise on architecture written between 27 and 17 BC had a profound effect on classical architecture. It is the only surviving Roman work on architectural theory and a major influence on Renaissance architects. Book 8 of his treatise is devoted to water

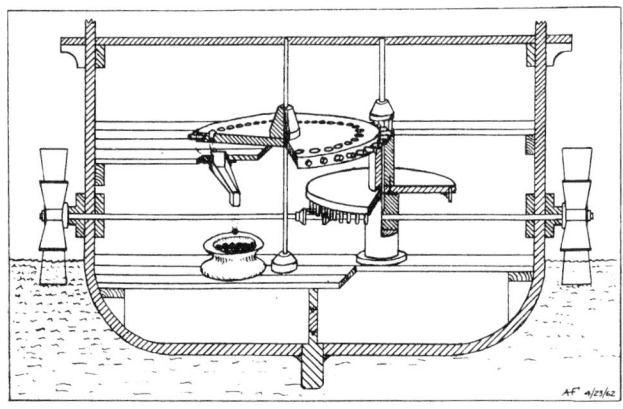

Figure H52 World's first mechanical ship's log as described by Vitruvius in the first century BC. (Drawn by A.H. Frazier.)

and ways of finding water and states that where surface water was not available, underground sources should be considered (Figure H51). He also indicated how a groundwater well should be designed and dug.

He was familiar with the meteorological writings of Aristotle and Theophrastus and had a clear concept of the hydrological cycle. He too believed that only the thinnest, or lightest, water was evaporated and the heaviest was left behind.

His treatise also dealt with aqueducts, pipes, wells and cisterns. Like other Romans, however, he was totally unaware that discharge depended on both velocity and cross-sectional area of flow, and it was general practice to calculate discharge from cross-sectional area only. He erred in his view about the source of the Nile, but the Nile was a constant wonder to the people of the ancient civilization and they kept speculating about its source and regularity of its inundation.

Although the current meter had to wait more than 1400 years to be invented, and in any case velocity of flow was not included in the discharge equation, Vitruvius invented a ships' log to determine the distance a ship had traveled on the sea (Figure H52). The final gear, as shown in Figure H52 had a series of holes through it and at regular

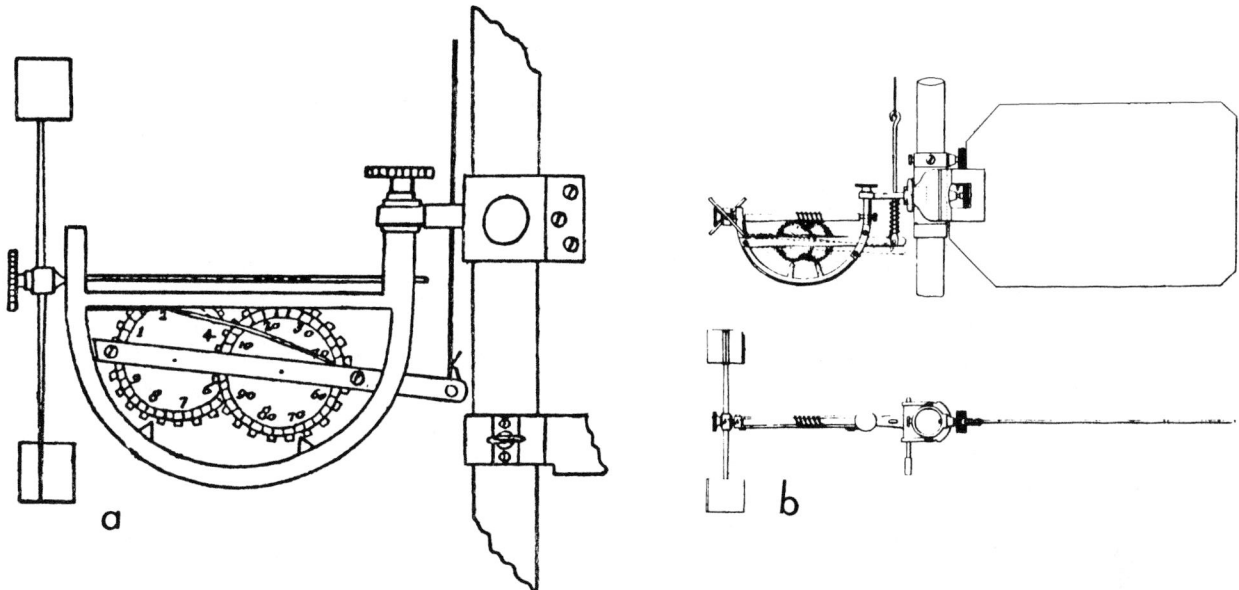

Figure H53 Woltman current meters. (a) Venturoli's version, 1818. (Redrawn by A.H. Frazier after Venturoli, *Element di Mechanica e Hydraulica*.) (b) Baumgarten's drawing, 1847. (From Baumgarten, *Sur le moulinet de Woltmann*, NMHT 314769, Smithsonian photo 73–777.)

intervals of time a single round stone would be placed in each of the holes. The gear reduction was so great that the ship had to travel a distance of 1 mile (1.6 km) to cause each successive stone to drop into a bronze container on the lower deck. Each mile of travel would be announced by a loud clang as the stone would strike the container and by counting the number of stones, the captain could determine the number of miles that the ship had traveled.

Reinhard Woltman (1757–1837)

Woltman was in charge of hydraulic operations at Cuhaven, Germany. He developed his current meter in 1790 and rated it successfully in still water (Figure H53). His book on the hydrometric current meter was published in 1790.

Sir Christopher Wren (1632–1723)

Wren is credited with the development of the first English raingauge which was of the recording type, although evidence is lacking as to its actual use in the field. Wren's second automatic raingauge was jointly developed with Robert Hooke. The gauge was a tipping bucket type provided with recording facilities.

R.W. Herschy

Sources

Biswas, A.K., 1970. *History of Hydrology*, North-Holland, Amsterdam

Frazier, A.H., 1974. Water current meters: in the Smithsonian collection of the National Museum of History and Technology, Smithsonian Institution, Washington, DC.

Chow, Ven Te, 1964. *Handbook of Applied Hydrology* McGraw-Hill, New York.

Bibliography

Adams, F.D., 1928. The origin of springs and rivers. *Fennia*, **50**.

Agricola, G., 1546. *De ortu et causis subterraneorum* lib. V. Basilae, H. Frobenium and N. Episcopium.

Agricola, G., 1950. *De re metalica*, translated by H.C. Hoover and L.H. Hoover. New York, Dover Publications Inc.

Alberti, L.B., 1955. *Ten books on architecture*, translated by J. Leoni. London, Alec Tiranti Ltd.

Anonymous, 1872–1873. Nathaniel Beardmore, Memoir. *Proceedings of the Institution of Civil Engineers*, **32**.

Aristotle, 1259. A, *Politics*, Book 1.

Aristotle, *De caelo* (On the heavens). Book 11, 294a.

Aristotle, *Metaphysica* (Metaphysics). Book 1, Chapter 3, 383b.

Aristotle, *Meteorologica*, Book 2, chapter 1.

Aristotle, 1952. *Meteorologica*, translated by E.W. Webster. Chicago, Encyclopaedia Britannica Inc.

Aristotle, 1952. *Degeneratione et corruptione*, translated by H.H. Joachim. *Great Books of the Western World*, Vol 8. Chicago, Encyclopaedia Britannica Inc.

Armitage, A., 1966. *Edmond Halley*. London, Thomas Nelson and Sons.

Bazin, H.E., 1898. *Etude d'une nouvelle formule pour calculer le débit des canaux découverts*. Paris, P. Vicq-Dunod (extract of *Annales des Ponts et Chaussées*, 1897).

Beardmore, N., 1862. *Manual of Hydrology*. London, Watelow & Sons.

Bernoulli, D., 1738. *Hydrodynamica*. Argentorati, J.R. Dulseckeri.

Biswas, A.K., 1965. Leonardo da Vinci and the hydrology cycle. *Civil Engineering, ASCE*, **35**.

Biswas, A.K., 1965. In defence of Plato. *Engineering*, **200**.

Biswas, A.K., 1965. Discussion of: Ground water management for the nation's future – ground water basin management. *Journal of the Hydraulics Division, ASCE*, **91**.

Biswas, A.K., 1966. The Nile: its origin and rise. Water and Sewage Works, **113**.

Biswas, A.K., 1967. The automatic rain gauge of Sir Christopher Wren. *Notes and Records of the Royal Society of London*, **22**.

Biswas, A.K., 1967. Hydrologic engineering prior to 600 BC. *Journal of Hydraulics Division, ASCE*, **93**. Discussions by G. Garbrecht.

Biswas, A.K., 1967. Hydrology during the Roman Civilisation. *Water and Sewage Works*, **114**.

Biswas, A.K., 1967. Development of rain gauges. *Journal of Irrigation and Drainage Division, ASCE*, **93**.

Biswas, A.K., 1967. Hydrology during the Hellenic Civilisation. *Bulletin, International Association of Scientific Hydrology*, **12**.

Biswas, A.K., 1969. A short history of hydrology. *Proceedings of the International Seminar for Hydrology Professors*, University of Illinois, Urbana.

Biswas, A.K., 1969. Experiments on atmospheric evaporation till the end of the eighteenth century. *Technology and Culture*, **10**.

Biswas, A.K., 1969. Development of hydrology in the nineteenth century. *Water Power*, **21**.

Biswas, A.K., 1969. Atmospheric evaporation: development of concept and measurement up to the end of the 18th century. *Technology and Culture*, **10**.

Biswas, A.K., 1970. *History of Hydrology*, North Holland, Amsterdam.

Botley, C.M., 1935. A founder of English meteorology. *Quarterly Journal of the Royal Meteorological Society*, **61**.

Burnet, J., 1914. *Greek philosophy: Thales to Plato*, Part 1, 1st edn. London, Macmillan & Co.

Castelli, B., 1628. *Della misura dell'acque correnti*. Roma, Nella Stamparia Camerale.

Castelli, B., 1661. *Of the mensuration of running waters*, translated by T. Salusbury. London, W. Leybourn.

Chapman, S., 1957. Edmond Halley, F.R.S. *Notes and Records of the Royal Society*, London, **12**.

Dalton, J., 1802. Meteorological observations. *Memoirs, Literary and Philosophical Society of Manchester*, **5**(2).

Darcy, H.P.G., 1856. *Les fontaines publiques de la ville de Dijon*. Paris, V. Dalmont.

Darcy, H.P.G. and H.E. Bazin, 1856. Recherches hydraulics. Paris.

Darmstaeder, L., 1927. The life of Edmé Mariotte. *Journal of Chemical Education*, **4**.

De Columbo, C.A., 1800. Expériences destinées a déterminer la cohérance des fluides et les lois de leur résistance dans les mouvements très lents. *Mémoires de l'Institut National des Sciences et Arts*, **3**.

De Frense, R., 1955. The life of Leone Battista Alberti in *Ten books of architecture*, translated by J. Leoni. London, Alec Tiranti Ltd.

De Lorenzo, G., 1920. *Leonardo da Vinci e la geologa*. Bologna.

Delorme, S., 1948. Piere Perrault, auteur d'un traité De l'origine des fontaines et d'une théorie de l'expérimentation. *Archives Internationales d'Histoire des Sciences*, **27**.

De Pitot, H., 1732. Description d'une machine pour mesurer la vitesse des eaux courantes et le sillage des vaisseuax. *Memoires de l'Académie Royale des Sciences*.

Derham, W., 1726. *Philosophical experiments and observations of the late eminent Robert Hooke*, London, W. and J. Innys.

De Selincourt, A., 1962. *The world of Herodotus*. London, Secker and Warburg.

De St Venant, J.C.B., 1851. Formules et tables nouvelles pour la solution de prolémes realtifs aux eaux courantes. *Anales des Mines*.

Dobson, D., 1777. Observations on the annual evaporation at Liverpool in Lancashire; and on evaporation considered as a test of the dryness of the atmosphere. *Philosophical Transactions of the Royal Soceity of London*, **67**.

Dooge, J.C.I., 1957. The rational method for estimating flood peaks. *Engineering*, **184**.

Dooge, J.C.I., 1959. Quantative hydrology in the 17th Century. *La Houille Blanche*, No. 6, 799–807 (in English and French).

Dooge, J.C.I., 1987. Manning and Mulvany, river improvement in 19th century Ireland, in *Hydraulics and hydraulic research, a historical review*, IAHR, Balkema, Rotterdam.

Dooge, J.C.I., 1989. The Manning formula in context, International Conference on Channel Flow and Catchment Runoff, Univ. Virginia, *Proc. Int. Conf. for Centennial of Manning's formula and Küchling's rational formula*, virginia, USA.

Dooge, J.C.I., 1996. *Robert Manning (1816–1897)*, University of Cork, Ireland.

Du Buat, P.L.G., 1786. *Principes d'hydraulique*, 2 vols, new edn. Paris.

Du Buat, P.L.G., 1822. *Chevalier Du Buat's principles of hydraulics*, translated by T.F. De Havilland, 2 vols. Madras, Asylum Press.

Duhem, P., 1906. *Etudes sur Léonard de Vinci*, vol. 1. Paris, A. Hermann.

Ellis, A.J., 1917. *The diving rod – a history of water witching*. Water-Supply Paper 416, US Geological Survey, US Government Printing Office, Washington, DC.

Fontana, G., 1640. *Dell' accrescimanto che hanno fatto li fiumi, torrenti, e fiossi che hanno causato l'inondation à Roma il natale*, reprinted at recommendation of Sig. Domenico Castelli. Gioiosi, Appresso Antonia Maria.

Faraday, M., 1832. *Philosophical Transactions of the Royal Society of London*, p. 75.

Fontana, G., 1696. *Utilissimo trattato dell'acque corrent*. Roma, G.F. Buagni.

Frazier, A.H., 1967. *William Gunn Price and the Price current meters*. Contributions from the Museum of History and Technology, Paper 70, Smithsonian Institution, Washington, DC.

Frazier, A.H., 1969. Robert Hooke and the Hooke current meter. *Journal of the Hydraulics Division, ASCE*, **95**.

Frazier, A.H., 1974. Water current meters in the Smithsonian collections of the National Museum of History and Technology, Smithsonian Institution, Washington, DC.

Frisi, P., 1818. *A treatise on rivers and torrents*, translation by J. Garstin. London, Longmans, Hurst, Rees, Orme and Brown.

Frisinger, H.H., 1965. Early thoeries on the Nile floods. Weather, **20**.

Ganguillet, E.O. and W.R. Kutter, 1877. *Versuch zur Aufstellung einer neuen allgemeinen Formel für die gleichförmige Bewegung des Wassers in Kanälen und Flüssen*. Bern, Druck von Lang.

Garbrecht, G., 1987. *Hydraulics and hydraulic research; a historical review (Ed)*. Rotterdam, Balkema.

Grossman, E., 1929. Frontinus, the water commissioner. *Journal of the Boston Society of Civil Engineers*, **16**.

Grover, N.C., 1938. Progress in branch activities to June 30, 1906. US Washington, DC, Geological Survey.

Guglielmini, D., 1690. *Aquarum fluentium mensura nova methodo inquisita*. Bononiae.

Guglielmimi, D., 1697. *Della natura de'fiumi, tratto fisico-mathematico*. Bolgna.

Guglielmimi, D., 1719. *Opera omnia mathematica, hydraulica, medica et physica*, Vol. 1, edited by J.B. Morgagni. Ginerva.

Halley, E., 1687. An estimate of the quantity of vapour raised out of the sea by the warmth of the sun. *Philosophical Transactions of the Royal Society of London*, **16**.

Halley, E., 1691. An account of the circulation of watry vapours of the sea, and of the cause of springs. *Philosophical Transactions of the Royal Society of London*, **16**.

Halley, E., 1694. An account of the evaporation of water. *Philosophical Transactions of the Royal Society of London*, **18**.

Halley, E., 1715. A short account of the cause of the saltness of the ocean. *Philosophical Transactions of the Royal Society of London*, **29**.

Hart, I.B., 1961. *The world of Leonardo da Vinci*. London, Macdonald & Co.

Heidel, W.A., 1921. Anaximander's book. The earliest known geographical treatise. *Proceedings of the American Academy of Arts, Sciences and Letters*, **56**.

Hellmann, G., 1890. *Die Anfäge der meteorologischen Beobachtungen und Instrumente*. Berlin, Druck von Wilhelm Gronau, 1024 pp. Also published in *Himmel und Erde*, vols 2–4.

Hellmann, G., 1901. Die Entwicklung der meteorologischen Beobachtungen bis zum Ende des XVII. Jahrhunderts. *Meteorlogische Zeitschrift*.

Hellmann, G., 1908. The dawn of meteorology. *Quarterly Journal of the Royal Meteorological Society*, **34**.

Herodotus, 1952. *The history of Herodotus*, translated by George Rawlinson. *Great Books of the Western World*, vol 6, book. Chicago, Encyclopaedia Britannica.

Herschel, C., 1887. The Venturi water meter: an instrument making use of a new method of gauging water. *Transactions, ASCE*, **17**.

Herschel, C., 1897. On the origin of the Chézy formula. *Journal of the Association of Engineering Societies*, **18**.

Herschel, C., 1899. *Frontinus and the water supply system of the city of Rome*. Boston, Dana Estes & Co.

Herschel, C., 1929. A farewell word on the Venturi meter. *Engineering News Record*, **102**, 636–637.

Hippocrates, 1952. *Hippocrates' writings*, translated by Francis Adams. *Great Books of the Western World*, Vol. 10. Chicago, Encyclopaedia Britannica Inc.

Hooke, R., 1695. An account of the quantities of rain fallen in one year in Gresham College, London, begun August 12 1695. *Philosophical Transactions of the Royal Society of London*, **19**.

Horton, R.E., 1919. The measurement of rainfall and snow. *Journal of the New England Water Works Association*, **33**.

Humphreys, A.A. and H.L. Abbot, 1861. *Report upon the physics and hydraulics of the Mississippi river*. Philadelphia.

Kent, W.G., 1912. *An appreciation of the two great works in hydraulics: Giovanni Batista Venturi and Clemens Herschel*. London, Blades East Blades.

Kirby, R.S., 1939. Henry de Pitot, pioneer in practical hydraulics. *Civil Engineering, ASCE*, **9**.

Kircher, A., 1656. *Itinerarium exstaticum, quo mundi opificium*, typis V. Mascardi Romae.

Kircher, A., 1665. *Mundus subterraneus*. Amstelodami, Apud J. Janssonium et E. Weyerstraten.

Kolupaila, S., 1933. Die bestimmung de Abflüsses des Memelströmes, Nemunas (1812–1932). IV Hydrologische Konferenz der Baltiscen Staten, Leningrad, September.

Kolupaila, S., 1933. Early history of hydrometry in the United States. *Journal of the Hydraulics division, ASCE*, **86**.

Kolupaila, S., 1961. *Bibliography of hydrometry*. Notre Dame, Indiana, University of Notre Dame Press.

Leliacsky, S., 1951. Historic development of the theory of the flow of water in the canals and rivers, no. 1. The Engineer, **91**.

Leonardo Da Vinci, 1643. *Manuscript volume*, edited by L.M. Arconati. Vatican Library. Later published as *De moto e misura dell'acqua*, E. Carusi and A. Favaro, Publicazioni dello Istituto de Studii Vinciani, Nuova Serie, Vol. 1. Bologna, 1923.

Leonardo Da Vinci, 1826. *Del moto e misura dell'acqua*, edited by Franceso Cardinali. In: Raccolta d'autori Italiani che trattano del moto del'acqua. Bologna.

Leonardo Da Vinci, 1939. The literary works of Leonardo da Vinci, ed J.P. Richer, Vol. 2. Oxford, University Press.

Leonardo Da Vinci, 1956. *The notebooks of Leonardo da Vinci*, ed. E. MacCurdy, Vol. 1. London, Jonathan Cape.

Leroux, D., 1927. *La Vie de Bernard Palissy*. Paris, Librairie Ancienne Honoré Champion.

Levie, Eenzo., 1995. *The science of water: the foundation of modern hydraulics*. Translated by Daniel E. Medina, ASCE, 649 pp.

Lucretus Carus, Titus, 1947. *De rerum natura*, Vol. 3, ed. C. Bailey. Oxford, Clarendon Press.

Maccurdy, E., 1956. *The notebooks of Leonardo da Vinci*, Vol. 2, new ed. London, Jonathan Cape.

Manley, G., 1968. Dalton's accomplishment in meteorology, in John Dalton and the progress of science, ed. D.S.L. Cardwell. Manchester, University Press.

Manning, R., 1851. Observations on subjects conected with arterial drainage. *Proceedings of the Institution of Civil Engineers of Ireland*, May.

Manning, R., 1890. On the flow of water in open channels and pipes. *Proceedings of the Institution of Civil Engineers of Ireland*, **20**.

Mariotte, E., 1686. *Philosophical Transactions of the Royal Society of London*, **16**.

Mariotte, E., 1686. *Traité du mouvement des eaux et des autres corps fluides*. Paris, E. Michallet.

Mariotte, E., 1718. *A treatise of the motion of water and other fluids with the origin of springs and cause of winds*, translation by J.T. Desaguliers. London.

Meinzer, O.E., 1934. The history and development of ground water hydrology. *Journal of the Washington Academy of Sciences*, **24**.

Meinzer, O.E., 1942. *Hydrology. Physics of the earth*, vol IX. New York, McGraw-Hill.

Morley, H., 1855. *The life of Bernard Palissey, of Saintes*, 2nd edn. London, Chapman & Hall.

Mouret, G., 1921. Antoine Chézy, histoire d'une formule d'hydraulique. *Annales des Ponts et Chaussées II*.

Mulvany, T.J., 1850–1851. On the use of self-registering rain and flood gauges in making observations of the relations of rainfall and flood discharges in a given catchment. *Proceedings of the Institution of Civil Engineers of Ireland*, **4**.

Palissy, B., 1876. *Resources: a treatise on water and springs*, translated by E.E. Willett. Brighton, D. O'Connor.

Palissy, B., 1580. *Discours admirable*. Paris, Martin Le Juene.

Palissy, B., 1957. *The admirable discourses*, translated by A. la Rocque. Urbana, University of Illinois Press.

Parker, P.A.M., 1949. *The control of water*. London, Routledge & Kegan Paul.

Partsch, J., 1909. Das Aristotles Buch 'Uber das steigen des Nil'. *Abhandlungen der Philologisch-Historischen Klasse de Königlichen Sächsischen Gesellschaft der Wissenschaften*. Leipzig, B.G. Teuber.

Perrault, P., 1678. *De l'origine des fontaines*. Paris, Pierre Le Petit.

Perrault, P., 1967. *Origin of fountains*, translated by A. La Rocque. New York, Hafner Publishing Co.

Plato, 1804. *The Critias or Altanticus. The works of Plato*, vol. 2, translated by T. Taylor and F. Sydenham, printed for T. Taylor by R. Wilks. London, Chancery-Lane.

Plato, 1892. *Critias*, translated by B. Jowett, Vol. 3, 3rd edn. Oxford, Oxford University Press.

Plato, 1892. *Phaedo. Dialogues of Plato*, translated by B. Jowett, Vol. 2, 3rd edn. Oxford, Oxford University Press.

Plato, 1892. *Timaeus*, translated by B. Jowett, Vol. 3, 3rd edn. Oxford, Oxford University Press.

Plato, 1916. *Critias. The works of Plato*, Vol. 2, translated by H. Davis. London, G. Bell and Sons.

Plato, 1929. *Critias*, translated by R.G. Bury. Loeb Classical Library. London, William Heinemann.

Plato, 1952. *Timaeus* in *The dialogues of Plato and the seventh letter*, translated by B. Jowett. Chicago, Encyclopaedia Brittanica Inc.

Pliny the Elder, 1855. *The natural history of Pliny*, translated by John Bostock and H.T. Riley. Bohn's Classical Library. London, Henry C. Bohn.

Plot, R., 1685. *De origine fontium tentamen philosophicum*. Oxonii.

Plot, R., 1686. *The natural history of Staffordshire*. Oxford, The Theatre.

Powell, R.W., 1968. The origin of Manning formula. *Journal of the Hydraulics Division*, ASCE, **94**.

Ray, J., 1691. *The wisdom of God manifested in the works of creation*. London.

Reymond, A., 1927. *Science in Greco-Roman antiquity*, translated by R. Gheury de Bray. London, Methuen & Co.

Rose, V., 1893. *Aristotle's pseudepigraphus*. Lipsiae.

Rouse, H. and S. Ince, 1957. *History of hydraulics*. Iowa City, Iowa Institute of Hydraulic Research.

Sarton, G., 1927. *Introduction to the history of science, from Homer to Omar Khayyam*, Vol. 1. Baltimore, William and Wilkins, pp. 135–136.

Sarton, G., 1953. A *history of science, ancient science through the Golden Age of Greece*. Cambridge, Harvard University Press.

Sheppard, T., 1917. Wiliam Smith; his maps and memoirs. *Proceedings of the Yorkshire Geographical Society*, **19**, 189–191 (published as a book in 1920 by A. Brown, Hull).

Smith, W., 1827. On retaining water in the rocks for summer use. *Philosophical Magazine, New Series* **1**.

Symons, G.J., 1891. A contribution to the history of rain gauges. *Quarterly Journal of the Royal Meteorological Society* **17**.

Symons, G.J., 1866. On the rainfall of the British Isles. *Report of the 35th Meeting of the British Association for the Advancement of Science*, Birmingham, Sept. 1865. London, John Murray.

Taylor, A.E., 1960. *Plato the man and his work*, 7th edn. London, Methuen and Co.

Thiem, A., 1881. Beitrtag zur Kenntnis der Grundwasserverhältnisse im nord-deutschen Tieflande. *Journal für Gasbeleuchtung and Wasserversorgung*, **24**.

Thiem, A., 1880. Der Versuchsbrunnen für die Wasserversorgung der Stadt München. *Journal für Gasbeleuchtung und Wasserversorgung*, **23**.

Thiem, A., 1876. Resultate des Versuchsbrunnens für die Wasserversorgung der Stadt Strassburg. *Journal Für Gasbeleuchtung und Wasserversorgung* **19**.

Thiem, A., 1888. Neue Messungsart natürlicher Grundwassergeschwindigkeiten. *Journal für Gasbeleuchtung und Wasserversorgung* **31**.

Thiem, A., 1887. Verfahren für Messung natürlicher Grundwassereschwindigkeiten. *Polytechnische Notizblatt*, **42**.

Townley, R., 1694. A letter from Richard Townley Esq., of Townley in Lancashire, containing observations on the quantity of rain falling monthly for several years successively. *Philosophical Transactions of the Royal Society of London*, **18**.

Tromp, S.W., 1972. Water divining, *Encyclopedia of Geochemistry and Environmental Science*. New York, Van Nostrand Reinhold.

Van Deman, E.B., 1934. *The building of Roman aqueducts*. Publication no 423, Carnegie Institute of Washington.

Venturi, G.B., 1826. *Experimental inquiries concerning the principle of the lateral communication of motion in fluids*, translation by W. Nicholson, 2nd edn. Tracts on hydraulics, ed. T. Tredgold, London, J. Taylor.

Venturi, G.B., 1797. *Recherches expérimentales sur le principe de la communication latérale du mouvement dans le fluides*. Paris.

Vitruvius, 1826. *The architecture of Marcus vitrivius Pollio* in 10 books, translated by J. Gwilt. London, Prieslty and Weale.

Vitruvius, 1926. *The architecture of Marcus Vitruvius Pollio* in 10 books, translated by J. Gwilt, book 8. London, Prieslty and Weale.

Vitruvius, 1914. *Ten books on architecture*, translated by M.H. Morgan, book 1. Cambridge, Harvard University Press.

Webster, C., 1966. Richard Townley, 1629–1707, and the Townley Group. *Transactions of the Historic Society of Lancashire and Cheshire*, **118**.

White, G., 1849. *The natural history of Selbourne*. Bohn's Illustrated Library. London, G. Bell and Sons.

Wolf, A., 1935. *A history of science, technology and philosophy in the 16th and 17th centuries*. London, George Allen and Unwin Ltd.

Woltman, R., 1790. *Theorie und Gebrauch des hydrometrischen Flugels*. Hamburg.

Yen, B.C., 1989. Catchment flow and catchment runoff, *Proc Int. Conf. for Centennial of Manning's Formula and Kuichling's Rational Formula*, University of Virginia, USA.

Cross references

Bernoulli energy equation
Chézy formula
Current metering
Float
Flow through weirs, flumes, orifices, sluices and pipes
Flow measurement: new technology
Flume
Groundwater
Hydrological cycle
Hydrology: lakes and reservoirs
Hydrometeorology
Precipitation
Perrault, Pierre (1611–1680)
Radar: precipitation measuring (weather) in Europe
Rain
Raingauge
Reynolds number

HYDROLOGY

There is no universally accepted definition of hydrology but the US Federal Council for Science and Technology (1962) stated that it is 'the science that treats of all the waters of the earth, their occurrence, circulation, and distribution, their chemical and physical properties, and their reaction with their environment including their relation to living things.' One of the best accounts is contained in Ward (1990).

Hydrology is one of the most comprehensive Earth sciences. In order to reduce the subject to manageable proportions, some scientific studies of water are conventionally, if rather arbitrarily, excluded. For example, although hydrology is concerned with the occurrence and distribution of all forms of precipitation on the Earth, aspects of atmospheric moisture fall mainly within the province of meteorology. Similarly, oceanic water comes within the science of oceanography and frozen water in the permanent ice sheets is primarily part of glaciology. Even so, a complete knowledge of hydrology requires a formidable breadth of expertise and many hydrologists have specialized interests that draw on a formal training gained previously in allied disciplines. Thus groundwater hydrology necessarily depends heavily on a knowledge of geology, soil moisture studies are founded on pedology and surfacewater hydrology requires a background in hydraulics and fluid mechanics. Other supporting sciences include physics, chemistry, biology, mathematics and statistics. Some hydrologists specialize in the study of water in particular physical environments, such as arid zones or karst areas, or particular landuse areas, such as forests or towns.

The scope of hydrology is further widened because it is not just a pure science, but it also has many important practical applications, especially in relation to the assessment, use and management of the world's freshwater resources. The applied aspects of hydrology embrace additional subjects such as agriculture, forestry and the branches of civil engineering dealing with water supply, irrigation, flood defense, water pollution control and wastewater disposal. In some of these areas hydrology begins to overlap with certain social and behavioral sciences such as economics and sociology. Despite detailed demarcation problems, a basic distinction can often be maintained between systematic hydrology, which is concerned with the basic science of moisture processes in the natural environment, and regional hydrology, which employs the drainage basin as the prime functional unit in hydrology and includes many practical studies relating to applied hydrology (Figure H54).

As with many other scientific disciplines, the most rapid strides in hydrological understanding have taken place in the twentieth century, when progress was particularly stimulated by the International Hydrological Decade (1965–1974). But hydrology also has a long history and, according to Biswas (1970), the first recorded evidence of water resources engineering work can be traced back to 3200 BC. Jiaqi (1987) has identified three separate stages of evolutionary development in hydrology which emphasise different aspects of the science, namely geographical hydrology, engineering hydrology and water resource hydrology.

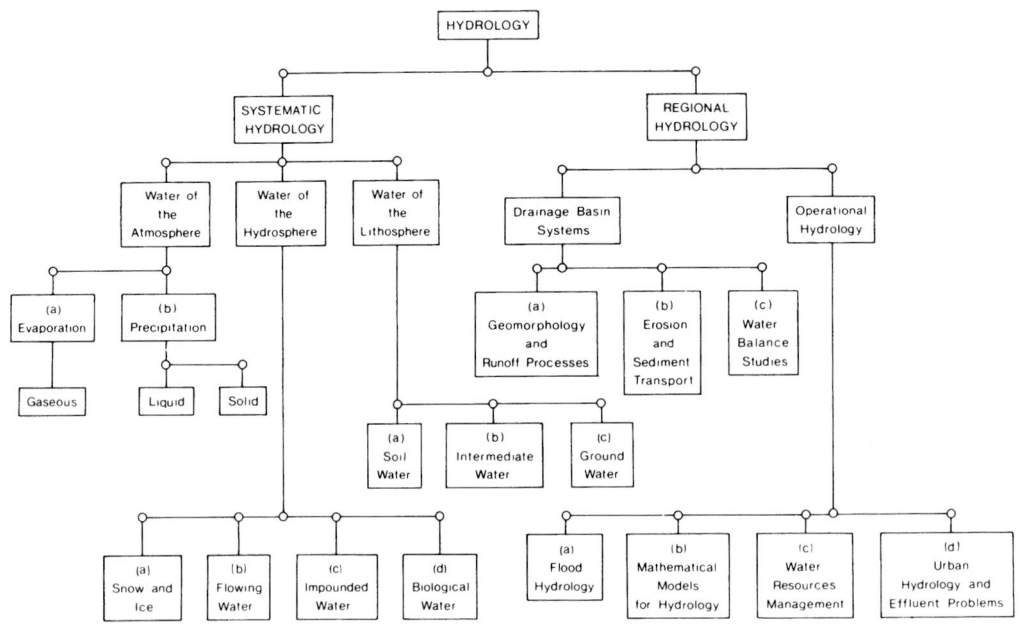

Figure H54 Components of hydrology. (After Meinzer, 1923.)

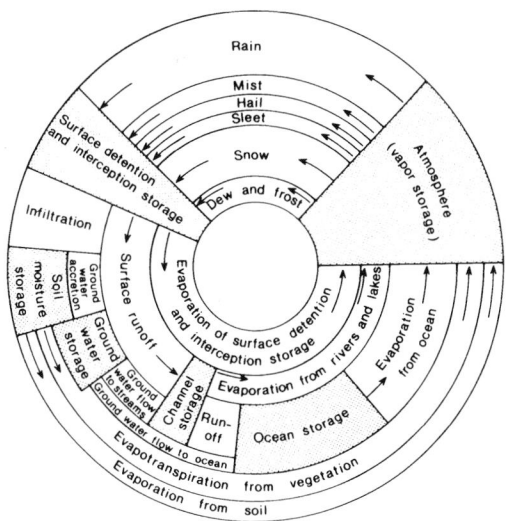

Figure H55 A diagrammatic representation of the hydrological cycle. (After Wisler and Brater, 1959.)

Hydrological cycle

Hydrology is organized around the concept of the hydrological cycle (q.v.) (Figure H55). This illustrates the dynamic nature of water movement. But the hydrological cycle also shows how water may occur in each of its three natural states as a liquid, a solid or a gas and how water is widely distributed through the Earth–atmosphere system in various storage phases. Solar energy powers the continuous circulation of moisture, which extends from a height of over 15 km in the atmosphere to an average depth of some 0.8 km in the lithosphere. There is no real beginning or end to the cycle, but the atmospheric storage phase is a convenient starting point since all water enters by the evaporative processes, which alone utilize 22% of the total solar radiation received at the top of the Earth's atmosphere, and all water exits through precipitation. When precipitation reaches the land areas of the Earth, it immediately begins its return journey to the atmosphere. Some precipitated water is evaporated as it falls, whilst intercepted precipitation is quickly re-evaporated from vegetation and building surfaces. Surface runoff carries water in the rivers directly back to the oceans, which in turn produce over 80% of the world's evaporation. Infiltrated water may be returned to the atmosphere by evapotranspiration from plants and the upper layers of the soil or may percolate to the water table. This groundwater will then move slowly underground to reappear at the surface either in springs or, via deeper seepage, in river channels or the ocean basins.

To some extent, the hydrological cycle concept implies a simplicity and smoothness which does not really exist. First, as indicated in Figure H55, there are numerous alternative pathways which water can follow, especially in the land-based part of the cycle, and there are, therefore, many hydrological cycles rather than just one. Second, the cycle is spasmodic. For example, although water is continuously being evaporated from the ocean surfaces, a very arid region may well experience a period of several years without precipitation. Similarly, the residence time of water in the various storage units varies enormously. Thus intercepted water is likely to be retained on vegetation for only a few hours before it is re-evaporated; the average residence time for a water molecule in the atmosphere is about 10 days, whilst other water molecules may be held for hundreds of years under virtually static conditions in ice caps and in deep aquifers.

Because the fundamental water transfer process is between the atmosphere and the earth, the hydrological cycle may also be seen as a downstroke or input which is dependent on precipitation, and an upstroke of output which results from evaporation. This interpretation has led to a systems view of the hydrological cycle. As a system, the hydrological cycle may be defined as a dynamic, physical and cascading system in which a chain of subsystems is linked together by the sequential transfer of water from one storage component to another. This sequential movement between the various storage subsystems of the vegetation, ground surface, soil moisture, channel

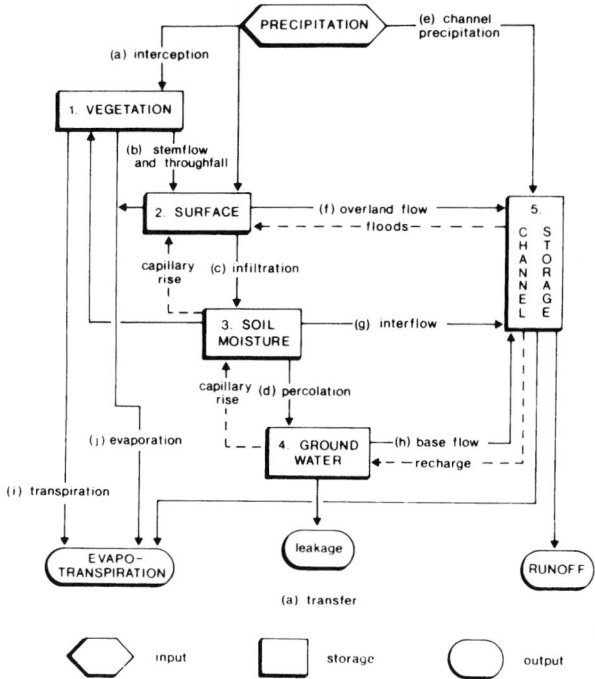

Figure H56 A systems representation of the natural hydrological cycle within a drainage basin. (After Ward, 1990.)

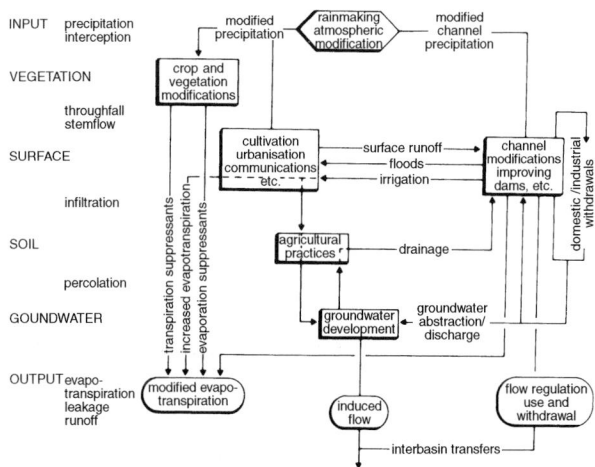

Figure H57 A systems representation of the basin hydrological cycle as modified by human activity. (After Ward, 1990.)

and groundwater is schematically illustrated in Figure H56. The system normally has a single input, precipitation, and two major outputs, evapotranspiration and runoff, but in detail the output from one subsystem becomes the input for the next.

The hydrological cycle is not only a highly complex natural system but it has also been progressively complicated by human interference, both deliberately and inadvertently. As shown in Figure H57, virtually every subcomponent of the system has been purposely modified in an attempt to secure a better temporal and spatial distribution of water resources, either by speeding up the circulation process through rain-making experiments or land drainage or by retarding the removal of the most useful water in the land-based phase of the cycle by the construction of storage dams or through the artificial recharge of groundwater reserves. Most of these changes are eventually reflected in streamflow and there can be few, if any, large rivers remaining in the world which still possess an entirely natural regime. Some of the modifications have been both unexpected and adverse, such as the waterlogging and salinization of soils which result from inadequate irrigation schemes.

Water balance

If the hydrological cycle is viewed as a series of storage units influenced by inputs and outputs, and if these factors can be quantified, it follows that water balances can be drawn up for different parts of the cycle. These balances are expressed mathematically by the hydrological or continuity equation which, in its simplest form, is written

$$I = O \pm \Delta S$$

where I is the input over any area during any given period, O is the output over the same area and period and S represents the resulting change in storage. The basic equation can, at least in theory, be applied to any of the storage subcompartments of the hydrological cycle, such as interception or groundwater, and can also be applied on virtually any scale of space or time. Thus, according to the nature of the problem to be solved, it could refer to a small experimental plot over a few hours, to a large drainage basin on a monthly basis or even to a continent in terms of seasonal or annual data. Generally speaking, the time scale increases with the areal dimensions in order to maintain a given level of resolution or accuracy.

Despite the fundamental simplicity of the water balance equation, it is often difficult to apply successfully. Essentially the hydrologist is interested in the water balance of the land areas, and the equation is normally solved on a drainage basin scale to obtain the relationships between the incoming precipitation and the outgoing evaporation and runoff. In this situation the equation would have to be rewritten as

$$P = E + R \pm \Delta S$$

where P is precipitation in all its forms, E is the total evapotranspiration process, R is runoff and S represents all the storage changes on the land surface such as interception, surface detention, soil moisture and groundwater. In practice, it is impossible to obtain uniformly reliable areal measurements for all these factors on a basin scale. Often the equation has been simplified to include commonly available data only, such as precipitation and runoff, and other values have been assumed rather than measured. One of the first reasonably complete water balance calculations on a basin scale was made in the UK by Penman (1950) when monthly values of precipitation, runoff and also evapotranspiration calculated from meteorological variables were used to estimate changes in groundwater storage over the chalk basin of the river Stour. It was then possible to compare the estimate of storage with monthly measurements of restwater levels in an observation well over a 16-year period, as shown in Figure H58. The seasonal fluctuations, and longer trends of estimated storage, are reflected quite satisfactorily in the well level fluctuations, although there is a difference in phasing owing to the delayed response of the

water table to surface events. Such lag effects are a common complicating factor in the water balance.

Alternatively, the water balance concept can be applied on a global scale. According to Nace (1969), the total capital stock of water in the Earth–atmosphere system is $1\,384\,000 \text{ km}^3 \times 10^{-3}$, of which 97.6% is located in the ocean basins. A further 1.9%, comprising almost 80% of the world's freshwater reserves, is locked up in glaciers and ice caps. The next largest storage unit is groundwater, which totals some 0.5% of all water. This represents by far the largest directly utilizable source available to humans and there are clearly finite limits to this resource (Postel et al., 1996).

Sellers (1965) has shown that there are large geographical variations in the water balance of the Earth's surface. As a whole, the oceans lose more water by evaporation than they gain by precipitation, the deficit being made good by runoff from those continents where precipitation is generally higher than evaporation. The Earth's wettest continental area is South America, which receives an average 1350 mm depth of precipitation annually, of which nearly two-thirds is regained to the atmosphere by evaporation from the land surface. In the dry continents of Australia and Africa this evaporative loss exceeds 75% of the annual rainfall.

Drainage basin hydrology

The most convenient areal unit for hydrological studies is the drainage basin or watershed. Its primary attribute is that of a drainage divide which, except in special geological circumstances, is assumed to coincide with the topographic perimeter so that all the precipitated water which is surplus to evaporation requirements, both surface and subsurface, eventually flows gravitationally to leave the basin through runoff at the lowest downstream point on the main river. Some water will be gained or lost to the basin by deep subterranean leakage, but in most cases that amount will be small in relation to other processes.

Most hydrological data collection systems throughout the world are still based on manually operated networks within drainage basins. In many studies streamflow measurements are the most valuable data. This is because rivers comprise that part of the hydrological cycle which is both most directly relevant to humans as a resource and is also most often modified, in quantity and quality, by human activity. In addition, measurements of river discharge provide a unique areal integration of basin hydrology since a point measurement of streamflow obtained at the basin outlet can readily be converted to a mean areal value of depth equivalent for comparison with other variables such as precipitation and evaporation. Fortunately the rate of flow in open channels can be measured relatively accurately using a variety of techniques (Herschy, 1978). The most widespread method is that adopted at velocity–area stations where the stream cross-section is divided into a number of parts for each of which the area, the velocity and the discharge are separately determined. The mean velocity is obtained by a current meter and, by adding together the partial discharges, a total is achieved for the stream. Other traditional methods include the use of gauging structures such as weirs and flumes, which employ a mathematical relationship between the height of water passing through a channel section of known artificial geometry and discharge, or chemical techniques whereby the downstream dilution of a tracer injected into the river at a known concentration provides a measure of streamflow. Newer techniques, especially for larger rivers, are based on electromagnetic and ultrasonic principles.

All other hydrological field measurements depend on point samples. The sample size is often small, for example in the case of an individual raingauge, and the spatial distribution of such measurements within the drainage basin is rarely satisfactory. Optimum sampling depends on the level of accuracy needed by the end user and on the spatial variation of the individual hydrological parameter. For example, the World Meteorological Organization has specified general minimum gauge densities for precipitation networks ranging from 25 km^2 per gauge in small mountainous islands with highly variable precipitation patterns to $1500–10\,000 \text{ km}^2$ per gauge for areas with arid or polar climates (WMO, 1981). Raingauges are also subject to systematic errors, and it has been known for over 200 years that standard gauges exposed above ground level catch less precipitation than actually reaches the ground. Point measurements of open water evaporation can be obtained directly by means of evaporation pans and values of evapotranspiration from vegetated surfaces can be derived from formulae, such as that originally devised by Penman

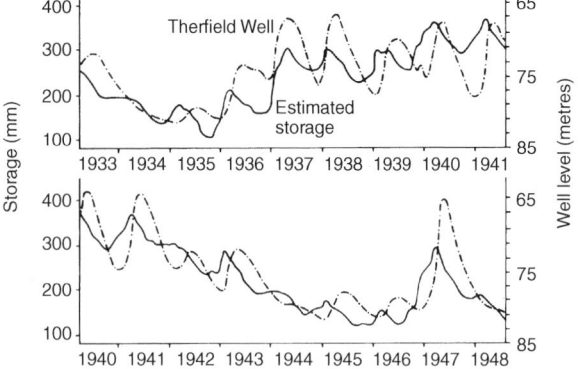

Figure H58 Estimated monthly changes in water storage over a drainage basin in eastern England compared with observed water-level changes in a well. The graph has been re-zeroed at the beginning of 1940 to reduce the significance of cumulative errors. (After Penman, 1950.)

(1948) and subsequently developed into the Penman–Monteith model (Monteith, 1981, 1985). Such evapotranspiration formulae require input measurements of solar radiation, air temperature, humidity and windspeed which are increasingly made within the drainage basin by automatic weather stations.

Technical advances in remote sensing have greatly increased the availability of all hydrological data. These advances include the long-distance interrogation of standard field instruments by telemetry and satellites, together with entirely novel methods of data collection, such as that for precipitation measurement by radar (Collier, 1987). Through the use of satellites, real-time hydrological data collection is now possible, even in remote areas (Engman and Gurney, 1991). Advances in data collection have been complemented by advances in data analysis and hydrological understanding. Empirical hydrological studies depend on monitoring the various processes within selected drainage basins, which are often designated either representative or experimental. Representative basins are intended to sample typical geographical assemblages of basin geometry, geology, climate, soils, vegetation and land use which remain largely constant over the observation period. Experimental basins, on the other hand, are established to record the hydrological consequences of deliberate land-use modifications over time, such as deforestation (Trimble *et al.* 1987) or urbanization (Hall, 1984).

The use of computers has stimulated a great deal of hydrological analysis based on the theoretical modeling of drainage basin behavior. Runoff forecasting has long been the principal objective in hydrology (Linsley, 1967) and many models are mathematical expressions of the basin response to rainfall input. The models are based on the physical laws of water movement and aim to simulate future runoff conditions, especially during flood and drought events. The most sophisticated approaches tend to rely on physically based distributed models, which treat the drainage basin as a spatially variable system, although for some practical purposes so-called lumped models, which aggregate processes over the basin, may be equally effective in transforming precipitation into runoff. One distributed model, widely used in Europe, is the SHE model which grids the drainage basin into squares (Institute of Hydrology, 1984).

The synergistic way in which improvements in data collection systems and drainage basin modeling have aided hydrological understanding over recent decades can be illustrated with reference to the advances documented by Clark (1994) for the USA:

- 1930s: Introduction of the unit hydrograph runoff concept and the start of rainfall–runoff model development
- 1940s: Initial establishment of regional River Forecast Centers responsible for issuing flow forecasts for major river basins.
- 1950s: Early computerization of flow forecasting techniques, e.g. the use of graphical antecedent precipitation indices for individual basins converted to mathematical algorithms.
- 1960s: Telemetry linkage of rainfall and river gauges provided rapid telephone access to hydrometeorological conditions at remote sites; the Stanford Watershed Model gave a better basin-scale representation of rainfall–runoff relations.
- 1970s: Satellite technology introduced, e.g. the GOES satellite launched in late 1960s, which gave fully remote collection and relay of realtime hydrometeorological data for large drainage basins.
- 1980s: Advent of Automated Local Evaluation in Real Time (ALERT) systems provided some improvement in flash flood forecasting for basins less than 250 km^2 in area.
- 1990s: New emphasis on quantitative precipitation forecasting (QPF) for periods from less than 24 h up to 10 days ahead using expanded computing capability and improved numerical models of storms.

Applied hydrology

From a practical viewpoint, the main uncertainties in hydrology center on the extremes of water availability. Much of applied hydrology is, therefore, concerned either with water resources studies, where the emphasis is on the nature and consequences of a deficiency in supply, or with drainage and flood studies where excess water is the problem. In both cases, one of the most important applications is in providing engineering design criteria for water control structures such as dams, storage reservoirs, sewers, bridges and irrigation systems. It has also become apparent that hydrologists increasingly recognize the importance of the quality dimension of water, as detailed in Gower (1980),

and water quality studies are now probably the fastest-growing area of applied hydrology.

In water resources studies a recurrent problem is to determine the safe yield of various supply sources, such as wells or reservoirs. The size of dam, and the subsequent capacity of any reservoir, is dependent on the degree to which natural fluctuations in streamflow can be modified by storage to correspond with water demands. An important hydrological factor is always the minimum annual or seasonal runoff which can be expected and, in the usual absence of long-term gauging records at the proposed site, estimates have to be prepared based on runoff simulation or other techniques. Low runoff may be associated with high evaporative demands, and in all such studies allowances must be made for evaporation and seepage losses from storage reservoirs. In arid regions, or during dry spells, crop-based agriculture relies heavily on irrigation. Successful irrigation is dependent on a clear understanding of water balance principles and ideal conditions occur when soil moisture is maintained within the range available to the crop rooting system, i.e. between field capacity and the wilting point.

Floods are the most obvious of all hydrological hazards. They can be controlled by engineering works in a way that extreme droughts, for example, cannot, and flood studies represent probably the most important single theme in applied hydrology. The ultimate aim of flood hydrology is to estimate the peak volume and maximum height of water in a river channel during a particular event, either retrospectively or predictively. The statistical concept of frequency occurrence is basic to all flood studies because, without some idea of the probability of a stated risk, engineering structures can be neither designed nor operated safely and economically. Some of the special features related to the statistical analysis of both floods and droughts were outlined in an key early paper by Gumbel (1958).

Since exceptional rainfall is the immediate cause of most floods, much emphasis is placed on an understanding of the hydrometeorological processes associated with storm rainfall. The characteristics of storm rainfall are often expressed in terms of frequency, intensity and areal extent since the intensity of precipitation increases as both the time duration and the spatial scale contract. These attributes reflect the importance of concentrated, short-lived convectional storms rather than the more widespread precipitation associated with frontal activity. In certain circumstances, such as the design of a major dam, it may be desirable to estimate the magnitude of the largest possible flood that can physically occur at that point. Such an extreme event, known as the probable maximum flood (PMF) represents the upper limit of flooding for a stated drainage basin size which the existing climatic regime can produce and will result from the maximum possible combination of precipitation and snowmelt together with minimum evaporative losses.

All hydrological activity, including the statistical analysis of floods and droughts, depends on the assumption of climatic stationarity, i.e. the belief that past atmospheric conditions and hydroclimatic events provide a reliable guide to the future. This assumption may not be valid if, as a result of increasing atmospheric concentrations of greenhouse gases, global warming, with all associated consequences, occurs over the coming decades. Concern about the regional implications of hydroclimatic change is already apparent in parts of the temperate latitudes (Arnell, 1992) but, on a global scale, the tropics appear especially vulnerable to climatic changes. Some of the poorest countries on earth are located in the tropics and the semiarid regions are already prone to seasonal, or longer periods, of drought whilst many water surplus areas of the tropics suffer from recurrent floods. Any changes in hydroclimate which increase the risk of droughts or floods will be most evident in the drainage basins of large, unregulated rivers in such areas. The less-developed countries lack the resources to cope with many potential impacts which may impinge adversely not only on water management but also on the quality of life and future development prospects.

Keith Smith

Bibliography

Arnell, N.W., 1992. Factors controlling the effects of climate change on river flow regimes in a humid temperate environment. *Journal of Hydrology*, **132**, 321–342.

Biswas, A.K., 1970. *History of Hydrology*, North-Holland Publishing Co., Amsterdam and London.

Clark, R.A., 1994. Evolution of the national flood forecasting system in the USA, in Rossi, G., Harmancioglu, N. and Yevjevich, V. (eds) *Coping with Floods*, Kluwer Academic Publishers, Dordrecht, pp. 437–444.

Collier, C.G., 1987. Accuracy of real-time radar measurements, in Collinge, V.K. and Kirby, C. (eds) *Weather Radar and Flood Forecasting*, John Wiley and Sons, Chichester, pp. 71–95.

Engman, E.T. and Gurney, R.J., 1991. *Remote Sensing in Hydrology*, Chapman & Hall, London.

Gower, A.M. (ed.), 1980. *Water Quality in Catchment Ecosystems*, John Wiley and Sons, Chichester.

Gumbel, E.J., 1958. Statistical theory of floods and droughts. *Journal of the Institution of Water Engineers*, **12**, 157–184.

Hall, M.J., 1984. *Urban Hydrology*, Elsevier, London.

Herschy, R.W. (ed.), 1978. *Hydrometry*, John Wiley and Sons, Chichester.

Institute of Hydrology, 1984. *Research Report 1981–1984*, Natural Environment Research Council, Wallingford.

Jiaqi, C., 1987. The new stage of development of hydrology – water resources hydrology, in *Water for the Future: Hydrology in Perspective*, International Association of Scientific Hydrology, Pub. 164, pp. 17–25.

Linsley, R.K., 1967. The relation between rainfall and runoff. *Journal of Hydrology*, **5**, 297–311.

Meinzer, O.E., 1923. *Outline of ground water hydrology*. Water Supply Paper 494, United States Geological Survey, Washington, DC.

Monteith, J.L., 1981. Evaporation and surface temperature. *Quarterly Journal of the Royal Meteorological Society*, **107**, 1–27.

Monteith, J.L., 1985. Evaporation from land surfaces: progress in analysis and prediction since 1948, in *Advances in Evaporation*, Association of the Society of American Engineers, pp. 4–12.

Nace, R.L., 1969. World water inventory and control, in Chorley, R.J. (ed.) *Water, Earth and Man*, Methuen, London, pp. 31–42.

Penman, H.L., 1948. Natural evapotranspiration from open water, bare soil and grass. *Proceedings of the Royal Society of London Ser. A*, **193**, 120–145.

Penman, H.L., 1950. The water balance of the Stour catchment area. *Journal of the Institution of Water Engineers*, **4**, 457–469.

Postel, S.L., Daily, G.C. and Ehrlich, P.R., 1996. Human appropriation of renewable fresh water. *Science*, **271**, 785–788.

Sellers, W.D., 1965. *Physical Climatology*, University of Chicago Press, Chicago and London.

Trimble, S.W., Weirich, F.H. and Hoag, B.L., 1987. Reforestation and the reduction of water yield on the Southern Piedmont since circa 1940. *Water Resources Research*, **23**, 425–437.

US Federal Council for Science and Technology, 1962. Scientific Hydrology *Ad Hoc Panel on Hydrology*, Washington, DC.

Ward, R.C., 1990. *Principles of Hydrology*, (3rd edn), McGraw-Hill, London.

Wisler, C.O. and Brater, E.F., 1959. *Hydrology*, (2nd edn), John Wiley and Sons, New York.

World Meteorological Organization (WMO), 1981. *Guide to Hydrological Practices, No. 168*, 4th edn, World Meteorological Organization, Geneva.

Cross references

HYDROLOGY: COASTAL TERRAIN

In coastal districts, the fresh water in the water table migrates slowly downhill to the sea. Because of their different densities, the fresh water and salt water do not generally mix, except in the ocean where the tides, waves and currents do the mixing. In the aquifers in coastal districts, the less dense fresh water tends to float on the more dense saline water just like an iceberg. Figure H59 shows the shape of the fresh water lens on a sandy island assuming that the fresh water is being replenished by rainfall. The relationship between the thickness of the freshwater body (a) and the depth of the lowest part of the freshwater body below sea level (b) is:

$$\frac{b}{a} = \frac{\text{Specific gravity of fresh water}}{\text{Specific gravity of seawater}} = \frac{40}{41}$$

Thus for every meter the fresh water stands above sea level, the surface of the salt water lies some 40 times as many meters below sea level. These are, of course, only approximate figures, depending mainly on the salinity of the seawater and the purity of the fresh water. Figure H60 shows the flow lines, i.e., the paths of water movement, for the fresh water contained within the lens. Both the lens and the underlying salt water will rise and fall with the tide unless there is a barrier between the underground water and the sea. The time of the peaks and troughs of the fluctuations becomes later as it is traced inland, just as the time of high and low tide becomes progressively later as it is traced up the tidal part of a river. The time between the peaks and troughs will remain the same, while the time lag will be constant for a given well.

Some mixing of the fresh and salt water does take place at the interface. Usually this is negligible, but it can reach appreciable proportions under certain favorable conditions. This produces a brackish water zone which may be quite thick. This zone occurs where there are considerable fluctuations in the level of the interface due to tidal action or irregular heavy rains. Thus a strong development of a brackish zone is found in the basalt aquifers along the coast of

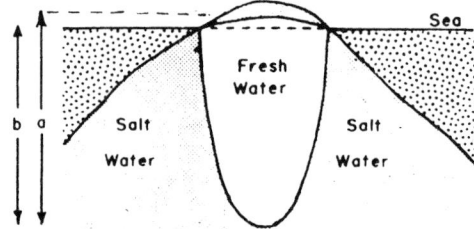

Figure H59 The characteristic shape of the freshwater lens in islands made of uniformly permeable materials in humid areas.

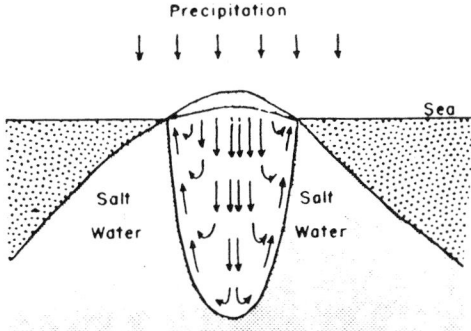

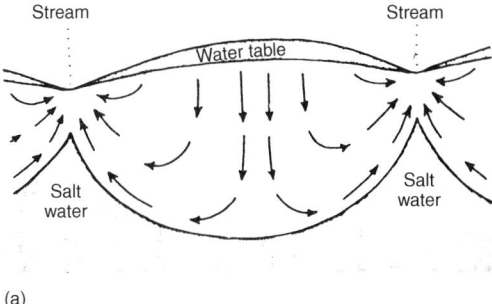

Figure H60 Flow lines within the freshwater lens on the island shown in Figure H59. The lens is recharged by infiltration of rain and snow into the ground but loses water by diffusion and by flowage into the sea.

Oahu in Hawaii (Visher and Mink, 1964). It is also increased by pumping the wells in these regions, as we shall see below. Like the fresh water, this brackish water lens moves slowly downslope.

Variations in water table pattern with climate and form of the aquifer

At coastal sites on permeable materials where several large rivers cross the area and there is continuous replenishment by precipitation, the floating fresh water and its flow lines are as in Figure H61a. With a seasonal reduction in the replenishment, seawater tends to enter the rivers in low-lying districts. Even in areas of tropical rainforest, seawater may penetrate nearly to the headwaters of coastal rivers in the drier seasons. Thus in May and early June, 1961, seawater reached some 187 km up the Barima River in Guyana, i.e. almost to its headwaters. In this case, the freshwater lens and its attendant flow lines take the form shown in Figure H61b. Tropical lowland rivers where this may occur are marked by a line of mangroves, usually *Rhizophora* species, lining the bank. Behind this line occurs the normal swamp or rainforest.

In desert areas on permeable materials where there is negligible recharge from local precipitation, the fresh water takes the form of a thin, almost horizontal sheet (Figure H61c).

This undergoes local modification where coastal dunes separate the sea from a zone below sea level as at Zaura in Libya. The relationship between the Mediterranean Sea, the dunes, the salty depression or sabkha, and the groundwater are shown in Figure H62. A freshwater lens occurs beneath the dunes and is replenished by rainfall. The water table slopes downhill inland to the sabkha. There the intense evaporation disposes of fresh water flowing towards the coast from farther inland, fresh water and brackish water flowing downslope from the dunes, and seawater which has traveled inland beneath the fresh water. Thus Zaura can only expand its freshwater supply by taking more water from beneath the dunes.

So far, we have assumed that all the rocks on the shore are permeable. This is often not the case. Where artesian basins dip seaward and end up beneath the sea, a special set of conditions applies. Figure H63 shows a typical example from the Baltimore area, Maryland. Where a well is sunk in the fresh water, the well is artesian, but where it penetrates the saline water, it is not. This is due to the different densities and hence piezometric surfaces of the two kinds of water.

Water tables and eustatic sea-level changes

The worldwide rise of sea level as the glaciers melt has an important effect on the water tables in coastal areas. As the sea-level rises, so does the water table and the saline water. This increases the tendency of tides to sweep saline water into rivers and it tends to push the fresh water shoreward. In the case of the Atlantic seaboard of the United States, the sea is rising at a rate equivalent to approximately 0.6 m per century. It has been estimated that this small rise will cause the fresh water in the artesian aquifer of New Jersey to recede inland at the rate of 0.6–2.5 km per century, depending on the dip of the aquifer (Long, in Parker, 1955). Thus a small rise in sea level can have quite a large effect.

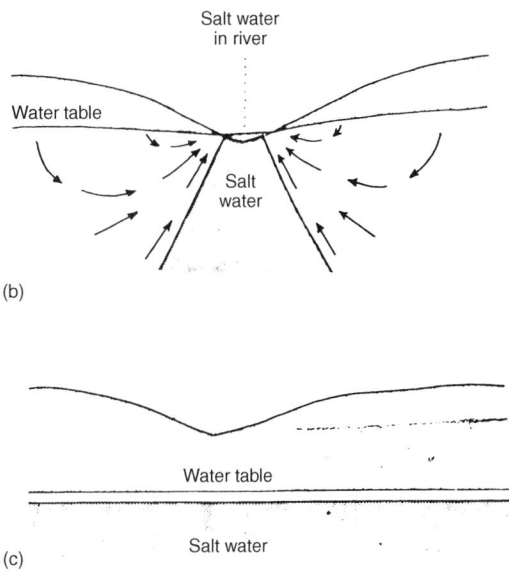

Figure H61 Idealized common situation on coasts. (a) A representation of a coastal area with uniformly permeable materials under a humid climate. (b) Shows what happens in a dry season when salt water backs up the rivers. (c) Shows the nature of the freshwater table along an arid coast. (Partly after Parker, 1955.)

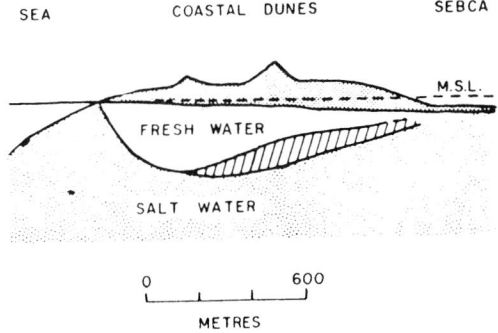

Figure H62 Cross-section of the groundwater zones below a sand dune belt at Zaura, Libya. Note the water table sloping inland, the brackish zone (cross-hatched) and the localized freshwater lens. Vertical exaggeration 1:10 000. (Adapted from Underhill and Atherton, 1964.)

Exploitation of fresh water

The thin layer of fresh water underlain by salt water means that great care must be taken in exploiting the fresh water. The usual method of exploitation today is by a well from which the water flows or is

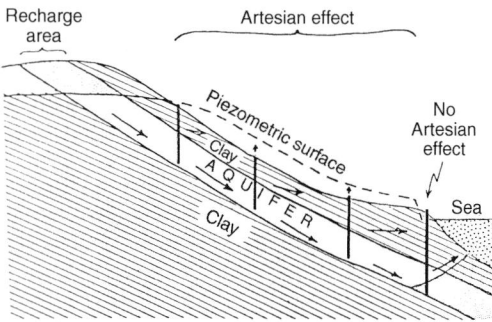

Figure H63 Idealized cross-section of a coast where an artesian aquifer outcrops some distance offshore as well as on the land surface. This applies to most of the Atlantic coast of the United States from Long Island to Florida, parts of the Gulf coast and Pacific coast of the United States, parts of the Hawaiian Islands, as well as to the Australian artesian basins. (After Parker, 1955.)

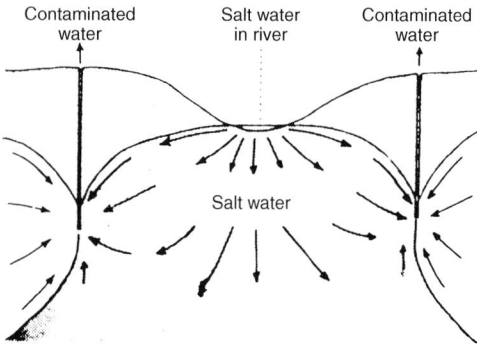

Figure H64 Effect of pumping wells along the margins of streams after seawater has backed up rivers as in Figure H61b. The contamination will last until fresh water again fills the river and the salt water has sunk to the position in Figure H61a.

pumped. When water is taken from a well, the surface of the water table is lowered close to the well by an amount depending on the output of the well and the porosity of the aquifer, among other factors. As we have already stated, the thickness of the fresh water below a point is some 41 times its height above sea level. Thus for every meter of lowering of the surface of the water table, the freshwater–saltwater boundary moves 40 m nearer the surface. Thus it does not take a great output of water to cause the bottom of the freshwater layer to rise to the bottom of the well. Thereafter the water produced by the well will be saline. By limiting the production, contamination can be stopped.

Particular care must be taken in the case of artesian basins in coastal regions. The quantity of fresh water that is stored is finite, and the amount of recharge is limited. Overpumping in such situations may cause the inflow of saline water and it may be necessary to resort to restricted pumping or artificial recharge or both (McCollum and Counts, 1964).

In the case of salt water backing up a river in the dry season, pumping water near the river produces the result in Figure H64. Once this has happened, the state cannot be altered until fresh water again fills the river. Then the salt water subsides to its former level and the wells can start producing fresh water once again.

This emphasizes the advantage of the Libyan collecting galleries used by the Romans. These are in a region where there is little precipitation and the freshwater layer is only a few inches deep. Any well would be contaminated by the saline waters underneath. However, the collecting galleries are about 400 m long and skim off the fresh water from a large area, thus producing a large quantity of fresh water with negligible draw down.

Compared with the collecting galleries, pumping out water always has the disadvantage of causing greater mixing of fresh and salt water (Figure H65). This is greater in the case of the standard single-

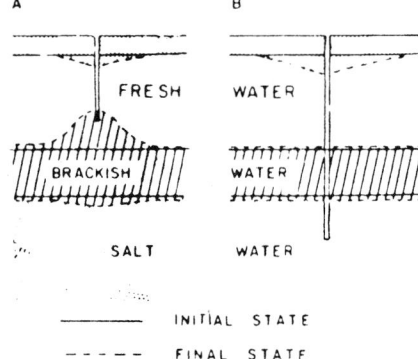

Figure H65 Results of intermittent pumping of a single pumped well (case A) contrasted with the effect of constant pumping of a double-pumped well (case B). In the latter case, water is abstracted from the brackish- and saltwater zones at a rate calculated to maintain a horizontal interface between the fresh water and brackish water. (After J.H. Edelman in Underhill and Atherton, 1964.)

pumped well than from the double-pumped well (Underhill and Atherton, 1964). In the latter case, water is pumped from the brackish and saline water zones in just the right quantities to keep the brackish water–freshwater interface horizontal. However, some means of disposing of the brackish and saline water must then be found, and this is not always easy.

Constantly pumped normal wells on shores such as at Oahu. Hawaii, show thickening of the brackish water lens on the seaward side only. Similar wells which are used intermittently produce a thickening of the brackish water lens at the expense of the freshwater lens on all sides of the well. This is due to the water movements involved in the alternate thickening and thinning of the freshwater lens.

Draining of swamps and marshes

Many tropical shores have large areas of freshwater swamps and marshes behind them. The shores of the Guianas (Guyana, Surinam and French Guiana) in South America are but one example. It is very tempting for agriculturists and land development corporations to try to drain these swamps and put them to some more profitable use. Unfortunately such swamps are the source of the fresh water in the coastal districts. Thus, when there was an attempt made to drain part of the Everglades behind Miami, the drains proved unsuccessful, but they caused a marked thinning of the freshwater layer. When salt water entered the drainage canals in the dry season, it promptly contaminated the freshwater lens upon which Miami relied for its water supply. Thus, if the city of Miami is to use groundwater for drinking purposes, the Everglades must remain a swampy wilderness.

In the case of the Guianas, the early Dutch settlers realized the problem, and instead of reclaiming the swamps, they reclaimed the intertidal zone. This required the building of hundreds of kilometers of embankments or polders, together with numerous tide-operated sluices so that the land lying below high-tide mark could be drained, thus ensuring an ample supply of irrigation water from the swamps behind the polders.

Stuart A. Harris

Bibliography

Bennett, R.R. and Meyer, R.R., 1953. Geology and ground-water resources of the Baltimore area, Md., *Maryland Bd. Nat. Resources Dept. Geol. Mines Water Resources Bull.*, **4**.
Cooper, H.H., Jr, 1964. Sea water in coastal aquifers, *US Geol. Surv., Water Supply Paper*, **1613-C**, 84 pp.
McCollum, M.J. and Counts, H.B., 1964. Relation of salt-water encroachment to the major aquifer zones, Savannah Area, Georgia and South Carolina, *US Geol. Surv., Water Supply Paper*, **1613-D**, 26 pp.
Parker, G.G., 1955. The encroachment of salt water into fresh, in *Water, Yearbook Agr. US Dept. Agr.*, 615–635.

Underhill, H.W. and Atherton, M.J., 1964. A coastal ground water study in the Libya and a discussion of a double pumping technique, *J. Hydrol.*, **2**, 52–64.

Visher, F.N. and Mink, J.F., 1964. Groundwater resources in Southern Oahu, Hawaii, *US Geol. Surv. Water Supply Paper*, **1778**, 133 pp.

Cross references

Aquifer
Drainage
Groundwater
Hydrology
Hydrology: subsurface waters
Water table

HYDROLOGY: LAKES AND RESERVOIRS

The hydrology of lakes and reservoirs is one of the principal areas of study in continental hydrology, relating to the study of water bodies with a slow water exchange. As elements of the geographic landscape, lakes and reservoirs comprise a specific group of water bodies, which differ greatly from rivers and from seas. In rivers the water flows mainly under the effect of gravity, but in lakes and reservoirs it flows under the effect of external forces, i.e. wind, atmospheric pressure, force of attraction, etc. In contrast to rivers, lakes and reservoirs have quite distinct basins which are gradually filled in by organic and inorganic matter during basin evolution.

The main difference between lakes and reservoirs compared to seas is the absence of a direct water exchange with the ocean. In addition, the shape and size of the lake or reservoir basin affect the development of its hydrological regime, which is not of importance in a sea. Each lake or reservoir originates and evolves in a certain geographic environment and is in constant contact with this environment. As the geographic environment is predominant in the development of a lake or reservoir ecosystem and in the formation of specific features of the hydrological regime of these water bodies, the hydrology of lakes and reservoirs belongs to geographical sciences. This does not exclude, however, the application of hydrophysical, hydrochemical, hydrobiological, hydrodynamic and other approaches within the framework of this discipline for a solution of particular problems.

The main objective of the hydrology of lakes and reservoirs, as a scientific discipline and an independent branch of knowledge, is to study the morphometric peculiarities of lake or reservoir basins, together with their bottom deposits, physical and chemical water properties and hydrological regimes. Therefore, the discipline of the hydrology of lakes and reservoirs comprises the following research areas:

- morphometric and morphological peculiarities of basins;
- dynamics of bottom deposits, sedimentation, formation of shores;
- water regime (sources of water supply, water balance, water level regime, runoff control);
- thermal regime (sources of energy, energy balance, heat storage, thermal stratification, freezing and ice break-up; dynamics of the ice cover);
- water motion (waves, currents, wind set up, seiches, water circulation and mixing);
- features of the hydrochemical regime.

Water bodies with slow water exchange, i.e lakes and reservoirs, are subdivided into those of natural origin and artificial ones. Natural water bodies are lakes, which range from small ones (less than 10 000 m^2 in area) up to very large lakes, often called seas (e.g. the Caspian Sea with an area of 400 000 km^2). Artificial water bodies comprise reservoirs and ponds, i.e. small artificial lakes (no more than 1 km^2 in area).

A lake together with its basin (drainage area) forms a natural complex which has been evolving for tens or hundreds or thousands of years. During its long interaction with the environment the lake attains the characteristic features of its hydrological regime which correspond closely to the physiographic features.

A reservoir, as a new artificial object, which appeared quite rapidly (from a historical viewpoint) and which has not had an evolution period as a single natural complex of a river basin, interacts with the environment quite intensively. In contrast to a lake, the reservoir–river basin system is a physiographic–technological complex within which the water exchange is formed during a complicated mutual interaction of an anthropogenic object (reservoir) and a physiographic object (river basin). Thus, when lakes are studied from the point of view of their hydrology, the main problem is in studying the lake regime and the establishment of the available interactions with the environment. However, when reservoirs are studied within this discipline, the major emphasis is on the interaction between this new water body and the environment under conditions of artificial control of the water regime in the reservoir. Since the extensive construction of reservoirs was initiated in the world in the second half of the twentieth century, the discipline of the hydrology of lakes and reservoirs reached its present form during the 1970s. It developed from limnology, i.e. a discipline which originated in the nineteenth century, which studied natural lakes.

When the hydrological peculiarities of lakes and reservoirs are studied, experimental and field research methods are applied. Field investigations are organized and made on particular water bodies with the use of land, airborne and satellite technical facilities. The experimental method is related to the study of hydrological peculiarities of lakes and reservoirs based on physical models in laboratories.

The hydrology of lakes and reservoirs, as a scientific discipline, is closely connected with the demands of the national economy. The research results in this field are widely applied for fisheries, power generation, navigation, water supply, salt yield and recreation. Besides, the results of studying the hydrological regime of lakes and reservoirs are used for the selection of optimal schemes of runoff control for power generation at hydroelectric power plants.

V.S. Vuglinsky

Bibliography

Anon., 1979. *Reservoirs of the World*. Moscow: Nauka, 287 pp. (in Russian).

Cole, G.A., 1983. *Textbook of Limnology*, 3rd edition. Saint Louis: C.V. Mosby, 231 pp.

Henderson-Sellers, B., 1984. Engineering Limnology. London: Pitman, 356 pp.

Hutchinson, G.E., 1957. *A Treatise on Limnology*, New York: John Wiley.

Matarzin, Yu M. (ed.), 1977–1979. *Water Storage Reservoir Hydrology*, in 4 volumes, Perm University Press (in Russian).

Wetzel, R.G., 1975. Limnology, Philadelphia. Pensylvania: W.B. Saunders, 743 pp.

Cross references

Aral Sea
Caspion Sea
Lake Balaton, Hungary
Lake Chad
Lakes
Lakes: lakewater
Lakes: largest worldwide
Limnology
Lough Neagh, UK
Okeechobee Lake, Florida, USA: human impacts, research and lake restoration
Quinghaihu Lake, China

HYDROLOGY: SUBSURFACE WATERS

Scope of topic

The study of subsurface water generally includes a consideration of its chemical and physical properties, geological environment, natural movement, recovery and utilization. More specifically, the hydrology of subsurface waters concentrates on the study of the laws of the occurrence and movement of subterranean waters (Meinzer, 1942).

History

Accounts of well water and well construction abound in ancient literature and are specially well known from the biblical record of *Genesis*.

Well construction in the Near East was by human and animal power aided by hoists and primitive hand tools, despite great difficulties. The Egyptians had perfected core drilling in rock as early as 3000 BC. The ancient Chinese people were able to achieve wells with depths up to 1500 m through sustaining a slow drilling rate over a period of years.

The greatest achievement in groundwater utilization by ancient peoples was in the construction of long infiltration galleries, or kanats, which collected water from alluvial fan deposits and soft sedimentary rock. These structures, commonly several kilometers long, collected water for both agricultural and municipal purposes. Kanats were probably first used more than 2500 years ago in Iran; however, the technique of construction spread eastward to Afghanistan and westward to Egypt. One extensive kanat system built about the year 500 BC in Egypt is said to have irrigated 3500 km² of fertile land west of the Nile (Tolman, 1937). Many kanats are still in use today in Iran and Afghanistan, the best known of which are in Iran on the alluvial fans of the Elburz Mountains.

Modern percussion methods of well drilling were developed more or less independently in Western Europe. The impetus for this development came largely from the discovery of flowing wells, first in Flanders about AD 1100, then a few decades later in eastern England and in northern Italy. One of the first wells was dug in AD 1126 by Augustinian monks from a convent near the village of Lillers (De Wiest, 1965). In Gonnehem, Flanders, near Bethune, four wells were drilled and were cased to nearly 3.5 m above ground level so that they were able to deliver water at sufficient height to drive a water mill. The wells were several tens or hundreds of meters deep and tapped water under pressure from a formation consisting of fractured chalk that had its outcrop area in the higher plateaus of the Province of Artois. These and other similar wells in the region of Artois became so famous that flowing wells were eventually called artesian wells after the name of the region.

The methods of drilling for water have improved rapidly since the end of the nineteenth century, partly owing to knowledge borrowed from oil and gas drilling. The most significant single advance in drilling techniques has been the development of hydraulic rotary methods. Early rotary drilling was carried out with the aid of an outer casing; however, in about 1890 thick mud was found to be sufficient for holding up the walls of the hole, and the outer casing was no longer used. With this new efficiency and with the successful drilling of the Spindle Top oil field in Texas in 1901 by rotary methods, rotary drilling has steadily gained in popularity. The perfection of the deep-well turbine pump in the years between 1910 and 1930 added a further stimulus to the well-drilling industry (Davis and De Wiest, 1966).

Founders of hydrology of subsurface waters

Although his scientific work was somewhat related to that by Hagen and Poiseuille, Henri Darcy (1803–58) was the first person to state clearly the mathematical law which governs the flow of groundwater. Darcy developed his formula as a result of experimentation with filter sands, and presented it in 1856 in an appendix of a report on the municipal water supply of Dijon, France. Jules Dupuit, of France, was the first scientist to develop a formula for the flow of water into a well. Modern methods of higher mathematics were first applied extensively to groundwater flow by Philip Forchheimer of Austria and C.S. Slichter of the United States.

Modern and most significant advances in groundwater hydraulics in the United States were made in the 1930–1940 period by Theis, Jacob, Muskat and Hubbert (Muskat, 1937; Hubbert, 1940; Todd, 1980).

Darcy's law

The hydrodynamic microscopic picture of the flow of groundwater is very complicated and not amenable to mathematical treatment. Indeed, for isothermal flow, the three velocity components, the pressure and the density at any point of the fluid are the five unknown quantities in problems of groundwater flow. Water is treated as incompressible, except in the calculation of the storage coefficient of water-bearing strata. If the density is assumed to be constant, it would theoretically be possible to solve for the unknown pressure and velocity, if equations of motion (of the Navier–Stokes type) and of conservation of mass were available. Hydrodynamically, such a problem would be tractable if the granular skeleton were a simple

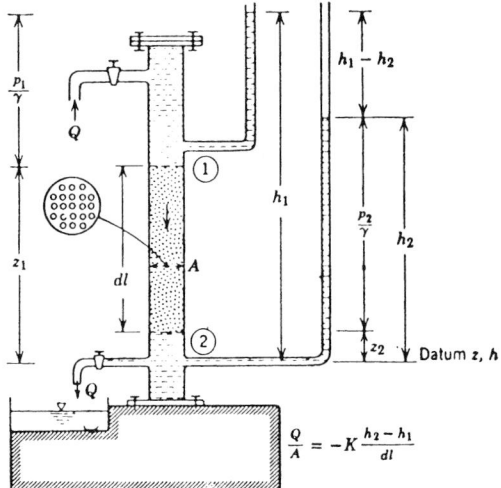

Figure H66 Apparatus to demonstrate Darcy's law. (Courtesy M.K. Hubbert.)

geometrical assembly of prismatic, unconnected tubes. The seepage path, far from being a prismatic channel however, is tortuous, branching into a multitude of tributaries and recombining several of them as the flow proceeds.

Darcy's law in its original form avoids the insurmountable difficulties of the hydrodynamic microscopic picture by introducing a doubly averaging microscopic concept. First, it considers a fictitious flow velocity, the Darcy velocity or specific discharge through a given cross-section A (Figure H66) of porous medium rather than the true velocity between the grains. Second, it treats average hydraulic values rather than local hydrodynamic values of this velocity. The basic reason for the introduction of this simplifying concept lies in the nature of Darcy's experiment: it utilized a sand-filled cylindrical pipe which permitted a measurement of only the average hydraulic values. The flow in the pipe of Figure H66 proceeds from higher to lower head h. The velocity head part in h is negligible so that

$$h = z + \frac{p}{\gamma} + \text{constant} \tag{H6}$$

in which z is the elevation head and p/γ is the pressure head, being the ratio of the water pressure in the pores to the unit weight of the fluid. The flow rate Q is proportional to the head loss, inversely proportional to the length of the flow path, and proportional to a coefficient K which depends on the nature of the sand and of the fluid. Darcy's law may be expressed as (Shaw, 1994).

$$Q = KA(h_1 - h_2)/dl = -KA\frac{dh}{dl} \tag{H7}$$

The coefficient K is the fluid conductivity and may be expressed as

$$K = k\frac{\gamma}{\mu} \tag{H8}$$

in which γ is the unit weight of the fluid, μ is its dynamic viscosity and

$$k = cd^2 \tag{H9}$$

is the permeability of the medium, where d is a characteristic length, say the average pore size of the sand, and c is a dimensionless constant or shape factor which takes into account effects of stratification, packing, arrangement of grains, size distribution and porosity.

Flow in confined aquifers

Aquifers are geological formations or strata containing water in their voids or pores that may be removed economically and used as a source of water. They are separated from each other by aquicludes or aquitards. Aquicludes are geological formations so impervious that for all practical purposes they completely obstruct the flow of

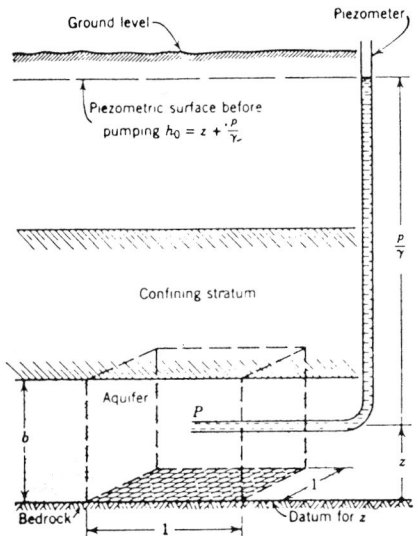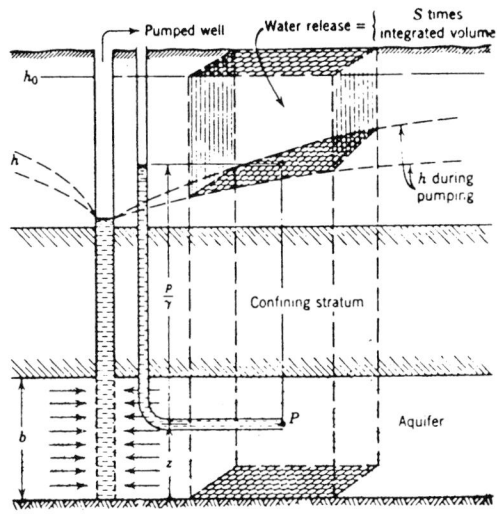

Figure H67 Physical interpretation of storage coefficient. (Courtesy R.J.M. de Wiest.)

groundwater (although they may be saturated with water themselves) and completely confine other strata with which they alternate in deposition. A shale in an example of an aquiclude. Aquitards are geological formations of a rather impervious and semi-confining nature which transmit water at a very slow rate compared to the aquifer. Over a large area of contact, however, they may permit the passage of large amounts of water between adjacent aquifers which they separate from each other. Clay lenses interbedded with sands, if thin enough, may form aquitards.

Consolidated sandstones are common examples of confined aquifers. The flow in such aquifers (Figure H67) is governed by the equation

$$\nabla^2 h = \frac{S}{T}\frac{\partial h}{\partial t} \tag{H10}$$

in which T is the transmissivity of the aquifer, the product of K and b, the thickness of the aquifer. S is the storage coefficient of the aquifer (Shaw, 1994; Todd, 1980):

$$S = \gamma b(\alpha + n\beta) \tag{H11}$$

in which α is the vertical compressibility of the granular skeleton of the medium, treated as a continuum, β is the compressibility of the fluid and n is the porosity of the medium. The storage coefficient is dimensionless, and it may be conceived of physically (Figure H67) as the amount of water in storage that is released from a column of aquifer with unit cross-sectional area and per unit decline of head (De Wiest, 1969). Its two parts may be interpreted as:

$\gamma b\alpha$ = water in storage released due to the compression of the intergranular skeleton per column with unit cross-section and per unit decline of head.

$\gamma bn\beta$ = water in storage released due to the expansion of the water per column with unit cross-section and per unit decline of head.

The coefficients T and S are called the formation constants of the aquifer and the knowledge of these coefficients is indispensable in the planning of the production of an aquifer. S and T are determined in the field by means of pumping tests. This proposed a solution of equation (H10) in the case of an aquifer of infinite extent, initially and uniformly under a constant head H, and then pumped for a constant flow rate Q. If s represents the drawdown $H - h$, the solution proposed by Theis is

$$S = \frac{Q}{4\pi T}W(u) \tag{H12}$$

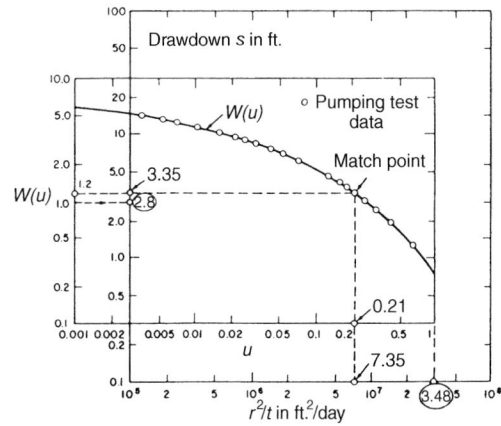

Figure H68 Theis' graphical method to determine S and T. (Courtesy R.J.M. de Wiest.)

in which

$$u = \frac{r^2 S}{4Tt} \tag{H13}$$

where r is the distance from the pumped well to the well where the drawdown is observed and t is the time since pumping started. The function $W(u)$ is tabulated as the exponential integral. S and T are determined by superimposing (Figure H68) a graph of $\log W(u)$ versus $\log u$ on a plot of $\log s$ versus $\log r^2/t$, so that the pumping data fit the tabulated data. A match point chosen on the overlapping portion of the sheets determines mutual values of s, $W(u)$, r^2/t and u which may be inserted in equations (H12) and (H13). The solution of equations (H12) and (H13) for S and T is then straightforward.

In unsteady flow, the drawdown s satisfies the equation

$$\nabla^2 s - \frac{s}{B^2} = \frac{S}{T}\frac{\partial s}{\partial t} \tag{H14}$$

in which $B = \sqrt{(Kbb'/K')}$ is the leakage factor, b' is the thickness of the semi-confining stratum, and the other symbols are as defined before.

The solution of equation (H14) for boundary conditions and initial condition as prevailing to obtain equation (H12) becomes

$$s = \frac{Q}{4\pi T} W\left(u, \frac{r}{B}\right) \tag{H15}$$

in which $W(u, r/B)$ is the well function for leaky artesian aquifers and u is defined by equation (H13). It is evident that the case of perfect confinement can be obtained from the more general case by making $B \to \infty$ in equations (H14) and (H15).

The formation constants S, T and B are determined graphically from pumping test data by a method similar to Theis' method.

Roger J.M. De Wiest

Bibliography

Davis, S.N. and De Wiest, R.J.M., 1966. *Hydrogeology*, New York, John Wiley & Sons, 463 pp.

De Wiest, R.J.M., 1965. *Geohydrology*, New York, John Wiley & Sons, 366 pp.

De Wiest, R.J.M., 1969. *Flow through Porous Media*, New York, Academic Press, 530 pp.

Hubbert, M.K., 1940. The theory of groundwater motion, *J. Geol.*, **48**(8), 785–944.

Meinzer, O., 1942. *Hydrology*, New York, Dover, 712 pp.

Muskat, M., 1937. The Flow of Homogeneous Fluids Through Porous Media, New York, McGraw-Hill Book Co., 763 pp.

Shaw, E.M., 1994. *Hydrology in Practice*, 3rd edn, London, Chapman & Hall, 569 pp.

Todd, D.K., 1980. *Groundwater Hydrology*, 2nd edn, New York & Sons, John Wiley, 535 pp.

Tolman, C.F., 1937. *Ground Water*, New York, McGraw-Hill Book Co.

Cross references

Groundwater
Groundwater: UK
Hydrology
Hydrology: coastal terrain
Ghyben–Herzberg theory
Water table

HYDROMECHANICS

The term 'hydromechanics' is generally applied to the fluid mechanics of incompressible flows. In the geological field, these flows include those of the oceans, rivers and groundwaters. Two phenomena are encountered in certain hydromechanical situations that are unique to flowing liquids: the existence of a free surface and the occurrence of cavitation, or low-pressure boiling.

Analyses of hydromechanical flows are based on the laws of conservation of mass, energy and momentum. These analyses can be found in the works of Lamb (1945), Milne-Thomson (1955), McCormick (1973), Schlichting (1960) and Valentine (1959). For reference, some of the equations needed in hydromechanical analyses are presented here. The reader should consult the listed references for more complete coverage.

Hydrostatics

As the term implies, hydrostatics refers to fluids at rest. Included in this area are problems in ship stability, pressure-hull design, and dam design (see Dams).

The basis equation of hydrostatics is the following (McCormick, 1973):

$$p = -\int_0^{-h} \gamma \mathrm{d}z = \begin{cases} \gamma h, & \gamma \text{ invariant} \\ F(h), & \gamma \text{ variable} \end{cases} \tag{H16}$$

where p is the static pressure, γ is the specific weight of the fluid, z is measured vertically upward from the free surface and h is the depth of the fluid as in Figure H69. In deep-ocean problems, γ of salt water does vary due to the enormous pressures experienced at great water depths.

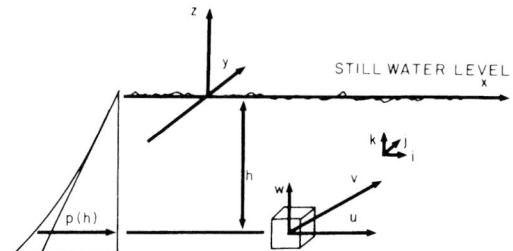

Figure H69 Notation for hydrostatics and hydrodynamics.

Hydrodynamics

As previously mentioned, hydrodynamic analysis involves the conservation of mass, momentum and energy. In two-dimensional flows, laws of conservation are expressed as follows.

The conservation of mass for an incompressible flow of velocity (Figure H69) is

$$v = ui + wk \tag{H17}$$

The conservation of mass is expressed as

$$\nabla \cdot v = 0 \tag{H18}$$

If the flow is irrotational, the velocity can be represented by a velocity potential ϕ as

$$v = \nabla(\phi) \tag{H19}$$

and equation (H17) can be expressed as

$$\nabla^2(\phi) = 0 \tag{H20}$$

which is Laplace's equation.

The general expression for the conservation of momentum of a flowing fluid, called the Navier–Stokes equation, is expressed as

$$\rho\left(\frac{\partial v}{\partial t} + v \cdot \nabla v\right) = -f - \nabla(p) + \mu\nabla^2(v) \tag{H21}$$

where ρ is the mass-density of the fluid, f is the resultant body where force on a fluid element, such as the gravitational force, and μ is the coefficient of viscosity. Equation (H21) has no general solution since it is non-linear due to the term $v \cdot \nabla v$. There are, however, many situations in which simplifications of equation (H21) can be made and solutions obtained (Schlichting, 1960).

If the viscosity can be neglected, then equation (H21) reduces to

$$\rho\left(\frac{\partial v}{\partial t} + v \cdot \nabla v\right) = f - \nabla(p) \tag{H22}$$

which is called Euler's equation.

If the non-linear term in equation (H22) is replaced by using the identity

$$v \cdot \nabla v = \nabla\left(\frac{v^2}{2}\right) - v(\nabla \cdot v) \tag{H23}$$

and the flow is assumed to be irrotational so that $\nabla \cdot v = 0$ and equation (H19) can be applied, then the resulting equation can be integrated to obtain Bernoulli's equation, which is the expression of the conservation of energy for an ideal flow:

$$\rho \frac{\partial \phi}{\partial t} zyx + \tfrac{1}{2}\rho v^2 + \rho gz + p = f(t) \tag{H24}$$

where the body force f of equation (H24) is assumed to be the gravitational force in the z-direction only.

Cavitation

When Bernoulli's equation is applied to flow situations in the ocean, the time function $f(t)$ in equation (H24) can be assumed to be zero simply by assuming that the still-water level – that is, $z = 0$ – is the energy datum. If the flow is steady, then $\partial\phi/\partial t = 0$ and equation (H24) can be rewritten as

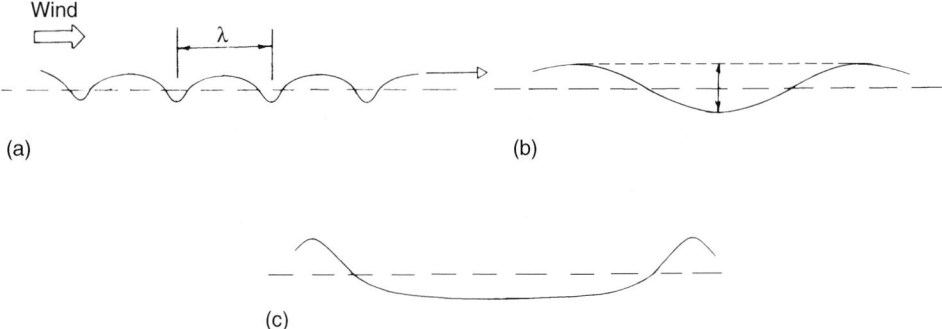

Figure H70 Three wind-generated water forms: (a) capillary wave; (b) sinusoidal wave; (c) non-linear wave.

$$p = -\rho g z - \rho v^2 \qquad \text{(H25)}$$

For a given value of z, say, $z = -d$, one sees that as the velocity, v, increases the pressure, p, decreases. It is possible, therefore, for the pressure to reduce to the vapor pressure, p_v, by increasing velocity. This situation is encountered on the tips of high-speed propellers.

A measure of cavitation susceptibility is given by the cavitation index or number:

$$\sigma = \frac{p - p_v}{\frac{1}{2}\rho v^2} \qquad \text{(H26)}$$

When $\sigma = 0$, cavitation or low-pressure boiling will occur.

Waves

Most waves on the surfaces of the oceans are caused by the wind and are thus called wind waves. The turbulent pressure fluctuation of the wind on the free surface of the water causes small deformations such as those sketched in Figure H70a. Note the narrow trough and broad crest of this wave. This wave profile is caused by the strong influence of surface tension. The wave, called a capillary wave, is first created by the wind. These small waves then grow due to the combined action of turbulence and viscous shear on the surface. Soon the profile resembles that of Figure H70b, that is, a sinusoid. The wave increases in height, H, and length, λ, and eventually has a profile like that sketched in Figure H70c. Note the difference between the profiles of Figures H70a and H70c.

The wave moves at a velocity, c, which is called phase velocity, or celerity. For an irrotational and sinusoidal wave, the celerity in deep water – that is, where $\lambda/2 < h$ is approximately

$$c = \left(\frac{g\lambda}{2\pi}\right)^{\frac{1}{2}} \qquad \text{(H27)}$$

and the wavelength is

$$\lambda = \frac{gT^2}{2\pi} \qquad \text{(H28)}$$

where T is the wave period.

The water particles in this wave travel in nearly circular orbits with the velocity components of

$$u = Hg/2c \cos(kx - wt)$$
$$\text{and } w = Hg/2c \sin(kx - wt) \qquad \text{(H29)}$$

where $k = 2\pi/\lambda$ is the wavenumber and $w = 2\pi/T$ is the circular frequency. When the horizontal particle velocity at a crest equals the celerity, then the wave is said to break, that is,

$$\frac{u}{\left(\frac{H}{2}\right)} = c \qquad \text{(H30)}$$

After the wave breaks, turbulence occurs, dissipating most of the wave energy.

Refer to Lamb (1945), Milne-Thomson (1955) and McCormick (1973), and the *Shore Protection Manual* (US Army Staff, 1974) for more complete discussions of the mathematical and physical descriptions of waves. Bascom (1980) gives an excellent non-mathematical discussion of waves.

Michael E. McCormick

Bibliography

Bascom, W., 1980. *Waves and Beaches*, Garden City, NY: Doubleday, 366 pp.
Lamb, H., 1945. *Hydrodynamics*. New York: Dover, 738 p.
McCormick, M.E., 1973. *Ocean Engineering Wave Mechanics*. New York: Wiley-Interscience, 179 pp.
Milne-Thomson, L., 1955. *Theoretical Hydrodynamics*. New York: Macmillan, 632 pp.
Schlichting, H., 1960. *Boundary Layer Theory*. New York: McGraw-Hill, 647 pp.
US Army Staff, 1974. *Shore Protection Manual*. Washington, DC: US Government Printing Office, 3 vols.
Valentine, H., 1959. *Applied Hydrodynamics*. London: Butterworths, 272 pp.

Cross references

Dams
Hydrodynamics: porous media

HYDROMETEOROLOGY

Hydrometeorology is the application of meteorology to hydrological problems. An alternative definition is to refer to hydrometeorology as the applied science which deals with interface problems between meteorology and hydrology. Whatever definition is used, it must recognize that the hydrological system is a totally integrated system, no one part being completely independent of other parts. It is therefore difficult to describe the interface between meteorology and hydrology in a way which is not dependent upon application.

C.G. Collier

Cross references

Hydrological cycle
Hydrology

HYDROPOWER AND ENERGY-RELATED PROJECTS

General

Energy is one of the most important commodities for the satisfaction of physical needs and for providing economic development of modern society. Energy needs are continually growing. To date, the world energy market has depended almost entirely upon the non-renewable, but lowcost, fossil fuels. Energy produced by hydroelectric developments throughout the world provides approximately one-fifth of the world's total electrical energy.

Electric power generation constitutes a major demand on water resources, so that hydrological data and information are essential to planning the exploitation of both renewable and non-renewable electrical energy sources. Although water is one of the two essential components in the production of hydroelectric energy, this is essentially a non-consumptive use as well as a non-polluting one. In the production of thermal-electric energy, water is required in practically all technical stages from the boring of test wells in oil and gas exploration to the transformation of fossil and nuclear fuels into electrical energy at thermal power stations – uses which are largely consumptive and/or polluting.

Recently, new problems have evolved with the exponential growth of electrical energy demands. These are the issues of water supply for energy production and the impact of energy developments on climate and the global environment.

Hydroelectric power

Hydropower is a source of electrical energy that is continually renewed and available in the runoff segment of the hydrological cycle. Energy from flowing water offers something unique to a nation's economic development – sustainability, which has been defined by the Bruntland Commission as 'economic activity that meets the needs of the present generation without jeopardizing the ability of future generations to meet their needs' (WCED, 1987). Hydroelectric schemes are diverse, not only as a result of the different natural conditions to which they may be adapted, but also because of the diversity of circumstances related to power demand and utilization. Hydroelectric power is frequently developed as part of a multipurpose project so that the project may involve the full range of water resources considerations, e.g. flood control, navigation, irrigation, municipal and industrial supplies, recreation, and fish and wildlife enhancement. Rarely does a project concern a local area only. Usually, an entire river basin is under investigation, which entails regional, national and even international considerations. In considering any magnitude of development, the planning phase must take into consideration all water resources needs of the region and the ways in which such needs are to be met. The effects of a proposed hydroelectric development on the resources and various needs in a region, and the capacity to meet those needs, must be carefully evaluated.

Although hydroelectric projects have become increasingly large during the past quarter-century, small hydroelectric plants of up to a few megawatts (MW) can economically exploit the energy at potential sites on small streams, or they can often be integrated into existing dams or artificial waterways.

Advantages

Although hydroelectric developments throughout the world provide a relatively small percentage of the global electrical energy demand, the importance of their outputs is proportionally greater than that from other sources. It is especially significant as an economic stimulus in developing countries and as an important part of complex power systems in more industrialized countries. Its importance will not diminish because

- hydroelectric energy is derived from a continuously renewable resource powered by the energy of the Sun, which sustains the hydrological cycle;
- hydroelectric energy is non-polluting – significant heat or noxious or greenhouse gases are not released in its production;
- hydroelectric plant efficiencies can be close to 90%, whereas fossil-fired thermal plants attain efficiencies of only 30–40%;
- hydroelectric plants have a long, useful life;
- hydroelectric technology is a mature technology offering reliable and flexible operation, and its equipment is readily adapted to site conditions;
- water in storage provides a means of storing energy and may be available for other purposes;
- hydroelectric plants are capable of responding, within seconds, to changes in electrical demands;
- hydroelectric generation has no fuel costs and, with low operating and maintenance costs, it is essentially inflation proof.

Of course, a potential hydroelectric development may be subject to geopolitical constraints, e.g. flooding of upstream areas, to create a head and/or storage, as well as environmental impacts, e.g. changing a riverine to a lacustrine ecology.

Site potential

Hydroelectric energy is developed by the transformation of the energy in water falling from a higher level to a lower level into mechanical energy on the turbine generator shaft, and thence into electrical energy through the generator rotor and stator. The power potential of a site, in kW, is:

$$P = 9.81\, Q\, h\, e \qquad\qquad (H26)$$

where Q is discharge in $m^3\, s^{-1}$, h is the net head (fall) in meters and e is the plant efficiency. The head to be utilized may result from control of the natural characteristics of the watercourse, such as steep gradients, rapids and falls, or it may be created artificially by the construction of a dam that may create a substantial storage or reservoir area that can be used to regulate or change the natural flow regime of the stream. The flow available for use is, allowing for losses, the flow of the watercourse on which the plant is constructed, but this flow may be modified in several ways:

- by regulation;
- by means of reservoirs;
- by diversions from contiguous river basins;
- by pumping to enable an upstream reservoir to be used for energy storage.

The amount of capacity to be installed at a potential site depends not only upon the magnitude and the regime of the streamflow and the available head or fall through which that flow may be utilized, but also upon the size of available storage capacity, the lengths of the waterways, operating limitations imposed in the interests of other water uses and, very importantly, upon the magnitude and characteristics of the power markets to be served.

The gross head on a hydroelectric plant is the difference between headwater (forebay) elevation and tailwater (tailrace) elevation when the plant is in operation. The gross head will vary with the magnitude of flow in the stream and the reservoir, or pond, water levels. In considering the economic feasibility of a project, it is important to know the average gross head that may be expected for computing average energy, as well as the minimum gross head to enable an estimate of the firm or dependable energy, i.e. the energy that can always be supplied to consumers on demand. The average gross head is dependent upon the flow in the river below the site. Under low-flow conditions, the tailwater level would be low so that the head is usually near its maximum value, whereas, under high-flow conditions, the tailwater level would be high so that the head would be in the minimum range. A low-head plant on a river subject to periodic, large flood flows may have such a minimum reduced gross head under flood conditions that operation of the plant would be impaired to the extent that it might have little or no firm (dependable) capacity, and its operation would have to be interrupted.

While most of the gross head can be utilized in producing electrical energy, there are hydraulic losses in the intake water passages which convey the water from the forebay to the turbines, as well as losses in the exit water passages, i.e. from the turbines through the draft tube into the tailrace. The draft tube is designed to regain most of the kinetic energy of the water at its exit from the turbine runner. Within limits, all of the foregoing losses are controllable because they decrease with the increasing size of water passages or with the type of design.

The net or effective head in a hydroelectric plant is the gross head minus all losses upstream of the entrance to the scroll cases that surrounds the reaction-type turbine and at the exit from the draft tube, or to the base of the nozzle in the case of the impulse-type turbine. Thus, the net head is a function of the gross head and discharge through the plant – because velocity-head losses increase approximately as the square of the discharge – as well as the flow in the river immediately below the power site.

In order to enable a reliable estimate to be made of the energy that can be generated at a selected site requires an adequate record of streamflow along with related information and hydrological data, as follows:

- daily and/or monthly streamflow data for an extended period of time, at least 10 years;
- streamflow diversions upstream from the dam or intake works;
- flow duration curves;
- drainage areas;
- evaporation losses from proposed reservoir surfaces;
- stage–discharge relationship immediately below proposed site;

- spillway design flood hydrograph;
- dam, spillway and outlet rating curves;
- project purposes, storage available, and operating rules;
- seepage losses, fish ladder requirements and diversions from storage;
- reservoir elevation-duration information;
- annual peak discharge data to assess risks associated with spillway design;
- minimum flow requirements downstream from the site.

Examples of techniques used in analyzing hydrological data to extract relevant information for design purposes are given in the *Hydrological Operational Multipurpose System (HOMS) Reference Manual* which also provides information on the availability of software packages for the application of these techniques.

Probably the most useful tool in hydropower feasibility studies is the flow duration curve. A flow duration curve based on the day as a unit will give a more accurate curve, particularly for those portions near each end of the curve, than one based on the month. The differences will be more obvious for streams with little or no natural surface storage. The area under the curve is equivalent to the mean flow of the stream for the period of the data. A very useful form of the flow duration curve is one that is dimensionless and can be obtained by expressing the ordinate as a ratio to the mean flow. Since the general form of this curve is much the same for different streams in a similar hydrological region, it provides a means of approximating a flow duration curve with only an estimate of the mean flow of a stream. This approach might be necessary if few or no streamflow records are available, but data on precipitation and general hydrological conditions are at hand to establish a reliable estimate of the mean yearly flow for the stream under study, and adequate flow records are available for the construction of flow duration curves for other streams in the region.

The capacity of the turbines to be installed in a development will usually be of a magnitude that could utilize the flow available 20–40% of the time. However, design plant capacity will be determined by comparing the cost of different sizes of plants, their resulting outputs and the value of this power to the system. The value of hydropower to a power system is dependent upon the cost of obtaining equivalent energy from an alternative source at that time. On the other hand, the cost of energy varies with time. At times of low energy demand, such as at night, only the most efficient plants of the system would be operating so that the incremental energy cost would be small, but under the much larger daytime demands, less efficient equipment would be used so that the incremental cost of energy would be larger during these periods.

The area under the flow duration curve up to the turbine discharge capacity will give the average yearly flow that can be utilized, which, with the net head, will provide the basis for computation of the available power from the proposed installation. The effect of storage (pondage) on a flow duration curve would be to raise the curve to the right of the mean and to lower it to the left of the mean flow. If it were possible to provide complete regulation, the duration curve below the storage reservoir would become a horizontal line corresponding to the mean flow of the stream.

Storage permits within-day and within-week fluctuations in output to respond to fluctuating demand. Peaking is a term used for within-day fluctuations. Hydroelectric plants are particularly well suited for peaking operations. Load changes may be handled in a matter of seconds by altering turbine gate openings. This capability can mean significant fuel savings as well as greater security of supply in a mixed hydro/thermal system. However, should the reservoir have sufficient capacity to meet not only daily and weekly fluctuations but also greater energy production during those seasons of the year when electrical energy is in greater demand, the project should be provided with seasonal storage. For example, in many regions, energy demands are greatest in winter when the river flow may be lowest. Storage of the generally high flows from snowmelt runoff during the spring season could then augment the winter flows. Occasionally, carryover storage can be provided for extended drought periods of 1 or more years.

The determination of storage–yield relationships is one of the basic hydrological analyses associated with the design of reservoirs. Physical constraints may include limits on the area that may be flooded, i.e. on the maximum water level, on the minimum level because of the location of the low level intakes, on the discharge capacity and on the downstream channel capacity. Constraints may

also require the maintenance of fisheries during spawning or for species of wildlife dependent on water levels for their survival. In northern regions, for example, the maintenance of an ice cover or the prevention of ice jams may require the establishment of appropriate discharge maxima during the winter period.

When the reservoir capacity is fixed by conditions at the site (it usually is), the firm flow which the runoff regime from the contributing area, together with available storage space, could sustain may be determined by a mass curve analysis. The firm yield is the sum of the usable storage in the reservoir and the usable inflow during the critical low-flow period. It may not always be a simple matter to select the critical-flow period. A combination of two moderately dry years in series may be more serious than a single, isolated, very dry year.

For the planning of hydropower utilization and for the design hydropower stations, a special duration curve should be derived, which is the resultant duration of the corresponding heads and discharges and corresponds to the planned plant efficiency. This output duration curve can be produced by the successive application of equation (H26) for selected corresponding points (Q, h) of their respective duration curves.

Another valuable product for estimating hydropower resources is the hydroenergetic longitudinal profile, which corresponds to the potential energy content of the river that can be generated in an average dry or wet year. This is based on the discharge records of the stream and on the corresponding water level slopes (or the energy line).

Rule curves

The water demands for conservation storage, in a multipurpose storage project, may be partially complementary or they might be competitive so that it is necessary to make proper capacity allocations among the competing demands by removing or resolving conflicts as far as possible. From the standpoint of power generation, it is desirable to use the water according to the electric power demand by maintaining the storage level as high as possible and by generating electrical energy under the resulting higher head. For irrigation, the required water is expected to be available during the irrigation season, and municipal and industrial water will be required throughout the year. In most instances, water used for power generation is discharged into the river and can be reused in the lower reaches. Thus there is no essential conflict between power generation and water supply but, among other purposes, basic conflicts can exist. If, for example, water is diverted from the reservoir by gravity canal for irrigation purposes, it would be in direct conflict with power generation. Even though the return flow from the irrigation project would eventually find its way back into the river, such use could be in direct conflict with downstream water supply and navigation because a substantial quantity of diverted water would have been consumed in the irrigation project and its quality may have deteriorated. Therefore, where multipurpose reservoirs are contemplated, reservoir capacity planning becomes much more complex. Each objective function adds to the complexity because each use must be evaluated and the final result must be obtained by a process of optimization.

Rule curves or operational criteria should be established to minimize the conflicts between purposes. A rule curve is a guideline for reservoir operation and is generally based on a detailed sequential analysis of various critical combinations of hydrological conditions and water demands. When hydroelectric energy is a principal output of the reservoir operation, a detailed flow analysis is required to coordinate energy production with other reservoir uses to determine the project's average energy output. This also will establish the firm power and energy over the critical period, particularly when the conservation storage is relatively large and the head can be expected to fluctuate over a fairly wide range. Various operational plans may be tried in an attempt to maximize energy output while meeting other uses. When the optimum output has been achieved, a rule for operation can be developed and tested during critical low- and high-flow periods.

A simple rule curve for power operation of a single-purpose storage project will show the reservoir elevation or storage volume required to assure the generation of firm power at any time of the year. Variations of the rule curve may be developed with an upper and lower curve corresponding to whether or not the reservoir supplies are above or below normal. Hydrological forecasts will assist in maximizing energy output by minimizing spillage of water. Flow restrictions may be necessary where there are downstream constraints due to flooding. The rule curve can be adapted to reflect such constraints. Probability analysis of supplies may be employed to guide the operator on whether or not to favour the upper or lower rule curve.

Water quality

Water quality is not usually a major concern for hydropower projects, either with respect to the inflow or the outflow. Current environmental considerations ensure that biomass degradation and reduction in flow aeration in the reservoir reach are minimized. In some rivers in tropical regions, the water may be dangerously acidic and corrosive to such an extent that it may attack the runner blades and other parts of the turbine machinery. The sediment load of a river may also be a factor in designing and in limiting the useful life of a reservoir, as well as of the embedded and moving parts of the hydraulic turbine.

Energy-related projects

Although the major use of water in electrical energy production is hydroelectric, it is also essential in the thermal production of energy and is necessary in practically all of its technological stages from the boring of test wells in oil and gas exploration to the transforming of fossil and nuclear fuels into electrical energy at thermal power stations. The following descriptions provide a guide to the quantity and quality of water required for processing and consumptive use and the quality of the effluent flow from such projects. Table H6 provides a summary of the general ranges of water requirements and consumption for a number of processes related to energy production.

Fossil fuel and nuclear power generation

Uses of water for electrical generation from fossil and nuclear fuels are similar. All such power plants use water for steam and condensate system make-up, general service and potable and miscellaneous water systems. The rate of use is dependent upon the condenser cooling and waste heat rejection systems. In the case of coal-fired generating stations, water is also needed for ash transport, which requires about $0.00095 \text{ m}^3 \text{ s}^{-1} \text{ MW}^{-1}$ and, where appropriate, flue-gas desulphurization requiring about $0.00019 \text{ m}^3 \text{ s}^{-1} \text{ MW}^{-1}$. However, water for condenser cooling is the single, most significant use, and the quantity required is typically in the range of $0.032–0.044 \text{ m}^3 \text{ s}^{-1} \text{ MW}^{-1}$ based on an 8°C temperature rise across the condenser. The principal waste heat rejection systems include once-through cooling, evaporative cooling towers, and dry cooling towers. The application of regulations controlling thermal pollution of water courses is resulting in a decline in the use of once-through cooling. Evaporative cooling towers are the largest water consumer and contributors of effluent water. Dry cooling towers dissipate waste head from a power plant directly to the atmosphere by means of air-cooled heat exchangers without the addition of heat to, or consumptive use of, natural bodies of water. Plants using this system, however, require increased fuel consumption and additional plant capital cost.

As is typical of any complex system, nuclear power plants are subject to a wide variety of unplanned occurrences that may interfere with their normal operation and, in extreme cases, endanger the health and safety of the public. The probability of occurrence of more serious accidents is undoubtedly quite small in view of the large factors of safety and safeguards that are an inherent part of the design of nuclear power plants. Volume II of the WMO (1951) *Meteorological and Hydrological Aspects of Siting and Operation of Nuclear Power Plants* describes the various types of nuclear power plants that are now part of many electric utility systems. It discusses hydrological and related water resources problems that may be encountered in planning, designing, operating and decommissioning nuclear power plants.

In view of the diversity and complexity of such problems, this publication provides some examples of techniques that could be useful in solving the most significant problems. Not only the very high but also very low flows carry a special significance for the operation and safety of a nuclear power plant. From a safety point of view, highly reliable water supplies are essential for the emergency core-cooling system, the cooling of the spent fuel, and for the ultimate heat sink (IAEA, 1980). Of particular importance is the requirement for protection against flooding from any conceivable source, because flooding may cause common-mode failure, i.e. failure of two or more systems, that could reduce the efficiency of system safety measures (IAEA, 1981). It is imperative, therefore, that the best available system of hydrological forecasting of the regime of water bodies affecting a nuclear power plant is applied and also that periodic reviews of the hydrological assumptions of the planning and design of the station are carried out.

In most energy-related projects, water quality considerations are not the determining factor in the viability of development, but they may be a contributing factor in the sizing, process design, economic siting or attractiveness of the project. The composition of water arising from different sources varies widely both in the amount of dissolved salts and in the dissolved gases that it contains. Surface waters usually contain suspended matter and often organic matter in solution or suspension, derived from either decayed plant material or sewage. The increasing use of synthetic detergents, some of which are not readily destroyed in sewage treatment processes, has resulted in measurable amounts of these chemicals being present, even in public water supplies. Rainwater in industrial areas and for considerable distances downwind of emission sources, such as coal- and oil-burning furnaces, may have a low pH and be potentially corrosive. Most waters, however, can be treated to make them suitable for

Table H6 Summary of water requirements for energy-related uses other than hydropower (source: Acres International, 1982)

Process	Process water consumption			Energy development water requirements	
	Standard unit (product)	Water consumption (m³ standard unit)	Comment	Standard production rate (product)	Water requirement (m³ standard production rate) (typical values)
Coal mining (surface and underground)[a]	tonnes	0.01–0.06	–	million t year^{-1}	0.0003–0.0019
Coal mining (hydraulic)[a]	tonnes	0.08–0.14	15% makeup	million t year^{-1}	0.0025–0.0044
Coal processing[a]	tonnes	0.4–1.5	10% makeup	million t year^{-1}	0.0127–0.0475
Coal slurry pipelines	tonnes	0.95	–	million t year^{-1}	0.0301
Coal liquefaction	tonnes	2.4–3.8	–	million t year^{-1}	0.0761–0.1204
Tar sands extraction	bbl	0.88	–	bbl day^{-1}	1.02×10^{-5}
Crude oil refining	bbl	0.163	–	bbl day^{-1}	1.88×10^{-6}
Fossil fuel power plant	MW h	0.9–5.4	–	MW	0.00025–0.0015
Nuclear power plant	MW h	1.5	–	MW	0.00043
Uranium milling					
Ontario and Newfoundland	kg	0.67	Low-grade ore	t year^{-1}	2.11×10^{-5}
Saskatchewan	kg	0.4	High-grade ore	t year^{-1}	1.27×10^{-5}
Methanol production (synthesis gas and biomass)	tonnes	1.75–3.5	14–25% make-up	t day^{-1}	2.03×10^{-5} 4.05×10^{-5}

[a] Coal mining is divided into mining and coal processing. Therefore, to determine the water requirements of a coal mine operation, surface and underground mining or hydraulic mining values must be added to coal processing values to determine the total water requirements of the developments.

condenser cooling, general service, ash transport and flue gas desulphurization. However, very pure water, containing no more than a trace of dissolved salt, is required for boiler feed make-up purposes. The cost of preparing this pure water will, in general, increase in proportion to the total dissolved salts that the natural water contains.

A fossil fuel thermal power plant generates a variety of wastewater streams, the most important of which are cooling water discharge and blowdown. The largest wastewater stream is cooling water from a once-through cooling system. For coal-fired power plants, approximately 6330 kJ of heat must be dissipated by means of cooling for every kilowatt-hour of electricity generated. Cooling water discharges are often 6–9°C higher than the temperature in the receiving stream. In recent years, cooling towers have become necessary in many installations to prevent thermal pollution of natural watercourses. The next largest waste stream in a fossil fuel power plant is the cooling tower blowdown of an evaporative cooling system. Blowdown water contains high dissolved amounts of calcium, magnesium, sodium, chloride and sulfate. It also contains other agents introduced for corrosion control.

Radioactive wastes are encountered in nuclear power generation and are due, to a large extent, to such factors as leakage, blowdown, maintenance, refuelling and other mechanisms. Circulating reactor water is used as a source of heat, and corrosion products formed in the system are the primary source of radioactive isotopes in the reactor water. It is mandatory that the water used for cooling purposes, as well as that used as the source for stream, is exceptionally pure because any salts or other impurities in the water may capture neutrons and become radioactive. Another potential source of radio-isotopes in the reactor water is the fission products formed within the fuel elements. The quantity of radioactive isotopes in the reactor water depends, therefore, on corrosion rates, frequency of failure of fuel element cladding and the rate of removal by condensate and reactor clean-up demineralizers. The possible presence of radioactive isotopes in the water necessitates waste treatment precautions. In the primary circulating system, great care is required to maintain the water at a high level of purity in order to minimize build-up of excessive radioactivity caused either by impurities or by corrosion products. No primary water is wasted but a portion is removed, purified and recirculated. The danger of stress corrosion requires that the boiler water contains very low concentrations of oxygen and chlorides. To achieve this, raw water is deaerated and evaporated to reduce oxygen and chloride levels to less than 0.03 and 0.3 mg l^{-1}, respectively.

Coal mining and processing

Very little water is used in either open-pit or underground mining for the extraction of coal. In fact, seepage water is usually a nuisance, and considerable effort and cost may be expended in removing it from mine workings. Coal preparation plants use large quantities of water to clean coal, but recycling systems are generally used with the result that about 10% make-up water is required.

Coal slurry technology has been available since just prior to the turn of the twentieth century. Slurry pipelines may be economical over certain large-volume, long-distance routes, but following separation of the pulverized coal product, the water must be treated prior to discharge into a natural watercourse. Effluent-treating facilities will depend on the quality of the coal (i.e. its sulfur content, ash and minerals) proposed for transport, the chemical additives required to inhibit corrosion in the pipeline and associated equipment, and the coagulating-agent chemicals used in dewatering.

Runoff from coal mining sites contains high levels of metals, suspended solids, and sulfate from pyrite and/or marcasite, which is commonly associated with coal, shale and sandstone deposits. Upon exposure to air, these minerals form both sulfuric acid and ferric hydroxide compounds. Acid mine drainage can result from tailings ponds, waste rock piles and wherever coal is stockpiled. Impacts to receiving waters can include high acidity (pH of 2–4) and high concentrations of aluminum, sulfate and iron, and trace levels of heavy metals.

The result of deforestation, establishing access roads and the mining process itself will create increased erosion, siltation, and nitrate and cation leaching into the receiving waters. Impacts are nutrient loading and increased turbidity in the receiving waters.

Uranium mining and processing

The use of water in both underground and open-pit uranium mining is generally small and is mostly required for potable supply. The total water usage during uranium milling is not large, with most of the water being used for wet grinding.

The processing of uranium ore in concentrating mills generates wastes and effluents that are both radioactive and non-radioactive. Solid, liquid and gaseous effluents are released into the environment to a greater or lesser extent, depending on the process control and waste management measures instituted.

Petroleum production

Water supply and availability, cost, energy conservation and environmental considerations have all had an impact on petroleum refining. Modern refineries are designed with the objective of reducing water intake to a magnitude of one-fiftieth of the older once-through systems. The emphasis is now on air cooling, rather than water cooling and multiple use of water (water recycling). The extent of water utilization depends on refinery complexity, which tends to be directly related to capacity, with the larger refineries being more complex. Unit water intake capacities can range from 0.1 to 3 m^3 bbl^{-1} depending on the size, complexity and design approach.

Discharges from petroleum production and refining operations require treatment prior to release into natural watercourses. These treatment processes typically constitute settling of solids and oil/water separation. Due to the large volumes of water required in some processes, recycling design is becoming essential in new refineries.

Methanol production

The conversion efficiency for producing methanol fuel from wood or natural gas is approximately 60%. Thus a large proportion of the heat content of the original carbon-rich source materials must be rejected during the process of converting them to methanol. Approximately half of the heat loss can be rejected via an evaporation cooler, requiring approximately 3 m^3 of water to be evaporated for every tonne of methanol produced. Alternatively, if direct cooling is used, and a 10°C temperature rise is permitted, then 170 m^3 of water would be passed through the heat exchanger to remove this heat with an induced evaporation loss of 1.5 m^3 tonne^{-1} of product. Clearly, if water is scarce or costly, the process designer must choose a water-conserving method of heat rejection.

By far the largest effluent stream in the manufacture of methanol from either natural gas or wood is cooling water. The degree of contamination of cooling water for these processes is minimal, and the main consideration in wastewater disposal is thermal pollution of the receiving waters.

Source

World Meteorological Organization, 1994. *Guide to Hydrological Practices*, 5th edn, WMO, Geneva.

Bibliography

International Atomic Energy Agency (IAEA), 1980. *Ultimate Heat Sink and Its Directly Associated Heat Transport Systems in Nuclear Power Plants: A Safety Guide*. Safety Series No. 50–S6-D6, Vienna.

International Atomic Energy Agency (IAEA), 1981. *Determination of Design Basis Floods for Nuclear Power Plants on River Sites: A Safety Guide*. Safety Series No. 50SG-S1OA, Vienna.

World Commission on Environment and Development (WCED), 1987. *Our Common Future*. Oxford University Press, Oxford.

World Meteorological Organization (WMO), 1981. *Meteorological and Hydrological Aspects of Siting and Operation of Nuclear Power Plants*. Volume II, Hydrological Aspects, Technical Note No. 170, WMO No. 550, Geneva.

Cross references

Water budget analysis
Water resources

HYDROSPHERE

More than 99% of the Earth's water is in the ocean, and its present volume appears to have been reached fairly early in planetary history. Small additional supplies of juvenile water may be added to that volume through steam from volcanoes or hydrothermal seeps at sea-floor spreading centers, but most of this apparently new water is probably recycled due to burial of crustal sediments and dehydration of sedimentary minerals.

The ocean, with its much larger thermal capacity and sluggish overturn, is often said to play the role of a regulating flywheel in the Earth's atmospheric heat engine. The rate of that overturn, shown by using radioactive tracers introduced by nuclear bomb testing, is of the order of 500–1000 years. However, that value is reduced to a few decades for the near-surface waters above the main thermocline (100–500 m). Thus climate variations of less than a century are registered by sea surface temperature (SST) and temperature indicators such as planktonic foraminifera, radiolaria, diatoms and Coccolithophoridae. The skeletons of these planktonic organisms accumulate on the deep-sea floor and can be analyzed (by $^{18}O/^{16}O$ ratios and distribution statistics). Unfortunately, in most areas the surface layers are disturbed or homogenized by bioturbation and it is found that in general climatic variations of less than 3000 years are smoothed out. On the other hand, the same process helps to standardize the evidence of fluctuations relating to the Milankovitch periodicities (Imbrie and Imbrie, 1979a,b). It was from this deep-sea sediment-core data that the positive proof of the Milankovitch theory was finally established.

The salinity of seawater (determined by the ratios of so-called conservative ions) remains more or less constant through time, but certain components such as CO_2, or the phosphate ion are highly variable and can be employed in the science of paleoceanography as climate indicators. CO_2 is highly soluble in seawater where there is 40 times more of it than in the atmosphere. Local disequilibrium is only of transient importance. Near-surface waters are commonly saturated with respect to $CaCO_3$ (both in calcite and aragonite phases), but this is not so for deep waters which, coming from polar latitudes, are cold and rich in CO_2. Thus the calcite or aragonite shells of planktonic organisms falling towards the bottom will gradually redissolve at great depths. The mean level of dissolution is known as the calcium carbonate compensation depth. It varies in different areas and over time. Its depth in any one place is thus a function of three things: state of mean sea level, atmospheric CO_2 concentration and supply of deep polar water. Sharp fluctuations in the calcium carbonate compensation depth through geologic time are a valuable sign of major climatic disturbance. Thus, for example, the late Eocene–Oligocene boundary and late Miocene (Messinian)–Pliocene boundary are two well-known eustatic fluctuations, when sea level (probably for tectonic reasons) dropped 100 m or more. Climatic cooling occurred, and p_{CO_2} rose in the deep water, resulting in a calcium carbonate compensation depth drop of over 1000 m (Van Andel, 1975, 1979).

The question of rising CO_2 in the atmosphere today is a matter of great concern, because of increasing amounts due to the unprecedented burning of fossil carbon (coal) and hydrocarbons (oil and natural gas) and to the large-scale clearing of forests, interfering with the usual removal of CO_2 during photosynthesis. CO_2, along with H_2O, SO_2 and so forth, are important greenhouse gases in the troposphere, and it is commonly assumed that a major rise of p_{CO_2} will cause a possibly runaway rise in temperature at the Earth's surface and then, due to eventual glacier melting, a rise of sea level (Barth and Titus, 1984). Some voices are heard to the contrary, however; for example, Newell and Hsiung (1984) argue that CO_2 rises during warm cycles because of climatic change and not vice versa, being a consequence of reduced upwelling and therefore of phytoplanktonic productivity. Analysis of gas bubbles in glacier ice from the late Pleistocene glacial maximum has shown that the p_{CO_2} at that time was only half its present value. The biological–oceanographic explanation of this paradox, pointed out by Newell and Hsiung, is a control by upwelling at low latitudes. Increased ice-age upwelling is due to steeper pole–equator temperature gradients and enhanced meridional (convective) winds. More nutrients are brought to the surface by the upwelling and increased metabolic activity raises the consumption of CO_2, leaving less in the atmosphere. Increased phosphate mobilization further encourages biosphere blooms and CO_2 intake, and further depletes the atmosphere (Broecker, 1981). During interglacials the warmer water and wider extent of the tropical belt decreases ocean

productivity and causes the CO_2 level to rise. The short-lived El Niño warming effect, which occurs on average about every 7 years, is in the nature of a mini-Hypsithermal Holocene (i.e. about 6000–8000 years ago).

A second climatic indicator in paleoceanography is phosphorite (Arthur and Jenkins, 1981). It forms discontinuously in the marine environment during times of high or rising sea level and warmer climates. Warm cycles increase chemical weathering on land, liberating phosphorus from igneous rocks. Poorly oxygenated shelf seas help retain organic phosphorus and have been extensive during the thalassocratic phases of Earth history. As Broecker (1981) showed, during glacial low sea-level stages, the shelves are uncovered, and the offshore water overturn favors phosphorus uptake by pelagic organisms.

A third paleoceanographic indicator is the oceanic anoxic event (OAE). These may be brief, lasting a few hundred or thousand years, as occurred episodically in the Mediterranean during the history of the Quaternary, or they may persist for several million years at a time as developed, for example, in North America during the Devonian. In either case there is an accumulation of black sapropelic mud (black shale, after diagenesis). They often contain beautifully preserved pelagic fossils, such as fish skeletons, which attest to the fact that the surface waters were oxygenated, whereas the bottom water was stagnant, anoxic, and free from bottom-scavenging organisms that would otherwise have destroyed the fish skeletons. The Mediterranean Quaternary examples appear to match times of sudden inrush of fresh water, principally from the Nile (Rossignol-Strick, 1985) but also from northern rivers carrying glacial meltwaters. In either case a density stratification is created that prevents the usual hydrodynamic overturn, thus reducing the bottom waters to anoxic stagnation (redox or euxinic condition).

The pre-Quaternary anoxic events of greater duration correlate not with brief interglacial transgressions but with extended periods of very warm climates (Schlanger and Cita, 1982). What were the oceanographic conditions that played a role in this relationship? The question is still open. It seems likely, as speculated long ago by Chamberlain (1909), that the density convection of present-day oceans might easily be reversed if the present polar bottomwater input were removed. Then the high equatorial evaporation would increase the density, so there would be an equatorial downwelling. From all indications, ^{18}O isotopes, for example, the Eocene (50 Ma BP) bottom temperatures were of the order of 15°C, in contrast to the present 1–2°C. Do the ancient black shales perhaps indicate a general bottom stagnation due to an extreme cloud cover that inhibited the usual evaporation? Not all ancient black shales were euxinic or anoxic, because tracks and trails of benthic organisms disclose a habitable condition (these are diagenetic), but mostly they reflect a high-H_2S poisonous environment, unfavorable to life except for sulfate-reducing bacteria.

Berry and Wilde (1978) postulated that the Paleozoic deep-sea environment was universally anoxic. Life was only possible on the broad continental shelves. Complete ventilation of this enormous redox reservoir only became possible toward the end of that era when cold, dense bottom waters of the Permo-Carboniferous glaciation led to overturning such as we see today.

In some of the black shale sequences there is a well-developed cyclicity, which suggests a climatic forcing. The cause of such redox cycles is controversial, however, and for cases studied by Dean and Gardner (pp. 70–75 in Schlanger and Cita 1982), a post-depositional diagenetic alteration of high- and low-organic layers is favored. A pulsating supply of organic debris can be easily explained by turbidity current, but even then a climatic forcing may be a key factor. However, a basin-wide production of rhythmic black shales is not the same as ocean-wide anoxic conditions, which represent a subject of ongoing research.

R.W. Fairbridge

Bibliography

Arthur, M.A. and H.C. Jenkins, 1981. Phosporites and paleoceanography, *Acta Oceanologica*, pp. 83–98.
Barth, M.C. and J.G. Titus (eds), 1984. *Greenhouse Effect and Sea Level Rise*. New York: Van Nostrand Reinhold.
Berner, E.K. and Berner, R.A., 1987. *The Global Water Cycle*. Englewood Cliffs, NJ: Prentice-Hall, 397 pp.
Berry, W.B.N. and P. Wilde, 1978. Progressive ventilation of the

oceans – an explanation for the distribution of the Lower Paleozoic black shales, *Am. Jr. Sci.*, **278**, 257–275.

Broecker, W.S., 1981. Glacial to interglacial changes in ocean and atmosphere chemistry, in *Climatic Variations and Variability: Facts and Theories*, A. Berger (ed.). Dordrecht: Reidel, pp. 111–121.

Chamberlin, T.C., 1909. Diastrophism as the ultimate basis of correlation, *J. Geology*, **17**, 685–693.

Drever, J.I., 1997. *The Geochemistry of Natural Waters*. Upper Saddle River, NJ: Prentice-Hall, 397 pp.

Imbrie, J. and K.P. Imbrie, 1979a. *Ice Ages: Solving the Mystery*. Short Hills, NJ: Enslow Publishing and New York: Macmillan.

Newell, R.E. and J. Hsuing, 1984. Sea surface temperature, atmospheric CO_2 and the global energy budget: Some comparisons between the past and present, in *Climatic Changes on a Yearly to Millennial Basis*, N.A. Mörner and W. Karlén (eds). Dordrecht: Reidel, pp. 533–561.

Rossignol-Strick, M., 1985. Mediterranean Quaternary sapropels, an immediate response of the African monsoon to variation in insolation, *Palaeogeography, Palaeoclimatology, Palaeoecology*, **49**, 237–263.

Schlanger, S.O. and M.B. Cita, 1982. *Nature and Origin of Cretaceous Carbon-rich Facies*. New York: Academic Press.

Van Andel, T.H., 1975. Mesozoic/Cenozoic calcite compensation depth and the global distribution of calcareous sediments, *Earth and Planetary Sci. Letters*, **26**, 187–194.

Van Andel, T.H., 1979. An eclectic overview of plate tectonics, paleogeography, and paleoceanography, in *Historical Biogeography, Plate Tectonics, and the Changing Environment*, J. Gray and A.J. Boucot (eds). Corvallis: Oregon State University Press, pp. 9–25.

Cross references

Atmospheric processes associated with water in the atmosphere
Climate and climate change
Greenhouse effect

INFILTRATION: INTRODUCTION

Infiltration is the process of water entry into the soil through the earth's surface. The water at the soil surface can originate from rain, snow-melt or anthropogenic activities (e.g. to regulate groundwater formation by artificial infiltration). As infiltration divides water resources into surface and subsurface water, it is a key process in the hydrological cycle. The infiltration depends on the availability of water at the soil surface and on soil characteristics which influence the water retention capacity and hydraulic conductivity. The movement of water into the soil is caused by gravitation and is affected by forces of soil particles on the water. As these forces depend mostly on the soil water content, infiltration is a non-linear time-dependent process. At the beginning the infiltration intensity of unsaturated soil is at its maximum as the gradient of water at the soil surface is maximal. Adsorption and capillary forces enhance the water movement into soil. If the upper soil becomes water saturated infiltration intensity decreases. The infiltration intensity narrows down to the hydraulic conductivity of saturated soil. During a rainfall event all water will infiltrate until the soil surface becomes saturated. After this ponding time the infiltration rate decreases exponentially. The infiltration is strongly influenced by vegetation which protects the soil surface against the kinetic energy of raindrops, which has impacts on the density of the upper soil through the roots and the macropore system and on the water storage at the soil surface. Bare soils usually have less infiltration than areas of the same soil type covered with vegetation.

Infiltration curves

Infiltration curves have the following features (refer to Figure I1):

1. the rainfall rate is high, the upper soil is saturated during the first time interval, the infiltration rate decreases;

2. the infiltration rate is equal to rainfall rate until the upper soil is saturated, after the ponding time the infiltration rate decreases until the hydraulic conductivity of saturated soil is reached;

3. the precipitation rate is less than the hydraulic conductivity of saturated soil, infiltration rate is equal to precipitation, and there is no ponding.

A.H. Schumann

Cross references

Drainage
Hydrological cycle
Hydrology
Infiltration
Water movement in unsaturated soils

INFILTRATION

Introduction

Infiltration is the process of entry into the soil, and subsequent movement, of water made available (under appropriately defined conditions) at its surface. This 'surface' may be the more or less horizontal upper surface of the soil, or it may be the bed and walls of a basin, pond, stream, ditch, irrigation bay or furrow, or, again, it may be the walls of a natural or artificial tunnel or cavity.

The process of infiltration is centrally important in the terrestrial sector of the hydrological cycle: it governs the fate of precipitated water reaching the land surfaces of the Earth. Some may be ponded temporarily, but ultimately (except for a generally minor evaporation loss) all the water either infiltrates into the soil or runs off. This represents a radical hydrological parting of the ways. The water which runs off will flow overland until it enters a surface channel which is a tributary of a branching system of streams generally discharging into the sea or a lake. With exceptions (which we shall note), the infiltrated water is lost to the surfacewater system. A large fraction of it is retained in the unsaturated zone between the soil surface and the groundwater (if any), ultimately returning to the atmosphere as transpiration from plants or as evaporation from base soil surfaces. Some part (very small in arid regions, but larger in humid lands) may join the groundwater and thereby ultimately enter surface streams, lakes or the seal and, in some circumstances, a small faction may take part in subsurface flow above an impermeable stratum, re-emerging from hillside springs or entering ponds or streams.

Beyond its central role in the natural hydrological cycle, infiltration is also of great importance to many forms of human intervention in the landscape. Irrigation, water distribution and delivery, water spreading, drainage and soil and water conservation are notable examples. Infiltration occurs also in connexion with the human misadventures of the spillage and leakage of pollutants.

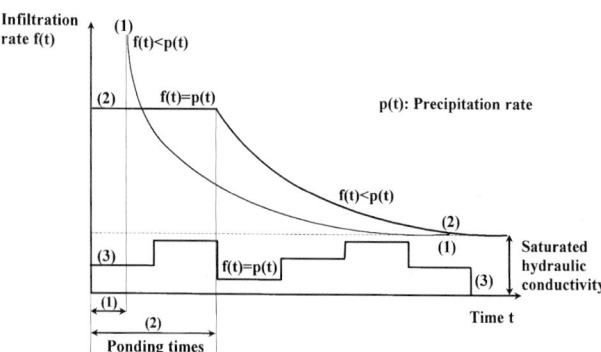

Figure I1 Features of infiltration curves.

In the natural hydrological context we are concerned usually with one-dimensional infiltration, with the relevant space coordinate vertically downward. Sloping topography requires some modification and we require a perturbation of the one-dimensional analysis. In some irrigation applications infiltration is one-dimensional, but in others and in many other human-made circumstances, the infiltration is two- or three-dimensional; and we need to understand the influence of dimensionality on the process.

In general, infiltration flows take place in unsaturated soils, and the physics of flow in such soils is central to understanding the infiltration process. Accordingly this article cross-refers heavily to the companion article on Water movement in unsaturated soils (q.v.). As in that article, we develop our theme in terms of the modern mathematical–physical approach to the analysis of water movement in unsaturated soils.

As we shall see, the dynamics of infiltration is the outcome of subtle interactions between gravity, capillarity and the geometry of the particular system. Note that here, and in what follows, we use 'capillarity' as shorthand for the totality of local soil–water interactions that include capillarity, physical adsorption and electrostatic effects (see Water movement in unsaturated soils).

We note that, consistent with the foregoing considerations, a fundamental point of departure for understanding the physics of infiltration is the study of absorption. This is the particular case of infiltration when the effect of gravity may be neglected, as in horizontal systems, in the early stages of infiltration, and in fine-textured soils where the influence of capillarity dominates that of gravity.

One-dimensional absorption and infiltration: potential boundary conditions

We begin by considering one-dimensional processes with water made available at the surface at constant potential, Ψ_1. In the natural hydrological context, with water supplied at essentially zero hydrostatic pressure, $\Psi_1 = 0$ and the surface value of the volumetric moisture content $\theta = \theta_1 = \theta(\Psi_1)$, the saturated moisture content.

Absorption

Absorption into an effective semi-infinite homogeneous system of uniform initial moisture content is described by equation (I1) subject to conditions (I2):

$$\frac{\partial \theta}{\partial t} = \frac{\partial}{\partial z}\left(D\frac{\partial \theta}{\partial z}\right) \tag{I1}$$

$$\begin{aligned} t = 0, \ z > 0, \ \theta = \theta_0 \\ t \geq 0, \ z = 0, \ \theta = \theta_1 \end{aligned} \tag{I2}$$

Equation (I1) is the one-dimensional absorption form of equation (W12). t is time, z is the space coordinate, and $D = K\mathrm{d}\Psi/\mathrm{d}\theta$ [equation (W11)] is the moisture diffusivity. The solution of equations (I1) and (I2) is

$$z(\theta, t) = \phi_1(\theta)t^{1/2} \tag{I3}$$

with $\phi_1(\theta)$ the solution of a non-linear ordinary equation which may be found by a simple and accurate numerical technique (Philip, 1955, 1957a). ϕ_1 may be obtained analytically for an indefinitely large class of functional forms of $D(\theta)$ (Philip, 1960). This similarity solution of the nonlinear diffusion equation (I1) (Boltzman, 1894) was first recognized in the soil-water context by Klute (1952).

Sorptivity

The foregoing is the simplest transient solution of the unsaturated flow equation. As we shall see, it plays a centre role in the study of more difficult problems involving the effect of gravity and/or geometrical complications. The early stages of all infiltration processes governed by boundary conditions like (I2) are virtually independent of both gravity and geometry, and are essentially the same as one-dimensional absorption. It follows that equation (I3) represents the leading term of perturbation-type analyses of more complicated problems. It is for this reason that the sorptivity, S, which arises naturally in the one-dimensional absorption problem and which we define below, is so important in the development of the theory of infiltration.

Reverting to the solution (I3), we denote by i (dimension [length], convenient unit m) the cumulative absorption into the soil at time t. It follows at once that

$$i = \int_{\theta_0}^{\theta_1} \phi\,\mathrm{d}\theta \cdot t^{1/2} = St^{1/2} \tag{I4}$$

where

$$S = \int_{\theta_0}^{\theta_1} \phi\,\mathrm{d}\theta \tag{I5}$$

Furthermore, we denote by v, the absorption (or infiltration) rate, which is also the flow velocity at $x = 0$. Then

$$v = \mathrm{d}i/\mathrm{d}t = \frac{1}{2}St^{-1/2} \tag{I6}$$

It is seen that the integral properties of the process [by which we mean functions such as $i(t)$, $v(t)$] are completely determined by the quantity S which we call the sorptivity of the soil (Philip, 1957d). Evidently, S embodies in a single parameter the influence of capillarity on the transient flow processes that follow a step-function change in θ (or, more precisely, the potential Ψ) at the surface of the soil. Strictly, we should write $S(\theta_0, \theta_1)$ or $S(\Psi_0, \Psi_1)$, since S has meaning only in relation to an initial state of the soil and an imposed boundary condition; however, we may usually omit the arguments of S without ambiguity. The dimensions of S are [length] [time]$^{-1/2}$, and we use the unit m s$^{-1/2}$. Compare the dimensions of conductivity K, [length] [time]$^{-1}$.

Philip (1957d) determined the variation of S with θ_0 both for the foregoing exact analysis and for the delta-function, or Green-Ampt, model discussed below. There is no special difficulty in calculating S in cases (such as when water is supplied under positive pressure) where the analysis must be made, at least in part, with Ψ as the dependent variable. Philip (1958a) studied the effect on S of the depth of the supplied water, and related the results to the delta-function model (Philip, 1958b).

Under the assumption that absorption is the consequence of viscous flow induced by capillarity (certainly the primary mechanism) we may define the intrinsic sorptivity, $\mathcal{S}$, by the equation

$$\mathcal{S} = (\mu/\sigma)^{1/2}S \tag{I7}$$

where μ is the dynamic viscosity and σ is the surface tension.

$\mathcal{S}$ is, then, like the intrinsic permeability κ, a characteristic of the internal geometry of the medium, and is independent of the properties of the absorbate. It is interesting to note that the dimensions of $\mathcal{S}$ are [length]$^{1/2}$, whereas those of κ are [length]^2.

Infiltration

When the system of the section on Absorption (above) is vertical, with gravity important, we replace equation (I1) by the vertical one-dimensional form of equation (W12)

$$\frac{\partial \theta}{\partial t} = \frac{\partial}{\partial z}\left(D\frac{\partial \theta}{\partial z}\right) - \frac{\mathrm{d}K}{\mathrm{d}\theta}\frac{\partial \theta}{\partial z} \tag{I8}$$

Here z becomes the vertical space coordinate, taken positive downward. Infiltration is then described by the solution of equation (I8) subject to conditions (I2). Philip (1954, 1957a,b) gave the required solution:

$$z(\theta, t) = \phi_1(\theta)t^{1/2} + \phi_2(\theta)t + \phi_3(\theta)t^{3/2} + \dots \tag{I9}$$

It is notable that ϕ_1 here is precisely that found for absorption, and that the functions ϕ_2, ϕ_3, etc., are the easily obtained solutions of linear ordinary equations. For the example depicted in Figure I2, computed for the Yolo light clay of Figures W14–W16, the first four terms of equation (I9) give an accurate solution ($\leq 0.5\%$) for t as large as 10^6 s.

This solution is supplemented (Philip, 1957c) by the asymptotic traveling-wave solution, valid for large t:

$$z(\theta, t) = (t - t_0)u + \zeta(\theta) \tag{I10}$$

with u, the downward velocity of the moisture profile, given by

$$u = [K(\theta_1) - K(\theta_0)]/(\theta_1 - \theta_0) \tag{I11}$$

ζ is found by a simple quadrature and t_0 from a matching procedure.

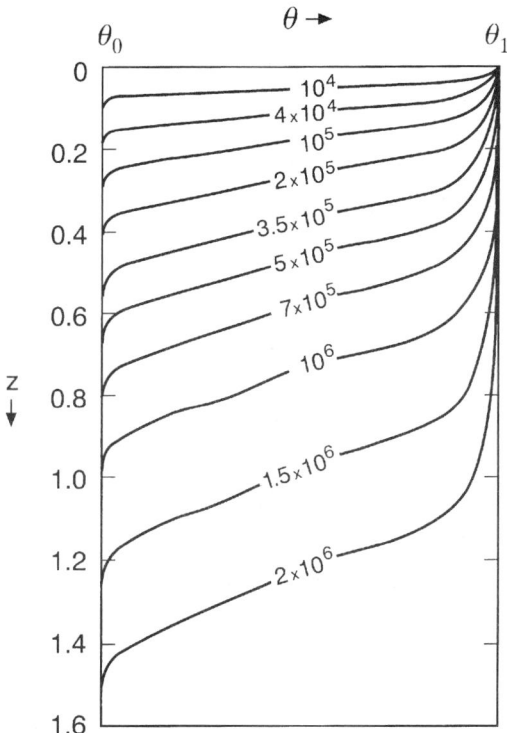

Figure I2 Computed moisture profiles for one-dimensional infiltration into the soil characterized by Figures I1, I2 and I3. Numerals on the profiles represent values of $t(s)$. Profiles for $t \leq 10^6$ calculated from first four terms of (I9); those for larger t are based on (I10). (After Philip, 1957b, c.)

Solutions (I9) and (I10) reveal the interesting evolution of the shape of the moisture profile: at small t we have, to a good approximation, the gradual absorption profile with its depth increasing as $t^{\frac{1}{2}}$; the relative stepness of the 'wetting front' gradually increases, until at large t a front of constant shape moves down into the soil with constant velocity u.

Cumulative infiltration and infiltration rate

Integrating equation (I9) with respect to θ gives the cumulative infiltration i as a series in powers of $t^{\frac{1}{2}}$. For small and intermediate times, we thus obtain (Philip, 1954, 1957d) the truncated two-term expression

$$i(t) = St^{\frac{1}{2}} + At, \quad v(t) = \frac{1}{2}St^{-\frac{1}{2}} + A \tag{I12}$$

The quantity A,

$$= \int_{\theta_0}^{\theta_1} \phi_2 \, d\theta$$

tends to be about $0.38K_1$ (Philip, 1987b).

On the other hand equation (I10) gives the large-time result

$$\lim_{t \to \infty} v(t) = K_1 \tag{I13}$$

The inconsistency between equation (I12) and (I13) is of academic rather than practical interest. See Philip (1987b) for a treatment of this 'infiltration joining problem'.

As we see from equation (I12), the infiltration rate is initially indefinitely large. This result sheds light on an important deficiency of the empirical Horton (1940) infiltration equation, which presupposes a finite 'initial infiltration rate'.

Quasi-analytical and analytical solutions

Brute-force use of high-speed computers can, in principle, solve any well-posed flow problem described by equation (W10), or any of its

special forms, subject to appropriate conditions for any properly characterized soil but this is not so fruitful a source of understanding as it might seem at first glance. The difficulty is that each of these non-linear problems requires for each soil its own *ad hoc* solution, so that the approach is necessarily piecemeal. Such calculations tell us relatively little about the fundamental structure of the solutions. On the other hand, through taking mathematical analysis as far as we can (without losing touch with physical reality), we are able to advance understanding of the mathematical shape and the physical structure of infiltration processes. Accordingly, this article deals primarily with quasi-analytical and analytical methods.

We describe as quasi-analytical solutions those in which the solution is taken a significant part of the way by mathematical analysis. The sections on Absorption and Infiltration (above) present typical quasi-analytical solutions.

We call solutions which may be found completely analytical solutions. Such solutions necessarily hold only for special forms of the functions characterizing the soil hydrologically. These vary in their ability to represent soils accurately, but matching procedures can often ensure acceptable estimates of integral properties of the solution (all that may be needed in practice).

One-dimensional infiltration: analytical solutions for potential boundary conditions

The foregoing quasi-analytical results (Philip, 1954, 1957b,c) preceded systematic consideration of the analytical solutions noted in this section.

Green–Ampt or delta-function soil

In a very early study the Australians Green and Ampt (1911) replaced the actual gradual wetting front by an infinitely sharp one, assigning a 'wetting front potential' to this putative front. They were thus able to describe one-dimensional infiltration by an ordinary differential equation. This was shown (Philip, 1957d, 1973b) to be equivalent to assigning to D in equation (I8) the value.

$$D(\theta) = \frac{1}{2}S^2(\theta_1 - \theta_0)^{-1}\delta(\theta_1 - \theta) \tag{I14}$$

with δ, the Dirac delta function (i.e. the diffusivity is essentially all concentrated at $\theta = \theta_1$).

Linear soil

The second analytical solution (Philip, 1966b, 1969a) was for the linear soil, for which

$$D(\theta) = \text{constant} = \frac{\pi S^2}{4(\theta_1 - \theta_0)^2}$$

$$\frac{dK}{d\theta} = \text{constant} = \frac{K_1 - K_0}{\theta_1 - \theta_0} \tag{I15}$$

Burgers or Knight soil

The third is for the Burgers or Knight soil (Philip, 1987b). Certain $D(\theta)$ and $K(\theta)$ functions reduce equation (I8) to Burgers' equation, which readily yields analytical solutions through a transformation (Philip, 1973a, 1974a; Clothier *et al.*, 1981). The specialized forms for initially relatively dry soil are

$$D(\theta) = \text{constant} = \frac{\pi S^2}{4(\theta_1 - \theta_0)^2}$$

$$\frac{dK}{d\theta} = 2(K_1 - K_0)\left[\frac{\theta - \theta_0}{\theta_1 - \theta_0}\right] \tag{I16}$$

The Knight soil convincingly represents the evolution of the traveling wave at large t. Neither the Green–Ampt nor linear soils can reproduce this feature of the full non-linear solution (Philip, 1987b, 1988).

Infiltration rate and infiltration capacity

In the natural hydrological context, and in artificial situations such as sprinkler irrigation, we must distinguish clearly between two quantities, the infiltration rate, v, and the infiltration capacity v_c.

For a given soil profile, the infiltration rate is the actual instantaneous rate of water entry, which may be limited by the rate of supply to the surface. On the other hand, for the same profile, the infiltration capacity is the instantaneous rate of water entry when water is available in excess at the surface.

The case of ponded infiltration with water available in excess, has been treated above. There we have used the term infiltration rate and the symbol v; but it will be understood that, in that case, $v_c = v$, and our expressions for v also give the infiltration capacity.

On the other hand, when the rate of supply (e.g. from rainfall or sprinkler irrigation), r, is limiting, we have $v = r < v_c$, and the distinction between actual rate and capacity becomes central.

Consider the partition of infiltration and runoff for a given rainstorm. Except in the unusual circumstance where the soil surface is initially saturated, we shall have, initially, that v_c is very large [cf. equation (I11)], exceeding any finite rainfall rate r. In consequence there will always be an initial storm period with no runoff, and with

$$v = r < v_c \tag{I17}$$

If r is large enough, the surface moisture content will increase with time, ultimately reaching its saturated value unless the profile is well-drained and $r < K_1$. At the instant of surface saturation, ponding will begin, the ponded infiltration regime will take over, and we shall have

$$K_1 < v = v_c < r \tag{I18}$$

The course of the subsequent ponded infiltration (until r becomes less than v_c) then follows dynamics similar to that found in the section on One-dimensional absorption and infiltration: potential boundary conditions (above), with one modification: although the second of conditions (I2) holds, the first does not. The moisture profile at the instant of ponding is non-uniform.

For the preponding stage, no analysis is needed to evaluate v: it is simply equal to r. On the other hand, analysis *is* needed to obtain the time-to-ponding, i.e. the t-value at which θ reaches its saturated value. Accordingly, we need solutions of equation (I8) subject to flux boundary conditions. We turn to this matter in the following section.

One-dimensional infiltration: flux boundary conditions

For flux boundary conditions, equation (I8) is subject to conditions such as

$$t = 0, \ z > 0, \ \theta = \theta_0$$
$$0 \le t \le t_p, \ z = 0, \ -D\frac{\partial\theta}{\partial z} - K = r(t) \tag{I19}$$

Here $r(t)$ is the (given) flux at the surface $z = 0$, and t_p is the time-to-ponding. For well-drained soils and $r < K$ for all $0 \le t \le \infty$, $t_p = \infty$.

Quasi-analytical solutions

The applicability of quasi-analytical solutions to this class of problem is very restricted (Philip, 1969a, 1988), so that relevant analytical solutions would be especially valuable. So far it has not proved possible, however, to match analytic solutions to physical reality as robustly as for ponded infiltration (Philip, 1988).

Green–Ampt soils

Mein and Larsen (1973) and Swartzendruber (1974) used Green–Ampt models of preponding infiltration. Unfortunately the real moisture profiles during this phase tend to be very gradual, giving maximum deviation from the step-function profiles of Green–Ampt, so that we cannot expect accurate results.

Linear soils

Braester (1973) analyzed preponding infiltration for the linear soil. The limitation here is that, since $D(\theta)$ varies over the moisture range typically by a factor of 1000, the use of a mean diffusivity grossly misrepresents the early stages of the process.

Burgers or Knight soil

Clothier *et al.* (1981) found preponding solutions for the Knight soil. These gave an improved representation of later stages of the process.

But for small values of t they suffer the same difficulty, arising from using an average value of D, as do the linear solutions.

Broadbridge–White soil

A valuable step forward was made when Broadbridge and White (1988) showed that, for the usefully realistic representations,

$$D(\theta) = a(b - \theta)^{-2}$$
$$K(\theta) = \beta + \gamma(b - \theta) + \lambda(b - \theta)^{-1} \tag{I20}$$

analytic solutions can be found, though only for constant flux conditions. This result holds for finite as well as semi-infinite regions (Broadbridge *et al.*, 1988).

Flux concentration method

Somewhat apart from the preceding approaches is the flux concentration method, a versatile, physically based, integral method.

Similarity methods reduce the number of independent variables by one. Where they do not apply, integral methods may achieve this same economy. A guessed 'shape' of the solution is put into an integral of the equation with respect to the variable to be eliminated. The shape may be a once-and-for-all guess, or it may be iteratively improved. The Green–Ampt model is the primitive integral method with the advancing moisture profile taken to be a step function.

The shape used in the flux concentration method is the dependence of flux density on θ, appropriately normalized. Unlike the moisture profile shape, the flux concentration function for infiltration problems is very conservative: it does not vary much with soil properties, nor with t during a particular unsteady flow process (Philip, 1973b).

Tight bounds can be established for it, and it can be guessed to good accuracy. Simple iterative procedures have been established for various processes (Philip and Knight, 1974). But non-iterated first approximations tend to be so good that iteration has not been much used to date. The method applies to processes involving a single space coordinate. Boundary conditions may be of either potential or flux type, and may be (monotonically) time dependent.

In a pioneering paper, Knight (1983) used the method to study the evolution of infiltration during the preponding and ponding phases. There remains much scope for more elaborate and more realistic analyses of the preponding/ponding complex using the flux concentration method.

Multidimensional infiltration

Quasi-analytical solutions

Series solutions like that for one-dimensional infiltration have been found for two- and three-dimensional axisymmetrical absorption, with the three-dimensional solution supplemented by an exact solution asymptotically valid for large values of t (Philip, 1969a). There are analogous solutions for the early stages of unsteady infiltration from cylindrical and spherical cavities (Philip, 1969b).

Analytical solutions

Solutions are available for unsteady multidimensional infiltration in the linear soil (Warrick, 1974; Lomen and Warrick, 1974; Philip, 1986c), and these can be matched at large values of t with the quasilinear solutions discussed below (Philip 1986c).

It is most desirable to establish more realistic analytical solutions, such as for the Knight soil. No transformation analogous to Hopf–Cole, however, exists to linearize Burgers' equation in two dimensions, according to Broadbridge (1986), and it is unlikely that one exists in three dimensions.

Quasi-linear analysis of multidimensional steady infiltration

Quasi-linear flow equation

As shown in the entry on Water movement in unsaturated soils, the exponential representation

$$K(\Psi) = K(\Psi_1)e^{\alpha(\Psi - \Psi_1)}, \ \alpha = \text{constant} > 0 \tag{I21}$$

reduces steady forms of the multidimensional flow equations [equations (W16) and (W17)] to the linear equation

$$\nabla^2\Theta = \alpha\partial\Theta/\partial z \tag{I22}$$

Θ is the Kirchhoff (1894) potential [equation (W18)]. The flow analysis based on equation (I22) is called quasi-linear because, although equation (I22) is linear, it embodies with reasonable accuracy the known strongly non-linear decrease of K as Ψ decreases through negative values.

The quantity $2\alpha^{-1}$ is the sorptive length and α is the sorptive number. Note that α is directly proportional to a characteristic pore radius R. In fact, $R = 7.4 \times 10^{-1}\alpha$ when R is in meters and α is in m^{-1} (Philip, 1987a, 1989). The value of α ranges from 0.2 m^{-1} or less in fine-textured soils to 20 m^{-1} or more in coarse ones.

The quasi-linear analysis needs only two parameters, K_1 and α. In many practical problems $\Psi_1 = 0$, so that K_1 is the saturated conductivity. The direct field determination of K_1 and α is under active study (Philip, 1987a, 1989; White, 1988).

Normalizing space coodinates with respect to a, a characteristic length of the water supply surface, reduces equation (I22) to

$$\nabla^2\theta = 2s\partial H/\partial z \qquad (I23)$$

with $s = \frac{1}{2}\alpha a$. The dimensionless parameter s is a measure of the relative importance of capillarity and gravity in producing the flow. As $s \to 0$, capillarity dominates the flow process. As $s \to \infty$, gravity dominates.

Quasi-linear analysis and arid hydrology

A significant virtue of the quasi-linear analysis is that it is well suited to investigation of multidimensional infiltration in arid and semiarid landscapes, where the water table may be nonexistent or at great depth. Many concepts and intuitions of soil and groundwater hydrology evolved in the well-watered regions of Europe and the eastern United States. This humid hydrology, worked out for usually moist environments with a water table near (and even at) the soil surface, has not always taken much account of unsaturated flow. It can, in fact, be quite unhelpful in drier regions such as Australia, Israel and much of the western United States. Thus the classical saturated flow approaches to problems of infiltration from finite water sources to a water table at effectively infinite depth (Kozeny, 1931; Vedernikov, 1940; Polubarinova-Kochina, 1962) are physically incorrect and practically misleading (Philip, 1989, 1990). Quasi-linear analysis of such problems is feasible and avoids the inherent deficiencies of the classical approach. There is a clear need for a soundly based arid soil hydrology and growing recognition that the quasi-linear analysis is an important tool towards this end.

Quasi-linear solutions

Numerous solutions of equations (I22) and (I23) have been established. Some are based on line and point source solutions (Philip, 1966a, 1969a), supplemented by the use of certain theorems (Raats, 1971; Philip, 1971). These source solutions were the earliest and simplest, and they serve to illustrate the general character of steady infiltration from two- and three-dimensional sources. Figure I3 shows how gravity distorts the infiltration wetted region much more strongly in two dimensions than in three.

Other solutions are based on separation of variables (Raats, 1970; Philip, 1984a,b). This technique yields solutions in series form to many important problems. The practical limit to direct summation of the series tends to be about $s = 10$. Much interest, however, attaches to solutions for large values of s, since gravity dominates in the limit as $s \to \infty$. This method thus left unanswered the question, important in many engineering contexts, of the role of capillary effects in infiltration and seepage problems with large length scales.

Scattering analog

This obstacle was removed by recognition of the exact analog between steady quasi-linear flows and the scattering of plane waves (Waechter and Philip, 1985). Many established results on wave scattering, and associated potent mathematical methods, become immediately available for the solution and understanding of problems in multidimensional steady infiltration (Philip, 1985, 1989). Of greatest significance are the asymptotic methods and results for large values of s.

To establish the analog, we use the substitution $H = He^{-sz}$ to transform equation (I23) into

$$\nabla^2 H = s^2 H \qquad (I24)$$

Equation (I24) is exactly the reduced wave (Helmholz) equation for imaginary wavenumber si. The boundary conditions on equation (I24)

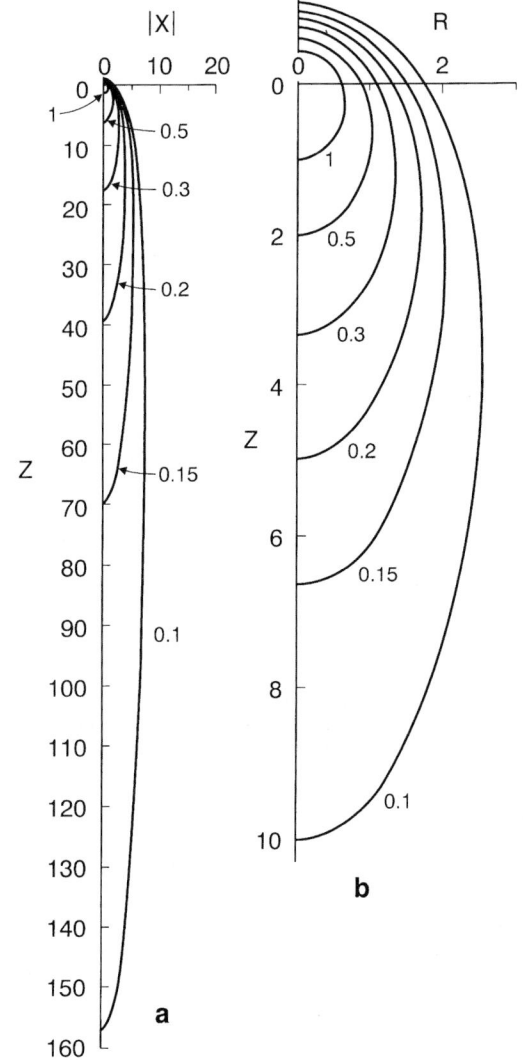

Figure I3 Dimensionless plots of quasilinear source solutions for (a) two-dimensional and (b) three-dimensional steady infiltration. Numerals on the curves are values of the dimensionless Kirchhoff potential (After Philip, 1969a.)

for steady infiltration from a cavity supplying water at constant potential Ψ_1 take the same form as those for scattering of a plane wave (wavenumber si) by an acoustically soft obstacle of the same shape. Hence the analog.

Many new results follow. An analog of the optical theorem in scattering enables the deduction of cavity discharge, and the far moisture field, from known scattering functions (Philip, 1985).

Singular character of gravity-dominated flows

It is customary to ignore capillary effects in seepage problems with large length scales or, more accurately, with large values of s. In terms of the optical analogy, ignoring capillarity is equivalent to using simple geometrical optics, in which an opaque obstacle illuminated by plane waves casts a coherent shadow projected to infinity without scattering. Note that for steady infiltration from a cavity, the 'gravity-only' solution consists simply of a saturated column extending to infinity vertically beneath the cavity, with no wetting outside the column.

The gravity-only solutions are singular in the sense that they give a wholly misleading picture of the extent and character of the wetted region (Figure I4). This result was established through the quasi-linear analysis, but it holds also for soils with $K(\Psi)$ general.

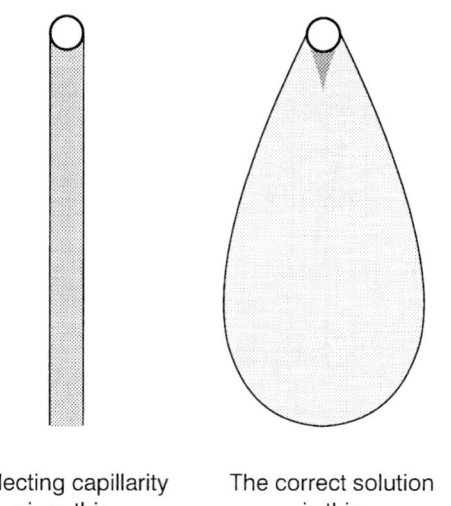

Neglecting capillarity   The correct solution
gives this   is this

Figure I4 Cartoon illustrating the singular character of the flow when gravity is dominant and capillarity is non-zero, i.e. the limit as $s \rightarrow \infty$. (After Waechter and Philip, 1985.)

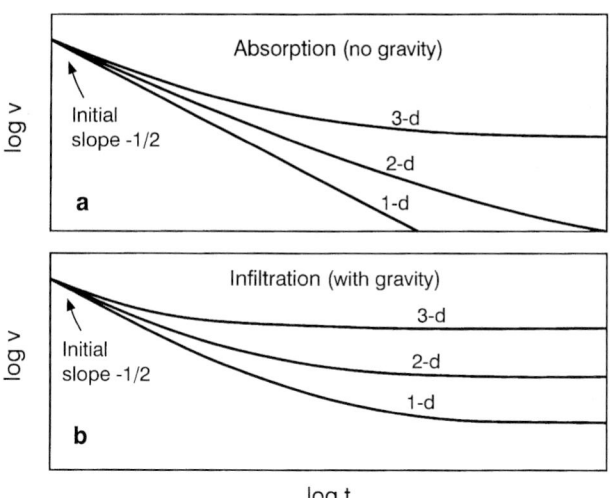

Figure I5 Schematic figure showing the effects of gravity and geometry on the dynamics of infiltration with water supplied at constant potential. For one- and two-dimensional absorption $v \rightarrow 0$ as $t \rightarrow \infty$. For three-dimensional absorption and all three infiltration processes, v approaches a non-zero value as $t \rightarrow \infty$. The steady state is approached more rapidly in three dimensions than two, and in two dimensions than one.

Infiltrometers and field permeameters

Important results of these studies of three-dimensional steady infiltration concern the performance and design of infiltrometers and field permeameters, such as rings set into the soil and boreholes. These instruments give a relatively rapid approach to a final constant rate of water loss, Q; and it has often been supposed that Q is proportional to the saturated conductivity, K_1.

The foregoing analyses of multidimensional steady infiltration show, however, that for a given instrument geometry Q is determined not only by K_1, but also by the unsaturated and capillary properties of the soil.

The scattering analog has been used to analyze steady infiltration from disc-shaped and spheroidal cavities (Philip, 1986a,b). The importance of capillarity relative to gravity in producing flow from a spheroidal source increases rapidly with aspect ratio. Design principles follow for permeameters used in unsaturated soils. Measurements with borehole permeameters with a large aspect ratio are confused by strong capillarity effects. On the other hand, these are minimal for surface disc permeameters with aspect ratio zero (Philip, 1986a; White, 1988; White and Sully, 1987, 1988; Perroux and White, 1988; White et al., 1992).

Effects of geometry and gravity on infiltration

The interactions between capillarity, gravity and geometry are complicated and subtle, and are central to the physical understanding of infiltration. Philip (1969a) compared one-, two- and three-dimensional infiltration in an infinite body of uniform soil, without and with the operation of gravity. Space considerations here allow only the summary offered by Table I1 and the schematic Figure I5. The reader is referred to the original article for a full explanation.

Extensions and limitations

In the preceding sections we have explored many aspects of infiltration *via* the basic fluid-mechanical theory of flow in un-

saturated soils (see Water movement in unsaturated soils). In this concluding section we briefly discuss further topics that demand attention. Some involve extensions of the foregoing analysis and others recognition of its limitations.

Hillslope infiltration: topographical effects

Most studies of infiltration involve the tacit assumption of a horizontal soil surface. In the last few decades, however, there has been much activity in hillslope hydrology, especially by geographers and foresters; but it is only recently that modern quantitative theory has been applied to hillslope infiltration.

Philip (1991a) solved the full non-linear flow equation for ponded infiltration on a long planar hillslope with slope angle γ. Infiltration normal to the slope follows dynamics similar to that given by equations (I9) and (I10) for a horizontal surface, except that the factor $(\cos\gamma)^{n-1}$ enters the nth term on the right of equation (I9) and the factors $\cos\gamma$ and $\sec\gamma$, respectively, enter the first and second terms on the right of equation (I10). Evidently infiltration normal to the slope is little affected so long as γ is small. Interesting aspects, however, are a time-independent total horizontal flow into the slope and a time-dependent total downslope flow which behaves as $t^{\frac{1}{2}}$ at small values t and as t at large values of t. Previously, such flows were attributed to soil anisotropy and layering, but these flows in a homogeneous isotropic soil are simple consequences of gravity and capillarity.

The analysis was extended (Philip, 1991b) to divergent and convergent slopes. The perturbations due to curvature of the contours were calculated, but need be taken into account only when contour radius is less than about $10 l_{grav}$, where l_{grav} is the characteristic infiltration length defined (Philip, 1991a) by

$$l_{grav} = \frac{S^2}{(K_1 - K_0)(\theta_1 - \theta_0)} \qquad (I25)$$

Table I1 Effect of geometry and gravity on infiltration

No. of dimensions	Absorption (without gravity)			Infiltration (with gravity)		
	Ultimate wetted region	Ultimate wetting	Steady state?	Ultimate wetted region	Ultimate wetting	Steady state?
1	Infinite	Complete	No	Infinite	Complete	Quasi-steady
2	Infinite	Complete	No	Finite	Incomplete	Yes
3	Finite	Incomplete	Yes	Finite	Incomplete	Yes

White and Sully (1987) showed that l_{grav} is of the same magnitude as the sorptive length $2x^{-1}$, and that both lengths tend to lie in the range 0.1–2 m.

The effect of slope convexity or concavity was also analyzed (Philip, 1991c). It was found that the perturbation to the long-slope solution is proportional to the total curvature of the surface, but may be neglected as long as this is greater than about $10l_{grav}$ (as it will usually be).

The long-slope study (Philip, 1991a) was much simplified by the assumption that, except in a small region near the slope crest (or a point of slope angle change), the solution is independent of the downslope coordinate. A linearized analysis of ponded infiltration near slope crests (Philip, 1992) showed that the crest effect is unimportant, and the long-slope solution is accurate, at downslope distances greater than a few times l_{grav}.

The foregoing studies for ponded infiltration were supplemented by linearized analyses for constant-rainfall infiltration (Philip, 1993a), which gave generally similar results. These various investigations of topographical effects on infiltration were reviewed in Philip (1995). In summary, the major topographical effect is due to slope angle, with the surface total curvature the second most significant topographical parameter.

Time condensation

The empirical idea of 'time condensation', and its name, are due to Sherman (1940, 1943; Sherman and Mayer, 1941). This intuitive notion proposes that, during infiltration from a rainstorm, $v_c(t)$ may be treated as a function solely of the cumulated infiltration $i(t) = \int_0^t v(t)\mathrm{d}t$, which serves as a time-like variable.

Evidently this notion cannot hold good, even approximately, for low-intensity storms ($r < K_1$); but, a priori, it seems possible that it may prove a useful approximation for storms with $r > K_1$ for at least the latter part of the preponding phase. In such cases we can eliminate t between $v(t)$ and $i(t)$ for ponded infiltration to secure an approximate $v_c(i)$ function. By illustration, we note that the two-parameter (small and moderate values of t) results [equation (I12)] then give

$$v_c = \frac{S(\sqrt{S_2 + 4Ai} + S)}{2i} + A$$

$$\sim S^2/i \text{ for } i \ll S^2/(4A) \tag{I26}$$

The $v_c(i)$ curve is asymptotic to a rectangular hyperbola in the limit as $i \to 0$.

An indication of the accuracy of time condensation follows from the work of White and Broadbridge (1989). These authors calculated for the Knight soil $v_c(i)$ in dimensionless form for three storm hyetographs: (1) for $r(t) \geq v_c(t)$, i.e. ponded infiltration; (2) $r(t) =$ constant; and (3) $r(t) \propto t$ (Figure I6).

These hyetographs are strongly contrasting, so the level of agreement of the three curves offers some support for the time condensation approximation. Note, however, that the curves diverge increasingly for normalized cumulative infiltration >1.5. For the White–Broadbridge normalization this value corresponds to $i \sim 3\alpha^{-1}(\theta_1 - \theta_0)$. We see that time condensation becomes increasingly unreliable the smaller the storm intensity, as we should expect.

Initial infiltration capacity: inertial and other effects

At various points we have emphasized that the initial infiltration capacity is indefinitely large. Strictly, of course, infinite flow velocities are physically impossible. An analysis of inertia and possible effects of turbulence and non-linear friction during the very early stages of absorption and infiltration (Philip, 1959), however, showed that these are negligible for all values of t greater than about 10^{-4} s. Equations such as (I6) and (I12) thus hold for all t – values of hydrological interest. See Water movement in unsaturated soils for further discussion of inertia.

Ponding depth effects: mixed saturated–unsaturated flow

To this point we have not discussed the effect of a definite ponded depth, or positive hydrostatic pressure, of water supplied in one-dimensional ponded infiltration. When, as is usual, the ponded depth is small compared with the sorptive length, there is no significant effect on infiltration dynamics. There is, however, no particular difficulty in taking water depth into account (see Philip, 1958a,b, for examples). In such circumstances a growing saturated zone develops

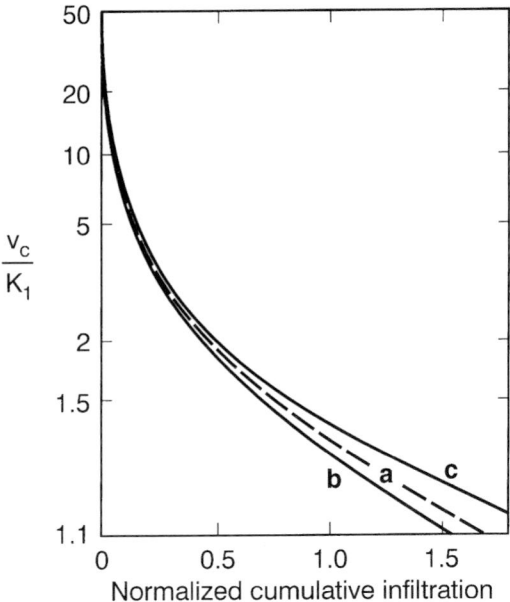

Figure I6 Three different time-condensation relations for the Knight soil: (a) for ponded infiltration, (b) for constant rainfall rates and (c) for rainfall rates proportional to t. (After White and Broadbridge, 1989.)

immediately beneath the soil surface, with an increase in sorptivity S being the principal effect on the dynamics.

The foregoing is an example of mixed saturated–unsaturated flow. More complicated mixed flows occur in multidimensional infiltration under positive hydrostatic pressure. Certain of these flows have been successfully analyzed, but there is much scope for further work. More difficult problems in mixed flow arise in cases where ephemeral perched water tables overlie impermeable strata.

Swelling soils

This article has treated infiltration in terms of the theory of flow in unsaturated non-swelling soils. The entry on Water movement in unsaturated soils develops the theory of flow and volume change in swelling soils. Here we do no more than emphasize that, because the overburden potential is an additional component of the total potential of water in swelling soils, the role of gravity in such soils is fundamentally different from that in non-swelling soils.

In mineral soils exhibiting normal swelling, for example, overburden potential essentially reverses the effect of gravity: it operates against infiltration rather than with it, so the dynamics resembles that of capillary rise in non-swelling soils (Philip, 1966c). Note that neglecting overburden potential can lead to the incorrect assumption that a swelling soil is highly impermeable: it may be a driving force which is lacking, not permeability.

On the other hand, in swelling peat soils, of specific gravity not much greater than 1, gravitational effects may be negligibly small, with infiltration dynamics indistinguishable from that of absorption.

Other complications

There are other complications we have not discussed, but which receive some attention in the entry on Water movement in unsaturated soils. We mention them here only briefly.

- Capillary hysteresis. This is of little significance in infiltration. It can emerge if rainfall rates decrease during preponding infiltration, but is of minor importance. (cf. Philip, 1993b).
- Soil air. Infiltration dynamics may be affected by air ahead of the wetting front which is entrapped or impeded in its escape.
- Hydrodynamic stability. The potential instability of infiltration flows must be recognized.
- Thermal effects. These are unlikely to be important in infiltration.
- Aggregated and cracked soils. Infiltration dynamics may be modified quantitatively, if not qualitatively, by soil aggregation. On

the other hand, gross cracking of swelling soils into deep monoliths may change completely the mechanics and dynamics of the infiltration process.

- Heterogeneity. Vertical variation of soil properties, either gradually or in layers, evidently affects infiltration dynamics.

J.R. Philip

Bibliography

Boltzmann, L., 1894) Zur Integration der Diffusionsgleichung bei variabeln Diffusionscoefficienten. *Ann. Physik*, **53**, 959–964.

Braester, C., 1973. Moisture variation at the soil surface and the advance of the wetting front during infiltration at constant flux. *Water Resour. Res.*, **9**, 687–694.

Broadbridge, P., 1986. Nonintegrability of nonlinear diffusion-convection equations in two spatial dimensions. *J. Phys. A Math. Gen.*, **19**, 1245–1257.

Broadbridge, P. and I. White, 1988. Constant rate rainfall infiltration: a versatile nonlinear model. 1. Analytic solution. *Water Resour. Res.*, **24**, 145–154.

Broadbridge, P., J.H. Knight and C. Rogers, 1988. Constant rate rainfall infiltration in a bounded profile: solutions of a nonlinear model. *Soil Sci. Soc. Am. J.*, **52**, 1526–1533.

Clothier, B.E., J.H. Knight and I. White, 1981. Burgers' equation: application to field constant flux infiltration. *Soil Sci.*, **132**, 255–261.

Green, W.H., and G.A. Ampt, 1911. Studies in soil physics. 1. The flow of air and water through soils. *J. Agric. Sci.*, **4**, 1–24.

Horton, R.E., 1940. Approach toward a physical interpretation of infiltration capacity. *Soil Sci. Soc. Am. Proc.*, **5**, 399–417.

Kirchhoff, G., 1894. *Vorlesungen über die Theorie der Wärme*. Leipzig: Barth.

Klute, A., 1952. A numerical method for solving the flow equation for water in unsaturated soils. *Soil. Sci.*, **73**, 105–116.

Knight, J.H., 1983. Infiltration functions from exact and approximate solutions of Richards' equation in *Advances in Infiltration*. St Joseph, Michigan: Am. Soc. Agric. Eng, pp. 24–33.

Kozeny, J., 1931. Grundwasserbewegung bei freiem Spiegel, Fluss und Kanalversickerung. *Wasserkraft Wasserwirt.*, **26**, 28–31.

Lomen, D.O. and A.W. Warrick, 1974. Time-dependent linearized infiltration. II. Line sources. *Soil Sci. Soc. Am. Proc.*, **38**, 568–572.

Mein, R.G., and C.L. Larsen, 1973. Modeling infiltration during a steady rain. *Water Resour. Res.*, **9**, 384–394.

Perroux, KM. and I. White, 1988. Designs for disc permeameters. *Soil Sci. Soc. Am. J.*, **52**, 1205–1215.

Philip, J.R., 1954. Some recent advances in hydrologic physics. *J. Inst. Engrs. Australia*, **26**, 255–259.

Philip, J.R., 1955. Numerical solution of equations of the diffusion type with diffusivity concentration-dependent. *Trans. Faraday Soc.*, **51**, 885–892.

Philip, J.R., 1957a. Numerical solution of equations of the diffusion type with diffusivity concentration-dependent. II. *Australian J. Phys.*, **10**, 29–42.

Philip, J.R., 1957b. The theory of infiltration: 1. The infiltration equation and its solution. *Soil Sci.*, **83**, 345–357.

Philip, J.R., 1957c. The theory of infiltration: 2. The profile at infinity. *Soil. Soc.*, **83**, 435–448.

Philip, J.R., 1957d. The theory of infiltration: 4. Sorptivity and algebraic infiltration equations. *Soil Sci.*, **84**, 257–264.

Philip, J.R., 1958a. The theory of infiltration: 6. Effect of water depth over the soil. *Soil Sci.*, **85**, 278–286.

Philip, J.R., 1958b. The theory of infiltration: 7. *Soil Sci.*, **85**, 333–337.

Philip, J.R., 1959. The early stages of absorption and infiltration. *Soil Sci.*, **88**, 91–97.

Philip, J.R., 1960. General method of exact solution of the concentration-dependent diffusion equation. *Australian J. Phys.*, **13**, 1–12.

Philip, J.R., 1966a. Absorption and infiltration in two- and three-dimensional systems, in *Water in the Unsaturated Zone*, Vol. 1. Paris: UNESCO, pp. 503–525.

Philip, J.R., 1966b. A linearization technique for the study of infiltration, in *Water in the Unsaturated Zone*, Vol. 1. Paris: UNESCO, pp. 471–478.

Philip, J.R., 1966c. The dynamics of capillary rise, in *Water in the Unsaturated Zone*, Vol. 1. Paris: UNESCO, pp. 559–564.

Philip, J.R., 1969a. Theory of infiltration. *Adv. Hydroscience*, **5**, 215–296.

Philip, J.R., 1969b. Early stages of infiltration in two- and three-dimensional systems. Australian J. Soil Res., **7**, 213–221.

Philip, J.R., 1970. Flow in porous media. *Anna. Rev. Fluid Mech.*, **2**, 177–204.

Philip, J.R., 1971. General theorem on steady infiltration from surface sources, with application to point and line sources. *Soil Sci. Soc. Am. Proc.*, **35**, 867–871.

Philip, J.R., 1973a. Flow in porous media, in *Theoretical and Applied Mechanics*. Berlin: Springer, pp. 279–294.

Philip, J.R., 1973b. On solving the unsaturated flow equation. 1. The flux-concentration relation. *Soil Sci.*, **116**, 328–335.

Philip, J.R., 1974a. Recent progress in the solution of nonlinear diffusion equations, *Soil Sci.*, **117**, 257–264.

Philip, J.R., 1974b. Fifty years progress in soil physics. *Geoderma*, **12**, 265–280.

Philip, J.R., 1984a. Steady infiltration from circular cylindrical cavities. *Soil Sci. Soc. Am. J.*, **48**, 270–278.

Philip, J.R., 1984b. Steady infiltration from spherical cavities. *Soil Sci. Soc. Am. J.*, **48**, 724–729.

Philip, J.R., 1985. Scattering functions and infiltration. *Water Resour. Res.*, **21**, 1025–1033.

Philip, J.R., 1986a. Steady infiltration from buried discs and other sources. *Water Resour. Res.*, **22**, 1058–1066.

Philip, J.R., 1986b. Steady infiltration from spheroidal cavities in isotropic and anisotropic soils. *Water Resour. Res.*, **22**, 1874–1880.

Philip, J.R., 1986c. Linearized unsteady multidimensional infiltration. *Water Resour. Res.*, **22**, 1717–1727. (Correction, **23**, 1710, 1987.)

Philip, J.R., 1987a. The quasilinear analysis, the scattering analog, and other aspects of infiltration and seepage, in *Infiltration Development and Application*. Honolulu: Water Resources Research Center, pp. 1–27.

Philip, J.R., 1987b. The infiltration joining problem. *Water Resour. Res.*, **23**, 2239–2245.

Philip, J.R., 1988. Quasianalytic and analytic approaches to unsaturated flow, in *Flow and Transport in the Natural Environment: Advances and Applications*. New York: Springer, pp. 30–47.

Philip, J.R., 1989. The scattering analog for infiltration in porous media. *Rev. Geophys.*, **27**, 431–448.

Philip, J.R., 1990. How to avoid free boundary problems, in *Free Boundary Problems: Theory and Applications*. London: Longman. *Research Notes in Mathematics*, **185**, 193–207.

Philip, J.R., 1991a. Hillslope infiltration: planar slopes. *Water Resour. Res.*, **27**, 109–117.

Philip, J.R., 1991b. Hillslope infiltration: divergent and convergent slopes. *Water Resour. Res.*, **27**, 1035–1040.

Philip, J.R., 1991c. Infiltration and downslope unsaturated flows in concave and convex topographies. *Water Resour. Res.*, **27**, 1041–1048.

Philip, J.R., 1992. A linearized solution of the slope crest infiltration problem. *Water Resour. Res.*, **28**, 1121–1132.

Philip, J.R., 1993a. Constant-rainfall infiltration on hillslopes and slope crests, in *Water Flow and Solute Transport in Soils: Developments and Applications*. Heidelberg: Springer. *Advanced Series in Agricultural Sciences*, **20**, 152–179.

Philip, J.R., 1993b. Variable-head ponded infiltration under constant or variable rainfall. *Water Resour. Res.*, **29**, 2155–2165.

Philip, J.R., 1995. Mathematical physics of infiltration on flat and sloping topography, in *Environmental Studies: Mathematical, Computational, and Statistical Analysis*. New York: Springer.

Philip, J.R. and J.H. Knight, 1974. On solving the unsaturated flow equation. 3. New quasianalytical technique. *Soil Sci.* **117**, 1–13.

Polubarinova-Kochina, P.Ya., 1962. *Theory of Ground Water Movement*. Princeton, NJ: Princeton University Press.

Pullan, A.J., 1990. The quasi-linear approximation for unsaturated porous medium flow. *Water Resour. Res.*, **26**, 1219–1234.

Raats, P.A.C., 1970. Steady infiltration from line sources and furrows. *Soil Sci. Soc. Am. Proc.*, **34**, 709–714.

Raats, P.A.C., 1971. Steady infiltration from point sources, cavities and basins. *Soil Sci. Soc. Am. Proc.*, **35**, 689–694.

Sherman, L.K., 1940. Derivation of infiltration-capacity (f) from average loss-rates (f_{av}). *Trans. Am. Geophys. Union*, **21**, 541–550.

Sherman, L.K., 1943. Comparison of *f*-curves derived by the methods of Sharp and Holtan and of Sherman and Mayer. *Trans. Am. Geophys. Union*, **24**, 465–467.

Sherman, L.K., and L.C. Mayer, 1941. Application of the infiltration-theory to engineering practice. *Trans. Am. Geophys. Union*, **22**, 666–677.

Swartzendruber, D.S., 1974. Infiltration of constant-flux rainfall into soil as analyzed by the approach of Green and Ampt. *Soil Sci.*, **117**, 272–281.

Vedernikov, V.V., 1940. Account of soil capillarity on seepage from a canal. *Prikl. Mat. Mekh.*, **28**(5.

Waechter, R.T. and J.R. Philip, 1985. Steady two- and three-dimensional flows in unsaturated soil: the scattering analog. *Water Resour. Res*, **21**, 1875–1887.

Warrick, A.W., 1974. Time-dependent linearized infiltration. I. Point sources. *Soil Sci. Soc. Am. Proc.*, **38**, 383–386.

White, I., 1988. Measurement of soil physical properties in the field, in *Flow and Transport in the Natural Environment: Advances and Applications*. New York: Springer, pp. 59–85.

White, I. and P. Broadbridge, 1989. Reply. *Water Resour. Res.*, **25**, 1054–1059.

White, I. and M.J. Sully, 1987. Macroscopic and microscopic capillary length and time scales from field infiltration. *Water Resour. Res*, **23**, 1514–1522.

White, I. and M.J. Sully, 1988. Field characteristics of the macroscopic capillary length or alpha parameter, in *Validation of Flow and Transport Models for the Unsaturated Zone: Conference Proceedings*. Las Cruces: Dept. of Agronomy and Horticulture, New Mexico State Univ., pp. 517–524.

White, I., M.J. Sully and K.M. Peroux, 1992. Measurement of surface–soil hydraulic properties: disk permeameters, tension infiltrometres, and other techniques, in *Advances in Measurement of Soil Physical Properties: Bringing Theory into Practice*. Madison: Soil Sci. Soc. Am., pp. 69–103.

Cross references

Drainage
Hydrological cycle
Hydrology
Water movement in unsaturated soils

INTERNATIONAL DATABASES

Hydrological Information Referral Service (INFOHYDRO)

The Hydrological Information Referral Service (INFOHYDRO) is a World Meteorological Organization (WMO) service for the dissemination of information on

1. national and international (governmental and non-governmental) organizations, institutions and agencies dealing with hydrology;
2. hydrological and related activities of the bodies mentioned in (1);
3. principal international river and lake basins of the world;
4. networks of hydrological observing stations of WMO members – number of stations and duration of records;
5. national hydrological data banks – status of collection, processing and archiving of data;
6. international data banks related to hydrology and water resources.

INFOHYDRO as a metadatabase does not contain or handle actual hydrological data, nor does it duplicate national referral systems. It is designed to facilitate the prompt dissemination of updated hydrological information to member countries, particularly for the benefit of their experts, agencies and enterprises engaged in water resources assessment, development and management requiring support from national, regional or international agencies dealing in operational hydrology. The information available in INFOHYDRO provides a good indication of water resources assessment activities of member countries.

The *INFOHYDRO Manual* (WMO, 1987) contains information concerning the entire database and its operation. It also contains all hydrological information available at present in INFOHYDRO. Thus

the *Manual* comprises, in a single volume, comprehensive information on the hydrological services of all countries and their data collection activities.

INFOHYDRO is maintained as a computerized database, in which data can be supplied on diskettes. Information can be supplied for particular countries or WMO Regions and can consist of any of the items described under 1 to 5 above. Requests should be addressed to WMO.

Global Runoff Data Centre (GRDC)

On 1 May 1987 a permanent Global Runoff Data Centre (GRDC) was established at the Federal Institute of Hydrology in Koblenz, Germany, under the auspices of WMO.

The GRDC operates for the benefit of WMO members and the international scientific community. It provides a mechanism for the international exchange of data pertaining to river flows and surface-water runoff on a continuous long-term basis. The GRDC receives data from many sources, principally through WMO. All data archived at the GRDC are available to users.

As of November 1991, the GRDC data bank consisted of flows for 2930 stations from 131 countries. Complete daily flows were available for 1478 stations, and daily flows for only a part of the record were available for a further 186 data series. Monthly flows were available for 1266 stations.

The core of the data bank is the daily flows for 1237 stations from 75 countries that were collected initially by WMO under the WMO/ICSU Global Atmosphere Research Programme (GARP) for use in validation of atmospheric general circulation models (GCMs), and later within the World Climate Programme (WCP). The first available year for this set of data was 1978, and there are data up to 1980 from nearly all of the stations. Data from 40 countries are also available up to 1982–1983 and from Australia up to 1984–1985. The database is being updated from time to time.

The stations have been selected according to the following criteria:

- uniform national geographical distribution (consistent with network conditions), with higher densities in areas of rapid variation of flow;
- coverage, to the greatest extent possible, of each type of homogeneous hydrological region of each country;
- relatively small river basins (up to about 5000 km^2, and in exceptional cases, up to 10 000 km^2);
- flow data representing natural river flow, i.e. they should have been corrected for any significant diversions, abstractions and redistributions by storage;
- good quality of records.

The GRDC has developed a set of programs to provide users with a selection of retrieval options to make the data and information readily accessible. The following retrieval options are currently available: tables of daily or monthly mean flows; hydrographs of daily or monthly mean flows; flow duration curves or tables; and station and catchment information.

Requests for data may be made in writing or by personal visit to the GRDC in Koblenz. Charges might be assessed to cover the costs of providing services to users (e.g. costs of tapes or diskettes, mailing and handling charges). The charges could be waived if the individual or institution were a contributor of data to GRDC.

World Climate Data Information Referral Service (INFOCLIMA)

The World Climate Data Information Referral Service (INFOCLIMA) is a service for the collection and dissemination of information on the existence and availability of climate data in the world. The information comprises, in particular:

- descriptions of available data sets, held at data centers and/or published;
- climatological and radiation station networks of the world and their histories;
- national climatological data banks including status of collection, processing and archiving of data.

The INFOCLIMA referral service is implemented by WMO under the World Climate Programme.

INFOCLIMA information is obtained from member countries of

WMO and, as regards data sets, also from contributions by individual data centers and international organizations. INFOCLIMA does not handle actual climate data but provides information on the existence and availability of climate data in the world. It is maintained as a computerized database.

The INFOCLIMA catalog contains descriptions of data sets that originate from a particular data collection or data-processing program. The information on data sets submitted by members and international centers is edited and entered into the INFOCLIMA computerized database in a standardized format after verification has taken place with the centers involved. Tape or diskette copies of portions of the database will soon be available upon request.

For practical purposes, climate data have been divided into a number of categories, namely upper-air data, surface climatological data, radiation (surface) data, maritime and ocean data, cryosphere data, atmospheric composition data, hydrological data and historical and proxy data.

World Meteorological Organization's Hydrological Operational Multi-purpose System (HOMS)

In recent decades, hydrological science and technology have made substantial progress, and significant contributions have been made in the development and management of water resources. The technology transfer system HOMS, developed by WMO and in operation since 1981, offers a simple but effective means of disseminating a wide range of proven techniques for the use of hydrologists.

Structure of HOMS

HOMS transfers hydrological technology in the form of separate components. These components can take many forms, such as sets of drawings for constructing (or instruction manuals for) hydrological equipment, reports describing a wide variety of hydrological procedures, and computer programs covering the processing, quality control and storage of hydrological data, as well as modeling and analysis of the processed data. About 400 components are available, which are operationally used by their originators, thus ensuring that each component is useful and that it actually works. To date, 35 countries have provided components to HOMS. Each component has a two-page summary description, written in a standard format, giving information on the content and applicability of the component package, together with details of the originator and available support. These descriptions are held in the *HOMS Reference Manual (HRM)* (WMO, 1988), a copy of which is kept in each country participating in HOMS. The *Manual* is divided into sections and subsections on the basis of subject matter, and the components are coded according to topic and complexity.

HOMS components can be grouped into sequences of compatible components that can be used to carry out larger tasks. The sequences also provide a means of accessing the component or components needed for some particular task.

Organization and operation of HOMS

HOMS is organized as a cooperative effort of WMO Members, with about 117 countries (February 1994) participating. Each participating country designates a HOMS National Reference Centre (HNRC), which is usually in some part of the national Hydrological Service. Regional focal points for specific areas have also been formed.

The functions of an HNRC include

- proposing suitable national components and sequences for use in HOMS;
- processing requests from other HNRCs for nationally supported components;
- obtaining components from abroad for national users;
- bringing HOMS to the attention of potential users in the country, and assisting with the selection and use of appropriate components.

The international activities of HOMS are supervised and coordinated by a steering committee that works within the framework of the WMO Commission for Hydrology. The HOMS Office in the WMO Secretariat keeps the HNRCs up-to-date by the provision of supplements to the *Reference Manual* containing details of new components and by publishing a newsletter on HOMS activities.

Hydrologists who wish to make use of HOMS components should contact the HNRC of their country, where they will be able to consult the *HOMS Reference Manual* (WMO, 1988). The HNRC will also be able to advise on the choice of components. Once it is decided which components are needed, the HNRC will be able to forward the formal requests to the HNRCs concerned. The HOMS Office monitors requests and is able to help with administrative formalities when required.

Source

World Meteorological Organization, 1994. *Guide to Hydrological Practices*, 5th edn, WMO, Geneva.

Bibliography

World Meteorological Organization (WMO), 1987. *Hydrological Information Referral Service – INFOHYDRO Manual*. Operational Hydrology Report No. 28, WMO No. 683, Geneva.
World Meteorological Organization, 1988. *Hydrological Operational Multipurpose System (HOMS) Reference Manual*, 2nd edn, Geneva.
World Meteorological Organization, 1989. *INFOCLIMA Catalog of Climate System Data Sets*. Hydrological Data Extract. WCDP-8, WMO/TD No. 343, Geneva.

Cross references

Hydrological services
Water agencies: national and international

INTERNATIONAL ORGANIZATIONS INVOLVED WITH HYDROLOGY AND WATER RESOURCES

General

This contribution provides an overview of the involvement of international organizations (governmental and non-governmental) in the field of water resources and of the various arrangements for system-wide and sectoral coordination and cooperation at the regional and global levels. It was prepared on the basis of information provided by the Secretariat of the Intersecretariat Group for Water Resources (ISGWR) of the UN Administrative Committee on Coordination (ACC) (United Nations 1982, 1992a).

Intergovernmental organizations (IGOs)

These are organizations, established by agreements, to which two or more States are party. Such organizations may be global or regional. A number of these organizations are active in some form in the field of water resources. Table I2 lists those United Nations organizations and specialized agencies active on a global level, while Table I3 provides information about the regional organizations of the United Nations and of other regional organizations. Both tables include the official acronyms and the addresses of the organizations.

Nature and interrelationship of the activities of the UN organizations in water resources development

The activities of the organizations of the United Nations system in the field of water resources are wide-ranging in scope and nature. Their involvement has grown during the past three decades, both in terms of the magnitude and complexity of the problems addressed. Table I4 presents a synoptic view of the involvement of the UN organizations with an indication of the main and applied areas of interest. The grouping has been done in accordance with the main areas of concern considered by the International Conference on Water and the Environment (United Nations, 1992b), namely:

1. water resources assessment and impacts of climate change on water resources;
2. protection of water resources, water quality and aquatic ecosystems;
3. water and sustainable urban development, and drinking water supply and sanitation in the urban context;

Table 12 Intergovernmental organizations dealing with hydrology and water resources – global[a]

Name	Abbreviation	Address
United Nations		
Department of Economic and Social Development	DESD	United Nations Headquarters, New York, NY 10017, USA
United Nations Children's Fund	UNICEF	Three United Nations Plaza, New York, NY 10017, USA
United Nations Development Programme	UNDP	One United Nations Plaza, New York, NY 10017, USA
United Nations Environment Programme	UNEP	PO Box 30552, Nairobi, Kenya
United Nations University	UNU	Toho Seimei Building, 15–1 Shibuya, 2-Chome, Shibuya-ku, Tokyo 150, Japan
World Food Programme	WFP	Via Cristoforo Colombo 426, 00145 Rome, Italy
United Nations Centre for Human Settlements	HABITAT	United Nations Office in Nairobi, PO Box 30030, Nairobi, Kenya
Department of Humanitarian Affairs-Office of the United Nations Disaster Relief Co-ordinator	DHA–UNDRO	Palais des Nations, CH-1211 Geneva 10, Switzerland
World Food Council	WFC	Via delle Terme di Caracalla, 00100 Rome, Italy
International Research and Training Institute for the Advancement of Woman	INSTRAW	PO Box 21747, Santo Domingo, Dominican Republic
Specialized agencies and other organizations		
International Labour Organization	ILO	4, route des Morillons, CH-1211 Geneva 22, Switzerland
Food and Agriculture Organization of the United Nations	FAO	Via delle Terme di Caracalla, 00100 Rome, Italy
United Nations Educational, Scientific and Cultural Organization	UNESCO	7, place de Fontenoy, 75700 Paris, France
World Health Organization	WHO	20, avenue Appia, CH-1211 Geneva 27, Switzerland
World Bank	IBRD	1818 H Street, NW, Washington, DC 20433, USA
World Meteorological Organization	WMO	PO Box 2300, CH-1211 Geneva 2, Switzerland
International Fund for Agricultural Development	IFAD	Via del Serafico 107, 00142 Rome, Italy
United Nations Industrial Development Organization	UNIDO	PO Box 300, Vienna International Centre, A-1400, Vienna, Austria
International Atomic Energy Agency	IAEA	PO Box 100, Vienna International Centre, A-1400 Vienna, Austria

[a] Status in 1992.

Table 13 Intergovernmental organizations dealing with hydrology and water resources – regional

Name	Abbreviation	Address
Organs of the United Nations		
Economic Commission for Africa	ECA	PO Box 3001, Addis Ababa, Ethiopia
Economic Commission for Europe	ECE	Palais des Nations, CH-1211 Geneva 10, Switzerland
Economic Commission for Latin America and the Caribbean	ECLAC	Casilla 179 D, Santiago, Chile
Economic and Social Commission for Asia and the Pacific	ESCAP	The United Nations Building, Reajadamnem Ave, Bangkok 10200, Thailand
Economic and Social Commission for Western Asia	ESCWA	PO Box 927 115, Amman, Jordan
United Nations Sudano-Sahelian Office (UNDP)	UNSO	One United Nations Plaze, Room DC 1100, New York, NY 10017, USA
Regional Commission on Land and Water Use in the Near east (FAO)	RNEA LWU	Via delle Terme di Caracalla, 00100 Rome, Italy
Others		
Arab Center for the Studies of Arid Zones and Drylands	ACSAD	PO Box 2440, Damascus, Syria
Caribbean Meteorological Organization	CMO	PO Box 461, Port of Spain, Trinidad
Comité inter-etats de lutte contre la sécheresse dans le Sahel	CILSS	BP 7049, Ouagadougou, Burkina Faso
Comité Regional para los Recursos Hídricos del Istmo Centroamericano	CRRH	c/o ICE, P.O. Box 10032, San José, Costa Rica[b]
Commission of the European Communities	CEC	200 rue de la Loi, Brussels 1040, Belgium
Council of Europe	CE	Avenue de l'Europe, 67 Strasbourg, France
Council for Mutual Economic Assistance	CMEA	Prospekt Kalinina 56, Moscow G-205, Russian Federation
Energy Organization of the Great Lakes Countries	CEPGL	BP 58, Gisenyi, Rwanda
European Space Agency	ESA	8–10 rue Mario Nikis, 75738 Paris, CEDEX 15, France
Comité interafricain d'études hydrauliques	CIEH	BP 369, Ouagadougou 01, Burkina Faso
Nordic Council	NC	Gamla Rigsdagshuset, Stockholm, Sweden
Organization of African Unity	OAU	PO Box 3243, Addis Ababa, Ethiopia
Organization of American States	OAS	Pan American Union Building, Washington, DC 20006, USA
Organization for Economic Co-operation and Development	OECD	Château de la Muette, 2 rue André Pascal, 75775 Paris, France

[a] Status in 1992.
[b] Rotational Secretariat.

Table 14 Involvement of organizations of the United Nations system in water resources development: indication of main and applied areas of interest[a]

Areas of concern	Organizations with main concern in indicated areas	Organizations with interest in applied aspects of indicated areas
1. Water resources assessment and impacts of climate change on water resources	WMO, UNESCO, DESD, FAO, IBRD, IAEA	WHO, UNDP, ECA, ECE, ECLAC, ESCAP, ESCWA, UNDRO
2. Protection of water resources, water quality and aquatic ecosystems	WHO, WMO, UNEP, DESD, ECE	All others
3. Water and sustainable urban development, and drinking water supply and sanitation in the urban context	IBRD, HABITAT, WHO, UNDP, UNICEF, INSTRAW	DESD, ECA, ECLAC, ESCAP, ESCWA, UNEP
4. Water for sustainable food production and rural development, and drinking water supply and sanitation in the rural context	FAO, IBRD, UNDP, WFP, WHO, UNICEF, DESD, HABITAT, INSTRAW, ILO	ECA, ECLAC, ESCAP, ESCWA
5. Integrated water resources management	DESD, ECA, ECE, ECLAC, ESCAP, INSTRAW, UNDP, IBRD	UNDRO, UNESCO, WMO, WHO, FAO

[a] Status in 1992.

Table 15 Involvement of the organizations of the United Nations system in the field of water resources[a]

Development and management functions	Agricultural water use	Drinking water supply	Industrial water use	Hydro-power	Navigation	Flood control	Drought management	Multi-purpose water use
1 Surfacewater hydrology	DESD, ECA, FAO, IBRD	DESD, UNICEF, ECA, ESCAP, ESCWA, INSTRAW, IBRD, HABITAT	DESD, ECA, IBRD, HABITAT	DESD, ECA, INSTRAW, UNESCO, IBRD	DESD, ECA, ESCAP, IBRD	DESD, ECA, ESCAP, ESCWA, UNESCO, FAO, WMO, IBRD, HABITAT	DESD, ECA, ESCAP, ESCWA, UNESCO, FAO, WMO, IBRD	DESD, ECA, ESCAP, ESCWA, IBRD, UNESCO, WMO, HABITAT
2 Groundwater hydrology	DESD, ECA, FAO, IBRD	DESD, ECA, INSTRAW, IBRD, HABITAT	ECA, IBRD, HABITAT			ECA, HABITAT, UNESCO, WMO	DESD, ECA, ESCAP, UNESCO, FAO, WMO, IBRD	DESD, ECA, ESCAP, ESCWA, HABITAT, UNESCO, WMO, IBRD
3 Surfacewater quality monitoring	DESD, ECA, FAO, WHO, IBRD	DESD, UNICEF, ECA, ESCAP, UNEP, WHO, IBRD, HABITAT	DESD, WHO, IBRD, HABITAT	ECA	ECA, ESCAP		DESD, ECA, UNESCO, FAO, WMO	DESD, ECA, ESCAP, UNEP, UNESCO, WHO, WMO, HABITAT
4 Groundwater quality monitoring	ECA, FAO, WHO, IBRD	UNICEF, ECA, ESCAP, UNEP, WHO, IBRD, HABITAT	WHO, IBRD, HABITAT				ECA, ESCAP, UNESCO, FAO, WMO	ECA, ESCAP, UNEP, UNESCO, WHO, WMO, HABITAT
5 Information on water use	ECA, ECE, ECLAC, ESCAP, INSTRAW, FAO, IBRD	UNICEF, ECA, ECE, ECLAC, ESCAP, INSTRAW, WHO, IBRD, HABITAT	ECA, ECE, ECLAC, ESCAP, IBRD, HABITAT	ECA, ECE, ECLAC, ESCAP, INSTRAW, IBRD	ECA, ECE, ECLAC, ESCAP	ECA, ECE, ECLAC, ESCAP, UNESCO, WMO	ECA, ECLAC, ESCAP, UNESCO, WMO, FAO, IBRD	ECA, ECE, ECLAC, ESCAP, INSTRAW, HABITAT, UNESCO, WMO, IRBD
6 Surfacewater development	DESD, ECA, ESCAP, FAO, WFP, IBRD	DESD, UNICEF, ECA, ESCAP, WHO, WFP, IBRD, HABITAT	DESD, ECA, IBRD, HABITAT	DESD, ECA, ESCAP, IBRD	ECA, ESCAP, IBRD	DESD, ECA, ESCAP, ESCWA, FAO, WFP, IBRD, HABITAT, UNESCO, WMO	DESD, ECA, ESCAP, UNESCO, WMO, FAO, WFP, IBRD	DESD, ECA, ECLAC, ESCAP, ESCWA, WFP, UNESCO, WMO, IBRD, HABITAT

Table 15 Continued

Development and management functions	Agricultural water use	Drinking water supply	Industrial water use	Hydro-power	Navigation	Flood control	Drought management	Multi-purpose water use
7 Groundwater development	DESD, ECA, ESCAP, FAO, WFP, IBRD	DESD, UNICEF, ECA, ESCAP, WHO, WFP, IBRD, HABITAT	ECA, IBRD, HABITAT				DESD, ECA, ESCAP, UNESCO, WMO, FAO, WFP, IBRD	DESD, ECA, ESCAP, ESCWA, WFP, IBRD, HABITAT, UNESCO, WMO
8 Wastewater reuse	DESD, ECA, ECE, FAO, IBRD	DESD, WHO	ECA, ECE, WHO, IBRD, HABITAT				ESCAP	DESD, ECA, ECE, ECLAC, ESCAP, ESCWA, HABITAT, UNESCO, WMO
9 Integrated water resources management	DESD, ECA, ECE, ESCAP, FAO, WHO, WFP, IBRD	DESD, ECA, ECE, ESCAP, WHO, WFP, IBRD, HABITAT	DESD, ECA, ECE, ESCAP, WHO, IBRD, HABITAT	DESD, ECA, ECE, IBRD	ECA, ECE, ESCAP, IBRD	DESD, ECA, ECE, ESCAP, ESCWA, WFP, IBRD, HABITAT, UNESCO, WMO	DESD, ECA, ECE, ESCAP, UNESCO, FAO, WMO, WFP, IBRD	DESD, ECA, ECE, ECLAC, ESCAP, ESCWA, WFP, IBRD, WMO, UNESCO, HABITAT
10 Water use management	DESD, FAO, ECA, ECLAC, ESCAP, IBRD	DESD, ECA, ECLAC, ESCAP, INSTRAW, WHO, IBRD, HABITAT	DESD, ECA, ECLAC, ESCAP, IBRD, HABITAT	ECA, ESCAP, ECLAC, INSTRAW, IBRD	ECA, ECLAC, IBRD	ECA, ECLAC, ESCAP, FAO, UNESCO, WMO, IBRD	ECA, ECLAC, ESCAP, FAO, IBRD, UNESCO, WMO	DESD, ECA, ECLAC, ESCAP, INSTRAW, UNESCO, WMO, IBRD, HABITAT
11 Waste water management	ECA, ECE, FAO, WHO, WFP	ECA, WHO, IBRD, HABITAT	ECA, ECE, WHO, IBRD, HABITAT					ECA, ECE, ECLAC, ESCAP, ESCWA, WMO, HABITAT, UNESCO
12 Strengthening of institutions	ECA, ECLAC, FAO, IBRD	UNICEF, ECA, ESCAP, ECLAC, WHO, IBRD, HABITAT	ECA, ECLAC, IBRD, HABITAT	ECA, ECLAC, IBRD	ECA, ESCAP	ECA, ECLAC, ESCAP, IBRD, HABITAT, UNESCO, WMO	ECA, ECLAC, UNESCO, FAO, WMO, IBRD	ECA, ECLAC, ESCAP, ESCWA, IBRD, HABITAT, UNESCO, WMO
13 Legislation	DESD, ECA, ECE, FAO	DESD, ECA, ECE, WHO, HABITAT	DESD, ECA, ECE	DESD, ECA, ECE	ECA, ECE, ESCAP	ECA, ESCAP, FAO, HABITAT	ECA, FAO	DESD, ECA, ECE, ECLAC, ESCAP, ESCWA, FAO, IBRD, HABITAT
14 Education and training	ECA, INSTRAW, FAO, WHO, WFP, IBRD	DESD, UNICEF, ECA, ESCAP, INSTRAW, WHO, IBRD, HABITAT	ECA, WHO, IBRD	ECA, INSTRAW, UNESCO, IBRD	DESD, ECA, ESCAP	ECA, ESCAP, WFP, UNESCO, WMO, IBRD, HABITAT	ECA, ESCAP, FAO, WFP, UNESCO, WMO, IBRD	ECA, ECLAC, ESCAP, INSTRAW, UNESCO, WMO, IBRD, HABITAT
15 Human resources development	ECA, INSTRAW, FAO, WHO, IBRD	DESD, UNICEF, ECA, ESCAP, INSTRAW, WHO, IBRD, HABITAT	ECA, IBRD, HABITAT	ECA, IBRD	ECA, ESCAP	ECA, ESCAP, IBRD, HABITAT, UNESCO, WMO	DESD, ECA, ESCAP, UNESCO, FAO, WMO, IBRD	ECA, ESCAP, INSTRAW, IBRD, HABITAT, UNESCO, WMO

[a] Status in 1992.

4. water for sustainable food production and rural development, and drinking water supply and sanitation in the rural context;
5. integrated water resources and management.

These areas of concern also correspond to those of Chapter 18 of Agenda 21 of the United Nations Conference on Environment and Development (UNCED; United Nations, 1992c). A sixth area considered at the conference, mechanisms for implementation and coordination at international, national and local levels, is relevant to the nature and scope of the activities of the organizations of the United Nations system, and to the means of coordination of these activities. Issues related to capacity building inevitably permeate all the areas depicted above.

Table I5 provides a more detailed view of the activities of the organizations. Each cell in the matrix shows which organizations are involved in development and management activities concerning specific water resources sectors. The development and management functions have been classified as follows:

1. surfacewater hydrology;
2. groundwater hydrology;
3. surfacewater quality monitoring;
4. groundwater quality monitoring;
5. information on water use;
6. surfacewater development;
7. groundwater development;
8. wastewater reuse;
9. integrated water resources management;
10. water use management;
11. wastewater management;

Table 16 International non-governmental organizations (NGOs) dealing with hydrology and water resources[a]

Name	Abbreviation	Address
International Association of Hydrogeologists	IAH	National Rivers Authority, 550 Steetsbrook Road, Solihul, West Midlands, B91 1QT, UK
International Association of Sedimentologists	IAS	Université de Liège, Place du Vingt-Aout 7, B-4000 Liege, Belgium
International Association of Theoretical and Applied Limnology	SIL	Sil Secretariat/Central Office, Department of Biological Sciences, University of Alabama, Tuscaloosa, Alabama 35487-0344, USA
International Association for Water Law	IAWL	Via Montevideo 5, I-00198 Rome, Italy
International Association for Water Quality	IAWQ	Alliance House, 29–30 High Holborn, London WC1V 6BA, UK
International Council of Scientific Unions	ICSU	Bd. de Montmorency 51, F75016 Paris, France
Committee on Space Research	COSPAR	See ICSU
Committee on Science and Technology in Developing Countries	COSTED	See ICSU
Committee on Data for Science and Technology	CODATA	see ICSU
Committee on Water Research (ICSU-UITA)	COWAR	CHO-TNO, PO Box 6067, 2500 JA, Delft, The Netherlands
Scientific Committee on Problems of the Environment	SCOPE	see ICSU
International Geographical Union (member of ICSU)	IGU	University of Alberta, Edmonton, Alberta, Canada T6G 2H4
International Institute for Applied Systems Analysis	IIASA	A-2361 Laxenburg, Austria
International Association on Water Pollution Research	IAWPRC	1 Queen Anne's Gate, London SW1H 9BT, UK
International Organization for Standardization	ISO	1, rue de Varembé, CH-1211 Geneva 20, Switzerland
International Society of Soil Science	ISSS	PO Box 353, 9 Duivendaal, 6700 AJ Wageningen, The Netherlands
International Training Centre for Water Resources Management	ITCWRM (CEFIGRE)	BP 13, Sophia Antipolis, F-06561 Valbonne CEDEX, France
International Union for Conservation of Nature and Natural Resources	IUCN	Avenue du Mont-Blanc, CH-1196 Gland, Switzerland
International Union of Geodesy and Geophysics (member of ICSU)	IUGG	Observatoire Royal, Avenue Circulaire 3, B-1180, Brussels, Belgium
International Association of Hydrological Sciences	IAHS	PO Box 6067, 2500 JA, Delft, The Netherlands
International Association of Meteorology and Atmospheric Physics	IAMAP	National Center for Atmospheric Research, PO Box 3000, Boulder, CO 80307 USA
International Union of Geological Science (member of ICSU)	IUGS	Maison de la Géologie, Rue Claude-Bernard 77, F-75005 Paris, France
International Water Resources Association	IWRA	University of Illinois, 205 North Mathews Avenue, Urbana, IL 61801 USA
International Water Supply Association	IWSA	1 Queen Anne's Gate, London SW1H 9BT, UK
Union of International Technical Associations	UITA	Unesco, 1 rue Miollis, F-75015 Paris, France
International Commission of Agricultural Engineering	CIGR	CHO-TNO, PO Box 6067, 2600 JA Delft, The Netherlands
International Union of Pure and Applied Chemistry	IUPAC	Bank Court Chambers, 2–3 Pound Way, Templars Square, Cowley, Oxford OX4 3YF, UK
International Association for Hydraulic Research	IAHR	Rotterdamseweg 185, PO Box 177, 2600 MH Delft, The Netherlands
International Commission on Large Dams	ICOLD	Bd. Haussmann 151, F-75008 Paris, France
International Commission of Irrigation and Drainage	ICID	48 Nyaya Marg, Chanakyapuri, New Delhi 110021, India
World Energy Conference	WBC	34 St James Street, London SW1A 1HD, UK
Permanent International Association of Navigation Congresses	PIANC	WTC-Tour 3, 26e étage, Boulevard S. Bolivar 30, B-1210 Brussels, Belgium

[a] Status in 1992.

12. strengthening of institutions;
13. legislation;
14. education and training;
15. human resources development.

The specific water resources sectors are as follows:

- agricultural water use;
- drinking water supply;
- industrial water use;
- hydropower;
- navigation;
- flood control;
- drought management;
- multipurpose water use.

Additional information as to the nature of the involvement of each organization, as well as a description of the scope and nature of water-related activities of the activities of the organizations of the United Nations system, with examples of typical projects executed by them, is provided in *The United Nations Organizations and Water* (United Nations, 1982) and in *The United Nations Organizations and Water: Briefing Notes on the Scope and Nature of the Activities of the Organizations of the United Nations System* (United Nations, 1992a).

Non-governmental organizations (NGOs)

These are international organizations that are not established by intergovernmental agreement. They include organizations that accept members designated by government authorities, provided that such membership does not interfere with the free expression of the views of the organization. The NGOs involved with hydrology and water resources are listed alphabetically in Table I6. They may pertain to any of the following categories:

- federations of international organizations;
- universal membership organizations;
- intercontinental membership organizations;
- regional membership organizations;
- semi-autonomous bodies;
- organizations of special form.

Table I6 is presented as follows:

1. Column (1). Organization name: the name of the organization is normally given in English.
2. Column (2). Acronym.
3. Column (3). Organization address: the address given is that of the international secretariat or principal contact as of 1992. Some secretariats rotate or move to another address depending on the changes in the composition of the governing bodies.

Institutionalized cooperation in international river and lake basins

There are many international agreements and treaties that concern the joint use of international rivers and boundary waters, and many of these agreements and treaties have resulted in institutionalized cooperation among the countries concerned. The list of the main international institutions of this kind can be found by WMO Region in the *INFOHYDRO Manual* (WMO, 1987).

Source

World Meteorological Organization, 1994. *Guide to Hydrological Practices*, 5th edition, WMO, Geneva.

Bibliography

United Nations, 1982. *The United Nations Organizations and Water*, 83–00237, New York.
United Nations, 1992a. *The United Nations Organizations and Water: Briefing Note on the Scope and Nature of the Activities of the Organizations of the United Nations System.*
United Nations, 1992b. *International Conference on Water and the Environment: Development Issues for the Twenty-first Century.* The Dublin Statement and Report of the Conference, 26–31 January (1992), Dublin, Ireland.
United Nations, 1992c. *Conference on Environment and Development (UNCED) – Agenda 21*, Rio de Janeiro, Brazil.

World Meteorological Organization (WMO), 1987. *Hydrological Information Referral Service – INFOHYDRO Manual.* Operational Hydrology Report No. 28, WMO No. 683, Geneva.

Cross references

Hydrological services
International databases
Water agencies: national and international

INTERNATIONAL RIVERS

More than 200 river systems, draining over half of the planet's area, are shared by two or more countries. Overpumping of groundwater aquifers that stretch under political borders also injects international

Table I7 Summary of the number of international continental river basins

(A)	Africa	57		
(B)	North and Central Americas	34[a]	Caribe	3
	Juradó {		USA–Mexico	43
			Central America	18
(C)	South America	36		
(D)	Asia	40		
(E)	Europe	48		
	Total	215[b]		

[a] Counted both in North and Central America.
[b] The actual total number is 214 because the Juradó river is included in both South America and Central America.

Table I8 International river and lake basins

1. Percentage of the aggregated areas of international river and lake basins, per 'continent':

	%
Africa	60
Asia	65
Europe	50
North and Central America	40
South America	60
Former USSR	33
The world	47

2. Inclusions and exclusions for each 'continent':

Africa	includes:	Sinai Peninsula, Madagascar.
Asia	includes:	Asian part of Turkey, Cyprus, Japan, Chinese islands, Asian part of Malaysia, Sri Lanka.
	excludes:	Philippines, Borneo I., Indonesia etc.
Europe	includes:	British Isles, Iceland, Svakvard, European part of Turkey.
North and Central America	includes:	all Canadian islands, the Caribbean Is.
	excludes:	Greenland, Hawaiian Is.
South America	includes:	Galapagos Is., Falkland Is. (Malvinas).
Former USSR	includes:	European and Asian parts, all Arctic and Pacific islands.
The world	excludes:	Antarctica.

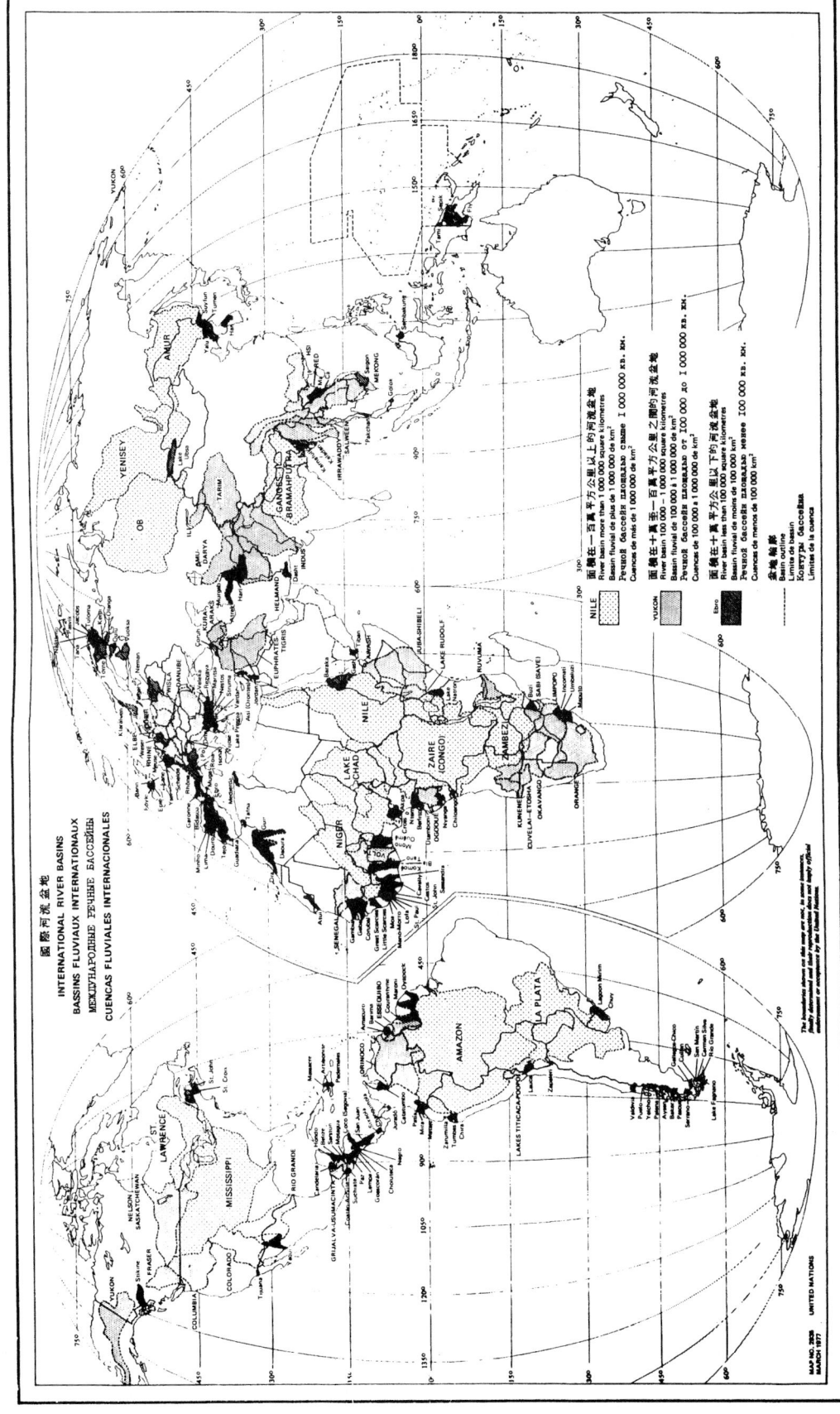

Figure 17 International river basins.

Table 19 List of separate river basins shared by two or more countries showing the share of the constituent countries

River basin	Area of basin (km²)	Constituent countries/ territories	Share per country (km²)	Share per country (%)	River basin	Area of basin (km²)	Constituent countries/ territories	Share per country (km²)	Share per country (%)
Africa					Ogooué	220 700	Gabon	195 000	88.4
							Congo Republic	19 300	8.7
Niger	2 200 000	Mali	620 000	28.2			Republic of Cameroon	4 400	2.0
		Nigeria	580 000	26.4			Equatorial Guinea	2 000	0.9
		Niger	490 000	22.3					
		Algeria	148 300	6.8	Kunene	112 000	Angola	101 000	90.0
		Guinea	95 000	4.3			Namibia	11 000	10.0
		Republic of Cameroon	90 000	4.1					
		Burkina Faso	81 700	3.6	Okavango	529 000	Botswana	194 500	36.8
		Benin	50 000	2.3			Namibia	144 500	27.3
		Ivory Coast	25 000	1.1			Angola	168 000	31.8
		Chad	20 000	0.9			Zimbabwe	22 000	4.1
Lake Chad	1 910 000	Chad	950 000	49.7	Limpopo	385 000	South Africa	180 000	46.8
		Niger	416 000	21.8			Botswana	73 000	19.0
		Central African Republic	214 980	11.3			Mozambique	71 000	18.4
		Nigeria	176 000	9.2			Zimbabwe	61 000	15.8
		Sudan	100 000	5.2					
		Republic of Cameroon	53 020	2.8	Sabi (Save)	103 000	Zimbabwe	73 000	71.0
							Mozambique	30 000	29.0
Nile	3 030 700	Sudan	1 900 000	62.7					
		Ethiopia	368 000	12.1	Ruvuma	166 500	Mozambique	103 000	62.0
		Egypt	300 000	9.9			United Republic of Tanzania	60 000	36.0
		Uganda	232 700	7.7					
		United Republic of Tanzania	116 000	3.8			Malawi	3 500	2.0
		Kenya	55 000	1.8	Juba-Shibeli	766 500	Ethiopia	333 500	43.5
		Congo, D.R.	23 000	0.8			Somalia	236 000	30.8
		Rwanda	21 500	0.7			Kenya	197 000	25.7
		Burundi	14 500	0.5	Lake Rudolf (L. Turkana)	203 300	Kenya	106 700	52.5
Congo	3 720 000	Congo, D.R.	2 310 000	62.1			Ethiopia	86 600	42.6
		Central African Republic	408 000	10.9			Sudan	6 700	3.3
		Angola	285 000	7.7			Uganda	3 300	1.6
		Congo Republic	255 000	6.9					
		Zambia	175 000	4.7	Awash	118 500	Ethiopia	115 000	97.0
		United Republic of Tanzania	170 000	4.5			Djibouti	3 500	3.0
		Republic of Cameroon	98 900	2.7	Medjerda	36 000	Algeria	18 700	52.0
		Burundi	13 300	0.4			Tunisia	17 300	48.0
		Rwanda	4 800	0.1					
					Tafna	8 800	Algeria	5 000	57.0
Zambezi	1 419 960	Zambia	577 600	40.7			Morocco	3 800	43.0
		Angola	260 000	18.3	Guir	98 500	Algeria	72 500	73.6
		Zimbabwe	226 360	15.9			Morocco	26 000	26.4
		Mozambique	161 000	11.4					
		Malawi	110 000	7.7	Daoura	65 700	Morocco	43 700	66.5
		Botswana	40 000	2.8			Algeria	22 000	33.5
		United Republic of Tanzania	28 000	2.0	Dra	80 500	Morocco	68 000	84.5
		Namibia	17 000	1.2			Algeria	12 500	15.5
Orange	950 000	South Africa	570 145	60.0	Atui	19 500	Western Sahara	12 500	64.0
		Namibia	250 000	26.3			Mauritania	7 000	36.0
		Botswana	99 500	10.5	Gambia	75 760	Senegal	50 000	66.0
		Lesotho	30 355	3.2			Guinea	15 500	20.0
							Gambia	10 200	13.5
Senegal	353 000	Mali	163 000	46.2	Geba	13 700	Guinea-Bissau	8 700	63.6
		Mauritania	93 000	26.4			Senegal	4 400	32.0
		Senegal	64 000	18.1			Guinea	600	4.4
		Guinea	33 000	9.3	Corubal	22 000	Guinea	14 000	63.6
Volta	379 000	Burkina Faso	172 500	45.5			Guinea-Bissau	8 000	36.4
		Ghana	159 000	42.0					
		Togo	19 000	5.0	Great Scarcies	8 500	Guinea	6 000	70.6
		Ivory Coast	13 000	3.4			Sierra Leone	2 500	29.4
		Benin	10 500	2.8	Little Scarcies	15 000	Sierra Leone	11 000	73.3
		Mali	5 000	1.3			Guinea	4 000	26.7

Table I9 Continued

River basin	Area of basin (km²)	Constituent countries/territories	Share per country (km²)	(%)
Moa	20 000	Guinea	10 500	52.5
		Sierra Leone	8 800	44.0
		Liberia	700	3.5
Mano-Morro	10 000	Liberia	7 500	75.0
		Sierra Leone	2 500	25.0
Lofa	12 000	Liberia	11 000	92.0
		Guinea	1 000	8.0
St Paul	18 000	Liberia	12 000	66.7
		Guinea	6 000	33.3
St John	13 500	Liberia	11 000	81.5
		Guinea	2 500	18.5
Cestos	12 500	Liberia	9 400	75.2
		Ivory Coast	2 500	20.0
		Guinea	600	4.8
Cavally	23 500	Ivory Coast	12 500	53.2
		Liberia	9 000	38.3
		Guinea	2 000	8.5
Sasasandra	77 500	Ivory Coast	70 000	90.4
		Guinea	7 500	9.6
Komoé	75 000	Ivory Coast	55 000	73.3
		Burkina Faso	20 000	26.7
Bia	13 100	Ghana	8 700	66.4
		Ivory Coast	4 400	33.6
Tano	14 000	Ghana	12 000	85.7
		Ivory Coast	2 000	14.3
Mono	25 600	Togo	21 800	85.2
		Benin	3 800	14.8
Ouémé	45 500	Benin	40 500	89.0
		Nigeria	2 500	5.5
		Togo	2 500	5.5
Cross	57 000	Nigeria	45 000	79.0
		Republic of Cameroon	12 000	21
Akpa	2 150	Nigeria	1 500	70.0
		Republic of Cameroon	650	30.0
Ntem	33 000	Republic of Cameroon	19 000	57.6
		Gabon	9 000	27.3
		Equatorial Guinea	5 000	15.1
Benito	15 900	Equatorial Guinea	14 000	88.1
		Gabon	1 900	11.9
Utamboni	12 800	Equatorial Guinea	7 000	54.7
		Gabon	5 800	45.3
Nyanga	26 000	Gabon	18 000	69.2
		Congo Republic	8 000	30.8
Chiloango	11 000	Congo, D.R.	6 800	61.8
		Cabinda (Angola)	3 000	27.3
		Congo Republic	1 200	10.9
Cuvelai-Etosha	126 000	Namibia	79 000	62.7
		Angola	47 000	37.3
Maputo	33 963	South Africa	18 500	54.5
		Swaziland	9 863	29.0
		Mozambique	5 600	16.5

River basin	Area of basin (km²)	Constituent countries/territories	Share per country (km²)	(%)
Umbeluzi	8 000	Swaziland	5 000	62.5
		Mozambique	1 800	22.5
		South Africa	1 200	15.0
Incomati	54 000	South Africa	34 000	63.0
		Mozambique	17 500	32.4
		Swaziland	2 500	4.6
Buzi	29 500	Mozambique	22 500	76.3
		Zimbabwe	7 000	23.7
Lake Natron	28 500	Kenya	16 800	59.0
		United Republic of Tanzania	11 700	41.0
Gash	32 000	Ethiopia	25 300	79.0
		Sudan	6 700	21.0
Baraka	66 200	Ethiopia	43 700	66.0
		Sudan	22 500	34.0
North and Central America				
St Lawrence	1 280 000	Canada	800 000	62.3
		United States of America	480 000	37.5
Mississippi	3 250 000	United States of America	3 180 000	97.8
		Canada	70 000	2.2
Yukon	765 000	United States of America	481 950	63.0
		Canada	283 050	37.0
Nelson-Saskatchewan	990 000	Canada	871 200	88.0
		United States of America	118 800	12.0
Fraser	260 000	Canada	249 600	96.0
		United States of America	10 400	4.0
Columbia	610 000	United States of America	506 300	83.0
		Canada	103 700	17.0
Colorado	615 000	United States of America	608 850	
		Mexico	6 150	
Rio Grande (Bravo del Norte)	550 000	United States of America	302 500	
		Mexico	247 500	
Stikine	56 700	Canada	51 600	
		United States of America	5 100	
St John	51 800	Canada	34 200	
		United States of America	17 600	
St Croix	8 100	United States of America	6 700	82.2
		Canada	1 400	17.3
Tijuana	1 635	Mexico	1 550	94.3
		United States of America	85	5.2
Grijalva-Usumacinta	120 000	Mexico	81 000	62.5
		Guatemala	39 000	32.5
Hondo	5 600	Belize	2 750	49.1
		Mexico	1 700	30.4
		Guatemala	1 150	20.5
Belize	6 960	Belize	5 040	72.4
		Guatemala	1 920	27.5
Sarstún	1 800	Guatemala	1 570	87.2
		Belize	230	12.8
Motagua	12 570	Guatemala	10 660	84.8
		Honduras	1 910	15.2

Table 19 Continued

River basin	Area of basin (km²)	Constituent countries/territories	Share per country (km²)	(%)
Coco (Segovia)	24 800	Nicaragua	20 590	
		Honduras	4 210	
San Juan	39 350	Nicaragua	26 780	68.1
		Costa Rica	12 570	31.9
Sixaola	3 300	Costa Rica	2 570	77.9
		Panama	730	22.1
Negro	2 830	Nicaragua	1 645	58.1
		Honduras	1 185	41.9
Goascorán	2 500	Honduras	1910	
		El Salvador	590	23.6
Lempa	9 870	El Salvador	5 855	59.3
		Honduras	3 555	36.0
		Guatemala	460	4.7
Paz	1 650	Guatemala	920	55.8
		El Salvador	730	44.2
Suchiate	1 840	Guatemala	1 310	71.2
		Mexico	530	28.8
Massacre	480	Dominican Republic	280	58.3
		Haiti	200	41.7
Artibonite	7 900	Haiti	5 250	66.5
		Dominican Republic	2 650	33.5
Pedernales	450	Dominican Republic	260	52.8
		Haiti	190	42.2
Yaqui	70 000	Mexico	67 000	95.7
		United States of America	3 000	4.3
Candelaria	10 800	Mexico	10 000	92.5
		Guatemala	800	7.4
Choluteca	6 260	Honduras	6 100	97.4
		Nicaragua	160	2.5
Changuinola	3 060	Panama	3 000	98.0
		Costa Rica	60	2.0
Coatán-Achute	1 250	Mexico	750	60.0
		Guatemala	500	40.0
Juradóa	620	Colombia	475	76.7
		Panama	145	23.4
South America				
Orinoco	966 000	Venezuela	626 000	64.8
		Colombia	340 000	35.2
Amazon	5 870 000	Brazil	3 715 000	63.3
		Peru	935 000	15.9
		Bolivia	700 000	11.9
		Colombia	340 000	5.8
		Ecuador	125 000	2.1
		Venezuela	50 000	0.9
		Guyana	5 000	0.1
La Plata	3 200 000	Brazil	1 425 000	44.5
		Argentina	990 000	
		Paraguay	407 000	12.7
		Bolivia	238 000	2.4
		Uruguay	140 000	4.4
Essequibo	147 000	Guyana	113 190	77.0
		Venezuela	33 810	23.0
Lakes Titicaca–Poopó system	114 000	Bolivia	59 300	52.0
		Peru	49 900	43.8
		Chile	4 800	4.2
Lagoon Mirim	55 700	Uruguay	32 500	58.3
		Brazil	23 200	41.7
Catatumbo	34 840	Colombia	19 300	55.4
		Venezuela	15 540	44.6
Amacuro	10 260	Venezuela	7 960	77.6
		Guyana	2 300	22.4
Courantyne	72 100	Surinam	36 900	51.2
		Guyana	35 200	48.8
Maroni	66 000	Surinam	37 000	56.1
		French Guiana	29 000	43.9
Oyapock	30 270	French Guiana	16 194	53.5
		Brazil	14 076	46.5
Mira	11 200	Ecuador	6 920	61.8
		Colombia	4 280	38.2
Mataje	870	Ecuador	630	72.4
		Colombia	240	27.6
Patia	22 540	Colombia	22 400	99.4
		Ecuador	140	0.12
Zarumilla	1 570	Ecuador	880	56.1
		Peru	690	43.9
Tumbes	4 655	Ecuador	2 640	56.7
		Peru	2 015	43.3
Chira	16 220	Peru	8 740	53.9
		Ecuador	7 480	46.1
Lauca	23 500	Bolivia	20 700	88.1
		Chile	2 800	11.9
Zapaleri	1 565	Chile	830	53.0
		Argentina	532	34.0
		Bolivia	203	13.0
Valdivia	11 280	Chile	10 070	89.3
		Argentina	1 210	10.7
Puelo	8 800	Argentina	5 610	63.8
		Chile	3 190	36.2
Yelcho	11 145	Argentina	6 815	61.2
		Chile	4 330	38.8
Palena	13 000	Chile	7 400	56.9
		Argentina	5 600	43.1
Aysen	15 300	Chile	14 400	94.1
		Argentina	900	5.9
Baker	25 700	Chile	20 350	79.2
		Argentina	5 350	20.8
Pascua	13 840	Chile	7 420	53.6
		Argentina	6 420	46.4
Serrano	9 100	Chile	6 960	76.5
		Argentina	2 140	23.5
Gallegos–Chico	12 240	Argentina	6 880	56.2
		Chile	5 360	43.8
Cullen	735	Chile	550	74.8
		Argentina	185	25.2
San Martin	370	Chile	340	91.9
		Argentina	30	8.1
Carmen Silva	1 620	Chile	980	60.5
		Argentina	640	39.5

Table 19 Continued

River basin	Area of basin (km²)	Constituent countries/territories	Share per country (km²)	(%)	River basin	Area of basin (km²)	Constituent countries/territories	Share per country (km²)	(%)
Río Grande	4 830	Chile	2 650	54.9	Red	169 600	China	90 000	53.1
		Argentina	2 180	45.1			Viet Nam[e]	78 000	46.0
Lake Fagnano	4 820	Argentina	3 930	81.5			Laos	1 600	0.9
		Chile	890	18.5	Hsi	436 000	China	419 000	96.1
Chuy	500	Brazil	390	78.0			Viet Nam[e]	17 000	3.9
		Uruguay	110	22.0	Coruh	21 000	Turkey	19 300	91.0
Juradó[b]	620	Colombia	475	76.6			Georgia	1 700	8.1
		Panama	145	25.4	Asi (Orontes)	13 300	Syrian Arab Republic	9 700	73.0
Barima	8 400	Guyana	7 500	89.3			Turkey	2 000	15.0
		Venezuela	900	10.7			Lebanon	1 600	12.0
Asia					Jordan	11 500	Jordan	6 200	53.9
Ob	3 010 000	Russia	2 955 000	98.2			Syrian Arab Republic	3 400	29.6
		China	55 000	1.8			Israel	1 200	10.4
Yenisey	2 530 000	Russia	2 200 000	87.0			Lebanon	700	6.1
		Mongolia	330 000	13.0	Tiban	6 200	Yemen	3 300	53.2
Amur	1 900 000	Russia	995 000	52.4			South Yemen	2 900	46.8
		China	845 000	44.5	Atrek	61 000	Iran	41 000	67.2
		Mongolia	60 000	3.1			Turkmenistan	20 000	32.8
Ganges–Brahmaputra	1 600 400	India	–	–	Hari	84 000	Afghanistan	37 000	44.0
		China	–	–			Turkmenistan	25 000	29.8
		Nepal	140 800	8.8			Iran	22 000	26.2
		Bangladesh	113 300	7.1	Murgab	73 000	Turkmenistan	43 000	58.9
		Bhutan	47 000	2.9			Afghanistan	30 000	41.1
Kura–Araks	225 000	Georgia	140 000	62.3	Dasht	36 000	Pakistan	29 000	80.6
		Turkey	57 000	25.3			Iran	7 000	19.4
		Iran	28 000	12.4	Karnafuli	10 500	Bangladesh	10 000	95.2
Euphrates–Tigris[c]	884 000	Iraq	362 500	41.0			India	500	4.8
		Iran	238 500	27.0	Kaladan	40 000	India	21 500	53.7
		Turkey	163 000	18.4			Myanmar	18 500	46.3
		Syrian Arab Republic	120 000	13.6	Lake Ubsa	67 000	Mongolia	49 000	73.1
Amu–Darya	653 000	Turkmenistan/Uzbekistan	503 000	77.0			Russia	18 000	26.9
		Afghanistan	150 000	23.0	Suyfun	16 500	China	9 400	57.0
Ili	176 000	Kazakhstan	111 000	63.1			Russia	7 100	43.0
		China	65 000	36.9	Tumen	34 400	China	24 000	69.7
Tarim	980 000	China	945 000	96.4			Democratic People's Republic of Korea	10 000	29.1
		Russia	35 000	3.6			Russia	400	1.2
Helmand	386 000	Afghanistan	300 000	77.7	Yalu	64 500	Democratic People's Republic of Korea	32 500	50.4
		Iran	78 000	20.2			China	32 000	49.6
		Pakistan	8 000	2.1	Han	34 700	Democratic People's Republic of Korea	28 500	82.1
Indus	980 000	Pakistan	–	–			Republic of Korea	6 200	17.9
		India	–	–	Ma	36 000	Viet Nam[e]	22 500	62.5
		Afghanistan	70 000	7.1			Laos	13 500	37.5
		China	–	–	Ca	28 500	Viet Nam[e]	20 200	70.9
Irrawaddy	396 000	Myanmar	345 000	87.1			Laos	8 300	29.1
		India	33 000	8.3	Saigon	44 000	Viet Nam[d]	35 000	79.5
		China	18 000	4.6			Cambodia	9 000	20.5
Salween	270 000	China	143 000	53.0					
		Myanmar	110 000	40.7					
		Thailand	17 000	6.3					
Mekong	786 000	Laos	199 500	25.4					
		Thailand	180 000	22.9					
		China	174 000	22.2					
		Cambodia	149 000	18.9					
		Viet Nam[d]	60 500	7.7					
		Myanmar	22 500	2.9					

Table 19 Continued

River basin	Area of basin (km²)	Constituent countries/ territories	Share per country (km²)	(%)	River basin	Area of basin (km²)	Constituent countries/ territories	Share per country (km²)	(%)
Pakchan	3 100	Myanmar	1 600	51.6	Jacobs	240	Norway	160	66.7
		Thailand	1 500	48.4			Russia	80	33.3
Golok	1 500	Malaysia	800	53.3	Pasvik	19 300	Finland	14 500	75.1
		Thailand	700	46.7			Norway	3 300	17.1
Sembakung	11 000	Indonesia	6 000	54.5			Russia	1 500	7.8
		Malaysia	5 000	45.5	Tuloma	23 600	Russia	20 600	87.3
Tami	4 600	Indonesia	4 100	89.1			Finland	3 000	12.7
		Papua New Guinea	500	10.9	Torne	32 400	Sweden	22 900	70.7
Sepik	71 000	Papua New Guinea	69 000	97.2			Finland	9 500	29.3
		Indonesia	2 000	2.8	Kemi	51 500	Finland	49 500	96.1
Fly	67 000	Papua New Guinea	64 000	95.5			Russia	2 000	3.9
		Indonesia	3 000	4.5	Olanga	20 700	Russia	15 300	73.9
							Finland	5 400	26.1
Europe					Oulu	25 000	Finland	22 000	88.0
Rhine	168 757	Germany	100 500	59.6			Russia	3 000	12.0
		Switzerland	27 500	16.3	Vuoksa	76 000	Russia	41 400	54.5
		France	23 700	14.1			Finland	34 600	45.5
		Netherlands	10 000	5.9	Klarälven	47 000	Sweden	38 600	82.1
		Austria	2 900	1.7			Norway	8 400	17.9
		Luxembourg	2 600	1.5	Neman	86 300	Lithuania	78 500	91.0
		Belgium	1 400	0.8			Poland	7 800	9.0
		Liechtenstein	157	0.1	Pregel	19 700	Poland	10 300	52.3
Meuse	41 400	Belgium	17 300	41.8			Russia	9 400	47.7
		France	12 300	29.7	Schelde	16 500	Belgium	9 700	58.8
		Netherlands	7 300	17.6			France	6 500	39.4
		Germany	4 500	10.9			Netherlands	300	1.8
Danubeᶠ	796 250	Romania	233 000	29.3	Yser	1 700	Belgium	900	52.9
		Yugoslavia	179 000	22.5			France	800	47.1
		Hungary	93 030	11.7	Bann	5 800	Northern Ireland (United Kingdom)	5 420	93.5
		Austria	79 549	10.0					
		Czech Republic and Slovakia	66 369	8.3			Ireland (Republic of)	380	6.5
		Germany	56 000	7.0	Foyle	4 000	Northern Ireland (United Kingdom)	2 760	69.0
		Bulgaria	42 500	5.3					
		Ukraine	41 500	5.2			Ireland (Republic of)	1 240	31.0
		Switzerland	2 788	0.4	Erne	4 750	Ireland (Republic of)	2 850	60.0
		Italy	2 000	0.3			Northern Ireland (United Kingdom)	1 900	40.0
		Poland	364	0.0					
		Albania	150	0.0	Fane	190	Ireland (Republic of)	165	86.8
Elbe	144 500	Czech Republic	49 700	34.4			Northern Ireland (United Kingdom)	25	13.2
		Germany	92 600	64.1					
		Austria	1 400	1.0	Rhône	95 300	France	88 000	92.3
		Poland	800	0.5			Switzerland	7 300	7.7
Oder	126 000	Poland	103 800	82.4	Garonne	53 425	France	52 900	99.0
		Germany	13 000	10.4			Spain	500	0.9
		Czech Republic and Slovakia	9 100	7.2			Andorra	25	0.1
Wisla (Vistula)	193 000	Poland	175 500	90.9	Ebro	84 440	Spain	83 600	99.0
		Ukraine	14 800	7.7			Andorra	440	0.5
		Czech Republic and Slovakia	2 700	1.4			France	400	0.5
Tana	13 000	Norway	6 900	53.1	Muga	1 300	Spain	1 100	84.6
		Finland	6 100	46.9			France	200	15.4
Näätämo	3 700	Finland	2 300	62.2					
		Norway	1 400	37.8					

Table 19 Continued

River basin	Area of basin (km²)	Constituent countries/ territories	Share per country (km²)	(%)	River basin	Area of basin (km²)	Constituent countries/ territories	Share per country (km²)	(%)
Bidasoa	820	Spain	640	78.0	Lake Prespa	1 400	Yugoslavia	800	57.1
		France	180	22.0			Greece	320	22.9
Minho	13 500	Spain	12 800	94.8			Albania	280	20.0
		Portugal	700	5.2	Vijosë	6 900	Albania	5 100	73.9
Lima	3 400	Spain	1 800	52.9			Greece	1 800	26.1
		Portugal	1 600	47.1	Vardar	24 000	Yugoslavia	19 300	80.4
Douro	94 500	Spain	78 600	83.2			Greece	4 700	19.6
		Portugal	15 900	16.8	Struma	14 500	Bulgaria	7 200	49.7
Tejo	82 000	Spain	56 500	68.9			Greece	5 500	37.9
		Portugal	25 500	31.1			Yugoslavia	1 800	12.4
Guadiana	61 400	Spain	53 800	87.6	Nestos	8 000	Bulgaria	4 400	55.0
		Portugal	7 600	12.4			Greece	3 600	45.0
Roia	750	France	500	66.7	Maritsa	56 000	Bulgaria	32 700	58.4
		Italy	250	33.3			Turkey	14 600	26.1
Po	74 300	Italy	70 600	95.0			Greece	8 700	15.5
		Switzerland	3 700	5.0	Rezvaya	1 100	Turkey	800	72.7
Isonzo	2 800	Croatia	1 450	51.8			Bulgaria	300	27.3
		Italy	1 350	48.2	Veleka	1 000	Bulgaria	800	80.0
Drin	17 100	Yugoslavia	9 500	55.6			Turkey	200	20.0
		Albania	7 600	44.4					

ᵃ Listed in South America.
ᵇ Listed also in North and Central America.
ᶜ The Government of Iraq proposed that the Euphrates and Tigris should be listed as separate basins and provided the Secretary-General with the following data:

			km²	%
Tigris	378 834 km²	Iraq	220 000	58
		Iran	110 000	28.8
		Turkey	48 000	13
		Syrian Arab Republic	834	0.2
Euphrates	400 000 km²	Iraq	240 000	60
		Turkey	105 000	26.3
		Syrian Arab Republic	55 000	13.7

ᵈ This part was formerly the Republic of Viet Nam.
ᵉ This part was formerly the Democratic People's Republic of Viet Nam.
ᶠ Recent data from the Danube Commission gives an area of 817 000 km² for the whole basin. Presumably it is based upon more detailed maps.

politics into the management of water scarcity. The international river basins are shown in Figure 17.

The list of international rivers which follows in Tables 17–19 (summarized in Tables 17 and 18) was prepared by the United Nations and published in 1978. It comprises all the world's separate basins shared by two or more countries. In some cases, most of the basin area is confined to one country and only a small part of it extends to one or more other countries. For example, in the case of the Mississippi and Colorado river basins the first is shared by Canada and the United States, but only 3% of the basin is within Canada and for practical purposes the river basin is considered a United States basin. In the case of the Colorado River, Mexico shared only 1% of the basin, but because it is the downstream country, the Colorado is considered to be a truly international river basin by both countries.

Measurements of areas in the UN survey were based on regional maps and taken with the aid of a planimeter. The reliability of the figures given varies according to the maps available at the UN Map Library and the possibility of cross-checking with other sources. Discrepancies between basin areas given in Table 17 and those in Rivers (q.v.: Showers, 1989) can be attributed to different interpretations of the location of the watershed. Such interpretation is especially difficult in relatively flat regions. In addition, topographic divides do not always coincide with groundwater flow divides, which implies that without investigations *in loco* an important source of river or lake basin waters may have been overlooked.

R.W. Herschy

Source

United Nations, 1978. *Register of International Rivers*, Pergamon Press, Oxford.

Cross references

Rivers
Water allocation and use
Water availability and river water quality
Water resources: Europe
Water use in the USA
World water balance

INTERNATIONAL STANDARDS IN FLOW MEASUREMENT

Introduction

The requirement for international standards is well established, and it is not appropriate to explain it here in detail. Suffice it to say that the flow measurement industry, like most other industries, recognized its need for international standards long ago, and has been active in their preparation and use for many years. Such standards, whilst not being mandatory, do promote consistency and common practice in some aspects of design, safety and use of many different types of flow-meters and techniques. This leads to advantages in interchangeability of equipment, adoption of minimum safety standards, and permits direct comparisons of flow measurements, regardless of where they are taken or who takes them.

Since international standards are written by industrial experts from countries around the world, their content therefore reflects the knowledge and experience of these experts who draft them, enhanced by those who subsequently submit comments on the drafts. Thus an international standard should represent the correct, best, up-to-date knowledge and experience possible, from experts worldwide, harmonized into a form which is acceptable to most – if not all – member countries.

Purpose and types of standards

The primary purpose of many standards is to promote consistent design of some equipment, by specifying, where appropriate, various dimensions or other design criteria, safety requirements or performance requirements. Other standards specify, or merely describe, test methods or measurement techniques, so that such procedures are conducted consistently throughout the flow measurement industry.

It follows, therefore, that there are several different types of international standard, depending on the intended purpose. The most obvious type is that of a specification which lays down well-defined rules, which have to be followed, on how something must be done, possibly for the construction of safe use of equipment. However, some standards give instructions rather than rules for, say, test methods, and others describe measurement techniques in detail. Still other standards give guidance and recommendations on good practice, and some are mere technical reports when their content is based on technologies which are still in a development stage and not yet sufficiently established to justify a firm standard.

The international scene

Standards fall into one or more groups, according to their applicability in the international scene. Many are not true International Standards, but are often mistakenly thought to be such because of their wide acceptance worldwide. The real international standards are prepared by the International Standards Organization (ISO) and are identified by their ISO number and date of publication. But there are some national standards e.g. BS, DIN (German) or ANSI (US) which are well known and often used as international references, and there are some specifically industrial standards e.g. IP (Institute of Petroleum) or API (American Petroleum Institute) which are used worldwide. There are also some purely regional standards which have only local status, but are well known throughout the industry e.g. CEN (European) Standards, which are prepared for use in the European Community but are widely used outside the EC.

There are many cases where standards on a particular subject have a common interest to two, or even to all three of these parties. This is illustrated in Figure I8, which shows that often, but not always, a national standard, or a regional standard, is common with an international standard, and in some cases there are common national, regional and international standards. In such cases, the text of the international standard is adopted and dual numbered, or even triple numbered, by the other parties. Thus it is quite possible for there to be a standard with a number such as ISO/EN/BS 1234.

There are some other series of documents which are widely recognized internationally, but do not have the status of true International Standards. Within this category is the International Organization for Legal Metrology (OIML) Recommendations, which are in fact legal requirements adopted by Governments in many countries for the purposes of legal metrology, and the WMO (World Meteorological Organization) Technical Regulations.

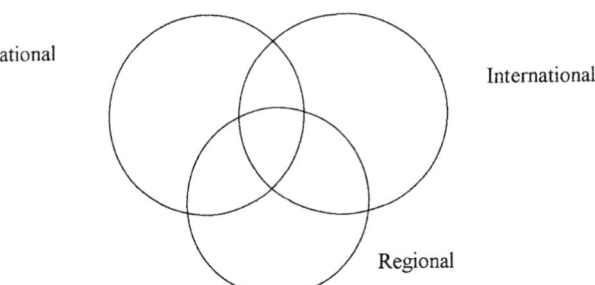

Figure I8 Common coverage of some standards.

Standards preparation

Drafting

Within the ISO organization, the initial drafting and development of international standards is the responsibility of a relevant technical committee. But the detailed work will probably be done in an appropriate sub-committee (SC), which may well establish a specific working group to undertake the initial draft preparation work. There are over 100 technical committees in ISO, each responsible for its own area of work. Any Technical committee might have six, or more, sub-committees to support it.

Some of the work is done at meetings of the appropriate group, but correspondence is circulated to all members whenever possible to avoid the requirement for too many costly meetings.

An ISO Standard is published in two languages – English and French. Often a new standard will be drafted in English, but the version which is circulated for final approval must be in both languages, usually one document in English and a separate document in French, although in some cases it is found to be more acceptable to present opposite pages or opposite columns in the two languages, within the one document.

Procedures

When the text of a new standard – or a revision of an existing standard – has been prepared, it is circulated to all member bodies of the committee, initially for comment and ultimately for approval. Thus all member bodies of a committee have an equal opportunity to contribute to the final version of the document. There is an established procedure, which is followed in all cases, which ensures that the membership of a committee reached 'consensus' (i.e. the lack of sustained objection), or at least a two-thirds majority approval, on any document before it is published by ISO.

The work of the technical committee is therefore more of a supervisory/management role than actual writing of standards. It is responsible for not only the preparation of new standards, but also for the review – and possible revision or amendment – of existing standards. A technical committee might meet once every $1\frac{1}{2}$ to 2 years, whereas a sub-committee might meet a little more frequently.

A proposal for a new standard can be made by any national member body of a committee, or by a committee itself. However, any proposal for a new project must be accompanied by at least an outline of the proposed standard, if not by a complete initial draft. Any new project must be supported by at least five committee member bodies who are willing to participate actively in its development, before the new project is approved and started.

Any queries about the content of a standard are resolved by the technical committee, or perhaps by the relevant sub-committee. It is not for any one person to express an interpretation of the content of a standard, without the approval of the committee as a whole. Thus any questions regarding specific points in a standard should be considered as part of the activity of the responsible committee, and not of any particular delegate to its meetings.

Committees

Members

The members of a technical committee, or of a sub-committee, are the national standards bodies who are members of ISO and have requested to be included in the membership of any specific technical Committee. Each of these national standards bodies can send delegates to any meeting of the technical committee or sub-committee. These delegates attend, not in a personal capacity or representing their employer or trade association, but purely as a delegate from a national standards body. Before attending a meeting of a committee, the delegates should have been briefed by their national body about what views to express and what attitudes to take.

Thus there are no personal members of any committee, although there are regular delegates who attend most, if not all, meetings of a committee (as a delegate of their national body). This does promote a measure of continuity in the discussions and the development of professional collaboration and liaison between member countries.

The only two permanent members of a technical committee or a sub-committee are its secretary and chairman.

Secretary

The secretariat of a committee is held by an appointed national member body. That national body is responsible for the provision of all the services required of the secretariat, and they usually appoint one person to undertake the required duties and to take all required actions. This person takes on the role of the committee secretary, but in fact again they are acting on behalf of their national standards body which is the member of ISO. A member body, once appointed, can retain the secretariat of a committee for as long as it chooses, but the person acting as the secretary will change from time to time for various reasons.

Chairman

The chairman of a committee is nominated by the secretary and approved by the membership, before being formally appointed by the parent committee. The chairman does not necessarily have to come from the same country as the secretariat, although this is often the case, and does help with communication between the two officers of the committee. The chairman serves for an initial period of 6 years, but may subsequently be reappointed for periods of 3 years. The chairman, of course, is an independent person, and does not represent his home country, who will send other delegates to the meetings.

Meetings

Meetings are held usually at the premises of the national standards body of one of the committee members. Thus there is no regular venue for meetings, and committee members tend to share the overall cost of hosting them, in different countries around the world.

Costs

The overall cost of standards work is high, due mainly to the effort put into it by the many experts worldwide who are involved in preparing and commenting on drafts, and acting as delegates to meetings. But this high proportion of the cost is usually paid by 'industry', specifically by the employers of those experts, since this part of the work is done voluntarily, and no payment is made for it by either ISO or the national standards organizations. Similarly, the chairman of a committee acts in a voluntary capacity, and receives no direct payment for his involvement. But, of course, the work of the secretariats is usually paid for by the relevant national standards organization, or sometimes by local industry.

Also, attendance by delegates to meetings is usually not paid for and is again on a voluntary basis, although in many cases a national standards organization will give assistance with travel expenses to international meetings.

When a standard is published, it is not distributed on free circulation, except for a nominal copy to each national standards organization involved. It is available to the worldwide market for sale, through the ISO Central Office, which often has sales arrangements with national standards bodies who can then offer the standards for sale through their own sales outlets. The income from such sales is held by the seller, and provides a contribution towards their overall costs.

ISO standards on flowmeters and flow metering

ISO standards on flowmeters and the measurement of flow (flow metering), and the work involved in their preparation, are divided into two distinct groups, relating to

- flow in closed conduits;
- open channel flow

There is one area which is at times of common interest to both committees. This relates to the measurement of flow in closed conduits which are only partially full. Since such conduits are often sewers or drainage pipes which are rarely full, the work is handled by the open channel committee.

Closed conduit flow

The work on flow in closed conduits (i.e. pipes) is the responsibility of ISO Technical Committee TC 30 – Measurement of fluid flow in closed conduits, running full. TC 30 is supported (currently) by five sub-committees, each of which deals with the detailed work of a specific section of the subject. The sub-committees of TC 30 were restructured in 1994, and now comprise:

SC 2   Pressure differential methods
SC 5   Velocity based methods
SC 7   Volume based methods
SC 9   General topics
SC 12 Mass flow methods

There are currently about 40 standards published as a result of past work by TC 30 and its sub-committees. These comprise standards on most common methods of flow measurement applied to closed conduits, or to the equipment used for those measurements and the methods and techniques involved.

The main current standards are:

ISO 4006 *Measurement of fluid flow in closed conduits – Vocabulary and symbols.*
ISO 5167 *Measurement of fluid flow by means of pressure differential devices – Orifice plates, nozzles and Venturi tubes in circular cross-section conduits.*
ISO 5168 *Measurement of fluid flow – Estimation of uncertainty of a flowrate measurement.*
ISO 3313 *Measurement of pulsating fluid flow.*
ISO 4064 *Measurement of water flow in closed conduits – Meters for cold, potable water.*
ISO 7858 *Measurement of water flow in closed conduits – Meters for cold, potable water – Combination meters.*
ISO 2975 *Measurement of water flow in closed conduits – Tracer methods.*

Other standards deal with flowmeter calibration techniques, or specific types of flowmeter, e.g. electromagnetic flowmeters, ultrasonic flowmeters or mass flowmeters.

Within this list, some standards are more important than others. Perhaps the most important standard is ISO 5167, which deals with orifice plates and other basic differential pressure devices of nozzles and venturis, which specifies the geometry and dimensions to be used, and describes installation and operating conditions. However, since ISO 5167 has grown in coverage over the years since it was first prepared, as the technology has become better understood, there is currently a proposal under consideration in TC 30 to divide ISO 5167 into four or five separate parts, dealing separately with orifice plates, Venturis, nozzles, Venturi nozzles and their installation.

However, it is acknowledged that there are known limits to the basic conditions which are specified in ISO 5167, and that there are times when practical conditions make it necessary to install or operate a flowmeter outside of these specified conditions. Therefore ISO Technical Report 12767 describes the sorts of effects and changes in performance which can be expected if the various basic conditions are surpassed. Also, ISO 15377 describes how orifice plates other than the basic designs can be used. ISO TR 3313 describes the effects of pulsating flow, and how this can be expected to influence the measurement of flow through an orifice plate.

Since the whole subject of flow metering by the use of differential pressure devices is complex, and the use of the ISO standards may not be easily understood by all who would like to use them, ISO 9464 is in preparation as a Code of Practice for the use of the devices specified in ISO 5167.

Whilst the above selection represents the standards relating to the basic differential pressure devices and their usage, there are of course other ranges of standards relating to other types of flow measurement. Thus ISO 4064 and ISO 7858 in their several parts relate to water meters of various types, specifying constructional features and connections, and describing test methods and equipment. Also, ISO 9951 deals with turbine meters, and ISO 4185, ISO 9368 and ISO 8316 deal with weighing and volumetric techniques, which may be considered as calibration techniques rather than industrial measurement techniques.

A new standard, ISO 10790, has recently been published, dealing with Coriolis mass flowmeters. Other new work is in hand to prepare new standards for vortex flowmeters and ultrasonic flowmeters, as well as bringing together all the different techniques for flowmeter calibration, and bringing in further new information relating to upstream pipe configurations and flow conditioners.

Open channel flow

The other area of work, on open channel flow, is the responsibility of ISO Technical Committee 113 – Hydrometry. This is obviously involved more in the flows in rivers, streams, culverts, flumes etc., but does also now include work on groundwater (flow from wells and boreholes), hydrometric data and its quality, as well as the uncertainties arising from the various measurements.

TC 113 is supported by seven sub-committees, comprising:

SC 1   Velocity–area methods
SC 2   Weirs and flumes
SC 3   Vocabulary and definitions
SC 4   Dilution methods
SC 5   Instruments and equipment
SC 6   Sediment transport
SC 8   Groundwater

There are currently nearly 60 standards published as a result of past work by TC 113 and its sub-committees. These comprise standards on most common methods of flow measurement in open channels, or to the equipment used for those measurements and the methods and techniques involved. The main standards are:

● ISO 772 – *Glossary of terms*, which lists and defines over 450 terms used in hydrology and hydrometry. The correct use of these terms is essential to ensure common understanding of the terminology and discussion and comparison of measurements taken worldwide.
● ISO 748 – *Velocity–area methods* and ISO 1100–1 and 1100–2 – *Establishment and operation of a guaging station*, and *The stage–discharge relation*. These standards describe the basic techniques which are common to many measurements of flow in open channels. But they are supported by a series of 15 other standards or technical reports explaining and specifying the correct techniques for many associated procedures, such as moving boat and ultrasonic and electromagnetic methods, and their application under various different conditions, such as the measurement of flood flows and under ice conditions. These are supported by other standards dealing with the data evolved, possible errors in it, corrections to be made, and guidance on the selection of the best or most appropriate method for a particular application.

Another series of standards relate to measurements using weirs and flumes. The principal standard in this series is ISO 1438, which describes and specifies thin-plate weirs. This also is supported by another series of 14 standards and technical reports dealing with other common types of weirs, flumes and gauging structures. ISO 8368 is a guide for the selection of measuring structures, giving guidance on the selection of an appropriate structure for a particular application.

Another series of standards describes and specifies many of the types of equipment which are used to make different measurements. Important amongst these is ISO 2537 – *Current meters*, but a series of 12 other standards specifies such equipment as water level measuring devices, cableway systems, hydrometric data transmission systems and the correct way to specify the performance of hydrometric equipment.

There are smaller series of standards relating to the measurement of suspended sediment transport and the sampling of bed materials, and also a few methods of dilution gauging with tracers.

All the above standards are complemented by ISO 5168 – *Estimation of the uncertainty of a flow rate method*, and ISO 7066

(Parts 1 and 2) describing linear and non-linear calibration relations respectively.

Apart from the above standards which are currently published (many of them for many years), further work is in progress to develop standards for techniques for the measurement of the volume in lakes and reservoirs, a mathematical model for unsteady flow, compound gauging structures, acoustic Doppler current profilers, automatic dilution gauging, suspended sediment in tidal channels and the test pumping of water wells.

European standards on flowmeters and flow metering

Whilst in the past, most of the activity on standards preparation has been within the relevant ISO technical committees and their sub-committees, the preparation of European Standards (ENs) through CEN committees is becoming increasingly important, and active. As in many other areas of work, this European activity is stimulated by the development of the European Market, and its associated European Directives.

In the past, CEN/TC 244 was responsible for the development of European standards on flow metering in closed conduits. However, there was little interest in this work, since there was no relevant European Directive to stimulate the work. Therefore, CEN/TC 244 has been declared dormant for some years past, but could be reactivated if and when required.

But work in CEN/TC 92 dealing with water meters has started again, after several years of inaction, because of their anticipation of a new European Directive which will probably be relevant to their interests.

A relatively new CEN committee, CEN/TC 318 – Hydrometry, is now developing its program of work, based on the adaptation (or adoption verbatim, if possible) of relevant ISO standards specifically into the European region. Whilst there is (as yet) no relevant European Directive, CEN/TC 318 is becoming involved in the development of relevant European standards, because of the growing interest in standards in hydrometry seen by the CEN members.

At times, the situation arises in which appropriate CEN and ISO committees share a common interest in certain subjects, and may wish to prepare similar standards. In order to avoid duplication of effort and ultimate publication of different, possibly conflicting standards, a formal agreement – known as 'The Vienna Agreement' – between the two organizations has been implemented. This Vienna Agreement sets out the liaison procedures to be followed such that the work is prepared in only one of the committees, but the other committee is involved in drafting, commenting, and voting on the subject document in parallel.

Summary

International standards on the measurement of flow, either in closed conduits or in open channels, have been developed and continue to be developed under the aegis of ISO, the International Standards Organization. There has in the past been some limited activity on the European scene, under the European Standards Organization CEN, but this is becoming more active, particularly in the fields of water meters and hydrometry.

G.G. Robson

Cross reference

Streamflow measurement

IRRIGATED LAND AREA: WORLD

Irrigation and drainage is by far the largest of the World Bank sectors, accounting for about 7% of Bank lending. From 1950 to the present the Bank lent some $21 billion for irrigation projects worldwide. Internationally about 70% of fresh water is used for irrigation, 23% for industry and 7% for domestic use. However, water-use pattern reflect the degree of industrialization and Africa, for example, uses 88% of its water resources for irrigation and only 5% for industry. Europe, on the other hand, uses 50% of its fresh water for industry and only 30% for agriculture, the remaining 20% being for domestic use.

Even though agricultural use of water has the lowest value per cubic meter, there is strong political oppositon to diverting water from

Table 110 Irrigated land area

Country	Irrigated land area (ha × 10³)			Country	Irrigated land area (ha × 10³)		
	1970	1980	1990		1970	1980	1990
Low-income economies				62 El Salvador	20	110	120
				63 Turkmenistan	–	927	1 240
1 Mozambique	26	65	115	64 Moldova	–	217	290
2 Ethiopia	155	160	162	65 Lithuania	–	–	–
3 Tanzania	38	120	150	66 Bulgaria	1 001	1 197	1 263
4 Sierra Leone	6	20	34	67 Colombia	250	400	520
5 Nepal	117	520	1 000	68 Jamaica	24	33	35
6 Uganda	4	6	9	69 Paraguay	40	60	67
7 Bhutan	–	–	–	70 Namibia	–	–	–
8 Burundi	27	56	72	71 Kazakhstan	–	1 961	2 300
9 Malawi	4	18	20	72 Tunisia	90	156	232
10 Bangladesh	1 058	1 569	2 936	73 Ukraine	–	2 013	2 600
11 Chad	5	6	10	74 Algeria	238	253	384
12 Guinea-Bissau	–	–	–	75 Thailand	1 960	3 015	4 300
13 Madagascar	330	645	920	76 Poland	213	100	100
14 Laos	–	–	–	77 Latvia	–	–	–
15 Rwanda	4	4	4	78 Slovak Republic	–	–	–
16 Niger	18	23	40	79 Costa Rica	26	61	118
17 Burkina Faso	4	10	20	80 Turkey	1 800	2 090	2 370
18 India	30 440	38 478	45 500	81 Iran, Islamic Rep.	5 200	4 948	5 750
19 Kenya	29	40	54	82 Panama	20	28	32
20 Mali	80	152	205	83 Czech Republic	–	–	–
21 Nigeria	802	825	870	84 Russian Federation	–	–	–
22 Nicaragua	40	80	85	85 Chile	1 180	1 255	1 265
23 Togo	4	6	7	86 Albania	284	371	423
24 Benin	2	5	6	87 Mongolia	10	35	77
25 Central African Rep.	–	–	–	88 Syrian Arab Rep.	451	539	693
26 Pakistan	12 950	14 680	16 960				
27 Ghana	7	7	8	*Upper-middle-income*			
28 China	37 630	44 888	47 403				
29 Tajikistan	–	617	690	89 South Africa	1 000	1 128	1 128
30 Guinea	5	8	25	90 Mauritius	15	16	17
31 Mauritania	8	11	12	91 Estonia	–	–	–
32 Sri Lanka	465	525	520	92 Brazil	796	1 600	2 700
33 Zimbabwe	46	157	220	93 Botswana	1	2	2
34 Honduras	70	82	90	94 Malaysia	262	320	342
35 Lesotho	–	–	–	95 Venezuela	70	137	180
36 Egypt, Arab Rep.	2 843	2 445	2 648	96 Belarus	–	163	149
37 Indonesia	4 370	5,418	8 177	97 Hungary	109	134	204
38 Myanmar	839	999	1 005	98 Uruguay	52	79	120
39 Somalia	95	105	118	99 Mexico	3 583	4 980	5 180
40 Sudan	1 625	1 770	1 900	100 Trinidad and Tobago	15	21	22
41 Yemen, Rep.	–	–	–	101 Gabon	–	–	–
42 Zambia	9	19	32	102 Argentina	1 280	1 580	1 680
				103 Oman	29	38	58
Middle-income economies				104 Slovenia	–	–	–
Lower-middle-income				105 Puerto Rico	–	–	–
				106 Korea, Rep.	1 184	1 307	1 345
43 Ivory Coast	20	44	64	107 Greece	730	961	1 195
44 Bolivia	80	140	165	108 Portugal	622	630	631
45 Azerbaijan	–	1 195	1 401	109 Saudi Arabia	365	555	900
46 Philippines	826	1 219	1 560	110 Ireland	–	–	–
47 Armenia	–	274	305	111 New Zealand	111	183	280
48 Senegal	110	170	180	112 Israel[a]	172	203	200
49 Cameroon	7	14	30	113 Spain	2 379	3 029	3 402
50 Kyrgyzstan Republic	–	955	1 030	114 Hong Kong[a]	8	3	2
51 Georgia	–	409	466	115 Singapore[a]	–	–	–
52 Uzbekistan	–	3 476	4 159	116 Australia	1 476	1 500	1 832
53 Papua New Guinea	–	–	–	117 United Kingdom	88	140	164
54 Peru	1 106	1 160	1 260	118 Italy	2 561	2 870	3 120
55 Guatemala	56	68	78	119 Netherlands	380	480	555
56 Congo Republic	1	3	4	120 Canada	421	596	860
57 Morocco	920	1 217	1 270	121 Belgium	–	–	–
58 Dominican Republic	125	165	225	122 Finland	16	60	64
59 Ecuador	470	520	552	123 United Arab Emirates[a]	–	–	–
60 Jordan	34	37	63	124 France	539	870	1 170
61 Romania	731	2 301	3 216	125 Austria	4	4	4

Table I10 Continued

Country	Irrigated land area (ha × 10³)		
	1970	1980	1990
126 Germany	284	315	332
127 United States	16 000	20 582	18 771
128 Norway	30	74	97
129 Denmark	90	391	430
130 Sweden	33	70	114
131 Japan	3 415	3 055	2 846
132 Switzerland	25	25	25

Selected economies not included in main WDI tables

Angola	–	–	–
Barbados	–	–	–
Cyprus	30	30	36
Fiji	1	1	1
Gambia, The	8	10	12
Guyana	115	125	130
Haiti	60	70	75
Iceland	–	–	–
Iraq	1 480	1 750	2 550
Kuwait	1	1	2
Liberia	2	2	2
Luxembourg	–	–	–
Malta	1	1	1
Surinam	28	42	59
Swaziland	47	58	62
Zaire	–	7	10

[a] Economies classified by the United Nations or otherwise regarded by their authorities as developing.

agriculture to other sectors. The result is that in many countries, industrial and developing alike, large volumes of water are used in irrigated agriculture, adding little economic value, whilst cities and industries, which would gladly pay more, cannot get enough. The mismatch is most striking in the areas around large cities. In the western United States, for example, farmers in Arizona pay less than 1 ¢ per cubic meter of water, while residents of Phoenix pay about 25 ¢. In the industrial heartland of China, around Beijing and Tianjin, 65% of water is used relatively inefficiently for low-value irrigation, while huge expenditure is contemplated to bring water from other river basins to the cities.

Table I10 shows the irrigated land areas of the world for the periods 1970, 1980 and 1990 based on four income categories. It can be seen that, with the exception of Sri Lanka, irrigation in low-income economies is increasing, and in the large countries by 10% or more per annum. In the high-income economies only the United States and Japan can be said to have reduced irrigation in 1990 compared to 1980.

R.W. Herschy

Source

The World Bank, *World Development Report*, 1994.

Cross references

Water allocation and use
Water availability and river water quality
Water resources: Europe
Water use in USA
World water balance

IRRIGATION AND DRAINAGE

Irrigation

A major objective of the management of an irrigation system is to maximize the crop yields per volume of water consumed by the system. In practice, four basic types of irrigation are used – surface, sprinkler, subsurface and drip or trickle. Where water is scarce and costly, the use of drip or trickle irrigation may become attractive.

The water consumed is needed for the following purposes.

- To meet the crop water requirement, which is defined by the Food and Agriculture Organization (FAO) (1975) as:

  the depth of water needed to meet the water loss through evaporation of a disease-free crop growing in large fields under non-restricting soil conditions including soil water and fertility and achieving full production potential under the given growing environment.

- To satisfy losses caused by
  - evaporation from weeds;
  - evaporation from the wet surfaces of vegetation and saturated soil;
  - evaporation from moist soil;
  - drainage of soil water;
  - seepage, leaks and evaporation from associated reservoirs and water-distribution canals.

Water management is directed toward ensuring that crop water requirements are met while minimizing other water losses.

Crop water requirements

Crop water requirements are usually estimated either from a knowledge of evaporation demand and crop characteristics or, more recently, from direct measurements of the soil water status or the physiological stress of the plants.

The Food and Agriculture Organization (1975) describes how the pan evaporation, radiation, Penman and Blaney–Criddle methods can be used to calculate a potential reference crop evaporation which, when multiplied by an appropriate crop coefficient, gives an estimate of the crop water requirement. In the former USSR, the heat–water balance method, based on data from a standard hydrometeorological network and developed by the State Hydrological Institute, has been applied widely (Kharchenko, 1975).

Blaney–Criddle method
Blaney–Criddle is one of the most widely used methods of estimating crop water requirements. An adaptation of this method is suggested in the FAO *Guidelines for Predicting Crop Water Requirements* (1975) to calculate the reference crop evapotranspiration for areas where only measured air temperature data are available. The original Blaney–Criddle approach involves temperature and percentage of daylight hours as climatic variables to predict the effect of climate on evapotranspiration. An empirically determined consumptive-use crop coefficient is then applied to establish the consumptive water requirement, which is defined as the amount of water potentially required to meet the evapotranspiration needs of vegetative areas so that plant production is not limited from lack of water.

However, crop water requirements will vary widely between climates having similar air temperature, for example, between very dry and very humid climates, or between generally calm and very windy conditions. Thus the effect of climate on crop water requirements is not fully defined by the temperature and day length. Consequently the consumptive-use crop coefficient will vary not only with the crop, but also with climatic conditions. Its value is both time and place dependent, and local field experiments are normally required for its determination.

The Blaney–Criddle consumptive-use factor f is calculated in the following way:

$$f = p\,(0.46t + 8.13) \qquad (I26)$$

where p is the mean daily percentage of annual daylight hours for a given month and latitude and t is the mean of the daily temperatures, in °C, during the month considered. The factor f is expressed in millimeters per day and represents the mean value for the given month. This adaptation of the Blaney–Criddle method should be used only when temperature data are the only weather data available. The empiricism involved in any evapotranspiration prediction from a single weather factor is inevitably high. Only for weather conditions similar in nature does a generally positive correlation seem to exist between f values and the reference crop evapotranspiration.

The use of the Blaney–Criddle method to calculate mean daily evapotranspiration should normally be applied for periods no shorter than 1 month. Unless verification of the general prevailing weather conditions can be obtained, for example, daytime minimum humidity, the ratio of actual to maximum possible sunshine hours or daytime wind conditions at 2 m height, predictions are highly questionable. Thus considerable care is needed in the use of this method because, for a particular month, actual sunshine hours may vary greatly from year to year. Hence it is suggested that evapotranspiration should be calculated for each calendar month for each year of record rather than by using mean temperatures based on several years' records.

This method should not be used in equatorial regions where temperatures remain relatively constant while other weather variables change. Nor should it be used for small islands where air temperature is generally a function of the surrounding sea surface temperature and shows little response to seasonal change in radiation. At high altitudes, the daytime radiation levels may be higher than under the conditions from which the method was derived. Also, in climates with a high variability in sunshine hours during transition months, e.g. monsoon climates and midlatitude climates during spring and fall, the method can be misleading.

Soil moisture
Tensiometer and neutron-probe soil moisture meters have been used to monitor soil moisture and to calculate application requirements based on the measurement of soil moisture deficit. They can also be used directly to schedule and control application amounts by turning irrigation on and off when predetermined levels of soil moisture have been reached (Richards and Marsh, 1961; Campbell and Campbell, 1982).

Water quality
Water for irrigation is required not only in sufficient quantities, but must also meet certain quality criteria. Plants are particularly sensitive to the level of dissolved salts in irrigation water. A high salt content in water and soil, and irregular irrigation can produce problems of salinization of irrigated lands, which are common in many regions of the world.

Water losses

Water losses are influenced by the particular configuration of an irrigation scheme and can be minimized by management practices that can result in considerable savings in the operational costs of the scheme. The drainage of soil water should not be regarded solely as a loss. A minimum drainage is required to remove salt accumulation from the soil. Where seasonal rainfall is insufficient to flush out accumulated salts, application rates should be increased to satisfy both the crop and the leaching requirements. Details on how this increment can be calculated from a knowledge of the quality of the irrigation water and the crop water requirement are provided in the FAO *Guidelines for Predicting Crop Water Requirements* (FAO, 1975).

The choice of method of irrigation will also influence evaporative losses. Overhead sprinkler systems will wet vegetation surfaces, and the loss rates of intercepted water are likely to be higher – in the case of tall crops this will be many times higher – than transpiration from dry crops.

Surface irrigation will result in evaporative losses from the wet soil surface, but this is not likely to be a significant proportion of the crop water requirement unless the surface is maintained wet for a significant proportion of the time, as in a rice paddy. It is of more concern that an oversupply of water may be necessary, in ridge and furrow application, to meet the crop water requirement, and the drainage losses may be large and spatially variable.

With drip or trickle irrigation, there is maximum potential for managing water applications to minimize both surface evaporation losses and drainage losses in excess of leaching requirements. By siting the drip points close to the crop, both weed growth and evaporative losses from the weeds can be minimized. Drip irrigation systems can be used in both large-scale irrigation schemes and on small-scale, gravityfed, smallholder irrigation schemes. Except for subsurface irrigation, which may not be cost effective in many circumstances, drip irrigation methods offer the highest potential for water-use efficiency (Hodnett et al., 1990).

Agricultural drainage

Definition

Agricultural drainage is the removal of excess groundwater or water from the land surface to create more favorable conditions for plant growth. Surface drainage can remove excess precipitation from the land surface at a rate that will prevent long periods of ponding or flooding without excessive erosion, so that pasture or crops will have the best possible moisture conditions.

Subsurface drainage lowers the water table so that it will not interfere with root development and it promotes leaching to maintain the proper salt balance in the soil. Detailed discussion on this subject is provided by Richards and Marsh (1961).

Factors affecting drainage

Agricultural drainage needs vary considerably because of differences in climate, geology, topography, soils, crops and farming methods. Visual evidence of inadequate drainage includes surface wetness, lack of vegetation, undesirable vegetation – such as marsh grass, sedge or swamp trees – crop stands of irregular color and growth, variations in soil color, and salt deposits on the surface of the ground. The topography, geology, human-made obstructions, or soils of a site and its surrounding area, may result in conditions that retard water movement and cause poorly drained sites. Site factors can be placed in several categories. These may exist separately or in various combinations. The following are some of the more important factors.

- Lack of a natural drainage way or depression to serve as an outlet. Such sites are common in glaciated and coastal plain areas where natural drainage systems are in the process of development.
- Lack of sufficient land slope to cause water to flow to an outlet. Such sites can be found in the irregular and pitted surfaces of glaciated land, above constrictions and natural barriers of valley floodplains, and above dams.
- Soil layers of low permeability that restrict the downward movement of water. Many soils have a heavy subsoil, rock formation or compact (hardpan) layer below the surface in the normal root zone of plants.
- Artificial obstructions, such as roads, fence rows, dams, dikes, bridges and culverts with insufficient capacity, which obstruct or limit the flow of water.
- Natural surface barriers that cause local concentrations of water in sufficient amounts to aggravate the drainage problem.
- Subsurface drainage problems in irrigated areas caused by deep percolation losses from irrigation and seepage losses from the system of canals and ditches serving the irrigated lands. Deep percolation losses from irrigation fall in the general range of 20–40% of the water applied. Seepage losses from canals and ditches vary widely and may be in the range of zero to 50% of the water applied.

Most soils in arid regions contain some salts, varying in concentration from slight to strong. High water table conditions caused by deep percolation from irrigation tend to concentrate salt accumulations in the root zone. One of the primary functions of subsurface drainage is to lower the water table and to keep the level of salt concentration below the root zone. Much of the subsurface drainage work in arid regions is actually for salinity control.

There is no danger of overdrainage of most soils with poor internal drainage. Close spacing of drains in soils in poor physical condition aids in the establishment and growth of vegetation needed for soil conditioning, even though this intensity of drainage may not be necessary in the same soil if it were in good physical condition. The removal of free water in the soil eliminates moisture in excess of that held by capillary action. Drainage does not remove the capillary water used by growing plants. The depth of the drains controls the height of the water table. If the water table is low in soils with low capillary suction, then moisture may not move upward into the root zone. This is a desirable condition in irrigated saline, saline-alkali and alkali soils.

There is a possibility of overdraining some extremely sandy soils and some peat and muck areas. These soils have a particular depth of water table that is best for plant growth that should be considered in designing the drainage system.

Benefits of farm drainage

The removal of free water, which promotes bacterial action in the soil that is essential for the manufacture of plant food, allows air to enter the soil. The roots of plants, as well as the soil bacteria, must have oxygen. Drainage accomplishes this by providing air space in the soil. Rainfall passing downward in the soil removes carbon dioxide and permits fresh air to infiltrate. Thus drainage provides needed soil aeration.

Surface drainage removes ponded water quickly, thereby allowing the remaining gravitational water to move through the soil.

The removal of free water by drainage allows the soil to warm more quickly because more heat energy is required to raise the temperature of wet soil. Soil warmth promotes bacterial activity, which increases the release of plant food and the growth of plants. Soils that warm up sooner in the spring can be planted earlier. Better germination conditions for seeds are also provided.

The removal of groundwater improves the conditions for root growth. For example, if free water is removed only from the top 25 cm of soil, crop roots will feed in this confined area, but if free water is removed from the top 1 m, this entire depth of soil is available as a root zone from which plants can obtain nutrients and moisture.

Basic types of drainage

Drainage is accomplished by establishing or accelerating gravity flow within the site, by diverting flow from the site, or by a combination of these two.

Relief drainage

Establishing or accelerating the flow of excess water within and from a site is referred to as relief drainage. Surface flows are removed by surface ditch systems and land grading. Subsurface flows are removed by relief drains, which are lateral drains located parallel, or approximately so, to the flow of groundwater.

Interception drainage

Interception ditches or drains located across the flow of groundwater (or seepage) are installed primarily for intercepting subsurface flow moving down a slope. While this type of drainage intercepts and diverts both surface and subsurface flows, the removal of surface water is generally referred to as diversion drainage and the removal of subsurface water by this method is referred to as interception drainage.

Methods of artificial drainage

Surface drainage

Surface drainage may be accomplished by open ditches and by shaping land surfaces for the movement of water to the disposal ditches. Drainage by this method applies to flat sites where

- soils are of low permeability throughout their profiles, e.g. low-permeability clays;
- soils are shallow (20–50 cm) over low-permeability subsoil or rock;
- soils would be responsive to subsurface drainage but lack a free subsurface outlet;
- subsurface drainage is not economically feasible;
- surface drainage supplements subsurface drainage.

Subsurface drainage

Subsurface drainage may be accomplished by various types of buried drains, by mole drains and open ditches. Subsurface drainage is applicable to saturated soil conditions where it is physically and economically feasible to use underdrains to remove free water from the root zone. The fertility of the soil must be such that sufficient drainage will result in additional yields of crops to justify the expense of installing the drains.

The need for, and the design of, subsurface systems are related to the amount of excess water entering the soil, the permeability of the soil and the underlying subsoil, and the crop requirements. In general, fine-textured soils have low permeabilities. In such soils, the pore spaces are so small and clogged with colloidal material that gravitational flow into the drain is obstructed, which restricts it to removal of the free water only from a limited area.

In some sandy, peat and muck soils, the pore spaces are large and the movement of water is rapid. Wetness occurs because of the high water table, particularly in the spring in non-irrigated areas and in the fall, or after the irrigation season, in irrigated areas. This must be corrected by drainage if maximum crop yields are to be produced. Soils of this type can be successfully drained, but many of them present installation and maintenance problems.

In some fine, sandy soils, there is insufficient colloidal material to hold the particles together, and there is danger of excessive movement of the sand particles into the drains. These soils require special precautions in drain construction. Open-ditch subsurface drainage may be practical.

In peat and muck soils there is some tendency for the fine soil particles to enter the drain, there is danger of a tile drain shifting due to the unstable nature of the soil, and there is a tendency for newly drained muck and peat soils to settle considerably. For these reasons it is recommended that drainage of these soils by buried drains should be delayed until initial settlement has taken place. Mole drainage and open ditches may be used for initial drainage of this type of soil.

In very porous soils, such as coarse sands and some peats, excessive lowering of the water table may cause a moisture deficiency during periods of drought. Such soils, being very porous, have low capillary suction and are unable to draw water up into the root zone of certain crops if the water table falls much below the root zone.

There are other soil conditions where drains are hazardous or impractical. In some soils, boulders or stones make drainage costs prohibitive. In other soils, the topsoil is satisfactory but it is underlaid with unstable sand at the depth where drains should be installed, thus making installation more difficult or impossible. In other soils (such as those containing glauconite, iron oxide or magnesium oxide), there is a tendency for the drain joints or perforations to seal due to chemical action.

Economic factors

Some soil can be drained satisfactorily, but the installation costs are so great that the benefits derived do not justify the expense. In most instances, drain spacing of less than 15 m for relief drainage cannot be justified unless high-value crops or substantial indirect benefits are involved. Indirect benefits should be considered when, for example, the drying out of orchards makes it possible for spray rigs to be used without bogging down.

Some soil can be drained satisfactorily, but inherent productivity is so low that yields do not justify the expense. Suitable outlets may not be available and can be obtained only at prohibitive cost.

In some cases, the financial ability of the farmer may not permit substantial indebtedness, even though returns from increased crop yields and reductions in the cost of production might pay for drain installation within a 5–10 year period.

Source

World Meteorological Organization, 1994. *Guide to Hydrological Practices* 5th edn, WMO, Geneva.

Bibliography

Campbell, G.S. and Campbell, M.D., 1982. Irrigation scheduling using soil moisture measurements: theory and practice. *Advances in Irrigation*, **1**, 25–42.

Food and Agriculture Organization, 1975. *Guidelines for Predicting Crop Water Requirements* (eds J. Doorenbros and W.O. Pruitt). Irrigation and Drainage Paper No. 24.

Hodnett, M.G., Bell, J.P., Ah Koon, P.D. *et al.*, 1990. The control of drip irrigation of sugarcane using 'index' tensiometer: some comparisons with control by the water-budget method. *Agricultural Water Management*, Special Issue.

Kharchenko, S.I., 1975. *Hydrology of Irrigated Lands*. Gidrometeoizdat, Leningrad.

Richards, S.J. and Marsh, A.W., 1961. Irrigation based on soil section measurements. *Proceedings of the Soil Science Society*, 65–69.

Cross references

Drainage
Irrigated land area: world

K

KARST HYDROLOGY

Introduction

Karst phenomena have long held a deep fascination for people who have been fortunate enough to live in one of the many limestone regions of the world. Mythological deities, particularly those in the Mediterranean region (Egyptian, Greek and Roman) are associated in many ways with karst features. The gods used caves as their sanctuary and were responsible for the origin of springs, the disappearance of rivers and the occurrence of droughts. Caves and caverns have been depicted as the entrance to Hades and the underworld and the place of purgatory and death. They also have been depicted as the source of life for water during droughts and designated as holy places for religious ceremonies.

Limestone and dolomite are common sedimentary rocks that are moderately soluble. These rocks crop out over approximately 12% of the land area of the Earth, and occur in all structural and climatic regions. Calcareous soils forms on them have been preferred for cultivation for thousands of years. This has led to a disproportionate concentration of settlement on the carbonate rocks, and it is now estimated that 25% of the world's population depends upon karst aquifers for their water supplies.

Karst terranes were among the first regions of the world to be adversely affected by the activities of humans. Carbonate rocks surround much of the Mediterranean Sea, and early civilizations located many of their settlements on slightly elevated limestone ridges and plateaus. These areas were more desirable than the low-lying wetlands near the rivers that were plagued with insects and disease. The early people were unaware of the consequences of inhabiting the karst areas. The one characteristic of karst regions that makes them particularly fragile environments and susceptible to environmental degradation is the erodability of the thin soils that develop atop limestone. As these areas became deforested through overgrazing or the use of trees for cooking fuel and later for charcoal in the smelting of copper, the protective vegetative mat that held the soil in place was destroyed. Consequently, the soil was removed either by sheet erosion or washed into solution cavities in the limestone. The loss of soil made it more difficult for the rainfall to infiltrate to the subsurface and recharge the groundwater. As a result, higher-level springs dried up and the flow of rivers decreased. Reforestation programs are now underway in many karst regions of the world to mitigate these problems.

There are many examples of how groundwater flow behaves differently in karst aquifers than in porous-media aquifers. One karst feature that set back understanding of the hydrological cycle for many centuries is the unique sinkhole on the island of Cephalonia, into which the water from the Ionian Sea drains at a rate of a few cubic meters per second. The early Greeks used this flow of water to generate energy with water wheels that came to be known as the sea mills of Cephalonia. This reverse flow of seawater prevented the early philosophers from recognizing the hydrological cycle. They observed the seawater going underground and believed it supplied high-elevation springs in the bordering mountains. They believed that rainfall did not supply the springs because they thought the carbonate rocks too impermeable for rainfall to infiltrate. The early philosopbers devised hypotheses that would remove the salts from the water and raise the fresh water to higher elevation of the mountains. They called upon such fallacious notions as the force of wind, condensation of moisture in the caves, siphon effects and even the curvature of the Earth to move fresh water to higher elevations.

Because limestone and other carbonate rocks have extremely low primary permeability, most of the occurrences and movement of groundwater are in fractures and dissolution channels. The principles of groundwater flow were developed through the scientific study of porous-media aquifers and must be greatly modified when applied to flow in fractures. Because of the enhanced permeability derived from dissolution of the rocks, greater understanding and knowledge of occurrence and movement of water have been gained in fractured and solutionally altered carbonates than in fractured rocks of other lithologies. Consequently, the study of fracture flow in carbonate terranes has transferred value to other terranes.

Most outcrops of carbonate rocks display karst dissolutional enlargements to some extent, and many are intensively karstified. The sedimentological and diagenetic environments are so varied that the number of distinctly different types of limestone and dolomite is greater than of all other types of consolidated sedimentary rocks. The fracturing effects of tectonic activity, together with the superposition of the episodes of paleokarstification on the initial lithological variety, produce a wide range of hydrological conditions in modern karst aquifers. Some karst aquifers can be treated as classical porous media aquifers with Darcy flow, others as fracture aquifers or as double-porosity (fracture–granular or wide fracture–narrow fracture) aquifers, others as purely cavernous conduit systems. The characteristics of many karst aquifers range between these differing conditions over distances as little as a few tens of meters.

The problems of predicting the behavior of karst aquifers and of managing and conserving them are correspondingly great. Because of the sensitive nature of limestone terranes, they provide an excellent long-term field laboratory to study human-induced changes in the environment. The effects of land-use practices can be easily observed in these fragile terranes and the understanding gained from studies of them may be applicable to more stable environments. An important goal of international karst studies is to learn to manage the local resources of limestone terranes and, in addition, to understand the consequences of their mismanagement. This experience could be used as a guide to use and manage the land, forest and water resources of non-karst terranes in a sustainable manner (see excellent discussion by LeGrand, 1984).

Occurrence and movement of water in karst terranes

In non-karstic terranes a clear distinction exists between surface water and groundwater, and between surface drainage basins and groundwater aquifers. Drainage basins, which determine the catchment and runoff characteristics of surface water are defined by drainage divides set by local topography. Drainage basin boundaries remain fixed in time, at least on the time scale of surface water runoff. Because of the slow rate of infiltrating surface water is often loosely coupled to the groundwater system, aquifers are usually continuous across surface divides, and may span many surface drainage basins. In unconfined aquifers, high water table elevation usually corresponds to high topography and thus to surfacewater divides; but in confined aquifers in structurally complex regions, groundwater gradients and resulting flow directions may be quite different from the direction of surface runoff.

In karst regions, distinctions between basins and aquifers are less because of the integrated system of conduits that carry water through the subsurface. The groundwater basins are often defined by the relationship between swallow holes of recharge areas and discharge points. The groundwater basins are closely related to the overlying surface basins because the flowpaths through the conduit system are alternative routes to the flow. In some cases, the boundaries of the surface and subsurface basins are identical and their discharge through the conduit serves as an underground bypass to the surface stream. In general, however, subsurface basins are not congruent with the surface basins.

Underground conduit systems, particularly those with low-gradients, may have more than one channel carrying water, with downstream discharges at more than one spring. The number of conduits carrying water and the details of the flowpath change depending on groundwater levels. Conduits that are flooded during high water levels may be dry cave passages when the water level is low.

The hydrological cycle is a fundamental concept of hydrology. It is a statement of mass conservation for the waters of the Earth. Water that falls as rain supplies both surface streams and groundwater. Groundwater returns to surface streams, which aggregate to large rivers, which drain to the ocean. Evaporation from the ocean provides the moisture that drifts over the continents to fall as rain, thus completing the cycle. A somewhat more complex hydrological cycle operates in karst terranes. The karst aquifer and associated groundwater basins may be in a larger basin containing non-karstic rocks. Runoff from the non-karstic parts of the basin can generate streams that flow onto the karst, some of which may remain on the surface across the karst, some of which may go underground and other streams from the borderlands may sink into swallow holes at the margin of the karst and immediately become part of the subsurface drainage system. Rainwater may penetrate the soils on the karst surface as diffuse infiltration through joints and fractures in the limestone. Little or no runoff occurs on karst surfaces. Overland flow disappears into sinkholes to enter the groundwater system as internal runoff. Surface and subsurface flows rejoin in the base-level surface streams to form total discharge that drains the region.

The sediment balance is important in the overall development of the conduit system because the amount of sediment transported varies with climatic regime, particularly with intensity and frequency of rainfall. Flow in conduit systems that is competent to carry the sediment load during high rainfall may not be competent to carry it during low rainfall. The continued deposition of sediments can fill the conduit system, and force the drainage back to the surface.

Unlike groundwater flow in other aquifers, much water in karstic aquifers flows in pipe-like solutions conduits, some of which are open cave stream channels, and others are openings such as fractures and pores. Thus theories of flow in porous media and flow in pipes and channels are applicable in karst aquifers. The conduit permeability of karst aquifers consists of integrated systems of openings ranging from solutionally widened joints and bedding plane partings to pipe-like passages many meters in diameter. The force of gravity is responsible for the flow of water in these natural systems of pipes and channels.

Karst aquifers are of diverse form and character, depending on their hydrogeological setting. Two extremes can be recognized: conduit aquifers, in which the groundwater throughput is completely dominated by the conduit system, and diffuse flow aquifers, in which conduit systems are either absent or so poorly integrated that they have little influence on the groundwater circulation. Many carbonate aquifers contain both elements.

The land surface can have three kinds of catchment areas: the karst surface itself, borderlands adjacent to the karst that may be the source for allogenic streams, and a caprock area standing above the karst, which may be the source of either sinking streams or direct vertical input into the karst aquifer below. Precipitation is partly lost by evapotranspiration; the remainder enters the aquifer as allogenic runoff, internal runoff and diffuse infiltration.

Sinking streams that channel the water collected from a considerable area to a single swallow hole usually continue underground in a conduit, which may appear as an open channel or as a flooded pipe. Water draining from surface catchments or perched aquifers above the limestone often takes a direct route through the vadose zone by means of vertical shafts. These may drain to an integrated conduit system or feed small springs directly.

Diffuse recharge moves uniformly downward through available joints, accounting for much of the infiltration from the karst surface. A few of the joints intersect conduits and discharge their water. Most seepage waters miss the conduit system completely and provide an important source of recharge to the more dispersed, deeper water body of phreatic storage. The balance between internal runoff and diffuse infiltration depends on soil permeability, depression density and the efficiency of the closed depression drains.

Below and beside the conduit system lies a region of bedrock with water stored in small channels, fractures, and primary pores. Water from phreatic storage supplies cave streams during periods of low recharge. Phreatic storage is also the source of water for most wells drilled in carbonate aquifers. The conduits create a low-gradient groundwater trough, so that water exchange between the cave stream–conduit system and the phreatic storage–diffuse flow system is both lateral and vertical. In the headwater reaches of the system, large conduits are more sparse and the diffuse flow system tends to dominate, whereas downstream, closer to the springs, the reverse is true.

The schematic diagram (Figure K1) can be adapted to many real aquifers by adjusting the mix of water traveling along the various flowpaths. For example, to obtain a diffuse flow aquifer, such as some of the principal dolomite aquifers, simply shut off the conduit system altogether. Seepage waters and some internal runoff make their way directly to the phreatic storage, from which they may be discharged through diffuse flow springs. The chemical and physical response of the springs tells much about the hydrogeological systems.

The scientific beginning of modern hydrogeochemistry occurred primarily in the study of karst aquifers. The application of principles of chemical thermodynamics was feasible in carbonate terranes because the dissolution and precipitation reactions are relatively rapid and because of the rather simple, nearly monomineralogic composition of carbonate rocks. The principles of chemical hydrogeology could not have been developed as readily in more complex terranes composed of silicate minerals.

Environmental fragility

Many environmental problems exist in karst regions, the majority of which are related to a combination of hydrological conditions and the long periods of inhabitation. The extremes in permeability in karst regions, described by LeGrand and LaMoreaux (1975, p. 11), range locally from large solution cavities that are capable of transmitting large quantities of water to dense rocks with almost no permeability; these extremes in permeability cause sensitive conditions with diverse responses within a karst region. These sensitive conditions often result in harmful environmental impact as a result of certain actions by humans, and problems may develop in places and at times that are not easily predicted. Many of the problems are economical and social in scope, some of which lead to legal involvement related to (1) pollution, (2) waste disposal and management, (3) mine hydrology, (4) drainage wells and (5) various aspects of water-level behavior (chiefly land subsidence; LeGrand, 1984).

Water in regions underlain by carbonate rocks is much more susceptible to contamination because of the characteristics of karstic terranes, for example (1) the lack of soil cover or permeable sediments precludes the normal filtering of water as it moves from land surface to the water table; (2) the rapid infiltration of water through solution channels does not provide enough time for degradation of organic material and elimination of bacteria; and (3) the high permeability of the rocks permits contaminants to move with

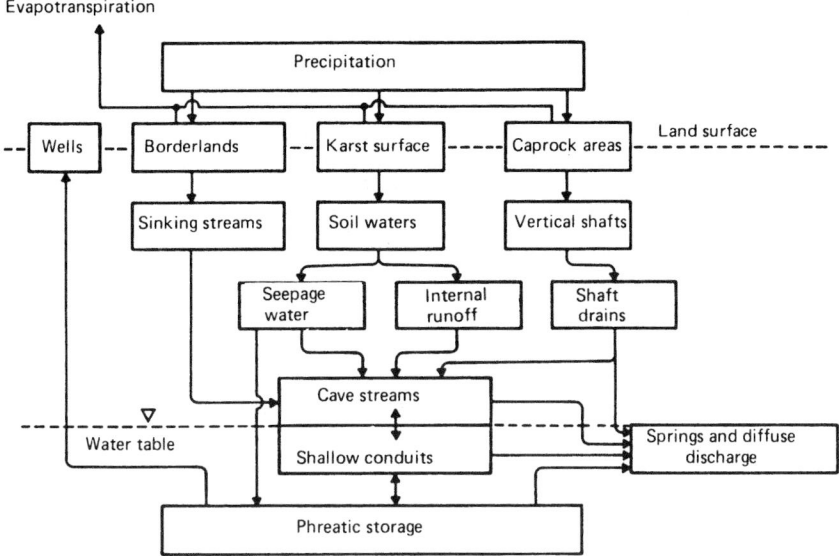

Figure K1 Internal flow system of a karst aquifer. (From White, 1988.)

relatively high velocity from the recharge area to the point of water collection. These factors preclude the completion of the biological processes that would provide natural self-purification. In addition, the great number of solution openings from the surface to the aquifer permits direct access of potential contaminants to the aquifer.

True understanding of a geological terrane ultimately comes about as the result of prolonged scientific study. The scientific challenges presented in the study of karst are many. The scientific problems associated with karst phenomena focus on understanding the physical, chemical and biological processes that control the hydrogeological behavior of karst terranes.

Natural resources

The occurrence of natural resources in carbonate terranes make the understanding of the functioning of karst hydrological and geological systems not only desirable but a necessity in order to use the resources in an effective manner (Back and Arenas, 1989). Karst terranes produce a variety of resources that have had a significant effect on the social and economic development of countries, the most important being water for agricultural, municipal and industrial use. Many of the world's major aquifers are in karstified limestone. Hydroelectric power has been extensively developed in karst regions, particularly Yugoslavia and Turkey. In these areas an interplay of hydrological factors makes development of hydroelectric power particularly desirable. The heavy precipitation at high elevations from orographic effects of the mountains, and the proximity of the high elevations to the ocean, causes a great drop instream gradient over a short distance that permits efficient operation of turbine engines.

One of the most important resources of karst areas is the tourism associated with the picturesque scenery and the numerous uses of caves. Tourist facilities in caves in many parts of the world include electric lights, walkways and boat trips, some of which are on flat-bottomed boats containing small musical bands or pianos for concerts and special theatrical performances. Because of stable temperature and hydrologic conditions within the caves both in the temperate and tropical climates, caves have other economic uses, such as for cultivation of mushrooms, aging of wines and cheeses, and storage of petroleum products. In many areas, calcite from dripstone deposits and bat guano, which is used for fertilizer, are common products obtained from caves.

In addition to the hydrology and geomorphology of limestone terranes, the mineralogy itself defines a resource. Limestone is used extensively as a building stone and provides a valuable source of lime, an essential ingredient for cement and agricultural soil improvement. Dolomite provides an agricultural source of magnesium. Karstified limestones are reservoirs for oil and gas and are host rocks for many ore deposits such as lead and zinc in the mid-continent area of the United States, bauxite in Jamaica and Hungary, and low-grade coal and manganese in many places.

The extensive surface and subsurface dissolution of limestone common in karst regions has resulted in unique engineering problems. These include damage to highways and buildings due to sinkhole collapse, the need for extensive grouting in the construction of dams, inadvertent drainage of reservoirs due to the opening of plugged solution channels, the occurrence of high water pressure in the construction of tunnels, and the challenges of constructing subterranean dams for additional water resources.

A challenge for the future

It is clear that engineers, geologists and other scientists in many parts of the world are gradually accumulating a body of scientific knowledge and engineering experience to understand the many phenomena associated with karst terranes. It is also equally clear that people living in karst terranes frequently tend to destroy the quantity and quality of land and water resources by not applying knowledge of the laws governing its complex hydrogeological behavior. This is especially critical now when the magnitude of the potential influence of human activities on the biosphere is largely known. The increased demands on karst terranes during the twentieth century because of population growth, water and land use, mining, energy production, industrialization and agriculture can combine to degrade the fragile karst environment.

It is a responsibility of scientists to educate members of society, such as managers, planners and politicians, of the consequences of activities undertaken in karst terranes. These consequences may be direct and immediate or may be indirect and gradual over a long period of time. As in other natural sciences, this challenge for the scientific community to provide usable information demands, more than ever, effective communication, meaningful cooperation and active collaboration among scientific researchers and technicians of all nations.

William Back

Bibliography

Back, W. and A.A. Arenas, 1989. *Karst terranes, resources and problems: Nature and Resources*, Special Issue, UNESCO, pp. 19–26.
Back, W., Herman, J.S. and H. Paloc, eds, 1992. *Hydrogeology of Selected Karst Regions*, International Association of Hydrogeologists, Vol. 13, 493 pp.
Brahana, J.V., Thrailkill, I., Freeman, T. and W.C. Ward, 1988. Carbonate rocks, in Back, W., Rosenshein, J.S. and P.R. Seaber,

eds, *Hydrogeology*, Boulder, Colorado: Geological Society of America, *The Geology of North America*, Vol. 0–2, 333 pp.

Burger, A. and L. Dubertret, eds, 1975. *Hydrogeology of Karstic Terranes*, International Union of Geological Sciences, Series B, No. 3, International Association of Hydrogeologists, Paris, 190 pp.

Burger, A. and L. Dubertret, eds, 1984. *Hydrogeology of Karstic Terranes, Case Histories*, Vol. 1, International Association of Hydrogeologists, Heise, Hannover, 264 pp.

Dreybrodt, W., 1988. *Processes in Karst Systems, Physics, Chemistry, and Geology*, New York: Springer-Verlag, 288 p.

Ford, D.C. and P.W. Williams, 1989. *Karst Geomorphology and Hydrology*, London: Unwin Hyman, 601 pp.

Günay, G., Johnson, A.I. and W. Back, eds, 1993. *Hydrogeological Processes in Karst Terranes*, IAHS Publication No. 207, Wallingford, UK: 412 pp.

Herak, M. and V.T. Stringfield, 1972. *Karst: Important Karst Regions of the Northern Hemisphere*, Amsterdam: Elsevier Publishing, 551 pp.

James, N.P. and P.W. Choquette, 1988. *Paleokarst*, New York: Springer-Verlag, 416 pp.

LeGrand, H.E. and P.E. LaMoreaux, 1975. Hydrogeology and hydrology of karst, Chapter 1, pp. 9–19, in *Hydrogeology of Karstic Terranes*, International Union of Geological Sciences, Series B., No. 3 (Published by International Association of Hydrogeologists), 190 pp.

LeGrand, H.E., 1984. Environmental problems in karst terranes, in Castany, G., Groba, E. and E. Romijn, eds, *Hydrogeology of Karstic Terranes: International Contributions to Hydrogeology*, Vol. 1, pp. 189–194.

Palmer, A.N., 1990. Groundwater processes in karst terranes, in Higgins, C.G. and D.R. Coates, eds, *Groundwater Geomorphology; The Role of Subsurface Water in Earth-Surface Processes and Landforms*, Boulder, Colorado: Geological Society of America Special Paper 252, pp. 177–209.

White, W.B., 1988. *Geomorphology and Hydrology of Karst Terranes*, New York: Oxford University Press, 464 pp.

White, W.B. and E.L. White, 1989. *Karst Hydrology: Concepts from the Mammoth Cave Area, Kentucky*, New York: Van Nostrand Reinhold, 346 pp.

Zoetl, J.G., 1989. Bibliography of the history of karst research, in LaMoreaux, P.E., ed., *Hydrology of Limestone Terranes, International Contributions to Hydrogeology*, Vol. 10, International Association of Hydrogeologists, Heise, Hannover, pp. 1–49.

Cross references

Groundwater
Hydrogeology
Hydrology
Hydrometeorology

KISSIMMEE RIVER, FLORIDA, USA

The River is located in central Florida, USA, between Lake Kissimmee and Lake Okeechobee. The river has a 6192 km² drainage basin consisting of approximately 50 lateral tributary watersheds and an interconnected chain of headwater lakes extending from the city of Orlando to Lake Kissimmee. The lower watersheds drain primarily cattle pasture; the headwater basin is moderately developed and populated, and includes the southern half of Orlando and several smaller municipalities.

The historical river channel was approximately 166 km long and meandered within a 1.6–3.2 km wide floodplain. To provide flood protection for the developing upper basin, the river was channelized between 1962 and 1971, and presently exists as a relatively straight, 90 km long, 9 m deep and 64–105 m wide canal, called C-38. The canal is divided by a series of dam-like water control structures into five pools with stepped water surface profiles. As a result of channelization, approximately 12 000 ha of the river's flanking floodplain wetlands were eliminated and historic fish and wildlife resources have diminished. A plan to restore 70 km of river channel and 11 000 ha of floodplain wetlands was implemented in 1996.

Louis A. Toth

Cross references

Okeechobee, Florida, USA: human impacts, research and lake restoration
Everglades, Florida, USA

L

LAKES

In the latter part of the nineteenth century Forel, the famous Swiss limnologist, defined a lake as a mass of still water situated in a depression in the ground without direct communication with the sea. Geologically speaking, they are temporary bodies of water which are usually formed rapidly and decay quickly, sometimes leaving only scant evidence of their existence in the geological record.

Today lakes are universally distributed; they are more abundant in high rather than low latitudes and are numerous in mountain regions. They are particularly abundant in those areas which have been recently glaciated. Lakes are common alongside rivers with low gradients and wide valley floors, where they are obviously connected with changing river channels. Low-lying areas bordering the sea may also have many lakes.

The waters of lakes may be fresh or saline, this condition being primarily controlled by climatic conditions. The source of lake waters today is direct precipitation and runoff; in some cases the water originally may have been seawater.

Lakes possess a marked individuality in physical, chemical and biological features; these are related to the isolation and position of any lake. Consequently, it is not possible to make too many generalities on this subject.

Classification of lakes

Any lake may be formed by one or a number of agents. Various authorities have classified lakes in different ways – Davis (1882) and Penck (1882) adopted a classification based on the agencies which may have produced basins and have grouped them into constructive, destructive or obstructive. Others have grouped lakes as rock basins, barrier basins and organic basins. Both these systems can be criticized because they cut across natural regional groupings. For the limnologist who is concerned with a group of lakes, it is better to consider the agencies which led to their formation. Hutchinson (1957) puts forward such a scheme, and a simplified version of his classification is presented here.

Origin of lakes

Tectonic basins

Lakes formed by gentle crustal movements
(1) Relict seas isolated from the sea by epeirogenetic earth movements, e.g. Caspian Sea and the Aral Sea. (2) Uplift of the sea floor allowing irregularities to become lakes, e.g. Lake Okeechobee, Florida. (3) The gentle tilting of an existing land surface leading to the reversal of drainage, e.g. Lake Kioga in East Africa (Figure L1).

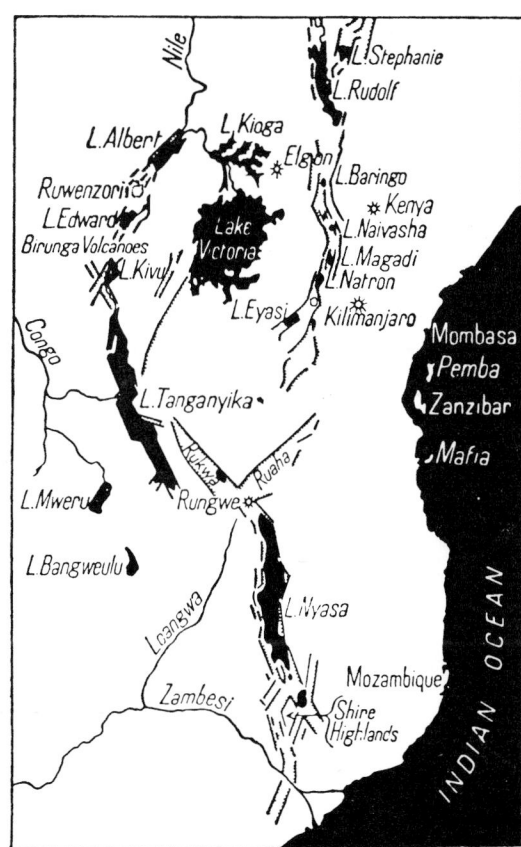

Figure L1 Lakes of East Africa, showing lake due to reversal of drainage (Lake Kioga), lake in depression due to upwarping of margins of a region (Lake Victoria), and lakes in grabens.

(4) The gentle upwarping of the margins of a region resulting in a central basin, e.g. Lake Victoria (Figure L1).

Lakes formed by uplift of peneplains during orogenic movements
Here peneplains have been uplifted and now appear as intermontane basins, and lakes may occupy part of these depressions. In some cases local faulting may delineate the limit of the lakes. Examples of lakes

of this type are found in the Altiplano of the Andes and include Lake Titicaca.

Lakes in basins formed by folding

A synclinal basin formed by folding is the site of Fählensee, Switzerland. In this case the synclinal basin is blocked by an anticline thrust across its lower end.

Lake basins formed by faulting

Lake basins formed by faulting and tilting, or by block faulting resulting in grabens, form an important category of lakes. Many of the world's great lakes fall into this category, e.g. Lake Baikal and Lake Balkhash of Central Asia, the lakes of the African Rift Valley (Figure L1), and the lakes of the Great Basin and the Sierra Nevada of North America.

Lakes associated with volcanic activity

Lakes in modified or partially modified craters

For example, the lake in Crater Butte, California and the maars of the Eifel District of Germany.

Lakes in calderas

For example, Crater Lake, Oregon, Medicine Lake, California, and Lago di Bolsena in the Roman Campagna.

Lakes in modified calderas where pre-existing faults are important

For example, Lake Toba, Sumatra.

Lakes on collapsed lava flows

For example, Myvatn, Iceland, and Yellowstone Lake, Yellowstone National Park.

Lakes formed by barriers of lava, volcanic mud or volcanoes

Lake Kivu, East Africa, lakes around Bandaisan, Japan, and Lake Bunyoni, East Africa, are respective examples.

Lakes formed by landslides

Lakes in this category are usually extremely short-lived. Lake Sarez in the Pamirs is held by a rock slide and is quite stable. Some lakes, e.g. Lac de St Andre, Mt Granier, develop in irregularities on the surface of landslides.

Lakes formed by glacial activity

Lakes formed by glacial agencies constitute a special category since they were formed during a very limited period of the Earth's history. However, the Pleistocene glaciation has produced far more lakes than all the other agencies combined.

Lakes held by ice

For example, Märjelensee, Switzerland, against the Aletsch Glacier, or some Greenland lakes which are held by the Greenland ice-sheet ('proglacial' or 'ice-dammed' lakes).

Lakes in glacial rock basins

(1) Cirque, Corrie or Cwm lakes formed at about the snow line in glaciated valleys, e.g. Llyn Glaslyn, Wales, Blea Tarn, English Lake District, Iceberg Lake, Glacier National Park. (2) Valley rock basins formed below the snow line by glacial erosion, for example the larger lakes of the English Lake District, the fjord lakes of Norway, such as the Nordfjord lakes, and many of the lakes in Alpine valleys. (3) Lakes in basins produced by continental ice sheets scouring on an enormous scale, e.g. the Great Slave Lake and the Great Lakes of North America. (4) Lake basins formed by glacial corrosion of planes of weakness on mature surfaces. Such lakes are usually small in size.

Lakes held by glacial deposits

Glacial moraines (often combined with glacial overdeepening) cause many lakes. Examples include the Italian lakes south of the Alps, Green Lake and the Madison lakes of Wisconsin which are held by terminal moraines, the Finger Lakes of New York State which are held by moraines at each end, and Lac de Barterand in the Jura which is held by a lateral moraine.

Drift basins

Kettle hole lakes, lakes filling depressions formed by the melting of blocks of ice trapped in glacial drift, are extremely common but are usually of small dimensions. Linsley Pond, Connecticut, is an example of this type of lake. Other basins in drift may have been caused by erosion of ground moraine by subglacial streams, e.g. Jelserseen, North Germany. In permafrost regions of the present day, areas of local melting give rise to small lake basins. Examples of this type of lake occur in Alaska.

Pingo lakes

Pingo depressions, roughly circular depressions surrounded by rings of unconsolidated glacial debris, may be filled with water and thus give rise to Pingo lakes and ponds. These depressions are formed by the repeated slipping of morainic material off a partially buried mass of stagnant ice. The slipped material gives rise to concentric rings of debris. With the final melting of the ice a central depression may be formed. Pingo depressions have been described in the Paris Basin, the Low Countries and Wales.

Solution lakes

(1) The solution of limestone by water can give rise to roughly circular depressions. These may be filled by water to give doline lakes, e.g. Deep Lake, Florida, while a fusion of dolines can give rise to uvala lakes, e.g. Muttensee, Switzerland.

(2) Lakes may be formed by solution in tectonically determined basins, i.e. polje lakes, e.g. Lake Scutari in the Karst region of Yugoslavia.

(3) Lakes formed by subsidence after natural solution underground of soluble salts, e.g. Mansfeldersee, Saxony, and possibly some of the lakes in the Landes region of France.

Lakes due to fluviatile action

Fluviatile erosion

For example Falls Lake and Castle Lake, which were plunge pools created by the falls in the Grand Coulee, Washington State, when it was a diversion channel in Pleistocene times.

Fluviatile deposition

(1) Alluvial fans and deltas divide existing lakes, e.g. Derwentwater and Bassenthwaite in the English Lake District and Loch Geal and Loch Lomond in Scotland. (2) Levees of a main river which dam tributaries, e.g. along the Yangtzekiang, the Danube or levees which create basins between the river and the scarp defining the floodplain, e.g. Catahoula Lake, Louisiana, or basins between levees in a delta, e.g. Lake St Catherine in the Mississippi Delta. (3) Basins formed by abandoned channels in mature floodplains, e.g. oxbow lakes.

Lakes associated with shorelines

Lakes are often formed behind bars, spits and tombolos. Their formation is favored by a rise in sea level, causing the drowning of estuaries, followed by a fall in sea level, thus stabilizing the bars. Examples include some of the lakes in the Landes region of France, and many lakes bordering the Gulf of Mexico.

Lakes formed by wind action

Lakes in basins dammed by windblown sand

For example Moses Lake, Washington.

Lakes between dunes

For example, lakes found in the Tarim Basin of Central Asia and in Cherry County, Nebraska.

Deflation basins formed by wind action

Found in arid, or previously arid regions; examples of lakes in this type of basin occur in Australia, northern Texas, South Africa and Egypt (Qattara Depression, and others).

Lakes formed by the accumulation of organic matter

This class of lake includes those dammed by dense growths of plants, such as Silver Lake, Nova Scotia, and the lake on Washington Island in the Central Pacific, which is in a coral atoll basin raised above sea level.

Lakes formed by meteorite impact

Ungava or Chubb Lake, Quebec, is believed to have been formed in this way. The 'Caroline bay' lakes of the southeast coast of North America have been claimed to have been formed in this way, but expert opinion is firmly opposed to this point.

Lakes produced by the complex behavior of higher organisms

This class of lake includes lakes caused by beaver dams and reservoirs, water-filled excavations and subsidence hollows created by humans.

Lakes below sea level

Many lake floors are far below sea level.

- Tectonic basins, e.g. Lake Baikal (1279 m), the Dead Sea (793 m) and Caspian Sea (972 m).
- Excavated basins, e.g. Hornindalsvatn (fjord lake), Norway (461 m) and Loch Morar (glacial basin), Scotland (301 m).
- Volcanic basins, e.g. Tazawako, Japan (175 m).

Some lake surfaces are below sea level, e.g. Dead Sea (396 m) and the Caspian Sea (25 m).

Shape of lake basins

The shapes of lakes vary considerably and are often related to their modes of origin. Some examples are:

- circular: volcanic crater, doline and wind formed;
- subcircular: cirque and kettle-hole basins;
- elliptical: possibly meteorite basins;
- rectangular and sub-rectangular: grabens and overdeepened valleys;
- dendritic: flooded valleys;
- lunate: oxbow lakes;
- triangular: drowned valleys behind bars;
- irregular: areas of glacial scouring and combined lakes.

Life history of lakes

From the moment of their formation all lakes begin to disappear. In humid regions this is brought about by the erosion of the barrier at the exit of the lake and the deposition of detrital and organic matter as deltas and bottom deposits. No marked change in the chemistry of the lake waters is likely in their short history. In arid regions lakes disappear because of evaporation and the deposition of wind- and waterborne sediments. Because of evaporation many lakes in arid regions become increasingly saline though the ancestral lake may have been fresh, e.g. the Great Salt Lake and possibly the Dead Sea. Lakes which are descended from trapped seas usually become more saline, e.g. the Aral Sea. The Caspian Sea is a special case because its waters have become less saline; the reason for this is that the Gulf of Karabogaz on the eastern side of the Caspian Sea is the site of intense evaporation and consequently has led to the removal of salts from the waters of the Caspian Sea.

Lake water

Composition of lake water

The amount of dissolved salts in lake waters varies considerably, e.g. the Great Salt Lake contains 238.12 g l^{-1} whereas Lake Geneva contains only 0.1775 g l^{-1}

The salts which are present in solution in lake waters are the result of the original composition of the ancestral lake, the salts in the waters entering the lake and the rate of evaporation (Table L1).

Movement of lake water

The movement of water in lakes is usually turbulent, i.e. at any point in a lake, in addition to any average directional flow, variable movements in any direction may be observed. The turbulence permits the transfer of material and heat in any direction and increases the apparent viscosity. This last point helps explain the importance of wind-driven currents. Without this viscosity such currents would only have a shallow effect. The main currents are (1) the movement of water from influents to effluent (this is rarely important); (2) tidal currents, as yet little studied and not believed to be important; (3) density currents – these are important bottom-sweeping currents caused by cold and/or sediment-laden waters moving under the main body of lake water; and (4) wind drift which leads to a piling of water at the down wind end of the lake and creates a return current. When the piling of water by wind ceases, the water flows back; since in the process the momentum is not lost, a new flow starts from what was the old upwind end. In this way in particular, a periodic rocking motion or seiche is produced. Uneven atmospheric pressure passing over a lake can also produce a seiche. The period of seiches is determined by the shape of a lake.

Temperature of lake water

The temperature of lake water varies with the season and from place to place in the lake. Insolation, atmospheric temperature, inflow of rivers and rain all play a part in controlling the temperature of a lake. The varying temperature of lake water can lead to the stratification of

Table L1 Analyses of lake waters

Constituent salt	1	2	3	4	5	6	7
CO_3	29.96	26.94	53.21	0.09	t[c]	0.66	–
SO_4	11.93	38.35	5.29	6.68	0.28	23.70	17.36
Cl	9.96	0.57	0.87	55.48	66.37[a]	41.67	53.28
Ca	19.15	29.65	33.74	0.16	4.37	3.01	–
Mg	1.32	2.27	1.99	2.76	13.62	5.97	16.01
Na	8.32	0.64	0.75	33.17	11.14	24.44	11.51
K	4.00	0.38	0.74	1.66	2.42	0.53	1.83
SiO_2	13.93	0.71	2.37	–	t[c]	0.02	–
$Al_2O_3 + Fe_2O_3$	1.45[b]	0.49[b]	1.04[b]	–	–	–	–
Others	–	–	–	–	1.80	–	–
Total	100.00	100.00	100.00	100.00	100.00	100.00	99.99
Salinity (ppm)	27	270.5	122	203 490	226 000	12 670	163 960

1. Lac de Champs, Switzerland: fresh water derived from igneous and metamorphic rocks (Clarke and Washington, 1924; recalculated from Boucart).
2. Lac Noir, Switzerland: fresh water derived from geosynclinal sediments (Clarke and Washington, 1924; recalculated from Boucart).
3. Lac Taney, Switzerland: fresh water derived from calcareous sediments (Clarke and Washington, 1924, recalculated from Boucart).
4. Great Salt Lake, Utah: saline (Hutchinson, 1957).
5. Dead Sea: saline (Hutchinson, 1957).
6. Caspian Sea: trapped sea water, reduced salinity (Russell, 1895).
7. Gulf of Karabogaz, Caspian Sea: saline water drawn from Caspian Sea (Russell, 1895).
[a] Includes 1.78% Br.
[b] Includes traces of Mn.
[c] t = trace.

the waters of a lake. (Stratification can also be caused by differences in the salinity and the amount of suspended sediment.) Lakes which are not holomictic (wholly circulating) are divided into an upper region of more or less uniformly warm circulating, fairly turbulent water, the epilimnion, and a lower relatively undisturbed region, the hypolimnion. The two regions are separated by a plane of rapid temperature change, the thermocline.

Stratification can be destroyed when the surface waters of the lake become cooler, and thus more dense, than the lower waters. When the cooled surface water sinks the lake waters are said to turn over. Turnovers tend to be seasonal. In those lakes where the temperature of the surface waters does not fall below 4°C (the temperature at which fresh water is most dense), there will be one turnover in the fall. In those lakes where the temperature of the surface waters falls below 4°C there is a possibility of two turnovers annually.

Sediments in lakes

Saline and freshwater lakes have markedly different deposits. Those of saline lakes may be predominantly evaporites interrupted by clastic sediments brought into the lake in periods of sudden flooding. The composition of the evaporites precipitated depends greatly upon the composition of the rocks in the catchment area, the climatic conditions and the degree of evaporation, and the position in the lake in relation to the inflow. Organically formed sediments may be rare in such lakes since very few forms of life can be supported in the saline waters.

The sediments of freshwater lakes vary considerably in response to a host of factors. These include the nature of the origin of the lake basin, the character of the rock and soil around the basin and in the drainage area, the size and depth of the basin, the extent of shallow water near the shorelines, the relief and amount and type of vegetational cover of the drainage area, the climatic conditions and the organisms dwelling in the lake. The deposits in any lake tend to be characteristic of that lake only; even closely adjacent freshwater lakes may have different deposits. Recent deposits in most freshwater lakes contain a high percentage of organic matter. Organic sediments range from recognizable plant and animal remains, sometimes called förna, down to finely divided plant and animal matter of colloidal dimensions, which, when composed of unstable protobitumins, may be called äfja. Gyttja is a deposit formed from äfja under oxidizing conditions. The terms förna, äfja and gyttja were proposed by Wasmund (1930). These highly organic deposits may accumulate rapidly in the still waters of some lakes. Over 5 m of such deposits have accumulated in Windermere, English Lake District, in about 8000 years. Many lake deposits can be dated accurately by a study of pollen and macroscopic plant remains trapped in the sediments. Such remains may also indicate changes in climatic conditions during the history of deposition in a lake. In recent years more accurate dating of organic sediments has been possible by the ^{14}C method.

Marked changes often occur in lake deposits soon after deposition, bacteria being particularly important in this respect. All newly deposited sediments are subjected to reworking by micro- and macroorganisms.

Studies of lake sediments have been largely confined to those in existing lakes. Some attention has been paid to sediments which predate the organic sediments described above. Particularly interesting are the varved sediments which occur in lakes which receive or have received at some time in their history, waters from melting ice sheets. A varve is the product of an annual cycle of sedimentation. The scientific study of lakes falls into the general discipline of limnology, which also includes rivers (Horne and Goldman, 1994). In ecological science, the lacustrine natural environment and its biota is known as a 'lotic ecosystem', being characterized by its generally still-water dynamics, although it is nevertheless subject to a certain amount of wave action and longshore currents as well as turbulent mixing (Fairbridge, 1998).

Alec J. Smith

Bibliography

Clarke, F.W. and Washington, H.S., 1924. The composition of the earth's crust, *US Geol. Surv. Profess. Paper*, **127**, 117 pp.

Davis, W.M., 1882. On the classification of lake basins, *Proc. Boston Soc. Nat. Hist.*, **21**, 315–381.

Fairbridge, R.W., 1968. *The Encyclopedia of Geomorphology*, New York, Reinhold Books Co.

Fairbridge, R.W., 1998. Lentic and lotic ecosystems, *The Encyclopedia of Environmental Sciences*, D.E. Alexander (ed.), London, Chapman and Hall.

Horne, A.J. and Goldman, C.R., 1994. *Limnology*, 2nd ed, New York, McGraw-Hill, 576 pp.

Hough, J.L., 1958. *Geology of the Great Lakes*, Urbana, University of Illinois Press, 313 pp.

Hutchinson, G.E., 1957. *A Treatise on Limnology*, Vol. 1, New York, John Wiley and Sons, 1015 pp.

Penck, A., 1882. *Die Vergletscherung der Deutschen Alpen, ihre Ursachen, periodische Wiederkekr und ihr Einfluss auf die Bodengestaltung*, Leipzig, J.A. Barth, 483 pp.

Russell, I.C., 1895. *Lakes of North America*, Boston and London, Ginn & Co., 125 pp.

Twenhofel, W.H. and McKelvey, V.E., 1941. Sediments of freshwater lakes, *Bull. Am. Assoc. Petrol. Geol.*, **25**, 826–849.

Wasmund, E., 1930. Bitumen, Sapropel and Gyttja, *Förh. Geol. Fören. Stockholm*, **52**, 315–350.

Cross references

LAKE BALATON, HUNGARY

This large lake ('Plattensee' in German) in western Hungary, has an area of 596 km^2, a length of 70 km and an average depth of 3.30 m. It was created by the interaction of several forces of nature. Its basin was formed by wind, alluvial deposits and volcanic activity after the recession of the sea. Towards the end of the glacial era, the lake basin developed in a tectonic rift.

The average annual precipitation on Lake Balaton and the surrounding area is 600–700 mm. In the driest year there was 450 mm and in the wettest one, 1100 mm. The shallow water follows rapidly the changes in air temperature. Thus Lake Balaton is pleasantly warm during the summer and freezes early in the winter. Owing to high water temperatures during the summer and frequent winds, there is substantial evaporation, amounting to an annual rate of 870–920 mm.

From the surrounding catchment of 5174 km^2 area a number of streams discharge into the lake (the most important is the Zala River with a mean discharge of 6.9 m^3 s^{-1}). Precipitation on the lake surface and inflow from the tributaries would raise the lake level by 890 mm in an average year. The excess water from the lake is drained to the Danube through the Sió Canal. The volume of water released annually corresponds to a uniform discharge of 12 m^3/s and to a lowering of the lake level by 630 mm over the year.

Since the lake level may be considered constant in the long-term, it may be assumed that outflow roughly equals precipitation over the lake surface and evaporation roughly equals the inflow. The water is completely renewed every 2 years. The lake volume is about 2000 million m^3.

Regulation of the lake level is accomplished by the sluice at Siófok. According to regulations a stage between 40 and 100 cm must be maintained, permitting only 60 cm changes during the year (referring to the zero point of the Siófok lake gauge at 104.09 m above sea level). Extreme climate conditions caused a maximum stage of 155 cm in 1947 and minimum of 22 cm in 1949. The present discharge capacity of the sluice, 80 m^3 s^{-1}, is adequate at full gate openings for lowering the lake level by about 12 mm day^{-1}.

The longitudinal setup due to southwesterly wind may cause level differences of as much as 120 cm between the end points of the lake. The wind-induced waves are short, steep and often create difficult conditions for sailing boats; waves may be as much as 2 m high.

Subsurface waters play an important role. Besides the shallow groundwaters controlled by the lake level, the water stored in the karstified rocks on the northern shore is of major importance. There are eight karstic springs and 14 mineral water springs. The only thermal water wells in the region are in the area of Hévíz, where there is a natural thermal lake, fed by thermal water of 30–40°C temperature.

Lake shore development is related mainly to the development of the recreation areas. The total length of the shoreline is 197 km, of which virtually no improvement is needed over 74 km. In order to control shore erosion and consequent silting, shore walls and linings were constructed.

In view of the recreational character of the region, passenger transport forms virtually the whole lake navigation based on 17 permanent and several temporary public ports. The lines – both across and along the lake – have a total length of 123.5 km. In the field of boating, sailing is the most important. Angling is also a popular activity.

From the point of water quality, Lake Balaton may be subdivided into four basins. Generally, the water is characterized by a magnesium and calcium hydrogencarbonate. The pH and the oxygen content are fairly uniform. The total content of nitrogen, phosphates, silicic acid and sulfate increases towards the Keszthely (most southerly) Bay. Variations in wind conditions also result in appreciable changes of water quality. The content of suspended matter may vary by a factor of two and the oxygen demand may double. Specific research into the process of eutrophication was started in 1968. The results of the Limnological Institute (Tihany) and the Water Resources Research Centre (Budapest) indicated the necessity of remedial measures: to avoid release of any sewage into the lake, and to construct a reservoir system upstream of the Zala mouth (Little Balaton Project). Large-scale models were also constructed to study silting of the Keszthely Bay. Plans were developed to protect lake water quality and they are being implemented, since the lake has a high recreational value both for Hungary and international tourism.

Problems of water supply in the region are solved substantially by a regional network surrounding the lake. The sources are the karstic water from the northern shores, the artesian water of deep aquifers and the treated Balaton water. Sewerage in the region has developed at a somewhat slower rate, and the final goals have not yet been achieved.

Literature published mainly in Hungarian and English, including international studies, is available. Some important literature in English is listed in the bibliography.

Ö. Starosolszky

Bibliography

Németh, Gy., 1988. *Hungary. A complete guide*, 3rd edn, Corvina, Budapest.

Somlyódy, L. *et al.*, 1983. *Eutrophication of shallow lakes: Modeling and management. The Lake Balaton case study*. IIASA Collaborative Proceedings Series, CP83-S3, Laxenburg.

Starosolszky, Ö., 1979. *Study trip to Lake Balaton Travel Guide for Hydraulic Engineers*, VIZDOK.

Cross references

Lakes
Lakes: largest worldwide
Limnology

LAKE CHAD

Lake Chad lies in the center of Africa, between 12°20' and 14°20'N and between 13° and 15°20'E. It is the vestige of a Paleochadian Sea which, during the Holocene, covered a very large area. It is the fourth largest lake in Africa, coming after lakes Victoria, Tanganyika and Nyasa, and is divided among four countries: Chad, Nigeria, Niger and Cameroon.

Situated in an endoreic basin, it is a very shallow lake, which explains the very rapid fluctuations in its area, ranging between 5000 and 20 000 km^2 at the present time, with evaporation losses of some 2.2 m year^{-1}.

Lake Chad's salt content is very low, as some of its annual ion inflow is trapped by the clay minerals of the lake bed and the rest seeps down to the groundwater supplies around the lake margins.

Since the discovery of Lake Chad by Europeans in 1823, scientists have described three phases in its fluctuations (according to Tilho):

- Great Chad phase, i.e. the lake as described by nineteenth century travelers and again in 1962–1965; the lake surface is 284 m above sea level, the estimated lake area is 25 000 km^2, navigable throughout, and the volume of water is over 100 km^3;
- Mid-Chad phase, with water surface 282 m above sea level and lake area of 15 000–20 000 km^2;
- Little Chad phase, with a water surface 280 m above sea level or less, the Grande Barrière above waterline; these conditions were observed in 1905, 1907, 1914 and were predominant from May 1973 to today.

In the Mid-Chad phase, one can distinguish between the lake's northern and southern pools, separated by a shallow zone called the Grande Barrière. The morphological features are very varied: freely navigable open water near the Chari Delta and south of the mouth of the Komadugu; a multitude of islands to the north and east formed by the dune tops of a submerged fixed erg; islets/banks that are also the summits of lakebottom dunes; and swamps or aquatic plant beds that can cover large areas.

With its arid and semiarid climate, Lake Chad receives an estimated mean annual rainfall of 270 mm, i.e. 13% of the basin's total yield (for the period 1932–1989), with extremes ranging from less than 100 mm (in 1972 and 1984) to 565 mm (1954). Over the past 20 years, however, owing to the persistence of drought in the Sahelian zone, the isohyet on the lake's southern shore has fallen from 550 to 400 mm.

Annual fluctuations in the lake level follow a regular pattern. About 82% of the inflow comes from the Chari in flood (5% coming from the El Beîd and the Komadugu Yobe), so that the lake fills up from its southern half. It reaches its maximum between late November and late January, and its minimum is in July.

The mean discharge of the River Chari over the observation period 1932–1994 was 33 km^3 (40 km^3 for 1932–1969 and 21.5 km^3 for 1970–1994). When mean discharge is below 15 km^3, only the southern pool fills. This was the case in 1984–85, when inflow from the Chari, at only 6.7 km^3, was the lowest at any time since measurements began.

Since its 1963 maximum, the lake level has been gradually falling. By 1973 its area had shrunk to one-third of the maximum and its volume to one-quarter; in 1985 the northern pool remained dry throughout the year.

The question of whether the lake may disappear had already been raised during the 1920s, and is still topical. However, it is unlikely to dry up entirely. First, the lake has a very short 'memory' (it only takes two consecutive years of abundant rainfall to fill it completely, as has occurred twice since the start of the twentieth century); and second, this could only occur if the regime of the River Chari were completely disrupted – a historically unprecedented event, although Africa's tropical rivers are currently undergoing a spectacular decline.

J.H. Sircoulon

Bibliography

Bouchardeau, A. and R. Lefèvre, 1957. *Monographie du lac Tchad*, Paris: Orstom, 122 pp.

Olivry, J-C. *et. al.*, 1996. *Hydrologie du lac Tchad*, Paris: Orstom, Col. Monographies Hydrologiques, No. 12.

Tilho, J., 1910. *Documents scientifiques de la mission Tilho (1906–1909)*, Paris: Imprimerie Nationale, 3 vols.

Touchebeuf de Lussigny, P. *et. al.*, 1969. *Monographie hydrologique du lac Tchad*, Paris: Orstom, 226 pp.

Cross references

Lakes
Lakes: largest worldwide
Limnology

LAKES: EFFECTS ON CLIMATE

Lakes, natural and artificial, modify the climates of their surrounding areas. These modifications are known collectively as 'lake effects' and extend from the microscale to the synoptic scale. Lake effects vary with the areal extent, depth and configuration of the lake, the velocity and direction of winds, the existence of winter ice cover and the general climatic environment within which the lake interacts. Thus large, deep lakes impose more pronounced modifications than small, shallow lakes, and the effects of lakes in areas of pronounced seasonal change are more significant than those in tropical areas.

The climatic effects of the Great Lakes of North America have been studied extensively, as have the effects of Lake Balaton in Hungary, Lake Baikal in the former Soviet Union, and the Masurian Lakes of Poland. Additional studies have examined the modifications made by many other lakes in the world, including artificial lakes such as Nassar and Karimba in Africa.

Reasons for climatic modifications by lakes

Essentially all lake effects stem from alterations of the boundary layer due to (1) the thermal lag of lake surface temperatures compared to the temperatures of surrounding land areas: (2) differences in the moisture budget of lakes compared to surrounding areas; (3) alteration of winds by lakes as a result of contrasts in surface roughness; and (4) interactions of (1), (2) and (3).

Thermal lag of lakes versus surrounding land areas

Net radiation, the difference between absorbed solar radiation and effective outgoing terrestrial radiation, is the primary source of energy to heat Earth surfaces. More net radiation is used for evaporation over lakes than over land. In addition, the heat capacity (Table L2) of water is greater than that of land, and more energy is required to heat water. Of much greater importance is the difference in the thermal conductivity (or temperature diffusivity) of water and land. Thermal conductivity expresses the rate of heat penetration downward. The thermal conductivity of land materials is small, consequently heat penetration goes on slowly by molecular exchange. For water bodies, the thermal conductivity is large and goes on rapidly due to mixing.

The temperature increase of water and land surfaces is inversely proportional to the conductive capacity (the heat capacity times the square root of the temperature conductivity; Table L2). The conductive capacity is much larger for water than for land, consequently water surfaces lag behind their surrounding shores during the spring warm-up period.

During the fall, as lake surface waters cool, they become denser (the maximum density of water is 4°C). Thus they sink, and are replaced by warmer, less dense water from below. The entire column of water must be cooled to maximum density before surface waters can be further cooled. Therefore, surface temperatures of large and deep lakes, such as Lake Superior and Lake Baikal, exhibit marked thermal lags and restricted seasonal changes as compared to temperatures of surrounding shores (Figure L2)

Contrasts in water budget

Evaporation over lakes may be high relative to land, particularly when lake temperatures are higher than that of overpassing air. The Great Lakes have high fall and early winter evaporation rates compared to surrounding land and are important sources of moisture during these months. In spring, evaporation rates over the Great Lakes are low, because of their cold temperatures. During some spring months, negative evaporation (condensation) may occur.

Table L2 Heating rates of various Earth surface materials (Pettersen, 1969, p. 63)

Material	Heat capacity[a] ρC (cal cm^{-3} °C^{-1})	Thermal (temperature) conductivity (cm^2 s^{-1}) K	Conductive capacity, $\rho C \sqrt{K}$	Diurnal penetration (m)
Ice	0.45	0.012	0.05	0.6
Dry sand	0.3	0.0013	0.011	0.2
Wet soil	0.4	0.01	0.04	0.5
Still water	1.0	0.0015	0.0015	0.2
Stirred water	1.0	50[b]	7[b]	40[b]

[a] Heat capacity equals the energy in calories required to heat 1 cm^3 by 1°C.
[b] Estimated and variable.

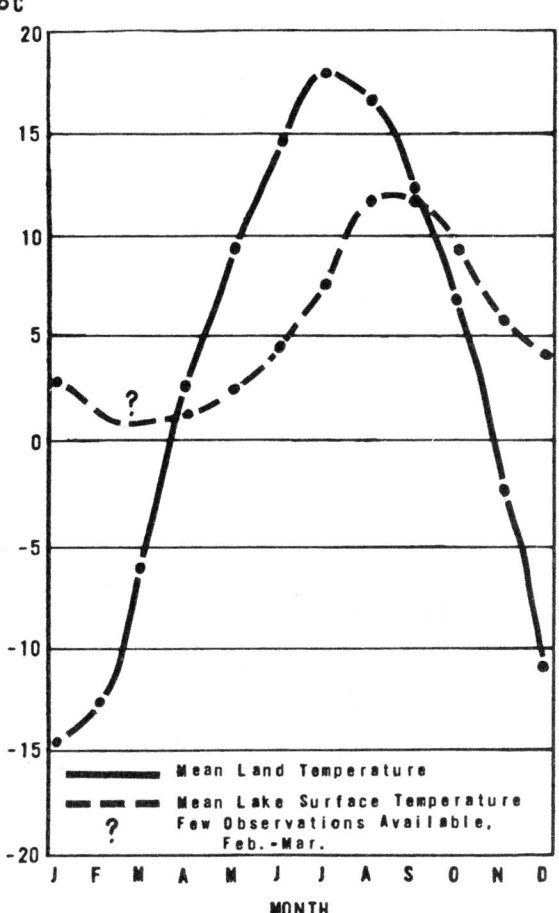

Figure L2 Mean temperatures over land and lake portions of Lake Superior basin. (Data from Phillips, 1978.)

Alterations of wind fields

Mechanical turbulence in the boundary layer induced by lakes is small relative to that induced by land. Thus winds increase in velocity over lakes, and directional changes occur as a result of changes of the Coriolis parameter. Divergence due to velocity increases occurs on upwind shores, whereas convergence occurs on downwind shores as wind speeds decrease. Additions of heat or cooling by the lakes may further alter wind fields.

Boundary layer modifications by lakes

Ship and aircraft measurements in the Great Lakes area have provided some data regarding heat and moisture fluxes between the lakes and the boundary layer. These measurements and estimates of energy transfer must be viewed as limited cases, although similar processes occur on different scales over other lakes. When the air is warmer than the lake, strong conduction inversions and mesoscale anticyclones may exist. Inversions typically may be 1 km in depth and restrict vertical turbulent energy transfer (Strong, 1972). Turbulent mixing is suppressed in overpassing air within 15–50 km from the upwind shore, and the inversion reduces heat, moisture, and momentum transfer from air to water (Bellaire, 1965). Condensation may occur on the lake surface and reduced momentum exchange minimizes wind and waves.

With air colder than the lake the flux of heat and moisture from the lake may be large. Vertical turbulent sensible heat fluxes of 5–10 m W cm^{-2} were measured along a flight path over Lake Michigan by Lenschow (1973) with fluxes of latent heat of 6–15 m W cm^{-2}. Over Lake Ontario, sensible heat fluxes of 75–135 W m^{-2} and latent heat fluxes of 210–300 W m^{-2} were measured during a cold surge by Grayson (1976), although great variation occurred.

Climatic modifications

Thermal modifications imposed by lakes

Figure L3 indicates the mean, maximum and minimum thermal impact of the Great Lakes. Modifications are more marked on downwind shores than on upwind shores. The mean daily range of temperature is reduced throughout the year. Mean daily temperatures are increased during the cool season and decreased during the warm season (Tables L3 and L4). In summer, the effects are most marked by a reduction of maximum temperature, but in winter minimum temperatures are most strongly affected. Autumn effects include increases in minimum temperatures, whereas spring witnesses large decreases of maximum temperatures with little effect on minimum temperatures.

Downwind from both Lake Ontario and Lake Superior, the largest thermal modifications occur with summer and spring maximum

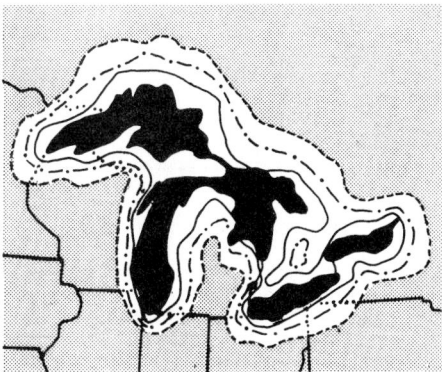

Figure L3 Thermal impact of the Great Lakes. (After Kopec, 1967.)

Table L3 Temperature modifications by Lake Superior and Lake Ontario (Phillips, 1978, p. 289)

	Lake Superior (upwind minus downwind)	Lake Ontario (upwind minus downwind)
Temperature (°C)		
Mean winter maximum temperature	−4.0	−1.4
Mean winter minimum temperature	−6.2	−2.3
Mean spring maximum temperature	8.9	5.1
Mean spring minimum temperature	1.3	0.0
Mean summer maximum temperature	11.0	3.7
Mean summer minimum temperature	4.1	−2.1
Mean fall maximum temperature	−3.8	1.1
Mean fall minimum temperature	−7.8	−4.0
Frost-free season (number of days)	−61	−33
Heating (degree days)	1275	306
Precipitation (mm)		
May to September total rainfall	50.8	30.7
November snowfall	−71	−41
Sunshine (h)		
May sunshine	−32	−
November sunshine	63	−
Wind (ms^{-1})		
Mean spring wind speed	−5.4	−2.4
Mean summer wind speed	−4.2	−2.9
Mean fall wind speed	−7.2	−4.3

Table L4 Weather modifications by Lake Michigan. Percentage difference between actual and predicted[a] weather conditions at downwind shoreline location[b] (after Gatz and Changnon, 1976, p. 29; Changnon and Jones, 1972, p. 369).

Weather conditions	Summer (June–August)	Winter (December–February)
Cloudy days	−15	+35
Solar radiation	?	?
Percent possible sunshine	+5	−30
Snowfall	−	+50
Precipitation	−10	+25
Days with ≥0.25 mm	−11	+45
Thunderstorm days	±10	+25 (fall)
Hail days	−33	+100
Surface wind speed	±5	+11
Surface wind direction	±14	+12
Mean maximum temperature	−3	+6
Days of ≥32.2°C	−50	−
Mean minimum temperature	−2	+15
Days of ≤0°C (minimum)		−6
Evapotranspiration	±3	+37
Surface dewpoint (1300)	0	+12
Days of heavy fog	+50	−28

[a] Predicted values determined by comparison with upwind shore values or by comparison with patterns constructed using no lake-effect stations beyond and around the lakes.
[b] Located in areas 1−40 km inland from lake shore and downwind.

temperatures. The modifications exerted by the large and deep Lake Superior are more marked than those imposed by Lake Ontario. It should be noted that warm-season modifications are confined close to the shorelines of the lakes, whereas cool-season modifications extend further inland.

The frost-free season is extended along the shores of the Great Lakes, chiefly due to retardation of the date of occurrence of the first fall frost. The frost-free season is extended by 2 months near the shores of Lake Superior and by 1 month near Lake Ontario.

Thermal modifications around the Great Lakes occurring with individual weather situations may greatly exceed the average values, particularly during winter when very cold air may move over relatively warm water surfaces. Nocturnal minima on downwind shores during cold air outbreaks may be 15–20°C warmer than those on upwind shores (Baker, 1976). As ice cover increases during late winter and the lake surface temperature drops, the modification decreases.

Smaller lakes also exert thermal modifications on surrounding shores. The effect of Big Bear Lake in California is most discernible on minimum temperatures during very cold days (Minnich, 1971). Lake Toya, a small lake (less than 200 km^2) in Japan, increases mean winter temperatures on its downwind shores by 2°C, with nocturnal increases of 5°C with moderate winds and cold air advection (Takahashi *et al.*, 1978). Lakes Inawashiro, Nakaumi and Kuttara in Japan exert a discernible influence on their shorelines, and investigations around Lakes Lema and Maggiore in Italy have shown complex thermal effects, governed by orientation of the wind. Even subtropical lakes, such as Lake Apapka in Florida, exert strong thermal modifications under certain conditions. Temperatures downwind of Lake Apapka may be higher than surrounding surface temperatures by as much as 5°C with moderate winds and cold air advection (Bill *et al.*, 1978).

Effects on winds

On the shores of the Great Lakes, lake breezes may occur during the warm season. When lake surface temperatures are cold relative to the land, a shallow conduction inversion exists, consisting of cold, dense air near the surface and warmer subsiding air aloft. The increase in density of the air caused by chilling from the lake develops a mesoscale anticyclone over the lake. During the day, heating of surrounding land areas establishes a pressure gradient of 2 mbar or more, thus setting air in motion from the higher pressure of the lake to lower pressure of the land (Fig. L4).

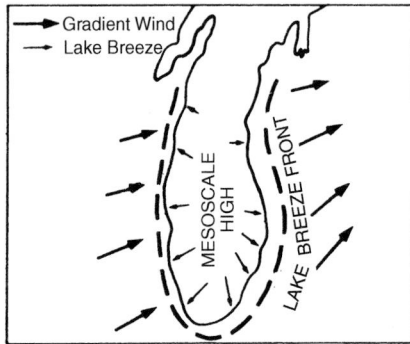

Figure L4 Mesoscale anticyclone and lake breeze over Lake Michigan. (After Lyons and Chandik, 1971.)

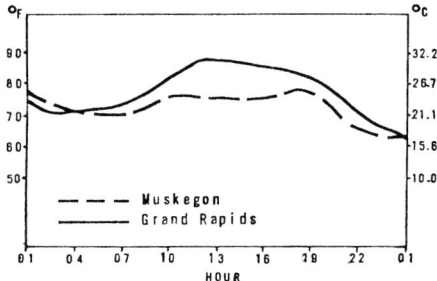

Figure L5 Diurnal temperatures for Muskegon and Grand Rapids, Michigan, on 26 June 1966, when a lake breeze occurred at Muskegon. (From Ressel, 1970.)

True lake breezes occur in opposition to the gradient wind. They may begin several hours after sunrise, and the leading edge (called the lake breeze front) may penetrate anywhere inland from several blocks to 40 km (Cole and Lyons, 1972), cooling the coastal zones (Figure L5). Lake breeze fronts are convergence zones and may be marked by increased cloud build-up.

Sunshine and cloud cover

When a lake is colder than the land, it can prevent convective cloud formation due to the suppression of vertical heat transport. Intense conduction inversions over the Great Lakes and subsidence aloft associated with mesoscale anticyclones may cause cloud-free zones along shorelines. The additional sunshine along the shores of the Great Lakes is an attractive feature for the summer visitor. The effects of added sunshine on the establishment of fruit belts on the downwind shores of the Great Lakes has been, for the most part, overlooked by researchers.

During the cool season, the Great Lakes increase cloudiness markedly due to heat and moisture transfer into overpassing cold air (Table L4). Large contrasts in cloudiness exist between upwind and downwind shores. Increased winter cloudiness is not confined to narrow shoreline zones but can extend far inland and plays a role in the thermal modification of these inland areas by increasing nocturnal temperatures.

Fogs

The shorelines of the Great Lakes have a higher incidence of fog during spring and summer, particularly in May and June when temperature contrasts between land and lake are large. Fog is particularly common along Lake Superior shorelines. Warm and moist air masses become chilled over colder water surfaces, thereby lowering temperatures to the dewpoint. Lake breezes may bring these fogs on shore, but generally only the immediate shore areas are affected as the fog dissipates on contact with warm land surfaces.

Advection-radiation fogs may also be common during the warmer season. These are primarily evening or night-time fogs. Air originating over water surfaces may be cooled by nocturnal radiation processes.

During the cool season, the Great Lakes tend to decrease the occurrence of fog on downwind shores (Table L4). However, over the lakes themselves, 'steam fog' may form as a result of water vapor transfer into cold air crossing warmer water. This type of fog is likely to dissipate downwind, giving way to cloudiness and snow shower activity.

Effects on humidity and precipitation

The measured effect of Lake Michigan on humidity is to increase mean annual dewpoints by 1–2°C (Gatz and Changnon, 1976), primarily as a result of fall and winter modifications. Increases of humidity have also been noted in the Masurian Lake district of Poland, on the shores of Lake Baikal, and around many other smaller lakes.

The flux of moisture and heat to and from the lakes also influences precipitation over and downwind from the lakes. Annual precipitation amounts increase between 1 and 20% in areas downwind from Lakes Michigan and Superior (Gatz and Changnon, 1976). Seasonal analyses indicate that the majority of the increase occurs during fall and winter, with spring showing the smallest influence by the lake. Effects on the diurnal frequencies of precipitation, number of precipitation days and on thunderstorm occurrences are also discernible.

Precipitation increases of 10–16% have also been observed on the southeast shores of Lake Baikal in November and December (Kornienko, 1969), and small influences dependent on wind direction, wind speed and lake water temperatures have been detected in the Masurian Lake area of Poland.

Lake-effect snowfall

The increase of snowfall on downwind shores of some large middle latitude lakes is striking. This snowfall (lake-effect snowfall) is a common occurrence in the Great Lakes area where it contributes 30–50% of seasonal snowfall on downwind shores (Eichenlaub, 1970). Downwind shores may receive amounts 50–100% larger than upwind shores (Table L4). Lake-effect snow causes snowbelts on the lee shores of the Great Lakes (Figure L6). These regions experience larger annual totals, more snowfall days, larger frequencies of heavy snows and greater snow depths than upwind areas and interior regions.

Lake-effect snowfall results from heat and moisture transfer into overpassing cold air. The intensity of the snowfall is determined by temperature contrasts between the lake surface and overpassing air, the fetch (or length of travel of air over water) and the synoptic situation. Lake-effect snowstorms frequently exhibit a banded structure. Consequently snowfall patterns may be sporadic and the boundaries sharply defined. Lake-effect snowfall is heaviest during fall and early winter, decreasing in intensity in late winter as lake temperatures decrease. Over the Great Lakes, larger interior displacement occurs during late fall and early winter, with a retreat towards shorelines of the zone of maximum intensity during late winter (Strommen and Harman, 1978).

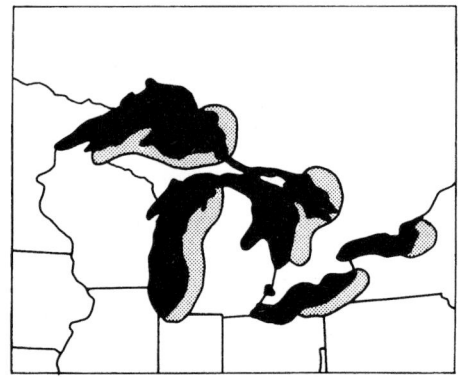

Figure L6 Snow belts of the Great Lakes. (After Eichenlaub, 1970.)

Inadvertent lake effects: role of dams and reservoirs in modifying climate

Some studies of actual and potential modifications occurring around reservoirs of various sizes are available. Data from shore and island stations around and in a midlatitude reservoir – the Rybinski reservoir in the Soviet Union – showed that the reservoir modulated summer temperatures with daytime temperatures 3°C less and night-time temperatures 2–4°C more than on surrounding shores (Guschina, 1967). An analysis of the Selset reservoir in the Tees Valley, UK, a water area of only 110 ha, showed only small variations in temperature, usually less than 0.5°C but rarely exceeding 2.5°C (Gregory, 1967). A tropical artificial lake – Lake Kariba in Africa – a large reservoir 32 km wide by 260 km long was found to have caused an increase in rainfall after 1960, the year in which the lake was formed (Hutchinson, 1973). It was postulated that the increase in precipitation was caused by the setting up of a lake breeze system. The potential climatic modification of the addition of a large lake to the Sahara was simulated through a general circulation model (Rapp and Warshaw, 1974). A significant change of precipitation and wind direction in the lower layers was observed.

Val L. Eichenlaub

Bibliography

Baker, D.G., 1976. Mesoscale temperature and dew-point fields of a very cold airflow across the Great Lakes, *Monthly Weather Rev.*, **104**(7), 860–867.

Bell, B. and T. Lajos (eds), 1974. *Climate of Lake Balaton, Hungary* (in Hungarian), Vol. 40. Országos Meteorológiai Szolgalat Budapest, Hivatalos Kiadványai, Budapest.

Bellaire, F.R., 1965. The modification of warm air moving over cold water in *Proceedings, Eighth Conference on Great Lakes Research*, Pub. No. 13, Ann Arbor: Great Lakes Research Division, pp. 249–256.

Bill, R.G. Jr *et al.*, 1978. Observations of the convective plume of a lake under cold-air advective conditions, *Boundary Layer Meteorol.*, **14**(4), 543–556.

Changnon, S.A. and D.M.A. Jones, 1972. Review of the influences of the Great Lakes on weather, *Water Resources Research*, **8**(2), 360–371.

Cole, H.S. and W.A. Lyons, 1972. The impact of the Great Lakes on the air quality of urban shoreline areas: Some practical applications with regard to air pollution control policy and environmental decision making in *Proceedings, Fifteenth Conference on Great Lakes Research*. Ann Arbor: International Association for Great Lakes Research, pp. 436–463.

Eichenlaub, V.L., 1970. Lake effect snowfall to the lee of the Great Lakes: Its role in Michigan, *Am. Meteorol. Soc. Bull.*, **51**(5), 403–412.

Eichenlaub, V.L., 1979. *Weather and Climate of the Great Lakes Region*. Notre Dame, IN: University of Notre Dame Press.

Gatz, D.F. and S.A. Changnon Jr, 1976. *Environmental Status of the Lake Michigan Region*, Vol. 8, Atmospheric Environment of the Lake Michigan Drainage Basin, ANL/ES-40. Argonne, IL: Argonne National Laboratory.

Grayson, T.H., 1976. *Analysis of Cool Season Lake-Related Mesoscale Phenomena Using Numerical Variational Analysis*. Silver Spring, MD: Techniques Development Lab, Systems Development Office, National Weather Service, National Oceanic and Atmospheric Administration.

Gregory, S., 1967. Local temperature and humidity contrasts around small lakes and reservoirs, *Weather*, **22**(12), 497–506.

Gushchina, L.A., 1967. Diurnal variation of air temperature and humidity on Rybinsk reservoir (in Russian), *Leningrad Glavnaia Geofizichezkaia Observatoriaa, Trudy*, **206**, 86.

Hutchinson, P., 1973. Increase in rainfall due to Lake Kariba, *Weather*, **28**(12), 499–504.

Kopec, R.J., 1967. Effects of the Great Lakes' thermal influence on freeze-free dates in spring and fall as determined by Hopkins' bioclimatic law, *Agric. Meteorol.*, **4**(4), 241–253.

Kornienko, V.I., 1969. The effect of evaporation from Lake Baikal on precipitation in adjacent regions (in Russian), *Glavnaia Geofizicheskaia Observatoriia, Trudy*, **247**, 122–126.

Lenschow, D.H., 1973. Two examples of planetary boundary layer modification over the Great Lakes, *J. Atmos. Sci.*, **30**(4), 568–581.

Minnich, R.A., 1971. Minimum temperatures in Big Bear Basin, California, *Pacific Coast Geog. Assoc. Yearbook*, **33**, 83–106.

Petterssen, S., 1969. *Introduction to Meteorology*, 3rd edn. New York: McGraw-Hill.

Phillips, D.W., 1978. Environmental climatology of Lake Superior, *J. Great Lakes Research*, **4**(3–4), 288–309.

Rapp, R.R. and M. Warshaw, 1974. *Some Predicted Climatic Effects of a Simulated Sahara Lake*, Rep. R-1415-ARPA. Santa Monica: Rand Corporation.

Strommen, N.S. and J.R. Harman, 1978. Seasonally changing patterns of lake-effect snowfall in western Lower Michigan, *Monthly Weather Rev.*, **106**(4), 423–427.

Strong, A.E., 1972. Influence of a Great Lake Anticyclone on the Atmospheric Circulation, *J. Appl. Meteorol.*, **11**(4), 598–612.

Takahashi, H. *et al.*, 1978. Local climate near the small lake, pt. 1, air temperature distributions near Lake Toya, Hokkaido, in winter, *Agric. Meteorol.*, **34**(2) 77–82.

Cross references

Climate and climate change
Lakes
Maritime climate, oceanicity

LAKES, LAKE WATER

Definition

A lake is a relatively isolated stretch of water which fills a ground depression. From this point of view, a lake consists of two distinct parts – the basin and the water body.

Lakes may have a very wide range of surface areas, from a few thousands of square meters to hundreds of thousands of square kilometers (e.g. the Caspian Sea, the largest lake in the world, 371 000 km²). Natural lakes cover some 2.7 million km² (1.8% of the Earth's surface), encompassing a water volume of about 176 400 km³ (0.013% of the planetary water volume).

In 1952 Welch suggested a definition of size distinguishing between lakes and ponds. According to his definition, the lake is 'an area of open, relatively deep water, sufficiently large to produce somewhere on its periphery, a barren wave-swept shore', while the pond is a 'very small, very shallow body of standing water in which quiet water and extensive occupancy by higher aquatic plants are common characteristics'.

In the author's opinion, the definition of a lake should not be subject to size, depth and annual variability, because these are characteristic features that shape the huge variety of lakes.

In terms of their formation, lakes may be grouped into natural lakes, with a great many genetic types of lake basin (formed by tectonic, volcanic, glacial, karst, fluviatile, marine and wind action, etc.), and artificial lakes, created to meet various needs (power generation, fresh water and industrial water supply, navigation, fish farming, flood control, etc).

Classifications of lake basins

Numerous classifications have been made by geomorphologists, geologists and biologists. W.M. Davis (1882), for instance, grouped lake basins according three natural processes – constructive, destructive and obstructive. A few years later, I.C. Russel (1895), considering Davis' classification incomplete, put forward another division 'based on material agencies which produce depressions in the earth's surface', depicting ten major natural agencies, nine geological and one organic. Much later, in 1957, E. Hutchinson recognized 11 major genetic processes which produce a total of 76 different lake types selected from all the continents of the Earth and provide the best international outlook on lake classification available. Space precludes the enumeration of the Hutchinson's 76 types; however, we shall review in brief 11 categories of natural processes involved in the formation of lake basins.

Tectonic lake basins

Faulting is a major result of tectonic activity and is responsible for the origin of important lake basins (e.g. lakes Baikal, the world's deepest lake, Tanganyika, Nyasa, Albert and Victoria). Submarine structural basins or depressions may become lakes when uplifted above sea level, or isolated from the sea (e.g. Caspian Sea, Aral, Chad, Eyre, Great Salt Lake, Balaton and Neuesiedler).

Lake basins produced by volcanic activity

There are three dominant types among the 13 volcanic groups: crater, caldera and dam lakes. The lakes formed in craters are small, because these depressions are changed (enlarged) by explosions, giving birth to calderas. The latter contain the great majority of lakes (e.g. Crater Lake, Oregon, which is a classical caldera). Many such lakes occur in the Central Massif of France, central Italy, Japan, the Philippines, Indonesia and New Zealand). Volcanic dam lakes are the work of lava barring some valleys (Chambon, d'Aydat, Cassière from the Central Massif in France).

Lake basins produced by landslides

These are formed in the valleys dammed by the superficial movement of earth materials, but since the dams are eroded by streamflow pressure, these lakes disappear rapidly. However, some still persist, e.g. the Sarez in the valley of the Murgab River, Pamir Mountains, and Lacul Roşu, on the Bicaz River, Romania.

Lake basins produced by the action of glaciers

These are of several types (in cirques, valleys, or on glaciers between moraines). They are the outcome of erosion and accumulation processes. There is a wide range of glacial lakes, from small ones in mountain glacial cirques to larger ones (the Great Lakes) on the old continental plains and blocks of the subpolar zone (Fennoscandia and Canada).

Lake basins produced by dissolution of bedrock

These occur mostly in limestone-based depressions (dolines, poljes), but also on salt deposits. A special case is chemical suffusion, a process developing in loess deposits and generating small, shallow depressions on the ground, in which water gathers. Their incidence is higher in plain areas, where such deposits do exist (e.g. the Romanian Plain).

Lake basins formed by fluviatile processes

These are found in the floodplains of large rivers, in meanders and abandoned channels (meanders, scrolls and oxbow lakes), wherever small tributaries become obstructed by the mainstream. Such lakes frequently occur on the lower course of lowland streams.

Lake basins formed by shoreline processes

Sea currents and sea waves dislodge sediments along the shore, forming bars which enclose smaller gulfs, turning them into lagoons, or small valleys and engendering fluvio-marine slugs. Examples of lagoon-rich coasts include the Gulf of Mexico between Yucatan and Florida, the Bay of Biscay and the Mediterranean Sea between the extremities of the Pyrenees and the Alps. The largest lagoon, however, is Maracaibo in Venezuela.

Lake basins formed by wind action

They are formed by the erosion processes of deflation, by the accumulation of windborne material, but are short-lived. They occur in arid areas after rainfalls.

Lake basins formed by organic accumulation

These include some lakes which emerge when a stream is dammed by vegetal remains or by the accumulation of corals on the shore of tropical and subtropical seas. Typical cases are the coral lagoons of the Pacific and Indian Oceans.

Lake basins formed by the impact of meteorites

These are crater-shaped depressions of various sizes. The best-known lake in this group is Chubb, situated in the Chubb crater, Ungava region, Canada (3350 m in diameter and 410 m deep; the lake itself is 251 m long). Other lakes of this type include Bosumtwi in Ghana and Lonar Deccan Tableland, India.

Lake basins formed by human activity

These go back over 4000 years. From ancient Egypt and Syria to this day, people have been building them to meet their various needs. Artificial lakes have a wide range of sizes and utilities, but they have also drastically changed some streamflows, producing ecological imbalances in their respective areas (for example, the Aswan Dam, Egypt and Sudan, has had the strongest ecological impact downstream on the Nile, with severe consequences for its delta, in particular).

Water balance

The lake water balance (WB) represents the quantity of inflow (I) and outflow (O) analyzed over a given time interval (day, month, season, year or several years) and put into a mathematical formula. WB is positive when $I > 0$, which means that a certain water volume accumulates in the lake ($+\Delta V$) and the water level rises; when $I < 0$, in which case the lake loses a certain volume of water ($-\Delta V$), and its level drops; and constant (neutral) when $I = 0$, and both the water volume and the level remain constant. The WB model depends on many natural or artificial components, in particular precipitation, evaporation and discharge of drainage basin. Quantitative WB values are related to the geographical (climatic) zone the lake lies in and the local characteristics of the area. In terms of water balance, lakes may be classified into permanent, temporary and ephemeral, with and without discharge (enclosed).

The water level of the lake is an accurate measure of water volume. The WB-governed level hydrograph is given by hypsographic curve changes. The water balance and water level variations are the main expression of climatic conditions, having facilitated medium- and long-term assessment of climate changes in certain regions of the world, following the pattern of some large lakes, especially enclosed lakes (without discharge).

Water chemistry and mineralization

Water chemical composition and mineralization depend on the rock composition of the lake basin and of the drainage basin, and on the climatic zone. The chemical composition accounts for the hydrochemical type, reflecting the proportion of major anions and cations. As a rule, three main anion-based hydrochemical types can be distinguished: Cl^- (chloride, characteristic of saline and brackish lakes); SO_4^{2-} (sulfate, specific to saline and brackish lakes); HCO_3^- (carbonate and hydrogencarbonate, peculiar to freshwater lakes). The main cations (N^+, K^+ Ca^{2+} and Mg^{2+}) are used to identify hydrochemical subtypes, e.g. sodium chloride; calcium hydrogencarbonate, etc.

According to their mineralization (quantity of salts in solution), lakes may be grouped into freshwater (up to 1 g l^{-1}; brackish (1–24.7 g l^{-1}); saline (24.7–50 g l^{-1}) and ultrasaline (over 50 g l^{-1}).

Thermal regime and classification

The thermal regime and structure of lakes depend on the climatic zone, lake basin pattern and size, water volume, type of water balance and mineralization. Thus there may be direct thermal stratification, with higher upper-layer (epilimnion) temperatures (temperature variations relate to the air temperature values); a sudden fall of temperature (thermal leap or thermocline) in the intermediate layer (metalimnion), as well as small, sometimes constant variation, close to 4°C, in the deep layer (hypolimnion). Alternatively, there may be indirect (reverse) stratification, with low epilimnion and high hypolimnion temperatures while in shallow lakes, thermal uniformity is induced mechanically (by wind) or by convection (in seasons of transition). This is the so-called homothermal phase.

Classifications of lakes into polar, temperate and tropical (A. Forel, 1901) and subsequently into intermediate divisions – subpolar and subtropical (S. Yoshimura, 1936) – were based on the criterion of water temperature passing through the maximum density threshold (4°C) in the course of one year. Hutchinson (1957) divides lake types according to the modality and number of thermal water mixtures: dimictic (two mixed phases in the temperate zone), amictic (no water mixing); oligomictic (low mixing), and polymictic (several mixtures).

Trophicity

In terms of the quantity of nutrients (mainly nitrates and phosphates), lakes are grouped into oligotrophic (nutrient poor, characteristic of polar, subpolar and high mountain zones), mesotrophic (nutrient rich) and eutrophic (nutrient excess, causing water blooming). Several intermediate groups besides these types are also reported.

Petre Gâştescu

Bibliography

Davis, W.M., 1882. *On the classification of lake basins, Proc. Boston Soc. Nat. Hist.*, **21**, 315–381.
Gâştescu, P., 1969. *Lacurile pe glob*, Editura ştiinţifică, Bucureşti.
Hutchinson, G.E., 1957. *A Treatise on Limnology*, John Wiley and Sons, New York.
Russel, I.C., 1895. *Lakes of North America*, Ginn and Company, Boston.
Welch, P.S., 1952. *Limnology*, 2nd edn, McGraw-Hill, New York.

Cross references

Lakes: largest worldwide
Limnology

LAKES: LARGEST WORLDWIDE

Table L5 gives details of location, area, elevation and greatest depth of all the world's natural lakes above 777 km^2 in area. Also included is a selection of other well-known natural lakes in certain countries but below 777 km^2 in area.

R.W. Herschy

Source

Showers, V., 1989. World Facts and Figures, John Wiley & Sons, New York.

Cross references

Lakes
Limnology

Table L5 Largest natural lakes, by continent

Latitude and longitude		Lake	Location	Area (km^2)	Elevation (m)	Greatest depth (m)
Africa						
1.00S,	33.00E	Victoria	Kenya–Tanzania–Uganda	62 940	1134	85
6.00S,	29.30E	Tanganyika	Burundi–Tanzania–Congo, D.R.–Zambia	32 000	774	1471
12.00S,	34.30E	Malawi (alt Niassa, Nyasa)	Malawi–Mozambique–Tanzania	22 490	475	706
13.20N,	14.00E	Chad (alt Tchad)	Cameroon–Chad–Niger–Nigeria	10 360/^{b} 25 900	240	11/4
3.30N,	36.00E	Rudolf (alt Turkana) (salt)	Ethiopia–Kenya	6 400	427	73
1.40N,	31.00E	Albert	Uganda–Congo, D.R.	5 590	617	60–50
2.00S,	18.20E	Mai-Ndombe (for Leopold II)	Congo, D.R.	2 070/ 8 210	340	12–10
11.05S,	29.45E	Bangweulu	Zambia	5 000	1140	5
1.30N,	33.00E	•Kyoga (alt Kioga)	Uganda	4 430	1036	8
9.00S,	28.45E	Mweru	Congo, D.R.–Zambia	4 350	922	3/2
12.10N,	37.20E	Tana (alt Tsana)	Ethiopia	3 600	1840	9
8.00S,	32.25E	Rukwa (salt)	Tanganyika, Tanzania	2 850	793	shallow
2.00S,	29.10E	Kivu	Rwanda–Congo, D.R.	2 220	1460	480
0.21S,	29.35E	Rutanzige (for Edward, Idi Amin Dada)	Uganda–Congo, D.R.	2 150	912	117
31.15N,	32.00E	•Manzala (off Manzilah) (salt, lag)	Egypt	1 360	0	shallow
6.20N,	37.50E	Abaya (for Margherita)	Ethiopia	1 160	1268	13
3.40S,	35.05E	Eyasi (salt)	Tanganyika, Tanzania	1 050	1030	
15.12S,	35.50E	Chilwa (alt Chirua, Shirwa) (salt)	Malawi–Mozambique	1 040	550	shallow
2.25S,	36.00E	Natron (salt)	Kenya–Tanzania	900	610	
11.10N,	41.47E	•Abbe (alt Abe; off Abhe) (salt)	Djibouti–Ethiopia	780		
5.15N,	3.14W	•Aby (salt, lag)	Ivory Coast	780	0	
16.45N,	3.54W	Faguibine	Mali	590		10
5.50N,	37.33E	Chamo (alt Chama; for Ruspoli)	Ethiopia	550	1235	10
8.36S,	26.26E	•Upemba	Congo, D.R.	530	1000	3
0.48S,	18.03E	Tumba	Congo, D.R.	500	340	12/10
8.00N,	38.50E	Ziway (alt Zeway, Zwai)	Ethiopia	430	1846	4
9.29N,	38.32E	Shala	Ethiopia	410	1567	250
14.38S,	35.12E	•Malombe	Malawi	390	470	6
Asia						
42.00N,	50.00E	Caspian (Sea) (salt)	Iran–Russia–Kazakhstan–Turkmenistan–Azerbaijan	378 400	−28	1025
45.00N,	60.00E	Aral (Sea) (off Aralskoye) (salt)	Kazakhstan–Uzbekistan	32 300	52	68
54.00N,	109.00E	Baykal (alt Baikal)	Russia	31 500	455	1741
46.00N,	74.00E	Balkhash (alt Balkash) (salt)	Kazakhstan	18 300	339	26
13.00N,	104.00E	•(Tonle) Sap	Cambodia	2 700/ 10 000		12
42.25N,	77.15E	Issyk(-Kul) (salt)	Kyrgyzstan	6 280	1609	702
37.40N,	45.30E	•Orumiyeh (for Rezaiyeh, Urmia) (salt)	Iran	3 880/ 5 960	1275	16
74.30N,	102.30E	Taymyr (alt Taimyr)	Russia	4 560	6	26
37.00N,	100.20E	Koko (alt Kuku, Chinghai, Tsinghai; off Qinghai) (salt)	Qinghai, China	4 460	3197	38
29,18N,	112.45E	Dongting (alt Tungting)	Hunan, China	3 100/ 5 200	11	10

Table L5 Continued

Latitude and longitude	Lake	Location	Area (km²)	Elevation (m)	Greatest depth (m)
38.33N, 42.46E	Van (salt)	Turkey	3 710	1646	25
45.00N, 132.24E	Khanka (alt Zinghai, Hsingkai)	China–Russia	3 030/ 4 190	68	11
29.00N, 116.25E	Poyang	Jiangxi, China	3 350	1800	20
50.20N, 92.45E	Uvs (alt Ubsa, Ubsu) (salt)	Mongolia	3 350	759	shallow
38.18N, 118.41E	Hongze (alt Hungtse, Hungtze)	Anhui-Jiangsu, China	2 700	15	4
46.10N, 81.50E	Ala(kol) (salt)	Kazakhstan	2 650	350	54
51,00N, 100.30E	Hovsgol (alt Hobsogol, Khubsugul, Kosogol)	Mongolia	2 620	1624	246
30.45N, 90.30E	Nam (alt Namu, Tengri) (salt)	Tibet, China	2 500	4627	
31.15N, 120.10E	Tai	Jiangsu–Zhejiang, China	2 210	12	5
54.50N, 77.30E	Chany (salt)	Russia	1 990	105	9
30.50N, 47.10E	•Hammar	Iraq	1 940		2
31.50N, 89.00E	•Siling (alt Chilin, Goring, Zilling) (salt)	Tibet, China	1 860	4495	8
48.00N, 92.10E	Har Us (alt Hara Usa, Khara-Us)	Mongolia	1 850	1153	4
48.00N, 84.00E	Zaysan (alt Zaisan)	Kazakhstan	1 800ᵃ	395	10
49.00N, 117.27E	•Hulun (alt Dalai) (salt)	Inner Mongolia, China	1 590	1275	2
40.30N, 90.30E	Lop (alt Lopu, Lob) (salt)	Xinjiang, China	0/3 010	768	2/0
38.45N, 33.25E	Tuz (salt)	Turkey	1 500	925	shallow
49.12N, 93.24E	Hyargas (alt Hirgis, Khirgis) (salt)	Mongolia	1 410	1028	80
31,00N, 86.22E	•Tangra (alt Tangkulayumu, Dangrayum) (salt)	Tibet, China	1 400	4724	
42,00N, 87.00E	Bosten (alt Possuteng, Baghrash, Bagrax) (salt)	Xinjiang, China	1 380	1038	
7.30N, 100.15E	•(Thale) Luang (alt Sap) (salt, lag)	Thailand	1 290	0	shallow
50.24N, 68.57E	Tengiz (salt)	Kazakhstan	1 160	304	8
2.35N, 98.40E	Toba	Sumatra I, Indonesia	1 150	906	529
44.55N, 82.55E	•Ebi(nur) (alt Aipi) (salt)	Xinjiang, China	1 070	213	15
19.45N, 85.25E	•Chilka (salt, lag)	Orissa, India	910/ 1 170	0	shallow
31.30N, 35.30E	Dead (Sea) (off Lut, Mayyit, Melah) (salt)	Israel–Jordan–West Bank	1 020	−393	433
34.35N, 117.13E	•Weishan	Jiangsu–Shandong, China	1 000		
31.31N, 117.33E	•Chao	Anhui, China	900		
14.23N, 121.15E	Bay (salt)	Luzon I, Philippines	890	2	6
28.35N, 90.20E	•Puma (alt Pomo, Pumuchang)	Tibet, China	880	4936	
47.20N, 87.10E	•Ulungur (alt Wulunku, Pulunto, Urungu) (salt)	Xinjiang, China	830	468	
68.20N, 91.00E	Khantayskoye	Russia	820	73	
31.06N, 85.35E	•Tielinanmu (alt Terinam, Tiehlinanmu) (salt)	Tibet, China	810	4684	
29.00N, 90.49E	•Yamzho (alt Yangchoyun, Yamdrok)	Tibet, China	800	4374	
34.45N, 51.36E	•Namak (salt)	Iran	750	300	shallow
53.15N, 73.15E	Seletyteniz (salt)	Kazakhstan	750	65	3
46.35N, 81.00E	Sasyk(kol) (salt)	Kazakhstan	740	347	5
53.00N, 79.36E	Kulunda (off Kulundinskoye)	Russia	730	98	5
69.45N, 87.45E	Pyasino	Russia	730	33	10
32.50N, 119.15E	•Gaobao (alt Kaoyu, Kaopao)	Anjui-Jiangsu, China	700		
35.15N, 136.05E	Biwa	Honshu I, Japan	690	87	96
31.10N, 88.15E	•Zhalin (alt Dzharing, Chalin)	Tibet, China	670	4708	
37.40N, 31.30E	Beysehir	Turkey	660	1121	9
34.55N, 98.00E	•Ngoring (alt Oling)	Qinghai, China	650	4270	
47.48N, 117.42E	Buyr (alt Buir, Peierh, Bor)	China–Mongolia	610	583	11
33.45N, 79.15E	•Pangong (alt Bangong, Pangkung) (salt)	China–Kashmir	600	4248	43
2.45S, 121.32E	Towuti	Celebes I, Indonesia	580	293	141
34.52N, 97.30E	•Gyaring (alt Chaling, Tsaring)	Qinghai, China	570	4270	
48.06N, 93.12E	Har (alt Hara, Khara)	Mongolia	570	1104	7
30.40N, 81.25E	•Mapam (alt Manasalowu, Manasarowar)	Tibet, China	520	4557	82
38.02N, 30.53E	Egridir (alt Egirdir)	Turkey	470	916	13
24.50N, 102.43E	Dian (alt Tien)	Yunnan, China	400	1950	
39.00N, 73.30E	Kara(kul)	Tadzhikistan	380	3914	238
1.52S, 120.35E	Poso	Celebes I, Indonesia	280	518	440
51.35N, 87.49E	Teletskoye [for Altyn(-Kol)]	Russia	220	436	325
2.28S, 121.20E	•Matana	Celebes I, Indonesia	190	382	590
32.48N, 35.35E	•Tiberias [alt (Sea of) Galilee; off Kinneret]	Israel	170	−209	48
38.13N, 72.50E	Sarez (off Sarezskoye)	Tadzhikistan	86	3239	505
51.27N, 157.05E	Kurile (off Kurilskoye)	Russia	77	104	306
42.45N, 141.20E	Shikotsu	Hokkaido I, Japan	77	248	363
40.28N, 140.55E	Towada	Honshu I, Japan	59	400	334
39.43N, 140.40E	Tazawa	Honshu I, Japan	26	249	425
1.01S, 100.43E	•Dibaruh	Sumatra I, Indonesia	11	1464	310

Table L5 Continued

Latitude and longitude	Lake	Location	Area (km²)	Elevation (m)	Greatest depth (m)
Europe					
61.00N, 31.30E	Ladoga (off Kadozhskoye)	Russia	17 700	4	230
61.30N, 35.45E	Onega (off Onezhskoye)	Russia	9 720	33	127
58.55N, 13.30E	Vanern (alt Vaner, Vener)	Sweden	5 580	44	98
57.19N, 30.52E	Peipus (off Chudskoye)	Estonia–Russia	3 550	30	15
58.24N, 14.36E	Vattern (alt Vatter, Vetter)	Sweden	1 910	88	128
61.15N, 28.15E	Saimaa (alt Saima)	Finland	1 760	76	82
55.00N, 21.00E	•Kurisches (Haff) (off Kurskiy) (salt, lag)	Lithuania–Russia	1 620	0	10
40.20N, 45.20E	Sevan (for Gokcha)	Armenia	1 360	1905	86
60.15N, 37.40E	White (off Beloye)	Russia	1 290	113	20
52.35N, 5.30E	•IJssel(meer) [for Zuider (Zee)]	Netherlands	1 210	8	shallow
59.30N, 17.12E	Malaren (alt Malar)	Sweden	1 140	0.3	64
63.40N, 34.40E	Vyg(ozero)	Russia	1 140	89	18
61.35N, 25.30E	Paijanne	Finland	1 090	78	93
69.00N, 28.00E	Inari (alt Enare)	Finland	1 000	114	60
65.40N, 32.00E	Top(ozero)	Russia	990	110	56
58.17N, 31.20E	Ilmen	Russia	980	18	10/4
53.46N, 14.14E	Oder(-Haff) [alt Szczecinski; for Stettiner (Haff)] (salt, lag)	Germany–Poland	900	0	9
64.20N, 27.15E	Oulu (alt Ule)	Finland	900	122	38
67.30N, 33.00E	Imandra	Russia	880	127	67
54.20N, 19.30E	•Vistula [for Frisches (Haff); off Vislinskiy, Wislany] (salt, lag)	Poland–Russia	860	0	5
63.15N, 29.40E	Pielinen	Finland	850	94	49
63.18N, 33.45E	Seg(ozero)	Russia	810	120	97
66.28N, 32.05E	Not(ozero)	Russia	740	80	
66.05N, 30.58E	Pya(ozero)	Russia	660	101	49
46.50N, 17.45E	Balaton (alt Platten)	Hungary	590	104	11
46.25N, 6.30E	Geneva (off Geneve, Ginevra, Leman)	France–Switzerland	580	372	310
47.35N, 9.23E	Constance [alt Boden(see), Costanza]	Austria–Switzerland–Germany	540	396	252
59.15N, 15.45E	Hjalmaren	Sweden	480	22	18
63.12N, 14.18E	Storsjon i Jamtland	Sweden	460	292	74
54.40N, 6.25W	•Neagh	Northern Ireland, UK	400	15	31
44.54N, 28.57E	Razelm (alt Razim) (salt)	Romania	390	3	3
42.10N, 19.,20E	Scutari (off Shkodres, Skadarsko) (salt)	Albania–Montenegro	380[b]	6	44
45.40N, 10.41E	•Garda	Lombardia–Trentino–Alto Adige–Veneto, Italy	370	65	346
60.40N, 11.00E	Mjosa	Norway	370	121	449
41.00N, 20.45E	Ohrid (off Ohridsko, Ohrit)	Albania–Macedonia	360[c]	695	286
47.50N, 16.45E	Neusiedler (alt Ferto)	Austria–Hungary	320	115	1
40.55N, 21.00E	Prespa (off Megal Prespa, Prespes, Prespansko)	Albania–Greece–Macedonia	280[d]	853	54
45.25N, 12.19E	•Venice (off Veneta) (salt, lag)	Veneto, Italy	280	0	shallow
66.14N, 17.30E	Hornavan	Sweden	230/280	425	221
46.54N, 6.53E	Neuchatel [alt Neuenburger-(see)]	Switzerland	220	429	153
45.57N, 8.39E	Maggiore [alt Langen(see), Majeur)]	Italy–Switzerland	210	193	372
53.26N, 9.14W	Corrib	Ireland	170	9	46
46.00N, 9.17E	•Como	Lombardia, Italy–Switzerland	150	199	412
60.02N, 10.08E	Tyrifjorden	Norway	130	63	295
53.25N, 12.42E	Muritz	Germany	120	62	33
47.00N, 8.28E	Lucerne (alt Lucerna, Vierwaldstatter)	Switzerland	110	434	214
53.46N, 21.44E	Sniardwy (for Spirding)	Poland	110	116	23
47.14N, 8.42E	Zurich(see) (alt Zurigo)	Switzerland	90	406	143
56.08N, 4.38W	•Lomond	Scotland, UK	70	8	190
57.18N, 4.27W	•Ness	Scotland, UK	57	16	230
59.54N, 8.55E	Tinnsjo	Norway	54	190	460
61.56N, 6.22E	Nornindals(vatnet)	Norway	51	53	514
59.06N, 8.12E	Fyres(vatn)	Norway	50	280	369
45.58N, 9.00E	Lugano [alt Ceresio, Luganer(see)]	Italy–Switzerland	49	270	288
64.43N, 11.40E	Sals(vatnet)	Norway	49	16	464
61.35N, 11.12E	Storsjoen	Norway	49	250	309
46.42N, 7.44E	Thun [alt Thuner(see)]	Switzerland	48	558	217
45.44N, 5.52E	Bourget	Rhone–Alpes, France	45	231	145
59.42N, 7.57E	Totak	Norway	38	685	306
59.35N, 6.45E	Suldals(vatnet)	Norway	29	68	376
59.24N, 8.15E	Bandak	Norway	26	72	325
56.57N, 5.43W	•Morar	Scotland, UK	26	9	310
58.22N, 6.36E	Lunde(Vatnet)	Norway	24	45	314

Table L5 Continued

Latitude and longitude	Lake	Location	Area (km^2)	Elevation (m)	Greatest depth (m)
North America					
48.00N, 88.00W	Superior	Canada–USA	82 260[e]	184	405
44.30N, 82.15W	Huron	Canada–USA	59 580[f]	177	229
44.00N, 87.00W	Michigan	Illinois–Indiana–Michigan–Wisconsin, USA	58 020	177	285
66.00N, 121.00W	Great Bear	NWT, Canada	31 150	156	413
61.30N, 114.00W	Great Slave	NWT, Canada	28 570	156	614
42.15N, 81.00W	Erie	Canada–USA	25 690[g]	174	64
52.00N, 97.30W	Winnipeg	Manitoba, Canada	24 390	217	28
43.45N, 78.00W	Ontario	Canada–USA	19 240[h]	75	244
11.30N, 85.30W	Nicaragua (alt Cocibolca)	Nicaragua	8 200	32	70
59.05N, 109.30W	Athabasca (alt Athabaska)	Alberta–Saskatchewan, Canada	7 940	213	124
57.15N, 102.40W	Reindeer	Manitoba–Saskatchewan, Canada	6 650	337	219
66.30N, 70.40W	Nettilling	NWT, Canada[i]	5 540	29	
52.30N, 100.00W	Winnipegosis	Manitoba, Canada	5 370	253	12
49.50N, 88.30W	Nipigon	Ontario, Canada	4 850	261	165
51.00N, 98.45W	Manitoba	Manitoba, Canada	4 660	248	28
41.10N, 112.30W	•Great Salt (salt)	Utah, USA	4 440	1280	15
49.15N, 94.45W	Woods	Canada–USA	4 100[j]	323	21
63.08N, 101.30W	Dubawnt	NWT, Canada	3 830	236	
65.00N, 71.00W	Amadjuak	NWT, Canada[3]	3 120	113	
53.45N, 59.30W	Melville (salt)	Newfoundland, Canada	3 070	0	256
58.15N, 103.20W	Wollaston	Saskatchewan, Canada	2 680	398	71
59.30N, 155.00W	Iliamna	Alaska, USA	2 590	15	299
51.00N, 73.30W	Mistassini	Quebec, Canada	2 340	372	183
60.30N, 99.30W	Nueltin	Manitoba–NWT, Canada	2 280	278	
57.10N, 98.40W	Southern Indian	Manitoba, Canada	2 250	255	18
54.00N, 64.00W	Michikamau	Newfoundland, Canada	2 030	460	80
64.10N, 95.20W	Baker	NWT, Canada	1 890	2	230
26.57N, 80.52W	Okeechobee	Florida, USA	1 810	6	6
63.15N, 116.55W	La Martre	NWT, Canada	1 780	265	
50.20N, 92.30W	Seul	Ontario, Canada	1 660	357	34
56.00N, 124.00W	Williston	British Columbia, Canada	1 660	664	168
30.13N, 90.07W	Pontchartrain (salt, lag)	Louisiana, USA	1 620	0	5
18.37N, 91.33W	•Terminos (salt, lag)	Campeche, Mexico	1 550	0	shallow
62.41N, 98.00W	Yathkyed	NWT, Canada	1 450	141	
58.35N, 112.05W	Claire	Alberta, Canada	1 440	213	2
57.30N, 106.30W	Cree	Saskatchewan, Canada	1 430	487	45
55.10N, 105.00W	La Ronge	Saskatchewan, Canada	1 410	364	41
56.00N, 74.30W	Eau Claire (for Clearwater)	Quebec, Canada	1 380	241	
54.00N, 100.10W	Moose	Manitoba, Canada	1 370	255	
53.20N, 100.00W	Cedar	Manitoba, Canada	1 350	253	
60.20N, 102.10W	Kasba	NWT, Canada	1 340	336	
44.35N, 73.20W	Champlain	Canada-USA	1 270	30	122
55.10N, 73.15W	Bienville (for Apiskigamish)	Quebec, Canada	1 250	427	
53.47N, 94.25W	Island	Manitoba, Canada	1 220	227	
57.56N, 156.23W	Becharof	Alaska, USA	1 190	4	
55.25N, 115.25W	Lesser Slave	Alberta, Canada	1 170	577	21
48.02N, 94.55W	Red	Minnesota, USA	1 170	358	9
42.28N, 82.40W	Saint Clair	Canada–USA	1 160[k]	175	6
54.45N, 94.00W	Gods	Manitoba, Canada	1 150	178	
20.15N, 103.00W	Chapala	Jalisco–Michoacan, Mexico	1 140	1525	13
15.23N, 83.55W	Caratasca (salt, lag)	Honduras	1 110	0	5
64.27N, 99.00W	Aberdeen	NWT, Canada	1 100	80	
45.50N, 60.50W	Bras d'Or (salt, lag)	Nova Scotia, Canada[l]	1 100	0	70
66.30N, 113.27W	Napaktulik (for Takiyuak)	NWT, Canada	1 080	381	
63.55N, 111.00W	MacKay	NWT, Canada	1 060	431	
12.21N, 86.21W	Managua (alt Xolotlan)	Nicaragua	1 040	37	80
48.35N, 72.05W	Saint-Jean (for Saint John)	Quebec, Canada	1 000	98	62
66.00N, 100.00W	Garry	NWT, Canada	980	148	
33.13N, 115.51W	Salton (Sea) (salt)	California, USA	970	−70	15
65.40N, 110.40W	Contwoyto	NWT, Canada	960	445	
48.42N, 79.45W	Abitibi	Ontario–Quebec, Canada	930	265	
65.04N, 118.29W	Hottah	NWT, Canada	920	180	
48.42N, 93.10W	Rainy	Canada–USA	910[m]	338	34
9.05N, 82.05W	•Chiriqui (salt, lag)[n]	Panama	900	0	deep
64.05N, 108.30W	Aylmer	NWT, Canada	850	375	
69.30N, 132.00W	Eskimo North	NWT, Canada	840	0.3	

Table L5 Continued

Latitude and longitude	Lake	Location	Area (km²)	Elevation (m)	Greatest depth (m)
46.17N, 79.45W	Nipissing	Ontario, Canada	830	196	22
70.35N, 153.26W	Teshekpuk	Alaska, USA	820	2	
62.40N, 109.30W	Nonacho	NWT, Canada	780	319	
55.55N, 108.44W	Peter Pond	Saskatchewan, Canada	780	421	24
59.30N, 133.45W	Atlin	British Columbia–Yukon Territory, Canada	770	668	283
54.47N, 97.22W	Cross	Manitoba, Canada	760	207	
57.30N, 75.00W	Minto	Quebec, Canada	760	168	
44.25N, 79.20W	Simcoe	Ontario, Canada	740	219	41
56.15N, 76.20W	Guillaume-Delisle (for Richmond G) (salt, lag)	Quebec, Canada	700	0	110
53.45N, 90.00W	Big Trout	Ontario, Canada	660	213	
53.52N, 98.05W	Playgreen	Manitoba, Canada	660	217	
54.46N, 107.17W	Dore	Saskatchewan, Canada	640	459	20
58.38N, 155.52W	Naknek	Alaska, USA	630	10	
52.45N, 66.15W	Ashuanipi	Newfoundland, Canada	600	529	
15.30N, 89.10W	Izabal	Guatemala	590	8	18
44.00N, 88.25W	Winnebago	Wisconsin, USA	560	228	7
49.00N, 57.20W	Grand	Newfoundland, Canada[o]	540	87	110
46.14N, 93.39W	Mille Lacs	Minnesota, USA	540	381	13
47.51N, 114.07W	Flathead	Montana, USA	510	881	67
18.27N, 71.39W	Enriquillo (salt)	Dominican Republic	500	−44	
39.06N, 120.02W	Tahoe	California–Nevada, USA	500	1899	501
19.55N, 101.95W	Cuitzeo	Guanajuato–Michoacan, Mexico	460	1821	3
47.09N, 94.24W	Leech	Minnestoa, USA	460	393	11
40.01N, 119.35W	Pyramid (salt)	Nevada, USA	440	1159	101
48.10N, 116.21W	Pend Oreille	Idaho, USA	380	629	366
42.24N, 121.54W	Upper Klamath	Oregon, USA	370	1262	14
40.12N, 111.48W	Utah	Utah, USA	360	1368	5
44.27N, 110.22W	Yellowstone	Wyoming, USA	350	2357	91
45.37N, 69.40W	Moosehead	Maine, USA	300	322	75
60.10N, 150.50W	Tustumena	Alaska, USA	300	27	
41.59N, 111.20W	Bear	Idaho–Utah, USA	280	1811	53
60.13N, 154.22W	Clarke	Alaska, USA	280		185
38.42N, 118.43W	Walker (salt)	Nevada, USA	280	1219	305
47.26N, 94.12W	Winnibigoshish	Minnesota, USA	280	396	8
53.18N, 126.42W	Eutsuk	British Columbia, Canada	270	859	323
60.18N, 163.43W	Dall	Alaska, USA	260		
41.55N, 120.25W	Goose (salt)	California–Oregon, USA	260	1437	7
52.33N, 120.59W	Quesnel	British Columbia, Canada	260	725	475
29.52N, 93.50W	Sabine (salt, lag)	Louisiana–Texas, USA	250	0	shallow
49.31N, 121.52W	Harrison	British Columbia, Canada	240	10	279
55.22N, 125.54N	Takla	British Columbia, Canada	240	689	287
30.15N, 90.30W	Maurepas (salt, lag)	Louisiana, USA	240	0	shallow
29.45N, 92.30W	White	Louisiana, USA	210	0.3	shallow
43.12N, 75.45W	Oneida	New York, USA	210	112	17
50.03N, 124.27W	Powell	British Columbia, Canada	190	53	358
43.37N, 71.21W	Winnipesaukee	New Hampshire, USA	190	154	52
42.40N, 76.41W	Cayuga	New York, USA	170	116	133
45.57N, 66.02W	Grand	New Brunswick, Canada	170	1	
42.39N, 76.53W	Seneca	New York, USA	170	136	188
59.59N, 158.59W	Nuyakuk	Alaska, USA	170		283
51.17N, 124.00W	Chilko	British Columbia, Canada	160	1177	366
47.50N, 120.01W	Chelan	Washington, USA	140	290	489
51.12N, 119.35W	Adams	British Columbia, Canada	130	413	457
14.42N, 91.12W	Atitlan	Guatemala	130	1563	320
43.37N, 73.33W	George	New York, USA	110	97	61
42.56N, 122.00W	Crater	Oregon, USA	54	1882	589

Australia and New Zealand

Latitude and longitude	Lake	Location	Area (km²)	Elevation (m)	Greatest depth (m)
28.30S, 137.20E	Eyre (salt)	South Australia, Australia	0/7 690	−12	1/0
31.00S, 137.50E	Torrens (salt)	South Australia, Australia	0/5 780	30	shallow
31.35S, 136.00E	Gairdner (salt)	South Australia, Australia	0/4 770	34	shallow
30.44S, 139.48E	Frome (salt)	South Australia, Australia	0/2 410	49	1/0
38.59S, 175.56E	Taupo	North I, New Zealand	610	357	159
35.26S, 139.10E	Alexandrina (lag)[p]	South Australia, Australia	570	0	0
45.12S, 167.48E	Te Anau	South I, New Zealand	340	207	276

Table L5 Continued

Latitude and longitude		Lake	Location	Area (km^2)	Elevation (m)	Greatest depth (m)
45.05S,	168.34E	Wakatipu	South I, New Zealand	290	310	378
44.30S,	169.08E	Wanaka	South I, New Zealand	190	279	331
43.48S,	172.25E	Ellesmere (salt, lag)	South I, New Zealand	180	0	2
44.07S,	170.10E	Pukaki	South I, New Zealand	170	494	
45.30S,	167.30E	Manapouri	South I, New Zealand	140	185	443
44.30S,	169.17E	Hawea	South I, New Zealand	140	345	392
South America						
9.40N,	71.30W	Maracaibo (salt, lag)	Venezuela	13 010	0	60
31.06S,	51.15W	•Patos (salt, lag)	Rio Grande do Sul, Brazil	10 140	0	5
15.48S,	69.24W	Titicaca	Bolivia–Peru	8 030	3809	304
32.45S,	52.50W	•Mirim (alt Merin) (salt, lag)	Brazil–Uruguay	2 970	0	10
46.30S,	72.00W	Buenos Aires (alt General Carrera)	Argentina–Chile	2 240	217	
30.42S,	62.36W	(Mar) Chiquita (salt)	Cordoba, Argentina	1 850	70	4/3
50.13S,	72.25W	Argentino	Santa Cruz, Argentina	1 410	200	300
18.45S,	67.07W	Poopo	Bolivia	1 340	3686	3
49.35S,	72.35W	Viedma	Santa Cruz, Argentina	1 090	250	
48.52S,	72.40W	San Martin (alt O'Higgins)	Argentina–Chile	1 010	200	170
45.30S,	68.48W	Colhue Huapi	Chubut, Argentina	800	265	4
41.08S,	72.48W	Llanquihue	Chile	800	52	350
54.38S,	68.00W	Fagnano (alt Cami)	Argentina–Chile	590	252	200
40.58S,	71.30W	Nahuel Huapi	Neuquen–Rio Negro, Argentina	550	767	438
48.55S,	71.15W	Cardiel	Santa Cruz, Argentina	460	270	
45.27S,	69.13W	Musters	Chubut, Argentina	430	271	100
40.14S,	72.24W	Ranco	Chile	400	70	80
10.11N,	67.45W	Valencia	Venezuela	370	406	40
39.15S,	72.06W	•Villarrica	Chile	170	230	

Source: Showers, V., 1969 *World Facts and Figures*, John Wiley and Sons, New York.
Abbreviations: •, unofficial data; alt, alternative name; for, former name; lag, lagoon; off, official name; salt, salt water; /, minimum or maximum value.
[a] This original area has been increased by damming to 5510 km^2.
[b] Average of areas officially given by Albania (370 km^2) and Yugoslavia (390 km^2).
[c] Average of areas officially given by Albania (370 km^2) and Yugoslavia (350 km^2).
[d] Average of areas officially given by Albania (280 km^2) and Yugoslavia (270 km^2).
[e] Average of areas officially given by Canada (82 100 km^2) and USA (82 410 km^2).
[f] Average of areas officially given by Canada (59 570 km^2) and USA (59 600 km^2).
[g] Average of areas officially given by Canada (25 670 km^2) and USA (25 720 km^2).
[h] Average of areas officially given by Canada (19 010 km^2) and USA (19 480 km^2).
[i] Baffin I.
[j] Average of areas officially given by Canada (4350 km^2) and USA (3850 km^2).
[k] Average of areas officially given by Canada (1110 km^2) and USA (1190 km^2).
[l] Cape Breton I.
[m] Average of areas officially given by Canada (940 km^2) and USA (890 km^2).
[n] Actually a bay.
[o] Newfoundland I.
[p] Fresh water.

LAND-USE CHANGE

Land-use determines many hydrological and biochemical processes. It affects water quality and quantity. Catchment characteristics which depend on vegetation, soil parameters and drainage networks are influenced. In general, land-use change has an impact especially on hydrological processes which depend on soil and vegetation, interception, infiltration, percolation and evapotranspiration.

The most important land-use changes with impacts on water quantity conditions within a catchment are: afforestation/deforestation, urbanization, agricultural intensification (mostly connected with irrigation or drainage).

The resulting effects on hydrology depend much on extent and intensity of land use changes in relation to the size and the hydrological characteristics of the catchment. For example, an increase of built-up areas as a result of urbanization will affect the total infiltration within a catchment in relation to the hydraulic conductivity of the soil and the relative size of newly built areas to the total catchment area. For this reason, in Table L6, only some generalized references to possible hydrological effects of land-use change on hydrological characteristics and no quantitative information on the impacts are given. Secondary effects of land-use changes on water

resources are caused by changes of climatological conditions and water quality.

Land-use change affects the fluxes of energy and water as the aerodynamic resistance at the land surface and the albedo is changed. Its climatic effects can be of local up to global importance. For example, the clear-cutting of a forest can have effects on the microclimate of a catchment, while the clear-cutting of tropical rainforests on a large scale can affect the global climate conditions. The climatic effects of land-use change can intensify the changes of hydrology, e.g. by urbanization the frequency of convective storms may increase and as a result also the flood frequency.

Some water quality aspects of land-use change are listed in Table L7. As their relative importance on water resources within a catchment depends greatly on the local conditions (chemistry of rain, chemical characteristics of soils) only some generalized references to possible water quality effects are given.

How to estimate the hydrological effects of land use change?

It is very common to make an assumption of hydrological effects of land-use change by hydrological modeling. This methodology includes many uncertainties as most hydrological models are not

Table L6 Impacts of land-use change on catchment hydrology

Land use change	Affected hydrological processes	Effects on hydrology
Afforestation (deforestation has converse effects)	Interception increased Transpiration increased Infiltration capacity increased (soil moisture deficit increased, soil porosity increased by macropores) Surface runoff reduced Snowmelt delayed	Annual flow reduced Low flow may be reduced Floods reduced
Urbanization	Infiltration and percolation reduced Surface runoff increased Runoff formation accelerated by anthropogenic drainage network Evapotranspiration reduced	Annual flow increased Low flow reduced Flood volume and peaks increased Groundwater formation reduced
Agricultural intensification	Infiltration may be reduced by changed soil characteristics and seasonal bare soils Surface runoff may increase Transpiration may be altered	Annual flows may change Floods may increase Low flow may be reduced Seasonal distribution of runoff may change
Drainage of wetlands	Water retention capacity reduced Evaporation reduced	Floods increased Annual flows may increase

Table L7 Impacts of land-use change on water quality

Land-use change	Effects on water quality
Afforestation (deforestation has converse effects)	Deposition of atmospheric pollutants increased, especially with mist or fog Leaching of nutrients is decreased by reduced erosion
Urbanization	Air pollution increased Water pollution with organic and inorganic substances
Agricultural intensification	Pesticides and fertilizers endanger groundwater quality Farm wastes

sufficiently physically based to describe changes in hydrological processes by variations of their parameters only without any calibration to the new conditions after changes. The most of these analyses can be seen as sensitivity analyses of the hydrological models instead of the hydrological conditions. To describe changing hydrological processes under land-use conditions with a model it should be applied first to catchments with different land-use characteristics without calibration of parameters which are related to land use to demonstrate its possibilities and limitations to describe land-use-related processes. Catchment experiments can be divided into two groups:

- the hydrological processes of a single catchment will be analyzed before and after land-use change;
- in two catchments which are similar in their characteristics, the hydrological processes are observed in parallel, but in only one of them does the land-use change happen.

The problem of the one-catchment experiment is that the hydrological conditions before and after land-use change are affected by the variability of hydrometeorological conditions. To obtain comparable,

significant and representative results, relatively long-time observations before and after land-use change are needed. The two-catchment experiment gives better results as the hydrological conditions of both catchments are comparable under any meteorological conditions. In many cases the relative size of changes in hydrological conditions are small in comparison to the errors of measurements of hydrological variables. To ensure that the error of measurement is smaller than the signal of changing hydrological conditions, in most cases special process-related data-collecting networks are needed. As some effects of land-use change cause contrary results, a prognosis of hydrological changes in summary is very difficult.

A.H. Schumann

Cross references

Agroclimatology
Climate and climate change
Drainage
Urban hydrology
Water quality for drinking: WHO guidelines
Water resources
Water resources: integrated river basin management

LIMNOLOGY: DEFINITION

There are two ways of interpreting the term 'limnology'. First, the science which studies lakes (its name comes from the Greek *limnos* which means 'lake', 'swamp' or 'pond'). Lakes can be classified according to lake basin genesis, the water thermal and chemical regime, the development of floral and faunal associations, relationships with the environment and usefulness for humans. Limnology is a border discipline between geography, hydrology and biology, and is also closely connected with other sciences, from which it borrows research methods. Physical limnology (the geography of lakes) studies lake biotopes, and biological limnology (the biology of lakes) studies lake biocenoses. The father of limnology is the Swiss scientist F.A. Forel, the author of a three-volume work entitled *Le Léman: monographie limnologique* (1892–1904), which focuses on the geology, physics, chemistry and biology of lakes. He was also author of the first textbook of limnology, *Handbuch der Seenkunde: allgemeine Limnologie* (1901).

Since both the lake biotope and its biohydrocenosis make up a single whole, the lake and lakes, respectively, represent the most typical systems in nature. They could be called limnosystems (lacustrine ecosystems), a microcosm in itself, as the American biologist St.A. Forbes put it (1887).

Second, limnology is the science which studies the biology of continental waters, that is of running waters, stagnant waters and springs, either fresh, brackish or saline. It is, in effect, a subdivision of hydrobiology. The promoters of this study trend were Aug. Thienemann and E. Naumann, who in 1922 had founded the Societas Internationalis Limnologiae (SIL) at Kiel in Germany. This society, which gathers remowned specalists in the field, organizes international congresses every 3 years.

Petre Gâştescu

Bibliography

Dussart, B., 1966. *Limnologie. L'étude des eaux continentals*, Gauthier-Villars, Paris.
Forel, F.A., 1892–1904. *Le Léman: monographie limnologique*, F. Rouge, Lausanne.
Forel, F.A., 1901. *Handbuch der Seenkunde: allgemeine Limnologie*, Stuttgart.
Forbes, St.A., 1887. *The Lake as a Microcosm*, Peoria, Historical Society, Illinois.
Gâştescu, P., 1971. *Lacurile din România*, Ed. Academiei Române, Bucharest.
Gâştescu, P., 1972. *Limnologia – ştiinţă de graniţă între geografie, hidrologie şi biologie*, 'Progresele ştiinţei', No. 3, Bucharest.
Gâştescu, P., 1979. *Lacurile Terrei*, Ed. Albatros, Bucharest.
Goldman, C.R. and Horne, A.J., 1983. *Limnology*, McGraw-Hill Book Company, New York.

Cross references

LIMNOLOGY

Limnology is the science of lakes, a synthesis of many disciplines, drawing its devotees from various scientific fields. It includes all aspects of the study of inland waters, although concerned largely with the physicochemical nature of lakes, their flora and fauna. Stream study has lagged behind lake investigation, although the ecological approach to rivers falls within the realm of limnology as now understood (Schwoerbel, 1987; Horne and Goldman, 1994).

F.A. Forel is considered by many to have been the first limnologist, and his first volume (1892) on Le Léman (Lake of Geneva) stands as a milestone. Because it dealt with environmental factors rather than the biota of the lake, subsequent studies on physicochemical aspects of lacustrine habitat have been termed Forelian limnology. In America, the pioneer work of E.A. Birge, C. Juday and their students in Wisconsin marked the onset of modern limnology and made conditions in Wisconsin lakes a yardstick for later studies in other regions (Mortimer, 1956). Birge, one of the first Americans to work with the microcrustaceans known as Cladocera, was led from what was essentially a biological study of their spatial and seasonal distribution in Lake Mendota, to a study of physical and chemical factors involved in puzzling fluctuations of cladoceran population – in other words, to Forelian limnology. Since then, limnological research has touched on countless facets of the lacustrine microcosm, but a unifying goal has linked most investigations. This involves the assay of productivity and the exploration of factors which interact to make a given lake more or less productive of lying material than its neighbor or some distant counterpart.

Volume I of *A Treatise on Limnology* by Hutchinson (1957) is an outstanding modern source for information on geological, physical, and chemical limnology. The original researches pertaining to most of the material presented below are cited in that volume and are not duplicated here.

Origin and fate of lakes

Lake basins owe their origins to diverse causes, many of which are geological 'accidents' or 'catastrophic.' Geologically, they are temporary phenomena in geomorphic evolution (see Lakes). Tectonic events have created some of the oldest and deepest lakes of the world: the African rift lakes and Lake Baikal of Siberia with an ancient, largely endemic fauna. Volcanism, glacial activities, solution of calcareous substrates, eolian forces, and even meteoritic impact have created lake basins. Hutchison (1957) has summarized 76 major categories and eight subdivisions of these events which have resulted in lake genesis.

Because of the nature of their origins, lakes often occur in definite geographic districts of related bodies of water, similar in age, geological history and surroundings. Comparison of lakes within districts, and between lake districts, has contributed much to understanding the factors which determine productivity and the fundamental trophic nature of a lake. Thus the importance of climate and of soils in influencing productivity has been revealed. Moreover, basin shape, often related to its origin, plays a role in determining the ultimate nature of a lake (Schiller *et al.*, 1991).

No matter what its origin, the lake is doomed to eventual extinction because of its concave nature and accumulation of autochthonous and allochthonous materials which gradually obliterate the depression. Thus a lake passes from a youthful stage to maturity, senility and extinction. The rate of succession depends on various factors; for example, introduction of domestic sewage enriches the lake and accelerates the aging process, Also, mismanagement of water resources over a long period reduces the lake capacity (Micklin, 1988). Although complete definitions of the following terms will be postponed, the youthful lake may be described as oligotrophic, the

mature lake as eutrophic. Many intermediate stages between extreme oligotrophy and extreme eutrophy occur, and the term mesotrophic can be applied to them. Senility is characterized by much shallower water and the conspicuous encroachment of large aquatic plants upon the open water. Extinction often involves a marshy meadow which is later colonized by plants typical of terrestrial situations. If drainage is poor and the lake is protected from wind, a floating bog-mat may close over and eventually obliterate the open water. Bog lakes are acid, or at least circum-neutral, and are typified by characteristic marginal vegetation contributing to the floating mat. When calcium content is low in bog-lake waters, decay of organic matter is reduced greatly. Plant fragments from the bog mat accumulate in flocculent layers, and the water may become tea colored from humic matter. Flocculated humic colloids contribute to bog sediments to form a characteristic deposit termed '*dy*' by Scandinavian researchers. Under such conditions, nutrients are not recycled by decay, and the lake approaches extinction as a dystrophic lake (McManus and Duck, 1993).

Study of sediment cores has given much information about lake succession and past conditions. The sequence of allochthonous pollen relics in lake deposits tells much of vegetational changes, and hence, ecological conditions of the past – paleoecology. The chemical and physical nature of sediments, and microfossils and plant pigments derived from the lake itself, bespeak lacustrine changes which occurred, and this is the theme of paleolimnology (Deevey, 1955; Bradley, 1962).

Morphometry

Limnological work on a body of water is enhanced if a bathymetric map is available, showing shoreline and subsurface contours. Important morphometric parameters derived from maps include area, volume, maximum and mean depth (volume/area). The form of the lake may be expressed as a shoreline index, comparing length of shoreline to that of a theoretical circle with the same area as the lake, and as volume development which compares basin shape to a cone with a height equal to the maximum depth, and basal area the same as that of the lake's surface. Related to area is the maximum length, which has particular significance when considered as a wind-effective dimension. Morphometric parameters often reflect the lake's origin, e.g. irregular where glacial scouring was involved, circular and with high volume development in lakes occupying caldera depressions, and high shoreline indices in lakes formed by damming of stream system (Schwoerbel, 1987).

Except for small lakes, there is generally an increase in depth as area increases and there is a relationship between depth and productivity. Lakes with mean depths greater than 20 m produce less per unit area than those with a shallower average depth, other conditions being equal (Rawson, 1955). A surprising number of lakes have maximum depths well below sea level, their basins being called crypto-depressions.

Physical limnology

Light

Radiant solar energy, directly or indirectly arriving at the lake surface, is considered the primary energy source for all productivity within, although in special cases the import of organic material may be significant. Common limnological usage expresses solar energy as total energy, g cal cm^{-2} of lake surface per day or some longer period of time.

In an ideal system of pure water, various light waves are absorbed at different rates and the penetration is selective. Blue light penetrates farthest because absorption of wavelengths of about 4700 Å is least. At the red end of the spectrum penetration is reduced greatly because most wavelengths above 7500 Å are absorbed and converted to heat within the first 1 m of water. This includes more than 50% of the total radiation falling on the surface. Optical properties of lakes vary because of differences in suspended or dissolved materials, with the latter particularly exhibiting selective absorption effects (Baker, 1994).

Subsurface photoelectric cells are used to determine the depths at which various percentages of surface radiation still persist, with 1% marking the lower limit of the euphotic zone. This is the lake region, somewhat arbitrarily chosen, in which photosynthesis can occur.

A standard method of comparing lake transparencies employs the Secchi disc, a white disc, usually of 20 cm diameter. The depth at which the disc first disappears from the observer's vision is the recorded transparency value. Secchi disc values of about 50 m represent extremes. Such great transparency goes hand in hand with intense blue color; waters poor in nutrients and organic productivity are usually blue. Some lakes are known in which Secchi-disc readings rarely surpass 1 m; such bodies of water are yellow-green to yellow because of abundant suspended microscopic organisms.

Thermics and lake classification

The distribution of heat in a lake, evidenced by vertical temperature profiles, is quite different from a curve constructed on data from the penetration of light. The difference is brought about by wind action, which distributes heat to greater depths than solar radiation and conduction could do alone. The work of the wind in producing the observed temperature distribution can be calculated; it varies from lake to lake, but a reasonable mean approximation is 0.1 g cm of work necessary to mix 1 cal of heat.

Seasonal warming and cooling of lake waters have profound effects because of concurrent density changes. In winter, an ice-covered lake may show an inverse temperature gradient with the warmest water at its greatest depths. This is because of the unique property of water; its greatest density is at 4°C. Just beneath the ice it is colder and therefore lighter than water at 4°C. Although the lake is protected from wind, and the period of ice cover is called the winter stagnation period, there are gentle currents beneath the ice. Early inferences about such currents were based on temperature changes, and these have been confirmed by introducing and tracing radio-sodium (^{24}Na) beneath the ice. Horizontal currents are especially striking. These are caused, in part, by a greenhouse effect of solar radiation through the ice especially at the lake margins. When water which is near 0°C is warmed, density increases and it flows down the basin slope.

Vernal ice melting results in a lake exposed to wind action, practically uniform in temperature and, therefore, in density. Wind-generated currents mix the lake thoroughly; this is the spring overturn. As days become warmer, heat accumulated in the upper strata is distributed downward by the wind, but there comes a time when the warm, upper waters differ so in density from the deeper waters that wind no longer mixes the lake completely. This results in thermal stratification and the period of summer stagnation. A nearly uniform layer, the epilimnion, mixed by wind, lies at the top; below, lies a region where temperature falls rapidly with depth. This is the thermocline or metalimnion, where density changes parallel, to some extent, the temperature changes. Recently the thermocline has been defined as a plane where the maximum change in temperature occurs, although once defined as the stratum where temperature drops at least 1 °C with each increase in depth of 1 m. Below this region lies a colder, dark region perhaps not much warmer than the temperature that prevailed during winter, and protected from wind action by the density barrier above. Thus the lake is divided into two parts separated by the metalimnion – the well-mixed epilimnion and the cold, stagnant hypolimnion.

An important point, however, is that density differences between adjacent temperatures are far greater at high than at low temperatures. Thus the density change between 24 and 25°C is 31 times that between 4 and 5°C. For this reason the thermocline, defined as a plane where the maximum temperature change occurs, may be less important than another plane where maximum density change occurs. The summer stability of a body of water, defined as work needed to mix the lake uniformly, destroying the stratification with no addition or loss of heat, might therefore be as great in a warm tropical lake with a gentle temperature profile, as in a colder lake with a pronounced temperature stratification (Figure L7).

In the fall as air temperatures fall, heat is lost from upper waters, the thermocline lowers, and eventually the lake is homothermal and of the same density. The stability is zero, and the wind mixes the entire lake so chemical and physical conditions throughout approach uniformity. This is the fall overturn. Cooling of the entire body of water continues until one calm, cold night a skim of ice forms and winter stagnation commences.

Seasonal temperature phenomena described above are typical of a dimictic lake, there being two periods of circulation separated by two periods when complete mixing is prohibited. Such lakes are common in temperate regions. However, basin shape and area, surrounding topography and climate, and their effects in diminishing or enhancing wind action, may modify this annual cycle. Some lakes are monomictic, mixing thoroughly but once a year. Cold monomictic lakes

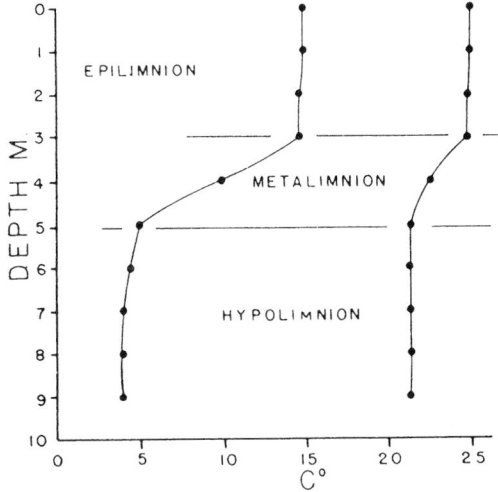

Figure L7 Vertical temperature profiles of a cold and warm lake. The vertical density-change relationships are the same in each curve.

with a winter stagnation period circulate freely during the rest of the year, if shallow enough and wind-exposed. Warm monomictic lakes are found at lower latitudes where ice cover does not occur and only summer stratification interrupts circulation. Polymictic lakes include small high-altitude lakes which stratify during warm days and become uniform in temperature at night because of surface cooling and ensuing convection currents. This term might apply also to lakes situated near the boundary of warm monomictic and dimictic lakes, where winter ice covers form and disappear several times with circulation occurring between each, and tropical lakes which frequently turn over and stratify. Some shallow lakes in mild climates never stratify and are best described on the basis of older terminology as third-class lakes. Conversely, in Antarctica permanently ice-covered lakes occur which must be defined as amictic.

All lakes described above except the amictic type, are holomictic, their entire water mass circulating at overturn periods. By contrast, meromictic lakes occur in which bottom waters are stabilized by the density imparted by dissolved matter and are permanently stagnant. Circulation is limited to the upper waters. Unusual factors combine to create meromixis. Certain plunge-basin lakes formed beneath old Pleistocene waterfalls are so deep relative to surface area that they cannot circulate completely after a critical amount of dissolved material accumulates in the deeps. In such cases, meromixis is caused, to a great extent, by unusual morphometric parameters.

Other meromictic lakes owe their character to events which introduce dense salty water into a freshwater lake, or conversely, introduce fresh water into a markedly saline lake (Micklin, 1988). In either case, a dense layer is overlain by superficial lighter water. In arid regions where evaporation rates are high, saline meromictic lakes are known which owe their stability to more complex processes. If ice cover occurs, freezing-out effects tend to concentrate various salts in deeper layers, and mirabilite precipitates if the solubility product of Na_2SO_4 is attained. In closed basins especially, these events combined with snow and runoff may effect meromictic conditions.

The upper layer of relatively light water, which may stratify thermally at times and circulate at others, is the mixolimnion. The deep layer of permanently stagnant water is the monimolimnion. Between the two is a region where salinity, and therefore density, increases markedly with depth. This is the chemocline, analogous to the metalimnion although the density gradient is caused by materials in solution rather than by temperature change.

Meromictic lakes usually reveal anomalous temperature profiles. Many show a marked low point within the curve with warm water lying beneath colder water, a condition in no way related to the normal temperature–density relationships that occur in the neighborhood of 4°C. For example, 25°C water may lie below 18°C water, which in turn lies below 20°C surface water. This condition is known as dichothermy. Another phenomenon, termed mesothermy, involves a maximum in the vertical temperature curve. The rare condition, in which one or more maxima and minima occur within the curve, is called poikilothermy. The apparent discrepancies in density and stability on the basis of temperature are compensated for by materials in solution which impart greater density to underlying warm waters.

The source of heat in the monimolimnia of meromictic lakes is not always obvious, although bacterial metabolism may account for slow heat accumulation which can be lost only through conduction and diffusion. In shallow meromictic lakes, solar radiation warms the saline bottom waters, and heat is stored as in certain solar furnaces.

The various types of annual temperature cycles, described above, used to categorize lake types are based on intensity factors. A further aspect of lacustrine thermal properties is the capacity factor – the heat gained or lost during a period of time. The annual heat budget is commonly calculated, although segments of it such as summer heat income can be considered alone. The former is the total amount of heat entering a lake between the times of its lowest and its highest heat content. Each extreme is determined by summing the products of temperature and volume of every stratum. Because total heat contents determined this way are a function of lake volume, comparisons between lakes of different sizes are validated by dividing total heat by surface area. The heat budget is, therefore, expressed in terms of calories per square centimeter.

In general, heat budgets are strongly correlated with mean depths, areas and volumes. Lake Baikal (q.v.), the world's deepest lake (1741 m), has an annual heat budget on the order of 60 000 cal cm^{-2}. This is two or three times that of the average dimictic lake in a temperate region. Tropical lakes, which show relatively little annual temperature variation, usually have low heat budgets. A notable exception is Lake Atitlán in Guatemala; exceptional wind action distributes an enormous quantity of heat throughout the lake during the summer warming period and the resulting heat income is some 22 000 cal cm^{-2}, a value much like those derived from many northern and alpine lakes in Europe.

Water movements

A further aspect of physical limnology deals with the movement of water, including various types of waves, currents and turbulence. The air–water interface is regarded as important in considering the motions within the water, for it is here that wind stress acts and radiant energy enters to produce different densities, which in turn may be modified by the action of wind. In addition, other phenomena are involved and interact with the current patterns: the Coriolis force and even tides can be measured in very large lakes; influent river water carrying dissolved and suspended materials, or of low temperature, may move through the lake as a so-called density current (Schiller et al., 1991).

A lake phenomenon of interest is the standing wave termed the seiche. This is periodic oscillation occurring when strong winds cease after having blown water to one side of the lake. The water flows back to the former windward side and then again to the downwind shore, rocking with a period determined by basin shape, length and depth. Surface seiches are especially obvious in large lakes with gentle marginal slopes, since large areas of the shallows may be alternately exposed and flooded with a seiche of relatively small amplitude. These, however, are far smaller than the internal seiche produced in stratified lakes; the accumulation of surface water depresses the thermocline and sets it rocking with an amplitude which surpasses that of the surface seiche above. Internal seiches are detected by the rise and fall of isotherms. These rhythmic currents, alternating in direction, are probably the most important deepwater movements in lakes, creating turbulence which distributes heat and dissolved and suspended materials both vertically and horizontally.

Chemical limmology

Inorganic ions and compounds

The commonest inorganic cations of inland fresh waters are generally these in sequence of abundance: $Ca^{2+} > Mg^{2+} > Na^+ > K^+$. Similarly, the anions are: CO_3^{2-} (usually as HCO_3^-) $> SO_4^{2-} > Cl^-$. There are many exceptions to such arrangements, but most dilute fresh waters are essentially calcium hydrogencarbonate waters (Baker, 1994). In some closed basins of arid regions, evaporation and concentration leads to the early precipitation of $CaCO_3$ and the relative enrichment of other ions. Further concentration will lead to the precipitation of gypsum ($CaSO_4 \cdot 2H_2O$), and the principal anion

becomes chloride. Such a sequence depends on the original composition of the water; if the calcium content is high, almost all the carbonate and sulfate will be removed as the solubility products of $CaCO_3$ and $CaSO_4$ are attained. Even after calcium is gone, sulfate may be removed by the precipitation of mirabilite ($Na_2SO_4 \cdot 10H_2O$). In some lakes, calcium precipitation has not removed the carbonate and the waters are now essentially soda waters, with $NaHCO_3$ and Na_2CO_3 as major constituents. These are less common than those saline lakes where sulfate and especially chloride predominate.

In oxygen-free hypolimnia of some eutrophic lakes and the monimolimnia of meromictic lakes, sulfate is reduced to hydrogen sulfide. This substance is poisonous to most organisms, but the sulfur bacteria require it as a hydrogen donor in chemosynthesis and a modified type of photosynthesis.

Dissolved oxygen and lake typology

Most dissolved oxygen, the distribution of which is of extreme importance to lacustrine fauna, arises as a by-product of algal photosynthesis or that of larger aquatic plants. The basic formulae for this synthesis of carbohydrate by green plants in the presence of light is:

$$6H_2O + 6CO_2 \rightarrow C_6H_{12}O_6 + 6O_2$$

Some dissolved oxygen may be derived through atmospheric exchange when the water surface is wind-disturbed, although diffusion of molecular oxygen through undisturbed water appears negligible.

During circulation periods, oxygen is distributed throughout, often approaching or achieving the saturation point. This is because its solubility increases inversely to water temperature, and overturn periods are usually cold; this applies especially to spring circulation following ice melt in a dimictic lake. At 25°C, 5 mg O_2/l would represent about 90% of saturation, but the same quantity at 5°C is less than 60% of the possible saturation. These data apply to sea level; correction factors are applied to account for altitude effects on solubility. During calm periods when intense photosynthesis prevails, upper waters may hold dissolved oxygen in excess of saturation; conversely, if respiration and decay are excessive, the waters may become understurated (Baker, 1994).

During summer the metalimnion serves as a barrier, prohibiting rapid penetration of wind-driven substances into the hypolimnion. Thus oxygen produced in the well-lit, turbulent epilimnion cannot be distributed to the deeps. In fact, oxygen and CO_2 increases in the stagnant hypolimnion as a result of decomposition, certain bacteria of decay utilizing oxygen and producing CO_2. The rate of hypolimnetic oxygen decrease depends on several factors, one of which is the amount of organic material sinking from the epilimnion, the so-called trophogenic zone where most living matter is produced. Lakes can be classified on the basis of their summer oxygen profiles. Fertile lakes with high rates of production show marked oxygen depletion below the thermocline; poorer producers in some instances show almost no decrease. Rich lakes, which show summertime clinograde oxygen stratification, are eutrophic. Thermally stratified lakes with abundant summer oxygen throughout reveal an orthograde oxygen curve, and are termed oligotrophic.

A method of comparing productivity of different lakes is based on observing decline in hypolimnetic oxygen between two or more dates and referring it to unit area of hypolimnion surface. Products of oxygen decrease at different depths and the volumes of the strata in which the decreases occur are summed to yield total oxygen lost. This is divided by hypolimnion surface area and the number of days involved in the decrease. One somewhat arbitrary scheme places the rate of hypolimnetic oxygen decrease in mesotrophic lakes between the upper limits for oliogotrophy and the lower limits for eutrophy, between 0.025 (mg O_2) cm^{-2} day^{-1} and 0.055 (mg O_2) cm^{-2} day^{-1}. The method applies best to lakes at least 20 m deep where light penetration and vertical turbulence play minimum roles in altering hypolimnion oxygen values.

In tropical lakes, a factor other than total organic productivity assumes importance in oxygen deficit and CO_2 production. Higher hypolimnion temperatures increase metabolic rates, and the oxygen curve may reflect this rather than the magnitude of organic productivity in the epilimnion.

The vertical oxygen profile may be modified by factors other than geological surroundings and climate which obviously govern fertility, productivity and rate of decomposition. These include morphometric parameters of the basin. A relatively large hypolimnion volume starts with a greater supply of oxygen to be utilized in metabolism than does a relatively smaller hypolimnion. Therefore, two lakes of identical surface area and productivity might show quite different rates of hypolimnetic oxygen depletion. The English Lake Windermere lies in a region of fertile soil and is productive; its hypolimnion volume is so great, however, that there is not a marked decrease of oxygen there, and its summertime oxygen profile resembles the oligotrophic type.

Unusual oxygen profiles prevail at times. More than 50 lakes are known which show plus-heterograde oxygen curves, characterized by a persistent metalimnetic maximum where oxygen is in excess of the theoretical 100% saturation existing at the spring overturn (Mosello *et al.*, 1994). In many such lakes the blue-green alga *Oscillatoria agardhii* produces dense blooms and high oxygen concentrations at the intermediate temperatures and low light intensities typical of the metalimnion.

Metalimnetic oxygen minima producing negative-heterograde curves are not easily explained. In some lakes, a concentration of non-migrating animals in the metalimnion may consume oxygen to such an extent that a minimum occurs within the curve.

An alternative to describing the trophic nature of a lake on the basis of intensity of hypolimnetic oxygen consumption is to measure the rate of CO_2 production. However, more CO_2 appears than can be accounted for by O_2 depletion alone. The CO_2 produced by aerobic oxidation of organic matter can be estimated within reason by multiplying the observed decline in oxygen by a factor of 0.85, a mean value for the respiratory quotient to acquatic organisms. But, in addition to aerobic bacterial action, anaerobic metabolism may occur in tropholytic waters. Anaerobic bacteria break down organic material without utilizing oxygen in the process. About 50% of the organic carbon thus decomposed becomes CO_2; the other half becomes methane, CH_4.

Hydrogen ion, CO_2 and alkalinity

For most lakes the hydrogen ion concentration, expressed conventionally as pH, ranges between 6.0 and 9.0, but remarkable exceptions occur. Some volcanic lakes are extremely acid with pH values below 2.0, and the pH of bog lakes is often in the neighborhood of 4.0. In these instances, H_2SO_4 is responsible for the acidity, and oxidation of pyrite is one common source of the acid. A further mechanism for acid production exists in bog lakes. It is well known that sphagnum peat softens the water percolated through it. This involves cation exchange; thus soluble sulfate compounds moving through peaty matter may lose metallic cations and gain hydrogen ions by base exchange. Rainwater alone could account for much of the sulfate ion because it contains a mean of about $2 \, mg \, l^{-1}$; furthermore, sulfur dioxide is atmospherically derived (Baker, 1994).

Persistent pH values of 10 or above are usually associated with high concentrations of sodium and magnesium in alkaline lakes, but in the average freshwater lake which show no extreme acid or alkaline qualities, pH is governed by a CO_2–hydrogencarbonate–carbonate buffering system.

Alkalinity is a rather poorly named component of water chemistry which is commonly determined in limnological procedure. Total alkalinity is acid-combining capacity, arrived at by titration with strong acid to the methyl orange or equivalent end-point at a pH of about 4.5. Under normal conditions, alkalinity is a measure of hydrogencarbonate and carbonate, although at a high pH some hydroxide may be present. Because calcium is one of the most abundant lake cations, and commonly exists as $Ca(HCO_3)_2$, alkalinity titrations in normal waters also reflect calcium content. Perhaps for this reason, total alkalinity serves as a rough index of productivity, those lakes with very low titers being less productive than those with moderate or high alkalinities. In the saline waters of 'soda lakes' the carbonates and hydrogencarbonates are compounds of sodium rather than calcium.

Close relationships exist between alkalinity, hydrogen ion content expressed as pH, and the gas CO_2 (Figure L8). It is customary to consider CO_2 as existing in three states; 'bound' as in CO_3, 'half-bound' as in HCO_3, and free as gas. In simple terms, CO_2 makes for lower pH and greater acidity as it combines with water to form carbonic acid, H_2CO_3. Limestone, or $CaCO_3$, is insoluble but in carbonic acid it dissolves as $Ca(HCO_3)_2$. Calcium hydrogencarbonate stays in solution only if there is CO_2 in excess of the equilibrium value. The excess CO_2 prevents the following reaction:

$$Ca(HCO_3)_2 \rightarrow CaCO_3 + H_2O + CO_2$$

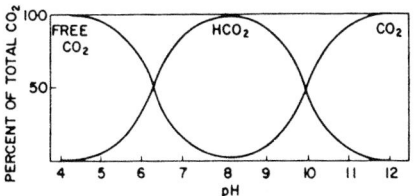

Figure L8 Relation of pH to the percentage of CO_2 in each of its states.

This explains events which occur as the issue from a subterranean spring moves downstream. In calcareous regions, spring water is high in $Ca(HCO_3)_2$, free CO_2 and hydrogen ion concentration. Warming and turbulence results in the loss of free CO_2 to the atmosphere, a rise, therefore, in pH and conversion of hydrogencarbonate to insoluble carbonate which precipitates as marl. Thus total alkalinity is less in the stream than at the spring source.

Biogenic marl precipitation results as photosynthetic plants remove CO_2 from water or, as most aquatic plants can do, remove CO_2 from the hydrogencarbonate radical, converting it to the insoluble carbonate. Marl formation is restricted largely to the shallows, because sinking carbonate comes in contact with carbonic acid in the tropholytic zone and goes into solution. As a result, hydrogencarbonate and, therefore, alkalinity increases in the stagnant hypolimnion.

The vertical distribution of pH in a stratified lake is explained by some of the above remarks. In the trophogenic epilimnion, photosynthesis removes free CO_2 as rapidly as it is formed. The result may be a pH value well in excess of 8.0. Below the thermocline in the tropholytic region, CO_2 is abundant and the pH is lower. Thus, in a eutrophic lake, the pH curve parallels, to some extent, the clinograde oxygen curve, and the CO_2 curve is the reverse of both. Fluctuations in free CO_2 have marked effects on the pH of soft-water lakes, but not so in well-buffered water with high alkalinity.

Iron cycle

The lacustrine iron cycle shows a close relation to oxygen, pH and CO_2 conditions. Ferric iron, the oxidized form, is insoluble and scarce in well-oxygenated waters. In the absence of oxygen and at a low pH, iron occurs in the ferrous reduced state which is soluble. At overturn periods with abundant oxygen at all depths, iron is scarce because it is precipitated as ferric phosphate and ferric hydroxide. An oxidized microzone is formed at the mud–water interface prohibiting diffusion of substances from the underlying sediments. In lakes with orthograde oxygen curves, this microzone may be permanent. In eutrophic lakes, however, reducing conditions prevail in the stagnant hypolimnion, and ferric compounds are converted to ferrous iron which diffuses into the hypolimnion. If it reaches oxygenated levels, it is converted to the ferric form and reprecipitated. For this reason, the hypolimnion acts as an iron trap and progressive enrichment occurs there. The disappearance of the oxidized barrier at the mud surface allows other substances to diffuse out, and there is not only a marked increase in hypolimnion iron, but in phosphates, hydrogen carbonates, silicates and ammonium nitrogen as well. These may be circulated at overturn periods and utilized by epilimnetic algae, accounting in part for the blooms which occur at the fall overturn.

Sometimes evidence of the relation of pH, CO_2, oxygen and iron is visible near springs. Flocculent masses of the reddish ferric hydroxide may be seen below, but not at the spring mouth. Ferrous iron carried in solution from underground sources is converted to ferric iron and precipitated when the water becomes oxygenated.

Under intense reducing conditions hydrogen sulfide appears in the hypolimnion and, with iron, forms ferrous sulfide. This tends to precipitate and accounts for the black color of the slimy sediment called sapropel, which is characteristic of meromictic lakes and others with long periods of anaerobiosis.

Nutrients and minor elements

Other chemical constituents of the aquatic environment are classified as nutrient, or in certain instances minor elements. Nitrogen and phosphorus are especially important. Compounds of the former are derived from atmosphere via rain, or from terrestrial decomposition. Furthermore, nitrogen fixation brought about by bacteria and blue-green algae is of some magnitude in lakes. Aquatic blue-greens of the family Nostocaceae, and especially species of *Anabaena*, are important in this fixation. Lake waters contain nitrogen in the combined forms – nitrate, nitrite and ammonia. Ammonium nitrogen, derived from protein breakdown, may be especially abundant in the hypolimnia of rich lakes during stagnation periods (Horne and Goldman, 1994).

Phosphorus is less abundant than nitrogen and is usually a limiting factor in fertility. It is derived largely from soil in the drainage area through weathering of phosphatic rocks, but phosphate compounds leach less readily than those of nitrogen. Experiments with radiophosphorus, ^{32}P, show that inorganic phosphate added to a lake disappears quickly from the water. It is taken up and stored by planktonic algae and sedimentation follows rapidly. Soluble, inorganic phosphate is often difficult to demonstrate in waters with high productivity because planktonic algae are remarkably efficient in taking it up. Soluble organic compounds of phosphorus are probably not useful, but bacteria rapidly reduce them to inorganic phosphates which may be utilized by autotrophic organisms.

Silica is taken up by diatomaceous algae for frustule construction. Its abundance dissolved in epilimnion water usually is negatively correlated with diatom population fluctuations, and it exhibits stratification.

Cobalt, copper, zinc, aluminum and other metallic elements occur in lake waters in trace quantities and presently are considered of minor importance. Further research will modify this outlook. Cobaltous ion, for example, is a component of vitamin B_{12}, and evidence is accumulating that this vitamin may control the species composition of planktonic algal populations. Also, molybdenum seems to affect general productivity, possibly through a role in governing nitrogen metabolism.

Primary productivity

The rate of photosynthesis carried on by aquatic plants may be referred quantitatively to oxygen produced, CO_2 utilized, carbon fixed or glucose produced per unit surface area over some span of time. In most lakes, the limnetic algae are the important primary producers. Various methods have been devised for determining their productivity (Ryther, 1956); most involve the use of replicate dark and light bottles. A water sample with its included planktonic populations is collected from some depth, placed in a clear bottle and a light-tight bottle and immediately returned to the original level where the bottles are suspended for a period of time. Oxygen determinations made on water from that level at this time serve as a base for the experiments. After 6–8 h, oxygen is determined in each bottle. Decrease in the dark bottle represents respiration of plankton and consumption of oxygen in bacterial decomposition of organic substances, and it is assumed the same amount of oxygen was utilized in each bottle. Oxygen increase in the light bottle represents net production for the period of exposure. Gross production is the sum of oxygen consumed in the dark bottle and the determined increase in the clear bottle.

Because the oxygen method is limited by the sensitivity of the iodometric Winkler test, some workers have determined CO_2 uptake as a measure of productivity. This may be done in two ways. First, initial pH and total alkalinity values are determined; from these data, free CO_2 may be calculated. At the end of the experiment, changes in pH reflect either net decrease in CO_2 because of assimilation or increase due to respiration in the dark bottle. Second, a known amount of radiocarbon, ^{14}C, is added to the light bottle whose initial free CO_2 and hydrogencarbonate content is known. Because ^{14}C is assimilated at a rate only slightly different from that of ^{12}C, later Geiger counter determination of the ^{14}C in algal cells filtered from the bottle water makes it possible to calculate total amount of carbon assimilated from the original $^{12}C/^{14}C$ ratio in the water. The radioactivity of filtered cells probably reflects carbon incorporated in the photosynthate minus that which was respired, and is therefore net productivity.

Obvious sources of error in the light–dark bottle method arise with differential increase of algae in the clear bottle making greater surface area available to bacteria, the extra bacterial activity, and any other discrepancy between the two bottles. To escape these inherent errors some workers have studied daily fluctuations in dissolved oxygen at different levels in the lake itself, multiplying by proper volumes to obtain total amounts. The hours of darkness represent the light-tight bottle; during daylight the lake is comparable to the clear bottle. Field measurements of oxygen changes used to calculate productivity are probably valid, if day and night respiration are the same and if atmospheric exchange is negligible. Similar methods have been employed in calculating primary productivity instream environments, where oxygen changes occur as water passes over beds of aquatic vegetation or substrates encrusted with photosynthetic algae.

A further method is based on the relationships between light intensities, the amount of chlorophyll *a* and photosynthesis. Chlorophyll *a* is determined colorimetrically following acetone extraction of residual cells filtered from a known water volume. Concurrent oxygen increases in the lake are observed and an assimilation number, the ratio of oxygen produced per milligram of chlorophyll, is established. Assimilation values are extremely variable, but useful means are determined empirically. Knowing the depth and volume of the euphotic zone and its chlorophyll content, the assimilation number is used to convert these data to rate of oxygen production or carbon fixation. Values from many lakes reveal mean assimilation rates of about 4.0 (mg C fixed) (mg chlorophyll $a)^{-1} h^{-1}$. This method assumes a fixed relation between light and photosynthesis and essentially ignores adaptations of algae to varying light intensities. Winter phytoplankton, for example, may include extreme 'shade species'; autotrophic assimilation under ice and snow cover has been demonstrated at 0.06% of incident radation (Horne and Goldman, 1994).

Habitats and communities

Thousands of species of plants and animals may occur in a lake, but relatively few have attracted the attention of limnologists. In general, they have been studied on a spatial community basis involving many taxonomic groups in each instance.

Plankton

The unique plankton community consists of many species, usually minute, which are found floating or feebly swimming in open waters, the limnetic zone. Plankton is collected by straining lake waters through nets of fine silk. Various devices, including the noteworthy Clarke–Bumpus sampler which records the volume of water strained, have been developed for quantitative collecting. Organisms captured by net are the net plankton. The nanoplankton includes diminutive algae and flagellates which pass through the interstices of the net. Centrifuging and membrane filtration are used for sampling nanoplankters, which surpass the net plankton in total mass and numbers. Freshwater plankton includes five main groups: the various plants known collectively as algae, a term which covers microscopic plants best assigned to several different phyla, and classed as phytoplankton; the Protozoa; the rotifers or Rotatoria; the cladoceran crustaceans; and the copepod crustaceans. The animals are the zooplankton.

To these five groups should be added the interesting dipterous larva, *Chaoborus*, termed the phantom larva. The adult resembles a mosquito and is assigned to a subfamily of the mosquitoes, the Chaoborinae. The larvae are usually found in eutrophic waters, spending the daylight hours on or near the sediment surface in the deeper portions of the lake. At night they rise to prey upon zooplankters, and are conspicuous in plankton collections taken at that time.

Plankton productivity has been examined in the search for indices of lake productivity, and there are many data on numbers, volumes and weights of plankton. A stumbling block has been the lack of precise knowledge concerning life histories and hence the span of time involved in a generation. For this reason, many published data refer only to standing crops; actual turnover rates, which would more closely approximate productivity, are not known.

Cladocera are extremely important members of most plankton communities. In recent years, attention has been focused on cladocerans of the family Chydoridae. Few live in the limnetic region, although at least one, *Chydorus sphaericus*, is a common, cosmopolitan plankter. Chydorid head shields remain well preserved in lake sediments and seem to be distributed more or less uniformly across the lake basin. The study of these relics in lake stratigraphy has become an important part of paleolimnology. Most planktonic Cladocera leave no well-defined microfossils, but species of *Bosmina* constitute an exception, and their carapaces and head shields, as in the chydorids, remain preserved in lacustrine deposits (Frey, 1964; Horne and Goldman, 1994).

Benthos

Thermal stratification makes possible definition of bottom regions, the benthic zones. The profundal zone is overlain by hypolimnion water and is composed of fine-grained sediments. In eutrophic lakes this is a planktogenic and coprogenic ooze, only partially oxidized or slightly reduced. This is a grey-brown material called gyttja. In oligotrophic lakes profundal sediments are often mineralized.

The littoral zone at the lake margins contains coarser sediments. It is often defined as the region from shore to the depth marking the deepest growth of the rooted aquatic plants. Between the littoral and profundal is the sublittoral zone.

The community of bottom-dwelling organisms is the benthos. In most instances, limnologists have studied the profundal benthos, neglecting the taxonomically richer communities at littoral and sublittoral depths. In eutrophic lakes with summertime oxygen deficits, the profundal benthos is impoverished, made up only of those species which can tolerate anaerobic conditions. Typically, certain larvae of the dipterous family, Chironomidae, occur here. These midge larvae, however, are largely so-called, 'red blood worms' of the *Chironomus plumosus* type which includes a handful of species. Evidence is accumulating that they feed largely on bacteria in the bottom deposits rather than on the organic sediments themselves, (McManus and Duck, 1993).

Tubificid oligochaete worms, tiny fingernail clams, especially species of *Pisidium*, and larval *Chaoborus* are also members of the profundal benthos in some eutrophic lakes. The scanty representation in this species list is largely a function of anoxic conditions.

In oligotrophic lakes the profundal fauna is enriched in species, but diminished in total biomass. Chironomid larvae of the *Tanytarsus* type occur in the profundal area where oxygen is adequate. Stratigraphic studies show that they are replaced by the *Chironomus* as eutrophication proceeds.

Quantitative benthic sampling utilizes dredges which cover known areas and take sediment samples to depths of 15–20 cm. The Ekman dredge, covering about 125 cm^2, is employed in most studies. The greatest standing crops of profundal benthos occurring in eutrophic lakes are near 500 kg fresh weight ha^{-1}, which is some 100 times greater than the mass reported from extremely oligotrophic lakes.

A largely neglected segment of the bottom fauna is the microbenthos, the assemblage of microscopic species which pass through the sieves normally employed in screening dredge samples. These include protozoans, nematodes, rotifers and various microcrustaceans such as ostracods.

Minor communities

The Aufwuchs or periphyton is the community encrusting submerged objects. It is comprised of algae and many taxonomic groups of animals.

In the interstitial water among beach sand grains, occurs another community, the psammon. Many microscopic forms are found here which are absent from the littoral sediments just a few centimeters away.

A minor community is the neuston, associated with the water–air surface film. Several plants and animals live either on the upper or under surface of this film.

The nekton is composed of powerful swimmers, an assemblage including fishes, some large crustaceans and insects. Salmonid fish are typical of oligotrophy or mesotrophy and disappear with eutrophy. In North America, the bass and sunfishes, centrarchids, replace them and typify eutrophic waters.

Community metabolism

Limnologists have been concerned with the quantitative flow of energy through lacustrine ecosystems since the important paper of Lindeman (1942) focused attention on this dynamic aspect of communities. The role of organisms as members of trophic levels, utilizing and passing on energy derived from lower levels is the essence of this approach.

Calories derived from solar radiation represent the original source of energy. The first trophic level includes autotrophic organisms of which the phytoplankters are most noteworthy: they are the producers. In spite of high annual production, they are relatively inefficient, converting less than 1% of the radiation to energy-containing substances.

The primary consumers are herbivorous animals which feed directly on plants, and fall prey to the carnivores, the secondary and tertiary consumers, which represent the third and fourth levels.

Each level contains less energy than the preceding one, the loss by respiration increasing progressively from lower to higher trophic levels. The efficiency of conversion of food energy to protoplasm, however, increases simultaneously. The simplicity of the scheme is misleading, because most carnivores are not restricted to feeding on organisms of the next lower level. Furthermore, detritus and associated bacterial flora are an important energy source.

The decomposers, mostly heterotrophic bacteria, break down organic substances in dead plants and animals to an inorganic state, thereby returning nutrients to the green producers for recycling.

Gerald A. Cole

Bibliography

Baker, L.A. (ed.), 1994. *Environmental Chemistry of Lakes and Reservoirs*. Washington, DC, American Chemical Society, 627 pp.

Bradley, W.H., 1962. Paleolimnology, in Frey D.G., ed., *Limnology in North America*, pp. 621–652, Madison, WI., University of Wisconsin Press.

Deevey, E.S., 1955. The obliteration of the hypolimnion, *Mem. Ist. Ital. Idrobiol., Suppl. 8*, 9–38.

Forel, F.A., 1892. *Le Léman: Monographie Limnologique*, Tome 1, *Géographie, Hydrographie, Géologie, Climatologie, Hydrologie*, Lausanne, F. Rouge, xiii + 543 pp.

Frey, D.G., 1964. Remains of animals in Quaternary lake and bog sediments and their interpretation, *Arch. Hydrobiol. Beih., Ergebn. Limnol.*, **2**, 1–114, 2 pl., 1 tab.

Horne, A.J. and Goldman, C.R., 1994. *Limnology*, McGraw-Hill, New York, 576 pp.

Hutchinson, G.E., 1957. *A Treatise on Limnology*, Vol. I, *Geography, Physics, and Chemistry*, New York, John Wiley & Sons, xiv + 1015 pp.

Lindeman, R.L., 1942. The trophic-dynamic aspect of ecology, *Ecology*, **23**, 399–418.

McManus, J. and Duck, R.W. (eds), 1993. *Geomorphology and Sedimentology of Lakes and Reservoirs*, New York, John Wiley and Sons, 278 pp.

Micklin, P.P., 1988. Desiccation of the Aral Sea: A water management disaster in the Soviet Union. *Science*, **241**, 1170–1175.

Mortimer, C.H., 1956. An Explorer of Lakes, in *G.C. Sellery, E.A. Birge – A Memoir*, pp. 163–211, Madison, Wisc., University of Wisconsin Press.

Mosello, R., Boggero, A., Carmine, M. *et al.*, 1994. Velbania Pallanza, Italy 436 pp.

Rawson, D.S., 1955. Morphometry as a dominant factor in the productivity of large lakes, *Verh. intern. Ver. Limnol.*, **12**, 164–175.

Ryther, J.H., 1956. The measurement of primary production, *Limnol. Oceanogr.*, **1**, 72–84.

Schiller, G., Lammela, R. and Spreafico, M., 1991. *Hydrology of natural and man-made lakes*, Report No. 206, IAHS, Wallingford, UK, 206 pp.

Schwoerbel, J., 1987. *Limnology*, New York, John Wiley and Sons, 228 pp.

Cross references

Caspian Sea
Lake Balaton, Hungary
Lakes
Lakes, lake water
Lakes: largest worldwide
Limnology
Lough Neagh
Okeechobee Lake, Florida, USA: human impacts, research and lake restoration

LOUGH NEAGH

This is the largest freshwater lake in the British Isles for every parameter except depth. Formed by glacial movement, creating glacial moraine against basaltic outcrop at its northern end, it is situated in Northern Ireland with a center at longitude 6°25'W latitude 54°38'N and it drowns out the original confluence of the Rivers Bann and Blackwater. With a surface area of 383 km², an area draining into the Lough of 4465 km², and a downstream river catchment of 925 km², the lough 'controls' 43% of the surface area of Northern Ireland. It has a volume of some 3.15 million Ml and a mean depth of 8.6 m, although there is a part that is some 30 m deep. It was in danger of becoming eutrophic due to pollution from human habitation and agricultural nutrients but action by the Government of Northern Ireland, and its successors, utilizing tertiary treatment of sewage, and agricultural fertilization and silage effluent control, has alleviated this danger, hopefully permanently and successfully. The Lough Neagh Basin is an important salmon fishery and an internationally famous source of eels. Public potable water supplies of some 200 Ml day^{-1} are drawn from the lough, about 120 Ml day^{-1} of which are for use in the Belfast conurbation. A number of sluices in the outflowing River Lower Bann control the flow regime through the system and, in normal circumstances, permit the control of the lough level within a 150 mm range, so that optimum use can be made of the peripheral arable agricultural land.

P.G. Holland

Cross references

Lakes
Lakes: largest worldwide
Limnology

M

MANNING FORMULA

Developed from seven different formulae by earlier scientists and engineers in various parts of Europe, in 1889 Robert Manning, from Ireland, introduced the following formula for uniform flow in open channels:

$$V = \frac{1.49}{n} R^{2/3} S^{1/2}$$

where V is the water velocity in feet per second, n is a roughness coefficient (non-dimensional), R is the hydraulic radius in feet and S is the energy gradient (non-dimensional).

The metric equation is:

$$V = \frac{1}{n} R^{2/3} S^{1/2}$$

where V is in meters per second and R is in meters.

Although similar to the **Chézy formula**, the Manning formula is simple and practical to apply to field work, so it remains a popular means of assessing the water velocities in rivers and streams. The key to the use of the formula is the determination of the roughness coefficient, known as 'Manning's n', but experienced engineers have produced listings of n for various types of conditions of river beds and banks and many users utilize a nomographic solution for the formula utilizing all the parameters of the formula.

P.G. Holland

Bibliography

BS 3680, Part 1, 1991. *Glossary of Terms*, British Standards Institution, London.

Cross reference

Hydrologists (600 BC–AD 1900)

MARITIME CLIMATE, OCEANICITY

The regional climatic type that is dominated by its propinquity to the sea is known as a maritime climate or, sometimes, oceanic climate or marine climate; the terms oceanic climate and marine climate, however, are not greatly favored because they may be misread to mean climates over the ocean rather than a land climate bordering on the ocean. A marine weather observation is one taken from a ship at sea; it has nothing to do with maritime weather. The technical measure of maritime character is oceanicity, the converse of continentality.

Maritime air and a maritime air mass refer to a type of air that has developed over or passed over an appreciable body of water, thereby obtaining a high moisture content. A secondary factor is that it has usually picked up a large quantity of salt nuclei from sea spray that will favor the nucleation of water droplets when onshore winds bring the clouds over hills or orographic barriers, causing uplift and precipitation.

Maritime climates are necessarily most marked where the prevailing winds are onshore. Most characteristic are the western shores of land masses in the westerly wind belts. Where offshore winds are the rule, there is generally a littoral desert, as in the Atacama Desert of South America, the Namib Desert of Southwest Africa, the desert coasts of Spanish Sahara and Mauritania, and parts of Baja California. In such places, a sea fog is sometimes the only maritime influence (Trewartha, 1961).

The most characteristic situations for maritime climates are on most oceanic islands, on the westerly sides of continents in the latitudes of the westerly winds fed by maritime polar air, and in certain areas of the monsoons, where the dry season wind reversal is locally abbreviated or suppressed for some reason.

The maritime influence may reach inland until it is orographically blocked by a climate divide; in regions of little relief, as in northwestern Europe the maritime time effects at times reach hundreds of kilometers inland. The daily temperature range is very small (1–1.5°C), while the rainfall is high and the weather often changes rapidly. In midlatitudes summers are cool and winters mild. The annual extremes are retarded as a rule for periods of 1–2 months after the solstices. This is in contrast to the high symmetry of continental climates. Typical maritime stations in the northern and Southern hemisphere westerlies are given in Table M1, although at Melbourne, temperature extremes are raised by periodic land winds in summer that briefly bring days of excessively high-temperatures such as are never recorded in Seattle, where any such tendencies are blocked by mountains. Although maritime temperature means are essentially equable, rainfall figures vary considerably, sometimes with

Table M1 Kerner's oceanicity index for selected stations (Landsberg, 1958, p. 291)

	T_o	T_a	T_w	T_c	A	O (%)
Verkhoyansk, Russia	−14.6	−12.1	16.5	−46.2	62.7	−4
Bismarck, ND	6.9	6.2	21.1	−12.7	33.8	2
Winnipeg, Manitoba	4.8	3.2	19.1	−19.9	39.0	4
Bergen, Norway	7.5	5.6	14.3	1.1	13.2	14
Tokyo, Japan	15.8	12.5	25.4	3.0	22.4	15
New York, NY	13.2	9.2	23.1	−0.8	23.9	17
San Diego, Calif.	17.3	14.6	20.2	12.2	8.0	34
San Francisco, Calif.	14.9	12.4	15.5	9.7	5.8	43
Honolulu, Hawaii	24.9	22.8	25.8	21.5	4.3	49

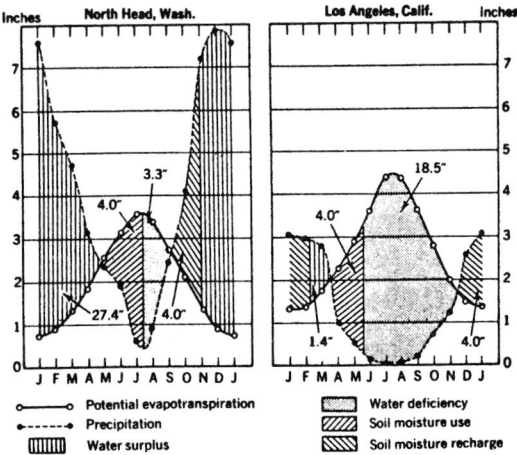

Figure M1 Potential evapotranspiration and precipitation at North Head, Washington, and Los Angeles, California. (From Thornthwaite, 1948.)

marked seasonality. Likewise, evapotranspiration shows big variations (Figure M1).

Oceanicity

The statistical determination of oceanicity (O) may be defined by what Kerner (1905) called the thermoisodromic ratio, thus

$$O = 100 \frac{T_o - T_a}{A}$$

where A is the annual range of temperatures (mean warmest to mean coldest month), and T_o, T_a are the monthly means for October and August. Typical examples are shown in Table M1, where Verkhoyansk (Siberia) in the Taiga climatic zone is selected as an extreme continental example (and thus shows negative oceanicity), whereas Honolulu, Hawaii, shows maximal oceanicity.

Ocean currents

Longshore oceanic currents also play an important role in determining the character of the maritime climate. Figure M2 illustrates a series of monthly mean temperatures for Barcelona, Spain, where for half the year the air is cooler than the ocean and the other half vice versa, thus keeping the annual range within 5°C (Landsberg, 1958). Another example, noted by Landsberg, is San Diego, California one of the mildest situations in the United States, where the mean annual range of the sea surface temperature is less than 6°C and its extreme range (for 30 years) is only 15°C; on the other hand, the situation is south of the westerlies and the maritime influence is limited to the coastal strip so that a few kilometers inland there is an annual temperature range of 41°C.

In the North Atlantic, because of its special configuration, the warm Gulf Stream and its extensions reaches far to the north of Scotland and Norway with its ameliorating effects.

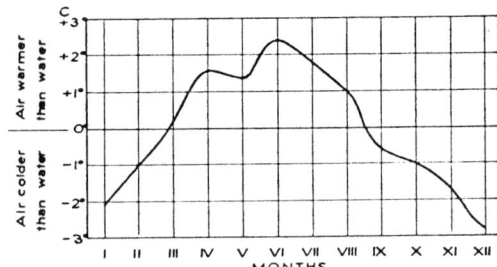

Figure M2 Difference between air temperature at the beach of Barcelona. (From Landsberg, 1958.)

Table M2 Diurnal variation of the precipitation at typical stations (Köppen, 1936)

Station	\multicolumn{6}{c}{Hours}
	0–4	4–8	8–12	12–16	16–20	20–24
Valencia, Ireland, 52°N						
A	18.1	18.3	16.0	14.9	16.2	16.5
D	18.6	21.3	15.4	12.8	14.5	16.4
I	38	33	41	46	44	40
San José, Costa Rica, 10°N						
A	1.3	0.6	4.4	34.2	48.5	8.4
D	12.6	5.6	5.4	22.4	34.3	19.1
I	35	31	23	108	131	40

Note: A = amount in percent of the total, D = duration in percent of the total duration, I = intensity in thousands of inches per hour of rain.

Precipitation

Maritime precipitation in the westerly belts is marked by the cyclonic activity characteristic of the polar fronts. These are most intense in the winter months, but the fall season is likely to be wettest. The amount of evaporation into the maritime air masses is related to both water and air temperature over the ocean. Accordingly, since the warmest conditions of the winter season occur in the fall, the latter is marked by the heaviest precipitation. In midwinter and spring, a cold continental air mass may become more important.

There is marked diurnal variation in maritime precipitation, having its maximum at night or in the early morning, in contrast to the continental type, which tends to occur in the afternoon or early evening (Haurwitz and Austin, 1944), although admittedly many variants are known. There may, for example, be a maritime type in winter and a convectional, continental type in summer (especially in the lower-latitude stations). In the tropics there is greater radiation over the ocean at night, although the sea surface remains warm; thus the night-time temperature gradient is steeper than that of the daytime, and convectional rain is common at night. A comparison between a typical westerly midlatitude station (Valencia, Ireland) and a tropical maritime station (San José, Costa Rica) shows this striking diurnal contrast (Table M2).

Humidity

Over the oceans humidity will reach a maximum in the afternoon and a minimum in the early morning, reflecting the diurnal air temperature cycle. The vapor pressure is controlled in part by the air temperature and in part by the turbulent transport that tends to disperse the humid layer upward, but the latter is less important over sea than over land, where a double cycle tends to develop over the 24 h period. Relative humidity is lowest in the high-pressure belts of each hemisphere (Figure M3); note the slight asymmetry, owing to the extensive water girdle of the Southern Ocean in latitudes 40–60°S. In the completely maritime stations (all-year) relative humidity is likely to be over 80% at all seasons.

Sea fog is commonly experienced along coasts of maritime climates, but more often along those of opposite conditions, for example where cold continental air flows out across a warm ocean (as in the evaporation fogs of 'Arctic sea smoke' or 'steam fog'), or where warm land air flows over a cold ocean current (as in the advection fogs of the Newfoundland Banks). The evaporation fogs

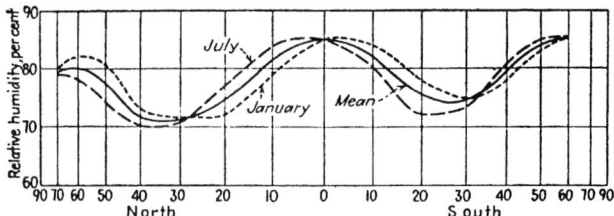

Figure M3 Zonal distribution of relative humidity. (From Haurwitz and Austin, 1944.)

tend to come more in the winter while the convection fogs come in summer when the water is cooler than the land; local geographic configurations play an important role.

Major lakes play a role analogous to the ocean (Landsberg, 1958) and are well illustrated by isotherms over the North American Great Lakes or Lake Baikal in Siberia.

In the middle to low latitudes, the maritime influence is reflected notably by the land and sea breezes, which lead to important amelioration of the midday and afternoon heat.

Rhodes W. Fairbridge and John E. Oliver

Bibliography

Anthes, R.A., H.A. Panofsky, S.J. Cahir and A. Rongo, 1978. *The Atmosphere*. Columbus, OH: Charles E. Merrill.

Currey, D.R., 1974. Continentality of extra-tropical climates. *Assoc. Am. Geog. Annals*, **64**, 268–280.

Critchfield, H.J., 1960. *General Climatology*. Englewood Cliffs. NJ: Prentice-Hall.

Haurwitz, B. and J.M. Austin, 1944. *Climatology*: New York: McGraw-Hill.

Kerner, F., 1905. Thermisodromen, versucheiner Kartographischen Dartstellung des jährlichen Ganges der Luftteperatur, *K.K. Geogr. Gesell. Wien* **6**(3).

Köppen, W., 1936. Das Geographische System der Klimate. *Handbuch der Klimatologie*, **1**, 1–44.

Landsberg, H., 1958. *Physical Climatology*, 2nd edn. Dubois, PA: Gray Printing Company.

Lockwood, J.G., 1974. *World Climatology*. New York: St Martin's Press.

Schroedor, M.J., 1967. Maritime air invasion of the Pacific Coast: A problem analysis, *Am. Meteorol. Soc. Bull.*, **48**, 802–808.

Thornthwaite C.W., 1948. An approach toward a rational classification of climate, *Geog. Rev.*, **38**, 55–94.

Trewartha, G.T., 1961. *The Earth's Problem Climates*. Madison: University of Wisconsin Press.

Trewartha, G.T. and L.H. Horn, 1980. *An Introduction to Climate*. New York: McGraw-Hill.

Cross references

Climate and climate change
Climate data: sources
Evapotranspiration
Lakes
Maritime zones

MARITIME ZONES

Maritime zones are areas of ocean or sea which are or will be subject to national or international authority. They are delimited as parts of the seabed, water column and sea surface, the subdivision being on the grounds of political jurisdiction relating to the use and ownership of marine resources.

Defining and delimiting maritime zones

Various maritime zones are claimed by coastal states seeking exclusive access to the resources of the sea and the seabed. In earlier centuries only two types of ocean space – territorial sea and high seas – were acknowledged. In the twentieth century, with the discovery of offshore oil and gas and the rapid development of technologies to explore and exploit these resources, the situation began to change. Post-World War II coastal states attempted to extend their jurisdiction beyond the territorial sea as interest in the resources of the sea and especially the seabed increased. Regularisation was attempted under the auspices of the United Nations, and the Conventions adopted by the 1958 United Nations Conference on the Law of the Sea created three types of ocean space:

- the sea surface and water column under national jurisdiction, this to comprise the territorial sea and a contiguous zone;
- the sea surface and water column beyond the limits of national jurisdiction, i.e. the high seas;
- the seabed under national jurisdiction – the so-called continental shelf.

The legal limits of these ocean spaces were, however, not clearly defined and were seen as somewhat open ended. This applied particularly to the continental shelf which was delimited as a national jurisdiction out to a line where the water depth reached 200 m but which could be, alternatively, legally extended as far as existing technology would allow resource exploitation. Inevitably there was an escalation of national claims and potential conflict between competing claims.

The current position is set out in the 1982 United Nations Convention on the Law of the Sea (UNCLOS) in which six types of maritime zone or ocean space are more clearly defined and delimited:

- the territorial sea;
- the contiguous zone;
- the exclusive economic zone;
- the continental shelf;
- the international seabed;
- archipelagic waters.

In addition to creating a new spatial order for the oceans, UNCLOS also provides the legal framework for the management and conservation of marine resources and the management of all major uses of the oceans, as well as creating new international institutions such as an International Seabed Authority and an International Tribunal of the Law of the Sea. Considerable disagreement exists, however, between many neighboring coastal states as to the delimitation of the various maritime zones, and certain countries, including the UK and USA, are not yet signatories to the 1982 Convention.

Baselines – the starting point for delimiting maritime zones

The current rules for drawing baselines from which maritime zones are measured were established in 1958 at the first United Nations Conference on the Law of the Sea. The low-water mark would normally determine the baseline around the coasts of any state, including the coasts of any islands which are part of the state. Along most coasts the low-water mark varies throughout the year, but the rules do not specify which low-water mark should be used. It is the responsibility of the government of the coastal state to select a low-water mark and indicate its location on a large-scale chart so that governments of other states are aware of the positions of the baselines. The conventional baseline corresponds to the mean low-water mark.

There are two circumstances in which deviations are permitted from the selected baseline:

- straight baselines may be drawn along short sections of coast, such as across river mouths or harbour entrances, to link up longer sections where the low-water mark is used; such straight baselines can be drawn across bays where the mouth of the bay does not exceed 24 nautical miles (44.5 km);
- straight baselines can be drawn along comparatively lengthy sections of coasts which are deeply indented or cut into or where there is a fringe of islands.

The drawing of straight baselines along indented and island-fringed coasts and across the mouths of bays and estuaries creates internal waters. Internal waters, lying on the landward side of the baseline, are not necessarily continuous and are legally treated as part of the land territory of the coastal state.

Coastal states may also draw a straight baseline across the mouth of a bay wider than 24 nautical miles (44.5 km) by claiming it as an historic bay. The historic bay concept was recognized but not defined at the 1958 UN Conference on the Law of the Sea. Bays of considerable size can therefore be claimed as historic waters providing those waters have been used exclusively for a long time by the claimant state, a formal claim to sovereignty has been made and that claim is accepted by other states. Where these three conditions are satisfied, an historic bay can be declared as internal waters and, in theory, the state's territorial sea is measured seaward from the baseline across the mouth of the bay. Historic bays are often disputed since the coastal state's claim to sovereignty is not accepted by all the other states.

Individual maritime zones

There are five maritime zones, comprising the sea surface and water column, whose extent is delimited seaward in relation to the baseline set by a coastal state.

Territorial sea

The territorial sea lies immediately offshore of a coastal state (and has also been referred to as the marine belt and marginal sea). From medieval times a distance of 3 nautical miles (*ca.* 5.5 km; based on the distance that could be defended by cannon from the shore) was generally accepted as the outer limit of a coastal state's territorial sea, but every coastal state now has the right to establish (codified in the 1982 UNCLOS Convention) a territorial sea which extends up to 12 nautical miles (22 km) from the coast or from baselines drawn in accordance with the agreed principles.

The territorial sea is treated as part of the national state, all national laws applying to the control of airspace, waters and seabed. The coastal state's exercise of sovereignty is absolute except that ships of other states are granted the right of 'innocent passage' in peacetime. Ships of other states could therefore pass through national waters so long as they did nothing to interfere with the rights or with the security of the coastal state, although the sovereign state may challenge such vessels at any time.

Contiguous zone

Located immediately seaward of the territorial sea the contiguous zone extends over an additional 12 nautical miles (22 km) i.e. its outer limit is 24 nautical miles (44.5 km) from the coast or baselines from which the territorial sea is measured.

Within the contiguous zone the coastal state exercises certain exclusive rights and controls to prevent infringement of its customs and fiscal, immigration and sanitary/health regulations with respect to its land territory or territorial sea or to punish infringement of those regulations committed within its territory or territorial sea.

Exclusive economic zone

Given the importance of marine resources beyond the territorial sea and contiguous zone, coastal states began making claims to more distant areas of sea or ocean. The first attempts to extend a coastal state's sovereignty beyond its territorial sea led to the designation of zones of exclusive fisheries as coastal states sought to protect their coastal fisheries as much as possible against foreign exploitation. Subsequently, zones of diffusion (ill-defined areas beyond the contiguous zone) were delimited by some coastal states in which those states claimed certain unilateral rights, such as weapon testing, fisheries and pollution control.

Such zones have now been subsumed into and formalized as the exclusive economic zone which extends 200 nautical miles (370 km) from the coast or baselines from which the territorial sea is measured. The exclusive economic zone is an ocean space *sui generis*, having in part qualities of both territorial sea and high seas but the status of neither. Thus the coastal state does not exercise sovereignty over this maritime zone in which other states enjoy many high-sea freedoms but it does have sovereign rights over all resources and economic uses of the zone. Foreign nationals cannot conduct any enterprise, such as fishing or mining, without the permission of the coastal state, although they do have rights to navigate vessels through the zone, to lay submarine cables and to overfly. The coastal state also exercises jurisdiction over marine scientific research in this zone and has responsibilities with regard to the protection of the marine environment and can control installations and creation of artificial islands.

Archipelagic waters

An archipelagic state, consisting of a group of islands which historically form one nation but whose components had in the past been separated by high seas to the detriment of national unity, have the right under international agreement to draw straight archipelagic baselines joining the outermost edges or fringes of their outermost islands, subject to four conditions:

- the pattern of straight baselines must not depart to any appreciable degree from the configuration of the archipelago;
- the ratio of water to land enclosed within the baselines must not exceed 9:1;
- no segment of straight baseline should be longer than 125 nautical miles (231 km);
- no more than 3% of the baseline segments may measure between 100 and 125 nautical miles (185–231 km).

Baselines drawn according to these rules enclose archipelagic waters. Beyond these baselines, archipelagic states may still claim a territorial sea, and exclusive economic zone, and a continental shelf, so bringing vast areas of ocean under some national jurisdiction.

Archipelagic waters bear a stronger similarity to the territorial sea than to internal waters since the archipelagic state must allow the traditional fishing rights of other states enclosed by the baselines as well as allowing states which have laid submarine cables through the area when it was part of the high seas access to repair those cables. The right of innocent passage also applies through archipelagic waters although the archipelagic state may designate shipping lanes through them which have to be observed by foreign vessels.

High seas

The high seas are the 'global commons', the rest of the ocean space lying beyond the exclusive economic zones of all coastal states. All states have equal rights to navigate, to overfly, to lay submarine cables, to construct artificial islands, to fish, and to conduct scientific research within the high seas.

The seabed is also subdivided on the basis of political jurisdiction relating to ownership of the marine resources. There are just two divisions – the continental shelf and the international or deep seabed: confusingly, whilst the terminology is geomorphological, the interpretation is legal.

Continental shelf

In geomorphological terms the continental shelf is the gently sloping offshore extension of a continent which is submerged by a shallow sea and which extends to the continental slope (marking, at a depth of about 200 m the transition zone between the continental shelf and the ocean depths). Continental shelves are structurally similar to their contiguous continents and can vary in width from virtually non-existent to over 350 nautical miles (650 km).

The political significance of the continental shelf is of recent origin, dating from the time when exploitation of underwater resources, other than fisheries, became a feasible proposition. Coastal states are now permitted to claim the seabed and substratal resources of the continental shelf which extends beyond their territorial seas throughout the natural prolongation of their land territory to the edge of the continental slope as a maritime zone. In certain cases the offshore extent of the continental shelf exceeds the 200 nautical mile outer limit of the exclusive economic zone and this prolongation of the shelf beyond the exclusive economic zone has been termed the continental margin.

Claim to the continental shelf by the coastal state is restricted to the mineral and other non-living resources of the seabed and substrata, along with living organisms belonging to sedentary species, which at the harvestable stage are either immobile on or under the seabed or are unable to move except in constant physical contact with it.

International seabed

The international seabed is the area beyond the limits of the continental shelf and is designated by the United Nations as part of 'the common heritage of mankind'. The intention is that the international or deep seabed comes under the jurisdiction of the International Seabed Authority which controls resource exploitation via a permit system and places royalties accruing to it into a UN fund for economic development.

In addition to these internationally recognized maritime zones there is a further theoretical zone advocated by the American geologist, H.D. Hedberg.

Hedberg zone

The Hedberg zone is a proposed maritime zone designed to allow each coastal state to claim the offshore area within which there is a reasonable chance of discovering hydrocarbon deposits. It is linked to the continental slope in that the line marking the junction between the continental slope and the continental rise is identified to serve as a baseline of a zone within which the outer edge of the continental margin should be set. There is no agreement on the width of the zone and, in many areas, there are problems locating the contact between the continental slope and rise.

Relationships between maritime zones

The spatial relationships between the various maritime zones are summarized in Figure M4 for a hypothetical coastal state. In the case

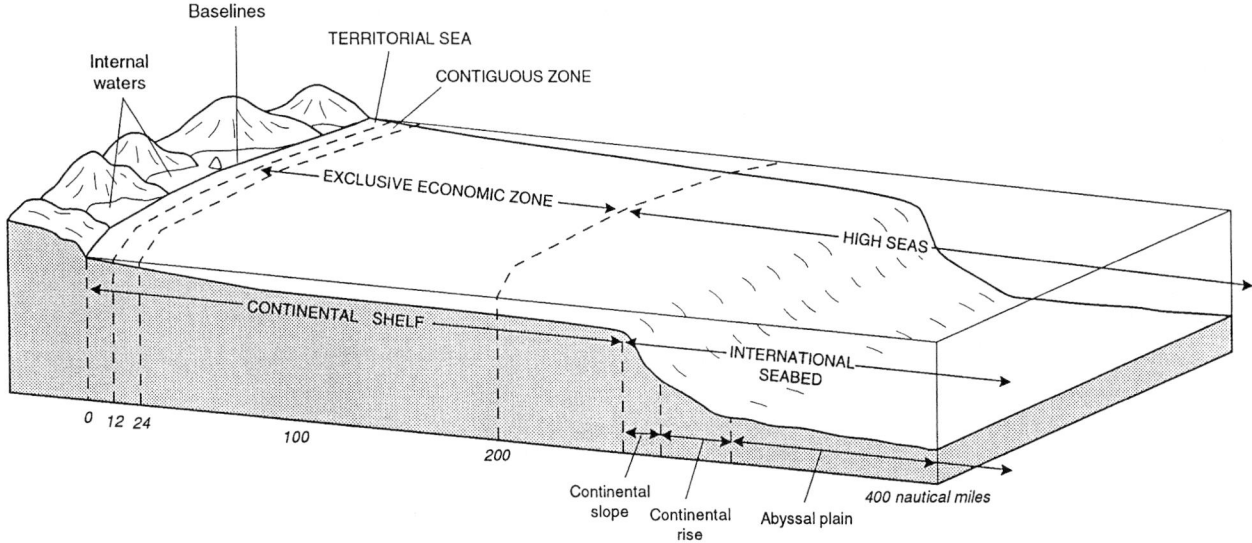

Figure M4 The spatial organization of maritime zones.

illustrated it is assumed that the continental shelf is wider than the exclusive economic zone but in many cases that relationship will be reversed, i.e. the exclusive economic zone is wider than the continental shelf.

Problems occur in determining the extent of territorial seas, contiguous zones and exclusive economic zones in shallow, semi-enclosed seas such as the North Sea where the claims of coastal states overlap. In such cases, median or dividing lines have to be agreed and where distances separating coastal states are small, e.g. the Straits of Dover, certain of the outer maritime zones cannot exist and even the territorial seas do not achieve their maximum permitted width.

The 1982 UNCLOS Convention, which has been called 'a constitution for the oceans', provided clear guidelines for the future use of oceans, both nationally and internationally. It clarified the delimitation of the political maritime zones and the powers vested in the coastal state with respect to the exploitation of marine resources. There were 'gainers' and 'losers' as a result of the changes initiated. Prominent amongst the gainers were large coastal states (many of them developed), archipelagic states and many island states (especially in the Pacific Ocean). As far as exploitation of resources is concerned the chief 'losers' are landlocked states and the so-called geographically disadvantaged states (with short coasts or other geographic impediments preventing them from acquiring a full-sized exclusive economic zone and/or continental shelf). About 35% of the global sea and ocean area now falls within maritime zones under the national jurisdiction of coastal states, with the rest – the high seas and international seabed – acknowledged as a common property resource. UNCLOS signatories accept an obligation to manage and protect the marine environment, but the Convention has yet to command universal support and there are disagreements between signatories as to the exact delimitation of certain boundary lines of maritime zones.

Brian Goodall

Cross reference

Maritime climate, oceanicity

MATHEMATICAL MODELS

Mathematical modeling is understood as the representation of selected aspects of performance of a system, process or object, with the help of mathematical equations. Among the variables occurring in a number of mathematical models, one may distinguish input, output and state variables and initial and boundary conditions.

Mathematical models are used in theoretical studies, aimed at increasing our understanding, or else they are aimed at applications

where they are applied for either simulation or forecasting purposes. Simulation is mimicking the behavior of the real system, for conditions which have actually occurred in the past, or for those chosen arbitrarily. Forecasting is estimating future values of variables of interest, based on the knowledge of past and present values of variables characterising the system. A mathematical model may describe only those aspects which are deemed essential in a particular application, ignoring or strongly simplifying those details which are not important; that is, there may exist a set of models of a particular object, differing in their structure, complexity, accuracy, data requirements, etc., which are used for various purposes.

Depending on the assumptions and the class of mathematical tools used, mathematical models can be static (if they express relationships between instantaneous or mean values of variables) and – more often – dynamic. In this latter case a model describes how the previous history of the object determines its present, or future, condition. Models can be stationary or non-stationary, depending on whether the properties of the object (e.g. parameters of model equations) change in time, or not. They can either be lumped, if spatial variations are not considered, or otherwise distributed. They can be linear or non-linear, depending on the equations used, while the linearity assumption which reduces the computational burden is frequently taken. Models may be deterministic or probabilistic (random). In the former case there would be the same result for a given set of input conditions, whereas in the latter case the results may differ. They may be discrete, i.e. describing the process in a number of discrete time instants usually with equal intervals (e.g. hour, day, 10 days, month, year), or continuous, i.e. representing the system behavior for any time instant.

The important class of dynamic mathematical models covers a spectrum of approaches. At the one extreme there are simple black box methods where modeling of relations between input and output signals is made, without the insight into the system. At the other extreme there are rigorous non-linear partial differential equations of mathematical physics, such as those expressing the laws of hydrodynamics governing the movement of water. Between these extremes there are a variety of conceptual models, where representation of some aspects of system performance is achieved with the help of simple conceptual elements.

Recent theoretical developments have led to a number of novel categories of mathematical models, such as fractals, deterministic chaos, fuzzy sets, pattern recognition and neural networks, which have been applied in hydrology and water resources research.

Mathematical models are the core of the software packages which are widely available and used in hydrology and water resources. Although many users of user-friendly software products may not be aware of the mathematical model implemented in the algorithm and

encoded in a programming language, mathematical models are still of
great interest to researchers and software developers.

Zbigniew W. Kundzewicz

Cross references

Modeling, of water resources systems
Model predictions: uncertainty in
Models: distributed models of catchment hydrology
Models: parameter estimation

MAXIMUM OBSERVED RAINFALLS

Some of the largest point rainfalls for selected durations that have
been observed are given in Table M3. These values, which approach
probable maximum precipitation magnitude, are enveloped by the
approximate equation (see Figure M5).

$$P = 422T^{0.475}$$

where P is rainfall in millimetres, and T is duration in hours. The
lower relation in Figure M5 is the UK relation: approximately
$P = 80T^{0.424}$.

R.W. Herschy

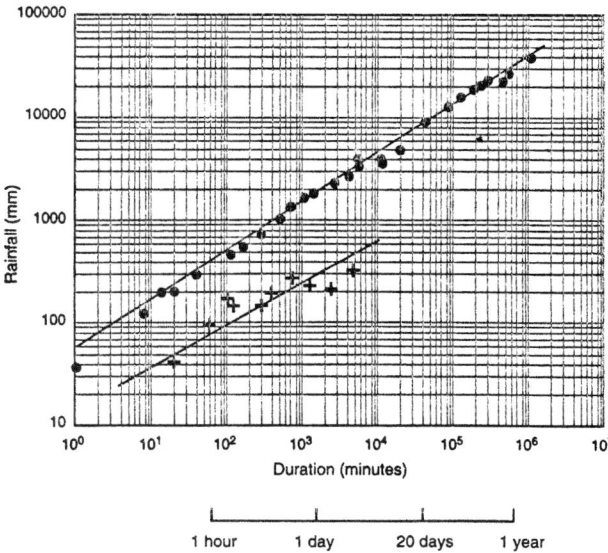

Figure M5 World's greatest observed point rainfalls (●) (adapted from
data in WMO, 1994), compared with greatest UK point rainfalls (+).

Table M3 World's greatest observed point rainfalls

Duration	Depth (mm)	Location	Date
1 min	38	Barot, Guadeloupe	26 November 1970
8 min	126	Fussen, Bavaria	25 May 1920
15 min	198	Plumb Point, Jamaica	12 May 1916
20 min	206	Curtea-de-Arges, Romania	7 July 1889
42 min	305	Holt, MO, USA	22 June 1947
1 h 00 min	401	Shangdi, Nei Monggol, China	3 July 1975
2 h 10 min	483	Rockport, WV, USA	18 July 1889
2 h 45 min	559	D'Hanis, TX, USA (27 km NNW)	31 May 1935
4 h 30 min	782	Smethport, PA, USA	18 July 1942
6 h	840	Muduocaidang, Nei Monggol, China	1 August 1977
9 h	1 087	Belouve, La Réunion	28 February 1964
10 h	1 400	Muduocaidang, Nei Monggol, China	1 August 1977
18 h 30 min	1 689	Belouve, La Réunion	28–29 February 1964
24 h	1 825	Foc Foc, La Réunion	7–8 January 1966
2 days	2 467	Aurere, La Réunion	7–9 April 1958
3 days	3 130	Aurere, La Réunion	6–9 April 1958
4 days	3 721	Cherrapunji, India	12–15 September 1974
5 days	4 301	Commerson, La Réunion	23–27 January 1980
6 days	4 653	Commerson, La Réunion	22–27 January 1980
7 days	5 003	Commerson, La Réunion	21–27 January 1980
8 days	5 286	Commerson, La Réunion	20–27 January 1980
9 days	5 692	Commerson, La Réunion	19–27 January 1980
10 days	6 028	Commerson, La Réunion	18–27 January 1980
11 days	6 299	Commerson, La Réunion	17–27 January 1980
12 days	6 401	Commerson, La Réunion	16–27 January 1980
13 days	6 422	Commerson, La Réunion	15–27 January 1980
14 days	6 432	Commerson, La Réunion	15–28 January 1980
15 days	6 433	Commerson, La Réunion	14–28 January 1980
31 days	9 300	Cherrapunji, India	1–31 July 1861
2 months	12 767	Cherrapunji, India	June–July 1861
3 months	16 369	Cherrapunji, India	May–July 1861
4 months	18 738	Cherrapunji, India	April–July 1861
5 months	20 412	Cherrapunji, India	April–August 1861
6 months	22 454	Cherrapunji, India	April–September 1861
11 months	22 990	Cherrapunji, India	January–November 1861
1 year	26 461	Cherrapunji, India	August 1860–July 1861
2 years	40 768	Cherrapunji, India	1860–1861

Revised: 29 November 1991, USA NWS; USA Bureau of Reclamation; Australian Bureau of
Meteorology.

Source

World Meteorological Organization, 1994. *Guide to Hydrological Practices*, 5th edn, WMO Geneva.

MEAN ANNUAL RUNOFF: CORRELATION WITH CATCHMENT CHARACTERISTICS

An estimate of mean annual runoff for a particular catchment may be derived by various methods, but the following approximate empirical relation is often used:

$$\bar{Q} = a\bar{S}^m \cdot \bar{P}^n / \bar{\theta}$$

where $\bar{Q}$ is the mean annual runoff, $\bar{S}$ is the mean slope, $\bar{P}$ is the mean annual rainfall (in the same units as $\bar{Q}$), $\bar{\theta}$ is the mean annual temperature (°C) and a, m and n are empirical constants for a given region; a also depends on the units used.

The average slope may be measured by one of several methods, which would influence the values of a and m. The constants should be derived empirically for catchments similar to the one to which they are to be applied.

R.W. Herschy

Source

World Meteorlogical Organization, 1974. *Guide to Hydrological Practices*, 3rd edn, WMO, Geneva.

MEDICINAL SPRINGS

The exploitation of medicinal or mineral springs with curative properties has long attracted attention, first by the Romans, and more recently, particularly in Central Europe during the nineteenth century.

The term 'spa' for a mineral spring or curative resort (taking its name from the Belgian village of Spa, near Liege) has been in the literature since the sixteenth century.

Balneo-geohydrology is a small and relatively new segment of geohydrology specializing in finding water sources characterized by unusual chemical composition and physical properties which can be exploited for medicinal purposes. Determination of the geological formations through which such water flows or is stored and evaluation of peloids found in some regions, such as peats, fango or muds, are other tasks of this branch of geology. In medical terminology, peloids are natural substances formed by a geological and/or biological process; they are classified according to origin, chemical composition and radioactivity. Fango is a type of peloid that is characterized by high inorganic content and low organic content, extensively used as a therapeutic mud in Italy.

Mineral waters are distinguished from potable (drinking) waters by a number of factors: (1) higher contents of mineral matter, (2) presence of gas, (3) presence of trace elements, (4) higher temperature, and (5) radioactivity. Various amounts and combinations of these factors can be present. As with other kinds of water, most mineral waters originate in the terrestrial water cycle. On its long upward journey from great depths, water leaches and dissolves many minerals present in the geological formations with which it comes in contact. Such mineral waters are called vadose. Primary volatiles from the Earth's mantle may sometimes become mixed with it. This apparently newly formed water is known as juvenile. Both vadose and juvenile waters may become heated by magmatic or geochemical reactions and emerge as hot springs.

Thermal springs producing water of 20–37°C are classified as warm and those discharging water above body temperature of 37°C are classified as hot springs. Some thermal springs contain relatively large amounts of dissolved mineral matter while others are poorly mineralized. Both may be radioactive to various degrees. It remains unproved but highly improbable that there are any mineral springs producing solely juvenile water. The temperature of water changes with the seasonal fluctuation of soil and air temperatures. The chemical composition of mineral springs is by no means constant as the concentration of single elements may vary for a number of reasons.

Mineral water appears on the surface of the Earth by way of natural springs, or it may be obtained artificially by sinking of deep wells. Drilling of wells is undertaken to supplement the occasionally insufficient quantity of water from existing mineral springs or to procure it in regions where no natural springs are present. In both instances, location of mineral water depends on a thorough geological survey prior to the contemplated drilling. On the other hand, temperature, radioactivity and the kind of minerals present in water can give important clues to the geological formations through which it passes.

The concentration of minerals in water is expressed in milligrams per liter or parts per million. In older analyses, the term grains per gallon was used. The composition of water is reported in the form of ions as cations (+) and anions (−). The undissociated combinations, e.g. silicic acid (H_2SiO_3), are listed separately. The following cations are frequently present: calcium (Ca^{2+}), magnesium (Mg^{2+}), sodium (Na^+), iron (Fe^{2+}), lithium (Li^+), aluminum (Al^{3+}) and potassium (K^+); fluoride (F^-), iodide (I^-), bromide (Br^-), chloride (Cl^-), hydrogenbicarbonates (HCO_3^-) and sulfates (SO_4^{2-}) appear as anions. Oxygen, carbon dioxide, hydrogen sulfide and radon are the usual gaseous constituents. In some localities, thermal springs carry radioactive elements in relatively high concentrations.

In addition to the dissolved mineral substances, gases and radioactive elements, temperature, taste, color, odor, turbidity and pH determine the medicinal usefulness of mineral water. Spectrographic examinations of mineral waters reveal the presence of many elements in extremely low amounts. The following oligoelements – copper, zinc, cobalt, molybdenum, boron, selenium, chromium and vanadium – seem to be of special therapeutic significance. However, the essentiality, the function and the physiological role of these elements are still not fully understood. The fact that some microelements ingested accidentally with food or water are deposited in animal and human tissues without having any noticeable effect or function makes this biological issue even more puzzling.

The 'safe levels' of radioactive constituents such as uranium, radium, thorium, radon, thoron, tritium, carbon (^{14}C) and potassium ($^{30,40,41}K$) in water suitable for internal or external use were tentatively established by health authorities and balneological organizations in Europe. The widely fluctuating and divergent views in regard to the safety and effectiveness of low doses of radiation prevents a uniform agreement. Thus considerably different levels are quoted in pertinent literature.

Only water containing more than $1\ g\ kg^{-1}$ of dissolved minerals is, in some European countries, classified as mineral water. Although a great many varieties of these waters are extensively used, some of lesser mineralization (the acratopegs) also enjoy a great popularity, probably due to the abundance of trace elements. According to the type, temperature and potability, mineral waters and peloids are employed therapeutically in many parts of the world. Chronic disorders of the integument, the musculoskeletal, gastrointestinal, biliary, respiratory and cardiovascular systems are the most common indications for spa therapy.

Since time immemorial, mineral springs were held in high esteem by all races and strata of the population. Mineral water is probably the only medicinal medium which has outlasted millennia of progress and is today, in spite of vast advances in the healing arts, in even greater demand than before. Keen observations extending over many centuries established empirically the therapeutic specificity of certain types of mineral waters. Depending on temperature, mineral contents, gases, taste and purity (bacteriological safety), mineral water is being used in Europe, Asia, Central and South America very successfully for bathing, drinking, inhalation, gargling, internal irrigation and for filling of therapeutic pools. Peloids are in great demand for treatment of arthritis, rheumatism and allied conditions and are dispensed as hot and mud baths or in the form of local applications. Selected kinds of pure or naturally carbonated mineral waters are bottled and widely distributed as table water.

In the UK the absence of recent volcanic activity has resulted in a corresponding dearth of hot springs; however, this has not eliminated all mineralized waters, such as Bath, which has exploited their health-giving properties since Roman times. In the United States, artificial baths have largely replaced or supplemented natural springs for physiotherapeutic purposes. In the high-heat flow areas of the Rocky Mountains, notably Wyoming and Colorado, the hot mineralized springs have long been a folk-remedy for various ailments, but they have never approached the success of the long-established balneology of central European establishments (e.g. Baden-Baden). Current

suspicion of polluted drinking water has created an almost worldwide
demand for bottled water (see Bottled water), some of which makes
claims for beneficial natural mineralization in varying degrees.

Certain kinds of peloids and mineral waters are important natural
assets. We feel that the pinpointing of their exact location and determination of their composition could be of great value in the future. It is
certainly not too far fetched to say that earth sciences may help medicine
to rediscover an excellent, but neglected, form of therapy.

I.H. Kornblueh and Rhodes W. Fairbridge

Bibliography

Amelung, A. and Evers, A. (eds), 1962. *Handbuch der Baeder und
Klimaheilkunde*, Stuttgart, W.K. Schattauer Verlag.
Archives of Medical Hydrology, Pisa, Italy, Nistri-Lischi Publishers
(official journal of the International Society of Medical Hydrology
and Climatology).
Cyclopedia of Medicine, Surgery, Specialties, 1959. Philadelphia, PA,
F.A. Davis.
Fairbridge, R.W. (ed.), 1972. *The Encyclopedia of Geochemistry and
Environmental Sciences*, New York, Van Nostrand Reinhold, (see
pp. 699–702, Medical geology – trace metals in mammals).
Fundamenta Balneo-Bioclimatologica, Stuttgart, F.K. Schattauer
Verlag (periodical in German).
Kuenen, P.H., 1955. *Realms of Water*, New York and London, John
Wiley & Sons.
Kurashova, C.V. *et al.*, 1962. *Kurorti USSR*, Moscow, State Publish-
ing Office for Medical Literature (in Russian).
*Kurortologia (Physical Therapy and Corrective Physical Culture,
Transactions)*, Moscow, Ministry of Health of USSR (periodical in
Russian).
Licht, Sidney (ed.), 1963. *Medical Hydrology*, New Haven, Co,
Elizabeth Licht Publ.
Therme (Review of European Thermalism), Italy Castrocaro Terme
(Forli) (periodical).
US Department of Agriculture, 1955. *Water, The Yearbook of
Agriculture*, Washington, DC.
Zetschrift für angewandte Bäder- und Klimalzeilkunde, Stuttgart, F.K.
Schattauer Verlag.

Cross references

Bottled water
Hydrology
Springs
Water: categories

MEDITERRANEAN CLIMATE

The climate of areas bordering the Mediterranean Sea is used as a
climate type that generally is described as humid due to winter rainfall
but that is water-deficient during part of its growing season. Areas of
southern Australia, the southern tip of Africa, and parts of the southwestern United States are examples of this type of climate. Although
classed as a single type, this climate has appreciable variations in
temporal distribution of rainfall.

Biseasonality is a distinctive characteristic of the Mediterranean-

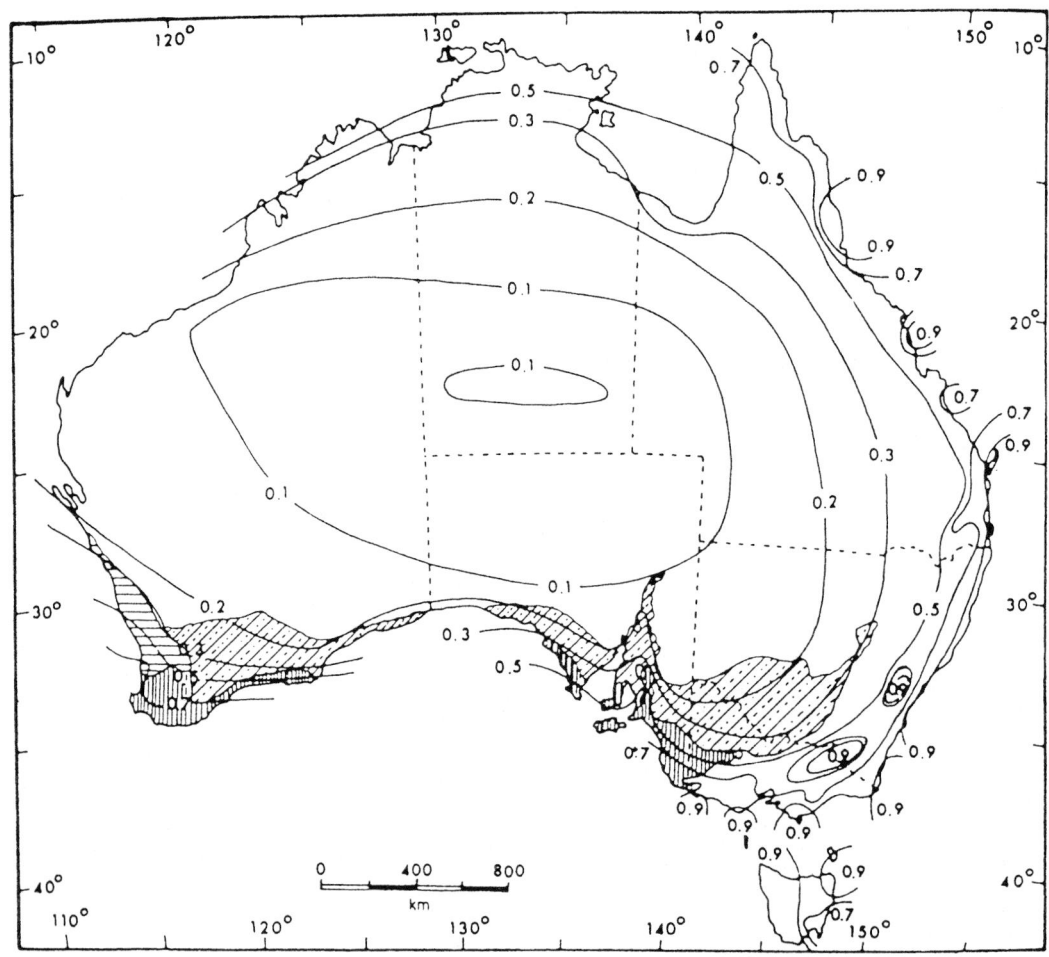

Figure M6 Distribution of Mediterranean climates (Köppen types Csa, horizontal hatching; Csb, vertical hatching; BSK, stippled) in Australia (after
Dick, 1975), superimposed with isolines of annual moisture index (after Specht and Moll, 1983).

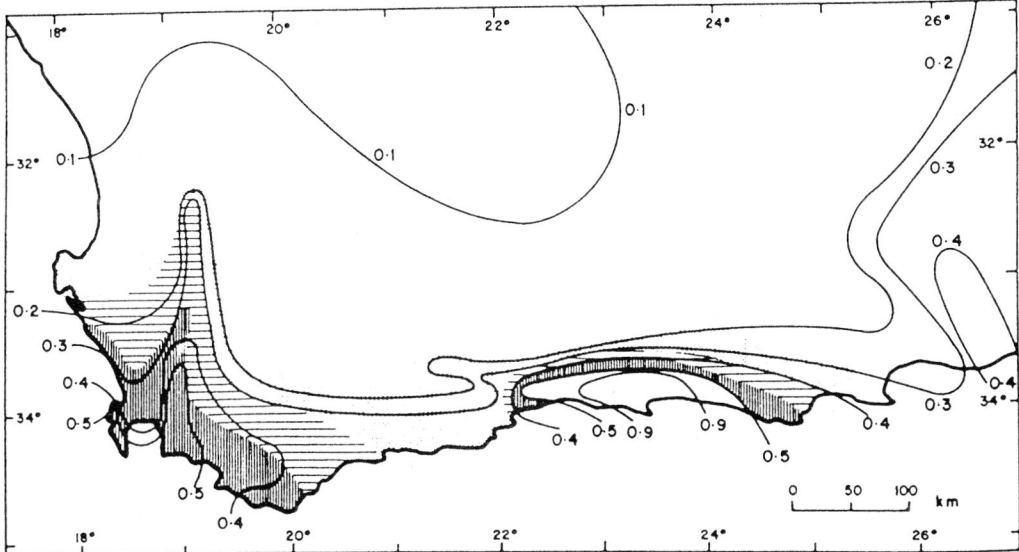

Figure M7 Distribution of Mediterranean climates (Köppen types Csa, horizontal hatching; Csb, vertical hatching; BSK, stippled, calculated according to Dick, 1975), in South Africa, superimposed with isolines of annual moisture index.

type climate, which has components of a hot, dry summer and a mild, rainy winter. Several approaches have been used to delimit this climate type. One approach delimits the Mediterranean climate by amounts of rainfall ranging between 275 and 900 mm, of which at least 65% falls in winter. Köppen (1936) defined the Mediterranean climate somewhat differently by distinguishing the Mediterranean regions (Cs) as a warm climate (C) with a mild, humid winter (at least 1 month >18°C) and a dry summer season(s). Various gradations in the intensity of the summer seasons (dry summer, mild summer, hot summer) have been recognized by the Köppen method.

Although the Köppen method emphasized the severity of summer drought in defining the Mediterranean climate, inland and more elevated localities within a Mediterranean-type region may experience severe winter frosts. Bioclimatologists believe that both the summer droughts and winter cold are independent stresses on the landscape morphology in such a region.

An analysis of several approaches of delimiting this climate type reveals that seasonality and not total precipitation amount is the most important factor. Therefore, many climatologists observe that there are not only the Mesic Mediterranean type climates but also Mediterranean semideserts and even extreme deserts. Figures M6 and M7 are examples of the extent of Mediterranean regions in Australia and South Africa.

Kamlesh P. Lulla

Bibliography

Dick, R.S., 1975. A map of the climates of Australia: According to Köppen's Principles of Definition, *Queensland Geog. J., Ser. 3:* **3**(3), pp. 3–69 and map.
Köppen, W., 1936. Das Geographische System der Klimate, in *Handbach der Klimatologue*, Vol. 1, pp. 1–44.
Mooney, H.A., 1977. *Covergent Evolution in Chile and California: Mediterranean Climate Ecosystems.* Stroudsburg, PA: Dowden, Hutchinson & Ross.
Orshan, G., 1964. Seasonal dimorphism of desert and Mediterranean chamaephytes and their significance as a factor in their water economy, in *Water in Relation to Plant*, A.F. Rutter and F.H. Whitehead (eds), Oxford: Blackwell, pp. 206–222.
Specht, R.S. and E.J. Moll, 1983. Mediterranean-type heathlands and sclerophyllous shrublands of the world: an overview, in *Mediterranean-Type Ecosystems*, F.J. Kruger, D.T. Mitchell and J.V.M. Jarvis (eds), New York: Springer-Verlag, pp. 41–65.
Walker, D.S., 1974. *The Mediterranean Lands.* New York: Wiley.

Cross references

Australia: climate
Climate data: sources
Europe: climate, a hydrological perspective
Precipitation distribution

METEOROLOGY

Meteorology is the science describing the structure and behavior of the atmosphere of the Earth, including the movement of water both as vapor and as liquid in the air. Over 97% of the Earth's water resides in the oceans, and only 0.001% of the total water supply of the world is in motion at any one time, yet associated with this small amount is all weather and river flow.

C.G. Collier

MIXING

Mixing is a process in which the particles of one or more materials are dispersed in another material. In the case of hydrology, the recipient is normally a water body (river, lake or groundwater), and solid, gaseous or fluid material, mostly an effluent, discharges into this recipient (Jolánkai, 1992). The water of the recipient transports the material which is the subject of the mixing process. Mixing is due to diffusion (molecular and/or turbulent), advection and dispersion. Advective transport is where a substance enters or leaves the water body with the flow of water, and dispersive transport is then the substance entering or leaving the water body due to the combined effect of molecular and turbulent diffusion, and the uneven flow velocity distribution.

The simple description of the diffusion is given by Fick's law: the mass transport in the direction of the concentration gradient is proportional to the concentration gradient and the diffusion coefficient. A more sophisticated description can be achieved by the three-dimensional transport and flow equations. In these transport equations the most important factor is the dispersion coefficient, which can be determined experimentally by a tracer test or can be calculated by empirical formulae, as a function of the local hydraulic conditions.

The phenomenon itself depends on the type of the water body, e.g. a river, lake or groundwater. In lakes, wave motion can also cause mixing, while in pervious strata the mixing is influenced by the pore sizes and their distribution.

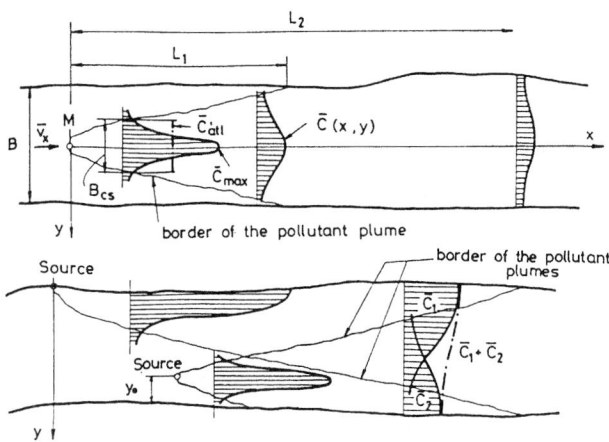

Figure M8 Transverse dispersion where $\partial\tau/\partial x$ approaches zero in the case of a continuous pollutant.

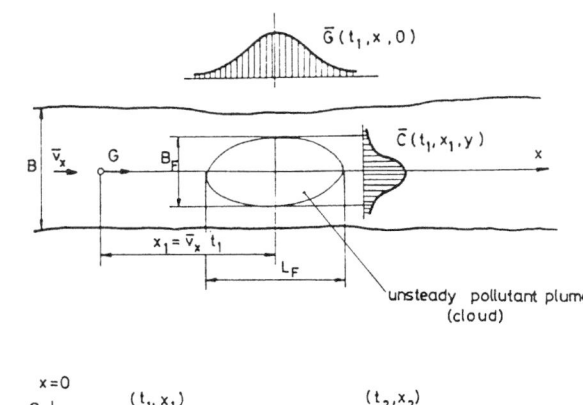

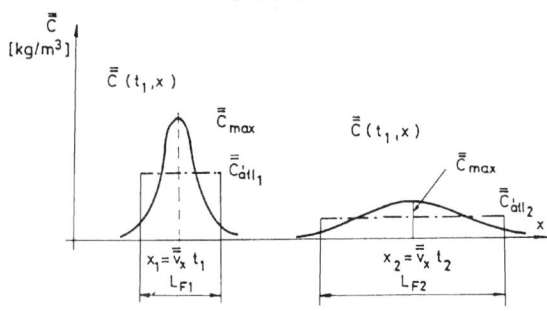

Figure M9 Variation of concentration with respect to time and location in longitudinal dispersion, with notation.

The rate of the mixing can be characterized by the spatial variation of the concentration. A forced mixing can be generated by a jet-type influence.

The mixing itself can be steady or unsteady. In steady mixing a more or less permanent plume can be distinguished which diminishes in the direction of the flow (Figure M8). Unsteady mixing can be caused by the variation of the confluence (Figure M9), or by the variation of the flow in the recipient (Starosolszky, 1987).

The variation of the concentration in the direction of the flow may be characteristic for the mixing process.

The contamination can be conservative, e.g. resistant to settlement, deposition, adhesion or any sort of chemical reaction. Non-conservative matter can change its substance, and in the transport equation this phenomenon should be reflected by source- or sink-type members.

Due to mixing, the longitudinal dispersion of matter may change, and the distribution of the concentration in the longitudinal profile is characteristic of the longitudinal dispersion coefficient (also determinable from a tracer test). Dispersion coefficients can be expressed by the equation

$$D_1 = A\frac{v^2B^2}{hu_x}$$

where v is the velocity of the flow, B is the channel width, h is the water depth and u_x is the shear velocity of the flow $u^x = (hSg)^{1/2}$. A is an empirical coefficient, S is the slope of the energy line in the river and g is the gravitational acceleration.

Numerical solutions for one- and two-dimensional cases have been derived (Starosolszky, 1987). Attempts have been made to develop numerically correct and stable solutions for three-dimensional flows. Analytical solutions are available for simplified boundary conditions, but the experimental coefficients should be estimated.

A special problem arises to forecast the concentration wave due to accidental water pollution (WMO, 1992). Warning systems for such conditions are necessary to prevent damage and health risk due to unexpected high concentrations at the water intakes.

The calculation of the concentration fields at outlets of sewage treatment plants, and a field survey of the effective mixing conditions is necessary. In certain cases without artificial mixing by special diffusers (point or line source), the required low level of the concentration cannot be solved. Standards for effluents can be useful measures against over concentration due to ineffective dilution of the contaminants.

The salt content of the discharges can be detected and the rate of mixing can be determined. If the salt content of the ambient or the effluent flow do permit the detection of the salt plume, artificial tracers can be used (Herschy, 1978). The concentration can be measured on the site or can be analyzed from water samples. The electrical conductivity of the water measured on the site can be the simplest detector of the salt content and the rate of mixing.

Water quality control in major rivers requires the expeditious determination of the mixing process, particularly in the vicinity of the confluences.

Ö. Starosolszky

Bibliography

Herschy, R., 1978. *Hydrometry, Principles and Practices*, John Wiley and Sons, Chichester.
Jolánkai, G., 1992. Hydrological, chemical and biological processes of contaminant transformation and transport in river and lake systems. IHP-IV. Project H-3.2. UNESCO, Paris.
Starosolszky, Ö., 1987. *Applied Surface Hydrology*, Water Resources Publications, Littleton, Colorado.
World Meteorological Organization, 1992. *Hydrological Aspects of Accident Pollution of Water Bodies*. Operational Hydrology Report, No 37. Geneva.

Cross references

Sanitation and clean water
Sewage treatment processes
Water resources: natural quality
Water resources: surface and groundwater

MODELING, OF WATER RESOURCES SYSTEMS

Introduction

Modeling of water resources systems is used extensively in the planning, development, management and operation of water resources systems. Whilst in the 1960s and 1970s there was a reluctance to translate systems analysis techniques from the research environment into practical use within water utilities and regulators, this reluctance has now been largely overcome as adequate data, more realistic models and improved computer facilities (in terms of both computational power and interface environments) have become more widely available. Nowadays, most water utilities and regulators have a whole range of techniques available to them, which often form part of sophisticated, integrated analytical packages of modeling techniques. These may often link in to remote data collection systems and telemetry and real-time forecasting systems. Others may be used more usually in an off-line mode, concentrating more on medium- to long-term operations, management and planning.

Data

Fundamental to the success of any modeling system is the need for data of adequate quality, whether this is to develop the model initially or to apply it in an operational context. Lack of data is an issue in the analysis of most water resources systems. This can either be because data is not measured at a site or it is not available with sufficient timeliness. For example, in a modeling system used in real-time for pump scheduling at a river abstraction site, river flow data available when downloaded from a logger 1 month later is unacceptable. In order to overcome the lack of data, modelers have to develop means of making up the shortfall, for example by deriving data from that collected at a nearby site or by regionalization. In other instances there is no option but to develop new data-gathering systems in conjunction with the modeling system. As bad as no data at all is poor quality-controlled data. This can especially be the case with tele-metered data. For the developer of a modeling system, it is important to identify when bad performance results from poor-quality data rather than any deficiency in the modeling itself. However, to a water resources manager, the cause of poor performance is immaterial, so there is a danger that the modeling system can face a credibility problem through no fault in the modeling.

Computational issues

Until recent years, a lack of computer power has held back the development of modeling systems and their use in day-to-day management of water resources systems. This was especially the case where systems were developed on PCs. Sometimes prototype systems work adequately in this environment. However, when extended up to the full-size problem, the performance of a PC may not be adequate. This issue is now less of an issue with computer technology developing such that PCs and workstations can have capabilities similar to those of mainframes in the 1970s. More relevant today is the need to have access to appropriate peripherals such as graphics screens and digitizing tablets and appropriate software tools and operating environment. Whilst the use of software tools and environments allows the more speedy and sophisticated development of modeling systems, they can create problems relating to transportability of software and forward compatibility. This is especially the case regarding graphics capabilities since there is still no unique graphics standard across the computer industry. These issues need to be considered in the development of any modeling system.

Successful implementation of water resources modeling

In developing a modeling system, there are several steps which a developer can take to ensure that implementation will be successful. Initially, formulation of the problem to be addressed should be jointly carried out by both the developer and the user of the system, taking into account not just the technical elements of the investigation, but also organizational issues such as the culture in which the modeling systems will ultimately rest. During development and implementation, there should be close liaison between the developer and user to ensure that any changes in requirements are quickly introduced, and so that the user gains familiarity in the use of the models and can gain confidence in the results being produced. This can be done by being able to compare results from the modeling system with those that would have obtained by traditional methods of analysis or where, if it is a new type of approach, the results are those that the user would have intuitively expected, or by using the model to recreate known past historic events. By vigorously testing a modeling system in parallel with development/implementation, the developer is able to ensure that the results produced are realistic and the solutions recommended are workable. This can only be done in close liaison with the user who is aware of the practical limitations of both the

water resources system and also the institutional framework in which the modeling system must work. This sort of approach also ensures that the modeling system provides answers to the questions being asked and at the appropriate level of complexity. By working together, the developer may be able to persuade the user that there may be other or more appropriate questions to be asked of the system. It is always easier to generate and discuss new ideas and developments round a partially working system rather than consider these issues in the abstract. Here, prototyping has role. However, the inherent dangers of producing a prototype which is valid for a small or simplified problem, yet is unsuited to the scale of the real working situation, should be considered.

Perhaps the most significant key to successful implementation is to ensure that the modeling system has easily understood results. It is essential that there is an appropriate user interface. What may be 'user friendly' to a novice user may be a source of frustration to an expert user. Ideally, the user interface should be matched with the users of the system, such that the same model is accessed by all users but the actual interface is different, depending not only on the expertise of the user but also the type of information required. For example, the information required for a board report would be rather different than that required by technical staff involved in operational control of a water resources system.

To ensure successful implementation, a modeling system should be flexible, so that the system is able to respond quickly and efficiently to the ever increasing variety of questions being asked of it. By providing more information, this in turn generates more potential uses of the system. Part of this flexibility lies in taking advantage of new modeling techniques as they become available.

Judging the success of a modeling system

A modeling system is likely to include two main components. First, a modeling tool will carry out the numerical analysis, for example simulate the operation of a water resources system. Second, an interface will allow the user to interact with the modeling tool by presenting it with data and instructions to initiate the run and by providing the results of the analysis, one hopes in a way which is easy for the user of the model to understand.

In judging the success or failure of the system, it is important to consider the achievement of success in both these areas. Volume of use of the system may not in itself be a good indicator of success. It is disconcerting to find that intrinsically sound models may not be well used if they do not have an acceptable user interface. Others are used extensively since their technical quality is sufficiently accepted by practicing engineers to outweigh any shortcomings in the user interface. More worrying is the fact that a scientifically poor model may achieve disproportionately high use purely because it has a good user interface.

An indicator of success of a modeling system is if it meets the user's needs. At its most basic, this may be judged by the user making no complaints to the developer and by there being a smooth implementation of the system culminating in its routine use in decision making. Success of a model is indicated by its accuracy, as measured by its ability to reproduce known expected results, its flexibility in being able to respond to changes in circumstances, and its robustness, but most of all by its acceptance by the organization using it. This requires a model to be not just good science, but also usable in a practical and timely way to meet user requirements.

Analytical techniques

A whole range of analytical techniques are available for the investigation of water resources systems. Many of these can be defined as 'modelling' or 'systems analysis' techniques. Such systems may be generalized or site- or problem-specific in approach. A generalized model is often costly and time consuming to develop initially, but is then easier and cheaper to apply to new problems in the future. In contrast, a site- or problem-specific model may be cheaper to develop for a first application, but requires a greater investment of effort to develop further models for each subsequent application.

Water resources modeling can be subdivided by type of application, for example models applied to the derivation of operating policies or used in new water resource development planning, or by analytical procedure, for example simulation compared with optimization using dynamic programming. The range of both analytical techniques and application areas are extensive, and therefore only brief examples are outlined below.

Feasibility studies

Water resources modeling based on simulation analyses are used extensively in feasibility studies of new or extended water resources developments. Investigations may be carried out using a site-specific model or a generalized model which allows the analysis of combination of reservoirs, pumping stations, river regulation control points, treatment works and aqueducts. Such models can be used to investigate future different resource alternatives to indicate the supply availability they provide, the impact on environmental indicators (for example, the pattern of reservoir levels and river flows) and operating costs. Sensitivity analysis can be carried out using simulation, such as to show the impact of changed assumptions regarding runoff into the system, together with demand profiles, operating policies and constraints placed on operation to protect the environment. An example of this approach is given in Smithers (1994).

Yield assessment

Assessment of yield and supply availability from different types of water resource are fundamental to the successful management of a water resources system. These assessments can involve the statistical generation of runoff data and probabilistic assessments of river flow and reservoir inflows. They may require the simulation of the water resources system. The output of such assessments is typically a yield–storage diagram which indicates the yield that can be supported by different sizes of useable storage in a resource system. A further assessment can be made of the average supply availability from the source and how this varies throughout the year and in relation to abstraction capacity. Such techniques are explained more fully in Walker (1991a).

Derivation of operating policies

Modeling is used extensively to derive operating policies for water resources systems. The models used range from the simple analysis of critical periods in the runoff record to more sophisticated approaches for complex, interconnected water resources systems that may involve optimization such as linear or dynamic programming techniques (Walker, 1991b). The objective function for optimization may be the minimization of operating costs, ensuring that a reliability criterion is met, or the maximization of the availability of flows for environmental purposes.

Such approaches typically use simulation analysis in support in order to test the robustness of the policy so derived and also to establish the impact of the policy on the environment, operating costs and risks to supply (Walker and Wyatt, 1989).

Budgetary control

Modeling can be used in the estimation and monitoring of financial budgets relating to the operation of water resources systems. Estimates of water demand over the coming financial year can be made and an appropriate operating policy introduced. By establishing the typical pattern of usage of the water resources, an estimate of the costs of operating the water resources system may be made. This can then be used to monitor against actual operation and any variance can be identified and explained.

Contingency planning

A key area in which modeling is used is in contingency planning and the operational management of water resources systems. An example of this is in drought management. Whilst these situations are, by their nature, rare, modeling techniques allow the management of such situations with the minimum impact on supply availability, whilst protecting the environment as far as is possible. Examples of such modeling techniques include risk assessment as a means of dynamically managing a drought (Walker et al., 1993) and the use of time of travel techniques (Beven et al., 1991) to manage pollution incidents effectively.

Water resources regulation

Water resources modeling can be used extensively in the regulation of water resources systems. Modeling can be used to mimic the behavior of an abstraction on a watercourse in order that the regulator can place appropriate conditions on the abstractor in order to protect the interests of the environment of other abstractors (Walker, 1993). Where overgenerous abstraction conditions have been applied in the past, to the detriment of the environment, modeling can be used to investigate more acceptable abstraction regimes in order that conditions on abstraction can be modified to the benefit of the environment.

Benefits of water resources modeling

By effectively using water resources modeling, the management of water resources systems may be improved by providing an appropriate balance between resource availability, risk and operating costs whilst meeting the reasonable demands for water supply, yet protecting the environment and any 'in-river needs'.

Susan Walker

Bibliography

Beven, K., Buckley, K.M. and Young, P.C., 1991. *ADZ – Analysis Manual*. CRES Report. TR 190, Lancaster University, Lancaster.
Smithers, H.A., 1994. Water resources management for an uncertain future in NW England, in *Advances in Water Resources Technology and Management*, eds Tsakiris, G. and Santos, M.A., Balkema, Rotterdam.
Walker, S. and Wyatt, T., 1989. The development and use of medium term policies for operation of a major regional water resource system. *Proceedings of the 2nd National Hydrology Symposium*, Sheffield, British Hydrological Society, Wallingford.
Walker, S., 1991a. The management of a major regional water supply network, in *Advanced Technology in Water Management*, Thomas Telford, London.
Walker, S., 1991b. Computer aided water resources planning and operational management in a water company in the UK, in *Decision Support Systems*, eds Loucks, D.P. and da Costa, J.R., Springer-Verlag, Berlin, pp. 305–330.
Walker, S., 1993. The systems approach to water resources management: a myth or reality, in *Information Technology for Civil and Structural Engineers*, eds Topping, B.H.V. and Khan, A.I., Civil Comp Press, Edinburgh.
Walker, S., Jowitt, P.W. and Bunch, A.H., 1993. Development of a decision support system for drought management within North West Water. *J. Instn. Wat. Environ. Management* **7**(2), 295–303.

Cross references

Mathematical models
Model prediction: uncertainty
Models: distributed models of catchment hydrology
Models: parameter estimation

MODEL PREDICTIONS: UNCERTAINTY

Uncertainty in model predictions results from many sources, including uncertainty in model structures, uncertainty in observations used as model inputs and boundary conditions, uncertainty in measurements of model outputs used in the calibration of parameter values and the nature of the calibration process itself. A very good review of the issues involved has been provided by Beck (1987) with specific relevance to water quality modeling. It is an area of growing interest, resulting from an increasing recognition of the limitations on predictive capability for many environmental systems, including hydrology. The proceedings of a symposium on new uncertainty concepts in hydrology and water resources (Kundzewicz, 1995) and a review of uncertainty estimation for hydrological modeling (Melching, 1992) have recently been published. A recent general review of the estimation of many different types of uncertainty intervals is given by Hahn and Meeker (1991). Predictive uncertainty is closely related to problems of parameter estimation and model sensitivity analysis (see Models: parameter estimation).

Until recently, very few hydrological studies attempted to estimate the uncertainty associated with their predictions, even where the statistical techniques for calculating uncertainty bounds or confidence limits in predictions were well established (such as for linear and multiple regression models). Similarly, values of hydrological quantities that depend on other measurements (such as hydraulic conductivities) have been rarely reported with any confidence limits or standard errors of estimation. The situation is worse in the case of predicted quantities based on non-linear hydrological models as the theory of uncertainty estimation for non-linear cases is not well developed. Resort must then normally be made to uncertainty estimation by simulation, using a variant on Monte Carlo simulation in which appropriately chosen random inputs are used to drive a model and the range of predictions is evaluated. In some cases, such as the estimation of the 100-year flood from a relatively short discharge record, the uncertainty in the estimate revealed in this way may be large and could have implications for risk analysis and decision-making processes. Freeze et al. (1990, 1992) have discussed a framework for including predictive uncertainty in decision analysis for subsurface contamination problems.

Traditional statistical approaches to uncertainty estimation have been based on an analysis of the variation in predictions around some best estimate, where that best estimate has been derived from fitting the parameters of a model. For linear models the theory is well developed but there is an assumption that the model, with its calibrated parameters, represents an optimal representation of the system within the limitations of the data available. Thus the uncertainty is estimated with respect to the variation in the model predictions around that optimal parameter set. Recent studies of hydrological models have revealed, however, that there may be many sets of parameters, across broad areas in the parameter space, that can give reasonable fits to the available data (e.g. Duan et al., 1992; Beven, 1993). This recognition, which has been called the equifinality problem by Beven (1993), introduces an additional aspect of uncertainty estimation that is only recently starting to be studied. An interesting perspective on this problem has been provided by Carrera (1993) who illustrates the limitations of current model conceptualizations for subsurface flow and transport prediction. Studies of multiple competing models are also starting in more traditional statistical theory (e.g. Draper, 1995).

This article summarizes briefly the main techniques in use for estimating uncertainty in model predictions, including first-order perturbation analysis, reliability analysis, stochastic differential equations, Monte Carlo techniques of different types, and fuzzy set theoretic methods.

First-order perturbation analysis

For the case of a smooth function or model $G(\theta)$ dependent on a vector of parameter values θ for which the covariance function is not large relative to any non-linearity in the model responses, approximate uncertainty bounds can be obtained by linearizing the parameter response surface using a Taylor series approximation (e.g. Benjamin and Cornell, 1970). Thus given some specific (perhaps 'optimum' in some sense) parameter vector θ^*, the mean and variance of a model output may be estimated from:

$$E[G(\theta)] = G^*$$

$$\mathrm{Var}[G(\theta)] = \sum_{i=1}^{N} \sum_{j=1}^{N} \frac{\partial G^*}{\partial \theta_i} \frac{\partial G^*}{\partial \theta_j} E[(\theta_i - \theta_i^*)(\theta_j - \theta_j^*)]$$

where G^* is $G(\theta)$ evaluated at θ^*, N is the number of parameters and the starred quantities are evaluated at the optimum. For linear models, it may be possible to evaluate the differential terms directly. This is not generally true for non-linear models when the differentials must be evaluated using small perturbations of each parameter in turn, requiring $(2N + 1)$ simulations. The expression for the variance is greatly reduced if the parameters can be considered to be independent, in which case:

$$\mathrm{Var}[G(\theta)] = \sum_{j=1}^{N} \left(\frac{\partial G^*}{\partial \theta_i}\right)^2 \sigma_i^2$$

Note that the covariance function on the parameters is usually assumed to be multivariate normal (perhaps after transformation of the parameters). The linearized equations then imply that the output quantity of interest will also be normally distributed. For non-linear models this latter assumption may not follow, even if it can be assumed that the uncertainty in the input parameters is normal.

Examples of such perturbation analyses have been given in rainfall-runoff modeling by Garen and Burges (1981), groundwater flow by Townley and Wilson (1985) and water quality modeling by Song and Brown (1990). Both Kuczera (1988) and Jones (1989) note the potential for error in the uncertainty estimates using this method for non-linear models when the input uncertainties are high.

A variation on first-order perturbation analysis is the method of Rosenblueth (1975), which has been used with a model of freezing soil by Guymon et al. (1981) and in rainfall-runoff modeling by Rogers et al. (1985) and Binley et al. (1991). In the Rosenblueth method, 2^N simulations are required in which each parameter is varied by plus or minus its estimated standard deviation.

First-order reliability analysis

The first-order reliability method (FORM) aims to improve on the methodology of perturbation analysis by changing the nature of the linearization by Taylor series. Rather than linearizing around central or optimal values of the parameters, the linearization point is varied to match the uncertainty quantile of the model output required. In that such a local linearization is concerned, the procedure may be less prone to error with non-linear models. However, this necessarily involves an iterative procedure to find the point required, and convergence of the procedure may be made more difficult by objective function surfaces in the parameter space that are locally not continuous or differentiable. Examples of the application of the FORM methodology can be found in rainfall–runoff modeling in Melching (1992) and in groundwater flow and transport in Cawlfield and Wu (1993).

Stochastic differential equations

Stochastic differential equation approaches aim to provide continuum descriptions for the evolution of the mean and standard deviation of quantities of interest, given information about the uncertainties or heterogeneities of parameter values and, in some cases, boundary conditions. To make this possible, they tend to require quite specific assumptions about the nature of the uncertainty or variability to allow closure of the system of equations at the second moment. This approach has been most often used in groundwater flow and transport problems (e.g. Dagan, 1986; Chrysikipoulous et al., 1990; Graham and McLaughlin, 1991; Cushman and Hu, 1995) and in water quality modeling (e.g. Zielinski, 1989; Curi et al., 1995). A stochastic analysis for unsaturated flow has been published by Yeh et al. (1985).

Monte Carlo methods

One method of uncertainty estimation that is very simple in concept, and can cope with non-linear models and high error variances, is the Monte Carlo method in which possible values of uncertain quantities are chosen randomly from specified distributions. Although in principle the Monte Carlo method is applicable to any uncertainty problem, in practice its use may be constrained by the computer demands of obtaining a large enough sample of the uncertain quantities to obtain good estimates of the uncertainty in the resulting predictions, particularly for any extreme quantities of interest. The sample required will increase significantly with an increasing number of uncertain input parameters or variables. A second limitation is properly specifying the distributions to use for the uncertainty in those parameters or variables and their interactions. The choice of one parameter value may need to be conditioned on the choice of another due to some real or apparent correlation between them. Simple distributions and correlation structures are readily handled in the Monte Carlo method. Analytical relationships for inversion of the cumulative distribution function, such that a random number is easily converted to a parameter or variable value, are helpful in this respect. The use of Monte Carlo methods in environmental modeling was pioneered in the generalized sensitivity analysis of Hornberger and Spear (1981) and by Gardner and O'Neill at the Oak Ridge National Laboratory (e.g. Gardner et al., 1980; O'Neill et al., 1982).

For a small number of uncertain variables (parameters or boundary conditions), the computer requirements may be reduced by treating the specified distributions in discrete form. The resulting latin hypercube technique then only evaluates model predictions at discrete increments along the variable axes, either regularly or irregularly spaced. One run is made for every combination of the increments.

Each prediction is then weighted by a probability density resulting from the product of the discrete probability densities associated with each variable, taking any correlations into account appropriately. If each variable is represented by at least ten increments, then the number of runs required by latin hypercube sampling grows as 10^N where N is the number of variables. The technique will normally therefore be limited to consideration of less than five variables, which would require 100 000 runs. This also gives an indication of the sampling problems associated with random sampling within the Monte Carlo procedure. Ten thousand runs may appear to be a great many, and will certainly give an indication of the range of behaviors to be expected from a model, but if five or more variables are being varied and the extremes of the model predictions are sensitive to extremes in the input variables being randomly chosen, the uncertainty in those extremes may not be well estimated.

Monte Carlo simulation can be used in estimating uncertainties of distributed models. In this case the number of parameter values that must be chosen randomly is generally large, but it may be possible to condition those choices on some observations. Delhomme (1979) and Clifton and Neumann (1982) give examples of the use of conditional simulation in groundwater simulation. Binley et al. (1991) have compared Monte Carlo and Rosenblueth uncertainty estimates for a low-parameter distributed model. Examples of the application of Monte Carlo methods in other related fields are Gilbert et al. (1995), Patwardhan and Small (1992) and Dilks et al. (1992) (see also the following sections).

Set membership methods

A significant problem with most of the approaches considered so far is that they depend upon the estimation of the parameters of the distributions of parameters or noise. This often requires simplifying approximations (such as restriction to Gaussian distributions of relatively small variances) that may be unrealistic. An alternative approach is based on the concept that the structure of the noise is unknown but is expected to be bounded. The identification problem then becomes one of identifying parameter sets that provide simulations which are consistent with the observations within the bounds set for the noise processes (see Keesman and van Straten, 1990, for an example within water quality modeling). The resulting ensemble of parameter sets will consist of those models considered to be possible or 'behavioral' simulators. The concept appears to have been first used in hydrologically related models by Hornberger and Spear (1981). The sample of parameter sets is usually chosen by a Monte Carlo procedure within some specified ranges. Following classification as behavioral or non-behavioural, each parameter set is given equal weight in estimating the range of the predictions, so that probabilistic measures of uncertainty are here replaced by the concept of possibility. Some refinements of this type of technique in which only sufficient runs are made to characterize the confidence region required are described by Klepper and Hendrix (1994). The set nature of this methodology is reflected in the name of set-theoretic or set-membership estimation that is sometimes used. The assumption of equal weights may be relaxed to be a fuzzy measure of set membership or a possibility measure (Klir and Folger, 1988; Bardossy and Duckstein, 1995).

Equifinality and uncertainty estimation

Recent studies involving extensive Monte Carlo explorations of the parameter space associated with particular models have revealed that parameter sets which produce reasonable simulations (given the limitations of both models and available data) may be widespread throughout the parameter space for many objective functions. There are a number of possible responses to this problem. One is to search for either a model structure (perhaps a reduced parameterization), objective function or parameter optimization technique that provides a more easily identified 'global optimum' parameter set. An alternative is to reject the idea that an optimal parameter set exists at all, but that all those parameter sets giving acceptable results are potential simulators of the future behavior of the system (unless there is some reason to believe otherwise, in which case it may be possible to reject some simulations on other grounds). There is some reason to believe that the latter strategy is both realistic and viable (see discussion in Beven, 1993), not least because of the different rankings of parameter sets that are often found when different periods of data are used in the calibration. The Monte Carlo-based generalized likelihood uncertainty

estimation (GLUE) technique of Beven and Binley (1992) provides one methodology for estimating the resulting predictive uncertainty in a way that allows Bayesian updating of parameter set likelihoods as more data or different types of data become available.

Uncertainty and model validation

The problem of model equifinality has obvious implications for concepts of model validation, which has been the subject of considerable recent discussion in the groundwater literature (e.g. Konikow and Bredehoeft, 1992; Oreskes et al. 1994). Oreskes et al. (1994) affirm that, within a critical rationalist philosophy, no model or theory can be ultimately verified or validated, but rather only falsified, and that the best that can be expected of models that perform acceptably (i.e. within some specified limits of uncertainty) in their predictions is a degree of confirmation of that model; that is, no more than to make it more probable. Confirmation is always a matter of degree of empirical adequacy (van Fraasen, 1980), and even that adequacy may be limited, requiring further tuning as more or different types of data become available. Such a view is consistent with the simulation of relative likelihood values or fuzzy possibility measures for different models or parameter sets where those deemed 'non-behavioral' are rejected. The estimation of such likelihoods is also consistent with a relativistic philosophical stance which does not require any necessary or strong correspondence between theory and reality. The latter view might be the more appropriate in most environmental modeling, since for most modeling problems of interest it is all too easy to falsify models on the basis of either their assumptions or their performance relative to observations. The modeling process is then saved by the adoption of less stringent criteria of acceptability or recourse to ancillary arguments which allow that it may not be possible to predict all the observations all of the time (arguments of scale, spatial heterogeneity, lack of time variability in parameter values, uncertainty in theoretical descriptions of the processes, etc).

Introducing the concept of falsifiability in this context, however, raises some interesting possibilities. It may be possible to design testable hypotheses and associated experiments that would allow model structures or parameter sets to be designated as non-behavioral, i.e. a certain class of models or parameter sets will be deemed falsified. The resulting studies might represent a very different approach from the experimental work associated with modeling work carried out at present in which the concern tends to be with the measurement of parameters or state variables, but at small (but manageable) scales. This may not be the most cost-effective approach to refining the likelihood associated with individual models and consequently to constraining the model uncertainty.

Keith Beven

Bibliography

Bardossy, A. and L. Duckstein, 1995. *Fuzzy rule-based modeling with applications to geophysical, biological and engineering systems,* CRC Press, Boca Raton, Florida, 232 pp.

Beck, M.B., 1987. Water quality modelling: A review of the analysis of uncertainty, *Water Resour. Res.,* **23**, 1393–1442.

Benjamin, J.R. and C.A. Cornell, 1970. *Probability, Statistics and Decision for Civil Engineers,* McGraw-Hill, New York, pp. 684.

Beven, K.J., 1993. Prophecy, reality and uncertainty in distributed hydrological modelling, *Adv. Water Resour.,* **16**, 41–51.

Beven, K.J. and A.M. Binley, 1992. The future of distributed models: model calibration and predictive uncertainty, *Hydrol. Process.,* **6**, 279–298.

Binley, A.M., K.J. Beven, A. Calver and L.G. Watts, 1991. Changing responses in hydrology: assessing the uncertainty in physically-based model predictions, *Water Resour. Res.,* **27**(6), 1253–1261.

Carrera, J., 1993. An overview of uncertainties in modelling ground-water solute transport, *J. Contam. Hydrol.,* **13**, 23–48.

Cawlfield, J.D. and M.-C. Wu, 1993. Probabilistic sensitivity analysis for one-dimensional reactive transport in porous media, *Water Resour. Res.,* **29**, 661–672.

Chrysikopoulous, C.V., Kitanidis P.K. and Roberts, P.V., 1990. Analysis of one-dimensional solute transport through porous media with spatially variable retardation factor, *Water Resour. Res.,* **26**, 437–446.

Clifton, P.M. and Neumann, S.P., 1982. Effects of kriging and inverse modelling on conditional simulation of the Avra River Valley in southern Arizona, *Water Resour. Res.*, **18**, 1215–1234.

Curi, W.F., Unny T.E. and Kay, J.J., 1995. A stochastic physical system approach to modelling river water quality, *Stochastic Hydrology and Hydraulics*, **9**, 117–132.

Cushman, J.H. and Hu, B.X., 1995. A resumé of non-local transport theories, *Stochastic Hydrology and Hydraulics*, **9**, 105–116.

Dagan, G., 1986. Statistical theory of groundwater flow and transport: pore to laboratory, laboratory to formation and formation to regional scale, *Water Resour. Res.*, **22**, 120S–134S.

Delhomme, J.P., 1979. Spatial variability and uncertainty in groundwater flow parameters: a geostatistical approach, *Water Resour. Res.*, **15**, 269–280.

Dilks, D.W., Canale R.P. and Maier, P.G., 1992. Development of Bayesian Monte Carlo techniques for water quality model uncertainty, *Ecological Modelling*, **62**, 149–162.

Draper, D., 1995. Assessment and propogation of model uncertainty, *J. Roy. Stat. Soc.*

Duan, Q., Sorooshian S. and Gupta, V. Effective and efficient global optimisation for conceptual rainfall–run-off models, *Water Resour. Res.*, **27**, 1253–1262.

Freeze, R.A., Massmann, J. Smith, L. *et al.*, 1990. Hydrogeological decision analysis. 1. A framework, *Ground Water*, **28**, 738–766.

Freeze, R.A., James, B., Massmann J. *et al.*, 1992. Hydrogeological decision analysis. 24. The concept of data worth and its use in the development of site investigation strategies, *Ground Water*, **30**, 574–588.

Gardner, R.H., Huff, D.D., O'Neill, R.V. *et al.*, 1980. Application of error analysis to a march hydrology model.

Garen, D.C. and Burges, S.J., 1981. Approximate error bounds for simulated hydrographs, *J. Hydraul. Div., ASCE*, **107**(HY11), 1519–1534.

Gilbert, R.O., Bittner E.A. and Essington, E.H., 1995. On the use of uncertainty analyses to test hypotheses regarding deterministic model predictions of environmental processes, *J. Environ, Radioactivity*, **27**, 231–260.

Graham, W.D. and McLaughlin, D.B., 1991. A stochastic model of solute transport in groundwater: application to the Borden, Ontario tracer test, *Water Resour. Res.*, **27**, 1345–1359.

Guymon, G.L., Harr, M.E., Berg, R.L. and Hromadka II, T.V., 1981. A probabilistic–deterministic analysis of one-dimensional ice segregation in a freezing soil column, *Cold Reg. Sci. Technol.*, **5**, 127–140.

Hahn, G.J. and Meeker, W.Q., 1991. *Statistical Intervals: A Guide for Practitioners*, Wiley, Chichester, 392 pp.

Hornberger, G.M. and Spear, R.C., 1981. An approach to the preliminary analysis of environmental systems, *J. Environ. Manag.*, **12**, 7–18.

Jones, L., 1989. Some results comparing Monte Carlo simulation and first order Taylor series approximation for steady groundwater flow, *Stochastic Hydrology and Hydraulics*, **3**(3), 179–190.

Keesman, K. and van Straten, G., 1990. Set membership approach to identification and prediction of lake eutrophication, *Water Resour. Res.*, **26**, 2643–2652.

Klepper, O. and Hendrix, E.M.T., 1994. A comparison of algorithms for global characterisation of confidence limit regions for nonlinear models, *Environ. Toxicol. Chem.*, **13**, 1887–1899.

Klir, G.J. and Folger, T.A., 1988. *Fuzzy Sets, Uncertainty and Information*, Prentice-Hall, New Jersey.

Konikow, L.F. and Bredehoeft, J.D., 1992. Groundwater models cannot be validated, *Adv. Water Resour.*, **15**, 75–83.

Kuczera, G., 1988. On the validity of first-order prediction limits for conceptual hydrologic models, *J. Hydrol.*, **103**, 229–247.

Kundzewicz, Z.W. (ed.), 1995. *New Uncertainty Concepts in Hydrology and Water Resources*, Cambridge University Press, Cambridge.

Melching, C.S., 1992. An improved first-order reliability approach for assessing uncertainties in hydrologic modelling, *J. Hydrol.*, **132**, 157–177.

O'Neill, R.V., Gardner R.H. and Carney, J.H., 1982. Parameter constraints in a stream ecosystem model: incorporation of *a priori* information in Monte Carlo error analysis, *Ecological Modelling*, **16**, 51–65.

Oreskes, N., Shrader-Frechette K. and Belitz, K., 1994. Verification, validation and confirmation of numerical models in the earth sciences, *Science*, **263**, 641–646.

Patwardhan, A. and Small, M.J., 1992. Bayesian methods for model uncertainty analysis with application to future sea-level rise, *Risk Analysis*, **12**, 513–523.

Rogers, C.C.M., Beven, K.J., Morris, E.M. and Anderson, M.G., 1985. Sensitivity analysis, calibration and predictive uncertainty of the Institute of Hydrology Distributed Model, *J. Hydrol*, **81**, 179–191.

Rosenblueth, E., 1975. Point estimates for probability moments, *Proc. Natl Acad. Sci. USA*, **72**(10), 3812–3814.

Song, B.Q. and Brown, L.C., 1990. DO model uncertainty with correlated inputs, *J. Environ. Eng., ASCE*, **116**, 1164–1180.

Townley, L. and Wilson, J.L., 1985. Computationally efficient algorithms for parameter estimation and uncertainty propagation in numerical models of groundwater flow, *Water Resour. Res.*, **21**, 1851–1860.

van Fraassen, B., 1980. *The Scientific Image*, Oxford University Press, Oxford.

Yeh, T.-C.J., Gelhar L.W. and Gutjahr, A.L., 1985. Stochastic analysis of unsaturated flow in heterogeneous soils, *Water Resour. Res.*, **21**, 447–471.

Zielinski, P.A., 1989. Stochastic dissolved oxygen model, *J. Environ. Eng., ASCE*, **114**, 74–90.

Cross references

Mathematical models
Modeling of water resources systems
Models: distributed models of catchment hydrology
Models: parameter estimation

MODELS: DISTRIBUTED MODELS OF CATCHMENT HYDROLOGY

Spatially distributed and probability distributed models

Distributed models of catchment hydrology attempt to model not only the output from a catchment area but also the variations in internal responses in space and time. There are two ways of trying to do this. The first is to try to model that variability in its correct spatial context using a model of one-, two- or three-dimensional components. A formulation for this type of distributed model, commonly referred to as a physically based distributed model, was first outlined by Freeze and Harlan (1969). Reviews of such models have been published by Freeze (1978) and Beven (1985). Examples are the SHE model (système hydrologique Européen, Abbott *et al.*, 1986) and the IHDM (Institute of Hydrology distributed model (Calver, 1988). This approach has the advantage that it can make direct use of spatial information about the catchment and its characteristics where it is available, and that any predictions are directly related to points in space. It has the disadvantage that the models tend to be computationally demanding and require large amounts of data files and parameter values to run.

The second approach attempts only to represent the variability in hydrological response in a probabilistic way. This has the advantage that the formulation may be less computationally demanding, since it is not necessary to predict the response for every point in the catchment; points that can be considered as responding in a hydrologically similar way can be treated in the same calculation. Examples of such models are TOPMODEL (Beven and Kirkby, 1979; Beven *et al.*, 1995) and the probability distributed model of Moore and Clarke (1981). The approach has the disadvantage that the model formulations tend to be more conceptual in nature since it is only possible to define hydrological similarity in an approximate or functional way. It is also more difficult to relate the predictions to specific points in space, although the topographic index of similarity used in TOPMODEL does allow a mapping of the predictions back into space (see below).

SHE model

Although three-dimensional formulations of distributed models of partially saturated flow domains have been used (e.g. Binley *et al.*, 1989), computational constraints have limited their application to small and generally hypothetical systems. For catchments of a scale for which predictions would be of more practical interest, it has to date been necessary to reduce the dimensionality of the solution to ease the computational requirements. Different models have done this in different ways. In the SHE model (Abbott et al., 1986), originally developed jointly by the Institute of Hydrology (UK), Danish Hydraulic Institute and SOGREAH (France), one-dimensional (vertical) unsaturated flow elements link two-dimensional (in plan) surface runoff and saturated-zone flow components. One-dimensional river channel elements complete the structure of the system. The different flow components are linked through common boundary conditions and fluxes. All the components are solved by means of finite difference approximations to the governing partial differential flow equations. The finite difference grid elements are generally square in plan, and in published applications have ranged in size from 250 m (Bathurst, 1986) in a catchment of less than 10 km^2 to 4 km on a side in a catchment of 1000 km^2 in India (Refsgaard et al., 1992; Jain et al., 1992). Soil and vegetation parameters can be varied from one grid element to another but must be assumed constant within each grid element.

Extensions of the SHE model (such as the MIKE SHE system of DHI and SHETRAN UK of the University of Newcastle) have been made by the original partners, including three-dimensional subsurface flow, sediment and solute transport components (e.g., Jensen et al., 1993; Bathurst and Purnama, 1991). These, of course, require additional parameter values to be specified, but make use of the flow simulation to provide information on distributed surface and subsurface velocities. Sophisticated graphical pre- and post-processing programs and user interfaces have been added to facilitate the input of parametric and boundary condition data and the interpretation of the very large quantity of results from the distributed simulations. Improvements in computer capabilities have meant that multiple-year catchment simulations can be run in a matter of hours on a dedicated workstation, although the grid sizes used still tend to be large. The cost of obtaining such a simulation is now increasingly dominated by gathering the data and measurements required to set up the model, rather than computer time.

TOPMODEL

TOPMODEL (Beven and Kirkby, 1979; Quinn and Beven, 1993; Beven et al., 1995) is an example of a probability distributed model, but one that is underlain by specific simplifying assumptions about the nature of the hydrological responses. In particular, it assumes that the saturated zone is always in a quasi-steady-state, with hydraulic gradients that can be approximated by the local surface slope and downslope soil transmissivities that are an exponential function of storage or water table depth. These assumptions allow a relationship to be built up between mean storage and local storage (or water table depth) that involves the local value of the topographic index ln(a/tan β) where a is the area draining through a point from upslope and tan β is the slope angle at that point. The ln(a/tan β) index then acts as an index of hydrological similarity and calculations are made using the distribution of the index within a catchment area which allows the prediction of variable saturated areas, either surface or subsurface, as the catchment wets and dries. The model is completed by simplified unsaturated zone and surface flow routing algorithms with a minimal number of parameters.

The nature of the basic assumptions means that the model is most appropriate to moderately sloping catchments with permeable layers that are shallow relative to slope lengths. The distribution of the index depends on slope forms within the catchment and can be calculated from digital terrain maps (e.g. Quinn et al., 1995). The index can be extended to take account of variability in soil transmissivities, the ln(a/T_0 tan β) index where T_0 is the transmissivity when the soil is just saturated, if such information is available. Other extensions have been proposed to allow different transmissivity profiles and hydraulic gradients that vary from the surface slope but can still be assumed constant such that deeper saturated zones can be treated in the same way.

The TOPMODEL concepts have now been applied to a variety of catchments in a range of different environments. One advantage of the approach is that because the distribution of the index is known in space, the predictions of the model can be mapped back into space for evaluation as to the realism of those predictions. This has, however, been done only in one or two studies, due to a general lack of spatial data sets on catchment responses (e.g. Burt and Butcher, 1986; Moore et al., 1988; Jordan, 1994), with varying degrees of success. The reproduction of spatially distributed observations will necessarily be approximate in such a model and the availability of such data poses some interesting questions in respect of parameter calibration which are discussed further below. The computational advantages of using a probability distributed model also mean that such models are much more easily used in a framework that considers uncertainty in the predictions. This generally requires many runs of the model with different parameter sets (e.g. Beven, 1993).

Problem of model structural definition

Both types of distributed models are associated with difficulties in model structural definition. Beven (1989a) has discussed the basis for the 'physically based' distributed models, such as SHE, and has suggested that the current generation of such models do not adequately represent the complexity of hydrological processes and that they should be considered to be lumped conceptual models at the model grid scale. There are processes such as preferential flow in unsaturated soil, which field studies suggest may be important in hydrological responses, but which are not represented in such models; indeed it may be impossible to do so in some physically based way due to the lack of knowledge of the soil structural parameters. Furthermore, even for those processes that are represented, it may be very difficult to estimate or measure the parameter values of the descriptive equations for each element of the discretization of the catchment used in the model. Such parameter values will also often change over time, exacerbating the parameter estimation problem. Since there may be multiple parameters for every grid element used in the spatial discretization of the catchment, a major problem of using this type of distributed model is that of parameter estimation and identifiability. Many of these parameters are, in principle, measureable in the field (although see the section below on scale problems) but this can clearly not normally be done for every parameter at every grid element. Thus, some parameter extrapolation or a priori estimation will be necessary. The origin of the parameters in physical process descriptions leads to an expectation that extrapolation or estimation techniques may be based on physical descriptions of soil, geology, vegetation and land-use characteristics. However, such an expectation remains a hope rather than a body of established techniques at the present time and it can be argued that it will always be necessary to resort to empirical, rather than theoretical, means of parameter estimation for such models.

A second problem, which cannot be considered completely solved, is the numerical problem of solving the descriptive equations of the model. These equations are non-linear partial differential equations for which there are no general proofs of stability and convergence of solutions. Solution techniques have improved dramatically over the last 20 years (e.g. Binley, 1992), but problems may still arise, even with the most sophisticated algorithms available today. Most distributed model codes have some statements to deal with the possibility of instabilities in the solution at occasional time steps, primarily induced when there are rapid and marked changes in the boundary conditions.

Probability distributed models were, in part, developed to offset some of the computational and parameter estimation problems of the physically distributed models. They are simpler on both counts, but as noted above rely on some theoretical or functional measure of hydrological similarity. A detailed conceptual description of catchment response based on probability distributions has been outlined by Beven (1989b), who shows that it is impossible to know all the distributions necessary given the limitations of current measurement techniques. Consequently, probability distributed models will always resort to some simpler conceptual analog of the variability in process response. In the model of Moore and Clarke (1981) this is a purely functional description with consequent advantages in parameter estimation. In TOPMODEL, as outlined above, it is based on a grossly simplified physical theory, but with a consequent reduction in the number of parameters to be estimated. There is, however, a resulting loss in the physical significance of those parameters (except perhaps those that can be directly related to lumped catchment responses, such as the recession curve) and in the potential for detailed spatial simulation. Moore (1984) has extended the probability distributed model to predict catchment sediment yield.

Scale problems: measurement techniques and model predictions

There is no doubt that some of the limitations of distributed models result directly from problems of scale and in particular from the differences in scale between the scale of measurements, the grid element scale at which parameter values are required in the model and the hillslope or catchment scale at which predictions are required. Most measurement techniques used in hydrology are point measurement techniques, sampling only a very small area or volume of the flow domain of interest. The continuum theory used in physically distributed models has been developed on the basis of such small-scale measurements and, nearly exclusively, tested in small-scale systems. When large numbers of such measurements are taken within even a grid element of a distributed model, great variability is revealed. This leads to an incommensurability between the variable and parameter values measured locally and the element scale variables and parameters of the model. There has not yet been a complete resolution of this problem, although some of the consequences are now being recognized. Scale dependence in observed conductivities and dispersion coefficients has been summarized for saturated subsurface flow by Neuman (1990) and Gelhar et al. (1992) and addressed theoretically by Dagan (1986). The problem is probably worse for unsaturated flow in structured soils and perhaps for routing surface runoff over uneven ground. The use of locally measured parameters to estimate the grid scale parameters of the model is really an act of faith such that variability is unimportant.

Problem of parameter estimation: using spatially distributed observations/remote sensing

An alternative approach is to determine parameters by calibration (changing the values to improve the fit between observed and predicted variables), using whatever data are available. Given the very large numbers of parameters that need to be calibrated in all physically distributed models, however, this process is fraught with difficulty. These models are totally overparameterized in a system identification sense, especially given the correlations and interactions to be expected between the effects of different parameters. Thus although such models have been fitted to discharge data alone in the past (e.g. Bathurst, 1986; Calver, 1988; Jain et al., 1992), the resulting parameter valuues should not be considered as in any way robustly estimated. In fact, it should be expected that there may be many sets of parameters that might give reasonable discharge simulations in calibration. These will, however, generally give different predictions when used for simulation (see Models: parameter estimation).

One way of ameliorating this situation is to add information into the calibration process. In particular, for a distributed model it would appear to be sensible to add distributed observations. It is surprising how rarely this appears to have been done in the past, not because of a lack of distributed models but much more because of a lack of spatial data sets. There is some evidence that adding spatial information does not necessarily help in distributed modeling, since the more that spatial observations are made available, the more complexity is revealed and the more difficult it is to model the variability in response (e.g. Stephenson and Freeze, 1974). Distributed measurements will reveal something of the nature of the local responses, but simulating those local responses will require local variations in parameter values. Thus increasing the number of measurements increases the dimensionality of the parameter space. This problem has been recognized for a long time in groundwater modeling where the 'inverse' problem of parameter calibration has been the subject of considerable research (see Models: parameter estimation). The problem is much worse when the additional nonlinearities of unsaturated and surface water flow are added.

The ultimate source of distributed data is from remote sensing. There has been some hope expressed in the past that remote sensing techniques might ease the problems of applying the distributed model (Moran et al., 1994). Experience to date, however, is not very promising. Certainly some catchment characteristics (topography and vegetation classification) can be obtained by remote sensing at scales of the order of the elements of a distributed model. The potential for determination of hydrological variables, however, has been limited to relatively coarse resolution of rainfall intensities using radar (with continuing calibration problems; e.g. French and Krajewski, 1994; Lord et al., 1995); surface soil moisture in relatively unvegetated terrain (e.g. Schmugge et al., 1994; Cognard et al., 1995) and the more indirectly useful thermal infrared measurement of vegetation or ground surface temperature which can be used in evapotranspiration estimation (e.g. Bastiaanssen et al., 1994). This is, however, an area of active research.

Uncertainty in the predictions of distributed models

From the discussion above it will be recognized that there are many potential sources for error in setting up and calibrating distributed models that have to do with limitations in the model process representations, spatial heterogeneity and scale problems in the determination of parameters, and problems in matching observed variables to those predicted by the model. For all these reasons it must be expected that the predictions of distributed models will be subject to uncertainty. There have been very few studies of the magnitude and nature of such uncertainty, in part due to the computational demands of distributed models, particularly the physically distributed models. It is, however, increasingly recognized that it may not be adequate to expect that parameter values can be specified accurately at every grid element. At best it may be possible to estimate some feasible range of values. Binley et al. (1991) for example, in a limited study, used Monte Carlo simulations of the IHDM implemented on a parallel computer, to evaluate the range of uncertainty in predictions of the effects of land-use change. This work was later extended by Beven and Binley (1992) using observed hydrographs to modify the uncertainties within a Bayesian framework. More detailed studies have been made using TOPMODEL (Beven, 1995). In each case, it has become clear that there may be many sets of parameter values that give reasonable simulations of observed discharge responses, undermining the idea that physical reasoning alone may be adequate to define parameter values.

Keith Beven

Bibliography

Abbott, M.B., J.C. Bathurst, J.A. Cunge et al., 1986. An introduction to the European Hydrological System – Système Hydrologique Européen 'SHE'. 1: History and philosophy of a physically based distributed modelling system, J. Hydrol., **87**, 45–59.

Bastiaanssen, W.G.M., D.H. Hoekman and R.A. Roebling, 1994. *A methodology for the assessment of surface resistance and soil water storage variability at mesoscale based on remote sensing measurements*, IAHS Special Publication No. 2, IAHS Wallingford.

Bathurst, J.C., 1986. Physically-based distributed modelling of an upland catchment using the Système Hydrologique Européen, J. Hydrol., **87**, 79–102.

Bathurst, J.C. and A. Purnama, 1991. Design and application of a sediment transport modelling system, in *Sediment and Stream Water Quality in a Changing Environment: Trends and Explanation*, IAHS Pub. No. 203, IAHS, Wallingford, pp. 305–313.

Beven, K.J., 1985. Distributed models in M.G. Anderson and T.P. Burt (eds), *Hydrological Forecasting*, Wiley, Chichester, pp. 405–435.

Beven, K.J., 1989a. Changing ideas in hydrology: the case of physically-based models. J. Hydrol., **105**, 157–172.

Beven, K.J., 1989b. Interflow, in H.J. Morel-Seytoux (ed.), *Unsaturated Flow in Hydrologic Modeling*, Kluwer, Dordrecht, 191–219.

Beven, K.J., 1993. Prophecy, reality and uncertainty in distributed hydrological modelling, *Adv, Water Resour.*, **16**, 41–51.

Beven, K.J. and M.J. Kirkby, 1979. A physically-based variable contributing area model of basin hydrology, *Hydrol. Sci. Bull.*, **24**, 43–69.

Beven, K.J. and A.M. Binley, 1992. The future of distributed models: model calibration and predictive uncertainty. *Hydrol. Process.*, **6**, 279–298.

Beven, K.J., R. Lamb, P. Quinn et al., 1995. TOPMODEL in V.P. Singh (ed.) *Computer Models of Watershed Hydrology*, Water Resource Publications, Colorado.

Binley, A.M., K.J. Beven and J. Elgy, 1989. A physically-based model of heterogeneous hillslopes. II. Effective hydraulic conductivities, *Water Resour. Res.*, **25**, 1227–1233.

Binley, A.M., K.J. Beven, A. Calver and L.G. Watts, 1991. Changing responses in hydrology: assessing the uncertainty in physically-based model predictions, *Water Resour. Res.*, **27**(6), 1253–1261.

Burt, T.P. and D.P. Butcher, 1986. Development of topographic indices for use in semidistributed hillslope runoff models, in D.

Baltenau and O. Slaymaker (eds), *Geomorphology and Land Management, Zits, Geomorph. Suppl. Band 58*, pp. 1–19.

Claver, A., 1988. Calibration, sensitivity and validation of a physically-based rainfall-runoff model, *J. Hydrol.*, **103**, 103–115.

Cognard, A.-L., C. Loumagne, M. Normand *et al.*, 1995. Evaluation of the ERS1/synthetic aperture radar capacity to estimate surface soil moisture: two-year results over the Naizin watershed, *Water Resour. Res.*, **31**, 975–982.

Dagan, G., 1986. Statistical theory of groundwater flow and transport: pore to laboratory, laboratory to formation and formation to regional scale, *Water Resour. Res.*, **22**, 120S–134S.

Freeze, R.A., 1978. Mathematical models of hillslope hydrology, in M.J. Kirkby (ed.), *Hillslope Hydrology*, Wiley, Chichester, pp. 177–226.

Freeze, R.A. and R.L. Harlan, 1969. Blueprint for a physically-based digitally simulated hydrologic response model, *J. Hydrol.*, **9**, 237–258.

French, M.N. and W.F. Krajewski, 1994. A model for real-time quantitative rainfall forecasting using remote sensing (2 parts), *Water Resour. Res.*, **30**, 1075–1097.

Gelhar, L.W., C Welty and K.R. Rehfeldt, 1992. A critical review of data on field-scale dispersion in acquifers, *Water Resour. Res.*, **28**, 1955–1974.

Jain, S.K., B. Storm, J.C. Bathurst *et al.*, 1992. Application of the SHE to catchments in India. 2. Field experiments and simulation studies with the SHE on the Kolar subcatchment of the Marmada river. *J. Hydrol.*, **140**, 25–47.

Jensen, K.H., K. Bitsch and P.J. Bjerg, 1993. Large-scale dispersion experiments in a sandy aquifer in Denmark: observed tracer movements and numerical analysis. *Water Resour. Res.*, **29**, 673–696.

Jordan, J.-P., 1994. Spatial and temporal variability of stormflow generation processes on a Swiss catchment. *J. Hydrol.*, **153**, 357–382.

Lord, M.P., C. Young and R.C. Goodhew, 1995. Adaptive radar calibration using raingauge data, in K.A. Tilford (ed.) *Hydrological uses of weather radar*, British Hydrological Society Occasional Paper No. 5, pp. 10–25.

Moore, I.D., G.J. Burch and D.H. Mackenzie, 1988. Topographic efects on the distribution of surface soil water and the location of ephemeral gullies, *Trans. Am. Soc. Agric. Eng.*, **31**, 1098–1107.

Moore, R.J., 1984. A dynamic model of basin sediment yield, *Water Resour. Res.*, **20**(1), 89–103.

Moore, R.J. and R.T. Clarke, 1981. A distribution function approach to rainfall–runoff modelling, *Water Resour. Res.*, **17**(5), 1367–1382.

Moran, M.S., T.R. Clarke, W.P. Kustas *et al.*, 1994. Evaluation of hydrologic parameters in a semi-arid rangeland using remotely sensed data, *Water Resour. Res.*, **30**, 1287–1298.

Neuman, S.P., 1990. Universal scaling of hydraulic conductivities and dispersivities in geologic media, *Water Resour. Res.*, **26**, 1749–1758.

Quinn, P.F. and K.J. Beven, 1993. Spatial and temporal predictions of soil moisture dynamics, runoff, variable source areas and evapotranspiration for Plynlimon, mid-Wales, *Hydrol. Process.*, **7**, 425–448.

Quinn, P.F., K.J. Beven and R. Lamb, 1995. The ln(a/tanβ) index: how to calculate it and how to use it in the TOPMODEL framework, *Hydrol. Process.*, **9**, 161–182.

Refsgaard, J.C., S.M. Seth, J.C. Bathurst *et al.*, 1992. Application of the SHE to catchments in India. 1. General results, *J. Hydrol.*, **140**, 1–23.

Schmugge, T., T.J. Jackson, W.P. Kustas *et al.*, 1994. Push broom microwave radiometer observations of surface soil moisture in Monsoon'90, *Water Resour. Res.*, **30**, 1321–1328.

Stephenson, G.R. and R.A. Freeze, 1974. Mathematical simulation of subsurface flow contributions to snowmelt runoff, Reynold's Creek Watershed, Idaho, *Water Resour. Res.*, **10**, 284–298.

Cross references

MODELS: PARAMETER ESTIMATION

The problem of model parameter estimation is always associated with a prior problem of model structure choice or identification (e.g. Beck *et al.*, 1990, for their discussion of identifiability in water quality modeling). The methods available depend on whether a linear or non-linear model structure is chosen. The theory of parameter estimation is well developed for linear systems but not for non-linear systems. However, although most hydrological systems are essentially non-linear, much has been achieved using the theories of linear systems analysis.

The essence of the parameter estimation problem is that, given some time series of measurements for the inputs (U) and outputs (Y) of the system of interest, and given some chosen model $G(U, \theta)$ where θ are parameters of the model, it is necessary to identify values of θ so as to minimize some criterion of error (index of goodness of fit, objective function or loss function) in reproducing the observations Y. The problem may be extended in some problems by the addition of additional observations on the boundary conditions of the flow, additional state variable observations to be reproduced, or multiple error criteria to be satisfied.

The most common error criteria used in hydrological modeling are variants on the sum of squared errors, S, defined by

$$S = \sum_{}^{N} \mathbf{e}^{T}(\mathbf{U}, \boldsymbol{\theta})\mathbf{W}^{-1}\mathbf{e}(\mathbf{U}, \boldsymbol{\theta})$$

Where e is a vector of errors dependent on the model $G(U, \theta)$, W is a matrix of weighting coefficients, and N is the number of time steps. If W is defined to be the identity matrix, then this criterion is identical to the leastsquare criterion and similar measures such as the modeling 'efficiency' of Nash and Sutcliffe (1970) defined by

$$E = 1 - \frac{\frac{1}{N}\sum^{N}\{e(U, \theta)\}^2}{\sigma_Y^2}$$

where σ_Y^2 is the variance of the observed outputs. The efficiency criterion has the advantage of a range between zero, where the model is no better than assuming that the mean output is known, and 1, when the model fit is perfect (negative values are also possible, but indicate a very poor fit). It is worth noting that these criteria based on the squared errors can be sensitive to timing errors between observed and predicted responses when the gradient of the response is steep, as in the rising limb of hydrographs. They also give weight to the largest errors, which tend to be around hydrograph peaks in rainfall–runoff modeling. If there is more interest in the prediction of longer-term recessions, then an alternative criterion such as the sum of the absolute errors might be more appropriate.

There are many other criteria that can be chosen, including those derived from maximum likelihood theory that stem from different assumptions about the nature and structure of the residuals (e.g. Sorooshian *et al.*, 1983). Once the choice of criterion is made, it is then traditional to proceed to estimate the 'optimum' parameter set, that is, the set which minimizes the error criterion. For linear models there are well-established techniques for finding the optimum and estimating the uncertainty in the evaluation of the parameter values (e.g. Young, 1984). Recursive methods, which step through the series updating the parameters at each time step, allow the changes in parameter values (and their estimation variances) to be followed and give some indication of the robustness of the parameter estimation for that model. Recursive estimation can be put to good effect in the initial study of non-linear systems by allowing the parameters of a linear model to be time variable using either the extended Kalman filter (EKF) or the less restrictive instrumental variable (IV) algorithms (Young, 1984; Beck, 1987). The IV approach requires much fewer assumptions about the nature of the uncertainties associated with the model parameters and residuals. Time-variable parameter estimation can give a good indication of the nature and strength of the non-linearity in the system response (e.g. Young and Beven, 1994).

In general linear systems analysis it may not always be clear how complex a model to use. This is the problem of model structure identification, but is closely related to the parameter identification problem since, in general, the more complex a model, the more parameters will need to be identified and the more degrees of freedom the model will have in fitting the data. Thus, although more complex models will commonly give better fits to the data, parameter estimation will often result in parameter values that have greater estimation variances. The error criteria defined above are not adequate to decide between competing models in this respect, and it is recommended that additional criteria, such as the Young information criterion (YIC; Young, 1989), which combines goodness of fit with a function of the estimation variances of the parameter values, should also be calculated. The YIC has a logarithmic scale and is normally negative for a well-fitting, well-estimated model. If a model has poorly estimated parameters, the value of the YIC will tend to grow rapidly, indicating that the model may be overparameterized.

For non-linear models, such as most rainfall–runoff models used in hydrology, there is no body of theory for parameter estimation. Early approaches were based on trial-and-error methods in which successive parameters are changed in value and the results compared with observations, either visually or with the help of quantitative error criteria. Parameters will be varied on the basis of physical reasoning or experience in use of the model. Such methods are still in use for models that are computationally expensive, limiting the number of runs that can be made (e.g. Bathurst, 1986). However, as soon as computers became more widely available in the 1960s studies of automatic optimization techniques were undertaken. These techniques are based on the use of multiple runs of the model to minimize the chosen error criterion by searching through the parameter space in some random or structured way. Examples of such 'hill-climbing' algorithms that do not require the calculation of function derivatives (difficult for most hydrological models) and which have been used quite widely include the Rosenbrock (1960) algorithm; the simplex algorithm of Nelder and Mead (1965) and the Powell algorithm (Powell, 1970). Example code for a number of different methods is provided by Press et al. (1989).

Early work showed that conceptual hydrological models based on combinations of linear and non-linear storage elements with threshold parameters controlling their responses were not well suited to automatic optimization techniques (e.g. Ibbitt and O'Donnell, 1974). Problems encountered include local minima in the error criterion response surface in parameter space, insensitive areas where the search procedure cannot easily find a way towards an optimum, and 'valley' areas resulting from interaction between parameters where many combinations of values may give similar degrees of fit, so that again it may be very difficult to find some global optimum. A good summary of the application of automatic optimization to lumped rainfall–runoff models is provided by Blackie and Eeles (1985). The attempt to produce a parameter estimation methodology that is robust to such problems has continued, with new methods, such as simulated annealing (e.g. Press et al., 1989; Tarantola, 1987), genetic algorithms (Wang, 1991; Ritzel et al., 1994) and shuffled complex evolution (Duan et al., 1992) providing new strategies for optimization procedures.

The parameter estimation problems of distributed models are considerably greater (see Models, distributed models of catchment hydrology) since the sheer number of parameter values involved is much greater. Such models are overparameterized in any systems identification sense, although this problem might be mitigated to some extent by basing parameter values on local measurements. The identification of distributed model parameter values has been extensively studied in the field of groundwater modeling where it is known as the inverse problem. A useful review of approaches to the inverse problem is given by Yeh (1986). Although saturated groundwater systems can be considered to be quasi-linear in their responses, it is clear that similar identification problems arise primarily because there may not be sufficient observational data available to distinguish adequately between different possible models or parameter sets. A recent study of modeling the Birmingham aquifer by Brooks et al. (1994), for example, based on optimal parameter searches from multiple starting points, concluded that the best two solutions found were virtually identical in terms of performance but were from very different parts of the parameter space.

The near equivalence of many different hydrological models and parameter sets in reproducing the available observations has been highlighted by Beven (1993), who suggests that the concept of an 'optimal' parameter set should be rejected and replaced by multiple competing models, each of which is associated with some likelihood of being an acceptable simulator of the system of interest. Such likelihoods can be based on error criteria (Beven and Binley, 1992; Romanowicz et al. 1994) and can be updated within a Bayesian framework as more data become available. The likelihoods can be the basis for estimating uncertainty in the predictions (see Model

predictions: uncertainty). Such an approach requires that many simulations are required both in calibration and prediction, but allows for the fact that sets of parameter values from different parts of the parameter space may provide equally good simulations. A similar approach based on a set membership framework has been outlined by Keesman and van Straten (1990; see Model predictions: uncertainty). They illustrate very nicely some of the problems of reproducing observations within the expected error of the measurements in a water quality modeling example. It is undoubtedly true that model structural errors continue to dominate errors in the prediction of response of environmental systems, resulting from lack of both theory and measurements at the appropriate scales of interest.

Keith Beven

Bibliography

Bathurst, J.C., 1986. Physically-based distributed modelling of an upland catchment using the Système Hydrologique Européen, J. Hydrol., **87**, 79–102.

Beck, M.B., 1987. Water quality modelling: A review of the analysis of uncertainty, Water Resour. Res., **23**, 1393–1442.

Beck, M.B., F.M. Kleiseen and H.S. Wheater, 1990. Identifying flowpaths in models of surface water acidification, Rev. Geophys., **28**, 207–230.

Beven, K.J., 1993. Prophecy, reality and uncertainty in distributed hydrological modelling, Adv. Water Resour., **16**, 41–51.

Beven, K.J. and A.M. Binley, 1992. The future of distributed models: model calibration and predictive uncertainty, Hydrol. Process., **6**, 279–298.

Blackie, J. and C. Eeeles, 1985. Lumped catchment models, in M.G. Anderson and T.P. Burt (eds), Hydrological Forecasting, Wiley, Chichester, pp. 311–345.

Brooks, R.J., D.N. Lerner and A.M. Tobias, 1994. Determining the range of predictions of a groundwater model which arises from alternative calibrations, Water Resour. Res., **30**, 2993–3000.

Duan, Q., S. Sorooshian and V. Gupta, 1992. Effective and efficient global optimisation for conceptual rainfall–runoff models, Water Resour. Res., **27**, 1253–1262.

Ibbitt, R. and T. O'Donnell, 1974. Designing conceptual catchment models for automatic fitting methods, IAHS Pubn., **101**, 461–475.

Keesman, K. and G. van Straten, 1990. Set membership approach to identification and prediction of lake eutrophication, Water Resour. Res., **26**, 2643–2652.

Nash, J.E. and J.V. Sutcliffe, 1970. River flow forecasting through conceptual models. 1. A discussion of principles, J. Hydrol., **10**, 282–290.

Nelder, J.A. and R. Mead, 1965. A simplex method for function minimisation, Computer J., **7**, 308–3.

Powell, M., 1970. A survey of methods for unconstrained optimisation, SIAM Rev., **12**, 79–97.

Press, W.H., B.P. Flannery, S.A. Teukolsky and W.T. Vetterling, 1989. Numerical Recipes: The Art of Scientific Computing, Cambridge University Press, Cambridge.

Ritzel, B.J., J.W. Eheart and S. Ranjithan, (1994) Using genetic algorithms to solve a multiple objective groundwater pollution containment problem, Water Resour. Res., **30**(5), 1589–1603.

Romanowicz, R., K.J. Beven and J. Tawn, 1994. Evaluation of predictive uncertainty in nonlinear models using a Bayesian approach, in V. Barnett and K.F. Turkman (eds), Statistics for the Environment II. Water Related Issues, Wiley, Chichester, pp. 297–317.

Rosenbrock, H.H., 1960. An automatic method of finding the greatest or least value of a function, Computer J., **3**, 175–184.

Sorooshian, S., V.K. Gupta and J.L. Fulton, 1983. Evaluation of maximum likelihood parameter estimation techniques for conceptual rainfall–runoff models: influence of calibration data variability and length on model credibility, Water Resour. Res., **19**, 251–259.

Tarantola, 1987. Inverse Problems.

Wang, Q.J., 1991. The genetic algorithm and its application to calibrating conceptual rainfall–runoff models, Water Resour. Res., **27**(9), 2467–2472.

Yeh, G.T., 1986. Review of parameter identification procedures in grouundwater hydrology – the inverse problem, Water Resour. Res., **22**, 95–108.

Young, P.C., 1984. *Recursive Estimation and Time-Series Analysis*, Springer-Verlag, Berlin, 300 pp.

Young, P.C., 1989. Recursive estimation, forecasting and adaptive control, in C.T. Leondes (ed.), *Control and Dynamic Systems*, Vol. 30, Academic Press, San Diego, pp. 119–165.

Young, P.C. and K.J. Beven, 1994. Data-based mechanistic modelling and the rainfall–flow nonlinearity, *Econometrics*, **5**, 335–363.

Cross references

Mathematical models
Modeling of water resources systems
Model predictions: uncertainty
Models: distribution models of catchment hydrology

MONSOON CLIMATES

The monsoons, large-scale seasonal winds, blow in response to the annual change in the difference in atmospheric pressure over land and sea that, in turn, results from the difference in temperature between land and sea. Where great continents border an ocean, temperature differences are large.

Summer versus winter circulation

When the Sun moves north of the equator in the northern hemisphere summer, the land mass of Asia, because of its relatively low heat capacity, is rapidly warmed. On the other hand, the northern Indian Ocean and the western Pacific Ocean store the Sun's heat within their deep surface layers. Consequently, as the land gives off heat more readily than the sea, air over the land becomes warmer and air pressure lower than over the neighboring ocean.

Thus, during summer; air flows from the Indian Ocean towards lower pressure over southern Asia, ascending as it is heated over the land until it reaches a level at which the pressure gradient is reversed, whereupon it flows on a return trajectory from land to sea. There it descends and is once more taken up by the landward-directed pressure gradient (Figure M10). As long as the land is significantly warmer than the sea this great circulation persists.

In winter the reverse occurs. The low heat capacity of Asia relative to the northern Indian and western Pacific Oceans ensures that air over the land is colder than over the sea. We observe, then, the typical winter monsoon in which, at low levels, air flows out from the continent over the sea where it rises and returns in the middle and higher layers of the troposphere to the land, sinks to the surface and resumes the cycle.

Monsoons over Africa have a somewhat different character from those over Asia. During the northern hemisphere summer, the desert areas of North Africa heat rapidly and pressure there falls in the same way as over Asia. South of the equator over Africa, during the southern hemisphere winter, cooling occurs, thus establishing a pressure gradient across Africa from south to north, which in turn sets up a massive flow of air, also from south to north, across the equator.

Because the deflecting force due to the Earth's rotation reverses its direction at the equator and is weak in equatorial regions, air flow is more directly from high to low pressure than is the case in higher latitudes. The influence of Africa on the atmospheric circulation extends 800 km east of the continent and merges with the influence of Asia farther north. In the northern hemisphere summer, wind circulates in a huge gyre from the southeast around the northern edge of the South Indian Ocean anticyclone and toward the coast of Africa near the equator, swinging into the south and then southwest to parallel the African, Arabian and Asian coasts, and finally sweeping across India, Burma and the Indochina–Thailand peninsula as the southwest or summer monsoon.

Six months later a complex reversal takes place. Northern Africa is cold and southern Africa is warm and so the winds blow from the north across the equator in the western Indian Ocean.

Climate in the vicinity of Australia is less intensely monsoonal. Although monsoon tendencies have been identified in many other regions, for example Texas, the Caspian Sea, and even parts of Europe, they by and large reflect annual variations in the tracks of moving surface pressure systems. Such systems are rare in the monsoon region proper, which extends from western Africa to

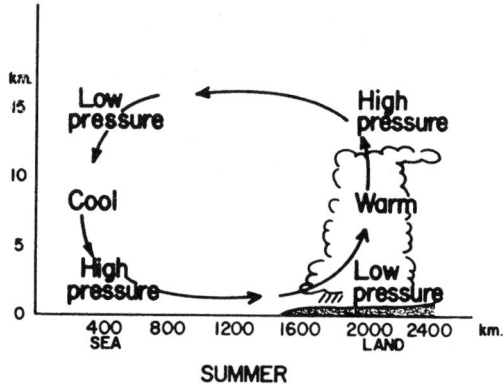

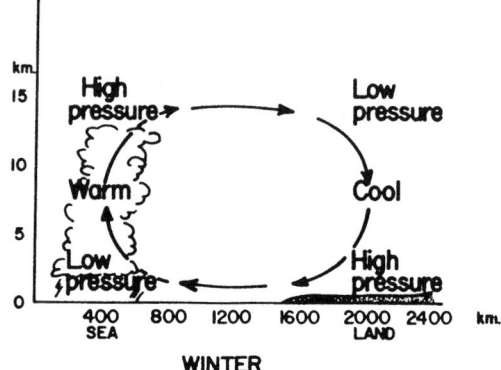

Figure M10 Schematic representation of the vertical circulations associated with the summer and winter monsoons.

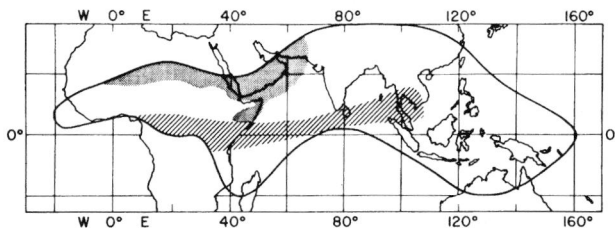

Figure M11 The area satisfying monsoon criteria in enclosed by the solid line. Within this area, deserts (annual rainfall < 250 mm) are stippled and the region with a bimodal annual rainfall variation is hatched. Except for a small portion of eastern peninsular India the remainder of the monsoon area experiences a summer or fall rainfall maximum.

Indonesia with southward protrusions to Madagascar and northern Australia (Figure M11).

Along the west coasts of northern and southern Africa and South America, equatorward-flowing air streams force cool water to upwell throughout the year. This, by ensuring a persistent seaward-directed temperature gradient, prevents monsoon development.

In the monsoon region a near-equatorial band possesses a double rainfall maximum (Figure M11) that is probably associated with temporary intensification of near-equatorial troughs during the transition seasons. Elsewhere in the region, more than 75% of the rain falls in summer or fall.

Monsoon modifications

In northwest India and west Pakistan, even the moist onshore winds of summer fail to disrupt the desert-like climate. Together with the deserts of Arabia and North Africa, this stems from the thermal–mechanical effect of the Himalaya Mountains–Tibet Plateau, causing

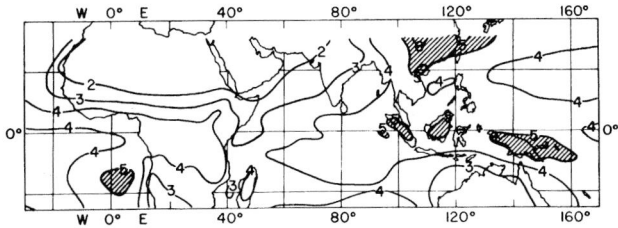

Figure M12 Mean annual cloudiness (in eighths of sky covered) based on 3 years of weather satellite photographs.

upper tropospheric convergence, thereby forcing air to subside throughout the year west of about 70°E. The opposite effect causes persistent cloudiness over China (Figure M12). Between 70°E and 100°E, at the longitude of India, subsidence prevails in winter and rising motion in summer with consequent enormous rainfall differences between the two seasons.

The chain of mountains extending from Turkey to west China, by protecting the land to the south from cold polar outbreaks, is also responsible for a sharp discontinuity in surface temperature characteristics along about 100°E. Thus in summer in the South Asian monsoon circulation is stronger (more intense heating) than the East Asian, whereas in winter the East Asian monsoon is the stronger (more intense cooling; Table M4).

The general character of the monsoons and their inter-regional variations reflect juxtaposition of continents and oceans. However, without the great mechanical and thermal distortions produced by the Himalayas and Tibet, the vast northern hemisphere deserts would be less desert-like, and central China would be much drier and no colder in winter than India.

The monsoons, as modified by the Himalaya Mountains–Tibet Plateau, set the stage for weather. Over and near the continents, the cold dry air of the winter monsoon usually predisposes weather to be unsettled. Nevertheless, despite the fact that during the monsoons moving surface disturbances are rare and surface winds are notably steady, rainfall, on scales from days to weeks, is surprisingly variable. The cause lies largely in synoptic changes brought about by moving disturbances that are strongest a few kilometers above the surface, or by intensification and decay of quasi-stationary bad weather systems.

March of the seasons

During spring and fall, the transition seasons between winter and summer monsoons, complex changes take place both in the large-scale circulation and in the types of synoptic systems. The changes differ within the monsoon area and also from one year to the next. The spring transition lasts about twice as long as the fall transition, because in spring the normal equator–pole global temperature gradient is being reversed, whereas the fall marks the return to the normal gradient. As might be expected, since the monsoons arise from the differing heat capacities of land and ocean, circulation in a transition season changes first in the surface layers. After several weeks the change extends throughout the troposphere and the transition season is over.

In the Indian Ocean–South Asia region, an east–west-oriented trough exists throughout the year in the tropics of each hemisphere. In the northern hemisphere summer the trough lies well away from the equator over southern Asia and there may be a very weak secondary trough close to the equator (Figure M13). As the year progresses from summer to winter, the primary trough weakens and dissipates and the secondary trough becomes predominant. The sequence reverses from winter to summer. This model satisfactorily removes two difficulties inherent in the old concept of a single trough undergoing an annual

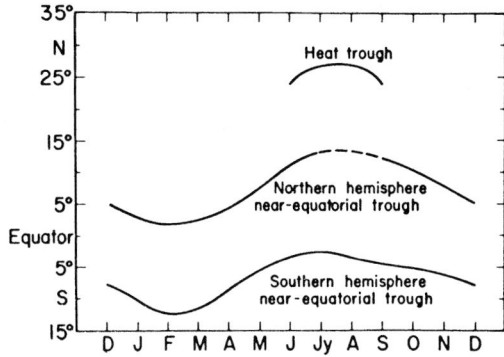

Figure M13 Annual latitudinal variation of lower tropospheric pressure troughs over the Indian Ocean.

oscillation between the summer tropics of both hemispheres – apparently discontinuous latitudinal jumps, and the usual absence of double rainfall maxima between 10 and 25°N. The transition season jumps are caused by rapid cooling in the fall and heating in spring of the deserts of northwest India and Pakistan.

South of the equator, despite considerable day-to-day fluctuations, a single trough follows the annual march of the sun.

Over Africa, a weak heat trough is sometimes found between 5–10° latitude in each hemisphere during the transition seasons. Then, as the year advances, the trough in the winter hemisphere disappears and the trough in the summer hemisphere intensifies and follows the sun poleward. The sequence then reverses through the following transition season.

The north Indian Ocean–South Asian regime is weakly duplicated in the northern Australian region, whereas over East Asia, absence of a vigorous heat trough leads to a relatively regular annual movement of the trough.

Over the ocean, troughs near the equator spawn weakly cyclonic bad weather systems; farther from the equator, tropical cyclones may develop in the troughs. When the midsummer monsoon trough of northern India extends into the northern Bay of Bengal, monsoon depressions develop there. Over land, west of 70°E, fine weather prevails in the troughs (heat troughs), but on their equatorward sides the weather is unsettled.

C.S. Ramage

Bibliography

Arakawa, H. (ed.), 1969. *Climates of Northern and Eastern Asia.* Amsterdam: Elsevier.
Chang, J.-H., 1967. The Indian summer monsoon, *Geog. Rev.*, **57**, 373–396.
Chang, J.-H., 1971. The Chinese monsoon, *Geog. Rev.*, **61**, 370–395.
Gentilli, J. (ed.), 1971. *Climates of Australia and New Zealand.* Amsterdam: Elsevier.
Griffiths, J.F. (ed.), 1972. *Climates of Africa.* Amsterdam: Elsevier.
Krishnamurti, T.N. (ed.), 1977. Monsoon meteorology, *Pure Appl. Geophys.*, **115**, 1087–1529.
Lighthill, J. and R.P. Pearce, 1981. *Monsoon Dynamics.* Cambridge: Cambridge University Press.
Ramage, C.S., 1971. *Monsoon Meteorology.* New York: Academic Press.
Ramage, C.S. and C.R.V. Raman, 1972. *International Indian Ocean Expedition Meteorological Atlas*, Vol. 2, *Upper Air.* Washington, DC: National Science Foundation.

Table M4 Mean monthly air temperature (°C) at Hong Kong (22°18'N; 114°10'E; 33 m above sea level) and Calcutta (22°32'N; 88°20'E; 6 m above sea level)

	Jan.	Feb.	Mar.	Apr.	May	June	July	Aug.	Sept.	Oct.	Nov.	Dec.	Year
Hong Kong	15.4	15.1	17.4	21.3	25.1	27.3	27.8	27.6	27.0	24.4	20.8	17.2	22.2
Calcutta	19.5	22.0	27.0	30.1	30.3	29.7	28.9	28.7	28.7	27.5	23.3	19.5	26.3

Rao, Y.P., 1976. *Southwest Monsoon*. New Delhi: India Meteoro-logical Department.
Thompson, B.W., 1965. *The Climate of Africa*. Nairobi: Oxford University Press.
van Loon, H. (ed.), 1983. *Climates of the Oceans*. Amsterdam: Elsevier.
Webster, P.J., 1981. Monsoons, *Sci. Am.*, August, 108–118.
Yoshino, M.M., 1971. *Water Balance of Monsoon Asia*. Tokyo: University of Tokyo.

Cross references

Asia: climate
Climate and climate change
Climate data: sources
Precipitation distribution

MULTI-PARAMETER DATA LOGGERS

The functional characteristics of multi-parameter data loggers can be partitioned into measurement, storage and control, and for many loggers, the telemetry of hydrological data. These three functions are reflected in the architecture of data loggers. As the name implies, multi-parameter data loggers are designed to integrate data from two or more measurement subsystems with a storage-and-control subsystem. The data logger must interact with other outside influences, such as sources of electrical power, the hydrological environment itself, data displays, and operators who may be initializing or making routine contact with the subsystem.

The function of a hydrological measurement subsystem is to sense a characteristic of water, and convert it to data form suitable to be displayed, recorded or processed. For example, mechanical water level measurements may be accomplished by use of a float to drive a pen on a recorder chart, or punch a paper tape, while microelectronic systems generate an electrical signal. The mechanical water level device may also have a display that allows the current value of water level to be observed directly by a visiting observer. More recently developed subsystems utilize other measurement technologies.

Storage-and-control subsystems of multi-parameter data loggers accept signals from two or more measurement subsystems, and store these signals in a form for later retrieval, analysis or telemetry. These signals may be relayed continuously or at fixed or irregular intervals of time. The command to transfer data may come from either side of the interfaces between subsystems. The communication of data across the interfaces must be clearly defined for each subsystem, and they must be compatible.

Many modern storage-and-control subsystems can perform complex analysis of data in real time and use such analyses to compute derived information, compact data or institute some action. For example, some subsystems can collect data on a rapidly changing condition, such as wind speed and direction (a highly variable set of parameters), and compute and store statistical data, rather than discrete data values.

The subsystem may take some automatic control action based on the value of the data received from the measurement technology. Modern subsystems are capable of initiating control signals to the measurement subsystem and causing additional data to be collected, or initiating control signals to the telemetry subsystem, resulting in the initiation of a warning or alert message to be transmitted.

Additionally, some multi-parameter data loggers that are equipped with telemetry can have their modes of operation remotely altered by interrogation through the telemetry subsystem.

Hydrological telemetry subsystems also consist of three elements, which are remote-site equipment, a communications medium, such as telephone lines or radio-communications links, and central receiving stations. The remote-site equipment is considered to be the multi-parameter data loggers defined earlier. The following discussion concentrates on the remote-site telemetry subsystem.

In some configurations, two-way communications between a remote hydrological station and a central receiving station has been provided in the system design. In other cases, the system may be only designed for one-way communications from the remote site to the central receiving station. In the former case, the remote station is generally interrogated and commanded to transmit its data. In the latter case, the remote site initiates a transmission after a specified elapsed time or because hydrological data have exceeded some threshold condition. A transmission after an elapsed time may occur after a fixed or random time interval.

Current hydrological telemetry subsystems rely on microwave, radio or telephone lines for communications. In microwave communication, the transmission is line of sight, while radio transmission may be line of sight or relayed via an intermediate medium. This medium may be a terrestrial relay link or Earth-orbiting satellites.

In telemetry subsystems there is requirement for the remote-site system to meet the communications standards of the communications medium. For example, the characteristics of particular grades of telephone lines can only support certain speeds of data communications and the telemetry subsystem must conform to these speeds. Similarly, the use of a satellite relay for telemetry of data requires the remote-site system to broadcast data within precisely defined limits of emitted power, bit rate and other communications standards. These of course are dictated by the operator of the satellite.

The key attributes of multi-parameter data loggers are their hardware, software and physical attributes of size, weight and electrical power.

The microprocessor, circuitry and other physical components of multi-parameter data loggers are known as hardware. The key component of microelectronic hardware is the microprocessor. The earliest microprocessors that were developed commercially were able to process four or eight bits of information simultaneously, and were known as 4-bit or 8-bit microprocessors, respectively. Subsequently, 16-bit and 32-bit microprocessors have been introduced.

Microprocessors in multi-parameter hydrological data loggers must be provided with a carefully defined sequence of instructions (software) to dictate the logger's operations. These instructions define many facets of the internal operation of the system, as well as how the microprocessor operates with other elements of hardware. Software determines how the microprocessor maintains track of time, how and at what frequency it provides data to data storage devices or to the telemetry subsystem, and a myriad of other tasks. Programming of the operation of the logger may be accomplished by means of a detachable device or by switches or a keyboard designed as integral parts of the storage-and-control subsystem.

Multi-parameter data loggers have become small and lightweight compared to the traditional hydrological data-collection instrumentation that they have replaced. Because of their small size and low electrical power requirements, they are usually battery operated and can be packaged in small environmentally protected cases. Many have displays which permit a visiting technician or hydrologist to assess the status of the logger and to review data that have been collected.

Source

World Meteorological Organization, 1994. *Guide to Hydrological Practices*, 5th edn, WMO, Geneva.

N

NATURAL DISASTERS

Some 300 million people annually are affected by natural disasters and probably 3 million people have lost their lives over the past 25 years. Most of the human losses and sufferings are geophysical, meteorological and hydrological in character or are weather related. Disasters caused by tropical cyclones (hurricanes and typhoons), floods, droughts, tornadoes, thunderstorms, squalls, lightning and hail are examples of weather-related events as are storm surges, snow storms, heat waves, landslides and avalanches.

Tropical cyclones in the western north Pacific and hurricanes in the Americas account for an annual average of about 20 000 deaths and some \$6 billion in damage. In 1993 the Philippines had a record number of 32 tropical cyclones and the Yangtze River floods in 1991 destroyed over 4 million dwellings while the damage caused by the 1993 Mississippi floods is estimated at over \$10 billion.

Table N1 Total number of events for each type of natural disaster, 1967–1991 (Obasi, 1994)

Type	Number	Percentage of total
Weather events		
Cyclone, hurricane, typhoon	894	15.1
Flood	1358	23.0
Storm	819	13.9
Cold and heat wave	133	2.3
Drought	430	7.3
Associated with weather events		
Avalanche	29	0.5
Landslide	238	4.0
Fire	729	12.4
Insect infestation	68	1.1
Famine	15	0.3
Food shortage	22	0.4
Epidemic	291	4.9
Geological		
Earthquake	758	12.8
Volcano	102	1.7
Tsunami	20	0.3

The statistics indicate that extreme meteorological and hydrological events account for 62% of all events recorded as natural disasters. If those associated with weather events are included, the percentage rises to 85%.

Table N2 Total number of people killed by each type of natural disaster, 1967–1991 (Obasi, 1994)

Type	Number killed
Weather events	
Cyclone	846 240
Hurricane	15 139
Typhoon	34 684
Flood	304 870
Storm	54 500
High wind	13 904
Cold and heat wave	4 926
Drought	1 33 728
Associated with weather events	
Avalanche	1 237
Landslide	41 992
Fire	81 970
Insect infestation	0
Famine	605 832
Food shortage	252
Epidemic	124 338
Geological	
Earthquake	646 307
Volcano	27 642
Tsunami	6 390

About 3.5 million people were killed by meteorological and hydrological events.

Droughts have persisted in many parts of Africa with catastrophic results, but at the same time floods persisted in other parts of the world.

Changing climate must be considered of major concern in the future, especially with respect to floods and droughts.

Table N1 presents the total number of events for each type of natural disaster over the period 1967–1991; Table N2 gives the total number of people who have lost their lives by each type and Table N3 shows the total number of people affected by each type over the same period. Table N1 shows that extreme meteorological and hydrological events account for some 85% of all events. Over the same period Tables N2 and N3 show that 3.5 million people lost their lives by meteorological and hydrological events, while some 2.8 billion were affected by them.

R.W. Herschy

Table N3 Total number of people affected by type of natural disaster, 1967–1991 (Obasi, 1994)

Type	Number affected
Weather events	
Cyclone	80 485 116
Hurricane	6 028 833
Typhoon	63 321 930
Flood	1 057 193 110
Storm	68 122 680
High wind	2 960
Cold and heat wave	71 000
Drought	1 426 239 250
Associated with weather events	
Avalanche	500 000
Landslide	3 603 580
Fire	814 341
Insect infestation	446 000
Famine	12 950 000
Food shortage	28 320 267
Epidemic	5 791 234
Geological	
Earthquake	42 943 009
Volcano	1 938 279
Tsunami	918

About 2.8 billion people were affected by meteorological and hydrological events.

Source

Obasi, G.O.P., 1994. WMO's role in the International Decade for Natural Disaster Reduction, *Bulletin of the America Meteorological Society*.

Cross references

Dams: failure
Drought
Drought in Western Europe, 1988–1992
Drought management
Droughts
Flood

NORTH AMERICA: CLIMATE

The climate of North America ranges from the frost-free tropical of southernmost Florida to the perennial ice and snow of the northern-most islands of the Canadian Archipelago, from the rain-drenched mountains of the northwest coast to the drought-ridden deserts of the southwestern United States. Within the limits set by these extremes is a variety of climates that results from the interplay of atmospheric systems with the complex geography of this large landmass. Its great size (about 25 million km^2), the span of latitudes from about 25°N to 80°N, the configuration and alignment of its shorelines, plus the distribution of major landform regions have generated a diversity of climates that very closely resemble those of the much larger continent of Eurasia within the same latitudes.

The scattering of persistent ice fields in the Canadian Arctic from the northern Ellesmere southward to Baffin Island may be viewed as a fragmented extension of the north polar ice cap that dominates most of Greenland's elevated mass to the east and the surface of the Arctic Ocean to the north. These ice fields are only a negligible component of North America's climatic array, but it is a reminder that past climates have permitted continental glaciers to spread over much of North America during earlier geological epochs.

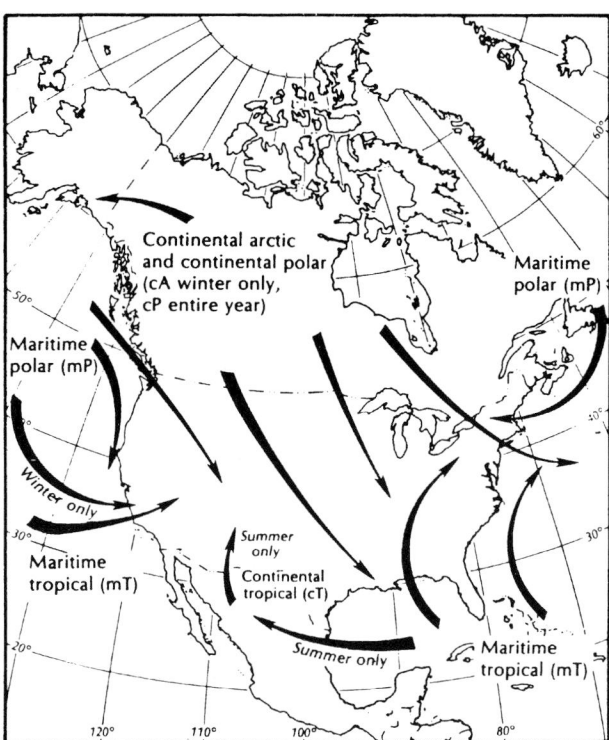

Figure N1 Source regions and paths of air masses influencing the North American continent.

Organization rationale

To deal adequately with the climates of North America in a limited number of pages requires careful selection of the material presented. Given the vast amount of climatic literature pertaining to the continent, it is possible to take a number of approaches to describe North American climates. The approach selected here is a descriptive appraisal using climate–vegetation relationships. This emphasis permits a rational description of the existing climates without undue analysis of causal factors.

One approach that assesses climates by cause draws upon air mass climatology (e.g. Oliver and Hidore, 1983). In this approach, source regions are identified (Figure N1) and their characteristics are related to prevailing conditions (Table N4). Air mass climatology is examined in related articles in this volume. Similarly, reasons for the distribution of temperature (Figure N2) and precipitation (Figure N3) and other climatic elements are also dealt with in other entries. It is also possible to delineate the climates according to a formal classification scheme, such as that of Köppen, and use the identified regions as the basis for description. This approach has been used in a number of textbooks (e.g. Trewartha and Horn, 1980). In all, a number of possible organizational bases exist, and that selected here represents but one approach. Figure N4 identifies the location of the regions described below.

Tundra climate

Northernmost of the continent's major climates is the seasonally ice-free tundra, a region of rolling, virtually treeless plains stretching eastward from the shores of the Bering Sea to the coast of Labrador. Long, severely cold winters alternate with short, cool summers. Sunless darkness prevails for several weeks before and after the winter solstice around 22 December. Lowest temperatures are normally reached in February, except near the sea after oceanic ice has attained its maximum extent, where March is the coldest month. February averages between −15 and −30°C. From time to time, values drop to −50°C or lower. At this season, precipitation diminishes over the interior and falls mainly in the form of fine, light, dry snow, brought by the fairly rapid passage of small, intense frontal storms from the North Pacific. Occasionally, when clear, transparent

Table N4 Characteristics of North American air masses (after Lutgens and Tarbuck, 1982)

Air mass	Source region	Temperature and moisture characteristics in source region	Stability in source region	Associated weather
cA	Arctic basin and Greenland ice cap	Bitterly cold and very dry in winter	Stable	Cold waves in winter
cP	Interior Canada and Alaska	Very cold and dry in winter Cool and dry in summer	Stable entire year	Cold waves in winter Modified to cPk in winter over Great Lakes bringing 'lake-effect' snow to leeward shores
mP	North Pacific	Mild (cool) and humid entire year	Unstable in winter Stable in summer	Low clouds and showers in winter Heavy orographic precipitation on windward side of western mountains in winter Low stratus and fog along coast in summer; modified to cP inland
mP	Northwestern Atlantic	Cold and humid in winter Cool and humid in summer	Unstable in winter Stable in summer	Occasional 'northeaster' in winter Occasional periods of clear, cool weather in summer
cT	Northern interior Mexico and southwestern USA (summer only)	Hot and dry	Unstable	Hot, dry and clear, rarely influencing areas outside source region Occasional drought to southern Great Plains
mT	Gulf of Mexico, Caribbean Sea, western Atlantic	Warm and humid entire year	Unstable entire year	In winter it usually becomes mTw moving northward and brings occasional widespread precipitation or advection fog. In summer, hot and humid conditions, frequent cumulus development and showers or thunderstorms
mT	Subtropical Pacific	Warm and humid entire year	Stable entire year	In winter it brings fog, drizzle and occasional moderate precipitation to NW Mexico and SW United States In summer it occasionally reaches western USA, providing moisture for infrequent conventional thunderstorms

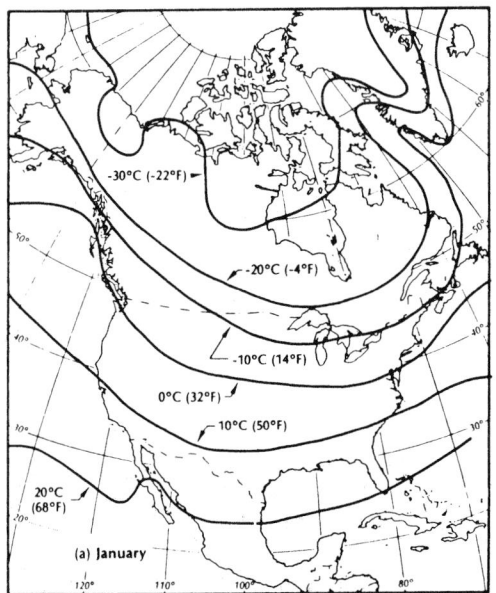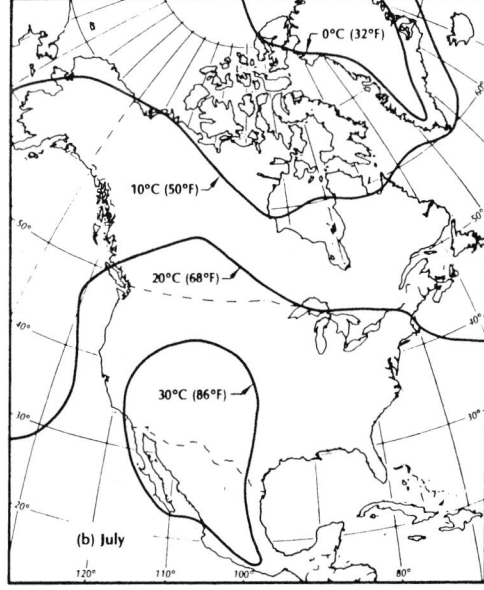

Figure N2 Mean January and July temperatures over North America.

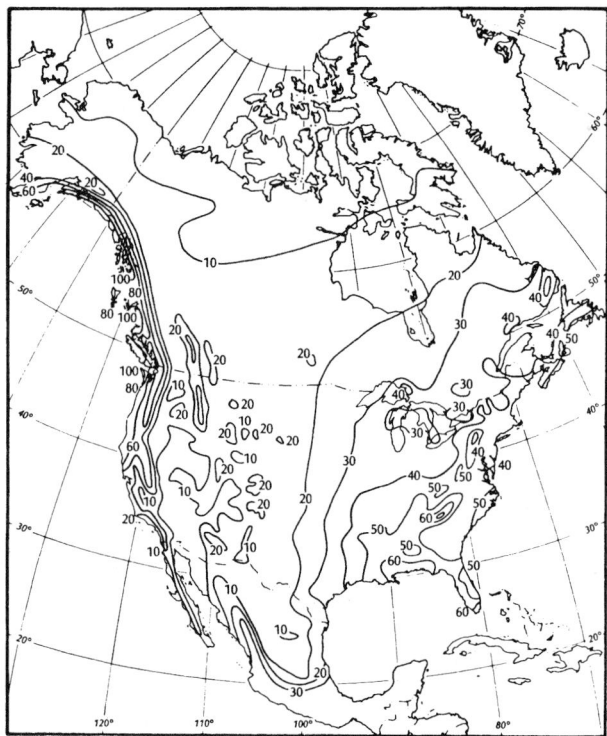

Figure N3 Distribution of precipitation in North America. (From Barry and Chorley, 1982.)

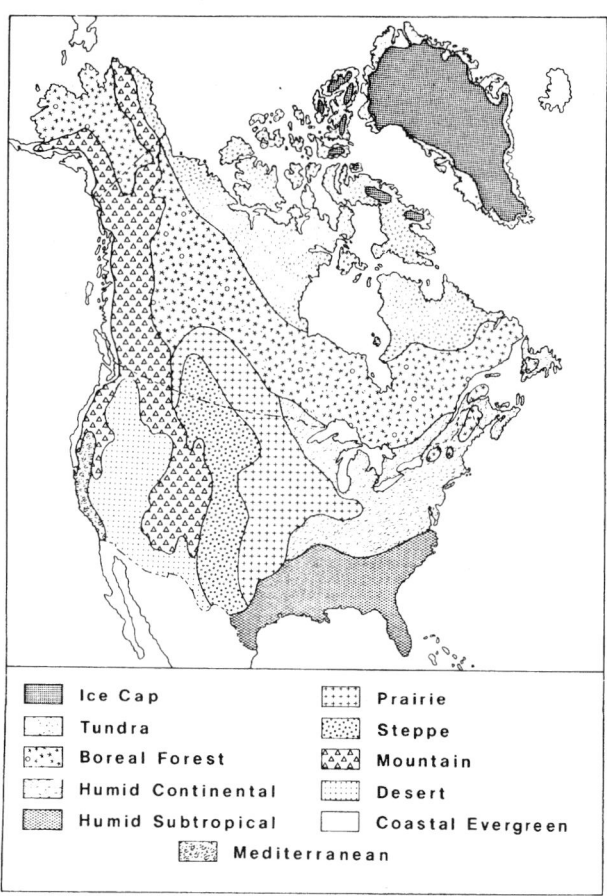

Figure N4 Climates of North America.

air is calm and intensely cold, showers of minute ice crystals fall from the cloudless sky. Normal yearly precipitation throughout the tundra is usually less than 250 mm, although coastal stations may average up to 500 mm or more.

Summer is a season of prolonged daylight when temperatures average above freezing for 3–4 months at most stations. The warmest month is commonly July when readings are above 5° but not above 10°C. Comparatively high temperatures are not unknown at numerous inland sites where values exceeding 30°C have been reached. However, it is the sporadic occurrence of sub-freezing temperatures from June to September that characterizes the luminous summer of the tundra. This is the season of maximum precipitation for most of the region, mainly falling as light, steady drizzling rain of frontal origin. Thunderstorms are rare. By midsummer, landscapes are released from the grip of frost, streams flow, lakes and ponds are ice free, and the low-lying plant life engages in the year's brief period of vegetative growth. Very few plant species tolerate the lengthy dormancy of winter and the short, cool, frost-threatened growing season. Mosses, lichens, a few flowering herbs and ligneous shrubs that spread laterally close to the surface, plus isolated clumps of hardy spruces, a few meters high in sheltered depressions, are the commonest plant forms of this severe climate.

Boreal forest climate

Along the southern margins of the tundra, treeless barrens give way gradually and unevenly to the boreal forest, also called the subpolar forest or subarctic forest. This is the northernmost forest climate and one in which only a very few species of coniferous trees, chiefly fir, spruce and larch are hardy enough to thrive despite the long, intensely cold winters and to form the dominant plant associations. A few robust, relatively short-lived broadleaf species of birch, poplar, cherry, and alder, along with evergreen jackpine, appear in transitory abundance throughout much of the forest.

Taiga, as the region is also called, stretches eastward from the Bering Sea to the Gulf of St Lawrence. Extending up Alaska's Yukon Valley, it spreads thinly across the lower slopes of the Canadian Rockies, reaching southward to the upper Great Lakes and in

separated segments to the Maritime Provinces and northern New England.

As in the tundra, temperature is the controlling climatic element. It is this climate that offers the widest annual temperature range in North America, inflicting the greatest thermal stress on all forms of life. The length and severity of winter are noteworthy characteristics. Deep cold settles over the land from October through to April. January is usually the coldest month when normal temperatures average below −12°C, and temperatures often plunge to less than −50°C. In northwest Canada, interior stations have recorded values below −60°C. The high frequency of anticyclonic weather is chiefly responsible for lengthy, intensely cold periods, when the calm, dry, crystal-clear atmosphere settles over the land, permitting rapid radiative heat loss to proceed from the surface with exceptional effect.

Winter is the season of diminished precipitation when most frontal disturbances move rapidly over the surface at about twice the speed of summertime, yielding snow that is light, fine and dry. Winter storms increase in frequency from west to east. The incidence of snowfall rises from about 50 days per year in the Yukon Valley to over 100 days in northern Quebec and Labrador. Mean yearly snowfall increases from around 1000 mm in central Alaska to over 5000 mm in easternmost Labrador.

Summer in the taiga brings a growing season of about 75 days. Although frost may occur sporadically, mean monthly temperatures are above 10°C for 3–4 months. July is commonly the warmest month with average readings above 13° but usually not above 18°C. Extreme temperatures in excess of 38°C have been recorded at some inland stations, although prolonged hot spells do not occur. This is the season of maximum precipitation when about two-thirds of the normal yearly amounts are received, except along the north shore of the Gulf of St Lawrence, which records a winter maximum. Most summer precipitation is produced by slow-moving frontal disturbances, yielding light, steady rains. Convective showers occur from time to time during invasions of warm, humid air from the south. Thunderstorms may develop on from 5 to 10 days of the summer period.

Midlatitude mixed forest climates

In eastern North America the thermally restrictive taiga yields to the more moderate forest climates of the midlatitudes. A yearly surplus of precipitation supports perennial lakes, streams and marshes and a self-regenerative mixed forest dominated by broadleaf deciduous trees. Some 200 species of prominent broadleaf and needleleaf trees indicate the transitional position of these climates between the species-poor taiga and species-rich forests of the tropics. Transitional also is the character of atmospheric behavior over this part of North America, where the interplay between synoptic systems of lower latitude origin and those of the higher latitudes is almost continuously in progress. Higher-latitude systems dominate in winter, lower-latitude systems in summer, resulting in four well-defined seasons within which the yearly program of weather events unfolds.

Southward from the Great Lakes–St Lawrence Valley to the Gulf of Mexico much climatic variety is encountered due to the influence of major water bodies and larger relief features such as the Adirondacks, Appalachians and the Ozarks on transient atmospheric systems. But two major climates are commonly identified, humid continental extending southward from the taiga approximately to the Chesapeake Bay, the Great Smokies and the Ozarks, beyond which humid subtropical reaches the Gulf of Mexico and the shores of peninsular Florida. The four seasons that characterize humid continental areas become much less distinct farther south, a result mainly of increasing nearness to the equator. Annual possible sunshine increases from about 40% in southeastern Canada to over 65% in Florida: mean daily solar radiation rises from less than 300 langleys in the north to more than 450 in the south, and the growing season lengthens from under 80 days to more than 350 days. The main seasonal change from north to south is in the length and severity of winter. January temperatures in the Great Lakes–St Lawrence Valley area average between −10 and −20°C and the thermometer averages below freezing for 4 or more months. South of the Chesapeake and the Great Smokies the coldest winter months remain substantially above freezing and January temperatures in southern Florida are between 18 and 20°C. Snow falls in northernmost areas on more than 80 days each winter and normal amounts for the season exceed 2500 mm, whereas throughout the subtropical region snow is expected on fewer than 10 days and does not normally exceed 100 mm.

Mean annual precipitation increases southward from about 800 mm near the taiga to over 1600 mm along the Gulf Coast and in southern Florida. Greater atmospheric moisture content is partly responsible for this, plus the increased vigor and frequency with which turbulent systems occur. Thunderstorms account for a much larger percentage of yearly precipitation, occurring on fewer than 30 days per year in the north but on over 100 days in southwest Florida. Thunderstorm activity reaches a maximum between June and October. This is also the peak season of tropical cyclones, which account for 15% or more of annual precipitation along the Gulf Coast and the Atlantic seaboard south of New England. Many tropical cyclones develop into hurricanes, by far the greater number of which deliver their destructive intensity and size to the humid subtropical south.

Grassland climates

Westward from the humid forests of eastern North America a gradual change takes place towards the desert regions beyond the Rocky Mountains. Between the regions of moisture surplus and moisture deficit are the transitional grassland climates of the interior plains. They occupy much of a broad corridor extending some 3000 km from the boreal forest in Alberta to the gulf of Mexico. From the foothills of the Rockies at an elevation of about 1500 m, the surface slopes gently eastward to about 150 m along the banks of the Mississippi River. The openness of the entire plains regions thus facilitates the constant interaction of atmospheric systems from alternately higher and lower latitudes, besides providing ease of eastward movement that contrasts sharply with the disruptive effect of rugged mountainous terrain to the west.

The grassland climates consist of two major divisions – tall-grass prairie in the east and short-grass steppe in the west. Prairie begins in central Illinois, where mean yearly precipitation is 750 mm or more, spreading westward across Iowa, much of Kansas, Nebraska and Oklahoma to about the 100th meridian. Here mean annual precipitation drops to around 500 mm and prairie gives way to steppe. From here to the foothills of the Rockies, precipitation decreases to less than 250 mm and scattered bunch grass replaces steppe.

The short-grass region is distinct from prairie not only because it receives less rainfall, but also because it has less cloud cover, lower atmospheric moisture content and a much higher variability of annual precipitation. Prairie stretches northwestward into southern Manitoba, Saskatchewan and Alberta from central Texas, a distance of nearly 3000 km, whereas drier steppe reaches some 2500 km from west Texas to southern Alberta.

The climates of the grasslands are distinctly continental in character, featuring four well-defined seasons, a large annual temperature range and a pronounced summer precipitation increase. From 65 to 80% of yearly precipitation occurs in the period from April through to September, maximum amounts commonly falling early in the growing season. Summer rains are mainly in the form of thermal convective showers of short duration and highly variable distribution. This results principally from the presence of warm, humid air from the Gulf of Mexico and the tropical Atlantic that dominates the growing season. Skies during summer are predominantly cloudless or partly cloudy, and from April through to September over 60% of possible sunshine is recorded, with a maximum of 70% in July and August. During July and August temperatures above 32°C occur on more than 15 days per month in the central grasslands, more than 25 days per month in Texas, but fewer than 10 days in the Canadian provinces.

West winds off the heights of the Rockies are felt frequently during summer, gaining heat as they move downward across the plains toward the Mississippi. They often accompany high temperatures and intervals of dry weather to impose the stress of severe drought. Frequent drought is much more extensive in the grasslands than elsewhere in North America. Evaporative moisture loss to the atmosphere is notably high in the grasslands, especially along the drier western margins, attaining values in excess of 2500 mm year^{-1} in southern sections and over 1400 mm year^{-1} in the north.

Summer storms are noteworthy in the grassland regions. Over 40 days with thunderstorms per year are commonly recorded, most occurring from April through September. During the same period, from 3 to 5 days with damaging hail may be expected. The grasslands of central North America are also notorious for the highest frequency and greatest destructive intensity of the atmosphere's most damaging storm, the tornado. Tornadoes may occur any month of the year, but do so most frequently between May and July in a zone of maximum incidence extending from southwest Texas to northern Illinois.

The growing season varies in length from above 240 days in central Texas to fewer than 120 days in the prairie provinces. July is the warmest month as a rule, averaging above 27°C from Kansas southward, diminishing to about 22°C in North Dakota, and to less than 18°C in Alberta.

Winter is dominated by air masses originating in the higher latitudes, most of which are continental in character and thus are often very cold and very dry. Moisture arrives mainly from the north Pacific, and frontal storms bring most of the season's precipitation. Snow, usually fine, light and dry and easily drifted, is the chief form, occurring on over 60 days per year in the north and amounting to about 1250 mm in depth. Snow is expected on fewer than 10 days per year in southern sections and rarely amounts to more than 25 cm. Freezing rain also occurs with notable frequency each winter, especially in a narrow zone from Missouri to Minnesota where the yearly incidence may reach 15 days. The cold wave, blizzard and chinook are also characteristic of winter in the grasslands. The cold wave arrives when a very cold, dry air mass moves rapidly southward, causing a temperature drop in 24 h of more than 12°C to a value of lower than −18°C. When deep cold, dry air moves rapidly into a developing low pressure system, a blizzard results. This is usually when wind speeds within the frontal system reach between 15 and 20 m s^{-1}. Accumulations of more than 600 mm from a single storm have been reported, with drifts to more than 3000 mm, the open plains allowing winds to attain speeds of up to 130 km h^{-1}. Much of the surface is laid bare by the winter's winds and ground frost may reach depths of 2500 mm in northern areas.

The chinook provides one of the more dramatic phenomena of winter's weather, particularly along the western margins of the drier grasslands. It is essentially a downslope wind, variable in strength and persistence, but capable of attaining speeds exceeding 130 km h^{-1}. Moving rapidly down the east slopes of the Rockies, it may bring a temperature rise of more than 20°C in 3 hours and usher in a period of higher than normal temperatures lasting for several days. Through sublimation, snow on the ground may diminish at the rate of 25 mm h^{-1}. The chinook may develop at any hour of the day or night throughout the year, but it mainly appears in January and March. January is the coldest month of winter and the chinook offers an ameliorating effect, especially since January temperatures average −20°C or lower in northern sections and may drop to −50°C.

Mountain climates

The western reaches of North America are dominated by high mountain ranges and intermontane plateaus and basins that create a climatic pattern of great complexity totally different from the rest of the continent. The Rocky Mountains extend from west Texas northwestward some 4500 km to Alaska. The Pacific mountain system is a series of narrower ranges reaching southward from the Alaska archipelago to include the Cascades of Washington and Oregon and the Sierra Nevada of eastern California. Less prominent coastal ranges overlook the Pacific from Mount Olympus in Washington to the peninsula of Baja California in Mexico. Between the Pacific system and the Rockies are broad tablelands like the Colorado Plateau and extensive elevated depressions like the Great Basin of Nevada and Utah.

Among the western mountains, maximum climatic diversity is encountered within short distances. Differences in altitude alone impose pronounced changes in atmospheric properties. With increasing altitude, moisture and dust content diminish, as well as atmospheric density and barometric pressure. The atmosphere's heat capacity also decreases, resulting in lower air temperatures. Along the Front Range in Colorado, for example, the mean annual temperature at Colorado Springs (el. 1860 m) is 8.9°C, whereas at Pike's Peak (el. 4300 m) about 15 km west, it is −7.2°C. As density decreases, there is an increase in the intensity of solar radiation, particularly in the ultraviolet range of the spectrum. Wind speed also increases with altitude as well as the proportion of precipitation that falls as snow.

Perhaps the chief effect on the atmosphere of mountain features is the barrier they interpose on the movement of airstreams over the Earth's surface. Airflow deflected upward is cooled adiabatically and encounters increasingly cooler atmospheric strata in the process. The visible result of these movements is the development of cloud formations along windward mountain slopes that offer one of the more striking features of the skies over mountain terrain. Precipitation is commonly produced by such deflection and is known as orographic rain or snow. Thus precipitation increases with altitude, reaching a maximum at heights that vary with latitude, but in western North America range from 1200 to 2400 m. Convective motions that generate the cumulus clouds of mountain regions are often intensified and prolonged sufficiently to produce spectacular cumulonimbus forms that yield extremely heavy showers. Such cloudbursts, added to meltwater runoff from higher snowfields, often cause sudden, destructive freshets to rush swiftly down through the lower stream courses with disastrous consequences to life and property.

With increasing altitude, most of the year's precipitation falls as snow. As latitude increases, snowfields are seen at ever lower elevations, approaching sea level in the coastal ranges of southern Alaska. Snow is without doubt one of the more significant products of atmospheric activity in mountain country. With its esthetic value is coupled its usefulness in winter sports and in the annual release of meltwater to mountain streams in spring and summer. Mountain runoff is indeed the chief source of water supply for streams in arid regions of the west. On the other hand, mountain snows can bring sudden death and destruction through rapid accumulation during unusually severe winter storms or through the sudden, swift downward rush of an avalanche. On mountain tops above 3000 m, snow has been observed during every month of the year.

The greatest amounts of snow are normally recorded in the coastal mountains of British Columbia, the Cascades and the Sierra Nevada. In the state of Washington, at levels from 1200 to 1700 m, normal wintertime snows amount to between 10 000 and 15 000 mm during the period from November to June. In Oregon, amounts average between 7600 and 14 000 mm at levels from 1400 to 1800 m, and in California mean amounts are around 1000 cm in the central Sierra Nevada between 1800 and 2400 m. Similarly high values are reported from British Columbia and are several times greater than the normal

amounts reported outside the mountainous west. Some exceptional amounts have been measured during a single winter season. At Tamarack, California (el. 2438 m) 12 000 mm fell during the winter of 1906–1907. At the Paradise Ranger station (el. 1676 m) on Mount Rainier in Washington, 25 400 mm was recorded in winter seasons 1953–1956. At Silver Lake (el. 2600 m) about 55 km northwest of Denver, 2200 mm of fresh snow fell in 27.5 h in April 1921.

Blizzards can arrive at any time during the winter months, sometimes even in spring, and in April 1997 a blizzard shut down much of the northern Plains with drifts up to 6 m closing many hundreds of kilometers of highways in Wyoming, the Dakotas, Nebraska and the eastern edge of Montana. The winter seasons 1995–1997 were particularly bad for snowfalls well above average, and especially in the north Plains where double the average was experienced (e.g. Maquette average 1995–1997 4 m, which was the actual 1996–1997 total depth).

Atmospheric turbulence is commonly intensified in mountain regions as a product of the barrier effect on normal air movement. Complex combinations of convective vorticity develop that add the risks of aircraft operation. Hazardous turbulence not infrequently occurs in clear, cloudless air over mountain terrain and is most noteworthy over middle and upper slopes. This is partly due to wind speeds increasing with altitude. The levels at which maximum wind speeds are attained vary widely, but one example suggests this common tendency. At the summit of Old Glory Mountain (el. 2347 m), British Columbia, average wind speeds during April have been 26 km h^{-1}, with a maximum for 1 h of 100 km. During the same period Carmi (el. 1245 m), average wind speed was 8 km h^{-1}, with a maximum for 1 h of 71 km.

The long-range effect of the change in atmospheric properties with altitude is readily perceivable in the vertical zoning plant formations. Much variety exists among the western mountains in the actual species associated with the area and the levels at which one stratum yields to another. However, where forest is present, as on the western slopes of the Sierra Nevada, a change from dominantly deciduous trees to conifers such as ponderosa pine, sugar pine and giant sequoia takes place at around 600 m. These in turn give way to shorter red fir and lodgepole pine beyond 1800 m. Fewer species of smaller conifers such as mountain hemlock and white bark pine appear at elevations above 2500 m, whereas only shrub forms appear at elevations of 3300 m and higher along with close-growing alpine grasses and flowering herbs.

Desert climates

The term desert traditionally has been applied to tracts of land afflicted by permanent drought. The capacity of a desert atmosphere to take up and retain moisture is overwhelmingly greater than its capacity to release it; the evaporation rate far exceeds precipitation. A desert area preserves an aspect of unrelieved aridity despite the occurrence of random rain or snow or even regular rains in small amounts. But a precise definition of desert as compared with near-desert, arid versus semiarid, is lacking.

Most North American deserts are in the elevated uplands between the Rockies and the Cascade–Sierra Nevada ranges. From north to south they include the desert of southeastern Washington and northern Oregon (el. 600 m), where the Snake River joins the Columbia; the Great Sandy Desert (el. 1500 m) in central Oregon; the Bighorn Basin (el. 1500 m) of central Wyoming; and Bridger Basin (el. 2100 m) in southwestern Wyoming. The largest desert region is the Great Basin, at a general elevation of more than 1500 m. It begins in southeastern Oregon and southwestern Idaho and extends into southern Nevada and eastward to the Salt Lake Desert in Utah. The arid region known as the Colorado Plateau spreads over western Colorado, southern Utah and northern Arizona and New Mexico at altitudes above 1500 m. At somewhat lower elevations in southern Arizona are the fringes of the Sonoran Desert and, in New Mexico, fringes of the Chihuahua Desert, both extensive arid regions of northern Mexico. The detailed distribution of actual desert climates is greatly complicated by the widespread scattering of grass-covered plains and the grassy foothills of forested mountain ranges. The boundaries between persistently arid land and the semiarid climate of grassland are repeatedly changing under the highly variable occurrence of effective precipitation.

Deserts of southern California are situated somewhat differently. The largest desert is the Mojave Desert north of Los Angeles, at an elevation of around 600 m. To the northeast is the structural basin of Death Valley with its lowest elevation at 85 m below sea level.

Southeast of Los Angeles is the Coachella Valley, which drops to 72 m below sea level, and south of that is the Imperial Valley, which extends into northern Mexico.

Clear, cloudless skies from dawn until dusk are the overriding atmospheric condition of the desert climate. In summer, most of the western deserts receive more than 80% of all possible sunshine from day to day; in southern California more than 90% is the rule. During the winter months clear skies are less frequent although even then southern deserts receive more than 70% of possible sunshine. In the northern deserts, frequent winter storms from the Pacific reduce the value to less than 40% in general and less than 40% in the Columbia Basin. Under clear skies the Sun's rays at dawn rapidly heat the bare ground and the overlying air begins to warm at once. Air temperatures rise to a maximum in mid-afternoon, falling more slowly with the approach of evening. At Yuma, Arizona, on a typical August day, air temperature reached a minimum of 26°C during the calm interval at dawn, but by 1000 h had risen to 32°C and by 1500 h reached a maximum of 44°C. The dry soil at the same location registered 64°C. Relative humidity was 8% at noon and remained below 10% during the early afternoon. High temperatures and low humidities after midday are strongly characteristic of desert climates.

Extremes of summer heat are well known in desert situations of low altitude. At Greenland Ranch (el. −54 ml) in Death Valley, the mean July temperature is 38°C, at Yuma (el. −61 m) it is 35°C and at Brawley (el. −30 m) in the Imperial Valley it is 34°C. Extreme summer temperatures are frequently above 38°C. At Greenland Ranch temperatures have reached 56.5°C, at Yuma 49°C and at Las Vegas, 47°C. In the deserts of higher altitude, summer temperatures are lower. At Redmond, Oregon (el. 884 m), July averages 19°C; at Idaho Falls (el. 1500 m), 20°C; at Deaver, Wyoming (el. 1250 m), 22°C; at Reno, Nevada (el. 1340 m), 21°C; and at Alamosa, Colorado (el. 2297 m), 18°C. Winter temperatures tend to remain above freezing except at higher elevations.

Mean yearly precipitation is generally less than 250 mm over most North American desert regions. Of far greater importance than the averages, however, are the degree of variability, the kinds of precipitation and the seasons of their occurrence. Over most of the northern regions and in southern California, precipitation falls mainly in winter and is produced by cyclonic storms from the Pacific. Over most of the Great Basin and the arid tracts of Wyoming and western Colorado, more than 30% of all precipitation falls as snow. In the southern deserts, snow falls infrequently and in trifling amounts. Here the bulk of winter precipitation falls as light, steady rain. Over the eastern deserts near the Rockies, from the Bighorn Basin southward, most precipitation falls during the summer. The main moisture source is the Gulf of Mexico and thermal convective showers are the chief rain-generating mechanisms. In the areas of summer maximum, departures from normal are over 20%. In winter precipitation regions, variability averages between 25 and 35%. The random scattering of precipitating clouds is a common cause of variability. A notable phenomenon, known as virga is the appearance of showers falling from a cloud base that fail to reach the surface.

West Coast climates

From southern Alaska to northwestern Mexico, the coastal climates of western North America are uniquely elongated and narrowly confined to within 80–160 km of the sea. They are dominated by atmospheric systems of the north Pacific. In summer, subtropical high-pressure fields expand northward, bringing the equatorward flow of their clockwise circulation over the margins of the continent. These southward airstreams become increasingly warmer, their capacity to take up moisture rises and precipitating processes are inhibited. Storm tracks are also displaced poleward, frontal systems develop less frequently, and are less vigorous, and precipitation is substantially reduced. In winter the anticyclonic circulation contracts and moves southward. Precipitating disturbances develop much more frequently, and are usually larger and more highly intensified. They pursue paths that lie much farther south than in summer and dispense the greater part of the year's precipitation. This is the only part of North America where winter precipitation is such a pronounced climatic characteristic.

Coastal evergreen forest

The most extensive climatic region is the coastal evergreen forest that spans a distance of 3500 km from Cook Inlet (latitude 60°) to San Francisco Bay (latitude 38°). Largely a coniferous forest of unusually tall trees, it is remarkable for the uniformity of its appearance. This reflects the dominant climatic role of maritime air off the unfrozen sea, bringing abundant precipitation, persistently high humidity and cool to mild temperatures. From Kodiak Island to the Columbia River, western hemlock, Sitka spruce and western red cedar form the leading association. Douglas fir, attaining heights of more than 60 m, outnumbers other species from there to northern California, and from there to San Francisco Bay, numerous stands of redwood, growing up to 90 m, predominate. This is a temperate climate, despite the high latitudes into which it extends. Cool, damp summers alternate with moderate, cloudy winters with much fog and frequent rain and snow.

In summer, temperatures for the warmest month are nearly uniform at coastal stations throughout the region. In July, at Kodiak and Yakutat monthly means are 12°C, at Sitka and Eureka, California, 13°C. Extreme maxima have rarely exceeded 30°C. Inland, sheltered stations are somewhat warmer. Anchorage, at the head of Cook Inlet, records a July mean of 14°C, Vancouver 18°C, Seattle 17°C and Portland 19°C. In the Puget Sound–Willamette Valley most stations have recorded extreme maxima of more than 38°C. The growing season is about 150 days along the Alaskan coast, increasing to 240 days from Vancouver to Eureka.

Summer is the season of least rainfall. In August storm tracks reach their northernmost position, and cyclonic activity is at a yearly minimum in the Gulf of Alaska. But some rain falls in every month of summer. During the three months June through to August, normal amounts increase northward, with less than 25 mm at Eureka, Portland 66 mm, Seattle 70 mm, Vancouver 140 mm, Sitka 370 mm and Yakutat 570 mm.

The increased frontal activity of winter brings a higher number of depressions to the Gulf of Alaska than anywhere else in the northern hemisphere at any time of the year. December is thus the wettest month at nearly all stations in the region. Repeated changes of wind direction arise from the passage of cyclonic systems, but the prevailing counterclockwise flow is predominantly parallel to the coast from Oregon to the Aleutians, consisting of south, southeast and easterly winds from December through to February. Although most of the year's precipitation occurs in winter, the actual percentage decreases northward. From October through to March, Eureka receives 83%, Portland 77%, Vancouver 73%, Sitka 60%, Yakutat 59% and Kodiak 53%. Snowfall accounts for an increasing proportion of winter's precipitation. However, at sea level stations it is usually less than 250 mm in southern sections rising to 2000 mm farther north. Greater amounts of precipitation fall on higher elevations. In general, drizzling rain is the chief source of annual precipitation, which averages over 2000 mm from northern California to southern Alaska. On seaward slopes just a short distance inland, amounts may range from 2500 to 5000 mm. The annual departure from normal averages about 15%.

Winter temperatures are unusually high for the latitudes, and Eureka averages 8°C, Sitka 0°C and Yakutat −2°C. Occasionally in winter a mass of cold, dry continental air from the deep-frozen interior spills out over the sea, producing very low thermometer readings at low-lying stations. Portland has reported −19°C, Vancouver −17°C, Sitka −21°C and Kodiak −24°C.

Mediterranean scrub woodland

Coastal evergreen forest merges with a strikingly different climate in California. Beginning in the northern reaches of the central valley near Mt Shasta, Mediterranean landscapes extend southward along the lower slopes of the Sierra Nevada and the interior valleys of the coastal ranges. South of San Francisco Bay, the Mediterranean climate dominates the coastal valleys and hillsides to the shores of the Pacific, continuing southward into Baja California, a total distance of nearly 1000 km. The visible expression of this climate is an open woodland of broadleaved evergreen and deciduous trees and a dense shrub vegetation known as chaparral. Evergreen oaks are the dominant tree forms, and although some may rise to heights of 23 m, most are less than 10 m, have short, heavy trunks, thick bark and foliage resembling the leaves of holly. The southern central valley is occupied to a large extent by steppe grassland reaching south to the Mojave Desert.

The Mediterranean climate is one of warm to hot, dry summers and mild, wet winters, with an abundance of cloud-free, sunny skies. Mean annual precipitation averages between 400 and 900 mm. An important climatic characteristic is that about 95% of the normal yearly amount falls during the 7 month period from October to April. Most stations record only negligible amounts the rest of the year. In southern districts the rainy season lasts for only 5 months. The region can be seen as clearly transitional between the coastal evergreen forest and the arid lands of southernmost California, sharing its winters with well-watered forest and its summers with drought-ridden desert.

Winter precipitation is produced almost entirely by frontal disturbances off the Pacific. Thunderstorms are rare, seldom occurring more than three or four times yearly, and are usually small, weak and of short duration. On the upper slopes of the Sierra Nevada however, they develop 10–15 times per year, and lightning at times ignites forest fires of destructive severity and size. Winter rains ordinarily begin before the end of October and increase to a maximum for the season in December at most stations, although February is the wettest month around Los Angeles. Average amounts for December are largest toward the north. Redding, far up the Sacramento Valley, averages 200 mm, the Russian River Valley, north of San Francisco, averages between 200 and 230 mm, and Placerville (el. 575 m) in the Sierra Nevada foothills also records 200 mm. Elsewhere, normal December amounts are about 100 mm. From year to year the total amount of winter's precipitation may vary widely around the arithmetic mean. Red Bluff, in the northern interior, averages 640 mm, but during a recent year received 1750 mm, followed a few years later by a total of 260 mm. San Francisco records a normal of 520 mm, but has received as little as 230 mm and as much as 900 mm. Variability rises to more than 40% in southern sections.

Winter rains are often heavy and prolonged, when slow-moving frontal storms persist for several days. From 1861 to 1960 severe flooding occurred in 18 winter seasons in southern California, attended by numerous local landslides and widespread property damage. From time to time, prolonged dry weather prevails when a large, intense and dry continental air mass lingers over most of the southwestern interior from the Rockies to the Pacific. For 60 consecutive days from 16 November 1876 to 16 January 1877, no rain whatever fell on San Francisco and in the winter of 1850–1851 only 190 mm were recorded at San Francisco and 130 mm at Sacramento. Snow seldom falls in the mediterranean region, except on the coastal mountains where it is a common occurrence at altitudes above 1200 m. At Sandberg (el. 1377 m) less than 80 km northwest of Los Angeles, winter snowfall averages about 700 mm. High in the Sierra Nevada, however, heavy snows accumulate each winter, creating an annual reserve of water that is released in spring and summer, adding meltwater to the rain-fed streamflow.

Despite the high frequency of precipitating disturbances during winter, the sun shines much of the time. About one-third of the days of each winter month are sunny, dry and pleasant. Under clear skies, nocturnal temperatures sometimes drop well below freezing. Red Bluff in the north has recorded −8°C in both December and January, Sacramento −6°C in January, and San Diego −2°C. January is without exception the coldest month and mean values are usually well above 2°C in the north and above 4°C elsewhere in the region. In February winter begins to wane as frontal activity diminishes and precipitating storms retreat northward.

Summer weather is controlled for the most part by the expanded clockwise circulation of the oceanic high-pressure fields; Northwest winds increase in frequency after the vernal equinox, providing the dominant airstreams until October. Except for the coast, most of mediterranean California is under clear, dry, cloudless skies from May to October. More than 90% of possible sunshine prevails from June to August. Temperatures rise rapidly after sunrise each day, reaching close to 30°C by mid-afternoon. July is the warmest month, when mean temperatures are largely in the mid to upper 20s Celsius; Red Bluff averages 28°C, Sacramento 24°C, and Bakersfield 29°C. The temperature rises above 32°C in July on 28 days at Red Bluff, 20 days at Sacramento and 30 days at Bakersfield. Relative humidity by mid-afternoon during July drops to low values, usually less than 30%. The average at Red Bluff is 18% and at Sacramento 28%. In the drier southern sections of the central valley, under the desiccating influence

of hot, dry winds, values often drop to between 5 and 10%. The period from May through to October accounts for over 75% of annual evaporation, which amounts to about 180 cm in the north and over 230 cm in the south. The growing season at interior locations averages between 240 and 270 days, whereas along the coast it increases to more than 330 days.

The summer weather of coastal California is cooler, more cloudy and more humid than areas that are sheltered from chilly air off the Pacific, even though many are only a few kilometres inland. The south-flowing California Current, coupled with upwelling deep water, brings sea surface temperatures of about 15°C past San Francisco, and up to 20°C south of Los Angeles, from June through to August. The warmest month is usually either August or September at most coastal stations. San Francisco averages 17°C in September, Santa Cruz 17°C from July through September and Santa Barbara 19°C in July and August.

Warm air off the open Pacific, 130–160 km offshore, passing over the cool coastal waters, produces frequent low stratus clouds and dense fog. Only about 60% of possible sunshine is recorded at shoreline stations, and most observe more than 60 days per year with dense fog. Cool, humid maritime air penetrates inland through the Golden Gate, providing lower temperatures and higher humidities over much of the San Francisco Bay area. A similar effect is seen in the Los Angeles Basin, which is exposed to the sea. In both areas combustion effluents frequently combine with stable, humid air to develop into California's notorious smog.

George R. Rumney

Bibliography

Barry, R.G. and R.J. Chorley, 1982. *Atmosphere, Weather and Climate*, 4th edn. London: Methuen.

Bryson, R.A. and F.K. Hare (eds), 1974. *Climates of North America*. New York: Elsevier.

Guttman, N.B., 1996. North America, in *The Encyclopedia of Climate and Weather* (ed. S.H. Schneider). Oxford: Oxford University Press.

Hare, F.K., 1950. Climate and zonal divisions of the Boreal Forest in eastern Canada, *Geog. Rev.*, **40**, 615–635.

Houghton, J.G., 1969. *Characteristics of Rainfall in the Great Basin*. Reno: University of Nevada Press.

Kendrew, W.G., 1961. *Climates of the Continents*, 6th edn. Oxford: Clarendon.

Lutgens, F.K. and E.J. Tarbuck, 1982. *The Atmosphere*, 2nd edn. Englewood Cliffs, NJ: Prentice Hall.

Oliver, J.E. and J. Hidore, 1983. *Climatology: An Introduction*. Columbus, OH: Merrill.

Pyke, C.B., 1971. *Some Meteorological Aspects of the Distribution of Precipitation in the Western United States and Baja California*, Water Resources Contrib. No. 139. Davis: University of California Desert Research Center.

Rumney, G.R., 1968. *Climatology and the World's Climates*. New York: Macmillan.

Thomas, M.K., 1964 *Snowfall in Canada*. Toronto: Department of Transport, Meteorological Branch.

Trewartha, G.T. and L.H. Horn, 1980. *An Introduction to Climate*, 5th edn. New York: McGraw-Hill.

US Weather Bureau, 1968. *Climatic Atlas of the United States*. Washington, DC: Environmental Sciences Services Administration.

Visher, S.S., 1954. *Climatic Atlas of the United States*. Cambridge, MA: Harvard University Press.

Cross references

Atmospheric processes associated with water in the atmosphere
Climate and climate change

O

OKEECHOBEE LAKE, FLORIDA, USA: HUMAN IMPACTS, RESEARCH, AND LAKE RESTORATION

Lake Okeechobee (26°58'N, 80°50'W, Florida, USA) is the central feature of the interconnected Kissimmee River–Lake Okeechobee–Everglades ecosystems, and often is referred to as the 'liquid heart' of south Florida. This large (1730 km²), shallow (mean depth of 2.7 m), subtropical lake supplies water to the remnant Everglades and Florida Bay, and provides water, flood protection and recreational benefits to a population exceeding 3.5 million people (Aumen, 1995). The lake's 12 000 km² watershed is primarily to the north of the lake, and the principal land uses are dairy farming and cattle ranching (Flaig and Havens, 1995). An ecologically diverse littoral zone occupies 25% of the lake's surface area, and is important habitat for wading birds, including threatened and endangered species.

Prior to human impacts, the lake and its littoral zone were larger, there were few natural outlets and water exited as wide sheet flow to the south and east (Figure O1). Extensive loss of life and property damage from hurricane-related flooding in the early twentieth century led to the construction of a levee around the lake. Today, almost all of the surfacewater inflows and outflows are controlled by gates and locks. Unnatural connections also have been made between the lake and the estuaries on the east and west coasts of Florida. These connections serve as major outlets for regulatory flood-control releases of water from the lake. When this occurs, there are sometimes harmful impacts on the estuarine biota, which are not adapted to fresh water.

Human activities in the basins upstream of Lake Okeechobee have resulted in high nutrient exports to the lake, and concerns over coincident declines in the lake's water quality and ecological health (Aumen, 1995). The lake is considered to be naturally eutrophic;

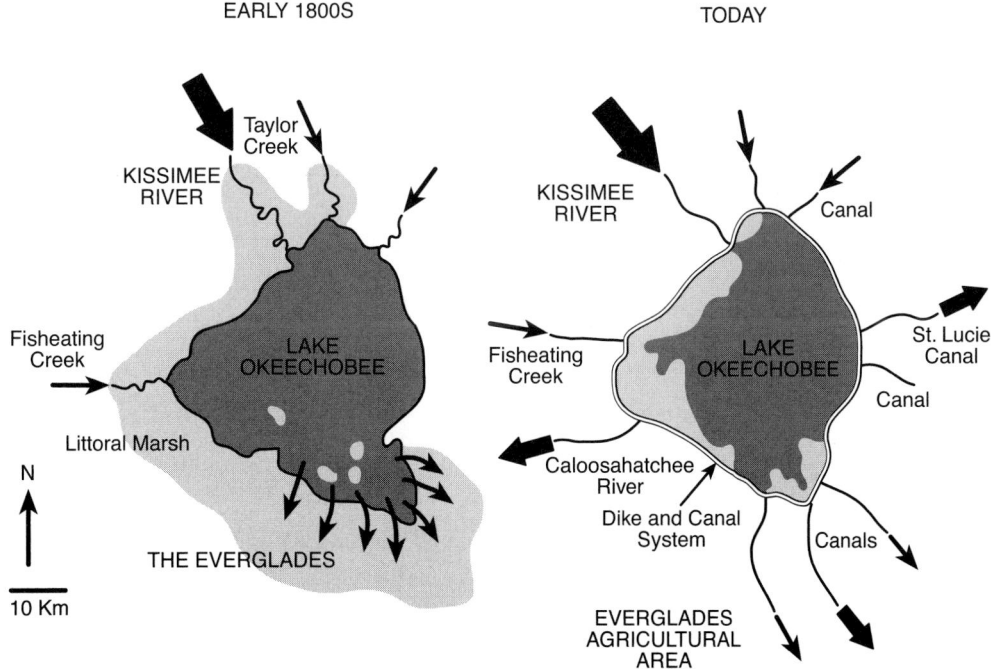

Figure O1 A representation of Lake Okeechobee prior to and subsequent to human impacts, highlighting the changes in the spatial extent of the lake and its littoral zone, and the construction of the levee, canals and water control structures. The arrows and their relative sizes indicate the changes in the general directions and magnitudes of surfacewater flows.

however, the concentration of total phosphorus in the lake doubled from about 50 μg l^{-1} in the early 1970s to about 100 μg l^{-1} in the early 1980s (James *et al.*, 1995). Nitrogen concentrations also increased during the same time period. These trends were attributed to both the increased external loads and to a reduction in the lake sediment's ability to assimilate nutrients from the water column. The main sources of external nutrient inputs are animal waste and fertilizers applied to food and forage crops. Basin land use and lake management programs have been implemented in response to these trends, including design and implementation of agricultural Best management practices (BMPs) for the dairy industry. Phosphorus loading to the lake has declined by nearly 200 tonnes per year in response to these programs, but the average loads still exceed a legislatively mandated target (based on a modified Vollenweider model) that averages 397 tonnes per year (Chapter 373.451–373.4595, Florida Statutes). Total phosphorus concentrations in the lake's water column stabilized after 1984, but unlike the external loads, the concentrations are not declining. This situation is probably due to internal loading from the lake's sediments. It has been estimated that internal loading now equals external loading on a yearly basis (Reddy *et al.*, 1995), and this will probably delay recovery of the lake from cultural eutrophication.

In contrast to phosphorus, the in-lake total nitrogen concentrations have declined in response to reductions in external loading. The net result of the nitrogen and phosphorus concentration trends is a 50% reduction in the ratios of total nitrogen to total phosphorus in the water column. The low ratios have led to lake-wide nitrogen limitation of algal growth where phosphorus limitation was once the norm (Havens, 1995), and conditions that are very favorable for the proliferation of bloom-forming blue-green algae (Smith *et al.*, 1995a). Some very large algal blooms occurred in the 1980s, with a bloom in summer 1986 covering over 40% of the pelagic surface area (Jones, 1987). Given the high levels of nutrients in the lake water, the warm water (up to 30°C during summer) and the high inputs of solar radiation, algal blooms will probably be a common feature of this eutrophic lake for many years to come.

In addition to accelerated eutrophication, Lake Okeechobee's water level is regulated to meet often-competing flood control, water supply and environmental objectives. Lake levels are controlled predominantly by climatic patterns (over 50% of water inputs are from rainfall onto the lake surface; and over 70% of water losses are by evapotranspiration), but large regulatory releases can be made to the coastal estuaries to prevent the lake from exceeding safe levels. In 1978 the lake's 'regulation schedule,' a United States Army Corps of Engineers (USACE) schedule of lake levels at which water can and cannot be released from the lake during particular times of the year, was modified in order to obtain greater water storage capacity. The maximum water level in the schedule was increased from 4.4–4.9 m to 4.7–5.3 m NGVD (national geodetic datum). This modified schedule remained in effect until 1991. There is scientific evidence that it resulted in wading bird declines, principally due to reduction in nesting and foraging habitat (Smith *et al.* 1995b). Sustained high lake levels also are related to declines in native vegetation, dominance by inundation-tolerant plants such as cattail, decreased macroinvertebrate diversities and lower gamefish abundance within the native vegetation (Aumen and Gray, 1995). Public pressure and these research results resulted in the adoption of a schedule having a slightly lower range of lake levels. This new schedule also incorporates a 'pulse release' discharge regime to the estuaries that reduces environmental damage from regulatory freshwater discharges by more closely mimicking natural discharge events.

A third problem facing the Lake Okeechobee ecosystem is the introduction and spread of exotic plants. These include melaleuca, torpedo grass and hydrilla. Melaleuca was planted on the levee by the USACE in order to reduce erosion. Shortly after introduction, the plant spread into the littoral zone, where it now covers several hundred square kilometers. Melaleuca forms a dense monoculture that chokes out native plants. It also is a poor habitat for native wildlife. Therefore, the South Florida Water Management District (SFWMD), along with other state and federal agencies, have launched major programs for melaleuca control. In Lake Okeechobee over 2 million melaleuca trees have been killed, but many more remain.

Torpedo grass and hydrilla are submerged and emergent exotic plants, respectively. They also have grown over large areas of the lake. At present, torpedo grass control efforts primarily utilize controlled fires, but herbicide treatments are under consideration. The SFWMD and USACE have active programs for hydrilla control by

herbicide spraying. However, this effort is restricted to waterways within the lake, and is carefully controlled in order to avoid damage to native plants.

Ecosystem-level research has shown that Lake Okeechobee consists of five distinct ecological zones, including four open-water zones and a littoral zone (Phlips *et al.*, 1993). The littoral zone is highly dynamic, responding dramatically and rapidly to disturbances such as fire, drought, freezing, and lake level fluctuations. The documentation of four distinct open-water zones has important management implications, because the lake had previously been considered as a well-mixed homogeneous unit (e.g. Canfield and Hoyer 1988). Current research and modeling efforts are being conducted in the context of the ecological zones. This change in approach should enhance our ability to predict responses of the lake to management actions.

A greater understanding of the Lake Okeechobee ecosystem has resulted from monitoring, ecological research and modeling. Efforts to enhance or restore Lake Okeechobee are designed around the results of this research, and are implemented through state and federal programs such as the Rural Clean Water Program (RCWP) and Florida's Surface Water Improvement and Management (SWIM) Act. This tight coupling of research and management is critical to success in protecting this vital natural resource.

Nicholas G. Aumen and Karl E. Havens

Bibliography

Aumen, N.G., 1995. The history of human impacts, lake management, and limnological research on Lake Okeechobee, Florida (USA). *Arch. Hydrobiol. Beih Ergebn. Limnol.*, **45**, 1–16.

Aumen, N.G. and S. Gray, 1995. Research synthesis and management recommendations from a five-year, ecosystem-level study of Lake Okeechobee, Florida (USA). *Arch. Hydrobiol. Beih Ergebn. Limnol.*, **45**, 343–356.

Canfield, D.E. and M. Hoyer, 1988. The eutrophication of Lake Okeechobee. Lake Reserv. Manage, **4**, 91–99.

Flaig, E.G. and K.E. Havens, 1995. Historical trends in the Lake Okeechobee ecosystem I. Land use and nutrient loading. *Arch. Hydrobiol. Suppl.* **107**, 1–24.

Havens, K.E., 1995. Secondary nitrogen limitation in a subtropical lake impacted by nonpoint-source agricultural pollution. *Environ. Pollut.*, **89**.

James, R.T., V.H. Smith and B.L. Jones, 1995. Historical trends in the Lake Okeechobee ecosystem III. Water quality. *Arch. Hydrobiol. Suppl.*, **107**, 49–69.

Jones, B.L., 1987. Lake Okeechobee eutrophication research and management. Aquatics, **9**, 21–26.

Phlips, E.J., F.J. Aldridge, P. Hansen *et al.*, 1993. Spatial and temporal variation of trophic state parameters in a shallow subtropical lake (Lake Okeechobee, Florida, USA). *Arch. Hydrobiol.*, **128**, 437–458.

Reddy, K.R., Y.P. Sheng and B.L. Jones, 1995. *Lake Okeechobee Phosphorus Dynamics Study: Summary Volume I.* Report, South Florida Water Management District, West Palm Beach, FL, 33416–4680, USA.

Smith, V.H., V.J. Bierman, B.L. Jones and K.E. Havens, 1995a. Historical trends in the Lake Okeechobee ecosystem. IV. Nitrogen: phosphorus ratios, cyanobacterial dominance, and nitrogen fixation potential. *Arch. Hydrobiol., Suppl.*, **107**, 71–88.

Smith, J.P., J.R. Richardson and M.W. Collopy, 1995b. Foraging habitat selection among wading birds (Ciconiiformes) at Lake Okeechobee, Florida in relation to hydrology and vegetative cover. *Arch. Hydrobiol. Beih Ergebn. Limnol.*, **45**, 247–285.

Cross references

Lakes
Limnology

OROGRAPHIC PRECIPITATION

When mountains of uplands act as barriers to air flow, forcing the air to ascend, moist air moving upslope cools adiabatically, producing clouds and precipitation. Showers and thunderstorms usually occur when the air is unstable; but if it is stable, precipitation will be more

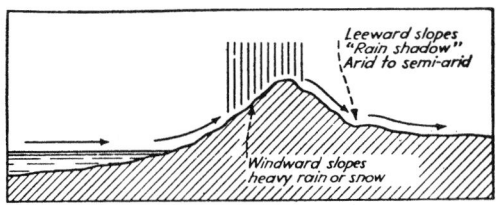

Figure O2 When moist winds are forced to cross mountain barriers, heavy precipitation falls on the windward slopes, but leeward slopes in the rain shadow are relatively dry.

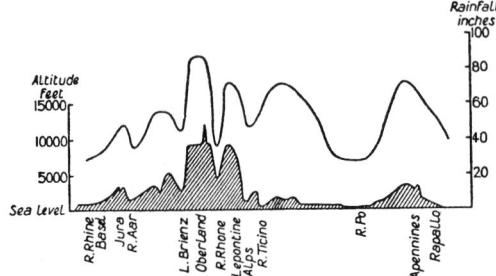

Figure O3 The control of rainfall by topography. The upper curve shows the mean annual rainfall along the section of the Alpine region from the Gulf of Genoa to Basel. The proximity of the Mediterranean gives the Apennines a heavy rainfall for the altitude. The upper Rhone Valley (762 m) has almost as little as the Po Plains (76 m) owing to the high surrounding ranges.

general and steady. The clouds and precipitation will be concentrated on the windward side of mountains or upland slopes (Figure O2).

In orographic precipitation, the important fact is that a landform acts as a wedge or barrier over which moist air is lifted and cooled to the dew point, and clouds release some moisture as precipitation. Mountains also cause precipitation by slowing the air flow and both causing convergence and retarding the air speed of storm systems. The barriers further enhance differential heating, which creates mountain breezes at night and valley breezes during the daytime.

These combined effects account for the high precipitation amounts along the windward slopes of topographic barriers, and a close relationship between topography and rainfall amounts is often found (Figure O3). Orographic precipitation may occur in downpours, but it usually takes the form of general rains. Since mountains are fixed in position, successive rains may yield substantial totals giving rise to high annual precipitation. For example, a station at Mt Waialeale, Hawaii, located on the windward coast of Kauai at an elevation of 1523 m, averages 11 680 mm of annual rainfall. Cherrapunji, India, at an elevation of 1313 m on the south slopes of the Himalaya Mountains averages 11 430 mm of rainfall per year.

As the air moves over the mountain crest and downslope, it is warmed adiabatically. Since the moisture-holding capacity is increased, precipitation ceases and clouds usually dissipate. The result is often a rain shadow desert. Some of the world's deserts that lie in the rain shadow of mountains are the Gobi Desert of Mongolia, the Atacama Desert of Peru and Chile, the Patagonia Desert of Argentina and the Mojave Desert of the United States.

James L. Guernsey

Bibliography

Dickson, D.R., 1982. *Weather and Flight*. Englewood Cliffs. NJ: Prentice-Hall.
Lutgens, F.K. and E.J. Tarbuck, 1983. *The Atmosphere*, 2nd edn. Englewood Cliffs, NJ: Prentice-Hall.
Miller, A., J.C. Thompson, R.E. Peterson and D.R. Haragan, 1983. *Elements of Meteorology*, 4th edn. Columbus, OH. Charles E. Merrill.

Cross references

Dew
Dewpoint
Dew pond
Fog and mist
Precipitation distribution

P

PALEOHYDROLOGY

Paleohydrology is the study of the waters of the Earth, their composition, distribution and movement on ancient landscapes from the first occurrence of precipitation to the beginning of hydrological record keeping. It provides a link between the hydrology of the present and the sciences concerned with Earth history and past environments (Gregory, 1983).

Paleohydrological speculations and conclusions depend on existing knowledge of the effects of climatic, vegetational and geological controls on runoff, sediment yield, sediment concentration, water chemistry, groundwater and lake levels, and the nature of floods. These relationships, however, are not directly applicable to the remote geological past when vegetation was absent or when it was evolving towards its present condition and distribution. This important paleohydrological consideration has been discussed by Cayeux, Russell, Schwarzback, and Tricart and Callieux (see references in Schumm, 1968). Because of this uncertainty concerning the basic interaction among components of the hydrological cycle in the remote geological past, most paleohydrological research focuses on the Quaternary.

In theory, paleohydrology is concerned with all components of the hydrological cycle. In practice, most paleohydrology research focuses on estimation of (1) river channel and discharge characteristics, (2) fluctuations in lake levels, (3) fluctuations in groundwater levels and isotope chemistry, and (4) proxy indicators of past precipitation characteristics, such as tree rings, ice cores, pollen or soils. This article will focus on estimation of Quaternary river and lake paleohydrology, but we begin with a discussion of how recent fluvial hydrological conditions may be used to infer pre-Quaternary conditions.

Data on runoff and sediment movement from a drainage basin, when averaged for a period of years, provide information on the hydrological and climatic regime of the drainage system. Therefore, relationships developed among modern data for average precipitation, temperature, runoff, and sediment yield permit estimates to be made of regional paleohydrology when paleoclimatic information is available. In addition, relationships that have been developed between channel morphology, sediment load and runoff provide a basis for the estimation of paleochannel discharge.

Relationships that exist among mean annual temperature, precipitation, runoff and sediment yield, based on data obtained for the United States, are shown in Figures P1 and P2. The curves show in a general way what average hydrological differences can be expected among the climatic regions. Using a somewhat different line of reasoning, the curves can also be used to demonstrate the hydrological effects of a climate change (Schumm, 1965).

Only the curve for an average temperature of 10°C will be discussed, but the principles apply to all curves of Figures P1 and P2. Sediment yield from a drainage basin will increase from a minimum in a region of no precipitation to a peak at between 250 and 380 mm of precipitation. The decrease in sediment yield beyond this peak is a

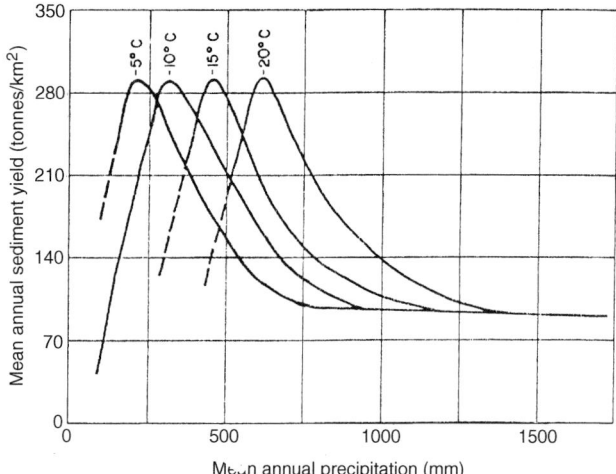

Figure P1 Curves illustrating the effect of average temperature on the relations between mean annual sediment yield and mean annual precipitation. (Source: Schumm, 1965.)

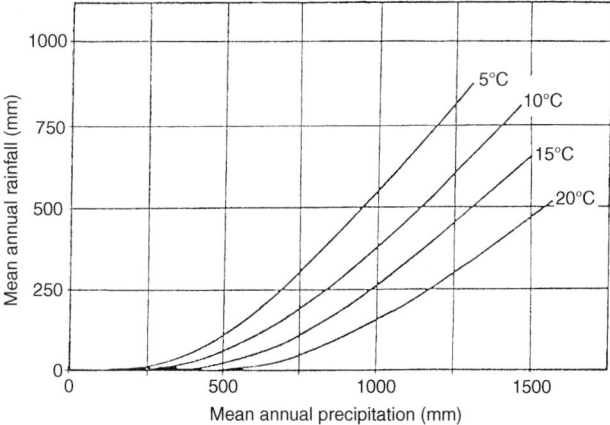

Figure P2 Curves illustrating the effect of average temperature on the relation between mean annual runoff and mean annual precipitation. (Source: Langbein et al., 1949.)

result of the increased effectiveness of vegetation in protecting the soil and in retarding erosion, i.e. a transition from desert shrubs to grassland. Sediment yield rates decrease further to lower values in forested regions of high rainfall. The curves of Figure P2. show how runoff will increase with increased precipitation for uniform monthly precipitation at different mean annual temperatures. For the 10°C curve, below about 500 mm of precipitation, runoff increases slowly with increased precipitation because of relatively high water losses to evaporation and infiltration. The quantity of water lost through transpiration and interception by vegetation increases, as precipitation increases, to about 1000 mm of precipitation when the water requirements of infiltration, evaporation and transpiration have been met; above 1000 mm of precipitation, runoff increases directly with precipitation (Langbein et al., 1949).

It should also be recognized that highly seasonal precipitation should increase sediment yields, as suggested by Fournier and Douglas (see references in Schumm, 1968). In fact, some information on sediment yields from regions of tropical monsoon climates suggest that the sediment yield curve will rise again as rainfall increases above 1300 mm; however, this increase may, in fact, be due to the rugged topography and intensive land use of the areas from which the data used to establish the relationship were obtained.

The relationships of Figures P1 and P2 demonstrate that, depending on the climate before a climate change, increased precipitation will increase runoff, but sediment yield rates may increase or decrease. In very arid regions an increase in precipitation will increase sediment yield because the vegetational cover will not improve sufficiently to retard erosion effectively whereas, under an initially semiarid climate, an increase of precipitation will improve the vegetational cover sufficiently to cause a decrease in sediment yield rates. Therefore, a change in the quantity of sediment moving out of a drainage system cannot be considered as a function of precipitation and runoff alone. The vegetational influence must also be evaluated, but this becomes increasingly difficult as the more remote geological periods are considered.

Four major divisions of geological time are paleohydrologically significant: (1) the time prior to evolution of vascular land plants, from the Precambrian through to the Silurian; (2) the time period during colonization of alluvial areas by primitive vascular vegetation (Silurian to Cretaceous); (3) the time period during colonization of interfluves by flowering plants (Cretaceous to Miocene); and (4) the time following the appearance of grasses (Miocene to Recent).

It can be assumed that the relationship in Figure P1 is not valid for periods prior to the Miocene, and that in pre-Silurian time a direct relationship existed between precipitation and sediment yield which was limited only by the erodability of the bedrock (Schumm, 1968). The sediment size made available to river channels would have also differed through time due to the protective effects of the vegetation. River sediment load will be finer under recent conditions, whereas fluvial sediment during pre-vegetation time and during colonization of the land by primitive vegetation should have been dominated by coarse-grained material because complete weathering was precluded by rapid erosion. This is not universally true, however, since pre-Silurian clay rich soils ('Green Clays') with deep weathering profiles are known to occur (Retallack and Mindszenty, 1994). The influence of vegetation in modern environments also helps to maintain channel stability and a single channel in planform. Those effects would be lost in the pre-Silurian, and minimized in pre-Miocene times. Therefore it may be expected that the very different hydrological characteristics of the geological past should be reflected in the characteristics of terrestrial sedimentary deposits and paleochannels, such as an increase in the number of braided rivers in the past due to the increased sediment supply, larger sediment size and lack of stabilizing bank influences. For example, a survey of over 100 published descriptions of fluvial deposits (Cotter, 1978) showed that almost all pre-Silurian rivers were braided, while younger geological periods show evidence for both braided and meandering rivers.

Quaternary fluvial paleohydrology

Fluvial paleohydrological studies of the recent geological past may be divided into those which seek to estimate (1) the basic fluvial characteristics (both channel geometry and flow regime) for relic channels preserved in the sedimentary record, and (2) the characteristics of flow regime along an active channel prior to the period when systematic records were kept.

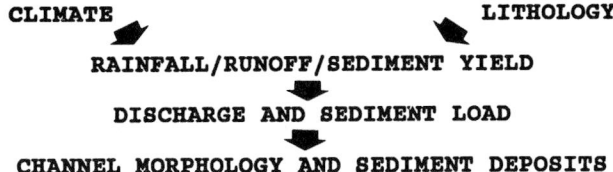

Figure P3 The primary variables that determine river behaviour and subsequent fluvial sedimentary deposits.

Studies of relic channels (which can be done for sediments and sedimentary rocks of any age) generally focus on some aspect of channel form (Ethridge and Schumm, 1978) such as meander geometry (Dury, 1976; Schumm, 1968, 1972; Wohl and Georgiadi, 1994), channel width/depth ratio (Schumm, 1960, 1969; Bridge, 1985), or the grain size of bedforms in sedimentary deposits. Using empirical relationships between these measureable features the flow regime, discharge and hydraulics of the relict channels may then be inferred, with the emphasis on mean, rather than extreme, flow conditions. The chain of inference may then be extended from discharge to rainfall–runoff regime (Figure P3). These studies are often hampered by environmental misapplication, inadequate preservation of the channel features, or the inadequacy and improper use of the algebraic equations describing channel form and dynamics.

Studies attempting to infer prehistoric flow regime along an active channel may use (1) historical information in the form of marked high-water levels, or written records such as journals, (2) sediment–flow regime relations, where the grain size, quantity or arrangement of sediment deposited by a large flood is used to infer flow regime, (3) botanical records in the form of corrasion scars, tree-ring anomalies, stand age, or adventitious sprouts, all of which are affected by flow regime if the plant grows close to a channel (Hupp, 1986, 1988) and (4) paleostage indicators in the form of silt lines, scour lines, lichen limits, debris accumulations, slackwater deposits or eroded landforms (O'Conner et al., 1986; Kochel and Baker, 1988). Historical information is restricted to the time span of human occupation within the drainage basin, and tends to be subject to human perception of extreme high or low flows. Sediment-based inferences from smaller (sand- and gravel-sized) sediments generally focus on mean flow conditions and are made using assumptions about river capacity, while coarser clasts may be used to infer the largest flows along a channel by assuming they represent the maximum flow competence. Botanical features record both mean and extreme flow regimes, and provide chronologically precise data. Inferences from such features are of course limited to the time span of the oldest vegetation along the channel. Paleostage indicators best record large-magnitude floods along stable channels with high-magnitude flow variability and a cross-sectional geometry that enhances stage change. Paleostage indicators may record flow during at least the past 20 000 years (Jarrett and Malde, 1987; O'Conner, 1993). Table P1 shows a comparison between prehistoric and modern examples of flooding.

Quaternary lacustrine paleohydrology

Although lakes cover only 1% of the Earth's continental surface, and contain only 0.02% of the water in the hydrosphere, they are sensitive to hydroclimatic fluctuations. Closed lake basins that have no outlet are particularly useful paleohydrological indicators. Variations in lake water volume, surface area, depth and lake chemistry are assumed to reflect variations in regional hydrological balance. The lake variations are in turn recorded by the elevations of shore features, and by the characteristics of organic and inorganic lake deposits and biota. Paleohydrological data from the shore and nearshore environment provide episodic records of lake area and elevation. Data from deeper-water environments may provide continuous records of lake chemistry.

Data from the shore environment generally consist of stratigraphic relationships, age and elevations of shorelines, wave-cut cliffs and beach ridges (Benson et al., 1990). Relationships of lake deposits to nearby alluvial fans, dunes, lunettes, tufa deposits and archeological sites may help to interpret lake fluctuations (Enzel et al., 1989, 1992). Lake level fluctuation may also be recorded by the presence or absence of peat, gytta, evaporites, marl, playa muds, lacustrine clays, silt and carbonates, and by sedimentary structures such as unconformities and mud cracks (Teller and Last, 1990).

Table P1 Comparison of some of the world's largest recorded flood events (Vaughn, 1984, p. 191)

Flood	Major source	Date	Volume (km³)	Average discharge (m³ s⁻¹)	Maximum discharge (m³ s⁻¹)	Average velocity (m³ s⁻¹)	Duration
Channelled scablands	Lake Missoula	18 000–13 000 BP	1563.000	–	21 296 640	9.0–18.0	7–14 days
Wabash River[a]	Lake Maume	14 000 BP	71.000	310 000	79 000	3.5	21 days
Snake River	Lake Bonneville	13 500 BP	1604.000	1 463 578	6 060 480	4.6–7.3	42 days
Bireh Ganga River	Lake Gohna	AD 1894	0.470	17 558	100 000	7.0	4.5 h
Indus River	Landslide-formed lake	AD 1840	1.300	14 160	–	–	24 h
Teton River	Lake Reservoir	AD 1976	0.321	–	42 500	–	5 h
St Francis Dam	Lake Reservoir	AD 1928	0.050	–	11 000 to 14 000	8.0	>12 h
Gros Ventre River	Landslide-formed lake	AD 1927	0.050	1 700	–	–	>12 h

[a] Discharge and velocity values for the upper Wabash Valley only.

Vertical and lateral variability in the geochemistry and mineralogy of lacustrine sediments may record changes in lake salinity and alkalinity via changes in mineralogy or in the isotopic composition of carbonates (McKenzie and Eberli, 1987), and micro- and macrofauna shells such as diatoms and ostracods (Battarbee, 1986; Forester, 1987).

Groundwater

Relatively little research has been devoted to the paleoclimatic and paleohydrological records that may be extracted from groundwater. Fluctuations of the water table, as recorded in soils and stratigraphy, may be used to infer changes in the precipitation regime through time. Mean annual paleotemperature may be derived from noble gases dissolved in radiocarbon-dated groundwater (Andrews and Lee, 1979; Phillips et al., 1986; Stute et al., 1995), and groundwater isotopes of oxygen, carbon and hydrogen may be used to infer recharge temperatures, as well as patterns and mechanism of precipitation transport (Dutton, 1995).

Deborah Anthony and Ellen Wohl

Bibliography

Andrews, J.N. and Lee, D.J., 1979. Inert gases in groundwater from the Bunter Sandstone of England as indicators of age and paleoclimate trends. *Journal of Hydrology*, **41**, 233–252.
Battarbee, R.W., 1986. Diatom analysis, in B.E. Berglund (ed.), *Handbook of Holocene Palaeoecology and Palaeohydrology*, John Wiley, 527–570.
Benson, L.V., Currey, D.R., Dorn, R.I. et al., 1990. Chronology of expansion and contraction of four Great Basin Lake systems during the past 35,000 years. *Palaeogeography, Palaeoclimatology, Palaeoecology*, **78**, 241–286.
Bridge, J.S., 1985. Paleochannel patterns inferred from alluvial deposits: A critical evaluation. *Journal of Sedimentary Petrology*, **55**, 579–589.
Cotter, E., 1978. The evolution of fluvial style, with special reference to the Central Appalachian Paleozoic, in A.D. Miall (ed.), *Fluvial Sedimentology*, Canadian Society of Petroleum Geologists, Memoir 5, 361–381.
Dury, G.H., 1964. *Principles of underfit streams*. US Geological Survey Professional Paper 452-A.
Dury, G.H., 1976. Discharge prediction present and former, from channel dimensions. *Journal of Hydrology*, **30**, 219–245.
Dutton, A.R., 1995. Groundwater isotopic evidence for paleorecharge in US. High Plains aquifers. *Quaternary Research*, **43**, 221–231.
Enzel, Y., Cayan, R.D., Anderson, R.Y. and Wells, S.G., 1989. Atmospheric circulation during Holocene lake stands in the Mojave Desert: evidence of a regional climatic change. *Nature*, **341**, 44–48.
Enzel, Y., Brown, W.J., Anderson, R.Y. et al., 1992. Short-duration Holocene lakes in the Mojave River drainage basin, southern California. *Quaternary Research*, **38**, 60–73.
Ethridge, F.G. and Schumm, S.A., 1978. Reconstructing paleochannel morphologic and flow characteristics: Methodology, limitations and assessment, in A.D. Miall (ed), *Fluvial Sedimentology*, Calgary, Canadian Society of Petroleum Geologists, pp. 703–721.
Forester, R.M., 1987. Late Quaternary paleoclimate records from lacustrine ostracodes, in W.F. Ruddiman and H.E. Wright Jr (eds),

North America and Adjacent Oceans During the Last Deglaciation, Boulder, Colorado, The Geological Society of America, pp. 261–276.
Gregory, K.J. (ed.), 1983. *Background to Paleogeohydrology*, New York, Wiley-Interscience.
Hupp, C.R., 1986. Botanical evidence of floods and paleoflood frequency. *International Symposium on Flood Frequency and Risk Analysis*, Baton Rouge, Louisiana.
Hupp, C.R., 1988. Plant ecological aspects of flood geomorphology and paleoflood history, in V.R Baker, R.C. Kochel and P.C. Patton (eds), *Flood Geomorphology*, New York, Wiley, pp. 335–356.
Jarrett, R.D. and Malde, H.E., 1987. Paleodischarge of the late Pleistocene Bonneville Flood, Snake River, Idaho, computed from new evidence. *Geological Society of America Bulletin*, **99**, 127–134.
Koehel, R.C. and Baker, V.R., 1988. Paleoflood analysis using slackwater deposits, in V.R. Baker, R.C. Kochel and P.C. Patton (eds), *Flood Geomorphology*, New York, Wiley, pp. 357–376.
Langbein, W.B. et al., 1949. *Annual Runoff in the United States*, US Geological Survey Circular 52, Washington, DC, US Government Printing Office.
McKenzie, J.A. and Eberli, G.P., 1987. Indications for abrupt Holocene climatic change: Late Holocene oxygen isotope stratigraphy of the Great Salt Lake, Utah, in W.H. Berger and L.D. Labeyrie (eds), *Abrupt Climatic Change, Evidence and Implications*, Dordrecht, D. Reidel, pp. 127–136.
O'Connor, J.E., 1993. *Hydrology, hydraulics, and geomorphology of the Bonneville Flood*. Geological Society of America Special Paper 274.
O'Connor, J.E., Webb, R.H. and Baker, V.R., 1986. Paleohydrology of pool-and-riffle pattern development: Boulder Creek, Utah. *Geological Society of America Bulletin*, **97**, 410–420.
Phillips, A.R., Peeters, L.A., Tansey, M.K. and S.N. Davis, 1986. Paleoclimatic inferences from an isotopic investigation of groundwater in the central San Juan Basin, New Mexico. *Quaternary Research*, **26**, 179–193.
Retallack, G.J. and Mindszenty, A., 1994. Well preserved late Precambrian paleosols from northwest Scotland, *Journal of Sedimentary Research*, **A64**(2), 264–281.
Schumm, S.A., 1960. *The shape of alluvial channels in relation to sediment type*. US Geological Survey Professional Paper 352-B.
Schumm, S.A., 1965. Quaternary paleohydrology, in H.E. Wright, Jr and D.G. Frey (eds), *The Quaternary of the United States*, Princeton, Princeton University Press, pp. 783–794.
Schumm, S.A., 1968. Speculations concerning paleohydrologic controls of terrestrial sedimentation, *Geological Society of America Bull.*, **79**, 1573–1588.
Schumm, S.A., 1969. River metamorphosis. *Journal of the Hydraulics Division, Proc. ASCE*, **95**, 255–273.
Schumm, S.A., 1972. Fluvial paleochannels, in J.K. Rigby and W.K. Hamblin (eds), *Recognition of Ancient Sedimentary Environments*, Society of Economic Paleontologists and Mineralogists Special Publication 16, 98–107.
Stute, M., Clark, J.F., Schlosser, P., Broccker, W.S. and Bonani, G., 1995. A 30,000 year continental paleotemperature record derived from noble gases dissolved in groundwater from the San Juan Basin, New Mexico. *Quaternary Research*, **43**, 209–220.
Teller, J.T. and Last, W.M., 1990. Paleohydrological indicators in playa and salt lakes, with examples from Canada, Australia, and

Africa. *Palaeogeography, Palaeoclimatology, Palaeoecology*, **76**, 215–240.

Vaughn, D., 1984. Paleohydrology and geomorphology of selected reaches of the upper Wabash Valley, Indiana. PhD dissertation, Terre Haute, Indiana State University.

Wohl, E.E. and Georgiadi, A.G., 1994. Holocene palaeomeanders along the Sejm River, Russia. *Zeitschrift für Geomorphologie*, **38**, 299–309.

Cross references

PERMAFROST

Introduction

The long, cold winters and short, cool summers in the polar regions result in the formation of a layer of frozen ground that does not completely thaw during the year. This perennially frozen ground, known as permafrost, affects many human activities in the Arctic, as well as in the Subarctic and at high altitudes, and causes problems that are not experienced elsewhere.

Permafrost is a naturally occurring material that has a temperature below 0°C continuously for 2 or more years (Muller, 1943, p. 3). This layer of frozen ground is designated exclusively on the basis of temperature. Part or all of its moisture may be unfrozen, depending upon the chemical composition of the water or depression of the freezing point by capillary forces. For example, permafrost with saline soil moisture, such as that found under the ocean immediately off the arctic shores, might be colder than 0°C for several years but would contain no ice and thus would not be firmly cemented. Most permafrost is consolidated by ice; permafrost with no water, and thus no ice, is termed dry permafrost. The upper surface of permafrost is called the permafrost table. In permafrost areas, the surficial layer of ground that freezes in the winter (seasonally frozen ground) and thaws in summer is called the active layer. The thickness of the active layer under most circumstances depends mainly on the moisture content; it varies from 10–20 cm in thickness in wet organic sediments to 2–3 m in well-drained gravels. Permafrost is a widespread phenomenon in the northern part of the northern hemisphere, underlying an estimated 20% of the land surface of the world (Figure P4).

Although the existence of permafrost was known to the inhabitants of Siberia for centuries, not until 1836 did scientists of the Western world take seriously isolated reports of thick frozen ground existing under northern forest and grasslands. Then, Alexander Theodor von Middendorff measured temperatures to depths of approximately 107 m in permafrost in the Shargin shaft, an unsuccessful well dug for the governor of the Russian-Alaskan Trading Company at Yakutsk, and estimated that the permafrost was 214 m thick (Péwé, 1974).

For the last 100 years, scientists and engineers in the former Soviet Union have been pioneers actively studying permafrost and applying results to development of the northern country (Melnikov, 1984). Similarly, prospectors and explorers have been aware of permafrost in the northern part of North America for many years.

In the early part of the twentieth century, a few railroads were built in permafrost terrain in North America, mainly the Alaska Railroad and the railroad to Fort Churchill on the south shore of Hudson Bay, Canada. Although permafrost was and is the predominant and most serious cause of engineering problems that affect the northern part of these railroads, no systematic studies of the perennially frozen ground were made prior to and during construction. In the 1950s, detailed engineering geology studies were made by scientists of the US Geological Survey in critical areas of permafrost trouble along the Alaska Railroad (Péwé, 1949; Wahrhaftig and Black, 1958, Péwé and Paige, 1963; and Fuglestad, 1986).

Prior to World War II, the industry in North America most concerned with problems created by permafrost was placer gold mining in the interior of Alaska and in the Yukon Territory of Canada. It was a small operation until the 1920s; then large-scale dredging for gold was undertaken in the perennially frozen deposits. The US Smelting, Refining and Mining Co. (Boswell, 1979), operating widely throughout central and western Alaska, studied the distribution and engineering characteristics of permafrost in detail in

its operations. This organization kept meticulous records and drawings of its experiences with permafrost and how they successfully solved engineering geology problems with frozen ground long before the word 'permafrost' was invented and the subject became a national concern (Boswell, 1979; Péwé, 1975b, footnote 1). For example, they were the first company in North America to attempt to delineate the distribution of permafrost using geophysical techniques (Joesting, 1941). Unfortunately, almost all of their records concerning permafrost, collected over 40 years, are proprietary or destroyed (Péwé, 1975a).

During World War II, permafrost became a national and international engineering geology concern in North America with the construction of the Alaska Highway in Canada and Alaska, Army air bases in Alaska and Canada (Barnes, 1946; Wilson, 1948), and the oil pipeline from Norman Wells south to the Alaska Highway near Whitehorse in Canada (Hemstock, 1949a,b). Millions of dollars were lost in trying to overcome engineering problems created by thawing of ice-rich permafrost (Péwé, 1948). The US Smelting, Refining and Mining Co. provided the US Army with data concerning their experiences in engineering on permafrost. S.W. Muller of the US Geological Survey compiled for the US Army Corps of Engineers the now classic first book in English on permafrost and related engineering problems (1943), which was to remain the textbook on permafrost in North America for the next 20 to 30 years. The word '*permafrost*', coined by Muller in 1943, is now in international use.

Immediately after the end of World War II, widespread systematic studies of permafrost and seasonal frost were undertaken by many scientists and engineers of North America, mainly those associated with the US Geological Survey, the Geological Survey of Canada, the US and Canadian military forces, the US National Academy of Sciences–National Research Council and the National Research Council of Canada (US Army Corps of Engineers, 1946, 1947, 1949, 1951, 1954, 1956; Black, 1950; Péwé, 1948; Hardy and D'Appolonia, 1946; Highway Research Board, 1948; Johnson, 1952). Such studies and many others were the vanguard of those that now permit more intelligent planning and construction in permafrost areas of the world.

With continued expansion of the utilization of areas in the high latitudes and high altitudes, the understanding of permafrost and related engineering problems has indeed become of international concern (Figure P4). This has given rise to important and constructive international conferences to discuss and publish advances and problems of the subject: United States, 1963; Yakutsk, Russia, 1973; Canada, 1978; Alaska, 1983; and Norway, 1988 (National Academy of Sciences–National Research Council, 1966, 1973, 1978, 1983a, 1984; National Research Council of Canada, 1978; French, 1982). Major textbooks on various aspects of permafrost are now available in North America (for instance, French, 1976; Washburn, 1980; Johnston, 1981). A milestone of the last decade in the science and engineering of permafrost research and applications was the establishment of the International Permafrost Association in 1983. After 10 years of preparation, the four founding countries – Canada, China, the United States, and the USSR – established the association at the Fourth International Permafrost Conference at Fairbanks, Alaska. The association now has adhering national bodies from 15 countries.

Permafrost profoundly affects human activities in the Arctic and Subarctic and requires that conventional engineering construction techniques and design be modified at additional costs (Corte, 1969). Agriculture, mining, water supply, sewage disposal and construction of all types are seriously affected by subsidence of the ground surface caused by thawing of the perennially frozen ground; furthermore, additional problems are brought on by the associated soil flowage and frost action. Unless planners, engineers and builders have a thorough understanding of the geological, thermal and mechanical problems unique to permafrost, impassable roads and railroads, unusable airstrips and destroyed or abandoned buildings and pipelines may result.

Origin and thermal regime of permafrost

In areas where the mean annual air temperature drops below 0°C, some of the ground frozen in the winter will not be completely thawed in the summer; therefore, a layer of permafrost will form and continue to grow downward in small increments from the seasonally frozen ground. The permafrost layer will become thicker each winter; the thickness is controlled by the thermal balance achieved between the

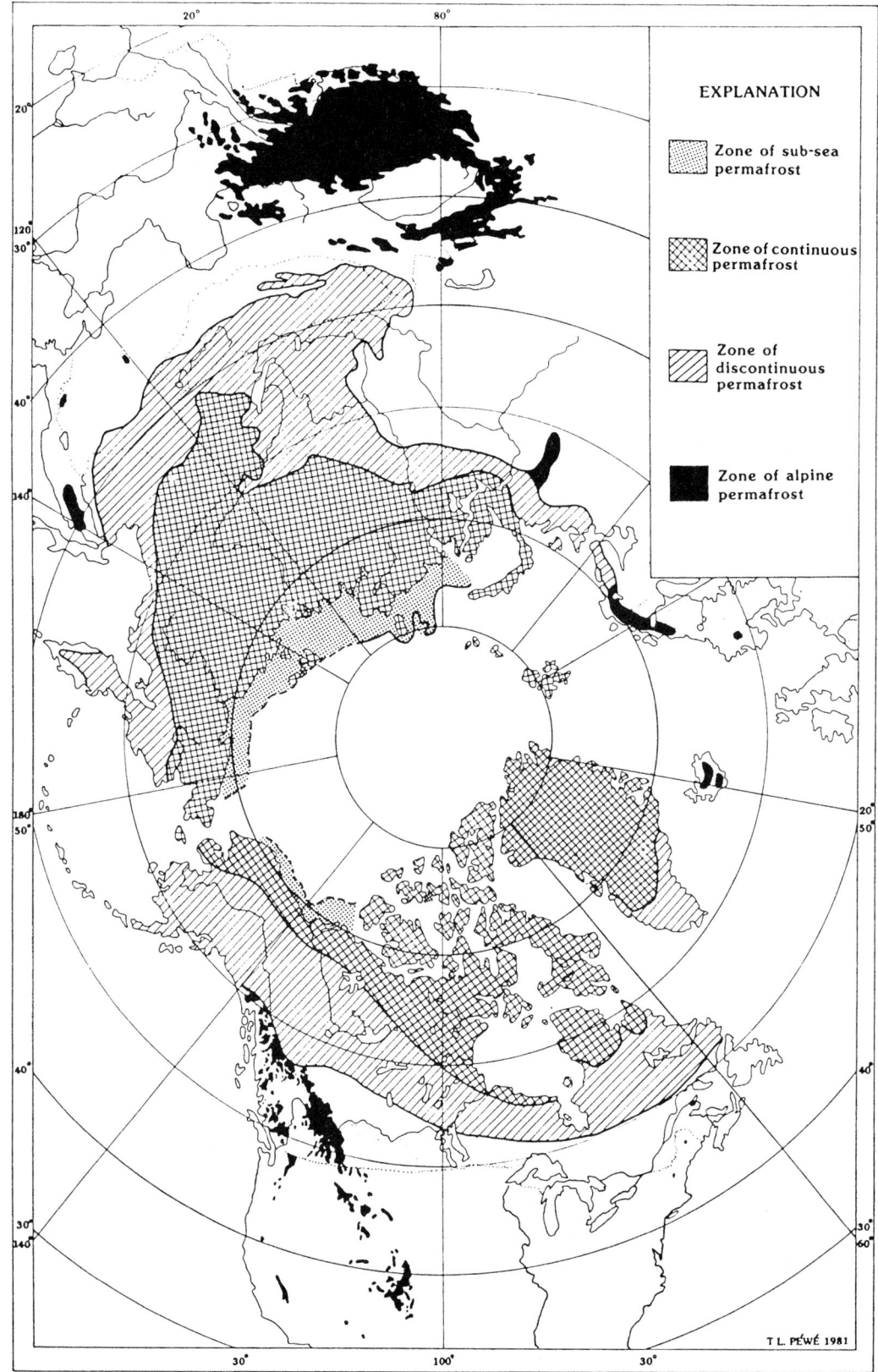

Figure P4 Distribution of permafrost in the northern hemisphere. Isolated areas of alpine permafrost not shown on the map exist in the high mountains and outside the map area in Mexico, Hawaii, Japan and Europe. Submarine permafrost is also reported beneath the Barents Sea. (From Péwé, 1983.)

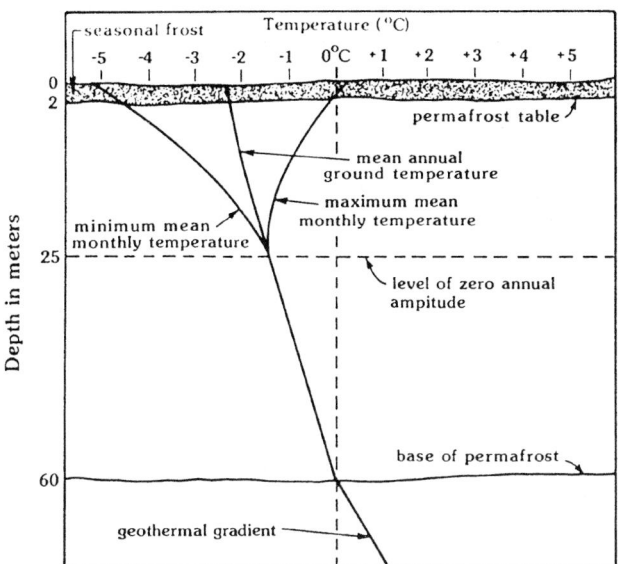

Figure P5 Hypothetical example of a temperature profile and thickness of permafrost in central Alaska. (from Péwé, 1975b.)

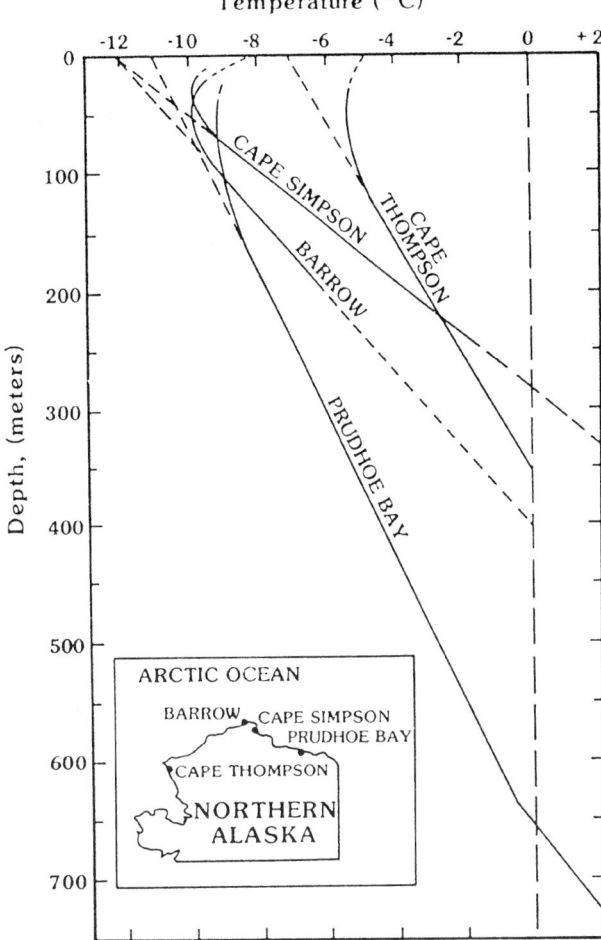

Figure P6 Generalized profiles of measured temperature on the Alaskan Arctic coast (solid lines). Dashed lines represent extrapolations. (From Lachenbruch et al., 1982.)

heat flowing upward from the Earth's interior and that flowing outward into the atmosphere – a balance that depends upon the mean annual air temperature and geothermal gradient. The average geothermal gradient is about 1°C increase in the temperature of the Earth for every 30–60 m of depth. Eventually the thickening permafrost layer reaches an equilibrium depth at which over several years the same amount of geothermal heat reaching the permafrost is lost into the atmosphere. A state of equilibrium takes thousands of years to be reached where permafrost is hundreds of meters thick.

An example of the change of temperature of frozen ground with depth and the upper and lower limit of permafrost is illustrated in Figure P5. The annual fluctuation of air temperature from winter to summer is reflected in a subdued manner in the upper few meters of the ground. This fluctuation diminishes rapidly with depth; it is only a few degrees at 8 m and is barely detectable at 15 m. The level at which the fluctuations are hardly detectable (10–15 m) is termed the level of zero amplitude. Below this depth the temperature increases steadily under the influence of geothermal heat. The temperature of permafrost at the depth of minimum annual seasonal change varies from near 0°C at the southern limit of permafrost to −10°C in northern Alaska and −13°C in northeastern Siberia. In the continuous zone, the temperature of the permafrost is less than −5°C. Permafrost is the result of present climate; however, many temperature profiles show that permafrost is not in equilibrium with the present climate at the sites of measurement, and in such areas, much of the permafrost is a product of a colder past climate.

Characteristics of permafrost

Distribution and thickness

Permafrost is essentially a phenomenon of the polar regions. It occurs in half of the former Soviet Union and Canada, in 85% of Alaska (Ferrians et al., 1969) (Figure P4), 20% of China and probably all of Antarctica. In the Northern Hemisphere permafrost is more widespread and extends to greater depths in the northern than in the southern regions. It is 740 m thick in northern Alaska, 1600 m thick in northern Siberia, and thins progressively towards the south. Permafrost is generally differentiated into two broad zones on land in the polar areas of the northern hemisphere: the continuous and the discontinuous. In the continuous zones (Figure P4), permafrost is nearly everywhere except under the large lakes and rivers that do not freeze to the bottom. The discontinuous zone includes numerous permafrost-free areas that increase progressively in size and number from the north to the south.

The thickness and areal distribution of permafrost are directly affected by natural surface features such as snow and vegetation cover, topography and bodies of water, in addition to the Earth's interior heat and the temperature of the atmosphere. The most conspicuous change in thickness of permafrost is related to climate. If the mean annual air temperature is the same in two areas, the permafrost will be thicker where the conductivity of the ground is higher and the geothermal gradient is less. Lachenbruch and others (1982) report an interesting example from northern Alaska. The mean annual air temperatures at Cape Simpson and Prudhoe Bay are similar, but permafrost thickness is 305 m at Cape Simpson and about 740 m at Prudhoe Bay because rocks at Prudhoe Bay are more siliceous and therefore have a higher conductivity and a lower geothermal gradient than do the rocks at Cape Simpson (Figure P6). Lachenbruch (1968) has graphically illustrated that bodies of water – lakes, rivers and the sea – have a profound effect on the distribution of permafrost (Figure P7). Inasmuch as south-facing slopes of hills receive more incoming solar energy per unit area than other slopes, they are warmer; permafrost is generally absent on these in the discontinuous zone (Figure P8) and is thinner in the continuous zone (Péwé, 1982).

The main role of vegetation in permafrost areas is to shield perennially frozen ground from solar energy. Snow cover also influences heat flow between the ground and the atmosphere and therefore affects the distribution of permafrost. Thus permafrost is not present in areas of the world where great snow thicknesses persist throughout most of the winter.

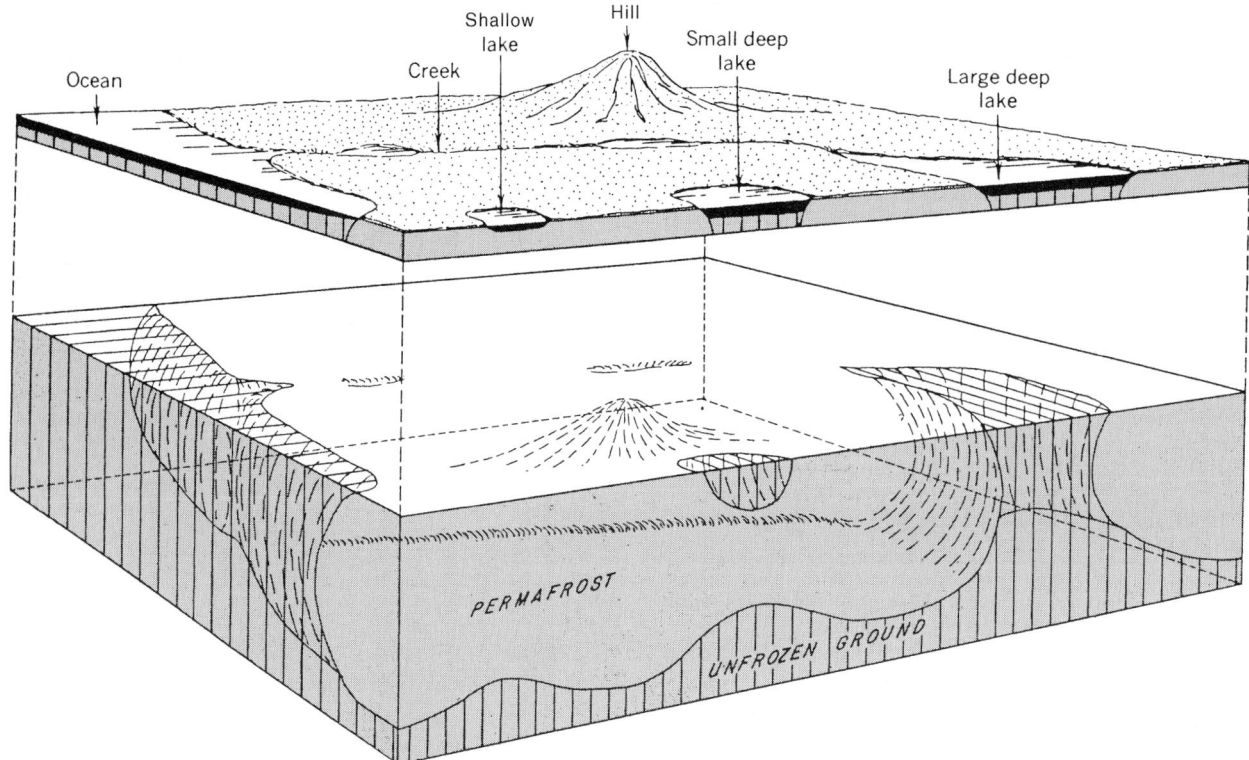

Figure P7 Cutaway block diagram with surface lifted, showing the effect of surface features on the distribution of permafrost in the continuous permafrost zone. (After Lachenbruch, 1968).

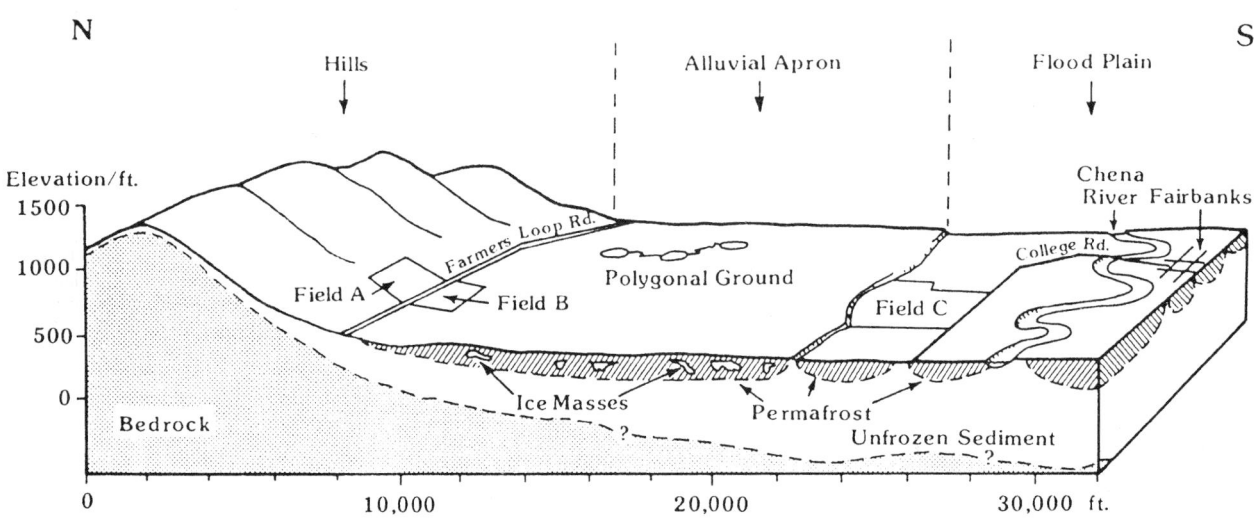

Figure P8 Character and distribution of permafrost in Fairbanks, Alaska, area. (From Péwé, 1954.)

Alpine permafrost

In addition to the widespread perennially frozen ground in the polar areas of the Earth, permafrost also exists at high altitudes in the lower latitudes and has been termed alpine permafrost (Figure P4). Although information about permafrost in the polar areas has been systematically accumulating for many years, data about frozen ground in high plateaus and mountains are sparse (Péwé, 1983).

In the contiguous United States, alpine permafrost is almost entirely limited to the high mountains of the western United States. About 100 000 km² is known to exist (Péwé, 1983). Permafrost occurs as low as 2500 m elevation in the northern states and about 3500 m elevation in Arizona. The largest area of alpine permafrost in the world is in western China (Figure P4), where 1 500 000 km² of permafrost is known (Péwé, 1981; Tong, 1981; Péwé, 1986).

The best evidence for the existence of perennially frozen ground in alpine areas is temperature measurements taken below the active layer that indicate temperatures of 0°C or colder for two or more years. Such temperature measurements are relatively rare. Active ice-cemented (lobate) rock glaciers (Haeberli, 1985), ice wedges, active cryoplanation terraces (Reger and Péwé, 1976), pingos, ice lenses, and pore ice are all evidence of the presence of perennially frozen ground in alpine areas. Moreover, the distribution of alpine permafrost is affected by the interaction of solar radiation and elevation, snowfall, orientation and slope of land, and vegetation; the thickness, in addition, is affected by the thermal conductivity of the rock (Péwé, 1983; King, 1984).

Subsea permafrost
One of the most active and exciting areas of current permafrost research concerns the distribution and properties of permafrost under the Arctic Ocean – on the continental shelf – termed subsea or offshore permafrost. The occurrence is unique and has no real analog on land (Sellmann and Hopkins, 1984). Because of the great hydrocarbon resources on the arctic continental shelves, investigations into subsea permafrost have progressed rapidly in the last 30 years (generally much faster than permafrost investigations on land).

Knowledge of the distribution, type, and water or ice content of subsea permafrost is critical for planning petroleum exploration, locating production structures, burying pipelines, and driving tunnels beneath the sea bed. Furthermore, the temperature of the sea bed must be known in order to predict potential sites for accumulation of gas hydrates or areas in which ground water or artesian pressures are likely. In addition, knowledge of the distribution of the subsea permafrost permits a thorough interpretation of the regional geological history and the position of ancient sea levels.

The following scenario suggests that the origin, distribution, and characteristics of subsea permafrost have a rather simple explanation. At the height of the glacial epochs, especially about 20 000 years ago, most of the continental shelf in the Arctic Ocean was exposed to the polar climates for thousands of years; inner parts were exposed longer than the outer parts. The climate caused cold permafrost to form to depths of more than 700 m (Hopkins *et al.*, 1977; Hunter *et al.*, 1976). Subsequently, within the last 10 000 years, the level of the Arctic Ocean rose and the sea advanced over a frozen landscape to produce a degrading relict subsea permafrost. The perennially frozen ground is no longer exposed to a cold atmosphere, and the saline water causes a reduction in strength and consequent melting of the ice-rich permafrost bonded with freshwater ice. The temperature of subsea permafrost, near −1°C, is no longer as cold, and therefore is sensitive to the warming from internal geothermal heat and encroachment activities of humans (Lachenbruch *et al.*, 1982; Sellmann and Hopkins, 1984).

Ground ice

The ice content of permafrost is probably the most important feature relevant to human life in the North. Ice in perennially frozen ground exists in various sizes and shapes, with definite distribution characteristics grouped into five main types: pore ice, segregated or Taber ice, foliated or ice-wedge ice, pingo and buried ice.

When investigators consider the subject of ground ice, the question is generally raised as to how much ice actually exists in the ground. Because such information would be interesting and valuable from a historical standpoint as well as essential in solving engineering problems posed by permafrost (Lachenbruch, 1970), the question is being considered. Estimates of the volume of worldwide ground ice range from 0.2 to 0.5 million km^3, less than 1% of the total volume of the Earth (Shumskiy and Vtyurin, 1966; Shumskiy *et al.*, 1964, p. 433).

On the basis of an examination of ice in the ground in many bore holes near Barrow, Alaska, and the extrapolation of this borehole information to the rest of the coastal plain, Brown (1967) estimated that 10% by volume of the upper 3.5 m of permafrost of the coastal plain of Alaska is composed of ice wedges (foliated ground ice). Taber ice is the most extensive type of ground ice, in places representing 75% of the ground by volume. Brown (1967) calculated that the pore and Taber ice content in the depth between 0.5 and 3.5 m (surface to 0.5 m is seasonally thawed) is 61%, and between 3.5 and 8.5 m, 41% by volume. The total amount of pingo ice is less than 0.1% of the permafrost. The total amount of perennial ice in the permafrost of the Arctic coastal plain of Alaska is estimated to be 1500 km^3, and below 8.5 m most of that is present as pore ice.

Ice wedges

The most conspicuous and controversial type of ground ice in permafrost is the large ice wedges or masses characterized by parallel or subparallel foliation structures. Most foliated ice masses occur as wedge-shaped, vertical, or inclined sheets or dikes 1 cm to 3 m wide and 1 to 10 m high where seen in transverse cross section (Péwé, 1991). Some masses seen on the faces of frozen cliffs may appear as horizontal bodies a few centimeters to 3 m in thickness and 0.5 to 15 m in length. The true shape of these ice wedges can be seen only in three dimensions. Ice wedges are parts of a polygonal network of ice enclosing cells of frozen ground 3–30 m or more in diameter. They reflect a polygonal microrelief pattern on the surface of the ground that is a characteristic feature of permafrost terrain (Péwé, 1991).

The origin of ice wedges is now generally accepted as being explained by the thermal contraction theory (Leffingwell, 1915, 1919). During the cold winter, polygonal thermal contraction cracks about 1–2 cm wide and 2–3 m deep form in the frozen ground. In early spring, water from the melting snow runs down these tension cracks and freezes and, with accumulating hoar frost, produces a vertical vein of ice that penetrates permafrost. When the permafrost warms and re-expands during the following summer, horizontal compression results in the upturning of the frozen sediment by plastic deformation. During the next winter, renewed thermal tension reopens the vertical ice-cemented crack, which may be a zone of weakness. Another increment of ice is added in the spring when meltwater enters and freezes. Over the centuries the vertical wedge-shaped mass of ice is produced (Lachenbruch, 1962).

Ice wedges require a more rigorous climate to grow than does permafrost. The mean annual air temperature alone does not control the formation of ice wedges (Péwé, 1966, 1975b); rather, the ground cracks when the temperature of the upper part of the permafrost is perhaps colder than −15°C and a winter 'cold snap' occurs, rapidly cooling the permafrost further and causing it to crack. Regions of the world where ice wedges are actively growing over widespread areas have a mean annual air temperature of about −6 to −8°C or colder (northern Alaska, for example). Ice wedges occasionally form in restricted areas (Péwé, 1966) or local cold spots or during colder periods of a few years' duration, in regions with a general mean annual temperature listed as slightly warmer than −6°C.

Impact on engineered works

Development of the polar regions demands understanding and the ability to cope with problems of the environment dictated by permafrost. The most dramatic, widespread and economically important examples of the influence of permafrost on life in the North deal with construction and maintenance of roads, railroads, airfields, bridges, buildings, dams, sewers, oil and gas pipelines, and communication lines (Péwé, 1991).

Principles of land use

A thorough study of frozen ground should be included in the planning of any project in the North. Except in cases of very thin permafrost where thawing is a possibility, it is generally best to disturb frozen ground as little as possible in order to maintain a stable foundation for engineering structures. Construction techniques that preserve permafrost are referred to as the passive method; those that destroy permafrost are termed the active method (Muller, 1943).

Muller (1943, pp. 85, 86) stated 'Once the frozen ground problems are understood and correctly evaluated, a successful solution is, for the most part, a matter of common sense, whereby the frost forces are utilized to play the hand of the engineer and not against it.' With few exceptions, all building, railroad and highway construction in permafrost areas has involved installation of a pad or gravel fill before construction (Figure P6). Dry gravel is generally a better conductor of heat than the underlying silt or vegetation mat. In the cooler parts of the Arctic, it is possible to install a layer of gravel thick enough (about 2 m) to contain seasonal freezing and thawing of the ground. In these instances there is no thawing of the underlying permafrost, and in some cases the permafrost table moves upward into the pad. However, the deep active layer in Subarctic areas precludes containment of seasonal freezing and thawing in a gravel fill. Alternate procedures are necessary, including placement of insulating materials under fill.

Four fundamental types of permafrost-related, land-use problems are (1) thawing of ice-rich permafrost with subsequent surface subsidence under unheated structures such as roads, airfields, agricultural fields and parks, (2) ground subsidence under heated structures, (3) frost action, generally intensified by poor drainage caused by permafrost and (4) freezing of buried sewer, water and oil lines.

Ground subsidence

The most ubiquitous and unique geological hazard of Arctic and Subarctic regions results from the thawing of ice-rich permafrost. This thawing promotes a loss of bearing strength, high moisture content and subsidence of the ground surface. Melting of large ground-ice masses produces dramatic differential settlement and can result from human disturbance of the thermal equilibrium of the ground or from climatic change.

Permafrost has adversely affected agricultural development in many parts of the Subarctic by influencing water supply, soil drainage, the stability of roads and buildings, and especially the topography of cultivated land (Péwé, 1954). The destructive effects of permafrost on cultivated fields result from the thawing of large ground-ice masses. Care must be used in selecting areas for cultivation, because thawing permafrost may force abandonment or modification to pasturage only a few years after clearing (Péwé, 1954). Although fields containing thermokarst mounds and pits can be utilized with repeated grading, excess time, money, and soil are expended in the struggle (Péwé, 1982).

Railroads

Serious construction and maintenance problems created by thawing of ice-rich permafrost affect railroads in Russia, Canada, China and Alaska. The northern part of the Alaska Railroad has been plagued with construction and maintenance problems in dealing with relatively warm permafrost in the discontinuous zone (Péwé, 1949; Péwé and Paige, 1963; Wahrhaftig and Black, 1958; Péwé and Bell, 1975). Differential settlement of the roadbed is extreme when the sensitive permafrost thaws. For example, in the 1960s, the annual maintenance cost of the railroad grade in a 50 km section west of Fairbanks averaged about $200 000. Because of permafrost thawing, an exceptionally wide roadbed was constructed on a thick layer of gravel over frozen ice-rich silt, but it has not prevented the track from settling up to 0.6 m year^{-1}, even after 50 years of construction (Péwé, 1982).

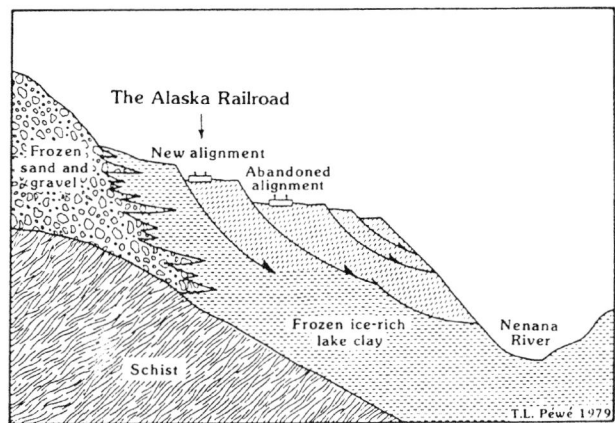

Figure P9 Diagrammatic sketch of landslides along the Alaska Railroad in perennially frozen lake clay. After the vegetation is removed, the heat that is absorbed from the channelled surface-drainage water thaws the lake clay. As the clay thaws, large blocks of sediment slip towards the Nenna River canyon, and the railroad track must be realigned.

The most spectacular engineering geology permafrost problem dealing with railroads in the United States is in the Nenana River Gorge (Healy Canyon) of the Alaska Range in central Alaska (Wahrhaftig and Black, 1958; Fuglestad, 1986). For 16 km the river flows through a two-story canyon. The outer canyon is a glacial valley with flaring walls that rise 770 m above the 0.7–1.2 km wide canyon floor. At the beginning of the canyon and again on its last 8 km, the river flows in a 150 m wide postglacial inner gorge that has walls 60–90 m high (Fuglestad, 1986). In the gorge are deposits of a glacial

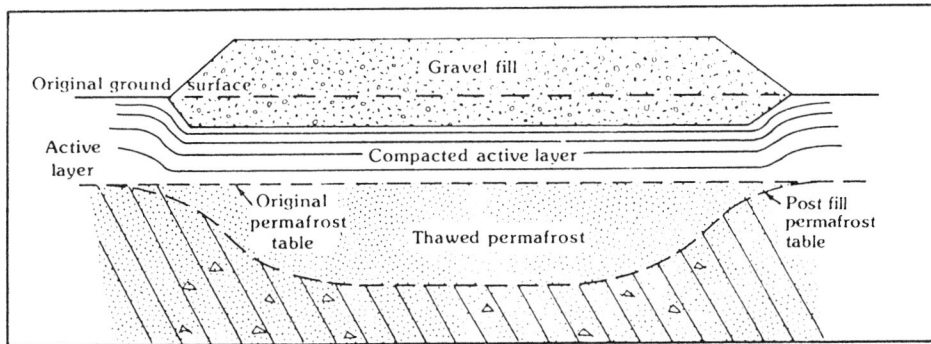

A. Effect on permafrost table if insulating of fill plus compacted active layer is less than the insulating effects of original active layer.

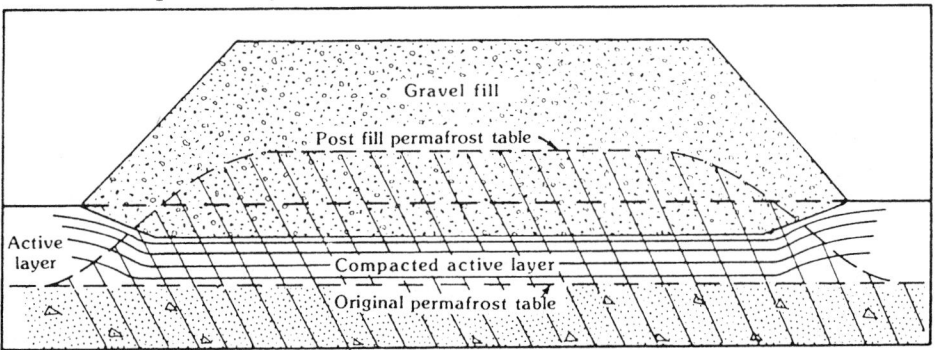

B. Effect on permafrost table if insulating effect of fill and active layer is greater than the insulating effect of the original active layer.

Figure P10 Diagram showing the effect of different thicknesses of gravel fill upon the thermal regime of the ground and the modification of the level of the permafrost table under the fill. (From Ferrians et al., 1969.)

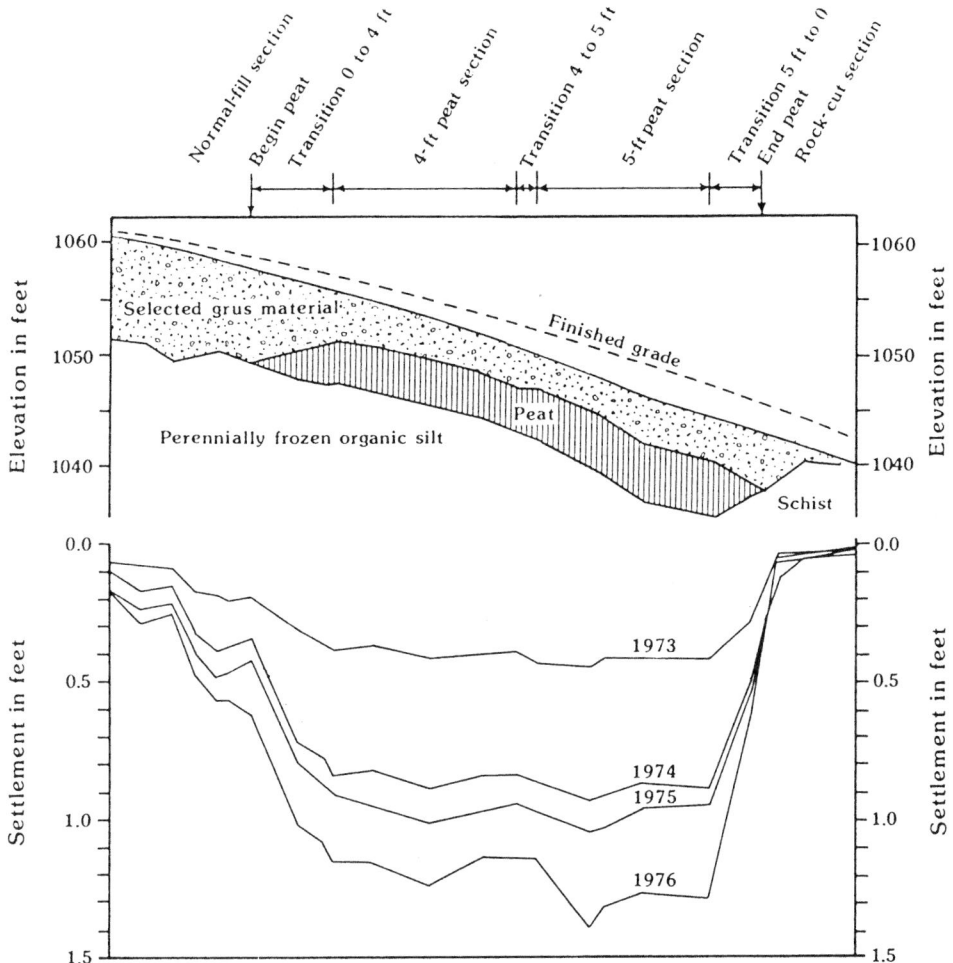

Figure P11 Comparison of longitudinal (centerline) profile of peat test section and amounts of settlement after 4 years (1973–1977), Mile 300.7, Richardson Highway, Alaska. (From Péwé and Reger, 1983.)

lake clay that are perennially frozen and ice rich. Numerous unstable bedrock slopes, and the thawing of frozen lake clays along with the downcutting and eroding of the river, create landslide conditions (Figure P9) that are exacerbated by railroad construction and maintenance that greatly disturbs the thermal equilibrium of the permafrost.

Railway service has been delayed for several days at a time as huge slides up to 1 km wide and 610 m long occur. The track is continually moved downward and laterally towards the river. Slumping has been as much as 1.2 m day^{-1}, the most critical area being the Moody landslide area (Péwé, 1991).

Roads, highways and airfields
Engineered works such as roads, highways and airfields are especially susceptible to deterioration and eventual destruction by thawing of ice-rich permafrost. Tremendous strides have been made in the last 20 years in utilizing detailed permafrost mapping to aid design prior to construction (Esch, 1984). Numerous techniques are being developed to prevent heat flow from the warm atmosphere in the summer into the cold, underlying ground (Figure P10). These efforts have met with considerable success; however, it has not been economically feasible as yet to insulate great distances, or great areas, by artificial means. One test in central Alaska (Péwé and Reger, 1983) along the Richardson Highway demonstrated that if prethawed and preconsolidated peat were placed beneath the road fill, it would delay the thawing of permafrost for some time and require a thinner thickness of road fill (Figure P11). The test section freezes much sooner and the underlying permafrost chills much colder than other sections.

An example of high-cost road maintenance in ice-rich permafrost environment occurred at the eastern entrance of the University of Alaska at Fairbanks. In 1962 a road was rerouted across ice-rich peat and organic silt at the east entrance. This 0.25 km rerouting cost $170 000. Shortly after the subgrade was paved, differential settlement occurred because of thawing of the underlying ground ice and ice-rich peat and subsequent compacting of the peat. Releveling of the road shoulders and similar annual maintenance, including leveling with asphalt, has been necessary since 1974.

Another example at the University of Alaska–Fairbanks has been called the Farmers Loop 'sinkhole' because of the tremendous amount of thaw-subsidence that has occurred during the past few years. An accumulated pavement thickness of up to 1.8 m has been formed from patches in limited areas (Péwé, 1991). To further the ongoing investigation, the road was painted white in 1983 in order to promote reflection of solar radiation. Thermal heat tubes were also installed at the west edge of the road in an attempt to freeze the ground beneath the highway and prevent further thawing and subsidence.

Destruction of highways from ice-rich permafrost continues to be a serious problem in the polar areas. The best defense against this natural hazard is preconstruction geological investigation. Subsequently a relocation of roads to non-frozen areas, or areas of sand and gravel containing scant ground ice, may be advisable. If these alternatives are not practical or economical, the placement of insulation between the gravel overlay and the subgrade may reduce the problem, but special maintenance may be necessary (Figure P10).

Buildings

Heated buildings introduce more heat into the ground and cause more thawing of ice-rich permafrost in a shorter time than a highway, railroad, cleared field or other area where the natural surface has been disturbed. Instead of gradual subsidence over a large area, there may be local differential lowering of the ground surface 12–25 cm or more a year.

Such distortion and destruction of structures are generally spectacular and emphasize the hazards of thawing permafrost and its effect on people's budgets. Most people are only vaguely aware of the maintenance or replacement costs of deformed highways and airfields. However, the cost and importance of the ice-rich permafrost hazard are dramatically realized when one observes tilted homes, abandoned cabins and ill-placed split-level homes (Péwé, 1991).

It is disheartening to notice the deformation and destruction of homes in certain areas of the world because of improper construction on ice-rich permafrost. Even though detailed information about local permafrost has been available from local and federal agencies since the 1940s – and earlier from some mining companies – buildings continue to be constructed in a manner that will result in deformation, increased maintenance costs and perhaps abandonment. Obviously, the ideal solution is to construct in permafrost-free areas or on permafrost with a relatively low ice content.

If construction must proceed in areas of ice-rich permafrost, special engineering techniques can be employed to preserve the permafrost intact so that building heat does not enter the ground. Placement on piles is one common solution; another is to provide openings under the building for cold air circulation and building-heat dispersion (Péwé, 1991).

Sufficient scientific and engineering expertise is now available for the trouble-free construction of engineered works on ice-rich permafrost. Unfortunately, frequency and avoidance costs of permafrost problems are higher in areas of 'warm' sensitive permafrost, such as Fairbanks, Alaska, than in areas where permafrost is colder and less sensitive.

Effects on buried pipelines

General statement

The extraction and transportation of commercial large-scale oil and natural gas in the Arctic have introduced a new set of permafrost-related geological, engineering and environmental problems. Although natural gas and oil have been extracted in the Arctic for many years, production and transportation have been relatively limited. Small, temporary pipelines for crude oil and refined products were built in northwestern Canada (Norman Wells) and Alaska (along the Alaska Highway) during World War II and shortly thereafter.

The most favored means of transporting petroleum products is by large-diameter, above- or below-ground pipeline. Natural gas may assume the temperature of its surroundings, or it may even be chilled and thus cause little thawing of the permafrost; nevertheless, the burial of a pipeline, with the stripping of surface vegetation, causes a disturbance of the thermal equilibrium of the frozen ground. Oil pipelines can cause a major disruption of the thermal equilibrium, because the flowing oil is warmer than permafrost. The basic problems involved with pipeline construction in permafrost regions are the thawing of permafrost if the pipe is buried, and frost action of the supporting piles if built above ground.

A few short- to long-distance natural-gas pipelines exist in permafrost regions of world. These pioneer lines are mostly above ground, even though the cold gas would not significantly thaw the permafrost. More importantly, the difficulty of excavation in permafrost and the resulting disturbance of the thermal equilibrium have favored above-ground construction. Such pipelines are exposed and subject to expansion and contraction, weathering and damage by humans and natural elements, especially wind. Also, the above-ground line serves as a barrier for humans and animals and must be locally buried or elevated.

Major natural gas pipelines in the north are known only in Russia. A 72 cm diameter unshielded line from Messoyakha to Norilsk for 300 km is about 1 m or more above ground and supported either by logs lying on the ground or bents supported by piles of either wood, concrete or steel (Péwé, 1991). Locally the line is about 3–4 m above the ground on longer piles to create artificial underpasses. The first line was built during 1967 and 1968, and the new line, incorporating more advanced designs, between 1971 and 1973. It is reported that wind caused the destruction of a part of the older lines.

A gas pipeline about 30 cm in diameter is buried 1 m in loess at a point 4 km west of Yakutsk, Siberia. Where this line crosses low terraces of the Lena River near Yakutsk, it divides into two smaller lines, and they are elevated on wood piles to form an underpass. These piles, and those on the Norilsk line, are subject to extreme frost heaving and differential jacking that deform the pipelines and destroy the continuity of supports (Péwé, 1976).

Crude oil transported in pipelines is warmer than the temperature of permafrost. The consequence, that initial heat of the oil plus frictional heating of the pipe will thaw the surrounding perennially frozen ground, is the basic problem of oil pipeline construction in permafrost regions. The rate and amount of thaw are dictated by the temperature of the oil and permafrost, if other parameters are stable.

Trans-Alaska pipeline system

The long-anticipated, vast petroleum potential of northern Alaska was in part realized with the discovery of oil at Prudhoe Bay in 1968. The 9.6 billion barrels of proven reserves suggested that one-third of the oil reserves of the United States in the 1970s were in northern Alaska. The warm crude oil was transported from the North Slope to an ice-free port at Valdez (Figure P12) by construction of a 1.2 m diameter pipeline that was 1285 km long. The Trans-Alaska Pipeline System (TAPS), a remarkable construction achievement in a permafrost environment, was completed in 1977. The pipeline is designed to transport as many as 2 million barrels per day. The total cost of constructing TAPS was about $8 billion. Of this, some $1 billion was spent to investigate and understand the characteristics of perennially and seasonally frozen ground, demonstrating the impact frozen ground had on the cost of construction.

The pipeline was originally designed for burial in permafrost along most of the route. With the oil temperature (at full production) estimated at 70–80°C, such an installation would have thawed the adjacent permafrost. Thawing of the widespread ice-rich permafrost by a warm-oil line can cause liquefaction, loss of bearing strength and soil flow. Differential settlement of the line can occur, and mudflows of thawed soil may form on slopes. The greatest differential settlement could occur in areas of ice wedges, where troughs could form and deflect surface water into the trenches, causing erosion and more thawing.

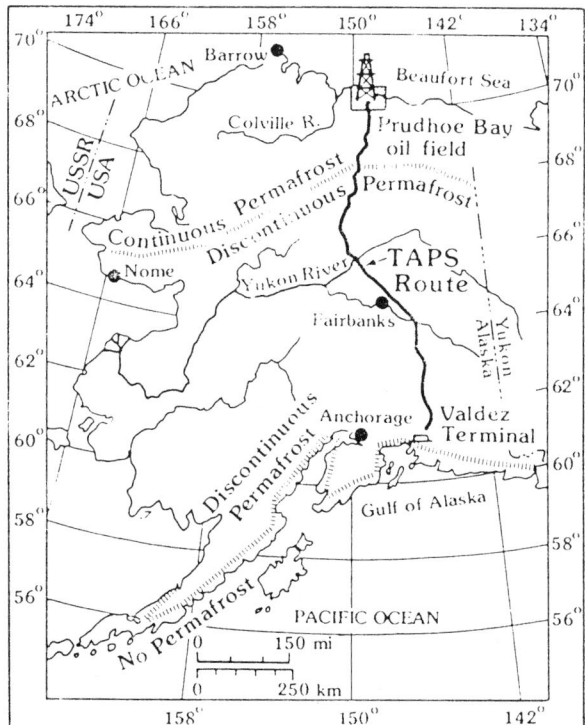

Figure P12 Index map of Alaska indicating the route of the Trans-Alaska Pipeline system (TAPS). (From Péwé, 1982.)

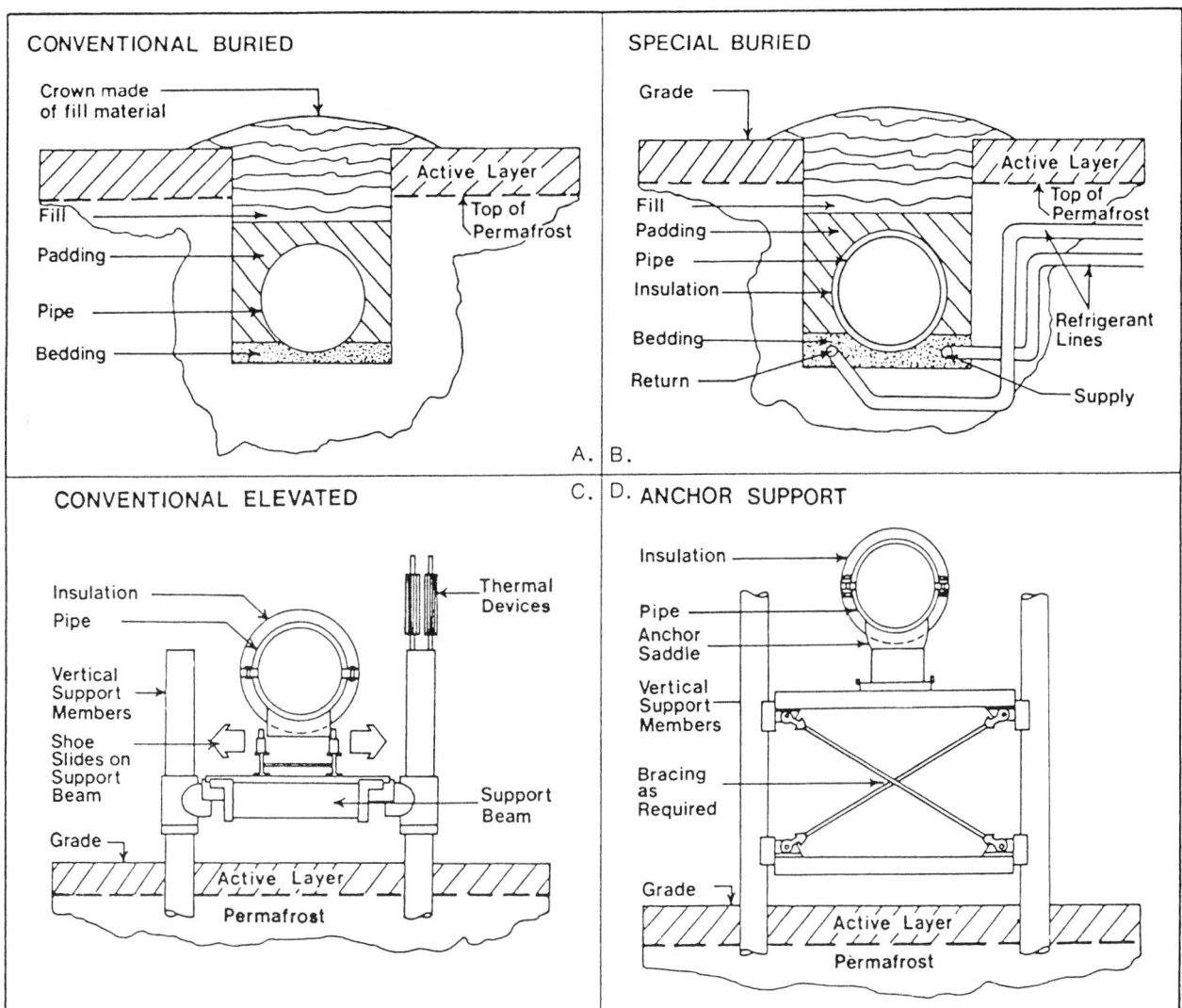

Figure P13 Different construction modes used for the building of the Trans-Alaska Pipeline by the Aleyeska Pipeline Service Company. (A) Conventional buried, (B) special buried, (C) conventional elevated and (D) anchor support. (Diagram courtesy Aleyeska Pipeline Service Company.)

The pipeline was built in three construction modes, depending on the environment, terrain and permafrost conditions (Figure P13). Oil is pumped through the pipe at temperatures up to 63°C, depending on production rates and the heat generated by pumping and friction with the pipe. Consequently, the potential effects of the heat on the specific frozen ground along the route determined the mode of pipeline installation. For example, in areas where the ice content is very low or absent or where no permafrost exists (some 658 km), the pipe is buried in the conventional manner, as it would be in most areas of the world (Figure P13A). In contrast, seven short sections of the pipeline (total 11.2 km) were buried and then frozen into the ground (Figure 10B). These sections can provide crossings for caribou and other animals and are located in both ice-poor and ice-rich permafrost environments. The temperature of the permafrost in which the pipe is buried is maintained by pumping refrigerated brine through pipes emplaced beneath the pipeline.

About half the pipeline (615 km) is elevated above ground with anchor support because of the presence of ice-rich permafrost (Figure P13D). Although an above-ground pipeline successfully discharges its heat into the air and does not directly affect the underlying permafrost, other problems caused by permafrost and associated phenomena must be considered. For example, to eliminate frost heaving of the 120 000 vertical support members (VSM) (Figure P13C, D), each VSM is frozen firmly into the permafrost using a special thermal

device as shown on Figure P13C. To compensate for expansion of the above-ground pipe caused by the warm oil, and contraction caused by extremely cold air temperatures in winter, the line is built in a flexible zigzag configuration (Péwé 1991), which converts expansion of the pipe into lateral movement and likewise accommodates pipe motions induced by earthquakes (Figure P13C). As the line expands or contracts, the pipe slides across the beam; the pipe is anchored on the crossbeams at the end of each zigzag configuration (every 240–540 m).

Utility lines

Water, steam, gas, sewer and other utility lines are commonly buried in the permafrost areas. To prevent freezing, such lines are usually placed in underground boxes, called 'utilidors', from 0.3 m to more than 2 m wide (Figure P14). Construction of buried utilidors in ice-rich permafrost generally creates problems similar to those of a heated building, triggering thawing of permafrost and subsidence of the utilidor. This causes pipe breakage and the eventual destruction of the system (Figure P14).

Dams

In permafrost areas the design and construction of dams are more complex because of the thawing effect of the impounded water on the perennially frozen ground underlying the reservoir and structures

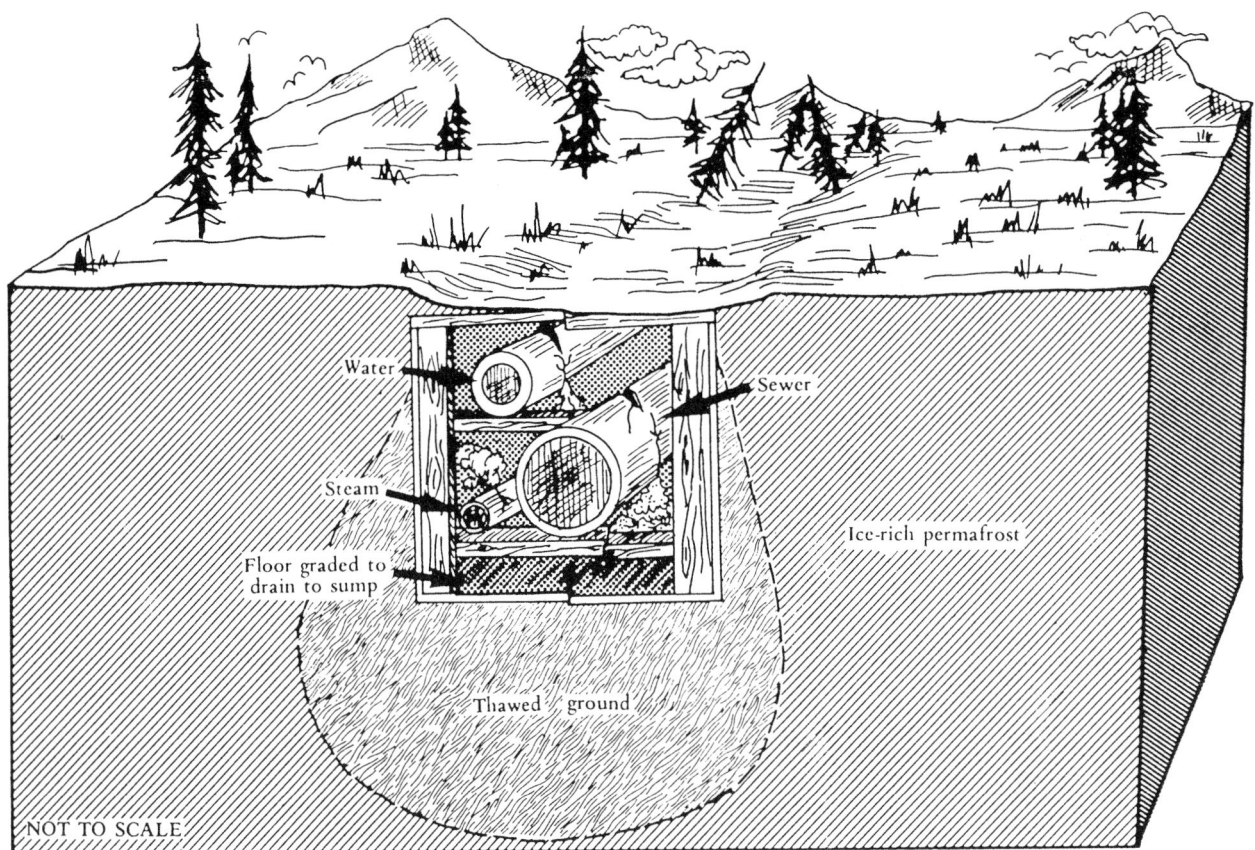

Figure P14 Diagrammatic sketch (exaggerated) of a utilidor buried in ice-rich permafrost. Heat from the utilidor has thawed the ground, and allowed the irregular subsidence of the utilidor that resulted in broken pipes. This is not an uncommon occurrence in permafrost areas, especially when the ice content is not uniformly distributed. (From Péwé, 1982.)

(Johnston, 1981; Sayles, 1987). Thawing may cause differential movement and seepage. Most dams are earthfill structures in permafrost areas, and two types may be constructed: impervious or semi-pervious. The former, designed to maintain the foundation of embankment in a frozen condition, is best in the continuous permafrost zone. The latter allows for thawing under the structure.

Only a few small structures have been built on permafrost in North America, and they are, for the most part, in the discontinuous zone (Rice and Simoni, 1966). Several large and small earthfill dams have been built in Russia in both the discontinuous and continuous zones (Figure P4; Johnston, 1981).

The largest dam built on permafrost (in 1968) is reported by Péwé (1973) to be on the Vilyuy River in Siberia at the Vilyuyskaya hydroelectric power station near the village of Chernysheviskiy. The dam is an earth-fill structure 75 m high and 600 m long and contains about 5 000 000 m^3 of material. It is built on jointed diabase that contains ice in the joints. The underlying bedrock has been thawing at a rate of 4–8 m year^{-1}; it is grouted as the ice melts.

Groundwater and permafrost

Groundwater flow systems in permafrost regions are greatly modified. The frozen ground acts as an impermeable layer that (1) restricts recharge, discharge, and movement of groundwater (2) acts as a confining layer and (3) limits the amount of liquid water that can be stored in unconsolidated sediments and bedrock (Williams, 1970). Groundwater may occur above, within and below permafrost.

Research investigations on groundwater in permafrost regions in North America essentially began immediately after World War II, even though some observations were reported much earlier (Tyrell, 1903). Perhaps the first systematic studies were in connection with sanitation engineering by Alter (1950a,b) in Alaska. Both the US Geological Survey and the Canadian Geological Survey began detailed studies some of which continue on a reduced scale (Cederstrom,

1952; Hopkins *et al.*, 1955; Cederstrom and Péwé, 1961; Williams, 1966, 1970; Brandon, 1965; and many others).

The earlier International Permafrost Conferences contained entire sections devoted to the subject of the science and engineering of groundwater problems (Brandon, 1966; Williams and Waller, 1966; Williams and Van Everdingen, 1973). Ground water basin studies were expounded in regard to investigations for pipeline projects and petroleum production installations. Hydrological basin model studies are continuing in Alaska and Canada (Williams and Van Everdingen, 1973). The most detailed studies concerned with groundwater movement and the origin of pingos are those by MacKay (1966, 1968, 1973, 1979, 1986).

Outlook for permafrost research: an assessment of future needs

Preliminary statement

After World War II, permafrost research was well established, but in the 1970s a new stimulus appeared: the discovery of petroleum in northern Alaska. In the 1970s major advances occurred in the understanding of permafrost (National Academy of Sciences–National Research Council, 1983b), especially the design and structure of the Trans-Alaska oil pipeline across about 1285 km of permafrost terrain. At that time the National Academy of Sciences–National Research Council spearheaded special studies to outline trends and needs for permafrost research: (1) basic research (1974), (2) research associated with the Trans-Alaska Pipeline (1975) and (3) problems and priorities in offshore permafrost (1976). With the economic impact of permafrost finally becoming better known, it is anticipated that in the next decade scientists and engineers will be called on to solve problems related to onshore and offshore oil and gas explorations in polar areas, transportation across permafrost terrain, and use of alpine permafrost areas for recreation and military facilities. The challenges of sea ice and related phenomena to offshore exploration are discussed in Weeks and Brown (1991).

Recommendations for permafrost research in the immediate future

The National Academy of Sciences–National Research Council (1983b) recognized a serious need to improve our ability to detect and determine the configuration of ground-ice masses by drilling, geophysics and applied sciences because of the serious engineering – and environment-related problems caused by the degradation of ice-rich permafrost. Furthermore, it recognized the need for a much improved understanding and knowledge of the origin of all types of ground ice.

The report concluded that to advanced engineering technology, it is necessary to (1) carefully monitor prototype existing facilities constructed in permafrost and (2) develop improved methods to predict heat and mass transport within permafrost and across its boundaries with adjacent media. Furthermore, research must be continued and expanded to determine the physical, chemical, mechanical, thermal and electrical properties of permafrost, especially saline permafrost.

Subsea permafrost investigations also remain high on the list of subjects that need increased knowledge. To attain this, both industry and government should continue to collect and analyze data concerning distribution, thickness, and properties of permafrost and also continue refinement of theoretical models of subsea permafrost, critical to predict the depth and thickness of ice-bonded permafrost.

There is now a unique opportunity for scientists and engineers to work together, collecting new data and analyzing and developing a better understanding of processes that affect permafrost and its long-term behavior. There is a clear relationship between the protection of permafrost environments and the long-term performance of artificial structures, which must be considered by designers (National Academy of Sciences–National Research Council, 1983b).

Source

Péwé, T.L., 1991. Permafrost, in Kiersch, G.A., ed., *The heritage of engineering geology, the first hundred years*. Geological Society of America Centennial Special Volume 3.

Bibliography

Alter, A.J., 1950a. *Arctic sanitary engineering*, Washington, DC, Federal Housing Administration, 106 pp.
Alter, A.J., 1950b. Water supply in Alaska, *American Water Works Association Journal*, **42**, 519–532.
Barnes, L.C., 1946. Permafrost; A challenge to engineers, *Military Engineering*, **38**, 9–11.
Black, R.F., 1950. Permafrost, in Trask, P.D., ed., *Applied sedimentation*. New York, John Wiley and Sons, pp. 247–275.
Boswell, J.C., 1979. *History of Alaskan operations of US Smelting, Refining, and Mining Company*. Fairbanks, University of Alaska, 126 pp.
Brandon, L.V., 1965. *Groundwater hydrology and water supply in the District of MacKenzie, Yukon Territory, and adjoining parts of British Columbia*. Canadian Geological Survey Paper 64–39, 102 pp.
Brandon, L.V., 1966. Evidences of groundwater flow in permafrost regions, in *International Conference on Permafrost, Lafayette, Indiana, 1963, Proceedings*, Washington, DC, National Academy of Sciences, National Research Council Publication 1287, pp. 176–177.
Brown, J., 1967. *An estimation of the volume of ground ice, coastal plain, northern Alaska*, US Army Materiel Command Cold Regions Research and Engineering Laboratory Technical Note, 22 pp.
Cederstrom, D.J., 1952. *Summary of ground-water development in Alaska, 1950*. US Geological Survey Circular 169, 37 pp.
Cederstrom, D.J. and Péwé, T.L., 1961. *Ground-water data, Fairbanks area, Alaska*, Alaska Department of Health and Welfare Hydrological Data Report 9, 28 pp.
Corte, A.E., 1969. Geocryology and engineering, in Varnes, D.J. and Kiersch, G.A., eds, *Reviews in engineering geology II*. Boulder, Geological Society of America, pp. 119–185.
Esch, D.C., 1984. Design and performance of road and railway embankments on permafrost, in *Permafrost, Fourth International Conference, Final Proceedings*, Fairbanks, Alaska. Washington, DC, National Academy of Science, pp. 25–30.
Ferrians, O.J. Jr., Kachodoorian, R. and Green, G.W., 1969. *Permafrost and related engineering problems in Alaska*, US Geological Survey Professional Paper 678, 37 pp.
French, A.M., 1976. *The periglacial environment*. London, Longman, 309 pp.
French, A.M., 1982. *Proceedings of the Fourth Canadian Permafrost Conference; The Roger J.E. Brown Memorial Volume*. Ottawa, National Research Council of Canada, 591 pp.
Fuglestad, T.C., 1986. *The Alaska Railroad between Anchorage and Fairbanks, guidebook to permafrost and engineering problems*. The Alaska Division of Geological and Geophysical Surveys, Guidebook no. 6, 82 pp.
Haeberli, W., 1985. *Creep of mountain permafrost; Internal structure and flow of alpine rock glaciers*. Zurich, Mittgeilugen der Versuchsanstalt fur Wasserbau Hydrologie und Glaziologie, no. 77, 142 pp.
Hardy, R.M. and D'Appolonia, E., 1946. Permanently frozen ground and foundation design. *Engineering Journal of Canada*, **29**, 1–11.
Hemstock, R.A., 1949a. *Engineering in permafrost in Canada's Mackenzie Valley*. National Research Council of Canada Technical Memoir 13, 3 pp.
Hemstock, R.A., 1949b. *Permafrost at Norman Wells; NWT*. Calgary, Canada, Imperial Oil Limited, 100 pp. (pub. in 1953).
Highway Research Board, 1948. *Bibliography on frost action in soils*. Washington, DC, Highway Research Board Bibliography no. 3, 47 pp.
Hopkins, D.M., Karlstrom, T.N.V. *et al.*, 1955. *Permafrost and groundwater in Alaska*. US Geological Survey Professional Paper 214-F, p. 113–146.
Hopkins, D.M. *et al.*, 1977. Earth science studies, in *Beaufort Sea synthesis report*, National Oceanic and Atmospheric Administration Arctic Special Bulletin 15, Fairbanks, University of Alaska, pp. 43–72.
Hunter, J.A.M., Judge, A.S., MacAualy, H.A. *et al.*, 1976. *Permafrost and frozen sub-sea bottom materials in the southern Beaufort Sea* Canada, Department of the Environment, Beaufort Sea Project report 22, 174 pp.
Joesting, H.R., 1941. Magnetometer and direct-current resistivity studies in Alaska *American Institute of Mining and Metallurgical Engineers Transactions*, **164**, 66–87.
Johnson, A.W., 1952. *Frost action in roads and on fields; Review of the literature 1765–1951*. Highway Research Board Special Report 1, 287 pp.
Johnston, G.H., ed., 1981. *Permafrost; Engineering design and construction*. New York, Wiley and Sons, 340 pp.
King, von L., 1984. *Permafrost in Scandinavia*. Heidelberg, Heidelberger Geographische Arbeiten no. 76, 174 pp.
Lachenbruch, A.H., 1962. *Mechanics of thermal contraction cracks and ice-wedge polygons in permafrost*. Geological Society of America Special Paper 70, 69 pp.
Lachenbruch, A.H., 1968. Permafrost, in Fairbridge, R.W. ed., *The Encyclopedia of Geomorphology*. New York, Reinhold Publishing Corporation, pp. 833–839.
Lachenbruch, A.H., 1970. *Some estimates of the thermal effects of a heated pipeline in permafrost*. US Geological Survey Circular 632, 23 pp.
Lachenbruch, A.N., Sass, J.H., Marshall, B.V. and Moses, T.H. Jr, 1982. Permafrost, heatflow, and the geothermal regime at Prudhoe Bay, Alaska. *Journal of Geophysical Research*, **84**(11), 9301–9316.
Leffingwell, E. de K., 1915. Ground-ice wedges, the dominant form of ground-ice on the north coast of Alaska. *Journal of Geology*, **23**, 635–654.
Leffingwell E. de K., 1919. *The Canning River region, northern Alaska*. US Geological Survey Professional Paper 109, 251 pp.
Mackay, J.R., 1966. Pingos in Canada, in *International Conference on Permafrost, Lafayette, Indiana, 1963, Proceedings*. Washington, DC, National Academy of Sciences, National Research Council Publication 1287, pp. 71–76.
Mackay, J.R., 1968. Discussion of the theory of pingo formation by water expulsion in a region affected by subsidence. *Journal of Glaciology*, **7**, p. 346–351.
Mackay, J.R., 1973. The growth of pingos, western Arctic coast, Canada. *Canadian Journal of Earth Science*, **10**, p. 979–1004.

Mackay, J.R., 1979. Pingos of the Tuktoyaktuk Peninsula area, Northwest Territories. *Geographie Physique et Quaternaire*, **33**, p. 3–61.

Mackay, J.R., 1986. Growth of Ibyuk Pingo, western Arctic coast, Canada, and some implications for environmental reconstructions. *Quaternary Research*, **26**(1), p. 68–80.

Melnikov, P.I., 1984. Major trends in the development of Soviet permafrost research, in *Permafrost, Fourth International Conference, Final Proceedings*, Fairbanks, Alaska. Washington, DC, National Academy of Sciences, pp. 163–166.

Muller, S.W., 1943. *Permafrost or permanently frozen ground and related engineering problems*. US Army, Office, Chief of Engineers, Strategic Engineering Study Special Report 62, 231 pp.; 1945, second printing with corrections; reprinted in 1947, Ann Arbor, MI, J.W. Edwards, Inc.

National Academy of Sciences–National Research Council, 1966. *International Conference on Permafrost, 1963, Proceedings*, Lafayette, Indiana. Washington, DC, National Research Council Publication 1287, 563 pp.

National Academy of Sciences–National Research Council, 1973. *Permafrost, the North American contribution to the Second International Conference*, Yakutsk, Siberia, USSR. Washington, DC, National Research Council, 783 pp.

National Academy of Sciences–National Research Council, 1974. *Priorities for basic research on permafrost*. Washington, DC, National Research Council, 54 pp.

National Academy of Sciences–National Research Council, 1975. *Opportunities for permafrost-related research associated with the Trans-Alaska Pipeline System*. Washington, DC, National Research Council, 37 pp.

National Academy of Sciences–National Research Council, 1976. *Problems and priorities in offshore permafrost research*. Washington, DC, National Research Council, 43 pp.

National Academy of Sciences–National Research Council, 1978. *Permafrost; the USSR contribution to the Second International Conference*, Yakutsk, Siberia, USSR. Washington, DC, National Research Council, 866 pp.

National Academy of Sciences–National Research Council, 1983a. *Permafrost; Fourth International Conference, Proceedings*, Fairbanks, Alaska. Washington, DC, National Research Council, 1524 pp.

National Academy of Sciences–National Research Council, 1983b. *Permafrost research; An assessment of future needs*. Washington, DC, National Research Council, 103 pp.

National Academy of Sciences–National Research Council, 1984. *Permafrost; Fourth International Conference, Final Proceedings*, Fairbanks, Alaska. Washington, DC, National Research Council, 413 pp.

National Research Council of Canada, 1978. *Proceedings; Third International Conference on Permafrost*, Edmonton. Ottawa, National Research Council of Canada, 1, 947, pp.; vol. 21, 255 pp.

Péwé, T.L., 1948. *Terrain and permafrost of the Galena Air Base, Galena, Alaska*. US Geological Survey Permafrost Program Progress Report 7, 52 pp.

Péwé, T.L., 1949. *Preliminary report of permafrost investigations in the Dunbar area Alaska*. US Geological Survey Circular 42, 3 pp.

Péwé, T.L., 1954. *Effect of permafrost on cultivated fields, Fairbanks area, Alaska*. US Geological Survey Bulletin 989-F, pp. 315–351.

Péwé, T.L., 1966. Ice wedges in Alaska; Classification, distribution, and climatic significance, in *International Conference on Permafrost, Lafayette, Indiana, 1963, Proceedings*. Washington, DC., National Academy of Sciences, National Research Council Publication 1287, pp. 76–81.

Péwé, T.L., 1973. Permafrost Conference in Siberia. *Geotimes*, December, 23–26.

Péwé, T.L., 1974. Permafrost. *Encyclopaedia Britannica*, vol. 14, pp. 89–95.

Péwé, T.L., 1975a. *Quaternary stratigraphic nomenclature in unglaciated central Alaska*. US Geological Survey Professional Paper 862, 32 pp.

Péwé, T.L., 1975b. *Quaternary Geology of Alaska*. US Geological Survey Professional Paper 835, 145 pp.

Péwé, T.L., 1976. Permafrost in *1976 Yearbook of Science and Technology*. New York, McGraw-Hill Book Co., pp. 32–47.

Péwé, T.L., 1981. Tibetan science updated. *Geotimes*, **26**(1), 16–20.

Péwé, T.L., 1982. *Geological hazards of the Fairbanks area, Alaska*. Division of Geological and Geophysical Survey Special Report 15, 109 pp.

Péwé, T.L., 1983. Alpine permafrost in the contiguous United States; A review. *Arctic and Alpine Research*, **15**(2), 145–156.

Péwé, T.L., 1986. China expands research in frozen ground; A report of the Third Chinese Conference on Permafrost. *Zeitschrift für Gletscherkunde und Glazialgeologie*, **22**, 1–7.

Péwé, T.L., 1991. Permafrost, in Kiersch, G.A., ed., *The heritage of engineering geology, the first hundred years*, Geological Society of America Centennial Special Volume 3.

Péwé, T.L. and Bell, J.W., 1975. Map showing distribution of permafrost in the Fairbanks D2 NW Quadrangle, Alaska. US Geological Survey Map FM 688 A, scale 1:24 000.

Péwé, T.L. and Paige, R.A., 1963. *Frost heaving of piles with an example from the Fairbanks area, Alaska*. US Geological Survey Bulletin 1111–I, pp. 333–407.

Péwé, T.L. and Reger, R.D., 1983. *Richardson and Glenn Highways, Alaska; Guidebook to permafrost and Quaternary geology*. Alaska Division of Geological and Geological Survey, Guidebook no. 1, 263 pp.

Reger, R.D. and Péwé, T.L., 1976. Cryoplanation terraces; Indicators of a permafrost environment. *Quaternary Research*, **6**, 99–109.

Rice, E.F. and Simoni, O.W., 1966. The Hess Creek dam, in *International Conference on Permafrost, Lafayette, Indiana, 1963, Proceedings*. Washington, DC, National Academy of Sciences, Natural Research Council Publication no. 1287, pp. 436–439.

Sayles, F.H., 1987. *Embankment dams on permafrost*. US Army Corps of Engineers, Cold Regions Research and Engineering Laboratory Special Report 87–11, 109 pp.

Sellmann, P.V. and Hopkins, D.M., 1984. Subsea permafrost distribution on the Alaskan shelf, in *Permafrost; Fourth International Conference, Final Proceedings*, Fairbanks, Alaska. Washington, DC, National Academy of Sciences, pp. 75–82.

Shumskiy, P.A. and Vtyurin, B.I., 1966. Underground ice, in *Permafrost International Conference, Lafayette, Indiana, 1963, Proceedings*. Washington, DC, National Academy of Science–National Research Council Publication 1284, pp. 108–113.

Shumskiy, P.A., Krenke, A.N. and Zotikov, I.A., 1964. *Ice and its changes in solid earth and interface phenomena*. Vol. 2, *Research and geophysics*. Cambridge, Massachusetts Institute of Technology Press, pp. 425–460.

Tong, Boling, 1981. Some features of permafrost on the Qinghai-Xizang Plateau and factors influencing them, in Liu Dongsheng, ed., *Geological and ecological studies of Qinghai-Xizang Plateau; Vol. 2, Environment and ecology of Qinghai-Xizang Plateau*. Beijing, Science Press, pp. 1795–1801.

Tyrell, J.B., 1903. A peculiar artesian well in the Klondike. *Engineering Mining Journal*, **75**, 188.

US Army Corps of Engineers, 1946. Airfield pavement design; Frost conditions, in *An interim engineering manual for War Department construction*. US Army Corps Engineers, Pt. 12, Ch. 4, 10 pp.

US Army Corps of Engineers, 1947. *Report on frost investigation 1944–1945*, US Army Corps Engineers, 66 pp.

US Army Corps of Engineers, 1949. *Addendum No. 1, 1945–1947, report on frost investigation 1944–45*. US Army Corps Engineers, 50 pp.

US Army Corps of Engineers, 1951. *et seq. Bibliography on snow, ice and permafrost*, vols 1–16, US Army SIPRE; vols 7–20, US Army CRREL; *Bibliography on snow, ice, and frozen ground, with abstracts*, Vols 21–22; *Bibliography on cold regions science and technology*, Vols 23–40.

US Army Corps of Engineers, 1954. Arctic and subarctic construction building foundations, in *Engineering manual for War Department construction*. US Army Corps Engineers, Pt. 15, Ch. 4, 14 pp.

US Army Corps of Engineers, 1956. Engineering problems and construction in permafrost regions, in *The Dynamic North*. US Office Naval Operations [Polar Projects] Bk. 2 (OPNAY P03–17), 53 pp.

Wahrhaftig, C. and Black, R.F., 1958. Engineering geology along part of the Alaska Railroad, in Wahrhaftig, C. and Black, R.F., eds., *Quaternary and engineering geology in the central part of the Alaska Range*. US Geological Survey Professional Paper 293, pp. 69–119.

Washburn, A.L., 1980. *Geocryology*. New York, John Wiley and Sons, 406 p.

Weekes, A. and Brown, J., 1991. The challenge of sea ice and related phenomena, in Kiersch, G.A., ed., *The heritage of engineering geology, the first hundred years*. Geological Society of America Centennial Special Volume 3.

Williams, J.R., 1966. *Ground water in permafrost regions; An annotated bibliography*. US Geological Survey Water-Supply Paper 1792, 294 pp.

Williams, J.R., 1970. *Ground water in the permafrost regions of Alaska*. US Geological Survey Professional Paper 696, 83 pp.

Williams, J.R. and Van Everdingen, R.O., 1973. Ground water investigations in permafrost regions of North America; A review, in *Permafrost, The North American contribution to the Second International Conference*, Yakutsk, Siberia, USSR. Washington, DC, National Academy of Science, pp. 435–446.

Williams, J.R. and Waller, R.M., 1966. Groundwater occurrence in permafrost regions of Alaska, in *International Conference on Permafrost, Lafayette, Indiana, 1963. Proceedings*. Washington, DC, National Academy of Sciences, National Research Council Publication 1287, pp. 159–164.

Wilson, W.K. Jr, 1948. The problem of permafrost. *Military Engineering*, **40**, pp. 162–164.

Cross references

Dams
Groundwater
Soil and water management

PERRAULT, PIERRE (1611–1680)

Pierre Perrault, the founder of scientific hydrology, was the second of five sons, nearly all of whom were destined for illustrious careers. Two are still famous today: Claude (1613–1688) for his plans of the Louvre facade and the building of the Paris Observatory and, above all, Charles (1628–1703) for his *Fairy Tales*.

Pierre was a lawyer until 1654, when he purchased the office of Receiver-General of Finances for Paris. Louis XIV remitted the account for 10 years, but Perrault had the unfortunate idea of recovering his costs by dipping into the receipts for 1664. Unmasked by Superintendent Colbert, he was obliged to sell his office at a loss and found himself in very straitened circumstances.

He then turned to literature and science – with unequal success. He amused himself by translating a now wholly forgotten work, Tassoni's *La Secchia Rapita*, from the Italian; and he drafted two dialogues, one a heavy-handed critique of Don Quixote de la Mancha and the other a tedious comparison of the *Iphigenias* of Racine and Euripides.

His scientific work was of a much higher order. *On the Origin of Springs* (*De l'origine des fontaines*), published anonymously in Paris in 1674, is regarded by UNESCO and the World Meteorological Organization as the starting point of scientific hydrology. It was indeed a major breakthrough: for the first time in history, it was demonstrated quantitatively that 'rain and snowmelt are adequate to feed springs and rivers'. This question had been debated since the ancient Greeks – Aristotle, seeking to account for the abundance of spring discharge, had argued that springs were partly fed from the Earth's interior. *On the Origin of Springs* was attributed to various writers, including Edmé Mariotte, before Pierre Perrault was identified as its author (Dooge, 1959; LaRocque, 1967).

To counter Aristotle's arguments, Perrault drew up the first known water budget, taking the Seine basin at Aignay-le-Duc (118 km²). He calculated precipitation from 3 years' measurements taken between 1668 and 1674 (at an unknown site) and obtained a mean value of 224 million muids (518 mm). He estimated the basin's runoff by taking a discharge 24 times greater than that of the Gobelins River near Versailles – 36.5 million muids or 9.5 million m³. In this way he showed that his basin's annual runoff amounted to one-sixth of the precipitation.

Perrault's calculation is very crude and the units of measurement he used are very complicated, but the order of magnitude he arrived at is not unreasonable for this part of France. Edmé Mariotte, in a book that is a landmark in the history of hydraulics, arrived at very similar values for the Seine basin upstream of Paris.

Self-taught but inspired, Pierre Perrault was the first to open the way to modern hydrology, taking an approach that was quite new for his day. While his enterprise did not regain him the favor he had lost, posterity has paid him a belated homage.

J.H. Sircoulon

Bibliography

Dooge, J.C.I., 1959. Quantitative hydrology in the 17th Century. *La Houille Blanche*, pp. 799–807, Paris.

LaRocque, A., 1967. *On the Origin of Springs, by Pierre Perrault*. Hafner Publishing Co., New York and London.

Mariotte, E., 1686. *Traité du mouvement des eaux et des fluides*, Jean Jombert, Paris.

Perrault, P., 1674. *De l'origine des fontaines*. Pierre le Petit imprimeur, Paris.

Sircoulon, J., 1990. Pierre Perrault, précurseur de l'hydrologie moderne. *Europe*, no. 739–740, 40–47.

Cross references

Drainage
Hydrologists (600 BC–AD 1900)

POPULATION DISPLACEMENT DUE TO DAM CONSTRUCTION

Population displacement caused by dam construction is the most serious counter-development social consequence of water resources development.

Displacement in water resources development occurs as a result of three categories of dams, irrigation dams, hydropower dams and dams and reservoirs for potable water supplies (multipurpose dams combine two or all three functions), by adversely affecting people whose lands or house, or both, are submerged (in full or in part), or people whose place of employment disappears. For example, in Indonesia the dams recently built at Saguling, Kedung Ombo and Cirata have dislocated some 65 000, 30 000 and 57 000 people, respectively. In China the Gezhouba Dam displaced more than 20 000 and the Dienjangkou dam over 383 000 people. In Brazil, the Sobradinho and the Itaparica reservoirs dislocated 65 000 and 40 000 people, respectively. In the building of the Three Gorges Dam in China over a million people will be relocated.

Like Brazil and China, India is among the developing countries with the largest dam construction programs, and consequently with massive involuntary displacement operations. the series of dams being built under the Gujarat medium irrigation projects I and II displaced about 150 000 people, Srisailam dam alone has dislocated some 100 000 people, the ongoing Andhra Pradesh irrigation project II has to resettle 80 000 people, and the Upper Krishna irrigation II project in Karnataka, building two major dams, assumes the gigantic task of compensating or relocating up to 240 000 people.

Africa has also many reservoir-related displacements. In Togo the Nangbeto Dam displaced some 12 000 people which, considering the size of the country, was significant.

Large water resource related displacements have occurred, or are occurring in Ghana, Mali, Sudan, Cote d'Ivoire, Zaire and elsewhere.

Not every large dam, however, displaces significant numbers of people, but on average the numbers remain high. A study of 39 dam projects approved for financing by the World Bank during 1979–1986 found that these projects entailed the involuntary relocation of about 750 000 people in 27 countries. In China the water conservancy projects constructed in the past 30 years caused the evacuation of over 10 million people. In India it is estimated that up to 20 million people might have been adversely affected by water resources (and other) development programs over the last four decades.

Underestimation of the scale of dislocation may have tragic consequences, both on the people affected and on the economics of the projects. In Kenya's Tana River basin, the Kiambere dam was reported during the feasibility and appraisal analysis to displace 1000 people, but during implementation it turned out to dislocate some 6000 people. For the Baardhere Dam project in Somalia, the borrower's assessment indicated only 400 families (2000 people) in the reservoir area; however, the study required by the World Bank and conducted by an independent team concluded that the reservoir's indigenous population numbered at least 15 000, excluding people now living in the refugee camps in the future reservoir. Perhaps the record of underestimation is held by the preparation report for the Ruzizi II hydropower project, started in 1983. While it was initially estimated by planners that the project would affect about 200 people, in reality the reservoir, power plant and transmission line ended up affecting 16 000 people, who lost either lands, houses or both. The engineering design of dams sometimes neglects the social problems and their solution. It is reported by the World Bank that a case in point is the new Kalabagh dam being considered in Pakistan, where an international team of consultants produced a detailed feasibility report of nine massive volumes, yet failed to include the displacement of some 80 000 people. Population displacement and its social consequences is now a major consideration in river basin development.

R.W. Herschy

Source

The World Bank.

Bibliography

Cernea, M.M., 1994. *Impoverishment risks from Population Displacement in Water Resources Development.* World Bank Reprint Series Number 476, Washington, DC.

Cross reference

Dams

PRANDTL–VON KARMAN EQUATION

Theodor von Karman produced an equation in 1930 for free flow in pipes and open channels which was modified by Prandtl in 1935 to the generalized form:

$$\frac{1}{\sqrt{(f)}} = 2 \log{(R\sqrt{(f)}} + 0.4$$

where R is the Reynolds number and and f is a friction factor.

The f-R relationship can be important in hydraulic calculations, especially for flow in pipes, and the Prandtl–von Karman equation is one of many such types of equation which are popular in use because of their simplicity. The key to its successful use is the assessment of the Reynolds number.

P.G. Holland

Bibliography

BS 3680, Part I, 1991. *Glossary of Terms*, British Standards Institution, London.

PRECIPITATION

Precipitation represents any form of water, liquid or solid, falling to the ground from the atmosphere. All forms of water or ice in the atmosphere are classified as hydrometeors, but only those that fall to the ground are classed as precipitation.

Types of precipitation

Liquid precipitation is subdivided into two categories: drizzle and rain. The major difference between drizzle drops and raindrops is one of size. Water drops with diameters less than 0.5 mm are regarded as drizzle. Drops of greater size are considered to be raindrops.

Drizzle usually falls from stratus clouds or fog. The droplets are fairly uniform in size and quite numerous, but because of their smallness they produce only small quantities of rainfall. On the other hand, raindrop sizes may range from 0.5 to greater than 5 mm in diameter. They most often fall from nimbostratus and cumulonimbus clouds.

Frozen precipitation comes in a variety of types: ice crystals, snow, snow grains, snow pellets, ice pellets and hail. Single crystals are usually in the form of flat six-sided plates, hexagonally branched dendrites, ice needles or hexagonal prisms. The dendritic crystals are sometimes composed of extremely beautiful and intricate patterns with nearly perfect hexagonal symmetry. At other times it is difficult to discern the hexagonal structure. The structure of the ice crystal is determined to a large extent by the temperature and water vapor content of the air in which it has grown.

When ice crystals collide and stick to one another, snow is produced. At times, each snowflake is composed of few individual crystals. On occasion, especially when temperatures at the ground are just below 0°C, many ice crystals may combine to form very large snowflakes.

Snow grains are small, white, opaque particles of ice. They generally are flattened and elongated in shape and have diameters less than 1 mm. They may be regarded as the frozen equivalent to drizzle and usually fall from stratus clouds or fog on days with sub-freezing temperatures.

Snow pellets are roughly spherical or sometimes conical particles of white, opaque ice having diameters of about 2–5 mm. When snow pellets strike a hard surface, they bounce and sometimes break up. In these respects they differ from snow grains, which neither bounce nor break when hitting the ground. Snow pellets usually fall in showers of brief duration, often at the same time as snow.

Ice pellets are composed of transparent or translucent ice particles with diameters less than 5 mm. Their shapes may be spherical, conical or irregular. They are harder than snow pellets, being mostly solid ice, and they bounce readily when striking a hard surface. Ice pellets may form from the same types of clouds as those that produce rain.

Hail particles are composed of ice in the form of spheres, ellipsoids, cones or irregular masses. They differ from ice pellets because of size, ranging in diameter from 5 mm to over 10 cm. A characteristic feature of hailstones, particularly the large ones, is that they are composed of alternate layers of opaque and clear ice. Hail is nearly always produced in cumulonimbus clouds.

Precipitation formation processes

Virtually all precipitation-producing clouds are formed as a result of upward motion of air. The ascending air expands and cools as it moves to levels of lower pressure. When the air has cooled to temperatures below the dewpoint, water vapor condenses on tiny particles in the air called condensation nuclei, and cloud droplets are formed. These nuclei are mostly sulfate and sea salt particles. Typical cloud droplet diameters range from about 2 to 50 μm. The concentrations are of the order of 100 cm^{-3}. The condensation mechanism in a rising region of air acts to make the drop size spectrum narrow since small droplets grow faster than large ones.

As noted earlier, precipitation particles range in size from several hundred micrometers (drizzle) to several centimeters (hail). A typical raindrop may be 2 mm in diameter and have a mass of the order of 10^5–10^6 times that of a cloud droplet. In order to produce a precipitation particle, a process is needed other than condensation that produces the cloud particles.

Ice crystals generally are produced by the direct deposition of water vapor on minute ice embryos. The latter may originate on solid particles called ice nuclei. The temperature at which an ice nucleus produces an ice crystal varies from one substance to another. Certain types of powdered soil can cause crystals to form at temperatures between about −10 and −15°C. Such substances as silver iodide and lead iodide initiate nucleation at about −4°C. When a cloud, in an ascending current, for example, is supersaturated with respect to ice, existing crystals can grow rapidly as water molecules diffuse directly toward the crystals and are deposited on them (sometimes called deposition).

In order to explain precipitation rates and total quantities of the magnitudes commonly observed, it is essential that there be another process besides condensation and deposition. The additional process is known as coalescence. As the name implies, coalescence is the merging of existing particles. This can take place when some of them move faster than others under the influence of gravity. Table P2 shows the variation of terminal velocity, i.e., the equilibrium velocity

Table P2 Terminal velocities of water particles of various sizes

Diameter (cm)	0.01	0.05	0.10	0.20	0.30	0.40	0.50
Terminal velocity (m s^{-1})	0.27	2.06	4.03	6.49	8.06	8.83	9.09

in still air, of water particles of various sizes (at $P = 760$ mm Hg, $T = 20°C$, relative humidity = 50%).

It is clear that with small particles, the terminal velocity increases rapidly with drop diameter, but as diameters approach 5 mm, the velocity levels off. As diameters increase, the drop shape deviates from sphericity. It is found that with larger drops, the equilibrium shape is that resembling a hemisphere with the flat face downward. Also, drops commonly change their shape by oscillating along a vertical axis. When diameters exceed about 5 mm, the droplets become quite unstable and break up into many smaller drops.

Rain may be produced by a combination of the condensation and coalescence processes provided some of the cloud droplets are several times larger than the preponderance of droplets. For example, one droplet in 10^6 having a diameter of 100 µm could lead to rainfall. The large droplets may be grown as a result of condensation on relatively rare, large, sea salt particles (called giant salt nuclei) or perhaps by chance collisions between existing smaller droplets. Once the larger cloud droplets have developed they fall at sufficient speeds relative to the smaller ones that they collide and coalesce. The number of collisions depends on the relative sizes and numbers of droplets.

It is known that over tropical latitudes, particularly over the oceans, the condensation–coalescence mechanism plays a vital role in the initiation and production of precipitation. It is also quite important at higher latitudes, especially in convective clouds, but its importance, in quantitative terms, has not yet been assessed.

Rain is also produced as a result of melted ice particles. It is common for clouds composed of water droplets to ascend to sub-freezing regions of the atmosphere while the droplets remain in the liquid state. The term supercooled is used to describe such a cloud. Supercooling to a level of $-10°C$ is often observed. At times it occurs at much lower temperatures. According to laboratory experiments and theory, providing the droplet diameters are greater than about 1 µm, all the droplets will freeze at temperatures above about $-40°C$.

When ice nuclei are introduced into a supercooled cloud, they lead to the formation and growth of ice crystals that grow rapidly by deposition at the expense of the supercooled droplets. This comes about because at temperatures below $0°C$, air saturated with respect to water is supersaturated with respect to ice. This physical difference between water and ice is most pronounced at about $-12°C$.

When an ice crystal reaches a diameter of several hundred micrometers, it begins to fall with respect to the supercooled cloud droplets and smaller ice crystal. In addition, the horizontal movement of a flat crystal as it 'flutters' towards the ground gives it a horizontal motion relative to the smaller particles. Both the vertical and horizontal relative motions lead to collisions. When ice crystals collide, they may stick together and form snowflakes. With increasing collisions, the snowflakes become larger as they fall toward the ground at terminal speeds of about $1–2$ m s^{-1}.

When ice crystals or a group of aggregated ones fall into a region of supercooled droplets, the captured water droplets freeze and may act to 'bind' the crystals together and produce snow pellets. If large quantities of supercooled droplets are encountered, the entire particle may be encased in a thin layer of ice. In this case an ice pellet is formed.

When the temperature near the ground is warm, the falling ice particles begin melting after falling through the level where the temperature is $0°C$. If the so-called 'melting level' is high enough, the particles reach the ground as liquid drops and are called rain.

On some occasions, a layer of air a few thousand meters above the ground may have temperatures above freezing, even though temperatures near the earth are below freezing. In such a circumstance, the precipitation may reach the ground at temperatures close to $0°C$. This produces a phenomenon called freezing rain. When the drops strike solid objects at sub-freezing temperatures, the water freezes and produces a layer of clear ice.

On some occasions the cooled raindrops refreeze before they reach the earth and produce solid, transparent ice particles. In this case, the precipitation is in the form of ice pellets.

Hail may be thought of as a special type of ice pellet. It requires the presence of deep layers of supercooled clouds and a strong current of upward moving air (called an updraft). Such conditions are found only in thunderclouds of great vertical extent.

In order to produce large hailstones, i.e. stones with diameters of several centimeters, it is essential that the initial ice pellet collides with large numbers of supercooled drops. The characteristic layers of clear and opaque ice in hailstones can be explained in terms of the rates at which the growing stones accumulate supercooled water.

When a hailstone passes through regions of the cloud with high water contents, the water swept out by the stone freezes slowly and air bubbles are forced out of the ice, leaving clear ice. With small water contents, freezing proceeds rapidly and air bubbles are trapped in the ice and give it a milky appearance.

The largest stones require very strong updrafts, tens of meters per second, in order that the growing ice mass can be held in the supercooled cloud for sufficient time to accrete the required super-cooled water. This might occur in a cloud with an essentially vertical updraft. However, some observational evidence suggests that many hail-producing thunderstorms have updrafts that are tilted off the vertical. In such clouds it is possible for some hailstones to make several up and down traverses through the supercooled parts of the thunderstorm before falling out.

The rate of fall of hailstones depends on their shapes and densities. A spherical ice sphere having a diameter of about 5 cm may fall at about 30 m s^{-1}. When hailstones fall below the $0°C$ level, they begin to warm and melt. However, the percentage change of particle diameter decreases as the diameter increases. Large stones, i.e. greater than 1 or 2 cm in diameter melt little even in falling 3 km from the melting level to the ground. In order for an ice particle to completely melt after a 3 km descent, its diameter would have to be less than about 3 mm.

Louis J. Battan

Bibliography

Dinsman, S.L., 1994. *Physical Hydrology*. London: Macmillan, 575 pp.
Fletcher, N.F., 1962. *The Physics of Rainclouds*. Cambridge: Cambridge University Press.
Mason, B.J., 1971. *The Physics of Clouds,* 2nd edn. London: Oxford University Press.
Mason, B.J., 1975. *Clouds, Rain and Rainmaking,* 2nd edn. Cambridge, England: Cambridge University Press.
Rogers, R.R., 1976. *A Short Course in Cloud Physics*. New York: Pergamon.
WMO, 1994. *Guide to Hydrological Practices*, 5th edn. Geneva: World Meteorological Organization, WMO No. 168, 735 pp.

Cross references

Acid rain
Atmosphere
Atmospheric process associated with water in the atmosphere
Atmospheric water vapour
Clouds (cloud seeding)
Hydrometeorology
Rain

PRECIPITATION DISTRIBUTION

The amount of precipitation received at any place on the Earth's surface and over any given time period depends on a variety of factors and results from several complex causes and interacting processes that are themselves time and space dependent. Atmospheric water vapor, the basic element in precipitation, varies geographically and temporally. For example, the mean precipitable water content of the atmosphere at any given moment is 25 mm with a maximum near the equator of 44 mm and a minimum in the polar regions of 2–8 mm depending on the season. Between 40 and 50° latitude, precipitable water ranges above 20 mm in the summer and drops to 20 mm in the winter. However, the presence of atmospheric moisture is a necessary but not a sufficient condition for precipitation. Precipitation totals depend not only on amount of precipitable water in the atmosphere but also on lifting mechanisms that cause the moisture to precipitate. Except for irregularities induced by rugged topography, the greatest amount of precipitation occurs along portions of the intertropical convergence zone (ITCZ). Any discussion of precipitation distribution, therefore, must of necessity address variations both in time and space. This can be accomplished by examining the major features of global patterns of precipitation and the factors that combine to influence those patterns, by looking at the nature and causes of seasonal, diurnal and other temporal variations in precipitation

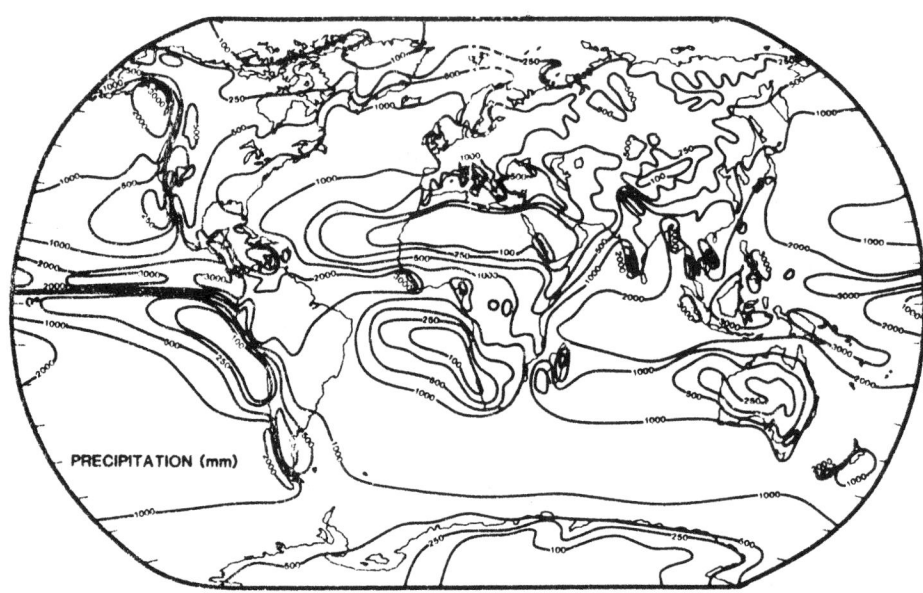

Figure P15 Mean annual precipitation (mm) over the Earth.

distribution to discern major temporal regimes and their forcing mechanisms, and finally, by examining distribution from a statistical perspective rather than from a purely climatological one. In this last case, distribution characteristics such as variability, persistence, frequency and intensity become important.

The distribution pattern of precipitation is considerably more complex than either temperature or global radiation because of the influences of several factors which can be classified into those influencing vertical motion in the atmosphere and those relating to the nature of the atmosphere itself. Of the latter group, the stability and instability of the atmosphere and its thermal and moisture characteristics, which in turn are determined by the nature of the source region of the air masses and their subsequent trajectory, are particularly noteworthy. Practically all precipitation results from the adiabatic cooling of ascending moist air masses. The areas of most frequent and rapid air mass ascent are the zones of convergent horizontal air flow in the equatorial belt and areas affected by the midlatitude cyclonic disturbances as well as along the windward side of high mountain ranges adjacent to extensive sources of moisture (Figure P15). Additionally, the distribution patterns of precipitation may be influenced by the distribution and density of the data collection network and the accuracy of the precipitation measuring instruments. Most measuring stations tend to be in highly populated areas with few stations over the oceans and in sparsely peopled regions. Moreover, gauge measurements tend to be underestimates of the true precipitation, largely because of wind-induced turbulence at the gauge orifice and wetting losses on the internal walls of the gauge. Monthly estimates of the bias often vary from 5 to 40%, are larger in winter than in summer, and over land masses in the northern hemisphere, increases to the north due largely to the deleterious effects of wind on snowfall.

Geography of annual distributions

The mean annual precipitation for the Earth is about 1000 mm but its spatial distribution is variable (Figure P15). Several factors combine to influence the global distribution of precipitation. Mean annual totals are greater in the equatorial belt, where the general circulation of the atmosphere is characterized by the rapid ascent of warm moist air in the tropical zone of converging winds, than they are in the subtropical area of dominant subsidence and high surface pressure which inhibit precipitation and lead to much lower totals. The tropical convergence zone with its widespread uplift of warm, humid air masses is the most effective precipitation producing part of the general circulation of the atmosphere. From this zone of heavy precipitation, amounts decrease irregularly toward the poles, and at high latitudes the lower moisture-carrying capacity of the colder

atmosphere, fewer incursions of moist warm air and less active thermal convection result in limited precipitation. The minimum in precipitation at latitudes between 20° and 30° is caused by subsidence in the high-pressure zones and the adjacent stable portions of the trade wind inversion. In midlatitudes, precipitation increases again because of synoptic storm systems there. The forced ascent of moist surface air in midlatitude cyclones and orographic uplift in the westerly flow give rise to heavy precipitation. Although the distribution of annual precipitation shows a strong latitudinal zonation, there are non-latitudinal differences, especially in the tropics. Mt Waialeale on the island of Kauai, the northwesternmost of the large islands of the Hawaiian chain, averages 11 675 mm of rain annually; Cherrapunji, at an elevation of 1313 m on the southern slope of the Khasi Hills in northeastern India, averages 11 419 mm; and Debundscha, near the base of the Cameroon Mountain just north of the equator in western Africa, averages 10 279 mm annually. Areas with the least rainfall are Arica in northern Chile, which over a 59-year period averaged 0.75 mm and Wadi Haifa, Sudan, averaged 3.0 mm over 39 years. Most of the coast of Peru and northern Chile averages less than 25 mm, as do portions of the southwest coast of Africa. In some places years may pass with no measurable precipitation and then a rare shower will account for the long-term average.

In middle and high latitudes the interannual differences in precipitation distribution patterns that lead, for example, to extensive areas of drought existing concurrently with other and equally extensive areas of severe flooding can be traced to interannual differences in the configuration and intensity of the mid-tropospheric jet stream that is responsible for steering storms and storm systems through a series of troughs and ridges, the wavelengths and amplitudes of which vary from year to year and place to place. This upper-air steering mechanism is directly related to the magnitude of the internal energy of the atmospheric system, which in turn is related to the amount of solar energy absorbed by radiatively active surfaces in the boundary layer and its apportionment into sensible heat energy, latent heat energy and energy stored in the submedia modulated by the heat capacity and the response time of each storage medium. Interannual variability of precipitation over a large part of the tropics and subtropics is related to the El Niño/Southern Oscillation (ENSO) phenomenon.

The distribution of land and water greatly influences the precipitation patterns in middle and high latitudes. Regions in the interior of continents remote from oceanic sources of moisture receive less precipitation than coastal areas. Windward coasts tend to receive more precipitation than leeward coasts in regions where favorable prevailing winds transport moisture towards the land. Ocean currents or drifts may introduce further modifications since warm currents such as the North Atlantic Drift increase the moisture content of the air in contact with it and in some cases induce atmospheric instability that may lead to precipitation. Cold currents, on the other hand, decrease precipitation over the adjacent land areas by cooling the lowest layers of the atmospheric column, increasing atmospheric stability, inhibiting vertical mixing and hence precipitation. The cooling process in the lower atmosphere may lead to fog or low stratus but no precipitation. Large lakes and lake systems in the middle and high latitudes produce extensive lake-effect snowfall on their downwind edge in winter (Dewey, 1970; Brahan and Dungey, 1984). Cold moist air, in its trajectory over the lake, becomes unstable as a result of heating from below. This instability and uplift, accentuated by convergence in the lower layers as the air mass moves from an aerodynamically smooth lake surface to a comparatively rough land surface, result in comparatively higher snowfall totals from given air masses on the downwind edge of the lake.

Forced ascent of moist air by topographic barriers concentrate precipitation on the windward slopes but create a rain shadow to leeward, by causing descending motion and hence heating of the already wrung-out air. Apart from tropical regions, the wettest places are on the windward side of mountains exposed to winds with long fetches across warm ocean surfaces. Mountains therefore tend to contribute to the spatial inequality in precipitation distribution. Examples of topographically induced inequalities in precipitation distribution are mountainous islands in the trade wind belt, the southern Andes along the Chilean–Argentine border, the Southern Alps of New Zealand and the coastal ranges of western North America (Figure P16). A cogent example is afforded by the state of Washington in the western United States (Figure P17). Along the Pacific coast where topographic variations are small the rainfall averages about 1525 mm; with increasing topographic variations annual totals may increase by 500 mm to totals of 2050 mm. On the western slopes of the Olympic Mountains up to 3800 mm are recorded. However just to the east in the Puget Sound–Willamette Valley region, totals average around 1020 mm. Farther east, the western slopes of the Cascades receive amounts of 1525–1780 mm. Across the Cascade Divide there is a sharp transition spanning the range from 180 mm in Yakima to about 510 mm around Spokane (Figure P17). Orographic barriers also affect the vertical distribution of precipitation such that totals tend to increase with elevation on the windward slopes up to a certain level. Very high totals due to orographic lifting also occur in monsoonal areas and where tropical cyclones occur frequently, such as the western coast of India, Burma, Sumatra and Borneo in the first case and in the Caribbean region in the second case.

The major features of global patterns of annual precipitation distribution ascertainable from the preceding discussion and Figure P15 are (1) large totals in the equatorial zone and moderate to heavy amounts in the middle latitudes; (2) comparative dryness in the subtropical belts and in the regions around the poles; (3) within the subtropics, the west coasts of continents tend to be dry whereas east coasts tend to be wet; (4) in higher latitudes, west coasts tend to be wetter than east coasts; (5) in mountainous regions, especially where the mountains are transverse to prevailing winds, precipitation tends to be abundant on windward slopes but sparser on leeward slopes; and (6) maritime locations receive more precipitation than interiors of continents that are remotely located with respect to sources of moisture in the oceans, except perhaps in innermost Amazonia where prevailing winds, extensive horizontal convergence, large evapotranspiration rates and orographic effects combine to produce large amounts in a continental interior (Lettau et al., 1979).

Seasonal distribution of precipitation

Variations in the spatial distribution of annual precipitation and the accumulated totals over a given year are quite important climatically, but total annual precipitation is an insufficient measure of moisture availability because it does not take into account how precipitation is distributed over time. Equally significant from the point of view of agriculture, hydroelectric power generation and water resources planning are the seasonality, the interannual dependability, month to month variations and the frequency and intensity. Precipitation inducing mechanisms and processes do not operate in the same way throughout any given year. Belts of convergence shift poleward in summer and equatorward in winter, causing a shift in the associated precipitation belts. In these convergence belts, rainfall is comparatively well distributed throughout the year, but a few degrees of latitude poleward, where alternating convergence and subsidence

Precipitation (mm)

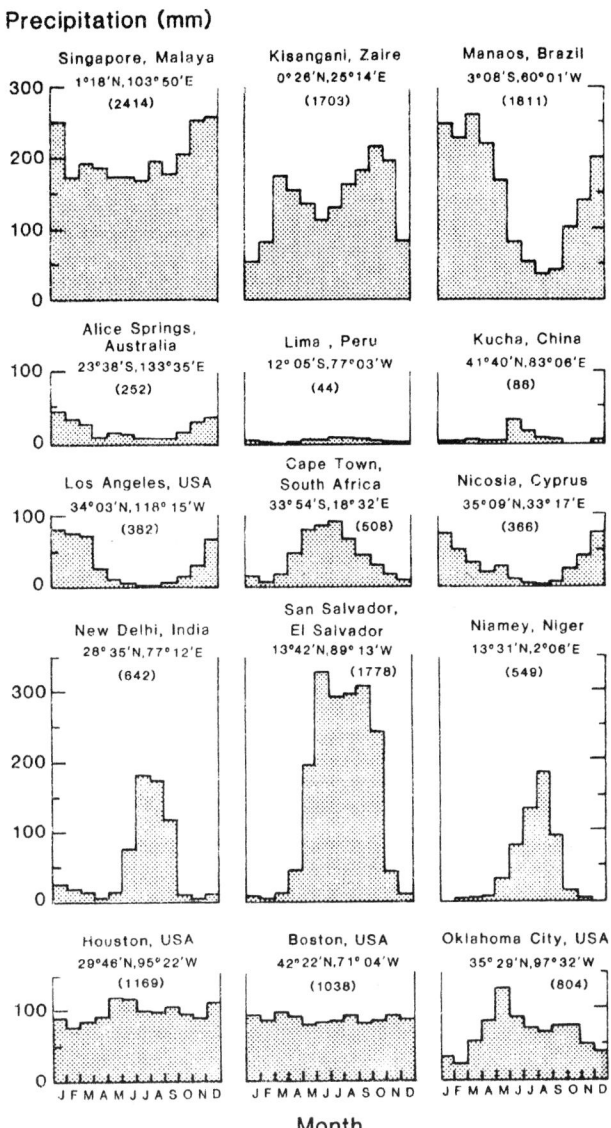

Figure P16 Seasonal precipitation distribution at selected groups of stations.

predominate, marked dry and wet seasons exist. Near the equator two precipitation maximums are discernible, one occurring approximately 1 month after the equinoxes. With increasing distance from the equator the two maxima move closer together in time, reducing one minimum in length while the other dry season becomes longer and more intense. The total amount of annual rainfall is less here than closer to the equator. Between 15 and 20° latitude the short dry season disappears completely and the two maximums merge. North and south of this wet–dry precipitation regime, the influence of the subtropical quasi-stationary high leads to dry summers, whereas wet winters generally follow with the increasing influence of the westerlies and migrating cyclones and frontal disturbances. In India and Southeast Asia, monsoonal variations result in even more striking differences in the seasonal distribution of precipitation. Monsoons are seasonally prevailing winds that blow from one direction in summer and an approximately opposite direction in winter. These winds are best developed in India and Southeast Asia and are the result of differences in heat capacities of land and water, seasonally unequal heating and the pressure differentiation it produces, and the influence of mid-tropospheric circulation patterns, especially the subtropical jet and its interaction with the Himalayas. Northern Australia experiences a similar, though less well-developed monsoon circulation, which is

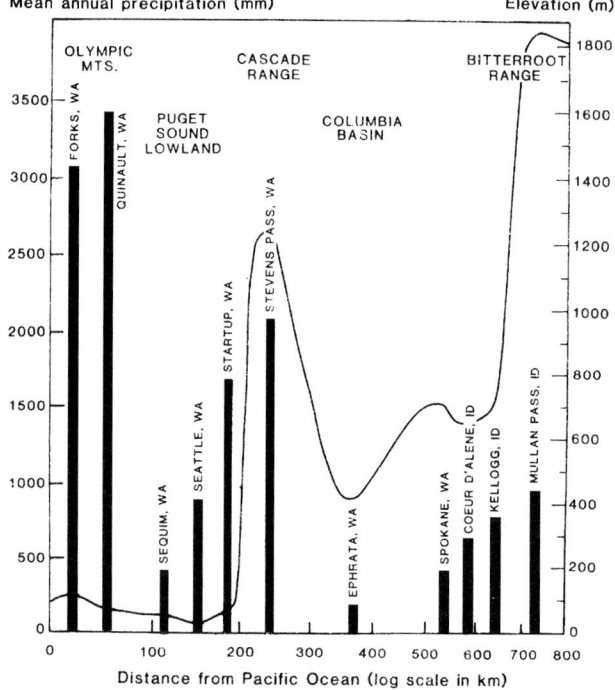

Figure P17 Orographic influences on precipitation exemplified by a transect of mean annual precipitation along the 48°N parallel in western North America. (Data from National Oceanic and Atmospheric Administration.)

also related to seasonal migration of components of the general circulation. Along the east coast of Asia and across central Africa there are monsoon effects that merge into other shifting patterns of the general circulation. The southeast monsoon of India, which blows from the Arabian Sea across much of India from June to October, brings with it general rains that, along the west coast and the foothills of the Himalayas, are among the heaviest on Earth. In monsoon regions there is usually a marked summer maximum in precipitation (Figure P16), and delays in the onset of the wet monsoon season have disastrous consequences for agriculture and water supplies. In East Africa, agricultural practices are predicated on two maxima, the larger of which is monsoonal whereas the smaller is due to the shifting tropical convergence zone. Over the tropical continents, seasonal rainfall distributions are further complicated by intense convectional processes that increase the amounts of rainfall locally.

The seasonal distribution of precipitation is as important as the total distribution over large portions of the Earth's surface. In many parts of the tropics precipitation is strongly seasonal and the times of initiation, duration and end of the rainy season control agricultural activities and water resources allocations. In the tropics, seasonal rainfall distribution is so important that it forms the basis of most classifications of tropical climates. This does not imply, however, that the seasonal distribution of precipitation is not important in middle latitudes; tropical rainfall is effective for plant growth no matter what time of year it falls, but in middle latitudes only that part of the annual precipitation that falls during the non-freezing season may be effective. In middle latitudes there is a fairly pronounced summer maximum of convective activity in continental interiors, where the summer maximum tends to lag later into summer the farther poleward one goes because of the lateness of snowmelt and ground thaw that delay convective activity.

In general terms, it is possible to discern the following seasonal regimes (Figure P16): (1) abundant rainfall, largely convectional, occurring year round in equatorial regions; (2) largely convectional rainfall occurring mainly in summer months; (3) low rainfall totals in all seasons; (4) mainly cyclonic precipitation in midlatitudes occurring in winter whereas summers are dry; (5) cyclonically induced precipitation more abundant in winter than in summer; (6) mainly

summer precipitation in continental regions; (7) abundant precipitation on east coasts due to onshore winds in low latitudes and inflowing warm moist air masses in summer and traveling cyclones in winter in middle latitudes; and (8) low levels of precipitation in polar regions with a maximum in summer.

Diurnal distribution

Diurnal rainfall distribution is generally not as important as seasonal distribution except in low latitudes. The diurnal regime influences transportation systems, many outdoor activities and evapotranspiration losses. Daytime precipitation input is subject to very heavy evaporation losses. Diurnal precipitation distributions are more regular in low latitudes than in higher latitudes because of the differences in the dominant rain-producing mechanisms. In the tropics, the convectional processes that predominate are forced by the diurnal march of global radiation input and are therefore considerably more regular than the midlatitude disturbances and fronts responsible for a large proportion of the diurnal variations in middle and high latitudes, especially in winter and spring. The nocturnal maximum in the midwestern United States as compared with the late afternoon maximum in the eastern part of the country indicate that even within midlatitude regions there can be distinguishable diurnal variations, particularly in summer. At almost all latitudes diurnal regimes are also influenced by the effects of local factors, such as topography, on small-scale atmospheric processes, such as thunderstorms, and by the form and physiographic configuration of the coastline. Two types of diurnal regimes are recognizable in low latitudes: (1) the inland or continental type characterized by a late morning or afternoon maximum attendant on convection initiated by surface heating of the land, and (2) the maritime or coastal type with a maximum during the night or early hours of the morning caused by severe night-time instability and convection predicated on the steepened lapse rate attendant on radiative cooling of the upper troposphere while the lower layers remain warm due to its thermal coupling with the warm ocean surface (Kraus, 1963; Malkus, 1964). Ramage (1952) questioned the validity of this explanation but substituted no generally accepted theory. In the Caribbean region both types coexist, leading to the expectation of showers soon after solar noon, especially during the wet season. Equally expected, although not with the same regularity, are nocturnal severe thunderstorms occurring late at night or in the early hours of the morning. Some of the most intense precipitation, rivaled only by that accompanying hurricanes, occurs in these nighttime episodes.

The manner in which precipitation falls is also important. If it falls in infrequent cloudbursts it may cause floods and heavy erosion. On the other hand, prolonged drizzle effectively moistens the soil and provides high-humidity conditions. Europe receives less than 750 mm of precipitation per year but in many areas this is spread over more than 150 days. Under the cool cloudy, humid conditions that prevail there, the rainfall is quite adequate. In some parts of the world precipitation falls nearly every day of the year. Bahia Felix, Chile, averages 325 days per year with measurable precipitation, so that there is an 0.89 probability that it will rain or snow on any given day of the year. Buitenzorg, Java, averages 322 days per year with thunderstorms. Conversely, Arica, Chile, averages about one rain day annually; at nearby Iquique, Chile, there was no measurable rain between 1899 and 1913.

Statistical distribution

It is evident from the foregoing that the character of the spatial and temporal distribution of precipitation can be very important and enlightening but, in order to adequately analyze past sets of precipitation measurements so that conclusions about the future can be drawn and in order to test the significance of physical experiments and check the validity of hypotheses concerning the distribution of precipitation, we need to know something about the statistical characteristics of the precipitation data sets. Statistical theory is based on the concept that a set of realizations is assumed to be representative of their population and therefore deductions can be made from that sample concerning the nature of the population. In order to draw inferences about a population, the variates within the data sample must be random, independent and homogeneous. The precipitation data sets we work with, however, rarely meet all these criteria at once. In the first place, there can be very little control over the selection of a sample and whatever nature provides over a period of time must be used; second, the degree of independence varies with the precipitation-inducing mechanisms, season and location, and, in some cases and places, persistence is prevalent. In many regions, sample data lack homogeneity because station sites have shifted due to urbanization or development and because separate and distinct causal mechanisms contribute to the precipitation sample. Moreover, precipitation is not an infinite continuum in both directions; it cannot be less than zero. Therefore, variations on the dry side of the average cannot be as great in magnitude as variations on the wet side, and a few wet years can balance a great number of comparatively dry years. Despite all this, precipitation samples tend to follow one of several types of statistical distributions. The most important of these is the normal or Gaussian distribution, which is a bell-shaped curve that is symmetrical about the mean value for the sample. If the sample is not normal it can be normalized by various methods such as using the logarithms or cube roots of the variates in the sample. Other types of distributions include the gamma distribution, the Pearson types I, II and III distributions and, as already alluded to, logarithmically transformed lognormal, logextremal and truncated lognormal distributions. Each of these distributions has special characteristics that allow researchers to extract the maximum amount of information from the many and diverse precipitation records available throughout the world.

The characteristics of the statistical distribution of a precipitation sample are described by statistical parameters. The most important of these relate to measures of central tendency, variability and skewness. The parameters that generally represent measures of central tendency are the mean, the median and the mode. The mean can be arithmetic, geometric or harmonic. The arithmetic mean is the most familiar and generally used. The geometric mean is the nth root of the product of n terms, so that if one takes the logarithm of the individual values of a precipitation record and takes the mean, the logarithm of the geometric mean will result. The harmonic mean is the reciprocal of the mean value of the reciprocals of the individual values (Chow, 1964; Munn, 1970; Panofsky and Brier, 1968). The arithmetic mean is used more often than any other measure of central tendency because of its computational simplicity and, in general, its greater sampling stability. However, in extremely skewed precipitation distributions such as would be found in samples taken in arid and semiarid regions, the mean may be misleading because, although it is based on sound mathematics, weight is given to each occurrence according to its magnitude so that extreme values are excessively stressed in comparison with middling values. In a distribution that approximates the normal, this is of minor importance at most. The median is the middle value or the variate that divides the frequencies in a distribution into two equal parts so that all variates greater or less than the median always occur half the time. In the distribution of discrete variables the variate that occurs most frequently is the mode. In a distribution of continuous variables, the mode is the variate with the maximum probability density. In many stations, particularly in the tropics, precipitation distributions are not unimodal but bimodal and in that case it becomes difficult to decide exactly where the mode is, especially since the actual value arrived at may in part result from a subjective choice of groupings. Furthermore, the mode does not possess any true mathematical quality having at best only a generalized relationship to the average.

The parameters representing variability or dispersion are the standard deviation, the variance, the range and the coefficient of variation. At many stations the frequency distribution of precipitation totals for individual years is positively skewed – the negative departures from the annual mean are more numerous than the positive ones. Such skewness is generally strongest where rainfall totals are low and where correct estimation of the rainfall that can be expected is most critical but where the annual mean is inflated by a few very high annual totals. For predictive purposes, medians and other probabilistic estimates such as quartiles, quintiles and deciles are more reliable despite the computational complexities and the requirement of long records. The standard deviation is the square root of the mean squared deviation of individual measurements from their mean, whereas the variance is the square of the standard deviation. The curve representing a particular normal or Gaussian distribution is completely determined by its standard deviation. Generally, for such a curve the standard deviation is the distance on either side of the center of the point where the slope is steepest. The coefficient of variation is the standard deviation divided by the mean, whereas the range is the difference between the largest and smallest values. Finally, skewness is a measure of the lack of symmetry of a distribution. The simplest measure of variability is the range but, because it is based on two observations only, there are large variations from sample to sample making the range very unstable.

Precipitation intensity, which relates the total amount of precipitation to its duration over some specified finite time period, is another descriptor of precipitation. It may be taken as the amount of precipitation occurring on an average rainy day of a certain period. It is an important measure because it controls the probability and seriousness of local floods and is critical in planning dams, reservoirs and drainage canals. In addition, when intensity exceeds the maximum infiltration rate of the soil, surface runoff results so that intensity can be related to erosion and sedimentation rates in lakes and reservoirs. In addition to intensity, the frequency distribution of different amounts of precipitation is also important. For example, in the tropical belt, in the wet season as well as in the dry season, the frequency distribution is represented surprisingly enough by a decay curve, demonstrating that days with no or very light rain are the most frequent and days with increasing amounts occur more rarely. Also, in the dry period nearly 85% of all days and even in the wet season 35% of all days have no or little rain.

The frequency distribution is used to organize precipitation data so their characteristics may be easily and quickly summarized. Also, an estimate of the frequency with which a given magnitude of precipitation may be exceeded in the future is based on the frequency with which it has been exceeded in the past. The parameters of the frequency distribution can be used in a probabilistic framework to calculate recurrence intervals of specified magnitudes of precipitation events with varying durations – hourly, daily or seasonal.

Precipitation distribution can be viewed from many different perspectives, each of which possesses both diagnostic and prognostic components. The time–space nature of that distribution and the many and varied causal mechanisms and relationships render the topic extremely complex, but at the same time immensely useful for an understanding of climate and its spatial and temporal variability and how these interact with the human-use system. At a time when the activities of human societies are modifying the climate system and reducing the margin between water supply and demand for agriculture, hydroelectric power generation and domestic use by an ever expanding population in many parts of the world, it is imperative that we not only know where the areas of precipitation deficits and surpluses are and the time periods in which they are most likely to exist, but also how the distribution patterns change over longer time periods.

Precipitation distribution in a warmer world

There is strong evidence in the instrumental records that global temperatures have been increasing over the last century and that the increase has been largest in the last two decades. The consensus, based on the results of general circulation models and higher-resolution regional models, is that most of that warming is due to increasing concentrations of trace gases in the atmosphere, and at the present rates of greenhouse warming, significant global environmental change will ensue. Among these changes are increased air and ocean surface temperatures, sea level rise, an increase in the intensity of tropical storms, and a probable increase in the frequency of extreme climatic events. If temperatures change because of trace gas-induced warming, changes in other elements will occur also. Precipitation patterns are likely to be altered. Total precipitation over the globe will increase as the water-holding capacity of the air and evapotranspiration increase, but it is difficult to predict precisely how the additional moisture will be distributed either spatially or temporally. The IPCC estimates that summer precipitation will decline by 5–10% and soil moisture by 15–20%. The American Mid-West will become drier in summer. Confidence in these estimates are low, however (Mitchell et al., 1990). Precipitation would be less frequent over most of Europe in summer and fall and throughout the year in the south. According to some projections, more rainfall is possible in parts of Africa and Southeast Asia. In parts of Africa such as the Sahel, winter rainfall will decline by 5–10% and summer precipitation will increase by as much as 5% (Wigley et al., 1986; Kellogg, 1987). Hurricane intensity is expected to increase, as is the precipitation and the flooding associated with such storms. Whatever the magnitude and specific distribution of precipitation that a warmer atmosphere and ocean will foster, global warming is sure to change the distribution of precipitation in ways that may be beneficial to some regions of the world but detrimental to others. Finally, the knowledge about and understanding of precipitation and its distribution should culminate in an improvement in our predictive capability if it is to serve as a useful input into decision-making processes in agriculture, water resources and precipitation-related disaster mitigation, if only in a probabilistic form.

Orman E. Granger

Bibliography

Braham, R.R. Jr and J. Dungey, 1984. Quantitative estimates of the effect of Lake Michigan on snowfall, *J. Clim. Appl. Meteorol.*, **23**, 940–949.

Chow, V.T. (ed.), 1964. *Handbook of Applied Hydrology*. New York: McGraw-Hill.

Dewey, K.F., 1970. An analysis of lake-effect snowfall, *Illinois Geog. Soc. Bull.*, **12**, 27–42.

Gilman, C.S., 1964. Rainfall, in *Handbook of Applied Hydrology*, V.T. Chow (ed.). New York: McGraw-Hill, pp. 955–956.

Kellogg, W.W., 1987. Mankind's impact on climate: the evolution of an awareness. *Climate Change*, **10**, 113–136.

Kraus, E.B., 1963. The diurnal precipitation change over the sea, *J. Atmos. Sci.*, **20**, 551–556.

Lettau, H., K. Lettau and L.C. Molion, 1979. Amazonia hydrologic cycle and the role of atmospheric recycling in assessing deforestation effects, *Monthly Weather Rev.*, **107**(3), 227–238.

Malkus, J.S., 1964. Tropical convection: Progress and outlook, in *Proceedings of the Symposium on Tropical Meteorology*. Boston: American Meteorological Society, pp. 247–277.

Mitchell, J.F.B., S. Manabe, T. Tokioka and V. Melishko, 1990. Equilibrium climate change, in J.T. Houghton, G.J. Jenkins and J.J. Ephraums (eds), *Climate Change: The IPCC Scientific Assessment*. Cambridge: Cambridge University Press, pp. 131–172.

Munn, R.E., 1970. *Biometeorological Methods*. New York: Academic Press.

Panofsky, H.A. and G.W. Brier, 1968. *Some Applications of Statistics to Meteorology*. University Park, PA: Pennsylvania State University.

Ramage, C.S., 1952. Diurnal variation of summer rainfall over East China, Korea, and Japan, *J. Meteorology*, **9**, 83.

US Department of Commerce, National Oceanic and Atmospheric Administration, *Climatological Data: Annual Summary*. Asheville, NC: National Climatic Center.

Wigley, T.M.L., P.D. Jones and P.M. Kelly, 1986. Empirical climate Studies, in B. Bolin, B.R. Doos, J. Jager and R.A. Warrick (eds), *The Greenhouse Effect, Climate Change and Ecosystems*, SCOPE 29, New York: John Wiley and Sons.

Cross references

Arid climates
Climate and climate change
Climate data: sources
Clouds (cloud seeding)
Hydroclimatology
Hydrometeorology
Paleohydrology
Precipitation
Water budget analysis
World water balance

PRECIPITATION: SOURCE

Water may take a number of different forms in the atmosphere. These forms are collectively termed 'precipitation', which includes rain, drizzle, sleet (partly melted snowflakes, or rain and snow falling together), snow and hail. The intensity and duration of precipitation are extremely variable in most areas of the world.

The source of precipitation is water vapor, which is always present in the atmosphere in varying amounts, although it makes up less than 1% by volume. However, the water vapor in the air must be cooled to allow water to be condensed into cloud droplets. These droplets then grow to form precipitation particles. The mass of water in the atmosphere in both liquid and vapor form is around 1.3×10^{16} kg compared with the mass of water in the oceans of around $1.3 \times$

10^{21} kg (Nace, 1967). Nevertheless this water is distributed very unevenly and is transported by the circulation of the atmosphere.

C.G. Collier

Bibliography

Nace, R.L., 1967. Water resources: a global problem with local roots, *Environ. Sci. Technol.*, **1**, 550–560.

Cross reference

Precipitation distribution

PROBABILITY OF A HYDROLOGICAL EVENT OCCURRING IN A GIVEN TIME INTERVAL

The return period (T_r) of an event (X_{T_r}) established by frequency analysis indicates only the average interval between events equal to or greater than X_{T_r}. The probability (Pr) that X_{T_r} will occur in any one year is given by

$$Pr = \frac{1}{T_r} \tag{P1}$$

The probability (Pr_t) that X_{T_r} will occur at least once in the first t years of a project's life is given by

$$Pr_t = 1 - \left(1 - \frac{1}{T_r}\right)^t \tag{P2}$$

The probability (Pr_{T_r}) that X_{T_r} will occur in the first T_r years of a project's life, i.e in a period equal to the return period, is thus

$$Pr_{T_r} = 1 - \left(1 - \frac{1}{T_r}\right)^{T_r} \tag{P3}$$

For large values of T_r, the value of Pr_{T_r} approaches $1 - e^{-1} = 0.63$. Table P3 shows some values of Pr_{T_r}.

It may happen, however, that during a definite period of T_r years, precipitation of the magnitude $P \geq P_{T_r}$ does not occur at all, or that it occurs several times. The probability that, during a given period of t years, a respective phenomenon will occur n times, is equal to

$$Pr_{n/t} = \left(\frac{t!}{n!(t-n)!}\right) p^n (1-p)^{t-n} \tag{P4}$$

where $p = 1/T_r$. Assuming, for example, that $t = T_r = 100$ years, then the probabilities for various values of n are:

n	0	1	2	3	4	5
$Pr_{n/100}$	0.366	0.370	0.185	0.061	0.015	0.003

Design return period

It is seen from Table P3 that for values of $T_r \geq 10$ years, the probability that X_{T_r} will occur within the next T_r years is about 65%.

Table P3 Probability (Pr_{T_r}) that an event X with return period T_r will occur within the next T_r years

T_r	Pr_{T_r}
2	0.750
5	0.672
10	0.651
20	0.642
50	0.636
100	0.634
500	0.633
∞	0.632

For project design purposes it is preferable to specify the acceptable probability, or calculated risk, that an event will occur within the life of the project and to compute the required design return period (T_d).

If a project has a desired lifetime N, and U is the calculated risk that failure will occur within a lesser interval, then

$$U = 1 - \left(1 - \frac{1}{T_d}\right)^N \qquad (P5)$$

A useful approximation to equation (P5) is

$$T_d = N\left(\frac{1}{U} - \frac{1}{2}\right) \qquad (P6)$$

where T_d is the design return period for calculated risk U and desired project life N years.

Table P4 shows values of the design return period (T_d) for calculated risk U that failure will occur during a project's desired life of N years. To illustrate the use of Table P4, assume that a proposed dam has a desired life of 50 years and that the designer wishes to take only a 10% calculated risk that the dam will be overtopped within this 50-year period. From Table P4 for $N = 50$ and $U = 0.1$, the design

Table P4 Design return period T_d for calculated risk U and project life N

Calculated risk of failure U	Desired life of project N (years)			
	2	10	50	100
0.01	198	996	4975	9953
0.10	20	95	475	950
0.50	3.4	15	73	145
0.75	2	7	36	73

return period (T_d) is found as 475. The dam must be designed to withstand the event of return period 475 years or probability 0.002 of occurring in any one year.

Source

World Meteorological Organization, 1974. Guide to Hydrological Practices, 3rd edn, WMO, Geneva.

Q

QINGHAIHU LAKE, CHINA

Survey of the lake

Qinghaihu is situated at the northeast end of the Qinghai–Xizang Plateau, between 97°50' and 101°20'E longitude and 36°15' and 38°20'N latitude, in the northwest of China. It was given its name because of its expanse of water and the azure color; in Chinese, *Qing* means azure, *Hai* means open sea and *Hu* means lake (Figure Q1). The lake surface has the shape of a pear, with a circumference of about 360 km, covering a length of some 109 km from east to west and a width of 65 km from south to north. Its annual water stage (in 1988), measured at the Shatuosi Hydrological Station on the north shore, is 3193.59 m, with a surface area of 4282.3 km² and a capacity of 73.88 billion m³. It is therefore the largest lake in China (Figure Q2).

Due to the secular degradation of water level, parts of the marginal waters have become separated from the main body and form some sublakes. They are Gahai, Erhai, Haiyanhu, Shadaohu and Yilangjianxiaohu, the area of these sublakes totaling about 200 km². After their separation, these sublakes underwent their own development processes which were different from each other, caused by the sudden changes to their respective hydrological conditions. Both Gahai and Erhai were formed at an early stage but, due to the lack of inlet feed, the former not only shrank in area but the water quickly became saline; the latter, however, remains stable in these two respects thanks to the continuous feeding by the Daotanghe River.

Figure Q1 The situation of Qinghaihu in China.

The biggest island in Qinghaihu is Haixinshan Mountain, which is located in the middle of the lake, having an area of 1.14 km², with an elevation of 3266 m at its highest point. The second island, situated in the southwest part of the lake, is Guchashan Mountain (meaning 'three stones'), which is made up of dense limestone-ledge rocks. The famous Bird Island, originally a small island close to the western shore, became connected with the shore in 1978 as a result of the drop in level and sediment being constantly carried in by the chief inflow river, Buhahe. Shadao Island, originally the largest island at the northeast part of the lake, has now become joined to the shores of the lake and its newly formed sublakes. All these islands are ideal dwelling and breeding places for migratory birds. Every summer, Bird Island alone can attracts several tens of thousands of migrants, including almost 30 varieties. Birds such as speckledhead geese, brownhead gulls, fish gulls and cormorants breed on the island. There are almost a thousand adult birds per mu ($\frac{1}{15}$ ha) of the nesting area, and this is a density seldom seen at other places in the world. Rare birds such as swans, blackneck cranes, etc, are to be found, mainly on the marshlands near the influxes of the west shore. In Qinghaihu there are six fish types, among which the naked carp is the only economic fish of any importance. According to a calculation made by the Northwest Plateau Organism Research Institute of the Academy of Sciences of China, the lake basin possessed a naked carp reserve of about 50 000 tonnes in about 1975, but the expected yearly catch now is about 4790 tonnes.

The Qinghaihu watershed is encircled by mountains such as Riyueshan, Qinghainanshan, Xijunshan, Hekashan, Datongshan, and most of them are some 4000 m above sea level. The peak Gangge, erxiaoheli (a transliteration from Tibetan meaning 'fairy maiden peak'), in the western section of Datongshan, is 5291 m in elevation, its summit being the highest point in the watershed. While extending through numerous river valleys, between continuous mountains and uplands, the watershed slopes from northwest to southeast, forming a basin at its southeast end by integrating the lake surface with littoral plains. Here, the watershed closes to join the lake water.

The watershed abounds with rivers. The rivers scattered over the western and northern parts are large and long, while the rivers over the eastern and southern parts are not only small and short, but obviously seasonal. There are more than 20 small freshwater lakes on the high-altitude areas of the western and northern parts. At the source of Buhahe there are 22 active glaciers which cover an area 1329 km², containing 0.59 billion m³ of ice.

On account of the low rainfall and cold weather, most of the vegetation in the watershed is prata and grass, bushes, marshes and deserts; forests are few with a total coverage of no more than 2 km². Relevant to such a vegetation cover, most of the watershed soil is made up of alpine prata soil, grass soil and chernozem soils, and also karaburan soil which makes up about 12% of the total coverage. Alpine areas experience permafrost layers, the thicknesses of which range from 0.5 m to several tens of meters, being 60 m at their thickest.

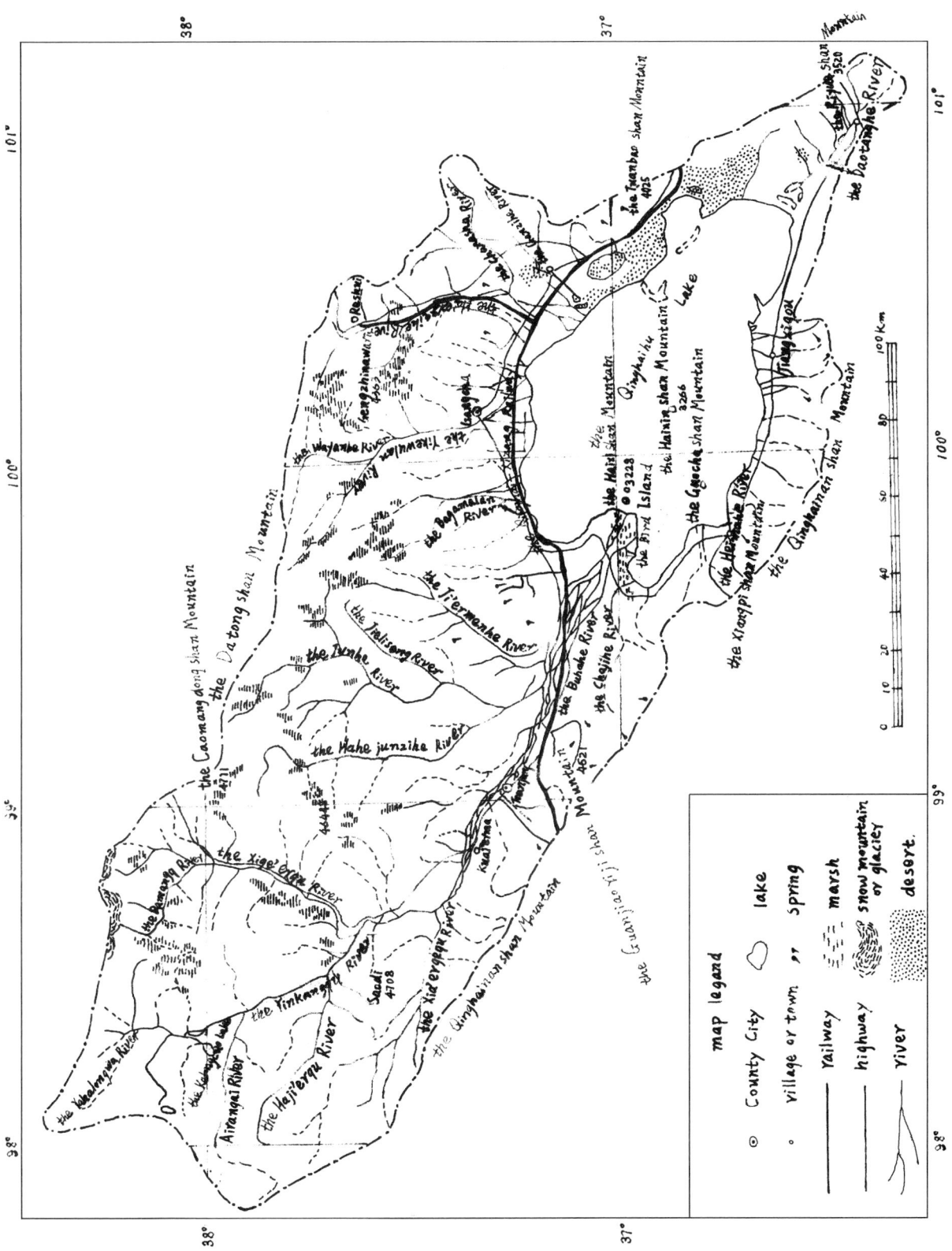

Figure Q2 The Qinghaihu watershed.

Formation and development

Qinghaihu is a tectonic sag lake. From the Tertiary period, due to the multiple tectonic movements and the repeated intense block fluctuations, the unified ancient planation surface of the Qinghaihu area was faulted, and the surrounding mountain ridges, flights of terraces and cryptodepressions appeared. The lake took its form at the initial stage of the Early Pleistocene period and came to its prime at a later stage of the same epoch. At that time, however, Qinghaihu was still connected to the Yellow River, with which it shared an expansion of waters. With the beginning of the Mid-Pleistocene period, the uplift of the surrounding mountains intensified, and the mountains situated at the southeastern part of the lake area were forced up, with their peaks reaching above the snow line. It happened that this bulging was forced into the second glacial period. The advent of this period not only brought about the decrease of inflow, the fall in water level and the shrinkage of water surface, but finally caused the separation of Qinghaihu from the Yellow River, and thus the formation of a closed lake was realized.

At the end of the Pleistocene (about 20 000 years ago), influenced by the aridly cold weather caused by the last glacial period, the lake surface then was lower than at present. With the coming of the Holocene epoch (about 12 000 years ago), the continuous uplift of the circumambient mountains and the alternation of warmth and coldness in climate caused the lake level to fluctuate. In the period of 7000 to 5000 years ago, the lake level maintained a fluctuating rise, for the watershed then possessed a climate which was relatively warm and damp. In the light of the various Chinese historical records about Qinghaihu, inference can be made that the lake level was relatively stable during the period from about 3000 to 1400 years ago; from 1400 years ago to the end of the nineteenth century, the lake level was continuously falling, with the yearly average fall being 0.9 cm.

After the dawn of the twentieth century, the lake level began to fall faster. A calculation made by Shiyafeng, an academician of the Academy of Sciences of China, on the basis of some relevant data, shows that, from 1908 to 1957, the rate of fall of the lake level was 17.2 cm year^{-1}. According to the observed data obtained from 1959 to 1972 at the Shatuosi Hydrological Station on the north shore of Qinghaihu, the lake level kept falling from 3196.55 m to 3193.69 m, the annual average fall being 8 cm.

It is mainly because of the rotation of warmth, coldness, dryness and dampness in the climate that the rise or fall in lake level and the expansion or contraction of lake surface come about.

Climate features

Nestling in the hinterland of the mid-Asian continent, the Qinghaihu watershed is at the intersection of the perennial westerly belt, the southeast monsoon region and the Qinghai–Xizang Plateau monsoon region. Influenced by the land configuration, altitude and atmospheric circulation, the watershed possesses such climatic features as dryness, intense solar radiation, cold winters and cool summers, great daily amplitude in temperature and frequent squalls.

The average air temperature of the watershed is between −1.5 and 1.5°C. The average monthly air temperature is between 16.0 and 20.0°C at its highest and between −18.0 and −23.0°C at its lowest, with the extreme lowest temperature being −35.8°C. The territorial distribution of air temperature shows that temperature is high in the southeastern parts and the lake basin area, and low in the northwestern parts and the hilly areas. The temperature variation within a year assumes a peak-to-valley shape, with the peak being in July and the valley in January. There is very little interannual change in air temperature. Since the 1950s the yearly average air temperature has shown a tendency of becoming warmer, with an obvious rise in winter and a slight drop in summer, thus the annual amplitude is gradually getting less.

The average wind velocity of the watershed is 3.2–4.4 m s^{-1}. The maximum instantaneous wind velocity is more than 30 m s^{-1}. Tianjun and Gangcha, the two counties located at the western and northern parts of the watershed, respectively, experience the highest number of days per year with gales, 97 days at Tianjun and 47 days at Gangcha. Gales are frequent in the months from March to May, and together with the gales there are always sandstorms and a drop in temperature.

The watershed enjoys plenty of sunshine, the total annual hours of which are 2430–3330 h with an annual insolation of 56–76%. The littoral plain and broad valleys enjoy more hours of sunshine while the hilly areas receive less. In a year, summer has most hours of sunshine, winter and spring see less, and the fall the least.

The vapor over the watershed comes mainly from the humid warm air flow stemming from the Bay of Bengal and the southeast coastal areas. Because of the situation of the watershed deep within an inland plateau, the humid warm air flow coming from afar has first to come across many obstructions and interceptions on its way, and consequently, when it arrives at the watershed, there is little vapor left, which accounts for the deficiency of rainfall. Being a semiarid area, the mean annual rainfall of the watershed is between 250 and 580 mm. The northern hilly areas see rather more rain, and in a wet year the rainfall is more than 700 mm. The western area and the southeast parts of the lake see less rain, the rainfall in a dry year being only about 210 mm. There is little interannual change, the coefficient of variation C_v for most of the various rainfall stations being around 0.20, and inter-regional C_v range being from 0.13 to 0.25. Within the territory of the lake, the rainfall in mid-lake is lower than that on the lake shores (see Figure Q3). Cloudbursts are seldom and they hold little rainfall, if any. The maximum rainfall in 24 h never exceeds 50 mm, and rains often occur during summer nights. Hail is more frequent on the north shore area than in any other areas. In Gangcha County, for example, the annual average number of days of hailfall is 15.3 days. More than 85% of the rainfall is concentrated in the months from May to September.

Under the combined actions played by the above-mentioned climatic features, water surface evaporation is intense and the annual evaporation is about 700–1200 mm. The territorial distribution of evaporation shows that it is greater in the lake basin than in the hilly areas, and in the south rather than in the north. In a year, the distribution is similar in various areas, that is, the months from June to September account for more than 60% of the evaporation. There is little interannual change, with the ratio of the annual maximum evaporation to the annual minimum being only about 1.5.

Rivers and their hydrological features

There are 48 direct inflow rivers, each of which has more than 5 km^2 of collecting area. Among these rivers, the biggest, with a drainage area of 14 385 km^2 and a length of 286 km, is Buhahe. It flows over the northwestern part of the watershed and flows into the lake on the western shore. The mean annual runoff of this river is 0.818 billion m^3, taking up 56.7% of the inflow mean annual runoff of the whole watershed. The second largest rivers are Yikewulanhe and Ha'ergaihe, the former with a drainage area of 1442 km^2, having a mean annual runoff of 0.246 billion m^3, and the latter with a drainage area of 1425 km^2, having a mean annual runoff of 0.242 billion m^3. The fourth largest rivers is the Wuha'alanhe. All of these rivers are distributed over the western and northern parts of the watershed. They are longer and larger in view of the drainage areas, branching because of their tributaries, and form the main body of the overland runoff inflow. There are also many inflow rivers over the eastern and southern parts of the watershed, but most of them are not only short in length, they have small drainage areas, and are also seasonal. Among these rivers, the Heimahe, with a drainage area of 112 km^2, is the largest.

The mean annual overland runoff of the Qinghaihu watershed is 1.612 billion m^3. The annual runoff depths differ at different areas, and at its maximum it amounts to 175 mm in the northern mountainous area, and at its minimum it is only 50 mm in the eastern part to the lake. Most of the rivers are fed by rain and snow. The months May to September, owing to the larger rainfall and the thawing of alpine ice and snow caused by higher temperature, produce more than 85% of the annual runoff, and most of the flood peaks appear during this period. There is a greater interannual change in the annual runoff; the ratio of the maximum to minimum annual runoff is 5.5, and the coefficient of variation C_v between the various rivers is about 0.40–0.65.

Having a larger catchment area, many tributaries and a longer path, the Buhahe has a smaller gradient in its lower reaches and thus, when any flood peak appears, it lasts a longer time (1–2 days) and assumes a horizontal crest. The other rivers are all mountain rivers, running through abrupt slopes with torrents, so when any flood peak appears in them, it is of short duration (from 0.5 to several hours) and assumes a steep shape. Larger flood peaks appear frequently in July and August, and in April and May, as a result of thawing, minor flood peaks (freshets) are experienced. This kind of flood peak often

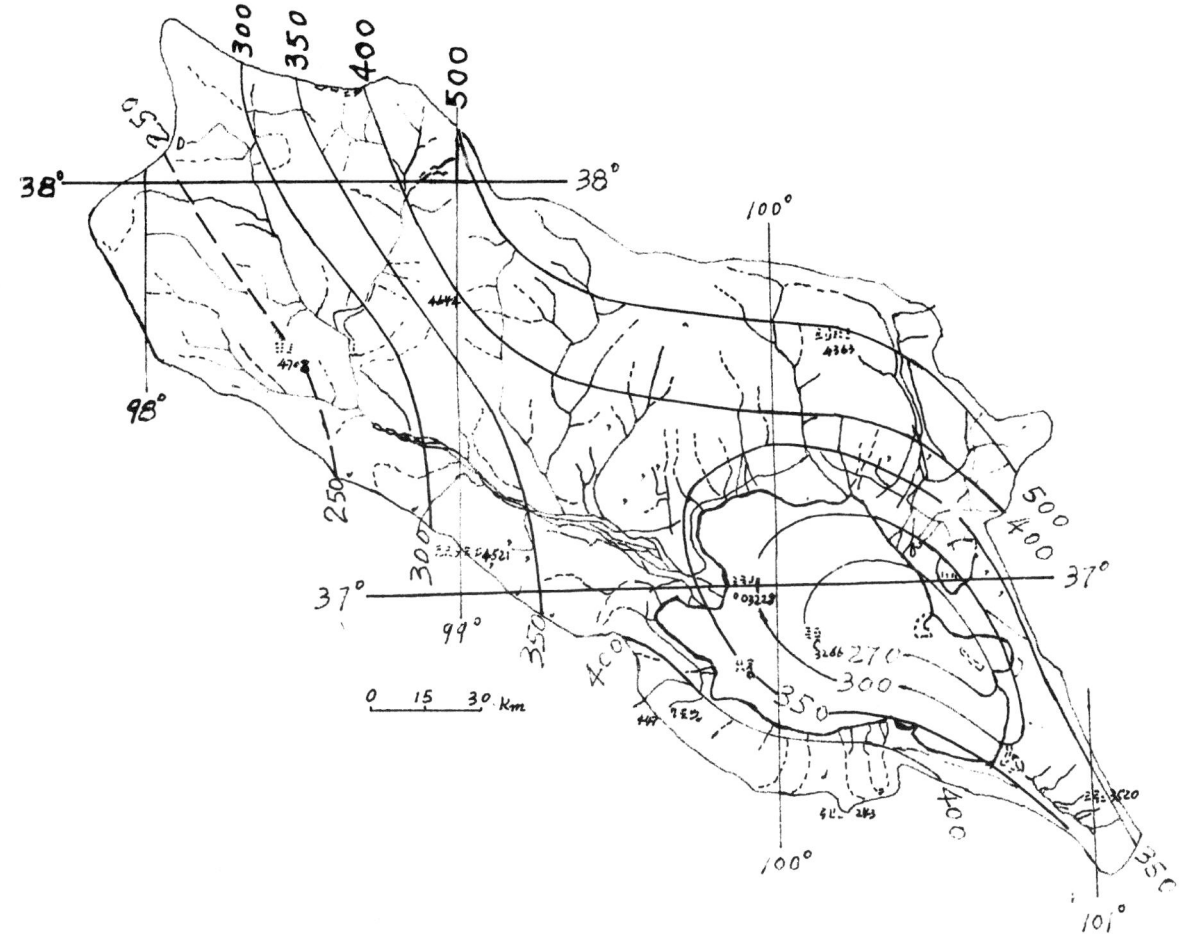

Figure Q3 Annual precipitation in the Qinghaihu watershed (in millimeters).

Table Q1 Runoff of the main rivers in the Qinghaihu watershed

River	Hydrological station name	Drainage area (km²)	Mean annual runoff (billion m³)	Maximum runoff		Minimun runoff		Maximum/ minimum ratio
				Runoff billion m³	Year	Runoff billion m³	Year	
Buhahe	Buhahekon	14 337	0.785	1.663	1967	0.1986	1973	8.4
Shallube	Gangcha	1 442	0.246	0.392	1988	0.1054	1979	3.7
Ha'ergaihe	Ha'ergal	1 425	0.242	0.335	1988	0.196	1960	1.7
Wuha'alanhe	Shatuosi	567	0.0546	0.0856	1988	0.0079	1960	10.8
Helmahe	Helmahe	107	0.0109	0.0407	1967	0.00157	1978	25.9

appears once a day for about 2 weeks with the flood hydrograph taking on a shape of serration.

The mean annual runoff of the main rivers in the Qinghaihu watershed are given in Table Q1. Figures Q4 and Q5, show the annual distribution and interannual variation of the Buhahe typical of these rivers.

The territorial distribution of sediment in suspension is the same as that of runoff, i.e. the main sediment contributing area is in the northwestern part of the watershed. The annual sediment discharge is unevenly distributed, and the discharge in the months from May to September makes up more than 95% of that for a year. There is a large interannual change, for instance the mean annual sediment discharge of the Buhahe is 484 700 tonnes, with the maximum annual

discharge being 862 000 tonnes, and a minimum of 231 000 tonnes, the ratio of these being 37.3.

The annual distribution and the interannual change in sediment discharge of the Buhahe, is shown in Figures Q6 and Q7.

There are broad and thick diluvial fans at the mouths of the rivers, caused by the accumulation of the large amount of bedload carried by the rivers while running through the mountainous areas during flood periods.

Because of the cold weather in the Qinghaihu watershed, the cryogenic period of the rivers lasts 6–7 months. The ice conditions are complicated, with ice phenomena such as shore ice, drift ice, complete freezing, ice damming and blocking, running water on ice, and floating ice, all being experienced. The farther upstream, the more

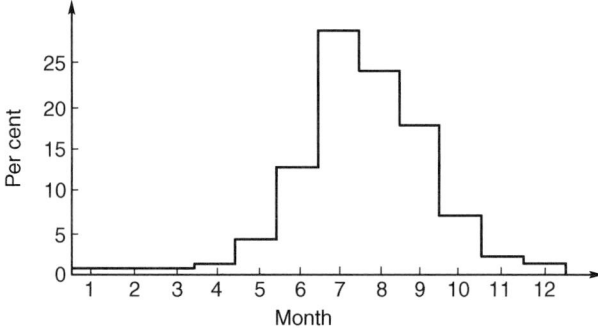

Figure Q4 Annual distribution of the Buhahe runoff.

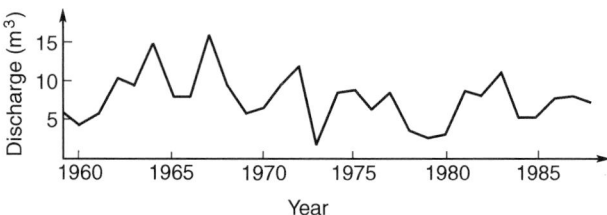

Figure Q5 Hydrograph of the interannual variation o the Buhahe runoff, 1959–1988.

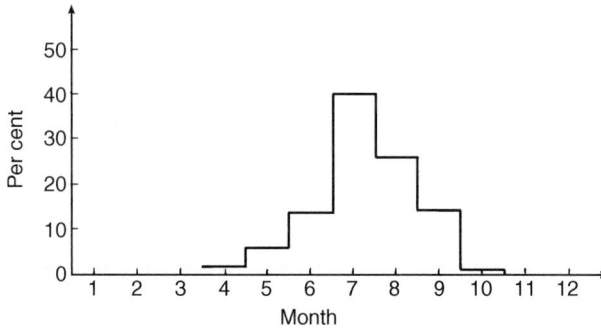

Figure Q6 Annual distribution of the sediment discharge of the Buhahe,

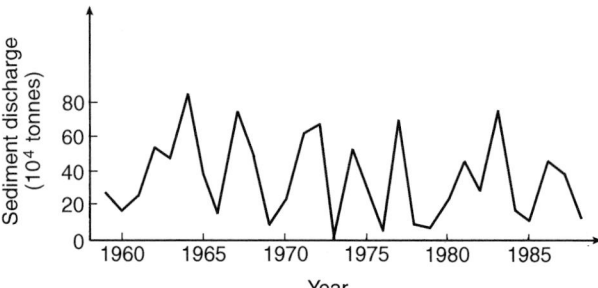

Figure Q7 Hydrograph of the interannual variation of sediment discharge of the Buhahe, 1959–1988.

complicated are the ice conditions. In the cryogenic period, river flow is low with obvious daily variations: after sunrise, with the gradual rise in air temperature, river flow gradually increases, and reaches its maximum at 3–5 pm; after sunset, the variation of river flow begins to reverse.

Hydrological balance

On the basis of the hydrological, meteorological and geological data collected for more than 30 years, the hydrological equation for Qinghaihu can be established in the light of water balance principle:

$$\Delta V = Q + R + W - D - E$$

Where ΔV is the annual loss and gain of water by Qinghaihu (billion m^3), Q is the surface runoff inflow (billion m^3), R is the groundwater runoff inflow (billion m^3), W is the precipitation on the lake surface (billion m^3), D is the the overlapping amount of surface and groundwater (billion m^3) and E is the evaporation amount from the lake surface (billion m^3).

Each factor in the formula can be estimated as follows.

- Q The mean annual runoff amount can be calculated using the contour line of the mean annual runoff depth in the watershed.

Using the method of hydrologic analogy, that is, first divide the rivers which are controlled by the hydrologic stations according to their coverage, and compute their respective runoff amounts. Then, with the guidance of the 'similarity principle', divide the other rivers into small divisions according to their natural geographic environments, the underlying conditions and runoff contributing systems, and thus the runoff amounts of these divisions with no observed data can be computed by making use of an analogy method. The addition of these derived data to the observed data provided by the hydrological stations throw light on the surface runoff amount for the whole watershed. The use of the above two methods has brought about similar findings. The mean annual runoff of the whole watershed is 1.612 billion m^3 using the hydrological analogy method. The following table (Table Q2)

Table Q2 Surface runoff in the Qinghaihu watershed, 1959–1988

Year	Runoff (10^9 m^3)
1959	1.441
1960	1.043
1961	1.277
1962	1.964
1963	1.895
1964	2.724
1965	1.576
1966	1.611
1967	2.876
1968	1.804
1969	1.230
1970	1.402
1971	1.922
1972	2.354
1973	0.654
1974	1.797
1975	2.027
1976	1.405
1977	1.481
1978	0.867
1979	0.681
1980	0.861
1981	1.896
1982	1.636
1983	2.210
1984	1.271
1985	1.316
1986	1.611
1987	1.629
1988	1.852

Table Q3 Annual precipitation amount on the lake surface, 1959–1988

Year	Precipitation on surface (10^9 m^3)
1959	1.614
1960	1.283
1961	1.524
1962	1.114
1963	1.334
1964	1.535
1965	1.710
1966	1.670
1967	2.375
1968	1.228
1969	1.394
1970	1.346
1971	1.635
1972	1.381
1973	1.109
1974	1.519
1975	1.590
1976	1.622
1977	1.268
1978	1.452
1979	1.271
1980	1.201
1981	1.782
1982	1.507
1983	1.617
1984	1.349
1985	1.801
1986	1.604
1987	1.624
1988	1.820

Table Q4 Evaporation from the lake surface, 1959–1988

Year	Surface evaporation (10^9 m^3)
1959	4.311
1960	4.247
1961	4.191
1962	4.233
1963	4.466
1964	3.961
1965	4.126
1966	4.330
1967	3.733
1968	4.052
1969	4.741
1970	4.460
1971	4.492
1972	4.407
1973	4.501
1974	4.071
1975	3.938
1976	4.077
1977	3.435
1978	3.952
1979	4.492
1980	4.398
1981	4.164
1982	3.801
1983	3.459
1984	3.910
1985	3.625
1986	3.778
1987	3.793
1988	3.836

shows the year-to-year watershed runoff by using this method. As shown in Table Q2, there is a greater interannual change in runoff, and the ratio of the maximum to the minimum is 4.3.

- R The hilly areas and the plain areas of the watershed have been separately calculated. The mean annual runoff of hilly areas is 0.935 billion m^3, but most of this overlaps with surface water, and it therefore must be excluded while performing water balance accounting. The mean annual runoff of the plain areas is 0.615 billion m^3. This is derived by adding up the bed undercurrent amount, the piedmout outflow amount, precipitation, infiltration and the canal system seepage supply, and then substracting the groundwater evaporation, river course infiltration and groundwater extraction. This amount is included in water-balance accounting.
- W Based on the precipitation data provided by the lake shore hydrometeorological stations, the value of this factor can be computed by using two methods, namely the arithmetic mean and coverage weighting. The use of these two methods has led to similar figures. The mean annual precipitation on the lake surface is 1.509 billion m^3 according to the figures derived by using the method of weighting. Table Q3 shows the year-to-year precipitation on the lake surface from 1959 to 1988. As seen from Table Q3, there is little interannual change in precipitation, the ratio of the maximum to the minimum being only 2.14.
- D While calculating the surface and ground runoff, the river course infiltration, etc. have been brought into the calculating process. Therefore, a deduction should be effected while calculating the water balance accounting. Hence the mean annual runoff is 0.093 billion m^3.
- E The yearly intensities of evaporation at the various hydrological stations can be computed by using the year-to-year lake surface evaporation data provided by these lake-surrounding stations. Then, using the Thiessen method, the average intensity of evaporation on the lake surface can be calculated and further, using the dependence of the lake level Z and the lake coverage F, the lake coverage corresponding to each year can be looked up. The yearly

intensity of evaporation multiplied by the corresponding coverage of each year gives the yearly lake surface evaporation amount. The quantities are shown in Table Q4. The mean annual amount is 4.10 billion m^3.

Apart from the five above-mentioned factors, there are other factors such as farm irrigation and industrial consumption, but they consume such a small amount of water that they are not significant. Therefore they have not been taken into consideration.

The ΔV values from 1959 to 1988 (30 years), have been calculated year to year as shown in Table Q5; the mean annual amount is −0.457 billion m^3. However, the yearly loss and gain rates can also be found from the dependence of the lake level Z and the lake capacity V, also given in Table Q5. According to the latter, the lake level fell by 2.96 m during the specified 30 years, and it had lost by 12.957 billion m^3 of water at an average annual rate of 0.432 billion m^3. It is obvious, therefore, that the use of the two methods has produced similar results.

The general trend of the Qinghaihu water level is that of a decline, which corresponds to the tendency of high consumption of the water. Totalling the ΔV values year by year, and taking this as ordinate, a hydrographic chart of the yearly average water level can be plotted. The two variation hydrographs are identical (Figure Q8).

Physicochemical properties of the lake water

With the long-standing depletion of the lake water, the degree of mineralization has been gradually increasing each year: 1962, 12.49 g l^{-1}; 1986, 14.37 g l^{-1}; 1987, 14.70 g l^{-1}; 1990, 15.30 g l^{-1}. Because the inflow rivers are mainly on the western and northern sides of the lake, the distribution of the degrees of mineralization is such that it is high in the eastern and southern parts of the lake, and low in the western and northern parts, and the minimum degree of mineralization appears at the influxes of large rivers such as the Buhahe.

The degrees of mineralization of all the inflow rivers are far lower than that of the lake water. For example, in the Buhahe it is 33 times lower than that of the lake water.

Table Q5 Water balance accounting of the lake, 1959–1988

Year	Water balance accounting			ΔV from the dependence of level and capacity (10^9 m³)
	$Q + R + W$ (10^9 m³)	$D + E$ (10^9 m³)	ΔV (10^9 m³)	
1959	3.670	4.424	−0.754	−0.507
1960	2.941	4.340	−1.339	−1.446
1961	3.416	4.284	−0.868	−0.632
1962	3.693	4.326	−0.633	−0.613
1963	3.844	4.559	−0.715	−0.635
1964	4.874	4.064	−0.820	0.452
1965	3.901	4.219	−0.318	−0.497
1966	3.896	4.423	−0.527	−0.497
1967	5.866	3.826	2.040	2.214
1968	3.647	4.145	−0.498	−0.181
1969	3.239	4.834	−1.595	−1.536
1970	3.363	4.553	−1.190	−0.761
1971	4.172	4.585	−0.413	0.044
1972	4.350	4.500	−0.150	−0.438
1973	2.378	4.594	−2.216	−1.576
1974	3.931	4.164	−0.233	−0.219
1975	4.232	4.031	0.201	0.315
1976	3.687	4.170	−0.483	−0.263
1977	3.364	3.528	−0.164	−0.569
1978	2.934	4.045	−1.111	−0.876
1979	2.567	4.585	−2.018	−1.795
1980	2.677	4.491	−1.814	−1.444
1981	4.293	4.257	0.036	0.087
1982	3.758	3.894	−0.136	−0.043
1983	4.442	3.552	0.890	0.831
1984	3.235	4.003	−0.768	−1.050
1985	3.732	3.718	0.014	−0.671
1986	3.830	3.871	−0.041	0
1987	3.868	3.886	−0.018	−0.449
1988	4.287	3.929	0.358	0.040

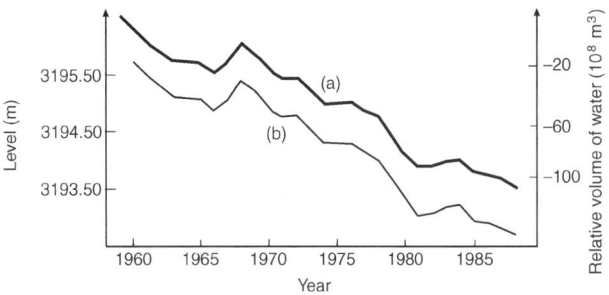

Figure Q8 (a) Variation in the water level and (b) annual change in the volume of Qinghaihu, 1959–1988.

There are three observation stations which have collected the offshore superficial water temperature for more than 30 years. Over this period the annual mean water temperature is 5–6°C with the maximum being 25.3°C and the minimum −1.1°C. In winter, the superficial water temperature is often negative, and the temperature gradually increases from the surface downward, with the temperature difference between the surface and the bottom being 3–4°C.

Ice on the lake appears somewhat later than that on the rivers. As a rule, they often appear between the middle of September and early November. At first, there is only some thin ice, then gradually increasing and drifting on the lake with winds accumulating to form shore ice in the meantime. With a gradual drop in temperature, the shore ice becomes widened and thickened, but stormy waves repeatedly break the ice, compelling the ice to drift on the lake. After several such repetitions, the lake can be completely covered with ice overnight if the weather is cold and windless, but this first closure of ice is easily broken up by storm waves, then windless cold weather brings about another closure, often needing several such repetitions before the steady closure of ice comes. More often than not, this steady closure of ice appears at the beginning of December, coming to its thickest in January and February. Generally, the thickness of ice layer is about 50 cm, but it can be 69 cm at the thickest. In spring, the ice layer begins to thaw. From the latter part of March to the middle of April, gales cause the ice layer to break, and as a result the lake opens.

The inflow of the Buhahe causes a principal circulation which circumambulates clockwise around the north side of Haixinshan Mountain. In addition to this, local circulations are formed within small areas near the influxes of the bigger rivers on the north side. Under the action of water temperature variation and evaporation, the specific densities of the lake water at various depths change continuously, causing the vertical movement of water. However, waves raised by winds exert an influence on both the circulation and the vertical movements of the water.

Forecast for the future

Due to the fact that inflow is smaller than consumption, the Qinghaihu water level has been falling for a long period of time. If the climatic conditions of the watershed do not change dramatically, according to the principle of water balance, the lake level will begin to maintain a fluctuating state of relative stability, when the lake surface shrinkage caused by the fall of water level reaches a point where the evaporation and the inflow begin to keep a balance. A calculation made by the Hydrological Service of Qinghai Province shows that this water level will be 3186.5 m, with the corresponding surface coverage being 3550 km² and the capacity being 46.49 billion m³. Compared with the state in 1988, the lake level will have fallen by 7.09 m, the area will have decreased by 732.3 km², and the capacity will have been reduced by 27.39 billion m³. It is estimated that this process will take about 130 years.

Most of the foreign and Chinese experts, while making forecasts about the climatic changes at the Qinghaihu watershed and its adjoining areas in the coming century, have come to the conclusion that it will tend to be warm and humid. If this is reliable, the state in which the Qinghaihu water level becomes relatively steady will appear ahead of the estimated time.

Jing Yueling and Cui Dewei

Cross references

Lakes
Lakes: largest worldwide
Limnology

R

RADAR: PRECIPITATION-MEASURING (WEATHER) RADAR IN EUROPE

The application of radar technology in meteorology has been the subject of over 50 years of development. Most applications originate from the ability of radar to detect cloud particles, raindrops, snow

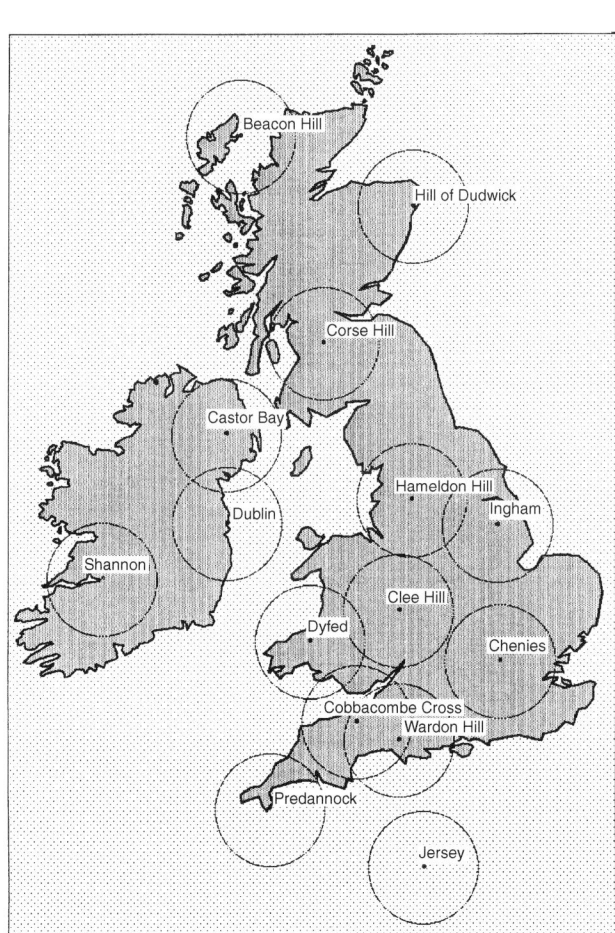

Figure R1 Weather radar network in the British Isles, February 1993 (circles correspond to 75 km range).

falling and ice particles. Commercial development of precipitation measuring (weather) radar in the UK, however, commenced in the 1950s with an X-band Type 40 storm warning radar having a 3 cm wavelength produced by the Decca Company, while later types were based on S-band (10 cm) and C-band (5.6 cm). In 1966 Plessey Radar (who had taken over the Decca Company), introduced a new S-band radar Type 43S with a beam width of 2° and with quantitative coverage of 75 km radius. Qualitative information, however, could be obtained up to 200 km radius. The first trial of the application of weather radar to hydrology and water resources started in 1967 at a site in North Wales and was sponsored by the Water Resources Board, the Meteorological Office, the Plessey Company and the Dee and Clwyd River Authority. It continued over a 10-year period (research was also being carried out in the US and Switzerland). The success of this project – the Dee Weather Radar Project – led to the eventual installation of a network of radars covering the British Isles (Figure R1). Western Europe followed and during a 14-year project (1979–1992) under the aegis of the European Commission's COST program, 16 countries participated and a network of some 100 precipitation measuring radars were installed covering most of Western Europe.

R.W. Herschy

Bibliography

Browning, K.A., 1978. *The short-period weather forecasting pilot project*, Meteorological Office RRI, Research Report No. 1.

Browning, K.A., 1978. Meteorological application of radar, *Rep. Prog. Phys*, **41**(5), 761–806.

Browning, K.A., 1987. Towards the more effective use of radar and satellite imagery in weather forecasting, in *Weather Radar and Flood Forecasting* (Collinge, V.K. and Kirby, C., eds), John Wiley & Sons, Chichester.

Bulman, P.J. and Browning, K.A., 1971. *Report, National Weather Radar Network*, Royal Radar Establishment, Malvern, UK, 17 pp.

Central Water Planning Unit, 1977. *Dee Weather Radar and real time hydrological forecasting project*. Report by Steering Committee, HMSO, London.

Collier, C.G., 1987. Accuracy of real-time radar measurements, in Weather Radar and Flood Forecasting (Collinge, V.K. and Kirby, C., eds), John Wiley & Sons, Chichester.

Collier, C.G. (ed.), 1992. *International Weather Radar Networking*, Kluwer Academic Publishers, Dordrecht.

Collier, C.G. and Chapuis, M. (eds), 1990. *Weather Radar Networking*, Kluwer Academic Publishers, Dordrecht.

Collinge, V.K. and Kirby, C. (eds), 1987. *Weather Radar and Flood Forecasting*, John Wiley & Sons, Chichester.

Newsome, D.H. (ed.), 1992. *Weather Radar Networking*, Kluwer Academic Publishers, Dordrecht.

Newsome, D.H., 1987. COST 72 and Weather radar in Western Europe, in *Weather Radar and Flood Forecasting* (Collinge, V.K. and Kirby, C., eds), John Wiley & Sons, Chichester.

Ryder, P. and Collier, C.G., 1987. Future development of the UK weather radar network, in *Weather Radar and Flood Forecasting* (Collinge, V.K. and Kirby, C., eds), John Wiley & Sons, Chichester.

Water Resources Board, 1973. Dee Weather Radar Project, Report by the operation systems group on the use of a radar network for the measurement and quantitative forecasting of precipitation, 21 pp and 9 appendices.

RAIN

The condensation of water vapor in the atmosphere, brought about by the movement of air upwards, provides water droplets or ice crystals in clouds. Such precipitation particles are denser than the air surrounding them, and therefore they begin to fall at a rate of a few centimeters per second. However, these particles will either evaporate in unsaturated air below the cloud, or will be held suspended by vertical currents within the cloud. They will only be able to reach the ground as precipitation if they become large enough to stand evaporation losses and overcome upwards air motions.

Over 100 years ago it was recognized that processes other than condensation were needed to cause the growth of precipitation particles in the observed time scales. In 1911 Wegener proposed that rain was formed by the melting of ice particles. Ice crystals grow by sublimation when they exist in cloud together with supercooled water droplets. As water vapor is removed from the air by this process, the air becomes unsaturated with respect to water so the droplets evaporate. This continues until either all the droplets have been evaporated or the ice crystals become so large that they fall from the cloud. This process takes from 10 to 30 min, and as the ice crystals fall they may melt to form rain (droplets with radii of about 20 μm) which can reach the ground. A theory of this process was derived by Bergeron in 1935, and observations confirming the theory were made by Findeisen in 1939. Hence this process of rain formation became known as the Wegener–Bergeron–Findeisen process, sometimes shortened to the Bergeron process.

Although this explains the formation of precipitation in midlatitudes where clouds usually extend well above the 0°C level, hence it is referred to as the cold rain process, it is observed that warm clouds with tops below 0°C also produce rain. Indeed, such clouds occur in midlatitudes as well as in the tropics, and a warm rain process is required. It was discovered that cloud particles normally begin as cloud condensation nuclei (CCN), which consist of partially or completely soluble aerosol particles. As the cloud particles grow by condensation or sublimation of the CCN, they begin to fall and collect other particles. The type of precipitation which is formed by such collisions depends upon the types of cloud particles present. If the cloud contains only water then rain is formed and the process is known as coalescence. However, if only ice crystals exist then snow results and the process is called aggregation.

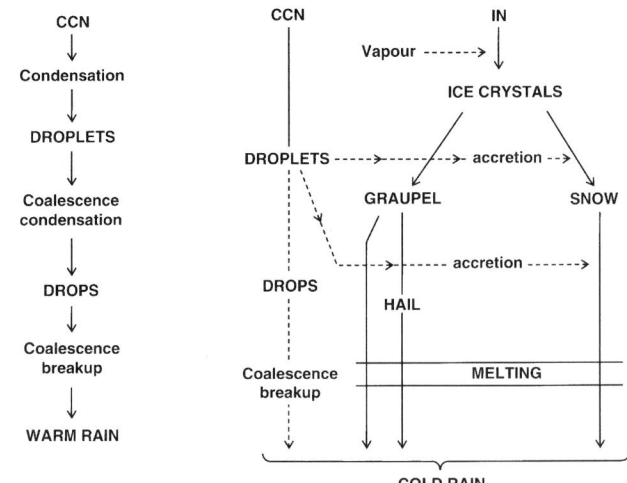

Figure R2 The evolution of warm and cold rain starting from cloud condensation nuclei (CCN) and ice nuclei (IN). (From List, 1977.)

If both water droplets and ice crystals exist, then ice or snow pellets (graupel) or hail may form and the process is known as accretion. Eventually the precipitation particles may reach such a size that they break up beginning the process again. Detailed descriptions of these processes may be found in Mason (1971). They are sometimes collectively referred to as the Langmuir process in recognition of the work of Langmuir in 1948. This process may exist together with the Bergeron process. A summary is given in Figure R2.

Rain systems

We have seen that water vapor in the atmosphere may be condensed to form clouds when the air is lifted. Cloud particles may then grow by various processes to form precipitation. The atmospheric motions which produce the lifting necessary to trigger these processes are organized on various scales as shown in Table R1. Although mesoscale precipitation areas and rain bands are entered in this table as separate phenomena, in fact they occur within the larger systems, and represent organization on an intermediate scale of localized convection. They will therefore be discussed within a description of the larger phenomena. It is not the intention here to provide detailed dynamical and physical descriptions of each system. Comprehensive reviews have been provided by Houze and Hobbs (1982) and Browning (1983a). The aim is to note the temporal and spatial distribution of precipitation, and the likely intensity extremes which are associated with each system.

Table R1 Scales of precipitation systems (partly after Browning, 1983a)

Precipitation system	Description	Horizontal scale (km)	Vertical velocity (cm s^{-1})
Localized convection	Precipitation from single clouds in the lower atmosphere or from convective generating cells within larger-scale systems. They may occupy a large depth of the atmosphere, in which case they are referred to as thunderstorms	$\sim10^{0.5}$	$\sim10^2$
Mesoscale precipitation area (MPA)	Cluster of convective cells	$\sim10^{1.5}$	~10
Mesoscale rainband	Convective cells occurring in lines, sometimes almost two dimensional	Width $\sim10^{1.5}$ Length $\sim10^2$	~10
Synoptic (or large) scale precipitation systems	Midlatitude depressions	Width $\sim10^2$ Length $\sim10^3$	~1
Tropical storms	Known as hurricanes in the southern part of the North Atlantic, cyclones in the northern part of the Indian Ocean, typhoons in the south western part of the North Pacific, and Willy Willys in the northeast of Australia	Radius $\sim10^{1.5}$	$\sim10^2$

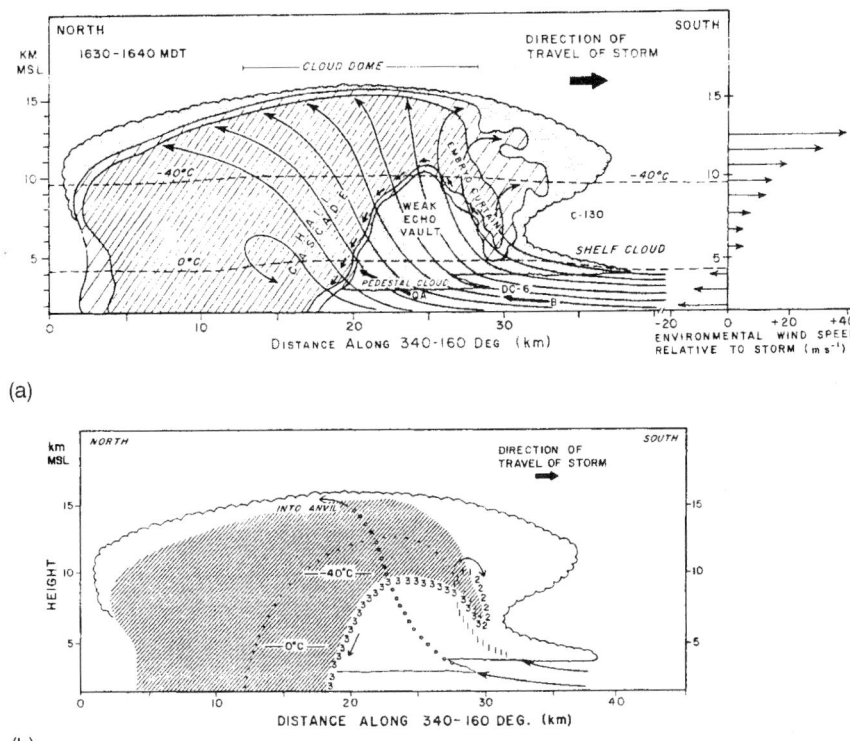

(a)

(b)

Figure R3 (a) Vertical section showing features of the visual cloud boundaries of the Fleming, northeastern Colorado, USA, storm at 1630–1640 MDT, 21 June 1972, superimposed on the pattern of radar echo. The outer contour derived from an airborne radar corresponds to a reflectivity of 0 dBz at a distance of 10 km and −20 dBz at 1 km. The inner contour is 16 dBz higher. The section is oriented in the direction of travel of the storm. Areas of cloud devoid of detectable echo are shown stippled. The location of four instrumented aircraft are indicated by C-130, QA, DC-6 and B. Bold arrows denote wind vectors in the plane of the diagram. Short thin arrows skirting the boundary of the vault represent a hail trajectory. The thin lines are streamlines of airflow relative to the storm. To the right of the diagram is a profile of the wind component along the storm's direction of travel, derived from a Sterling sounding 50 km south of the storm. (b) Hail trajectories in a vertical section along the direction of travel of the same storm as (a). Trajectories 1, 2 and 3 represent the three stages in the growth of large hailstones. The transition from stage 2 to 3 corresponds to the re-entry of a hailstone embryo into the main updraft prior to a final up-and-down trajectory during which the hailstone may grow large. (From Browning and Foote, 1976.)

Localized convection

As the lower layers of the atmosphere are warmed by the sun, or cold air passes over the warm sea, the air becomes less dense and rises. If the atmosphere is unstable, then vertical motion will occur giving rise to condensation of water vapor and the formation of cloud. The same effect may be produced by air forced to rise over hills. Such clouds grow to a height of several kilometers in 20–40 min producing rain of intensity which is dependent upon the depth of the cloud.

The vertical velocity depends upon the difference between the environmental lapse rate and the saturated adiabatic lapse rate. If this difference is large, then the vertical motion will be large, and a small shower cloud may develop rapidly to a height of around 10 km, becoming a thunderstorm. Electricity is generated by processes which are still not clearly understood, producing lightning. Within thunderstorms, hail may be produced by precipitation particles sweeping up supercooled droplets in an updraft until they are of such a size that they fall out of the updraft. However, the hailstones may be caught in the updraft at a lower level as it is not usually vertical, and therefore they can be recycled several times before falling to the ground. This gives rise to a layered structure to the hail caused by growth in different parts of the cloud.

In recent years a considerable amount of research has been carried out into the structure of thunderstorms. Two main types have been identified: (1) multi-cell storms and (2) supercell storms (Figure R3a). Most storms fall into the multicell category. They consist of several cells (clouds) at different stages of development at any one time. New cells may form where the outflow regions from other cells intersect (Purdom, 1976). The supercell storms have a structure which consists of a single storm-scale circulation comprising one giant updraft–downdraft pair. Either of these types of storm may produce a tornado, a region a few kilometers wide or less of high rotation (winds in excess of 320 km h^{-1}) and low pressure visualized as a funnel cloud. Rainfall may be very heavy, although usually of short duration. Thunderstorms occur in temperate and tropical regions throughout the world, and some examples of extreme storm rainfall totals are given in Table R2.

Mesoscale precipitation systems

Several thunderstorms may be grouped together within what Maddox (1980) has referred to as 'mesoscale convective complex' (MCC). Indeed, it has been suggested that 50–60% of the summer rainfall in the Great Plains and the mid-western United States is due to MCC systems.

The precipitation area is continuous, and larger than precipitation from any individual storm, and this has been cited as evidence for an organized circulation system. In the early stages of development the MCC is very convective and the rainfall is dominated by that produced by the individual cells. However, as the MCC develops, lifting on a larger scale occurs, although the actual mechanism is not yet clearly understood. Such systems have also been identified in the UK (Browning and Hill, 1984). Rainfall totals are similar to those from individual thunderstorms, although large storm totals may accrue. The MCC which caused severe flooding at Johnstown, Pennsylvania, USA on 19–20 July 1977 produced a storm total of over 300 mm in a 9 h period.

Table R2 Examples of rainfall totals from intense thunderstorms occurring worldwide

Location	Date	Duration (h)	Rainfall total (mm)
Middle Knoll, Dunsop Valley, Lancashire, UK	8 August 1967	1.5	117
Hampstead, London, UK	14 August 1975	2.5	170
Manchester, UK	6 August 1981	1	83
Big Thompson Canyon, Colorado, USA	31 July 1976	4	250
Strongstown, Pennsylvania, USA	20 July 1977	1	71
Rapid City, South Dakota, USA	10 June 1972	4	~305
Sydney, Australia	11 November 1976	11 min	100
Kyushu, Japan	27 June 1972	1	100
		10 min	30

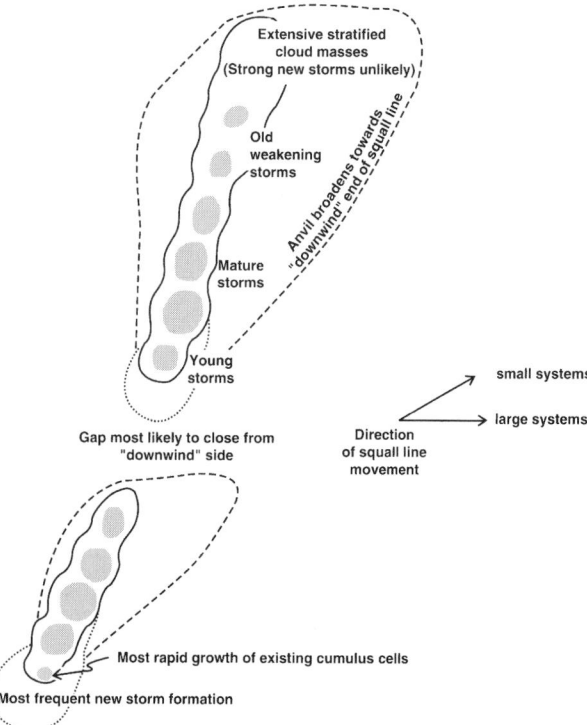

Figure R4 Schematic diagram of one squall line configuration. (After Newton and Frankhouser, 1964; Houze and Hobbs, 1982.)

A similar type of system occurs in tropical areas as a cloud cluster (e.g. Leary and Houze, 1979), and both systems may develop into squall lines, the configuration of which vary depending upon the environmental wind shear (Figure R4). At all latitudes the squall line tends to travel by a combination of translation and discrete propagation as new cells form on the leading edge of the system.

Midlatitude depressions

Although thunderstorms produce large amounts of rainfall in midlatitudes at any one place, they are quite rare. Most of the rainfall originates from low-pressure systems termed 'depressions'. With the invention of the barometer in the seventeenth century it became possible to measure the pressure of the atmosphere frequently, but it was not until the nineteenth century that particular atmospheric storms were studied using observations made at more than one place. The first theory involved the idea of a 'cyclone', a revolving cylinder of air with a more or less upright axis. No sharp discontinuities or 'fronts' were admitted.

This concept held sway until the early twentieth century when the idea of fronts within a rotating baroclinic fluid (that is, a fluid with both temperature and density gradients) was developed by what became known as the Bergen School in Norway (Bjerknes and

Solberg, 1922). They considered a wave developed on a boundary between warm and cold air, giving rise to a warm front and cold front structure as shown in Figure R5. Rain from stratiform (layered) cloud is associated with the warm front, and rain from convective cloud and thunderstorms is associated with the cold front. The basis of a dynamical description of the depressions is the release of energy in sloping (baroclinic) connection.

This model of a depression has formed the basis of practical weather forecasting until the present day. However, with the increasing availability of observations from radars and satellites it became clear that, in practice, the structure of a depression was much more complex than that depicted in Figure R5. Indeed, many small-scale features were evident, and a new model, shown in Figure R6, has been formulated to provide a framework for the development of these features (Browning, 1971; Harrold, 1973; Carlson, 1980).

In this model the distribution of rainfall is explained by reference to two main flows termed the warm conveyor belt (WCB) and the cold conveyor belt (CCB). The WCB originates at low levels to the south (in the northern hemisphere) of the depression, and ascends ahead of the cold front, turning southwards above the cold air ahead of the surface warm front. The CCB originates to the northeast of the depression at low levels, ascending ahead of the surface warm front beneath the WCB, and then turning to run parallel but below the WCB.

The dry air behind the cold front often overrides the WCB in the region AB shown in Figure R6. This causes the air to become unstable, if it is ascending, with the result that convection is released in small cells at mid-levels known as generating cells. These cells tend to occur in clusters and give rise to mesoscale precipitation areas (MPAs) which can produce moderate or heavy rainfall. This rainfall is produced by what is known as a 'seeder-feeder' mechanism, (Bergeron, 1950). Ice particles from the generating cells fall into lowerlevel frontal cloud, and act as natural seeding particles causing a rapid growth of precipitation particles. This mechanism, together with dynamical processes, is important in producing a number of different types of rain band associated with the depression. Figure R7 summarizes the types of rain bands which have been identified. Rainfall totals from midlatitude depressions are not generally large, although heavy rain may occur from thunderstorms developing on the cold front. Nevertheless, the precipitation often occurs steadily over long periods, particularly over hilly areas. In such cases, or when a system becomes stationary or slow moving, storm totals of over 100 mm in 24 h may occur. For example, 211 mm of rain fell on 10–11 November 1929 over the Rhondda Valley, South Wales, UK, and 229 mm at Richmond, Natal, South Africa, on 17–18 May 1959.

One special form of depression occurs in deep polar air over the northeast Atlantic, and is known as a polar low. These systems are small and resemble small tropical cyclones. It has been suggested that either they arise from the instability produced by the cold air moving over a relatively warmer ocean, or that baroclinic instability occurs in a shallow layer (Harrold and Browning, 1969). Such systems may produce large amounts of snow or rain over short time periods.

Tropical storms

Tropical cloud clusters were described above. One favored area for the formation of these clusters is the intertropical convergence zone (ITCZ) located close to the equator. Here the northeast trade winds of the northern hemisphere converge with the southeast trade winds of

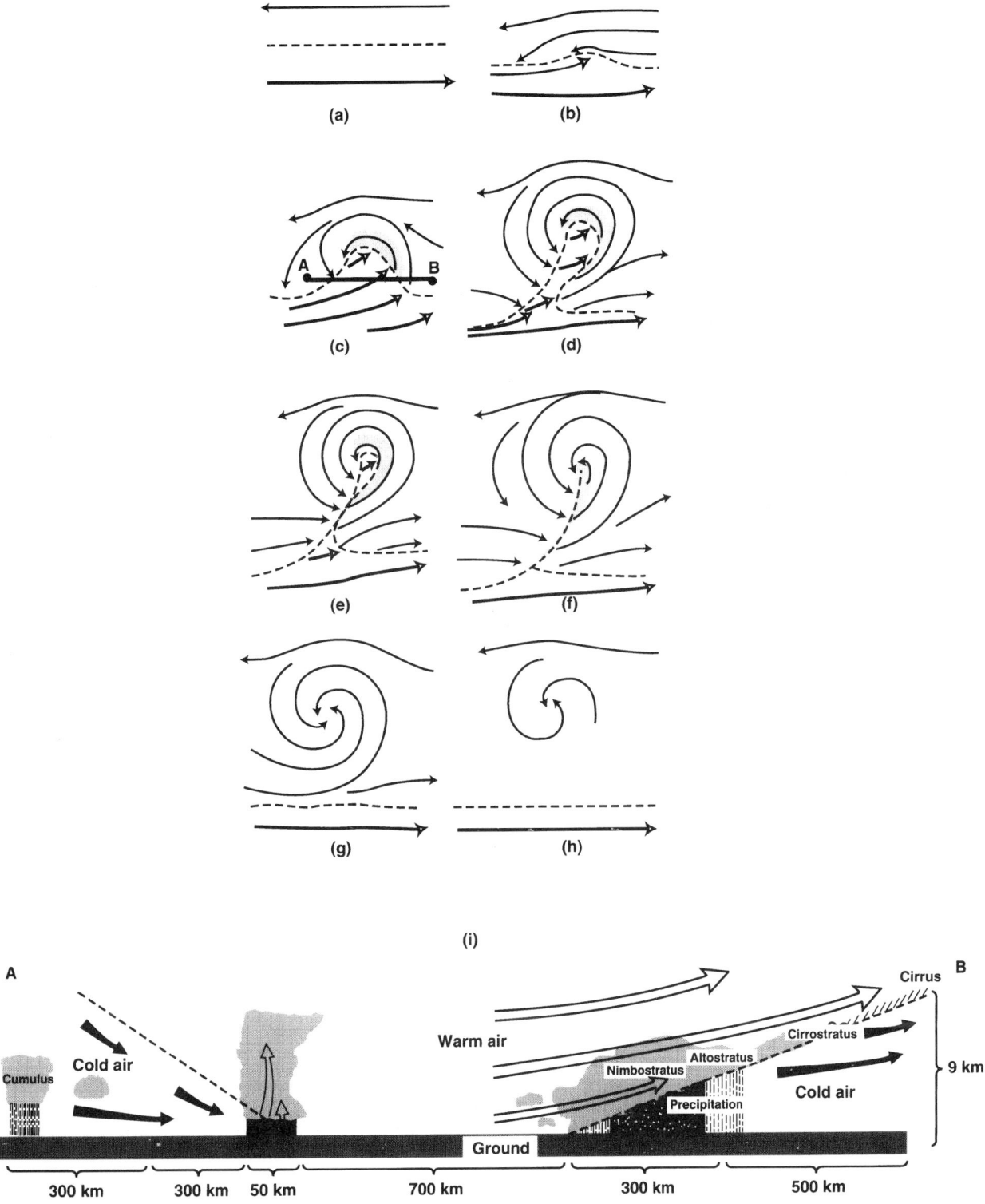

Figure R5 Evolution of cyclone and its fronts, according to Bjerknes and Solberg (1922) (from Ludlam, 1966). Rain areas are shaded; (i) is a cross-section along AB in (c) showing the cloud and air motion associated with the fronts (from Houze and Hobbs).

the southern hemisphere. A small fraction of clusters contain the necessary circulation structure to enable them to develop into hurricanes (cyclones or typhoons; Table R1). The upper clouds become circular and concentrated in, usually, several bands around the storm center or 'eye'. Heavy precipitation occurs in the main cloud band, known as the eyewall band, and other bands, although lighter precipitation does occur between them (for a review, see Anthes, 1982).

Hurricanes may be symmetric, having a closed eyewall band, or asymmetric, having an eyewall band which is not closed. In a

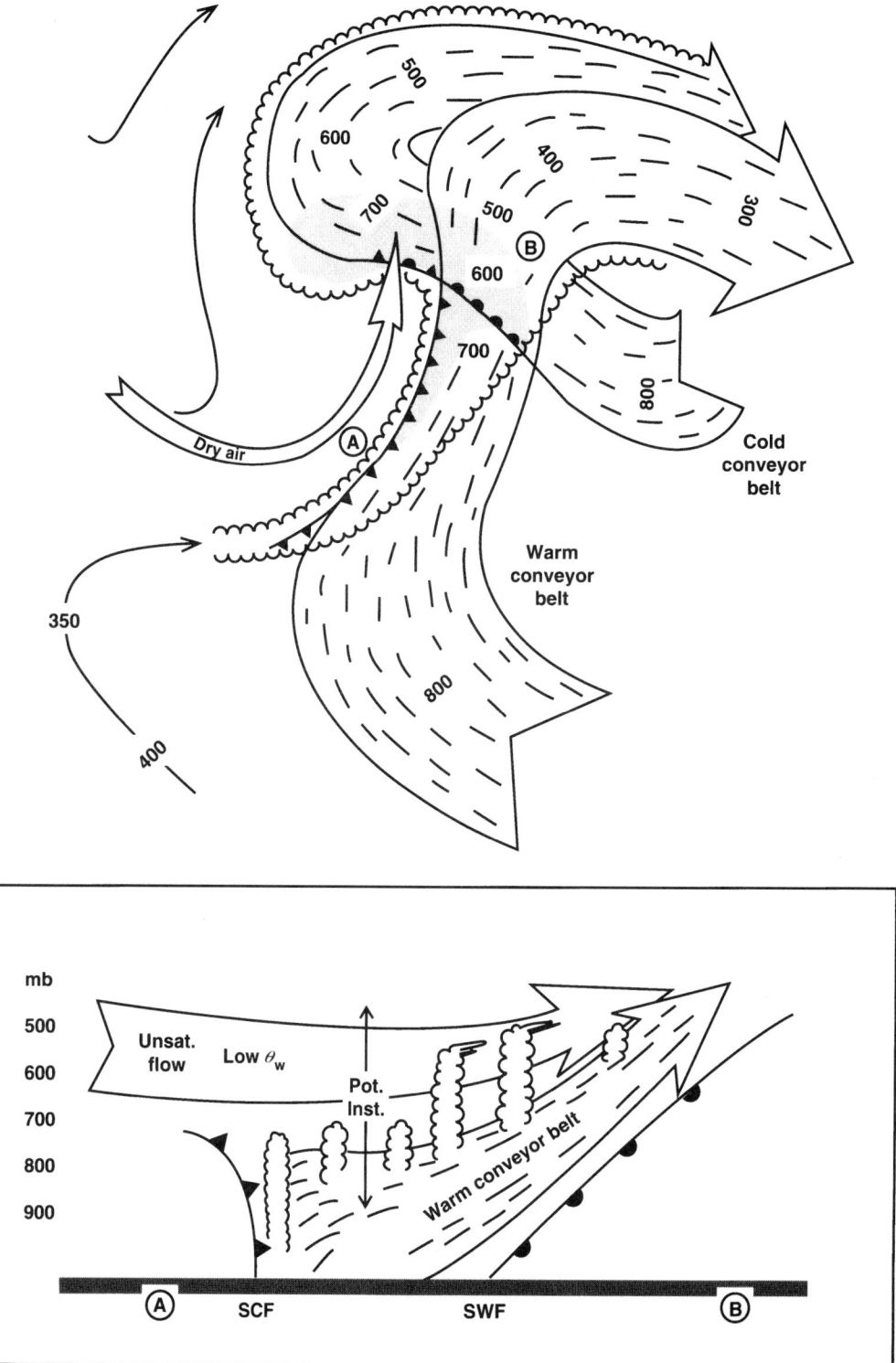

Figure R6 Model depicting the main features of the large-scale flow that determines the distribution of cloud and precipitation in a midlatitude depression. The arrows represent flow, the height of which is labeled in millibars. The scalloped line represents the outline of the cloud pattern and the dotted shading represents the extent of the surface precipitation. (From Browning and Carlson, 1980; after Carlson, 1980.)

symmetric system the center of the wind circulation (counter-clockwise in the northern hemisphere and clockwise in the southern hemisphere) is located in the center of the circle defined by the eyewall boundary. The eyewall band often contracts as the storm develops. Eventually the eyewall band vanishes, and is replaced by a new eyewall band at a radius of 50–150 km. This process continues and each time the eyewall band is replaced, the central pressure of the storm rises and then proceeds to fall. Asymmetric storms do not

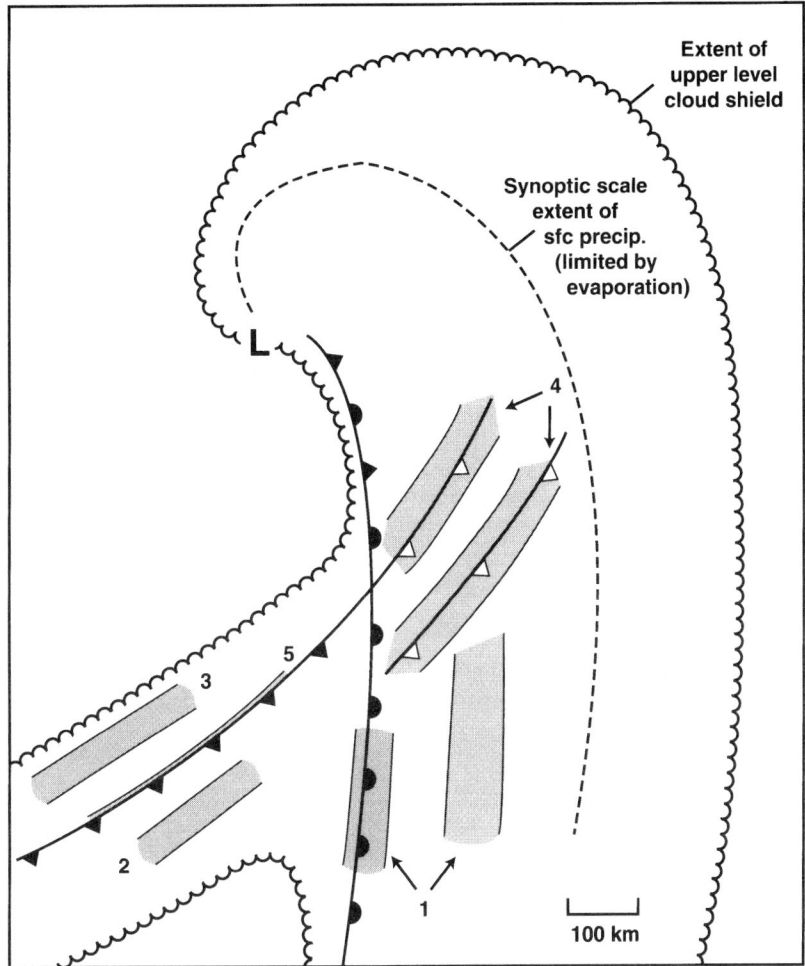

Figure R7 Major types of rain bands (stippled) observed in midlatitude depressions. Type 1, warm front bands; type 2, warm sector bands; type 3, cold front bands; type 4, upper-level cold surge bands (situated along the leading edge of cold air over-running the warm front). (From Browning and Mason, 1980).

Table R3 Examples of storm rainfall totals from tropical storms occurring worldwide

Location	Date	Duration (h)	Rainfall total (mm)
Jamaica, West Indies	4–7 October 1963	24	508
Virginia, USA	20 August 1969	8	710
Horian Province, China	5–7 August 1975	72	1605
NW Australia	20 December 1976	6.33	356
Kyushu, Japan	12–13 September 1976	36	1950
Masirah, Arabian Gulf	12–13 June 1977	24	431
Hope, Hong Kong	2–4 August 1979	72	287
Commerson, Reunion, SW Indian Ocean	14–28 January 1980	360 (15 days)	6433
Grand-Ilet Reunion, SW Indian Ocean	14–28 January 1980	1	110
		12	1170
		24	1742

appear to follow this cycle. Although hurricanes form over the tropical oceans, the rainfall from them, when they reach land, may be very large indeed as shown by the examples in Table R3.

Orographic effects on precipitation distribution

Orography can affect the distribution of rainfall by forcing air to ascend. The ascent may lead to condensation and rainfall (Figure R8a), or may result in convection, leading to rainfall (Figure R8b). A third process is the enhancement of rain over hills from pre-existing

clouds as a result of a natural seeding process (Figure R8c). Mechanisms (a) and (c) in Figure R8 occur in frontal rainfall situations. The natural seeding process was first proposed by Bergeron (1950). Detailed observations in the UK (Browning, 1980) have established that this mechanism can substantially enhance rainfall amounts over hills of only small size, as shown in Figure R9.

The increase in rainfall over hilly areas is very evident on average annual rainfall maps in midlatitudes for areas exposed to moist air moving from the sea. Indeed, there is an almost linear relationship between elevation and average annual rainfall amount. In the lee of ranges of hills, descending air (or subsidence) may lead to reduced rainfall in areas known as rain shadows.

Topographical effects on precipitation distribution

Precipitation amount may be increased in coastal areas by differences in the convergence of air caused by increased friction over the land over that experienced over the sea. Figure R10 shows such an effect as reflected in monthly rainfall totals over the coastal area of the Netherlands.

Differences between land and sea can cause fronts associated with midlatitude depressions to become stationary just offshore, which can lead to very large falls of precipitation in coastal regions, particularly if the land is hilly (Bosart, 1975, 1981; Browning, 1983a). The sea or land breezes associated with air circulation caused by temperature differences between the land and the sea may organize convection such that precipitation amounts can be very much enhanced. Such effects are evident in winter on the Great Lakes in North America (e.g. Passarelli and Braham, 1981).

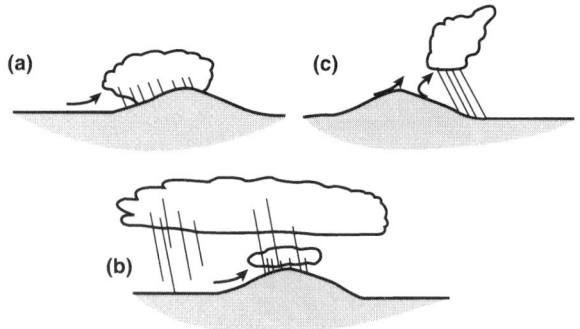

Figure R8 Mechanism of orographic rain generation. (From Smith, 1979.)

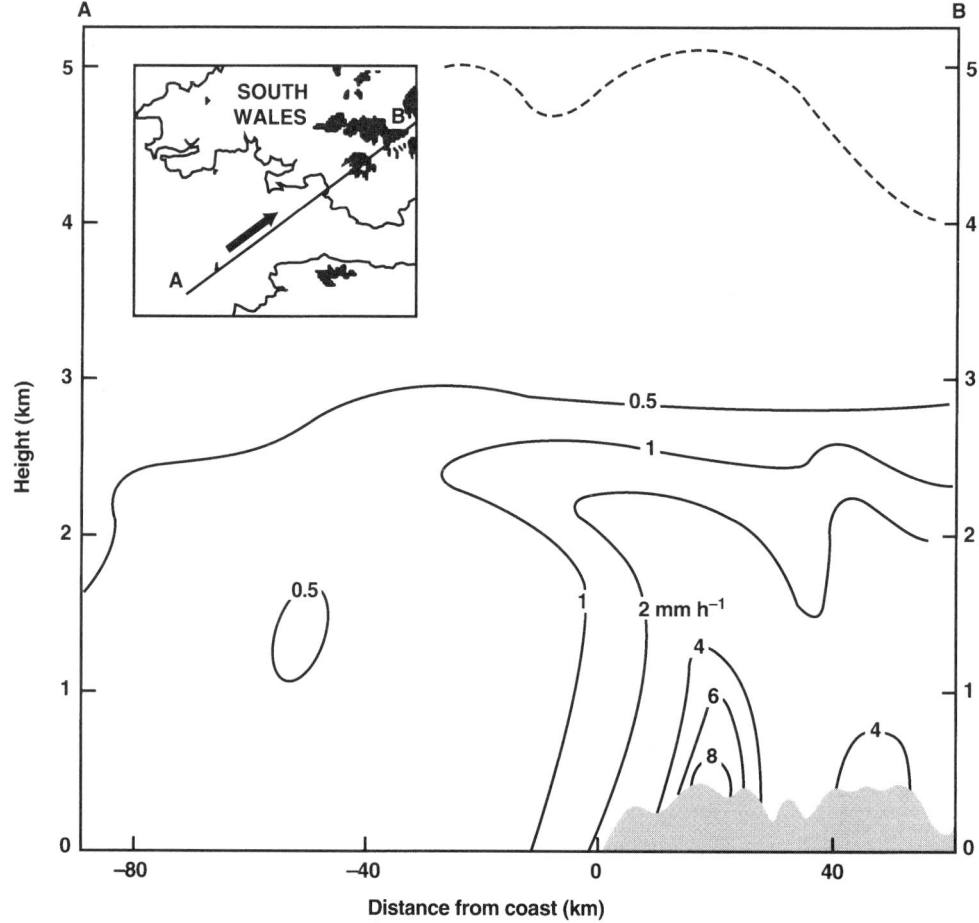

Figure R9 Distribution of mean rainfall intensity (mm h^{-1}) within a vertical section (AB) along the direction of motion of rain traveling from the sea over the hills of South Wales, UK, during a 5 h period of warm-sector rain. The inset shows the orientation of the section AB in relation to the coastline and hills (> 400 m). (From Browning, 1980.)

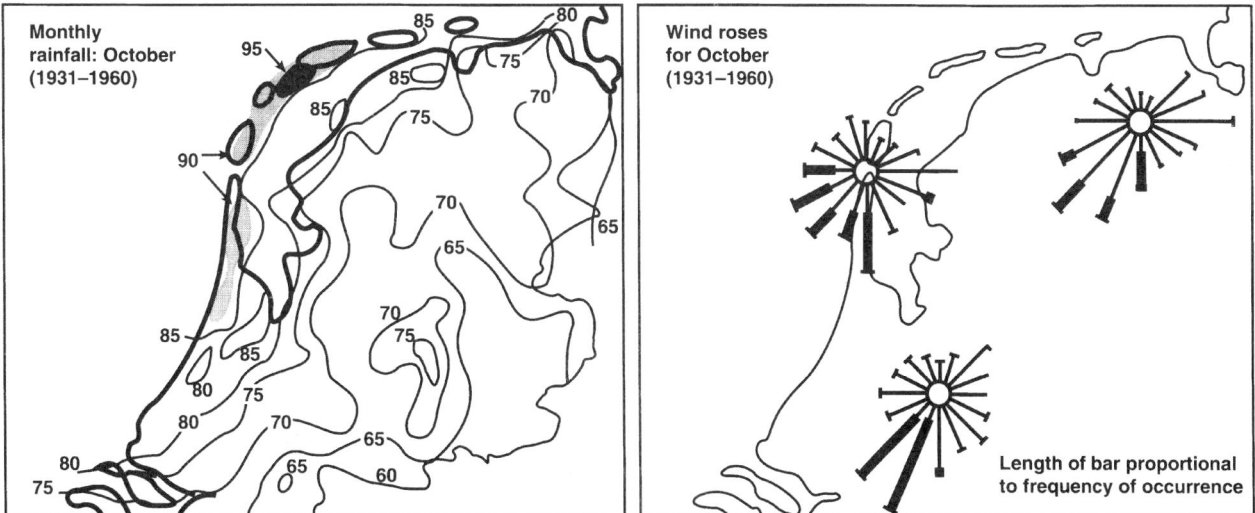

Figure R10 Coastal effect on precipitation as illustrated by the monthly mean rainfall (mm) for October over the coast of the Netherlands. Wind roses for three stations are also shown. Note the coastal convergence produced by predominant southwesterly winds inland compared with the significant occurrence of more westerly winds over the coast. (From *Klimaatatlas van Nederland*, KNMI, 1972.)

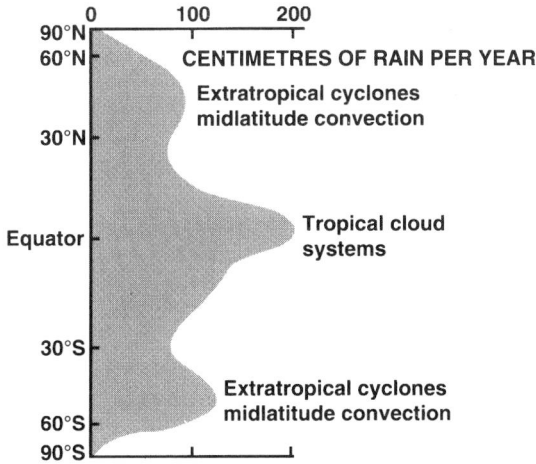

Figure R11 Globally averaged annual precipitation. (From House, 1981, adapted from Sellers, 1965.)

Global atmospheric circulation

So far we have discussed the way in which atmospheric water vapor may be condensed to form precipitation. The average time that a molecule of water is resident in the atmosphere as vapor or within a cloud is 10–12 days (Miller, 1977). During this period the molecule may travel a considerable distance within the atmospheric circulation before being returned to the surface of the Earth as precipitation. Indeed, it is the global circulation of the atmosphere as a whole which dictates the occurrence of the atmospheric systems outlined earlier, and hence the global precipitation distribution. The climatological distribution of annual rainfall is summarized in Figure R11.

The precipitation maxima in midlatitudes are associated with depressions, showers and thunderstorms. In the tropics the rainfall maximum is associated with cloud clusters and tropical storms. The water which passes from the atmosphere to the surface of the Earth is replenished by the processes of evaporation and transpiration.

C. G. Collier

Bibliography

Anthes, R.A., 1982. Tropical cyclones: their evolution, structure and effects. *Meteorol. Monog.*, **19**, No 41.

Bergeron, T., 1935. On the physics of cloud and precipitation, in *Proc. 5th Assembly Int. Union Geodesy Geophys.*, 1933, Vol. 2, pp. 156–161.

Bergeron, T., 1950. Uber den mechanisms der ausgiebigen nieder-schläge, *Ber. Deutsch Wetterdienst*, **12**, 225–232.

Bjerknes, J. and Solberg, H., 1922. Life cycle of cyclones and the polar front theory of atmospheric circulation, *Goefys. Publikasjoner*, **3**(1), 3–18.

Bosart, L.F., 1975. New England coastal frontogenesis, *Quart. J. R. Meteorol. Soc.*, **101**, 957–978.

Bosart, L.F., 1981. The President's Day snowstorm of 18–19 February 1979: a subsynoptic-scale event, *Mon. Wea. Rev.*, **109**, 1542–1566.

Browning, K.A., 1971. Radar measurements of air motion near fronts, *Weather*, **26**, 320–340.

Browning, K.A., 1980. Structure, mechanism and prediction of orographically enhanced rain in Britain, in *Orographic effects in planetary flows* (ed. R. Hide and P.W. White). GARP Publ. Series No. 23, pp. 85–114.

Browning, K.A., 1983a. *Mesoscale structure and mechanisms of frontal precipitation systems.* Course on Mesoscale Meteorology, May 30 to June 10, Pinnarpsbaden, Sweden, Lecture notes II, No. 17, SMHI.

Browning, K.A., 1983b. Air motion and precipitation growth in a major snowstorm, *Quart. J. R. Meteorol. Soc.*, **109**, 225–242.

Browning, K.A. and Foote, G.B., 1976. Airflow and hail growth in supercell storms and some implications for hail suppression, *Quart. J. R. Meteorol. Soc.*, **102**, 499–533.

Browning, K.A. and Hill,   F., 1984. Structure and evolution of a maritime mesoscale convective system, *Quart. J. R. Meteorol. Soc.*, **110**, 897–913.

Browning, K.A. and Mason, B.J., 1980. Air motion and precipitation growth in frontal system, *Pageoph*, **119**, Birkhauser Verlag, Basel, 1–17.

Browning, K.A., Frankhauser, J.C., Chalon, F.P. *et al.*, 1976. Structure of an evolving hailstorm, Part V. Synthesis and implications for hail growth and hail suppression, *Mon. Wea. Rev.*, **104**, 603–610.

Carlson, T.N., 1980. Airflow through mid-latitude cyclones and the comma cloud pattern, *Mon. Wea. Rev.*, **108**, 1498–1509.

Findeisen, W., 1939. Zur frage der regentropfendilbung in reinen wasserwolken, *Meteorol. Z.*, **56**, 365.

Harrold, T.W., 1973. Mechanisms influencing the distribution of precipitation within baroclinic disturbance, *Quart. J. R. Meteorol. Soc.*, **99**, 232–252.

Harrold, T.W. and Browning, K.A., 1969. The polar low as a baroclinic disturbance, *Quart. J. R. Meteorol. Soc.*, **95**, 719–730.

Hide, R. and Mason, P.J., 1975. Sloping convection in a rotating fluid, *Adv. Phys.*, **24**(1), 47–100.

Houze, R.A., 1981. Structure of atmospheric precipitation systems – A global survey, *Radio Science*, **16**(5), 671–689.

Houze, R.A. and Hobbs, P.V., 1982. Organisation and structure of precipitation cloud systems, *Advances in Geophysics*, **24**, 225–315.

Langmuir, I., 1948. The production of rain by a chain reaction in cumulus clouds at temperatures above freezing, *J. Meteorol.*, **5**, 175–192.

Leary, C.A. and Houze, R.A., 1979. The structure and evolution of convection in a tropical cloud cluster, *J. Atmos. Sci.*, **36**, 437–457.

List, R., 1977. The formation of rain, *Trans. R. Soc. Canada, Ser. IV*, **XV**, 333–347.

Mason, B.J., 1971. *The Physics of Clouds*, Clarendon Press, Oxford, 671 pp.

Miller, D.H., 1977. *Water at the Surface of the Earth. An introduction to Ecosystem Hydrodynamics*, Academic Press, New York, 557 pp.

Newton, C.W. and Frankhauser, J.C., 1964. On the movements of convective storms with emphasis on size discrimination in relation to water-budget requirements, *J. Appl. Meteorol.*, **3**, 651–688.

Passarelli, R.E. and Braham, R.R., 1981. The role of the winter land breeze in the formation of Great Lake snow storms, *Bull. Am. Meteorol. Soc.*, **62**(4), 482–491.

Purdom, J., 1976. Some uses of high resolution GOES imagery in mesoscale forecasting of convection and its behaviour, *Mon. Wea. Rev.*, **104**, 1474–1483.

Rasmussen, E., 1979. The polar low as an extratropical CISK disturbance, *Quart. J. R. Meteorol. Soc.*, **105**(445), 531–549.

Sellers, W.D., 1965. *Physical Climatology*, Univ. Chicago Press, Chicago, 272 pp.

Smith, R.B., 1979. The influence of mountains on the atmosphere, in *Adv. Geophys.*, **21**, 87–230, Academic Press, New York.

Wagner, A., 1911. *Thermodynamik der Atmosphäre*, Barth, Leipzig.

Cross references

Maximum observed rainfalls
Precipitation
Precipitation distribution
Precipitation: source

RAINGAUGE

A raingauge is the most basic hydrological instrument. In its simplest form the gauge may be a bucket collecting falling rain or snow. This is manually emptied and the depth of water that has accumulated since it was last emptied is noted. The earliest raingauges probably date back to almost 2500 years ago in India, and gauges were reported in Europe from about the seventeenth century.

Despite this long history it is still not possible to measure the amount of precipitation with a known degree of accuracy. Some sources of error are easily recognized and dealt with, such as sheltering by buildings or overhanging trees, or disturbance by humans or animals. Evaporation of the collected water can be minimized by funneling water through a narrow-bore tube from the raingauge into a storage container, and splash in or out of the gauge can be reduced by proper siting of the gauge and design of the collector. More difficult to deal with is the problem that wind turbulence around the gauge may cause it to undercatch. This is because the gauge itself presents an obstacle to the airflow, resulting in an increase in wind speed and turbulent eddies over the gauge which may carry falling precipitation across and clear past its orifice, leading to a systematic underestimate. This error is more serious for taller gauges, at windy locations or times of the year, for storms with smaller (hence lighter) raindrops, and for frontal rainfall rather than convective storms. Some of the earliest measurements of rainfall in Europe were made using gauges on secure sites such as the roof of a house, although it was noted by Benjamin Franklin as early as 1771 that higher gauge rims caught less rainfall. A few years earlier, Heberden (1769) had noted the lower catch of a raingauge on the roof of Westminster Abbey in England than that of a similar gauge sited on the ground.

Empirical relations have been derived to correct for a number of sources of raingauge error (Sevruk, 1982), of which the most important is wind speed. When the loss of catch is compared to a reference gauge (such as one with its rim at ground level) errors of perhaps 2–7% in annual catches have been noted in temperate latitudes. The smallest errors occur in equatorial and tropical countries where rainfall intensities and drop sizes are large, whilst much higher errors have been recorded in areas prone to snow. Due to their large surface area and low density, snowflakes are easily swept across raingauges, or even out of the collector. Gauge undercatches during snowfall may be 50% or more.

In principle it should be possible to apply empirical corrections for these errors, but this is very difficult in practice since it would require very detailed concurrent meteorological information, including the changing wind speed and precipitation type (intensity, drop size and whether liquid or solid).

In addition to considering the gauge location, there have been many efforts to reduce the wind-induced error by changes to the gauge itself. Various attempts have been made to reduce the turbulence around gauges; these include shields, fences and turf ridges. However, to date, no gauge has yet been designed that completely eliminates catch errors due to wind.

The minimum disturbance to airflow would be achieved by installing a raingauge in a pit, so that its rim could be level with the ground surface. However, such a solution is not suitable during periods of blowing snow, when lying snow may be raised into the air and blown into the gauge and pit, which would then fill up to the level of snow on the surrounding ground however often the gauge was emptied.

Different countries have developed different designs for raingauges. There are over 40 designs of raingauge, which vary with regard to their height, the diameter of their opening and whether or not they have an associated wind shield. In general, taller gauges with larger orifices are preferred in areas where snow is common. In the UK, for example, the standard gauge has a diameter of 126 mm with a rim height at 0.305 m (1 ft) above the ground surface, as a compromise between splash into the gauge and wind turbulence. In countries prone to snow, gauge heights of 1–2 m may be used. In Sweden, for example, gauges are 1.5 m high with an opening of 357 mm diameter.

The use of different types of gauges in different countries creates problems when making comparisons across national boundaries (Table R4). Even within countries there may be inconsistencies, both

Table R4 Different types of widely used storage raingauges

Country of origin	Gauge name	Orifice area (cm²)	Rim height above ground (m)	Estimated number worldwide
UK	MK2/Snowdon	127	0.3	17,800
France	SPIEA	400	1.0	1,800
France	Association	400	1.0	3,000
Germany	Hellmann	200	1.1	30,100
Russia	Tretyakov	200	2.0	13,500
USA	Weather Bureau	324	1.1	11,300
China	Chinese	314	0.7	19,700
India	Indian	200	0.3	11,000
Australia	Australian	324	0.3–1.3	7,600

Source: based on data in Sevruk (1982) and Sevruk and Klemm (1989)

between individual regions and over time as a result of changes to the type of instrument in use.

In addition to storage gauges, there are also recording gauges using float systems, weighing systems or tipping buckets to provide a record of the time distribution of rainfall. Sir Christopher Wren developed several automatic raingauges in England during the seventeenth century (Biswas, 1970).

Mark Robinson

Bibliography

Biswas, A.K., 1970. *History of Hydrology*, North Holland Publishing Co., London and New York, 336 pp.
Sevruk, B., 1982. *Methods of correction for systematic error in point precipitation measurement for operational use*. Operational Hydrology Report 21, World Meteorological Organization, Geneva.
Sevruk, B. and Klemm, S., 1989. *Catalogue of national standard precipitation gauges*. Instruments and Observing Methods Report No 39, World Meteorological Organization, Geneva, 50 pp.
Ward, R.C. and Robinson, M., 1990. *Principles of Hydrology*. McGraw-Hill Book Company, New York.

Cross references

Precipitation
Precipitation distribution
Rain

RAIN SHADOW: GENERAL

More precipitation falls on the windward side of hills and mountains than on the leeward side, which is in the rain shadow of the mountains. The cause for this difference is the orographic rainfall generation process: moisture-bearing air is transported by wind and forced to lift up as it flows over hills and mountains. As a result of this uplift the air pressure on the wet air is reduced, the air expands and cools down. As cool air stores less humidity than warm air, the saturation point can be reached and water condenses, causing rainfall. On the leeward side of the mountain the air humidity is less since surplus humidity is given off. As a result the rain amount at this site is less than on the luff windward site. The orographic rainfall is proportional to the wind speed up the slope and the amount of moisture in the air. The rain-shadow effect is especially marked as only few rainfall-producing circulation patterns exist in a particular region. This is the case mostly in maritime climates, e.g. in Western Europe.

A.H. Schumann

Cross references

Precipitation
Precipitation distribution
Rain
Rain shadow

RAIN SHADOW

The leeside of mountain ranges is a preferred location for arid climates. This is a result of the rain-shadow effect, the role mountains play in promoting dry conditions. Many of the world's deserts owe their existence to this effect. The rain shadow also occurs on a small scale, producing a mosaic of wet and dry conditions on the lee side of many small peaks and ranges.

There are several reasons why rainfall is suppressed in the lee of mountains. First, the air generally is dry, because it loses its moisture through precipitation on the windward side or on the peaks. This precipitation is a result of the air mass ascending as it encounters the high terrain. After this dry air mass traverses the mountain barrier and reaches the leeside, it subsides (i.e. descends). This subsidence heats the air, enhancing its dryness and possibly producing a temperature inversion and stable conditions. The streamlines of the flow also

diverge during descent. The subsidence, divergence, stability and dryness in the lee all act to suppress precipitation.

The deserts resulting primarily from rain-shadow effect are the desert in the Great Basin of western North America and the Patagonian desert of South America. The semiarid Great Plains region of the United States is also a consequence of the rain shadow of the Rockies. This effect produces particularly dry conditions in low-lying basins surrounded by high mountains. It is thus a major factor in the creation of the deserts of eastern Asia, particularly the Gobi and Takla-Makan. A rain-shadow effect probably also enhances the aridity in some coastal deserts, like the Namib.

The contrast in precipitation between the windward and leeward sides is dramatic. Over the Sierra Nevada in the western United States, mean annual rainfall is characteristically about 1200 mm on the peaks, but in the valleys of the leeward rainshadow it falls to 200–300 mm. Similarly, over Asia rainfall can be as low as 10 mm per year in depressions such as the Tarim Basin, but several hundred to more than 1000 mm per year on the surrounding slopes and highlands.

This rain-shadow effect produces wide variation of rainfall conditions within a relatively short distance. In the lee of the Cascades in the state of Washington, for example, mean annual rainfall in many sectors changes from 1600 mm to 200 mm within less than 80 km. This produces a virtual mosaic of small desert basins interspersed with lush green slopes, and separated by only small distances. Such effects are also seen in the high mountains of Asia, in the Andes and in the highlands of Ethiopia.

Sharon E. Nicholson

Bibliography

Nicholson, S.E., 1994. What is a Desert? in *The Illustrated Library of the Earth: Deserts* (M.K. Seely, ed.), Sydney: Weldon-Owen, pp. 14–25.

Cross references

Precipitation
Precipitation distribution
Rain

RAZIM–SINOIE LAKE COMPLEX, ROMANIA

The Razim–Sinoie lake complex is situated on the northwestern coast of the Black Sea, Romania. The surface area is 863.5 km², the maximum depth is 3.5 m. The lake is of the genetic basin-lagoonal type. It is the largest lake complex of the whole Black Sea coast, located south of the Danube Delta (approximately 29°E long. and 45°N lat.). The complex encompasses the following lakes: Razim, which is the largest (415 km²), Sinoie (171.5 km²), Golovița (118.7 km²), Zmeica (54.6 km²), Babadag (23.7 km²), Nuntași–Tuzla (10.5 km²) and Istria (5.6 km²), as well as several smaller lakes (see Figure R12).

Basin origin

About 2000 years ago the area was a sea gulf known then as Halmyris. The gulf appears to have been much deeper, because the then city of Histria (between the seventh century BC and the fifth century AD), used to be a stopover on the route of ancient Greek ships. The remains of Histria can still be seen on a crystalline promontory on the left bank of Lake Sinoie, inside the complex. The Halmyris Gulf was separated from the sea by several marine levees formed by Danube-carried sediments and by local organic material brought by sea currents.

Drainage basin

There are several smaller streams (including the Taița, Telița, Slava, Beidaud and Săruri), totalling 1725 km². However, the climate being semiarid continental (precipitation 350–400 mm year^{-1}) they supply only 1.44 m^3 s^{-1} to the lake discharge.

Human-induced changes

When this lacustrine environment was in a natural condition, that is, before human intervention, the lake complex was connected to the

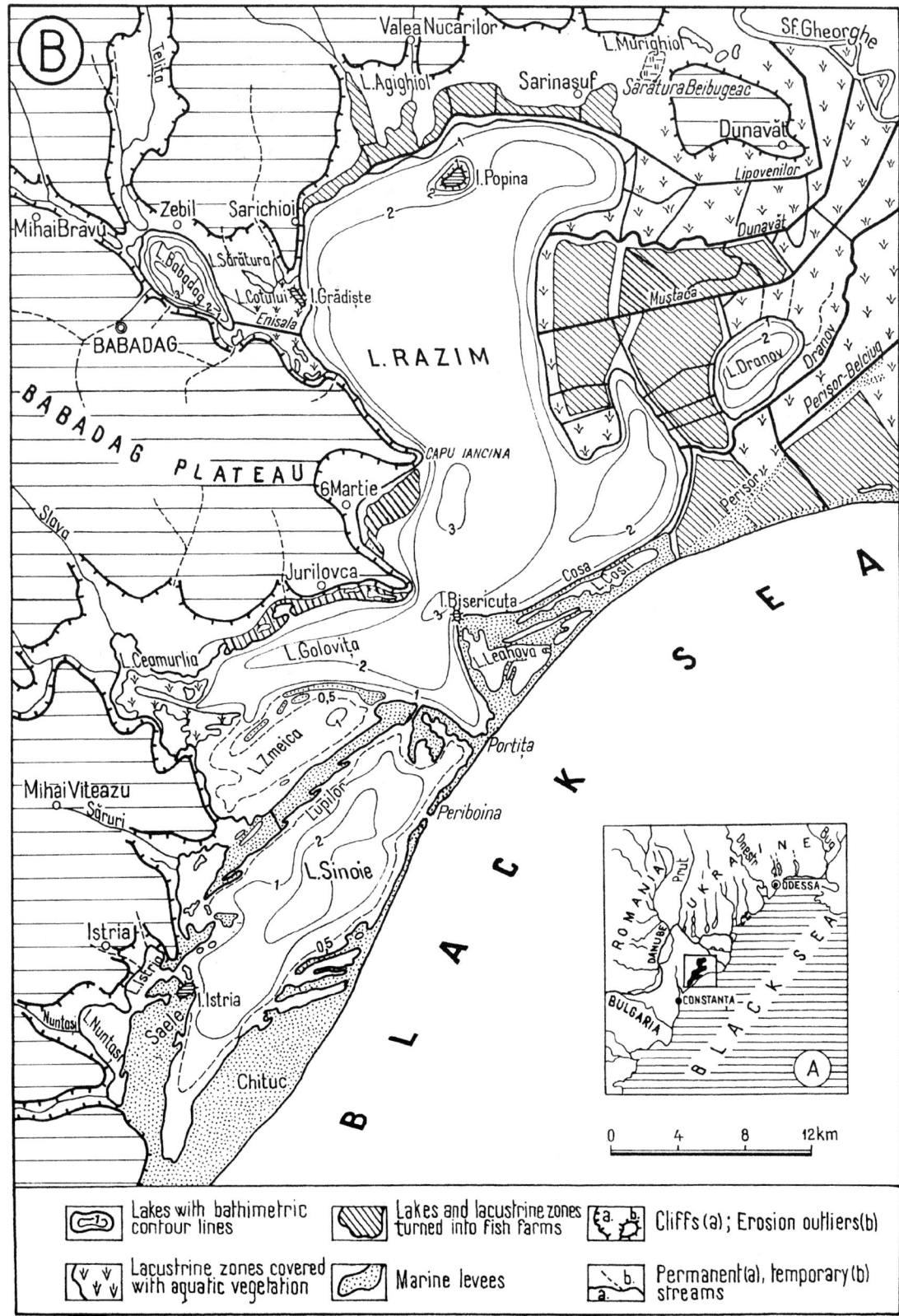

Figure R12 Razim–Sinoie lake complex. (A) Location on Black Sea coast, (B) morphohydrographical sketch map.

Black Sea through a breach in the marine levee, a kind of gateway ('periboine' and 'portiţe'). A major link between the Black Sea and Lake Razim was Gura Portiţçei, which enabled a permanent exchange between fresh and sea waters, with lake and sea levels coming very close. The dam built in 1970 destroyed the link, and Lake Razim remained a fresh waterbasin used for the irrigation of 1600 km² in the adjoining area. Irrigation water is supplied by the Danube through the Sfântu Gheorghe Arm. At present, the only connection between the lake complex and the Black Sea is Lake Sinoie in the south.

These human-induced changes (water coming from the Danube in the north – Lake Razim – and from the sea in the south – Lake Sinoie) have altered the mineralization, i.e. fresh water turning to brackish and saline in the more isolated lakes of the south (Nuntași–Tuzla). At the same time, the area was put to new uses, e.g. fish farming, irrigation, spas and tourism.

Water balance

The anthropically modified water balance is

$$P + Y_B + Y_D - E_V - I_R - Y_M = \pm \Delta V$$

where P is the precipitation, Y_B is the supply from a drainage basin, Y_D is the supply from the Danube via canals, E_V is the surface lake water evaporation, I_R is the water volume for irrigation, Y_M is the discharge into the Black Sea and $\pm \Delta V$ is the lake complex water volume stocked or evacuated within a certain time interval.

When the irrigation systems functioned normally (1984–1987), the water balance annual means were as follows (mm year⁻¹): $P = 328$ (13.7%), $Y_B = 50.7$ (2.1%), $Y_D = 2014$ (84.2%), $E_V = 742$ (35.1%), $I_R = 608$ (28.8%), $Y_M = 760$ (36.1%), $\pm \Delta V = 280.7$.

The great diversity of the Razim–Sinoie lake complex ecosystem, as well as the vicinity of the Danube Delta, caused the authorities to declare the area a nature reserve in 1990 – the Danube Delta Biosphere Reserve – which encompasses both geographical units.

Petre Gâștescu

Bibliography

Gâștescu, P., 1993. The Danube Delta Biosphere Reserve. Present-day Conditions and Ecological Recovery, *Geojournal*, **29**(1), 57–67
Gâștescu, P. and Breier, A., 1976. *Le complexe lacustre Razim–Sinoie (Roumanie)*, Memorie della Societa Geografica Italiana, Scritti geografici in onore di Riccardo Riccardi, Roma, pp. 247–269

Cross references

Lakes
Lakes, lake water
Lakes: largest worldwide
Limnology

RELATIVE HUMIDITY

Relative humidity (or simply humidity) is the ratio of atmospheric vapor pressure to saturation vapor pressure. The World Meteorological Organization prefers to define relative humidity as the ratio of specific humidity (or mixing ratio) to saturation mixing ratio, expressed in percent. It can be computed from wet-bulb temperature and related psychrometric data. Relative humidity (U, in percent) is expressed as:

$$U = 100 \frac{e'}{e_w},$$

Table R5 Relative humidity (%) from temperature and temperature-dewpoint depression

Dewpoint depression (°C)	Temperature (°C)																	
	40	35	30	25	20	15	10	5	0	−5	−10	−15	−20	−25	−30	−35	−40
1	95	95	94	94	94	94	94	93	93	93	92	92	92	91	91	91	90
2	90	89	89	89	88	88	87	87	86	86	85	85	84	83	83	82	81
3	85	85	84	83	83	82	82	81	80	79	79	78	77	76	75	74	73
4	81	80	79	78	78	77	76	75	74	73	73	71	70	69	68	67	66
5	76	75	75	74	73	72	71	70	69	68	67	66	64	63	62	60	59
6	72	71	70	69	68	67	66	65	64	63	61	60	59	57	56	54	53
7	68	67	66	65	64	63	62	61	59	58	57	55	54	52	51	49	47
8	64	63	62	61	60	59	57	56	55	53	52	50	49	47	46	44	42
9	61	60	59	57	56	55	54	52	51	49	48	46	45	43	41	39	38
10	58	56	55	54	53	51	50	48	47	45	44	42	41	39	37	35	34
11	54	53	52	50	49	48	46	45	43	42	40	39	37	35	33	32	27
12	51	50	49	47	46	45	43	41	40	38	37	35	34	32	30	28	24
13	48	47	46	44	43	42	40	38	37	35	34	32	30	29	27	25	21
14	46	44	43	41	40	39	37	36	34	32	29	29	28	26	24	23	19
15	43	42	40	39	37	36	34	33	31	30	28	27	25	23	22	20	16
16	40	39	38	36	35	33	32	30	29	27	26	24	23	21	20	16	
17	38	37	35	34	32	31	29	28	27	25	23	22	21	19	18	14	
18	36	34	33	32	30	29	27	26	24	23	21	20	19	13	16	12	
19	34	32	31	30	28	27	25	24	22	21	20	18	17	15	14	11	
20	32	30	29	28	26	25	23	22	21	19	18	16	15	14	12	09	
21	30	28	27	26	24	23	22	20	13	17	16	15	14	12	09		
22	28	27	25	24	23	21	20	19	17	16	15	13	12	11	08		
23	26	25	24	22	21	19	13	17	13	13	13	12	11	10	07		
24	25	23	22	21	19	18	17	16	14	13	12	11	10	09	06		
25	23	22	21	19	18	17	16	14	13	12	11	10	09	08	06		
26	22	20	19	18	17	16	14	13	12	11	10	09	08	05			
27	20	19	18	17	16	14	13	12	11	10	09	08	07	05			
28	19	18	17	15	14	13	12	11	10	09	08	07	06	04			
29	18	17	15	13	13	12	11	10	09	08	07	06	06	04			
30	17	16	14	13	12	11	10	03	08	03	03	06	03	03			

Note: computed from saturation vapor pressures (with respect to water surface) given in *Smithsonian Meteorological Tables* (6th rev. edn; Washington DC, 1951).

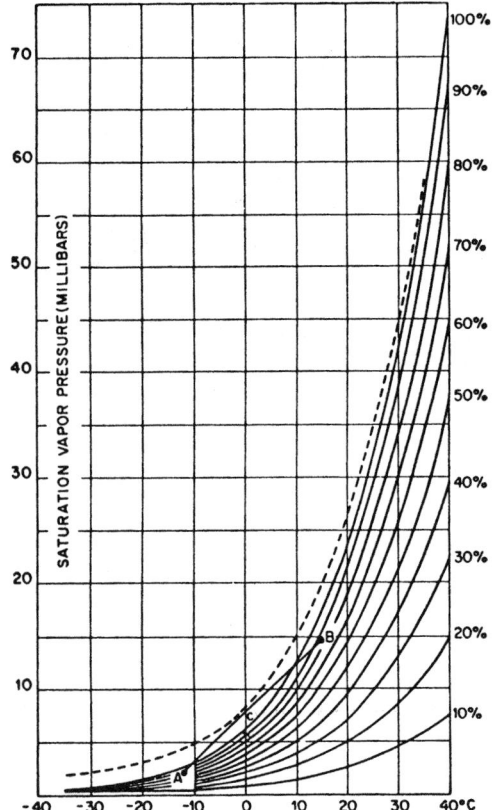

Figure R13 Relation between vapor pressure and air temperature for standard values of relative humidity. The continuous curves represent saturation with respect to water. The dashed curve indicates supersaturation corresponding to 0.5 g of liquid water per cubic meter. For explanation of A, B and C, see text. (From Petterssen, 1956.)

or

$$U = \frac{\text{absolute humidity}}{\text{saturation absolute humidity}} \times 100$$

where e' is the actual vapor pressure of the air and e_w' is the saturation vapor pressure at the same pressure and temperature. The mixing ratio (r) is defined as the ratio of mass of water vapor (m_v) to mass of dry air (m_a), thus $r = m_v/m_a$. Saturation occurs when moist air at pressure p and temperature T may exist in equilibrium with pure water or ice at the same temperature and pressure. In terms of mixing ratio (r) and saturation mixing ratio (r_w), it may be expressed as:

$$U = 100 \frac{r}{r_w} \times \frac{0.62197 + r_w}{0.62197 + r}$$

Table R5 provides relative humidity values derived, in this example, from temperature and temperature-dewpoint depression.

Relative humidity generally decreases with altitude, like many other atmospheric characteristics, but it will increase just below clouds since in these the relative humidity is generally 100%. It also has a marked diurnal oscillation opposite in phase to temperature, with a daily maximum at dawn and a minimum in the afternoon. In addition, in certain climatic belts it has an annual variation, also in contrasting phase to temperature. In temperate regions it is generally above 60–80%, in arid regions 20–40%.

In the event of mixing of air masses of differing relative humidity (A, B), a resultant midpoint (C) can be established (Figure R13). If C falls above the 100% saturation curve, then condensation occurs.

Rhodes W. Fairbridge

Bibliography

McIntosh, D.H. (ed.), 1963. *Meteorological Glossary*. London: Her Majesty's Stationery Office.
Meteorological Office, 1978. *Elementary Meteorology*. London: Her Majesty's Stationery Office.
Petterssen, S., 1956. *Weather Analysis and Forecasting*. Vol. 2. New York: McGraw-Hill.
Wallace, J.M. and P.V. Hobbs, 1977. *Atmospheric Sciences: An Introductory Survey*. New York: Academic Press.

Cross references

Dew
Dewpoint

REMOTE SENSING (1)

Introduction

Hydrological variables are generally measured as point values whilst their use for modeling purposes is as an areal average or as spatially varying values. Many attempts have been made to provide spatial estimates of hydrological variables using factors such as topography, land cover, soil type and underlying geology. As an alternative, remotely sensed images offer the possibility of extrapolating point values to the area of interest.

In the past, the use of remotely sensed images by the hydrological community has been limited. There are a number of reasons for this:

- inadequate temporal, spatial and/or spectral resolution;
- the cost of acquiring and analyzing the images;
- problems with cloud cover;
- the need to correct for atmospheric effects;
- the need to provide calibration or 'ground-truth' data.

Recent advances in sensor design and image processing algorithms have ensured that data obtained from analyzing remotely sensed imagery are being increasingly used to help solve hydrological problems. A number of platform/sensor combinations are available, the most relevant depending on the particular application. For studying small areas, ground-based systems on stationary tripods or mobile vehicles may be used. For larger areas, aircraft or satellite imagery will be more appropriate. For the latter, the ground resolution may vary between 10 m and 2.5 km. Similarly, satellite imagery has a wide range of temporal resolutions, ranging from about 30 min to 26 days. Inevitably, the choice of satellite imagery to be used is a compromise, with the temporal resolution being reduced as the spatial resolution increases.

Remote sensing applications relevant to hydrology

Rainfall

Traditionally, areal rainfall has been estimated using networks of raingauges. Values from individual gauges have been used to give estimates of areal rainfall using topographic considerations (Spreen, 1947), by regression analysis (Rodda, 1962) or by determining the area of influence of each gauge in the network (Thiessen, 1911). However, the cost of installation and maintenance, and the acquiring and processing of data from the dense networks required, together with uncertainties in estimating areal inputs from the networks, have resulted in alternative methods being employed. Remotely sensed images, with their spatial attributes, are ideal for such an application. Whilst the various types of remotely sensed images give a qualitative distribution of rainfall, networks of raingauges are still required to provide a quantitative calibration. A detailed description of the various remote sensing techniques employed is given in Kuittinen (1988).

Weather radar

Ground-based radar has been used operationally for areal rainfall detection for a number of years. The technique is based on the fact that microwaves are attenuated by precipitation particles – raindrops, snowflakes and ice particles. Thus a ground station, comprising a combined emitter/receiver, can provide a measure of rainfall, given as the intensity of the radar echo, over the area of a circle around the station. The radius of the circle depends on the wavelength employed. For most applications, C-band radar (5 cm) is used. This will cover an area of approximately 100 km radius around the station. A detailed description of the design of weather radar systems is given in Lawler (1989).

The intensity of the radar echo Z is related to the rainfall rate R by a function of the type:

$$Z = aR(\exp b)$$

where a and b are constants. It has been found that the values of a and b depend on the nature of the particle size distribution of raindrops, snowflakes and hailstones (Battan, 1973), and it is useful to identify the type of precipitation being observed before applying parameter values to the above equation.

Other corrections applied to radar-derived rainfall estimates include (Collier, 1986)

- an empirically derived range-dependent correction to allow in part for increasing underestimation of rainfall at far ranges, due to incomplete filling of the radar beam, or to failure of the radar to observe low-level rainfall;
- areas of permanent clutter (reflections from hills, buildings, etc.) are identified using a 'mask' and are ignored; data for these areas are estimated directly from raingauge observations;
- signal enhancement by melting snow, the 'bright band' effect, is corrected using estimates of reflectivity profile, derived from past observations.

The accuracy of rainfall estimates by weather radar depends on a number of factors, notably the type of precipitation being measured and the distance from the radar. In general, radar rainfall estimates are expected to vary between 0.5 and 2.0 times the 'true' rainfall for about 75% of the time (Dalezois, 1988). Typical errors of radar rainfall estimates for various worldwide studies are given in Collier (1989a, Table 3.2, p. 68) from which further details of all aspects of weather radar systems may also be obtained.

Techniques for the determination of precipitation rates using weather radar have improved to a stage where they are now being used operationally in over 100 countries throughout the world. Applications include flash flood warnings, severe weather warnings including wind, hail and tornadoes, and air terminal surveillance. Individual radars are often networked to provide complete coverage of the area of interest. Descriptions of the use of weather radar in various countries of the world are given in COST73 (1990).

Satellite imagery
Operational satellite based techniques for estimating precipitation rates rely on images in the visible and infrared regions of the electromagnetic spectrum. Algorithms have been developed to process images from geostationary satellites such as METEOSAT and GOES for use in weather forecasting. In addition, algorithms are being developed using microwave images, although these cannot, as yet, be regarded as operational.

Techniques to estimate rainfall rates from visible/infrared imagery can broadly be divided into two types – cloud characteristics techniques and life history methods – although as development has progressed, the difference between the techniques has become less distinct.

The estimation of rainfall by the identification of cloud characteristics (Barrett and Martin, 1981, pp. 43–64) was the first technique to be developed, and is based on developing regressions of rainfall totals against remotely sensed variables such as the brightness temperature of the cloud tops and the stage of development of clouds (O'Sullivan et al., 1990). Initially, the techniques utilized either the visible or infrared regions of the electromagnetic spectrum, though many current procedures utilize both regions. The methods are the least dependent on sophisticated software and hardware systems. However, they involve some degree of subjective interpretation, and the resulting rainfall estimates are not available in real time. They are mainly used to provide long-term (>1 day) estimates of rainfall over large areas. A detailed review is given in D'Souza (1988).

Cloud indexing methods, utilizing the visible bands, were first investigated in the late 1960s (Barrett, 1970). Basically, the area of interest is divided into a number of grid squares or cells. Remotely sensed images are used to identify the types of clouds and percentage areas within each cell. Look-up tables give probabilities of rainfall and intensities of rainfall for each cloud type. The rainfall for a given period over a grid square is given by the sum of cloud area times the values in the look-up tables for each cloud type.

Cloud-top temperatures, measured using satellite thermal infrared images, can also be used to estimate precipitation rates (Lethbridge, 1967). The approach adopted is to define temperature thresholds at which rainfall at different intensities occurs. This forms the basis of the derivation of a global precipitation index from geostationary meteorological satellites.

Both of the above techniques have limitations. For the cloud indexing method, it has been found that clouds with similar appearances do not always precipitate equally, whilst the same could be said of clouds with similar top temperatures. This led to the development of techniques using both visible and infrared imagery (O'Sullivan et al., 1990).

Life-history methods (Barrett and Martin, 1981, pp. 65–81) are based on the premise that most precipitation originates from convective clouds which can be distinguished from other types of clouds using satellite imagery. The cloud locations are determined using visible and/or infrared images, and their development over time is monitored using sequential images. To do this, the interval between consecutive images must be short compared with the lifetimes of precipitating clouds. The rainfall patterns observed may be used for short-term applications such as flood warning.

Snow

The mapping of snow extent from satellite images has been shown to be feasible to the extent that it is one of the few hydrological parameters which is measured by satellites on an operational basis. This is particularly so for those areas of the world (e.g. the Rocky mountains, the Himalayas and Scandinavia) which experience substantial precipitation in the form of snow, and where over 50% of annual catchment runoff originates from melting snow (Bergstrom, 1975). In these areas, an assessment of the extent, depth, water equivalent, and onset and rate of melting is crucial for flood alleviation, water resource evaluation and hydroelectric power schemes.

The most widely used images for snow monitoring are the visible and infrared bands on the Landsat (Baumgartner et al., 1986) and NOAA satellites (Collin and Carlisle, 1989). The Advanced Very High Resolution Radiometer (AVHRR) onboard the NOAA series of satellites, providing four images per day in 4–5 channels at a ground resolution of 1.1 km at nadir (Lauriston et al., 1979) is particularly suited for snow mapping in large areas (>1000 km^2; Rango and Martinec, 1986). For smaller areas, or where snow cover is highly variable, it is more appropriate to use images from the Landsat series of satellites.

Very little information may be obtained about quality aspects of snow in the visible and infrared bands. For this, microwave images may be used, although the analysis involved is complex and at present at an experimental stage. At the present stage of knowledge it is suggested that a combination of microwave sensors be used (Matzler et al., 1982).

Soil moisture

All the available remote sensing wavebands can provide information of value to the interpretation of soil moisture availability, but it is essential that the physical basis behind the use of each waveband is understood to enable their advantages and limitations to be carefully matched to their application.

In the visible and near-infrared wavelengths, the albedo of bare soil is an indicator of soil moisture (Idso et al., 1975), though the relationships vary with soil type and roughness (Evans, 1979). Also, the rapid drying of the thin surface crust (Hoffer and Johannsen, 1969) limits the applications to local qualitative comparisons. Vegetation reflectance provides a more practical indication of soil moisture, since it responds to water availability within the whole of the root zone rather than a very thin surface layer as in the case of bare soil. Studies have been conducted over a number of crop types (see, for example, Richardson and Everitt, 1987, for grass; Collier, 1989b, for cotton and Danson et al., 1990, for sugar beet) and, although encouraging results have been obtained, care must be taken since other factors such as fertilizer applications can produce changes in leaf reflectance similar to changes of soil moisture. Also, unless a complete canopy cover exists, reflectance from underlying soil will have a major effect on the overall signal.

Middle infrared radiation is strongly absorbed by the presence of moisture in soil and vegetation, with water absorption at 1.45 and

1.95 μm. However, absorption by atmospheric water prevents the use of these wavelengths for remote sensing, so the intervening regions are used, the most common being those selected for Landsat TM bands 5 (1.55–1.75 μm) and band 7 (2.08–2.35 μm).

For bare soil, Musick and Pelletier (1986) found a good relationship between soil moisture and the ratio of Landsat bands 5 and 7. Further work showed that the effect of different soil types was largely eliminated by expressing soil moisture relative to field capacity (Musick and Pelletier, 1988). For vegetated surfaces, laboratory measurements by Ripple (1986) showed that mid-infrared reflectance was governed primarily by leaf moisture content, but he confirmed that soil reflectance was an important factor when leaf cover was less than 100%. Whilst the results are encouraging, further work is required before mid-infrared–vegetation–soil relationships are better understood.

Thermal infrared measures surface temperature which can best be related to soil moisture via the thermal inertia method. Wet soil has a higher thermal capacity than dry soil so it exhibits a smaller diurnal range of temperature, appearing cooler during the day and warmer at night. Models to derive surface soil moisture and evaporation over bare earth and short grass were developed during the European TELLUS project to test data from the Heat Capacity Mapping Mission satellite (De Paratesi, 1981). Nieuwenhuis (1986) calculated evaporation of crops from aircraft thermal data and derived soil moisture via the SWATRE soil water model. As evaporation is crop dependent, coincident visible/near-infrared data was used for crop classification. This appears to be the best approach for obtaining a quantitative measure of soil moisture using thermal remote sensing.

Active microwave or radar has great potential for providing soil moisture distribution (Blyth, 1993). Because of their cloud-penetrating capability, repetitive monitoring enables soil moisture changes to be studied whilst unwanted effects, such as those caused by variations in surface roughness, are reduced as they are largely time invariant. Although in a very early stage of development, this is an area of great potential as a number of satellite radars will be launched within the next 5 years.

Surface temperatures

Two regions of the electromagnetic spectrum have been used for the determination of surface temperatures; passive microwave and thermal infrared. Whilst the microwave method overcomes the problem of cloud cover, complications due to the presence of snow and water bodies limit its applicability (McFarland et al., 1990), particularly for the detection of frozen ground surfaces. In addition, the poor ground resolution associated with passive microwave imagery from satellites (typical pixel size 50 km) limits its application to global studies.

The thermal infrared technique for determining surface temperatures utilizes images in the 10.5–12.5 μm range. The most suitable images are those provided by the Advanced Very High Resolution Radiometer (AVHRR) sensor on-board the NOAA series of satellites.

Estimated land surface temperatures have been used in a number of relevant applications. These include the detection of heat losses from urban areas (Birnie et al., 1984), water availability for agricultural purposes (Cihlar, 1980), the detection of frost hollows in fruit-growing areas (Caselles and Sobrino, 1989), areal estimates of evaporation (Carlson and Buffum, 1989), the detection of icy road conditions (Thornes, 1989) and the mapping of areas of very low temperatures (Collier et al., 1989).

There are two problems associated with the determination of land surface temperature using thermal infrared data. The first concerns the elimination of atmospheric effects. These are caused by absorption and scattering of radiation by particulate matter or gases within the layer of atmosphere between the emitting surface and the sensor. One method of correction is based on the fact that these atmospheric effects vary with wavelength. In the AVHRR sensor on the latest NOAA satellites, the original 10.5–12.5 μm channel has been separated into a 10.5–11.5 μm and a 11.5–12.5 μm channel. Atmospheric correction by the use of images in two such bands is known as the 'split window' technique. For the Along Track Scanning Radiometer (ATSR) on board the ERS-1 satellite, atmospheric correction may be done by making use of the fact that different atmospheric path lengths are associated with nadir and forward look images (Mutlow et al., 1994).

The second problem is the effect of emissivity. Unlike water bodies, for which the emissivity is close to 1.00 with little variation,

land surfaces exhibit large variations in emissivity (Griggs, 1968). Neglecting this variation may result in appreciable errors in the estimates of land surface temperatures (Becker, 1987). Since the emissivity also varies with wavelength, the split window approach can give misleading results (Kerr et al., 1992). A number of theoretical formulations have been developed to solve the problem (e.g. Becker and Li, 1990; Wan and Dozier, 1989). Alternatively, derived emissivity values from ground-based radiometers for different land surfaces may be used to correct the observed brightness temperatures over the area of interest according to the distribution of land cover. Another approach is to correct the brightness temperatures observed by the satellite using measured ground or air temperatures (McClatchey et al., 1987). Both of the latter techniques suffer from the fact that, whilst the ground observations are point values, the brightness temperatures are averaged over a considerable area (1 km^2 in the case of the AVHRR or the ATSR sensors).

Land cover

The mapping of land cover is probably the most widely used remote sensing application. A land-use classification, perhaps in combination with other spatial data sets such as soils, topography and geology, has many possible hydrological applications. These include the identification of suitable areas of study or to judge the representativeness of existing study sites. Also, such a classification could be used to extrapolate results from small plot studies to the catchment scale. Such an approach for catchment water balance purposes has been demonstrated by Roberts and Roberts (1992), and could be extended to estimate chemical and sediment losses, provided that the required data from small plot studies were available. Another possibility is in the study of the effects of land use change, where the analysis of multitemporal images will give an indication of the size of the areas affected.

As with all remote sensing applications, the type of image used depends on the spatial and temporal requirements of the particular application. Aerial photography probably remains the most widely used remote sensing medium for land-use mapping. Such photographs are widely available, are economic to buy, and have a sufficiently fine spatial resolution to map small features, e.g. riparian vegetation, within the area of interest. Disadvantages of using aerial photography include the relative coarseness of the spectral resolution attainable by the use of optical filters, the difficulty of converting the image to digital form, and hence the effort entailed in transforming the data to map projections or as input to a geographic information system.

Imagery from multispectral scanners in digital form has a number of advantages over photographic products. In the first place, the scanners operate in distinct spectral regions designed to highlight differences in the reflectances of land cover types. Also, since the data are in digital form, they may be subject to detailed analysis using computer software written specifically for this purpose. This includes image enhancement and the combination of reflectances in different spectral bands. On this latter point, a very useful index for land classification purposes is the 'Normalized Difference Vegetation Index' (NDVI). This is basically the difference between reflectances in the near-infrared and red region of the electromagnetic spectrum. Green vegetation has a high reflectance in the infrared region and a low reflectance in the red; for non-vegetated surfaces the reverse is true, so the NDVI is an index related to crop cover.

There are two basic classification methods that can be used – unsupervised and supervised. In the former, a clustering algorithm is used to divide the total population of picture elements (pixels) in the area of interest according to their distribution in multispectral space (Belward et al., 1990), without any user input. Users must then decide how the resultant classes relate to their perception of 'ground truth'. In a supervised classification (Schowengerdt, 1983), the user defines representative pixels of known land class; their mean spectral response is then used to define the spectral properties of each class and 'unknown' pixels are matched and assigned to the most 'similar' class.

A particularly relevant aspect of land cover to hydrology is the extent of urbanization. The change in catchment response to rainfall inputs following urbanization has been the subject of much research and has led to the development of predictive rainfall–runoff models. These vary in their complexity from simple techniques (Lloyd-Davies, 1906) through hydrograph methods (Watkins, 1962) to the more advanced models that are in general use (Colyer and Petwick, 1976). Most of the latter include modules that calculate surface runoff for a given rainfall intensity and duration. This depends on several catchment characteristics including soil type, catchment wetness and, most importantly, the percentage of impervious surface within the catchment. This latter parameter has usually been assessed, for a given catchment, using large-scale maps, aided by on-the-ground surveys and the manual interpretation of aerial photographs. However, remotely sensed data have been used to assess this parameter and can be considered a viable operational technique (Finch *et al.*, 1989).

The application of remotely sensed data from satellites to mapping urban areas was an early use made of the data (Jackson *et al.*, 1976) when it first became available at the beginning of the 1970s with the launch of Landsat 1 carrying the Multispectral Scanner (MSS). The spatial resolution associated with the MSS (70 m) was found to be inadequate for urban mapping, and it was not until the launch of Landsat 4 with its Thematic Mapper (30 m) sensor and the SPOT (20 m) satellite in the early 1980s that the application became viable. Even then, problems were encountered with the fine spatial detail prevalent in urban areas. More accurate classifications are obtained using sensors mounted on aircraft (Finch *et al.*, 1989) or on satellites in low earth orbit (Van Genderen, 1989). Alternatively, algorithms have been developed to estimate the individual spectral contributions made by different land classes within mixed pixels (Fisher and Pathirana, 1990).

Water quality

A number of remote sensing techniques can be used to measure parameters that are directly or indirectly related to water quality. As with most remote sensing techniques, direct measurements are limited, and ground truths or predictive techniques have to be used to provide the required parameter. One exception to this is surface water temperature which can be derived from radiances measured in the thermal infrared. For airborne sensors, the use of internal temperature reference plates to calibrate the observed radiances will generally ensure relative accuracies of 0.1°C and absolute accuracies of 0.3°C. In contrast, satellite imagery has to be carefully corrected for atmospheric effects (see Surface temperature, above), and the resulting accuracies are likely to be lower. The mapping of surface water temperatures has many possible hydrological applications such as tracking sewage and power station discharges, the identification of groundwater sources and the mapping of lake or coastal currents.

No comprehensive ways have yet been found to measure the levels of chemical pollutants in water. However, images in the visible and near-infrared bands can be used to determine dissolved organic carbon, suspended mineral particles and the concentration of phytoplankton chlorophyll. These determinations are based on characteristic reflectances of the various components in natural waters. In the case of sediments, these reflectances are very dependent on the types of soil; information regarding this is required in order that accurate surveys may be made.

The simplest way of detecting algal growth by remote sensing is by utilizing the green chlorophyll reflectance peak. This technique can only be used for relatively high chlorophyll concentrations under clear sky conditions. Under more difficult conditions, a variety of multiband algorithms have been devised that predict water quality parameters from the relative change in water colour. Alternatively, tunable laser spectrometers have been used very successfully for identifying solute types (e.g. Fantasia and Ingrao, 1974).

Other hydrologically relevant parameters

Two vegetation parameters relevant in particular to estimates of evapotranspiration are albedo and leaf area index. Albedo is defined as the ratio of the reflected to incoming solar radiation, and its determination will enable an estimate of the amount of solar energy that is available for evapotranspiration. A number of investigations have been conducted, but there is as yet no effective way of remotely sensing albedo. There are three main problems.

- Remote sensors collect only a small portion of light in the specific electromagnetic bands of the sensor. For the determination of albedo it is necessary to consider radiation over all portions of the electromagnetic spectrum.
- Remote sensors collect radiation only in the field of view of the instrument, whereas light reflected in all directions needs to be estimated/measured.
- Corrections are required for the effects of the atmosphere on the reflectances recorded by the sensor.

Studies are being conducted on methods to overcome these difficulties (e.g. Brest and Goward, 1987).

Leaf area index is normally expressed as a function of the NDVI, the difference between measured reflectances in the near-infrared and red parts of the electromagnetic spectrum (Curran *et al.*, 1992). However, variations in a number of environmental factors – solar angle, understory vegetation, atmospheric conditions and phenology – cause variability in the relationship, and it is generally necessary to undertake ground validation and to develop site-specific relationships between these and the remotely sensed data.

A final hydrologically relevant parameter that may be obtained by remote sensing is altitude although, for most purposes, this information may be obtained from existing maps. When no maps are available, a number of techniques may be used. Conventionally, stereo aerial photography, taken from a height conducive to the surface detail requirement, would be analyzed to produce topographic heights and contours, from which ground slope and aspect can be measured much more quickly than from ground surveys. Alternatively, estimates may be obtained using stereo images from the SPOT satellite. The high spatial resolution of the scanner on board this satellite will provide values with errors in the range 5–20 m in both the vertical and horizontal planes. In areas experiencing cloud cover for long periods of the year, altimeters on board aircraft (Menenti and Ritchie, 1994) and satellites such as ERS-1 (e.g. Rapley *et al.*, 1992) will provide useful information.

Gareth Roberts

Bibliography

Barrett, E.C., 1970. The estimation of monthly rainfall from satellite data. *Monthly Weather Rev.*, **98**, 322–327.

Barrett, E.C. and Martin, D.W., 1981. *The Use of Satellite Data in Rainfall Monitoring*. Academic Press, London.

Battan, L.J., 1973. *Radar Observation of the Atmosphere*. University of Chicago Press, Chicago, 324 pp.

Baumgartner, M.F., Seidel, K., Haefner, H. *et al.*, 1986. Snow cover mapping for runoff simulations based on Landsat-MSS data in an alpine basin. *Proc. Workshop on Hydrologic Applications of Space Technology*, IAHS Publ. 160, 190–199.

Becker, F., 1987. The impact of spectral emissivity on the measurement of land surface temperature from a satellite. *Int. J. Remote Sensing*, **8**, 1509–1522.

Becker, F. and Li, Z.L., 1990. Toward a local split window method over land surfaces. *Int. J. Remote Sensing*, **11**, 363–393.

Belward, A.S., Taylor, J.C., Stuttard, M.J. *et al.*, 1990. An unsupervised approach to the classification of semi-natural vegetation from Landsat Thematic Mapper data. A pilot study on Islay. *Int. J. Remote Sensing*, **11**(3), 429–445.

Bergstrom, S., 1975. The development of a snow routine for the HBV-2 model. *Nordic Hydrol.*, **6**, 73–92.

Birnie, R.V., Ritchie, P.F.S., Stone, G.C. and M.J. Adams, 1984. Thermal infrared survey of Aberdeen City: data processing, analyses and interpretation. *Int. J. Remote Sensing*, **5**, 47–63.

Blyth, K., 1993. The use of microwave remote sensing to improve spatial parameterization of hydrological models. *J. Hydrol.*, **152**, 103–129.

Brest, C.L. and Goward, S.N., 1987. Deriving surface albedo measurements from narrow band satellite data. *Int. J. Remote Sensing*, **8**(3), 351–367.

Carlson, T.N. and Buffum, M.J., 1989. On estimating daily evapotranspiration from remote surface temperature measurements. *Remote Sensing Environ.*, **29**, 197–207.

Caselles, V. and Sobrino, J.A., 1989. Determination of frosts in orange groves from NOAA-9 data. *Remote Sensing Environ.*, **29**, 135–146.

Cihlar, J., 1980. Soil water and plant canopy effects on remotely measured surface temperatures. *Int. J. Remote Sensing*, **1**, 167–173.

Collier, C.G., 1986. Accuracy of rainfall estimates by radar. Part 1: Calibration by telemetring raingauges. *J. Hydrol.*, **83**, 207–233.

Collier, C.G., 1989a. *Applications of Weather Radar Systems – A Guide to Uses of Radar Data in Meteorology and Hydrology*. Ellis Horwood Ltd., 294 pp.

Collier, P., 1989b. Radiometric monitoring of moisture stress in irrigated cotton. *Int. J. Remote Sensing*, **10**, 1445–1450.

Collier, P., Runacres, A.M.E. and McClatchey, J., 1989. Mapping very low surface temperatures in the Scottish Highlands using NOAA AVHRR data. *Int. J. Remote Sensing*, **10**, 1519–1529.

Collin, R.L. and Carlisle, P.J., 1989. Snow assessment in small catchments – the operational context in England and Wales. *Proc. 15th Annual Conference of the Remote Sensing Society*, Nottingham, 69–75.

Colyer, P.J. and Petwick, R.W., 1976. *Storm drainage design methods – a literature review*. Hydraulics Research Center Internal Report, INT 154, Wallingford.

COST73, 1990. *Weather Radar Networking*. Seminar on COST Project 73 (eds C.G. Collier and M. Chapuis). Commission of the European Communities, 567 pp.

Curran, P.J., Dungan, J.L. and Gholz, H.L., 1992. Seasonal LAI in slash pine estimated with Landsat TM. *Remote Sensing Environ.*, **39**, 3–13.

Dalezios, N.R., 1988. Objective rainfall evaluation in radar hydrology. *J. Water Resourc. Plan. Manage.*, **114**(5), 531–546.

Danson, F.M., Steven, M.D., Matthews, T.J. and Jaggard, K.W., 1990. Spectral response of sugar beet to water stress. *Proc. 16th Annual Conference of the Remote Sensing Society*, Nottingham, 49–58.

De Paratesi, S.G., 1981. *Heat Capacity Mapping Mission*. Investigation No. 25 (Tellus Project). EEC Final Report to NASA, Washington, DC.

D'Souza, G., 1988. Mid- to long-term, objective rainfall estimation techniques, in Barrett, E.C., Power, C.H. and Micallef, A. (eds), *Satellite Remote Sensing for Hydrology and Water Management*. Gordon and Breach Science Publishers, London, pp. 47–72.

Evans, R., 1979. Air photos for soil survey in lowland England: factors affecting the photographic images of bare soil and their relevance to assessing soil moisture content and discrimination of soils by remote sensing. *Remote Sensing Environ.*, **8**, 39–63.

Fantasia, J.F. and Ingrao, H.C., 1974. Development of an experimental airborne laser remote sensing system for the detection and classification of oil spills. *Proc. 9th Int. Symp. on Remote Sensing of Environment*, Vol III, pp. 1711–1745, April 1974.

Finch, J.W., Reid, A. and Roberts, G., 1989. The application of remote sensing to estimate land cover for urban drainage catchment modelling. *J. Inst. Water and Environ. Manage.*, **3**, 558–563.

Fisher, P.F. and Pathirana, S., 1990. The evaluation of fuzzy membership of land cover classes in the suburban zone. *Remote Sensing Environ.*, **34**, 121–132.

Griggs, M., 1968. Emissivities of natural surface in the 8–14 micron spectral range region. *J. Geophys. Res.*, **73**, 7545–7551.

Hoffner, R.M. and Johannsen, C.J., 1969. Ecological potentials in spectra signature analysis, in Johnson, P.L. (ed.), *Remote Sensing in Ecology*, University of Georgia Press, pp. 1–16.

Idso, S.B., Jackson, R.D. and Reginato, R.J., 1975. Detection of soil moisture by remote surveillance. *Am. Sci.*, **63**, 549–557.

Jackson, T.J., Ragan, R.M. and Shubinski, R.P., 1976. Flood frequency studies on ungauged urban watershed using remotely sensed data. *Proc. Natl Symp. on Urban Hydrology Hydraulics and Sediment Control*, University of Kentucky, Lexington, KY, pp. 31–39.

Kerr, Y.H., J.P. Lagouarde and J. Imbernon, 1992. Accurate land surface temperature retrieval from AVHRR data with use of an improved split window algorithm. *Remote Sensing Environ.*, **40**, 1–20.

Kuittinen, R., 1988. *Application of satellite data for estimation of precipitation*. World Meteorological Organization Technical Report to the Commission for Hydrology No. 26. WMO/TD No. 300. Secretariat of the World Meteorological Organization, Geneva, Switzerland.

Lauriston, L., Nelson, G.J. and Porto, F.W., 1979. *Data extraction and calibration of TIROS-N/NOAA Radiometers*. NOAA Technical Memorandum NESS 107, National Oceanic and Atmospheric Administration, Washington, DC.

Lawler, K.P., 1989. Weather radar system design. *Weather Radar and the Water Industry. Opportunities for the 1990s*. British Hydrological Occasional Paper No. 2, 14–24.

Lethbridge, M., 1967. Precipitation probability and satellite radiation data. *Monthly Weather Rev.*, **95**, 487–490.

Lloyd-Davies, D.E., 1906. The elimination of storm water from sewerage systems. *Proc. Inst. Civ. Eng.*, **164**, 41–67.

McClatchey, J., Rumacres, A.M.E. and Collier, P., 1987. Satellite images of extremely low temperatures in the Scottish Highlands. *Meteorol. Mag.*, **116**, 376–386.

McFarland, M.J., Miller, R.L. and Neale, C.M.U., 1990. Land surface temperatures derived from the SSM/I passive microwave brightness temperatures. *IEEE Trans. Geosci. Remote Sensing*, **28**, 839–845.

Matzler, C., Schanda, E. and Good, W., 1982. Toward the definition of optimum sensor specifications for microwave remote sensing of snow. *IEEE Trans. Geosci. Remote Sensing*, **GE-20**, 57–66.

Menenti, M. and Ritchie, Jerry, C., 1994. Estimation of effective aerodynamic roughness of Walnut Gulch watershed with laser altimeter measurements. *Water Resour. Res.* **30**, 1329–1337.

Musick, H.B. and Pelletier, R.E., 1986. Response of some Thematic Mapper band ratios to variation in soil water content. *Photogram. Eng. Remote Sensing*, **52**, 1661–1668.

Musick, H.B. and Pelletier, R.E., 1988. Response to soil moisture of several indexes derived from bidirectional reflectance in Thematic Mapper Wavebands. *Remote Sensing Environ.*, **25**, 167–184.

Mutlow, C.T., Llewellyn-Jones, D.T., Zavody, A.M. and Barton, I.J., 1994. The Along Track Scanning Radiometer on ESA's ERS-1 satellite – early results and performance. *J. Geophys. Res. (Oceans)*.

Nieuwenhuis, G.J.A., 1986. Integration of remote sensing with a soil water balance simulation model (SWATRE) *Proc. Workshop on Hydrologic Applications of Space Technology*. IAHS Publ. 160, 119–140.

O'Sullivan, F., Wash, C.H., Stewart, M. and Motell, C.E., 1990. Rain estimation from infrared and visible GOES satellite imagery. *J. Appl. Meteorol.*, **29**, 209–223.

Rango, A. & Martinec, J., 1986. The need for improved snow-cover monitoring techniques. *Proc. Workshop on Hydrologic Applications of Space Technology*. IAHS Publ. 160, 173–179.

Rapley, C.G., Baker, S.G., Birkett, C.M. *et al.*, ERS-1 altimetry of inland water and land. *Proceedings of the First ERS-1 Symposium – Space at the Service of the Environment*, Cannes, France, 4–6 November (1992), pp. 539–542.

Richardson, A.J. & Everitt, J.H., 1987. Monitoring water stress in buffelgrass using hand-held radiometers. *Int. J. Remote Sensing*, **8**, 1797–1806.

Ripple, W.J., 1986. Spectral reflectance relationships to leaf water stress. *Photogram. Eng. Remote Sensing*, **52**, 1669–1675.

Roberts, G. and Roberts, A.M., 1992. Computing the water balance of a small agricultural catchment in southern England by consideration of different land-use types. II. Evaporative losses from different vegetation types. *Agric. Water Manage.*, **21**, 155–166.

Rodda, J.C., 1962. An objective method for the assessment of areal rainfall amounts. *Weather*, **17**, 54–59.

Schowengerdt, R.A., 1983. *Techniques for Image Processing and Classification in Remote Sensing*. Academic Press, London, pp. 143–145.

Spreen, W.C., 1947. Determination of the effect of topography upon precipitation. *Trans. Am. Geophys. Union*, **28**, 285–290.

Thiessen, A.H., 1911. Precipitation averages for large areas. *Monthly Weather Rev.*, 39, 1082–1084.

Thornes, J.E., 1989. A preliminary performance and benefit analysis of the UK national ice prediction system. *Meteorol. Mag.*, **118**, 93–99.

Van Genderen, J.L., 1989. High-resolution satellite data for urban monitoring. *Int. J. Remote Sensing*, **10**, 257–258.

Wan, Z. and Dozier, J., 1989. Land-surface temperature measurements from space: physical principles and inverse modelling. *IEEE Trans. Geosci. Remote Sensing*, **27**, 268–278.

Watkins, L.H., 1962. *The design of urban sewer systems*. Road Research Technical Paper No. 55, DSIR.

Cross references

REMOTE SENSING (2)

Introduction

Generally remote sensing is considered as a set of techniques making possible the acquisition of information about an object without physical contact.

The development of remote sensing can be divided in two periods, prior to and after 1959. Until 1959, aerial photography was the only system used. After 1959 the technological development of remote sensing began systematically with Earth orbital observation. In 1958 the first space photography of the Earth was transmitted by Explorer 6, and TIROS 1, the first meteorological satellite, was launched in 1960.

Remote sensing is based on the detection of electromagnetic radiation emitted or reflected by the Earth's surface, between ultraviolet and very low frequencies, representing only a limited part of the electromagnetic spectrum. It is used by scientists and engineers to observe the Earth's surface, particularly its natural and cultural resources, using aircraft, shuttles and satellites; to extract available information concerning a given theme such as cartography, geology, hydrology, pedology, agronomy and botany, land use and oceanography; and to present acquired data in its most usable form – analog or digital – providing information, for a specific area, which can be incorporated in decision-making processes and systems.

Electromagnetic radiation is measured in analog (photographic) or digital (imaging) forms. The electromagnetic energy received by the sensor depends on several factors:

- the nature and state of the radiating object, for a given wavelength;
- the terrestrial environment of the radiating object;
- the propagation through the atmosphere.

Remote sensing and the electromagnetic spectrum

In theory it is possible, in the remote sensing domain, to use continuously a large part of the electromagnetic spectrum, i.e. the transmission bands. Until now only experimental airborne spectroimaging systems have had this capacity, and operational remote sensing is actually based on the use of a few narrow bands, as with Landsat MSS, TM and SPOT from space, and Daedalus and other systems from aircraft.

The most important bands unaffected by atmospheric absorption are as follows.

- The visible (0.4–0.75 μm) and near-infrared (0.75–1.1 μm). The radiation, detected by passive mode, is induced by solar energy reflected by terrestrial objects. The sensors used are photographic cameras and multispectral scanners (Landsat MSS, SPOT HRV).
- The mid-infrared (1.55–2.35 μm). The reference sensors used are Landsat TM (operational) and JERS OPS (experimental).
- Thermal infrared including two regions: 3–5 μm, where reflected solar energy is dominant, and 6–14 μm, where emitted solar energy is more important. In these two thermal bands the energy radiated by the terrestrial objects is a function of their temperature, emissivity and surface conditions. The sensors used are scanning devices equipped with detection cells sensitive to thermal IR radiations. The reference sensors are HVHR (NOAA meteorological satellite), HCMR (HCMM experimental satellite), TM (Landsat TM operational satellite) and airborne systems.
- The microwave region (1 cm up to several meters). The sensors used are of two basic types, active and passive. In active microwave sensors, the radar (synthetic aperture radar or SAR and sidelooking airborne radars or SLAR) illuminates the surveyed area and registers the backscattering beams, the intensity of which is a function of dielectric constant and surface roughness. They are operating from aircraft (Goodyear, Motorola, and Intera), shuttles (Seasat, SIRA, B and C) and satellites (ERS1, JERS1 and ALMAZ), with different wavelengths X, C, S and L respectively 3, 6, 10 and 25 cm. Passive microwave sensors detect radiation naturally emitted at these wavelengths. They are still at the experimental stage.

Data processing

Whatever the spectral bands used, the acquired data generally constitute a digital image, a sampled and quantized numeric representation of a particular scene. This image also has spectral and temporal dimensions if viewed in many spectral bands and at different times. To manipulate or operate upon this matrix of numbers for a particular objective, digital image processing is a basic and necessary function which is performed using a computer. Image processing can be broadly classifed into three categories: image correction, image enhancement and information extraction.

Image correction concerns both geometric and radiometric errors. These errors require corrections to support image analysis and interpretation. The degree of correction varies with the application. Cartography and land use require a high geometric accuracy or fidelity. Agricultural and hydrological applications involving the detection of change are based on subtraction and addition, and the combination of multiple and diverse data sets requires images which are geometrically registered with respect to each other.

Image enhancement processing methods increase the information content of images to be used for visual interpretation. Several techniques have been developed since 1972, the launch of Landsat 1 (ERTS) providing a tremendous stimulus to digital processing of Earth observation data. They are presented in different publications (e.g. Prat, 1978; Colby et al., 1981). Some of these techniques are used routinely: contrast stretching, color composition, ratioing and principal component. Past experience shows that interactive manipulation in conducting processing is a necessary principle for the different applications.

Processing adapted to the extraction of information contained in digital images enables spectral and textural classification, and also, to a different extent, pattern recognition. Their respective accuracy depends on the type of application. Spectral classification is commonly used for agricultural surveys, the accuracy of the mapping result being considered high. Therefore, to become operational, textural classification requires more experimentation particularly at a large scale, and pattern recognition is still at the research stage in remote sensing. It should be noted that pattern recognition now benefits from the actual development of three-dimensional space acquisition (SPOT).

Principles of remote sensing mapping

To extract information specific to a given theme from remote sensing data, three main methods are generally followed: visual image interpretation, digital interactive image interpretation and spectral classification.

Visual image interpretation was first developed for various applications of aerial photographs. It is based on the experience and thematic qualification of the interpreter who can, to some extent, directly identify the objects, when they are quite visible (roads, houses or rivers) or delineate targets to be controlled on the ground. Generally the information, mainly in geology, is deduced from the landscape through global interpretation based on the analysis of different parameters, spectral, textural, structural and geomorphological.

The spectral parameter is defined as the energy emitted (or reflected) by the constituents of the Earth's surface under specific conditions, as a function of wavelength. In theory the spectral intensity of an object is characteristic of that object in a given state and should permit its identification and mapping. In geology and pedology, the spectral properties of rocks and soils have been studied in laboratory conditions in the optical domain (Hunt and Salisbury, 1971) in the mid and thermal infrared region (Salisbury et al., 1988) and for microwaves (Ulaby et al., 1982). Practically, the spectral characteristic is used in visual interpretation to separate and delineate the different objects.

Texture in the image interpretation method corresponds to all the signals, points and traces which constitute the scene. It is evidently related to the grain of emulsion (photography) or the size of the pixel (digital image). Textural elements are gathered according to certain geometric rules, straight lines, squares and circles, and the mode of distribution, number and density, constituting what is known as a structure.

Textures and structures induce specific forms and networks, the recognition of which is an important parameter is visual image interpretation, which to a certain extent can be operated by digital methods: textural classification and pattern recognition.

Geomorphology, or study of the forms induced by relief is the basis of stereo geological photo interpretation. Landforms are the result of various interactions of morphogenetic factors on a particular type of material under definite climatic conditions. Three main factors are generally considered: the nature of the rocks (mechanical and physical

Table R6 Data for Earth observation satellites

Satellite	Swath width (km)	Spectral bands	Spatial resolution (m)	Information registered (MB)
Landsat MSS	185	4	76	16
Landsat TM	185	7	30 (120 m for thermal band)	230
SPOT XS	60	4	20	27
SPOT Panchro	60	1	10	36

properties), external action (wind, rain, frost and sea) and internal action (mechanical, thermal and chemical action). Thus in many cases landform analysis permits the differentiation of various types of exposed rocks, without direct identification, in general terms such as hardness, permeability, plasticity and anisotropy.

Therefore, to perform a more complete and accurate landform analysis and then to obtain a correct geological interpretation, it is recommended – as it is with aerial photographs – to study stereo pairs of images. Until the launch of SPOT, satellite stereoscopy was only feasible through aerial surveys. Since 1986 SPOT has been widely used in the field of geomorphology, making possible stereo pair acquisition and interpretation using a mirror stereoscope (Scanvic, 1993), digital elevation model calculation by correlation methods, and the development of quantitative analysis for slope, slope exposure, drainage and finally pattern recognition.

Digital interactive image interpretation is a compromise between visual and digital methods. It takes advantage of the potential of the human eye and experience on one hand, and software capacity on the other. With this method the interpreter works directly on the data processing system screen, the cartographic observation being digitized in real time, and takes benefit from all the available functions such as image enhancement, classification, measurements and stereoscopy.

Spectral classification is a digital method to determine automatically the information class of each distinct region of the ground. It is based on a pixel's spectral information. If the pixel's information consists only of sensor measurements from one observation time, the classification is called multispectral. If information is provided for more than one observation time, for the same ground area, the classification is called multispectral–multitemporal (Haralick and King-Sun Fu, 1981).

To maximize automatic spectral pattern recognition and classification, it is first necessary to define the type of objects to discriminate, using training data (supervised method). Then spectral pattern recognition is done by processing each pixel information separately over the entire image and each measurement is assigned to a category, including an unknown category if necessary. However, as there is not a unique measurement assigned to a specific category, but rather a probability of distribution indicating a relative frequency of occurrence, the likelihood that it corresponds to a true category is estimated.

At last the procedure has become relatively simple: the training data are used to determine the class sample mean and the measurements are partitioned so that each class associates with it all the spectral measurements closest to it.

Until recently the quality of results obtained by using the different methods of interpretation have varied from one method to another and varied as a function of the application domain and development status.

When objects observed from the sensor are quite visible on the Earth surface, classification presents a high degree of accuracy, crop mapping for instance. When parcels are partly canceled and mapping is deduced through landscape units, as is quite often in the case of geology, image interpretation is the more useful method.

Therefore in both cases digital interactive image interpretation should rapidly become an optimal method for all areas of application, when the necessary development has been completed in two directions, namely improvement of systems and an adequate methodology. Whatever the methodology used, the global performance still depends on several sensor specifications: spectral and spatial resolution, swath width, repetitivity and stereo capacity.

Spectral resolution corresponds to sensor characteristics, Landsat TM or SPOT HRV for instance. It is defined by two parameters, the location on the electromagnetic spectrum and the width of the spectral band. Each selected band has been specified according to spectral reflectance or emittance of the characteristic Earth constituents,

detection of which is an objective of the mission satellite, e.g. water, crops and rocks. It enables image reconstruction through digital processing based on a selection of three of the different spectral bands acquired. Therefore the display colour does not always correspond to the acquired wavelengths. Thus SPOT images are generally displayed so that green (XS1) becomes blue, red (XS2) becomes green and near-infrared (XS3) becomes red: it is a false colour composite image enabling near-infrared – which is normally invisible – photographic reconstruction.

The spatial resolution induces the choice of optimal enlargement and the associated image scale. Specific digital processing and resampling enables spatial resolution improvement: the visibility of objects is better but unregistered information due to acquisition conditions is not yet displayed.

Swath width determines the surface area observed from the sensor and influences the synoptic view, an important parameter in global landscape image interpretation. Swath width, spatial resolution and the number of spectral bands influence the quantity of registered information, as indicated for some Earth observation satellites (Table R6).

Repetitivity is the time required for complete Earth coverage (for a Sun-synchronous satellite) or for a portion of the Earth (geostationary satellite). It conditions the revisiting of a particular area, in similar conditions (local time and orbit), several times a year. It is only feasible on a regular and economic basis through satellites.

The repeat cycle is 18 days for Landsat, 25 days for SPOT (using the routine method) or 5 days (programmed for a specific purpose). It is a valuable parameter for studying vegetation changes, soil moisture conditions, land use evolution and, to a certain extent, natural hazards and disasters.

Stereoscopy is the capacity of the sensor to acquire a pair of images, making it possible to observe the landscape in relief. Until 1986 this was only feasible with aircraft camera or by digital simulation. In 1986 the launch of the SPOT satellite has opened the way to stereoscopy. Therefore with SPOT 1, 2 and 3, stereo images are not acquired on the same orbit and this is sometimes a handicap when climatic conditions quickly change. The JERS1 experimental satellite has demonstrated the feasibility of simultaneous stereo acquisition and SPOT 4 (1998) should operate on this basis.

Remote sensing hydrology

Hydrology studies circulation of water in the three parts of the Earth system: the atmosphere (gaseous envelope), hydrosphere (water covering the surface) and lithosphere (the solid rocks). Remote sensing from Earth observation satellites is *a priori* adapted to the survey of those of the terrestrial hydrologic process which are phenomena that vary rapidly in space and time (hydrosphere) or influence the landscape (lithosphere). Its capacity to provide synoptic observation with high observational density over a relatively large area and on a regular revisiting basis makes it possible to assume that remote sensing can be used (and it has been used) effectively in hydrology in different ways (Jackson *et al.*, 1983):

- simple qualitative observations guiding the location of *in situ* observations, assisting in interpolation and extrapolation;
- detection and mapping of features such as land occupation, quantitative analysis of drainage, location of fractures, faults, quantification of biomass which influences evapotranspiration, and soil moisture storage.

This remote sensing capacity is widely exploited and makes a large contribution to operational hydrology as an effective complement to others sources of information. Maps, an effective form of presenting hydrological data, have been prepared for numerous surveys either on their own or by integration into a geographic information system (GIS), making possible a statistical approach according to rules established by specialists in the various fields concerned. This particularly concerns research for areas favorable to percolation prospecting, for areas favorable for water resources, the identification of drilling sites, the study of drainage and irrigation networks, cartography, the delimitation of drainage basins, predictive soil erosion vulnerability mapping, the evaluation of biomass and the location of karstic phonomena to prevent pollution. It now widely contributes to the topographic and geological preparation of the different hydrological maps, as proposed by IASH (1962).

Remote sensing is also used for the direct estimation of hydrological parameters through the development of correlative models between the remotely sensed observations and corresponding *in situ* measurements. Such research is now extensively developed on the basis of active microwave data, provided by the ERS1 satellite, in which the moisture content of rocks and soils plays an important role in determining the dielectric constant value registered by the sensor. In this, the optical teletransmission of *in situ* measurements, through satellite (Argos system) should be a valuable source of improvement.

Conclusions

Remote sensing is now the source of positive results in hydrology. The utility of remote sensing is in facilitating the hydrologist's task of observation, measurement, description and comprehension, making available information which is inaccessible through other ways. Therefore if the analysis of satellite imageries alone brings pertinent informations it is in combination with observations provided by the mean of diverse methods and techniques that space data are most useful.

Jean-Yves Scanvic and Philippe Dutartre

Bibliography

Colby, C., Murphrey, S.W. and Snyder J., 1981. Image geometry and rectification, in *Manual of remote sensing*. 2nd edn, vol. 1, chapter 21. American Society of Photogrammetry, pp. 873–922.
Haralick, R.M. and King-Sun Fu, 1981. Pattern recognition and classification, in *Manual of remote sensing*, 2nd edn, vol. 1, chapter 18. American Society of Photogrammetry, USA, pp. 793–805.
Hunt, E.B. and Salisbury, J.W., 1971. Visible and near infrared spectra of minerals and rocks. *Modern Geol.* **1–2**, 1–95.
IASH, 1972. Hydrogeology: 'a legend for hydrogeological maps'. *The encyclopedia of geochemistry and environment*, Colombia University, New York, pp. 521–530.
Jackson, T., Lucas, J., Moore, G. *et al.*, 1981. Water resources assessment, in *Manual of remote sensing*. 2nd edn, vol. 2, chapter 29. American Society of Photogrammetry, pp. 1497–1570.
Prat, W.K., 1978. *Digital image processing*. John Wiley and Son, New York, 170 pp.
Salisbury, J.W., Walter, L.S., Vergo, N. and Arin, D.M., 1992. Infrared (2.5 to 13.5 μm) spectra of minerals, in *The John Hopkins study of Earth and space sciences*. University Press, Baltimore, London, pp. 1–108.
Scanvic, J.Y., 1993. *Télédétection aérospatiale et informations géologiques*. Manuels et méthodes no. 24. Editions BRGM Orléans, France, 284 pp.
Ulaby, F.T., Moore, R.K. and Fung, A.K., 1982. *Microwave remote sensing, Active and passive*, vol. 2. University of Michigan, Ann Arbor, USA, pp. 456–1064.

Cross references

Remote sensing (1)
Remote sensing data: use in hydrological modeling
Satellite hydrology: present trends and outstanding needs

REMOTE SENSING DATA: USE IN HYDROLOGICAL MODELING

Different studies have shown that several surface parameters, important for the monitoring of the soil water budget, may be inferred from remote sensing. For example, land use and vegetation fractional cover can be estimated from visible and near infrared reflectance. Thermal infrared and radar measurements may also be used to obtain evapotranspiration and soil moisture at the regional scale. In the thermal infrared, the methods are based on the use of the surface radiative temperature to estimate soil parameters and surface fluxes. These variables may be assimilated in hydrological models to improve the simulation of the exchanges between the soil and the atmosphere and finally the water flow over the catchments.

Already tested in the framework of the HAPEX-MOBILHY experiment on the Adour river basin (Ottlé and Vidal-Madjar, 1994), this methodology is in the process of being applied on a large number of agricultural river basins. These catchments are situated in Brittany (France) where a large dataset of AVHRR/NOAA, SAR/ERS1 images and ground truth coincident measurements have been compiled for a 2.5-year period (from winter 1992 to winter 1994) and for which a conceptual hydrological model has been developed (Loumagne *et al.*, 1995). The methods to infer vegetation parameters from remote sensing data, the calibration of the SAR data in terms of soil moisture and the methodology for estimating soil moisture in the surface layer from thermal infrared data are presented. Then, an example of their introduction in a hydrological model of the Adour river basin is shown.

Introduction

Remote sensing may be applied in different domains of hydrological modeling. The determination of surface characteristics, for example land use, pedology and snow cover, may be estimated using satellite data and are already used in hydrological models. Evapotranspiration and soil moisture may also be estimated at the mesh scale of hydrological models and used to control the model simulations. In fact, most of the hydrological models are based on a mass balance between the input of the system (the precipitation) and one of the outputs (streamflow). Generally, part of the input is stored in a surface reservoir and this amount of water controls the distribution between runoff and infiltration. This surface reservoir should evolve like soil moisture. Then, regular measurements of soil water content could be allowed to control the temporal evolution of the surface reservoir and then could improve forecasts of water flow. The advantage of remote sensing data compared to ground-truth data is that it provides us with repetitive areal measurements that can be assimilated in regional models.

In this contribution, two studies where remote sensing data have been used in hydrological models are presented. The first one was done in the framework of the HAPEX-MOBILHY experiment which took place in southwestern France in 1985–1986. In this study, optical and thermal infrared data have been used to infer land uses and the annual cycle of vegetation and soil moisture in the root zone for assimilation in a distributed hydrological model (Ottlé and Vidal-Madjar, 1994). The second work is still in progress and consists of the synergistic use of radar, optical and thermal infrared data for hydrological modeling of small agricultural watersheds in Brittany, France (Cognard *et al.*, 1995). About 30 basins have been selected at the crossing of ascending and descending tracks of the ERS1 satellite, leading to the acquisition of SAR images at least once every 3 days during orbiting phases B and D (winter 1992 and winter 1994). Among these basins, the Naizin watershed is a long-term test site for hydrological studies and has been selected to evaluate the SAR/ERS1 capacity to estimate soil moisture (Cognard *et al.*, 1995). Ground-truth measurements of soil moisture, vegetation cover and other atmospheric and hydrological parameters have been made during 2 years, 1992–1993. The first results are presented here. The spatial tool is first described and the methods to infer surface parameters from satellite imagery are briefly presented.

Satellite imagery available for hydrological studies

The spatial and temporal variability of surface hydrological parameters and variables require the use of data from operational meteorological satellites because of their good repetitivity. The polar-orbiting satellites like the NOAA/TIROS-N series have a repetitivity of 1 day, a 1 km resolution and carry many instruments. Among them, the Advanced Very High Resolution Radiometer (AVHRR) measures the radiances emitted or reflected by the Earth's surface at different wavelengths.

In the visible and near infrared domains, this instrument measures the solar radiation reflected by the surface. The reflectivity depends on the soil texture and on the vegetal coverage. Because the chlorophyll responsible for plant photosynthesis reflects near infrared radiation more than visible radiation, the comparison of the two signatures is related to the vegetation amount. It is then possible to define vegetation indices combining these two reflectances and taking into account approximate soil, atmospheric and geometric effects. The most simple and widely used one, is the Normalized Difference Vegetation Index (NDVI, Deering *et al.*, 1975) which is the ratio of the difference of the sum of the reflectances in the visible and near infrared domains. This index is a good tool to estimate the vegetation density within the satellite pixel.

In the thermal infrared domain, the radiometer measures the energy emitted from the Earth's surface which is the sum of the surface and atmospheric emittances. Different methods allow to correct the atmospheric effects and to retrieve the surface temperature from such measurements. The most popular is the split-window method (Prabhakara *et al.*, 1974; Anding and Kauth, 1979; Deschamps and Phulpin, 1980). It is based on the combination of the radiances measured at two particular wavelengths carefully chosen to minimize the water vapor atmospheric absorption and the influence of the other constituents.

Microwave instruments may also provide soil parameters which are interesting for hydrological studies. These techniques are based on the large contrast between the dielectric properties of liquid water and dry soil. Passive instruments are still not adapted for hydrological purposes because of their low spatial resolution. Active techniques are more suitable. Over bare soil, the radar backscattered signal is related to soil moisture and soil roughness. Different approaches (theoretical and empirical) may be followed to separate the effects of roughness from soil moisture. In the case of vegetated areas, the relation is more complex because the vegetation scattering and absorption must be taken into account but some models are under development.

The Synthetic Aperture Radar (SAR) aboard the experimental European Remote Sensing Satellites (ERS 1 and 2) is equipped with a C-band suitable for the retrieval of soil moisture. The swath width is about 100 km and the resolution of the standard products is 20×20 m^2. The orbits of ERS1 were divided into two phases. The first one, known as the 'ice phase' (or phases B and D) has been implemented to have a repetition period of 3 days, but no global coverage of the Earth's surface. This is particularly interesting for hydrological studies since the characteristic time for surface soil moisture changes is a few days. Phase B was active from the end of 1991 to the end of March 1992 and phase D from the beginning of January 1994 to the end of March 1994. Between these two periods, the satellite was in phase C with a global coverage but a repetition period of 35 days. This configuration is unfortunately less useful for hydrology.

Methods to infer hydrological surface parameters from remote sensing data

Vegetation coverage

Reflectances in the visible and near infrared domains are good indicators of the vegetation amount. The follow-up of indices based on measurements such as the NDVI (Deering *et al.*, 1975), combined with ground-truth information, permits land-use classification and estimation of the vegetation fractional cover. This parameter, which is necessary to partition direct evaporation from evapotranspiration, can be deduced from the value of the vegetation index compared to its maximum and minimum values. In the Hapex study, all the cloudless NOAA/AVHRR images available over the 2-year period have been processed and a rough classification of the vegetation has been performed through the study of the NDVI. It has been possible to separate forested areas, and regions of summer and winter crops. Moreover, the annual variation of this index allowed to identify the different phenological periods of the vegetation. As an illustration, Figure R14 presents the time variation of NDVI over summer crops situated within the HAPEX-MOBILHY region between January 1985 and January 1987. The periods of bare soil, corresponding to minimum values of the NDVI are clearly differentiated from the periods of growing of the vegetation when the NDVI rises to its maximum value, from the period of maturity characterized by a plateau and finally from the period of senescence corresponding to the rapid decrease of the index. It is then possible from these plots to estimate roughly not only the state of maturity of the vegetation but also the fraction of vegetation within a pixel. The vegetation

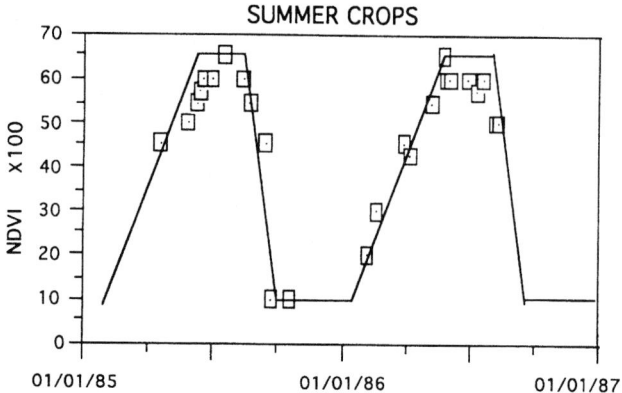

Figure R14 Temporal variation of the NDVI of summer crops over the Hapex–Mobilhy region.

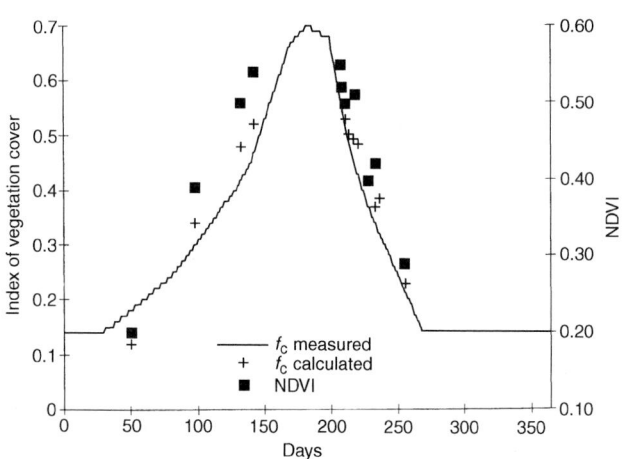

Figure R15 Fraction of vegetational cover f_c over the Naizin watershed calculated from ground-truth measurements compared to the one deduced from the NDVI values plotted on the same graph.

fractional cover f_c can be deduced from the value of the NDVI compared to its minimum and maximum values, weighted by its maximum value, to take into account the permanent non-vegetated areas:

$$\frac{f_c}{f_{c\,max}} = \frac{NDVI - NDVI_{min}}{NDVI_{max} - NDVI_{min}} \quad (R1)$$

where $NDVI_{min}$ and $NDVI_{max}$ are the minimum and maximum values observed during the whole year.

The same work has been conducted over the Brittany watersheds (Cognard *et al.*, 1995) and particularly over the Naizin river basin where a precise vegetation survey was done in 1992. Figure R15 shows the NDVI values computed with NOAA/AVHRR images after correction of the atmospheric effects using the 5S model (Tanré *et al.*, 1990). On the same graph is plotted the fraction of vegetation f_c calculated using equation (R1) (crosses) and the one measured from the ground truth vegetation survey for the whole year 1992 (full line). A value of 0.7 has been assumed for $f_{c\,max}$ which corresponds to the averaged maximum occupation rate of the region. Unfortunately, no cloudless images were available between Julian days 150 and 200, leaving an uncertainty in the exact location of the maximum. Nevertheless, the plot shows the good agreement between f_c calculated and measured values, demonstrating the capacity of the NDVI to monitor the vegetation cover.

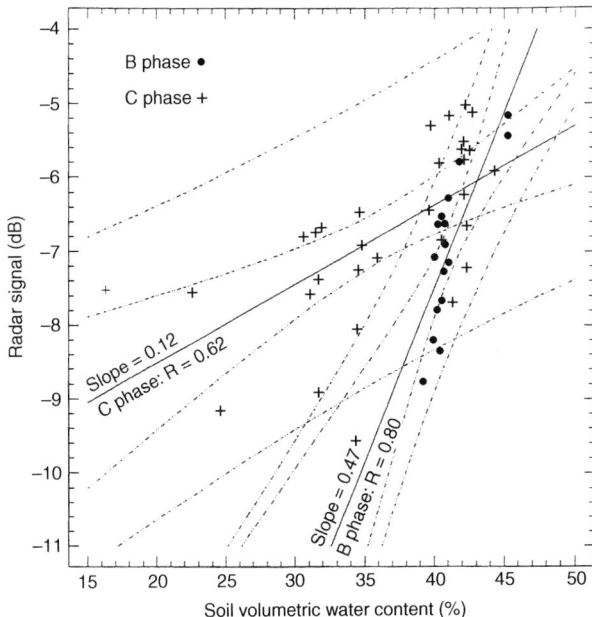

Figure R16 Correlation between soil moisture measured by an automatic probe and mean radar signal over the entire Naizin watershed.

Surface soil moisture

One of the possible applications of satellite radar remote sensing is the estimation of surface soil moisture. The Naizin watershed is a good site to evaluate the SAR/ERS1 capacity in this field. All possible SAR images were acquired together with two types of ground truths: intensive ground measurements during 14 field campaigns and automatic point measurements over the entire period. All the results are presented in Cognard et al. (1995). They have shown that at the field scale, the relation between the radar backscattering and the soil moisture depends on the nature of the surface. At a larger scale (a few square kilometers), during the fall, winter and mid-spring periods (phase B and D), the correlation is high but is lost at the end of spring and in summer, corresponding to the period of denser vegetation (phase C), as shown in Figure R16. It may then be possible to define a hydrological state index from ERS1/SAR images at least during the periods of low vegetation density. Work is underway to propose an algorithm to correct the C-band radar signal from the vegetation attenuation (Taconet et al., 1995).

Root zone soil moisture

Various studies have shown that the surface temperature estimated from space can be used to infer surface fluxes or variables such as evapotranspiration and soil moisture by a surface model describing the exchanges between the soil, the vegetation and the atmosphere (SVAT; e.g. Carlson et al., 1981; Taconet et al., 1986). The one which is used here is a one-dimensional soil–vegetation–atmosphere model described in several papers (Taconet et al., 1986; BenMehrez et al., 1992a,b; Ottlé and Vidal-Madjar, 1994). The model calculates the surface fluxes, the surface temperature and the surface soil moisture by solving simultaneously the energy budget equation at the soil surface level and above the canopy. An adequate partition of the incident energy fluxes between the vegetation and the soil is assumed, as well as a knowledge of the daily variation of the atmospheric forcing and the characteristics of the soil and vegetation. The soil is represented by a twolayer system derived from Deardorff (1978) and modified by Bernard et al. (1986). The vegetation parameters may be derived from spatial vegetation indices.

The SVAT model can be used in the direct way to calculate the fluxes and the surface radiative temperature, if the soil and vegetation parameters, the soil humidity and the atmospheric forcing are known. The atmospheric input data necessary to run the interface model comprise the daily variation of the incoming solar radiation, the air temperature and humidity, and the wind speed at 2 m.

Moreover, it can be used in the inverse manner. If the radiative surface temperature and the functional parameters are known, the model may be inverted to retrieve one unknown variable, which could be the soil moisture in one of the two layers. Over bare soil, as the surface temperature is very sensitive to soil evaporation, which is mostly determined by soil humidity in the surface layer W_s, the surface temperature allows the estimation of W_s. In the same way, over dense vegetation cover, since the evapotranspiration is strongly related to the soil humidity in the root zone, the foliage temperature permits the estimation of soil moisture of the bulk layer W_b. Over partial covers, where the soil and the vegetation both contribute to the total evapotranspiration, it is possible to estimate either of the two soil moisture measures, W_s or W_b, assuming one is known. This methodology has been applied to estimate the bulk soil moisture at a regional scale, using AVHRR thermal infrared data over the Adour river basin and over the Brittany basins.

Application to hydrological models

Methodology

Remotely sensed variables are of great interest for physically based hydrological models because their introduction can improve the simulation of the water exchanges at the surface. The following describes how soil moisture, land use and some vegetation characteristics inferred from space measurements may be used in hydrological models. From the knowledge of the daily values of precipitation and potential evaporation, hydrological models simulate the daily water flows at the outlets of the riverbasin. Every time a satellite image is available, the thermal and radar data place controls an estimate of the soil humidity in the surface layer and in the root zone. The optical data are used to adjust vegetation parameters. This can be done for all the meshes of the hydrological model if it is a distributed one.

Application to the Adour riverbasin hydrological modeling

Vegetation parameters and root zone soil moisture inferred from AVHRR data have been used in a distributed hydrological model implemented on the Adour river basin in the framework of the HAPEX–MOBILHY experiment. The land-use classification allowed to assign to each grid mesh the dominant type of vegetation and its annual variation, namely the evolution of the vegetation fractional cover and the dates of the different steps of maturity. This information is used in the hydrological model to improve the calculation of the total daily evapotranspiration. The soil humidity inferred from the surface temperature measurements has also been introduced in the hydrological model to correct the simulated value on each mesh, on each day that a cloudless satellite image was available. The methodology and the results of this work are presented in the next sections.

Description of the hydrological model

The HAPEX–MOBILHY experiment (André et al., 1988) took place in southwestern France during a period of 2 years in 1985–1986. The chosen site is watered by the Adour river system. The upper part of the region is covered by the Landes forest and the rest of the square by mixed crops. To simulate the streamflows over the Adour river basin, the hydrological model of Girard (1974) and Ledoux (1980) has been implemented (Girard and Boukerma, 1985). The model is described in Ottlé and Vidal-Madjar (1994). It is a deterministic model, composed of two coupled domains, the surface and the underground. These two domains have been discretized over square meshes with sides ranging from 1.25 to 5 m. The primary objective of this work was first to improve the simulation of the water budget, especially the estimation of the actual evapotranspiration, and to assimilate the soil water content of soil layers contributing to the evaporation. For this purpose it has been necessary to separate the surface reservoir into two layers: the superficial soil layer corresponding to the first 10 cm of soil and the root zone extending approximately through the first 1 m. Consequently, we have introduced in the hydrological model the surface parameterization already used in the (surface flux/surface temperature) inversion model, and the calculation of the evapotranspiration has been modified by taking into account the vegetation.

To improve the calculation of the evapotranspiration, the evaporation of the bare soil has been separated from the transpiration, according to Deardorff (1978) and BenMehrez *et al.* (1992a,b). They have shown the importance of taking into account the nature of the surface and its vegetal covering. The total evapotranspiration can be calculated as the sum of the contributions of the vegetation and of the bare surface with respect to the fraction of vegetal covering. In the hydrological model, the total evapotranspiration of the surface layer is given by the following equation:

$$E_b = f_c E_v + (1 - f_c) E_{bs} \qquad (R2)$$

where E_v is the daily evapotranspiration and E_{bs} is the daily bare soil evaporation. The fraction of vegetation f_c is estimated from the NDVI annual variations.

Since the input data of the hydrological model is the decade potential evaporation, it has been necessary to parameterize E_{bs} and E_v in terms of the potential evaporation and of the soil moisture in the respective soil layers. For the evaluation of E_v, the state of maturity of the vegetation has been taken into account because it influences strongly the stomatal resistance to the evapotranspiration. For more details, see Ottlé and Vidal-Madjar (1994).

Use of the inferred soil moisture in the hydrological model and results

One of the objectives of this study was to use the estimation of the soil moisture, deduced from the energy budget observed from space, to correct the simulated content of the hydrological model's surface reservoir. For that purpose, only the images corresponding to dense covers with an LAI greater than 2 have been kept, where it has been demonstrated that the radiative surface temperature is more sensitive to root zone humidity than to superficial humidity.

For each of the days for which all data and satellite images are available, the soil–vegetation–atmosphere interface model can be run with the hydrological model surface moisture as an initial value on all the meshes. The soil and vegetation parameters of the interface model have been calibrated for each mesh and each day considered, according to the surface characteristics estimated by combination of ground-truth measurements and NDVI data. The atmospheric parameters necessary to run the model, have been obtained by interpolation of the PATAC meteorological network to the center of each mesh. The soil humidity W_b is modified until the difference between the modeled surface radiative temperature and the observed one is less than 1.5 K. If the variation of the soil moisture has no effect on the surface simulated temperature, the model is not inverted.

The difference between the soil moisture inferred in this way and the initial one, simulated by the hydrological model, is then introduced in the hydrological model for correcting the modeled soil moisture, W_b. The model is then run until a new satellite image is available. This work has been done for 24 images acquired mostly during the growing period of the vegetation between May and September for the 2 year duration of the experiment.

It can be noticed that when the actual soil moisture is obtained on each mesh by inversion of the interface model, the evapotranspiration is also retrieved and could also be incorporated in the hydrological model. But, since it varies strongly in time, it is not relevant to correct it for the few days where cloudless satellite images are available. As a result, this information has not been introduced in the hydrological model. The results have been presented by Ottlé and Vidal-Madjar (1994), but only those concerning the evapotranspiration will be presented here.

Two simulations of the hydrological model in the years 1985–1986 will be compared: the first one, called hereafter 'Run 1', was run with the initial version of the hydrological model (Girard, 1974) using surface parameterization; the second one, called 'Run 2', was run with the second version of the model in which we have adjusted the vegetation parameters from the visible imagery and assimilated the soil moisture inferred from the thermal infrared imagery.

An important point is the estimation of regional evapotranspiration by the hydrological model. The SAMER stations give an estimate of evapotranspiration at the field scale which is representative of a larger region. André *et al.* (1990) have shown that the SAMER measurements are well correlated with the aircraft flux measurements integrated over a larger region.

Evapotranspiration as measured by the SAMER stations during the 5 months of measurements of the Special Observing Period (SOP) in 1986 has been compared to the one simulated over the corresponding meshes. Figure R17 presents monthly averaged values. The results show a better agreement of Run 2 with the observations, especially for the months of July and August when Run 1 simulates very low and

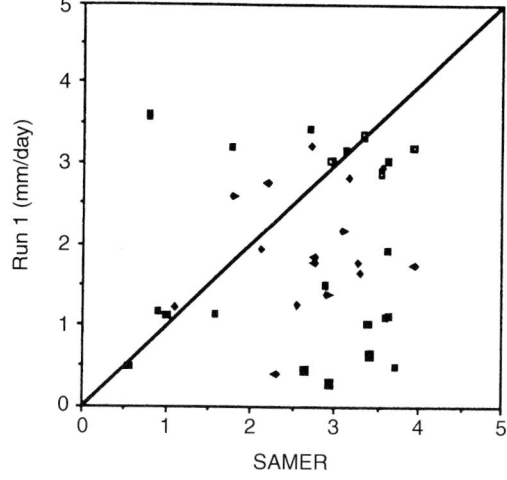

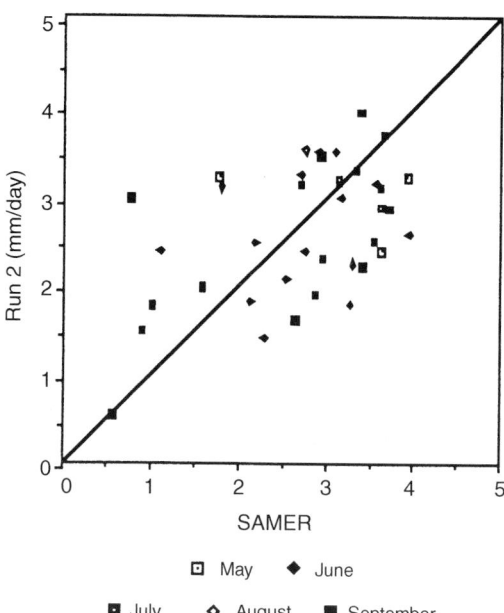

Figure R17 Monthly averaged evapotranspiration simulated by run 1 and run 2 compared to the SAMER measurements.

unrealistic values. The standard deviation is 1.1 mm for Run 1 against SAMER measurements and 0.7 mm for Run 2.

Application to hydrological modeling of Brittany river basins

The same methodology is underway on the 30 small river basins situated in central Brittany, where SAR/ERS1 and AVHRR/NOAA11 data have been acquired over a period of more than 2 years.

Description of the hydrological model

The GRHUM model used in this study (Loumagne *et al.*, 1995), is based on a daily conceptual rainfall–runoff model called GR4 (Edijatno and Michel, 1989; Nascimento and Michel, 1992). This model has already been successfully applied to 120 French catchments with different characteristics (soil, vegetation and climate; Edijatno, 1991).

The upper soil surface is modeled as a two-layer system: a surface superficial layer representing about the first 10 cm of the soil and the bulk layer representing the root zone (about the first 1 m of soil). This depth depends on the particular soil properties and vegetation type of the basin under study. The exchanges between the two layers depend on the soil moisture conditions (Chkir, 1994). Rainfall is divided into

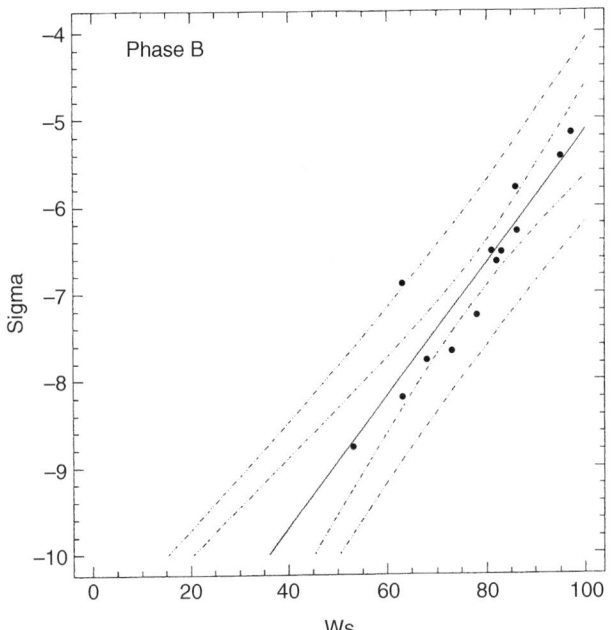

Figure R18 Correlation between relative soil humidity simulated by the hydrological model and the mean radar signal over the entire Naizin watershed.

two terms: the first feeds the soil reservoir and the second is routed directly to the outlet of the catchment through a transfer function. This partition depends on the moisture level in the soil bulk layer. The NDVI index calculated from AVHRR data is used to estimate the fractional vegetation cover (following a methodology similar to the one used in the HAPEX work) which partitions soil evaporation from transpiration.

A model validation has also been performed on the 36 basins of central Brittany. These basins have been chosen because hydrological data (streamflow and rainfall) are available for more than 6 years.

The catchments range in area from 20 to 500 km^2 and most of them have a Brioverian schist substratum. This region is characterized by intensive agricultural exploitation. One of them, the Naizin experimental watershed, has been chosen as a permanent test site for ground-truth measurements dedicated to the assessment of the ability of the ERS1/SAR radar in evaluating the soil surface water content (Cognard *et al.*, 1995).

The accuracy of the model in reproducing discharges observed at the outlet of the catchment and its aptitude to simulate soil moisture in a surface soil layer and in a bulk layer have been checked.

Preliminary results

A preliminary study on the Naizin watershed has shown good correlation between the backscattering coefficient averaged over the whole river basin and the soil moisture (the first few centimetres) modeled by GRHUM during the periods of bare soil in 1992 and 1993. Figure R18 shows the correlation between the relative soil humidity simulated by GRHUM over the Naizin watershed in the first 10 cm and the mean radar signal over the basin (in decibels). The measurements correspond to phase B where the Naizin basin is mostly covered by bare soils or permanent grass (fraction of vegetation f_c is less than 0.2). The good correlation (R = 0.93) shows that it may be possible to use the radar signal to define a surface soil moisture index for bare soils.

The next step is the treatment of the thermal infrared data for the estimation of root zone soil moisture and the definition of a second soil moisture index for the root zone in vegetated areas. The final step is the introduction of these indices in the hydrological model.

Conclusion

These two attempts at using optical, thermal infrared and radar remote sensing in hydrological models are very encouraging. They show

particularly that the temporal evolution of soil moisture may be assimilated for the control of the surface reservoirs. It has been shown in the framework of the HAPEX–MOBILHY experiment that better simulation of water exchange at the land–atmosphere interface leads to major improvements in the simulation of regional evapotranspiration and water flow. The synergy radar/thermal infrared technique opens new perspectives. These results ensure confidence that in the very near future it will be possible to improve hydrological modeling of gauged and ungauged basins with remote sensing data.

Catherine Ottlé

Bibliography

Anding, D. and R. Kauth, 1979. Estimation of sea surface temperature from space. *Remote Sensing Environ.*, **1**, 217–220.

André, J.C. *et al.*, 1988. HAPEX-MOBILHY: First results from the Special Observing Period. *Ann. Geophys.*, **6**, 477–492.

André, J.C., P. Bougeault and J.P. Goutorbe, 1990. Regional estimates of heat and evaporation fluxes over non-homogeneous terrain. Examples from the HAPEX-MOBILHY program. *Boundary-Layer Meteorol.*, **50**, 77–108.

BenMehrez, M., O. Taconet, D. Vidal-Madjar and Y. Sucksdorff, 1992a. Calibration of a fluxes model over bare soils during HAPEX-MOBILHY experiment. *Agric. For. Meteorol.*, **58**, 257–283.

BenMehrez, M., O. Taconet, D. Vidal-Madjar and C. Valencogne, 1992b. Estimation of canopy stomatal resistance and canopy evaporation during HAPEX-MOBILHY experiment. *Agric. For. Meteorol.*, **58**, 285–313.

Bernard, R., J.V. Soares and D. Vidal-Madjar, 1986. Differential bare field drainage properties from airborne microwave observation. *Water Resour. Res.*, **22**(6), 869–875.

Carlson, T.N., J.K. Dodd, S.G. Benjamin and J.N. Cooper, 1981. Satellite estimation of the surface energy balance, moisture availability and thermal inertia. *J. Appl. Meteorol.*, **20**, 67–87.

Chkir, N., 1994. Mise au point d'un modèle hydrologique conceptuel intégrant l'état hydrique du sol dans la modélisation pluie-débit. Thèse de L'Ecole Nationale des Ponts et Chaussées de Paris (France), Sciences et Techniques de l'Environnement.

Cognard, A.L., C. Loumagne, M. Normand *et al.*, 1995. Evaluation of the ERS1/synthetic aperture radar capacity to estimate surface soil moisture: two years results over the Naizin watershed., *Water Resour. Res.*, **31**(4), 975–982.

Deardorff, J.W., 1978. Efficient prediction of ground surface temperature and moisture, with inclusion of layer of vegetation. *J. Geophys. Res.*, **83**(C4), 1889–1903.

Deering, D.W., J.W. Rouse, R.H. Haas and J.A. Schall, 1975. Measuring forage production of grazing units from Landsat MSS data. *Proceedings of the 10th International Symposium on Remote Sensing of the Environment*, Ann Arbor, Michigan, p. 1169.

Deschamps, P.Y. and T. Phulpin, 1980. Atmospheric correction of infrared measurements of sea surface temperature using channels at 3.7, 11 and 12 μm. *Boundary Layer Meteorol.*, **18**, 131–143.

Edijatno, M.C., 1989. Un modèle pluie-débit journalier à trois paramètres. *La Houille Blanche*, no. 2, 113–121.

Edijatno, M.C., 1991. Mise au point d'un modèle élémentaire pluie-débit au pas de temps journalier, Thèse de doctorat, Université de Strasbourg, CEMAGREF, 625 pp.

Girard, G., 1974. Modèle global ORSTOM. Première application du modèle journalier à discretisation spatiale sur le bassin versant de la crique Grégoire en Guyane, in *Atelier hydrologique sur les modèles mathématiques*, ORSTOM, Paris.

Girard, G. and B. Boukerma, 1985. *Projet HAPEX–MOBILHY, Calage du Modèle Hydrologique*. Report LHM/RD/85/110 (available from CIG, 35 rue Saint-Honoré, 77305 Fontainebleau Cédex, France).

Ledoux, E., 1980. Modèlisation intégrée des écoulements de surface et des écoulements souterrains sur un bassin hydrologique. Thèse de Docteur-Ingénieur, ENSMP et Université P.M. Curie, Paris, France.

Loumagne, C., M. Normand and C. Michel, 1991. Etat hydrique du sol et prévision des débits. *J. Hydrol.*, **123**, 1–17.

Loumagne, C., N. Chkir, M. Normand *et al.*, 1995. Introduction of the soil–vegetation–atmosphere continuum in a rainfall–runoff model for remote sensing data assimilation. *IAHS Review*, **41**(6), 889–902.

Ottlé, C. and D. Vidal-Madjar, 1994. Assimilation of soil moisture inferred from infrared remote sensing in a hydrological model over the HAPEX–MOBILHY region. *J. Hydrol.*, **158**, 241–264.

Prabhakara, C., G. Dalu and V.G. Kunde, 1974. Estimation of sea surface temperature from remote sensing in the 11–13 μm window region. *J. Geophys. Res.*, **79**, 5039–5044.

Taconet O., D. Vidal-Madjar, Ch. Emblanch and M. Normand, 1995. Taking into account vegetation effects to estimate soil moisture from C-band radar measurements. *Remote Sensing Environ.*

Taconet, O., R. Bernard, and D. Vidal-Madjar, 1986. Evapotranspiration over an agricultural region using a surface flux/temperature model based on NOAA–AVHRR data. *J. Climate Appl. Meteorol.*, **25**, 284–307.

Tanré, D., C. Deroo, P. Duhaut *et al.*, 1990. Description of a computer code to simulate the satellite signal in the solar spectrum: the 5S code. *Int. J. Remote Sensing*, **11**, 659–668.

Cross references

Remote sensing (1)
Remote sensing (2)
Satellite hydrology: present trends and outstanding needs

RESERVOIR CAPACITY ESTIMATION

General

The examination of the natural variability of streamflow may indicate if a stream will frequently be deficient with respect to the estimated water requirements of a particular management project. Low flows may be augmented by reservoir storage. The effectiveness of a reservoir depends primarily upon the rate of withdrawal, called reservoir draft D, the reservoir storage capacity S, and the time-series structure of the streamflows. The relationship between reservoir storage capacity, draft, and the resulting reliability R of the water supply is called the storage equation (WMO, 1973). In reservoir design, any two of these three variables can be taken as independent and assigned specific values. The value of the third variable may be computed from the storage equation for the given hydrological regime. There are several methods for the solution of storage equations that depend on the inflow regime representation, e.g. a historical or a synthetic streamflow record, probabilistic properties of the inflow process, or inflow duration curves.

The usual task in the design of a low-flow augmentation reservoir is to find the reservoir storage capacity necessary to maintain a given rate of draft with a given reliability, i.e. to solve the storage equation of the form

$$S = f(D, R) \qquad (R3)$$

where R is expressed either as a percentage of time of non-failure operation, as a risk that a failure in operation will occur within 1 year or any specified period, or as the amount of water actually supplied in the long run to the consumer, expressed as a percentage of the demand.

The simplest solution is possible if the reliability for the design period can be considered to be 100%, i.e. if no failures in water supply are allowed. Although such a case never occurs in practice, because of its simplicity, it is often used in preliminary computations.

A more realistic case, in which reliability is less than 100% within the design period, cannot be solved directly in the form of equation (R3). The usual method of solution is one of successive approximations of the storage equation in the form $R = f(D, S)$. The solution proceeds in such a way that, keeping one of the independent variables constant, the other is adjusted until the desired value of R is obtained.

In many cases, the value of R is not specified explicitly, but another criterion of reservoir performance is used, e.g. the maximum expected economic gain, the minimum expected economic regret. However, problems of this category are beyond the scope of this article. Approaches to their solutions are presented by Maass *et al.* (1962).

Estimation of water losses from surfacewater systems

Nature of losses

Water losses, e.g. evaporation and seepage, occur under pre-project conditions and are reflected in the streamflow records used for estimating water supplies. The construction of new reservoirs and canals is often accompanied by additional evaporation and infiltration. Estimation of these losses may be based upon measurements at existing reservoirs and canals. The measured inflows, outflows and rates of change of storage are balanced by the computed total loss rate.

In terms of a water balance, seepage and infiltration do not constitute a loss because they contribute to groundwater recharge or to discharge downstream from a river-control structure. However, they do constitute a loss as far as the primary purpose of the project is concerned, e.g. the water lost through seepage is lost for power generation or for withdrawal for water supply. Thus the term loss is to be understood in the water management, rather than the hydrological, sense.

The depth of water evaporated annually from a reservoir surface may vary from about 400 mm in cool, humid climates to more than 2500 mm in hot, arid regions. Therefore evaporation is an important consideration in many projects and deserves careful attention.

Seepage loss from reservoirs and irrigation and navigation canals may be significant if these facilities are located in an area underlain by permeable strata. The reduction of seepage losses can be expensive, and the technical difficulties involved can render a project unfeasible.

Losses from irrigated areas

Water losses in an irrigation system may be several times greater than the water actually utilized by the crops. These losses consist of excess water drained from the land surface or percolating to the groundwater, evaporation from the soil, transpiration by undesirable vegetation, and seepage and evaporation from canals. From 20 to 60% of the water diverted for irrigation may appear as return flow and may contribute either to streamflow or to groundwater recharge. Chemical constituents in irrigation return flows are usually more concentrated than in the original water and may contain additional undesirable elements.

Losses through evaporation from water surfaces and soils may be estimated by various methods, which are reviewed by WMO (1958) Seepage and percolation may be estimated from field observations of groundwater levels, on small pilot or experimental irrigation systems established within the area of interest, or by the water budget method.

Evaporation from reservoirs

Methods for estimating reservoir evaporation include those from pan observations and from meteorological data. In the absence of pan evaporation or other appropriate meteorological observations at or near the reservoir site, regional estimates of these quantities are used to assess reservoir evaporation.

Prior to its submersion by a reservoir, land loses water by evapotranspiration. For reservoir design purposes, it is desirable to estimate increased water loss from the reservoir area due to formation of a lake, i.e. the difference between estimated reservoir evaporation and estimated pre-project evapotranspiration from the land area to be covered by the reservoir. Direct measurement of evapotranspiration presents a number of largely unsolved problems (WMO 1955, 1958a,b, 1983).

Whenever practicable, the minimum-storage surface area per unit volume of storage should be sought in the selection of the dam sites. Extensive research has been conducted into evaporation suppression by the spreading on water surfaces of monomolecular films, but major practical problems in the application of these techniques to large storages still remain unsolved (Fiering, 1967). Thermal stratification in reservoirs and the temperature difference between inflow and outflow can have a significant impact on reservoir evaporation.

Seepage from reservoirs

Seepage from reservoirs depends on the structure and permeability of the underlying strata and on local conditions. An estimate of expected seepage may be derived from evaluation of seepage at existing reservoirs, from geological investigations at the site and from the area–depth relationship for a planned reservoir. Following the construction of a reservoir and after collection of flow and other data, seepage or total losses may be evaluated and a seepage–depth relationship can be derived.

Influence of reservoir location

When the reservoir is to be located at or near the point of diversion or utilization, the estimated flow available for the project may be considered to be the reservoir inflow, while the estimated water requirements and losses may be considered to be the required draft from storage. If, however, the reservoir is to be located some distance upstream, the flow entering the reservoir will represent only a portion of the total flow available. The storage capacity of the reservoir should be based on projected water requirements reduced by the amount that can be supplied from uncontrolled runoff from the drainage area between the reservoir and the point of demand.

The procedures described below apply specifically to the case where storage is located at the point of utilization, but with suitable adaptations these procedures may also be applied in most other cases.

Influence of sedimentation

Sedimentation results in a continual reduction of the reservoir storage capacity. If the rate of sedimentation is small with respect to the capacity of the reservoir, a mean annual volume of sediment can be considered as a constant yearly reduction of storage in the computations. If a large volume of sediment occurs, the decrease in the capacity should be related to the annual streamflow or to each major flood event.

Sequential analysis

The time series used for storage reservoir design can be either historical streamflow records, a historical streamflow record extended by synthesis from another streamflow or precipitation record, or a synthesized reservoir inflow series. Most frequently, the computations are based on time series of mean monthly flows, mean 10-day flows or mean daily flows.

Numerical procedure

The numerical procedure is best organized in a tabular form as shown in Table R7. The computation can be carried out either for a prescribed initial amount of water in storage, e.g. starting with an empty or a full reservoir, or for the so-called steady-state condition, where the initial storage is equal to the final storage at the end of the design period. In the steady-state case, the computations are first carried out with an arbitrary initial storage and are then repeated with the initial storage set equal to the final storage obtained in the first run. The results from the second run represent the steady-state situation.

Table R7 shows a segment of computations for a case where the storage capacity $S = 300 \times 10^6 \, \text{m}^3$ is between a minimum $S_{min} = 2.0 \times 10^6 \, \text{m}^3$ and a maximum $S_{max} = 2.3 \times 10^6 \, \text{m}^3$. It may be assumed that the minimum storage is required, for example, to facilitate

navigation in the reservoir, while the minimum storage must not be exceeded because of the danger of damage to shore property. Thus, whenever the release of the full draft would require the storage to drop below $2 \times 10^6 \, \text{m}^3$, the outflow must be reduced to a rate that prevents a violation of this constraint. Likewise, should the release of the draft cause the storage to rise above $2.3 \times 10^6 \, \text{m}^3$, the outflow must be increased to prevent such a rise.

In the present example, the rate of draft varies with the season of the year (column 7 of Table R7). The reservoir inflow is represented by a series of monthly inflow totals (column 6). Also given are monthly precipitation and evaporation totals in millimeters (columns 2 and 4).

The volumes of precipitation P and evaporation E (columns 3 and 5, respectively) for a given month are computed by using the reservoir surface area at the end of the previous month (column 11). Each row of the table represents the reservoir water balance for one month, i.e. a solution to the storage equation:

$$S_i = S_{i-1} + I_i + P_i - E_i - O_i = S_{i-1} + \Delta S_i \qquad (R4)$$

where the release O_i equals the water demand D_i, subject to the constraint $S_{min} \leq S_i \leq S_{max}$ (the values of P and E are taken from columns 3 and 5).

The violation of the lower constraint is prevented by a reduction of the outflow by the amount $S_{min} - S_i$, which is registered as a water deficit (e.g. column 13 for November). If the constraint S_{min} is removed, the reservoir becomes semi-infinite in the sense of being bottomless. Such an assumption is used to determine the storage capacity required to prevent shortages in water supply for the duration of the inflow series, i.e., for $R = 100\%$. The design storage capacity would equal the maximum storage depletion recorded during the design period. The violation of the upper constraint is prevented by increasing the outflow by an amount, $S_i - S_{max}$, registered as a spill (column 12 for May).

After completing the computations, the water deficits are used to compute the reliability R. The value of R, together with the original values of S and D, represent one solution of the storage equation for the specified input series. Since the same pair of S and D will lead to different values of R for different input series, the value of R which is obtained from the historical record may not be representative of a future period. It is, therefore, advantageous to perform the computations of R for a number (at least 50) of synthetic input series and to take their average as the design value. When confidence limits on R are desired, at least 1000 values should be computed. In such cases, the volume of computations is large, and the tabular form should be computerized. The ease of computerization makes the numerical time series approach the most flexible and powerful tool for reservoir design (WMO 1973; Fiering 1967).

Table R7 Sequential computations of reservoir operation

Year, Month	Precipitation, P		Evaporation, E		Inflow, I	Desired draft, D	Outflow, O	Storage change, ΔS	Storage at end of month, S	Reservoir area	Spills	Water deficits
	(mm)	($10^3 \, \text{m}^3$)	(mm)	($10^3 \, \text{m}^3$)	($10^3 \, \text{m}^3$)	($10^3 \, \text{m}^3$)	($10^3 \, \text{m}^3$)	($10^3 \, \text{m}^3$)	($10^3 \, \text{m}^3$)	(km^2)	($10^3 \, \text{m}^3$)	($10^3 \, \text{m}^3$)
1	2	3	4	5	6	7	8	9	10	11	12	13
1954												
:												
:									2254	0.56	0	0
Sept.	40	22	90	50	20	150	150	−158	2096	0.54	0	0
Oct.	50	27	70	38	20	80	80	−71	2025	0.53	0	0
Nov.	40	21	50	26	20	50	40	−25	2000	0.52	0	10
Dec.	50	26	30	16	30	30	30	+10	2010	0.52	0	0
1955												
Jan.	40	21	30	16	20	20	20	+5	2015	0.53	0	0
Feb.	60	32	40	21	30	20	20	+21	2036	0.53	0	0
Mar.	80	42	50	26	50	20	20	+46	2082	0.54	0	0
Apr.	90	49	70	38	160	20	20	+51	2233	0.55	0	0
May	70	38	90	50	140	20	61	+67	2300	0.56	41	0
:												
:												

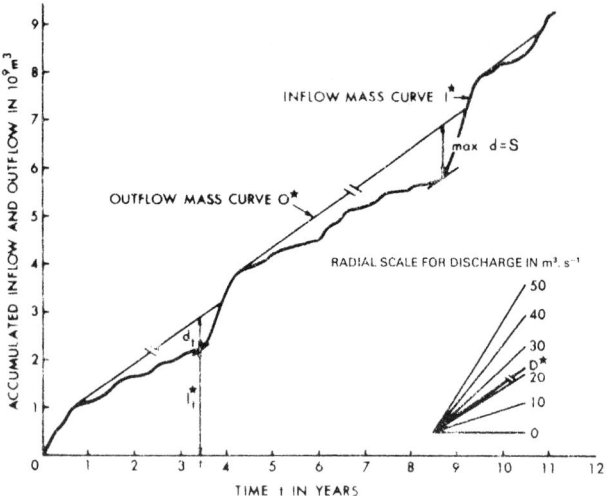

Figure R19 Mass curve approach to the determination of storage capacity S required for supplying constant draft D.

Graphical procedure

In a reservoir subject to an inflow I and an outflow O, the storage S at time t is mathematically defined as

$$S_t = S_o + \int_0^t (I - O)\mathrm{d}\tau = S_o + \int_0^t I\mathrm{d}\tau - \int_0^t O\mathrm{d}\tau = S_o + I_t^* - Q_t^* \quad \text{(R5)}$$

Plots of the cumulative sums I^* and O^* represent the inflow and outflow mass curves, respectively, with S_o being the initial reservoir storage. It can be seen in equation (R3) that the reservoir storage capacity is the difference between the input and output mass curves. An example of this technique is shown in Figure R19 in which the reservoir storage capacity S is determined for a constant draft D with the constraint that no failure is allowed during the design period. The procedure employs the concept of a semi-infinite (bottomless) reservoir and yields the required storage capacity as the minimum storage depletion recorded in an initially full reservoir during the design period. The following graphical procedure applies. The constant draft corresponds to a constant slope of the draft mass curve D^*. A line, parallel to D^*, is drawn through each peak on the inflow mass curve I^*. The design storage capacity S is the maximum vertical distance between any point on I^* and any of the lines that are parallel to D^*.

Probabilistical approach

If streamflow is interpreted as a stochastic process, each realization of this process, i.e. each time series, is governed by the probabilistical properties pertinent to the given process. If these properties are relatively simple, then it is possible to use them directly for reservoir analysis without generating time series. Methods employing this approach are called rigorous and methods ignoring the sequential structure of the process and making use only of its probabilistical distribution are called approximate. Sometimes these processes are called stochastic and probabilistic, respectively (Chow, 1964).

Rigorous methods

Only two stochastic methods have been developed to simulate the reservoir inflow process – the random and the first-order autoregressive inflow models (Lloyd, 1967). The problem has to be posed in the form $R = f(D, S)$. The general procedure is as follows: the storage capacity is divided into k intervals ΔS_i, $i = 1, 2, \ldots, k$, so that $S = \Sigma \Delta S_i$. Each interval, or storage layer, is represented by one value S_i, usually the interval midpoint. The objective of the computations is to find the probability distribution of storage states $S_1, S_2, \ldots, S_k$ for the given values of S and D. The reliability is obtained from this distribution. In finding this distribution, advantage is made of the fact that, whatever the initial storage, the distribution soon reaches an equilibrium or steady state that is independent of the initial storage.

Hence it is sufficient to find the distribution of storage states for any instant t_m of the equilibrium state of the process by first establishing the conditional probability that the reservoir will be in a given state S_i if it had been in state S_j at the time t_{m-1} for each storage state at time t_m. This probability is called the transition probability p_{ij} and can be determined from the probability distribution of the inflow and the value of D. After knowing all the transition probabilities, it is then possible to compute the probability $P_i^{(m)}$ of reservoir state S_i at time t_m from the equation

$$P_i^{(m)} = P_1^{(m-1)} p_{i1} + P_2^{(m-1)} p_{i2} + \ldots + P_k^{(m-1)} p_{ik} \quad \text{(R6)}$$

Such an equation is written for each storage state $i = 1, 2, \ldots, k$. This form of analysis can be employed only for a representation of the inflow process by series of annual, seasonal or monthly flows that can be assumed to be either a random or a first-order autoregressive process.

Approximate methods

The design of storage reservoirs intended for delivering relatively low rates of draft must often be based on a detailed time representation of the inflow process. This representation is typically a series of daily flows with a complex stochastic structure that prevents the use of rigorous probabilistical methods. In such cases, the only alternative to the time-series approach is the use of the approximate probabilistical methods that are applied in the preliminary stages of planning and design. The most common of the approximate methods employ low-flow discharge and volume frequencies and are based on the assumption that sufficient dry periods are separated by wet periods to fill the reservoir before the next dry period starts.

The fact that the lowest flows often occur during one period of the year in sequences seldom interrupted by high flows makes it possible to disregard the actual sequential order and to find the approximate storage capacity necessary for raising the lowest flows to some constant level D from the duration curve. The storage capacity is defined as a volume represented by the area of a wedge bounded by that part of the duration curve for which the flow is less than D and a line defined by flow equal to D. A risk assessment of the storage required for seasonal drought conditions can be made by using low-flow frequency curves. For example, if a mean daily discharge lower than 5 m^3 s^{-1} is permitted on the average only once every 4 years, then the required storage can be determined from curves of the type of Figure R19.

Duration curves and low-flow frequency curves should not be used for ephemeral streams. For these streams and for high-flow perennial streams, the discharge duration frequency curves are more convenient. As an example, if a draft of 3 m^3 s^{-1} is desired from the river, the available volume of 80% of the years can be computed from Figure R20.

The volume available for draft at a given discharge is most conveniently determined from volume duration frequency curves. This method can also be used for a preliminary estimate of storage capacity of a flood-control reservoir. Thus, for example, if mean daily flows higher than 100 m^3 s^{-1} are permitted on the average only once every 5 years (i.e. with a 20% exceedance probability), then the required detention storage can be determined by means of Figure R21.

Storage–draft reliability relationship

To facilitate comparisons and economic evaluations of various alternatives of reservoir design, it is convenient to solve storage equation (R3) for a wide range of the three variables involved and represent the solution in a concise manner. The most usual representation for a given site on a stream is to plot the draft as a function of reservoir storage capacity, by using the reliability characteristic as a parameter. Such relationships can be further generalized by expressing both the draft and the storage capacity as ratios to the mean annual inflow (Figure R22).

Multipurpose reservoirs

Most storage reservoirs serve a number of purposes. It is usually not practicable to allocate a fixed portion of storage for each purpose. In most cases, such an allocation is restricted to emergency purposes. For example, a buffer zone is often created immediately above the dead storage zone and is reserved for use in exceptional circumstances, such as flushing the downstream river section in case of some

The present

Since ancient times, technical advances have provided engineers with the potential to modify significantly the characteristics of a river to meet the demands of society. The economies of many developed and developing countries have been built by exploiting their river systems.

In the USA, for example, the country has benefited considerably from the development of the Mississippi system (*Horizon*, 1994). Here navigation was improved by increasing channel depth to a minimum of 2.7 m, enabling paddle steamers to transport crops from the interior to sea ports. This was achieved by encouraging the channel to self-scour through the construction of groynes to restrict its flow area. The success of this led to the development of the river corridor and the desire to build on the floodplains. With the river in its natural state these are inundated each spring. Successful development therefore required the prevention of this by the construction of flood protection works. The first stage was the construction of levees on the river banks to prevent overtopping. When this proved insufficient, over 200 dams and flood control reservoirs were constructed on the river's tributaries (*Horizon*, 1994).

However, river systems have a strong desire to return to their natural state, the consequence of which is that the long-term effect of large-scale engineering intervention has not always been adequately anticipated. This has resulted in an estimated expenditure of $180 million per year to maintain the Mississippi system (*Horizon*, 1994).

In the developing world, Pakistan provides a good example of the advantages and disadvantages of basing economic development on the exploitation of a river system. Since independence in 1947, it has striven to attain greater material prosperity (Meadows *et al.*, 1994). Success relies heavily on the use of water from the Indus for both irrigation and power generation. After the partition of India at independence the water supply in the head waters of the Chenab, Ravi and Sutlej rivers was diverted into India, depriving large areas of agricultural land of irrigation water. A solution in the form of the Indus Water Treaty was reached some 10 years later. Essentially, this was an aid package from the World Bank and Western governments to enable Pakistan to re-engineer the system. Two large dams were constructed, Mangela on the Jhelum River and Tarbela on the Indus River, along with 50 barrages and around 1000 km of link canals (Pearce, 1992). These divert water across the Punjab to fill the irrigation canals. While solving the problem in terms of water availability, the treaty has increased Pakistan's dependence on large dams. Mangela and Tarbela now provide 45% of the nation's electricity and 50% of its water (Meadows *et al.*, 1994). In common with all large reservoirs, sedimentation is causing a loss of storage capacity. Tarbela is worst affected and is estimated to be losing around 21% of its storage capacity each year. The sediment delta is progressing at an alarming rate, and if not stopped is in danger of encroaching on the intakes for the hydroelectric generators within the next few years. To compensate for this, a new large dam will be required; however, good sites are scarce and those that do exist are likely to meet with strong environmental objections.

In other countries excessive development of their river systems has led to unmitigated environmental disasters. The development of the Volga basin by the former USSR, principally for irrigation, has resulted in a 40% reduction in the surface area of the Aral Sea and has decimated the fishing industry once centred on the town of Muynak (Pearce, 1992; O'Grady, 1995). The exposed salt bed is now the source of dust storms that annually deposit an estimated 75 million tonnes of salt on the surrounding landscape, causing severe problems for agricultural development.

The future

These long-term problems coupled with society's increasing awareness of environmental issues has resulted in new responsibilities for river engineers. This situation was reinforced by the principal outcome of the United Nations Conference on Environment and Development (the Earth Summit) held in Rio de Janeiro in 1992, which was that sustainable development should play a central role in exploitation of the Earth's natural resources.

One noticeable trend (in the developed world) is that of river rehabilitation where rivers that have suffered environmental degradation are restored to a condition close to their natural condition. The most obvious improvement for many developed rivers is in water quality through the control of discharges and the construction of treatment works. Other initiatives however, include engineering environmental improvement schemes, such as meander reinstatement and fish habitat construction (Hoey *et al.*, 1995; O'Grady, 1995; RSPB, 1994).

The need to balance economic development with environmental protection will be a key river engineering issue in future years.

G. Pender

Bibliography

Hansen, V.E., Israelsen, O.W. and Sringham G.E., 1980. *Irrigation Principles and Practices*, 4th edn, John Wiley & Sons, Chichester.

Hoey, T.B., Smart, D.W.J., Pender, G. and Metcalf, N., 1995. *Alternative Methods for River Management for Scottish Rivers*, Scottish Natural Heritage Report SNH/045/95 EBS.

Horizon, 1994. *After the Flood*, Text adapted from the television program transmitted on 18 April 1994, London, British Broadcasting Corporation.

Meadows, P.S., Pender, G. and Meadows, A., 1994. The Indus River and Pakistan's Economy: Energy, Environmental Resources, and Developmental Policy, *Science, Technology and Development*, **12**(2), 40–50, Frank Cass, London.

Micklin, P.P., 1994. The Aral Sea problem, *Proceedings of the Institution of Civil Engineers*, **102**(3), 114–121.

O'Grady, M., 1995. The enhancement of salmonid rivers in the Republic of Ireland, *Journal of Water and Environmental Management* **9**(2), 164–172.

Pearce, F., 1992. *The Dammed*, London, The Bodley Head.

Royal Society for the Protection of Birds (RSPB), National Rivers Authority and Royal Society for Nature Conservation, 1994. *The New Rivers and Wildlife Handbook*

RIVERS

Table R9 gives details of the world's rivers. The location is given of the mouth of the river, the outflow (sea, lake, river, etc.) and the country in which it is located, the total length of the watercourse, and the average discharge. The discharge is given of the main stream averaged over several recent years and measured at the gauging station giving the greatest discharge (or nearest the mouth of the river). Included in the table are rivers with a continuous watercourse of at least 800 km (500 miles) in Africa, Asia and South America, and 480 km (300 miles) in Europe, North America and Oceania. Also included are other well-known rivers, including the longest in certain countries and border or frontier rivers of strategic importance. Non-tributary rivers in the United States of at least 400 km are also given. The lengths of rivers are rounded to the nearest 10 km, drainage basins to the nearest 1000 km^2 and the average discharge to the nearest 10 m^3 s^{-1} (1 m^3 s^{-1} when less than 1000 m^3 s^{-1}).

R.W. Herschy

Source

Showers, V, 1989. *World Facts and Figures*, John Wiley and Sons, New York.

Cross reference

International rivers

Table R9 Longest rivers, by continent (see footnotes for abbreviations)

Latitude and longitude	River	Outflow and location	Length (km)	Drainage basin (1000 km²)	Discharge rate (m³ s⁻¹)
Africa					
31.32N, 31.51E[a]	•Nile (off Nil)–Kagera–Ruvuvu–Luvironza	Mediterranean Sea, Egypt	6670	3349	2 830
6.04S, 12.24E	Congo (alt Kongo)–Lualaba	Atlantic O, Angola–Congo, D.R.	4630	3822	39 000
4.20N, 6.00E	•Niger	G of Guinea, Nigeria	4100	2092	5 700
18.50S, 36.17E	•Zambezi (alt Zambesi, Zambeze)	Mozambique Channel, Mozambique	2650	1331	7 070
0.30S, 17.42E	•Ubangi (alt Oubangui)–Uele–Kibali	Congo R; Congo–Congo, D.R.	2460	773	7 500
28.38S, 16.27E	Orange (alt Oranje)	Atlantic O, Namibia–South Africa	2250	855	215
3.02S, 16.57E	•Kasai (alt Cassai)	Congo R, Congo, D.R.	1930	904	9 950
0.12N, 42.45E	•Shebele (alt Shabale, Shabeelle, Shibeli: for Scebeli)	Balli Swamp, Somalia	1930	200	320
15.48N, 16.32W	Senegal–Bafing	Atlantic O, Mauritania–Senegal	1700	440	815
18.53S, 22.24E	•Okovanggo (alt Cubango, Okavango)	Okovango Basin, Botswana	1610	785	255
5.47N, 0.43E	•Volta-Black Volta (alt Volta Noire)	G of Guinea, Ghana	1600	398	1180
25.12S, 33.32E	•Limpopo (alt Crocodile)	Indian O, Mozambique	1590	412	169
0.15S, 42.38E	•Juba (alt Ganana Jubba; for Giuba)–Ganale–Dorya (alt Genale)	Indian O, Somalia	1560	196	200
6.46S, 26.58E	•Luvua–Luapula–Chambezi	Lualaba R, Congo, D.R.	1500	250	
15.38N, 32.31E	•Blue Nile (off Abay,Azraq)	Nile R, Sudan	1450	331	1620
0.46N, 24.16E	•Lomami	Congo R, Congo, D.R.	1450	110	
12.58N, 14.31E	•Shari (alt Chari)–Sara (alt Ouham)	Chad L, Cameroon–Chad	1450	700	1230
8.00S, 39,20E	•Rufiji–Luwegu	Indian O, Tanganyika, Tanzania	1400	178	973
1.13S, 16.49E	•Sangha–Kadei	Congo R, Congo Republic	1400	181	1800
7.50N, 6.50E	•Benue (alt Benoue)	Niger R, Nigeria	1300	337	3 170
1.13N, 23.36E	•Aruwimi–Ituri	Congo R, Congo, D.R.	1290	116	
0.05N, 18.17E	•Ruki-Busira-Tshuapa	Congo R, Congo, D.R.	1290	174	
0.49S, 9.00E	•Ogooeu (alt Ogowe)	G of Guinea, Gabon	1210	216	4 670
4.17S, 20.25E	•Sankuru–Lubilash	Kasai R, Congo, D.R.	1210	156	2 500
29.04S, 23,38E	•Vaal	Orange R, Cape, South Africa	1210	155	36
11.35N, 41.38E	Awash (alt Hawash)	Abbe L, Ethiopie	1200	55	160
13.27N, 16.37E	Gambia (alt Gambie)	Atlantic O, Gambia	1130	182	
3.14S, 17.22E	•Kwango (alt Cuango)	Kasai R, Congo, D.R.	1130	263	2 700
17.40N, 33.58E	•Atbarah (alt Atbara)–Takaze (off Satit, Tekeze)	Nile R, Sudan	1120	69	389
14.30N, 4.12W	Bani-Bagoe	Niger R, Mali	1110	130	796
3.01S, 16.58E	•Fimi–Lukenie	Kasai R, Congo, D.R.	1070	132	
3.22S, 17.22E	•Kwilu (alt Cuilo)	Kwango R, Congo, D.R.	1050	90	71
13.49N, 10.50W	•Bakoye–Baoule	Senegal R, Mali	1000	95	
17.47S, 25.10E	•Chobe (alt Linyanti)–Kwando (alt Cuando)	Zambezi R, Botswana–Namibia	1000	56	
15.56S, 28.55E	•Kafue	Zambezi R, Zambia	970	150	
28.31S, 20.13E	•Molopo	Orange R, Cape, South Africa	970		
9.19S, 13.08E	Cuanza	Atlantic O, Angola	960	156	835
12.06N, 15.02E	•Logone–Mbere	Shari R, Chad	960	76	403
17.20S, 11.50E	Cunene (alt Kunene)	Atlantic O, Angola–Namibia	940	83	215
3.35N, 9.38E	Sanaga–Lom	Bight of Bonny of G of Guinea, Cameroon	920	135	2 190
5.02S, 21.07E	•Lulua	Kasai R, Congo, D.R.	890	65	
9.10N, 1.15W	•White Volta (alt Volta Blanche)	Volta R, Ghana	890	105	240
3.17N, 9.54E	•Nyong	Bight of Bonny of G of Guinea, Cameroon	860	26	125
0.43N, 18.23E	•Lulonga–Lopori	Congo R, Congo, D.R.	820	66	
4.28S, 11.41E	•Kouilou–Niari	Atlantic O, Congo Republic	810	56	940
5.10N, 5.00W	•Bandama	G of Guinea, Ivory Coast	800	60	327
25.46S, 32.43E	•Komati (alt Incomati)	Delagoa B of Indian O, Mozambique	800	46	73
0.14S, 20.42E	•Lomela	Busira R, Congo, D.R.	800		
15.36S, 30.25E	•Luangwa (alt Aruangua)	Zambezi R, Zambia	800	145	
4.31N, 35.59E	Omo	Rudolf L, Ethiopia	760	67	
9.22N, 31.33E	•Sobat–Baro	Nile R, Sudan	740	245	
10.29S, 40.28E	Ruvuma (alt Rovuma)	Indian O, Mozambique–Tanzania	730	155	
4.08N, 22.27E	•Bomu (alt Mbomou)	Ubangi R, Central African Republic–Congo, D.R.	720	250	
2.32S, 40.31E	•Tana	Indian O, Kenya	710	42	151
5.12N, 3.44W	•Comoe (alt Komoe)	G of Guinea, Ivory Coast	700	74	274
6.29N, 2.32E	•Oueme	Porto-Novo Lagoon, Benin	700	40	170

Table R9 Continued

Latitude and longitude		River	Outflow and location	Length (km)	Drainage basin (1000 km²)	Discharge rate (m³ s⁻¹)
36.02N,	0.08E	•Sheliff (alt Cheliff; off Shalaf)	Mediterranean Sea, Algeria	700	35	17
11.24N,	4.07E	Sokoto	Niger R, Nigeria	630	83	71
21.00S,	35.02E	Sabi (alt Save)	Mozambique Channel, Mozambique	610	88	158
13.31S,	40.32E	Lurio	Mozambique Channel, Mozambique	600	61	232
33.19N,	8.20W	•Oum er Rbia	Atlantic O, Morocco	600	34	130
4.58N,	6.05W	•Sassandra–Tienba	G of Guinea, Ivory Coast	600	50	325
17.42S,	35.19E	•Shire (alt Chire)	Zambezi R, Mozambique	600	130	589
11.45N,	15.35W	•Geba-Corubal (alt Koliba)	Atlantic O, Guinea-Bissau	560	8	
21.29S,	43.41E	•Mangoky	Mozambique Channel, Madagascar	560	50	459
29.14S,	31.30E	•Tugela	Indian O, Natal, South Africa	560	29	149
7.48N,	0.08E	•Oti	Volta R, Ghana	550	73	500
35.06N,	2.20W	•Moulouya	Mediterranean Sea, Morocco	520	54	44
4.22N,	7.32W	•Cavally (alt Cavalla)	Atlantic O, Ivory Coast–Liberia	510	30	
6.16N,	1.49E	Mono	Bight of Benin of G of Guinea, Benin–Togo	500	25	96
34.16N,	6.41W	•Sebou	Atlantic O, Morocco	450	39	200
8.30N,	13.15W	•Sierra Leone–Rokel	Atlantic O, Sierra Leone	440		
37.07N,	10.13E	•Medjerda (off Majardah)	G of Tunis of Mediterranean Sea, Tunisia	360	23	19

Asia

Latitude and longitude		River	Outflow and location	Length (km)	Drainage basin (1000 km²)	Discharge rate (m³ s⁻¹)
31.48N,	121.10E	Yangtze (off Chang)	East China Sea, Jiangsu, China	5980	1827	32 190
71.50N,	84.40E	Yenisey (alt Yemisei)–Angara[b]	Yenisey G of Kara Sea, Russia	5870	2580	17 600
52.65N,	141.10E	Amur (alt Heilong, Heilung)–Argun (alt Ergun, Oerhkuna)–Kerulen (alt Kolulun; off Herlen)	Tatar Strait, Russia	5780	1855	9 860
66.45N,	69.30E	Ob-Irtysh[c] (alt Irtish)	G of Ob of Kara Sea, Russia	5410	2990	12 400
37.32N,	118.19E	Huang (alt Hwang, Yellow)	G of Chihli of Yellow Sea, Shandong, China	4840	771	1530
72.25N,	126.40E	Lena	Laptev Sea, Russia	4400	2490	16 600
10.15N,	105.55E	•Mekong (alt Khong, Lancang, Lantsand, Mekongk, Tien Giang)	South China Sea, Vietnam	4180	811	14 200
61.04N,	68.52E	Ob (upper)–Katun	Ob R, Russia	3180	765	4 920
46.03N,	61.00E	Syr (Darya)-Naryn–Bolshoy Naryn	Aral (Sea) L, Kazakhstan	3020	219	730
65.48N,	88.04E	LowerTunguska (off Nizhnyaya Tunguska)	Yenisey R, Russia	2990	473	3 300
24.20N,	67.47E	Indus (alt Yintu)	Arabian Sea, Pakistan	2880	1165	6 640
22.50N,	90.50E[d]	Brahmaputra (alt Tsangpo, Yarlung Zangbo, Yalutsangpu)	B of Bengal, Bangladesh	2840	580	19 200
16.31N,	97.37E	•Salween (alt Khong, Lu, Nu)	G of Martaban of Andaman Sea, Myanmar	2820	324	10 000
41.05N,	86.40E	Tarim (alt Talimu)–Yarkant (alt Yeherhchiang; for Yarkand)	Tarim Basin, Xinjiang, China	2750	447	146
64.24N,	126.26E	Vilyuy (alt Vilyui)	Lena R, Russia	2650	454	1470
43.40N,	59.01E	Amu (Darya) (alt Oxus)–Panj (alt Pyandzh)–Vakhan	Aral (Sea) L, Uzbekistan	2540	309	2 000
21.55N,	88.05E[e]	•Ganges (alt Ganga)–Bhagirathi	B of Bengal, Bangladesh–India	2510	952	11 650
69.30N,	161.00E	Kolyma (alt Kolima)–Kulu	East Siberian Sea, Russia	2510	647	2 250
57.42N,	71.12E	Ishim	Irtysh R, Russia	2450	177	55
29.57N,	48.34E	(Shatt al) Arab–Euphrates (off Firat, Furat)–Kura Su	Persian G, Iran–Iraq	2430	1105	2 860
47.00N,	51.48E	Ural	Caspian (Sea) L, Kazakhstan	2430	237	360
63.28N,	129.35E	Aldan	Lena R, Russia	2270	729	5 010
73.00N,	119.55E	Olenek	Laptev Sea, Russia	2270	219	820
57.43N,	83.51E	Chulym (alt Chulim)–Belyy lyus	Ob R, Russia	2020	134	773
15.50N,	95.06E	•Irrawaddy (off Iyawadi)–Nmai	Andaman Sea	1990	409	12 660
59.26N,	112.34E	Vitim-Vitimkan	Lena R, Russia	1980	225	1520
70.48N,	148.54E	Indigirka-Khastakh	East Siberian Sea, Russia	1970	360	1570
22.45N,	113.37E	Zhu (alt Canton, Chu, Pearl, Yueh)-Xi (alt Hsi, Si, West)–Hongshui (alt Hungshui)–Nanpan	South China Sea, Guangdon, China	1960	426	12 500
61.36N,	90.18E	Stony Tunguska (off Podkamen–naya Tunguska)	Yenisey R, Rusia	1860	240	1690
47.42N,	132.30E	Sungari (alt Sunghua; off Songhua)	Amur R, Heilongjiang, China	1860	524	2 450
31.00N,	47.25E	Tigris (off Dicle, Dijlah)	(Shatt al) Arab R, Iraq	1850	373	1250
72.56N,	106.00E	Khatanga-Kotuy	Khatanga G of Laptev Sea, Russia	1640	364	3 320
58.10N,	68.12E	Tobol	Irtysh R, Russia	1630	426	802
58.06N,	93.00E	Yenisey (upper)	Yenisey R, Russia	1630	299	2 910

Table R9 Continued

Latitude and longitude	River	Outflow and location	Length (km)	Drainage basin (1000 km²)	Discharge rate (m³ s⁻¹)
58.55N, 81.32E	Ket	Ob R, Russia	1620	94	445
70.51N, 153.34E	Alazeya–Kadylchan	East Siberian Sea, Russia	1600	65	300
53.20N, 121.26E	Shilka–Onon	Amur R, Russia	1590	206	532
25.30N, 81.53E	•Yamuna (alt Jamna)–Chambal	Ganges R, Uttar Pradesh, India	1530	359	
28.57N, 70.30E	Panjnad–Sutlej	Indus R, Pakistan	1520	533	3 080
30.34N, 114.17E	Han (Shui)	Yangtze R, Hubei, China	1500	174	1200
71.31N, 136.32E	Yana–Sartang	Laptev Sea, Russia	1490	238	935
62.38N, 134.32E	Amga	Aldan R, Russia	1460	69	170
17.00N, 81.45E	Godavari	B of Bengal, Andhra Pradesh, India	1460	313	3 180
45.24N, 74.08E	Ili-Tekes	Balkhash L, Kazakhstan	1440	140	470
60.22N, 120.42E	Olekma	Lena R, Russia	1440	210	1000
15.57N, 80.59E	Krishna (alt Kistna)	B of Bengal, Andhra Pradesh, India	1400	259	1990
67.32N, 78.40E	Taz	G of Ob of Kara Sea, Russia	1400	150	905
57.47N, 67.16E	Tavda–Lozva	Tobol R, Russia	1360	88	429
40.39N, 122.12E	Liao	G of Liadong of Yellow Sea, Liaoning, China	1340	215	624
58.06N, 94.01E	Taseyeva–Chuna (alt Uda)	Andara R, Russia	1320	128	736
26.37N, 101.48E	Yalong (alt Yalung)	Yangtze R, Sichuan, China	1320	144	2 500
21.38N, 72.36E	Narmada (alt Narbada)	G of Cambay of Arabian Sea, Gujarat, India	1310	93	1290
45.00N, 67.44E	Chu–Dzhuvanaryk	Sauma(kol) L, Kazakhstan	1300	62	61
50.15N, 127.35E	Zeya	Amur R, Russia	1240	233	1790
41.45N, 35.59E	Kizil (Irmak)	Black Sea, Turkey	1180	77	133
63.28N, 118.50E	Markha	Vilyuy R, Russia	1180	99	375
45.26N, 124.39E	Nen (alt Nonni, Nun)	Sungari R, Heilongjiang, China	1170	244	700
29.23N, 71.02E	Chenab–Chandra	Panjnad R, Pakistan	1160	138	2 050
59.34N, 69.17E	Demyanka	Irtysh R, Russia	1160	35	
64.54N, 176.13E	Anadyr (alt Anadir)	G of Anadyr of Bering Sea, Russia	1150	191	1680
37.24N, 60.38E	Hari (Ryud) (alt Tedzhen)	Kara (Kum) Desert, Turkmenistan	1150	71	24
20.17N, 106.34E	Red (alt Coi, Koi; off Hong, Yuan)	G of Tonkin of South China Sea, Vietnam	1150	120	630[1]
29.26N, 113.08E	Siang (off Hsiang)	Tungting L, Hunan, China	1150	100	2 500
50.21N, 106.05E	Orhon (alt Orkhon)	Selenga R, Mongolia	1120	133	120
31.12N, 61.34E	•Helmand (alt Helmund, Hirmand)	Seistan Basin, Afghanistan–Iran	1110	259	
68.42N, 158.36E	Omolon	Kolyma R, Russia	1110	113	
60.40N, 69.46E	Konda	Irtysh R, Russia	1100	73	270
33.12N, 118.33E	Hwai (off Huai)	Hungtze L, Anhwei, China	1090	210	1090
54.59N, 73.22E	Om	Irtysh R, Russia	1090	53	61
63.46N, 121.35E	Tyung	Vilyuy R, Russia	1090	50	
25.47N, 84.37E	Ghaghara (alt Gogra, Kauriala)	Ganges R, Bihar-Uttar Pradesh, India	1080	127	
59.07N, 80.46E	Vasyugan	Ob R, Russia	1080	62	343
60.55N, 73.40E	Bolshoy Yugan	Ob R, Russia	1060	35	138
30.35N, 71.49E	•Ravi	Chenab R, Pakistan	1060		250
54.30N, 134.38E	Maya	Aldan R, Russia	1050	171	1170
40.30N, 80.48E	•Khotan (off Hotien)–Kara-Kash (off Kalakashih)	Tarim R, Sinkiang, China	1030		
57.12N, 66.56E	Tura	Tobol R, Russia	1030	80	174
67.31N, 77.55E	Pur-Pyakupur	G of Ob of Kara Sea, Russia	1020	112	875
57.43N, 95.24E	Biryusa (alt Ona)	Taseyeva R, Russia	1010	56	344
0.25S, 109.40E	Kapuas (alt Kapuas-Besar; for Kapoeas)	South China Sea, Borneo I, Indonesia	1010	102	
41.48N, 86.47E	•Konche (Darya) [alt Kuruk (Darya); off Kungchiao]-Khaydyk(gol) (off Kaitu)	Tarim Basin, Sinkiang, China	1010		
37.23N, 50.11E	•Safid-Qezel Owzan	Caspian (Sea) L, Iran	1000	58	130
29.34N, 106.35E	Kialing (off Chialing)	Yangtze R, Szechwan, China	1000	160	2 500
13.32N, 100.36E	Chao Phraya (alt Menam)-Nan	G of Siam, Thailand	990	150	883
38.18N, 61.12E	Murgab (alt Morghab)	Kara (Kum) Desert, Turkmenia	980	47	52
50.30N, 69.59E	Nura	Tengiz L, Kazakhstan	980	61	18
38.57N, 117.43E	Hai-Pai (alt Pei)	G of Chihli of Yellow Sea, Hopeh, China	970	208	234
60.45N, 76.45E	Vakh	Ob R, Russia	960	77	448
73.08N, 113.36E	Anabar	Laptev Sea, Russia	940	100	498
60.22N, 120.50E	Chara	Olekma R, Russia	920	86	626
71.54N, 102.06E	Kheta	Khatanga R, Russia	920	120	

Table R9 Continued

Latitude and longitude	River	Outflow and location	Length (km)	Drainage basin (1000 km²)	Discharge rate (m³ s⁻¹)
29.43N, 107.24E	Wu	Yangtze R, Szechwan, China	920	88	1500
48.28N, 135.02E	Ussuri (alt Wusuli)–Ulakhe	Amur R, China–Russia	910	187	953
56.36N, 66.24E	Iset-Miass	Tobol R, Russia	900	82	69
28.46N, 104.38E	•Min-Tatu	Yangtze R, Szechwan, China	900	134	
66.30N, 87.12E	Kureyka (alt Kureika)	Yenisey R, Russia	890	45	710
39.20N, 119.10E	Luan	G of Chihli of Yellow Sea, Hopeh, China	880	46	140
39.32N, 63.45E	Zeravshan	Kyzyl(kum) Desert, Uzbekistan	880	18	162
51.19N, 106.59E	Khilok	Selenga R, Russia	870	38	105
29.12N, 116.00E	Kan-Kung	Poyang L, Kiangsi, China	860		2 500
34.41N, 110.10E	Wei	Yellow R, Shensi, China	860	63	
28.58N, 111.49E	Yuan	Tungting L, Hunan, China	860		2 500
52.52N, 83.36E	Aley (alt Alei)	Ob R, Russia	860	21	34
20.19N, 86.45E	Mahanadi	B of Bengal, Orissa, India	860	132	2 120
30.25N, 48.12E	•Karun	(Shatt al) Arab R, Iran	850	61	522
21.26N, 95.15E	•Chindwin	Irrawaddy R, Myanmar	840	114	
75.41N, 99.20E	Taymyra (alt Taimyra)	Kara Sea, Russia	840	124	988
56.50N, 84.27E	Tom	Ob R, Russia	830	62	1120
68.44N, 103.42E	Moyyero (alt Moyero)	Kotuy R, Russia	820	31	
73.50N, 87.10E	Pyasina	Kara Sea, Russia	820	182	2 600
41.07N, 30.39E	Sakarya	Black Sea, Turkey	820	55	257
64.10N, 65.28E	Severnaya Sosva–Bolshaya Sosva	Ob R, Russia	820	98	860
48.57N, 104.48E	Tuul (alt Tola)	Orhon R, Mongolia	820	50	
59.08N, 135.06E	Yudoma–Nitkan	Maya R, Russia	820	44	
48.54N, 93.23E	Dzavhan (alt Dzabkhan)	Ayrag L, Mongolia	810	71	60
58.44N, 81.35E	Parabel–Kenga	Ob R, Russia	810	25	90
56.42N, 74.36E	Tara	Irtysh R, Russia	810	18	41
58.48N, 130.35E	Uchur	Aldan R, Russia	810	113	1300
39.55N, 124.20E	Yalu (alt Amnok)	Korea B of Yellow Sea, China–North Korea	810	63	1040
21.15N, 105.20E	Black (alt Lihsien; off Da, Lixian)	Red R, Vietnam	800		
62.54N, 111.06E	Chona	Vilyuy R, Russia	800	41	109
25.32N, 83.10E	•Gomati (alt Gumti)	Ganges R, Uttar Pradesh, India	800	19	
1.16S, 104.05E	•Hari (for Djambi)	Berhala Strait, Sumatra I, Indonesia	800	53	1500
11.09N, 78.52E	•Kaveri (alt Cauvery)	B of Bengal, Tamil Nadu, India	800	80	934
64.57N, 124.36E	Linde	Lena R, Russia	800	20	
60.32N, 116.14E	Nyuya	Lena R, Russia	800	38	115
48.36N, 52.30E	Uil	Aralsor L, Kazakhstan	800	31	
68.23N, 145.50E	Uyandina–Irgichyan	Indigirka R, Russia	800	41	128
Europe					
45.45N, 47.52E	Volga	Caspian (Sea) L, Russia	3530	1360	8 060
45.20N, 29.40E	Danube (off Donau, Duna, Dunaj, Dunarea, Dunav, Dunay)	Black Sea, Romania, Russia	2860	816	6 250
46.30N, 32.18E	Dnieper (off Dnepr)	Black Sea, Ukraine	2200	504	1610
47.04N, 40.30E	Don	Sea of Azov of Black Sea, Russia	1870	422	873
64.32N, 40.30E	Northern Dvina (off Severnaya Dvina)–Vychegda (alt Vichegda)	White Sea, Russia	1860	357	3 400
68.13N, 54.15E	Pechora (for Petchora)	Barents Sea, Russia	1810	322	4 000
55.25N, 50.40E	Kama	Volga R, Russia	1800	507	2 800
56.20N, 43.59E	Oka	Volga R, Russia	1500	245	1300
55.54N, 53.33E	Belaya	Kama R, Russia	1430	142	845
39.24N, 49.19E	Kura	Caspian (Sea) L, Azerbaijan	1360	188	570
46.18N, 30.17E	Dniester (off Dnestr)	Black Sea, Ukraine	1350	72	293
51.47N, 4.10Eᶠ	Rhine (off Rhein, Rhin, Rijn)	North Sea, Netherlands	1320	252	2 490
55.36N, 51.30E	Vyatka (alt Viatka)	Kama R, Russia	1310	129	866
54.21N, 18.56E	Vistula (for Visla, Weichsel; off Wisla)–Bug (alt Zapadnyy Bug)	G of Danzig of Baltic Sea, Poland	1200	194	1040
53.50N, 9.00E	Elbe (alt Labe)	North Sea, Niedersachsen–Schleswig-Holstein, Germany	1160	144	703
50.33N, 30.32E	Desna	Dnieper R, Ukraine	1130	89	346
39.56N, 48.20E	Araks (alt Aras)	Kura R, Azerbaijan	1070	102	285
47.35N, 40.54E	Donets (alt Northern Donets; off Severnyy Donets)	Don R, Russia	1050	99	153

Table R9 Continued

Latitude and longitude	River	Outflow and location	Length (km)	Drainage basin (1000 km²)	Discharge rate (m³ s⁻¹)
47.16N, 2.11W	Loire	B of Biscay, Pays de la Loire, France	1020	120	871
57.00N, 24.00E	Western Dvina (alt Daugava; for Duna; off Zapadnaya Dvina)	G of Riga of Baltic Sea, Latvia, Russia	1020	88	603
38.40N, 9.24W	Tagus (off Tajo, Tejo)	Atlantic O, Portugal	1010	81	128
59.57N, 30.20E	Neva–Volkhov–Lovat	G of Finland of Baltic Sea, Russia	1000	281	2 480
45.30N, 28.12E	Prut (alt Prutul; for Pruth)	Danube R, Romania–Russia	990	27	70
49.36N, 42.19E	Khoper	Don R, Russia	980	61	151
66.11N, 43.59E	Mezen	White Sea, Russia	970	78	642
45.15N, 20.17E	Tisza (alt Tisa, Tissa; for Theiss)	Danube R, Serbia, Yugoslavia	970	157	844
51.47N, 4.10Eᵍ	Meuse (alt Maas)	North Sea, Netherlands	950	49	269
53.32N, 14.38E	Oder (alt Odra)–Warta (for Warthe)	Baltic Sea, Germany–Poland	950	119	560
55.18N, 21.23E	Neman (alt Nemunas; for Memel, Niemen)	Baltic Sea, Lithuania	940	98	578
44.50N, 20.28E	Sava (alt Save; for Sau, Szava)	Danube R, Serbia, Yugoslavia	940	96	1700
54.40N, 56.00E	Ufa	Balaya R, Russia	920	53	390
40.43N, 0.54E	Ebro	Mediterranean Sea, Cataluna, Spain	910	85	173
45.20N, 37.22E	Kuban	Sea of Azov of Black Sea, Rusia	910	58	425
41.08N, 8.40W	Douro (alt Duero)	Atlantic O, Portugal	890	98	312
56.18N, 46.24E	Vetluga	Volga R, Russia	890	39	231
56.06N, 46.00E	Sura	Volga R, Russia	840	67	207
64.08N, 41.54E	Pinega–Belaya	Northern Dvina R, Russia	820	42	357
43.20N, 4.50E	Rhone	Mediterranean Sea, Languedoc–Provence–Cote d'Azur, France	810	99	1500
46.59N, 31.58E	Southern Bug (off Yuzhnyy Bug)	Dniepe R, Ukraine	810	64	83
44.55N, 46.32E	Kuma	Caspian (Sea) L, Russia	800	33	11
46.15N, 20.12E	Maros (alt Mures, Muresul)	Tisza R, Hungary	800	30	154
51.46N, 55.01E	Sakmara	Ural R, Russia	800	30	133
47.31N, 40.45E	Sal	Don R, Russia	800	21	12
37.14N, 7.22W	Guadiana	G of Cadiz of Atlantic O, Portugal–Spain	780	68	91
49.26N, 0.26E	Seine	English Channel, Haute–Normandie, France	780	79	272
51.10N, 30.30E	Pripet (off Pripyat)	Dnieper R, Ukraine	770	114	370
51.27N, 32.34E	Seym (alt Seim)	Desna R, Ukraine	750	27	103
49.35N, 42.41E	Medveditsa	Don R, Russia	740	35	71
43.43N, 24.51E	Olt (for Aluta)	Danube R, Romania	740	24	160
53.32N, 8.34E	Weser–Werra	North Sea, Niedersachsen, Germany	730	46	334
57.42N, 11.52E	Gota-Kiar	Kattegat, Sweden	720	50	640
52.35N, 14.39E	Oder (upper)	Oder R, Poland	720	54	
49.01N, 33.32E	Psel	Dnieper R, Ukraine	720	23	54
45.33N, 18.55E	Drava (alt Drau, Drave)	Danube R, Croatia	710	40	611
45.24N, 28.01E	Siret (for Sereth; off Seret, Siretul)	Danube R, Romania	710	48	400
51.30N, 53.22E	Ilek–Zharyk	Ural R, Russia	700	41	36
59.27N, 28.02E	Narva–Velikaya	G of Finland of Baltic Sea, Estonia	700	56	415
56.10N, 42.58E	Klyazma	Oka R, Russia	690	42	202
52.31N, 21.05E	Vistula (upper)	Vistula R, Poland	680	85	
52.01N, 47.24E	Bolshoy Irgiz	Volga R, Russia	670	24	24
52.08N, 27.17E	Goryn	Pripet R, White Russia	660	28	88
36.47N, 6.22W	Guadalquivir	G of Cadiz of Atlantic O, Andalucia, Spain	660	57	182
54.44N, 41.53E	Moksha	Oka R, Russia	660	51	154
65.57N, 56.55E	Usa–Bolshaya Usa	Pechora R, Russia	660	94	1030
45.35N, 1.03W	Girdone–Garonne (alt Garona)	B of Biscay, Aquitaine–Poitou–Charentes, France	650	85	590
51.57N, 30.48E	Sozh	Dnieper R, Ukraine–Belarus	650	42	207
43.44N, 46.33E	Terek	Caspian (Sea) L, Russia	620	43	224
44.57N, 12.04E	Po	Adriatic Sea, Veneto, Italy	620	75	1540
52.33N, 30.14E	Berezina	Dnieper R, White Russia	610	24	127
59.12N, 10.57E	Glomma (off Glama)	Skagerrak, Norway	600	42	720
60.30N, 32.48E	Svir-Suna	Ladoga L, Russia	600	84	617
58.13N, 56.22E	Chusovaya	Kama R, Russia	590	48	226
53.10N, 50.04E	Samara	Volga R, Russia	590	46	47
62.48N, 4.56E	Vaga	Northern Dvina R, Russia	570	45	396
60.45N, 46.20E	Yug	Northern Dvina R, Russia	570	36	300
65.48N, 24.08E	Torne (alt Tornio)–Muonio–Konkama (alt Kongama)	G of Bothnia of Baltic Sea, Finland–Sweden	570	39	366
60.46N, 46.24E	Sukhona	Northern Dvina R, Russia	560	50	452

Table R9 Continued

Latitude and longitude	River	Outflow and location	Length (km)	Drainage basin (1000 km²)	Discharge rate (m³ s⁻¹)
52.53N, 11.58E	Havel–Spree	Elbe R, Germany	550	24	90
46.41N, 32.50E	Ingulets	Dnieper R, Ukraine	550	14	9
65.47N, 24.30E	Kemi	G of Bothnia of Baltic Sea, Finland	550	51	578
50.22N, 7.36E	Moselle (alt Mosel)	Rhine R, Rheinland-Pfalz, Germany	550	28	292
44.43N, 21.03E	Morava (off Velika Morava)–Southern Morava (off Juzna Morava)	Danube R, Serbia, Yugoslavia	540	37	253
65.19N, 52.54E	Izhma	Pechora R, Russia	530	31	196
57.20N, 43.08E	Unzha–Kema	Volga R, Russia	530	27	176
50.00N, 8.18E	Main	Rhine R, Hessen, Germany	520	27	100
48.49N, 2.24E	Marne	Seine R, Region Parisienne, France	520	14	98
60.38N, 17.27E	Dal	G of Bothnia of Baltic Sea, Sweden	520	29	333
51.31N, 39.05E	Voronezh–Polnoy Voronezh	Don R, Russia	520	22	71
48.35N, 13.28E	Inn	Danube R, Bayern, Germany	510	26	735
65.36N, 44.35E	Peza-Rochuga	Mezen R, Russia	510	15	130
54.54N, 23.53E	Viliya (alt Neris; for Wilja)	Neman R, Lithuania	510	25	188
39.09N, 0.14W	Jucar	Mediterranean Sea, Valencia, Spain	500	21	60
55.05N, 38.51E	Moscow (off Moskva)	Oka R, Russia	500	18	64
40.52N, 26.12E	Maritsa (alt Evros, Meric)	Aegean Sea, Greece–Turkey	490	35	
44.18N, 0.20E	Lot	Garonne R, Aquitaine, France	480	10	128
52.26N, 20.42E	Narew (alt Narev)	Bug R, Poland	480	75	316
45.44N, 4.50E	Saone	Rhone R, Rhone–Alpes, France	480	30	424
45.02N, 0.35W	Dordogne	Garonne R, Aquitaine, France	470	23	286
63.47N, 20.16E	Ume	G of Bothnia of Baltic Sea, Sweden	460	27	450
48.50N, 34.05E	Vorskla	Dnieper R, Ukraine	460	15	30
58.50N, 37.11E	Mologa	Volga R, Russia	460	30	314
62.48N, 17.56E	Angerman	G of Bothnia of Baltic Sea, Sweden	450	32	448
65.35N, 22.03E	Lule	G of Bothnia of Baltic Sea, Sweden	450	25	444
58.5N, 31.20E	Msta	Ilman L, Russia	440	23	161
50.45N, 21.51E	San	Vistula R, Poland	440	17	131
46.18N, 16.55E	Mur (alt Mura)	Drava R, Croatia	440	14	166
51.57N, 11.55E	Saale (alt Sachsische Saale)	Elbe R, Germany	430	24	105
51.22N, 4.15E	Scheldt (aolt Escaut, Schelde)	North Sea, Netherlands	430	20	155
48.07N, 22.20E	Somes (off Somesul, Szamos)	Tisza R, Romania	430	9	80
50.22N, 14.28E	Vltava (alt Moldau)	Elbe R, Bohemia, Czech Republic	430	28	145
63.58N, 38.02E	Onega	Onega B of White Sea, Russia	420	57	493
40.35N, 22.50E	Vardar (alt Axios, Vardaris)	G of Salonika of Aegean Sea, Greece	420	28	135
45.10N, 12.20E	Adige (for Etsch)	Adriatic Sea, Veneto, Italy	410	15	262
44.42N, 27.51E	Ialomita	Danube R, Romania	410	12	70
41.44N, 12.14E	Tiber (off Tevere)	Tyrrhenian Sea, Lazio, Italy	410	17	239
47.55N, 18.00E	Vah (for Vag, Waag)	Danube R, Slovakia	390	11	158
52.30N, 9.55W	Shannon	Atlantic O, Ireland	370	16	198
70.30N, 28.23E	Tana	Tana Fjord of Barents Sea, Norway	360	16	190
5.25N, 3.00W	Severn	Bristol Channel, England, UK	340	21	62
53.32N, 0.08E	•Humber–Trent	North Sea, England, UK	330	23	198
38.06N, 0.38W	Segura	Mediterranean Sea, Valencia, Spain	320	16	6
51.30N, 0.45E	Thames	North Sea, England, UK	320	16	67
41.52N, 8.51W	Mino (alt Minho)	Atlantic O, Portugal–Spain	310	18	276
47.36N, 8.13E	Aare (alt Aar)	Rhine R, Switzerland	290	18	552
41.45N, 19.34E	Drin	Adriatic Sea, Albania	280	6	290
43.31N, 10.17E	•Arno	Ligurian Sea, Toscana, Italy	240	8	140
63.47N, 20.48E	•Thjorsa	Atlantic O, Ireland	230	7	395
56.22N, 3.21W	Tay	North Sea, Scotland, UK	190	6	156
56.29N, 10.13E	Gudena	Kattegat, Denmark	160	3	16

North America

29.02N, 89.15W	Mississippi–Missouri[h]–Jefferson–Beaverhead–Red Rock	G of Mexico, Louisiana, USA	5970	3230	15 040
69.15N, 134.08W	Mackenzie–Slave–Peace–Finlay	Beaufort Sea, NWT, Canada	4240	1787	8 940
48.09N, 67.10W	Saint Lawrence (alt Saint-Laurent)–(Great Lakes)–Saint Louis	G of Saint Lawrence, Quebec, Canada	3320[i]	1424	10 050
62.32N, 163.54W	Yukon-Lewes-Teslin-Nisutlin	Bering Sea, Alaska, USA	3180	850	6 310
25.58N, 97.09W	Rio Grande (alt Bravo)	G of Mexico, Mexico–USA	3030	460	34
57.04N, 92.30W	Nelson-Saskatchewan–South Saskatchewan-Bow	Hudson B, Manitoba, Canada	2570	1132	2 270
33.47N, 91.04W	Arkansas	Mississippi R, Arkansas, USA	2350	417	1160

Table R9 Continued

Latitude and longitude	River	Outflow and location	Length (km)	Drainage basin (1000 km²)	Discharge rate (m³ s⁻¹)
31.54N, 114.57W	Colorado	G of California, Baja California–Sonora, Mexico	2330	640	42
29.53N, 91.28W	Atchafalaya–Red	G of Mexico, Louisiana, USA	2260	246	6 990[j]
46.15N, 124.03W	Columbia–Snake[k]	Pacific O, Oregon–Washington, USA	2240	668	5 490
28.52N, 95.22W	Brazos	G of Mexico, Texas, USA	2110	120	217
36.59N, 89.08W	Ohio–Allegheny	Mississippi R, Illinois–Kentucky, USA	2100	528	7 710
58.47N, 94.12W	Churchill–Beaver	Hudson B, Manitoba, Canada	2100	298	996
38.49N, 90.07W	Mississippi (upper)	Mississippi R, Missouri, USA	1880	446	2 900
41.03N, 95.53W	Platte–North (Platte)	Missouri R, Nebraska, USA	1590	233	181
29.42N, 101.2W	Pecos	Rio Grande R, Texas, USA	1490	99	2
35.27N, 95.03W	Canadian	Arkansas R, Oklahoma, USA	1460	124	143
37.04N, 88.34W	Tennessee–Holston	Ohio R, Kentucky, USA	1450	106	1850
46.12N, 119.02W	Columbia (upper)	Columbia R, Washington, USA	1430	251	3 400
28.36N, 95.59W	Colorado	Matagorda B of G of Mexico, Texas, USA	1390	109	42
49.04N, 123.07W	Fraser	Strait of Georgia, British Columbia, Canada	1370	233	3 410
53.15N, 105.05W	North Saskatchewan	Saskatchewan R, Saskatchewan, Canada	1290	133	242
45.25N, 74.00W	Ottawa (alt Outaouais)	Saint Lawrence R, Quebec, Canada	1270	146	1970
30.41N, 88.00W	Mobile-Alabama–Coosa–Etowah	Mobile of G of Mexico, Alabama, USA	1260	113	1940
35.16N, 95.31W	North Canadian	Canadian R, Oklahoma, USA	1260	39	18
58.40N, 110.50W	Athabasca (alt Athabaska)	Athabasca L, Alberta, Canada	1230	163	681
38.11N, 109.53W	Green	Colorado R, Utah, USA	1170	122	178
48.03N, 106.19W	Milk	Missouri R, Montana, USA	1170	58	19
60.05N, 162.25W	Kuskokwim	Kuskokwim B of Bering Sea, Alaska, USA	1170	127	1160
37.09W, 88.24W	Cumberland	Ohio R, Kentucky, USA	1160	47	1090
33.57N, 91.05W	White	Mississippi R, Arkansas, USA	1160	73	646
29.45N, 94.43W	Trinity	Galveston B of G of Mexico, Texas, USA	1150	47	202
42.52N, 97.18W	James (alt Dakota)	Missouri R, South Dakota, USA	1140	57	11
39.07N, 94.37W	Kansas (alt Kaw)–Smoky Hill	Missouri R, Kansas, USA	1140	159	200
50.24N, 96.48W	Red-Assiniboine	Winnipeg L, Manitoba, Canada	1140	287	218
36.07N, 96.30W	Cimarron	Arkansas R, Oklahoma, USA	1120	49	33
61.51N, 121.18W	Liard	Mackenzie R, NWT, Canada	1120	280	2 350
18.24N, 92.38W	Usumacinta–Chixoy	B of Campeche of G of Mexico, Tabasco, Mexico	1110	103	1730
47.58N, 103.59W	Yellowstone	Missouri R, North Dakota, USA	1110	181	370
65.09N, 151.57W	Tanana–Chisana	Yukon R, Alaska, USA	1060	115	674
50.37N, 96.20W	Winnipeg–English	Winnipeg L, Manitoba, Canada	1050	135	835
49.53N, 97.07W	Red (upper)–Otter Tail	Red R, Manitoba, Canada	1030	124	150
32.43N, 114.33W	Gila	Colorado R, Arizona, USA	1010	150	11
21.36N, 105.26W	Santiago (alt Grande de Santiago)–Lerma	Pacific O, Nayarit, Mexico	1010	125	363
34.08N, 96.36W	Washita	Red R, Oklahoma, USA	1010	21	39
52.17N, 81.31W	Albany–Cat	James B of Hudson B, Ontario, Canada	980	134	1020
56.02N, 87.36W	Severn–Black Birch	Hudson B, Ontario, Canada	980	101	713
67.15N, 95.15W	Back (for Great Fish)	Arctic O, NWT, Canada	970	107	470
31.16N, 91.50W	Black–Ouachita	Red R, Louisiana, USA	970	47	496
29.59N, 93.47W	Sabine	Sabine L, Louisiana–Texas, USA	930	27	210
47.36N, 102.25W	Little Missouri	Missouri R, North Dakota, USA	900	23	17
64.16N, 96.05W	Thelon	Baker L, NWT, Canada	900	155	757
53.50N, 79.00W	Grande (alt Fort George)	James B of Hudson B, Quebec, Canada	890	98	1700
39.03N, 96.48W	Republican–Arikaree	Kansas R, Kansas, USA	890	65	24
58.30N, 68.10W	Koksoak–Caniapiscau (alt Kaniapiskau)	Ungava B of Hudson Strait, Quebec, Canada	870	137	2 420
33.07N, 79.17W	Santee-Wateree-Catawba	Atlantic O, South Carolina, USA	870	39	95
53.20N, 60.20W	Churchill (for Hamilton)–Ashuanipi	Atlantic O, Newfoundland, Canada	860	96	1750
40.23N, 91.25W	Des Moines	Mississippi R, Iowa–Missouri, USA	860	41	165
44.41N, 101.18W	Cheyenne	Missouri R, South Dakota, USA	850	66	24
37.48N, 88.02W	Wabash	Ohio R, Illinois–Indiana, USA	850	86	783
67.49N, 115.04W	Coppermine	Arctic O, NWT, Canada	840	40	102

Table R9 Continued

Latitude and longitude	River	Outflow and location	Length (km)	Drainage basin (1000 km²)	Discharge rate (m³ s⁻¹)
29.43N, 84.58W	Apalachicola–Chattahoochee	G of Mexico, Florida, USA	840	51	834
64.33N, 100.06W	Dubawnt	Beverly L, NWT, Canada	840	69	329
64.55N, 157.32W	Koyukuk	Yukon R, Alaska, USA	840	84	411
31.08N, 87.57W	Tombigbee	Mobile R, Alabama, USA	840	52	878
43.42N, 99.27W	White	Missouri R, South Dakota, USA	820	27	15
49.00N, 117.36W	Pend Oreille–Clark Fork	Columbia R, British Columbia, Canada	810	67	772
62.12N, 159.43W	Innoko	Yukon R, Alaska, USA	800		
38.35N, 91.58W	Osage–Marais des Cygnes	Missouri R, Missouri, USA	800	39	289
30.11N, 89.31W	Pearl	G of Mexico, Louisiana–Mississippi, USA	790	26	280
32.2N, 90.54W	Yazoo–Tallahatchie	Mississippi R, Mississippi, USA	790	23	292
46.44N, 105.26W	Powder	Yellowstone R, Montana, USA	780	34	17
49.15N, 117.39W	Kootenay (alt Kootenai)	Columbia R, British Columbia, Canada	780	50	798
51.25N, 78.55W	Nottaway-Bell-Megiscane	James B of Hudson B, Quebec, Canada	780	66	1040
17.55N, 102.10W	Balsas	Pacific O, Guerrero–Michoacan, Mexico	770	112	439
51.15N, 78.32W	Eastmain	James B of Hudson B, Quebec, Canada	760	46	909
51.30N, 78.48W	Rupert–Temiscamie	James B of Hudson B, Quebec, Canada	760	43	875
15.00N, 83.10W	•Coco (alt Segovia)	Caribbean Sea, Honduras–Nicaragua	750	27	500
52.57N, 82.18W	Attawapiskat	James B of Hudson B, Ontario, Canada	750	50	411
28.27N, 96.47W	Guadalupe	San Antonio B of G of Mexico, Texas, USA	740	28	59
35.48N, 95.18W	Neosho (alt Grand)	Arkansas R, Oklahoma, USA	740	33	214
64.03N, 95.35W	Kazan	Baker L, NWT, Canada	730	74	413
31.20N, 81.20W	Altamaha–Ocmulgee–South	Atlantic O, Georgia, USA	720	36	388
46.09N, 107.28W	Bighorn-Wind	Yellowstone R, Montana, USA	720	58	111
55.16N, 77.48W	Grande (Riviere) de la Baleine (alt Great Whale	Hudson B, Quebec, Canada	720	43	547
66.34N, 145.19W	Porcupine	Yukon R, Alaska, USA	720	120	337
50.56N, 109.54W	Red Deer	South Saskatchewan R, Saskatchewan, Canada	720	47	70
49.39N, 99.34W	Souris	Assiniboine R, Manitoba, Canada	720	62	13
41.07N, 100.41W	South Platte	Platte R, Nebraska, USA	710	63	13
39.32N, 76.04W	Susquehanna	Chesapeake B of Atlantic O, Maryland, USA	710	71	1190
60.51N, 115.44W	Hay	Great Slave L, NWT, Canada	700	49	100
18.36N, 92.39W	Grijalva (alt Mezcalapa)	B of Campeche of G of Mexico, Tabasco, Mexico	700	52	200
33.22N, 79.16W	Pee Dee–Yadkin	Winyah B of Atlantic O, South Carolina, USA	700	23	281
48.10N, 69.45W	Saguenay–Peribonca	Saint Lawrence R, Quebec, Canada	700	88	1180
21.45N, 105.30W	San Pedro (alt Mezquital)	Pacific O, Nayarit, Mexico	700	18	110
69.42N, 129.01W	Anderson	Wood B of Beaufort Sea, NWT, Canada	690	62	162
38.41N, 85.11W	Kentucky	Ohio R, Kentucky, USA	690	18	235
42.46N, 98.03W	Niobrara	Missouri R, Nebraska, USA	690	33	44
42.59N, 91.09W	Wisconsin	Mississippi R, Wisconsin, USA	690	31	246
42.29N, 96.27W	Big Sioux	Missouri R, Iowa–South Dakota, USA	680	23	28
51.58N, 98.04W	Dauphin–Fairford–Red Deer	Winnipeg L, Manitoba, Canada	680	83	62
38.58N, 90.28W	Illinois–Kankakee	Mississippi R, Illinois, USA	680	72	628
67.41N, 134.32W	Peel–Ogilvie	Mackenzie R, NWT, Canada	680	71	768
34.37N, 90.35W	Saint Francis	Mississippi R, Arkansas, USA	680	22	149
45.51N, 116.47W	Salmon	Snake R, Idaho, USA	680	36	323
27.37N, 110.39W	Yaqui–Bavispe	G of California, Sonora, Mexico	680	66	108
45.15N, 66.04W	Saint John	B of Fundy of Atlantic O, New Brunswick, Canada	670	55	849
36.56N, 76.27W	James–Jackson	Chesapeake B of Atlantic O, Virginia, USA	670	23	214
41.16N, 72.20W	Connecticut	Long Island Sound of Atlantic O, Connecticut, USA	660	28	469

Table R9 Continued

Latitude and longitude	River	Outflow and location	Length (km)	Drainage basin (1000 km²)	Discharge rate (m³ s⁻¹)
35.56N, 76.42W	Roanoke	Albermarle Sound of Atlantic O, North Carolina, USA	660	26	229
63.18N, 139.24W	Stewart	Yukon R, Yukon Territory, Canada	640	51	451
39.15N, 75.20W	Delaware	Delaware B of Atlantic O, Delaware–New Jersey, USA	630	30	332
38.50N, 82.08W	Kanawha–New	Ohio R, West Virginia, USA	630	31	425
70.01N, 126.42W	Horton	Amundsen G of Arctic O, NWT, Canada	620		
38.00N, 76.23W	Potomac	Chesapeake B of Atlantic O, Maryland–Virginia, USA	620	38	326
51.20N, 80.24W	Moose–Abitibi	James B of Hudson B, Ontario, Canada	610	109	793
62.47N, 137.20W	Pelly	Yukon R, Yukon Territory, Canada	610	52	382
38.03N, 121.56W	Sacramento	Suisun B of Pacific O, California, USA	610	70	702
16.30N, 97.31W	•Atoyac (alt Verde)	Pacific O, Oaxaca, Mexico	600	19	90
70.27N, 150.07W	Colville	Beaufort Sea, Alaska, USA	600	62	
29.35N, 104.25W	•Conchos	Rio Grande R, Chilhauhua, Mexico	590	64	24
37.54N, 87.30W	Green	Ohio R, Kentucky, USA	580	24	318
41.10N, 91.10W	Iowa–Cedar	Missouri R, Iowa, USA	580	34	198
37.16N, 110.26W	San Juan	Colorado R, Utah, USA	580	60	72
32.02N, 80.53W	Savannah–Seneca	Atlantic O, Georgia–South Carolina, USA	580	26	343
54.09N, 130.05W	Skeena	Chatham Sound of Pacific O, British Columbia, Canada	580	55	910
52.43N, 108.15W	Battle	North Saskatchewan R, Saskatchewan, Canada	570	31	16
39.04N, 113.07W	Sevier	Sevier L, Utah, USA	570	16	5
41.27N, 112.15W	Bear	Great Salt L, Utah, USA	560	18	49
25.45N, 109.22W	•Fuerte–Verde	G of California, Sinaloa, Mexico	560	36	171
58.50N, 66.10W	George	Ungava B of Hudson Strait, Quebec, Canada	560	42	740
45.40N, 100.45W	Grand	Missouri R, South Dakota, USA	560	16	7
39.06N, 84.30W	Licking	Ohio R, Kentucky, USA	560	10	117
49.10N, 68.15W	Manicouagan	Saint Lawrence R, Quebec, Canada	560	46	871
67.00N, 162.30W	Noatak	Kotzebue Sound of Chukchi Sea, Alaska, USA	560	33	
44.30N, 68.48W	Penobscot	Atlantic O, Maine, USA	560	26	338
46.21N, 72.31W	Saint–Maurice	Saint Lawrence R, Quebec, Canada	560	43	703
38.04N, 121.51W	San Joaquin	Suisun B of Pacific O, California, USA	560	36	136
61.03N, 123.22W	South Nahanni	Liard R, NWT, Canada	560	34	403
35.48N, 95.19W	Verdigris	Arkansas R, Oklahoma, USA	560	20	105
27.50N, 97.29W	Nueces	Nueces B of G of Mexico, Texas, USA	550	49	24
54.45N, 114.17W	Pembina	Athabasca R, Alberta, Canada	550	13	38
50.39N, 59.29W	Petit Mecatina (alt Little Mecatina)	G of Saint Lawrence, Quebec, Canada	550	20	495
41.07N, 96.18W	Elkhorn	Platte R, Nebraska, USA	540	18	34
56.31N, 132.24W	Stikine	Stikine Strait, Alaska, USA	540	51	1550
38.25N, 87.44W	White	Wabash R, Indiana, USA	540	31	330
32.03N, 91.04W	Big Black	Mississippi R, Mississippi, USA	530	8	103
51.10N, 79.45W	Harricana (alt Harricanaw)	James B of Hudson B, Ontario, Canada	530	29	134
44.54N, 93.09W	Minnesota	Mississippi R, Minnesota, USA	530	44	103
33.23N, 112.18W	Salt-Black	Gila R, Arizona, USA	530	18	27
33.53N, 78.01W	Cape Fear–Deep	Atlantic O, North Carolina, USA	520	16	164
59.33N, 124.01W	Fort Nelson–Sikanni Chief	Liard R, British Columbia, Canada	520	47	334
47.01N, 96.49W	Sheyenne	Red R, North Dakota, USA	520	26	5
37.58N, 89.57W	Kaskaskia	Mississippi R, Illinois, USA	510	16	112
22.16N, 97.47W	Panuco–Santa Maria	G of Mexico, Tamaulipas–Veracruz, Mexico	510	66	548
67.27N, 133.45W	Arctic Red	Mackenzie R, NWT, Canada	500	19	153
25.45N, 102.50W	•Nazas	Mayran L, Coahuila, Mexico	500	36	95
49.05N, 68.23W	Outardes	Saint Lawrence R, Quebec, Canada	500	19	385
43.49N, 117.02W	Owyhee	Snake R, Oregon, USA	500	31	12
50.18N, 63.48W	Romaine	G of Saint Lawrence, Quebec, Canada	500	14	315
40.42N, 74.01W	Hudson	New York B of Atlantic O, New Jersey–New York, USA	490	35	388
56.11N, 117.19W	Smoky	Peace R, Alberta, Canada	490	55	367
50.14N, 121.34W	Thompson–North Thompson	Fraser R, British Columbia, Canada	490	57	786
39.35N, 96.34W	Big Blue	Kansas R, Kansas, USA	480	25	61
35.53N, 84.29W	Clinch	Tennessee R, Tennessee, USA	480	9	132

Table R9 Continued

Latitude and longitude		River	Outflow and location	Length (km)	Drainage basin (1000 km²)	Discharge rate (m³ s⁻¹)
58.46N,	70.05W	Feuilles (alt Leaf)	Ungava B of Hudson Strait, Quebec, Canada	480	43	587
39.23N,	93.06W	Grand	Missouri R, Missouri, USA	480	18	110
57.00N,	92.15W	Hayes	Hudson B, Manitoba, Canada	480	108	650
36.12N,	111.48W	Little Colorado	Colorado R, Arizona, USA	480	70	7
51.03N,	80.55W	Moose (upper)–Mattagami	Moose R, Ontario, Canada	480	65	420
30.23N,	88.37W	Pascagoula–Chickasawhay	Mississippi Sound of G of Mexico, Mississippi, USA	480	18	283
41.9N,	90.37W	Rock	Mississippi R, Illinois, USA	480	28	173
28.48N,	111.49W	•Sonora	G of California, Sonora, Mexico	480	29	
55.16N,	85.05W	Winisk	Hudson B, Ontario, Canada	470	67	484
44.26N,	102.18W	Belle Fourche	Cheyenne R, South Dakota, USA	470	20	10
39.59N,	118.36W	Humboldt	Humboldt L, Nevada, USA	470	44	7
41.24N,	97.19W	Loup–Middle Loup	Platte R, Nebraska, USA	470	39	18
45.18N,	100.43W	Moreau	Missouri R, South Dakota, USA	470	16	6
47.21N,	107.57W	Musselshell	Missouri R, Montana, USA	470	23	8
32.32N,	87.51W	Black Warrior	Tombigbee R, Alabama, USA	460	16	382
53.56N,	122.42W	Nechako	Fraser R, British Columbia, Canada	460	43	300
35.38N,	91.19W	Black	White R, Arkansas, USA	450	21	244
51.21N,	78.53W	Broadback	James B of Hudson B, Quebec, Canada	450	21	314
39.19N,	92.57W	Chariton	Missouri R, Missouri, USA	450	6	35
60.18N,	145.03W	Copper	G of Alaska, Alaska, USA	450	63	1060
45.44N,	120.39W	John Day	Columbia R, Oregon, USA	450	21	59
66.54N,	160.38W	Kobuk	Kotzebue Sound of Chukchi Sea, Alaska, USA	450	31	425
29.58N,	93.51W	Neches	Sabine L, Texas, USA	450	21	147
35.06N,	76.29W	Neuse	Pamlico Sound of Atlantic O, North Carolina, USA	450	8	82
59.03N,	158.23W	Nushagak-Mulchatna	Bristol B of Bering Sea, Alaska, USA	450	37	641
31.58N,	82.32W	Oconee	Altamaha R, Georgia, USA	450	13	142
18.42N,	95.38W	Papaloapan–Tuxtepec	Alvarado Lagoon, Veracruz, Mexico	440	23	1240
30.24N,	81.24W	Saint Johns	Atlantic O, Florida, USA	440	23	157
33.44N,	80.38W	Congaree–Broad	Santee R, South Carolina, USA	440	21	266
58.15N,	67.38W	Baleine (alt Whale)	Ungava B of Hudson Strait, Quebec, Canada	430	32	537
32.30N,	86.16W	Tallapoosa	Alabama R, Alabama, USA	430	12	139
45.39N,	122.46W	Willamette	Columbia R, Oregon, USA	430	29	970
30.57N,	84.34W	Flint	Apalachicola R, Georgia, USA	430	21	197
38.40N,	91.33W	Gasconade	Missouri R, Missouri, USA	430	9	73
46.25N,	105.52W	Tongue	Yellowstone R, Montana, USA	430	16	12
43.03N,	86.15W	Grand	Michigan L, Michigan, USA	420	13	102
41.17N,	91.21W	Iowa (upper)	Iowa R, Iowa, USA	420	11	81
61.15N,	150.36W	Susitna	G of Alaska, Alaska, USA	420	50	1400
45.38N,	120.55W	Deschutes	Columbia R, Oregon, USA	400	28	166
41.33N,	124.05W	Klamath	Pacific O, California, USA	400	31	513
31.50N,	81.03W	Ogeechee	Atlantic O, Georgia, USA	400	8	66
29.17N,	83.10W	Suwannee	G of Mexico, Florida, USA	400	26	301
44.06N,	77.34W	Trent-Otonabee-Irondale	Ontario L, Ontario, Canada	400	12	137
40.04N,	109.40W	White	Green R, Utah, USA	400	13	18
34.06N,	98.10W	Wichita	Red R, Texas, USA	400	10	8
40.32N,	108.59W	Yampa	Green R, Colorado, USA	400	10	44
45.27N,	75.40W	Gatineau	Ottawa R, Quebec, Canada	390	24	360
20.17N,	75.56W	Cauto	G of Guacanayabo of Caribbean Sea, Cuba	370	11	
43.58N,	69.52W	Androscoggin–Magalloway	Atlantic O, Maine, USA	360	10	174
13.14N,	88.49W	Lempa	Pacific O, El Salvador	320	18	377
17.32N,	88.14W	•Belize	Caribbean Sea, Belize	290		
19.15N,	72.47W	Artibonite	G of Gonave of Caribbean Sea, Haiti	280	10	
19.51N,	71.41W	Yaque del Norte	Atlantic O, Dominican Republic	280		49
46.03N,	73.08W	Richelieu	Saint Lawrence R, Quebec, Canada	170	24	360

Oceania

35.22S,	139.22E	Murray-Darling[l]–Culgoa–Balonne–Condamine	Indian O, South Australia, Australia	3750	1057	326
34.07S,	141.55E	Murray (upper)	Murray R, New South Wales, Australia	1750	267	168
29.56S,	146.20E	Barwon–Macintyre–Dumaresq–Severn	Darling R, New South Wales, Australia	1580	225	60

Table R9 Continued

Latitude and longitude		River	Outflow and location	Length (km)	Drainage basin (1000 km²)	Discharge rate (m³ s⁻¹)
34.43S,	143.12E	Murrumbidgee	Murray R, New South Wales, Australia	1580	97	73
34.21S,	143.57E	Lachlan	Murrumbidgee R, New South Wales, Australia	1480	85	17
8.25S,	143.10E	•Fly–Strickland	G of Papua of Coral Sea, Papua New Guinea	1290	64	4 450
3.51S,	144.34E	•Sepik	Pacific O, Papua New Guinea	1130		
23.32S,	150.52E	•Fitzroy–Dawson	Pacific O, Queensland, Australia	1110	143	191
30.07S,	147.24E	Macquarie	Barwon R, New South Wales, Australia	950	47	25
30.00S,	148.07E	Namoi	Barwon R, New South Wales, Australia	850	43	21
17.36S,	140.36E	Flinders	G of Carpentaria, Queensland, Australia	840	108	16
24.52S,	113.37E	Gascoyne	Indian O, Western Australia, Australia	820	80	17
29.57S,	146.21E	•Bogan	Barwon R, New South Wales, Australia	720	26	2
19.39S,	147.30E	Burdekin	Pacific O, Queensland, Australia	710	131	287
1.26S,	137.53E	Mamberamo–Taritatu (for Idenburg)	Pacific O, Irian Jaya, Indonesia	670		
29.27S,	149.48E	Gwydir	Barwon R, New South Wales, Australia	670	26	25
15.12S,	129.43E	•Victoria	Timor Sea, Northern Territory, Australia	650	78	87
4.02S,	144.40E	•Ramu (for Ottilien)	Pacific O, Papua New Guinea	640		
30.12S,	147.32E	Castlereagh	Barwon R, New South Wales, Australia	550	18	5
7.07S,	138.42E	Digul (for Digoel)	Arafura Sea, Irian Jaya, Indonesia	540		
33.30S,	151.10E	Hawkesbury	Pacific O, New South Wales, Australia	470	22	66
32.50S,	151.42E	Hunter	Pacific O, New South Wales, Australia	470	20	30
37.23S,	174.42E	Waikato	Tasman Sea, North Island, New Zealand	420	14	334
29.25S,	153.22E	Clarence	Pacific O, New South Wales, Australia	390	23	116
46.21S,	169.48E	Clutha–Makarora	Pacific O, South Island, New Zealand	320	22	651

South America

Latitude and longitude		River	Outflow and location	Length (km)	Drainage basin (1000 km²)	Discharge rate (m³ s⁻¹)
0.01S,	49.00W	•Amazon (off Amazonas)–Ucayali–Tambo–Ene–Apurimac	Atlantic O, Amapa–Para, Brazil	6570	6150	175 000
35.00S,	57.00W	Plata–Parana–Grande	Atlanic O, Argentina–Uruguay	4880	3100	22 900
3.22S,	58.45W	•Madeira–Mamore–Grande (alt Guapay)	Amazon R, Amazonas, Brazil	3200	1200	21 800
2.37S,	65.44W	•Jurua (alt Yurua)	Amazon R, Amazonas, Brazil	3000	240	4 000
3.42S,	61.28W	•Purus	Amazon R, Amazonas, Brazil	3000	240	12 600
10.30S,	36.24W	Sao Francisco	Atlantic O, Alagoas–Sergipe, Brazil	2780	623	2 890
1.00S,	48.30W	Para–Tocantins	Atlantic O, Para, Brazil	2750	836	8 630
27.18S,	58.38W	Paraguay (alt Paraguai)	Parana R, Argentina–Paraguay	2600	1100	4 400
3.08S,	64.46W	•Caqueta (alt Japura, Yapura)	Amazon R, Amazonas, Brazil	2280	310	7 000
2.24S,	54.41W	•Tapajos–Juruena	Amazon R, Para, Brazil	2220	463	6 000
5.21S,	48.41W	•Araguaia	Tocantins R, Para, Brazil	2200	320	6 140
34.12S,	58.18W	Uruguay (alt Uruguai)–Canoas	Plata R, Argentina–Uruguay	2200	307	5 500
8.37N,	62.15W	Orinoco	Atlantic O, Venezuela	2140	880	25 200
1.30S,	51.53W	•Xingu	Amazon R, Para, Brazil	2100	450	2 060
3.08S,	59.55W	•Negro (alt Guainia)	Amazon R, Amazonas, Brazil	2000	1000	35 000
3.07S,	67.58W	•Putumayo (alt Ica)	Amazon R, Amazonas, Brazil	2000	112	5 000
11.54S,	65.01W	•Guapore (alt Itenez)	Mamore R, Bolivia–Brazil	1800	600	2 000
3.00S,	41.50W	•Parnaiba	Atlantic O, Maranhao–Piaui, Brazil	1700	350	2 400
11.06N,	74.51W	Magdalena	Caribbean Sea, Colombia	1540	260	8 000
31.42S,	60.44W	Salado (alt Salado del Norte)	Parana R, Santa Fe, Argentina	1500	800	38
1.24S,	61.51W	•Branco–Uraricoera	Negro R, Roraima, Brazil	1470	195	5 400
39.50S,	62.08W	Colorado–Salado–Desaguadero–Bermejo	Atlantic O, Buenos Aires, Argentina	1430	110	133
4.30S,	73.27W	Maranon	Amazon R, Peru	1410		
7.21S,	58.03W	•Teles Pires (alt Sao Manuel, Tres Barras)	Tapajos R, Mato Grosso–Para, Brazil	1400		1750
8.54N,	74.28W	Cauca	Magdalena R, Colombia	1350	63	2 200
4.03N,	67.44W	Guaviare	Orinoco R, Colombia	1350		
3.55N,	67.42W	Inirida	Guaviare R, Colombia	1350		
25.36S,	54.36W	Iguacu (alt Iguazu; for Iguassu)	Parana R, Argentina–Brazil	1320	62	1750
5.07S,	60.24W	•Aripuana–Roosevelt	Madeira R, Amazonas, Brazil	1290		
10.23S,	65.24W	•Beni–Madre de Dios	Madeira R, Bolivia	1290	69	2 310
20.07S,	51.05W	Paranaiba	Parana R, Mato Grosso do Sul–Minas Gerais, Brazil	1270		1500
41.02S,	62.47W	Negro–Neuquen	Atlantic O, Buenos Aires–Rio Negro, Argentina	1210	125	1010
1.23S,	69.25W	Apaporis	Caqueta R, Colombia	1200		
2.52S,	44.12W	•Itapecuru	Sao Jose B of Atlantic O, Maranhao, Brazil	1200	45	

Table R9 Continued

Latitude and longitude	River	Outflow and location	Length (km)	Drainage basin (1000 km²)	Discharge rate (m³ s⁻¹)
2.43S, 66.57W	•Jutai	Amazon R, Amazonas, Brazil	1200	31	500
4.21S, 70.02W	Javari (alt Yacarana, Yavari)	Amazon R, Brazil–Peru	1180	91	120
5.10S, 75.32W	Huallaga	Maranon R, Peru	1140	95	3 500
21.37S, 41.03W	Paraiba do Sul (alt Paraiba)	Atlantic O, Rio de Janeiro, Brazil	1140	57	331
20.40S, 51.35W	•Teite	Parana R, Sao Paulo, Brazil	1130	72	378
3.04S, 44.35W	•Mearim	Sao Marcos B of Atlantic O, Maranhao, Brazil	1100	100	
25.21S, 57.42W	Pilcomayo	Paraguay R, Paraguay	1100	192	197
15.51S, 38.53W	•Jequitinhonha	Atlantic O, Bahia, Brazil	1090	62	557
26.52S, 58.23W	Bermejo	Paraguay R, Chaco–Formosa, Argentina	1060	94	325
19.37S, 39.49W	•Doce	Atlantic O, Espirito Santo, Brazil	1000	83	969
3.52S, 52.37W	•Iriri	Xingu R, Para, Brazil	1000		
6.12N, 67.28W	Meta	Orinoco R, Colombia–Venezuela	1000	104	2 500
0.02N, 67.16W	•Vaupes (alt Uaupes)	Negro R, Amazonas, Brazil	1000		
6.58N, 58.23W	•Essequibo	Atlantic O, Guyana	970	69	2 190
7.37N, 66.25W	Apure-Uribante	Orinoco R, Venezuela	960	130	1890
8.21N, 62.43W	Caroni	Orinoco R, Venezuela	920	95	4 750
38.49N, 64.57W	Colorado (upper)–Grande	Colorado R, La Pampa, Argentina	920	25	
1.29S, 48.30W	•Guama–Capim	Para R, Para, Brazil	900		
22.40S, 53.09W	•Paranapanema	Parana R, Parana–Sao Paulo, Brazil	900	56	348
3.35S, 64.47W	•Tefe	Amazon R, Amazonas, Brazil	900		
6.25N, 58.37W	•Mazaruni–Cuyuni	Essequibo R, Guyana	880	21	1150
3.20S, 72.40W	•Napo	Amazon R, Peru	880		
10.44S, 73.45W	Urubamba	Ucayali R, Peru	860		
6.15S, 42.52W	•Caninde	Parnaiba R, Piaui, Brazil	860		
23.18S, 53.42W	•Ivai	Parana R, Parana, Brazil	860	36	340
4.26S, 74.05W	Tigre	Maranon R, Peru	840		
7.24N, 66.35W	Arauca	Orinoco R, Venezuela	810	18	393
43.20S, 65.03W	Chubut	Atlantic O, Chubut, Argentina	810	31	49
10.25S, 58.20W	•Arinos	Juruena R, Mato Grosso, Brazil	800		1280
10.58S, 66.09W	•Beni (upper)	Beni R, Bolivia	800		
3.41S, 44.48W	•Grajau	Mearim R, Maranhao, Brazil	800		
1.13S, 46.06W	•Gurupi	Atlantic O, Maranhao–Para, Brazil	800	61	
11.47S, 37.32W	•Itapicuru	Atlantic O, Bahia, Brazil	800	39	17
11.45S, 50.44W	•Mortes	Araguaia R, Mato Grosso, Brazil	800		
33.24S, 58.22W	•Negro	Uruguay R, Uruguay	800	70	637
15.39S, 38.57W	•Pardo	Atlantic O, Bahia, Brazil	800	45	62
12.28S, 64.24W	•Itonamas–San Miguel	Guapore R, Bolivia	760		
17.13S, 44.49W	•Velhas	Sao Francisco R, Minas Gerais, Brazil	760		
8.17N, 76.58W	Atrato	G of Uraba of Caribbean Sea, Colombia	750	35	4 900
7.38N, 64.53W	Caura-Merevari	Orinoco R, Venezuela	720	50	2 700
5.43N, 53.58W	•Maroni (alt Marowyne)	Atlantic O, French Guiana–Suriname	680	62	1850
5.55N, 55.10W	•Suriname	Atlantic O, Suriname	600	16	440
1.55S, 55.35W	•Trombetas	Amazon R, Para, Brazil	550	124	1500
47.49S, 73.37W	Baker	G of Penas of Pacific O, Chile	440	25	600
6.00N, 57.04W	•Courantyne (alt Corantijn)	Atlantic O, Guyana–Suriname	440	67	2 000
21.26S, 70.04W	Loa	Pacific O, Chile	440	34	2
36.49S, 73.10W	Bio-Bio	Arauco G of Pacific O, Chile	380	24	1000

Abbreviations: •, unofficial data; alt, alternative name; B, Bay; for, formerly; G, Gulf; I, Island; L, Lake; O, Ocean; off, official name; R, River.

[a] For Damietta (off Dumyat) R distributary; for Nile R proper: 30.10N, 31.06E.

[b] Data for Angara-Selenga-Ider: 3410 km; 1 039 000 km²; 5080 m³ s⁻¹.

[c] Data for Irtysh: 4250 km; 1 643 000 km²; 2150 m³ s⁻¹.

[d] For Meghna R distributary; for Brahmaputra R proper: 24.02N, 90.59E.

[e] For Hooghly R distributary; for Ganges R proper: 23.22N, 90.32E.

[f] For Haringvliet Estuary distributary; for Rhine R proper: 51.52N, 6.02E.

[g] For Haringvliet Estuary distributary; for Meuse R proper: 51.49N, 5.01E.

[h] Data for Missouri–Jefferson–Beaverhead–Red Rock: 4080 km; 1 370 000 km²; 2290 m³ s⁻¹.

[i] For Saint Lawrence R proper: 960 km.

[j] Inc approximately 25% of the Mississippi R flow, diverted to control flooding; the average discharge rate of the Mississippi at Vicksburg, above the point of diversion, is 16 400 m³ s⁻¹. The average discharge rate of the Red R is 875 m³ s⁻¹.

[k] Data for Snake: 1670 km; 282 000 km²; 1600 m³ s⁻¹.

[l] Data for Darling-Culgoa-Balonne-Condamine: 2910 km; 640 000 km²; 225 m³ s⁻¹.

RIVER POLLUTION PREVENTION: HISTORICAL

Until the beginning of the Industrial Revolution, significant pollution of rivers in the UK occurred only in the largest towns, and then the pollution occurred mainly as a result of dumping refuse in rivers. The Industrial Revolution permitted a great increase in factory development and output. Where the industries were water users, for example in textile manufacture, or productive of foul chemical wastes as in the early gasworks processes, gross river pollution developed. The concentration of population in the new industrial towns, coupled with the advent of piped water supplies and water-carriage sanitation systems, exacerbated river pollution and the spread of waterborne disease. The first widespread and evil consequences of inadequate environmental sanitation and water pollution were demonstrated in the massive cholera outbreaks which developed in almost every town. The first steps were then taken in water pollution control by switching sources of public water supply from local polluted river sources, and less rigorous but genuine, if somewhat ineffective, endeavors were made to control river pollution.

The earliest statute on pollution prevention was apparently enacted during the reign of Richard II (in 1388), which prohibited the dumping of refuse into ditches and rivers in or near to towns. During the reign of Henry VIII a number of statutes were enacted. A Bill of Sewers was passed in 1531 enabling the establishment of Commissioners of Sewers with land drainage, tidal defence and public health responsibilities. Three statutes relating to pollution of harbors and navigations by solid refuse were also enacted during Henry's reign, and two further similar measures were enacted in 1745 and 1814.

The Lighting and Watching Act of 1833 included provision to control stream pollution arising from gasworks waste. In 1847 the first extensive legislation was passed for prevention of pollution of inland waters. The Cemeteries Clauses Act, the Gasworks Clauses Act, the Harbours Docks and Piers Clauses Act, all passed in 1847, prohibited pollution of inland waters, harbors and docks, and waterworks reservoirs and aqueducts by drainage from gasworks and cemeteries and other offensive matter.

Also in 1847, the first of a successive series of measures, relating to the public health aspects of sewerage and sewage disposal, was passed in the Towns Improvement Clauses Act. There followed the first Public Health Act in 1848, the Local Government Acts of 1858 and 1861, the Sewage Utilization Act of 1865, and various additions and amendments of these measures were made, culminating in the Public Health Act of 1875, all directed at the provision of sewerage and sewage disposal arrangements and the control of sewage pollution.

The first public body set up for river pollution prevention was the Thames Conservancy in 1857. The first national legislation for river pollution prevention generally was the Rivers (Prevention of Pollution) Act 1876. This Act prohibited pollution caused by dumping of refuse, or by discharge or deposit of sewage matter, industrial waste and mining waste. By and large, this Act was a sound legal measure, but it had two characteristics which greatly diminished the effectiveness of its enforcement. The first of these shortcomings, which could hardly be avoided having regard to the embryonic state of waste disposal technology of the times, was that pollution could not be penalized if the offenders could show that they had used the best practicable and available means to render the polluting matter harmless. The second, and major, shortcoming of the 1876 Act was that it was to be enforced by the local sanitary authorities, who were themselves responsible for sewage disposal and hence for much of the pollution occurring.

The Local Government Act of 1888 enabled county councils, and joint committees of county councils set up specifically for pollution prevention purposes, to enforce the Rivers (Prevention of Pollution) Act of 1876.

In 1898 a Royal Commission on Sewage Disposal was appointed. This Commission published a series of reports between 1898 and 1915. Many of these are still accepted today. For example, the well-known requirement that sewage effluents should have a BOD of not more than 20 mg l^{-1}, a suspended solids content of not more than 30 mg l^{-1} and should be diluted with at least 8 volumes of clean river water with a BOD of not more than 2 mg l^{-1} under dry-weather flow conditions.

The Salmon & Freshwater Fisheries Act 1923 set up Fishery Boards to protect fisheries, and although their powers included provision to take action against pollution harmful to fish, these Boards were inadequately financed and indeed could do nothing about pollution of waters which had already been rendered fishless.

The first effective provision for river pollution control in England and Wales was made when 32 River Boards were set up over the period 1950–52 following enactment of the River Boards Act of 1948. These River Boards took over the existing land drainage authorities (Rivers Catchment Boards), the few existing, specialist pollution prevention authorities, with the existing Fisheries Boards, and were given combined responsibilities for land drainage, salmon and freshwater fisheries and pollution prevention. These river boards, until their demise in 1965, were given new powers for the prevention of river pollution, as specified in the Rivers (Prevention of Pollution) Act of 1951, the Clean Rivers (Estuaries and Tidal Waters) Act of 1960, and the Rivers (Prevention of Pollution) Act of 1961. The Thames Conservancy and the Lee Conservancy Catchment Board were also empowered to operate the 1951 and 1961 Acts in addition to their private acts (they had no tidal waters in their areas to which the 1960 Act was applicable). In 1963 the 32 river boards were superseded by 27 river authorities, and given additional powers for water conservation, by virtue of the Water Resources Act of 1963. The Thames and Lee Boards remained in existence and were empowered to operate the 1963 Act.

The Control of Pollution Act of 1974 extended control of pollution powers to the discharge of trade and sewage effluents into rivers and coastal waters and sanitary appliances on vessels. In 1989 the Water Act established the National Rivers Authority (NRA) and ten water services companies covering the whole of England and Wales. The former included in its terms of reference all hydrometric matters as well as pollution control of rivers, and the water services companies became responsible for both water supply and sewage purification. In 1990 the introduction of the Environment Protection Act set up Her Majesty's Inspectorate of Pollution (HMIP) which controlled prescribed emissions to land, water and air. Finally, the Environment Protection Act of 1995 brought together the NRA, the HMIP and the regional Waste Regulatory Authorities under one authority – the Environment Agency with effect from 1 April 1996. In Scotland the Scottish Environment Protection Agency was launched on the same date. It combines the Scottish River Purification Boards, Her Majesty's Industrial Pollution Inspectorate and the Waste Regulation and Air Pollution functions of local councils.

R.W. Herschy

Source

Fish, H. 1973. *Principles of Water Quality Management*, Thunderbird Enterprises, London.

S

SAHEL REGION: CLIMATES AND HYDROLOGY

The 'Sahel' as a concept is of recent origin and not very precise. Etymologically, it means an interface between two environments: the Arab word *sahil* means 'shore'. The interface in question is the transition zone between desert Africa and humid Africa north of the equator, and scientists generally apply the term to a belt about 600 km wide, stretching from the Atlantic to the Red Sea. However, the width and boundaries of the Sahel vary considerably according to the criteria used to define it. For botanists, its northern limit is the beginning of the grassland zone, where rainfall is 100–150 mm per year, and its southern limit is the line where shea butter trees (*Butyrospermum parkii*) begin, at 700 mm (27") per year. For climatologists, broadly speaking, the Sahelian climate lies between the 100 and 600 mm isohyets, while hydrologists consider that the Sahel proper covers the belt between 300 mm, where generalized runoff begins, and 750 mm, where hydrographic degradation stops. By extension, hydrologists tend also to include the sub-desert climate zone between 100 and 300 mm per year. None of these defining boundaries are geographically stable; however, in the space of a few decades, the mean isohyets can shift several hundred kilometers north, as they did in the 1950–1965 wet period or even farther towards the south (1966–1995).

General climatology

The Sahelian climate is conditioned by the annual oscillation of the meteorological equator – the zone of contact between the dry, continental air over the Sahara (the northeasterly harmattan winds) and the moist air of the southwesterly monsoon winds that blow in from the Atlantic. The line of contact between these two air masses at ground level used to be called the intertropical front or ITF, but this designation is now somewhat outdated and has been more or less replaced by the term intertropical convergence zone (ITCZ), which refers both to the ground trace and to the zonal alignment of cloud formations in the middle strata. It is also the upflowing side of the Hadley cell. In the lower strata, where the monsoon winds blow, the annual migration of the ITCZ over the land is closely linked to ground temperature; this is why the zone migrates much farther in East Africa, which is entirely continental, than it does in West Africa, which is bathed by the Atlantic Ocean on its southern side.

Weather and rainfall patterns are linked to the position of the convergence zone. It is at the beginning and end of the rainy season that the most violent storms, the tornadoes, generally occur; these are isolated occurrences of limited area. In the middle of the wet season, when the ITCZ reaches its most northerly position, the dominant pattern is one of squall lines following easterly air waves that generally come from Sudan. These squall lines can travel several thousand kilometers, regenerating themselves as they go, and can be observed locally for a day or two at a time. They coincide with the African easterly jet (AEJ), which occurs during the northern summer, at 650 hPa. There is also a second jet stream during the northern summer, higher in the atmosphere, at 200 hPa; this is the tropical easterly jet (TEJ) and it comes from the high plateaus of Tibet. These two jet streams fulfill a very important role, supplying energy for the development and maintenance of the precipitation systems. Annual precipitation in a given place will be only partly determined by the duration of the influence of the monsoon and the vertical depth of this airstream; other factors also play a part, the behavior of the jet streams especially. For example, a weak TEJ over the African continent, combined with an AEJ strong enough to cut through the upcurrents, will be a factor for drought.

Rainfall regime and characteristics

The regular annual behavior pattern of these air masses explains why, especially in West and Central Africa, the typical features of the Sahelian precipitation regime follow a very simple pattern as one moves northward along a meridian: the lower the annual rainfall, the greater the interannual variations. Except in the highlands, the interannual isohyets line up with the parallels, with an average increase of 1 mm km^{-1}.

There is a single rainy season, always centered on the month of August. At the 200 mm isohyet, the rainy season lasts only 6 weeks, while at 600 mm it lasts at least 3 months. The number of rain days per year ranges from 20 to 60, while mean daily and 10-day rainfall amounts are, respectively, 35 and 70 mm in the north and 60 and 100 mm in the south. The tornadoes take the form of brief convective rainstorms lasting 15 min to 2 h; the peak intensity of such a storm can be as much as 150 mm h^{-1} in 5 min (at an annual total of 200–300 mm). These rains generally cover several tens of square kilometers, and rarely more than 100–200 km^2. With the squall lines, rainfall is somewhat less heavy and the hyetogram can show several peaks of intensity.

East Africa

The above description is valid from Mauritania to the Sudan. Farther east, the pattern is more complex owing to the major orographic impact of the highlands and the dual influence of monsoonal airstreams from the Congo Basin and Atlantic Ocean on the one hand and airstreams from the Indian Ocean on the other. Thus northern and central Ethiopia have a bimodal rainfall regime, with a subsidiary peak in April due to the Indian monsoon, and a main peak in August due to the Atlantic or Congolese monsoon. In addition, ENSO (El Niño/Southern Oscillation), which has a limited effect on West and Central Africa, has a significant influence from Ethiopia to the Horn of Africa. This dual influence on rainfall dynamics explains why, in some years, Ethiopia has a different rainfall pattern to that of the 'Sahel' proper.

Available data

A few raingauge stations were set up at the beginning of the twentieth century or even earlier, e.g. the St Louis station in Senegal, which was opened in 1851, but is not very representative as it is located in the coastal zone. However, there were no proper, organized pluviometric networks until the 1920s. Although the stations are too few and not sufficiently uniform, the data gathered do allow global assessment of annual, monthly or daily trends and fluctuations in the rainfall regime since the start of the twentieth century. At the level of the individual rainfall event, however, knowledge of characteristics such as intensity, duration and spatial spread is sparse and inadequate.

To advance our knowledge in this sphere, two crucial experiments have recently been conducted in Niger. The first, EPSAT-NIGER, was designed to improve evaluation of precipitation systems and was fully operational from 1990 to 1993, with 93 rain recorders in a 16 000 km^2 area and a C-band meteorological radar. The other experiment was HAPEX-SAHEL, conducted in 1992. This was designed to improve the performance of the general circulation models (GCMs) used in this climatic zone, by improving our knowledge of continent–atmosphere interactions at the scale of 1 degree square.

Climate trends

Since instrumental observations began, i.e. since the start of the twentieth century, rainfall has fluctuated considerably in the Sahelian zone, with drought in 1913, a wet period from 1950 to 1964, etc. But the drought that has now persisted since 1968 has been exceptionally severe and long-lasting, and also exceptional in its geographical extent during the peak periods of 1972–1973 and 1983–1984, when its effects were felt as far south as the equatorial zone. 1984 is considered to have been the driest year in the twentieth century on the African continent, in both hemispheres. These rainfall deficits are expressed as regional indices (Lamb, Nicholson), or are calculated at the most representative stations. The severity of the drought over several decades is illustrated by the southward migration of the mean isohyets, as mentioned above, and by variations in the rainfall averages in the Sahel, the 1961–1990 averages are in many places 20–30% lower than the 1931–1960 averages.

Surfacewater hydrology

Knowledge of the Sahel's hydrological characteristics is recent, since the earliest flood studies only date from the early 1950s, with a series of surveys in the subdesert zone and the gradual equipping of some 30 representative basins, monitored for two or three years. The permanent stations in the Sahelian networks only monitor water courses south of the 400 mm isohyet. Fewer than 30 are active throughout the zone, giving an observation coverage of 30–35 years at most (and with gaps in many cases).

It is very difficult to assess water resources in the Sahelian zone because precipitation varies so widely from place to place. There are also practical reasons why little or no observation data are available for some regions:

- sparsely scattered measuring networks, poor access and widespread insecurity make it difficult, and not always possible, to record measurements regularly;
- measurement may be difficult or dangerous where runoff is brief or floods sudden and violent;
- staff may be poorly skilled, too few in number and often ill equipped.

Characteristics of runoff in the Sahel

Runoff is characteristically intermittent and very unevenly spread over time and space. Hydrographic degradation is also typical.

Intermittence

In the mountains and foothills of the subdesert zone, from July to the end of August, there can be sudden, brief floods. In the Aïr mountains, for example, at the 150 mm interannual isohyet, local observation shows that there are 50 days of runoff in the wettest years and 25 in an average year, whereas in 1984 there were only 35 h. South of the 400 mm isohyet there is a series of less violent floods lasting about 3 months in the year, often with continuous streamflow between floods. In small basins, current drought conditions seem not to be having too much impact on total annual duration of runoff; in large basins, on the other hand, runoff duration is far more sensitive to climatic variations; in the very large basins it has decreased spectacularly. The river Komadugu, for example, which is fed by well-supplied aquifers and,

in an average year, flows into Lake Chad for 9 months of the year, flowed for only 4 months in 1984.

Non-uniformity in time and space

Rainfall patterns (see above) are the primary factor in the variability of runoff, the second being physio-geographical conditions (soil type, nature of bedrock and vegetation, relief).

As the rains are localized, their runoff coefficients are naturally very variable; in the event of heavy rain, these can very well vary from 5% on granitic sand to 75% on clayey regs (hence the usefulness of knowing which parts of a basin are 'active'). Moreover, the permeability of a soil can change considerably over the course of a wet season, depending on the degree of saturation, vegetation growth and surface crusting. Of course, specific discharge falls rapidly as basin size increases, and hydrographic degradation makes for even greater unevenness by causing a rapid loss of runoff by evaporation proceeding downstream.

In addition to spatial variability, there is of course considerable irregularity from year to year. Take, for example, the 12 km^2 granitic Abou Goulem Basin in Chad. Here, in a median year with 400 mm of rainfall, the runoff coefficient (K_r) is 3% i.e. a runoff depth (L_e) of 12 mm. It is estimated that once every 100 years, an L_e of 102 mm (i.e. $K_r = 13\%$ and $P = 780$ mm) should occur. But once every 20 years on average, runoff must be strictly zero for the whole year.

Hydrographic degradation

This generalized phenomenon occurs as the slope of a watercourse begins to flatten out, and is caused by several factors:

- the length of the dry season, which tends to denude the soil;
- sporadic floods that cannot evacuate all the eroded materials from the stream bed,
- wide, gently sloping expanses downstream, enabling the runoff to spread out and gradually evaporate (endoreic conditions).

Calculating baseline parameters

The two hydrological parameters which it is most important to estimate are the annual volume of runoff (depth of runoff) and flood characteristics (base time, increase time, peak flow and form). Publications are now available with charts which, despite gaps and imperfections, can be used to calculate runoff volumes and flood data, for different probability frequencies, in basins that are poorly known or have not been measured. However, to avoid calculation errors of more than 100% it is essential to follow checklist instructions carefully. One needs to know, for example, that in 10 km^2 basins observed in West and Central Africa, the maximum 10-year probability flood varies between 4 and 160 m^3 s^{-1} depending on the type of basin, i.e. a ratio of 1:40.

For a given surface area and rainfall, such a questionnaire would cover the permeability of the basin, soil compactness, relief (slope) and vegetation, but also the degree of hydrographic degradation, the stream network pattern, hydraulic conditions, etc.

Changes in Sahelian watercourses since 1960

For the drought period from 1966 to the present, the hydrological behavior recorded by permanent measuring stations on Sahelian basins of less than several thousand square kilometers has been in sharp contrast with that of the region's major tropical rivers, the Senegal, Niger and Logone-Chari. For the small basins, studies of recent floods show no reduction in maximum peak discharge. This information is of capital importance for calculating engineering projects; moreover, calculations of water budget factors for several Sahelian basins in Burkina Faso, such as the Gorouol at Dolbel, show that, although annual rainfall has fallen by 20–40% over the past 20 years, annual runoff over the same period has not fallen, since the annual runoff coefficients have gradually increased (and in some cases doubled). These findings are corroborated by experiments run on a number of basins or plots in Burkina Faso. These show that, in Sahelian zones, surface runoff and infiltration depend almost exclusively on surface state. The present drought has considerably favored the development of eroded and crusted areas, and hence a gradual increase in impermeability and fewer obstacles to runoff.

In basins of more than 10 000 km^2, however, the combined effects of hydrographic degradation, evaporation in areas where surface runoff spreads (2.5–3 m per year) and the absence of infiltration in the dry season, drastically reduce annual runoff, by 40–60% or even, as in 1984, by 90%.

Causes of the present drought

Nowhere in the world has drought persisted for nearly 30 years as it now has in the Sahel. A great deal of research has been done to identify possible causes. At present, there is a consensus of opinion that the causes are planetary, and that the local impact of human activity in hastening desertification (overgrazing, deforestation, etc.), or the role of albedo (Charncy's theory) with a feedback effect on land surface states, are of lesser importance. Researchers have found clear evidence of the influence of certain planetary anomalies, especially the role of sea surface temperatures. Thus a year with high sea surface temperatures in the Southern Ocean and Indian Ocean, combined with cooling of the waters of the North Atlantic, leads to a drier than usual rainy season; trial short-term rainy season forecasts based on this have already given promising results. Atmospheric circulation also plays a major part in generating rainfall deficits: this is so if the ITCZ migrates late and covers only a short distance, if the West African Walker cell weakens, or if a week TEJ combines with a strong AEJ (see above). Nonetheless, many aspects are still not understood and current multidisciplinary research is fully justified.

J.H. Sircoulon

Bibliography

Folland, C.K., Palmer, T.N. and Parker, D.E., 1986. Sahel rainfall and worldwide sea temperatures, 1901–1985, *Nature*, **320**, 602–607.
Hulme, M., 1992. Rainfall changes in Africa: 1931–60 to 1961–90, *International Journal of Climatology*, **12**, 685–699.
Nicholson, S.E., Kim, J. and J. Hoopingarner, 1988 *Atlas of African Rainfall and its Interannual Variability*. Dept. of Meteorology, Florida State University, Tallahassee, Florida.
Rodier, J.A., 1982. Evaluation of annual runoff in tropical African Sahel, in *Travaux et Documents de l'ORSTOM* no. 145, Paris, 211 pp.
Rodier, J.A., 1985. Aspects of arid zone hydrology, in J.C. Rodda (ed.), *Facets of Hydrology*, vol. II. John Wiley and Sons, Chichester, pp. 205–247.
Rodier, J.A. and Roche, M., 1978. River flow in arid regions, in R.W. Herschy (ed.) *Hydrometry, principles and practices*, John Wiley and Sons, Chichester, pp. 453–472.
Sircoulon, J., 1992. Caractéristiques des ressources en eau de surface en zones arides de l'Afrique de l'Ouest. Variabilité et évolution actuelle, in *L'Aridité une contrainte au développement*. Coll. Didactiques ORSTOM (ed.), pp. 53–68.

Cross references

Climate and climate change
Climate data: sources
Desert hydrology
Desertification
Deserts
Drought
Droughts

SALTWATER WEDGE

A saltwater wedge is a wedge-like intrusion of a large mass of salt (saline) water flowing in from the sea on a flood (inflowing) tide under the fresh water in a tidal waterway. The wedge occurs because of the differing densities of the saline and the fresh water and the difference may be exacerbated by differing temperatures of the two types of water. Consideration of the effects of a saltwater wedge is important in relation to sedimentation, the conveyance of solids in the waters and matters of water quality. In recent years, many coastal sewage outfalls have had to be modified, rebuilt or even abandoned because of a lack of previous understanding of the effects of saltwater wedges.

P.G. Holland

Bibliography

BS 3680, Part 1, 1991. *Glossary of Terms*, British Standards Institution, London.

SAND DUNES

Dunes are eolian features of the desert surface, regular patterns in the surface topography that result from the mobilization and deposition of sand. Sand dunes occupy about 15–20% of the surface area of the world's deserts. A hierarchy of dune types and a variety of forms exist, based on size and on dune morphology and its relationship to the sand-transporting winds. Dune form and orientation provide much information about the prevailing wind regime which produced them.

The largest dunes, sometimes termed draa, are features on the scale of kilometers. In the Kalahari–Namib Desert and parts of Asia these attain heights of 200–250 m or more. More commonly, dunes are of the order of tens to hundreds of meters. These smaller dunes may rest on the surface of the larger draa.

Satellite imagery has shown that a great variety of dune forms exist. Classically, however, two main types of dunes are distinguished: longitudinal (or linear) dunes, which are roughly parallel to the prevailing winds, and transverse dunes, which are aligned more or less normal to the wind (Figures S1 and S2).

The longitudinal dunes are arranged in long, linear, parallel ridges, with an interdune spacing of the order of 1-3 km, somewhat smaller in

Figure S1 Linear dunes of the Namib Desert. These form in a bimodal wind regime, i.e. winds concentrated in two prevailing directions.

Figure S2 Transverse dunes along the coast of the Namib Desert. These form in a bimodal wind regime in which the prevailing winds come from one general direction with relative constancy.

Figure S3 Barchan dune in the southern Namib. Like the transverse dunes of Figures S2, these form in a unimodal wind regime, but under conditions of limited sand supply. The slight back-tilt of the crest is deformation resulting from an easterly berg wind, a disruption of the normal wind patterns in the region.

width, and up to 20 km or more in length. They comprise about 50% of the areas of dunes worldwide. Simple, compound and complex linear dunes are also distinguished on the basis of dune morphology. Simple linear dunes consist of a single ridge with a narrow, single crest, while compound dunes have multiple, narrow ridges along the crest. Complex linear dunes, also called star dunes or rhourds, have regularly shaped peaks along a main ridge and frequently secondary dunes form at oblique angles to the orientation of the primary ridge.

Transverse dunes are likewise parallel ridges, but roughly perpendicular to the wind. The crescent-shaped barchan dunes are also oriented normal to the sand-moving winds, with the horns of the crescent facing upwind. Barchans are quite rare, occupying only about 0.01% of the area of dunes worldwide (Figure S3).

Dunes arise within vast sand seas, the formation of which is reasonably well established. Sand is transported from areas with a high wind energy and relatively constant direction, such as seashores, and accumulates in areas of weaker and more variables winds. Such conditions exist in deserts like the western Sahara and the Kalahari, which lie in the vicinity of the subtropical high-pressure belt. The mechanisms by which dunes arise within the sand seas are not fully understood.

In general, the control of dune formation is closely linked to sand supply and wind regime. The transverse and barchan dunes are associated with high-energy, unidirectional wind, with barchans forming in areas where sand supply is limited. The linear dunes form where the wind regime is bimodal, with one primary direction, or where it is predominantly from one sector but with much variation of direction within that sector. The compound and complex dunes result from multidirectional wind regimes.

The mechanism for barchan formation is well established: winds elongate a mass of sand then flow accelerates around this sandy 'obstacle', stretching the horns in a downwind direction. This, together with a lee vortex transporting sand back upwind between the ridges, produces the characteristic crescent shape. The shape rapidly becomes exaggerated, since the horns are narrower than the bulk of the dune and hence response more readily to the winds. Barchans are relatively mobile, advancing downwind as much as 10–50 m per year. One type of linear dune, the seif dune with a sinuous crest, results from an elongation of barchans when these advance into a bidirectional wind regime.

The formation of other linear dune types and of transverse dunes appears to be linked to atmospheric vortices. Bagnold (1941) explained transverse dunes as a consequence of atmospheric Kelvin–Helmholtz waves sculpting the surface at preferred intervals, corresponding to their wavelength. He postulated that simple linear dunes results from helical roll vortices in the atmosphere blowing essentially along the ridges and piling up sand where they converge.

Alternate mechanisms of linear dune formation, offered recently by Tsoar (1983) and Livingstone (1990), are more widely accepted.

These involve vortices produced as oblique winds traverse a dune crest. The flow is deflected and blows parallel to the dune on the leeward side, transporting material and elongating the dune.

Sharon E. Nicholson

Bibliography

Bagnold, R.A., 1941. *The Physics of Blown Sand and Desert Dunes*, London: Chapman & Hall.
Cooke, R.U. and A. Warren, 1975. *Geomorphology in Deserts*, London: B.T. Batsford.
Livingstone, I., 1990. Desert sand dune dynamics: review and prospect, in *Namib Ecology* (M.K. Seely, ed.), Pretoria: Transvaal Museum, pp. 47–53.
Tsoar, H., 1983. Dynamic processes acting on a longitudinal (seif) dune, *Sedimentology*, **30**, 567–578.

Cross references

Desert hydrology
Desertification
Deserts
Savanna

SANITATION AND CLEAN WATER

Introduction

Although the provision of clean water and sanitation is often omitted from the list of priority environmental challenges, in many parts of the developing world it ranks at the top. Two environmental issues are involved: the costs to human health and productivity of polluted water and inadequate sanitation, and the stresses placed on water resources by rapidly growing human demands for water.

Water supply and sanitation as environmental priorities

Inadequate sanitation is a major cause of the degradation of the quality of groundwater and surface water. Economic growth leads to larger discharges of wastewater and solid wastes per capita. Inadequate investments in waste collection and disposal mean that large quantities of waste enter both groundwater and surfacewater. Groundwater contamination is less visible but often more serious because it can take decades for polluted aquifers to cleanse themselves and because large numbers of people drink untreated groundwater.

More environmental damage occurs when people try to compensate for inadequate provision. The lack or unreliability of piped water causes households to sink their own wells, which often leads to overpumping and depletion. In cities such as Jakarta, where almost two-thirds of the population relies on groundwater, the water table has declined dramatically since the 1970s. In coastal areas this can cause saline intrusion, sometimes rendering the water permanently unfit for consumption. In, for example, Bangkok excessive pumping has also led to subsidence, cracked pavements, broken water and sewerage pipes, intrusion of seawater, and flooding.

Inadequate water supply also prompts people to boil water although in many cases fuel is only available for cooking and the family drink polluted unboiled water. The practice is especially common in Asia. In Jakarta more than $50 million is spent each year by households for this purpose – an amount equal to 1% of the city's GDP. Investments in water supply can therefore reduce fuelwood consumption and air pollution.

Effects on health

The health benefits from better water and sanitation are significant. When services were improved in the industrial countries in the nineteenth and twentieth centuries, the impact on health was revolutionary. Today, adequate water and sanitation services are just as vital: diarrheal death rates are typically about 60% lower among children in households with adequate facilities than among those in households without such facilities.

Specific investments that matter for health

The potential health benefits from improved water and sanitation services are huge.

Water quality

Contrary to common belief, contamination of water in the home is relatively unimportant. What matters is whether the water coming out of the tap or pump is contaminated. In most developing countries the imperative is to get from 'bad' quality (say, more than 1,000 fecal coliforms per 100 ml) to 'moderate' quality (less than 10 fecal coliforms per 100 ml), not necessarily to meet the stringent quality standards of industrial countries.

Water availability

As long as families have to go out of the yard to collect water, the quantities used will remain low (typically between 15 and 30 l per capita per day). The use of water for personal hygiene usually increases only when availability rises to about 50 liters per capita per day and generally depends on getting the water delivered to the yard or house.

Excreta disposal

It is necessary to distinguish between the effects on the household and on the neighborhood. For the household, the health impacts of improved sanitation facilities depend only on getting the excreta out of the house and are thus similar whether family members use an improved pit latrine, a cesspool overflowing into a street drain, or a conventional sewerage system. For the neighborhood, the key is the removal of excreta, a task done well by a wide range of technologies but badly by many commonly used systems (such as nightsoil collection and unemptied septic tanks). Because all the fecal–oral transmission routes are much more important when people live in close proximity to each other, the ill effects of poor environmental sanitation are greatest in high-density urban settlements.

Effects on productivity

Improved environmental sanitation has economic benefits. Consider the case of sewage collection in Santiago, Chile. The principal justification for investments was the need to reduce the extraordinarily high incidence of typhoid fever in the city. A secondary motive was to maintain access to the markets of industrial countries for Chile's increasingly important exports of fruit and vegetables. To ensure the sanitary quality of these exports, it was essential to stop using raw wastewater in their production. In the light of the recent cholera epidemic in Latin America, this reasoning was prescient. In just the first 10 weeks of the cholera epidemic in Peru, losses from reduced agricultural exports and tourism were estimated at $1 billion – more than three times the amount that the country had invested in water supply and sanitation services during the 1980s.

Improved access to water and sanitation also yields direct economic benefits. For many rural people, obtaining water is time consuming and heavy work, taking up to 15% of women's time. Improvement projects have reduced the time substantially. In a village on the Mueda Plateau in Mozambique, for instance, the average time that women spent collecting water was reduced from 120 to 25 min per day. Family well being was thus improved, as the time saved could be used to cultivate crops, tend a home garden, trade in the market, keep small livestock, care for children or even rest. Because users clearly perceive these time savings, they are willing to pay substantial amounts (as discussed below) for easier access.

In the absence of formal services, people have to provide their own services, often at high cost. In Jakarta, for instance, about 800 000 households have installed septic tanks, each costing several hundred dollars (not counting the cost of the land). In many cities and towns large numbers of people buy water from vendors. A review of vending in 16 cities shows that the unit cost of vended water is always much higher than that of water from a piped city supply – from 4 to 100 times higher, with a median of about 12. The situation in Lima is typical; although a poor family uses only one-sixth as much water as a middle-class family, its monthly water bill is three times as large. Consequently, in the slums around many cities water costs the poor a large part of household income – 18% in Onitsha, Nigeria, and 20% in Port-au-Prince, for example.

The economic costs of compensating for unreliable services – by building in-house storage facilities, sinking wells, or installing booster pumps (which can draw contaminated groundwater into the water distribution system) – are substantial. In Tegucigalpa, for example, the sum of such investments is so large that it would be enough to double the number of deep wells providing water to the city. The costs of compensating for poor water quality are great, too. In

Bangladesh boiling drinking water would take 11% of the income of a family in the lowest quartile. With the outbreak of cholera in Peru the Ministry of Health urged all residents to boil drinking water for 10 min. The cost of doing so would amount to 29% of the average household income in a squatter settlement.

Investments in sanitation and water offer high economic, social and environmental returns. Universal provision of these services could become a reality in the coming generation. But the next four decades will see urban populations in developing countries rise threefold and domestic demand for water increase fivefold. Current approaches will not meet these demands, and there is a real possibility that the numbers unserved could rise substantially, even while aquifers are depleted and rivers degraded.

Managing water resources better

When there was little competition for water, it was (correctly) used in large quantities for activities in which the value of a unit of water was relatively low. In many countries irrigated agriculture became the dominant 'high-volume, low-value' user. Today about 70% of all water withdrawals (and higher proportions of consumptive use) are for irrigation. This share is even higher in low-income countries, as shown in Table S1. In most countries this water is provided at heavily subsidized prices, with users seldom paying more than 10% of operating costs.

As demand by households, industries and farmers increases, governments find it hard to change existing arrangements. The allocation of water in all countries is a complex issue and is governed by legal and cultural traditions. Users typically have well-established rights. Reallocation is a contentious and ponderous process that generally responds to changes in demand only with long lags.

Paradoxically, there is good news in these distortions. Their very size indicates that urban shortages could be met with only modest reallocation. In Arizona, for instance, the purchase of the water rights from just one farm is sufficient to provide water for tens of thousands of urban dwellers. Because of the low value of water in irrigated agriculture, the loss of this marginal water has little overall effect on farm output. To help transfers, new market-driven methods for reallocation have been developed. When a recent drought dangerously reduced available water, the State of California set up a voluntary 'water bank' that purchased water from farmers and sold it to urban areas. The farmers made a profit by selling the water for more than it was worth to them, while the cities got water at a cost well below that of other sources of supply.

In developing countries, too, a start is being made in applying innovative methods for managing water resources. China's State Science and Technology Commission found that the economic rate of return to a cubic meter of water used for agriculture was less than 10% of the return to municipal and industrial users. Once agricultural and urban users accepted that they had to look at water as an economic commodity with a price, progress – including reallocation – was possible. In addition, Jakarta has been reasonably successful in reducing the overpumping of its aquifers by registering groundwater users (especially commercial and industrial establishments) and by introducing a groundwater levy.

The striking features of these 'market-based' reallocation methods are that they are voluntary, they yield economic benefits for both buyers and sellers, they reduce the environmental problems caused by profligate use of water in irrigation and they lessen the need for more dams.

Without effective management of water resources, the cost of supplying water to cities will continue to rise. The most dramatic examples will be in large and growing urban areas. In Mexico City, where much water is used for irrigation, the city has to contemplate

Table S1 Sectoral water withdrawals, by country income group (source: World Resources Institute, 1990)

Income group	Annual withdrawals per capita (m³)	Withdrawals, by sector (%)		
		Domestic	Industry	Agriculture
Low-income	386	4	5	91
Middle-income	453	13	18	69
High-income	1167	14	47	39

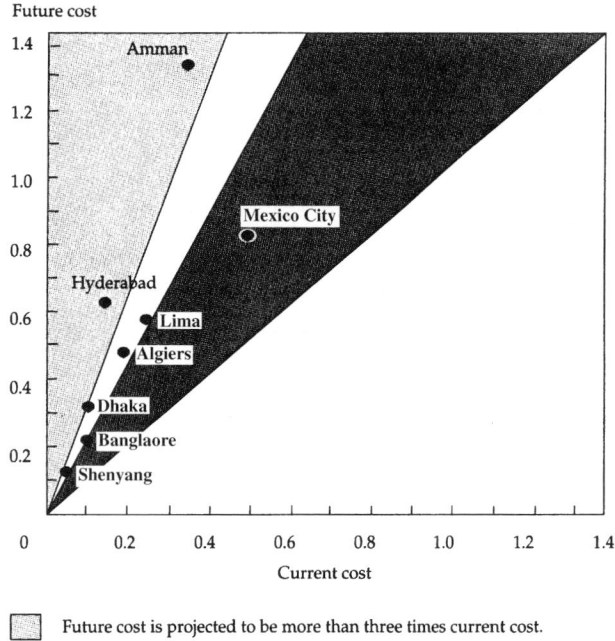

Future cost

☐ Future cost is projected to be more than three times current cost.

☐ Future cost is projected to be more than twice current cost.

■ Future cost is projected to be more than current cost.

Figure S4 The increasing cost of supplying water. (Source: World Bank, 1992d.)

pumping water over an elevation exceeding 1000 m into the Valley of Mexico; in Lima upstream pollution has increased treatment costs by about 30%; in Shanghai water intakes have already been moved upstream more than 40 km at a cost of about $300 million; and in Amman the most recent works involve pumping water up 1200 meters from a site about 40 km from the city. A recent analysis of the costs of raw water for urban areas in World Bank-financed projects (Figure S4) shows that the unit cost of water would more than double – and in some cases more than triple – under a new water development project.

Industries and households also need to be given incentives to use water efficiently. Cities, like farmers, have tended to take demand as given and to see their task as increasing supplies to meet it. As was the case with energy 20 years ago, little attention is paid to conservation and demand management in the water sector. This is both economically and environmentally unsound. Consider the case of Washington, DC. In the 1960s the US government concluded that 16 dams and more than $400 million were required to meet the water needs of the metropolitan area. Because of resistance from environmentalists to the construction of the dams, the plan had to be reconsidered. Eventually the number of dams was reduced to one and the total cost of the scheme to $30 million. The key changes were a revised plan for managing demand during droughts and more efficient operating rules. This illustrates once again that better economics and a better environment are compatible.

Experience in industrial and developing countries alike shows the potential for using water more cost effectively in industry. In the United States withdrawals of fresh water by manufacturing industries are expected to be 62% less in 2000 than in 1977, primarily because of the increased costs industries have to pay for disposing of industrial wastewater. In São Paulo, Brazil, the imposition of effluent charges induced three industrial plants to reduce their water demand by between 42 and 62%.

A particularly important conservation alternative is reclamation of wastewater. The reclamation of water for urban, industrial and agricultural use is attractive both for improving the environment and for reducing the costs of water supply. Reclaimed wastewater has been used for many years for flushing toilets in residential and commercial buildings in Japan and Singapore. A recent reclamation scheme in the Vallejo area of Mexico City illustrates the great

potential, both economic and environmental, of wastewater reuse and, to anticipate a theme developed below, the scope for the private sector.

At present, in most countries management of water resources is fragmented (industrial users, for example, do not have to take account of the costs that their use and pollution of water imposes on domestic users downstream) and is done by 'command and control' (most allocations are set by administrative fiat). The challenge is to replace this system with one that recognizes the unitary nature of the resource and its economic value and that relies heavily on price and other incentives to encourage efficient use of water.

Environmental improvement, management of water resources, and the private sector in Mexico

In 1989, faced with rising water prices and potential water shortages, a group of companies in the Vallejo area of Mexico City sought an alternative to water supplied by the public agency. At about the same time, the Mexican government decided to involve the private sector in water supply and wastewater treatment.

The industrialists realized that if sewage flows could be adequately treated, this could provide a cost-effective and reliable source of industrial water (and, incidentally, could improve the environment by treating wastes and reducing the need for new water supplies). Twenty-six Vallejo companies organized a new for-profit firm, Aguas Industriales de Vallejo (AIV), to rehabilitate an old municipal wastewater treatment plant. Each shareholder company contributed equity on the basis of its water requirements, with total equity amounting to $900,000.

AIV operates the plant under a 10-year concession from the government. The plant now provides $60 \, l \, s^{-1}$ to shareholders and $30 \, l \, s^{-1}$ to the government as payment for the concession. The concession agreement gives AIV the right to withdraw up to $200 \, l \, s^{-1}$ of wastewater from the municipal trunk sewer. AIV plans to double the plant's capacity within 5 years at an estimated cost of $1.5 million. The firm provides treated water to shareholder companies at a price equivalent to 75% percent of the water tariff charged by the government (currently, $0.95 \, m^{-3}$).

Providing services that people want and are willing to pay for

During the United Nations Drinking Water and Sanitation Decade of the 1980s, coverage increased. But about 1 billion people still lack an adequate water supply, and about 1.7 billion people do not have adequate sanitation facilities. The quality of service often remains poor. In Latin America, for example, levels of leakage and pipe breakage are, respectively, four times and 20 times higher than is normal in industrial countries. In Lima 70% of the water distribution districts provide inadequate water pressure. In Mexico 20% of the water supply systems have unreliable chlorination facilities.

Developing countries cannot afford to provide all people with in-house piped water and sewerage connections. The policy has usually been to concentrate primarily on the (subsidized) provision of water, often through house connections for the better-off and standpipes or handpumps for the poor.

Consumers in most industrial countries pay all of the recurrent costs (operations, maintenance and debt service) of both water and sewerage services. They also pay most of the capital costs of water supply and a large (typically over half) and rising portion of the capital costs of sewerage. In developing countries, by contrast, consumers pay far less. A recent review of World Bank-financed projects showed that the effective price charged for water is only about 35% of the average cost of supplying it. The proportion of total project financing generated by utilities points in the same direction: internal cash generation accounts for only 8% of project costs in Asia, 9% in Sub-Saharan Africa, 21% in Latin America and the Caribbean, and 35% in the Middle East and North Africa.

In urban areas there is abundant evidence that most people want on-plot water supplies of reasonable reliability and are willing to pay the full cost of these services. In some areas this standard solution will have to be adjusted and special efforts made to accommodate poor people. In Latin America and, more recently, in Morocco, utilities have helped poor families to install a connection and in-house plumbing by giving them the option of paying over several years. Another option is a 'social tariff' whereby the better-off cross-subsidize the poor. Properly executed, such policies are both sensible

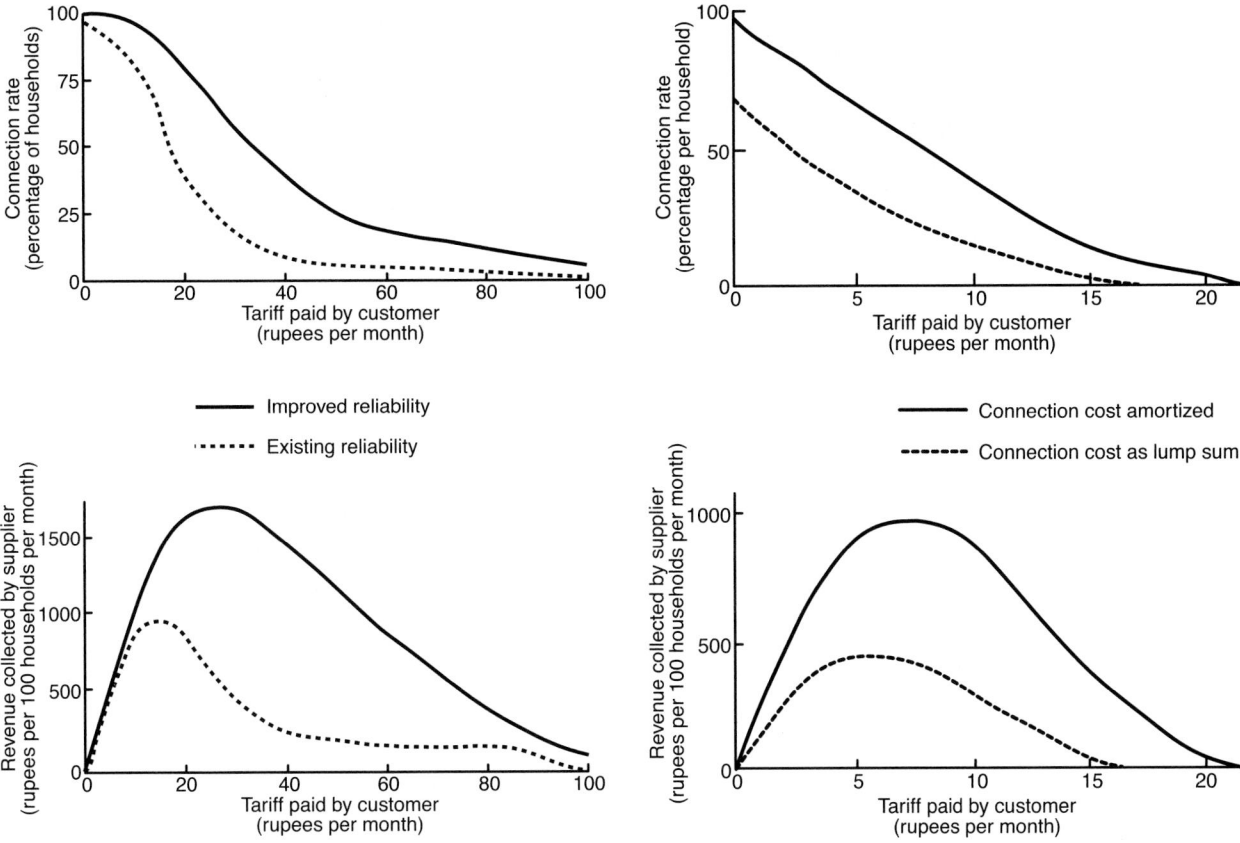

Figure S5 How reliability of supply affects willingness to pay for piped water: Punjab, Pakistan. (Source: World Bank Water Demand Research Team.)

Figure S6 How spreading connection costs over time affects willingness to pay for piped water: Kerala, India. (Source: World Bank Water Demand Research Team.)

(since the poor use relatively little water) and compassionate. But there are dangers. Social tariffs can lead to a general spread of subsidies. In addition, the assignment of non-commercial objectives to a public enterprise generally has as insidious effect on the achievement of all its objectives, commercial and non-commercial alike.

It is widely assumed that the demand situation in rural areas is quite different, that there people have only a 'basic need' which can be met with a public tap or handpump. But a recent multi-country study by the World Bank of rural water demand found that most rural people want and are willing to pay for a relatively high level of service (yard taps). As shown in Figure S5, they will pay substantially more if that service is reliable. As Figure S6 illustrates, more people will make use of improved water supplies if innovative financing mechanisms are employed.

Willingness to pay for water in rural areas

The World Bank, in conjunction with other agencies, completed a study of rural water demand in Brazil, Haiti, India, Nigeria, Pakistan, Tanzania and Zimbabwe. The study suggests that where water demand is concerned, there are four broad categories of rural community.

Type I: willingness to pay for private connections is high and willingness to pay for public water points is low. Communities in this group offer exciting possibilities because people want and are willing to pay the full costs of a reliable water service delivered by way of private metered connections into the house or yard. The availability of free public taps (for the poor) will not appreciably affect the demand for private connections. The appropriate strategy is to offer private connections and even encourage them (specifically, by amortizing connection costs in monthly water bills), to recover all costs through the tariff and to deliver a reliable service. A striking finding from the World Bank study is that this category is larger than is commonly

assumed; it probably includes many communities in Southeast Asia, South Asia, Latin America, the Middle East and North Africa.

Type II: only a minority of households are willing to pay the full costs of private connections, but most households are willing to pay the full costs of public water points. Although overall willingness to pay for improved water service is considerable in type II communities, users vary greatly in their willingness to pay for different levels of service. In these villages the provision of free public water points (such as standpipes, wells or boreholes) would significantly reduce the demand for private connections. When there is heavy reliance on public water points, some charge must be levied on water from these sources in order to finance the system. Here the greatest challenge is to devise revenue collection systems that are sensitive to people's preferences about when they want to buy water and how they want to pay for it. Kiosks appear to be an attractive and flexible option for many households. Those who wish to have house connections should be able to do so but must have metered connections and must pay the full cost. Many of the better-off communities in Sub-Saharan Africa and poorer communities in Asia and Latin America probably fall into this category.

Type III: households' willingness to pay for improved service is high but not high enough to pay the full costs of an improved service. This group typically includes poor communities in arid areas in South Asia and Sub-Saharan Africa. As in type II villages, people are willing to pay a relatively large share of their income for an improved water service. The distinction is that the costs of supply are so high, as a result of a combination of aridity and low population densities, that improved systems will not be built and operated without subsidies. Given the high priority that people give to improved water supply, if transfers were available from central government or from foreign donors, households would typically choose to spend the funds on an improved water supply. The primary service offered in such communities would be public taps, wells or boreholes, although in piped systems metered yard taps should be allowed, with tariffs set to recover full costs.

Type IV: willingness to pay for any kind of improved service is low. This group typically includes poor communities in which (1) traditional water supplies are considered more or less satisfactory by the population or (2) water supply is seen as the financial responsibility of the government. In such communities self-financed improved water supplies are not feasible. Given the low priority accorded improved water supply, available subsidies could be better used in providing other, more highly valued infrastructural services. For the time being, the appropriate rural water supply policy in such cases is simply to do nothing. For the second category, once government paternalism ceases, communities may express a willingness to pay and will become type II communities.

Breaking out of the 'low-level equilibrium trap' in northeast Thailand

A well-documented case in northeast Thailand, covering a 20-year period, demonstrates the importance of discovering what users of rural water services want rather than making assumptions about the answers.

Since the people in the area were poor, the initial project was intended to provide protected water at the lowest possible cost. Because groundwater is abundant in the region, the technology chosen was handpumps. After 5 years most of the handpumps were not working, and water-use habits were largely unchanged. In a follow-up phase, motor pumps provided piped water at community standpipes. Again, the project failed. Five years after implementation 50% of the systems were not working at all, and another 25% operated intermittently.

As was consistent with conventional assumptions, the failures were attributed to technologies that were too complex to maintain and to the inability of the villagers to pay for improved supplies. Gradually, however, it became apparent that the main problem was not the capabilities of the villagers but the fact that the service being offered was not what they wanted. They did not want handpumps, which were not considered an improvement over the traditional rope-and-bucket system. Also, standpipes, being no closer than their traditional sources, offered no obvious benefits. Only piped water to yardtaps could meet people's aspirations.

In the next project yardtaps were allowed, with the users paying the full costs of connection. Five years later the verdict was in: 90% of the systems were functioning reliably, 80% of the people were served by yardtaps, meters had been installed and locally adapted charging systems had been developed. Not only were the systems well maintained, but because the service was so popular, many systems had extended distribution lines to previously unserved areas.

In other words, in terms of the typology discussed above when these (poor) people were treated as 'type IV' cases, the result was the familiar low-level equilibrium trap. When they were treated as 'type I' communities, the cycle was broken and a high-level equilibrium was established.

Twenty years of experience with the provision of water in rural Thailand shows how it is possible to break out of a 'low-level equilibrium trap' (in which a low level of services is provided, thus willingness to pay and thus revenues are low, and the operation consequently deteriorates) to a 'high-level equilibrium' in which users get a high level of service, pay for it and maintain the desired system.

Increasing investments in sanitation

Public investment in water supply and sanitation accounts for 10 % of total public investment in developing countries, or about 0.6 % of GDP. Spending on sewerage and sanitation accounts for substantially less than one-fifth of lending in World Bank-financed projects. Most of this has been for sewage collection, with little spent on treatment. An indication of the huge underinvestment in treatment is that only 2 % of sewage in Latin America is treated. Similarly, only a small proportion (typically 5 % in developing countries, compared with 25 % in industrial countries) of all spending on solid wastes is directed to their safe disposal.

There is abundant evidence that urban families are willing to pay substantial amounts for the removal of excreta and wastewater from their neighborhoods. People want privacy, convenience and status; polluted water smells unpleasant and fosters mosquitos; and the installation of sewers typically increases property prices. As with water supply, so with sanitation: where public provision is absent, people pay significant amounts for privately provided services. Even in poor cities the amounts paid are considerable. In Kumasi, Ghana, for example, the use of public latrines and bucket latrines accounts for large recurrent expenditures – about 2.5 and 1 %, respectively, of family income. In Kumasi and in Ouagadougou families are willing to pay about 2 % of household income for an improved sanitation system. This is roughly the amount paid for water and for electricity. The following examples of northeast Brazil and of Orangi, Pakistan, show the willingness of households to pay for having wastewater carried out of the neighborhood (by means of a low-cost sewer).

Innovative sewerage in northeast Brazil: the condominial system

The condominial system is the brainchild of José Carlos de Melo, a socially committed engineer from Recife. The name 'condominial' was chosen for two reasons. First, a block of houses was treated like a horizontal apartment building – or *condominiais*, in Portuguese. Second, 'Condominial' was the title of a popular Brazilian soap opera and so was associated with the best in urban life. The result is a layout radically different from the conventional system, with a shorter grid of smaller and shallower 'feeder' sewers running through backyards and with the effects of shallower connections to the mains rippling through the system. These innovations cut construction costs to between 20 and 30% of those of a conventional system.

The more fundamental and radical innovation, however, is the active involvement of the population in choosing the level of service and in operating and maintaining the 'feeder' infrastructure. Families can choose to continue with their current sanitation system, to connect to a conventional waterborne system (which usually means a holding tank discharging into an open street drain), or to connect to a 'condominial' system.

If a family chooses to connect to a condominial system, it has to pay a connection charge (financed by the water company) of, say, X cruzados, and a monthly tariff of Y cruzados. If it wants a conventional connection, it has to pay an initial cost of about $3X$ cruzados and a monthly tariff of $3Y$ cruzados, reflecting the higher capital and operating costs of the conventional system.

Families are free to continue with their current system. In most cases, however, those families that initially choose not to connect eventually change their minds. Either they succumb to heavy pressure from their neighbors, or they find the build-up of wastewater in and around their houses intolerable once the (connected) neighbors fill in the rest of the open drain.

Individual households are responsible for maintaining the feeder sewers, with the formal agency tending only to the trunk mains. This has several related positive results. First, it increases the communities' sense of responsibility for the system. Second, the misuse of any portion of the feeder system (by, say, putting solid wastes down the toilet) soon shows up as a blockage in the neighbor's portion of the sewer. The consequence is rapid, direct and informed feedback to the misuser. This virtually eliminates the need to 'educate' the users of the system about do's and don'ts and results in fewer blockages than in conventional systems. Third, because of the greatly reduced responsibility of the utility, operating costs are much lower.

The condominial system is now providing service to hundreds of thousands of urban people in northeast Brazil. The danger is that the clever engineering may be seen as 'the system.' Where the community and organizational aspects have been missing, the technology has worked poorly (as in Joinville, Santa Catarina) or not at all (as in the Baixada Fluminense in Rio de Janeiro).

Innovative sewerage in a Karachi squatter settlement: the Orangi Pilot Project

In the early 1980s Akhter Hameed Khan, a world-renowned community organizer, began working in the slums of Karachi. He asked what problem he could help resolve and was told that 'the streets were filled with excreta and wastewater, making movement difficult and creating enormous health hazards.' What did the people want, and how did they intend to get it? he asked. What they wanted was clear – 'people aspired to a traditional sewerage system . . . it would be difficult to get them to finance anything else.' How they would get it, was also clear – they would have Dr Khan persuade the Karachi

Development Authority (KDA) to provide it free, as it did (or so the poor perceived) to the richer areas of the city.

Dr Khan spent months going with representatives of the community to petition the KDA to provide the service. When it was clear that this would never happen, Dr Khan was ready to work with the community to find alternatives. (He would later describe this first step as the most important thing he did in Orangi – liberating, as he put it, the people from the immobilizing myths of government promises.)

With a small amount of core external funding, the Orangi Pilot Project (OPP) was started. It was clear what services the people wanted; the task was to reduce the costs to affordable levels and to develop organizations that could provide and operate the systems. On the technical side, the achievements of the OPP architects and engineers were remarkable and innovative. Thanks partly to the elimination of corruption and the provision of labor by community members, the costs (for an in-house sanitary latrine and house sewer on the plot and underground sewers in the lanes and streets) were less than $50 per household.

The related organizational achievements are equally impressive. OPP staff members have played a catalytic role: they explain the benefits of sanitation and the technical possibilities to residents, conduct research and provide technical assistance. The OPP staff never handle the community's money. (The total costs of the OPP's operations amounted, even in the project's early years, to less than 15% of the amount invested by the community.) The households' responsibilities include financing their share of the costs, participating in construction and electing a 'lane manager' who typically represents about 15 households. Lane committees, in turn, elect members of neighborhood committees (typically representing about 600 houses), which manage the secondary sewers.

The early successes achieved by the project created a 'snowball' effect, in part because of the increased value of properties with sewerage systems. As the power of the OPP-related organizations increased, they were able to put pressure on the municipality to provide funds for the construction of trunk sewers.

The Orangi Pilot Project has led to the provision of sewerage services to more than 600 000 poor people in Karachi and to recent initiatives by several municipalities in Pakistan to follow the OPP method and, according to OPP leader Arif Hasan, 'have government behave like an NGO.' Even in Karachi the mayor now formally accepts the principle of 'internal' development by the residents and 'external' development (including trunk sewers and treatment) by the municipality.

Expanding the range of supply options

A vital element of a demand-driven sanitation strategy is to expand the menu of services from which users can choose.

In city centers there is no alternative to costly waterborne systems. But even in relatively poor cities the difficulties are not insoluble. In Fortaleza, a poor city in northeast Brazil, developers of all high-rise buildings are required to, and do, install package sewage collection and treatment systems. The point here is not that this is a good technical solution but that even in a relatively poor city, developers can easily absorb such costs and pass them on to those who purchase units in the buildings.

Beyond the urban core, however, conventional sewerage systems (with average household costs anywhere from $300 to $1,000) are too expensive for most developing countries. In recent decades efforts have been made to develop technological alternatives. Most of this work has concerned the on-site disposal of excreta. Pour-flush latrines and ventilated improved pit (VIP) latrines are often the technologies of choice – they provide good service (privacy and few odors) at reasonable cost (typically about $100–200 per unit), and their installation and functioning does not depend on the municipality or other organization. At even lower cost, there are yet simpler improvements, such as the latrine slab program that proved successful in Mozambique.

For a variety of reasons – high housing densities, impermeable soils and the need to dispose of considerable quantities of domestic wastewater – on-site solutions do not function well in many urban areas. Sewage and wastewater collect in the streets and in low-lying areas, creating serious aesthetic and health problems. In many settings people aspire to 'the real thing', i.e. waterborne sewerage.

- Effluent sewerage is a hybrid between a septic tank and a conventional sewerage system. Its distinctive feature is a tank, located between the house sewer and the street sewer, that retains the solids, thereby allowing smaller sewers to be laid at flatter gradients and with fewer manholes. Such systems have been widely used in small towns in the United States and Australia and in Argentina, Brazil, Colombia, India, Mozambique and Zambia. The (limited) cost data suggest that solids-free sewerage costs about 20% less than conventional sewerage.
- Simplified sewerage, developed in São Paulo, allows smaller, shallower, flatter sewers with fewer manholes. This simplified design works as well as conventional sewerage but costs about 30% less. It is now routinely used in Brazil.
- The condominial system described above has been developed and applied in northeast Brazil. It comprises shallow, small-diameter backyard sewers laid at flat gradients and costs about 70% less than a conventional system.
- The Orangi Pilot Project in Karachi (see above) adapted the principles of effluent sewerage and simplified sewerage to the realities of a hilly squatter settlement in Karachi. The result – not just the result of clever engineering – was a drastic reduction in the cost of sewers, from the $1,000 per household that was standard in Karachi to less than $50 per household (excluding the cost of the trunk sewers). The achievement is extraordinary – about 600 000 people in Orangi are now served with self-financed sewers.

Investing in waste disposal

There is an important difference between 'private goods' (including water supply and even wastewater and solid waste collection), in which the primary benefits accrue to individual households, and waste treatment and disposal, in which the benefits accrue to the community at large. In the first case, willingness to pay is an appropriate guide to the level of service to be provided, and the main source of finance should be direct charges to the users. In the case of waste disposal, however, public financing is essential. Governments that subsidize 'private' water supply and wastewater collection services are left with less money to finance treatment and disposal services.

No developing country, however, will have the luxury of collecting and treating wastewater from all households. Because the costs of meeting such goals are extremely high, even in industrial countries the full population is not served by wastewater treatment facilities; coverage is only 66% in Canada and 52% in France. In making the inevitable choices, the best ratio of benefits to costs will usually be achieved by concentrating most public funds on waste treatment in large cities, especially those that lie upstream from large populations.

In recent decades some important advances have been made in innovative sewage treatment processes. At the lower end of the spectrum is the stabilization pond, a technology that has proved robust, easy to operate and (where land is not costly) relatively inexpensive. A promising intermediate (in both cost and operational complexity) is the upflow anaerobic sludge blanket, which has performed well in Brazil and Colombia. The point is the importance of developing technical solutions that are adapted to the climatic, economic and managerial realities of developing countries.

Rethinking institutional arrangements

A comprehensive review of 40 years of World Bank experience in water and sanitation pinpoints 'institutional failure' as the most frequent and persistent cause of poor performance by public utilities. This section deals with the key areas for institutional reform. A World Bank review of more than 120 sector projects over 23 years concludes that only in four countries – Botswana, Korea, Singapore and Tunisia – have public water and sewerage utilities reached acceptable levels of performance. A few examples illustrate how serious the situation is:

- In Accra only 130 connections were made to a sewerage system designed to serve 2000 connections.
- In Caracas and Mexico City an estimated 30% of connections are not registered.
- Unaccounted-for water, which amounts to 8% in Singapore, is 58% in Manila and about 40% in most Latin American cities. For Latin America as a whole, such water losses cost between $1 billion and $1.5 billion in revenue forgone every year.
- The number of employees per 1000 water connections is between two and three in Western Europe and about four in a well-run developing country utility (Santiago), but between 10 and 20 in most Latin American utilities.

Financial performance is equally poor. A recent review of Bank projects found that borrowers often broke their financial performance covenants. A corollary is that the shortfalls have to be met through large injections of public money. In Brazil, from the mid-1970s to mid-1980s about $1 billion a year of public monies was invested in the water sector. The annual federal subsidy to Mexico City for water and sewerage services amounts to more than $1 billion a year, or 0.6% of national GDP.

Public utilities play a dominant role in the provision of water and sanitation services throughout the world. There are many examples of such utilities working effectively in industrial countries and, as described above, a few cases in developing countries. An essential requirement for effective performance is that both the utility and the regulatory body (essential for such natural monopolies) must be free from undue political interference. In the case of the utility the vital issue is managerial autonomy, particularly as regards personnel policies; in the case of the regulatory body, it is the setting of reasonable tariffs. Although this recipe is simple and has been well tested in many industrial countries, it has been extraordinarily difficult to implement in developing countries other than those with high levels of governance. Sometimes utilities and regulators are nominally autonomous, but usually key policies (on investments, personnel policies and tariffs, for instance) are effectively made by government and heavily influenced by short-term political considerations.

Many projects financed by external agencies have addressed the problems of public water utilities through sizeable action plans, technical assistance components and conditionality. Some of these efforts, such as that undertaken recently by Sri Lanka's National Water Supply and Drainage Board, have led to significant improvements in performance. As with public enterprises in other sectors, however, most of these efforts failed because, in the words of a recent Bank review, 'public enterprises . . . are key elements of patronage systems, . . . overstaffing is often rife, and appointments to senior management positions are frequently made on the basis of political connections rather than merit.' In addition, things have been getting worse rather than better. The achievement of institutional objectives in World Bank-financed water and sanitation projects fell from about two in three projects in the late 1970s to less than one in two projects 10 years later.

Improving the performance of public utilities nevertheless remains an important goal, for two reasons. First, in the medium term, public utilities will continue to provide services to many. Second, improvement in the performance of public utilities is often a precondition if private operators are to be induced to participate.

Experience in industrial countries shows that a central problem in improving environmental quality is that the public sector acts both as supplier of water and wastewater services and as environmental regulator – it is both gamekeeper and poacher. The results of this conflict of interest are similar throughout the world. In England and Wales prosecutions of those responsible for sewage treatment were rare when the river basin authorities were responsible for water resource management, environmental protection and services. In 1989 private companies were given responsibility for the delivery of water and sewerage services (with public agencies retaining regulatory authority). Since then, fines have been increased substantially and violators have been prosecuted. The other side of the separation of powers is that service delivery agencies are, in the process, liberated from serving multiple tasks and can pursue well-defined and specific objectives.

Expanding the role of the private sector

Increased private sector involvement is warranted in two areas. One is in services to public utilities. In industrial countries the engineering of public works is dominated by private firms, which depend for their survival on their reputation for performance and which assume legal liability for the consequences of any professional negligence. These factors provide powerful incentives for supplying cost-effective, high-quality services and concurrently furnish a stringent environment for the supervised apprenticeship training that is a required part of professional certification in these countries. By contrast, in many developing countries (particularly in Asia and Africa) the engineering of public works is dominated by large public sector bureaucracies. Employment security is total, promotion is by seniority alone, good work goes unrecognized, poor work is not subject to sanctions and an atmosphere of lethargy prevails. The direct consequence is the construction of high-cost, low-quality facilities; the indirect effects

include a weak professional labor force. The obvious answers are, first, to decrease the direct involvement of the government in public works and, second, to nurture a competitive engineering consultancy sector.

More private involvement in the operation of water, sewerage and solid waste companies is also warranted. Many industrial countries have found it difficult to reform public enterprises, except as part of a move to privatize them. Indeed, privatization is increasingly seen as a way not only to effect performance improvements but also to lock in the gains.

In developing countries there has been some experience with private sector operation of water and sanitation utilities. Côte d'Ivoire has been a pioneer: SODECI, in Abidjan, is considered one of the best-run utilities in Africa. After Macao's water utility was privatized in 1985, performance improved dramatically; the percentage of unaccounted-for water fell by 50% over 6 years. Guinea, which recently let a lease contract for supplying water to its principal cities, experienced dramatic improvements in the financial condition of the utility in just the first 18 months as a result of raising the efficiency of bill collection from 15 to 70%.

Other countries have taken more incremental approaches. EMOS, the utility serving Santiago, has used private contracts for functions such as meter reading, pipe maintenance, billing and vehicle leasing. As a result, it has a high staff productivity rate, three to six times higher than that for other companies in the region. Many other countries, faced with persistently poor performance by their public utilities, are seriously considering greater private sector involvement following, in general, variations of the French model. For example, in Latin America concession contracts are currently being let for the supply of water and sewerage services in Buenos Aires and Caracas.

Private involvement in the sector is not a panacea and is never simple. In the UK water privatization is generally considered the most complex of all privatizations undertaken. In developing countries there are formidable problems. For the private operator the risk involved is typically high. In addition to the obvious political and macroeconomic risks, knowledge about the condition of the assets is usually only rudimentary and there is uncertainty about the government's compliance with the terms of the contract. Groups such as existing agencies and labor unions that stand to lose from greater private sector involvement often strongly oppose privatization.

For the government, too, there are problems. Because of economies of scale, it is virtually impossible to have direct competition among suppliers in a specific area. Countries have tried a variety of solutions: in France, there is periodic competition for markets, and in England and Wales, economic regulators reward efficiency by comparing the relative performance of different companies (a practice that is unlikely to be applicable elsewhere). In addition, in many developing countries it is often difficult to attract private sector interest. Only a handful of firms compete internationally for such contracts.

The case for private sector involvement is stronger still in the solid waste collection business. Whereas foreign control of water supply is often perceived to involve losing sovereignty over a strategic sector, nobody cares if foreigners pick up the garbage. In addition, for populations of more than about 50 000 there are no economies of scale and thus no natural monopoly. Experience in many countries, including Argentina, Brazil, Canada, Chile, Colombia, Japan, Switzerland and the United States, has shown that the private sector almost invariably collects solid wastes more efficiently than municipalities. Unit costs for public systems are 50–200% higher, with the private sector efficiency gains apparently greatest in the developing countries listed.

Community groups and other NGOs also have an important role to play in the supply of water and sanitation services and the collection of wastes. In the urban fringe the most productive relationship between community groups and the formal sector is that of partnership, with the formal sector responsible for the 'external' or 'trunk' infrastructure and the community paying for, providing and managing the 'internal' or 'feeder' infrastructure.

Because many water and sanitation services are monopolies, consumers cannot force suppliers to be accountable by giving their business to a competitor. To give consumers a voice in the political process, consumers' associations and ratepayers' boards are vital. Paradoxically, because there is such an obvious need for oversight of the activities of a private operator of a natural monopoly, greater private sector involvement stimulates greater consumer involvement.

In the UK, for example, water users have had a much greater say in running the industry since privatization.

In recent years external agencies and governments alike have become aware that in rural areas involvement of the users is essential if water supplies are to be sustained. Generally it has been assumed that support to rural communities – in the form of information, motivation and technical assistance – will come from the government. The difficulty is that governments, especially in rural areas, are often weak, and their officials rarely have an incentive to provide support. Here the private sector (including NGOs) may be able to help.

Several promising examples of the involvement of small-scale private operators in developing countries have emerged:

- In rural Pakistan about 3 million families have wells fitted with pumps, many of which are motorized. The water supplies are paid for in full by the families, and all the equipment is provided and serviced by a vibrant local private sector industry.
- In Lesotho the government trained bricklayers to build improved pit latrines. Government banks also provided (unsubsidized) credit to finance the latrines. The program has been a singular success, thanks mainly to the aggressive role of the bricklayers in expanding their markets (and providing services as well).
- In West Africa a private handpump manufacturer has developed a 'Sears Roebuck'–type scheme whereby purchase of a pump comes with 5 years of support, including training and the provision of spare parts. Later on, the community will be able to maintain the pump and will purchase the necessary spare parts from local traders. Because the private sector agent has clear incentives for providing services effectively, this arrangement may work better than government support to the communities.

Finally, women have a central role to play in these reforms. In most countries the collection of water has been considered 'women's work' (except where the water is sold). Only recently, however, have systematic efforts been made to involve women in project identification, development, maintenance and upkeep. The results have generally been encouraging. In an urban slum in Zambia a women's organization improved drainage around public taps. Women have been trained as caretakers for handpumps in Bangladesh, India, Kenya, Lesotho and Sudan. In Mozambique women engineers and pump mechanics perform alongside, and as effectively as, their male counterparts. In Sri Lanka women's cooperatives have been set up to assemble and maintain a locally manufactured handpump. Women's cooperatives manage communal standpipes and collect money to pay for metered supplies in Honduras, Kenya and the Philippines. Women who are trained to manage and maintain community water systems often perform better than men because they are less likely to migrate, more accustomed to voluntary work and better trusted to administer funds honestly.

It can be seen that massive improvements can be made in health, economic efficiency and equity through better provision of sanitation and water. The key is firmly in the hands of governments, for the single most important factor needed is political will. Where there are long-established and deeply entrenched traditions of sound governance (as in Botswana, Korea and Singapore), it is evident that autonomous, accountable public sector agencies can provide efficient and equitable service. For many countries, however, such levels of governance are not attainable in the short run, so that greater involvement of the private sector and NGOs will be crucial to the provision of accountable and efficient services.

To allow helpful change to occur, the government must concentrate on the things that it, and only it, can do. Its job is to define and enforce an appropriate legal, regulatory and administrative framework. This includes tasks as fundamental and diverse as rewriting legislation so that water markets can come into existence, rewriting contract laws so that the private sector can participate with confidence, building a capacity for environmental and, where appropriate, economic regulation, developing financial mandates for utilities that encourage conservation, and setting and enforcing quality standards for equipment. The government must also create conditions under which others – the private sector, NGOs, communities and consumers – can play their parts.

What might be accomplished?

More than 1 billion people are still without access to safe water and 1.7 billion people are without access to adequate sanitation facilities. Elementary calculations show that an 'unchanged practices' or 'business-as-usual' scenario would lead to a rise in the number of people without service in the coming decades (the top curves in Figure S7). This is a result of rising unit costs, as well as unprecedented increases in population. If the shares of total investment allocated to sanitation (currently 0.6% of gross investment) and to water supply (currently 1.7%) were raised by, say, 50 and 30%, respectively, the numbers unserved might still rise, although not as

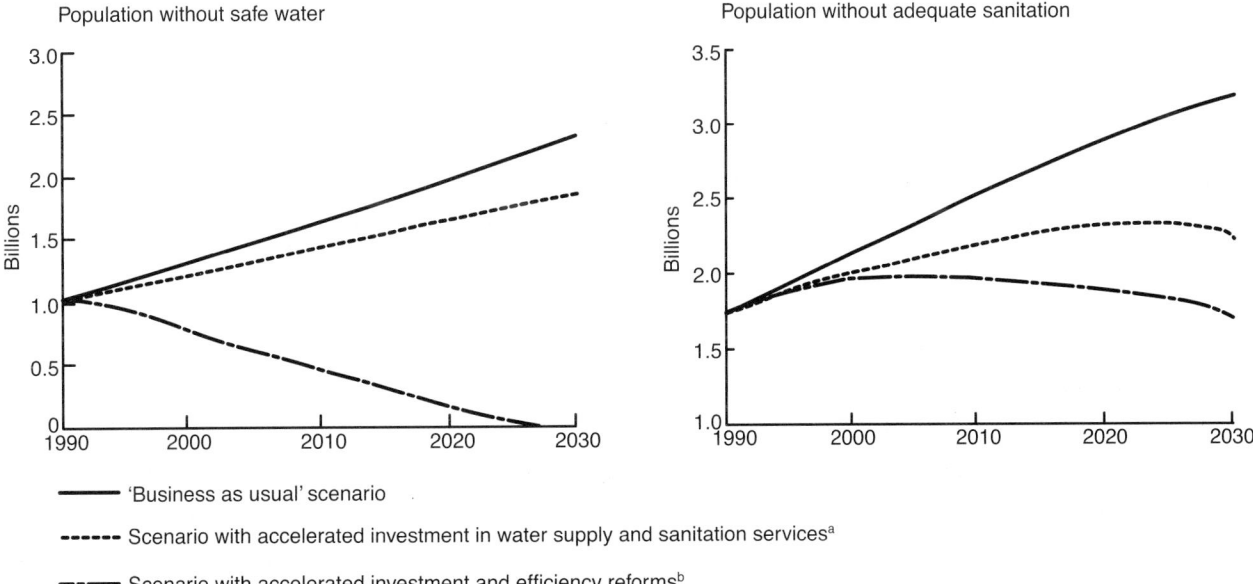

Figure S7 Safe water and adequate sanitation: three scenarios, 1990–2030. Assumptions: per capita income elasticity 0.3; price elasticity, −0.25; initial prices 60% of marginal costs, gradually rising to efficiency levels over a 25-year period; initial supply costs 50% higher than with good practices (due to managerial inefficiencies), gradually being reduced in step with price efficiency reforms; and marginal costs rising at 3% per year. [a]Investment in water supply increases 30%, and investment in sanitation services increases 0% over the period. [b]To realize this scenario in low-income countries, efficiency reforms – and the resulting increase in investment shares – would need to be greater than average. (Source: World Bank estimates.

much (the middle curves in the figure). Far more important (as shown by the bottom curves) is the combination of policy reforms and accelerated investment. By attracting financial, managerial and skilled labor into the sector and by freeing enterprises to invest more and improve maintenance, this new approach, which is already being adopted in some countries, could bring about dramatic increases in access to sanitation and clean water within the next generation.

R.W. Herschy

Source

The World Bank, 1992. *World Development Report.*

Bibliography

Brazil, Direccao Nacional de Aguas e Energia, 1992. Projeto Rio Doce. Brasilia.

DANIDA (Danish International Development Agency). (1991) *The Copenhagen Report, Implementation Mechanisms for Integrated Water Resources Development and Management.* Copenhagen: Ministry of Foreign Affairs, Copenhagen.

Davezies, L. and Remy Prud'homme, 1993. The economics of public–private partnership in infrastructure, in Claude Martinand, ed., *Private Financing of Public Infrastructure: The French Experience.* Paris: Ministry of Public Works, Transportation and Tourism.

de Melo, J.C., 1985. Sistemas condominiais de Esgotos, *Engenharia Sanitaria* **24**(2), 237–38.

Garn, H.A., 1987. *Patterns in the data reported on completed water supply projects.* World Bank, Infrastructure and Urban Development Department, Water Supply and Sanitation Division, Washington, DC.

Garn, H.A., 1990. *Financing Water Supply and Sanitation Services.* Prepared for the Collaborative Council as a background paper for the Delhi Conference. World Bank, Infrastructure and Urban Development Department, Water Supply and Sanitation Division, World Bank, Washington, DC.

Hasan, A., 1986. Innovative sewerage in a Karachi squatter settlement: the low-cost sanitation programme of the Orangi Pilot Project and the process of change in Orangi. Orangi Pilot Project, Karachi.

Hasan, A., 1990. Community groups and NGOs in the urban field in Pakistan. *Environment and Urbanization*, **2**, 74–86.

India, Planning Commission, 1992. Water supply and sanitation, in *The Eighth Five Year Plan (1992–97).* New Delhi.

India, Ministry of Urban Development, 1993. *Proceedings of National Conference of Urban Water Supply and Sanitation Policy.* New Delhi.

International Conference on Water and the Environment, 1992. *The Dublin Statement and Report of the Conference.* Dublin.

Petrei, A.H., 1989. *El Gasto Publico Social y sus Efectos Distributivos.* Santiago: ECIEL.

Ruhrverband, 1992. *Tasks and Structure.* Essen.

Singh, Bhanwar, Radhika Ramasubban, Ramesh Bhatia, John Briscoe, Charles C. Griffin, and Chongchun Kim, 1993. Rural water supply in Kerala, India: how to emerge from a low-level equilibrium trap. *Water Resources Research*, **29**(7), 1931–42.

United Nations Commission on Sustainable Development, 1994. *Financing of Freshwater for Sustainable Development.* Background Paper 5 for the Intersessional Ad Hoc Working Group on Finance. New York.

US National Research Council, 1994. *Wastewater Management for Coastal Urban Areas.* Water Science and Technology Board. Washington, DC.

Warning on costs of European Union 'Green' Law, *Financial Times*, 21 February 1994, p. 8.

Watson, Gabrielle, 1992. Water and Sanitation in Sâo Paulo, Brazil; Successful Strategies for Service Provision in Low-Income Communities. Master's thesis in city planning, Massachusetts Institute of Technology, Cambridge, MA.

World Bank, 1988. *FY88 Annual Sector Review – Water Supply and Sanitation.* World Bank, Infrastructure and Urban Development Department, Water and Sanitation Division, Washington, DC.

World Bank, 1991a. *FY90 Sector Review – Water Supply and Sanitation.* World Bank, Infrastructure and Urban Development Department, Water and Sanitation Division, Washington, DC.

World Bank, 1991b. *FY91 Sector Review – Water Supply and Sanitation*, World Bank, Infrastructure and Urban Development Department, Water and Sanitation Division, Washington, DC.

World Bank, 1992a. *FY92 Sector Review – Water Supply and Sanitation*, World Bank, Infrastructure and Urban Development Department, Water and Sanitation Division, Washington, DC.

World Bank, 1992b. *Utility Reform and Environmental Clean-up in Formerly Socialist Countries: Report of a Workshop on the Baltic Sea.* Water and Sanitation Utilities Partnership Report 3, World Bank, Infrastructure and Urban Development Department, Water and Sanitation Division, Washington, DC.

World Bank, 1992c. *Water Supply and Sanitation Projects: The Bank's Experience, 1967–1989*, World Bank, Operations Evaluation Department, Washington, DC.

World Bank, 1992d. *World Development Report 1992: Development and the Environment.* New York: Oxford University Press.

World Bank, 1993a. *Water Resources Management.* A World Bank Policy Paper. Washington, DC.

World Bank, 1993b. *World Development Report 1993: Investing in Health.* New York: Oxford University Press.

World Bank, 1994. *World Development Report 1994: Investing in Infrastructure.* New York: Oxford University Press.

World Bank, 1995. *Towards Sustainable Management of Water Resources.* Ismail Serageldin, Washington, DC.

World Bank, 1995. *Water Supply, Sanitation and Environmental Sustainability.* Ismail Serageldin, Washington, DC.

World Bank Water Demand Research Team, 1993. The Demand for Water in Rural Areas: Determinants and Policy Implications. *World Bank Research Observer*, **8**(1), 47–70.

Yepes, G., 1991. *Water Supply and Sanitation Sector Maintenance: The Costs of Neglect and Options to Improve It.* World Bank, Latin America and Caribbean Region Technical Department, Washington, DC.

Cross references

Water allocation and use
Water availability and river water quality
Water charges: UK
Water inventory
water quality for drinking: WHO guidelines
Water resources: introduction

SATELLITE HYDROLOGY: PRESENT TRENDS AND OUTSTANDING NEEDS

Introduction

Although no dedicated hydrological satellites have yet been operated, interest in satellite hydrology has been growing rapidly in recent years. Studies have been undertaken to identify observational requirements in hydrology and water management, and systems and methods have been developed by which such requirements might be addressed. At present, however, some can, whilst others cannot. Recent trends are reviewed leading to today's capabilities, and future needs for satellite remote sensing in support of both scientific research and operational methods in the hydrological and associated arenas. For a recent general review of this field, see Engman and Gurney (1991).

Present trends in satellite hydrology

The past and present of satellite hydrology may be summarized through a number of trends, amongst which the following may be considered of special importance.

Maturing of 'original' methods and algorithms for satellite data analysis

The first satellite-based algorithms to be developed for use in hydrology were relatively simple, particularly in terms of their reliance on data from a single channel. Many such techniques have now been developed to a quite mature stage, as evidenced by the regular generation of product data sets from such algorithms, either in the context of ongoing research, and/or in the context of operations in hydrology and water management. Examples of these are to be found in satellite rainfall monitoring, which has been tested and developed since 1970, based at first on either visible and/or infrared data, and more recently on passive microwave data. Sophisticated examples of such methods include that used in the 'IFFA' (the Interactive Flash Flood Analysis system of NOAA/NWS), whose interactive rainfall algorithm assumes that local, high-intensity rainfall is a function of several factors, including meteorology, cloud-top temperature, growth rates of cloud tops, and enhancement factors. Different versions of this approach are in operational use in the USA, namely for summer convection, winter convection, hurricanes, winter extratropical storms and warm-top systems. This methodology is applied every day in the USA to provide near real-time estimates of high-intensity rains on a county basis, and in collaboration with a simple moisture model, short term forecasts of flash floods (Borneman, 1988). On a different scale, and for different purposes, the 'B4' methodology provides for 5×5 km or 25×25 km rainfall estimates every day over selected regions in the tropics (including West Africa, East Africa, Brazil and parts of Southeast Asia) in support of crop monitoring and river flow prediction. This approach utilises frequent geostationary infrared imagery and interprets this in terms of spatially and temporally varying rain:no-rain thresholds, rain cloud durations, and rain rates, calibrated both by climatological 'background fields' and meteorological synoptic station reports (Barrett, Beaumont and Todd, 1994). In the passive microwave area, techniques have been developed to generate global rainfall, monthly at a 0.5° grid square resolution, using different passive microwave frequencies, polarizations and combinations of these, in addition to masks for difficult types of ambiguous areas (e.g. areas of ice and snow), thereby providing good quality products for climatological applications over both land and water.

Regarding solid-state water, e.g. snow on the ground, mature methodologies have been developed based on multi-temporal and multispectral analyses of NOAA-AVHRR data, evidencing the snow line location, snow area, partial or complete snow cover, and snow condition (e.g. melting or accumulating).

Further incremental improvements in the above are likely to be relatively small.

Growth of international projects

Increasing attention has been paid in recent years to the global and continuous natures of the hydrological cycle and its associated processes, as recognized by umbrella programs like the well-known GEWEX (the Global Energy and Water Cycle Experiment) of WCRP (World Climate Research Programme). Under the auspices of GEWEX, a number of regional and global projects and products have been developed, for example the global rainfall products of the Global Precipitation Climatology Project (GPCP).

Meanwhile, international cooperation in Europe has taken a somewhat different form, notably under the auspices of the Research Training and Development (RTD) programs of the European Commission. It is now commonplace for groups of laboratories, perhaps even several in number, to collaborate on multi-year projects across existing national boundaries. The STORM Project, involving nine institutes in the UK, Italy and Spain, is an example from the hydrological field (Barrett and Cheng, 1996).

Last but not least, the opportunities provided by computer networking have led to projects designed to focus the efforts of scientists in different laboratories and even different countries upon common data sets and related analytical problems. The WetNet Project led by NASA from 1990 to 1995 is a good example. Using the Marshall Space Flight Center in Huntsville, AL, as its hub, WetNet has integrated the activities of scientists in the USA, Australia, Japan and several European countries in respect to the utility of data from the US DMSP satellite series primarily for water and water-related applications (Dodge and Goodman, 1994). More projects of a similar nature are expected as the Internet becomes more widely accessed and appreciated.

Integration of remotely sensed and collateral data sets via GIS

As data become available in a digital form from an increasingly wide number of sources – both conventional and remote sensing – it has become necessary to utilize geographic information systems to organize and interrelate data of many different types. An early example of this relates to the Universal Soil-loss Equation Database

(Pelletier, 1985). More recently, GIS has been used in support of the snow measurement and monitoring method described above, involving terrain effects (especially elevation, aspect and slope), vegetation effects (e.g. forests are often dark in AVHRR band 1, heathland and scrub are brighter, and arable and pasture land are bright), and for the extrapolation of snow areas beneath clouds (e.g. in relation to well-defined valley features). In these ways information from historic data sets can be combined with the more dynamic remotely sensed observations to provide improved end products for hydrology, pure or applied.

'Community' efforts via algorithm intercomparison projects

It is in the area of satellite remote sensing of rainfall that most progress has been made with algorithm development and improvement as a result of intercomparison activities (Barrett, 1993). These have included intercomparisons of regional infrared-based techniques by the FAO over the Sahel (1987); the regionally based AIP (Algorithm Intercomparison Project) series over Japan (1990), Northwest Europe (1992–1993), and the Western Tropical Pacific (1994–1995) involving infrared, passive microwave and radar and raingauge data sets; and the PIP (Precipitation Intercomparison Project) series of WetNet, applied globally in the case of PIP-1 and PIP-3, and to event case studies in PIP-2. These have helped advance the satellite science, and pinpoint problems with existing data sets, for example from raingauges (whose distribution is still inadequate even for climate purposes, particularly over the global oceans) and radar data (as discussed below).

These intercomparison projects have underlined the need for fewer, good algorithms, rather than more algorithms, many of which 'reinvent the wheel.'

Development of new satellite sensor systems

Much systematic and concerted effort has been put, and is being put, into the development of new satellite systems to support Earth system science, particularly under the 'Earth Observation System' concept. Many of the new satellites and sensors currently under development or being built will have a large part to play in further satellite studies in the mid-term future (CEOS, 1995). Amongst the most important new satellites will be the Canadian Radarsat, the Japanese ADEOS, the US/Japanese TRMM, and the 'Polar Platforms' of the USA, ESA and Japan. In such regards, a significant trend has been towards the development of larger and more expensive satellite/sensor systems. Not surprisingly, some of these are running into budgetary problems, and 'downscaling' has become alarmingly widespread.

Interest in 'smallsats'

Largely because of the problems of escalating satellite size and associated costs, magazine articles have begun to appear entitled 'Small is Beautiful: mint and micro sats find international favor'. Small satellites offer much reduced costs of construction and operation in orbit, and are allowing developing countries and small laboratories to develop their own space programs. Even more major space powers are thinking in this direction too. One example of such a system is the Portugese PoSat-1, weighing only 50 kg, and costing only £1.2 million ($1.8 million; less than 1% of the cost of an average environmental satellite). Other small satellites are being actively considered, or constructed, by Brazil, China, Finland, Germany, Ireland, Israel, Italy/Spain/Greece, Sweden and the UK, with other countries joining the list quite frequently. Most recently, the European Space Agency has begun to debate the possible 'Earth Explorer Missions', amongst which some may focus upon hydrologically significant targets, including precipitation, atmospheric chemistry or surface characteristics.

New analytical approaches, especially multisource, and 'artificial intelligence'

Reference was made earlier to the maturing of the pioneering types of algorithms of interest to satellite hydrology and water management. In recent years, an increasing number of laboratories have sought to improve the performance of their algorithms by including data from more than one sensor or even satellite, along with collateral data in a GIS context (Xu et al., 1993). For example, AVHRR methods for snow monitoring may be supplemented usefully with information from passive microwave satellite imagery, which evidences snow beneath most types of cloud, and provides evaluations of additional parameters, e.g. the vitally important snow water equivalent (SWE). The STORM Project (see above) focused on the design of schemes for high-intensity rainfall recognition and prediction over parts of southern Europe using Meteosat infrared imagery, DMSP passive microwave imagery, radar data and raingauge observations, all within a GIS context, and taking cognizance also of synoptic weather analyses and forecast weather situations.

For water balance modeling, it has been suggested that a variety of new data types, including those from Synthetic Aperture Radar (SAR) and expected from HIRIS (High Resolution Infrared Spectrometer) data should all be beneficial.

So far as new analytical techniques are concerned, these are desperately needed in view of the practical difficulties of capitalizing upon many different data sources simultaneously. In order to do this efficiently, new mathematical approaches are necessary, amongst which the use of 'artificial intelligence' and/or 'neural networks' appear to be the most promising. Much progress may be expected in these directions in the near future, and as a result much progress may be expected with multichannel and multisource algorithms for hydrology and water management.

Outstanding needs in satellite hydrology

Following from the points summarized above in respect of the past and present, it is possible to suggest a number of ways in which satellite hydrology and water management might be advanced in the foreseeable future. These include the following.

Better surface (*in situ*) calibration opportunities

Reference was made earlier to the several precipitation inter-comparisons which have been undertaken, including those of the PIP series. Through PIP-1 it was confirmed that surface observations of rainfall are inadequate, over both land and sea. Over land surfaces, less than half the 2.5° grid squares over the world's continents were found to contain two or more raingauges, even for the year 1987 (for which the best climate data set had been prepared); meanwhile, over the world's oceans, the situation was – and is – even worse: over most of the world's oceans we simply do not know with reasonable accuracy the monthly rainfalls which might be expected in an average year. Indeed, herein is one of the greatest single needs in respect of the global hydrological cycle at the present time: even modest, but reliable, calibration and validation data for rainfall estimation over oceans. Various experimental methods have been proposed for non-satellite rainfall monitoring over sea areas, and of these optical or acoustical raingauges seem to promise most (Thiele, 1992). However, problems are associated with both of these innovations, and more development and testing is necessary before sets of such gauges could be placed in transects across zones of particularly high rainfall gradients over ocean areas.

Similar, and arguably even greater, practical problems exist with other parameters of high hydrological significance, including evapo-transpiration and soil moisture.

Improved satellite data resolution (especially spatial and temporal)

Whilst reasonably good spatial and temporal resolution is obtainable already for some applications (e.g. convective cloud monitoring over the tropics using geostationary satellite infrared data), and Earth resources satellite data (e.g. for monitoring long-term changes in stream channels and other drainage features over land from Landsat or SPOT), for many hydrological applications present time and space resolutions are not yet good enough to satisfy many potential end users. Passive microwave satellite data are superior to infrared data in many ways, not least for rainfall monitoring because of their cloud penetration capabilities, but their spatial resolutions may be even orders of magnitude less good than those of infrared sensors; and polar orbiting satellites may provide only one or two overpasses per day, compared with hourly or more frequent imagery from geo-stationary platforms. New sensing systems are needed, e.g. to improve the spatial and temporal resolution of passive microwave satellite data. A system has been proposed recently to be carried on a geostationary platform, albeit with some changes in the wavelengths investigated (Savage *et al.* 1995). Even communications satellites could be used as carriers of such instruments, which therefore do not need to wait for a further generation of meteorological satellites to be designed. Such imagination and innovation may yield big dividends if vigorously pursued.

Concerted efforts in respect of particularly elusive parameters

Key parameters for hydrology and water management include evaporation/evapotranspiration and soil moisture, but extended period monitoring is required in respect of all parameters. As suggested earlier, some hydrological parameters have proved particularly elusive to remote sensing, as they were to *in situ* monitoring before it. Such problems will not simply go away: they need to be addressed head on. Increasing activity and sensitivity in instrument design are therefore necessary.

A related issue is that concerted efforts are needed to ensure that satellite algorithms are tested more fully than before, because case studies can be most misleading, and it is only through long-period analyses that the true qualities of existing algorithms can be determined, and selected aspects of our environment can be identified and properly understood. Using the 'B4' range of hydrological monitoring methods described above, it has been established that the optimum rain:no rain temperature threshold in infrared data is not constant, as has been assumed in some widely-used methods of this kind, but highly variable: by up to or even more than 50 K year^{-1}, and by as much over horizontal distances of merely a few hundred kilometres at certain times of the year. Clearly it is very important that such variations should be recognized, modeled and allowed for in the design of operational techniques.

Agreement on standards for satellite hydrology

Agreement on standards is vital in respect of sensors, algorithms, processing systems, etc. It is well known that different types of instruments provide different values for single parameters in small site experiments. The difficulties such variations in performance engender are now becoming evident on much larger scales, e.g. through differences in raingauge design, and in the performance and calibration of neighboring radars. Such problems may become evident only when data are aggregated over significant periods of time, e.g. with respect to the UK FRONTIERS radar system: the application of standard procedures to the data from different radars would help to reduce both general local effects and related problems.

Other difficulties which are becoming increasingly apparent in satellite hydrology are related to hardware (where virtually no two computer systems are ever the same), software (frequent upgrades of which are a continual concern for laboratory managers), algorithm parameters (e.g. the rain:no rain boundary, which may be set variously at 0, 0.1, 0.5, 1.0 mm h^{-1} or more), user requirements (some countries still expect system outputs in non-SI units) and storage (e.g. computer-compatible magnetic tapes of different densities and dimensions, magneto-optical disks and CD-ROMs). Of special concern is the need to copy archived data sets frequently from one tape to another, or from one type of medium to another, simply in order to preserve the data already archived. In all probability, environmental data are being lost at present at a greater rate than ever before in the history of the world.

Dedicated satellites for hydrological applications: 'Hydrosat' or 'Watersat'?

Earlier studies were undertaken to identify the needs of hydrology and water management for satellite data in respect of a wide range of parameters (Herschy *et al.*, 1984). Specifications and justifications for dedicated hydrological satellite systems have been drawn up to meet those needs. However, we still await the first dedicated hydrological satellite, and none is yet in sight. Until such a system is in operation, hydrologists and water managers – perhaps more than any other broad church of environmental data users – will have to continue to rely on data obtained primarily for other purposes. For us in the hydrology and water management community, this is a much less than ideal situation.

More stable political climates for scientific R&D

Many natural environmental cycles are irregular and uncertain. Unfortunately for those of us who are concerned about them, political and economic circumstances can change at least as rapidly, and maybe even more predictably. At present the North American scientific community, and others, are suffering because of the political decision to balance the US Budget within 7 years. Many environmental monitoring programs, including some satellite observations projects, are suffering as a result. Greater stability in both programs and budgets is essential for satisfactory and cost-effective science.

Meanwhile, in Europe and elsewhere the ravages of privatization are also becoming obvious. One very recent example is the initiation of encryption of Meteosat satellite data transmissions to local reception stations, as from 4 September 1995. It seems unlikely that much income will be generated for EUMETSAT by this process, but it is certain that much science, and dependent operational usage of Meteosat data, will be adversely affected thereby.

More rapid uptake of satellite hydrological methods and products by 'operational' hydrologists and water managers

Disappointingly in view of the recent trends outlined above, but not surprisingly in view of the other present needs, relatively little satellite activity has found its way into operational hydrology and water management, particularly in the realms of monitoring and forecasting, as opposed to analysis and mapping. Some noteworthy examples do exist however, including the Africa Real Time Environmental Monitoring using Imaging Satellites (ARTEMIS) of FAO, providing rainfall and vegetation index products for large areas of the African continent in support of famine relief and food security endeavours. Meanwhile, few water authorities and similar bodies utilize remote sensing on a regular basis, despite the established benefits (e.g. in respect of satellite snow monitoring) of combining information from the remotely sensed source with other data already available and in operational use. More dialogue between the scientists and the end users is necessary to increase the uptake of data and techniques in operational contexts: few would now argue that benefits might not be expected, but concerted effort is still required to bring such advances about.

Conclusions

Great strides have been made in the uses of satellite data in areas of immediate concern to hydrologists and water managers, but much remains to be done before satellites can be thought to have taken their rightful place alongside *in situ* sensing systems as parts of integrated operational monitoring and management methods.

It is important that we continue to press for satellites and sensors which are appropriate to our needs, and the requirements of the applications community we seek to serve.

Eric C. Barrett

Bibliography

Barrett, E.C., 1993. Precipitation measurement by satellites: towards community algorithms, *Advances in Space Research*, **13**(5), 119–136.
Barrett, E.C. and Cheng, M.C., 1996. The identification and evaluation of moderate to heavy precipitation areas using IR and SSM/I satellite imagery over the Mediterranean region for the STORM Project, in E.C. Barrett (ed.), *Remote Sensing Reviews*.
Barrett, E.C., Beaumont, M.J. and Todd, M., 1994. Rainfall monitoring by Meteosat over West Africa: results from an extended study, and their wider implications. *Proceedings of Tenth Meteosat Scientific Users Meeting*, Cascais, Portugal, 3–7 September 1994, 299–307.
Barrett, E.C., Power, C.H. and Micallef, A. (eds), 1990. *Remote Sensing for the Hydrology of Mediterranean Coastlands and Island*. Gordon and Breach (London), 338 pp.
Borneman, R., 1988. Satellite rainfall estimating program of the NOAA/NESDIS Synoptic Analysis Branch. *National Weather Digest*, **13**(2), 7–15.
CEOS, 1995. Coordination for the next decade *1995 CEOS Yearbook*, ESA, Paris, 133 pp.
Dodge, J.C. and Goodman, H.M., 1994. The WetNet Project, *Remote Sensing Reviews*, **11**, 5–21.
Engman, E.T. and Gurney, R.J., 1991. Remote sensing, in *Hydrology, Remote Sensing Applications*, Chapman & Hall, 223 pp.
Herschy, R.W., Barrett, E.C. and Roozekrans, J.N., 1984. *The World's Water Resources: a Major Neglect*, ESA BR-40, 41 pp.
Pelletier, R.E., 1985. Evaluating non-point pollution using remotely sensed data in soil erosion models, *Journal of Soil Water Conservation*, **40**, 332–335.
Savage, R.C., Smith, E.A. and Mugnai, A., 1995. Concepts for a Geostationary Microwave Imaging Sounder (GeoMIS), *International Geoscience and Remote Sensing Symposium*, **I**, 10–14 July, Firenze, Italy, 652–654.
Thiele, O.W., 1992. Ground truth for rain measurement from space, *Proceedings of the International Symposium on Aqua and Planet*, Tokyo, 25–26 June 1990, 245–260.
Xu, H, Bailey, J.O., Barrett, E.C. and Kelly, R.J., 1993. Measurements of snow area and depth by remote sensing and their integration with GIS, *International Journal of Remote Sensing*, **14**, 3259–3268.

Cross references

Remote sensing (1)
Remote sensing (2)
Remote sensing data: use in hydrological modeling

SAVANNA

The savanna is a vegetation formation which characteristically includes grasses, trees and shrubs in varying proportions. It is generally a transitional landscape, located between the forests and the deserts and marking the transition from humid to dry climates. Savannas occupy about 23 million km^2, or about 20% of the global land surface. They are most abundant in tropical and subtropical latitudes of Africa, Australia and South America. Large expanses of savanna have been converted to croplands and settlements. Those that remain are home to a multitude of grazing animals and to herds of carnivorous game.

The main feature of the savanna climate is a highly seasonal rainfall regime. A wet season occurs during the high-sun period (i.e., summer) and a dry season, with a deficiency of available soil moisture, prevails during the cooler, low-sun months. This seasonal moisture shortage is the prime factor in the existence of the savannas. However, other aspects of the physical environment, such as soils, nutrients, drainage, geology and environmental history, also play a role in their formation. Some savannas are an end product of human activities, such as setting fires to clear the land at the end of the dry season.

The savanna environment includes a number of diverse landscapes. Their commonalities are a distinctive seasonal cycle of growth in response to the seasonal availability of moisture and vegetation characteristics that permit tolerance of the seasonal drought. Since the savanna is the habitat of grazing animals, many species of trees and shrubs possess structures which protect them from the foragers. Thorns are common to many species, particularly the ubiquitous *Acacia*.

Two broad classes of savanna are distinguished on the basis of climate. The 'wet' savanna woodlands consist primarily of trees which are relatively closely spaced; these correspond roughly to areas of wet–dry tropical climates. The 'dry' savannas consist mainly of low trees and shrubs, well spaced with a continuous grass cover in between. These are generally areas of semiarid or semidesert tropical climates. In most regions a continuum exists along the rainfall gradient, with a gradual transition between wet savanna woodlands and low tree and shrub savannas.

Mean annual rainfall in the world's savannas ranges from about 250 mm in those along the desert margins to about 1500 mm or more in the savanna woodlands along the forest margins. The transition between low tree and shrub savannas and woodlands occurs at around 500 mm. The dry season is of varying length and intensity, ranging from about 3 or 4 months in the wetter savannas to about 8 or 9 months in the drier ones. Most of the rainfall is derived from a small number of brief but intense showers.

The savannas occupy relatively low latitudes, so solar insolation and temperature are high for most of the year. Consequently, water demand (i.e. potential evapotranspiration or PET) is high, of the order of 1000–1500 mm annually. Mean maximum temperatures of the warmest months are generally 30–35°C near the desert margin and 25–30°C at the forest boundary. Temperatures of the cooler months are of the order of 13–18°C and 8–13°C along these same boundaries.

Figure S8 Typical thorn tree and grass savanna of Namibia.

Figure S10 The savannas were occupied by huge herds of animals, but humans have drastically reduced their numbers in recent times.

ago, savanna occupied most of the present-day Sahara Desert. Figures S8–S10 show typical savanna of Africa.

Sharon E. Nicholson

Bibliography

Bourlière, F. (ed.), 1983. *Tropical Savannas, Ecosystems of the World*, vol. 13, Amsterdam: Elsevier.
Cole, M.M., 1986. *The Savannas*, London: Academic Press.
Cole, M.M., 1987. The savannas, *Progress in Physical Geography*, **11**, 334–356.

Cross references

Desert hydrology
Deserts
Semiarid regions

Figure S9 Dwellings in the African savanna are generally constructed with thatched roofs and walls of woven mats, features which are well adapted to the peculiarities of the savanna climate. The roof both effectively sheds the rainfall and prevents the water from penetrating through to the interior. The weave of the mats expands and contracts with atmospheric humidity, being more open in the dry season to allow ventilation but close and nearly impermeable during the wet season.

The highest temperatures often occur just before the start of the rainy season, when the sun is high in the sky and the sparse vegetation cover and dry soils promote intense heating. The annual range of temperatures tends to be relatively small, being of the order of 3–10°C in Africa and South America. However in Australia and Asia it can reach 15–20°C.

Because rainfall occurs during the high-sun season, much of it is lost through evaporation and is unavailable for vegetation growth. In savannas with a wet–dry tropical climate, rainfall exceeds PET during at least several months and there is a substantial water surplus (the excess of rainfall over evapotranspiration and soil moisture storage) during the wet season. During the dry season, water deficits (amount by which potential evapotranspiration exceeds rainfall plus soil moisture storage) are of the order of 200 mm or more. In the dry savannas, rainfall may exceed PET in only 1 or 2 months and it may suffice only to saturate the soil, producing no water surplus.

The interannual variability of rainfall is high in most savannas. These regions are therefore prone to severe drought and to longer-term climatic fluctuations. During past centuries and millenia, climatic fluctuations have influenced the savannas, so that their extent and the rainfall conditions in them have changed over time. Towards the last ice age maximum about 18 000 years ago, the low-latitude deserts encroached onto many of the savannas, contracting them and/or displacing them equatorward. After the ice age ended, the deserts retreated and savannas expanded into these regions. About 5000 years

SEA LEVEL: MEAN

To provide a datum for heights shown on topographical maps, a reference point for many coastal investigations and engineering works, and for a variety of other purposes, knowledge of mean sea level is required (Doodson, 1960). The level of the sea along a coast varies over time. There are the regular fluctuations of marine tides with, most often, a semidiurnal periodicity, on which are superimposed fortnightly and monthly terms; there are fluctuations of an annual nature related to changes of wind and atmospheric pressure; and there are local and irregular variations caused by storms (onshore winds will temporarily raise local sea level), sea surges, or flooding in river estuaries. Additionally, there may be longer-term changes in the relative levels of land and sea, of eustatic or isostatic origin, acting over thousands of years (Mörner, 1980). Mean sea level is determined by a tide gauge, in which the movement of a float is either mechanically recorded by pencil on a revolving drum or digitally registered on magnetic tape. Analysis of the curve so recorded can then yield mean sea level over any desired period of time. An approximate value will be obtained from hourly water heights observed over 1 month, but such a value may be misleading because of abnormal weather conditions and because it ignores the annual tide. If a high degree of accuracy is needed, several years' observations must be obtained. The siting of a tide gauge is an important consideration: both estuaries and highly exposed coasts should be avoided, and the tide gauge must be constructed on the most stable foundations possible, ideally hard bedrock. Long-term tide-gauge records can provide evidence of changes in relative land and sea level (Fairbridge, 1960; Rossiter, 1967): mean sea level computed annually at a given place may reveal a consistent positive or negative shift. For example, the tide gauge at Felixstowe in eastern England showed a relative rise of sea level averaging 1.7 mm year^{-1} between 1918 and 1950; the reverse tendency has been recorded at Aberdeen, where relative sea level fell at an average rate of 0.5 mm year^{-1} between 1862 and 1913 (Valentin, 1952, 1953). The longest record of mean sea level in Britain is that given by the Newlyn tide gauge in Cornwall, which commenced operation in 1916 and provides the present datum for the Ordnance Survey. Much longer, however, is the record of the Amsterdam tide gauge that operated from 1682 to 1930, when it was interrupted by closure of the Zuider Zee. Until 1930 the record showed subsidence of Amsterdam relative to sea level averaging 0.7 mm year^{-1}. Other long records in Europe include those from the Baltic, where approximate measurements of water level span three centuries; more precise observations have been kept since 1839. North American records date mostly from the twentieth century (Gutenberg, 1941).

Clifford Embleton

Bibliography

Doodson, A.T., 1960. Mean sea level and geodesy, *Bull. Géodeseque*, **55**, 69–77.

Fairbridge, R.W., 1960. The changing level of the sea, *Sci. Am.*, **202**, 70–79.

Gornitz, V., 1995. A comparison of differences between recent and late Holocene sea-level trends from eastern North America and other selected regions, in Holocene Cycles, C.W. Finkl (ed.), Fort Lauderdale: *Jour. Coast. Res.*, Special issue, **17**, 287–298.

Gutenberg, G., 1941. Changes in sea level, post-glacial uplift, and mobility of the Earth's interior, *Geol. Soc. Am. Bull.*, **52**, 721–772.

Mörner, N.A., 1980. Eustasy and geoid changes as a function of core/mantle changes, in *Earth Rheology, Isostasy, and Eustasy*, N.A. Mörner (ed.), New York: Wiley, pp. 535–553.

Rossiter, J.R., 1967. An analysis of annual sea-level variations in European waters, *Royal Astron. Soc. Geophys. Jour.*, **12**, 259–299.

Valentin, H., 1952. *Die Küsten der Erde*. Gotha: Justus Perthes.

Valentin, H., 1953. Present-day vertical movements of the British Isles, *Geog. J.*, **119**, 229–305.

SEICHE

A seiche is an oscillation of an open water surface, such as a lake surface, caused mainly by winds and variations of atmospheric pressure. A seiche is normally only discernible over a large open surface, say in excess of some 250–300 km^2 when geodesic (geodetic) influences are prominent enough to have an effect. It is a cyclic type of oscillation, with a cycle period of between 18 and 23 min in most parts of the world, and it manifests itself in the form of a slowly changing swell with little or no surface overbreak into waves. Waves onto a beach (called surfbeat) and wind-induced waves appear to be the cause of visible wave formation, but there is always a seiche component in the wave formation, and it is often the major component of the wave formation.

P.G. Holland

Bibliography

BS 3680, Part 1, 1991. *Glossary of Terms*, British Standards Institution, London.

SEMIARID REGIONS

Introduction

The semiarid regions of the Earth's surface occur as transition zones between the arid deserts and the subhumid belts. Water movement will shape the landscape, according to its geology and past topography, and will work in conjunction with wind erosion, solar insolation, temperature changes and soils (stable or in movement), as well as with the vegetation and the animals that live thereon, to produce an ecological balance of all factors, either in a temporary or a permanent sense.

Such a balance will incorporate past balances, achieved under past conditions, and may preserve some of them as fossil types; the balance itself will have been changed by human activities, living on the water,

the vegetation and the animals. Under the rigorous conditions of life in the semiarid regions, successful human intervention has generally tilted the balance in the direction of diminishing returns from those products on which humans can live (Haragan, 1990).

Semiarid regions of the earth

Strict definitions of climatic regions vary according to the purpose for which they are made; plant ecology, faunal distribution, living conditions for humans and domesticated flocks impose different boundaries for the semiarid regions of the world. However, the controlling factors in the arid zones are clearly low precipitation and the amount of heat, and these have been the basis of many scientific classifications since that of Martonne and Aufrère in 1925. The usual classification today is based on Thornthwaite (1948) and has been modified by Meigs (1952) who produced maps of the distribution of the arid homoclimates (Figure S11).

In defining arid and semiarid areas, Meigs used only three factors: humidity, season of precipitation, and temperature. Extremely arid (E) is defined as an area with at least one entirely rainless twelve months; arid (A) and semiarid (S) are regions where precipitation is less than potential evaporation after the 1948 Thornthwaite formulae. Precipitation may be irregular and non-seasonal (a); it may occur in summer (b), as on the savanna and pampas under a monsoon or tropical semiarid climate; or it may occur in winter (c), as on the steppe or even tundra (not further considered here) under the Mediterranean semiarid climate. Finally, Meigs gives indices for the temperatures of the coldest and hottest month, with (0) for $-0°C$; (1) for $0–10°C$; (2) for $10–20°C$; (3) for $20–30°C$; (4) for $+30°C$.

Thus '$S.c$.1.3' would indicate a semiarid area, precipitation occurring in a cold winter, with the coldest month in the $0–10°C$ average range and the hottest month in the $20–30°C$ average range; this would be a typical Mediterranean semiarid climate, occurring in Morocco, Algeria, Lebanon, northern Iran and also on the western coast of the United States around latitude 35°N.

Arid and semiarid lands account for over one-third of the land surface of the Earth, whereas cultivated lands account for but one-tenth of the whole. The greatest belt of arid and semiarid regions extends across North Africa as the Sahara, through the Arabian Peninsula with the 'Empty Quarter' of extreme aridity, into the Salt Desert of Iran, and the Takla Makan of Central Asia. In North America, the Great American Desert has thus been described by Shreve (1942) in plant ecological terms:

> Beginning in the north, it will be seen that the desert extends southwards from central and eastern Oregon, embracing nearly all of Nevada and Utah except the higher mountains, into southwestern Wyoming and western Colorado, reaching westwards into California to the eastern bases of the Sierra Nevada, San Bernadino and Cuyamaca mountains. From southern Utah the desert extends into northeastern Arizona at the same time that it occupies the western and southwestern parts of that State. On the highlands of southeastern Arizona and southern New Mexico the continuity of the desert is broken by 'desert grassland transition.' At a slightly lower elevation, it reappears in the valleys of the Rio Grande and Pocos rivers, extending as far east in Texas as the lower courses of Devil's River. In the north, an isolated area of desert occupies part of the Columbia River basin in eastern Washington.

Hydrology

In connection with the International Hydrological Decade of 1966–1976, hydrology has been defined as 'the science which deals with the waters of the Earth, their occurrence, circulation and distribution on the planet, including their response to human activity' (White, 1963).

A stricter definition is that given in the *Journal of the Institution of Water Engineers* (White, 1963, p. 381): 'Hydrology is the science of the occurrence and movement of water over and under the Earth's surface from the moment of precipitation to the moment of entry into the ocean or of evaporation into the atmosphere.'

Hydrological cycle

The limited amounts of water that enter and circulate within the semiarid and arid regions are the main subject of this presentation. But to clarify the position, it is desirable to glance at the general circulation of water that takes place above, on, or at limited depths below the surface of the Earth.

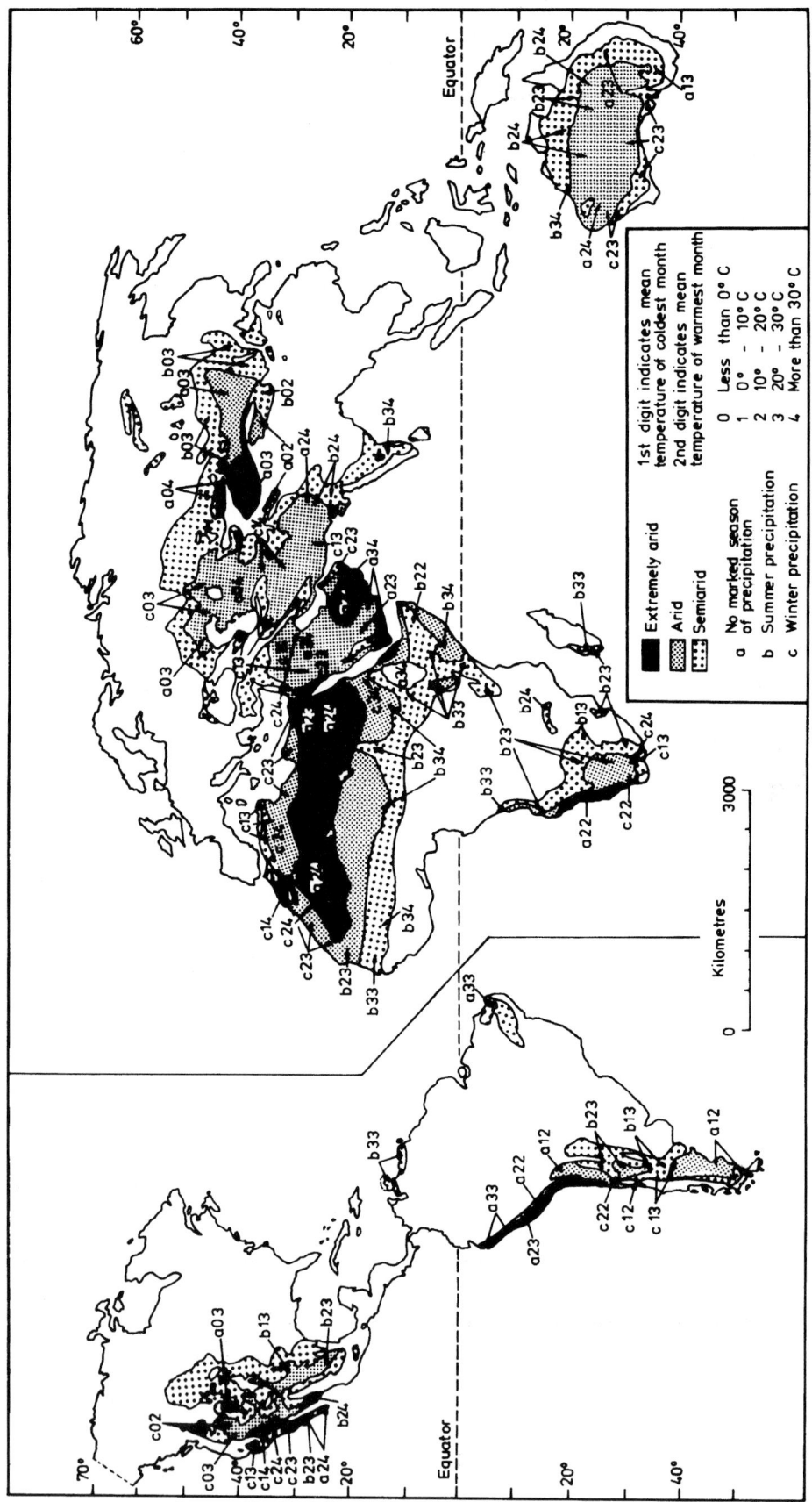

Figure S11 World distribution of arid and semiarid climates. (After Meigs, 1952; White, 1961.)

The hydrologic cycle describes the circulation of water in nature. It traces the water from its evaporation from the surface of the Earth, through its movement as water vapor in the atmosphere to its condensation as clouds and mist. If the minute droplets of this initial condensate can grow by coalescing, then precipitation will fall as rain, hail or snow, according to initial and subsequent temperature conditions. Of the water that returns to the surface of the Earth, much will be evaporated or transpired through the vegetation almost at once, depending on the heat available for such vaporization. Precipitation that falls on the sea short-circuits much of the hydrological cycle (Dinsman, 1994).

The non-evaporated portion will first sink into the soil and fill its pores. Thereafter, the remaining non-evaporated water may run off on surface or sink underground to reach the porous and permeable rocks, known as aquifers, which can transmit and store water. Surface runoff will form streams, rivers, and lakes; the main river of the region may flow to the sea or evaporation may be so great and the quantity of water so small that no outlet to the sea can be achieved. Then the surface drainage takes place in a closed basin, very common in the semiarid regions of the world. The underground water will move through the aquifers according to hydraulic conditions, remaining free of evaporation losses, until it reaches the springs or seepages through which it is discharged from the aquifer. It then once more forms part of surface flow and is again subject to evaporation.

Water balance

A watershed is an area of land that sheds or discharges all its surface runoff through a common point, which is often the estuary of the main river draining the watershed to the sea. If there is no movement of groundwater into or out of the watershed beneath its encircling water divides, then the watershed forms a hydrological unit for which the following balance holds for any unit period of time, which is usually the hydrological year of 12 months:

> Total precipitation on the watershed = evapotranspiration to the atmosphere + surface runoff (including spring discharge) + infiltration to the groundwater ± changes in water storage (surface and underground)

If all the items in this balance sheet can be measured directly, then the amount of errors incurred can be determined insofar as the equation does not balance. If some major constituent is not, or cannot be, measured, then its size is found by difference, and all errors of measurement tend to become indeterminate. Very often, evapotranspiration is determined by difference, since it is difficult to measure directly; since evapotranspiration is of major importance in the semiarid and arid zones, its determination by difference greatly reduces the accuracy of the water balance sheet (Figure S12).

Heat balance

The Sun supplies heat to the surface of the Earth, and this heat is again dissipated, so that over periods of say a year, the surface of the Earth neither gains nor loses heat; radiation maintains this thermal balance. The heat reaching the Earth from the Sun amounts to $2 \, g \, cal \, cm^{-2} \, min^{-1}$ and in the arid zones more than 80% of this heat reaches the surface of the Earth – 89% at Yuma, Arizona, and 84% at Helwan in Egypt (Thornthwaite, 1958).

In order to avoid heating up, the Earth's surface must return this incident heat to the atmosphere. The rate of return is determined by the reflectivity and emissivity of the different surfaces – bare soil, vegetation, water and others. The ability of a surface to reflect incident heat is known as its albedo; it is high in desert surfaces and accounts for the brilliant light and high air temperatures.

Water is a good conductor of heat, and the wetting of a dry soil will double its heat capacity and increase tenfold its heat conductivity. Thus the intake of heat into a wet soil is high, yet the heat is used to vaporize part of the water; due to the high latent heat of vaporization of water, such water takes back much heat to the atmosphere. Where the heat supplied is generally greater than the heat required to vaporize all water on the surface of the area (including water rising by capillary action), there will be found the arid and semiarid zones of the Earth (Elbaz, 1984).

Collection of hydrological data in semiarid regions

Conditions within the arid and semiarid zones are generally unfavorable to the collection of hydrologic data. The lack of permanent

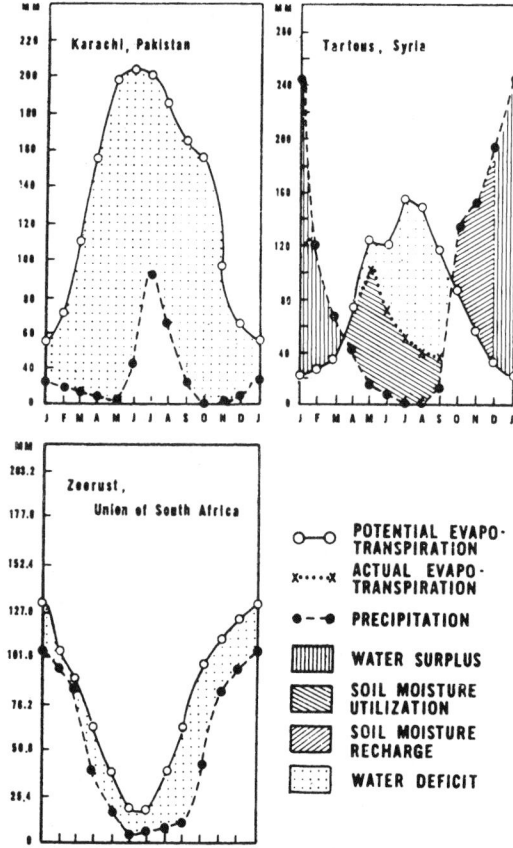

Figure S12 Water balance diagrams for Karachi, Tartous and Zeerust. (These averages were obtained using the Thornthwaite's method for computation; see White, 1961).

settlements in the past has resulted in almost no long-term records, whereas the scattered settlements of today provide insufficient meteorological posts for wide regions of sharply differing precipitation.

Within the semiarid regions of the eastern Mediterranean, for example, the longest record of precipitation is that maintained by the American Colony in Jerusalem; this record commences in 1844 and gives some idea (Burdon, 1959) of precipitation variations over the past 150 years. Such a glance into the past is valuable, but other indicators, such as the geomorphology, plant ecology and historical accounts of floods and other natural phenomena, must be used when information on earlier climates is required.

Instruments

The difficulties regarding the use of instruments to measure precipitation, heat, humidity, wind strength and direction, as well as dew and dewpoint and evaporation from measuring pans, are clear. There are too few permanent habitations at which to establish stations, still fewer trained and conscientious people to monitor such stations. For simple precipitation stations there is also a lack of interest; rain may fall only 10–20 days a year. The application of radar to determine areal rainfall (Kessler, 1966) and the use of photographs from satellites to show cloud cover, areas wetted by precipitation and vegetative growth offer new possibilities of overcoming some of the data collection difficulties in arid regions and are now being tested.

Gauging stations for surface runoff present different problems. The presumed river bed often is only a wide expanse of boulders with numerous channels, down any one of which the water may flow. Such flow may be lost over quite short distances by infiltration into the gravels. To choose a gauging site is difficult but not impossible. Seldom can the flow of the river at the gauging station be calibrated, so that a height–discharge relationship is known; each flow must be measured for depth and speed, possibly over the first 20 years. So a person must be at the gauge station when the river flows, and such flow may, like the rain, occur but a few times each winter. In the rare cases where reservoirs have been constructed, they can be used to measure runoff whose existence, but not amount, was known when the dam was planned (Rodier and Roche, 1978).

Measurements of the amount of groundwater in storage can be achieved by recorders or depth measurements at wells or boreholes; such points are often centers of settlement. Soil moisture measurements are seldom possible on a permanent basis but are made for specific investigations. This also applies to the composition of the waters, though regular chemical sampling of groundwater is feasible.

Ecological conditions, past and present

In the northern temperate zones, the present ecology has been greatly influenced by the glacial and interglacial periods of the Quaternary era. Farther south, the glacials were represented by arid periods, more or less alternating with the Pluvials, during which precipitation was greater over what are now the semiarid and arid zones; the present deserts may have greatly narrowed or disappeared, though some investigators hold that they shifted toward the equator and remained as deserts. The last Pluvial began some 15 000 years ago, with a small increase in pluviosity from 5000 to 3000 BC. There is much evidence for such increased pluviosity in the past, including the deeply incised wadis of the Syrian, Arabian and North African deserts, the types of soil that have developed, many under light forests, as well as the great quantities of groundwater stored in their aquifers, not a little of which may be fossil water, infiltrated in periods of much higher precipitation (Rowland, 1993).

Attempts have been made to date this presumed fossil groundwater by ^{14}C methods; the age as so determined shows a curious and indeed suspicious convergence on an age of $\pm 25\ 000$ years. Such convergence suggests an equilibrium of ^{14}C and ^{12}C in the water-aquifer system, thus a false dating by the ^{14}C method.

Again, geomorphological studies can reveal much of the past climates of the dry regions of the Earth: the detection of lost rivers and river channels, determination of places where floods have occurred and may occur again, and the origin of piedmont deposits where vanished rivers once lost their carrying power as they swept sediment down from the hills and deposited it at their feet.

Since the fundamental concept in plant ecology is that the vegetation of an area taken as a whole is the result of a long period of adaptation to the whole environment – rainfall, evaporation, dew deposition, wind, insolation, temperatures, soil, etc, it follows that present-day plant ecology will not only give data on present-day climate but will also assist in understanding the past climates that brought about the present ecological pattern (Weischet and Caviedes, 1993). For the plant ecology of the arid and semiarid zones of North America, a summary has been made by McGinnies (1955) showing the relationships between plant associations and climates and indeed defining some climatical areas in terms of plant ecology.

Presentation of data

Hydrological data is mainly presented in tables and on maps; most atlases cover the main features of the arid and semiarid zones. The special maps prepared by Meigs (1952) have already been mentioned as defining and classifying these zones.

Also of special interest in defining the water balance for the semiarid and arid zones of the eastern Mediterranean regions are the maps prepared by Carter (1957). These show, for the regions bordering the Red Sea and the Persian Gulf, the mean annual potential evapotranspiration, water deficiency and water surplus.

Many more types of maps exist, each showing different aspects of the semiarid and arid zones. As more data are accumulated and analyzed, so will more detailed maps reveal present position and future development potentials for these zones where life has been difficult in the past but where modern development methods should produce great improvements in the future.

Precipitation in semiarid regions

Precipitation in the semiarid regions is restricted and kept low by the inability of moisture-bearing winds to penetrate into, and cool down within, such regions. Zones of high pressure may prevent the entry of winds, and the great desert areas are mainly associated with this meteorological phenomenon. Such winds as do enter arid and semiarid regions may have had no opportunity to acquire moisture by passage over oceans or sea, or they may have been forced to lose their moisture in passing over high mountains, as in the rain-shadow deserts of Imperial Valley and the Jordan–Syrian steppe. Again, lack of orogenetic effects within the regions, combined with high heat reflected from the ground, may prevent cooling of the incoming winds so that no moisture condenses to form clouds or precipitation, as in the coastal deserts of Chile, southern California, Morocco and western Australia.

Average figures are misleading, but the semiarid regions of the world tend to receive between 400 and 100 mm of precipitation; below 100 mm, the regions become truly arid. Within the Mediterranean semiarid climatic zone, a well-distributed cold-weather precipitation of 250 mm is sufficient to mature a crop of winter-sown barley; an additional 100 mm is required for a good wheat crop. However, precipitation within the semiarid zones is distinguished by great variations from year to year, and the intensities of such rains as occur tend to be high; light showers may be evaporated before they reach the ground. As one goes from semiarid to arid regions, the regular seasonal precipitation tends to be entirely replaced by irregular showers. They may be of high intensity and very restricted extent, and often occur when cold air overlies warm air and some factor (orographic or meterological) permits a sudden uprush of warm air that cooling condenses its moisture to produce a cloudburst.

Condensation over the seas and coasts tends to take place mainly on nuclei of sea salts as well as volcanic and cosmic dust. Over deserts and semideserts, eolian erosion processes introduce much dust into the atmosphere – the famous red sunsets and dust storms of the deserts and wrongly cultivated marginal lands – which form condensation nuclei. In these zones, precipitation tends to contain material of continental as well as marine origin.

Almost all precipitation in the semiarid and arid regions occurs as rain, though snow is not unknown, especially on ground over 1000 m in elevation. Dew and even hoar frost are also of importance and are due to the great differences between day and night temperatures. Where infiltration conditions are good, as over coastal sand dunes, or suitable vegetation exists, such dew may make a permanent addition to the useful water resources of the area. Some plants, such as tomatoes, appear to be able to take in dew directly through the stomata on the leaves (White, 1961, p. 50); such intake by desert vegetation is under study.

Wind-wells, mounds of high-albedo stones through which the wind can blow, are reported by many authors from areas such as the Crimea and the south of France. They supply water by night condensation on the rapidly cooling stones. In the Negev of Sinai, it was found that the Byzantines had irrigated vines by planting them at the base of an octahedron of open stones, the upper pyramid above ground surface to condense moisture and the lower inverted pyramid leading the condensate down to the vine root (Hinman & Hinman, 1992).

Evaporation and transpiration

High evaporation – and high transpiration in vegetated areas – is the dominant hydrological characteristic of the semiarid and arid zones. Both transpiration and evaporation are high because abundant heat energy is supplied to change the limited amounts of liquid water into water vapor, either directly or through biological processes. In this way the heat balance of the area is maintained.

Evaporation may be considered as involving three dynamic processes that occur simultaneously (King, 1961, p. 55):

- a flow of water vapor by turbulent and molecular diffusion from the evaporating surface to the atmosphere;
- a flow of heat by radiation, convection and conduction to the evaporating surface and its removal therefrom as latent heat of evaporation;
- a flow of water through the soil and plants to the evaporating surface.

King has presented his conception of these processes in terms of complex equations and has simplified them in diagrams, e.g. Figure S13. Note that the evaporating surface extends from the top of the vegetation down to the surface of the soil; evaporation may also take place within the pores of the soil, but a cover of dust will develop and is a good retarder of evaporation.

With diffuse rainfall, as much as 100% evaporation loss may occur, so that there is no surface runoff, no replenishment of the groundwater and indeed no increase even in the soil moisture; the amount of heat

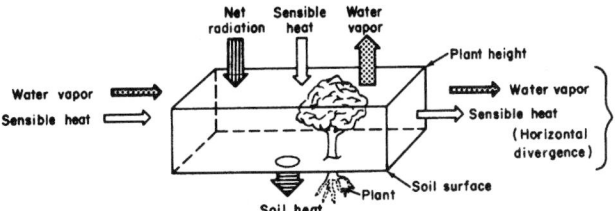

Figure S13 Complete energy balance for a land surface. (After King, 1961.)

stored in the ground and supplied during and immediately after precipitation has been sufficient to vaporize all the water supplied by the precipitation. Since the heat supply is greater during summer (the hot season), precipitation of an amount and intensity that would be completely evaporated in summer may produce increased soil moisture, infiltration and even surface runoff in the cold winter season. Night-time rainfall may suffer less evaporation than daytime rainfall, though such diurnal changes are less important than seasonal effects.

On the other hand, the intensity of the precipitation may enable much more water to be supplied to a surface than can be evaporated by the heat then available. If the water were to remain *in situ* just as it fell, then subsequent heat supplies would, almost by definition within the semiarid zone, evaporate all the precipitation. In nature, however, the non-evaporated precipitation is first used to increase soil moisture or flow off at the surface as overland flow. Where the soil cover is nil or negligible, infiltration takes place directly into the pores and cracks of the rock, or overland flow occurs over impermeable outcrops.

The water stored as soil moisture remains within the influence of the heat supply; if the soil is planted, this heat can be used beneficially for plant growth, but if the soil is barren, most of the water is evaporated. However, if the soil is deep, a dry dust cover can develop, with an insulating effect that reduces or stops evaporation from the deeper soil layers; the effect of capillary rise of water is of particular importance in this stage of the hydrological cycle.

The problems of evaporation are not difficult to describe in general terms, but in detail they are extremely complex and indeed have not been solved insofar as either the design of instruments to collect the necessary data is concerned or with regard to the different theoretical thermodynamical approaches possible. Much detailed information will be found in Deacon *et al.* (1958), but the position can only be summed up in the words of Hare (1961): 'We have woken up to the fact that natural evaporation is one of the most bafflingly difficult processes to study, and that we get nowhere unless we consider the evidence gathered by five or six different groups of workers – hydrologists, meteorologists, soil scientists, plant physiologists, agronomists, and others.' So, evapotranspiration still remains a major problem in the hydrology of the semiarid regions of the Earth (Dinsman, 1994).

Surface runoff

The water that runs off on the surface also tends to remain within the influence of surface heat and thus evaporates. In the immediate sense, this leads to rivers with dwindling volumes of water or to temporary lakes found only after rains and that are dry through most of the year. Evaporation from the surface of such bodies of standing water will be very high, and it is for this reason that artificial surface storage in ponds and reservoirs is subject to many disadvantages in semiarid regions. Attempts are being made to reduce evaporation by covering the water surface with a monomolecular layer of chemicals (fatty alcohols), but such layers can be broken by wind or contaminated by dust and so rendered impermanent (see Control and storage of surface runoff, below).

The inability of surface waters to maintain themselves against evaporation has a long-term effect in that it permits the formation of basins of inland drainage. Any basin of closed drainage will cease to exist if the average annual storage of surface water exceeds evaporation from its central lake system, for then the lake will rise and spread each year till it overtops the lowest point of the encircling water divide, over which it will discharge to the ocean level and also cut its way down so as to reduce the size of the lake. Such a general

conception may be nullified by heavy infiltration (as in the closed karst basins of humid and other regions) or by rising mountain barriers and other geogenetic phenomena, which may produce closed basins of surface drainage; but where such extraneous factors are not present, basins of closed drainage are characteristic of the semiarid lands of the world. The ability of the Nile, the Euphrates–Tigris, the Indus, the Colorado and similar rivers to keep open their basins is due to the fact that the amount of incoming surface waters (originating in non-arid regions) exceeds the evaporation losses.

Infiltration and groundwater

Finally, turning to the water that infiltrates to the groundwater, this is the only water that places itself effectively beyond the reach of further heat supply and thus further evaporation. Within the surface zone it would appear that heat changes are sufficient to produce some evaporation and condensation, as in certain sand dunes in eastern Saudi Arabia. However, in general, evaporation losses are nil from water stored underground in the rock aquifers, whose temperatures are unaffected by surface insolation, though the geothermal gradient in the Earth's crust (and other causes) will heat up groundwaters to much higher temperatures than that of the original infiltration.

Since precipitation stored in the rock aquifers by natural infiltration is immune from evaporational losses and is often cool, it is of particular importance in the semiarid zone; in the past, it has been tapped by wells and galleries (kherazes, khanats, fogarras, chains-of-wells, etc.) of numerous types and is now exploited from boreholes. Extraction by modern methods has often exceeded the small natural replenishment, and this has led first to the idea of inducing additional infiltration or recharge to depleted aquifers and recently to the more positive approach of storing water underground, just as in many places natural gas is stored underground.

The different types of groundwater that can occur in the aquifers need to be distinguished carefully in the semiarid zones, since only truly meteoric groundwater is a renewable resource, and extraction of other types is equivalent to the mining of a non-renewable mineral. Groundwater originating by current infiltration of precipitation is known as meteoric water, whereas groundwater of similar origin, but which has formed in the past and is no longer forming today, is known as fossil groundwater. The latter may have formed during the pluvial periods of the Quaternary, but attempts to date it by the time of decay of radioactive constituents have not yet proved completely reliable. Water that was entrapped in the sediments at the time of formation is known as connate groundwater; in well-flushed aquifers, it is not of importance, but low water throughput under semiarid conditions may permit saline connate water to affect the composition of the normal meteoric groundwater. Finally, there is magmatic or juvenile groundwater, said to consist of water given off from cooling magma; current thinking minimizes the amount of such water and emphasizes the heating of meteoric water that comes into the zone of high geothermal gradients associated with recent granitic and volcanic activities.

The temperature of groundwater is influenced by the temperature of the infiltration, by the geothermal gradient in the aquifer, by the depth and time the groundwater takes to pass through the aquifer, and by certain chemical reactions between the water and the aquifer, of which the most important are exothermic oxidation processes, such as the conversion of sulfides (marcasite, pyrite, etc.) to sulfates. In certain areas, such as the rift valley of Jordan, warm and hot springs at Hamme in the Yarmouk tributary must owe some of their heat to the effect of recently intruded and extruded basaltic magma.

Hydrochemistry and soil chemistry

Precipitation is never pure H_2O, but contains salts and gases in solution. The salts are dissociated into cations, mainly Ca, Mg, Na and K, while the anions are HCO_3, Cl and SO_4; carbon dioxide is the main dissolved gas. These elements in solution in precipitation may be of marine or terrestrial origin. It is estimated that the annual precipitation of sea salts is perhaps $3\ kg\ ha^{-1}$ for the drier steppe regions south of the Sahara, $2\ kg\ ha^{-1}$ in the Kalahari and $1\ kg\ ha^{-1}$ for the high plateaus of Iraq and Iran. Full evaporation of the water which carries these salts will result in their deposition more or less where they fall; surface runoff will concentrate them in the central evaporating pans in basins of closed drainage, while infiltration to the aquifers may be with water which already is far from pure. Thus, in Syria, 'the chemical composition of the precipitation may be changed from an initial figure of some 20 ppm to concentrations of from 100–200 ppm by evapotranspiration and leaching of precipitates in the zones of precipitation; thus recharge waters to aquifers in Syria may contain from 50–200 ppm of total soluble salts' (Burdon and Mazloum, 1959, p. 87).

Precipitation of salts from immediate, full evaporation of precipitation can, over long periods, produce an appreciable *in situ* salt content in the soil. Soil formation itself in the semiarid zone is often more mechanical than chemical or organic; hence, primary soils reflect the composition of the underlying rock without much change by leaching. Where rainfall is higher however, solution effects become more important. Thus in limestone regions, soil formation is almost nil with very low rainfall; solution and redeposition (caliche, havara and nari) occur in regions of intermediate rainfall, while under strong rainfall (upper limit of arid zone) solution is dominant and karst topography, with residual terra rossa soils, marks the true Mediterranean climatic regions.

However, it is in areas of closed drainage where waters have been concentrated by surface flow, and subsequently their salts have been precipitated by evaporation, that the main effects of chemical precipitation can be seen. In this way are formed the salt pans and salt lakes of Australia and South Africa, the salinas of North America and the sabkhas of North Africa and the Middle East. The floors are true solonchak soils, with salts flocculating the silt and clay fractions into grains readily moved by saltation and so producing erosion.

While the composition of the infiltrating waters has a strong influence on the composition of underground water, nevertheless the composition of the aquifer, as well as the volume of water and its rate of movement, will dominate the chemical composition of the underground waters in the semiarid zone.

The solution of low-soluble carbonates by conversion to hydrogencarbonates will continue as long as there is some carbon dioxide in solution in the groundwater; such action will produce hydrogencarbonate waters with up to 600 ppm of total dissolved salts. Thereafter, mineralization of the groundwater will continue by hydration of sulfates, halite (NaCl) and other soluble salts in the aquifers. In some aquifers, such as the Fars Formation of Iraq–Syria (of lagoonal facies), such soluble salts will be very abundant, while in other aquifers, such as the continental, arkosic sandstones of the Sahara and Arabia, soluble minerals are almost completely absent. When the amount of groundwater flowing through the aquifer is large, such soluble salts tend to be removed and the aquifer flushed and cleaned out; likewise, fast-moving groundwater will flush an aquifer quicker than slow-moving water. Since the amount (and sometimes even the rate of movement) of groundwater in the semiarid zone tends to be small, mineralization by dissolution of the aquifer tends to be high.

At the point of natural discharge from aquifers, springs or marshy ground occur. If the spring is large, a perennial river carries off the discharge, and an oasis is formed, or else a great city such as Damascus (fed by the Barada River flowing mainly from Ain Figeh) comes into existence. If the discharge is small or diffuse, a saline marsh tends to form, of which one of the greatest is the Qatarra Depression in Egypt, the probable discharge zone for the sandstone aquifer of the Western Desert of Egypt.

Water development in the semiarid zones

Studies of the hydrology of the arid and semiarid zones are of much interest to pure science, but they are undertaken mainly to facilitate the proper development of the most critical resource of such regions – the water. Accordingly, a few notes are given on some of the lines along which development of water is taking place – for domestic supplies, including urban development and tourism, for watering the pasturing animals, for irrigation and for industry.

Surface management

Surface management in the semiarid zones is, from the water viewpoint, directed to making use of the water before it is lost by evaporation. It is directed to increasing transpiration through useful vegetation. Thus it covers the maintenance of the optimum plant cover over the region, and in this sense optimum plant cover is that which gives maximum grazing to the pasturing herds and flocks. In addition to transpiring much of the direct precipitation, such vegetative cover reduces soil erosion while its shade conserves moisture in the soil. Its root system will help infiltration and generally build up the soil structure.

Control and storage of surface runoff

When surface runoff occurs, it is liable to be intense but short-lived; the amount will be in excess of the requirements of the vegetation at that time. One method of control is to spread the water over large areas, by diverting it from the wadi or stream bed and controlling it behind earth banks in such a way that its flow velocity is never sufficient to erode the retaining structures. In order to control the flood above such water-spreading areas, conditions may warrant the construction of a spate-breaker, a dam with a permanent but restricted underflow, whose reservoir will fill only temporarily after floods. In such schemes, the surface runoff is mainly stored in the soil, from which it can be extracted by the vegetation; recharge to aquifers may occur.

Water may be stored for longer periods of time in surface reservoirs, but in such cases the water surface is subject to heavy evaporation losses. The stored water may be for watering animals, for domestic supplies, for irrigation or for power generation; the longer it is stored, the more is lost. To reduce such losses it is possible to cover the water surface with a protective layer or film one molecule thick, whose action McArthur (1959) has described as follows: 'When a long-chain fatty alcohol such as hexadecanol is placed on the surface of the water, because the work of adhesion between the hydroxyl group of the molecule and the water is greater than the work of cohesion between alcohol molecules, the latter separate from the mass and move across the water surface to form an oriented surface film one molecule thick.' This film substantially reduces water evaporation, but is liable to be destroyed by dust, broken by wind and in need of constant replenishment.

Control of aquifers

Aquifers may be controlled in many ways, of which the most obvious are extraction from galleries, wells and boreholes. But loss of water from the shallow portion of aquifers can be reduced by the control of phreatophytes whose long roots can reach down over 20 m to tap the groundwater. Some phreatophytes such as alfalfa are beneficial and should be encouraged; others, such as mesquite, cottonwood, etc., are of little value yet use up considerable quantities of valuable groundwater.

Seepages and springs may be developed so as to minimize evaporation losses and make the full discharge available for beneficial use. When the maximum discharge of a spring occurs when its waters are least required, it may be possible to overpump the aquifer when water is required so that all recharge is stored below the spring discharge level and there is no flow to waste during the period when water is not required.

Underground storage of groundwater

While engineering works have been carried out to use surface waters to recharge aquifers, the approach has been negative in that it is used to correct the ill-effects of overpumping from the aquifers. There is now a tendency to take a more positive approach and to attempt to store surplus water underground, in the way in which surplus natural gas, and even oil, is stored. Water so stored is free of evaporation losses, though the location and the use of such underground reservoirs call for detailed geotechnical investigations.

Such storage has not yet been attempted in many places, but in Morocco it has been tried successfully at Tafilalet, in the Dra Valley and in the Souss Valley.

Weather modification

Attempts to increase precipitation by various forms of cloud seeding (silver iodide, dry ice and even water drops) have been made over many of the semiarid regions of the world, often with what appears to be success. However, all such methods of increasing precipitation depend for true success on the ability of the atmospheric circulation to bring more moisture into the region; in this, cloud seeding does not appear to be very effective.

Consideration has also been given to modifying climate by allowing large surfaces of water to form upwind of the semiarid area in question. Would the introduction of the Mediterranean Sea to the Qatarra depression increase precipitation along the Alexandrian coast in addition to generating power from a fall of some 70 m?

Desalination of brackish waters

The desalination of brackish waters is of much importance to the semiarid regions, for not a little of their existing water resources are rendered unusable due to salinities lying above the limits of human, irrigation or animal use, yet well below the salinity of seawaters. Solar energy is generally available for the desalting of such waters, but efficiency is low and costs are comparatively high. On some oil fields, such as those of Libya and Kuwait, waste gases offer a source of energy which would otherwise go to waste.

In considering modern engineering methods, it must not be forgotten that natural adaptation processes have evolved vegetation and animals which are able to convert brackish waters into liquids and foods which can be readily consumed. Dates flourish in high saline water; some rices will grow almost in the sea. Again, the halophyte vegetation of the coasts and salt marshes can be eaten by grazing animals, while waters with salinities of up to 10 000 ppm are efficiently converted by sheep into nourishing milk, one of the mainstays of life for many types of nomad.

David J. Burdon

Bibliography

Arnon, I., 1992. *Agriculture in Dry Lands: Principles and Practice*. Amsterdam: Elsevier, 979 pp.

Burdon, D.J., 1959. *Handbook of the Geology of Jordan*. Amman, Jordan: Government Printer.

Burdon, D.J. and S. Mazloum, 1959. Some chemical types of groundwater from Syria, in *Salinity Problems in the Arid Zones*, UNESCO – Arid Zone Research XIV. Paris: UNESCO, pp. 73–90.

Carter, D.B., 1957. *World Climatic Atlas – Special Sheets for Red Sea and Persian Gulf*. Centerton, NJ: Laboratory of Climatology.

Deacon, E.L., C.H.B. Priestley and W.C. Swinbank, 1958. Evaporation and the water balance, in *Climatology: Review of Research*, UNESCO – Arid Zone Research X. Paris, UNESCO, pp. 9–34.

Dinsman, S.L., 1994. *Physical Hydrology*, London: Macmillan, 575pp.

Elbaz, F. (ed.), 1984. *Deserts and Arid Lands*. The Hague: Elsevier.

Hare, F.K., 1961. *Summing up of Symposium on Evaporation*, Cat. No. R32–361/2. Ottawa: The Queen's Printer.

Haragan, D., 1990. *Human Intervention in the Climatology of Arid lands*. Albuquerque, NM.

Hinman, C.W. and Hinman, J.W., 1992. *The Plight and Promise of Arid Land Agriculture*. Cambridge: Cambridge University Press.

Kessler, E., 1996. Radar measurements for the assessment of areal rainfall: Review and outlook, *Water Resources Research*, **2**, 413–425.

King, K.M., 1961. Evaporation from land surfaces, in *Proceedings of Hydrology Symposium No. 2: Evaporation*, Cat. No. R32–361/2. Ottawa: The Queen's Printer, pp. 55–82.

McArthur, I.K.H., 1959. *Control of Evaporation Losses from Water Surfaces*. New Delhi, India: International Committee on Irrigation and Drainage.

McGinnies, W.G., 1955. Plant ecology – The United States and Canada, in *Review of Research*, UNESCO – Arid Zone Research VI. Paris: UNESCO, pp. 250–301.

Meigs, P., 1952. World distribution of arid and semi-arid homo-climates, in *Review of Research on Arid Zone Hydrology*, UNESCO – Arid Zone Research I. Paris: UNESCO, pp. 208–214.

Rodier, J. and Roche, M., 1978. River flow in arid regions, in *Hydrometry* (ed. R.W. Herschy). Chichester: John Wiley, 511 pp (2nd edn in press).

Rowland, J.R.J., 1993. *Dryland Farming in Africa*, London: Macmillan, 336 pp.

Shreve, F., 1942. The desert vegetation of North America, *Botan. Rev.*, **8**, 195–246.

Thornthwaite, C.W., 1948. An approach towards a rational classification of climate, *Geog. Rev.*, **38**, 55–94.

Thornthwaite, C.W., 1958. *Introduction to Arid Zone Climatology*, UNESCO – Arid Zone Research XI. Paris: UNESCO, pp. 15–22.

Webster, C.C. and Wilson, P.N., 1989. Harlow: Longman, 640 pp.

Weischet, W. and Caviedes, C.N., 1993. *The Persisting Ecological Constraints of Tropical Agriculture*. Harlow: Longman, 319 pp.

White, G.F., 1961. *Science and the Future of Arid Lands*. Paris: UNESCO.

White, G.F., 1963. *Preparatory Meeting on the Long-Term Programme of Research in Scientific Hydrology*. Paris: UNESCO.

White, G.F., 1963. The education and training of hydrologists, *J. Inst. Water Eng.*, **17**, 381–391.

Cross references

Albedo and reflectivity
Arid climates
Aridity indices
Arid lands
Arid zone hydrology
Atmosphere
Desalination
Desertification
Evapotranspiration
Precipitation
Streamflow measurement
Water budget analysis

SEWAGE TREATMENT: GENERAL INTRODUCTION

The principles behind sewage treatment are basically separation and settlement. The processes may vary from site to site, but the principle is very much the same, from a high-technology sewage treatment plant to the most basic of rural treatment plants.

The first stage is preliminary treatment, in which gross solids are removed from the flow of sewage by one of several methods, or sometimes a combination of the methods. The most commonly used method is the removal of the solids by the use of screens. The screens are either manually raked or automatically raked with the debris being removed off-site for disposal. An alternative to screens is the use of a macerator which shreds the solids and either puts them back into the flow to settle out later or removes them for off-site disposal. A screenings compacter may be introduced to compact macerated or non-macerated screenings on-site prior to disposal.

The second phase of sewage treatment is the removal of grit. This is usually done in a detritor or constant-velocity grit channel. The speed of the flow of sewage is sufficiently slowed (about 0.3 m s^{-1}) to allow particles of grit to settle out. It is fed into a hopper and pumped to a classifier where it is then swept into a holding unit prior to disposal.

Screened and degritted sewage then passes to primary treatment, where the matter held in suspension is settled out in primary settlement tanks to form raw sludge, which is either treated on site or sometimes tankered to a larger site for further treatment. On entering the primary settlement tanks the velocity of the sewage is slowed down dramatically to allow suspended solids to settle to the bottom of the tank, where it is again swept into a central hopper and pumped away.

The effluent from the primary settlement tanks then passes to secondary treatment which is a variation of the same form of treatment, i.e. biological treatment. The most prevalent form of secondary treatment is biological filtration where the effluent is passed over filter beds containing a media of blastfurnace slag or similar material covered in a film of bacteria. The bacteria break down the colloidal material in the effluent as it passes over the media. This mixture then passes into humus tanks where the mixture is allowed to settle to form a clarified effluent, which is usually fit to discharge to a receiving watercourse, and a humus sludge, which is treated with the raw sludge. The alternative to filter treatment is the oxidation of the primary tank effluent with the bacteria in a large oxidation tank. Oxygen is introduced to this mixture and kept in an agitated state for around 4 h. This mixture is called a mixed liquor. Again the mixed liquor passes to humus tanks where the bacteria and clarified effluent separate. The effluent passes to the watercourse if there is no tertiary treatment and the bacteria ('returned activated sludge') is returned to the start of the aeration process where some is drained off and the remainder is used again.

Many works may have tertiary treatment or nitrifying filters which convert ammonia to nitrite and nitrate, which are not harmful to fish in the watercourse. Other forms of tertiary treatment include pebble bed clarifiers and land treatment areas. These are used to polish the final effluent prior to discharge.

The raw sludge formed from the primary treatment stage is treated in various ways. It can be heated in digestors to form a digested sludge, which is fairly innocuous, or dewatered and treated with lime and then pressed to produce a cake. In some cases the sludge is burned in an incinerator to produce ash. The disposal of treated sludge is dependent upon the nature of the sewage it originated from. Heavy metals and chemicals, amongst other things, contaminate industrial sewage and these sludges are disposed of. Rural sludges are usually returned to the land as fertilizer.

Diane Ireland

Cross references

Sewage treatment processes
Activated sludge process

SEWAGE TREATMENT PROCESSES

In describing the treatment of sewage it is conventional to refer to the processes applied in terms of the type of treatment plant involved as the sewage passes through the treatment works.

The three basic processes of sewage treatment are as follows:

- removal of solids from sewage, comprising
  - removal of gross solids and grit contained in crude sewage;
  - removal of finer solids from crude sewage by gravity settlement;
  - removal of solids from bio-oxidized effluents by settlement;
  - polishing processes for removal of solids from settled, bio-oxidized effluents;
- biochemical purification comprising:
  - bio-oxidation and associated processes.
  - bio-reduction processes;
- separation of water from sludges, comprising
  - thickening processes;
  - conditioning processes prior to dewatering;
  - drying processes;
  - mechanical dewatering processes.

In contrast the conventional description of sewage treatment processes is as follows:

- preliminary treatment for the removal of gross solids and grit;
- primary settlement in tanks for the removal of finer suspended solids.
- secondary treatment, in the form of bio-oxidation (aerobic breakdown) of the settled sewage by means of percolating filtration or activated sludge treatment, followed by removal of secondary solids by settlement;

- tertiary or polishing treatment;
- sludge treatment by a variety of means including digestion.

To identify better the stated basic processes with the conventionally described processes, Table S2 links the various processes between the two types of description.

After treatment of sewage there remains to be disposed of the effluent, being the water fraction of the sewage flow plus any polluting matter remaining therein, and the solids removed during treatment. The means of disposal of effluent usually employed is to discharge this to water resources, while sludge disposal usually takes the form of dispersal on land, dumping at sea (this must cease by 31 December 1998 in the European Community) or incineration.

Figure S14 shows aerial views of the Davyhulme Sewage Treatment Works in the UK which treats the sewage from Manchester. The Davyhulme works is one of the largest in Europe, serving a population of some 660 000 and a total population equivalent of 1 386 000. The works extend to 103 ha. The works were originally built in 1894, treating sewage by chemical precipitation and land filtration. The present flow is $360 \, \text{ml day}^{-1}$.

Removal of solids from sewage

Removal of gross solids and grit from crude sewage

On the average the content of suspended matter in crude sewage is about $350 \, \text{mg} \, \text{l}^{-1}$. In sewages which are stronger than average, the concentration of suspended solids may be as high as $600 \, \text{mg} \, \text{l}^{-1}$ whereas a weak sewage may contain only $150 \, \text{mg} \, \text{l}^{-1}$ suspended solids. Of these suspended solids, some are in the form of large solids which can be readily removed by screening, and some are in the form of relatively heavy inorganic solids such as sand or grit (detritus), which can be readily settled out of the sewage flow. The remainder of the suspended solids present consists about equally of solids which can be induced to settle out of the sewage flow, and of solids which cannot be removed in this way.

In the screening of large solids from sewage, vertical-bar screens, manually or mechanically raked, are the most used. Separated screenings at small works are usually disposed of by burial on the sewage works site, but the clearing and handling of rags, paper and fecal matter separated on screens is an unpleasant job. To avoid this on many of the larger works, mechanically separated screenings are subjected to maceration and are returned to the sewage flow, or the larger solids are not screened out of the sewage flow but chopped up mechanically by comminutors and remain in the sewage flow.

Grit removal at smaller works is normally achieved in specially designed channels, at least two in number, of parabolic cross-section, which regulate the velocity of sewage flow to about $0.3 \, \text{m} \, \text{s}^{-1}$, permitting grit, reasonably free of organic matter, to settle out of the

Table S2 Descriptions of sewage treatment processes

Conventionally described processes	Type of treatment given	Nature of process	Basic treatment process
Preliminary treatment	Screening and grit removal	Physical	Removal of solids (gross) from sewage
Primary settlement	Settlement in tanks	Physical	Removal of solids (fine) from sewage
Secondary treatment	Percolating filtration or Activated sludge treatment	Bio-oxidation	Biochemical (bio-oxidation) purification
	Settlement in tanks	Physical	Removal of solids (fine) from effluent
Tertiary treatment	Sand filtration or microstraining or land treatment or lagooning	Mainly physical	Removal of solids (fine) from effluent (with minor other effects)
Sludge treatment	Digestion	Bio-reduction	Biochemical (bio-reduction) Purification
	Thickening Conditioning Drying Mechanical dewatering	Physical Chemical or physical Physical Physical	Separation of water from sludges

Figure S14 Aerial views of the Davyhulme Sewage Treatment Works, UK.

sewage flow. The settled grit is removed manually or mechanically for subsequent disposal on the sewage works site. However, at the larger works the removal of grit is achieved in specially designed tanks, known as detritors, working on the usual principle of maintaining a constant flow velocity through the tanks. The compactness of such tanks, and their better adaptability to mechanical removal of settled grit, are the main advantages when compared to grit removal channels.

Table S3 Some design criteria for primary settlement tanks

Type of tank	Detention time at dry weather flow (h)	Maximum rate of upward flow at 3 × dry weather flow (m h⁻¹)	Maximum surface loading at 3 × dry weather flow (m³ m⁻² day⁻¹)
Rectangular tank	8–12	–	20–40
Pyramid (upward flow) tanks	–	1.2–1.6	–
Circular tanks	6–8	–	20–40

The quantity of grit arriving at sewage works is small, except from partially separate and combined sewerage systems when rainfall wash-off of grit from paved surfaces and roofs can be considerable. The removal, or break-up, of large solids and removal of detritus from the sewage flow at the preliminary stage is done more to protect subsequent mechanical plant and pumps from blockage or excessive grit abrasion, etc., than to achieve purification of the sewage.

Settlement of fine solids from crude sewage

Settlement of the finer, lighter solids from sewage is an essential part of treatment. This settlement is achieved by passing the sewage, after screening and grit removal, through tanks of a capacity and design appropriate to permit the optimum removal of solids. These primary settlement tanks may be of a rectangular cross-section, with an average depth of about 2 m and at least twice as long as it is broad, through which the sewage flow is effectively horizontal. The more modern settlement tanks at smaller works have an inverted pyramid shape through which the sewage flow is effectively vertical. At the larger modern works, primary settlement is achieved in circular tanks, through which the sewage flow is effectively radial.

The essential objective in the use of these tanks is to cause as much suspended matter as practicable to settle out of the sewage onto the floor of the tanks, for subsequent removal manually or by means of mechanical scrapers, with withdrawal of the accumulated liquid sludge under the hydrostatic head of the sewage in the tanks. Floating grease and scum accumulating at the surface of the tanks is retained by means of scum boards and removed manually or mechanically along with the sludge. The liquid sludge is pumped to sludge treatment processes, while the supernatant settled sewage leaving the tanks passes on to bio-oxidation treatment.

At some of the larger sewage works, circular primary settlement tanks incorporate a center section which is stirred gently by means of power-driven paddles. This stirring of the sewage encourages flocculation of fine solids into larger particles which settle more easily. Similar flocculation can be achieved by gentle stirring of the sewage in a separate tank before primary sedimentation.

Continuous-flow settlement of sewage solids is an empirically derived process, in which factors such as the nature and concentration of suspended solids, rates of sewage flow, the effects of wind on the tank contents and the necessary compromises between construction costs, tank depth and sludge condition militate against precision in derivation of design criteria and in consistency of performance. Typical design criteria for the various types of primary settlement tank are given in Table S3. Primary settlement usually removes between 60 and 70% of the suspended solids present in the tank inflow and about 40% of the BOD_5.

Various possibilities exist for improvement, or even elimination, of some of the current primary settlement arrangements, but in the immediate future no marked shift from these arrangements seems likely. The use of chemical coagulants such as lime, aluminum sulfate or chlorohydrate, and polyelectrolytes for aiding the settlement of sewage solids is practised at a few sewage works.

Settlement of solids from bio-oxidized effluents

Mention has already been made of the two basic processes used for the further purification of settled sewage, namely percolating filtration and activated sludge treatment. While further reference will be made to these processes later, it can be noted now that the effluents from both of these processes contain suspended solids which need removal by settlement in tanks.

Percolating-filter effluents contain between 50 and 200 mg l⁻¹ of suspended solids (humus solids) according to the precise nature of the process, the concentration of suspended solids passing onto the filters from primary settlement, and the season of the year. These solids settle about twice as readily as the solids present in sewage after screening and grit removal and consequently the settlement tank capacity required for removal of these solids is about half that required for primary settlement. These secondary or humus tanks may be of the rectangular, horizontal-flow type or the upward-flow pyramid type at the smaller works, and of the circular, radial-flow type at the larger works. Apart from the lower capacity of these humus tanks, the settlement process and provision for sludge removal are substantially the same as in primary settlement tanks.

However, the effluents from activated sludge treatment processes contain a high concentration of solids, up to 6000 mg l⁻¹ or more, because these processes depend upon the maintenance, by recirculation of sludge, of a high concentration of activated sludge solids in the treatment process. When the activated sludge plant is working properly, the suspended solids in the plant effluent readily flocculate in settlement tanks, and continuous removal of sludge from the settlement tanks is necessary. Accordingly, circular radial-flow, settlement tanks, which are mechanically scraped and with a capacity of about one-half of that of primary settlement tanks, are normally used, although there are installations using upward-flow pyramid tanks. Part of the sludge removed from these tanks is returned to the activated sludge treatment process and the balance passed to sludge treatment.

The effluents from secondary settlement should contain less than 30 mg l⁻¹ of suspended solids. This result is more consistently achieved in the activated sludge processes than in percolating filter processes. Where it is necessary to ensure that the effluent contains considerably less than 30 mg l⁻¹ suspended solids, the use of a 'polishing' process for further removal of solids is required.

Removal of residual solids from settled bio-oxidized effluents

The polishing processes in current use for reducing the content of suspended solids in the effluents from secondary tanks can be grouped into

- straining processes;
- flocculative removal;
- broad irrigation on grassland;
- lagooning.

Table S4 gives an indication of the relevant design criteria and performance of these processes.

The straining processes in current use are sand filtration and microstraining. Both of these processes are derivations of processes used in the treatment of public water supply. Sand filtration is of the slow sand-filtration type. Essentially a slow sand filter consists of a shallow tank, containing a layer of special sand about 0.1 m deep overlying a similar layer of graded gravel resting on underdrains. The tank is flooded with effluent from secondary tanks and in passage down through the filter, solids are strained out on the upper layers of sand. From time to time the filter is drained down and dried out, the top layer of sand and solids is removed and replaced with clean sand, and the filter brought back into use.

At larger sewage works, rapid-gravity sand filters are used, fitted with backwashing arrangements. These filters are 1–1.2 m deep, again consisting of a layer of special sand overlying graded gravel, with an underdrainage system serving for effluent collection, and provision for upward-flow backwashing of the filter. The effluent to be treated is pumped into the top of the filter, and is strained of solids in its downward passage through the filter. When the filter is becoming clogged, as indicated by a loss of hydrostatic head of about 2 m through the filter, it is pressure backwashed with filter effluent and given a scouring with compressed air delivered to the base of the filter. The backwashings are returned to the primary settlement tanks. Upward-flow filters, with both the filtration and the filter wash

Table S4 Polishing processes for removing solids from effluents

Type of process	Rate of treatment	Purification achieved (% removal)	
		Suspended solids	BOD
Broad irrigation on grassland	100–2000 m^3 ha^{-1} day^{-1}	40–60	40–50
Lagoons	3000–5000 m^3 ha^{-1} day^{-1}	45–75	50–70
Slow sand filters	2–3 m^3 (m^{-2} filter surface) day^{-1}	50–70	40–50
Rapid gravity sand filters	150–200 m^3 m^{-2} day^{-1}	70–85	50–70
Upward-flow sand filters	300–400 m^3 m^{-2} day^{-1}	60–80	50–60
Microstrainers	200–300 m^3 m^{-2} day^{-1}	40–70	30–50
Upward-flow clarifiers	15–20 m^3 m^{-2} day^{-1}	40–60	25–40

proceeding in an upward direction, may also be used at sewage works. These have the advantage over downward-flow filters of providing for greater filtration depth in the filter bed and thus giving longer filter runs. In the downward-flow filter, most of the solids removal occurs in the upper layers of sand, whereas in the upward-flow filter solids are removed in the lower layers of gravel and coarse sand, and finally strained out in the topmost layers of sand. Sand filters do in fact give rather more effluent purification than simple straining and this aspect of additional purification will be referred to later.

Microstrainers consist essentially of a large wide drum, closed at one end, and covered with a very finely woven, stainless steel fabric, rotating slowly on a horizontal axis. The drum is partially submerged in a tank divided into two compartments, one containing the liquid to be treated and the other the microstrained effluent. The influent passes through the open end of the drum and passes out radially through the fabric. The strained-out solids collecting on the inside of the fabric are washed into a collecting trough within the drum by pressure jets of strained effluent and returned to the primary settlement tanks. To control the growth of bacterial slime on the fabric it is irradicated with ultraviolet light, and at intervals of a few weeks the microstrainer is put out of action and cleansed by chlorination. Microstrainers have only the one treatment action, that of physically straining solids out of the secondary tank effluent.

Flocculative removal of the residual solids in the effluents from secondary tanks at smaller sewage works is often achieved by means of upward-flow, pebble-bed clarifiers. The clarifier consists of a shallow bed of pea-gravel, supported on a perforated floor, located in the upper part of a humus tank or similar compartment. The effluent leaving the tank must pass upwards through the gravel, causing flocculation of solids and settlement of these on the surface of the pea-gravel. At about weekly intervals the tank is drained down and the gravel cleansed by pressure spraying with a hose. The bed of pea-gravel is sometimes omitted, and the supporting floor constructed of closely spaced wedge-wire, or perforated tiles, the apertures of these structures causing the solids in the upward-flowing liquid to be flocculated and to settle out.

Chemical flocculation can be achieved in secondary settlement tanks, by adding a coagulant such as aluminum sulfate to the incoming liquid, to improve the removal of solids in the settlement tank; this process is not used to any significant degree at present. Mechanical flocculation arrangements can be made in secondary settlement tanks in much the same way as in primary tanks, but this process is little used at present.

Irrigation of secondary tank effluent on grassland is frequently used at small to medium-sized works for effluent polishing. Grass plots with a steady gradient of about 1 in 100 are fed with the secondary tank effluent through a feeder channel, and the liquor flows through the grass to collecting channels, being strained of suspended solids in the process. Other, mainly bio-oxidative, purification is achieved in some degree during this irrigation process. The grass on the plots need not be cut, but eventually (after many years in most cases) each plot in turn has to be put out of use for cleansing, cultivation and reseeding. Sometimes the distribution of secondary tank effluents onto the land is done by spray-irrigation equipment.

Lagooning of secondary-tank effluents in one or more small lakes about 2 m deep, and with a storage capacity of between 5 and 17 days, permits further settlement of solids to occur. However, such lagooning permits other biochemical processes to occur, some advantageous and others disadvantageous. In particular, the disadvantage of development of algal blooms in summer, causing the lagoon effluent to

contain more suspended solids than the influent, has to be taken into account.

Biochemical purification of sewage

The biochemical processes occurring in sewage purification are many and varied, and current understanding of the detailed nature of these complex processes is far from complete. From the moment that domestic sewage enters a sewer, these essentially natural processes begin to take place, but at sewage works, treatment plant is provided to concentrate and accelerate certain of these processes. The main processes utilised are the bio-oxidative processes proceeding in percolating filter and activated sludge installations, and the bio-reduction process of sludge digestion. However, these and associated processes also occur in a limited or subsidiary way in other treatment units. These will be considered later. The bio-oxidative processes themselves are a complex of certain major, and a number of minor yet important processes. In contrast, the bio-reduction process of sludge digestion (and septic tank treatment) is less diverse, but it is not the only bio-reductive process which proceeds in sewage treatment.

Bio-oxidation and associated processes

In the larger sewage works it is usual to treat the sewage for removal of suspended solids by screening, grit separation and primary settlement in the ways already described before subjecting the sewage to bio-oxidation. However, processes of 'extended aeration' of sewage have been developed for small sewage works over the last 15 years, which operate successfully on crude sewage.

No matter what method of achieving bio-oxidation is involved, the overall processes depend upon provision of means whereby sewage can absorb oxygen from air and a suitable concentration of aerobically growing microorganisms can remain for a suitable time in contact with the sewage and dissolved oxygen. The basic processes utilized are

- growth and reproduction of the microorganisms;
- oxidation of carbonaceous organic matter by the microorganisms to carbon dioxide and water (plus oxidation of nitrogenous matter to nitrate where required).

These basic processes are essentially natural ones, proceeding in rivers and the soil wherever sufficient oxygen is available, and the means provided at sewage works for bio-oxidation of sewage are simply devices for intensifying the operation of these processes.

There are two basic methods of achieving bio-oxidation at sewage works.

- Those in which a structure is provided on which the microorganisms grow as a film and to which the sewage is brought. These structures are usually referred to as 'filters', although they are not at all intended to achieve filtration in the sense of trapping solids from the sewage flow.
- Those in which the microorganisms are kept in suspension in a sludge, and the microorganisms in the sludge are brought to the sewage. This method is usually referred to as an 'activated sludge' or 'bio-aeration' process.

There are many variations of these two basic methods, and the principles of the basic methods and their variants are now considered. Percolating filters of the traditional type consist of a bed of clinker, gravel, stone or slag about 2 m deep laid over underdrains, the pieces of 'medium' being nominally about 25–75 mm in size depending on how the filtration process is to be operated. The filter beds are circular or rectangular in plan. Settled sewage is distributed over the surface of the medium by revolving distributor arms in the case of circular filters, or by distributors traversing forwards and backwards over the length of rectangular beds. Revolving distributors may be operated by the reaction of jets sprinkling the feed liquid from holes in the distributors arms, by an electric motor drive or by a simple water-wheel drive in some smaller units. Distributors on rectangular beds are usually operated by power-driven rope haulage. The sewage passes downwards through the medium and out of the filter via the underdrains, while natural ventilation upwards (and downwards) through the filter provides the air flow for the supply of oxygen essential to achieve bio-oxidation. As the flow of settled sewage trickles over the medium, a film of bacterial and fungal slime (filter film or zoogloeal film) accumulates.

Since the average time of retention of sewage in a percolating filter is about 20 min or less, depending on the rate of application of sewage, it is evident that in the main the organic matter in the sewage is absorbed and/or adsorbed on the film and oxidized there by respiration of bacteria. The oxygen necessary is taken up from any present in the sewage, and from filter air present in the space between the pieces of media, onto the film while the carbon dioxide end product (and other intermediate products) of the oxidation diffuses from the film back into the sewage flow, and into the spaces between the pieces of medium. In this process bacterial film is being continually produced as the bacteria grow and die, and but for the effect of 'scouring organisms' the filter would clog as film blocks up the spaces between the pieces of medium. The open packing of the medium in the filter is a very suitable habitat for fly larvae, worms and a host of other animals which break up and graze on filter film, or consume one another, grow, excrete and die, and in effect prevent blockage of the filter. The broken-up film, and the excreta, corpses and remains of these organisms pass out of the filter with the effluent as 'humus' solids, for removal in the humus tanks or secondary settlement tanks already described. This grazing activity of filter animals is seasonal, vigorous in the warmer months like all other aquatic and terrestrial animals and slow in winter, the latter giving a tendency for the sewage applied to the filter surface to 'pond' as film builds up and interferes with flow through the filter.

The oxidation of carbonaceous matter is not the only bio-oxidative process proceeding in a percolating filter; the oxidation of ammonia present in the sewage passing through the filter will occur when the demand of the carbonaceous-oxidizing bacteria for oxygen diminishes and an oxygen–carbon dioxide balance in the air-space–film–sewage complex becomes more favorable to the growth of nitrifying bacteria. In effect, in the upper layers of filter medium bio-oxidation of carbonaceous matter dominates because the availability of such matter is high, and the bacteria carry out the oxidation growth at a much faster rate than the nitrifying bacteria. In the lower regions of a reasonably loaded filter, nitrification of ammonia can proceed almost to completion under favorable circumstances. Both types of oxidation are temperature dependent, as might be expected having regard to their bio-chemical nature, proceeding faster in summer than in winter.

The degree of hydraulic loading onto the filter, the strength and nature of the sewage applied, the nature, size and shape of the filter medium, and the degree of exposure of the location of the filter to cold wind, are all factors which have a bearing on the efficiency and effectivenes of a percolating filter.

For example, the greater the hydraulic load on a filter, the less favorable the interior of the filter becomes for the mixed fauna of grazing animals, and the more the filter will tend to pond in cold weather unless the medium is larger in size. The larger the medium, the less the available surface on which film can grow, and so the lower the degree of purification, yet the filter will tend to clog and pond less in cold weather; and the lower the rentention time in the filter and the greater the degree of hydraulic scouring through the medium. A filter receiving a sewage rich in carbohydrates, from industrial effluent discharges to sewers such as milk or cannery wastes, will tend to grow mats of fungus (fusarium, etc.) on the filter surface. A heavy loading of organic matter on the filter usually results in little or no nitrification occurring, because the whole filter depth becomes dominated by the carbonaceous oxidation process. Honeycomb medium such as clinker or slag presents a high surface area to the sewage and hence achieves better purification than, say, gravel or stone, and the colder a filter becomes due to exposure to the weather, the less efficient its operation.

The normal basis of loading of a percolating filter operating on single filtration, that is, a filter expected to produce an effluent which after settlement will usually contain less than 30 mg l^{-1} suspended solids and 20 mg l^{-1} BOD$_5$, is 0.35 m^3 of average strength settled sewage per cubic meter of medium per day, equivalent to a BOD$_5$ loading of 0.09 (kg BOD$_5$) m^{-3} day^{-1} of medium. Even at this relatively low rate of loading the filter performance is likely to be below standard during very cold weather, and in spring when 'sloughing' of humus solids occurs as the filter animals surge into break-up and grazing of winter-accumulated film. Despite these shortcomings, the process is simple, requiring little skilled supervision, and robust in its reaction to variations in sewage quality resulting from variations in industrial waste content of the sewage flow. Many attempts have been made to improve the efficiency of percolating filters and they deserve brief mention.

Following World War II, use of pumped recirculation of settled filter effluent became fashionable as a means of permitting almost double the rate of BOD$_5$ loading on filters. This was advantageous insofar as this rendered the effluent from overloaded filters better in BOD$_5$ quality, reduced fly swarming from filters, and spring sloughing of humus solids, by modifying the activity of grazing animals. However, the total benefits of simple recirculation may be marginal unless the degree of filter loading is kept within reasonable bounds, or effluent quality requirements are not stringent.

Double filtration of settled sewage, with intermediate humus settlement, is generally advantageous, whether this is associated with alternating double filtration (ADF) or high-rate filtration. ADF involves essentially changing over the primary and secondary filters every week or so, thereby controlling film growth which otherwise would be excessive due to the degree of organic load applied. The primary filter grows heavy film, and when this filter is switched to the secondary role, the much weaker feed it receives causes much of the film formed on primary feed to become starved, and it breaks up, clearing much of the accumulated film before the filter is switched back into the primary role. High-rate double filtration is aimed essentially at producing a high rate of carbonaceous oxidation on the heavily loaded primary filter, packed with large-sized medium to minimize ponding, while the secondary filter carries out a clean-up of carbonaceous oxidation and a high degree of nitrification. Both ADF and high-rate filtration permit more than a doubling of the rate of loading possible on single filtration [1 m^3 m^{-3} day^{-1} for average sewage or 0.25 (kg BOD) m^{-3} day^{-1} overall] to produce approximately the same quality of effluent.

The use of specially shaped plastic media packings for percolating filters has been developed in recent years for a variety of purposes. One successful development has been that of the tower filter, packed with relatively widely spaced, preformed plastic sheets, dosed at a high rate to remove carbonaceous BOD as a 'roughing' process before bio-oxidation in conventional filters. The application of this technique has been even more successful in the 'pretreatment' of very strong organic industrial wastes.

There have been recent developments of moving 'filters', where spaced large-diameter discs or large-diameter drums containing plastic media rotate partially submerged in tanks into which settled sewage is fed. The part of the disc or drum rotating through the air permits oxygen take-up by the bacterial film which, as the discs or drum rotate, then enters and passes through the sewage permitting take-up of organic matter, in total producing the same effect as a percolating filter but using less area, and not producing fly swarms and sloughing of humus solids.

Activated sludge processes utilize aeration tanks or channels into which settled sewage and activated sludge pass, and the 'mixed liquor' is given aeration by means of air delivered to the bottom of the tanks through numerous air diffusers or jets, or by means of mechanical agitation of the surface of the liquor using vertically rotating bladed cones or horizontally rotating cylindrical-shaped brushes or multi-bladed rotors. The treatment process depends on the presence of an adequate concentration of activated sludge in intimate contact with the settled sewage, for an appropriate period of time, in the presence of dissolved oxygen. The latter requirement is precise, but the concentration of activated sludge and the retention time in the aeration tanks may be varied, depending on the degree of bio-oxidation required.

The basic oxidation processes are the same as those which proceed in a percolating filter. The activated sludge, carrying a dense and highly active population of bacteria and other micro-organisms, first oxidizes the carbonaceous matter in the sewage and then the nitrogenous matter. The activated sludge does not contain animals larger than protozoa, and these play an essential role in feeding on bacteria suspensions and other colloidal matter developing or present in the mixed liquor. There is a continuous production of new activated sludge in the aeration tanks as bacteria and protozoa grow and die. After the prescribed period of aeration, the mixed liquor passing out of the aeration tanks is settled in tanks in the manner already described, and part of the settled sludge is returned to the inlet of the aeration tanks to maintain the appropriate concentration of solids in the mixed liquor, while the balance of sludge is passed to sludge treatment and disposal processes. Frequently, activated sludge returned to aeration tanks is given reaeration with blown air to recondition it before further use.

At this point it is appropriate to compare the production and removal of excess activated sludge with the production and removal of bacterial film in filters. In the former case the sludge is kept in suspension by the mixing caused by air injection or mechanical-splash aeration until it leaves the aeration tanks, when it is settled out and removed. In the case of the percolating filter, the removal of excess filter film is achieved partly by the grazing of insects and other macrofauna and partly by hydraulic scouring of the downward flow of sewage – with the exception that in alternating double-filtration processes, the periodic alternation of feed strength to the filters additionally causes film breakdown by periodic 'starving'.

There are a variety of ways of operating activated sludge processes, and each different technique utilizes variations in tank shape in plan, in detention time, in the ways in which sludge return is carried out and in the means of aeration. Yet all of these variations can be classified according to the type and objective of treatment. The treatment may be at a high rate to achieve quick clarification and removal of carbonaceous matter (bioflocculation), at a normal rate to achieve a high degree of purification including very good nitrification, or 'extended aeration' or 'contact stabilization' to give good purification, and partial oxidation of sludge, at small plants without skilled supervision. Table S5 relates the variations in technique to the basic classification of treatment type and purification objectives, and gives an indication of design criteria. In addition to these solely activated sludge processes, the use of high-rate activated sludge followed by percolating filtration is used on a number of the larger sewage works, particularly where the proportion of the trade effluent in the sewage flow is high.

Despite the facts that activated sludge treatment processes, in comparison to percolating filter processes, on a large scale require greater skill in supervision and effective operation, and are more prone to upset by trade effluent discharges, it seems clear that they will increasingly displace percolating filter processes. The activated sludge processes require less land, do not give odor or fly nuisance, are cheaper, and have greater flexibility in producing ranges of purification. More important, if properly loaded, and protected by good trade effluent control and employing an aeration time of around 10 h with a high ($\sim$6000 mg l^{-1}) mixed-liquor solids concentration, they will achieve more complete and consistent purification than percolating filters. Because of the controlled rate of aeration achieved in all but the small installations, they are less susceptible to reduced purification in very cold weather.

In lagoons, land irrigation and sand filtration installations used for the 'polishing' of settled, bio-oxidized effluents, the normal processes of oxidation of carbonaceous matter as ammonia, are continued and other associated processes occur. Continuation of bio-oxidation of carbonaceous matter is slow because the residual of such matter in the settled effluents from percolating filter and activated sludge installations is low. The concentration of ammonia in these effluents may be high, up to 35 mg l^{-1} if little nitrification has occurred in the preceding treatment, and polishing processes may favor considerable further oxidation of ammonia to nitrate. The slow sand filter may oxidize up to 30% of the ammonia applied to it, while rapid sand filters, lagoons and land treatment usually achieve less than this.

Lagoons are, as already mentioned, prone to produce prolific blooms of algae during the warmer months of the year, the appearance of such blooms (and heavy crops of minute animals such as *Daphnia* and *Cyclops*) being stimulated by the residuals of organic matter, nitrate and phosphate, and the relatively high concentration of carbon dioxide present in sewage effluents. While the formation of such blooms amounts in effect to a disadvantageous synthesis of suspended organic matter by photosynthesis, the daytime production of oxygen by such algae when actively growing is advantageous in favoring further bio-oxidation of carbonaceous matter and ammonia. No

significant algal growths are produced in sand filters and land-treatment polishing processes.

However, there is one further useful purification phenomenon which occurs in lagoons, land treatment areas and sand filters. This is the reduction of numbers of bacteria and viruses of fecal origin present in the effluents from percolating filter and activated sludge processes. While these latter processes can bring about a 90% reduction in the fecal bacteria present in sewage (*E. coli*), the effluents still contain many thousands of such bacteria per 100 ml. Lagooning of the final effluents from sewage works for only 5 days will in summer reduce the numbers of these bacteria to very low levels, while it has been shown that storage of effluent in lagoons for 17 days brings about elimination of *E. coli* and certain viruses. Land treatment and sand filtration of sewage effluents are not as effective as lagooning in this respect.

The reduction of bacterial and virus numbers in lagoons is brought about by predation by larger microorganisms and by the sterilizing effect of sunlight.

Bio-reduction processes

The bio-reduction processes used currently in sewage treatment are in sludge digestion and in the simple septic tank. The latter is an uncontrolled and mixed process, while sludge digestion is specifically aimed at diminution of sludge bulk by the reduction of organic matter in the sludge, in the absence of oxygen, to methane and carbon dioxide. This basic digestion process proceeds slowly at ambient temperature in any sludge deposits accumulated in tanks or lagoons, and in such uncontrolled circumstances the occurrence of side reactions causes the development of foul odors.

Sludge digestion installations at sewage works are operated at a temperature in the sludge of between 30 and 35°C to accelerate the process to yield the optimum results in terms of maximum gas production and minimum odor nuisance. This digestion is carried out in heated, enclosed, primary digestion tanks giving a nominal detention period of between 15 and 30 days. The sludge withdrawn from primary digestion is usually passed into secondary, unheated open tanks giving a detention time of 20–50 days, where some further digestion occurs, the sludge solids consolidate and much sludge liquor separates. By decanting this separated liquor, the volume of the original sludge may be reduced by about 65% and the weight of sludge solids by about 40%. The sludges from both primary settlement tanks and from secondary settlement of bio-oxidized effluents are usually subjected to digestion, mixing of the sludges being necessary since secondary sludges alone do not digest readily.

The essential requirements for efficient operation of primary digesters are

- exclusion of oxygen from the digestion tanks;
- maintenance of a temperature within the digesting sludge of around 30°C;
- frequent additions of small quantities of fresh sludge to the digestion tanks and good mixing of this and digesting sludge within the digestion tanks;
- maintenance of a pH within the digesting sludge in the range 6.8–7.8;
- exclusion of materials derived from trade effluents discharged to sewers at concentrations which are toxic to, or inhibitive to the growth of, the anaerobic bacteria performing the digestion.

Given these requirements, the daily gas yield from the primary digesters will be about 0.025 m^3 per head of population, the gas consisting of a mixture of about 70% methane, 30% carbon dioxide and small quantities of nitrogen and hydrogen sulfide.

Primary digestion tanks are usually circular in plan, with a sloping conical floor and either a floating gasholder-type roof or a fixed roof, and a capacity of about 0.05 m^3 per head of population served. The sludge should be fed into the digester at a rate of about 2.5 kg organic matter per m^3 of digester capacity per day. Mixing and heating of the sludge within the digester can be achieved in a variety of ways, the most usual being by recirculating the sludge from the digester through external heat exchangers, supplied with heat from boilers fired by sludge gas or the waste heat from gas-operated power-generation engines. The development of too low a pH in the digester, as a result of excess accumulation of fatty acids from the organic matter being digested, is usually corrected with small additions of lime.

A variety of substances inhibit sludge digestion. The accumulation of excess concentrations of synthetic detergents (greater than 750 mg l^{-1}) interferes with digestion, but this can be corrected by

Table S5 Activated sludge processes

Type of process	Objective	Variations in technique					Design criteria	
		Tank shape	Nominal aeration time (h)	Concentration of mixed liquor solids (mg/l)	Means of aeration	Sludge return system	Daily loading in kg (kg BOD$_5$ m^{-3} aeration tank capacity)	Air supply or aeration power required
High rate	Bio-flocculation for partial purification or to precede percolating filtration	Circular or square tanks 2.5–3.5 m deep or long rectangular tanks 2–4 m deep, or shallow channels 1.2 m wide by 1.2 m deep	2–6	750–1500	Diffused or coarse bubble air or cone, brush or paddle surface aeration	With or without re-aeration	1–1.5	30–50 m^3 kg^{-1} BOD removed
Normal rate	Full treatment including good nitrification	Square tanks or long rectangular tanks	8–12	3000–6000	As above	Often with separate sludge re-aeration before return	0.4–0.6	60–100 m^3/kg BOD removed 0.6 k W h kg^{-1} BOD removed
Extended aeration	Full treatment and reduction of sludge bulk in small plants without skilled supervision	Circular or rectangular special design tanks or a ring channel 5 m wide by 1 m deep	24–48	3000–8000	As above	Usually a simple built-in sludge return system, but excess removed periodically	0.15–0.25	150–300 m^3/kg BOD loading 1.2 k W h kg^{-1} BOD removed
Contact stabilization	As above but sometimes used to give little or no nitrification	Circular or rectangular tanks of special design	6–12 with most aeration time in sludge re-aeration	3000–6000	Usually coarse bubble air	Separate sludge re-action	0.4–0.6	150 m^3 kg^{-1} BOD removed

additions of stearine amine acetate. Of the inhibiting substances
derived from trade effluents, the chlorinated hydrocarbons such as
chloroform and trichloroethane are among the worst, concentrations
of 0.1 and 0.01 mg l^{-1} of the former and latter respectively in sewage
causing significant reduction of gas production in digestion. A variety
of other organic compounds and toxic metals also exert inhibitory
effects.

Because of inhibition difficulties of these kinds and the decreasingly favorable overall economics of power production from digestion, the process has lost some of its attraction in sewage treatment in
recent years. However, with good trade effluent control, which is
necessary anyway for effective bio-oxidation of sewage, digestion
still has much to commend it.

In septic tanks the reduction processes are aimed at minimizing the
volume of sludge accumulating in the tanks, but the reduction
processes also proceed within the overlying sewage, giving rise to
odor nuisance mainly as a result of reduction of sulpfate in the sewage
to hydrogen sulpfide. Overall, a septic tank produces an effluent of
considerably worse organic quality than the influent. For these reasons
the use of septic tanks is limited to very small installations serving
single or small groups of dwellings, where open settlement tanks,
requiring regular desludging, are neither desirable nor practicable.

The denitrification (reduction) of nitrate by bacteria to nitrogen gas
in bio-oxidized effluents, both in the sludges in aeration tanks and in
secondary settlement tanks where dissolved oxygen may fall to very
low levels, frequently occurs. However, at the present time this
phenomenon is mainly a nuisance at sewage works because it may
result in 'rising sludge' in secondary tanks, thereby reducing
settlement efficiency. This process is likely to be developed for
general use in removing nitrate from sewage effluents discharged to
lakes, or to rivers drawn on for public supply.

Separation of water from sludges

One of the biggest problems in sewage disposal is the economical
disposal of sludges. Since primary and secondary sludges as run from
tanks usually contain 97–99% water, it is of great importance that the
optimum removal of water from sludges, to reduce their bulk, should
take place before disposal by any means is attempted, the separated
water being passed back to the primary settlement tanks for
purification along with the incoming sewage.

The sludges from both primary and secondary settlement tanks are
colloidal in character and do not readily dewater on standing,
although primary sludges are better in this respect. Primary sludge as
drawn from tanks contains 95–98% water according to the nature of
the settlement process, the thicker sludge coming from horizontal
flow tanks where desludging is carried out periodically.

Regarding secondary sludges, humus sludge from percolating
filters contains 94–97% water while activated sludge contains
98–99.5% water. Digested sludge contains between 92 and 95%
water. The sludges from high-rate bio-oxidation processes are usually
thin and difficult to dewater.

The ease (or difficulty) with which various sludges can be
dewatered can be best assessed by comparison of their individual
'specific resistance to filtration'. This parameter can be derived from
the results of measurement of the capillary suction time of the sludge
through a standard filter paper, and can be used in conjunction with
other empirically derived data to determine optimum design and
operational requirements in sludge dewatering installations.

The various processes used for separation of water from sludges
are

● thickening processes;
● conditioning processes used to prepare sludge for dewatering;
● drying processes;
● mechanical dewatering processes.

A variety of special equipment is available to perform, or facilitate
performance of, most of these processes.

Thickening processes

These are directed at causing water, which is in loose colloidal
binding with the particles of sludge solids, to separate. One process,
that of thickening by storage for 24 h in tanks, with or without gentle
stirring, facilitates the consolidation of sludge solids while liquor
separates at the surface for removal. Another process, that of flotation
thickening, involves the introduction of a mixture of compressed air
and recirculated liquor into the bottom of long, narrow tanks

containing the sludge being treated. The fine bubbles of air rising
upwards through the sludge accumulate sludge solids and lift this to
the tank surface where it is removed.

Conditioning processes

These processes are directed at modifying or substantially destroying
the colloidal structure of the sludge–water mixture, enabling the finer
particles of sludge to coalesce or coagulate, thereby facilitating
subsequent mechanical filtration of the conditioned sludge. Chemical,
biochemical or physical processes of conditioning are used.

Chemical conditioning involves the addition of lime and ferrous
sulfate, aluminum chlorohydrate or polyelectrolytes to the sludge.
These chemicals alter the state of electrical charge on the sludge
particles in colloidal suspension in the sludge water, resulting in
coagulation of these particles and separation of water. Sometimes,
chemical conditioning is preceded by elutriation, a process whereby
treated sewage effluent is brought into contact with the sludge,
washing out materials which tend to inhibit coagulation, and then
separating the elutriating effluent. In this way the quantities of
chemicals required for conditioning can be reduced. The biochemical
processes of sludge conditioning are the anaerobic (bio-reduction)
process of sludge digestion and the aerobic (bio-oxidation) digestion
of activated sludge as used in extended aeration treatment systems
both of which processes are carried out by bacteria.

The physical processes of sludge conditioning involve pressure
cooking or freezing of the sludge, which substantially destroys the
colloidal structure of the sludge. In the Porteus heat treatment process,
sludge is heated to around 180–200°C with steam under pressure for
about half an hour, following which rapid separation of water as a
very strong liquor occurs. This process has the advantage of
sterilizing the sludge, which is of public health significance, in
relation to final disposal of the sludge on agricultural land. However,
the cooking of the sludge transfers some of the organic matter in the
sludge into the sludge water, thereby reversing in some small degree
the objective of the settlement and bio-oxidative processes of sewage
treatment. In the Zimpro process of wet air oxidation, the sludge is
heated to about 200°C under pressure with air to condition and
partially oxidize the sludge. Freezing of sludge followed by thawing
partially destroys the colloidal structure of the sludge and results in
separation of water.

Drying of sludges

These processes are intended to remove water from sludges by
atmospheric or heated evaporation. Sludge, with or without prior
conditioning is dried in lagoons or drying beds, while the heated
processes, for obvious reasons of heat economy, are normally used to
dry further the sludges which have already been subjected to
substantial dewatering. The sludge lagoon is a very slow and
obnoxious arrangement for the drying of sludges, and serves much
more as a space for dumping sludge rather than as a treatment
process. Drying beds serve two functions in permitting a kind of
dewatering filtration of sludge and in atmospheric drying of sludge. A
typical drying bed consists of a layer of clinker or gravel about 0.3 m
deep over underdrains, topped with fine clinker, pea-gravel or sand.
When sludge is run onto the bed to a depth of about 15 cm (or up to
0.3 m if the drying of earlier-filled beds is slow), drainage of sludge
liquor through the bed occurs rapidly at first and then falls off to zero,
and thereafter further dewatering occurs by evaporation. Lifting of the
dried sludge by mechanical means has now replaced manual lifting at
most of the larger sewage works. Rotary kiln dryers and flash dryers
are specially designed installations for transforming sludge cakes
from drying beds or mechanical dewatering processes into powder
containing about 10% moisture.

Mechanical dewatering processes

These processes mainly involve subjecting sludge to considerable
mechanical stress by pressure filtration, vacuum filtration or centrifuging, but one process, the rotary sludge concentrator, is different in
operation.

Filter presses have been in operation for sludge dewatering for
many years. The process consists of pumping chemically conditioned
sludge at high pressure through filter cloths covering rectangular,
recessed cast iron or steel plates, set up in a multiple sandwich in the
press installation. The press is closed hydraulically, sludge is pumped
into the press, passes out through holes in the center of the plates into

the spaces between filter cloths, water is forced through the cloths and the sludge solids are retained as a cake. Pressing time varies from a few hours to more than 24 h depending on the nature of the sludge. The press is then opened and the sludge cakes containing 50–70% moisture are discharged. The sludge cake can be handled readily, and if stacked in the open it will break down into a friable state. The process is a batch one which has now been automated to a considerable extent.

Vacuum filtration has been developed as an automated continuous process, in which a horizontally mounted segmented drum, 1.3–1.7 m in diameter and covered with a filter cloth, coil springs or a specially formed 'precoat', rotates into a trough of chemically conditioned sludge. Water is drawn into the drum from the sludge by vacuum, and a layer of sludge cake forms on the filtering medium, to be lifted off automatically as the drum rotates to re-enter the sludge trough. A mobile disk filter has also been developed for vacuum filtration of sludge. This is a segmented disk, 1.8 m in diameter, covered on both sides with filter cloth, rotating through a sludge trough. Vacuum applied as the disk passes through the trough draws water through the filter cloth, which picks up a layer of sludge. Before the disk re-enters the trough, the layer of sludge cake formed on the filter cloth is blown off by air under pressure. Sludge cakes from vacuum filtration contain around 70–75% water.

Processes of filter-belt pressing and of centrifuging of sludge are new developments now being installed at sewage works. The filter-belt press consists of an endless filter belt and an endless pressure belt running together slowly through rollers. Conditioned sludge delivered onto the filter belt is dewatered to 70–80% water content as it passes under increasing pressure through the machine. Centrifuges of the continuously operated, solid bowl type can produce sludge cake containing 65–85% water.

The *rotary sludge concentrator* consists of a hollow drum about 1 m in diameter and 0.35 m wide covered with filter cloth. Sludge is passed into the drum, and as this dewaters by drainage through the cloth it accumulates as a rolling plug of sludge at the bottom of the drum. As the plug of sludge increases in size it drops off the edges of the drum. The system does not work well on all sludges, but primary sludge and mixtures of some primary and secondary sludges can be dewatered to around 70–75% water without prior chemical conditioning.

A characteristic of all processes of sludge filtration is that the separated liquors contain variable proportions of sludge fines, depending on the nature of the sludge and the process used. The concentrations of solids and BOD in such liquors returned to primary treatment are matters which cannot be ignored in respect of the additional load thereby passed to the sewage treatment processes. The efficiency of any particular installation for sludge filtration in terms of cake yields is likely to be affected by a number of factors of operational detail.

Disposal of sludges

Sewage sludge is disposed of by return to the land as a medium-quality fertilizer and soil conditioner, by tipping along with domestic refuse, by dumping at sea (to cease as of 31 December 1998), and by incineration.

The traditional, and normally the best method, is to return it to agricultural land. Liquid sludge, in particular digested sludge, is increasingly being disposed of by tankers onto grassland, while disposal on land of the less friable sludges from drying beds and some mechanical dewatering processed is decreasing, because of handling and spreading difficulties. The potential for disposal of heat-treated and filter-pressed sludges, after stacking at sewage works and on farms, is considerable because the sludge cake is sterile and readily spreadable after breakdown in stacks. Heat-dried and powdered sludges, and composted mixtures of liquid sewage sludge and household refuse, are of course very suitable for spreading on land. However, care must be exercised in disposing of sludges derived from sewage containing much industrial waste, because the levels of toxic materials, particularly the toxic metals lead, cadmium, chromium, nickel, and zinc, in the sludge may result in damage to plants and animals. Sewage sludge has a reasonable fertilizer value, containing in dry sludge about 2.5% nitrogen, 1.5% phosphorus (P_2O_3) and 0.3% potassium (K_2O), and it contains a high proportion of organic matter.

Disposal of liquid sludge at sea by tanker ships is practiced in London, Manchester and other coastal towns in the UK. Provided the sludge does not contain excessive concentrations of toxic substances,

and is disposed of by dispersal well out to sea, this method of disposal is satisfactory since the capacity of the sea to oxidize organic matter is massive. As noted above, this procedure ceases in 1999.

Incineration of sewage sludge, alone or with household refuse, is being increasingly practiced, using multiple-hearth, rotating drum and fluidized-bed furnaces. Sludge when incinerated alone, without added heat, should not contain much more than 50% water in order to be completely burned. A temperature of around 700°C should be achieved in the furnace to avoid odor nuisance from flue gases. Scrubbing of the flue emissions from incinerators is always desirable, and necessary in the case of some sludges and some furnace installations.

New sewage treatment processes

Because of the heightening concern regarding the effects of increasing discharge of sewage effluents in increasing the nitrate, phosphate and residual organic matter content of surface waters drawn on, or expected to be drawn on in the foreseeable future, for public water supply, attention has been directed to the development of new processes, or to adaptation of existing water treatment processes, for giving additional purification of sewage effluents. These developments can be divided into two categories:

- those aimed at removing more impurity from sewage than the existing processes;
- those aimed at removing most of the water from sewage effluents in a very clean state.

The removal of phosphate from sewage effluents to levels less than 1 mg l^{-1} can be done readily by simple precipitation and settlement in tanks using lime or aluminum sulfate. The removal of ammonia and phosphate from screened and de-gritted sewage can be achieved by adding lime to the sewage to achieve a pH value greater than 10.5, which precipitates the phosphate as calcium phosphate as well as flocculating much of the colloidal matter present in the sewage. Ammonia present in the lime-treated and settled sewage can then be partially removed by air stripping, that is, the sewage flows down a packed tower through which air is blown upwards. Unfortunately this latter process is inhibited by cold weather, and suffers from scale formation in the tower. Also, if in the sewage being treated with lime, the urea present has not already been hydrolyzed to ammonia, further hydrolysis is halted when the pH of the sewage is raised by lime addition. An alternative process would be to take the well-nitrified sewage effluent after normal treatment, and to bring about denitrification by adding a suitable carbon source (waste activated sludge or methanol) to reduce the nitrate present to nitrogen gas, and following this with phosphate precipitation.

Good rates of removal of ammonia, nitrate, phosphate and other unwanted impurities including organic matter, can be achieved by subjecting a normal sewage to specific ion exchange processes, but these are relatively expensive. The use of activated carbon by additions of powdered carbon or filtration through filters packed with granular activated carbon, for removal of residual organic matter in sewage effluents, has been demonstrated to be a feasible and not a highly expensive process. The use of ozone for destroying the color of, and sterilizing, effluents has also considerable potential.

Of the processes which are suitable for use in the water supply industry for desalinating water, electrodialysis, distillation, freeze-desalination and reverse osmosis, only the latter shows promise of being of value in removing clean water from sewage effluents. The other three processes will be expensive and inhibited in their efficiency because of the effects of contaminants in the sewage effluent. Reverse osmosis is a process which operates by forcing the water in sewage effluent through a semi-permeable membrane under very high pressure, while most of the impurities cannot pass the membrane. Problems of membrane fouling have yet to be overcome before the process becomes economically feasible on a large scale.

Storm overflows at sewage works

It is common practice at sewage works to make provision for overflow of wet-weather sewage flows to prevent excess hydraulic loading of sewage purification plant. Usually, these overflows are set such that 3 × DWF (dry-weather flow) goes forward to the normal treatment processes while excess flows are shed to storm sewage treatment facilities. It was also quite common for a normal storm overflow to be provided at the sewage works inlet to ensure that no more than $6 \times$ DWF was passed to treatment ($3 \times$ DWF to normal treatment processes and $3 \times$ DWF to storm sewage treatment facilities). The general principle of good practice is that the carrying capacity of sewers leading to sewage works should be used to the full, and that the only separation of storm flow should be between the $3 \times$ DWF given full treatment and the excess flow which is passed to storm sewage treatment. The necessity for pumping might call for compromise.

At sewage works having separated storm flows in excess of $3 \times$ DWF, usually by means of side-weir overflows located downstream of screens and flow measuring devices, these excess flows are passed into storm tanks. These tanks, traditionally of capacity equal to 6 h DWF, are normally used to store the first, most polluted, sewer-cleansing flush of storm flow, and to provide settlement of solids from the storm flow in excess of tank capacity. In many cases where at least two tanks are provided, one or more tanks are used as 'blind' tanks without an overflow to contain completely the first flush of flow, while remaining tankage is used as settlement tanks which overflow when full. At some works all the tanks are filled in succession and when these are full overflow from the tank occurs. The former arrangement is probably the better one at the smaller sewage works. When storm flows subside, the storm tank contents are drained off into the normal flow being given full treatment. Quite often any overflow from the storm tanks is given further purification by broad irrigation on land before discharge to stream. Small sewage works, serving substantially separate sewerage systems, are not provided with storm overflows or storm tanks, but are made capable of giving full treatment to all flows received, up to a maximum of $6 \times$ DWF.

The capacity to be provided in storm tanks may be designed to be equal to 70 l per head of population served by combined and/or partially separate sewerage draining to the sewage works.

The quality of discharges of storm sewage from storm sewage treatment arrangements at sewage works depends mainly on the nature of the normal sewage flow and on the treatment arrangements made. Where the storm tankage provision is on the basis of 70 l per head, and is operated such that the first flush of storm flow separated at the sewage works is retained in the tank(s), and any excess flow is given settlement before discharge, the discharge should normally not have a suspended solids content in excess of 150 mg l^{-1} nor have a BOD$_5$ greater than 150 mg l^{-1} and on average the discharge will be of much better quality than this. An overflow set at $6 \times$ DWF could be expected to discharge, on average, a sewage with a suspended solid content of about 400 mg l^{-1}, a BOD$_5$ of about 200 mg l^{-1}, an ammonia nitrogen content of about 4 mg l^{-1}, and a very high content of bacteria of fecal origin. The first discharge of such an overflow during a summer storm after a dry period of several days could be expected to be very much worse in quality, while the peak discharge during a spell of winter rainfall could be more like normal surfacewater runoff in quality. Where industrial discharges are passed to sewers in flushes, near to storm overflows, the discharges from these overflows may contain appreciable quantities of toxic and other highly polluting materials derived from such industrial discharges.

The polluting effects of any given storm sewage discharge on the receiving stream depend on the quality of the discharge at a particular time, and the quality and flow of the receiving stream. Discharges of storm sewage arising from local summer storms often occur when the flows of the receiving streams are low and are not being appreciably augmented by clean inflow from the local storm. This is often the root of storm sewage pollution – the discharges are at their strongest when the receiving streams are at their lowest flows.

At pumping stations receiving the flows from combined or partially separate sewers, it is common practice to provide a storm overflow upstream of the station to bring the rate of flow of sewage to be pumped down to the practicable level. At pumping stations which are dependent solely on mains power supply, it has been common practice to provide an emergency overflow set at a high level in the pumping well to prevent flooding of the station in the event of mains power failure. The occurrence of industrial disputes and power shut-off in the electricity supply industry has highlighted the pollution risks which can arise from overflows of sewage in dry weather at pumping stations wholly dependent on mains power supply. In consequence, in the catchment areas of rivers used for water supply and on reaches of rivers and streams of high amenity value, steps have been taken to install standby pumping capability at the larger pumping stations to prevent sewage overflow in an emergency. For use at the smaller stations during power shut-off or failure, where the cost of provision

of standby equipment would be prohibitive, mobile pumping equipment has been provided.

In some rural localities, where storm overflow from sewers is necessary yet no significant stream pollution from such overflows can be permitted, the overflowing sewage is passed to small areas of land, thereby giving a reduction in the volume, and a marked improvement in the quality, of the overflowing sewage reaching the stream. Similar arrangements are made from time to time for dealing with any emergency overflows from small pumping stations.

R.W. Herschy

Source

Fish, H., 1973. *Principles of Water Quality Management*, Thunderbird Enterprises.

Cross references

Activated sludge process
Sewage treatment: general introduction

SINK, SINKHOLE, SWALLOW HOLE

By far the most common feature of limestone solution in karst landscapes, sink or sinkhole applies to any depression ranging from a shallow saucer shape, where runoff water quickly sinks into the ground, to a funnel-shaped or cylindrical pipe that normally gives access to underground caves. Where a stream disappears into the hole, it is usual to call it a swallow hole or *ponor* in Serbo-Croat. A deep karst pipe leading down into a cavern system is termed a pothole in England and a *jama* in Yugoslavia. Explorers of such features are popularly known as 'potholers.' The depression may contain a pond or lake. Malott (1945) estimated that there are about 300 000 sinkholes within the area of the southern Indiana karst, covering up to 500 km^{-2}.

Thornbury (1954) distinguishes between a depression, lowered by solution beneath a soil cover, ultimately to form a dolina, and the undermining of a cavern that leads to a collapsed sink; this definition seems to run counter to the usual understanding that a collapsed sink is a dolina, and further collapse and extension provide a series: sink–dolina–uvala–polje.

Large sinkholes in Florida and Yucatan, today half-filled with water, were formed during the Pleistocene low sea level stages. In Yucatan these 'cenotes' were regarded as sacred by the Mayas in this otherwise waterless limestone platform (Shrock, 1946; Termier and Termier, 1963). Drowned karst holes and pipes are commonly seen in Pacific and Indian Ocean coral reefs that predate the last glacial drop of sea level, when the groundwater table was lowered by over 100 m (Fairbridge, 1948). They are often seen in calcareous eolianite rocks in Bermuda, Bahamas, South Africa, Western Australia and elsewhere (Fairbridge, 1950). Where large deep examples are found by sounding offshore, as in the Bahamas, they are known as 'oceanholes' or 'blue holes,' in contrast to the green shallows. Small ones in the Bahamas on the land often contain rich soil and are called 'banana holes' for that is what is usually planted there. In some early papers, these pipes were mistaken for fossil palmetto trunks. Certainly they are often associated with fossil roots, 'rhizomorphs' or 'rhizoconcretions' (Northrop, 1890).

Sinks also form sometimes in other soluble rocks such as salt, gypsum and anhydrite.

Rhodes W. Fairbridge

Bibliography

Fairbridge, R.W., 1948. Notes on the geomorphology of the Pelsart Group of the Houtman's Abrolhos Islands, *J. Roy. Soc. W. Australia*, **33** (for 1946–47), 1–43.
Fairbridge, R.W., 1950. The geology and geomorphology of Point Peron, Western Australia, *J. Roy. Soc. W. Australia*, **34**, (for 1947–48), 35–72.
Malott, C.A., 1945. Significant features of the Indiana karst, *Proc. Indiana Acad. Sci.*, **54**, 8–24.

Northrop, J.I., 1890. Notes on the geology of the Bahamas, *Trans. N.Y. Acad. Sci.*, **10**, 4–22.

Shrock, R.R., 1946. Karst features in Maya region of Yucatan peninsula, Mexico, *Proc. Indiana Acad. Sci.*, **45**, 111–116.

Termier, H. and Termier, G., 1963. *Erosion and Sedimentation*, London and Princeton, NJ, D. Van Nostrand Co., 433 pp. (translated by D.W. and E.E. Humphries).

Thornbury, W.D., 1954. *Principles of Geomorphology*, New York, John Wiley & Sons, 618 pp.

Cross reference

Karst hydrology

SNOW MEASUREMENT

Introduction

Snow and other forms of solid precipitation pose greater problems of measurement than liquid precipitation. Snow is hard to catch, it is readily redistributed on the ground, it melts differentially and its presence restricts access to measurement points.

For the majority of hydrological purposes the most important information required is the volume of snow as a water equivalent – the depth of water that would result from melting – and the rate of melt. These are equivalent to the more readily measured volume and rate of rainfall. However, to ensure that these can be quantified it is often necessary to track the progress of accumulation through the snow season. For many other purposes such as transport management, recreational use of snow and avalanche prediction, it is the measurement of falling snow and the properties of the snow cover through the snow season which are of greater importance.

Measurement of snowfall

In countries such as the UK where snow contributes a comparatively small proportion of total precipitation, standard storage and recording raingauges are also used to measure falling snow. Where daily storage gauges are used, snow in the collecting funnel is melted either by bringing indoors or by adding a known quantity of warm water. Recording gauges may be equipped with a heating element, sufficient to protect the collecting mechanism of tipping bucket or siphon gauges but less frequently to melt snow in the funnel.

Such conventional gauges suffer seriously from the effects of wind eddies created by nearby obstructions and the gauge itself, by interception of drifting snow, by snow bridging the gauge orifice and occasionally by complete burial in heavy snow. Wilson (1954) demonstrated the effect of wind speed on gauge catch and showed, for example, that with a wind speed of 11 m s^{-1} the deficiency is of the order of 60%.

In countries where a substantial proportion of the annual precipitation falls as snow, special snow gauges have been devised. Such gauges are raised above the maximum expected snow depth and special shields are fitted around them to minimize the effects of turbulent eddies. Most widely used are the flexible Alter shield (Figure S15a) developed in the United States, the similar Tretyakov shield in Russia and the solid Nipher shield (Figure S15b) deployed in Canada. Even with these modifications, investigations show that measured snowfall is generally deficient. Goodison and McKay (1978) illustrate the extent of the deficiency (Figure S16), and demonstrate the superior performance of Nipher shielded gauges with respect to the effects of wind speed. It is, however, more prone to bridging.

Even if it were possible to measure snowfall accurately, the values obtained are of indirect hydrological use, since the amount and distribution of the snowpack at the final onset of melting may be quite different from the distribution of snowfall. The snow may be redistributed by wind, reduced by melting at lower elevations whilst continuing to accumulate at higher levels. It is therefore usually more profitable to measure the properties of the snow cover directly.

Snow cover at a point

Depth is the most obvious and most commonly measured property of the snowpack, but it is not in itself a useful hydrological measure, as

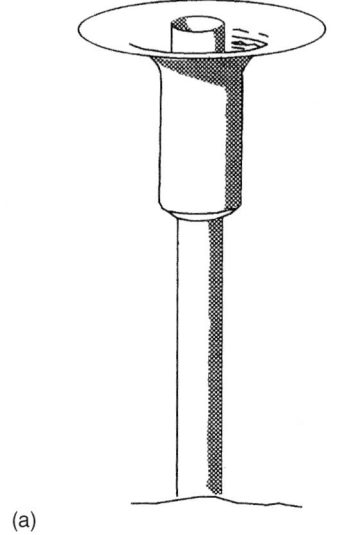

(a)

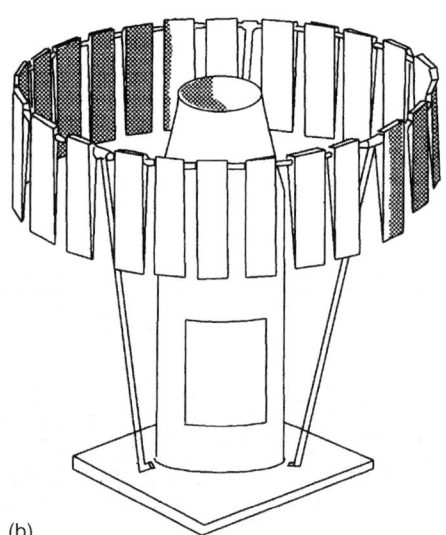

(b)

Figure S15 Shielded gauges for measurement of snowfall: (a) Nipher shield and (b) Alter shield.

density may vary from as little as 50 kg m^{-3} for fresh dry snow to more than 500 kg m^{-3} for old compacted or wet snow. Snow density can be considered as the ratio of the volume of meltwater from a snow sample to the original volume of snow or, more simply, as the ratio of their depths.

Snow depth is measured with a ruler for shallow packs and by using a series of fixed graduated stakes for deeper snowpacks as part of a snow course. In remote areas these may be adapted for aerial survey. Ultrasonic distance-measuring sensors placed over the snow pack have been developed for continuous recording and transmission of depth at remote sites (e.g. Bergman, 1989).

Snow water equivalent at a point is most commonly measured using a snow core sampler. The sampler is pushed or driven vertically through the snowpack and the weight of the snow core is taken as the difference between the weight of the sampler with and without snow. Weight and water equivalent are directly related given the diameter of the sampler, and indeed, the Mount Rose snow sampler used in the United States has a scale directly calibrated in water equivalent units. Small-diameter samplers (e.g. Mount Rose, 3.78 cm), with a cutting edge to break through ice layers, are preferred in deep packs. Larger

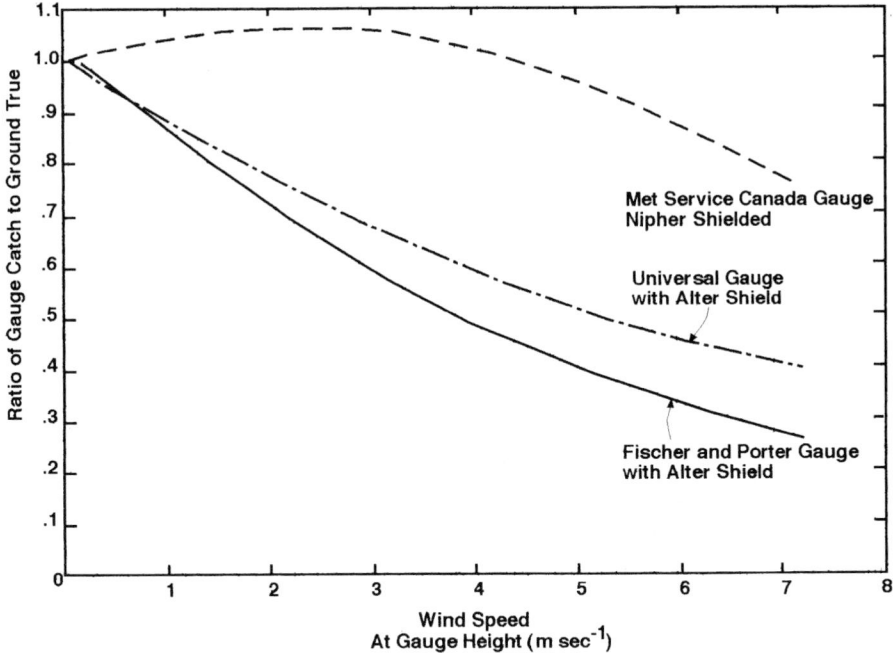

Figure S16 Gauge catch as a function of wind speed for shielded precipitation gauges. (After Goodison and McKay, 1978.)

diameters are required for improved sensitivity in shallow or low-density packs and in the UK a 10.16 cm tube is used (Figure S17) with a central spindle and impeller, which is rotated through the snow core to the base, to hold the snow in place whilst weighing (Johnson, 1975).

Whilst the snow core sampler has advantages of mobility and low cost, it has the disadvantage that later measurements cannot be made at precisely the same point for comparison, as the pack is disturbed. Also, it cannot be adapted for remote and automatic readings, and for this purpose a number of devices have been developed.

The snow pressure pillow is such a device. It is a bladder of butyl rubber or similar material, 1.5–4 m in diameter and 100–250 mm in depth, filled with an antifreeze solution (Figure S18). When snow accumulates on the pillow, the pressure is transmitted via a connecting pipe to a stilling well. Changes in level (millimeters) in the well directly reflect the water equivalent of the snow with a small adjustment for the specific gravity of the pillow solution. Such measurements are readily adapted for telemetry. Differences in sequential measurements indicate rates of snow accumulation and ablation, and have been put to practical use in river flow and flood forecasting (Farnes, 1984).

There are some problems, however. In shallow packs, albedo and roughness differences from surrounding vegetation may produce unrepresentative measurements and in deeper packs ice lenses may form bridges over the pillow, reducing accuracy. Small-scale fluctuations also arise from temperature and other meteorological conditions. The pillow is vulnerable to animal damage, although protective measures reduce the risk.

Radioisotope snow gauges have been developed for deep snow covers in remote locations. They are based on the attenuation of radiation by the snow cover. A radiation source, usually ^{60}Co, is mounted above the snow cover and a receiver, usually a Geiger–Muller tube, is placed on the ground below. The gamma radiation passing through the snow is absorbed according to the equation

$$I_w = I_0 e^{-\mu w} \qquad (S1)$$

where I_0 is the intensity of incident radiation, I_w is the intensity after passing through the snowpack, μ is the linear absorption coefficient for water (cm^{-1}) and w is the water equivalent of the snowpack.

Whilst such vertical gauges give the total water equivalent of the snowpack, horizontal profiling gauges (Figure S19) give additional information on variations of density with depth in the pack, by having a source and detector which can be moved synchronously through

Figure S17 Using a Johnson snow core sampler in shallow snow packs in England.

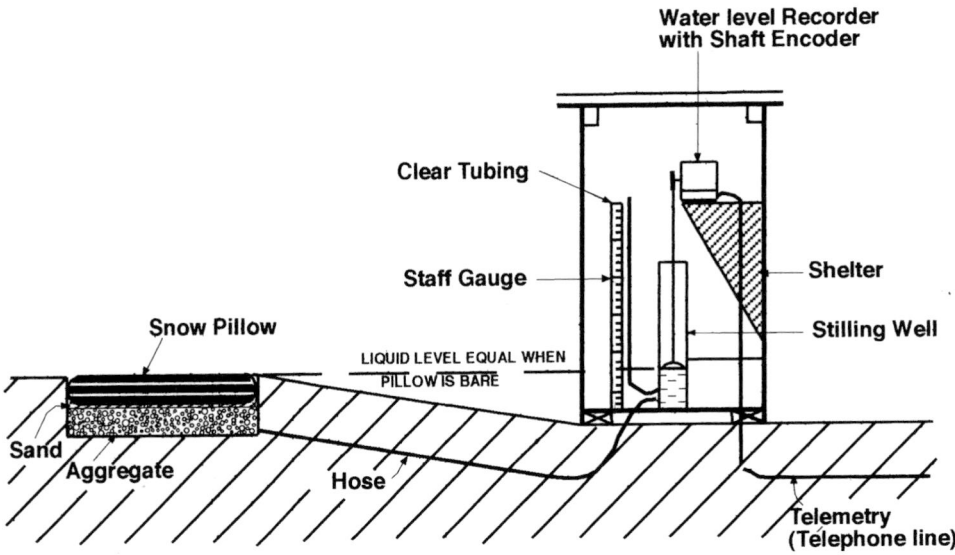

Figure S18 Snow pressure pillow.

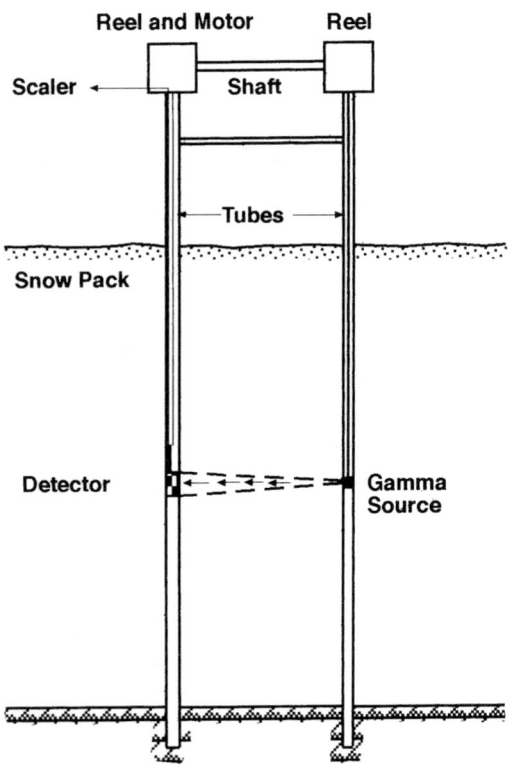

Figure S19 Profiling radioactive snow gauge. (After Martinec, 1976.)

vertical tubes about 0.6 m apart (Randolph *et al.*, 1972). Radioisotope gauges are expensive to install and maintain and are primarily used for research.

Areal snow assessment

From the foregoing it is evident that a range of instruments is available for measurement at a point. However, because of the heterogeneous nature of the snowpack and its redistribution on a local as well as a basin-wide scale, formidable difficulties exist in translating these measurements into areal values. Ground-based snow

measurement networks are designed to sample local variability of water equivalent over a snow course, where repeated measurements of water equivalent are made along a line of stakes at almost the same locations. Broader basin variability is generally sampled by locating snow courses within unit source areas, especially with respect to elevation, aspect and vegetation cover. Nevertheless, even where many basin measurements are made, the resulting areal estimate is still generally regarded as an index rather than an absolute estimate of basin snow cover water equivalent. Such indices can, however, be usefully employed for runoff volume and flood prediction.

Martinec (1976) points out that sampling efficiency can be improved by use of the conservative nature of snow density, which varies spatially much less than snow depth on a given date. Thus there may be little loss of accuracy in assessing areal water equivalent, by reducing the number of density and water equivalent measurements so long as the number of simpler depth measurements is maintained.

Airborne and satellite remote sensing techniques have been applied to obtain direct areal measurements of the snow cover. A method based on the attenuation of natural gamma radiation from the soil by a snow cover has been used operationally in many countries including Russia (Dmitriev *et al.*, 1973), Canada and the United States (Peck *et al.*, 1971). The magnitude of attenuation is related to the mass of water between soil and detector [c.f. equation (S1)]. Airborne surveys are carried out at a flight elevation below 150 m and water equivalent is determined as the average of a band about 300 m wide beneath the flight path. Corrections must be made for soil moisture and background atmospheric radiation, but in favorable conditions measurement of water equivalent is possible to within 10–25 mm. Use of the method in rugged terrain is limited by the low flight elevation.

Satellite remote sensing techniques have been widely used to assess snow cover extent (Hall and Martinec, 1985). Usage of these techniques may be limited by pass frequency – 16 days for Landsat, and daily for AVHRR (Advanced High Resolution Radiometry) satellites – or by spatial resolution (Landsat, 30×30 m; AVHRR, 1.1×1.1 km). Pass frequency and spatial resolution are inversely related. Cloud cover also restricts the use of images in the visible range, although techniques have been developed to distinguish automatically between snow and cloud surfaces in partly cloudy images (Lucas, 1990). Satellite remote sensing is particularly useful in assessing snow cover changes over broad areas with a prolonged snowpack and protracted depletion.

The assessment of snow depth based on snow albedo has been widely investigated and early work suggested that reasonable results could be achieved up to a limit where underlying vegetation is totally masked. However, snow surface conditions, the effect of shadows and atmospheric contamination also affect albedo and limit the operational usefulness of the results.

Passive microwave radiometry offers potential for the assessment of snow water equivalent in both cloudy and cloud-free conditions. Although the imagery is of lower spatial resolution than visible and infrared imagery, techniques are giving moderate success in dry snow conditions over relatively flat continental areas using brightness temperature difference algorithms. (Goodison and McKay, 1990).

In spite of improving ground-based and remote sensing technologies, the measurement of snow remains an important research and development area in hydrology.

David Archer

Bibliography

Bergman, J.A., 1989. An evaluation of the acoustic snow depth sensor in a deep Sierra Nevada snowpack. *Proc. Western Snow Conf.*, **57**, 126–129.

Dmitriev, A.V., Kogan, R.M., Nikoforov, M.V. and Fridman, Sh.D., 1973. The experience and practical use of aircraft gamma-ray survey of snow cover in the USSR. *UNESCO/WMO/IAHS Symposia on The Role of Snow and Ice in Hydrology*, Banff, IAHS Publ. 107. Vol. 1, 702–712.

Farnes, P.E., 1985. Predicting time of peak snowmelt runoff from snow pillow runoff. *Proc. Western Snow Conf.*, **52**, 132–138.

Goodison, B.E. and McKay, D.J., 1978. Canadian snowfall measurements; some implications for the collection and analysis of data from remote stations. *Proc. Western. Snow. Conf.*, **46**, 48–57.

Goodison, B.E., Walker, A.E. and Thirkettle, F.W., 1990. Determination of snow cover on the Canadian prairies using passive microwave data. *Proc. International Symposium on Remote Sensing and Water Resources*. Enschede, Netherlands, pp. 127–136.

Hall, D.K. and Martinec, J., 1985. *The Remote Sensing of Snow and Ice*. Chapman & Hall, London.

Johnson, P., 1975. Snowmelt, in *Proceedings of the Flood Studies Conference*, Instn of Civil Engineers, pp. 5–21.

Lucas, R.M., 1990. *Development of satellite techniques for operational snow monitoring in the United Kingdom*. Report to the Dept. of the Environment, Remote Sensing Unit, University of Bristol.

Martinec, J., 1976. Snow and Ice, in *Facets of Hydrology* (ed. J.C. Rodda), John Wiley, pp. 85–118.

Peck, E.L., Bissel, V.C., Jones, E.B. and Burge, D.L., 1971. Evaluation of snow water equivalent by airborne measurement of passive terestrial gamma radiation. *Water Resources Research*, 7, 1151–1159.

Randolph, P.D., Coates, R.A., Killian, E.W. *et al.*, 1972. A network of telemetered profiling isotopic snow gauges. *UNESCO/WMO/IAHS* Symposia on the Role of Snow and Ice in Hydrology, Banff, IAHS Publ. 107, Vol. 1, pp. 688–701.

Wilson, W.T., 1954. Discussion of Precipitation at Barrow Alaska, greater than recorded (by R.F. Black), *Trans. Am. Geophys. Union*, **35**, 203–207.

Cross references

Snow
Snowfall and snow cover: recent variations in northern high latitudes, present status of research and relevance to the climate change problem

SNOW

As warm moist air ascends, water vapor begins to condense to form cloud. When the cloud temperatures drop below freezing, conditions are suitable for forming snow. However, several processes are involved which govern the different types of snow which may be produced. Figure S20 summarizes the processes involved. At about $-5°C$ the aerosol nuclei (size $0.01–1\ \mu m$) which are always present in the atmosphere form small (diameters $< 75\ \mu m$) crystals by ice nucleation. These crystals have simple shapes, but may continue to grow by sublimation to form snow crystals which often have very complex shapes. A number of snow crystals together form a snowflake. Sometimes snow crystals pass through parts of the cloud which have many cloud droplets (size $10–40\ \mu m$), and therefore riming (droplets freeze when they come into contact with the crystals) occurs if the crystals are larger than about $300\ \mu m$. This occurs at temperatures from -5 to $-20°C$. If riming continues for a significant time then snow pellets (graupel) may be formed.

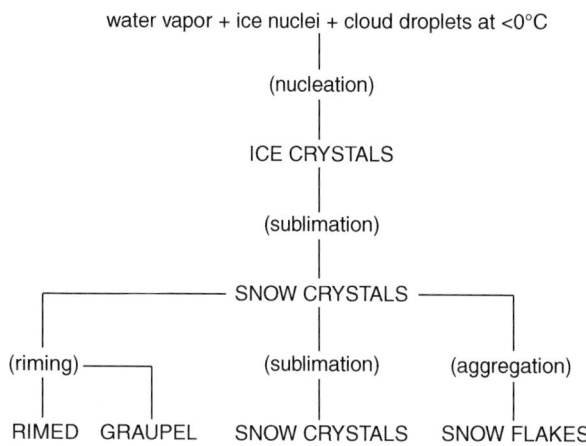

Figure S20 Flow diagram of the formation of different types of snow. (Source: Gray and Male, 1981.)

At the temperatures appropriate for the formation of snow, a cloud may only be slightly supersaturated with respect to water, but 10–20% supersaturated with respect to ice. Hence there is a transfer of water vapor from the cloud droplets to the ice crystals, which consequently grow. This is a Bergeron process. The shape of an ice crystal depends upon the temperature at which it grows, but its rate of growth and secondary crystal features depend upon the degree of supersaturation. The range of shapes are: 0 to $-4°C$, plates; -4 to $-10°C$, prism-like crystals, scrolls, sheaths and needles; -10 to $-20°C$ thick plates, dendrites and sector plates; and -20 to $-35°C$ sheaths and hollow columns. Figure S21 show examples of the structure of snowflakes. The rate of increase with time t of a mass m of a crystal through diffusion of water vapor onto its surface is

$$dm/dt = 4\prod C\,D\,F\,A\,(P_{oo} - P_o) \tag{S2}$$

where C is the shape factor, D is the diffusivity of water vapor in air, F is the ventilation factor depending on the relative motion of the crystal with respect to the air, A is the function of crystal size, P_{oo} is the vapor density (mass of water vapor per unit volume of moist air) at a large distance from the crystal and P_o is the vapor density at the surface of the crystal.

The mass growth rate due to riming depends upon the fall speed of the ice crystals relative to the cloud droplets, and the efficiency with which the droplets freeze and remain attached to the crystals. Hence

$$dm/dt = \prod r^2 ab w W \tag{S3}$$

where r is the radius of the ice crystal, a is the adhesion efficiency, b is the collision efficiency ($a\,b$ is between 0.1 and 1), w is the fall speed of the ice crystal relative to the droplets ($< 5\ m\ s^{-1}$) and W is the liquid water content of the cloud.

Formation of snow cover and its effects on the atmosphere

Several conditions have been found to be necessary, though not sufficient, for the occurrence of significant snowfall, these being

- sufficient moisture and aerosol nuclei for the formation and growth of ice crystals;
- sufficient depth of cloud to permit snow crystal growth;
- temperatures below 0°C in most of the layer through which the snow falls;
- sufficient moisture and aerosol nuclei to replace losses caused by precipitation.

Operational weather forecasts tend to use the temperature of the lower part of the atmosphere from 500 to 1000 hPa, known as the thickness temperature, as an indicator of likely snowfall. However, the other factors are very important, and currently numerical prediction models cannot represent the processes involved accurately enough to provide

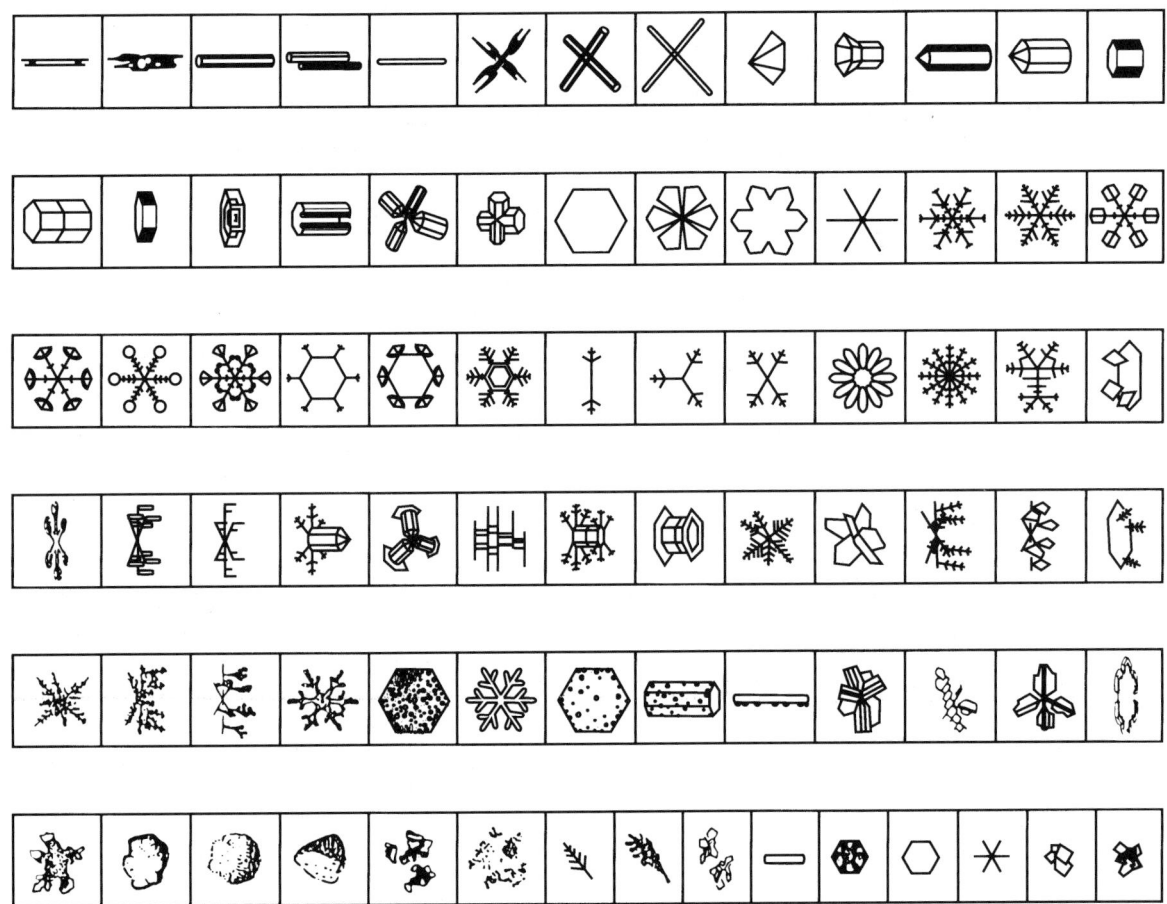

Figure S21 Snowflake structures. (From Gray and Male, 1981.)

completely reliable guidance. Fulkes (1935) indicated the rate of precipitation obtained from adiabatically ascending air for a 100 m layer with a vertical velocity of 1 m s^{-1} as shown in Figure S22. Such information is useful, but the vertical velocity of the air is determined by the nature of the atmospheric system, (e.g. Browning and Mason, 1981), orographic effects (Browning, 1983) and topographic effects (Layoie, 1972). Hence the estimation of likely snowfall on particular occasions is a very difficult problem.

Heavy falls of snow are not always associated with high latitudes, although clearly the lower temperatures experienced in these regions are conducive to snowfall for much of the year. Other areas where very cold air occasionally crosses relatively warm stretches of water are also likely to be subjected to heavy snow. Hence in winter, snow cover can persist for some time in midlatitudes. The Rocky Mountains, the Alps and the Himalayas all retain snow cover at the highest elevations all year round. Even the Scottish Highlands, only about 1 km high, lose their snow cover for just 1 or two months each year.

Both land and sea (the Arctic) areas are covered by snow. The areas involved vary with the seasons, with night and day, and with the day-to-day weather over particular areas. Hence it is not surprising to discover that snow cover is an important factor in determining climate and climatic change. Snow cover influences the atmospheric circulation by interacting with and affecting overlying air masses, which results in either the amplification or stabilization of circulation anomalies that often cause the weather.

Snow strongly reflects visible and near-infrared light, that is, it has a high albedo. Consider Table S6 which compares the albedo of snow with that of other natural surfaces. The albedo of snow varies with the age of the snow, but is considerably higher than those of most other natural surfaces. Hence, because of the seasonal changes in the extent of the snow cover, the albedo of the surface of the Earth varies from season to season. Since snow also has a high thermal emissivity, a low

thermal conductivity and a low water vapor pressure, the energy balance within the atmosphere will change as the seasons change. As melting occurs, the water vapor pressure and the thermal conductivity of the snow increase, and the latent heat of the snow must be taken into account in the energy balance.

Since the snow cover at higher latitudes persists for much of the year, less solar energy is absorbed at these latitudes, causing a profound effect upon the meridional flux of energy. Higher latitudes act as source regions for cold air masses which move to midlatitudes, resulting in the atmospheric systems (Walsh, 1984). Deviations from the regular seasonal changes in snow cover may be one factor in triggering climatic change.

Formation of ice sheets

As the surface of an area of water cools, the cooler, more dense water near the surface sinks and is replaced by less dense water from below. For temperatures less than about 4°C, the less dense, cooler water at the surface begins to freeze even though the water below it is relatively warmer. The initial orientation of the ice crystals is random, but as the ice thickens some crystals grow more quickly downwards than others. In sea ice more horizontally orientated crystals are favored, whereas for ice on lakes this is not the case (e.g. Hobbs, 1974). The water below the ice continues to lose heat by conduction through the ice sheet, albeit much more slowly, and as a consequence the ice thickens downwards. If snowfall is low then the layer of ice may be over 3 m thick and is know as 'black ice'.

As snow accumulates on the ice its density increases with depth, and the existing black ice is depressed below the original water level. The temperature within the lower layers of the snow cover may rise slightly, so that sublimation occurs and a layer of 'white ice' is formed on the black ice separated by a layer in which some melting occurs due to the flooding caused by the snow-cover depression. This

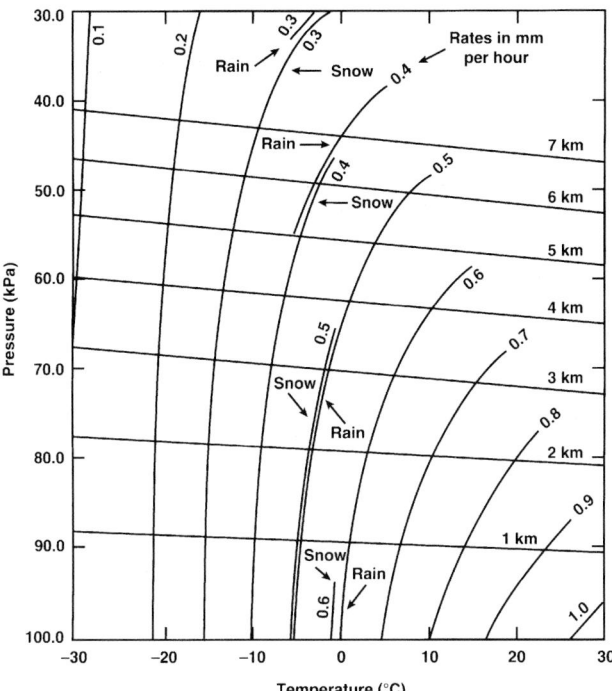

Figure S22 Rates of precipitation from adiabatically ascending air for a 100 m layer with a vertical velocity of 1 m s⁻¹. (Source: Fulkes, 1935.)

Table S6 Percentage of incident shortwave radiation reflected by some natural surfaces (from Gray and Male, 1981)

Surface	Albedo (%)
Fresh snow	75–95
Old snow	40–70
Sea ice	30–40
Desert	24–30
Water	5–30
Bare fields	15–25
Field crops	3–25
Tundra	15–20
Forests	3–20

area of 'slush' may also be caused by warmer water moving upwards by capillary action through cracks in the black ice, but which quickly refreezes.

Like a snow-cover, the ice on lakes delays runoff from a catchment area and therefore may have a significant impact upon hydrological forecasting. Melting will require appropriate modeling. Extensive sea ice affects the weather on a much larger scale in the same way that extensive snow fields influence atmospheric systems.

Characteristics of snow cover

The characteristics of snow cover are determined by atmospheric conditions and the condition of the land surface. The rate of precipitation, deposition and condensation with the magnitude of the turbulent transfer of heat and moisture, radiative exchange and air movement all contribute to the form of the snow cover. In addition, the features of the land may modify the atmospheric conditions and therefore the snow cover, and the amount of moisture retained by the land will also have its affect.

Snow depth will be increased as temperatures fall with increasing land elevation. Indeed, snow cover may not be established below

Table S7 Densities of snow cover (from Gray and Male, 1981)

Snow type	Density (kg m⁻³)
Wild snow	10–30
Ordinary new snow immediately after falling in still air	50–65
Settling snow	70–90
Very slightly toughened by wind immediately after falling	63–80
Average wind-toughened snow	280
Hard wind slab	350
New firn-snow[a]	400–550
Advanced firn snow	550–650
Thawing firn snow	600–700

[a] Firn-snow consolidated partly into ice (after Seligman, 1962).

particular elevations dependent on season and type of weather (e.g. Minnich, 1986). Wet snow, which is not easily blown by the wind, often occurs when air temperatures are near 0°C, such as may occur when air flows over large bodies of water (Passarelli and Braham, 1981). Terrain slope and aspect also have their effect. There are large differences between snowfall on the windward and leesides of hills and over steep and gentle slopes. The roughness of the land surface causes turbulent flow in the lowest layers of the atmosphere. This turbulence influences the snow cover distribution and properties, mainly density. The density increases as wind causes drifting. Table S7 shows the range of densities measured in snow cover. Condensation and melting also affect density.

The rate of transport of snow depends upon the wind speed, the nature of the terrain and the type of climatic region. In the Rocky Mountains, Martinelli (1973) measured a transport flux rate of 136 tonne m⁻¹, whereas along the Arctic coast Mikhell et al. (1989) measured a rate of 907 tonne m⁻¹. The snow will be deposited where the wind speed decreases markedly, often over rough ground. The result will be large drifts if upwind of the deposition area there is a long fetch covered with loose snow, and a high wind speed has been sustained for a considerable time. The presence of forests often causes drifting on the upwind edge, provided the forest is reasonably dense. Since forests also modify atmospheric conditions and flow, they also change the snow cover within them relative to that in non-forested areas, sometimes in quite complex and poorly understood ways.

Melting caused by changing atmospheric conditions will clearly modify the snow cover significantly. However, before major melting begins, the characteristics of the snow will be modified by radiative fluxes. The amount of solar (shortwave) radiation absorbed by the snow cover, and therefore available for melting, is governed by the albedo of the snow. In cloudy conditions or at night, longwave radiation is lost from the snow cover. Whilst this is usually not significant in the melting process, Olyphant (1986) has demonstrated that the surrounding rockwalls in alpine snowfields cause energy reflection which may reduce this loss of longwave radiation by about 50%, and so reduce the rate of melting. For example, the snow fields on Ben Nevis in Scotland, which are sheltered by steep slopes, and may occasionally persist throughout the year, are one of the very few locations in the UK where this may occur.

C.G. Collier

Bibliography

Browning, K.A., 1983. Air motion and precipitation growth in a major snowstorm, *Quart. J. R. Meteorol. Soc.*, **108**, 225–242.

Browning, K.A. and Mason, B.J., 1981. Air motion and precipitation growth in frontal systems, *Pure Appl. Geoph. Basle*, **118**, 577–593.

Fulkes, J.R., 1935. Rate of precipitation from adiabatically ascending air, *Mon. Wea. Rev.*, **63**, 291–294.

Gray, D.M. and Male, D.H. (eds), 1981. *Handbook of Snow. Principles processes, management and use*, Pergamon Press, Toronto, Oxford.

Hobbs, P.V., 1974. *Ice Physics*, Clarendon Press, Oxford, 837 pp.

Layoie, R., 1972. A mesoscale numerical model and lake-effect storms, *J. Atmos. Sci.*, **29**, 1025–1040.

Martinelli, M. Jr, 1973. Snow fences for influencing snow accumulation, in *Proc. Banff Symp. on The Role of Snow and Ice in Hy-*

drology, Sept. 1972, UNESCO–WMO–IASH, Geneva-Budapest-Paris, Vol. 2, 1394–1398.

Mikhil', V.M., Rudneva, A.V. and Lipôuskaya, V.I., 1969. *Snowfall and snow transport during snowstorms over the USSR*, (English Transl. 1971 Israel Prog. Sci. Transl, Transl. No. 5909).

Minnich, R.A., 1986. Snow levels and amounts in the mountains of southern California, *J. Hydrol*, **89**, 37–58.

Olyphant, G.A., 1986. Longwave radiation in mountainous areas and its influence on the energy balance of Alpine snowfields, *Water Resour. Res.*, **22**, 62–66.

Passarelli, R.E. Jr and Braham, R.R., 1981. The role of the winter land breeze in the formation of Great Lake snow storms, *Bull. Am. Meteorol. Soc.*, **62**, 482–491.

Seligman, G., 1962. *Snow Structure and Ski Fields* (reprint of 1936 edition) publ. by Jos. Adam, Brussels.

Walsh, J.E., 1984. Snowcover and atmospheric variability, *Am. Sci.*, **72**(1), 50–57.

Cross references

Snow measurement

Snowfall and snow cover: recent variations in northern high latitudes, present status of research and relevance to the climate charge problem

SNOWFALL AND SNOW COVER: RECENT VARIATIONS IN NORTHERN HIGH LATITUDES: PRESENT STATUS OF RESEARCH AND RELEVANCE TO THE CLIMATE CHANGE PROBLEM

Introduction

Frozen precipitation and snow on the ground play an important role in the hydrological cycle and ecology of high latitudes. They also significantly affect global climate variability due to a snow-cover feedback to the planetary heat balance.

Snow accumulated during the cold season (up to 10 months in the Arctic environment) provides a major contribution to runoff and feeds numerous swamps in tundra lowlands. It affects an important loop of the global hydrological cycle, fresh water input into the Arctic Ocean. The latter is often considered a key element in the ocean thermohaline circulation (Broecker, 1991).

Insulation properties of snow define and secure a fragile ecosystem in the Arctic. Snow cover protects the upper soil layer from freezing (or reduces the occurrence and degree of freezing) and saves vegetation and fauna from perishing during severe winter conditions.

The most renowned effect of the presence of snow on the ground is the change of surface albedo, the portion of solar radiation that is reflected by the surface. Albedo increases up to 0.85 for freshly fallen snow from its usual values of 0.15–0.30 for different types of soils. This significantly (on average by 5–6°C) decreases near-surface air temperatures (Voeikov, 1889; Gray and Male, 1981) and in turn sup-

ports (due to lower temperatures) the established snow cover. The last effect is often named snow-cover feedback. This feedback enhances any climate changes, being a positive feedback, and thus has a noticeable impact on global climate variability (Cess *et al.*, 1991, Randall *et al.*, 1994, Groisman *et al.*, 1994a,b).

These three functions of snow cover in high latitudes are currently being affected by ongoing climatic change. It appears that snow cover itself is affected by climatic changes and thus can feed back to the above-mentioned processes, changing them significantly.

Recent findings

During the past 10 years several works have been completed that show a noticeable increase in high latitudinal precipitation, particularly in the cold season, over the past 100 years (Bradley *et al.*, 1987; Groisman, 1991; Karl *et al.*, 1993, Groisman and Easterling, 1994; Dahlström, 1994; Hanssen-Bauer and Førland, 1994; Brown and Goodison, 1996). Groisman and Legates (1996) summarized these findings (Table S8) and reached the following conclusions.

- There is strong evidence that precipitation (including specifically its frozen form) has increased during the past 100 years over the northern extratropical land areas, encompassing roughly regions to the north of 45°N in North America and 50–55°N in Eurasia (on average 10% per 100 years). In some regions with a shorter period of instrumental record, e.g. northern Canada, these changes have occurred during the past four decades.
- The introduction of a new generation of instrumentation for measuring precipitation may inadvertently affect our ability to monitor and detect subtle but important climatic changes in precipitation. In high latitudes the issue is especially important because prominent changes in Arctic precipitation have already occurred, and because severe environmental conditions dictate an urgent need to automate the observational network.

Satellites have become an important tool for monitoring Arctic snow cover (Robinson *et al.*, 1993; Ferraro *et al.*, 1994). While winter snow-cover extent (SE) variations are not relevant to the Arctic studies (regions of winter SE variations are located in middle latitudes: in Central Asia and the contiguous USA) the changes in continental snow cover in intermediate seasons, especially in late spring, are very important. During the past 20 years, spring SE retreated by 10%, and these changes are closely related to temperature increase along the snow line (Groisman *et al.* 1994a,b; Figures S23 and S24). If this is a manifestation of ongoing global warming, the consequences of this change may be foreseen (up to a certain limit) and taken into account in future strategies of regional development.

Snow cover water-equivalent is being monitored by *in situ* methods (snow courses and less reliable snow-stick observations), by airborne gamma radiation measurements (Carroll, 1995), and by microwave satellite observations (Chang *et al.*, 1981). In Arctic conditions, remote methods of observations are preferable, but the current status of these observations shows that further studies are required in this direction.

Table S8 References for the increase of terrestrial precipitation in the extratropical regions during the period of instrumental observations; the unusually warm (globally) 1980s gave record high decadal precipitation totals in several regions of North America (Karl *et al.* 1993) and Eurasia (Georgievsky *et al.* (1995), thus somewhat exaggerating the trend estimates, making them 'visible' above the level of natural variability (adapted from Groisman and Legates 1996)

Authors	Region	Type of precipitation		Period (years)
Bradley *et al.* 1987	35–70°N	Annual indices		120
Diaz *et al.* 1989	25–60°N 25–60°S	Seasonal indices		97
Groisman 1991	35–70°N except northern Canada and China	Annual totals		98
Karl *et al.* 1993	Canada	Annual totals	55–70°N,	41
			<55°N	100
Hannsen-Bauer and Førland, 1994	Norway	Annual totals		90
Dahlström, 1994	Northern Europe	Annual totals		100
Groisman and Easterling 1994	North America	Annual and cold season totals	45–55°N,	100
			55–70°N	43
Georgievsky *et al.* 1995	European part of the former USSR	Annual and warm season totals		41

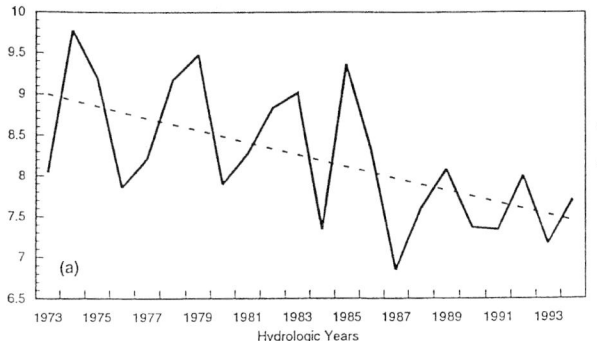

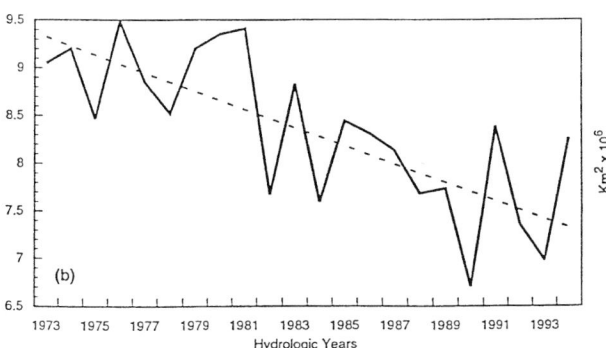

Figure S23 Spring snow cover extent over (a) North America and (b) east Asia and their linear trends (Groisman *et al.*, 1994b.)

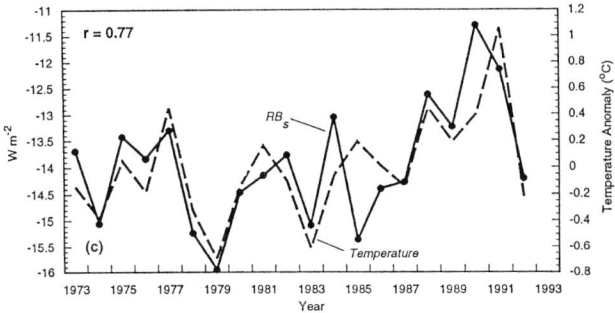

Figure S24 Spring (April to May) impact of snow cover on the radiative balance (RB_s) and surface air temperature anomalies over the northern extratropical land area. (From Groisman *et al.*, 1994b.)

Ongoing studies

There are two important projects that are currently being completed.

- The 20-year work on documentation of the world cryosphere (*World Atlas of Snow and Ice Resources*, chief editor: V.M. Kotlyakov, 1996–1997) has been completed by a large group of scientists from several institutes of Russia. It accumulates current knowledge on the present and past cryosphere, and will serve as the starting point for assessing the state of the environment in the 1970s and 1980s, and for comparison with the present global warming and its consequences.

- A reassessment of the amount of solid precipitation in high latitudes using the results of the WMO Solid Precipitation Intercomparison Project and (more numerous) field experiments carried on by national weather services of northern countries in Scandinavia, USA, Russia, and Canada is well underway (Goodison *et al.*, 1992; Golubev *et al.*, 1995).

Intercomparisons of *in situ*, airborne and satellite snow cover and snow water equivalent measurements have been intensively conducted during the past 10 years. While the problem of documentation of snow cover extent seems to be nearly resolved, snow cover water-equivalent (SWE) measurements from satellite are still unreliable and require further work (Armstrong and Brodzik, 1995). Promising results are delivered by the airborne gamma measurements of SWE (Carroll, 1995). However, expansion of this method to a global scale (i.e. to the entire Arctic Basin) would be quite expensive. Currently it is used only over North America with the most complete coverage in agricultural regions of the USA.

Need for further research

There is strong evidence that snow cover in high latitudes, being a function of snowfall and surface temperature, is quite sensitive to ongoing climatic changes. Modern climate models and empirical evidence indicate that in high latitudes these two elements (especially temperature) will change in the process of global warming by a higher degree than over other parts of the globe (IPCC 1995). Therefore, studies of snow cover variations and their relationship with other components of the climatic system should be a priority. These studies must be based on reliable observational data and implement all knowledge about snow-related processes, including those from global climate models. The most important snow-cover characteristic, its water equivalent, has to be monitored carefully for this purpose, but current observational systems are not yet capable of this monitoring. Most promising is the use of airborne gamma methods for SWE monitoring. This may require significant investments and/or special studies devoted to selection of a cost-effective program of airborne SWE monitoring.

Pavel Ya. Groisman

Bibliography

Armstrong, R.L. and M.J. Brodzik, 1995. An earth-gridded SSM/I data set for cryospheric studies and global change monitoring, *Advances in Space Research*, **16**, 155–163.

Barry, R.G., R.L. Armstrong, A.N. Krenke and T. Kadomtseva, 1994. *Cryospheric indices of global change, Final Report*. NSF Grant SES-91-12420, Boulder, CO, 25 pp.

Basist, A. and N.C. Grody, 1994. Identification of snow cover using SSM/I measurements, *in Proc. 6th Conference on Climate Variations*, Boston: American Meteorological Society, pp. 252–256.

Bradley, R.S., H.F. Diaz, J.K. Eischeid *et al.*, 1987. Precipitation fluctuations over Northern Hemisphere land areas since the mid-19th century, *Science*, **237**, 171–275.

Broecker, W.S., 1991. The great ocean conveyor, *Oceanography*, **4**, 79–89.

Brown, R.D. and B. Goodison, 1996. Interannual variability in Canadian snow cover, *J. Climate*.

Brown, R.D., M.G. Hughes and D.A. Robinson, 1995. Characterizing the long-term variability of snow-cover extent over the interior of North America, *Annals of Glaciology*, **21**, 45–50.

Carroll, T.R., 1995. GIS used to derive operational hydrologic products from in situ and remotely sensed snow data, in A. Carrara and F. Guzzetti (eds) *Geographical Information Systems in Assessing Natural Hazards*, Netherlands: Kluwer Academic Publ., pp. 335–342.

Chang, A.T.C., J.L. Foster, D.K. Hall *et al.*, 1981. *Snow water equivalent determination by microwave radiometry*, NASA Technical Memorandum 82074, Goddard Space Flight Center, 18 pp.

Cess, R.D. and 32 others, 1991. Interpretation of snow-climate feedback as produced by 17 general circulation models, *Science*, **253**, 888–892.

Dahlström, B., 1994. Short term fluctuations of temperature and precipitation in Western Europe, in *Climate Variations in Europe*, Proceedings of the European Workshop on Climate Variations, Kirkkonummi, Finland, 15–18 May 1994: Publ. of the Academy of Finland, pp. 30–38.

Diaz, H.F., R.S. Bradley and J.K. Eischeid, 1989. Precipitation fluctuation over global land areas since the late 1800s, *J. Geophys. Res.*, **94**, 1195–1240.

Ferraro, R., N. Grody, D. Forsyth *et al.*, 1994. Microwave measurements produce global climatic, hydrologic data, *EOS*, **75**(30), 337–338 and 343.

Georgievsky, V.Yu., S.A. Zhuravin and A.V. Ezhov, 1995. Assessment of trends in hydrometeorological situation on the Great Russian Plain under the effect of climate variations, in *Proc. of the AGU 15th Annual Hydrology Days*, Atherton, California, USA: Hydrology Days Publ., pp. 47–58.

Goodison, B.E., V.S. Golubev, T. Gunter and B. Sevruk, 1992. Preliminary results of the WMO Solid Precipitation Measurement Intercomparison, in *Proc. of WMO Technical Conference on Instruments and Methods of Observation*, 10–15 May 1992, Vienna, Austria, pp. 81–85.

Golubev, V.S., V.V. Koknaeva and A.Yu. Simonenko, 1995. Results of atmospheric precipitation measurements by national standard gauges of Canada, USA, and Russia, *Meteorologia i Gydrologia*, No. 2, 102–110 (in Russian).

Gray, D.M. and D.H. Male, 1981. *Handbook of Snow*, New York: Pergamon Press, 776 pp.

Groisman, P.Ya., 1991. Data on present-day precipitation changes in the extratropical part of the Northern hemisphere, in Schlesinger M.E. (ed.) *Greenhouse-Gas-induced Climatic Change: A Critical Appraisal of Simulations and Observations*, Amsterdam: Elsevier, pp. 297–310.

Groisman, P.Ya. and D.R. Easterling, 1994. Variability and trends of precipitation and snowfall over the United States and Canada, *J. Climate*, **7**, 184–205.

Groisman, P.Ya., T.R. Karl and R.W. Knight, 1994a. Observed impact of snow cover on the rise of continental spring temperatures, *Science*, **263**, 198–200.

Groisman, P.Ya., T.R. Karl, R.W. Knight and G.L. Stenchikov, 1994b. Changes of snow cover, temperature, and the radiative heat balance over the Northern Hemisphere, *J. Climate*, **7**, 1633–1656.

Groisman, P.Ya. and D.R. Legates, 1996. Documenting and detecting long-term precipitation trends: where we are and what should be done, *Climatic Change*.

Hanssen-Bauer, I. and E.J. Førland, 1994. Homogenizing long Norwegian precipitation series, *J. Climate*, **7**, 1001–1013.

Intergovernmental Panel on Climatic Change (IPCC), 1995. *Climate Change, Second Scientific Assessment*.

Karl, T.R., P.Ya. Groisman, R.R. Heim Jr and R.W. Knight, 1993. Recent variations of snow cover and snowfall in North America and their relation to precipitation and temperature variations, *J. Climate*, **6**, 1327–1344.

Leathers, D.J., A.W. Ellis and D.A. Robinson, 1995. Characteristics of temperature depressions associated with snow cover across the Northeast United States, *J. Appl. Meteorol.*, **34**, 381–390.

Legates, D.R., 1987. A climatology of global precipitation, *Publ. Climatol.*, **40**(1), 84 pp.

Randall, D.A. and 26 others, 1994. Analysis of snow cover feedbacks in 14 general circulation models, *J. Geophys. Res.*, **99**(D10), 20757–20771.

Robinson, D.A., K.F. Dewey, and R.R. Heim Jr, 1993. Global snow cover monitoring: an update, *Bull. Am. Meteorol. Soc.*, **74**, 1689–1696.

Rudolf, B., H. Hauschild, M. Reiss, *et al.*, 1992. Contributions to the Global Precipitation Climatology Centre, *Meteorologische Zeitschrift*, Neue Folge, Heft 1, pp. 7–84 (in German, with English summaries).

Salomonson, V.V., D.K. Hall and J.Y.L. Chien, 1995. Use of passive microwave and optical data for large-scale snow-cover mapping, in *Proc. 2nd Topical Symp. on Combined Optical-Microwave Earth and Atmosphere Sensing*, 3–6 April, 1995, Atlanta, GA, pp. 35–37.

Sevruk, B., 1982. Methods of correction for systematic error in point precipitation measurement for operational use. *Oper. Hydrol. Rep.*, **21**, Publ. 589, Geneva, Switzerland: World Meteorological Organization 91 pp.

Sevruk, B., (ed.), 1992. Snow cover measurements and areal assessment of precipitation and soil moisture, *Oper. Hydrol. Rep.*, **35**, Publ. 749, Geneva, Switzerland: World Meteorological Organization, 283 pp.

Voeikov, A.I., 1889. Snow cover, its effects on soil, climate, and weather and methods of investigations, *Notes of Russian Geographical Society on General Geography*, **18**, No. 2 (in Russian).

World Water Balance and Water Resources of the Earth (WWB), 1974–1978. Leningrad: Gidrometeoizdat (1974 in Russian, 1978 in English).

Cross references

Snow
Snow measurement

SOIL AND WATER MANAGEMENT

Soil erosion by water

The widespread and sometimes devastating effects of soil erosion by water, commonly induced or accelerated by unwise agricultural practices, are well documented. Such erosion, which was a factor in the downfall of some empires (Stallings, 1957; Hudson, 1981), continues to be a serious problem even in technologically advanced parts of the world, and is one of the chief obstacles to agricultural efficiency, and hence to economic progress, in most developing countries. It tends to be a regional rather than a localized problem because accelerated erosion not only denudes the affected areas but produces much detritus deposited by streams at lower levels, where it impairs soil drainage and fertility, infills stream channels and reservoirs, and damages roads, buildings and other structures. This erosion and related damage is enormously expensive (Stallings, 1957; Stall, 1973).

The main agents of soil erosion by water are raindrop splash and runoff moving over the surface as either sheet flow or channelized flow in rills and gullies, and subsurface throughflow which removes soil in solution and suspension, and erodes pipes which frequently collapse to initiate gullies. Throughflow is also a major source of runoff when it emerges at the surface as return flow. Numerous studies have demonstrated that the rate and amount of soil lost due to these agents vary according to slope gradient and length, and are also influenced by slope curvature, contour curvature, surface microrelief and slope aspect. The steeper the slope, the greater the erosion for various reasons: there is more splash downhill, there will be more runoff and it will flow faster. The amount of this erosion is not simply proportional to the slope gradient, but rises rapidly as the gradient increases. The length of slope has a similar effect on soil loss. Thus the amount, velocity and depth of runoff are greatly increased on a longer slope. This produces scour erosion, which would not occur on a shorter slope, or where the effective slope for runoff is reduced, as between terraces. Values derived for these exponential relationships are summarized by Smith and Wischmeier (1962) and Hudson (1981). (See also Stallings, 1957; Carson and Kirkby, 1972; Young, 1972; Strahler, 1973; Kirkby and Morgan, 1980; Morgan, 1986; Hudson, 1987; models of overland flow processes and soil erosion are reviewed by Gerits *et al.* (1990).)

Soil erosion is a function of erosivity and erodibility. **Erosivity** is governed by rainfall and so cannot be controlled. **Erodibility** is the susceptibility of a soil to erosion. It depends on many factors, primarily including the nature of the soil itself, ground configuration and vegetation cover. Vegetation is especially important because its disturbance or removal reduces the proportion of rainwater that infiltrates the soil, permits splash erosion, increases runoff and hence leads to accelerated erosion by surface flow. Such disturbance is inevitable in land use, however, and so the amount of subsequent erosion depends largely on the quality of management.

Control of runoff and erosion

The term management is used here to include land management and crop management. These correspond respectively with the two main kinds of erosion control measures: mechanical measures dealing with the ground itself and non-mechanical measures concerned with crops and animals. Mechanical measures are essentially applied geomorphology in that they consist of reshaping surface form to manipulate the associated geomorphological processes. In particular, they aim to modify the gradient, length, and microrelief of slopes in order to regulate the concentration and velocity of surface water flow. Terraces are one of the most widely used of these mechanical measures. They consist of bench-like earthworks with banks along their downslope margins, constructed at right angles to the direction of maximum gradient, that is, along the contour. In this way the original slope is divided into several small catchments, corresponding to a number of short slope segments, each with a gentler gradient than originally. On each terrace, or catchment, this gentler gradient and shorter slope length not only decrease the amount of runoff by increasing the proportion of rainfall that infiltrates the soil, but also, by lowering the velocity of runoff, reduce soil loss and cause more flowing water to be absorbed as it moves slowly over the ground.

Terraces may be divided into two categories according to their primary function: level, or absorption, terraces to conserve moisture, and graded, or dispersion, terraces for the orderly disposal of water during periods of excess rainfall. Level terraces are common in drier areas where rainfall is inadequate for maximum crop growth, and graded terraces are suited to more humid conditions. A level terrace is level along the contour in order to hold rainwater so that it infiltrates the soil. In contrast, a graded terrace is constructed as a very shallow channel with a longitudinal gradient slightly oblique to the contour so that surplus runoff flows away at non-erosive velocities to a place where it can be safely discharged. Level terraces may be built either by excavation or, in drainage floors, by impounding natural sedimentation. Although approximately level along the contour, the shelf of a level terrace may have a slight downslope gradient – that is, in the direction of the original slope – or a reverse gradient in the opposite direction to that of the original slope. The bank along the downslope edge of such a level terrace is relatively steep and is commonly, although not always, capped by a raised lip of earth to prevent water on the shelf from overflowing the bank. The downslope margin of a graded terrace, however, normally consists of a gentler-sided, levee-like embankment constructed to impound water flowing in the shallow channel immediately upslope.

A simpler form of earthwork designed to intercept surface runoff on gentle slopes is the contour bund. This is a low ridge of soil thrown up along the contour, commonly planted with grass or shrubs to stabilize it and also to assist in trapping silt washed downslope by runoff. In many areas the chief function of contour bunds is to hold runoff until it infiltrates the soil, and hence they are especially common in semiarid areas. Other forms of such artificial microrelief include tied ridging, and ridges and furrows. Tied ridging consists of ridges closely spaced in two directions at right angles so that the ground is covered with small rectangular depressions intended to prevent runoff and wash erosion. However, the well-known ridge-and-furrow method is used primarily to improve surface drainage, but it also affords some measure of erosion control.

As graded terraces are intended to protect arable land by leading away surplus runoff, they are commonly designed in conjunction with two other components of such an artificial geomorphological system, stormwater drains and grass waterways. A stormwater drain is a ditch constructed along the contour upslope from the terraces to intercept runoff, which would otherwise flow down from higher ground on to the arable land. The runoff in both graded terraces and stormwater drains must be led away from the area, and so the grass waterway is designed to fulfill that function. It is excavated down the slope with discharge inlets from the terraces and their stormwater drain.

Other control measures that are adapted to surface configuration, but intended to modify its character, include contour plowing and contour strip cropping. Contour plowing is one of the most effective control measures for cultivated cropland, because it increases the proportion of rain that infiltrates the soil, disrupts surface flow and reduces its velocity by increasing surface roughness. Deeper plowing is also a feature of semiarid areas, in order to impede as much runoff as possible by creating pronounced microrelief, and to facilitate infiltration by breaking surface crusts and hardpans, thus enabling the soil to absorb and retain more water. Contour strip cropping involves alternating strips of cultivated row crops and close-growing or sod crops. The row crops induce erosion, of course. However, the sod crops protect the underlying soil from erosion by both raindrop splash and surface flow. Moreover, during all but extremely heavy or prolonged rains, a sod strip can absorb the water flowing from the row crop immediately upslope, and thus prevent run-on to the adjoining row crop downslope. This increases the infiltration rate and reduces the overall amount and erosive capacity of runoff.

The design of these various management measures is largely determined by geomorphological considerations. Thus terraces are planned as small drainage catchments that can be handled through one outlet or system of outlets and can accommodate the anticipated rainfall and runoff intensities without overtopping or breaching of their banks. The uppermost terrace of a sequence is especially important here because its capacity must not be overtopped by runoff, or the whole system will be at risk. Therefore it must be built with a suitably small drainage catchment near the crest of the slope or be protected by diversion ditches at its upslope margin. These considerations – including the construction of channels to collect, divert or lead

away water – are described by Stallings (1957), Hudson (1981), Bennett and Chapline (1973) and Morgan (1986) who give standard design formulae in which the governing factors for terraces are the slope and configuration of the ground, and for channels, the size, shape, gradient and bed roughness of the proposed structure in relation to estimated discharge. Similarly, the planning and effectiveness of contour cultivation and strip cropping are strongly influenced by the gradient and length of slope as well as by soil type (Food and Agriculture Organization, 1965).

Traditional applications

Such methods of soil and water management, and geomorphological controls in their location and design, have received much attention since about the beginning of the twentieth century (Glenn, 1973). Geomorphology has been applied in this way since time immemorial, however, and many twentieth-century methods are refinements of long-established techniques. These traditional applications of geomorphology characterize agricultural land use in the semiarid province of Murcia in southeast Spain, for example. The district around the town of Mula is typical. Except on steeper slopes the landscape has been completely transformed by terraces, bunds and related structures for soil and water management. These artificial landforms are closely attuned to their geomorphological setting and reflect the farmers' traditional understanding of the associated natural processes. The main aim is to exploit all gently or moderately sloping surfaces (gradients up to about 12%) receiving run-on and influent seepage from upslope, mostly by building level terraces and bunds. On interfluves these surfaces occur chiefly on footslopes adjoining low hills, cuestas and tablelands, and on the flanks of low rises in undulating plains. In their original form they comprised concave slopes up to about 9%. However, gradients were commonly 5–6% on the shorter, more curved footslopes of hills and flanks of rises, whereas longer, near-planar surfaces of about 4–5% characterized the footslopes of cuestas and tablelands.

These morphological differences are reflected in terraces design. Thus hillfoot terraces are more closely spaced (10–15 m) and have relatively steep reverse gradients to impound as much water as possible. In contrast, terraces on the longer scarpfoot slopes are more widely spaced (25–30 m), with slight downslope gradients to ensure that water will flow gently over the whole surface. Terraces are relatively uncommon on hillslopes with gradients greater than about 18% because of severe erosion problems. On less steep hillslopes these problems are countered by using narrow terraces, with shelves less than 10 m across, and reducing runoff by keeping a vegetated strip of the original surface immediately upslope from each terrace shelf.

Here as elsewhere, drainage floors have the greatest concentration of run-on and seepage waters. Therefore, all floors throughout the area have been converted for cultivation by using terraces and bunds. The floors comprise drainage trenches and drainage zones. The former include trunk drainage floors along the main rivers where irrigated terraces have been constructed. They also include the floors of tributary drainage trenches occurring especially in 'badlands,' which periodically have large amounts of inflowing waters from densely branching networks of tributary gullies and their steep side slopes. The trenches receive much fine-grained sediment brought in by these waters and by mass movements. Hence terrace construction here simply involves harnessing these geomorphological processes by building dams to impede sediment transport and thus infill the floor behind each dam. In this way the farmers not only construct water-retaining structures but make deep soil where none existed before. To illustrate the magnitude of these transformations, in the Mula River floodplain, southeastern Spain, a badland trench almost 1500 m long has been completely infilled and terraced except for a narrow gorge section in upper sectors. The character of such a badland floor is reflected in terrace design. Hence in steeper upper sectors (original longitudinal gradients about 12%), terraces are more closely spaced, with shelves having pronounced reverse gradients and the highest banks; in mid-sectors (original gradients 6–7%) they are more widely spaced, with slight down-valley gradients; and in the lowest sectors (original gradients 2–3%) they are most widely spaced, nearly flat or with gentle reverse gradients.

The area also includes extensive branching networks of very shallow, unchanneled drainage 'zones' occurring on undissected interfluve plains and footslopes. These were probably sites of natural infilling by colluvium and alluvium, in contrast to the eroding nature of tributary trenches, and phased downcutting and alluviation in the main trenches. Original gradients ranged from about 2% in the trunkline drainage zones to about 5% in tributaries. Their surfaces are now characterized by bunds as well as terraces, which are more closely spaced in tributary zones, where greater runoff and colluvation lead to soil accumulation immediately upslope from each bund. In many parts the very shallow depressions corresponding to these drainage zones may not be perceived by visual inspection. However, detailed ground measurement and air photo analysis reveal that throughout the whole area their occurrence and boundaries have been identified with great precision by the local farmers in constructing these bunds. Both terraces and bunds are built in association with various forms of microrelief designed to maximize infiltration and minimize erosion. Surfaces around tree crops are kept rough and broken, young trees are grown in circular depressions until they are properly established, and contour plowing is practiced even on the most gently inclined terrace surfaces.

The foregoing discussions of natural hazards, soil erosion and water management underline the central role of water in human activities. Water is not only a most essential need, together with food, but also commonly a limiting factor in food production, a major hazard when out of control, and a key element in such contrasted problems as environmental pollution and landslide prevention. It is the single most specific determinant of many physical processes in the natural environment and is indispensable to all biological processes. Lack of space prevents consideration of these diverse questions here. However, further examples to illustrate the primary importance of applied geomorphology in manipulating surficial water systems include gully control (Jepson, 1973), irrigation (Stallings, 1957; Schwab *et al.*, 1966; Askochensky, 1973; Withers and Vipond, 1974; Currey, 1977), groundwater extraction, groundwater recharge (Freeman, 1974), land drainage (Schwab *et al.*, 1966), disposal of waste and refuse (Schneider, 1975; Foose and Hess, 1976), and river impoundment (Petts, 1984).

Robert L. Wright

Bibliography

Askochensky, A.N., 1973. Basic trends and methods of water control in the arid zones of the Soviet Union, in D.R. Coates (ed.), *Environmental Geomorphology and Landscape Conservation, III: Non-Urban Regions*. Stroudsburg, PA: Dowden, Hutchinson & Ross, pp. 207–216.

Bennett, H.H. and Chapline, W.R., 1973.. Soil erosion – a national menace. Part I: Some aspects of the wastage caused by soil erosion, in D.R. Coates (ed.), *Environmental Geomorphology and Landscape Conservation, III: Non-Urban Regions*. Stroudsburg, PA: Dowden, Hutchinson & Ross, pp. 57–83.

Carson, M.A. and Kirkby, M.J., 1972. *Hillslope Form and Process.* Cambridge, UK: Cambridge University Press, 475 pp.

Currey, D.T., 1977. The role of applied geomorphology in irrigation and groundwater studies, in J.R. Hails (ed.), *Applied Geomorphology*. Amsterdam: Elsevier, 51–83.

Dinsman, S.L., 1994. *Physical Hydrology.* London: Macmillan, 575 pp.

Food and Agriculture Organization, 1965. Soil erosion by water – some measures for its control on cultivated lands, *Agricultural Development Paper 87*, 284 pp.

Foose, R.M. and Hess, P.W., 1976. Scientific and engineering parameters in planning and development of a landfill site in Pennsylvania, in D.R. Coates (ed.), *Geomorphology and Engineering*, Stroudsburg, PA: Dowden, Hutchinson & Ross, pp. 289–312.

Foth, H.D., 1990. *Fundamentals of Soil Science*, New York.

Freeman, V.M., 1974. Water spreading as practiced by the Santa Clara water-conservation district, Ventura County, California, in D.R. Coates (ed.), *Environmental Geomorphology and Landscape Conservation, II: Urban Areas*. Stroudsburg, Pa.: Dowden, Hutchinson & Ross, 111–117.

Glenn, L.C., 1973. Denudation and erosion in the southern Appalachian region and the Monongahela basin, in D.R. Coates (ed.), *Environmental Geomorphology and Landscape Conservation, III: Non-Urban Regions*. Stroudsburg, PA: Dowden, Hutchinson & Ross, 36–56.

Gerits, J.J.P., Lima, J.L.M.P. and van den Broek, T.M.V., 1990. Overland flow and erosion, in M.G. Anderson and T.P. Burt (eds), *Process Studies in Hillslope Hydrology*, Chichester: Wiley, pp. 173–214.

Hudson, N., 1971. *Soil Conservation*. London: Batsford, 320 pp.

Hudson, N.W., 1981. *Soil Conservation*. London: Batsford, 324 pp.

Hudson, N.W., 1987. *Soil Conservation in Semi-arid Areas, FAO Soils Bull.*, **57**, Rome: UN Food and Agriculture Organization, 172 pp.

Jepson, H.G., 1973. Prevention and control of gullies, in D.R. Coates (ed.), *Environmental Geomorphology and Landscape Conservation, III: Non-Urban Regions*, Stroudsburg, PA: Dowden, Hutchinson & Ross, 283–298.

Kirkby, M.J. and Morgan, R.P.C. (eds), 1980. *Soil Erosion.* Chichester, UK: Wiley, 312 pp.

Morgan, R.P.C., 1986. *Soil Erosion and Conservation*. Harlow, UK: Longman, 298 pp.

Petts, G.C., 1984. *Impounded Rivers: Perspectives for Ecological Management*. Chichester, UK: Wiley, 326 pp.

Schneider, W.J., 1975. Hydrologic implications of solid-waste disposal, in G.D. McKenzie and R.O. Utgard (eds.), *Man and His Physical Environment: Readings in Environmental Geology* 2nd edn. Minneapolis: Burgess, pp. 125–134.

Schwab, G.O., Frevert, R.K., Edminster, T.W. and Barnes, K.K., 1966. *Soil and Water Conservation Engineering*, 2nd edn. New York: Wiley, 683 pp.

Smith, D.D. and W.H., Wischmeier, 1962. Rainfall erosion, *Advances in Agronomy*, **14**, 109–148.

Stall, J.B., 1973. Man's role in affecting the sedimentation of streams and reservoirs, in D.R. Coates (ed.), *Environmental Geomorphology and Landscape Conservation, III: Non-Urban Regions*. Stroudsburg, PA: Dowden, Hutchinson & Ross, pp. 103–119.

Stallings, J.H., 1957. *Soil Conservation*. Englewood Cliffs, NJ: Prentice-Hall, 575 pp.

Strahler, A.N., 1973. The nature of induced erosion and aggradation, in D.R. Coates (ed.), *Environmental Geomorphology and Landscape Conservation, III: Non-Urban Regions*. Stroudsburg, PA: Dowden Hutchinson & Ross, pp. 18–35.

Withers, B. and Vipond, S., 1974. *Irrigation Design and Practice*. London: Batsford, 306 pp.

Young, A., 1972. *Slopes*. London: Longmans, 288 pp.

Cross references

Alluvial valley engineering
Deltaic plains
Weathering

SOUTH AMERICA

Covering 17.8 million km^2, the South American continent has a meridional extent of 7150 km and measures 5150 m in the longitudinal direction of the equatorial and tropical belt, which takes in 80% of the total area, while temperate to subarctic climates are limited to the south. Its length of shoreline is about 26 000 km. Its topography is dominated by the Andes, along the western side, with many peaks over 6000 m (highest Mt Ilyampu, 7014 m), and embracing a series of high intermontane plateaus, the Altiplano (3500–4100 m), which cover 220 000 km^2 and are marked largely by endoreic drainage (8% of the total); Lake Titicaca is a prominent feature. Some 85% of the drainage area is to the Atlantic, with only 7% to the Pacific.

South America's three major rivers, the Amazon (the world's largest in terms of discharge), the Orinoco and Parana/Uruguay/LaPlata, all drain to the east. There are three semiarid to hyperarid regions: the littoral Atacama Desert of northern Chile – southern Peru which comes under the influence of the north-setting Peru ('Humboldt') current, the rain-shadow semi-desert of Patagonia, and the periodically dry Nordeste province of Brazil. Both littoral Peru and the Nordeste come under the cyclical influence of the El Niño (2–7 years), when heavy precipitation often takes place.

Most of the continent comes under the effects of the monsoonal and trade wind circulation and associated precipitation, although the Amazon Basin is characterized by convectional rains with 23% involving no moisture import or export. On the western (Pacific) slope of the Andes the equatorial monsoon brings very heavy precipitation; in fact, Serrania-de-Baudo, with 9000 mm annually, is one of the rainiest spots on Earth. The southern part is affected by moisture transport from the South Pacific westerlies and on the east by frontal systems from the South Atlantic. The snow-line ranges from above 6000 m in the Central Andes down to 500 m in Tierra del Fuego.

Table S9 South America's river systems

Name	Discharge (m³ s)	Discharge (km³ year⁻¹)	Area (km²)
Amazon	220 000	6 930	6 915 000
Orinoco	29 100	917	1 000 000
Parana/Uruguay	23 000	724	2 970 000
Magdalena	8 200	258	240 000
Sao Francisco	2 000	63	600 000
Others (Atlantic)	–	1 538	4 875 000
Others (Pacific)	–	1 330	1 200 000
Total		11 760	

A review of South America's climate is provided in *The Encyclopedia of Climatology* (Oliver and Fairbridge, 1987) and of the hydrology in the UNESCO volume *World Water Balance and Water Resources of the Earth* (USSR Committee, 1978). The number of hydrometric (river discharge) stations established by 1994 was 7924 and the number of rain gauging stations was about 19 000.

From Table S9 it may be seen that more than 88% of the discharge is to the Atlantic (two-thirds by the Amazon) and 11% to the Pacific. Some 58.7 km³ of the flow is endoreic. Average precipitation for the entire continent is 661 mm. More than 50% of the total is rather accurately monitored. Interannual fluctuations are quite large, ranging from the maximum (1934) of 12 870 km³ year⁻¹ to a minimum (1936) of 10 570 km³ year⁻¹. Maximum flow to the Atlantic occurs from April to August, but to the Pacific it is from February to June.

Water resources

The total resources of fresh or only slightly mineralized water in South America is calculated to be 3 010 000 km³, but 99.6% of this is geologically stored underground (to depths of up to 2000 m), and only 0.4% is recycled annually; 0.033% is in lakes and < 0.002% in glaciers. Very little of the geological reserve (essentially nonrenewable) has been exploited to date. The average residence time for river water is 35 days; that is, it is renewed more than 10 times each year. Per unit area, South America occupies first place among all the continents for its water abundance.

There has been relatively little development of dams in South America, for hydroelectric potential or irrigation purposes, with the notable exception of those on the Paraná River (Iguaçu Falls) and on the São Francisco both in Brazil. Besides these major works, ten of the larger reservoirs in South America conserve about 286 km³ of water, of which 123 km³ is renewed annually. In addition there are the natural reserves of 'once-only' river channel storage for the continent which amounts to about 1000 km³. The total amount of annually renewed water for the continent comes to 4300 km³, or 2300 m³ for every human inhabitant.

Rhodes W. Fairbridge

Bibliography

Oliver, J.E. and Fairbridge, R.W., eds, 1987. *The Encyclopedia of Climatology*. New York: Van Nostrand Reinhold, 986 pp.
USSR Committee for I.H.D., 1978. *World Water Balance and Water Resources of the Earth*. Moscow and Paris: UNESCO, 663 pp.

Cross references

Arid climates
Arid land
Arid zone hydrology
Climatic data: sources
Monsoon climates
Rivers

SPRINGS

A spring is a place where, without human intervention, water flows from a rock or soil upon the land or into a body of surface water

(Price, 1996). Throughout the world, springs are important sources of domestic and stock water, and some are large enough to supply cities, industries or irrigation projects or to form attractive recreational areas. Their flow, along with groundwater seepage too diffuse to be called a spring, is the principal source of dry-weather flow in streams.

Springs occur in a variety of forms and in varying circumstances, and it has proved difficult to agree a comprehensive classification. A number of groupings have been suggested over the years on the basis of individual characteristics (Ward, 1967) such as genesis, rock structure, discharge, temperature and variability, the most generally accepted classification being that of Bryan (1919). Bryan divided springs into (1) those whose water is of meteoric origin and flows under the influence of gravity, and (2) those whose water is of deep origin (including connate, juvenile and deeply circulating meteoric water) and, in part, is brought to the land surface by forces in addition to gravity. He divided the gravity springs into four groups; as redefined by Meinzer (1942, p. 419), these are (Figure S25)

- depression springs, which are due to the water table intersecting the land surface in permeable rocks;
- contact springs, which are due to an out-crop of permeable water-bearing rock overlying relatively impermeable rock.
- artesian springs, which are due to water rising from a permeable water-bearing bed, confined between relatively impermeable beds, either at an outcrop or through a fissure penetrating the upper confining bed;
- springs that issue from tubular openings or fractures in otherwise impermeable rocks.

The springs of deep origin Bryan divided into two groups:
- volcanic springs, which are due to or associated with volcanism or volcanic rocks;
- fissure springs, which are due to fractures extending deep into the Earth's crust.

As a result of the size of springs, Meinzer (1923, p. 53) proposed a classification based on discharge (Table S10).

In the United States, according to a study by Meinzer (1927), about 65 springs of the first magnitude are known. Of these, 38 rise in volcanic rocks or associated gravel, 24 rise in limestone and 3 rise in sandstone. Among the largest springs, according to records of the United States Geological Survey, are Malad Springs, Idaho (average discharge 35.4 m³ s⁻¹ in 1956); Thousand Springs, Idaho (average discharge 35.4 m³ s⁻¹ in 1956); Big Springs, Missouri (average discharge 12.3 m³ s⁻¹ for 41 years of record); Comal Springs, Texas (average discharge 8.2 m³ s⁻¹ for 30 years of record); Silver Springs, Florida (average discharge 23.1 m³ s⁻¹ for 30 years of record); and Giant Springs, Montana (discharge about 17 m³ s⁻¹). There are innumerable springs of lesser magnitude, including perhaps hundreds of second magnitude and thousands of third magnitude.

In Europe, the largest known karst spring is reputed to be at Trebisnjica in the former Yugoslavia with a mean annual discharge of 41 m³ s⁻¹ (Bourdon, 1966).

The flow of springs fluctuates. Some springs are perennial; others are intermittent; a geyser, for example, is a special kind of intermittent spring whose discharge is caused by the expansive force of highly heated steam. Some springs fluctuate seasonally or over prolonged periods of time in accordance with seasonal and long-term variations in groundwater recharge (Todd, 1980). Some are periodic, i.e. they fluctuate at intervals related not to variations in recharge but, rather, to such causes as diurnal changes in transpiration and evaporation, variations in atmospheric pressure, tidal fluctuations and perhaps even special local conditions that produce an intermittent siphon action. Some fluctuate, or are permanently changed, as a consequence of developments by humans. For example, Kissengen Spring, formerly one of the large springs of the Florida Peninsula, ceased to flow in 1950 owing to the heavy draft from wells in the surrounding region. In contrast, a large increase – from about 108 m³ s⁻¹ in 1902 to about 167 m³ s⁻¹ in 1956 – in the discharge of springs to the Snake River, Idaho, between Milner and Bliss, was due to the increase in recharge from irrigation on the Snake River Plain.

Springs are classified also as thermal and non-thermal (Meinzer, 1923, pp. 54–55). Thermal springs are those whose water has a temperature noticeably above the mean annual air temperature for their localities. They may be divided into hot springs, whose water has a higher temperature than that of the human body, and warm springs, whose water has a lower temperature than that of the human body.

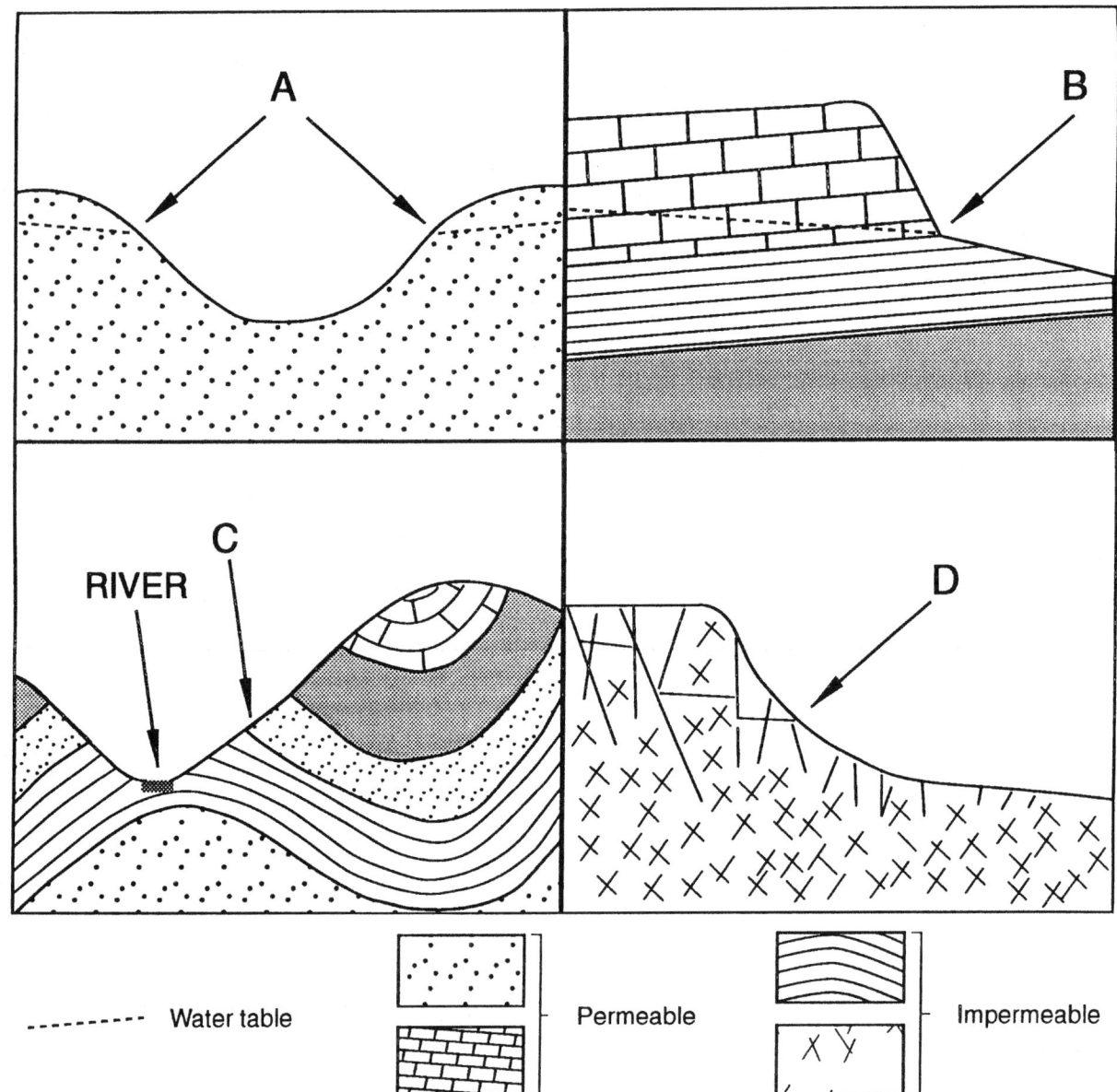

Figure S25 Some types of gravity springs. A, depression spring; B, contact spring, C, artesian spring; D, fracture spring.

Table S10 Classification of springs based on discharge (Meinzer, 1923)

Magnitude	Discharge
First	100 cfs 3 m³ s⁻¹
Second	10–100 cfs (0.3–3 m³ s⁻¹)
Third	1–10 cfs (0.03–0.3 m³ s⁻¹)
Fourth	100 gal min⁻¹–1 cfs (0.007–0.031 m³ s⁻¹)
Fifth	10–100 gal min⁻¹ (0.7–7 l s⁻¹)
Sixth	1–10 gal min⁻¹ (0.07–0.7 l s⁻¹)
Seventh	1 pint–1 gall min⁻¹ (0.01–0.07 l s⁻¹)
Eighth	≤1 pint min⁻¹ (≤0.01 l s⁻¹)

Thermal springs are commonly associated with areas of present or geologically recent volcanic activity, such as Yellowstone National Park in Wyoming; the lava areas of Idaho, eastern Oregon and northern California; the lava of the Auvergne region in France and of areas of volcanic rocks in central Europe, Italy and the circum-Pacific belt, notably Japan, New Zealand and Chile. They are also associated with areas in which the rocks have been intensely folded and faulted, such as the Alps and Pyrenees and the mountains of the western United States (Waring, 1965). More than 1000 thermal springs are known in the United States, and several thousand in the world.

The principal mineral substances in water from springs are generally the same as those in other natural groundwater. Springs whose water contains unusually large quantities of mineral salts in solution are known as mineral springs. Many of these are thermal, and some have gained widespread notice because of the therapeutic value attributed to their water. Some present a striking appearance because of their relationship to colorful terraces formed by minerals precipitated from the spring water, for example, at Mammoth Hot Springs in Yellowstone National Park.

O.M. Hackett

Bibliography

Bourdon, D.J., 1966. The largest karst spring? *J. Hydrol.*, **4**, 104.

Bryan, Kirk, 1919. Classification of springs, *J. Geol.*, **27**, 522–561.

Meinzer, O.E., 1923. Outline of ground-water hydrology, with definitions, *US Geol. Surv. Water Supply Paper* **494**, 48–56 (71 pp).

Meinzer, O.E., 1927. Large springs in the United States, *US Geol. Surv. Water Supply Paper* **557**, 94 pp.

Meinzer, O.E. (ed.), 1942. *Hydrology*, in *Physics of the Earth Series*, Vol. 9, pp. 416–432, New York, McGraw-Hill Book Co., 712 pp.

Price, M., 1996. *Introducing Groundwater*, 2nd edn, Chapman & Hall, London.

Stearns, N.D., Stearns, H.T. and Waring, G.A., 1937. Thermal springs in the United States, *US Geol. Surv. Water Supply Paper*, **679B**, 59–206.

Todd, D.K., 1980. *Groundwater Hydrology*, John Wiley, Chichester, 2nd, edn, 535 pp.

Tolman, C.F., 1937. *Ground Water*, pp. 435–466, New York, McGraw-Hill Book Co., 593 pp.

Ward, R.C., 1967. *Principles of Hydrology*, McGraw-Hill, London, 403 pp.

Waring, G.A., 1965. Thermal springs of the United States and other countries of the world, revised by R.R. Blankenship and R. Bentall, *US Geol. Surv. Profess. Paper* **492**, 383 pp.

White, D.E. and Brannock, W.W., 1950. The sources of heat and water supply of thermal springs, with particular reference to Steamboat Springs, Nevada, *Am. Geophys. Union Trans.*, **31**, 566–574.

Cross references

Groundwater
Medicinal springs
Water: categories
Water table

STAGE–DISCHARGE RELATION

General

Both the stage and the discharge of a stream vary most of the time and, in order to obtain a continuous record of discharge, the stage is recorded and the discharge computed from a correlation of stage and discharge. This correlation, or calibration, is known as the stage–discharge relation.

The operations necessary to develop the stage–discharge relation at a gauging station include making a sufficient number of discharge measurements and developing a rating curve by plotting the measured discharges against the corresponding stages and drawing a smooth curve of the relation between the two quantities. Discharge measurements are carried out over the range of stage variation in order to establish the rating curve as quickly as possible. Normally the lower and medium stages present little difficulty, but discharges at the higher stages may take some time and it may be necessary to resort to careful extrapolation until such time as the higher discharges are available to be measured.

If the channel is stable, comparatively few measurements may be required, although very few rivers have completely stable characteristics. The calibration, therefore, cannot be carried out once and for all, but has to be repeated as frequently as required by the rate of change in the stage–discharge relation.

It is, therefore, the stability of the stage–discharge relation that governs the number of discharge measurements that are necessary to define the relation at any time. In order to define the relation in sand-bed channels, for example, several discharge measurements a month may be required because of random shifts in the stream geometry.

Sound hydrometric practice requires that the discharge curve is determined as rapidly as possible after the establishment of a new station. Unless this is done, and the curve maintained, the record of stage cannot be converted into a reliable record of discharge.

Further information can be found in Herschy (1993, 1995).

The station control

An analysis of the stage–discharge relation and the construction of the rating curve requires an appreciation of the functioning of the channel control.

In order to have a permanent and stable stage–discharge relation, the stream channel at the gauging station must be capable of stabilizing and regulating the flow past the station so that for a given stage the discharge through the measuring section will always be the same. The shape, reliability and stability of the stage–discharge relation are normally controlled by a section or reach of channel at, or downstream from, the gauging station, known as the station control. The geometry of the station control eliminates the effects of all other downstream features on the discharge at the measuring section. The channel characteristics forming the control include the cross-sectional area and shape of the stream channel, the channel sinuosity (meanders and loops), the expansions and restrictions of the channel, the stability and roughness of the stream bed and banks, and the vegetation cover, all of which collectively constitute the factors determining the channel conveyance.

The simple stage–discharge curve

The general procedure in establishing the stage–discharge curve is as follows. The discharge measurements are plotted on arithmetic graph paper with discharge on the horizontal scale (abscissa) and the corresponding gauge height on the vertical scale (ordinate). If a measurement of discharge was not made at steady stage, the mean gauge height during the measurement is used. The plotted observations are labeled in chronological order, and rising and falling stages during the measurement are indicated by distinguishing symbols if necessary (Figure S26).

The relation should be defined by a sufficient number of measurements suitably distributed throughout the range in stage, taking into account the shape of the stage–discharge relation. Ideally, the number and spacing of the observations are made to conform to the relative frequency of flow at the various stages; that is, the number of observations at various subranges is in proportion to the probable occurrence of discharge at these same ranges, covering the whole range of discharge for which the relation is plotted.

Nevertheless, in practice, it is desirable to have as many observations as possible at the extreme ranges, both at the low-flow and at the high-flood stages.

The curve of relation, the rating curve, is drawn evenly and smoothly through the scatter of plotted data points.

Although all current-meter discharge measurements are checked and considered correct before plotting, observations that plot more than, say, 5% in discharge off the curve should again be checked for possible uncertainties. Particular attention is paid to the need to adjust or weight the gauge height, to the correct current-meter rating and to errors in the computation. With respect to the latter, it is useful to make a plot of the cross-sectional area of flow and the mean velocity against gauge height for each discharge measurement. Such plots reveal the presence of an error and where it is located in the computation, either in the velocity or in the cross-sectional area. If no apparent error is found to be caused by the above, the condition of the control is investigated before the measurement is discarded or a shift correction applied if applicable. From the above, it is evident that a copy of the stage–discharge curve is conveniently available in the instrument house.

Fitting the stage–discharge curve: visual estimation

There are several methods of fitting a median curve to observed or measured data points. This can be performed, however, quite satisfactorily by visual estimation of the plot with the aid of drafting curves, which usually are designed to conform to parabolic equations. Very often the trend of discharge measurements plotted on graph paper follows a particular drafting curve owing to the fact that stream discharge tends to vary as some power of the depth of flow.

The criterion used when fitting a median curve to observations by visual estimation is that there are about the same number of plus and minus deviations. A deviation is considered negative for a measurement lying above the curve and positive when lying below the curve.

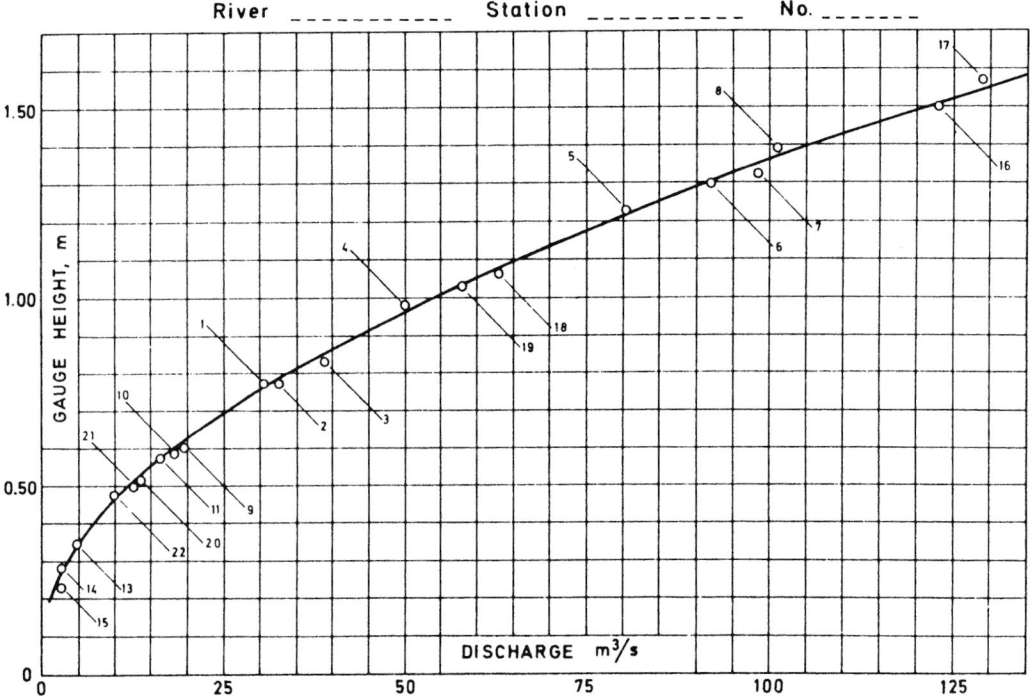

Figure S26 Stage–discharge curve. (Courtesy O.A. Tilrem.)

Logarithmic method

In many cases the stage–discharge curve may be established by plotting the logarithms of stage against the logarithms of discharge. The use of logarithmic graph paper obviates the necessity of computing the logarithms and the plotting of the observations is performed in the same manner as before. There are certain important advantages in using the logarithmic method.

- The logarithmic form of the rating curve can be developed into a straight line, or straight-line segments, by adding or subtracting a constant value (datum correction) to the gauge height logarithmic scale.
- The straight-line graph can be described by a simple mathematical equation that is easily handled by pocket calculator or computer.
- The straight-line graph may be conveniently analyzed for uncertainties.
- A percentage distance off the curve is always the same regardless of where it is located. Thus a measurement that is 10% off the curve at high stage will be the same distance away from the curve as a measurement that is 10% off at low stage.
- It is easier to identify the range in stage for which different controls are effective.
- The gauge height scale may be conveniently altered by halving, doubling or adding a percentage to the scale. The curve will merely shift position but retain the same shape.
- The curve can be extrapolated at either the top or the bottom and if the curve is a single segment and the control is stable, then extrapolation may be performed with more confidence than if the curve is made up of several segments.

Theory of the logarithmic curve (Herschy, 1993, 1995)

The stage-discharge relation may be expressed by an equation of the form

$$Q = C(h + a)^n \tag{S5}$$

which is the equation of a parabola where Q is the discharge, h is the gauge height, C and n are constants, and a is the stage at zero flow (datum correction). This equation may be transformed by logarithms to

$$\log Q = \log C + n \log(h + a) \tag{S6}$$

which is in the form of the equation of a straight line

$$y = n'x + C' \tag{S7}$$

where n' is the gradient and C is the intersection of the line on the y axis. By plotting Q against $(h + a)$ therefore, on double logarithmic graph paper, a straight line is obtained.

Often two or more straight lines may be required to fit the data, and it is usually possible, initially, to decide on the approximate location of the break points of each range by a careful investigation of the controls. The actual break points may be determined by solving the two equations concerned for Q and h or by purely graphical means. For very irregular channels, or for non-uniform flow, equation (S5) cannot be expected to apply throughout the whole range of stage. Sometimes the curve changes from a parabolic to a complex curve or vice versa, and sometimes the constants and exponents vary throughout the range.

The logarithmic rating equation, therefore, is seldom a single straight line or a gentle curve throughout the entire range of stage at a gauging station. Even if the same channel cross-section is the control for all stages, a sharp break in the contour of the cross-section causes a break in the slope of the rating curve. Also, the other constants C and a in equation (S5) are related to the physical characteristics of the stage–discharge control.

If the control section changes at various stages, it may be necessary to fit two or more equations, each corresponding to the portion of the range over which the control is applicable. If, however, too many changes in the parameters are necessary in order to define the relation, it is possible that the logarithmic method may not be suitable and a curve fitted by visual estimation can be employed as previously described.

Normally, in graphical analysis the dependent variable, Q in equation (S5), would be plotted on the vertical axis and the independent variable, h, plotted on the abscissa. It has been a tradition in stream gauging, however, that this procedure is reversed while still retaining Q as the dependent variable and taking n, the slope, as the cotangent instead of the tangent.

The geometry or shape of the channel section is reflected in the slope n of the stage–discharge equation (S5). This property is a useful indicator when carrying out a preliminary survey at a new site. The following are approximate relations between n and channel sections:

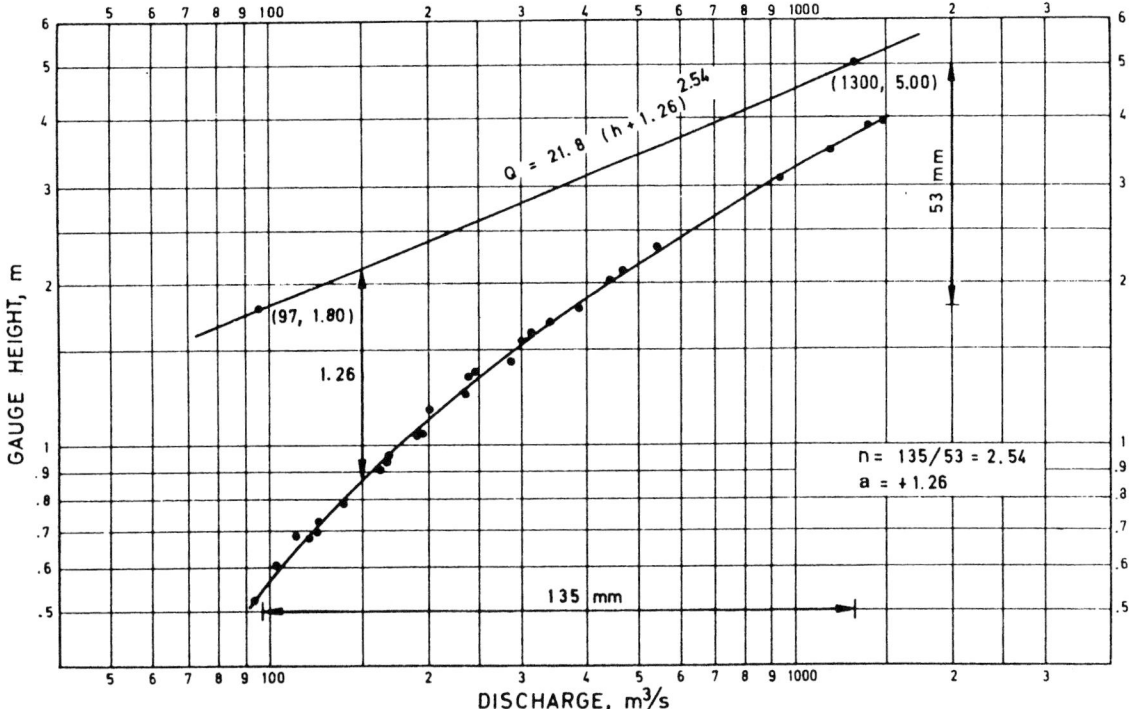

Figure S27 Stage–discharge curve established by the logarithmic method. (Courtesy O.A. Tilrem.)

- for a rectangular channel section, $n = \frac{3}{2}$;
- for a concave section of parabolic shape, $n = 2$;
- for a triangular or semicircular section, $n = \frac{5}{2}$.

Changes in channel resistance and slope with stage, however, will affect the exponent n. The net result of these factors is that the exponent in equation (S5) for relatively wide rivers with channel control will generally vary from about 1.3 to 1.8. For relatively deep, narrow rivers with section control, the exponent n will almost always be greater than 2 and may often exceed a value of 3.

The stage of zero flow

The datum correction a is the value of the stage at zero flow and corresponds to the lowest point on the low-water control. It is defined as the gauge height at which water ceases to flow over the control. Usually, this stage does not coincide with the zero of the gauge unless the zero of the gauge is specifically set to the lowest level of an artificial control or the crest of a measuring structure. The point of zero flow is, therefore, easily determined for artificial controls and in those cases where the control is well defined by a rock ledge.

The stage of zero flow is determined by subtracting the depth of water over the lowest point on the control from the stage indicated by the gauge reading. If the gauge is at some distance from the control, an adjustment is made for the slope. The difficulty in determining the point of zero flow is in establishing the lowest point on the control, as not all controls are easily identified. Generally, a cross-section is surveyed across the stream at the first complete break in the slope of the water surface below the measuring section. This is usually the location of the upstream lip of the low-water control. For a channel-controlled gauging station, the maximum depth directly opposite the gauge will give a reasonable approximation of the depth to be subtracted from the gauge reading in order to obtain the stage of zero flow.

The position of the point of zero flow is best determined at a time of low water when rivers can often be waded. In those cases, however, where the control section is difficult to identify, it may be located by surveying a close grid of spot levels or by running a sufficient number of cross-sections over the area of the assumed control section or reach.

It should be noted that when a quantity has to be added to the gauge heights in order to obtain a straight line, a is taken as positive, that is,

the zero of the gauge is in this case positioned at a level above the point of zero flow. Conversely, when a quantity has to be subtracted from the gauge heights, a is taken as negative, and in this case the zero of the gauge is positioned at a level below the point of zero flow. When the zero of the gauge coincides with the level of the point of zero flow, then a is zero.

Trial-and-error procedure of estimating the datum correction a

All discharge measurements are plotted on double logarithmic graph paper ('log–log paper') and a median line is drawn through the scatter of observations. Usually, this line will be a curved line. Various trial values, one value for each trial, are added or subtracted to the gauge heights of the measurements until the plot obtained forms a straight line. The trial value forming the straight line is the required value of a (Figure S27). All the plotted observations may be used in the trial operation. However, it is better to use only a few points selected from the median line first fitted to the points.

Computer program procedure

If automatic data processing by computer is available, the datum correction can be found conveniently by a computer program, which is designed essentially to carry out an iteration procedure until the best straight line is obtained. In the iteration program, different values of a are tested against the correlation coefficient for the line of best fit. The value of a giving the maximum correlation coefficient is the value selected. Usually, about 20 or more iterations are necessary. This is a purely mathematical procedure and probably gives the best results, but every endeavor is made to ensure that the result can be confirmed by a site investigation.

Estimating the constants C and n

After a straight-line plot of the discharge measurements on log–log graph paper has been obtained, the constants C and n of the rating equation (S5) can be computed by the least-squares procedure or graphically.

The stage–discharge relation is first analyzed from a plot on log–log graph paper in order to establish whether the rating curve is composed of one or more straight-line segments, each having its own

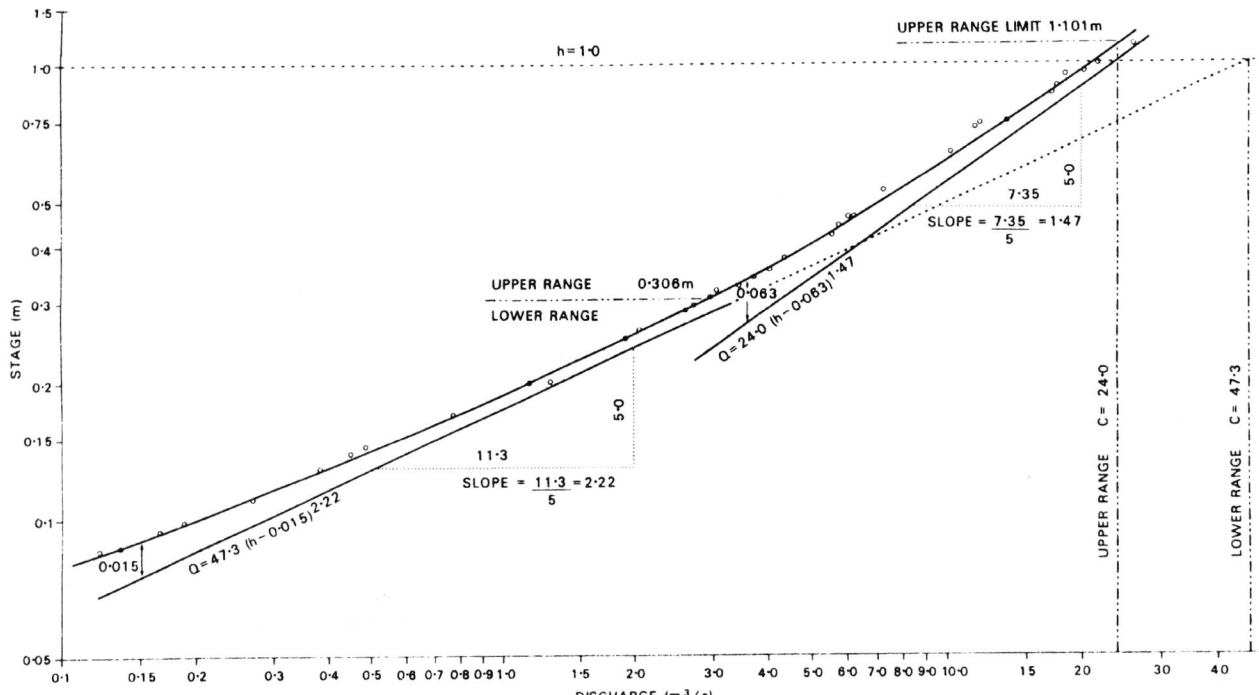

Figure S28 Example of logarithmic stage–discharge curve having lower and upper segments.

constants C and n. The constants for each separate segment are calculated separately.

Graphical procedure

The graphical method of determining C and n is simple and normally as effective in giving good results as the two methods described above.

The value of n is found by scaling the horizontal and vertical projections of the line and calculating this ratio. In Figure S27 the horizontal projection has been scaled as 135 mm and the vertical projection as 53 mm. The value of n is therefore calculated as

$$n = \frac{135}{53} = 2.54$$

The value of C is given by the numerical value of the discharge when $(h + a) = 1$ since [equation (S6)]

$$\log Q = \log C + n \log(h + a)$$

and when $(h + a) = 1$, $n \log(h + a) = 0$ and therefore $Q = C$.

Referring to Figure S27, it will be observed that the logarithmic stage–discharge curve does not go through the line $(h + a) = 1$. If the line is projected, however (not shown in Figure S27), it is found that when $(h + a) = 1$, $Q = 21.9$ and therefore $C = 21.9$. Otherwise, knowing n, C may be found by solving the rating equation (S5) for C as follows:

$$Q = C(h + a)^{2.54}$$

when $Q = 355$ and $(h + a) = 3.00$ and

$$C = \frac{355}{3^{2.54}} = 21.79$$

Therefore

$$Q = 21.79(h + 1.26)^{2.54}$$

Figure S28 shows another example of a stage–discharge curve. From an examination of this curve and an inspection in the field, it was clear that the high-water control became operative at a gauge height of 0.306 m. The curve, therefore, had two segments. The computer program procedure gave values of a for the lower range of 0.015 m and for the upper range 0.063 m (the trial-and-error procedure would give similar results). These values of a are scaled off graphically in Figure S28 and two straight lines of relation established as shown. The ordinate scale for these lines now becomes $(h - a)$ (both a values are negative). The value of the slope n of each line is scaled off as shown and the C values obtained on the abscissa at $(h - a) = 1$. It will be observed that in order to obtain C for the lower line, the line requires to be projected to $(h - a) = 1$.

The equations of the lines of relation are therefore

$$Q = 47.3(h - 0.015)^{2.22} \tag{S9}$$

for values of h up to 0.306 m, and

$$Q = 24.0(h - 0.063)^{1.47} \tag{S10}$$

for larger values of h.

R.W. Herschy

Source

Flow Measurement and Instrumentation, Oxford: Butterworth-Heinemann, **4**(1), 11–15.

Bibliography

Herschy, R.W., 1995. *Streamflow Measurement*, Chapman & Hall, London and New York.
Herschy, R.W., 1993. The Stage–discharge relation. *Flow Meas. Instrum*, **4**(1), 11–15.

STORMS, BRIEF DEFINITIONS

Tropical cyclones

Defined as being made up of countless thunderstorms of rotating spiral systems which rise over warm oceans. At their greatest intensity they are known as hurricanes, typhoons or cyclones, depending on the country or region in which they occur.

Hurricanes

Arise from tropical cyclones that spiral off the west coasts of Africa and Central America, usually in late summer. The hurricane's central 'eye', a calm, sunlit cylinder often as much as 100 km across, experiences winds swirling at speeds as much as 300 km h^{-1}.

Typhoons

Tropical cyclones along the northwest Pacific Ocean coasts which can be experienced at any time and often kill 100 000 people at one time. In India and Australia typhoons are known as cyclones.

Storm surges

Storm surges from oceans are associated with typhoons and hurricanes and often cause flooding of low-lying areas. Storm surges may cause a temporary increase in sea level of as much as 6 m.

Tornadoes

Produced by severe thunderstorms and more compact than hurricanes; many of these could fit into a hurricane's 'eye'. They are caused by a constant infusion of warm, moist air.

Water spouts

Weak tornadoes which form over open water, although they often whirl to shore causing damage to boats, docks, etc.

Land spouts

Small-sized tornadoes.

Dust devils

Found in deserts with wind speeds exceeding as much as 100 km h^{-1} and formed when rising heat encounters isolated pockets of rolling air.

R.W. Herschy

Bibliography

Time International, 1996. Unravelling the mystery of twisters, Vol. 147, No. 21.

STREAMFLOW MEASUREMENT

Field measurement

Streamflow is the combined result of all climatological and geographical factors that operate in a drainage basin. It is the only phase of the hydrological cycle in which the water is confined in well-defined channels which permit accurate measurements to be made of the quantities involved. Other measurements of the hydrological cycle are point measurements for which the uncertainties, on an areal basis, are difficult, if not impossible, to estimate.

Good water management is founded on reliable streamflow information and the final reliability of the information depends on the initial field measurements. The hydrologist making these measurements therefore has the responsibility of ensuring that raw data of acceptable quality are collected. The application, processing and publication of the data depend largely on the quality of the field measurements.

Objectives of a streamflow program

There are many different uses of streamflow data within the broad context of water management, such as water supply, pollution control, irrigation, flood control, energy generation and industrial water use. The importance placed on any one of these purposes may vary from country to country. In India and China, for example, emphasis may be placed upon irrigation and flood control whereas in the UK water supply may be given priority. The emphasis for any one need may also change over short or longer periods of time. What appears to be axiomatic, however, is that none of these needs can be met without reliable streamflow data being available at the right time, the right place and the right quality.

Categories of streamflow data

The type of streamflow information required may be classified into two distinct categories. The first is that required for planning and design while the second is that required for current use, i.e. operational management.

Data for planning and design may not necessarily have an immediate use but are valuable in the long term for civil engineering works of various types and for flood forecasting and control. Planning and design data are also used to examine long-term trends as are data on the stream environment.

Current use data have an immediate high return value since the data are invariably required initially for operation and control. Current use streamflow stations are operated for as long as the need remains.

Designers of water control and water-related facilities increasingly use the statistical characteristics of streamflow rather than flow over specific historic periods. The probability that the historical sequence of flow history at a given site will occur again is remote. Indeed, when a hydrologist makes just one measurement of discharge, it is probable that the exact conditions under which the discharge occurred may rarely happen again.

It is often desirable to consider the future, not in terms of specific events but in terms of probability of occurrence over a span of years. For example, many highway bridges are designed on the basis of the flood that will be exceeded on the average only once in 50 years. Storage reservoirs are designed on the basis of the probability of failure of a particular capacity to sustain a given draft rate. The water available for irrigation, dilution of waste or other purposes may be stated in terms of the mean flow, or probability of flow magnitudes, for periods of a year, season, month, week or day. In addition there is a trend towards flow simulation based on statistical characteristics, such as the mean, standard deviation and skew. To define statistical characteristics, a record of at least 30 years is desirable for reliable results.

Cost effectiveness

In most countries the cost effectiveness of streamflow data collection is an important consideration; this is particularly the case where streamflow is included in the budget for water management. Cost effectiveness may be measured by the benefit/cost ratio, but to estimate this ratio for streamflow is difficult, mainly due to the problems associated with assessing the benefits accruing. This problem sometimes leaves the hydrological service at a disadvantage in bidding for funds.

The wide variety of uses of streamflow data also makes the estimation of national benefits difficult. The question of marginal gains through network changes is therefore not straightforward. It is, however, a fact that costs have risen sharply in providing gauging stations, and in data capture and publication. The gains, on the other hand, are not easily quantified and each use of streamflow data may demand different and perhaps sophisticated analysis before benefits of streamflow data collection can be realized. This, however, is not usually the case in developing countries where the gain from a flood control scheme or an irrigation scheme may be enough to cover the cost of the entire network many times over. Benefit/cost ratios in these circumstances may be as high as 50 or more.

In other countries, however, a period of years may elapse before a useful record is generated to quantify the benefits of current data, and even then any satisfactory assessment is complicated.

The first objective, therefore, is to develop a suitable method to identify potential quantifiable benefits to various types of data user. Such benefits are usually to be found in data required for reservoir design, water abstraction, flood warning, flood control including flood proofing, irrigation, highway bridge design, hydroelectric power generation, river pollution control, sewage purification and so on.

The costs of providing these services are quantified over a defined period and the benefits accruing from streamflow data are calculated for each. More often the benefits may have to be determined from an agreed percentage of the cost of the services.

Flood proofing, for example, reduces the cost of flood damage and this figure can be conveniently quantified from flood damage records. If no flood damage records exist, a percentage benefit based on the cost of the scheme can usually be calculated. The benefits are calculated for each use to which the streamflow data are put and totalled. This total is divided by the cost of obtaining the streamflow data. This is usually the cost of the operation of the gauging stations and processing the data or, more conveniently, the sum of capital and staff costs.

In a benefit/cost study of the UK streamflow network (1989) carried out for the Department of the Environment, the annual benefits were found to be in the range \$16.5–90 million depending on how these were quantified, the best estimate being \$31.5 million. The annual cost of operating the network, including overheads, management, data processing, etc., was \$13.5 million. The benefit/cost ratio was therefore in the range 1.2–7 with a best estimate of 2.3. It was concluded, therefore, that even at the lowest level of benefit/cost ratio, the UK streamflow network represents a sound economic investment. The ratio would have been higher if some of the intangibles (e.g. consents for discharge effluents) could have been quantified. Not surprisingly, only small annual economic benefits could be quantified from flood forecasting, flood warning or flood alleviation.

International standards in stream gauging

Water in a stream in a specific locality knows no jurisdictional boundaries, local or national. That same water may eventually move to any other part of the Earth through the hydrological cycle. Streamflow data are therefore needed from all parts of the Earth to enable hydrologists to discover the quantity of the Earth's water resources on a comprehensive and continuous basis. Streamflow records that have been gathered by non-standard methods may be suspect. For this and other reasons, the International Organization for Standardization (ISO) set up, in 1964, a technical committee on streamflow measurement. This committee, known as TC113, has produced a number of international standards on streamflow which are now used worldwide. Of the 89 ISO member countries, representing some 95% of the world's population, some 26 are members of TC113.

In addition, the World Meteorological Organization (WMO) publishes guides and technical reports on stream gauging and selected ISO Standards, in the form of Technical Regulations which are circulated to some 185 WMO member countries.

Standardization activity at the European level is the responsibility of CEN (European Committee for Standardization – Comité Européen de Normalisation) and CENELEC (European Committee for Electrotechnical Standardization). Together these bodies make up the Joint European Standards Institution (ESI). The aim of European standardization is the harmonization of standards on a Euro-wide basis in order to facilitate the exchange of goods and services by eliminating barriers to trade which might result from requirements of a technical nature. The national standardization institutes of 18 countries support CEN (15 EU and three EFTA countries). In addition, 11 other European countries have Affiliate status. Streamflow is under TC318 'Hydrometry', formed in 1994.

Summary of methods

A summary of the methods of streamflow measurement follows.

Velocity–area method

The discharge is derived from the sum of the products of stream velocity, depth and distance between verticals (Figure S29), the stream velocity usually being obtained by a current meter. For a continuous record of discharge in a stable prismatic open channel with no variable backwater effects, a unique relation exists between water level (stage) and discharge. Once established, this stage–discharge relation is used to derive discharge values from recordings of stage. With the exception of the dilution method, which is a direct method, it could be inferred that all methods of streamflow measurement are based on the velocity–area principle.

Float gauging

The water velocity is measured by recording the time taken for a float to travel a known distance along the channel. Observations are made using floats at different positions across the channel and discharge is derived from the sum of the products of velocity, width and depth.

Generally, this method is used only when the flow is either too fast or too slow to use a current meter or where ice floe would cause damage to the meter.

Slope–area method

The discharge is derived from measurements of the slope of the water surface and the cross-section of the channel over a fairly straight reach, assuming a roughness coefficient for the channel boundaries.

Stage–fall–discharge method

In a stable open channel affected by backwater, a relation is established between fall (slope) and discharge.

Weirs and flumes

The relation between stage (or head) and discharge over a weir or through a flume is established from laboratory (or field) calibration. The discharge is subsequently derived from this rating equation.

Dilution method

A tracer liquid is injected into the channel and the water is sampled at a point further downstream where turbulence has mixed the tracer uniformly throughout the cross-section. The change in concentration between the solution injected and the water at the sampling station is converted into a measure of the discharge.

Moving boat method

A current meter is suspended from a boat which traverses the channel normal to the streamflow. The component of the velocity in the direction of the stream is computed from the resultant velocity and the angle of this resultant. The discharge is the sum of the products of the stream velocity, depth and distance between observation points.

Ultrasonic method

The velocity of flow is measured by transmitting an ultrasonic pulse diagonally across the channel in both directions simultaneously. The

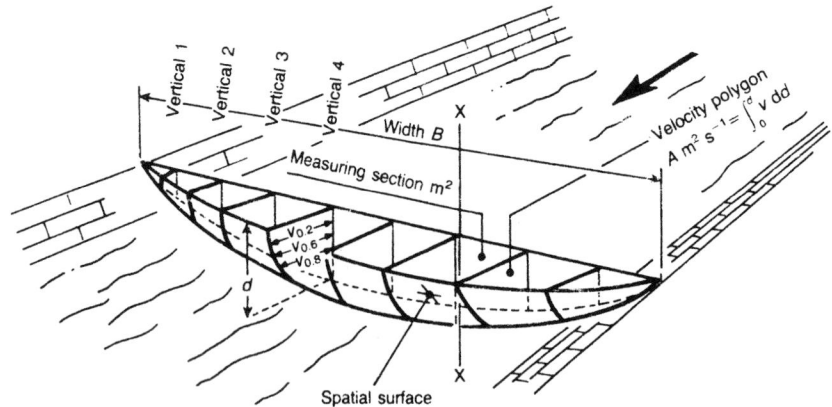

Figure S29 The measuring section. The volume of water is bounded by the measuring section, the water surface, the bed and the spatial surface as shown schematically. At any section XX, the area of the velocity polygon is the integral vdd (with limits from 0 to d) and equal to A m^2 s^{-1}. The volume of water passing per second is then found from the integral of Adb (with limits from zero to b) and equal to the integral of $vdd\ db$ (with limits from 0 to d and 0 to b) which is equal to the total flow Q in m^3 s^{-1}.

difference in time transits is a measure of the velocity which has to be multiplied by the cross-sectional area to derive discharge. The ultrasonic method therefore also follows the principles of velocity–area measurements. Ultrasonic meters are also used in pipes and in sewage works channels.

Electromagnetic method

The discharge is found by measuring the electromotive force (emf) produced by a moving conductor (the flowing water) through a magnetic field produced by a coil placed either below or above the open channel. The emf is proportional to the discharge. Electromagnetic meters are also used in pipes to measure flow rate.

Accuracy

Considerable research into uncertainties in streamflow measurement over recent years has led to the publication of several international standards.

Flow in pipes (closed conduit flowmeters)

Differential pressure meters are based on the principle that when water passes through a contraction in a pipe, it accelerates. The resulting increase in kinetic energy is balanced by a decrease in the static pressure at that point in the pipe and the pressure drop caused by the contraction is proportional to the square of the flow rate for a given flowmeter.

R.W. Herschy

Source

Herschy, R.W., 1995. *Streamflow Measurement*, 2nd edn. Chapman & Hall, London, 524 pp.

Bibliography

Ackers, P., White, W.R., Perkins, J.A. and Harrison, A.J.M., 1978. *Weirs and Flumes for Flow Measurement*. John Wiley and Sons, Chichester.
Bos, M.G., 1976. *Discharge Measurement Structures*. Publication No. 161, Delft Hydraulics Laboratory, Delft.
Department of the Environment, 1989. *The Benefit Cost of Hydrometric Data: River Flow Gauging*. Report by CNS, Reading, UK. The Foundation for Water Research, UK.

International Standards on liquid flow measurement in open channels, International Organization for Standardization, Geneva, Switzerland.

ISO 748–1997. *Velocity–Area Methods.*
ISO 772–1996. *Vocabulary and Symbols.*
ISO 1070–1992. *Slope–Area Method.*
ISO 1088–1973. *Velocity–Area Methods – Collection of Data for Determination of Errors in Measurement.*
ISO 1100/1–1996, Part 1. *Establishment and Operation of a Gauging Station.*
ISO 1100/2–1998, Part 2. *Determination of the Stage–Discharge Relation.*
ISO 1438/1–1980, Part 1. *Thin Plate Weirs.*
ISO 2425–1998. *Measurement of Flow in Tidal Channels.*
ISO 2537–1988. *Cup-type and Propeller-type Current Meters.*
ISO 3454–1983. *Direct Sounding and Suspension Equipment.*
ISO 3455–1976. *Calibration of Rotating Element Current Meters in Straight Open Tanks.*
ISO 3716–1977. *Functional Requirements and Characteristics of Suspended Sediment Load Samplers.*
ISO 3846–1989. *Rectangular Broad Crested Weirs.*
ISO 3847–1977. *End-depth Method (Rectangular Channels).*
ISO 4359–1983. *Flumes.*
ISO 4360–1984. *Triangular Profile Weirs.*
ISO 4362–1992. *Trapezoidal Profile Weirs.*
ISO 4363–1993. *Methods of Measurement of Suspended Sediment.*
ISO 4364–1977. *Bed Material Sampling.*
ISO 4365–1984. *Determination of Concentration and Particle Size.*
ISO 4366–1979. *Echo Sounders for Water Depth Measurements.*
ISO 4369–1979. *Moving Boat Method.*
ISO 4371–1984. *End Depth Method (Non-rectangular Channels).*
ISO 4373–1995. *Water Level Measuring Devices.*
ISO 4374–1989. *Round-nose Horizontal Crest Weirs.*
ISO 4375–1979. *Cableway System for Stream Gauging.*
ISO 4377–1989. *Flat V Weirs.*
ISO 5168–1978. *Estimation of Uncertainty of Flow-rate Measurement.*
ISO 6416–1992. *Ultrasonic Method.*
ISO 6419/1–1984. *Hydrometric Data Transmission Systems – General.*
ISO 6419/2–1992. *Hydrometric Data Transmission Systems – Specification.*
ISO 6420–1984. *Position Fixing Equipment for Hydrometric Boats.*
ISO 6421–1984. *Sediment Accumulation in Reservoirs.*
ISO 7066/1–1985. *Uncertainty in Linear Calibration Curves.*
ISO 7066/2–1985. *Uncertainty in Non-linear Calibration Curves.*
ISO 7178–1983. *Investigation of the Total Error in Measurement of Flow by Velocity–Area Methods (TR).*
ISO 8333–1985. *V-shaped Broad Crested Weirs.*
ISO 8363–1997. *Guide for the Selection of Method.*
ISO 8368–1985. *Guide for the Selection of Gauging Structures.*
ISO 9123–1986. *Stage–Fall Discharge Method.*
ISO 9195–1992. *Gravel Bed Material Sampling and Analysis.*
ISO 9196–1992. *Measurement Under Ice Conditions.*
ISO 9209–1989. *Wet-line Correction.*
ISO 9210–1992. *Measurement in Meandering Rivers.*
ISO 9212–1992. *Bed Load Discharge TR.*
ISO 9213–1992. *Electromagnetic Method (TR).*
ISO 9555/1–1994. *Dilution Methods – General.*
ISO 9555/2–1992. *Dilution Methods – Radioactive Tracers.*
ISO 9555/3–1992. *Dilution Methods – Chemical Tracers.*
ISO 9555/4–1992. *Dilution Methods – Fluorescent Tracers.*
ISO 9823–1990. *Three Verticals Method.*
ISO 9824/1–1990. *Free Surface Flow in Closed Conduits (Methods) (TR).*
ISO 9824/2–1990. *Free Surface Flow in Closed Conduits (Equipment) (TR).*
ISO 9825–1994. *Measurement of Flood Flows TR.*
ISO 9826–1992. *Parshall and Saniiri Flumes.*
ISO 9827–1994. *Streamlined Triangular Profile Weirs.*
ISO 11328–1994. *Equipment for the Measurement of Discharge under Ice Conditions.*
ISO 11330–1997. *Volume of Lakes and Reservoirs.*
ISO 11332–1998. *Flow in Unstable Channels and Ephemeral Streams.*
ISO 11627–1998. *Model for Unsteady Flow.*
ISO 11655–1995. *General Performance Specification for Hydrometric Equipment.*
ISO 11656–1993. *Mixing Length of a Tracer in Open Channels.*
ISO 11974–1998. *Electromagnetic Current Meters (TR).*

Cross references

Flow measurement: new technology
Stage–discharge relation
Water resources: dictionary of basic terms

SURFACEWATER YIELD CALCULATION

The traditional definition of surface water yield is that output which a water resource system could sustain under a specified design runoff criteria. For a river abstraction this is a function of the minimum flow under the design runoff and any restrictions on abstraction such as licence conditions relating to a prescribed or maintained flow. For a reservoir system, the yield would be a function of the storage availability as well as runoff. The full usable capacity typically contributes to the calculation of the yield. This is combined with a cumulative runoff sequence of either historic or specified severity. All possible combinations of consecutive cumulative months' runoff from the record are considered to establish the amount which could theoretically be abstracted from the reservoir in each case. The minimum value represents the gross yield of the reservoir. The yield available for supply purposes would be this gross value minus any compensation water or other releases. Similar analyses, but more complex, are possible for river regulation reservoirs, but would typically involve simulation of the operation of the reservoir to meet a downstream prescribed or maintained flow in order that the fluctuating daily pattern of releases, taking into account natural river flows, can be considered.

The critical period of the source is that period over which the yield can be sustained, with the reservoir storage falling from full to just empty. As such, yield is a design term since in practice a source would not be operated in this hypothetical way. It does, however, provide a mechanism for comparing the theoretical resource availability between water resources systems. Depending on the ratio of runoff to storage capacity, a reservoir source may be single, two-season or multi-season critical. Here, 'season' relates to the susceptibility to a period of drawdown; that is, a single-season source could maintain its design yield over one drawdown period, i.e. from reservoir full, typically in the early months of the year, to reservoir empty in the summer or fall of that same year. A two-season critical source could be critical over two drawdown seasons, i.e. it would drawdown in the first year, would fail to refill fully in the intervening winter and would just empty during the following summer/fall.

The runoff criteria used to calculate yield can be based on either a minimum historic or probabilistic basis, but any quotation of yield should make it plain as to the criterion used. The advantage of using the minimum runoff pattern in the historic runoff record is that the time-series data are often more readily available than probabilistic records, especially over the long periods which are required. Whilst cumulative probabilistic monthly records can be used in the assessment of yield for reservoirs, a daily runoff pattern is ideally required for the assessment of yield involving either river abstractions or regulation releases. Such derived time series are often difficult to produce, especially for those sites with short runoff records. The advantage of probabilistic methods, however, is the compatibility of yield calculations from different sites, making quantification of yields across a region or the country more consistent. Yields calculated on a minimum historic basis are hard to compare on a region-wide basis because of the susceptibility of resource systems in different locations to different historic droughts. Such yield assessments are greatly influenced by not only the length of historic runoff record but also by the inclusion within the period of extreme events of the length of critical period to the resource.

In recent years there has been a move away from this traditional concept of yield. These developments have taken two forms. First, there has been a move towards the calculation of yield on an operational basis rather than in a design context. Such estimates are useful in day-to-day management, particularly during droughts. In these cases, rather than providing a yield based on the reservoir starting full and an associated runoff pattern, estimates are based on current reservoir contents and a forecast of runoff from the present. Such estimates are used in real-time to establish the supply which can be supported by the resource. In this way, yield can actually be perceived as a dynamic rather than static concept that can be used to manage actively the resource system.

A second development has been the need to consider operating practices and constraints and levels of service in yield calculations. Such approaches recognise that the supply which can be supported from a water resources system, particularly for a complex, interlinked water resources system reflects as much infrastructure and operating practices as the hydrology of the situation. This philosophy also recognizes that a water resources system operates to meet a varying demand profile rather than a fixed rate of abstraction. In addition, levels of service are in place which recognize that a water resources system would not, in practice, be allowed to empty as assumed in the design yield concept. Measures such as hosepipe bans and even standpipes would be introduced to restrict demand should a water resources system be in danger of failure. Whilst traditional estimates of yield ignore such measures, more recent approaches assume that such actions would be introduced with a specified frequency (level of service) usually based on reservoir storage and take this into account in yield calculation. Such approaches require the use of simulation involving the analysis of varying demand patterns and the introduction of cutback regimes. Simulation would typically be carried out using a time series based on either a historic or probabilistic runoff record.

Whilst the assessment of yield, taking into account varying demand profiles and levels of service, is more realistic than design concepts, its major disadvantage is that it becomes more difficult to compare yields between water resources systems. Even if runoff patterns of similar severity are used in each case, the pattern of demand and the criteria on which level of service restrictions are imposed are likely to vary between resource systems. Also, the susceptibility of the different resource systems to runoff events is very much a function of the degree of connectivity within the water resources system which allows the movement of supplies from zones of plenty to those under greater stress.

Susan Walker

Cross references

Reservoir capacity estimation
Water resources

SUSPENDED SEDIMENT MONITORING: USE OF ACOUSTIC DOPPLER CURRENT PROFILER

Introduction

During recent years, increasing attention has been focused on the mechanism of suspended sediment transport in both fluvial and marine environments. The need for understanding suspended sediment transport processes has been brought about by a wide range of problems and phenomena which are related to the erosion, transport and deposition of fine material. Quantitative and qualitative aspects of suspended sediment transport can be compared with qualitative aspects being strongly related to the quantitative ones. The following examples are arbitrarily taken from a long list of problems associated with suspended sediment transport, to give an impression of the problems being addressed:

- the impact of huge river reservoirs on suspended sediment transport;
- the deposition of fine materials eroded from agricultural areas along the longitudinal profile of rivers;
- the deposition and erosion of fine materials in inundated areas and wetlands during flood events;
- long-term changes in estuaries due to changes of suspended sediment transport;
- the impact of suspended sediment transport due to coastal currents, tides and waves on coastal morphology and coastal structures.
- surfacewater quality, which is strongly related to small-sized suspended sediments that are potent concentrating media for agricultural chemicals, heavy metals and other pollutants.

In both fluvial and marine environments, suspended sediment transport is characterized by a high stochastic component in the temporal and spatial pattern, respectively, of the various complexly linked impacts. Thus, reliable suspended sediment monitoring demands a fine temporal and spatial resolution. Applying traditional monitoring methodology, the acquisition of representative data requires considerable time and human resources. The requirement for time- and cost-saving methods for *in situ* measuring methods have therefore been considered. However, new methodical approaches for suspended sediment monitoring have to be judged not only on their ability to supply information about parameters such as particle concentration and particle size distribution but also for their contribution to solving monitoring problems.

Acoustic Doppler current profilers (ADCPs) are state-of-the-art instruments in oceanography and hydrometry for current velocity and discharge measurements. They transmit short acoustic pulses along narrow, nearly vertical beams and process the echo from scatterers in the sonicated water column. In addition to the data needed for calculating velocity and discharge, ADCPs provide the so-called backscatter intensity which is related to the amount and the properties of scatterers in the ensonified water column. Thus efforts are made to derive characteristics of suspended particles from the ADCP data. This entry describes the advantages and limitations of applying ADCPs for suspended sediment monitoring. Special attention is given to the demands of suspended sediment monitoring and the state-of-the-art monitoring methods which are available.

Standard methods of suspended sediment monitoring

Sampling and sample processing

Until now the procedure most often used for suspended sediment monitoring is first to take water samples, sometimes several hundred litres, then dewater the suspended sediments by centrifugation or filtering and finally, perform the necessary analysis in the laboratory. Standard parameters to describe suspended sediment transport quantitatively are the concentration of suspended sediments and the particle size distribution. Recent techniques for suspended sediment monitoring are summarized in UNESCO (1983). The most recent compilation of available methods for particle size analysis is given in Syvitski (1991).

Although being state of the art, the available methods of taking samples, dewatering the samples and performing laboratory analysis have many disadvantages.

Sampling

A comparative study of three standard methods for water sampling (Verhoeven et al., 1989) including the bottle sampler, the pump sampler and the XRB van Dorn water sampler, shows remarkable differences in measuring concentration and their influences on particle size distribution. Gardner (1977) proved the influence of standard bottle samplers on suspended particle concentration and particle size distribution. Gibbs (1981) demonstrated that sediment aggregates are readily disrupted by their passage through the pumping systems usually used for particle sampling purposes.

Dewatering

Errors caused by using a continuous-flow centrifuge for dewatering suspended sediments from large volumes of river water were evaluated by Rees et al. (1991). In addition to negative effects on the concentration measurement by losses of material due to cleaning processes, changes of the particle size distribution resulting from particle fractionation within the centrifuge bowl may be significant.

Pretreatment

The effects of sample pretreatment on size analysis are summarized in Matthews (1991a). The disturbances most often observed are fractionation of particle aggregates by the use of ultrasonics, and chemical cracking by reagents.

Analysis

Finally, the laboratory analysis involves different problems and errors of its own. The automated determination of particle size distribution is generally based on recording parameters which are related to the particle size, such as the settling velocity, laser diffraction or changes in electroresistance of particles passing through an electric field. To obtain particle size distributions, assumptions have to be made about the shape, density and/or surface of the particles. Thus, strictly speaking, all available methods for routine analysis do not provide a real particle size distribution but an equivalent particle size distribution based on the assumption of spherical particles. The effects of particle shape and density on size measurement are given in Matthews

(1991b). Comparative studies between standard methods of particle size analysis, including the impact of particle properties, were conducted by Swift et al. (1972), Fadout (1973), Harfield and Miller (1984) and McCave and Syvitsky (1991).

In situ measurements

Acoustic measurement devices

Acoustic measurements of suspended particles in water are based on two approaches. The first method is to measure the acoustic attenuation due to the suspended particles of an acoustic pulse passing through the water column (Figure S30). By making acoustic spectroscopy measurements on the received signal, the particle size distribution and the concentration within the water column can be derived (Schaafsma, 1992). However, estimates of the distribution of these parameters as a function of depth cannot be made.

The second approach is by interpreting backscatter, which is the scattering by the suspended particles back to the transducer (Figure S31). Anticipating the more detailed discussion in the next section, it can be stated that interpreting the backscatter provides information about suspended sediment parameters as a function of the acoustic path, i.e. in most cases as a function of depth. The method of interpreting the backscatter is well established for near-bed investigations to measure a concentration profile of well-sorted sand close to the bottom, typically using a resolution of around 100 depth cells with a height of 1.0–2.0 cm per cell (Young et al., 1982; Hess and Bedford, 1985; Varadan and Varadan, 1985; Hanes et al., 1988; Vincent et al., 1991).

Submersible one-point measurement devices

Based on the same physical laws of attenuation and backscatter which are used for acoustic measurement devices, optical sensors are now being applied (Hanes et al. 1988; Fahrentholz and Koske, 1992). In contrast to the remote sensing acoustic devices, they are generally designed to measure at the very point where they are positioned, 'point' meaning a 1–30 cm transmission path for optical sensors. Thus, if information along a measurement profile is necessary, they have to be monitored continuously. The attainable information about suspended particles is subject to the same restrictions as that given for monofrequency acoustic devices described below.

The most recent efforts using submersible prototypes of commercially available particle sizers are described in Bale and Morris (1987, 1991). This approach offers the opportunity to measure the particle size distribution directly. Similar results can be obtained by combining underwater photography with image processing, whereby both systems provide information about the position where they are moored.

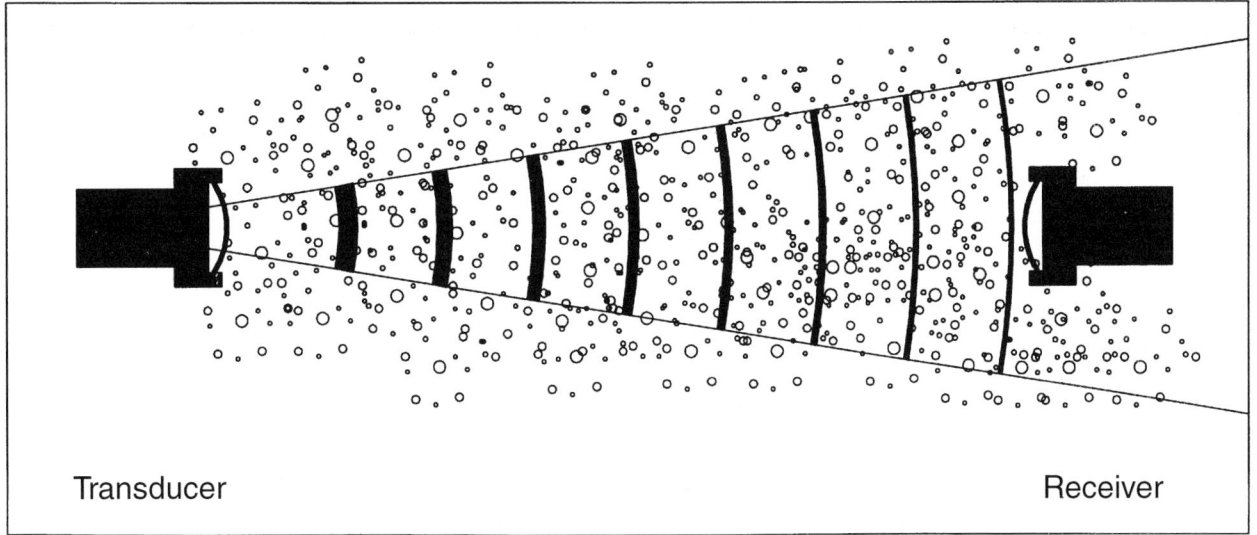

Figure S30 The use of acoustic attenuation for measuring suspended sediment.

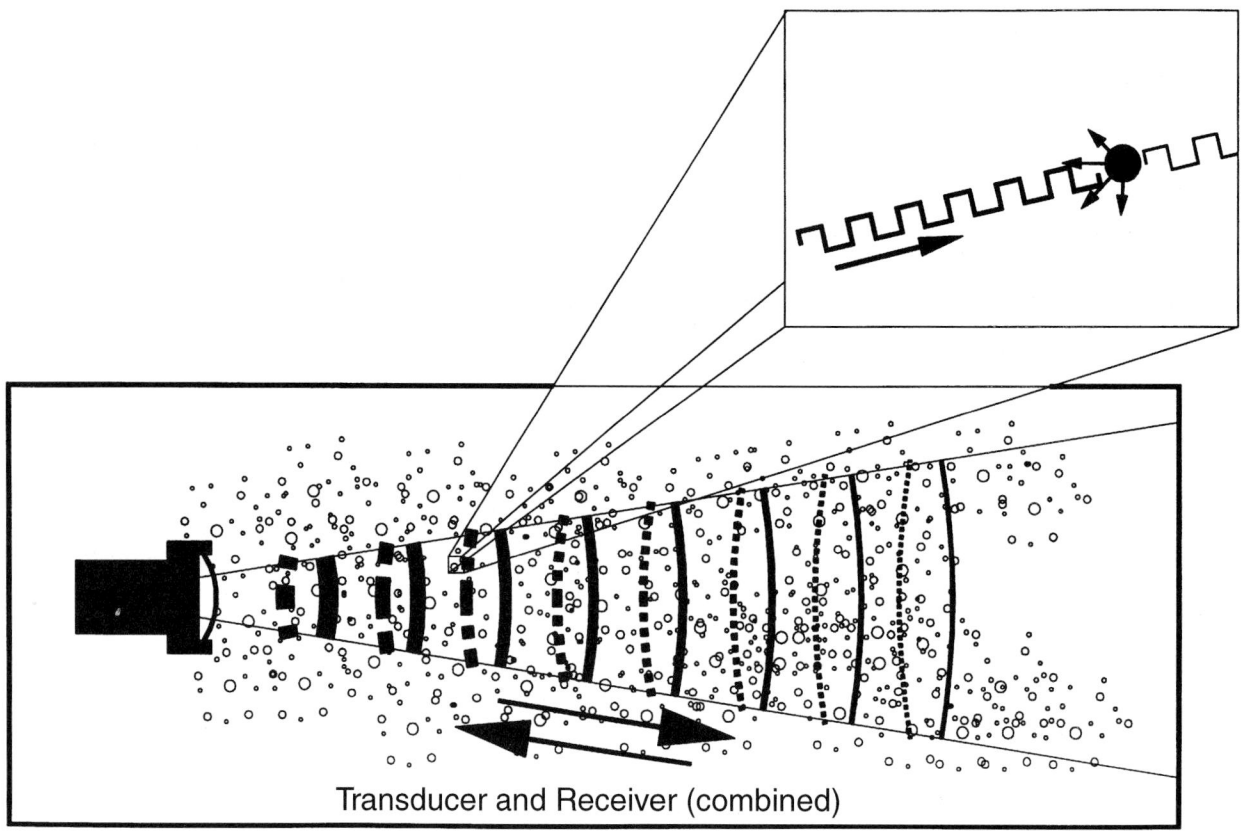

Figure S31 The use of acoustic backscattering for measuring suspended sediment.

ADCP methodology

Acoustic principles using acoustic backscatter

Active sonar equation
The evaluation of the acoustic backscattering for measuring suspended sediments is based on the active sonar equation (Urick, 1973):

$$EL = SL - 2TL - TS \text{ [dB]} \tag{S11}$$

where EL is the echo level (intensity of received signal), SL is the source level (intensity of emitted signal), TL is the transmission loss and TS is the target strength (intensity of signal echoed by a target).

The source level SL is a parameter determined by the equipment and therefore generally known or measurable. The transmission loss TL describes the weakening of sound on its path between the source of sound and a given distance in the medium, i.e. the distance between the source of sound and the backscattering particles. Transmission loss can be considered as the sum of a loss due to spreading and attenuation. Additionally, other losses, for example due to the shadowing of ensonified particles, may occur, but for both fresh water and seawater they are much smaller than the losses due to spreading and attenuation (Urick, 1973). Thus, as far as general problems of suspended sediment monitoring are concerned, additional losses can be neglected. General standard values to estimate the transmission loss are available in literature. To obtain more reliable results, however, calibration processes *in situ* are necessary.

Calculating the target strength
The real interest in suspended particles monitoring is focused on the target strength TS, which is the intensity of the signal echoed and scattered back to the source of sound by a target, i.e. by the particles. TS is a function of the concentration and the size distribution of the backscattering particles, of their properties – such as shape, density, compressibility and rigidity – and of the particle size to acoustic wavelength ratio. As long as the concentration of backscattering particles remains below a limiting concentration, the backscatter of

each individual particle can be considered to be independent from all the other particles (single-scattering theory). Ma *et al.* (1983) showed that this is suitable for a kaolin–water suspension up to about 8% volume concentration of kaolin. Up to the limit concentration the backscatter of an ensonified water column can be considered as the sum of the backscattering TS_i of the individual particles:

$$TS_i = Ts_i(d_i, \; d_i/\lambda_{sound}, \; a_1, a_2, a_3, \ldots) \tag{S12}$$

$$TS = \sum_{i=1}^{n} TS_i \tag{S13}$$

where TS_i is the backscatter of an individual particle, d_i is the diameter of an individual particle, λ_{sound} is the acoustic wavelength, $a_1, a_2, a_3,$ ... are the properties of individual particles such as shape, density, compressibility and rigidity, n is the number of particles within the ensonified water column and TS is the total backscattering of the ensonified water column.

Mono- and multifrequency approaches
The acoustic backscatter received by the measuring instrument represents an integrated signal for all scattering particles. Thus, as long as an acoustic device with a constant frequency is used (monofrequency approach), it is not possible to distinguish between the portion of the signal due to the particle concentration and the portion of the signal due to the particle size (respectively the particle size distribution). The interpretation of the data collected by monofrequency instruments depends greatly on *a priori* information about the scattering particles.

Considerable more information about the scattering particles is attainable when multifrequency techniques are applied. Since the acoustic response (target strength) of the particles is heavily dependent on the ratio of wavelength to particle diameter, it is advantageous to ensonify the water column with pulses of different frequencies. The particle size and the particle concentration can be obtained by solving the inverse problem for the backscatter signal recorded for all pulses. Relevant approaches are given in Ma *et al.* (1983) and Greenlaw and Johnson (1983).

ADCP measurement techniques

Acoustic Doppler current profilers (ADCPs) measure the three-dimensional components of the current velocity with high spatial and vertical resolution (Gordon, 1989). The ADCP transmits short acoustic pulses along narrow beams at a fixed frequency (available with a working frequency of 75–2400 kHz, depending on the transducer head). It receives and processes the echoes from successive volumes along the acoustic beams and determines the changes in frequency. Based on the Doppler effect frequency, shifts are used to calculate the relative velocity between ADCP and the particle scattering in the sonicated water column. As each beam measures only the velocity component parallel to the beam, the calculation of the three-dimensional vector of the current velocity requires multiple beams. Through range gating it is possible to compute the velocity $v(x, y, z_i)$ in a series of discrete ranges along each beam (depth cells), with x, y being the ADCP position, z_i the mean depth of a depth cell i, and i being the index of the available depth cells. The standard beam geometry as used in this study is given in Figure S32.

As a by-product of the current velocity measurements, the backscatter intensity (corresponding to the echo level EL) for each beam is recorded with the same resolution (depth cells) as the velocity data. By adding the terms for the transmission loss, a normalized 'range-corrected backscatter intensity' is obtained, which represents the integrated response of all particles within the single-depth cells. Due to the practically fixed frequency – the frequency can only be altered

by exchanging the transducers – the ADCP works as a mono-frequency instrument.

Calculating backscatter intensity for given suspended sediment data

In the literature, modeling approaches of different complexity are described to calculate the target strength (backscatter intensity) of particles within the sonicated water column. The selection of the model to be applied has to be based on the ratio of particle diameter to acoustic wavelength, on the properties of the scattering particles and, above all, on the data being available. Ma *et al.* (1983) state that, as far as field measurements are concerned, it is practically impossible to measure all particle characteristics needed to establish a deterministic relationship between target strength and particles. Thus, when more complex modeling approaches are used, an extensive database is needed for calibration, and vice versa, when the properties and characteristics of the scattering particles are not known in detail, the application of more complex models does not provide more distinct information than the application of simple models.

The simplest approach to calculating the theoretical backscatter intensity for given suspended particles is based on an elementary Rayleigh scattering model (Urick, 1973). This model is a rough, elementary approach which implies several deficiencies. Rayleigh scattering is based on the assumption of small, fixed and rigid spheres (Urick, 1973). Additionally, the Rayleigh scattering model is restricted to particles whose ratio of circumference to wavelength is much smaller than unity. Using a 1.2 MHz ADCP implies that the Rayleigh scattering model is applicable for particles with a diameter much smaller than 400 μm.

The Rayleigh scattering model considers the target strength TS_R as a function of the acoustic wavelength λ, the distribution function of particle size $F(d)$ and the particle concentration c:

$$TS_R = TS_R(\lambda, F(d), c) \tag{S14}$$

$$TS_R = I_0 \frac{\pi^2}{\lambda^4} \left(\frac{5}{2}\right) \sum_{i=1}^{n} \left[\frac{4}{3}\left(\frac{d_i}{2}\right)^3 \pi\right]^2 \tag{S15}$$

where I_O is the intensity of the incident signal at the location of the scatterers, d_i is the diameter of an individual particle i, λ is the acoustic wavelength and n is the number of particles. The intensity I_O can be calculated by subtracting the transmission losses due to spreading and adsorption from the ADCP's source level.

In the following, the theoretical backscatter intensity is exemplified by reflection for six hypothetical suspended sediment scenarios (Table S11; Figure S33). The scenarios were established by Reichel *et al.* (1993) considering limits and measurement results available in literature for the River Danube. They cover a range from low flow to flood conditions concerning typical suspended sediment concentrations and particle size distributions. According to laboratory analysis results, the particle sizes were assumed to be lognormally distributed (Table S11; Figure S33). The theoretical backscatter intensity TS_R was calculated based on the Rayleigh scattering model according to equation (S15).

Two parameters are used to judge the backscatter signal. The first represents the total backscatter signal denoted as TS_R. The range of the expected total backscatter signal for all scenarios compared to the dynamic range of the ADCP transducer shows whether the ADCP's measuring range covers all the suspended sediment situations of

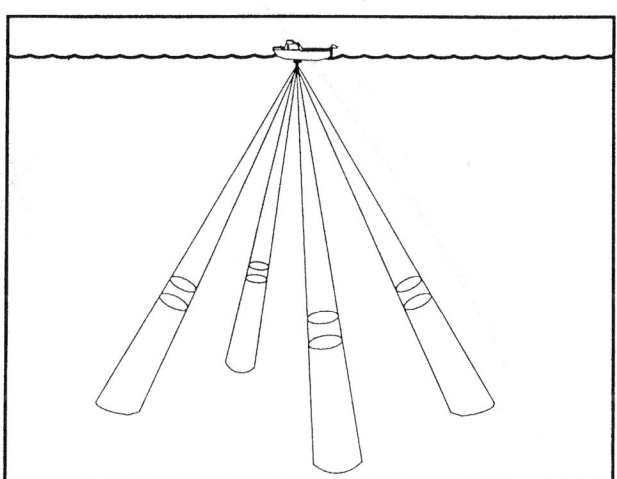

Figure S32 ADCP standard beam geometry and location of one depth cell. (Source: Gordon, 1989.)

Table S11 Presumed characteristics of six scenarios of suspended sediment and results of the computation of the theoretical backscatter signal

Scenario	d_{50} = median (μm)	$d_{97.5}$ (μm)	d_{99} (μm)	Presumed concentration (mg l⁻¹)	Backscatter intensity (dB)	d_{-3dB} (μm)	V_{-3dB} (%)
HW, flood event	13.5	305	550	15 000	111.7	146	2.2
NW1, low flow	4.5	48	75	3.0	63.7	98	0.4
NW2, low flow	2.7	30	46	2.0	60.4	70	0.2
SS001, low flow	1.6	26	42	5.0	75.6	87	<0.1
SS039, mid-flow	2.7	48	80	38.1	67.6	123	0.4
SS066, mid-flow	2.8	65	112	3.8	2.8	130	0.3

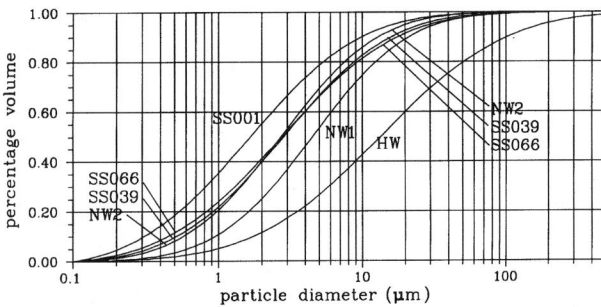

Figure S33 Presumed particle size distributions for six scenarios.

interest. The second parameter, denoted as V_{-3dB} serves to illustrate the dependence of the backscatter signal on larger particles. It is based on the fact that a 3 dB change of acoustic signal is equivalent to a doubling or halving of the signal's intensity. Thus, to obtain the second parameter V_{-3dB} the diameter d_{-3dB} was calculated such that the signal of all particles smaller than d_{-3dB} equals the signal of all particles larger than d_{-3dB}. The parameter V_{-3dB} represents the percentage volume of all particles larger than d_{-3dB}. The results of these calculations are summarized in Table S11.

It appears that the span between the lowest and the highest backscatter signal is much smaller than the dynamic range of the ADCP transducers (55 dB versus 90 dB given in RDI, 1991). Thus, when the uncertainties induced by the scattering model are also included, the ADCP should cover all suspended sediment situations of interest.

The percentage V_{-3dB}, however, appears to be very small. It is less than 0.4% for all low and mid-flow scenarios and 2.2% for the flood scenario (Table S11). The meaning of this can be illustrated by a simple example. The doubling of the concentration of particles smaller than d_{-3dB} (this is nearly equivalent to a doubling of the suspended sediment concentration) results in the same changes of the signal as a doubling of the concentration of the particles larger than d_{-3dB}, which is equivalent to an increase of suspended sediment concentration of several per cent.

Example of ADCP field measurements

In the expectation of the ADCP being a suspended sediment measurement tool, the results of the theoretical considerations presented above induce some uncertainty. Regarding the ADCP as an instrumental monitoring tool, however, the following results prove its suitability.

The data were collected during a research project in a reservoir in the Austrian section of the River Danube (Müller *et al.*, 1992). The aim of this project was to investigate the impact of this reservoir on the suspended sediment transport. During the project, water samples of 50–120 l were collected during three campaigns. Suspended particle concentrations were obtained by drying and weighing. The particle size distribution was determined by wet sieving with a 60/40 μm ultrasonic sieve and applying a Sedigraph 5000ET on particles smaller than 40 μm.

In addition to taking water samples, ADCP measurements were performed in selected river cross-sections. Figure S34 shows one of five recordings of the backscatter data collected within a river cross-section (at km 1982.0). The time required to record the data was around 3 min. The total range of the backscatter signal was 12 dB. Clearly, a stratification appears showing the highest backscatter signal in the northern parts of the transect (left side) and an increase of the backscatter intensity as a function of depth. This structure was

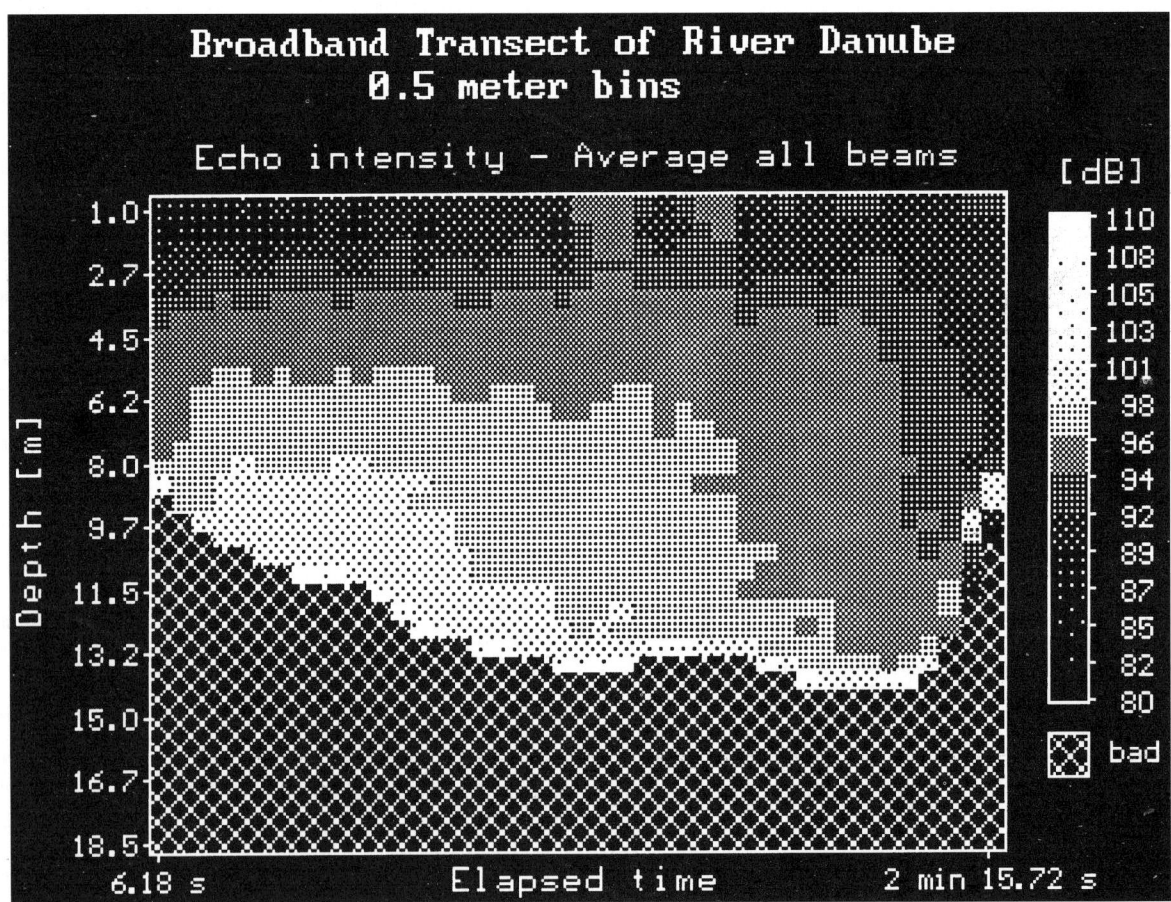

Figure S34 Recorded backscatter signal for a river transection (north to south).

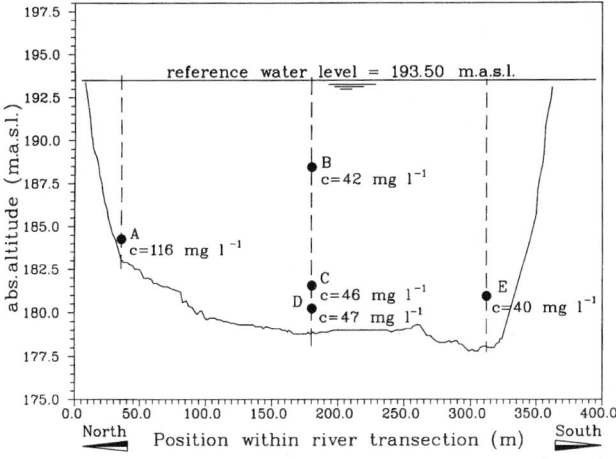

Figure S35 Concentration of suspended sediments within the river transection shown in Figure S34. (Concentrations were obtained by centrifugation of 50–120 l samples.)

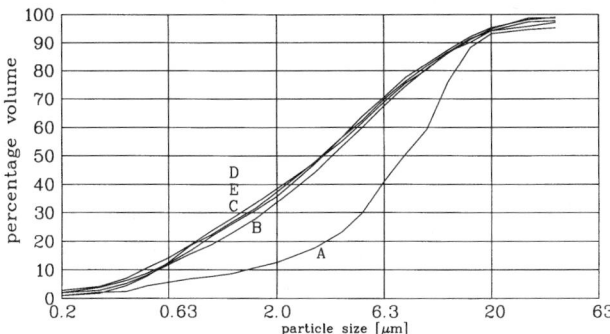

Figure S36 Particle size distribution for water samples from the river transection shown in Figures S34 and S35.

Table S12 Suspended sediment characteristics and ADCP backscatter intensity within the cross-section (see Figures S35 and S36)

Sample location	Suspended sediment concentration (mg l⁻¹)	$d_{50} =$ median (μm)	d_{95} (μm)	Backscatter intensity (dB)
A	116	5.16	40.26	99
D	47	3.23	26.39	97
C	46	2.87	26.38	96
B	42	3.09	25.72	94
E	40	2.01	24.57	94

consistent during the whole period of 30 min when the five data sets were recorded within this transect.

The suspended sediment concentration and the particle size distribution for five water samples corresponding to the backscatter data presented in Figure S34 are given in Figure S35 and S36. The suspended sediment concentration at the northern, near-bottom sample position is 2.5–3 times higher than at the other positions. In addition, the particle size distribution at this position shows a significant proportion of larger particles (Figure S36, graph A). As a conclusion, a good correlation between backscatter data and the sample processing results was found, whereby the changes of the backscatter signal result from both concentration and particle size effects (Table S12).

Summary and conclusions

The ADCP is designed as a remote-measuring current velocity meter. As a by-product the ADCP also yields information for suspended sediment monitoring. Working as a monofrequency instrument, the ADCP shows substantial limitations as far as suspended sediment measuring is concerned. The backscatter intensity represents the integrated response of all particles within the ensonified depth cell. Effects due to the particle concentration and the particle size distribution cannot be separated in principle. A monofrequency measurement can only yield a coarse or very qualitative feature of the size distribution of particles and cannot separate parameters such as grain size distribution and concentration. The interpretation of the data using this method will depend greatly on additional information about the scattering particles.

Due to the wide range of particle sizes characterizing most suspended sediments, only a very small percentage of the particles dominate the backscatter signal. Thus, even if the theoretical distribution function underlying the particle size and the concentration is known, it is almost impossible to estimate reliable parameters of the particle size distribution.

In comparison with standard methods for suspended sediment measurements, the ADCP shows essential advantages. While methods which require sampling and processing of water samples are demanding in both time and personnel, the ADCP provides a prompt qualitative survey of the suspended sediment transport with high spatial and temporal resolution. In addition, the ADCP functions by remote sensing applicable under extreme hydrological conditions such as flood events.

The sonicated water volume represented by the backscatter signal varies as a function of instrument settings and evaluated data (one to four beams). Volumes may be varied within the range of 2.5–50 l whereby greater volumes are equivalent to averaging smaller volumes. Due to this averaging the ADCP provides more representative information than small bottle samples.

Compared with existing field measurements, the ADCP backscatter signal is in good agreement with laboratory analysis results. This means that changes in the backscatter signal indicate changes of the particle concentration and/or the particle size distribution. Thus the ADCP can serve as a valuable tool for detecting and monitoring spatial and temporal variabilities in the suspended sediment transport and for determining sample sites. It can be instrumental in temporal interpolation and regionalization between point values of parameters obtained by standard monitoring methods. In addition, when detailed information about the particles being monitored is available (i.e from calibration processes) and when the size distribution of these particles does not change, the ADCP's backscatter signal can be processed to derive reliable solids concentrations. A software package for data collection, calibration and visualization of suspended monitoring for such conditions is under development at present.

Günther Reichel

Bibliography

Bale, A.J. and A.W. Morris, 1987. *In situ* measurements of particles in estuarine waters. *Estuarine, Coastal and Shelf Sciences*, **24**, 253–263.

Bale, A.J. and A.W. Morris. 1991. *In Situ* measurements of suspended particles in estuarine and coastal waters using laser diffraction, in Syvitski, J.P. (ed.), *Principles, methods, and application of particle size analysis*, Cambridge University Press, Cambridge, pp. 197–208.

Fahrentholz, B. and P. Koske, 1992. A multi frequency ultrasonic acoustic scatterometer for the detection of suspended matter in coastal waters, in *International Coastal Congress*, Kiel, pp. 173–176.

Faudot, 1973. Etude de la finesse des poudres. *Informations Chimie*, no. 136.

Gardner, W.D., 1977. Incomplete extraction of rapidly settling particles from water suspensions. *Limnologie and Oceanographie*, **22**, 764–768.

Gibbs, R.J., 1981. Floc breakage by pump. *Journal of Sedimentary Petrology*, **52**, 657–670.

Gordon, R.L., 1989. Acoustic measurement of river discharge. *Journal of Hydraulic Engineering*, **115**(7), 925–936.

Greenlaw, C.F. and R.K. Johnson, 1983. Multiple-frequency acoustic estimation. *Biological Oceanography*, **2**(2–4), 227–252.

Hanes, D.M., C.E. Vincent, D.A. Huntley and T.L. Clarke, 1988. Acoustic measurements of suspended sand concentration on the C2S2 experiment at Stanhope Lane, Prince Edward Island. *Marine Geology*, **81**, 185–196.

Harfield, J.G. and B.V. Miller, 1984. Improved agreement between Coulter Counter and sedimentation for porous particles, in *Fine Particle Society, Annual Meeting*, Orlando.

Hess, F.R. and K.W. Bedford, 1985. Acoustic backscatter system (ABSS): The instrument and some preliminary results. *Marine Geology*, **66**, 357–379.

Ma, Y., V.K. Varadan, V.V. Varadan and K.W. Bedford, 1983. Multifrequency remote acoustic sensing of suspended material in water. *Journal of the Acoustical Society of America*, **74**(2), 581–585.

Matthews, M.D., 1991a. The effect of pretreatment on size analysis, in Syvitski, J.P.M. (ed.), *Principles, methods and application of particle size analysis*, Cambridge University Press, Cambridge pp. 34–42.

Matthews, M.D., 1991b. The effect of grain shape and density on size measurement, in Syvitski, J.P.M. (ed.), *Principles, methods and application of particle size analysis*, Cambridge University Press, Cambridge pp. 22–33.

McCave, I.N. and J.P.M. Syvitski, 1991. Principles and methods of geological particle size analysis, in Syvitski, J.P.M. (ed.), *Principles, methods and application of particle size analysis*, Cambridge University Press, Cambridge pp. 4–21.

Müller, H.W., H.P. Nachtnebel, B. Schwaighofer and G. Reichel, 1992. Schwebstoffanalyse und -bilanz in Flußstauhaltungen (Suspended sediments in river reservoirs: analysis and balance). Österreichische Verbundgesellschaft, Vienna.

RDI, 1991. *Direct-reading broadband acoustic Doppler current profiler*, Technical Manual, San Diego.

Rees, T.F., J.A. Leenheer and J.F. Ranville, 1991. Use of a single-bowl continuous-flow centrifuge for dewatering suspended sediments: effects on sediment physics and chemical characteristics. *Hydrological Progresses*, **5**, 201–214.

Reichel, G., H.P. Nachnebel and P. Schreiner, 1993. Zur Anwendung eines BB-ADBP in flachen Fließgewässern und Stauräumen (Applying a BB-ADBP in shallow rivers and river reservoirs). *Österreichische Wasserwirtschaft*, **45**(1–2), 24–35.

Schaafsma, A.S., 1992. *In situ* acoustic attenuation spectroscopy of sediment suspensions, in *International Coastal Congress*, Kiel, pp. 177–180.

Swift, D.J.P., J.R. Schubel and R.W. Sheldon, 1972. Size analysis of fine-grained suspended sediments: a review. *Journal of Sedimentary Petrology*, **42**(1), 122–134.

Syvitski, J.P. (ed.), 1991. *Principles, methods, and application of particle size analysis*, Cambridge University Press, Cambridge.

UNESCO, 1983. *Study of the relationship between water quality and sediment transport*. Technical Papers in Hydrology, UNESCO, Paris.

Urick, R.J., 1973. *Principles of underwater sound*, 3rd edn, McGraw-Hill, New York.

Varadan, V.V. and V.K. Varadan, 1985. Theoretical analysis of the acoustic response of suspended sediment of HEBBLE. *Marine Geology* **66**, 267–276.

Verhoeven, R., P. Verdonck, D., Fransaer and J. Van Rensbergen, 1989. A comparative study of three methods of sediment transport measurements, in *Workshop on Instrumentation for Hydraulic Laboratories*, Canada, pp. 259–266.

Vincent, C.E., D.M. Hanes and A.J. Bowen, 1991. Acoustic measurements of suspended sand on the shoreface and the control of concentration by bed roughness. *Marine Geology*, 1–18.

Young, R.A., J.T. Merrill, T.L. Clarke and J.R. Proni, 1982. Acoustic profiling of suspended sediments in the marine bottom boundary layer. *Geophysical Research Letters*, **9**, 175–178.

Cross references

Flow measurement: new technology
Streamflow measurement

T

THAMES BARRIER

The River Thames flooding has had a long established history. Notable floods include those of 1663, 1791, 1834, 1852, 1874, 1875, 1881, 1928 and 1953. During the 1953 tidal surge, walls and embankments failed in the outer Thames estuary and along the North Sea coast causing a massive 'escape' of water heading for London. Sadly 300 people in East Anglia, Essex and Kent lost their lives during this event. If this flood had reached the capital's highly populated low-lying areas ($1\frac{1}{4}$ million people over 324 km^2) the result could have been catastrophic. The reasons for Thames flooding risk are that surge tides, which are the special threat, occur under certain meteorological conditions. When a trough of low pressure moves eastwards across the Atlantic towards the British Isles, the sea beneath it rises above the normal level, thus creating a 'hump' which moves eastwards with the depression. If the depression passes the north of Scotland and veers southwards into the North Sea, extremely dangerous conditions may be created. A surge happens when this mass of water coming from the deep ocean reaches the relatively shallow southern part of the North Sea. The height of the surge may be further increased by strong northerly winds.

A second, less serious threat arises when a depression travels eastwards up the English Channel. This kind of surge is smaller than one that comes from the north of Scotland.

If a high surge coinciding with a high 'spring' tide (spring tides occur twice in each month) reaches the bottleneck of the Straits of Dover and enters the Thames Estuary, then there could be a real flood danger along most of the tidal Thames. The adverse trend in water levels steadily increases this possibility of flooding.

Such a flood in London could have paralyzed the central part of the London Underground, knocked out freshwater and sewer systems, power, gas and vital telephone and data services and severely hit thousands of homes, shops, factories, businesses and buildings. It could have taken months to get London functioning normally again. The cost of a major flood could have been enormous – easily more than $2000 million and that is not counting the sheer human misery, suffering and loss of life.

The risk of flooding, however, is slowly and steadily increasing due to a combination of facts. There is a steady increase in tide levels caused by a combination of factors including tilting of the British Isles, with the southeastern corner tipping downwards, the settlement of London on its bed of clay, increasing mean sea levels, increasing tidal amplitude, and because of events linked to global warming. As a result high tide levels are rising in central London by about 75 cm per century.

The traditional solution has been to raise and strengthen river walls and embankments. Following the Thames Flood Act of 1879, long stretches of the river bank were so treated, and again after the 1928 flood the banks were raised between 1930 and 1935. This was the last major wall raising until 1971, and because of the factors already described the risk was by then already greater than in the 1930s.

Wall raising has many advantages as a solution. Walls are permanent, easy to maintain and not likely to fail through human error. On the other hand, to build walls of ever mounting height would increasingly shut out the Thames from the view of Londoners and mar the beauty of London's river line for the 27 million tourists who use the river to explore London each year.

Public feeling is very much in favor of improving the Thames and the riverside areas, and although it would be possible to redevelop all the sites along the river with high defences and still preserve views and local beauty spots, it was calculated that this piece by piece solution would have taken many years.

The comprehensive flood prevention strategy therefore went beyond wall raising and had as a key feature a movable flood barrier. The years of research led to the conclusion that it should be a rising-sector gate barrier and that it should be sited across the Thames in the Woolwich Reach (Figure T1). This solution, combined with carefully designed bank raising where necessary, gave the desired protection, could be built in the shortest time and would cause relatively little interference with river traffic.

Royal Assent to the Act giving power to the GLC and Water Authorities to go ahead was given in August 1972. Some 32 km of flood defences were built downstream of the Barrier, with bank levels 2 m higher than previously existed. To improve London's defences against flooding whilst the main defences were being constructed, 102 km of interim bank raising from Putney to Purfleet were carried out in 1971–1972. These included the 60 m high Barking Barrier. This has a drop gate which is held out of the water when not in use, to allow uninterrupted passage by commercial shipping using Barking Creek.

Defences upstream of Putney on the south bank and Hammersmith on the north were also raised to give the same degree of protection as in Central London and to reduce the risk of flooding when high upland flows coincide with high tides which are not sufficiently large to warrant a Barrier closure.

The building of attractive river walks as part of the permanent bank reconstruction, and similar riverside improvements, were constantly in mind as details of work were planned and carried out. Particular attention was also given to ensuring the continued potential of existing riverside commerce and industry.

Basically, the rising-sector gate Barrier is a series of separate moveable gates positioned end-to-end across the river (Figure T2). Each gate is pivoted and supported between concrete piers which house the operating machinery and control equipment.

Closing the Barrier when required seals off part of the upper Thames from the sea. When not in use, the gates rest out of sight in curved recesses in concrete sills in the river bed, allowing free passage of river traffic through the openings between the piers (Figure T2).

If a dangerously high tidal surge threatens, the gates swing up through about 90° from their river-bed position, forming a continuous steel wall facing down-river ready to stem the tide. Further rotation of the gate to the maintenance position renders every part accessible for maintenance (Figure T2).

Figure T1 The Thames Barrier: view looking upstream.

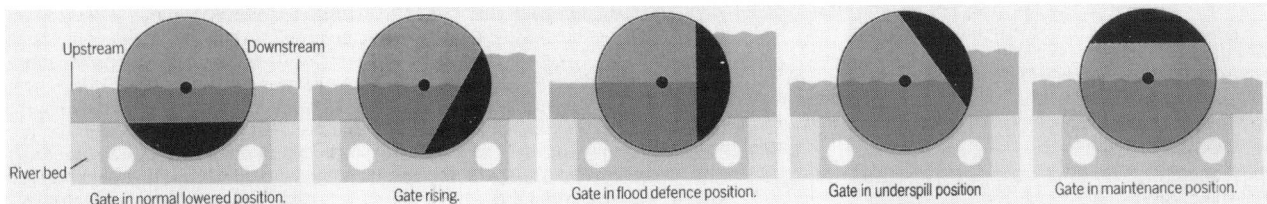

Figure T2 Operating gate positions of the Barrier.

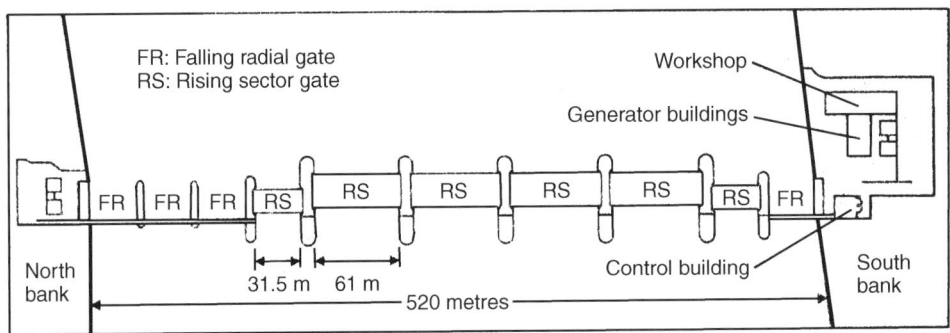

Figure T3 Plan view of the Barrier showing layout of gates.

The width of the Barrier from bank to bank is about 520 m with four main openings each having a clear span of 61 m (Figure T3). The four main gates are massive. Each gate is designed as a hollow steel-plated structure over 20 m high and weighing, with counterweights, about 3700 tonnes. Each is capable of withstanding an overall load of more than 9000 tonnes.

A total of 65 staff operate and maintain the Barrier. Dangerous conditions can be forecast about 12 h ahead. The decision to close the Barrier is taken by the Barrier Controller. This is derived from the predicted height of the incoming tide estimated by the East Coast Storm Tide Warning Service (ECSTWS) based at Bracknell, together with information from the Barrier's own sophisticated computer analysis of the situation. The ECSTWS monitors tides along the east coast and issues warnings of dangerously high waters. These estimates are based on meteorological information from oil rigs in the North Sea and the land-based meteorological stations. They also receive tide level readings from recorders as far away as Stornoway in the Western Isles and Wick in the north of Scotland.

Barrier closure will probably take place about 1 h after low water or about 4 h before the peak of the incoming surge tide reaches the site. Before closing the Barrier, staff inform the Port of London Authority navigation service. They in turn inform shipping by radio, and in addition notice boards upstream and downstream of the site are illuminated. Lights on the Barrier itself also indicate that closure is to take place. The gates, controlled and powered from the southern shore take only 30 min to close against the threatened surge. The cost of operating and maintaining the Barrier is estimated to be $5 million per year.

The barrier and its associated works formed a massive project. The final cost of the flood defences within the Thames Water area of responsibility was in the order of \$800 million (1984). Taxpayers (central government through the Ministry of Agriculture, Food and Fisheries) met 75% of the approved costs and ratepayers were responsible for the remainder.

The Barrier project took 10 years to complete and was undertaken by the GLC Department of Public Health Engineering with consulting engineers Rendel Palmer and Tritton responsible for the design of the Barrier and for the supervision of construction. The main civil engineering contractor for the Barrier river works was the Costain, Tarmac, Hollandsche Beton Maatschappij Joint Venture. The contractor for the steel gates and operating machinery was the Davy Cleveland Barrier Consortium. The shore works contractor was Sindall Construction Ltd and the main services contractor Balfour Kilpatrick Ltd.

The Department itself designed and supervised some 23.5 km – 18.5 km downstream and 5 km upstream – of bank raising works with construction being carried out by a number of contractors. The Barrier was opened by Her Majesty the Queen on 8 May 1984, and since 1 April 1996 responsibility has passed to the new Environment Agency. Since its opening, and up to April 1997, the Barrier has been closed on 30 occasions, other than for testing. The most recent closure and highest tide in the Barrier's history was on 29 October 1996 when hurricane Lili threatened to drive the tide to within 200 mm of wall tops in Central London.

R. W. Herschy

Source

The National Rivers Authority, Thames Region, 1995. *The Thames Barrier, A Visitor's Brochure*.

Bibliography

Gilbert, S. and Horner, R., 1984. *The Thames Barrier*, Thomas Telford, London.
The Thames Barrier Design, 1978. The Institution of Civil Engineers, London.

Cross references

Dams
Floods
Flood studies for the British Isles

THAMES, RIVER

The River Thames commands a unique place in the history and culture of the UK and, despite its modest size, it is also one of the most intensely studied of the world's rivers. The Thames catchment has contributed substantially to the understanding of human impact on hydrological processes and the development and management of complex water resources systems.

The traditional source of the Thames is near Cirencester in the Cotswold Hills. It flows 382 km to its tidal limit at Teddington Weir and forms an estuary from the Thames Barrage (flood barrier) to the North Sea. Average rainfall over the catchment above Teddington

(9950 km^2) is about 710 mm, distributed fairly evenly throughout the year but with a slight tendency towards a maximum in the fall. Evaporation losses, which on average account for around 65% of the rainfall, are concentrated in the April–September period and impose a marked seasonality on the flow regime, the average January flow being over four times the August mean. Throughout the catchment, relief is relatively subdued but outcrops of major aquifers – the Inferior and Great Oolite and the Chalk and Upper Greensand – contribute to the characteristic scarp-and-vale topography. Groundwater outflows provide, on average, about half the total flow of the Thames; this baseflow sustains the river through dry summers and more sustained droughts.

Historically, the Thames was a major source of power for the many mills along its length. Few remain but today the river is heavily utilized for water supply purposes. The river contributes over half of the water used in the Thames Valley and accounts for around 70% of London's needs. On average, the volume of water abstracted from the lower Thames is the equivalent of the August gauged flow. This water is transferred to major bankside reservoirs which provide a substantial measure of security against drought. This security has been increased by the linkage and joint management of these reservoirs with those in the neighboring River Lee catchment, and by the greater operational flexibility in meeting water demands resulting from the completion of the London Ring Main. The Thames with its tributaries is one of the most intensively used river systems in the world with over 3500 abstractions and about 300 sewage treatment works, supporting a population of about 13 million (Sexton, 1988).

The evolving pattern of water utilization has inevitably had a profound impact on the flow regime. Abstractions from the lower Thames have increased from around 3 m^3 s^{-1} in the 1880s to over 24 m^3 s^{-1} now. Groundwater abstraction and a mid-catchment pumped storage reservoir provide (via effluent returns) a measure of flow regulation. Land-use changes, especially urbanization and changes in agricultural practice, impinge rather more subtly on the regime. Agricultural drainage changes (particularly in the nineteenth century) and river management changes have affected the average stream velocity and the river's character. In terms of impact on the community, however, perhaps the most significant historically was the discharge of untreated sewage into the tideway following the invention of the water closet. This resulted in epidemics of cholera and typhoid, during the 1850s especially, and stimulated the introduction of several pioneering conservation and pollution legislative procedures, paving the way for further measures and better management practices which ultimately allowed salmon to return to the Thames in the 1970s after a gap of 150 years.

Flows on the Thames, initially based on float velocities, have been measured since the 1850s but a daily hydrometric record did not commence until Teddington Weir was rebuilt in 1883. Although this usefully increased the weir's capacity, its principal function was to maintain levels for navigation purposes; accuracy, especially in the high and low flow ranges, was limited. Over the ensuing 70 years the measuring structure was subject to a number of important changes which contributed to more precise flow measurement. Nonetheless, with any structure as complex as Teddington Weir – it was a barrage of sluices and gates over 200 m in total length – leakage could be very substantial and further losses occurred through lockages and via the lock gates themselves. In the 1970s an ultrasonic gauging station was successfully commissioned in the Kingston reach, 3 km upstream of

Table T1 Naturlised flows of the River Thames

9–month period ending month/year	Mean flow (m^3 s^{-1})	% lta[a]	18-month period ending month/year	Mean flow (m^3 s^{-1})	% lta[a]	24–month period ending month/year	Mean flow (m^3 s^{-1})	% lta[a]
12/1921	19.1	24	11/1934	24.6	31	05/1935	38.7	50
11/1934	21.7	28	10/1944	25.2	32	04/1903	40.2	51
10/1944	21.9	28	11/1902	29.4	38	08/1891	40.9	52
09/1976	23.7	30	09/1891	33.7	43	03/1992	4.81	53
11/1898	28.6	37	11/1976	34.5	44	06/1945	42.5	54
02/1902	28.9	37	10/1992	35.6	46	07/1923	43.0	55
12/1893	30.0	38	10/1991	39.6	51	09/1906	44.1	56
12/1990	30.6	39	11/1948	40.0	51	09/1949	45.5	58

[a] Mean flow as a percentage of the long-term average (78 m^3 s^{-1}).

Teddington, and in 1986, when a multipath version was installed, the ultrasonic station became the principal monitoring site.

In order to provide a more objective basis on which to develop management procedures, assess human impact on the flow regime and identify any significant trends, a 'naturalized' flow series has been maintained which takes account of the major abstractions upstream of London. Neither the long-term catchment rainfall nor naturalized runoff plots for the Thames show any overall trend, but the series are characterized by several extended periods of below, or above, average runoff.

Towards the end of the exceptionally intense drought of 1975–1976, flows over Teddington Weir effectively ceased; there is no known historical precedent for such an occurrence. During the more recent protracted drought of 1988–1992, minimum flows remained substantially higher but accumulated runoff totals were extremely depressed (Marsh et al., 1994). Comparisons with historical drought and flood events need to be undertaken with caution: the Thames is very different in character from 200 years ago. Nonetheless, as a consequence of the underestimation of low flows throughout most of the pre-1951 record (Littlewood and Marsh, 1995), the significance of these recent minima is certainly greater than is implied by the rankings presented in Table T1. The recent drought served to highlight the problems associated with the over abstraction of groundwater in parts of the Thames Basin; the shrinkage in the stream network was without modern parallel and helped initiate the investigation of alleviation strategies whereby reduced groundwater abstraction in sensitive headwater areas would contribute to increased spring flows and the re-creation of a healthy aquatic environment.

Difficulties also attend comparisons between the impact of contemporary and historical high-flow events. Washlands have been drained and the ancient bridges and weirs which restricted flows in the Thames have been superseded as part of an ongoing improvement to arterial drainage. Most significantly, during the twentieth century there has been increasing development on the floodplain; this both restricts the storage area available for overbank flows and, as a result of urban and commercial drainage networks, speeds storm runoff to the main channel. During the 1987–1995 period there was a cluster of notable winter spates on the Thames, but most of these have been only around half of the 1947 March peak when, after a lengthy cold spell with substantial snowfall, rainfall triggered a thaw on frozen ground causing very severe flooding. In the gauged flow record this event is eclipsed only by November 1894, but flood marks on bridges and riverside building testify to a number of extreme events, notably in 1768, 1774, 1809 and 1821.

An important consequence of floodplain development is that the number of properties, and the scale of economic activity, at risk is far greater than even 50 years ago. Sophisticated flood warning measures are being continually refined and the National Rivers Authority's planned Maidenhead Flood Relief Scheme will be one of the largest river engineering projects in the UK, providing a pioneering example of a modern, more environmentally sympathetic approach to flood alleviation.

T.J. Marsh

Bibliography

Littlewood, I.G. and Marsh, T.J., 1995. A re-assessment of the monthly naturalised flow record for the River Thames at Kingston since 1883 and the implications for the relative severity of historical droughts, in *Regulated Rivers: Research and Management*.
Marsh, T.J., Monkhouse, R.A., Arnell, N.W. *et al.*, 1994. *The 1988–92 Drought*. Hydrological Data UK Series, Institute of Hydrology, Wallingford, UK, 84 pp.
Sexton, J.R., 1988. Regulation of the River Thames, a case study on the Teddington flow proposal. *Regulated Rivers: Research and Management*, **2**, 323–333.

Cross references

Flood studies in the British Isles
Floods
Floods: world's maximum observed
Thames Barrier

THAMES WATER RING MAIN

The Thames Water Ring Main was built to improve the reliability of potable water supplies to some 6 million Londoners across an area of 1500 km². The capacity of the 2.5 m average diameter main is 130 Ml day⁻¹. The main has 21 shafts of which five take water from water treatment works and 11 are pumping stations transferring water into local mains supply networks. The main, which was completed in 1994 at a cost of £250 million ($375 million), is 80 km long – the longest water tunnel ever built in the UK – with an average depth of 40 m (Figure T4).

Thames Water, the largest water company in the UK, supplies 2600 million liters of treated water daily to over 7 million customers and treats sewage from nearly 12 million. 190 000 dry tonnes of sewage sludge per annum is utilized or dispersed and 60% of this is recycled to agriculture land as fertilizer. Renewable energy generated from sludge gas provides over 200 GWh or 25% of the annual electricity consumption.

In 1994 a record salmon run and 115 fish species were recorded, making the Thames one of the cleanest metropolitan rivers.

R.W. Herschy

Bibliography

Dickens, W.J. and Bensted, I.H., 1988. The London Water Ring Main, *Proc. Inst. Civ. Eng. Part 1*, **84**, 445–474.
The London Water Ring Main, 1994. Institution of Civil Engineers, Special Issue 2, **102**, suppl., 82 pp.

THIESSEN POLYGON

The Thiessen polygon (Figure T5) is a commonly used methodology for computing the mean areal precipitation for a catchment from raingauge observations which was presented by A.H. Thiessen (1911). The Thiessen method is based on the assumption that measured amounts at any station can be applied halfway to the next

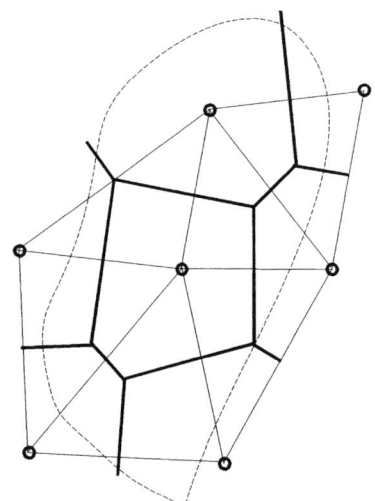

   o   rain gauges

------- catchment boundary

—— lines joining nearby stations

—— Thiessen Polygon network

Figure T5 Example of a Thiessen polygon.

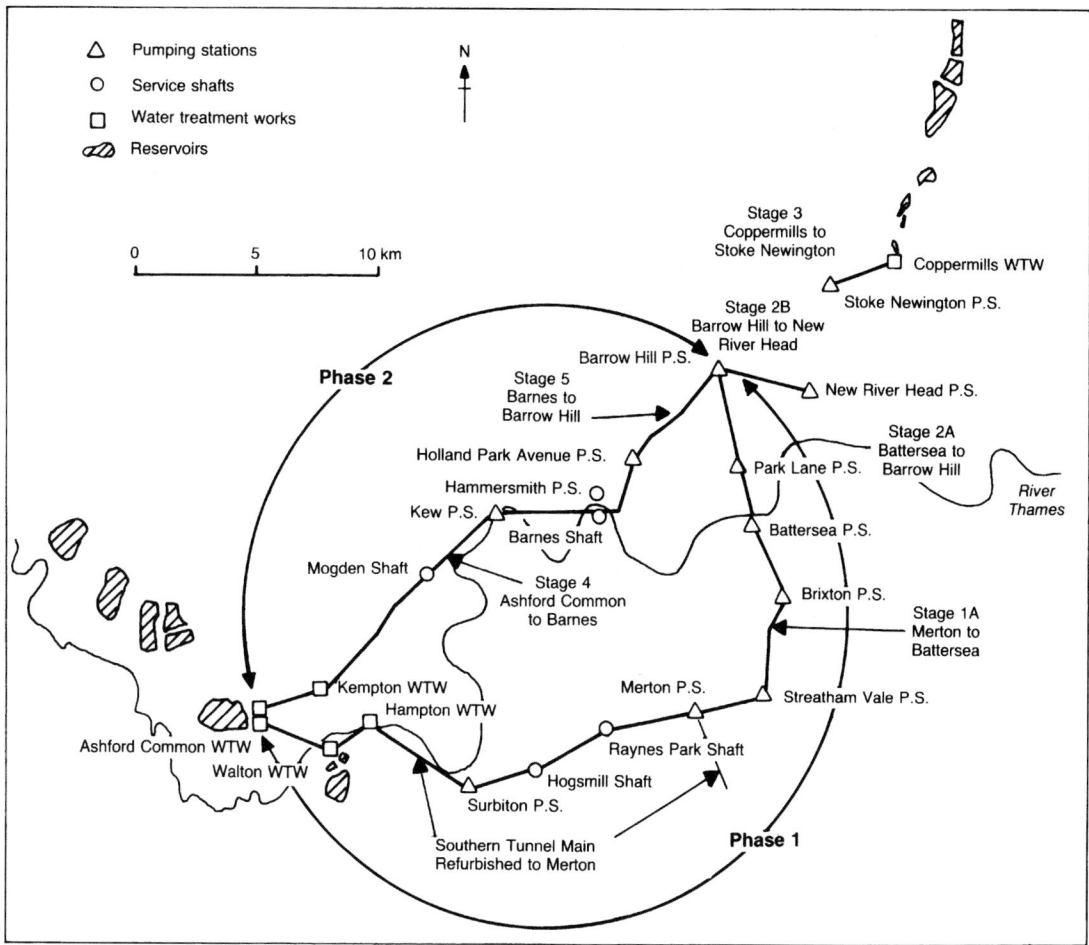

Figure T4 Route plan of the Thames Water Main. (From Dickens and Bensted, 1988.)

station in any direction, which means that for any point rainfall is equal to the observed rainfall at the closest gauge. The weights of the rain gauges are computed by their relative areas, which are estimated with the Thiessen polygon network. The polygons are formed by the perpendicular bisectors of the lines joining nearby stations. The area of each polygon is used to weight the rainfall amount of the station in the center of the polygon. If the amount for any station is missing, the polygon must be changed. The Thiessen method is unable to consider orographic differences in rainfall distributions.

A.H. Schumann

Bibliography

Thiessen, A.H., 1911. Precipitation for large areas, *Monthly Weather Review*, **39**, 1082–1084.

Cross references

Precipitation
Rain
Raingauge

TRAFALGAR SQUARE BOREHOLE, LONDON, UK

The Trafalgar Square borehole penetrates the Chalk and Upper Greensand aquifer in central London (UK). Groundwater levels have been monitored since the borehole was commissioned in 1844 – sporadically over the early part of the record – and the historical series provides a singularly important record of human impact on groundwater resources.

The Chalk and Upper Greensand aquifer forms part of the London Basin syncline outcropping extensively along the western and northwest southern boundaries of the basin. To the east, the Chalk dips below progressively younger strata and at Trafalgar Square is confined beneath 60 m of Lower London Tertiary deposits, London Clay and superficial deposits. In the unconfined Chalk, the water table fluctuates seasonally in response to natural recharge, but variations reduce gradually away from the outcrop and below London water level changes are dominated by the effect of groundwater pumping.

An artesian situation existed throughout much of the London Basin prior to the sinking of the first deep boreholes in the eighteenth century. The burgeoning growth of the metropolis was accompanied by increased domestic and industrial demand for water. In addition, major land drainage projects and the the loss of large tracts of open land to accommodate the urban sprawl reduced the potential for significant local recharge. By the 1820s the rapidly growing demand for groundwater exceeded the natural replenishment to the Chalk below London. The consequent decline in groundwater levels was first noted by Clutterbuck (1848). His exhortations to exploit more prudently what was widely regarded as an inexhaustible resource had little impact and the severe pollution of the Thames in the mid-nineteenth century provided a further impetus to groundwater development. London's demand for groundwater increased over sevenfold in the 100 years to 1920 by which time the water table had fallen around 45 m.

The decline in groundwater levels continued through World War II during which extensive damage was inflicted on both the wells and boreholes throughout London and the water supply network they serviced. This, together with the increased pumping costs associated with raising water from ever-increasing depths, stimulated a switch to surface water supplies drawn from major reservoir developments in the lower Thames and lower Lee valleys. The trend towards reduced groundwater abstraction was encouraged by the borehole licencing provisions introduced in the Water Act 1945 and by the migration of industry away from London. Aquifer development came to a virtual stop after the war.

The growing preference for surface water supplies again shifted the balance between groundwater abstraction and replenishment. Gradually, the decline in groundwater levels eased and then, in the 1960s, reversed. Over a 25-year period, the annual rise averaged 1.5 m and the water table had, by 1995, returned to a level last recorded around 1900. The steady and proctracted recovery caused a number of geotechnical difficulties providing a foretaste of the problems more widely encountered in a number of major conurbations throughout the world. In London the engineering difficulties include those associated with the construction of foundations in waterlogged conditions, the flooding of basements and tunnels and settlement associated with the rewatering of the clay. Also, water quality problems have been encountered as groundwater levels rise into contaminated surface horizons. On the other hand, the water table rise has provided the option for re-exploitation of the aquifer for water supply purposes and, in so doing, affect a degree of control over future levels.

T.J. Marsh

Bibliography

Clutterbuck, J.C., 1848. On the periodical alternations and progressive permanent depression of the chalk water-level under London. *Min. Proc. Instn Civ. Engr.*, **9**, 151–180.

Marsh, T.J. and Davies, P.A., 1983. The decline and partial recovery of groundwater levels below London: *Proc. Inst. Civ. Eng., Part 1*, **74**, 263–276.

Water Resources Board, 1972. *Artificial recharge of the London Basin, I: Hydrogeology*. HMSO, London.

Cross references

Chilgrove House well, UK
Groundwater

TREE RINGS IN HYDROLOGICAL STUDIES

Introduction

Instrumental records are often too short to characterize accurately even basic statistical parameters such as the annual mean, variance or serial correlation of climatic and hydrological data (Rodriguez-Iturbe, 1969), particularly the frequency and magnitude of droughts and floods. Estimates of hydrological parameters can be improved from documents and anecdotal observations of natural phenomena, but data may still be insufficient to describe the long-term variability needed for flood control measures, bridge construction and water-use planning, to name but a few concerns. The growth increments (rings) of trees, however, form annually in temperate and boreal regions and thus can be dated precisely. The width and morphology of rings are controlled in part by hydroclimatic factors, most notably precipitation and temperature, resulting in the usefulness of ring series as proxy records to estimate past hydrological conditions over hundreds or thousands of years. Long-lived trees suitable for hydrological inference are available almost everywhere that trees grow, with the possible exception of parts of the tropics (Stockton *et al.*, 1985).

Dendrochronologists have demonstrated that tree growth is correlated with diverse hydroclimatic phenomena, including lake levels (Stockton and Fritts, 1973; Clague *et al.*, 1982; Begin and Payette, 1988; Meko and Stockton, 1988), streamflow or runoff (Stockton, 1975; Holmes *et al.*, 1979; Stockton and Boggess, 1980; Yanosky, 1982, 1983, 1984; Cook and Jacoby, 1983; Phipps, 1983; Cleaveland and Stahle, 1989), snowpack (Tunnicliff, 1975), precipitation (Stahle and Cleaveland, 1992), temperature (Briffa *et al.*, 1990; Cook *et al.*, 1991), sea level and upper atmospheric pressure patterns (Fritts *et al.*, 1979; Briffa *et al.*, 1986), sea surface temperatures (Douglas, 1976), the Southern Oscillation (Lough and Fritts, 1985; Cleaveland *et al.*, 1992), sunshine duration (Stahle *et al.*, 1991) and drought indices (Stockton and Meko, 1975; Stahle *et al.*, 1988; Cleaveland and

Duvick, 1992). Tree-ring data are held by the International Tree-Ring Data Bank, maintained by the National Oceanic and Atmospheric Administration at the World Data Center-A for Paleoclimatology, National Geophysical Data Center, Boulder, Colorado 80303 USA. Of the approximately 1300 chronologies on hand, 700 are from North America; the remainder are from South America, Eurasia and Africa.

This article considers tree-ring studies applied directly to specific aspects of hydrology, most notably long-term variations in streamflow, flooding and contaminant hydrology. Numerous studies unreported here also relate tree-rings and environmental variables that affect the hydrological cycle, but these latter studies are general in scope and were not designed to investigate specific hydrological phenomena. Three general categories of tree-ring methodologies are presented.

- Hydrological reconstructions from tree-rings. Transformed ring widths correlated with hydrological variables are used to generate proxy data in annual series.
- Tree-ring anatomy. The proportions and dimensions of cells within individual rings are used to infer the occurrence of hydrological events during the interval of ring formation.
- Elemental analysis of tree-rings. Element concentrations within rings are used to infer the contamination history of hydrological systems.

Hydrological reconstructions from tree-rings

The application of tree-rings to hydrological problems depends on construction of long, well-replicated and environmentally sensitive tree-ring chronologies from collections of trees of the same species. Small cylindrical samples are removed with an increment borer (Figure T6) from opposite sides of each tree (Stokes and Smiley, 1968; Phipps, 1985; Swetnam *et al.*, 1985), glued into mounts and sanded until cellular structures are distinguishable (Figure T7) at 10–50× magnification under a dissecting microscope. All rings are carefully dated and measured to the nearest 0.01 mm. The resulting numerical series are transformed to improve their statistical properties and enhance the environmental signals they contain.

Several basic principles guide the compilation of chronologies (Schulman, 1941; Fritts, 1976).

- Dendrochronologists carefully select trees with variable ring-width patterns indicating 'sensitivity' to factors that affect annual growth. Trees with 'complacent' (nearly uniform) rings are generally less useful for reconstructions because radial growth cannot be correlated with hydrological and other environmental variables. Trees growing on marginal sites near ecotones (for example, the Arctic tree line or the prairie margin), or near the distributional limit of a species, typically produce the most sensitive series (Fritts, 1976; Stahle and Hehr, 1984). The sensitivity of upland trees is generally greatest on steep slopes with shallow, well-drained soils. However, some trees growing on wet bottomlands are sensitive to drought (Phipps, 1982; Stahle and Cleaveland, 1992). A mean sensitivity

Figure T6 An increment borer removes a core sample 4–5 mm in diameter from a tree. The cutting bit and hollow shaft are bored into a tree and the sample is removed with a grooved extractor spoon. (Photo from Phipps, 1985.)

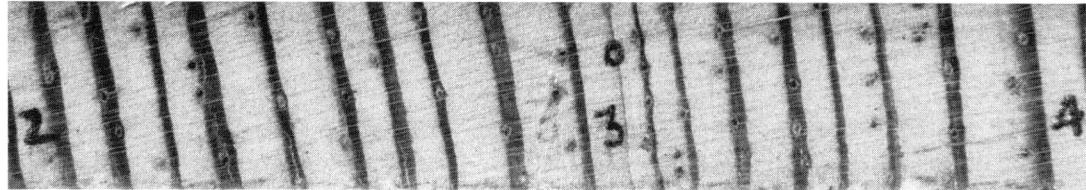

Figure T7 Part of a core from a pine (magnification ×8). Each ring consists of a light band ('earlywood') and a dark band ('latewood'). Numbers placed on the core are decadal indicators. Note the unusually thin latewood zone of the 1930 ring and the small 'o' indicating that the 1931 ring is missing. (Photo from Phipps, 1985.)

value (Fritts, 1976) is calculated for each core and is a practical statistic even if reconstructions are not desired. For example, Yanosky (1982) used mean sensitivity values to show that the growth of floodplain trees was limited more by flood intensity than by flood frequency alone.

- Patterns of annual growth must carefully be compared within and among chronologies to detect missing and false rings (Douglass, 1941; Schulman, 1941; Stokes and Smiley, 1968; Swetnam *et al.*, 1985). Occasionally a ring may not form along the entire circumference of the lower trunk, resulting in an incomplete ring series if the tree is cored at that point. Conversely, a false ring is a cellular feature that can be mistaken for an actual ring. Both missing and false rings must be detected by 'cross-dating' (matching climatically controlled ring-width patterns) to ensure the exact chronological placement of each ring in a series. Cross-dating is an absolute necessity for the application of tree-rings to high-resolution hydrological reconstructions.

- Growth trends unrelated to environmental factors must be removed by 'standardization' prior to the construction of chronologies. Trees typically form wider rings near the center and progressively narrower rings near the bark (Figure T8A). Standardization transforms the ring series to dimensionless indices by dividing the raw value of each ring by a corresponding value generated by some mathematical function, such as a smoothing spline (Figure T8B). Each ring series transformed to indices (Figure T8C) has a mean of 1.0, thus giving it equal weight when averaged, and homogeneous average and variability from beginning to end (Fritts, 1976; Graybill, 1982; Cook, 1987). Indices represent the climatic (including the hydrological) component of ring width, although some climatic information is necessarily lost because the indexing process is not perfect. However, the serious problem created by biological trends in ring width rarely leaves an alternative to standardization. Detrended indices compensate for differences in tree age and growth rates, and hence are more useful than raw ring widths for climatic or ecological studies.

The standardized index value of each ring in a core is averaged with the corresponding indices of all other cores in the collection, resulting in a single combined standard chronology for the site. A natural characteristic of the standard chronology is that the successive annual values are not completely independent, owing to growth carry-over effects from year to year (Fritts, 1976). This positive persistence is measured by serial correlation, that is, the relation between a series and itself lagged one or more years. Positive persistence may arise from soil moisture storage, carbohydrate reserves, root mass, internal phenomena such as fruiting cycles, and needles that continue to function beyond the year of their formation, thus providing a type of biological inertia (Fritts, 1976; LaMarche, 1974). A persistence-free, or 'residual' chronology constructed by removing serial correlation with autoregressive modeling (Box and Jenkins, 1976; Meko, 1981; Cook, 1985) is useful for statistical analysis because the annual values are independent. After estimating past hydrological conditions from the residual chronology, persistence in the instrumental record can be added to the reconstruction to align its time-series properties with those of the actual series (Meko, 1981).

Reconstructions require a transfer function derived by statistically calibrating tree-ring chronologies with hydroclimatic data. Transfer functions often are coefficients from linear regression, although more complex functions involving both multiple predictors and predictands have been used (Fritts, 1976, 1991; Fritts and Swetnam, 1989). When applied to an entire ring series, the transfer function produces a reconstruction, or estimate, of the calibration variable. Part of the available observed environmental data is withheld in order to validate

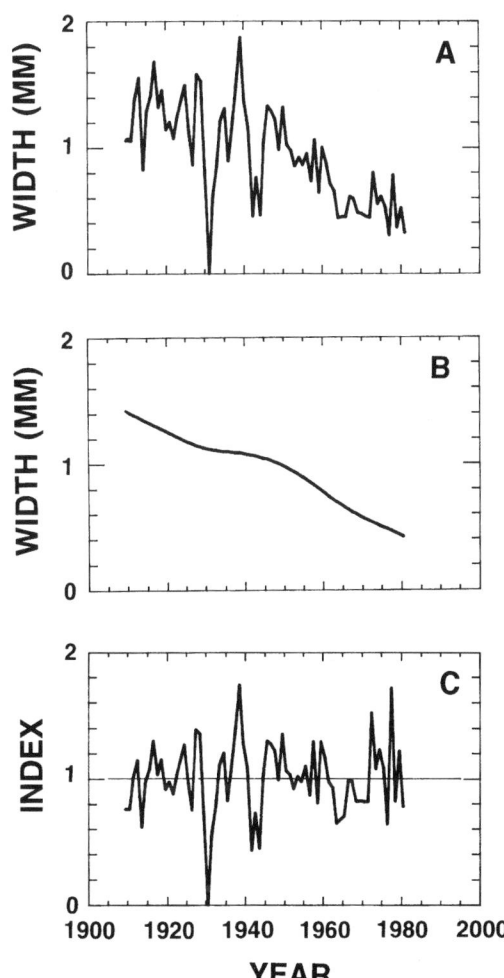

Figure T8 Graphs derived from ring width measurements of a single core.(A) Raw ring widths. (B) Growth trend curve derived by applying cubic spline to raw ring width data. (c) Standardized indices of ring width derived by dividing raw ring widths by corresponding values from the growth trend curve. (From Phipps, 1985.)

the reconstruction (Fritts, 1976, 1991). Although most reconstructions use indices derived from ring widths, X-ray densitometric data from coniferous rings have also been used to generate indices for reconstructions (Briffa *et al.*, 1988). Similarly, indices can be constructed from ring brightness values generated by reflected light imaging techniques (Yanosky *et al.*, 1987; Sheppard *et al.*, 1996), but widespread application awaits further technical development.

Tree-ring chronologies provide a source of seasonal to annual climatic information, but their use as proxy hydrological variables is limited. The relation between ring growth and environmental factors may be difficult to determine (Phipps, 1972), and a substantial fraction of climatic or hydrological variance (typically 20–50%) may remain unexplained. Annual ring widths reflect prevailing hydroclimate over the growing season and longer, and typically do not reflect short-term meteorological or hydrological events (i.e. high-intensity precipitation and flash flood events). The climatic response of trees also is limited by long periods of dormancy each year. The accuracy of regional reconstructions can be improved by sampling a mix of species, each responding somewhat differently to climatic variables or possessing different rates or durations of annual radial growth. Despite these caveats, reconstructions contain useful paleohydrological information if growth responses are robust and associated errors are random in sign and magnitude (Fritts, 1976, 1991; Fritts and Swetnam, 1989).

Reconstructions of drought, and thus of low streamflow, are generally more feasible than those of sustained periods of abundant soil moisture because radial growth is limited severely by soil moisture depletion. Conversely, during the absence of moisture deficits, maximum rates of radial growth may not be achieved owing to microsite conditions or competition, nor are rates necessarily linear during the wettest periods. The growth of bottomland trees tolerant of extensive flooding sometimes increases during droughts as soils become more aerated, producing an inverse relationship between ring width and soil moisture (Phipps, 1972; Stockton and Fritts, 1973; Stahle and Cleaveland, 1992). Thus in most cases it seems likely that reconstructions of low streamflow are more accurate than those of floodflows.

Proxy data reconstructed from tree-ring series have been used to investigate numerous specific hydrological problems. When the Canadian government wanted to maintain a natural regime of water level variability in Lake Athabasca, for example, trees with an inverse growth relationship to water levels provided 158 additional years of information, nearly five times the length of the observed record (Stockton and Fritts, 1973). Lake levels also can be inferred from dating driftwood on exposed strand lines (Clague *et al.*, 1982). Ditching altered hydrological conditions within a wetland and resulted in subsequent changes in tree-growth responses to temperature and precipitation (Phipps *et al.*, 1979).

Tree-ring data from the Colorado Plateau, USA, indicate that long-term mean annual discharge of the Colorado River at Lee Ferry is as much as 2.5 km^3 less than the 19.4 km^3 now contractually divided between upper and lower basin users by the Law of the River (Stockton and Jacoby, 1976; Stockton *et al.*, 1985). The gauging data available when the compact was negotiated spanned a period of 'anomalously persistent high runoff from the Colorado River basin, and . . . [this anomalous runoff episode] apparently was the greatest and longest high-flow period within the last 450 years' (Stockton and Jacoby, 1976).

In the eastern United States, Cook and Jacoby (1983) reconstructed annual summer flows of the Potomac River since 1730 and found that the period of observed streamflow was not representative of the entire series; flows exceeded the long-term median for extended periods during the gauged record. Reconstructed summer streamflow of the Occoquan River, Virginia, USA, indicated that critical low flows occurred more frequently during the reconstructed period than during the gauged period (Phipps, 1983). A reconstruction of runoff in the White River, Arkansas, USA (Cleaveland and Stahle, 1989), suggested that longer periods of consecutive deficit and surplus runoff occurred between 1700 and 1900 than during the 1900–1980 gauged record, and that the extreme reconstructed and gauged runoff values contained more interannual persistence than would be expected by chance. In addition, tree-ring chronologies facilitated double mass analysis (Kohler, 1949) that discovered homogeneity anomalies in discharge measurements of the White River, thus providing a new hydrological application of tree-ring data (Cleaveland and Stahle, 1989).

Brinkman (1987) used 16 tree-ring chronologies representing five different species to reconstruct June–July net basin supplies (NBS) to the Great Lakes. Reconstructions varied in length for different lakes because the shortest chronology limited the separate reconstructions. The lengthened NBS records improved estimates of the independence of summer water supplies to the lakes. The NBS were intercorrelated to about the same degree as presently observed, with Lake Superior showing the lowest correlation (and the most independence). The reconstructions also showed that large water supply anomalies frequently affected the entire region. Non-random basin-wide drought occurred more frequently than did high NBS, most probably resulting from long-term climatic trends.

Tree-ring anatomy

Tree-rings are composed of a system of conducting, strengthening and storage tissues. Differences in the size, proportion and arrangement of cells comprising these tissues confer many of the physical properties of wood and permit identification to genus or species. Although it has been long speculated that general wood architectural patterns are evolutionary responses to climate (Carlquist, 1975), the individual cells comprising rings demonstrate a range of morphologies correlated with short-term environmental variations during their growth and maturation. Anatomical variations within individual rings are easiest to detect when rings are wide, such as those produced during vigorous juvenile growth or in complacent trees (Yanosky, 1984). Thus complacent trees sometimes preserve a record of environmental variation not detected in the narrow rings of trees typically used in ring-width reconstructions. Unlike studies based strictly on the widths of rings along single core radii, anatomical studies are best performed using cross-sectional disks that permit detailed observations along the entire stem circumference. In addition to direct microscopic examinations, digital image analysis has been used to facilitate the detection and description of ring anatomical features (Yanosky and Robinove, 1986; Yanosky *et al.*, 1987).

A striking anatomical feature is the production of callus (scar) tissue following cambial damage by abrasion or burning. Callus is an undifferentiated mass of cells produced near the margins of the damaged zone. New cambium eventually grows over the damaged area and restores typical ring growth, but the scar persists for the remainder of the tree's life. Thus counting the number of rings between the bark and the scar affords a way to determine when the tree was damaged. For example, flood-induced scars in riparian trees preserve evidence of floods (Sigafoos, 1964; Hupp, 1987). The magnitude of large, infrequent floods can be estimated by detecting scars that form along the upper bole of tall trees (Harrison and Reid, 1967), or by recovering evidence from trees growing at successively higher floodplain elevations. Similarly, the magnitude of ice jamming along northern rivers (Parker *et al.*, 1973) and lakes can be determined by recovering scar data from shoreline trees. Miller (1960) used stem scarring to estimate the size of a giant wave generated in 1853–1954 in Lituya Bay, Alaska. The tilting of stems by floods also may result in the production of asymmetric ring-width growth (reaction wood) along directly opposite radii, even if scarring is not evident (Phipps, 1970; Hupp, 1988).

The morphology of vessels sometimes changes following an extreme hydrological event, thus preserving evidence of its occurrence. Vessels are conducting cells typical of hardwoods, and in ring-porous species are larger in the inner (earlywood) than outer part (latewood) of each ring. Sigafoos (1964) found that the trunk of an ash (*Fraxinus pennsylvanica*) buried rapidly by flood-borne sediments produced earlywood and latewood vessels more like those of roots than of stems; similarly, he reasoned, roots excavated rapidly by flood waters would produce vessels similar to those of stemwood. Yanosky (1983) detected anomalous rings of vessels in ash trees damaged by floods generated during the growing season. Trees with leaves stripped by flood waters subsequently produced a second leaf set, and with it tangential rows of vessels within the latewood that more typically resembled those of the earlywood. Yanosky (1983) concluded that flood timing could be estimated from the radial position of anomalous vessels relative to total ring width, and that the stage of large floods could be estimated by detecting flood-induced anomalies in tall trees near the channel or in small trees on terraces.

Periodic flooding during the growing season sometimes alters the dimensions of tracheids or fibrous, non-conducting cells in some trees. Controlled flooding resulted in increased fiber diameters in *Alnus japonica* seedlings (Yamamoto *et al.*, 1995) and the production of false rings in baldcypress (*Taxodium distichum*; Young *et al.*, 1993). It may be that periodic flooding alters hormonal dynamics, which in turn may affect the rate of cambial activity and the subsequent development of the anatomical features of wood (Yanosky, 1982; Yamamoto *et al.*, 1995). For example, ash trees flooded for several days by non-damaging floodflows abruptly produced large, thin-walled fibers compared to those produced prior to inundation and in unflooded trees nearby (Yanosky, 1984). Growth responses were particularly sensitive to flooding in late summer, when streamflow is typically low. In some cases, evidence was preserved of more than one small-magnitude flood within a single growth year. Wood fiber evidence of large floods was found in trees that grew where the probability of high-velocity floodflow damage was low, such as along the inside of channel bends and at high floodplain levels (Yanosky, 1984).

Numerous developmental anatomical features are related at least in part to climatic variation, and thus directly or indirectly to hydrological phenomena. For example, Woodcock (1989) was able successfully to reconstruct precipitation from the average diameters of latewood vessel of oak. Hill (1982, 1983) showed that anatomical features within two species of hickory seemed related to climate. Models of coniferous cell numbers and dimensions (Frutts et al., 1991) suggest that these features correlate with short-term climatic variables. As understanding improves of factors controlling ring morphologies, it seems likely that anatomical studies will have greater applicability to hydrological problems.

Elemental analysis of tree-rings

Few studies have investigated the relation between element concentrations within tree-rings (dendrochemistry) and hydrological phenomena. These studies are based on the premise that the uptake, transport and deposition of elements within tree-rings at least to some extent are proportional to element availabilities in soils. Thus elements sorbed to sediments deposited during periodic flooding would be expected to furnish an exchangeable pool of elements potentially available for uptake by bottomland trees. Similarly, trees with roots in close proximity to shallow aquifers might form wood preserving the relative elemental composition of groundwater. Element concentration trends within rings might be useful to monitor historical changes in water quality and to indicate spatial gradients of hydrological contaminants.

Element concentrations are measured either simultaneously within core fragments comprising ring aggregates, or within individual rings. The former generally involves cutting the core sample into the desired number of fragments and preparing them in the appropriate manner for analysis by inductively coupled plasma atomic emission spectrometry, atomic absorption spectroscopy or neutron activation analysis. The analysis of individual rings can be performed by non-destructive methods such as proton-induced X-ray emission (PIXE) and X-ray fluorescence. In PIXE, for example, a standard 5 mm core is mounted in a lucite clamp and scraped to a flat surface with a high-quality stainless steel scalpel. A proton beam that induces X-rays from the constituent elements within rings is directed in turn at each desired target ring. X-ray spectra are used to determine simultaneously the relative concentration of 12–20 elements. If necessary, the detection limit of some elements can be lowered by increasing the time of irradiation. The analysis of individual rings potentially increases the likelihood of detecting sudden changes in radial concentrations of an element and thus may be particularly suited to short-term monitoring.

The utility of dendrochemical studies is complicated by a poor understanding of the fate of elements absorbed into the transpiration stream, and in particular by the radial movement (translocation) of some elements across rings. In other words, an element taken up during any given year may not necessarily remain within that year's ring. Although elements within the non-functional heartwood are seemingly immobile, those within the outer zone of rings (sapwood) containing some living cells may be subject to translocation into newer or older sapwood rings. For example, phosphorus is typically most highly concentrated within the outermost sapwood near the cambial zone, not because phosphorus availability was greater during than before the outermost sapwood formed, but rather because phosphorus accumulates there at the expense of metabolic energy. The extent to which an element is translocated seems to be species related and perhaps also depends on its relative uptake by plants; in some instances, concentrations seemingly in excess of physiological requirements are translocated from the outer to inner sapwood and thus ultimately become permanent constituents of the heartwood. Further discussions of the assumptions, limitations and potential applications of dendrochemical studies to environmental problems are summarized by Cutter and Guyette (1993) and Hagemeyer (1993).

Apparently the first attempt to relate dendrochemical data to surface water quality was in 1970 by Ishizaki and colleagues (Hagemeyer, 1993). They found higher concentrations of zinc and cadmium in rings of Cryptomeria exposed to river water contaminated with these elements than in nearby uncontaminated trees. However, element concentration trends within contaminated trees were not synchronized sufficiently to permit estimates of the historical onset of contamination. Positive correlations between the metal

concentrations in tree-rings and those of sediments were found by Sheppard and Funk (1975), and peak concentrations of iron and chromium in rings coincided to some extent with mining histories if allowances were made for transport delays of sediments. Similarly, Hupp et al. (1993) found that tree-ring concentrations of zinc, copper, nickel and lead varied among trees but generally were proportional to concentrations in sediments at a site; high concentrations of zinc within the innermost rings sampled suggested that zinc sediment burdens may have been greater in the past than at present.

A study of element concentrations within rings of Taxodium provided spatial and historical information concerning saltwater flooding within an estuary (Yanosky et al., 1995). Trees exposed abruptly to increased tidal flooding caused by channel modifications contained larger concentrations of sodium, chloride and bromide than did nearby trees in freshwater reaches of the estuary. Correlations among concentrations of the three elements strengthened the hypothesis that saline flooding was the source of elements detected in rings. Each element was translocated laterally within the sapwood of trees, thus obscuring a direct relationship between ring chemistry and element availability; however, it was possible to estimate the approximate onset of saline flooding from lagged element concentrations within heartwood rings.

The dynamics of groundwater contaminant hydrology have been inferred from element concentrations within trees growing over shallow aquifers. Vroblesky and Yanosky (1990) found that concentrations of chloride and iron in Liriodendron seemed to be related to the disposal history and attempted remediation of a hazardous waste site in which groundwater was contaminated. The transport velocity of iron in groundwater was estimated from iron concentrations of rings in trees at successive distances from the waste site. Similarly, the transport velocity of the groundwater was estimated from ring concentrations of chloride, which generally is transported at the same rate as groundwater. They concluded that velocity data derived from dendrochemical data could have utility for estimating horizontal hydraulic conductivity values. Vroblesky et al. (1992) also used concentration patterns of potassium in trees to determine the spatial extent of potassium contamination within the underlying aquifer. They found that trees growing over contaminated groundwater apparently absorbed large concentrations of potassium and deposited excess amounts within inner sapwood rings and thus, ultimately, into the heartwood. Trees growing where potassium was not elevated, in contrast, contained only minimal concentrations of potassium in heartwood rings.

Concentrations of nickel were studied in several tree species growing over shallow aquifers in which nickel concentrations were elevated (Yanosky and Vroblesky, 1992). At one site, trees were established before the onset of contamination and contained abrupt increases in ring nickel concentrations believed to document the local onset of nickel contamination; at another site, trees probably became established after the onset of nickel contamination, but contained generally high concentrations of nickel within their rings. An additional study at this site (Yanosky and Vroblesky, 1995) suggested that evidence for the movement of a petroleum plume floating on the aquifer might have been preserved in trees along the flowpath as abrupt increases in the concentrations of iron and manganese. Dendrochemical studies at an abandoned creosote works detected high concentrations of some elements in trees at the site but concluded that contaminated groundwater flow had not reached a nearby forest (Yanosky and Carmichael, 1993).

Thomas M. Yanosky and Malcolm K. Cleaveland

Bibliography

Begin, Y. and S. Payette, 1988. Dendroecological evidence of lake-level changes during the last three centuries in subarctic Quebec, Quaternary Research, **30**, 210–220.

Box, G.E.P. and G.M. Jenkins, 1976. Time Series Analysis: Forecasting and Control. Oakland, California: Holden-Day, 575 pp.

Briffa, K.R., P.D. Jones and F.H. Schweingruber, 1988. Summer temperature patterns over Europe: A reconstruction from 1750 A.D. based on maximum latewood density indices of conifers, Quaternary Research, **30**, 36–52.

Briffa, K.R., P.D. Jones and T.M.L. Wigley, 1986. Climate reconstruction from tree rings: part 2, spatial reconstruction of summer mean sea-level pressure patterns over Great Britain, Journal of Climatology, **6**, 1–15.

Briffa, K.R., T.S. Bartholin, D. Eckstein *et al.*, 1990. A 1,400-year tree-ring record of summer temperatures in Fennoscandia, *Nature*, **346**, 434–439.

Brinkmann, W.A.R., 1987. Water supplies to the Great Lakes – reconstructed from tree rings, *Journal of Climate and Applied Meteorology*, **26**, 530–538.

Carlquist, S., 1975. *Ecological Strategies of Xylem Evolution.* Berkeley: Univ. of California, 259 pp.

Clague, J., L.A. Jozsa and M.L. Parker, 1982. Dendrochronological dating of glacier-dammed lakes: an example from Yukon Territory, Canada, *Arctic and Alpine Research*, **14**, 301–310.

Cleaveland, M.K. and D.N. Duvick, 1992. Iowa climate reconstructed from tree rings, A.D. 1640 to 1982, *Water Resources Research*, **28**, 2607–2615.

Cleaveland, M.K. and D.W. Stahle, 1989. Tree ring analysis of surplus and deficit runoff in the White River, Arkansas, *Water Resources Research*, **25**, 1391–1401.

Cleaveland, M.K., E.R. Cook and D.W. Stahle, 1992. Secular variability of the Southern Oscillation detected in tree-ring data from Mexico and the southern United States, in H.F. Diaz and .V. Markgraf, eds, *El Nino: The Historical and Paleoclimatic Record of the Southern Oscillation*, Cambridge: Cambridge University Press, pp. 271–291.

Cook, E.R., 1985. A Time Series Analysis Approach to Tree-Ring Standardization. PhD dissertation. Tucson: University of Arizona, 171 pp.

Cook, E.R., 1987. The decomposition of tree-ring series for environmental studies, *Tree-Ring Bulletin*, **47**, 37–59.

Cook, E.R. and G.C. Jacoby, 1983. Potomac River streamflow since 1730 as reconstructed by tree rings, *Journal of Climate and Applied Meteorology*, **22**, 1659–1672.

Cook, E.R., T. Bird, M. Peterson *et al.*, 1991. Climatic change in Tasmania inferred from a 1089-year tree-ring chronology, *Science*, **253**, 1266–1268.

Cutter, B.E. and R.P. Guyette, 1993. Anatomical, chemical, and ecological factors affecting tree species choice in dendrochemistry studies, *Journal of Environmental Quality*, **22**, 611–619.

Douglas, A.V., 1976. Past Air–Sea Interactions over the Eastern North Pacific Ocean as Revealed by Tree-Ring Data. PhD dissertation. Tucson: University of Arizona, 196p.

Douglass, A.E, 1941. Crossdating in dendrochronology, *Journal of Forestry*, **39**, 825–831.

Fritts, H.C., 1976. *Tree Rings and Climate*. London: Academic Press. 567 pp.

Fritts, H.C., 1991. *Reconstructing Large-Scale Climatic Patterns from Tree-Ring Data*. Tucson: University of Arizona Press, 286 pp.

Fritts, H.C., and T.W. Swetnam, 1989. Dendroecology: a tool for evaluating variations in past and present forest environments, in M. Begon, A.H. Fitter, E.D. Ford and A. MacFadden, eds, *Advances in Ecological Research, vol. 19*, New York: Academic Press, pp. 111–188.

Fritts, H.C., G.R. Lofgren and G.A. Gordon, 1979. Variations in climate since 1602 as reconstructed from tree rings, *Quaternary Research*, **12**, 18–46.

Fritts, H.C., E.A. Vaganov, I.V. Sviderskaya and A.V. Shashkin, 1991. Climatic variation and tree-ring structure in conifers: empirical and mechanistic models of tree-ring width, number of cells, cell size, cell-wall thickness and wood density, *Climate Research*, **1**, 97–116.

Graybill, D.A., 1982. Chronology development and analysis, in M.K. Hughes, P.M. Kelly, J.R. Pilcher and V.C. LaMarche Jr, eds, *Climate from Tree Rings*, Cambridge: Cambridge University Press, pp. 21–31.

Hagemeyer, J., 1993. Monitoring trace metal pollution with tree rings: A critical reassessment, in B. Markert, ed., *Plants as Biomonitors. Indicators for Heavy Metals in the Terrestrial Environment*, New York: VCN Weinheim, pp. 541–563.

Harrison, S.S. and J.R. Reid, 1967. A flood-frequency graph based on tree-scar data, *Proceedings, North Dakota Academy of Science*, **21**, 23–33.

Hill, J.F., 1982. Spacing of parenchyma bands in wood of *Carya glabra* (Mill.) Sweet, pignut hickory, as an indicator of growth rate and climatic factors, *American Journal of Botany*, **69**, 529–537.

Hill, J.F., 1983. Relationship among vessel diameter, vessel frequency, and spacing of parenchyma bands in wood of *Carya tomentosa* Nutt., mockernut hickory, *American Journal of Botany*, **70**, 934–939.

Holmes, R.L., C.W. Stockton and V.C. LaMarche Jr, 1979. Extension of river flow records in Argentina from long tree-ring chronologies, *Water Resources Bulletin*, **15**, 1081–1085.

Hupp, C.R., 1987. Botanical evidence of floods and paleoflood history, in V.P. Singh, ed., *Regional Flood Frequency Analysis*, Boston: D. Reidel, pp. 355–369.

Hupp, C.R., 1988. Plant ecological aspects of flood geomorophology and paleoflood history, in V.R. Baker, R.C. Kochel and P.C. Patton, eds., *Flood Geomorphology*, New York: Wiley, pp. 335–356.

Hupp, C.R., M.D. Woodside and T.M. Yanosky, 1993. Sediment and trace element trapping in a forested wetland, Chickahominy River, Virginia, *Wetlands*, **13**, 95–104.

Kohler, M.A., 1949. On the use of double-mass analysis for testing the consistency of meteorological records and for making required adjustments, *Bulletin of the American Meteorological Society*, **30**, 188–189.

LaMarche, V.C. Jr, 1974. Paleoclimatic inferences from long tree-ring records, *Science*, **183**, 1043–1048.

Lough, J.M. and H.C. Fritts, 1985. The Southern Oscillation and tree rings: 1600–1961, *Journal of Climate and Applied Meterorology*, **24**, 952–966.

Meko, D.M., 1981. Applications of Box–Jenkins Methods of Time Series Analysis to the Reconstruction of Drought from Tree Rings. PhD dissertation. Tucson: University of Arizona, 149 pp.

Meko, D.M. and C.W. Stockton, 1988. Tree-ring inferences on historical changes in the level of Great Salt Lake, in P.A. Kay and H.F. Diaz, eds, *Problems and Prospects for Predicting Great Salt Lake Levels*, Salt Lake City: University of Utah, pp. 63–76.

Miller, D.J., 1960. Giant waves in Lituya Bay Alaska, *US Geol Survey Prof. Paper 354-C*, pp. 51–86.

Parker, M.L., L.A. Jozsa and R.D. Bruce, 1973. *Dendrochronological investigations along the Mackenzie, Liard, and South Nahanni Rivers, Northwest Territories, part II: using tree-ring analysis to reconstruct geomorphic and climatic history, Technical Report to Glaciology Division, Water Resources Branch, Environment Dept.* Vancouver: Forintek Canada Corp. Western Forest Products Lab, 74 pp.

Phipps, R.L., 1970. The potential use of tree rings in hydrologic investigations in eastern North America with some botanical considerations, *Water Resources Research*, **6**, 1634–1640.

Phipps, R.L., 1972. Tree rings, stream runoff, and precipitation in central New York – a reevaluation, *US Geol. Survey Prof. Paper 800-B*, pp. B259–B264.

Phipps, R.L., 1982. Comments on interpretation of climatic information from tree rings, Eastern North America, *Tree-Ring Bulletin*, **42**, 11–22.

Phipps, R.L., 1983. Streamflow of the Occoquan River in Virginia as reconstructed from tree-ring series, *Water Resources Bulletin*, **19**, 735–743.

Phipps, R.L., 1985. Collecting, preparing, crossdating and measuring tree increment cores, *US Geol. Survey Water Resour. Inv. Rpt. 85–4148*, 48 pp.

Phipps, R.L., D.L. Ierley, and C.P. Baker, 1979. Tree rings as indicators of hydrologic change in the Great Dismal Swamp, Virginia and North Carolina, *US Geol. Survey Water Resour. Inv. Rpt. 78–136*, 26 pp.

Rodriguez-Iturbe, I., 1969. Estimation of statistical parameters of annual river flows, *Water Resources Research*, **5**, 1418–1421.

Schulman, E., 1941. Some propositions in tree-ring analysis, *Ecology*, **22**, 193–195.

Sheppard, J.C. and W.H. Funk, 1975. Trees as environmental sensors monitoring long-term heavy metal contamination of the Spokane River, Idaho, *Environmental Science and Technology*, **9**, 638–642.

Sheppard, P.R., L.J. Graumlich and L.E. Conkey, 1996. Reflected-light image analysis of conifer tree rings for reconstructing climate, *The Holocene*, **6**, 62–68.

Sigafoos, R.S., 1964. *Botanical evidence of floods and flood-plain deposition*, US. Geol. Survey Prof. Paper 485-A, 35 pp.

Stahle, D.W. and M.K. Cleaveland, 1992. Reconstruction and analysis of spring rainfall over the southeastern U.S. for the past 1000 years, *Bulletin of the American Meteorological Society*, **73**, 1947–1961.

Stahle, D.W. and J.G. Hehr, 1984. Dendroclimatic characteristics of post oak across a precipitation gradient in the southcentral United States, *Annals of the Association of American Geographers*, **74**, 561–573.

Stahle, D.W., M.K. Cleaveland and R.S. Cerveny, 1991. Tree-ring reconstructed sunshine duration over the central United States, *International Journal Climatology*, **11**, 285–295.

Stahle, D.W., M.K. Cleaveland and J.G. Hehr, 1988. North Carolina climate changes reconstructed from tree rings: A.D. 372 to 1985, *Science*, **240**, 1517–1519.

Stockton, C.W., 1975. *Long-Term Streamflow Records Reconstructed from Tree Rings. Laboratory of Tree-Ring Research Papers No. 5.* Tucson: University of Arizona Press, 111 pp.

Stockton, C.W. and W.R. Boggess, 1980. Augmentation of hydrologic records using tree rings, in *Proceedings of the Engineering Foundation Conference on Improved Hydrologic Forecasting – Why and How*, New York: American Society of Civil Engineers, pp. 239–265.

Stockton, C.W. and H.C. Fritts, 1973. Long-term reconstruction of water level changes for Lake Athabasca by analysis of tree rings, *Water Resources Bulletin*, **9**, 1006–1027.

Stockton, C.W. and G.C. Jacoby, 1976. *Long-Term Surface-Water Supply and Streamflow Trends in the Upper Colorado River Basin. Lake Powell Research Project Bulletin No. 18.* Los Angeles: University of California Institute of Geophyics and Planetary Physics, 70 pp.

Stockton, C.W. and D.M. Meko, 1975. A long-term history of drought occurrence in western United States as inferred from tree rings, *Weatherwise*, **28**, 244–249.

Stockton, C.W., W.R. Boggess and D.M. Meko, 1985. Climate and tree rings, in A.D. Hecht, ed., *Paleoclimate Analysis and Modeling*, New York: John Wiley and Sons, pp. 71–161.

Stokes, M.A. and T.L. Smiley, 1968. An *Introduction to Tree-Ring Dating*. Chicago: University of Chicago Press, 73 pp.

Swetnam, T.W., M.A. Thompson and E.K. Sutherland, 1985. *Using dendrochronology to measure radial growth of defoliated trees*, Agriculture Handbook No. 639, US Dept. of Agriculture Forest Service, Cooperative State Research Service, 39 pp.

Tunnicliff, B.M., 1975. The Historical Potential of Snowfall as a Water Resource in Arizona. M. Sci. thesis. Tucson: University of Arizona, 137 pp.

Vroblesky, D.A. and T.M. Yanosky, 1990. Use of tree-ring chemistry to document historical ground-water contamination events, *Ground Water*, **28**, 677–684.

Vroblesky, D.A., T.M. Yanosky and F.R. Siegel, 1992. Increased concentrations of potassium in the heartwood of trees in response to ground-water contamination, *Environmental Geology and Water Science*, **9**, 71–74.

Woodcock, D.W., 1989. Climate sensitivity of wood anatomical features of a ring-porous oak (*Quercus macrocarpa*). *Canadian Journal of Forest Research*, **19**, 639–644.

Yamamoto, F., T. Sakata and K. Terazawa, 1995. Growth, morphology, stem anatomy, and ethylene production in flooded *Alnus japonica* seedlings, *International Association of Wood Anatomists Journal*, **16**, 47–59.

Yanosky, T.M., 1982. Hydrologic inferences from ring widths of flood-damaged trees, Potomac River, Maryland, *Environmental Geology*, **4**, 43–52.

Yanosky, T.M., 1983. Evidence of floods on the Potomac River from anatomical abnormalities in the wood of flood-plain trees, *US Geol. Survey Prof. Paper 1296*, 42 pp.

Yanosky, T.M., 1984. Documentation of high summer flows on the Potomac River from the wood anatomy of ash trees, *Water Resources Bulletin*, **20**, 241–250.

Yanosky, T.M. and Carmichael, J.K., 1993. *Element concentrations in growth rings of trees near an abandoned wood-preserving plant site at Jackson, Tennessee*, US Geol. Survey Water Res. Inv. Rpt., 93–4223, 68 pp.

Yanosky, T.M. and C.J. Robinove, 1986. Digital image measurement of the area and anatomical structure of tree rings, *Canadian Journal of Botany*, **64**, 2896–2902.

Yanosky, T.M. and D.A. Vroblesky, 1992. Relation of nickel concentrations in tree rings to groundwater contamination, *Water Resources Research*, **28**, 2077–2083.

Yanosky, T.M. and Vroblesky, D.A., 1995. Element analysis of tree rings in ground-water contamination studies, in T.E. Lewis, ed., *Tree Rings as Indicators of Ecosystem Health*, Boca Raton, Florida: CRC Press, pp. 177–205.

Yanosky, T.M., Hupp, C.R. and Hackney, C.T., 1995. Chloride concentrations in growth rings of *Taxodium distichum* in a saltwater-intruded estuary, *Ecological Applications*, **5**, 785–792.

Yanosky, T.M., C.J. Robinove and R.G. Clark, 1987. Progress in the image analysis of tree rings, in G.C. Jacoby and J.W. Hornbeck, eds, *Proceedings of International Symposium on Ecological Aspects of Tree-Ring Analysis*, Publication CONF 8608144, Washington, DC: US Dept. of Energy, pp. 658–665.

Young, P.J., J.P. Megonigal, R.R. Sharitz and F.P. Day, 1993. False ring formation in baldcypress (*Taxodium distichum*) saplings under two flooding regimes, *Wetlands*, **13**, 293–298.

Cross references

Tree rings: Fortingall yew tree, UK

TREE RINGS: FORTINGALL YEW TREE, UK

The Fortingall yew tree, near Aberfeldy in Scotland, is reputed to the earliest known planting of a surviving tree in Europe (Figure T9).

The earliest written account of the tree is by William Pennant. Pennant inspected the tree in 1769 on a tour of Scotland and stated (1771):

> In the churchyard of Fortingall near the foot of Glenlyon there is the remains of a prodigious yew tree fifty-six and a-half [ft.] in circumference. The middle part of it is now decayed to the ground but within memory was limited to the height of three feet. Captain Campbell of Glenlyon having assured me that when a boy he had often clambered or rode over the connecting part.

The Hon. Davies Barrington, a barrister, in a letter published in the *Royal Society Transactions* of 1769 wrote:

> I measured the circumference of this yew tree twice and therefore cannot be mistaken when I inform you that it amounted to fifty-two

Figure T9 The Fortingall yew tree, 1996.

feet. Nothing scarcely remains now but the outward bark which hath been separated by the centre of the tree decaying within these twenty years. What still appears however is thirty four feet in circumference.

Both Pennant and Barrington corresponded with Gilbert White of Selborne.

In 1831 De Candolle, the Swiss botanist, estimated the age of the tree to be 2500–2600 years in 1770, but commended further studies. Sir Robert Christison in his paper 'The Fortingall Yew' (1876–1879), gave an account of his measurement of the growth of trees and concluded:

It is better to use the general rules formerly arrived at according to which the tree in the first place may be assumed to have attained a girth of 22 feet in a thousand years. After that age no information yet got warrants a rate of more than one inch in thirty five years. Taking the lowest measurements of Barrington at 52 feet the difference will thus add 2000 years to the age of the Fortingall Yew making it in all 3000 years when measured in 1768–69. The result is startling but not so improbable as may at first be thought, if it is to be considered that several English yews of scarcely half the girth are not without good reason held to surpass materially a thousand years of age, yet still appear to be in vigorous health and steadily increasing.

A. Mitchell created a National Tree Register of the British Isles (1995) which became the major part of his work. In an article in *The Field* (1991), he wrote:

Britain's oldest tree could date back 5000 years. There are now 93 000 trees of 1730 species and 1200 varieties in the register. All the oldest trees in Britain are yews, some dating back at least 1000 years. Their exact ages are a matter for speculation and circumstantial evidence. As all the trees (yew) are hollow no early growth exists to be tested so no ring counts reach back to their youth. The biggest and oldest yews were there long before the Saxon churches which were built near them to appropriate the ancient pagan sanctity of the ground . . . the oldest by general consent is the Fortingall Yew . . . it was thought to be 2500 years old but recent studies suggest it could be over 5000 years old.

Meredith (1988), a contemporary worker of Mitchell, says that before World War II there were possibly 1000 ancient yew trees aged 1000 years or more in Britain. Now he estimates there may only be some 450, the best example being found at Fortingall. The obstacles to measurements he notes as:

- ring counts are difficult to make and incomprehensible when the tree is old;
- the increase of girth sometimes stops for a time;
- old yew trees are hollow;
- there is irregular and asymmetrical growth due to differences in climate.

On the other hand, other known facts may help corroborate ages from tree-rings; for example, the vaults of a church in Surrey dating back 1000 years which appear to have been built over the roots of a tree that still stands. Further updating is helped by the examination of old trees destroyed in the storm of 1987. Meredith concludes: 'The Fortingall yew is certainly over 5000 years old, of this I have no doubt.' A ring count of 32 trees of girth 3–36 ft (0.9–9 m) gives Fortingall an age of 9000 years.

Dickson (1982) noted that the age of the Fortingall yew varies with the investigator and agreed with Christison and Mitchell that it was the oldest surviving tree in Britain. He added that confirmation of the status of ancient yew trees in Britain was not disputed since yew was proved by radiocarbon dating to be part of English limestone woodlands 7000 years ago.

For comparison, the world's oldest living vegetation is the Huon pine in Mount Read in Tasmania, reputed to be some 10 500 years old and with a spread of some 1.5 ha. The Californian bristlecone pine named 'Methuselah', previously thought to be the oldest tree, is reputed to be over 4700 years old. Tree-ring analysis enables the study of climate reconstruction over a period of many hundreds of years (see Tree-rings in hydrological studies). In dendroclimatological work by Briffa and Wigley (1985) successful reconstruction of local area-average temperature and precipitation have been carried out for the UK and in a study in northern Fennoscandia temperature changes have been estimated from AD 500 (Briffa et al., 1992). This latter study showed that the sixth and seventh centuries were somewhat cool, particularly the 530s and 540s. Most of the eighth century was warm, especially the 750s; 870–900, 920–940 and 970–1000 were warm periods and between 970 and 1120 it was consistently warm, warmer than the 1930s and 1940s. At about 1100, however, the warmth terminated abruptly, marking a major oscillation from warm to cold (1100–1150) and back to warm (1150–1190) conditions. Summers from 1200 to 1360 were generally near the long-term mean. Temperatures thereafter rose again and general warmth prevailed through much of the fifteenth and early sixteenth centuries. The two decades centred on 1500 in July and August were as warm as any period and noticeably warmer than the warmth of the early twentieth century. Summers thereafter remained generally cold until the middle of the eighteenth century. In second half of the eighteenth and the first half of the nineteenth century temperatures oscillated around the long term-mean, with the 1750s, 1760s and 1820s being warm. Finally, it is estimated that the last 100 years have been the warmest since the eleventh century.

R.W. Herschy

Bibliography

Barrington, D., 1769. Letter to Royal Society, *Phil. Trans. R. Soc. London*, **VI**.

Briffa, K.R. and Wigley, T.M.L., 1985. Tree rings and soil moisture, *Climate Monitor*, **14**(4).

Briffa, K.R., Jones, P.D., Bartholin, T.S. *et al.*, 1992. Fennoscandian summers from AD 500 temperature changes in short and long timescales, in *Climate Dynamics*, Vol. 7, Springer-Verlag, Berlin, pp. 111–119.

De Candolle, 1831. The Fortingall Yew Tree, *Bib. Univ.*, ii66.

Christison, R., 1876–1879. The Fortingall Yew Tree, *Trans. Bot. Soc.*, XIII, London.

Dickson, J., 1991. The yew trees – *Taxus baccata* – in Scotland – Native or early introduction of both? M. of A.

Keddie, J.B., 1996. Sources of Information on the yew tree at Fortingall, private communication.

Meredith, A., 1988. *The Yew Tree Campaign*.

Mitchell, A., 1991. The National Tree Register, *The Field*, London.

Mitchell, A., 1995. *The National Tree Ring Register*, Royal Horticultural Society, London.

Pennant, W., 1771. *Tour of Scotland, 1769*.

Cross reference

Tree rings in hydrological studies

U

UNITED STATES GEOLOGICAL SURVEY: NATIONAL WATER-USE INFORMATION PROGRAM

Background

The United States possesses abundant water resources and has developed and used those resources extensively. The future health and economic welfare of the nation's population depend on a continuing supply of uncontaminated fresh water. Many existing sources of water are being stressed by withdrawals from groundwater wells and diversions from rivers and reservoirs to meet the needs of homes, cities, farms and industries. Increasing requirements to leave water in the streams and rivers to meet environmental, human and recreational needs further complicate the problem. Recent drought in some areas has accentuated the need to balance water demand with supply.

Traditionally, water management in the United States has focused on manipulating the country's supplies of fresh water to meet the needs of users. A number of large dams were built during the early twentieth century to increase the supply of fresh water at any given time. The era of building large dams and conveyance systems to meet water demand in the United States is drawing to a close, as shown in Figure U1; in the twenty-first century, the limited water supply and established infrastructure will require that demand is managed effectively within the available supply.

'New' water supplies likely will be from conservation, recycling, reuse and improved water-use efficiency rather than from large development projects. It is apparent that the nation can no longer meet insatiable water demands by continuously expanding a supply that has physical, ecological and economic limits. The transition is well underway to an era of 'integrated water resources management' in which traditional supply-management options will be balanced with demand-management options.

The Governors of the Western States recently issued a policy statement calling for sharply enhanced efficiency in water use, and the President signed into law the 1994 Energy Policy Bill, which calls for government agencies to take the lead in water-use efficiency measures and sets new standards for water-conserving plumbing fixtures. Water resources planners, managers and hydrologists need comprehensive, credible and reliable water-use data to assess the impacts and efficiency of demand management strategies and to balance the competition between traditional uses and new recreational and environmental uses.

National Water-Use Information Program

The US Geological Survey's (USGS) National Water-Use Information Program is a cooperative program with state and local governments and is designed to collect, store, analyze and disseminate water-use information nationally and locally to a wide variety of government agencies and private organizations. The program was begun in 1978 to meet the need for a single source of uniform information on water use and to serve as the focal point for water-use information. The Water-Use Program is financed through the Federal–State Cooperative Program of the USGS with Federal–State co-operative matching funds available to support water-use information activities in the 50 States and Puerto Rico. By law, these Federal funds must be matched with equal amounts of State or local agency funds.

Why is the Program in the USGS?

The USGS has compiled estimates of water use every 5 years since 1950. However, before the National Water-Use Information Program was started, these estimates were derived from many sources and were based on a variety of methods of data collection and analysis that differed in accuracy. Therefore, the available information fell short of providing a national database that was current, readily accessible and reliable.

Because the USGS has offices in each state, it is in a good position to operate and manage the National Water-Use Information Program. The Program compiles data that are used to estimate water use at the State level on the basis of nationally consistent guidelines and procedures, and uniform quality assurance for each state.

Objectives

The objectives of the National Water-Use Program are to

- determine on local and national levels how much fresh and saline ground and surface water is withdrawn and for what purposes;
- develop and refine computerized systems to store and retrieve water-use information at state and national levels;
- devise and apply new standards, methods and techniques to improve the collection and analysis of water-use information;
- disseminate water-use information at state and national levels.

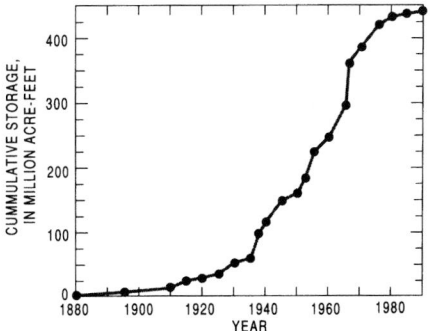

Figure U1 Cumulative reservoir storage in the United States.

Accomplishments

Through the cooperative water-use program, standardized collection and analysis methods are developed that allow evaluations on the basis of similar assumptions and comparable data, site-specific water-use databases are being developed to help ensure effective, efficient communication and data handling among other federal agencies, the states and the USGS, and the data are aggregated at 5-year intervals and published in the USGS circular series *Estimated Use of Water in the United States* to respond to interstate, regional and national water-use data needs. In addition, many States publish their own water-use reports in cooperation with the USGS.

Information on water use in the United States is now available on the Internet through the US Geological Survey Water Resources Information home page. The Universal Resource Locator (URL) for the National Water-Use Home Page is:

http://h2o.usgs.gov/public/watuse/

It may be accessed by using a number of Web browser software packages.

Instream flow requirements

Instream use is defined as any water use that occurs within the stream channel for such purposes as hydroelectric power generation, navigation, fish and wildlife preservation, water quality improvement and recreation. The National Water-Use Program compiles information for hydroelectric power. Quantitative estimates for most other instream uses are difficult to quantify and compile on a national scale. However, because such uses compete with traditional uses such as domestic, industrial and irrigation, effective water resources management requires methods and procedures to be devised to enable instream uses to be assessed quantitatively.

Regulatory instream flow requirements are increasing at an unprecedented rate. Current activities in many states calls for substantial increases in instream flows to meet a variety of human and environmental needs. For example:

- recent legislation in Florida requires substantial increases in instream flow deliveries to Everglades National Park;
- releases from Jordanelle Reservoir to support fish habitat in Utah are six times greater than called for in the original 1960s design;
- efforts are underway in Nevada to convert irrigation water rights to instream flow rights for Stillwater wetlands;
- the economic market associated with whitewater rafting in Virginia has led the US Army Corps of Engineers to investigate alternative release schedules;
- numerous projects in California call for increased instream flows for fish and wildlife preservation.

Water use in the United States

Approximately 339 000 million gallons per day (1 283 000 million liters per day) of fresh water (about one-quarter of the national renewable supply) was withdrawn from streams, reservoirs and wells during 1990 for use in the nation's homes, cities, farms and industries. After use, about 220 000 million gallons per day (832 700 million liters per day) was returned to the natural system. The following sections describe some prominent water issues and how the water is used in selected high-intensity use areas.

Idaho

Idaho ranks second in the irrigation of crops, which dwarfs all other water uses in the state. Idaho is a leading national producer of several crops including potatoes, barley, sugarbeets, hops, mint, onions, sweet corn, dried beans, alfalfa, hay and wheat. Agriculture predominates in the south-central and southeastern parts of the state along the Snake River and in tributary valleys adjacent to the Snake River Plain. Irrigation steadily increased by 37% from 1970 to 1985 and decreased 9% from 1985 to 1990. Idaho uses more water for fish farms than any other state. The fish farms are conveniently located in spring outlets along the Snake River.

California

The Central Valley has long been one of the premier agricultural areas of California, as well as the nation. It produces crops of the widest diversity and highest value of any comparable region in the world. Withdrawals for irrigation in California account for more than 20% of all irrigation in the United States, with most of the irrigation occurring in the Central Valley. Major irrigated crops include alfalfa, cotton, pasture, grapes and wheat, as well as a wide assortment of nuts, fruit and vegetables. The intensive agricultural development in the Central Valley is supported by groundwater pumpage and large imports of surface water. During the 1960s and the 1970s, withdrawals from wells caused water levels to decline hundreds of feet in the southern and western parts of the San Joaquin Valley. Overall, the net loss of aquifer storage resulted from inelastic compaction of clay beds that produced the largest volume of land subsidence in the world. However, in recent years, importing surface water, changing crop patterns, and decreasing groundwater withdrawals have controlled land subsidence in the seriously affected areas.

The Imperial Irrigation District, which is located southeast of the Central Valley, is also a productive agricultural area and receives water from the Colorado River. The total irrigated acreage in California has remained constant since 1975, and water withdrawals declined from 1975 to 1990 because of active conservation programs, irrigation drainage problems and changes in crop types.

The High Plains

The High Plains aquifer provides groundwater to irrigate millions of hectares and to support large livestock populations in the High Plains, which stretches south from the Dakotas to Texas and is bordered by the Rocky Mountains and the Mississippi River. Major crops include wheat, corn, sorghum and cotton. Irrigation has declined in recent years because of substantial decreases in groundwater levels. Depletion of the High Plains aquifer system threatens the livelihood of millions of Americans and the existence of many small mid-America farm communities.

Changes in irrigation practices, such as from flood irrigation to more efficient sprinkler irrigation, and the education of farmers on water-saving techniques may slow the rate of groundwater depletion. Accurate estimates of pumpage throughout the High Plains will assist in determining the effectiveness of conservation programs and in assessing the viability of this major aquifer system.

Arizona

The Central Arizona Project (CAP), which was created to mitigate groundwater overdraft in central Arizona, conveys water up to 542 km away from its source – the Colorado River. The CAP deliveries are used primarily for irrigation and are the source for about 10% of the state's total water withdrawals. The principal crops in Arizona are cotton, hay and grains.

Because detailed water-use information is required to monitor the effectiveness of conservation requirements and management plans, the Arizona Legislature passed the Ground-Water Management Act in 1980. This Act established the Arizona Department of Water resources, which administers the water law and manages the groundwater resources in designated Active Management Areas and Irrigation Non-Expansion Areas.

The northeast

Excluding thermoelectric power withdrawals, public supply and domestic and commercial withdrawals dominate water use in the major urban centers in the Northeast, led by New York City and Philadelphia. Many of the major public supply utilities have implemented active water-conservation programs to increase dramatically the reliability of their water systems. Industries have implemented new water-conserving processes and increased recycling to decrease their wastewater discharges.

Illinois/Indiana

Public supply withdrawals in Illinois rank fourth in the USA, and Indiana ranks first in industrial withdrawals. Water for public supplies in the Chicago and the East St Louis areas is withdrawn primarily from surface water to serve nearly one-half of the population in Illinois. Industries in the area are predominantly oil and ore refineries, and chemical and steel plants. Surfacewater quality is a major concern because it is being degraded by sewage and industrial wastewater discharges in densely urbanized areas. Because of advances in water treatment technology, the use of surface water has not yet been restricted.

Florida

Within an area that crosses the state from Tampa to Daytona Beach, large withdrawals satisfy demands for public supply, industrial and irrigation water. The nine counties in the area are home to nearly 4 million people and to millions of vacationers that visit the area annually. Also, demands for water to support the Everglades National Park are at their highest.

Irrigation of sugarcane and citrus dominates water use in the six counties that comprise the Everglades Agricultural Area and the Indian River Citrus Area in Florida. About 1 million acres (404 700 ha) are irrigated with more than 2 billion gallons (7.5 billion liters) of water per day.

R.W. Herschy

Source

Solley, W.B., 1995. United States Geological Survey, Reston, Virginia.

Cross reference

Water use in USA

UNIT HYDROGRAPH METHOD IN UK FLOOD STUDIES

This method for deriving values of Q_T for various return periods is the best when only 1–3 years of records are available. During such a short period, there is usually a sufficient number of significant storms from which the rainfall–runoff relationship for a catchment can be found. The unit hydrograph method is also the recommended procedure to follow when estimates of the probable maximum flood and the shape of the flood hydrograph are required.

Application of the unit hydrograph to obtain a flood flow hydrograph assumes that the storm rainfall is uniformly distributed over the catchment. The extent of areally uniform high intensity rainfall is limited and thus the analysis of records to obtain means of deriving synthetic unit hydrographs has been restricted to catchments with areas less than 500 km². However, the unit hydrograph method can be used for catchments up to 1000 km² provided that the rain storms are spatially uniform over the area. To obtain design floods for large rivers, flood flows must be determined separately on the major tributaries and routed downstream to the required location.

Unit hydrograph derivation

(1) Several large events (>5), with the storm rainfall well spread over the catchment area, should be selected from the short-period record. The recorded hydrographs and corresponding rainfalls are abstracted and an average unit hydrograph calculated.

(2) For ungauged catchments, <500 km², a synthetic triangular unit hydrograph (Figure U2) may be constructed from catchment characteristics using relationships derived from analysis of countrywide data. The triangular unit hydrograph is defined by three parameters: time to peak, T_p, in h, the peak flow, Q_p, in m³ s⁻¹ (10 mm)⁻¹ and the time base, TB, in h. In the UK, from FSSR16 (NERC, 1985), it is

recommended that T_p is found via $T_p(0)$ the time to peak of the instantaneous unit hydrograph given by

$$T_p(0) = 283.0 \, \text{S}1085^{-0.3}(1 + \text{URBAN})^{-2.2}\,\text{SAAR}^{-0.54}\,\text{MSL}^{0.23}$$

where MSL is the main stream length (km), SAAR is the standard average annual rainfall (mm) and the other variables are as defined previously. A convenient data interval is chosen such that $t \sim T_p(0)/5$. Then $T_p(t) = T_p(0) + t/2$.

If there are some rainfall and stage data for the catchment, it is more reliable to use

$$T_p(0) = 0.604 \, \text{LAG}^{1.44}$$

where LAG is defined here as the time (h) from the centroid of effective rainfall to the peak runoff.

$$Q_p = 220/T_p[\text{m}^3 \, \text{s}^{-1}(100 \, \text{km}^2)^{-1}(10 \, \text{mm})^{-1}]$$

and

$$\text{TB} = 2.52 \, T_p$$

For moderately sized catchments of 200–500 km², an interval t of 1 h may be expected.

Design storm

The return period of a possible design flood is chosen at this stage and a corresponding design storm assessed. As a result of varying storm rainfall profiles combined with differing initial catchment wetness design storms of a given return period do not produce floods having the same frequency of occurrence. It requires a storm of lesser frequency (longer return period) to produce a flood of a given return period.

Storm duration

A design storm duration D can be obtained by rounding to the nearest odd integer multiple of the unit hydrograph data interval t, the estimate for D from

$$D = (1.0 + \text{SAAR}/1000)T_p$$

where SAAR is the standard average annual rainfall (1916–1950). This ensures that the storm peak is at a central data point. The design storm duration will normally be less than 48 h.

Storm depth

The evaluation of the design storm requires data from maps and diagrams given in Vols II and V of the *Flood Studies Report*. The Meteorological Office devised a series of relationships between storm rainfall depth, duration and frequency from a wide variety of data from 6000 raingauges in the UK. The design task requires a $MT - D$ h rainfall, i.e. the total D h storm rainfall with a T year return period (or $MT - d$ min).

The following are the required mapped data:

- SAAR: the standard average annual rainfall (1916–1950);
- $M5$–60 min rain: the depth of rain falling in 60 min once in 5 years;
- $M5$–2 day rain: the depth falling in 2 days once in 5 years.

For the catchment location, the rainfall values for 60 min and 2 days for $M5$ return period are taken from the relevant maps and the ratio r, $M5$–60/$M5$–2 day rainfalls calculated as a percentage. For this value of r, the appropriate percentages of $M5$ rainfall are applied for d min or D h from a standard table. A table of growth factors, $MT/M5$, relating the desired T year return period fall to the 5-year fall, are then applied ($MT/M5 \times M5 - D$ h), to give the required design storm rainfall, $MT - D$ h.

It must be noted that this is only a point rainfall estimate, and to give the rainfall over the catchment, an areal reduction factor (ARF) must be applied. Thus the catchment storm rainfall for return period T years and duration D h is given by

$$P(\text{mm}) = \text{ARF} \times MT - D \text{ h}$$

Antecedent catchment condition and percentage runoff

Although the wetness of a catchment will change during a period of rainfall, it is assumed that a catchment wetness index (CWI) derived

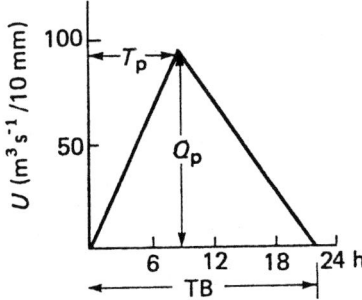

Figure U2 A synthetic unit hydrograph (10 mm effective rain).

from the average annual rainfall will be applicable throughout the storm duration. A standard percentage runoff factor (SPR) is calculated from the SOIL and then two dynamic components of percentage runoff are evaluated as follows:

$$\text{DPR}_{\text{CWI}} = 0.25 \, (\text{CWI} - 125)$$

and for larger rainfall events

$$\text{DPR}_{\text{RAIN}} = 0.45 \, (P - 40)^{0.7} \, (P > 40 \text{ mm})$$

Then an estimate of

$$\text{PR}_{\text{RURAL}} = \text{SPR} + \text{DPR}_{\text{CWI}} + \text{DRP}_{\text{RAIN}}$$

Adding an URBAN component and the percentage runoff PR for the design storm P mm is then given by

$$\text{PR} = \text{PR}_{\text{RURAL}} \, (1.0 - 0.3 \text{ URBAN}) + 70(0.3 \text{ URBAN})$$

This revised estimate of percentage runoff for a storm of P mm from FSSR16 (NERC, 1985) gives a soil type weighting to impervious areas.

Storm rainfall profile

In order to apply the P mm of rainfall in D h to the triangular unit hydrograph, a distribution of the rain depth during the rainfall period, a storm profile, is required. A family of storm profiles for different seasons in the year has been compiled from the autographic records from many stations. These all assume a symmetrical shape with the peak intensities occurring in the middle of the storm. That peakiness of the storm tends to be greater in the summer than in the winter owing to the greater likelihood of short intense thunderstorms. The distribution of the rainfall within the storm duration is represented in a series of cumulative percentage graphs related to the storm center.

Design hydrograph

By applying the percentage runoff, PR, uniformly to the rainfall amounts throughout the storm, the values of effective rain to be applied to the unit hydrograph are obtained. The convolution of the effective rainfall with the unit hydrograph gives the surface runoff hydrograph. The resultant hydrograph is an estimate of the design flood hydrograph, from which can be obtained the peak flow and the total volume of surface runoff. If relevant, an estimated baseflow component, ANSF, can be added; this may be negligible compared with the flood discharges:

$$\text{ANSF} \, (\text{m}^3 \, \text{s}^{-1} \, \text{km}^{-2}) = [33(\text{CWI} - 125) + 3.0 \, \text{SAAR} + 5.5] \times 10^{-5}$$

A software package, Micro-FSR, incorporating the latest improvements for PCs is available from the Institute of Hydrology.

Elizabeth M. Shaw

Source

Shaw, E.M., 1994. *Hydrology in Practice*, Chapman & Hall, London.

Bibliography

Natural Environment Research Council (NERC), 1985. Flood Studies Supplementary Report No. 16, Institute of Hydrology, UK.

Cross references

UNSTEADY FLOW MODEL FOR COMPUTING STREAMFLOW

Definitions

Boundary condition

A boundary condition is a condition that a dependent variable of a differential equation must satisfy along the boundary of the model domain. Boundary conditions for the dependent variables must be specified at the physical extremities of the modeled region for the duration of model application.

Courant condition

The usual condition for the numerical stability of the explicit formulation of a numerical scheme which requires that the ratio of the propagation speed of a physical disturbance to that of a numerical signal should not exceed unity is a Courant condition.

Explicit finite difference numerical scheme

Explicit numerical schemes convert either the characteristic equations or the governing equations to a system of linear algebraic equations from which the unknowns may be solved directly (explicitly) without iterative computations. Dependent variables on the advanced time level are determined one point at a time from known values and conditions at the present or previous time levels. Explicit schemes are only conditionally stable, meaning that errors may grow as the solution progresses, and the errors are a function of the time and distance finite-difference step sizes. Explicit schemes are generally stable when the Courant condition is met, which results in limitations on the maximum time and distance steps which can be used.

Gradually varied, unsteady flow

This is generally non-uniform flow in which there are no abrupt changes in depth along the longitudinal axis of the channel, and in which depth (and velocity and discharge) change with time.

Hydrograph

This is a relation in graphical, equational or tabular form between time and flow variables such as discharge, depth, velocity and stage. Stage and discharge hydrographs are typically used for open channel flows.

Implicit finite difference numerical scheme

Implicit numerical schemes convert either the characteristic equations or the governing equations to a system of non-linear algebraic equations from which the unknowns must be solved iteratively. All of the unknowns within the model domain are determined simultaneously, rather than point by point as with explicit methods. Implicit methods are generally stable, and are more computationally efficient than explicit schemes, but implicit schemes require more complex computer algorithms than explicit schemes.

Initial conditions

These are a description of the dynamic conditions (typically, discharge and depth of flow for unsteady flow models) in the model domain at some specified time, usually the beginning of the simulation period. For all subsequent times, the governing equations and the boundary conditions describe the state of the system.

Method of characteristics

The method of characteristics is a mathematical approach for solving boundary-value problems by transforming the original partial differential equations representing the physical system into corresponding characteristic equations. The characteristic equations are ordinary differential equations and generally are more amenable to numerical solution than are the original partial differential equations.

Momentum coefficient

The momentum coefficient, also known as the Boussinesq coefficient, quantifies the deviation of the velocity at any point in a cross-section from a uniform velocity distribution in the same cross-section. A value of unity indicates that a uniform velocity distribution is present in the cross-section. The momentum coefficient generally varies between about 1.01 and 1.12 for fairly straight, prismatic channels; coefficients are typically smaller for large, deep channels than for small channels.

Principles of unsteady flow models

Governing equations

The foundations for the fundamental derivation of the governing one-dimensional unsteady flow equations were laid by the nineteenth century hydraulicians Coriolis, Boussinesq and Saint Venant. The governing equations are the one-dimensional, cross-sectionally averaged expressions for (1) the conservation of mass (or equation of continuity),

$$\frac{\partial A}{\partial t} + \frac{\partial Q}{\partial x} = q \tag{U1}$$

and, (2) conservation of linear momentum

$$\frac{\partial Q}{\partial t} + \frac{\partial}{\partial x}\left(\beta\frac{Q^2}{A}\right) + gA\frac{\partial z}{\partial x} + gA(S_f - S_o) = qu' \tag{U2}$$

where A is the cross-sectional area of the channel, and varies with x, t and z; t is time; Q is the discharge, and varies with x, t and z; u' is longitudinal component of the lateral inflow velocity, and varies with x and t; x is the longitudinal position along the channel axis; z is the depth of flow, and varies with x and t; g is the acceleration of gravity; β is the momentum coefficient, and varies with x, z, and t; q is the lateral inflow per unit length of channel, and varies with x and t; S_o is the bed slope, and varies with x; and S_f is the friction slope, and varies with x, t and z.

The momentum coefficient may be computed as

$$\beta = \frac{\int u^2 dA}{U^2 A} \tag{U3}$$

where u is the velocity in some elemental area dA, and U is the mean velocity in the same cross-section having a total area A.

The friction slope S_f accounts for the resistance due to external boundary stresses. The friction slope is generally written as

$$S_f = \frac{Q|Q|n^2}{AR^{4/3}} \tag{U4}$$

where R is the hydraulic radius and n is the Manning coefficient.

Both R and n can vary as a function of x, z and t. Equation (U4) is based on the assumption that the Manning equation for steady, uniform flow provides a reasonable approximation for S_f in unsteady, non-uniform flow.

Equation (U2) can be modified to include a term accounting for the momentum imparted to the water by a temporally and spatially varying wind. Equations (U1) and (U2) also can be written with (1) depth and velocity, (2) stage and velocity or (3) stage and discharge as the dependent variables.

Equations (U1) and (U2) apply to the unsteady, spatially varied, turbulent free-surface flow of an incompressible, viscous fluid in an open channel of arbitrary cross-section and alignment. The equations are solved simultaneously for the unknowns z (depth of flow) and Q (discharge) as a function of time (t) and longitudinal position (x).

Assumptions upon which governing equations are based

Equations (U1) and (U2) are derived from first principles, and may be obtained directly from the three-dimensional equation of mass continuity and the Navier–Stokes equations, which are generally three-dimentional statements of the conservation of momentum for any fluid flow. A number of assumptions are required to derive equations (U1) and (U2). An unsteady flow model which is based on equations (U1) and (U2) should generally be applied to those conditions in which none of the major assumptions is severely violated. The assumptions are as follows:

- the flow is approximately one-dimentional, meaning that the predominant spatial variation in dynamic conditions (discharge, velocity, and stage) is in the longitudinal direction;
- the fluid density is homogeneous throughout the modeled reach;
- vertical accelerations are negligible (the hydrostatic pressure distribution is applicable);
- velocity is uniformly distributed in a given cross-section. Inclusion of the momentum coefficient in equation (U2) allows this assumption to be violated somewhat, but there should be no flow separation, and streamlines should not be highly curvilinear;

- neither aggradation nor degradation of the flow channel occurs;
- turbulence and energy dissipation can be described by resistance laws formulated for steady, uniform flow [required for equation (U4)];
- there are no abrupt changes in channel shape or alignment;
- the velocity is zero at the channel boundary;
- there is no superelevation of the water level at any cross-section;
- surface tension and the density of air at the free surface are negligible.

Simplified models

A number of techniques have been used to simplify equations (U1) and (U2) to provide approximate unsteady flow models. These simplified models generally provide results with less computational effort and fewer data than are required for solution of the full equations. However, the models have limited applicability, and it is more appropriate to use a general unsteady flow model based on equations (U1) and (U2) to obtain reliable records of discharge under a wide range of conditions. A brief summary of simplified models follows.

Empirical models
Empirical models are based on observations of past flood events. These models are limited to applications in which sufficient observations of inflows and outflows of a river section are available to calibrate essential empirical relations or routing coefficients. These models are typically applied to slowly fluctuating rivers with negligible lateral inflows and backwater effects.

Hydrological models
Hydrological models are based on the continuity equation written as

$$I - O = dS/dt \tag{U5}$$

where I is the inflow to the modeled river section, O is the outflow from the section and dS is the change in storage within the section during the time interval dt.

The storage is generally assumed to be related to the inflow or outflow by some empirically determined storage constant. Hydrological models are limited to applications in which the stage–discharge relation is single-valued, and are not applicable to flows having backwater effects, significant lateral inflows or looped stage–discharge relations. Difficulties in solving equation (U5) are often encountered when flows are changing rapidly with time.

Linearized models
Linearized models are derived from equations (U1) and (U2) by ignoring or linearizing non-linear terms in the equations. The linearized equations can then be analytically integrated with less computational effort than is required for numerical integration of equations (U1) and (U2). The most common simplifying assumptions for these models are

- the acceleration term (second term) in the momentum equation [equation (U2)] is negligible;
- the cross-sectional area (A) and channel bottom slope (S_o) are constant;
- the friction slope (S_f) is linearized with respect to discharge and depth;
- there is no lateral inflow;
- the routed flood wave has a simple shape described by an analytical expression.

These assumptions severely limit the applicability of linearized models.

Kinematic wave model
The kinematic wave model is derived by assuming that all terms in the momentum equation are negligible relative to the friction slope S_f and the bed slope S_o, and that there is no lateral inflow, so that

$$S_f = S_o \tag{U6}$$

As a consequence of equation (U6), the discharge for a kinematic flow is equal to the normal discharge. This means that the momentum of the unsteady flow is described by an expression, such as the Manning or Chézy equations, in which flow is a single-valued function of depth of flow. Moreover, kinematic waves travel without attenuation of the peak flow, but the shape of the flood wave is modified as the wave is translated downstream. The kinematic wave model allows only the downstream propagation of flow disturbances, so that backwater and tidal effects cannot be modeled. Numerous analytical solutions exist for applications of the kinematic wave model to specific flow geometries, and these models are most widely used in the routing of overland flow of precipitation runoff.

Diffusion analogy model
The diffusion analogy model is obtained by assuming that the channel is prismatic, that the local and convective acceleration terms in the momentum equation are negligible and that there is no lateral inflow. The continuity and momentum equations may then be combined to form a single parabolic partial differential equation, which is in the form of the so-called convective diffusion equation with the single unknown of discharge. The local and convective acceleration terms, the first two terms in equation (U2), are often small in steep streams.

The diffusion analogy model can be used to compute flows affected by backwater conditions. However, the diffusion model is limited to applications in which flows change relatively slowly, and in which the channel has a rather uniform geometry throughout the modeled reach.

Numerical techniques for solution of governing equations

No known analytical solutions exist for equations (U1) and (U2). Consequently, numerical techniques are used to convert equations (U1) and (U2) into algebraic equations that may be solved for z and Q at finite, incremental values of x and t. This solution depends on the proper description of the cross-sectional area as a function of x and t, and on the availability of accurate boundary condition data.

A variety of numerical techniques have been proposed and used to solve the unsteady flow equations. Although finite element methods may be used to solve the equations, finite difference techniques generally are more appropriate for the solution of the one-dimensional partial differential equations describing unsteady open channel flow. The three broad categories of numerical techniques are (1) method of characteristics, (2) explicit finite difference methods and (3) implicit finite difference methods. Numerous variations of each of these general categories of techniques exist. The methods are briefly reviewed to provide some perspective on the advantages and disadvantages of each method.

Method of characteristics
The method of characteristics is a mathematical approach for solving boundary value problems by transforming the original partial differential equations representing the physical system into corresponding characteristic equations. In this context, the characteristic is the speed of a wave relative to a stationary observer. Characteristic equations are ordinary differential equations and are generally more amenable to numerical solution than the original partial differential equations. The characteristic equations are solved using either explicit or implicit finite difference methods.

The method of characteristics can be used with a curvilinear grid or a rectangular grid in the x–t domain. The curvilinear grid is not generally used for solution of the unsteady flow equations in natural open channels. The nature of characteristics is such that the wave trains in the x–t domain usually are not orthogonal, so that solutions of the characteristic equations typically do not coincide with a point on the rectangular grid representing the natural system. Consequently, an interpolation scheme is required to transfer results from the characteristic network to the rectangular grid representing the flow system. The accuracy of the interpolation scheme plays a major role in determining the performance of the method of characteristics in solving the governing equations.

Explicit finite difference methods
Explicit numerical schemes convert either the characteristic equations or the governing equations to a system of linear algebraic equations from which the unknowns may be solved directly (explicitly) without iterative computations. Dependent variables of the advanced time level are determined one point at a time from known values and conditions at the present or previous time levels. Explicit schemes are only conditionally stable, which means that errors may grow as the solution progresses, and the errors are a function of the time and distance finite-difference step sizes. Explicit schemes are generally stable when the Courant condition is met, which results in limitations on the maximum time and distance steps which can be used.

In order to meet numerical stability requirements, the computational time step must decrease as the hydraulic depth increases.

Consequently, computational time steps may be required to be of the order of a few minutes for unsteady flow models of large rivers, which makes the models computationally somewhat inefficient. Explicit finite difference schemes also require the computational distance steps to be equal throughout the model domain, which may be a disadvantage for some systems.

Implicit finite difference methods

Implicit numerical schemes convert either the characteristic equations or the governing equations to a system of non-linear algebraic equations from which the unknowns must be solved iteratively. Consequently, a system of $2N$ algebraic equations is generated for a model having N cross-sections along the x axis. All of the unknowns within the model domain are determined simultaneously, rather than point-by-point as with explicit methods.

Weighting factors are typically required in the application of implicit schemes. These factors determine the time between adjacent time levels at which (1) the spatial derivatives and (2) functional quantities are evaluated. Functional quantities are features such as cross-sectional area, top width and hydraulic radius, all of which are functions of the computed depth of flow. Some judgement is required in selecting these weighting factors, and the weighting factors are often adjusted as part of the model calibration process. The accuracy of the numerical scheme generally decreases as the factor approaches one, where the terms in the governing equations are expressed entirely in terms of the future time step.

Fewer numerical stability problems are encountered with implicit schemes than with explicit schemes. Numerical instabilities can occur when modeling rapidly varying flows if the time step is large and if the spatial derivatives are not sufficiently weighted towards the future time step. Non-linearities caused by irregular cross-sections having widths that vary rapidly along the channel or with depth also can cause numerical instabilities in implicit models.

Data requirements

Data are required to construct, calibrate, test and apply unsteady flow models. Referenced International Standards for the measurement of velocity and discharge, and for collection of water-level and discharge records should be followed.

In general, data are required at model boundaries for the entire period for which flow is to be computed using the unsteady flow model. Short-term records and discrete measurements are needed at locations within the model domain for the period which is used for model calibration and testing.

Selection of model boundaries

Reliable, accurate and appropriate boundary condition data are required for the successful computation of streamflow using an unsteady flow model. Model boundaries must be selected prior to the installation of data collection instrumentation. Boundaries should be in locations where there are fewest flow disturbances, such as sharp bends, rapid changes in cross-sectional geometry and major inflows. The modeled reach length also should be sufficiently long to permit accurate determination of the longitudinal water surface slope so that adverse effects of measurement errors are minimized. Moreover, as subsequently discussed, discharge records should be used as the upstream boundary condition if at all possible. For these reasons it may be expedient to extend the model domain beyond the reach for which streamflow computations are actually needed in order to obtain the necessary data for model boundary conditions.

Stage data

Stage data are required at all external boundaries of the modeled system in order to specify boundary conditions. (Stage is typically required even if discharge is used as the upstream boundary condition because stage is generally needed in the computation of discharge, whether from a stage–discharge relation or *in situ* velocity meters.) Multiple-channel systems require stage data at upstream and downstream external boundaries of each channel. Stage data are not required at internal junctions, where channels join within the model domain. Stage also should be measured in at least one, and preferably three, locations within the model domain to provide data for model calibration and testing.

It is critically important that all stage measurements be referenced to a common datum. Errors in gauge datum translate into errors in water surface slope, which greatly affects computed streamflow through the third term in equation (U2). Correct datums are particularly important (and perhaps more difficult to obtain) for low-gradient channels, where unsteady flow models are often applied. The use of discharge as the upstream boundary condition removes much of the uncertainty associated with potential errors in gauge datums.

Except in very large rivers (widths of several hundred meters or more), stage should be measured at intervals of 15 min or less for reliable modeling of flow transients. If possible, the stage measurement interval should be a whole multiple of the computational interval, and should be no more than about five times the computational interval.

Synchronous measurement of stage at all recorders is also required for the application of unsteady flow models for computing streamflow. Asynchronous measurements, like datum errors, translate into errors in water surface slope and, hence, errors in computed streamflow.

Stage is measured following procedures outlined in ISO 1100/1 using equipment described in ISO 4373.

Velocity data

Measurements of velocity in the study reach are required to (1) evaluate the assumption of one-dimensional flow and (2) compute the momentum coefficient [equation (U3)]. Stream velocities obtained during discharge measurements are generally adequate for these purposes.

Velocities are measured following procedures outlined in ISO 748 or ISO 2425 using equipment described in ISO 2537 and ISO 3454.

Discharge data

Discharge data are required for model calibration, and may be needed as boundary data. Discharge data may be either (1) a continuous time series obtained from a stage–discharge rating or by using the *in situ* velocity meters, such as ultrasonic velocity meters, or (2) discrete measurements. Time series of discharge are generally required only at the upstream boundaries. Discrete measurements are made within the model domain for the purposes of model calibration and testing.

Discharge is measured using methods described in ISO 555, ISO 748, ISO 1070, ISO 1100/1 and 1100/2, ISO 2425 and ISO 6418 (see International standards in flow measurement).

Lateral inflows and withdrawals

Time-varying records of major inflows into and withdrawals from the modeled reach must be included in the model to maintain mass balance. Inflows or losses which are relatively constant throughout the modeled reach, such as groundwater inputs or losses, can be lumped into a few discrete points in the reach or can be included at each computational node. Inflows from major streams must be gauged. Inflows from minor streams and local areas along the channel can be estimated using data from nearby gauged streams and drainage area ratios.

Channel cross-section data

Cross-sectional data consists of a set of longitudinal, lateral and vertical coordinates which describe the location and configuration of the cross-section. A measured cross-section is typically required at each computational node within the model. The exact spacing of the computational nodes, however, is not known until convergence testing has been completed and stability criteria have been evaluated.

Numerical solutions of the governing equations generally use the average of the measured cross-sections at the upstream and downstream ends of a reach to represent the cross-sectional geometry of the entire reach bounded by the two measured sections. Consequently, measured cross-sections should be fairly representative of conditions upstream and downstream of the measurement. The measured sections also should be spaced sufficiently close so that large changes in channel geometry do not occur between the sections.

The longitudinal distance between cross-sections can be determined from a map having a scale of 1:24 000 or larger. Longitudinal distances should be measured along the centerline of the channel.

Cross-sections are measured in accordance with ISO 748 or by using a recording fathometer. It is generally not practical to run levels from an established benchmark to each cross-section. Consequently, cross-sections can be measured during steady flow, when water levels are not changing, and the stage at the measurement site can be interpolated from measured stage at upstream and downstream stage recorders. Cross-sections must be referenced to the same datum as the stage records.

Calibration and testing of unsteady flow models

Modeling is based on the abstraction of a physical system to a mathematical expression and replication of the system using these expressions and appropriate field data. The analyst must identify the important features of the flow system and ensure that those features are reflected in the model which is selected for application to the study reach. Important general model attributes include (1) the ability to simulate a wide range of flow conditions; (2) the ability to represent a range of complex channel conditions and geometries; (3) a stable, numerically convergent, efficient computational scheme and (4) a system for processing model input data and output simulation results.

Preliminary tests

In many cases it is appropriate to conduct preliminary tests using simplified channel geometry and boundary conditions with the unsteady flow model. Tests should be conducted if the model is poorly documented, or if the user is unfamiliar with the model.

Tests should be conducted using a channel which has a uniform, rectangular cross-section, and with the model configured for the study reach. Tests using the rectangular cross-section model could include the following:

- no inflow and no bed slope – no flow should be generated within the model domain;
- steady inflow – mass should be conserved;
- unsteady inflow – mass should be conserved;
- triangular-shaped inflow hydrograph in channel with no bed slope – peak flow should not be significantly attenuated.

After the model for the study reach has been constructed using the measured cross-sectional data, tests which should be performed include

- no inflow or water surface slope – no flow should be generated in the model domain; this test also determines if unintentional openings in the boundaries are present;
- steady and unsteady flow – mass should be conserved;
- rapid change in inflow boundary conditions – no numerical instabilities should be generated;
- change in boundary conditions from one steady flow to another flow – the amount of time required for all flows within the model domain to reach the new steady-state condition is an indication of how long the initial conditions persist within the model domain; model results are generally not accepted until the effects of the initial conditions are transported out of the model domain so that the model is responding to boundary conditions only.

Other tests may be performed as needed, but these simple tests can be used to document general model performance and should allow users to gain a better understanding of model capabilities and limitations prior to application to the study reach.

Computational grid and time step

The computational grid is used to represent the physical system in the unsteady flow model. Junctions, inflows and outflows must be represented by the computational grid.

The channel system is subdivided into a number of finite segments for solution of the numerical approximations of the governing equations. The solution points are either at the ends or the midpoint of each segment. The computational grid should be established such that computations of stage and discharge coincide with locations of data collection, or at locations where computed data are required. Measured cross-sectional data should be available at the ends of each computational segment or grid cell.

Some models allow non-uniform segment lengths, but others require all segments to have the same length. Segment lengths should be at least three times greater than the width of the channel, and are frequently five to ten times the channel width. Exact segment length is determined during convergence testing.

The representation of overbank areas, or the floodplain, in the model must be done with caution. For relatively narrow floodplains, in which the flow length of the floodplain is very nearly equal to the flow length of the channel, model assumptions of one-dimensional flow are not violated. Broad floodplains, which bound a sinuous channel, may have flow lengths distinctly different than that of the channel. Moreover, filling and draining of the floodplain may lag the rise and fall of water in the channel. In these cases, the one-dimensional flow model described in this standard is not appropriate for application.

In some instances, off-channel storage reservoirs have been included in one-dimensional flow models to account for the storage and release of water from the floodplain. These formulations account for the slow storage and release of floodplain waters. However, such a model design also converts a process which occurs throughout some finite reach of the channel into a single-point process.

The computational time step should be sufficiently small to represent accurately the flow transients which occur in the modeled system. Generally, the computational time step is reduced to meet stability criteria rather than adjusting the spatial discretization interval. As with the computational grid, convergence testing helps define the maximum time step which can be used.

Convergence testing

A finite-difference solution to a partial differential equation is spatially convergent if the numerical solution approaches the true solution of the differential equation as the finite-difference spatial discretization approaches zero. Spatial convergency can be tested by repeatedly applying the model with a fixed set of boundary conditions for successively smaller computational discretizations. The model is spatially convergent if no further change in model results is observed as the spatial step is refined. Likewise, a model is temporally convergent if model results remain substantially unchanged as the computational time step is decreased. To determine the effects of spatial discretization and time step on model results, convergence testing should be conducted prior to model calibration.

Boundary and initial conditions

Two initial conditions (Q and z for the formulation of the unsteady flow equations used in this standard) are required at each computational node in the model domain. For the initial application of the model to a study reach, common initial conditions are a steady flow, equal to the initial boundary condition flow, and a water surface which slopes linearly from a measured upstream stage to a measured downstream stage. Output from a previous unsteady flow model application may also be used to determine initial conditions for a simulation which follows sequentially in time.

The model may only be applied for periods which have measured boundary conditions. Boundary conditions include a time series of measured stage at the downstream boundaries, measured stage or discharge at the upstream boundaries, and measured lateral inflows. As previously discussed, better results are often obtained when discharge is used as the upstream boundary condition.

Calibration

Model calibration is required to adapt a general unsteady flow model to the specific application for which streamflows are to be computed. Calibration is accomplished by adjusting the model parameters until model results agree with observations Essentially all components of the model are subject to adjustment during model calibration. Components that are directly measurable and physically well defined, however, are typically less subject to adjustment than are those that might not be directly measured. Measures for quantifying the calibration are discussed under Validation, below.

Initially, the resistance coefficient, momentum coefficient and weighting coefficients for the numerical scheme should be varied, because these parameters cannot be measured. Boundary gauge datums may be adjusted slightly if there is some uncertainty about the accuracy of the datum. Cross-sectional geometry may also be adjusted during the calibration process. The adjustment of channel geometry is justified because the measured cross-sections are used to represent the average conditions within a computational segment, rather than the actual conditions at the measured cross-section.

It is entirely possible to achieve a well-calibrated model with empirical coefficients which bear little resemblance to those justified by the physics and setting of the study reach. In such a case, the application of the model to other conditions is not justified without another calibration.

Because of the important assumptions of the governing equations, that the channel is stable and that most natural channels undergo continuous change, model calibrations should be repeated at least annually. The amount of data required for the subsequent model calibrations should not be as great as for the initial calibration, unless the user suspects dramatic changes in the stream geometry.

Validation

Validation refers to the comparison of model results to measured stage and discharge. Calibration data should not be used for validation.

Graphical comparisons of measured and simulated information are often used, but may be misleading. Quantitative measures of validation include measures of deviation and testing of the statistical significance of the deviation. Measures of deviation between model results and data include absolute and relative error, and root mean square deviations.

Statistical tests of significance are necessary to determine whether deviations are meaningful or whether they are simply related to the variability in the data. However, statistically independent data are needed to test for significance. Consequently, data and model results must be sampled at intervals greater than the correlation time scale before applying certain statistical tests, such as the t-test.

If possible, the model should be validated over the range of flow conditions for which the model is to be applied. The model may be applied to flows reasonably outside the range for which the model was tested, as long as conditions do not change appreciably. For example, a model calibrated for flows less than bankfull should not be applied to simulate overbank flows.

Greater emphasis should be placed on validating against discharge rather than stage, because the purpose of the model is to compute streamflow. However, in most cases, stage data are much easier and less expensive to obtain than discharge data, and more stage data will be available for use in calibration, validation and testing.

Sensitivity testing

Sensitivity testing consists of evaluating the sensitivity of model results to changes in selected model parameters. Parameters, or conditions, which should be included in the sensitivity testing include the resistance coefficient, the weighting coefficients used in the numerical scheme, the momentum coefficient, boundary gauge datums (particularly if stage is used as the upstream boundary condition) and channel geometry. The usual procedure is to increment successively the parameters by small amounts, apply the model and compare the results with results from the calibrated model. A model which is highly sensitive to small changes in one or more parameters may become unstable for conditions outside those used for model calibration, and greater care must be taken when applying such a model.

Uncertainties

Governing equations

Equations (U1) and (U2) strictly apply to an infinitesimally small volume at an instant in time. For the development and application of unsteady flow models, the equations are assumed to apply to some finite volume, which may have a length of the order of hundreds of meters, a width of tens of meters and a depth of several meters; the equations also are assumed to apply to some finite duration, which may be as much as 1 h. Uncertainties associated with the extrapolation of the differential equations to these finite volumes and times exist and are difficult to quantify exactly. However, the previously described convergence testing does provide an indication, along with the satisfaction of numerical stability criteria, that the appropriate time and distance discretization has been selected.

Flow in open channels is three-dimensional in nature, but is approximated by the governing equations as having only variations in the longitudinal direction. In addition to assumption of one-dimensional flow, a number of other assumptions were used in the development of the unsteady flow equations. Some of these assumptions are restrictive and deviations from the assumption (such as no channel geometry change and no density gradients) will probably produce erroneous simulations. Other assumptions, such as negligible surface tension effects, will not be violated in open channel flow. All deviations from model assumptions should be clearly documented, and model testing should be used to assist in quantifying the effects of the deviations on flow computations.

Numerical approximations to governing equations

Equations (U1) and (U2) are simplified mathematical expressions for the complex three-dimensional turbulent flow field in an open channel. The expressions are further simplified by making numerical approximations of the partial differential equations. These approximations introduce further uncertainty into the problem. The uncertainty, which is difficult to quantify for any given application, can be evaluated through testing and documentation of the numerical scheme.

Clear documentation of the numerical scheme, including equations, discretization and results of tests of the scheme, should be available. Certain numerical schemes, including variations of the method of characteristics, explicit and implicit schemes for one-dimensional unsteady flow modeling, have general testing and documentation that is widely available. However, many schemes which have been devised for specific problems may require extensive evaluation by model users.

The documentation of a numerical scheme should include a discussion of the relations among grid size, time step, and the stability and accuracy of the scheme. Numerical dispersion introduced by the scheme, which can result in the damping of a steep wave front, should be quantified.

Model parameters

A number of physical processes which are not explicitly expressed in the momentum equation [equation (U2)] are combined into the channel rugosity coefficient (Manning n, in this case). The Manning n is an empirically derived coefficient which is an approximation of channel rugosity in a steady uniform flow. The parameter in equation (U2), however, is applied to unsteady flow and, from the physical perspective, includes the effects of turbulent dissipation of energy by several processes.

The Manning n value, as well as weighting factors used in certain numerical schemes, cannot be measured, and these parameters are adjusted during model calibration to obtain agreement between model results and prototype measurements. Consequently, these model parameters also may include the effects of (1) the deviation of the modeled system from model assumptions, (2) the effects of numerical approximations to the governing equations and (3) errors in field measurements. The uncertainties associated with combining these processes into a few model parameters is difficult, if not impossible, to quantify. However, model parameters selected during the calibration and testing phase of model development should be within the range of previously published values. Otherwise, the unrealistic model parameter values may be masking some serious errors within the model – errors which may result in poor simulations of streamflow for conditions which were not evaluated during model testing.

Data for model development, testing and application

The collection of hydraulic field data is subject to a number of uncertainties. The quantification of uncertainties associated with the measurement of stage, velocity, discharge and cross-sectional data has been discussed in previously published International Standards.

Uncertainties associated with the estimation of lateral inflows and withdrawals should be minimized by accurately measuring as many of the inputs and losses as possible.

The effects of uncertainties associated with errors in gauge datums can be minimized by using discharge as an upstream boundary condition rather than stage.

Source

United States Geological Survey, 1996. Draft, Technical Report for ISO/TC113/SC1.

Bibliography

Daugherty, R.L. and Franzini, J.B., 1965. *Fluid mechanics with engineering applications*. New York, McGraw-Hill Book Company, 578 pp.

DeLong, L.L., 1989. Mass conservation – I-D open channel flow equations. *Journal of Hydraulic Engineering, American Society of Civil Engineers*, **115**(2), p. 263–269.

Ditmars, J.D., Adams, E.E., Bedford, K.W., and Ford, D.E., 1987. Performance evaluation of surface transport and dispersion models. *Journal of Hydraulic Engineering, American Society of Civil Engineers*, **113**(8), 961–980.

Ervine, D.A., Willetts, B.B., Sellin, R.H.J. and Lorena, M., 1993. Factors affecting conveyance in meandering compound flows. *Journal of Hydraulic Engineering, American Society of Civil Engineers*, **119**(12), 1383–1399.

Fread, D.L., 1988. *The NWS DMBRK model – theoretical background and user documentation*. Silver Spring, Maryland, Office of Hydrology, National Weather Service, 123 pp. + app.

Henderson, F.M., 1966. *Open Channel Flow*. New York, Macmillan Publishing Company, 522 pp.

ISO 9555, 1993. *Liquid flow measurement in open channels – Dilution methods for measurements of steady flow*.

ISO 748, 1997. *Liquid flow measurement in open channels – Velocity–area methods*.

ISO 772, 1996. *Liquid flow measurement in open channels – Vocabulary and symbols*.

ISO 1070, 1992. *Liquid flow measurement in open channels – Slope–area method*.

ISO 1100/1, 1996. *Liquid flow measurement in open channels – Part 1: Determination of the stage-discharge relation*.

ISO 2425, 1998. *Measurement of flow in tidal channels*.

ISO 4360, 1979. *Measurement of liquid flow in open channels – Moving-boat method*.

ISO 4373, 1996. *Measurement of liquid flow in open channels – Water-level measurement devices*.

ISO 6416, 1992. *Measurement of liquid flow in open channels – Measurement of discharge by the ultrasonic (acoustic) method*.

ISO 11332, 1998. *Flow in unstable channels and ephemeral streams*.

Lee, J.K., 1989. The one-dimensional equations of unsteady open-channel flow, in Schaffranek, R.W., ed., *Proceedings of the Advanced Seminar on One-Dimensional, Open-Channel Flow and Transport Modeling*. US Geological Survey Water-Resources Investigations Report 89–4061, pp. 6–10.

Overton, D.E. and Meadows, M.E., 1976. *Stormwater Modeling*. New York, Academic Press, 358 pp.

Roache, P., 1982. *Computational Fluid Dynamics*. Albuquerque, New Mexico, Hermosa Publishers, 446 pp.

Schaffranek, R.W., 1989. Appendix I – glossary of technical terminology, in Schaffranek, R.W., ed., *Proceedings of the Advanced Seminar on One-Dimensional, Open-Channel Flow and Transport Modeling*. US Geological Survey Water-Resources Investigations Report 89–4061, pp. 77–98.

Schaffranek, R.W., Baltzer, R.A. and Goldberg, D.E., 1981. A model for simulation of flow in singular and interconnected channels: Chapter C3, Book 7, *Automated Data Processing and Computations, Techniques of Water-Resources Investigations of the United States Geological Survey*, 110 pp.

Strickland, A.G. and Bales, J.D., *Simulation of unsteady flow in the Roanoke River from near Oak City to Williamston, North Carolina*. US Geological Survey Water-Supply Paper 2408, Chapter A, 34 pp.

Strelkoff, T., 1969. One-dimensional equations of open-channel flow. *Journal of the Hydraulics Division, American Society of Civil Engineers*, **95**(HY3), 861–876.

Weinmann, P/E. and Laurenson, E.M., 1979. Approximate flood routing methods – a review. *Journal of the Hydraulics Division, American Society of Civil Engineers*, **105**(HY12), 1521–1536.

Xia, R. and Yen, B.C., 1994. Significance of averaging coefficients in open-channel flow equations: *Journal of Hydraulic Engineering, American Society of Civil Engineers*, **120**(2), 169–190.

Yen, B.C., 1973. Open-channel flow equations revisited: *Journal of the Engineering Mechanics Division, American Society of Civil Engineers*, **99**(EM5), 979–1009.

Cross references

Modeling, of water resource systems
Model predictions: uncertainty
Models: distributed models of catchment hydrology
Models: parameter estimation

Streamflow measurement
Water resources: dictionary of basic terms

URBAN HYDROLOGY

Urban hydrology is not a new field (Leopold, 1968; Johnson, 1971). Urbanization results in radical changes in land use and in the interaction between land and water. Hence the hydrology of urbanized areas differs notably from that of the same land in its preceding rural condition. Urban hydrology is the scientific application of hydrological principles and knowledge to the planning and management of urban areas and their surroundings. It embraces all aspects of the interactions between humanity and water in occupancy of land and includes the special hydrological studies needed to accomplish these ends. Urban hydrology deals with minimizing the adverse effects of the human use of land and water and with maximizing the effective use of the available water resources (see Hydrogeology).

More than two-thirds of the US population is urban, occupying less than 10% of the land. This majority may well control the destiny not only of urban areas, but also of the remaining 90% of the land. Population trends in the United States suggest that the urban population in the year 2000 will be three-quarters of the total population of the nation and more than the total population of today (1996). Sound management of total resources, and especially of urban areas, will be increasingly urgent.

As urbanization proceeds, an increasing proportion of the total land area becomes covered with impermeable surfaces such as roofs and pavement. Rainfall, which formerly trickled slowly through vegetated areas or soaked into the ground, now runs quickly over the surface to streams. This creates one of the major problems of urban hydrology – flooding – or the safe transmission and rapid disposal of greatly increased surface runoff. One role of the urban hydrologist is to estimate the quantities of maximum runoff and, on this basis, to design storm sewers, discharge channels and disposal and treatment facilities. Impervious surfaces also reduce infiltration, soil moisture and groundwater recharge. When combined with the increased groundwater withdrawal that often accompanies urbanization, this often results in lowered water tables and decreased groundwater yields (Rantz, 1970; Schneider, 1970; Kuo, 1993).

Human activity in an urban environment produces large quantities of wastes that can find their way into and degrade the quality of the natural waters of the area. Surface streams receive both solid particles (sediment) and dissolved matter. Groundwater normally receives only dissolved substances. The control of contaminants and/or the correction of pollution is a major function of urban hydrology. Another important facet of urban hydrology is the provision of adequate quantities of water for a concentrated population. If sources of water are not available in the immediate vicinity, a search in more distant areas becomes necessary. Thus urban hydrology involves the planning and development of public water supplies and of the works for the disposal of liquid wastes (Schneider et al., 1973; Thanh and Biswas, 1990).

Urban hydrology is concerned with the management of floodplains under urban conditions. Many urban areas have been built on floodplains because these areas are level, easily built on and usually dry. A floodplain is a strip of relatively smooth land adjacent to a river channel, constructed by the present river in its existing regimen, and covered with water when the river overflows its banks at times of high water. Buildings and other objects on the floodplain tend to be submerged or swept away at times of high water. Measures to prevent or reduce the loss of life and property on urbanized parts of floodplains thus become an important part of urban hydrology. The choice of measures to accomplish this includes structural measures, such as levees, dikes, channel improvement and upstream storage, and non-structural measures such as zoning and building regulations. The best solution in any given situation depends on local circumstances, and a combination of structural and non-structural measures often produces the best results (Bue, 1967; McHarg, 1969).

Floods are the most destructive of all geological hazards and take tolls on lives and damage to property. Yet flooding streams serve an important, increasingly recognized function in the local and regional environmental balance. Human development of drainage basins has a profound effect on drainage, flooding and erosion, as well as water quality and aquatic life. Flood control has become intimately entangled with water-use rights, and this has made flood control the most fiercely politicized issue of land-use planning and environmental management (Nuhfer et al., 1993).

The avoidance or abandonment of large tracts of land during urban growth and development is usually neither practical nor necessary. The mapping of geological hazards defines specific areas to be avoided for particular types of development. Otherwise hazardous sites may make ideal parks or green belt space. Towns such as Jamesville, Wisconsin, have created beautiful parks in areas zoned as floodplains, thus avoiding placing expensive structures where flooding will cause destruction (Nuhfer et al., 1993).

The use and control of water for the aesthetic enhancement of the urban environment, for example watering of desirable vegetation, judicious draining of urban swamps to increase their beauty and utility, and providing parkland lakes and watercourses, also fall within the purview of urban hydrology. Closely allied to this function is the provision of water and the adaptation of natural bodies of water in urban localities to serve best the recreational needs of a crowded populace. This may range from the provision of a sprinkler on a city fire hydrant on a hot summer day to the construction of lakes in city parks that are used for rowing, fishing, skiing or swimming. It includes the provision of public swimming pools, and the adaptation of the shores of major bodies of water to varied recreational uses.

Thus we see that urban hydrology covers a wide range of water management and control functions within and around cities. Where water is a threat or inconvenience, it must be controlled. Where it can serve a useful or satisfying purpose, it must flow in appropriate channels. Where human ignorance or greed results in structures that impinge on the natural channels or reduce the utility or quality of water, restrictions must be imposed so that we can live safely and harmoniously with the water that is an essential and inescapable part of our environment and that serves many of our needs.

Urban hydrology is related to numerous other sciences and disciplines, and when applied in conjunction with them, affords wise planning and management of the entire urban environment (see Hydrology). The goal of such management should be to permit the efficient use of the hydrological regime, and to preserve this vital resource for the benefit of future generations.

Philip E. LaMoreaux and Lois D. George

Bibliography

Abdullaev, K.M., Malakhov, I.A., Poletaev, L.N. and Sobol, A.S., 1992. *Urban Waste Waters: Treatment for Use in Steam and Power Generation*, Ellis Horwood Limited, West Sussex, UK.

Beck, B.F., ed., 1993. *Karst Geohazards, Engineering and Environmental Problems in Karst Terrane*, Proceedings of the Fifth Multidisciplinary Conference on Sinkholes and the Engineering and Environmental Impacts of Karst, Gatlinburg, Tennessee, 2–5 April 1993, A.A. Balkema, Rotterdam, Netherlands, 581 pp.

Beck, B.F., ed., 1993. *Applied Karst Geology*, Proceedings of the Fourth Multidisciplinary Conference on Sinkholes and the Engineering and Environmental Impacts of Karst, Panama City, Florida, 25–27 January 1993, A.A. Balkema, Rotterdam, Netherlands, 295 pp.

Bernknopf, R.L., Brookshire, D.S., Soller, D.R. et al., 1993. *Societal Value of Geologic Maps*, US Geol. Survey Circ. 1111, 53 pp.

Bolt, B.A., 1975. *Geologic Hazards*, Springer-Verlag, New York, NY.

Bue, Conrad D., 1967. *Flood information for flood plain planning, US Geol. Survey Cir. 539*, 10 pp.

European Commission, 1995. *Cost action 65 – Hydrogeological aspects of groundwater protection in karstic areas. Final report*, Luxembourg: Office for Official Publications of the European Communities, Belgium, 446 pp.

Griggs, G.B., 1983. *Geologic Hazards, Resources, and Environmental Planning*, Wadsworth Publishing Company, Belmont, CA.

Johnson, J.H., 1971. *Urban Geology – An Introductory Analysis*. Oxford: Pergamon Press, 188 pp.

Kuo, C.Y., ed., 1993. *Engineering Hydrology*. Proceedings of the ASCE Symposium, San Francisco, CA, July 25–30, 1993, American Society of Civil Engineers, New York, NY.

Legget, R.F., 1973. *Cities and Geology*, McGraw-Hill Book Company, New York, NY.

Leopold, L.B., 1968. *Hydrology for urban land planning – a guidebook on the hydrologic effects of urban land use*, US Geol. Survey Circ. 554, 18 pp.

Mathewson, C.C. and Keaton, J.R., 1988. *Flood Hazard Recognition and Mitigation on Alluvial Fans*, ASCE Proceedings of Hydraulic Engineering Symposium, Colorado Springs, CO.

McHarg, I., 1969. *Design with Nature*, American Museum of Natural History, Natural History Press, Garden City, NY, 197 pp.

National Research Council, Panel on Land Subsidence, 1991. *Mitigating Losses from Land Subsidence in the United States*, National Academy Press, Washington, DC, 58 pp.

Nuhfer, E.B., Proctor, R.J. and Moser, P.H. (eds), 1993. *The Citizens' Guide to Geologic Hazards*, American Institute of Professional Geologists, Arvada, CO 80003, 134 pp.

Rahn, P.H., 1986. *Engineering Geology – An Environmental Approach*: Elsevier, New York.

Rantz, S.E., 1970. *Water in the urban environment: urban sprawl and flooding in southern California*, US Geol. Survey Circ. 601-B, 11 pp.

Schneider, W.J., 1970. *Water in the urban environment: hydrologic implications of solidwaste disposal*, US Geol. Survey Circ. 601-F, 9 pp.

Schneider, W.J., Rickert, D.A. and Spieker, A.M., 1973. *Water in the urban environment: role of water in urban planning and management*, US Geol. Survey Circ. 601-H, 10 pp.

Thanh, N.C. and Biswas, A.K., eds, 1990. *Environmentally-Sound Water Management*, Oxford University Press, Oxford. 276 pp.

Utgard, R.O., ed., 1978. *Geology in the Urban Environment*, Burgess Publishing Company, Minneapolis, MN.

Cross references

Alluvial valley engineering
Hydrology

V

VELOCITY–AREA METHOD

General

The velocity–area method for the determination of discharge in open channels consists of measurements of stream velocity, depth of flow and distance across the channel between observation verticals. The velocity is measured at one or more points in each vertical by current meter and an average velocity determined in each vertical. The discharge is derived from the sum of the product of mean velocity, depth and width between verticals. The discharge so obtained is normally used to establish a relation between water level (stage) and streamflow. Once established, this stage–discharge relation is used to derive discharge values from records of stage at the gauging station.

Not all current-meter measurements, however, are made to establish a stage–discharge relation, and individual determinations or 'spot measurements' are very often required for management functions. Such measurements may not require the measurement of stage, but otherwise the method of measurement is the same. At some stations a record of stage only may be required for purposes such as flood warning. At most gauging stations, however, both stage and discharge are measured to establish a relation between these two variables.

Further details may be found in Herschy (1993, 1995).

Spacing of verticals

In order to describe the bed shape and the horizontal and vertical velocity distributions completely, an infinite number of verticals would be necessary; for practical reasons, however, only a finite number is possible. In practice, therefore, the cross-section is divided into segments by spacing verticals at a sufficient number of locations across the channel to ensure an adequate sample of both velocity distribution and bed profile. The spacing and number of verticals are crucial for the accurate measurements of discharge, and for this reason between 20 and 30 verticals are normally used. This practice applies to rivers of all widths except where the channel is so narrow that this number of verticals would be impracticable. Uncertainties in streamflow measurement are expressed as percentages. The percentage uncertainty, therefore, for using, say, 20 verticals is of the same order for all widths of river notwithstanding the width of the segments (in absolute terms the uncertainty will increase as the width of segment increases).

Verticals may be spaced on the basis of the following criteria:

- equidistant;
- segments of equal flow;
- bed profile.

The choice will depend largely on the flow conditions, the geometry of the cross-section and the width of the river. For very wide rivers (over 300 m), for example, it is sometimes convenient to make the verticals equidistant; for rivers having an asymmetrical horizontal velocity distribution, or a significant variation in the horizontal velocity distribution, it is normally advisable to space the verticals in such a manner as to achieve segments of equal flow over the required range; for rivers having abnormalities in the bed profile, the verticals are spaced so as to make allowance for depressions or obtrusions and general irregularities of the bed. A general rule, however, for current-meter measurements is to make the width of segments less as the depth and velocities become greater.

Irrespective of which criterion is followed, the spacing of the verticals is arranged so that no segment contains more than 10% of the total flow. The best measurement is normally one having no segment with more than 5% of the total flow.

Computation of current-meter measurements

Mid-section method

In the mid-section method of computation it is assumed that the velocity sampled at each vertical represents the mean velocity in a segment. The segment area extends laterally from half the distance from the preceding vertical to half the distance to the next, and from the water surface to the sounded depth, as shown by the hatched area in Figure V1. The segment discharge is then computed for each segment, and these are summed to obtain the total discharge. Referring to Figure V1, which shows diagrammatically the cross-section of a stream channel, the discharge passing through segment 5 is computed as

$$q_5 = \bar{v}_5 \left(\frac{(b_5 - b_4) + (b_6 - b_5)}{2} \right) d_5 \qquad (\text{V1})$$

$$= \bar{v}_5 \left(\frac{b_6 - b_4}{2} \right) d_5 \qquad (\text{V2})$$

where q_5 is the discharge through segment 5, $\bar{v}_5$ is the mean velocity in vertical 5, b_4, b_5, b_6 are distances from an initial point on the bank to verticals 4, 5 and 6 and d_5 is the depth of flow at vertical 5.

For the end segment, 1, shown hatched, the discharge may be computed as

$$q_1 = \bar{v}_1 \left(\frac{b_2 - b_1}{2} \right) d_1 \qquad (\text{V3})$$

and for the other end segment, n, as

$$q_n = \bar{v}_n \left(\frac{b_n - b_{n-1}}{2} \right) d_n \qquad (\text{V4})$$

Mean-section method

Segment discharges are computed between successive verticals. An example of one such segment is shown hatched in Figure V2. The

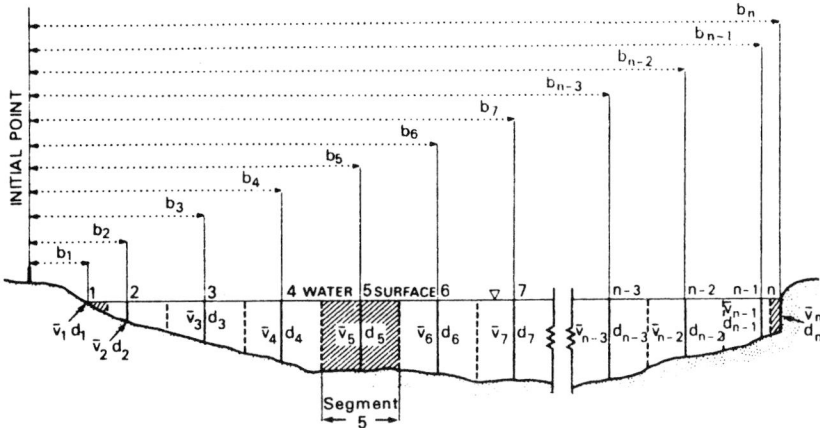

Figure V1 The mid-section of computing current-meter measurements: 1, 2, 3, ..., n, number of verticals; b_1, b_2, b_3, ..., n, distance from initial point; d_1, d_2, d_3, ..., n, depth of flow at verticals; v, average velocity in verticals.

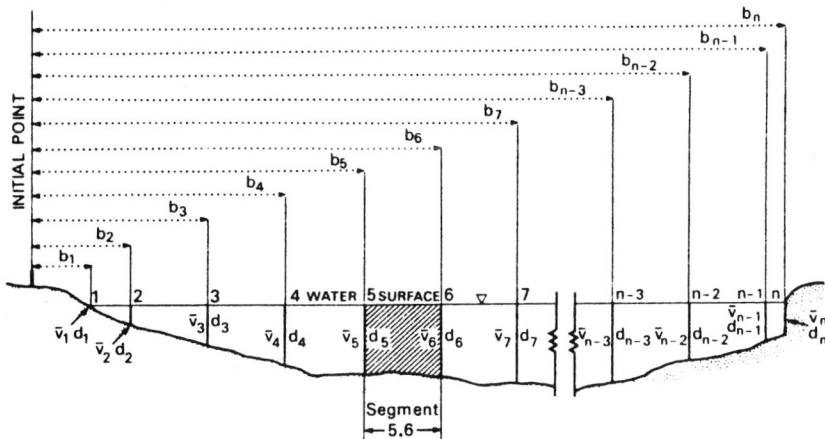

Figure V2 The mean-section of computing current-meter measurements: 1, 2, 3, ..., n, number of verticals; b_1, b_2, b_3, ..., n, distance from initial point; d_1, d_2, d_3, ..., n, depth of flow at verticals; v, average velocity in verticals.

velocities and depths for successive verticals are each averaged, the segment discharge being the product of the two averages.

Referring to Figure V2, the discharge passing through segment 5–6 is computed as

$$q_{5-6} = \left(\frac{\bar{v}_5 + \bar{v}_6}{2}\right)\left(\frac{d_5 + d_6}{2}\right)(b_6 - b_5) \qquad \text{(V5)}$$

where q_{5-6} is discharge through segment 5–6, $\bar{v}_5$, $\bar{v}_6$ are the mean velocities in verticals 5 and 6, d_5, d_6 are the depths of flow at verticals 5 and 6, b_5, b_6 are distances from an initial point on the bank to the verticals 5 and 6.

It will be noted that the depth of flow at vertical 1 is zero and the problem of computing the flow in the hatched end segment does not arise in this method, nor does it arise when the bank is vertical and if the velocity can be taken as approximately zero at the end vertical. The computation is therefore carried out for the end segments in exactly the same way as for the other segments. Nevertheless, this facility does not give the mean-section method an overall advantage over the mid-section method, the latter being simpler to compute and therefore quicker if the calculations are being performed manually. There is little difference in time, however, if a pocket calculator is employed for the calculation.

Measurement of velocity

The mean velocity in each vertical is determined by current-meter observations by any of the following methods.

Velocity distribution method

In this method velocity observations are made in each vertical at a sufficient number of points distributed between the water surface and bed to define effectively the vertical velocity curve, the mean velocity being obtained by dividing the area between the curve and the plotting axes by the depth. The number of points required depends on the degree of curvature, particularly in the lower part of the curve, and usually varies in the range 6–10. Observations are normally made at 0.2, 0.6 and 0.8 of the depth from the surface, so that the results from the vertical velocity curve can be compared with various combinations of reduced points methods, and the highest and lowest points should be located as near to the water surface and bed as possible.

This method is the most accurate if done under ideal, steady-stage conditions, but is not considered suitable for routine gauging owing to the length of time required for the field observations and for the ensuing computation. It is used mainly for checking velocity distribution when the station is first established and for checking the accuracy of the reduced points methods.

The velocity curve may be extrapolated to the bed by the use of the equation

$$v_x = v_a \left(\frac{x}{a}\right)^{1/c} \qquad \text{(V6)}$$

where v_x is the point velocity required in the extrapolated zone at distance x from the bed [note: if $x = 0$ (bed level), $v_o = 0$], v_a is the velocity at the last measuring point on the velocity curve at distance *a* from the bed, and *c* is a constant varying from 5 for coarse beds to 7 for smooth beds and generally taken as 6.

An example of the use of equation (V6) is as follows. In a velocity distribution measurement the lowest observation in the vertical was at a point 0.25 m from the bed. The value of the velocity at this point was 0.15 m s^{-1}. Find the approximate velocity at a point 0.1 m from the bed in order to complete the vertical velocity curve.

From equation (V6)

$$v_x = 0.15 \left(\frac{0.10}{0.25}\right)^{1/6} = 0.13 \text{ m s}^{-1}$$

An alternative method of obtaining the velocity in the region beyond the last measuring point, and so completing the vertical velocity curve, is based on the assumption that the velocity for some distance up from the bed may often be taken as being proportional to the logarithm of the distance *x* from the bed. If the observed values of velocities, therefore, are plotted against corresponding values of log *x*, the best-fitting straight line through these points can be extended to the bed. The required velocities close to the bed may then be read directly from the graph.

The 0.6 depth method

Velocity observations are made at a single point at 0.6 of the depth from the surface and the value obtained is accepted as the mean for the vertical. This assumption is based both on theory and on results of analysis of many vertical velocity curves, which showed that in the majority of cases the 0.6 method produced results of acceptable accuracy. The value of the method is its essential reliability, the ease and speed of setting the meter at a single point, and the reduced time necessary for completion of a gauging.

The 0.2 and 0.8 depth method

Velocity is observed at two points at 0.2 and 0.8 of the depth from the surface and the average of the two readings is taken as the mean for the vertical. Here again, this assumption is based on theory and on the study of vertical velocity curves, and experience has confirmed its essential accuracy. Generally the minimum depth of flow should be about 0.75 m when the 0.2 and 0.8 depth method is used.

Six-point method

Velocity observations are made by taking current-meter readings on each vertical at 0.2, 0.4, 0.6 and 0.8 of the depth below the surface and as near as possible to the surface and bed. The mean velocity may be found by plotting in graphical form and using a planimeter, or from the equation

$$\bar{v} = 0.1(v_{\text{surface}} + 2v_{0.2} + 2v_{0.4} + 2v_{0.6} + 2v_{0.8} + v_{\text{bed}}) \quad \text{(V7)}$$

where *v* is the velocity.

Five-point method

Velocity observations are made by taking current-meter readings on each vertical at 0.2, 0.6 and 0.8 of the depth below the surface and as near as possible to the surface and bed. The mean velocity may be

found by plotting in graphical form and using a lanimeter, or from the equation

$$\bar{v} = 0.1(v_{\text{surface}} + 3v_{0.2} + 3v_{0.6} + 2v_{0.8} + v_{\text{bed}}) \quad \text{(V8)}$$

Equations (V7) and (V8) are established from the area of a plane surface by a simple arithmetic procedure. In the six-point method, for example, the surface area of the curve ($\bar{v}D$) is approximately

$$(v_1 \times 0.1D + v_2 \times 0.2D + v_3 \times 0.2D + v_4 \times 0.2D + v_5 \times 0.2D + v_6 \times 0.1D) \text{ m}^2 \text{ s}^{-1}$$

and the average velocity is found by dividing by *D*, the total depth, giving

$$0.1(v_1 + 2v_2 + 2v_3 + 2v_4 + 2v_5 + v_6) \text{ m s}^{-1}$$

Similarly, equation (V8) is established from the surface area of the curve, giving

$$(v_1 + 0.1D + v_2 \times 0.3D + v_3 \times 0.3D + v_4 \times 0.2D + v_5 \times 0.1D) \text{ m}^2 \text{ s}^{-1}$$

and dividing by *D* gives the average velocity as

$$0.1(v_1 + 3v_2 + 3v_3 + 2v_4 + v_5) \text{ m s}^{-1}$$

Three-point method

Velocity observations are made by taking current-meter readings on each vertical at 0.2, 0.6 and 0.8 of the depth below the surface. The average of the three values may be taken as the mean velocity in the vertical. Alternatively, the 0.6 measurement may be weighted and the mean velocity obtained from the equation

$$\bar{v} = 0.25(v_{0.2} + 2v_{0.6} + v_{0.8}) \quad \text{(V9)}$$

The origin of the average velocity occurring at 0.6 of the depth and also at 0.2 and 0.8 of the depth from the surface is based essentially on the theoretical distribution of velocity in an open channel. For the condition of turbulent flow over a rough boundary, the vertical velocity curves have approximately the form of a parabola whose axis, coinciding with the filament of maximum velocity, is parallel with the surface and is in general situated between the surface and one-third of the depth of the water from the bed. As the depth and velocity increase, however, the curve approaches a vertical line in its limiting position (this fact is used to advantage in the moving boat method where the current meter is located at approximately 1 m from the surface).

R.W. Herschy

Bibliography

Herschy, R.W., 1993. The velocity–area method, *Flow Meas. Instrum.*, **4**(1), 7–10.
Herschy, R.W., 1995. *Streamflow Measurement.* Chapman & Hall, London and New York.

Cross references

Streamflow measurement
Water resources: dictionary of basic terms

W

WATER, ON EARTH AND OTHER PLANETS

Water, probably the most critical component in the geochemistry of planet Earth, only makes up about 0.02% of the near-surface mass of the Earth. As one of the essentials for self-reproducing life, it is a substance of supreme importance to humankind, for survival, for the environment and for the economy (Kuenen, 1955).

In its pure form as 'distilled water', water is colorless, odorless and tasteles. This oxide of hydrogen (H_2O) freezes at 0°C and boils at 100°C (at standard atmospheric pressure). It has a high specific heat, is a poor conductor of electricity and is very slightly compressible. Its maximum density is reached at 4°C, so that in natural environments such as lakes, ice forms at the surface while the bottomwater often remains warm (for encyclopedia citations with references, see Fairbridge, 1966, 1967b, 1972).

Water in one form or another is probably present in every planet except Mercury and in many of their satellites (but not the Moon). Nevertheless, on Earth it is unique. This is due to the Earth's approximately 1.5×10^6 km distance from the Sun. A little closer, and a Venusian hothouse would result (no liquid phase). A little farther away, and a Martian cold world (ice phase) would exist.

Over most of its existence, the Earth has enjoyed a global mean temperature in the range of 20 ± 5°C. Evidence for this generalization is based on (1) sedimentary products of running water (former sands, gravels), (2) biological evolutionary continuity and fossils such as bacteria and algal mats (stromatolites several billion years old have present-day equivalents) and (3) geochemical equilibria (e.g. crystal forms of aragonite versus calcite, and isotopic ratios). This uniformitarian deduction leads to the assumption that the liquid water phase had been continuously dominant on the Earth's surface since its earliest Precambrian existence. Repeated ice ages generated glaciers that probably have never reached more than 45° from the poles, and in spite of evidence of periodic biological crises, no major phyla have ever become extinct.

The continued presence of liquid water on Earth implies that the Earth's position (actually that of the Earth–Moon pair) must have remained very close to its present orbit for at least 4 billion years and that the effect of incoming solar radiation (terrestrial insolation) has not materially increased or decreased. Nevertheless, solar evolution models predict a 30% increase in luminosity. This paradox presents important cosmological constraints from the planetary science viewpoint.

As a natural solvent on planet Earth, water in unequalled (see Water: substance and solvent; Langmuir and Barnes in Fairbridge, 1972, p. 1244). It has a weak tendency to become ionized as H^+ and OH^- which form complexes with solute species. CO_2 is readily soluble in water creating a weak acid, which is generated during rainfall. Humid environments cause weathering of common surface minerals such as calcite in limestones or feldspars in granites or basalts. In contrast, in semiarid regions, soils and ephemeral lake waters become enriched by alkaline products, notably the carbonates and sulfates of calcium and sodium. In these high-pH environments, silica (SiO_2, the principal component of quartz sand, sandstone and granite) hydrates to form a weak acid, H_4SiO_4, represented by H^+ ions plus anionic species such as $H_3SiO_4^-$ and $H_2SiO_4^{2-}$. Differences in regional climates, through water's solvent behavior, are responsible for a large amount of geochemical rearrangement at the Earth's surface. Thanks to stratigraphic burial and geologists' ability to date ancient deposits, a paleoclimatic record is thus established which contains built-in environmental data spanning almost the whole of geological time.

Water in planet Earth is distributed in the successive spheres: the atmosphere, hydrosphere and lithosphere. In the atmosphere its mass is 1.3×10^{13} tonnes, $\ll 1\%$ of the total; in the hydrosphere 1.4×10^{18} tonnes, 37%; and in the crust or outer lithosphere, $2.2–2.6 \times 10^{18}$ tonnes, 63%.

In the atmosphere, water is almost entirely in the vapor phase. Most of the water is concentrated in the troposphere with very little in the stratosphere; 90% is in the lowest 5 km. Atmospheric water plays a key role in the modulation of the Earth's daily climate. The principal dynamic focus is the equatorial/intertropical zone where water, evaporated from the sea surface and humid forest lands, is convected up to the lower stratosphere and redistributed polewards (a 'Hadley circulation'). This circulation is affected by the terrestrial rotation, the distribution of land and sea, and the related atmospheric dynamic systems. On an annual basis, this circulation is almost subject to seasonal redistribution. Climatic trends of monthly, yearly or decadal significance are constrained, not by the atmosphere (which has a 'short memory'), but by the hydrosphere, where the ocean plays the role of 'the giant flywheel of the atmospheric heat engine' (Hastenrath, 1991; Peixoto and Oort, 1992; Leroux, 1996).

The distribution of water in the hydrosphere is overwhelmingly (97%) in the oceans (1.34×10^{18} tonnes; a volume of 1.34×10^9 km^3 or 1.34×10^{21} l). Fresh water constitutes $\sim 29 \times 10^6$ km^3 of glacial ice, with $\sim 360\,000$ km^3 in lakes and rivers; and 23×10^6 km^3 in groundwater (shallow and deep) and sedimentary rocks. During the last 2 million years, the successive glacial/interglacial cycles (about 100 000 years) have left a eustatic signal of sea-level fluctuation, indicating expansion and contraction of the ocean by about 3% (by volume) with a corresponding vertical rise and fall of 100–135 m. If all the world's present glacier ice melted, seal level would rise 66 m.

Fresh water at the Earth's surface is thus severely restricted (only about 2.52% of the total water). The main difference of the rest is that the ocean is salty, on average about 3.5% or 35 (33–38) parts per mille. (The salts are largely the chlorides and sulfates of sodium, magnesium and calcium, but including up to about 100 minor constituents. The corresponding mass of the dissolved solids is 4.9×10^{16} tonnes). Because of this salinity, seawater is a good electrolyte (in contrast to fresh water) and conducts electricity at about $4\,\Omega\,m^{-1}$; its conductivity increases with temperature.

Biologically important components of seawater include dissolved CO_2 and $H_2CO_3 + Ca^2 \rightleftharpoons Ca_2(HCO_3)$ (calcium hydrogencarbonate) which provide respectively the basis for photosynthesis and the fixation of carbonate skeletal materials. Marine populations tend to be constrained by the supply of silicon, phosphorus and vanadium (each averaging $\sim$3 mg l^{-1}). Dissolved oxygen has slowly increased through geological time ($>4 \times 10^9$ years).

A long-term balance of salinity in the oceans is maintained. Through geological time, salts have been added to the ocean through weathering on land and transport via the hydrological cycle while salts are subtracted through evaporation in rift basins and lagoons and subsequent burial by sediments. Recycling occurs at spreading rifts. In addition, life forms withdraw salts from the sea, notably $CaCO_3$ (by nannoplankton, foraminifera, molluscs and corals) and SiO_2 (by diatoms and radiolaria). In this complex history, ocean salinity has likely been maintained approximately at the present level at least for the last 1 billion years.

Average seawater has a density of 1.025 g cm^{-3} and in an estuary, the river's fresh water tends to 'float' above the seawater wedge below it. Hydrostatic pressure rises with depth at about 1 atm (10.1 m)$^{-1}$; in SI units, expressed as N m^{-2}, the rise is $\sim$1 $\times$ 10^6 N m^{-2} (100 m)$^{-1}$, and at 5000 m (mean depth for 27% of the Pacific Ocean) the pressure is $\sim$49 $\times$ 10^6 N m^{-2}.

Water in the Earth's lithosphere is in the soil, in porous sedimentary formations and in chemically combined form in the various minerals. According to Russian analyses, carried out on behalf of a UNESCO project (USSR Committee, 1978) the total water reserves of the three outer spheres of the planet Earth are approximately 1386 million km^3, of which 23.4 million km^3 (1.7%) are classified as groundwater.

The average global thickness of the lithosphere, down to the Mohorovičić discontinuity (with an area of 510 million km^2) is 17 km. Beneath the continents its mean thickness is 35 km; of this only an upper 1 km, on average, consists of potentially porous sediments. Free water (not chemically bonded to minerals) in the lithosphere is classified as gravitational and capillary, i.e. sinking down from surface precipitation, and rising up from water droplets trapped by sedimentation.

Other water resources in the lithosphere are called 'pellicular', i.e. chemically bound in minerals such as gypsum ($CaSO_4 \cdot 2H_2O$). Under slightly elevated temperature and/or pressure, gypsum will dehydrate to anhydrite ($CaSO_4$), liberating 'bound water'. The reaction is reversible, so that buried anhydrite when exposed to weathering hydrates to gypsum. Hydration–dehydration reactions are common among the evaporite minerals (halides), the clays and micas (Fairbridge, 1983). Hot aqueous fluids triggered by igneous activity often reach the surface in volcanic regions, creating hot springs and geysers which liberate large quantities of steam to the atmosphere. On average, the rocks of the lithosphere have been estimated to contain 3.56% combined water, which would suggest that the whole lithosphere contained some 84.2 million km^3 of water (USSR Committee, 1978).

Besides the atmosphere, hydrosphere and lithosphere, there are two special types of 'sphere' that are superimposed on all three of the above. First is the biosphere, the overall habitat of organized life. For particular species, that habitat may range up to the higher troposphere (where migratory birds are observed flying over the Himalayas at elevations of 8000–10 000 m) or reach down to ocean depths exceeding 10 000 m. Microscopic organisms are found in soils or porous sedimentary rocks, to considerable depths ($>$10 km).

Second, there is the cryosphere, the base of which is the boundary between average freezing and thawing, a critical constraint for most inhabitants of the biosphere. The cryosphere boundary on planet Earth is generally stated in its relation to mean sea level (m.s.l.). Poleward of about 70° latitude, the cryosphere is generally below m.s.l.; the ocean there is regularly frozen over (shifting seasonally). Near the equator it is around 6000 m elevation. During glacial periods, the cryosphere boundary falls dramatically. Around 20 000 years ago in the northern hemisphere it dropped below m.s.l. at about 45°N, and in mountainous regions of low latitudes it was about 1000 m lower (in exceptional spots like New Guinea near 5°S, glaciers almost reached present m.s.l.). On planet Mars, with its major orbital variations, the cryosphere boundary undergoes a dramatic seasonal (polar) reversal.

Evidence of water from outside of planet Earth is not lacking. Ordinary chondritic meteorites carry 10–1000 ppm water, and water is always present in the 'dirty snowballs' of comets. If these indicators are taken as representing the material of the proto-Earth planet starting with a mass of 5.98×10^{27} g, if its water content were 0.5%, we would have 2.99×10^{25} g or approximately 30 billion km^3. Some

water has been lost to outer space by photodissociation of H_2O in the outer atmosphere, through geological time.

Most of the other planets display some evidence of water, either spectroscopically or through water's physiographic behavior. There is no water on the Moon or Mercury, even in mineral-bound form; on Venus it is identified in cloud droplets. On Mars, geomorphological evidence shows that some times in the past, liquid water existed in sufficient amounts to erode dendritic drainage areas and carve out giant canyons. Because of long-term orbital changes, Mars is probably subject to climate cycles which mobilize water that is at present stored in permafrost form beneath the surface regolith. In the outer planets, water can only be indicated spectroscopically, so far only on Jupiter, and on natural satellites of the outer planets (Shirley and Fairbridge, 1997).

Rhodes W. Fairbridge

Bibliography

Back, W. and Freeze, R.A. (eds), 1983. *Chemical Hydrogeology*. Stroudsburg: Hutchinson Ross Publ. Co. (Benchmark Papers in Geology, vol. 73), 416 pp.

Fairbridge, R.W., 1960. The changing sea level of the sea. *Scientific American*, **292**(5), 70–79.

Fairbridge, R.W. (ed.), 1966. *The Encyclopedia of Oceanography*. New York: Reinhold, 1021 pp.

Fairbridge, R.W., 1967a. Carbonate rocks and paleoclimatology in the biogeochemical history of the Earth, in Chilinger, G.V. *et al.* (eds), *Carbonate Rocks*. Amsterdam: Elsevier, pp. 399–432.

Fairbridge, R.W. (ed.), 1967b. *Encyclopedia of Atmospheric Sciences and Astrogeology*. New York: Reinhold Publ. Co., 1200 pp.

Fairbridge, R.W. (ed.), 1972. *The Encyclopedia of Geochemistry and Environmental Sciences*. New York: Van Nostrand Reinhold Co., 1321 pp.

Fairbridge, R.W., 1983. Syndiagenesis–anadiagenesis–epidiagenesis: phases in lithogenesis, in Larsen, G. and Chilinger, G.V. (eds), *Diagenesis in Sediments and Sedimentary Rocks*, vol. 2, Amsterdam: Elsevier, pp. 17–113.

Hastenrath, S., 1991. *Climate Dynamics of the Tropics*. Dordrecht: Kluwer, 488 pp.

Holland, H.D., 1972. The geologic history of sea water – an attempt to solve the problem. *Geochim. Cosmochim. Acta*, **36**, 637–651.

Kuenen, P.H., 1955. *Realms of Water*, New York: John Wiley & Sons, 327 pp.

Peixoto, J.P. and Oort, A.H., 1992. *Physics of Climate*. New York: American Institute of Physics, 520 pp.

Shirley, J.H. and Fairbridge, R.W. (eds), 1997. *The Encyclopedia of Planetary Sciences*. London: Chapman & Hall, 1040 pp.

Short, N.M. 1975. *Planetary Geology*. Englewood Cliffs: Prentice-Hall, 361 pp.

Tardy, Y., 1986. *La Cycle de l'Eau: Climats, Paléoclimats et Géochimie Globale*. Paris: Masson, 338 pp.

USSR Committee for IHD, 1978. *World Water Balance and Water Resources of the Earth*. Paris: UNESCO, 663 pp.

Wyllie, P.J., 1971. Role of water in magma generation and initiation of diapiric uprise in the mantle. *J. Geophys. Res.*, **76**, 1328–1338. (Reprinted in Garfunkel, Z., 1985. *Mantle Flow and Plate Theory*, New York: Van Nostrand Reinhold, pp. 130–140).

Cross references

WATER AGENCIES: NATIONAL AND INTERNATIONAL

Development of water agencies

Fresh water is a basic human need, both for drinking and for food production. Where available in the form of streams, rivers and lakes it has also been used from earliest times for the disposal of waste, for power production and for transport. The provision of a reliable source of good-quality fresh water has always been one of the major factors permitting or limiting socio-economic development, even population growth, in any region of the world. An excess of fresh water, on a continuous basis in the form of swamps, or in frequent flooding, has also been a factor in the location of human settlements and the use of land for agricultural purposes.

As human settlements grew in size, so did the challenge of providing an adequate supply of fresh water and protecting the population from drought and from floods. As villages grew to become towns and then cities, the original sources of water became inadequate to the task and also became polluted and it was necessary to construct even larger works to bring water from more distant sources, to treat it and provide for removal of used water. Urban development encroached onto wetlands and the floodplains which had been left vacant in earlier times. This called for the development of drainage works and flood protection schemes. Add to this the increased per capita demand for fresh water, brought about by an increase in the standard of living. The simple solutions provided by the village well, the river-side pump, the earthen levee bank and the drainage ditch could be handled by small communities from their own resources, using local expertise. The towns and then cities had to develop specialized teams that grew into Departments of Public Works which could undertake the projects needed by the larger communities.

Urban populations grew into the hundreds of thousands and the development of hydropower and irrigation schemes called for the damming of major rivers and the construction of large aqueducts. The communities concerned established authorities to oversee the construction and management of the hydraulic works involved, either under a single agency or under a series of agencies with greater or lesser independence. Universities and technical colleges taught water engineering as a separate subject and water research institutes were established both off and on campus.

As demands continued to grow, there was a natural tendency towards the establishment of more specialized agencies, in parallel with the amalgamation or absorption of smaller agencies in neighboring areas, leading to administrative structures involving

- departments of public works (flood control, road drainage);
- departments of public health (pollution control);
- departments of agriculture (rural water supply and irrigation);
- water supply authorities (urban water supply);
- drainage boards (land drainage, flood control);
- highway and railway departments (road and rail drainage);
- universities and colleges specializing in water sciences and engineering (education and research);
- water research institutes (research);
- sewage treatment authorities (sewage treatment);
- flood protection boards (flood control);
- electric power companies (hydropower production);
- parks and recreation departments (public parks and water-based recreation).

National and state (provincial) policy governed the extent to which these agencies (1) were administered at national, state or local community level, and (2) were funded from local or national taxation, or could raise their own income by charging for water supplied or licences issued. They undertook their own construction and maintenance work by employing their own staff, or contracted the work to private companies. They also established themselves as private or semi-private companies operating under various degrees of public oversight and legislation.

No dates have been associated with the above description because these developments have taken place at different times and at different rates in various regions of the world. It is clear that even the ancient civilizations of the Americas, the Indian subcontinent and of Rome and China all evolved specialized water agencies at national or regional level which disintegrated when the civilizations themselves collapsed. It is only in the second half of the twentieth century that

one can talk of world-wide movements in the management of water affairs.

Until the 1950s, the general trend had been towards greater specialization, and hence the spawning of new independent or semi-independent agencies, usually at ever higher levels of administration. One major controlling factor in federal states, however, was whether water affairs were primarily the responsibility of the federal or of state governments. If all authority lay with the federal government, then federal agencies were the norm, operating through state branches. If the national constitution gave state governments the authority, then a full set of agencies was commonly established in each state, leaving the federal authorities with only a coordinating role. This coordination could be very powerful, however, especially where it involved the distribution of federal funds in support of state activities.

In the 1970s and 1980s there were two important developments which have had a major influence on the structure of water agencies in many countries. The first development resulted from the environmental movement. Water was identified as part of the natural environment which required monitoring and protection. Departments of the environment were established. Some operated in parallel with departments of public works, some absorbed the latter. In general the effect was to add new agencies, or at least new subsidiary bodies, to the structure, the purpose of which was to collect and analyze data on the availability and quality of fresh waters and advise on the protection of these waters.

The second development was a wider recognition of the river basin and/or large aquifer as the natural hydrological unit. Given that the freshwater resources of a basin or aquifer are a single unit, it was seen as logical that they should be administered as such, and new basin authorities were set up by absorbing various of the functions of the previous national or state authorities having responsibilities in the freshwater sector. Rarely, if ever, were all the functions assigned to the new authorities, however.

A further development took place during the 1980s and 1990s: the privatization of various public bodies or the commercialization of certain of their activities. City watersupply companies or entire river basin authorities were sold into private ownership, public authorities retaining varying degrees of oversight and control. Even where agencies remained in public hands, they were required to charge for more of their services and to do so increasingly at commercial rates. This applied not only to agencies which supplied products such as potable water and hydropower, but also those which provided services such as water pollution studies and the training of specialists.

These various developments have led to a great variety of structures of water agencies within the countries of the world, and one that is constantly in a state of flux. Geophysical, economic and political reasoning each lead to quite different 'optimal' solutions and few countries have maintained the same administrative structure for more than 10 or 15 years at the most. It is doubtful, given the many forces and factors involved, whether a perfect structure exists which is applicable to all or even a few countries. What is certain is that, while the changes of the past 20 to 30 years have led in many countries to more effective and logical systems of administration, in others they have given rise to complex structures that are almost impossible to untangle and which lead to departmental rivalry and duplication of effort which can threaten even the most dedicated attempts to solve the countries' water problems.

International organizations involved in freshwater resources

Fresh water is a national responsibility, except where the river basin or aquifer concerned lies within the territory of more than one country. It is not surprising therefore that worldwide organizations concerned with freshwater affairs are comparatively recent in origin. The International Meteorological Organization (IMO), the direct predecessor of the present World Meteorological Organization (WMO), established a Commission for Hydrology back in the 1940s but its activities lapsed during World War II and were not taken up again by WMO until the late 1950s. However, hydrology as a geophysical science, to be taught and in which to conduct research, knows no national boundaries. The non-governmental scientific and technical communities, therefore, predate the intergovernmental agencies in their establishment and the implementation of truly international programs. Particular mention can be made in this regard of the International Association of Hydrological Sciences (IAHS), founded as the International Association for Scientific Hydrology in 1922, and the International Commission on Large Dams (ICOLD), founded in 1928.

The family of independent organizations which was established in the 1940s and 1950s to form the intergovernmental United Nations system reflected very much the division of responsibilities between agencies found at national level at that time. Once established, with their own constitutions, country membership, governing bodies and budgets, there have been few changes in the division of responsibility between them for water-related matters. This is not to say that the individual organizations within the family have not restructured their activities internally. The most important changes have occurred within the United Nations Organization (UNO) itself which, while being one organization, implements a range of programs, some as large if not larger than other organizations of the family. For example, in the early 1970s, the UNO established the United Nations Environment Programme (UNEP) in response to increasing concerns over environmental matters. More recently, the UNO has moved major responsibility for its water-related activities to UNEP and to the UN's Regional Economic and Social Commissions.

The present (1995) structure of UN organizations and programs involved in freshwater issues may be summarized as follows. In the United Nations Office the following organizations are involved:

- Department of Economic and Social Development (DESD);
- United Nations Children's Fund (UNICEF);
- United Nations Development Programme (UNDP);
- United Nations Environment Programme (UNEP);
- United Nations University (UNU);
- Economic Commission for Africa (ECA);
- Economic Commission for Europe (ECE);
- Economic Commission for Latin America and the Caribbean (ECLAC);
- Economic and Social Commission for Asia and the Pacific (ESCAP);
- Economic and Social Commission for Western Asia (ESCWA);
- United Nations Centre for Human Settlements (HABITAT);
- United Nations Disaster Relief Coordinator (UNDRO);
- International Research and Training Institute for the Advancement of Women (INSTRAW);
- World Food Programme (WFP).

The following specialized agencies of the United Nations and related organizations are involved:

- Food and Agriculture Organization of the United Nations (FAO);
- United Nations Educational, Scientific and Cultural Organization (UNESCO);
- World Health Organization (WHO);
- World Bank (WB);
- World Meteorological Organization (WMO);
- United Nations Industrial Development Organization (UNIDO);
- International Atomic Energy Agency (IAEA).

The various organizations within the UN system are coordinated through the Administrative Committee on Coordination (ACC) on which sits the Secretary-General of the UNO and the Executive Heads of the Specialized Agencies and related organizations. Water-related activities are coordinated at working level by the ACC Sub-committee on Water Resources.

Also important among the international agencies in fresh water are the various international river basin authorities, such as the Rhine and Danube Commissions, the Mekong Committee, the River Niger Basin Authority and the Permanent Joint Technical Committee for Nile Waters. Some have the power to enforce regulations on water use or transport agreed between the countries concerned. Others have little executive power, but provide the framework within which the countries seek to manage the water resource concerned in a manner which is fair to all concerned. Some are very limited in their mandates, some very broad. There are cases, such as on the Rhine, where there is more than one international agency for a single body of water, each concerned with a different aspect of management.

A major new development is the growing interest of regional politico-economic bodies in water issues, because of the importance of these issues to the environment and to sustainable economic development. For example, the Southern African Development Community (SADC) supports work in this field and the European Union has established the European Environment Agency. The latter has a mandate that covers at regional level many of the activities assigned to national water agencies at national level and to the United Nations

at international level. Such agencies are certain to increase in number and importance in the future and their rôle in relation to existing national and international structures will no doubt evolve with time.

The non-governmental organizations with an interest in water matters have increased over the years. Since the UN Conference on Environment and Development, held in Rio de Janeiro in 1993, they have become increasingly recognized as full members of the international water community with a valuable role to play. Some are members of unions, others are independent. A basic, but not exhaustive, list includes:

- International Council of Scientific Unions (ICSU):
  - International Union of Geodesy and Geophysics (IUGG);
  - International Union of Geological Sciences (IUGS);
  - International Association of Hydrological Sciences (IAHS): a member of IUGG;
  - International Association of Hydrogeologists: affiliated to IUGS;
  - Scientific Committee for the International Geosphere–Biosphere Programme (IGBP);
  - Scientific Committee on Water Research (SCOWAR);
  - International Association on Water Quality (IAWQ);
- International Union of Technical Associations and Organizations (UATI):
  - International Commission of Irrigation and Drainage (ICID);
  - International Association for Hydraulic Research (IAHR);
- Other bodies:
  - International Organization for Standardization (ISO);
  - International Water Resources Association (IWRA);
  - International Institute for Applied Systems Analysis (IIASA);
  - International Commission on Large Dams (ICOLD);
  - European Committee for Standardization (CEN);
  - International Oceanographic Foundation (IOF);
  - International Desalination Association (IDA).

The ICSU and UATI water-related associations used to coordinate their activities through the Committee on Water Resources (COWAR). This was disbanded in 1994 and its role has since been assumed, at least in part, by the International Water-related Association's Liaison Committee (IWALC).

Arthur J. Askew

Cross reference

International organizations involved with hydrology and water resources

WATERAID

WaterAid was created in 1981 by the British water industry in response to the UN-sponsored International Drinking Water Supply and Sanitation Decade to do grassroots work among some of the world's poorest communities. The aim then, as now, was to tackle one of the root causes of poverty through the provision of safe water and effective sanitation.

The need is immense. Two billion people lack access to potable water supplies within reasonable walking distance of their homes. More than 3 billion lack any form of sanitation (Figures W1 and W2). The result is that hundreds of millions of women and children walk long distances every day to fetch and carry water, frequently from polluted sources (Figure W3). Quite apart from the appalling burden and drudgery involved, the price paid is enormous: 9 million children die every year from water-related illnesses (one child every 3 s); 3 million of these are under the age of 5 (one child every 10 s).

The experience of the world's richer countries shows that something can be done to improve water and sanitation. The issue for organizations like WaterAid is how such improvements can be made in an affordable, lasting and sustainable manner.

WaterAid's approach has been to focus its efforts in a very particular way. Projects are implemented not by WaterAid itself but by local organizations. These may be non-governmental organizations (NGOs), including churches and women's groups; sometimes they are departments of central or local government. WaterAid has a member of its salaried staff based in each country where it is working. He or she supervizes funding, and where necessary gives technical or management advice to strengthen these partner organizations.

Figure W1 All members of the village take part in the construction of the well, Gabulig Tingit, Bolgatanga, Ghana. (Photo: WaterAid/Caroline Penn.)

Figure W2 An open sewer running past people's homes in the slums of Addis Ababa, Ethiopia. (Photo: WaterAid/Steve Morgan.)

Figure W3 Water carriers, India.

All projects require major and active participation on a self-help basis by those who will benefit from them.

WaterAid is a small player on the global stage. With an annual income of some $10.5 million, its direct support can only ever have a modest influence. However, the style and type of work funded can have a radical impact on the policies of other much larger organizations. Funding for projects comes from a variety of sources. Though the British water industry has always been a major supporter, it was never intended that WaterAid should be under its control. From the outset, other people of relevant interest and experience were drawn into its governance and day-to-day running, and WaterAid became a self-governing, not-for-profit organization registered in the UK as a charity.

A network of supporters throughout the UK helps to raise awareness of developing country needs and raise money for projects supported by WaterAid. In each UK region, the company or public authority with overall responsibility for water and sewerage has established a WaterAid committee initiating and coordinating a wide variety of educational and fund-raising activities. About one-third of current income results directly from initiatives by the water industry and its staff. More than 15 000 of those staff give personal support by regular weekly or monthly deduction from pay. More than 20 million WaterAid appeal leaflets are mailed each year to customers in England and Wales with their water bills.

This has led to growing support from the general public through donations, covenants and other means, and from various trusts and community groups, now in total producing a further one-third of all income. The remaining one-third is provided by the British government and the European Union from their co-funding schemes.

To date, most of WaterAid's experience has been gained in rural areas where the majority of people in the developing world have lived. However, that is changing; by the year 2000 half the 6.2 billion population of the planet will be urban.

An important shift is taking place in the Third World, where urban populations are growing at fantastic speed – much faster than the parallel shift in the industrial world. By the year 2000, 23 cities will have populations of at least 10 million people and 18 of these cities will be in developing countries.

The unplanned and often chaotic growth of these mega-cities will mean that at least half of their inhabitants will live in crowded tenements, shanty-towns and slums without basic amenities. Overcrowding, filth and squalor will encourage the spread of infectious diseases and pose massive threats to urban health.

Similar organizations are involved in other parts of the world, particularly in the United States under the name 'Water for People' operating out of Denver, Colorado, and also in Australia, Canada (Water Can) and New Zealand (Water for Survival) as well as others.

R.W. Herschy

Source

Lane, J., 1996. WaterAid: Doing the almost impossible, in *New World Water*, Sterling Publications, London.

Cross references

Drinking water and sanitation
Environment priorities for development: water
River pollution prevention: historical
Sanitation and clean water
Water availability and river water quality – world
Water allocation and use
Water resources: quality assessment

WATER ALLOCATION AND USE

Competition between farmers and cities for water supplies is already constraining many countries' development strategies. The problem will grow as populations increase and economies expand. The large fixed costs associated with water distribution, uncertainties about the physical availability of water from year to year, and widely held cultural and religious proscriptions against treating water as a commodity are likely to compel governments to continue to allocate water administratively.

The largest single demand for water comes from irrigation. Inefficient use of irrigation water puts pressure on other users and imposes environmental costs. About 85% of irrigated land relies on traditional surface systems based on canals and gravity flow. Their design is often too inflexible to provide water with the timeliness and predictability that farmers desire as they adopt improved crop varieties and turn to intensified and diversified cropping systems. Instead, water is delivered on arbitrary schedules and for limited periods of time, with incentives for use further distorted by subsidized prices. Farmers respond by taking as much water as possible while they can. The results are often wasted water, waterlogging, leaching of soil nutrients and excessive runoff of agricultural chemicals with drainage water.

It is often better to improve existing systems than to build new ones. Lining canals reduces water losses, and installing drainage helps combat salinization and waterlogging. But modernizing installed designs is generally more expensive than achieving comparable gains through improved management.

Better pricing of water (and of electricity used to pump groundwater) to reflect its scarcity and the environmental costs of overuse is fundamental to better management. Governments often worry that reducing subsidies will hurt poor farmers and will be unacceptable if water delivery is unpredictable. Implementing improved pricing is difficult. Water flows are hard to measure in the open canal systems that characterize most irrigation systems. Closed-pipe conveyance systems are best for charging by water volume, but unless there is good communication between farmers and the delivery agency, they are vulnerable to tampering and damage to volumetric gauges.

A number of countries are finding that progress is possible. In China financially semi-autonomous water supply agencies sell water wholesale to water users who are grouped by village or township, partly on the basis of volume. These user groups in turn collect fees from their members, typically on the basis of the area irrigated or, less frequently, the volume of water used. Although the charges are generally set well below real costs, the link to quantities used encourages savings. Moreover, the system reinforces financial responsibility at each level because the fees collected remain in the irrigation budgets. Tighter overall budgets in other countries have prompted increases in water fees from the subsidized rates.

Additional public investment in surface irrigation must take account of increasing infrastructural costs, low commodity prices and environmental costs. Some developments will be ruled out by the environmental consequences of reservoir inundation, water diversion, increased water pollution from non-point agricultural sources and alteration of hydrological systems.

New techniques such as drip and sprinkler systems can use water more efficiently and deliver water when farmers need it. Although they are unlikely to supplant the large surface irrigation systems for grain crops, these techniques will become more important for future expansion of irrigation, partly because they can be employed with high-value crops grown on unleveled land and permeable soils where traditional surface irrigation is impossible. They are already spreading in developing countries, especially in North Africa and the Middle East, China and Brazil.

The spreading of these irrigation techniques will require a change in the traditional role of governments in irrigation. The new techniques work on a far smaller scale than traditional surface irrigation, and the source of water is usually a privately owned tubewell rather than a publicly managed dam. Manufacturers can be relied on to promote the systems because more marketable equipment is involved than in surface canal systems. Any price distortions that affect investment decisions by farmers must be corrected, since the farmers, rather than direct public investment, will be the main agents of expansion. Governments must also monitor aggregate use of groundwater and regulate tubewell pumping to prevent excessive drawdown of aquifers.

If the potential efficiency gains from these technologies are to be realized, the new methods must be integrated into a broader approach to the interactions among water, plants, soils, nutrients and other farm inputs. Farmers will need research and extension support to acquire new management skills, credit to enable them to afford mechanical equipment, and secure legal rights to water to encourage them to invest in new technology.

R.W. Herschy

Source

The World Bank, 1992. *World Development Report*.

Cross references

Access to and accountability of water resources
Drinking water and sanitation
Irrigated land area: world
Water charges (UK): abstractions and discharges

WATER AVAILABILITY AND RIVER WATER QUALITY

Availability

The Département Hydrogéologie in Orléans, France, compiles water resource and withdrawal data from published documents, including national, United Nations and professional literature. The Institute of Geography at the National Academy of Sciences in Moscow also compiles global water data on the basis of published work and, where necessary, estimates water resources and consumption from models that use other data, such as area under irrigation, livestock populations and precipitation. These and other sources have been combined by the World Resources Institute to generate the data for Table W1. Data for small countries and countries in arid and semiarid zones are less reliable than are those for larger countries and those with higher rainfall.

Annual internal renewable water resources refer to the average annual flow of rivers and of aquifers generated from rainfall within the country. The regional and income group totals presented here are compiled from data that are not strictly additive, since they are based on differing sources and dates. In addition, annual country data may conceal large seasonal, year-to-year and long-term variations.

For each region or income group, annual withdrawal as a share of water resources refers to total water withdrawal as a percentage of internal renewable water resources. Withdrawals include those from non-renewable aquifers and desalting plants but do not include evaporative losses.

Per capita figures are calculated using 1990 population estimates. Withdrawals can exceed 100% renewable supplies when extractions from non-renewable aquifers or desalting plants are considerable or if there is significant water reuse.

Sectoral withdrawal is divided into three categories: agriculture (irrigation and livestock), domestic (drinking water, private homes, commercial establishments, public services and municipal use or provision), and industry (including water for cooling thermoelectric plants). The sectoral proportions are based on national reports and models that use estimates from other data and thus should be inter-

preted with care. Numbers may not sum to 100% because of rounding.

Generally, countries with an annual water availability of less than 1000 m^3 per capita face chronic water scarcity, while those with less than 2000 m^3 face water stress and major problems in drought years.

River water quality

The global water quality monitoring project (GEMS/Water) was established in 1976 as part of the Global Environment Monitoring System (GEMS). In 1990 there were a total of 488 reporting stations in 64 countries. Water quality data are available from 1979 to the present. Data shown in Table W2 comprise two of the 50 indicators of water quality that are reported within the GEMS system and have been made available by the Canada Centre for Inland Waters, which acts as the global data center. Not all stations collect all data, and the frequency and physical accuracy of measurement vary among stations. Four-year periods are used in the table to minimize seasonal and year-to-year variability and to emphasize general trends, if any.

Dissolved oxygen is a critical factor in the health of aquatic organisms. In general, for life, growth and reproduction, values must exceed 5.5 mg l^{-1} for warm-water habitats and 6.5 mg l^{-1} for cold-water habitats. Lower values of dissolved oxygen endanger stocks of fish and other oxygen-dependent organisms.

Fecal coliforms are most commonly associated with animal and human feces. This measure is used as a sentinel indicator for the presence or potential presence of many other pathogenic organisms that are more difficult to observe and measure. Water for human consumption should usually contain zero fecal coliforms per 100 ml sample, and bathing water and water used for irrigation should contain less than 1000 per 100 ml sample.

R.W. Herschy

Source

The World Bank, 1992. *World Development Report*.

Table W1 Water availability

Country group	Total annual internal renewable water resources (km^3)	Total annual water withdrawal (km^3)	Annual withdrawal as a share of total water resources (%)	Per capita annual internal renewable water resources, 1990 (m^3)	Per capita annual water withdrawal, year of data (m^3)	Sectoral withdrawal as a share of total water resources		
						Agriculture	Domestic	Industry
Low income	14 272	1 257	9	4 649	498	91	4	5
China and India	4 650	840	18	2 345	520	90	5	6
Other low-income	9 622	417	4	8 855	460	95	3	2
Middle income	13 730	492	4	12 597	532	69	13	18
Lower-middle income	6 483	290	4	10 259	550	71	11	18
Upper-middle income	7 247	202	3	15 824	508	66	16	18
Low and middle income	28 002	1 749	6	6 732	507	85	7	8
Sub-Saharan Africa	3 713	55	1	7 488	140	88	8	3
East Asia and the Pacific	7 915	631	8	5 009	453	86	6	8
South Asia	4 895	569	12	4 236	652	94	2	3
Europe	574	110	19	2 865	589	45	14	42
Middle East and North Africa	276	202	73	1 071	1 003	89	6	5
Latin America and the Caribbean	10 579	173	2	24 390	460	72	16	11
Other economies	4 486	375	8	13 976	1 324	66	6	28
High income	8 368	893	11	10 528	1 217	39	14	47
OECD members	8 365	889	11	10 781	1 230	39	14	47
Other	4	4	119	186	372	67	22	12
World	40 856	3 017	7	7 744	676	69	9	22

Table W2 Selected water quality indicators for various rivers

Country	River, city	Dissolved oxygen Annual mean concentration (mg l^{-1}) 1979–82	1983–86	1987–90	Average annual growth rate for series (%)	Fecal coliform Annual mean concentration (number per 100 ml sample) 1979–82	1983–86	1987–90	Average annual growth rate for series (%)
Low income									
Bangladesh	Karnaphuli	5.7	6.1	–	−1.1 (5)	–	–	–	− (3)
Bangladesh	Meghna	6.5	7.0	–	2.6 (5)	3 133	700	–	−35.1 (5)
China	Pearl, Hong Kong	7.6	7.8	7.8	0.4(11)	519	563	174	−14.4(10)
China	Yangtze, Shanghai	8.3	8.3	8.2	−0.1(11)	316	464	731	10.6(11)
China	Yellow, Beijing	9.8	9.7	9.8	−0.1(11)	711	1 337	1 539	9.8(11)
India	Cauveri, d/s from KRS Reservoir	7.2	7.6	7.3	0.8 (9)	51	681	445	63.8 (9)
India	Cauveri, Satyagalam	7.0	7.3	7.5	1.1 (9)	10	684	920	121.8 (9)
India	Godavari, Dhalegaon	6.5	6.6	6.7	0.3 (9)	–	–	–	− (0)
India	Godavari, Mancherial	8.0	8.0	7.3	−1.1 (9)	5	5	8	19.7 (7)
India	Godavari, Polavaram	7.2	7.2	6.9	0.0 (8)	4	2	4	−3.8 (7)
India	Sabarmati, Dharoi	9.4	9.1	8.9	0.0 (9)	248	222	220	−15.4 (8)
India	Subarnarekha, Jamshedpur	8.0	7.9	7.5	−0.2 (9)	659	4 513	5 2800	89.0 (9)
India	Subarnarekha, Ranchi	6.7	4.0	5.3	−6.2 (9)	1 239	7 988	3 100	70.5 (9)
India	Tapti, Burhanpur	7.5	6.9	6.1	−2.3 (9)	–	110	130	−23.2 (4)
India	Tapti, Nepanagar	7.2	7.0	7.0	−0.6 (9)	–	19	163	76.0 (4)
Pakistan	Chenab, Gujra Branch	6.2	6.8	7.1	1.8(10)	436	463	446	−1.7(10)
Pakistan	Indus, Kotri	7.6	7.2	2.6	−13.6(11)	105	121	78	−3.4(11)
Pakistan	Ravi, d/s from Lahore	6.8	5.7	6.3	−1.4(12)	378	746	555	−2.4(10)
Pakistan	Ravi, u/s from Lahore	7.2	6.7	7.0	−0.8(12)	275	392	249	−6.6(10)
Sudan	Blue Nile	7.3	8.2	–	3.3 (7)	–	–	–	− (0)
Middle income									
Argentina	de la Plata, Buenos Aires	7.6	7.5	–	0.0 (8)	828	230	–	−23.1 (8)
Argentina	Paraná Corrientes	8.1	8.0	8.1	0.1(10)	185	146	111	−6.6(10)
Brazil	Guandu, Tomada d'Agua	8.1	7.8	7.7	−0.7(11)	1 202	2 452	6	−47.0 (8)
Brazil	Paraiba, Aparecida	6.0	6.1	6.0	−0.4 (7)	13 950	9 800	6 075	−11.5 (7)
Brazil	Paraiba, Barra Mansa	7.4	7.6	7.8	0.4(11)	8 003	8 100	8	−33.4 (7)
Chile	Maipo, el Manzano	12.9	13.2	10.8	−1.4(10)	871	705	775	5.3 (8)
Chile	Mapocho, Los Almendros	11.8	12.1	10.0	−1.7(10)	2	2	5	8.0 (8)
Colombia	Cauca Juanchito	–	5.2		1.0 (5)	–	10 000	10 000	0.0 (4)
Ecuador	San Pedro	7.7	7.8	–	−0.1 (5)	80 000	30 603	–	−31.5 (4)
Fiji	Waimanu	7.6	7.8	8.0	0.5 (9)	600	1 605	–	8.1 (7)
Hungary	Danube	9.4	10.4	9.9	1.7(10)	3 419	3 075	3 750	1.2(10)
Korea	Han	–	10.5	10.4	−0.2 (8)	–	8	12	14.4 (8)
Malaysia	Kinta	6.8	7.5	8.3	2.9 (7)	–	–	–	− (0)
Malaysia	Klang	3.0	3.3	2.8	−1.1 (9)	–	–	–	− (1)
Malaysia	Linggi	3.4	3.6	3.7	0.9(10)	–	–	–	− (0)
Malaysia	Muda	7.3	7.2	6.3	−1.3 (8)	–	–	–	− (0)
Mexico	Atoyac	3.5	1.7	0.3	−47.5 (9)	157 500	105 000	916 667	23.9 (7)
Mexico	Balsas	7.6	6.3	6.8	−1.9(10)	1 558	26 333	130 000	95.4 (8)
Mexico	Blanco	5.0	3.4	4.1	−3.7 (9)	21 717	39 500	12 150	1.8 (8)
Mexico	Colorado	7.9	8.7	8.2	1.4 (9)	227	58	37	−28.7 (7)
Mexico	Lerma	0.3	0.4	0.5	−18.6(10)	192 250	165 000	67	5.7 (7)
Mexico	Panuco	7.7	8.1	8.3	0.7(11)	110	201	–	−27.8 (6)
Panama	Aguas Claras	7.9	8.2	–	0.4 (7)	219	143	–	−14.4 (6)
Panama	San Felix	8.2	8.0	–	−1.0 (7)	850	753	–	−6.2 (6)
Philippines	Cagayan	7.8	7.9	8.1	0.3(11)	–	–	–	− (3)
Portugal	Tejo, Santarem	8.9	8.6	8.4	−0.7 (9)	2 252	4 163	4 225	24.6 (9)
Thailand	Chao Phrya, d/s from Nakhon Sawan	6.3	6.3	–	0.2 (8)	1 093	1 745	–	47.7 (7)
Thailand	Prasak, Kaeng Khoi	6.6	7.7	–	8.0 (5)	596	2 724	–	9.9 (8)
Turkey	Porsuk, Agackoy	9.0	9.,1	9.2	0.7 (9)	–	–	–	− (1)
Turkey	Sekarya, Adetepe	9.2	8.7	8.9	−0.3 (8)	–	–	–	− (1)
Uruguay	de la Plata, Colonia	–	–	–	7.1 (3)	–	453	93	54.6 (4)
Uruguay	Uruguay Bella Union	–	7.9	8.4	−1.4 (4)	–	200	1 100	66.9 (4)
High income									
Australia	Murray	10.0	9.4	9.1	1.0 (6)	–	–	–	− (0)
Australia	Murray, Mannum	7.1	8.2	8.6	2.4 (8)	33	103	80	15.8 (8)
Belgium	Escaut, Bleharies	5.7	6.2	5.9	1.1(11)	76	579	867	40.8(11)
Belgium	Meuse, Heer/Agimont	10.5	10.8	11.3	0.8(11)	30	1 391	1 700	69.7(11)
Belgium	Meuse, Lanaye Ternaaien	9.2	8.4	8.9	−0.7(11)	147	5 233	7 100	78.2(11)
Japan	Kiso, Asahi	10.0	10.6	11.7	1.7(11)	300	400	216	−4.1(11)

Table W2 Continued

Country	River, city	Dissolved oxygen				Fecal coliform			
		Annual mean concentration (mg l^{-1})			Average annual growth rate for series (%)	Annual mean concentration (number per 100 ml sample)			Average annual growth rate for series (%)
		1979–82	1983–86	1987–90		1979–82	1983–86	1987–90	
Japan	Kiso, Inuyama	10.8	10.5	10.8	−0.2(10)	610	491	600	−2.0(10)
Japan	Kiso, Shimo-Ochiai	11.2	11.1	11.4	0.3(10)	546	443	353	−6.0(10)
Japan	Shinano, Zuiun Bridge	10.1	10.3	10.3	0.2(10)	290	346	193	−3.0(10)
Japan	Tone, Tone-Ozeki	10.0	9.9	10.4	0.5(10)	521	593	618	3.7(10)
Japan	Yodo, Hirakata Bridge	8.7	8.4	8.4	−0.4(11)	72 000	70 333	–	9.3 (7)
Netherlands	Ijssel (arm of Rhine)	8.7	7.9	–	−3.3 (6)	9 833	2 050	–	−43.0 (5)
Netherlands	Rhine, German frontier	8.5	8.0	–	−2.6 (6)	17 633	10 500	–	−11.8 (5)
United Kingdom	Thames	9.9	10.3	9.1	0.2 (8)	–	–	–	– (0)
United States	Delaware, Trenton, N.J.	11.1	10.6	–	−2.5 (7)	74	197	–	−4.0 (7)
United States	Hudson, Green Island, N.Y.	9.8	12.1	–	4.2 (7)	941	792	–	−7.4 (7)
United States	Mississippi, Vicksburg, Miss.	8.4	8.3	–	−0.2 (7)	435	1 473	–	40.2 (7)

Note: d/s, downstream; u/s upstream. Numbers in parentheses denote the number of years of observations. Data have been presented only when they are available for 4 or more years.

Cross references

Access to and accountability of water resources
Drinking water and sanitation
Sanitation and clean water
Sewage treatment: general introduction
Sewage treatment processes
Water allocation and use
Water quality for drinking: WHO guidelines

WATER BALANCE

As defined by Thornthwaite and Mather (1957), '. . . the term water balance refers to the balance between the income of water from precipitation and the outflow of water by evapotranspiration.'

By comparing monthly or seasonal values of precipitation with evapotranspiration, other associated moisture parameters such as water surplus, water deficit, soil moisture storage and water runoff may be measured.

One can distinguish between the water balance at a locality and the water balance of the world. The latter will be discussed in the last section of this article.

Seasonal change of evapotranspiration and precipitation

The Thornthwaite system of determining water balance represents a valuable contribution to the field of climatology. His method begins with a monthly accounting of the precipitation and potential evapotranspiration (*PE*). When precipitation exceeds *PE*, there is a net gain in soil moisture for that month. If the soil is at field capacity (i.e. its saturation limit), then the difference between excess precipitation and *PE* is water runoff. As long as the soil remains at field capacity, evapotranspiration will continue at the potential rate.

Information necessary to determine the water balance

The water balance at a locality may be determined when the following information is available:

- mean monthly or daily air temperature;
- mean monthly or daily precipitation;
- conversion and computational tables to reduce the complication in the relationship between evaporation, temperature, latitude and length of day;
- the depth of the root zone of the soil (and thus the water holding capacity, which may vary with soil type and vegetation).

Information on the last item is the most difficult to establish, as the water holding capacity of the soil depends on (1) soil type and structure, and (2) the type of vegetation growing on the surface. For example, a sandy soil will hold only 1–2 cm of moisture per 30 cm depth of soil while a silt or clay may hold 10 cm of water in the same depth. Also, the amount of water in the root zone of a soil at field capacity can vary from a few milimeters on shallow sand to over 400 mm on a deep, well-aerated silt loam (Thornwaite and Mather, 1955).

Thornwaite's bookkeeping procedures

Sample monthly water balance computation for Seabrook, NJ and Bismarck, ND, are indicated in Table W3. The bookkeeping procedures will be briefly discussed line by line. Note that PE is potential evapotranspiration.

Line 1. *T*: °C. Record the mean monthly air temperature on line 1.

Line 2. *I*: heat index. Obtain the heat index *I* from tables and record on line 2. Summation of the 12 monthly values yields the index *I*. Note that *I* is zero when the mean monthly temperature is below 0°C, as the logarithm of a negative number is indeterminate.

Line 3. Unadjusted *PE*. Obtain the unadjusted daily *PE* from tables and record on line 3. Note that *PE* is zero below 0°C.

Line 4. *PE*. Multiply unadjusted *PE* by the appropriate month and day length correction factor (obtained from tables) for the station's latitude. Record this new adjusted monthly value of *PE* on line 4.

Line 5. Precipitation. Record the mean monthly precipitation on line 5.

Line 6. Precipitation (*P*). Determine the difference between precipitation and *PE* and record on line 6. A negative value indicates a period of moisture deficiency, while a positive value indicates water available for soil moisture recharge and runoff. Note that Seabrook has an annual moisture excess of 352 mm whereas Bismarck has an annual moisture deficiency of 178 mm.

Line 7. Accumulated potential water loss. The accumulated sum of the negative precipitation – *PE* values are recorded on line 7. Since the annual value of precipitation – *PE* is positive for Seabrook, the value of accumulated potential water loss with which one starts accumulating the negative values of line 6 is zero. In the case of Bismarck, the annual value of precipitation – *PE* is negative. Therefore, it is necessary to refer to the tables and by a series of successive approximations determine the accumulated potential water loss.

Line 8. *ST*: storage. Values of soil moisture storage are recorded on line 8 after they are determined by reference to tables and lines 1, 6 and 7.

Line 9. Δ*ST*: change in soil moisture. Determine the difference in soil moisture storage from one month to the next and record on line 9.

Table W3 Sample monthly balance computations (all values except *T* and *I* in mm)

Seabrook, New Jersey (water-holding capacity in root zone of soil is 300 mm)

	Jan.	Feb.	Mar.	Apr.	May	June	July	Aug.	Sept.	Oct.	Nov.	Dec.	Year
1. $T°C^a$	0.9	1.2	5.9	11.3	17.5	22.3	24.7	23.7	20.2	14.0	7.6	2.3	
2. I	0.07	0.12	1.29	3.44	6.66	9.62	11.23	10.55	8.28	4.75	1.89	0.31	58.21
3. Unadjusted PE	0.1	0.1	0.6	1.3	2.5	3.5	4.1	3.9	3.1	1.8	0.8	0.1	
4. PE	3	2	19	43	93	131	156	138	97	52	20	2	756
5. P	87	93	102	88	92	91	112	113	82	85	70	93	1108
6. $P-PE$	84	91	83	45	−1	−40	−44	−25	−15	33	50	91	352
7. Accumulated potential WL					−1	−41	−85	−110	−125				
8. ST	300	300	300	300	299	261	225	207	197	230	280	300	
9. ΔST	0	0	0	0	−1	−38	−36	−18	−10	+33	+50	+20	
10. AE	3	2	19	43	93	129	148	131	92	52	20	2	734
11. D	0	0	0	0	0	2	8	7	5	0	0	0	22
12. S	84	91	83	45	0	0	0	0	0	0	0	71	374
13. RO	59	76	79	62	31	15	8	4	2	1	1	36	374
14. $SMRO$	0	0	0	0	0	0	0	0	0	0	0	0	
15. Tot. RO	59	76	79	62	31	15	8	4	2	1	1	36	374
16. DT	360	375	379	362	330	277	233	211	199	231	280	335	

(Snow 0)

Bismarck, North Dakota (water-holding capacity in root zone of soil is 200 mm)

	Jan.	Feb.	Mar.	Apr.	May	June	July	Aug.	Sept.	Oct.	Nov.	Dec.	Year
1. $T°C^a$	−13.4	−12.1	−4.3	5.6	12.5	17.6	21.0	19.6	14.5	7.2	−1.9	−9.6	
2. I	0	0	0	1.19	4.00	6.72	8.78	7.91	5.01	1.74	0	0	35.35
3. Unadjusted PE	0	0	0	0.9	2.0	2.9	3.5	3.3	2.4	1.1	0		
4. PE	0	0	0	31	78	115	140	121	76	31	0	0	592
5. P	11	11	23	39	59	85	57	46	31	24	14	14	414
6. $P-PE$	11	11	23	8	−19	−30	−83	−75	−45	−7	14	14	−178
7. Accumulated potential WL				(−116)	−135	−165	−248	−323	−368	−375			
8. ST	69	80	103	111	101	87	57	39	31	30	44	58	
9. ΔST	11	11	23	8	−10	−14	−30	−18	−8	−1	14	14	
10. AE	0	0	0	31	69	99	87	64	39	25	0	0	414
11. D	0	0	0	0	9	16	53	57	37	6	0	0	178
12. S	0	0	0	0	0	0	0	0	0	0	0	0	
13. RO	0	0	0	0	0	0	0	0	0	0	0	0	
14. $SMRO$	0	0	0	0	0	0	0	0	0	0	0	0	
15. Tot. RO	0	0	0	0	0	0	0	0	0	0	0	0	
16. DT	69	80	103	111	101	87	57	•39	31	30	44	58	

(Snow 73 mm)

[a] Abbreviations: *T*, mean air temperature; *I*, heat index; Unadjusted PE, unadjusted potential evapotranspiration; *PE*, potential evapotranspiration; *P*, precipitation; *P−PE*, precipitation minus the potential evapotranspiration; Accumulated potential WL, accumulated potential water loss (accumulated sum of the negative *P−PE* values); *ST*, storage; ΔST, change in soil moisture; *AE*, actual evapotranspiration; *D*, moisture deficit; *S*, moisture surplus; *RO*, water runoff; *SMRO*, snow melt runoff; Totl. *RO*, total runoff; *DT*, total moisture detention.

Line 10. AE: actual evapotranspiration. When the precipitation is greater than *PE*, the soil remains at field capacity and *AE* approximates *PE*. When precipitation is less than *PE*, the soil begins to dry out and $AE < PE$. In those months AE = precipitation + $|\Delta ST|$. *AE* is recorded on line 10.

Line 11. D: moisture deficit. $D = PE - AE$ and is recorded on line 11.

Line 12. S: moisture surplus. Any excess precipitation after the soil reaches field capacity is recorded on line 12 as moisture surplus which can become runoff.

Line 13. RO: water runoff. For large catchment areas, only about 50% of the surplus water will run off.

Line 14. SMRO: snow melt runoff. Since ST is less than the water-holding capacity in the case of Bismarck, that is no *SMRO* as it is assumed it will go into soil storage.

Line 15. Tot. RO: total runoff. Tot. $RO = \Sigma RO + \Sigma SMRO$.

Line 16. DT: total moisture detention. The moisture detention represents the total of the water stored in the soil, the snow remaining on the soil surface and the surplus water which has been detained for a month and is in the process of running off.

Note A: a daily water balance can also be determined for a station by computing procedures similar to the one already discussed for the monthly water balance.

Note B: Seabrook, NJ, and Bismarck, ND, represent stations with one 'wet' and one 'dry' season. A monthly water balance can also be computed for stations with multiple wet and dry seasons by more detailed procedures described in Thornthwaite and Mather (1957).

Water balance of the Earth

Although precipitation equals evaporation on a worldwide basis, the amounts vary considerably from region to region (Gentilli, 1958).

Table W4 indicates that the area poleward of 40°N has precipitation exceeding evaporation by 25 900 km³ year⁻¹. Most of this precipitation is caused by the polar front between 40 and 60°N. In contrast, the area between 10 and 40°N has evaporation exceeding precipitation by 42 300 km³. The belt between the equator and 10°N gains 19 300 km³ of water, the greatest amount of any 10° latidudinal belt on Earth.

Thus, the middle and middle–low latitudes of the northern hemisphere (10–40°N) lose a tremendous amount of water to the higher latitudes and to the intertropical front. The greater land masses in the northern hemisphere cause the intertropical front to be generally located north of the equator. Consequently, the northern hemisphere obtains approximately 3000 km³ of water from the southern hemisphere.

As shown in Table W4, the area between the equator and 30°S loses 32 700 km³ of water per year, 30 000 of which go to the higher latitudes with the westerlies, the other 3000 crossing the equator northward.

The prevailing westerlies of both hemispheres transport the net vapor equivalent of nearly 56 000 km³ of water toward the polar fronts. A tremendous amount of heat is needed to evaporate this water into water vapor. When the water vapor condenses and returns to the earth in liquid form as precipitation, a corresponding amount of heat is liberated. Thus, water in the air plays a large role in the heat transfer of the Earth.

Table W4 Water balance of the earth[a]: precipitation minus evaporation

Latitude	Oceans		Continents		World	
	(cm)	(km³ × 10⁻³)	(cm)	(km³ × 10⁻³)	(cm)	(km³ × 10⁻³)
	(cm)	$(km^3 \times 10^{-3})$	(cm)	$(km^3 \times 10^{-3})$	(cm)	$(km^3 \times 10^{-3})$
90–80°N	(+10)	(+0.3)	(+29)	(+0.1)	(+12)	(+0.4)
80–70°N	(+20)	(+1.7)	(+17)	(+0.6)	(+20)	(+2.3)
70–60°N	+36	+2.0	(+23)	(+3.1)	(+27)	(+5.0)
60–50°N	+56	+6.0	+14	+2.1	+31	+8.1
50–40°N	+47	+7.1	+18	+2.9	+32	+10.1
40–30°N	−45	−9.3	+14	+2.2	−20	−7.1
30–20°N	−93	−23.4	+29	+4.3	−48	−19.0
20–10°N	−58	−18.1	+16	+1.8	−38	−16.2
10–0°	40	+13.5	+57	+5.8	+44	+19.3
0–10°S	−19	−6.2	+59	+6.1	0	−0.2
10–20°S	−54	−17.9	+20	+1.8	−37	−16.0
20–30°S	−61	−18.7	+23	+2.2	−42	−16.5
30–40°S	−1	−0.2	+6	+0.2	0	+0.1
40–50°S	+34	+10.3	+37	+0.4	+34	+10.7
50–60°S	+47	+11.9	+82	+0.2	+47	+12.0
60–70°S	(+20)	(+3.5)	(+20)	(+0.1)	(+19)	(+3.6)
70–80°S	(+10)	(+0.3)	(+25)	(+2.2)	(+19)	(+2.5)
80–90°S	(0)	(0.0)	(+25)	(+1.0)	(+25)	(+1.0)
World	−10.0	−37.1	+24.9	+37.1	0	0.0

[a] After Wüst (1922); less reliable estimates in brackets.

It is more difficult to determine the water balance of the tropical and equatorial regions from Table W4. Vertical air movements predominate over horizontal movements in low latitudes. Therefore, enormous quantities of water vapor which are absorbed by the tropical easterlies over the oceans return to earth within a narrow latitudinal belt along the intertropical front, and also where a coastline intercepts the easterlies.

Oceanic evaporation is much more important than evaporation from continents in terms of the contribution to the total moisture balance of the atmosphere. Precipitation exceeds evaporation over the continents by 37 100 km³ of water a year. This amount is then returned to the oceans as runoff. Note also the considerable transfer of heat from the oceans to the continents by the transfer of vapor which condenses later over the land.

Robert M. Hordon

Bibliography

Gentilli, J., 1958. *A Geography of Climate*, 2nd edn revised, Perth, Australia, University of Western Australia Press.

Prescott, J.A., 1958. Climatic indices in relation to the water balance, *Climatology and Microclimatology* (Canberra Symposium), Paris, UNESCO, pp. 48–51.

Thornthwaite, C.W. An approach toward a rational classification of climate, *Geog. Rev.*, **38**, 55–94.

Thornthwaite, C.W. and Mather, J.R,. 1955. The water balance, *Johns Hopkins Univ., Laboratory in Climatology, Publ. in Climat.*, **8**, No. 1.

Thornthwaite, C.W. and Mather, J.R., 1957. Instructions and tables for computing potential evapotranspiration and the water balance, *Johns Hopkins Univ., Laboratory in Climatology, Publ. in Climat.*, **10**, No. 3.

Wüst, G., 1922. *Z. Ges. Erdk., Berlin*, 35.

Cross references

Evapotranspiration
Hydrological cycle
Hydrology
Precipitation
Water budget analysis
World water balance

WATER BUDGET ANALYSIS

Water budget analyses represent an environmental systems approach to the hydrological cycle, with emphasis on the transport, storage and utilization of water at the Earth's surface. The geographical scales of analyses range from global water budgets down to studies of the income, outflow and storage of water from small tanks set in the soil, known as lysimeters or evapotranspirometers. Time scales range from average water budgets derived from climatic data averaged over a number of years (typically 30 or more) to continuous daily water budgets within a real-time framework. This overview of water budget methodology begins with a description of the climatic water budget, with applications organized within a time scale beginning with long-term averages and closing with daily water budgets.

Mather (1974, 1978) has published comprehensive compilations of water budget procedures, analyses and applications; Muller (1982) has published an overview of his water budget approaches to runoff and river regimen; and Miller (1977) has published a detailed survey of interactions of water with ecosystems at the surface of the Earth.

Climatic water budget

The climatic water budget is usually developed for a place, with the data inputs based on mean monthly temperature and precipitation derived from a climatic station or array of stations. The climatic water budget was introduced into the literature by Thornthwaite and his colleagues (Thornthwaite, 1948; Thornthwaite and Mather, 1955), initially for analyses of global and regional climatic classification in terms of the interactions of energy and moisture in the various regions. Focus was directed towards the determination of humid and dry climatic realms, identification of subhumid climatic regions neither humid nor dry, and establishment of an orderly or regular structure of climatic types based solely on climatic parameters rather than vegetation characteristics (Carter and Mather, 1966). Numerous applications for environmental monitoring and analyses were developed at about the same time, and professional interest in applications quickly surpassed the initial studies of climatic classification (Thornthwaite and Mather, 1955).

Another key feature of the climatic water budget is that it serves as an excellent instructional tool for environmental managers and decision makers, while providing a working foundation of many climatological principles for students interested in water resources. The various components of the water budget are easy to understand, and the bookkeeping methodology provides a straightforward assessment of the distribution of moisture in the environment.

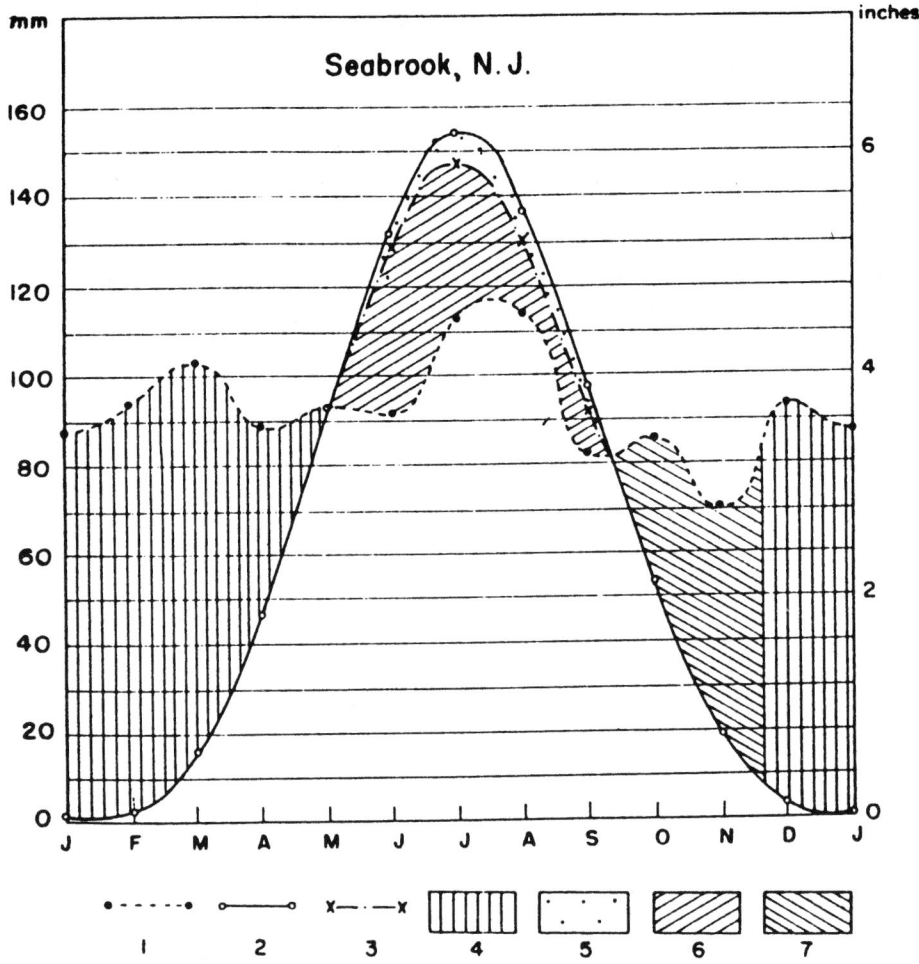

Figure W4 Traditional climatic water-budget graph. (1) P; (2) PE; (3) AE; (4) S; (5) D; (6) soil moisture utilization; (7) soil-moisture recharge (Source: Thornthwaite and Mather, 1955.)

The climatic water budget for Seabrook, New Jersey (Figure W4), is used to illustrate components of the budget. Thornthwaite developed and refined many of his water budget concepts at the Laboratory of Climatology at Seabrook Farms, and this average water budget for Seabrook appeared in the initial monograph about the water budget and its applications (Thornthwaite and Mather, 1955).

Potential evapotranspiration

Potential evapotranspiration (PE), the basic building block of the water budget, was developed as a concept by Thornthwaite in the mid-1940s. Potential evapotranspiration is defined as the amount of water that would evaporate and transpire from a landscape fully covered by a homogeneous stand of vegetation without any shortage of soil moisture within the rooting zone. Another specification is that estimates of PE are determined for a large area with similar vegetation and soil moisture conditions so that advection effects are eliminated.

Thornthwaite used measurements of water use in irrigation districts (primarily from the United States) and evapotranspirometer data to derive a complex set of empirical equations for estimation of monthly PE based solely on temperature data and a latitude factor to adjust for day length. The calculation procedures were set out in an instruction manual (Thornthwaite and Mather, 1957), so that climatologists rather quickly calculated PE and average climatic water budgets for most climatic regions.

PE can be measured on a daily basis by means of moist evapotranspirometers or weighing lysimeters. These devices are large tanks filled with soil set into the ground with the top flush with the soil surface. Plants are grown in the evapotranspirometers, with a large

buffer zone of similar plants around the tanks, to reflect natural field conditions.

The evapotranspirometer can be treated as a water budget system in terms of transport and storage of water. Precipitation or irrigation represents income, and measured percolation to subsurface drainage tanks represents one component of outflow. If the soil moisture in the tank is maintained close to capacity, the difference between total income (precipitation and irrigation water) and percolation represents an estimate of PE.

Weighing lysimeters are much more elaborate and expensive, but these instruments provide much more information. Decreases in weight represent evapotranspiration losses, after the other incomes and outflows of water are taken into account. Hence, weighing lysimeters have the capability even of estimating the diurnal regimes of PE rates, but it is difficult to maintain representativeness to the surrounding landscape. In the United States weighing lysimeters have been limited mostly to drier western states where the economy depends on the availability of inexpensive water from well-watered uplands.

PE can be thought of as an energy supply term and, at the same time, as the climatic demand for water from the landscape. PE became the keystone of the water budget analysis; it represented an approximate estimate of potential or optimum water demand in the landscape that could be met by current precipitation and soil moisture utilization. It provided the basis for the budgeting procedure.

At the same time the PE concept became controversial. For some investigators PE estimates were great improvements over pan evaporation data; pans behave much like wet wicks in dry landscapes, and pan evaporation data needed to be adjusted downward by geographical coefficients to represent water loss from ponds and small lakes (Farnsworth *et al.*, 1982). At the same time, the PE concept for a dry region is not entirely satisfying. This dilemma revolves around the dimensions of irrigated buffer zones necessary to eliminate advection effects for the measurement of PE in dry landscapes.

Thornthwaite also recognized that PE estimates based on temperature and daylength were necessary surrogates for measurements of surface energy budgets needed for rigorous estimates of PE, and he justified the use of empirical equations because of the unavailability of routine energy budget data. Over the ensuing decades more rigorous estimates of PE based on solar and long-wave radiation, humidity and wind have been developed, with the most frequently used estimates based on modifications of formulations originally developed by Penman (1948) in the UK and Jensen and Haise (1963) in the United States. Nevertheless, the unavailability of energy budget data on a routine basis has restricted use of the more rigorous PE estimates mostly to scientific investigations, with the standard Thornthwaite PE estimates used for regional analyses. In general, the error rates for regional PE estimates are considered to be rather minimal, particularly in most midlatitude environments; Brutsaert (1982) has published a comprehensive analysis of evaporation, including the PE concept.

Precipitation

Precipitation (P) represents the atmospheric delivery of moisture for the fundamental interactions of energy and moisture at the surface. Unfortunately, most national and regional networks of precipitation gauges represent minuscule samples of precipitation. Some networks are reasonably adequate for water budget analyses, especially on an average annual basis. However, standard precipitation data are clearly inadequate in areas marked by large spatial and temporal variability in rainfall, over many sparsely settled areas and especially in mountainous regions with complex patterns of orographic precipitation and rain-shadow valleys. Furthermore, there are few long-term observational series over the world's oceans, which occupy more than two-thirds of the surface areas of the Earth, and remote sensing techniques have not as yet been refined to fill in data gaps between standard gauges satisfactorily. Of particular significance is the generally agreed upon undercatch of standard rain gauges, especially during windy storms, as opposed to the true precipitation delivered to global landscapes (Mather, 1974); indeed, in river basin analyses, precipitation may be more in error than PE.

Soil moisture storage

Soil moisture storage (ST) represents water available within rooting zones of the plants for transpiration and evaporation. In classical terms, it represents the differences between field capacity and the wilting point. In the water budget framework, soil moisture is generally expressed in units equivalent to precipitation. On an average worldwide basis, Thornthwaite originally estimated the available soil moisture storage capacity to approximate 100 mm, but 300 mm was later used for global comparisons (Carter and Mather, 1966). In the United States the modern county soil surveys published by the Natural Resources Conservation Service include useful information for estimating available soil moisture capacities for combinations of geological substrate, soil types, slopes, land use and vegetation covers.

Within the water budget model, it is assumed that soil moisture will be utilized by plants to meet the demands of PE when precipitation is less than PE. For the calculation of soil moisture depletion, Thornthwaite originally suggested that plants could draw on soil moisture equally as needed until the available soil moisture within the rooting zone was exhausted; this approach became known as the equal-availability model (Figure W5, curve A). A few years later, Thornthwaite and his colleagues introduced the decreasing-availability model (Figure W5, curve B), which stated that the plants would withdraw soil moisture to meet the PE demand in proportions relative to the availability of soil moisture within the rooting zone. For example, if available soil moisture within the rooting zone amounted to only 60% of capacity, the plants would withdraw 60% of the PE demand, but if available soil moisture were only 20% of capacity, plants would be able to withdraw only 20% of the demand (Mather, 1974). Figure W5 also illustrates various other soil moisture depletion models either proposed or in use.

Most investigators prefer some form of the decreasing-availability model over the equal-availability model. A later innovation has been the introduction of a two-layer or even multiple-layer accounting

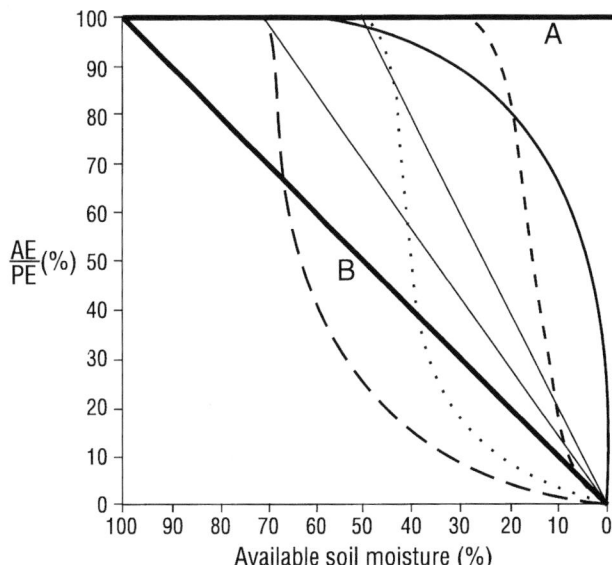

Figure W5 Selected soil moisture depletion curves. Curve A represents equal availability, and B decreasing availability. (After Mather, 1974.)

system for soil moisture storage and depletion. The most commonly used approach is to set up a two-layer accounting system, with an upper shallow layer in which moisture is treated as equally available and given the first priorities for soil moisture depletion and recharge. Soil moisture in the thicker lower layer is treated as decreasingly available and second in the priority system of depletion and recharge. Multiple-layer systems are most commonly used with daily or weekly water budgets, and are particularly well-adapted to computer modeling.

Actual evapotranspiration

Actual evapotranspiration (AE) in Thornthwaite's average water budget models represents precipitation and soil moisture withdrawals actually used by the plants to try to meet the energy demand represented by PE. Actual evapotranspiration, then, can be equal to but never greater than PE; there are, of course, places and times when AE is less than PE. Thornthwaite's adoption of the term actual evapotranspiration was unfortunate, however, because it implies empirical measurements of evapotranspiration, rather than calculations within the model framework. Most researchers in the agricultural sciences have substituted evapotranspiration, or simply ET, for actual evapotranspiration (AE), regardless of whether ET is measured or calculated.

No distinction is made between transpiration by plants and evaporation from soil surfaces in the Thornthwaite water budget models. Thornthwaite pointed out that most water loss from fully vegetated surfaces was transpiration by plants rather than small amounts of evaporation from shaded soil surfaces; linking of the two pathways of water vapor to the atmosphere as evapotranspiration, either potential or actual, was the cornerstone of water budget analyses as applied to landscapes, river basins and regions.

In most situations, ET represents water passing through plant systems. Indeed a number of measures of biological activity, plant development and growth, crop yields and even physical and chemical processes in the soils are tied to climatic processes and are indexed effectively from a climatic perspective by AE or ET.

Moisture deficit and surplus

The moisture deficit (D), often simply termed the deficit, is defined as the difference between PE and ET. The deficit represents the additional water that would have been used by plants if it were available, and can serve as a measure of irrigation potential of places and regions; thus, D is also a measure of potential increases in plant growth and crop yields.

The moisture surplus (S) represents precipitation not used for evapotranspiration or soil moisture recharge, and therefore water available for surface runoff to lakes and streams or for percolation to groundwater tables. The term surplus betrays an agricultural bias; surplus becomes the input to the surface and groundwater systems studied by hydrologists and engineers. In water budget climatology, surplus water is just as much a part of regional climates as clouds and precipitation are for meteorologists and atmospheric scientists. The surplus is useful for estimating runoff, streamflow, groundwater recharge and physical or chemical processes in the soil related to downward migration of water.

Figure W4 shows the standard graphical format of the average climatic water budget components of a place. Seabrook, New Jersey, is used here to represent typical annual regimes for a humid climate region in middle latitudes. This average monthly budget is based on an available soil moisture storage capacity of 300 mm and the decreasing-availability model of soil moisture depletion (Figure W5 curve B). Figure W4 emphasizes the seasonality of water budget components, soil moisture withdrawals and deficits during summer, soil moisture recharge during fall, and production of surplus water during winter and early spring.

The average climatic water budget model can be expressed in two interrelated fundamental equations expressing water movement and storage (1) $P = ET + S$; and energy demands (2) $PE = ET + D$, with the terms as defined in the text.

Average water budgets

In recent decades there has been considerable interest in global water budgets, in which both storage and exchanges of water in the atmosphere, on the continents in terms of surface waters, soil moisture, groundwater, snow and glacial ice, and ocean water are evaluated. These analyses indicate, in part, the very small storage of water vapor in the atmosphere at any one time, estimated to average only about 25 mm of equivalent precipitation on a global basis, the large magnitude of evapotranspiration on the continents relative to runoff to the oceans, the importance of evaporation from the oceans and the dominance of maritime sources of atmospheric moisture for precipitation over most regions of the continents (Mather, 1974; Baumgartner and Reichel, 1975). Figure W6 is a schematic representation of the hydrological cycle in terms of the average annual water budget of the Earth.

The water budget graph for Seabrook, New Jersey (Figure W4) is a representation of the average water budget of a place. Thornthwaite and his associates have calculated average climatic water budgets for thousands of places using a soil moisture storage capacity of 300 mm and the decreasing-availability model of soil moisture depletion; these average budgets have been published by continents in *Publications in Climatology* (Mather, 1962–1965). A moisture index (I_m) was also derived from the average water budget calculations; it represents a comparative index of the degree to which mean annual precipitation meets the climatic demands for water (PE) at each place. The index was used as part of an overall climatic classification having five basic types from wettest to driest: per-humid, humid, subhumid, semiarid and arid (Carter and Mather, 1966). Detailed maps of each of the

continents are available, but there has been less interest in classification than applications.

Components of the average water budget, especially PE, AE, and I_m, have been used to illustrate regional and global patterns of natural vegetation (Mather and Yoshioka, 1968) and the basic types of the United States comprehensive soil classification (Mather, 1978). Actual evapotranspiration (ET) has also been used to describe global patterns of a host of biological interactions with climate including primary productivity of land plants (Leith and Box, 1972), the decomposition of organic debris and forest fire hazards (Meentemeyer, 1978), and plant litter production (Meentemeyer *et al.*, 1982). Carter *et al.* (1972) prepared a summary statement setting out environmental relationships and responses to components of the average budget, extending relationships to weathering and geomorphology.

Monthly water budgets

For study and management of environmental resources in most climatic regions, average water budget components suggest an unrealistic seasonal stability of moisture conditions. Although the variability of energy availability from one year to another on a given month (expressed by PE) is relatively small, the variability of precipitation for a given month over a series of years can be very large. In addition, the use of average monthly precipitation in average water budget calculations masks the effects of precipitation variability on evaporation, deficits and surpluses through time.

As an example, Table W5 shows mean annual water budget components for Baton Rouge, Louisiana (based on the period 1961–1990), calculated first by means of the average water budget, and second by means of a continuous monthly water budget procedure. In the continuous monthly water budget procedure, components for each month are calculated before proceeding on to the following month through the period of interest. Monthly precipitation variability is taken into account, and a deficit, for example, will be calculated for a very dry month that on average is very wet. When monthly means of continuous water budgets are calculated and summed on an average annual basis, Table W5 shows that average annual ET (AE) decreased at Baton Rouge from 1045 to 946 mm, or a decrease of nearly 100 mm. At the same time, deficits and surpluses increased to 105 mm and 601 mm, respectively. Mean annual totals based on continuous monthly water budget analyses are more representative of the interannual variability of environmental conditions than components of the average water budget.

Figure W7 illustrates the monthly water budget components at Baton Rouge during a recent period of years. This graphical style does not show monthly precipitation, but emphasizes the seasonality of generation of surplus water for runoff in humid subtropical and midlatitude climatic regimes. The effects of climatic variability also are illustrated in Figure W8, which shows seasonal and annual values of the AE/PE ratio plotted against PE for four places on the United States beginning with New Brunswick, New Jersey, in the humid continental climatic region. Despite the designation as a humid climate, dry years at New Brunswick are much drier in terms of the AE/PE ratio than wetter years in the subhumid climate at Dodge City, Kansas, where the variability is very great. Bradford, in the mountains of Pennsylvania, is almost always wet in these terms, and Alamosa, located on a rain-shadow valley of Colorado, is always dry.

The Palmer drought index series (Palmer, 1965) and the crop moisture index are developed from continuous monthly and weekly water budgets using a two-layer soil moisture system. These two indices represent unique adaptations of water budget components for

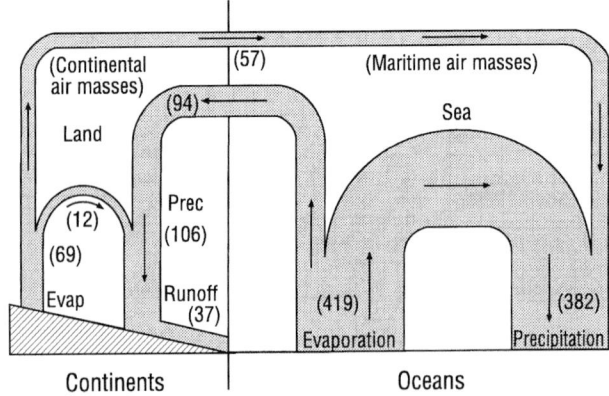

Figure W6 Schematic representation of the global hydrological cycle. Widths are proportional to volumes of water; values are mean annual water volumes in thousands of cubic kilometers. (After Mather, 1974.)

Table W5 Comparison of average annual and continuous monthly water budgets, Baton Rouge, Louisiana, 1961–1990 (mm)

Water budget component	Average annual budget	Means of continuous monthly budgets
PE	1064	1051
P	1546	1546
AE	1045	946
D	20	105
S	502	601

Source: Louisiana Office of State Climatology, Southern Regional Climate Center, Louisiana State University.

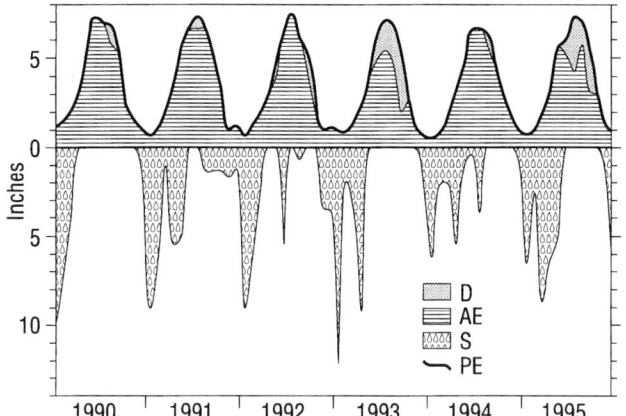

Figure W7 Modeled continuous monthly water budget components for Baton Rouge, Louisiana, 1990–1995.

the identification of periods of significant departures from normal of seasonal moisture characteristics for each place or region. The Palmer indices and the crop moisture index are both prepared on a near real-time basis by the Joint Agricultural Weather Facility (JAWF) of the National Weather Service and the US Department of Agriculture for climatic monitoring across the United States; these products are available in the *Weekly Weather and Crop Bulletin*, published by JAWF.

An especially interesting application of continuous monthly water budget procedures has been the study of relationships of land use and land cover to runoff and streamflow. For a given climatic regime, does a parcel of land or an entire drainage basin yield more or less water now than, say, 50 years ago because of changes in the use of the land and because of engineering modifications of the drainage systems? These relationships are difficult to evaluate because the investigator is usually searching for relatively small differences among quantities that are difficult to estimate well for entire drainage basins, with climatic variation adding to the complexity of the problem.

The first water budget analysis of this problem was an investigation of the effects of reforestation on water yield from small drainage basins on the Allegheny Plateau in New York (Muller, 1966). Monthly water budget calculations were used to estimate runoff and streamflow due directly to month-by-month variations of PE and precipitation. Monthly estimates of streamflow were then compared to measured streamflow, which is the consequence of climatic variability and changes in land cover. In these drainage basins the land-use changes were mostly replacement of pastures with conifer reforestation blocks. The differences between calculated and measured runoff (Figure W9) were assumed to represent the effects of reforestation, and this study indicated that reforestation significantly reduced annual runoff for a number of years.

Water budget analyses have also been used to show some of the effects of urbanization on runoff. Figure W10 shows the results of a simulation of the monthly surplus or runoff for undeveloped land in Middlesex County, New Jersey, and for the same land and years as if this land had been developed for typical middle-class subdivisions (Muller, 1969). The figure shows extra water coming off the subdivisions; most of the supplementary runoff is generated during the summer and fall when undeveloped land normally produces little to no runoff. Examples of other hydrologic applications of continuous monthly water budget procedures include, for example, the effects of land-use changes on river-basin regimen for water resources optimization (Shelton, 1981), comparisons of drainage basin modeled and measured runoff (Rohli and Grymes, 1995) and long-term trends of precipitation and modeled runoff (Keim *et al.*, 1995). Other applications include monitoring salinity regimes in estuaries (Mather *et al.*, 1972), analyses of water budget components of a large lake system (Sanderson, 1966), relationships among water budget components and crop yields (Chang, 1968) and evaluations of the relationships of pine bark beetle outbreaks to deficits and surpluses (Kalkstein, 1981).

Weekly and daily water budget analyses

The annual sums of estimated evapotranspiration, deficits and surpluses are changed if weekly or daily water budget calculations are used to replace continuous monthly calculations. In humid and subhumid climatic regions, daily and weekly variability of precipitation results in smaller seasonal or annual sums of ET and correspondingly larger sums for D and S. Weekly or daily water budget data are even more representative of real-world climate and environmental interactions than continuous monthly water budgets. As a result of the increased demand for daily and weekly water budgets, there are now a number of computer programs available for long-term analysis.

Daily water budgets, in which daily values of PE are compared to available soil moisture, are being developed extensively in the agricultural sciences. Basically, daily budgets are being used to model crop development and growth, and to study the water economies of commercial crops. Examples include the quantification of relationships between ET and productivity (Hank, 1974), development of specialized water budget models for agriculture (Ritchie, 1972) and simulation models for growth of corn and sorghum (Stapper and Arkin, 1980). Daily water budget models also have been used to analyze differential stress in terms of deficits on crop production at

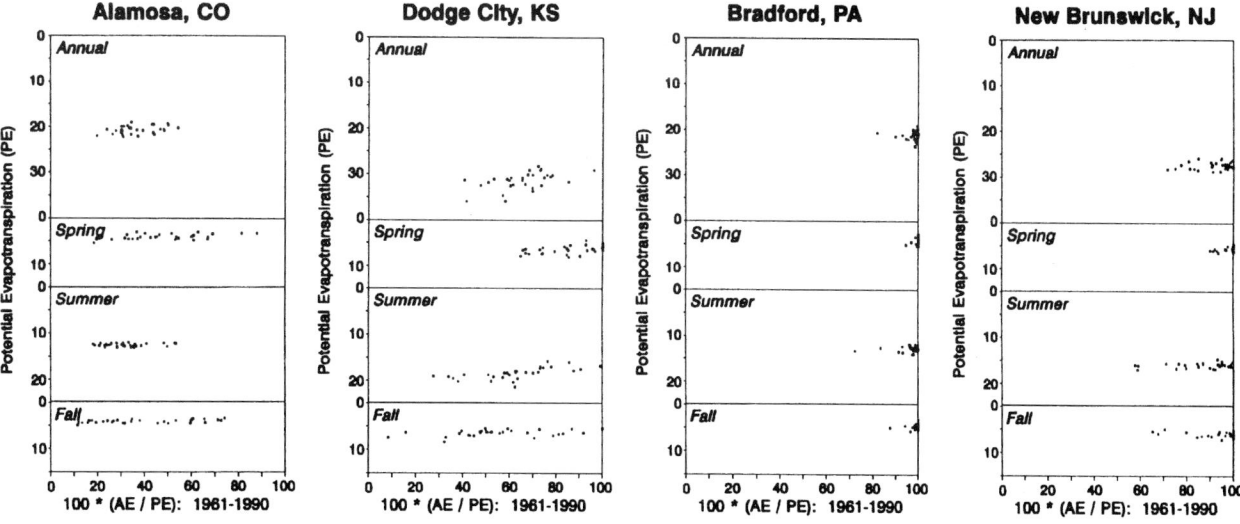

Figure W8 Seasonal and annual ratios of AE/PE plotted against PE for the humid East (New Brunswick, New Jersey, and Bradford, Pennsylvania), the subhumid Great plains (Dodge City, Kansas), and a high rainshadow desert (Alamosa, Colorado).

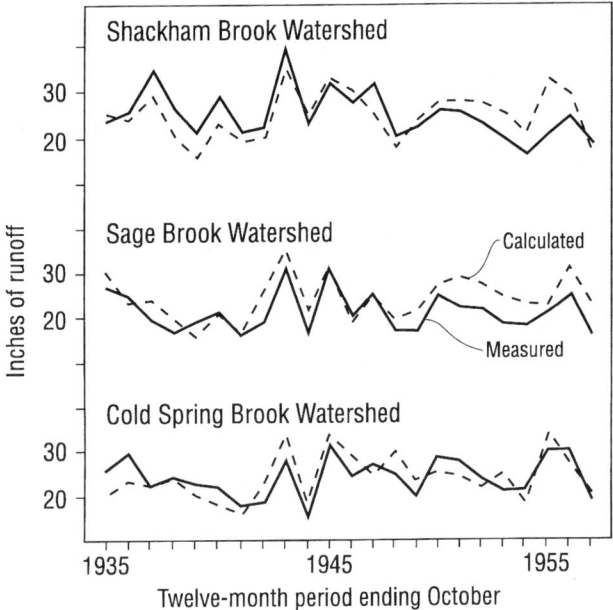

Figure W9 Modeled and measured water yield from partially reforested watersheds on the Allegheny Plateau, New York. (After Muller, 1966.)

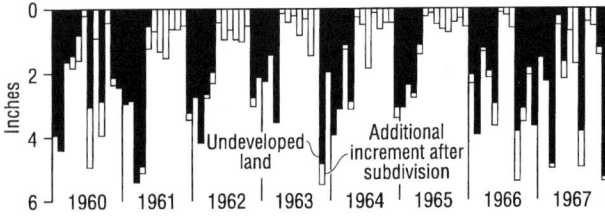

Figure W10 Water budget simulation of surplus water generated from undeveloped land and residential subdivisions in Middlesex County, New Jersey. (After Muller, 1982.)

various physiological growth stages in corn and soybeans (Sudar *et al.*, 1981). Daily water budget models also have been incorporated by the National Weather Service in their River Forecast Centers for real-time prediction of river stages and floods on selected drainage basins (Peck, 1976).

Robert A. Muller and John M. Grymes III

Bibliography

Baumgartner, A. and E. Reichel, 1975. *The World Water Balance: Mean Annual Global, Continental and Maritime Precipitation, Evaporation and Runoff*, R. Lee (translation). Amsterdam: Elsevier.

Brutsaert, W., 1982. *Evaporation into the Atmosphere: Theory, History, and Applications*. Dordrecht, Netherlands: Reidel.

Carter, D.B. and J.R. Mather, 1996. Climatic classification for environmental biology. *Publ. Climatology*, **19**, 305–395.

Carter, D.B., T.H. Schmudde and D.M. Sharpe, 1972. *The Interface as a Working Environment: A Purpose for Physical Geography*. Technical Paper No. 7. Washington, DC: Commission on College Geography, Association of American Geographers.

Chang, J.H., 1968. *Climate and Agriculture: An Ecological Survey*. Chicago: Aldine.

Denmead, O.T. and R.H. Shaw, 1960. The effects of soil moisture stress at different stages of growth on the development and yield of corn. *Agron. J.*, **45**, 385–390.

Farnsworth, R.K., W.S. Thompson and E.L. Peck, 1982. *Evaporation Atlas for the Contiguous 48 United States*. NOAA Technical Report NWS 33. Washington, DC: US Government Printing Office.

Hank, R.J., 1974. Model for predicting plant growth as influenced by evapotranspiration and soil water. *Agron. Journal*, **66**, 35–41.

Jensen, M.E. and H.R. Haise, 1963. Estimating evapotranspiration from solar radiation. *Am. Soc. Civil Engineers Proc. J. Irrigation and Drainage Div.*, **89**, 15–41.

Kalkstein, L.S., 1981. An improved technique to evaluate climate-southern pine beetle relationships. *Forest Science*, **27**, 579–589.

Keim, B.D., G.E. Faiers, R.A. Muller *et al.*, 1995. Long-term trends of precipitation and runoff in Louisiana, U.S.A. *Int. J. Climatology*, **15**, 531–541.

Leith, H. and E.O. Box, 1972. Evapotranspiration and primary productivity. *Publ. Climatology*, **25**, 37–46.

Mather, J.R. (ed.), 1962–1965. Average climatic water balance data of the continents. *Publ. Climatology*, **15–18**.

Mather, J.R., 1974. *Climatology: Fundamentals and Applications*. New York: McGraw-Hill.

Mather, J.R., 1978. *The Climatic Water Budget in Environmental Analysis*. Lexington, MA: Lexington Books.

Mather, J.R. and G.A. Yoshioka, 1968. The role of climate in the distribution of vegetation. *Annals Assoc. Am. Geog.*, **58**, 29–41.

Mather, J.R., F.J. Swaye Jr and B.J. Hartmann, 1972. The influence of the climatic water balance on conditions in the estuarine environment. *Publ. Climatology*, **25**, 1–41.

Meentemeyer, V., 1978. Microclimate and lignin control of litter decomposition rates. *Ecology*, **59**, 465–472.

Meentemeyer, V., E.O. Box and R. Thompson, 1982. World patterns and amounts of terrestrial plant litter production. *Bioscience*, **32**, 125–128.

Miller, D.H., 1977. *Water at the Surface of the Earth: An Introduction to Ecosystem Hydrodynamics*. New York: Academic Press.

Muller, R.A., 1966. The effects of reforestation on water yield – a case study using energy and water balance models for the Allegheny Plateau, New York. *Publ. Climatology*, **19**, 251–304.

Muller, R.A., 1969. Water balance evaluations of the effects of subdivisions on water yield in Middlesex County, New Jersey. *Proc. Assoc. Am. Geog.*, **1**, 121–125.

Muller, R.A., 1970a. Frequency of moisture deficits and surpluses in the humid subtropical climatic region of the United States. *Southeastern Geog.*, **10**, 30–40.

Muller, R.A., 1970b. Frequency analyses of the ration of actual to potential evapotranspiration for the study of climate and vegetation relationships. *Proc. Assoc. Am. Geog.*, **2**, 118–122.

Muller, R.A., 1982. The water budget as a tool for inventory and analysis of factors affecting variability and change of river regimen, in *The Environment: Chinese and American Views*, L.J.C. Ma and A.G. Noble (eds). New York: Methuen, pp. 171–186.

Muller, R.A. and P.B. Larimore Jr, 1975. Atlas of seasonal water budget components of Louisiana. *Publ. Climatology*, **28**, 1–19.

Palmer, W.C., 1965. *Meteorological Drought*. Weather Bureau Research Paper No. 45. Washington, DC: US Department of Commerce.

Peck, E., 1976. *Catchment Modeling and Initial Parameter Estimation for National Weather Service River Forecast System*. NOAA-NES-Hydro 31. Washington, DC: US Government Printing Office.

Penman, H.L., 1948. Natural evaporation from open water, bare soil, and grass. *Proc. Royal Soc. London*, **A193**, 120–145.

Ritchie, J.T., 1972. Model for predicting evaporation from a row crop with incomplete cover. *Water Resources Res*, **8**, 1204–1213.

Rohli, R.V. and J.M. Grymes III, 1995. Differences between modeled surplus and USGS-measured discharge in Lake Pontchartrain basin, U.S.A. *Water Resources Bull.*, **31**, 97–107.

Sanderson, M., 1966. A climatic water balance of the Lake Erie Basin. *Publ. Climatology*, **19**, 1–87.

Shelton, M.L., 1981. Runoff and land use in the Deschutes Basin. *Annals Assoc. Am. Geog.*, **71**, 11–27.

Stapper, M. and G.F. Arkin, 1980. *Dynamic Growth and Development Model for Maize*. Texas Agr. Exp. Sta. Report 80–2.

Sudar, R.A., K.F. Saxton and R.E. Spomer, 1981. A predictive model of water stress in corn and soybeans. *Trans. Am. Soc. Agric. Engineers*, **24**, 97–102.

Thornthwaite, C.W., 1948. An approach toward a rational classification of climate. *Geog. Review*, **38**, 55–94.

Thornthwaite, C.W. and J.R. Mather, 1955. The water balance. *Publ. Climatology*, **8**, 9–86.

Thornthwaite, C.W. and J.R. Mather, 1957. Instructions and tables for computing potential evapotranspiration and the water balance. *Publ. Climatology*, **10**, 185–311.

Cross references

Agroclimatology
Aridity indices
Climate data: sources
Evaporation: measurement
Evapotranspiration
Hydroclimatology
Precipitation distribution
Semiarid regions
Water balance
Water resources: introduction
World water balance

WATER: CATEGORIES

The water of the planet Earth is conveniently classified into a number of categories, defined either by origin or the place where it is found, which may be in the atmosphere, hydrosphere or lithosphere. Seawater, which constitutes by far the bulk of all the Earth's water, is treated separately. The most commonly identified categories are as follows.

Meteoric water

Meteoric water is derived from the atmosphere, precipitating as rain or snow, and participating in the hydrological cycle (q.v.). Rainwater, being largely derived from the ocean by evaporation (also from lakes and rivers, and from land surfaces immediately following precipitation), is strongly modified by the distance it has traveled from the source area; rains near the sea coast are enriched geochemically with 'cyclic salts', i.e. sea salts carried into the clouds from the bubbles of breaking waves.

Surface water

The term surface water includes all those waters associated with the land surface, as in streams, rivers, ponds, lakes and inland seas. Their geochemistry is not only modified by the character of the rainwater, but also by the soils, biology and geology of the various regions.

Groundwater

Groundwater is part of the subsurface or underground water that is present below the water table, in the zone of saturation, as well as in the underground streams of karst (limestone) terrane. In porous aquifers it can reach depths of 10 km or more. It may be tapped economically in artesian wells or in places it may escape in natural springs, even offshore where it can bubble up to the surface, e.g. off eastern Florida or in the Persian Gulf where it has been exploited by the Gulf Arabs for thousands of years. The geochemistry of groundwater is dominated by that of descending meteoric water and surface waters, but at depth and with rising levels of ambient heat flow the mineral solutions increase, as does contamination by mixing with connate water (see below). In deep oil wells temperatures commonly exceed 100°C.

Phreatic water

Phreatic water is often used as a synonym of groundwater (Vollmer, 1967; Bates and Jackson, 1997), but it was originally intended for unconfined groundwater, that is, in the upper part of the zone of saturation only (Meinzer, 1923, 1939). The word 'phreatic' is from the Greek, pertaining to an artificial well. It would exclude deep groundwater in unconnected pore spaces (Davis and de Wiest, 1966, p. 39). A phreatic cycle is seen in the rise and fall of the water table (q.v.), which may be annual or longer, or in coastal regions may be diurnal, following the luni-solar tides in the nearby ocean. A phreatic or phreatomagmatic explosion occurs during a volcanic eruption when the hot, rising magma encounters an abundant aquifer or surface water body (such as a crater lake).

Vadose water

The term 'vadose' (first used between 1890 and 1895) comes from the Latin root *vadosus*, meaning shallow and having a link with the English word to wade (i.e. in shallow water). Vadose water is found in the zone of saturation (hence vadose zone) near the surface of the ground (Meinzer, 1923) and embraces all water above the phreatic zone, i.e. the groundwater. The boundary between them is the water table (Davis and de Wiest, 1966, p. 39). The upper part of the vadose zone is called soil water and the basal part may be referred to as capillary water, both having seasonal variability. The expression 'suspended water' is used by Hem (in Fairbridge, 1972) as a synonym for vadose water, but it is not recommended in Jackson (1997).

Connate water

Water that is trapped in marine sediments at the time they are laid down in the sea is commonly called connate water (so named by Lane, 1906). As the term implies, connate water is produced at the same time as the rock and constitutes a sort of fossil seawater (Hem, in Fairbridge, 1972). When marine deposits are raised above sea level they are subjected to slow downwards movement by meteoric (ground) waters, which results in mixing and leaching. In large sedimentary basins that are still actively accumulating and subsiding, the basal loading increases and compaction steadily rises to expel connate waters which join a slow upward circulation. This may leave evidence in the form of diagenetic minerals such as dolomite ('anadiagenesis') and authigenic feldspars. Both quartz and calcite are often found in joints and along fault planes. White (1957) has suggested criteria for distinguishing connate water by means of ratios of concentrations of certain of the dissolved ions to one another. In most connate water the ratios of bromide and iodide to chloride are relatively high and ratios of potassium and lithium to sodium are low. Analyses of a variety of connate brines have been published by White *et al.* (1963). During sedimentation, burial, lithification and subsequent emergence and weathering, a complex history of diagenesis and authigenesis leads to loss or release of groundwater to the chemically bound (crystalline) state in certain minerals, so-called 'pellicular water', such as hydrous clay minerals and the gypsum–anhydride reactions.

Juvenile water

Juvenile water refers to water which has been derived from the interior of the Earth, and at the time of its appearance in the circulating water of the hydrosphere it represents an accretion to the available water supply. Juvenile water is therefore water which has not previously been a part of the hydrosphere (Hem, in Fairbridge, 1972).

Hem points out that it is extremely difficult to define juvenile water in a geochemical sense, because over long periods of time the surface rocks (with their connate waters) are believed to be subject to a secular recycling into the mantle during subduction and convective overturn. However, the theory of mantle plumes rising from great depth suggests that some, at least, of the water contained in these 'basaltic' plumes represents the planetesimals or 'dirty snowballs' that go back to the primeval development of the solar system. Thus the plumes that penetrate only the oceanic crust and appear in the axes of sea-floor spreading zones may contain some traces of the original mantle.

Another source of genuinely extraterrestrial water is to be found in the comet tails and annual influx to the Earth's atmosphere of 'dirty snowballs' associated with meteor showers. It has been estimated that 40 tonnes of water-ice from this source joins the hydrosphere each year.

Magmatic water

The water derived from the melting of rocks, or which is present in rock melt, is termed magmatic water. A major part of the gaseous exudations in volcanic regions is water vapor. White *et al.* (1963) reported that steam generally constitutes more than 90% of the volatile material coming from volcanoes. However, there is usually considerable opportunity for meteoric water to become incorporated in volcanic gases, so not all of this can be attributed to magmatic sources. There is a strong probability that a portion of the water issuing from hot springs in volcanic regions is of magmatic origin.

However, White (1957) suggested that the proportion of magmatic water in such solutions was probably not generally more than 5% (Fyfe *et al.*, 1978). The magmas referred to in most cases are those associated with thick continental crust, as distinct from the thin oceanic crust.

Magmatic water may be partly juvenile, that is, water which has never been in the hydrosphere before. However, many magmas have probably had opportunity to pick up water from the hydrosphere at some time in their past history.

Magmatic water, according to White (1957) can contain considerable quantities of some of the more volatile elements and may often be high in chloride, fluoride or boron, and have a high ratio of lithium to sodium. The water liberated from a magma in the late stage of its cooling is called deuteric water, being strongly reactive, and it then attacks the minerals already formed. Rising upward, it often fills cracks and fissures with hydrothermal deposits, some of which constitute valuable ore deposits.

Rhodes W. Fairbridge

Bibliography

Berner, E.K. and Berner, R.A., 1987. *The Global Water Cycle*, Englewood Cliffs, NJ: Prentice-Hall., 397 pp.
Craig, H., Boato, G. and White, D.E., 1946. Isotopic geochemistry of thermal waters, *Proc. 2nd Conf. Nuclear Processes in Geologic Settings, Publ. 400*, Natl. Acad. Sci., Natl Res. Coun., pp 29–38.
Davis, S.N. and de Wiest, R.J.M., 1966. *Hydrogeology*. New York: Wiley and Sons.
Drever, J.I., 1997. *The Geochemistry of Natural Waters*, Upper Saddle River, NJ: Prentice-Hall, 397 pp.
Fairbridge, R.W., 1972. *The Encyclopedia of Geochemistry and Environmental Sciences*. New York: Van Nostrand Reinhold, 1321 pp.
Fyfe, W.S., Price, N.J. and Thompson, A.B., 1978. *Fluids in the Earth's Crust*, New York: Elsevier, 383 pp.
Gorham, E., 1961. Factors influencing supply of major ions to inland waters, with special reference to the atmosphere, *Geol. Soc. Am. Bull.*, **72**, 795–840.
Hem, J.D., 1985. *Study and Interpretation of the Chemical Characteristics of Natural Water*, US Geological Survey Water Supply Paper 2254, 263 pp.
Jackson, J.A., 1997. *Glossary of Geology* (4th edn). Alexandria, VA: Am. Geol. Inst.
Lane, A.C., 1906. The chemical evolution of the oceans. *J. Geol.*, **14**, 221–225.
Meinzer, O.E., 1923. *Outline of Ground water Hydrology, with Definitions*. US Geological Survey Water Supply Paper, 494, 71 pp.
Rona, P., Bostrom, K., Laubier, J. and Smith, S., 1983. *Hydrothermal Processes at Sea Floor Spreading Centers*. New York: Plenum Press.
Vollmer, E., 1967. *Encyclopedia of Hydraulics, Soil and Foundation Engineering*. New York: Elsevier, 219 pp.
White, D.E., 1957. Magmatic, connate and metamorphic waters, *Geol. Soc. Am. Bull*, **68**, 1659–1682.
White, D.E., Hem, J.D. and Waring, G.A., 1963. Chemical composition of subsurface waters. US Geol. Surv. Profess. Paper 440F, F30–F39.

Acknowledgement

Kindly reviewed and improved by Robert Horden, Rutgers University, NJ.

Cross references

Acid rain
Groundwater
Hydrological cycle
Medicinal springs
Water on Earth and other planets
Water resources: natural quality

WATER CHARGES (UK): ABSTRACTIONS AND DISCHARGES

Under Act of Parliament the Environment Agency is required to raise, through charges, sufficient funds to cover the costs relating to issuing and monitoring discharge consents and their impact on receiving waters. The charges are raised from persons or industry making either an abstraction or discharge or both. The principle behind the scheme of charges is that the Environment Agency costs should be recovered from those making the abstraction or discharge rather than the general taxpayer. About 70% of the funding of the Agency is raised through water abstraction charges, levies on local authorities for flood defense work, pollution control and fisheries, navigation and recreation charges. The remainder is in the form of grants from central government.

Abstraction charges (1996)

The annual abstraction charge is calculated as follows:

$$\text{annual abstraction charge} = V\,A\,B\,C\,U \text{ (pounds sterling)} \quad (W1)$$

where V is the annual licenced volume ($m^3 \times 10^3$), A is the source factor (normally unity), B is the season factor, C is the loss factor and U is the unit charge based on region in pounds sterling per 1000 m^3.

A (the source factor) is normally unity unless the abstraction is made from certain rivers (3.0) or from tidal rivers (0.2). B (the season factor) is unity if the abstraction is made all year, 1.6 for summer-only abstraction and 0.16 for winter-only abstraction. C (the loss factor) is based on the purpose of the abstraction and has four categories as follows:

- high loss: water abstracted but not returned (e.g. spray irrigation);
- medium loss: public and private water supply, etc.;
- low loss: (e.g. vegetable washing);
- very low loss: (e.g. power generation, effluent dilution).

U is the standard unit charge which is assessed for each of the ten regions in England and Wales as given in Table W6.
Example calculations:

- A licence in Anglian region authorizes 100 000 m^3 per annum abstraction for spray irrigation during the months from April to August.
  Therefore from equation (W1):

$$\text{annual charge} = 100 \times 1.0 \times 1.6 \times 1.0 \times 13.94$$
$$= \pounds2230.40 \ (\$3345)$$

- A licence in the Thames region authorizes 10 000 000 m^3 per annum abstraction for public water supply.
  From equation (W1):

$$\text{annual charge} = 10\,000 \times 1.0 \times 1.0 \times 1.6 \times 7.95$$
$$= \pounds47\,700 \ (\$71550)$$

In addition to the above a sum of £100 ($150) is payable for a licence to abstract. Abstractions of less than 20 m^3 per day are exempt.

Table W6 Standard unit charges for abstraction for each water authority in England and Wales

Region	Standard unit charge (pounds sterling per 1000 m^3)
Anglian	13.94
Northumbria	16.22
North West	7.98
Severn Trent	8.44
Southern	10.28
South West	12.50
Thames	7.95
Welsh	7.76
Wessex	11.00
Yorkshire	6.29

Table W7 Volume factors V for discharges

Volume (m^3 day^{-1})		
More than	Up to and including	Factor
	5	0.3
5	20	0.5
20	100	1.0
100	1 000	2.0
1 000	10 000	3.0
10 000	50 000	5.0
50 000	150 000	9.0
150 000		14.0

Discharge charges (1996)

The annual charge for discharging to rivers is calculated from the following formula:

$$\text{annual discharge charge} = V\,C\,W\,R \text{ (pounds sterling)} \quad (W2)$$

where V is the volume factor based on cubic meters per day, C is the contents factor, W is the receiving water factor and R is the unit rate factor (£401 for 1995–1996). V is found from Table W7. C (the contents factor) is based on seven bands of effluent discharge depending on the type of effluent as follows.

- Band A (factor 14.0), the most critical, includes trade or sewage effluent containing pesticides, fungicides, herbicides, biphenyls, hydrocarbons, alcohols, phenol compounds, etc.
- Band B (factor 5.0) includes trade or sewage effluent containing metals, cyanides, sulpfides, ethanol, etc.
- Band C (factor 3.0) includes sewage effluent and organic trade effluents.
- Band D (factor 2.0) includes discharges which do not fall under A, B or C and trade effluent not specified in E, F or G.
- Band E (factor 1.0) includes site drainage, storm discharges and cooling water.
- Band F (factor 0.5) includes surface water and trade effluent with due regard to suspended solids, iron, pH and chlorine.
- Band G (factor 0.3) includes trade effluent of direct cooling water with regard to volume, temperature, pH and chlorine.

W, the receiving water factor, may have the following values:

Groundwater or land	0.5
Coastal	0.8
Surface water	1.0
Estuarial water	1.5

R, the financial factor or unit rate for 1995–1996 was £401 ($600)

Example calculations:

- A consent allows for a discharge of 8000 m^3 per day of organic sewage effluent to a river. Equation (W2) is:

$$\text{annual discharge charge} = VCWR \text{ (pounds sterling)}$$
$$= 3 \times 3 \times 1 \times 401$$
$$= £3609 (\$5413)$$

- A consent allows for a discharge of 50 m^3 per day of site drainage from trade premises to coastal waters.

$$\text{annual discharge charge} = 1 \times 1 \times 0.8 \times 401$$
$$= £320.80 (\$480)$$

Charges are subject to $17\tfrac{1}{2}\%$ tax. No charge is made for discharge of sewage effluent of 5 m^3 or less per day.

R.W. Herschy

Source

UK Environment Agency, Annual Charges 1995/96.

Cross reference

Water allocation and use

WATER DIVINING (1)

Water divining (or dowsing), known in the USA as water witching, is a mystical, non-scientific technique which exponents claim enables them to locate auspicious places to construct new wells. Dowsing is also used as a wider term referring to the application of these methods to locate any mineral including oil, gold and metal ores, or buried objects such as pipes, and even missing persons or the bodies of murder victims. Water divining is found in some form in all countries.

The most commonly used technique in water divining (Ellis, 1917; Todd, 1980; Randi, 1991) involves holding a forked stick (hazel and willow being the most favored) which is gripped by the two forks with the elbows held to the sides, and with the bottom end of the stick pointing away from the body. The dowser then walks over the local area until the end of the stick is twisted downwards, which the dowser interprets as indicating the presence of groundwater. Another popular method uses two metal rods of copper or aluminum each about 60 cm long. These are bent in a right angle at one end to form a handle. The rods are held one in each hand so that they are parallel and project out from the body. They swing either together or apart to signify the apparent presence of groundwater. Randi (1991) comments that the way in which the forked stick or rods are held ensures that they are in very unstable equilibrium, so that the slightest shift of the hand causes them to move violently, providing the effect used by the dowser as evidence for the presence of water.

A lack of scientific evidence in its favor has done little to reduce a popular belief in most countries that water divining is the only reliable way to locate a new well. Consequently, despite official advice to the contrary from authorities such as the US Geological Survey, water diviners continue to be used in siting new water wells. So far, water divining (or any other form of dowsing) has not been demonstrated to work when subjected to the scrutiny of scientific objectivity. Several large-scale experiments have been carried out (Lehr, 1985; Randi, 1991) where practicing dowsers were asked to locate buried pipes, either empty or containing water. In each instance, it is stated, the successes were as may be predicted by chance alone. It has been suggested that the reputation for success of many water dowsers may result from them operating in localities underlain by water-bearing rocks where a successful well may be drilled almost anywhere (Brassington, 1995).

Vogt and Hyman (1959), an anthropologist and a psychologist, carried out a survey to review the practice of water divining across the USA. They found that 0.018% of the population practiced water divining, with greater concentrations in rural areas and where it is difficult to locate the site for successful wells. They suggested that water divining is a 'magical divination', that is, an irrational system of decision making in which the signs have no demonstrable connection to the anticipated outcome. Many commentators on both sides of the debate on the success of water divining regard all forms of dowsing as supernatural phenomena akin to religious experiences, extrasensory perception (ESP) and magic (not conjuring). Devotees in all these areas share a conviction (or faith) in the truth of their beliefs. In these circumstances, the success of scientific proof of whether or not water divining really does detect water is irrelevant to the majority of water diviners and explains the continuing widespread conviction that water divining actually works, and perhaps the apparent indifference of dowsers to participate in scientific investigations as reported by Randi (1991).

It has been suggested that the continued use of water diviners to search for water rather than use the scientific approach adopted by hydrogeologists is likely to lead to serious problems. Price (1985) argues that the water diviner's role in locating the presence of water is only a small part of what is required for successful groundwater development. In contrast, the hydrogeologist must determine the long-term effects of the abstraction, such as the impact on springs and other parts of the water environment or whether the water quality will change with time. Lehr (1985) makes similar points and forcefully argues that water diviners should not be regarded as experts in the groundwater field. Water diviners tend to explain their activities and the way in which water flows through the ground in terms which are completely incompatible with hydrogeological theory. Lehr fears that serious environmental damage could result if the water diviners' concepts of groundwater flow are used by individuals and companies to decide policies on activities which may pollute groundwater, such as waste disposal, the storage of fuel oils and industrial chemicals and the use of agricultural chemicals.

F.C. Brassington

Bibliography

Brassington, R., 1995. *Finding Water*, 2nd edn. John Wiley and Sons, Chichester, 271 pp.

Ellis, A.J., 1917. *The divining rod – a history of water witching*. US Geological Survey Water Supply Paper 416, 59 pp.

Lehr, J.H., 1985. Tolerence of water witching rhetoric is reprehensible. *Well Water Journal*, June, 8–11.

Price, M., 1985. *Introducing Groundwater*. George Allen & Unwin, London, 195 pp.

Randi, J., 1991. *James Randi: Psychic Investigator*, Boxtree Ltd, London, 159 pp.

Todd, D.K., 1980. *Groundwater Hydrology*, 2nd edn. John Wiley and Sons, New York, 535 pp.

Vogt, E.Z. and R. Hyman, 1959. *Water Witching USA*. University of Chicago Press, Chicago, 248 pp.

Cross references

Groundwater
Water divining (2)

WATER DIVINING (2)

Historical

The origin of water divining is lost in antiquity. Students of the subject have discovered in ancient literature many references to the rod, and although it is certain that divining (or dowsing) rods were in use among ancient civilizations for forecasting events or searching for lost objects, little is known of the manner in which they were used or what relationship, if any, they may have had to the modern devices in use today. The 'rod' is mentioned many times in the bible, especially in the books of Moses (*Numbers* XX, 9–11) also in *Hosea* IV, 12 and *Ezekiel* XXI, 21.

Various origins have been ascribed to the word 'dowse', and as far back as 1692 the philosopher John Locke referred to the forked stick as a 'dowsing rod' or *virgula divina* (*virgula* being the Latin word for twig). Water divining is one of the many forms of dowsing, the word dowsing having a much wider significance in matters of search and detection, involving water, minerals, oils, cavities, pipes, persons and animals, and embracing a wide medical field and animal husbandry.

What is believed to be the first published description of the divining rod is Georgius Agricola's *De re Metallica* published in 1556. There were numerous publications, however, during the seventeenth and eighteenth centuries, but the first publication which referred specifically to water divining was probably by Claude Galien (1630) on the discovery of the Chateau Thierry mineral waters by Baroness Beausoleil. From about this time the divining rod was used in southern Europe as much in the search for water as in the location for minerals, although it was not used in England for water divining until near the end of the eighteenth century. However, it is almost certain that the divining rod came into common use in Germany in the latter part of the sixteenth century as a means for locating mines and for the discovery of buried treasure. German miners were imported to England during the reign of Elizabeth (1558–1603) to lend an impetus to the industry in Cornwall. By the end of the seventeenth century the divining rod had spread through most of the countries of Europe. The first publication in England was probably Robert Fludd's *Philosophia Moysaica*, in Latin, in 1638.

Sir William Barrett, however, is accredited the honor of being one of the first distinguished scientists of the twentieth century who paid serious attention to water divining. Barrett's research was conducted on behalf of the Society for Psychical Research at the end of the nineteenth century. Barrett was Professor of Physics in the Royal College of Science for Ireland, and in addition to publishing his results in two large volumes (1897, 1901) he joined with Theodore Besterman to augment his work. The results were published by Besterman (1926). Barrett and Besterman listed more than 600 references which they claimed comprised no more than a quarter of the publications known to them, some dating back to the fifteenth

century. They noted, however, that the published record, either as books, papers or paintings, was very much older, going back to the menhirs and stone circles of Neolithic times (megaliths) and into a remoteness of time beyond that. Recent publications in the UK are contained in the *Journal of the British Society of Dowsers*, first published in 1933. The Society now has a membership of some 1200. Publications in English are also available in the USA and Canada, the Society of Dowsers of the former having a membership of 5000. Publications in French and German are held by the respective societies in these countries.

On the basis of the objects used in water divining the subject is classified into two major fields:

- Rhabdomancy, using different types of rods. The name is derived from the Greek words *rhabdos* = rod and *manteia* = divination.
- Radiesthesia or pallomancy, using different types of pendulums. The first name indicates the sensitivity to radiations: pallomancy derives from the Greek words *pallo* = to shake and *manteia* = divination.

Forms of divining rod

Experienced dowsers have their favourite tools but the most common form consists of forkshaped rods made usually of wood (not necessarily hazelwood), plastic or whalebone, (or two metal rods one rod held in each hand); (Figure W11). The divining rod is held with one branch of the fork in each hand and directly in front of the dowser's body as shown in Figure W12. The principle is that, given a flow of underground water, the rod will rise, or fall, gently and turn, or tend to turn, as the dowser crosses the flow. Expert dowsers claim that they are able to locate groundwater precisely and provide an estimate of the actual depth to water level and probable flow with an accuracy far beyond chance and frequently against findings of geologists. Yields of some 180 000 l h^{-1} and depths of up to 2700 m have been recorded. Table W8 shows dowsing results by Maby (1947–1948) in which predicted yields and depths are compared with actual yields and depths. In addition to divining rods, dowsers often use gyrating pendulums both in the field and in map studies (map dowsing) (Figure W13).

C.V. Beadon (personal communication, 1995) has shown that, if a peridot (gemstone) is placed on the center of the flow line of an underground source of water found by dowsing, a circle can be marked off around the center of the source of radius equivalent to the depth of the source. Using fragments of different minerals and a pair of peridots on a flat surface, Beadon found that radii of 1 m, one megalithic yard (0.829 m), one Royal cubit (0.524 m), one Egyptian remen (0.370 m), one Greek foot (0.308 m) and one phi (1.618 m) can

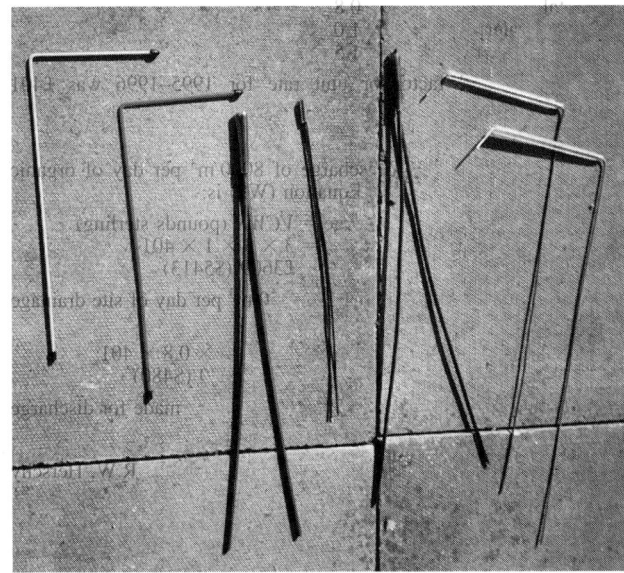

Figure W11 A selection of dowsing rods: from left to right, steel L-shaped; plastic rod; whalebone rod; and steel rod with plastic-covered handles.

Figure W12 Demonstrating of how the rod is held.

be marked off. [The megalithic yard was the standard unit of measurement in the building of megaliths in the British Isles and Brittany, *c.* 2000 BC. The Royal cubit was the unit used in building the pyramids of Egypt and was equal to $\sqrt{2}$ Egyptian remens (Ivimy, 1974). Phi (ϕ) is the limiting value of the ratio between successive numbers in the Fibonacci series (a series of numbers beginning 0, 1, 1, 2, 3, 5, 8, 13, 21) in which each term is the sum of the two previous terms. As the number in the series get progressively larger, so the rates between each term and its immediate predecessor approximates closely to phi.] The minerals used by Beadon to calibrate these units of measurement are, respectively, silver, gold, galena, iron, tin and copper. If a droplet of water is used instead of a mineral, a radius of exactly 0.314 m, or $\pi/10$ m can be marked off.

To perform this experiment, a pendulum is allowed to gyrate freely over one of the peridots and the mineral in question; the second peridot moves until the pendulum ceases to gyrate and the distance is measured. This distance represents one of the above units depending on the mineral used.

Beadon's theory, based on these experiments, postulates that these ancient units of measurements were chosen because they could be shown to be repeatable constants of linear distance under all laboratory conditions and that the architects or engineers involved were in fact either dowsers or had an intimate knowledge of dowsing. They were therefore capable of measuring off the various unit lengths and applying them as required for the builders or surveyors to use.

Many dowsers use an understanding of color as an aid to positive recognition. For example, a dowser may be shown three similar gem stones, all three having the same color visually, of which two are identical twins and the third is a lookalike of a different mineral background. Without touching the stones, the dowser can work out their dowsing colors (four) for each stone and compare the results. Two would have the same four-color sequence, while the third would

be the odd one out, with different colors (C.V. Beadon, personal communication, 1995). Apparently color can be used successfully in water divining and map dowsing by means of the socalled Mager color wheel. However, it is clear that expert dowsers will normally have a knowledge of the geological structure and types of strata with which they have to work in order to obtain the best results.

After the dowser has marked the possible location of underground water, the drilling contractor is required to drill with great care, since a fissure can be missed by a few centimeters or a sluggish vein can easily be blocked by drill action. The contractor may accidentally run right through the water-bearing stratum and line it off with steel tubing, or the drill is stopped short of a likely aquifer.

Understanding of dowsing

After some 5000 years of dowsing the whole subject is still not understood – the simple 'why' and 'how' has defeated scientific explanation. It is doubtful whether so much investigation and discussion have been bestowed on any other subject with such absolute lack of positive explanation. Indeed the subject is still deeply veiled in so much mystery that the term 'water witch' was often used in the past in the United States and water divining became known as 'water witching'.

Nevertheless water diviners often claim a 95% success rate, although it is difficult to obtain comparable results with those of geologists, there being a paucity of such data. However, Table W9 gives a set of results for Tanzania in which the successes of the Water Development Department (WDD) are compared with those of dowsers (Williamson, 1963). It can be seen that the dowser had a much better success rate than the Water Department's professional geologists. Table W10 gives dowsing results for East Africa from 1957 to 1962 (Williamson, 1963) showing actual yields and depths.

Although no satisfactory explanation as to the 'why' and 'how' of dowsing has yet been accepted, dowsers agree that the rod will not move nor the pendulum gyrate without the operator first being motivated and that the movement of either is created out of muscular action. It is the operator that matters and not what is held in the hands. Dowsing appears to be a highly individual art and most operators have their own particular way of interpreting the information which reaches them according to their degree and type of extrasensory perception. It is perhaps unfortunate that some dowsers surround the subject with a supernatural atmosphere. Such attitudes tend to create a secrecy which ultimately generates suspicion. The only real test of dowsers' capabilities is to compare their results with those obtained by geologists and geophysicists working under similar conditions, preferably in situations where the more orthodox means of finding a water source has been deemed improbable (C.V. Beadon, personal communication, 1995). Nevertheless, many scientific ideas have been advanced as an explanation for the phenomenon including the following.

Human senses

The use of our five senses enables us to explore the environment and permits us to assess environmental qualities directly through neurophysical responses to gases, solids and liquids with which we have immediate contact. They can make us aware of more distant features through the patterns of energy propagation associated with them.

Electromagnetic radiation

All bodies with temperatures above absolute zero generate and emit energy in radiant form. Each radiation source, or radiator, emits a characteristic array of radiation waves assessed in terms of wavelengths and intensities. Hence a spectral signature may be obtained for each type of body (e.g. water). This phenomenon is applied successfully in remote sensing from aircraft or satellites, the principle being that electromagnetic energy is transmitted from an object with a specific spectral signature to a sensor.

Electromagnetic induction

The Faraday principle of electromagnetic induction is well known and states that a conductor (the water) moving through the Earth's magnetic field induces an electromotive force which, when sensed by a sensor, is found to be directly related to the flow. Many rivers in the UK are measured by this principle except that, to amplify the

Table W8 Some dowsing results in the UK (Source: Cecil Maby and W.G. Lines)

Site	Yield (gal h⁻¹)		Depth (ft)	
Forecasts and results of bores (Lines) 1947–1948	Prediction (for the vein)	Actual total (all bds)	Prediction (for the vein)	Actual w.b.s.[a] (as reported)
Leominster RDC				
Luston	1200 (Feb.)	1200 (June)	26–30 and 99–108	40–45 and 70–76
Kyre Bank	400 (Sept.)	500 (July)	72–90	70-100
Richards Castle	780 (March)	1000 (May)	33–38	25–40
Wigmore	1300 (April)	90 (Aug.)	45 and 94–106	45
Brimfield	320 (March)	540 (June)	129–138	135–140
Monkland	600 (May)	600 (June)		
	1600 (Feb.)		43–48 and 72	42–47 and 68–75
Hereford RDC				
Breinton Court	320 (Sept.)	350 (Sept.)	60–75	52–61 and 72
Burghill[a]	250 (July)	350 (Oct.)	80–85	40–45 and 70
Burghill[b]	600 (Aug.)	300 (Nov.)	95–109	52–110
Preston Wynne	600 (Oct.)	600 (Nov.)	165–172	58–110 and 120
Dormington	1100 (Oct.)	1080 (Feb.)	25–28 and 135–141	5–6 and 26–54 and 120–140
Sutton St. Nicholas	620 (Aug.)	250 (Feb.)	140–160	20–97 (end of bore)
Holme Lacey	500 (July)	150 (April)	128–130	120–140
Stoke Edith	2000 (Oct.)	2100 (May)	138 and 184	30–36 and 100
Pershore RDC				
Fladbury	1100 (Oct.)	15(?) (Sept.)	305–320	310
Kington	1100 (May)	800 (July)	90–100	18–24 and 48–52
Norton	780 (Feb.)	500 (Jan.)	150–165	110(?)
Cropthorne	1100 (April)	150 (Aug.)	60–66	16–18 and 43(?)

[a] w.b.s. = waterbearing stratum.
[b] Possible variations of yield from forecasts with season and with additional static water from upper levels in certain cases.
[c] Depth of flowing vein only usually predicted, which may come in lowest w.b.s. True actual levels sometimes dubious, owing to upper w.b.s. and standing water in bore.

Figure W13 Demonstration of map dowsing using color wheel and gyrating pendulum.

Table W9 Summary of results of dowsing and a comparison with the results of the Water Development Department of Tanganyika (Tanzania) in similar formations (Source: Williamson, 1963)

1	2	3	4	5	6	7
			Average	Average	Holes exceeding	
		No. of	yield per	depth hole	3 000 (gal h^{-1})	Successes
Sites selected by	Formation	sites	hole (gal h^{-1})	(ft)	(%)	(%)
A.C.W. (dowser)	Tertiary–recent volcanics	25	2140	333	26	76
W.D.D.		19	480	278	0	30
A.C.W. (dowser)	Basement complex	23	1700	286	23	87
W.D.D.		212	980	253	6	79

Column

4. The Dodoma Town water supply from the W.D.D. Site No. 10 59 yielded 16 000 gal h^{-1} when using special air-life compressor pumps. This gives a generous boost to the W.D.D. average yield per hole. It is very likely that some of the dowser's sites shown as 3 000 gal h^{-1} plus could compete with this performance if high-powered pumps were employed.
5. It will be noted that the average drilling depths of the dowser's sites in both classes of formations are greater: 55 ft and 33 ft respectively greater.
6. This column, together with column 4, show significant comparative figures.
7. This is based on yields of 100 gal h^{-1} and more. This may seem small but low yields may be very successful as small domestic supplies in many Masailand villages in very arid areas. Site No. 50 is a case in point: a domestic supply was desperately needed and the yield of 100 gal h^{-1} was considered a blessing, particularly as the dowser predicted a low yield before drilling started (see Table W10).

Table W10 Dowser's drilling results from 1957 to March 1962, East Africa (Source: Williamson, 1963)

Site No.	Results gal h^{-1}	Depth (ft)	Location	Drillers	Remarks
Basement formation: Granites, gneiss, quartz, etc.					
7	2 500	300	Mtindiro Est, Tanga, Tanganyika	Mowlems Ltd	
8	240	360	Lugongo Est, Tanga, Tanganyika	Mowlems Ltd	
20	50	373	Seronero Camp, Tanganyika	Mowlems Ltd	
23	Nil	490	Ulu Est., Konza, Kenya	Mowlems Ltd	
24	800	694	Ulu Est., Konza, Kenya	Mowlems Ltd	
27	1 400	501	Kapiti Plains Est., Konza, Kenya	Mowlems Ltd	Over 100 ft of clay
29	800	300	Yoani Est., Kima, Kenya	Mowlems Ltd	
32	20	520	Tsavo West Park, Kenya	Mowlems Ltd	
46	240	342	Kianzabe Est., Thika, Kenya	Mowlems Ltd	
26	600	300	Kalonzoni Est., Konza, Kenya	Richardson	
50	100	318	Matua Est., Thika, Kenya	Mowlems Ltd	Domestic supply
68	2 000	100	Kibaranga Est., Tanga, Tanganyika	Adsco Ltd	
64	3 800+	117	Mnazi Est., Mkomasi, Tanganyika	Adsco Ltd	
90	1 100	152	Kyampisi Est., Uganda	Grue & Co.	Hard-branded granite prevented deeper drilling
91	240	105	Kyampisi Est., Uganda	Grue & Co.	
92	150	135	Kyampisi Est., Uganda	Grue & Co.	
82	1 100	117	Lugazi Est., Lugazi, Uganda	Grue & Co.	
72	10 000+	300	Mazinde Est., Mombo, Tanganyika	Adsco Ltd	
60	4 800+	140	Mtibwa Est., Turiani, Tanganyika	Adsco Ltd	Shallow alluvium over basalt
97	3 000	250	Khumbara Est., Mombo, Tanganyika	Adsco Ltd	
62	3 000+	231	Msowero Est., Kimamba, Tanganyika	Adsco Ltd	Artesian 5 ft head
96	1 300	240	Hale Est. (Bird & Co.), Muheza, Tanganyika	Adsco Ltd	
73	1 400	300	Mazinde Est., Mombo, Tanganyika	Adsco Ltd	
Tertiary–recent volcanics					
1	750	145	Nyati Est., Sanya Juu., Tanganyika	A.J. Holmes	
2	1 000	184	Msingi Est., Sanya Juu., Tanganyika	Mowlems Ltd	
3	5 000+	114	Kifaru, Est., Sanya Juu, Tanganyika	Mowlems Ltd	
5	1 800	180	Magadini Est., Sanya Juu, Tanganyika	Mowlems Ltd	
11	2 500	273	Weru Weru, Moshi, Tanganyika	Mowlems Ltd	
12	8 000+	118	Riverside Farm, Sanya Juu, T.T.	Mowlems Ltd	
15	650	285	Rongai Ranch, Sanya Juu, Tanganyika	Mowlems Ltd	
18	2 000	295	Sangiti Est., Moshi, Tanganyika	Mowlems Ltd	
21	150	474	Mountainside Farm, Ol Molog, T.T.	Mowlems Ltd	
25	3 000+	350	Pongo Est., Sanya Juu, Tanganyika	Mowlems Ltd	
36	400	354	Kiyungi Est., Moshi, Tanganyika	Mowlems Ltd	
45	2 400	279	Sangiti Est., Moshi, Tanganyika	Mowlems Ltd	
56	960	408	Gatua Nyanga, Thika, Kenya	Mowlems Ltd	Clay 105 ft
59	7 200+	329	Selian Est., Arusha, Tanganyika	Mowlems Ltd	

Table W10 Continued

Site No.	Results gal h^{-1}	Depth (ft)	Location	Drillers	Remarks
19	360	300	Laki Laki Est., Arusha, T.T.	A.J. Holmes	
67	6 000	240	Riverside Farm, Sanya Juu, T.T.	Mowlems Ltd	
28	2 400	400	Mezi Est., Moshi, Tanganyika	Craelius Ltd	
66	4 600+	139	Shah Plantations, Oldeani, T.T.	Craelius Ltd	
100	880	403	Uru Est., Moshi, Tanganyika	Craelius Ltd	
109	50	683	Uru Est., Moshi, Tanganyika	Craelius Ltd	Clay 267 ft
110	–	510	Tipperary Estate, Oldeani, T.T.	Craelius Ltd	
113	–	537	Nalendo Estate, Arusha, T.T.	Craelius Ltd	Clay 29 ft
114	–	600	Mondul Estate, Monduli, T.T.	Craelius Ltd	Clay 97 ft
117	–	402	New Brandon Est., Oldeari, T.T.	Craelius Ltd	
124	3 500+	330	Laki Laki Est., Arusha, T.T.	A.J. Holmes	

electromagnetic signals, a coil installed under the river is employed instead of the Earth's field.

Radiesthesia

Radiesthesia may be defined as sensitivity to radiation, or the ability of the dowser to react to radiation emitted from the Earth. It follows the pattern of electromagnetic radiation from water bodies. It is a term much used by the dowsing fraternity and was devised early in the twentieth century by the famous French dowser, the Abbe Alexis Bouly (Beadon, personal communication, 1995) under the idea that the phenomenon observed in the act of dowsing was due to some form of radiation.

At present two branches of radiesthesia are known:

- Teleradiesthesia or intuitive or psychical radiesthesia. The persons using this method claim that with a pendulum above a map or photograph they can indicate the location of water, ore deposits and certain diseases, hundreds of kilometers away. However, in no critical tests has it been possible to confirm these claims.
- Physical radiesthesia. In this case the dowser operates in the field in direct relation to a geophysical discontinuity in the soil. The rotation or swinging of the pendulum is explained by physical and physiological laws.

Geopathy

Dowsing has also been used to check houses for geopathic stress fields generated from underground streams, large mineral deposits, or faults in the substrata. These energy fields radiate from deep under the earth (Alexander, 1995). It is possible that geopathic stress affects the health of many people worldwide. It has also been suggested that geopathic stress could have some bearing on cot deaths (Beadon, personal communication, 1995). The phenomenon of geopathy has been researched in Germany for many years and is taken very seriously, many building sites being tested for geopathic stress prior to house building. With the aid of the divining rod (Beadon, personal communication, 1995), these energy lines which may cross inside houses and cause stress and restlessness can be recognized and returned to their natural pattern of energy from which they originally came.

Perhaps the answer to the 'why' and 'how' lies somewhere between all of these but the research for a solution continues. The history of water divining is long, but the validity of the techniques has been questioned by some engineers and scientists. Nevertheless there are numerous documented cases where water divining has proved to be successful. Its use might therefore be taken into consideration when contemplating investigations into groundwater resources.

R.W. Herschy

Bibliography

Alexander, J., 1995. Dowsing may help detect the Earth's harmful rays, *Daily Mail*, May 6, London.

Barrett, W.F., 1897, 1900. On the so-called divining rod or *virgula divina*, *Soc. Physical Res. Proc. London*, **13** (1897), **15** (1900).

Beadon, C.V., 1977. Planning well extensions, *J. Br. Soc. Dowsers*, **XXV**(175), 242–248, Ashford, UK.

Beadon, C.V., 1979. Downsing at home and abroad, *J. Br. Soc. Dowsers*, **XXVII**(184), 2–13, Ashford, UK.

Bent, C., 1968. Water diving experiences, *J. Br. Soc. Dowsers*, **XXI**(142), 26–31, Ashford, UK.

Besson, J., 1569. *The art and science of finding water and fountains hidden under ground* (translated from the French).

Besterman, T., 1926. *The divining rod*, Methuen, London.

Boyle, R., 1661. Tentamina quaedam physiologica (some physiological experiments) *Phil. Trans. London*.

Burridge, G., 1953. Primary water, *J. Br. Soc. Dowsers*, **XI**(82), 194–197, Ashford, UK.

Burridge, G., 1959. Steam and treasure, *J. Br. Soc. Dowsers*, **XIV**(99), Ashford, UK.

Burridge, G., 1959. Geothermic steam, *J. Br. Soc. Dowsers*, **XV**(106), 213–215, Ashford, UK.

Cameron, V.L., 1954. Hot and cold springs and geysers in the USA, *J. Br. Soc. Dowsers*, **XII**(86), 100–102, Ashford, UK.

Chavairox, M.F., 1879. *Die Wunschelrute* (the divining rod), Leipzig.

Cookworthy, W., 1751. Observations on the properties and use of the virgula divina, *Gentleman's Mag. and Hist. Chronicle*, London.

Creke, R., 1934. Water table and fissure streams, *J. Br. Soc. of Dowsers*, 1(5), 92–102, Ashford, UK, 1934.

Creke, R., 1960. The point depth method, *J. Br. Soc. Dowsers*, **XV**(107), 255–257, Ashford, UK.

de Beaucorps, A. and F., 1900. *Empirical study, by means of the divining rod, of the underground source of the rivers of the Loiret. Project for obtaining water for the city of Paris. History of the divining rod, theory of its use, application to the valley of the Loire and the springs of the Loiret*, Orleans.

de France, H., 1930. *The modern dowser* (translated from the French by A.H. Bell).

Donaldson, J.E., 1974. Water find in Yorkshire, *J. Br. Soc. Dowsers*, **XXIV**(164), 50, Ashford, UK.

Edney, A.J., 1936. Dowsing, *J. Br. Soc. of Dowsers*, 11(13), 236–238, Ashford, UK.

Ellis, A.J., 1917. *The divining rod: a history of water witching (with bibliography 1500–1916)*, United States Geological Survey Water Supply Paper 416, Washington, DC.

Elliot, J.S., 1973. So far so good, *J. Br. Soc. Dowsers*, **XXIV**(163), 2–10, Ashford, UK.

Endriss, K., 1910. The divining rod and professional water finders, Berlin.

Fenwick, C.D.A., 1965. Dowsing for water, *J. Br. Soc. Dowsers*, **XVIII**(127), 256–262, Ashford, UK.

Galien, C., 1630. *The discovery of the mineral waters of Chateau-Thierry and their properties*, Paris (translated from the French).

Harpers Weekly, 1913. The divining rod in Germany, **57**, Feb. 22, p. 22, New York.

Herschy, R.W., 1963. The use of water for agriculture, in *Conservation of Water Resources in the UK Proc. Instn. Civ. Eng.*, No. 68.

Independent, 1913. The mystery of the divining rod, **76**, Oct. 9, 64–65, New York.

Invimy, J.W.L., 1974. *The Sphinx and the Megaliths*, Turnstone Books, London.

Kniepf, A., 1906. Radioactivity and the divining rod, *Die Gegenwart*, **70**, 166–169, Berlin.

Lang, A., 1883. The divining rod, *Cornhill Magazine*, **47**, 83–91, London.

Latimer, C., 1876. *The divining rod: virgula divina baculus divinatoribus, water witching*, Cleveland.

Le Grand, J.P., 1939. Underground water supplies, *Proc. Inst. San. Eng.*, 1959 (now CIWEM) reprinted in *The Surveyor*, March 1939, *J. Br. Soc. Dowsers*, **III**(20), 362–367, Ashford, UK.

Lines, W.G., 1948. Rural district water supplies and the dowser, *J. Br. Soc. Dowsers*, **VIII**(60), 16–20, Ashford, UK.

Lines, W.G., 1956. Experiences of a practical water dowser, *J. Br. Soc. Dowsers*, **XIII**(90), 348–354, Ashford UK.

Maby, J.C., 1939. Some personal recollections of radiesthesia, *J. Br. Soc. Dowsers*, **XVII**(117), 85–92, Ashford, UK, 1962; British Society of Dowsers, Investigation Committee on radiesthesia; *The physics of the divining rod*, G. Bell & Sons, London.

Maby, J.C., 1948. Dowsing and contracting for water supplies in Britain, *J. Br. Soc. Dowsers*, **VIII**(61), 64–66, Ashford, UK.

Mager, H., 1910. *For the discovery of springs, mines and treasures by means of the divining rod and various scientific apparatus*, Paris.

Mager, H., 1910. The radiation of the earth, and experiments adapted to prove the causes of the movements of certain divining rods, in *First Cong. Exper. Psychology Rept.* Paris, pp. 196–206.

Mager, H., 1913. *Water witches and their methods: the divining rod and the magic pendulum*, Paris.

Mager, H., 1913. *Communication on the lines of force capable of influencing man and of being registered by a simple divining rod*, Academy of Science, Paris.

Mantel, E.A., 1913. Report on the congress of diviners at Paris presented to the special commission of underground hydrology of the Ministry of Agriculture, Paris.

Merrylees, K.W., 1959. A suggested explanation, *J. Br. Soc. Dowsers*, **XV**(106), 212–221, Ashford, UK.

Merrylees, K.W., 1973. Water supplies in England, *J. Br. Soc. Dowsers*, **XXIII**(161), 242–246, Ashford, UK.

Moffat, W.S., 1979. Wells and well drilling, *J. Br. Soc. Dowsers*, **XXVI**(183), 298–306, Ashford, UK.

Morin, H., 1960. History of radiesthesia, *J. Br. Soc. Dowsers*, **XVI**(110), 99–107, Ashford, UK.

Palen, L.S., 1954. Letter on primary water, *J. Br. Soc. of Dowsers*, **XI**(84), 349, Ashford, UK.

Phippen, F., 1853. *Narrative of practical experiments proving to demonstrate the discovery of water, coal and minerals in the earth by means of the dowsing fork or divining rod*, London.

Platters, G., 1639. *A discovery of subterraneall treasure*, London.

Pogson, C.A., 1940. Dowsing over 'still' water, *J. Br. Soc. Dowsers*, **IV**(30), 196–197, Ashford, UK.

Taylor, E., 1974. Twenty-five years of practical dowsing, *J. Br. Soc. Dowsers*, **XXIV**(165), 123–129, Ashford, UK.

Thouvence, P., 1781. *Physical and medical memoir showing the evident relations between the phenomena of the divining rod and animal magnetism and electricity*, Paris and London.

Trinder, A.A., 1933. Some hints for beginners, *J. Br. Soc. Dowsers*, **1**(1), 15–17, Ashford, UK.

Tromp, S.W., 1949. *Physical physics*, Elsevier, Amsterdam.

Tromp, S.W., 1972. Water divining (dowsing), in *The Encyclopedia of Geochemistry and Environmental Sciences* (Ed. R.W. Fairbridge), Van Nostrand Reinhold, New York.

Voll, A., 1910. *The divining rod and the sidereal pendulum. An attempt at a scientific study*, Leipzig.

Wentheimer, J., 1910. Experiments with water finders, *Royal Soc. Arts J.*, **59**, 384–389, London.

Williamson, A.C., 1963. Dowsing in East Africa, *J. Br. Soc. Dowsers*, **XVII**(119), 167–180, Ashford, UK.

Cross references

Groundwater
Water divining (1)

WATERFALLS

Table W11 presents details of the world's highest waterfalls (individual leaps) above 91 m (300 ft). Also included are other well-known waterfalls in certain countries but below this height. The flow of the greatest waterfalls above 566 m^3 s^{-1} (20 000 ft^3 s^{-1}) is also given.

R.W. Herschy

Source

Showers, V., 1989. *World Facts and Figures*, John Wiley & Sons, New York.

Table W11 Highest waterfalls (individual leaps), by continent

Latitude and longitude	Waterfall	River and location	Height (m)	Average flow (m^3 s^{-1})
Africa				
28.45S, 28.56E	Tugela: highest fall	Tugela R, Natal, South Africa	411	
10.12S, 27.27E	Kaloba (alt Lofoi)	Lofoi R, Congo, D.R.	340	
18.36S, 32.43E	Mtarazi	Mtarazi R, Mozambique–Zimbabwe	305	
18,25S, 32.47E	Pungwe	Pungwe (alt Pungoe, Pungue) R, Zimbabwe	277	
8.36S, 31.14E	Kalambo	Kalambo R, Tanzania–Zambia	215	
29.52S, 28.04E	Maletsunyane	Maletsunyane R, Lesotho	192	
9.12N, 37.56E	Finchaa	Finchaa R, Ethiopia	155	
28.35S, 20.23E	Aughrabies (alt King George's)	Orange (alt Oranje) R, Cape, South Africa	147	
5.28N, 40.12E	Baratieri	Ganale–Dorya (alt Genale) R, Ethiopia	140	
31.26S, 29.38E	Magwa	Magwa R, Cape, South Africa	137	
8.51S, 31.02E	Izi (alt Chirombo)	Izi R, Zambia	134	
0.49S, 36.46E	Kitaru	Tana R, Kenya	134	
14.35S, 29.07E	Lunsemfwa	Lunsemfwa R, Zambia	122	
20.21S, 57.27E	Tamarin	Tamarin R, Mauritius	122	
31.15S, 28.57E	Tsitsa	Tsitsa R, Cape, South Africa	114	
17.23S, 14.15E	Ruacana	Cunene (alt Kunene) R, Angola–Namibia	107	
29.29S, 30.14E	Howick	Umgeni R, Natal, South Africa	95	
17.55S, 25.51E	Victoria (alt Mosi-oa-Tunya)	Zambezi (alt Zambesi, Zambeze) R, Zambia–Zimbabwe	92	1 090
20.26S, 57.23E	Chamarel	Cap R, Mauritius	91	
9.06S, 15.57E	Dianzundu (alt Duque de Braganca)	Lucala R, Angola	61	
11.29N, 37.35E	Tisissat (off Tis Isat)	Blue Nile (Abay, Azraq) R, Ethiopia	43	
2.17N, 31.41E	Kabalega (for Murchison)	Nile (alt Nil) R, Uganda	40	1 200

Table W11 Continued

Latitude and longitude		Waterfall	River and location	Height (m)	Average flow (m³ s⁻¹)
Asia					
25.10N,	91.45E	Mawsmai	Sohryngkew R, Meghalaya, India	350	
6.04N,	116.29E	Kalapis	Kalapis R, Sabah, Malaysia	335	
25.30N,	91.40E	Thylliejlongwa	Umngi R, Meghalaya, India	304	
14.14N,	74.50E	Gersoppa (alt Jog): highest fall	Sharavati R, Karnataka–Maharashtra, India	253	
25.34N	91.50E	Nohkalikai	Umtru R, Meghalaya, India	198	
7.05N,	80.05E	Kurundu Oya	Kurundu R, Sri Lanka	189	
6.44N,	81.02E	Diyaluma	Punagala R, Sri Lanka	171	
1.19N,	124.54E	Tondano	Manado (alt Menado) R, Sulawesi, Indonesia	150	
45.31N,	148.53E	Ilya Muromets	? R, Iturup I, Russia	141	
6.46N,	80.50E	Bambarakanda: lower fall	? R, Sri Lanka	141	
35.05N,	133.41E	Kamba	Asahi R, tributary, Honshu I, Japan	140	
11.27N,	77.41E	Bhavani	Bhavani R, Tamil Nadu, India	137	
33.40N,	135.53E	Nachi	Nachi R, Honshu I, Japan	133	
6.54N,	80.30E	Laksapana	? R, Sri Lanka	115	
24.32N,	81.18E	Bihar	Bihar R, Madhya Pradesh, India	113	
7.22N	80.55E	Ratna Ella	Ratna R, Sri Lanka	111	
6.38N,	80.34E	Kirinidi Ela	Kirindi R, Sri Lanka	106	
7.04N,	80.42E	Ramboda	Panna (alt Puna) R, Sri Lanka	100	
12.15N,	77.10E	Kaveri (alt Cauvery)	Kaveri (alt Cauvery) R, Karnataka–Tamil Nadu, India	98	934
36.44N,	139.27E	Kegon	Daiya R, Honshu I, Japan	97	
13.56N,	105.56E	Khone	Mekong (alt Khong, Lancang, Lantsang, Mekongk, Tien Giang) R, Cambodia–Laos	21	11 610
Europe					
62.28N,	7.54E	Monge	Rauma R, Norway	774	
		Ormeli	? R, Norway	563	
62.05N,	6.73E	Tusse	Tussa L. Norway	533	
62.34N,	8.11E	Vestre Mardals (alt Western Mardals)	Eikesdals L., Norway	468	
42.42N,	0.00	Gavarnie: highest fall	Pau R, Midi–Pyrenees, France	422	
62.21N,	8.04E	Verma	Verma R, Norway	381	
60.50N,	7.30E	Austerbo	? R, Norway	380	
46.44N,	8.01E	Giessbach	Giess(bach) Creek, Switzerland	350	
60.31N,	7.15E	Rembesdals: highest fall	Rembesdals L., Norway	300	
60.29N,	7.15E	Skykkjedals: highest fall	Skykkjua R, Norway	300	
60.07N,	6.48E	Tyssetrengene: highest fall	Tysso R, Norway	300	
46.36N,	7.54E	Staubbach	Staub(bach) Creek, Switzerland	299	
62.34N,	8.11E	Austre Mardals (alt Eastern Mardals): upper fall	Eikesdals L, Norway	297	
46.24N,	12.55E	Farina del Diavolo	? R, Friuli–Venezia Giulia, Italy	280	
61.22N,	7.55E	Vettis: highest fall	Morkedola R, Norway	275	
60.22N,	7.08E	Valur	Veig R, Norway	272	
69.20N,	21.50E	Molli	Molles R, Norway	269	
67.21N,	15.47E	Austerkrok: highest fall	Austerkrok R, Norway	257	
46.15N,	11.15E	Stuls: highest fall	Cascata R, Trentino–Alto Adige, Italy	230	
60.45N,	7.07E	Kjos	Flams R, Norway	225	
62.34N,	8.11E	Austre Mardals: lower fall Rogaland: highest fall	Eikesdals L, Norway	220	
58.13N,	4.52W	Eas Coul Aulin	Glencoul L, Scotland, UK	201	
61.23N,	7.26E	Feigum: highest fall	Feigum R, Norway	200	
61.27N,	7.59E	Maradals	Maradals Glacier, Norway	200	
61.53N,	8.21E	Aurstaupet: highest fall	Aura R, Norway	193	
		Sote: highest fall	? R, Norway	176	
46.05N,	10.04E	Serio: highest fall	Serio R, Lombardia, Italy	166	
61.07N,	10.30E	Mesna	Mesna R, Norway	160	
46.25N,	9.24E	Pianazzo	Scalcoggia R, Lombardia, Italy	160	
60.70N,	6.46E	Skjeggedals	Tysso R, Norway	160	
45.40N,	6.59E	Rutor: highest fall	Rutor R, Valle d'Aosta, Italy	150	
60.26N,	7.15E	Vorings: highest fall	Bjoreia R, Norway	145	
46.24N,	8.24E	Frua (alt Toce)	Toce R, Piemonte, Italy	143	
		Hundkastet: highest fall	? R, Norway	140	
47.12N,	12.10E	Krimmler: lower fall	Krimmler R, Austria	140	
47.12N,	12.10E	Krimmler: upper fall	Krimmler R, Austria	140	
60.44N,	6.53E	Rjoande: highest fall	Rjoand(ani) R, Norway	140	
60.50N,	6.40E	Stalheims	Jordals R, Norway	126	
57.16N,	5.18W	Glomach	Glomach R, Scotland, UK	113	
67.20N,	15.40E	Fagerbakk: highest fall	Fager(bak) Creek, Norway	112	
41.58N,	12.48E	Tivoli: highest fall	Aniene R, Lazio, Italy	108	

Table W11 Continued

Latitude and longitude		Waterfall	River and location	Height (m)	Average flow (m³ s⁻¹)
59.52N,	8.34E	Rjukan	Mane R, Norway	105	
		Heis: highest fall	? R, Norway	100	
47.12N,	12.10E	Krimmler: middle fall	Krimmler R, Austria	100	
46.10N,	10.44E	Nardis	Nardis R, Trentino-Alto Adige, Italy	100	
59.49N,	6.48E	Novle: highest fall	Stor R, Norway	100	
41.50N,	13.29E	Zompo lo Schioppo	Romito R, Abruzzi e Molise, Italy	100	
46.43N,	8.12E	Reichenbach	Reichen(bach) Creek, Switzerland	91	
42.33N,	12.43E	Marmore: highest fall	Velino R, Umbria, Italy	90	
47.07N,	13.08E	Gastein: lower fall	Gasteiner Ache R, Austria	85	
47.41N,	8.37E	Rhine (alt Rhein, Rhin)	Rhine (alt Rhein, Rhin) R, Switzerland	21	700
North America					
37.41N, 119.39W		Ribbon	Ribbon Creek, California, USA	491	
49,15N, 125.45W		Della	Tofino Creek, British Columbia, Canada	440	
37.45N, 119.36W		Yosemite: upper fall	Yosemite Creek, California, USA	436	
51.30N, 116.29W		Takakkaw: highest fall	Yoho R tributary, British Columbia, Canada	366	
37.42N, 119.40W		Silver Strand (alt Widow's Tears)	Meadow Brook, California, USA	357	
28.13N, 108.14W		Basaseachic	(Arroyo) Basaseachic Creek, Chihuahua, Mexico	311	
51.33N, 116.33W		Twin	Twin Fall Creek, British Columbia, Canada	274	
52.17N, 125.46W		Hunlen	Atnarko Creek, British Columbia, Canada	253	
46.47N, 121.42W		Fairy	Stevens Creek, Washington, USA	213	
39.34N, 121.17W		Feather	Fall R, California, USA	195	
37.43N, 119.39W		Bridalveil	Bridalveil Creek, California, USA	189	
52.11N, 117.03W		Panther	Nigel Creek, Alberta, Canada	183	
37.43N, 119.32W		Nevada	Merced R, California, USA	181	
45.34N, 122.06W		Multnomah: highest fall	Multnomah R, Oregon, USA	165	
37.43N, 119.36W		Sentinel: lower fall	Sentinel Creek, California, USA	152	
51.57N, 120.11W		Helmcken	Murtle R, British Columbia, Canada	137	
49.11N, 121.44W		Bridal Veil	Bridal Creek, British Columbia, Canada	122	
37.43N, 119.34W		Illilouette	Illilouette Creek, California, USA	113	
46.47N, 121.47W		Comet	Van Trump Creek, Washington, USA	98	
37.45N, 119.36W		Yosemite: lower fall	Yosemite Creek, California, USA	98	
37.44N, 119.33W		Vernal	Merced R, California, USA	97	
44.43N 110.28W		Yellowstone: lower fall	Yellowstone R, Wyoming, USA	94	
10.44N, 61.24W		Maracas	Caraguate R, Trinidad I, Trinidad and Tobago	91	
46.47N, 121.43W		Sluiskin	Paradise R, Washington, USA	91	
61.38N, 125.42W		Virginia	South Nahanni R, NWT, Canada	90	
46.55N, 71.10W		Montmorency	Montmorency R, Quebec, Canada	83	
47.33N, 121.49W		Snoqualmie	Snoqualmie R, Washington, USA	82	
35.39N, 85.22W		Fall Creek	Fall Creek, Tennessee, USA	78	
53.30N, 64.10W		Churchill (for Grand)	Churchill (for Hamilton) R, Newfoundland, Canada	75	1390
42.33N, 76.34W		Taughannock	Taughannock Creek, New York, USA	66	
42.36N, 114.26W		Shoshone	Snake R, Idaho, USA	59	
43.04N, 79.04W		Niagara	Niagara R, Canada–USA	57	5830
Oceania					
21.10N, 156.49W		Kahiwa	Wailau (Stream), Molokai I, Hawaii, USA	533[a]	
21.10N, 156.48W		Papalaua	Kawainui (Stream), Molokai I, Hawaii, USA	366[a]	
8.54S, 140.06W		Ahui	? R, Nuku-Hiva I, French Polynesia	350	
21,01N, 156.37W		Honokohau	Honokohau (Stream), Maui I, Hawaii, USA	341[a]	
30.32S, 152.03E		Wollomombi: highest fall	Wollomombi R, New South Wales, Australia	334	
20.06N, 155.36W		Hiilawe	Wailoa (Stream), Hawaii I, Hawaii, USA	305	
34.38S, 150.34E		Belmore: three falls	Barrengarry Creek, New South Wales, Australia	300	
		Cannabullen	Cannabullen Creek, Queensland, Australia	300	
33.39S, 150.16E		Horseshoe	Govetts Leap Creek, New South Wales, Australia	300	
18.17S, 146.03E		Wallaman (alt Stony Creek)	Stony Creek, Queensland, Australia	296	
17.43S, 145.35E		Elizabeth Grant	Tully R, Queensland, Australia	274	
45.28S, 167.10E		Helena	Helena R, South I, New Zealand	253	
44.48S, 167.44E		Sutherland: upper fall	Arthur R, South I, New Zealand	248	
22.06N, 159.40W		Waipoo: two falls	Waimea R, Kauai I, Hawaii, USA	244	
15.50S, 145.39E		Barron	Barron R, Queensland, Australia	235	
44.48S, 167.44E		Sutherland: middle fall	Arthur R, South I, New Zealand	229	
13.55S, 171.45W		Tiavi	Vaisigano R, Upolu I, Samoa	183	
17.47S, 145.35E		Tully: highest fall	Tully R, Queensland, Australia	168	
44.40S, 167.55E		Bowen	Bowen R, South I, New Zealand	158	
44.36S, 167.52E		Stirling	Stirling R, South I, New Zealand	146	

Table W11 Continued

Latitude and longitude	Waterfall	River and location	Height (m)	Average flow (m^3 s^{-1})
19.51N, 155.09W Akaka		Kolekole (Stream), Hawaii I, Hawaii, USA	135	
34.39S, 150.29E Fitzroy: highest fall		Shoalhaven R tributary, New South Wales, Australia	122	
33.43S, 150.23E Wentworth: upper fall		Wentworth R, New South Wales, Australia	110	
44.48S, 167.44E Sutherland: lower fall		Arthur R, South I, New Zealand	103	
20.09N, 155.38W Waiilikahi		Waiilikahi (Stream), Hawaii, I, Hawaii, USA	98	
South America				
5.57N, 62.30W Angel: upper fall		Churun R, Venezuela	807	
23.07S, 48.36W Itatinga		Itatinga R, Sao Paulo, Brazil	628	
27.12S, 49.21W Pilao		Itajai R, Santa Catarina, Brazil	524	
4.00N, 62.50W Montoya		Porah–Pi R, Venezuela	505	
5.46N, 61.08W Great (alt King George VI)		Kamarang R. Guyana	488	
5.13N, 60.51W Cuquenan (Alt Kukenaam): highest fall		Cuquenan (alt Kukenaam) R, Guyana–Venezuela	317	
5.22N, 72.45W Candelas		Cusiana R, Colombia	300	
12.15S, 73.45W Sewords		Cutibireni R, Peru	267	
5.43N, 59.38W Tiboku (alt King Edward VIII)		Semang R, Guyana	256	
5.09N, 59.29W Kaieteur		Potaro R, Guyana	226	650
20.20S, 46.22W Casca d'Anta		Sao Francisco R, Minas Gerais, Brazil	203	
5.57N, 62.30W Angel: lower fall		Churun R, Venezuela	172	
4.35N, 74.18W Tequendama		Bogota (alt Funza) R, Colombia	157	
6.32N, 60.50W Sakaika: highest fall		Ekreku R, Guyana	140	
6.38N, 60.44W Wakowaieng		Morong R, Guyana	134	
22.31S, 43.10W Fagundes		Piabanha R, Rio de Janeiro, Brazil	126	
17.12S, 54.07W Itiquira		Itiquira R, Mato Grosso, Brazil	120	
5.25N, 59.30W Marina: highest fall		Ipobe R, Guyana	110	
13.02S, 58.17W Utiariti		Saueruina (alt Papagaio) R, Mato Grosso, Brazil	107	
2.22N, 52.40W Mano (alt Manaua)		Oyapock (alt Oiapoque) R, Brazil–French Guiana	105	
5.03N, 65.41W Quenque (alt Tencua)		Manapiare R, Venezuela	100	
9.24S, 38.13W Paulo Afonso		Sao Francisco R, Alagoas–Bahia, Brazil	80	2 890
25.41S, 54.26W Iguacu (alt Iguazu; for Iguassu)		Iguacu (alt Iguazu; for Iguassu) R, Argentina–Brazil	70	1 700
24.02S, 54.16W Sete Quedas (alt Guaira)[b]		Parana R, Brazil–Paraguay	65	8 260
20.18S, 49.10W Maribondo (alt Marimbondo)		Grande R, Minas Gerais–Sao Paulo, Brazil	35	1 500
31.14S, 57.55W Grande		Uruguay (alt Uruguai) R, Argentina–Uruguay	23	4 500
20.36S, 51.33W Urubupunga		Parana R, Mato Grosso do Sul–Sao Paulo, Brazil	9	2 750

Abbreviations: alt, alternative name; I, Island(s); off, official name; R, River; ?, unknown.
[a] Total fall.
[b] The Sete Quedas waterfall was totally submerged after the completion of the Itaipu dam in 1982.

WATER INVENTORY

This term usually implies a national system of data collection, processing and delivery to the user on the regime and resources of surface and subsurface waters, their quality and utilization. These systems operate in many countries of the world for the provision to a wide variety of users of official information on the state of continental waters. Continental waters which are usually presented in a system of water inventory embrace rivers, lakes, reservoirs, canals, underground water and glaciers.

Depending on the functional features, the following three stages may be selected in the system.

The first stage involves data collection from the hydrological network, which is the main source for obtaining information on continental waters; the main objective of the network is to make observations of all basic characteristics of the hydrological regime of water bodies. In many countries the hydrological network is subdivided into basic and special networks. The basic network provides observations on rivers, lakes and reservoirs according to standard programs including base hydrological regime components, i.e. water discharge, sediment yield, water levels, and ice events in rivers, lakes and reservoirs, as well as accompanying meteorological components. A hydrological station organized for direct measurements of a particular water body is the primary structural unit of the basic network. These are stations operated manually by an observer or

completely automatic stations, information from which is transmitted to the processing center by telemetry. Several stations make up a complete hydrological station, which is a structural unit of a higher order; it supervises the operation of stations and provides data collection from the area.

The special network is organized for the observation of individual characteristic components of the hydrological regime, or for multipurpose hydrometeorological observations on special catchments or particular kinds of water bodies. The special network may consist of points for observations of evaporation from soil, water or snow, river channel changes and the deformation of shores in reservoirs, ice jams and ice dams, etc. In addition, the special network embraces water balance stations and experimental plots where detailed observations of the water balance components of river basins are made by special programs, as well as swamp or wetland stations and stations where the hydrological regime is studied, and providing points for mudflow studies, etc. Observation points from the special network can be combined with hydrological stations of the basic network, while observation programs at the special network may comprise measurements of basic characteristics of the hydrological regime.

Observations in hydrological networks (basic and special ones) are made by standard instruments and equipment according to national standard methodologies. There are international recommendations on the design of hydrological networks and on the methods for hydrological parameters measurements, e.g. the WMO *Guide to Hydrological Practices* (WMO, 1994). Observations are made either by observers or by automatic systems. In a number of developed countries, autonomous hydrological stations operate in the automatic mode, with data storage in the computer memory and data transmission to the data processing center, either through direct communication channels or via satellites.

The second stage of the water inventory system consists of the processing of information received at the first stage. It may be either concentrated or scattered. In the case of the concentrated variant, which is quite usual for small countries, a single center for all hydrological data processing is established. This center receives data directly from hydrological stations by communication channels (by mail, teletype, electronic mail, etc.) where the data are processed. In large countries a scattered processing of hydrological information is usually practiced.

The primary processing may be made at the hydrological station so that the processed data may be compiled for the area within the responsibility of the station. Further processing may be made in some regional centers or in the single national center to get the generalized information on particular regions or on the country as a whole. The processing of information is aimed at gathering generalized characteristics of the hydrological regime in standard formats on diskettes or as publications (yearbooks, reports, bulletins, etc.). The processing of hydrological information is mainly done by modern computers; moreover, the primary processing at the hydrological station is made by personal computers (PCs). Large computers are used in national centers for processing the hydrological information from the territories of large countries.

The third stage of the system involves data storage and its delivery to the users. These functions are usually realized in each structural unit of the system, from the individual hydrological station up to the national center for water inventory. Moreover, data are stored and delivered to the users at hydrological stations within the area of the station responsibility; regional centers deliver data for the appropriate region, and the national center provides generalized data storage and delivery for the whole country. Archives or databases are usually made at hydrological stations for their appropriate areas; these stations provide the users with standard formats of data output (tables, graphs, diagrams, etc.). Hydrological data banks operate at national centers of water inventory, which comprise automatic data archives, technical, language, programming and methodological facilities, providing a centralized storage of data on diskettes and the joint multipurpose use of data (including processing by special request, the delivery of design hydrological parameters, preparation of different generalized data on water body regimes, etc.). Databases are usually placed in PCs, while banks of data are stored in large computers. In many countries automatic information systems (AIS) of water inventory are in operation which provide a successive automatic processing of information at each of the three stages of the system by computers. They may differ in their organization structure, the composition of the data to be processed, the technical facilities applied, the types of processed and distributed data, but the purpose of the AIS is always the same, i.e. the provision of operational data collection, processing and distribution of multi-aspect information on continental water for a rational development and protection of water resources of individual states and of large physiographic regions.

V.S. Vuglinsky

Bibliography

Laura, D., 1984. Water resources assessment – the United States experience. Proceeding of the scientific session. *International management of water resources*, Paris Press, pp. 73–110.

Moss, M.E. *et al.*, 1982. *Design of surface-water data network for regional formation*, Washington DC: US Government Printing Office.

National Water Summary, 1984, 1985. *Hydrological events, selected water-quality trends and ground water resources*, US Geological Survey Water-Supply Paper No. 2275, Washington, DC: US Government Printing Office, 467 pp.

Surface Water Data – Reference Index. Canada, 1975. Ottawa: Thorn Press Limited, 276 pp.

Vuglinsky, V.S., Kolobaev, N.N. and L.S. Yazvin, 1990. Problems of improvement of hydrological observation system and national water resources inventory, *Proc. of Vth All-Union Hydrological Congress*, Vol. I, Leningrad: Gidrometeoizdat, pp. 81–98 (in Russian).

Vuglinsky V.S. and Gusev S.I., 1989. National water cadastre system (stages of development, present state, prospects), in *Problemy sovremennoi gidrologii* (Problems of Modern Hydrology), Leningrad: Gidrometeoizdat, pp. 97–108 (in Russian).

WMO, 1994. *Guide to Hydrological Practices*, 5th edn, WMO No. 168, 736 pp.

Cross reference

Hydrological year books

WATER MOVEMENT IN UNSATURATED SOILS

Introduction

In its natural state, the soil is normally unsaturated; that is, it contains both water and air. Most of the water involved in the terrestrial hydrological cycle is located in unsaturated soil between the time of its arrival as rain at the soil surface and that of its return to the atmosphere. A small fraction of precipitation does not enter the soil, but moves over the land surface directly into streams or lakes, and a second small fraction percolates downward through the unsaturated zone and joins the groundwater. In Australia, for example, about 93% of the precipitation enters the soil, and of this 92% returns directly to the atmosphere, only about 1% reaching the groundwater.

The questions of just how water is held and moves in unsaturated soil are thus central to the scientific study of the land sector of the hydrological cycle and in the related problems of irrigated and dryland agriculture, of plant ecology and of the biology of soil flora and fauna. These questions are, in addition, of great significance in connection with the transport through the soil of materials in solution, such as natural salts, fertilizers, and urban and industrial wastes and pollutants.

Among the processes of water movement in unsaturated soil of great concern to hydrologists are infiltration (the entry into the soil of water arriving at its surface), drainage and retention of water in the soil strata, extraction of soil water by plant roots and its subsequent transpiration; and evaporation of water directly from the soil.

The character of these everyday, but all-important, processes depend on the physical uniqueness of water. The latent heat of evaporation of water is greater than that of any other substance, and this produces the tight linkage between the water balance and the energy balance at the Earth's surface which is central to our understanding of natural evaporation. Equally significant is the very great surface tension of water and related properties of the water molecule. These generate very large capillary (and other surface) forces in the soil water. Because of these the soil can hold appreciable quantities of water against gravity. This is a great advantage to plant life and, moreover, makes possible the whole range of moisture conditions at the surface: in the absence of the surface tension (and related properties) of water, the land surface would be either desert or swamp.

Occupying a central place in this article is the mathematical–physical approach to the study of water movement in unsaturated soils which has evolved, with sundry fits and starts, over the twentieth century. See Philip (1974), Gardner (1986), and Sposito (1987) for historical accounts. The physical and mathematical foundations of the basic approach are described in the following paragraphs, including its limited generalization to flow and volume change in swelling soils, and finally the various limitations and extensions are recognized.

Flow in unsaturated non-swelling soils

Darcy's law for saturated soils

Saturated soil is free of air, the pores being completely filled with water. We may write Darcy's law for the flow of water in the liquid phase through saturated soils in the form

$$U = -K\nabla\Phi \qquad (W3)$$

where U is the vector flow velocity, Φ is the total potential and K is the hydraulic conductivity of the medium.

U has the character of a macroscopic mean. It is a flow rate per unit cross-section formed by averaging over an area large compared with the cross-section of individual pores and grains of the medium. U has the dimensions [length][time]$^{-1}$ and the specific units m s^{-1}.

Φ is the potential defined by the equation

$$\Phi = (P/\rho g) + \Omega \qquad (W4)$$

where P is the pressure, ρ the density of water, g the acceleration due to gravity and Ω the potential of the external forces per unit weight of water. Like U, P is a macroscopic quantity and is the outcome of smoothing the actual microscopic distribution of pressure in the water over a volume rather larger than that of the individual pores. Usually the only external force on the system is that of gravity, and then Ω equals the vertical space coordinate. As we have defined it, Φ has the dimensions [length], and the specific unit m.

We employ this definition of Φ, which corresponds to the hydraulic head used by engineers in their flow calculations. The formalism could be developed equally well in terms of a potential based on unit mass. It is an elementary matter to transpose our various results into this form – at the expense of the intrusion of the factor ρg into various expressions, and the use of a more cumbersome set of units.

It then follows that K has the dimensions [length][time]$^{-1}$, with the specific units m s^{-1}. For isotropic soils K is a scalar. For anisotropic soils, on the other hand, K is a second-order symmetric tensor.

A great deal has been written on the theoretical basis of Darcy's law, although much of it seems to constitute self-education rather than serious research. It is sufficient to remark that there are both hydrodynamic and statistical elements to the basis of Darcy's law. The hydrodynamic one is as follows: the Navier–Stokes equation governing the flow of a viscous, incompressible, Newtonian fluid is linear in the limit of small Reynolds number. We have hinted already at the statistical one in defining U and P: it is that, although it is not feasible to know details of the internal geometry of the medium and of the microscopic distribution of velocity and potential, local mean quantities such as U and Φ can be defined and exist.

Darcy's law for unsaturated soils

Buckingham (1907) and Richards (1931) assumed, and Childs and Collis-George (1950a) and others confirmed experimentally, that Darcy's law holds for the flow of liquid water in unsaturated media in a modified form in which K is a function of the volumetric moisture content, θ. The theoretical validity of this concept depends on the (usually very reasonable) assumption that the drag at the air–water interfaces in the soil is negligibly small (Philip 1957a, 1972; Mahony, 1972). The general behavior of the $K(\theta)$ function is now fairly well established, thanks to the work of Richards (1931), Moore (1939), Childs and Collis-George (1950a) and other workers in soil physics and petroleum engineering research. K is found to decrease very rapidly as θ decreases from its saturation value. This is not surprising, for the following reasons (Philip, 1957b).

- The total cross section available for flow decreases with θ.
- The largest pores are emptied first as θ decreases. Since the contribution to K per unit area varies roughly as the square of pore 'radius', K can be expected to decrease much more rapidly than θ.
- As θ decreases, the probability increases that water will occur in pores and wedges isolated from the general three-dimensional network of water films and channels. Once continuity fails, there can be no flow in the liquid phase, other than flow through liquid 'islands' in series–parallel with the vapor system (Philip and de Vries, 1957). (Flow of this type is usually negligible in the absence of temperature gradients, and need not detain us here.)

K may, in fact, vary through six or more decades over the range of interest of θ (Gardner, 1960). Figure W14 shows a typical $K(\theta)$ relationship.

Although the possibility of hysteresis in $K(\theta)$, analogous to that in the moisture characteristic, discussed below, cannot be excluded *a priori*, the indications are that it is relatively unimportant (e.g. Topp and Miller, 1966).

We thus express Darcy's law for the flow of water in unsaturated soils in the modified form of equation (W3):

$$U = -K(\theta)\nabla\Phi \qquad (W5)$$

Total potential and moisture potential of water in unsaturated soils

A keen disciple of Willard Gibbs, Buckingham (1907) proffered Gibbs' thermodynamic insights to soil-water physics between three and seven decades before their central relevance to water transfer in many cognate technologies had begun to be understood.

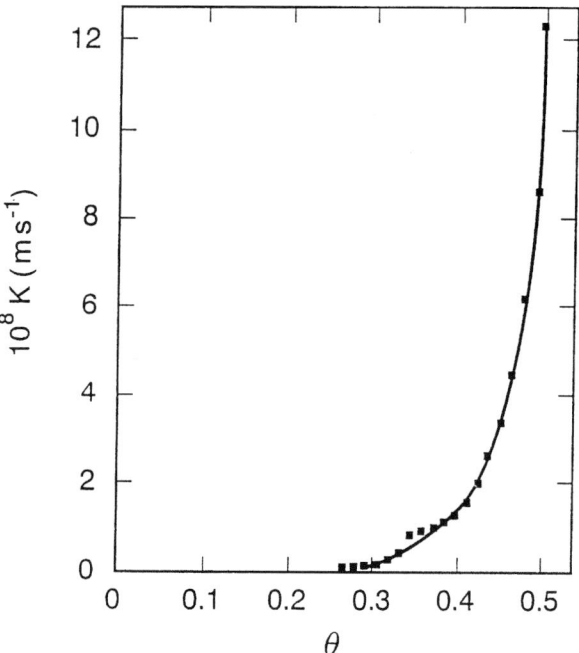

Figure W14 Relationship between hydraulic conductivity K and moisture content θ for Yolo light clay. (Source: Moore, 1939.)

In (water wet) unsaturated soils the water is not free in the thermodynamic sense because of capillarity, physical adsorption and, in colloidal soils, electrostatic effects. Capillarity is dominant in wet, coarse-textured soils and adsorption assumes its greatest importance in dry soils. Buckingham (1907) was the first to appreciate that the conservative forces governing the equilibrium and movement of water in soil are amenable to treatment through their associated scalar potentials.

We define such potentials relative to the reference state of free water at atmospheric pressure and datum elevation $z = 0$. Here z is the vertical space coordinate, conveniently taken to be positive downward. Then, for a nonswelling medium the *total potential* is

$$\Phi = \Psi - z \qquad (W6)$$

The *moisture potential* Ψ is the potential of the forces arising from local interactions between soil particles and water.

Buckingham (1907) called $-\Psi$ the 'capillary potential'. Other names for Ψ or $(-\Psi)$ include 'moisture tension', 'moisture suction', 'negative pressure', 'pressure head', 'matric potential' (Richards, 1949; Childs and Collis-George, 1950a,b; Miller and Klute, 1967; Rose, 1966) and, in swelling soils 'swelling pressure' (Bolt, 1956). Petroleum engineers use 'capillary pressure'. The functional dependence of Ψ on the volumetric moisture content $\Psi(\theta)$ is called the moisture characteristic.

It is not essential either to know or to specify these interaction forces in detail: it suffices that Ψ can be measured by well-established techniques (Croney *et al.*, 1952; Richards, 1965; Holmes *et al.*, 1967; Klute, 1986; Bruce and Luxmoore, 1986; Cassell and Klute, 1986; Reeve, 1986; Rawlins and Campbell, 1986; Campbell and Gee, 1986).

With potentials expressed per unit weight, Ψ has the dimensions [length] and the specific unit m. The gravitational contribution to the total potential is $-z$: see equation (W6). In water-wet non-swelling soils, $\Psi = 0$ at saturation, and $\Psi < 0$ in unsaturated soils, decreasing with the volumetric moisture content θ to very large negative values (typically -10^4 m) at the dry end of the moisture range of interest. Figure W15 depicts the $\Psi(\theta)$ relation for the soil for which $K(\theta)$ is given in Figure W14.

The partial volumetric Gibbs free energy associated with the local solid–water interaction is $\rho g\Psi$. In the absence of solutes the liquid and vapor systems at $z = 0$ are connected by the relation

$$H = \exp g\Psi/RT \qquad (W7)$$

with H being the relative humidity, R the gas constant for water vapor

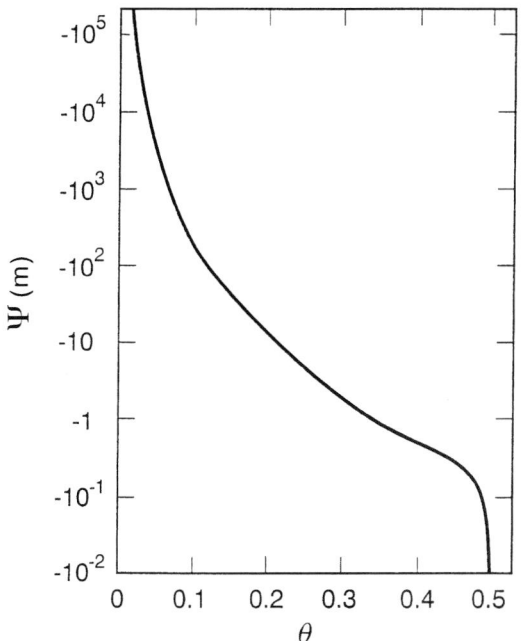

Figure W15 Relationship between moisture potential Ψ and moisture content θ for the soil of Figure W14 (Source: Moore, 1939).

($= 461.5 \text{ J kg}^{-1} \text{ K}^{-1}$), and T the absolute temperature. The function $\Psi(\theta)$ is thus a restatement of its adsorption isotherm for water.

In the absence of solutes $\rho g \Psi$ is the activity or chemical potential of the water. The matter becomes more complicated, however, in the presence of solutes: $\rho g \Psi$, the mechanically operative component of the activity, remains the determinant of liquid phase flow (in the absence of semi-permeable barriers), but it is the total activity, including the component due to solutes, which now fixes H and governs equilibrium and flow in the vapor (and any absorbed) phase.

As the link between liquid and vapor phases, equation (W7) (modified where required to embrace solutes) is important to the study of the evaporation of soil water, of its transport under temperature gradients, and of its isothermal transport in very dry soils.

General partial differential equation of soil-water movement

Combining equation (W5) with the requirement of continuity that

$$\partial\theta/\partial t = -\nabla \cdot U \tag{W8}$$

we obtain the equation

$$\partial\theta/\partial t = \nabla \cdot (K \nabla \Phi) \tag{W9}$$

Here t denotes time (s). Substituting equation (W6) in equation (W9), we then have

$$\partial\theta/\partial t = \nabla \cdot (K \nabla \Psi) - \partial K/\partial z \tag{W10}$$

Equations (W9) and (W10) hold quite generally in the sense that they apply equally to both homogeneous and heterogeneous soils, and do not depend on any requirement that relations between K, Ψ and θ should be single-valued. For a homogeneous soil, however, we may express equation (W10) in more tractable form in the two following ways. If K and $d\Psi/d\theta$ are single-valued functions of θ, we may introduce the quantity D, also a single-valued function of θ, such that

$$D = K \, d\Psi/d\theta \tag{W11}$$

Then (W10) may be written as

$$\frac{\partial\theta}{\partial t} = \nabla \cdot (D \nabla \theta) - \frac{dK}{d\theta} \frac{\partial\theta}{\partial z} \tag{W12}$$

The coefficient $dK/d\theta$ is evidently a function of θ. Alternatively, if K and θ are single-valued functions of Ψ, we may take Ψ as the dependent variable, replacing equation (W12) by

$$\frac{d\theta}{d\Psi} \frac{\partial\Psi}{\partial t} = \nabla \cdot (K \nabla \Psi) - \frac{dK}{d\Psi} \frac{\partial\Psi}{\partial z} \tag{W13}$$

The coefficients $d\theta/d\Psi$ and $dK/d\Psi$ are functions of Ψ.

Equations (W12) and (W13) are not wholly equivalent. Equation (W13) may still apply when Ψ exceeds the air-entry value [the value at which air enters initially saturated soil (Luthin and Miller, 1953)] or is positive (as it may be when a depth of free water is ponded over the soil), whereas equation (W12) cannot. On the other hand, whenever a solution of equation (W12) may be found, a corresponding solution of equation (W13) may be found also.

Equation (W12) is a non-linear Fokker–Planck or convection–diffusion equation. Linear forms of equation (W12) arise in physical phenomena such as heat conduction in moving media (Carslaw and Jaeger, 1947) and diffusion under an external force field (e.g. sedimentation with Browian motion), in the theory of Markov processes in mathematical probability (e.g. Chandrasekhar, 1943; Bailey, 1964), and certain studies of turbulent diffusion (Mahony and Philip, 1967; Philip, 1968a).

Only since 1950 have non-linear forms of equation (W12) been much studied, and mainly in the soil-water context. Work on the closely related non-linear diffusion equation

$$\partial\theta/\partial t = \nabla \cdot (D \nabla \theta) \tag{W14}$$

however, goes back at least to Boltzmann (1894). Equation (W12) reduces to equation (W14) for horizontal systems and in other instances where gravity may be neglected.

We see that equation (W12) is of the diffusion, or heat conduction, form but with two complications that add greatly to the difficulty of its solution. First, the equation contains on its right side the first-order term representing the influence of gravity on the process. Second, the coefficients D and $dK/d\theta$ are both markedly dependent on θ. D may vary typically through three or more decades (and $dK/d\theta$ through several more) between the wet and dry ends of the moisture range. Usually neither complication can be ignored, so quantitative study of the physics of soil-water movement depends centrally on solving non-linear diffusion and Fokker–Planck equations.

D is called the moisture diffusivity. It has the dimensions $[\text{length}]^2[\text{time}]^{-1}$, and the unit $\text{m}^2 \text{ s}^{-1}$ is appropriate. Figure W16 is a graph of $D(\theta)$ for the soil with the $K(\theta)$ and $\Psi(\theta)$ functions of Figures W14 and W15.

The basic concepts leading to equation (W12) were implicit in Buckingham (1907). Richards (1931) developed equations (W10) and (W13). Childs and George (1948) recognized the diffusion character of the form of equation (W14) which applies to one-dimensional horizontal systems. Klute (1952) explicitly derived equation (W12). Philip (1954, 1955, 1957c) extended the approach to include moisture transfer in the vapor and adsorbed phases in the same mathematical

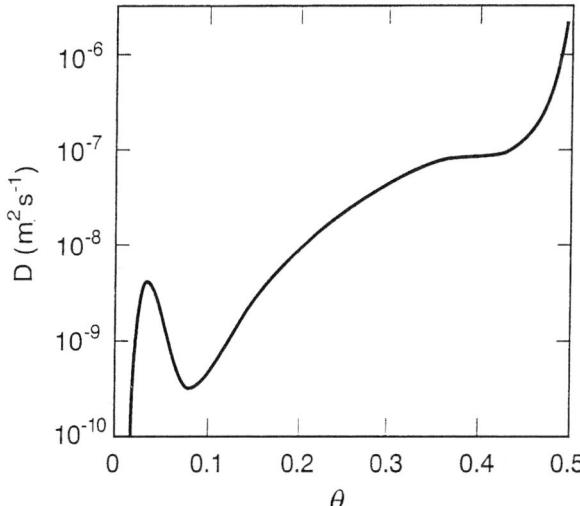

Figure W16 Relationship between moisture diffusivity D and moisture content θ for the soil of Figure W14. (Note that D for $\theta \leq 0.06$ includes dominant contribution in vapor phase. Source: Philip, 1955.)

formulation. This made use of equation (W7) and the chain rule for differentiation.

Although the analysis developed above is relatively elaborate, it depends on various simplifying assumptions. These are justified in many circumstances. In some where they are not, extensions of the analysis may be made. We address these matters below.

Applications to hydrological processes

Solutions of equations (W12)–(W14) satisfying appropriate initial and boundary conditions accord with experiment and offer insight into the mechanics of many hydrological soil-water processes. Examples are infiltration (q.v.), capillary rise (Philip, 1966), evaporation from bare soil (Philip 1954, 1957a), drainage and retention of soil water (Staple and Lehane, 1954; Day and Luthin, 1956; Philip, 1957a; Whisler and Watson, 1968), and extraction of soil water by plant roots (Philip, 1957b; Gardner, 1960; Cowan, 1965).

These citations center on early pioneering studies of these processes. By now there is a vast literature of solutions of these equations relevant to hydrological processes. We do not attempt a comprehensive bibliography here.

Steady flows

Insight into processes may sometimes be gained through solutions of the simpler steady (time-independent) equations. The steady forms of equations (W10), (W12) and (W13) are, respectively,

$$\nabla \cdot (K \nabla \Psi) = \partial K / \partial z \qquad (W15)$$

$$\nabla \cdot (D \nabla \theta) = \frac{dK}{d\theta} \frac{\partial \theta}{\partial z} \qquad (W16)$$

$$\nabla \cdot (K \nabla \Psi) = \frac{dK}{d\Psi} \frac{\partial \Psi}{\partial z} \qquad (W17)$$

The Kirchhoff (1894) transformation:

$$\Theta = \int_{\Psi_0}^{\Psi} K d\Psi = \int_{\theta_0}^{\theta} D d\theta \qquad (W18)$$

reduces the steady form of equation (W14) to the linear and well-known Laplace equation:

$$\nabla^2 \Theta = 0 \qquad (W19)$$

For one-dimensional systems equations (W15)–(W17) and (W19) become readily soluble ordinary differential equations.

The exponential representation

$$K(\Psi) = K(0)e^{x\Psi}, \ \alpha = \text{constant} > 0 \qquad (W20)$$

combined with equation (W18), reduces both equations (W16) and (W17) to the linear form

$$\nabla^2 \Theta = \alpha \frac{\partial \Theta}{\partial z} \qquad (W21)$$

This was first noticed by Gardner (1958). Quasi-linear analysis, based on equation (W19), has yielded a large body of solutions for steady two- and three-dimensional unsaturated soil-water flows. See Philip (1989) and Pullan (1990) for reviews.

Flow and volume change in saturated and unsaturated swelling soils

One important limitation of the analysis developed above warrants special attention in this section. The non-swelling soils we have considered behave as rigid matrices with interconnected pores in which the volume of water changes as Ψ changes. On the other hand, many soils of high colloid content, and peats, may change bulk volume as water content, and Ψ, change. The energetics and hydrodynamics in such swelling soils differ fundamentally from those in non-swelling soils.

It is therefore necessary to modify and extend the foregoing analysis. We describe here the generalization for one-dimensional systems only. Further extension requires incorporation into the analysis of an adequate macroscopic representation of stress–strain relations in multi-dimensional systems. A further limitation is that irreversible structural changes, which may occur in the presence of

large enough stresses, lie outside the following analysis. Nevertheless it has important consequences: it reveals that many classical concepts of soil- and groundwater hydrology, based on the behavior of non-swelling soils, fail completely for swelling soils.

Extension to swelling soils

At the outset we note that, since K, Ψ and the *moisture ratio* ϑ are all free to vary in both saturated and unsaturated swelling soils, both are amenable to the same general formulation. ϑ is the ratio of the volume occupied by water to that occupied by particles, and is equal to $\theta(1 + e)$, where e is the void ratio, defined as the ratio of void volume to particle volume. For swelling media it is usually more convenient to work with ϑ rather than θ.

The analysis is simpler and more straightforward for saturated or two-component (soil, water) systems, since they necessarily exhibit 'normal' volume change (Keen, 1931; Marshall; 1959) and $e = \theta$. Unsaturated or three-component (soil, water, air) systems exhibit 'residual' volume change, with e dependent both on ϑ and on the normal stress P [not to be confused with use of P for pressure in equation (W4)]. Note that both the particles and the water are taken to be incompressible.

Three new basic elements enter the extension of flow theory to swelling soils.

- In unsteady swelling systems, the soil particles are, in general, in motion, so that it must be recognized (Gersevanov, 1937) that Darcy's law applies to flow relative to the particles. We therefore replace equation (W5) by

$$U_r = -K(\vartheta)\nabla\Phi \qquad (W22)$$

with U_r being the vector flow velocity in the local rest frame of the particles.

- For one-dimensional systems involving self-weight and/or surface loading, equation (W6) must be generalized to include the overburden potential Ω (Philip 1969b), so that it becomes

$$\Phi = \Psi + \Omega - z \qquad (W23)$$

It is convenient to take Ψ as the 'unloaded' moisture potential, and Ω is then the contribution to Φ due to the normal stress, P. The measured water pressure in such systems is $\Psi + \Omega$. We discuss overburden potential further under Hydrostatics in swelling media (below).

- Whereas $K(\theta)$ and $\Psi(\theta)$ provide a sufficient hydrodynamic characterization of non-swelling soils, we require for unsaturated systems $K(\vartheta)$, $\Psi(\vartheta)$, and $e(\vartheta, P)$, as well as particle specific gravity γ_s. Figure W17 shows $e(\vartheta, P)$ for an illustrative swelling soil. For mineral soils $\gamma_s \sim 2.7$. Saturated systems need much less elaborate characterization. They require merely $K(\vartheta)$, $\Psi(\vartheta)$ and γ_s.

Unsteady swelling systems may be subjected either to Eulerian analysis (Prager, 1953; Philip 1968b) in the physical space coordinate, or to Lagrangian analysis (Hartley and Crank, 1949; McNabb, 1960; Smiles and Rosenthal, 1968) in material coordinate m such that

$$\frac{dm}{dw} = (1 + e)^{-1} \qquad (W24)$$

Here w is the one-dimensional space coordinate and the datum of m is in a plane where particles are stationary or which lies outside the soil mass. The Lagrangian analysis is much simpler and more useful than the Eulerian one. Recognition that the particles move in unsteady systems is no mere exercise in pedantry: it is demonstrable that the mass flow with the particles can be of the same order of magnitude as the Darcy flow relative to the particles (Philip, 1968b).

Hydrostatics in swelling media

The overburden potential Ω enters the analysis of vertical and loaded systems. The separation of the non-gravitational contributions to the total potential (i.e. $\Phi + z$) into Ψ and Ω is, in a sense, arbitrary; but it enables us to identify, study and calculate the separate contributions to water pressure arising from local interaction with the particles (Ψ) and from the normal stress produced by the weight of overlying strata and/or surface loading (Ω). It has been shown by Bolt (Philip, 1970b; Groenevelt and Bolt, 1972) that

$$\Omega = P, \text{ with } \alpha(\vartheta, P) = P^{-1}\int_0^P \frac{\partial}{\partial\vartheta}[e(\vartheta, P)]dP \qquad (W25)$$

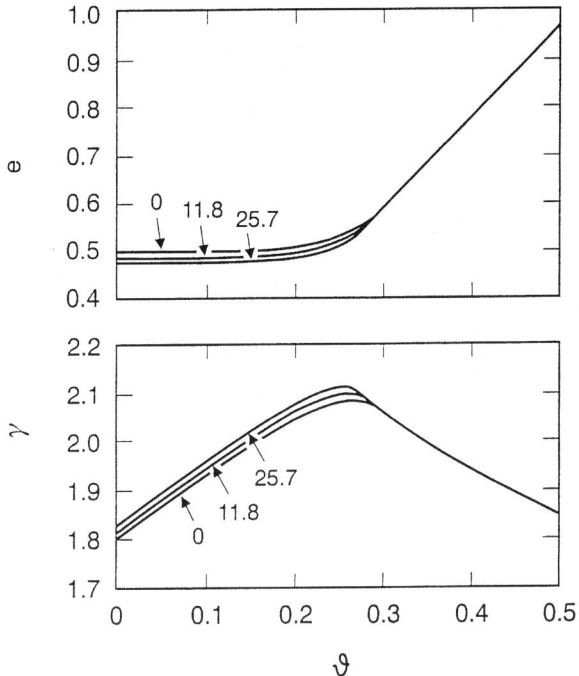

Figure W17 The functions $e(\vartheta, P)$ and $\gamma, (\vartheta, P)$ for an illustrative swelling soil. e is the void ratio, γ the apparent wet specific gravity and P the normal stress. Numerals on the curves denote values of $P(m)$. (Source: Philip, 1971.)

For a vertical column at equilibrium, then, the use of equation (W25) in equation (W23) yields

$$\Phi = \Psi - z + \alpha\left[P(0) + \int_0^z \gamma \, dz\right] = \text{constant} = -Z \quad (W26)$$

Here $P(0)$ is the vertical stress due to loading at the upper surface, $z = 0$ and γ is the apparent wet specific gravity, defined by

$$\gamma = (\vartheta + \gamma_s)/(1 + e)$$

Figure W17 shows γ for the illustrative swelling soil. Z is the water table depth. With γ_s and characterizing functions $\Psi(\vartheta)$ and $e(\vartheta, P)$ known, equation (W26) may be solved to yield the equilibrium moisture profile corresponding to a given value of Z. Note that α here should not be confused with the α in equation (W20).

There are three quite distinct types of equilibrium profiles: the soil moisture may increase, decrease or remain constant with depth (Philip, 1969b). Equilibrium moisture distributions in swelling soils are thus quite different from those of non-swelling soils.

There was earlier recognition of overburden effects in hydrology. Schofield (1935b) saw that Φ should include an overburden contribution. Coleman and Croney (1952) introduced a 'compressibility factor' α, though their definition, evaluation and use of α differ from those here (see also Collis-George, 1961; Rose *et al.*, 1965).

Steady vertical flows

Combining equations (W22) and (W26) gives the equation for vertical flow. The resulting steady flow equation is a linear ordinary differential equation. Steady vertical flows prove possible only for certain combinations of the value of ϑ at the surface and at large depths. The details are complicated (Philip, 1969b).

Unsteady flows

The equation for unsteady flow and volume change in vertical swelling systems is most efficiently expressed with m and t as independent variables, and ϑ as dependent variable. For two-component systems with $e = \vartheta$, it is a non-linear Fokker–Planck equation similar to the one-dimensional vertical form of equation (W12). It is analogous, but more elaborate, for three-component systems.

Horizontal, unloaded, unsteady flows obey a non-linear diffusion equation like the one-dimensional form of equation (W14). It is of interest that the classical theory of consolidation in soil mechanics (Terzaghi, 1923) arrived at an Eulerian linear diffusion equation. That work took K and $d\Psi/d\vartheta$ as constant and ignored both gravity and mass flow.

Unsteady non-linear solutions have been found by Smiles and Rosenthal (1968), Philip (1968b) and Philip and Smiles (1969). There is much scope for further studies.

Hydrological consequences

Although further work is needed, it is clear that hydrologic theory based on equations (W22), (W23) and (W25) differs profoundly from the classical theory which takes no account of swelling and neglects the contribution to Φ of the overburden potential Ω. The simplest general statement one can offer on the influence of swelling is this: the net effect of gravity on the equilibrium and flow of water in swelling media is approximately $(1 - \gamma\alpha)$ times that in non-swelling ones. For a mineral soil this factor is about -1 in the normal range, increasing to 0 when γ is maximum, and approaching $+1$ as $\alpha > 0$ at small values of ϑ. Various conventional intuitions of the hydrologist are therefore invalidated.

We have seen that equilibrium moisture distributions in swelling soils differ fundamentally from those in non-swelling soils. The same holds also for vertical flows. The course of infiltration in a swelling soil evidently has analogies with that of capillary rise in a non-swelling one, and vice versa: and evaporation from an initially wet swelling soil does not necessarily exhibit the sharp transition between constant-rate and falling-rate phases characteristic of non-swelling soils (Philip 1954, 1957d).

Limitations and extensions

The basic fluid-mechanical theory of water movement in unsaturated soils has been described earlier. In this section we identify limitations to this theory and extensions needed to cope with certain complications. The basic theory is not nullified by such extensions, but is subject to more or less severe perturbation. We discuss the limitations under eight headings, but the complications are such that there is overlap between many of them.

Local thermodynamic disequilibrium; aggregated and cracked soils

The moisture potential Ψ is basically a concept of equilibrium thermodynamics, yet it is used in the analysis of transient flows. The large body of experimental data on unsteady unsaturated flows suggests, however, that the deviations of $\Psi(\theta)$ from its equilibrium value are negligible in unaggregated soils. On the other hand, for soils containing microporous aggregates that form the 'pains' of a macroporous (cracked) structure, local disequilibrium between Ψ values in the microporosity and in the macroporosity may become important during transient processes. Especially when the aggregates (or monoliths between cracks) are large and relatively impermeable, more elaborate flow models are required, along lines typified by Philip (1968c,d).

Capillary hysteresis

An important limitation to the applicability of equations (W12) and (W13) arises from hysteresis in $\Psi(\theta)$, first discussed by Haines (1930). Hysteresis does not, however, invalidate the use of equations (W12) and (W13) so long as the $\Psi(\theta)$ relation adopted is appropriate to the phenomenon under study. A wetting curve is thus the $\Psi(\theta)$ function appropriate to the analysis of absorption or infiltration into a homogeneous soil of uniform initial moisture content; and conversely, a drying curve should be used in the analysis of suitably uniform processes of removal of soil water. As Childs first pointed out (see Philip, 1955), however, it is not in general useful to treat in this way phenomena, such as redistribution and drainage after infiltration ceases, wherein wetting and drying occur simultaneously in the one soil mass. In general, the 'diffusion' analysis fails in such cases: different points in the soil mass follow different hysteresis scanning curves, so that there is no definite relationship between gradients of Ψ and gradients of θ. In such cases, knowledge of the complete family of scanning curves within the main $\Psi(\theta)$ hysteresis loop is prerequisite to quantitative treatment of the transfer process. (Some special cases of simultaneous wetting and drying to which the diffusion analysis is, nevertheless, relevant have been treated by Philip, 1957b, 1991, and Youngs, 1958.)

Collis-George (1955) suggested that the independent domain model of hysteresis due to Everett (1954) might be applicable to capillary hysteresis. Miller and Miller (1956) gave an original and perceptive treatment of capillary hysteresis, but not a quantitative one. Poulovassilis (1962) gave the first published account of capillary hysteresis in terms of the independent domain model and presented experimental evidence for its validity.

The model is equivalent to the supposition that any infinitesimal volume element of the pore space may be characterized by two values of Ψ: the arithmetically smallest value for which the element may be occupied by air, and the arithmetically largest value for which it may be occupied by water. The bivariate distribution density function of these values then completely defines the hysteresis behavior of the soil. Similarity hypotheses about this function (Philip, 1964; Mualem, 1973) greatly simplify the treatment. More complicated representations have been proposed, but their use may not be warranted in hydrological applications. It is of interest that Philip (1991) showed that, although neglecting hysteresis produced serious errors in details of the moisture profiles, it had little effect on the total rate of water transfer.

Soil air

The analyses in previous sections neglect any effect of pressure differences in the soil air. We may develop a set of equations similar to the foregoing, which embrace flows of both soil water and soil air. This slightly more general two-component formalism is in fact used in petroleum engineering. In most applications to soil-water movement this elaboration is unnecessary, as the pressure differences within the soil atmosphere are trivially small. We discuss some instances where this is not so, and other complications that may arise from interactions between the soil air and the soil water.

Air escape impeded or prevented
The dynamics of processes such as the absorption and infiltration of water may be modified significantly if the air initially in the soil is not free to escape as it is displaced by water. The process may be studied, to some extent, by use of the two-component analysis mentioned above (e.g. Elrick, 1961). However, air pressures may increase to the point where the pressure build up is partly relieved by a bubbling air escape, with disruption of soil structure (Peck, 1965a,b). Such a phenomenon is, of course, beyond the scope of a Darcy-type analysis.

Limits to air escape are usually unimportant in the field, but they can affect infiltration into large inundated areas. Soil-air pressures may become great enough to lift highway pavements passing through flooded regions.

Entrapped air
Air is more likely to be trapped in pores remote from the water supply surface than in those near the surface. Such spatial differences in air entrapment will be reflected in spatial differences in the $\Psi(\theta)$ relation. The effect is thus essentially one of spatial heterogeneity but it is apparently of minor significance. It is a possible explanation of the transition zone which is sometimes observed in moisture profiles during infiltration.

The temperature dependence of the pressure and volume of entrapped air bubbles complicates the effect of temperature on the $\Psi(\theta)$ relation (e.g. Peck, 1960). This influences the thermal effects considered below.

Deviations from Darcy's law

We discuss two possible causes of deviations from Darcy's law.

Inertial effects
The Navier–Stokes equation underlying Darcy's law ceases to be linear when the 'inertia terms' are important. For soil-water flows, this has the consequence that Darcy's law fails when the Reynolds number of the flow (based on $|U|$ and on a characteristic pore dimension) exceeds a value that tends to lie in the range 1 to 10. It is a non-linear regime, not a turbulent one, which emerges as the Reynolds number increases. The Reynolds numbers characteristic of steady soil-water flows are very small indeed, and the possibility of the failure of

Darcy's law due to inertial effects cannot be entertained in such cases.

The matter is more complicated for transient flows, but analyses indicate that deviations from Darcy's law due to inertia are unlikely to be of much significance during transient flows in either saturated (Philip, 1957e) or unsaturated (Philip, 1959) soils.

Non-Newtonian behavior
The applicability of the Navier–Stokes equation depends on the soil water behaving as a Newtonian fluid (in which the stresses are proportional to the rates of strain). Bulk water is a Newtonian fluid, and it would seem that the flow behavior of soil water can differ significantly from that of bulk water only if a large fraction of the soil water is subject to surface forces capable of non-linear modification of its rheology. There are claims in the literature for 'non-Darcy' behavior of water in clays and colloid pastes, but it seems that at least some results are artifacts of poor experimental techniques.

A more significant issue is the behavior of non-Newtonian solutions in soils and other porous media. Some provisional modifications of the analysis presented above have been applied to this problem.

Hydrodynamic stability

The problems of hydrodynamic stability are well known in general fluid mechanics. The process of fingering has been recognized in petroleum engineering for many decades, but attention to stability problems in soil-water movement is relatively recent.

Instability in soils is typically manifested in a downward vertical flow from a horizontally uniform source into a horizontally uniform soil. One would expect the flow also to be uniform in the horizontal, with a one-dimensional flow equation applicable. Under certain conditions, however, the flow becomes unstable and three-dimensional, breaking into advancing fingers of water separated by unwetted soil. We cannot offer a full discussion here. The reader is referred to the review of hydrodynamic stability in porous media by Homsey (1987), and to the enumeration of the extensive literature on instability in unsaturated soils by Iwata et al. (1995).

The onset of instability is relatively well understood. In soil-water flows it occurs under various circumstances, such as penetration of the flow to a non-wetting stratum, compression of air ahead of the advancing flow, redistribution following infiltration, and increase of conductivity with depth. On the other hand, there seems as yet to be no convincing physical explanation of the observed stable persistence of fingers after they have developed.

Colloidal behavior

The flow analysis for soils subject to colloidal swelling and shrinkage has been described above. It applies for a fixed electrolyte concentration in the soil water. If electrolyte concentration varies during a process, any electrolyte dependence of the structure and hydraulic conductivity (Quirk and Schofield, 1955) must be incorporated into this analysis.

Thermal effects

The analyses of earlier sections are for isothermal systems. Extensions to non-isothermal systems have been made by Philip and de Vries (1957; de Vries, 1958; de Vries and Philip, 1959). Analysis of such systems involves, in general, the solution of simultaneous equations for both heat and moisture transfer. Later work has related the analysis to the thermodynamics of irreversible processes. Philip and de Vries (1957; Philip 1957d) have shown that thermally induced moisture movement tends to be most important in rather dry systems where vapor diffusion is a dominant mechanism.

Experimental (Anderson and Linville, 1962; Anderson et al., 1963) and theoretical (Philip 1960a,b) studies indicate that thermal effects on unsaturated flow due to heat of wetting are quite unimportant.

Heterogeneity

Often soils change in properties with depth, either gradually or in layers. These modes of heterogeneity, on the local scale, are more important for unsaturated flow processes than is the larger-scale horizontal heterogeneity that may arise in the saturation flow problems of regional groundwater hydrology.

Provided the spatial variation is known, soil-water processes in these systems may be analyzed through extensions of the approaches above. Important elements not present in analogous homogeneous situations may arise, such as unstable flows, perched water tables and hysteretic behavior. Complications may also be produced by micropores (e.g. wormholes and old root holes) and cracks. When their disposition is known, their influence may be estimated using models of aggregated or cracked soils. White (Perroux and White, 1988) has observed the useful simplification that, even when macropore variability produces large heterogeneities in saturated permeability, such macropores are empty of water at small negative values of Ψ, so that the heterogeneity may have little effect on the hydraulics of the unsaturated soil.

Obstacles to analyzing soil-water behavior in insaturated field soils arise in two ways: first, the magnitude of the task of measuring adequately the spatial variation of hydraulic properties, and second, the magnitude of the task of performing the consequently elaborate calculations (even if a proper specification of properties were to hand).

Many workers have pursued stochastic methods of evading these obstacles; see Jury *et al.* (1991) for an account of concepts arising in the statistical characterization of soil hydrologic properties. The logical difficulty of stochastic approaches is that they depend on assumptions of stationarity and ergodicity. Stationarity can seldom be verified; and ergodicity demands an unbounded ensemble of realizations, where there is not even a finite ensemble, simply one reality (Philip, 1986).

Clearly there are deep methodological difficulties; and it may be that some questions put to soil-water hydrologists will be recognized ultimately as 'trans-scientific' in the sense of Weinberg (1972): although they may be posed as scientific questions, precise answers lie beyond the practical power of natural science.

J.R. Philip

Bibliography

Anderson, D.M. and A. Linville, 1962. Temperature fluctuations at a wetting front: I. Characteristic temperature time curves. *Soil Sci. Soc. Am. Proc.*, **26**, 14–18.

Anderson, D.M., G. Sposito and A. Linville, 1963. Temperature fluctuations at a wetting front: II. The effect of initial water content of the medium on the magnitude of the temperature fluctuations. *Soil. Sci. Soc. Am. Proc.*, **27**, 367–369.

Bailey, N.J.T., 1964. *The Elements of Stochastic Processes*. New York: Wiley.

Bolt, G.H., 1956. Physico-chemical analysis of the compressibility of pure clays. *Geotechnique*, **6**, 86–93.

Boltzmann, L., 1894. Zur Integration der Diffusionsgleichung bei variabeln Diffusionscoefficienten. *Ann. Physik*, **53**, 959–964.

Bruce, R.R. and R.J. Luxmoore, 1986. Water retention: field methods. *Methods of Soil Analysis*, Part 1. Madison: American Society of Agronomy, pp. 663–686.

Buckingham, E., 1907. *Studies on the movement of soil moisture. US Dept. Agric. Bull* 38.

Campbell, G.S. and G.W. Gee, 1986. Water potential: miscellaneous methods. *Methods of Soil Analysis*, Part 1. Madison: American Society of Agronomy, pp. 619–633.

Carslaw, H.S. and J.C. Jaeger, 1947. *Conduction of Heat in Solids*. Oxford: Clarendon Press.

Cassell, D.K. and A. Klute, 1986. Water potential: tensiometry, in *Methods of Soil Analysis*, Part 1. Madison: American Society of Agronomy, pp. 563–596.

Chandrasekhar, S., 1943. Stochastic problems in physics and astronomy. *Rev. Mod. Phys.*, **15**, 1–89.

Childs, E.C. and N. Collis-George, 1948. Soil geometry and soil–water equilibria. *Discussions Faraday Soc.*, **3**, 78–85.

Childs, E.C. and N. Collis-George, 1950a. The permeability of porous materials. *Proc. R. Soc. London*, **A201**, 392–405.

Childs, E.C. and N. Collis-George, 1950b. The control of soil water. *Adv. Agron.*, **2**, 232–272.

Colman, J.D. and D. Croney, 1952. *The estimation of the vertical moisture distribution with depth in unsaturated cohesive soils*. Road Res. Lab. Note RN/1709/JDC DC.

Collis-George, N., 1955. Hysteresis in moisture content-suction relationships in soils. *Proc. Natl Acad. Sci. India*, **A24**, 80–85.

Collis-George, N., 1961. Free energy considerations in the moisture profile at equilibrium and effect of external pressure. *Soil Sci.*, **91**, 306–311.

Cowan, I.R., 1965. Transport of water in the soil–plant–atmosphere system. *J. Appl. Ecology*, **2**, 221–239.

Croney, D., J.D. Coleman and P.M. Bridge, 1952. *The suction of moisture held in soil and other porous materials*. Road Res. Tech. Paper 24.

Day, P.R. and J.N. Luthin, 1956. A numerical solution of the differential equation of flow for a vertical drainage problem. *Soil Sci. Soc. Am. Proc.*, **20**, 443–447.

de Vries, D.A., 1958. Simultaneous transfer of heat and moisture in porous media. *Trans. Am. Geophys. Union*, **39**, 909–916.

de Vries, D.A. and J.R. Philip, 1959. Temperature distribution and moisture transfer in porous materials. *J. Geophys. Res.*, **64**, 386–388.

Elrick, D.E., 1961. Transient two-phase capillary flow in porous media. *Phys. Fluids*, **4**, 572–575.

Everett, D.H., 1954. A general approach to hysteresis, 3. *Trans. Faraday Soc.*, **51**, 1077–1096.

Gardner, W.H., 1986. Early soil physics into the mid-20th century. *Adv. Soil Sci.*, **4**, 1–101.

Gardner, W.R., 1958. Some steady-state solutions of the unsaturated moisture flow equation with application to evaporation from a water table. *Soil Sci.*, **85**, 228–232.

Gardner, W.R., 1960. Dynamic aspects of water availability in plants. *Soil Sci.*, **89**, 63–73.

Gersevanov, N.M., 1937. *The Foundations of Dynamics of Soils*, 3rd edn. Moscow, Leningrad: Stroiizdat.

Groenevelt, P.H. and G.H. Bolt, 1972. Water retention in soil. *Soil Sci.*, **113**, 238–245.

Haines, W.B., 1930. Studies in the physical properties of soil. V. The hysteresis effect of capillary properties, and the modes of moisture distribution associated therewith. *J. Agric. Sci.*, **20**, 97–116.

Hartley, G.S. and J. Crank, 1949. Some fundamental definitions and concepts in diffusion processes. *Trans. Faraday Soc.*, **45**, 801–818.

Holmes, J.W., Taylor, S.A. and Richards, S.J., 1967. Measurement of soil water, in *Irrigation of Agricultural Lands*. Madison: American Society of Agronomy, pp. 547–579.

Homsey, G.M., 1987. Viscous fingering in porous media. *Ann. Rev. Fluid Mech.*, **19**, 271–311.

Iwata, S., Tabuchi, T. and Warkentin, B.P., 1995. *Soil–Water Interactions*, 2nd ed. New York: Dekker.

Jury, W.A., Gardner, W.R. and Gardner, W.H., 1991. *Soil Physics*, 5th edn. New York: Wiley.

Keen, B.A., 1931. *The Physical Properties of the Soil*. London: Longmans.

Kirchhoff, G., 1894. *Vorlesungen über die Theorie der Wärme*. Leipzig: Barth.

Klute, A., 1952. A numerical method for solving the flow equation for water in unsaturated soils. *Soil. Sci.*, **73**, 105–116.

Klute, A., 1986. Water retention: laboratory methods. *Methods of Soil Analysis*, Part 1. Madison: American Society of Agronomy, pp. 635–662.

Luthin, J.N. and Miller, R.D., 1953. Pressure distribution in soil columns draining into the atmosphere. *Soil Sci. Soc. Am. Proc.*, **17**, 329–333.

McNabb, A., 1960. A mathematical treatment of one-dimensional soil consolidation. *Quart. Appl. Math.*, **17**, 337–347.

Mahony, J.J., 1972. Three-dimensional Stokes flow. *Bull. Australian Math. Soc.*, **7**, 77–89.

Mahony, J.J. and Philip, J.R., 1967. Equations modelling spatially variable stochastic processes. *Phys. Fluids*, **10**, 1403–1405.

Marshall, T.J., 1959. *Relations between Water and Soil*. Farnham Royal: Comm. Agr. Bureau.

Miller, E.E. and Klute, A., 1967. The dynamics of soil water. Part I – mechanical forces, in *Irrigation of Agricultural Lands*. Madison: American Society of Agronomy, pp. 209–244.

Miller, E.E. and Miller, R.D., 1956. Physical theory for capillary flow phenomena. *J. Appl. Phys.*, **27**, 324–332.

Moore, R.E., 1939. Water conduction from shallow water tables. *Hilgardia*, **12**, 383–426.

Mualem, Y., 1973. Modified approach to capillary hysteresis based on a similarity hypothesis. *Water Resour. Res.*, **9**, 1324–1331.

Peck, A.J., 1960. Change of moisture tension with temperature and air pressure: theoretical. *Soil. Sci.*, **9**, 303–310.

Peck, A.J., 1965a. Moisture profile development and air compression during water uptake by bounded porous bodies: 2. Horizontal columns. *Soil Sci.*, **99**, 327–334.

Peck, A.J., 1965b. Moisture profile development and air compression during water uptake by bounded porous bodies: 3. Vertical columns. *Soil Sci.*, **100**, 44–51.

Perroux, K.M. and White, I., 1988. Designs for disc permeameters. *Soil Sci Soc. Am. J.*, **52**, 1205–1215.

Philip, J.R., 1954. Some recent advances in hydrologic physics. *J. Inst. Eng. Australia*, **26**, 255–259.

Philip, J.R., 1955. The concept of diffusion applied to soil water. *Proc. Natl Acad. Sci. India*, **A24**, 93–104.

Philip, J.R., 1957a. Remarks on the analytical derivation of the Darcy equation. *Trans. Am. Geophys. Union*, **38**, 782–784.

Philip, J.R., 1957b. The physical principles of soil water movement during the irrigation cycle, in *Proc. 3rd Intern. Congr. Irrigation Drainage*, San Francisco, pp. 8.125–8.154.

Philip, J.R., 1957c. The theory of infiltration: 1. The infiltration equation and its solution. *Soil Sci.*, **83**, 345–357.

Philip, J.R., 1957d. Evaporation, and moisture and heat fields in the soil. *J. Meteorol.*, **14**, 354–366.

Philip, J.R., 1957e. Transient fluid motions in saturated porous media. *Australian J. Phys.*, **10**, 43–53.

Philip, J.R., 1959. The early stages of absorption and infiltration. *Soil Sci.*, **88**, 91–97.

Philip, J.R., 1960a. Energy dissipation during absorption and infiltration: 1. *Soil Sci.*, **89**, 132–136.

Philip, J.R., 1960b. Energy dissipation during absorption and infiltration: 2. *Soil Sci.*, **89**, 353–358.

Philip, J.R., 1964. Similarity hypothesis for capillary hysteresis in porous materials. *J. Geophys. Res.*, **69**, 1553–1562.

Philip, J.R., 1966. The dynamics of capillary rise, in *Water in the Unsaturated Zone*, Vol. 1, Paris: UNESCO, pp. 559–564.

Philip, J.R., 1968a. Diffusion by continuous movements. *Phys. Fluids*, **11**, 38–42.

Philip, J.R., 1968b. Kinetics of sorption and volume change in clay-colloid pastes. *Australian J. Soil Res.*, **6**, 249–267.

Philip, J.R., 1968c. The theory of absorption in aggregated media. *Australian J. Soil Res.* **6**, 21–30.

Philip, J.R., 1968d. Diffusion, dead-end pores, and linearized absorption in aggregated media. *Australian J. Soil Res.* **6**, 21–30.

Philip, J.R., 1969a. Theory of infiltration. *Adv. Hydroscience*, **5**, 215–296.

Philip, J.R., 1969b. Hydrostatics and hydrodynamics in swelling soils. *Water Resour. Res*, **5**, 1070–1077.

Philip, J.R., 1970a. Flow in porous media. *Ann. Rev. Fluid Mech.*, **2**, 177–204.

Philip, J.R., 1970b. Reply to note by E.G. Youngs and G.D. Towner on 'Hydrostatics and hydrodynamics in swelling soils'. *Water Resour. Res.*, **6**, 1248–1251.

Philip, J.R., 1971. Hydrology of swelling soils, in *Salinity and Water Use*. London: Macmillan, pp. 95–107.

Philip, J.R., 1972. Flows satisfying mixed no-slip and no-shear conditions. *ZAMP*, **23**, 353–372.

Philip, J.R., 1973. Flow in porous media, in *Theoretical and Applied Mechanics*. Berlin: Springer, pp. 279–294.

Philip, J.R., 1974. Fifty years progress in soil physics. *Geoderma*, **12**, 265–280.

Philip, J.R., 1978. Water on the Earth, in *Water, Plants and People*. Canberra: Australian Academy of Science, pp. 35–59.

Philip, J.R., 1986. Issues in flow and transport in heterogeneous porous media. *Transp. Porous Media*, **1** 319–338.

Philip, J.R., 1989. The scattering analog for infiltration in porous media. *Rev. Geophys.*, **27**, 431–448.

Philip, J.R., 1991. Horizontal redistribution with capillary hysteresis. *Water Resour. Res.*, **27**, 1459–1469.

Philip, J.R., 1992. Flow and volume change in soils and other porous media, and in tissues, in *Mechanics of Swelling: from Clays to Living Cells and Tissues*. Berlin: Springer, pp. 3–31.

Philip, J.R. and D.A. de Vries, 1957. Moisture movement in porous materials under temperature gradients. *Trans. Am. Geophys. Union*, **38**, 222–232.

Philip, J.R. and D.E. Smiles, 1969. Kinetics of sorption and volume change in three-component systems. *Australian J. Soil Res.*, **7**, 1–19.

Poulovassilis, A., 1962. Hysteresis of pore water, an application of the concept of independent domains. *Soil Sci.*, **93**, 405–412.

Prager, S., 1953. Diffusion in binary systems. *J. Chem. Phys.*, **21**, 1344–1347.

Pullan, A.J., 1990. The quasilinear approximation for unsaturated porous medium flow. *Water Resour. Res.*, **26**, 1219–1234.

Quirk, J.P. and R.K. Schofield, 1955. The effect of electrolyte concentrations on soil permeability. *J. Soil Sci.*, **6**, 163–178.

Rawlins, S.L. and G.S. Campbell, 1986. Water potential: thermocouple psychrometry, in *Methods of Soil Analysis*, Part 1. Madison: American Society of Agronomy, pp. 597–618.

Reeve, R.C., 1986. Water potential: piezometry, in *Methods of Soil Analysis*, Part 1. Madison: American Society of Agronomy, pp. 545–561.

Richards, L.A., 1931. Capillary conduction of liquids through porous mediums. *Physics*, **1**, 318–333.

Richards, L.A., 1949. Methods of measuring soil moisture tension. *Soil Sci.*, **68**, 95–112.

Richards, L.A., 1965. Water conducting and retaining properties of soils in relation to irrigation, in *Desert Research*, Res. Council Israel Spec. Report 2, pp. 523–546.

Rose, C.W., 1966. *Agricultural Physics*. London: Pergamon.

Rose, C.W., W.R. Stern and J.E. Drummond, 1965. Determination of hydraulic conductivity as a function of depth and water content for soil *in situ*. *Australian J. Soil Res.*, **2**, 1–9.

Schofield, R.K., 1935. The pF of water in soil, in *Trans. 3rd Cong. Soil Sci.*, Vol. 2, pp. 37–48.

Smiles, D.E. and M.J. Rosenthal, 1968. The movement of water in swelling materials. *Australian J. Soil Res.*, **6**, 237–248.

Sposito, G., 1987. The 'physics' of soil water physics. *History of Geophysics*, **3**, 93–98.

Staple, W.J. and J.J. Lehane, 1954. Movement of moisture in unsaturated soils. *Can J. Soil Sci.*, **42**, 247–253.

Terzaghi, K., 1923. Die Berechnung der Durchlassigkeitsziffer des Tones aus dem Verlauf der hydrodynamischen Spannungserscheinungen. *Sitsb. Akad. Wiss. (Wien) Abt. 2a*, **132**, 125–138.

Topp, G.C., and E.E. Miller, 1966. Hysteretic moisture characteristics and hydraulic conductivities for glass-bead media. *Soil Sci. Soc. Am. Proc.*, **30**, 156–162.

Weinberg, A.M., 1972. Science and trans-science. *Minerva*, **10**, 209–222.

Whisler, F.D. and K.K. Watson, 1968. One-dimensional gravity drainage of uniform columns of porous materials. *J. Hydrol.*, **6**, 277–296.

Youngs, E.G., 1958. Redistribution of moisture in porous materials after infiltration: 1. *Soil Sci.*, **84**, 283–290.

Cross references

Water resources
Water use
World water balance

WATER QUALITY FOR DRINKING: WHO GUIDELINES

The WHO recommended guideline values are set at a level to protect human health; they may not be suitable for the protection of aquatic life. The guidelines apply to bottled water and ice intended for human consumption but do not apply to natural mineral waters, which should be regarded as beverages rather than drinking water in the usual sense of the word.

The recognition that fecally polluted water can lead to the spread of microbial infections has led to the development of sensitive methods for routine examination to ensure that water intended for human consumption is free from fecal contamination. Although it is now possible to detect the presence of many pathogens in water, the methods of isolation and enumeration are often complex and time consuming. It is therefore impracticable to monitor drinking water for every possible microbial pathogen. A more logical approach is the detection of organisms normally present in the feces of humans and other warm-blooded animals as indicators of fecal pollution, as well as of the efficacy of water treatment and disinfection.

Thousands of organic and inorganic chemicals have been identified in drinking-water supplies around the world, many in extremely low concentrations. The chemicals selected for the development of guideline values include those considered potentially hazardous to human health, those detected relatively frequently in drinking water, and those detected in relatively high concentrations.

Table W12 Orally transmitted waterborne pathogens and their significance in water supplies

Pathogen	Health significance	Persistence in water supplies[a]	Resistance to chlorine[b]	Relative infective dose[c]	Important animal reservoir
Bacteria					
Campylobacter jejuni, C. coli	High	Moderate	Low	Moderate	Yes
Pathogenic *Escherichia coli*	High	Moderate	Low	High	Yes
Salmonella typhy	High	Moderate	Low	High[d]	No
Other salmonellae	High	Long	Low	High	Yes
Shigella spp.	High	Short	Low	Moderate	No
Vibrio cholerae	High	Short	Low	High	No
Yersinia enterocolitica	High	Long	Low	High(?)	Yes
Pseudomonas aeruginosa[e]	Moderate	May multiply	Moderate	High(?)	No
Aeromonas spp.	Moderate	May multiply	Low	High(?)	No
Viruses					
Adenoviruses	High	?	Moderate	Low	No
Enteroviruses	High	Long	Moderate	Low	No
Hepatitis A	High	?	Moderate	Low	No
Enterically transmitted non-A, non-B hepatitis viruses, hepatitis E	High	?	?	Low	No
Norwalk virus	High	?	?	Low	No
Rotavirus	High	?	?	Moderate	No(?)
Small round viruses	Moderate	?	?	Low(?)	No
Protozoa					
Entamoeba histolytica	High	Moderate	High	Low	No
Giardia intestinalis	High	Moderate	High	Low	Yes
Cryptosporidium parvum	High	Long	High	Low	Yes
Helminths					
Dracunculus medinensis	High	Moderate	Moderate	Low	Yes

?, not known or uncertain.
[a] Detection period for infective stage in water at 20°C: short, up to 1 week; moderate, 1 week to 1 month; long, over 1 month.
[b] When the infective stage is freely suspended in water treated at conventional doses and contact times. Resistance moderate, agent may not be completely destroyed; resistance low, agent completely destroyed.
[c] Dose required to cause infection in 50% of healthy adult volunteers; may be as little as one infective unit for some viruses.
[d] From experiments with human volunteers.
[e] Main route of infection is by skin contact, but can infect immunosuppressed or cancer patients orally.

Table W13 Bacteriological quality of drinking water[a]

Organisms	Guideline value
All water intended for drinking	
E. coli or thermotolerant coliform bacteria[b,c]	Must not be detectable in any 100 ml sample
Treated water entering the distribution system	
E. coli or thermotolerant coliform bacteria[b]	Must not be detectable in any 100 ml sample
Total coliform bacteria	Must not be detectable in any 100 ml sample
Treated water in the distribution system	
E. coli or thermotolerant coliform bacteria[b]	Must not be detectable in any 100 ml sample
Total coliform bacteria	Must not be detectable in any 100 ml sample. In the case of large supplies, where sufficient samples are examined, must not be present in 95% of samples taken throughout any 12-month period

[a] Immediate investigative action must be taken if either *E. coli* or total coliform bacteria are detected. The minimum action in the case of total coliform bacteria is repeat sampling; if these bacteria are detected in the repeat sample, the cause must be determined by immediate further investigation.
[b] Although *E. coli* is the more precise indicator of fecal pollution, the count of thermotolerant coliform bacteria is an acceptable alternative. If necessary, proper confirmatory tests must be carried out. Total coliform bacteria are not acceptable indicators of the sanitary quality of rural water supplies, particularly in tropical areas where many bacteria of no sanitary significance occur in almost all untreated supplies.
[c] It is recognized that, in the great majority of rural water supplies in developing countries, fecal contamination is widespread. Under these conditions, the national surveillance agency should set medium-term targets for the progressive improvement of water supplies, as recommended in Volume 3 of *Guidelines for drinking-water quality*.

Some potentially hazardous chemicals in drinking water are derived directly from treatment chemicals or construction materials used in water supply systems. Such chemicals are best controlled by appropriate specifications for the chemicals and materials used. For example, a wide range of polyelectrolytes are now used as coagulant aids in water treatment, and the presence of residues of the unreacted monomer may cause concern. Many polyelectrolytes are based on acrylamide polymers and copolymers, in both of which the acrylamide monomer is present as a trace impurity. Chlorine used for disinfection has sometimes been found to contain carbon tetrachloride. This type of drinking-water contamination is best controlled by the application of regulations governing the quality of the products

Table W14 Chemicals of health significance in drinking water

	Guideline value (mg l^{-1})	Remarks
Inorganic constituents		
Antimony	0.005 (P)[a]	
Arsenic	0.01[b] (P)	For excess skin cancer risk of 6×10^{-4}
Barium	0.7	
Beryllium		NAD[c]
Boron	0.3	
Cadmium	0.003	
Chromium	0.05 (P)	
Copper	2 (P)	ATO[d]
Cyanide	0.07	
Fluoride	1.5	Climatic conditions, volume of water consumed, and intake from other sources should be considered when setting national standards
Lead	0.01	It is recognized that not all water will meet the guideline value immediately; meanwhile, all other recommended measures to reduce the total exposure to lead should be implemented
Manganese	0.5 (P)	ATO
Mercury (total)	0.001	
Molybdenum	0.07	
Nickel	0.02	
Nitrate (as NO$_3^-$)	50	The sum of the ratio of the concentration of each to its respective guideline value should not exceed 1
Nitrate (as NO$_2^-$)	3 (P)	
Selenium	0.01	
Uranium		NAD
Organic constituents		
Chlorinated alkanes		
Carbon tetrachloride	2	
Dichloromethane	20	
1,1-Dichloroethane		NAD
1,2-Dichloroethane	30[b]	For excess risk of 10^{-5}
1,1,1-Trichloroethane	2000 (P)	
Chlorinated ethenes		
Vinyl chloride	5[b]	For excess risk of 10^{-5}
1,1-Dichloroethene	30	
1,2-Dichloroethane	50	
Trichloroethene	70 (P)	
Tetrachloroethene	40	
Aromatic hydrocarbons		
Benzene	10[b]	For excess risk of 10^{-5}
Toluene	700	ATO
Xylenes	500	ATO
Ethylbenzene	300	ATO
Styrene	20	ATO
Benzo[a]pyrene	0.7[b]	For excess risk of 10^{-5}
Chlorinated benzenes		
Monochlorobenzene	300	ATO
1,2-Dichlorobenzene	1000	ATO
1,3-Dichlorobenzene		NAD
1,4-Dichlorobenzene	300	ATO
Trichlorobenzenes (total)	20	ATO
Miscellaneous		
Di(2-ethylhexyl)adipate	80	
Di(2-ethylhexyl)phthalate	8	
Acrylamide	0.5[b]	For excess risk of 10^{-5}
Epichlorohydrin	0.4 (P)	
Hexachlorobutadiene	0.6	
Edetic acid (EDTA)	200 (P)	
Nitrilotriacetic acid	200	
Dialkyltins		NAD
Tributyltin oxide	2	
Pesticides		
Alachlor	20[b]	For rxcess risk of 10^{-5}
Aldicarb	10	
Aldrin/dieldrin	0.03	
Atrazine	2	

Table W14 Continued

	Guideline value (mg l^{-1})	Remarks
Bentazone	30	
Carbofuran	5	
Chlordane	0.2	
Chlorotoluron	30	
DDT	2	
1,2-Dibromo-3-chloropropane	1 ^{b}	For excess risk of 10^{-5}
2,4-D	30	
1,2-Dichloropropane	20 (P)	
1,3-Dichloropropane		NAD
1,3-Dichloropropene	20 ^{b}	For excess risk of 10^{-1}
Ethylene dibromide	NAD	
Heptachlor and heptachlor epoxide	0.03	
Hexachlorobenzene	1 ^{b}	For excess risk of 10^{-5}
Isoproturon	9	
Lindane	2	
MCPA	2	
Methoxychlor	20	
Metolachlor	10	
Molinate	6	
Pendimethalin	20	
Pentachlorophenol	9 (P)	
Permethrin	20	
Propanil	20	
Pyridate	100	
Simazine	2	
Trifluralin	20	
Chlorophenoxy herbicides other than 2,4-D and MCPA		
2,4-DB	90	
Dichlorprop	100	
Fenoprop	9	
MCPB		NAD
Mecoprop	10	
2,4,5-T	9	
Disinfectants		
Monochloramine	3	
Di- and trichloramine		NAD
Chlorine	5	ATO. For effective disinfection there should be a residual concentration of free chlorine of $\geqslant$0.5 mg l^{-1} after at least 30 min contact time at pH <8.0
Chlorine dioxide		A guideline value has not been established because of the rapid breakdown of chlorine dioxide and because the chlorite guideline value is adequately protective for potential toxicity from chlorine dioxide
Iodine		NAD
Disinfectant by-products		
Bromate	25 ^{b}(P)	for 7 $\times$ 10^{-5} excess risk
Chlorate		NAD
Chlorite	200 (P)	
Chlorophenols		
2-Chlorophenol		NAD
2,4-Dichlorophenol		NAD
2,4,6-Trichlorophenol	200 ^{b}	For excess risk of 10^{-5}, ATO
Formaldehyde	900	
MX		NAD
Trihalomethanes		The sum of the ratio of the concentration of each to its respective guideline value should not exceed 1
Bromoform	100	
Dibromochloromethane	100	
Bromodichloromethane	60 ^{b}	For excess risk of 10^{-5}
Chloroform	200 ^{b}	For excess risk of 10^{-5}
Chlorinated acetic acids		
Monochloroacetic acid		NAD
Dichloroacetic acid	50 (P)	
Trichloroacetic acid	100 (P)	
Chloral hydrate (trichloroacetaldehyde)	10 (P)	
Chloroacetone		NAD

Table W14 Continued

	Guideline value (mg l⁻¹)	Remarks
Halogenated acetonitriles		
Dichloroacetonitrile	90 (P)	
Dibromoacetonitrile	100 (P)	
Bromochloroacetonitrile		NAD
Trichloroacetonitrile	1 (P)	
Cyanogen chloride (as CN)	70	
Chloropicrin		NAD

ᵃ (P), Provisional guideline value. This term is used for constituents for which there is some evidence of a potential hazard but where the available information on health effects is limited; or where an uncertainty factor greater than 1000 has been used in the derivation of the tolerable daily intake (TDI). Provisional guideline values are also recommended: (1) for substances for which the calculated guideline value would be below the practical quantification level, or below the level that can be achieved through practical treatment methods; or (2) where disinfection is likely to result in the guidelines value being exceeded.

ᵇ For substances that are considered to be carcinogenic, the guideline value is the concentration in drinking water associated with an excess lifetime cancer risk of 10⁻⁵ (one additional cancer per 100 000 of the population ingesting drinking water containing the substance at the guideline value for 70 years). Concentrations associated with estimated excess lifetime cancer risks of 10⁻⁴ and 10⁻⁶ can be calculated by multiplying and dividing, respectively, the guideline value by 10.

In cases in which the concentration associated with an excess lifetime cancer risk of 10⁻⁵ is not feasible as a result of inadequate analytical or treatment technology, a provisional guideline value is recommended at a practicable level and the estimated associated excess lifetime cancer risk presented.

It should be emphasized that the guideline values for carcinogenic substances have been computed from hypothetical mathematical models that cannot be verified experimentally and that the values should be interpreted differently than TDI-based values because of the lack of precision of the models. At best, these values must be regarded as rough estimates of cancer risk. However, the models used are conservative and probably err on the side of caution. Moderate short-term exposure to levels exceeding the guideline value for carcinogens does not significantly affect the risk.

ᶜ NAD, No adequate data to permit recommendation of a health-based guideline value.

ᵈ ATO, Concentrations of the substances at or below the health-based guideline value may affect the appearance, taste or odor of the water.

Table W15 Chemicals not of health significance at concentrations normally found in drinking water

Chemical	Remarks
Asbestos	U
Silver	U
Tin	U

U, It is unnecessary to recommend a health-based guideline value for these compounds because they are not hazardous to human health at concentrations normally found in drinking water.

Table W16 Radioactive constituents of drinking water

	Screening value (Bq l⁻¹)	Remarks
Gross alpha activity	0.1	If a screening value is exceeded, more detailed radionuclide analysis is necessary. Higher values do not necessarily imply that the water is unsuitable for human consumption
Gross beta activity	1	

themselves rather than the quality of the water. Similarly, strict national regulations on the quality of pipe material should avoid the possible contamination of drinking water by trace constituents of plastic pipes. The control of contamination of water supplies by *in situ* polymerized coatings and coatings applied in a solvent requires the development of suitable codes of practice in addition to controls on the quality of the materials used.

The human pathogens that can be transmitted orally by drinking-water are listed in Table W12, together with a summary of their health significance and main properties. Those that present a serious risk of disease whenever present in drinking-water include Salmonella spp.,

Shigella spp., pathogenic *Escherichia coli*, *Vibrio cholerae*, *Yersinia enterocolitica*, *Campylobacter jejuni* and *Campylobacter coli*, the viruses listed in Table W12, and the parasites *Giardia* spp., *Cryptosporidium* spp., *Entamoeba histolytica* and *Dracunculus medinensis*. Most of these pathogens are distributed worldwide. However, outbreaks of cholera and infection by the guinea worm *D. medinensis* are regional. The elimination of all these agents from water intended for drinking has high priority. Eradication of *D. medinensis* is a recognized target of the World Health Assembly (World Health Assembly resolution WHA44.5, 1991).

Other pathogens are accorded moderate priority in Table W12 or are not listed, either because they are of low pathogenicity, causing disease opportunistically in subjects with low or impaired immunity, or because, even though they cause serious diseases, the primary route of infection is by contact or inhalation, rather than by ingestion.

Opportunistic pathogens are naturally present in the environment and are not formally regarded as pathogens. They are able to cause disease in people with impaired local or general defence mechanisms, such as the elderly or the very young, patients with burns or extensive wounds, those undergoing immunosuppressive therapy, or those with acquired immunodeficiency syndrome (AIDS). Water used by such patients for drinking or bathing, if it contains large numbers of these organisms, can produce various infections of the skin and the mucous membranes of the eye, ear, nose and throat. Examples of such agents are *Pseudomonas aeruginosa* and species of *Flavobacterium*, *Acinetobacter*, *Klebsiella*, *Serratia*, *Aeromonas* and certain 'slow-growing' mycobacteria.

Certain serious illnesses result from inhalation of water in which the causative organisms have multiplied because of warm temperatures and the presence of nutrients. These include Legionnaires' disease (*Legionella* spp.) and those caused by the amoebae *Naegleria fowleri* (primary amoebic meningoencephalitis) and *Acanthamoeba* spp. (amoebic meningitis, pulmonary infections).

Schistosomiasis (bilharziasis) is a major parasitic disease of tropical and subtropical regions, and is primarily spread by contact with water during bathing or washing. The larval stage (cercariae) released by infected aquatic snails penetrates the skin. If pure drinking water is readily available, it will be used for washing, and this will have the benefit of reducing the need to use contaminated surface water.

It is conceivable that unsafe drinking water contaminated with soil or feces could act as a carrier of other parasitic infections, such as balantidiasis (*Balantidium coli*), and certain helminths (species of *Fasciola*, *Fasciolopsis*, *Echinococcus*, *Spirometra*, *Ascaris*, *Trichuris*,

Table W17 Substances and parameters in drinking water that may give rise to complaints from consumers

	Levels likely to give rise to consumer complaints	Reasons for consumer complaints
Physical parameters		
Color	15 TCU[b]	Appearance
Taste and odor	–	Should be acceptable
Temperature	–	Should be acceptable
Turbidity	5 NTU[c]	Appearance; for effective terminal disinfection, median turbidity $\leqslant 1$ NTU, single sample $\leqslant 5$ NTU
Inorganic constituents		
Aluminum	0.2 mg l^{-1}	Depositions, discoloration
Ammonia	1.5 mg l^{-1}	Odor and taste
Chloride	250 mg l^{-1}	Taste, corrosion
Copper	1 mg l^{-1}	Staining of laundry and sanitary ware (health-based provisional guideline value 2 mg l^{-1})
Hardness	–	High hardness: scale deposition, scum formation Low hardness: possible corrosion
Hydrogen sulfide	0.05 mg l^{-1}	Odor and taste
Iron	0.3 mg l^{-1}	Staining of laundry and sanitary ware
Manganese	0.1 mg l^{-1}	Staining of laundry and sanitary ware (health-based provisional guideline value 0.5 mg l^{-1})
Dissolved oxygen	–	Indirect effects
pH	–	Low pH; corrosion High pH: taste, soapy feel, preferably <8.0 for effective disinfection with chlorine
Sodium	200 mg l^{-1}	Taste
Sulfate	250 mg l^{-1}	Taste, corrosion
Total dissolved solids	1000 mg l^{-1}	Taste
Zinc	3 mg l^{-1}	Appearance, taste
Organic constituents		
Toluene	24–170 μg l^{-1}	Odor, taste (health-based guideline, value 700 μg l^{-1})
Xylene	20–1800 μg l^{-1}	Odor, taste (health-based guideline, value 500 μg l^{-1})
Ethylbenzene	2–200 μg l^{-1}	Odor, taste (health-based guideline value 300 μg l^{-1})
Styrene	4–2600 μg l^{-1}	Odor, taste (health-based guideline value 20 μg l^{-1})
Monochlorobenzene	10–120 μg l^{-1}	Odor, taste (health-based guideline value 300 μg l^{-1})
1,2-Dichlorobenzene	1–10 μg l^{-1}	Odor, taste (health-based guideline value 1000 μg l^{-1})
1,4-Dichlorobenzene	0.3–30 μg l^{-1}	Odor, taste (health-based guideline value 300 μg l^{-1})
Trichlorobenzenes (total)	5–50 μg l^{-1}	Odor, taste (health-based guideline value 20 μg l^{-1})
Synthetic detergents	–	Foaming, taste, odor
Disinfectants and disinfectant by-products		
Chlorine	600–1000 μg l^{-1}	Odor, taste (health-based guideline value 5 mg l^{-1})
Chlorophenols		
2-Chlorophenol	0.1–10 μg l^{-1}	Taste, odor
2,4-Dichlorophenol	0.3–40 μg l^{-1}	Taste, odor
2,4,6-Trichlorophenol	2–300μg l^{-1}	Taste, odor (health-based guideline value 200 μg l^{-1})

[a] The levels indicated are not precise numbers. Problems may occur at lower or higher values, according to local circumstances. A range of taste and odor threshold concentrations is given for organic constituents.
[b] TCU, time color unit.
[c] NTU, nephelometric turbidity unit.

Toxocara, Necator, Ancylostoma, Strongyloides and *Taenia solium*). However, in most of these the normal mode of transmission is ingestion of the eggs in food contaminated with feces or fecally contaminated soil (in the case of *Taenia solium*, ingestion of the larval cysticercus stage in uncooked pork) rather than ingestion of contaminated drinking water.

Blooms of *Cyanobacteria* (commonly called blue-green algae) occur in lakes and reservoirs used for potable supply. Three types of toxin can be produced, depending upon species:

- hepatotoxins, produced by species of *Microcystis, Oscillatoria, Anabaena* and *Nodularia*, typified by microcystin LR:R, which induce death by circulatory shock and massive liver hemorrhage within 24 h of ingestion;
- neurotoxins, produced by species of *Anabaena, Oscillatoria, Nostoc, Cylindrospermum* and *Aphanizomenon*;
- lipopolysaccharides.

There are a number of unconfirmed reports of adverse health effects caused by algal toxins in drinking water, including an epidemiological study of mild, reversible liver damage in hospital patients receiving drinking water from a reservoir with a very large toxic bloom of *Microcystis aeruginosa*. Only activated carbon and ozonation appear to remove or reduce toxicity; however, knowledge is impeded by the lack of suitable analytical methods. There are insufficient data to allow guidelines to be recommended, but the need to protect impounded surface water sources from discharges of nutrient-rich effluents is emphasized.

WHO tables of guideline values

The Tables W12–W17 present a summary of guideline values for microorganisms and chemicals in drinking water. Individual values should not be used directly from the tables. The guideline values must be used and interpreted in conjunction with the information contained in the text of Volume 1, *Guidelines for drinking-water quality – Recommendations* (WHO, 1993) and in Volume 2, *Health criteria and other supporting information.*

The 1980 European Union Drinking Water Directive, based largely on the WHO guidelines, is currently being revised. The Directive is mandatory to all 15 countries of the European Union, although cases against several countries were brought by the European Commission in the European Court for non-compliance.

R.W. Herschy

Source

World Health Organization, 1993. *Guidelines for Drinking-Water Quality*, Vol. 1, *Recommendations*, 2nd edn.

WATER RESOURCES: INTRODUCTION

Fresh water: a vital environmental resource

Various other entries in this encyclopaedia have described the hydrological cycle and the freshwater environment (see Evaporation, measurement of; Groundwater; Lakes; Rainfall; Rivers; Springs). Non-saline (i.e. 'fresh') terrestrial waters are vital to a host of living creatures. They sustain numerous beneficial uses to human populations, as follows:

- consumptive uses (in which the water is not necessarily returned to source):
  - river water abstraction and storage in reservoirs (for public, private and industrial water supplies);
  - abstraction from groundwaters (for same water supply purposes);
  - irrigated agriculture;
  - watering of livestock;
- non-consumptive uses:
  - recreational enjoyment (fishing, walking, sightseeing, bathing);
  - water abstracted for cooling purposes (e.g. power stations and industrial refrigeration plant);
  - navigation in rivers and canals;
  - dilution and dispersion of suitably treated effluents from sewage treatment works, farms and industry;
  - fisheries, including natural waters used for angling, and fish farms and commercial fishing;
  - hydropower generation.

Sources of water

Referring to Figures W18 and W19, one can distinguish the following types of water source:

- Natural lakes. These usually offer a limited supply, since their range of acceptable drawdown is severely constrained.
- Natural rivers. Their flows vary seasonally and in upland areas streamflow varies wildly ('flashy discharge') and is commonly contained by impoundment in reservoirs (see below). Where a river receives groundwater inputs, via springs, its flow will be more steady. Lowland rivers are partially sustained by effluents of used water, for example from sewage treatment works and from the return of irrigation drainage water.
- Reservoirs. By creating an artificial lake behind a dam, a reservoir of water may be stored for later use. In climates where winter rainfall is assured, water stored in winter may be withdrawn for use in the following summer. But in many parts of the world reservoirs have to cater for a long series of dry years. Reservoirs may be of the following types.
  - Direct supply reservoir. Most upland reservoirs supply water by pipeline, often by gravity flow (i.e. pumps are not required to feed the water) and offer water of excellent quality.
  - River-regulating reservoirs, which are major water reserves for sending down a river, to sustain its discharge in conditions of exceptional low natural flow.
  - Pumped storage, in which river water is pumped into an off-channel storage, formed by embankments, for subsequent use with improved quality, by direct supply or by return to the river for regulation of low flows.
  - Estuary barrage, which impounds an area that otherwise would be tidal. This is the ultimate means of trapping fresh water

before it enters the sea; it entails making provision for the to-and-fro passage of migratory fish (e.g. salmon and sea trout) and providing locks for shipping.
- Boreholes (pumped) are the main sources of groundwater, usually with electric submersible pumps or with mechanical shaft-drive pumps (including wind-pumps); they range in yield from 1 to 10^4 m^3 day^{-1}, depending on the pump capacity and on available resources in the aquifer.
- Open wells (bucket and winch, or hand-pumped) are small sources in rural areas.
- Springs. These occur where aquifers crop out above impermeable rock; depending very much on local hydrogeological conditions, springs may be perennial, affording a reliable minor source of supply, or only seasonal, with no year-round availability.
- Saline waters. Brackish waters, which have a salt content >0.1% are amenable to desalination by reverse osmosis systems; the process is expensive, but is applied extensively in arid areas (e.g. the Arabian/Persian Gulf) where fuel is cheap. Even seawater can be treated thus (e.g. in Malta) or it can be desalinated by distillation processes.

Water quality

Sources differ widely in quality, i.e. untreated 'raw' water can have a great range of chemical and microbiological content.

Rivers range from clear mountain streams, with as few as 50 ppm of dissolved solids and low counts of harmful bacteria, to turbid lowland rivers, carrying over 700 ppm of dissolved matter, including agricultural, industrial and sewage effluents (the latter not necessarily harmful) and considerable counts of fecal bacteria, cysts and other potentially harmful microorganisms. Within limits, most raw waters can be satisfactorily treated to create a potable (safely drinkable) water supply, but the expense of doing this increases greatly as the quality of the raw water decreases. Effluent discharges and treatment standards are therefore set, not just for the wellbeing of the river environment (notably for fish, but including human activities and wild flora and fauna), but also as part of a water resources management strategy, to provide suitable water quality for downstream abstractors. A constant alert has to be kept against accidental pollution of rivers, due to industrial or transport mishaps, which can very quickly jeopardize the good quality of treated water.

Lakes and reservoirs allow the settlement of suspended matter that enters, either from natural inflow or by pumped input. As water typically remains in storage for several months, the interaction of several biological processes causes a gradual improvement in the bacteriological quality. Unicellular plants called algae grow in the water, if it has suitable nutrients, as will generally be the case; a succession of different algal species usually occurs in springtime. The abundance of algal populations depends on many factors, but tends to be greatest where high sunshine, nitrate and phosphate levels occur. If algae become too abundant, the lake or reservoir may have to be temporarily abandoned for a few weeks, due to water treatment and taste problems.

Groundwaters usually have a better bacteriological quality than surface waters, but this cannot be taken for granted, especially when drawn from shallow open wells. The recharge of groundwater, by infiltration of rainwater though the ground, creates a naturally filtered water, but in its passage through the soil and subsoil the water increases its content of soluble matter. In calcareous soils and rocks, such as the Chalk, the Jurassic Oolites and the Carboniferous Limestone, groundwaters have 'lime-hardness' (calcium hydrogen-carbonate) and in certain Triassic rocks the gypsum present gives rise to 'permanent hardness' (sulfates in solution). Where grassland has been ploughed and where nitrogeneous fertilizers have been liberally applied to arable crops. groundwaters are liable to develop excessive nitrate concentrations, rendering them unfit for potable supply, unless diluted or specially denitrified. Pollution of groundwaters has to be guarded against, the greatest hazards being:

- unlawful disposal of dangerous chemicals on land (e.g. sheep dip; industrial solvents);
- landfill leachates (i.e. contaminated liquors that develop if domestic and industrial waste is percolated by rainfall);
- spills of oils and hydrocarbon fuels (e.g. tanker accidents on roads, railways and airfields);
- pesticide residues (from spraying of roadside verges and railways; or from agriculture).

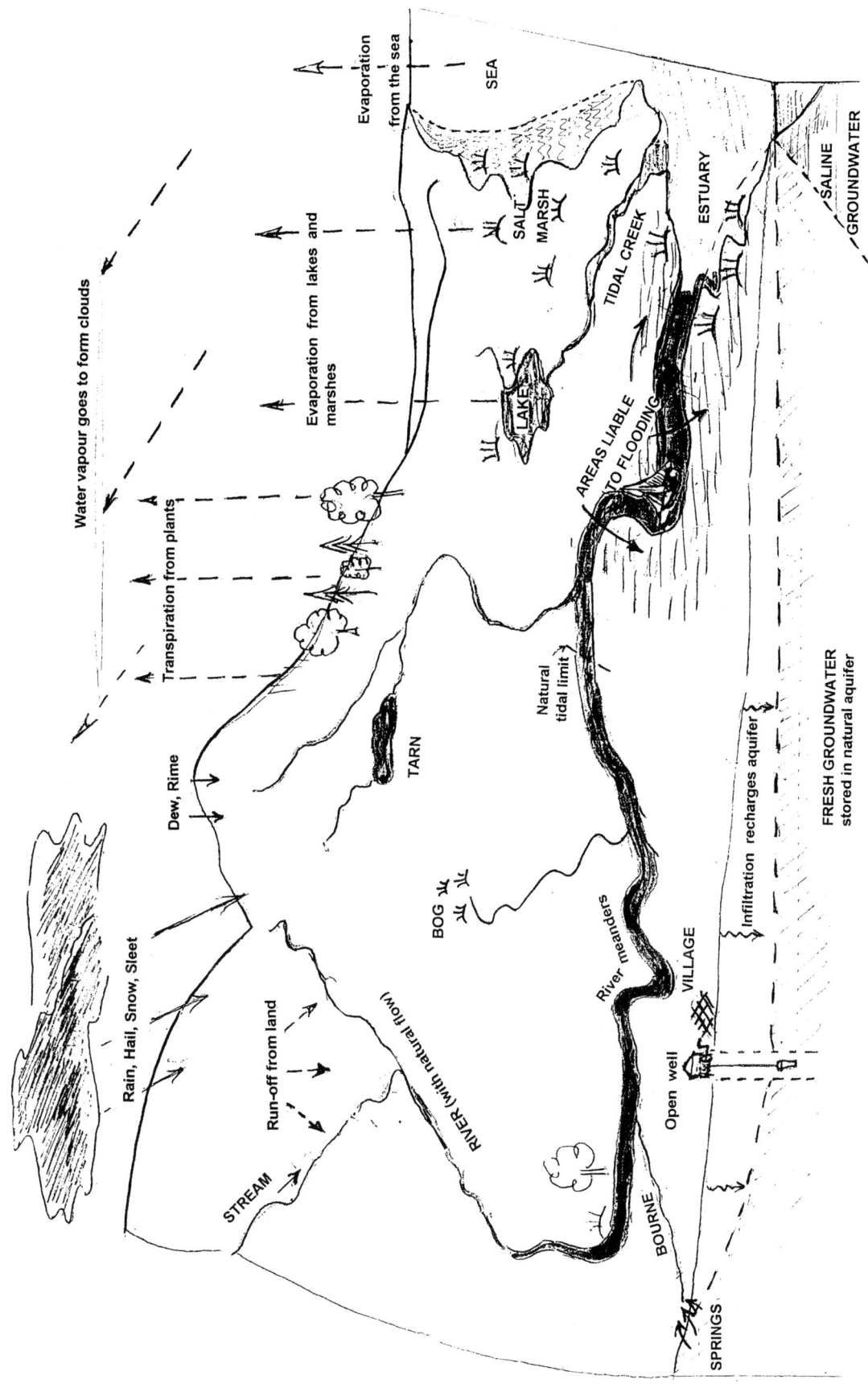

Figure W18 The natural hydrological cycle.

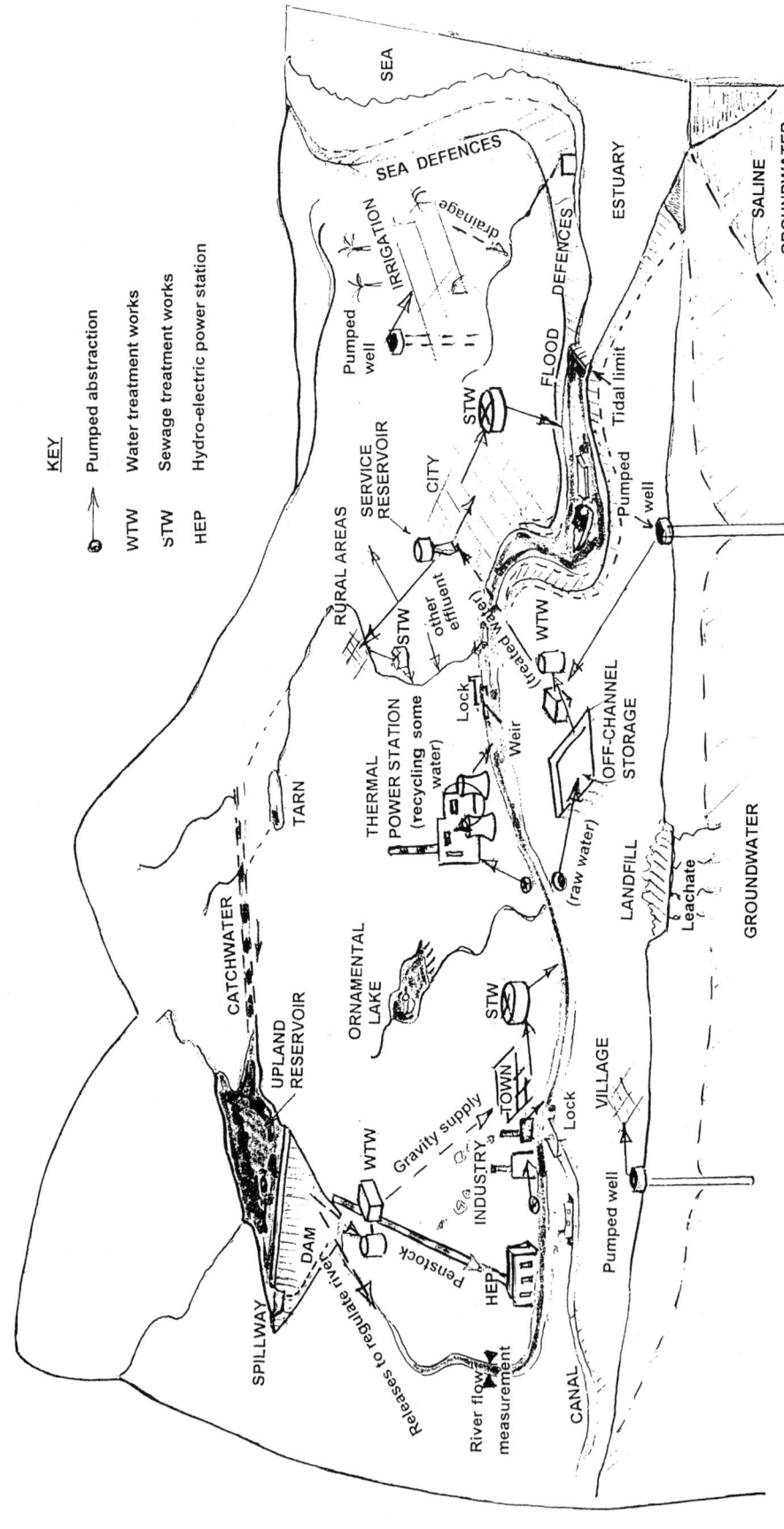

Figure W19 Human influence upon the hydrological cycle.

Planning and operation of regional water resources

Water resources are finite – they are exhaustible if overused, with reservoirs going empty, springs and rivers drying up and groundwater sinking below pumping levels. Hydrologists use river flow and rainfall data to evaluate the amount that each source can produce regularly, even under moderate drought conditions: this amount is termed the reliable yield of the source, and is expressed as a rate of delivery (e.g. $m^3\,day^{-1}$) that will be sustainable most years (e.g. 49 years in every 50).

Regional demands for water vary gradually with time, depending considerably on the level of commercial and other economic activity, such as manufacturing and agriculture. Purely domestic needs are typically $0.2\,m^3\,day^{-1}$ per head; the figure has risen over the years with the greater use of domestic appliances, such as automatic laundry machines and dishwashers. At present the cost of water hardly affects people's use of it. In fact there are several moves afoot in the UK to restrain demand. These include engineering means (e.g. pressure reduction) and restriction of use (e.g. hosepipe bans), but also control by cost (e.g. metering water use and imposing a variable tariff that penalizes excessive use). Good maintenance of water-distribution mains and of domestic pipework is also vital, otherwise as much as 30% of supplies can be lost through leakage.

The allocation of sources to meet present and future needs is a politically guided process, as regards setting priorities. The starting point is normally that of pinpointing centers of demand (e.g. towns and industrial complexes) and currently available sources. With each demand center there is a list of the present-day elements of water supply, including leakage, firefighting and waterworks maintenance. From the present-day figure, projections are made of future needs, allowing for any population trends, industrial changes in the area and leakage control measures. The reliable yields of sources feeding each demand center are evaluated. It then becomes possible to identify surpluses and deficits in present-day and future supplies: zones of surplus can in some cases be used to meet deficits elsewhere, by transfer of water by canal, pipeline or regulated river. Where that course proves uneconomic, a combination of water-saving measures and new source development will be planned to meet the future needs.

Social and environmental factors

Social and environmental factors have to be considered, especially when new reservoir schemes are proposed; there has to be a careful reckoning of the potential loss of farm land, dwellings and of access to scenic areas, set against the gains in water supplies and any new recreational opportunities created by the future reservoir. A host of recreational uses can be cited for rivers, lakes and reservoirs:

- angling, from banks or boats;
- sailing, rowing and canoeing;
- power boating, water skiing and scuba diving;
- sightseeing, picnicking, rambling and painting;
- birdwatching and natural history.

Clearly not all of these can coexist at one place, but there are substantial pressures on water companies (and on the national regulatory bodies controlling natural water resources) to facilitate the public's wishes for recreational uses of surface waters. Many companies allow organized sailing clubs to use their reservoirs, but other water sports are usually not allowed.

Groundwaters also can give rise to conflicts of use, such as when water levels become lowered by abstraction, with the consequence that springs feeding watercress beds or sustaining a trout stream are depleted in flow or even dry up completely.

In the UK various national bodies [in England and Wales, the new Environment Agency; in Scotland, the Environment Protection Agency; and in Northern Ireland, the Department of the Environment (NI)] all act in their respective spheres as the planning and regulating authorities. Their licence is required for all significant abstraction, and their consent is required for the promotion of new water resource schemes, which can entail going before a public inquiry, with the final say in the hands of a UK Government minister.

Water budget

When a hydrologist studies the water resources of an area, one of his first acts is to assemble the available data for that area, of the measured hydrological variables, principally:

1. precipitation (rainfall and snow water equivalent);
2. stream and spring discharges (including calibration data and estimated accuracy of measured discharge);
3. climatological observations (wind speed, air temperature and humidity, net solar radiation);
4. soil data (moisture content and hydraulic properties);
5. groundwater data (water level or pressure head; aquifer boundaries and hydraulic properties).

Items (1) and (2) are those amenable to direct measurement and (3) allows evaporation and transpiration to be calculated; (3) and (4)

Table W18 Purposes of water abstraction

Purpose	Daily requirement	Water quality
Drinking water		
For humans	2.5 l per head	WHO standards apply: freedom from waterborne disease organisms and from toxic chemicals is essential
For livestock	20 l per head (cattle) or 4 l per head (sheep and goats)	Not as stringently limited as above, but same principles apply in avoiding disease and chemical toxicity
Other domestic uses		
Preparing and cooking food	10 l per head	As above
Sanitation, showering and laundry	25 l per head	As above
Industrial processes		
Steelmaking	100 000 l tonne^{-1} of steel	Each industry has own water quality requirements, ranging from ultrapure (e.g. for electronics manufacture) to the converse (e.g. cooling water in a steelworks)
Petrochemicals	500 000 l tonne^{-1} of product	
Other chemicals	>5000 l tonne^{-1} of product	
Breweries	>2500 l m^{-3} of beer	
Food processing	10,000 l tonne^{-1} of product	
Commercial users		
Laundries	500 l tonne^{-1} of throughput	
Hotels	200 l per resident	
Garages	500 l per employee	
Shops, stores	100 l per employee	
Agriculture		
Rice and cotton (in tropics)	10^4 m^3 ha^{-1} of crop	
Wheat and maize (in warm temperate climates)	10^3 m^3 ha^{-1} of crop	

Table W19 Diverse sources of water

Source type	Typical water supply user	Reliability of source
Wells (hand-dug) with bucket and winch	Household, farm or small village ($<50\,000$ l day^{-1})	Depends on groundwater levels; source remains available as long as water table stays above base of well
Borehole (drilled) with submerged electric or mechanically driven pump	Farmstead, small town, factory (mostly $10\,000$–10^6 l day^{-1} but exceptionally 10^7 l day^{-1})	As above, but also constrained by avoiding adverse influence on adjoining abstractions
Springs captured at emergence	Household, village or small industrial user	Flow diminishes as groundwater levels recede, and can dry up after long periods of zero recharge
Rainwater storage tanks	Individual households, mainly in tropics	Good reliability dependent on having storage volume to cater for longest realistic dry season
Natural lakes and impounding reservoirs created by damming	Towns and whole regions; irrigation districts; independent industrial firms	Very dependent on the active volume–yield relationship, which is governed by local hydrology

together allow the infiltration of water into underground aquifers to be estimated.

Water needs and sources

Water is abstracted for a variety of requirements, as illustrated in Table W18. The situation can and often does arise where the demands for water by large industrial and agricultural interests tend to overshadow those of the human population. Such conflict is resolvable by an orderly planning process, which allows water uses to be assigned priorities and a reasonable apportionment to be arrived at, via the route of consultation and political representation. Hence the importance of setting up administrative arrangements.

The sources of water are very diverse, as indicated in Table W19.

J.A. Cole

Bibliography

General planning topics

Department of the Environment, 1992. *Using water wisely*. London: DoE consultation document.

Harris, R. (ed.), 1994. *Who's Who in European Water 1994/95*. London: Sterling Publications, 224 pp.

Rees, Y.J. and Zabel, T.F., 1995. *Eurowater: institutional mechanisms for water management in the context of European environmental policies – vertical report on the UK*. National Rivers Authority: R&D Note 422, 207 pp.

Wiseman, R., 1994. *Water Management in Europe 1994/95*. London: Sterling Publications, 180 pp.

Water resources topics

Cole, J.A., 1975. Assessment of surface water sources, in *Engineering Hydrology Today*, Institution of Civil Engineers, London, pp. 113–125.

Maas, A., Dorfman, G., Fair, G. and Hufschmidt, M., 1963. *The Design of Water Resource Systems*. London: Macmillan.

National Rivers Authority, 1994. Water, nature's precious resource: an environmentally sustainable water resources development strategy for England and Wales. London: Her Majesty's Stationery Office, 94 pp.

Plester, H.R.F. and Binnie, C.J.A., 1995. The evolution of water resources development in Northern Ireland. *J. CIWEM*, **9**(3), 272–280.

The Scottish Office, 1993. *Public Water Supplies 1991–1992, Water Resources Survey*. Edinburgh: Scottish Office Environment Directorate, 23 pp + 2 figs.

Susani, L., 1995. Reservoir dogged. *Water Services*, **99**(1189), 14–16.

Water Authorities Association, 1985. Water Facts. London: Water Authorities Association, 47 pp.

Water Resources Board, 1973. *Water Resources in England and Wales*. London: Her Majesty's Stationery Office; Vol. 1, 67 pp + 5 maps; Vol. 2, 56 pp + diagrams.

Wright, P., 1995. Water resources management in Scotland. *J. CIWEM*, **9**(2), 153–163.

Demand forecasting

Daniels, B.W., Achttienribbe, G.E. and Schoot, A.J.M., 1994. Forecasting domestic water demand and the effect of economy measures. *H$_2$O*, **27**(25), 736–739 (in Dutch).

Hall, M.J., Postle, S. and Hooper, D., 1989. A data management system for demand forecasting. *Water Resources Development*, **5**(1), 3–10.

Herrington, P., 1995. *Climate change and the demand for water*. Leicester: University of Leicester final report to the Institute of Hydrology and the Department of the Environment, under NERC Contract F3CR05-C1-37-03, 163 pp.

Merlo, D., 1989. The future development of water demand and problems connected with water demand forecasts. *Gas, Wasser, Abwasser*, (5), 25–31.

Merlo, D., White, R.J., Hu, M.S., *et al.*, 1990. Statistics for water demand forecasts. *Water Supply*, **7**(2–3), SS1, 1–13

Pygall, A., Turton, P. and Smith, R.J., 1980. *The impact of social and economic factors on consumption from public supplies*. Reading: Central Water Planning Unit, Tech. Note No. 29, 16 pp.

Thackray, J.E. and Archibald, G.E., 1981. The Severn–Trent studies of industrial water use. *Proc. Inst. Civ. Eng., Part 1*, **70**, 403–432.

Weber, J.A., 1989. *Forecasting demand and measuring price elasticity*. J. Am. Water Works Assoc. **81**(5), 57–65.

Winje, D., 1983. Development of domestic, industrial water consumption in the public network *Gas, Wasser und Warme*, **37**(4), 114–119. (in German).

Water requirements of specific industries

Nelson, W.L., 1963. Clean-water needs of refineries. *Oil and Gas J.*, **61**(3), 80.

Cross references

Drinking water and sanitation
Groundwater
River pollution prevention: historical
Rivers
Sanitation and clean water
Sewage treatment processes
Water availability and river water quality
Water balance
Water quality for drinking: WHO guidelines
Water resources: Europe

WATER RESOURCES: INTERNATIONAL

The water volume in any water body or within any territory may be considered in two ways, either (1) as definite water storage, which is important in an assessment of its contribution to the total water circulation of the Earth, or (2) as a definite moisture resource from the

Table W20 Water resources of selected rivers, lakes and reservoirs

Water body	Continent	Water resources (km³)
Rivers		
Amazon	South America	6 940
Congo	Africa	1 450
Yangtze	Asia	1 070
Yenisei	Asia	630
Mississippi	North America	600
Parana	South America	550
Lena	Asia	530
Lakes		
Caspian	Asia	78 200
Baikal	Asia	23 000
Great Lakes	North America	22 700
Tanganyika	Africa	17 800
Nyasa	Africa	8 400
Isyk-Kul	Asia	1 730
Reservoirs		
Owen Falls	Africa	205
Bratskoje	Asia	169
Cariba	Africa	160
Naser	Africa	157
Volta	Africa	148
Daniel-Johnson	North America	142

viewpoint of its opportunity to be used for the needs of the national economy. Since any water, either on the surface or subsurface, may be used (either at present or in the future, if appropriate technologies are developed), then all water in the hydrosphere including rivers, lakes, reservoirs, seas and oceans, underground water and glacial water, may form water resources in the wide understanding of the term. According to the *International Glossary of Hydrology* (1992), water resources is defined as 'Water available for use in sufficient quantity and quality at a location and over a period of time appropriate for an identifiable demand'. Water resources serve as a natural resource and may be subdivided into real and potential. Real water resources comprise water which may be used for the needs of the national economy at the specified stage of production development. Potential water resources comprise water which is required for industrial production but for some reason (e.g. because of inadequate technology) may be used only in the future. Real and potential resources together make up the total water resources.

Unlike other natural resources, water, with some exceptions, is not transformed into new matter but taken from the water cycle. During utilization, water is either (1) not changed quantitatively (e.g. hydro-power generation, fisheries or recreation, etc.) or (2) some portion is diverted from the source (for irrigation, industrial and municipal water supply, reservoir filling, etc.), thus causing irretrievable losses from a particular region or water source, but without changing the Earth's total water storage. Due to a great variety of methods of water utilization, including those which lead to decrease in water storage, it is not possible to estimate accurately the amount of water source available for present or for future use both for a particular water body and for different regions. Therefore the assessment of real and potential water resources is difficult.

In hydrological practice, the term 'water resources' means the average water storage in those water bodies or within areas which are subject to human impact. As water storage in rivers, lakes and reservoirs, and groundwater are used most intensively, the term 'water resources' is mainly applicable to these sources. It is most often applied to mean long-term volumes of water in the water body or within some region, irrespective of the fact that the relative water body is recharged by stationary or dynamic water. Stationary water storage only recovers slowly and is concentrated in seas, glaciers, lakes and other water bodies with a slow water exchange. Dynamic water storage is formed annually during a continuous water cycle and arrives on land as river runoff. When water resources are defined, these two categories of water storage are considered. Hence the water resources of a river is its mean long-term volume of annual runoff under natural conditions; that is, the river source of the water

resources is an annually recovered dynamic water storage. The water resources of a water body is the water volume within it, which corresponds to the mean long-term water level position under natural conditions. In this case the water storage of the water body consists of (1) slowly recovered stationary water storage, (2) some portion of the dynamic storage of river water and (3) some portion of the dynamic moisture storage in the atmosphere which arrives as precipitation (minus evaporation). The ratio between these three categories of water storage is determined by the rate of water exchange in the water body, its geographic situation and the character of anthropogenic use.

The water resources of reservoirs are usually identified as water storage at full reservoir level. Groundwater water resources are estimated by the amount of mean long-term water storage in some particular deposit of underground water.

Water resources of any region consists of the total amount of water storage in water bodies within the region. In hydrological practice, the regional water resources indicate mean long-term annual runoff in all rivers within the region. Some data on the water resources of large rivers, lakes and reservoirs of the world are given in Table W20.

V.S. Vuglinsky

Bibliography

International Glossary of Hydrology, 1992. WMO, Geneva.
Mints, A.A., 1972. *Economic assessment of natural resources*, Moscow: Mysl, 304 pp. (in Russian).
Spoehr, A., 1965. Cultural differences in the interpretation of the natural resources, in *Man's role in changing the face of Earth*, Chicago, pp. 48–59.
Vuglinsky, V.S., 1991. On a more accurate definition of the term 'water resources'. *Trudy GGI*, **352** (in Russian).

Cross references

Lakes: largest worldwide
Water balance
Water resources: introduction
Water resources: Europe
Water resources: natural quality
World water balance

WATER RESOURCES: DICTIONARY OF BASIC TERMS

Closed conduit flow

Constant level tank A tank, the level of liquid in which is controlled for example by a weir, the length of which should be as long as possible to ensure stable flow conditions in the circuit being supplied with liquid.

Differential pressure The pressure difference generated by the primary device, when there is no difference in datum level between the upstream and downstream pressure tappings.

Differential pressure devices A device inserted in a conduit to create a pressure difference whose measurement, together with a knowledge of the fluid conditions and of the geometry of the device and the conduit, enables the flow rate to be calculated.

Discharge coefficient Coefficient given by the formula

$$C = \frac{\alpha}{E}$$

Dynamic pressure For an element of fluid in a conduit, the dynamic pressure is the increase in pressure above the static pressure which would result from the complete isentropic transformation of the kinetic energy of the fluid into pressure energy. It is equal to the product $\frac{1}{2}\rho v^2$, where ρ is the fluid mass density and v is the velocity of the element of fluid.

The mean dynamic pressure in a cross-section of a straight flow is equal to the ratio of the power which flows through the cross-section in a kinetic energy form to the volume flow-rate. It can be expressed by the formulae $\alpha \times \frac{1}{2}\rho U^2$ where U is the discharge velocity in the cross-section and where the kinetic energy coefficient α is equal to

$$\frac{1}{S}\iint_S \left(\frac{v}{U}\right)^3 dS,$$

dS being a surface element and S the total area of the cross-section. (In the flow usually occurring in practical installations, α may vary between approximately 1 and 1.2.)

Effective pressure The difference between the local absolute pressure of the fluid and the atmospheric pressure at the place and time of the measurement.

Electromagnetic flowmeter A flowmeter which creates a magnetic field perpendicular to the flow so enabling the flow rate to be deduced from the induced electromotive force (emf) produced by the motion of a conducting fluid in the magnetic field. The electromagnetic flowmeter consists of a primary device and a secondary device.

Flow coefficient Coefficient given in the case of a flow of fluid considered as incompressible by the formula

$$\alpha = \frac{q_m}{\frac{\pi}{4}d^2 \sqrt{(2\,\Delta p \rho_1)}}$$

where q_m is the mass flow-rate, d is the diameter of the primary device, Δp is the differential pressure and ρ_1, is the mass density of the fluid upstream of the device.

Flow rate of a fluid through a cross-section of a conduit The amount of fluid flowing through the cross-section of a conduit in unit time.

Fully developed velocity distribution A velocity distribution that does not change between two cross-sections of a flow. It is generally obtained at the end of a sufficient straight length of a conduit.

Laminar flow Flow under conditions where forces due to viscosity are significant in comparison to the forces due to inertia. Laminar flow may be unsteady but is completely free from turbulent mixing. Poiseuille flow is an example of steady laminar flow in a circular pipe.

Mass flow rate through a cross-section of a conduit The mass of fluid flowing through the cross-section of a conduit in unit time.

Nozzle Convergent device having a curved profile without discontinuities which may blend into a cylindrical throat.

Orifice plate A plate having a hole through it conforming to certain specifications.

Pressure loss (caused by a primary device) The irrecoverable pressure loss caused by the presence of a primary device in the conduit.

Primary device A device which generates a signal enabling the flow rate to be determined. According to the principle used, the primary device can be internal or external to the conduit. This device contains the following elements: an electrically insulated meter tube through which the conductive fluid to be metered flows, a pair of diametrically opposed meter electrodes across which the signal generated in the fluid is measured, and an electromagnet for producing a magnetic field in the meter tube. The primary device develops a signal proportional to the flow rate and in some cases the reference signal.

Pulsating flow of mean constant flow rate Flow in which the flow rate in a measuring section is a function of time but has a constant mean value when averaged over a sufficiently long period of time. Two types of pulsating flow are found: periodic pulsating flow and fluctuating (random) pulsating flow.

Regular velocity distribution The distribution of velocities which sufficiently approaches that established in a long straight length of the closed conduit to permit an accurate measurement of the flow rate to be made.

Secondary device A device which receives from the primary device a signal and displays, records, transforms and/or transmits it as a measure of the flow rate. This equipment contains the circuitry which extracts the flow signal from the electrode signal and converts it to a standard output signal directly proportional to flow rate. This equipment may or may not be mounted on the primary device.

Square-edged thin-orifice plate Plate the thickness of which is small compared with the diameter of the measuring conduit, the orifice of which is circular, concentric with the conduit axis, and sharp and square on the upstream edge.

Static pressure Pressure measured in a fluid in such a way that the velocity of this fluid has no effect on the measurement.

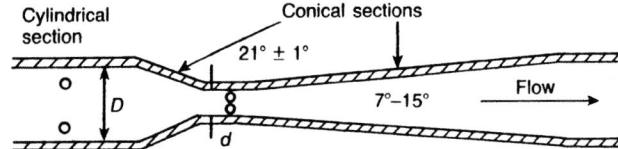

Figure W20 Classical Venturi tube.

Steady flow Flow in which the flow rate through a measuring section does not vary significantly with time. The steady flows observed in conduits are in practice flows in which quantities such as velocity, pressure, mass density and temperature vary in time about mean values independent of time; these are actually 'statistically steady flows'.

Swirl remover Device inserted in a conduit to eliminate or reduce circumferential velocity components which produce swirl.

Total pressure The sum of the effective pressure and of the dynamic pressure. For an element of fluid at rest, the effective pressure and the total pressure have the same numerical value.

Transition flow Flow lying between a laminar flow and a turbulent flow. As a guide, the Reynolds number for the transition flow of a Newtonian fluid, when referred to the conduit diameter, is generally between a lower limit of 2000 and upper limit which varies between 7000 and 12 000 according to the conduit roughness and other factors.

Turbulent flow Flow under conditions where forces due to viscosity are small in comparison to the forces due to inertia. It is a flow in which irregular (random) velocity fluctuations in time and space are superimposed on the main flow.

Unsteady flow Flow which may be laminar or turbulent, in which the flow rate in a measuring section fluctuates randomly with time. The time interval being considered is to be long enough to exclude from this definition the random components of the turbulent flow itself.

Velocity of approach factor Coefficient given by the formula

$$E = (1 - \beta^4)^{-1/2} = \frac{D^2}{\sqrt{(D^4 - d^4)}}$$

where β is the diameter ratio $\frac{d}{D}$, D is the upstream internal diameter of the conduit and d is the diameter of the primary device orifice or throat.

Venturi tube Device consisting of

- an entrance cylinder;
- a convergence (converging section);
- a throat (cylindrical portion);
- a divergence (diffuser or expanding section).

See Figure W20.

Volume flow rate through a cross-section of a conduit The volume of fluid flowing through the cross-section of a conduit in unit time at the conditions of pressure and temperature pertaining to that section.

Volumetric method Method of measurement, principally used for liquids, in which the flow is directed during a certain time into a calibrated volumetric tank, the capacity of which is known as a function of the level to which the tank is filled.

Coefficients and losses

Coefficient of contraction Ratio of the *vena contracta* of a jet discharged under pressure from an orifice to the area of the orifice.

Coefficient of discharge Ratio of the actual discharge of liquid over or through a weir pipe or orifice, to the theoretical discharge.

Coefficient of permeability Ratio of velocity of flow of water through soil to the hydraulic gradient causing the flow.

Coefficient of velocity Ratio of the actual discharge velocity to the theoretical discharge velocity of liquid flowing through an orifice.

Friction coefficient Factor that is a measure of the frictional resistance to flow in a conduit dependent upon the material, dimensions and conditions.

Friction loss (friction head deprecated) Head loss due to friction between a liquid and the internal surface of a conduit.

Head loss Reduction in head when fluid flows from one point to another.

Vena contracta Cross-section of minimum area in a jet of water beyond the orifice or notch through which it emerges.

Dams: types

Arch buttress dam Buttress dam in which the upstream part comprises a series of arches.

Arch dam Dam that obtains its resistance to the thrust of water from horizontal arching action, the thrust being transmitted to its abutments.

Buttress dam Dam that consists of a watertight wall supported at intervals on the downstream side by a series of buttresses.

Constant angle arch dam Arch dam in which the angle subtended by any horizontal section is constant throughout the height of the dam.

Constant radius arch dam Arch dam in which every horizontal section of the dam has approximately the same radius of curvature.

Crib dam Gravity dam built using crossed timber members and fill.

Double-curvature arch dam Arch dam that is curved vertically as well as horizontally.

Earth dam Embankment dam constructed of soil.

Embankment dam Dam formed as an embankment.

Fabridam Dam made of flexible membrane, anchored to the river bed, that can be inflated with water or air, or both, and is completely collapsible.

Fish pass Artificial passage, down which water flows, to enable migratory fish to surmount an obstruction such as a weir or a dam.

Flat slab buttress dam Buttress dam in which the upstream part is a relatively thin inclined flat slab usually made of reinforced concrete.

Gabion dam Gravity dam built using gabions.

Gravity arch dam Dam that obtains its resistance to the thrust of water both from horizontal arching action and from its own weight.

Gravity dam Dam that relies on its weight for stability.

Hydraulic fill dam Embankment dam constructed of hydraulic fill.

Prestressed dam Dam the stability of which depends in part on the tension in vertical steel wires, cables or rods that pass through the dam and are anchored into the foundation rock.

Rock fill dam Embankment dam constructed mainly of rock.

Solid head buttress dam Buttress dam in which the upstream ends of all buttresses are enlarged symmetrically to meet those adjacent, thereby forming a continuous structure.

Underground dam An impervious barrier (usually clay) emplaced in the sand and gravel bed of an ephemeral stream, e.g. in the monsoonal Sahel belt of West Africa.

Dams: parts

Core Zone of material of low permeability in an embankment dam that extends from the foundation to near the top of dam and that inhibits seepage in a broadly horizontal direction.

Core wall Wall of material of low permeability in an embankment dam.

Cut-off trench Trench excavated below the general foundation level to a layer of low permeability and then filled with a material of low permeability.

Cut-off wall Wall of material of low permeability built into the foundation to reduce seepage.

Heel of dam Junction of the upstream face of a gravity dam with the ground surface.

Toe of dam Junction of the downstream face of a gravity dam or of the face of a shoulder with the ground surface.

Top of dam (crest of dam deprecated) Uppermost surface of a dam, usually a road or walkway excluding any parapet wall, railings, etc.

Erosion prevention

Apron, (discharge apron, anti-scour apron, deprecated) Hard surface or layer of suitable material placed adjacent to a structure, such as an outfall, or on a bed or bank to prevent scour.

Dutch mattress Mattress formed mainly of reeds.

Bank protection Construction works to protect bank of a canal, river or reservoir from erosion.

Faggoting Revetment of river banks constructed using fascines.

Fascine (faggot deprecated) Long cylindrical bundle of brushwood.

Gabion Rectangular container made of wire or plastics mesh, filled with stones and placed to form retaining walls or provide protection against erosion.

Groyne Wall or embankment built out from the shore to reduce littoral transport.

Pitching Bank protection formed of hand-placed stones of similar size or concrete blocks.

Pole-wharfing (pole-warping deprecated) Bank protection formed of longitudinal timber poles and short piles used as toe support at the bottom of the banks of tidal lengths of river.

Reno mattress Mattress formed of stones contained in a cage.

Revetment Construction that comprises one or more layers of material to provide protection to a slope against erosion.

Riprap Bank protection formed of large uncoursed stones, broken rock or precast blocks placed in random fashion.

Flow control structures

Headwall Retaining wall at the end of a culvert or pipe.

Head water level Water level upstream of a hydraulic structure.

Tail water level Water level downstream of a hydraulic structure.

Gates

Drum gate Spillway gate or gate of a barrage that consists of a long hollow drum that is held in its raised position by the water pressure in a flotating chamber beneath the drum and that rises with the reservoir water level and lowers when overtopped by floods, usually automatically.

Radial gate Gate with a curved upstream plate supported by radial arms.

Roller drum gate Spillway gate in the form of a hollow cylinder, for regulating the flow at a dam spillway, carried at each side on large toothed wheels that mesh with steeply inclined racks up which the gate moves when being opened.

Sector gate (1) Gate with a curved upstream plate supported by several radial arms pivoted at the downstream rim of a recess into which the gate retracts to allow increased flow over the gate.

Sector gate (2) Roller drum gate in which the roller is not cylindrical but a sector of a circle.

Sluice gate Rectangular gate that moves vertically between guides.

Spillway gate Gate on the crest of a spillway to control overflow or reservoir water level.

Vertical lift gate Gate, large and mechanically operated, that moves vertically within guides.

Hydrology

Abstraction Removal of water from any source, either permanently or temporarily.

Aquifer Water-bearing formation of permeable rock, sand or gravel capable of yielding significant quantities of water.

Base flow (dry-weather flow deprecated) That part of the discharge in a watercourse not directly derived from runoff.

Borehole Hole, usually vertical, bored to determined ground conditions, for extraction of water or measurement of groundwater level.

Catchment area Area of land that drains naturally to a given point on a river.

Creek Small inlet on coast or estuary.

Deep well Well equal to or more than 15 m deep.

Design flood Flood parameters adopted for the design of water engineering construction works.

Design storm Rainstorm parameters adopted for the design of water engineering construction works.

Estuary (regional terms: firth, frith, kyle) Mouth of a river connected to the sea where the tide meets the river current.

Flood routing Determining at successive points along a river the timing and form of the hydrograph of a flood.

Free water Groundwater in interconnected interstices in saturation zone, that extends down to the first impervious barrier and moves under the influence of gravity in the direction of the slope of the water table.

Frazil ice (slush ice deprecated) Ice formed, as colloidal crystals, in a watercourse when the temperature of the entire body of water flowing along it is reduced to 0°C.

Groundwater Water within saturation zone.

Hydrograph Graph that shows the relationship, with time, of level, discharge or velocity of water in a river or channel.

Hydrography Applied science concerned with the study and measurement of seas, lakes and other waters where volume is more important than the rate of flow.

Hydrology Science of the occurrence and movement of water over and below the surface of the Earth from the moment of precipitation to the moment of entry into the ocean or of evaporation into the atmosphere.

Hydroscopic water Water in soil that is in equilibrium with atmospheric water vapor pressure and that is essentially water in which attraction between water molecules can hold against evaporation.

Infiltration Passage of water through the soil surface into the soil or from the ground into a conduit.

Intensity of rainfall Rate of precipitation expressed as a depth in unit time, for example millimeters per hour.

Interception Process by which precipitation is retained by and later evaporated from a structure or vegetation and is thus prevented from reaching the ground.

Isohyet Line on a map joining places with equal rainfall.

Mass diagram Graph that shows cumulative flow quantities, such as the integration of a time–flow curve.

Pellicular water (adhesive water deprecated) Water retained in soil by attraction between water and soil molecules and which forms a coating around the particles that may move from one particle to another.

Phreatic surface Surface to which groundwater would rise in an open ended pipe.

Phreatic water Water in the zone below the water table.

Phreatic zone Zone below the phreatic surface.

Precipitation Water derived from atmospheric vapour and deposited on a surface as mist, rain, hail, sleet and snow or dew.

Recharge Flow of water to groundwater storage from precipitation, infiltration from surface streams and other sources.

River Water flow along a large watercourse.

Safe yield Maximum rate at which water may be extracted from an aquifer over a period without depleting the supply or causing a deterioration in quality.

Salting (outmarsh, saltmarsh deprecated) Area of land periodically covered by saline water and that usually supports vegetation.

Saturation zone Zone below water table and above rock flow zone throughout which all fissures are filled with water under hydrostatic pressure.

Sewage catchment Area of land that drains to a single sewage treatment works or outfall.

Shallow well Well less than 15 m deep.

Spate Sudden short-lived increase in discharge.

Specific retention Ratio of the volume of water retained against gravity by unit volume of rock or soil that has been saturated and allowed to drain completely to a remote body of mobile water by way of continuous capillary interstices to unit volume.

Specific yield Ratio of the volume of water yielded by unit volume of permeable rock or soil when drained by gravity under specified conditions after being saturated to unit volume.

Specific storage Ratio of the volume of water that unit volume of a vertical column of an aquifer releases from storage as the head within the column declines unit distance to unit volume.

Storage coefficient Ratio of the volume of water that a vertical column of an aquifer of unit cross-sectional area releases from storage as the head within the column declines unit distance to unit volume.

Stream (regional terms: beck, bourne, brook, burn, gill, rhine, rife, rill, runnel, spur) Water flow along a small watercourse.

Time of concentration Time taken for a droplet of surface water that falls on the most remote part of a catchment area to flow to a given point under investigation.

Transmissivity Rate of flow of groundwater through a vertical strip of unit width of an aquifer, the strip extending the full saturated depth, under unit hydraulic gradient at a fixed temperature.

Underflow Groundwater movement in an aquifer.

Unit hydrograph Theoretical hydrograph that would result from a storm of unit intensity of rainfall and unit duration.

Water cycle Complete natural cycle of water circulation from the atmosphere to the earth and return to the atmosphere through various stages or processes such as precipitation, runoff and evaporation.

Watershed Boundary between catchment areas.

Water table Surface of groundwater in saturated soil or rock.

Hydropower works

Headrace Free-flow conduit that conveys water from a forebay to a water turbine or waterwheel.

Hydroelectric power station Power station in which the energy of falling water is used to drive water turbines coupled to electrical generators.

Millrace Channel that carries water from a river, stream or millpond to a waterwheel.

Overshot waterwheel Waterwheel in which the water strikes the buckets at the top of the wheel and passes over the wheel.

Pumped storage Means of storing energy by pumping water up to a high level reservoir and releasing it at suitable times to generate electricity in a hydroelectric power station.

Power tunnel Tunnel that carries water to a hydroelectric power station.

Tailrace Conduit that conveys the discharge from a waterwheel or water turbine.

Undershot waterwheel Waterwheel in which the water strikes the paddles at the bottom of the wheel.

Waterwheel Wheel, with buckets or paddles on the rim, which is rotated by water about a horizontal axis to provide power.

Irrigation

Basin irrigation Surface irrigation of orchards by which each area or group of trees is surrounded by a bund.

Border irrigation Flood irrigation in which land is divided into strips and water is delivered into each strip from an irrigation canal.

Check irrigation Flood irrigation in which fields are divided into rectangles by low bunds.

Continuous flow irrigation Method of delivery of irrigation water by which irrigators receives their allotted quantity of water at a continuous rate.

Flood irrigation Irrigation in which water is made to cover the surface of the land to such a depth as to cause saturation for a considerable time.

Furrow irrigation Surface irrigation in which water is run in furrows between crops.

Gravity irrigation Irrigation in which the flow of water is maintained by gravity alone.

Lift irrigation Gravity irrigation in which water is lifted by a pump or other device.

Pumped irrigation Irrigation in which water is delivered by pumping through pipes.

Spate irrigation Surface irrigation for which earth diversion embankments are built across normally dry watercourses to divert spate water into canals leading to bunded fields where the water is ponded until absorbed.

Sprinkler irrigation Irrigation in which water is conducted in pipes to sprinklers that distribute water like rain.

Sub-irrigation Irrigation in which water is applied below the ground surface.

Supplemental irrigation Irrigation practice whereby water is supplied for comparatively short and usually irregular periods of drought during the crop growing season so as to overcome the shortage of soil moisture.

Surface irrigation Irrigation in which water is directed across the land surface.

Microorganisms found in water

Achorutes subviaticus See *Hypogastrura viatica*.

Aelosoma A member of the family Aelosomatidae of the oligochaete worms. Found in aquatic habitats and in activated sludges.

Amoeba A protozoan of the class Rhizopoda, which moves by the formation of temporary protrusions (pseudopodia).

Arcella A shelled rhizopod protozoan.

Aspidisca A hypotrichous ciliate found in activated sludge and indicative of good conditions.

Ascaris A parasitic nematode worm of humans (human roundworm), the eggs of which can be disseminated by way of sewage effluents.

Asellus aquaticus **(the water hog louse)** An isopod crustacean found in the zone of recovery from organic pollution in a receiving water and therefore a useful indicator organism.

Baetis rhodani A species of mayfly (Ephemeroptera) commonly present in rivers and tolerant of reduced dissolved oxygen levels, in contrast to many other species of mayflies which are intolerant of such conditions. Consequently often present under mildly organically polluted conditions.

Beggiatoa A filamentous autotrophic bacterium capable of oxidizing sulfur compounds to elemental sulfur, according to the equation

$$H_2S + \tfrac{1}{2}O_2 \rightarrow S + H_2O \text{ and energy}$$

Bodo A non-pigmented flagellate protozoan found in activated sludge and usually indicative of unsatisfactory conditions.

Caenis A genus of mayfly which lives on a muddy substratum.

Callitriche An aquatic macrophyte, commonly known as starwort.

Campylobacter First recognized as a source of self-limiting diarrheal disease in the early 1970s and later shown to be the commonest reported cause of acute diarrheal illness in the UK.

Carchesium A colonial attached peritrichous ciliate which is found in activated sludge and in biological-filter slimes, as well as in streams and ponds, where it is a constituent of 'sewage fungus'.

Chlamydomonas A green flagellate, found in nutrient-rich waters.

Cladophora A filamentous green alga found commonly in waters. It may cause a nuisance by producing excessive growths in nutrient-enriched waters, e.g. downstream of an oxidized sewage effluent discharge or in the final recovery zone in organically polluted receiving waters. It is then known as 'blanket weed'.

Clostridium A genus of spore-forming anaerobic bacteria. *Clostridium perfringens* can form resistant spores which survive in water much longer than other fecal indicators. Its presence implies remote or intermittent fecal pollution. Some species of *Clostridium* can reduce sulfate to sulfide.

Colpidium A ciliate protozoan found in activated sludge and biological-filter slimes, where it is indicative of inferior conditions.

Cryptosporidium A protozoan parasite found in humans, other mammals, birds and reptiles. *Crytosporidium parvum* (*C. parvum*) is the species of *Cryptosporidium* believed to be capable of causing disease in humans and livestock.

Cyclops A copepod crustacean, common in zooplankton.

Cysticercosis The bovine effect of infection by *Taenia*.

Cysticercus bovis A parasitic beef tapeworm *Taenia saginata*.

Daphnia A genus of small crustacea of the sub-order Cladocera and commonly known as 'water fleas'. Common member of the zooplankton of lakes and reservoirs, and sometimes used as test animals in toxicity tests.

Dendrocoelum lacteum A flatworm (Turbellaria) of the phylum Platyhelminthes. Found in streams and used as an indicator organism.

Desulfovibrio desulfuricans A bacterium which reduces sulfate to hydrogen sulfide, thus:

$$H_2SO_4 + 8H \rightarrow H_2S + 4H_2O$$

Entamoeba histolytica A parasitic ameboid protozoan which causes dysentery in humans and is spread by contact with polluted waters.

Enterobius A nematode worm, parasitic in humans, the eggs of which can be found in sewage.

Enteromorpha A tubular thalloid green alga related to sea lettuce *Ulva*. Found in brackish and polluted waters, and especially abundant in polluted estuaries.

Enterovirus A subgroup (genus) of the Picornaviridae family of viruses. Over 70 different types have been isolated from humans. Their normal site of infection is the intestinal tract but other organs may become involved. Diseases caused range from the trivial to the fatal, including infectious hepatitis, poliomyelitis and meningitis.

Epistylis A colonial peritrichous ciliate commonly occurring in activated sludge or biological-filter film.

Eristalis tenax A dipterous fly with a bee-like appearance, the larva of which (known as the rat-tailed maggot) is aquatic and which can, by breathing atmospheric air through a telescopic tail, exist in most organically polluted waters and sludges when these are not too deep. Its presence is indicative of organic pollution.

Erpobdella A genus of freshwater leeches, commonly found in streams, which are useful indicator organisms.

Escherichia coli A bacterium living in the alimentary tract of humans and other mammals. As it is passed out with feces in large numbers its presence in water is indicative of fecal contamination and the possible presence of pathogenic organisms of enteric origin; it is not itself normally pathogenic. Also known as *E. coli*, *Esch. coli* or *Bact. coli*.

Fusarium aquaeductum A common biological-filter fungus having sickle-shaped spores and often having a pink coloration.

Geotrichum A fungus commonly occurring on biological filters.

Giardia *Giardia intestinalis* (*lamblia*) is a protozoan parasite capable of infecting humans and domestic animals. It causes acute diarrheal illness.

Glossiphonia A genus of leeches common in freshwater streams, useful as indicator organisms.

Gammarus pulex A freshwater shrimp which may be common in streams and is a useful indicator organism.

Helobdella stagnalis A species of leech, useful as an indicator organism.

Hydrobaenus A genus of chironomid midge, an active grazer in some biological filters. Also termed '*Spaniotoma*'.

Hydropsyche A genus of Trichoptera or caddis flies. In their larval stages they do not build cases like most caddis but catch their food in nets which they spin in flowing water. Useful indicator organisms.

Hypogastrura viatica A species of the primitive insect order Collembola, characterized by having a simple life cycle and by not having wings. A common member of the grazing fauna of many biological filters. Also termed '*Achorutes subviaticus*'.

Legionella Organisms of the genus *Legionella* are classified as Gramnegative bacteria of which there are at least 30 species, over half of which have been implicated in human disease. *L. pneumophila* is the causative agent of Legionnaires' disease, an acute pneumonic infection of the lungs, and greater than 85% of cases are caused by *L. pneumophila* serogroup 1.

Leptomitus lacteus A non-septate aquatic fungus found in certain organically enriched waters.

Leptospira icterohemorrhagiae A pathogenic spirochaete which may invade the blood causing leptospiral jaundice (Weil's disease), transmitted by sewer rats and their urine.

Leptothrix A filamentous genus of the iron bacteria. Chlamydobacteriacae, found in organically polluted waters.

Leuctra A genus of Plecoptera (stone-flies) having aquatic nymphs in flowing waters; indicative of well-aerated water.

Limnaea A genus of freshwater snail commonly found in fresh waters and tolerant of a wide range of conditions.

Lumbricillus Or *Pachydrilus*, a genus of the family Enchytraeidae. Small worms of a pale colour which are common grazers in biological filters.

Lumbricus A genus of worm of the family Lumbricidae, which includes the common earth worm. Other species are grazers in biological filters.

Litonotus Or Lionotus, a genus of free swimming ciliate protozoa common in activated sludge.

Nitrobacter A bacterium which converts nitrite to nitrate.

Nitrosomonas A bacterium capable of oxidizing ammonia to nitrite.

Nitzschia palea A common species of diatom of value as an indicator organism of organic pollution.

Opercularia A genus of colonial peritrichous ciliate protozoan common in activated sludge and used as an indicator of its condition.

Pachydrilus Or *Lumbricillus*; worms of the family Enchytraeidae, commonly found as a grazer in biological filters.

Paramecium A genus of free-swimming ciliate protozoa common in activated sludge and a useful indicator organism.

Phormidium A blue-green alga the filaments of which intertwine to form a sheet. Commonly on the surface of biological filters where it may cause ponding.

Psychoda A genus of small moth-like flies the larvae of which are 'grazers' in biological filters.

Potamogeton Pondweed, which is an aquatic macrophyte with many species growing in ponds and rivers.

Salmonella A group of enteropathogenic bacteria responsible for typhoid fevers and food poisoning. Common in polluted waters.

Sphaerotilus A filamentous bacterium of the order Chlamydobacteriaceae, which may form plumose growths in polluted waters known as 'sewage fungus' and when present in activated sludge is associated with bulking.

Streptococcus fecalis A bacterial indicator of fecal contamination of water.

Sylvicola fenestralis A dipterous fly whose larvae are common grazers in biological filters. The adults may cause a nuisance when they leave the filters. Formerly *Anisopus fenestralis*.

Taenia A genus of parasitic tapeworm of the class Cestoda, the eggs of which are dispersed through sewage and polluted waters.

Tanypus A genus of chironomid fly the larvae of which are aquatic.

Thiobacillus thiooxidans A sulfur-oxidizing bacterium which, when present in sewers, converts sulfur into sulfuric acid (which in turn attacks concrete) according to the equation:

$$2S + 3O_2 + 2H_2O \xrightarrow[thiooxidans]{Thiobacillus} 2H_2SO_4$$

The bacterium grows best in a strongly acid medium (pH 2.0–6.0) and can also oxidize thiosulfate to sulfate, thus:

$$Na_2S_2O_3 + 2O_2 + H_2O \xrightarrow[thiooxidans]{Thiobacillus} Na_2SO_4 + H_2SO_4$$

Ulothrix A filamentous green alga, common in fresh waters and on the surface of biological filters.

Vaucheria A filamentous green algae, common in streams.

Vorticella A genus of peritrichous ciliate, common in activated sludge.

Zoogloea ramigera A bacterium which grows embedded in a common gelatinous matrix, forming a bacterial slime. A member of the family Pseudomonadaceae. Common as a slime in biological filters in flocs of activated sludge.

Regime

Barrage Structure built across a river to regulate the upstream level that comprises a series of gates that when fully open allow flood water to pass without appreciably raising the head water level.

Barrier Gated structure, located in a tidal waterway, which can be closed to prevent exceptionally high tides reaching inland areas vulnerable to flooding.

Cross-regulator Barrage across an artificial channel.

Flood channel Channel that carries excess water during flood conditions.

Flood plain Area of land that borders a river and is partly or wholly covered with water during floods.

Headwater Source or upper reach of a stream or river.

Low flow channel (dry-weather channel deprecated) That part of a multistage channel that is designed to carry baseflow.

Reach Stretch of water between two features alongside or across a river or stream.

Regime (1) Dynamic equilibrium between accretion and erosion in an open channel system or a coastal environment where the balance existing between the two may vary over a time span from days to tens or even hundreds of years.

Regime (2) State in which a river or canal has adjusted its slope and cross-section to an equilibrium condition.

Riffle A sector or reach of a stream with rough water created by boulders in the stream bed.

Surge Difference between actual and predicted water levels due to meteorological and or tidal effects.

Two-stage channel Channel in which normal flow is contained in the lower part of a wider main channel.

Washland Low land adjacent to a river or drain used for storage of floodwater.

Wind setup Superelevation of water surface over its normal elevation due to wind action alone.

Reservoirs

Active storage (usable storage, useful, storage, working storage deprecated) Volume of water in a reservoir that is available for its intended use.

Bank storage (ground storage deprecated) Water that has infiltrated from a reservoir into the surrounding ground where it remains until water level in the reservoir is lowered.

Capacity curve Graph that shows the volume of water in a reservoir or tank at any given water level.

Dam freeboard (gross freeboard deprecated) Vertical distance between top water level (1) or (2) and top of dam.

Dead storage Volume of water in a reservoir measured below the invert level of the lowest outlet.

Drawdown Lowering or drop in water level of a reservoir.

Drawdown range Difference between top water level (2) and the level of the top of the inactive storage.

Drawdown zone Area of the bank of a reservoir and dam uncovered by drawdown.

Flood surcharge Maximum rise of still water level above top water level (1) or (2) during a design flood.

Freeboard Vertical distance between water level and lowest level of a water retaining structure at which uncontrolled overflow would occur.

Impounding reservoir (conservation reservoir, deprecated) Reservoir, usually in the upper reaches of a watercourse, that collects surface water from an open area of land.

Inactive storage Volume of water in a reservoir measured between invert level of lowest outlet and minimum operating level.

Live storage Sum of active storage and inactive storage.

Maximum water level Highest still water level in a reservoir during the probable maximum flood.

Millpond Impounding reservoir that serves a watermill.

Net freeboard (dry freeboard, flood, freeboard wave deprecated) Vertical distance between maximum water level and top of dam or other water retaining structure.

Pumped storage reservoir High-level reservoir in a pumped storage scheme.

Reservoir capacity (gross storage, gross capacity of reservoir deprecated) Gross capacity of a reservoir from the bottom of the reservoir up to top water level (1) or (2).

Retention water level Maximum level to which water rises under normal operating conditions, exclusive of any provision for flood surcharge.

Scour outlet (bottom outlet deprecated) Structure in bed of reservoir at upstream end of scour pipe or scour tunnel.

Scour pipe Outlet pipe from a reservoir, usually an impounding reservoir, at the lowest possible level for discharging water.

Scour sluice Sluice gate that controls flow through a scour outlet.

Scour tunnel (scour culvert deprecated) Outlet tunnel for discharging water and sediment from a reservoir, usually an impounding reservoir, at the lowest possible level.

Top water level (1) Lowest overflow level of crest of a spillway.

Top water level (2) (normal top water level, full supply level deprecated) Highest water level in a reservoir in which the overflow level is controlled, exclusive of any provision for flood storage.

Spillways and energy-dissipating works

Bellmouth spillway (shaft spillway, morning glory spillway deprecated) Spillway that consists of a vertical shaft with a bellmouth inlet.

Siphon spillway Spillway that operates on the siphon principle.

Ski jump spillway Spillway with a chute at the bottom of which there is a flip bucket that throws water up into the air to dissipate energy at a safe distance from the bottom of the spillway.

Baffle block (impact block deprecated) Block constructed in a channel or stilling basin to increase turbulence and thereby dissipate the energy of water flowing at high velocity.

Chute block Baffle block constructed in the chute of a spillway.

Energy dissipator Device constructed in a waterway to reduce substantially the kinetic energy of fast-flowing water.

Plunge pool Deep water-filled basin for dissipating the energy of a free-trajectory jet entering it from a spillway.

Stilling basin (water cushion, stilling pool deprecated) Basin constructed downstream of a structure to dissipate the energy of fast-flowing water and to protect the bed and banks from erosion.

Streamflow

Afflux The rise in water level in an open channel immediately upstream of and due to an obstruction. More usually the obstruction takes the form of a weir or a gate structure.

Air line correction See main entry Current metering.

Approach channel The reach of an open channel upstream of a gauging station in which suitable flow conditions have to be established to ensure correct stream gauging.

Backwater The afflux upstream from a given location on an open channel resulting from any impedances offered to flow. It is caused by channel storage for which the reservoir properties vary with the depth of flow at the given location.

Backwater curve The profile of a water surface in an open channel which is concave upwards along an open channel from the raised surface at an obstruction or confluence to the point upstream at which the flow is at normal depth. Surface profiles which are non-uniform with respect to distance upstream or downstream of a given location sometimes are referred to as backwater curves but this use of the term is deprecated. (See also drawdown curve).

Baffle A straightening vane or a guide vane in an open channel. Also a wall, or blocks, placed in an open channel downstream of a structure to dissipate energy or to improve the stream velocity distribution.

Bankfull stage The stage at which the water flowing in an open channel just begins to overflow the tops of the banks which contain the water flow. Quite often, the bankfull stage is the point at which the water flow becomes a flood.

Bed load See Sediment transport.

Bed material load See Sediment transport.

Bed slope The difference in elevation of a channel bed per unit horizontal distance in the direction of the water flow. Also known as the bottom slope, the slope usually is shown mathematically as positive downwards in the direction of the water flow.

Bottom slope See Bed slope.

Braided stream A stream pattern characterized by a wide and shallow open channel in which the flow passes through a number of small interlaced channels separated by shoals. Generally there is little or no meandering of the main channel, but meandering in the minor channels is usual. Erosion of the main channel banks is not usual but changes in the shoal profiles commonly occur. Characteristic of bed overload in strongly seasonal or semi-arid climates.

Cableway See Current metering.

Calibration tank A tank containing still water through which one or more current meters, suspended from a travelling gantry, are moved at a constant velocity for calibration of the meters. This method mirrors the field use of current meters where flowing water velocities are measured by means of statically held current meters. Trials are in hand for holding current meters statically in a calibration tank and allowing water to flow through the tank to activate the meters. This could be a more accurate method of calibration but there are uncertainties about the methods of assessing the water velocity through the tank. Also known as a rating tank.

Channel storage The volume of water flow contained in an open channel at a given instant. It is represented specifically by the mean depth of water in a defined reach of an open channel at the given instant.

Concentration time The time period over which sediment concentration is assessed in an open channel.

Contracted weir See main entry, Weir, flow measurement.

Control The physical properties of a cross-section or a reach of a natural or an artificial open channel, that govern the relation between stage and discharge at a location in the open channel.

Coriolis energy coefficient See Energy correction factor.

Crest The stage at which the free surface of water flowing in an open channel just begins to flow over a weir placed in the open channel. The term is used also to describe that part of the weir which allows the first flow of the water over the weir.

Crest stage gauge A gauge placed in an open channel to indicate peak stage.

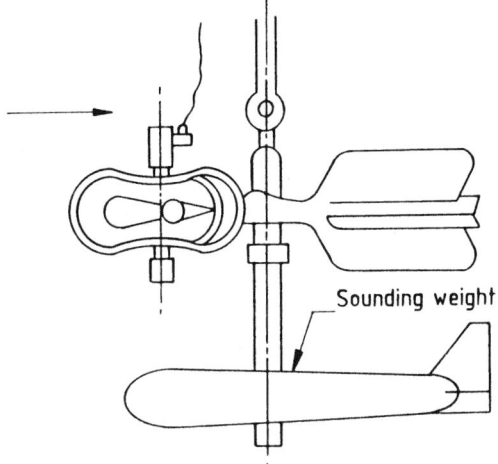

Key

→ is the direction of the flow.

Figure W21 Cup-type current meter.

Critical flow See Flow.

Current meter An instrument used for measuring water velocity. Basically the instrument is placed in the stream so that the velocity of the water passing the current meter is sensed by mechanical or electromagnetic means. Typical meters are shown in Figures W21 and W22, but there are design differences according to differing manufacturers. Some form of weighted sinker is incorporated into, or attached to, current meters used in open channel conditions in order to stabilize the position of the current meter in the stream.

Darcy–Weisbach formula A formulae devised for pipeflow considerations by the German Engineer Julius Weisbach in 1845 and modified by the French Scientist H.P.G. Darcy in 1854. This formulae can be used for open channel flow when it is necessary to consider channel friction coefficient and the Reynolds number for assessing the conveyance, the degree of rugosity being an important factor in some channel flow calculations. The user is referred to standard hydraulics texts for details of the Darcy–Weisbach formulae and its applications.

Discharge The volume of water flowing through an open channel cross section in unit time. It must not be confused with flow. In an open channel the determination of discharge, almost invariably, is based on the application of the relationship $Q = AV$ where Q is the volume of discharge ($m^3 s^{-1}$), 2A is the area of the wetted cross section at the point of measurement (m^2) and 2V is the mean stream velocity through the measuring cross section ($m s^{-1}$).

Double float See main entry Float.

Drawdown See Surface drawdown.

Drawdown curve The profile of the water surface in an open channel when the surface slope exceeds the bed slope due to surface

drawdown. An increase in the bed slope, or an abrupt drop in the bed level, will cause the drawdown curve to peter out further downstream but the surface profile along the open channel, from the point of increase or drop at the bed, will be convex upwards in an upstream direction and concave upwards in a downstream direction.

Drowned flow See Flow.

Duration curve A graphical representation of the percentage of a finite time period during which particularly measured or assessed values of a specific parameter are exceeded. For open channels the most commonly used duration curve is the flow duration curve.

Energy correction factor A factor proposed by the French scientist G. Coriolis in 1836 to take account of velocity head computational inaccuracies caused by non-uniform velocities in open channels.

Energy gradient The slope of the energy head line along an open channel. It is of greatest significance when the bed slope is steep.

Error Fundamentally, an error can be a random error or a systematic error. A random error is the component of an error of measurement which, in the course of a number of measurements of the same measurand varies in an unpredictable way; it is not possible to correct for a random error. A systematic error is the component of the error which remains constant or which varies in a predictable way; the error and its cause may be known or unknown. Also, there may be spurious errors which invalidate a measurement. Generally they are caused by the incorrect recording of one or more significant digits or by instrumental malfunction. Errors must not be confused with uncertainties (see Uncertainties in measurement, below).

Float gauge See main entry, Gauge.

Flow In the context of water, flow is the movement of a substance that is neither a solid nor a gas, which is practically incompressible, which offers insignificant resistance to change of shape and which flows freely. It can be anything from water that is sensibly pure to water with sediment. It is important that the word flow used on its own describes movement without any quantitative or qualitative connotation. In hydrometry the types of flow encountered most commonly are as follows.

- Free surface flow, which is gravitational and occurs under open atmospheric conditions;
- Critical flow in open channels, which occurs when the specific energy is a minimum for a given discharge and, under this condition, the Froude number is unity and small surface disturbances cannot travel upstream. As critical flow is a given state during discharge, there is an associated critical velocity. When the Froude number is greater than unity, supercritical flow pertains. When the Froude number is less than unity, subcritical flow pertains (in which case, small surface disturbances can travel upstream).
- Steady flow in open channels, which is the condition when the discharge does not change in magnitude with respect to time. If there is a change, then unsteady flow pertains.
- Uniform flow in open channels, which pertains when the water depth and the water velocity remain constant along the open channel: this can only happen in open channels of constant cross-section.
- Modular flow in open channels, which occurs when a structure is placed in the open channel and the upstream level of the water flow through or over the structure is independent of the downstream water level for a given discharge. If the downstream water level

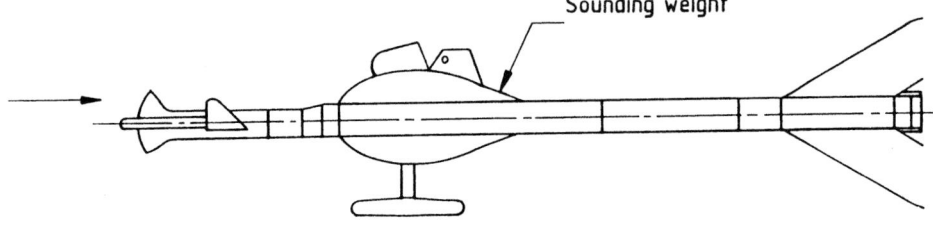

Key

→ is the direction of the flow.

Figure W22 Propeller-type current meter.

does affect the flow at the structure, the flow is known as non-modular or drowned flow. The state of flow when non-modular conditions begin to exist is known as the modular limit.
See also Viscosity.

Flow duration curve A graphical representation of the percentage of the time that discharge exceeds a given value at a given open channel location. The usual notation is for the percentage time to be on the horizontal axis of the graph and the discharge values to be in ascending order of magnitude on the vertical axis. Strictly speaking, it should be called a 'discharge duration curve' but the historically used 'flow duration curve' is accepted and understood by users.

Flume See main entry, Flume.

Free surface flow See Flow.

Friction coefficient A coefficient used to calculate the energy gradient caused by friction in open channel flow.

Guide vane See Baffle.

Hagan formula In 1876 G.H.L. Hagan, a German engineer, promulgated a formula based on the Chézy formula. However, although a scientifically sound formula, it was superseded in 1889 by the Manning formula.

Hook gauge See main entry, Gauge.

Hydraulic jump The sudden transition in an open channel from supercritical flow to subcritical flow. Upstream of the jump the velocity and the depth are respectively greater than and less than the critical depth. Downstream of the jump the velocity and the depth are respectively less than and greater than the critical depth.

Hydraulic mean depth A numerical value, expressed as a length, of the area of the cross section of water flowing in an open channel divided by the width of the open channel at the water surface. Not to be confused with hydraulic radius.

Hydraulic radius A numerical value, expressed as a length, of the area of the cross section of water flowing in an open channel divided by the length of the wetted perimeter at that cross section. Not to be confused with hydraulic mean depth.

Hydrometry The science of the measurement of water including the methods, techniques and instrumentation used. Effectively, this is the measurement of any or all of the water elements contained in the hydrological cycle including precipitation (rainfall and snowfall), evaporation, transpiration, soil moisture, ground water and runoff. The use of the term 'hydrometry' is not related in any way to the hydrometer used for measuring the relative densities of liquids.

Inclined gauge See Gauge.

Mass discharge curve A graphically plotted curve in which the cumulative volume of flow mass in an open channel is shown against time. Known also as a cumulative volume curve.

Mean section segment The area in an open channel cross section bounded by two consecutive verticals the bed and the water surface (Figure W23).

Mid-section segment In an open channel cross section the area at a vertical (Figure W23) multiplied by one-half of the distance between the preceding and succeeding verticals.

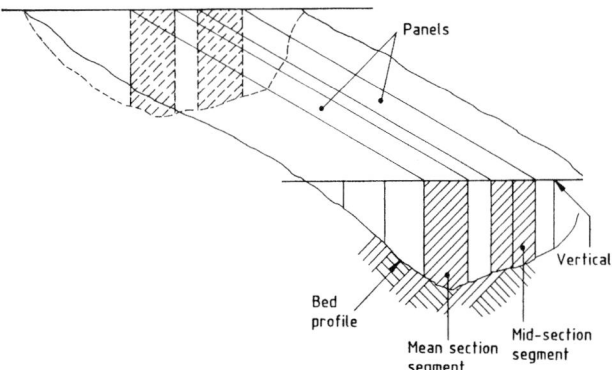

Figure W23 Geometrical definitions.

Modular flow See Flow.

Modular limit See Flow.

Moving boat method See main entry, Current metering.

Nappe The jet formed by the flow of a weir.

Non-modular flow See Flow.

Non-uniform flow See Flow.

Normal depth The depth from the water surface in an open channel to the thalweg in conditions of uniform flow, depending upon the bed slope, the roughness, the channel geometry and the discharge.

Open channel The longitudinal boundary surface in a waterway consisting of the bed and banks within which the water flows with a free surface. The implication is that the water flows throughout at atmospheric pressure unless the depth is sufficiently large to affect significantly the pressure at the lower depths.

Peak stage The maximum instantaneous stage in an open channel during a given time period. Usually it is the stage at which maximum discharge occurs but there can be circumstances when this is not so.

Piezometric head In an open channel this is the elevation of the free surface plus any pressure head. Thus, at any cross-section, it is the total head (see Energy head) above a given datum minus the velocity head at that cross section.

Point gauge See Gauge.

Potential head Head at a specified datum equal to the elevation of the liquid above that datum.

Pressure head Head due to pressure, expressed as the ratio of the pressure to the density of the liquid.

Radioactive tracer See Tracer.

Ramp gauge See main entry, Gauge.

Random error See error.

Rating curve See Stage–discharge relation.

Rating tank See Calibration tank.

Reach A length of open channel between two defined cross-sections.

Remote sensing In a general sense, remote sensing is the acquisition of data or information on some property of an object or a phenomenon by a sensor which is significantly remote from the object or a phenomenon. The remoteness is a matter of degree according to the parameters under consideration, but in hydrometry the term implies that the sensor is mounted in an aircraft or in a space vehicle; the use of the term is not recommended for the use of a sensor which, merely, is not in contact with the object or the phenomenon.

River See Stream.

Rod float See main entry, Float.

Roughness coefficient See Rugosity coefficient.

Rugosity coefficient A coefficient that characterizes the roughness of the wetted perimeter of a water-carrying channel or pipe and which is taken into account when computing the resistance to flow of the channel or pipe. For open channels Manning's n is an example of a rugosity coefficient, derived from the use of the Manning Formula.

Saltwater wedge The wedge-like intrusion of a large mass of salt (saline) water flowing in from the sea on a flood (inflowing) tide under the fresh water in a tidal waterway. The wedge occurs because of the differing densities of the saline and the fresh water and the difference may be exacerbated by differing temperatures of the two types of water. Consideration of the effects of a saltwater wedge is important in relation to sedimentation, the conveyance of solids in the waters and matters of water quality. In recent years, many coastal sewage outfalls have had to be modified, rebuilt or even abandoned because of a lack of previous understanding of the effects of saltwater wedges.

Sediment Solid particles, formed as a result of erosion, that are, or have been transported by water flow in open channels.

Sediment concentration The proportion by mass or volume of the dry sediment in a water-sediment mixture to the total mass or volume

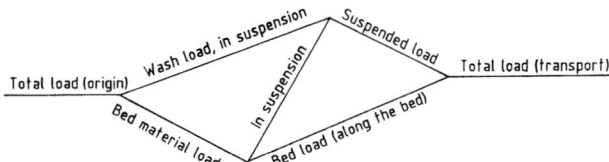

Figure W24 Definition sketch.

of the mixture. Usually it is quoted as a weight-to-volume ratio averaged over a specified time; for example, mg l^{-1} day^{-1}.

Sediment transport The movement of solids transported in any way by a flowing liquid. For water in open channels the progression of the original total sediment load to the total sediment load transported is via bed material load, bed load along the bed, wash load in suspension and suspended load, as shown in (Figure W24).

Sensor A device that responds to a physical or chemical stimulus. The output is often in the form of a numerical measurement. Not to be confused with a transducer.

Sill Sometimes spelled as 'cill', in open channels. This is the effective bed surface or flat invert of a structure, such as a broad-crested weir, built into the open channel. Its existence provides a stable bed datum for depth and head measurements and also a clearly defined lowest point of flow through the structure.

Single-gauge station See main entry, Gauging station.

Slope–area method An indirect method of discharge determination in a reach which is based on the surface slope, the reach roughness, the wetted perimeters and the flow areas of the various cross-sections in the reach.

Specific energy In an open channel this is the sum of the elevation of the free surface above the channel bed and the velocity head at a cross-section based on the mean velocity at that section. It differs from energy head in that it is the summation of measurable quantities without taking into account any energy correction consideration.

Speed of flow The ratio of the distance covered by a body of water, moving in a specified direction, to the time taken to cover that distance (not to be confused with velocity).

Spurious error See Error.

Staff gauge See main entry, Gauge.

Stage The elevation of the free surface of a stream, lake or reservoir relative to a specified datum. For practical purposes it can be known as the water level.

Stage-discharge relation A curve, equation or table that expresses the relation between the stage and the discharge in an open channel at a given cross-section, for a given condition of steady, rising or falling stage (Figure W25).

In determining the discharge values using observations of stage and velocity – usually at a gauging station – the meticulous building up and monitoring of the stage–discharge relation is critical to accurate assessments of the discharges in a watercourse, over a period of time. The discharge values are determined using the equation $Q = AV$ where Q is the discharge (m^3 s^{-1}), A is the wetted cross-sectional area (m^2) and V is the mean velocity in the cross-section (m s^{-1}). Values of A are calculated using measurements of the cross-sectional profile and of stage, and values of V are determined using current metering or other means.

The calculation of the stage–discharge relation can be arithmetic, graphical, computer-aided or by least-squares techniques.

Stage recorder See Water level recorder.

Static head Pressure head of a stationary liquid.

Steady flow See Flow.

Stilling well See main entry, Gauging station.

Straightening vane See Baffle.

Stream The water flowing in an open channel. A river is the stream of water in a natural open channel. These definitions are the strictest ones in a hydraulics sense, but it is quite common for a stream to be considered as a small river; however, this is not truly correct. See also Watercourse and main entry, Hydrology.

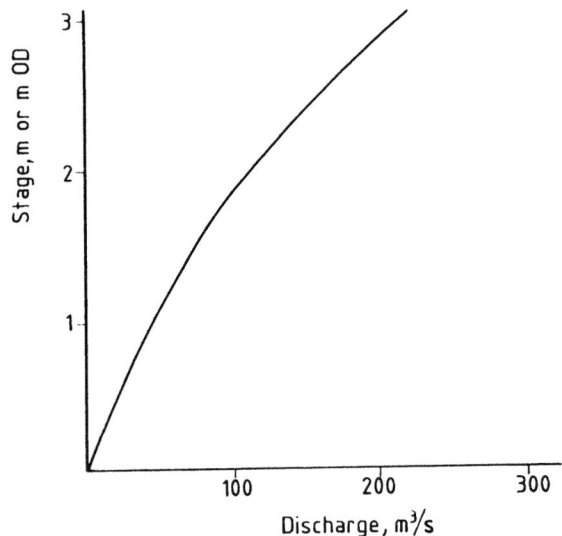

Figure W25 Stage–discharge relation.

Streamflow This is a term used often to describe the movement of water but flow is a quite sufficient term.

Strickler formula This is one of several formulae setting out a modification of the Manning formula. It was promulgated by the German Engineer, A. Strickler, in 1923 but it never succeeded in ousting the simple, down-to-earth, Manning formula in general usage.

Sub-surface float See main entry, Float.

Subcritical flow See Flow.

Supercritical flow See Flow.

Surface drawdown The local lowering of the water surface in an approach channel caused by the acceleration of the flow passing over an obstacle or though a control.

Surface float See main entry, Float.

Surface runoff This is the proportion of precipitation in the form of rainfall or ultimately melted snow, which runs off over ground via a watercourse to some outlet such as the sea. It does not include rainwater which percolates through the solid surface to find its way into ground storage, but it does include percolated water which eventually finds its way into a watercourse.

Surface slope The difference in elevation of a water surface in an open channel per unit horizontal distance along a length of the channel, measured in the direction of flow.

Systematic error See Error.

Thalweg (Sometimes spelled 'TALWEG'). There are two interpretations of this German term:

● A line in plan joining the deepest points of a stream bed, a channel or a valley: thus a line on a plan;

● the line of greatest depth, and thus the lowest water thread, along the stream channel: thus a line that can be shown on a plan or as a line on a longitudinal section.

Total head Sum of velocity head, pressure head and potential head.

Total load See Sediment transport.

Tracer An ion, compound or radionuclide introduced into an open channel flow system to follow the behavior of some component of that system. Used in dilution gauging, tracers can be of several forms with elemental or proprietary names, but they fall into three main groupings: chemical, fluorescent and radioactive. Potential users of tracers are advised – in some cases, obliged – to check that a tracer which they intend to use is not a prohibited substance in the geographical location of intended use.

Transducer A device that responds to a phenomenon and produces a signal which is a function of one or more characteristics of the phenomenon. Not to be confused with a sensor.

Transition flow See Flow.

True value The value which characterizes a quantity perfectly defined in the conditions which exists when the quantity is considered. It is an ideal value which can be determined only if all causes of measurement error are eliminated.

Turbulent flow See Flow.

Twin-gauge station See main entry, Gauging station.

Uniform flow See Flow.

Unsteady flow See Flow.

Velocity The speed of flow past a point in a specified direction (not to be confused with speed of flow without a direction qualification).

Velocity head This is the kinetic energy of water flow and it is the head, due to velocity, equal to the vertical height through which a fall under the influence of gravity alone would give the water a velocity equal to its actual velocity, expressed as the square of the mean velocity divided by twice the acceleration due to gravity: that is

$$\frac{\bar{V}^2}{2g}$$

Velocity rod See main entry, Float.

Vertical In an open channel the vertical line in which velocity measurements or depth measurements are made.

Vertical gauge See main entry, Gauge.

Viscosity In layman's term this is the degree of 'thickness' and 'stickyness' of water. The extent of viscosity in an open channel, together with gravitational forces, relative to inertial forces, is a factor in determining the type of flow. Laminar (or smooth) flow pertains when the viscosity is strong enough to overcome inertial forces. Conversely, turbulent flow pertains when the inertial forces outweigh the viscosity. The flow pattern between laminar and turbulent is known as transitional flow. The Reynolds number can give an evaluation of the kinematic viscosity in a particular state of flow, but viscosity can be present in a dynamic form. In general, kinematic viscosity is equal to the dynamic viscosity divided by the mass density of the water.

Wash load See Sediment transport.

Water level See Stage.

Water level recorder A device that records automatically, either continuously or at regular time intervals, the water level as sensed by a sensor.

Watercourse A general term, used often in statutory or legal documents, given to describe a natural or artificial open channel, a pipe, a conduit, a river, a canal or other type of continuous water-carrying feature or structure.

Weir See main entry, Weir flow measurement.

Wetted perimeter This has two interpretations:

- The length of wetted contact between a stream of flowing water and its containing open channel, at a cross section, measured in a direction normal to the flow.
- The wetted boundary of an open channel at a specified section.

Wire weight gauge See main entry, Gauge.

Tidal

Density current The phenomenon of gravity flow of a liquid relative to another liquid, or of relative flow within a liquid medium due to difference in density.

Diurnal inequality

1. The difference in heights and durations of the two successive high waters or two successive low waters of each day.
2. The difference in speed and direction of the two flood currents or the two ebb currents of each day.

Duration of tide The time taken for the completion of one tidal cycle; usually 12.42 h for a semi-diurnal tide or 24.84 h for a diurnal tide.

Ebb current The seaward movement of water along a tidal channel.

Ebb tide The occurrence of falling water surface of a tide.

Ebb volume The total discharge of an ebb tide.

Flood current The landward movement of water along a tidal channel.

Flood tide The occurrence of rising water surface of a tide.

Flood volume The total discharge of a flood tide.

High water The state of tide when water is highest for any given tidal cycle.

Low water The state of the tide when water is lowest for any given tidal cycle.

Neap tide Tide of small amplitude occurring twice during a lunar month, near the time of quadrature of the Moon with the Sun, i.e. when the resultant tractive force acting upon the Earth is at a minimum.

Spring tide Tide of large amplitude occurring twice during the lunar month when the resultant tractive force of the Sun and the Moon acting upon the Earth is at a maximum.

Tidal amplitude One half of the difference in height between consecutive high water and low water, hence half the tidal range.

Tidal cycle A period that includes a complete set of tide conditions or characteristics, such as a tidal day.

Tidal day The interval between two upper transits of the Moon over a local meridian, approximately 24.84 h.

Tidal limit (of a river) The location beside a river at which the rise and fall of water at equinoctial spring tide are just perceptible. If there is a dam or sluice then this may be the tidal limit.

Tidal prism The volume of water that flows into a tidal channel and out again during a complete tide, with the movement of the tide, excluding any upland discharges.

Tidal range The difference in level between high water and low water of a tide. The range is specific to a particular tide if consecutive high water and low water are used; otherwise, the range can refer to extremes of high water and low water over any specified period of time.

Tidal water Any part of the sea or of a river water within the ebb and flow of the equinoctial spring tides.

Tide The periodic rise and fall of water due principally to the gravitational attraction of the Sun and the Moon.

Turbines

Francis turbine Reaction turbine in which water enters the turbine runner radially inwards and exists axially.

Impulse turbine Water turbine in which kinetic energy only is converted into rotary energy.

Kaplan turbine Propeller turbine with vertical shaft and variable pitch blades.

Mixed flow turbine Inward flow reaction turbine in which the water acts on the turbine runners both radially and axially.

Propeller turbine Reaction turbine in which water enters and leaves the turbine axially.

Pelton wheel Impulse turbine with a turbine runner that comprises a set of double cup-shaped buckets rotated by the impact of a jet or jets of water on the buckets.

Reaction turbine Water turbine in which both pressure and kinetic energy are converted into rotary energy.

Uncertainties in measurement

Absolute error of measurement The result of a measurement minus the conventional true value of the measurand.

The term relates equally to

- the indication;
- the uncorrected result;
- the corrected result.

The known parts of the error of measurement may be compensated by applying appropriate corrections. The error of the corrected result can only be characterized by an uncertainty.

'Absolute error', which has a sign, should not be confused with 'absolute value of an error', which is the modulus of an error.

Accuracy A qualitative expression for the closeness of a measured value to the true value. The quantitative expression of accuracy should be in terms of uncertainty. Good accuracy implies small random errors and systematic errors.

Arithmetic weighted mean; weight average ($\bar{x}_w$) The sum of the products of each value and its weight of measurement divided by the sum of the weights of measurement (which can be positive or zero). It is given by the following formula:

$$\bar{x}_w = \frac{\sum\limits_{i=1}^{n} w_i x_i}{\sum\limits_{i=1}^{n} w_i}$$

Average value ($\bar{x}$) Arithmetic mean of n readings of the value x. The average value is calculated using the following formula:

$$\bar{x} = \frac{1}{n} \sum_{i=1}^{n} x_i$$

Calibration The process of comparing the response of a measuring device with a calibrator or a measuring standard over the measurement range.

Calibration hierarchy The chain of calibrations which link or trace a measuring device to a primary standard.

Confidence level The probability that the true value will lie between the specified confidence limits, assuming negligible systematic error. This is generally expressed as a percentage.

Confidence limits The lower and upper limits within which the true value is expected to lie with a specified probability, assuming negligible systematic error.

Deviation The difference between the value of a quantity and a standard of reference value. Particularly in statistics, the reference value is frequently the arithmetic mean of a series of measurements.

Elemental error A random error or a systematic error associated with a single source or process in a chain of sources or processes.

Experimental standard deviation (s) For a series of n measurements of the same measurand, the parameter characterizing the dispersion of the results and given by the following formula:

$$s = \sqrt{\frac{\sum\limits_{i=1}^{n} (x_i - \bar{x})^2}{n - 1}}$$

where x_i is the result of the ith measurement and $\bar{x}$ is the arithmetic mean of the n results considered.

The experimental standard deviation should not be confused with the population standard deviation σ of a population of size N and of mean m, given by the following formula:

$$\sigma = \sqrt{\frac{\sum\limits_{i=1}^{n} (x_i - m)^2}{N}}$$

If the series of n measurements is considered to be a sample of a population, s is an estimate of the population standard deviation.

Experimental standard deviation of the mean $s(x)$ Estimate of the standard deviation of the arithmetic mean $\bar{x}$ with respect to the mean m of the overall population. It is given by the following formula:

$$s(\bar{x}) = \frac{s(x)}{\sqrt{n}}$$

Experimental variance (s^2) A measure of the scatter or spread of a distribution. It is estimated by calculating the sum of the squares of deviations of measurements about the mean, divided by the number of degrees of freedom, i.e.

$$s^2 = \frac{\sum\limits_{i=1}^{n} (x_i - \bar{x})^2}{n - 1}$$

Frequency distribution The relationship between the measured values of a variable and their frequencies of occurrence.

Method of least squares A technique used to compute the coefficients of the equation when a particular form of equation is chosen for fitting a curve to data. The principle of the method of least squares is the minimization of the sum of squares of deviations of the data from the curve.

Normal distribution; Laplace-Gauss distribution The probability distribution of a continuous random variable x such that the probability density is

$$f(x) = \frac{1}{\sigma\sqrt{(2\pi)}} \exp\left[-\frac{1}{2}\left(\frac{x - m}{\sigma}\right)^2\right]$$

where m is the arithmetic mean and σ is the standard deviation of the normal distribution.

Number of degrees of freedom (v) In general, the number of observations minus the number of parameters. For example, the standard deviation is said to have $(n - 1)$ degrees of freedom because, for the estimation of the mean, it is necessary to use one degree of freedom.

Outlier An observed value in a set of data which appears to be inconsistent with the remainder of the set of data.

Population Totality of items under consideration.

Precision The closeness of agreement between the results obtained by applying the experimental procedure several times under prescribed conditions. The smaller the random part of the experimental errors which affect the results, the more precise is the procedure.

Random error The component of the error of measurement which, in the course of a number of measurements of the same measurand, varies in an unpredictable way. It is not possible to correct for random error (Figure W26).

Random uncertainty (U_r) The component of uncertainty associated with a random error. Its effect on the mean value can be reduced by taking many measurements. (Figure W26).

Regression The process of quantifying the dependence of one variable on one or more other variables. Regression is a procedure for determining the unknown constants of a proposed model in such a manner that predictions from the model are as close as possible to the data in some way. Often 'as close as possible' is taken to mean that the sum of squares of the deviations is a minimum.

Many of the available computer programs suitable for curve fitting have the word 'regression' in the title. For the purposes of this contribution 'regression' and 'least squares' may be regarded as synonyms.

Residual variance (s_R^2) The square of the standard error of estimate.

Resolution A quantitative expression of the ability of an indicating device to distinguish meaningfully between closely adjacent values of the quantity indicated.

Sample One or more items taken from a population and intended to provide information on the population, and possibly to serve as a basis for a decision concerning the population or the process which produced it.

Sample size (n) The number of items which are to be included in the sample.

Sensitivity coefficient, θ_x; influence coefficient, θ_x The ratio of the change in a result R to a change in an input parameter x, i.e.

$$\theta_x = \frac{\Delta R}{\Delta x}$$

In relative terms, this becomes

$$\theta_x = \frac{\Delta R}{R} \bigg/ \frac{\Delta x}{x}$$

Spurious errors Errors which invalidate a measurement. They generally have a single cause, such as the incorrect recording of one or more significant digits or the malfunction of instruments (Figure W26).

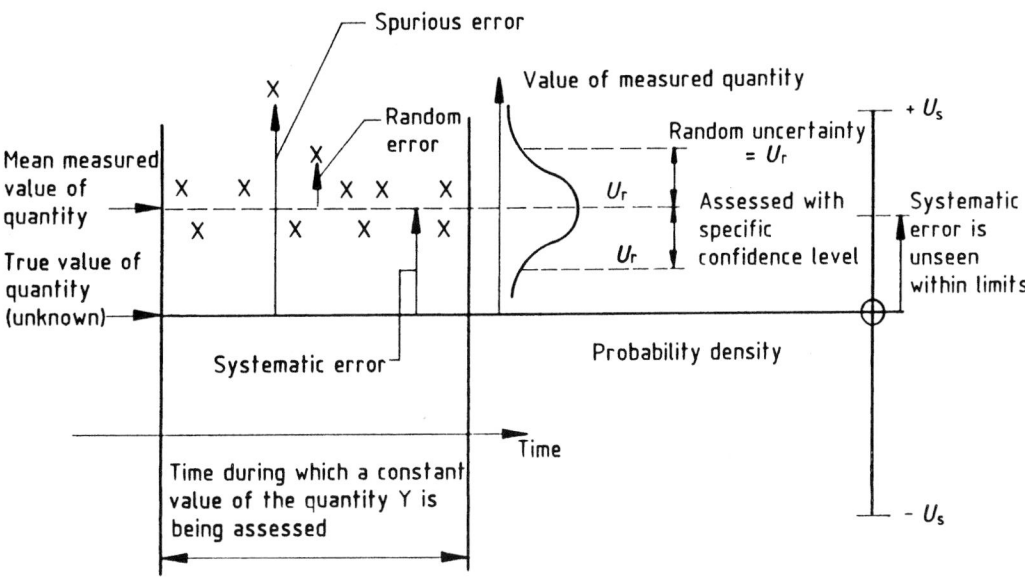

Figure W26 Diagram illustrating the terms relating to errors and uncertainties.

Standard error of estimate; residual standard deviation (s_R) The measure of dispersion of the dependent variable (output) about the least-squares line obtained by curve fitting or regression analysis. For a curve based on n data points and for which the equation has a number, k, of coefficients, the standard error of estimate is calculated as follows:

$$s_R = \sqrt{\frac{\sum\limits_{i=1}^{n} (y_i - \bar{y})^2}{n - k}}$$

This equation is similar to the expression for standard deviation except that the curve-fit value y_i replaces the mean value $\bar{y}$ and k replaces 1.

Student's t distribution The distribution of the deviations of the mean values of the samples from the population mean expressed as a proportion of the sample standard deviation (the samples being taken from normal distributions). It is used to set the confidence limits of the population mean, in particular in cases where the mean has been estimated from small samples. It is obtained from tables giving the number of degrees of freedom and the confidence level, where

$$t = \frac{\bar{x} - \mu}{s/\sqrt{n}}$$

where μ is the population mean.
Example:

$$(U_r)_{95} = t_{95}s$$

where $(U_r)_{95}$ is the random uncertainty at the 95% confidence level and t_{95} is the appropriate value of Student's t.

Systematic error The component of the error of measurement which, in the course of a number of measurements of the same measurand, remains constant or varies in a predictable way. Systematic errors and their causes may be known or unknown (Figure W26).

Systematic uncertainty (U_s) The component of uncertainty associated with systematic error. Its effect cannot be reduced by taking many measurements.

True value The value which characterizes a quantity perfectly defined in the conditions which exist when that quantity is considered. It is an ideal value which can be determined only if all causes of measurement error are eliminated.

Uncertainty (U) The interval within which the true value of a measured quantity can be expected to lie with a stated probability: it is given as $\pm ts_Y$, with the value of t equal to that corresponding to the chosen probability (Figure W26).

Weight of measurement (W_i) The number which expresses the degree of confidence in the result of a measurement of a certain quantity in comparison with the result of another measurement of the same quantity.

Waste water and water supply

Activated sludge Biological mass produced from sewage by growth of bacteria and other microorganisms in the presence of dissolved oxygen.

Activated sludge process Biological sewage treatment process in which a mixture of sewage and activated sludge is subjected to aeration.

Absorption Process in which molecules of a fluid are taken up by capillary, osmotic, chemical or solvent action.

Adsorption Process in which a solid takes up a fluid by surface adhesion.

Aerobic action Biological process in the presence of oxygen.

Anaerobic action Biological process in the absence of oxygen.

Biochemical oxygen demand (BOD abbrev.) Mass concentration of dissolved oxygen consumed under specified conditions by biological oxidation of matter in water.

Biological filter Bed of inert material that promotes or assists natural aerobic degradation of sewage.

Chemical oxygen demand (COD abbrev.) Mass concentration of oxygen equivalent to the amount of chemical oxidant consumed under specified conditions by dissolved or suspended organic matter.

Chlorination Process of adding either chlorine or chlorine compounds to water or sewage.

Combined system Drainage system or sewerage system in which foulwater, wastewater and surface water are conveyed by the same pipeline.

Compensation water Statutory amount of water discharged to a river or stream from an impounding reservoir, borehole or other abstraction point.

Consent standard Limitations on physical, biological and chemical content of final effluent, trade effluent and sometimes on rate of discharge, included in a discharge consent.

Desalination Treatment of seawater or brackish water to remove salts.

Digested sludge Sludge from sewage that has been subjected to aerobic action or anaerobic action to reduce concentration of organic matter and pathogens.

Discharge consent Statutory approval to discharge final effluent, trade effluent, surface water or stormwater to inland or tidal waters.

Dry-weather flow (DWF abbrev.) Rate of flow foulwater, wastewater and infiltration, if any, in a sewer in specified dry weather conditions.

Effluent Liquid discharged from a given process.

Eutrophication Enrichment of water by nutrients, especially compounds of nitrogen and phosphorus, which accelerates growth of algae and higher forms of plant life.

Final effluent (sewage effluent deprecated) Effluent that is discharged from a sewage treatment works after completion of treatment of sewage.

Fluoridation Addition of compound that contains fluorine to a drinking water supply to maintain the fluoride ion concentration within agreed limits.

Gross solids Solids in raw sewage of such a size that they affect treatment adversely if not disintegrated or removed at the start of the process.

High-rate biological filter (high-rate filter deprecated) Biological filter that consists of coarse inert material with a hydraulic loading that exceeds $3 \, m^3 \, m^{-3} \, day^{-1}$ or an organic loading that exceeds 2.0 kg of biochemical oxygen demand per m^3 per day.

Influent Liquid received for a given process.

Ion exchange softening (base exchange softening deprecated) Water softening in which a material is used that is able to exchange the soluble salts of calcium and magnesium for another salt, usually of sodium.

Land treatment Treatment of final effluent by irrigation.

Leachate Liquid, usually polluted with soluble substances as a result of percolating through or being exuded from solid waste.

Microstrainer Cup screen that removes fine screenings from water.

Oxidation ditch Endless circuit in which sewage is treated by an activated sludge process.

Oxygen sag curve Graphical representation of the concentration of dissolved oxygen along a river or estuary.

Ozonation (ozonization deprecated) Process of adding ozone to water, wastewater or gases.

Preliminary treatment (of sewage) The removal or disintegration of gross sewage solids and the removal of grit. It can include the removal of grease and oil from sewage prior to sedimentation, and pre-aeration and neutralization.

Primary treatment (of sewage) The stage of treatment usually involving the removal of the bulk of solids capable of settling. In the case of sewage it follows immediately after preliminary treatment.

Rapid gravity filter Construction for removing impurities from water that comprises a tank containing filter media through which water flows vertically by gravity at rates between 3 and $10 \, m \, h^{-1}$ with associated pipework and facilities for backwashing.

Reverse osmosis Flow of water through a membrane from a more concentrated to a less concentrated solution as a result of application of pressure to the former in excess of the osmotic pressure difference between the two.

Schmutzdecke Biological layer formed on sand in a filter by algae and microorganisms.

Secondary treatment Treatment of sewage by biological processes, such as biological filtration and settlement, or activated sludge, as distinct from preliminary treatment (grit separation, comminution, etc.), primary treatment (primary sedimentation) and tertiary treatment (effluent polishing by sand filtration, microstraining, etc.).

Slow sand filter Construction for removing impurities from water that comprises an open tank containing filter media, on which matter is retained by a schmutzedecke, through which water flows vertically by gravity at a rate of the order of $0.3 \, m \, h^{-1}$ into an under drain collection system, and from which the filter media is periodically removed for cleaning.

Sludge dewatering Process whereby wet sludge, usually conditioned by a coagulant, has its water content reduced by physical means.

Stormwater (storm sewage deprecated) Surface water from heavy rainfall combined with wastewater diverted from a sewer by a storm overflow.

Stormwater overflow Device, on a combined or partially separate sewerage system or at a sewage treatment works, which relieves the system of excess flow.

Stormwater retention tank Tank or reservoir for storage of stormwater that cannot be accepted immediately by a sewage system, sewage treatment works or watercourse.

Suspended solids (SS abbrev.) Solids in suspension in a liquid measured under specified conditions.

Tertiary treatment (polishing deprecated) Third stage of sewage treatment that further reduces suspended solids and biochemical oxygen demand.

Top water level (TWL abbrev.) Highest level of a body of water, in or behind a water retaining structure, before it overflows or ceases to enter.

Water conservation Preservation, control and development of water resources by storage and other means, and prevention of pollution.

Water hardness Property of water that manifests itself by resistance to the development of lather with soap, due mainly to the presence of calcium and magnesium ions.

Water quality

Algae A large group of single- or many-celled organisms, including so-called cyanobacteria, which usually contain chlorophyll or other pigments. They are usually aquatic and capable of photosynthesis.

Alpha factor In an activated sludge plant, the ratio of the oxygen transfer coefficient in mixed liquor to the oxygen transfer coefficient in clean water.

Bacteriophages A group of particular viral agents whose life cycle occurs in specific bacterial hosts.

Bacteria A large group of microscopic, metabolically active, single-cell organisms with a dispersed (not discrete) nucleus, mostly free living, and usually multiplying by binary fission.

Beta factor In an activated sludge plant, the ratio of the oxygen saturation value in mixed liquor to the oxygen saturation value in clean water at the same temperature and atmospheric pressure.

Biodegradability The susceptibility of an organic substance to biodegradation.

Biofilm (of a sand filter) The film, consisting of living, dead or moribund organisms, that forms on the surfaces of the medium in a slow sand filter or other biological filter.

Bioassay A technique for evaluating the biological effect, either qualitatively or quantitatively, of various substances in water by means of changes in a specified biological activity.

Biomass The total mass of living material in a given body of water.

Black water Wastewater and excreta from water closets excluding wastewater from baths, showers, handbasins and sinks.

Carbon adsorption/chloroform extraction (CCE) (obsolete) A procedure in which materials, predominantly organic, are absorbed from water onto activated carbon under specified conditions, and are subsequently extracted into chloroform, prior to analysis.

Coliform organisms A group of aerobic and facultatively anaerobic Gram-negative, non-spore-forming, lactose-fermenting bacteria which typically inhabit the large intestine of man and animals. Generally, apart from *E. coli*, many of them are able to survive and multiply in the natural environment.

Cyprinid Fish belonging to the family *Cyprinidae*, for example roach, rudd and carp, sometimes used as a biological indicator of water quality.

Dissolved organic carbon; DOC That part of the organic carbon in a water which cannot be removed by a specified filtration process.

Ecosystem A system in which, by the interaction of the different organisms present and their environment, there is a cyclic interchange of materials and energy.

Elutriation A conditioning process by which sludge is washed with either fresh water or plant effluent to reduce the alkalinity of the sludge, particularly by removing ammoniacal compounds, thereby reducing the amount of coagulant required.

Enteroviruses; enteric viruses A group of viruses which can multiply in the gastrointestinal tract of humans and other animals.

Equilibrium pH The thermodynamically stable pH value of a solution, or body of water, when equilibrium is attained, not only within the aqueous phase itself, but also between it and any other phases with which it may be in contact.

Escherichia coli (E. coli) An aerobic and facultatively anaerobic thermotolerant coliform organism which ferments lactose (or mannitol) at a temperature of 44°C with the production of both acid and gas, and which also produces indole from tryptophan. Its normal habitat is the large intestine of humans and warm-blooded animals. *E. coli* is usually not able to multiply in wastewater and polluted surface water.

Eukaryotic Descriptive of organisms whose cells have a visible and definite nucleus.

Euphotic zone The upper layer of a body of water where light penetration is sufficient to support effective photosynthesis.

Fecal streptococci Various aerobic and facultatively anaerobic species of streptococci which possess Lancefield's Group D antigen and which normally inhabit the large intestine of humans and/or animals. Their presence in water, even in the absence of *E. coli*, indicates fecal pollution.

Fungi A large group of heterotrophic organisms which usually form spores and have well defined nuclei, but lack photosynthetic material such as chlorophyll. Yeasts are single-celled fungi which reproduce by budding. Other fungi are multicellular and filamentous, e.g. *Fusarium* species which cause ponding on biological filters, and *Geotrichum* species, which cause bulking of activated sludge.

Halocline A layer in a stratified body of water in which the salinity gradient is at a maximum.

Gray water (sullage) Wastewater from household baths and showers, handbasins and kitchen sinks, but excluding wastewater and excreta from water closets.

Haloforms; triholomethanes (THM) Compounds in which three of the hydrogen atoms of the methane molecule have been substituted by chlorine, bromine or iodine atoms. They may be formed from organic matter in water which has been treated or disinfected by halogens (excluding fluorine or oxidants capable of releasing halogens).

Hazen number A number used to indicate the intensity of colour of water, the standard unit being the colour produced by a solution containing 1 mg of platinum per liter [in the form of hydrogen hexachloroplatinate(IV)], in the presence of 2 mg of cobalt(II) chloride hexahydrate per liter.

Heterotrophic bacteria Bacteria which require organic matter as a source of energy, in contrast to autotrophic bacteria.

Imhoff cone Conical transparent vessel, usually of capacity 1 l and graduated near its apex, used for determining the volume of settleable matter in waters.

Ionic balance The algebraic sum of the product of the molar concentration and ionic charge of each cationic and anionic species present. In all waters this sum must be equal to zero. Any deviation from zero, of the balance calculated from the actual analytical results, is an indication either of the incompleteness of the determination (some ions not determined) or errors in analysis.

Isokinetic sampling A technique in which the sample from a water stream passes into the orifice of a sampling probe with a velocity equal to that of the stream in the immediate vicinity of the probe.

Langelier index The value obtained by subtracting the saturation pH (pHs) from the measured pH of a water sample. pHs is the calculated pH that would be obtained if the water were in equilibrium with solid calcium carbonate.

Macrophytes Large water plants.

Mass balance The relationship between input and output of a specified substance in a defined system, for example in a lake, river or sewage treatment works, taking into account the formation or decomposition of that substance in the system.

Mesophilic microorganisms Those microorganisms whose optimum temperature for growth lies between about 20 and 45°C.

Mesotrophic water A water of intermediate nutrient status, naturally occurring or due to nutrient enrichment, between *oligotrophic* and *eutrophic* states.

Methemoglobinemia A condition of the blood which occurs in infancy due to methemoglobin excess when nitrites, formed in the gut mainly by bacterial reduction of ingested nitrates, become attached to hemoglobin, and interfere with oxygen uptake and transport, thus causing cyanosis.

Night soil Human wastes accumulated in a container and removed periodically. These used to be removed at night, hence the name.

Oligosaprobic A description of the zone in running water where mineralization is complete. The zone has abundant dissolved oxygen and can support a wide range of plants and animals, primarily photoautotrophic plants and oxygenous animals.

Oligotrophic Description of a body of water, poor in nutrients and containing many species of aquatic organisms, each of which is present in relatively small numbers. This body of water is characterized by high transparency, a high concentration of oxygen in the upper layer, and by bottom deposits which are usually colored in shades of brown and contain only small amounts of organic matter.

Organoleptic Descriptive of those attributes of water, for example color, taste, odor and appearance, which are perceptible by the sense organs.

Oxygen sag curve The curve resulting from plotting the concentration of dissolved oxygen against distance or time of flow in a river downstream from a source of pollution that has an oxygen demand.

Pathogen An organism capable of producing disease in a susceptible plant or animal, including humans.

Photoautotrophic bacteria Bacteria which obtain their energy from light, and whose sole source of carbon is inorganic, i.e. CO_2.

Photosynthesis The synthesis of organic matter from carbon dioxide and water in the presence of light by living organisms, employing photochemically reactive pigments, such as chlorophyll.

Phytoplankton Plants present in *plankton*.

Plankton Organisms drifting or suspended in water, consisting chiefly of minute plants or animals, but including larger forms having only weak powers of locomotion.

Plate count; colony count An estimate of the numbers of viable microorganisms (comprising bacteria, yeasts and moulds) in a given volume of water, obtained from the number of colonies which form in, or on, a given culture medium under specified conditions.

Plumbo-solvent Descriptive of a water which is able to dissolve lead from pipes and fittings.

Polychlorinated biphenyls; PCB By convention, a collective term for biphenyls having chlorine substituents. In practice, it also includes monochlorinated biphenyls. Many polychlorinated biphenyls are persistent in nature and accumulate in the food chain. Some of them have long-term adverse effects on living organisms.

Polylectrolytes Polymers having ionized groups, some types of which are used for coagulating colloidal particles and/or flocculating suspended solids.

Polynuclear aromatic hydrocarbons; PAH Organic compounds composed of two or more benzene rings where the adjacent rings share two carbon atoms; non-aromatic rings may also be present.

Protozoa A phylum of unicellular eukaryotic animals varying from simple uninucleate organisms to cell colonies or highly organized structures and with a considerable diversity of forms and nutrition.

Pycnocline A layer in a stratified body of water in which the density gradient is at a maximum.

Salinity (absolute); absolute salinity (S_a) The ratio of mass of dissolved material in seawater to the mass of seawater. In practice, this quantity cannot be measured directly and a practical salinity is defined for reporting oceanographic observations.

Salinity (practical); practical salinity (S) A dimentionless value which, for the purposes of checking water quality, may be regarded as an estimate of the concentration, in grams per kilogram, of the dissolved salts in seawater. It is defined algorithmically, in terms of the ratio (K_{15}) of the electrical conductivity of the sample, at 15°C and 1 atm, to that of a defined potassium chloride solution (32 436 6 g kg^{-1} of sample) at the same temperature and pressure.

***Salmonella* species** A group of aerobic and facultatively anaerobic, Gram-negative, non-spore-forming bacteria which can cause intestinal infections in humans and animals. *Salmonella* species are excreted in the feces of cases and carriers in human and animals, and may, therefore, occur in sewage and farm wastes. They are a common cause of food poisoning in human.

Salmonid (fish) Fish belonging to the family Salmonidae, for example Atlantic salmon, brown trout and char, often used as biological indicators of water quality.

Self-purification The natural processes of purification in a polluted body of water.

Sewage fungus An adherent growth, consisting of communities of filamentous bacteria (for example *Sphaerotilus natans*) and fungi (for example *Fusarium aqueductum*) and other species, with protozoa, which may occur in sewage treatment plants or in streams as a result of the discharge of incompletely treated sewage, effluent or industrial wastewater.

Sludge cake Dewatered sewage sludge discharged from a filter press or similar device, usually containing about 25 to 35% (*m/m*) dry solids.

Sulfite-reducing clostridia A large group of Gram-positive, anaerobic bacteria which form spores. Their natural habitat is the soil or the large intestine of humans and animals. Most species are saprophytic organisms in the soil. Their spores can survive for long periods in feces, soil, dust and water. Their presence in water can be used to detect remote or intermittent fecal pollution. They are able to reduce sulfites to sulfides.

Sullage Household wastewater, other than fecal and urinary wastes (see also Gray water).

Synergism The increase in intensity of an effect (chemical or biological) by one substance or organism, due to the presence of another substance or organism; the combined effect is greater than the additive effects of the separate substances or organisms.

Thermal water Water of a hot or warm spring.

Thermotolerant/fecal coliform organisms Coliform organisms which can grow and which have the same fermentative and biochemical properties at 44°C as they have at 37°C (see *Escherichia coli*).

Turbidity Reduction of transparency of a liquid caused by the presence of undissolved matter.

Viruses A large group of ultramicroscopic agents (20–300 nm in diameter) which essentially consist of nucleic acid encased in a protein shell. They reproduce only in living cells. Viruses can pass through filters which retain bacteria.

Zooplankton Animals present in *plankton*.

P.G. Holland and R.W. Herschy

Bibliography

BS EN 24006, 1993. *Flow in Closed Conduits, Vocabulary and Symbols*, British Standards Organization.
BS 6100, 1989–1992. *Glossary of Building and Civil Engineering Terms*, British Standards Organization.
CIWEM, 1993. *Handbook of UK Wastewater Practice: Glossary*, CIWEM.
ISO 772, 1996. *Hydrometry, Glossary of Terms*, International Standards Organization.
ISO 1000, 1992. *The Use of SI Units*, International Standards Organization.
ISO 4006, 1997. *Flow in Closed Conduits, Vocabulary and Symbols*, International Standards Organization.
ISO 6107, 1986–1993. *Water Quality: Glossary (in 8 parts)*, International Standards Organization.

WATER RESOURCES: EUROPE

In order to identify the most prominent European environmental problems, a meeting of European environment ministers was held at Dobris Castle near Prague in 1991. As a result of this meeting, work began by making a survey of the literature and analyzing international initiatives by, among others, the EU, UNECE, UNEP, OECD and the US Environment Protection Agency. This resulted in a list of 56 environmental problems in Europe of which the following 12 problems were considered the most significant.

- the management of water;
- waste management;
- climate change;
- stratospheric ozone depletion;
- urban stress;
- forest degradation;
- chemical risks;
- acidification;
- major accidents;
- the loss of biodiversity;
- tropospheric ozone and other photochemical oxidants;
- coastal zone threats and management.

Only the first of these problems, management of water, is briefly discussed here from information presented in the report of the study by the European Environment Agency.

Introduction

Human health and development are threatened in many places because of insufficient or poor-quality water. Flooding is a serious problem in some countries, and destruction of aquatic habitats by channelization, rough maintenance schemes and damming of rivers can lead to an overall impoverishment of native plant and animal species.

By tradition, most evaluations of the quality of the aquatic environment have been based on measurements of a set of concentrations, speciations and physical partitions of inorganic or organic substances in the water. Other elements such as the amount of water and the physical conditions of the water body have only recently been considered to be of equal importance to the water quality *per se* for determining the ecological quality of the aquatic environment.

Demand for water of good quality has increased with the advent of industrialization and rapid population growth. This trend has continued over time and has become more widespread geographically. In addition to domestic and industrial use of water, other requirements have become increasingly important. These include improved personal hygiene, agricultural irrigation and livestock supply, hydropower generation, cooling water for power plants and industry, as well as recreational purposes such as boating, swimming and fishing. Each of these intentional water uses affects, more or less, the quality of the water. Together with increased intensity of water use, discharge of untreated domestic and industrial wastes, excessive application of fertilizers and pesticides in agriculture and accidental spills of harmful substances (including radioactive substances) have led to increasing pollution of many European water bodies: groundwaters, rivers, lakes, coastal areas and seas. Figure W27 illustrates in chronological order the sequence of water pollution problems found in European fresh waters since 1850. Most pollution problems have evolved unrecognized over time until they have become apparent and measurable. Recognition of a problem, therefore, took a considerable time and control measures took, in most cases, even longer.

Not all water quality problems are due solely to human impacts. Locally, natural geochemical conditions may cause a high content of reduced iron (e.g., in the Russian Federation and Denmark), fluoride (in Moldova and Bavaria, Germany), arsenic and strontium (in some mountainous countries) and salts in the groundwater, reducing its use as a source of drinking water. Natural events like volcanic eruptions and subsequent mud flows, floods and droughts can lead to serious local and regional deterioration of the aquatic environment. The impact of some of these events, however, can be made worse by human activities; for example, by land-use changes, deforestation and river channelization.

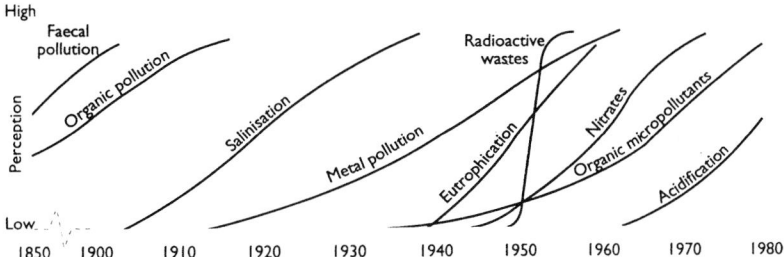

Figure W27 Consecutive occurrence and perception of freshwater pollution problems in industrialized countries. (Source: Meybeck and Helmer, 1989.)

Water pollution used to be primarily a local problem, with identifiable sources of pollution by liquid waste. Up to a few decades ago most of the wastes discharged to waters came from animal and human excreta and other organic components from industry. In areas with low population density without sewerage systems, such problems are to a great extent alleviated by the natural self-purification capacity of the receiving water. However, with the increasing urbanization of the nineteenth and early twentieth centuries, and subsequent expansion of sewerage systems without any or adequate treatment, liquid waste loads have become so large that the self-purification capacity of receiving waters downstream of large human settlements can no longer prevent adverse effects on water resources. The results of discharges of such materials include dying fish, offensive smells and the risk of infection. In addition, the widespread channelization of rivers that took place over this period contributed significantly to the reduction of the natural self-purification capacity of rivers.

Over the years, the pollution load of most receiving waters has further increased. In addition to impacts from point sources, pollution from non-point (diffuse) sources, for example leaching and runoff from agricultural areas and long-range transported air pollutants, have become increasingly important. Consequently, the associated problems are no longer just local or regional, but have become continental in scope.

For Europe, no general overview of water quality exists. Although the information available in many cases is anecdotal, an evaluation of water quality trends, and the present state of the water bodies has been related to natural processes, and to human activities in the catchments overlying the aquifers or draining to the water bodies in question.

The water resource

A thorough understanding of water in the hydrosphere is necessary if we are to appreciate its role in the Earth system and provide a solid basis for rational water management to meet human demands, while at the same time preserving the integrity of the environment. The International Hydrological Programme and the Hydrology and Water Resources Programme of UNESCO and WMO, respectively, have provided important contributions to knowledge about the hydrology and water resources of the Earth and its continents. Other significant contributions to the understanding of global water resources have been provided by Lvovich (1973), Baumgartner and Reichel (1975) and Shiklomanov (1991).

Hydrological cycle

The Earth's salt water and freshwater have been formed in the course of the evolution of the planet as a by-product of numerous chemical processes transforming rock matter at large depths. It penetrated to the surface of the Earth as water vapor in particular as a result of volcanic activity, having leaked from the interior of the Earth.

In total, it has been estimated by UNEP (1991) that the Earth's total water amount equals approximately 1360 million km^3 of which less than 3% is present as liquid fresh water or locked up temporarily in ice caps, 8 million km^3 are present as groundwater, 0.2 million km^3 as fresh surface water, and 29 million km^3 form the polar ice caps and minor glaciers elsewhere. For Europe the approximate figures are, according to original estimates and UNESCO (1978): 1 million km^3 of groundwater (very approximate), 2580 km^3 surface water (131 km^3 in rivers, 2027 km^3 in lakes and 422 km^3 in reservoirs) and 4090 km^3 of water locked up in glaciers.

Although the total amount of water on Earth is fixed on a short timescale, the physical state of the water is continuously changing between the three phases (solid in ice, liquid, and as atmospheric water vapor), circulating through the different environmental compartments (ocean, atmosphere, glaciers, rivers, lakes, soil moisture and groundwater) and renewing the resources. The typical average renewal rates of water resources show considerable differences, ranging from thousands of years for the ocean and polar ice down to fortnightly or even weekly renewal of water in rivers and the atmosphere.

Average annual global evaporation from the ocean is six times higher than the evaporation from land (0.43 versus 0.07 million km^3), whereas the ratio between precipitation over ocean and land is 3.5 (0.39 against 0.11 million km^3), showing that approximately 40 000 km^3 of water each year is transported from the ocean via the atmosphere to renew the freshwater resources. This amount of water constitutes the natural water resource that appears as global river runoff and which is potentially available for consumption each year.

Water resources

The annual average renewable water resources per capita show very large variability between different geographical regions in Europe. The populations of Nordic countries have in general between six and eight times as much water available for consumption per capita than the population of the other three geographical regions: Eastern, Southern and Western Europe.

Regional averages, however, provide only a very general measure of the adequacy of water availability, and obscure any variation between and within countries. It is interesting to examine the per capita distribution of available water resources in Europe, which is highly uneven, as shown in Figure W28.

Low, very low or extremely low water availabilities are found in 36% of the European countries, in particular in the Southern countries, with Malta having the lowest amount of water available of all European countries – 80 m^3 per capita per year. Densely populated Western countries with moderate precipitation (e.g., Belgium, Germany, Denmark and the UK) also fall in these categories. Water availability is also low in some Eastern countries (Moldova and Ukraine), mainly due to low precipitation (Tables W21 and W22).

About 32% of European countries fall in the category of medium availability. For Europe as a whole, above medium, high or very high water availabilities are found in another 32% of the countries. Plenty of water is found either in sparsely populated countries where precipitation is very high, like the Nordic countries (Iceland, for example, has more than 600 000 m^3 of water available per year for each of its citizens), or in countries with large transboundary rivers running through them. For example, the per capita water availability in Spain is only one-third of that for Portugal, which receives 48% of its water in transboundary rivers from Spain. Many other countries are dependent on external contributions of water for meeting their demands. The importance of transboundary rivers as water suppliers is summarized in Figure W29 for those countries which receive a significant proportion (more than 10%) of their water from other countries.

In this context, 20 European countries receive a high proportion of their water from transboundary rivers. Five countries – Hungary, Moldova, Romania, Luxembourg and The Netherlands – are heavily dependent on external inputs, receiving more than 75% as external contributions, while three countries – Latvia, Ukraine and former Czechoslovakia (CSFR) – receive between 50 and 75% of their

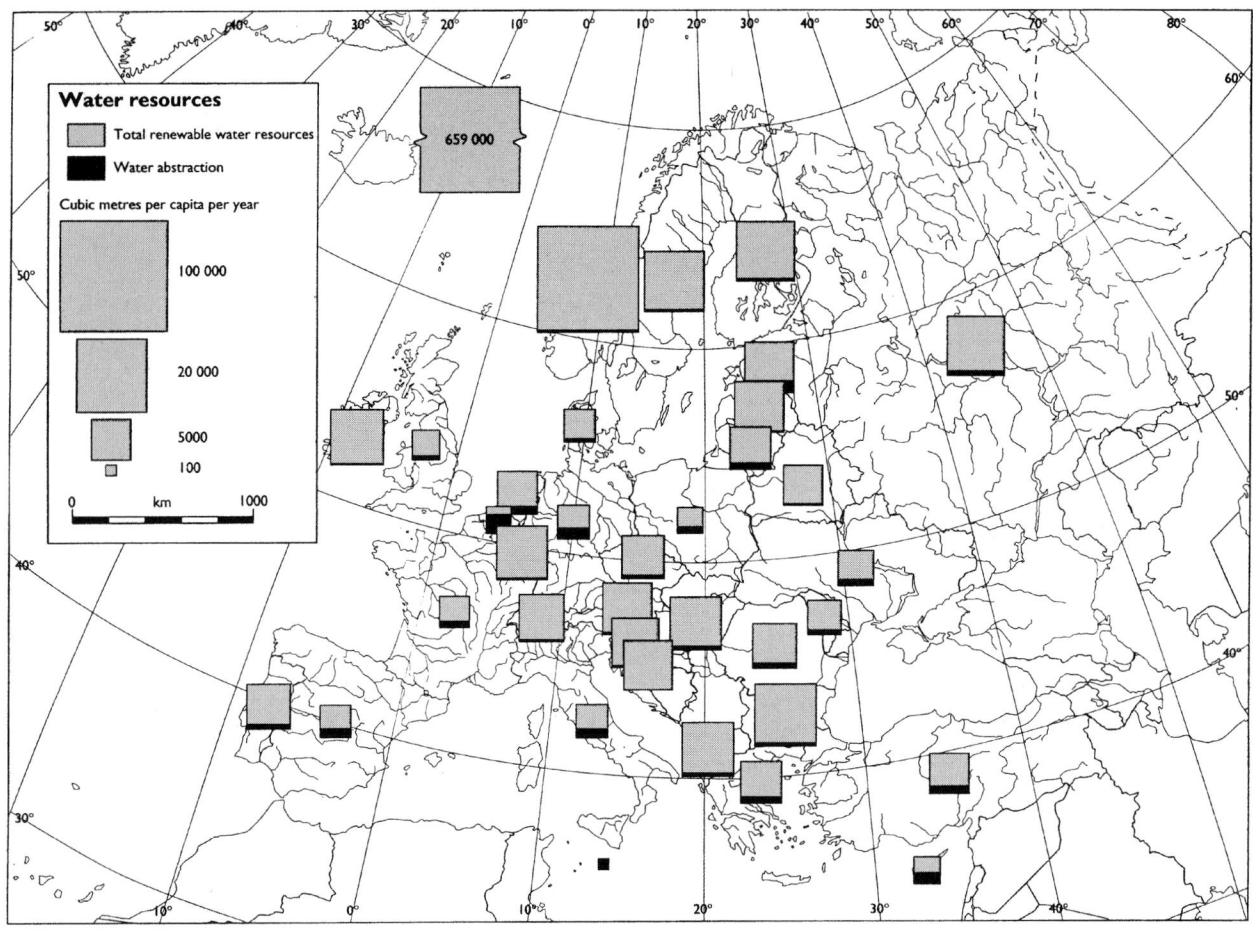

Figure W28 Per capita available water resources (total renewable water resource) and water abstractions for each European country. (Source: multiple sources compiled by EEA-TF and Eurostat.)

renewable water resources from abroad. Bulgaria and Romania share the water resources of their boundary river, the Danube.

Several large European rivers have catchment areas shared between many countries (e.g. the Danube, Rhine, Elbe, Neman and Zapadnaya Dvina), and in some cases collaboration between riparian countries has been initiated to protect water resources. As examples of the importance for national water resources, The Netherlands receives 85% of its water from transboundary rivers (e.g. Rhine, Maas). Hungary receives an annual equivalent of nearly 1200 mm over the country from the Danube, whereas only 65–70 mm of runoff is actually generated within Hungary (Figure W29). There are some 110 bilateral and multilateral agreements to regulate water quality, use and economy in Europe, thus helping to reduce international dispute over water supply sharing. Other international agreements, such as the EC directives on water quality, also contribute to safeguarding European water resources for the future.

In recognition of the basin-wide, upstream–downstream and riparian interdependencies in the development, use and protection of the waters (particularly transboundary rivers), their basins and eco-systems, increasing effort in recent years has been devoted to integrated river basin management to ensure their sustainable development.

Under the terms of the Bucharest Declaration signed in 1985, the countries of the Danube Basin began a program of international joint sampling and analysis of water quality at border crossing points on the River Danube. This initiative was followed by the formulation of the Environmental Programme for the Danube River Basin, with the overall aim of strengthening collaboration between countries on the examination and development of solutions to environmental problems in the region. It was established for a three-year period at a 1991 meeting in Sofia of riparian countries, international and non-governmental organizations, and certain G-24 countries as financing

partners, with the aim of developing a strategic action plan for the basin. This was signed by ministerial declaration in 1994.

The necessity for transboundary management of the River Rhine was recognized for shipping as early as the end of the eighteenth century. The transboundary pollution aspect was given attention in 1932, when the Dutch government protested against emissions of residual salt into the French part of the river. After World War II, the pollution of the river increased, and in 1950 the International Rhine Commission started to study wastewater and water quality problems. The international cooperation against pollution was strengthened by the agreement, signed in Berne in 1963, regarding the International Commission for the Protection of the Rhine against Pollution. This was followed by Council Decision (77/586/EEC), to conclude the Convention for the Protection of the Rhine against chemical pollution to the Berne agreement of 1963. Especially to accelerate ecological improvements in the Rhine, ministers from riparian countries decided in 1986 to establish the Rhine Action Programme, with the following aim:

> The ecosystem of the Rhine must become a suitable habitat to allow the return to this great European river of the higher species which were once present here and have since disappeared (such as salmon).

In order to improve water quality ecological conditions in the Elbe (one of the most impacted larger rivers in Europe) and its tributaries, an agreement (Vereinbarung über die Internationale Kommission zum Schutz der Elbe – IKSE) was signed in 1990 by riparian countries and the EC.

Extensive international collaboration is now also taking place on the North and Baltic seas and the Mediterranean to reduce riverine emissions of nutrients and dangerous substances. Nevertheless,

Table W21 Water resources of European countries

	Renewable water resources		Total water abstraction			
	(Mm^3 year^{-1})	(m^3 year^{-1} per capita)	Reference date	(Mm^3 year^{-1})	(m^3 year^{-1} per capita)	Water use intensity (%)
Albania	41 000	12 615	1989	2 970	928	7
Austria	92 000	11 929	1989	2 120	278	2
Belarus	58 000	5 654	1989	3 000	293	5
Belgium	12 500	1 254	1980	9 030	917	72
Bosnia–Herzegovina	–	–	–	–	–	
Bulgaria	190 000	21 088	1988	11 000	1 225	6
Croatia	47 600	10 160	1990	562	120	1
CSFR (former)	95 000	6 066	1990	5 786	369	6
Cyprus	900	1 282	1989	380	547	42
Denmark	13 000	2 529	1988	1 200	234	9
Estonia	15 000	9 476	1989	3 300	2 098	22
Finland	108 000	21 661	1989	3 001	605	3
France	198 000	3 490	1990	37 730	665	19
Germany	171 000	2 156	1990	58 852	742	34
Germany (east Germany)	34 000	2 092	1990	11 345	698	33
Germany (west Germany)	162 000	2 568	1990	47 507	753	29
Greece	58 650	5 794	1980	6 945	720	12
Hungary	120 000	11 371	1990	6 263	593	5
Iceland	168 000	658 824	1985	100	415	0
Ireland	50 000	14 273	1980	793	233	2
Italy	175 000	3 035	1990	56 200	980	32
Latvia	32 000	11 994	1988	640	240	2
Lithuania	23 000	6 718	1989	4 400	1 192	19
Luxembourg	5 000	13 089	1989	59	156	1
Malta	30	80	1987	39	113	130
Moldova	13 000	2 980	1990	1 655	379	13
Netherlands	91 000	6 086	1986	14 481	994	16
Norway	392 000	92 409	1983	2 025	491	1
Poland	59 000	1 545	1990	15 097	395	26
Portugal	73 000	7 398	1989	7 288	737	10
Romania	219 000	9 440	1989	20 340	879	9
Russian Federation	1 500 000	15 238	1989	106 227	721	7
Slovenia	18 672	9 585	1980	495	254	3
Soviet Union (former)	4 739 000	16 421	1989	358 103	1 249	8
Spain	117 000	3 003	1990	36 900	947	32
Sweden	168 000	19 612	1990	2 932	343	2
Switzerland	54 000	8 045	1989	1 166	175	2
Turkey	234 000	4 179	1990	30 600	521	13
Ukraine	212 700	4 103	1990	33 029	637	16
United Kingdom	120 000	2 090	1990	14 237	248	12
Yugoslavia (former)	265 000	11 130	1980	8 767	393	3

limited water resources, industrial pollution, mining activities, and/or dam construction plans regularly create some international tension. Two examples are disputes between Portugal and Spain on the sharing of the water of the Tajo River, and between Hungary and the Slovak Republic on the construction of the Gabcikovo Dam on the Danube.

For countries with major rivers running through them, estimates of total renewable water resources tend to overestimate sustainable water resources. In humid, temperate climates, a good estimate of a country's sustainable freshwater resources is the long-term average river runoff generated within the country itself. Therefore, consideration of the spatial and temporal variability of river flows is of utmost interest for the assessment and management of European water resources.

River flow characteristics in Europe

River runoff is one of the main sources of freshwater which the various water demands are satisfied.

There is considerable spatial and temporal variation in river flow across Europe. The seasonal flow regime of the Mediterranean catchment, with its winter high flows, dry periods in summer and frequent flash floods, is very different from that of an English lowland

chalk stream, where flows vary relatively little over the year. They both have a very different seasonal regime from a catchment in Poland, which has a maximum flow following the spring snowmelt. These factors, climatic and physical properties of the catchment, lead to such variations in seasonal flow regimes. In addition, European rivers also experience considerable variation from year to year. Finally river flow regimes are also affected by human activities, such as river impoundment and land-use changes.

Average annual runoff

The average annual runoff in Europe follows very closely the pattern of average annual rainfall – and topography. Annual runoff is greater than 4500 mm in western Norway, and decreases to less than 25 mm in parts of Spain, central Hungary and eastern Romania, and in large regions of Ukraine and the southern part of the Russian Federation. The greater variation of runoff in Western Europe, compared with Eastern Europe, reflects the greatest variability in topography, and hence rainfall.

Across most of lowland Europe, between 25 and 45% of rainfall runs off into water bodies. In high-rainfall areas, such as the Alps, western Norway and western Scotland, over 70% of the rainfall may

Table W22 Inland water abstraction for European countries in 1990

	Total water abstraction (Mm^3 year^{-1})	Groundwater abstraction as % of total water abstraction
Albania	2 970	–
Austria	2 120	53
Belarus	3 000	40
Belgium	–	–
Bosnia–Herzegovina	–	–
Bulgaria	11 000	15
Croatia	562	–
CSFR (former)	5 786	28
Cyprus	380	–
Denmark	1 200	–
Estonia	3 300	15
Finland	3 001	8
France	37 730	16
Germany	58 852	13
Germany (east Germany)	11 345	14
Germany (west Germany)	47 507	13
Greece	–	–
Hungary	6 263	16
Iceland	–	–
Ireland	–	–
Italy	56 200	–
Latvia	640	47
Lithuania	4 400	14
Luxembourg	59	46
Malta	39	69
Moldova	1 655	19
Netherlands	–	–
Norway	–	–
Poland	15 097	16
Portugal	7 288	42
Romania	20 340	11
Russian Federation	106 227	–
Slovenia	495	22
Soviet Union (former)	358 103	10
Spain	36 900	15
Sweden	2 932	20
Switzerland	1 166	81
Turkey	30 600	21
Ukraine	33 029	13
United Kingdom	14 237	19
Yugoslavia (former)	–	–

become runoff. In drier regions, particularly southern Spain, runoff may amount to less than 10% of the annual rainfall.

The annual average runoff for Europe is estimated at 3100 km^3 over a territory of 10.2 million km^2, that is, 304 mm year^{-1}, or 9.6 l s^{-1} km^{-2}. This is the equivalent of 4560 m^3 year^{-1} per capita for a population of 680 million. Compared to the present average total water abstraction of 700 m^3 year^{-1} per capita (1920 l day^{-1} per capita) in Europe, the immediate conclusion would be that Europe faces no water shortage problems.

Even when average runoff, if evenly distributed, would suffice to cover the needs of water users, the natural variation from one season to another, or from year to year, can set constraints. The part of runoff which can serve as a continuous supply source, even under drought conditions, is obviously much less than average runoff, and depends primarily on the local hydrological regime. For most of Western Europe, it has been estimated that this runoff, taken as the mean annual minimum 10-day flow, is between 1 and 3 l s^{-1} km^{-2}.

Seasonal river flow regimes

Variations in the distribution of river flows through the year define different flow regimes. All are based, to a large extent, on the timing of maximum and minimum flows, the number of flow seasons and the origin of the river flows (Gottschalk et al., 1979; Krasovskaia and

Gottschalk, 1992). Climate is usually the dominant influence on river flow regimes, but there are cases where the effect of physical characteristics of the catchment can dominate. Such 'atypical' regimes, with very little variability in flow from month to month, include those of catchments with very large lakes, or where river runoff is controlled by seepage from an aquifer.

River flow regimes vary considerably across Europe, but insufficient information is available to produce maps showing their geographical distribution.

The river flow regimes of large catchments (Figure W30) can be different from those of small catchments. In particular they are much less variable because they integrate runoff over a large area, and can include subcatchments with very different characteristics.

The differences in river flow regimes are apparent between Western Europe (where flows are at a minimum in summer and late fall), the mountain-fed catchments (where flows are at a maximum in summer), and Eastern and Northern Europe (where most runoff occurs during the spring snow-melt period).

Interannual variability

The droughts of the early 1990s, which affected several parts of Europe, illustrate clearly the variability in river runoff from year to year. From an analysis of runoff data from 14 gauging stations with long records (between 40 and 150 years) across Europe, three important conclusions can be drawn.

- Consistent patterns in the variability of river flow regimes from year to year exist across Europe. Droughts, for example, tend to affect large parts of Europe at the same time. The regional consistencies and differences between wide areas reflect the correlation with large-scale weather patterns.

- It appears that extreme years tend to cluster. Several low-flow years occurred in succession in the early 1970s in Western Europe, followed by a period in the 1980s where several years were above average.

- There is no evidence of any consistent trend in river flow characteristics. Some records show upward or downward trends for a part of the record, but in each case these periods are followed by periods with no trend, or an opposing trend.

Human influences on river flow regimes

Many European river flow regimes are heavily affected by human activities, ranging from direct abstraction of water through regulation of flow regimes by reservoirs, to changes in catchment land use. It is, however, difficult to generalize about the impacts of human activities on flow regimes. However, it is important to recognize that these impacts are not limited to areas with high population densities (examples are reservoirs and river channelization).

Main sources of water for abstraction

Surface water is the main source for water abstraction by all utilization sectors in Europe. On average, 70% of total abstraction is drawn from this source, but with large variation between countries. Groundwater is by far the next most important source, and other sources include desalinated seawater.

Countries such as Spain, Belgium, The Netherlands, Finland and Moldova, with insufficient groundwater supplies, abstract more than 90% from surfacewater sources. However, in Cyprus, Switzerland, Slovenia, Iceland and Denmark, which are countries with extensive groundwater reservoirs, more than 70% is drawn from groundwater.

The public water supply systems serve primarily domestic users and some industrial demands, and the source of water for this sensitive use is predominantly groundwater. As a source of public drinking water, groundwater is of enormous importance. For Europe as a whole, about 65% of the public supply is provided from groundwater (Zektser et al., 1992) which is normally of a better quality than surface water. On average, the total water abstraction in Europe amounts to approximately 480 km^3 year^{-1} (or 700 m^3 year^{-1} per capita), ranging from below 200 m^3 year^{-1} per capita in Luxembourg, Malta and Switzerland to above 1000 m^3 year^{-1} per capita in Bulgaria, Estonia, Lithuania and Spain.

Over the last two decades total water abstraction has in general increased in Europe. This trend, however, masks great variability between countries. Abstraction increases have been particularly marked in Southern European countries, but also in the majority of countries in Eastern and Western Europe more water was abstracted

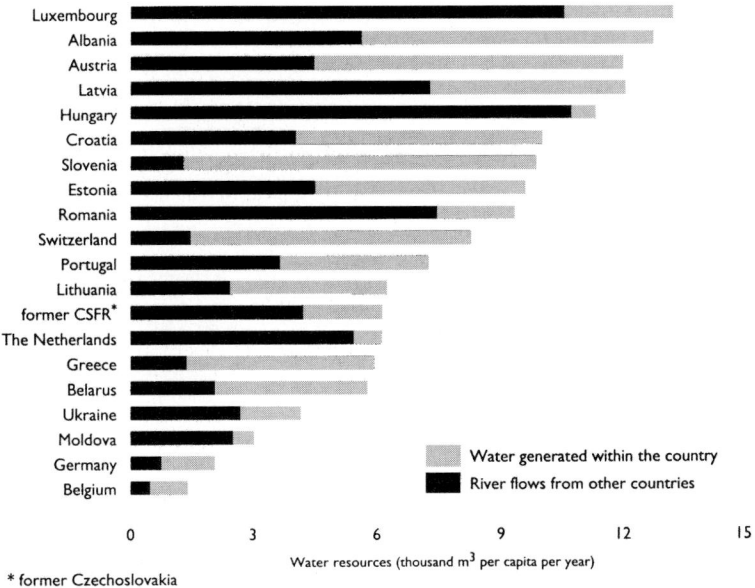

Figure W29 Internal and external contributions to the total renewable water resource in selected European countries. (Source: multiple sources compiled by EEA-TF and Eurostat. See OECD, 1992; Arnell *et al.*, 1993.)

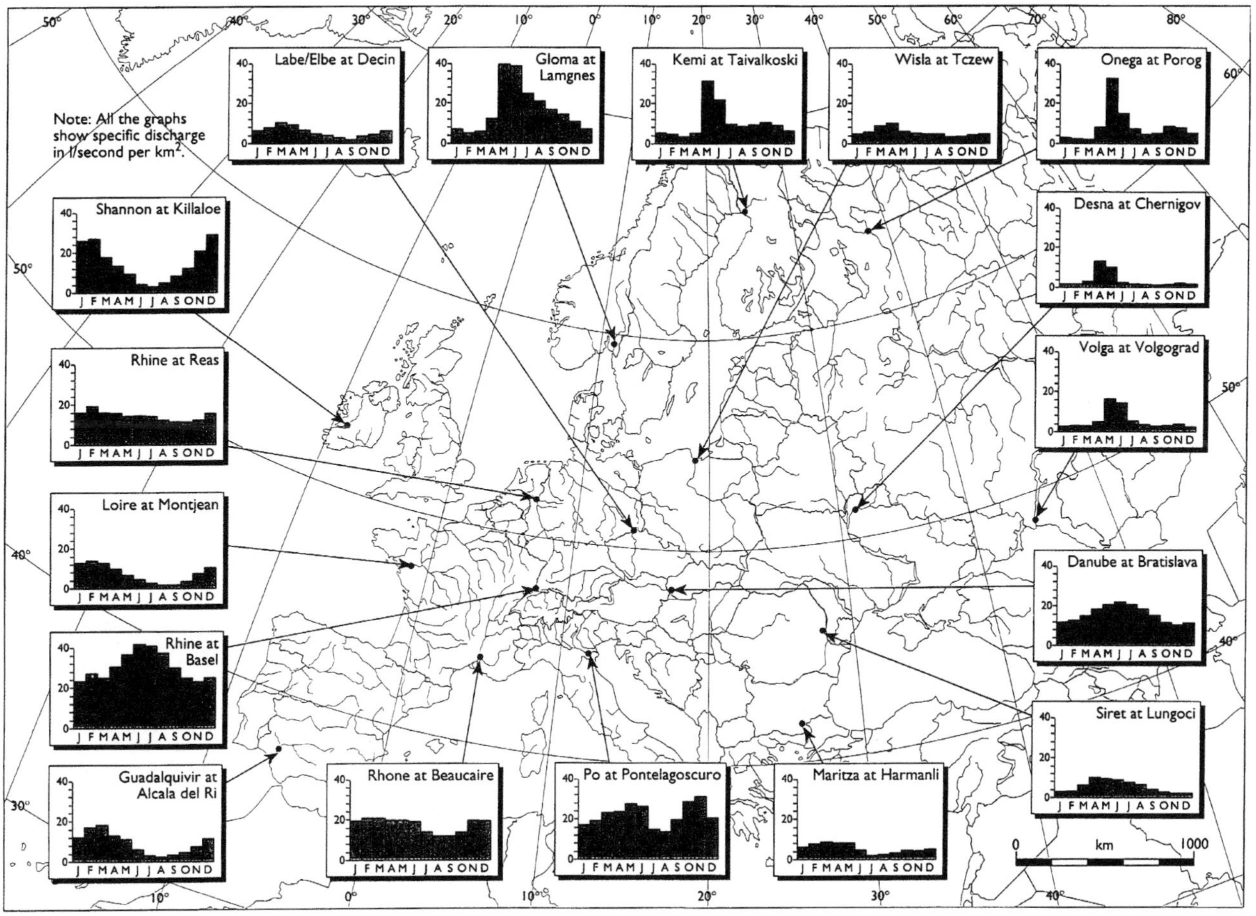

Figure W30 Mean monthly flow for several major river catchments in Europe. All graphs show specific discharge in l s^{-1} km^{-2}. (Source: Arnell *et al.*, 1993.)

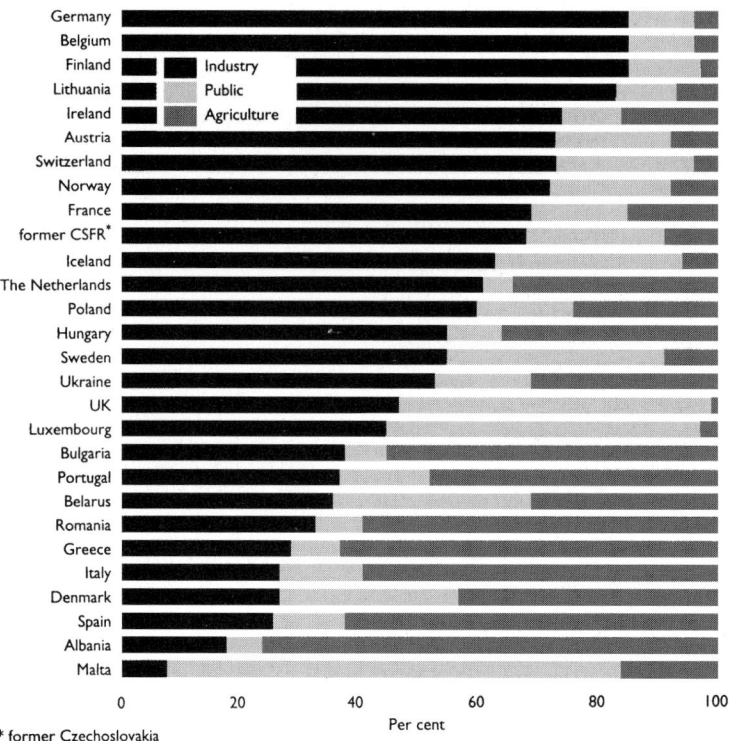

Figure W31 Total water abstraction (surface- and groundwater) by economic sectors in European countries. (Sources: if not otherwise specified, data are from WRI, 1990, as presented by Veiga da Cunha, 1993. Data from Germany, UK, Belarus, Ukraine and Lithuania are extracted from Umweltbundesamt, 1992; UK DoE, 1992.)

in the late 1980s than before. Stabilization or even an abstraction decrease has occurred in some of these countries, including Austria, Bulgaria, The Netherlands and Switzerland. This is also the case in the Nordic countries: Sweden and Finland.

For Europe as a whole, 53% of the abstracted water (surface and groundwater) is used for industrial purposes, 26% in agriculture and only 19% for domestic purposes (WRI, 1990).

There is a large variability in sectoral water abstraction between countries (Figure W31). Interpretations of such statistics, however, should be carried out with caution because it is not always clear how they have been derived and what they comprise. Water use in the industrial sector is particularly difficult to evaluate, since it is not always clear whether cooling water and water used for power generation is included in the industrial share. Water used for these purposes often accounts for about 70–80% of industrial water use, and most of this is used in power generation.

Agriculture's share is highest (30–70%) in a number of Southern and Eastern European countries with low net precipitation, and in Denmark and The Netherlands where agriculture is very intensive. Water use in agriculture is mainly for irrigation and livestock.

Water abstraction for public use shows less variability between European countries than other sectoral abstractions, ranging between 15 and 25% of total abstractions in the majority of countries to more than 40% in Luxembourg and the UK.

The intensity of water use in a country can, according to the OECD (1992), be represented as the percentage abstraction of total available water resources, which includes internally generated water and inputs from neighboring countries via transboundary rivers (Figure W30).

The calculated water-use intensity indicators cover a very wide spectrum, from less than 0.1% in Iceland to over 70% in Belgium, with a general average of around 15%. If this indicator were calculated by relating abstractions exclusively to internal resources, its values would be appreciably higher in those countries where total resources are, to a considerable extent, accounted for by external resources.

This is particularly true for Belgium, Bulgaria, Hungary, Moldova and The Netherlands, where water abstraction approximately equals total internal resources, or even exceeds it, as in the case of The

Netherlands. These countries, as well as Romania, Lithuania, Portugal and Ukraine, are therefore particularly sensitive to upstream impacts affecting the quantity and quality of the water they receive through transboundary rivers.

Groundwater

Characteristics and distribution

Groundwater is an important element in the Earth's hydrological cycle. It remains one of the least studied and most difficult water resources to determine. Natural groundwater resources are stored in aquifers, which are permeable rock formations or unconsolidated deposits, chiefly gravels, sands and silts.

The main characteristics of groundwater systems are

- invisible and relatively inaccessible location;
- very low flow rates, resulting in long residence times and a slow reaction to changes on the surface;
- huge extent of aquifers.

In spite of its non-visibility, groundwater has very important functions, including economic, ecological and those relating to public health – which are not always fully recognized. For example, groundwater is an important source for drinking water. Human activity, however, can have a great effect on quantity and quality of the available groundwater resources. Due to the aforementioned characteristics, groundwater systems are normally very stable, in both quantity and quality. However, the effects of pollution and over-exploitation will accumulate over time. In general, the periods of recovery will be centuries and decades, respectively.

The availability of groundwater (as a natural resource) is limited by three factors:

- the total amount of recharge (renewal of groundwater), resulting from precipitation, evapotranspiration, infiltration and seepage from rivers and lakes;
- the quality of the recharged water;
- the properties of the soil and aquifer (permeability, porosity, etc).

As with surface water, there is an uneven distribution of natural groundwater resources within Europe.

Recharge and loads

The renewal of groundwater resources occurs through natural and artificial recharge. The natural recharge is the amount of water available for percolation into the aquifer as a result of excess precipitation and runoff. The precipitation excess is the amount of precipitation minus evapotranspiration.

Artificial recharge is the result of excess irrigation or of human-induced recharge (forced feeding) of aquifer systems. Although irrigation is common in large parts of Europe, systematic data on groundwater sources and quantities used are not available. As a result, it is not possible to quantify the effects of irrigation on the groundwater systems. The essential objective of forced-feeding recharge is to transform surface water with a periodically unreliable quality into a safe source for water supply. It can be done by applying natural gravity forces, for example, by infiltration canals, or by damming rivers, or via injection wells. Forced feeding of aquifers is currently being used in a number of European countries (e.g. river bank infiltration in Germany and The Netherlands, and surfacewater infiltration in the Uppsala area of Sweden, and in dunes in The Netherlands). In Germany, for example, these techniques together account for 15.7% of public water supply (Umweltbundesamt, 1992).

All groundwater contains natural chemical constituents in solution. During its passage underground, the water dissolves and deposits various substances, while other solutes are being transformed or degraded. Most of these changes tend to be slow but long lasting. Pollution of the percolating water due to human activities sometimes has a severe impact on the constituent load reaching the groundwater.

In many places, the natural quality of groundwater is degraded because of human activities. Basically, two different sources of groundwater pollution can be recognized: non-point (diffuse) sources and point sources. The diffuse sources are mainly from agriculture (e.g., irrigation excess, animal wastes, fertilizers and pesticide residues), urban areas (storm drainage) and contaminant deposition from the atmosphere. The point sources comprise municipal and industrial activities, such as surface disposal of liquid and solid waste, improper storage of materials used in manufacture, sewer, tank and pipeline leakages, and activities related to mining (especially mine tailings) and oil exploitation.

Groundwater quantity

During recent centuries, human-made changes in the hydrological cycle have had enormous impacts on groundwater levels and flows. These have changed for many reasons, including pumping of groundwater for domestic, industrial and agricultural water supply, provision of cooling water, mine pit drainage, intensified and deeper drainage of agricultural lands, changing natural land into agricultural and urban areas, and regulation of surface waters.

These changes have important economic and ecological effects. Economic effects can include crop and industrial production as well as the production of drinking water. Groundwater-related terrestrial ecosystems are notably affected by changing groundwater levels, while at the same time wetlands are becoming more scarce. Moreover, groundwater is the last remaining water source for rivers and small surface waters during dry periods. If the baseflow in the rivers is not maintained, severe damage can be caused to aquatic ecosystems, and the economic functions of the rivers, such as navigation and water supply, will also suffer.

Abstractions

In many areas, groundwater abstractions exceed the recharge, and the aquifer becomes overexploited, leading to a systematic and in many cases ongoing lowering of the groundwater levels. This results in a wide range of problems: damage to wetlands, drying up of springs and upper river reaches, reduction of river flows, saltwater intrusion (the flow of salt water into an aquifer) and settling phenomena and damage to buildings.

Reported cases of overexploitation include abstraction for drinking and industrial water supply, irrigation water and drainage of mines. Drainage of waterlogged agricultural soils and some river engineering works also lead to lowering of the groundwater level. Many cases of lowering of the groundwater table up to several tens of meters have been reported, and in a number of cases the lowerings are in the order

of hundreds of meters, affecting the groundwater level in an area up to 500 km away from the overexploited site. Major lowerings of the groundwater table caused by mining activities have, for example, been reported for the area around Kharkov (Ukraine), Lille (France) and in the Ruhr Basin (Germany).

However, due to their requirement for water, most overexploited sites are found in or near large urban and industrial centers. Thus, about 60% of the European cities with more than 100 000 inhabitants (or a total of approximately 140 million people) are located in or near areas with groundwater overexploitation. The consequences of this can be far reaching, as shown by the recent severe problems in Spain and Greece, where cities have been given restrictions on the use of water. Rivers and marshes also have dried out, leading to loss of wetlands. An estimated 6% of the area with aquifers suitable for abstractions is presently overexploited.

Saltwater intrusion in aquifers may result from groundwater overexploitation along the coast. Since urban, towns and industrial centers are commonly located in this zone, intrusion of salt water is a problem of many coastal regions, especially along the Mediterranean, Baltic and Black Sea coasts.

A systematic European inventory of human-induced lowerings of groundwater levels and other changes in the groundwater systems is not available.

Potential wetland damage

Wetlands, or wet ecosystems, are considered very important nature reserves, with a high ecological value. Many wetlands are located in areas prone to flooding, such as river floodplains, and the systems are highly dependent on the shallow depth of the water table, making them very sensitive to minor changes in the groundwater level. A tragic example is the Table de Daimiel nature reserve in Spain (Custodio, 1991). The two main rivers feeding the area are fully dependent on groundwater but the groundwater resources in the area have been severely overexploited for irrigation, and at the beginning of 1993 the nature reserve had almost completely dried out.

Raising of groundwater levels

The water table has been rising due to reduced groundwater abstractions by industry in some cities and industrial areas (e.g. Paris, London, Birmingham and Liverpool). Rises in groundwater level can cause problems such as flooding of tunnels and basements, chemical attack on structural materials and increased humidity in houses (UK DoE, 1992). Also, increased volumes in sewers may decrease their effective capacity. Similar effects may result from large hydrotechnical constructions changing the groundwater level in the drainage basin (e.g. in the Russian Federation and Ukraine) and from artificial recharge and irrigation. Another example can be found in the Russian Federation around the Caspian Sea, where the rise in groundwater level was caused by a sea level rise of a few meters. In Ukraine rising groundwater level, in an area of approximately 25 000 km^2 in the northern and central parts of the country, has caused problems in 2000 towns and settlements. The groundwater rise, which was caused by excess irrigation, produces unfavorable geological processes such as displacements, landslides and soil salinization.

Groundwater pollution

The main pollution threats to groundwater are

- a wide range of inorganic and organic contaminants from point sources in urban, industrial, mining, military and landfill areas;
- leaching of nitrates;
- leaching of pesticides;
- acidification.

Point sources

Soil pollution on urban, industrial, mining, military and landfill areas can affect the quality of groundwater, either directly or after a delay in time. Severe problems of groundwater pollution have already occurred (Brömssen, 1986).

Actual data on polluted groundwater sites in Europe are very scarce. The area of groundwater potentially polluted by industry, mining, military activities and landfills can be estimated using various assumptions and by extrapolating available data. It is thus estimated that potential groundwater pollution by point sources occurs in less than 1% of the European territory. Although this seems to be a rather insignificant portion, it should be borne in mind that most ground-water abstraction takes place in the vicinity of the areas potentially contaminated, posing a threat to the public water supply.

Nitrate leaching

The use of manure and fertilizers can lead to leaching of nitrate, ammonium, sulfate, potassium and, to a lesser extent, phosphorus into the groundwater, and hence into surface water. Because nitrate in water above certain concentrations can be a danger to human health and adversely affects the stability of aquatic ecosystems, this report focuses only on this contaminant. Whether or not nitrate will leach to the groundwater depends primarily on the time and rate of applica-tion, crop uptake, the soil type and climatological conditions.

Several processes influence the fate of nitrate as it percolates from the root zone to the groundwater. Denitrification is particularly important in this respect, leading to a decrease of nitrate concentra-tions between the root zone and groundwater level as well as in the deeper groundwater, provided that (bio)chemical conditions are favorable.

The physical processes affecting the fate of nitrate involve mixing of 'old' resident groundwater with percolating 'young' water from upper soil layers. As a result of mixing of two types of water with different nitrate concentrations (lower in groundwater than in the percolating water), the nitrate concentrations in abstraction wells and seepage areas supporting wetlands, rivers and lakes are generally well below the leachate concentrations and/or their increase is retarded in time.

For large parts of Europe there is a complete lack of monitoring data on groundwater pollution by nitrate. A standardized groundwater monitoring network across Europe (based on unified and agreed principles for network design, site selection, sampling frequency, choice of determinants, etc.) would enable a much clearer picture of the problems.

Pesticides

Approximately 600 different pesticides (as broadly defined to include herbicides as well as fungicides and insecticides) are applied in European agriculture, silviculture and horticulture. On their passage through the subsurface environment, these 600 active ingredients are transformed in an (often) unknown number of degradable products (residues). The effects of active ingredients and their residues on non-target terrestrial organisms (side effects), their fate in the soil and their undesired effects in groundwater and surface waters are far from being known with certainty for every substance.

The extent to which pesticides in groundwater have reached concentrations limiting the use of groundwater for drinking-water abstraction is essentially unknown. Only a few pesticide measure-ments in groundwater are available for a restricted number of con-stituents. The information about pesticide use per country (amounts of pesticide) available for assessment is also limited.

There is an almost complete lack of systematic monitoring data for pesticides in Europe. Where measurements are available, they are mostly incidental (not part of a systematic monitoring network) and the density of monitoring points is very low. Often no specifications are given about the exact location of the measurement site and the monitoring depth. In general, only a few compounds (mostly fewer than ten) have been included in monitoring programs. The main reason is the high cost of chemical analyses, combined with the high number of chemical compounds. Out of approximately 600 pesticides, only about 30 have been monitored. These 30 pesticides are neither the most mobile nor representative of all pesticides. The availability of suitable analytical methods is also an important limiting factor.

Acidification

Acidification of the soil is a well-documented and serious problem in large parts of Europe. In addition to natural acidification, sandy and poorly buffered (low alkalinity) soils are subject to enhanced acidification caused by atmospheric deposition of sulfur and nitrogen compounds, fertilizer application and land drainage. Because water must filter through the soil before it reaches the aquifer, it is likely that groundwater below those areas may also have become acidified. Unlike surface waters there are very few data available to show the present extent of groundwater acidification and its trends in Europe, and only a few examples of documented groundwater acidification are available.

Acidification of forest soils is, in particular, a well-described and generally recognized problem in Northern and Central Europe (Last

and Watling, 1991). It is caused primarily by atmospheric deposition of acidifying compounds emitted from human activities. As a result, increased concentrations of aluminum, sulfate and hydrogen ions and sometimes nitrate are reported in the upper groundwater layer below sandy soils in Norway, Sweden, Denmark, Finland, the Czech Republic, the Slovak Republic, Germany and The Netherlands.

In Sweden, for example, a national groundwater survey showed a severe influence of acid inputs on wells in forested areas (Bernes and Grundsten, 1992). Thus more than 50% of surveyed wells in 14 municipalities (of a total of 52 municipalities surveyed) showed very strong or strong acidification effects in the mid-1980s.

Acidification of deeper groundwater under agricultural land has also been reported (Targbill, 1986; Brömssen, 1986; Overgaard, 1986), but in these cases acidification cannot be explained solely by atmospheric deposition. Other processes (such as oxidation of ammonia to nitrate by nitrification) can also contribute to acidification (Rebsdorf et al., 1991), as can denitrification if nitrate is reduced by ferrous sulfide.

In Denmark, for example, an analysis of a 30-year time series (period 1952–1982) of pH, alkalinity and nitrate in groundwater from unconfined aquifers below predominantly agricultural land in the western, sandy part of the country showed that pH and alkalinity had declined steadily over time, whereas nitrate had increased. The average yearly decrease was 0.023 units for pH and 33 μmol l^{-1} for alkalinity. For nitrate the increase was 14 μmol l^{-1} (Rebsdorf et al., 1991). If alkalinity continues to decrease at the present rate, the groundwater in that part of Denmark would completely lose its buffering capacity within 20 to 25 years. This would make the groundwater unsuitable for human consumption and almost all other purposes, unless expensive water treatment processes are intro-duced.

The most serious consequences of acidification of groundwater are the increased mobilization of trace elements, especially aluminum, and the increased solubility of some metals in water distribution systems, both resulting from a lowering of the pH.

Rivers, reservoirs and lakes

Considerable environmental information on rivers and lakes is currently collected and reported by various regional or national authorities. However, as the European continent covers about 10 mil-lion km^2 and there are several million kilometers of flowing waters and more than a million lakes, this information is very heterogeneous, and therefore difficult to collate on a pan-European basis. The primary focus is on frequently measured water quality parameters (e.g. organic matter in rivers, nutrients in rivers and lakes, and acidification of rivers and lakes) since the wide geographical coverage makes these variables well suited to illustrate the general environmental state of European inland surface waters.

A river system comprises both the main course and all the tributaries that feed into it. The main characteristic of rivers is their continuous one-way flow in response to gravity. In addition, because of changes in physical conditions such as slope and bedrock geology, rivers are dynamic and may change nature several times during their course (e.g. from a fast-flowing mountain stream to a wide, deep, slowly flowing lowland river).

When assessing river characteristics and water quality, it is important to bear in mind that a river comprises not only the main course, but also a vast number of tributaries. Thus, although the main course of Europe's largest river, the Volga, is 3500 km long, it receives water from ten tributaries each longer than 500 km, and more than 151 000 tributaries each longer than 10 km (Fortunatov, 1979).

Rivers are greatly influenced by the characteristics of the catchment area. The climatic conditions influence the water flow, as does bedrock geology and soil type. The latter also affects the mineral content of the river water. Human activity affects river systems in numerous ways, for example through afforestation or deforestation, urbanization, agricultural development, land drainage, pollutant dis-charge and flow regulation (dams, channelization, etc.). The lakes, reservoirs and wetlands in a river system attenuate the natural fluctuation in discharge and serve as settling tanks for material transported by the rivers. For example, whereas the water of the Rhine is very muddy and turbid when entering the Bodensee, it is clear and transparent when leaving. Water flow and water quality are therefore the net result of the various characteristics of the catchment.

Lakes are bodies of standing water that are usually fresh, but which may also be brackish. Although lakes may be characterized by physical features of the lake basin, such as lake area and water depth, the characteristics of the catchment are important when describing the lake environment. Nutrient loading of a lake is determined not only by the bedrock geology and soil type in the catchment, but also by the human activity.

Reservoirs are human-made lakes created to serve one or more purposes. As their water residence time is generally relatively short, and as the water level fluctuates much more widely and frequently than in natural lakes, they can be regarded as hybrids between rivers and lakes.

European rivers

On average, European rivers discharge a total of 3100 km^3 of fresh water to the sea each year, about 8% of total world discharge. Because Europe has a temperate humid climate and a high percentage of limestone in the surface rock, the weathering rate is the highest of all the continents; as a result, 12.6% of all dissolved solids discharged to the oceans are derived from Europe (Kempe et al., 1991). That Europe is relatively densely populated and has a high proportion of agricultural areas also affects the concentration of dissolved substances in river water; thus the median nitrate level is 1.8 (mg N) l^{-1} in European rivers as compared with only 0.25 (mg N) l^{-1} in non-European rivers (Meybeck and Helmer, 1989).

Major European river catchments

In proportion to its land area, Europe has the longest coastline of all continents. As it is a relatively young and structured continent, geologically, river catchments are numerous but relatively small and rivers are short. About 70 European rivers have a catchment area exceeding 10 000 km^2, and only rivers arising deep inside the continent are relatively large. The three largest rivers in Europe, the Volga, the Danube and the Dnepr, drain one-quarter of the continent, but are only small by world standards, their catchments ranking 14th, 29th and 48th, respectively (see Rivers).

The 31 largest European rivers, all of which have catchments exceeding 50 000 km^2, drain approximately two-thirds of the continent. More than half of these rivers have their catchment area in the European part of the former USSR. The major rivers flowing north into the Barents Sea and the White Sea are the Severnaya (Northern) Dvina and the Pechora. The Volga and the Ural, which flow south, and the Kura, which flows east, drain into the Caspian Sea while the Dnepr and the Don drain south into the Black Sea. The largest river to discharge into the Black Sea is the Danube, which has its catchments in 16 countries of Central Europe and the Balkans. The main rivers to discharge into the Baltic Sea are the Neva, the Wisla, the Oder and the Neman. Ten rivers with catchments larger than 50 000 km^2 drain into the Atlantic and the North Sea, with the Rhine, the Elbe, the Loire and the Douro being the largest. The European rivers that drain into the Mediterranean are relatively small, the Rhone, the Ebro and the Po being the largest. Nevertheless, since the damming of the Nile, the Rhone has become the Mediterranean's most important freshwater source (Kempe et al., 1991).

Major rivers in European countries

Countries whose coastline is long in relation to their area, for example Iceland, the UK, Ireland, Norway, Sweden, Denmark, Italy and Greece, are usually characterized by having a large number of relatively small river catchments and short rivers, the three to four largest of which drain only 15–35% of their area (Table W23). The population tends to congregate in towns along the coastline, and wastewater is consequently discharged directly into coastal areas rather than into the river systems.

Many European countries are drained by only a few river catchments; thus the Wisla and Oder drain more than 95% of Poland, and the Danube drains most of Hungary, Romania and Slovenia (Table W23).

European lakes and reservoirs

Many natural European lakes appeared 10 000–15 000 years ago, being formed or reshaped by the last glaciation period, the Weichsel. The ice sheet covered all of Northern Europe, but in Central and Southern Europe it was restricted to the mountain ranges. As a rule the regions that have many natural lakes are those that were affected by the Weichsel ice. Norway, Sweden, Finland and the Karelo-Kola part of the Russian Federation have numerous lakes that account for approximately 5–10% of national surface area. Large numbers of

lakes were also created in the other countries around the Baltic Sea, as well as in Iceland, Ireland and the northern and western parts of the UK. In Central Europe most natural lakes lie in mountain regions, those at high altitudes being relatively small and those in the valleys being the largest, for example Lac Léman, Bodensee, Lago di Garda, Lago di Como and Lago Maggiore in the Alps and Lake Prespa and Lake Ohrid in the Dinarian Alps. Exceptions are the two large lakes, Lake Balaton and Neusiedler See, which lie on the Hungarian Plain.

In contrast to glaciation, processes such as tectonic and volcanic activity have played only a minor role in the formation of European lakes. Numerous lakes have been created by natural damming of rivers and coastal areas, however.

Countries that were little affected by the glaciation period, such as Portugal, Spain, France, Belgium, southern England, central Germany, the Czech Republic, the Slovak Republic and the Central European part of the Russian Federation, have few natural lakes. In these areas human-made lakes such as reservoirs and ponds are often more frequent than natural lakes. Many river valleys have been dammed to create reservoirs, and a large number have been built in mountain ranges for use by the hydroelectric industry. In several countries, for example The Netherlands, Germany, France and the former Czechoslovakia, numerous artificial small lakes have been created by other human activities such as peat and sand quarrying, and for use as fish ponds.

Natural lakes in Europe

There are more than 500 000 natural lakes larger than 0.01 km^2 (1 ha) in Europe; of these about 80–90% are small, with a surface area between 0.01 and 0.1 km^2, and only about 16 000 have a surface area exceeding 1 km^2. Three-quarters of the lakes are located in Norway, Sweden, Finland and the Karelo-Kola part of the Russian Federation.

The approximate number and size distribution of natural lakes is shown for each country in Table W24; however, the number of small lakes is somewhat uncertain, and the figures given are generally minimum estimates.

Human-made lakes

Reservoirs are the most important human-made lakes in Europe, there being more than 10 000 major reservoirs covering a total surface area of more than 100 000 km^2. The numbers of relatively large reservoirs are greatest in the Russian Federation (ca. 1250), Spain (ca. 1000), Norway (ca. 810) and the UK (ca. 570). Other countries with a large number of reservoirs are Hungary (ca. 300), Italy (ca. 270), France (ca. 240) and Sweden (ca. 225). Many European countries have numerous smaller human-made lakes, for example Latvia, Bulgaria and Estonia, which have about 800, 500 and 60, respectively.

Large lakes and reservoirs in Europe

There are 24 natural lakes in Europe that have a surface area larger than 400 km^2, the largest being Lake Ladoga, which covers an area of 17 670 km^2. The latter is located in the northwestern part of the Russian Federation, together with Lake Onega, the second largest lake in Europe. Both are considerably larger than other European lakes and reservoirs, but nevertheless rank only 18th and 22nd in world order (Herdendorf, 1982). The third largest European freshwater body is the 6450 km^2 Kuybyshevskoye reservoir on the Volga. Another 19 natural lakes larger than 400 km^2 are found in Sweden, Finland, Estonia and the northwestern part of the Russian Federation, and three in Central Europe – Lake Balaton, Lac Léman and Bodensee, the surface areas of which are 596, 584 and 540 km^2, respectively.

The six largest reservoirs are located in the Volga river system in the Russian Federation, the two largest being the 6450 km^2 Kuybyshevskoye and the 4450 km^2 Rybinskoye reservoirs. Of the 13 European reservoirs with an area exceeding 1000 km^2, only the Dutch reservoir Ijsselmeer lies outside the Russian Federation and Ukraine.

Deep and shallow lakes and reservoirs

Lake water depth is an important parameter with which to characterize the lake environment. It is determined largely by the surrounding topography, lakes in mountainous regions generally being deeper than those in lowland areas. In two lowland countries, Finland and Poland, most lakes have a mean depth of 3–10 m; lakes with a mean depth greater than 10 m are rare. In Austria and Switzerland, in contrast, large shallow lakes are virtually absent, and most lakes have a mean

Table W23 The major European rivers, their catchment areas, and the percentage of the country area drained (see also Rivers, table of length, drainage basin and discharge)

Country	River	Catchment Area (10^3 km^2)	Catchment Per cent of country
Albania	Drin	7.0	23
	Seman	5.6	20
	Vijosë	4.5	16
Austria	Danube	80.6	96
	Inn	15.9	19
	Rhine	2.4	3
Belarus	Dnepr	130.0	63
	Neman	45.5	22
	Zapadnaya Dvina	30.0	14
Belgium	Meuse	13.5	44
	Schelde	10.0	33
Bosnia-Herzegovina	Sava	37.5	74
	Neretva	10.0	20
Bulgaria	Danube	48.2	43
	Maritza	21.1	19
	Struma	10.8	10
Croatia	Danube	37.0	65
	Sava	24.5	44
	Drava	8.0	14
Czech Republic	Elbe	51.4	64
	Morava	25.0	30
	Oder	4.7	6
Denmark	Gudenå	2.6	6
	Skjern Å	2.3	5
	Storå	1.1	3
Estonia	Narva	20.0	45
	Pärnu	6.9	15
	Jägala	1.5	3
Finland	Vuoksa	61.1	18
	Kemijoki	51.1	15
	Kymijoki	37.2	12
	Kokemäenjoki	27.0	9
Former Yugoslavic Republic of Macedonia	Vardar	20.5	81
	Drin	3.5	13
France	Loire	117.5	21
	Rhone	85.6	16
	Garonne	85.0	16
	Seine	79.0	14
Georgia	Kura	42.5	61
	Rioni	13.4	19
	Alazani	7.5	11
Germany	Rhine	102.1	29
	Elbe	97.0	27
	Danube	59.6	17
	Wesser	45.8	13
Greece	Aliákmon	9.5	7
	Piniós	7.1	5
	Strimon	6.0	5
Hungary	Danube	93.0	100
	Tisza	44.6	48
	Drava	6.2	7
Iceland	Jökulsá-á-Fjöllum	7.8	8
	Thjórsá	7.5	7
	Ölfusa	6.1	6
Ireland	Shannon	14.0	20
	Barrow	5.5	8
	Suir	3.6	5
Italy	Po	69.0	23
	Tevere	17.2	6
	Adige	12.2	4
Latvia	Daugava	23.6	37
	Lielupe	8.8	14
	Gauja	7.9	12
Lithuania	Neman	46.6	71
	Vilnya	13.8	71
	Venta	5.2	8

Table W23 Continued

Country	River	Catchment Area (10^3 km²)	Per cent of country
Luxembourg	Süre	2.0	77
	Mosel	0.5	19
Moldova	Dnestr	18.0	53
	Prut	12.0	36
	Kogel'nik	3.9	11
The Netherlands	Rhine	25.0	60
	Meuse	6.0	14
Norway	Glomma	41.4	13
	Drammens-elva	17.1	5
	Tana	10.9	3
Poland	Wisla	191.8	61
	Oder	114.2	37
	Rega	2.6	1
Portugal	Douro	20.0	22
	Tajo	18.0	20
	Guadiana	14.0	15
Romania	Danube	232.2	98
	Muresul	27.8	12
	Oltul	24.0	10
Russian Federation	Volga	1360.0	35
	Don	380.0	10
	Severnaya Dvina	357.0	9
	Pechora	322.0	8
	Neva	220.0	6
	Ural	110.0	3
	Dnepr	105.0	3
Serbia-Montenegro	Danube	95.0	93
	Sava	20.0	19
	Drin	4.5	4
Slovak Republic	Danube	46.5	97
	Váh	17.5	37
	Tisza	16.0	33
Slovenia	Sava	10.8	53
	Drava	5.0	24
Spain	Ebro	84.2	17
	Douro	78.0	15
	Tajo	62.0	12
	Guadiana	58.0	11
Sweden	Göta älv	41.0	9
	Torne älv	34.1	8
	Ångermanälven	30.6	7
	Dalälven	29.0	6
Switzerland	Rhine	28.0	68
	Aare	17.8	43
	Rhone	10.4	25
Ukraine	Dnepr	293.0	48
	Yuzhnyy Bug	63.7	11
	Dnestr	52.7	9
United Kingdom	Thames	15.0	6
	Severn	11.6	5
	Trent	10.5	4

Note: same rivers with different names: Strimon/Struma; Daugava/Zapadnaya Dvina.

depth greater than 25 m. As with natural lakes, the deepest reservoirs are located in mountainous regions of countries such as Norway, Spain, France, Scotland and Greece. Examples are the 190 m deep Spanish reservoir Almendra, the 132 m deep Greek reservoir Kremasta, and the 125 m deep Norwegian reservoir Blåjø (these are maximum depths).

Regulation of European rivers

River regulation is a general term describing the physical changes that people impose on watercourses. Many of the rivers in Europe have now been regulated; in some countries there are very few unregulated rivers (Brookes, 1987; Petts, 1988; Garcia De Jalon, 1987). River regulation has been undertaken to the greatest extent in Western and Southern Europe. Thus in countries such as Belgium, England, Wales and Denmark, the percentage of river reaches that are still in a natural state is low, typically 0–20%. By contrast, in countries such as Poland, Estonia and Norway, many rivers still have 70–100% of their reaches in a natural state.

River regulation often causes major changes in river processes, primarily the flow regime and the transport of dissolved and particulate matter. The effects are seen not just locally, but may be extensive, with downstream reaches nearly always being affected, and upstream reaches and the surrounding areas often being affected as well. Some types of river regulation, for example land drainage, may affect much if not all of a catchment area, but that which has the most widespread

Table W24 Number of lakes in different European countries

Country	0.01–0.1	0.1–1	1–10	10–100	>100
	Surface area (km²) Number of lakes				
Albania	–	–	–	>3	3
Austria	some 100s		19	7	2
Bulgaria	53	175	288	14	0
Croatia	–	1	3	0	0
Denmark	354	256	74	6	0
England and Wales	1665		50	2	0
Estonia	750	209	41	1	3
Finland	40 309	13 114	2283	279	47
France	–	128	23		1
Georgia	799	58	21	14	0
Germany	–	–	~100	~20	2
Greece	–	–	–	>16	1
Hungary	–	–	–	2	2
Iceland	~7000	1650	176	17	0
Ireland	–	–	~100	14	3
Italy	–	>168	>82	13	5
Latvia	2164	740	122	20	0
Moldova	>3300	48	30	6	0
The Netherlands	–	–	–	47	3
Norway	208 000		2000	450	7
Poland	6050	2627	545	32	2
Russian Federation	471 000		4626	412	51
Spain	–	–	–	800	
Sweden	59 500	19 374	3990	358	22
Switzerland	1300		10	15	5
Ukraine		950		>4	2
Former Yugoslavia	>200		>10	15	4

and marked effect is the construction of reservoirs. Nevertheless, many river systems have also been changed by channelization, especially those in lowland areas.

Reservoirs

Because reservoirs usually have a relatively short water residence time – often less than a year and sometimes just a few days – they can be regarded as a hybrid between a river and a lake that is internally divided in three zones: a river-like zone at the inflow end, a lake-like zone at the outflow end and a transitional zone in between. Another prominent feature of many reservoirs is that the water level fluctuates, so that the littoral zone is biologically poorly developed where the fluctuations exceed those of natural lakes in the region.

Reservoirs have been constructed in Europe for thousands of years; the earliest were relatively small, and used mainly for domestic water supply and crop irrigation. During the last two centuries there has been a marked increase in both reservoir size and number, with large storage capacity reservoirs constructed in many countries, especially the former USSR. Thus there are currently about 3900 large reservoirs with dams higher than 15 m in Europe (not including the territory of the former USSR), half of which have been built since 1961 (Boon, 1992). To this must be added the many large reservoirs in the European part of the former USSR, and the thousands of smaller reservoirs and ponds spread throughout Europe.

The development of reservoir construction in Europe can be illustrated using the UK and Spain as examples. In the UK the number of reservoirs grew rapidly during the second half of the nineteenth century from about 50 to about 200; from then until the mid-1970s, the rate was about six new reservoirs per year (Boon, 1992). In Spain the number of reservoirs grew at the rate of about two per year between 1900 and World War II, but at a rate of about 20 per year in the post-war period from 1950 to 1980 (Garcia De Jalon, 1987; Riera *et al.*, 1992). Reservoir construction in Europe has now fallen off and growth in total reservoir area seems to be stagnant. This is mainly because of the lack of suitable sites (Williams and Musco, 1992).

Reservoir usage
Reservoirs are usually built to serve several purposes, the primary uses typically being the generation of hydroelectric power, irrigation,

flood control, and domestic and industrial water supply. Other uses are commercial fisheries and various recreational activities.

Environmental problems related to reservoirs
Reservoir construction leads to a number of environmental problems, both during the building phase and following completion. As the water level in the reservoir rises upon the closing of the dam, major changes often take place in the area to be inundated; farmland can be lost, settlements flooded and the groundwater table elevated. Once the reservoir has been established, the environmental problems can be divided in two groups: those that render the reservoir unsuitable for its purpose, for example algae and toxic substances in reservoirs used for drinking water, and those that induce ecological deterioration of the river system, especially downstream of the reservoir.

The water quality of a reservoir, as reflected by the content of pathogens, toxic chemicals and poisonous algae, is of primary concern when the reservoir is used for drinking water, commercial fisheries, industrial processes, and recreational activities such as bathing and water sports. At present it is not possible to give a general overview of the extent to which European reservoirs are contaminated with toxic substances. However, contamination of reservoirs with oil, organic solvents, heavy metals and radionuclides has been reported to be a problem in the Russian Federation and in Ukraine (Mnatsakanian, 1992; Gavrilov *et al.*, 1989). Many of the environmental problems to which reservoirs are subject, such as eutrophication and heavy metal, organic and thermal pollution, also occur in natural lakes and rivers.

Since reservoirs interrupt the natural continuity of a river, the ecological consequences can be manifold. Access to spawning sites for migratory fish is prevented, the problem being especially acute for fish such as salmon, trout and sturgeon. As reservoirs trap the suspended matter flowing into them, they reduce the suspended matter load to downstream reaches; in contrast, because of their high biological productivity, the organic load to the downstream reaches may increase. Since reservoirs regulate the water flow, sudden flow fluctuations may occur downstream of reservoirs used for hydro-electric power generation. In other cases reservoirs may have a stabilizing effect on the downstream flow regime, thereby benefitting downstream floral and faunal communities. Still other reservoirs warm up the river water, thereby elevating the downstream water temperature. Finally, the water flowing from reservoirs with bottom-water outlets comes from the deeper layers and is consequently colder and, due to poor mixing and decomposition of organic matter, occasionally low in oxygen content. The changes in flow regime and water temperature detrimentally affect the downstream aquatic community. To quote Casado *et al.* (1989), these are 'a reduction of macrophytes, a reduction in faunistic richness both of fish and invertebrates, and a reduction of fish biomass, density and growth'.

During the last 10–15 years there has been growing public opinion against the construction of dams and reservoirs. For example, in Norway the Lappish population protested against the damming of the river Alta, and in Spain, a 101 m high dam built on the river Esla in 1970 to form the Riano reservoir remains unused because the inhabitants of the valley to be expropriated have refused to move (Garcia De Jalon, 1987). A more recent example is the construction of the Gabcikovo hydroelectric power plant on the Danube; although both Hungary and the former Czechoslovakia were jointly involved in the original project, environmental groups in Hungary forced the Hungarian government to withdraw. The Hungarians are now calling attention to an environmental catastrophe in the area.

River channelization

The objective when modifying the course of a river is to improve certain features, for example flood control, drainage of the surrounding land, navigation and erosion prevention. River channelization comprises a number of physical measures, each of which is related to hydrological parameters; hence straightening changes the slope, dredging changes the depth and width, and dredging and weed cutting change the roughness. Other more radical methods of river channelization are culverting, lining and piping.

In many countries where there is intensive agricultural production, many of the rivers have been regulated. In Denmark, for example, 85–98% of the total river network has been straightened (Brookes, 1987; Iversen *et al.*, 1993).

Physical and biological effects

Channelization has a great impact on a river because it disrupts the existing physical equilibrium of the watercourse; to compensate for the alteration in one or more of the hydraulic parameters, and to establish a new, stable equilibrium, other parameters will change. Because straightening of a river increases its slope, the energy in the moving water has to be dispersed over a smaller surface; as a result the water is able to move larger particles and sediment discharge increases through bank erosion. If the river is not repeatedly manipulated or stabilized by culverting, lining, etc., this will eventually lead to widening of the river channel and to a subsequent reduction in water velocity. River channelization generally changes a heterogeneous system into a homogeneous one such that flow becomes uniform, pools are lost and the substrate becomes uniform throughout the channel.

Channelization can also have great impact on riparian vegetation; trees are often logged to allow channel maintenance (e.g. machine dredging) and scrub is cut to ensure sufficient drainage. This increases solar radiation at the stream surface, thereby increasing the water temperature, reducing the concentration of dissolved oxygen and increasing the in-stream primary production. In nutrient-rich watercourses this results in enhanced growth of benthic algae and macroalgae.

Another effect of the channelization of rivers and drainage of wetlands may be increased nutrient and organic matter loading of rivers and the marine environment. The reason is that while the annual nitrogen removal capacity of wetlands and natural rivers can be as much as several hundred kilograms per hectare, that of channelized rivers and drained wetlands may be negligible. Naturally meandering riparian zones alongside rivers may therefore play an important role in balancing intensive agricultural and ecological interests.

The velocity of river water is one of the major factors regulating the structure of riverine plant and animal communities (Westlake, 1973; Brookes, 1988). The uniform and often unstable sediment found in channelized watercourses is suitable for few, if any, plant species. Furthermore, as the uniform water flow precludes areas with little or no flow, resting sites for fish and invertebrates are virtually absent. The general effect of channelization is therefore a reduction in habitat number and diversity and a consequent reduction in species number and diversity. The latter may be further reduced by the above-mentioned decrease in oxygen concentration. Hence the biomass of organisms such as fish and invertebrates is usually lower in channelized watercourses.

It is not only animals and plants living within the watercourse that are affected by channelization, however. Thus animal species which depend on the bank for foraging and/or breeding decline in number, with the consequence that species diversity on the river banks also decreases. In addition, a number of plant species that are confined to the more or less water-saturated soil adjacent to the river are also affected.

Organic pollution of rivers

Organic matter derived from diverse human activities is a major source of pollutant discharge to rivers. The decomposition and breakdown of this organic matter is mediated by microorganisms and takes place mainly at the surface of the sediment and vegetation in smaller rivers, and in the water column in larger rivers. Since the process requires the consumption of oxygen, severe organic pollution may lead to rapid deoxygenation of the river water and hence to the disappearance of fish and aquatic invertebrates. The habitat then becomes uniform with only a few robust species able to tolerate the low oxygen concentration. Decomposition of organic waste also results in the release of ammonium which, although not in itself toxic, may, depending on the pH and temperature of the water, be converted to ammonia, which is poisonous to fish.

The most important sources of the organic waste feeding into rivers are domestic and industrial sewage. Immediately downstream of a sewage effluent, organic matter decomposition reduces the oxygen content of the water and results in the release of ammonium (Figure W32). Further downstream the concentration of organic matter decreases as a result of dilution and continuing decomposition. As the distance from the effluent increases, bacteria oxidize the ammonium to nitrate, and oxygen enters the water via the water surface, thereby increasing its oxygen content. Eventually the levels of organic matter, oxygen and ammonium reach those present immediately upstream of the sewage effluent; this process of recovery is called self-

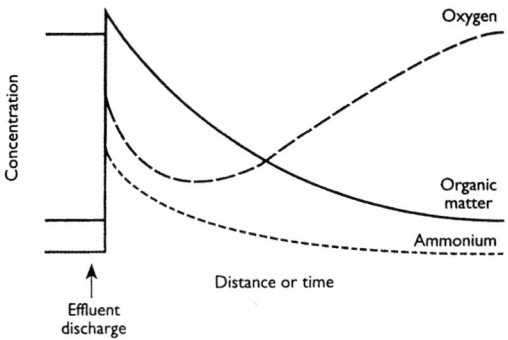

Figure W32 Impact of an organic matter effluent on river concentrations of organic matter, oxygen and ammonium.

purification. An example is the Danube, which is already polluted by organic matter when it enters Hungary. As it winds its way through the country the river receives large amounts of organic matter from tributaries and cities, especially from Budapest. However, by the time it leaves the country and enters Croatia, an amount of organic matter equal to that discharged in Hungary has been decomposed (Varga *et al.*, 1990; Benedek and Major, 1992). Nevertheless, this does not imply that rivers can take up an unlimited amount of organic matter without suffering as a result; the pollution may be so severe, widespread and long lasting that self-purification is insufficient. Thus the Danube is still polluted when it leaves Hungary, and the Rhine was polluted with such excessive amounts of organic matter between World War II and the early 1970s that there was very serious oxygen depletion in its middle and lower course and the river virtually died (Friedrich and Müller, 1984).

Organic matter content in European rivers

Because decomposition of organic matter requires oxygen, the amount of organic matter in a river can be measured in terms of the biochemical oxygen demand (BOD) or the chemical oxygen demand (COD), the units of which are (mg O_2) l^{-1}. River reaches little affected by human activities generally have a BOD below 2 (mg O_2) l^{-1} whereas a BOD exceeding 5 (mg O_2) l^{-1} generally indicates pollution. Measurement of BOD is the most widespread method in Europe, but many countries also measure COD, and some use only COD. Although both BOD and COD indicate the potential oxygen demand of the organic matter in the water, there is not necessarily a correlation between the two measurements. In large rivers suffering from severe eutrophication, elevated BOD values can occur due to decomposition of phytoplankton, and in these cases high BOD values are not necessarily indicative of organic pollution.

BOD, COD and oxygen content data from a large number of river stations in 33 European countries are collated in Table W25. Median BOD, COD and oxygen are 2.8, 14.5 and 9.7 (mg O_2) l^{-1} respectively. As can be seen, annual mean BOD was below 5 (mg O_2) l^{-1} and the oxygen content above 8 (mg O_2) l^{-1} at more than 75% of the river stations. Extremely high BOD is generally seen only in smaller rivers polluted with raw sewage or animal slurry, a BOD exceeding 500 (mg O_2) l^{-1} then being possible.

In Iceland, Norway, Sweden and Finland organic matter content is measured only as COD. In these countries discharge into rivers of organic waste derived from human activity is negligible and COD levels therefore are generally low. In terms of BOD, rivers in Ireland, Georgia, Estonia, Latvia, Austria, Switzerland, The Netherlands, the UK, Denmark and Croatia are least affected, with less than 25% of the rivers having a BOD exceeding 3.5 (mg O_2) l^{-1}. In Hungary, Lithuania, Portugal, France, Ukraine, Germany, Slovenia and Italy the rivers are moderately affected, with less than 25% having a BOD exceeding 5 (mg O_2) l^{-1}. More affected rivers are found in Albania, Poland, the Czech Republic, Moldova, the Russian Federation and Spain, where more than 25% of the rivers have a BOD exceeding 5 (mg O_2) l^{-1}. BOD is highest in Bulgaria, Belgian Flanders and Romania, exceeding 5 (mg O_2) l^{-1} in 60, 69 and 80% of the rivers, respectively.

Table W25 Descriptive statistics for annual mean organic matter and oxygen at European river stations

	Number of river stations	Mean	Percentage of river stations with organic matter and oxygen concentrations (not exceeding (mg O_2) l^{-1})				
			10%	25%	50%	75%	90%
All rivers							
BOD	645	4.5	1.4	1.9	2.8	4.7	7.9
COD	470	18.5	4.5	7.8	15.0	25.0	36.6
Oxygen	620	9.4	6.4	8.4	9.7	10.7	11.6
Near-pristine rivers							
BOD	11			1.2	1.6	2.7	
COD	23			5.1	13.3	29.9	
Oxygen	8			10.2	10.6	11.1	

Table W26 Descriptive statistics of annual mean physical and chemical variables at European river stations

	Number of river stations	Percentage of river stations with concentrations not exceeding		
		25%	50%	75%
pH	717	7.5	7.8	8.0
Total alkalinity (meq l^{-1})	274	1.0	2.5	4.0
Chloride (mg Cl l^{-1})	442	17.3	26.5	68.3
Organic matter and oxygen level (mg O_2 l^{-1})				
Biochemical oxygen demand	645	1.9	2.8	4.7
Chemical oxygen demand	740	7.8	15.0	25.0
Dissolved oxygen	620	8.4	9.7	10.7
Nitrogen (mg N l^{-1})				
Ammonium	580	0.1	0.2	0.4
Nitrate	654	0.7	1.8	3.9
Total nitrogen	329	0.8	2.1	4.5
Phosphorus (μg P l^{-1})				
Dissolved orthophosphate	412	45.0	124.0	286.0
Total phosphorus	546	59.0	170.0	366.0
Heavy metal (μg l^{-1})				
Copper	192	1.0	4.0	8.0
Zinc	176	5.0	10.0	

Table W27 Quality of European river reaches

	Water quality (%)			
	Good[a]	Fair[b]	Poor[c]	Bad[d]
Austria, 1991	14	82	3	1
Belgium, 1989–1990	17	31	15	37
Bulgaria, 1991	25	33	31	11
Croatia	15	60	15	10
Czech Republic, 1990	12	33	27	28
Denmark, 1989–1991	4	49	35	12
England and Wales, 1990	65	23	10	2
Finland, 1989–1990	45	52	3	0
Germany, 1985	44	40	14	2
Iceland	99	1	0	0
Ireland, 1987–1990	77	12	10	1
Italy	27	31	34	8
Latvia	10	70	15	5
Lithuania	2	97	1	0
Luxembourg	53	19	17	11
Netherlands, 1990	5	50	40	5
Northern Ireland, 1990	72	24	4	0
Poland, 1990	10	33	29	28
Romania	31	40	24	5
Russian Federation	6	87	5	2
Scotland, 1990	72	24	4	0
Slovenia, 1990	12	60	27	1

[a] River reaches with nutrient-poor water, low levels or oganic matter; saturated with dissolved oxygen; rich invertebrate fauna; suitable spawning ground for salmonoid fish.
[b] River reaches with moderate organic pollution and nutrient content; good oxygen conditions; rich flora and fauna; large fish population.
[c] River reaches with heavy organic pollution; oxygen concentration usually low; sediment locally anaerobic; occasional blooming of organisms insensitive to oxygen depletion; small or absent fish population; periodic fish kill.
[d] River reaches with excessive organic pollution; prolonged periods of very low oxygen concentration or total deoxygenation; anaerobic sediment, severe toxic input; devoid of fish.

Human activities and organic matter

The organic matter naturally occurring in rivers originates from soil erosion and from dead plants and animals, and is normally relatively insoluble and only slowly decomposed. In contrast, organic matter derived from human activities is generally soluble, finely divided and rapidly decomposed, the result being a marked and abrupt increase in oxygen consumption in the river.

As the population density in catchments increases, the level of organic matter in the rivers generally increases, and the oxygen content decreases. Thus, whereas BOD concentration is lower than 2 (mg O_2) l^{-1} in catchments with fewer than 15 inhabitants km^{-2}, it generally exceeds 5 (mg O_2) l^{-1} in catchments with more than 100 inhabitants km^{-2}. The great variation found in extensively populated areas is attributable mainly to variation in the extent of wastewater treatment – well-functioning treatment plants can decompose up to 90% of the organic matter in the wastewater.

Descriptive statistics for frequently measured physical and chemical variables in European rivers are given in Table W26.

Assessment of river quality

In most European countries the environmental state of the rivers has been monitored for many years. However, as pollution has both physicochemical and biological effects on the receiving water, the quality of the water can be assessed in many different ways, and there are numerous methods in use throughout Europe.

The quality of European rivers is summarized in Table W27. About a quarter of the river reaches are classified as having poor or bad water quality. Most of the countries classify 50% or more of their river reaches as having good or fair quality. Iceland, Scotland, Northern Ireland and Ireland have the highest proportion of rivers classified as having good water quality, while the Russian Federation, Finland, England and Wales, West Germany, Austria, Croatia, Lithuania and Latvia classify more than 75% of their rivers as having good or fair water quality. More than 25% of the rivers have poor or bad water quality in Bulgaria, Romania, the Czech Republic, Poland,

Denmark, The Netherlands, Luxembourg, Slovenia and Italy. Furthermore, the percentage of river reaches classified as having bad water quality is highest in the Czech Republic and Poland (28%) and in Belgian Flanders (37%).

The greater the amount of organic matter present in river water, the lower the oxygen concentration, and the higher the ammonium concentration. A low oxygen concentration influences the river fauna. To maintain a salmon or trout population in a river, a minimum oxygen concentration of 6 (mg O_2) l^{-1} is necessary (Council Directive 78/659/EEC). To reduce the risk of fish kill, the ammonium concentration should not exceed 1 (mg N) l^{-1}. Although ammonium is not in itself toxic to fish, it becomes toxic when converted to ammonia. Thus if the ammonia concentration exceeds 0.025 mg l^{-1}, trout growth is prevented, and if it exceeds 0.25 mg l^{-1}, the trout die. Rivers with large sewage discharges exceed these limits.

Trends in organic matter discharge to rivers

After World War II riverine discharge of organic waste increased in many European countries with resultant severe oxygen depletion. During the last 15–20 years, however, biological treatment of domestic and industrial wastewaters has intensified, and organic matter loading of rivers has consequently decreased in many parts of Europe, the result being that many rivers are now fairly well oxygenated.

An example is the River Rhine. Rebuilding of industry after World War II led to high loads of poorly or untreated sewage being discharged into the river. This caused oxygen contents of the water to fall considerably, and long stretches became biologically devastated.

R.W. Herschy

Source

Stanners, D. and Bordeau, P. (eds), 1995. *Europe's Environment: The Dobris Assessment*, The European Environment Agency, Copenhagen.

Bibliography

The water resource

Arnell, N., Oancea, V. and Oberlin, G., 1993. *European river flow regimes*. Report to the European Environment Agency Task Force. Institute of Hydrology, Wallingford, and CEMAGREF, Lyons.

Baumgartner, A. and Reichel, E., 1975. *The World Water Balance*. R. Oldenburg Verlag, Munich.

Bernes, C. and Grundsten, C., 1992. *The environment national atlas of Sweden*. The National Environmental Protection Agency, Stockholm.

Boon, P., 1992. Essential elements in the case of river conservation, in Boon, P.J., Calow, P. and Petts, G.G. (eds) *River Conservation and Management*, pp. 11–33, Wiley, New York.

Brömssen, U. von, 1986. *Acidification of drinking water–groundwater*, pp. 251–61. Elsevier, Amsterdam.

Custodio, E., 1991. *Characterization of aquifer overexploitation: comments on hydrogeological and hydrochemical aspects: The situation in Spain*. IAH XXIII International Congress, Puerto de la Cruz, Tenerife, Spain.

Gottschalk, L. Jensen, J.L., Lundquist, D. *et al.*, 1979. Hydrological regions in the Nordic Countries. *Nordic Hydrology*, **10**, 273–86.

Krasovskaia, I. and Gottschalk, L., 1992. Stability of river flow regimes. *Nordic Hydrology*, **23**, 137–54.

Last, F.T. and Watling, R. (eds), 1991. *Acidic deposition: its nature and impacts*. Proceedings of the international symposium held in Glasgow, Scotland, 16–21 September 1991. Royal Society of Edinburgh.

Lvovich, M.I., 1973. The global water balance. *Transactions of the American Geophysical Union*, **54**, 28–42.

OECD, 1992. *Environmental indicators. (In-depth analysis of individual indicators: intensity of use of water resources)*. OECD Secretariat, Group of the State of the Environment, Paris.

Shiklomanov, I.A., 1991. The world's water resources. In: *Proc International Symposium to Commemorate the 25 Years of IHD/IHP*, pp. 93–126. UNESCO, Paris.

Umweltbundesamt, 1992. *Daten zur Umwelt*. Erich Schmidt Verlag, Berlin.

UK DoE, 1992. *The UK Environment*. UK Department of the Environment HMSO, London.

UNEP, 1991. *Freshwater pollution*. UNEP/GEMS Environment Library, No. 6, Nairobi.

UNESCO, 1978. *World Water Balance and Water Resources of the Earth*. UNESCO Series Studies and Reports, No. 25, UNESCO, Paris.

Veiga da Cunha, L., 1993. Water resources in Europe, in Sesimbra (ed.) *The European common garden. Based on reports prepared for the 'East–West Interparliamentary Meeting'. Strasbourg, 17–20 May 1992*, pp. 159–204, Sesimbra, Brussels.

WRI, 1990. *World resources 1990–91*. Oxford University Press, World Resources Institute, Oxford and New York.

Zektser, I.S., Lorne, G.E. and Cullen, S.J., 1992. Groundwater pollution: an international perspective. *European Water Pollution Control*, Vol. 2 (6).

Rivers, reservoirs and lakes

Benedek, P. and Major, V., 1992. In: *Point source pollution in the Danube Basin. County technical reports. Bulgaria, the CSFR, Hungary, and Romania. Volume III*. Water and Sanitation for Health Project, USAID, Washington DC.

Boon, P.J., 1992. Essential elements in the case for river conservation. In: Boon, P.J., Calow, P. and Petts, G.E. (eds) *River conservation and management*, pp. 11–33, Wiley, New York.

Brömssen, U. von, 1986. *Acidification of drinking water-groundwater*, pp. 251–61. Elsevier, Amsterdam.

Brookes, A., 1987. The distribution and management of channelized streams in Denmark. *Regulated Rivers*, **1**, 3–16.

Brookes, A., 1988. *Channelized rivers: perspectives for environmental management*. Wiley, New York.

Casado, C. de Jalon, D.G., del Olmo, C.M., 1989. The effect of an irrigation and hydroelectric reservoir on the downstream communities. *Regulated Rivers*, **4**, 275–84.

Fortunatov, M.A., 1979. Physical geography of the Volga basin, in Mordukhai-Boltovskoi, E.D. (ed) *The River Volga and its life*, Junk Publishers, The Hague, 1–29, Monographiae Biologicae 33.

Friedrich, G. and Müller, D., 1984. Rhine, in Whitton, B.A. (ed.) *Ecology of European rivers*, pp. 265–315. Blackwell, Oxford.

Garcia De Jalon, D., 1987. River regulation in Spain. *Regulated Rivers*, **1**, 343–8.

Gavrilov, V.H., Gritchenko, Z.G., Ivanova, L.M. *et al.*, 1989. Strontium-90, caesium-134, and caesium-137 in water reservoirs of the Soviet Unions' Baltic region (1986–1988), in Three years observation of levels of some radionuclides in the Baltic Sea after the Chernobyl accident. Seminar on the radionuclides in the Baltic Sea. 29 May 1989. Rostock-Warnemünde, German Democratic Republic. *Baltic Sea Environment Proceedings*, **31**, 62–80.

Herdendorf, C.E., 1982. Large lakes of the world. *Journal of Great Lakes Research*, **8**, 379–412.

Iversen, T.M. Kronvang, B. Madsen, B.L. *et al.*, 1993. Re-establishment of Danish streams, restoration and maintenance measures, *Aquatic Conservation*, **3**, 1–20.

Kempe, S. Pettine, M. and Cauwet, G., 1991. Biogeochemistry of European rivers, in Degens, E.T. Kempe, S. and Richey, J.E. (eds) *Biochemistry of major world rivers*, pp. 169–211. SCOPE 42, Wiley, New York.

Meybeck, M. and Helmer, R., 1989. The quality of rivers: from pristine stage to global pollution. *Palaeogeography, Palaeoclimatology, Palaecology*, **75**, 283–309.

Mnatsakanian, R.A., 1992. *Environmental legacy of the former Soviet Republics*. Centre for Human Ecology, University of Edinburgh, Edinburgh.

Overgaard, K., 1986. *Nitrate and pH in drinking water*. Miljøprojekt 75, National Agency for Environmental Protection, Copenhagen.

Petts, G.E., 1988. Regulated rivers in the United Kingdom, *Regulated Rivers*, **2**, 201–20.

Rebsdorf, A. Thyssen, N. and Erlandsen, M., 1991. Regional and temporal variation in pH, alkalinity and carbon dioxide in Danish streams, related to soil type and land use. *Freshwater Biology*, **25**, 419–35.

Riera, J.L. Jaume, D, de Manuel, J. *et al.*, 1992. Patterns of variation in the limnology of Spanish reservoirs: a regional study. *Limnetica* **8**, 111–23.

Targbil, G., 1986. *The Norwegian monitoring programme for long range transport of air pollutants, results 1980–1984*. The Norwegian State Pollution Control Authority.

Varga, P., Abraham, M. and Simor, J., 1990. Water quality of the Danube in Hungary and its major determining factors. *Water Science and Technology*, **22**, 113–18.

Westlake, D.F., 1973. Aquatic macrophytes in rivers: a review. *Polskie Archivum Hydrobiologii*, **20**, 31–40.

Williams, D. and Musco, H., 1992. *Research and technological development for the supply and use of freshwater resources – SAST Project 6.* Strategic Dossier. Report published by the Commission of the European Communities, DG XII, Brussels, Luxembourg.

Cross references

Drinking water and sanitation
Groundwater
River pollution prevention
Rivers, table of length, drainage area and discharge
Sanitation and clean water
Sewage treatment
Water availability and quality, world
Water balance
Water pollution, types, causes and effects
Water quality for drinking, WHO guidelines
Water treatment, potable water
World water balance

WATER RESOURCES: INTEGRATED RIVER BASIN MANAGEMENT

What is integrated river basin management?

Integrated river basin management (IRBM) is now an important concept in the strategic planning of water resources development. It is essential in catchments where the water in the rivers is used intensively, i.e. where it is used for many different, and often conflicting, purposes. For example, a river may be required as a source of raw water for potable supplies and also as an effluent carrier. A compromise is thus required to cater for the needs of both type of use and a balance has to be struck between the volume of water abstracted for treatment and its associated cost before being put into supply, and the cost and volume of wastewater treatment necessary to provide river water of a quality suitable for abstractions downstream.

This entry summarizes data requirements and evolutionary phases of every river system leading to the introduction of IRBM methodology.

Data requirements

IRBM's use depends on the construction of an accurate mathematical model of the whole river system and its catchment area. It uses three kinds of data, namely static, (catchment/subcatchment areas, slopes, details of water and wastewater treatment plants, etc.), dynamic (precipitation, stream and river flows, water quality data, etc.), and socio-economic data (population, its size and growth rate, food and fiber requirements).

The three categories of data are set out because they require fundamentally different methods of acquisition and are subject to different, but related, evolutionary stages. The static and near-static data, although essential, involve only initial effort and thereafter minimal updating is required. This type of data will not be discussed further.

Perfection cannot, of course, be achieved immediately; the data collection system grows and changes in sympathy with, and in response to, the changing requirements for information. There is, however, a common thread which runs through the evolutionary pattern which can be divided into three stages:

- Stage 1 is the acquisition of basic knowledge, the accumulation of records for planning, research and other *non-operational* purposes;
- Stage 2 occurs with the introduction of operational management practice which, initially, is mainly concerned with water resources development and management;
- Stage 3 occurs with the introduction of integrated river basin management.

Stage 1

Assuming that no information about a river basin exists, Stage 1 has as its main objectives:

- the acquisition and recording of basic knowledge of the catchment area;
- the behavior of its drainage system.

For these purposes, both static and dynamic data, must be collected, but the collection of socio-economic data is limited to an examination of the proposed development plans for the catchment.

The static data are gathered principally by surveys to establish the catchment area and characteristics, for example its boundary, geomorphology, geology, vegetation cover and land use, and putting them all into a reference framework, e.g. a national grid referencing system or GIS.

Collecting dynamic data involves the sustained measurement and recording of all surface- and groundwater parameters, from which the limits of variation of the key parameters, such as flow and quality and the pattern of variation within those limits, can be determined.

An important activity of Stage 1 is the production of codes of practice for data collection and operating procedures. These not only enable the Stage 1 measurement scheme to be operated consistently and effectively, but will guide further development.

A less obvious, but nevertheless important function of Stage 1, is to prepare for the subsequent evolutionary stage, which will be almost certainly different in character from Stage 1. For instance, important areas of the catchment must be identified. These may be areas where the catchment is to be considered for agricultural, industrial, urban or even water resources development. 'Sensitive' areas should also be identified. These include areas where flooding may occur, or where water quality may be adversely affected to an unacceptable degree by planned developments.

The data collected in Stage 1 are not normally used for operational purposes; they are used for retrospective analysis. In addition, it will be discovered subsequently that too many data have been collected because, until the important interrelationships of the catchment are known, it is not possible to define data requirements precisely.

It is important to realize that, although Stage 1 may take several years to complete, it is finite. It will be complete when it is considered that a working knowledge has been obtained of the system under study under normal conditions. Knowledge about extreme events, floods and droughts (which, by their nature, occur much less frequently), must continue to be gained at every possible opportunity.

Stage 2

Whilst there is no clear-cut boundary between Stages 1 and 2, the latter implies the adoption by management of an overall concept for the future development of operational schemes and the data and information systems which are an inherent part of them. Thus the dynamic data should be collected in a timely fashion using telemetry systems to enable operational management decisions to be made as and when required.

Stage 3

In certain cases, where integrated river basin management becomes necessary because of the need to use the water resources of the catchment intensively, or where positive control is required for other reasons, for example recreational or amenity requirements, Stage 3 will be entered.

In this stage, systems for the collection of dynamic data have to be progressively refined and will closely resemble those used in process control applications. Socio-economic data have to be collected to ascertain how many people are using the river for recreational purposes (boating, fishing, swimming, etc.) and the value that they place upon it; how many enjoy informal recreation (walking, picnicking, etc.) and the 'general environmental value' of having a pleasant river in the area.

Evolutionary phases of a river system

It may be of interest to note that every river system, whose catchment is being developed, passes through three distinct phases in the deterioration of the river water quality.

- Phase 1. Before any development occurs, any population is thinly spread over the catchment and the water in the river system is of its natural quality. Phase 1 commences when the population increases and congregates in settlements. Waste is difficult to dispose of, so it is thrown into the river. This is called the stage of pathogenic pollution and is responsible for outbreaks of conventional water-borne diseases such as dysentery, cholera and typhoid. It can be overcome by adequate wastewater treatment and the cessation of dumping of waste into the river.

- Phase 2. As development continues in the catchment, industrialization develops and wastes with different characteristics are discharged to the river. This is a period in which the river suffers from high biochemical oxygen demand (BOD), and consequently low dissolved oxygen (DO) and high suspended solids (SS). The flora and fauna suffer badly and the river may, indeed, become 'dead'. It is called the phase of gross pollution and can be overcome by sufficient expenditure on wastewater treatment plant.
- Phase 3. Finally, and possibly the most dangerous phase of all, is when massive production of new and highly sophisticated synthesized chemicals and trace organics (micropollutants such as herbicides and pesticides) are manufactured or applied within the catchment, the effluents from which drain into the river. It is known as the chemical pollution phase. It is very expensive and difficult to overcome completely; in many cases it is more cost effective to prohibit the discharge of the pollutant into the receiving water.

In some developing countries, advances in industrialization take place at such a rate that it is possible to discover river systems where all three evolutionary phases are present at the same time.

The severity of the pollution of the river system is kept in check by legislation which seeks to minimize it. The legislation must keep in step with the technical capability to treat the pollution. Together with a suitable institutional framework, the use of the river system can be optimized to meet the needs of the different users of the river water.

Also, by the use of the mathematical model, the effects of a proposed new discharge, or a new abstraction, can be assessed before a consent to discharge – or abstract – is granted without making any actual quality or quantity changes to the river. Thus discharge consent conditions can be derived which minimize the effect of a proposed new discharge on other downstream users; for example, dischargers can be asked to site the discharge elsewhere in the catchment or, indeed, they may be refused permission to discharge at all.

It should be noted that one of the consequences of river basin development in any river system is the production of waste from sewage, industrial and agricultural processes. Research must be initiated into the constructive use of these wastes. This is vital if we want to preserve the environment and not leave a legacy of pollution for subsequent generations.

IRBM is not a technique which is applied once and all consents to abstract and discharge are then fixed for all time; it has the virtue of being extremely flexible in operation, provided that the timely data collection system is maintained, reviewed and refined on an ongoing basis. Should the overall objectives of the river basin be changed, the model of the catchment can be reoptimized to meet these new objectives whilst minimizing the effect on abstractors and dischargers.

Integrated river basin management is the key to overall planning policy and, in order to achieve it efficiently, some amalgamation of existing organizations dealing with water conservation and resources development, treatment and supply, wastewater treatment and disposal, river management and recreational uses of the river system is essential.

David H. Newsome

Bibliography

Newsome, D.H., 1975. Integrated river basin management, *Mitteilungen aus dem Institut für Wasserwirtschaft, Hydrologie und Landwirtschaftlichen Wasserbau der Technischen Universität Hannover*, pp. 419–437.
Seventh River Basin Management Conference, 1995. Kruger National Park, South Africa, May 15–19 1995, *International Association on Water Quality, Water Science and Technology*, **32**(6).

Cross references

Drought
Floods
Water resources: introduction

WATER RESOURCES: NATURAL QUALITY

The Earth's water resources consist of the oceans and seas, polar icecaps and glaciers, underground waters, and the waters in streams, lakes and rivers. Of this so-called hydrosphere, which together with atmospheric water and the water held in living things completes the hydrological cycle, about 97% consists of seawater and about 2% is ice located mainly at the poles. The remaining 1% consists of fresh and brackish inland waters.

Very nearly all of these waters were substantially of natural quality until the Industrial Revolution initiated the growth of modern technology, and human actions began to cause serious local pollution of inland waters and estuaries. Now we have reached the stage where the entire hydrological cycle is contaminated, and apparently no water of absolutely natural quality exists. Fortunately, this global contamination of water is very low, and serious pollution exists only in the inland and coastal waters of the developed countries. All the technologically developed nations are now seeking to manage properly the quality of their inland and coastal waters, and attempts are being made to establish control of pollution of seas on an international basis. While the latter is still very much in an embryonic state, the organization and techniques of management of inland and coastal water quality are well advanced.

In considering the management of inland and coastal water resources, it is essential to establish first some notion of what the natural quality of these waters should be, if only for the purpose of specifying broadly the natural background, or baseline, of quality management operations.

Unfortunately quality, even in relation to water, is an abstract term. Other abstract terms such as pure, clean, wholesome, dirty and polluted are commonly used to identify the various quality states of water. But these only become universally meaningful if they are related to the use to which the water is put. For example a water clean enough to support many species of freshwater fish may be too polluted to be safe for public bathing, or a water wholesome enough for human drinking may not be sufficiently pure for use in the manufacture of electronic equipment.

To decide whether a water is of a quality suitable for a particular purpose, its quality must be specified in scientific terms. Largely as a result of empirical observations, the physical, chemical and biological characteristics (or quality parameters) of waters have been quantified in relation to various uses. This process is continuously being revised and extended, and various national and international publications of water quality criteria relative to water uses have been published.

Using such physical, chemical and biological assessments as appropriate, the natural quality of inland and coastal waters can be described.

R.W. Herschy

Source

Fish, H., 1973. *Principles of Water Quality Management*, Thunderbird Enterprises, UK.

Cross references

Water resources: Europe
Water resources: surface and groundwater
World water balance

WATER RESOURCES: QUALITY ASSESSMENT

The assessment of water resources quality involves the streams, rivers, canals, groundwaters, lakes and tidal waters forming water resources, and the human-made factors affecting the quality of these resources. The purpose of assessment is to establish the present position, how this changes under the influence of controllable and uncontrollable events in the short and long term, what control should be exercised and subsequently modified, and the effectiveness of the control applied.

Essentially the assessment process consists of two parts – the making of observations and the interpretation of these observations. The term 'monitoring' is frequently used to mean the making of observations, or the interpretation of observations, or both. So, to minimize confusion, the term will not be used in the present context.

The methods of observation of resources quality and the factors affecting this fall into three categories:

- simple observation by means of sight, smell and hearing;
- field sampling and subsequent analysis;
- automatic instrumental measurement and recording.

The importance of personal observation must be strongly emphasized. First, because in this electronic age little credence is given to human efficiency and effectiveness, and second because it is a simple, proven fact that a well-trained and experienced person can both observe and assess many aspects of water quality and related matters with great efficiency and effectiveness. Much information about the physical and biological quality of a water can be deduced from the appearance of the water and a brief, but careful, examination of the plant and animal life present. Where the observer has had the opportunity of relating such observations to results of chemical analysis of waters, from experience it will be possible to deduce a fair amount of information regarding the chemistry of the water using judgment of its physical and biological quality. Much the same applies, in the chemical sense, to observation of effluent discharges. The question may be asked how the sense of hearing contributes to observation of water quality. Quite often the rise of a fish, the plunge of a water vole, the flap or click of a moorhen, or the splash of an outfall pipe or tributary inflow are sounds which give indirect clues as to the quality of a water – these things may appear to be of low scientific relevance, but they are things that every true water scientist would be glad to hear, and take in.

The sampling of waters, discharges, deposits and other materials, and their subsequent chemical analysis, in the field or in the laboratory, will of course provide information on the chemical quality of the samples according to the extent and limits of detectability of the analyses carried out. Field estimations of dissolved oxygen saturation (by chemical means or electronic meter), pH value (by pH paper or meter) and ammonia content (by chemical or electronic means) will yield very useful information on the spot. The support of a mobile laboratory, particularly at times when serious pollution arises, can be invaluable in giving quick answers to control problems. Otherwise, analysis of samples in a well-equipped laboratory is the proper, main source of information on chemical quality derived from samples. Biological assays in the field undoubtedly have their value in formulation of a general opinion, particularly when related to the physical characteristics of the water. The taking of samples for assay in a distant laboratory, coupled with observations of the depth, velocity of flow, degrees of shading of the water, and mapping of the plant growth, are of great value. Only when the latter observations are coupled with adequate chemical, biological and bacteriological examinations of samples in laboratories can a really sound observation of water quality be made. Problems of representative sampling arise in both the chemical and biological contexts, with the latter predominant. Observation of the fish life in rivers, particularly by electrical fishing, trapping or netting, biological and chemical examination of fish samples, and live toxicity tests are part of the direct biological and indirect chemical observation of water quality. The taking of automatic samples of waters over a period of a few hours or 2 or 3 days greatly extends the degree of observation of chemical quality. This job can of course be done manually but not as easily or as economically.

The automatic measuring and recording of chemical water quality, using portable or fixed instrumentation, is playing an increasing role, limited in both cases by the availability of sensors and, in the case of the fixed, permanent installations, by sensor fouling and the reliability of electronics. Recording of measured data can be done at the site, by telemetry to a distant point or by both means.

Specialists in water resources quality have their own preferences and judgments in interpreting the data produced from observations of water quality. Some specialists tend to adhere fairly rigidly to the few criteria which have been generally adopted for interpreting chemical and biological data relating to water resources, while others are more

Table W28 Classification of Royal Commission on Sewage Disposal of river water quality according to BOD_5

Classification	BOD_5 value (mg l^{-1})
Very clean	1.0
Clean	2.0
Fairly clean	3.0
Doubtful	5.0
Bad	10.0

flexible in their approach, relying in greater or lesser degree on their own practical experience based on personal observations.

Of the general criteria available for assessment of water resources quality, those produced in the UK by the Royal Commission on Sewage Disposal classifying river water quality according to BOD_5 levels, set out in Table W28, were the earliest. Various attempts to classify river water quality in terms of the biology of the water have been made, of which Table W29 is a good example. Both of these criteria are now recognized as inadequate as general classifications, because they ignored the effect of the physical characteristics of waters on its chemistry and biology. The latest set of criteria produced for general assessment of river quality is given in the *Quality of Rivers and Canals in England and Wales 1990–1992* (National Rivers Authority, 1994) (Table W30). This chemical grading in Table W30 defines six grades (denoted by A to F) on the basis of the concentrations of biochemical oxygen demand (BOD), total ammonia and dissolved oxygen. Grades A and B represent water of 'good' chemical quality, whilst grades C and D together equate to 'fair' quality and grades E and F represent 'poor' and 'bad' quality respectively. Under their assessment scheme the chemical quality of the total length of rivers and canals in England and Wales in 1991–1992 was as follows:

A. 21%
B. 32%
C. 23%
D. 12%
E. 11%
F. 2%

The Environment Agency's tentative biological assessment scheme, or score sheet, is shown in Table W31, which assigns points to particular taxa according to their sensitivity to pollution. The most sensitive taxa, such as stone flies, mayflies and caddis larvae score 10, the less sensitive taxa, like freshwater shrimps and water beetles, score 5, molluscs score 3 while the most pollution insensitive oligochaete worms score 1. The score for a site is the sum of all the scores of the taxa found in standard samples.

Note: taxon (plural taxa) is the term used for a subunit of the taxonomic classification system applied to living organisms. Organisms may be grouped together in, for example, classes, families, genera and species. Any of these groupings may be termed a taxon.

Assessment programs

The basic problem of setting up a program for assessing water quality and the human-made factors affecting this, is one of deciding what observations should be made, and where, how often and at what times of the day, week, month and year they should be made. A consequent problem is to decide how the data produced are to be processed, interpreted, presented and used for quality control purposes.

The main factors to be taken into account in setting up a program for assessing water resources quality are as follows:

- There are four basic reasons for assessing water resources quality:
  - to establish what the quality of resources is, and how it is changing over the longer term;
  - for establishing the quality of anthropogenic additions of polluting matter and flow augmentation water, the effect of these and of other quality-changing actions and whether requirements are being met, and if not with what divergence;
  - to give warning of abnormal changes in water quality so that any action possible may be taken to control these and to avoid damage to water abstraction interests;
  - for research and special investigation purposes.

Table W29 Classification of water quality according to biotic index (source: Sir Hugh Fish)

			Total number of groups of animals present				
			0–1	2–5	6–10	11–15	16+
Clean							
	Plecoptera nymph present	More than one species	–	7	8	9	10
		One species only	–	6	7	8	9
	Ephemeroptera nymph present	More than one species	–	6	7	8	9
		One species only	–	5	6	7	8
Organisms in order of tendency to disappear as degree of pollution increases	Trichoptera larvae present	More than one species	–	5	6	7	8
		One species only	4	4	5	6	7
	Gammarus present	All above species absent	3	4	5	6	7
	Asellus present	All above species absent	2	3	4	5	6
	Tubificid worms and/or red chironomid larvae present	All above species absent	1	2	3	4	–
Polluted	All above types absent	Some organisms not requiring dissolved oxygen may be present (e.g. *Eristalis tenax*)	0	1	2	–	–

A 'Group' means any one of the species included in the following list of organisms:

Platyhelminthes (flatworms)
Annelida (worms) except *Nais*
Hirudinae (leeches)
Mollusca (snails)
Crustacea (hog-louse, shrimps)
Plecoptera (stone fly)
Ephemeroptera (mayfly) except *Baetis rhodani*
Baetis rhodani (mayfly)

Trichoptera (caddis fly)
Neuroptera larvae (alder fly)
Chironomidae (midge larvae) except *Ch. thummi*
Chironomous *Ch. thummi* (blood worms)
Simulidae larvae (black fly)
Other fly larvae
Coleoptera (beetles)
Hydracarina (water mites)

Table W30 Chemical grading for rivers and canals (Source: NRA)

Water quality	Grade	Dissolved oxygen (% saturation) 10 percentile	Biochemical oxygen demand (ATU[a]) (mg l^{-1}) 90 percentile	Ammonia (mg N) l^{-1}), 90 percentile
Good	A	80	2.5	0.25
	B	70	4	0.6
Fair	C	60	6	1.3
	D	50	8	2.5
Poor	E	20	15	9.0
Bad	F[b]	–	–	–

[a] As suppressed by adding allyl thio-urea.
[b] Quality which does not meet the requirements of grade E in respect of one or more determinands.
Note: The grades are defined in terms of the 90 percentile for BOD and ammonia and the 10 percentile for dissolved oxygen; in other words, the river reach should contain less than the specified levels of BOD and ammonia for at least 90% of the time, whilst the level of dissolved oxygen must not fall below the prescribed level for more than 10% of the time.

- Assessments are required for management of resources quality on three scales of command, relating to the river basin, the subcatchment area and the local scale.
- The range and frequency of observations of river quality required varies directly with the existing use value and environmental importance of the water resources in question, and directly with the degree to which the existing use value approaches the total potential use value.
- The quality of water resources, even when wholly natural, is not static. It varies from year to year, season to season, and so on down to hour to hour, and these variations have to be taken into account in planning an assessment program. The position is even more complex when quality is artificially affected. Furthermore, the effects of differences of geography and geology on water quality ensure that resources in different localities have different variations in quality.

The sorting out of all these variables to give the optimum quality assessment program is clearly a very complex matter, requiring a computerized, systems analysis approach. While this could be justified for research purposes, off which would spin much useful guidance for resources managers, general applications cannot yet be justified, and rough approximations have to be made from experience and commonsense guided by such general findings as result from research.

To assist the reader, a number of general rules can be quoted, which eliminate, or at least sequester, much of the complex variability in the factors enumerated.

- The quality assessment of long-term changes needs be done at only a few strategically located stations on water resources. On a river, such stations would be basically near the headwaters around mid-catchment, and at the tidal limit. Other stations may be needed, for example where a massive effluent discharge or water abstraction occurs, and upstream of the entry of major tributaries. Massive and major in these contexts relates to the effect of these additions or abstractions on the river in its more valuable reaches. On tidal waters, the upper, middle and lower estuary would be appropriate stations, with additional stations up and down the coast and at points remote from estuaries and the coast to give an assessment of the sea position. On a lake, stations near the inlet, midway and at the outlet should normally suffice.

Table W31 Allocation of biological scores (Source: NRA, 1994)

Taxa	Score
Siphlonuridae Heptageniidae Leptophlebiidae Ephemerellidae Potamanthidae Ephemeridae Taeniopterygidae Leuctridae Capniidae Perlodidae Perlidae Chloroperlidae Aphelocheiridae Phryganeidae Molannidae Beraeidae Odontoceridae Leptoceridae Goeridae Lepidostomatidae Brachycentridae Sericostomatidae	10
Astacidae Lestidae Agriidae Gomphidae Cordulegasteridae Aeshnidae Corduliidae Libellutidae Psychomyiidae (Ecnomidae) Philopotamidae	8
Caenidae Nemouridae Rhyacophilidae (Glossosomatidae) Polycentropodidae Limnephilidae	7
Neritidae Viviparidae Ancylidae (Acroloxidae) Hydroptilidae Unionidae Corophiidae Gammaridae (Crangonyctidae) Platycnemididae Coenagriidae	6
Mesoveliidae Hydrometridae Gerridae Nepidae Naucoridae Notonectidae Pleidae Corixidae Haliplidae Hygrobiidae Dytiscidae (Noteridae) Gryrinidae Hydrophilidae (Hydraenidae) Clambidae Scirtidae Dryopidae Elmidae Hydropsychidae Tipulidae Simuliidae Planariidae (Dogesiidae) Dendrocoelidae	5
Baetidae Sialidae Piscicolidae	4
Valvatidae Hydrobiidae (Bithyniidae) Lymnaeidae Physidae Planorbidae Sphaeriidae Glossiphoniidae Hirudinidae Erpobdellidae Asellidae	3
Chironomidae	2
Oligochaeta	1

- Assessment of short-term changes will be possible from the observations at the long-term stations, provided the frequency of observation is sufficient. Other additional points will be needed close upstream of important points of water use and of high environmental value (sometimes observations downstream of major abstractions will be required), and upstream and downstream of most effluent discharges.
- Observations of quality for warning purposes, where these are needed, bearing in mind the value of fish as pollution indicators, should as far as possible be integrated with observations at long- or short-term change observation stations.
- Where observations need to be made for river management, effluent disposal and water abstraction purposes in one locality, then it should be possible for the same range and frequency of observations at one key site – normally the water abstraction site – to serve all purposes.
- Minor streams need relatively little observation unless they are subject to major, human-made, quality changes. Large streams, which are mainly of natural quality and are not likely to suffer accidental pollution, need be subject to no more observation than is necessary for assessment of long-term changes. On the other hand, large streams which are subject to much use for effluent disposal need close observation, and where simultaneously such streams are subject to much abstraction for public supply, observation requirements reach the maximum.

- The frequency of observation required depends substantially on the purpose of the observations and the variability of the quality involved. The optimum frequency of observation at any point to give a specific precision and confidence in the results can be selected with the aid of statistical considerations. Warning stations need observations at about 1–2 h intervals on large streams, 0.5 h on smaller streams and more frequent observation on alarm. Clearly such frequencies are only achievable over more than a few days by using automatic, remote observation. Observations of long-term changes do not need to be made with very high frequency, while observations of short-term changes need to be made with a frequency matching the circumstances and needs. It follows that the installation of permanent automatic quality observation stations can be fully justified for the combined purposes of observing long-term and short-term changes and for warning purposes. Without the latter requirement to protect an important use, the justification for a permanent station will often remain. Otherwise, automatic observation as and when required by means of temporary stations, in support of manual sampling, will usually be the appropriate provision to be made.
- The observation of effluent quality as discharged to resources is essentially a matter for observation of conformity of the effluent quality to a required standard, and it is difficult at present to imagine this being done in any way other than by manual sampling. However, the future may require much better than this, and automatic observation of effluent quality is already necessary at the more important wastewater outfalls on rivers subject to much water reuse in public supplies. On such rivers the effluent outfalls should properly be regarded as the sources of water supply, and subject to considerable and even violent changes in quality at any time.
- Regarding the quality parameters which should be observed, a brief summary of the considerations involved in the observation of the most widely used parameters is set out in Table W32. It is obviously impossible to specify in general terms the rules indicating what parameters of quality should be observed and with what frequency. Each case must be judged on its merits.

It is of course of great importance that observed data should be adequately processed to facilitate interpretation, presentation and use, but often this is not done and large quantities of data lie unused. Frequently this situation may arise wholly or partly because the data collected are unwanted or of little value. However, where an assessment program is expected to produce a considerable volume of data, an appropriate scheme of data processing should be provided. The most usual form of presentation of processed data is in the form of mean monthly values of the various quality parameters together with maximum and minimum values, and percentile values for the year.

Objectives of control

The objective of controlling water resources quality is to secure the maximum benefit from that resource in terms of use and environmental value. This is not a simple matter because the use of water resources for wastewater disposal is a legitimate, necessary and valuable use, which in many respects can and does depreciate the value of waters for other uses and in environmental terms, while giving the advantage of water reuse. The economist simplifies the position by stating that the control objective is to strive towards the optimum cost–benefit position; that is, the quality of a water should be controlled at the level where the marginal costs of improving that quality equal the marginal benefits to be derived from such improvement. Unfortunately even this pragmatic approach is difficult because many of the aesthetic benefits to be derived from maintenance or improvement of water resources quality cannot be quantified in terms of cash value. Yet in practical terms there are only four broad control objectives reasonably available on non-tidal surface waters:

- maintenance of very clean and beautiful waters in that condition;
- maintenance of a water in a condition fit for all normal uses, including abstraction for public supply;
- maintenance of a water as an environmental and amenity asset but not suitable for abstraction for public supply;
- maintenance of a water as a drainage channel free of nuisance in terms of environmental hygiene.

Table W32 Practical aspects of observation of parameters of water resources quality

Parameters	Practical aspects of observation of parameter
Suspended solids	Required for most quality assessments. Very frequent observations required where: • The stream is essentially a drainage channel • At tidal limit • Over short periods during floods Can be automatically observed in the field. Not normally of value as a 'warning' parameter
Temperature	Required for all observations of dissolved oxygen saturation. Otherwise very frequent observation necessary only where problems of heating of water resources exist. Can be automatically observed in the field. Not normally a 'warning' parameter
pH value	Required for most quality assessments. Very frequent observation not necessary except where serious industrial effluent or eutrophication problems exist. Can be automatically observed in the field. Not normally a 'warning' parameter
Conductivity	Of doubtful general value *per se* except on cleanest waters, but of particular value in relation to assessments of warnings of marked changes of other quality parameters. Can be automatically observed in the field
Dissolved oxygen	Required for most quality assessments. Very frequent observations required only where oxygen balance is critical in relation to fisheries and environmental hygiene requirements. Then a necessary warning parameter. Can be electronically observed in the field
Carbonaceous BOD_5 (ATU modified test)	Required for most quality assessments but very frequent measurements seldom justified. Of little value unless related to dissolved oxygen concentration. Not a 'warning' parameter. Not automatically observed in the field. The unmodified BOD_5 test is no longer considered appropriate for general use
Permanganate value (4 h)	Superseded by COD. Not measurable automatically in the field
Chemical oxygen demand (COD)	An important measure of organic matter content of water. Should be observed for long-term changes and short-term changes. Not measurable automatically in the field
Total organic carbon (TOC)	Will replace COD when easy electronic measurement practicable, and when automatically measurable in the field it will become an important long-term change, short-term change, warning parameter
Ammonia	Required for most quality assessments. Very frequent observations required if river is much used for effluent disposal, for protection of waterworks intakes and fisheries. An important 'warning' parameter. Can be automatically observed in the field
Nitrate or Nitrite–nitrate	Required for most quality assessments. Very frequent observations not required except in relation to public supply sources subject to wide fluctuations in nitrate content. Can be automatically observed in the field
Chloride	Required for virtually all quality assessments. Very frequent observations required only when it serves as a warning parameter or is to be used in flow estimations. Can be observed automatically in the field
Phosphate	Required for many quality assessments, but very frequent observations seldom justifiable at present. Can be automatically observed in the field
Sulfate	Not required for many surfacewater quality assessments, but usual in groundwater quality assessments. Very frequent observations seldom required. Can be automatically observed in the field
Anionic Synthetic Detergents	Required for most quality assessments, but very frequent observation seldom required. An important indicator of sewage pollution. Cannot readily be observed automatically in the field
Hardness Alkalinity Acidity Calcium Magnesium Sodium Potassium	Required for many quality assessments, but very frequent observations seldom required
Zinc Lead Copper Cadmium Mercury Nickel Chromium Arsenic	Of importance in assessments of toxic characteristics of waters draining old mining areas and industrial conurbations. Very frequent observations seldom required
Cyanide Sulphide Free Chlorine Chloramines	Normally observed in special circumstances where contaminants may be present. Very frequent observations of free chlorine and chloramines required in connection with control of chlorination/dechlorination used for biota control in pipelines in water transfer schemes. Can be automatically observed in the field

Table W32 Continued

Parameters	Practical aspects of observation of parameter
Pesticides Polychlorinated biphenyls Other trace contaminants	Observed as required in laboratories, often requring sophisticated analytical techniques
Bacteriological and virological examinations	Plate counts at 22°C and *E. coli* estimations usually required in environmental hygiene and raw water supply checks. Very frequent observations seldom required. Routine virological examinations barely established at present. Cannot be observed automatically in the field
Biological examinations	Counts of species of fauna, examinations of fish, weed-growth mapping and algal counts usually necessary at selected points on surface waters. Frequency of observation normally low, often restricted to seasonal observations unless special investigations involved. Each observation of fauna and flora must be related to chemical and physical parameters
Discharge or flow	Desirable at most points of observation of chemical or biological quality, and essential at stations for observation of long-term changes and at warning stations. Hourly observations of flow satisfactory at most permanent flow gauging stations. Instantaneous observations of flow often adequate. Flow estimations by chemical dilution techniques, and of time of travel by chemical means, often valuable mapping
Other physical characteristics	Mainly a matter for personal observation and recording, particularly in connection with biological assessments and maintenance and improvements of fisheries

In respect of tidal waters there are two broad objectives possible:

- maintenance of an estuary in a condition appropriate for the safeguard of migratory fisheries, coupled with avoidance of local nuisance;
- maintenance of non-objectionable conditions, from the environmental hygiene aspect, in any tidal water, and avoidance of toxic hazards to sea fish, or through sea fish to humans.

In relation to groundwaters there is only one broad control objective which can be reasonably applied and this is maintenance of the groundwater of a quality fit for abstraction for public supply.

It is of course manifest that many waters will have to be considerably, indeed in some cases markedly, improved in quality to meet desired control objectives. However, improvement targets are of no meaning if it is not intended to maintain the target quality once improvement has taken place.

Controlling the controllable

It has already been stated that the control of the quality of water resources can never be complete. Indeed there are many potential sources of accidental or unintentional pollution which are substantially uncontrollable. Thus any serious attempt to control the quality of water resources consists in part of setting up a system for controlling the controllable factors affecting water quality and in part of ensuring that the uncontrollable factors are dealt with as and when they cause interference with water resources quality.

Considering first the control system proper, the factors to be controlled are

- abstractions of water, additions of flow-augmentation water and physical alterations of stream channels;
- discharges of sewage and trade effluents to surface waters;
- other contaminations of surface waters;
- disposals of refuse and other solid or liquid wastes on tips and otherwise into the ground.

In the case of inland surface waters, the setting of a general policy for controlling physical changes of flow and regime to meet specified control objectives should not present great difficulty. From flow measurements and records of existing abstractions and returns of water, the minimum likely flows at various key points can be assessed, and the reserves of water available for abstraction, if any, irrespective of quality considerations, can be calculated. Applying, then, the selected quality control objective, the required general policy will emerge. For example, in the case of the stream maintained solely as a drainage channel where the quality objective is simply to avoid nuisance, the possible general policies will be either

- limitation of further loss of water through abstraction only to the degree necessary to ensure that the stream meets existing abstraction needs and continues to flow at all points; or

- limitation of loss of water through abstraction to the degree necessary to ensure that a minimum specified clean-water dilution is available for effluents.

The control of abstractions and physical changes of regime on tidal waters is not generally necessary on quality grounds, but in some circumstances of safeguarding migratory fisheries and environmental requirements in the upper reaches of estuaries, control may well be necessary on terms designed to meet the circumstances arising. In the case of groundwaters, the control necessary will relate to safeguarding groundwater levels for the maintenance of spring yields to streams to meet the objectives in those streams. Additionally, control may be necessary to limit infiltration of salt water underground in coastal areas or the induced movement of groundwaters otherwise of low quality.

Clearly, the major control problem arises in connection with effluent discharges to surface waters. The control tool is the river authority's specification of the permissible volumes and qualities of discharges. The greatest control difficulty arises where an inland stream is to be maintained for all normal use purposes, including public supply abstraction, while at the same time it is also extensively used for effluent disposal.

In applying this control objective, there are two states of the river concerned which are critical to the uses being protected:

- the late summertime lowest-flow period;
- the low-flow period during a severe winter cold spell.

Generally, if control of the quality of the river water is maintained adequately to protect all river uses during these critical periods, then protection at all other times will be achieved. The characteristics of water quality at these two critical periods are peculiar. In the summertime drought the water and air temperatures are high, biochemical oxidation of organic matter and ammonia proceeds at the maximum rate, diurnal variations of dissolved oxygen due to plant photosynthesis are at the maximum, the oxygen-holding and absorbing capacity of the water is at its minimum, while the clean-water dilution available from effluents is also at the minimum. At this time the local effects of effluent discharges in physical terms, in terms of toxic effects on fisheries and other fauna, and in terms of oxygen balance are at the maximum. Provided these local effects are adequately controlled, the river basin as a whole will be in a satisfactory state. If the total volume of effluent discharged is high relative to the drought flow of the river and is increasing, then control of the content of residual non-degradable organic matter, of toxic metal residuals, nitrate, phosphate, and perhaps chloride, in the river may be necessary for the protection of public water supply abstractions. The need for this control can be deferred wherever the summertime low flow of the river can be augmented by addition of water of low effluent content by inter-river transfer, by discharge from a winter-storage impoundment, or by discharge of groundwater from a groundwater development or recharge scheme. All these physical means of aiding quality control would of course be done primarily to meet quantitative demands for water at the intakes, giving quality protection as a secondary benefit. However, simple and attractive though this approach may appear, it can produce its problems if care is not exercised.

In contrast, during a prolonged cold spell in winter, the river flow may well be low, but not as low as the drought flow, due to ground freezing, effluent purification plants (except the very large, well-designed, high-quality treatment plants) will be operating at lowest efficiency due to low temperatures, and biochemical oxidation in the river will be at a minimum for the same reason, while the oxygen-holding capacity of the water will be at a maximum. In these circumstances parameters of quality such as BOD_5, ammonia and phosphate will be conservative – they will change little on passage down-river. Because of low oxygen demands on the river and high dissolved oxygen levels in the water from reaeration, the fair (relative to drought flow conditions) dilution with clean water available, and low rates of metabolism in fish and other fauna, local quality control problems are not likely to be a problem. Where an effluent discharge is greatly fouled, as a result of the effect of bad weather on bad works design and management, there might well be problems but these should not be permitted to occur twice. The main quality control problems involved will be river-basin ones of ensuring that ammonia levels at waterworks intakes are kept at reasonable levels.

Thus, in the type of river which must be kept satisfactory to meet all normal uses, the quality control to be applied to effluent discharges will relate essentially to ensuring the following.

- That oxygen demand and toxic effects local to individual outfalls do not become excessive during the lowest summer flow (including any flow augmentation).
- That loadings of substantially non-degradable residuals during lowest summer flow do not become excessive at any point of concern. In this context the point of discharge of the river to its tideway is a point of concern from the point of view of estuary and sea pollution.
- That the total loadings of ammonia passed to the river upstream of each waterworks intake during prolonged cold spells are kept as low as possible.

The permissible standards of quality for each effluent can thus be calculated on the basis of need with an accuracy related to the accuracy with which the critical river flows can be estimated or have been measured, and with which the rates of oxidation of organic matter and of ammonia, of phosphate precipitation or uptake by plants, of precipitation of heavy metals, and so on, in the different localities have been established or estimated. This approach ignores of course the currently accepted position that sewage effluents should normally be purified to at least a standard of $30\ \mathrm{mg\ l^{-1}}$ suspended solids and $20\ \mathrm{mg\ l^{-1}}\ BOD_5$. It should be borne in mind, however, that if the river needs in respect of any particular outfall indicate that a discharge of low quality is acceptable, the discharge will need to be made at such point and in such a manner that it does not cause any noticeable polluting effect at the outfall. In short, relaxations of effluent standards calculated as acceptable in the ways indicated are done so in terms of immediate mixing of the effluent with the total river flow, and, within the limits of reasonable practicability, the premise of the calculations should be realized in practice.

However, it should also be borne in mind that if the calculations of effluent quality are based on summer and winter critical flows which occur, say, once in 20 years, then it is certain that for much of the time the effluents concerned will be purified to a degree greater than strictly necessary. This overprovision of effluent treatment can be reduced by the setting of summer and winter standards of quality, or even relating standards of effluent quality to river flow rates. This only becomes really worthwhile where treatment to very high standards of quality by highly controllable techniques is demanded at critical times, and less high standards at less critical times can be readily produced at considerable cost saving.

Turning now to consider the position on a polluted river which is to be improved in quality, once the improvement target has been specified, the effluent standards necessary can be determined in the same way as described in the case of the river subject to all normal uses. Not every reach of the river need be subject to the same improvement target, and where a variability of targets is selected this obviously adds to the complexity of the calculations of required standards. Furthermore, along such polluted rivers, there are usually

severe problems related to discharges of storm sewage from combined sewerage systems which add further complications. Where a reach of river is not to be restored to the state where it can be described as an environmental and amenity asset, consideration can properly be given, on economic grounds, to the possibility of artificially increasing the self-purification capacity of the river itself to reduce the provision of treatment plant at sewage works. This would normally be done by providing mechanical aeration of the river water or by forming self-purification lakes on the river system. In making such decisions, however, there must be a reasonable certainty that demands for upgrading the quality of the river, in the selected treatment reach, will not have to be met long before the capital value of the river treatment system has been realized. In other words, care must be taken to ensure that economic gains calculated on a long-term basis do not in the practical turn of events prove to be short-term losses.

Whatever effluent quality program is decided upon, it must cover a reasonable period of time, and take into account the best obtainable estimates of the changes expected in water consumption, population, industrial output, river use and environmental and amenity demands over the control period. This quality control plan should cover, say, 30 years and should specify in outline the river quality targets, and the standards required of the major effluent discharges, in relation to time. From this outline plan, and subsequent amendments made as circumstances change from those forecasted, 7-year program plans, rolling on annually, can be produced specifying in greater detail the required targets, the standards required and the timing of achievement of these.

In controlling effluent discharges to estuaries, apart from differences in targets there are differences in circumstances from those relating to inland waters. In seeking to prevent the discharge of quantities of non-degradable toxic substances in excess of the limits required for fisheries protection, estimations can be made of the total load in estuarine discharges, in excess of river discharge loads, permissible under the worst expected circumstances of river discharge and clean seawater influx. This criticality would be estimated at the most unfavorable point in the estuary if the whole of this is to be maintained as a fishery, or at the seaward point where fisheries are to be maintained.

Calculations of the loads of degradable substances permissible in effluent discharges is a complex matter in the case of the heavily industrialized estuary, but not particularly difficult for other estuaries. Where the loadings of organic matter and ammonia are high, computer programs have to be used to determine effluent quality standards. Factors such as the total oxygen demand of effluents, usually referred to as the ultimate oxygen demand (UOD), which can be regarded as a 50-day BOD rather than the normal 5-day BOD, and the degree of clean seawater influx, have to be taken into account. Where estuaries are stratified in salinity, additional complications become involved. Again, from such calculations and forecast of expected changes over 30 years, a quality control program can be set up.

A particular problem of bacterial contamination by sewage effluent discharges arises in respect of the protection of shell fisheries. In these circumstances sewage effluents may need to be disinfected by chlorination or by storage in lagoons before discharge in the vicinity of shellfish beds.

The control of sewage discharges to coastal waters, and to estuaries of high amenity value, is mainly a matter of public health protection in terms of avoidance of contamination of bathing beaches. Where adequate protection cannot be achieved by construction of long outfalls giving a high degree of dispersion of the sewage in the tideway and rapid death of bacteria, full treatment and disinfection of the sewage may be necessary.

In seeking to control contaminations of surfacewater discharges running off surfaces subject to pollution, such as industrial yards and garage forecourts, the degree of control required depends on the quality target applicable to the receiving stream. On the river used for all normal purposes, a strict control is necessary and is often only achievable by prohibiting or minimizing the contamination, usually by arranging for the offending surface to be drained to soil sewers, to a treatment plant or to a storage pond. Only a limited control may be necessary on a stream serving solely as a drainage channel. For the protection of the cleanest streams the runoff from motorways often needs passage through a storage pond before discharge to stream.

The use of deicing chemicals such as alcohol–glycol mixtures and urea on airfield runways, and of salt and other materials on roads,

needs careful watching, and in some circumstances positive regulation. Where positive regulation is required for the protection of river
uses, the likely pollution consequences of the deicing operations have
to be examined in respect of local effects of individual applications,
and in respect of the maximum quantity of polluting matter likely to
reach key points of the river system during particular freeze-up
periods. On this basis, reasonably adequate control arrangements can
be developed and applied with the cooperation of airfield authorities,
and if and when necessary with the highway authorities.

Control of the use of chemicals in watercourses for weed control
and in cultivation of watercress is another necessity in many areas of
the UK. The strictest control is exercised where the river waters are
greatly used for abstraction for public supply and crop irrigation.

Control of the discharge of sewage from craft navigating inland
waters is variable according to the area concerned. On the River
Thames, for example, the control is prohibitive, strict and effective
over some 12 000 registered launches by virtue of regulatory byelaws
made under the provision of the Thames Conservancy Acts. However,
in some areas no control exists and sewage matter is discharged crude
to rivers and canals.

The dumping of refuse in inland waters by individuals is the one
aspect of water pollution which in theory is controlled, but in practice
cannot be controlled. Many people dump beds, prams, cars, drums
and garden rubbish in rivers and streams, but very few do this when
and where they can be observed, reported and prosecuted for doing
so. The only prospect of reducing the great amount of rubbish dumped
in streams is by means of education and persuasion of the public not
to do this, and the provision of effective alternative, and free, means
whereby people can dispose of the rubbish properly.

The control of groundwater contamination is not always as
complex a procedure as might appear at first sight, particularly if the
correct control target – that of ensuring that use of the groundwater
for public supply is not adversely affected – is applied. The main
danger of groundwater pollution, of course, is that once it has
happened, it may be substantially irreversible and remain for years.

For purposes of waste disposal into it, underground strata can be
divided into three types, permeable, semi-permeable and impermeable, and clearly it is the free-draining permeable strata which has the
greatest attraction for waste disposal, the greatest value in terms of
yield of groundwater supplies and is exposed most to pollution.

The main permeable strata, fissured chalk and limestone, sandstones and gravels, will pass through them the liquid fraction of, and
the rainfall percolating through, wastes dumped on or into the strata,
and any such disposal of waste which is not entirely solid and inert
must be considered as presenting an *a priori* probability of causing
contamination of underlying groundwater. However, depending on
the nature of the wastes disposed of, the nature of the stratum and the
nature of the groundwater flow, the resulting groundwater contamination may or may not be of significance.

Different wastes passed to different strata will produce different
results in terms of groundwater contamination on differing time
scales. For example, all non-inert wastes disposed of into chalk or
limestone stratum containing large and deep fissures will quickly lead
to contamination appearing in the underlying groundwater. If the
groundwater flow is artificially increased by pumping of water from
the aquifer close to the disposal point, contamination of the pumped
water will become evident in a matter of hours. Through less heavily
fissured chalk, the time of travel of contamination will be less rapid,
permitting greater opportunity for the contamination to be absorbed,
adsorbed, precipitated, exchanged or otherwise removed from the
percolating liquid, or contaminated rainfall derived from the waste to
be disposed of. In chalk where fissures are small, it may take years for
polluting matter to reach any borehole or spring in the underlying
aquifer, and the polluting matter may be only a fraction of the original
contamination, much of it having been held in the chalk. For example,
most of the toxic metals draining from disposal of waste will be
precipitated as hydroxides and held in the chalk. Zinc, being soluble
in alkaline water, will not be so held and will be conveyed through the
chalk.

Sandstone strata will normally behave in a similar way to finely
fissured chalk, and many gravels in much the same way, although
some sandstones and some gravels can be relatively impervious and
others seemingly as permeable as heavily fissured chalk or limestone.

The disposal of degradable wastes into permeable underground
strata may be an effective method of disposal – anaerobic decomposition of the portion of degradable waste passing through the stratum

occurring first, and then dilution and aerobic decomposition of the
remaining contamination proceeding in the groundwater flow to give
no significant final contamination. However, at the groundwater table
there may not be sufficient dilution with water containing dissolved
oxygen to permit final removal of degradable waste, and the whole
groundwater flow may be rendered anaerobic. The disposal of
nitrogen-rich wastes into permeable strata may result in the groundwater being fouled with ammonia, or rendered less suitable for public
supply, because of its nitrate content derived from the underground
oxidation of ammonia.

There are many other considerations likely to arise in respect of the
disposal of particular wastes into particular strata in particular ways at
particular points. Nevertheless, the control of pollution of underground strata in, or not securely separated from, hydraulic connection
with groundwaters of value, must be based on sound principles.

In the case of many semi-permeable strata, for example mixtures of
gravel and clay, brick-earths, clayey sands and so on, these overlie
impermeable clays and the disposal of wastes into these strata may
result only in polluted liquors and rainfall leaching out, or springing
out, at the ground surface at lower levels to pollute surface waters.
Careful control must be exercised of any waste disposal permitted,
and generally the use of such sites for waste disposal cannot be
permitted unless leaching of liquors from the site can be contained
and controlled satisfactorily. Where such semi-permeable strata are in,
or are only lightly separated from, hydraulic connection with
groundwaters of value, the position must be judged as similar to, but
perhaps less critical than, that in permeable strata.

Coping with the uncontrollable

There are many ways in which polluting matter sporadically and
unintentionally enters surface waters and underground strata, but the
major unintentional pollutions, uncontrollable by river authorities,
occur as a result of the following.

- Spillages and leakages of oils and chemicals by road accident,
  failure of storage installations and failure of pipelines and mechanical plant.
- Spillages of oils and chemicals as a result of human failure or
  neglect. Oil spillages are the major source of unintentional
  pollution.

The practical action that an authority takes regarding these
occurrences involves

- immediate investigation of reports of spillages, assessment of the
  consequences, taking of action to mitigate the consequences, giving
  of warnings as necessary to river users, establishing the causes and
  ensuring that all practicable steps are taken to prevent a recurrence;
- instituting legal proceedings in cases of gross negligence or where
  advice previously given, of the precautionary measures necessary,
  has been ignored.
- clearing up any fish loss and ensuring that the fishery is restored.

Some authorities have highly developed arrangements for dealing
with accidental pollutions. For example, some authorities have made
arrangements with police authorities for reports of pollution to be
made via the emergency call system at any time, and for police and
fire authorities to notify the river authority of all calls they receive
involving spillages of materials on roads so that pollution control
measures may be instituted forthwith.

In the case of some pollution which occurs, for example from
soluble toxic material such as cyanide, phenol, pesticides, etc., there is
nothing that can be done to remove the polluting matter, but
frequently steps can be taken to manage the river to minimize the
extent of the pollution. For example, discharges of water can
sometimes be made from reservoirs to dilute polluting matter in a
river, water levels can sometimes be built up at sluices to delay the
travel of polluting matter, water reaeration can be attempted where
appropriate, and distressed fish can be rescued and moved to clean
water. Nevertheless, despite all such endeavors, a gross accidental
pollution on a large river can be disastrous in its consequences to river
life and river users, and the wise river authority has plans and
organization prepared for dealing with such emergencies.

R.W. Herschy

Source

Fish, H. 1973. *Principles of Water Quality Management*, Thunderbird Enterprises, UK.

Cross references

Drinking water and sanitation
Groundwater
River pollution prevention: historical
Sanitation and clean water
Water quality for drinking, WHO guidelines
Water treatment: potable water

WATER RESOURCES: SURFACE AND GROUNDWATER

Rainfall is the starting point of the terrestrial section of the hydrological cycle. For obvious reasons of its origin, and its short period of existence before arrival on the Earth's surface, rainfall, though not sterile, contains no significant flora or fauna. The physical characteristics of natural rainfall are readily identified. Rainfall is naturally clear and colorless, with a temperature varying from ambient according to the height of formation of the rain. The chemical quality of rainfall is less readily defined.

Rainwater is not pure water. It contains impurities dissolved from the atmosphere. First, it contains gases, mainly oxygen, nitrogen and carbon dioxide dissolved from the air. Second, it contains solid matter dissolved from impurities in the atmosphere derived from the Earth's surface. Third, it contains traces of materials derived from chemical reactions of natural materials in the atmosphere, initiated by solar and cosmic radiation. It also contains impurities dissolved from anthropogenic atmospheric pollutions, but this aspect will be considered later.

The major source of natural impurity in rainwater is natural atmospheric contamination from the Earth's surface. The breaking of waves on the sea, and sea spray, cause seawater to enter the atmosphere and to be dissolved in rain. Generally, the farther away from the coast the rain falls, the less its contamination with seawater chemicals. The main impurities from this source are of course the principal constituents of seawater: chloride, sodium, sulfate, magnesium, calcium and potassium. Volcanic eruptions and winds carry dust into the atmosphere. Calcium, magnesium, sodium and potassium are the principal impurities dissolved in rain from this source.

Thus, apart from the effects of atmospheric pollutions, and the effects of chemical reactions in the atmosphere which are not of marked significance except where atmospheric pollution is considerable, the natural quality of rainfall is determined by seawater and dust take-up. This will vary according to weather and locality. Rainfall naturally contains around 10–20 mg l^{-1} of dissolved solids, and often more in coastal areas and duststorm regions. It is somewhat acid in character due to solution of carbon dioxide from the air, with a pH value around 5. However, rainwater on the ground may be more contaminated than rainfall. Even under natural circumstances the particulate impurities carried in the atmosphere 'fallout' in dry weather onto vegetation and the ground surface, as well as being precipitated in rainfall. Thus rainfall arriving on the ground may pick up such accumulated fallout.

On reaching the ground surface, rainfall will either run off the surface stratum if the underlying stratum is impermeable, or pass underground if the subsurface stratum is permeable. Depending of course on the nature of the substratum, the permeability of the substratum and water levels therein, and the intensity of rainfall, both runoff and percolation into the substratum may occur at the same time. This differentiation of the route taken by rainwater in commencing its journey to the sea is of great importance in its natural effects on freshwater quality. For a number of reasons, rainwater running off the surface stratum is usually of very different quality from that which percolates through underground strata, and it is appropriate to deal separately with the quality of the waters following the two different routes.

Surface runoff

Although the term 'surface runoff' is usually meant to include stream and river flow as well as runoff discharging thereto, it is far more convenient here to consider the quality of surface runoff as it first discharges into a stream, leaving the greater complexities of the quality of stream and river waters, which often include groundwater discharged from springs, to be dealt with later. From this restricted viewpoint, it is clear that most of the natural surface runoff will be from areas of substantially impermeable rock, the igneous and metamorphic rocks, and the clays.

Mechanical and chemical weathering of these rocks by rainfall results in the runoff containing materials both in suspension and solution. Additionally, as the water passes through soil and vegetation cover, matter may be taken up in suspension and solution. Mechanical weathering carries sediment in suspension. Chemical weathering, proceeding whenever mechanical weathering exposes new rock surface to the chemical action of rainwater, results in the take-up of soluble material.

The mineral silicates in volcanic rocks are broken down by the action of the acid rainwater (derived from carbon dioxide in solution in the rainwater). The harder and more compact the rock, the less the extent of this effect. Generally, the runoff from volcanic rocks contains a low concentration of dissolved solids, consisting mainly of calcium, magnesium, sodium, potassium and carbonates, with chloride and sulfate contents not significantly different from that of the original rainfall. Other impurities will be present according to the chemistry of the rock. However where the rock is covered with a peaty soil and associated vegetation, the take-up of organic matter from this is often heavy. The runoff from peat areas is usually colored yellow-brown, highly so in times of heavy runoff, and of low pH derived from that of the rainfall and additionally from solution of organic matter from the peat. A similar situation applies to the runoff from woodland areas, where ground cover with leaf-mould produces much the same effect. The organic matter producing this effect, apart from organic particles in suspension, consists of humic and fulmic acids produced in the peat or leaf-mould by decay of vegetable matter. Such waters show a high chemical oxygen demand, as measured by the dichromate COD (chemical oxygen demand) test or the 'oxygen absorbed' test, due to the organic matter present. A typical runoff from such peat or woodland areas would have a pH in the range 3 to 6, a color of between 100 and 200 hazen units, and a COD between 20 and 60. Such waters also contain traces of a variety of substances, but have a content of sodium, calcium, magnesium, sulfate and chloride very similar to that of rainfall.

The runoff from clay areas is characterized by its content of calcium and sulfate derived from solution of these materials naturally present in clays. This is shown up, on analysis of the water, by the presence of relatively high concentrations of calcium sulfate and consequently of non-carbonate (permanent) hardness. Considerable concentrations of nitrates are also often present due to the action of bacteria in the soil in transforming (fixing) the nitrogen in the air to nitrate. Turbidity of the water is often evident due to the presence of colloidal clay particles. The chloride content of the runoff may be three or more times in excess of that of the rainfall onto the land. Analysis of a typical water running off a clay catchment is shown in Table W33.

Between these two extremes of runoff quality from differing, impermeable rock substrata, there is a seemingly infinite variety of runoff qualities derived from different (and less impermeable) substrata. For example, the runoff from a chalk or limestone area during very wet weather, when the excess of rainfall over the water-

Table W33 Typical analysis of runoff from a Boulder Clay catchment area (mg l^{-1} unless otherwise stated)

Ammonium nitrogen	0.20
Oxygen absorbed from permanganate (4 h at 27°C)	3.0
pH	7.8
Color (Burgess scale)	20
Chloride (Cl)	50
Nitrate (N)	6
Total hardness (CaCO$_3$)	410
Non-carbonate hardness (CaCO$_3$)	250
Magnesium (Mg)	8.0
Phosphate (PO$_4$)	0.6
Sulfate (SO$_4$)	140
Silicate (SiO$_2$)	6
Iron (Fe)	0.05

absorptive capabilities of these rocks results in such runoff, could be expected to contain a considerably higher content of calcium hydrogencarbonate than the original rainfall. However, because natural surface runoff of the type under consideration here has a short period of contact with surface strata, it is generally much less mineralized than waters which pass underground through permeable rocks.

Groundwaters

The chemistry of groundwaters from the more important aquifers is an extensive and well-documented subject. The physical characteristics of most groundwaters in the UK are that the water is usually very clear and colorless with a temperature between 10–14°C, although some groundwaters may have a turbidity due to their peculiar chemistry. Obviously, because of their history, groundwaters have no significant flora or fauna and their bacteriological quality, particularly of the deeper-traveling groundwaters, is extremely high, very few bacteria being present.

When rainwater percolates into permeable strata it may accumulate impurities from overlying soil and vegetation in much the same way as surface runoff. It will accumulate organic matter, color, nitrate, minerals, additional carbon dioxide and so on. However, the water will have been filtered free of most of its suspended matter in percolating through the surface soil. As this water then penetrates the permeable underlying strata it will then change in quality in a variety of ways, according to its rate of travel, the nature of the substrata, and the chemical reactions proceeding as the water travels on. The nature of the more important of these processes can be best explained by brief reference to particular types of aquifers.

In unconfined chalk, that is, chalk with no overlying stratum other than surface soil, the water is characterized by its content of calcium hydrogencarbonate, derived from the action of carbon dioxide, dissolved in the percolating water, on the calcium carbonate of the chalk. It is of course through fissures in the chalk, large and small, initially produced by folding of the chalk during natural upheavals, that the bulk of water passes. Solid chalk is relatively impermeable to water, and it is the fissuring of the chalk which makes it an important aquifer.

The calcium hydrogencarbonate content of the water in unconfined chalk gives the water a temporary hardness, that is, hardness which can be precipitated by boiling of around 200–350 mg l^{-1} depending on locality. A typical analysis is shown in Table W34. The major factors determining the concentration of calcium hydrogencarbonate in unconfined chalk waters are first the amount of carbon dioxide accumulated in the rainwater as it percolates through the surface soil into the chalk, and second the length of time the water has been in the chalk. Where clay strata overlies the chalk the quality of the water in the chalk undergoes considerable change. Runoff from the periphery of the clay cover passes into the chalk, containing carbon dioxide from rain and soil, and often sulfate, calcium and magnesium. Further solution of calcium carbonate from the chalk under the action of

Table W34 Typical analysis of water in unconfined chalk, UK (mg l^{-1} unless otherwise stated)

Ammonium nitrogen	0.00
Oxygen absorbed from permanganate (4 h at 27°C)	0.03
pH	7.2
Turbidity (units)	<0.5
Color (Burgess scale)	0.0
Chloride (Cl)	8.5
Nitrate (N)	6.2
Alkalinity (CaCO$_3$)	275
Free carbon dioxide	38
Total hardness (CaCO$_3$)	303
Non-carbonate hardness (CaCO$_3$)	28
Magnesium (Mg)	1.2
Calcium (Ca)	119
Sodium (Na)	6.2
Potassium (K)	1.2
Phosphate (PO$_4$)	0.09
Sulfate (SO$_4$)	2.5
Silicate (SiO$_2$)	11
Iron (Fe)	<0.01

Table W35 Analysis of water in chalk underlying London Clay, UK (mg l^{-1} unless otherwise stated)

Ammonium nitrogen	0.48
Oxygen absorbed from permanganate (4 h at 27°C)	0.06
pH	7.5
Turbidity (units)	0.0
Color (Burgess scale)	0.0
Chloride (Cl)	80.0
Nitrate (N)	0.0
Alkalinity (CaCO$_3$)	230.0
Free carbon dioxide	12.0
Total hardness (CaCO$_3$)	212.0
Non-carbonate hardness (CaCO$_3$)	0.0
Magnesium (Mg)	26.0
Calcium (Ca)	42.5
Sodium (Na)	80.0
Potassium (K)	10.0
Phosphate (PO$_4$)	<0.02
Sulfate (SO$_4$)	46.5
Silicate (SiO$_2$)	17
Iron (Fe)	0.06

dissolved carbon dioxide increases the calcium hydrogencarbonate content of the groundwater, and the calcium sulfate derived from the clay may increase the permanent (non-carbonate) hardness. Where the clay cover is thick, and sodium-bearing minerals exist, ion exchange reactions may occur which result in the calcium of the groundwater being substantially replaced by sodium, and in the hardness of the water being drastically reduced thereby. Where the source of sodium is sodium chloride, then the chloride content of the chalk groundwater may be increased considerably. A typical analysis is shown in Table W35. Chalk water underlying boulder clay deposits may contain 500 mg l^{-1} or more of calcium sulfate. Clays overlying chalk often contain sodium chloride naturally present from inclusion of seawater when the deposits were formed, and this adds to the chloride content of the chalk groundwater in greater or lesser degree according to circumstances.

Within the chalk aquifers lying deep under overlying clay, where downward percolation of rainfall is very small, the groundwater may lose all the oxygen originally dissolved in it as it percolated through the surface outcrop of the chalk. This is brought about by the breakdown of organic matter dissolved in the water by the bacteria of decay, an oxygen-consuming and carbon dioxide-producing process. The latter product permits of course greater solution of calcium hydrogencarbonate in the water. In such circumstances of deoxygenation, other natural bacteria, the nitrate-reducing bacteria, may break down nitrate present in the groundwater mainly to nitrogen gas, causing a reduction of the nitrate content of the water. However, this process is of greater significance in the human-made circumstances where nitrate contamination of the surface soil occurs, as a result of use of nitrogenous fertilizers in agriculture or of discharge of sewage effluents in the ground. When the nitrate content of the groundwater has been substantially destroyed, another group of bacteria, the sulfate reducers, may begin to break down sulfate in the water to hydrogen sulfide.

In summary it can be said that the chalk groundwaters in the downland areas of chalk outcrop are naturally slightly alkaline and contain in the main calcium hydrogencarbonate, giving a carbonate hardness usually in the range 200–350 mg l^{-1}. In contrast, in the chalk aquifers underlying superficial clay deposits the quality of the water is more variable, depending on the nature of the overlying clay and the location of the groundwater beneath the clay cover. Such waters have considerable carbonate and non-carbonate hardness due to the presence of calcium hydrogencarbonate and calcium sulfate. They may be quite soft because they contain in the main sodium hydrogencarbonate and sodium sulfate as a result of natural ion exchange. They may contain very little nitrate because of nitrate reduction, and have a very low concentration of dissolved oxygen. They may be completely deoxygenated and contain hydrogen sulfide as a result of sulfate reduction.

In the sandstone aquifers, much the same type of situation exists as in the chalk aquifers, on a somewhat different scale. In the unconfined, outcropping sandstone aquifers, the water is of the calcium hydrogencarbonate type with a carbonate hardness in the range 50–175 mg l^{-1}. This hardness is derived from the percolating water dissolving calcium carbonate in the natural binding material between the sand grains of the rock. Where the aquifer lies under the Keuper Marl, the quality of the water changes with depth, the calcium sulfate content of the water increasing its permanent hardness. The deeper waters are usually soft, containing mainly sodium hydrogencarbonate, sodium chloride and sodium sulfate. At the lowest levels the waters may be highly saline, containing 15 000 mg l^{-1} or more of chloride. Both the Lower Greensand and limestone aquifers show similar general characteristics. The groundwaters of the Coal Measures and Millstone Grit are of very variable quality, some fairly hard, some soft, and others very saline. Similar waters exist very deep in nearly all of the major aquifers.

The waters contained in gravel aquifers are very variable because of the wide variety of the types of rock forming the gravel deposit. Waters from river gravels may contain iron and manganese in solution, and otherwise are often similar in quality to the rivers and streams to which they have an underground connection. Such iron-rich waters when they emerge from springs turn turbid and brown, because on aeration the iron in solution in the water is oxidized and precipitated as ferric hydroxide.

Glacial sands and gravels also give waters of varying quality, including both hard waters of the calcium hydrogencarbonate or the calcium sulfate type, and soft mineralized waters. The spring discharges from some gravels may contain high concentrations of carbon dioxide (up to 90 mg l^{-1}), giving them an acid character toxic to fish as a result of solution of carbon dioxide from overlying soil in the rainfall percolating into the gravel.

The natural flow path of these groundwaters is generally to emerge as springs discharging to streams and rivers, but in the deep aquifers, connecting eventually with the sea, a considerable proportion of the aquifer flow passes direct to the sea. The flow rates of spring discharges are not subject to significant short-term variation because the aquifers yielding the water even out the day-to-day variations in the amount of percolating rainfall. The degree to which the spring outflows reflect seasonal and longer-term differences in rainfall varies according to the natural circumstances prevailing, but of course human actions in ground-surface development and in abstracting water bring about marked changes in spring yields.

Inland surface waters

It is from surface runoff (and rainfall), and from groundwater springs, that the waters of streams and ponds, rivers and lakes are naturally formed, and it is thus axiomatic that the quality of these bodies of water will depend mainly on the nature and relative contribution of water from these sources. Yet a new and important factor applies to modify the chemical and physical qualities of inland surface waters – the influence of living things, particularly plants growing in the water. The biology of any surface water is determined substantially by the chemistry and physics of the water. It follows that the biology of any surface water provides a living expression of the chemistry and physics of that water. This relationship between the chemistry, physics and biology of a water is such that if any two of the three factors are broadly known, it is possible to predict roughly the third factor. But it is important to bear in mind that the physics of water in this context includes factors such as velocity of flow, depth and degree of shading of the water by trees. In passing, it should be noted that this fundamental relationship between aquatic chemistry, physics and biology also holds good under the artificial circumstances created by human actions, such as those arising from pollution, construction of weirs, flow diminution by abstraction and so on. For this reason, before considering the natural qualities of stream waters in broad categories it is necessary to examine the natural processes which underlie the relationship between the physical, chemical and biological characteristics of stream waters.

The natural shape of any stream channel is determined essentially by the volume of stream flow in unit time (the discharge of the stream), and the nature of the terrain through which the stream passes. It is these factors which primarily determine the gradient of the stream, its velocity of flow, width, depth, location of shoals and pools, and so on. The velocity of streamflow at any point determines whether suspended material is transported or deposited.

Contact of the streamflow with the atmosphere will result in closer adjustment of the stream water temperature to ambient. For example, the even temperature of a chalk spring discharge will be warmer than ambient in winter and cooler in summer, and when flowing in a stream channel this will slowly warm or cool as the case may be, by heat transfer to and from the air. At the same time any degree of unsaturation of the stream water, with oxygen in solution, will be corrected by absorption from the atmosphere. The weight of oxygen in saturated solution in water is about 14 mg l^{-1} at 1.5°C, 10 mg l^{-1} at 15°C, and 8 mg l^{-1} at 26.5°C; that is, the warmer the water, the less oxygen it can take up from the air. The more turbulent the streamflow, the faster it will absorb oxygen from the air until the saturation value is reached. A stream containing oxygen at less than saturation value, with a velocity of 0.24 m s^{-1} will absorb oxygen from the air (reaerate) three times faster than water with the same oxygen deficiency at a velocity of 0.06 m s^{-1}. At waterfalls the spill of water entrains large numbers of air bubbles which accelerate the rate of oxygen uptake. These physical phenomena have their bearing on the chemical quality of water.

The eroding effect of rainfall runoff has already been referred to. The eroding effect of the streamflow itself may affect the chemical quality of the stream water. For example, the erosion of a peat bank will add organic matter in solution and suspension to the water, while erosion of a clay bank will add clay solids in suspension and clay minerals in solution.

Where the clarity of the stream water, its depth and the absence of tree shading permit sufficient sunlight to penetrate to the stream bed, the growth of rooted submerged weed will be favored. If the stream bed is unstable because of the lift-and-drag effect of high-flow velocities, rooted weed growth will not be favored. If the stream bed is of clay and the water is slow flowing and turbid, then the growth of rooted emergent weed such as reeds, rushes and surface-floating types will be favored. But the species and density of aquatic weed growth is also dependent on the chemical quality of the water, submerged weeds being more susceptible to water quality than emergent ones. Algae, both benthal (on the stream bed), and suspended in the water in single cells, colonies or filaments, may grow wherever low flow velocities and the chemical quality of the water permit.

All of these green plants are autotrophic, that is, they create their substance from carbon dioxide (or hydrogencarbonate) dissolved in the water, and from the water itself, under the influence of sunlight. This photosynthesis yields oxygen as a waste product which is absorbed by the water, and may lead to the water becoming supersaturated with oxygen. At night the same plants respire, using up oxygen dissolved in the stream water and releasing carbon dioxide to the water. Thus any marked plant growth in the water will cause the oxygen content of the water to be higher during the day than during the night. This diurnal variation of dissolved oxygen content of the water, due to plant photosynthesis, is not of much significance under wholly natural circumstances, unless the streamflow is fairly sluggish and the weed growth is exceedingly prolific, as for example in a marsh ditch or fenland dike. Plant photosynthesis is of course the biosynthetic section of the well-known, natural carbon cycle.

The decay of weed and algae can add much organic matter to a stream, most of which will be in suspension, and some in solution. Leaf fall from trees also adds organic matter. Organic matter in suspension in the stream, from runoff, erosion, weed growth and leaf fall will be deposited at points of sluggish flow. This organic detritus is the primary food of the bacteria of decay, and of most of the grazing animals on the stream bed and on weeds, which in turn form much of the food of the higher aquatic animals including fishes.

Within the accumulations of organic silt on the stream bed, so long as oxygen is present the bacteria of decay feed on the organic matter, breaking this down to carbon dioxide and water, and using up oxygen in the process. This process also proceeds in any organic matter in solution or fine suspension in the streamflow. Within the deposits of organic silt, when no oxygen is present, or diffusing therein from the overlying water, breakdown of organic matter by anaerobic bacteria commences, yielding methane (marsh gas) and simpler chemical substances than the initial organic matter. Within the silt beds, detritus-consuming burrowing animals proliferate. Their action in stirring up the silt, and the escape of methane gas, release the anaerobic decomposition products to the overlying water where normal aerobic breakdown can proceed. Thus under stream conditions both these degradative parts of the carbon cycle exert a demand for oxygen on the streamflow, which can only be made good by the natural reaeration of the flowing water.

There are other natural elemental cycles of consequence to stream water quality. Ammonia present in the streamflow, the end product of the bacterial breakdown of proteinaceous organic matter, can be oxidized to nitrite by the bacterium *Nitrosomonas* and thence to nitrate by *Nitrobacter*. Again this process uses up dissolved oxygen. Yet if the dissolved oxygen present in the water drops to less than 10% of saturation, another group of bacteria will reduce the nitrate to nitrogen gas, provided sufficient organic matter is present to feed the bacteria. In the complete absence of dissolved oxygen and nitrate/nitrite, a group of bacteria known as the sulfate reducers will produce hydrogen sulfide by reduction of sulfate. Hydrogen sulfide, rotten-egg gas, may be changed to water and elemental sulfur, or sulfuric acid, by specific bacteria in the presence of oxygen, and it will form black iron sulfide with any iron dissolved in the water.

All of these important biochemical processes proceed at rates which are temperature dependent. In a stream which is naturally and heavily silted, supporting a prolific weed growth, and of very sluggish flow, during the warmer months of the year the dissolved oxygen content of the water may be reduced to a low level, as a combined result of bioxidation (aerobic breakdown) and benthal decomposition (bioreduction or anaerobic breakdown) of organic matter, the oxidation of ammonia present in the water, and plant respiration during the night. But such gross effects on the oxygen balance of the water do not happen in most streams under wholly natural circumstances.

As might be expected the natural, biological production of a stream can produce its feedback effect on the physical characteristics of a stream. Prolific weed growth may reduce the cross-sectional area of the stream through which water can flow, to such a degree that water levels are raised and flow rates through clear channels between the weed beds are increased, causing new bed movement and silt transport. Also, the growth of algae in suspension in a stream water, accelerated by the natural presence of nitrate and phosphate, may become sufficient to give such a turbidity and color to the water that submerged weed growth is inhibited by lack of light penetration.

Turning now to consider the quality of particular types of streams, a rough classification into chalk (and limestone) streams, streams draining clay catchments and streams draining upland and woodland areas will be used.

The flow of a typical chalk (or limestone) stream naturally consists mainly of groundwaters discharged from springs. Certainly this applies to its upper and middle reaches, while the flow of its lower reaches may be influenced by runoff from impermeable areas. It will be recalled that the groundwater discharged from and in the vicinity of the unconfined chalk strata will be very clear and colorless, free of suspended matter and large numbers of bacteria, of a fairly steady temperature around 10–14°C, and slightly alkaline due to its predominant content of calcium hydrogencarbonate. The first flow of the stream will consist of this water.

Provided the water is not heavily shaded by trees, it will immediately grow an abundance of submerged aquatic plants in which water crowfoot (*Ranunculus*) and water celery (*Apium*) will be typical. The clarity and mineral content of the water are major factors in producing this prolific plant growth. In these plants, and in the surface layers of the stream bed, a prolific fauna will also develop in which shrimp (*Gammarus*) and the larvae of mayflies (*Ephemera*) and caddis (*Agapetus*) will be predominant. Again the abundance of calcium and general mineralization of the water are important factors in the abundance of these animals. The first fish, minnows, bullhead, small trout and grayling, feeding on the aquatic organisms, will appear. The plant growth will begin to obstruct the free flow of water raising the water level and altering its pattern of flow.

Further down the course of the chalk stream, where the chalk strata is usually covered with impermeable strata such as London or Boulder Clays, the emissions from springs will diminish. Where breaks in this impermeable cover occur and groundwater levels permit, springs producing water of a character in the range between the calcium hydrogencarbonate type and the calcium sulfate/sodium carbonate types may discharge. Tributaries containing waters of these kinds have a different biota to the typical chalk stream, and the difference will be most marked the higher the concentration of sodium sulfate or carbonate in the water – in the extreme such water would not support much growth of submerged plants, and the fauna species and numbers would be restricted, mainly because of the low concentration of calcium in the water. Any influx of surface runoff from clay-covered areas would of course add turbidity, organic matter, calcium sulfate and nitrate to the stream flow, giving the receiving streams partially the character of streams draining clay catchments.

For a number of reasons, of which nature of the surface stratum is the dominant factor, few significant streams draining clay catchment areas now retain much of their natural quality. Such catchments are now either well developed for agricultural purposes or urbanized, and such changes from natural have a marked effect on the quality of the surface runoff, which in turn affects the quality of the streams draining these areas. However the flow of most streams draining clay catchments does not naturally consist only of surface runoff; minor aquifers in gravel deposits usually contribute to the dry-weather flow of the streams, and as already indicated the quality of the discharges from such aquifers is variable.

The main natural physical characteristics of streams draining clay catchments are low dry-weather flows of fairly clear water, and much higher flows of turbid water in wet weather, at a temperature varying according to the ambient temperature. The natural chemical characteristics of the water flow in dry weather are of a slightly alkaline water, containing mainly calcium hydrogencarbonate and calcium sulfate (with the latter often predominating) and a chloride concentration of around 35–50 mg l^{-1}. The presence of nitrate of around 5 mg l^{-1}, and a content of organic matter corresponding to about 3 mg l^{-1} 'oxygen absorbed', is typical. During periods of high surface runoff, the water is turbid with clay solids in suspension and slightly colored with organic matter in solution. The concentration in mineral matter, chloride and nitrate is variable according to flow. The first flow addition from surface runoff in wet weather is little different in total calcium content, but with higher chloride and nitrate concentrations, when compared to the dry-weather flow. During peak runoff the water contains less impurity in solution, but greater turbidity than under dry-weather flow conditions.

Because of the turbidity of wet-weather flows, and the general clayey nature of the stream bed, the flora is predominantly of emergent plants such as water lily (*Nuphar*) and reed mace (*Typha*), or submerged pondweeds, except in the riffles of the dry-weather flow channel where submerged weeds such as water celery (*Apium*) and starwort (*Callitriche*) occur but not abundantly. In the riffles the fauna is similar to that of a chalk stream, but less diverse and abundant. In summertime the water may produce noticeable growths of phytoplankton (*diatoms* and *algae*) mainly because of low velocities of flow. Because of the low productivity of submerged weeds the stream does not produce a great deal of natural organic silt. Despite this, the diurnal variation of dissolved oxygen concentration in the water under natural circumstances is about the same as that of a chalk stream. In this, the lesser effect of lower organic silt content on oxygen balance is by and large countered by the lower rates of dry-weather flow and hence of reaeration from the atmosphere.

The streams draining upland peaty and general woodland catchments derive their flow mainly from surface runoff, but some woodland areas have sandy or gravelly surface strata which yield reasonable spring flows even in dry weather. In upland peat areas, in dry weather the flow may disappear or almost so. Nevertheless, in the main the streams draining peat and woodland areas are subject to marked flow increases in wet weather. Thus the water in a stream in an upland peat area is always likely to be acid and colored with natural organic matter, but the degree of mineralization is low. Woodland stream waters are often similar in quality, but the dry-weather flow may be reasonable and the water less colored with natural organic matter. Both types of streams are similar in quality during wet weather when runoff is high, but often the stream draining the peat area is more colored with natural organic matter than the woodland stream. A typical analysis of a stream draining a woodland area is shown in Table W36.

The biology of both types of water is usually very limited in terms of numbers and diversity of species. The acidity, color and low mineralisation of the peaty water are the main causes of this. The same applies, but often to a lesser degree, to the woodland stream, yet

Table W36 Analysis of water in a stream draining a woodland area (mg l^{-1} unless otherwise stated)

Ammonium nitrogen	<0.01
pH	5.9
Chemical oxygen demand (COD)	19
Chloride (Cl)	13
Nitrate (N)	0.1
Total hardness (CaCO$_3$)	24
Non-carbonate hardness (CaCO$_3$)	13
Phosphate (PO$_4$)	<0.1
Iron (Fe)	<0.70

the diminution of sunlight by tree shading in woodland areas is significant in inhibiting growth of aquatic plants.

Therefore in considering the natural quality of any river water at any point at any time, in relation to proper management of uses of it, it is necessary to assess and appreciate (among other things) not only the quality of the water *per se*, but also the likely various origins of the water at that point and time.

In considering the quality of the water in lakes and ponds, much depends on the physical circumstances in individual cases. A lake or pond which is a part of a stream, or receives much spring discharge from underground strata, is likely to be considerably different in quality to an enclosed one maintained solely by rainfall and surface runoff. In the former case the basic chemical quality of the water will reflect that of the through-flowing water which may have originated from a distant source, whereas in the latter case local rainfall and local runoff will dominate the water chemistry. Lakes which contain water which is well mineralized, such as water from the chalk or from a clay catchment area, are referred to as 'eutrophic'. Lakes containing water from upland runoff and rainfall are scantily mineralized and are known as 'oligotrophic'. The currently fashionable use of the word 'eutrophication' to denote anthropogenic pollutive mineralization is a loose use of the word.

The factors regulating oxygen balance in a lake are complex. First, without any flow turbulence, reaeration of the water from the atmosphere will be slow except when winds agitate the surface. Water mixing will also be slow except under the influence of wave action. It is generally reckoned that a wave 0.3 m high will mix water down to a depth of 10 m. Without any wind effect, the transfer of oxygen, taken up from the air at the water surface, to the deeper levels will depend entirely on diffusion – a comparatively slow process. The growth of rooted submerged and emergent weeds will only occur normally in depths of less than about 2 m. Growths of suspended algae and diatoms (phytoplankton) in the surface layers of water will be prolific if the water is well mineralized, and contains appropriate residuals of the plant nutrients nitrate and phosphate. Thus a shallow lake may virtually be covered with rooted weed growth in summer, or it may take on the appearance of pea soup if phytoplankton growths are heavy in sunny weather. Indeed any well-mineralized shallow lake will, over a period of several hundred years, produce such an amount of silt from weed growth that the lake will fill up to resemble a bog.

In lakes of depth exceeding 10 m or so, stratification of the water may occur. The well-known 'autumn or fall turnover' phenomenon may well become an unfailing annual event. The stratification arises because in the warmer months of the year, the surface layers of water are warmed by solar radiation and thus are less dense and stay floating on top of the colder and denser water below. This effectively segregates the warmer well-mixed upper epilimnion from the colder unmixed hypolimnion. In the lower hypolimnion benthal decomposition on the lake bed may result in deoxygenation of the water, with resultant release of methane, chemical residuals such as ammonia, hydrogen sulfide and trace elements from the bed silt into the water. In the fall the upper water cools as air temperatures fall and increases its density. This upper water then sinks, pushing water to the surface which had hitherto been at lower depths, reducing dissolved oxygen levels and causing natural contamination, such as manganese, hydrogen sulfide, phosphate and ammonia to be present in appreciable quantity at the water surface. In severe weather, the icing up of lakes and ponds can prevent solution of oxygen from the atmosphere, and if prolonged can result in substantial oxygen depletion in the whole lake. The numbers of carp and bream killed in small lakes and ponds in the UK by this cause in the severe winter of 1962–63 were massive, but of course most of these were also subject to the exacerbating effects of human-made depreciations of the natural quality of the waters.

Tidal waters

Regarding tidal water resources, the water quality position in estuaries is more complex than in the case of coastal waters, because of the influence of freshwater discharges on estuaries. The uppermost part of any estuary is essentially similar in natural quality to the discharging river. Moving down-estuary from the tidal limit, the substantially freshwater zone gives way to an intertidal zone of fluctuating extent according to variations of tidal level and freshwater discharge. In this zone the number and diversity of species of flora and fauna are very

Table W37 Principal constituents of seawater (parts per thousand)

Sodium	10.8
Magnesium	2.0
Calcium	0.4
Potassium	0.4
Chloride	19.4
Sulfate	2.7

limited because of the variability of salinity. Down estuary of the intertidal zone the biota assumes an essentially marine character.

Two new physical factors have to be taken into account regarding the oxygen balance of estuarine waters. First, the fact that the degree of solution of oxygen into saline water decreases with salinity. Second, the residence time of the water in the estuary is affected by the two-way flow.

While estuarial waters are often somewhat turbid as a result of the stirring up of bottom silt at the higher tidal flow velocities, some are exceedingly so in consequence of the unusual silt load they carry. Coastal waters are of course seawaters and have the general quality shown in Table W37. The coastal waters close to estuaries may reflect in minor degree the chemical quality of the freshwater discharges from those estuaries. In both estuaries and tidal waters the essential natural biochemical phenomena produced by bacteria apply. For obvious reasons of the broad waters and seas involved, the biochemical and biophysical influences of biota are of no great consequence in terms of secondary effects on water chemistry or physics under natural circumstances. In the human-made circumstances now pertaining the position may be quite different.

From the foregoing brief outline of the factors affecting the quality of natural waters, it is clear that the diversity and dynamic balance of the living things in waters is determined primarily by the physical and chemical characteristics of the waters. At the same time the living processes of the biota exert in turn a secondary or 'feedback' influence on the chemical and physical characteristics of the water. All these things in total can be summarized as the 'ecology' of the water. It is thus manifest that any human-made alterations of the chemical and physical characteristics of the water will produce their consequences in changing the ecology of the waters.

It should be equally apparent that the proper assessment of water quality, even in the relatively simple natural circumstances, involves much more than perusing the results of chemical analyses. It even involves more than perusing the results of both chemical analyses and biological examinations of waters. In addition, or should it be said initially, it is necessary to inspect the water on site, and to assess the physical factors involved to enable the results of this to be applied in a total consideration of the quality factors involved. Then and only then, having established some reasonable notion of the background or baseline of water quality that nature intended, it is possible to assess how humans have changed this and how their actions need regulation to prevent further damaging change, and as and when possible to reverse past changes. But the first priority must be to prevent further damage.

R.W. Herschy

Source

Fish H., 1973. *Principles of Water Quality Management*, Thunderbird Enterprises, UK.

Cross references

Precipitation
Precipitation distribution
Precipitation: source
Rain
Water resources: natural quality
Water quality for drinking, WHO guidelines

WATER: SUBSTANCE AND SOLVENT

Structure

Water, present in the geological environment as solid, liquid, vapor and supercritical fluid, is one of the most important components of Earth's fluid envelope. Without water and its special properties, Earth would be a vastly different planet in many ways. Water plays critical roles in atmospheric processes, including the absorption of radiant energy and the redistribution of energy through weather mechanisms, thus fundamentally influencing climate. In the lithosphere, water serves as an essential solvent and transport medium for many of the elements, a modifier of rock physical properties, and a lubricant for tectonic processes. The biosphere also depends on water for its solvent properties and, in photosynthesis, as an electron-donating nutrient.

The two protons of each water molecule (molecular weight, 18.015) are deeply embedded in the oxygen atom as shown in Figure W33. Ice, up to about 2 kbar, has an open structure of tetrahedrally coordinated H_2O molecules, with a density of 0.9167 g cm^{-3} at 0°C and 1 atm. On melting (heat of fusion, 1.436 kcal mol^{-1} at 0°C), expanded, ice-like clusters of H_2O molecules form surrounded by unordered monomeric molecules. Dodecahedral cavities in the clusters are sufficiently large to accept an interstitial H_2O molecule which accounts for water having a greater density (0.999841 g cm^{-3}) than ice at 0°C and 1 atm. The increase with temperature of the fraction of cavities filled by interstitial molecules counteracts thermal expansion of the clusters and leads to an increase in density to a maximum at 3.98°C and 1 atm. Above 3.98°C about half the cavities remain filled

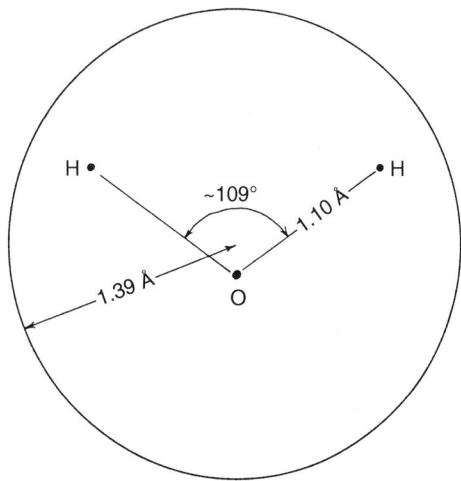

Figure W33 Geometry of a water molecule in the liquid at 25°C (schematic) showing the atomic nuclei and the 'hard-sphere' radius which is nearly identical to that of oxygen. (Data from Narten *et al.*, 1967)

and normal thermal expansion causes a decrease in water density with increasing temperature (Berner and Berner, 1987).

The abnormally wide range of thermal stability for liquid water at 1 atm, and its high surface tension (71.97 dyn cm^{-1} at 25°C) reflect the strength of hydrogen bonding between water molecules in the clusters. These bonds are continuously being broken and re-established so that the average lifetime of a cluster is about 10^{-10} s. The concentration and distribution of cluster sizes are uncertain (Kavanau, 1964) but both are affected by solutes and temperature. Pressure may have only a minor effect on water structure because of water's low compressibility (4.57 × 10^{-11} cm^2 dyn^{-1} at 25°C). Some properties of water along the liquid-vapor curve are given in Table W38. $P - V - T$ properties at other pressures and temperatures are tabulated in Clark (1966, p. 371). For properties at 25 and 100°C, see Table W39.

Solutes affect the stability of liquid water. Solids are generally more soluble in water than in ice or water vapor and therefore tend to increase water stability, shifting the ice–liquid curve to lower temperatures and pressures and the liquid–vapor curve to higher temperatures and pressures. Because gases are more soluble in water vapor than in liquid water, and in liquid water than in ice, gases in solution tend to shift both the liquid–vapor and ice–liquid curves to lower temperatures and pressures. Consistent with this behavior, salts generally increase the critical temperature (T_c) and decrease the ice point of water, whereas gases decrease T_c and lower the temperature of the ice point. For example, an NaCl solution saturated at 25°C (6.18 molal, 26.4 wt%) has a T_c of 700°C (1237 bar), whereas T_c for a 27 mol% CO_2 solution is 275°C (885 bar).

Water as a solvent

The unequalled ability of water as a natural solvent for ionic solutes reflects its dipolar nature and strong molecular polarizability, and in acid or alkaline solutions, the capacity of H^+ or OH^- ions to form complexes with solute species. The dielectric constant of water, ϵ (Table W38), is a measure of the combined effects of its dipolar and polarizing character, and is higher than for any other inorganic liquid. The dependence of ionic solubilities on ϵ in dilute solutions is given by the Debye–Hückel limiting law (Robinson and Stokes, 1959, p. 230), which for an ionic species i at constant temperature and ionic strength may be written

$$\log (1/\gamma_i) = k\epsilon^{3/2}$$

where k is a constant, and γ_i is the individual ion activity coefficient, a factor which is an inverse function of the solubility of salts forming solute ions. The force of attraction between two ionic species in solution is proportional to the reciprocal of the dielectric constant. Thus forces between solute ions in water are relatively small because of the ease of polarization of water molecules about an ion, an effect causing ion hydration and complexing. The dielectric constant of water decreases with rising temperature (Table W38), but increases with increasing pressure or density below the critical point.

A second important mechanism of solution in water is complex formation and ion association between H^+ and OH^- and solute species. This mechanism is, of course, more important in acidic or alkaline solutions. Most minerals, including those of the feldspar and carbonate groups, are salts of weak acids and strong bases. Their attack by water produces cations, OH^- ions and other species.

Table W38 Properties of liquid water in equilibrium with vapor

	Temp (°C)	Pressure (bar)	Specific volume (cm^3 g^{-1})	Dielectric constant	Log K_w[a]
Ice point	0	0.006107	1.0002	87.74	14.944
Triple point[b]	0.01	0.006112	1.0002	87.74	14.944
Maximum density point	3.98	0.008129	1.0000	86.20	14.793
	25	0.031663	1.0030	78.30	13.997
Boiling point	100	1.01325	1.0435	55.32	12.254
	200	15.551	1.1565	34.51	11.254
	300	85.92	1.4036	19.55	11.034
	350	165.37	1.741	12.55	11.422
Critical point	374.15	221.2	3.17	9.01	11.997

[a] $k_w = (a_{H^+}) \cdot (a_{OH^-})$.
[b] A standard point on the temperature scale.

Table W39 Properties of water[a]

Property	At 25°C	At 100°C
Dipole moment (Debye units)	1.87	–
Enthalpy (kcal mol^{-1})	−68.315	−67.747
Entropy (cal mol^{-1} k^{-1})	16.71	20.76
Gibbs free energy (kcal mol^{-1})	−56.688	−53.820
Heat capacity (cal mol^{-1} k^{-1})	17.995	18.15
Heat of vaporization (kcal mol^{-1})	–	9.717
Index of refraction, Na light:		
Absolute	1.33287	1.31819
In air	1.33251	1.31783
Resistivity, intrinsic (ohm)	18.24×10^6	(4.97×10^6 at 55°C)
Specific volume (cm^3 g^{-1}):		
1 bar	1.0030	1.0435
1000 bar	1.0000	0.9999
10 000 bar	0.807	0.8389
Surface tension, in air (dyne cm^{-1})	71.97	58.9
Thermal conductivity (cal cm^{-1} s^{-1} C^{-1})	0.00145	0.00160
Viscosity (poises)	0.008904	0.002790

[a] At 25 and 100°C and 1 bar (0.986923 atm), except where otherwise noted.

Consequently they are most soluble under acid conditions. In contrast, the solubility of silica, which hydrates to form the weak acid H_4SiO_4, increases under alkaline conditions by the formation of H^+ ions plus anionic species such as $H_3SiO_4^-$ and $H_2SiO_4^{2-}$. A number of multivalent metal ions, such as Fe^{3+} and Al^{3+} which form relatively insoluble oxyhydroxides in near-neutral solutions, are amphoteric, exhibiting high cationic solubilities in acid waters and high anionic solubilities in alkaline waters.

The acidity or alkalinity of an aqueous solution at any temperature or pressure is measured by its pH, defined as the log of the reciprocal of the hydrogen ion activity. The pH scale is temperature dependent as is evident from the increase in the activity product of water with temperature (Table W38); the scale is also slightly pressure dependent. The neutral point, where the activities of H^+ and OH^- are equal, decreases from pH = 7.00 at 25°C to pH = 5.52 at 300°C for solutions along the liquid–vapor curve. Barnes and Ellis (in Barnes, 1967) give pH data for other conditions to 700°C and over 2000 bar.

At a given temperature in dilute solutions, the solubility of minerals which dissolve to form ions increases with ionic strength, i.e. on the addition of other ions. The solubility of gases and solids which form molecular species in solution is controlled by the hydration process and generally decreases slightly with increasing ionic strength (Ben-Naim, 1974).

The effects on solubility of pressure and temperature variations are complex. Increasing pressure favors solubility to a minor extent at low temperatures for salts where ΔV_{soln} and solution compressibility are small, and to a major extent for gases at higher temperatures where both V_{soln} and solution compressibility are large. In the subcritical region, minerals which react with a gas on dissolving, such as calcite, with CO_2, usually decrease in solubility with increasing temperature; exceptions are the solubilities of cinnabar and acanthite in solutions with H_2S, where solubilities of both minerals pass through a maximum at about 100°C. Minerals such as halite or quartz, which do not involve gas species in their solution reactions, generally increase in solubility with rising temperature in subcritical solutions; however, some minerals, such as anhydrite and fluorite show decreasing solubility. In the supercritical region, solubility is often a direct function of solution density. Summaries of high-temperature solubilities are given in Barnes (1967, Chapters 8, 9 and 11).

Geological distribution

The water content of the crust of the Earth is $2.2–2.6 \times 10^{18}$ tonnes distributed among the atmosphere, 1.3×10^{13} tonnes in the hydrosphere, and 1.4×10^{18} tonnes in the lithosphere (Poldervaart, 1955). Water of the hydrosphere and atmosphere is believed to have originated by the continuous dehydration through geological time of the hydrous minerals of crustal and subcrustal igneous rocks (Rubey, 1955), and from extraterrestrial sources such as comets and meteors. Igneous rocks contain an average of 1.15% water by weight, primarily

as structural water of hydration in mineral groups such as the amphiboles and micas. The extent of hydration of crustal rocks depends on the thermodynamic water pressure or fugacity of water, a property which is a function of temperature, pressure and water purity. Fugacity data for water are presented by Anderson (in Barnes, 1967).

Hydrological cycle

In the environment, water is highly abundant and mobile. The exchanges of water among different parts of the environment are summarized in the hydrological cycle (Figure W34). Over 99% of Earth's modern inventory of water is in the oceans or in ice. During periods of extensive glaciation, some of the oceanic inventory is shifted to ice. Most of the remaining global inventory is in groundwater. The division between shallow and deep groundwater is somewhat arbitrary but indicates a distinction between groundwater that is readily available as a resource (although not necessarily available with suitable chemical composition) and groundwater that is not easily extracted. The illustration also shows that the water content of the crust diminishes with depth. Even so, evidence from metamorphic rocks indicates that liquid water exists at depths of at least 20 km in the crust.

Most marine precipitation is derived from marine evaporation, and most continental evaporation is returned to the continents as precipitation. On the other hand, marine evaporation exceeds marine precipitation, while continental precipitation exceeds continental evaporation. The system is balanced by transport of water vapor from the marine domain to the continental and the drainage of liquid water from the continents to the oceans. Aside from its effect on the global water budget, the net transport of water vapor to the continents is a significant part of the climatic moderating influence associated with marine air masses (as noted above, vapor transport = latent heat transport).

Composition of natural waters

Average rain or snow contains about 10 ppm dissolved solids, chiefly as Na^+, Ca^{2+}, Cl^- and SO_4^{2-}, and has a pH of about 5.5 due to carbonic acid formed by solution of atmospheric CO_2. Rain falling near ocean coastlines or saline deposits occasionally has more than 100 ppm dissolved solids due to solution of windblown salts (Carroll, 1962).

Ca^{2+} and HCO_3^- are the predominant ions in most surface- and groundwaters. The pH of most natural waters is between 6 and 8, buffered by the dissociation of carbonic acid and hydrolysis reactions with carbonate and silicate minerals. The lowest pHs, near 1, occur in surface waters which contain H_2SO_4 or rarely HCl. High pHs of 9–10, or exceptionally 12, may be found in alkali lakes and springs, in the presence of carbonate and evaporate minerals.

The dissolved solids content of surface waters is highly variable, ranging from less than 50 ppm in small mountain streams on siliceous

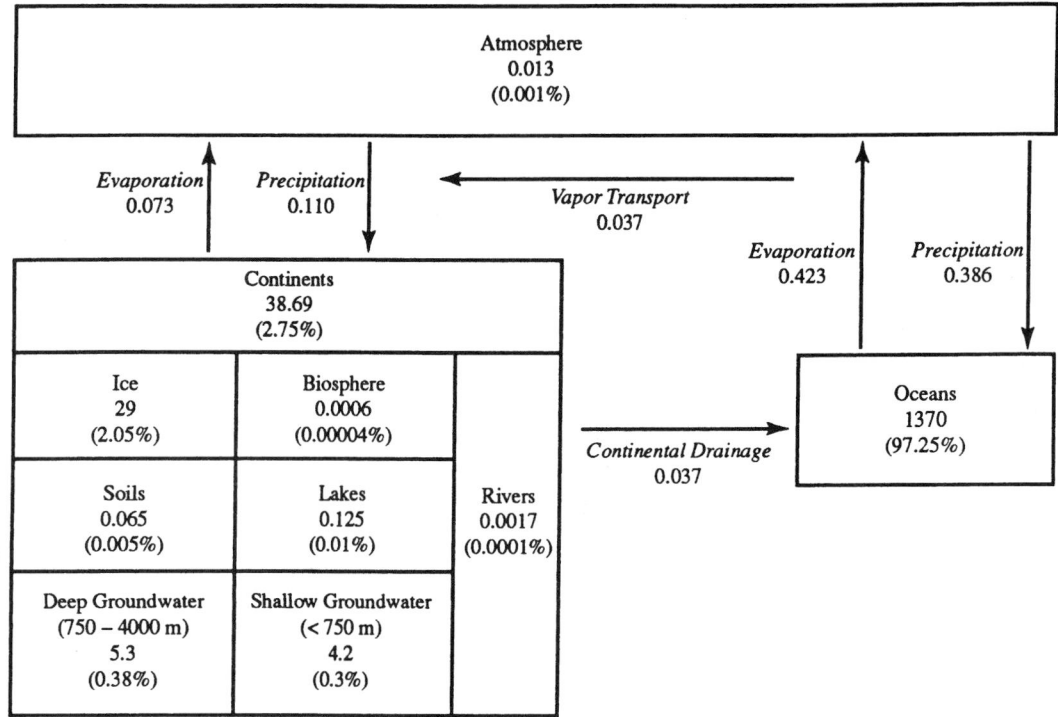

Figure W34 The hydrological cycle. Reservoir inventories are 10^6 km^3 (with percentage of global water inventory), and fluxes are 10^6 km^3 year^{-1}. (Data from Berner and Berner, 1967.)

rocks, to as much as 50 000 ppm in streams flowing over saline deposits, and 400 000 ppm in some closed soda alkali lakes. Average river water contains 120 ppm solutes, including 58.4 HCO_3^-, 11.2 SO_4^{2-}, 7.8 Cl^-, 1 NO_3^-, 15 Ca^{2+}, 4.1 Mg^{2+}, 6.3 Na^+, 2.3 K^+, 0.67 total iron and 13.1 SiO_2 (Livingstone, 1963). Ocean water contains 34 500 ppm solutes as follows: 18 980 Cl^-, 10 560 Na^+, 2650 SO_4^{2-}, 1270 Mg^{2+}, 400 Ca^{2+}, 380 K^+, 140 HCO_3^-, 65 Br^-, and others less than 10 ppm (Lide and Kehiaian, 1995).

The dissolved solids content of groundwater generally exceeds that of surface water (White *et al.*, 1963), and ranges from less than 100 ppm in shallow groundwaters from siliceous rocks in areas of high rainfall, to 300 000–600 000 ppm in some saline groundwaters. Compositions of thermal springs, which vary even more widely, have been compiled by Waring (1965).

Hardness of water reflects chiefly its Ca^{2+} and Mg^{2+} content. Generally, waters with dissolved solids less than about 120 ppm are considered soft; from 120 to 350 ppm moderately hard to hard; and greater than 350 ppm very hard.

Potable waters in the United States are usually expected to contain less than 500 ppm dissolved solids, although in some western states where supplies are high in solutes, they contain 500–1000 ppm. Half the United States population served by public supplies receives water containing less than 150 ppm, whereas 90% receives water with less than 550 ppm (Durfor and Becker, 1964).

The US Public Health Service states that a water is toxic and a health hazard, and should not be used for drinking if it contains substances in excess of the following mg l^{-1} concentrations: As 0.05, Ba 1.0, Cd 0.01, Cr^{6+} 0.05, CN 0.2, F 1.5 (approx.), Pb 0.05, Se 0.01 and Ag 0.05. The Health Service further recommends that when more suitable supplies are available, water should not be used if concentrations in mg l^{-1} exceed the following: alkyl benzenesulfonate (from detergents) 0.5, Cl^- 250, Cu 1.0, Fe 0.3, Mn 0.05, NO_3^- 45, phenols 0.001, SO_4^{2-} 250, total dissolved solids 500 and Zn 5.0 (Lide and Kehiaian, 1995).

Donald Langmur and H.L. Barnes

Bibliography

Barnes, H.L., (ed.), 1967. *Geochemistry of Hydrothermal Ore Deposits*, New York, Holt, Rinehart and Winston, Inc., 670 pp.

Ben-Naim, A., 1974. *Water and Aqueous Solutions*, New York, Plenum, 474 pp.

Berner, E.K. and Berner, R.A., 1987. *The Global Water Cycle*, Englewood Cliffs, NJ, Prentice-Hall, 397 pp.

Carroll, D., 1962. Rainwater as a chemical agent of geologic processes – a review, *US Geol. Surv. Water-Supply Paper*, **1535G**, 18 pp.

Clark, S.P. Jr. (ed.), 1966. Handbook of Physical Constants, *Geol. Soc. Am. Mem.*, **97**, 587 pp.

Durfor, C.N. and Becker, E., 1964. Chemical quality of public water supplies of the United States and Puerto Rico, 1962, *US Geol. Surv. Hydrologic Investig. Atlas*, **HA-200**.

Kavanau, J.L., 1964. *Water and Solute-Water Interactions*, San Francisco, Holden-Day, Inc., 101 pp.

Lide, D.R. and Kehiaian, H.V., 1995. *CRC Handbook of Thermophysical and Thermochemical Data*, Boca Baton, CRC Press, 518 pp.

Livingstone, D.A., 1963. Chemical composition of rivers and lakes, in Fleischer, M., ed. *Data of Geochemistry, US Geol. Surv. Profess. Paper*, **440-G**, 64 pp.

Narten, A.H. *et al.*, 1967. X-ray diffraction study of liquid water in the temperature range 4–200°C, *Discussions Faraday Soc.*, **43**, 97–107.

Poldervaart, A., 1955. Chemistry of the earth's crust, in Poldervaart, A., ed., *Crust of the Earth, Geol. Soc. Am. Spec. Paper*, **62**, 119–144.

Robinson, R.A. and Stokes, R.H., 1959. *Electrolyte Solutions*, 2nd edn., New York, Academic Press, 559 pp.

Rubey, W.W., 1955. Development of the Hydrosphere and Atmosphere, in Poldervaart, A., ed., *Crust of the Earth, Geol Soc. Am. Spec. Paper*, **62**, 631–650.

Waring, G.A., 1965. Thermal Springs of the United States and other countries of the world – a summary, *US Geol. Surv. Prof, Paper* **492**, 383 pp. (revised by R.A. Blankenship and R. Bentall).

White, D.E., 1965. Saline waters of sedimentary rocks, in Fluids in subsurface environments – A symposium, *Am. Assoc. Petrol. Geol. Mem.*, **4**, 342–366.

White, D.E. *et al.*, 1963. Chemical composition of subsurface waters, in Fleischer, M., ed., *Data of Geochemistry, US Geol. Surv. Prof. Paper* **440-F**, 67 pp.

Cross references

Atmosphere
Groundwater
Hydrological cycle
Hydrosphere
Precipitation
Rain
Water resources: quality assessment
Water quality for drinking: WHO guidelines
Water treatment: potable water

WATER TABLE

The water table is the surface in an unconfined aquifer at which the pressure of the water in the void spaces is exactly equal to atmospheric pressure. This surface is usually taken as the boundary between the saturated zone beneath and the unsaturated zone above. It should be noted that small voids in the aquifer will usually be completely filled with water for some distance above the water table, the water being held in place by capillary forces; this water will be at a pressure less than that of the atmosphere, and occupies a zone usually referred to as the capillary zone or capillary fringe.

The word table suggests something that is flat and static. The water table is neither. It usually follows the topography in a subdued way, so that it is at or near the surface in valleys, and deep below the surface beneath hills. The gradient of the water table is an indicator of the hydraulic gradient that causes groundwater movement, with flow occurring away from regions where the water table is high to regions where it is low. At any point, the water table will usually rise and fall through the year, being high after recharge and falling as recharge ceases and groundwater discharges from the aquifer to springs and rivers. The height of the water table above a datum such as sea level is thus an indicator of the amount of water in storage in an aquifer.

Michael Price

WATER TREATMENT: POTABLE WATER

It is estimated 25 000 people a day die as a consequence of waterborne and related diseases such as typhoid, cholera and dysentery. These deaths are virtually all associated with the underdeveloped, underprivileged and poorer nations.

Present water treatment technology and operational expertise can prevent the transfer of disease and panders to organoleptic standards now expected by the Western world. The World Health Organization (WHO), an international body, provides guidelines for drinking water quality; however, individual and groups of countries have developed their own quality criteria based on these guidelines. Examples are the US Environmental Protection Agency Standards (Safe Drinking Water Act, 1974), and the European Drinking Water Directive, 1980.

Contaminants

Water may naturally contains many substances harmful and or unpalatable to humans. In addition, humans discharge many substances into their environment which cause contamination of potential water sources.

To provide acceptable drinking-water quality it is necessary to identify potentially harmful and unwanted substances and their concentration, and decide upon acceptable concentrations that will protect the user from their possible harmful or unpleasant effects. Contaminants can best be categorized into three groups:

- organoleptics, being those substances pertaining to the senses, i.e. sight, taste and odor;
- chemical, inorganic and organic substances which are potentially toxic, aesthetically unwanted or interfere with treatment processes and water storage and transfer;
- bacteriological, harmful microorganisms normally associated with the waste products of humans and other animals.

Organoleptics

In assessing drinking water quality, consumers will respond to their natural senses i.e. visual, smell and taste. It is therefore necessary that these parameters are considered and reduced to an acceptable level.

The visual problems are color, turbidity and particulate matter. Color can be the consequence of organic acids associated with decayed vegetation i.e. peat, or the presence of iron and manganese which can form chemical complexes with the organic materials mentioned.

Turbidity can be due to the suspension of fine particulate matter such as algae or the more problematic emulsions typically produced by clays.

Water contains an assortment of solids which must be removed, not only for aesthetic reasons, but also for the protection of water mains from organisms which can utilize residual organic material as a source of food.

Taste and odor can result as consequence of chemical contamination, algal decay products and bacteria by-products producing chemicals that are detectable by taste at very low concentrations. Phenolic derivatives produced as a result of disinfection with chlorine are very taste and odor sensitive at very low concentrations.

Chemicals

The most commonly encountered are the metals iron, manganese and aluminum with lead, copper, cadmium and silver also occasionally present; the non-metals, nitrate, fluoride and phenols are also occasionally present. Depending on the concentrations and the standards imposed, these waters require treatment before they can be used for drinking purposes.

In addition to those found naturally, humans have enhanced existing levels of certain chemicals and introduce new contaminants by their pursuits and activities.

Nitrate is a good example of chemical enhancement as a result of intensive farming techniques, but the greatest potential danger are the chemically synthesized compounds which have the potential to enter the food chain and accumulate in animal tissue, e.g. certain pesticides, herbicides and PCBs (polychlorinated biphenyls).

Microorganisms

Bacteria can very simply be described as non-photosynthetic organisms which reproduce by fission, i.e. the dividing of cells. Under ideal conditions they can reproduce in less than 15 min, they are microscopic in size and exist as spheres, rods or spirals.

Whilst the vast majority of bacteria are harmless and in fact essential to life, some forms and groups termed pathogens are responsible for illnesses from relatively minor ones to potentially life-threatening diseases. Fortunately the once-common waterborne diseases cholera, typhoid fever and bacillary dysentery are now rare in the more developed countries.

Viruses are a group of pathogenic organisms which are smaller than bacteria and are harder to control. Unlike bacteria they can only reproduce within a host cell and are parasitic in nature. The more common viral diseases are poliomyelitis and infectious hepatitis.

The protozoan waterborne parasites *Cryptosporidium* and *Giardia*, whilst not life-threatening to a healthy person, are proving to be a serious problem in the Western world, causing a severe form of gastroenteritis.

Potable water sources

Raw water sources are conveniently segregated into four categories: sea, ground, upland and lowland. Each category has typical characteristics which are convenient in deciding the treatment process necessary to produce a wholesome quality of water. Table W40 provides a comparison of the properties of borehole water, upland water and lowland water. Seawater is not often used as a source of potable water due to the high cost of desalination either by distillation or membrane filtration technology. Both methods involve high capital and operating costs and would normally only be considered in the absence of other sources.

Ground water

An aquifer is a natural underground reservoir of water contained in the voids and pores of rock formations. Rainwater which does not evaporate or run off to surface streams or rivers penetrates permeable

Table W40 Comparison of properties of borehole water, upland water and lowland water

Parameter	Borehole water	Upland water	Lowland water
Temperature	Relatively constant	Varies with the seasons and depth of abstraction	Varies with the seasons
Turbidity	Low	Normally low but can develop some turbidity as a result of algal growth	Variable; moderate to high
Color	Naturally low but can develop color due to iron and manganese	Moderate to high due to the presence of organic acids	Moderate to low
Mineral content	Generally high but constant	Consistently low	Usually moderate to high but varies with catchment
Soluble Fe and Mn	Normally present, concentrations can vary significantly	Normally present at moderate to high concentrations and often complexed with organic compounds	Usually low
Carbon dioxide	Often present in high concentrations	Low levels	Low levels
Dissolved oxygen	Low to zero	Near saturation	Near saturation
Ammonia	Very low	Very low	Naturally low but can be influenced by effluent discharges
Nitrates	Often present in high concentrations	Expected to be low	Expected to be low; can be affected by effluent discharges
Living organisns	Expected to be very low; ferrobacteria occasionally found	Always present and will vary considerably with the nature of the catchment and temperature	Always present and will vary considerably with the nature of the catchment and temperature
Organic matter	Very low	Moderate	Moderate to high
Pesticides and herbicides	A potential problem in agricultural areas	Normally low	Often present and a potential problem

surfaces, usually soil, and infiltrates the ground by gravitational forces. The rate of infiltration (hydraulic conductivity) is governed by two major factors: 'porosity', the volume available within the material, and 'permeability', the rate at which the material will allow water to pass through it. The movement of infiltrated water percolates through the upper non-saturated soil and gravel layers until it reaches the water table where it will saturate the underlying porous rock. Excess water moves very slowly towards the sea due to gravitational forces.

A borehole is a shaft sunk into the ground to a depth that will pierce an aquifer. The depth of the aquifer will obviously vary. It will essentially be of adequate depth to protect it from direct surface contamination but reasonably accessible to make it physically and economically attractive to abstract water by pumping. As water is abstracted the borehole is constantly replenished by the natural movement of water through the porous rock and fissures, provided the pumping does not exceed the rate of replenishment.

Borehole water is described as sparkling, without taste and of consistent quality making it a very popular choice with users. However, it is vulnerable to contamination from intensive agricultural practices and industrial effluent and solid waste disposal. The major pollutant from agriculture is nitrate, but pesticides and herbicides must be considered as a major potential threat. Should an aquifer become polluted or contaminated, it will often be abandoned as its recovery can take tens or hundreds of years to restore it to potable quality.

Water abstracted from boreholes or springs often requires only disinfection to render it safe for drinking. The degree of disinfection employed will depend on the bacteriological quality and the risk of contamination. It is the practice in some countries not to disinfect bacteriologically reliable groundwater sources.

Iron, manganese, aluminum, carbon dioxide and nitrate are commonly present in groundwaters, occasionally in concentrations that require their reduction.

Upland water

Upland can be taken as a relative term in that it is used to describe a water from a catchment above the general lie of the land, which does not necessarily have to be a mountainous area, the nature of the collection grounds being the main criteria. The explanation is that the rain collected on upland gathering grounds is relatively uncontaminated compared to the types of pollution encountered in lowland waters. As a consequence they are not normally associated with eutrophication and its related problems; however, there is normally adequate background nutrient to allow the seasonal development of algal growths which can result in complaints concerning the taste associated with the decay products of the algae.

Upland sources have a tendency to be acidic and undergo seasonal color variation, the acidity and color originating from peat and other vegetable residues draining from the surrounding catchment area. Upland waters are generally soft and unstable with respect to pH, which may require the addition of alkalinity prior to treatment. However, there are impounding reservoirs in chalk and clay areas which are neutral in reaction and relatively hard in character.

The presence of iron and manganese is a common feature of upland water, making treatment more difficult because of their ability to form complexes with organic materials. Deep lakes are susceptible to thermal stratification which can result in the seasonal mixing of the separate layers, producing significant variation in water quality particularly in respect of iron, manganese and turbidity.

Lowland waters

Potable lowland waters are characterized by their ability to change rapidly in quality and nature, influenced by rainfall, drought and temperature. These variations can affect the color, turbidity, taste and odor, metal concentrations and other parameters. Color may originate from decaying vegetation, algal activity or possibly iron and manganese. Turbidity results from the movement of sediment, algal activity or emulsified particles such as clay. Sediment will either be in suspension or lie on the bed of the river dependent upon the velocity of flow and the density of the particle, while algae may be present as an emulsion or suspension. Taste and odor problems can result as a consequence of fungal and bacteriological activity.

Lowland waters commonly receive sewage effluent and storm overflows from sewerage systems, industrial discharges, farm waste and fertilizer runoff, and are susceptible to spillage on roads and industrial areas. A treatment plant has to be capable of treating all the above variations.

A river may be described as 'flashy', meaning that it is susceptible to significant flow variation as a result of rainfall. Rivers are also vulnerable to drought conditions which are combated by compensation water from impounding and storage reservoirs. The use of lowland waters, because of the problems related above, may in certain circumstances require storage prior to treatment to accommodate unacceptable quality conditions in the river.

Processes employed in the treatment of potable water (Figure W35)

Storage

When water undergoes a period of storage in an open reservoir there is normally a significant reduction in the number of bacteria, significantly those of intestinal origin. Associated pathogens and other harmful microorganisms are similarly reduced and can be totally removed. This method of treatment is usefully employed in the warmer climates where land is plentiful and funds are minimal. Ultraviolet light, which is a natural biocide, sedimentation and various biotic agencies account for the significant improvement in water quality. Table W41 indicates the order of improvement that may be achieved from a poor-quality source by suitable storage.

Screening of raw water

Screening is a preliminary treatment stage to protect subsequent treatment stages. Screening is necessary for upland water to prevent debris and fish entering the treatment process and is normally satisfactorily achieved by conventional bar screens. Lowland rivers, on the other hand, provide a rubbish collection site for nature and humans. Consequently an abstraction has to accommodate the associated problems and requires a far more sophisticated arrangement which is an essential stage of the treatment process. The purpose of screens is to remove the larger-sized debris, to prevent the entry of fish, particularly game fish, and to protect pumping equipment. The screens need to be robust to withstand large objects while being

Table W41 Improvement achieved from a poor-quality source by suitable storage

	% Reduction
Color	67
Turbidity	51
Organic nitrogen	57
Iron	85
Manganese	67
Coliform	99
E. coli	99

capable of preventing smaller objects from causing damage to pumping equipment, yet of adequate width to prevent blinding. A screen will often be protected by a floating boom to prevent floating materials, especially oil and immiscible solvents, and floating algae, from entering the abstraction. The introduction of a compressed air screen prior to the boom further improves the effectiveness of the boom and discourages fish passage.

In the absence of a raw-water storage reservoir, abstraction screening requires additional process screening to remove smaller debris which may interfere with or affect the efficiency of the consequent treatment processes. Band- or drum-type screens are successfully employed for this purpose.

A band screen is a vertically operated screen, with 5–6 mm diameter holes or slots, installed in a channel requiring all the flow to pass through. It rotates on a vertical plane, introducing a clean screen operated either by a time or level control or both. To maintain a continuous process, the screen will have a cleaning mechanism which can be a mechanical rake, a brush system or, probably most commonly, a water jet which washes the screenings into a channel which is then returned to the river. A dry brush system of solids removal can be employed to exclude the need for a river discharge.

A drum screen is, as the name implies, a large drum, typically size 2 m diameter and 3 m long with the curved surface perforated with 6 mm diameter apertures. It rotates slowly on an axis through its center and raw water is fed into the internal section of the drum and flows from inside to outside, the solids being retained inside the drum. The solids are removed as a continuous process by fine jets of water directed onto the top of the filter, washing the solids into a trough, which are then discharged back to the river.

pH adjustment

Water may be acidic due to organic acids or carbon dioxide or alkaline due to dissolved base metals. pH adjustment is sometimes the only treatment required for borehole water and some natural spring sources, whilst upland and lowland waters invariably require pH adjustment as a prerequisite to further treatment.

Lime in the form of a slurry is the most common choice of alkali because of its low cost; however, it does cause operating difficulties. Caustic solutions are easier to handle and control but are not cost effective, and consequently tend to be employed only on small plants. Both hydrochloric and sulfuric acid are employed for neutralizing drinking water, the choice depending on cost and availability.

Solids removal

The removal of fine solids is achieved mainly by sand filtration or microstraining. Sand filtration is mainly employed as the final stage of a multistage process and is explained in some detail later.

Microstrainers are revolving drums of stainless steel with a very fine mesh surrounding the drum (23–35 mm pore size), constructed from mild steel, fabric or a synthetic material designed to remove small particles. Raw water is fed to the center of the drum filtering to the outside, whilst the drum is continuously rotating. To prevent the screen blocking, it has to be continuously cleaned. The conventional method is to employ a high-pressure water jet to dislodge any material into a channel for disposal. As high as 1% of the plant output can be required to maintain this system. In addition to the physical washing of the screen it is often necessary to remove growths of algae by ultraviolet light.

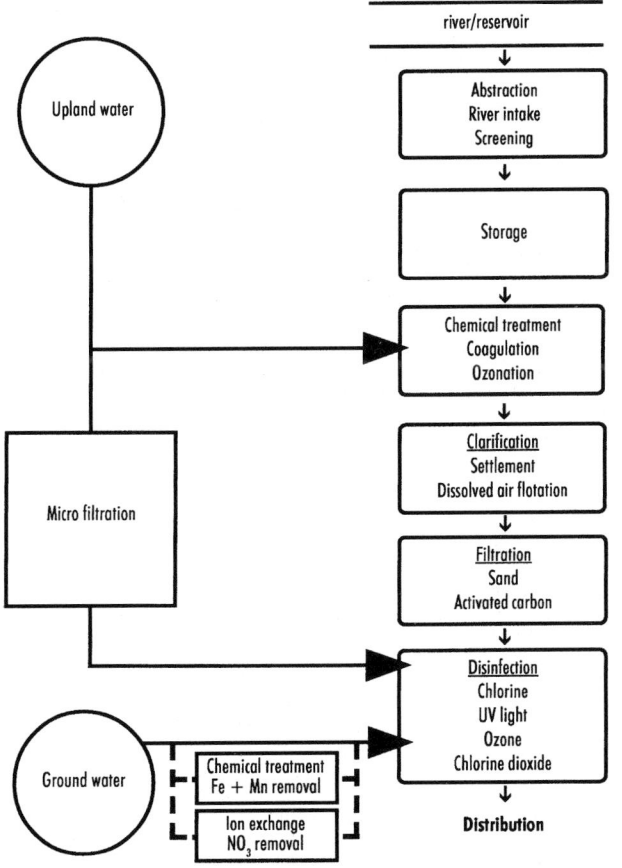

Figure W35 Schematic diagram of potable water treatment processes.

The loss of head through a unit varies from 150–450 mm and a single drum of size 3 m × 3 m can treat up to 1.35 Ml h^{-1}.

Microstrainers are extremely efficient in the removal of plankton, algae and other microscopic particles but do not remove color or finely divided particles such as clay.

Coagulation

Coagulation is the process of forming and developing a precipitate, which promotes the destabilization of suspended particles and the precipitation of soluble particles. To ensure maximum efficiency and effect the coagulant must be dispersed rapidly within the stream of water requiring treatment. The pH at which coagulation takes place is critical and will vary for different types of water and with variation in water quality. Typical coagulants include the following.

- Aluminum sulfate, $Al_2(SO_4)_3 \cdot 18H_2O$, which is acidic and will require chemical neutralization. The most effective pH for coagulation is 5.8–6.2, outside this range the precipitation of the hydroxide salt is incomplete and will cause residual aluminum to bypass the process. This pH restriction is not normally a problem except if the water has a high color, when a lower pH is more effective.
- Ferric salts have the ability to flocculate at a lower pH than alum and perform effectively at pH 4.5–5.0. This is the ideal level for the coagulation of humic substances which impart problematic color into the water.
- Ferric sulfate, $Fe_2(SO_4)_3 \cdot 9H_2O$, is strongly acidic and like alum requires suitably protected storage tanks to prevent corrosion.

The coagulation process will remove natural color derived from peat and other organic substances such as humic fulvic acids and other, more complex compounds found as true solutions or colloidal suspensions. Coagulation also removes animal and vegetable debris, plankton and other organic material that may be present.

Flocculation

Flocculation is the term describing the development of coagulated floc particles. The floc particles are encouraged to coalesce to a size and density which effects efficient clarification of the treated water. These particles can be separated from the bulk of the water by settlement or flotation.

Coagulation aids

Whilst referred to as coagulation aids, these are in effect flocculation aids acting as bridging or linking molecules for adjacent particles. The water industry employs mainly polyelectrolytes, which are synthetic, organic, water-soluble and high molecular weight polymers. They are expensive, but the cost can be offset by the improved efficiency of the solids removal process. Their dose rates are low compared to the coagulant dose and are typically 0.1 mg l^{-1}.

Clarification following the coagulation stage

The coagulation process adds chemicals resulting in the water being visibly more contaminated; however, it is now more amenable to further treatment and improvement. The type of raw water, or specifically the nature of the solids in solution and suspension, will normally decide which method of clarification is employed. A typical soft upland water encountering seasonal fluctuations in color and algal blooms lends itself to a flotation process, whilst a lowland river susceptible to intermittent high suspended solids of an inorganic nature may best be accommodated by a settlement process.

Settlement and sludge blanket clarification

Suspension/settlement perhaps best describes the removal of solids by this method. Circular or square tanks of conical shape are normally employed. Regardless of shape, the flow is introduced at the base of the tank and the treated water flows vertically through a concentrated suspension of suspended solid material referred to as the sludge blanket. The sludge blanket is a concentrated suspension of coagulant and particles produced by the coagulation process.

The sludge blanket acts as a barrier or filter to the incoming chemically treated water. The blanket is kept at a predetermined level and allows the clarified water to travel vertically up the tank through the sludge blanket and to overflow a weir.

Dissolved air flotation

This is achieved by introducing water supersaturated with air at 7 bar pressure to the water immediately following the coagulation and flocculation stage as it enters the dissolved air flotation tank. The carrier water for the dissolved air is treated water and represents between 10 and 20% of the total treatment.

The air-supersaturated water is released from a series of submerged header pipes situated at the inlet to the clarification tank, injecting a mass of microbubbles which attach themselves to the floc particles causing them to rise to the surface and form a thick scum at the surface of the water. A slow-moving scum board continuously moves the sludge from the top of the tank to a trough positioned at the outlet end of the tank, whilst the clarified water overflows a weir beyond the solids removal system.

Due to high separation velocities, the throughput of the system is much faster than the more conventional upward flow clarifiers and requires less space and considerably less civil engineering construction work. Disadvantages are the reliance on power and the relatively high energy consumption to maintain the compressors, recirculation and sparge pumps, etc. The plant also has to be covered to protect it from the elements, especially wind, but temperature does not impair the performance and at temperatures below 4°C it is a far better means of clarification than sedimentation.

Filtration is necessary after the coagulation stage to remove the carryover of fine solids and to combat the occasional periods of floc breakthrough.

Physical removal of solids

Sand filtration

Sand is a cheap and very effective material employed in the treatment of drinking water for the removal of fine solids either in a raw water or a partially treated water (Table W42).

Its objectives are

- removal of suspended solids from the raw or treated water;
- to achieve some biological treatment (?);
- reduction of bacteria and viruses (?).

It is achieved by

- physical straining;
- adsorption;
- biological activity;

There are several types of sand filters:

- slow sand filters;
- rapid gravity sand filters;
- upward flow filters;
- dual media filters;
- pressure filters.

Slow sand filters were first introduced as a filter of relatively good-quality water and was the first type of media filter employed on a large scale. The filters are essentially a layer of sand varying from 0.5 to 1.0 m in depth with some form of underdrainage. Water percolates through the sand leaving particulate matter deposited both on the surface and within the body of the sand filter. These filters were

Table W42 Sand filter media and loadings guide

Type	Application	Media	Flow rate (m h^{-1})
Slow sand	Color (slight) turbidity, organic (some)	Fine sand (0.1–0.4 mm)	0.1–0.5
Rapid gravity	Turbidity	Medium sand (0.5–1.0 mm)	5–10
Rapid gravity, high rate	Turbidity, Fe and Mn	Coarse sand (1–3 mm)	10–20
Rapid GAC filters	Taste and odor	Granular activated carbon (0.5–1 mm)	5–10

described as slow filters because the sand would become partially blinded with solid material, resulting in a relatively slow throughput. Consequently they required large areas and a large labor force to maintain them. A significant advantage of the slow sand filter was its ability to allow a layer of alluvial mud, organic and bacterial matter to form on the surface of the sand. Whilst not initially appreciated, this layer, later known as the schmutzdecke, plays an important role in the effectiveness of the slow sand filter in reducing harmful organisms and compounds which may be the precursors to taste-forming compounds and toxic chemical production when chlorine is employed as a disinfectant.

Slow sand filters were inevitably replaced by rapid gravity sand filters which are much more compact, less labor intensive and lend themselves to automation. A rapid gravity filter is a vertical column of sand through which water is passed at the greatest possible rate to achieve an acceptable quality of filtrate. The filter has the facility to backwash the sand of the collected debris at a frequency required to achieve maximum overall throughput.

A filter would normally vary with respect to sand quality, sand size and sand depth and be designed to match the particular requirements of the water to be treated. Filters can employ several different grades of sand or anthracite to improve the efficiency of its operation and are described as dual media filters.

Upward flow filters were again a further development of the rapid gravity filter to permit increased flow rates.

Pressure filters are totally enclosed filters constructed of stainless steel and operated under pressure, normally employing the natural head of the raw water source to force the water downwards through the media.

Granulated activated carbon filters (GAC)

Carbon has been employed in the treatment of potable water for many years in the form of a very fine black powder, primarily for the treatment of taste and odor. Over recent years its replacement of the conventional sand filter has become a well-proven and accepted method of combating a wide range of undesirable organic compounds, e.g. herbicides, pesticides and naturally forming chemical compounds.

The soluble impurities in the water are removed by 'physical' straining and surface adsorption and are retained in the pores of the carbon by weak molecular van der Waals forces. The efficiency of the adsorption depends on the size of the molecular particles and the size of the pores in the carbon. Since the pore size varies with the raw carbon source, it is important to match the carbon with the likely impurities to be encountered. Activated carbon can purify potable water beyond levels normally attainable by conventional biological and physicochemical treatment.

Activated carbon filters are manufactured from peat, wood, coal or coconut fiber by heating them to a very high temperature, often in the presence of steam. This causes the volatile constituents to be driven from the carbon leaving a network of crosslinked micropores of molecular dimensions. These pores can give the activated carbon an enormous internal surface area, typically $1000 \text{ m}^2 \text{ g}^{-1}$. It is a product of high density, durability and granularity capable of withstanding the abrasion and damage associated with repeated reactivation, hydraulic transport, air scouring, backwashing and mechanical handling.

Disinfection

Disinfection is the most important process employed in the treatment of drinking water, but it must not be considered in isolation since a water has to be of a good quality, i.e. free from toxic substances, low levels of suspended and dissolved or emulsified organic material to achieve effective disinfection.

Quality standards require that the indicator organisms fecal streptococci and the coliform group of bacteria, specifically fecal *E. coli*, should be absent. The standard is based on the fact that coliform bacteria are present in abundance in animal excreta, e.g. humans have of the order of 5×10^6 *E. coli* per gram of excreta. Therefore, in the absence of coliform and streptococcal bacteria, the water is assumed to be free of other waterborne pathogenic organisms associated with humans and other animals. This hypothesis does produce considerable debate and will no doubt continue to do so, inevitably resulting in some additional bacteriological analysis to help improve quality assurance further.

Chlorine and chlorine compounds are the most widely used disinfectants employed in treatment of drinking water. Ozone and ultraviolet light are more powerful disinfectants but neither have the residual powers of chlorine which protects water from contamination during its transfer from treatment to the user. Small-scale water treatment filters utilize the sterilizing properties of silver at concentrations of 0.02 mg l^{-1} to disinfect drinking water, which is employed in survival kits.

Chlorination

Chlorine in its gaseous state or in solution is brought into contact with the drinking water, producing the very effective disinfectant hypochlorous acid, HOCl. The concentration, expressed as chlorine, is dependent on the quality of the water and is typically $0.5-1.0 \text{ mg l}^{-1}$, but some surface water may require as much as 5 mg l^{-1} to ensure complete pathogen kill. Groundwater normally requires little disinfection, only to safeguard its transfer from source to user. The required concentration is dependent on both the natural chlorine demand and the pH of the water. Chlorine must be allowed contact with the water for some 30 min to ensure complete destruction of harmful organisms. Excess chlorine will remain in solution and continue to combat any bacteriological growth or contamination during its transfer.

A disadvantage of chlorine is its ability to react readily with other substances and produce compounds that result in taste and odor problems and in some circumstances compounds considered to be toxic. Of major concern is the formation of trihalomethanes compounds, which are the products of the reaction of chlorine with naturally derived organic compounds associated with surface waters. Strict operational control of a treatment plant will minimize the formation of these compounds.

Ozone

Ozone is a very powerful disinfectant requiring a relatively low contact time of 5 min or less and low concentrations to achieve effective disinfection. However, it does requires relatively sophisticated contact arrangements as the gas is extremely toxic, and the dosing and contact tank must be sealed and carefully monitored.

Unlike chlorine, ozone can considerably enhance the organoleptic qualities of water by removing or converting organic matter which may have the potential to impart taste or color to the raw water.

Ozone is particularly effective for removing phenols, detergents, polycyclic hydrocarbons and certain pesticides, such as aldrin. Ozone also removes numerous sapid organic compounds but can, unless properly managed, produce only partial degradation resulting in the production of taste-forming intermediaries such as ketones and aldehydes.

Typical dosing rates vary from 0.1 to 0.8 mg O_3 per mg total organic carbon for lowland waters with a contact time of 5 min. The dosage rate prior to using GAC filters is lower than for raw water treatment because of the reduced degradable material present. It is reported that ozone addition prior to coagulation promotes flocculation.

The combining of ozone and activated carbon treatment produces an effective and reliable treatment process optimizing the advantages of both systems. By improving the biodegradability of the water and the breaking down of higher molecular weight compounds, the work load of the activated carbon is significantly decreased. This can reduce the period between regeneration by between 20 and 50%.

An important additional benefit of ozone is its ability to inactivate viruses and its reported effectiveness against *cryptosporidium* and *Giardia* oocysts.

Ultraviolet light

Ultraviolet light is an natural source of energy emitted by the Sun's rays. They are of relatively short wavelength, $<400 \text{ nm}$ as compared with $400-700 \text{ nm}$ for visible light. The ultraviolet spectrum has very powerful biocidal properties which are reproduced by passing an electrical current through mercury vapor. Ultraviolet light instantaneously penetrates the cell walls of microorganisms and disrupts the DNA, thereby destroying the cell.

UV lamps are thin quartz tubes (25 mm diameter) filled with a mixture of argon and mercury vapor. The vapor pressure in a 'low-pressure' lamp is of the order 10^{-3} mbar, and these are the ones most commonly employed in the water industry. 'Medium-pressure' lamps, with a vapor pressure of 10^3 mbar, are commercially available but less economic to operate.

Treated water is passed across a series of tubes designed to ensure complete irradiation of all the flow. Provided the water is of good clarity and low organic content, reliable disinfection will be achieved.

Peter Price

Cross reference

Water quality for drinking, WHO guidelines

WATER USE

Table W43 shows the recent growth in water consumption in Germany and is typical of developed countries, where most of the population is supplied with potable tap water. Table W44 shows a similar trend in England and Wales. These trends are typical of developed countries where demand for water increased rapidly during the 1960s, rose at a slower rate in the 1970s and have been fairly

Table W43 Water supply in Germany (data for Federal Republic of Germany only; source: Wasserstatistik)

Year	Household (Ml day^{-1})	Industry (Ml day^{-1})	Other (Ml day^{-1})	Total (Ml day^{-1})
1970	5175	2893	641	8 709
1975	6324	2323	720	9 367
1980	7148	2082	786	10 086
1985	7414	1836	762	10 010
1990	8057	1929	762	10 748
1992	8088	1748	789	10 625

Table W44 Water supply in England and Wales (source: Waterfacts, OFWAT)

Year	Unmeasured (Ml day^{-1})	Measured (Ml day^{-1})	Non-potable (Ml day^{-1})	Total (Ml day^{-1})
1964	7 415	4034	320	11 769
1969	8 480	4600	505	13 585
1974	9 555	4801	548	14 904
1979	10 961	4490	643	16 094
1984–85	11 956	3990	558	16 504
1989–90	12 424	4135	714	17 273
1994–95	12 347	4138	603	17 088

static since the early 1980s. This has been mainly due to a fall in industrial demand for water as industry has become more efficient in its water use and there has been a move away from processes involving heavy use of water.

Table W45 illustrates how this water delivered to customers is allocated in England and Wales. As a result of the privatization of water and increased regulation there has been a move to measure the components of water use more carefully in order to allocate charges fairly between the various classes of customers and assess the economics of metering and leakage. Each component has been rigorosly defined, especially to ensure that the leakage element does not include other 'unaccounted for' components such as fire fighting and mains flushing. Recently, better information has emerged on the division of leakage between the companies' mains (distribution losses) and the customers' supply pipes (total leakage). Leakage control is increasingly being regarded as a source of water to be judged in economic terms against new resources.

The usual initial division of how water is used is the separating out of household (plus an element of commercial and small business use

Table W45 Water delivered by category[a] (source: OFWAT)

Year	Unmeasured Non-household (Ml day^{-1})	Unmeasured Household (Ml day^{-1})	Measured Non-household (Ml day^{-1})	Measured Household (Ml day^{-1})	Miscellaneous Water taken unbilled (Ml day^{-1})	Miscellaneous Operational use (Ml day^{-1})	Distribution Distribution losses (Ml day^{-1})	Distribution Distribution input (Ml day^{-1})	Distribution Total leakage (Ml day^{-1})	Unmeasured Per capita household (l day^{-1} per head)
1992/93	599	7800	3898	231	74	54	3612	16 268	4639	141
1993/94	547	7767	3736	298	95	47	3665	16 155	4644	142
1994/95	513	7938	3776	362	134	47	3715	16 485	4816	147

Table W46 Water supplied to households and industry by country (source: IWSA)

Country	Household and small business 1980 (l day^{-1} per head)	Household and small business 1993 (l day^{-1} per head)	Industry and other 1980 (l day^{-1} per head)	Industry and other 1993 (l day^{-1} per head)	Total 1980 (l day^{-1} per head)	Total 1993 (l day^{-1} per head)
Australia	–	316	–	163	–	479
Austria	155	170	100	92	252	262
Belgium	104	120	59	37	163	157
Denmark	165	155	96	74	261	229
France	109	157	58	58	167	215
Germany	137	136	74	41	211	177
Hungary	110	121	107	63	217	184
Italy	211	251	69	78	280	329
Luxemburg	183	178	76	83	259	261
Netherlands	142	171	37	32	179	203
Norway	154	180	247	340	401	520
South Africa	–	276	–	267	–	543
Spain	157	210	58	90	215	300
Sweden	195	203	120	73	315	276
Switzerland	229	242	163	120	392	362
United Kingdom	154	231	100	100	254	331

Table W47 Age and water use in the Netherlands (l day⁻¹ per head; source: VEWIN, 1992)

Water use	Age (years)						
	<24 (l day⁻¹ per head)	25–34 (l day⁻¹ per head)	35–44 (l day⁻¹ per head)	35–54 (l day⁻¹ per head)	55–64 (l day⁻¹ per head)	>65 (l day⁻¹ per head)	Average (l day⁻¹ per head)
Bath	3.90	6.70	9.30	12.70	6.26	9.53	8.24
Shower	54.14	50.12	35.38	32.13	13.05	28.56	39.46
Wash basin	3.80	3.00	3.57	3.88	4.23	4.64	3.69
Toilet	32.97	40.51	40.51	45.88	49.13	49.43	42.70
Hand washing	2.63	1.83	0.80	1.71	3.94	5.31	2.57
Washing machine	22.22	27.26	23.38	23.97	23.59	16.91	23.23
Hand dishwashing	6.52	5.54	0.95	7.84	13.15	17.15	8.78
Dishwasher	0.17	0.63	0.95	0.76	0.93	0.51	0.70
Other	5.90	5.90	5.90	5.90	5.90	5.90	5.90
Total	132.25	141.5	125.83	134.26	140.18	137.94	135.27

Table W48 Family size and water use in the Netherlands (l day⁻¹ per head; source: VEWIN, 1992)

Water use	1 person l/head	2 persons l/head	3 persons l/head	4 persons l/head	5 persons + l/head	Average l/head
Bath	4.60	8.66	9.18	9.54	13.26	9.05
Shower	40.21	40.21	45.36	35.10	38.99	39.97
Wash basin	3.49	3.91	3.53	3.85	3.71	3.69
Toilet	43.45	43.84	43.28	40.68	34.82	41.22
Hand washing	5.03	2.23	1.60	1.14	0.46	2.09
Washing machine	20.63	24.25	26.71	23.14	22.86	23.52
Hand dishwashing	12.80	9.53	6.68	5.53	3.94	7.70
Dishwasher	0.43	0.79	0.79	0.79	0.96	0.75
Other	5.90	5.90	5.90	5.90	5.90	5.90
Total	136.62	139.32	143.03	125.68	124.90	133.91

Table W49 Household water use by appliance in the UK (source: UK, Anglian Water plc, 1992 and 1993)

Appliance	%
Bath	13
Shower	4
Kitchen sink	16
Toilet	33
Wash basin	9
Washing machine	21
Dishwasher	1
Outside taps	3

Table W50 Household water use in the UK by day of week and size of household (l day⁻¹ per head; source: UK, Anglian Water plc, data for 1992 and 1993)

By day of week							
Sunday	Monday	Tuesday	Wednesday	Thursday	Friday	Saturday	Average
151	145	138	137	138	142	152	145

By household size								
1	2	3	4	5	6	7	8	Average
220	160	137	125	110	112	117	60	145

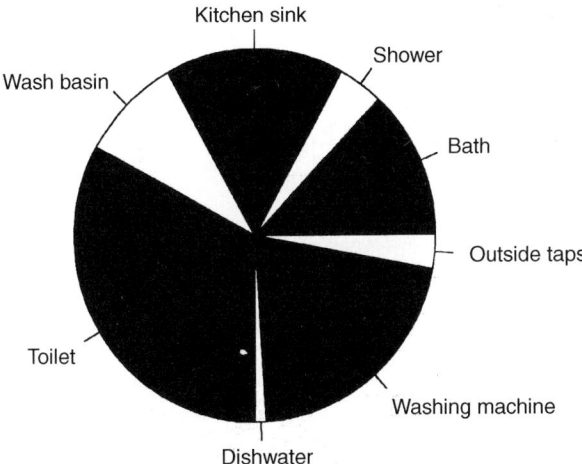

Figure W36 Household water use in UK.

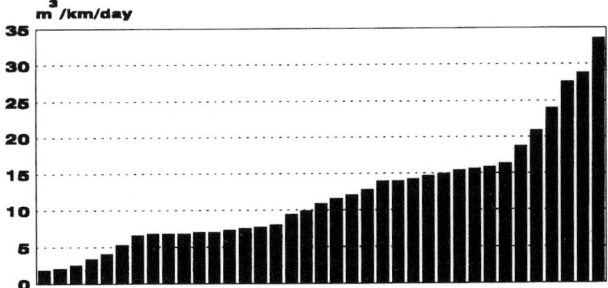

Figure W38 Leakage levels in the 37 largest cities in Germany.

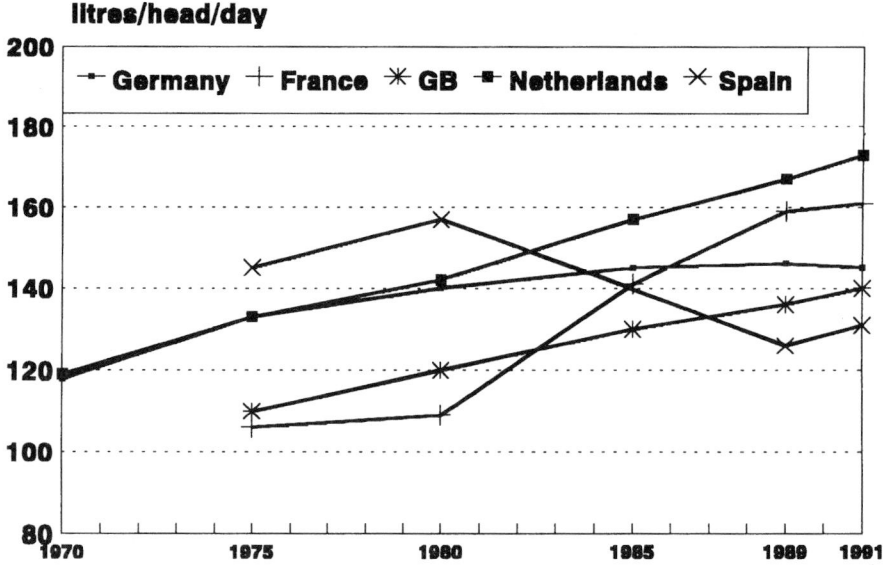

Figure W37 Per capita household consumption in Europe. (Source: Wasserstatistik 1982, BGW.)

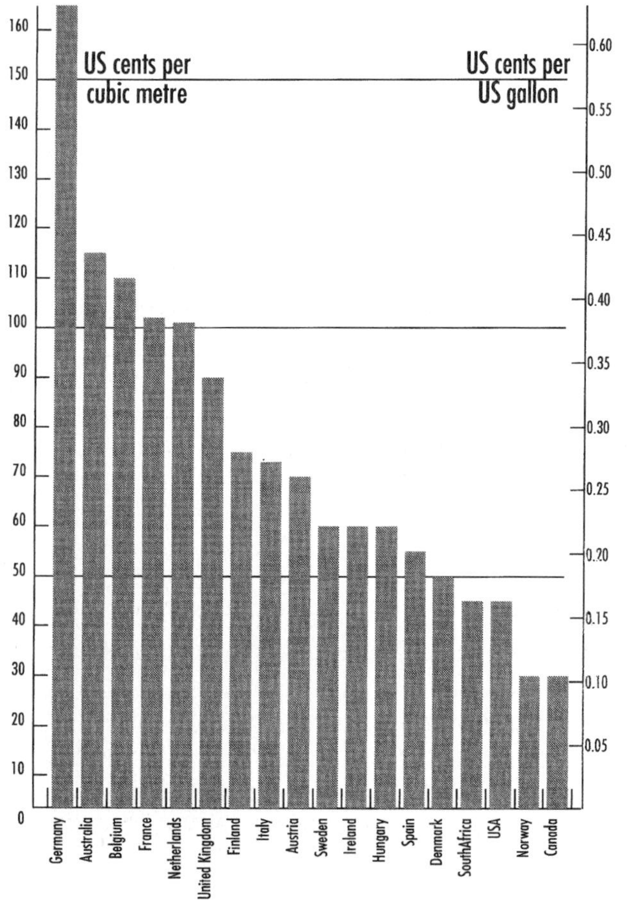

Figure W39 International water prices in US cents per cubic meter.

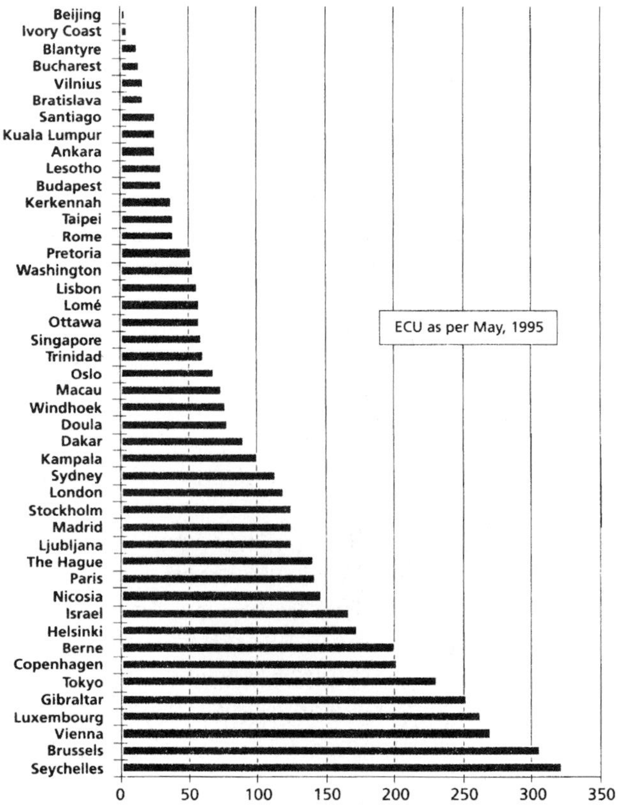

Figure W40 Comparison of water prices (annual) for capital and major cities, in ECU (1 ECU = $1.2). (Source: IWSA.)

Table W51 Leakage levels in England and Wales

Company	Total losses as percentage of distribution input (%)	Distribution losses l h^{-1} per property	Total underground supply pipe losses (l h^{-1} per property)	Distribution losses per km of main (m^3 km^{-1} day^{-1})	Distribution losses per km of main and communication pipe (m^3 km^{-1} day^{-1})
Hartlepools	15.6	5.3	2.0	8.1	6.6
Wrexham	16.9	3.2	2.5	3.0	2.6
East Surrey	17.1	3.6	2.1	4.5	3.8
Bournemouth and W Hants	17.1	5.3	2.4	7.1	5.9
Sutton	17.3	3.1	1.5	7.2	5.3
Bristol	18.2	4.7	1.9	6.4	5.3
Portsmouth	18.1	3.3	2.5	5.8	4.6
Tendring Hundred	18.9	2.9	1.7	4.1	3.4
Anglian	20.3	4.5	2.3	4.6	4.0
Essex and Suffolk	20.4	4.7	2.2	8.0	6.4
Cambridge	21.0	5.0	1.9	5.3	4.6
Southern	21.1	4.6	2.4	6.7	5.5
Chester	21.8	4.4	2.3	7.5	6.0
North Surrey	22.0	5.6	1.8	9.0	7.3
York	22.4	4.5	2.3	7.2	5.8
Three Valleys	22.7	5.7	2.4	10.1	8.0
South East	22.8	4.6	2.7	5.5	4.6
Folkestone and Dover	23.9	6.1	2.7	8.4	6.9
North East	24.0	5.1	2.4	7.2	5.9
Northumbrian	24.5	6.9	2.8	8.9	7.4
Mid Kent	24.7	6.7	2.1	7.4	6.3
South Staffs	26.6	6.7	2.4	12.5	9.7
Mid Southern	28.9	9.0	2.7	11.1	9.4
Thames	29.1	9.4	2.2	20.2	15.3
South West	30.5	8.4	2.3	7.4	6.5
Severn Trent	31.0	8.4	2.3	12.7	10.4
Wessex	33.2	11.6	2.5	10.6	9.3
Yorkshire	36.7	11.2	2.6	15.8	13.0
North West	36.9	11.9	2.6	17.8	14.5
Dwr Cymru (Welsh)	37.8	11.9	3.8	11.7	10.2
Average	29.0	8.3	2.4	11.7	9.6

[a] Assumes an average of 3 m of communication pipe per property.
[b] Source: OFWAT, Water Delivered Report 1994/95.

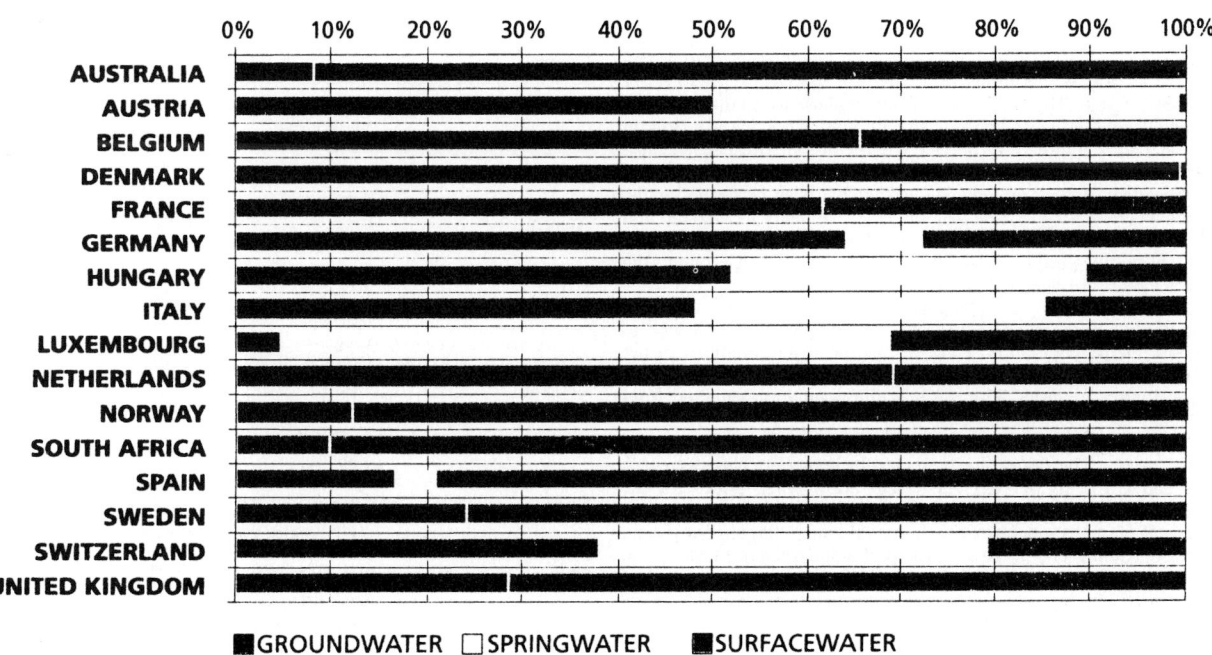

Figure W41 Percentage of sources from groundwater spring water and surface water in 16 selected countries. (Source: IWSA.)

Table W52 Water use in the Far East (source: Asian Development Banks; most data for 1992)

Country	City supply	Population (10^6)	Per capita (l day^{-1} per head)	Annual water use (10^3 Ml)	Domestic (%)	Industrial and commercial (%)	Other (%)	Unaccounted for water (%)
Bangladesh	Dhaka	6.0	44	205	32	2	4	62
China	Beijing	5.7	149	636	47	20	5	28
China	Shanghai	7.5	193	1540	34	40	1	25
China	Tianjin	4.6	125	547	36	50	2	12
Fiji	Suva	0.2	203	0.3	50	8	6	36
Hong Kong	Hong Kong	5.7	111	876	27	35	12	26
India	Bombay	9.8	–	894	42	10	24	24
India	Calcutta	4.4	213	376	48	16	–	36
India	Delhi	9.4	257	830	54	7	9	30
India	Madras	3.8	–	120	–	–	–	–
Indonesia	Bandung	1.8	96	37	42	8	8	42
Indonesia	Jakarta	8.4	148	321	27	8	8	57
Indonesia	Medan	2.0	153	70	48	10	8	34
Laos	Vientiane	0.4	140	21	33	10	24	33
Malaysia	Kuala Lumpur	1.1	222	131	32	21	10	37
Malaysia	Penang	1.2	203	182	50	28	–	22
Myanmar	Mandalay	0.6	153	19	–	67	–	33
Myanmar	Yangon	3.3	120	141	–	87	–	13
Nepal	Kathmandu	0.4	97	23	–	55	–	45
Pakistan	Karachi	9.1	172	584	58	12	–	30
Pakistan	Lahore	4.5	–	–	–	–	–	–
Philippines	Cebu	1.1	139	29	51	8	3	38
Philippines	Manila	7.9	133	909	27	11	4	58
Singapore	Singapore	2.8	168	434	43	27	22	8
Sri Lanka	Colombo	1.4	168	131	28	9	12	51
South Korea	Seoul	10.9	180	1799	40	15	3	42
South Korea	Ulsan	0.7	166	74	54	15	1	30
Taiwan	Taipei	3.7	281	675	57	19	–	24
Thailand	Bangkok	5.6	217	1047	34	25	10	31
Thailand	Chiang Mai	0.2	172	17	34	5	22	39
Vietnam	Hanoi	0.9	157	117	30	17	–	53
Vietnam	Ho Chi Minh	3.9	131	252	49	9	1	41

– but usually domestic-related use) and industry. Table W46 shows this for a selection of worldwide countries in per capita terms. There will inevitably be an element of leakage in some of the figures, often depending on where the water meters readings are taken.

There are few published studies on how water is used in the home. Tables W47 and W48 shows the results of a Dutch investigation and Tables W49 and W50 summarize a study in the UK Anglian Water Company region. The differences from country to country can be substantial, relating particularly to the amount of outside use (low in the Netherlands and the UK) and the volume of toilet flushes (see also Figures W36 and W37).

A great deal of work on reducing leakage has been performed in the UK and Table W51 gives the data by water company in England and Wales. Few countries have collected their leakage data centrally. Figure W38 gives information from 37 major cities in Germany; the average values are very similar to the UK.

Table W52 shows recent water-use figures from the Pacific region. A substantial number of these countries supply potable water to 99–100% of the population. However, as the wide variations in per capita consumption figures and 'unaccounted for' water show, many are developing countries with much of the population supplied from communal sources or local wells.

Figure W39 gives a selection of prices for water from 18 countries at 1994 prices. The prices are presented in both US cents per cubic meter and US cents per US gallon and are average prices for the countries concerned.

Table W53 gives a comparison of annual water charges based on a family consuming 200 m^3 year^{-1}. The prices are given in ECUs where 1 ECU = $1.2 approximately. Figure W40 gives a comparison of prices (in ECUs) for capital or major cities, while Figure W41 gives the percentage of groundwater, springwater and surface water sources for 16 selected countries.

Philip Turton

Table W53 Comparison of annual water charges for a family living in a house consuming 200 m^3 year^{-1} (IWSA)

Country	Town or city	GDP per capita (ECU)	Annual charge (ECU)
Australia	Sydney	13 028	118.4
	Melbourne		126.6
	Perth		78.1
	Brisbane		110.6
	Adelaide		111.7
Austria	Linz	20 817	172.6
	Salzburg		212.3
	Vienna		271.4
Belgium	Antwerp	18 658	140.6
	Brussels		306.7
	Liege		229.0
Cameroon	Doula	365	78.6
Canada	Ottawa	13 756	58.9
	Winnipeg		86.7
	Edmonton		98.9
	Vancouver		55.6
	Toronto		57.2
Chile	Santiago	2 496	26.5
	Valparaiso		40.1
	Concepcion		31.8
China	Beijing	238	5.4
	Shanghai		7.2
	Guangzhou		9.0
Cyprus	Nicosia	7 163	147.3
	Limassol		58.2
	Larnaca		106.2

Table W53 Continued

Country	Town or city	GDP per capita (ECU)	Annual charge (ECU)
Denmark	Copenhagen	23 581	203.0
	Aarhus		119.8
	Odense		143.5
Finland	Helsinki	16 485	174.1
	Tampere		143.2
	Turku		254.0
France	Ban.d.Paris	19 052	265.5
	Lyon		268.1
	Marseille		256.4
	Nice		254.2
	Paris		143.6
Gibraltar		12 303	253.1
Hungary	Budapest	2 056	31.4
	Miskolc		79.0
	Pecs		60.6
Israel		7 812	167.1
Italy	Bologna	11 878	82.2
	Milan		21.4
	Naples		90.8
	Rome		39.7
	Turin		37.1
Ivory Coast		353	7.1
Japan	Nagoya	33 931	144.9
	Osaka		135.1
	Sapporo		247.7
	Tokyo		231.1
	Yokohama		148.2
Lesotho		230	31.2
Lithuania	Vilnius	563	17.8
	Kaunas		9.5
	Klaipeda		9.8
Luxembourg	Luxembourg	25 914	264.0
Macau		6 125	75.3
Malaysia	Kuala Lumpur	2 415	27.3
	Ipoh		28.2
	Georgetown		13.5
Malawi	Blantyre	49	14.1
Namibia	Keetmanshoop	1 120	54.4
	Swakopmund		69.8
	Windhoek		77.4
Netherlands	Amsterdam	18 109	153.3
	The Hague		241.6
	Utrecht		123.0
Netherlands Antilles		7 603	3.0
Norway	Oslo	20 685 75.1[a]	73.5[b]
	Bergen		197.8[b]
	Trondheim	96.0[b]	120.5[b]
Portugal	Lisbon	5 954	57.5
	Porto		97.5
	Coimbra		109.4
Romania	Bucharest	333	15.0
Senegal	Dakar	319	90.6
	Saint-Louis		90.6
Seychelles		5 244	323.5
Singapore		16 829	60.8
Slovakia	Bratislava	1 937	18.0
	Kosice		13.7
	Trnava		14.4
Slovenia	Ljubljana	5 261	125.5
	Maribor		62.4
	Celje		59.6
South Africa	Bloemfontein	2 337	51.5
	Cape Town		44.9
	Durban		53.2
	Johannesburg		46.5
	Pretoria		52.8

Table W53 Continued

Country	Town or city	GDP per capita (ECU)	Annual charge (ECU)
Spain	Barcelona	9.555	119.8
	Madrid		125.5
	Sevilla		52.2
Sweden	Stockholm	17 238	125.1
	Goteborg		80.6
	Malmö		133.3
Switzerland	Berne	33 022	201.3
	Geneva		327.1
	Zurich		348.3
Taiwan	Taipei	8 404	39.4
	Taichung		33.3
	Kaoshung		33.3
Togo	Lomé	212	58.6
	Dapaong		58.6
	Sokode		58.6
Trinidad and Tobago		2 579	62.0
Tunisia	Kerkennah	1 449	37.9
Turkey	Anakara	565	27.4
	Canakkale		30.8
	Eskisehir		29.7
United Kingdom	Bristol	13 178 153[a]	110.7[b]
	Newcastle	275[b]	108.3[b]
	Manchester	184[a]	121.7[b]
	London	140[a]	119.3[b]
	Cardiff	221[a]	156.9[b]
Uganda	Kampala	149	100.9
United States	Los Angeles	18 723	96.0
	Washington		53.7
	New York		53.7
	Miami		48.4
	Des Moines		81.7

The annual charges are quoted using the charging base on which most customers are charged. This is on a measured basis for all countries except Norway and the UK. In these countries in places where some customers are charged on a measured basis, the annual metered charge is included as well (1 ECU = $1.2).
[a] Measured.
[b] Unmeasured.

Sources

Achtienribbe, G., Horner, V., Papp, E. and Weiderkehr, 1992. International comparison of drinking water prices. *J. Water SRT Aqua*, **41**(6).
International Water Price Survey, 1994. National Utility Services, UK.
International Water Supply Association UK (IWSA), 1995. Statistics and Economics Committee, Published for the IWSA Congress, Durban, South Africa.

WATER USE IN THE USA

Introduction

Water use as reported by the US Geological Survey is subdivided into offstream use, instream use and wastewater release. The difference among these types of use is explained below.

Offstream use is a water use that depends on water being diverted or withdrawn from a surface- or groundwater source and conveyed to the place of use. To determine the total quantity of water used (self-supplied withdrawals and public supply deliveries), five subtypes of use are evaluated, as explained below and shown in Figure W42.

- Withdrawal, the quantity of water diverted or withdrawn from surface- or groundwater (A in Figure W42).
- Delivery/release, the quantity of water delivered at the point of use (B) and the quantity released after use (C).

Table W54 Total offstream water use in the USA by state in 1990 (1 gal US = 0.0038 m³; figures may not add to totals because of independent rounding)

State	Population (10³)	Per capita use, freshwater (gal day⁻¹)	Withdrawals including irrigation conveyance losses (Mgal day⁻¹) By source and type									Reclaimed waste water (Mgal day⁻¹)	Conveyance losses (Mgal day⁻¹)	Consumptive use, fresh water (Mgal day⁻¹)
			Ground water			Surface water			Total					
			Fresh	Saline	Total	Fresh	Saline	Total	Fresh	Saline	Total			
Alabama	4 041	2 000	394	9.1	403	7 680	0.0	7 680	8 080	9.1	8 090	0.0	0.0	454
Alaska	550	517	64	48	112	221	308	529	284	357	641	0	0.1	26
Arizona	3 665	1 790	2 740	5	2 740	3 830	0.6	3 830	6 570	1.2	6 570	183	1 010	4 350
Arkansas	2 353	3 330	4 710	0	4 710	3 130	0	3 130	7 840	0	7 840	0	368	4 140
California	29 760	1 180	14 600	310	14 900	20 500	11 400	31 900	35 100	11 700	46 800	133	1 560	20 900
Colorado	3 294	3 850	2 770	30	2 800	9 910	0	9 910	12 700	30	12 700	3.7	2 990	5 250
Connecticut	3 287	325	165	0	165	902	3 780	4 680	1 070	3 780	4 840	0	0	103
Delaware	666	1 540	89	0	89	939	339	1 280	1 030	339	1 370	0	0	59
DC	607	15	1.0	0	1.0	8.0	0	8.0	9.0	0	9.0	0	0	16
Florida	12 938	582	4 660	0	4 660	2 870	10 400	13 200	7 530	10 400	17 900	170	64	3 130
Georgia	6 478	816	996	0	996	4 290	65	4 360	5 290	65	5 350	36	0	822
Hawaii	1 108	1 070	589	0.6	590	600	1 550	2 150	1 190	1 550	2 740	6.2	127	627
Idaho	1 007	19 600	7 590	0	7 590	12 100	0	12 100	19 700	0	19 700	0	7 160	6 090
Illinois	11 431	1 570	920	25	945	17 100	0	17 100	18 000	25	18 000	0	0	750
Indiana	5 544	1 700	621	0	621	8 810	0	8 810	9 430	0	9 430	0	0	451
Iowa	2 777	1 030	495	0	495	2 370	0	2 370	2 860	0	2 860	0	0	271
Kansas	2 478	2 460	4 360	0	4 360	1 720	0	1 720	6 080	0	6 080	6.0	146	4 410
Kentucky	3 685	1 170	247	0	247	4 070	0	4 070	4 320	0	4 320	0	0.5	309
Louisiana	4 220	2 200	1 340	0.6	1 340	7 950	67	8 010	9 290	67	9 350	0	90	1 590
Maine	1 228	433	85	0	85	446	609	1 060	532	609	1 140	0	0	51
Maryland	4 781	306	240	0	240	1 220	4 950	6 170	1 460	4 950	6 410	63	0	126
Massachusetts	6 016	338	338	0	338	1 690	3 490	5 180	2 030	3 490	5 520	0	0	195
Michigan	9 295	1 250	703	4.6	707	10 900	0	10 900	11 600	4.6	11 600	0	0	738

Minnesota	4 375	748	797	0	797	2 480	0	2 480	3 270	0	3 270	0	0	872
Mississippi	2 573	1 290	2 670	0	2 670	648	316	963	3 320	316	3 640	1.0	188	1 800
Missouri	5 117	1 150	727	0.1	728	5 150	1 060	6 200	5 870	1 060	6 930	0	0	529
Montana	799	11 600	205	13	218	9 100	0	9 100	9 300	13	9 320	0	4 620	2 090
Nebraska	1 578	5 660	4 790	4.7	4 800	4 150	0	4 150	8 940	4.7	8 940	0	2 160	4 230
Nevada	1 202	2 780	1 060	12	1 070	2 280	0	2 280	3 340	12	3 350	11	615	1 690
New Hampshire	1 109	378	64	0	64	356	894	1 250	420	894	1 310	0	0	26
New Jersey	7 730	287	566	0.2	566	1 650	10 600	12 200	2 220	10 600	12 800	0	0	211
New Mexico	1 515	2 300	1 760	0	1 760	1 720	0	1 720	3 480	0	3 480	0	590	2 060
New York	17 990	583	839	1.5	840	9 650	8 490	18 100	10 500	8 490	19 000	0	0	562
North Carolina	6 629	1 350	435	0	435	8 500	5.5	8 510	8 940	5.5	8 940	17	0	390
North Dakota	639	4 190	141	0	141	2 540	0	2 540	2 680	0	2 680	0	5.9	228
Ohio	10 847	1 080	904	0	904	10 800	0	10 800	11 700	0	11 700	0	0.1	901
Oklahoma	3 146	452	662	243	905	760	0	760	1 420	243	1 670	12	5.4	659
Oregon	2 842	2 970	767	0	767	7 660	0	7 660	8 430	0	8 430	0	1 270	3 160
Pennsylvania	11 882	827	1 020	0	1 020	8 810	0	8 810	9 830	0	9 830	0	0	581
Rhode Island	1 003	132	25	0	25	108	393	501	133	393	526	14	0	18
South Carolina	3 487	1 720	282	0	282	5 720	0	5 720	6 000	0	6 000	0	0	293
South Dakota	696	851	251	0	251	341	0	341	592	0	592	0.7	62	345
Tennessee	4 877	1 880	503	0	503	8 690	0	8 690	9 190	0	9 190	56	0	252
Texas	16 986	1 180	7 380	492	7 880	12 700	4 610	17 300	20 100	5 100	25 200	39	660	9 020
Utah	1 723	2 540	964	7.2	971	3 410	93	3 510	4 380	100	4 480	0	624	2 230
Vermont	563	1 120	45	0	45	587	0	587	632	0	632	0	0	29
Virginia	6 187	762	443	0	443	4 270	2 150	6 420	4 710	2 150	6 860	0	3.6	224
Washington	4 867	1 630	1 450	0	1 450	6 460	36	3 490	7 910	36	7 940	0	997	2 830
West Virginia	1 793	2 560	728	0	728	3 860	0	3 860	4 580	0	4 580	0	0	509
Wisconsin	4 892	1 330	681	0	681	5 830	0	5 830	6 510	0	6 510	0	0	461
Wyoming	454	16 700	384	19	403	7 200	0	7 200	7 580	19	7 600	0	2 150	2 730
Puerto Rico	3 522	163	157	0	157	419	2 470	2 880	576	2 470	3 040	0	17	199
Virgin Islands	102	91	1.9	1.2	3.1	7.4	153	160	9.3	154	164	0	0	1.5
Total	252 336	1 340	79 400	1 220	80 600	259 000	68 200	327 000	339 000	69 400	408 000	750	27 500	94 000

Table W55 Total water withdrawals in the USA by water-use category and state, 1990 (Mgal day^{-1}; 1 gal US = 0.0038 m³; Figures may not add to totals because of independent rounding)

State	Public supply Fresh	Domestic Fresh	Commercial Fresh	Irrigation Fresh	Livestock Fresh	Industrial Fresh	Industrial Saline	Mining Fresh	Mining Saline	Thermoelectric Fresh	Thermoelectric Saline	Total Fresh	Total Saline
Alabama	707	28	3.5	94	141	784	0	11	9.1	6 310	0.0	8 080	9.1
Alaska	92	6.9	18	0.6	0.6	111	0	25	357	31	0	284	357
Arizona	707	32	18	5 300	89	163	0	156	0.7	103	0.4	6 570	1.2
Arkansas	309	51	222	5 250	189	177	25	2.5	310	1 640	0	7 840	0
California	5 830	318	234	27 900	411	129	25	20	30	246	11 400	35 100	11 700
Colorado	650	19	8.6	11 600	162	118	0	54	30	114	0	12 700	30
Connecticut	374	46	18	15	1.5	80	68	2.2	0	530	3 710	1 070	3 780
Delaware	85	11	1.8	32	2.4	65	6.0	0	0	831	333	1 030	339
DC	0	0	0	0	0	0.5	0	0.5	0	8.0	0	9.0	0
Florida	1 930	299	52	3 730	78	403	56	315	0	732	10 300	7 530	10 400
Georgia	963	100	42	441	46	657	33	12	0	3 030	33	5 290	65
Hawaii	238	9.9	40	755	7.2	43	0.6	1.4	0	95	1 550	1 190	1 550
Idaho	201	48	16	18 700	560	196	0	8.4	0	6.1	0	19 700	0
Illinois	1 860	115	173	78	63	464	0	69	25	15 200	0	18 000	25
Indiana	604	118	63	51	46	2 480	0	97	0	5 960	0	9 430	0
Iowa	322	45	27	23	121	219	0	34	0	2 070	0	2 860	0
Kansas	373	25	6.2	4 190	114	53	0	26	0	1 300	0	6 080	0
Kentucky	427	56	13	12	33	313	0	18	0	3 440	0	4 320	0
Louisiana	619	50	13	708	551	2 360	67	37	0	4 950	0	9 290	67
Maine	106	49	34	1.8	1.7	254	0	3.7	0	82	609	532	609
Maryland	798	70	26	29	20	70	379	28	21	421	4 550	1 460	4 950
Massachusetts	714	37	74	100	1.7	87	0	5.0	0	1 010	3 490	2 030	3 490
Michigan	1 400	123	35	240	29	1 680	3.7	55	0.9	8 060	0	11 600	4.6
Minnesota	515	168	71	195	67	154	0	220	0	1 860	0	3 270	0

State													
Mississippi	320	33	16	1 880	411	269	0	3.4	0	386	316	3 320	316
Missouri	677	62	22	371	55	85	0	25	0.1	4 580	1 060	5 870	1 060
Montana	135	16	0	9 000	52	57	0	6.2	13	33	0	9 300	13
Nebraska	301	47	0.2	6 100	139	41	0	131	4.7	2 180	0	8 940	4.7
Nevada	385	9.8	23	2 820	5.6	10	0	49	12	34		3 340	12
New Hampshire	95	27	0.6	0.9	1.0	37		2.8	0	255	894	420	894
New Jersey	1 040	68	17	58	2.1	326	1 020	110	0	599	9 550	2 220	10 600
New Mexico	273	24	17	3 010	22	6.3	0	80	0	50		3 480	0
New York	2 910	120	61	54	26	274	0	45	16	6 990	8 740	10 500	8 490
North Carolina	805	103	17	114	201	390	5.5	96	0	7 210	0	8 940	5.5
North Dakota	76	12	0.1	164	24	8.8	0	3.7		2 390	0	2 680	0
Ohio	1 300	134	36	15	34	354	0	243	243	9 550	0	11 700	243
Oklahoma	515	41	6.3	601	131	35	0	2.8	0	89	0	1 420	0
Oregon	470	64	711	6 860	21	284	0	1.5	0	15	0	8 430	0
Pennsylvania	1 730	141	24	14	53	1 870	0	252	0	5 750	393	9 830	393
Rhode Island	102	4.9	5.6	2.1	0.3	12	0	6.8	0	0		133	0
South Carolina	352	103	2.1	55	25	632	0	12	0	4 820	0	6 000	0
South Dakota	76	8.8	17	392	43	15	0	38	0	3.2	0	592	0
Tennessee	695	59	55	38	49	882	0	90	0	7 320	0	9 190	0
Texas	3 090	93	62	8 490	228	884	1 460	139	491	7 130	3 150	20 100	5 100
Utah	508	6.1	4.2	3 590	34	106	2.3	41	98	87	100	4 380	100
Vermont	39	17	3.8	0.5	6.1	44	66	3.7	0	519	0	632	0
Virginia	709	113	35	36	29	495	36	91	0	3 210	2 080	4 710	2 150
Washington	875	104	27	6 030	30	501	0	3.0	0	334	0	7 910	36
West Virginia	160	49	3.1	0	4.8	132	0	527	0	3 710	0	4 580	0
Wisconsin	595	90	11	151	99	468	0	0.2	0	5 100	0	6 510	0
Wyoming	88	8.5	1.6	7 160	27	16	0	101	19	184	0	7 580	19
Puerto Rico	404	6.7	0.1	140	8.3	11	0	2.6	0	2.6	2 470	576	2 470
Virgin Islands	6.4	1.6	0.7	0	0.5	0.1	51	0	0	0	103	9.3	154
Total	38 500	3 390	2 390	137 000	4 500	19 300	3 270	3 310	1 650	131 000	64 500	339 000	69 400

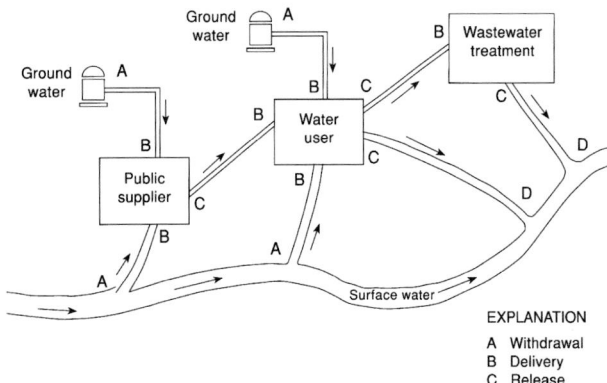

EXPLANATION

A Withdrawal
B Delivery
C Release
D Return flow

Figure W42 Relationships between subtypes of water use.

- Conveyance loss, the quantity of water that is lost in transit, for example from point of withdrawal to point of delivery (A–B), or from point of release to point of return (C–D).
- Consumptive use, that part of water withdrawn that is evaporated, transpired or incorporated into products or crops. In some instances, consumptive use will be the difference between the volume of water delivered and the volume released (B–C).
- Return flow, the quantity of water that is discharged to a surface- or groundwater source (D) after release from the point of use and thus becomes available for further use.

Self-supplied withdrawals, deliveries from public suppliers (where applicable) and consumptive use estimates are given for seven categories of offstream use: domestic, commercial, irrigation, livestock, industrial, mining and thermoelectric power. For the public supply category, in addition to withdrawals, Tables W54 and W55 also gives water delivered to domestic, commercial, industrial and thermoelectric power users.

Each category of use typically has different effects on the reuse potential of return flows. Reuse potential reflects the quality and the quantity of water available for subsequent use; for example, irrigation return flow may be contaminated by pesticides and fertilizers and, because of the high consumptive use of water during irrigation, the mineral content of the return flow is often substantially greater than that of the water applied. Consequently, irrigation return flow frequently has little reuse potential. This is a significant contrast to the reuse potential of most water discharged from thermoelectric plants, where the principal change in the water is an increase in temperature.

Instream use is a water use that takes place without the water being diverted or withdrawn from surface- or groundwater sources. Examples of instream uses are hydroelectric power generation, navigation, freshwater dilution of saline estuaries, maintenance of minimum streamflow to support fish habitat and the assimilation of wastewater.

Quantitative estimates for most instream uses are difficult to compile on a national scale. However, because such uses compete with offstream uses and affect the quality and quantity of water resources for all uses, effective water resources management requires methods and procedures to be devised to enable instream uses to be assessed quantitatively.

The only instream use estimates compiled are for hydroelectric power generation. Unlike other instream uses, the water used for hydroelectric power generation is a measurable quantity because the amount of water passed through the plant can be documented. Consumptive use in actual hydroelectric power generation (as opposed to evaporation from impoundments created by hydroelectric dams) is generally negligible.

Wastewater release refers to water released from private and public wastewater treatment facilities. Information is provided on the number of publicly and privately owned wastewater treatment facilities and on releases from only the public wastewater treatment facilities. The releases can be either returned to the natural environment or reclaimed for beneficial uses, such as irrigation of golf courses and parks.

Offstream use

The total fresh and saline withdrawals during 1990 were an estimated 408 000 million US gallons per day (Mgal day^{-1}; 1 gal = 3.78 l) for all offstream water-use categories (public supply, domestic, commercial, irrigation, livestock, industrial, mining and thermoelectric power), or 2% more than the withdrawals estimated for 1985. Average per-capita use was 1620 gal day^{-1} of fresh water and saline water and 1340 gal day^{-1} of fresh water. Total surfacewater withdrawals were an estimated 327 000 Mgal day^{-1} during 1990, or 1% more than during 1985. About 68 200 Mgal day^{-1} of surface water withdrawn (21%) was saline water. Total groundwater withdrawals were an estimated 80 600 Mgal day^{-1} or 9% more than during 1985. About 99% of groundwater withdrawn was fresh water. The use of reclaimed wastewater averaged about 750 Mgal day^{-1} or 30% more than during 1985.

The coastal regions (New England, Mid Atlantic, South Atlantic–Gulf, Pacific Northwest and California) accounted for nearly one-half of the total water withdrawn in the United States. About 54% of the nation's total withdrawals were in the East (water resources regions east of and including the Mississippi regions). These regions account for about one-third of the nation's land area.

A comparison of total withdrawals by State (Table W54) indicates that California accounted for the largest withdrawals, 46 800 Mgal day^{-1} more than the total withdrawn in Texas and Idaho, the next largest users. Some 20 States and the District of Columbia had less water withdrawn for offstream uses during 1990 than during 1985.

Irrigation is the largest category of freshwater use and thermoelectric power is the largest category of freshwater and saline water use. The California and Missouri Basin water resources regions accounted for 21% of total freshwater withdrawals during 1990. In these water resources regions, 73% of the withdrawals were for irrigation. The State of California accounted for the most freshwater withdrawn for public supply, domestic and irrigation, and the most saline water withdrawn for thermo-electric power (Table W55). Largest surfacewater withdrawals occurred in the Mid-Atlantic region, which is fifteenth out of 21 regions in land area. Of the 45 000 Mgal day^{-1} withdrawn in the Mid Atlantic region, 56% was saline water used for thermoelectric power plants. The State of California led the nation in both freshwater and saline surfacewater withdrawals. Five water resources regions, the Lower Mississippi, Missouri Basin, Arkansas–White–Red, Pacific Northwest and California, accounted for 75% of the nation's irrigation groundwater withdrawals. The State of California accounted for 18% of total groundwater withdrawals. Irrigation was the predominant use of groundwater in 22 states, most located in the West.

Freshwater consumptive use in the East was about 12% of the fresh water withdrawn in the East and accounted for 21% of the nation's freshwater consumptive use. By comparison, freshwater consumptive use in the West was about 44% of freshwater withdrawals. The higher consumptive use in the West is attributed to the fact that 90% of the total water withdrawn for irrigation occurred in the West and irrigation accounts for the largest part of consumptive use. California accounted for the largest consumptive use.

The distribution of per–capita freshwater withdrawals by State is shown in Figure W43 and Table W55. High per-capita values are characteristic of thinly populated states having large areas of irrigated land, such as Idaho, Montana and Wyoming. In contrast, Figure W44 shows the intensity of freshwater withdrawals by state in million gallons per day per square mile. The smaller states in the northeast show the most intense withdrawals by area.

For an overview of how the 339 000 Mgal day^{-1} of fresh water withdrawn during 1990 was used, the eight offstream categories mentioned above have been combined into five major categories: public supply, domestic and commercial, irrigation and livestock, industrial and mining and thermoelectric power. The source (withdrawals), use (withdrawals and deliveries), and disposition of fresh water for each category of use are summarized in Figure W45. The source column shows the proportion of withdrawals by source and the distribution of withdrawals by water-use category. Source data indicate, for example, that surface water was the source of 259 000 Mgal day^{-1} of freshwater, or 76.5% of total freshwater withdrawals in the United States. Of the 259 000 Mgal day^{-1} of surface water, 50.2% was withdrawn directly for thermoelectric power. Public supply is considered a source of water and Figure W44 shows the total quantity of water withdrawn by public supply, the percentage of surface- and groundwater withdrawn, and the percentage of water delivered to the

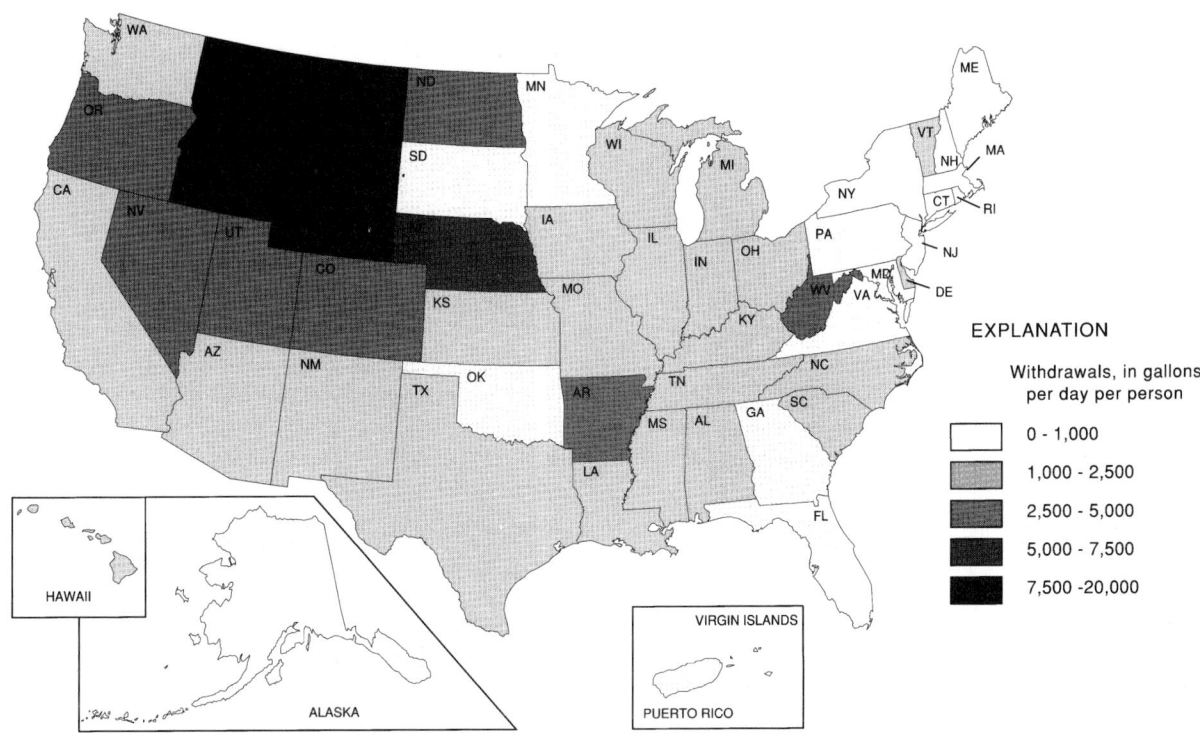

Figure W43 Intensity of freshwater withdrawals per capita by state, 1990.

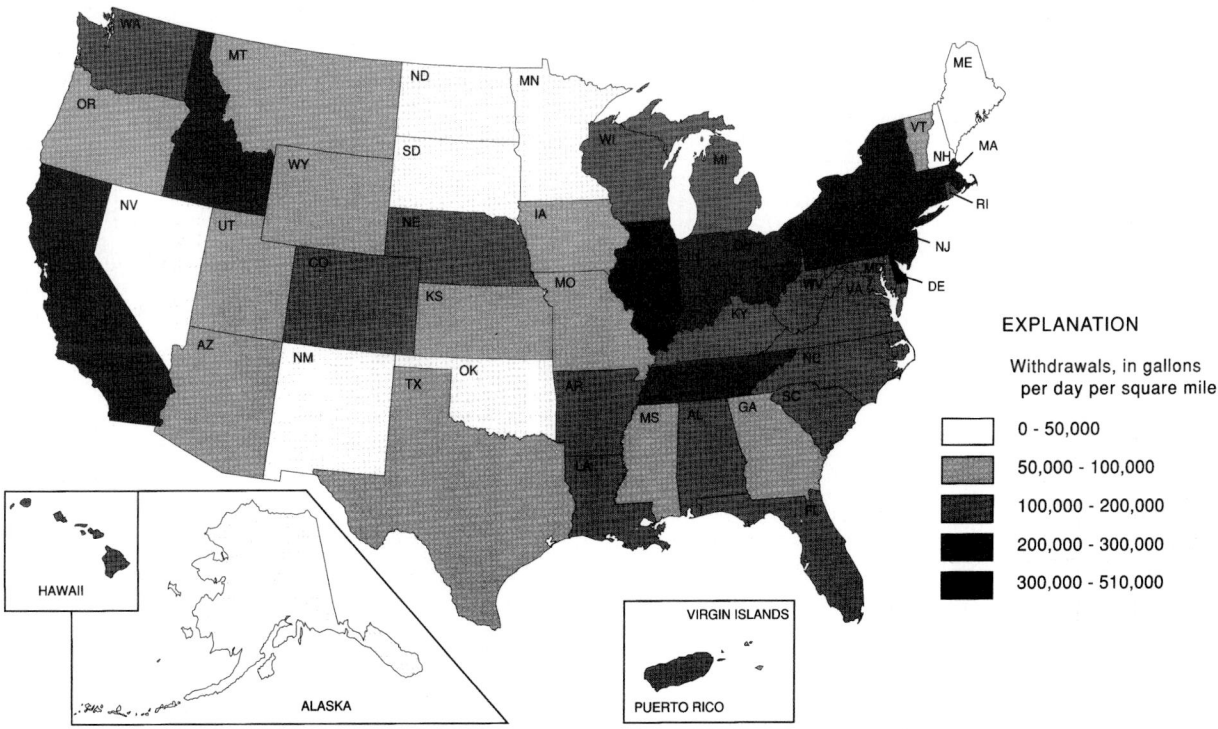

Figure W44 Intensity of freshwater withdrawals per area by state in 1990.

other water-use categories. The use column shows total freshwater use for each category and the percentage each category represents of total offstream water use. In addition, the use column shows the proportion of the source (surface water, groundwater and public supply) and disposition (consumptive use or return flow) for each category. The use data indicate, for example, that domestic and commercial use totaled 39 100 Mgal day^{-1} (including losses in the public supply distribution system), or 11.5% of the nation's total freshwater

SOURCE   USE   DISPOSITION

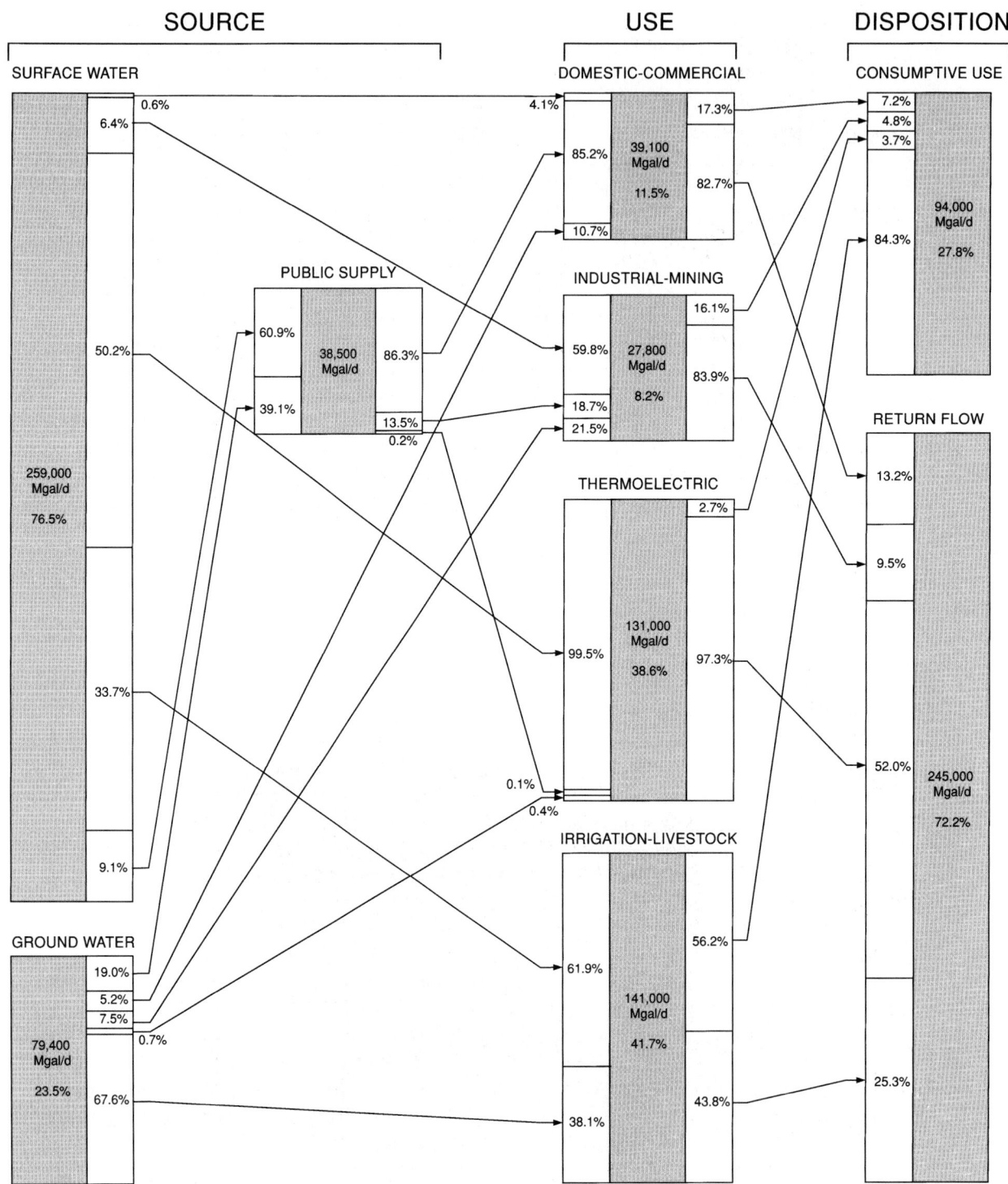

Figure W45 Source, use, and disposition of fresh water in the United States, 1990. For each water-use category, this diagram shows the relative proportion of water source and disposition and the general distribution of water from source to disposition. The lines and arrows indicate the distribution of water from source to disposition for each category; for example, surface water was 76.5% of total fresh water withdrawn, and going from 'Source' to 'Use' columns, the line for the surfacewater block to the domestic and commercial block indicates that 0.6% of all surface water withdrawn was the source for 4.1% of total water (self-supplied withdrawals, public supply deliveries) for domestic and commercial purposes. In addition, going from the 'Use' to 'Disposition' columns, the line from the domestic and commercial block to the consumptive use block indicates that 17.3% of the water for domestic and commercial purposes was consumptive use; this represents 7.2% of total consumptive use by all water-use categories.

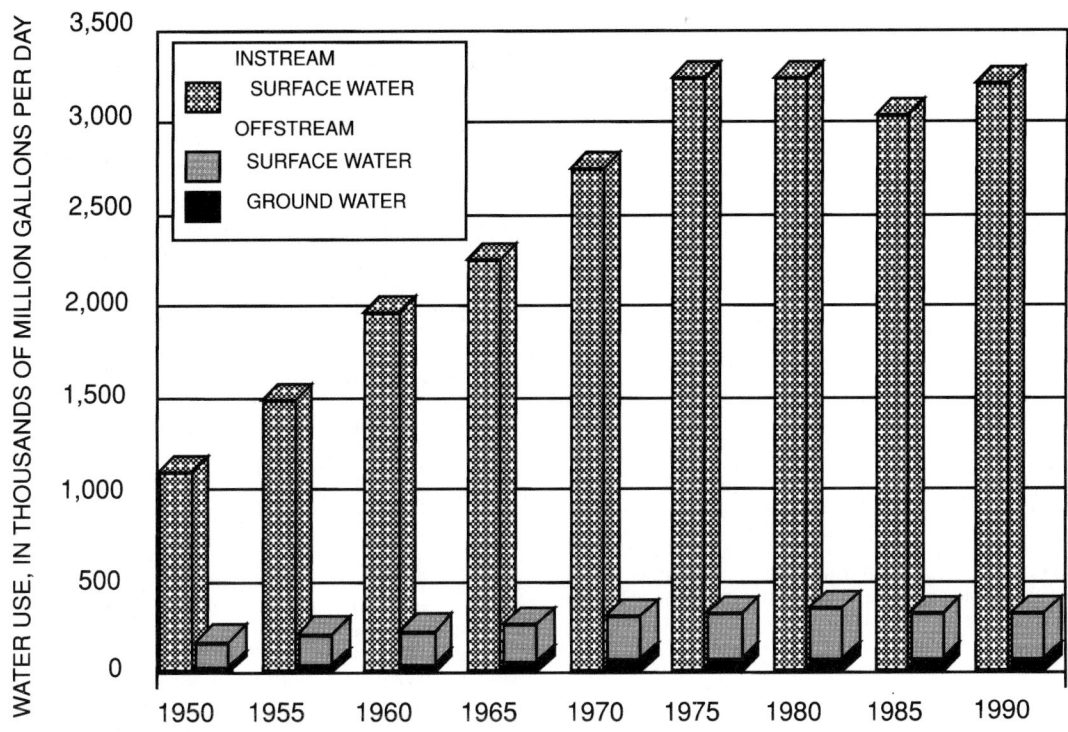

Figure W46 Trends in offstream and instream water uses, 1950–1990.

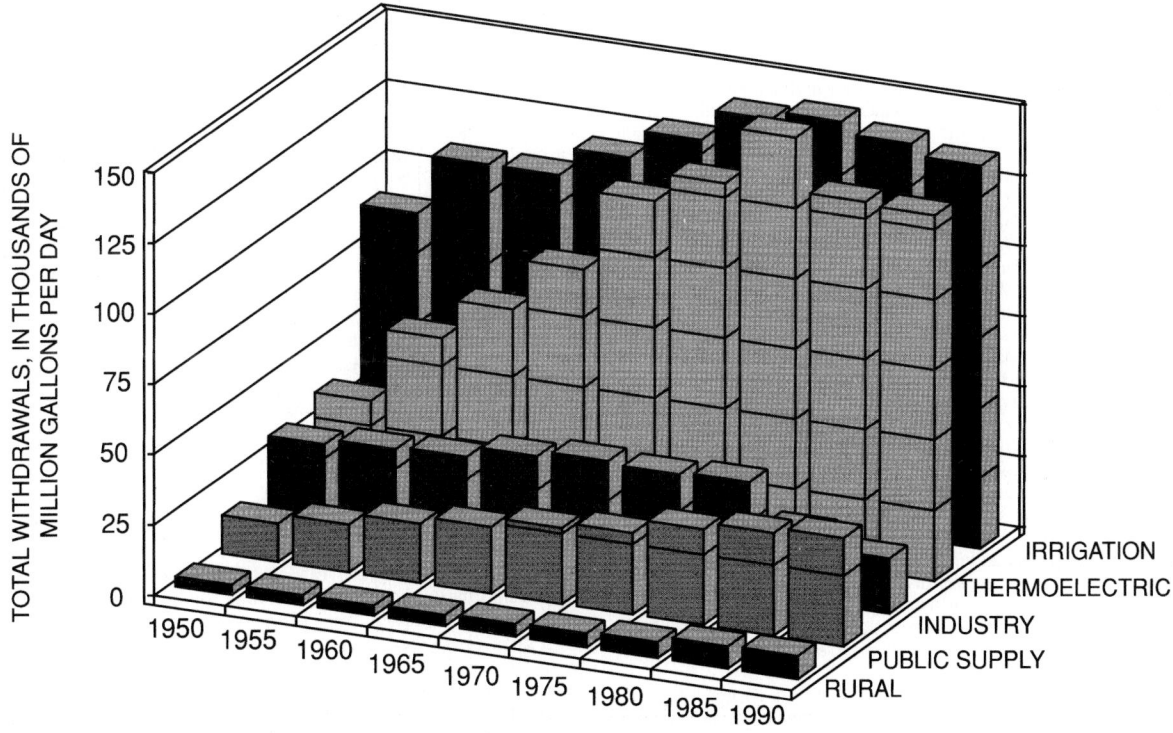

Figure W47 Trends in freshwater withdrawals by water-use category for rural , public supply, industry, thermoelectric and irrigation, 1950–1990.

withdrawals. Of this 39 100 Mgal day^{-1}, 85.2% was supplied by public supply systems, and 82.7% was returned to a surface- or groundwater source after use. The disposition column shows the quantity of consumptive use and return flow after use. The disposition column shows that of the total freshwater withdrawn, consumptive use was 94 000 Mgal day^{-1}, or 27.8%, and return flow was 245 000 Mgal day^{-1}, or 72.2% (including 27 500 Mgal day^{-1} of irrigation conveyance losses). Irrigation accounted for 84.3% consumptive use and thermoelectric power accounted for 52.0% of return flow. Trends in US water use for 1950–1990 are shown graphically in Figures W46 and W47.

R.W. Herschy

Source

Solley, W.B., Pierce, R.R. and Perlman, H.A., 1993. *Estimated use of water in the United States in 1990*. US Geological Survey Circular 1081, Washington, DC.

Cross reference

United States Geological Survey: National Water-Use Information Program

WEATHERING

Weathering may be defined as any geological destructive process that tends to weaken or disintegrate the solid materials of a planetary surface (Fairbridge, 1968). It is a strongly climate-dependent phenomenon (Büdel, 1981). While first recognized on planet Earth, the process is also to be expected on all hard planetary surfaces at varied levels of intensity depending on the local conditions. Those variables include factors such as the sun, atmospheric chemistry and wind velocities.

Two fundamental categories of weathering are (1) physical, and (2) chemical weathering.

Physical weathering

Physical weathering is carried out by a number of processes, notably thermal and abrasive. Thermal action results in brittle fracture and is usual diurnal or seasonal in its attack (Robinson and Williams, 1994). It involves rock fracture and is thus called thermoclastic. It is well displayed on planet Earth in desert regions, both hot ones such as the Sahara or cold ones such as the 'dry valleys' of East Antarctica. Heating of an outer layer of rock is achieved under solar radiation or 'insolation', causing a scale-like surface layer to expand ('exfoliation'); abrupt cooling at night causes the scale to crack and split (which we may call 'thermoclastic weathering'). In the Sahara some areas covering many hundreds of square kilometers are covered by sharp rock fragments ('thermoclasts') measuring about 5–15 cm across. The optimum size is physically determined by the linear expansion coefficients of brittle solids.

In high latitudes and in high mountains a wide range of weathering phenomena are observed, a study organized under the label 'geocryology'. A standard work on the subject (Washburn, 1980) cites no less than 2200 references to its various manifestations. The freezing process is often accompanied by the presence of water (or melted snow). This soaks into a crack and on freezing the force of crystallization (an expansion by 1 in 9) forces the crack more open until the rock surface or boulder splits into pieces. Mountain sides, escarpments and hill slopes in subglacial (and former periglacial) regions are very commonly covered by a talus (or scree) of fractured rock. The process is variously known as 'cryofracture' (from the Greek), and gelifracture or congelifraction (from the French *gel* for frost). The individual fragment may be called a 'gelifract' or 'congelifract'. Anglo-Saxon terms for this process are 'frost riving' or 'frost splitting'. 'Gelivity' is the susceptibility of a rock to cryogenic activity; thus a well-cleaved or tectonically jointed rock is far more susceptible to penetration by meltwater than is a massive rock (which, in contrast, is more susceptible to thermal expansion and contraction).

Abrasion, or abrasive action is, in contrast to the abrupt fracturing disintegration described above, an extremely gradual wearing down, scratching, scraping or scouring induced by friction which creates microfractures and is somewhat comparable to the role of sandpaper on a plank of wood (Hopwood, 1996). The energy for abrasion is provided by wind (eolian abrasion), moving water (fluvial or wave action) or moving ice (glacial action). However, neither wind nor moving water nor glaciers can act as an erosional force without the 'armament' of an abrasive agent such as boulders, sand grains or dust. The arming material itself is abraded and polished in the process;

this is particularly clear with the boulders that arm a glacier, the often rounded blocks becoming distinctively faceted (flattened and grooved) on the side in contact with the bedrock. The finest particles are easily dissolved, and can be detected by changes in glacial stream chemistry as an 'abrasion pH'.

The term 'corrasion' is virtually synonymous, but certain authors use it to distinguish the mechanical erosion by water from that by wind action.

On planet Mars wind action is widespread and persistent; thus it may be anticipated that eolian abrasion is widespread. Both thermoclastic and cryoclastic weathering are to be expected. Dust and larger particles, the armament for eolian abrasion, are known to be abundant on the Martian surface and are presumptive evidence for chemical weathering (see below).

Chemical weathering

Chemical weathering is carried out in the presence of water or moist air, principally in the form of rain or groundwater, by a number of chemical reactions. These include hydrolysis, hydration, oxidation, carbonation, ion exchange and solution (Barshad, 1972; Colman and Dethier, 1986). These may take place almost anywhere on the land surfaces, but mainly in the temperate and tropical regions marked by regular precipitation, most of the reactions being favored by rising temperature (except for carbonate solution inasmuch as the solubility of CO_2 in water is inversely related to temperature rise, but positively related to rising pressure, as in the deep ocean).

In contrast, the hot deserts like the Sahara, Atacama and Namib deserts have very little rainfall and almost negligible groundwater; chemical weathering in such places is almost exclusively limited to dew. Due to diurnal accumulation and reprecipitation, respectively, during the night-time cooling and the solar warming the next day, dew is a major cause of the 'desert varnish' or patina that forms as a thin veneer on desert thermoclasts; their color is dark red to black thanks to the concentration of iron and manganese oxides in part derived from their long-distance transport as dust. Dew was much more important in most terrestrial deserts during the glacial intervals of the last few million years than at the present time due to lower ambient temperatures (Colman and Dethier, 1986).

In hot deserts, apart from sand dune areas, there are pervasive red colors in the local rocks and soils. These are iron oxides generated by seasonal rains (generally monsoonal) that may only infrequently reach the area, or the red soils may be inherited from a more humid period. Over most of the Sahara, for example, there were widespread summer rains in the interval from 13 500 to 5500 BP (Fairbridge, 1968).

Cold deserts, as in polar and subpolar regions, normally inhibit chemical weathering but commonly enjoy a few weeks, even months, when daytime temperatures rise above freezing and with daylight persisting up to 24 h. At such times winter snow and permafrost melt, the latter superficially, but the meltwater, usually acidified by lush tundra vegetation, provides a powerful solvent for most minerals, except for quartz (SiO_2); as a result, tundra soils are usually podsols (leached soils, predominantly of quartz), with a strong organic component.

In certain places moving water dissolves out caverns in the ice of the upper permafrost, but beneath a carpet of soil and peat. These caverns and hollows are known as 'thermokarst', by analogy with the limestone karst of lower latitudes. Thermokarst caverns proved to be lethal traps for late glacial-age mammoths that migrated northwards during the summer season (about 12 000–6000 BP) but frequently strayed into thermokarst areas, becoming trapped when the soft roofs of the caverns collapsed. Tens of thousands of such trapped cadavers have been discovered, themselves preserved in part in the renewed permafrost (a number are to be seen in the Natural History Museum in St Petersburg). On planet Mars the strongly seasonal climate, with long-term cycles, makes it highly probable that widespread thermokarst will be discovered.

Chemical weathering also takes place in the marine realm. With the abrupt change of pH from river waters (often 5–7) to the ocean water (often >8), sediment particles, especially in the clay fractions, begin a rapid set of changes, sometimes called 'submarine weathering' or 'halmyrolysis' (Fairbridge, 1983; Julian, 1995). Many rock types (limestones, basalts and granites) are very slowly soluble in seawater.

Biological weathering

On planet Earth a special weathering form, unique to this planet, is often recognized: biological weathering. This action is often a modifier for each of the two principal categories. The physical biological process may be exemplified by the splitting of rocks and boulders by the slow but inexorable expansion of a growing root or by the action of cattle or sheep hooves in wearing down a friable rock surface or on hillsides creating 'terracettes' (also known as 'lynchets' in the UK). The chemical biological (or biochemical) process is much more pervasive, being most active at the microbial level or at the root hairs of a growing plant; in both cases the principal reaction involves the production of CO_2 which in the presence of water creates a weak acid, H_2CO_3, which has a slow but highly destructive effect on many rocks, notably limestones ($CaCO_3$). Biologically acidified groundwaters are thus able to dissolve channels and caverns in limestone down to many hundreds of meters beneath the surface, creating underground landscapes (known as 'karst', from the classic area of that name, 'kras' in Bosnia and Herzegovina).

Rhodes W. Fairbridge

Bibliography

Barshad, I., 1972. Weathering – chemical, in Fairbridge, R.W. (ed.). *The Encyclopedia of Geochemistry and Environmental Sciences.* New York: Van Nostrand Reinhold, pp. 1264–1269.
Büdel, J., 1981. *Klima-Geomorphologie*, 2nd edn. Berlin, Stuttgart: Gebr. Borntraeger, 304 pp.
Colman, S.M. and Dethier, D.P, 1986. *Rates of Chemical Weathering of Rocks and Minerals*, New York: Academic Press, 603 pp.
Fairbridge, R.W., 1963. *The Encyclopedia of Geomorphology.* New York: Van Nostrand Reinhold, 1295 pp.
Fairbridge, R.W., 1983. Syndiagenesis–anadiagenesis–epidiagenesis: phases in lithogenesis, in Larsen, G. and Chilinger, G.V. (eds), *Diagenesis in Sediments and Sedimentary Rocks*, Vol. 2, Amsterdam: Elsevier, p. 17–113.
Hopwood, J., 1996. In Schneider, S.H. (ed.), *Encyclopedia of Climate and Weather.* Oxford: Oxford University Press.
Julian, P.Y., 1995. *Erosion and Sedimentation.* Cambridge: Cambridge University Press, 280 pp.
Robinson, D.A. and Williams, R.B.J. (eds), 1994. *Rock Weathering and Landform Evaluation.* New York: Wiley, 519 pp.

Washburn, A.L., 1980. *Geocryology: a Survey of Periglacial Processes and Environments.* New York: Halsted Press, 406 pp.

Cross references

Acid rain
Arid climates
Arid zone hydrology
Climate and climate change
Karst hydrology

WEBER NUMBER

The Weber number (W_e) is a dimensionless parameter that represents the ratio of the inertia forces to the surface tension forces. For water in open channels the Weber number is given by the expression:

$$W_e = \frac{\rho \bar{V}^2 \bar{D}}{\sigma}$$

where ρ is the density of the liquid, $\bar{V}^2$ is the squared value of the mean velocity of the liquid, $\bar{D}$ is the mean depth of the cross-section and σ is the surface tension of the liquid.

P.G. Holland

Bibliography

BS 3680, Part 1, 1991. *Glossary of Terms.* British Standards Institution, London

WEIR: FLOW MEASUREMENT

Weir

A weir is an overflow structure that may be used for controlling upstream surface level or for measuring discharge or for both. The general equation takes the form (see Figure W48)

$$Q = \left(\frac{2}{3}\right)^{3/2} c\, b\sqrt{(g)} H^{3/2} \text{ m}^3 \text{ s}^{-1}$$

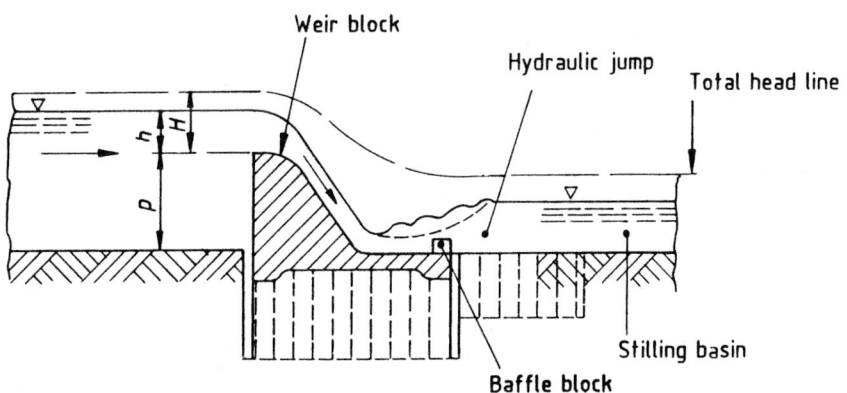

Key

▽ is the level of the water;
→ is the direction of flow;
H is the *total head*;
h is the head of *liquid level*;
p is the height of the *weir* crest above the bed of the *approach channel*.

Figure W48 Weir.

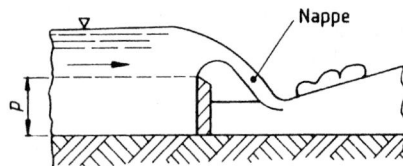

Key

∇ is the level of the water;

$\rightarrow$ is the direction of the flow;

p is the height of the *weir* crest above the bed of the *approach channel*.

Figure W49 Thin-plate weir.

Thin-plate weir

A thin-plate weir is a weir constructed of a vertical thin plate in such a manner that the nappe springs clear of the crest when operated within specified limits (Figure W49).

Thin-plate notch weir

A thin-plate notch weir is a weir, the crest of which is a notch cut in a thin plate (Figure W50).

Broad-crested weir

A broad-crested weir is a weir of such crest length in the direction of flow that critical flow occurs on the crest of the weir (Figure W51).

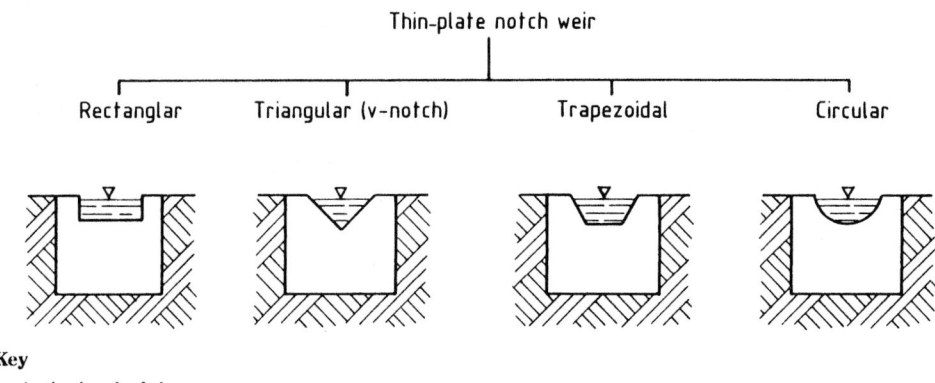

Key

∇ is the level of the water.

Figure W50 Thin-plate notch weirs.

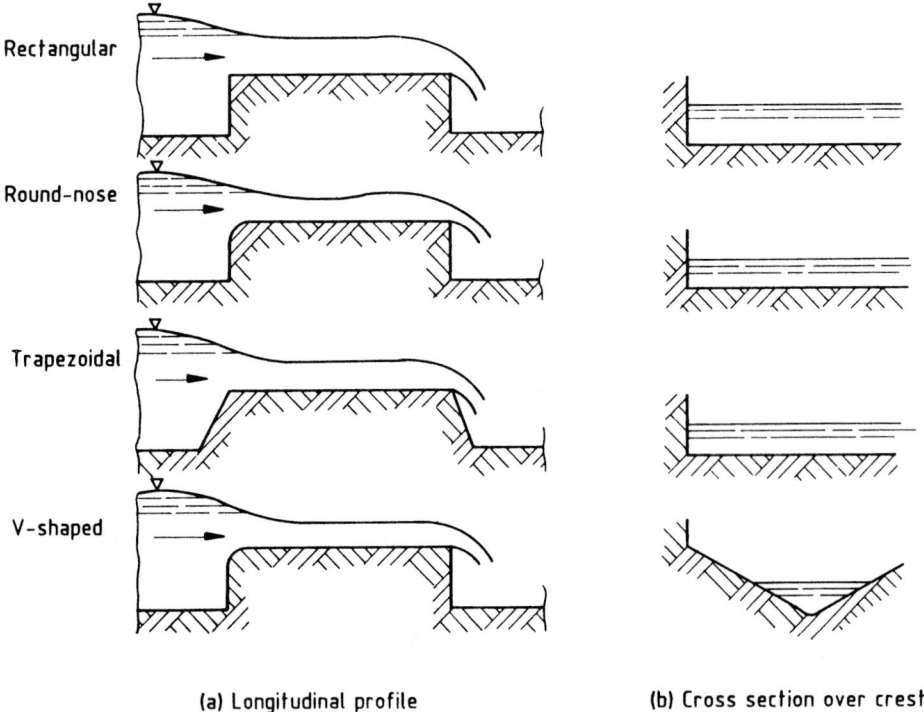

Key

∇ is the level of the water;

$\rightarrow$ is the direction of the flow.

Figure W51 Broad-crested weirs.

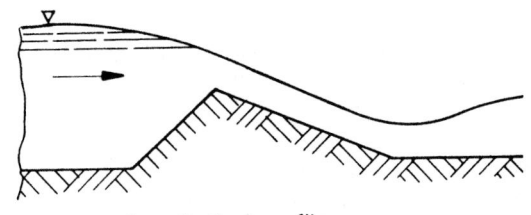

Longitudinal profile

Key

∇ is the level of the water;

→ is the direction of the flow.

Figure W52 Triangular-profile weir.

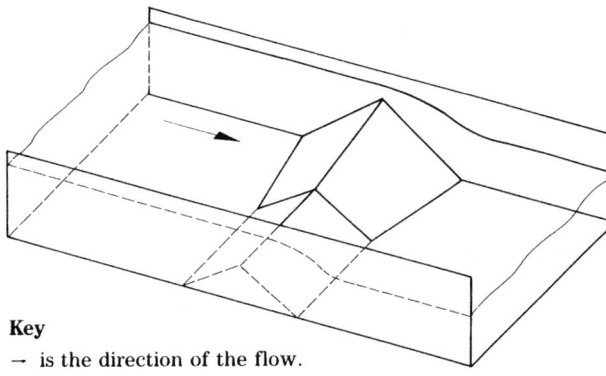

Key

→ is the direction of the flow.

Figure W53 Flat-V weir.

Triangular-profile weir

A triangular-profile weir is a long-base weir with a triangular longitudinal profile (Figure W52).

Flat-V weir

A flat-V weir is a weir in which the crest takes the form of a shallow V when viewed in the direction of flow (Figure W53).

P.G. Holland

Bibliography

BS 3680, Part 4A, 1981. *Thin-plate weirs*, British Standards Institution, London.
BS 3680 Part 4B, 1986. *Triangular-Profile Weirs*, British Standards Institution, London.
BS 3680 Part 4D, 1989. *Compound Gauging structures*, British Standards Institution, London.

BS 3680 Part 4E, 1990. *Rectangular broad-crested weirs*, British Standards Institution, London.
BS 3680 Part 4F, 1990. *Round-nose horizontal weirs*, British Standards Institution, London.
BS 3680 Part 4G, 1990. *Flat-V weirs*, British Standards Institution, London.
BS 3680 Part 4H, 1986. *Guide for the selection of measuring structure*, British Standards Institution, London.
BS 3680 Part 4I, 1986. *V-shaped broad-crested weirs*, British Standards Institution, London.
Herschy, R.W., 1995. *Streamflow Measurement*, 2nd edn, Chapman & Hall, London and New York.
ISO 1438, 1980. *Thin-plate weirs*, International Standards Organization, Geneva.
ISO 4360, 1984. *Triangular-profile weirs*, International Standards Organization, Geneva.
ISO 3846, 1989. *Rectangular broad-crested weirs*, International Standards Organization, Geneva.
ISO 4374, 1989. *Round-nose horizontal weirs*, International Standards Organization, Geneva.
ISO 4377, 1989. *Flat-V weirs*, International Standards Organization, Geneva.
ISO 8368, 1985. *Guide for the selection of measuring structure*, International Standards Organization, Geneva.
ISO 8333, 1985. *V-shaped Broad-crested weirs*, International Standards Organization, Geneva.
ISO 14139, 1998. *Compound gauging structures*, International Standards Organization, Geneva.

Cross reference

Flow through weirs and flumes, orifices, sluices and pipes

WORLD WATER BALANCE

Figure W54 shows diagrammatically the hydrological (water) cycle with major reservoirs. The values shown are approximate and may differ depending on how they are calculated. Table W56 gives the breakdown by continent (per annum) and Figure W55 shows diagrammatically a summary of the world water availability. It can be seen that, of the total volume of water, only 0.6% is available for use, and of this only 1% is available from rivers and lakes (i.e. about 0.006%).

Of the 0.6% of fresh water available some 99% is groundwater of which nearly 50% is below 1 km.

The general pattern of the world's annual rainfall and runoff are shown in Figures W56 and W57, respectively.

R.W. Herschy

Bibliography

Doxiadis, C.A., 1967. *Water and Environment*, International Conference on Water for Peace, Washington, DC.
Lvovitch, M.I., 1958. Streamflow Formation Factors, *IASH*, **3**(45), 122–132.
Lvovitch, M.I., 1973. *EOS*, **54**(1), American Geophysical Union.

Table W56 World water balance: breakdown by continent (km³)

	Europe	Asia	Africa	North America	South America	Australia	Total land area
Precipitation	7 162	32 590	20 780	13 810	29 255	6 405	110 000
Total river runoff	3 110	14 190	4 295	5 960	10 480	1 965	40 000
Groundwater runoff	1 065	3 410	1 465	1 740	3 740	465	11 885
Evaporation	4 055	18 500	16 455	7 850	18 800	4 340	70 000
Relative values							
Groundwater runoff as percentage of total runoff	34	26	35	32	36	24	31
Coefficient of groundwater discharge into rivers	0.21	0.15	0.08	0.18	0.16	0.10	0.14
Coefficient of runoff	0.43	0.40	0.23	0.31	0.35	0.31	0.36

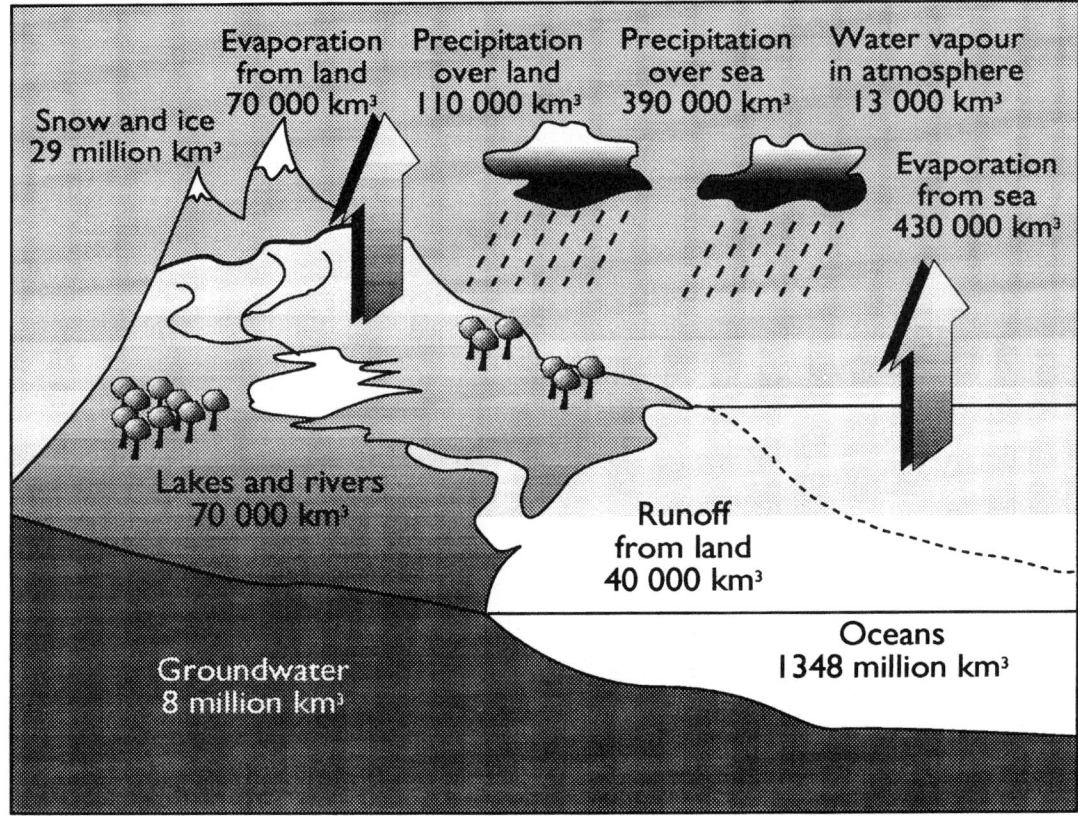

Figure W54 The hydrological cycle with major reservoirs. The fluxes of evaporation, precipitation and runoff are in km^3 year^{-1}. (Source: UNEP, 1991.)

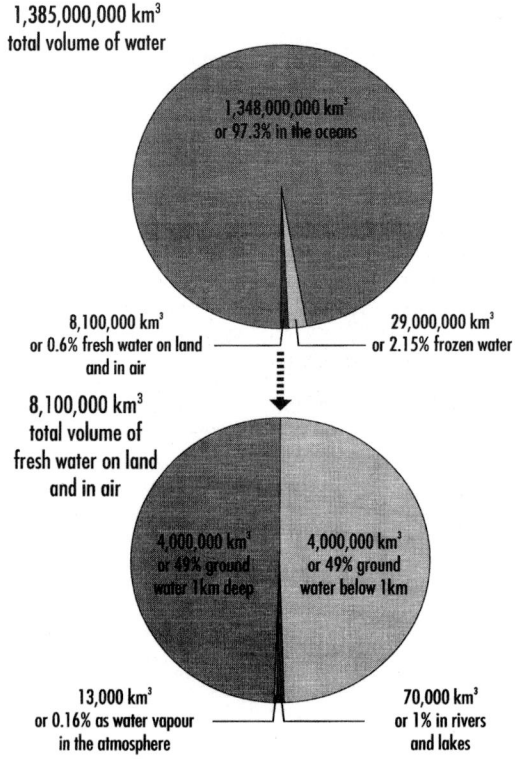

Figure W55 World water availability.

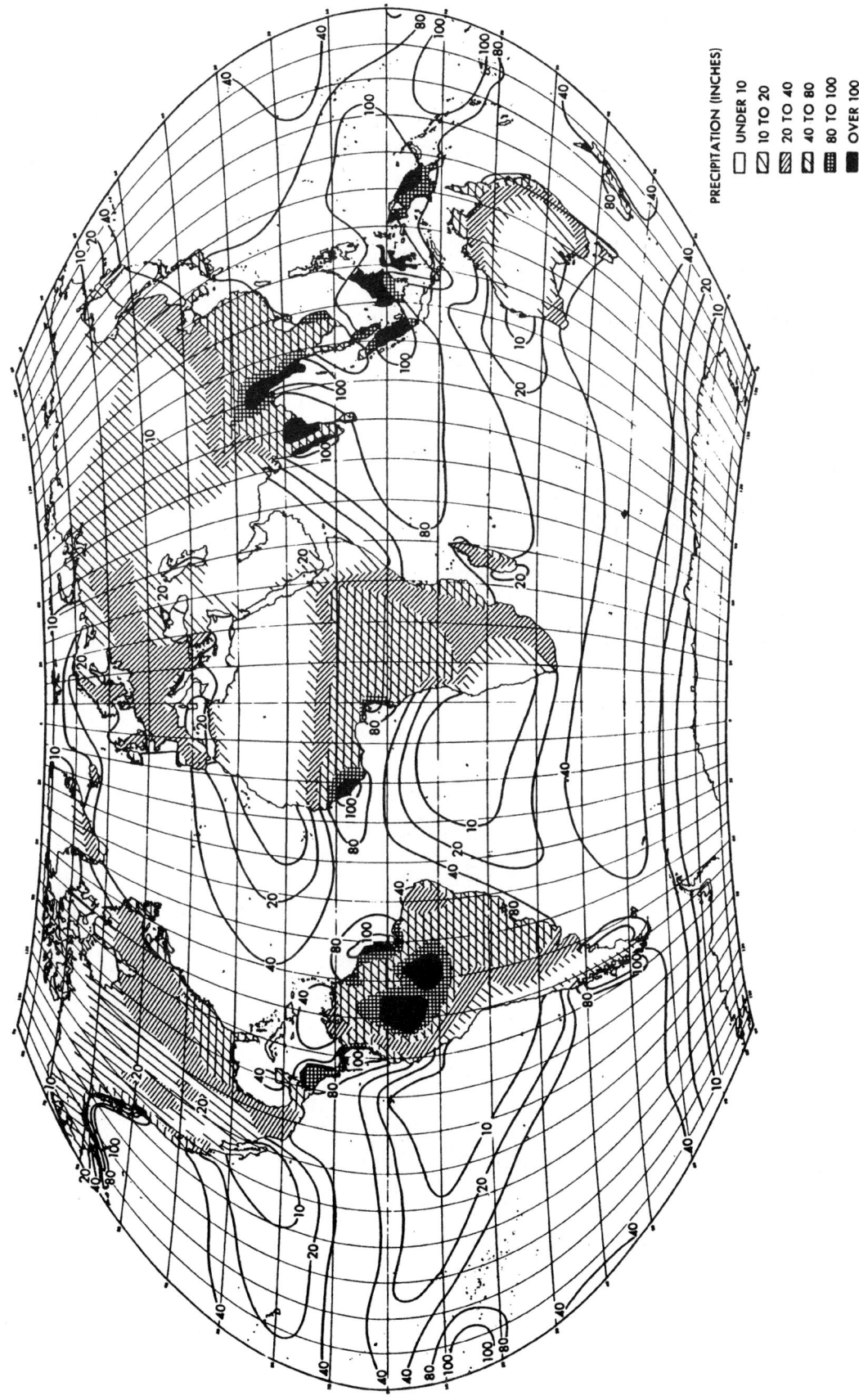

Figure W56 Annual world precipitation. Source: Environmental Science Service: Administration, *Climates of the World*, 1969.)

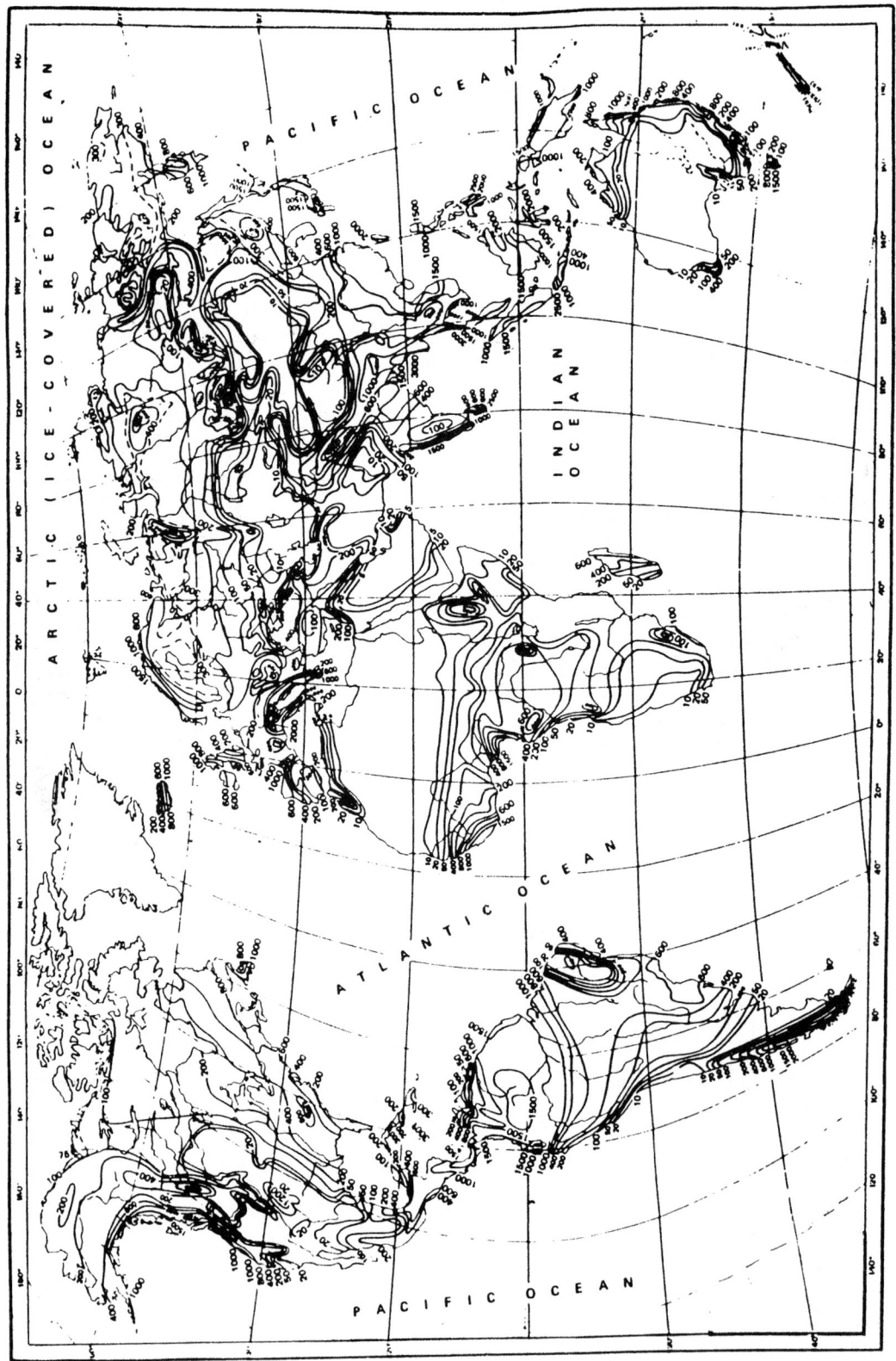

Figure W57 Annual world river runoff (mm), including groundwater discharge to rivers. (Source: Lvovitch, 1973.)

Y

YANGTZE THREE GORGES DAM, CHINA

Work has started on what is claimed by China to be the world's largest civil engineering project – the Yangtze River's Three Gorges Dam. When completed in the year 2009 at an estimated cost of $26 billion the dam will be 176 m high, 2.4 km long and will produce 85 billion kwh of electricity, more than 10% of China's annual consumption, for the benefit of the 200 million inhabitants of the Yangtze River basin. A series of locks will enable ships to negotiate the dam as well as two ship elevators. To construct the dam some 1.3 million people from 300 towns and villages will be relocated. The reservoir formed behind the dam will be nearly 640 km long and stretch upstream as far as Chungking, China's largest city with a population of 15 million and 4000 km from the coast, which will eventually be reached for the first time by oceangoing ships of 10 000 tonnes. The dam will also mitigate flooding in the Yangtze where in 1954 33 000 people were drowned in one of China's catastrophic floods. China has also experienced dam collapse due to flooding, the worst being the Banqiao and the Shimantan on the Huai River in Henan province in 1975 when 230 000 people lost their lives.

R.W. Herschy

Bibliography

The Times, October 28, 1995.

Cross references

Dams
Dams: failure
Dams: world

Author index

Subject index